| *Family name* | *Functional group structure*[a] | *Simple example* | *Name ending* |
|---|---|---|---|
| Sulfone | 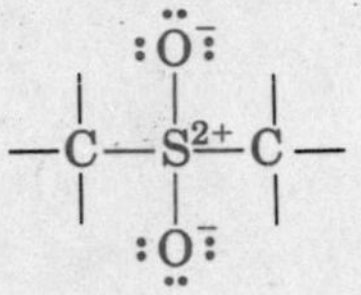 $-\underset{\mid}{\overset{\mid}{C}}-\underset{:\ddot{O}:^{-}}{\overset{:\ddot{O}:^{-}}{S^{2+}}}-\underset{\mid}{\overset{\mid}{C}}-$ | 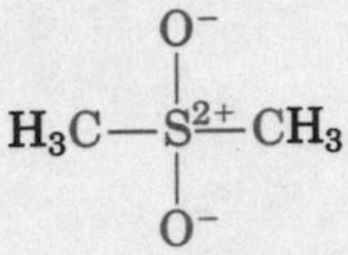 $H_3C-\underset{O^-}{\overset{O^-}{S^{2+}}}-CH_3$ | *sulfone*<br>Dimethyl sulfone |
| Thiol | 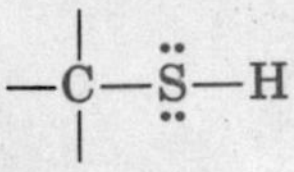 $-\underset{\mid}{\overset{\mid}{C}}-\ddot{\underset{..}{S}}-H$ | $H_3C-SH$ | *-thiol*<br>Methanethiol |
| **Carbonyl,**  $-\overset{:O:}{\overset{\parallel}{C}}-$ | | | |
| Aldehyde | 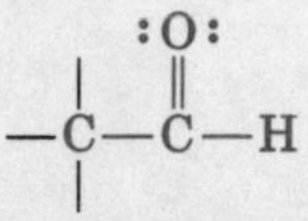 $-\underset{\mid}{\overset{\mid}{C}}-\overset{:O:}{\overset{\parallel}{C}}-H$ | 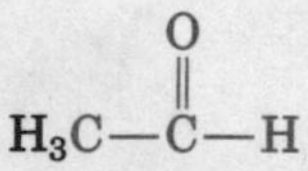 $H_3C-\overset{O}{\overset{\parallel}{C}}-H$ | *-al*<br>Ethanal<br>(Acetaldehyde) |
| Ketone | 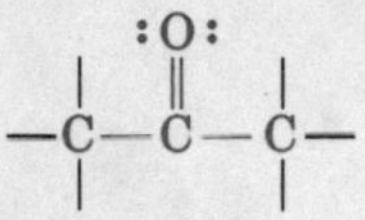 $-\underset{\mid}{\overset{\mid}{C}}-\overset{:O:}{\overset{\parallel}{C}}-\underset{\mid}{\overset{\mid}{C}}-$ | 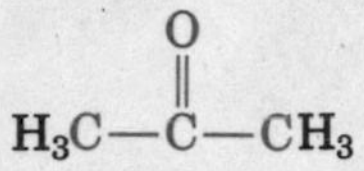 $H_3C-\overset{O}{\overset{\parallel}{C}}-CH_3$ | *-one*<br>Propanone<br>(Acetone) |
| Carboxylic acid | 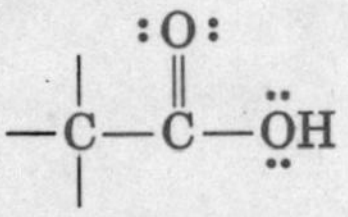 $-\underset{\mid}{\overset{\mid}{C}}-\overset{:O:}{\overset{\parallel}{C}}-\ddot{\underset{..}{O}}H$ | 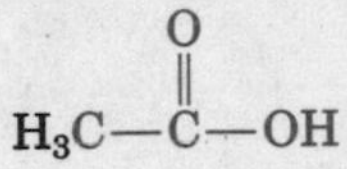 $H_3C-\overset{O}{\overset{\parallel}{C}}-OH$ | *-oic acid*<br>Ethanoic acid<br>(Acetic acid) |
| Ester | 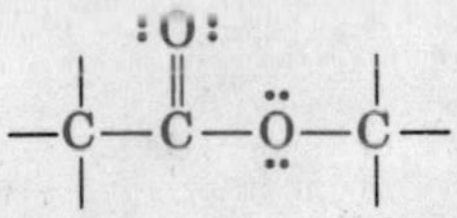 $-\underset{\mid}{\overset{\mid}{C}}-\overset{:O:}{\overset{\parallel}{C}}-\ddot{\underset{..}{O}}-\underset{\mid}{\overset{\mid}{C}}-$ | 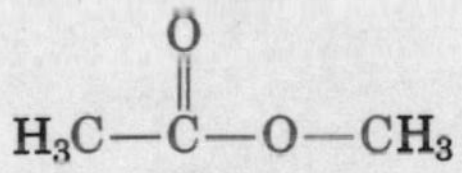 $H_3C-\overset{O}{\overset{\parallel}{C}}-O-CH_3$ | *-oate*<br>Methyl ethanoate<br>(Methyl acetate) |
| Amide | 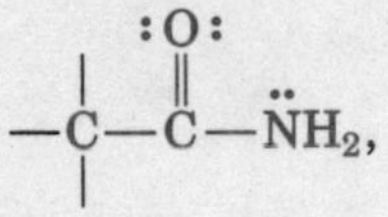 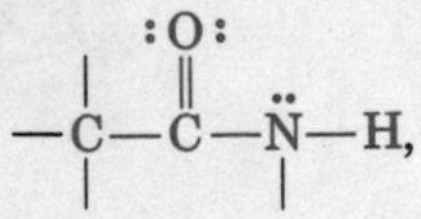 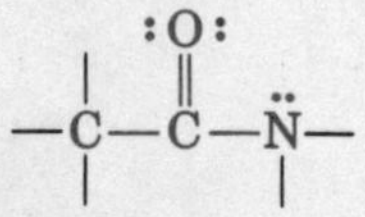 $-\underset{\mid}{\overset{\mid}{C}}-\overset{:O:}{\overset{\parallel}{C}}-\ddot{N}H_2,$<br>$-\underset{\mid}{\overset{\mid}{C}}-\overset{:O:}{\overset{\parallel}{C}}-\underset{\mid}{\ddot{N}}-H,$<br>$-\underset{\mid}{\overset{\mid}{C}}-\overset{:O:}{\overset{\parallel}{C}}-\underset{\mid}{\ddot{N}}-$ | 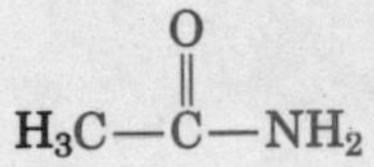 $H_3C-\overset{O}{\overset{\parallel}{C}}-NH_2$ | *-amide*<br>Ethanamide<br>(Acetamide) |
| Carboxylic acid chloride | 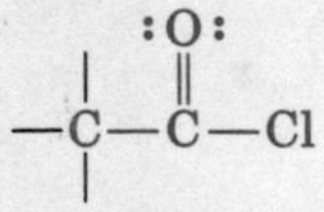 $-\underset{\mid}{\overset{\mid}{C}}-\overset{:O:}{\overset{\parallel}{C}}-Cl$ | 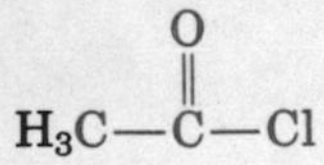 $H_3C-\overset{O}{\overset{\parallel}{C}}-Cl$ | *-oyl chloride*<br>Ethanoyl chloride<br>(Acetyl chloride) |
| Carboxylic acid anhydride | 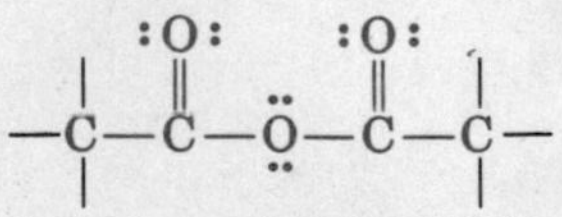 $-\underset{\mid}{\overset{\mid}{C}}-\overset{:O:}{\overset{\parallel}{C}}-\ddot{\underset{..}{O}}-\overset{:O:}{\overset{\parallel}{C}}-\underset{\mid}{\overset{\mid}{C}}-$ | 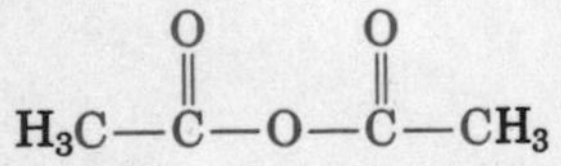 $H_3C-\overset{O}{\overset{\parallel}{C}}-O-\overset{O}{\overset{\parallel}{C}}-CH_3$ | *-oic anhydride*<br>Ethanoic anhydride<br>(Acetic anhydride) |

SECOND EDITION

# Organic Chemistry

**John McMurry**
Cornell University

BROOKS/COLE PUBLISHING COMPANY
Pacific Grove, California

**Brooks/Cole Publishing Company**
A Division of Wadsworth, Inc.



Printed in the United States of America

10 9 8 7 6 5 4 3

**Library of Congress Cataloging-in-Publication Data**

McMurry, John.
Organic chemistry.

Includes index.
1. Chemistry, Organic. I. Title.
QD251.2.M43 1988 547 87-15835

ISBN 0-534-07968-7
International Student Edition ISBN 0-534-98107-0

SPONSORING EDITOR: Sue Ewing
EDITORIAL ASSISTANT: Lorraine McCloud, Heidi Wieland
PRODUCTION COORDINATORS: Phyllis Niklas, Joan Marsh
MANUSCRIPT EDITOR: Patricia Cain
PERMISSIONS EDITOR: Carline Haga
INTERIOR DESIGN: Stan Rice, Janet Bollow
COVER DESIGN: Janet Bollow
COVER ILLUSTRATION: Vantage Art
INTERIOR ILLUSTRATION: Vantage Art
TYPESETTING: Jonathan Peck Typographers, Ltd.
COVER PRINTING: The Lehigh Press, Inc.

# Preface

In the first edition of this text, I stated that my goal was to write a lucid, readable, and effective *teaching* text, while providing an accurate and up-to-date view of organic chemistry. The reception that edition received from both students and faculty suggests that the goal was largely met.

This second edition of *Organic Chemistry* builds on the strengths of the first. All the features that made the first edition a success have been improved, and many new ones have been added:

- The writing, already clear and accessible to students, has been reworked to enhance understanding further. Particular attention has been paid to such traditionally difficult subjects as stereochemistry and nucleophilic substitutions.
- Much of the artwork has been redrawn with the added use of color and with the addition of computer-generated structures to ensure accuracy.
- The summaries and problem sets have been expanded. Also, many carefully worked-out sample problems have been added to the text.
- Other features of this second edition include expanded treatments of acid/base chemistry, resonance, and isomerism, and more detail in the coverage of spectroscopy.

## Organization

This book uses a dual functional-group/reaction-mechanisms organization. The primary organization is by functional group, beginning with the simple (alkenes) and progressing to the more complex. Within this primary organization, however, heavy emphasis is placed on explaining the fundamental mechanistic similarities of reactions. This is particularly evident in the coverage of carbonyl-group chemistry (Chapters 19–23), where mechanistically related reactions like the aldol and Claisen condensations are treated together.

Insofar as possible, topics have been arranged in a modular way. Thus, the chapters on simple hydrocarbons are grouped together (Chapters 3–8), the chapters on spectroscopy are grouped together (Chapters 12–14), and the chapters on carbonyl-group chemistry are grouped together (Chapters 19–23). This organization brings to these subjects a cohesiveness not found in other texts and allows the instructor a flexibility to teach in an order different from that used in the book.

## The Lead-Off Reaction: Addition of HBr to Alkenes

Many students attach great importance to the lead-off reaction in a text, because it is the first reaction they see and it is discussed in such detail. I have chosen a simple polar reaction—the addition of HBr to an alkene—as the lead-off to illustrate the general principles of organic reactions. This choice has the advantage of relative simplicity (no prior knowledge of chirality or kinetics is required), yet it is also an important polar reaction on a common functional group. As such, I believe that this choice is a more useful introduction to functional-group chemistry than a lead-off such as free-radical alkane chlorination.

## Coverage

The coverage in this book is up-to-date, reflecting the most important advances of the last decade. For example, $^{13}C$ NMR is introduced as a routine spectroscopic tool; the advantages of using lithium diisopropylamide as a base for carbonyl-group alkylations are discussed; selenoxide elimination as a method of introducing double bonds is covered; the chemistry of DNA sequencing and synthesis is treated; and entire chapters are devoted to polymer chemistry and to pericyclic reactions.

## Organic Synthesis

Organic synthesis is emphasized as a teaching device to help students learn to organize and work with the large body of factual information that makes up organic chemistry. Two sections, the first in Chapter 8 (Alkynes) and the second in Chapter 16 (Chemistry of Benzene), clearly explain the thought processes involved in working synthesis problems. The value of starting from what is known and logically working backward one step at a time is emphasized. Also important in this respect are several summaries included in the appendixes. A summary of functional-group reactions, another on functional-group preparations, and a third on reagents used in organic synthesis are enormously useful when working problems.

## Pedagogy

Every effort has been made to make this second edition as effective, clear, and readable as possible—to make it easy to learn from. In addition to features already mentioned, a wide assortment of additional pedagogical devices are used:

Paragraphs start with summary sentences.

Transitions between paragraphs and between topics are smooth.

New concepts are introduced only as needed and are immediately illustrated with concrete examples.

Extensive use is made of computer-generated three-dimensional art and carefully rendered stereochemical formulas.

Extensive cross-referencing to earlier material is used.

Numerous summaries are included, both within chapters and at the ends of chapters.

More than 1400 problems of varying difficulty are included within the text and at the ends of chapters.

Use of an innovative vertical format to explain reaction mechanisms, introduced so successfully in the first edition, is expanded. Mechanisms are printed vertically, while explanations of the changes taking place in each step are printed next to the reaction arrow. This format allows the reader to see easily what is occurring at each step in a reaction without having to jump back and forth between the text and structures.

## Appendixes

The back of this book contains a wealth of material helpful for learning organic chemistry. Included are a large glossary of more than 250 terms, an explanation of how to name polyfunctional organic compounds, a list of reagents used in organic chemistry, a list summarizing functional-group reactions, and a list summarizing functional-group preparations.

## Study Guide and Solutions Manual

A carefully prepared *Study Guide and Solutions Manual* accompanies this text. Written by Susan McMurry, this companion volume answers all in-text and end-of-chapter problems and explains in detail how the answers are obtained. In addition, the following supplemental materials are included: a list of study goals and additional problems for each chapter, a list of suggested readings, a table of acidities, a table of bond-dissociation energies, a list of important organic compounds used in industry, and tables of spectroscopic information.

## Acknowledgments

It is a great pleasure to thank the many people whose help and suggestions were so valuable in preparing this second edition. Foremost is my wife, Susan, who read, criticized, and improved the manuscript, and who was my constant companion throughout all stages of this book's development. Among the reviewers providing valuable comments were Ronald Caple (University of Minnesota, Duluth), John Cawley (Villanova University), Clair Cheer (University of Rhode Island), George Clemans (Bowling Green State University), Dennis Davis (New Mexico State University), John Hogg (Texas A&M University), Paul Hopkins (University of Washington), John Huffman (Clemson University), Glen Kauffman (East Mennonite College), Joseph Lambert (Northwestern University), Thomas Livinghouse (University of Minnesota), Clarence Murphy (East Stroudsburg University), Oliver Muscio (Murray State University), Russell Petter (University of Pittsburgh), Neil Potter (Susquehanna University), Naser Pourahmady (Southwest Missouri State University), Jan Simek (California Polytechnic State University), Marcus W. Thomsen (Franklin & Marshall College), Daniel Weeks (Northwestern University), and David Wiemer (University of Iowa). In addition, special thanks are due William Russey (Juniata College), for his thoughtful page-by-page comments on the entire book.

Of course, the advice of the reviewers of the first edition continues to be invaluable: Robert A. Benkeser (Purdue University), Donald E. Bergstrom (University of North Dakota), Weston T. Borden (University of Washington), Larry Bray (Miami Dade Community College), William D. Closson (State University of New York, Albany), Paul L. Cook (Albion College), Otis Dermer (Oklahoma State University), Linda Domelsmith, David Harpp (McGill University), David Hart (Ohio State University), Norbert Hepfinger (Rensselaer Polytechnic Institute), Werner Herz (Florida State University), Paul R. Jones (North Texas State University), Thomas Katz (Columbia University), Paul E. Klinedinst, Jr. (California State University, Northridge), James G. Macmillan (University of Northern Iowa), Monroe Moosnick (Transylvania University), Harry Morrison (Purdue University), Cary Morrow (University of New Mexico), Wesley A. Pearson (St. Olaf College), Frank P. Robinson (University of Victoria), Neil E. Schore (University of California, Davis), Gerald Selter (California State University, San Jose), Ernest Simpson (California State Polytechnic University, Pomona), Walter Trahanovsky (Iowa State University), Harry Ungar (Cabrillo College), Joseph J. Villafranca (Pennsylvania State University), Daniel P. Weeks (Northwestern University), Walter Zajac (Villanova University), and Vera Zalkow (Kennesaw College).

Finally, I would like to thank Phyllis Niklas for her fine work as production editor; and Sue Ewing, Joan Marsh, Mike Needham, and the entire staff at Brooks/Cole for a thoroughly professional job. It was a pleasure working with them all.

# A Note for Students

We have similar goals: Yours is to learn organic chemistry, and mine is to do everything possible to help you learn. It's going to require some work on your part, but the following hints should prove helpful:

*Don't read the text immediately.* As you begin each new chapter, look it over first. Read the introductory paragraphs, find out what topics will be covered, and then turn to the end of the chapter and read the summary. You'll be in a much better position to understand new material if you first have a general idea of where you're going.

*Work the problems.* There are no shortcuts here; working problems is the only way to learn organic chemistry. The practice problems show you how to approach the material, the in-text problems at the ends of most sections provide immediate practice, and the end-of-chapter problems provide both additional drill as well as some real challenges. Full answers and explanations for all problems are given in the accompanying *Study Guide and Solutions Manual.*

*Use the study guide.* The *Study Guide and Solutions Manual* that accompanies this text gives complete solutions to all problems, an outline and list of study goals for each chapter, tables of supplementary material, and additional unsolved problems.

*Use the appendixes.* The back of this book contains a wealth of supplementary material, including a large glossary of terms, an explanation of how to name polyfunctional organic compounds, and extensive lists summarizing the reagents commonly used in organic chemistry, the ways in which different functional groups are prepared, and the reactions that functional groups undergo. This material can be extremely useful when you're studying for an exam and for getting an overview of organic chemistry. Find out what's there now, so you'll know where to find it when you need help.

*Ask questions.* Faculty members and teaching assistants are there to help you. Most of them will turn out to be genuinely nice people with a sincere interest in helping you learn.

*Use molecular models.* Organic chemistry is a three-dimensional science. Although this book uses many careful drawings to help you visualize molecules, there's no substitute for building a molecular model and turning it around in your hands.

Good luck. I sincerely hope you enjoy learning organic chemistry and come to see the beauty and logic of its structure. I heard from many students who used the first edition of this book and would be glad to receive more comments and suggestions from those who use this new edition.

# Brief Contents

# Contents

## 16 Chemistry of Benzene: Electrophilic Aromatic Substitution 522



## Organic Reactions: A Brief Review 571



## 17 Alcohols and Thiols 580



## 18 Ethers, Epoxides, and Sulfides 620

CHAPTER 1

# Structure and Bonding

What is organic chemistry? The answer is all around. The proteins that make up our hair, skin, and muscles; the nucleic acids, RNA and DNA, that control our genetic heritage; the foods we eat; the clothes we wear; and the medicines we take—all are organic chemicals.

The foundations of organic chemistry were built in the mid-eighteenth century as chemistry was evolving from an alchemist's art into a modern science. At that time, unexplainable differences were noted between substances derived from living sources and those derived from minerals. Compounds from plants and animals were often difficult to isolate and purify. Even when pure, these compounds were difficult to work with and were more sensitive to decomposition than compounds from mineral sources. In 1770, the Swedish chemist Torbern Bergman first expressed this difference between "organic" and "inorganic" substances, and the phrase **organic chemistry** soon came to mean the chemistry of compounds from living organisms.

To many chemists at the time, the only explanation for the difference in behavior between organic and inorganic compounds was that organic compounds contained a peculiar and undefinable "vital force" as a result of their coming from living sources. One consequence of the presence of this vital force, chemists believed, was that organic compounds could not be prepared and manipulated in the laboratory as could inorganic compounds.

Although the vitalistic theory was believed by many influential chemists, its acceptance was by no means universal, and it's doubtful that the development of organic chemistry was much delayed. As early as 1816, the theory received a heavy blow when Michel Chevreul[1] found that soap, prepared by the reaction of alkali with animal fat, could be separated into several pure organic compounds, which he termed "fatty acids." Thus, for

[1]Michel Eugène Chevreul (1786–1889); b. Angers, France; Paris, Muséum d'Histoire Naturelle, professor of physics, Lycée Charlemagne (1813); professor of chemistry (1830).

the first time, one organic substance (fat) had been converted into others (fatty acids plus glycerin) without the intervention of an outside vital force.[2]

$$\text{Animal fat} \xrightarrow[H_2O]{NaOH} \text{Soap} + \text{Glycerin}$$

$$\text{Soap} \xrightarrow{H_3O^+} \text{“Fatty acids”}$$

A little more than a decade later, the vitalistic theory suffered still further when Friedrich Wöhler[3] discovered in 1828 that it was possible to convert the “inorganic” salt ammonium cyanate into the previously known “organic” substance urea.

$$\underset{\text{Ammonium cyanate}}{NH_4^+ \ ^-OCN} \xrightarrow{\text{Heat}} \underset{\text{Urea}}{H_2N{-}\overset{\overset{\displaystyle O}{\|}}{C}{-}NH_2}$$

By the mid-nineteenth century, the weight of evidence was clearly against the vitalistic theory. In 1848, William Brande[4] wrote in a paper that: “No definite line can be drawn between organic and inorganic chemistry . . . any distinctions . . . must for the present be merely considered as matters of practical convenience calculated to further the progress of students.” Chemistry today is unified; the same basic scientific principles that explain the simplest inorganic compounds also explain the most complex organic molecules. *The only distinguishing characteristic of organic chemicals is that all contain the element carbon.* Nevertheless, the division between organic and inorganic chemistry, which began for historical reasons, maintains its “practical convenience . . . to further the progress of students.”

Organic chemistry, then, is the study of carbon compounds. Carbon, which has atomic number 6, is a second-period element whose position in an abbreviated periodic table is shown in Table 1.1. Although carbon is the principal element in organic compounds, most also contain hydrogen, and many contain nitrogen, oxygen, phosphorus, sulfur, chlorine, and other elements.

**Table 1.1** An abbreviated periodic table

| | *Group* | | | | | | | | |
|---|---|---|---|---|---|---|---|---|---|
| *Period* | IA | IIA | | IIIA | IVA | VA | VIA | VIIA | 0 |
| 1 | H | | | | | | | | He |
| 2 | Li | Be | | B | C | N | O | F | Ne |
| 3 | Na | Mg | | Al | Si | P | S | Cl | Ar |
| 4 | K | Ca | Sc Ti V Cr Mn Fe Co Ni Cu Zn | Ga | Ge | As | Se | Br | Kr |

[2]In the equations that follow, a *single* arrow is used to indicate an actual reaction. Later in this book you will also see forward and backward arrows, ⇄, indicating equilibrium, and a double arrow, ⇒, indicating a multistep transformation whose individual steps aren't specified.

[3]Friedrich Wöhler (1800–1882); b. Escherheim; studied at Heidelberg (Gmelin); professor, Göttingen (1836–1882).

[4]William Thomas Brande (1788–1866); b. London; lecturer in chemistry, London (1808); Royal Institution (1813–1854).

Why is carbon special? What is it that sets carbon apart from all other elements in the periodic table? The answers to these questions are complex but have to do with the unique ability of carbon atoms to bond together, forming long chains and rings. Carbon, alone of all elements, is able to form an immense diversity of compounds, from the simple to the staggeringly complex: from methane, containing one carbon, to DNA, which can contain hundreds of *billions*. Nor are all carbon compounds derived from living organisms. Chemists in the past 100 years have become extraordinarily sophisticated in their ability to synthesize new organic compounds in the laboratory. Medicines, dyes, polymers, plastics, food additives, pesticides, and a host of other substances—all are prepared in the laboratory, and all are organic chemicals. Organic chemistry is a science that touches the lives of all; its study can be a fascinating undertaking.

## 1.1 The Nature of Atoms: Quantum Mechanics

Throughout the nineteenth century, and into the twentieth, scientists sought to understand the nature of the atom and the nature of the forces holding atoms together in molecules. A major breakthrough occurred in 1926 when the theory of **quantum mechanics** was proposed independently by Paul Dirac, Werner Heisenberg, and Erwin Schrödinger.[5] All three formulations are mathematical expressions that describe the electronic structure of atoms, but Schrödinger's is the one most commonly used by chemists.

The Schrödinger equation offers a detailed description of the electronic structure of atoms. It says that the motion of an electron around the nucleus can be described mathematically by what's known as a **wave equation**—the same kind of expression that's used to describe the motion of waves in a fluid.

The solution to a wave equation is called a **wave function**. If we could determine the wave function for every electron in an atom, we would have a complete electronic description of that atom. In practice, wave equations are mathematically so complex that only approximate solutions can be obtained, even with the fastest computers now available. These approximate solutions agree so well with experimental facts, however, that quantum mechanics is a universally accepted theory for understanding atomic structure.

## 1.2 The Nature of Atoms: Atomic Orbitals

How can we interpret quantum mechanical wave functions in terms of physical reality? A good way of viewing a wave function is to think of it as an expression that predicts the volume of space around a nucleus where an electron can be found. Though we can never know the *exact* position of an electron at a given moment, the wave function tells us where we would be most likely to find it.

[5]Erwin Schrödinger (1887–1961); b. Vienna, Austria; University of Vienna (1910); assistant, University of Vienna (1910); assistant to Max Wein, University of Stuttgart, Germany (1920); professor of physics, University of Zurich, Berlin, Graz, Dublin; Nobel prize in physics (1933).

The volume of space around a nucleus in which an electron is most likely to be found is called an **orbital**. It's often helpful to think of an orbital as a kind of time-lapse photograph of an electron's movement around the nucleus. Such a photograph would show the orbital as a blurry cloud indicating where the electron has recently been. This electron cloud doesn't have a discrete boundary, but for practical purposes we can set the limits of an orbital by saying that it represents the space where an electron spends *most* (90–95%) of its time.

What do orbitals look like? The exact shape and size of an electron's orbital depend on its energy level. Electrons can be thought of as belonging to different layers, or **shells**, around the nucleus, where each shell contains different numbers and kinds of orbitals. For example, the first shell (the one nearest the nucleus) has only one orbital, called a 1*s* orbital. The second shell has four orbitals, one 2*s* and three 2*p*; the third shell has nine orbitals, one 3*s*, three 3*p*, five 3*d*; and so on. Relative energy levels of the different kinds of atomic orbitals are shown in Figure 1.1.

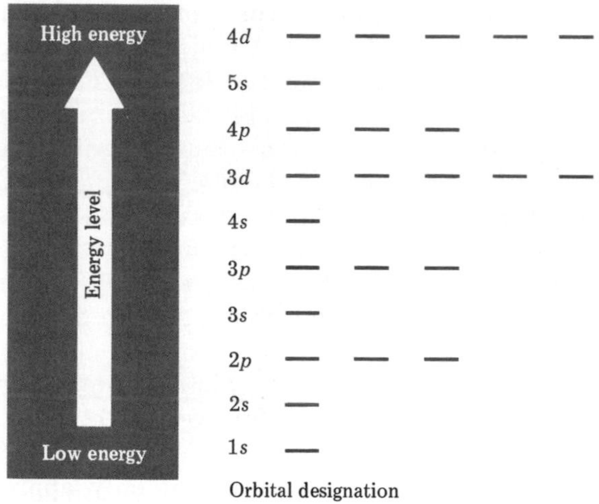


**Figure 1.1** Relative energy levels of atomic orbitals

The lowest-energy electrons occupy the 1*s* orbital. The *s* atomic orbitals are spherical and have the nucleus of the atom at their center, as shown in Figure 1.2. Next in energy after the 1*s* electrons are the 2*s* electrons. Because they are higher in energy, 2*s* electrons are farther from the positively charged nucleus on average, and their spherical orbital is somewhat larger than that of 1*s* electrons.

The 2*p* electrons are next higher in energy. As Figure 1.3 indicates, there are three 2*p* orbitals, each of which is roughly dumbbell shaped. The three 2*p* orbitals are equal in energy and are oriented in space such that each is perpendicular to the other two. They are denoted $2p_x$, $2p_y$, and $2p_z$, to show on which axis they lie. Note that the plane passing between the two lobes of a *p* orbital is a region of zero electron density. This **nodal plane** has certain consequences with respect to chemical bonding that will be taken up in Chapter 30.

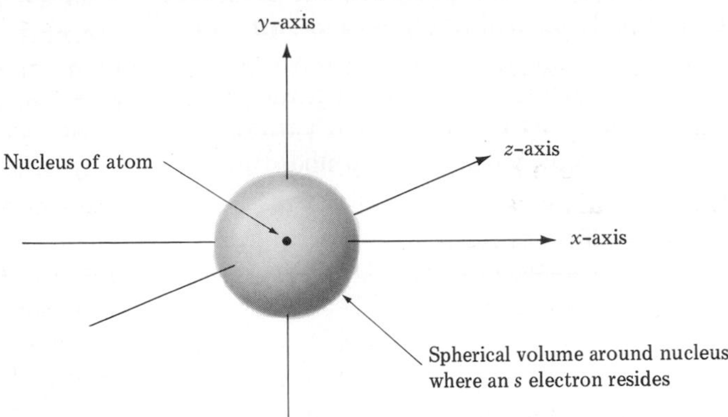


**Figure 1.2** The spherical shape of an *s* atomic orbital

A $2p_x$ orbital

A $2p_y$ orbital

A $2p_z$ orbital

A $2p$ orbital and its nodal plane

Three mutually perpendicular $2p$ orbitals

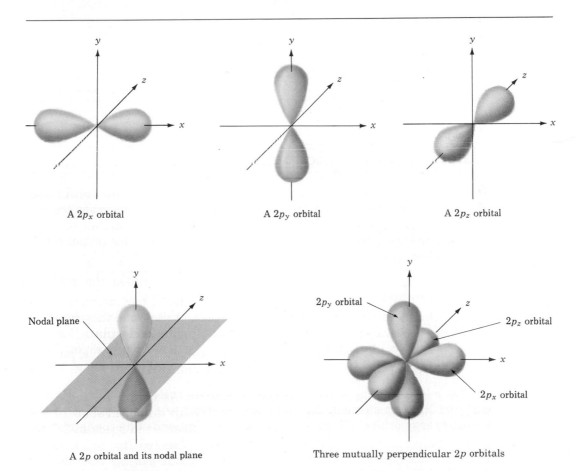

**Figure 1.3** Shapes of the $2p$ orbitals: Each of the three dumbbell-shaped orbitals has a nodal plane passing between its two lobes.

Still higher in energy are the 3*s* orbital (spherical), 3*p* orbitals (dumbbell shaped), 4*s* orbital (spherical), and 3*d* orbitals. There are five 3*d* orbitals of equal energy, one of which is shown in Figure 1.4. The *d* orbitals don't play as important a role in organic chemistry as the *s* and *p* orbitals do, and we won't be concerned with them. Note, however, that the 3*d* orbital shown has four lobes and two nodal planes.

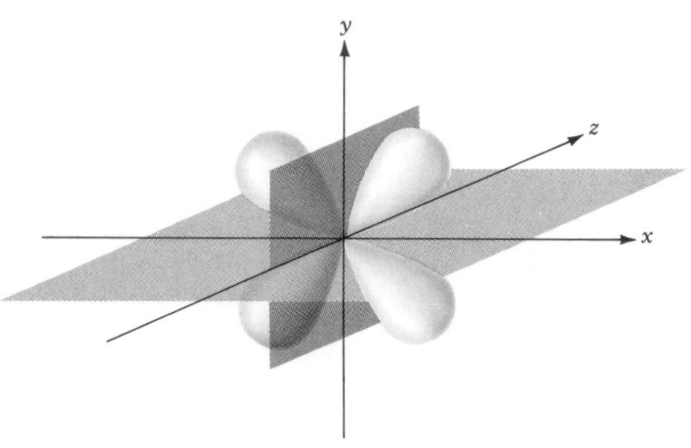


**Figure 1.4** A 3*d* orbital

## 1.3 The Nature of Atoms: Electronic Configurations

The lowest-energy arrangement, or **ground-state electronic configuration**, of an atom is a description of what orbitals the atom's electrons occupy. This arrangement can be found by using our knowledge of atomic orbitals and their energy levels. Available electrons are assigned to the proper orbitals by following three rules:

1. Always fill the lowest-energy orbitals first (called the **aufbau principle**).
2. Only two electrons can be put into each orbital, and they must be of opposite spin[6] (called the **Pauli exclusion principle**).
3. If two or more empty orbitals of equal energy are available, put one electron in each until all are half-full (called **Hund's rule**).

Let's look at some examples to see how these rules are applied. Hydrogen, the lightest element, has only one electron, which we assign to the lowest-energy orbital. This gives hydrogen a 1*s* ground-state configuration.

[6]For the purposes of quantum mechanics, electrons can be considered to spin around an axis in much the same way that the earth spins. This spin can have two equal and opposite orientations, denoted as up, ↑, and down, ↓

Carbon has six electrons, and a ground-state configuration $1s^22s^22p_x2p_y$ is arrived at by applying the three rules.[7] These and other examples are shown in Table 1.2.

**Table 1.2** Ground-state electronic configurations of some elements

| *Element* | *Atomic number* | *Configuration* | *Element* | *Atomic number* | *Configuration* |
|---|---|---|---|---|---|
| Hydrogen | 1 | 1s ↑ | Lithium | 3 | 2s ↑<br>1s ↑↓ |
| Carbon | 6 | 2p ↑ ↑ —<br>2s ↑↓<br>1s ↑↓ | Neon | 10 | 2p ↑↓ ↑↓ ↑↓<br>2s ↑↓<br>1s ↑↓ |
| Sodium | 11 | 3s ↑<br>2p ↑↓ ↑↓ ↑↓<br>2s ↑↓<br>1s ↑↓ | Argon | 18 | 3p ↑↓ ↑↓ ↑↓<br>3s ↑↓<br>2p ↑↓ ↑↓ ↑↓<br>2s ↑↓<br>1s ↑↓ |

PROBLEM

**1.1** Give the ground-state electronic configuration for these elements:
(a) Boron (b) Phosphorus (c) Oxygen (d) Chlorine

PROBLEM

**1.2** How many electrons does each of these elements have in its outermost electron shell?
(a) Potassium (b) Aluminum (c) Krypton

## 1.4 Development of Chemical Bonding Theory

By the mid-nineteenth century, with the vitalistic theory of organic chemistry dead and with the distinction between organic and inorganic chemistry nearly gone, chemists began to probe the forces holding molecules together. In 1858, August Kekulé[8] and Archibald Couper[9] independently proposed that, in all organic compounds, carbon always has four "affinity units." That is, carbon is *tetravalent*; it always forms four bonds when it joins other elements to form compounds. Furthermore, said Kekulé, carbon atoms can bond to each other to form extended chains of atoms linked together.

[7] A superscript is used here to represent the number of electrons at a particular energy level. For example, $1s^2$ indicates that there are two electrons in the 1s orbital. No superscript is used when there is only one electron in an orbital.

[8] Friedrich August Kekulé (1829–1896); b. Darmstadt; University of Giessen (1847); studied under Liebig, Dumas, Gerhardt, and Williamson; assistant to Stenhouse, London; professor, Heidelberg (1855), Ghent (1858), and Bonn (1867).

[9] Archibald Scott Couper (1831–1892); b. Kirkintilloch, Scotland; studied at the universities of Glasgow and Edinburgh (1852) and with Würtz in Paris; assistant in Edinburgh (1858).

Shortly after the tetravalent nature of carbon was proposed, extensions to the Kekulé–Couper theory were made when the possibility of *multiple* bonding between atoms was suggested. Emil Erlenmeyer[10] proposed a carbon-to-carbon triple bond for acetylene, and Alexander Crum Brown[11] proposed a carbon-to-carbon double bond for ethylene. In 1865, Kekulé provided another major advance in bonding theory when he postulated that carbon chains can double back on themselves to form rings of atoms.

Perhaps the most significant early advance in understanding bonding in organic molecules was the contribution made independently by Jacobus van't Hoff[12] and Joseph Le Bel.[13] Although Kekulé had satisfactorily described the tetravalent nature of carbon, chemistry was viewed in an essentially two-dimensional way until 1874. In that year, van't Hoff and Le Bel added a third dimension to our conception of molecules by proposing that the four bonds of carbon have specific spatial direction. Van't Hoff went even further and correctly proposed that the four atoms to which carbon is bonded sit at the corners of a tetrahedron, with carbon in the center. A representation of a tetrahedral carbon atom is shown in Figure 1.5. Note carefully the conventions used in Figure 1.5 to show three-dimensionality: Heavy wedged lines represent bonds coming out of the page toward the viewer; normal lines represent bonds in the plane of the page; and dashed lines represent bonds receding back behind the page, away from the viewer.

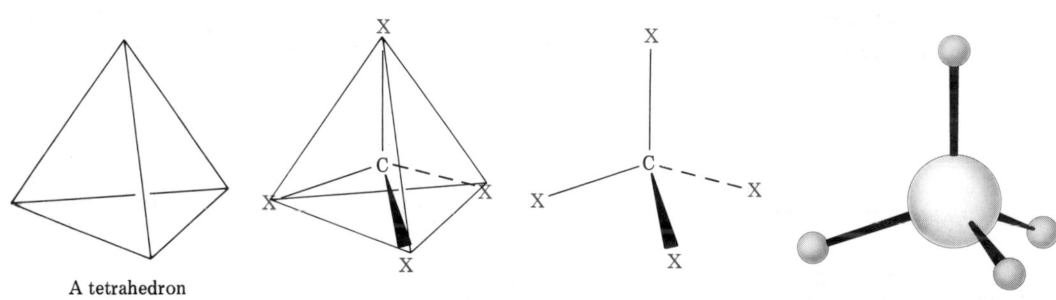


**Figure 1.5** Van't Hoff's tetrahedral carbon atom: The heavy wedged line comes out of the plane of the paper; the normal lines are in the plane; and the broken line goes back behind the plane of the page.

PROBLEM

**1.3** Draw a molecule of chloroform, $CHCl_3$, using wedged, normal, and dashed lines to show its tetrahedral geometry.

[10]Richard A. C. E. Erlenmeyer (1825–1909); b. Wehen, Germany; studied in Giessen and in Heidelberg; professor, Munich Polytechnicum (1868–1883).

[11]Alexander Crum Brown (1838–1922); b. Edinburgh; studied at Edinburgh, Heidelberg, and Marburg; professor, Edinburgh (1869–1908).

[12]Jacobus Henricus van't Hoff (1852–1911); b. Rotterdam; studied at Polytechnic at Delft, Leyden, Bonn, and Paris, and received doctorate at Utrecht (1874); professor, Utrecht, Amsterdam (1878–1896), Berlin; Nobel prize (1901).

[13]Joseph Achille Le Bel (1847–1930); b. Péchelbronn, Alsace; studied in the École Polytechnique, and at the Sorbonne; industrial consultant.

## 1.5 Modern Picture of Chemical Bonding: Ionic Bonds

What is the modern picture of chemical bonding? Why do atoms bond together, and how does the quantum-mechanical view of the atom describe bonding? The *why* question is relatively easy to answer: Atoms form bonds because the compound that results is more stable (has less energy) than the alternative arrangement of isolated atoms. Energy is always *released* when a chemical bond is formed. The *how* question is more difficult. To answer it, we need to know more about the properties of atoms.

We know through empirical observation that eight electrons (an *octet*) in the outermost electron shell impart special stability to the inert-gas elements in Group 0: Ne (2 + *8*); Ar (2 + 8 + *8*); Kr (2 + 8 + 18 + *8*). We also know that the chemistry of many elements with *nearly* inert-gas configurations is dominated by attempts to achieve the stable inert-gas electronic makeup. The alkali metals in Group I, for example, have single *s* electrons in their valence shells. By losing this electron, they can achieve an inert-gas configuration.

The amount of energy it takes to pull an electron away from an atom is called the **ionization energy (IE)** of the element. Alkali metals, at the far left of the periodic table, give up an electron easily, have low ionization energies, and are thus said to be **electropositive**. Elements at the middle and far right of the periodic table hold their electrons more tightly, give them up less readily, and therefore have higher IE's. In other words, a low IE corresponds to the ready loss of an electron, and a high IE corresponds to the difficult loss of an electron. Table 1.3 lists some ionization energies.

$$\text{Atom} + \text{Energy (IE)} \longrightarrow \text{Atom}^+ + \text{(Electron)}$$

Just as the electropositive alkali metals at the left of the periodic table have a tendency to form *positive* ions by *losing* an electron, the halogens

**Table 1.3** Ionization energies of some elements[a]

| *Element (electronic configuration)* | | *Cation (electronic configuration)* | *Ionization energy (kcal/mol)* |
|---|---|---|---|
| Li ($1s^22s$) | $\xrightarrow{-e^-}$ | $Li^+$ ($1s^2$—same as He) | 125 |
| Na ($1s^22s^22p^63s$) | $\xrightarrow{-e^-}$ | $Na^+$ ($1s^22s^22p^6$—same as Ne) | 118 |
| K ($\ldots 3s^23p^64s$) | $\xrightarrow{-e^-}$ | $K^+$ ($\ldots 3s^23p^6$—same as Ar) | 100 |
| C ($1s^22s^22p^2$) | $\xrightarrow{-e^-}$ | $C^+$ ($1s^22s^22p$) | 259 |
| $C^+$ ($1s^22s^22p$) | $\xrightarrow{-e^-}$ | $C^{2+}$ ($1s^22s^2$) | 562 |
| F ($1s^22s^22p^5$) | $\xrightarrow{-e^-}$ | $F^+$ ($1s^22s^22p^4$) | 401 |
| Ne ($1s^22s^22p^6$) | $\xrightarrow{-e^-}$ | $Ne^+$ ($1s^22s^22p^5$) | 497 |

[a]Organic chemists have been slow to adopt SI (Système International) units, preferring to use kilocalories (kcal) as a measure of energy, rather than kilojoules (kJ, pronounced kilojools). This book will use dual units, with values shown both in kcal/mol and kJ/mol, where 1 kilocalorie = 4.184 kilojoule, or 1 kcal/mol = 4.184 kJ/mol.

(Group VIIA elements) at the right of the periodic table have a tendency to form *negative* ions by *gaining* an electron. By so doing, the halogens can achieve an inert-gas configuration. The measure of this tendency to gain an electron is called the **electron affinity (EA)**. Energy is released when an electron is added to most elements, and EA's are therefore negative numbers.

Elements on the right of the periodic table have a much greater tendency to add an electron than elements on the left side and are said to be **electronegative**. Thus, the halogens release a large amount of energy when they react with an electron and have much larger negative electron affinities than the alkali metals. Table 1.4 lists the electron affinities of some common elements.

$$\text{Atom} + (\text{Electron})^- \longrightarrow \text{Atom}^- + \text{Energy (EA)}$$

**Table 1.4** Electron affinities of some elements

| *Element (electronic configuration)* | | *Anion (electronic configuration)* | *Electron affinity (kcal/mol)* |
|---|---|---|---|
| Li $(1s^22s)$ | $\xrightarrow{+e^-}$ | $Li^-$ $(1s^22s^2)$ | −13.6 |
| Na $(1s^22s^22p^63s)$ | $\xrightarrow{+e^-}$ | $Na^-$ $(1s^22s^22p^63s^2)$ | −5.0 |
| C $(1s^22s^22p^2)$ | $\xrightarrow{+e^-}$ | $C^-$ $(1s^22s^22p^3)$ | −28.9 |
| F $(1s^22s^22p^5)$ | $\xrightarrow{+e^-}$ | $F^-$ ($1s^22s^22p^6$—same as Ne) | −79.6 |
| Cl $(\ldots 3s^23p^5)$ | $\xrightarrow{+e^-}$ | $Cl^-$ ($\ldots 3s^23p^6$—same as Ar) | −83.2 |

The simplest kind of chemical bonding is that between an electropositive element (low IE) and an electronegative element (large negative EA). For example, when sodium metal [IE = 118 kcal/mol (494 kJ/mol)] reacts with chlorine gas [EA = −83.2 kcal/mol (−348 kJ/mol)], sodium donates an electron to chlorine forming positively charged sodium ions and negatively charged chloride ions. The product, sodium chloride, is said to have **ionic bonding**. That is, the ions are held together purely by electrostatic attraction between the two unlike charges. A similar situation exists for many other metal salts such as potassium fluoride ($K^+F^-$), lithium bromide ($Li^+Br^-$), and so on. This picture of the ionic bond, first proposed by Walter Kössel[14] in 1916, satisfactorily accounts for the chemistry of many inorganic compounds.

## 1.6 Modern Picture of Chemical Bonding: Covalent Bonds

Elements on the left and right sides of the periodic table form ionic bonds by gaining or losing an electron to achieve an inert-gas configuration. How,

[14]Walter Ludwig Julius Paschen Heinrich Kössel (1888–1956); b. Berlin; assistant in physics in Heidelberg (1910) and Munich (1913); professor of physics, Kiel (1921), Danzig (Poland) (1932–1945), and Tübingen (1947).

though, do elements in the middle of the periodic table form bonds? Let's look at the carbon atom in methane, $CH_4$, as an example. Certainly the bonding in methane isn't ionic, since it would be very difficult for carbon ($1s^22s^22p^2$) either to gain or to lose *four* electrons to achieve an inert-gas configuration.[15] In fact, carbon bonds to other atoms, not by donating electrons, but by *sharing* them. Such shared-electron bonds, first proposed in 1916 by G. N. Lewis,[16] are called **covalent bonds**. The covalent bond is the most important bond in organic chemistry.

A simple shorthand way of indicating covalent bonds in molecules is to use what are known as **Lewis structures**, or **electron-dot structures**. In this method, the outer-shell electrons of an atom are represented by dots. Thus, hydrogen has one dot representing its 1*s* electron, carbon has four dots ($2s^22p^2$), oxygen has six dots ($2s^22p^4$), and so on. A stable molecule results whenever the inert-gas configuration is achieved for all atoms, as in the following examples:

$$\cdot\dot{\underset{\cdot}{C}}\cdot + 4\,H\cdot \longrightarrow \begin{array}{c} H \\ H:\ddot{\underset{\cdot\cdot}{C}}:H \\ H \end{array}$$

Methane ($CH_4$)

$$2\,H\cdot + \cdot\ddot{\underset{\cdot}{O}}: \longrightarrow \begin{array}{c} H:\ddot{\underset{\cdot\cdot}{O}}: \\ H \end{array}$$

Water ($H_2O$)

$$2\,H\cdot + \cdot\ddot{\underset{\cdot\cdot}{O}}\cdot + H^+ \longrightarrow \begin{array}{c} + \\ H:\ddot{\underset{\cdot\cdot}{O}}:H \\ H \end{array}$$

Hydronium ion ($H_3O^+$)

$$3\,H\cdot + \cdot\dot{\underset{\cdot\cdot}{N}}\cdot \longrightarrow \begin{array}{c} H \\ H:\ddot{\underset{\cdot\cdot}{N}}:H \end{array}$$

Ammonia ($NH_3$)

$$3H\cdot + \cdot\dot{\underset{\cdot}{C}}\cdot + \cdot\ddot{\underset{\cdot}{O}}: + H\cdot \longrightarrow \begin{array}{c} H \\ H:\ddot{\underset{\cdot\cdot}{C}}:\ddot{\underset{\cdot\cdot}{O}}: \\ H\ \ H \end{array}$$

Methanol ($CH_3OH$)

Lewis structures are valuable because they make electron "bookkeeping" possible and constantly remind us of the number of outer-shell electrons (**valence electrons**) involved. Simpler still is the use of **"Kekulé" structures**, also called **line-bond structures**, in which a two-electron covalent

[15]The electronic configuration of carbon can be written either as $1s^22s^22p^2$ or as $1s^22s^22p_x2p_y$. Both notations are correct, but the latter is more informative since it indicates which of the three equivalent *p* orbitals are involved.

[16]Gilbert Newton Lewis (1875–1946); b. Weymouth, Mass.; Ph.D. Harvard (1899); professor, Massachusetts Institute of Technology (1905–1912), University of California, Berkeley (1912–1946).

bond is indicated simply by a line drawn between atoms. Pairs of nonbonding valence electrons are often ignored when drawing line-bond structures, but you must still be mentally aware of their existence. It's useful when starting out always to include them. Some of the molecules already considered are shown in Table 1.5.

**Table 1.5** Lewis and Kekulé structures of some simple molecules

| *Name* | *Lewis structure* | *Kekulé structure* |
|---|---|---|
| Water ($H_2O$) | H:Ö:<br>H | H—O<br>\|<br>H |
| Ammonia ($NH_3$) | H<br>H:N:H | H<br>\|<br>H—N—H |
| Methane ($CH_4$) | H<br>H:C:H<br>H | H<br>\|<br>H—C—H<br>\|<br>H |
| Methanol ($CH_3OH$) | H<br>H:C:Ö:<br>H H | H<br>\|<br>H—C—O<br>\| \|<br>H H |

PROBLEM

**1.4** Write Lewis structures for these molecules:

(a) $CHCl_3$, chloroform
(b) $H_2S$, hydrogen sulfide
(c) $CH_3NH_2$, methylamine
(d) $BH_3$, borane
(e) NaH, sodium hydride
(f) $CH_3Li$, methyllithium

PROBLEM

**1.5** Which of these molecules would you expect to have covalent bonds and which ionic bonds? Explain.

(a) $CH_4$
(b) $CH_2Cl_2$
(c) LiI
(d) KBr
(e) $MgCl_2$
(f) $Cl_2$

## 1.7 Molecular Orbital Theory

How can we describe the covalent bond in electronic terms? The most generally satisfactory method for dealing with organic compounds is **molecular orbital (MO) theory**. The major postulate of molecular orbital theory states that covalent bonds are formed by an *overlapping* of atomic orbitals. For example, we can describe the hydrogen molecule (H—H) by imagining what might happen if two hydrogen atoms, each with a 1*s* atomic orbital, meet and join together. As the two spherical atomic orbitals approach each other and combine, a new egg-shaped orbital results. This new orbital is called a **molecular orbital** because it belongs to the entire $H_2$ *molecule*, rather than to one of the individual atoms. The molecular orbital is filled by two electrons, one from each hydrogen:

H• + •H ⟶ H:H = H:H

1s atomic orbital

1s atomic orbital

$H_2$ molecular orbital

The new arrangement of electrons in the hydrogen molecule is considerably more stable than the original arrangement in individual atoms. During the reaction 2 H• → $H_2$, 104 kcal/mol (435 kJ/mol) of energy is *released*. Since the product $H_2$ molecule has 104 kcal/mol *less* energy than the starting 2 H•, we say that the product is more stable than the starting material, and that the new H—H bond has a **bond strength** of 104 kcal/mol. In other words, we would have to put 104 kcal/mol of energy (heat) *into* the H—H bond in order to break the hydrogen molecule apart into two hydrogen atoms. Figure 1.6 shows the relative energy levels of the different orbitals.

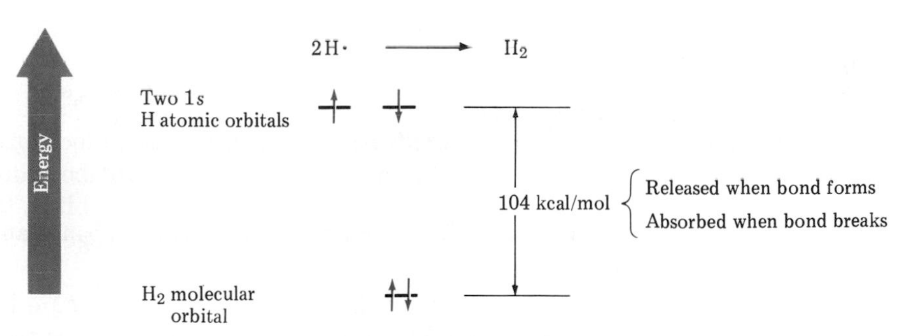


**Figure 1.6** Energy levels of $H_2$ orbitals

How close are the two nuclei in the hydrogen molecule? If the two positively charged nuclei are *too* close together, they will repel each other electrostatically; yet if the nuclei are too far apart, they won't be able to share the bonding electrons adequately. Thus, there is an optimum distance between the two nuclei that leads to maximum stability. This optimum distance, called the **bond length**, is 0.74 angstrom (Å) in the hydrogen molecule.[17] Every covalent bond has both a characteristic bond strength and bond length.

One further point that should be mentioned in this description of the hydrogen molecule is that an orbital seems to have disappeared. We began forming the hydrogen molecule by combining *two* atomic orbitals, each of which, if filled, could have held two electrons, for a total of four. We ended up, however, with what seems to be *one* molecular orbital, which can hold only two electrons. In fact, an orbital hasn't disappeared; we simply haven't paid it much attention. When we combine a *pair* of atomic orbitals, a *pair* of molecular orbitals is produced. One of the molecular orbitals is lower in energy than the starting atomic orbitals, and the other molecular orbital is correspondingly higher in energy, as shown in Figure 1.7.

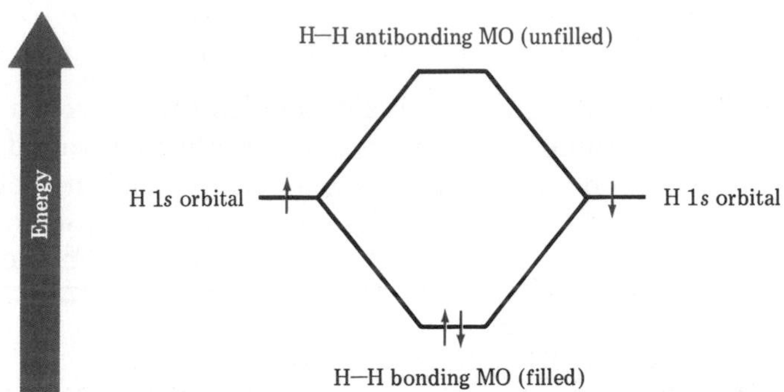


**Figure 1.7** Molecular orbitals of $H_2$: The combination of two hydrogen 1*s* atomic orbitals leads to the formation of two molecular orbitals. The lower-energy (bonding) molecular orbital is filled, and the higher-energy (antibonding) molecular orbital is unfilled.

The lower-energy orbital is called a **bonding MO**, whereas the higher-energy orbital is called an **antibonding MO**. The two electrons occupy the low-energy bonding orbital, and the high-energy antibonding orbital is unfilled. (If electrons were to occupy this high-energy orbital, the molecule would be higher in energy than the two isolated hydrogen atoms, and no bond could result—thus the term *antibonding*.)

The bonding molecular orbital in the hydrogen molecule has the elongated egg shape that we might get by pressing two spheres together. If an imaginary plane were to pass through the middle of the orbital, the intersection of the plane and the orbital would look like a circle. In other words, the H—H bond is *cylindrically symmetrical*, as shown in Figure 1.8.

Bonds that have circular cross-sections and are formed by head-on overlap of two atomic orbitals are called **sigma ($\sigma$) bonds**. Although sigma bonds are the most common kind, there are other types of bonds as well.

[17]The angstrom (Å; 1 Å = $10^{-10}$ m) is a convenient unit of measure still used by many chemists, even though it has been replaced in SI units by the picometer (pm, pronounced **pea-co**-meter) where 1 pm = $10^{-12}$ m and 1 Å = 100 pm. Because of the easy decimal conversion, only angstrom measurements will be given in this book.

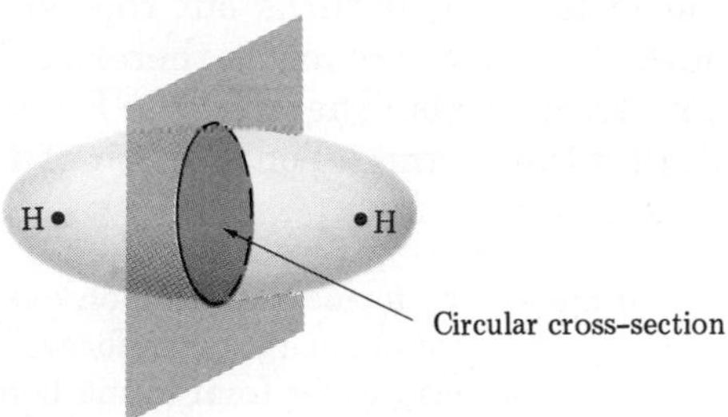


**Figure 1.8** The cylindrical symmetry of the H—H bond: The intersection of a plane cutting through the orbital looks like a circle.

Let's consider the fluorine molecule, $F_2$. A fluorine atom has seven outer-shell electrons and the electronic configuration $1s^22s^22p^5$. By bonding together, two fluorine atoms can each achieve stable outer-shell octets:

$$:\ddot{\underset{..}{F}}\cdot + \cdot\ddot{\underset{..}{F}}: \longrightarrow :\ddot{\underset{..}{F}}:\ddot{\underset{..}{F}}:$$

Unlike the situation in the hydrogen atom, a fluorine atom has an unshared $2p$ electron rather than a $1s$ electron. How can two $p$ orbitals come together to form a bond?

The general answer to this question was provided by Linus Pauling[18] in 1931 when he stated the **principle of maximum orbital overlap**. According to this principle, the strongest bond will be formed when the two orbitals achieve maximum overlap. There are two geometric possibilities for $p$ orbital overlap in the fluorine molecule: The $p$ orbitals can be oriented in a head-on fashion to form a sigma bond, or they can overlap in a sideways manner to form what is called a **pi ($\pi$) bond**, shown in Figure 1.9.

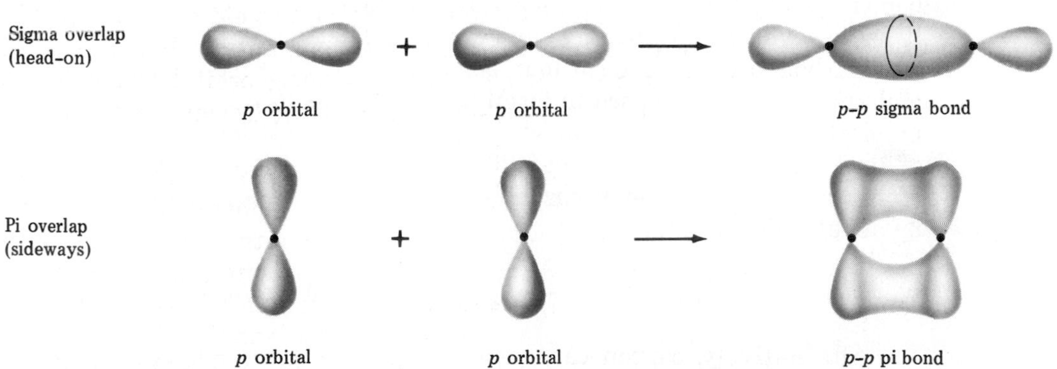


**Figure 1.9** The formation of sigma and pi bonds by overlap of $p$ orbitals

[18]Linus Pauling (1901– ); b. Portland, Oregon; Ph.D. California Institute of Technology (1925); professor, California Institute of Technology (1925–1967); University of California, San Diego; Professor Emeritus, Stanford University (1974– ); Nobel prize (1954, 1963).

Though it's difficult to predict which kind of bonding leads to maximum overlap and is favored, it turns out that sigma bonding is usually more efficient than pi bonding. Fluorine therefore forms a sigma molecular bond between two 2*p* orbitals. The new F—F bond has a bond strength of 38 kcal/mol (159 kJ/mol) and a bond length of 1.42 Å.

PROBLEM

**1.6** We said that sigma bonds, formed by head-on overlap of orbitals, have cylindrical symmetry. Draw a cross section of a pi bond, formed by sideways overlap of *p* orbitals. How does pi bond symmetry differ from sigma bond symmetry?

## 1.8 Hybridization: *sp*³ Orbitals and the Structure of Methane

The bonding in both the hydrogen molecule and the fluorine molecule is fairly straightforward. The situation becomes more complicated, however, when we turn to organic molecules with tetravalent carbon atoms. Let's start with the simplest case and consider methane, $CH_4$. Carbon has the ground-state electronic configuration $1s^22s^22p_x2p_y$. The outer shell has four electrons, two of which are paired in the 2*s* orbital, and two of which are unpaired in different 2*p* orbitals:

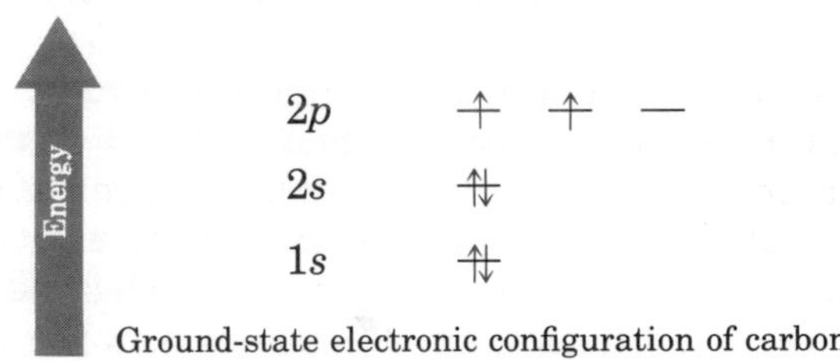

Ground-state electronic configuration of carbon

The first question we face is immediately apparent: How can carbon form *four* bonds if it has only *two* unpaired electrons? Why doesn't carbon bond to *two* hydrogen atoms to form $CH_2$? In fact, $CH_2$ is a known compound. It is, however, highly reactive and has only fleeting existence. We can see why carbon prefers to form four bonds instead of two by looking at the amount of energy released in forming $CH_2$ versus forming $CH_4$. By experimental measurement, we know that a typical C—H bond has a strength of approximately 100 kcal/mol (420 kJ/mol). Thus, the reaction of a carbon atom with two hydrogen atoms to form $CH_2$ should be energetically favored by about 200 kcal/mol:

$$\cdot\ddot{C}\cdot + 2\,H\cdot \longrightarrow H:\ddot{C}:H + \sim 200 \text{ kcal/mol}$$

Alternatively, carbon can adopt an electronic configuration *different* from the ground-state configuration. If one electron is promoted from the 2*s* orbital into the vacant $2p_z$ orbital, carbon can achieve the new configuration $1s^22s2p_x2p_y2p_z$. This new electronic arrangement is called an **excited-state configuration**; 96 kcal/mol (402 kJ/mol) of energy is required to accomplish the electron promotion from lower-energy ground state to higher-energy excited state.

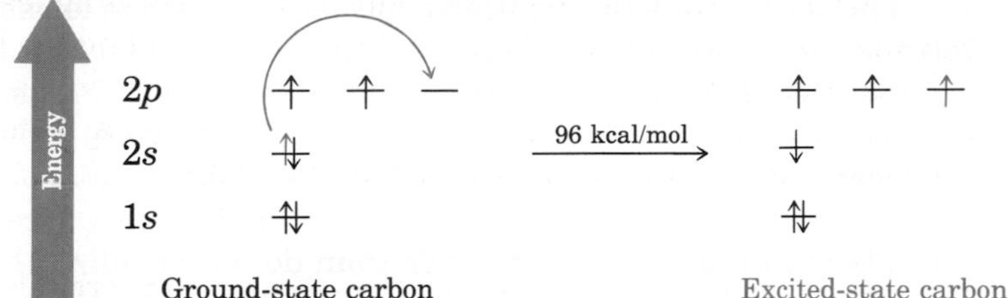


In the excited state, carbon has four unpaired electrons and can form four bonds to hydrogens. Although 96 kcal/mol is required to promote the 2*s* electron to a 2*p* orbital, this energy loss is more than offset by the formation of four stable C—H bonds, rather than two. Approximately 300 kcal/mol of energy is released in forming $CH_4$, versus the 200 kcal/mol of energy released in forming $CH_2$. In Lewis structures:

$$\cdot\ddot{C}\cdot \xrightarrow{96\ \text{kcal/mol}} \cdot\dot{\underset{\cdot}{C}}\cdot \xrightarrow{4\ H\cdot} H{:}\underset{H}{\overset{H}{\ddot{\underset{\cdot\cdot}{C}}}}{:}H + {\sim}400\ \text{kcal/mol}$$

$$\text{Net energy change} = (400 - 96)\ \text{kcal/mol} \approx 300\ \text{kcal/mol}$$

What are the four C—H bonds in methane like? Since excited-state carbon uses *two* kinds of orbitals for bonding purposes, we might expect methane to have *two* kinds of C—H bonds. In fact this is not the case; a large amount of evidence shows that all four C—H bonds in methane are identical. How can we explain this?

The answer was provided in 1931 by Linus Pauling, who showed that an *s* orbital and three *p* orbitals can mathematically mix or **hybridize** to form four equivalent new atomic orbitals that are spatially oriented toward the corners of a tetrahedron. These new tetrahedral orbitals, shown in Figure 1.10, are called ***sp*³ hybrids**,[19] since they are mathematically constructed from three *p* orbitals and one *s* orbital.

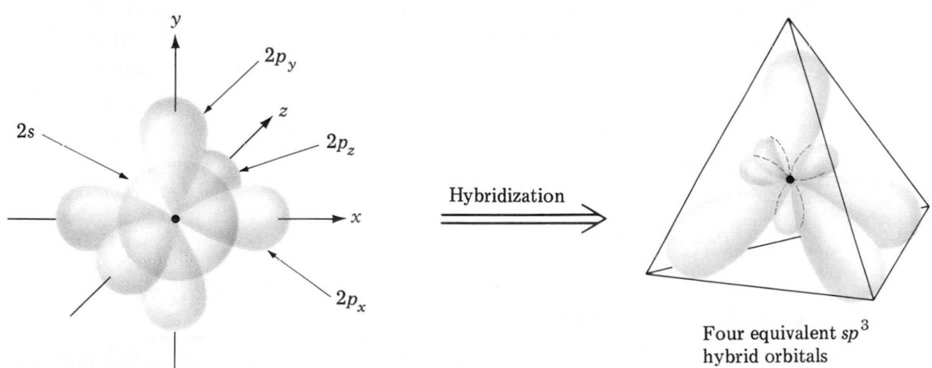


**Figure 1.10** The formation of four $sp^3$ hybrid orbitals by combination of an atomic *s* orbital and three atomic *p* orbitals

[19]Note that the superscript used to identify an $sp^3$ hybrid orbital tells how many of each type of atomic orbital combine to form the hybrid; it doesn't tell how many electrons occupy that orbital.

The concept of hybridization explains *how* carbon forms four equivalent tetrahedral bonds but doesn't answer the question of *why* it does so. Viewing a cross section of an $sp^3$ hybrid orbital suggests the answer. When an *s* orbital hybridizes with three *p* orbitals, the resultant hybrids are unsymmetrical about the nucleus. One of the two lobes of an $sp^3$ orbital is much larger than the other, as shown in Figure 1.11. As a result, $sp^3$ hybrid orbitals form much stronger bonds than do unhybridized *s* or *p* orbitals.

This lack of symmetry in $sp^3$ orbitals arises because of a property of orbitals that we've not yet considered. When the wave equation for a *p* orbital is solved, the two lobes have opposite algebraic signs, + and −. Thus, when a *p* orbital hybridizes with an *s* orbital, one lobe is *additive* with the *s* orbital, but the other lobe is *subtractive*. The resultant hybrid orbital (Figure 1.11) is therefore strongly oriented in one direction.

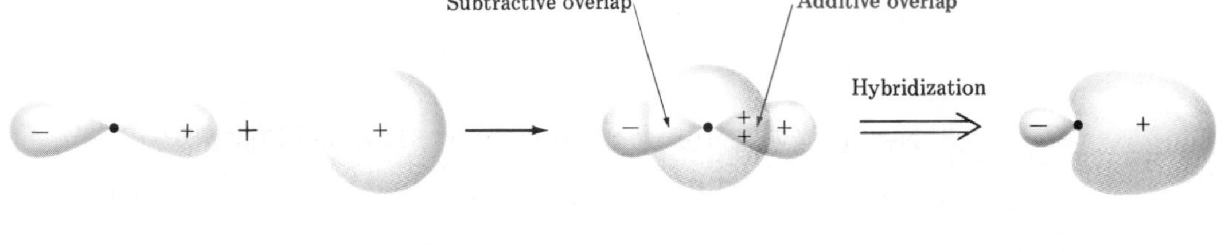


**Figure 1.11** The formation of an $sp^3$ hybrid orbital by overlap of a *p* orbital with part of an *s* orbital: Overlap of the *s* orbital with the positive *p* lobe is additive, but overlap with the negative *p* lobe cancels out. The resultant hybrid orbital is strongly oriented in one direction.

We describe the $sp^3$ hybrid as a *directed* orbital, and we find that it is capable of forming very strong bonds by overlapping the orbitals of other atoms. For example, the overlap of a carbon $sp^3$ hybrid orbital with a hydrogen 1*s* orbital gives a strong C—H bond (Figure 1.12).

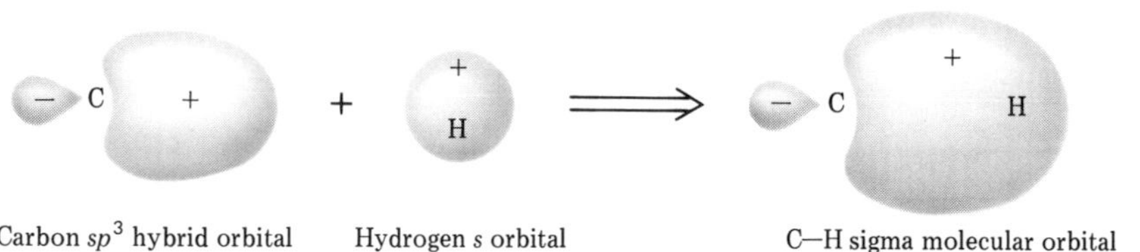


**Figure 1.12** The formation of a C—H bond by head-on (sigma) overlap of a carbon $sp^3$ hybrid orbital with a hydrogen 1*s* orbital

The C—H bond in methane has a measured bond strength of 104 kcal/mol (435 kJ/mol) and a bond length of 1.10 Å. Since the four bonds have a specific geometry, we can also define a third important physical property of pairs of bonds, called the **bond angle**. The angle formed by each H—C—H is exactly 109.5°, the tetrahedral angle. Methane therefore has the structure shown in Figure 1.13.

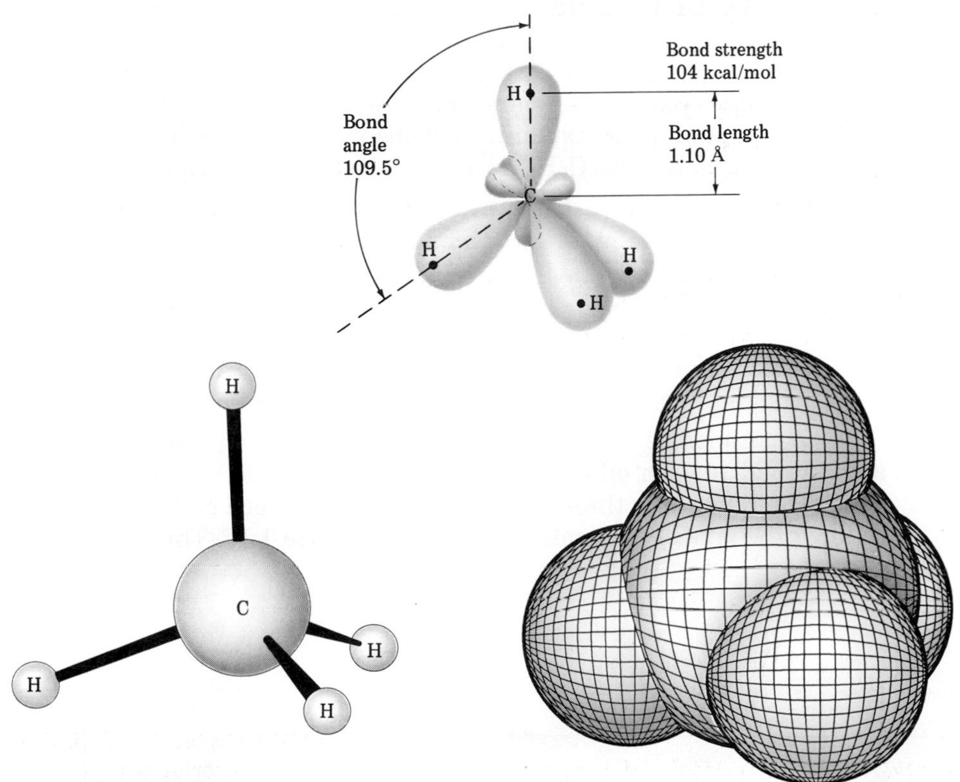


**Figure 1.13** The structure of methane: The two lower drawings are computer generated.

Before ending this discussion of the methane structure, it should again be pointed out that no orbitals are "lost" during hybridization and C—H bond formation. An $sp^3$-hybridized carbon has four hybrid orbitals and four electrons. Four hydrogen atoms also provide four 1*s* orbitals and four electrons. Methane therefore has four bonding C—H molecular orbitals, which are filled, and four antibonding C—H orbitals, which are unfilled. All eight MO's are shown in Figure 1.14.

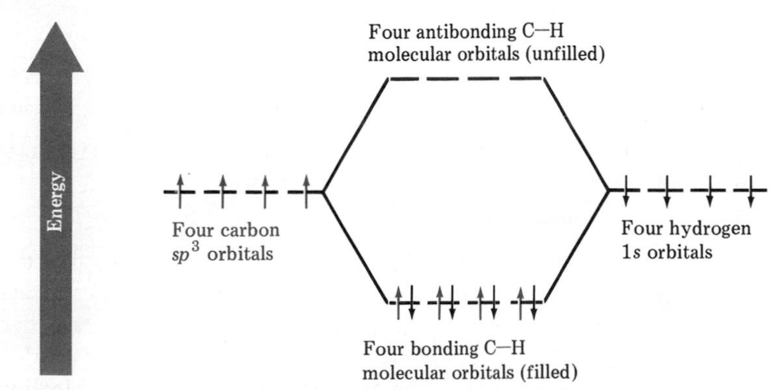


**Figure 1.14** Molecular orbitals of methane

## 1.9 The Structure of Ethane

A special characteristic of carbon is that it can form stable bonds to other carbon atoms. Exactly the same kind of hybridization that explains the methane structure also explains how one carbon atom can bond to another to form a chain. Ethane, $C_2H_6$, is the simplest molecule containing a carbon–carbon bond:

```
                      H   H
   H H                |   |
H:C:C:H            H—C—C—H            CH3CH3
   H H                |   |
                      H   H
```

Some representations of ethane

We can picture the ethane molecule by assuming that the two carbon atoms bond to each other by sigma overlap of an $sp^3$ hybrid orbital from each. The remaining three $sp^3$ hybrid orbitals on each carbon are then used to form the six C—H bonds, as shown in Figure 1.15. The C—H bonds in

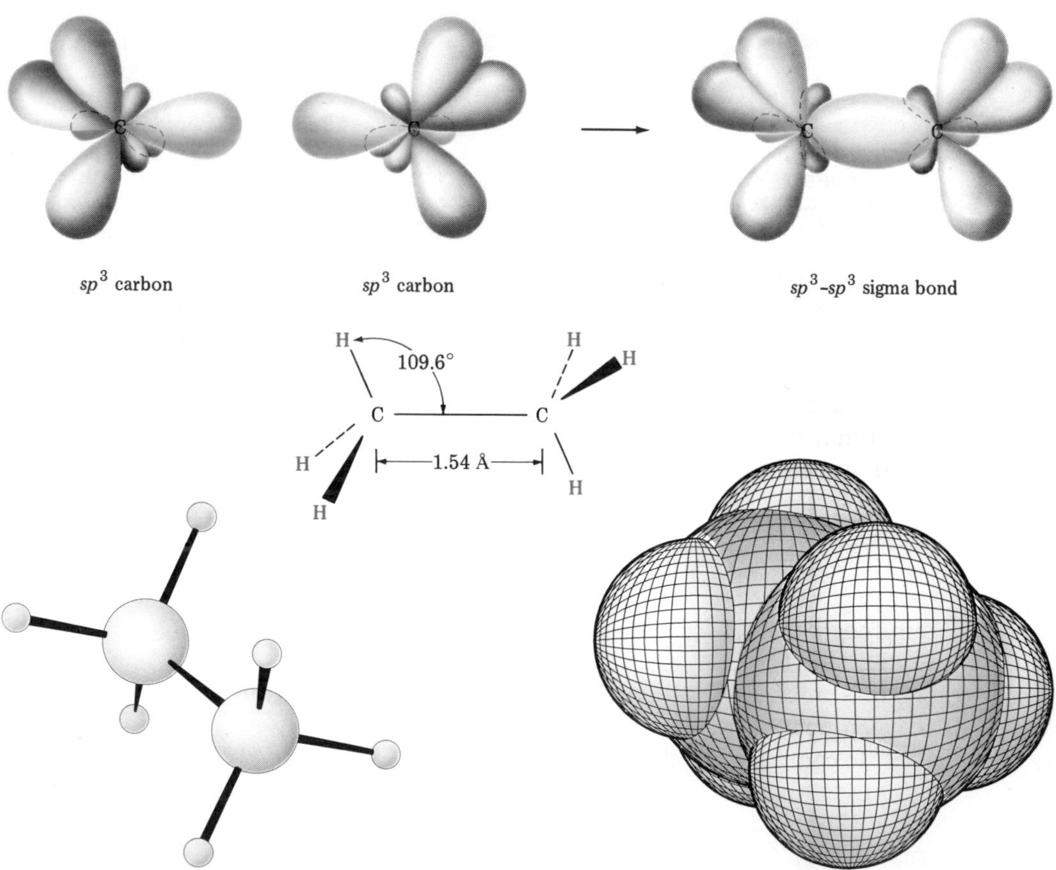


**Figure 1.15** The structure of ethane: The carbon–carbon bond is formed by sigma overlap of two carbon $sp^3$ hybrid orbitals.

ethane are similar to those in methane, though a bit weaker (98 kcal/mol for ethane versus 104 kcal/mol for methane). The C—C bond is 1.54 Å long and has a strength of 88 kcal/mol (368 kJ/mol). All the bond angles of ethane are very near the tetrahedral value, 109.5°.

## 1.10 Hybridization: $sp^2$ Orbitals and the Structure of Ethylene

Although $sp^3$ hybridization is the most common electronic state of carbon found in organic chemistry, it's not the only possibility. For example, let's look at ethylene, $C_2H_4$. It was recognized over 100 years ago that ethylene carbons can be tetravalent only if the two carbon atoms are linked by a *double* bond. How can we explain formation of the carbon–carbon double bond in molecular orbital terms?

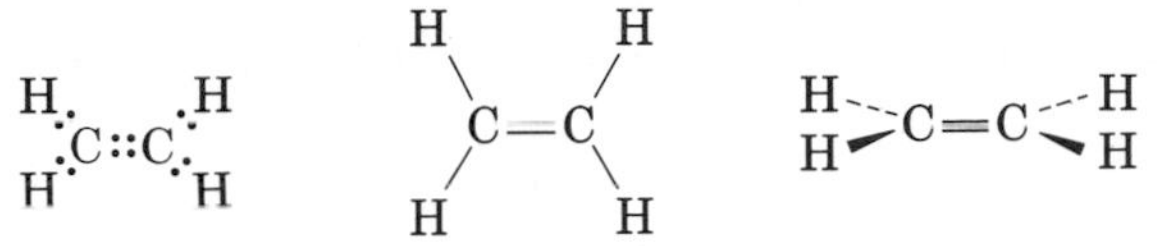

Top view

Side view

Ethylene

When we formed $sp^3$ hybrid orbitals to explain the bonding in methane, we first promoted an electron from the 2*s* orbital of ground-state carbon to form excited-state carbon with four unpaired electrons. We then mathematically mixed the four singly occupied atomic orbitals to construct four $sp^3$ hybrids. Imagine instead that we mathematically combine the 2*s* orbital with only *two* of the three available 2*p* orbitals. Three hybrid orbitals called **$sp^2$ hybrids** result, and one unhybridized 2*p* orbital remains unchanged. The three $sp^2$ orbitals lie in a plane at angles of 120° to each other, with the remaining *p* orbital perpendicular to the $sp^2$ plane, as shown in Figure 1.16 on page 22.

As with $sp^3$ hybrid orbitals, $sp^2$ hybrids are strongly oriented in a specific direction and can form strong bonds. If we allow two $sp^2$-hybridized carbons to approach each other, they can form a strong sigma bond by $sp^2$–$sp^2$ overlap. When this occurs, the unhybridized *p* orbitals on each carbon also approach each other with the correct geometry for sideways overlap to form a pi bond. The combination of $sp^2$–$sp^2$ sigma overlap and 2*p*–2p pi overlap results in the net sharing of four electrons and the formation of a carbon–carbon double bond (Figure 1.17, page 22).

To complete the structure of ethylene, we need only allow four hydrogen atoms to sigma bond to the remaining $sp^2$ orbitals. Thus, ethylene should be a planar (flat) molecule with H-C-H and H-C-C bond angles of approximately 120°, a geometry that has been verified by experimental observation. Ethylene is indeed flat, with H-C-H bond angles of 116.6° and H-C-C bond angles of 121.7°. Each C—H bond has a length of 1.076 Å and a strength of 103 kcal/mol (431 kJ/mol), as shown in Figure 1.18 on page 23.

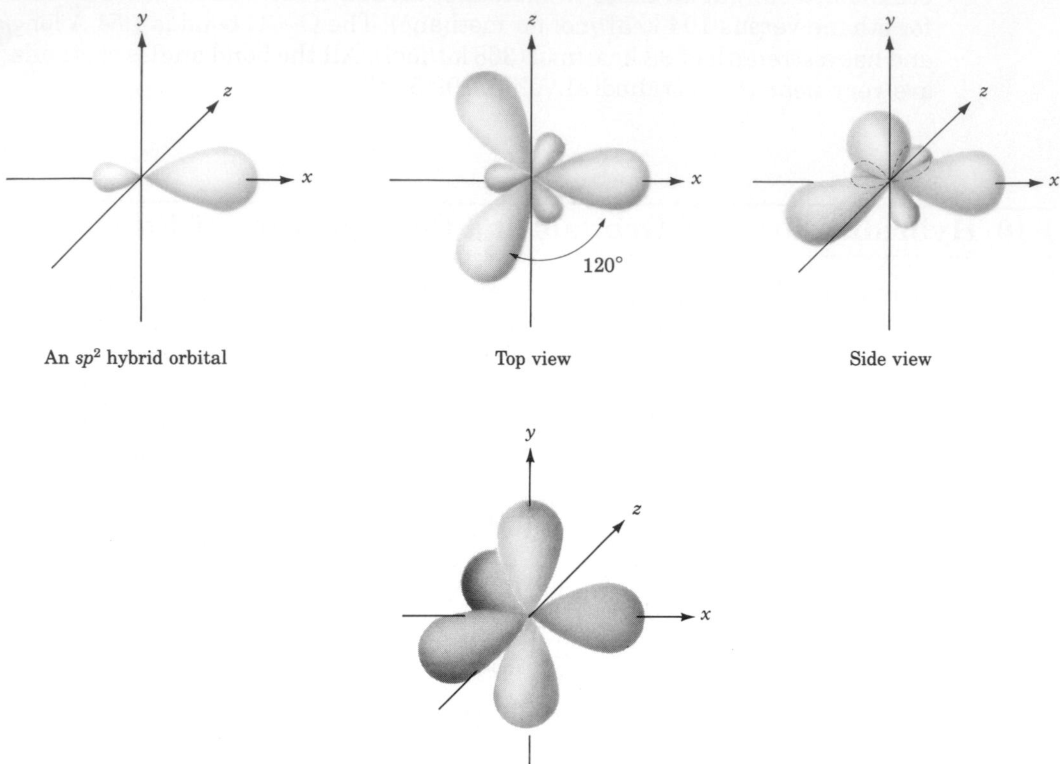


**Figure 1.16** An $sp^2$-hybridized carbon

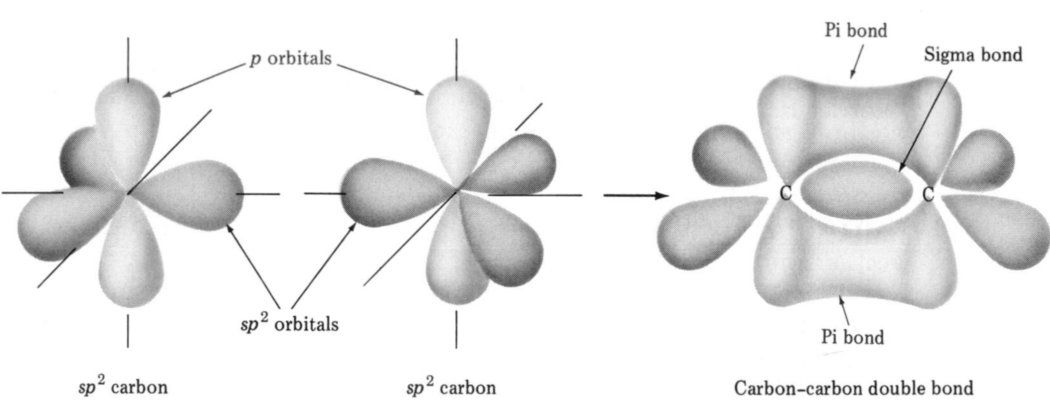


**Figure 1.17** Orbital overlap in a carbon–carbon double bond

We might also expect that the central carbon–carbon double bond in ethylene should be both shorter and stronger than the ethane single bond, an expectation that has also been verified. Ethylene has a C—C bond length of 1.33 Å and a bond strength of 152 kcal/mol (636 kJ/mol). Note, however, that the strength of the carbon–carbon double bond is not *exactly* twice as large as that of the corresponding C—C single bond in ethane (152 kcal/mol

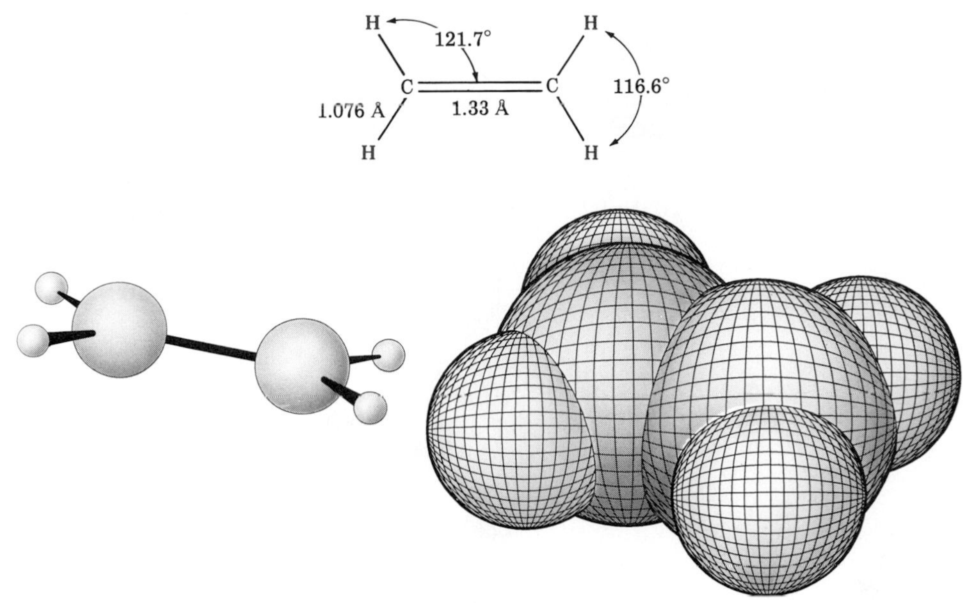


**Figure 1.18** The structure of ethylene

versus 88 kcal/mol, respectively), because overlap in the pi part of the double bond is not as effective as overlap in the sigma part.

PROBLEM

**1.7** Draw all of the bonds in propene, $CH_3CH{=}CH_2$. Indicate the hybridization of each carbon, and predict the value of each bond angle.

PROBLEM

**1.8** Carry out an analysis of 1,3-butadiene, $H_2C{=}CH{-}CH{=}CH_2$, similar to the one you carried out in Problem 1.7.

## 1.11 Hybridization: *sp* Orbitals and the Structure of Acetylene

In addition to being able to form single and double bonds, carbon can form a third kind of bond. Acetylene, $C_2H_2$, can be satisfactorily pictured only if we assume that it contains a carbon–carbon *triple* bond. We must construct yet another kind of hybrid orbital, an ***sp* hybrid**, to explain the bonding in acetylene:

$$H:C:::C:H \qquad H{-}C{\equiv}C{-}H$$

Acetylene

Imagine that, instead of combining with two or three *p* orbitals, a carbon 2*s* orbital hybridizes with only a single *p* orbital. Two *sp* hybrid orbitals result, and two *p* orbitals remain unchanged. The two *sp* orbitals are linear (180° apart on the *x*-axis), whereas the remaining two *p* orbitals are perpendicular on the *y*-axis and the *z*-axis, as shown in Figure 1.19 (page 24).

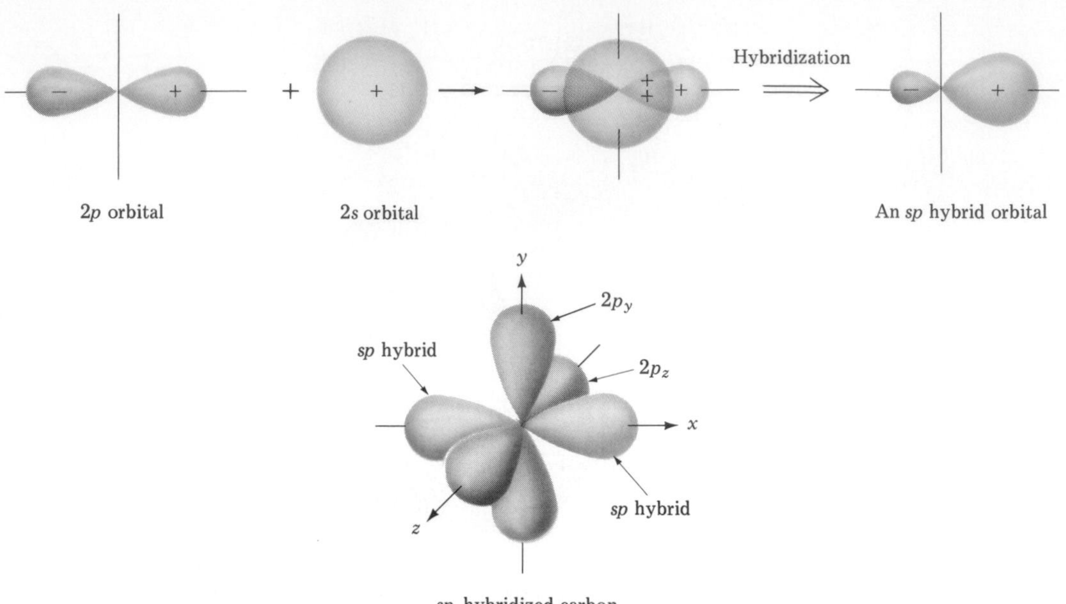


**Figure 1.19** An *sp*-hybridized carbon atom

If we allow two *sp*-hybridized carbon atoms to approach each other, *sp* orbitals from each carbon can overlap head-on to form a strong *sp–sp* sigma bond. In addition, the $p_z$ orbitals from each carbon can form a $p_z$–$p_z$ pi bond by sideways overlap, and the $p_y$ orbitals can overlap similarly to form a $p_y$–$p_y$ pi bond. The net effect is formation of one sigma bond and two pi bonds—that is, a carbon–carbon triple bond. The remaining *sp* hybrid orbitals can each form a sigma bond to a hydrogen 1*s* orbital to complete the acetylene molecule (Figure 1.20).

Because of *sp* hybridization, acetylene is a linear molecule with H-C-C bond angles of 180°. The carbon–hydrogen bond in acetylene has a length of 1.06 Å and a strength of 125 kcal/mol (523 kJ/mol). The carbon–carbon bond length is 1.20 Å, and its strength is 200 kcal/mol (837 kJ/mol). It's not surprising to find that the triple bond is so short and so strong; in fact, these values (1.20 Å and 200 kcal/mol) are the shortest and strongest known for any carbon–carbon bond. A comparison of *sp*, $sp^2$, and $sp^3$ hybridization is given in Table 1.6.

**Table 1.6** Comparison of carbon–carbon and carbon–hydrogen bonds

| *Molecule* | *Bond* | *Bond strength (kcal/mol)* | *Bond length (Å)* |
|---|---|---|---|
| Methane, $CH_4$ | $C_{sp^3}$—$H_{1s}$ | 104 | 1.10 |
| Ethane, $CH_3CH_3$ | $C_{sp^3}$—$C_{sp^3}$ | 88 | 1.54 |
| | $C_{sp^3}$—$H_{1s}$ | 98 | 1.10 |
| Ethylene, $H_2C{=}CH_2$ | $C_{sp^2}{=}C_{sp^2}$ | 152 | 1.33 |
| | $C_{sp^2}$—$H_{1s}$ | 103 | 1.076 |
| Acetylene, $HC{\equiv}CH$ | $C_{sp}{\equiv}C_{sp}$ | 200 | 1.20 |
| | $C_{sp}$—$H_{1s}$ | 125 | 1.06 |

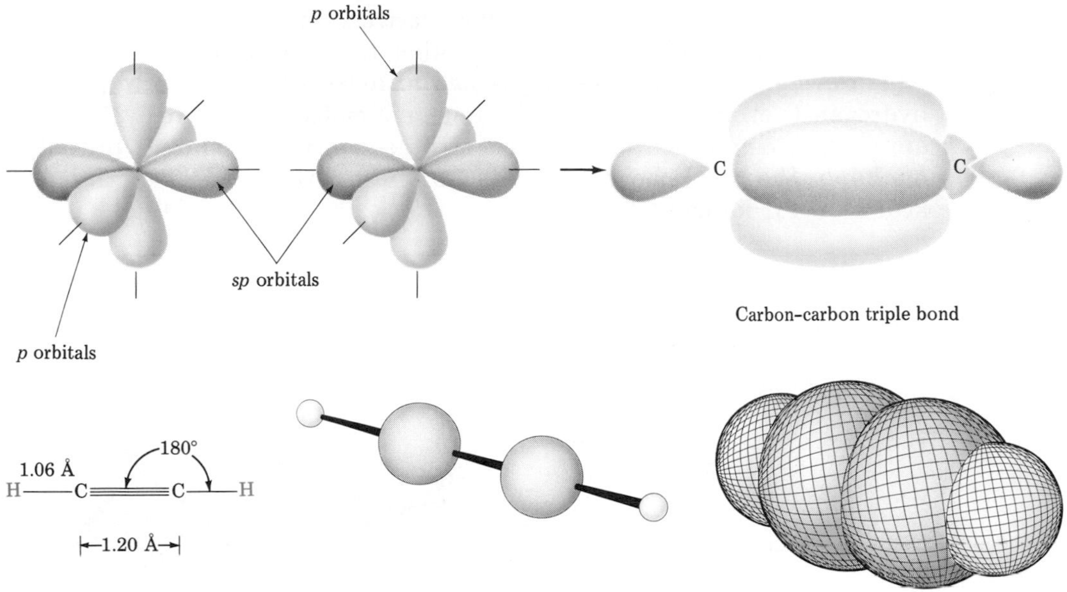


Figure 1.20 The carbon–carbon triple bond in acetylene

PROBLEM

**1.9** Draw all the bonds in propyne, $CH_3C{\equiv}CH$. Indicate the hybridization of each carbon, and predict a value for each bond angle.

## 1.12 Hybridization of Other Atoms: Nitrogen

The description of covalent bonding we've developed up to this point isn't restricted just to carbon compounds. All covalent bonds formed by other elements in the periodic table also can be described in terms of hybrid orbitals. The situation becomes more complex when elements heavier than carbon are involved, but the general principles remain the same.

Let's look at ammonia, $NH_3$, as an example of covalent bonding involving nitrogen. A nitrogen atom has the ground-state electronic configuration $1s^2 2s^2 2p_x 2p_y 2p_z$, and we might therefore expect nitrogen to combine with three hydrogen atoms:

$$\cdot\ddot{N}\cdot + 3\,H\cdot \longrightarrow H:\ddot{\underset{\cdot\cdot}{N}}:H \quad \text{or} \quad H{-}\ddot{N}{-}H$$

(with a third H below the N in each product structure)

What plausible geometry might ammonia have? Since the three unpaired electrons of nitrogen occupy half-filled $2p$ orbitals, one possibility is that hydrogen $1s$ orbitals might overlap the three nitrogen $2p$ orbitals to form three sigma bonds. If this occurred, ammonia would have H-N-H bond angles of 90°. [*Remember:* The $2p$ orbitals are at right angles to each other.] In fact, this picture is wrong. The experimentally measured H-N-H bond angle in ammonia is 107.1°, nearly the tetrahedral value (109.5°).

In fact, nitrogen hybridizes to form four $sp^3$ orbitals, *exactly as carbon does*. Since nitrogen has five outer-shell electrons, one of the four $sp^3$ orbitals is occupied by two electrons and the other three each have one electron. Sigma overlap of these three half-filled nitrogen $sp^3$ hybrid orbitals with hydrogen 1*s* orbitals completes the ammonia molecule (Figure 1.21). Thus, ammonia is a tetrahedral molecule with geometry very similar to that of methane. The N—H bond length is 1.01 Å, and the bond strength is 103 kcal/mol (431 kJ/mol).

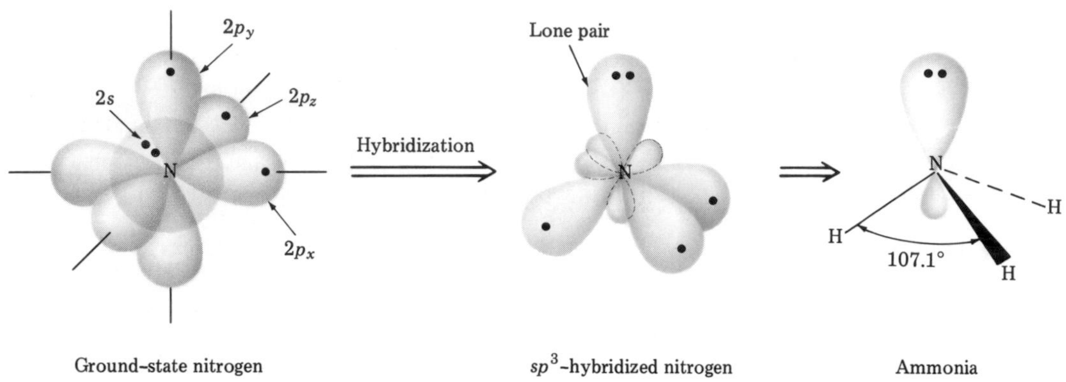


**Figure 1.21** Hybridization of nitrogen in ammonia: The nitrogen atom is $sp^3$ hybridized, just like the carbon atom in methane.

Ammonia is tetrahedral because of stability—such an arrangement of bonds is the lowest-energy form of all possible alternatives. The energy required to hybridize the nitrogen from the ground-state configuration to $sp^3$ configuration is more than offset by the added strength gained by bonding to $sp^3$ orbitals (strongly directed, good overlap) versus bonding to *p* orbitals (poorly directed, poor overlap).

Note that an unshared electron pair is present in nitrogen, occupying an $sp^3$ orbital. This **lone pair** of electrons occupies nearly as much space as an N—H bond and is very important in the chemistry that ammonia exhibits.

PROBLEM

**1.10** What kind of bonding do you think is present in the nitrogen molecule, $N_2$? (Both nitrogens are *sp* hybridized.)

## 1.13 Hybridization of Other Atoms: Oxygen and Boron

We saw in ammonia that nonbonding lone-pair electrons can occupy hybrid orbitals just as bonding electron pairs can. The same phenomenon is seen again in the structure of water, $H_2O$. Ground-state oxygen has the electronic configuration $1s^2 2s^2 2p_x^2 2p_y 2p_z$, and oxygen is therefore divalent; that is, it forms two bonds.

$$2\,\text{H}\cdot + \cdot\ddot{\underset{..}{\text{O}}}\cdot \longrightarrow \text{H}:\ddot{\underset{..}{\text{O}}}:\text{H}$$

We can imagine several hypothetical models for the bonding in water:

1. Perhaps oxygen uses two unhybridized *p* orbitals to overlap with hydrogen 1*s* orbitals. The two oxygen lone pairs would then remain in a 2*s* and a $2p_x$ orbital.
2. Perhaps oxygen undergoes *sp* hybridization and uses the two *sp* hybrid orbitals for bonding. The lone pairs would then both remain in the two unhybridized *p* orbitals.
3. Perhaps oxygen undergoes $sp^3$ hybridization and uses two $sp^3$ hybrid orbitals for bonding. The lone pairs would then occupy the remaining two $sp^3$ orbitals.

Only the third model, the hybridization of oxygen into $sp^3$ orbitals, allows strong bonds and maximum distance between the outer-shell electrons. The oxygen in water is therefore $sp^3$ hybridized, as illustrated in Figure 1.22.

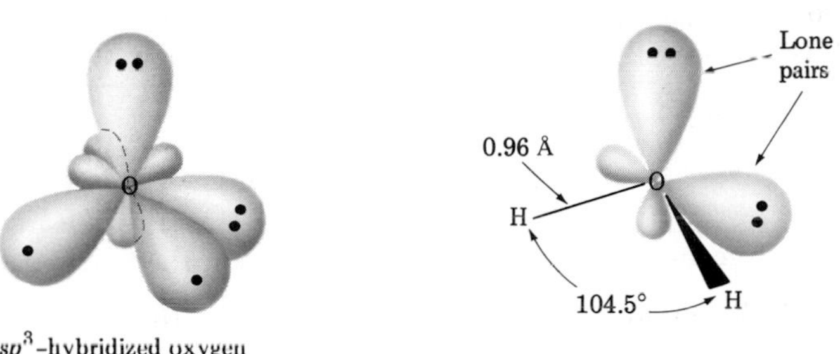


**Figure 1.22** The structure of water: The oxygen atom is $sp^3$ hybridized like the carbon atom in methane.

Measurements on water indicate that the oxygen doesn't have perfect $sp^3$ hybrid orbitals; the actual H-O-H bond angle of 104.5° is somewhat less than the predicted tetrahedral angle. We can explain this bond angle difference by assuming that there is a repulsive interaction between the two lone pairs that forces them apart and thus compresses the H-O-H angle.

One final example of orbital hybridization that we'll consider is found in molecules like boron trifluoride, $BF_3$. Since boron has only three outer-shell electrons ($1s^2 2s^2 2p_x$), it can form a maximum of three bonds. Even though we can promote a 2*s* electron into a $2p_y$ orbital and then hybridize in some manner, there is no way to complete a stable outer-shell electron octet for boron.

$$3\,:\ddot{\underset{..}{F}}\cdot + \cdot\dot{B} \longrightarrow \begin{matrix} :\ddot{F}: \\ :\ddot{F}:\ddot{B} \\ :\ddot{F}: \end{matrix} \quad \text{or} \quad \begin{matrix} & F & \\ & | & \\ & B & \\ F & & F \end{matrix}$$

Since boron has no lone-pair electrons to take into account, we might predict that it will hybridize in such a way that the three B—F bonds will be as far away from one another as possible. This prediction implies $sp^2$ hybridization and a planar structure for $BF_3$ in which each fluorine bonds to a boron $sp^2$ orbital, with *the remaining p orbital on boron left vacant.* Boron trifluoride has exactly this predicted structure (Figure 1.23).

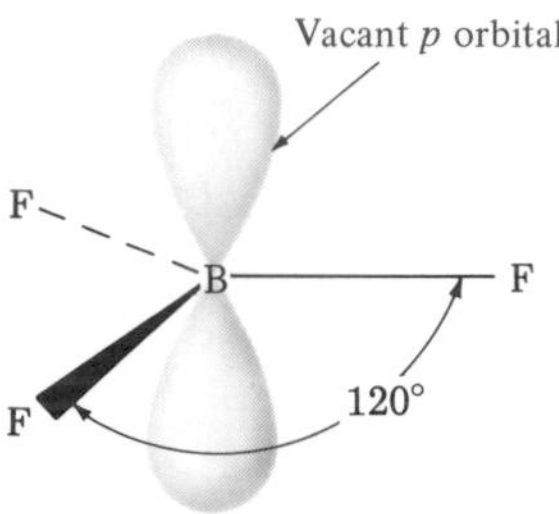


**Figure 1.23** The structure of boron trifluoride: The boron atom is $sp^2$ hybridized and has a vacant $p$ orbital perpendicular to the $BF_3$ plane.

PROBLEM

**1.11** What geometry would you expect for each of the following?

(a) The oxygen atom in methanol, $H_3C—\ddot{\underset{\cdot\cdot}{O}}—H$

(b) The nitrogen atom in trimethylamine, $H_3C—\ddot{N}(CH_3)—CH_3$

(c) The phosphorus atom in $PH_3$ [*Hint:* How many outer-shell electrons does phosphorus have? What element is it similar to?]

## 1.14 Summary and Key Words

**Organic chemistry** is the study of carbon compounds. For historical reasons, a division into organic and inorganic chemistry occurred, but there is no scientific reason behind the division.

The electronic structure of any atom can be described mathematically by the Schrödinger **wave equation**, in which electrons are considered to occupy **orbitals** centered around the nucleus. Different orbitals have different energy levels and different shapes. For example, $s$ orbitals are spherical and $p$ orbitals are dumbbell shaped. The **electronic configuration** of an atom can be found by assigning electrons to the proper orbitals, beginning with the lowest-energy ones.

There are two fundamental kinds of chemical bonds—**ionic bonds** and **covalent bonds**. Ionic bonds are based on the electrostatic attraction of unlike charges and are commonly found in inorganic salts. Covalent bonds are formed when an electron pair is shared between two atoms. This electron sharing occurs by **overlap** of two atomic orbitals to give a new **molecular orbital**. Bonds that have a circular cross-section and are formed by head-on overlap are called **sigma ($\sigma$) bonds**; bonds formed by sideways overlap of two *p* orbitals are called **pi ($\pi$) bonds**.

In order to form bonds in organic molecules, carbon first hybridizes to an **excited-state configuration**. When forming only single bonds, carbon is $sp^3$ hybridized; it has four equivalent $\boldsymbol{sp^3}$ **hybrid orbitals** with tetrahedral geometry. When forming double bonds, carbon is $sp^2$ hybridized; it has three equivalent $\boldsymbol{sp^2}$ **hybrid orbitals** with planar geometry and one unhybridized *p* orbital. The carbon–carbon double bond is formed when two $sp^2$-hybridized carbon atoms bond together. When forming triple bonds, carbon is *sp* hybridized; it has two equivalent ***sp*** **hybrid orbitals** with linear geometry and two unhybridized *p* orbitals. The carbon–carbon triple bond results when two *sp*-hybridized carbon atoms bond together.

Other atoms such as nitrogen, oxygen, and boron also hybridize in order to form stronger bonds. The nitrogen atom in ammonia and the oxygen atom in water are $sp^3$ hybridized; the boron atom in boron trifluoride is $sp^2$ hybridized.

## WORKING PROBLEMS

**There's no surer way to learn organic chemistry than by working problems** Learning organic chemistry requires familiarity with a large number of facts. Each page in this book presents new factual information that has to be digested and correlated with what has come before. Although careful reading and rereading of this text is important, reading alone isn't enough. In addition, you must be able to work with the information you've read and be able to use your knowledge in new situations. Working problems gives you the opportunity to do this.

Each chapter in this book provides many problems of different sorts. The in-chapter problems are placed for immediate reinforcement of new ideas just presented. The end-of-chapter problems provide additional practice and are of two types: drill and thought. Early problems are primarily of the drill type, providing an opportunity for you to practice your command of the fundamentals. Later problems tend to be more thought provoking, and many are real challenges to your depth of understanding.

As you study organic chemistry, take the time to work the problems. Do the ones you can, and ask for help on the ones you can't. If you're stumped by a particular exercise, check the accompanying *Study Guide and Solutions Manual* for an explanation that will help clarify the source of difficulty. Working problems takes effort, but the payoff in knowledge and understanding is immense.

## ADDITIONAL PROBLEMS

**1.12** How many outer-shell (valence) electrons does each of the following atoms have?
(a) Magnesium (b) Sulfur (c) Bromine

**1.13** Give the ground-state electronic configurations of the following elements. For example, carbon is $1s^2 2s^2 2p^2$.

(a) Sodium (b) Aluminum (c) Silicon (d) Calcium

**1.14** Write Lewis (electron-dot) structures for these molecules:

(a) $H{-}C{\equiv}C{-}H$ (b) $AlH_3$ (c) $CH_3{-}\ddot{\underset{..}{S}}{-}CH_3$

(d) $H_2C{=}CH\ddot{\underset{..}{Cl}}{:}$ (e) $H_2C{=}CH{-}CH{=}CH_2$ (f) $CH_3{-}\overset{\overset{\ddot{O}}{\|}}{C}{-}\ddot{\underset{..}{O}}{-}H$

**1.15** Write a Lewis (electron-dot) structure for acetonitrile, $H_3C{-}C{\equiv}N$. How many electrons does the nitrogen atom have in its outer shell? How many are used for bonding, and how many are not used for bonding?

**1.16** Fill in any unshared electrons that are missing from the following line-bond structures:

(a) $CH_3{-}O{-}CH_3$ (b) $CH_3{-}\overset{\overset{O}{\|}}{C}{-}CH_3$ (c) $CH_3{-}\overset{\overset{O}{\|}}{C}{-}NH_2$ (d) $CH_2ClF$

**1.17** Convert the following Kekulé (line-bond) structures into molecular formulas. For example,

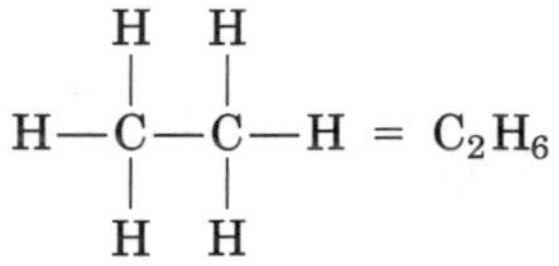

$$H{-}\underset{\underset{H}{|}}{\overset{\overset{H}{|}}{C}}{-}\underset{\underset{H}{|}}{\overset{\overset{H}{|}}{C}}{-}H = C_2H_6$$

(a) 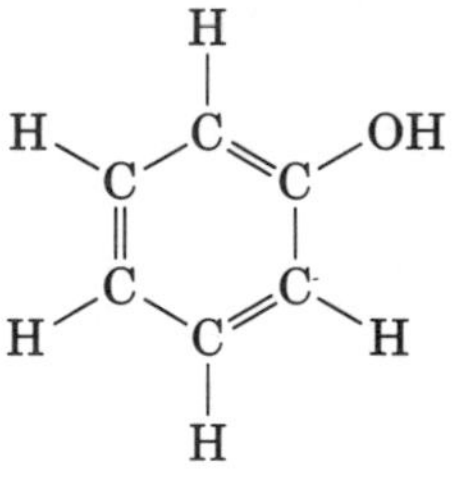 Phenol

(b) Aspirin

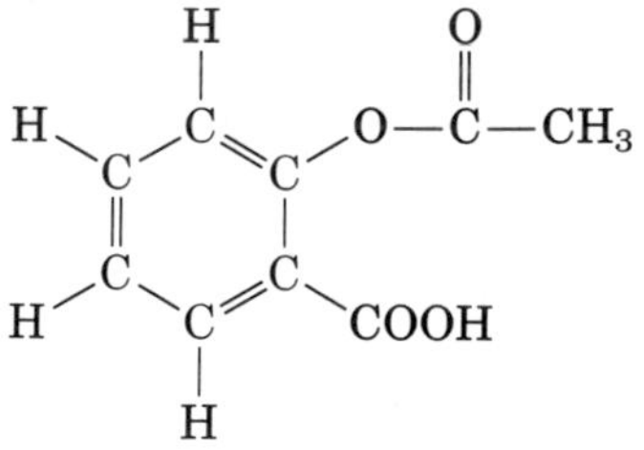

(c) Vitamin C

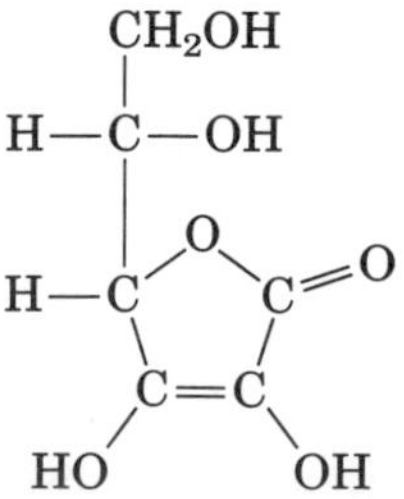

(d) Nicotine

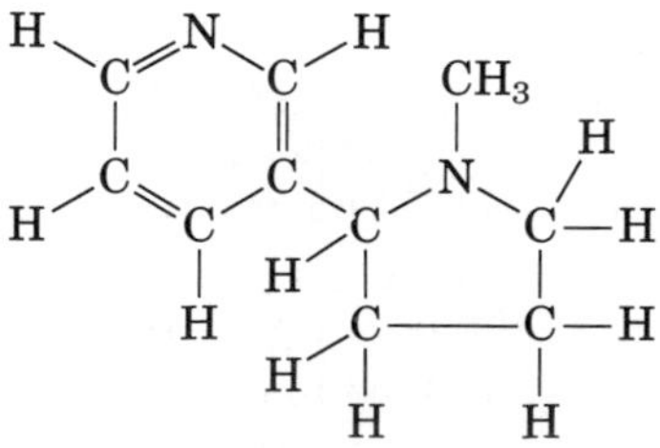

(e)

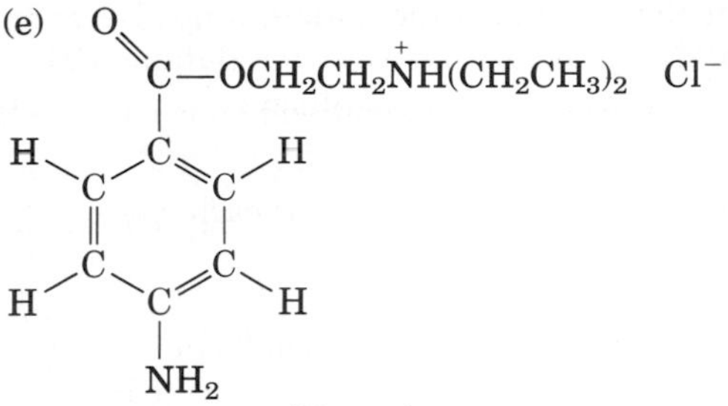

```
     O
      \\         +
       C—OCH2CH2NH(CH2CH3)2  Cl−
       |
 H    C    H
  \ //  \ /
   C     C
   ||    |
   C     C
  / \  // \
 H    C    H
      |
     NH2
```

Novocain

(f)

```
           CH2OH
            |
            C——O
     H     /|    \     H
      \   / H     \   /
       C          C
      / \ OH   H /  \
    HO   \|    |/    OH
          C————C
          |    |
          H    OH
```

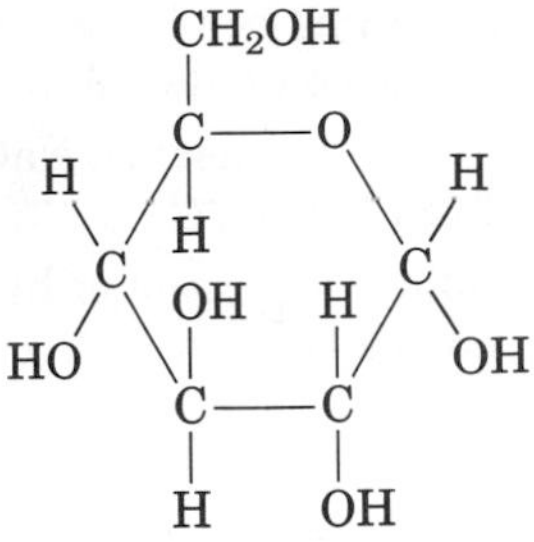

Glucose

**1.18** Convert the following molecular formulas into Kekulé structures that are consistent with valence rules:

(a) $C_3H_8$ (b) $CH_5N$

(c) $C_2H_6O$ (2 possibilities) (d) $C_3H_7Br$ (2 possibilities)

(e) $C_2H_4O$ (3 possibilities) (f) $C_3H_9N$ (4 possibilities)

**1.19** Indicate the kind of hybridization you might expect for each carbon atom in these molecules:

(a) Propane, $CH_3CH_2CH_3$ (b) 2-Methylpropene, $(CH_3)_2C{=}CH_2$

(c) 1-Buten-3-yne, $H_2C{=}CH{-}C{\equiv}CH$ (d) Cyclobutene,

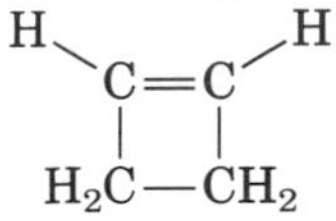

```
H       H
 \     /
  C = C
  |   |
 H2C—CH2
```

(e) Dimethyl ether, $CH_3OCH_3$

**1.20** What is the hybridization of each carbon atom in benzene? What overall shape would you expect benzene to have?

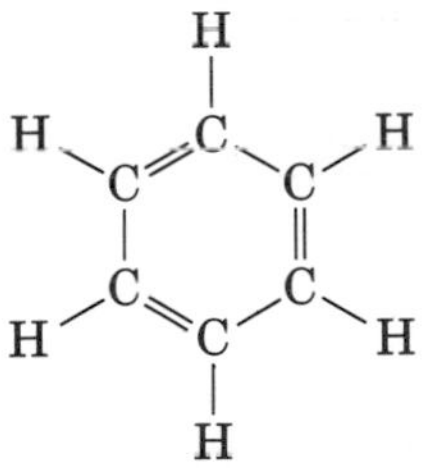

```
         H
         |
   H     C     H
    \  //  \  /
     C       C
     |       ||
     C       C
    /  \\  /  \
   H     C     H
         |
         H
```

**1.21** What kind of hybridization would you expect for the following?

(a) The oxygen in dimethyl ether, $CH_3{-}O{-}CH_3$

(b) The nitrogen in dimethylamine, $CH_3NHCH_3$

(c) The boron in trimethylborane, $(CH_3)_3B$

**1.22** On the basis of your answers to Problem 1.21, what bond angles would you expect for the following?

(a) The C-O-C angle in $CH_3{-}O{-}CH_3$ (b) The C-N-C angle in $CH_3NHCH_3$

(c) The C-N-H angle in $CH_3NHCH_3$ (d) The C-B-C angle in $(CH_3)_3B$

**1.23** What shape would you expect these species to have?

(a) The ammonium ion, $NH_4^+$ (b) Trimethylborane, $(CH_3)_3B$

(c) Trimethylphosphine, $(CH_3)_3P$ (d) Formaldehyde, $H_2C{=}O$

**1.24** Draw a three-dimensional representation of the oxygen-bearing carbon atom in ethanol, $CH_3CH_2OH$, using the standard convention of normal, heavy wedged, and dashed lines.

**1.25** Consider the molecules $SO_2$ and $SO_3$ and the ion $SO_4^{2-}$.

(a) Write Lewis structures for each. (b) Predict the shape of each.

**1.26** Draw line-bond structures for these covalent molecules:
(a) Acetonitrile, $CH_3CN$ (b) Ethanol, $CH_3CH_2OH$ (c) Butane, $CH_3CH_2CH_2CH_3$

**1.27** Sodium methoxide, $NaOCH_3$, contains both covalent and ionic bonds. Which do you think is which?

**1.28** Indicate the kind of hybridization you might expect for each carbon atom in these molecules:

(a) Acetic acid,

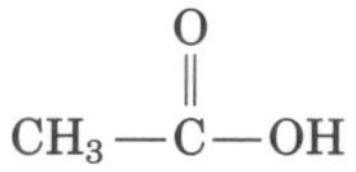

$$CH_3-\overset{\overset{\displaystyle O}{\|}}{C}-OH$$

(b) 3-Buten-2-one,

$$H_2C{=}CH-\overset{\overset{\displaystyle O}{\|}}{C}-CH_3$$

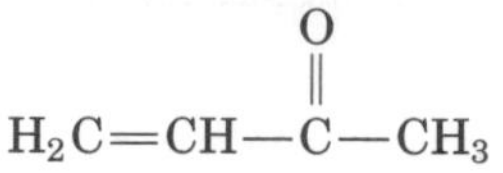

(c) Acrylonitrile,

$$H_2C{=}CH-C{\equiv}N$$

(d) Benzoic acid,

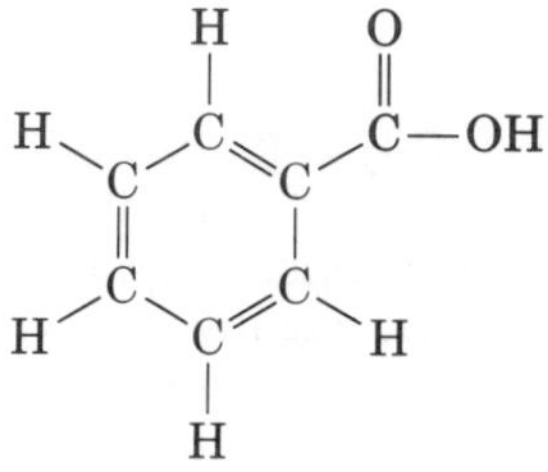

H, H, H, H, H ring carbons C; C—OH with C=O

**1.29** What kind of hybridization would you expect for the following?
(a) The nitrogen in aniline,

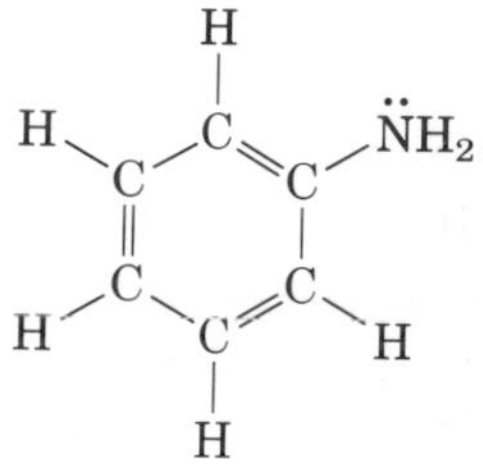

H, H, H, H, H ring carbons C; $\ddot{N}H_2$

(b) The nitrogen in pyridine,

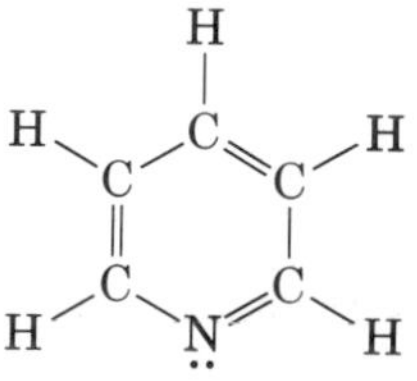

H, H, H, H, H ring carbons C; $\ddot{N}$

(c) The beryllium in dimethylberyllium,

$$CH_3-Be-CH_3$$

(d) The phosphorus in trimethylphosphine,

$$(CH_3)_3P:$$

**1.30** On the basis of your answers to Problem 1.29, what bond angles do you expect for the following?
(a) The C-N-H angle in aniline (b) The C-Be-C angle in $(CH_3)_2Be$
(c) The C-P-C angle in $(CH_3)_3P$

**1.31** Identify the bonds in these molecules as either ionic or covalent:
(a) NaCl (b) $CH_3Cl$ (c) $Cl_2$ (d) HOCl

**1.32** Allene is an unusual molecule that has the structure $H_2C{=}C{=}CH_2$. Draw a picture of the hybrid orbitals in allene. Is the central carbon atom $sp^2$ or $sp$ hybridized? What about the hybridization of the terminal carbons? What shape would you predict for allene?

**1.33** Allene (Problem 1.32) is related structurally to carbon dioxide, $CO_2$. Draw an orbital picture of $CO_2$ and identify the hybridization of carbon.

**1.34** Although almost all stable organic species have tetravalent carbon atoms, species with trivalent carbon atoms are known to exist. *Carbocations* are one such class of compounds.

(a) If a neutral carbon atom has 8 valence electrons associated with it (2 from each of 4 bonds), how many valence electrons does the positively charged carbon atom have?
(b) What hybridization might you expect this carbon atom to have?
(c) What geometry does the carbocation have?
(d) What relationship do you see between a carbocation and a trivalent boron compound such as $BF_3$?

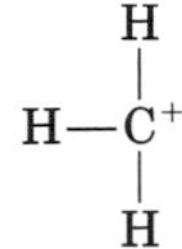

A carbocation

**1.35** A *carbanion* is a species that contains a negatively charged trivalent carbon atom.
(a) How many valence electrons does the negatively charged carbon atom have?
(b) What hybridization might you expect this carbon atom to have?
(c) What geometry does the carbanion have?
(d) What relationship do you see between a carbanion and a trivalent nitrogen compound such as $NH_3$?

H<br>|<br>H—C:$^-$<br>|<br>H

A carbanion

**1.36** Divalent species called *carbenes* are known to be capable of fleeting existence. For example, methylene, $:CH_2$, is the simplest carbene. The two unshared electrons in methylene can be either spin-paired in a single orbital or unpaired in different orbitals. Predict the type of hybridization you would expect carbon to adopt in singlet (spin-paired) methylene and triplet (spin-unpaired) methylene. Draw pictures of each, and indicate the types of carbon orbitals present.

CHAPTER 2

# Bonding and Molecular Properties

## 2.1 Drawing Chemical Structures

In the Kekulé structures we've been using, a line between atoms represents the two electrons in a covalent bond. These structures have served chemists well for many years and comprise a universal chemical language. A chemist in China and a chemist in England may not speak each other's language, but a chemical structure means the same to both of them.

Most organic chemists find themselves drawing many structures each day, and it would soon become awkward if every bond and atom had to be indicated. For example, vitamin A, $C_{20}H_{30}O$, has 51 different chemical bonds uniting the 51 atoms. Vitamin A can be drawn showing each bond and atom, but doing so is a time-consuming process, and the resultant drawing is cluttered (see Table 2.1).

Chemists have therefore devised a shorthand way of drawing structures that greatly simplifies matters. The rules for this shorthand are simple:

*Rule 1* Carbon atoms are not usually shown. Instead, a carbon atom is simply assumed to be at each intersection of two lines (bonds) and at the end of each line. Occasionally, a carbon atom might be indicated for emphasis or for clarity.

*Rule 2* Hydrogen atoms bonded to carbon are not shown. Since carbon always has a valence of 4, we mentally supply the correct number of hydrogen atoms to fill the valence of each carbon.

*Rule 3* All atoms other than carbon and hydrogen are indicated.

Table 2.1 gives examples of how these rules are applied in specific cases.

**Table 2.1** Kekulé and shorthand structures for several compounds

| *Compound* | *Kekulé structure* | *Shorthand structure* |
|---|---|---|
| Butane, $C_4H_{10}$ | 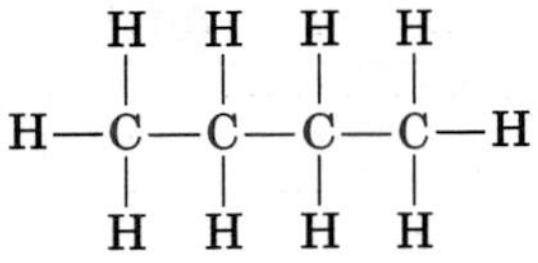 | 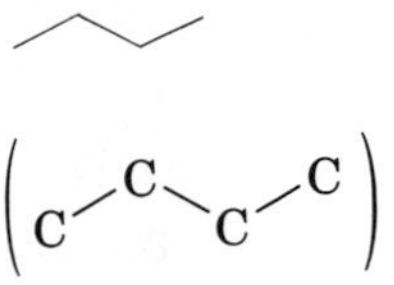 |
| Chloroethylene (vinyl chloride), $C_2H_3Cl$ | 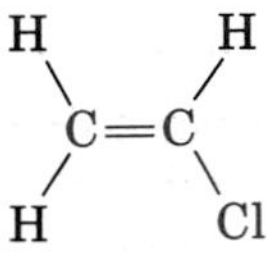 | 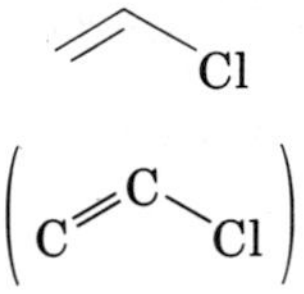 |
| 2-Methyl-1,3-butadiene (isoprene), $C_5H_8$ | 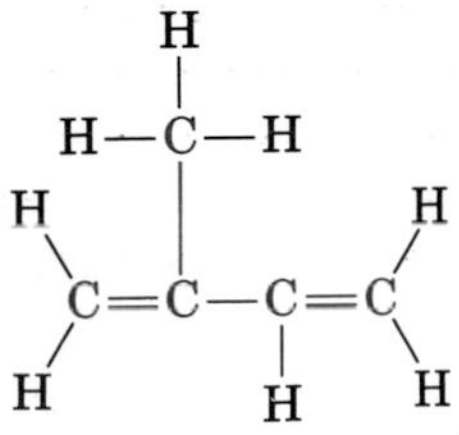 | 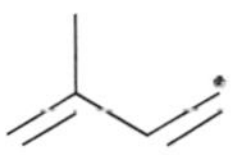 |
| Cyclohexane, $C_6H_{12}$ | 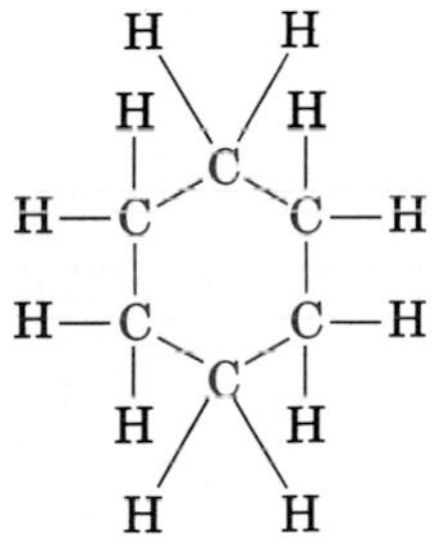 | 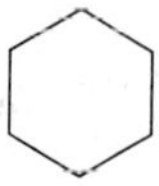 |
| Vitamin A, $C_{20}H_{30}O$ | 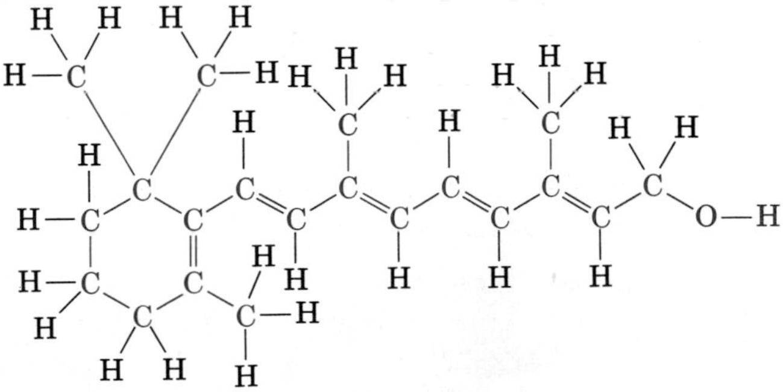 | 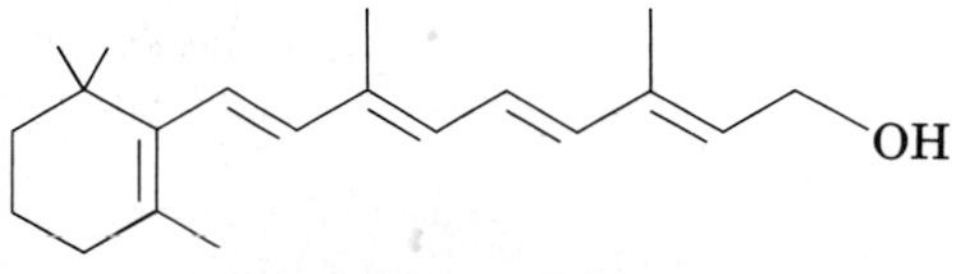 |

PROBLEM

**2.1** Convert these shorthand structures into molecular formulas:

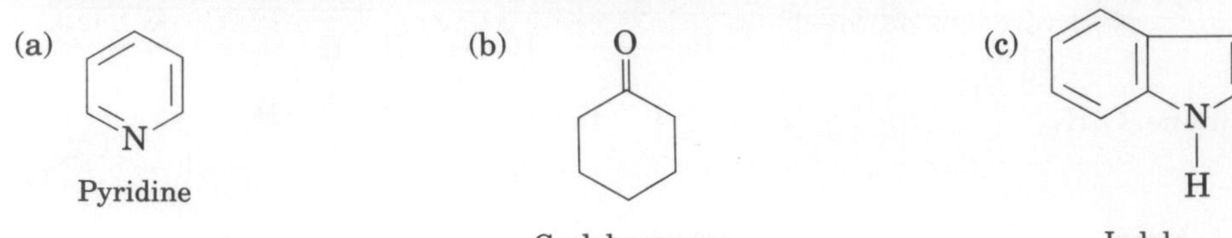


PROBLEM

**2.2** Propose shorthand structures for compounds that satisfy these molecular formulas (there is more than one possibility in each case):
(a) $C_5H_{12}$ (b) $C_2H_7N$ (c) $C_3H_6O$ (d) $C_4H_9Cl$

## 2.2 Molecular Models

Organic chemistry is a three-dimensional science, and molecular shape often plays a crucial role in determining the chemistry a compound undergoes. One very easy technique that simplifies the learning of organic chemistry is to use **molecular models**. With practice, you can learn to see many spatial relationships even when viewing two-dimensional drawings, but there is no substitute for building a molecular model and turning it in your hands to get different perspectives.

Many kinds of models are available, some at relatively modest cost, and every student should have ready access to a set of models while studying this book. Research chemists generally prefer to use either space-filling models such as Corey–Pauling–Koltun (CPK™) Molecular Models, or skeletal models such as Dreiding Stereomodels™. Both are quite expensive but are precisely made to reflect accurate bond angles, intramolecular distances, and atomic radii. CPK models are generally preferred for examining the degree of crowding within a molecule, whereas skeletal models allow the user to measure bond angles and interatomic distances more readily. For student use, ball-and-stick models are generally the least expensive and most durable. Figure 2.1 shows two kinds of models of acetic acid, $CH_3COOH$.

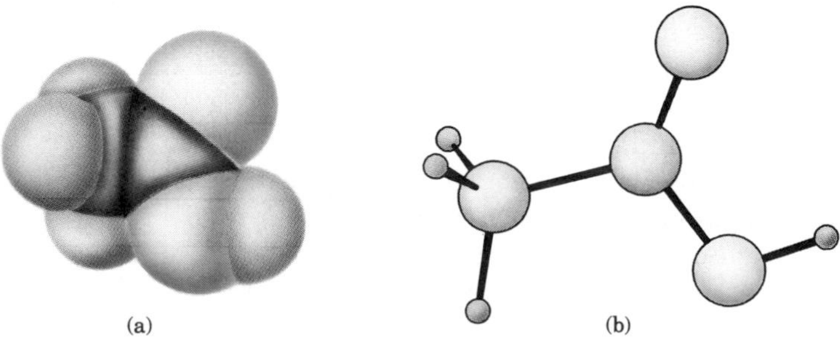


**Figure 2.1** Molecular models of acetic acid, $CH_3COOH$: (a) space-filling; (b) ball-and-stick

**PROBLEM**

**2.3** Build a molecular model of ethane, $H_3C—CH_3$, and look at the relationships between hydrogens on the different carbons by sighting along the C—C bond.

## 2.3 Formal Charges

Most organic molecules can be accurately represented by the Kekulé line-bond structures we saw in the last chapter. Sometimes, though, electron bookkeeping requires that we attach **formal electrical charges** to specific atoms within a molecule. For example, nitromethane ($CH_3NO_2$) must be represented as having a positive charge on nitrogen and a negative charge on the singly bound oxygen:

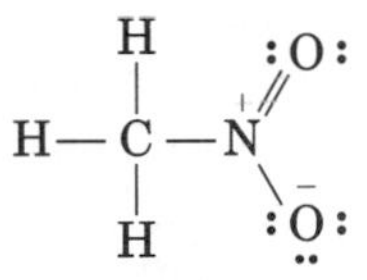

Nitromethane

Let's see why it is necessary to show these charges. In the normal covalent bond, each atom donates one electron. Although the bonding electrons are shared by both atoms, each atom can still be thought of as "owning" one electron for bookkeeping purposes. In methane, for example, there are four C—H bonds with a total of eight electrons. Carbon donated four of these bonding electrons and may still be considered to own four. Since a neutral, isolated carbon atom has four valence electrons and since the carbon atom in methane still owns four, the methane carbon is electrically neutral—it has neither gained nor lost electrons.

The same is true for ammonia, which has three covalent N—H bonds. Atomic nitrogen has five valence electrons, and the ammonia nitrogen also has five (one from each of three shared N—H bonds plus two in the lone pair). Thus, the nitrogen atom in ammonia is electrically neutral.

The situation is different for nitromethane. Atomic nitrogen has five valence electrons, but the nitromethane nitrogen has only *four* (one from the C—N bond, one from the N—O single bond, and two from the N=O double bond). Thus, the nitrogen has lost an electron and must therefore have a positive charge. A similar calculation for the singly bound oxygen atom shows that it has gained an electron and must have a negative charge. (Atomic oxygen has six valence electrons, but the singly bound oxygen in nitromethane has *seven*—one from the O—N bond and two from each of three lone pairs.)

It's a good idea to work out the formal charges in a molecule in a logical manner to make sure that you understand the reasons behind the answers. To express the calculations in a general way, however, we can say that every atom in a molecule can be assigned a formal charge, which is equal to the number of valence electrons in a neutral, isolated atom minus the number of electrons still owned by that atom in the molecule:

$$\textbf{Formal charge} = \begin{pmatrix}\text{Number of}\\ \text{valence electrons}\\ \text{in free atom}\end{pmatrix} - \begin{pmatrix}\text{Number of}\\ \text{valence electrons}\\ \text{in bound atom}\end{pmatrix}$$

$$= \begin{pmatrix}\text{Number of}\\ \text{valence}\\ \text{electrons}\end{pmatrix} - \begin{pmatrix}\text{Half of}\\ \text{bonding}\\ \text{electrons}\end{pmatrix} - \begin{pmatrix}\text{Number of}\\ \text{nonbonding}\\ \text{electrons}\end{pmatrix}$$

For the methane carbon,

$$\cdot\dot{\underset{\cdot}{C}}\cdot + 4\,H\cdot \longrightarrow H:\overset{H}{\underset{H}{\ddot{\underset{\cdot\cdot}{C}}}}:H$$

Carbon valence electrons = 4
Carbon bonding electrons = 8
Carbon nonbonding electrons = 0

$$\text{Formal charge} = 4 - \tfrac{8}{2} - 0 = 0$$

For the ammonia nitrogen,

$$\cdot\ddot{\underset{\cdot}{N}}\cdot + 3\,H\cdot \longrightarrow H:\ddot{\underset{\underset{H}{\cdot\cdot}}{N}}:H$$

Nitrogen valence electrons = 5
Nitrogen bonding electrons = 6
Nitrogen nonbonding electrons = 2

$$\text{Formal charge} = 5 - \tfrac{6}{2} - 2 = 0$$

For the nitromethane nitrogen,

$$CH_3NO_2 = H:\overset{H}{\underset{H}{\ddot{\underset{\cdot\cdot}{C}}}}:N\begin{matrix}\vdots\ddot{O}\colon\\ \cdot\ddot{O}\colon\end{matrix}$$

Nitrogen valence electrons = 5
Nitrogen bonding electrons = 8
Nitrogen nonbonding electrons = 0

$$\text{Formal charge} = 5 - \tfrac{8}{2} - 0 = +1$$

For the singly bound nitromethane oxygen,

| | | |
|---|---|---|
| Oxygen valence electrons | = | 6 |
| Oxygen bonding electrons | = | 2 |
| Oxygen nonbonding electrons | = | 6 |

$$\text{Formal charge} = 6 - \frac{2}{1} - 6 = -1$$

Molecules such as nitromethane that are neutral overall but have charges on individual atoms are called **dipolar molecules**. Dipolar character in molecules often has important chemical consequences, and it's important to be able to identify and calculate these charges correctly.

**PROBLEM**

**2.4** Dimethyl sulfoxide, a common solvent, has the structure indicated. Show by calculations why dimethyl sulfoxide must have formal charges on S and O.

$$\begin{array}{c} :\ddot{O}:^- \\ | \\ H_3C-\underset{\cdot\cdot}{S^+}-CH_3 \end{array}$$

Dimethyl sulfoxide

**PROBLEM**

**2.5** Calculate formal charges for the atoms in these molecules:

(a) Diazomethane, $H_2C{=}N{=}\ddot{N}:$ (b) Acetonitrile oxide, $CH_3-C{\equiv}N-\ddot{\underset{\cdot\cdot}{O}}:$

(c) Methyl isocyanide, $H_3C-N{\equiv}C:$

## 2.4 Polar Covalent Bonds: Electronegativity

Thus far, we have viewed chemical bonding in an either/or manner: A given bond is either covalent or ionic. A more accurate view, however, is to look at bonding as a continuum of possibilities between a perfectly covalent bond with a symmetrical electron distribution on the one hand, and a perfectly ionic bond between positive and negative ions on the other (Figure 2.2).

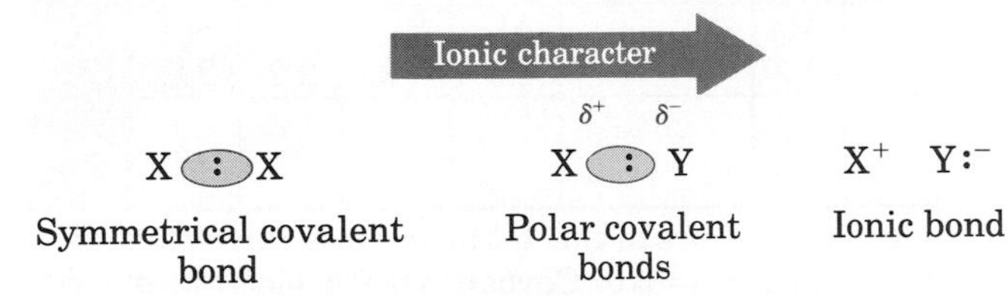


**Figure 2.2** The continuum in bonding from covalent to ionic as a result of unsymmetrical electron distribution: The symbol $\delta$ (Greek delta) means *partial* charge, either positive ($\delta^+$) or negative ($\delta^-$).

The carbon–carbon bond in ethane, for example, is electronically symmetrical and therefore perfectly covalent; the two bonding electrons are equally shared between the two equivalent carbon atoms. The bond in sodium chloride, by contrast, is purely ionic; positively charged sodium ions and negatively charged chloride ions are held together by electrostatic attraction. In between these two extremes lie the great majority of chemical bonds, in which the electrons are attracted *somewhat* more strongly by one atom than by the other. Such bonds are **polar**. A **polar covalent bond** is one in which the bonding electrons are unequally shared by two nuclei. Thus, the electron distribution in polar covalent bonds is unsymmetrical.

Bond polarity is due to **electronegativity**—the intrinsic ability of an atom to attract electrons. As shown in the electronegativity table (Table 2.2), carbon and hydrogen have similar electronegativities, and C—H bonds are therefore relatively nonpolar. Elements on the *right* side of the periodic table, such as oxygen, fluorine, and chlorine, are more electronegative than carbon; that is, they attract electrons more strongly than carbon. When carbon bonds to one of these elements, the bond is polarized so that the bonding electrons are drawn more toward the electronegative atom than toward carbon. This leaves carbon with a *partial positive charge* (denoted by $\delta^+$; $\delta$ is the Greek letter delta) and the electronegative atom with a *partial negative charge* ($\delta^-$). For example, the C—Cl bond in chloromethane is a polar covalent bond:

$Cl^{\delta-}$
$C^{\delta+}$
H H H

Chloromethane

**Table 2.2** Relative electronegativities of some common elements[a]

| | Group | | | | | | | |
|---|---|---|---|---|---|---|---|---|
| *Period* | IA | IIA | | IIIA | IVA | VA | VIA | VIIA |
| 1 | H 2.2 | | | | | | | |
| 2 | Li 1.0 | Be 1.6 | | B 2.0 | **C 2.5** | N 3.0 | O 3.4 | F 4.0 |
| 3 | Na 0.9 | Mg 1.3 | | Al 1.6 | Si 1.9 | P 2.2 | S 2.6 | Cl 3.1 |
| 4 | | | | | | | | Br 3.0 |

[a]Electronegativity values in this table are on an arbitrary scale, with H = 2.2 and F = 4.0. Carbon has an electronegativity value of 2.5. Any element more electronegative than carbon has a value greater than 2.5, and any element less electronegative than carbon has a value less than 2.5.

An arrow $\leftrightarrow$ is used to indicate the direction of polarity. By convention, *electrons move with the arrow*. The tail of the arrow is electron-poor ($\delta^+$), and the head of the arrow is electron-rich ($\delta^-$).

Metallic elements on the left side of the periodic table are less electronegative than carbon and attract electrons less strongly. Thus, when carbon bonds to one of these elements, the bond is polarized so that carbon bears a partial negative charge and the other atom bears a partial positive charge. Organometallic compounds such as tetraethyllead, the "lead" in gasoline, provide good examples of this kind of polar bond:

$$\begin{array}{c} \overset{\delta^-}{C}H_2CH_3 \\ | \\ CH_3\overset{\delta^-}{C}H_2 \overset{\delta^+}{—} Pb — \overset{\delta^-}{C}H_2CH_3 \\ | \\ {}_{\delta^-}CH_2CH_3 \end{array}$$

Tetraethyllead

When we speak of an atom's ability to polarize a bond, we use the term **inductive effect**. An inductive effect is simply the shifting of electrons in a bond in response to the electronegativity of nearby atoms. Electropositive elements such as lithium and magnesium inductively *donate* electrons, whereas electronegative elements such as oxygen and chlorine inductively *withdraw* electrons. Inductive effects play a major role in chemical reactivity and will be encountered many times throughout this text to explain a wide variety of chemical phenomena.

PROBLEM

**2.6** Which element in each of the following pairs is more electronegative?
(a) Li or H (b) Be or Br (c) Cl or I (d) C or H

PROBLEM

**2.7** Use the $\delta^+/\delta^-$ convention to indicate the direction of expected polarity for each of the bonds indicated.
(a) $H_3C—Br$ (b) $H_3C—NH_2$ (c) $H_3C—Li$ (d) $H_2N—H$
(e) $H_3C—OH$ (f) $H_3C—MgBr$ (g) $H_3C—F$

## 2.5 Polar Covalent Bonds: Dipole Moment

Since individual bonds are often polar, molecules as a whole are often polar also. Overall molecular polarity results from the summation of all individual bond polarities, formal charges, and lone-pair contributions in the molecule. The measure of this net molecular polarity is a quantity called the **dipole moment**.

Dipole moments can be viewed in the following way: Assume that there is a "center of gravity" of all positive charges (nuclei) in a molecule. Assume also that there is a center of gravity of all negative charges (electrons) in the molecule. If these two centers don't coincide, then the molecule is electrically unsymmetrical and has a net polarity. The dipole moment, $\mu$ (Greek mu), is defined as the magnitude of a unit charge $e$ times the distance $r$ between the centers, and is expressed in debye[1] units (D):

$$\mu = e \times r \times 10^{18}$$

where

$$e = \text{Electric charge in electrostatic units (esu)}$$
$$r = \text{Distance in centimeters (cm)}$$

For example, if one proton and one electron (charge $e = 4.8 \times 10^{-10}$ esu) are separated from each other by 1.0 Å ($10^{-8}$ cm), then the dipole moment is 4.8 D:

$$\mu = (4.8 \times 10^{-10}\ \text{esu})(10^{-8}\ \text{cm})\left(10^{18}\ \frac{\text{D}}{\text{esu cm}}\right) = 4.8\ \text{D}$$

It is relatively easy experimentally to measure dipole moments, and examples are given in Table 2.3. Once the dipole moment is known, it's then possible to work backward to get an idea of the amount of charge separation in a molecule. Let's take chloromethane, $\mu = 1.87$ D, as an example. If we assume that the contributions of the nonpolar C—H bonds are small, then most of the chloromethane dipole moment is due to the C—Cl bond. Since the C—Cl bond distance is 1.78 Å ($1.78 \times 10^{-8}$ cm), the dipole moment of chloromethane would be $1.78 \times 4.8$ D = 8.5 D if the C—Cl bond were ionic. But because the measured dipole moment is 1.87 D, the C—Cl bond is only about 1.87/8.5 = 20% ionic. Thus, the chlorine atom in chloromethane has an excess of about 0.2 electron, whereas the carbon atom has a deficiency of about 0.2 electron.

**Table 2.3** Dipole moments of some compounds

| *Compound* | *Dipole moment (D)* | *Compound* | *Dipole moment (D)* |
|---|---|---|---|
| NaCl | 9.0 | $NH_3$ | 1.47 |
| $CH_3—\overset{+}{N}(=O)—O^-$ Nitromethane | 3.46 | $CH_4$ | 0 |
| | | $CCl_4$ | 0 |
| | | $CH_3CH_3$ | 0 |
| $CH_3Cl$ | 1.87 | Benzene | 0 |
| $H_2O$ | 1.85 | | |
| $CH_3OH$ | 1.70 | $BF_3$ | 0 |
| $H_2C{=}\overset{+}{N}{=}N^-$ Diazomethane | 1.50 | | |

[1]Peter Joseph Wilhelm Debye (1884–1966); b. Maastricht, Netherlands; Ph.D. Munich (1910); professor of physics, Zurich, Utrecht, Göttingen, Leipzig, Berlin; professor of chemistry, Cornell University (1936–1966); Nobel prize (1936).

Sodium chloride (NaCl) has an extraordinarily large dipole moment because it is 100% ionic. Nitromethane ($CH_3NO_2$) also has a large dipole moment because it has formal charges on two atoms (that is, it is dipolar). Water and ammonia (Figure 2.3) also have large dipole moments, and this too is easily explained. The electronegativity table (Table 2.2) shows that both oxygen and nitrogen are electron-withdrawing relative to hydrogen. In addition, the lone-pair electrons normally make large contributions to overall dipole moments since they have no atom attached to them to "neutralize" their negative charge.

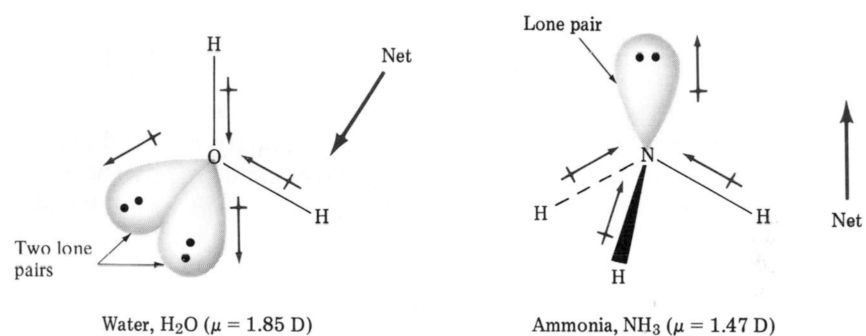


**Figure 2.3** Dipole moments of water and ammonia: The lone-pair electrons make large contributions to the dipole moments of both molecules.

By contrast, methane, tetrachloromethane, and ethane have zero dipole moments. Because of their symmetrical structures, the individual bond polarities in these molecules exactly cancel each other.

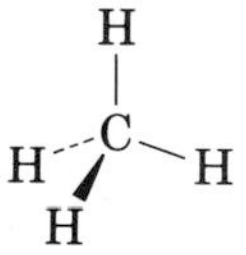

| Methane ($\mu$ = 0 D) | Tetrachloromethane ($\mu$ = 0 D) | Ethane ($\mu$ = 0 D) |
|---|---|---|

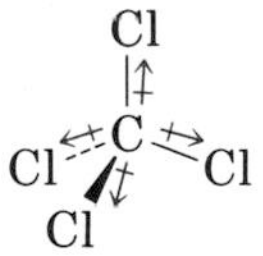

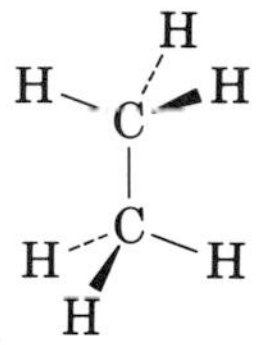

PROBLEM

**2.8** Account for the observed dipole moments of
(a) Methanol, $CH_3OH$ (1.70 D) (b) Benzene,  (0 D)

PROBLEM

**2.9** Make three-dimensional drawings of these compounds and predict whether each has a dipole moment. If a dipole moment is expected, show its direction.
(a) $H_2C{=}CH_2$ (b) $CHCl_3$ (c) $CH_2Cl_2$ (d) $H_2C{=}CCl_2$

## 2.6 Acids and Bases: The Brønsted–Lowry Definition

**Acidity** and **basicity** are related to the concepts of electronegativity and polarity just described, and it's a good idea to review these important topics. We'll soon see that the acid–base behavior of organic molecules helps explain much of their chemistry.

According to the **Brønsted–Lowry definition**, an acid is a substance that donates a proton (hydrogen ion, $H^+$), and a base is a substance that accepts a proton. For example, when HCl gas dissolves in water, an acid–base reaction occurs. Hydrogen chloride acts as an acid to donate a proton, and water acts as a base to accept the proton. The products of the reaction are $H_3O^+$ and $Cl^-$.

The species that results when an acid donates a proton is called the **conjugate base** of the acid; thus, $Cl^-$ is the conjugate base of HCl. The species that results when a base accepts a proton is called the **conjugate acid** of the base; thus, $H_2O$ is the conjugate acid of $HO^-$. Other common mineral acids such as sulfuric acid, nitric acid, and hydrogen bromide behave similarly, as do organic carboxylic acids such as acetic acid, $CH_3COOH$ (Section 20.3).

In a general sense:

$$\underset{\text{An acid}}{\mathrm{H{-}A}} + \underset{\text{A base}}{:\mathrm{B}} \rightleftharpoons \underset{\text{Conjugate base}}{\mathrm{A}:^-} + \underset{\text{Conjugate acid}}{\mathrm{H{-}B^+}}$$

For example:

$$\underset{\text{Acid}}{\mathrm{H{-}\ddot{\underset{..}{Cl}}:}} + \underset{\text{Base}}{:\ddot{\mathrm{O}}\mathrm{{-}H}\ (\mathrm{|H})} \rightleftharpoons \underset{\text{Conjugate base}}{:\ddot{\underset{..}{\mathrm{Cl}}}:^-} + \underset{\text{Conjugate acid}}{\mathrm{H{-}\ddot{O}^+{-}H}\ (\mathrm{|H})}$$

$$\underset{\text{Acid}}{\mathrm{H{-}\ddot{\underset{..}{O}}{-}H}} + \underset{\text{Base}}{\mathrm{H{-}\ddot{N}{-}H}\ (\mathrm{|H})} \rightleftharpoons \underset{\text{Conjugate base}}{\mathrm{H{-}\ddot{\underset{..}{O}}:^-}} + \underset{\text{Conjugate acid}}{\mathrm{H{-}N^+{-}H}\ (\mathrm{H \text{ above}, H \text{ below}})}$$

In all reactions of Brønsted–Lowry acids, a proton is transferred from the acid to the base. Note that water can act *either* as an acid or as a base. In its reaction with HCl, water accepts a proton to give the hydronium ion, $H_3O^+$; in its reaction with ammonia, water donates a proton to give the ammonium ion ($NH_4^+$) and hydroxide ion, $HO^-$.

Acids differ in proton-donating ability. Strong acids such as HCl react almost completely with water, whereas weaker acids such as acetic acid

($CH_3COOH$) react only slightly. Since all acid–base reactions are equilibrium processes, we can describe them using **equilibrium constants, $K_{eq}$**:

$$HA + H_2O \rightleftharpoons H_3O^+ + A^-$$

$$K_{eq} = \frac{[H_3O^+][A^-]}{[HA][H_2O]}$$

where HA represents any acid.[2]

In the customary dilute aqueous solution used for measuring $K_{eq}$, the concentration of water, $[H_2O]$, remains nearly unchanged at approximately 55.5*M*. We can therefore rewrite the equilibrium expression using a new term called the **acidity constant, $K_a$**. The acidity constant for any generalized acid, HA, is simply the equilibrium constant multiplied by the molar concentration of water:

$$HA + H_2O \rightleftharpoons H_3O^+ + A^-$$

$$K_a = K_{eq}[H_2O] = \frac{[H_3O^+][A^-]}{[HA]}$$

Strong acids have their equilibria toward the *right*, and have large acidity constants. Weaker acids have their equilibria toward the *left*, and have smaller acidity constants. We normally express acid strengths by quoting $pK_a$ values, where the $pK_a$ is equal to the negative logarithm of the acidity constant:

$$pK_a = -\log K_a$$

A strong acid (large acidity constant, $K_a$) has a *low* $pK_a$; conversely, a weak acid (small $K_a$) has a *high* $pK_a$. Table 2.4 lists the $pK_a$'s of some common acids in order of their strength.

**Table 2.4** Relative strength of some common acids and their conjugate bases

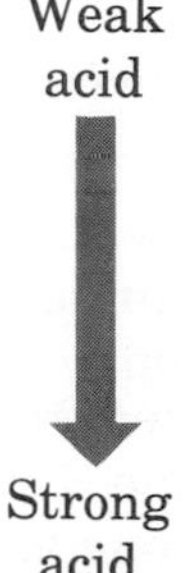

| | *Acid* | *Name* | $pK_a$ | *Conjugate base* | *Name* | |
|---|---|---|---|---|---|---|
| Weak acid | $CH_3CH_2OH$ | Ethanol | 16.00 | $CH_3CH_2O^-$ | Ethoxide ion | Strong base |
| | $H_2O$ | Water | 15.74 | $HO^-$ | Hydroxide ion | |
| | HCN | Hydrocyanic acid | 9.2 | $CN^-$ | Cyanide ion | |
| | $CH_3COOH$ | Acetic acid | 4.72 | $CH_3COO^-$ | Acetate ion | |
| | HF | Hydrofluoric acid | 3.2 | $F^-$ | Fluoride ion | |
| Strong acid | $HNO_3$ | Nitric acid | −1.3 | $NO_3^-$ | Nitrate ion | Weak base |
| | HCl | Hydrochloric acid | −7.0 | $Cl^-$ | Chloride ion | |

[2]Recall that brackets, [ ], refer to the concentration of the enclosed species expressed in moles per liter.

Although we have considered only acids thus far, base strength can be viewed in a similar manner. Thus, the conjugate base of a strong acid must be a weak base, since it has little affinity for protons. Similarly, the conjugate base of a weak acid must be a strong base, since it has a high affinity for protons. For example, chloride ion (the conjugate base of the strong acid HCl) is a weak base, since it has little affinity for a proton; acetate ion (the conjugate base of the weaker acid $CH_3COOH$) is a stronger base, with a modest affinity for a proton; and hydroxide ion (the conjugate base of the weak acid $H_2O$) is a still stronger base with a high affinity for a proton. Table 2.4 also shows the relative strengths of several common conjugate bases.

PROBLEM

**2.10** Formic acid, HCOOH, has $pK_a = 3.7$; and picric acid, $C_6H_3N_3O_7$, has $pK_a = 0.3$. Which is the stronger acid?

PROBLEM

**2.11** Amide ion, $H_2N^-$, is a much stronger base than hydroxide ion, $HO^-$. Which would you expect to be a stronger acid, $H_2N$—H (ammonia) or HO—H (water)? Explain.

## 2.7 Predicting Acid–Base Reactions from p$K_a$ Values

Compilations of p$K_a$ values such as those in Table 2.4 are extremely useful for predicting whether or not a given acid–base reaction will take place to any great extent. In general, an acid will react with (donate a proton to) the conjugate base of any acid with a higher p$K_a$. Conversely, the conjugate base of an acid will abstract a proton from any acid with a lower p$K_a$. For example, the data in Table 2.4 indicate that hydroxide ion will react with acetic acid, $CH_3COOH$, to yield acetate ion, $CH_3COO^-$, and water:

$$CH_3COOH + {}^-\!:\ddot{O}H \rightleftharpoons CH_3COO^- + H_2O$$

Acetic acid (p$K_a$ = 4.72) | Hydroxide ion | Acetate ion | Water (p$K_a$ = 15.74)

According to the p$K_a$ data in Table 2.4, water (p$K_a$ = 15.74) is a weaker acid than acetic acid (p$K_a$ = 4.72). Thus, hydroxide ion has a greater affinity for a proton than acetate ion has, and the reaction of hydroxide ion with acetic acid will occur.

Another way for predicting acid–base reactivity is simply to remember that the products must be more stable than the reactants in order for reaction to occur. In other words, the *product* acid and base must be weaker and less reactive than the *starting* acid and base. For example, in the reaction of acetic acid with hydroxide ion, the product base (acetate ion) is weaker than the starting base (hydroxide ion), and the product acid (water) is weaker than the starting acid (acetic acid):

$$\underset{\text{Stronger acid}}{CH_3COOH} + \underset{\text{Stronger base}}{HO^-} \rightleftarrows \underset{\text{Weaker acid}}{H_2O} + \underset{\text{Weaker base}}{CH_3COO^-}$$

**PRACTICE PROBLEM**

Water has $pK_a = 15.74$ and acetylene has $pK_a = 25$. Which of the two is more acidic? Would you expect hydroxide ion to react with acetylene?

$$H{-}C{\equiv}C{-}H + H{-}\ddot{\underset{\cdot\cdot}{O}}{:}^- \xrightarrow{?} H{-}C{\equiv}C{:}^- + H{-}\ddot{\underset{\cdot\cdot}{O}}{-}H$$

*Solution* In comparing two acids, the one with the lower $pK_a$ is stronger. Thus, water is a stronger acid than acetylene. Since water gives up a proton more easily than acetylene, the $H{-}O^-$ ion must have less affinity for a proton than the $H{-}C{\equiv}C{:}^-$ ion. In other words, the anion of acetylene is a stronger base than hydroxide ion, and the reaction will not proceed as written.

**PROBLEM**

**2.12** Is either of the following reactions likely to take place, according to the $pK_a$ data in Table 2.4?

(a) $H{-}CN + CH_3COO^-\ Na^+ \xrightarrow{?} Na^+\ ^-CN + CH_3COO{-}H$

(b) $CH_3CH_2O{-}H + Na^+\ ^-CN \xrightarrow{?} CH_3CH_2O^-\ Na^+ + H{-}CN$

**PROBLEM**

**2.13** Ammonia, $H_2N{-}H$, has $pK_a \approx 36$ and acetone has $pK_a \approx 20$. Will the following reaction take place?

$$H_3C{-}\overset{\overset{\displaystyle O}{\|}}{C}{-}CH_3 + {}^-{:}\ddot{N}H_2 \xrightarrow{?} H_3C{-}\overset{\overset{\displaystyle O}{\|}}{C}{-}\ddot{C}H_2 + {:}NH_3$$

## 2.8 Acids and Bases: The Lewis Definition

The Brønsted–Lowry concept of acidity is a useful one that can be extended to all compounds containing hydrogen. Of even more use, however, is the **Lewis definition** of acids and bases: A **Lewis acid** is a substance that accepts an electron pair; a **Lewis base** is a substance that donates an electron pair.

The Lewis definition of acidity is much broader than the Brønsted–Lowry definition. Lewis acids include not only proton donors but many other species as well. For example, a proton (hydrogen ion, $H^+$) is a Lewis acid because it has a vacant *s* orbital and needs a pair of electrons to fill its empty valence shell. In addition, compounds such as $BF_3$ and $AlCl_3$ are Lewis acids because they also have vacant orbitals that can accept electron pairs from Lewis bases, as shown in Figure 2.4. Both the boron atom in $BF_3$ and the aluminum atom in $AlCl_3$ have only six electrons in their outer shells.

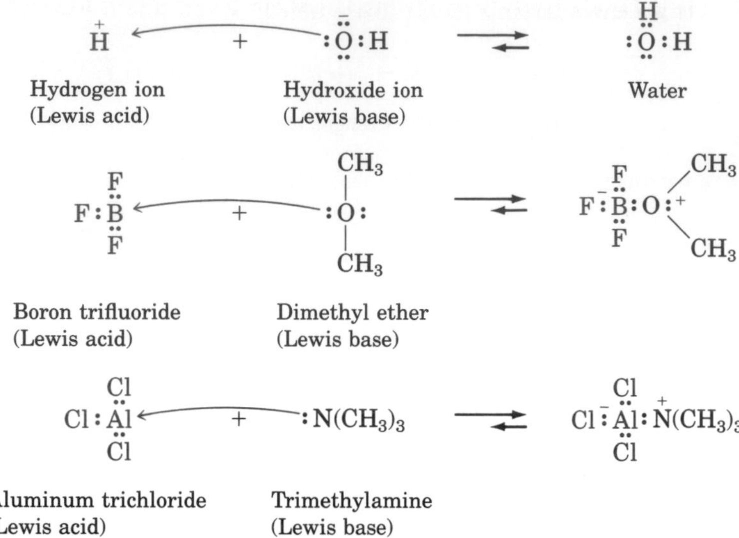


**Figure 2.4** The reactions of some Lewis acids with some Lewis bases: The Lewis acids all have a vacant orbital that can accept an electron pair; the Lewis bases all have a pair of nonbonding electrons. Note how the flow of electrons from the Lewis base to the Lewis acid is indicated by the curved arrow.

Less obvious is the fact that many transition-metal compounds, such as $TiCl_4$, $ZnCl_2$, $FeCl_3$, and $SnCl_4$, are excellent Lewis acids. The bonding in such compounds is complex, since it involves *d* orbital hybridization, but all of these transition-metal compounds have a vacant valence orbital that can accept an electron pair.

Note particularly how the acid–base reactions in Figure 2.4 are shown. A curved arrow indicates the direction of electron pair flow from the Lewis base (electron-rich) to the Lewis acid (electron-poor). This kind of arrow is used extensively in organic chemistry and *always* has the same meaning—a pair of electrons moves from the atom at the tail of the arrow to form a bond with the atom at the head of the arrow.

The Lewis definition of basicity—a compound that can donate an electron pair—is quite similar to the Brønsted–Lowry definition: A Lewis base has a lone pair of electrons that it can donate to a Lewis acid for use in forming a new bond. Thus, $H_2O$, with its two lone pairs of electrons on oxygen, serves as a Lewis base by donating an electron pair to a proton in forming the hydronium ion, $H_3O^+$:

$$\text{:}\ddot{\text{Cl}}\text{—H} + \text{:}\ddot{\text{O}}(\text{H})\text{—H} \rightleftharpoons \text{H—}\ddot{\text{O}}^{+}(\text{H})\text{—H} + \text{:}\ddot{\text{Cl}}\text{:}^{-}$$

Acid    Lewis base    Hydronium ion

In a more general sense, most oxygen- and nitrogen-containing organic compounds are good Lewis bases because they have lone pairs of available electrons. For example, dimethyl ether, ethyl alcohol, and acetone are Lewis bases because their oxygen atoms each have two lone pairs; trimethylamine is a Lewis base because its nitrogen atom has a lone pair:

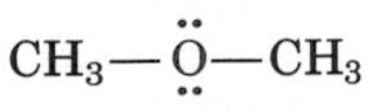

$$CH_3-\ddot{\underset{..}{O}}-CH_3 \qquad CH_3CH_2-\ddot{\underset{..}{O}}-H$$

Dimethyl ether 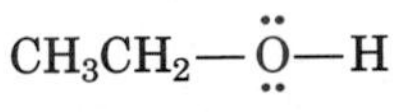  Ethyl alcohol

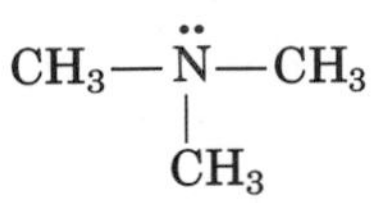

$$CH_3-\ddot{N}(CH_3)-CH_3 \qquad CH_3-C(=\ddot{O}:)-CH_3$$

Trimethylamine 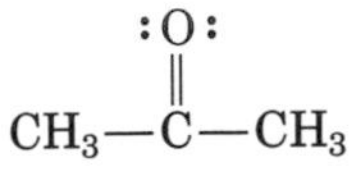  Acetone

PROBLEM

**2.14** Which of the following are Lewis acids and which are Lewis bases?

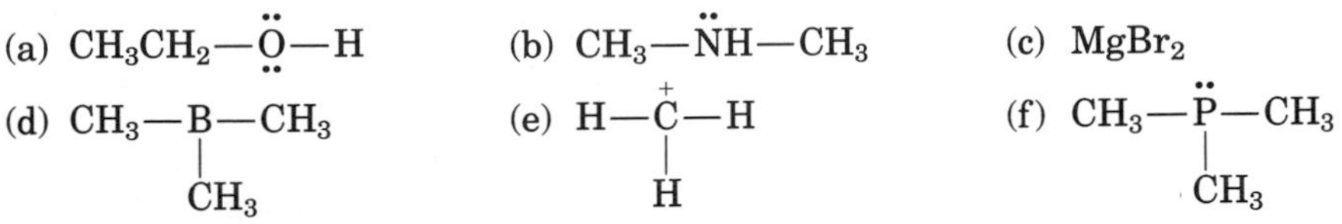

(a) $CH_3CH_2-\ddot{\underset{..}{O}}-H$ (b) $CH_3-\ddot{N}H-CH_3$ (c) $MgBr_2$

(d) $CH_3-B(CH_3)-CH_3$ (e) $H-\overset{+}{C}(H)-H$ (f) $CH_3-\ddot{P}(CH_3)-CH_3$

PROBLEM

**2.15** Explain by formal-charge calculations why the following acid–base complexes have the charges indicated:

(a) $F_3\overset{-}{B}-\overset{+}{O}(CH_3)-CH_3$ (b) $Cl_3\overset{-}{Al}-\overset{+}{N}(CH_3)_2-CH_3$

PROBLEM

**2.16** Boron trifluoride reacts with formaldehyde to give an acid–base complex. Which partner is the acid and which is the base?

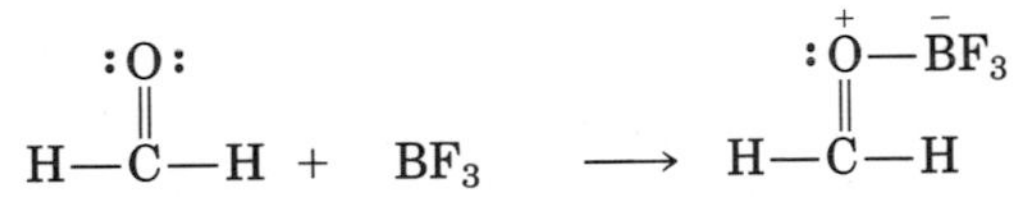

$$H-C(=\ddot{O}:)-H + BF_3 \longrightarrow H-C(=\overset{+}{\dot{O}}:-\overset{-}{B}F_3)-H$$

Formaldehyde  Boron trifluoride

## 2.9 Analysis of Organic Compounds

In the late eighteenth century, only 30 or so elements were known, and chemists faced with a newly isolated compound had great difficulty even identifying the elements present. A series of experiments on combustion carried out in 1772–1777 by Antoine Lavoisier[3] provided the first real breakthrough in the analysis of organic compounds. Lavoisier's techniques, though suitable for determining the *identity* of elements present in organic compounds, were by no means accurate enough to determine the relative proportions of the elements.

[3]Antoine Lavoisier (1743–1794); b. Paris; studied at College Mazarin; considered the founder of modern chemistry; guillotined during French Revolution.

The second breakthrough came in 1831 when Justus von Liebig[4] devised the method of organic analysis that is still used today. The key to Liebig's method was his recognition that organic compounds are efficiently burned on contact with red-hot copper oxide. For example, oxidation of benzene, $C_6H_6$, proceeds according to the following equation:

$$C_6H_6 + 15\ CuO \xrightarrow{900^\circ C} 6\ CO_2 + 3\ H_2O + 15\ Cu$$

The water produced by the combustion is swept by a stream of oxygen gas into a tube filled with calcium chloride, where it is retained. By weighing the tube before and after combustion, the amount of water formed can be accurately determined. The $CO_2$ produced passes through the $CaCl_2$ tube into a separate tube containing potassium hydroxide, KOH, where it is absorbed. Again, the amount of $CO_2$ present can be determined by weighing the tube before and after combustion, and the percentage composition of carbon and hydrogen present in the original sample can be determined.

The Liebig technique can't directly determine the amount of oxygen present in a compound. If, however, no other elements are detected and the combined percentages of carbon and hydrogen do not total 100, then the percentage of oxygen present is taken as the difference. Let's assume, for example, that we have analyzed a 0.55 gram (g) sample of a colorless organic liquid obtained by the distillation of wine. On weighing the $CaCl_2$ and KOH tubes, we find that 0.66 g $H_2O$ and 1.037 g $CO_2$ have been formed. We can then calculate the percentages of carbon and hydrogen in the unknown sample.

First, we find the weight of hydrogen in the sample by finding how much water is produced. We then calculate the percent hydrogen in the sample by dividing the hydrogen weight by the sample weight:

$$\text{Weight of H in sample} = \text{Weight of } H_2O \times \frac{\text{Molecular weight of } H_2 \text{ (2.016)}}{\text{Molecular weight of } H_2O \text{ (18.016)}}$$

$$= (0.66)(0.112) = 0.074 \text{ g H}$$

$$\%\text{ H in sample} = \frac{\text{Weight of H}}{\text{Weight of sample}} = \frac{0.074}{0.55} = 13.44\%$$

In a similar manner, we calculate the percent carbon in the sample by first finding the weight of carbon from the amount of $CO_2$ produced and then dividing by the sample weight:

$$\text{Weight of C in sample} = \text{Weight of } CO_2 \times \frac{\text{Molecular weight of C (12.01)}}{\text{Molecular weight of } CO_2 \text{ (44.01)}}$$

$$= (1.037)(0.273) = 0.283 \text{ g C}$$

$$\%\text{ C in sample} = \frac{\text{Weight of C}}{\text{Weight of sample}} = \frac{0.283}{0.55} = 51.47\%$$

[4]Justus von Liebig (1803–1873); b. Darmstadt; Ph.D. at Erlangen in 1822; professor, Giessen (1824–1852), Munich.

Since the percentages of carbon and hydrogen add up to only 64.91%, we can assume that the sample also contains 35.09% oxygen. The next step is to determine the *atomic ratios* by dividing the percentage of each element by its atomic weight:

$$\text{C:}\quad \frac{51.47\%}{12.01} = 4.29$$

$$\text{H:}\quad \frac{13.44\%}{1.008} = 13.33$$

$$\text{O:}\quad \frac{35.09\%}{16.00} = 2.19$$

The atomic ratio of elements in our sample is C, 4.29:H, 13.33:O, 2.19, which reduces to C, 1.95:H, 6.1:O, 1 when the numbers are divided by 2.19 (the lowest number in the series). Rounding off these ratios gives the **empirical formula** $C_2H_6O$. *Analysis gives only atomic ratios.* To determine the **molecular formula**, which may be a multiple of the empirical formula, we also need to determine the molecular weight. In the present case, though, any higher multiple of $C_2H_6O$ would be impossible by the rules of valency. Since carbon has a valence of four, an organic compound with $n$ carbons can have no more than $2n + 2$ hydrogens; that is, two hydrogen atoms per carbon plus one hydrogen at each end of the chain:

$$\text{H}\left(\begin{array}{c}\text{H}\quad\text{H}\qquad\quad\text{H}\\ |\quad\;\; |\qquad\quad\; |\\ \text{C}-\text{C}\;\;\cdots\;\;\text{C}\\ |\quad\;\; |\qquad\quad\; |\\ \text{H}\quad\text{H}\qquad\quad\text{H}\end{array}\right)\text{H}$$

$$C_nH_{2n+2}$$

A formula such as $C_4H_{12}O_2$ for our unknown would be impossible, and the liquid (since it comes from wine) is ethyl alcohol, $CH_3CH_2OH$.

The Liebig method of analysis for carbon and hydrogen, and a similar method introduced by Jean Dumas[5] in 1830 for the analysis of nitrogen, were remarkable achievements at the time. They were, however, limited in their usefulness by the large sample sizes required. Often, it is practically impossible to obtain more than milligram amounts of new compounds, and an analysis that destroys $\frac{1}{2}$ gram of material at a time is unthinkable.

The major limiting factor in the Liebig and Dumas analyses was the accuracy of the analytical balances used to weigh the samples and the collection tubes. As the science of chemistry developed, however, scientific instrumentation became more sophisticated. Under the leadership of Fritz Pregl,[6] a microbalance of great precision was developed, allowing highly accurate weighing of submilligram amounts. Pregl further refined all aspects of the Liebig method and, in 1911, introduced a method of microanalysis that could be carried out on 5–10 milligram (mg) samples. For his accomplishments, he received the 1923 Nobel prize in chemistry.

---

[5]Jean Baptiste André Dumas (1800–1884); b. Alais, France; professor, École Polytechnique (1835), École de Médecine, Sorbonne.

[6]Fritz Pregl (1869–1930); b. Laibach, Austria; Ph.D. Graz (1893); professor, Innsbruck, Graz; Nobel prize in chemistry (1923).

Today, microanalysis of organic compounds is still carried out by the methods pioneered by Liebig, Dumas, and Pregl, although the techniques have become highly automated. Modern chemists do not consider a new compound to be fully characterized until accurate combustion analyses have been carried out, and many chemical journals still require such data before they publish new work.

PROBLEM

**2.17** Calculate the percentage of each element in these molecular formulas:
(a) Benzene, $C_6H_6$
(b) Laetrile, $C_{14}H_{15}NO_7$
(c) Quinine, $C_{20}H_{24}N_2O_2$
(d) Diethylstilbestrol, $C_{18}H_{20}O_2$

PROBLEM

**2.18** The chemical responsible for the odor of lemon is a substance named *citral*. Combustion analysis shows citral to contain 78.9% C and 10.6% H. Assuming that the remainder is due to oxygen, what is the empirical formula of citral? If citral has a molecular weight of 152, what is its molecular formula?

PROBLEM

**2.19** A sample of squalene, isolated from shark oil, was submitted for combustion analysis. An amount weighing 8.00 mg gave 25.6 mg $CO_2$ and 8.75 mg $H_2O$. Calculate the empirical formula of squalene. If squalene has a molecular weight of 410, what is its molecular formula?

## 2.10 Summary and Key Words

Plus (+) and minus (−) signs are used to indicate the presence of **formal charges** on atoms. Assigning formal charges to specific atoms in neutral compounds is a bookkeeping technique that allows us to keep track of the valence electrons in an atom:

$$\textbf{Formal charge} = \begin{pmatrix}\text{Number of valence electrons}\\ \text{in the free atom}\end{pmatrix} - \begin{pmatrix}\text{Number of electrons}\\ \text{in the bound atom}\end{pmatrix}$$

Organic molecules often have **polar covalent bonds**. Bond polarity is a result of unsymmetrical electron sharing due to the intrinsic **electronegativity** of atoms. For example, a carbon–chlorine bond is polar because chlorine attracts the shared electrons more strongly than carbon does. Carbon–metal bonds, however, are usually polarized in the opposite sense, because carbon attracts electrons more strongly than most metals. Carbon–hydrogen bonds are relatively nonpolar. Many molecules as a whole are also polar owing to the cumulative effects of individual polar bonds, formal charges, and electron lone pairs. The polarization of a molecule is measured by its **dipole moment**.

Acidity and basicity are related to polarity and electronegativity. A **Brønsted–Lowry acid** is a compound that can donate a proton (hydrogen ion, $H^+$); a **Brønsted–Lowry base** is a compound that can accept a proton.

The exact strength of a Brønsted–Lowry acid or base is expressed in terms of acidity constants, $K_a$. More useful is the Lewis definition of acids and bases. A **Lewis acid** is a compound that has a low-energy unfilled orbital and can accept an electron pair; $BF_3$, $AlCl_3$, and $H^+$ are examples. A **Lewis base** is a compound that donates an unshared electron pair; $NH_3$ and $H_2O$ are examples. Many organic molecules that contain oxygen and nitrogen are weak Lewis bases.

The analysis of organic compounds can be carried out accurately on milligram amounts of sample. The organic material is burned, and the combustion products are weighed to give information that can be used to establish the empirical formula of an unknown.

Chemists normally draw **line-bond structures** using a shorthand method in which carbons and most hydrogen atoms are not indicated. A carbon atom is assumed to be at the ends and at the intersections of lines (bonds), and the correct number of hydrogens is mentally supplied. For example:

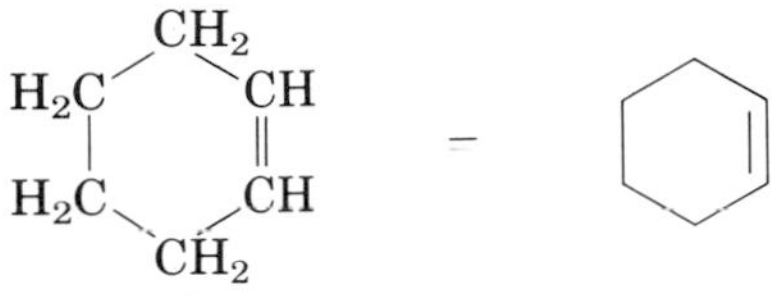

Cyclohexene

## ADDITIONAL PROBLEMS

**2.20** Convert the following structures into shorthand drawings:

(a) Naphthalene

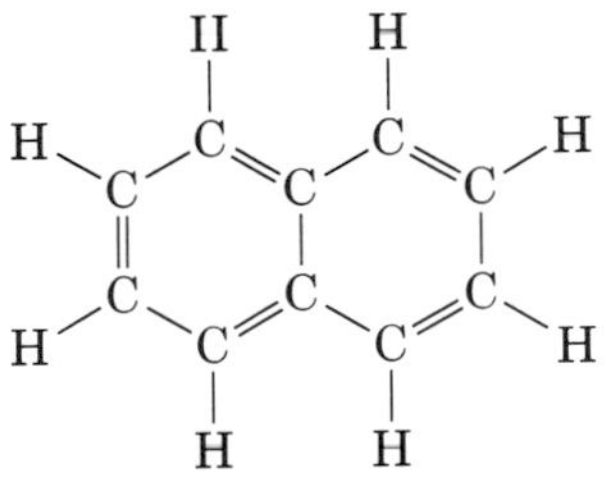

(b) 1,3-Pentadiene

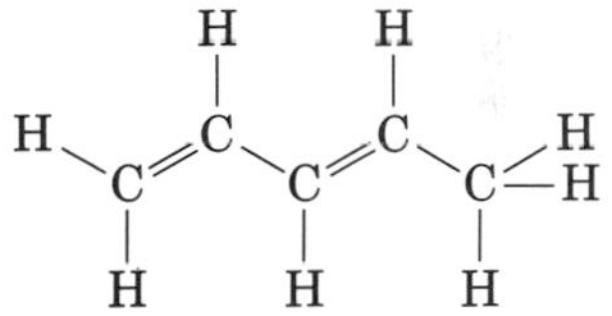

(c) 1,2-Dichlorocyclopentane

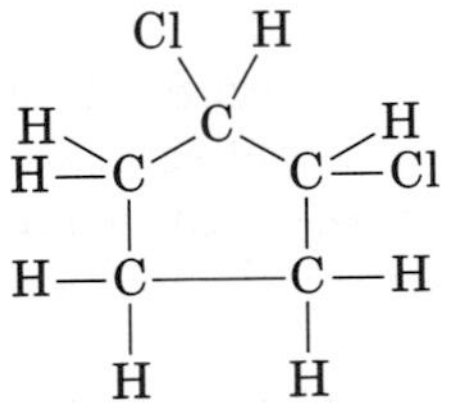

(d) Quinone

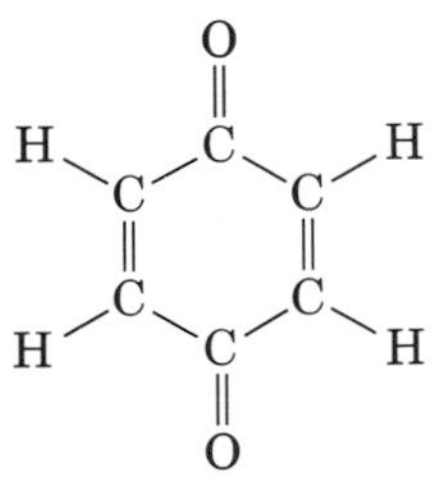

**2.21** Convert these shorthand drawings into Kekulé structures that show all carbons and hydrogens:

(a)

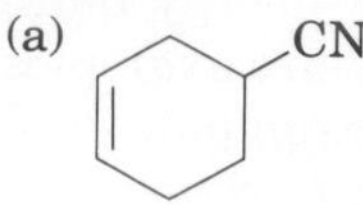


(b) 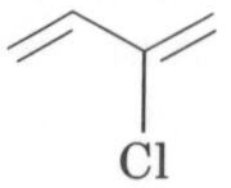 Cl

(c) 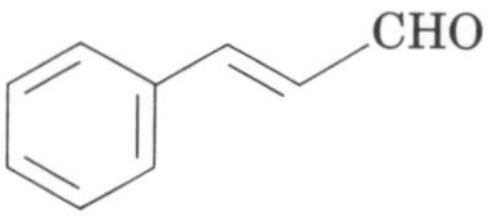 CHO

(d) 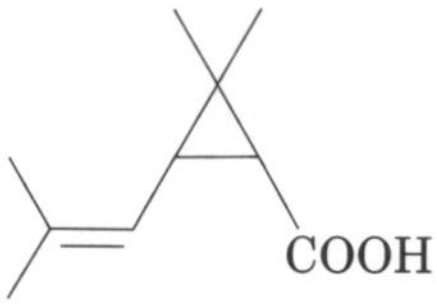 COOH

**2.22** Calculate the formal charges on the atoms indicated:

(a) $(CH_3)_3\ddot{O}\ BF_4$

(b) $H_2\ddot{C}—N{\equiv}N{:}$

(c) $H_2C{=}N{=}\ddot{N}{:}$

(d) $:\ddot{O}{=}\ddot{O}—\ddot{\underset{\cdot\cdot}{O}}{:}$

(e) 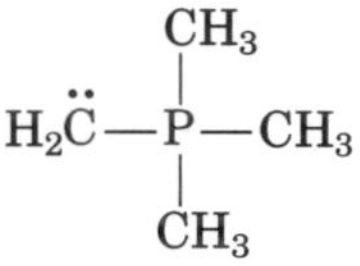 $H_2\ddot{C}—P(CH_3)_3$ (P bearing three $CH_3$ groups)

(f) 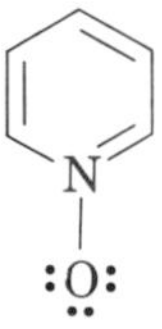 N :O: (pyridine N-oxide)

**2.23** Use the electronegativity table (Table 2.2) to predict which bond in each of the following sets is more polar:

(a) $H_3C—Cl$ and $Cl—Cl$

(b) $H_3C—H$ and $H—Cl$

(c) $HO—CH_3$ and $(CH_3)_3Si—CH_3$

(d) $H_3C—Li$ and $Li—OH$

**2.24** Indicate the direction of bond polarity for each compound in Problem 2.23.

**2.25** Which of these molecules have dipole moments? Indicate the expected direction of each.

(a) 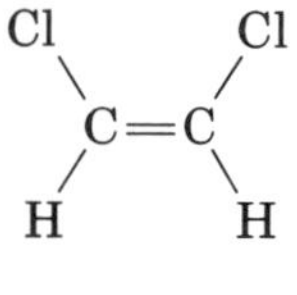 Cl Cl / C=C / H H

*cis*-1,2-Dichloroethylene

(b) 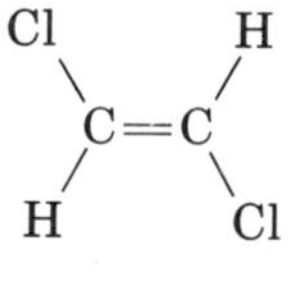 Cl H / C=C / H Cl

*trans*-1,2-Dichloroethylene

(c) LiH

(d) $F_3B—N(CH_3)_3$

(e) 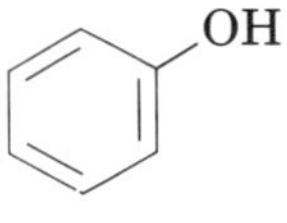 OH

(f) 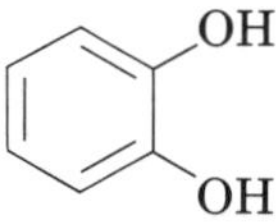 OH OH

(g) 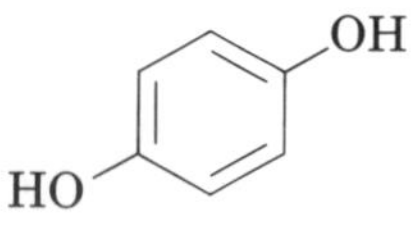 OH HO

(h) 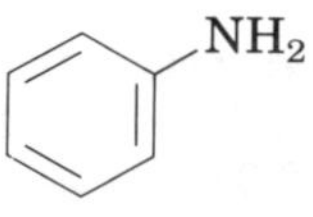 $NH_2$

**2.26** How can you explain the fact that the O—H hydrogen in acetic acid is more acidic than any of the C—H hydrogens? [*Hint:* Consider bond polarity.]

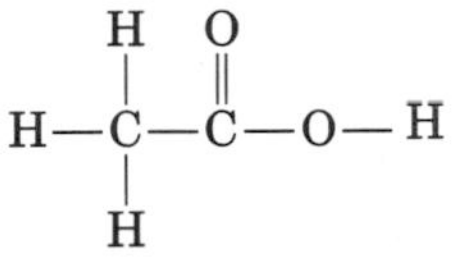

```
    H  O
    |  ‖
H—C—C—O—H
    |
    H
```

Acetic acid

**2.27** Classify the following reagents as either Lewis acids or Lewis bases:
(a) $AlBr_3$ (b) $CH_3CH_2NH_2$ (c) $BH_3$
(d) HF (e) $CH_3—S—CH_3$ (f) $TiCl_4$

**2.28** Draw Lewis electron-dot structures for each of the molecules in Problem 2.27. Make sure that you indicate the unshared electron pairs where present.

**2.29** Assign formal charges to these molecules:

(a) 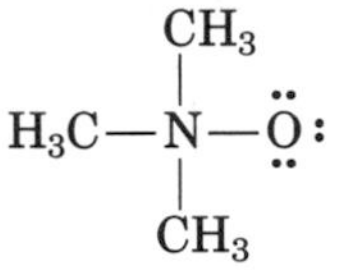
```
        CH3
        |       ..
H3C—N—O:
        |       ..
        CH3
```

(b) 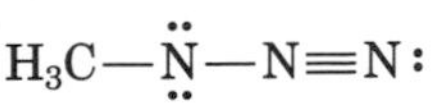 $H_3C—\ddot{N}—N≡N:$ (N with two lone pairs)

(c) 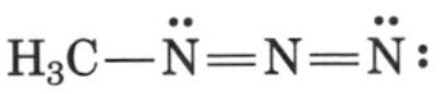 $H_3C—\ddot{N}=N=\ddot{N}:$

**2.30** Rank the following four substances in order of increasing acidity:

(a) 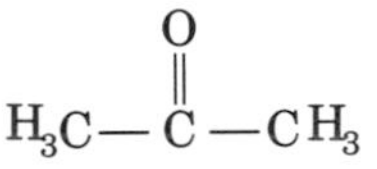
```
       O
       ‖
H3C—C—CH3
```
Acetone
(p$K_a$ ≈ 20)

(b) 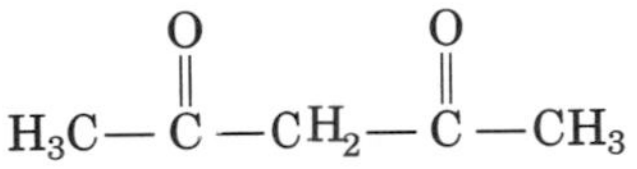
```
       O            O
       ‖            ‖
H3C—C—CH2—C—CH3
```
2,4-Pentanedione
(p$K_a$ ≈ 9)

(c) 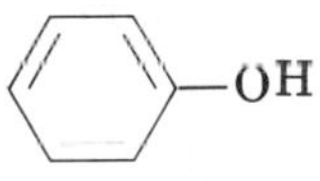 Phenol ($C_6H_5$—OH)
(p$K_a$ ≈ 10)

(d) 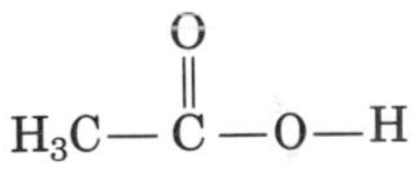
```
       O
       ‖
H3C—C—O—H
```
Acetic acid
(p$K_a$ ≈ 4.7)

**2.31** Which, if any, of the four substances in Problem 2.30 are strong enough acids to react completely with NaOH? (The p$K_a$ of $H_2O$ is 15.7.)

**2.32** Is *tert*-butoxide anion a strong enough base to react with water? In other words, will the following reaction take place as written? (The p$K_a$ of *tert*-butyl alcohol is approximately 18.)

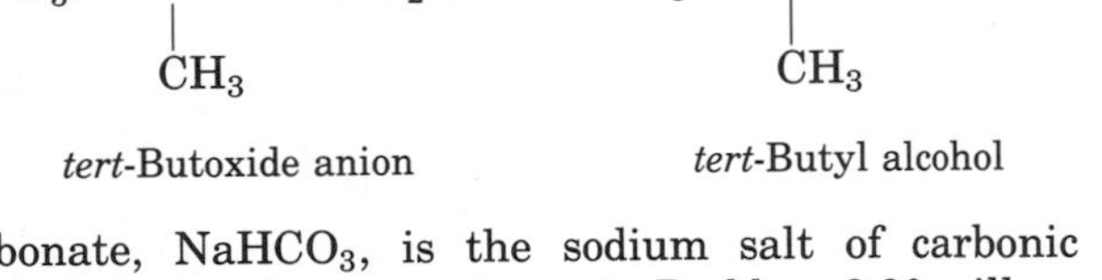

$$(CH_3)_3C—O^- + H_2O \xrightarrow{?} (CH_3)_3C—OH + HO^-$$

*tert*-Butoxide anion | *tert*-Butyl alcohol

**2.33** Sodium bicarbonate, $NaHCO_3$, is the sodium salt of carbonic acid ($H_2CO_3$), p$K_a$ ≈ 6.4. Which of the substances shown in Problem 2.30 will react with sodium bicarbonate?

**2.34** Assume that you have two unlabeled bottles, one of which contains phenol (p$K_a$ ≈ 10) and one of which contains acetic acid (p$K_a$ ≈ 4.7). In light of your answer to Problem 2.33, propose a simple qualitative way for telling what is in each bottle.

**2.35** Identify the acids and bases in these reactions:

(a) $CH_3OH + H^+ \longrightarrow CH_3\overset{+}{O}H_2$

(b) $CH_3OH + {}^-NH_2 \longrightarrow CH_3O^- + NH_3$

(c) $CH_3\overset{O}{\overset{\|}{C}}CH_3 + TiCl_4 \longrightarrow CH_3{-}\overset{\overset{+}{O}{-}\overset{-}{Ti}Cl_4}{\overset{\|}{C}}{-}CH_3$

(d) 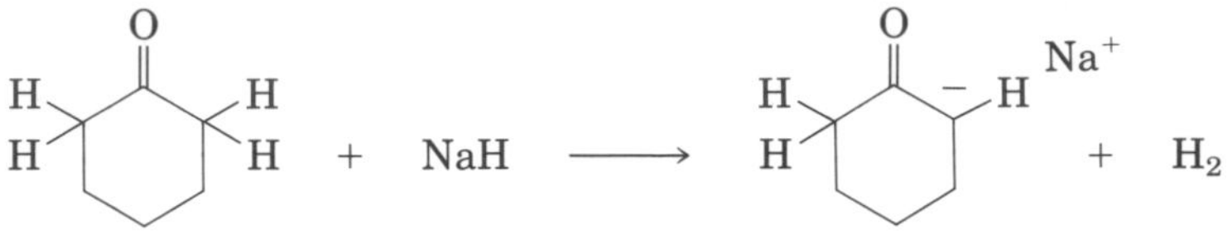 2,2,6,6-H cyclohexanone + NaH ⟶ enolate (carbanion, $Na^+$) + $H_2$

(e) 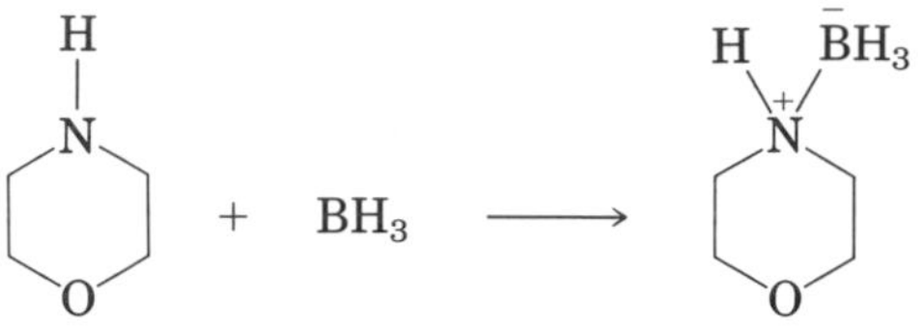 morpholine (N–H) + $BH_3$ ⟶ morpholine–$\overset{+}{N}$–$\overset{-}{B}H_3$ adduct

(f) 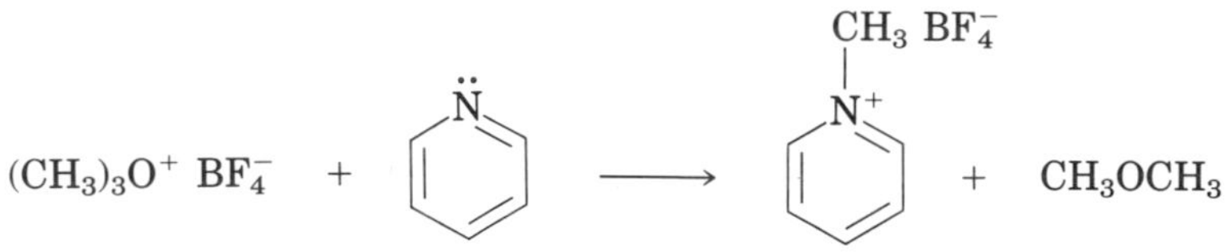 $(CH_3)_3O^+ \; BF_4^-$ + pyridine ⟶ N-methylpyridinium ($CH_3$–$N^+$) $BF_4^-$ + $CH_3OCH_3$

**2.36** Calculate the percentage of each element in the following formulas:
(a) Aspirin, $C_9H_8O_4$
(b) Muscone (musk oil), $C_{16}H_{30}O$
(c) Morphine, $C_{17}H_{19}NO_3$
(d) Strychnine, $C_{21}H_{22}N_2O_2$

**2.37** $\alpha$-Pinene, the main constituent of turpentine, has been shown by mass spectroscopy to have a molecular weight of 136. Combustion analytical data indicate that $\alpha$-pinene contains 88.3% C and 11.6% H by weight. What is the molecular formula of $\alpha$-pinene?

**2.38** Jasmone, an odoriferous compound isolated from the jasmine flower, is valued for its use in perfume. A pure sample of jasmone is analyzed and found to be 80.7% C and 9.7% H. Its molecular weight is 164. What other element is probably present in jasmone? What is the molecular formula of jasmone?

**2.39** Progesterone, the so-called pregnancy hormone, was isolated by Adolf Butenandt in 1934. By extracting the ovaries of 50,000 sows, Butenandt was able to isolate 20 mg of the pure hormone. Microanalysis of a 0.005 g sample by the Pregl techniques leads to the production of 0.0147 g $CO_2$ and 0.0041 g $H_2O$. What is the empirical formula for progesterone? Since we now know the molecular weight of progesterone to be 314, what is the molecular formula?

**2.40** The Dumas method of analysis for nitrogen content in a molecule involves measuring the amount of nitrogen gas that is produced in a complex degradation reaction. Application of the gas laws tells us that 28 mg $N_2$ (1.0 millimole, mmol) has a volume of 22.4 milliliters (mL). By measuring the amount of $N_2$ derived from a sample of known weight, we can arrive at a value for the percentage of nitrogen in the sample.

Cadaverine, an aptly named amine with a molecular weight of 102, was analyzed for C, H, and N. Analysis of a 0.040 g sample yielded 0.086 g $CO_2$, 0.051 g $H_2O$, and 8.6 mL $N_2$ gas collected at standard temperature and pressure. How many grams of $N_2$ gas were produced? What is the percentage of N in the 0.040 g sample? Using this information, derive the molecular formula for cadaverine. What olfactory properties would you expect this molecule to possess?

**2.41** Dimethyl sulfone has a high dipole moment ($\mu = 4.4$ D). Calculate the formal charges present on oxygen and sulfur, and suggest a geometry that is consistent with the observed dipole moment.

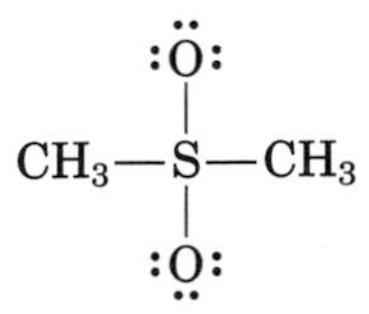

```
        :Ö:
         |
CH3—S—CH3
         |
        :Ö:
```

Dimethyl sulfone

CHAPTER 3

# The Nature of Organic Compounds: Alkanes and Cycloalkanes

According to *Chemical Abstracts*, the invaluable publication that abstracts and indexes the chemical literature, there are more than 8 million known organic compounds. Each of these compounds has its own physical properties such as melting point and boiling point, and each has its own chemical reactivity.

Chemists have learned through many years of experience that organic compounds can be classified into families according to their structural features, and that the chemical reactivity of the members of a given family is often predictable. Instead of 8 million compounds with random reactivity, there are several dozen general families of organic compounds whose chemistry is roughly predictable.

## 3.1 Functional Groups

The structural features that allow us to class compounds together by reactivity are called **functional groups** (see Table 3.1). A functional group is a part of a larger molecule; it is composed of an atom or group of atoms whose bonds have a characteristic chemical behavior. Chemically, a given functional group behaves approximately the same in every molecule it's a part of. For example, one of the simplest functional groups is the carbon–carbon double bond. We saw in Section 1.10 that a carbon–carbon double bond consists of two parts—a sigma bond formed by head-on overlap of an $sp^2$ orbital from each carbon, and a pi bond formed by sideways overlap of a $p$ orbital from each carbon (Figure 3.1).

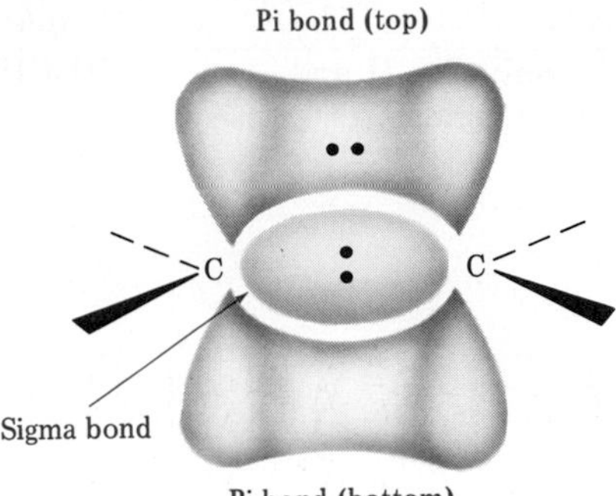


**Figure 3.1** An orbital picture of a carbon–carbon double bond: The bond has two parts—a sigma bond formed by head-on overlap of $sp^2$ orbitals and a pi bond formed by sideways overlap of $p$ orbitals.

Since the electronic nature of the carbon–carbon double bond remains essentially the same in all molecules where it occurs, its chemical reactivity also remains the same. Ethylene, the simplest compound with a double bond, undergoes reactions that are remarkably similar to those of cholesterol, a much more complicated molecule. Both, for example, react with bromine to give products in which a bromine atom has *added* to each of the double-bond carbons (Figure 3.2).

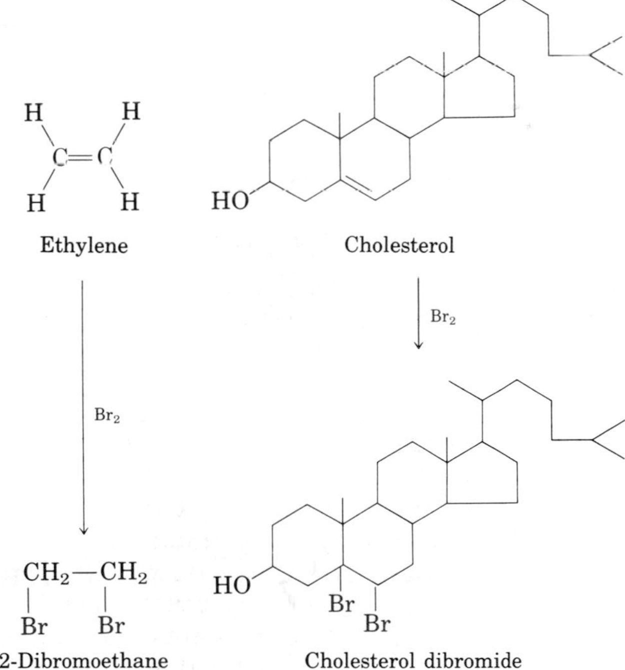


**Figure 3.2** The reactions of ethylene and cholesterol with bromine: In both cases, bromine reacts with the C=C double-bond functional group in exactly the same way. The size and nature of the remainder of the molecule is irrelevant.

Table 3.1 Structures of some important functional groups

| Family name | Functional group structure[a] | Simple example | Name ending |
|---|---|---|---|
| Alkane | (Contains only C—H and C—C single bonds) | $CH_3CH_3$ | *-ane* Ethane |
| Alkene | $>C=C<$ | $H_2C=CH_2$ | *-ene* Ethene (Ethylene) |
| Alkyne | —C≡C— | H—C≡C—H | *-yne* Ethyne (Acetylene) |
| Arene | six-membered ring of alternating C=C and C—C bonds, one open bond on each C | $C_6H_6$ (an H on each ring C) | None Benzene |
| Halide | $-\overset{\vert}{\underset{\vert}{C}}-\ddot{\underset{\cdot\cdot}{X}}:$ (X = F, Cl, Br, I) | $H_3C—Cl$ | None Chloromethane |
| Alcohol | $-\overset{\vert}{\underset{\vert}{C}}-\ddot{\underset{\cdot\cdot}{O}}-H$ | $H_3C—O—H$ | *-ol* Methanol |
| Ether | $-\overset{\vert}{\underset{\vert}{C}}-\ddot{\underset{\cdot\cdot}{O}}-\overset{\vert}{\underset{\vert}{C}}-$ | $H_3C—O—CH_3$ | *ether* Dimethyl ether |
| Amine | $-\overset{\vert}{\underset{\vert}{C}}-\ddot{N}(H)-H$, $-\overset{\vert}{\underset{\vert}{C}}-\ddot{N}(—)-H$, $-\overset{\vert}{\underset{\vert}{C}}-\ddot{N}(—)-$ | $H_3C—NH_2$ | *-amine* Methylamine |
| Nitrile | $-\overset{\vert}{\underset{\vert}{C}}-C\equiv N:$ | $H_3C—C\equiv N$ | *-nitrile* Ethanenitrile (Acetonitrile) |
| Nitro | $-\overset{\vert}{\underset{\vert}{C}}-\overset{+}{N}(=\ddot{O}:)-\ddot{O}:^{-}$ | $H_3C—\overset{+}{N}(=O)—O^{-}$ | None Nitromethane |
| Sulfide | $-\overset{\vert}{\underset{\vert}{C}}-\ddot{\underset{\cdot\cdot}{S}}-\overset{\vert}{\underset{\vert}{C}}-$ | $H_3C—S—CH_3$ | *sulfide* Dimethyl sulfide |
| Sulfoxide | $-\overset{\vert}{\underset{\vert}{C}}-\underset{\cdot\cdot}{S^{+}}(-\ddot{O}:^{-})-\overset{\vert}{\underset{\vert}{C}}-$ | $H_3C—S^{+}(O^{-})—CH_3$ | *sulfoxide* Dimethyl sulfoxide |

[a]The bonds whose connections aren't specified are assumed to be attached to carbon or hydrogen atoms in the rest of the molecule.

| Family name | Functional group structure[a] | Simple example | Name ending |
|---|---|---|---|
| Sulfone | 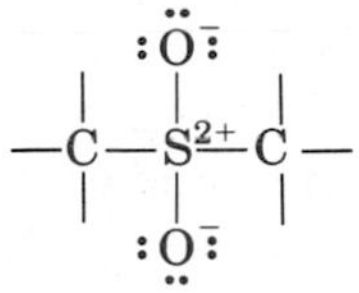 $-\overset{\vert}{\underset{\vert}{C}}-\overset{:\ddot{O}^{-}:}{\underset{:\ddot{O}^{-}:}{S^{2+}}}-\overset{\vert}{\underset{\vert}{C}}-$ | 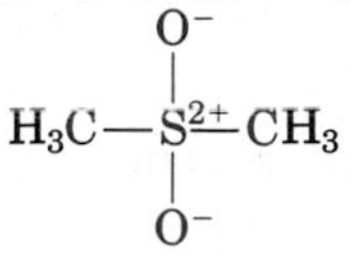 $H_3C-\overset{O^-}{\underset{O^-}{S^{2+}}}-CH_3$ | *sulfone*<br>Dimethyl sulfone |
| Thiol | 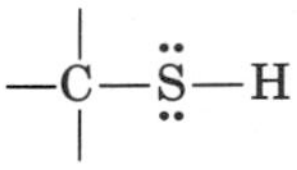 $-\overset{\vert}{\underset{\vert}{C}}-\ddot{\underset{\cdot\cdot}{S}}-H$ | $H_3C-SH$ | *-thiol*<br>Methanethiol |
| **Carbonyl,** 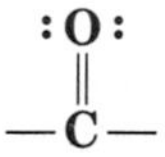 $-\overset{:O:}{\overset{\Vert}{C}}-$ | | | |
| Aldehyde | 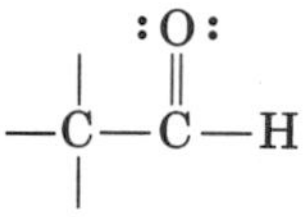 $-\overset{\vert}{\underset{\vert}{C}}-\overset{:O:}{\overset{\Vert}{C}}-H$ | 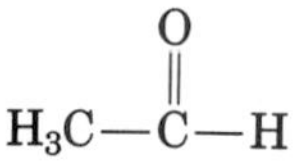 $H_3C-\overset{O}{\overset{\Vert}{C}}-H$ | *-al*<br>Ethanal<br>(Acetaldehyde) |
| Ketone | 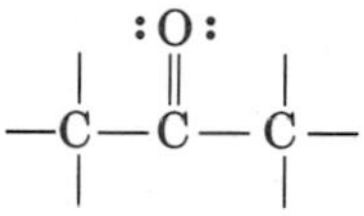 $-\overset{\vert}{\underset{\vert}{C}}-\overset{:O:}{\overset{\Vert}{C}}-\overset{\vert}{\underset{\vert}{C}}-$ | 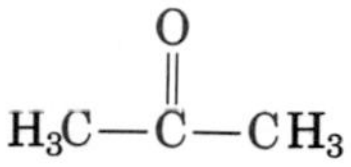 $H_3C-\overset{O}{\overset{\Vert}{C}}-CH_3$ | *-one*<br>Propanone<br>(Acetone) |
| Carboxylic acid | 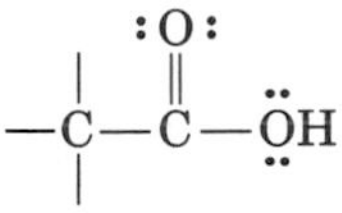 $-\overset{\vert}{\underset{\vert}{C}}-\overset{:O:}{\overset{\Vert}{C}}-\ddot{\underset{\cdot\cdot}{O}}H$ | 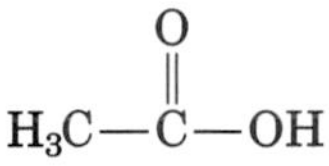 $H_3C-\overset{O}{\overset{\Vert}{C}}-OH$ | *-oic acid*<br>Ethanoic acid<br>(Acetic acid) |
| Ester | 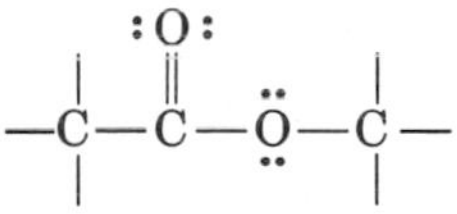 $-\overset{\vert}{\underset{\vert}{C}}-\overset{:O:}{\overset{\Vert}{C}}-\ddot{\underset{\cdot\cdot}{O}}-\overset{\vert}{\underset{\vert}{C}}-$ | 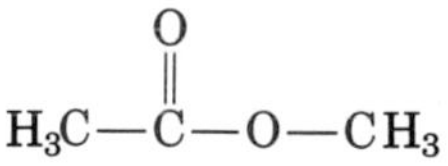 $H_3C-\overset{O}{\overset{\Vert}{C}}-O-CH_3$ | *-oate*<br>Methyl ethanoate<br>(Methyl acetate) |
| Amide | 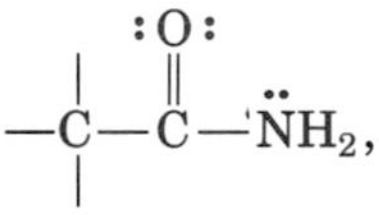 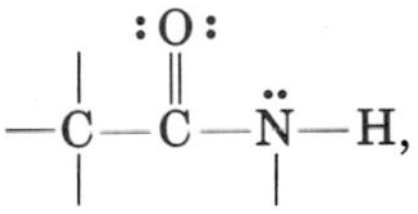 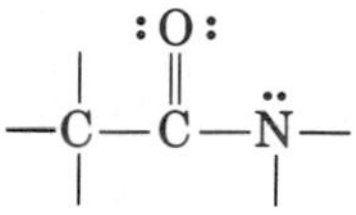 $-\overset{\vert}{\underset{\vert}{C}}-\overset{:O:}{\overset{\Vert}{C}}-\ddot{N}H_2$,<br>$-\overset{\vert}{\underset{\vert}{C}}-\overset{:O:}{\overset{\Vert}{C}}-\underset{\vert}{\ddot{N}}-H$,<br>$-\overset{\vert}{\underset{\vert}{C}}-\overset{:O:}{\overset{\Vert}{C}}-\underset{\vert}{\ddot{N}}-$ | 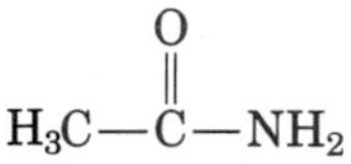 $H_3C-\overset{O}{\overset{\Vert}{C}}-NH_2$ | *-amide*<br>Ethanamide<br>(Acetamide) |
| Carboxylic acid chloride | 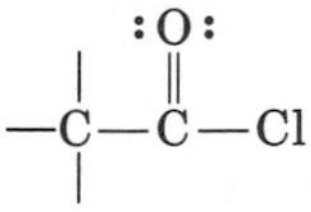 $-\overset{\vert}{\underset{\vert}{C}}-\overset{:O:}{\overset{\Vert}{C}}-Cl$ | 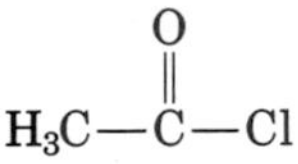 $H_3C-\overset{O}{\overset{\Vert}{C}}-Cl$ | *-oyl chloride*<br>Ethanoyl chloride<br>(Acetyl chloride) |
| Carboxylic acid anhydride | 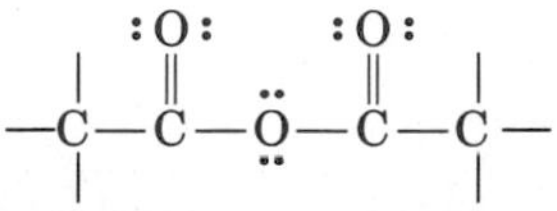 $-\overset{\vert}{\underset{\vert}{C}}-\overset{:O:}{\overset{\Vert}{C}}-\ddot{\underset{\cdot\cdot}{O}}-\overset{:O:}{\overset{\Vert}{C}}-\overset{\vert}{\underset{\vert}{C}}-$ | 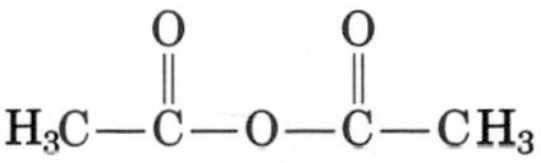 $H_3C-\overset{O}{\overset{\Vert}{C}}-O-\overset{O}{\overset{\Vert}{C}}-CH_3$ | *-oic anhydride*<br>Ethanoic anhydride<br>(Acetic anhydride) |

The example shown in Figure 3.2 is typical: The chemistry of all organic molecules, regardless of size and complexity, is determined by the functional groups they contain. Table 3.1 lists many of the common functional groups and gives simple examples of their occurrence. Look carefully at this table to see the many types of functional groups found in organic compounds. Some functional groups, such as alkenes, alkynes, and aromatic rings, have only carbon–carbon double or triple bonds; others have a halogen; and still others have oxygen, nitrogen, or sulfur. Much of the chemistry you'll be studying in the remainder of this book is the chemistry of these functional groups.

It's a good idea at this point to memorize the structures of the functional groups shown in Table 3.1 so that they will be familiar when you see them again. They can be grouped into several categories to aid memorization.

## FUNCTIONAL GROUPS WITH CARBON–CARBON MULTIPLE BONDS

Alkenes, alkynes, and arenes (aromatic compounds) all contain carbon–carbon multiple bonds. Alkenes have a double bond, alkynes have a triple bond, and aromatic rings have three alternating double and single bonds in a six-membered ring of carbon atoms. Because of their structural similarities, these compounds also have some chemical similarities:

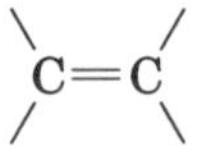

$$\begin{array}{ccc} >C{=}C< & -C{\equiv}C- & \text{(six-membered ring of C atoms with alternating double and single bonds)} \\ \text{Alkene} & \text{Alkyne} & \text{Arene (aromatic ring)} \end{array}$$

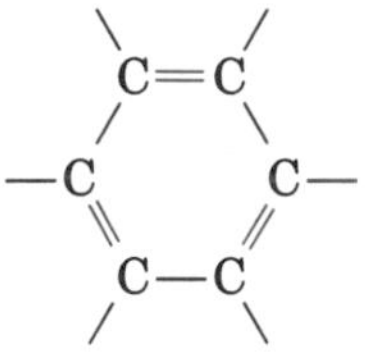

## FUNCTIONAL GROUPS WITH CARBON SINGLY BONDED TO ELECTRONEGATIVE ATOMS

Alkyl halides, alcohols, ethers, amines, sulfides, thiols, and several others all have a carbon atom singly bonded to an electronegative atom—a halogen, an oxygen, a nitrogen, or a sulfur. Alkyl halides have a carbon atom bonded to halogen, alcohols have a carbon atom bonded to a hydroxyl (—OH) group, ethers have two carbon atoms bonded to the same oxygen, amines have a carbon atom bonded to a nitrogen, thiols have a carbon atom bonded to an —SH group, and sulfides have two carbon atoms bonded to the same sulfur. In all cases, the bonds are polar, with the carbon atom bearing a slight positive charge ($\delta^+$) and the electronegative atom bearing a slight negative charge ($\delta^-$):

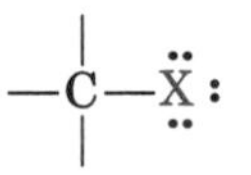

$$\begin{array}{ccc} -\overset{|}{\underset{|}{C}}-\ddot{\underset{\cdot\cdot}{X}}: & -\overset{|}{\underset{|}{C}}-\ddot{\underset{\cdot\cdot}{O}}-H & -\overset{|}{\underset{|}{C}}-\ddot{\underset{\cdot\cdot}{O}}-\overset{|}{\underset{|}{C}}- \\ \text{Alkyl halide} & \text{Alcohol} & \text{Ether} \end{array}$$

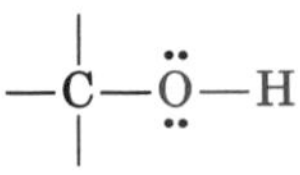

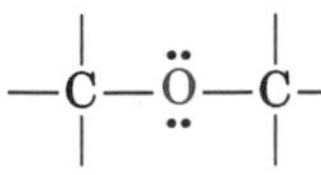

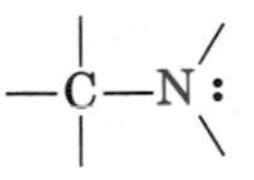

—C—N: Amine  —C—S—H Thiol  —C—S—C— Sulfide

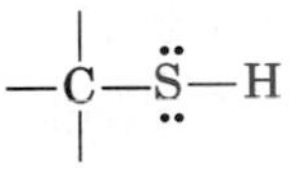

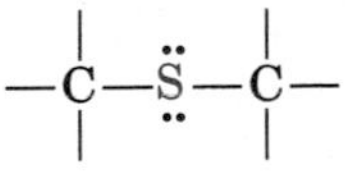

## FUNCTIONAL GROUPS WITH A CARBON–OXYGEN DOUBLE BOND (CARBONYL GROUPS)

Note particularly in Table 3.1 the different families of compounds that contain the **carbonyl group, C=O** (pronounced car-bo-**neel**). Carbon–oxygen double bonds are present in some of the most important compounds in organic chemistry. These compounds are similar in many respects but differ depending on the identity of the atoms bonded to the carbonyl-group carbon. Aldehydes have one carbon and one hydrogen bonded to the C=O, ketones have two carbons bonded to the C=O, carboxylic acids have one carbon and one —OH group bonded to the C=O, esters have one carbon and one ether-like oxygen bonded to the C=O, amides have one carbon and one amine-like nitrogen bonded to the C=O, acid chlorides have a chloro group bonded to the C=O, and so on:

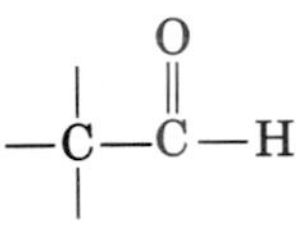

—C—C(=O)—H Aldehyde  —C—C(=O)—C— Ketone  —C—C(=O)—O—H Carboxylic acid

—C—C(=O)—O—C— Ester  —C—C(=O)—N Amide  —C—C(=O)—Cl Acid chloride

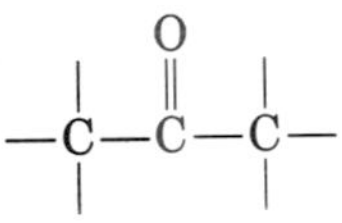

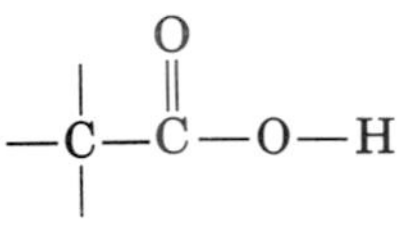

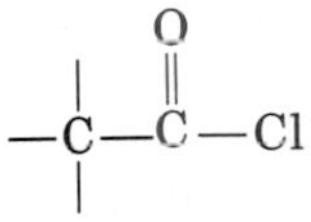

**PROBLEM**

**3.1** Circle and identify the functional groups present in each of the following molecules:

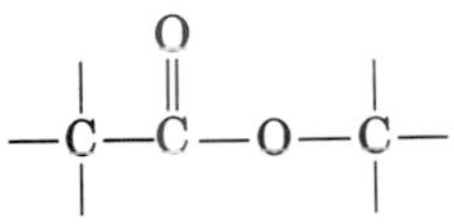

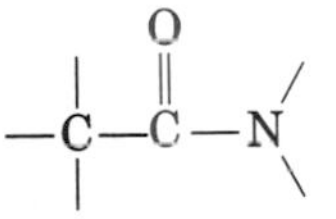

(a) 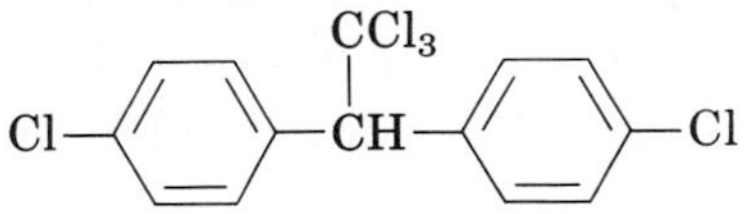 Cl—$C_6H_4$—CH($CCl_3$)—$C_6H_4$—Cl
DDT

(b) 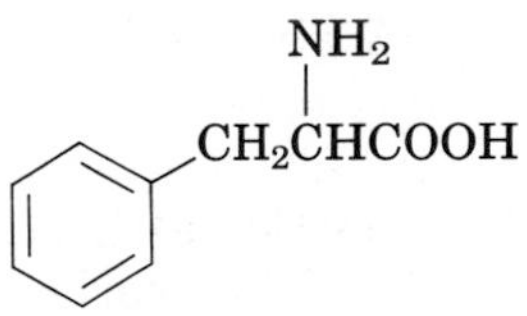 $C_6H_5$—$CH_2CH(NH_2)COOH$
Phenylalanine

(c) 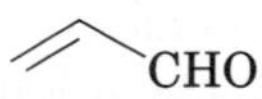 $CH_2$=CH—CHO
Acrolein

(d) 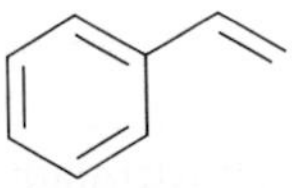 $C_6H_5$—CH=$CH_2$
Styrene

PROBLEM

**3.2** Propose structures for simple molecules that contain these functional groups:
(a) Alcohol (b) Aromatic ring (c) Carboxylic acid
(d) Amine (e) Both ketone and amine (f) Two double bonds

## 3.2 Straight-Chain Alkanes and Alkyl Groups

We saw earlier (Section 1.9) that carbon–carbon single bonds result from overlap of two carbon $sp^3$ orbitals. If we imagine joining three, four, five, or even more carbon atoms together, it's possible to generate a large number of compounds of *increasing chain length*, as shown in Figure 3.3.

```
     H             H   H               H   H   H               H   H   H
     |             |   |               |   |   |               |   |   |
  H—C—H   <   H—C—C—H   <   H—C—C—C—H   <   H—C—C—C—   —H
     |             |   |               |   |   |               |   |   |
     H             H   H               H   H   H               H   H   H

  Methane        Ethane              Propane                  Butane

                                            H   H   H   H     H
                                            |   |   |   |     |
                                      < H—C—C—C—C. . .C—H
                                            |   |   |   |     |
                                            H   H   H   H     H

                                                 CnH2n+2
```

**Figure 3.3** The increasing chain length of normal alkanes, $C_nH_{2n+2}$

Compounds generated in this way are called **straight-chain alkanes** or **normal alkanes**. All have the general formula $C_nH_{2n+2}$, where $n$ is any integer. Alkanes are often called **saturated hydrocarbons**—*saturated* because they have the maximum possible number of hydrogens per carbon and *hydrocarbons* because they contain only hydrogen and carbon. They are also occasionally referred to as being **aliphatic**, derived from the Greek word *aleiphas*, "fat." As we'll see in Chapter 28, animal fats do indeed contain long carbon chains similar to alkanes.

A given alkane can be depicted arbitrarily in a great many ways. For example, butane, the straight-chain alkane with four carbons, can be represented by *any* of the structures in Figure 3.4. These structures don't imply any particular geometry for butane; they indicate only that butane has a chain of four carbons. In practice we soon tire of drawing bonds at all and refer to butane as $CH_3CH_2CH_2CH_3$ or simply as $n$-$C_4H_{10}$, where $n$ signifies *normal*, straight-chain butane.

With the exception of the first four compounds—methane, ethane, propane, and butane—whose names have historical roots, alkanes are named from Greek numbers according to how many carbons are present (Table 3.2).

$$CH_3{-}CH_2{-}CH_2{-}CH_3 = CH_3{-}CH_2{-}CH_2(CH_3) = CH_3{-}CH_2(CH_2{-}CH_3) = CH_3(CH_2)_2CH_3$$

$$= CH_2(CH_3){-}CH_2(CH_3) = CH_3\,CH_2\ CH_2\,CH_3 = \text{(line structure)}$$

**Figure 3.4** Some representations of butane, $C_4H_{10}$: The molecule is the same, regardless of how it's drawn.

**Table 3.2** Alkane names

| *Number of carbons (n)* | *Name* | *Formula* ($C_nH_{2n+2}$) | *Number of carbons (n)* | *Name* | *Formula* ($C_nH_{2n+2}$) |
|---|---|---|---|---|---|
| 1 | Methane | $CH_4$ | 11 | Undecane | $C_{11}H_{24}$ |
| 2 | Ethane | $C_2H_6$ | 12 | Dodecane | $C_{12}H_{26}$ |
| 3 | Propane | $C_3H_8$ | 13 | Tridecane | $C_{13}H_{28}$ |
| 4 | Butane | $C_4H_{10}$ | 14 | Tetradecane | $C_{14}H_{30}$ |
| 5 | Pentane | $C_5H_{12}$ | 20 | Icosane | $C_{20}H_{42}$ |
| 6 | Hexane | $C_6H_{14}$ | 21 | Henicosane | $C_{21}H_{44}$ |
| 7 | Heptane | $C_7H_{16}$ | 22 | Docosane | $C_{22}H_{46}$ |
| 8 | Octane | $C_8H_{18}$ | 30 | Triacontane | $C_{30}H_{62}$ |
| 9 | Nonane | $C_9H_{20}$ | 40 | Tetracontane | $C_{40}H_{82}$ |
| 10 | Decane | $C_{10}H_{22}$ | 50 | Pentacontane | $C_{50}H_{102}$ |

The suffix *-ane* is added to the end of each name to indicate that the molecule identified is an alkane. Thus, pent*ane* is the five-carbon alkane, hex*ane* is the six-carbon alkane, and so on. The names of at least the first ten should be memorized.

If one hydrogen atom is removed from an alkane, the part of the molecule that remains is called an **alkyl group**. Alkyl groups are named by replacing the *-ane* ending of the parent alkane by an *-yl* ending. For example, removal of a hydrogen from methane, $CH_4$, generates the **methyl group**, $CH_3$—. Similarly, removal of a hydrogen atom from an end carbon of any *n*-alkane produces the series of straight-chain *n*-alkyl groups shown in Table 3.3. The combination of an alkyl group with any of the functional groups listed earlier allows us to generate and name many hundreds of thousands of compounds.

| Methane | A methyl group[1] | Methylamine | Methyl alcohol |
|---|---|---|---|
| $H{-}CH_2{-}H$ | $H{-}CH_2{-}\xi$ | $H{-}CH_2{-}NH_2$ | $H{-}CH_2{-}OH$ |

[1]The symbol $\xi$ will be used throughout this book to indicate that the partial organic structure shown is bonded to another, unspecified group

**Table 3.3** Some straight-chain alkyl groups

| *Alkane* | *Alkyl group* | *Example* |
|---|---|---|
| $CH_4$ | $CH_3—$ | $CH_3—OH$ |
| Methane | Methyl (abbreviated Me) | Methyl alcohol |
| $CH_3CH_3$ | $CH_3CH_2—$ | $CH_3CH_2—NH_2$ |
| Ethane | Ethyl (abbreviated Et) | Ethylamine |
| $CH_3CH_2CH_3$ | $CH_3CH_2CH_2—$ or $n$-$C_3H_7$ | $CH_3CH_2CH_2—Li$ |
| Propane | Propyl (abbreviated Pr) | Propyllithium |
| $CH_3CH_2CH_2CH_3$ | $CH_3CH_2CH_2CH_2—$ or $n$-$C_4H_9$ | $CH_3CH_2CH_2CH_2—Br$ |
| Butane | Butyl (abbreviated Bu) | Butyl bromide |

## 3.3 Branched-Chain Alkanes: Isomers

Methane and ethane each have only one kind of hydrogen. That is, the four hydrogens in $CH_4$ are all equivalent, and the six hydrogens in $C_2H_6$ are all equivalent. No matter which of the four methane hydrogens or six ethane hydrogens we remove, only one kind of methyl group and one kind of ethyl group result. The situation is more complex with higher alkanes, however. Propane and butane each have two kinds of hydrogens, pentane and hexane each have three kinds of hydrogens, and so on, as shown in Figure 3.5.

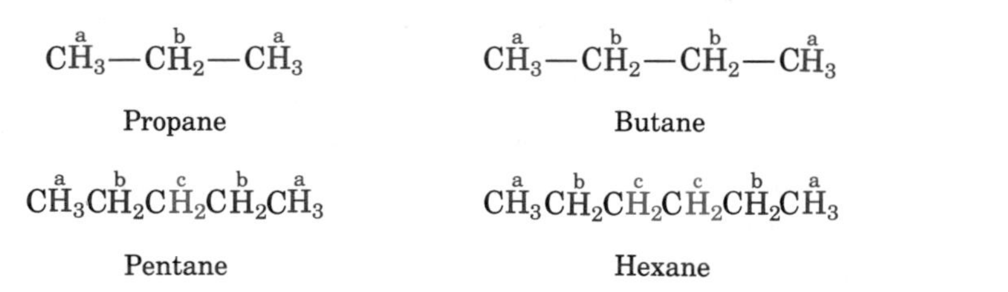


**Figure 3.5** Kinds of hydrogen atoms in propane and some higher alkanes: The superscript letters denote the different kinds of hydrogens in each molecule.

The simplest way to tell how many kinds of hydrogens a compound contains is to replace each hydrogen, one at a time, by some other atom such as Cl and see how many different compounds result. For example, replacement of any methane hydrogen by Cl would yield only one compound, $CH_3Cl$; replacement of one hydrogen at a time in propane, however, would lead to two compounds, $CH_3CH_2CH_2Cl$ and $CH_3CHClCH_3$.

We generated the series of straight-chain alkanes in Table 3.2 by successively replacing a terminal hydrogen of a lower alkane with a methyl group. It's equally possible to imagine replacing *internal* hydrogen atoms with alkyl groups and to generate thereby a vast number of **branched-chain alkanes**.

Beginning with propane, for example, we can replace a terminal hydrogen by a methyl group to generate butane, or we can replace an internal hydrogen by a methyl group to generate the branched four-carbon alkane, 2-methylpropane or *isobutane*. Although butane and isobutane have the same formula, $C_4H_{10}$, they have clearly different structures.

```
                                  H   H   H   H
                                  |   |   |   |
                          ---> H—C—C—C—C—H
                         /        |   |   |   |
    H   H   H           /         H   H   H   H
    |   |   |          /
 H—C—C—C—H                         Butane
    |   |   |          \
    H   H   H           \                 H
                         \                |
                          \      H   H—C—H   H
                           \     |        |       |
                            ---> H—C———C———C—H
                                 |        |       |
                                 H        H       H

                                  2-Methylpropane
                                    (Isobutane)
```

Compounds that have the same formula but different chemical structures are called **isomers** (from the Greek *isos* + *meros*, "made of the same parts"). Isomers are indeed made of the same parts: They have the same numbers and kinds of atoms, but their atoms are arranged differently. Compounds such as butane and isobutane, which have their atoms connected in a different order, are called **constitutional isomers**. We'll see shortly that other kinds of isomerism are also possible, even among compounds whose atoms are connected in the same order.

There are an enormous number of possibilities for branching in the alkane series. Although there is only one methane, one ethane, and one propane, there are two butane isomers, three pentane isomers, five hexane isomers, and so on. As Table 3.4 on page 68 shows, there are more than 62 trillion possible isomers of $C_{40}H_{82}$!

Constitutional isomerism is not limited to alkanes—it occurs widely throughout organic chemistry. It's often useful to classify constitutional isomers into three subgroups—skeletal, functional, and positional. **Skeletal isomers** are compounds such as butane and isobutane that have the same formula but have different carbon skeletons. The differences are usually obvious:

Skeletal isomers (different carbon skeletons)

```
       CH3
       |
CH3—CH—CH3      and      CH3—CH2—CH2—CH3
```

Table 3.4 Numbers of possible alkane isomers

| *Formula* | *Isomers* | | |
|---|---|---|---|
| $C_1H_4$ | $CH_4$ | | |
| $C_2H_6$ | $CH_3CH_3$ | | |
| $C_3H_8$ | $CH_3CH_2CH_3$ | | |
| $C_4H_{10}$ | $CH_3CH_2CH_2CH_3$ | $CH_3$<br>\|<br>$CH_3CHCH_3$ | |
| $C_5H_{12}$ | $CH_3CH_2CH_2CH_2CH_3$ | $CH_3$<br>\|<br>$CH_3CHCH_2CH_3$ | $CH_3$<br>\|<br>$CH_3—C—CH_3$<br>\|<br>$CH_3$ |
| $C_6H_{14}$ | $CH_3CH_2CH_2CH_2CH_2CH_3$ | $CH_3$<br>\|<br>$CH_3CHCH_2CH_2CH_3$ | $CH_3$<br>\|<br>$CH_3CH_2CHCH_2CH_3$ |
| | $CH_3$ $CH_3$<br>\| \|<br>$CH_3CH—CHCH_3$ | $CH_3$<br>\|<br>$CH_3—C—CH_2CH_3$<br>\|<br>$CH_3$ | |

| *Formula* | *Number of isomers* | *Formula* | *Number of isomers* |
|---|---|---|---|
| $C_7H_{16}$ | 9 | $C_{12}H_{26}$ | 355 |
| $C_8H_{18}$ | 18 | $C_{15}H_{32}$ | 4,347 |
| $C_9H_{20}$ | 35 | $C_{20}H_{42}$ | 366,319 |
| $C_{10}H_{22}$ | 75 | $C_{30}H_{62}$ | 4,111,846,763 |
| $C_{11}H_{24}$ | 159 | $C_{40}H_{82}$ | 62,491,178,805,831 |

**Functional-group isomers** are compounds that have the same formula but have different functional groups. For example, ethyl alcohol and dimethyl ether both have the formula $C_2H_6O$, but one contains an alcohol functional group (—OH) and the other contains an ether functional group (C—O—C). Again, the differences between functional-group isomers are usually obvious:

Functional-group isomers (different functional groups)

$CH_3—CH_2—OH$ and $CH_3—O—CH_3$

Ethyl alcohol — Dimethyl ether

**Positional isomers** are compounds that have the same formula, carbon skeleton, and functional groups but have the functional groups located at different positions along the carbon skeleton. For example, 1-propanol and 2-propanol are both straight-chain alcohols, but one has the —OH group attached to C1 and the other has the —OH group attached to C2:

Positional isomers (different position of functional group)

$CH_3—CH_2—CH_2—OH$ and

OH<br>
|<br>
$CH_3—CH—CH_3$

1-Propanol — 2-Propanol

PROBLEM

**3.3** There are seven constitutional isomers with the formula $C_4H_{10}O$. Draw as many as you can.

PROBLEM

**3.4** Draw structures of the nine isomers of $C_7H_{16}$.

PROBLEM

**3.5** Propose structures that meet the following descriptions:
(a) Two isomeric esters with formula $C_5H_{10}O_2$
(b) Two isomeric nitriles with formula $C_4H_7N$

PROBLEM

**3.6** How many isomers are there that have the following structures?
(a) Alcohols with formula $C_3H_8O$ (b) Bromoalkanes with formula $C_4H_9Br$

## 3.4 Nomenclature of Alkanes

In earlier times when relatively few pure organic chemicals were known, new compounds were named at the whim of their discoverer. Thus, urea ($CH_4N_2O$) is a pure crystalline substance isolated from urine; morphine ($C_{17}H_{19}NO_3$) is an analgesic (painkiller) isolated by Sertürner[2] in 1805 from the opium poppy and named after Morpheus, the Greek god of dreams; and barbituric acid is a tranquilizing agent named by its discoverer in honor of his friend Barbara.

As the science of organic chemistry slowly grew in the nineteenth century, so too did the need for a systematic method of unambiguously naming organic compounds. The system of **nomenclature** we'll use in this book is that devised by the International Union of Pure and Applied Chemistry (IUPAC—usually spoken as **eye**-you-pac). IUPAC rules are available for dealing with all functional groups and for unambiguously naming all but the most complex structures.

A chemical name has three parts in the IUPAC system: prefix, parent, and suffix. The **parent** tells how many carbon atoms are in the main chain and indicates the overall size of the molecule, the **suffix** identifies the functional groups present in the molecule, and the **prefix** specifies the location of the functional groups and other substituents on the main chain:

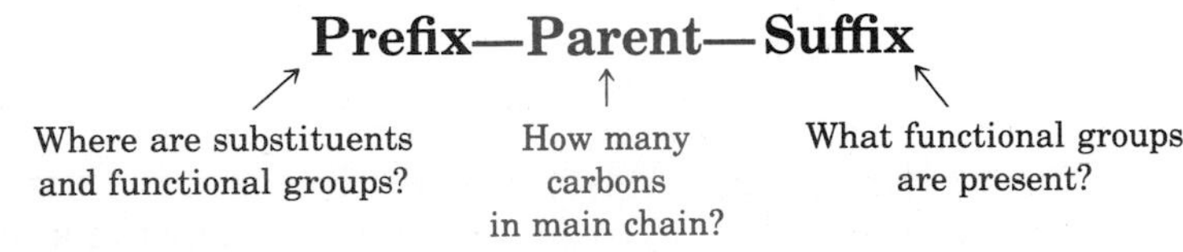


As we cover new functional groups in later chapters, the applicable IUPAC rules of nomenclature will be given. In addition, Appendix A at the end of this book shows how compounds that contain more than one functional group can be named. For the present, though, let's see how we can name branched-chain alkanes.

[2]Friedrich Wilhelm Adam Sertürner (1783–1841); b. Neuhaus, Germany; apothecary in Paderborn, Eimbeck, and Hameln.

According to IUPAC rules, most branched-chain alkanes can be named by following four steps. For a few very complex alkanes, a fifth step is needed.

*Step 1* Find the parent hydrocarbon.

a. Find the *longest continuous carbon chain* present in the molecule and use the name of that chain as the parent name. The longest chain may not always be apparent from the manner of writing; you may have to "turn corners."

```
              CH2CH3
              |
CH3CH2CH2CH—CH3                Named as a substituted hexane

           CH3
           |
           CH2
           |
   CH3—CHCH—CH2CH3             Named as a substituted heptane
              |
              CH2CH2CH3
```

b. If two different chains of equal length are present, select the one with the larger number of branch points as the parent:

```
     CH3
     |
CH3CHCHCH2CH2CH3               Named as a hexane with two
       |                       substituents
       CH2CH3

     CH3                       NOT
     |
CH3CH—CHCH2CH2CH3              as a hexane with one
       |                       substituent
       CH2CH3
```

*Step 2* Number the atoms in the main chain.

a. Beginning at the end *nearer the first branch point*, number each carbon atom in the longest chain you have identified:

```
        1 CH3                                 7 CH3
          |                                     |
        2 CH2                                 6 CH2
          |                                     |
  CH3—CHCH—CH2CH3          NOT         CH3—CHCH—CH2CH3
      3  |4                                5  |4
         CH2CH2CH3                            CH2CH2CH3
          5   6   7                            3   2   1
```

The first branch occurs at C3 in the proper numbering system but at C4 in the improper system.

b. If there is branching an equal distance away from both ends of the parent chain, begin numbering at the end nearer the *second* branch point:

```
      9  8
      CH3CH2          CH3  CH2CH3
         |             |    |
CH3—CHCH2CH2CH—CHCH2CH3
     7  6   5   4     3  2   1
```

*NOT*

```
      1  2
      CH3CH2          CH3  CH2CH3
         |             |    |
CH3—CHCH2CH2CH—CHCH2CH3
     3  4   5   6     7  8   9
```

*Step 3* Identify and number the substituents.

a. Using the numbering system you've decided is correct, assign a number to each substituent according to its point of attachment to the main chain:

```
      9  8
      CH3CH2          CH3  CH2CH3
         |             |    |
CH3—CHCH2CH2CH—CHCH2CH3        Named as a nonane
     7  6   5   4     3  2   1
```

Substituents: On C3, $CH_2CH_3$ (3-ethyl)
On C4, $CH_3$ (4-methyl)
On C7, $CH_3$ (7-methyl)

b. If there are two substituents on the same carbon, assign them both the same number. There must always be as many numbers in the name as there are substituents:

```
                CH3
                 |
CH3CH2—C—CH2CHCH3          Named as a hexane
 6   5     4|  3   |2 1
               CH2     CH3
                 |
                CH3
```

Substituents: On C2, $CH_3$ (2-methyl)
On C4, $CH_3$ (4-methyl)
On C4, $CH_2CH_3$ (4-ethyl)

*Step 4* Write out the name as a single word, using hyphens to separate the different prefixes and using commas to separate numbers. If two or more *different* substituents are present, cite them in alphabetical order. If two or more *identical* substituents are present, use one of the prefixes *di-*, *tri-*, *tetra-*, and so forth. Don't use these prefixes for alphabetizing purposes, however.

Full names for some of the examples we have been using follow:

$$\begin{array}{l} \phantom{CH_3CH_2CH_2}\overset{2}{C}H_2\overset{1}{C}H_3 \\ \phantom{CH_3CH_2CH_2}| \\ \overset{6}{C}H_3\overset{5}{C}H_2\overset{4}{C}H_2\overset{3}{C}H-CH_3 \end{array}$$

3-Methylhexane

$$\begin{array}{c} {}^{1}CH_3 \\ | \\ {}^{2}CH_2 \\ | \\ CH_3-\underset{3}{C}H\underset{4}{C}H-CH_2CH_3 \\ | \\ \underset{5}{C}H_2\underset{6}{C}H_2\underset{7}{C}H_3 \end{array}$$

4-Ethyl-3-methylheptane

$$\begin{array}{l} CH_3 \\ | \\ \underset{1}{C}H_3\underset{2}{C}H\underset{3}{C}H\underset{4}{C}H_2\underset{5}{C}H_2\underset{6}{C}H_3 \\ \phantom{CH_3CH}| \\ \phantom{CH_3CH}CH_2CH_3 \end{array}$$

3-Ethyl-2-methylhexane

$$\begin{array}{c} CH_3 \\ | \\ \overset{6}{C}H_3\overset{5}{C}H_2-\overset{4}{C}-\overset{3}{C}H_2\overset{2}{C}H\overset{1}{C}H_3 \\ | \qquad\qquad | \\ CH_2 \qquad CH_3 \\ | \\ CH_3 \end{array}$$

4-Ethyl-2,4-dimethylhexane

$$\begin{array}{l} \overset{9}{C}H_3\overset{8}{C}H_2 \qquad\qquad CH_3 \quad CH_2CH_3 \\ \qquad | \qquad\qquad\qquad | \qquad\quad | \\ CH_3-\overset{7}{C}H\overset{6}{C}H_2\overset{5}{C}H_2\overset{4}{C}H-\overset{3}{C}H\overset{2}{C}H_2\overset{1}{C}H_3 \end{array}$$

3-Ethyl-4,7-dimethylnonane

Application of the preceding four steps allows us to name many thousands of organic compounds. In some particularly complex cases, however, a fifth step is necessary. It occasionally happens that a substituent of the main chain has sub-branching:

$$\begin{array}{l} \qquad\qquad\qquad\qquad\qquad\qquad\qquad\quad CH_3 \\ \qquad\qquad\qquad\qquad\qquad\qquad\qquad\quad | \\ \overset{1}{C}H_3\overset{2}{C}H-\overset{3}{C}H\overset{4}{C}H_2\overset{5}{C}H_2\overset{6}{C}H-CH_2CH-CH_3 \\ \qquad | \qquad | \qquad\qquad\quad | \\ \quad CH_3 \;\; CH_3 \qquad\quad \underset{7}{C}H_2\underset{8}{C}H_2\underset{9}{C}H_2\underset{10}{C}H_3 \end{array}$$

Named as a 2,3,6-trisubstituted decane

In this case, the substituent at C6 is a four-carbon unit with a sub-branch. To name the compound fully, we must first name the complex sub-branched substituent.

*Step 5* Name the complex substituent. A complex substituent is named by applying the four primary steps exactly as if it were a compound itself. In the present case, the complex substituent is a substituted propyl group:

$$\begin{array}{l} \qquad\qquad\qquad\qquad\qquad\quad CH_3 \\ \qquad\qquad\qquad\qquad\qquad\quad | \\ \text{Molecule}-\!\!\wr\!-\underset{1}{C}H_2-\underset{2}{C}H-\underset{3}{C}H_3 \end{array}$$

We begin numbering *at the point of attachment* to the main chain and find that the complex substituent is a 2-methylpropyl group. To avoid confusion, this group name is set off in parentheses when the name of the complete hydrocarbon is given:

```
                                   CH3
 1  2      3  4   5   6             |
CH3CH—CHCH2CH2CH—CH2CH—CH3
      |      |            |
     CH3  CH3       CH2CH2CH2CH3
                          7   8   9  10
```

2,3-Dimethyl-6-(2-methylpropyl)decane

As a further example:

```
                              CH3  CH3                    CH3  CH3
                               |    |                      |    |
CH3CH2CH2CH2CH—CH—CHCH3               ~CH—CHCH3
 9   8   7   6  5|                                 1     2  3
              4 CH2    CH3                 5-(1,2-Dimethylpropyl)-
                 |      |
                CH2—CH—CH3
                 3    2    1
```

2-Methyl-5-(1,2-dimethylpropyl)nonane

**PRACTICE PROBLEM**

What is the IUPAC name of the following alkane?

```
      CH2CH3            CH3
       |                 |
   CH3CHCH2CH2CH2CHCH3
```

*Solution* The molecule has a chain of eight carbons (octane) with two methyl substituents. Numbering from the end nearer the first methyl substituent indicates that the methyls are at C2 and C6, giving the name 2,6-dimethyloctane.

**PROBLEM**

**3.7** Provide proper IUPAC names for these compounds:

(a) The three isomers of $C_5H_{12}$

(b)
```
           CH3
            |
   CH3CH2CHCHCH3
              |
             CH2CH3
```

(c)
```
                CH3
                 |
   (CH3)2CHCH2CHCH3
```

(d)
```
                   CH3
                    |
   (CH3)3CCH2CH2CH
                    |
                   CH2CH3
```

PROBLEM

**3.8** Draw structures corresponding to these IUPAC names:
(a) 3,4-Dimethylnonane (b) 3-Ethyl-4,4-dimethylheptane
(c) 2,2-Dimethyl-4-propyloctane (d) 2,2,4-Trimethylpentane

PROBLEM

**3.9** The following names are incorrect. Draw the structures they represent, explain why the names are incorrect, and give correct names.
(a) 1,1-Dimethylpentane (b) 3-Methyl-2-propylhexane
(c) 4,4-Dimethyl-3-ethylpentane (d) 5-Ethyl-4-methylhexane
(e) 2,3-Methylhexane (f) 3-Dimethylpentane

## 3.5 Nomenclature of Alkyl Groups

Earlier in this chapter, we saw that straight-chain alkyl groups are formed by removal of a terminal hydrogen atom from straight-chain alkanes. It's equally possible to generate an enormous number of *branched* alkyl groups by removing *internal* hydrogen atoms from alkanes. For example, there are two possible three-carbon alkyl groups and four possible four-carbon alkyl groups (Figure 3.6). The possibilities expand at an enormous rate as the number of carbon atoms increases.

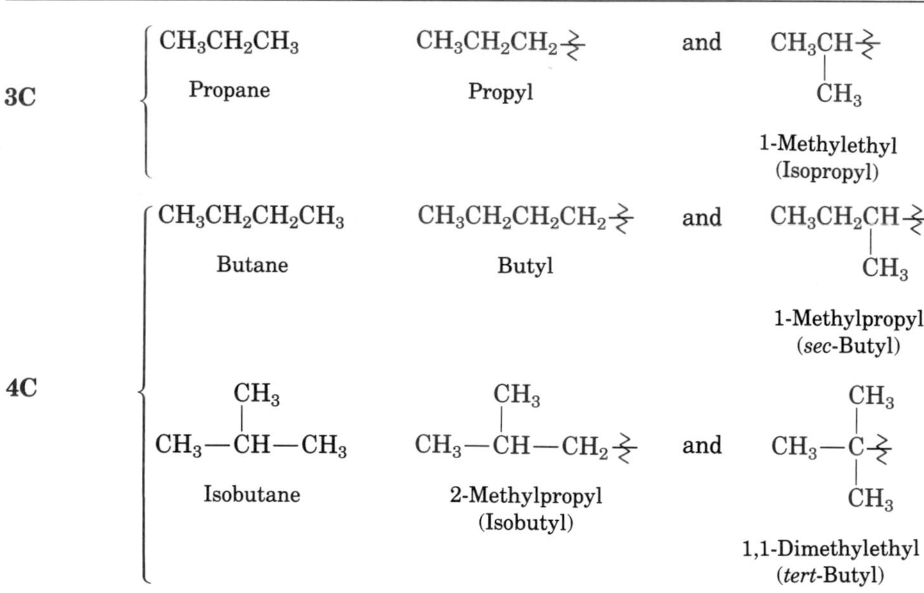


**Figure 3.6** Generation of straight- and branched-chain alkyl groups from *n*-alkanes

Branched-chain alkyl groups can be systematically named as discussed earlier in step 5 of Section 3.4. These groups are always numbered so that the point of attachment to the rest of the molecule is C1, with the longest continuous chain *beginning from the point of attachment* taken as the parent.

For historical reasons, some of the simpler branched-chain alkyl groups also have nonsystematic or *common* names.

1. Three-carbon alkyl group:

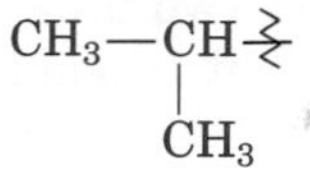

$$\begin{array}{l} CH_3-\underset{\displaystyle CH_3}{\underset{|}{CH}}- \end{array}$$

Isopropyl (abbreviated *i*-Pr)

2. Four-carbon alkyl groups:

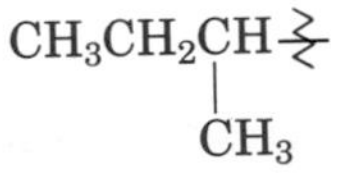

$$CH_3CH_2\underset{\displaystyle CH_3}{\underset{|}{CH}}- \qquad CH_3\underset{\displaystyle CH_3}{\underset{|}{C}}HCH_2- \qquad CH_3-\overset{\displaystyle CH_3}{\overset{|}{\underset{\displaystyle CH_3}{\underset{|}{C}}}}-$$

*sec*-Butyl (*sec*-Bu) (for *secondary*)

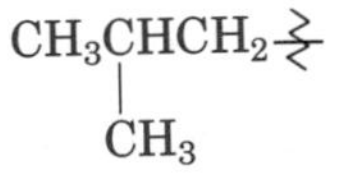

Isobutyl (*i*-Bu)

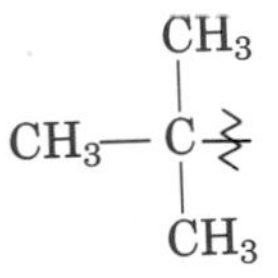

*tert*-Butyl or *t*-Butyl (*t*-Bu) (for *tertiary*)

3. Five-carbon alkyl groups:

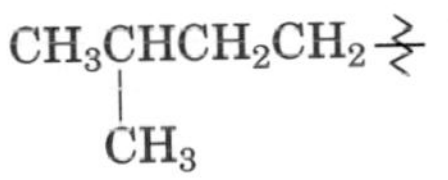

$$CH_3\underset{\displaystyle CH_3}{\underset{|}{C}}HCH_2CH_2- \qquad CH_3-\overset{\displaystyle CH_3}{\overset{|}{\underset{\displaystyle CH_3}{\underset{|}{C}}}}-CH_2- \qquad CH_3CH_2-\overset{\displaystyle CH_3}{\overset{|}{\underset{\displaystyle CH_3}{\underset{|}{C}}}}-$$

Isopentyl, also called Isoamyl (*i*-Amyl)

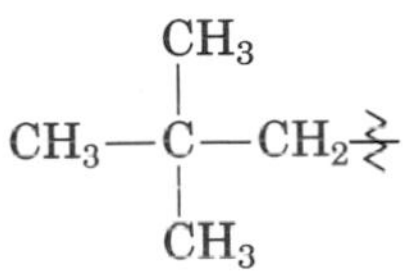

Neopentyl

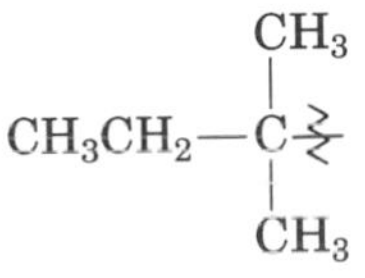

*tert*-Pentyl, also called *tert*-Amyl (*t*-amyl)

The common names of these simple alkyl groups are so well entrenched in the chemical literature that the IUPAC rules make allowance for them. Thus, the following compound may be properly named *either* 4-(1-methylethyl)heptane or 4-isopropylheptane. There's no choice but to memorize these common names; fortunately, there aren't very many of them.

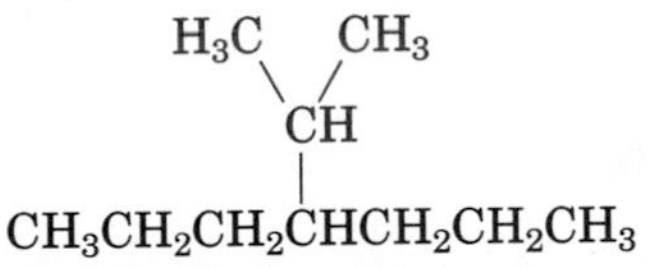

$$\begin{array}{c} H_3C \quad CH_3 \\ \diagdown \; \diagup \\ CH \\ | \\ CH_3CH_2CH_2CHCH_2CH_2CH_3 \end{array}$$

4-(1-Methylethyl)heptane or 4-Isopropylheptane

When writing an alkane name, the prefix *iso-* is considered to be part of the alkyl group name for alphabetizing purposes, but the hyphenated prefixes *sec-* and *tert-* are not. Thus, isopropyl and isobutyl are listed alphabetically under *i*, but *sec*-butyl and *tert*-butyl are listed under *b*.

One further word of explanation about branched-chain alkyl groups is necessary: The prefixes *sec-* (for secondary) and *tert-* (for tertiary) refer to the *degree of alkyl substitution* at the carbon atom in question. A *primary* carbon is directly bonded to one other carbon atom, a *secondary* carbon is directly bonded to two other carbons, and so on. By extension, a *primary hydrogen* is one that is bonded to a primary carbon, a *secondary hydrogen* is bonded to a secondary carbon, and so on. There are four possible substitution patterns for carbon:

$$\begin{array}{c} H \\ | \\ R-C-H \\ | \\ H \end{array}$$

*Primary* carbon (1°); directly bonded to one other carbon

$$\begin{array}{c} H \\ | \\ R-C-H \\ | \\ R \end{array}$$

*Secondary* carbon (2°); directly bonded to two other carbons

$$\begin{array}{c} R \\ | \\ R-C-H \\ | \\ R \end{array}$$

*Tertiary* carbon (3°); directly bonded to three other carbons

$$\begin{array}{c} R \\ | \\ R-C-R \\ | \\ R \end{array}$$

*Quaternary* carbon (4°); directly bonded to four other carbons

The symbol **R** is used here and throughout this text to represent a generalized organic portion of the molecule—methyl, ethyl, propyl, or any of an infinite number of others. You might think of **R** as representing the **R**est of the molecule, which we aren't bothering to specify because it is not important.

The terms *primary, secondary, tertiary,* and *quaternary* are routinely used in organic chemistry, and their meanings must become second nature. For example, if we were to say "The product of the reaction is a primary alcohol," we would be talking about the *general class of compounds* that has an alcohol functional group (—OH) bonded to a primary carbon atom, $RCH_2OH$:

**R** represents **R**est of molecule

$$\begin{array}{c} H \\ | \\ R-C-OH \\ | \\ H \end{array}$$

General class of primary alcohols, $R-CH_2OH$

$CH_3-CH_2OH$

$(CH_3)_2CHCH_2-CH_2OH$

(cyclohexane ring)$-CH_2-OH$

Specific examples of primary alcohols

PROBLEM

**3.10** There are eight five-carbon alkyl groups (pentyl isomers). Draw them and assign systematic names.

PROBLEM

**3.11** Draw and name alkanes that meet the following descriptions:
(a) An alkane with three tertiary carbons
(b) An alkane with two isopropyl groups
(c) An alkane with one quaternary and one secondary carbon
(d) A secondary alcohol

PROBLEM

**3.12** Identify the kinds of carbon atoms in these molecules as primary, secondary, tertiary, or quaternary:
(a) $(CH_3)_2CHCH_2C(CH_3)_3$
(b) $(CH_3)_2CHCH(CH_3)CH_2CH_2CH_3$
(c) $CH_3CH_2C(CH_3)_2C(CH_3)_2CH_2CH_3$

## 3.6 Occurrence of Alkanes: Petroleum

Many alkanes occur naturally in the plant and animal world. For example, the waxy coating on cabbage leaves contains nonacosane ($n$-$C_{29}H_{60}$), and the wood oil of the Jeffrey pine, common to the Sierra Nevada mountains, contains heptane. Beeswax contains, among other things, hentriacontane ($n$-$C_{31}H_{64}$).

By far the major sources of alkanes are the world's natural gas and petroleum deposits. Laid down eons ago, these natural deposits are derived from the decomposition of marine organic matter. **Natural gas** consists chiefly of methane, but ethane, propane, butane, and isobutane are also present. These simple hydrocarbons are used in great quantities to heat our homes, cook our food, and fuel some of our industries. **Petroleum** is a highly complex mixture of hydrocarbons that must be *refined* into different fractions before it can be used.

Refining begins by **fractional distillation** of crude oil into three principal cuts: straight-run gasoline (bp 30–200°C), kerosene (bp 175–300°C), and gas oil (bp 275–400°C). Finally, distillation under reduced pressure gives lubricating oils and waxes, and leaves an undistillable tarry residue of asphalt (Figure 3.7, page 78).

Simple distillation of crude oil is only the first step in gasoline production. It turns out that straight-run gasoline is a rather poor fuel because of the phenomenon of *engine knock*. In the normal internal-combustion engine, the downward intake stroke of the piston causes a carefully metered mixture of air and fuel to be drawn into the cylinder. The piston then compresses this mixture on its upward stroke and, at a certain point just before the end of compression, the spark plug ignites the mixture.

Not all fuels burn equally well. When poor fuels are used, combustion can be initiated in an uncontrolled manner by a hot surface in the cylinder before the spark plug fires. This *preignition*, detected as an engine knock or ping, can destroy the engine in short order by putting excessive and irregular forces on the crankshaft and by raising engine temperature.

| | | |
|---|---|---|
| Petroleum | Asphalt<br>Lubricating oils<br>Waxes | |
| | Gas oil<br>(bp 275–400°C) | $C_{14}$–$C_{25}$ hydrocarbons |
| | Kerosene<br>(bp 175–300°C) | $C_{11}$–$C_{14}$ hydrocarbons |
| | Straight-run gasoline<br>(bp 30–200°C) | $C_5$–$C_{11}$ hydrocarbons |
| | Natural gas | $C_1$–$C_4$ hydrocarbons |

**Figure 3.7** The products of petroleum refining

The fuel **octane number** is the measure by which the antiknock properties of fuels are judged. It was recognized long ago that straight-chain hydrocarbons are far more prone to induce engine knock than are highly branched compounds. Heptane, a particularly bad fuel, is assigned a base value of 0 octane number; 2,2,4-trimethylpentane (commonly known as isooctane), which has excellent antiknock characteristics, is given a rating of 100.

$CH_3CH_2CH_2CH_2CH_2CH_2CH_3$

Heptane
Octane number = 0

$$CH_3{-}\overset{\displaystyle CH_3}{\underset{\displaystyle CH_3}{C}}{-}CH_2{-}\overset{\displaystyle CH_3}{CH}{-}CH_3$$

2,2,4-Trimethylpentane
(Isooctane)
Octane number = 100

Since straight-run gasoline has a high percentage of unbranched alkanes and is therefore a poor fuel, petroleum chemists have devised sophisticated methods for producing higher-quality fuels. Two such methods are known as **catalytic cracking** and **catalytic reforming**. Although the actual chemistry taking place is extremely complex, catalytic cracking involves taking the high-boiling kerosene cut ($C_{11}$–$C_{14}$) and "cracking" it into smaller molecules suitable for use in gasoline. The process takes place on a silica-alumina catalyst at temperatures of 400–500°C, and the major products are light hydrocarbons in the $C_3$–$C_5$ range. These small hydrocarbons are then catalytically recombined to yield useful $C_7$–$C_{10}$ alkanes. Fortunately, the $C_7$–$C_{10}$ molecules that are produced are highly branched and are perfectly suited for use as high-octane fuels.

Catalytic reforming is the process by which straight-chain alkanes present in straight-run gasoline are converted into aromatic molecules such as toluene and benzene. Aromatics have high octane ratings and are therefore desirable components of gasoline.

## 3.7 Properties of Alkanes

Alkanes are often referred to as **paraffins**, a name derived from the Latin *parum affinis* ("slight affinity"). This term aptly describes their behavior: Alkanes show little chemical affinity for other molecules and are chemically inert to most reagents normally used in organic chemistry. Alkanes do, however, react with oxygen under appropriate conditions.

Reaction with oxygen occurs during combustion in an engine or furnace when the alkane is used as a fuel. Carbon dioxide and water are formed as products, and a large amount of heat is released. For example, methane (natural gas) reacts with oxygen according to the equation

$$CH_4 + 2\,O_2 \longrightarrow CO_2 + 2\,H_2O + 213 \text{ kcal/mol } (890 \text{ kJ/mol})$$

Alkanes show regular increases in both boiling point and melting point as molecular weight increases (Table 3.5), a regularity that's also reflected in other properties. For example, the average carbon–carbon bond parameters are nearly the same in all alkanes, with bond lengths of 1.54 ± 0.01 Å and bond strengths of 85 ± 3 kcal/mol (355 ± 10 kJ/mol). The carbon hydrogen bond parameters are also nearly constant, at 1.09 ± 0.01 Å and 95 ± 3 kcal/mol (400 ± 10 kJ/mol).

**Table 3.5** Physical properties of some alkanes

| *Number of carbons* | *Alkane* | *Melting point (°C)* | *Boiling point (°C)* | *Density (g/mL)* |
|---|---|---|---|---|
| 1 | Methane | −182.5 | −164.0 | 0.5547 |
| 2 | Ethane | −183.3 | −88.6 | 0.509 |
| 3 | Propane | −189.7 | −42.1 | 0.5005 |
| 4 | Butane | −138.3 | −0.5 | 0.5788 |
| 5 | Pentane | −129.7 | 36.1 | 0.6262 |
| 6 | Hexane | −95.0 | 68.9 | 0.6603 |
| 7 | Heptane | −90.6 | 98.4 | 0.6837 |
| 8 | Octane | −56.8 | 125.7 | 0.7025 |
| 9 | Nonane | −51.0 | 150.8 | 0.7176 |
| 10 | Decane | −29.7 | 174.1 | 0.7300 |
| 20 | Icosane | 36.8 | 343.0 | 0.7886 |
| 30 | Triacontane | 65.8 | 450.0 | 0.8097 |
| 4 | Isobutane | −159.4 | −11.7 | 0.579 |
| 5 | Isopentane | −159.9 | 27.85 | 0.6201 |
| 5 | Neopentane | −16.5 | 9.5 | 0.6135 |
| 8 | Isooctane | −107.4 | 99.3 | 0.6919 |

Table 3.5 also shows that increased branching has the effect of lowering an alkane's boiling point. Thus, pentane boils at 36.1°C, isopentane (2-methylbutane) has one branch and boils at 27.85°C, and neopentane (2,2-dimethylpropane) has two branches and boils at 9.5°C. Similarly, octane boils at 125.7°C, whereas isooctane (2,2,4-trimethylpentane) boils at 99.3°C. This effect can be understood by looking at what occurs during boiling.

Nonpolar molecules such as alkanes are weakly attracted to each other by intermolecular **van der Waals forces**. These forces, which operate only over very small distances, result from induced polarization of the electron clouds in molecules. Although the electron distribution in a molecule is uniform on average over a period of time, the distribution at any given instant is not uniform. One side of a molecule may, by chance, have a slight excess of electrons relative to the opposite side. When that occurs, the molecule has a temporary dipole moment. This temporary dipole in one molecule causes a nearby molecule to adopt a temporarily opposite dipole, with the result that a tiny attraction is induced between the two molecules (Figure 3.8).

Temporary dipoles have a fleeting existence and are constantly changing, but the cumulative effect of an enormous number of these interactions produces attractive forces sufficient to cause the molecules to stay in the liquid state rather than the gaseous state. Only when sufficient heat energy is applied to overcome these forces does the liquid boil.

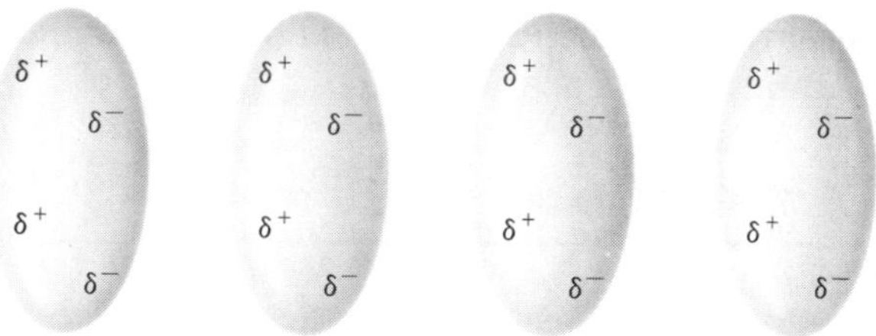


**Figure 3.8** Attractive van der Waals forces caused by temporary dipoles in molecules

You might expect that van der Waals forces would increase as molecule size increases, an expectation that is borne out in the alkane series. Although other factors are also involved, at least part of the increase in boiling point on going up the alkane series is due to increased van der Waals forces.

The effect of branching on boiling points can also be explained by invoking van der Waals forces. Branched alkanes are more nearly spherical than straight-chain alkanes. As a result, they have smaller surface areas, fewer van der Waals forces, and consequently lower boiling points.

## 3.8 Cycloalkanes

Though we've discussed only open-chain alkanes up to this point, chemists have known for over 100 years that compounds with rings of carbon atoms also exist. Such compounds are called **cycloalkanes** or **alicyclic compounds** (**ali**phatic **cyclic**).

Alicyclic compounds with many different ring sizes abound in nature. For example, *chrysanthemic acid* contains a three-membered ring (cyclopropane). Various esters of chrysanthemic acid occur naturally as the active insecticidal constituents of pyrethrum flowers.

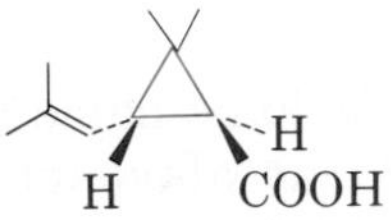

H COOH H

Chrysanthemic acid

*Prostaglandins*, such as $PGE_1$, contain a five-membered ring (cyclopentane). Prostaglandins are potent hormones that control a wide variety of physiological functions in humans, including blood platelet aggregation, bronchial dilation, and inhibition of gastric secretions.

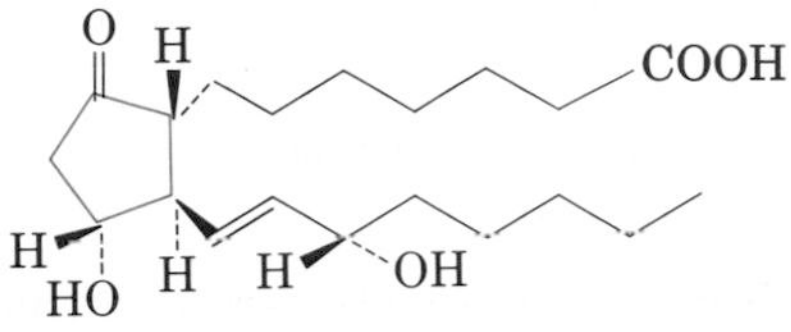

O H COOH H HO H H OH

Prostaglandin $E_1$ ($PGE_1$)

*Steroid hormones* such as cortisone contain four rings—three six-membered (cyclohexane) and one five-membered (cyclopentane)—all joined together.

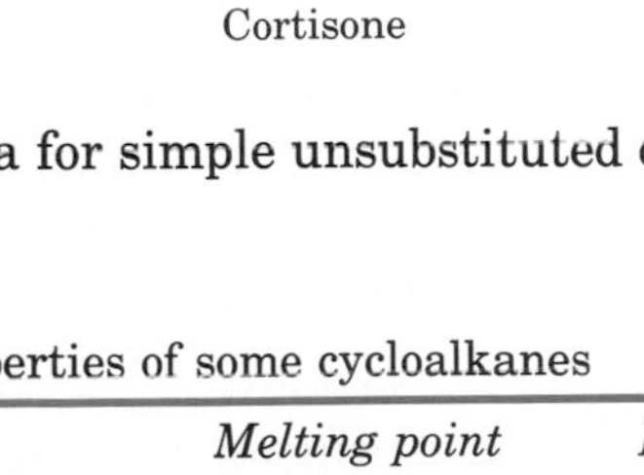

$HOH_2C$ O $H_3C$ O OH $H_3C$ H H H O

Cortisone

Some physical data for simple unsubstituted cycloalkanes are given in Table 3.6.

**Table 3.6** Physical properties of some cycloalkanes

| *Name* | *Formula* | *Melting point (°C)* | *Boiling point (°C)* | *Density (g/mL)* |
|---|---|---|---|---|
| Cyclopropane | $C_3H_6$ | −127.6 | −32.7 | |
| Cyclobutane | $C_4H_8$ | −50.0 | 12.0 | 0.720 |
| Cyclopentane | $C_5H_{10}$ | −93.9 | 49.3 | 0.7457 |
| Cyclohexane | $C_6H_{12}$ | 6.6 | 80.7 | 0.7786 |
| Cycloheptane | $C_7H_{14}$ | −12.0 | 118.5 | 0.8098 |
| Cyclooctane | $C_8H_{16}$ | 14.3 | 148.5 | 0.8349 |

## 3.9 Nomenclature of Cycloalkanes

The systematic naming of cycloalkanes follows directly from the rules given previously for open-chain alkanes. For the majority of cases, there are only two rules:

1. Use the cycloalkane name as the base name. Compounds are normally named as alkyl-substituted cycloalkanes, rather than as cycloalkyl-substituted alkanes. The only exception to this rule occurs when the alkyl side chain contains a greater number of carbons than the ring. In such cases, the ring is considered to be a substituent on the parent open-chain alkane. For example,

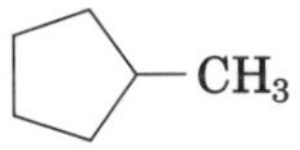

$-CH_3$ *BUT* $-CH_2CH_2CH_2CH_3$

3 carbons 4 carbons

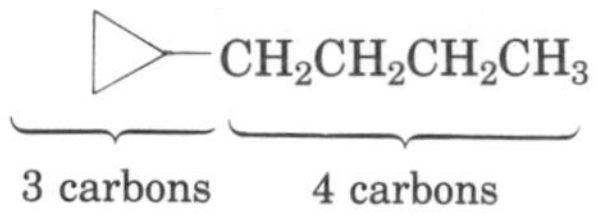

Methylcyclopentane 1-Cyclopropylbutane

2. Number the substituents on the ring so as to arrive at the lowest sum:

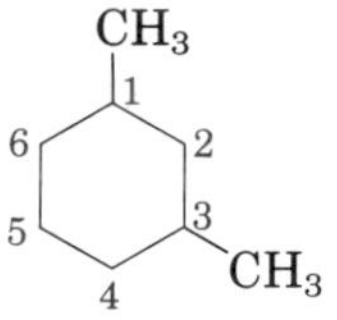

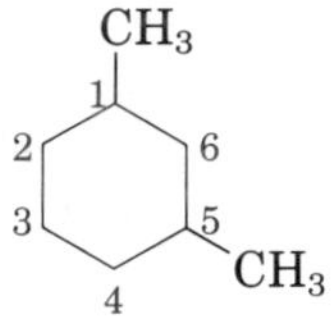

1,3-Dimethylcyclohexane *NOT* 1,5-Dimethylcyclohexane

a. When two or more alkyl groups are present, they are numbered alphabetically:

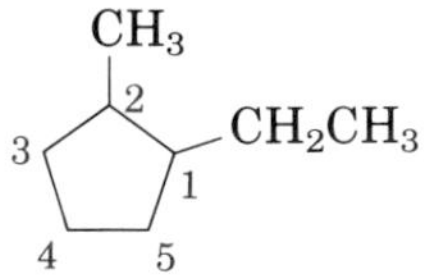

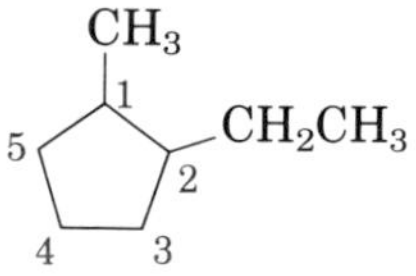

1-Ethyl-2-methylcyclopentane *NOT* 2-Ethyl-1-methylcyclopentane

b. Halogen substituents, if present, are treated exactly like alkyl groups:

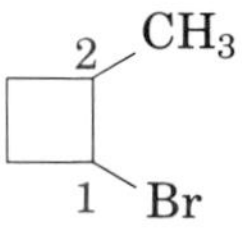

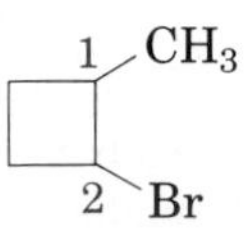

1-Bromo-2-methylcyclobutane *NOT* 2-Bromo-1-methylcyclobutane

Some additional examples follow:

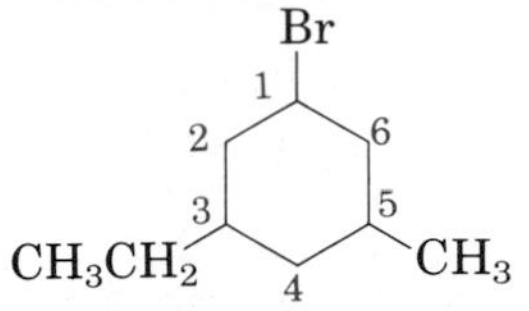

Br, 1, 2, 3, 4, 5, 6, $CH_3CH_2$, $CH_3$

1-Bromo-3-ethyl-5-methylcyclohexane

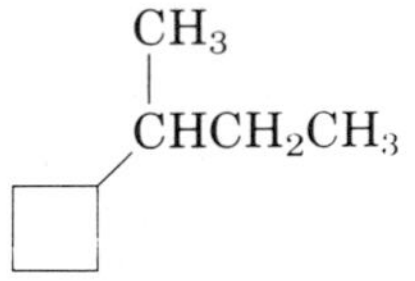

$CH_3$, $CHCH_2CH_3$

(1-Methylpropyl)cyclobutane
(or *sec*-Butylcyclobutane)

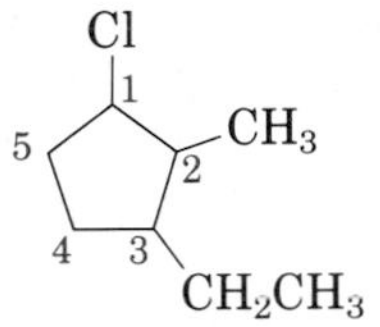

Cl, 1, 2, 3, 4, 5, $CH_3$, $CH_2CH_3$

1-Chloro-3-ethyl-2-methylcyclopentane

PROBLEM

**3.13** Give IUPAC names for the following cycloalkanes:

(a) $CH_3$, $CH_3$

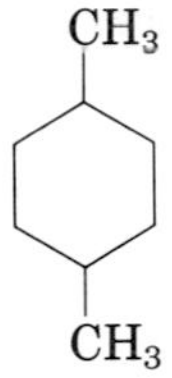

(b) $CH_2CH_2CH_3$, $CH_3$

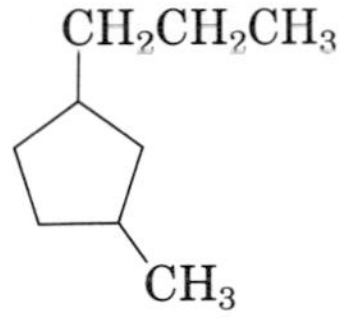

(c)

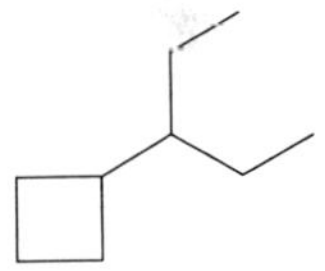

(d) $CH_2CH_3$, Br

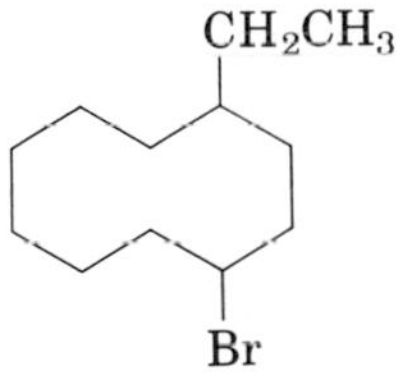

(e) $CH_3$, $CH(CH_3)_2$

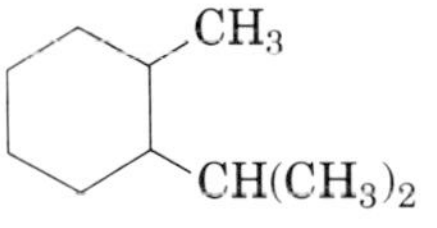

(f) Br, $CH_3$, $C(CH_3)_3$

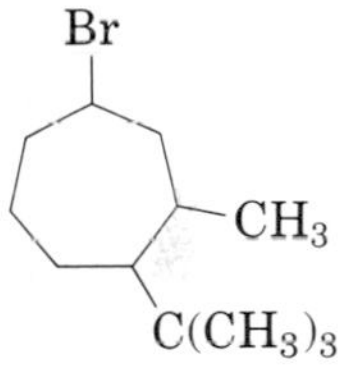

PROBLEM

**3.14** Draw structures corresponding to these IUPAC names:
(a) 1,1-Dimethylcyclooctane
(b) 3-Cyclobutylhexane
(c) 1,2-Dichlorocyclopentane
(d) 1,3-Dibromo-5-methylcyclohexane

## 3.10 Cis–Trans Isomerism in Cycloalkanes

In many respects, the chemistry of the cycloalkanes mimics that of open-chain (acyclic) alkanes. Both classes of compounds are relatively nonpolar and are chemically inert to most reagents. There are, however, some marked differences between cyclic and acyclic alkanes.

One difference is that cycloalkanes have less flexibility than their open-chain relatives. To see what this means, think for a minute about the nature of carbon–carbon single bonds. We know from Section 1.7 that sigma bonds are cylindrically symmetrical. In other words, the intersection of a plane cutting through a carbon–carbon single-bond orbital looks like a circle (Figure 3.9).

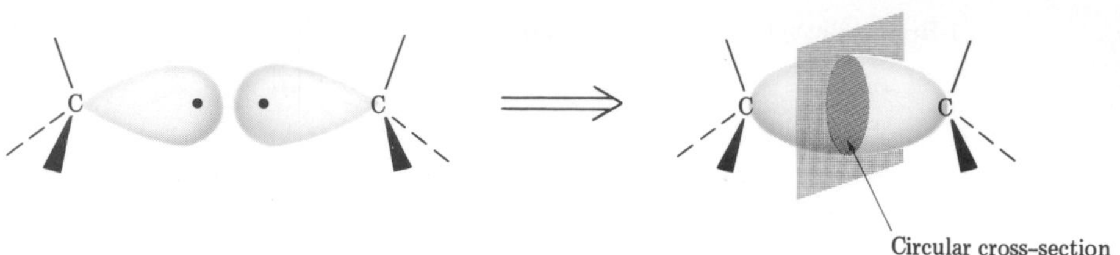


**Figure 3.9** The cylindrical symmetry of carbon–carbon single bonds

Because of the cylindrical symmetry of sigma bonds, there is **free rotation** around carbon–carbon bonds in open-chain molecules. In ethane, for example, C—C bond overlap is exactly the same for all geometric arrangements of the hydrogens (Figure 3.10).

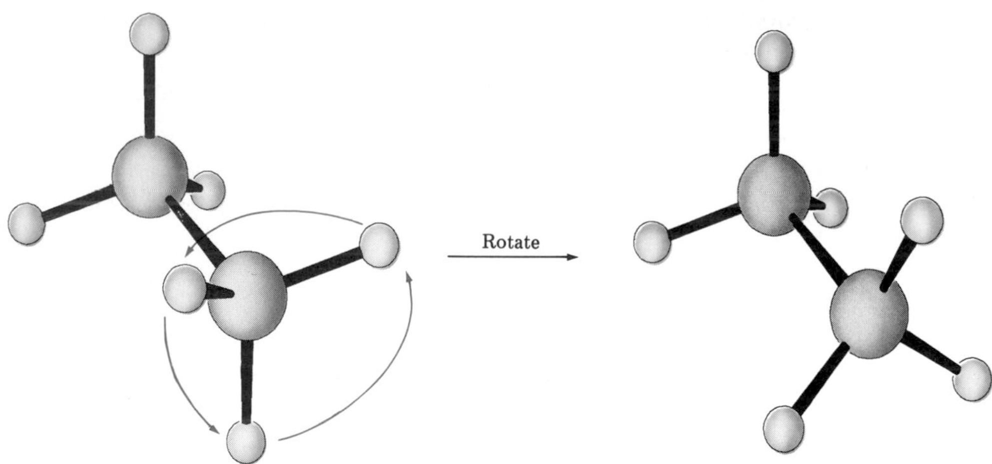


**Figure 3.10** Free rotation can occur around the carbon–carbon single bond in ethane as a consequence of sigma bond symmetry.

In contrast to the free rotation possible around single bonds in open-chain alkanes, there is much less freedom in cycloalkanes, which are geometrically constrained. For example, cyclopropane *must* be a flat, planar molecule with a rigid structure (three points define a plane). No bond rotation around a cyclopropane carbon–carbon bond is possible without breaking the ring (Figure 3.11).

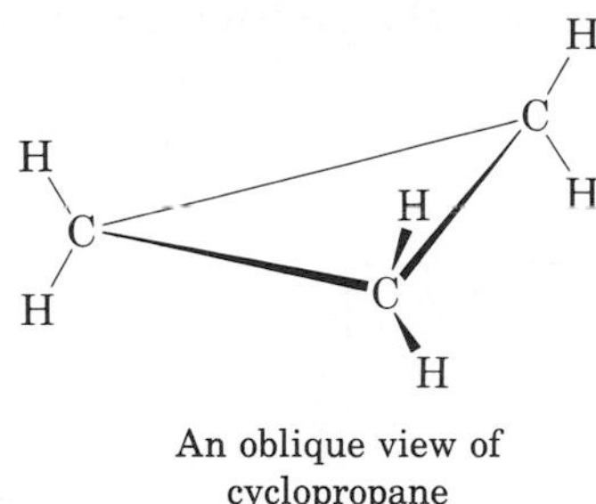


An oblique view of cyclopropane

**Figure 3.11** The structure of cyclopropane: No rotation is possible around the carbon–carbon bonds without breaking open the ring.

Higher cycloalkanes have increasingly more freedom, and the very large rings ($C_{25}$ and up) are so floppy that they are nearly indistinguishable from open-chain alkanes. The common ring sizes ($C_3$, $C_4$, $C_5$, $C_6$, $C_7$), however, are quite restricted in their molecular motions.

The most important consequence of their cyclic structure is that cycloalkanes have two distinct sides, a "top" side and a "bottom" side, leading to the possibility of isomerism in substituted cycloalkanes. For example, two different 1,2-dibromocyclopropane isomers exist. One isomer has the two bromines on the same side of the ring, and one isomer has them on opposite sides. The two isomers can't be interconverted without breaking and reforming chemical bonds; both are stable compounds that can be isolated. Make molecular models to prove this to yourself.

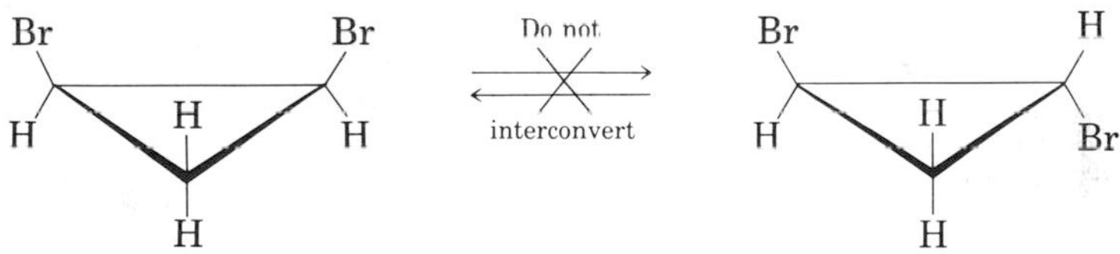


*cis*-1,2-Dibromocyclopropane (bromines on same side of ring)

*trans*-1,2-Dibromocyclopropane (bromines on opposite sides of ring)

Unlike the *constitutional isomers* butane and isobutane (Section 3.3), which have their atoms connected in a different order, the two 1,2-dibromocyclopropanes have the *same* order of connection but differ in the spatial orientation of their atoms. Compounds that have their atoms connected in the same order but that differ in three-dimensional orientation are called **stereoisomers**.

Constitutional isomers (different connections between atoms)

$$CH_3-\underset{\displaystyle |}{\overset{\displaystyle CH_3}{CH}}-CH_3 \quad \text{and} \quad CH_3-CH_2-CH_2-CH_3$$

Stereoisomers (same connections but different three-dimensional geometry)

H₃C CH₃ / H H and H₃C H / H CH₃

The 1,2-dibromocyclopropanes are special kinds of stereoisomers called **cis–trans isomers**. The prefixes *cis-* (Latin, "on the same side") and *trans-* (Latin, "across") are used to distinguish between them. Cis–trans isomerism is a common occurrence in substituted cycloalkanes.

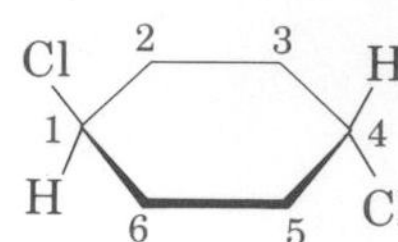

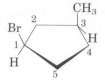

*cis*-1,3-Dimethylcyclobutane    *cis*-1-Bromo-3-methylcyclopentane    *trans*-1,4-Dichlorocyclohexane

**PROBLEM**

**3.15** Draw the structures of these molecules:
(a) *trans*-1-Bromo-3-methylcyclohexane (b) *cis*-1,2-Dimethylcyclopentane
(c) *trans*-1-*t*-Butyl-2-ethylcyclohexane

## 3.11 Summary and Key Words

A **functional group** is a part of a larger molecule. It is composed of an atom or group of atoms that has characteristic chemical reactivity. Because functional groups behave approximately the same in all molecules where they occur, the chemical reactions an organic molecule undergoes are largely determined by the functional groups it contains.

**Alkanes** are a class of **hydrocarbons** with the general formula $C_nH_{2n+2}$. They contain no functional groups, are chemically rather inert, and may be either straight-chain (***normal*** **alkanes**) or branched. All alkanes can be uniquely named by a series of **IUPAC rules**.

Compounds that have the same chemical formula but different structures are called **isomers**. More specifically, compounds such as butane and isobutane that differ in the order of their connections between atoms are called **constitutional isomers**. Constitutional isomers are conveniently classified into three subgroups: **Skeletal isomers** have different carbon skeletons; **functional-group isomers** have different functional groups; and **positional isomers** have their functional groups in different positions along the carbon chain.

**Cycloalkanes** contain rings of carbon atoms and have the general formula $C_nH_{2n}$. In comparison with open-chain alkanes, conformational mobility is greatly reduced in cycloalkanes. Although rotation is possible around carbon–carbon single bonds in open-chain molecules, rotation around carbon–carbon bonds in cycloalkanes is not possible. Disubstituted cycloalkanes can therefore exist as **cis–trans** isomers. The cis isomer has both substituents on the same side of the ring; the trans isomer has substituents on opposite sides of the ring. Cis–trans isomers are just one kind of **stereoisomers**—compounds that have the same order of connection between atoms but differ in their three-dimensional arrangements.

## ADDITIONAL PROBLEMS

**3.16** Locate and identify the functional groups present in these molecules:

(a) OH

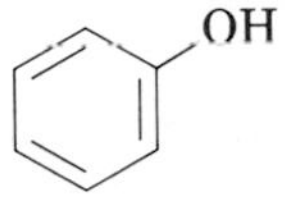

Phenol

(b) O

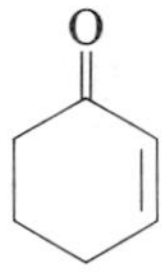

2-Cyclohexenone

(c) $NH_2$

$CH_3CHCOOH$

Alanine

(d) $NHCOCH_3$

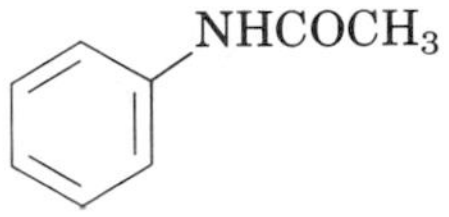

Acetanilide

(e) O

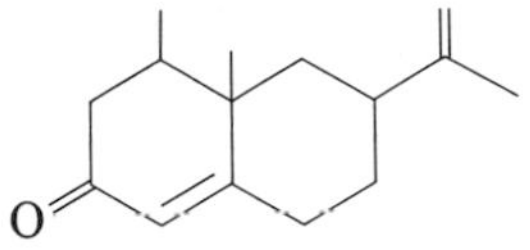

Nootkatone (from grapefruit)

(f)

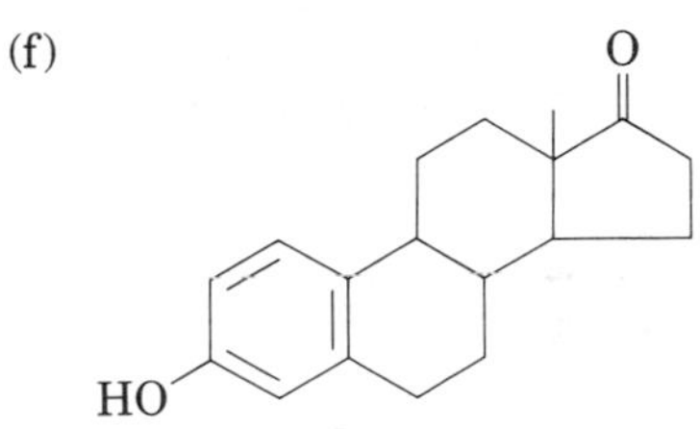


Estrone

(g) OH

HO

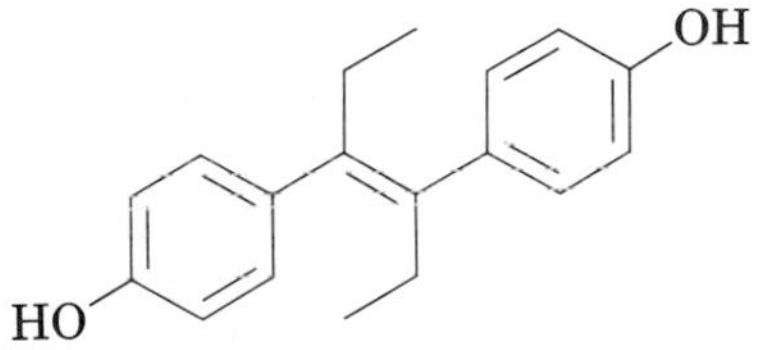

Diethylstilbestrol

(h)

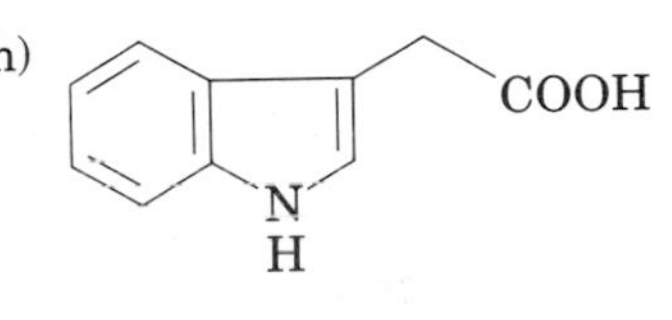


3-Indoleacetic acid

**3.17** Propose structures that meet these descriptions:
(a) A ketone with five carbons
(b) A four-carbon amide
(c) A five-carbon ester
(d) An aromatic aldehyde
(e) A keto ester
(f) An amino alcohol

**3.18** Propose suitable structures for the following:
(a) A ketone, $C_4H_8O$
(b) A nitrile, $C_5H_9N$
(c) A dialdehyde, $C_4H_6O_2$
(d) A bromoalkene, $C_6H_{11}Br$
(e) An alkane, $C_6H_{14}$
(f) A cycloalkane, $C_6H_{12}$
(g) A diene (dialkene), $C_5H_8$
(h) A keto alkene, $C_5H_8O$

**3.19** How many compounds can you write that fit these descriptions?
(a) Alcohols with formula $C_4H_{10}O$
(b) Amines with formula $C_5H_{13}N$
(c) Ketones with formula $C_5H_{10}O$
(d) Aldehydes with formula $C_5H_{10}O$
(e) Esters with formula $C_4H_8O_2$
(f) Ethers with formula $C_4H_{10}O$

**3.20** Draw compounds that contain the following:
(a) A primary alcohol
(b) A tertiary nitrile
(c) A secondary bromide
(d) Both primary and secondary alcohols
(e) An isopropyl group
(f) A quaternary carbon

**3.21** Draw and name all monobromo derivatives of pentane.

**3.22** Draw and name all monochloro derivatives of 2,5-dimethylhexane.

**3.23** What hybridization would you predict for the carbon atom in these functional groups?
(a) Ketone (b) Nitrile (c) Carboxylic acid (d) Ether

**3.24** Draw structural formulas for the following:
(a) 2-Methylheptane (b) 4-Ethyl-2,2-dimethylhexane
(c) 4-Ethyl-3,4-dimethyloctane (d) 2,4,4-Trimethylheptane
(e) 3,3-Diethyl-2,5-dimethylnonane (f) 4-Isopropyl-3-methylheptane

**3.25** Draw a compound that:
(a) Has both primary and tertiary (but no secondary) carbons
(b) Has no primary carbons
(c) Has four secondary carbons

**3.26** For each of the following compounds, draw an isomer having the same functional groups:

```
        CH3
        |
(a) CH3CHCH2CH2Br
```

(b) $OCH_3$

(c) $CH_3CH_2CH_2C{\equiv}N$

(d) OH

(e) $CH_3CH_2CHO$

(f) $CH_2COOH$

**3.27** Which of the following Kekulé structures represent the same compound, and which represent different compounds?

```
                                        H
                                        |
                                      H—C—H
      H  H  H  H                 H  H   |                 H  H  H
      |  |  |  |                 |  |   |                 |  |  |
(a) H—C—C—C—C—H              H—C—C—C—H             H—C—C—C—H
      |  |  |  |                 |  |   |                 |  |  |
      H  H  H  H                 H  H   H                 H  |  H
                                                          H—C—H
                                                             |
                                                             H

                                                           H
                                                           |
                                                         H—C—H
      H  H  Br H                 H  Br H  H                |  H  H
      |  |  |  |                 |  |  |  |                |  |  |
(b) H—C—C—C—C—H              H—C—C—C—C—H           H—C—C—C—H
      |  |  |  |                 |  |  |  |                |  |  |
      H  H  H  H                 H  H  H  H                H  Br H

          CH3                CH3
          |                  |
(c) CH3CHBrCHCH3          CH3CHCHBrCH3          (CH3)2CHCHBrCH2CH3
```

(d) 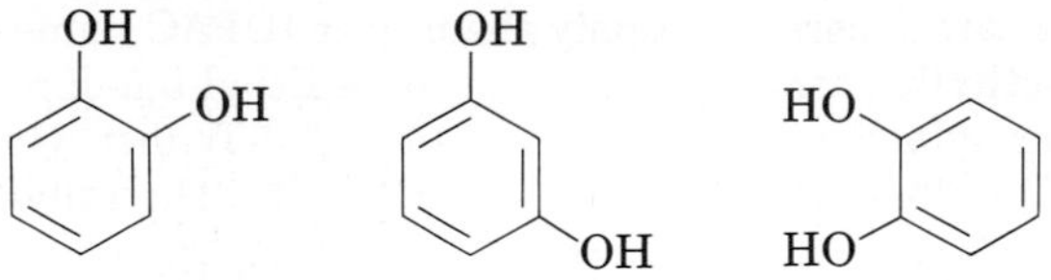
OH
OH
OH
OH
HO
HO

(e)

$$\begin{array}{c} CH_3 \\ | \\ HOCH_2CHCHCH_3 \\ | \\ CH_3 \end{array} \qquad \begin{array}{c} CH_3 \\ | \\ CH_3CH_2CHCH_2CHCH_2OH \\ | \\ CH_3 \end{array} \qquad \begin{array}{c} HOCH_2CHCH(CH_3)_2 \\ | \\ CH_3 \end{array}$$

**3.28** Draw structures for these compounds:
(a) *trans*-1,3-Dibromocyclopentane
(b) *cis*-1,4-Diethylcyclohexane
(c) *trans*-1-Isopropyl-3-methylcycloheptane
(d) Dicyclohexylmethane

**3.29** Identify the kinds of carbons (1°, 2°, 3°, or 4°) in these molecules:

(a)
$$\begin{array}{c} CH_3 \\ | \\ CH_3CHCH_2CH_3 \end{array}$$

(b) $(CH_3)_2CHCH(CH_2CH_3)_2$

(c)
$$\begin{array}{r} CH_3 \\ | \\ (CH_3)_3CCH_2CH_2CH \\ | \\ CH_3 \end{array}$$

(d) 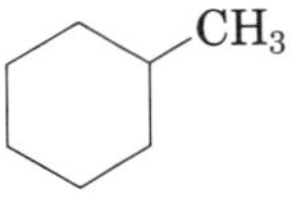
CH3

(e) 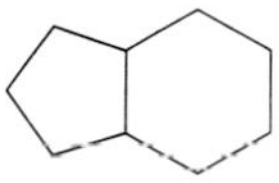

(f) 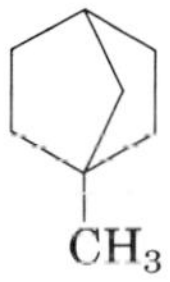
CH3

**3.30** Supply proper IUPAC names for these compounds:

(a)
$$\begin{array}{c} CH_3 \\ | \\ CH_3CHCH_2CH_2CH_3 \end{array}$$

(b) $CH_3CH_2C(CH_3)_2CH_3$

(c) $(CH_3)_2CHC(CH_3)_2CH_2CH_2CH_3$

(d)
$$\begin{array}{c} \quad CH_2CH_3 \quad CH_3 \\ | \qquad\quad | \\ CH_3CH_2CHCH_2CH_2CHCH_3 \end{array}$$

(e)
$$\begin{array}{c} CH_3 \quad CH_2CH_3 \\ | \qquad | \\ CH_3CH_2CH_2CHCH_2CCH_3 \\ | \\ CH_3 \end{array}$$

(f) $(CH_3)_3CC(CH_3)_2CH_2CH_2CH_3$

(g)
$$\begin{array}{c} CH_2CH_2CH_3 \\ | \\ CH_3CHCH_2CCH_2CH_3 \\ | \qquad | \\ CH_3CH_2 \quad CH_3 \end{array}$$

**3.31** Draw and name the five isomers of $C_6H_{14}$.

**3.32** The following names are incorrect. Supply the proper IUPAC names.
(a) 2,2-Dimethyl-6-ethylheptane
(b) 4-Ethyl-5,5-dimethylpentane
(c) 3-Ethyl-4,4-dimethylhexane
(d) 5,5,6-Trimethyloctane
(e) 2-Isopropyl-4-methylheptane
(f) *cis*-1,5-Dimethylcyclohexane

**3.33** Propose structures and give the correct IUPAC names for the following:
(a) A dimethylcyclooctane
(b) A diethyldimethylhexane
(c) A cyclic alkane with three methyl groups
(d) A (3-methylbutyl)-substituted alkane

**3.34** Malic acid, a compound of formula $C_4H_6O_5$, has been isolated from apples. Since this compound reacts with 2 equivalents (equiv) of base, it can be formulated as a dicarboxylic acid.
(a) Draw at least five possible structures.
(b) This compound can be shown to be a secondary alcohol. What is the structure of malic acid?

**3.35** Cyclopropane was first prepared by reaction of 1,3-dibromopropane with sodium metal. First formulate the cyclopropane-forming reaction and then predict the product that you might obtain from the following reaction. What geometry would you expect for the product?

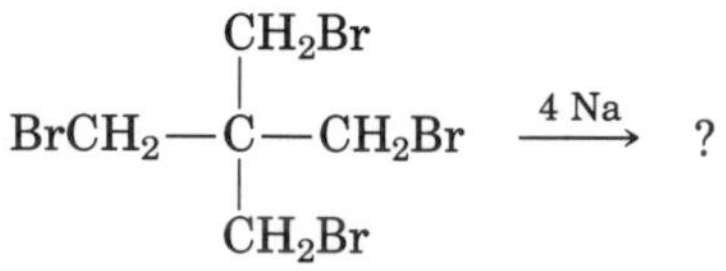

$$\mathrm{BrCH_2{-}C(CH_2Br)_2{-}CH_2Br} \xrightarrow{4\ \mathrm{Na}} ?$$

**3.36** The compound $\alpha$-methylenebutyrolactone is a skin irritant that has been isolated from the dogtooth violet. What functional groups does it contain?

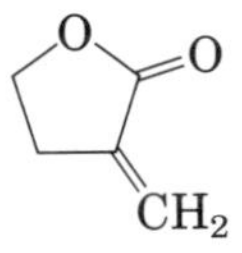

O, O, $CH_2$

$\alpha$-Methylenebutyrolactone

**3.37** Formaldehyde, $H_2C{=}O$, is a simple compound known to all biologists because of its usefulness as a tissue preservative. When pure, formaldehyde *trimerizes* to give trioxane, $C_3H_6O_3$. Trioxane, surprisingly enough, has no carbonyl groups. Only one monobromo derivative of trioxane is possible. Propose a structure that fits these data.

CHAPTER 4

# Stereochemistry of Alkanes and Cycloalkanes

Up to this point, we've been primarily concerned with the general nature of organic molecules. We've viewed most molecules in a two-dimensional way and have given little thought to any chemical consequences that might arise from the spatial arrangement of atoms in molecules. Now it's time to add a third dimension to our study. **Stereochemistry** is the branch of chemistry concerned with the three-dimensional aspects of molecules.

## 4.1 Conformations of Ethane

We know that an $sp^3$-hybridized carbon atom has tetrahedral geometry and that the carbon–carbon bonds in alkanes result from sigma overlap of two carbon $sp^3$ orbitals. Let's now look into the three-dimensional consequences of such bonding. What are the spatial relationships between the hydrogens on one carbon and the hydrogens on the other carbon?

We saw in the previous chapter (Section 3.10) that there is *free rotation* around carbon–carbon single bonds in open-chain molecules such as ethane, as a consequence of sigma bond symmetry. Bond overlap is exactly the same for all geometric arrangements of the atoms (Figure 4.1). The different arrangements of atoms caused by rotation about a single bond are called **conformations**, and a specific structure is called a **conformer** (**confor**mational iso**mer**). Unlike other kinds of isomers, however, different conformers cannot usually be isolated, because they interconvert too rapidly.

Chemists have adopted two ways of representing conformational isomers. **Sawhorse representations** view the carbon–carbon bond from an oblique angle and indicate spatial orientation by showing all the C—H

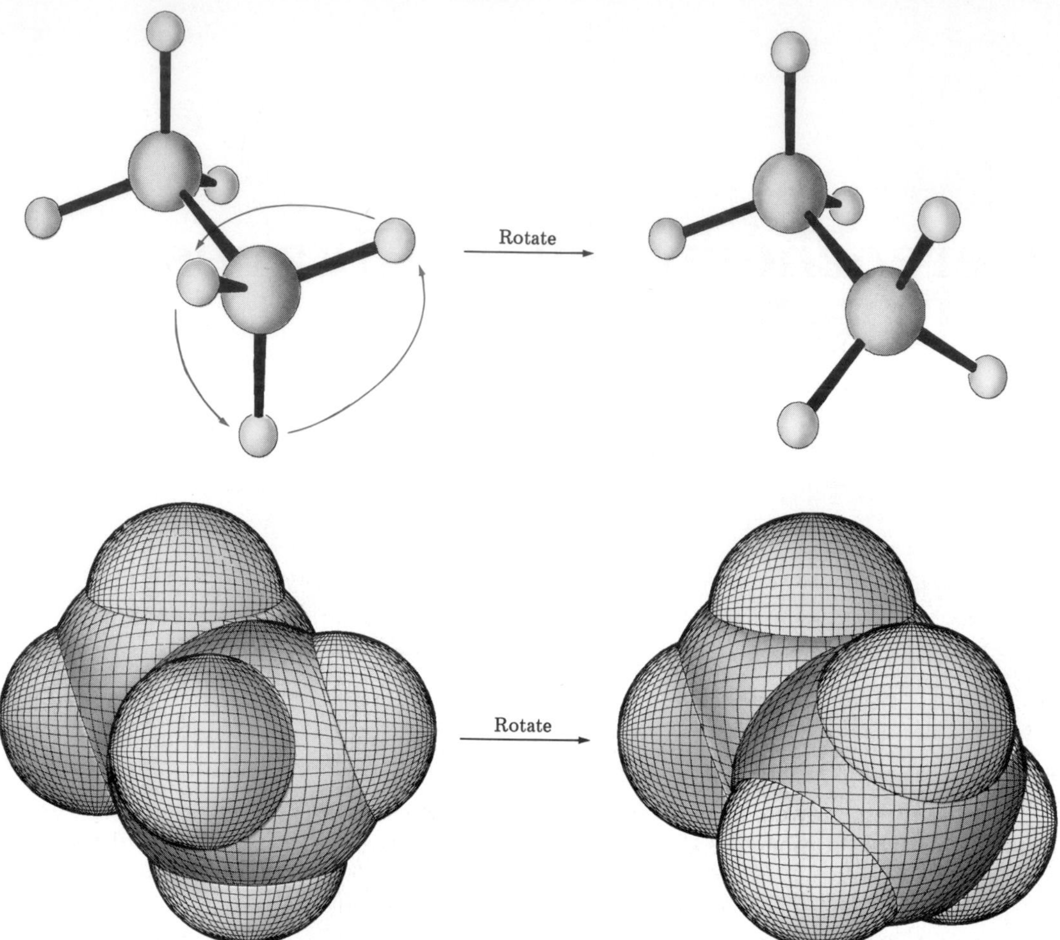


**Figure 4.1** Some conformations of ethane: Rapid rotation around the carbon–carbon single bond interconverts the different forms. The drawings are computer generated.

bonds. **Newman[1] projections** view the carbon–carbon bond directly end-on and represent the two carbon atoms by a circle. Substituents on the front carbon are represented by lines going to the center of the circle, and substituents on the rear carbon are represented by lines going to the edge of the circle. The advantage of Newman projections is that the relationships among substituents on the different carbon atoms are easily seen (Figure 4.2).

In view of the symmetry of the sigma bond, it is perhaps surprising that we don't observe *perfectly* free rotation in ethane. Experiments show that there is a slight [2.9 kcal/mol (12 kJ/mol)] barrier to rotation and that some conformations are more stable (have less energy) than others. The lowest-energy, most stable conformation is the one in which all six carbon–hydrogen bonds are as far away from each other as possible (**staggered**

[1]Melvin S. Newman (1908– ); b. New York; Ph.D. Yale University (1932); professor, Ohio State University.

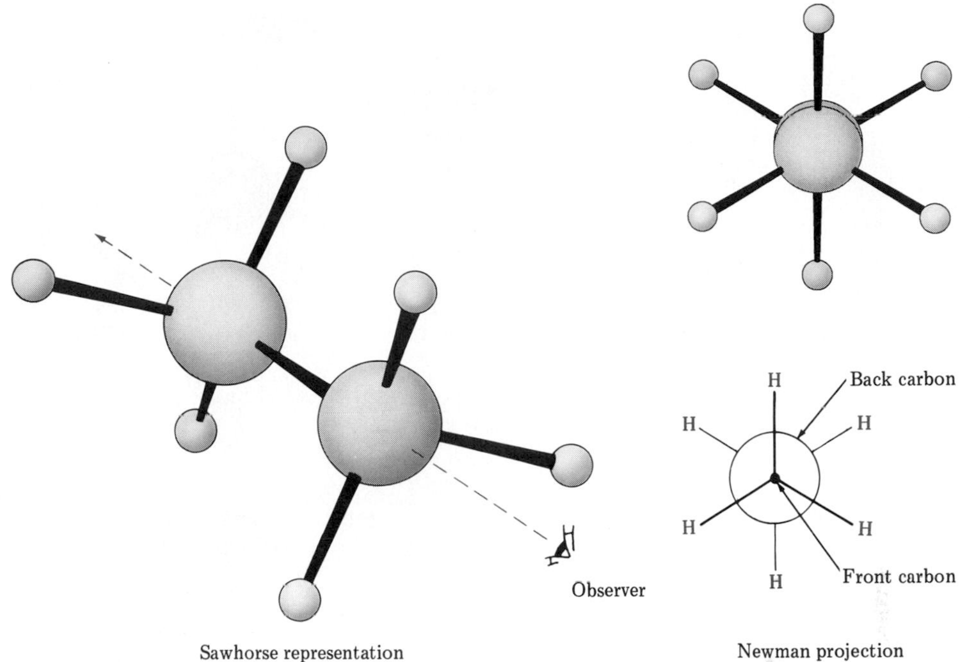


**Figure 4.2** A sawhorse representation and a Newman projection of ethane

when viewed end-on in a Newman projection). The highest-energy, least stable conformation is the one in which the six carbon–hydrogen bonds are as close as possible (**eclipsed** in a Newman projection). (In between these two limiting conformations are, of course, an infinite number of other possibilities.) Since the barrier to rotation is 2.9 kcal/mol, and since the barrier is caused by three equal hydrogen–hydrogen eclipsing interactions, we can assign a value of approximately 1 kcal/mol (4 kJ/mol) to each single interaction.

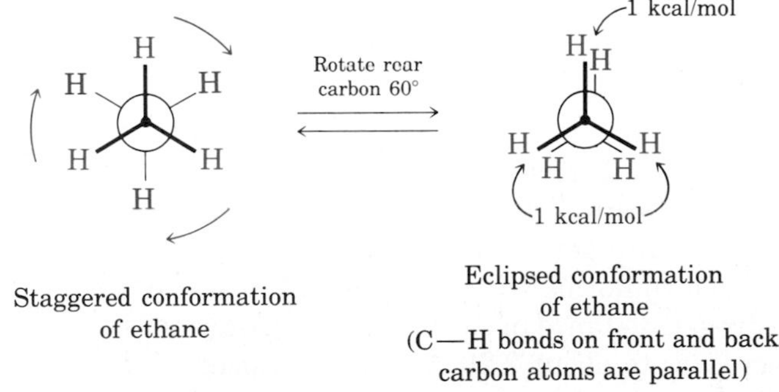


The 2.9 kcal/mol of **strain energy** present in the eclipsed conformation of ethane is called **torsional strain**. The barrier to rotation that results from torsional strain can be represented on a graph of potential energy versus degree of rotation, where the angle between C—H bonds on front

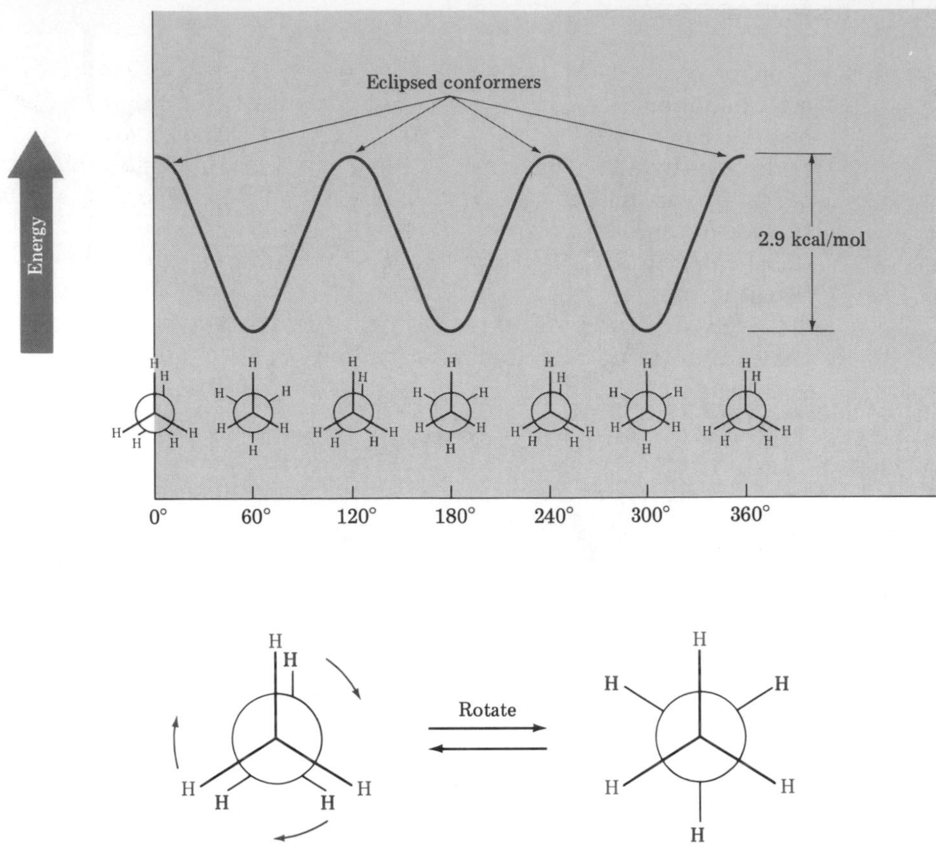


**Figure 4.3** A graph of potential energy versus bond rotation in ethane: The staggered conformers are 2.9 kcal/mol lower in energy than the eclipsed conformers.

and back carbons (the **dihedral angle**) goes full circle from 0° to 360° when seen end on. Energy minima occur at staggered conformations; energy maxima occur at eclipsed conformations, as shown in Figure 4.3.

To what is torsional strain due? The reasons for its existence have been the subject of some controversy, but most theoretical chemists now believe that the strain is due to the slight repulsion between electron clouds in the carbon–hydrogen bonds as they pass by each other at close quarters in the eclipsed conformer. Calculations indicate that the internuclear hydrogen–hydrogen distance in the staggered conformer is 2.55 Å but that this distance decreases to about 2.29 Å in the eclipsed conformer.

**PROBLEM**

**4.1** Build a molecular model of ethane and examine the interconversion of staggered and eclipsed forms.

## 4.2 Conformations of Propane

Propane is the next higher member in the alkane series, and we again find a torsional barrier that results in slightly hindered rotation about the carbon–carbon bonds. In the eclipsed conformer there are two ethane-type hydrogen–hydrogen interactions and one additional interaction between a carbon–hydrogen bond and a carbon–carbon bond. A slightly higher barrier to rotation is found in propane [3.4 kcal/mol (14 kJ/mol)] than in ethane (2.9 kcal/mol). Since we said that each eclipsing hydrogen–hydrogen interaction has an energy "cost" of 1.0 kcal/mol, we can assign a value of $3.4 - (2 \times 1.0)$ kcal/mol = 1.4 kcal/mol (6 kJ/mol) to the interaction between the carbon–methyl bond and the carbon–hydrogen bond (Figure 4.4).

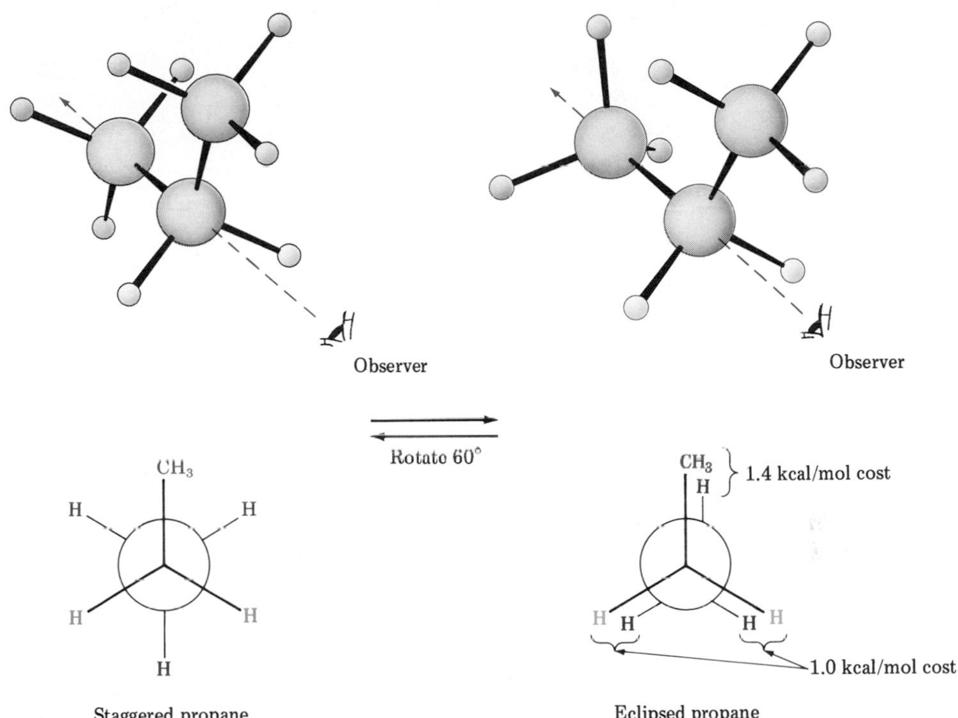


**Figure 4.4** Newman projections of propane showing staggered and eclipsed conformations: The staggered conformer is lower in energy by 3.4 kcal/mol.

PROBLEM

**4.2** Construct a graph of potential energy versus angle of bond rotation for propane. Assign quantitative values to the energy maxima.

## 4.3 Conformations of Butane

The conformational situation becomes more complex for higher alkanes. In butane, for example, a plot of potential energy versus rotation about the C2—C3 bond is shown in Figure 4.5 (page 96).

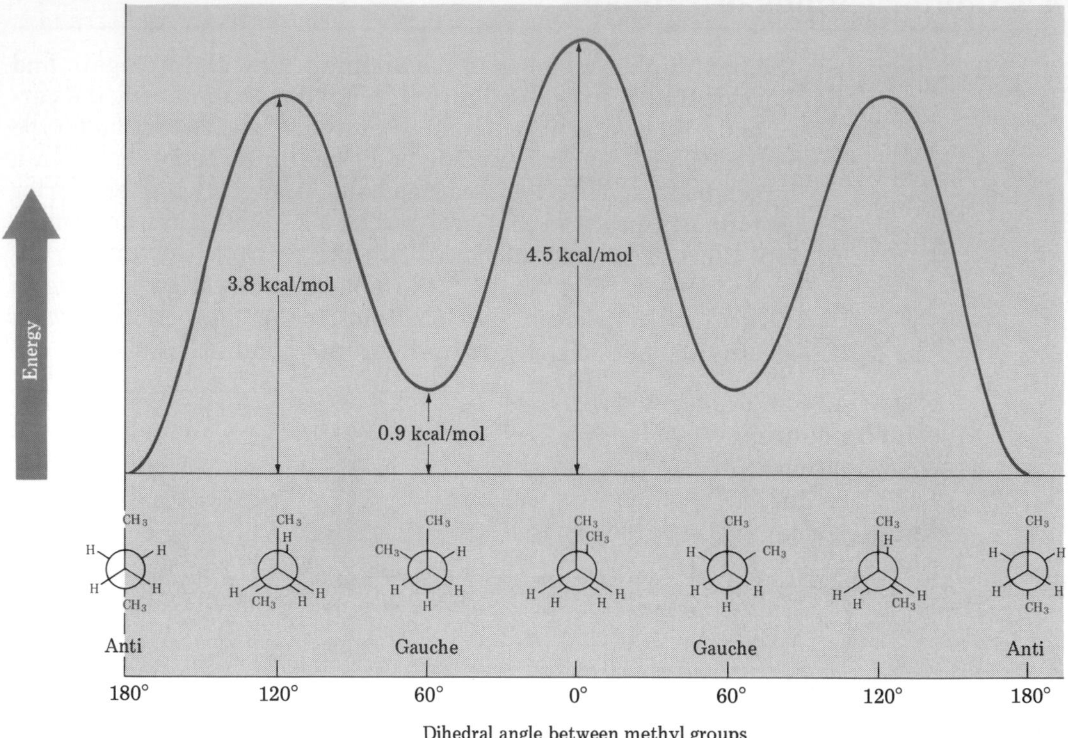


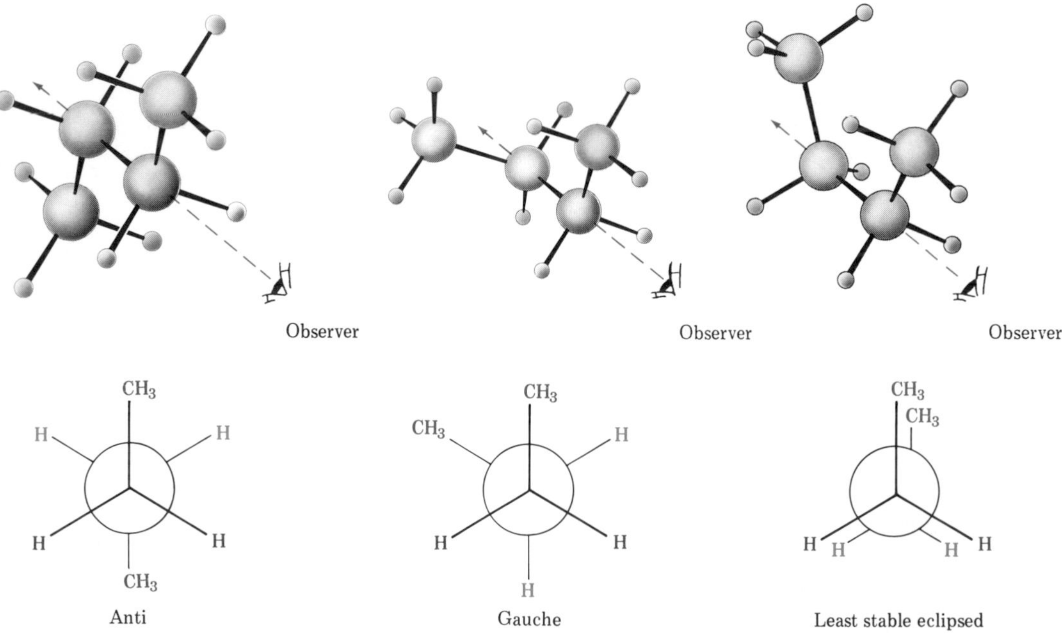


**Figure 4.5** A plot of potential energy versus rotation for the C2—C3 bond in butane: The energy maximum occurs when the two methyl groups eclipse each other.

Not all the staggered conformations of butane have the same energy, and not all eclipsed conformations are the same. The lowest-energy arrangement, called the **anti conformation**, is the one in which the two large groups (methyls) are as far apart as possible—that is, 180°. As rotation around the C2—C3 bond occurs, an eclipsed conformation is reached in which there are two methyl–hydrogen interactions and one hydrogen–hydrogen interaction. If we assign the energy values for eclipsing interactions that were previously derived from ethane and propane, we predict that this eclipsed conformation is more strained than the anti conformation by 2 × 1.4 kcal/mol (two methyl–hydrogen interactions) plus 1.0 kcal/mol (one hydrogen–hydrogen interaction), or a total of 3.8 kcal/mol (16 kJ/mol). This is exactly what is observed.

When the rotation is continued, an energy minimum is reached at the staggered conformation where the methyl groups are 60° apart. Called the **gauche conformation**, it lies 0.9 kcal/mol higher in energy than the anti conformation *even though it has no eclipsing interactions*. This energy difference is due to the fact that the large methyl groups are near each other in the gauche conformation, resulting in **steric strain**. Steric strain is the repulsive interaction that occurs when two groups are forced to be closer to each other than their atomic radii allow; it is the result of trying to force two objects to occupy the same space (Figure 4.6).

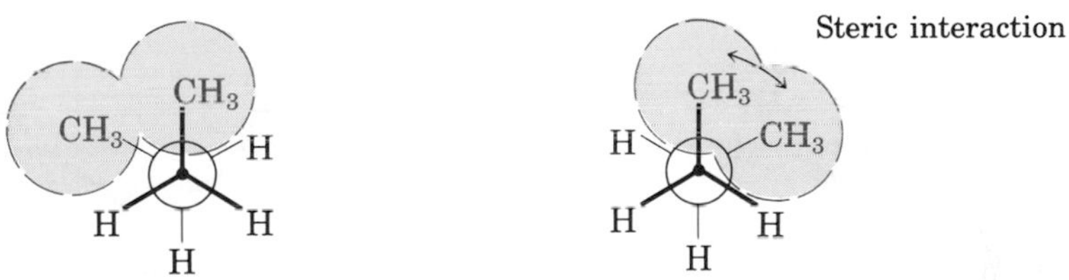


**Figure 4.6** The spatial interaction between two methyl groups in gauche butane: Steric strain results when the two methyl groups are too close together.

As the dihedral angle between the methyl groups approaches 0°, the energy maximum is reached. Since the methyl groups are forced even closer together than in the gauche conformation, a large amount of both torsional and steric strain is present. A total strain energy of 4.5 kcal/mol (19 kJ/mol) has been estimated for this conformation, allowing us to calculate a value of 2.5 kcal/mol (11 kJ/mol) for the methyl–methyl eclipsing interaction [total strain (4.5 kcal/mol), less strain of two hydrogen–hydrogen eclipsing interactions (2 × 1.0 kcal/mol), results in 2.5 kcal/mol].

After 0°, the rotation becomes a **mirror image** of what we've already seen. Thus, another gauche conformation is reached, another eclipsed conformation, and finally a return to the anti conformation occurs.

The concept of assigning definite energy values to specific interactions within a molecule is a very useful one. A summary of what we've found thus far is given in Table 4.1.

**Table 4.1** Energy costs for interactions in alkane conformers

| *Interaction* | *Cause* | *Energy cost* | |
|---|---|---|---|
| | | *(kcal/mol)* | *(kJ/mol)* |
| H—H eclipsed | Torsional strain | 1.0 | 4 |
| H—$CH_3$ eclipsed | Mostly torsional strain | 1.4 | 6 |
| $CH_3$—$CH_3$ eclipsed | Torsional plus steric strain | 2.5 | 11 |
| $CH_3$—$CH_3$ gauche | Steric strain | 0.9 | 4 |

The same principles just developed for butane apply to pentane and to all higher alkanes. The most favored conformation for any alkane is the one where all carbon–carbon bonds have staggered arrangements and where large substituents are arranged anti to each other. A generalized alkane structure is shown in Figure 4.7.

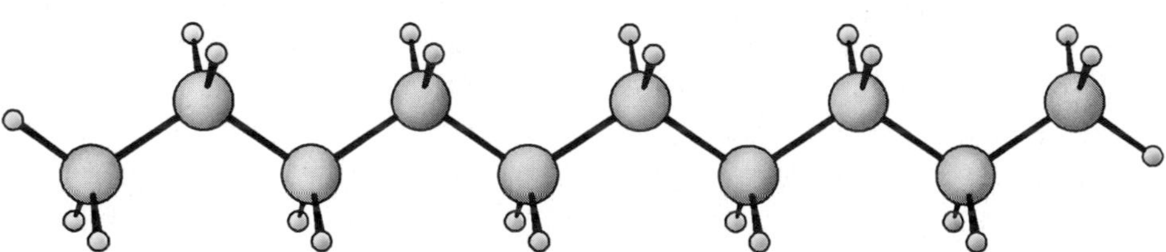

**Figure 4.7** The most stable alkane conformation is the one where all substituents are staggered and where the carbon–carbon bonds are arranged anti, as in this computer-generated structure of decane.

**One final point** It is important to remember that, when we speak of a particular conformer as being "more stable" than another, we don't mean the molecule in question adopts and maintains *only* the more stable conformation. At room temperature, enough thermal energy is present to ensure that rotation around sigma bonds occurs very rapidly and that all possible conformers are in a fluid equilibrium. At any given instant, however, a larger percentage of molecules will be found in a more stable conformation than in a less stable one.

PROBLEM

**4.3** Sight along the C2—C3 bond of 2,3-dimethylbutane and draw a Newman projection of the most stable conformation.

PROBLEM

**4.4** Consider 2-methylpropane (isobutane). Sighting along the C1—C2 bond:
(a) Draw a Newman projection of the most stable conformation.
(b) Draw a Newman projection of the least stable conformation.
(c) Construct a qualitative graph of energy versus angle of rotation about the C1—C2 bond.
(d) Since we know that a hydrogen–hydrogen eclipsing interaction costs 1.0 kcal/mol and a hydrogen–methyl eclipsing interaction costs 1.4 kcal/mol, assign quantitative values to your graph.

## 4.4 Conformation and Stability of Cycloalkanes: The Baeyer Strain Theory

Chemists in the late 1800s had accepted the idea that cyclic molecules exist, but the limitations on feasible ring sizes were unclear. Numerous compounds containing five-membered and six-membered rings were known, but smaller and larger ring sizes had not been prepared. For example, no cyclopropanes or cyclobutanes were known, despite numerous efforts to prepare them.

A theoretical interpretation of this observation was proposed in 1885 by Adolf von Baeyer.[2] Baeyer suggested that, if carbon prefers to have tetrahedral geometry with bond angles of 109°, ring sizes other than five and six may be too *strained* to exist. Baeyer based his hypothesis on the simple geometric notion that a three-membered ring (cyclopropane) must be an equilateral triangle with bond angles of 60°, a four-membered ring (cyclobutane) must be a square with bond angles of 90°, a five-membered ring (cyclopentane) must be a regular pentagon with bond angles of 108°, and so on. According to this analysis, cyclopropane, with a bond angle compression of 109° − 60° = 49°, has a large amount of **angle strain** and must therefore be highly reactive. Cyclobutane (109° − 90° = 19° angle strain) must be similarly reactive, but cyclopentane (109° − 108° = 1° angle strain) must be nearly strain-free. Cyclohexane (109° − 120° = −11° angle strain) must be somewhat strained, but cycloheptane (109° − 128° = −19° angle strain) and higher cycloalkanes must have bond angles that are forced to be too large. Carrying this line of reasoning further, Baeyer suggested that very large rings should be impossibly strained and incapable of existence.

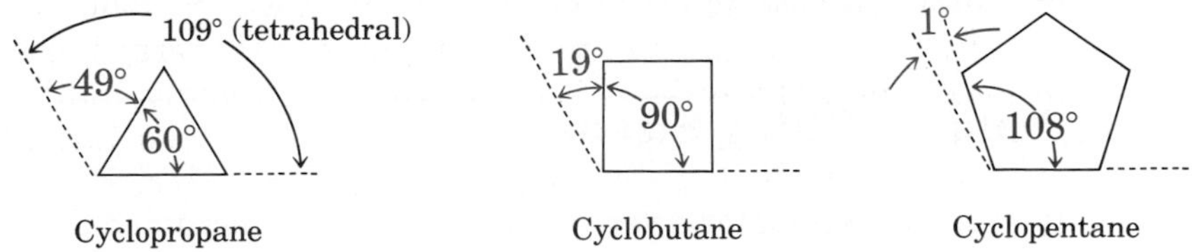


[2]Adolf von Baeyer (1835–1917); b. Berlin; Ph.D. Berlin (1858); professor, Berlin, Strasbourg (1872–1875), Munich (1875–1917); Nobel prize (1905).

Although there is some truth to Baeyer's assertions about angle strain in small rings, he was incorrect in his belief that small and large rings are too strained to exist. We know today that rings of all sizes from 3 through 30 and beyond can be prepared. Nevertheless, the concept of *angle strain*—the resistance of a bond angle to compression or expansion from the ideal tetrahedral angle—is a very useful one. Let's see what the facts are.

## 4.5 Heats of Combustion of Cycloalkanes

How can the amount of strain in a compound be measured? To do this, we must first measure the total amount of energy in the compound and then subtract the amount of energy in a hypothetical strain-free reference compound. The difference between the two values should represent the amount of extra energy possessed by the molecule due to strain.

The simplest way to determine strain energies is to measure heats of combustion of the cycloalkanes. The **heat of combustion** of a compound is the amount of heat (energy) released when the compound burns completely with oxygen:

$$—(CH_2)_n— + n\tfrac{3}{2}\,O_2 \longrightarrow n\,CO_2 + n\,H_2O + \text{Heat}$$

The more energy (strain) the sample contains, the more energy (heat) that's released on combustion. If we compare the heats of combustion of two isomeric substances, more energy is released during combustion of the more strained substance because that compound has a higher energy level to begin with.

Since the heat of combustion of a hydrocarbon depends on its molecular weight, it's more useful for comparison purposes to look at heats of combustion per $CH_2$ unit. In this way, the size of the hydrocarbon is not a factor, and we can compare cycloalkane rings of different sizes to a standard, strain-free, acyclic alkane. Table 4.2 and Figure 4.8 show the results of this comparison. Total strain energies are calculated by taking the difference between sample heat of combustion per $CH_2$ and reference heat of combustion per $CH_2$, and multiplying by the number of carbons, $n$, in the sample ring.

The data in Table 4.2 and the graph in Figure 4.8 show clearly that Baeyer's theory is not fully correct. Cyclopropane and cyclobutane are indeed quite strained, just as predicted, but cyclopentane is more strained than predicted, and cyclohexane is perfectly strain-free. For rings of larger size, there is no regular increase in strain, and rings having more than 14 members are again strain-free. Why is Baeyer's theory wrong?

## 4.6 The Nature of Ring Strain

Baeyer was wrong for a very simple reason—he assumed that all rings were flat. In fact, most cycloalkanes are not flat; they adopt puckered, three-dimensional conformations that allow bond angles to be nearly tetrahedral. Nevertheless, the concept of angle strain is a valuable one that goes far toward explaining the reactivity of three- and four-membered rings.

Table 4.2 Heats of combustion of cycloalkanes

| *Cycloalkane* $(CH_2)_n$ | *Ring size,* $n$ | *Heat of combustion (kcal/mol)* | *Heat of combustion per* $CH_2$ *(kcal/mol)* | *Total strain energy (kcal/mol)* |
|---|---|---|---|---|
| Cyclopropane | 3 | 499.8 | 166.6 | 27.6 |
| Cyclobutane | 4 | 655.9 | 164.0 | 26.4 |
| Cyclopentane | 5 | 793.5 | 158.7 | 6.5 |
| Cyclohexane | 6 | 944.5 | 157.4 | 0 |
| Cycloheptane | 7 | 1108 | 158.3 | 6.3 |
| Cyclooctane | 8 | 1269 | 158.6 | 9.6 |
| Cyclononane | 9 | 1429 | 158.8 | 12.6 |
| Cyclodecane | 10 | 1586 | 158.6 | 12.0 |
| Cycloundecane | 11 | 1742 | 158.4 | 11.0 |
| Cyclododecane | 12 | 1891 | 157.6 | 2.4 |
| Cyclotridecane | 13 | 2051 | 157.8 | 5.2 |
| Cyclotetradecane | 14 | 2204 | 157.4 | 0 |
| Alkane (reference) | | | 157.4 | 0 |

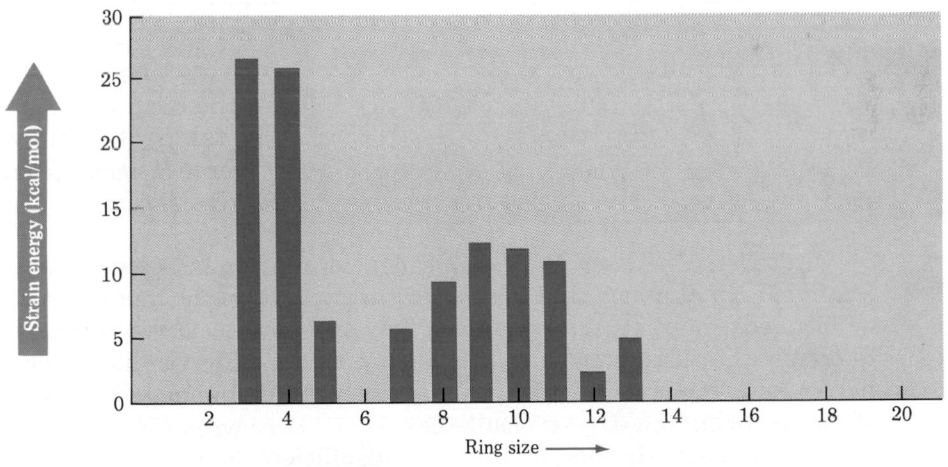


**Figure 4.8** Cycloalkane strain energy as a function of ring size

Several factors in addition to angle strain are important in determining the shape and total strain energy of rings. One factor is **torsional strain** (also called **eclipsing strain**). Torsional strain was encountered earlier in our discussion of alkane conformations (Section 4.1), in which we said that open-chain alkanes are most stable in the staggered conformation and least stable in the eclipsed conformation. A similar conclusion holds for cycloalkanes—torsional strain is present in cycloalkanes unless all the bonds have a staggered arrangement. For example, cyclopropane must have considerable torsional strain (in addition to angle strain), because C—H bonds on neighboring carbon atoms are eclipsed (Figure 4.9). Larger cycloalkanes attempt to minimize this strain by adopting puckered, nonplanar conformations.

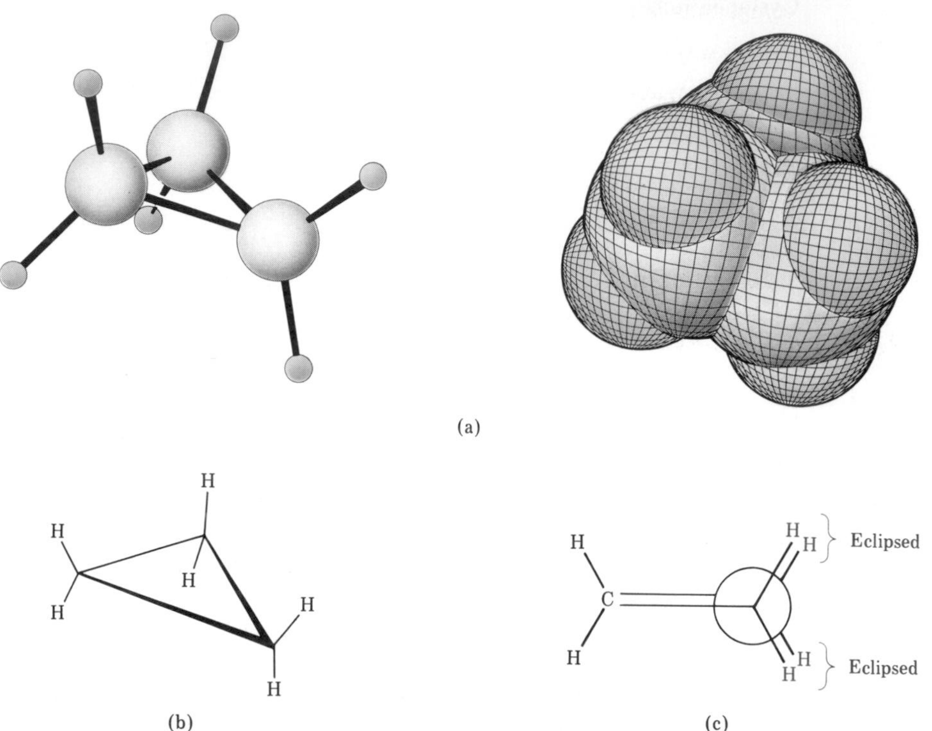


**Figure 4.9** The conformation of cyclopropane: Part (c) is a Newman projection along a C—C bond, showing the eclipsing of neighboring C—H bonds.

**Steric strain** (Section 4.3) is a third factor that contributes to the overall strain energy of a molecule. As we saw in gauche butane (Section 4.3), two nonbonded groups repel each other if they approach too closely and attempt to occupy the same point in space. Such nonbonded steric interactions are particularly important in determining the minimum-energy conformations of medium-ring ($C_7$–$C_{11}$) cycloalkanes.

In summary, the Baeyer theory is insufficient to explain the observed strain energies and geometries of cycloalkanes. Cycloalkanes adopt their minimum-energy conformations for a combination of three reasons:

1. **Angle strain**, the strain due to expansion or compression of bond angles
2. **Torsional strain**, the strain due to eclipsing of neighboring bonds
3. **Steric strain**, the strain due to repulsive interaction of atoms approaching each other too closely

PROBLEM

**4.5** We saw in Section 4.1 that each hydrogen–hydrogen eclipsing interaction in the eclipsed conformation of ethane "costs" about 1.0 kcal/mol. How many such eclipsing interactions are present in cyclopropane? What fraction of the overall 27.6 kcal/mol (115 kJ/mol) strain energy of cyclopropane can be ascribed to eclipsing strain?

PROBLEM

**4.6** *cis*-1,2-Dimethylcyclopropane has a higher heat of combustion than *trans*-1,2-dimethylcyclopropane. How can you account for this difference? Which of the two compounds is more stable?

PROBLEM

**4.7** If both propane and cyclopropane were equally available and equally priced, which would be the more efficient fuel? Explain.

## 4.7 Cyclopropane: An Orbital View

Cyclopropane is a colorless gas (bp −33°C) that was first prepared by reaction of sodium with 1,3-dibromopropane:

$$BrH_2C{-}CH_2{-}CH_2Br \xrightarrow{2\ Na} \underset{\text{Cyclopropane}}{(CH_2)_3} + 2\ NaBr$$

1,3-Dibromopropane — Cyclopropane

Since three points (the carbon atoms) define a plane, cyclopropane *must* be a flat, symmetrical molecule and must have C—C—C bond angles of 60°. How, though, can molecular orbital theory account for this great distortion of the bonds from the normal 109° tetrahedral angle?

The answer is that cyclopropane is best thought of as having *bent bonds*. In an unstrained alkane, maximum bonding efficiency is achieved when two atoms are located so that their overlapping orbitals point directly toward each other. In cyclopropane, however, the orbitals can't point directly toward each other; rather, they must overlap at a slight angle (Figure 4.10). The result of this poor overlap is that cyclopropane bonds are weaker and more reactive than normal alkane bonds.

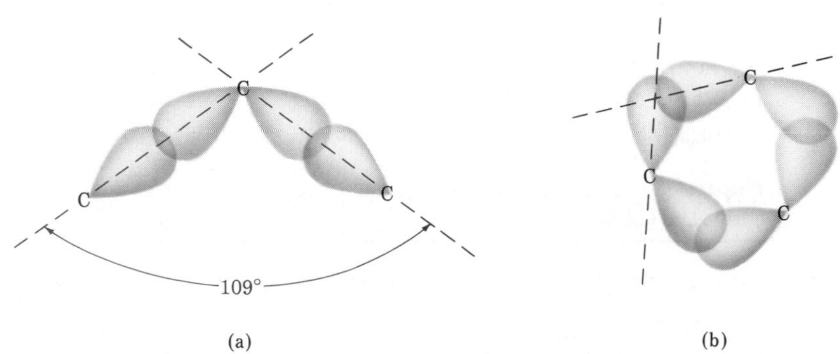


**Figure 4.10** An orbital view of bent bonds in cyclopropane: (a) Normal alkane C—C bonds have good overlap; (b) cyclopropane bent bonds have poor overlap.

## 4.8 Conformations of Cyclobutane and Cyclopentane

Cyclobutane has nearly the same total amount of strain [26.4 kcal/mol (110 kJ/mol)] as cyclopropane [27.6 kcal/mol (115 kJ/mol)]. The reason for this similarity is that cyclobutane has more eclipsing strain than cyclopropane by virtue of its larger number of ring hydrogens, even though it has less angle strain. Spectroscopic measurements indicate that cyclobutane is not quite flat but is slightly bent so that one carbon atom lies about 25° above the plane of the other three (Figure 4.11). The effect of this slight bend is to *increase* angle strain but to *decrease* eclipsing strain, until a minimum-energy balance between the two opposing effects is achieved.

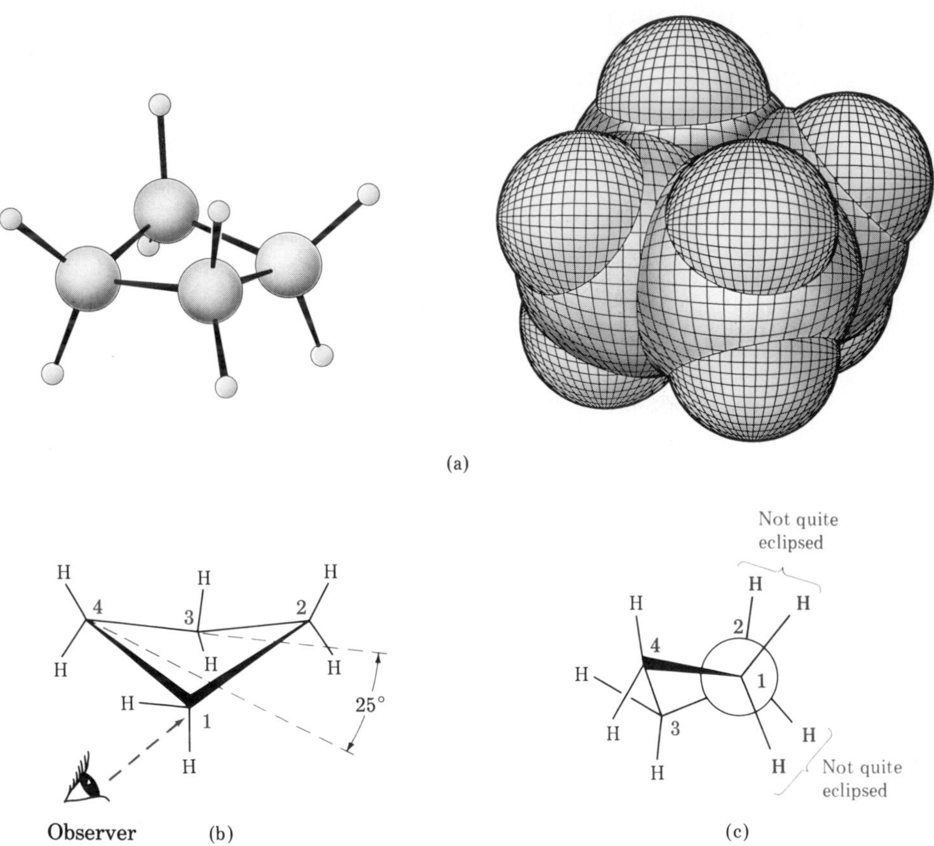


**Figure 4.11** Conformation of cyclobutane: Part (c) is a Newman projection along the C1—C2 bond.

Cyclopentane was predicted by Baeyer to be nearly strain-free, but heat-of-combustion data indicate that this is not the case. Although cyclopentane has practically no angle strain, the 10 pairs of neighboring hydrogens introduce considerable eclipsing strain into a planar conformation. Cyclopentane therefore adopts a puckered, out-of-plane conformation that strikes a balance between increased angle strain and decreased eclipsing strain. Four of the cyclopentane carbon atoms are in approximately the same plane, with the fifth bent out of the plane. Most of the hydrogens are nearly staggered with respect to their neighbors (Figure 4.12).

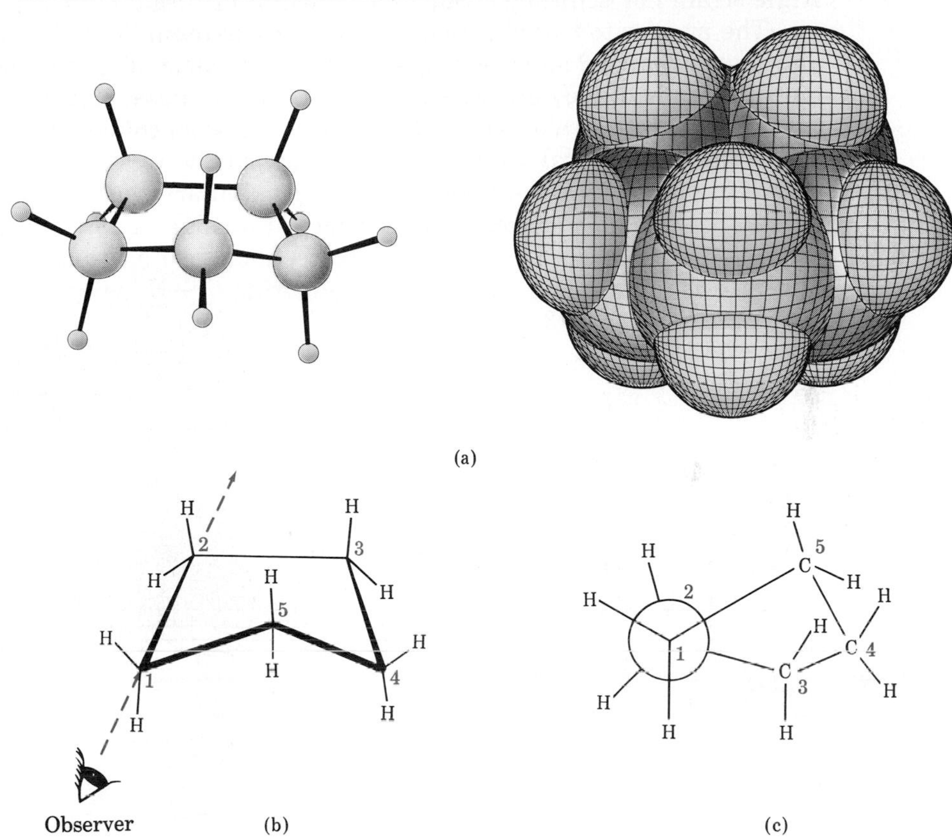


**Figure 4.12** Conformation of cyclopentane: Carbons 2, 3, 4, and 5 are nearly planar, but carbon 1 is out of the plane. Part (c) is a Newman projection along the C1—C2 bond.

**PROBLEM**

**4.8** *cis*-1,2-Dimethylcyclobutane is less stable than its trans isomer, but *cis*-1,3-dimethylcyclobutane is more stable than its trans isomer. Explain.

**PROBLEM**

**4.9** Draw the most favored conformation of *cis*-1,3-dimethylcyclobutane (Problem 4.8).

**PROBLEM**

**4.10** How many hydrogen–hydrogen eclipsing interactions would be present if cyclopentane were planar? Assuming an energy cost of 1.0 kcal/mol for each eclipsing interaction, how much total strain would you expect planar cyclopentane to have? How much of this strain is relieved by puckering if the measured total strain of cyclopentane is 6.5 kcal/mol?

## 4.9 Conformation of Cyclohexane

Cyclohexane compounds are the most important of all cycloalkanes because of their wide occurrence in nature. A large number of compounds, including many important pharmaceutical agents, contain cyclohexane rings. Experimental data show that cyclohexane rings are strain-free; they have neither angle strain nor eclipsing strain. Why should this be?

The answer to this question was first suggested in 1890 by H. Sachse and later expanded on by Ernst Mohr.[3] Cyclohexane rings are not flat as Baeyer assumed; they are puckered into a three-dimensional conformation that relieves all strain. The C-C-C angles of cyclohexane can reach the strain-free tetrahedral value if the ring adopts the **chair conformation** shown in Figure 4.13, so-called because of its similarity to a lounge chair—a back, a seat, and a footrest. Furthermore, if we sight along any one of the carbon–carbon bonds in a Newman projection, we find that chair cyclohexane has no eclipsing strain; all neighboring C—H bonds are perfectly staggered (Figure 4.13).

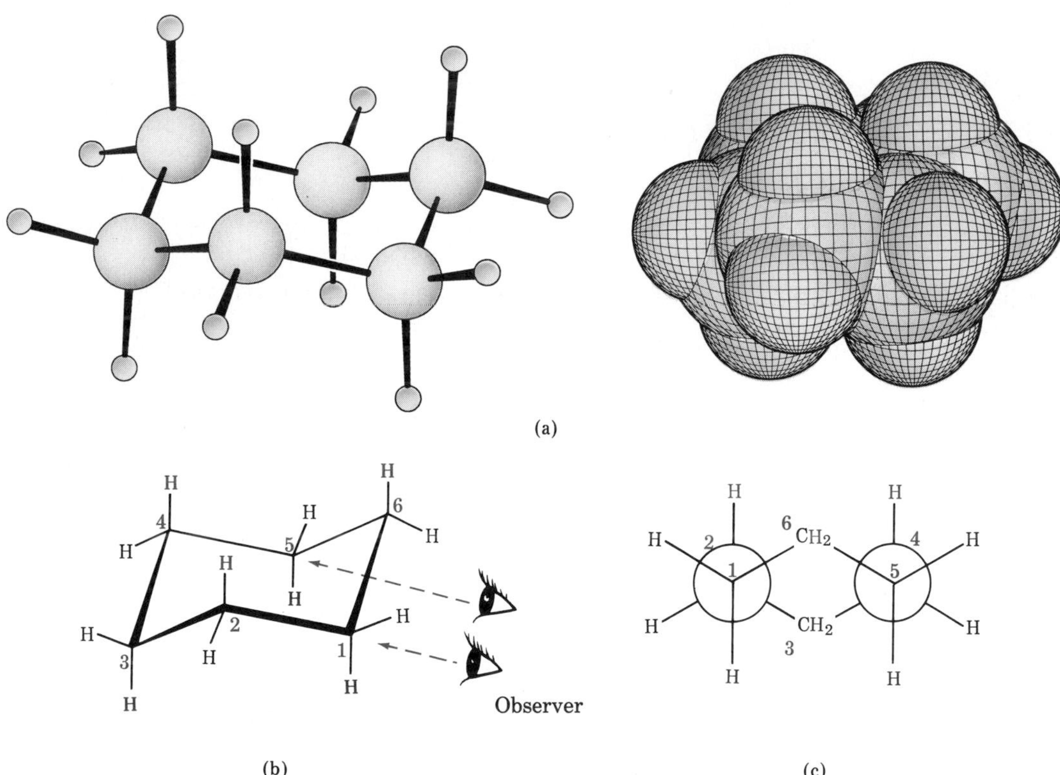


**Figure 4.13** The strain-free chair conformation of cyclohexane: All C-C-C bond angles are 111.5°, close to the ideal 109.5° tetrahedral angle.

[3]Ernst Mohr (1873–1926); b. Dresden; Ph.D. Kiel (1897); professor, University of Heidelberg.

The simplest way to visualize strain-free chair cyclohexane is to build and examine a molecular model. Two-dimensional drawings such as Figure 4.13 are useful, but there is no substitute for holding, twisting, and turning a three-dimensional model in your hands.

The chair conformation of cyclohexane is of such great importance that all organic chemists must learn how to draw it properly. The best way to do this is to follow the three steps shown in Figure 4.14.

1. Draw two parallel lines, slanted downward and slightly offset from each other. This means that four of the cyclohexane carbon atoms lie in a plane.

2. Locate the topmost carbon atom above and to the right of the plane of the other four and connect the bonds.

3. Locate the bottommost carbon atom below and to the left of the plane of the middle four and connect the bonds. Note that the bonds to the bottommost carbon atom are parallel to the bonds to the topmost carbon.

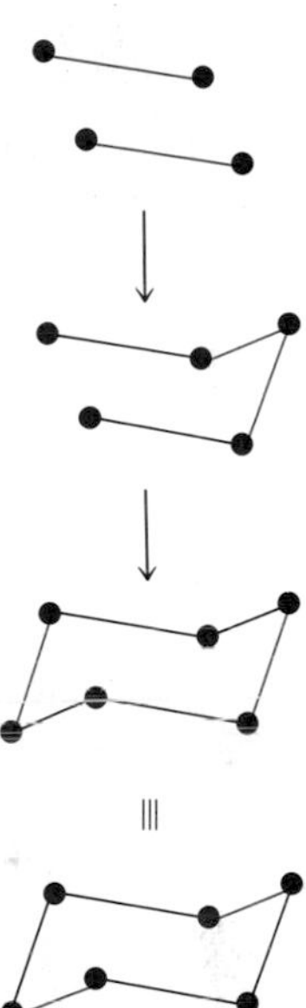

**Figure 4.14** How to draw the cyclohexane chair conformation

It's important to remember when viewing chair cyclohexane that the lower bond is in front and the upper bond is in back. If this convention isn't defined, an optical illusion can make it appear that the reverse is true. For clarity, the cyclohexane rings drawn in this book will have the front (lower) bond heavily shaded to indicate its nearness to the viewer.

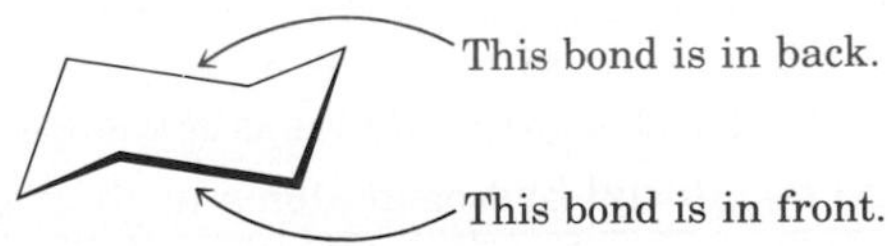
This bond is in back.
This bond is in front.

## 4.10 Axial and Equatorial Bonds in Cyclohexane

The chair conformation of cyclohexane has many consequences. For example, we'll see later that the chemical behavior of substituted cyclohexanes is intimately involved with their conformation. Another consequence of the chair cyclohexane conformation is that there are two kinds of hydrogen atoms on the ring—**axial hydrogens** and **equatorial hydrogens** (Figure 4.15).

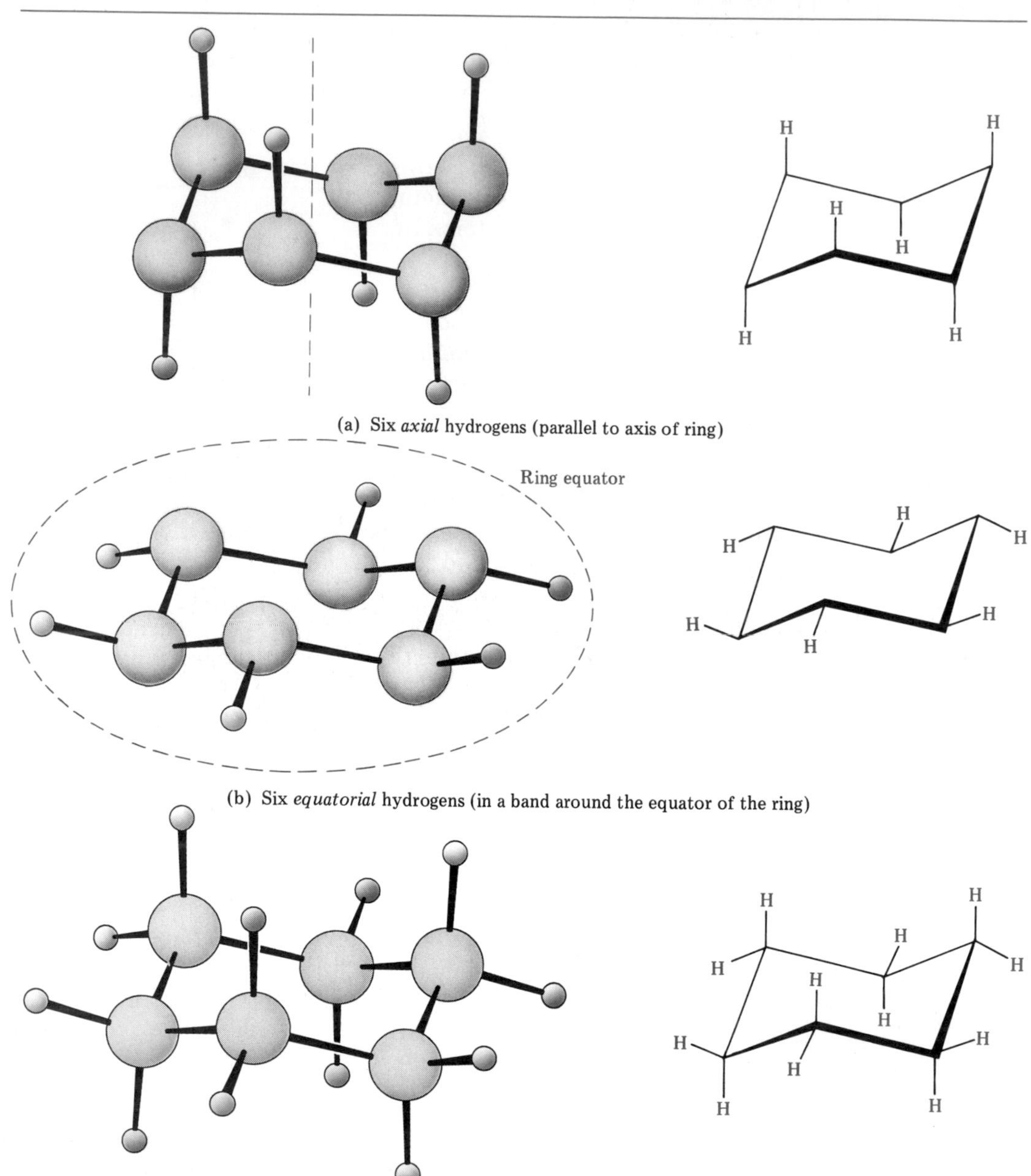


(a) Six *axial* hydrogens (parallel to axis of ring)

(b) Six *equatorial* hydrogens (in a band around the equator of the ring)

(c) Chair cyclohexane with all its hydrogen atoms

**Figure 4.15** Axial and equatorial hydrogen atoms in chair cyclohexane

As Figure 4.15 indicates (and molecular models illustrate much better), chair cyclohexane has six axial hydrogens that are perpendicular to the ring (parallel to the ring *axis*) and six equatorial hydrogens that are more or less in the rough plane of the ring (around the ring *equator*).

Look carefully at the disposition of the axial and equatorial hydrogens in Figure 4.15. Each carbon atom in cyclohexane has one axial and one equatorial hydrogen. Furthermore, every cyclohexane ring has two sides, or "faces"—a top face and a bottom face. Each face has both axial and equatorial hydrogens in an alternating axial–equatorial–axial–equatorial arrangement. For example, the top face of the cyclohexane ring shown in Figure 4.16 has axial hydrogens on carbons 1, 3, and 5 but has equatorial hydrogens on carbons 2, 4, and 6. Exactly the reverse is true for the bottom face: Carbons 1, 3, and 5 have equatorial hydrogens, but carbons 2, 4, and 6 have axial hydrogens.

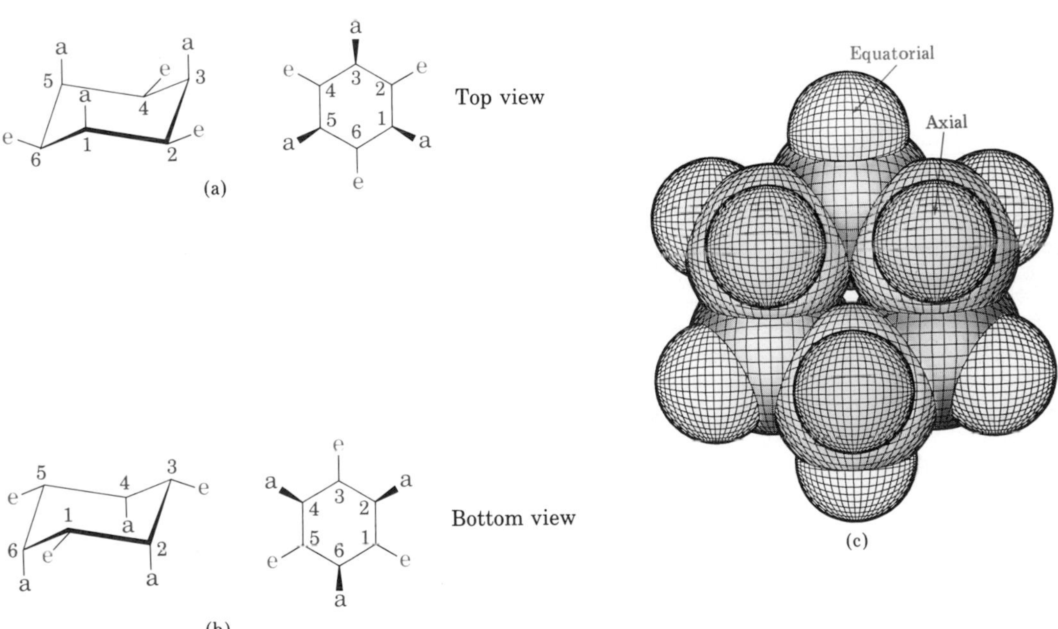


**Figure 4.16** Alternating axial and equatorial positions in chair cyclohexane. Each carbon atom has both an axial and an equatorial position, and each face has alternating axial–equatorial positions.

Note that we haven't used the words *cis* and *trans* in this discussion. Two hydrogens on the same face of the ring are always cis, regardless of whether they are axial or equatorial and regardless of whether they are adjacent. Similarly, two hydrogens on opposite faces of the ring are always trans, regardless of whether they are axial or equatorial.

It's important to practice drawing axial and equatorial bonds, a skill most easily learned using the following guidelines (illustrated in Figure 4.17). Look at a molecular model as you practice.

1. *Axial bonds* The six axial bonds (one on each carbon) are parallel and have an alternating up–down relationship.
2. *Equatorial bonds* The six equatorial bonds come in three sets of two parallel lines. Each set is also parallel to two ring bonds. (Parallel bonds are shown in red in Figure 4.17.)

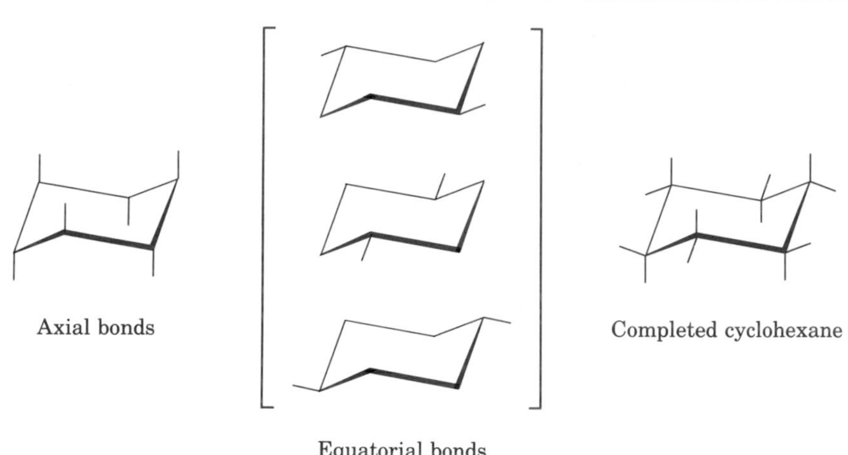


**Figure 4.17** How to draw axial and equatorial chair cyclohexane bonds

## 4.11 Conformational Mobility of Cyclohexane

In light of the fact that chair cyclohexane has two kinds of positions, axial and equatorial, we might expect a monosubstituted cyclohexane to exist in two isomeric forms. In fact, though, this expectation is wrong. At room temperature, there is only *one* methylcyclohexane, *one* bromocyclohexane, *one* cyclohexanol, and so on.

The explanation of this paradox is that cyclohexane is *conformationally mobile*. Different chair conformations can readily interconvert, with the result that axial and equatorial positions become interchanged. The interconversion of chair conformations, usually referred to as a **ring-flip**, is shown in Figure 4.18. Molecular models show the process more clearly, and you should practice with models while studying this material.

We can mentally ring-flip a chair cyclohexane by holding the middle four carbon atoms in place and folding the two ends in opposite directions. The net result of carrying out a ring-flip is the interconversion of axial and equatorial positions; an axial position in one chair form is equatorial in the ring-flipped chair form, and vice versa. For example, axial methylcyclohexane becomes equatorial methylcyclohexane after ring-flip. Spectroscopic

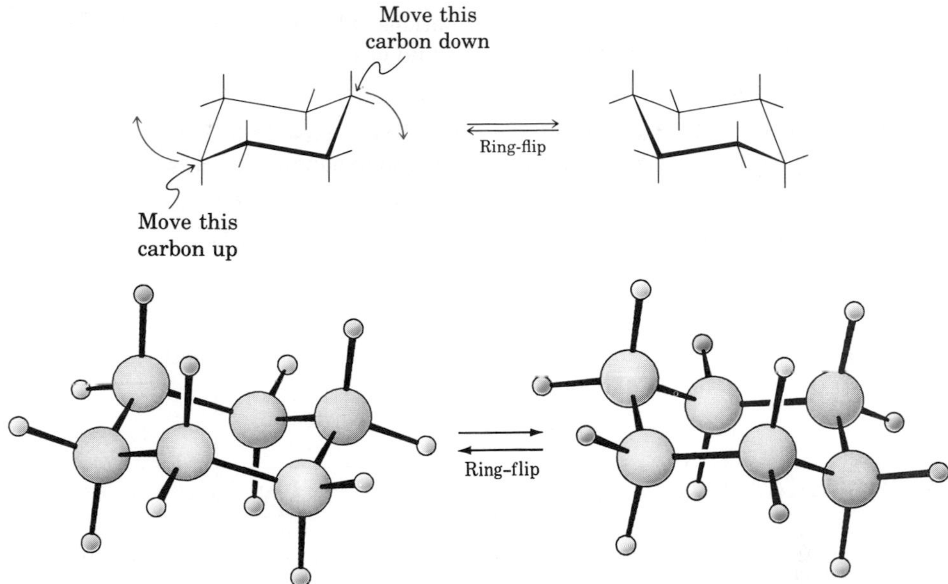


**Figure 4.18** A ring-flip in chair cyclohexane interconverts axial and equatorial positions.

measurements indicate that the energy barrier to chair–chair interconversions is about 10.8 kcal/mol (45 kJ/mol), a value low enough to make the process extremely rapid at room temperature. We therefore see only a single, rapidly interconverting average structure, rather than distinct axial and equatorial isomers.

Axial
$CH_3$
H
Ring-flip
(rapid)
Equatorial
$CH_3$
H

**PRACTICE PROBLEM**

Draw 1,1-dimethylcyclohexane, indicating whether each methyl group is axial or equatorial.

*Solution* First draw a chair cyclohexane ring and then put two methyl groups on the same carbon. The methyl group in the rough plane of the ring must be equatorial and the other (above or below the ring) must be axial.

Axial methyl group
$CH_3$
$CH_3$
Equatorial methyl group

PROBLEM

**4.11** Draw two different chair conformations of bromocyclohexane, showing all hydrogen atoms. Identify each position as axial or equatorial.

PROBLEM

**4.12** Explain why a 1,2-cis disubstituted cyclohexane such as *cis*-1,2-dichlorocyclohexane *must* have one group axial and one group equatorial.

PROBLEM

**4.13** Explain why a 1,2-trans disubstituted cyclohexane must have either both groups axial or both groups equatorial.

PROBLEM

**4.14** Draw two different chair conformations of *trans*-1,4-dimethylcyclohexane and label all positions as axial or equatorial.

## 4.12 Conformation of Monosubstituted Cyclohexanes

Let's look at the consequences of axial and equatorial cyclohexane bonds on substituted cyclohexanes. We've said that cyclohexane rings rapidly flip conformations at room temperature. At any given instant, however, the substituent of a monosubstituted cyclohexane is either axial or equatorial. The two conformers are not equally stable; the equatorially substituted conformer is more stable than the axially substituted conformer. In methylcyclohexane, for example, the energy difference between axial and equatorial methyl is 1.8 kcal/mol (8 kJ/mol).

It's possible to calculate the exact percentages of any two isomers at equilibrium, as shown in Table 4.3. For example, since the energy difference between axial and equatorial methylcyclohexane isomers is 1.8 kcal/mol, Table 4.3 shows that, at any given instant, about 95% of methylcyclohexane molecules have the methyl group equatorial, and only 5% have the methyl axial.

**Table 4.3** The relationship between stability and isomer percentages at equilibrium[a]

| *More stable isomer (%)* | *Less stable isomer (%)* | *Energy difference (25°C)* | |
|---|---|---|---|
| | | *(kcal/mol)* | *(kJ/mol)* |
| 50 | 50 | 0 | 0 |
| 75 | 25 | 0.651 | 2.72 |
| 90 | 10 | 1.302 | 5.45 |
| 95 | 5 | 1.744 | 7.29 |
| 99 | 1 | 2.722 | 11.38 |
| 99.9 | 0.1 | 4.092 | 17.11 |

[a]The values in this table are calculated from the equation $K = e^{-(\Delta E/RT)}$, where $K$ is the equilibrium constant between isomers; $e \approx 2.718$ (the base of natural logarithms); $\Delta E$ = energy difference between isomers; $T$ = absolute temperature (in kelvins); and $R = 1.986$ cal/mol · $K$ (the gas constant).

The reason for the energy difference between axial and equatorial conformers is steric strain due to **1,3-diaxial interactions**. The axial methyl group on C1 is too close to (suffers steric interference from) the axial hydrogens three carbons away on C3 and C5 (Figure 4.19).

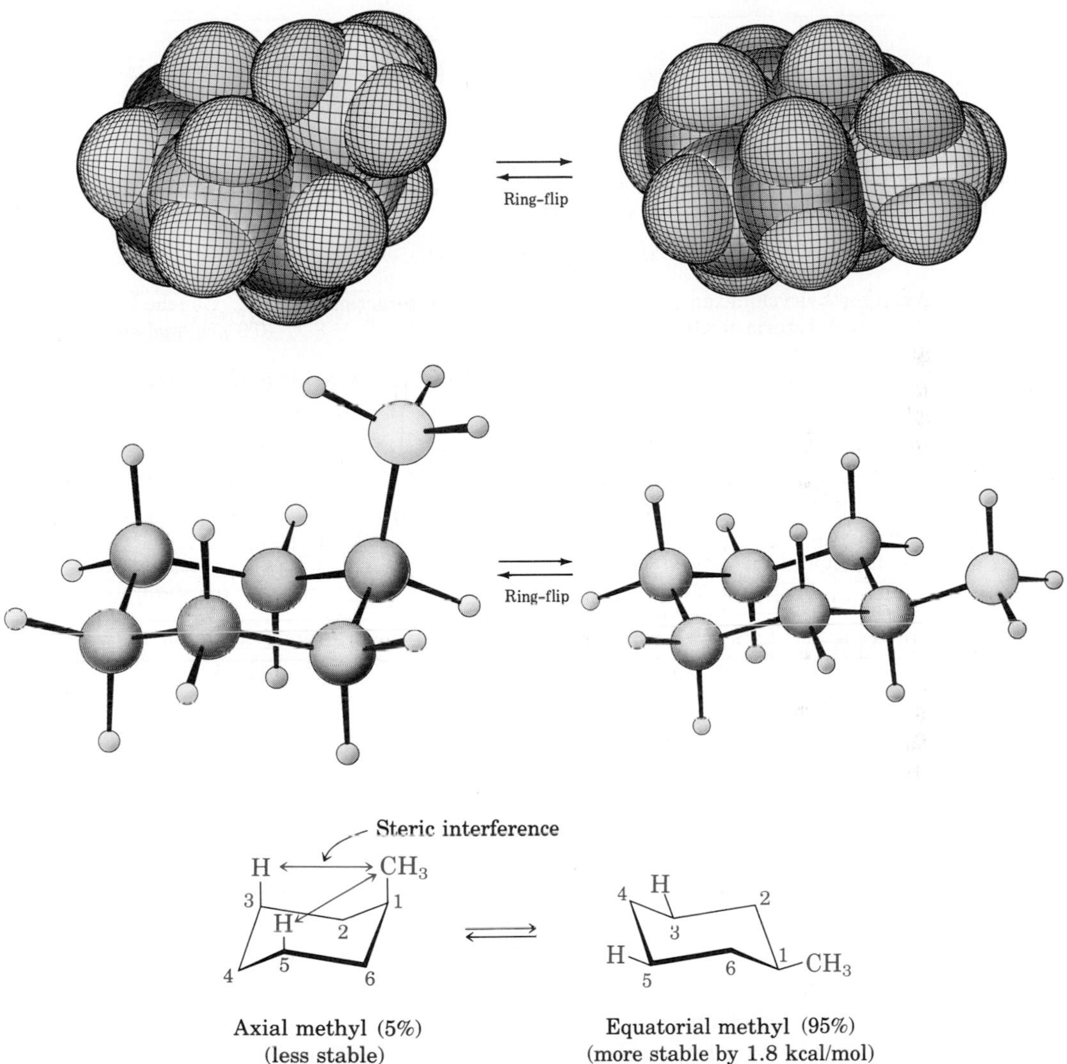


**Figure 4.19** Interconversion of axial and equatorial methylcyclohexane: The equatorial conformation is more stable by 1.8 kcal/mol.

1,3-Diaxial steric strain is familiar to us; we've seen it before as the steric strain between methyl groups in gauche butane (Section 4.3). Recall how we remarked, during the discussion of alkane conformations, that gauche butane is less stable than anti butane by 0.9 kcal/mol because of steric interference between hydrogen atoms on the two methyl groups. If we look at a four-carbon fragment of axial methylcyclohexane as a gauche butane, it's apparent that the steric interaction is the same. Since methylcyclohexane has two such interactions, however, there is $2 \times 0.9 = 1.8$ kcal/mol steric strain. The origin of 1,3-diaxial steric strain is shown in Figure 4.20.

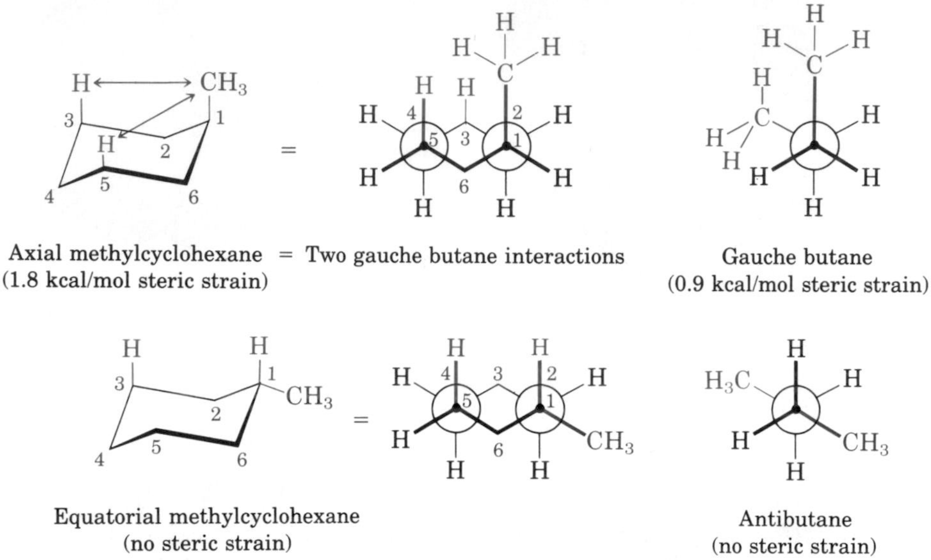

Axial methylcyclohexane (1.8 kcal/mol steric strain) = Two gauche butane interactions

Gauche butane (0.9 kcal/mol steric strain)

Equatorial methylcyclohexane (no steric strain)

Antibutane (no steric strain)

**Figure 4.20** Origin of 1,3-diaxial cyclohexane interactions

Sighting along the C1—C2 bond of axial methylcyclohexane shows that the axial hydrogen at C3 has a gauche butane interaction with the axial methyl group at C1. Sighting similarly along the C1—C6 bond shows that the axial hydrogen at C5 has a gauche butane interaction with the axial methyl group at C1. Both of these interactions are absent in equatorial methylcyclohexane, and we therefore predict an energy difference of 1.8 kcal/mol (8 kJ/mol) between the two forms. Experiment agrees perfectly with this prediction.

What is true for methylcyclohexane is also true for all other monosubstituted cyclohexanes: A substituent is almost always more stable in an equatorial position than in an axial position. The exact amount of 1,3-diaxial steric strain in a specific compound depends, of course, on the nature and size of the axial group. Table 4.4 lists some values for common substituents. As you might expect, the amount of steric strain increases through the series $H_3C$— < $CH_3CH_2$— < $(CH_3)_2CH$— ≪ $(CH_3)_3C$— in parallel with the increasing bulk of the successively larger alkyl groups. Note that the values in Table 4.4 refer to 1,3-diaxial interactions of the indicated group with a *single* hydrogen atom. These values must therefore be doubled to arrive at the amount of strain in a monosubstituted cyclohexane.

Table 4.4 Steric strain due to 1,3-diaxial interactions

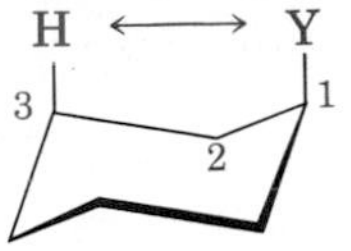

| Y | *Strain of one* H—Y *1,3-diaxial interaction* (*kcal/mol*) | (*kJ/mol*) |
|---|---|---|
| —F | 0.12 | 0.5 |
| —Cl | 0.25 | 1.4 |
| —Br | 0.25 | 1.4 |
| —OH | 0.5 | 2.1 |
| $—CH_3$ | 0.9 | 3.8 |
| $—CH_2CH_3$ | 0.95 | 4.0 |
| $—CH(CH_3)_2$ | 1.1 | 4.6 |
| $—C(CH_3)_3$ | 2.7 | 11.3 |
| $—C_6H_5$ | 1.5 | 6.3 |
| —COOH | 0.7 | 2.9 |
| —CN | 0.1 | 0.4 |

PROBLEM

**4.15** How can you account for the fact (Table 4.4) that an axial *tert*-butyl substituent has much larger 1,3-diaxial interactions than isopropyl, but isopropyl is fairly similar to ethyl and methyl? Use molecular models to help with your answer.

PROBLEM

**4.16** Why do you suppose an axial cyano substituent causes practically no 1,3-diaxial steric strain (0.1 kcal/mol)?

PROBLEM

**4.17** Look at Table 4.3 and estimate the percentages of axial and equatorial conformers present at equilibrium in *tert*-butylcyclohexane.

## 4.13 Conformational Analysis of Disubstituted Cyclohexanes

Monosubstituted cyclohexanes almost always prefer to have the substituent equatorial. In disubstituted cyclohexanes, however, the situation is more complex because the steric effects of both substituents must be taken into account. All of the steric interactions in the possible conformations must be analyzed before deciding which conformation is more favorable.

Let's look first at 1,2-dimethylcyclohexane. There are two isomers, *cis*-1,2-dimethylcyclohexane and *trans*-1,2-dimethylcyclohexane, and we have to consider them separately. In the cis isomer, both methyl groups are on the same side of the ring, and the compound can exist in either of the two chair conformations shown in Figure 4.21, page 116. (Note that it's often easier to see whether a compound is cis- or trans-disubstituted by first drawing a *flat* representation and then converting to a chair conformation.)

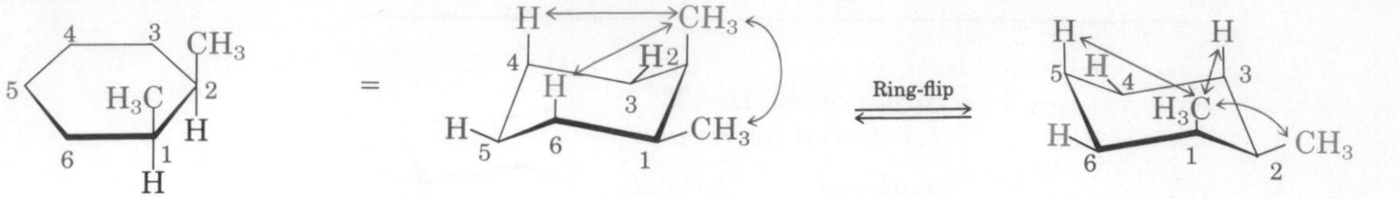

*cis*-1,2-Dimethyl-cyclohexane

One gauche interaction between $CH_3$ groups = 0.9 kcal/mol plus two $CH_3$—H 1,3-diaxial interactions; total strain = 0.9 + 1.8 = 2.7 kcal/mol

One gauche interaction between $CH_3$ groups = 0.9 kcal/mol plus two $CH_3$—H 1,3-diaxial interactions; total strain = 0.9 + 1.8 = 2.7 kcal/mol

**Figure 4.21** Conformations of *cis*-1,2-dimethylcyclohexane: The two conformations are equal in energy.

Both conformations of *cis*-1,2-dimethylcyclohexane in Figure 4.21 have one methyl group axial and one methyl group equatorial. The conformation on the left has an axial methyl group at C2 that has 1,3-diaxial interactions with hydrogens on C4 and C6. The ring-flipped conformation on the right has an axial methyl group at C1 that has 1,3-diaxial interactions with hydrogens on C3 and C5. In addition, both conformations have gauche butane interactions between the two methyl groups. *The two conformations are exactly equal in energy*, with a total steric strain of 2.7 kcal/mol (11.3 kJ/mol).

In the trans isomer, the two methyl groups are on opposite sides of the ring, and the compound can exist in either of the two chair conformations shown in Figure 4.22. The situation here is quite different from that of the cis isomer. The trans conformation on the left in Figure 4.22 has both methyl groups equatorial and therefore has one gauche butane interaction but no 1,3-diaxial interactions. The conformation on the right, however, has *both methyl groups axial*. The axial methyl group at C1 interacts with axial hydrogens at C3 and C5, and the axial methyl group at C2 interacts with axial hydrogens at C4 and C6. These four 1,3-diaxial interactions make the diaxial conformation 3.6 − 0.9 = 2.7 kcal/mol less favorable than the diequatorial conformation. We therefore predict that *trans*-1,2-dimethylcyclohexane will exist almost exclusively (> 99%) in the diequatorial conformation.

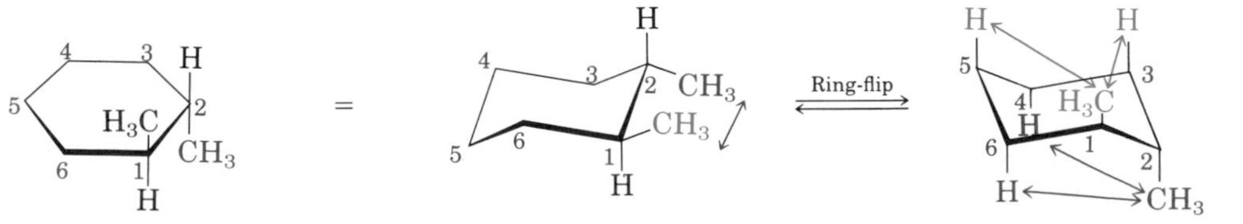

*trans*-1,2-Dimethylcyclohexane

One gauche interaction between $CH_3$ groups = 0.9 kcal/mol

Four 1,3-diaxial interactions = 3.6 kcal/mol

**Figure 4.22** Conformations of *trans*-1,2-dimethylcyclohexane: The conformation with both methyl groups equatorial is favored by 2.7 kcal/mol.

The same kind of **conformational analysis** just carried out for 1,2-dimethylcyclohexane can be carried out for any substituted cyclohexane. For example, let's look at *cis*-1-*tert*-butyl-4-chlorocyclohexane in the following practice problem.

**PRACTICE PROBLEM**

What is the most stable conformation of *cis*-1-*tert*-butyl-4-chlorocyclohexane, and by how much is it favored?

*Solution* First draw the two chair conformations of the molecule. *cis*-1-*tert*-Butyl-4-chlorocyclohexane can exist in either of the two chair conformations indicated:

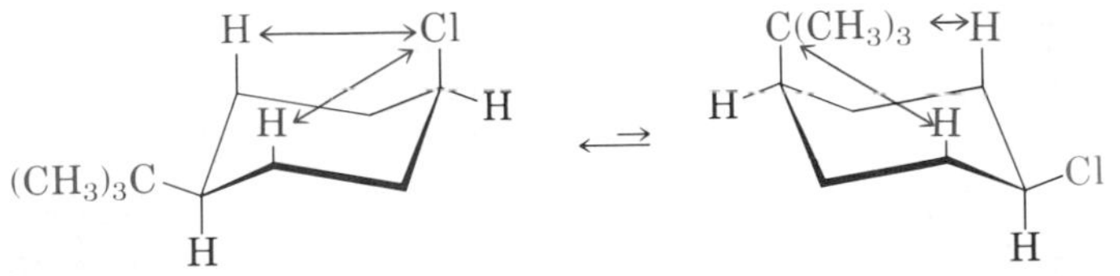

$2 \times 0.25 = 0.5$ kcal/mol steric strain $\qquad$ $2 \times 2.7 = 5.4$ kcal/mol steric strain

In the left-hand conformation, the *tert*-butyl group is equatorial and the chlorine is axial. In the right-hand conformation, the *tert*-butyl group is axial and the chlorine is equatorial. These conformations aren't of equal energy because an axial *tert*-butyl substituent and an axial chloro substituent produce different amounts of steric strain. Table 4.4 shows that a single *tert*-butyl–hydrogen 1,3-diaxial interaction costs 2.7 kcal/mol, whereas a single chlorine–hydrogen 1,3-diaxial interaction costs only 0.25 kcal/mol. An axial *tert*-butyl group therefore induces $(2 \times 2.7) - (2 \times 0.25) = 4.9$ kcal/mol more steric strain than an axial chlorine, and the compound therefore adopts the conformation having the chlorine axial and the *tert*-butyl equatorial.

The extremely large amount of steric strain caused by an axial *tert*-butyl group effectively holds the cyclohexane ring in a single conformation. We can sometimes take advantage of this steric locking if we wish to study the chemical reactivity of an immobile cyclohexane ring.

**PROBLEM**

**4.18** Draw the most stable chair conformation of these molecules and estimate the amount of 1,3-diaxial strain in each.

(a) *trans*-1-Chloro-3-methylcyclohexane
(b) *cis*-1-Ethyl-2-methylcyclohexane
(c) *cis*-1-Bromo-4-ethylcyclohexane
(d) *cis*-1-*tert*-Butyl-4-ethylcyclohexane

## 4.14 Boat Cyclohexane

Molecular models indicate that, in addition to chair cyclohexane, a second conformation known as **boat cyclohexane** is free of angle strain. We haven't paid it any attention thus far, however, because boat cyclohexane is much less stable than chair cyclohexane. As Figure 4.23 (page 118) shows, boat cyclohexane has a high degree of both steric strain and eclipsing strain.

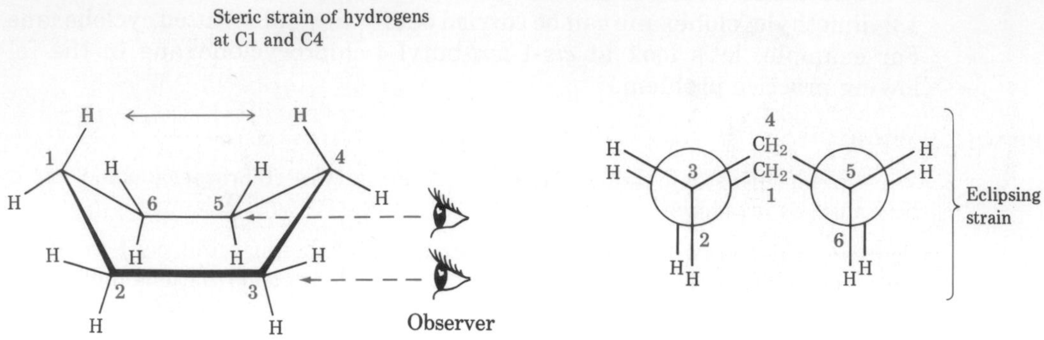

**Figure 4.23** The boat conformation of cyclohexane: There is no angle strain in this conformation, but there are large amounts of both steric strain and eclipsing strain.

There are two kinds of carbon atoms in boat cyclohexane. Carbons 2, 3, 5, and 6 lie in a plane, with carbons 1 and 4 above the plane. The inside hydrogen atoms on carbons 1 and 4 approach each other closely enough to cause considerable steric strain, and the four pairs of hydrogens on carbons 2, 3, 5, and 6 are eclipsed. The Newman projection in Figure 4.23, obtained by sighting along the C2—C3 and C5—C6 bonds, shows this eclipsing clearly.

Spectroscopic measurements indicate that boat cyclohexane is approximately 7.0 kcal/mol (29 kJ/mol) less stable than chair cyclohexane, although this value is reduced to about 5.5 kcal/mol by twisting slightly, thereby relieving some eclipsing strain (Figure 4.24). Even the **twist-boat conformation** is still far more strained than the chair conformation, and molecules adopt this geometry only rarely and under special circumstances.

**PROBLEM**

**4.19** There is good evidence for believing that *trans*-1,3-di-*tert*-butylcyclohexane exists largely in a twist-boat conformation. Why might this be so? Draw the likely twist-boat conformation.

## 4.15 Conformation of Polycyclic Molecules

The last point we'll consider about cycloalkane stereochemistry is what happens when two or more cycloalkane rings are fused together to construct **polycyclic molecules** such as decalin.

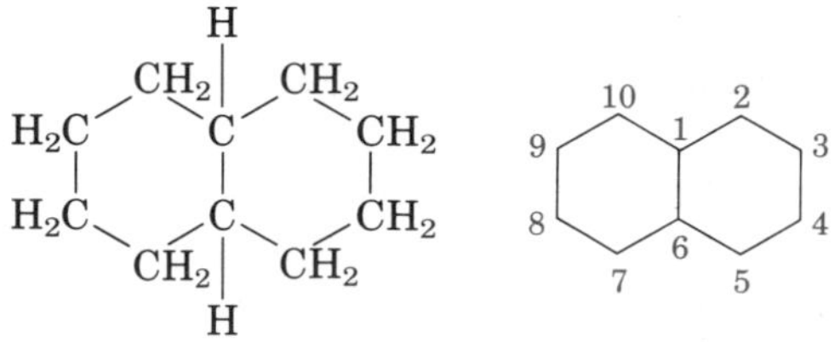

Decalin (two fused cyclohexane rings)

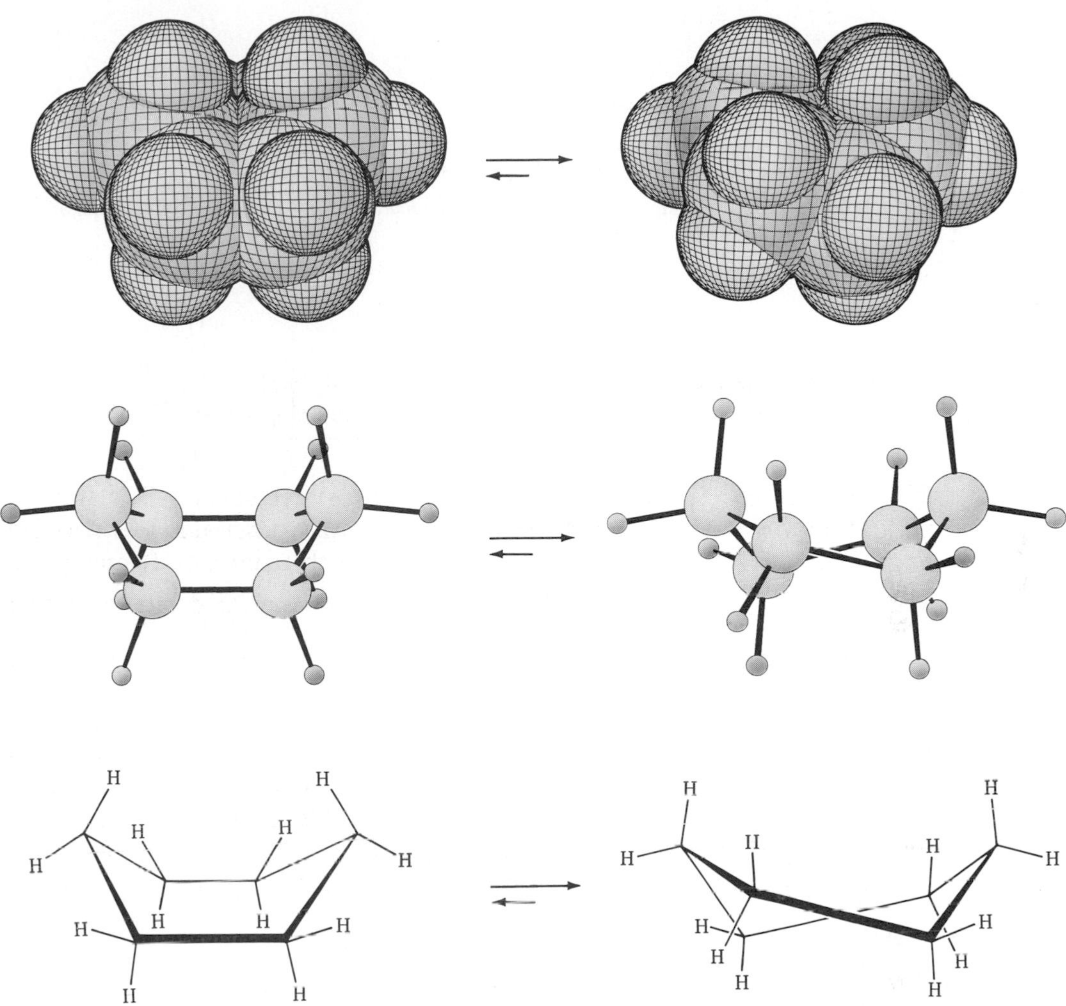


**Figure 4.24** Boat and twist-boat conformations of cyclohexane: The twist-boat conformation is lower in energy than the boat conformation by 1.5 kcal/mol. Both conformations are much more strained than chair cyclohexane.

Decalin consists of two cyclohexane rings joined together to share two carbon atoms (the **bridgehead** carbons, C1 and C6) and a common bond. Decalin can exist in either of two isomeric forms, depending on whether the rings are trans fused or cis fused. In *trans*-decalin, the hydrogen atoms at the bridgehead carbons are on opposite sides of the rings; in *cis*-decalin, the bridgehead hydrogens are on the same side. Figure 4.25 shows how both compounds can be represented using chair cyclohexane conformations.

Note that *trans*- and *cis*-decalin are not interconvertible by ring-flips or other rotations. They are stereoisomers (Section 3.10) and bear the same relationship to each other that *cis*- and *trans*-1,2-dimethylcyclohexane do.

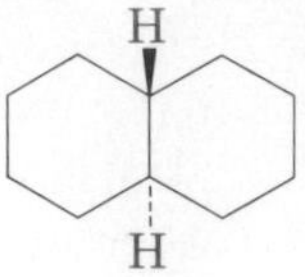

*trans*-Decalin

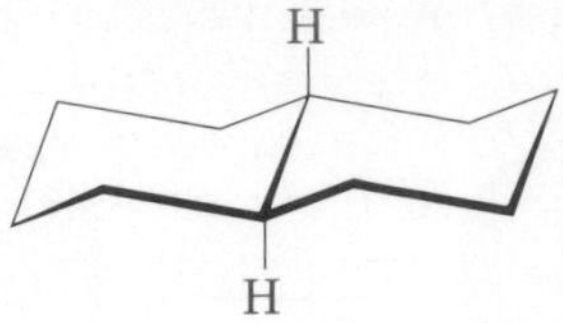

Bridgehead hydrogens on opposite sides of cyclohexane rings

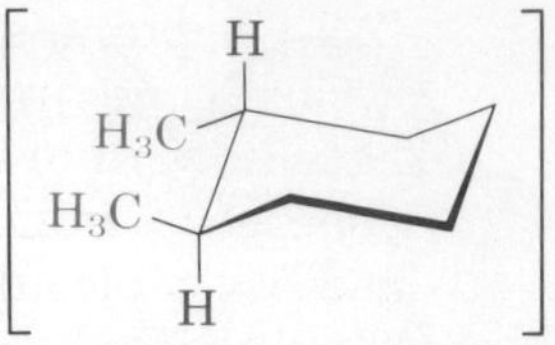

*trans*-1,2-Dimethylcyclohexane

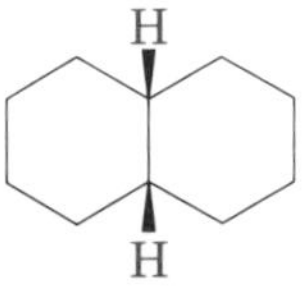

*cis*-Decalin

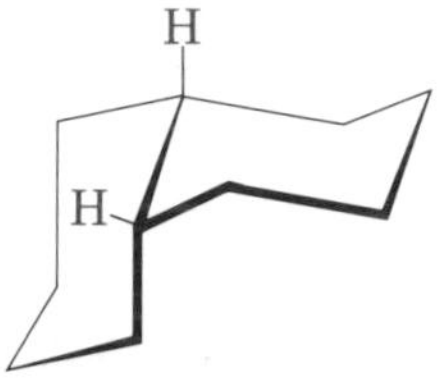

Brideghead hydrogens on same side of cyclohexane rings

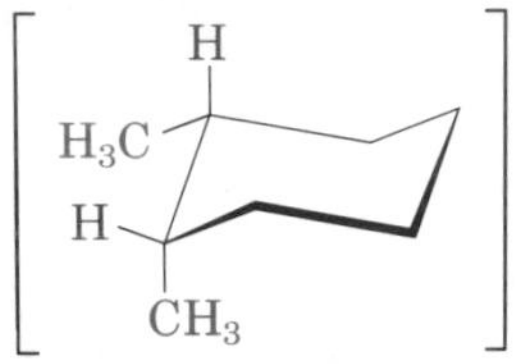

*cis*-1,2-Dimethylcyclohexane

**Figure 4.25** Representations of *trans*- and *cis*-decalin, compared with *trans*- and *cis*-1,2-dimethylcyclohexane

Polycyclic compounds are of great importance, and many valuable substances have fused-ring structures. For example, the steroids, such as cholesterol, have four rings—three six-membered and one five-membered—fused together. Though steroids look quite complicated in comparison to cyclohexane or decalin, they consist simply of chair cyclohexane rings locked together. The same principles that we applied to the conformational analysis of simple cyclohexane rings apply equally well (and often better) to steroids.

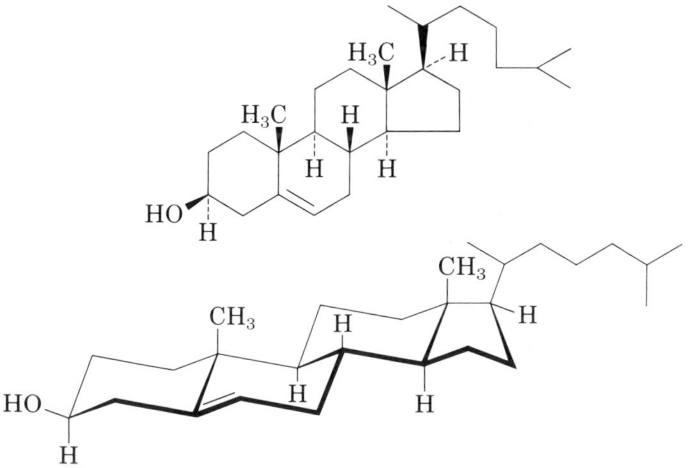

Cholesterol (a steroid)

Another common fused-ring system is the norbornane or bicyclo-[2.2.1]heptane structure. Norbornane has a conformationally locked boat cyclohexane ring in which carbons 1 and 4 are joined by an extra $CH_2$ group. Note how, in drawing this structure, we show one bond crossing in front of another by indicating a break in the rear bond. It's particularly helpful to make a molecular model when trying to see the three-dimensionality of norbornane.

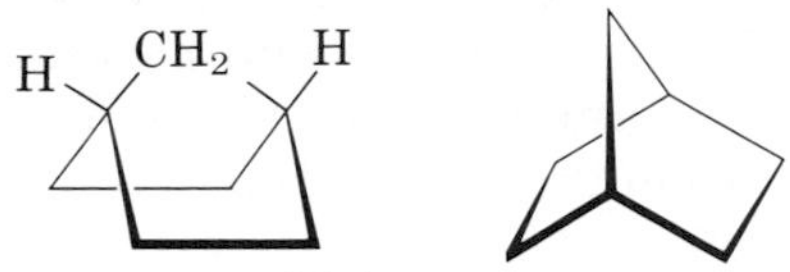

H $CH_2$ H

Norbornane
Bicyclo[2.2.1]heptane

Substituted norbornanes, such as camphor, are found widely in nature, and many have played an important historical role in developing organic structural theories.

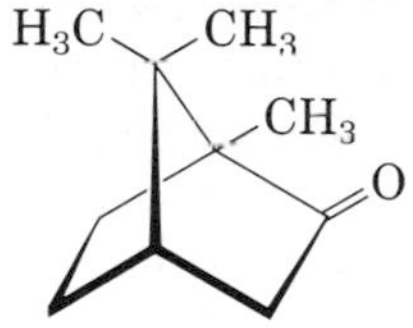

$H_3C$ $CH_3$ $CH_3$ O

Camphor

PROBLEM

**4.20** Which isomer do you think is the more stable, *trans*-decalin or *cis*-decalin? Explain your answer in terms of the steric interactions present in the two molecules.

## 4.16 Summary and Key Words

Carbon–carbon bonds in alkanes are formed by sigma overlap of two carbon $sp^3$ hybrid orbitals. Rotation occurs around sigma bonds as a result of their cylindrical symmetry, and alkanes therefore have a large number of rapidly interconverting **conformations**. **Newman projections** allow us to visualize the spatial consequences of bond rotation by sighting directly along a carbon–carbon bond axis. The **staggered** conformation of ethane is 2.9 kcal/mol (12 kJ/mol) more stable than the **eclipsed** conformation as a result of **torsional strain**. In general, any alkane is most stable when all of its bonds are staggered.

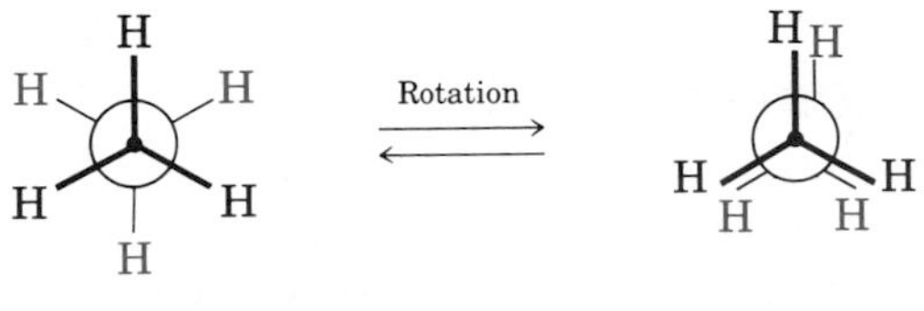

Rotation

Staggered ethane

Eclipsed ethane

Most organic molecules contain saturated rings. These **alicyclic** (aliphatic cyclic) molecules have special characteristics that affect the chemistry they undergo. Data obtained from studies on heats of combustion indicate that not all cycloalkanes are equally stable. Cyclopropane and cyclobutane are the least stable rings, whereas cyclopentane and the medium-sized rings (cycloheptane through cyclotridecane) all have varying degrees of strain. Cyclohexane and large rings (cyclotetradecane and above) are virtually strain-free.

Three different kinds of strain contribute to the overall energy level of a cycloalkane: (1) **angle strain**, the resistance of a bond angle to compression or expansion from the normal 109° tetrahedral value; (2) **eclipsing strain** (torsional strain), the energy cost of having neighboring C—H bonds eclipsed rather than staggered; and (3) **steric strain**, the result of the repulsive van der Waals interaction that arises when two groups try to occupy the same spatial position.

Cyclopropanes are highly strained (27.6 kcal/mol) because of both angle strain and eclipsing strain. Cyclobutane is also highly strained (26.4 kcal/mol) because of both angle strain and eclipsing strain. Cyclopentane is free of angle strain but suffers from a large number of eclipsing interactions. Both cyclobutane and cyclopentane pucker slightly away from planarity in order to relieve eclipsing strain.

Cyclohexane rings are the most important of all ring sizes because of their wide occurrence. Cyclohexane is strain-free by virtue of its puckered **chair conformation**, in which all bond angles are near 109° and all neighboring C—H bonds are staggered. Chair cyclohexane has two kinds of hydrogens—**axial** and **equatorial**. Axial hydrogens are directed up and down, parallel to the ring axis, whereas equatorial hydrogens lie in a belt more or less along the equator of the ring. Each carbon atom has one axial and one equatorial hydrogen.

Chair cyclohexanes are conformationally mobile; they can undergo a **ring-flip** that interconverts axial and equatorial positions:

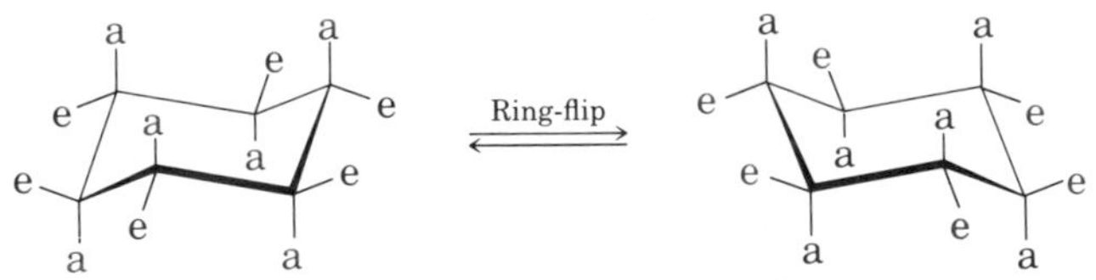

Substituents on the ring are more stable in the equatorial position, since axial substituents cause **1,3-diaxial steric strain**. The amount of 1,3-diaxial strain caused by an axial substituent depends on its bulk. The stereochemistry of cyclohexane and its derivatives is best learned by using molecular models to examine conformational relationships.

## ADDITIONAL PROBLEMS

**4.21** Define these terms in your own words:
(a) Angle strain (b) Steric strain (c) Torsional strain
(d) Heat of combustion (e) Conformation (f) Staggered
(g) Eclipsed (h) Gauche butane

**4.22** Consider 2-methylbutane (isopentane). Sighting along the C2—C3 bond:
(a) Draw a Newman projection of the most stable conformation.
(b) Draw a Newman projection of the least stable conformation.
(c) If a $CH_3$—$CH_3$ eclipsing interaction costs 2.5 kcal/mol and a $CH_3$—$CH_3$ gauche interaction costs 0.9 kcal/mol, construct a quantitative diagram of energy versus rotation about the C2—C3 bond.

**4.23** What are the energy differences between the three possible staggered conformations around the C2—C3 bond in 2,3-dimethylbutane?

**4.24** Construct a qualitative potential-energy diagram for rotation about the C—C bond of 1,2-dibromoethane. Which conformation would you expect to be more stable? Label the anti and the gauche conformations of 1,2-dibromoethane.

**4.25** Which conformation of 1,2-dibromoethane (Problem 4.24) would you expect to have the larger dipole moment? The observed dipole moment is $\mu = 1.0$ D. What does this tell you about the actual structure of the molecule?

**4.26** The barrier to rotation about the C—C bond in bromoethane is 3.6 kcal/mol (15 kJ/mol).
(a) What energy value can you assign to an H—Br eclipsing interaction?
(b) Construct a quantitative diagram of potential energy versus rotation for bromoethane.

**4.27** Define these terms:
(a) Axial bond (b) Equatorial bond
(c) Chair conformation (d) 1,3-Diaxial interaction

**4.28** Draw a chair cyclohexane ring and label all positions as axial or equatorial.

**4.29** Why is a 1,3-cis disubstituted cyclohexane always more stable than its trans isomer?

**4.30** Why is a 1,2-trans disubstituted cyclohexane always more stable than its cis isomer?

**4.31** Which is more stable, a 1,4-trans disubstituted cyclohexane or its cis isomer?

**4.32** *N*-Methylpiperidine has the conformation shown. What does this tell you about the relative steric requirements of a methyl group versus an electron lone pair?

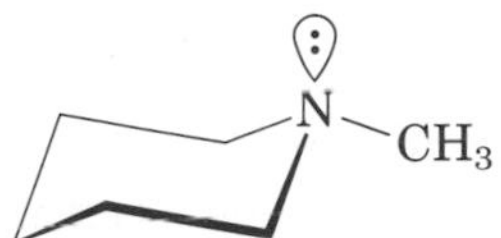


*N*-Methylpiperidine

**4.33** Draw the two chair conformations of *cis*-1-chloro-2-methylcyclohexane. Which is more stable and by how much?

**4.34** Draw the two chair conformations of *trans*-1-chloro-2-methylcyclohexane. Which is more stable and by how much?

**4.35** $\beta$-Glucose contains a six-membered ring in which all of the substituents are equatorial. Draw $\beta$-glucose in its more stable chair conformation.

HO O $CH_2OH$
HO OH
OH

Glucose

**4.36** From the data in Tables 4.3 and 4.4, calculate the percentages of molecules that have their substituents in an axial orientation for these compounds:
(a) Isopropylcyclohexane (b) Bromocyclohexane
(c) Cyclohexanecarbonitrile, $C_6H_{11}CN$ (d) Cyclohexanol, $C_6H_{11}OH$

**4.37** Assume that you have a variety of cyclohexanes substituted in the positions indicated. Identify the substituents as either axial or equatorial. For example, a 1,2-cis relationship means that one substituent must be axial and one equatorial, whereas a 1,2-trans relationship means that both substituents are axial or both are equatorial.
(a) 1,3-Trans disubstituted (b) 1,4-Cis disubstituted
(c) 1,3-Cis disubstituted (d) 1,5-Trans disubstituted
(e) 1,5-Cis disubstituted (f) 1,6-Trans disubstituted

**4.38** Draw the two possible chair conformations of *cis*-1,3-dimethylcyclohexane. The diaxial conformation is approximately 5.4 kcal/mol (23 kJ/mol) less stable than the diequatorial conformation. Can you suggest a reason for this large energy difference?

**4.39** Approximately how much steric strain does the 1,3-diaxial interaction between the two methyl groups (Problem 4.38) introduce into the diaxial conformation?

**4.40** In light of your answer to Problem 4.39, draw the two chair conformations of 1,1,3-trimethylcyclohexane and estimate the amount of strain energy in each. Which conformation is favored?

**4.41** We'll see in Chapter 11 that alkyl halides undergo an elimination reaction to yield alkenes on treatment with strong base. For example, chlorocyclohexane gives cyclohexene on reaction with $NaNH_2$.

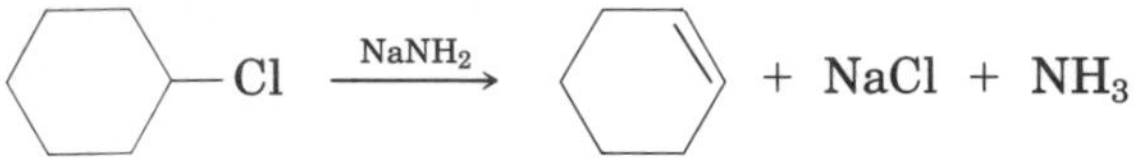

$$C_6H_{11}Cl \xrightarrow{NaNH_2} C_6H_{10} + NaCl + NH_3$$

If axial chlorocyclohexanes are generally more reactive than their equatorial isomers, which do you think would react faster, *cis*-2-chloro-*tert*-butylcyclohexane or *trans*-2-chloro-*tert*-butylcyclohexane? Explain.

**4.42** We saw in Problem 4.20 that *cis*-decalin is less stable than *trans*-decalin. Assume that the 1,3-diaxial interactions in *trans*-decalin are similar to those in axial methylcyclohexane (one $CH_3$–H interaction costs 0.9 kcal/mol) and calculate the magnitude of the energy difference between *cis*- and *trans*-decalin.

**4.43** Using molecular models as well as structural drawings, explain why *trans*-decalin is rigid and cannot ring-flip, whereas *cis*-decalin can easily ring-flip.

**4.44** How many geometric isomers of 1,2,3,4,5,6-hexachlorocyclohexane are there? Draw the structure of the most stable isomer.

**4.45** Propose an explanation of the observation that the all-cis isomer of 4-*tert*-butylcyclohexane-1,3-diol reacts readily with acetone and an acid catalyst to form an acetal, but that other stereoisomers do not react.

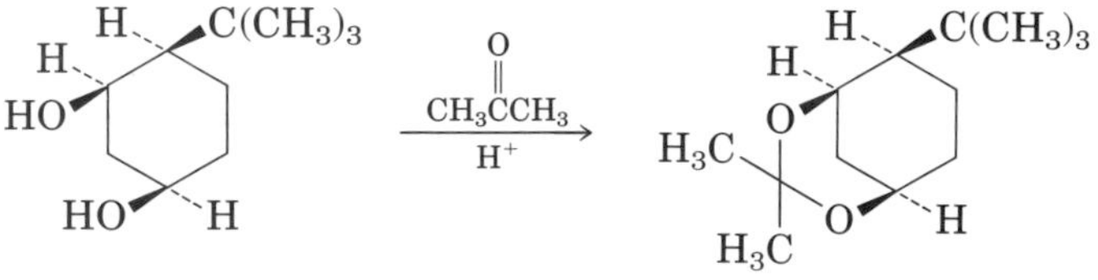

H, $C(CH_3)_3$, H, HO, HO, H $\xrightarrow[H^+]{CH_3\overset{O}{\overset{\|}{C}}CH_3}$ H, $C(CH_3)_3$, H, O, $H_3C$, $H_3C$, O, H

An acetal

In formulating your answer, draw the stable chair conformations of all four stereoisomers and of the product acetal. Use molecular models for help.

**4.46** Increased substitution around a bond always leads to increased strain. Take the four substituted butanes listed below, for example. For each compound, sight along the C2—C3 bond and draw Newman projections of the most stable and least stable conformations. Use the data in Table 4.1 to assign strain energy values to each conformation. Which of the eight conformations is most stable and which is least stable?

(a) 2-Methylbutane
(b) 2,2-Dimethylbutane
(c) 2,3-Dimethylbutane
(d) 2,2,3-Trimethylbutane

CHAPTER 5

# An Overview of Organic Reactions

When first approached, organic chemistry often seems like a bewildering collection of facts—a collection of millions of compounds, dozens of functional groups, and a seemingly endless number of reactions. With study, though, it becomes evident that there are only a few fundamental concepts that underlie *all* organic reactions.

Far from being a collection of isolated facts, organic chemistry is a beautifully logical subject unified by a few broad themes. When these themes are understood, learning organic chemistry becomes much easier and rote memorization can be avoided. The aim of this book is to point out the themes and to clarify the patterns that unify organic chemistry. We'll begin by taking an overview of the fundamental kinds of organic reactions that take place and seeing how they can be described.

## 5.1 Kinds of Organic Reactions

Organic chemical reactions can be organized in two ways—by *what kinds* of reactions occur, and by *how* reactions occur. When beginning a study of the subject, it's easier to look first at the kinds of reactions that take place. There are four particularly important kinds of organic reactions—additions, eliminations, substitutions, and rearrangements.

**Addition reactions**, as their name implies, occur when two reactants add together to form a single new product with no atoms "left over." We might generalize the process as:

These reactants add together . . . $A + B \longrightarrow C$ . . . to give this single product.

As an example of an important addition reaction that we'll be studying soon, alkenes such as ethylene react with acids such as HBr to yield alkyl halides:

These two reactants . . . H—Br + $H_2C{=}CH_2$ ⟶ $H{-}CH_2{-}CHBr{-}H$ . . . add to give this product.

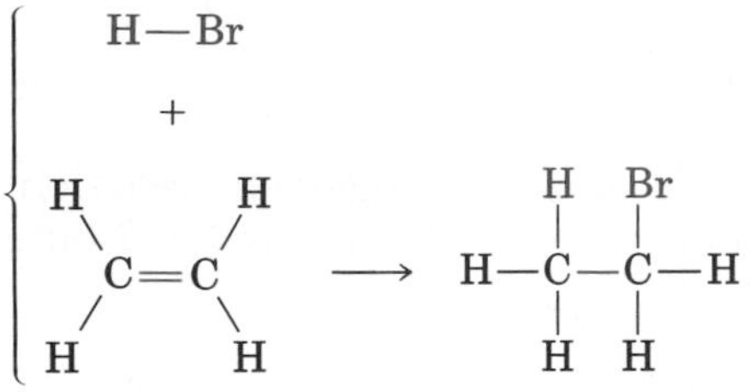

Ethylene (an alkene)

Bromoethane (an alkyl halide)

**Elimination reactions** are, in a sense, the opposite of addition reactions. Eliminations occur when a single reactant splits apart into two products, a process we can generalize as

This one reactant . . . $A \longrightarrow B + C$ . . . splits apart to give these two products.

As an example of an important elimination reaction, alkyl halides such as bromoethane split apart into an acid and an alkene when treated with base:

This one reactant . . . $H{-}CH_2{-}CHBr{-}H \xrightarrow{NaOH} H_2C{=}CH_2 + H{-}Br$ . . . gives these two products.

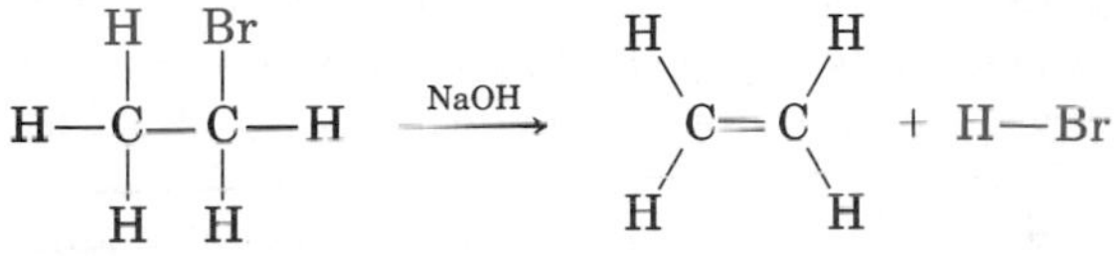

Bromoethane (an alkyl halide)

Ethylene (an alkene)

**Substitution reactions** occur when two reactants exchange parts to give two new products, a process we can generalize as

These two reactants exchange parts . . . $A{-}B + C{-}D \longrightarrow A{-}C + B{-}D$ . . . to give these two new products.

As an example of a substitution reaction, alkanes such as methane react with chlorine gas in the presence of ultraviolet light to yield alkyl halides. A —Cl group from chlorine replaces (substitutes for) the —H group of methane, and two new products result:

These two reactants . . . $H{-}CH_2{-}H + Cl{-}Cl \xrightarrow{Light} H{-}CH_2{-}Cl + H{-}Cl$ . . . give these two products.

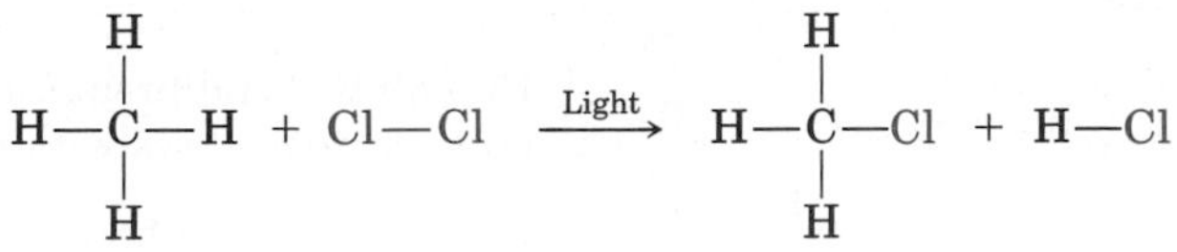

Methane (an alkane)

Chloromethane (an alkyl halide)

**Rearrangement reactions** occur when a single reactant undergoes a reorganization of bonds and atoms to yield a single isomeric product, a process we can generalize as

This single reactant . . . A $\longrightarrow$ B . . . gives this isomeric product.

As an example of a rearrangement reaction, the alkene 1-butene can be converted into its constitutional isomer, 2-butene, by treatment with an acid catalyst:

$$CH_3CH_2CH{=}CH_2 \xrightarrow{\text{Acid catalyst}} CH_3CH{=}CHCH_3$$

1-Butene 2-Butene

PROBLEM

**5.1** Classify these reactions as additions, eliminations, substitutions, or rearrangements:

(a) $CH_3Br + KOH \longrightarrow CH_3OH + KBr$

(b) $CH_3CH_2OH \longrightarrow H_2C{=}CH_2 + H_2O$

(c) $H_2C{=}CH_2 + H_2 \longrightarrow CH_3CH_3$

## 5.2 How Organic Reactions Occur: Mechanisms

Having looked at the organization of organic chemistry according to the kinds of reactions that take place, let's now see the organization according to how reactions occur. An overall description of how a specific reaction occurs is called a **reaction mechanism**. A mechanism describes in detail exactly what takes place at each stage of a chemical transformation. It describes which bonds are broken and in what order, which bonds are formed and in what order, and what the relative rates of each step are. A complete mechanism must also account for all reactants used, all products formed, and the amounts of each.

All chemical reactions involve bond breaking and bond making. When two reactants come together, react, and yield products, specific chemical bonds in the starting materials are broken and specific chemical bonds in the products are formed. Fundamentally there are only two ways that a covalent two-electron bond can break: A bond can break in an electronically *symmetrical* way such that one electron remains with each product fragment, or a bond can break in an electronically *unsymmetrical* way such that both bonding electrons remain with one product fragment, leaving the other fragment with an empty orbital. The symmetrical cleavage is called a **homolytic** process, and the unsymmetrical cleavage is called a **heterolytic** process.

$$A : B \longrightarrow A\cdot + \cdot B$$

Homolytic bond breaking (radical)
(one electron stays with each fragment)

$$A : B \longrightarrow A^+ + :B^-$$

Heterolytic bond breaking (polar)
(two electrons stay with one fragment)

Conversely, there are only two ways that a covalent two-electron bond can form: A bond can form in an electronically symmetrical (**homogenic**) way when one electron is donated to the new bond by each reactant, or a bond can form in an electronically unsymmetrical way (**heterogenic**) when both bonding electrons are donated to the new bond by one reactant.

A· + ·B ⟶ A : B — Homogenic bond making (radical) (one electron donated by each fragment)

$A^+$ + :$B^-$ ⟶ A : B — Heterogenic bond making (polar) (two electrons donated by one fragment)

Those processes that involve symmetrical bond breaking and bond making are called **radical reactions**. A **radical** (sometimes called a "free radical") is a species that contains an *odd* number of valence electrons and thus has a single, unpaired electron in one of its orbitals. Those processes that involve unsymmetrical bond breaking and bond making are called **polar reactions**. Polar reactions always involve species that contain an *even* number of valence electrons. Polar processes are the more commonly encountered reaction type in organic chemistry, and a large part of this book is devoted to their description.

In addition to polar and radical reactions, there's a third, less commonly encountered type called **pericyclic reactions**. Rather than explain pericyclic reactions now, though, we'll study them in more detail at a later point.

## 5.3 Radical Reactions and How They Occur

Radical reactions aren't as common as polar reactions, but they are nevertheless quite important in organic chemistry, particularly in certain industrial processes. Let's see how they occur.

Although most radicals are electrically neutral, they are highly reactive because they contain an odd number of electrons (usually seven) in their outer shell rather than a stable inert-gas configuration (an octet). They can achieve the desired octet in several ways. For example, a radical can abstract an atom from another molecule, leaving behind a new radical. The net result is a radical *substitution* reaction:

Rad· (Unpaired electron) + A : B ⟶ Rad : A + ·B (Unpaired electron)

Reactant radical; Substitution product; Product radical

Alternatively, a reactant radical can add to an alkene, taking one electron from the alkene double bond and yielding a new radical. The net result is a radical *addition* reaction:

Unpaired electron — Unpaired electron

$$\text{Rad}\cdot + \mathrm{C{=}C} \longrightarrow \text{Rad}{-}\mathrm{C{-}C}\cdot$$

Reactant radical — Alkene — Addition product radical

Let's look at a specific example of a radical reaction to see its characteristics. The details of the reaction aren't important at this point; it's only important to see that radical reactions involve odd-electron species.

## 5.4 An Example of a Radical Reaction: Chlorination of Methane

The chlorination of methane is a typical **radical substitution reaction**. Although inert to most reagents, alkanes react readily with chlorine to give chlorinated alkane products:

$$CH_4 + Cl{-}Cl \xrightarrow{\text{Light}} CH_3Cl + H{-}Cl$$

Methane — Chlorine — Chloromethane

A more detailed discussion of this radical substitution reaction is given in Chapter 10. For the present, it's only important to know that careful studies have shown alkane chlorination to be a multistep process involving radicals. Radical substitution reactions normally require three kinds of steps—an initiation step, propagation steps, and termination steps.

1. *Initiation step* The initial production of radicals: The initiation step begins the reaction by producing reactive radicals. In the present case, the relatively weak chlorine–chlorine bond is homolytically broken by irradiation with ultraviolet light. Two reactive chlorine radicals are produced, and further chemistry ensues.

$$:\ddot{\mathrm{Cl}}:\ddot{\mathrm{Cl}}: \xrightarrow{\text{Light}} 2\,:\ddot{\mathrm{Cl}}\cdot$$

2. *Propagation steps* Radicals undergo substitution reactions: Once chlorine radicals have been produced in small amounts, propagation steps take place. When a highly reactive chlorine radical collides with a methane molecule, it abstracts a hydrogen atom to produce

HCl and a methyl radical ($\cdot CH_3$). This methyl radical reacts further with $Cl_2$ in another substitution step to give the new products chloromethane and chlorine radical. The chlorine radical then cycles back into the first propagation step. Once the sequence has been initiated, it becomes a self-sustaining cycle of repeating steps 2a and 2b, making the overall process a *chain reaction.*

a. $:\ddot{\underset{..}{Cl}}\cdot + H:CH_3 \longrightarrow H:\ddot{\underset{..}{Cl}}: + \cdot CH_3$

b. $\cdot CH_3 + :\ddot{\underset{..}{Cl}}:\ddot{\underset{..}{Cl}}: \longrightarrow :\ddot{\underset{..}{Cl}}:CH_3 + :\ddot{\underset{..}{Cl}}\cdot$

c. Repeat steps a and b over and over.

3. *Termination steps* The chain is broken when two radicals combine: Occasionally, two radicals might collide and combine to form a stable product in a termination step. When this happens, the reaction cycle is broken and the chain is ended. Such termination steps occur infrequently, however, because the concentration of radicals in the reaction at any given moment is very small. Thus, the likelihood that two radicals will collide is also small.

$$\left.\begin{array}{l} :\ddot{\underset{..}{Cl}}\cdot + \cdot\ddot{\underset{..}{Cl}}: \longrightarrow :\ddot{\underset{..}{Cl}}:\ddot{\underset{..}{Cl}}: \\ :\ddot{\underset{..}{Cl}}\cdot + \cdot CH_3 \longrightarrow :\ddot{\underset{..}{Cl}}:CH_3 \\ H_3C\cdot + \cdot CH_3 \longrightarrow H_3C:CH_3 \end{array}\right\} \text{Possible termination steps}$$

Alkane chlorination is not a generally useful reaction because most alkanes (other than methane and ethane) have several different kinds of hydrogens, and *mixtures* of chlorinated products usually result in such cases. Nevertheless, radical chain reactions constitute a basic reaction type of considerable importance.

The radical substitution reaction just discussed is only one of several different processes that radicals can undergo. The fundamental principle behind all radical reactions is the same, however: All bonds are broken and formed by reaction of odd-electron species.

PROBLEM

**5.2** When a mixture of methane and chlorine is irradiated, reaction commences immediately. When irradiation is stopped, the reaction gradually slows down but does not stop immediately. How do you account for this behavior?

PROBLEM

**5.3** Draw and name all monochloro products you would expect to obtain from reaction of 2-methylpentane with chlorine.

PROBLEM

**5.4** Radical chlorination of pentane is a poor way to prepare 1-chloropentane, $CH_3CH_2CH_2CH_2CH_2Cl$, but radical chlorination of neopentane, $(CH_3)_4C$, is a good way to prepare neopentyl chloride, $(CH_3)_3CCH_2Cl$. How do you account for this difference?

## 5.5 Polar Reactions and How They Occur

**Polar reactions** occur as the result of attractive forces between positive and negative charges on molecules. In order to see how these reactions take place, we need first to recall our previous discussion of polar covalent bonds and then to look more deeply into the effects of bond polarity on organic molecules.

Most organic molecules are electrically neutral; that is, they have no net charge, either positive or negative. We saw in Section 2.4, however, that specific bonds within a molecule, particularly the bonds in functional groups, are often polar. Bond polarity is a consequence of unsymmetrical electron-density distribution in the bond and is due to the electronegativity of the atoms involved. A table of electronegativities (Table 5.1, which repeats Table 2.2 for convenience) shows that atoms such as oxygen, nitrogen, fluorine, chlorine, and bromine are more electronegative than carbon. A carbon atom bonded to one of these electronegative atoms has a partial positive charge ($\delta^+$), and the electronegative atom has a slight negative charge ($\delta^-$). For example,

$$\overset{\delta^+}{\text{—C}}\text{—}\overset{\delta^-}{\text{Y}} \qquad \text{where Y = O, N, F, Cl, or Br}$$

**Table 5.1** Relative electronegativities of some common elements

| Period | Group | | | | | | | |
|---|---|---|---|---|---|---|---|---|
| | IA | IIA | | IIIA | IVA | VA | VIA | VIIA |
| 1 | H 2.2 | | | | | | | |
| 2 | Li 1.0 | Be 1.6 | | B 2.0 | C **2.5** | N 3.0 | O 3.4 | F 4.0 |
| 3 | Na 0.9 | Mg 1.3 | | Al 1.6 | Si 1.9 | P 2.2 | S 2.6 | Cl 3.1 |
| 4 | | | | | | | | Br 3.0 |

When carbon forms bonds to atoms that are *less* electronegative, polarity in the *opposite* sense results. For example, the reaction of bromomethane with magnesium metal yields methylmagnesium bromide—a so-called **Grignard reagent**. In Grignard reagents, and in most species that contain carbon–metal (**organometallic**) bonds, the carbon atom is *negatively* polarized with respect to the metal:

$$H_3C\text{—Br} + Mg \longrightarrow H_3\overset{\delta^-}{C}\text{—}\overset{\delta^+}{MgBr}$$

A Grignard reagent

The polarity patterns of some common functional groups are shown in Table 5.2.

**Table 5.2** Polarity patterns in some common functional groups

| *Compound type* | *Functional group structure* | *Compound type* | *Functional group structure* |
|---|---|---|---|
| Alcohol | $\geq\overset{\delta^+}{C}-\overset{\delta^-}{O}H$ | Carbonyl | $>\overset{\delta^+}{C}=\overset{\delta^-}{O}$ |
| Alkene | $>C=C<$ Symmetrical, nonpolar | Carboxylic acid | $-\overset{\delta^+}{C}(=\overset{\delta^-}{O})-OH^{\delta^-}$ |
| Alkyl halide | $\geq\overset{\delta^+}{C}-\overset{\delta^-}{X}$ | Carboxylic acid chloride | $-\overset{\delta^+}{C}(=\overset{\delta^-}{O})-Cl^{\delta^-}$ |
| Amine | $\geq\overset{\delta^+}{C}-\overset{\delta^-}{N}H_2$ | Aldehyde | $-\overset{\delta^+}{C}(=\overset{\delta^-}{O})-H$ |
| Ether | $\geq\overset{\delta^+}{C}-\overset{\delta^-}{O}-\overset{\delta^+}{C}\leq$ | Ester | $-\overset{\delta^+}{C}(=\overset{\delta^-}{O})-\overset{\delta^-}{O}-C$ |
| Nitrile | $-\overset{\delta^+}{C}\equiv\overset{\delta^-}{N}$ | | |
| Grignard reagent | $\geq\overset{\delta^-}{C}-\overset{\delta^+}{Mg}Br$ | | |
| Alkyllithium | $\geq\overset{\delta^-}{C}-\overset{\delta^+}{Li}$ | Ketone | $-\overset{\delta^+}{C}(=\overset{\delta^-}{O})-C$ |

This discussion of bond polarity is oversimplified in that we've considered only bonds that are *inherently* polar due to electronegativity effects. Polar bonds can also result from the interaction of functional groups with solvents and with acids. For example, the polarity of the carbon–oxygen bond in methanol is greatly enhanced by protonation:

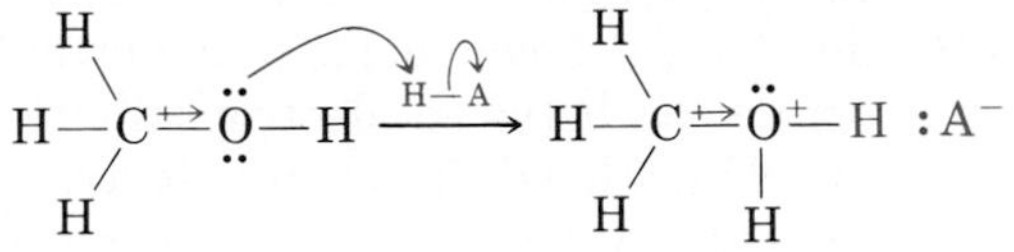

$$H_3C \overset{\leftrightarrow}{-} \ddot{\underset{..}{O}}-H \xrightarrow{H-A} H_3C \overset{\leftrightarrow}{-} \ddot{O}^{+}(H)-H \;\; :A^-$$

Methanol
(weakly polar C—O bond)

Protonated methanol cation
(strongly polar C—O bond)

In neutral methanol, the carbon atom is somewhat electron-poor because the electronegative oxygen attracts carbon–oxygen bond electrons. In the protonated methanol cation, a full positive charge on oxygen *strongly* attracts electrons in the carbon–oxygen bond and makes the carbon much more electron-poor.

Yet a further consideration is the **polarizability** (as opposed to polarity) of an atom. As the electric field around a given atom changes due to changing interactions with solvent or with other polar reagents, the electron distribution around that atom also changes. The measure of this response is the polarizability of the atom. Larger atoms with more loosely held electrons are more polarizable than smaller atoms with tightly held electrons. Thus, iodine is much more polarizable than fluorine. This means that the carbon–iodine bond, although electronically symmetrical according to the electronegativity table (Table 5.1), can nevertheless react as if it were polar.

What does functional-group polarity mean with respect to chemical reactivity? *Since unlike charges attract, the fundamental characteristic of all polar organic reactions is that electron-rich sites in the functional groups of one molecule react with electron-poor sites in the functional groups of another molecule.* Bonds are made when the electron-rich reagent donates a *pair* of electrons to the electron-poor reagent; conversely, bonds are broken when one of the two product fragments leaves with the electron *pair*.

As we saw earlier in Section 2.8, chemists normally indicate the electron pair movement that occurs during polar reactions by using **curved arrows**. *A curved arrow indicates the motion of electrons, not atoms.* It means that an electron pair has moved from the tail to the head of the arrow during the reaction. In referring to this fundamental polar process and to the species involved, chemists have coined the words *nucleophile* and *electrophile.* A **nucleophile** is a reagent that is "nucleus-loving"; nucleophiles have electron-rich sites and can form a bond by donating a pair of electrons to an electron-poor site. Nucleophiles are often negatively charged, though this is not always the case. An **electrophile**, by contrast, is "electron-loving"; electrophiles have electron-poor sites and can form a bond by accepting a pair of electrons from a nucleophile.

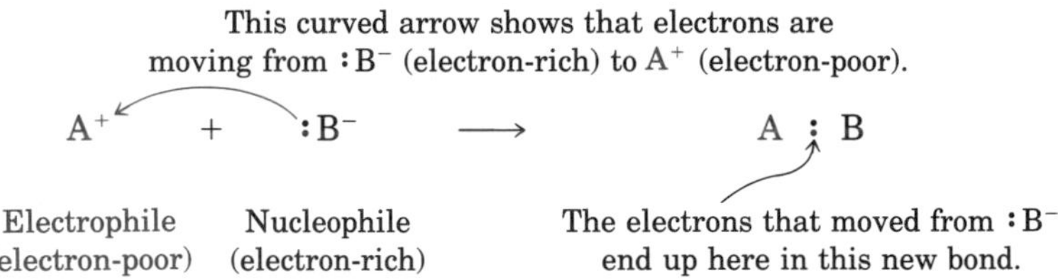


If the definitions of nucleophiles and electrophiles sound similar to those given in Section 2.8 for Lewis acids and Lewis bases, that's because there is indeed a correlation between electrophilicity–nucleophilicity and Lewis acidity–basicity. Thus, Lewis bases are electron donors and usually behave as nucleophiles, whereas Lewis acids are electron acceptors and usually behave as electrophiles. The major difference, however, is that the terms *electrophile* and *nucleophile* are usually used only when bonds to *carbon* are involved. We'll explore these ideas in much more detail in Chapter 11.

PROBLEM

**5.5** Identify the functional groups present in these molecules and show the direction of polarity in each.

(a) Acetone, $CH_3\overset{O}{\overset{\|}{C}}CH_3$

(b) Ethyl propenoate, $H_2C{=}CH\overset{O}{\overset{\|}{C}}OCH_2CH_3$

(c) Chloroethylene, $H_2C{=}CHCl$

(d) Tetraethyllead, $(CH_3CH_2)_4Pb$ (the "lead" in gasoline)

PROBLEM

**5.6** Which of the following would you expect to behave as electrophiles and which as nucleophiles?

(a) $H^+$ (b) $H\ddot{\underset{..}{O}}:^-$ (c) $:\ddot{\underset{..}{Br}}^+$

(d) $:NH_3$ (e) $CO_2$ (f) $Mg^{2+}$

## 5.6 An Example of a Polar Reaction: Addition of HBr to Ethylene

Let's look in detail at a typical polar process, the addition reaction of ethylene with HBr that we saw earlier. When ethylene is treated with hydrogen bromide at room temperature, bromoethane is produced. Overall, the reaction can be formulated as follows:

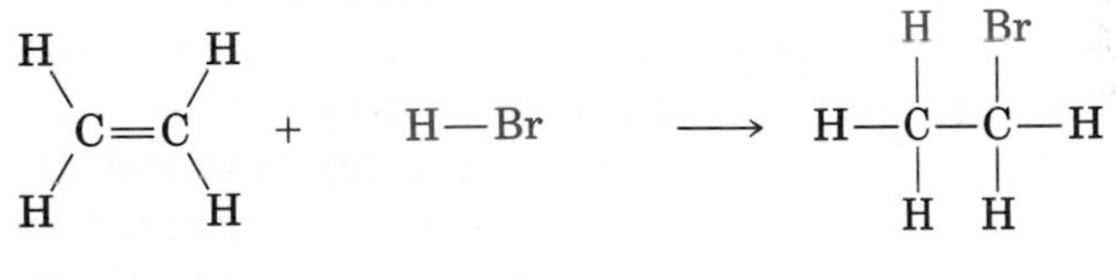

$$H_2C{=}CH_2 \quad + \quad H{-}Br \quad \longrightarrow \quad H{-}\underset{H}{\overset{H}{C}}{-}\underset{H}{\overset{Br}{C}}{-}H$$

Ethylene (nucleophile) — Hydrogen bromide (electrophile) — Bromoethane

This reaction, an example of a general polar reaction type known as an **electrophilic addition**, can be understood in terms of the general concepts just discussed. We'll begin by looking at the nature of the two reactants.

What do we know about ethylene? We know from Section 1.10 that a carbon–carbon double bond results from orbital overlap of two $sp^2$-hybridized carbon atoms. The sigma part of the double bond results from $sp^2$–$sp^2$ overlap, whereas the pi part results from *p*–*p* overlap.

What kind of chemical reactivity might we expect of carbon–carbon double bonds? We know that *alkanes*, such as ethane, are rather inert, since all valence electrons are tied up in strong, relatively nonpolar, carbon–carbon and carbon–hydrogen bonds. Furthermore, we know that the bonding electrons in alkanes are inaccessible to external reagents since they are sheltered in sigma orbitals between nuclei.

The electronic situation in ethylene and other alkenes is quite different, however. For one thing, double bonds have greater electron density than single bonds—four electrons in a double bond versus only two electrons in a single bond. Equally important, the electrons in the pi bond are accessible to external reagents because they are located above and below the plane of the double bond, rather than being sheltered between the nuclei (Figure 5.1).

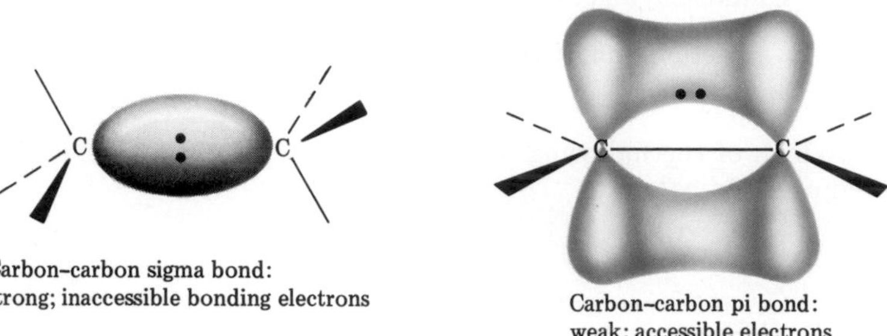


**Figure 5.1** A comparison of carbon–carbon single and double bonds: A double bond is both more electron-rich (nucleophilic) and more accessible than a single bond.

Both electron richness and electron accessibility lead us to predict high reactivity for carbon–carbon double bonds. In the terminology of polar reactions just introduced, we might predict that carbon–carbon double bonds should behave as *nucleophiles*. That is, the chemistry of alkenes should be dominated by reaction of the electron-rich double bond with electron-poor reagents. This is exactly what we find: The most important reaction of alkenes is their reaction with electrophiles.

What about HBr? As a strong mineral acid, HBr is a powerful proton ($H^+$) donor. Since a proton is positively charged and electron-poor, it is a good electrophile. Thus, the reaction between $H^+$ and ethylene is a typical electrophile–nucleophile combination, characteristic of all polar reactions. (Although chemists often talk about "$H^+$" when referring to acids, in fact there is really no such species. Protons are always associated with another molecule for stability, for example in $H_3O^+$.)

Although we'll see more details about alkene electrophilic addition reactions shortly, for the present we can look at the reaction as taking place by the pathway shown in Figure 5.2. The reaction begins when the alkene (ethylene) donates a pair of electrons from its C—C double bond to form a new single bond to $H^+$. (Note that the curved arrow in the first step of Figure 5.2 shows an electron pair moving out from the double bond to form a single bond to hydrogen.)

One of the double-bond carbons forms a new bond to the incoming hydrogen, but the other carbon atom, having lost its share of the double-bond electrons, is now trivalent, has only six valence electrons, and is left

The electrophile, HBr, is attacked by the pi electrons of the alkene, and a new C—H sigma bond is formed. This leaves the other carbon atom with a + charge and a vacant *p* orbital.

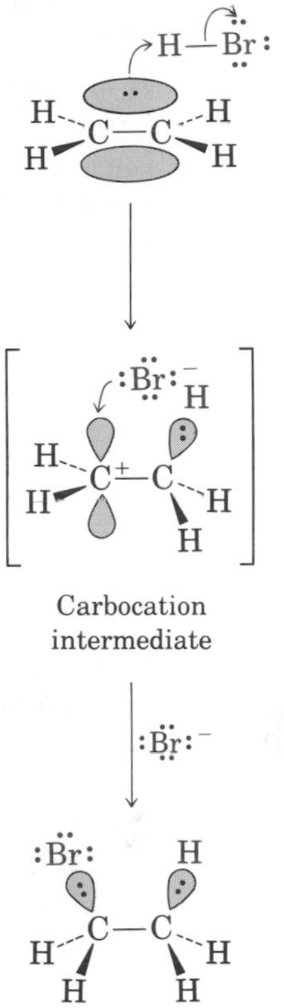

Carbocation intermediate

:Br:⁻ donates an electron pair to the positively charged carbon atom, forming a C—Br bond and yielding a neutral addition product.

**Figure 5.2** The polar addition reaction of HBr and ethylene: The reaction takes place in two steps, both of which involve electrophile–nucleophile interactions.

with the positive charge. This positively charged species, a carbon cation or **carbocation**, is itself an electrophile that can accept an electron pair from nucleophilic bromide anion to form a C—Br bond and give neutral bromoethane addition product. Once again, a curved arrow in Figure 5.2 is used to show the electron pair movement from bromide anion to carbon.

The electrophilic addition of HBr to ethylene is only one example of a polar process; there are many other types that we'll study in detail in later chapters. Regardless of the details of individual reactions, all polar processes can be accounted for in the general terms just presented. To repeat: Polar reactions take place between an electron-poor site and an electron-rich site and involve the donation of an electron pair from a nucleophile to an electrophile.

PROBLEM

**5.7** What product would you expect from reaction of HBr with cyclohexene?

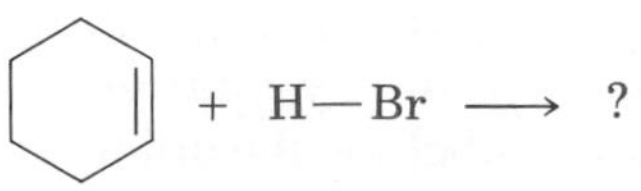

+ H—Br ⟶ ?

## 5.7 Describing a Reaction: Rates and Equilibria

In principle, all chemical reactions can be written as equilibrium processes. Starting materials can react to give products, and the products can revert back to starting materials. We can express a chemical equilibrium by an equation in which $K_{eq}$, the equilibrium constant (Section 2.6), is equal to the product concentrations multiplied together, divided by the concentrations of the starting materials multiplied together. For the reaction

$$a\mathrm{A} + b\mathrm{B} \rightleftharpoons c\mathrm{C} + d\mathrm{D}$$

we have

$$K_{eq} = \frac{[\text{Products}]}{[\text{Reactants}]} = \frac{[\mathrm{C}]^c[\mathrm{D}]^d}{[\mathrm{A}]^a[\mathrm{B}]^b}$$

This equation tells us the position of the equilibrium—that is, which side of the reaction arrow is energetically more favored. If $K_{eq}$ is larger than 1, then the product concentrations $[\mathrm{C}]^c[\mathrm{D}]^d$ are larger than the reactant concentrations $[\mathrm{A}]^a[\mathrm{B}]^b$, and the reaction proceeds as written from left to right. Conversely, if $K_{eq}$ is smaller than 1, the reaction does not take place as written.

What the equilibrium equation doesn't tell us is the *rate* of reaction. How fast is the equilibrium established? Some reactions are extremely slow even though they have highly favorable equilibrium constants. For example, gasoline is stable indefinitely when stored, because the rate of its reaction with oxygen is slow under normal circumstances. Under other circumstances, however—contact with a lighted match, for example—gasoline reacts rapidly with oxygen and undergoes complete conversion to the equilibrium products water and carbon dioxide. Rate (*how fast* a reaction occurs) and equilibrium (*how much* a reaction occurs) are entirely different.

What determines the equilibrium position of a reaction? Let's again look at the addition reaction of ethylene with HBr as an example. We can write the reaction as an equilibrium and can determine experimentally that the equilibrium constant is approximately $10^8$:

$$H_2C{=}CH_2 + HBr \rightleftharpoons CH_3CH_2Br + \text{Energy}$$

$$\frac{[CH_3CH_2Br]}{[H_2C{=}CH_2][HBr]} = K_{eq} \approx 10^8$$

Since $K_{eq}$ is relatively large, the reaction proceeds as written and energy is given off. Although we often speak of such reactions as "going to completion," this is imprecise terminology since in practically no reaction does *every* molecule react. For practical purposes, though, equilibrium constants of greater than $10^3$ can be considered to indicate "complete" reaction, since the amount of reactant left will be barely detectable (less than 0.1%).

The total amount of energy change during a reaction is called the **standard Gibbs free-energy change, $\Delta G°$**. (The Greek letter delta, $\Delta$, is the mathematical symbol for the difference between two numbers—in this case, the difference between the free energy of the products and the free energy of the starting materials.) For a reaction where $\Delta G°$ is negative, energy is released; for a reaction where $\Delta G°$ is positive, energy must be added.

Because the equilibrium constant, $K_{eq}$, and the free-energy change, $\Delta G°$, both measure whether or not a reaction is favored, they are mathematically related:

$$\Delta G° = \text{Free energy of products} - \text{Free energy of reactants}$$

$$\Delta G° = -RT \ln K_{eq} \quad \text{or} \quad K_{eq} = e^{-\Delta G°/RT}$$

where

$$R = 1.986 \text{ cal/mol} \cdot \text{K (the gas constant)}$$

$$T = \text{Absolute temperature (in kelvins)}$$

$$e \approx 2.718 \text{ (the base of natural logarithms)}$$

$$\ln K_{eq} = \text{Natural logarithm of } K_{eq}$$

As an example of how this mathematical relationship can be used, the reaction of ethylene with HBr has $K_{eq} \approx 10^8$, and we can therefore calculate $-12$ kcal/mol (50 kJ/mol) at room temperature (300 K).

To what is the energy change during a reaction due? The Gibbs free-energy change is attributable to a combination of two factors, an **enthalpy** factor, $\Delta H°$, and an **entropy** factor, $\Delta S°$:

$$\Delta G° = \Delta H° - T\Delta S°$$

where $T$ is the absolute temperature.

The enthalpy term, $\Delta H°$, called the **standard heat of reaction**, is a measure of the change in total bonding energy during a reaction. If $\Delta H°$ is negative, heat is evolved and the reaction is said to be **exothermic**. If $\Delta H°$ is positive, heat must be added and the reaction is said to be **endothermic**. For example, if a certain reaction breaks reactant bonds with a total strength of 100 kcal/mol and forms new product bonds with a total strength of 120 kcal/mol, then $\Delta H°$ for the reaction is $-20$ kcal/mol and the reaction is exothermic. [*Remember:* Breaking bonds takes energy, and making bonds releases energy.]

| | | |
|---|---|---|
| Strength of bonds broken (positive $\Delta H°$) | = | 100 kcal/mol |
| Strength of bonds formed (negative $\Delta H°$) | = | $-120$ kcal/mol |
| Net change | = | $-20$ kcal/mol |

The entropy term, $\Delta S°$, is a measure of the amount of "disorder" caused by a reaction. To illustrate, in an elimination reaction of the type

$$\text{A} \longrightarrow \text{B} + \text{C}$$

there is more freedom of movement (disorder) in the products than in the reactant because one molecule has split into two. Thus, there is a net gain of entropy during the reaction, and $\Delta S°$ has a positive value.

On the other hand, for addition reactions of the type

$$\text{A} + \text{B} \longrightarrow \text{C}$$

exactly the opposite is true. Because such reactions restrict the freedom of movement of two molecules by joining them, the products have *less* disorder than the reactants, and $\Delta S°$ has a negative value. The reaction of ethylene and HBr is one such example.

Of the two terms that make up $\Delta G°$, the enthalpy term ($\Delta H°$) is usually larger and more important than the entropy term ($T\Delta S°$) at normal reaction temperatures. Furthermore, enthalpy changes during reactions are relatively easily measured, and large compilations of data are available. For these reasons, the entropy contribution, $T\Delta S°$, is sometimes ignored when making thermodynamic arguments, and chemists often make the simplifying assumption that $\Delta G° \approx \Delta H°$.

In summary, the standard Gibbs free-energy change, $\Delta G°$, is a measure of the overall amount of energy change during a reaction. It is the net result of two contributing terms, one term that deals with changes in bond strengths between reactants and products ($\Delta H°$), and one term that deals with the amount of disorder caused by the reaction ($\Delta S°$). Table 5.3 describes these terms more fully.

**Table 5.3** Explanation of thermodynamic quantities: $\Delta G° = \Delta H° - T\Delta S°$

| *Term* | *Name* | *Explanation* |
|---|---|---|
| $\Delta G°$ | Gibbs free-energy change (kcal/mol) | Overall energy difference between reactants and products. When $\Delta G°$ is negative, a reaction can occur spontaneously. $\Delta G°$ is related to the equilibrium constant by the equation $\Delta G° = -RT \ln K_{eq}$ |
| $\Delta H°$ | Enthalpy change (kcal/mol) | Heat of reaction; the energy difference between strengths of bonds broken in a reaction and bonds formed |
| $\Delta S°$ | Entropy change (cal/degree · mol) | Overall change in freedom of motion or "disorder" resulting from reaction; usually much smaller than $\Delta H°$ |

PROBLEM

**5.8** Which reaction is more favored, one with $\Delta G° = -11$ kcal/mol or one with $\Delta G° = +11$ kcal/mol?

PROBLEM

**5.9** Which reaction is more exothermic, one with $K_{eq} = 1000$ or one with $K_{eq} = 0.001$?

## 5.8 Describing a Reaction: Bond Dissociation Energies

We've just seen that energy is released (negative $\Delta H°$) when a bond is made, and energy is consumed (positive $\Delta H°$) when a bond is broken. The measure of the energy change on bond making or bond breaking is a quantity called the **bond dissociation energy**. Bond dissociation energy is defined as the amount of energy required to homolytically break a given bond into two radical fragments when the molecule is in the gas phase at 25°C.

$$A : B \xrightarrow[\text{energy}]{\text{Bond dissociation}} A\cdot + \cdot B$$

Each specific bond has its own characteristic strength, and extensive tabulations of bond strength data are available (Table 5.4).

**Table 5.4** Bond dissociation energy data for the reaction A—B → A· + B·

| *Bond* | $\Delta H°$ *(kcal/mol)* | *Bond* | $\Delta H°$ *(kcal/mol)* | *Bond* | $\Delta H°$ *(kcal/mol)* |
|---|---|---|---|---|---|
| H—H | 104 | $(CH_3)_3C$—I | 50 | $CH_3$—$CH_3$ | 88 |
| H—F | 136 | $H_2C{=}CH$—H | 108 | $C_2H_5$—$CH_3$ | 85 |
| H—Cl | 103 | $H_2C{=}CH$—Cl | 88 | $(CH_3)_2CH$—$CH_3$ | 84 |
| H—Br | 88 | $H_2C{=}CHCH_2$—H | 87 | $(CH_3)_3C$—$CH_3$ | 81 |
| H—I | 71 | $H_2C{=}CHCH_2$—Cl | 69 | $H_2C{=}CH$—$CH_3$ | 97 |
| Cl—Cl | 58 | $C_6H_5$—H | 112 | $H_2C{=}CHCH_2$—$CH_3$ | 74 |
| Br—Br | 46 | $C_6H_5$—Cl | 97 | $C_6H_5$—$CH_3$ | 102 |
| I—I | 36 | $C_6H_5CH_2$—H | 85 | $C_6H_5$—$CH_2$—$CH_3$ | 72 |
| $CH_3$—H | 104 | $C_6H_5CH_2$—Cl | 70 | $CH_3C(=O)$—H | 86 |
| $CH_3$—Cl | 84 | $C_6H_5$—Br | 82 | HO—H | 119 |
| $CH_3$—Br | 70 | $C_6H_5$—OH | 112 | HO—OH | 51 |
| $CH_3$—I | 56 | HC≡C—H | 125 | $CH_3O$—H | 102 |
| $CH_3$—OH | 91 | | | $CH_3S$—H | 88 |
| $CH_3$—$NH_2$ | 80 | | | $C_2H_5O$—H | 103 |
| $C_2H_5$—H | 98 | | | $CH_3C(=O)$—$CH_3$ | 77 |
| $C_2H_5$—Cl | 81 | | | $CH_3CH_2O$—$CH_3$ | 81 |
| $C_2H_5$—Br | 68 | | | $NH_2$—H | 103 |
| $C_2H_5$—I | 53 | | | H—CN | 130 |
| $C_2H_5$—OH | 91 | | | | |
| $(CH_3)_2CH$—H | 95 | | | | |
| $(CH_3)_2CH$—Cl | 80 | | | | |
| $(CH_3)_2CH$—Br | 68 | | | | |
| $(CH_3)_3C$—H | 91 | | | | |
| $(CH_3)_3C$—Cl | 79 | | | | |
| $(CH_3)_3C$—Br | 65 | | | | |

Let's look at methane as an example. Methane has a measured bond dissociation energy $\Delta H° = +104$ kcal/mol (435 kJ/mol), meaning that 104 kcal/mol is required to break a C—H bond of methane to give the two radical fragments $\cdot CH_3$ and $\cdot H$. Conversely, 104 kcal/mol of energy is *released* when a methyl radical and a hydrogen atom combine to form methane.

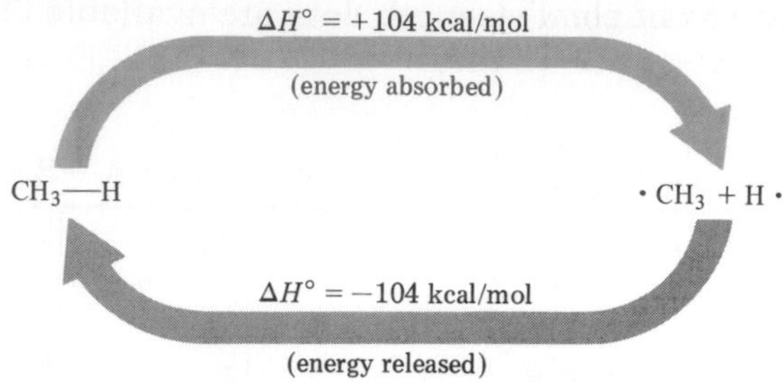


The data given in Table 5.4 are extremely useful—if enough bond strengths were known, it would seem possible to turn organic chemistry into a more quantitative science. Ideally, if we want to know whether a predicted reaction could occur, we could calculate $\Delta H°$ for the process and avoid a lot of time-consuming work in the laboratory. Unfortunately, there are two problems. The first is that the calculation says nothing about the probable rate of reaction; a reaction may have a favorable $\Delta H°$ and still not take place. The second problem is that bond dissociation energy data refer only to reactions occurring in the gas phase; the data aren't directly relevant to solution chemistry.

In practice, the vast majority of organic reactions are carried out in solution, and solvent molecules can interact strongly with dissolved reagents (**solvation**). Solvation effects can weaken bonds and cause large changes in the value of $\Delta H°$ for a given reaction. The entropy term, $\Delta S°$, can also be affected by solvent molecules, since the solvation of polar reagents by polar solvents causes a certain amount of orientation (reduces the amount of disorder) in the solvent. Although we can often use bond dissociation energy data to get a rough idea of how thermodynamically favorable a given reaction might be, we have to keep in mind that the answer is only approximate.

To take a familiar example, what do the thermodynamics look like for the reaction of gaseous HBr with gaseous ethylene? By totaling the energy released in making new bonds and subtracting the energy required to break old bonds, we can calculate $\Delta H°$ for the overall reaction:

$$H_2C{=}CH_2 + H{-}Br \longrightarrow H{-}\underset{H}{\overset{H}{C}}{-}\underset{H}{\overset{Br}{C}}{-}H$$

| *Bonds broken* | | | *Bonds formed* | | |
|---|---|---|---|---|---|
| H—Br | $\Delta H°$ = | 88 kcal/mol | C—H | $\Delta H°$ = | −98 kcal/mol |
| $\frac{1}{2}$ C=C | $\Delta H°$ = | 64 kcal/mol | C—Br | $\Delta H°$ = | −68 kcal/mol |
| | $\Delta H°$ = | 152 kcal/mol | | $\Delta H°$ = | −166 kcal/mol |

*Net change:* $\Delta H°$ = −14 kcal/mol

We calculate that the gas-phase reaction of HBr with ethylene is favorable by approximately −14 kcal/mol (59 kJ/mol), and the reaction might therefore occur under suitable conditions. We can't be *certain* that the reaction will actually take place, since the calculation says nothing about reaction rates. In fact, though, the reaction does occur as written.

We've just seen an example of how bond dissociation energy data can be used to calculate the energetics of reaction taking place in the gas phase. It has to be emphasized again, though, that the majority of organic reactions take place in solution. In practice, many reactions, particularly those that take place through polar mechanisms, are often highly susceptible to solvent influences. Thus, the accuracy of bond energy calculations is correspondingly lower. We'll see examples of solvent effects on reactions at numerous places in later chapters.

**PROBLEM**

**5.10** Calculate $\Delta H°$ for each step in the radical substitution reaction of chlorine with methane (Section 5.4):

(a) $Cl_2 \longrightarrow 2\ Cl\cdot$

(b) $CH_4 + Cl\cdot \longrightarrow \cdot CH_3 + HCl$

(c) $\cdot CH_3 + Cl_2 \longrightarrow CH_3Cl + Cl\cdot$

What is the overall $\Delta H°$ for the reaction? Consider only the propagation steps (b) and (c) in deciding your answer.

**PROBLEM**

**5.11** Calculate $\Delta H°$ for these polar reactions:

(a) $CH_3CH_2OCH_3 + HI \longrightarrow CH_3CH_2OH + CH_3I$

(b) $CH_3Cl + NH_3 \longrightarrow CH_3NH_2 + HCl$

## 5.9 Describing a Reaction: Energy Diagrams and Transition States

In order for a reaction to take place, reactant molecules must collide, and reorganization of atoms and bonds must occur. As an example, let's again look at the addition reaction of HBr and ethylene:

$$H_2C{=}CH_2 \xrightarrow{H-Br} \left[ H{-}CH_2{-}CH_2^{+} \right] \xrightarrow{:\ddot{Br}:^-} H{-}CH_2{-}CH_2{-}\ddot{Br}:$$

Carbocation

As the reaction proceeds, ethylene and HBr must approach each other, the ethylene pi bond and H—Br bond must break, a new carbon–hydrogen bond must form in the first step, and a new carbon–bromine bond must form in the second step. Over the years, chemists have developed a method for graphically depicting the energy changes that occur during a reaction using

**reaction energy diagrams** of the sort shown in Figure 5.3. The vertical axis of the diagram represents the total energy of all reactants, whereas the horizontal axis represents the progress of the reaction from beginning (left) to end (right). Let's take a careful look at the reaction, one step at a time, and see how the addition of HBr to ethylene can be described on a reaction energy diagram.

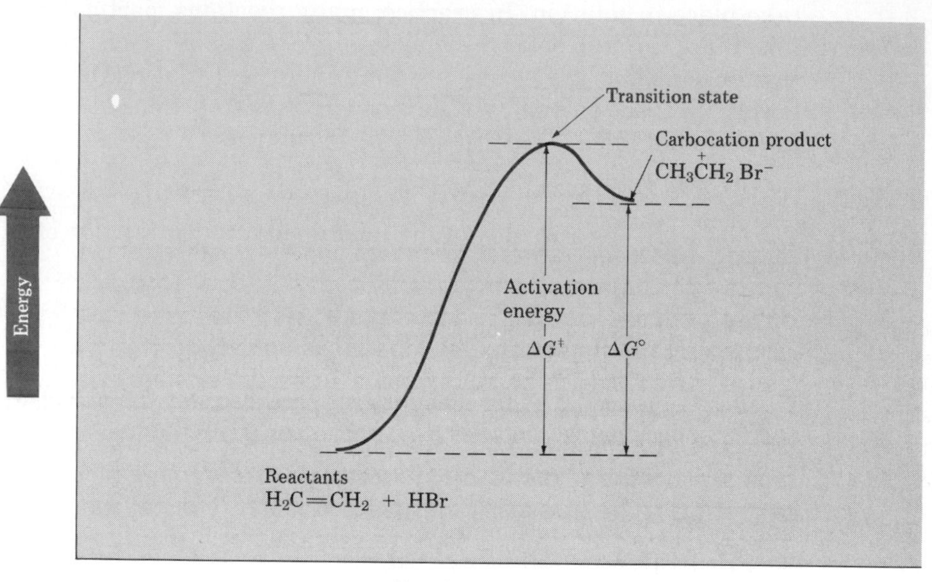


**Figure 5.3** A reaction energy diagram for the first step in the reaction of ethylene with HBr: The energy difference between reactants and transition state, $\Delta G^{\ddagger}$, controls the reaction rate. The energy difference between reactants and carbocation product, $\Delta G^{\circ}$, controls the position of the equilibrium.

At the beginning of the reaction, ethylene and HBr have the total amount of energy indicated by the reactant level on the far left side of the diagram. As the two molecules approach each other and reaction commences, a *repulsive* interaction occurs and the energy level therefore rises. This repulsive interaction is due to the steric strain introduced by crowding the reactants too closely together. In electronic terms, the electron clouds of the two reactants approach and repel each other. If the collision has occurred with sufficient force and proper orientation, however, the reactants continue to approach each other until the new carbon–hydrogen bond starts to form. At some point, a structure of maximum energy is reached, a structure we call the **transition state**.

Since the transition state represents the *highest*-energy structure involved in the step, it is unstable and can't be isolated. We can get no direct information about the exact nature of the transition-state structure, but we can imagine it to be a kind of activated complex of the two reactants in which the carbon–carbon pi bond is partially broken and the new carbon–hydrogen bond is partially formed (Figure 5.4).

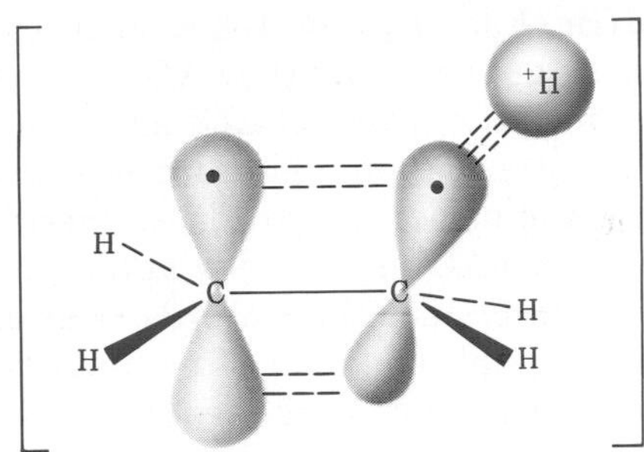


**Figure 5.4** A hypothetical transition-state structure for the first step of the reaction of ethylene with HBr: The C—C pi bond is just beginning to break, and the C—H bond is just beginning to form.

The energy difference between reactants and transition state, called the **activation energy, $\Delta G^{\ddagger}$**, determines how rapidly the reaction occurs. (The double dagger superscript, ‡, is always used to refer to the transition state.) A large activation energy, corresponding to a large energy difference between reactants and transition state, results in a slow reaction because few reacting molecules collide with enough energy to climb the high barrier. A small activation energy, however, results in a rapid reaction since almost all reacting molecules are energetic enough to climb to the transition state.

The situation of reactants needing enough energy to climb the barrier from starting material to transition state may be likened to the situation of hikers who need enough energy to climb over a mountain pass. If the pass is a high one, the hikers need a lot of energy and will surmount the barrier very slowly. If the pass is low, however, the hikers need less energy and will reach the top quickly.

Although it's difficult to generalize accurately, most organic reactions have activation energies in the range of 10–35 kcal/mol (40–150 kJ/mol). Reactions with activation energies less than 20 kcal/mol take place spontaneously at room temperature or below, whereas reactions with higher activation energies normally require heating. Heat provides the energy necessary for the reactants to climb the activation barrier.

Once the high-energy transition state has been reached, the reaction can either continue on to give the carbocation or revert back to starting materials. Since both choices are energetically "downhill" from the highpoint of the transition state, both are equally likely. If reversion to starting materials occurs, of course, no net change in the system is observed. If, however, the reaction continues on to give carbocation, energy is released as the new C—H bond forms fully, and the curve on the reaction energy diagram therefore turns downward until it reaches a minimum.

This minimum point represents the energy level of the carbocation product of the first step. The energy change, $\Delta G^{\circ}$, between starting materials and carbocation is simply the difference between the two levels on the diagram. Since the carbocation is less stable than the starting alkene, the first step is endothermic, and energy is absorbed.

Not all reaction energy diagrams are like that for the reaction of ethylene and HBr. Each specific reaction has its own specific energy profile. Some reactions are very fast (low $\Delta G^{\ddagger}$) and some are very slow (high $\Delta G^{\ddagger}$); some have a negative value of $\Delta G°$ and some have a positive value of $\Delta G°$. Figure 5.5 illustrates some different possibilities for energy profiles. Note in this figure the use of the words *exothermic* and *endothermic* to refer to reactions in which $\Delta G°$ is negative and positive, respectively. This usage is not strictly correct, since these words refer only to $\Delta H°$ and not to $\Delta G°$. As stated earlier, though, chemists often make the simplifying assumption that $\Delta G°$ and $\Delta H°$ are approximately equal.

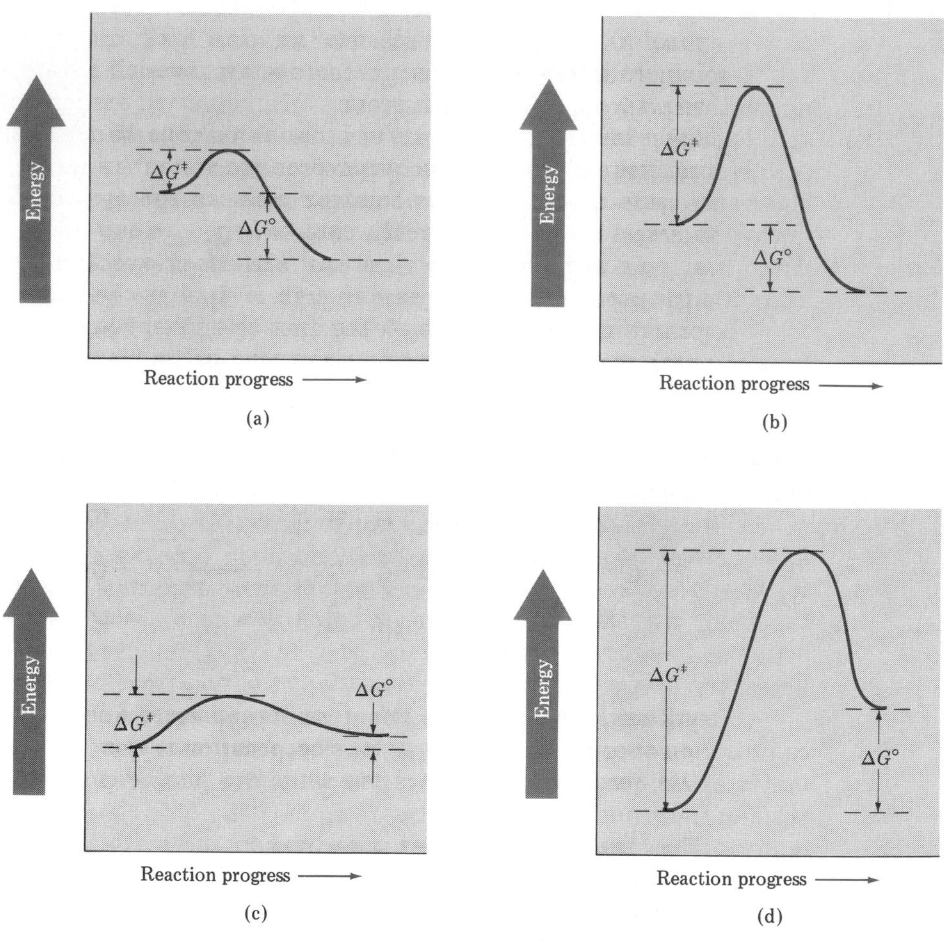


**Figure 5.5** Hypothetical reaction energy diagrams: (a) A fast exothermic reaction (small $\Delta G^{\ddagger}$, negative $\Delta G°$); (b) a slow exothermic reaction (large $\Delta G^{\ddagger}$, negative $\Delta G°$); (c) a fast endothermic reaction (small $\Delta G^{\ddagger}$, small positive $\Delta G°$); (d) a slow endothermic reaction (large $\Delta G^{\ddagger}$; positive $\Delta G°$)

It should be emphasized again that the overall energy change occurring during a reaction is measured by $\Delta G°$, not $\Delta H°$. Calculations of changes in bond dissociation energies ($\Delta H°$), such as that done earlier for the reaction of ethylene and HBr (Section 5.8), are useful, but can give only an *indication* as to whether or not a given reaction will have a favorable equilibrium constant. Such calculations don't take solvent effects or entropy factors into account.

**PROBLEM**

**5.12** Which reaction is faster: one with $\Delta G^{\ddagger} = 45$ kJ/mol or one with $\Delta G^{\ddagger} = 70$ kJ/mol? Is it possible to predict which of the two has the larger $K_{eq}$?

## 5.10 Describing a Reaction: Intermediates

How can we describe the carbocation structure formed in the first step of the reaction of ethylene with HBr? The carbocation is clearly different from the starting materials, yet it's not a transition state and it's not a final product.

$$H_2C{=}CH_2 \xrightarrow{H{-}Br} \left[ H{-}\overset{H}{\underset{H}{C}}{-}\overset{H}{\underset{H}{C^{+}}} \right] \xrightarrow{:\ddot{Br}:^{-}} H{-}\overset{H}{\underset{H}{C}}{-}\overset{H}{\underset{H}{C}}{-}Br$$

Reaction intermediate

We call the carbocation, which is formed briefly during the course of the multistep reaction, a **reaction intermediate**. As soon as the intermediate is formed in the first step by reaction of ethylene with $H^+$, it reacts further with bromide ion in a second step to give the final product, bromoethane. This second step has its own activation energy, $\Delta G^{\ddagger}$, its own transition state, and its own energy change, $\Delta G°$. We can view the second transition state as an activated complex between the electrophilic carbocation intermediate and nucleophilic bromide anion, in which the new C—Br bond is just starting to form as bromide ion donates a pair of electrons to the positively charged carbon atom.

A complete energy diagram for the overall reaction of ethylene with HBr is shown in Figure 5.6. In essence, we draw diagrams for each of the individual steps and then join them in the middle so that the *product* of step 1 (the carbocation) serves as the *starting material* for step 2. As indicated in Figure 5.6, the reaction intermediate lies at an energy minimum between steps 1 and 2. Since the energy level of this intermediate is considerably higher than the level of either starting material (ethylene and HBr) or product (bromoethane), the intermediate is highly reactive and can't be isolated. It is, however, more stable than either of the two transition states that surround it.

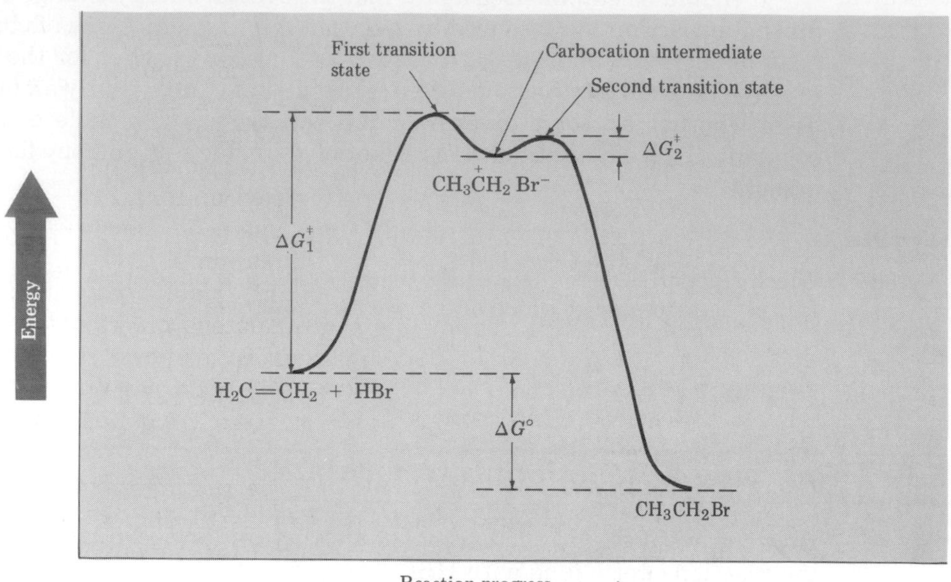


**Figure 5.6** Overall reaction energy diagram for the reaction of ethylene with HBr: Two separate steps are involved, each with its own transition state. The energy minimum between the two steps represents the carbocation reaction intermediate.

In general, each individual step in a multistep process can always be considered separately. Each step has its own $\Delta G^\ddagger$ (rate) and its own $\Delta G°$ (energy change). The *overall* $\Delta G°$ of the reaction, however, is the energy difference between initial reactants (far left) and final products (far right). This is always true regardless of the shape of the reaction energy curve. Figure 5.7 illustrates some different possible cases.

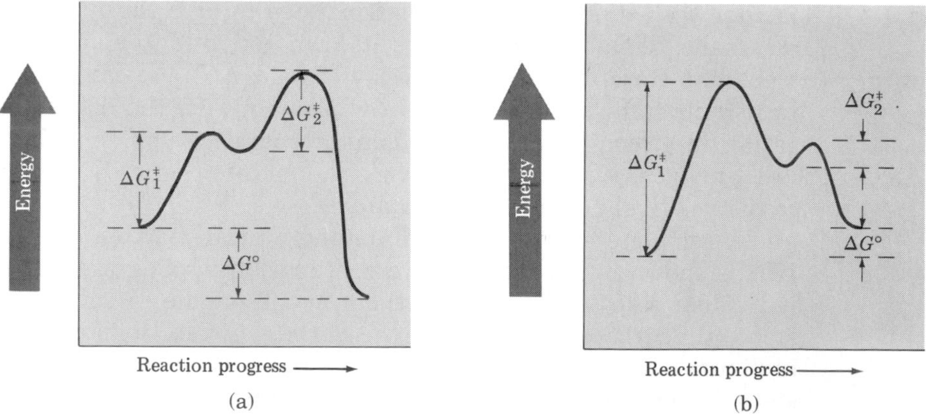


**Figure 5.7** Hypothetical reaction energy diagrams for some two-step reactions: The overall $\Delta G°$ for any reaction, regardless of complexity, is simply the energy difference between starting materials and final product. Note that reaction (a) is exothermic, whereas reaction (b) is endothermic.

PROBLEM

**5.13** Draw a single reaction energy diagram for the two propagation steps in the radical reaction of chlorine with methane. Is the overall $\Delta G°$ for this reaction positive or negative? Label the parts of your diagram corresponding to $\Delta G°$ and $\Delta G^{\ddagger}$.

$$Cl_2 + CH_4 \xrightarrow{\text{Light}} CH_3Cl + HCl$$

## 5.11 Summary and Key Words

Organic chemistry can be organized either according to the kinds of reactions that take place or according to how reactions take place. There are four important kinds of reactions—additions, eliminations, substitutions, and rearrangements. **Addition reactions** take place when two reactants add together to give a single product; **elimination reactions** take place when one reactant splits apart to give two products; **substitution reactions** take place when two reactants exchange parts to give two new products; and **rearrangement reactions** take place when one reactant undergoes a reorganization of bonds and atoms to give a new product.

| | |
|---|---|
| Addition | $A + B \longrightarrow C$ |
| Elimination | $A \longrightarrow B + C$ |
| Substitution | $A{-}B + C{-}D \longrightarrow A{-}C + B{-}D$ |
| Rearrangement | $A \longrightarrow B$ |

A full description of how a reaction occurs is called its **mechanism**. There are two kinds of mechanisms by which reactions can take place—**radical** mechanisms and **polar** mechanisms. Polar reactions, the most common type, occur as the result of attractive interactions between an electron-rich site (**nucleophile**) in the functional group of one molecule and an electron-poor site (**electrophile**) in the functional group of another molecule. Bonds are formed in polar reactions when the nucleophile donates an electron pair to the electrophile. This movement of electrons is indicated by a **curved arrow** showing the direction of electron travel from the nucleophile to the electrophile. Radical reactions involve odd-electron species; bonds are formed when each reactant donates one electron to the new bond.

Polar $B{:}^- + A^+ \longrightarrow A : B$

Nucleophile Electrophile

Radical $B\cdot + \cdot A \longrightarrow A : B$

The energy changes taking place during reactions can be described by considering both rates (how fast reaction occurs) and equilibria (how much reaction occurs). The position of a chemical equilibrium is determined by the magnitude of the **Gibbs free-energy change** ($\Delta G°$) that takes place during reaction. The free-energy change is composed of two parts, $\Delta G° = \Delta H° - T\Delta S°$. The **enthalpy** term ($\Delta H°$) corresponds to the net change in strength of chemical bonds broken and formed during reaction. The **entropy** term ($\Delta S°$) corresponds to the change in disorder during reaction. Since the enthalpy term is usually larger and more important than the entropy term, chemists often make the assumption that $\Delta G° \approx \Delta H°$.

Reactions can be described pictorially using **reaction energy diagrams** that follow the reaction course from starting material through transition state to product. The **transition state** is an activated complex occurring at the highest-energy point during reaction. The amount of energy needed by reactants to reach this high point is the **activation energy**, $\Delta G^\ddagger$. It is the magnitude of $\Delta G^\ddagger$ that determines the rate of the reaction: The higher the activation energy, the slower the reaction. Most organic reactions have activation energies in the range of 10–35 kcal/mol.

Many reactions take place in more than one step and involve the formation of **reaction intermediates**. An intermediate is a species that lies at an energy minimum between steps on the reaction curve and is formed briefly during the course of a reaction.

## ADDITIONAL PROBLEMS

**5.14** Identify the functional groups present in these molecules and predict the direction of polarity in each.

(a) $CH_3CN$

(b) 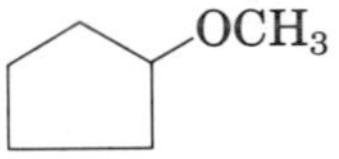 [cyclopentane with $OCH_3$ substituent]

(c) $CH_3\overset{O}{\overset{\|}{C}}CH_2\overset{O}{\overset{\|}{C}}OCH_3$

(d) 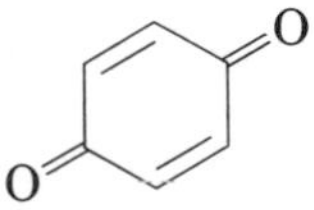 [quinone structure, O= ... =O]

A quinone

**5.15** Identify these reactions as being additions, eliminations, substitutions, or rearrangements:

(a) $CH_3CH_2Br$ + NaCN $\longrightarrow$ $CH_3CH_2CN$ (+ NaBr)

(b) 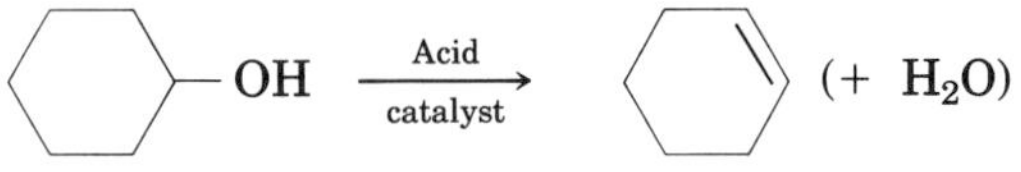 [cyclohexane]—OH $\xrightarrow[\text{catalyst}]{\text{Acid}}$ [cyclohexene] (+ $H_2O$)

(c) 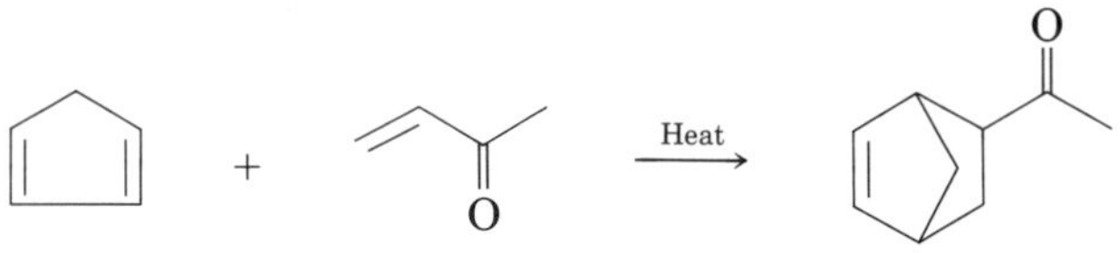 [cyclopentadiene] + [methyl vinyl ketone] $\xrightarrow{\text{Heat}}$ [bicyclic ketone product]

(d) 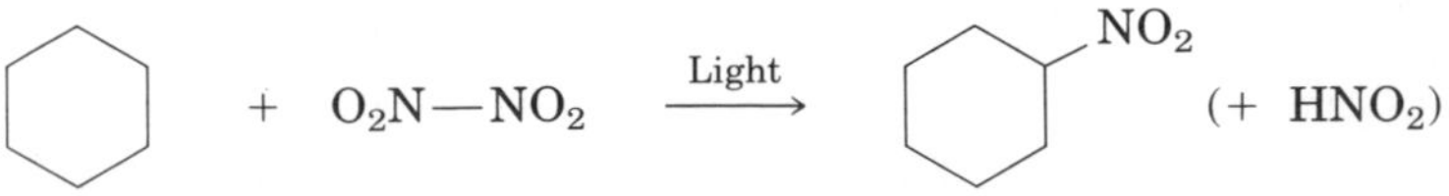 [cyclohexane] + $O_2N$—$NO_2$ $\xrightarrow{\text{Light}}$ [cyclohexane with $NO_2$] (+ $HNO_2$)

**5.16** Explain the differences between addition, elimination, substitution, and rearrangement reactions.

**5.17** Define the following:
(a) Polar reaction
(b) Heterolytic bond breakage
(c) Homolytic bond breakage
(d) Radical reaction
(e) Functional group
(f) Polarization

**5.18** Give an example of each of the following:
(a) A nucleophile
(b) An electrophile
(c) A polar reaction
(d) A substitution reaction
(e) A heterolytic bond breakage
(f) A homolytic bond breakage

**5.19** Which of these would you classify as nucleophiles, and which as electrophiles?

(a) $:\ddot{\underset{..}{Cl}}:^-$ (b) $BF_3$ (c) $H\ddot{\underset{..}{O}}:^-$ (d) $CH_3\ddot{N}H_2$

**5.20** Draw a reaction energy diagram for a one-step endothermic reaction. Label the parts of the diagram corresponding to reactants, products, transition state, $\Delta G°$, and $\Delta G^‡$. Is $\Delta G°$ positive or negative?

**5.21** Draw a reaction energy diagram for a two-step exothermic reaction. Label the overall $\Delta G^‡$ and $\Delta G°$, transition states, and reaction intermediate. Is $\Delta G°$ positive or negative?

**5.22** Describe the difference between a transition state and a reaction intermediate.

**5.23** Draw a reaction energy diagram for a two-step exothermic reaction for which the second step is faster than the first step.

**5.24** Draw a reaction energy diagram for a reaction with $K_{eq} = 1$. What is the value of $\Delta G°$ in this reaction?

**5.25** Consider the reaction energy diagram shown here and answer the following questions:

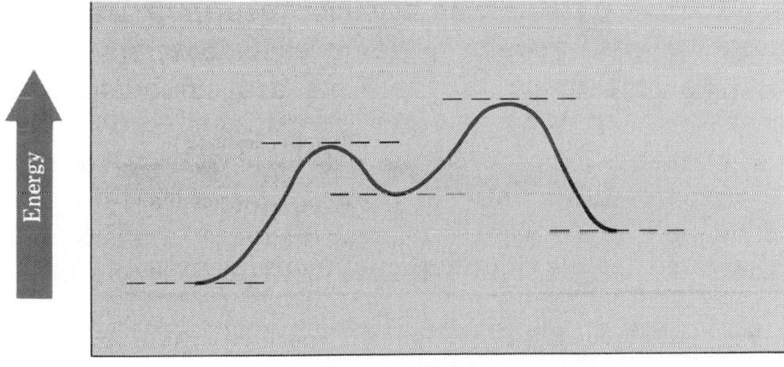


(a) Indicate $\Delta G°$ for the reaction. Is it positive or negative?
(b) How many steps are involved in the reaction?
(c) Which step is faster?
(d) How many transition states are there? Label them.

**5.26** Use the information in Table 5.4 to calculate $\Delta H°$ for these reactions:

(a) $CH_3OH + HBr \longrightarrow CH_3Br + H_2O$

(b) $CH_3CH_2OH + CH_3Cl \longrightarrow CH_3CH_2OCH_3 + HCl$

**5.27** Use the information in Table 5.4 to calculate $\Delta H^\circ$ for the reaction of ethane with chlorine, bromine, and iodine:

(a) $CH_3CH_3 + Cl_2 \longrightarrow CH_3CH_2Cl + HCl$

(b) $CH_3CH_3 + Br_2 \longrightarrow CH_3CH_2Br + HBr$

(c) $CH_3CH_3 + I_2 \longrightarrow CH_3CH_2I + HI$

What can you conclude about the relative ease of chlorination, bromination, and iodination?

**5.28** An alternative course for the reaction of bromine with ethane would result in the formation of bromomethane:

$$H_3C—CH_3 + Br_2 \longrightarrow 2\ CH_3Br$$

Calculate $\Delta H^\circ$ for this reaction, and comment on how it compares with the value you calculated in Problem 5.27 for the formation of bromoethane.

**5.29** Radical chlorination of alkanes is of interest mechanistically, but is of little general utility because mixtures of products usually result when more than one kind of C—H bond is present in the substrate. Calculate approximate $\Delta H^\circ$ values for the possible monochlorination reactions of 2-methylbutane. You should use the bond dissociation energies measured for $CH_3CH_2—H$, $H—CH(CH_3)_2$, and $H—C(CH_3)_3$ as representative of typical primary, secondary, and tertiary C—H bonds.

**5.30** Provide IUPAC names for each of the products formed in Problem 5.29.

**5.31** Despite the limitations of radical halogenation of hydrocarbons, the reaction is still useful for synthesizing certain halogenated compounds. For which of the following compounds does radical halogenation give single monohalogenation products?

(a) $C_2H_6$

(b) $(CH_3)_2CH_2$

(c) 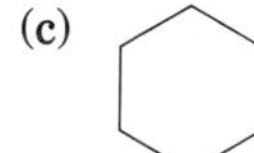

(d) $(CH_3)_3CCH_2CH_3$

(e) 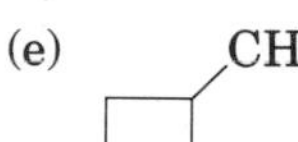


(f) $CH_3C{\equiv}CCH_3$

**5.32** We've said that the chlorination of methane proceeds by the following steps:

(a) $Cl_2 \xrightarrow{\text{Light}} 2\ Cl\cdot$

(b) $Cl\cdot + CH_4 \longrightarrow HCl + \cdot CH_3$

(c) $\cdot CH_3 + Cl_2 \longrightarrow CH_3Cl + Cl\cdot$

Alternatively, one might propose a different series of steps:

(d) $Cl_2 \longrightarrow 2\ Cl\cdot$

(e) $Cl\cdot + CH_4 \longrightarrow CH_3Cl + H\cdot$

(f) $H\cdot + Cl_2 \longrightarrow HCl + Cl\cdot$

Calculate $\Delta H^\circ$ for each individual step in both possible routes. What insight does this provide into the relative merits of each route?

**5.33** When isopropylidenecyclohexane is treated with strong acid at room temperature, isomerization occurs to yield 1-isopropylcyclohexene by the mechanism shown below:

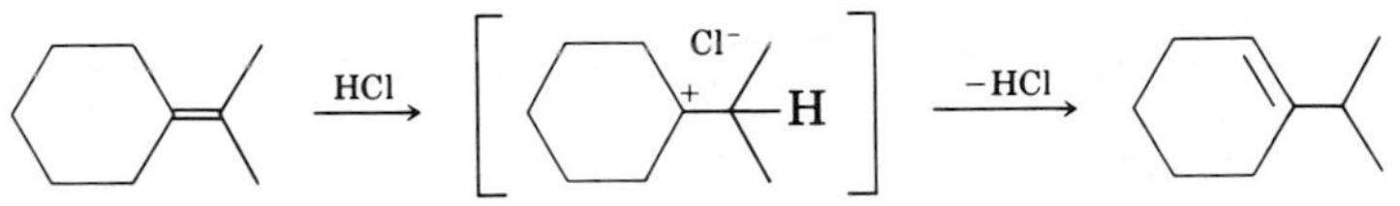

$Cl^-$ HCl → [ + H ] −HCl →

Isopropylidenecyclohexane

1-Isopropylcyclohexene

At equilibrium, the product mixture contains about 30% isopropylidenecyclohexane and about 70% 1-isopropylcyclohexene.

(a) What kind of reaction is occurring? Is the mechanism polar or radical?
(b) Draw curved arrows to indicate electron flow in each step.
(c) Calculate $K_{eq}$ for the reaction.
(d) Use Table 4.3 to calculate $\Delta G°$ for the reaction.
(e) Since the reaction occurs slowly at room temperature, what is its approximate $\Delta G^{\ddagger}$?
(f) Draw a quantitative energy diagram for the reaction.

CHAPTER 6

# Alkenes: Structure and Reactivity

**A**lkenes are hydrocarbons that contain a carbon–carbon double-bond functional group. The word **olefin** is often used as a synonym in the chemical literature, but alkene is the generally preferred term. Alkenes occur abundantly in nature, and many have important biological roles. For example, ethylene is a plant hormone that induces ripening in fruit, and α-pinene is the major component of turpentine.

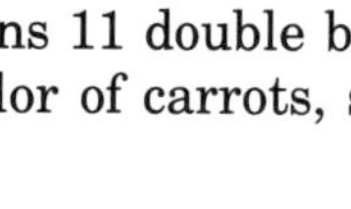

$$H_2C{=}CH_2$$

Ethylene

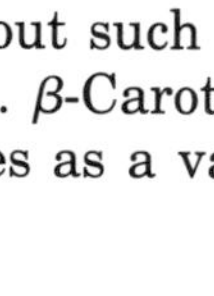

α-Pinene

Life itself would be impossible without such alkenes as β-carotene, a compound that contains 11 double bonds. β-Carotene, the orange pigment responsible for the color of carrots, serves as a valuable dietary source of vitamin A.

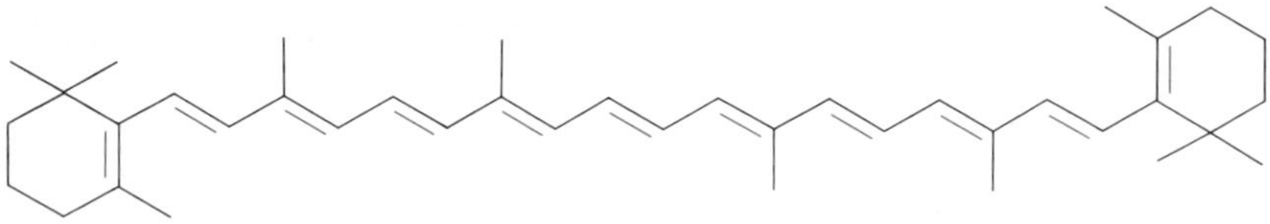

β-Carotene
(orange pigment and vitamin A precursor)

## 6.1 Industrial Preparation and Use of Alkenes

Ethylene (ethene) and propylene (propene), the simplest alkenes, are the two most important organic chemicals produced industrially. More than 30 billion lb of ethylene and almost 15 billion lb of propylene are produced each year in the United States for use in the synthesis of polyethylene, polypropylene, ethylene glycol, acetic acid, acetaldehyde, and a host of other raw materials (Figure 6.1).

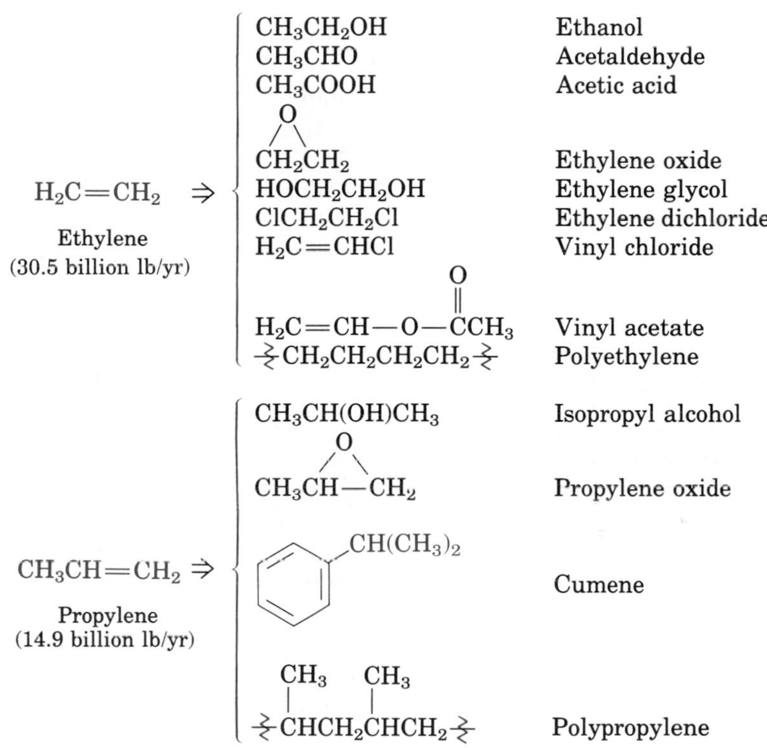


**Figure 6.1** Compounds derived industrially from ethylene (ethene) and propylene (propene)

Ethylene, propylene, and butene are synthesized industrially by thermal cracking of both natural gas ($C_1$–$C_4$ alkanes) and straight-run gasoline ($C_4$–$C_8$ *n*-alkanes):

$$CH_3(CH_2)_nCH_3 \xrightarrow[\text{Steam}]{850\text{–}900°C} H_2 + CH_4 + H_2C{=}CH_2 + CH_3CH{=}CH_2 + CH_3CH_2CH{=}CH_2$$
$$n = 0\text{–}6$$

Thermal cracking, introduced in 1912, takes place in the absence of catalysts at extremely high temperatures up to 900°C. Although it undoubtedly involves radical reactions, the exact processes are complex. Evidently, the high-temperature reaction conditions cause spontaneous homolysis of carbon–carbon and carbon–hydrogen bonds, with resultant formation of smaller fragments:

$$CH_3CH_2{-}CH_2CH_3 \xrightarrow{900°C} 2\ \overset{\displaystyle H}{\overset{|}{C}}H_2{-}CH\cdot \longrightarrow 2\,H_2C{=}CH_2 + H_2$$

Thermal cracking is an example of a reaction whose energetics are dominated by entropy ($T\Delta S°$) rather than by enthalpy ($\Delta H°$) in the free-energy equation $\Delta G° = \Delta H° - T\Delta S°$. The large positive entropy change resulting from the fragmentation of one large molecule into several smaller pieces, together with the extremely high temperature, $T$, makes the $T\Delta S°$ term larger than the $\Delta H°$ term, thus favoring the cracking reaction.

## 6.2 Calculation of the Degree of Unsaturation

Because of their double bond, alkenes have fewer hydrogens than alkanes with the same number of carbons ($C_nH_{2n}$ for an alkene versus $C_nH_{2n+2}$ for an alkane). Thus, alkenes are often referred to as being **unsaturated**. Alkanes, by contrast, have as many hydrogens as possible and are **saturated**.

In general terms, each ring or double bond in a molecule causes a *pair* of hydrogens to be removed from the alkane formula, $C_nH_{2n+2}$. This knowledge is quite useful because it allows us to work backward from a molecular formula to calculate the **degree of unsaturation** of a molecule—the number of rings and/or pi bonds present in an unknown.

Let's assume, for example, that we want to find the structure of an unknown hydrocarbon. A molecular weight determination on the unknown yields a value of 82, which corresponds to a molecular formula of $C_6H_{10}$. Since the fully saturated $C_6$ hydrocarbon, hexane, has the formula $C_6H_{14}$, we can calculate that the unknown compound has two fewer pairs of hydrogens ($H_{14} - H_{10} = H_4 = 2\ H_2$). The unknown therefore contains two double bonds, one ring and one double bond, two rings, or one *triple* bond. There's still a long way to go to establish structure, but our simple calculation has told us a lot about the molecule.

Similar calculations can be carried out for compounds containing elements other than just carbon and hydrogen.

1. *Organohalogen compounds*, C, H, X, where X = F, Cl, Br, or I: Since a halogen substituent is simply a replacement for hydrogen (both are monovalent), we can *add* the number of halogens and the number of hydrogens to arrive at a base hydrocarbon formula from which the number of double bonds and/or rings can be found.

For example,

Replace 2 Br by 2 H

$$Br{-}CH_2CH{=}CHCH_2{-}Br = H{-}CH_2CH{=}CHCH_2{-}H$$

$$\underbrace{C_4H_6Br_2}_{\text{Add}} = \text{“}C_4H_8\text{”}$$ One unsaturation: one double bond

2. *Organooxygen compounds*, C, H, O: Since oxygen is divalent, it doesn't affect the formula of the parent hydrocarbon. The easiest way to convince yourself of this is to look at what happens when an oxygen atom is inserted into an alkane C—C or C—H bond: There is no change in the number of hydrogen atoms. For example,

O removed from here

$$H_2C{=}CHCH{=}CHCH_2{-}O{-}H = H_2C{=}CHCH{=}CHCH_2{-}H$$

$$C_5H_8O = \text{“}C_5H_8\text{”}$$ Two unsaturations: two double bonds

3. *Organonitrogen compounds*, C, H, N: Since nitrogen is trivalent, an organonitrogen compound has one more hydrogen than its base hydrocarbon. We therefore *subtract* the number of nitrogens from the number of hydrogens to arrive at a base hydrocarbon formula. Again, the best way to convince yourself of this is to see what happens when a nitrogen atom is inserted into an alkane bond: Another hydrogen atom is required to fill the third valency on nitrogen, and we must therefore mentally subtract this extra hydrogen atom to arrive at the corresponding base hydrocarbon formula. For example,

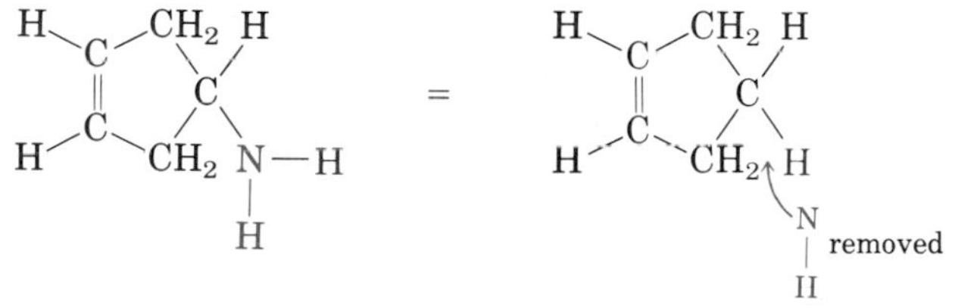

H, CH₂, H, C, C, C, H, CH₂, N—H, H = H, CH₂, H, C, C, C, H, CH₂, H, N, H removed

$$C_5H_9N = \text{“}C_5H_8\text{”}$$ Two unsaturations: one ring and one double bond

PROBLEM

**6.1** Calculate the degree of unsaturation in these hydrocarbons:

(a) $C_8H_{14}$ (b) $C_5H_6$ (c) $C_{12}H_{20}$

(d) $C_{20}H_{32}$ (e) $C_{40}H_{56}$, β-carotene

PROBLEM

**6.2** Calculate the degree of unsaturation in these formulas, and then draw as many structures as you can for each.

(a) $C_4H_8$ (b) $C_4H_6$ (c) $C_3H_4$

PROBLEM

**6.3** Calculate the number of pi bonds and/or rings in these formulas:

(a) $C_6H_5N$ (b) $C_6H_5NO_2$ (c) $C_8H_9Cl_3$
(d) $C_9H_{16}Br_2$ (e) $C_{10}H_{12}N_2O_3$ (f) $C_{20}H_{32}O_2$

## 6.3 Nomenclature of Alkenes

Alkenes are named systematically by following a series of rules similar to those developed for alkanes, with the suffix *-ene* instead of *-ane*. Three basic steps are used:

*Step 1* Name the parent hydrocarbon. Find the longest carbon chain containing the double bond, and name the compound accordingly, using the suffix *-ene*:

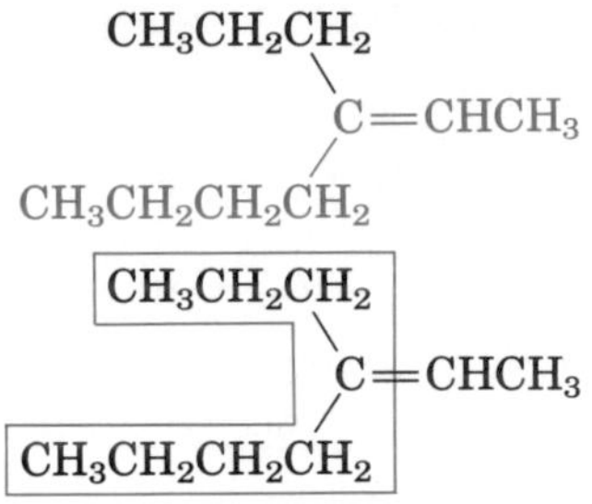

$CH_3CH_2CH_2$ \
C=$CHCH_3$ Named as a heptene
$CH_3CH_2CH_2CH_2$ /

*NOT*

$CH_3CH_2CH_2$ \
C=$CHCH_3$ as an octene, since the double bond is not contained in the eight-carbon chain
$CH_3CH_2CH_2CH_2$ /

*Step 2* Number the carbon atoms in the chain. Beginning at the end nearer the double bond, assign numbers to the carbon atoms in the chain. If the double bond is equidistant from the two ends, begin at the end nearer the first branch point. This rule assures that the double bond carbons receive the lowest possible numbers:

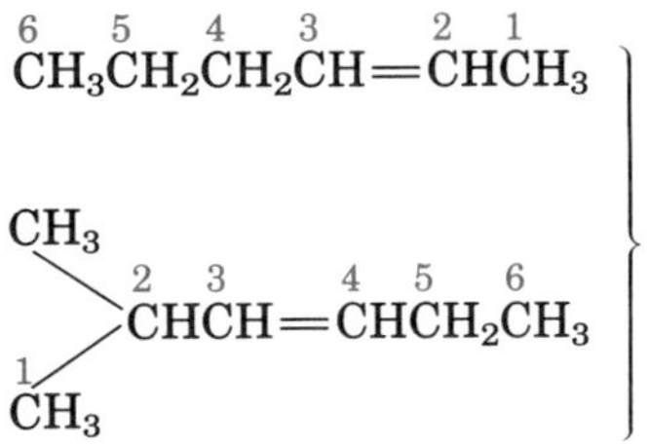

$\overset{6}{C}H_3\overset{5}{C}H_2\overset{4}{C}H_2\overset{3}{C}H{=}\overset{2}{C}H\overset{1}{C}H_3$

$CH_3$ (1) / $\overset{2}{C}H\overset{3}{C}H{=}\overset{4}{C}H\overset{5}{C}H_2\overset{6}{C}H_3$ / $CH_3$

Correct numbering

*Step 3* Write out the full name. Number the substituents according to their position in the chain and list them alphabetically. Indicate the position of the double bond by giving the number of the *first* alkene carbon. If more than one double bond is present, indicate the position of each and use the suffixes *-diene*, *-triene*, *tetraene*, and so on.

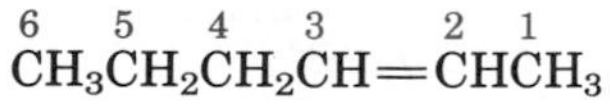

6 5 4 3 2 1
$CH_3CH_2CH_2CH{=}CHCH_3$

2-Hexene

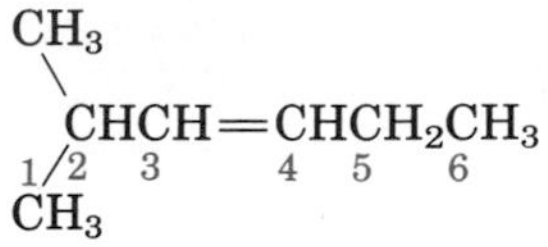

$CH_3$
\
$CHCH{=}CHCH_2CH_3$
1/2 3 4 5 6
$CH_3$

2-Methyl-3-hexene

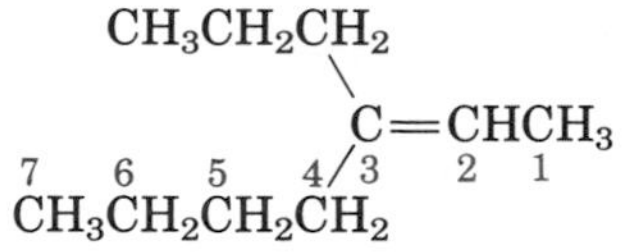

$CH_3CH_2CH_2$
\
$C{=}CHCH_3$
7 6 5 4/3 2 1
$CH_3CH_2CH_2CH_2$

3-Propyl-2-heptene

$CH_3$
|
$H_2C{=}C{-}CH{=}CH_2$
1 2 3 4

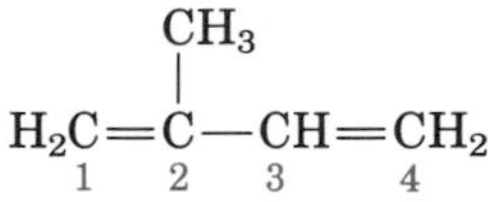

2-Methyl-1,3-butadiene

Cycloalkenes are named in a similar way. Since there's no chain end to begin from, though, we number the cycloalkene so that the double bond is between C1 and C2 and the first branch point has as low a value as possible:

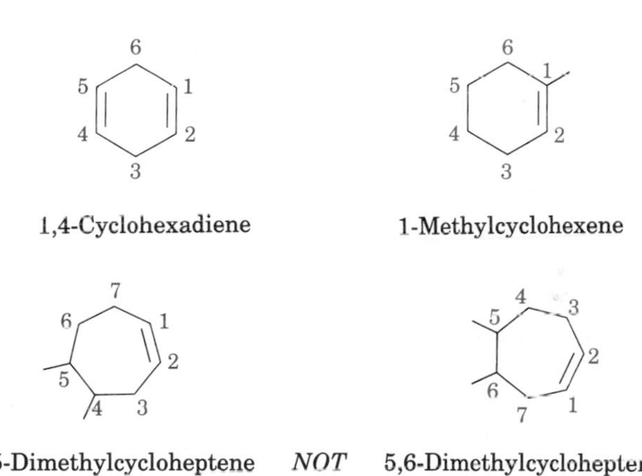

1,4-Cyclohexadiene 1-Methylcyclohexene

4,5-Dimethylcycloheptene *NOT* 5,6-Dimethylcycloheptene

A small number of alkenes have names that, though firmly entrenched in common usage, don't conform to strict rules of nomenclature. For example, the alkene derived from ethane should properly be called ethene, but the name ethylene has been used so long that it is accepted by IUPAC. Table 6.1 (page 160) lists several other common names that are often used and that are recognized by IUPAC. Note that an $H_2C{=}CH{-}$ substituent is called a **vinyl group**, and an $H_2C{=}CH{-}CH_2{-}$ substituent is called an **allyl group**:

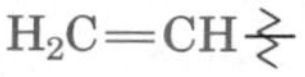

$H_2C{=}CH{-}$ $H_2C{=}CH{-}CH_2{-}$

A *vinyl* group 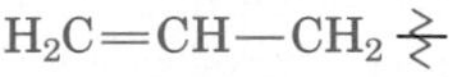 An *allyl* group

Table 6.1 Common names of some alkenes[a]

| *Compound* | *Systematic name* | *Common name* |
|---|---|---|
| $H_2C{=}CH_2$ | Ethene | Ethylene |
| $CH_3CH{=}CH_2$ | Propene | Propylene |
| $(CH_3)_2C{=}CH_2$ | 2-Methylpropene | Isobutylene |
| $H_2C{=}C(CH_3){-}CH{=}CH_2$ | 2-Methyl-1,3-butadiene | Isoprene |
| $CH_3CH{=}CHCH{=}CH_2$ | 1,3-Pentadiene | Piperylene |
| $H_2C{=}CH{-}$ | Ethenyl | Vinyl (an alkenyl group) |
| $H_2C{=}CH{-}CH_2{-}$ | 2-Propenyl | Allyl |
| $H_2C{=}$ | Methylene | |
| $CH_3CH{=}$ | Ethylidene | |

[a]Both common and systematic names are recognized by IUPAC.

PROBLEM

**6.4** Give proper IUPAC names for these compounds:

(a) $H_2C{=}CHCH(CH_3)C(CH_3)_3$ (b) $CH_3CH_2CH{=}C(CH_3)CH_2CH_3$

(c) $CH_3CH{=}CHCH(CH_3)CH{=}CHCH(CH_3)_2$

PROBLEM

**6.5** Draw structures corresponding to these IUPAC names:
(a) 2-Methyl-1,5-hexadiene
(b) 3-Ethyl-2,2-dimethyl-3-heptene
(c) 2,3,3-Trimethyl-1,4,6-octatriene
(d) 3,4-Diisopropyl-2,5-dimethyl-3-hexene
(e) 4-*tert*-Butyl-2-methylheptane

PROBLEM

**6.6** Provide proper names for these cycloalkenes:

(a) 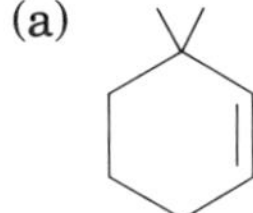

(b) 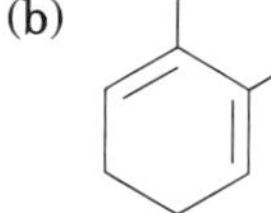

(c) 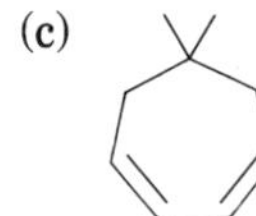

## 6.4 Electronic Structure of Alkenes

We saw earlier (in Section 1.10) that a carbon–carbon double bond consists of two parts—a sigma bond and a pi bond. The carbon atoms are $sp^2$ hybridized and have three equivalent orbitals directed to the corners of an equilateral triangle. The fourth carbon orbital is an unhybridized $p$ orbital, which is perpendicular to the $sp^2$ plane.

When two such carbon atoms approach each other, they form two kinds of bonds—a sigma bond formed by head-on overlap of $sp^2$ orbitals, and a pi bond formed by sideways overlap of $p$ orbitals. The doubly bonded carbons and the four groups attached to them therefore lie in a plane, with bond angles of approximately 120° (Figure 6.2). As you might expect, a carbon–carbon double bond is both stronger [152 kcal/mol versus 88 kcal/mol (636 kJ/mol versus 368 kJ/mol)] and shorter (1.33 Å versus 1.54 Å) than a carbon–carbon single bond. Table 6.2 (page 162) compares the experimentally determined bond parameters of ethylene and ethane.

$H_2C{=}CH_2$

Ethylene

$H_3C{-}CH_3$

Ethane

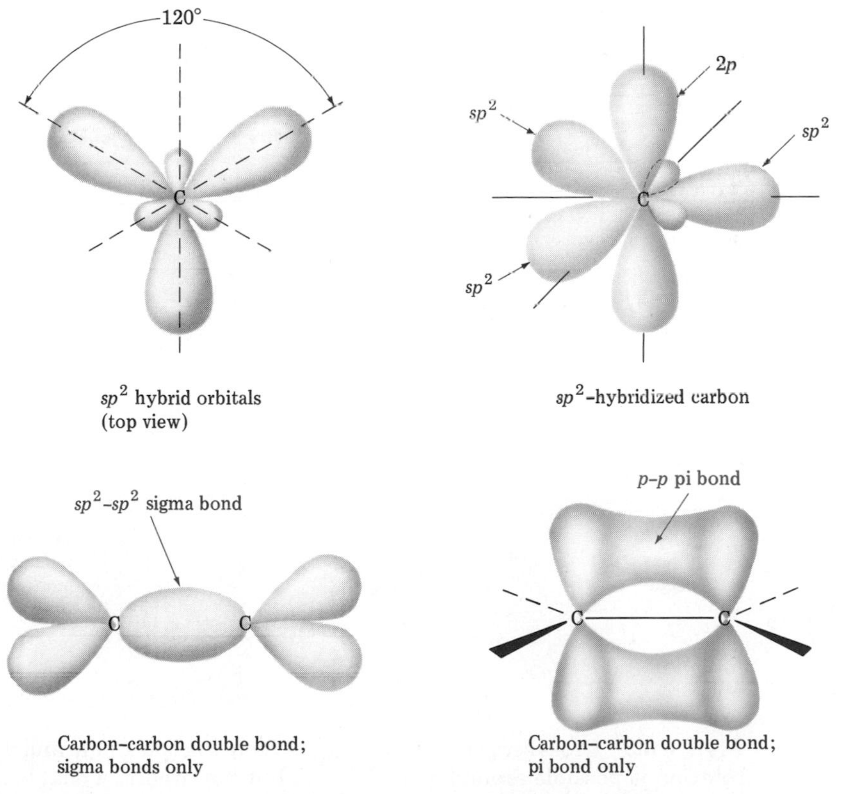


**Figure 6.2** An orbital picture of the carbon–carbon double bond

Table 6.2 Molecular parameters for ethylene and ethane[a]

| | *Ethylene* | *Ethane* |
|---|---|---|
| H—C—H bond angle (degrees) | 116.6 | 109.3 |
| H—C—C bond angle (degrees) | 121.7 | 109.6 |
| C—C bond strength (kcal/mol) | 152 | 88 |
| C—C bond length (Å) | 1.33 | 1.54 |
| C—H bond strength (kcal/mol) | 103 | 98 |
| C—H bond length (Å) | 1.076 | 1.10 |

[a]The double bond is both stronger and shorter than the single bond.

The presence of the double bond in alkenes has numerous consequences. One consequence is the phenomenon of **restricted rotation**. We know from Section 4.1 that relatively free rotation is possible around sigma bonds, and that open-chain alkanes such as butane therefore have an infinite number of rapidly interconverting conformations. The same is not true for double bonds. Carbon–carbon double bonds do not have circular cross-sections, and rotation can't occur freely (Figure 6.3).

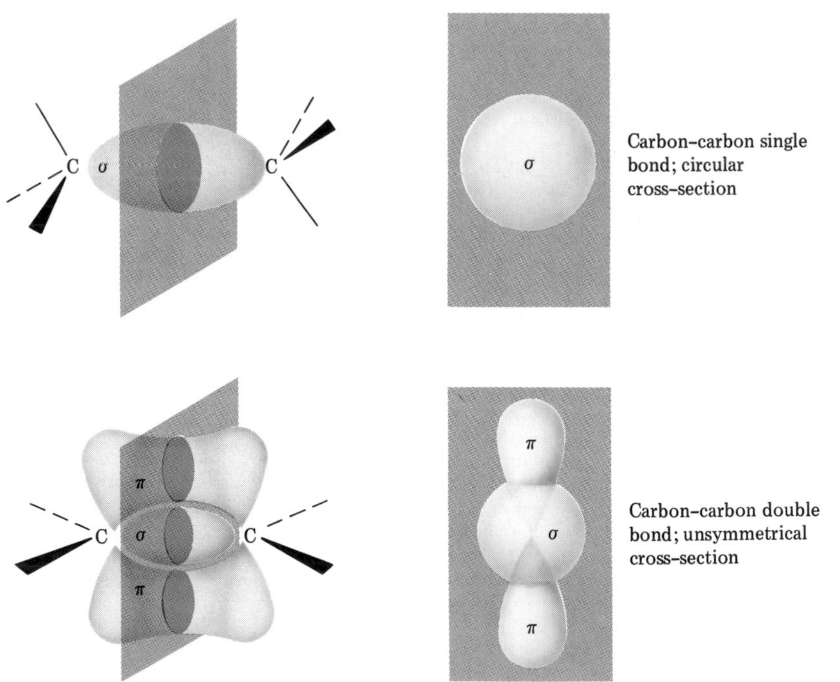


**Figure 6.3** Cross sections cut through carbon–carbon single and double bond: Free rotation is possible around a single bond but not around a double bond.

If we were to *force* rotation to occur, we would need to break the pi bond temporarily (Figure 6.4). Thus, the barrier to double-bond rotation must be at least as great as the strength of the pi bond.

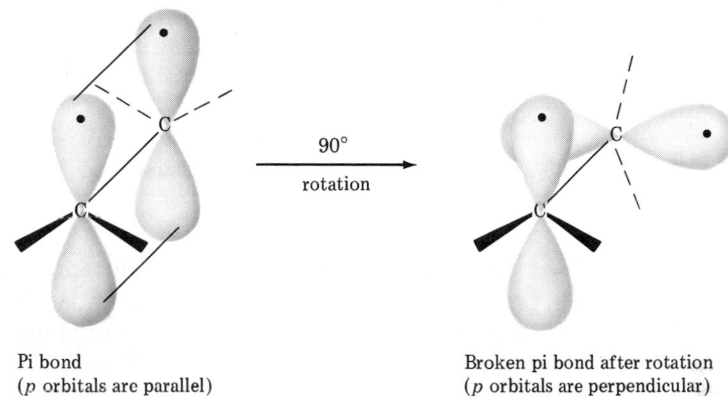


**Figure 6.4** The pi bond must break in order for rotation to take place around a carbon–carbon double bond.

Breaking a chemical bond normally requires a large amount of energy. We can make a rough estimate of how much energy is required to break the pi bond of an alkene by subtracting the value for the strength of an average carbon–carbon sigma bond from the total bond strength value for ethylene. This calculation predicts an approximate bond strength of 64 kcal/mol (268 kJ/mol) for the ethylene pi bond, and it is therefore clear why rotation can't occur. (Recall that the barrier to bond rotation in ethane is only 2.9 kcal/mol.)

| | |
|---|---|
| Ethylene C═C bond strength (sigma + pi) | 152 kcal/mol |
| Ethane C—C bond strength (sigma only) | 88 kcal/mol |
| Difference (pi bond only) | 64 kcal/mol |

## 6.5 Cis–Trans Isomerism in Alkenes

The lack of rotation around the carbon–carbon double bond is of more than just theoretical interest; it also has chemical consequences. Imagine the situation for a disubstituted alkene such as 2-butene. (*Disubstituted* means that there are two substituents other than hydrogen bonded to the double-bond carbons.) The two methyl groups in 2-butene can either be on the *same* side of the double bond or on *opposite* sides, a situation reminiscent of substituted cycloalkanes (Section 3.10). Figure 6.5 shows the two 2-butenes.

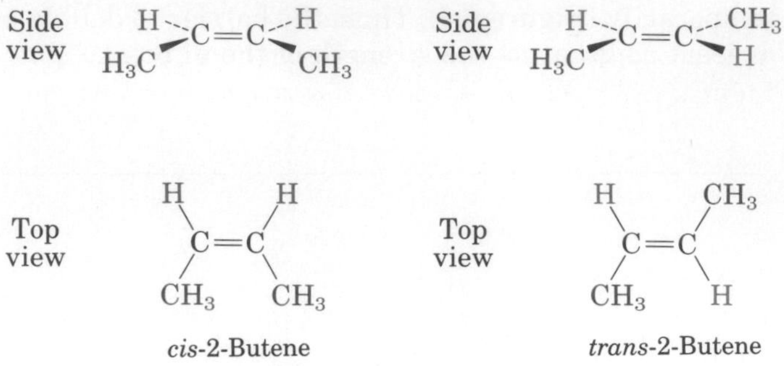

Side view $H$, $H_3C$ — $C{=}C$ — $H$, $CH_3$ Side view $H$, $H_3C$ — $C{=}C$ — $CH_3$, $H$

Top view

$$\begin{array}{ccc} H & & H \\ & C{=}C & \\ CH_3 & & CH_3 \end{array}$$

*cis*-2-Butene

Top view

$$\begin{array}{ccc} H & & CH_3 \\ & C{=}C & \\ CH_3 & & H \end{array}$$

*trans*-2-Butene

**Figure 6.5** The two cis–trans isomers of 2-butene

Since bond rotation can't occur, the two 2-butenes can't spontaneously interconvert. They are distinct, isolable compounds. As with substituted cycloalkanes (Section 3.10), we call such compounds **cis–trans stereoisomers** because they have the same formula and overall skeleton but differ in the spatial arrangement of atoms. The compound with substituents on the same side of the double bond is referred to as *cis*-2-butene; the isomer with substituents on opposite sides is *trans*-2-butene.

Cis–trans isomerism is a common feature of alkene chemistry and is not limited to disubstituted alkenes. Isomerism can occur whenever each of the double-bond carbons is attached to two different groups. If one of the double-bond carbons is attached to two identical groups, however, then cis–trans isomerism is not possible (Figure 6.6).

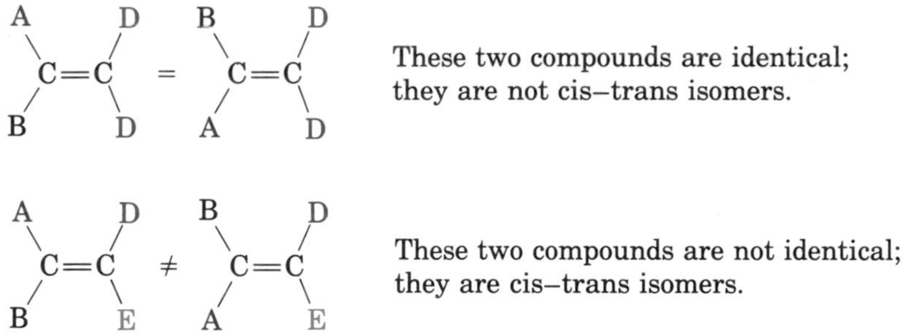

$$\begin{array}{ccc} A & & D \\ & C{=}C & \\ B & & D \end{array} \; = \; \begin{array}{ccc} B & & D \\ & C{=}C & \\ A & & D \end{array}$$

These two compounds are identical; they are not cis–trans isomers.

$$\begin{array}{ccc} A & & D \\ & C{=}C & \\ B & & E \end{array} \; \neq \; \begin{array}{ccc} B & & D \\ & C{=}C & \\ A & & E \end{array}$$

These two compounds are not identical; they are cis–trans isomers.

**Figure 6.6** The requirement for cis–trans isomerism in alkenes: Compounds that have one of their carbons bonded to two identical groups can't exist as cis–trans isomers.

PROBLEM

**6.7** Which of the following compounds can exist as pairs of cis–trans isomers? Draw each cis–trans pair and indicate the geometry of each isomer.

(a) $CH_3CH{=}CH_2$

(b) $(CH_3)_2C{=}CHCH_3$

(c) $CH_3CH_2CH{=}CHCH_3$

(d) $(CH_3)_2C{=}C(CH_3)CH_2CH_3$

(e) $ClCH{=}CHCl$

(f) $BrCH{=}CHCl$

PROBLEM

**6.8** How can you account for the observation that cyclohexene does not show cis–trans double-bond isomerism, whereas cyclodecene can exist in both cis and trans forms? Making molecular models should be helpful.

## 6.6 Sequence Rules: The *E,Z* Designation

In the previous discussion of isomerism in the 2-butenes, we used the terms *cis* and *trans* to denote alkenes whose substituents were on the same side and opposite side of a double bond, respectively. This cis–trans nomenclature is unambiguous and quite acceptable for all disubstituted alkenes. But how do we denote the geometry of trisubstituted (three substituents other than hydrogen) and tetrasubstituted (four substituents other than hydrogen) double bonds?

The answer is provided by the *E,Z* system of nomenclature, which uses a series of **sequence rules** to assign priorities to the substituent groups on the double-bond carbons. Considering each of the double-bond carbons separately, we use the sequence rules to decide which of the two groups on each carbon is higher in priority. If the higher-priority groups on each carbon are on the same side of the double bond, the alkene is designated *Z* (for the German word *zusammen*, "together"). If the higher-priority groups are on opposite sides, the alkene is designated *E* (for the German word *entgegen*, "opposite"). The easiest way to learn which is which is to think with a German accent: *Z* = groups on "ze zame zide" (*E* = the other guy). These assignments are shown in Figure 6.7.

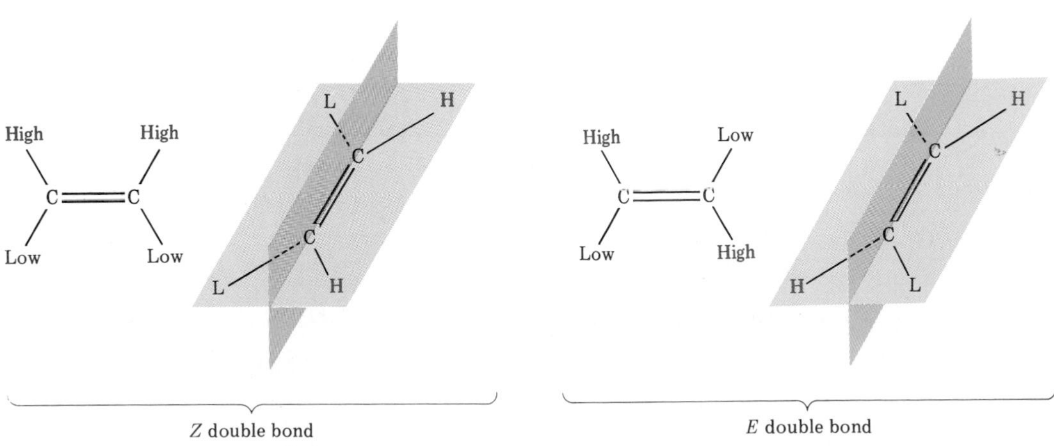


**Figure 6.7** The *E,Z* system of nomenclature for substituted alkenes

Introduced by Cahn, Ingold, and Prelog in 1964, the sequence rules have proven to be extraordinarily useful. The rules are as follows.

*Sequence rule 1* Look at the atoms directly attached to each carbon and rank them in order of decreasing atomic number. That is, an atom with higher atomic weight number receives higher priority than an atom with lower atomic weight. Thus, the common atoms that we might find attached to a double bond would be assigned the priority sequence Br > Cl > O > N > C > H. For example,

Low priority H, High priority Cl, High priority $CH_3$, Low priority $CH_3$ (on $C{=}C$)

(*E*)-2-Chloro-2-butene

Low priority H, Low priority $CH_3$, High priority $CH_3$, High priority Cl (on $C{=}C$)

(*Z*)-2-Chloro-2-butene

Since chlorine has a higher atomic number than carbon (17 versus 6), it receives higher priority than a methyl ($CH_3$) group. Methyl receives higher priority than hydrogen, however; and the left-hand isomer in the preceding display is therefore assigned *E* geometry (high-priority groups on opposite sides of the double bond). The right-hand isomer has *Z* geometry (high-priority groups on "ze zame zide" of the double bond).

*Sequence rule 2* If a decision can't be reached by considering the first atoms in the substituent (rule 1), look at the second, third, or fourth atoms away from the double-bond carbons until a difference is found. Thus, an ethyl substituent, $-CH_2CH_3$, and a methyl substituent, $-CH_3$, are equivalent by rule 1, since both have carbon as the first atom. By rule 2, however, ethyl receives higher priority than methyl since the *second* atoms in the group are one carbon and two hydrogens rather than three hydrogens. Look at the following examples to see how this rule is applied:

$-CH_3$ (Lower) vs. $-CH_2CH_3$ (Higher)

$-O-H$ (Lower) vs. $-O-CH_3$ (Higher)

$-CH(CH_3)CH_3$ (Higher) vs. $-CH_2CH_3$ (Lower)

$-CH(CH_3)NH_2$ (Lower) vs. $-CH_2Cl$ (Higher)

*Sequence rule 3* Multiple-bonded atoms are considered to be equivalent to the same number of singly bonded atoms. For example, an aldehyde substituent (—CH=O), which has a carbon atom *doubly* bound to *one* oxygen, is considered equivalent to a substituent having a carbon atom *singly* bound to *two* oxygen atoms:

—CH=O is equivalent to —CH(O)—O—C

This carbon is doubly bound to one oxygen

This oxygen is doubly bound to one carbon

This carbon is singly bound to two oxygens

This oxygen is singly bound to two carbons

As further examples, the following pairs are equivalent:

—CH=CH₂ same as —C(H)(C)—C(H)(C)—H

—C≡C—H same as —C(C)(C)—C(C)(C)—H

Taking all the sequence rules into account, we can assign the configurations shown in the following examples. Work through each one to convince yourself that the assignments are correct.

(*E*)-3-Methyl-1,3-pentadiene

(*E*)-1-Bromo-2-isopropyl-1,3-butadiene

(*Z*)-2-Hydroxymethyl-2-butenoic acid

**PRACTICE PROBLEM**

Assign *E* or *Z* configuration to the double bond in this compound:

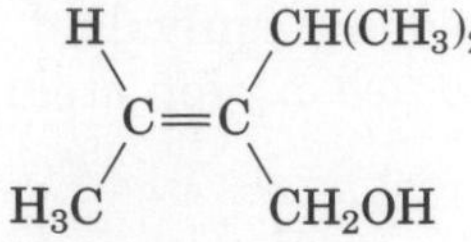

$$\begin{array}{ccc} H & & CH(CH_3)_2 \\ & C=C & \\ H_3C & & CH_2OH \end{array}$$

*Solution* Look at the two double-bond carbons individually. The left-hand carbon has two substituents, $—H$ and $—CH_3$, of which $—CH_3$ receives higher priority by sequence rule 1. The right-hand carbon also has two substituents, $—CH(CH_3)_2$ and $—CH_2OH$, but rule 1 doesn't allow a priority assignment to be made since both groups have carbon as their first atom. By rule 2, however, $—CH_2OH$ receives higher priority than $—CH(CH_3)_2$, since $—CH_2OH$ has an *oxygen* and two hydrogens as its second atoms, whereas $—CH(CH_3)_2$ has two *carbons* and one hydrogen as its second atoms. Thus, the two high-priority groups are on the same side of the double bond, and we assign *Z* configuration.

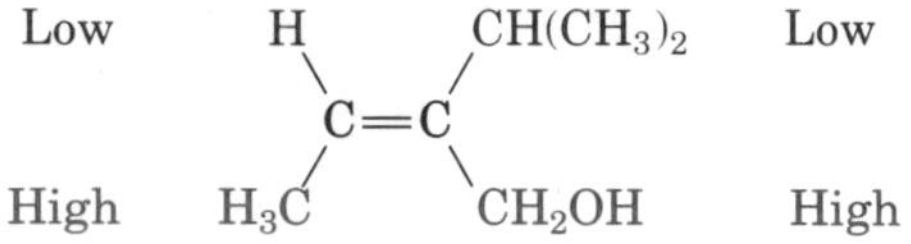

$$\begin{array}{ccccc} \text{Low} & H & & CH(CH_3)_2 & \text{Low} \\ & & C=C & & \\ \text{High} & H_3C & & CH_2OH & \text{High} \end{array}$$

*Z* configuration

**PROBLEM**

**6.9** Which member in each set is higher in priority?

(a) $—H$ or $—Br$ (b) $—Cl$ or $—Br$ (c) $—CH_3$ or $—CH_2CH_3$

(d) $—NH_2$ or $—OH$ (e) $—CH_2OH$ or $—CH_3$ (f) $—CH_2OH$ or $—CH{=}O$

**PROBLEM**

**6.10** Rank the sets of substituents in order of Cahn–Ingold–Prelog priorities:

(a) $—CH_3$, $—OH$, $—H$, $—Cl$

(b) $—CH_3$, $—CH_2CH_3$, $—CH{=}CH_2$, $—CH_2OH$

(c) $—COOH$, $—CH_2OH$, $—C{\equiv}N$, $—CH_2NH_2$

(d) $—CH_2CH_3$, $—C{\equiv}CH$, $—C{\equiv}N$, $—CH_2OCH_3$

**PROBLEM**

**6.11** Assign *E* or *Z* configuration to these alkenes:

(a)
$$\begin{array}{ccc} CH_3 & & CH_2OH \\ & C=C & \\ CH_3CH_2 & & Cl \end{array}$$

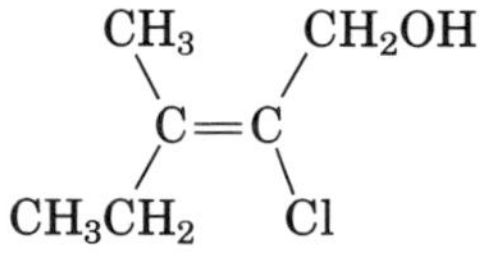

(b)
$$\begin{array}{ccc} Cl & & CH_2CH_3 \\ & C=C & \\ CH_3O & & CH_2CH_2CH_3 \end{array}$$

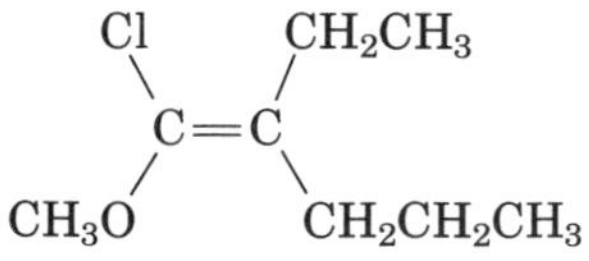

(c) 3-methylcyclopentyl ring carbon $C=C$ with $COOH$ and $CH_2OH$

$$\begin{array}{ccc} CH_3\text{-cyclopentylidene} & C=C & COOH \\ & & CH_2OH \end{array}$$

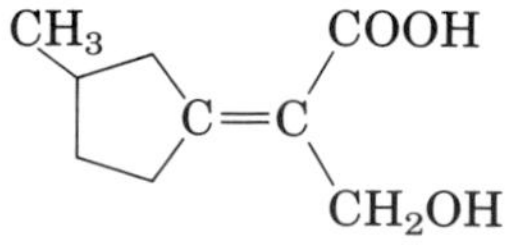

(d)
$$\begin{array}{ccc} H & & CN \\ & C=C & \\ CH_3 & & CH_2NH_2 \end{array}$$

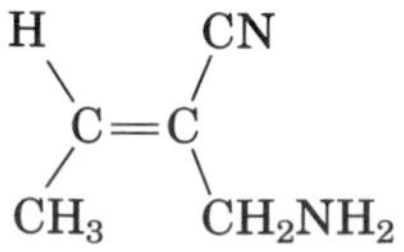

## 6.7 Alkene Stability

Although the cis–trans interconversion of alkene isomers does not occur spontaneously, it can be made to happen under appropriate experimental conditions (for example, on treatment with a strong acid catalyst). If we were to interconvert *cis*-2-butene with *trans*-2-butene and allow them to reach equilibrium, we would find that they are not of equal stability. At equilibrium, the ratio of isomers is 76% trans to 24% cis.

$$\underset{\text{Trans (76\%)}}{(H)(CH_3)C{=}C(CH_3)(H)} \underset{\text{catalyst}}{\overset{\text{Acid}}{\rightleftharpoons}} \underset{\text{Cis (24\%)}}{(CH_3)(H)C{=}C(CH_3)(H)}$$

Using the relationship between equilibrium constants and free-energy differences in Table 6.3 (which repeats Table 4.3 for convenience), we can calculate that *cis*-2-butene is less stable than *trans*-2-butene by 0.66 kcal/mol (2.6 kJ/mol) at room temperature.

**Table 6.3** The relationship between stability and isomer percentages at equilibrium

| *More stable isomer (%)* | *Less stable isomer (%)* | *Energy difference at 25°C* | |
|---|---|---|---|
| | | *(kcal/mol)* | *(kJ/mol)* |
| 50 | 50 | 0 | 0 |
| 75 | 25 | 0.65 | 2.72 |
| 90 | 10 | 1.30 | 5.45 |
| 95 | 5 | 1.74 | 7.29 |
| 99 | 1 | 2.72 | 11.38 |
| 99.9 | 0.1 | 4.09 | 17.11 |

It turns out to be a general phenomenon that cis alkenes are less stable than their trans isomers because of steric (spatial) strain between the two bulky substituents on the same side of the double bond. This is the same kind of steric interference that we saw previously in the axial conformation of methylcyclohexane (Section 4.12).

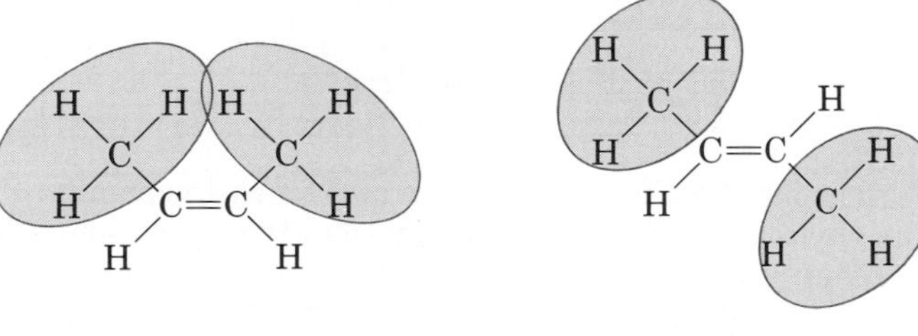

Steric strain in *cis*-2-butene

No steric strain in *trans*-2-butene

Although it is sometimes possible to obtain information about relative stabilities of alkenes by establishing a cis–trans equilibrium through treatment of the alkene with strong acid, there's an easier way to gain nearly the same information. One of the more important reactions that alkenes undergo is **catalytic hydrogenation**. In the presence of catalysts such as palladium or platinum, hydrogen adds to carbon–carbon double bonds to yield the corresponding saturated alkanes:

$$\underset{\textit{trans}\text{-2-Butene}}{\text{H(CH}_3\text{)C=C(H)CH}_3} \xrightarrow[\text{Pd}]{\text{H}_2} \underset{\text{Butane}}{\text{CH}_3\text{CH}_2\text{CH}_2\text{CH}_3} \xleftarrow[\text{Pd}]{\text{H}_2} \underset{\textit{cis}\text{-2-Butene}}{\text{CH}_3\text{(H)C=C(H)CH}_3}$$

Consider the hydrogenations of *cis*- and *trans*-2-butene, both of which react with hydrogen to give the *same* product, butane. Energy diagrams for the two reactions are shown in Figure 6.8. Since *cis*-2-butene is less stable than *trans*-2-butene by 0.66 kcal/mol, the energy diagram shows the cis alkene at a higher energy level. After reaction, however, both products are at the same energy level (butane). It therefore follows that $\Delta G°$ for reaction of the cis isomer must be larger than for reaction of the trans isomer. In other words, more energy is evolved in the hydrogenation of the cis isomer than of the trans isomer because there was more energy present in the cis isomer to begin with.

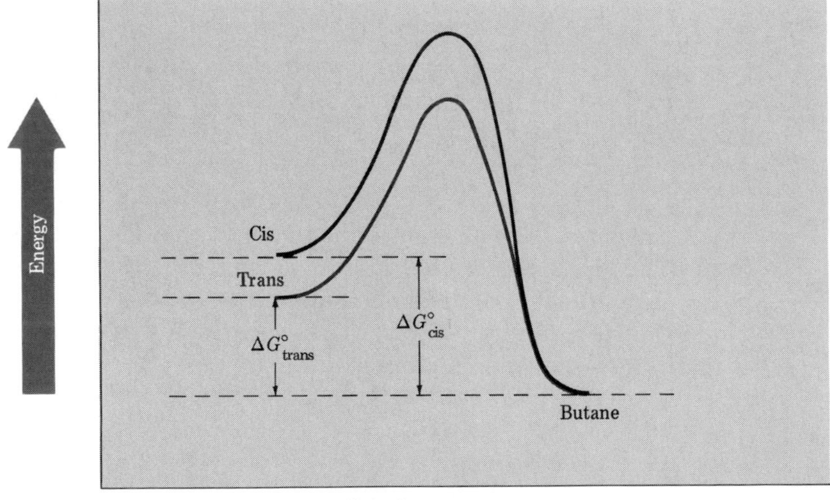


**Figure 6.8** Reaction energy diagrams for hydrogenation of *cis*- and *trans*-2-butene: The cis isomer is higher in energy than the trans isomer by about 1 kcal/mol and therefore gives off more energy in the reaction.

If we were to measure the heats of reaction for the two hydrogenations and find their difference, we could determine the relative stabilities of cis and trans isomers without having to measure an equilibrium position. A large number of such **heats of hydrogenation ($\Delta H^\circ_{hydrog}$)** have been measured, and the results bear out our expectation. For *cis*-2-butene, $\Delta H^\circ_{hydrog}$ = 28.6 kcal/mol (120 kJ/mol); for the trans isomer, $\Delta H^\circ_{hydrog}$ = 27.6 kcal/mol (116 kJ/mol). (Note that these $\Delta H^\circ$ values really should be negative numbers since heat is evolved during hydrogenation. The minus sign is usually dropped, however, to make comparisons easier.)

Trans isomer
$\Delta H^\circ_{hydrog}$ = 27.6 kcal/mol

Cis isomer
$\Delta H^\circ_{hydrog}$ = 28.6 kcal/mol

Although the energy difference in the heats of hydrogenation for the 2-butene isomers (1 kcal/mol) agrees reasonably well with the energy difference calculated from equilibrium data (0.66 kcal/mol), the two numbers aren't exactly the same. There are two reasons for this. The first is simply experimental error; heats of hydrogenation require considerable expertise and specialized equipment to measure accurately, and we are looking at a small difference between two large numbers. The second reason is that heats of reaction and equilibrium constants don't measure exactly the same quantity. Heats of reaction measure enthalpy changes, $\Delta H^\circ$, whereas equilibrium constants measure overall free-energy changes, $\Delta G^\circ$ ($\Delta G^\circ = \Delta H^\circ - T\Delta S^\circ$). We therefore expect a slight difference when comparing the two measurements.

Although heats of hydrogenation are not quite as accurate as we might like, we can nevertheless gain some useful and interesting information from them. Table 6.4 lists some representative data.

**Table 6.4** Heats of hydrogenation of some alkenes

| *Substitution* | *Alkene* | $\Delta H^\circ_{hydrog}$ (*kcal/mol*) | (*kJ/mol*) |
|---|---|---|---|
| | $H_2C{=}CH_2$ | 32.8 | 137 |
| Monosubstituted (one alkyl group next to double bond) | $CH_3CH{=}CH_2$ | 30.1 | 126 |
| | $CH_3CH_2CH{=}CH_2$ | 30.3 | 127 |
| | $(CH_3)_2CHCH{=}CH_2$ | 30.3 | 127 |
| | $(CH_3)_3CCH{=}CH_2$ | 30.3 | 127 |
| Disubstituted (two alkyl groups) | Cis $CH_3CH{=}CHCH_3$ | 28.6 | 120 |
| | Trans $CH_3CH{=}CHCH_3$ | 27.6 | 115 |
| | $(CH_3)_2C{=}CH_2$ | 28.4 | 119 |
| | $CH_3CH_2(CH_3)C{=}CH_2$ | 28.5 | 119 |
| Trisubstituted (three alkyl groups) | $(CH_3)_2C{=}CHCH_3$ | 26.9 | 113 |
| Tetrasubstituted (four alkyl groups) | $(CH_3)_2C{=}C(CH_3)_2$ | 26.6 | 111 |

The data in Table 6.4 show that alkenes become more stable with increasing substitution. For example, ethylene has $\Delta H^\circ_{\text{hydrog}} = 32.8$ kcal/mol (137 kJ/mol); but, when one alkyl substituent is attached as in 1-butene ($\Delta H^\circ_{\text{hydrog}} = 30.3$ kcal/mol), the alkene becomes approximately 2.5 kcal/mol more stable. Further increasing the degree of substitution leads to still further stability. As a general rule, alkenes follow the stability order

Tetrasubstituted > Trisubstituted > Disubstituted > Monosubstituted

$$R_2C{=}CR_2 \;>\; R_2C{=}CHR \;>\; RCH{=}CHR \;\approx\; R_2C{=}CH_2 \;>\; RCH{=}CH_2$$

The reasons for the observed stability order for substituted alkenes are not well understood. Several different explanations have been advanced, but none has received universal acceptance. Some chemists feel that the stability order is due to a phenomenon termed **hyperconjugation** (Figure 6.9). Hyperconjugation is a stabilizing effect due to orbital overlap between the carbon–carbon pi bond and a properly oriented carbon–hydrogen sigma bond on a neighboring substituent. The more substituents that are present, the more opportunities exist for hyperconjugation and the more stable the alkene.

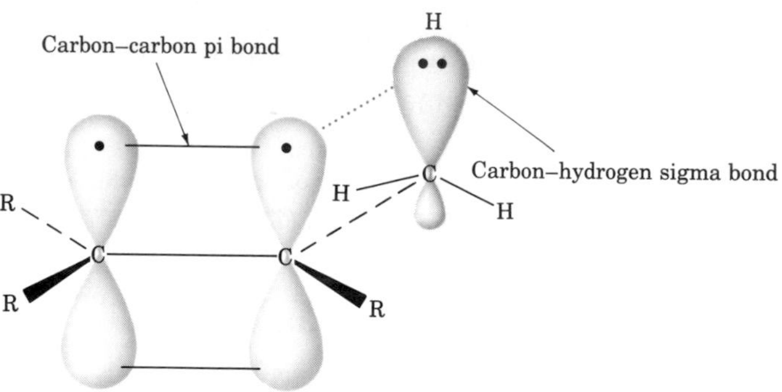


**Figure 6.9** Hyperconjugation—a stabilizing overlap between a *p* orbital and a neighboring C—H sigma bond orbital

Other chemists view hyperconjugation as unlikely and prefer a simple bond strength argument to account for the observed alkene stability order. A bond between an $sp^2$ carbon and an $sp^3$ carbon is somewhat stronger than a bond between two $sp^3$ carbons. Thus, in comparing the two isomers, 1-butene and 2-butene, the monosubstituted isomer has one $sp^3$–$sp^3$ bond and one $sp^3$–$sp^2$ bond, and the disubstituted isomer has two $sp^3$–$sp^2$ bonds.

Highly substituted alkenes always have a higher ratio of $sp^3$–$sp^2$ bonds to $sp^3$–$sp^3$ bonds than less substituted alkenes and are therefore more stable:

$$\underset{\substack{\text{2-Butene}\\\text{(more stable)}}}{CH_3\overset{sp^3-sp^2}{\overset{\downarrow}{-}}CH{=}CH\overset{sp^3-sp^2}{\overset{\downarrow}{-}}CH_3} \qquad \underset{\substack{\text{1-Butene}\\\text{(less stable)}}}{CH_3\overset{sp^3-sp^3}{\overset{\downarrow}{-}}CH_2\overset{sp^3-sp^2}{\overset{\downarrow}{-}}CH{=}CH_2}$$

Regardless of which explanation ultimately gains acceptance, this ongoing controversy underscores the fact that chemistry is a living science. There are many observations in organic chemistry that are still not well understood.

PROBLEM

**6.12** Which alkene in each of the following sets is more stable?
(a) 1-Butene or 2-methylpropene (b) (*Z*)-2-Hexene or (*E*)-2-hexene
(c) 1-Methylcyclohexene or 3-methylcyclohexene

PROBLEM

**6.13** The double bonds in small-ring cycloalkenes must have cis geometry, because a stable trans double bond is impossible within the confines of a five- or six-membered ring. At some point, however, a ring becomes large enough to accommodate a trans double bond. The following heats of hydrogenation have been measured:

| | $\Delta H^\circ_{hydrog}$ | |
|---|---|---|
| | *(kcal/mol)* | *(kJ/mol)* |
| *cis*-Cyclooctene | 23.0 | 96.2 |
| *trans*-Cyclooctene | 32.2 | 134.7 |
| *cis*-Cyclononene | 23.6 | 98.7 |
| *trans*-Cyclononene | 26.5 | 110.9 |
| *cis*-Cyclodecene | 20.7 | 86.6 |
| *trans*-Cyclodecene | 24.0 | 100.4 |

How do you explain these data? Make molecular models of the trans cycloalkenes to see their conformations.

## 6.8 Reactions of Alkenes

Before beginning a detailed discussion of alkene reactions, let's review briefly what was said in the previous chapter (Sections 5.6–5.10). At that time, we said that alkenes behave as nucleophiles. The carbon–carbon double bond is electron-rich and can donate a pair of electrons to an electrophile in polar reactions. For example, reaction of ethylene with HBr leads to the formation of bromoethane. Careful study of this and other **electrophilic**

**addition reactions** by Sir Christopher Ingold[1] and others many years ago has led to the generally accepted mechanism shown in Figure 6.10.

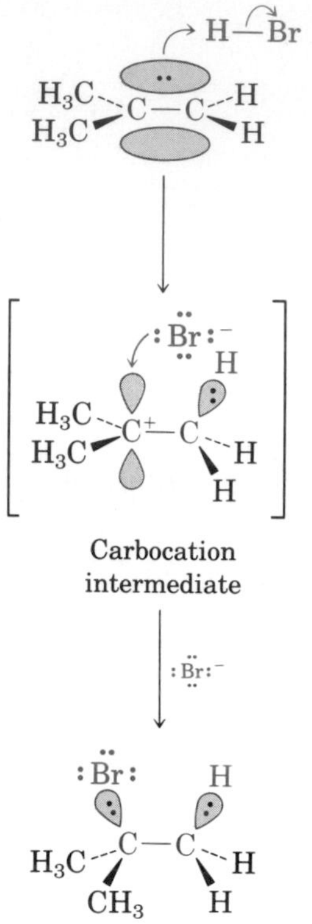

The electrophile, HBr, is attacked by the pi electrons of the alkene, and a new C—H sigma bond is formed. This leaves the other carbon atom with a $^+$ charge and a vacant *p* orbital.

Carbocation intermediate

:Br:$^-$ donates an electron pair to the positively charged carbon atom, forming a C—Br bond and yielding a neutral addition product.

**Figure 6.10** Mechanism of the electrophilic addition of HBr to ethylene: The reaction occurs in two steps and involves a carbocation intermediate.

The reaction begins with an attack on the electrophile, HBr, by the electrons of the nucleophilic pi bond. Two electrons from the pi bond form a new sigma bond between the entering hydrogen and an alkene carbon, as shown by the curved arrow at the top of Figure 6.10. The carbocation intermediate that results is itself an electrophile that can accept an electron pair from nucleophilic bromide ion to form a C—Br bond and yield a neutral addition product.

The reaction energy diagram for the overall electrophilic addition reaction, previously shown in Figure 5.6 and repeated in Figure 6.11, has two peaks (transition states) separated by a valley (carbocation intermediate). The energy level of this intermediate is higher than that of the starting alkene, but the reaction as a whole is exothermic (negative $\Delta G°$). The first step, protonation of the alkene to yield the intermediate cation, is relatively slow, but, once formed, the cation intermediate rapidly reacts further to yield the final bromoalkane product. The relative rates of the two individual steps are indicated in Figure 6.11 by the fact that $\Delta G_1^{\ddagger}$ is larger than $\Delta G_2^{\ddagger}$.

[1]Sir Christopher Ingold (1893–1970); b. Ilford, England; D.Sc., London (Thorpe); professor, Leeds (1924–1930), University College, London (1930–1970).

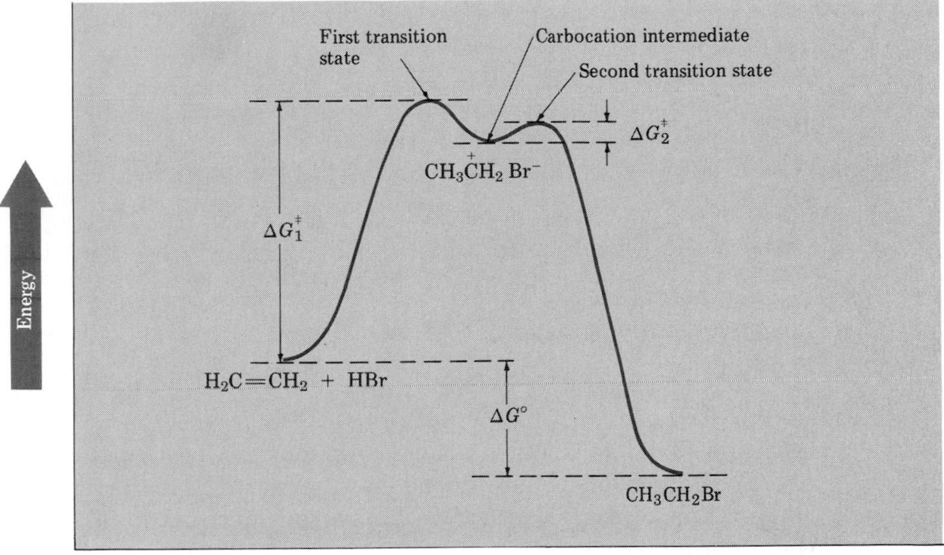


**Figure 6.11** Reaction energy diagram for the two-step electrophilic addition of HBr to ethylene: The first step is slower than the second step.

## 6.9 Addition of HX to Alkenes: Carbocations

Electrophilic addition of HX to alkenes is a general reaction that allows us to synthesize a variety of products.[2]

[2]Organic reaction equations may be written in different ways depending on the emphasis desired. For example, the reaction of 2-methylpropene with HCl might be written in the format A + B → C, emphasizing that *both* reaction partners are equally important for the purposes of the present discussion. The reaction solvent and notes about any other reaction conditions such as temperature or concentration can be noted either above or below the reaction arrow:

$$(CH_3)_2C{=}CH_2 + HCl \xrightarrow[25°C]{\text{Ether}} (CH_3)_3CCl$$

Alternatively, we might choose to write the same reaction in the format

$$A \xrightarrow{B} C$$

emphasizing that reagent A is the starting material whose chemistry is of greater interest. Reagent B is then noted above the reaction arrow, together with notes about solvent and reaction conditions:

$$(CH_3)_2C{=}CH_2 \xrightarrow[\text{Ether, 25°C}]{HCl} (CH_3)_3CCl$$

(HCl ← Reagent; Ether, 25°C ← Solvent)

Both reaction formats are used in this book, and it is important that the different roles of chemicals shown above or below the reaction arrow be understood. The only way to be sure whether the indicated substance is a reagent or a solvent is to look carefully at the transformation itself.

For example, addition of HCl and HBr is straightforward:

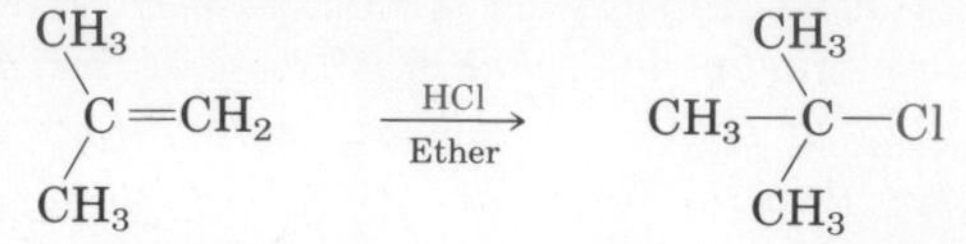

$$(CH_3)_2C{=}CH_2 \xrightarrow[\text{Ether}]{\text{HCl}} CH_3{-}C(CH_3)_2{-}Cl$$

2-Methylpropene → 2-Chloro-2-methylpropane (94%)

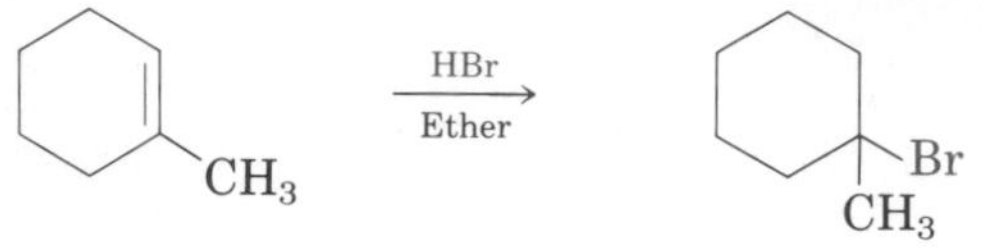

$$\text{1-Methylcyclohexene} \xrightarrow[\text{Ether}]{\text{HBr}} \text{1-Bromo-1-methylcyclohexane}$$

1-Methylcyclohexene → 1-Bromo-1-methylcyclohexane (91%)

Addition of HI to alkenes also occurs, but it's best to use a mixture of phosphoric acid and potassium iodide to generate HI in the reaction mixture, rather than to use HI directly. The overall mechanism is the same as for the other additions:

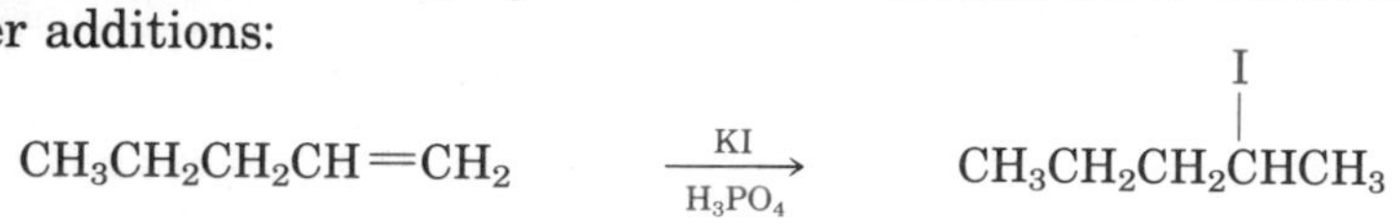

$$CH_3CH_2CH_2CH{=}CH_2 \xrightarrow[H_3PO_4]{KI} CH_3CH_2CH_2CHICH_3$$

1-Pentene (HI) 2-Iodopentane

## 6.10 Orientation of Electrophilic Addition: Markovnikov's Rule

Look carefully at the three reactions just shown. In all cases, an unsymmetrically substituted alkene has given a *single* addition product, rather than the mixture that might have been expected. For example, 2-methylpropene might have reacted with HCl to give 1-chloro-2-methylpropane (isobutyl chloride) in addition to 2-chloro-2-methylpropane, but it did not. We say that such reactions are **regiospecific** (pronounced **ree**-jee-oh-spe-cific), when only one of two possible directions of addition is observed.

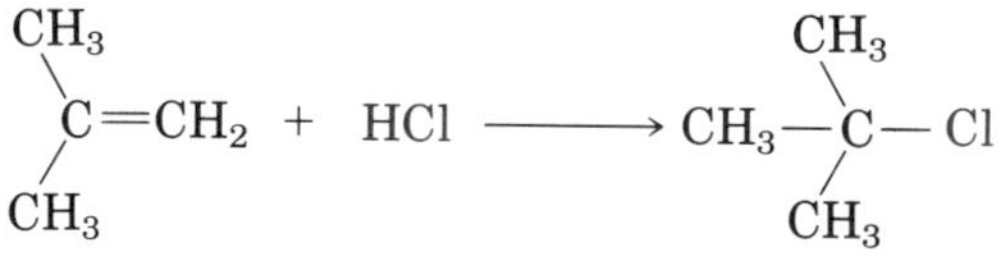

$$(CH_3)_2C{=}CH_2 + HCl \longrightarrow CH_3{-}C(CH_3)_2{-}Cl$$

2-Methylpropene

2-Chloro-2-methylpropane (sole product; a regiospecific reaction)

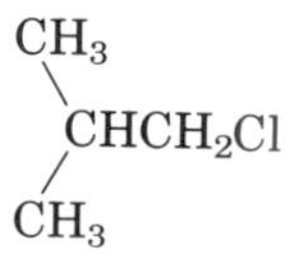

$$\left[ (CH_3)_2CHCH_2Cl \right]$$

1-Chloro-2-methylpropane (*not formed*)

By looking at the results of many such reactions and cataloging the results, the Russian chemist Vladimir Markovnikov[3] proposed in 1869 what has become known as **Markovnikov's rule**: *In the addition of HX to an alkene, the acid hydrogen becomes attached to the carbon with fewer alkyl substituents, and the X group becomes attached to the carbon with more alkyl substituents:*

$$(CH_3)_2C{=}CHCH_2CH_3 + HCl \xrightarrow{\text{Ether}} (CH_3)_2CClCH_2CH_2CH_3$$

2 alkyl groups on this carbon; 1 alkyl group on this carbon

2-Methyl-2-pentene → 2-Chloro-2-methylpentane

1-Methylcyclohexene (2 alkyl groups here; 1 alkyl group here) $\xrightarrow[H_3PO_4]{KI}$ 1-Iodo-1-methylcyclohexane

When both ends of the double bond have the same degree of substitution, a mixture of products results:

$$CH_3CH_2CH{=}CHCH_3 \xrightarrow[\text{Ether}]{HBr} CH_3CH_2CH_2CHBrCH_3 + CH_3CH_2CHBrCH_2CH_3$$

2-Pentene → 2-Bromopentane + 3-Bromopentane

Since carbocations are involved as intermediates in these reactions, another way to state Markovnikov's rule is to say that, in the addition of HX to an alkene, the more highly substituted carbocation is formed as an intermediate in preference to the less highly substituted one. For example, as shown on page 178, addition of $H^+$ to 2-methylpropene yields the intermediate *tertiary* carbocation rather than the primary carbocation, and addition to 1-methylcyclohexene yields a tertiary rather than a secondary cation. Why should this be so?

[3]Vladimir Vassilyevich Markovnikov (or Markownikoff) (1833–1904); b. Nijni-Novgorod, Russia; pupil of Butlerov, Erlenmeyer, Baeyer, and Kolbe; professor in Odessa (1871) and Moscow (1873).

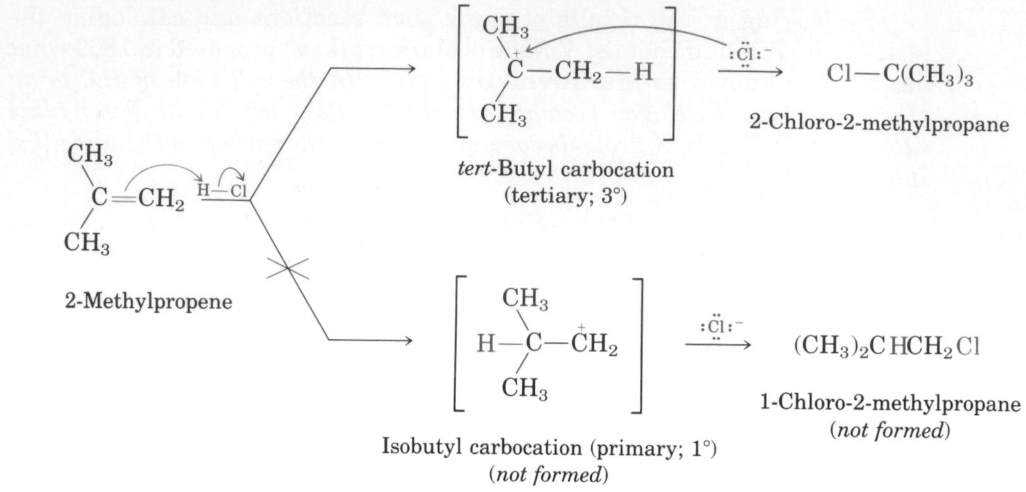
CH3
C=CH2
CH3
H—Cl
2-Methylpropene
CH3
C—CH2—H
CH3
tert-Butyl carbocation
(tertiary; 3°)
:Cl:−
Cl—C(CH3)3
2-Chloro-2-methylpropane
CH3
H—C—CH2
CH3
Isobutyl carbocation (primary; 1°)
(not formed)
:Cl:−
(CH3)2CHCH2Cl
1-Chloro-2-methylpropane
(not formed)

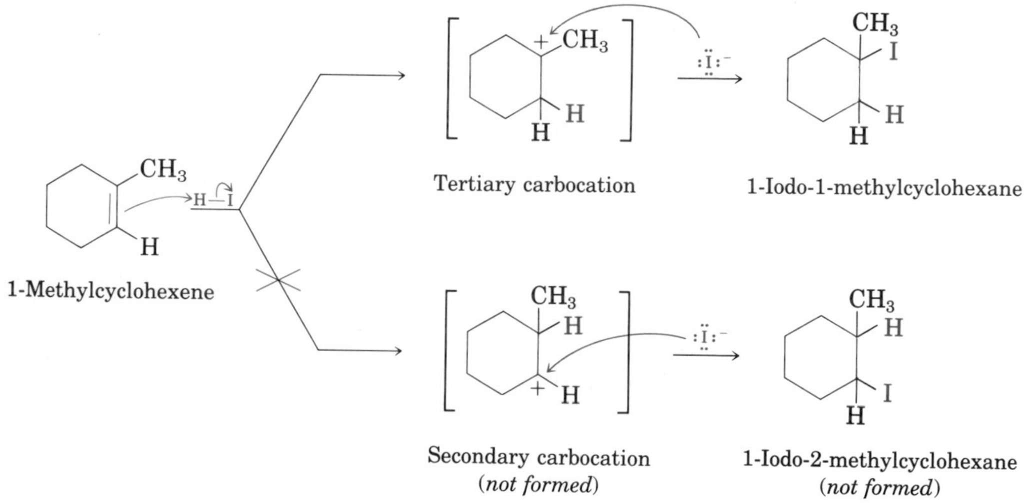
CH3
H—I
H
1-Methylcyclohexene
CH3
H
H
Tertiary carbocation
:I:−
CH3
I
H
H
1-Iodo-1-methylcyclohexane
CH3
H
H
Secondary carbocation
(not formed)
:I:−
CH3
H
I
H
1-Iodo-2-methylcyclohexane
(not formed)

PROBLEM

**6.14** Predict the products of these reactions:

(a) 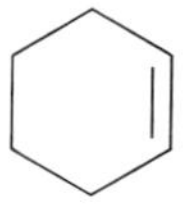 + HCl $\longrightarrow$

(b) $(CH_3)_2C{=}CHCH_2CH_3 \xrightarrow{HBr}$

(c) $CH_3CH_2CH_2CH{=}CH_2 \xrightarrow[KI]{H_3PO_4}$

(d) 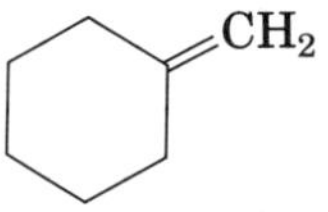
CH2
+ HBr $\longrightarrow$

PROBLEM

**6.15** What alkenes would you start with to prepare these alkyl halides?

(a) Bromocyclopentane

(b) $CH_3CH_2CHBrCH_2CH_2CH_3$

(c) 1-Iodo-1-ethylcyclohexane

## 6.11 Carbocation Structure and Stability

To understand the reasons for the Markovnikov orientation of electrophilic addition reactions, we need to learn more about the structure and stability of carbocations and about the general nature of reactions and transition states. The first point to explore involves structure.

A great deal of evidence points to the conclusion that carbocations are *planar*; the carbon is $sp^2$ hybridized and the three substituents are oriented to the corners of an equilateral triangle, as indicated in Figure 6.12. Since there are only six valence electrons on carbon, and since all six are used in the three sigma bonds, the $p$ orbital extending above and below the plane is unoccupied. [Note the electronic similarity of carbocations to trivalent boron compounds such as $BF_3$ (Section 1.13).]

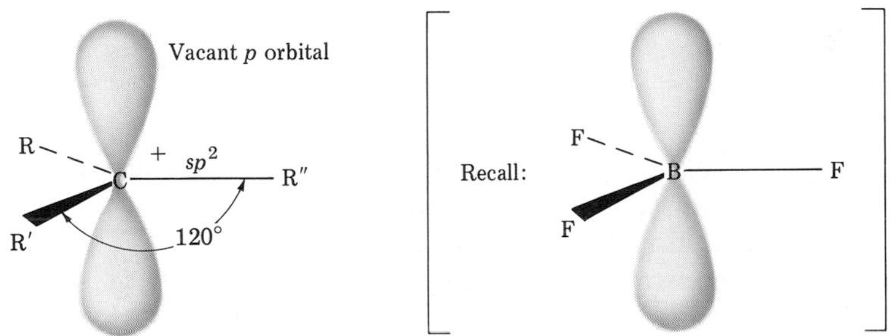


**Figure 6.12** The electronic structure of a carbocation: The carbon is $sp^2$ hybridized and has a vacant $p$ orbital.

The second point to explore involves carbocation stability. 2-Methylpropene might react with HCl to form a carbocation having three alkyl substituents (a tertiary ion, 3°), or it might react to form a carbocation having one alkyl substituent (a primary ion, 1°). Since the tertiary chloride, 2-chloro-2-methylpropane, is the only product observed, formation of the tertiary cation is evidently favored over formation of the primary cation. Thermodynamic measurements show that, indeed, the stability of carbocations increases with increasing substitution: More highly substituted carbocations are more stable than less highly substituted ones.

One way of determining carbocation stabilities is to measure the amount of energy required to form a carbocation from its corresponding alkyl halide:

$$R{-}X \longrightarrow R^+ + :X^-$$

Tertiary halides ionize to give carbocations much more readily than do primary ones (see Table 6.5, page 180).

**Table 6.5** Enthalpies of ionization[a] for alkyl halides in the gas phase: $R{-}Cl \rightarrow R^+ + :\ddot{Cl}:^-$

| Type | Reaction | ΔH° (kcal/mol) | ΔH° (kJ/mol) |
|---|---|---|---|
| Methyl | $CH_3Cl \longrightarrow CH_3^+ + :\ddot{Cl}:^-$ | 227 | 950 |
| Primary | $CH_3CH_2Cl \longrightarrow CH_3CH_2^+ + :\ddot{Cl}:^-$ | 195 | 816 |
| Secondary | $(CH_3)_2CHCl \longrightarrow (CH_3)_2CH^+ + :\ddot{Cl}:^-$ | 173 | 724 |
| Tertiary | $(CH_3)_3CCl \longrightarrow (CH_3)_3C^+ + :\ddot{Cl}:^-$ | 157 | 657 |

[a]Enthalpies are calculated in the following way:

$$
\begin{array}{lllll}
 & CH_3{-}Cl & \longrightarrow CH_3\cdot + Cl\cdot & \Delta H^\circ = 84 & \text{kcal/mol bond strength} \\
 & CH_3\cdot & \longrightarrow CH_3^+ + e^- & \Delta H^\circ = 226.7 & \text{kcal/mol ionization energy} \\
 & Cl\cdot + e^- & \longrightarrow Cl^- & \Delta H^\circ = -83.2 & \text{kcal/mol electron affinity} \\
\hline
\text{Net:} & CH_3{-}Cl & \longrightarrow CH_3^+ + Cl^- & \Delta H^\circ = 227 & \text{kcal/mol net ionization enthalpy}
\end{array}
$$

As Table 6.5 shows, there are large differences in the gas-phase stabilities of substituted carbocations. Trisubstituted (tertiary, 3°) carbocations are more stable than disubstituted (secondary, 2°) ones, which are more stable than monosubstituted (primary, 1°) ones:

Tertiary (3°) > Secondary (2°) > Primary (1°) > Methyl

$$(CH_3)_3C^+ > (CH_3)_2CH^+ > CH_3CH_2^+ > CH_3^+$$

Although the data in Table 6.5 are taken from measurements made in the gas phase, a similar carbocation stability order is found in solution. The values for ionization are much lower in solution since polar solvents can stabilize the ions, but the order of carbocation stability remains the same.

Why are more highly substituted carbocations more stable than less highly substituted ones? Most chemists think there are at least two reasons. Part of the answer has to do with hyperconjugation and part has to do with inductive effects. Hyperconjugation, discussed earlier (Section 6.7) in connection with the stability order of substituted alkenes, is the overlap of a *p* orbital and a neighboring C—H sigma bond orbital. In the present situation, hyperconjugation between the vacant carbocation *p* orbital and a neighboring C—H sigma bond acts to lower the energy level of the carbocation (Figure 6.13).

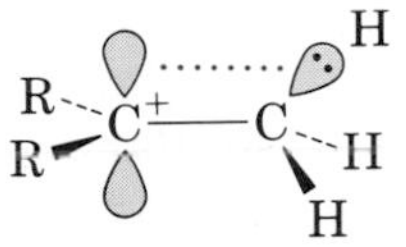

R R C+ C H H H

**Figure 6.13** Stabilization of a carbocation through hyperconjugation: The sigma electrons in the neighboring C—H bond help stabilize the positive charge.

The effect of hyperconjugation in carbocations is to allow neighboring sigma bond electrons to stabilize the positive charge by spreading the charge out, or *delocalizing* it, over a greater volume of space. We'll see repeatedly in later chapters that delocalizing charge over a greater volume invariably leads to greater stability. In the present instance, the more alkyl groups that are present on the carbocation, the more possibilities there are for hyperconjugation, and the more stable the carbocation—in other words, tertiary > secondary > primary > methyl.

Inductive effects, discussed earlier (Section 2.4) in connection with polar covalent bonds, result from the shifting of electrons in a bond in response to the electronegativity of a nearby atom. In the present instance, electrons from a relatively large and polarizable alkyl group can shift toward a neighboring positive charge more easily than electrons from an attached hydrogen can. Thus, the more alkyl groups there are attached to the positively charged carbon, the more electron density shifts toward the charge and the more inductive stabilization of the cation there is.

## 6.12 The Hammond Postulate

To summarize our knowledge of electrophilic addition reactions up to this point, we know two facts:

1. We know that electrophilic addition reactions to unsymmetrical alkenes involve the more highly substituted carbocation. A more highly substituted carbocation evidently forms faster than a less highly substituted one, and, once formed, rapidly goes on to give the final product.
2. We know that more highly substituted carbocations are more stable than less highly substituted ones. That is, the stability order of carbocations is tertiary > secondary > primary > methyl.

What we haven't yet seen is how these two pieces of information are related. Why is it that the *stability* of the carbocation intermediate determines the *rate* at which the carbocation is formed and thereby determines the structure of the final product? After all, carbocation stability is determined by $\Delta G^\circ$, but reaction rate is determined by $\Delta G^\ddagger$ (activation energy). The two quantities aren't directly related.

Although there is no precise thermodynamic relationship between the stability of a high-energy carbocation intermediate and the rate of its formation, there *is* an intuitive relationship between the two. It is generally true when comparing two similar reactions that the more stable intermediate usually forms faster than the less stable one. The situation is shown graphically in Figure 6.14, where the reaction energy profile shown in part (a) represents the usual situation rather than the profile shown in part (b). That is, the two curves remain on parallel courses rather than crossing over each other.

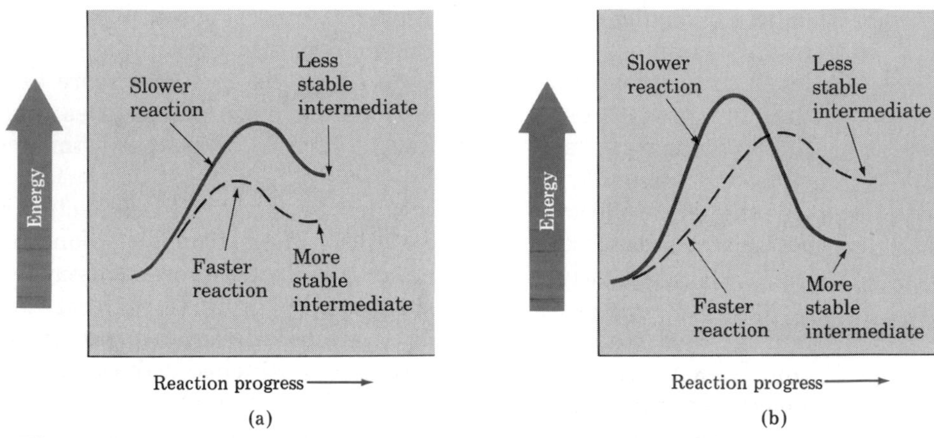


**Figure 6.14** Reaction energy diagrams for two similar competing reactions: In (a), the faster reaction yields the more stable intermediate. In (b), the slower reaction yields the more stable intermediate. The curve shown in (a) represents the usual situation.

An explanation of the relationship between reaction rate and intermediate stability was first advanced in 1955. Known as the **Hammond postulate**,[4] this explanation isn't a thermodynamic law; it is simply a reasonable account of observed facts. It intuitively links reaction rate and intermediate stability by looking at the energy level and structure of the transition state.

Transition states represent energy maxima. They are high-energy activated complexes that occur transiently during the course of a reaction and that immediately go on to a more stable species. Although we can't actually *observe* transition states, because they have no finite lifetime, the Hammond postulate says that we can get an *idea* of the structure of a particular transition state by looking at the structure of the nearest stable species. In terms of reaction energy diagrams, we can imagine the two cases shown in Figure 6.15. The reaction profile in part (a) shows the energy curve for an endothermic reaction step, and the profile in part (b) shows the curve for an exothermic step.

[4]George Simms Hammond (1921– ); b. Auburn, Maine; Ph.D. (1947), Harvard University; professor, Iowa State University; California Institute of Technology; University of California, Santa Cruz; Allied Chemical Company.

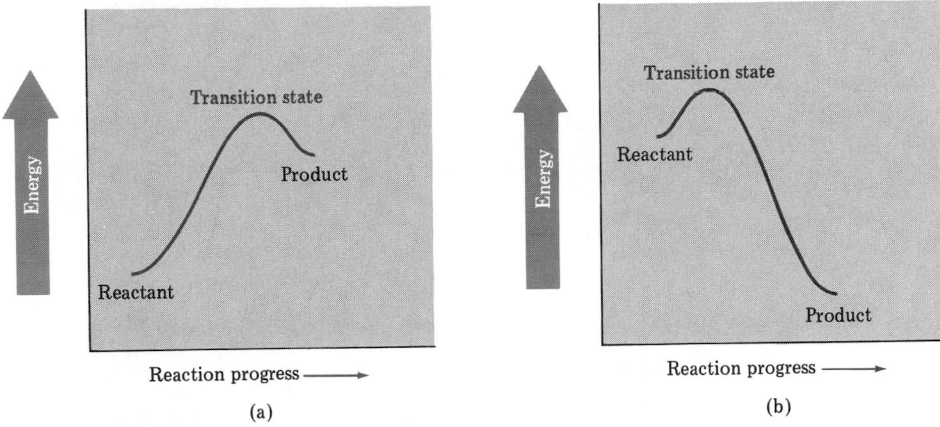


**Figure 6.15** Reaction energy diagrams for endothermic and exothermic steps: (a) In the endothermic step, the energy levels of transition state and *product* are similar. (b) In the exothermic step, the energy levels of transition state and *reactant* are similar.

In the endothermic reaction in Figure 6.15(a), the energy level of the transition state is much closer to that of the product than to that of the reactant. Since the transition state is closer *energetically* to the product, we make the natural assumption that it is also closer *structurally*. In other words, we can say that *the transition state for an endothermic reaction step structurally resembles the product.* Conversely, the transition state for the exothermic reaction in part (b) is much closer energetically to the reactant than to the product, and we say that *the transition state for an exothermic reaction step structurally resembles the reactant.*

How does the Hammond postulate apply to electrophilic additions to alkenes? We know that the formation of a carbocation by protonation of an alkene is an endothermic step. Therefore, the transition state for alkene protonation should structurally resemble the carbocation intermediate. Any factor that makes the carbocation product more stable should also make the nearby transition state more stable. Since increasing alkyl substitution stabilizes carbocations, it also stabilizes the transition states leading to those ions, thus resulting in faster reaction. More highly substituted carbocations form faster because their stability is reflected in the transition state that forms them. A hypothetical transition state for alkene protonation might be expected to look like that shown in Figure 6.16 on page 184.

Since the transition state for alkene protonation shown in Figure 6.16 resembles the carbocation product, we can imagine it to be a structure in which one of the alkene carbon atoms has almost completely rehybridized from $sp^2$ to $sp^3$ and in which the remaining alkene carbon bears a substantial portion of the positive charge. The positive charge in this transition state is delocalized and stabilized by hyperconjugation in the same way that the product carbocation is stabilized. The more alkyl groups that are present, the greater the extent of charge stabilization in the transition state and the faster it forms. Figure 6.17 summarizes the situation by showing competing reaction energy profiles for the reaction of 2-methylpropene with HCl.

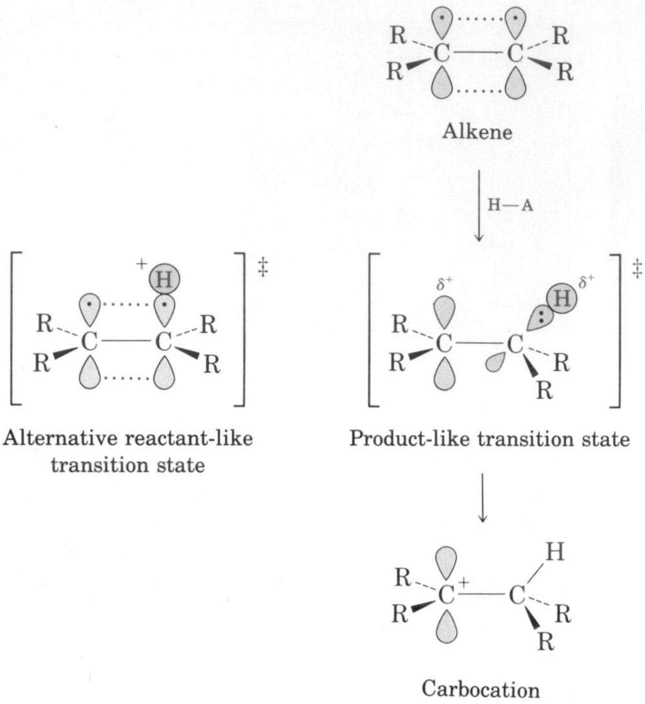

Alkene

H—A

Alternative reactant-like transition state

Product-like transition state

Carbocation

**Figure 6.16** The structure of a hypothetical transition state for alkene protonation: The transition state is closer in both energy and structure to the carbocation than to the reactant. Thus, an increase in carbocation stability (lower $\Delta G°$) also causes an increase in transition-state stability (lower $\Delta G^{\ddagger}$).

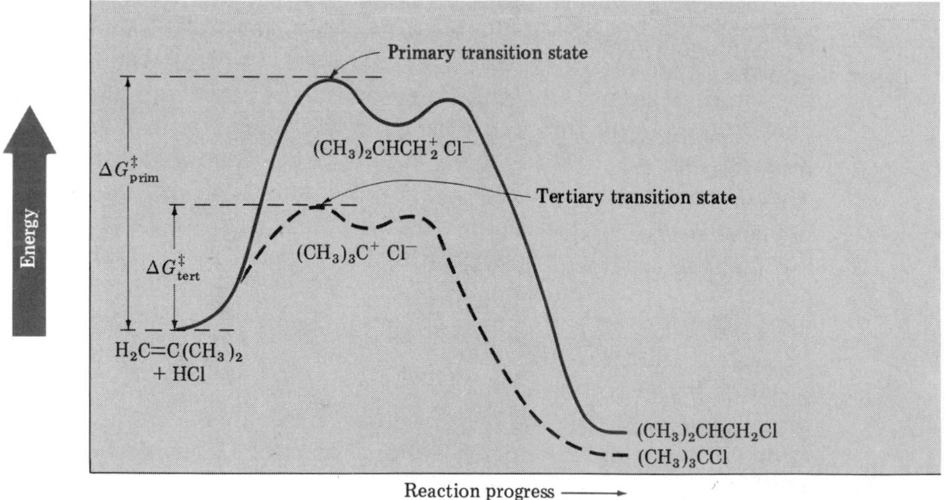


**Figure 6.17** A reaction energy diagram for the electrophilic addition of HCl to 2-methylpropene: The tertiary cation forms faster than the primary cation intermediate because it is more stable. The same factors that make the tertiary cation more stable also make the transition state leading to it more stable.

PROBLEM

**6.16** What about the second step in the electrophilic addition of HCl to an alkene—the reaction of chloride ion with the carbocation intermediate? Is this step exothermic or endothermic? According to the Hammond postulate, should the transition state for this second step resemble the reactant (carbocation) or product (chloroalkane)? Make a rough drawing of what you would expect the transition-state structure to look like.

## 6.13 Mechanistic Evidence: Carbocation Rearrangements

How do we know that the carbocation mechanism for addition of HX to alkenes is correct? The answer is that we *don't* know with absolute certainty. Although *incorrect* reaction mechanisms can be *disproved* by demonstrations that they don't satisfactorily account for observed data, correct reaction mechanisms can never be entirely proven. The best we can do is to show that a proposed mechanism is fully consistent with all known facts. If enough data are satisfactorily explained, then the mechanism is probably correct.

What evidence is there to support the two-step carbocation mechanism we've proposed for the reaction of HX with alkenes? How do we know that the two reactants, HX and alkene, don't simply come together in a single step to give the final product without going through a carbocation intermediate? One of the best pieces of evidence for a carbocation mechanism was discovered during the 1930s by F. C. Whitmore,[5] who found that structural **rearrangements** often occur during the reaction of HX with an alkene. For example, reaction of HCl with 3-methyl-1-butene yields a substantial amount of 2-chloro-2-methylbutane in addition to the "expected" product, 2-chloro-3-methylbutane:

$$\underset{\text{3-Methyl-1-butene}}{CH_3\overset{\overset{CH_3}{|}}{C}HCH{=}CH_2} + HCl \longrightarrow \underset{\substack{\text{2-Chloro-3-methylbutane}\\(\sim 50\%)}}{CH_3\overset{\overset{CH_3}{|}}{C}H\underset{\underset{Cl}{|}}{C}HCH_3} + \underset{\substack{\text{2-Chloro-2-methylbutane}\\(\sim 50\%)}}{CH_3\underset{\underset{Cl}{|}}{\overset{\overset{CH_3}{|}}{C}}CH_2CH_3}$$

How can the unexpected formation of 2-chloro-2-methylbutane be explained? If the reaction takes place in a *single* step, it would be hard to account for rearrangement, but, if the reaction takes place in *two* steps, rearrangement is more easily explained. Whitmore suggested that it is a carbocation intermediate that is undergoing rearrangement. For example, the secondary carbocation intermediate formed by protonation of 3-methyl-1-butene can rearrange to a more stable tertiary carbocation by a **hydride**

[5]Frank C. Whitmore (1887–1947); b. North Attleboro, Mass.; Ph.D. Harvard (E. L. Jackson); professor, Pennsylvania State University.

**shift**—the shift of a hydrogen atom and its electron pair (a **hydride ion, :H⁻**) from C2 to C1.

$$CH_3-\overset{\overset{H}{|}}{\underset{\underset{CH_3}{|}}{C}}-CH=CH_2 \xrightarrow{H-Cl} \left[ CH_3-\overset{\overset{H}{|}}{\underset{\underset{CH_3}{|}}{C}}-\overset{+}{C}H-\overset{\overset{H}{|}}{C}H_2 \underset{\text{shift}}{\overset{\text{Hydride}}{\rightleftarrows}} CH_3-\overset{+}{\underset{\underset{CH_3}{|}}{C}}-\overset{\overset{H}{|}}{C}H-CH_3 \right]$$

A 2° carbocation; A 3° carbocation

$$\downarrow :\ddot{Cl}:^- \qquad \qquad \downarrow :\ddot{Cl}:^-$$

$$CH_3-\overset{\overset{H}{|}}{\underset{\underset{CH_3}{|}}{C}}-\overset{\overset{Cl}{|}}{C}H-CH_3 \qquad CH_3-\overset{\overset{Cl}{|}}{\underset{\underset{CH_3}{|}}{C}}-\overset{\overset{H}{|}}{C}H-CH_3$$

Carbocation rearrangements can also occur by the shift of an *alkyl group* with its electron pair. For example, reaction of 3,3-dimethyl-1-butene with HCl leads to an equal mixture of 2-chloro-3,3-dimethylbutane and rearranged 2-chloro-2,3-dimethylbutane. In this instance, a secondary carbocation rearranges to a more stable tertiary carbocation by the shift of a methyl group:

$$CH_3-\overset{\overset{CH_3}{|}}{\underset{\underset{CH_3}{|}}{C}}-CH=CH_2 \xrightarrow{H-Cl} \left[ CH_3-\overset{\overset{CH_3}{|}}{\underset{\underset{CH_3}{|}}{C}}-\overset{+}{C}H-\overset{\overset{H}{|}}{C}H_2 \underset{\text{shift}}{\overset{\text{Methyl}}{\rightleftarrows}} CH_3-\overset{+}{\underset{\underset{CH_3}{|}}{C}}-\overset{\overset{CH_3}{|}}{C}H-CH_3 \right]$$

3,3-Dimethyl-1-butene; A 2° carbocation; A 3° carbocation

$$\downarrow :\ddot{Cl}:^- \qquad \qquad \downarrow :\ddot{Cl}:^-$$

$$CH_3-\overset{\overset{H_3C}{|}}{\underset{\underset{CH_3}{|}}{C}}-\overset{\overset{Cl}{|}}{C}H-CH_3 \qquad CH_3-\overset{\overset{Cl}{|}}{\underset{\underset{CH_3}{|}}{C}}-\overset{\overset{CH_3}{|}}{C}H-CH_3$$

2-Chloro-3,3-dimethylbutane (~50%); 2-Chloro-2,3-dimethylbutane (~50%)

Note the similarities of these two carbocation rearrangements. In both cases, a group ($:H^-$ or $^-:CH_3$) is moving to a positively charged carbon, *taking its electron pair with it*. In both cases also, a less stable carbocation is rearranging to a more stable ion. Rearrangements of the sort just shown are a common feature of carbocation chemistry. We'll see at numerous places in future chapters that their occurrence in a reaction provides strong mechanistic evidence for the presence of carbocation intermediates.

PROBLEM

**6.17** Propose a mechanism to account for the observation that reaction of vinylcyclohexane with HBr yields 1-bromo-1-ethylcyclohexane as the major product.

$$\text{Vinylcyclohexane} + HBr \longrightarrow \text{1-Bromo-1-ethylcyclohexane } (C_6H_{10}(CH_2CH_3)Br)$$

Vinylcyclohexane 1-Bromo-1-ethylcyclohexane

## 6.14 Summary and Key Words

**Alkenes** are hydrocarbons that contain one or more carbon–carbon double bonds. Because they contain fewer hydrogens than related alkanes, alkenes are often referred to as being **unsaturated**. A double bond has two parts: a sigma bond formed by head-on overlap of two $sp^2$ orbitals, and a pi bond formed by sideways overlap of two $p$ orbitals. The overall bond strength of an alkene double bond is greater than that of a carbon–carbon single bond, with the pi bond part estimated at 64 kcal/mol (268 kJ/mol).

Alkenes are named by IUPAC rules using the suffix *-ene*. Because rotation around the double bond is restricted, substituted alkenes can exist as a pair of **cis–trans stereoisomers**. The geometry of a double bond can be specified by application of the Cahn–Ingold–Prelog **sequence rules**, which assign priorities to double-bond substituents. If the high-priority groups on each carbon are on the same side of the double bond, the geometry is ***Z*** (*zusammen*, "together"); if the high-priority groups on each carbon are on opposite sides of the double bond, the geometry is ***E*** (*entgegen*, "apart"). The stability order of alkyl-substituted double bonds is:

$$\begin{array}{ccccccc} \text{Tetrasubstituted} & > & \text{Trisubstituted} & > & \text{Disubstituted} & > & \text{Monosubstituted} \\ R_2C{=}CR_2 & > & R_2C{=}CHR & > & RCH{=}CHR \approx R_2C{=}CH_2 & > & RCH{=}CH_2 \end{array}$$

The chemistry of alkenes is dominated by **electrophilic addition reactions**. When HX reacts with an unsymmetrically substituted alkene, **Markovnikov's rule** predicts that the hydrogen will add to the carbon having fewer alkyl substituents and that the X group will add to the carbon having more alkyl substituents. Electrophilic additions to alkenes take place through **carbocation** intermediates formed by reaction of the nucleophilic alkene pi bond with electrophilic HA. Carbocation stability follows the order

$$\begin{array}{ccccccc} \text{Tertiary (3°)} & > & \text{Secondary (2°)} & > & \text{Primary (1°)} & > & \text{Methyl} \\ R_3C^+ & > & R_2CH^+ & > & RCH_2^+ & > & CH_3^+ \end{array}$$

Markovnikov's rule can be restated by saying that, in the addition of HX to an alkene, the more stable carbocation intermediate is formed.

Although there is no mathematical reason why a more stable intermediate must be formed faster than a less stable one, the **Hammond postulate** provides an intuitive explanation. According to the Hammond postulate, the transition state of an exothermic reaction step structurally resembles the reactant, whereas the transition state of an endothermic reaction step structurally resembles the product. Since an alkene protonation step is endothermic, the stability of the more highly substituted carbocation is reflected in the stability of the transition state leading to its formation.

One of the best pieces of evidence in support of a carbocation mechanism for electrophilic addition reactions is the observation of structural **rearrangements** that sometimes occur during reaction. Rearrangements occur by shift of either a hydride ion, $:H^-$ (a **hydride shift**), or an alkyl group anion, $:R^-$, from a neighboring carbon atom to the positively charged carbon. The net result is isomerization of a less stable carbocation to a more stable one.

## ADDITIONAL PROBLEMS

**6.18** Calculate the number of double bonds and/or rings in these formulas:

(a) Benzene, $C_6H_6$ (b) Cyclohexene, $C_6H_{10}$ (c) Myrcene (bay oil), $C_{10}H_{16}$
(d) Lindane, $C_6H_6Cl_6$ (e) Pyridine, $C_5H_5N$ (f) Safrole, $C_{10}H_{10}O_2$

**6.19** Calculate the number of pi bonds and/or rings in these formulas and then draw five possible structures for each:

(a) $C_{10}H_{16}$ (b) $C_8H_8O$ (c) $C_7H_{10}Cl_2$
(d) $C_{10}H_{16}O_2$ (e) $C_5H_9NO_2$ (f) $C_8H_{10}ClNO$

**6.20** A compound of formula $C_{10}H_{14}$ undergoes catalytic hydrogenation but absorbs only two equivalents of hydrogen. How many rings does the compound have?

**6.21** Provide IUPAC names for these alkenes:

(a) $CH_3CH{=}CH\overset{\overset{\displaystyle CH_3}{|}}{C}HCH_2CH_3$

(b) $CH_3CH{=}CH\overset{\overset{\displaystyle CH_2CH_2CH_2CH_3}{|}}{C}HCH_2CH_2\underset{\underset{\displaystyle CH_3}{|}}{C}HCH_3$

(c) $H_2C{=}C(CH_2CH_3)_2$

(d) $H_2C{=}CH\overset{\overset{\displaystyle CH_3}{|}}{C}H\underset{\underset{\displaystyle CH_3}{|}}{C}HCH{=}CHCH_3$

(e) $CH_3CH_2\overset{\overset{\displaystyle CH_3}{|}}{C}{=}CHCH{=}CH_2$

(f) $H_2C{=}C{=}CHCH_3$

(g) $H_2C{=}CHC(CH_3)_3$

(h) $(CH_3)_3CCH{=}CHC(CH_3)_3$

**6.22** The triene ocimene is found in the essential oil of many plants. What is its correct IUPAC name?

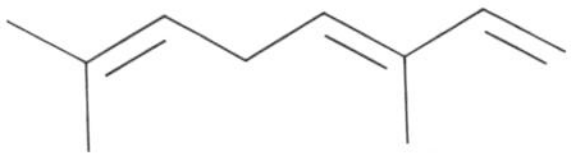

Ocimene

**6.23** α-Farnesene is a constituent of the natural wax found on apples. What is its IUPAC name?

α-Farnesene

**6.24** Draw structures corresponding to these systematic names:
(a) 2,4-Dimethyl-1,4-hexadiene
(b) 3,3-Dimethyl-4-propyl-1,5-octadiene
(c) 4-Methyl-1,2-pentadiene
(d) 2,6-Dimethyl-1,3,5,7-octatetraene
(e) 3-Butyl-2-heptene
(f) 2,2,5,5-Tetramethyl-3-hexene

**6.25** These names are incorrect. Draw structures and provide correct IUPAC names:
(a) 2-Methyl-2,4-pentadiene
(b) 3-Methylene-1-pentene
(c) 3,6-Octadiene
(d) 5-Ethyl-4-octene
(e) 3-Propyl-3-heptene
(f) 3-Vinyl-1-propene

**6.26** Draw and name the five possible pentene isomers, $C_5H_{10}$. Ignore *E,Z* isomers.

**6.27** Draw and name the 13 possible hexene isomers, $C_6H_{12}$. Ignore *E,Z* isomers.

**6.28** According to heat-of-hydrogenation data, *trans*-2-butene is more stable than *cis*-2-butene by 1 kcal/mol. Hydrogenation measurements also show that *trans*-2,2,5,5-tetramethyl-3-hexene is more stable than its cis isomer by 9.3 kcal/mol. Explain this large difference.

| | | $\Delta H^\circ_{hydrog}$ | |
|---|---|---|---|
| | | *(kcal/mol)* | *(kJ/mol)* |
| Cis | $CH_3CH{=}CHCH_3$ | 28.6 | 120 |
| Trans | $CH_3CH{=}CHCH_3$ | 27.6 | 116 |
| Cis | $(CH_3)_3CCH{=}CHC(CH_3)_3$ | 36.2 | 151 |
| Trans | $(CH_3)_3CCH{=}CHC(CH_3)_3$ | 26.9 | 112 |

**6.29** Allene (1,2-propadiene), $H_2C{=}C{=}CH_2$, has two adjacent double bonds. What kind of hybridization must the central carbon have? Sketch the bonding orbitals in allene. What shape do you predict for allene?

**6.30** 1,4-Pentadiene, a compound with two nonadjacent double bonds, has $\Delta H^\circ_{hydrog}$ = 60.8 kcal/mol, which is, as expected, approximately twice the value for propene ($\Delta H^\circ_{hydrog}$ = 30.1 kcal/mol). However, 1,2-propadiene has $\Delta H^\circ_{hydrog}$ = 71.3 kcal/mol. What does this tell you about the stability of 1,2-propadiene? What explanation might there be for these results?

**6.31** Predict the major product in each of these reactions:

(a) $CH_3CH_2CH{=}C(CH_3)CH_2CH_3$ + HCl ⟶
(b) 1-Ethylcyclopentene + HBr ⟶
(c) 2,2,4-Trimethyl-3-hexene + HI ⟶
(d) 1,6-Heptadiene + 2 HCl ⟶
(e) 3-methylcyclohexene ($CH_3$) + HBr ⟶

**6.32** Rank the following sets of substituents in order of priority according to the Cahn–Ingold–Prelog sequence rules:

(a) $—CH_3$, $—Br$, $—H$, $—I$

(b) $—OH$, $—OCH_3$, $—H$, $—COOH$

(c) $—COOH$, $—COOCH_3$, $—CH_2OH$, $—CH_3$

(d) $—CH_3$, $—CH_2CH_3$, $—CH_2CH_2OH$, $—\overset{O}{\overset{\|}{C}}CH_3$

(e) $—CH{=}CH_2$, $—CN$, $—CH_2NH_2$, $—CH_2Br$

(f) $—CH{=}CH_2$, $—CH_2CH_3$, $—CH_2OCH_3$, $—CH_2OH$

**6.33** Assign *E* or *Z* configuration to the following alkenes:

(a) 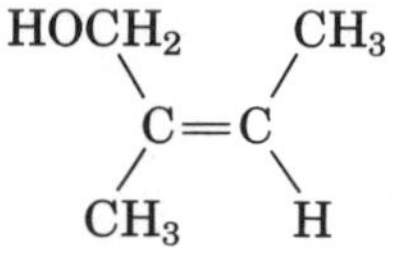 $HOCH_2$ CH$_3$ / C=C / $CH_3$ H

(b) HOOC H / C=C / Cl $OCH_3$

(c) NC $CH_3$ / C=C / $CH_3CH_2$ $CH_2OH$

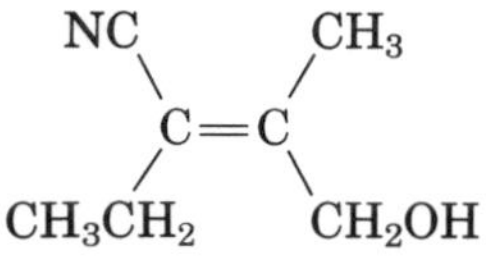

(d) 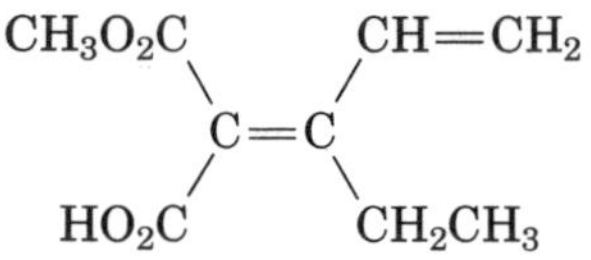 $CH_3O_2C$ $CH{=}CH_2$ / C=C / $HO_2C$ $CH_2CH_3$

**6.34** Name these cycloalkenes according to IUPAC rules:

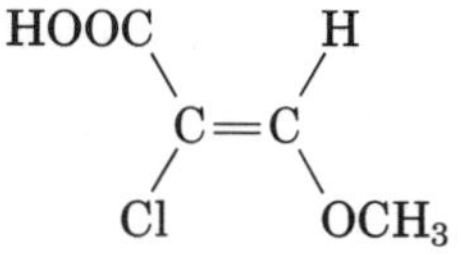

(a) 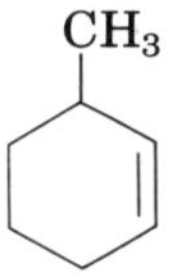 $CH_3$

(b)

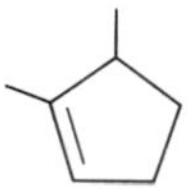

(c)

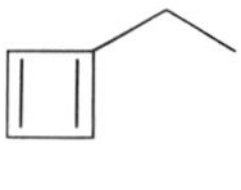

(d)

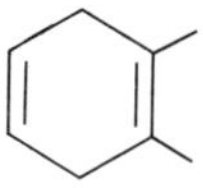

(e)

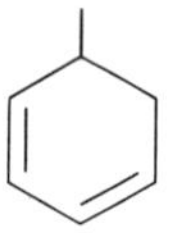

(f)

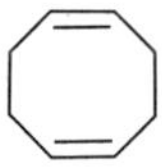

**6.35** Which of the given *E*,*Z* designations are correct, and which are incorrect?

(a) 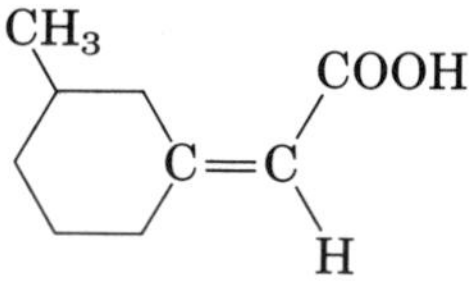 $CH_3$ COOH / C=C / H

*Z*

(b) 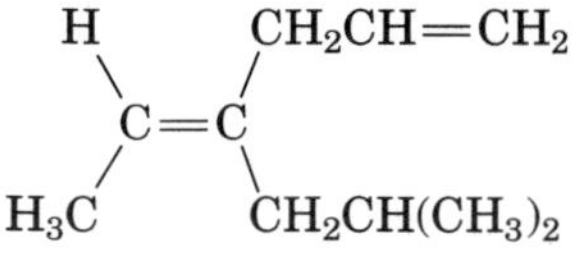 H $CH_2CH{=}CH_2$ / C=C / $H_3C$ $CH_2CH(CH_3)_2$

*E*

(c) 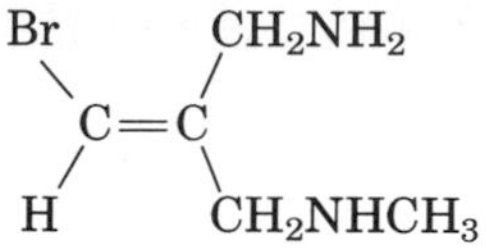 Br, H / C=C / $CH_2NH_2$, $CH_2NHCH_3$

*Z*

(d) 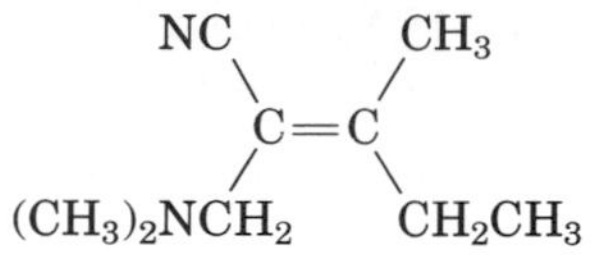 NC, $(CH_3)_2NCH_2$ / C=C / $CH_3$, $CH_2CH_3$

*E*

(e) 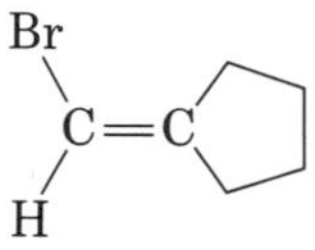 Br, H / C=C (cyclopentylidene)

*Z*

(f) 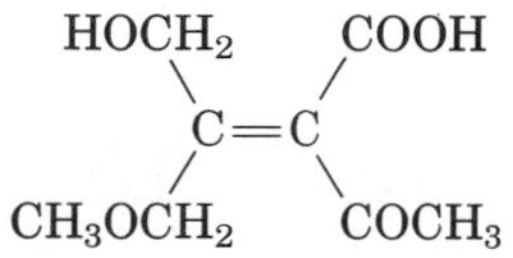 $HOCH_2$, $CH_3OCH_2$ / C=C / COOH, $COCH_3$

*E*

**6.36** Calculate the degree of unsaturation of these formulas:
(a) Cholesterol, $C_{27}H_{46}O$
(b) DDT, $C_{14}H_9Cl_5$
(c) Prostaglandin $E_1$, $C_{20}H_{34}O_5$
(d) Caffeine, $C_8H_{10}N_4O_2$
(e) Cortisone, $C_{21}H_{28}O_5$
(f) Atropine, $C_{17}H_{23}NO_3$

**6.37** Draw a reaction energy diagram for the addition of HBr to 1-pentene. Let one curve on your diagram show the formation of 1-bromopentane product and another curve on the same diagram show the formation of 2-bromopentane product. Label the positions for all reactants, intermediates, and products. Which curve has the higher-energy carbocation intermediate? Which curve has the higher first transition state?

**6.38** Make sketches of the transition-state structures involved in the reaction of HBr with 1-pentene (Problem 6.37). Identify each structure as resembling either starting material or product.

CHAPTER 7

# Alkenes: Reactions and Synthesis

We saw in the preceding chapter that the addition of electrophiles is one of the most important reactions of alkenes. Although we've studied only the addition of HX thus far, many other electrophilic reagents also add to alkenes.

## 7.1 Addition of Halogens to Alkenes

Bromine and chlorine both add readily to alkenes to yield 1,2-dihaloalkanes. For example, more than 5 million tons per year of 1,2-dichloroethane (ethylene dichloride) are synthesized industrially by the addition of $Cl_2$ to ethylene. The product is used both as a solvent and as starting material for use in the manufacture of poly(vinyl chloride), PVC.

$$H_2C{=}CH_2 \xrightarrow{Cl_2} H_2\overset{\displaystyle Cl}{\overset{|}{C}}{-}\overset{\displaystyle Cl}{\overset{|}{C}}H_2$$

Ethylene → 1,2-Dichloroethane (Ethylene dichloride)

The addition of bromine to an alkene also serves as a simple and rapid laboratory test for unsaturation. A sample of unknown structure is dissolved in tetrachloromethane (carbon tetrachloride, $CCl_4$) and placed in a test tube to which several drops of bromine in $CCl_4$ are added. Immediate disappearance of the reddish $Br_2$ color signals a positive test and indicates that the sample is an alkene.

Cyclopentene $\xrightarrow{Br_2 \text{ in } CCl_4}$ 1,2-Dibromocyclopentane (95%)

Fluorine tends to be too reactive and difficult to control for most laboratory applications, and iodine does not react with most alkenes.

A possible mechanism for the reaction of halogens with alkenes is shown in Figure 7.1. As a bromine molecule approaches a nucleophilic alkene, the Br—Br bond becomes polarized (Section 5.5). The pi electron pair of the alkene then attacks the positive end of the polarized bromine molecule, displacing bromide ion. The net result is that electrophilic $Br^+$ adds to the alkene in the same way that $H^+$ adds. The intermediate electrophilic carbocation then immediately reacts with nucleophilic bromide ion to yield the dibromo addition product.

The electron pair from the alkene pi bond attacks the positively polarized end of the bromine molecule, forming a C—Br sigma bond and causing the Br—Br bond to break. Bromide ion departs with both electrons from the Br—Br bond.

Bromide anion then uses an electron pair to attack the carbocation intermediate, forming a C—Br bond and yielding neutral dibromo addition product.

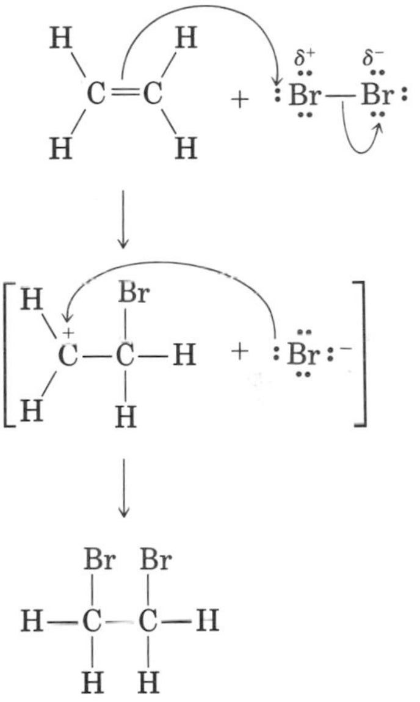

**Figure 7.1** A possible mechanism for the electrophilic addition of $Br_2$ to an alkene

Although the mechanistic description shown in Figure 7.1 for the addition of halogen to alkenes is consistent with what we've learned thus far, further examination shows that it is not *fully* consistent with known data. In particular, the proposed mechanism does not explain the *stereochemistry* of halogen addition. Let's look more closely at the addition reaction of bromine with the cyclic alkene cyclopentene to see what is meant by "stereochemistry of addition."

Let's assume that $Br^+$ adds to cyclopentene from the bottom face to form the carbocation intermediate shown in Figure 7.2. Since the positively charged carbon is planar and $sp^2$ hybridized, it could be attacked by bromide ion from either the top or the bottom to give a *mixture* of products. One product has the two bromine atoms on the same side of the ring (cis), whereas the other has the bromines on opposite sides (trans). We find, however, that only *trans*-1,2-dibromocyclopentane is produced. None of the cis product is formed.

Since the two bromine atoms add to opposite faces of the cyclopentene double bond, we say that the reaction occurs with **anti** stereochemistry. If the two bromines had added from the same face, the reaction would have had **syn** stereochemistry. Note that the word *anti* has a similar meaning in the present stereochemical context to the meaning it has in a butane conformational context (Section 4.3). In both cases, the two groups being referred to are 180° apart.

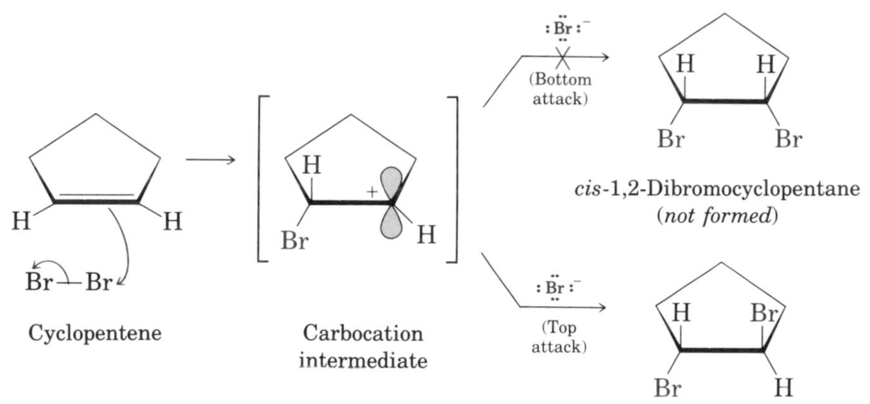


**Figure 7.2** Stereochemistry of the addition reaction of bromine with cyclopentene: Only the trans product is formed.

An explanation of the phenomenon of anti addition was suggested in 1937 by George Kimball and Irving Roberts, who postulated that the true reaction intermediate is not a carbocation, but a **bromonium ion**. A bromonium ion is a species containing a positively charged, divalent bromine, $R_2Br^+$. (A *chloronium ion*, similarly, contains a positively charged, divalent chlorine, $R_2Cl^+$.) In the present case, the bromonium ion is in a three-membered ring and is formed by the overlap of bromine lone-pair electrons with the vacant *p* orbital of the neighboring carbocation (Figure 7.3). Although Figure 7.3 depicts three-membered-ring bromonium ion formation as stepwise, this is done only for clarity. It's likely that the bromonium ion is formed in a single step by interaction of the alkene double-bond electrons with $Br^+$.

Alkene pi electrons attack bromine, pushing out bromide ion and leaving a bromo carbocation.

The neighboring bromo substituent stabilizes the positive charge by using two of its electrons to overlap the vacant carbon *p* orbital, giving a three-membered-ring bromonium ion.

Bromonium ion

**Figure 7.3** Formation of a bromonium ion intermediate by electrophilic addition of $Br^+$ to an alkene

How does bromonium ion formation account for the observed anti stereochemistry of addition to cyclopentene? If a bromonium ion is formed as an intermediate, we can imagine that the bromine atom might "shield" one face of the alkene double bond. Attack by bromide ion in the second step could then occur only from the opposite, unshielded face to give anti product.

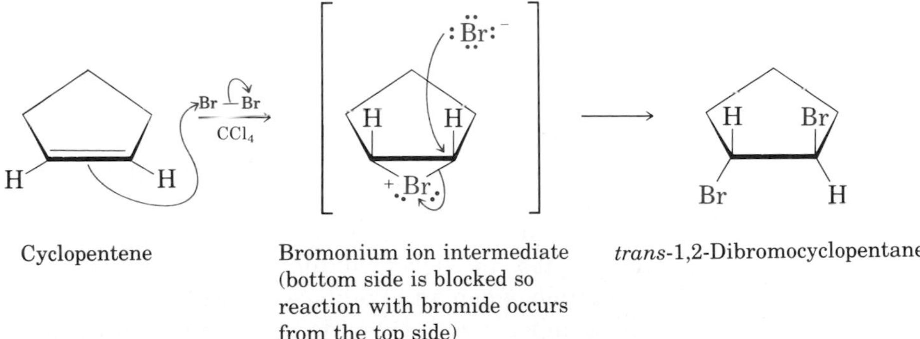


Cyclopentene

Bromonium ion intermediate (bottom side is blocked so reaction with bromide occurs from the top side)

*trans*-1,2-Dibromocyclopentane

The halonium ion postulate, made some 50 years ago to explain the stereochemistry of halogen addition to alkenes, is a remarkable example of the use of deductive logic in chemistry. Arguing from known experimental results, chemists were able to make a hypothesis about the intimate mechanistic details of alkene electrophilic reactions. Much more recently, strong evidence supporting the postulate has come from the work of George Olah, who has observed and studied specially prepared *stable* solutions of cyclic bromonium ions. Thus, it appears that bromonium ions are indeed a reality.

$$(CH_3)_2C(F)-CH(Br)CH_3 \xrightarrow[\text{Liquid } SO_2]{SbF_5} \text{cyclic bromonium ion } (H_3C)(CH_3)C-C(H)(CH_3)\ Br^+ \quad SbF_6^-$$

A bromonium ion stable in $SO_2$ solution

PROBLEM

**7.1** What product would you expect to obtain from addition of $Cl_2$ to 1,2-dimethylcyclohexene? Show the stereochemistry of the product.

PROBLEM

**7.2** Unlike the reaction in Problem 7.1, addition of HCl to 1,2-dimethylcyclohexene yields a mixture of two products. Show the stereochemistry of each and explain why a mixture is formed.

## 7.2 Halohydrin Formation

A great many different kinds of electrophilic additions to alkenes can take place. For example, alkenes can add HO—Cl or HO—Br under suitable conditions to yield 1,2-halo alcohols, or **halohydrins**. Halohydrin formation doesn't take place by direct reaction of an alkene with the reagents HOBr or HOCl. Rather, the addition is done indirectly by reaction of the alkene with either $Br_2$ or $Cl_2$ in the presence of water.

$$>C=C< \xrightarrow[H_2O]{X_2} >C(OH)-C(X)< + HX$$

An alkene A halohydrin

We've seen that, when a solution of bromine in carbon tetrachloride reacts with an alkene, the cyclic bromonium ion intermediate is trapped by the only nucleophile present, bromide ion. If, however, the reaction is carried out *in the presence of an additional nucleophile*, the intermediate bromonium ion can be "intercepted" by the added nucleophile and diverted to a different product. For example, when an alkene reacts with bromine in the presence of water, water competes with bromide ion as nucleophile and reacts with the bromonium ion intermediate to yield a mixture of dibromide and bromo alcohol (a **bromohydrin**). The net effect is addition of HO—Br to the alkene. The reaction takes place by the pathway shown in Figure 7.4.

In practice, few alkenes are soluble in water, and bromohydrin formation is often carried out in a solvent such as aqueous dimethyl sulfoxide (DMSO), using a reagent called *N*-bromosuccinimide (NBS) as a bromine source. (NBS is a stable, easily handled compound that slowly decomposes

Reaction of the alkene with $Br_2$ yields a bromonium ion intermediate.

Water acts as a nucleophile, using a lone pair of electrons to open the bromonium ion ring and form a bond to carbon. Since oxygen donates its electrons in this step, it now has the positive charge.

Loss of a proton ($H^+$) from oxygen then gives HBr and the neutral bromohydrin product.

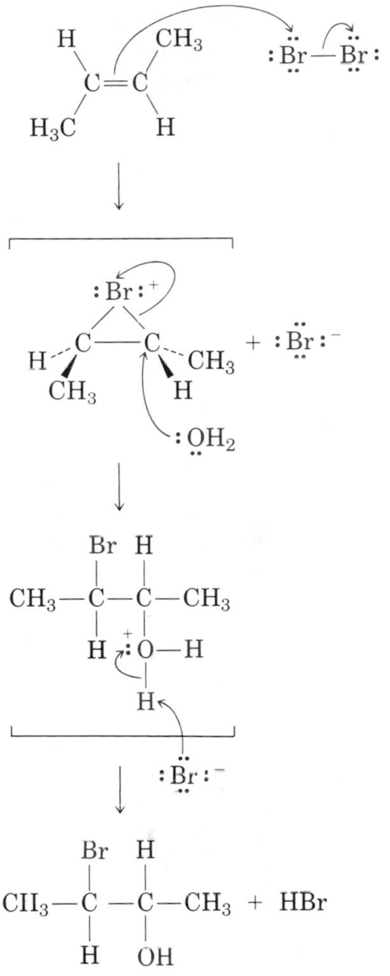

3-Bromo-2-butanol
(a bromohydrin)

**Figure 7.4** Mechanism of bromohydrin formation by reaction of an alkene with bromine in the presence of water

in water to yield $Br_2$ at a controlled rate. Bromine itself can also be used in the addition reaction, but it is much more dangerous and more difficult to handle.) Under these conditions, high yields of bromohydrins can be obtained. For example,

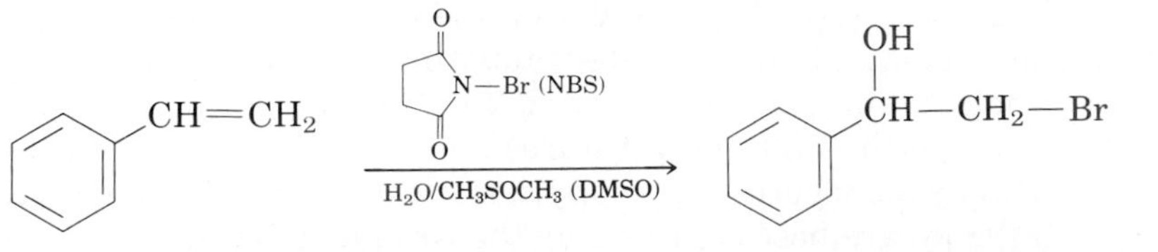

$$C_6H_5CH{=}CH_2 \xrightarrow[H_2O/CH_3SOCH_3\ (DMSO)]{\text{N-Br (NBS)}} C_6H_5CH(OH){-}CH_2{-}Br$$

Styrene

2-Bromo-1-phenylethanol (76%)

Note that the aromatic ring in the example is inert to bromine under the conditions used, even though it contains three carbon–carbon double bonds. Aromatic rings are a good deal more stable than might be expected, a property that will be examined in Chapter 15.

PROBLEM

**7.3** What product(s) would you expect from the reaction of cyclopentene with NBS and water? Show the stereochemistry.

PROBLEM

**7.4** When an unsymmetrical alkene such as propene is treated with *N*-bromosuccinimide in aqueous dimethyl sulfoxide, the major product has the bromine atom bonded to the less substituted carbon atom:

$$CH_3CH{=}CH_2 \xrightarrow[\text{DMSO}]{\text{NBS, } H_2O} CH_3\overset{\displaystyle OH}{\overset{|}{C}}HCH_2Br$$

How can you account for this result? Is this Markovnikov or non-Markovnikov orientation?

PROBLEM

**7.5** Iodine azide, I—$N_3$, adds to alkenes by an electrophilic mechanism similar to that of bromine. If a monosubstituted alkene is used, only one product results:

$$CH_3CH_2CH{=}CH_2 + I{-}N_3 \longrightarrow CH_3CH_2\overset{\displaystyle N_3}{\overset{|}{C}}HCH_2I$$

In light of this result, what is the polarity of the I—$N_3$ bond? Propose a mechanism for the reaction.

## 7.3 Hydration of Alkenes: Oxymercuration

Water can be added to simple alkenes such as ethylene and 2-methylpropene to yield alcohols. This **hydration** reaction takes place on treatment of the alkene with water and a strong acid catalyst (H—A) by a mechanism similar to that of HX addition. Thus, protonation of an alkene double bond yields a carbocation intermediate that then reacts with water as nucleophile to yield a protonated alcohol product ($ROH_2^+$). Loss of $H^+$ from this protonated alcohol gives the neutral alcohol and regenerates the acid catalyst (Figure 7.5).

Alkene hydration by the acid-catalyzed route is suitable for large-scale industrial procedures, and over one-half million tons per year of ethanol are manufactured in the United States by hydration of ethylene at 300°C. Nevertheless, the reaction is of little value for most laboratory applications because the high temperatures and strongly acidic conditions required are simply too vigorous for many organic molecules to survive. In practice, most alkenes are best hydrated by the **oxymercuration** procedure.

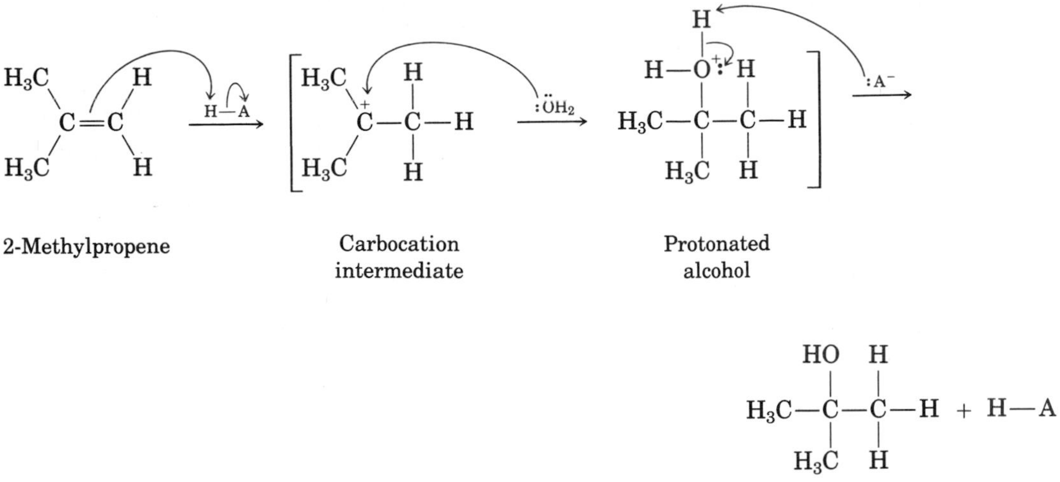

2-Methylpropene

Carbocation intermediate

Protonated alcohol

2-Methyl-2-propanol

**Figure 7.5** Mechanism of the acid-catalyzed hydration of an alkene

When an alkene is treated with mercuric acetate [$Hg(O_2CCH_3)_2$, usually abbreviated $Hg(OAc)_2$] in aqueous tetrahydrofuran (THF) solvent, electrophilic addition to the double bond rapidly occurs. The intermediate *organomercury* compound is then treated with sodium borohydride, $NaBH_4$, and an alcohol is produced. For example,

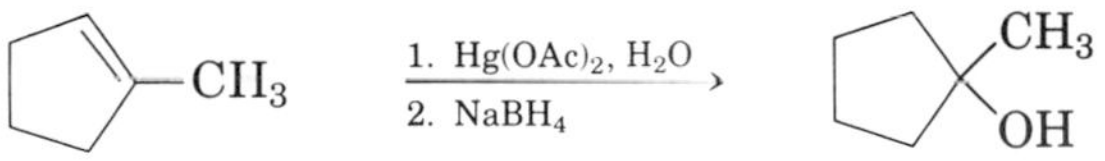

1. $Hg(OAc)_2$, $H_2O$
2. $NaBH_4$

1-Methylcyclopentene

1-Methylcyclopentanol (92%)

The oxymercuration reaction is initiated by electrophilic addition of mercuric ion to the alkene (Figure 7.6) to give an intermediate carbocation. Nucleophilic attack of water followed by loss of a proton then yields a stable organomercury addition product. The final step, reaction of the organomercury compound with sodium borohydride, is not well understood but appears to involve radicals as intermediates. Note that the regiochemistry of the reaction corresponds to Markovnikov addition of water; that is, the hydroxyl group becomes attached to the more highly substituted carbon atom and the hydrogen becomes attached to the less highly substituted carbon.

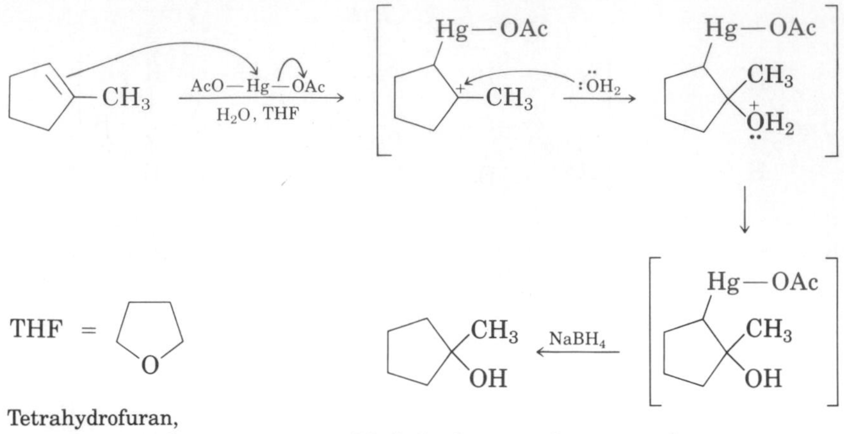

Hg—OAc Hg—OAc

CH$_3$ $\xrightarrow[\text{H}_2\text{O, THF}]{\text{AcO—Hg—OAc}}$ CH$_3$ $\xrightarrow{:\ddot{O}H_2}$ CH$_3$ $\overset{+}{O}H_2$

THF =

Tetrahydrofuran, a common organic solvent

CH$_3$ OH $\xleftarrow{\text{NaBH}_4}$ Hg—OAc CH$_3$ OH

1-Methylcyclopentanol (92%)

Organomercury intermediate

**Figure 7.6** Mechanism of the oxymercuration of alkenes to yield alcohols: This electrophilic addition reaction is similar to that of halohydrin formation.

**PROBLEM**

**7.6** What products would you expect from oxymercuration of these alkenes?
(a) $CH_3CH_2CH_2CH{=}CH_2$ (b) 2-Methyl-2-pentene

## 7.4 Hydration of Alkenes: Hydroboration

One of the most useful of all alkene additions is the **hydroboration** reaction first reported by H. C. Brown[1] in 1959. Hydroboration involves addition of a B—H bond of borane, $BH_3$, to an alkene to yield an **organoborane**:

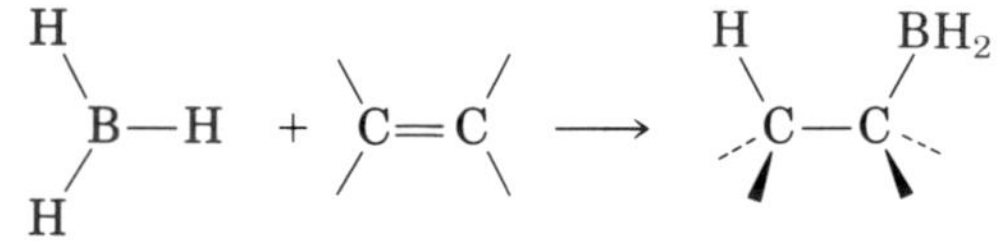

$$H_2B{-}H + C{=}C \longrightarrow H{-}C{-}C{-}BH_2$$

Borane

An organoborane

Borane is highly reactive by itself since the boron atom has only six electrons in its valence shell. In tetrahydrofuran (THF) solution, however, $BH_3$ accepts an electron pair from a solvent molecule to complete its octet and form a stable $BH_3$—THF complex. Although formal-charge calculations for this complex show that there must be a negative charge on boron and a positive charge on oxygen, for all practical purposes the $BH_3$—THF complex behaves chemically as if it were $BH_3$.

[1]Herbert Charles Brown (1912– ); b. London; Ph.D. (1938) University of Chicago (Schlessinger); professor, Purdue University (1947– ); Nobel prize (1979).

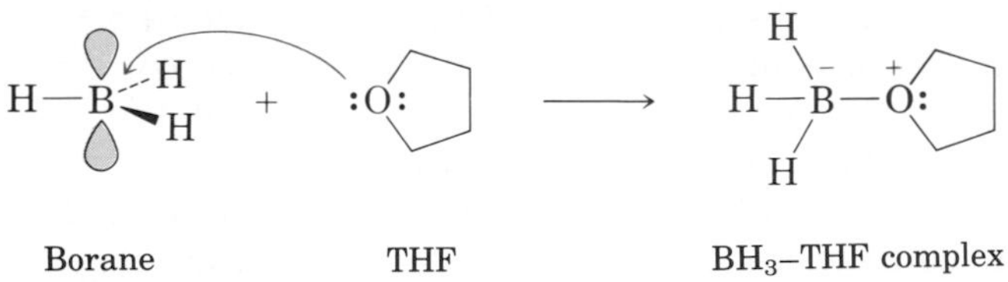

$$H-BH_2 + :O(CH_2)_4 \longrightarrow H_3\overset{-}{B}-\overset{+}{O}(CH_2)_4$$

Borane THF $BH_3$–THF complex

When an alkene reacts with $BH_3$ in tetrahydrofuran solution, rapid addition to the double bond occurs. Since $BH_3$ has three hydrogens, addition occurs three times and a *trialkylborane* product, $R_3B$ is formed. For example, one equivalent of $BH_3$ adds to three equivalents of ethylene to yield triethylborane (Figure 7.7).

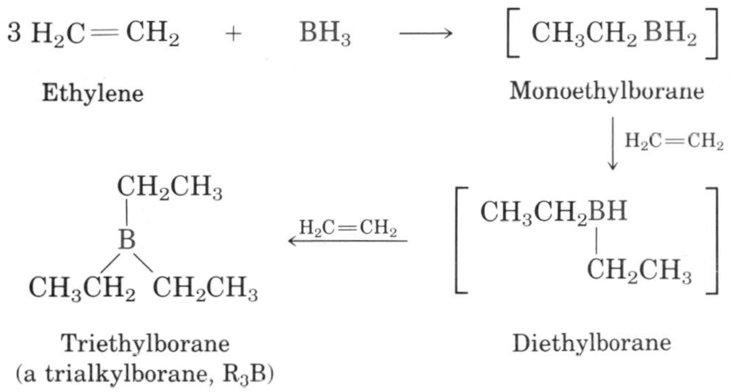

$$3\ H_2C{=}CH_2 + BH_3 \longrightarrow [CH_3CH_2BH_2] \xrightarrow{H_2C=CH_2} [(CH_3CH_2)_2BH] \xrightarrow{H_2C=CH_2} (CH_3CH_2)_3B$$

Ethylene; Monoethylborane; Diethylborane; Triethylborane (a trialkylborane, $R_3B$)

**Figure 7.7** The hydroboration of ethylene to yield triethylborane

Alkylboranes are of great value in synthesis because of the further reactions they undergo. For example, when tricyclohexylborane is treated with aqueous hydrogen peroxide ($H_2O_2$) in basic solution, an **oxidation** takes place. The carbon–boron bond is broken, a hydroxyl group is added, and three equivalents of cyclohexanol are produced. The net effect of the two-step hydroboration-plus-oxidation sequence is hydration of the alkene double bond.

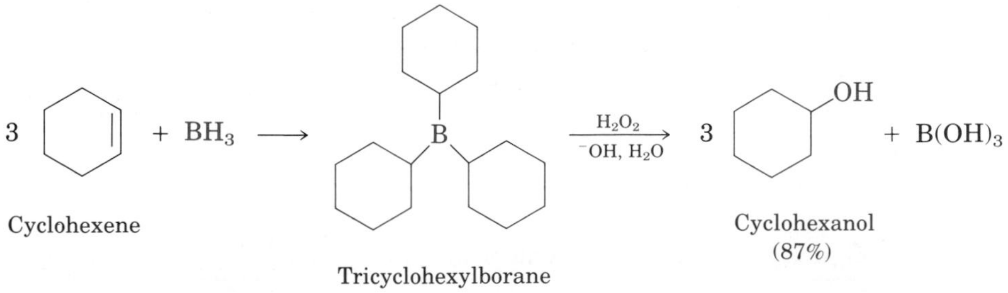

$$3\ C_6H_{10} + BH_3 \longrightarrow (C_6H_{11})_3B \xrightarrow[{}^{-}OH,\ H_2O]{H_2O_2} 3\ C_6H_{11}OH + B(OH)_3$$

Cyclohexene; Tricyclohexylborane; Cyclohexanol (87%)

One of the features that makes the hydroboration reaction so useful is the regiochemistry that results when an unsymmetrical alkene is hydroborated. For example, hydroboration–oxidation of 1-methylcyclopentene yields *trans*-2-methylcyclopentanol. Boron and hydrogen add to the alkene with syn stereochemistry, with boron attaching to the less highly substituted carbon. During the oxidation step, the boron is replaced by a hydroxyl with the same stereochemistry, resulting in an overall syn non-Markovnikov addition of water. This stereochemical result is particularly useful since it is *complementary* to the Markovnikov regiochemistry observed for oxymercuration.

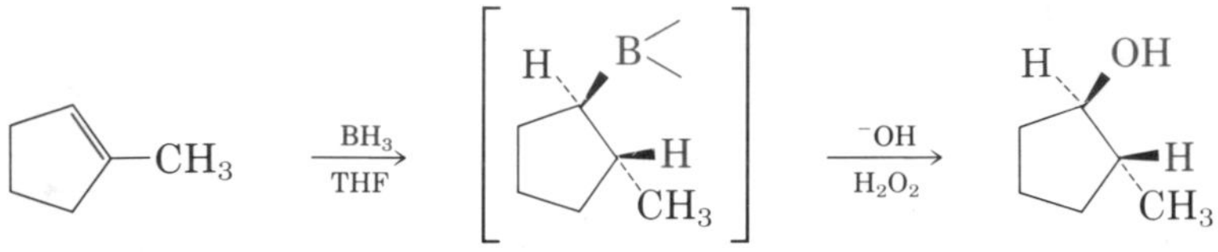

$CH_3$ $\xrightarrow[THF]{BH_3}$ [H, B, H, $CH_3$] $\xrightarrow[H_2O_2]{^-OH}$ H, OH, H, $CH_3$

1-Methylcyclopentene — Alkylborane intermediate — *trans*-2-Methylcyclopentanol (85%)

Why does alkene hydroboration take place with "non-Markovnikov" regiochemistry? Alkene hydroboration differs from other addition reactions in that it occurs in a single step; no carbocation intermediate is involved. We can view the reaction as taking place through a four-center cyclic transition state in which both carbon–hydrogen and carbon–boron bonds form at approximately the same time (Figure 7.8). Since both carbon–hydrogen

Addition of borane to the alkene pi bond occurs in a single step through a cyclic four-membered-ring transition state. The dotted lines indicate partial bonds that are breaking or forming.

A neutral alkylborane addition product is then formed when reaction is complete.

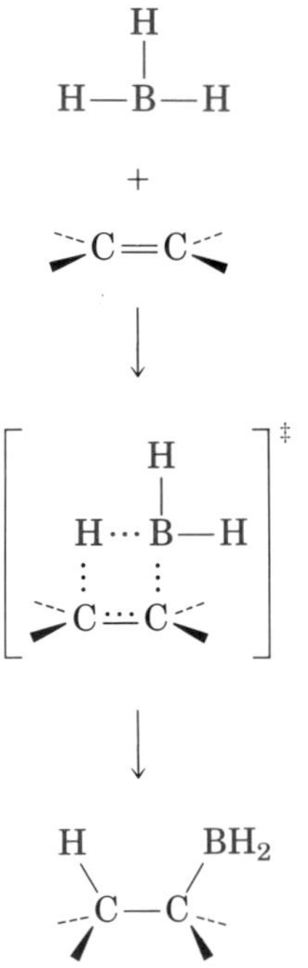

$BH_3$ + C=C → [H···B—H, C···C]‡ → H, $BH_2$, C—C

**Figure 7.8** Mechanism of alkene hydroboration

and carbon–boron bonds form simultaneously from the same face of the alkene, this mechanism satisfactorily explains why syn stereochemistry is observed.

The regiochemistry of the reaction can also be accounted for by the mechanism shown in Figure 7.8. Although hydroboration does not involve charged intermediates as other addition reactions do, the interaction of borane with an alkene nevertheless has a large amount of polar character to it. Borane, with only six valence electrons on boron, is a powerful electrophile because of its vacant *p* orbital. Thus, the interaction of $BH_3$ with a nucleophilic alkene must involve a partial transfer of electrons from the alkene to boron, with consequent buildup of polar character in the four-membered-ring transition state. Boron thus carries a partial negative charge ($\delta^-$) since it has gained electrons, and the alkene carbons carry a partial positive charge ($\delta^+$) since they have lost electrons.

In the addition of $BH_3$ to an unsymmetrically substituted alkene, there are two possible four-center transition states. From what we know about the nature of electrophilic additions, we would expect the transition state that places a partial positive charge on the more highly substituted carbon to be favored over the alternative that places the charge on the less highly substituted carbon. Thus boron tends to add to the less highly substituted carbon (Figure 7.9).

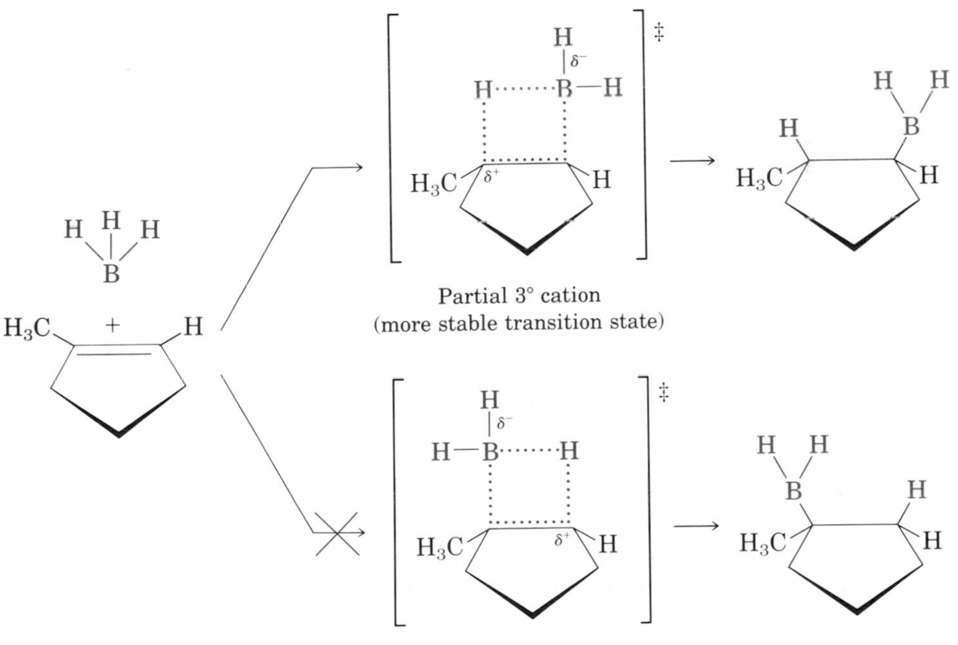

Partial 3° cation
(more stable transition state)

Partial 2° cation
(less stable transition state; *not formed*)

**Figure 7.9** Mechanism of the hydroboration of 1-methylcyclopentene: The favored transition state is the one that places the partial positive charge on the more highly substituted carbon.

In addition to electronic factors, a steric factor is probably also involved in determining the regiochemistry of hydroboration. Attachment of the boron group is favored at the less sterically hindered carbon atom of the alkene, rather than at the more hindered carbon, because there is less steric crowding in the resultant transition state:

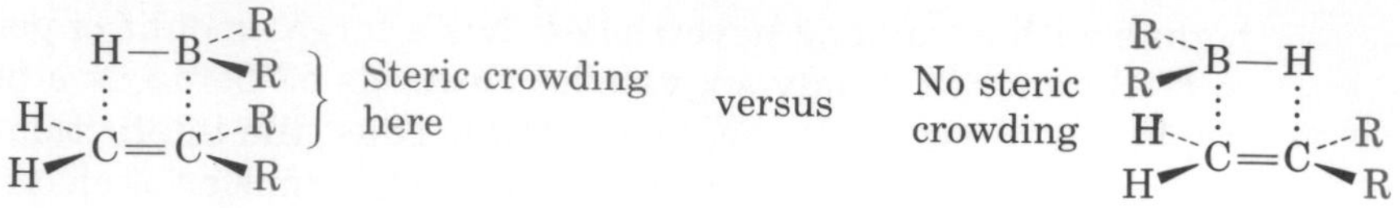

Steric crowding here versus No steric crowding

Both steric and electronic arguments predict the observed regiochemistry, and it is difficult to say which is more important. Evidence is accumulating, however, that suggests steric factors probably have greater influence than electronic factors.

PROBLEM

**7.7** What product will result from hydroboration–oxidation of 1-methylcyclopentene with deuterated borane, $BD_3$? Show both the stereochemistry (spatial arrangement) and regiochemistry (orientation) of the product.

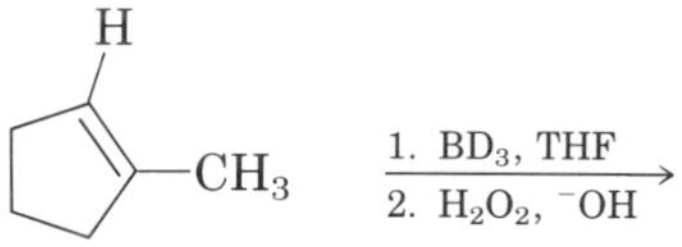

$$\xrightarrow[2.\ H_2O_2,\ ^-OH]{1.\ BD_3,\ THF}$$

PROBLEM

**7.8** What alkenes might be used to prepare these alcohols by hydroboration–oxidation?

(a) $(CH_3)_2CHCH_2CH_2OH$ (b) $(CH_3)_2CHCH(OH)CH_3$ (c) 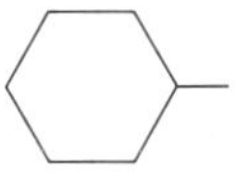 cyclohexyl–$CH_2OH$

## 7.5 A Radical Addition to Alkenes: HBr/Peroxides

The polar mechanism of electrophilic addition reactions is now well established. When HBr adds to an alkene under normal conditions, we know that an intermediate carbocation is involved and that Markovnikov orientation will be observed. This was not the case prior to 1933, however. In the 1920s the addition of HBr to 3-bromopropene had been studied several times by different workers, and conflicting results were reported in the chemical literature. The usual rule of Markovnikov addition did not appear to hold; *both* possible addition products were usually obtained, but in widely different ratios by different workers. For example,

$$\underset{\text{3-Bromopropene}}{H_2C{=}CHCH_2Br} \xrightarrow{\text{Liquid HBr}} \underbrace{\underset{\text{1,3-Dibromopropane}}{BrCH_2{-}CH_2CH_2Br} + \underset{\text{1,2-Dibromopropane}}{CH_3{-}CHBrCH_2Br}}_{\text{Product mixture}}$$

A careful examination of the reaction ultimately resolved the problem when it was found that HBr (but not HCl or HI) can add to alkenes by two entirely different mechanisms. It turns out that 3-bromopropene, the substrate many investigators had been using, is highly air-sensitive and absorbs oxygen to form *peroxides* (R—O—O—R, compounds with oxygen–oxygen bonds). Since the oxygen–oxygen bond is weak and easily broken, peroxides are excellent sources of radicals and serve to catalyze a *radical addition* of HBr to the alkene, rather than an electrophilic addition. In subsequent work, it was found that non-Markovnikov radical addition of HBr occurs with many different alkenes. For example,

$$H_2C{=}CH(CH_2)_8COOH \xrightarrow[\text{Peroxides}]{\text{HBr}} BrCH_2CH_2(CH_2)_8COOH$$

10-Undecenoic acid — 11-Bromoundecanoic acid

Peroxide-catalyzed addition of HBr to alkenes proceeds by a radical chain process involving addition of bromine radical, Br·, to the alkene. As with the radical chain mechanism we saw earlier for the light-induced chlorination of methane (Section 5.4), both initiation steps and propagation steps are required. In writing the mechanism of this radical reaction, note how a half-arrow or "fishhook" ⇀, is used to show the movement of a single electron, as opposed to the full arrow used to show the movement of an electron pair in a polar reaction.

*Initiation steps* The reaction is initiated in two steps. In the first step, light-induced homolytic cleavage of the weak oxygen–oxygen peroxide bond generates two alkoxy radicals, RO·. An alkoxy radical then abstracts a hydrogen atom from HBr in the second initiation step to give a bromine radical, Br·.

1. $$R{-}\ddot{\underset{..}{O}}{-}\ddot{\underset{..}{O}}{-}R \xrightarrow{\text{Light}} 2\,R{-}\ddot{\underset{..}{O}}\cdot$$

A peroxide — Alkoxy radicals

2. $$R{-}\ddot{\underset{..}{O}}\cdot + H{-}\ddot{\underset{..}{Br}}: \longrightarrow R{-}\ddot{\underset{..}{O}}{-}H + :\ddot{\underset{..}{Br}}\cdot$$

*Propagation steps* Once a bromine radical has formed in the initiation steps, a chain-reaction cycle of two repeating propagation steps begins. In the first propagation step, bromine radical adds to the alkene double bond, giving an alkyl radical. One electron from Br· and one electron from the pi bond are used to form the new C—Br bond, leaving an unpaired electron on the remaining double-bond carbon. In the second propagation step, this alkyl radical reacts with HBr to yield addition product plus a bromine radical to cycle back into the first propagation step and carry on the chain reaction.

3. $$:\ddot{\underset{..}{Br}}\cdot + H_2C{=}CHCH_2CH_3 \longrightarrow Br{-}CH_2{-}\dot{C}HCH_2CH_3$$

4. $$Br{-}CH_2{-}\dot{C}HCH_2CH_3 + H{-}\ddot{\underset{..}{Br}}: \longrightarrow Br{-}CH_2{-}\overset{H}{\overset{|}{C}}HCH_2CH_3 + :\ddot{\underset{..}{Br}}\cdot$$

According to this radical chain mechanism, the regiochemistry of addition is determined in the first propagation step when bromine radical adds to the alkene. For an unsymmetrical alkene such as 1-butene, this addition could conceivably take place at either of two carbons, to yield either a primary radical intermediate or a secondary radical. Experimental results, however, indicate that only the more substituted secondary radical is formed (Figure 7.10). How can we explain these results?

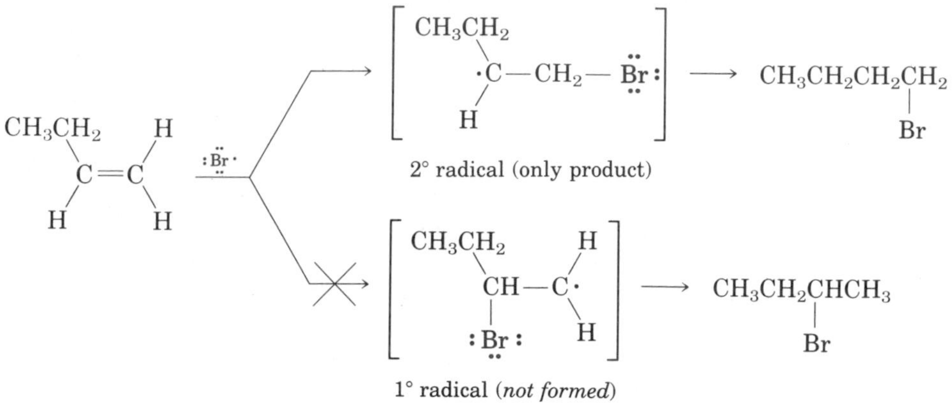

$$CH_3CH_2CH{=}CH_2 \xrightarrow{:\ddot{Br}\cdot} \left[CH_3CH_2\dot{C}H{-}CH_2{-}\ddot{Br}:\right] \longrightarrow CH_3CH_2CH_2CH_2Br$$

2° radical (only product)

$$CH_3CH_2CH{=}CH_2 \not\xrightarrow{:\ddot{Br}\cdot} \left[CH_3CH_2CH(\ddot{Br}:){-}\dot{C}H_2\right] \longrightarrow CH_3CH_2CH(Br)CH_3$$

1° radical (*not formed*)

**Figure 7.10** Addition of bromine radical to 1-butene: The reaction is regiospecific, leading exclusively to formation of the secondary radical.

## STABILITY OF RADICALS

In explaining the observed Markovnikov regiochemistry of polar electrophilic addition reactions (Section 6.12), we invoked the Hammond postulate to account for the fact that more stable carbocation intermediates form faster than less stable ones. In explaining the observed regiochemistry of the *radical* addition of HBr to alkenes, we need to invoke a similar argument. First, we need to compare the relative stabilities of substituted radicals by looking at the bond dissociation energies for different kinds of carbon–hydrogen bonds (Table 7.1).

**Table 7.1** Bond dissociation energies of different kinds of C—H bonds in alkanes

| *Bond broken* | *Radical products* | *Bond type*[a] | *Bond dissociation energy, ΔH°* (*kcal/mol*) | (*kJ/mol*) |
|---|---|---|---|---|
| $H{-}CH_3$ | $H\cdot + CH_3\cdot$ | Methyl | 104 | 435 |
| $H{-}CH_2CH_2CH_3$ | $H\cdot + CH_3CH_2CH_2\cdot$ | Primary | 98 | 410 |
| $CH_3\overset{H}{C}HCH_3$ | $H\cdot + CH_3\dot{C}HCH_3$ | Secondary | 95 | 397 |
| $H{-}C(CH_3)_3$ | $H\cdot + (CH_3)_3C\cdot$ | Tertiary | 92 | 385 |

[a]Tertiary C—H bonds are weaker than secondary ones, and secondary bonds are weaker than primary ones.

As we saw in Section 5.8, bond dissociation energy is the amount of energy that must be supplied to cleave a bond homolytically into two radical fragments. When $\Delta H^\circ$ is high, the bond is strong and there is a large difference in stability between reactant and products. When $\Delta H^\circ$ is low, the bond is weaker, and there is a smaller difference in stability between reactant and products. If the energy required to break a primary C—H bond of propane [98 kcal/mol (410 kJ/mol)] is compared with that required to break a secondary C—H bond of propane [95 kcal/mol (397 kJ/mol)], there is a difference of 3 kcal/mol. Since we are starting with the same reactant in both cases (propane), and since one product is the same in both cases (H·), the 3 kcal/mol energy difference is a direct measure of the difference in stability between the primary propyl radical ($CH_3CH_2CH_2\cdot$) and the secondary propyl radical ($CH_3\dot{C}HCH_3$). In other words, the secondary propyl radical is more stable than the primary propyl radical by 3 kcal/mol.

A similar comparison between a primary C—H bond of 2-methylpropane [98 kcal/mol (410 kJ/mol)] and the tertiary C—H bond in the same molecule [92 kcal/mol (385 kJ/mol)] leads to the conclusion that the tertiary radical is more stable than the primary radical by 6 kcal/mol (25 kJ/mol). We thus find that the stability order of radicals is as follows:

$$\begin{array}{ccccccc} \text{Tertiary} & > & \text{Secondary} & > & \text{Primary} & > & \text{Methyl} \\ R_3C\cdot & > & R_2\dot{C}H & > & R\dot{C}H_2 & > & \cdot CH_3 \end{array}$$

This stability order of radicals is identical to the stability order of carbocations, and for much the same reasons. Carbon radicals, even though they are uncharged, are electron-deficient like carbocations. Thus, the same substituent effects stabilize both.

## AN EXPLANATION OF THE REGIOCHEMISTRY OF RADICAL-CATALYZED ADDITION OF HBr TO ALKENES

Now that we know the stability order of radicals, we can complete an explanation for the observed regiochemistry of radical-catalyzed HBr additions to alkenes. Although the addition of bromine radical to an alkene is a slightly exothermic step, there is nevertheless a certain amount of developing radical character in the transition state. Thus, according to the Hammond postulate (Section 6.12), the same factors that make a secondary radical more stable than a primary one also make the transition state leading to the secondary radical more stable than the transition state leading to the primary one. In other words, the more stable radical forms faster than the less stable one.

We conclude that, although peroxide-catalyzed radical addition of HBr to an alkene gives an *apparently* non-Markovnikov-oriented product, the reaction in fact proceeds through the more stable intermediate just as a polar reaction does. A reaction energy diagram for the overall process is shown in Figure 7.11 on page 208.

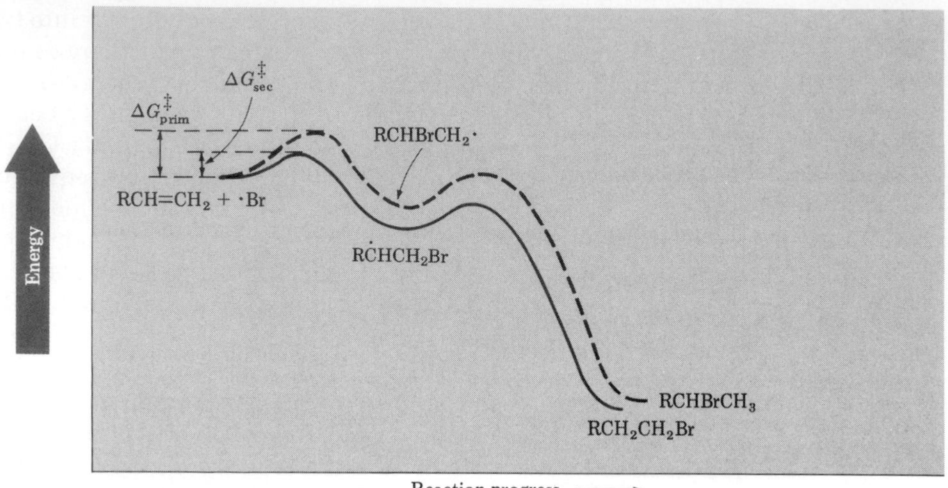


**Figure 7.11** Reaction energy diagram for the addition of bromine radical to an alkene: The more stable secondary radical forms faster than the less stable primary radical.

PROBLEM

**7.9** Show how you would synthesize the following compounds. Identify the alkene starting material and indicate what reagents you would use in each case.

(a) $CH_3CH_2C(CH_3)_2$ with Br on the C (i.e., $CH_3CH_2CBr(CH_3)_2$)    (b) $CH_3CH_2CH_2CH_2Br$    (c) $CH_3CH_2CHCHCH_2CH_3$ with $H_3C$ and Br substituents

PROBLEM

**7.10** Draw a reaction energy diagram for the radical addition of HBr to 2-methyl-2-butene (consider only the propagation steps). Construct your diagram so that reactions leading to the two possible addition products are both shown. Which of the two curves has the lower activation energy for the first step? Which of the two curves has the lower $\Delta G°$ for the first step?

## 7.6 Hydrogenation of Alkenes

Alkenes react with hydrogen in the presence of a suitable catalyst to yield the corresponding saturated-alkane addition products. We describe the result by saying that the double bond has been **hydrogenated**, or **reduced**. Although we looked briefly at catalytic hydrogenation as a method of determining alkene stabilities, the reaction also has enormous practical value.

$$C{=}C \quad + \quad H{-}H \xrightarrow{\text{Catalyst}} H{-}C{-}C{-}H$$

An alkene $\qquad$ An alkane

Either platinum or palladium is used as the catalyst for most alkene hydrogenations. Palladium is normally employed in a very finely divided state "supported" on an inert material such as charcoal to maximize surface area (Pd/C). Platinum is normally used as $PtO_2$, a reagent known as **Adams catalyst** after its discoverer, Roger Adams.[2]

Catalytic hydrogenation, unlike most other organic reactions, is a *heterogeneous* process rather than a homogeneous one. That is, the hydrogenation reaction occurs on the surface of solid catalyst particles rather than in solution. For this reason, catalytic hydrogenation has proven difficult to study mechanistically. Observation has shown, however, that hydrogenation usually occurs with syn stereochemistry; both hydrogens add to the double bond from the same face.

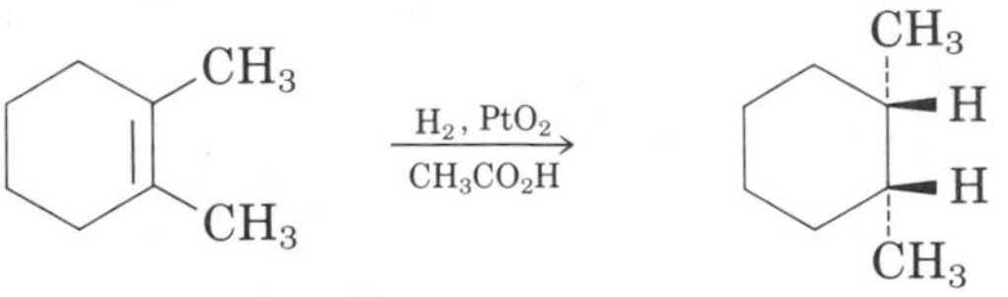

$CH_3$ $CH_3$ $\xrightarrow[CH_3CO_2H]{H_2,\ PtO_2}$ $CH_3$ H H $CH_3$

1,2-Dimethylcyclohexene

*cis*-1,2-Dimethylcyclohexane (82%)

The first step in the reaction is adsorption of hydrogen onto the catalyst surface. Complexation then occurs between catalyst and alkene by overlap of vacant metal orbitals with alkene pi electrons. In the final steps, hydrogen is inserted into the pi bond, and the saturated product diffuses away from the catalyst (Figure 7.12). It's clear from this picture why the stereochemistry of hydrogenation is syn: It *must* be syn because both hydrogens add to the double bond from the same catalyst surface.

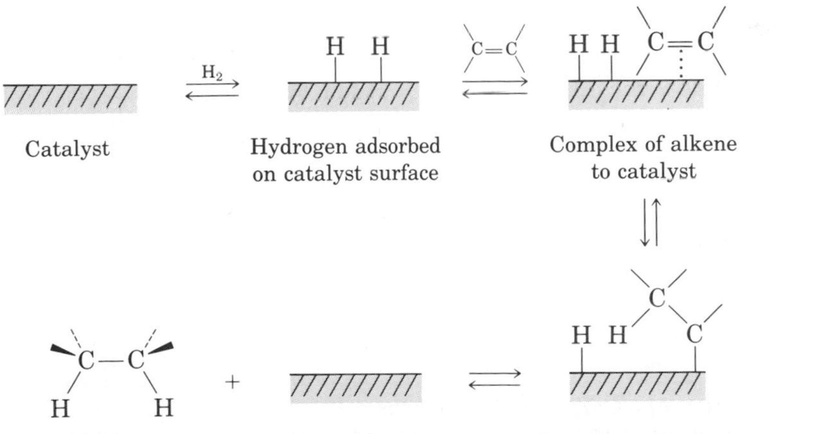

Catalyst $\xrightarrow{H_2}$ Hydrogen adsorbed on catalyst surface (H H) ⇌ Complex of alkene to catalyst (H H C=C)

Insertion of hydrogen into carbon–carbon double bond (H H C C) ⇌ Alkane product (C—C, H H) + Regenerated catalyst

**Figure 7.12** Mechanism of alkene hydrogenation: The reaction takes place with syn stereochemistry on the surface of catalyst particles.

[2]Roger Adams (1889–1971); b. Boston; Ph.D. (1912), Harvard (Torrey, Richards); professor, University of Illinois (1916–1971).

Alkenes are much more reactive than most other functional groups toward catalytic hydrogenation, and the reaction is therefore quite selective. Other functional groups such as ketones, esters, and nitriles survive normal alkene hydrogenation conditions unchanged. Reaction with these groups does occur under more vigorous conditions, however.

O O

$\xrightarrow[\text{Pd/C in ethanol}]{H_2}$

2-Cyclohexenone Cyclohexanone

Ketone not reduced

$CO_2CH_3$ $CO_2CH_3$

$\xrightarrow[\text{Pd/C in ethanol}]{H_2}$

Methyl 3-phenylpropenoate Methyl 3-phenylpropanoate

Benzene ring, ester not reduced

CN CN

$\xrightarrow[\text{Pd/C in ethanol}]{H_2}$

Cyclohexylideneacetonitrile Cyclohexylacetonitrile

Nitrile not reduced

Note that, in the hydrogenation of methyl 3-phenylpropenoate, the benzene ring functional group is not affected by hydrogen on palladium even though it contains three double bonds. This is yet another example of the remarkable unreactivity of aromatic rings.

In addition to its usefulness in the laboratory, catalytic hydrogenation is of great commercial value in the food industry. Unsaturated vegetable oils, which usually contain numerous double bonds, are catalytically hydrogenated on a vast scale to produce the saturated fats used in margarine and solid cooking fats.

## 7.7 Hydroxylation of Alkenes

**Hydroxylation** of an alkene—the addition of an —OH group to each of the two alkene carbons—can be carried out with reagents such as potassium permanganate ($KMnO_4$) and osmium tetraoxide ($OsO_4$). Since oxygen is added to the alkene during the reaction, we call this an **oxidation**. Both of these hydroxylation reactions occur with syn, rather than anti, stereochemistry and yield 1,2-dialcohols (**diols**, also called **glycols**).

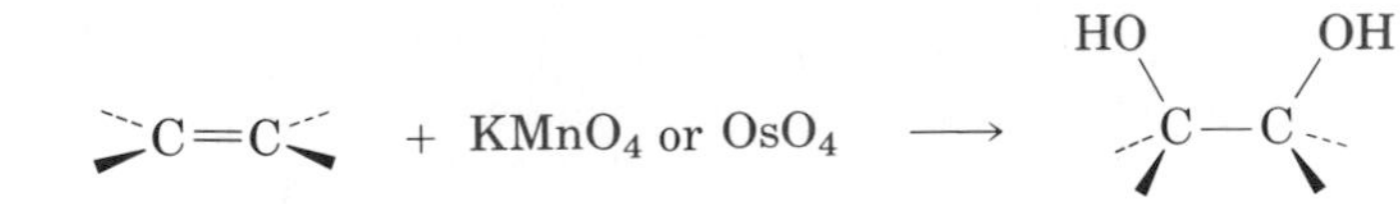

$$\text{C}{=}\text{C} + KMnO_4 \text{ or } OsO_4 \longrightarrow \text{HO–C–C–OH}$$

An alkene A 1,2-diol

Potassium permanganate is a common and inexpensive oxidant that reacts with alkenes under carefully controlled conditions in alkaline medium to yield 1,2-diols. Unlike many other addition reactions we've studied, hydroxylation does not involve a carbocation intermediate. Instead, the reaction occurs through an intermediate cyclic **manganate** species, formed in a single step by addition of permanganate ion to the alkene. This cyclic manganate is then hydrolyzed to give the diol product. For example, oxidation of cyclohexene, which was first carried out in 1879 by Vladimir Markovnikov, gives *cis*-1,2-cyclohexanediol in rather low yield. This result is typical: Syn hydroxylation is always observed, but yields are sometimes poor because of side reactions.

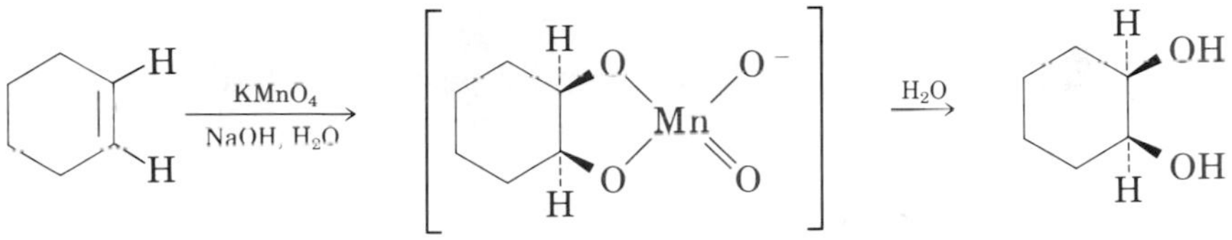

$KMnO_4$, NaOH, $H_2O$; $H_2O$

Cyclohexene — A cyclic manganate intermediate — *cis*-1,2-Cyclohexanediol (37%)

For small-scale laboratory preparations, osmium tetraoxide in place of potassium permanganate is much preferred. Although both toxic and expensive, $OsO_4$ reacts with alkenes to give high yields of cis 1,2-diols. As with permanganate hydroxylation, reaction with $OsO_4$ occurs through a cyclic intermediate, which is then cleaved to yield a cis diol. A cyclic **osmate** is less reactive than a cyclic manganate, however, and must be cleaved in a separate step. Aqueous sodium bisulfite, $NaHSO_3$, is often used to accomplish this cleavage:

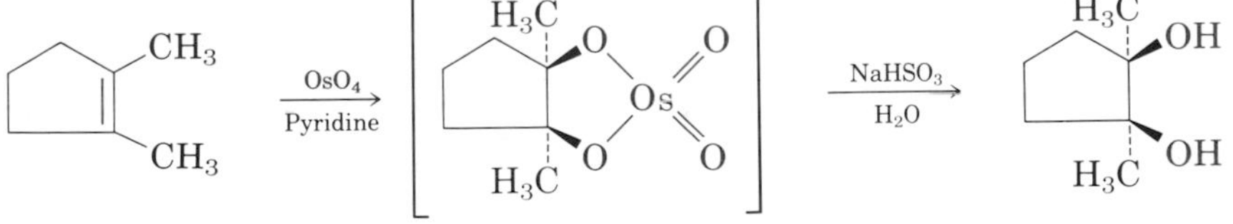

$OsO_4$, Pyridine; $NaHSO_3$, $H_2O$

1,2-Dimethylcyclopentene — Cyclic osmate intermediate — *cis*-1,2-Dimethyl-1,2-cyclopentanediol (87%)

**PROBLEM**

**7.11** How would you prepare the following compounds? Show the starting alkene and the reagents you would use.

(a) cyclohexane ring bearing $CH_3$, OH, OH, H

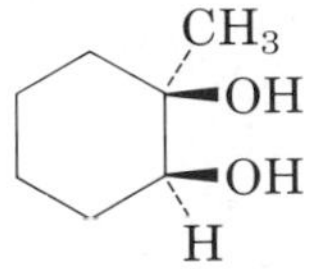

(b) $CH_3CH_2CH(OH)C(OH)(CH_3)_2$

(c) $CH_2(OH)CH(OH)CH(OH)CH_2OH$

PROBLEM

**7.12** Explain the observation that hydroxylation of *cis*-2-butene with $OsO_4$ yields a different product than hydroxylation of *trans*-2-butene. First draw the structure and show the stereochemistry of each product, and then make molecular models. (We'll explore the stereochemistry of the products in more detail in Chapter 9.)

## 7.8 Oxidative Cleavage of Alkenes

In all the alkene addition reactions we've seen thus far, the carbon skeleton of the starting material has been left intact. The carbon–carbon double bond has been converted into a new functional group (halide, alcohol, 1,2-diol) by adding different reagents, but no carbon bond framework has been broken or rearranged. There are, however, powerful oxidizing reagents that *cleave* carbon–carbon double bonds to produce two fragments.

Ozone ($O_3$) is the most useful double-bond cleavage reagent. Prepared conveniently in the laboratory by passing a stream of oxygen through a high-voltage electrical discharge, ozone adds rapidly to alkenes at low temperature to give cyclic intermediates called **molozonides**. Once formed, molozonides then rapidly rearrange to form **ozonides**. (We won't study the mechanism of this unusual rearrangement in detail. It involves the molozonide coming apart into two fragments, which then recombine in a new way.)

$$3\ O_2 \xrightarrow[\text{discharge}]{\text{Electric}} 2\ O_3$$

$$\mathrm{C{=}C} \xrightarrow[\mathrm{CH_2Cl_2,\ -78^\circ C}]{O_3} [\text{molozonide}] \longrightarrow \text{ozonide} \xrightarrow[\mathrm{CH_3COOH/H_2O}]{\mathrm{Zn}} \mathrm{C{=}O} + \mathrm{O{=}C}$$

A molozonide — An ozonide

Low-molecular-weight ozonides are highly explosive and are therefore not isolated. Instead, ozonides are usually further treated with a reducing agent such as zinc metal in acetic acid to convert them to carbonyl compounds. The net result of the **ozonolysis**–zinc reduction sequence is that the carbon–carbon double bond is cleaved, and oxygen becomes doubly bonded to each of the original alkene carbons. If an alkene with a tetrasubstituted double bond is ozonized, two ketone fragments result; if an alkene with a trisubstituted double bond is ozonized, one ketone and one aldehyde result, and so on.

$$\text{Isopropylidenecyclohexane (tetrasubstituted)} \xrightarrow[\text{2. Zn, } H_3O^+]{\text{1. } O_3} \text{Cyclohexanone} + CH_3\overset{O}{\overset{\|}{C}}CH_3 \text{ (Acetone)}$$

Isopropylidenecyclohexane (tetrasubstituted)

Cyclohexanone Acetone

84%; two ketones

$$CH_3(CH_2)_7CH{=}CH(CH_2)_7COOCH_3 \xrightarrow[\text{2. Zn, } H_3O^+]{\text{1. } O_3} CH_3(CH_2)_7CH{=}O + O{=}CH(CH_2)_7COOCH_3$$

Methyl 9-octadecenoate (disubstituted)

Nonanal + Methyl 9-oxononanoate

78%; two aldehydes

$$\beta\text{-Pinene} \xrightarrow[\text{2. Zn, } H_3O^+]{\text{1. } O_3} \text{Nopinone} + O{=}CH_2$$

β-Pinene (disubstituted)

Nopinone Formaldehyde

75%; one ketone, one aldehyde

Certain oxidizing reagents other than ozone also cause double-bond cleavage. For example, potassium permanganate in neutral or acidic solution cleaves alkenes, giving carbonyl-containing products in low to moderate yield. If hydrogens are present on the double bond, carboxylic acids are produced; if two hydrogens are present on one carbon, $CO_2$ is formed. Although we won't go into mechanistic details, the reaction involves formation and decomposition of the same kind of cyclic manganate shown previously for alkene hydroxylation (Section 7.7).

$$(CH_3)_2CHCH_2CH_2CH_2\overset{CH_3}{\overset{|}{C}}HCH{=}CH_2 \xrightarrow[H_2O]{KMnO_4} (CH_3)_2CHCH_2CH_2CH_2\overset{CH_3}{\overset{|}{C}}HCOOH + CO_2$$

3,7-Dimethyl-1-octene

2,6-Dimethylheptanoic acid (45%)

**PROBLEM**

**7.13** What products would you expect from reaction of 1-methylcyclohexene with these reagents?

(a) Aqueous acidic $KMnO_4$ (b) $O_3$, followed by Zn, $CH_3COOH$

PROBLEM

7.14 Propose structures for alkenes that yield the following products on reaction with ozone, followed by treatment with Zn.

(a) $(CH_3)_2C{=}O + H_2C{=}O$ (b) 2 equiv $CH_3CH_2CH{=}O$

## 7.9 Oxidative Cleavage of 1,2-Diols

1,2-Diols can be oxidatively cleaved by reaction with periodic acid ($HIO_4$) to yield carbonyl compounds in a reaction similar to the potassium permanganate cleavage of alkenes discussed in the previous section. The sequence of (1) alkene hydroxylation with $OsO_4$ followed by (2) diol cleavage with $HIO_4$ is often an excellent alternative to direct alkene cleavage with ozone or potassium permanganate.

If the two hydroxyls are on an open chain, two carbonyl compounds result. If the two hydroxyls are on a ring, a single dicarbonyl compound is formed. As indicated in the following examples, the cleavage reaction is believed to take place via a cyclic periodate intermediate.

$CH_3$ — 1. $OsO_4$, 2. $NaHSO_3$ → $H_3C$, OH, OH, H — $HIO_4$ → [$H_3C$, O, OH, I=O, O, O, H]

A 1,2-diol

Cyclic periodate intermediate

↓

O, $CH_3$, CHO

6-Oxoheptanal (86%)

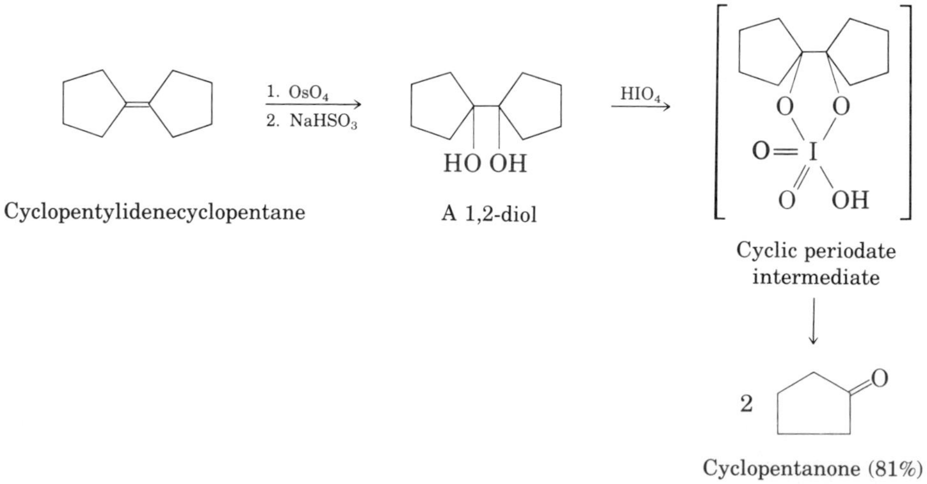

1. $OsO_4$, 2. $NaHSO_3$ → HO OH — $HIO_4$ → [O, O, O=I, O, OH]

Cyclopentylidenecyclopentane

A 1,2-diol

Cyclic periodate intermediate

↓

2 (O)

Cyclopentanone (81%)

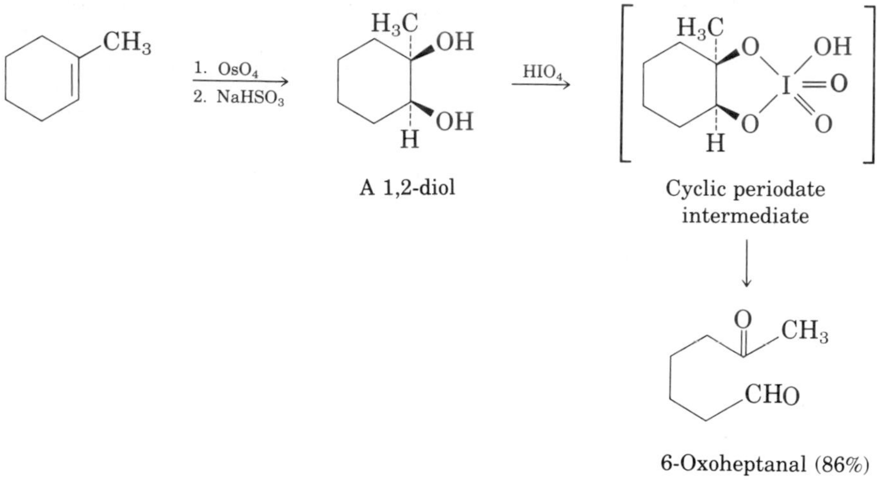

## 7.10 Addition of Carbenes to Alkenes: Cyclopropane Synthesis

The last alkene addition reaction we'll consider is the addition of a **carbene**, $\mathbf{R_2C:}$, to an alkene to yield a cyclopropane. A carbene is a neutral molecule containing a divalent carbon that has only six electrons in its valence shell. It is therefore highly reactive and can be generated only as a reaction intermediate, rather than as an isolable substance. Because they have only six valence electrons, carbenes are electron-deficient and behave as electrophiles. Thus, they react with nucleophilic carbon–carbon double bonds in a single step without intermediates.

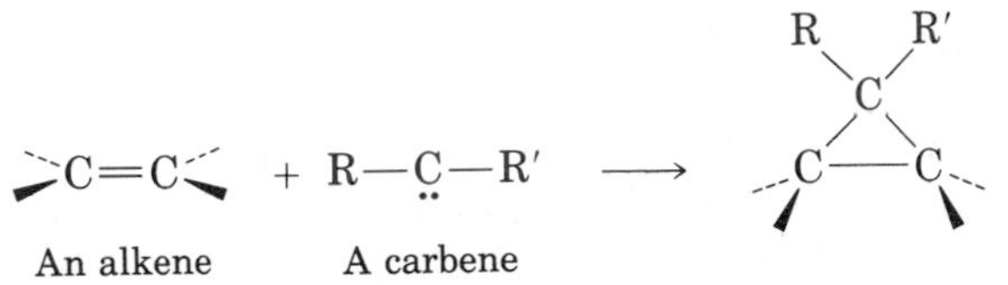

$$>C{=}C< \;+\; R{-}\ddot{C}{-}R' \longrightarrow \text{cyclopropane}$$

An alkene A carbene A cyclopropane

One of the best methods for generating a substituted carbene is by treatment of chloroform, $CHCl_3$, with a strong base such as potassium hydroxide. Loss of a proton from $CHCl_3$ generates the trichloromethanide anion, $^-:CCl_3$, which expels a chloride ion to give dichlorocarbene, $:CCl_2$ (Figure 7.13).

Strong base abstracts the chloroform proton, leaving behind the electron pair from the C—H bond and forming the trichloromethanide anion.

$$CHCl_3 \xrightarrow{K^+\,:\ddot{O}H^-} [\,Cl_3C:^-\,] + H_2O$$

Trichloromethanide anion

Loss of a chloride ion and associated electrons from the C—Cl bond yields the neutral dichlorocarbene.

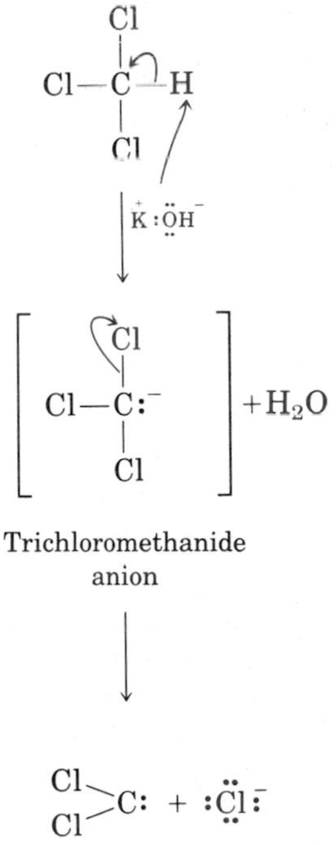

$$Cl_2C: \;+\; :\ddot{Cl}:^-$$

**Figure 7.13** Mechanism of the formation of dichlorocarbene by reaction of chloroform with strong base

The dichlorocarbene carbon atom is $sp^2$ hybridized, with a vacant $p$ orbital extending above and below the plane of the three atoms, and with an unshared pair of electrons occupying the third $sp^2$ lobe. Note that this electronic description of dichlorocarbene is similar to that for carbocations (Section 6.11), with respect both to hybridization of carbon and to the presence of a vacant $p$ orbital.

Vacant $p$ orbital

$sp^2$ orbital

Dichlorocarbene

A carbocation ($sp^2$ hybridized)

If dichlorocarbene is generated in the presence of an alkene, addition of the electrophilic carbene to the double bond occurs, and a dichlorocyclopropane is formed. As the reaction of dichlorocarbene with *cis*-2-pentene demonstrates, the addition is **stereospecific**, meaning that only a single stereoisomer is formed as product. Starting from the cis alkene, only cis-disubstituted cyclopropane is produced.

$$\text{cis-}CH_3CH_2CH{=}CHCH_3 + CHCl_3 \xrightarrow{KOH} \text{cis-1,1-dichloro-2-ethyl-3-methylcyclopropane} + KCl$$

*cis*-2-Pentene — 65%

$$\text{Cyclohexene} + CHCl_3 \xrightarrow{KOH} \text{7,7-dichlorobicyclo[4.1.0]heptane} + KCl$$

Cyclohexene — 60%

The best method for preparing nonhalogenated cyclopropanes is the **Simmons–Smith reaction**. This reaction, first investigated at the Du Pont company, does not involve a free carbene. Rather, it utilizes a **carbenoid**—a reagent with carbene-like reactivity. When diiodomethane is treated with a specially prepared zinc–copper alloy, (iodomethyl)zinc iodide, $ICH_2ZnI$, is formed. If an alkene is present, (iodomethyl)zinc iodide transfers a $CH_2$ group to the double bond and yields the cyclopropane. For example, cyclohexene reacts cleanly and in good yield to give the corresponding cyclopropane. Although we won't discuss the mechanistic details of these specific reactions, carbene addition to an alkene is an example of a general class of processes called **cycloadditions**, which we'll study more carefully in Chapter 30.

$$CH_2I_2 + Zn(Cu) \xrightarrow{\text{Ether}} I{-}CH_2{-}Zn{-}I = \text{“}:CH_2\text{”}$$

Diiodomethane — (Iodomethyl)zinc iodide (a carbenoid)

$$\text{Cyclohexene} + CH_2I_2 \xrightarrow[\text{Ether}]{Zn(Cu)} \text{Bicyclo[4.1.0]heptane} + ZnI_2$$

Cyclohexene

Bicyclo[4.1.0]heptane (92%)

**PROBLEM**

**7.15** What products would you expect from these reactions?

(a) Methylenecyclohexane $=CH_2 + CHCl_3 \xrightarrow{KOH}$

(b) $(CH_3)_2CHCH_2CH{=}CHCH_3 + CH_2I_2 \xrightarrow{Zn(Cu)}$

**PROBLEM**

**7.16** Simmons–Smith reaction of cyclohexene with diiodomethane gives a single cyclopropane product. Reaction with 1,1-diiodoethane, however, gives (in low yield) a mixture of two isomeric methylcyclopropane products. Formulate the reactions, and account for the formation of the product mixture.

## 7.11 Some Biological Alkene Addition Reactions

The chemistry of living organisms is a fascinating field of study—the simplest one-celled organism is capable of more complex organic synthesis than any human. Yet as we learn more, it becomes clear that the same principles applying to laboratory chemistry also apply to biochemistry.

Biological organic chemistry takes place in the aqueous medium inside cells rather than in organic solvents, and involves complex catalysts called **enzymes**. Nevertheless, the *kinds* of reactions are remarkably similar. Thus, there are many cases of biological addition reactions to alkenes. For example, the enzyme fumarase catalyzes the addition of water to fumaric acid much as sulfuric acid might catalyze the addition of water to ethylene on an industrial scale:

$$\text{HOOC(H)C{=}C(H)COOH} \xrightarrow[\text{H}_2\text{O, pH 7.4}]{\text{Fumarase}} \text{HOOC{-}CH}_2\text{{-}CHOH{-}COOH}$$

Fumaric acid — Malic acid

Enzymes can even catalyze the formation of bromohydrins from alkenes. For example, the enzyme chloroperoxidase can catalyze the addition of HO—Br to double bonds:

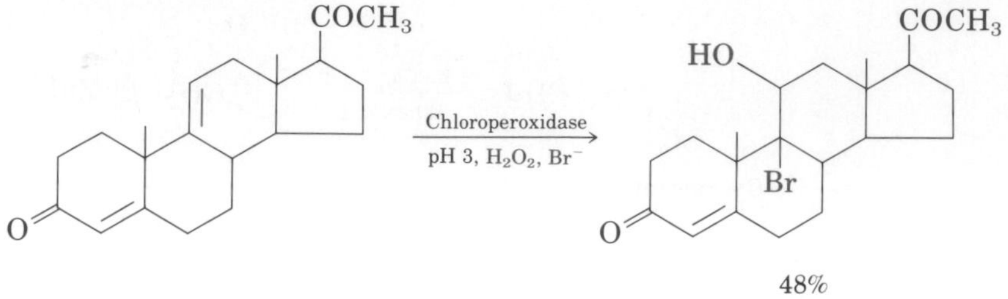

COCH$_3$ (steroid) $\xrightarrow[\text{pH 3, } H_2O_2\text{, } Br^-]{\text{Chloroperoxidase}}$ HO, Br, COCH$_3$ (bromohydrin) 48%

Enzyme-catalyzed reactions in organisms are usually much more chemically selective than their laboratory counterparts. Fumarase, for example, is completely inert toward maleic acid, the cis isomer of fumaric acid. Nevertheless, the fundamental processes of organic chemistry are the same in the living cell and in the laboratory.

## 7.12 Preparation of Alkenes: Elimination Reactions

Just as the chemistry of alkenes is dominated by addition reactions, the preparation of alkenes is dominated by **elimination reactions**. Additions and eliminations are in many respects two sides of the same coin. That is, an addition reaction might involve the *addition* of HBr or $H_2O$ to an alkene to form an alkyl halide or an alcohol, whereas an elimination reaction might involve the *loss* of HBr or $H_2O$ from an alkyl halide or alcohol to form an alkene.

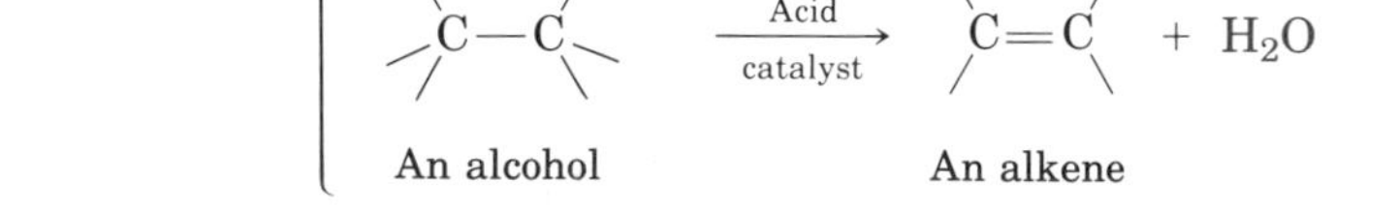

Elimination reactions

$$\text{H–C–C–X} \xrightarrow{\text{KOH}} \text{C=C} + \text{KX} + H_2O$$

An alkyl halide → An alkene

$$\text{H–C–C–OH} \xrightarrow[\text{catalyst}]{\text{Acid}} \text{C=C} + H_2O$$

An alcohol → An alkene

The two most common alkene-forming elimination reactions are **dehydrohalogenation**—the loss of HX from an alkyl halide—and **dehydration**—the loss of water from an alcohol. Dehydrohalogenation usually occurs by reaction of an alkyl halide with strong base such as KOH. For example, bromocyclohexane yields cyclohexene when treated with potassium hydroxide in ethanol solution:

$$\text{Bromocyclohexane} \xrightarrow[CH_3CH_2OH]{KOH} \text{Cyclohexene (81\%)} + KBr + H_2O$$

Bromocyclohexane Cyclohexene (81%)

Dehydration is often carried out by treatment of an alcohol with a strong acid. For example, loss of water takes place and 1-methylcyclohexene is formed when 1-methylcyclohexanol is warmed with aqueous sulfuric acid in tetrahydrofuran solvent:

$$\text{1-Methylcyclohexanol} \xrightarrow[50°C]{H_3O^+,\ THF} \text{1-Methylcyclohexene (91\%)} + H_2O$$

1-Methylcyclohexanol 1-Methylcyclohexene (91%)

Both kinds of elimination reactions are mechanistically complex enough, however, so that it is best to defer a detailed discussion until Chapter 11. For the present, it's sufficient just to realize that alkenes are readily available from simple precursors.

**PROBLEM**

**7.17** One of the problems in using elimination reactions is that mixtures of products are often formed. For example, treatment of 2-bromo-2-methylbutane with KOH in ethanol yields a mixture of two alkene products. What are their structures?

$$CH_3CH_2\overset{\overset{\displaystyle Br}{|}}{C}(CH_3)_2 \xrightarrow{KOH} ?$$

## 7.13 Summary and Key Words

**Electrophilic addition** is the most important reaction of alkenes. HCl, HBr, and HI add to carbon–carbon double bonds by a two-step mechanism involving initial reaction of the nucleophilic double bond with $H^+$ to form a carbocation intermediate, followed by attack of halide ion nucleophile on the cation intermediate. Bromine and chlorine add to alkenes via three-membered-ring **halonium ion** intermediates to give addition products having **anti stereochemistry**. If water is present during halogen addition reactions, a **halohydrin** is formed.

**Hydration** of alkenes (addition of water) is carried out by either of two procedures, depending on the product desired. **Oxymercuration** involves electrophilic addition of mercuric ion to an alkene, followed by trapping of the cation intermediate with water and treatment with $NaBH_4$. **Hydroboration** of alkenes involves addition of borane ($BH_3$) followed by oxidation of the intermediate organoborane with alkaline hydrogen peroxide. The two hydration methods are complementary: Oxymercuration gives the product of Markovnikov addition, whereas hydroboration–oxidation gives the product of non-Markovnikov syn addition.

HBr (but not HCl or HI) can also add to alkenes by a radical chain pathway to give the non-Markovnikov product. Radicals have the stability order:

$$\begin{array}{ccccccc} \text{Tertiary} & > & \text{Secondary} & > & \text{Primary} & > & \text{Methyl} \\ R_3C\cdot & > & R_2\dot{C}H & > & R\dot{C}H_2 & > & \cdot CH_3 \end{array}$$

Alkenes are **reduced** by addition of hydrogen in the presence of a catalyst such as platinum or palladium. **Catalytic hydrogenation** is a heterogeneous process that occurs on the surface of catalyst particles, rather than in solution. The reaction occurs with syn stereochemistry. Cis 1,2-diols can be made directly from alkenes by **hydroxylation** with osmium tetraoxide, $OsO_4$. Alkenes can also be cleaved to produce carbonyl-containing compounds by reaction with ozone, followed by reduction with zinc metal.

**Carbenes, $R_2C:$**, are neutral molecules containing a divalent carbon with only six valence electrons. Carbenes are highly reactive toward alkenes, adding to give cyclopropanes. Dichlorocarbene, usually prepared from $CHCl_3$ by reaction with base, adds to alkenes to give 1,1-dichlorocyclopropanes. Nonhalogenated cyclopropanes are best prepared by treatment of the alkene with $CH_2I_2$ and zinc–copper alloy—the **Simmons–Smith reaction**.

Methods for the preparation of alkenes generally involve elimination reactions. For example, treatment of an alkyl halide with a strong base effects elimination of HX (**dehydrohalogenation**) and yields the alkene. Alcohols undergo a similar elimination of water (**dehydration**) by treatment with strong acid.

## LEARNING REACTIONS

What's seven times nine? Sixty-three, of course. You didn't have to stop and figure it out; you didn't have to count your fingers and toes; you knew the answer immediately because you've memorized the multiplication tables. Learning the reactions of organic chemistry requires the same approach: Reactions have to be memorized for immediate recall if they are to be useful.

Different people take different approaches to learning reactions. Some people make flash cards; others find studying with friends to be helpful. To help guide your study, the chapters in this book end with a summary of the new reactions just presented. In addition, the accompanying *Study Guide and Solutions Manual* has several appendixes that organize organic reactions from other viewpoints. Fundamentally, though, there are no short cuts. Learning organic chemistry takes effort.

## 7.14 Summary of Reactions

1. Addition reactions of alkenes

a. Addition of HX, where X = Cl, Br, or I (Sections 6.9 and 6.10)

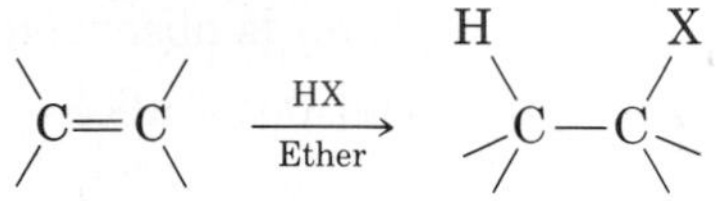

$$\mathrm{C{=}C} \xrightarrow[\text{Ether}]{\text{HX}} \mathrm{H{-}C{-}C{-}X}$$

Markovnikov regiochemistry is observed: H adds to the less substituted carbon, and X adds to the more substituted carbon.

b. Addition of halogens, where $X_2$ = $Cl_2$ or $Br_2$ (Section 7.1)

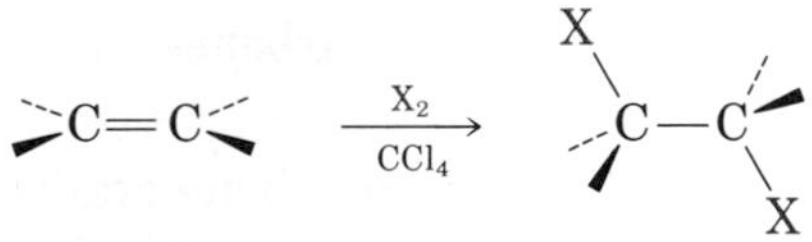

$$\mathrm{C{=}C} \xrightarrow[\mathrm{CCl_4}]{\mathrm{X_2}} \mathrm{X{-}C{-}C{-}X}$$

Anti addition is observed.

c. Halohydrin formation (Section 7.2)

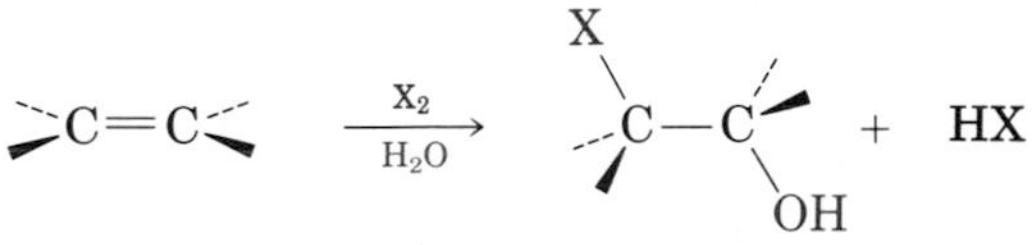

$$\mathrm{C{=}C} \xrightarrow[\mathrm{H_2O}]{\mathrm{X_2}} \mathrm{X{-}C{-}C{-}OH} + \mathrm{HX}$$

Markovnikov regiochemistry and anti stereochemistry are observed.

d. Addition of water by oxymercuration (Section 7.3)

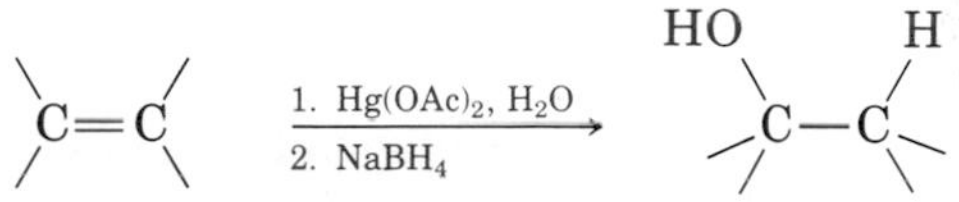

$$\mathrm{C{=}C} \xrightarrow[\text{2. NaBH}_4]{\text{1. Hg(OAc)}_2,\ \mathrm{H_2O}} \mathrm{HO{-}C{-}C{-}H}$$

Markovnikov regiochemistry is observed, with the OH group at the site of the more substituted carbon.

e. Addition of water by hydroboration–oxidation (Section 7.4)

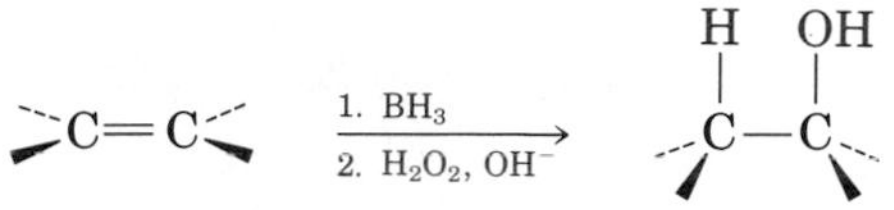

$$\mathrm{C{=}C} \xrightarrow[\text{2. H}_2\text{O}_2,\ \text{OH}^-]{\text{1. BH}_3} \mathrm{H{-}C{-}C{-}OH}$$

Non-Markovnikov syn addition is observed.

f. Radical addition of HBr to alkenes (Section 7.5)

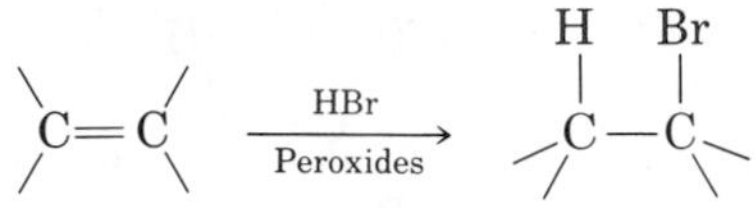

$$\mathrm{C{=}C} \xrightarrow[\text{Peroxides}]{\text{HBr}} \mathrm{H{-}C{-}C{-}Br}$$

Non-Markovnikov addition is observed.

g. Hydrogenation of alkenes (Section 7.6)

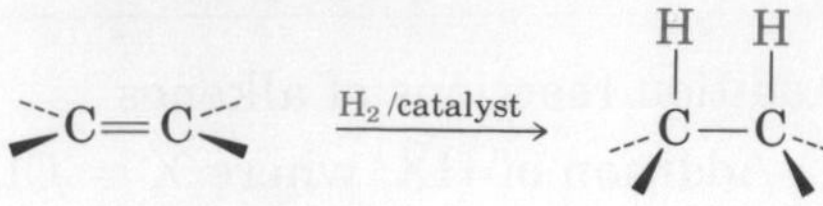

$$\text{C}=\text{C} \xrightarrow{H_2\text{/catalyst}} \text{H--C--C--H}$$

Syn addition is observed.

h. Hydroxylation of alkenes (Section 7.7)

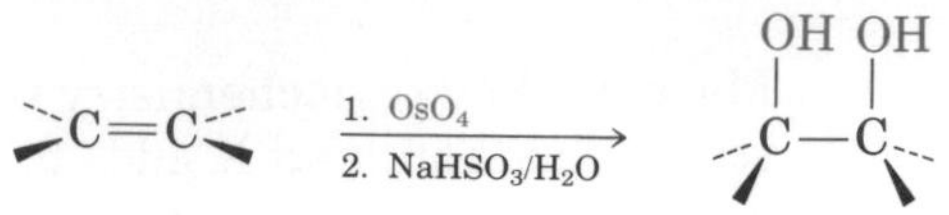

$$\text{C}=\text{C} \xrightarrow[\text{2. } NaHSO_3/H_2O]{\text{1. } OsO_4} \text{HO--C--C--OH}$$

Syn addition is observed.

i. Addition of carbenes to alkenes to yield cyclopropanes (Section 7.10)

(1) Dichlorocarbene addition

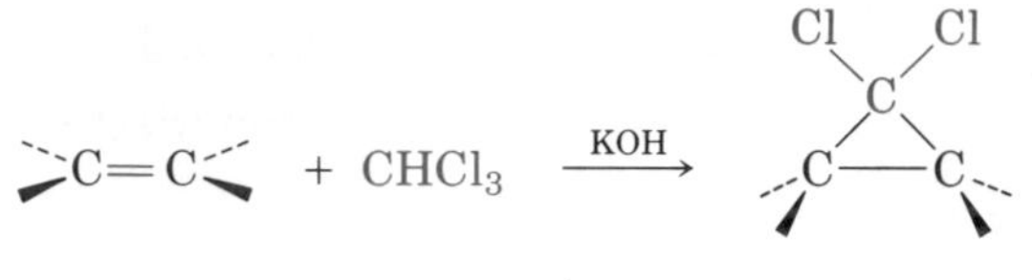

$$\text{C}=\text{C} + CHCl_3 \xrightarrow{KOH} \text{cyclopropane with } CCl_2$$

(2) Simmons–Smith reaction

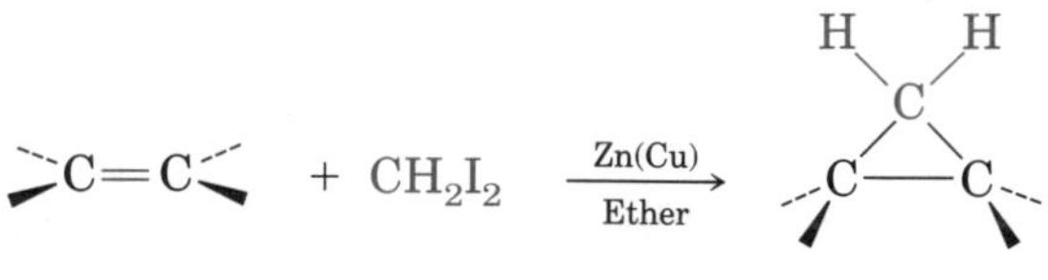

$$\text{C}=\text{C} + CH_2I_2 \xrightarrow[\text{Ether}]{Zn(Cu)} \text{cyclopropane with } CH_2$$

2. Oxidative cleavage of alkenes (Section 7.8)

a. Treatment with ozone, followed by zinc in acetic acid

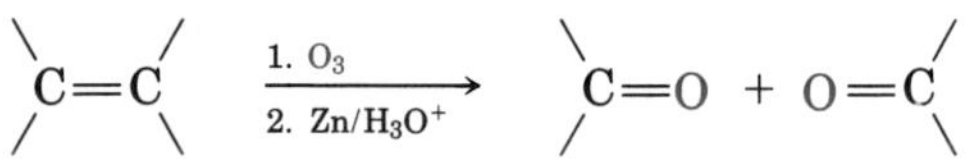

$$\text{C}=\text{C} \xrightarrow[\text{2. } Zn/H_3O^+]{\text{1. } O_3} \text{C}=\text{O} + \text{O}=\text{C}$$

b. Reaction with $KMnO_4$ in acidic solution

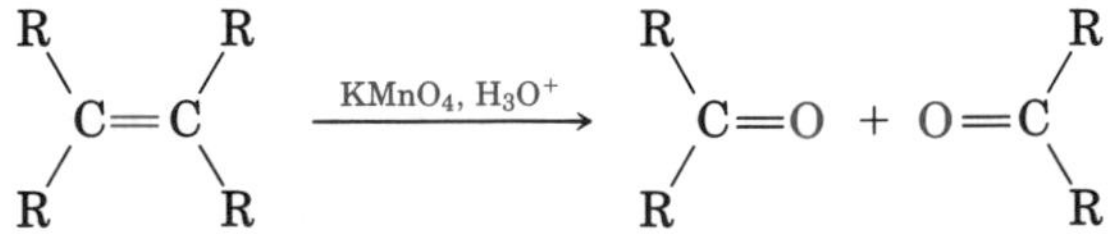

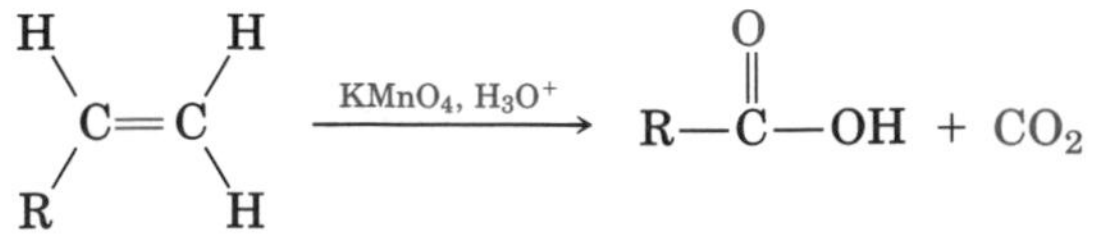

$$R_2C=CR_2 \xrightarrow{KMnO_4,\ H_3O^+} R_2C=O + O=CR_2$$

$$RHC=CH_2 \xrightarrow{KMnO_4,\ H_3O^+} R\text{—}\overset{O}{\overset{\|}{C}}\text{—}OH + CO_2$$

3. Oxidative cleavage of 1,2-diols (Section 7.9)

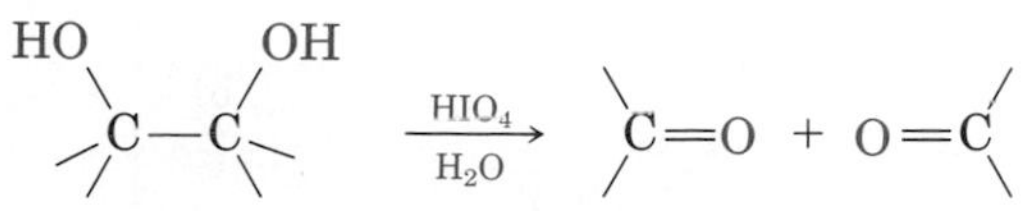

$$\text{HO–C–C–OH} \xrightarrow[H_2O]{HIO_4} \text{C=O} + \text{O=C}$$

4. Synthesis of alkenes (Section 7.12)
   a. Dehydrohalogenation of alkyl halides

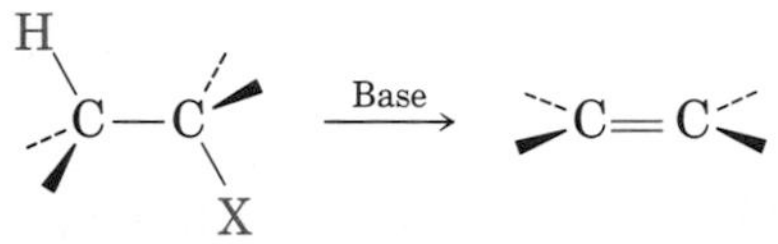

$$\text{H–C–C–X} \xrightarrow{\text{Base}} \text{C=C}$$

   b. Dehydration of alcohols

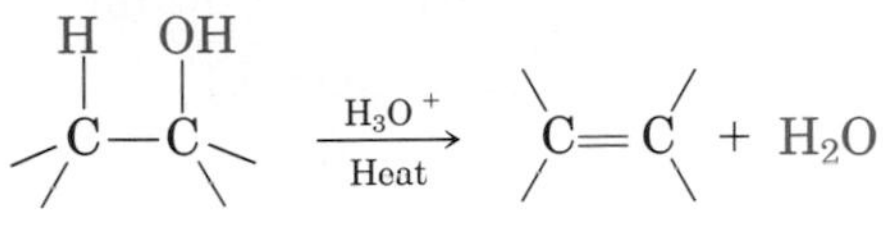

$$\text{H–C–C–OH} \xrightarrow[\text{Heat}]{H_3O^+} \text{C=C} + H_2O$$

## ADDITIONAL PROBLEMS

**7.18** Predict the products of the following reactions (the benzene ring is unreactive in all cases). Indicate regiochemistry when relevant.

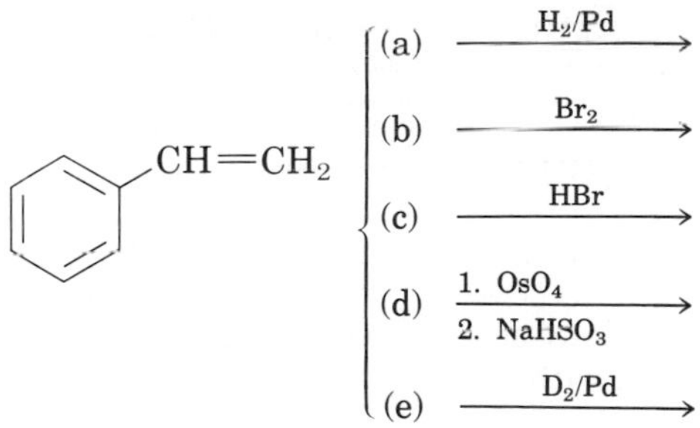

$C_6H_5CH{=}CH_2$

(a) $\xrightarrow{H_2/Pd}$

(b) $\xrightarrow{Br_2}$

(c) $\xrightarrow{HBr}$

(d) $\xrightarrow[\text{2. } NaHSO_3]{\text{1. } OsO_4}$

(e) $\xrightarrow{D_2/Pd}$

**7.19** Suggest structures for alkenes that give the following reaction products. There may be more than one answer for some cases.

(a) ? $\xrightarrow{H_2/Pd}$ 2-Methylhexane

(b) ? $\xrightarrow{H_2/Pd}$ 1,1-Dimethylcyclohexane

(c) ? $\xrightarrow{Br_2/CCl_4}$ 2,3-Dibromo-5-methylhexane

(d) ? $\xrightarrow[\text{2. } NaBH_4]{\text{1. } Hg(OAc)_2,\ H_2O}$ $CH_3CH_2CH_2CH(OH)CH_3$

(e) ? $\xrightarrow{\text{HBr/peroxides}}$ 2-Bromo-3-methylheptane

(f) ? $\xrightarrow{\text{HCl, ether}}$ 2-Chloro-3-methylheptane

**7.20** Predict the products of the following reactions, indicating both regiochemistry and stereochemistry where appropriate.

(c) 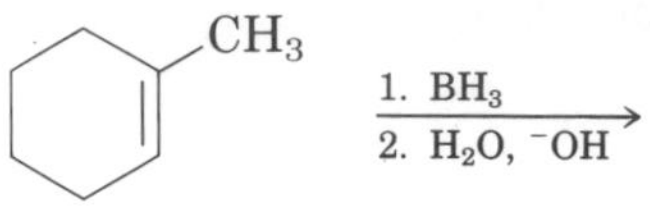 1-methylcyclohexene (CH3) $\xrightarrow[\text{2. } H_2O,\ ^-OH]{\text{1. } BH_3}$

(d) 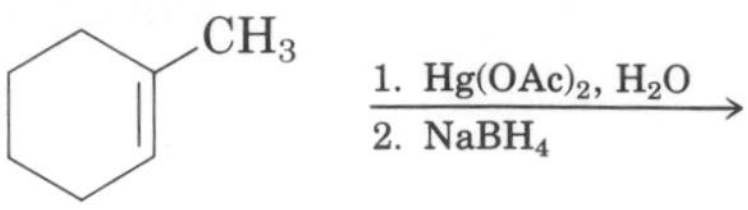 1-methylcyclohexene (CH3) $\xrightarrow[\text{2. } NaBH_4]{\text{1. } Hg(OAc)_2,\ H_2O}$

**7.21** How would you carry out the following transformations? Indicate the proper reagents.

(a) 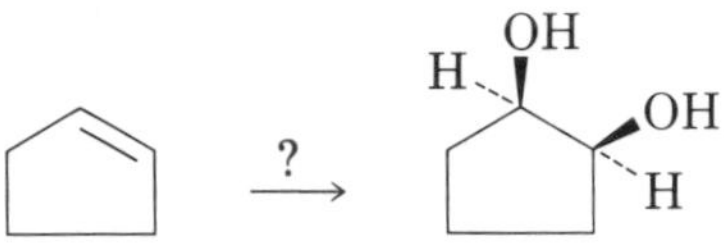 cyclopentene $\xrightarrow{?}$ cyclopentane-1,2-diol (OH, H; OH, H)

(b) 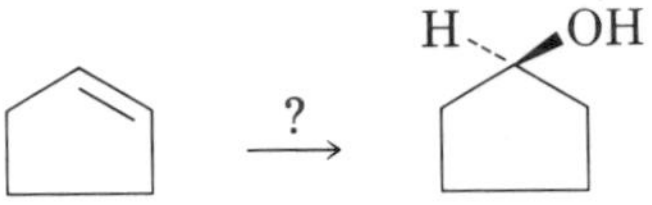 cyclopentene $\xrightarrow{?}$ cyclopentanol (H, OH)

(c) 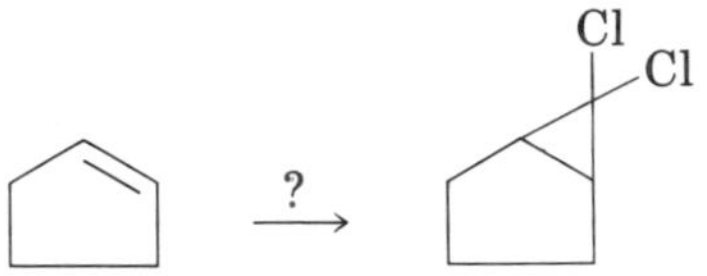 cyclopentene $\xrightarrow{?}$ dichlorocyclopropane-fused cyclopentane (Cl, Cl)

(d) $CH_3CH{=}CHCH(CH_3)_2 \xrightarrow{?} CH_3CHO + (CH_3)_2CHCHO$

(e) $H_2C{=}C(CH_3)_2 \xrightarrow{?} (CH_3)_2CHCH_2OH$

(f) 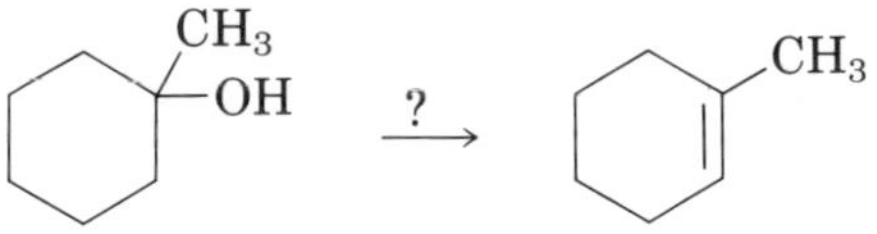 1-methylcyclohexanol (CH3, OH) $\xrightarrow{?}$ 1-methylcyclohexene (CH3)

**7.22** Give the structure of an alkene that provides only $(CH_3)_2C{=}O$ on ozonolysis followed by treatment with zinc.

**7.23** Draw the structure of a hydrocarbon that reacts with 1 mol equiv of hydrogen on catalytic hydrogenation and gives only pentanal, $n$-$C_4H_9CHO$, on ozonolysis. Write out the reactions involved.

**7.24** Show the structures of alkenes that give the following products on oxidative cleavage with $KMnO_4$ in acidic solution.

(a) $CH_3CH_2COOH + CO_2$ (b) $(CH_3)_2C{=}O + CH_3CH_2CH_2COOH$

(c) 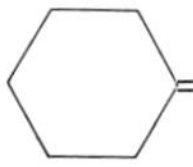 cyclohexanone ($=$O) $+ (CH_3)_2C{=}O$

**7.25** Compound A has the formula $C_{10}H_{16}$. On catalytic hydrogenation over palladium, it reacts with only 1 equiv of hydrogen. Compound A undergoes reaction with ozone, followed by zinc treatment, to yield a symmetrical diketone, B, with formula $C_{10}H_{16}O_2$.

(a) How many rings does A have? (b) What are the structures of A and B?

(c) Formulate the reactions.

**7.26** An unknown hydrocarbon, A, with formula $C_6H_{12}$, reacts with 1 equiv of hydrogen over a palladium catalyst. Hydrocarbon A also reacts with $OsO_4$ to give a diol, B. When oxidized with $KMnO_4$ in acidic solution, A gives two fragments. One fragment can be identified as propanoic acid, $CH_3CH_2COOH$, and the other fragment can be shown to be a ketone, C. What are the structures of A, B, and C? Write out all reactions, and show your reasoning.

**7.27** Using an oxidative cleavage reaction, explain how you would distinguish between the following two isomeric dienes:

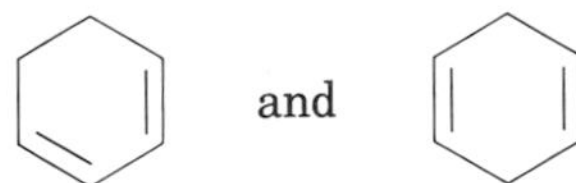


**7.28** Compound A, with formula $C_{10}H_{18}O$, undergoes reaction with dilute $H_2SO_4$ at 250°C to yield a mixture of two alkenes, $C_{10}H_{16}$. The major alkene product, B, gives only cyclopentanone after ozone treatment followed by reduction with zinc in acetic acid. Formulate the reactions involved and identify A and B.

**7.29** Which reaction would you expect to occur faster, addition of HBr to cyclohexene or to 1-methylcyclohexene? Explain your answer.

**7.30** Predict the products of the following reactions, and indicate regiochemistry if relevant.

(a) $CH_3CH{=}CHCH_3 \xrightarrow{HBr}$

(b) $CH_3CH{=}CHCH_3 \xrightarrow{BH_3} A \xrightarrow[^-OH]{H_2O_2} B$

(c) $(CH_3)_2C{=}CH_2 \xrightarrow[Peroxide]{HBr}$

(d) $CH_3CH{=}C(CH_3)_2 \xrightarrow[Peroxide]{HI}$

**7.31** Draw the structure of a hydrocarbon that absorbs 2 mol equiv of hydrogen on catalytic hydrogenation and gives only butanedial, $CHOCH_2CH_2CHO$, on ozonolysis.

**7.32** In planning the synthesis of one compound from another, it's just as important to know what *not* to do as to know what to do. The following proposed reactions all have serious drawbacks to them. Explain the potential problems of each reaction.

(a) $(CH_3)_2C{=}CHCH_3 \xrightarrow[Peroxides]{HI} (CH_3)_2CHCHICH_3$

(b) 1. $OsO_4$ 2. $NaHSO_3$ → H, OH, HO, H

(c) 1. $O_3$ 2. Zn → CHO, CHO

(d) $CH_3$ 1. $BH_3$ 2. $H_2O_2$, $^-OH$ → H, $CH_3$, OH, H

**7.33** Which of these alcohols could *not* be made selectively by hydroboration–oxidation of an alkene?

(a) $CH_3CH_2CH_2CH(OH)CH_3$

(b) $(CH_3)_2CHC(OH)(CH_3)_2$

(c) 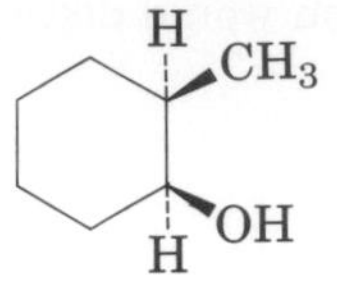 cyclohexane with H, $CH_3$ and H, OH

(d) 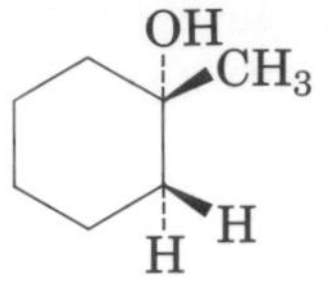 cyclohexane with OH, $CH_3$ and H, H

**7.34** What alkenes might be used to prepare these cyclopropanes?

(a)  cyclopropane–$CH(CH_3)_2$

(b) 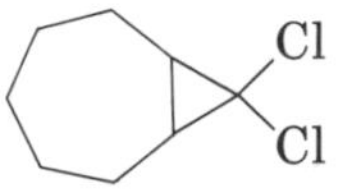 bicyclic cyclopropane fused to cycloheptane with Cl, Cl

**7.35** Predict the products of the following reactions. Don't worry about the size of the molecule; concentrate on the functional groups.

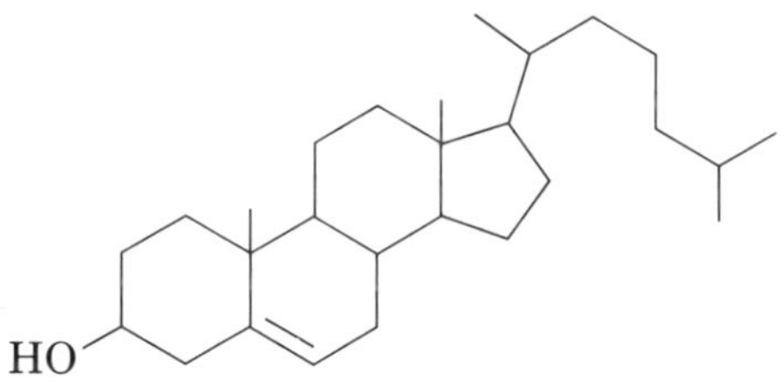

HO

Cholesterol

$\xrightarrow{Br_2}$ A

$\xrightarrow{HBr}$ B

1. $OsO_4$ 2. $NaHSO_3$ → C

1. $BH_3$, THF 2. $H_2O_2$, $^-OH$ → D

$CH_2I_2$ / Zu(Cu) → E

**7.36** Compound A has the formula $C_8H_8$. It reacts rapidly with $KMnO_4$ to give $CO_2$ and a carboxylic acid, B, with formula $C_7H_6O_2$, but reacts with only 1 equiv of $H_2$ on catalytic hydrogenation over a palladium catalyst. On hydrogenation under conditions that reduce aromatic rings, 4 equiv of $H_2$ are taken up, and hydrocarbon C, formula $C_8H_{16}$, is produced. What are the structures of A, B, and C? Formulate the reactions.

**7.37** Reaction of cyclohexene with mercuric acetate in methyl alcohol rather than water, followed by treatment with $NaBH_4$, yields cyclohexyl methyl ether rather than cyclohexanol. Suggest a mechanism to account for this ether synthesis.

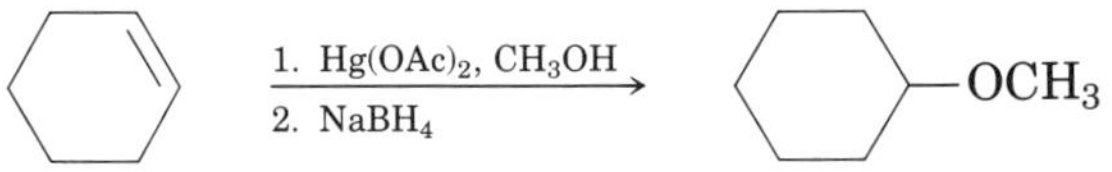

Cyclohexene $\xrightarrow[2.\ NaBH_4]{1.\ Hg(OAc)_2,\ CH_3OH}$ Cyclohexyl methyl ether ($-OCH_3$)

**7.38** When 4-penten-1-ol is treated with aqueous bromine, a cyclic bromo ether is formed, rather than the expected bromohydrin. Propose a mechanism for this transformation. [*Hint:* See Problem 7.37.]

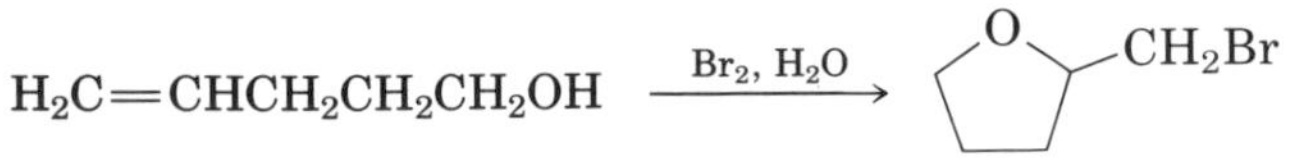

$H_2C{=}CHCH_2CH_2CH_2OH \xrightarrow{Br_2,\ H_2O}$ (O, $CH_2Br$)

4-Penten-1-ol → 2-(Bromomethyl)tetrahydrofuran

**7.39** How would you distinguish between the following pairs of compounds using simple chemical tests? Tell what you would do and what you would see.
(a) Cyclopentene and cyclopentane (b) 2-Hexene and benzene

**7.40** Ethylidenecyclohexane, on treatment with a strong acid, isomerizes to yield 1-ethylcyclohexene. Propose a mechanism for this reaction. Which is the more stable alkene, ethylidenecyclohexane or 1-ethylcyclohexene?

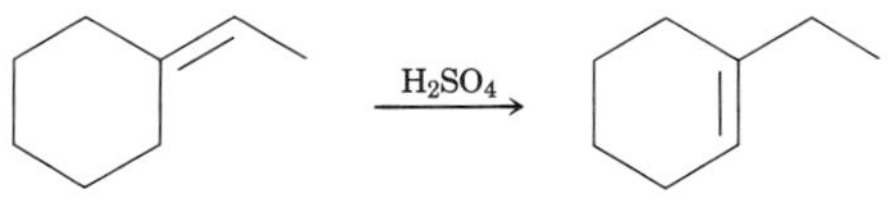

$$\xrightarrow{H_2SO_4}$$

Ethylidenecyclohexane 1-Ethylcyclohexene

**7.41** In addition to its preparation from $CHCl_3$, dichlorocarbene can also be generated by heating sodium trichloroacetate:

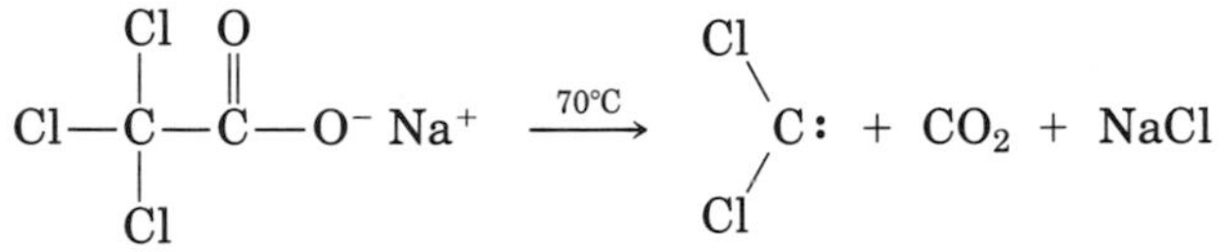

$$Cl_3C{-}CO{-}O^- \, Na^+ \xrightarrow{70°C} Cl_2C\colon + CO_2 + NaCl$$

Propose a mechanism to account for this reaction, and use curved arrows to indicate the movement of electrons. What relation does your mechanism bear to the base-induced elimination of HCl from chloroform?

**7.42** $\alpha$-Terpinene, $C_{10}H_{16}$, is a pleasant-smelling hydrocarbon that has been isolated from oil of marjoram. On hydrogenation over a palladium catalyst, $\alpha$-terpinene reacts with 2 mol equiv of hydrogen to yield a new hydrocarbon, $C_{10}H_{20}$. On ozonolysis, followed by reduction with zinc and acetic acid, $\alpha$-terpinene yields two products, glyoxal and 6-methyl-2,5-heptanedione.

How many degrees of unsaturation does $\alpha$-terpinene have? How many double bonds? How many rings? Propose a structure for $\alpha$-terpinene that is consistent with the foregoing data.

$OHC{-}CHO$ $CH_3COCH_2CH_2COCH(CH_3)_2$

Glyoxal 6-Methyl-2,5-heptanedione

**7.43** Evidence that cleavage of 1,2-diols by $HIO_4$ occurs through a five-membered cyclic periodate intermediate is based on *kinetic data*—the measurement of reaction rates. When diols A and B were prepared and the rates of their reaction with $HIO_4$ were measured, it was found that diol A cleaved approximately 1 million times faster than diol B. Explain these results by making molecular models of A and B and of potential cyclic periodate intermediates.

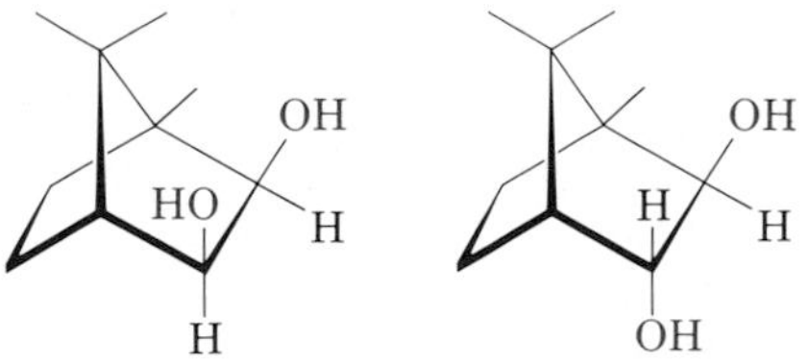

OH HO H H OH H H OH

A (cis diol) B (trans diol)

CHAPTER 8

# Alkynes

**Alkynes** (also called **acetylenes**) are hydrocarbons that contain a carbon–carbon triple bond. Acetylene itself (H—C≡C—H), the simplest alkyne, was once widely used in industry as the starting material for the preparation of acetaldehyde, acetic acid, vinyl chloride, and other high-volume chemicals, but more efficient routes using ethylene as starting material are now more common. Acetylene is still used in the preparation of acrylic polymers, however, and is prepared industrially by high-temperature decomposition (**pyrolysis**) of methane. This method is not of general utility in the laboratory.

$$2\,CH_4 \xrightarrow[1200^\circ C]{Steam} HC{\equiv}CH + 3\,H_2$$

Methane    Acetylene

A large number of naturally occurring acetylenic compounds have been isolated from the plant kingdom. For example, the following *triyne* from the safflower, *Carthamus tinctorius* L., evidently forms part of the plant's chemical defenses against nematode infestation:

$$CH_3{-}CH{=}CH{-}C{\equiv}C{-}C{\equiv}C{-}C{\equiv}C{-}CH{=}CH{-}CH{=}CH_2$$

A triyne

## 8.1 Electronic Structure of Alkynes

A carbon–carbon triple bond results from the overlap of two *sp*-hybridized carbon atoms (Section 1.11). Recall that the two *sp* hybrid orbitals of carbon adopt a linear geometry. They lie at an angle of 180° to each other along an axis that is perpendicular to the axes of the two unhybridized $2p_y$ and $2p_z$ orbitals. When two *sp*-hybridized carbons approach each other for bonding, the geometry is perfect for the formation of one *sp–sp* sigma bond and two *p–p* pi bonds—a net *triple* bond (Figure 8.1).

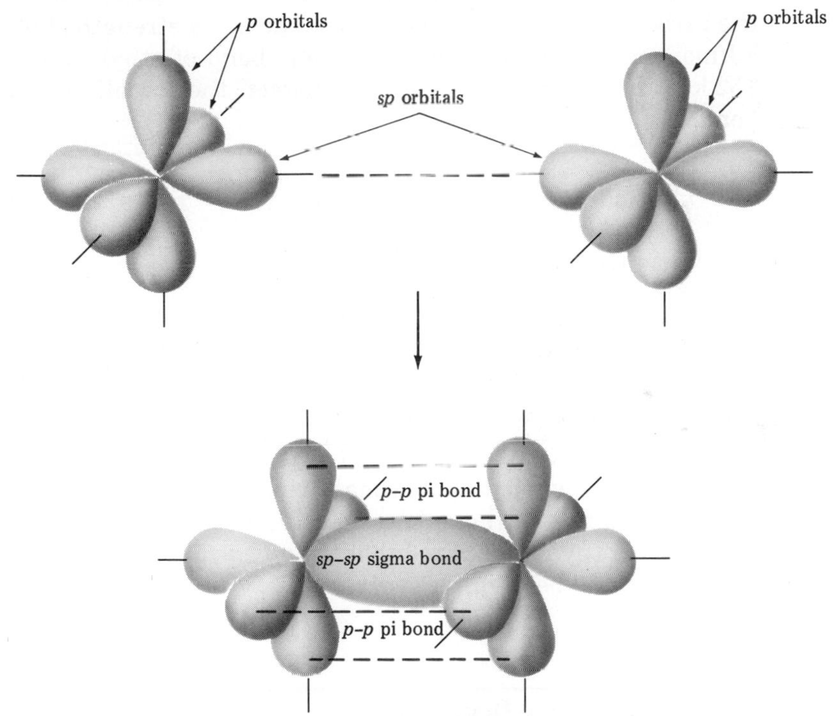


**Figure 8.1** Formation of a carbon–carbon triple bond by overlap of two *sp*-hybridized carbons

The two remaining *sp* orbitals form bonds to other atoms at an angle of 180° from the carbon–carbon bond. For example, acetylene, $C_2H_2$, has been shown by experimental measurement to be a linear molecule with H-C-C bond angles of 180° (Figure 8.2, page 230).

**Figure 8.2** The structure of acetylene: The H-C-C bond angles are 180°

The strength of the carbon–carbon triple bond is approximately 200 kcal/mol (837 kJ/mol), making it the strongest known carbon–carbon bond. The bond length is 1.20 Å. In a purely bookkeeping sense, we can assign bond strengths to each of the three triple-bond "parts." Since we know that the carbon–carbon *single* bond in ethane has a strength of 88 kcal/mol (368 kJ/mol) and the carbon–carbon *double* bond of ethylene has a strength of 152 kcal/mol (636 kJ/mol), we can "dissect" the overall carbon–carbon triple bond:

$$-C{\equiv}C- \longrightarrow -\dot{C}{=}\dot{C}- \qquad \Delta H^\circ = 200 - 152 = 48 \text{ kcal/mol}$$

$$>C{=}C< \longrightarrow >\dot{C}{-}\dot{C}< \qquad \Delta H^\circ = 152 - 88 = 64 \text{ kcal/mol}$$

$$-\overset{|}{\underset{|}{C}}-\overset{|}{\underset{|}{C}}- \longrightarrow -\dot{C}\ \ \dot{C}- \qquad \Delta H^\circ = 88 \text{ kcal/mol}$$

Although the two alkyne pi bonds are actually equivalent, our crude calculation shows that approximately 48 kcal/mol (200 kJ/mol) are needed to break the first of them. Since this value is 64 − 48 = 16 kcal/mol less than the energy required to break an *alkene* pi bond, we might predict that alkynes should be highly reactive.

## 8.2 Nomenclature of Alkynes

Alkynes follow the general rules of hydrocarbon nomenclature already discussed. The suffix *-yne* is substituted for *-ane* in the base hydrocarbon name to denote an alkyne, and the position of the triple bond is indicated by its number in the chain. Numbering begins at the chain end nearer the triple bond.

$$\overset{8}{C}H_3\overset{7}{C}H_2\underset{\underset{CH_3}{|}}{\overset{6}{C}}H\overset{5}{C}H_2\overset{4}{C}{\equiv}\overset{3}{C}\overset{2}{C}H_2\overset{1}{C}H_3$$

Begin numbering at the end nearer the triple bond

6-Methyl-3-octyne

Compounds with more than one triple bond are called *diynes, triynes,* and so forth; compounds containing both double and triple bonds are called *enynes* (not *ynenes*). Numbering of an enyne chain always starts from the end nearer the first multiple bond, even though this might give the triple bond the lower number. When there is a choice in numbering, however, double bonds receive lower numbers than triple bonds. For example,

$$\underset{7}{HC}\equiv\underset{6}{C}\underset{5}{CH_2}\underset{4}{CH_2}\underset{3}{CH_2}\underset{2}{CH}=\underset{1}{CH_2}$$

1-Hepten-6-yne

$$\underset{1}{HC}\equiv\underset{2}{C}\underset{3}{CH_2}\underset{4}{\overset{\overset{CH_3}{|}}{C}H}\underset{5}{CH_2}\underset{6}{CH_2}\underset{7}{CH}=\underset{8}{CH}\underset{9}{CH_3}$$

4-Methyl-7-nonen-1-yne

As was the case with hydrocarbon substituents derived from alkanes and alkenes, *alkynyl* groups are also possible:

| $CH_3CH_2CH_2CH_2$– | $CH_3CH_2CH=CH$– | $CH_3CH_2C\equiv C$– |
|---|---|---|
| Butyl (an alkyl group) | 1-Butenyl (a vinylic group) | 1-Butynyl (an alkynyl group) |

PROBLEM

**8.1** Provide correct IUPAC names for these compounds:

(a) $(CH_3)_2CHC\equiv CCH(CH_3)_2$

(b) $HC\equiv CC(CH_3)_3$

(c) $CH_3CH=CHCH=CHC\equiv CCH_3$

(d) $CH_3CH_2C(CH_3)_2C\equiv CCH_2CH_2CH_3$

(e) $CH_3CH_2C(CH_3)_2C\equiv CCH(CH_3)_2$

(f)

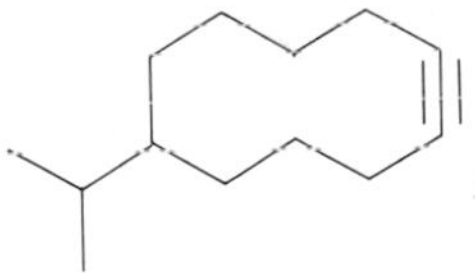

PROBLEM

**8.2** There are seven isomeric alkynes with the formula $C_6H_{10}$. Draw them and name them according to IUPAC rules.

## 8.3 Reactions of Alkynes: Addition of HX and $X_2$

You might expect, based on the electronic similarity of alkenes and alkynes, that the chemical reactivity of the two functional groups should also be similar. The pi part of the triple bond is weak (48 kcal/mol), and the electrons are readily accessible to attacking reagents. Alkynes do indeed exhibit much chemistry similar to that of alkenes, but there are also significant differences.

As a general rule, electrophilic reagents add to alkynes in the same way that they add to alkenes. With HX, for example, alkynes give the expected addition products. Although the reactions can usually be stopped after addition of 1 equiv of HX, an excess of acid leads to a dihalide product.

For example, reaction of 1-hexyne with 2 equiv HBr yields 2,2-dibromohexane. As the following examples indicate, the regiochemistry of addition follows Markovnikov's rule. Halogen adds to the more highly substituted side of the alkyne bond, and hydrogen adds to the less highly substituted side. Trans stereochemistry of H and X is normally (though not always) found in the product.

$$CH_3CH_2CH_2CH_2C{\equiv}CH \xrightarrow[CH_3COOH]{HBr} CH_3CH_2CH_2CH_2C(Br){=}CH(H)$$

1-Hexyne → 2-Bromo-1-hexene

$$\downarrow HBr$$

$$CH_3CH_2CH_2CH_2CBr_2{-}CH_2H$$

2,2-Dibromohexane

$$CH_3CH_2C{\equiv}CCH_2CH_3 \xrightarrow[CH_3COOH]{HCl,\ NH_4Cl} \begin{matrix} Cl & & CH_2CH_3 \\ & C{=}C & \\ CH_3CH_2 & & H \end{matrix}$$

3-Hexyne → (*Z*)-3-Chloro-3-hexene (95%)

The vinyl chloride (chloroethylene) used as starting material for preparation of poly(vinyl chloride) (PVC) was once produced on an immense industrial scale by mercuric chloride-catalyzed addition of HCl to acetylene. Now, however, other synthetic methods are used.

$$H{-}C{\equiv}C{-}H + HCl \xrightarrow{HgCl_2} H_2C{=}CHCl$$

Acetylene → Vinyl chloride (Chloroethylene)

Bromine and chlorine also add to alkynes to give addition products, and trans stereochemistry again results:

$$CH_3CH_2C{\equiv}CH \xrightarrow[CCl_4]{Br_2} \begin{matrix} CH_3CH_2 & & Br \\ & C{=}C & \\ Br & & H \end{matrix} \xrightarrow[CCl_4]{Br_2} CH_3CH_2CBr_2CHBr_2$$

1-Butyne → (*E*)-1,2-Dibromo-1-butene → 1,1,2,2-Tetrabromobutane

Although you might expect alkynes to be more reactive than alkenes in electrophilic additions because the triple bond has four pi electrons rather than two, the reverse is true. Triple bonds are *less* reactive toward electrophilic addition than double bonds. For example, ethylene reacts rapidly at room temperature with concentrated $H_2SO_4$, but acetylene is inert to this reagent. Why should this be?

The probable reason for the decreased reactivity of alkynes toward electrophilic reagents has to do with the mechanism of the reaction. When an electrophile such as HBr adds to an alkene (Sections 6.9–6.10), the reaction takes place in two steps and involves an alkyl carbocation intermediate. When HBr adds to an *alkyne*, however, a **vinylic carbocation** is formed as the intermediate. [*Remember: Vinylic* means on a double bond.]

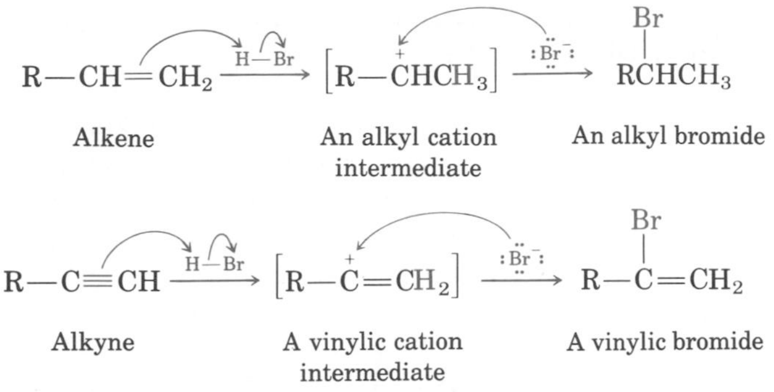

$$R{-}CH{=}CH_2 \xrightarrow{H{-}Br} \left[R{-}\overset{+}{C}HCH_3\right] \xrightarrow{:\ddot{Br}:^-} R\overset{Br}{\overset{|}{C}}HCH_3$$

Alkene — An alkyl cation intermediate — An alkyl bromide

$$R{-}C{\equiv}CH \xrightarrow{H{-}Br} \left[R{-}\overset{+}{C}{=}CH_2\right] \xrightarrow{:\ddot{Br}:^-} R{-}\overset{Br}{\overset{|}{C}}{=}CH_2$$

Alkyne — A vinylic cation intermediate — A vinylic bromide

Vinylic carbocations are generally less stable than similarly substituted alkyl cations. Thus, a *secondary* vinylic cation is only about as stable as a *primary* alkyl cation. The relative order of carbocation stability is as follows:

$$R_3\overset{+}{C} > R_2\overset{+}{C}H > R\overset{+}{C}{=}CH_2 \approx R\overset{+}{C}H_2 > RCH{=}\overset{+}{C}H$$

| Tertiary alkyl | Secondary alkyl | Secondary vinylic | Primary alkyl | Primary vinylic |
|---|---|---|---|---|

The reasons for the relative instability of vinylic carbocations are not fully understood but involve, at least in part, the lack of any stabilizing interaction with neighboring groups, such as occurs in alkyl cations. Although a vinylic cation might be either $sp^2$ hybridized or $sp$ hybridized, studies indicate that linear $sp$ hybridization is lower in energy. In this electronic configuration, the positively charged carbon atom has a vacant $p$ orbital perpendicular to the double bond (Figure 8.3).

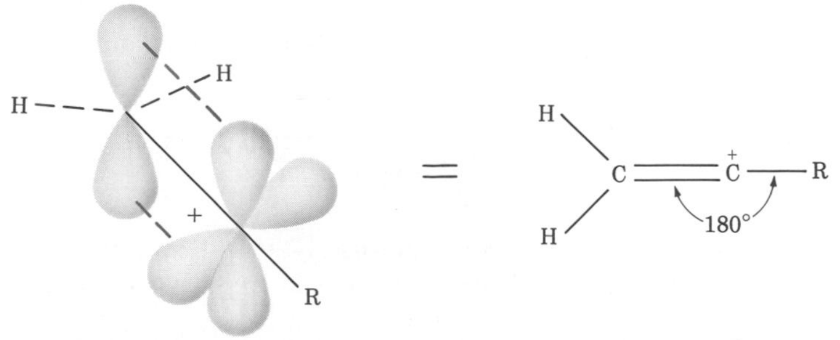

**Figure 8.3** The electronic structure of a vinylic carbocation: The cationic carbon atom is *sp* hybridized.

PROBLEM

**8.3** What products would you expect from these reactions?

(a) $CH_3CH_2CH_2C\equiv CH + 2\,Cl_2 \longrightarrow$ (b) 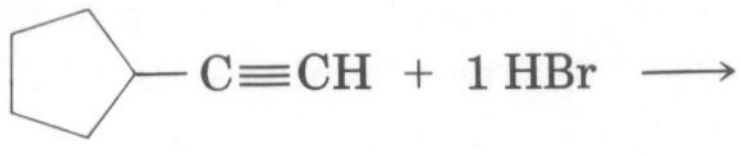 (cyclopentyl)$-C\equiv CH + 1\,HBr \longrightarrow$

(c) $CH_3CH_2CH_2CH_2C\equiv CCH_3 + 1\,HBr \longrightarrow$

## 8.4 Hydration of Alkynes

Alkynes cannot be hydrated as easily as alkenes because of their lower reactivity toward electrophilic addition. For example, aqueous sulfuric acid by itself has no effect on a carbon–carbon triple bond. In the presence of mercuric sulfate catalyst, however, hydration occurs readily:

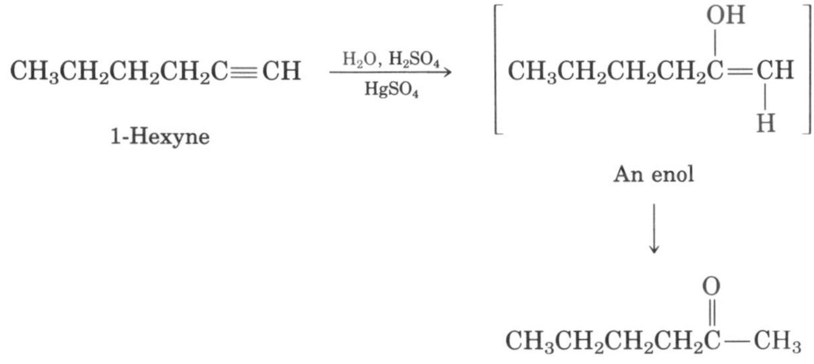

$$CH_3CH_2CH_2CH_2C\equiv CH \xrightarrow[HgSO_4]{H_2O,\ H_2SO_4} \left[ CH_3CH_2CH_2CH_2\overset{OH}{\overset{|}{C}}=\underset{\underset{H}{|}}{CH} \right] \longrightarrow CH_3CH_2CH_2CH_2\overset{O}{\overset{\|}{C}}-CH_3$$

1-Hexyne — An enol — 2-Hexanone (78%)

Markovnikov regiochemistry is observed for the hydration, but the product is not the expected vinylic alcohol or **enol** (*ene* + *ol*). Although the vinylic alcohol may well be an intermediate in the reaction, it immediately rearranges to a more stable isomer, a ketone ($R_2C=O$). It turns out that enols and ketones rapidly achieve an equilibrium—a process called **tautomerism**. **Tautomers** (pronounced **taw**-toe-mers) are special kinds of constitutional isomers that are readily interconvertible through a rapid equilibration, a process we will study in more detail in Section 22.1. With few exceptions, the **keto–enol tautomeric equilibrium** lies heavily on the side of the ketone. Vinylic alcohols are almost never isolated.

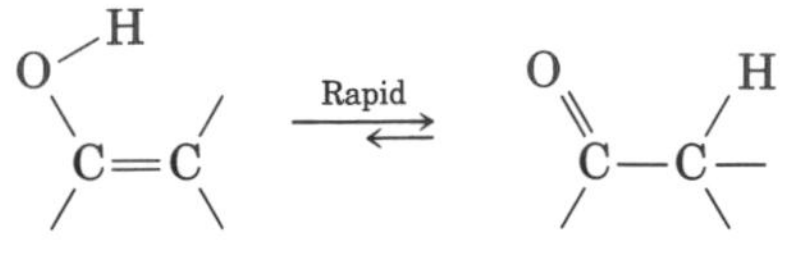

$$\overset{O-H}{\overset{\backslash}{}}C=C\!\!\diagup \;\underset{}{\overset{\text{Rapid}}{\rightleftarrows}}\; \overset{O}{\overset{\|}{C}}-\overset{H}{\overset{|}{C}}-$$

Enol tautomer (less favored) — Keto tautomer (more favored)

The mechanism of the mercuric ion-catalyzed alkyne hydration reaction is probably analogous to the oxymercuration reaction of alkenes (Section 7.3). Electrophilic addition of mercuric ion to the alkyne gives a vinylic cation, which reacts with water and loses a proton to yield an organomercury intermediate. In contrast to alkene oxymercuration, no treatment with $NaBH_4$ is necessary to remove the mercury. The acidic reaction conditions alone are sufficient to allow replacement of mercury by hydrogen (Figure 8.4).

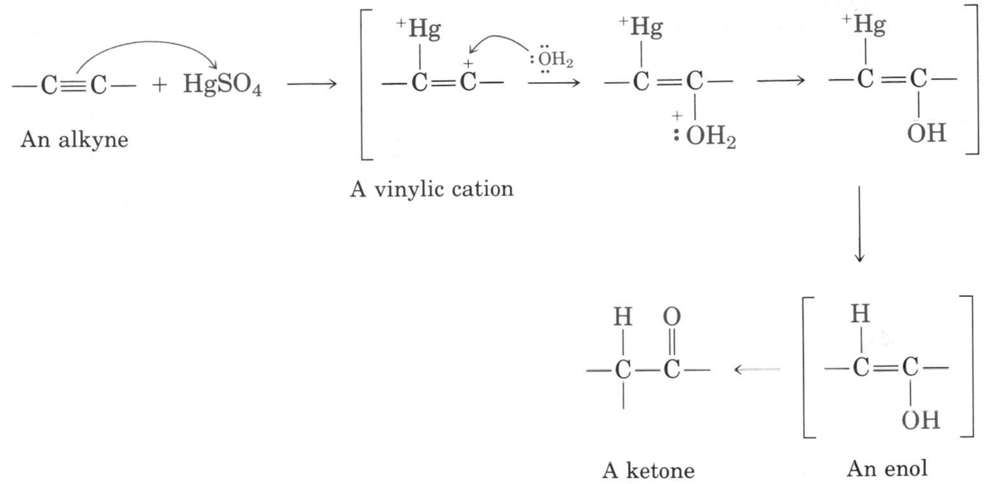

$$-C{\equiv}C- + HgSO_4 \longrightarrow \left[ \overset{^{+}Hg}{|}{-C{=}\overset{+}{C}-} \xrightarrow{:\ddot{O}H_2} {-\overset{^{+}Hg}{C}{=}\underset{\overset{+}{:}OH_2}{C}-} \longrightarrow {-\overset{^{+}Hg}{C}{=}\underset{OH}{C}-} \right]$$

An alkyne — A vinylic cation

$$\left[ {-\overset{H}{C}{=}\underset{OH}{C}-} \right] \longrightarrow {-\overset{H}{\underset{|}{C}}-\overset{O}{\overset{\|}{C}}-}$$

A ketone — An enol

**Figure 8.4** Mechanism of the mercuric ion-catalyzed hydration of an alkyne to yield a ketone: The reaction yields an intermediate enol that rapidly tautomerizes to give a ketone.

A mixture of both possible ketones results when an *internal* alkyne is hydrated, and the reaction is therefore most useful when applied to *terminal* alkynes (R—C≡C—H), since only methyl ketones are formed.

An internal alkyne

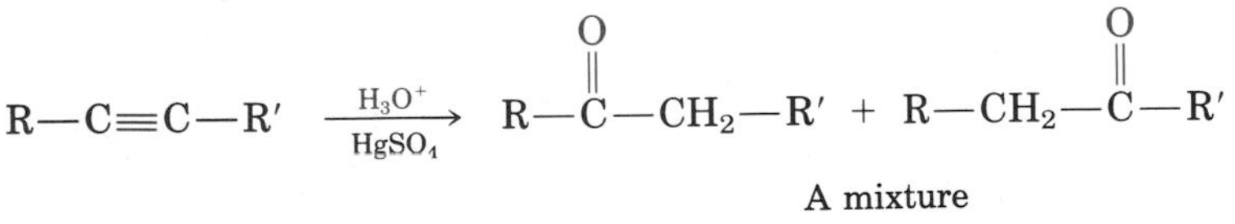

$$R-C{\equiv}C-R' \xrightarrow[HgSO_4]{H_3O^+} R-\overset{O}{\overset{\|}{C}}-CH_2-R' + R-CH_2-\overset{O}{\overset{\|}{C}}-R'$$

A mixture

A terminal alkyne

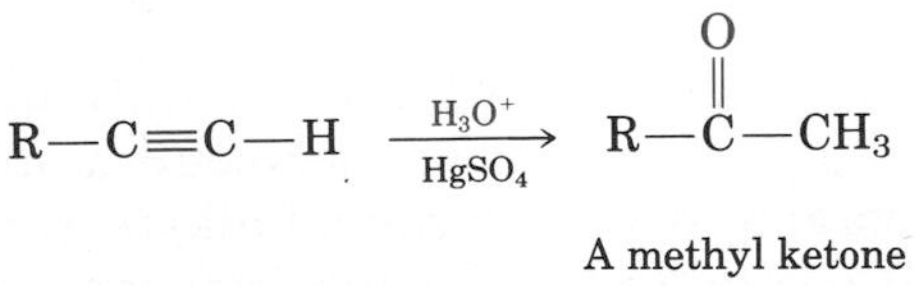

$$R-C{\equiv}C-H \xrightarrow[HgSO_4]{H_3O^+} R-\overset{O}{\overset{\|}{C}}-CH_3$$

A methyl ketone

**PROBLEM**

**8.4** What alkynes would you start with to prepare these ketones by a hydration reaction?

(a) 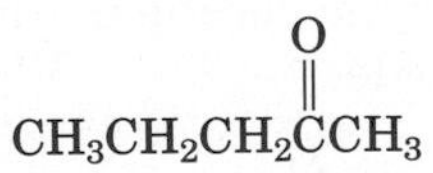 $CH_3CH_2CH_2\overset{O}{\overset{\|}{C}}CH_3$  (b) 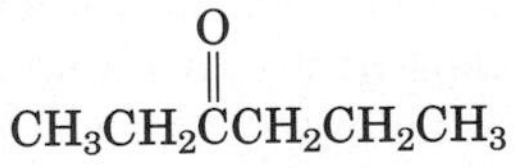 $CH_3CH_2\overset{O}{\overset{\|}{C}}CH_2CH_2CH_3$

## 8.5 Hydroboration of Alkynes

We saw in Section 7.4 that addition of $BH_3$ to an alkene takes place in a single step through a four-membered-ring transition state without involving a carbocation intermediate. Borane also adds rapidly to alkynes, and the resulting vinylic boranes are quite useful. Hydroboration of *symmetrically* substituted internal alkynes gives vinylic boranes that are oxidized by basic hydrogen peroxide to yield ketones (via enols). Hydroboration of *unsymmetrically* substituted internal alkynes gives a mixture of two ketones.

$$3\ CH_3CH_2C{\equiv}CCH_2CH_3 \xrightarrow[THF]{BH_3} \left[ CH_3CH_2\overset{\overset{\diagdown B \diagup}{\diagdown}}{C}{=}\overset{\overset{H}{\diagup}}{C}CH_2CH_3 \right]$$

3-Hexyne — A vinylic borane

$\downarrow$ $H_2O$, $HO^-$ | $H_2O_2$

$$3\ CH_3CH_2\overset{\overset{O}{\|}}{C}CH_2CH_2CH_3 \longleftarrow 3\ CH_3CH_2\overset{\overset{HO}{\diagdown}}{C}{=}\overset{\overset{H}{\diagup}}{C}CH_2CH_3$$

3-Hexanone — An enol

Terminal alkynes also react with borane, although in practice the reaction is difficult to stop at the vinylic borane stage. Under normal circumstances, a second addition of borane to the intermediate vinylic borane occurs:

$$R{-}C{\equiv}CH \xrightarrow[THF]{BH_3} [R{-}CH{=}CHBH_2] \xrightarrow[THF]{BH_3} R{-}CH_2{-}CH\begin{matrix} \diagup B{-} \\ \diagdown B{-} \end{matrix}$$

Terminal alkyne — Vinylic borane

To prevent this double addition, a bulky, sterically hindered borane such as bis(1,2-dimethylpropyl)borane (known commonly as disiamylborane) can be used in place of borane. When a terminal alkyne reacts with disiamylborane, addition of B—H to the C—C triple bond occurs with the expected non-Markovnikov regiochemistry. A second addition is rather slow, however, since the steric bulk of the large dialkylborane reagent makes approach to the double bond difficult. Oxidation of the vinylic borane intermediate then leads first to an enol and finally to an aldehyde.

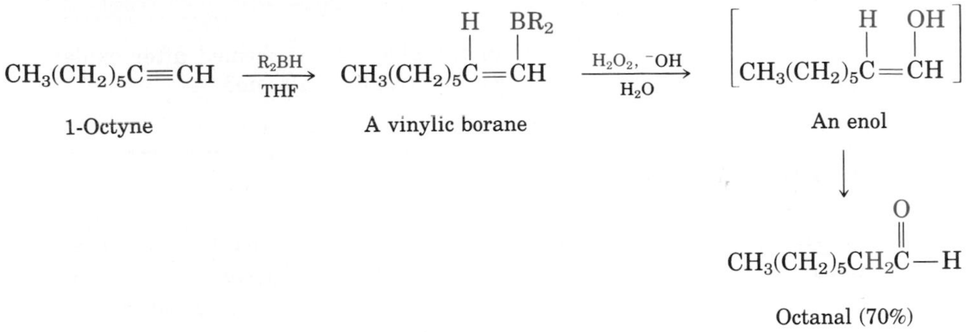

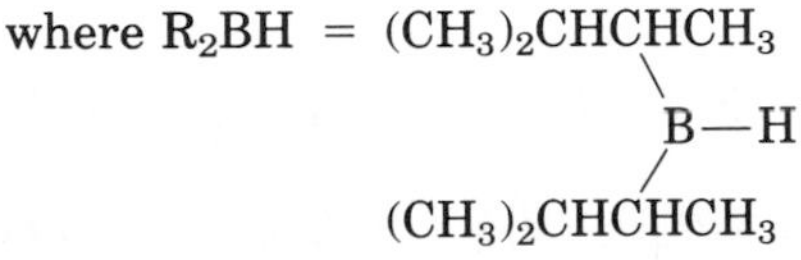

$$\underset{\text{1-Octyne}}{CH_3(CH_2)_5C{\equiv}CH} \xrightarrow[\text{THF}]{R_2BH} \underset{\text{A vinylic borane}}{CH_3(CH_2)_5\overset{H}{C}{=}\overset{BR_2}{CH}} \xrightarrow[H_2O]{H_2O_2,\ ^-OH} \underset{\text{An enol}}{\left[CH_3(CH_2)_5\overset{H}{C}{=}\overset{OH}{CH}\right]} \longrightarrow \underset{\text{Octanal (70\%)}}{CH_3(CH_2)_5CH_2\overset{O}{\overset{\|}{C}}{-}H}$$

where $R_2BH = $ $(CH_3)_2CHCHCH_3$ \ B—H / $(CH_3)_2CHCHCH_3$

Disiamylborane
[bis(1,2-Dimethylpropyl)borane]

Note that the hydroboration–oxidation sequence is *complementary* to the direct hydration reaction of alkynes, since different products result. Direct hydration of a terminal alkyne with aqueous acid and mercuric sulfate leads to a methyl ketone, whereas hydroboration–oxidation of the same alkyne leads to an aldehyde:

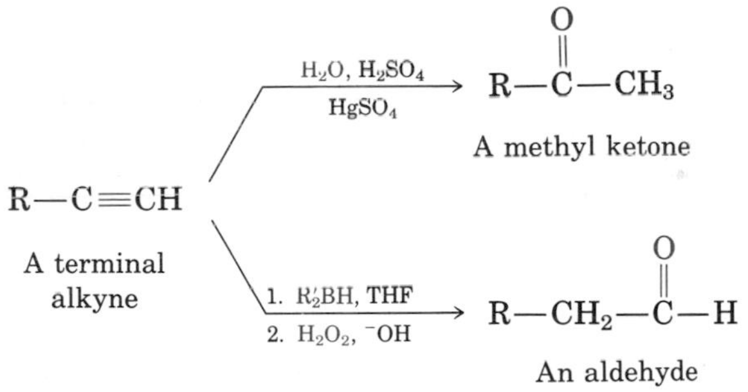

$$\underset{\text{A terminal alkyne}}{R{-}C{\equiv}CH} \xrightarrow[HgSO_4]{H_2O,\ H_2SO_4} \underset{\text{A methyl ketone}}{R{-}\overset{O}{\overset{\|}{C}}{-}CH_3}$$

$$\underset{\text{A terminal alkyne}}{R{-}C{\equiv}CH} \xrightarrow[2.\ H_2O_2,\ ^-OH]{1.\ R_2BH,\ THF} \underset{\text{An aldehyde}}{R{-}CH_2{-}\overset{O}{\overset{\|}{C}}{-}H}$$

**PROBLEM**

**8.5** What alkynes would you start with to prepare these compounds by a hydroboration–oxidation reaction?

(a) 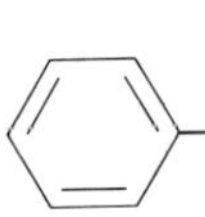 (benzene ring)—$CH_2CHO$

(b) $(CH_3)_2CHCH_2\overset{O}{\overset{\|}{C}}CH(CH_3)_2$

**PROBLEM**

**8.6** Disiamylborane is made by addition of $BH_3$ to 2 equiv of an alkene. What alkene is used?

$$[(CH_3)_2CH\underset{CH_3}{\underset{|}{CH}}]_2BH$$

Disiamylborane

PROBLEM

8.7 What organic product, other than the aldehyde, is formed after oxidation when disiamylborane is used to hydroborate a terminal alkyne?

## 8.6 Reduction of Alkynes

Alkynes are easily converted to alkanes by addition of hydrogen over a metal catalyst. Heat-of-hydrogenation data indicate that the first step in the reaction has a larger $\Delta H°$ than the second step, and we might therefore expect alkynes to reduce more readily than alkenes. This turns out to be the case.

$$HC{\equiv}CH \xrightarrow[\text{Catalyst}]{H_2} H_2C{=}CH_2 \qquad \Delta H°_{\text{hydrog}} = 42 \text{ kcal/mol}$$

$$H_2C{=}CH_2 \xrightarrow[\text{Catalyst}]{H_2} CH_3{-}CH_3 \qquad \Delta H°_{\text{hydrog}} = 33 \text{ kcal/mol}$$

Since alkynes reduce somewhat faster than alkenes, triple-bond hydrogenation can be stopped at the alkene stage if a suitable catalyst is used. The catalyst most often used for this purpose is the **Lindlar catalyst**, a finely divided palladium metal that has been precipitated onto a calcium carbonate support and then deactivated by treatment with lead acetate and quinoline, an aromatic amine. Since hydrogenation occurs with syn stereochemistry on the catalyst surface (Section 7.6), alkynes yield cis alkenes on reaction with 1 equiv of hydrogen. This reaction has been explored extensively by the Hoffmann-LaRoche pharmaceutical company, where it is used in the commercial synthesis of vitamin A.

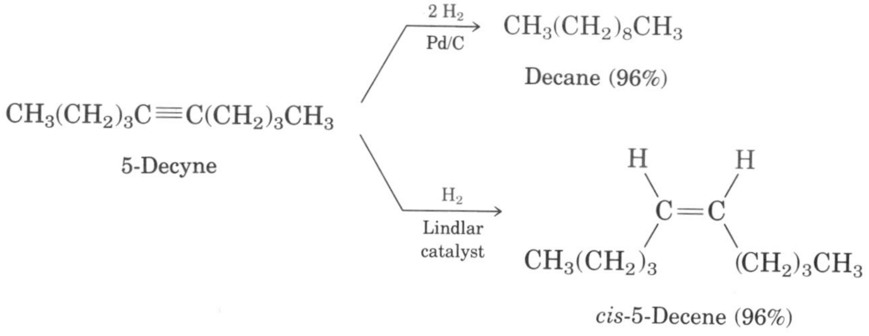
2 H2
Pd/C
CH3(CH2)8CH3
Decane (96%)
CH3(CH2)3C≡C(CH2)3CH3
5-Decyne
H2
Lindlar catalyst
H H
C=C
CH3(CH2)3 (CH2)3CH3
cis-5-Decene (96%)

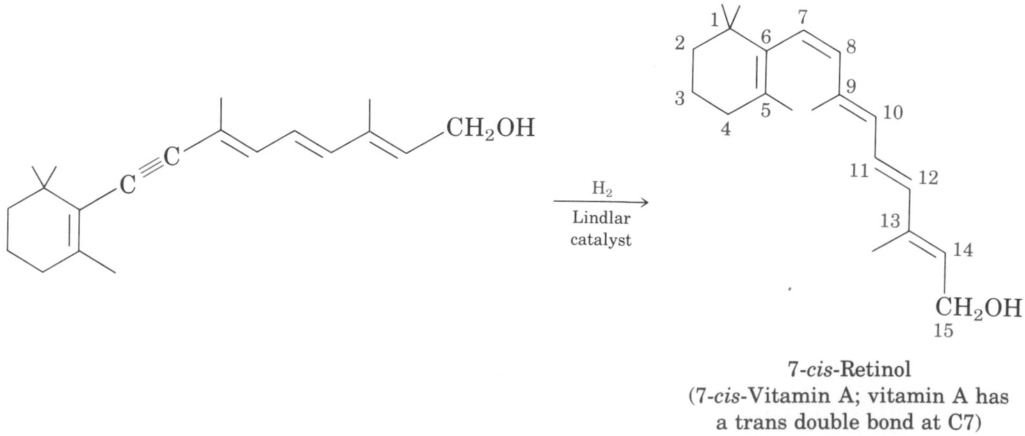
C
C
CH2OH
H2
Lindlar catalyst
1 2 3 4 5 6 7 8 9 10 11 12 13 14
CH2OH
15

7-*cis*-Retinol
(7-*cis*-Vitamin A; vitamin A has a trans double bond at C7)

A second method for the conversion of alkynes to alkenes employs sodium or lithium metal in liquid ammonia as solvent. This method is complementary to the Lindlar reduction, since it produces trans rather than cis alkenes. Remarkably, alkali metals such as lithium and sodium dissolve in pure liquid ammonia at −33°C to produce a deep blue solution. When an alkyne is added to this blue solution, reduction of the triple bond occurs and a trans alkene results. For example, 5-decyne gives *trans*-5-decene in good yield on treatment with lithium in liquid ammonia.

$$CH_3CH_2CH_2CH_2C{\equiv}CCH_2CH_2CH_2CH_3 \xrightarrow[2.\ H_2O]{1.\ Li/NH_3} (n\text{-}C_4H_9)HC{=}CH(C_4H_9\text{-}n)$$

5-Decyne → *trans*-5-Decene (78%)

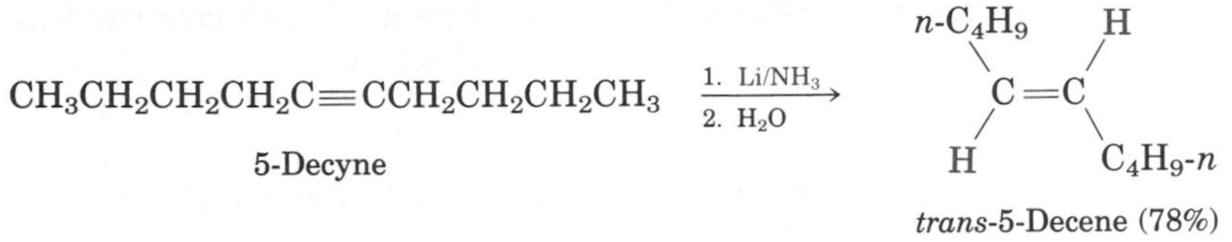

The mechanism of alkyne reduction by lithium in liquid ammonia involves donation of an electron to the triple bond to yield an intermediate **anion radical**—a species that is *both* a radical (has an odd number of electrons) and an anion (has a negative charge). This intermediate then takes a proton from ammonia to give a vinylic radical. Addition of a second electron to the vinylic radical gives a vinylic anion, which takes a second proton from ammonia to give trans alkene product (Figure 8.5).

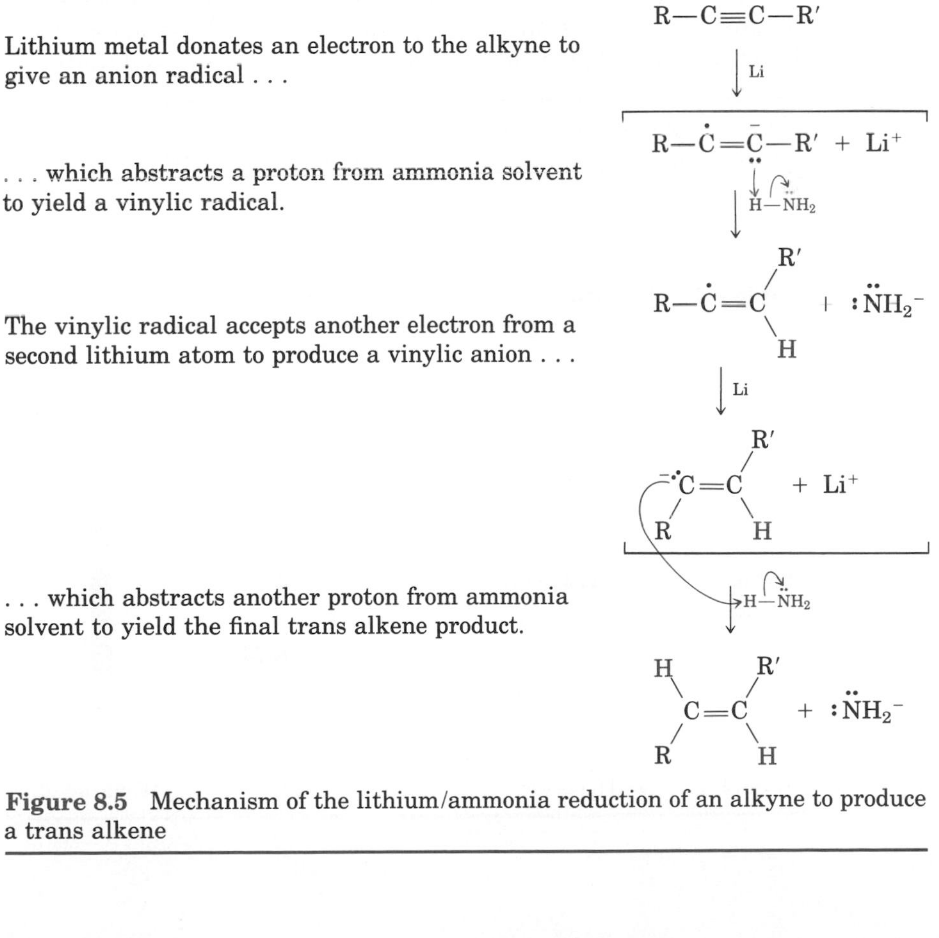

---

Lithium metal donates an electron to the alkyne to give an anion radical . . .

$$R{-}C{\equiv}C{-}R' \xrightarrow{Li} R{-}\dot{C}{=}\ddot{C}^{-}{-}R' + Li^+$$

. . . which abstracts a proton from ammonia solvent to yield a vinylic radical.

$$\xrightarrow{H{-}\ddot{N}H_2} R{-}\dot{C}{=}C(H)R' + :\ddot{N}H_2^-$$

The vinylic radical accepts another electron from a second lithium atom to produce a vinylic anion . . .

$$\xrightarrow{Li} R(\ddot{C}^{-}){=}C(H)R' + Li^+$$

. . . which abstracts another proton from ammonia solvent to yield the final trans alkene product.

$$\xrightarrow{H{-}\ddot{N}H_2} (H)(R)C{=}C(R')(H) + :\ddot{N}H_2^-$$

**Figure 8.5** Mechanism of the lithium/ammonia reduction of an alkyne to produce a trans alkene

The trans stereochemistry of the final alkene product is established during the second reduction step when the less hindered, trans vinylic anion is formed from the vinylic radical. Vinylic radicals undergo rapid cis–trans equilibration, but vinylic anions equilibrate much less rapidly. Thus, the more stable trans (rather than less stable cis) vinylic anion is formed and then protonated before it can equilibrate.

PROBLEM

**8.8** Starting from any alkyne you choose, how would you prepare these alkenes?
(a) *trans*-2-Octene (b) *cis*-3-Heptene (c) 3-Methyl-1-pentene

## 8.7 Alkyne Acidity: Formation of Acetylide Anions

The most striking difference between the chemistry of alkenes and that of alkynes is that *terminal alkynes are weakly acidic*. When a terminal alkyne is treated with a strong base such as sodium amide, $NaNH_2$, the terminal hydrogen is removed, and an **acetylide anion** is formed:

$$R—C\equiv C—H + :\ddot{N}H_2Na^+ \longrightarrow R—C\equiv C:^-Na^+ + :NH_3$$

Acetylide anion

According to the Brønsted–Lowry definition, an acid is a species that donates a proton. Although we usually think of oxyacids ($H_2SO_4$, $H_2O$) or halogen acids (HCl, HBr) in this context, *any* compound containing a hydrogen atom can be considered to be an acid under the proper circumstances. As discussed earlier (Section 2.6), we can establish an **acidity order** by measuring dissociation constants of acids and expressing the results as $pK_a$ values (Table 8.1). Recall from Section 2.6 that a low $pK_a$ corresponds to a strong acid, and a high $pK_a$ corresponds to a weak acid.

**Table 8.1** Strengths of some common acids ($pK_a$ values)

| *Acid* | *Formula* | $pK_a$ | |
|---|---|---|---|
| Sulfuric acid | $H—O—S(=O)_2—O—H$ | ~−9 | Strong acid |
| Acetic acid | $CH_3C(=O)—O—H$ | 4.72 | ↑ |
| Water | H—O—H | 15.74 | |
| Ethanol | $CH_3CH_2O—H$ | 16 | |
| Acetylene | $H—C\equiv C—H$ | 25 | |
| Ammonia | $H_2N—H$ | 35 | Weak acid |

At one end of the scale, the mineral acids such as sulfuric acid are very strong. Carboxylic acids such as acetic acid are moderately strong, water and alcohols such as ethanol are intermediate, and ammonia is a very weak acid. Since a stronger acid donates its proton to the anion of a weaker acid in an acid–base reaction, a rank-ordered list allows us to know what bases are needed to deprotonate what acids. For example, since acetic acid ($pK_a$ = 4.72) is a stronger acid than ethanol ($pK_a$ = 16), we know that the anion of ethanol (ethoxide ion, $CH_3CH_2\ddot{O}\!:^-$) will remove a proton from acetic acid. Similarly, amide ion ($^-\!:\!NH_2$), the anion of ammonia ($pK_a$ = 35), will remove a proton from ethanol ($pK_a$ = 16).

$$CH_3CH_2\ddot{O}:^- + H-\ddot{O}\overset{O}{\overset{\|}{C}}CH_3 \rightleftharpoons CH_3CH_2\ddot{O}-H + {}^-:\ddot{O}\overset{O}{\overset{\|}{C}}CH_3$$

$$H_2\ddot{N}:^- + H-\ddot{O}CH_2CH_3 \rightleftharpoons H_2\ddot{N}-H + {}^-:\ddot{O}CH_2CH_3$$

| Stronger base | Stronger acid | Weaker acid | Weaker base |
|---|---|---|---|

Where do hydrocarbons lie on the acidity scale? As the data in Table 8.2 indicate, both methane ($pK_a$ = 49) and ethylene ($pK_a$ = 44) are very weak acids; for all practical purposes, neither can be deprotonated by a base. Acetylene, however, has a $pK_a$ of 25 and is thus quite susceptible to deprotonation by a sufficiently strong base. The anion of any acid whose $pK_a$ is greater than 25 will abstract a proton from acetylene.

**Table 8.2** Acidity of simple hydrocarbons

| *Type* | *Example* | | | | $K_a$ | $pK_a$ | |
|---|---|---|---|---|---|---|---|
| Alkyne | $HC\equiv C-H$ | $\xrightleftharpoons{H_2O}$ | $HC\equiv C:^-$ | $+ H_3O^+$ | $10^{-25}$ | 25 | Stronger acid |
| Alkene | $H_2C=CH-H$ | $\xrightleftharpoons{H_2O}$ | $H_2C=\ddot{C}H^-$ | $+ H_3O^+$ | $10^{-44}$ | 44 | ↑ |
| Alkane | $CH_3-H$ | $\xrightleftharpoons{H_2O}$ | $:CH_3^-$ | $+ H_3O^+$ | $10^{-49}$ | 49 | Weaker acid |

The acidity data in Table 8.2 indicate that alkynes are much more acidic than either alkanes or alkenes. Any base whose conjugate acid has a $pK_a$ greater than 25 should be able to effect acetylide formation. Amide ion, $NH_2^-$, for example, is the conjugate base of ammonia ($NH_3$; $pK_a$ = 35) and is therefore able to abstract a proton from terminal alkynes.

What makes terminal alkynes so acidic? The best intuitive explanation is simply that acetylide anions have more "*s* character" in the orbital containing the electron pair than do alkyl or vinylic anions. [Alkane anions are $sp^3$ hybridized, so the negative charge resides in an orbital that has one-quarter *s* character and three-quarters *p* character; vinylic anions are $sp^2$ hybridized and therefore have one-third *s* character; and acetylide anions (*sp*) have one-half *s* character.] Since *s* orbitals are lower in energy and are nearer the positively charged nucleus than *p* orbitals, a negative charge is stabilized to a greater extent in an orbital with high $s$ character than in an orbital with low $s$ character (Figure 8.6).

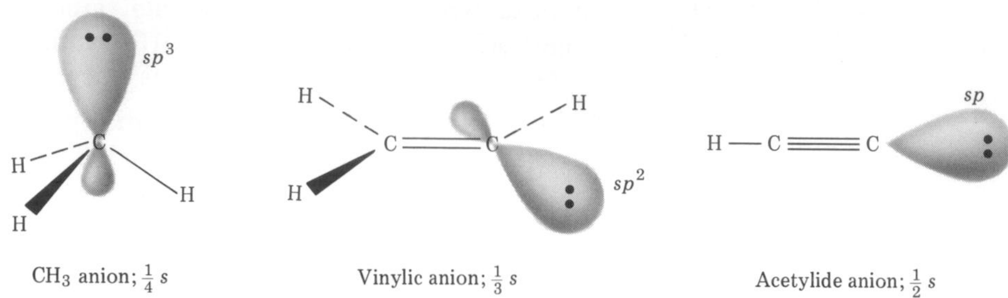


$CH_3$ anion; $\frac{1}{4}$ $s$ Vinylic anion; $\frac{1}{3}$ $s$ Acetylide anion; $\frac{1}{2}$ $s$

**Figure 8.6** A comparison of alkyl, vinylic, and acetylide anions: The acetylide anion is more stable.

PROBLEM

**8.9** The p$K_a$ of acetone, $CH_3COCH_3$, is 20. Which of the following bases are strong enough to deprotonate acetone?

(a) KOH (p$K_a$ of $H_2O$ = 15.7)

(b) $Na^+$ $^-C{\equiv}CH$ (p$K_a$ of $C_2H_2$ = 25)

(c) $NaHCO_3$ (p$K_a$ of $H_2CO_3$ = 6.4)

(d) $Na^+$ $^-OCH_3$ (p$K_a$ of $CH_3OH$ = 15.6)

## 8.8 Alkylation of Acetylide Anions

The chemistry of acetylide anions is both interesting and useful. For example, the presence of an unshared electron pair makes acetylide anions strongly nucleophilic. When treated with an alkyl halide such as bromomethane, acetylide anions substitute for halogen and bond to the alkyl-group carbon:

$$H{-}C{\equiv}C{:}^- \ Na^+ + H{-}\underset{H}{\overset{H}{C}}{-}Br \longrightarrow H{-}C{\equiv}C{-}\underset{H}{\overset{H}{C}}{-}H + NaBr$$

Although we won't study the details of this substitution reaction until Chapter 11, we might picture it as happening by the pathway shown in Figure 8.7. The nucleophilic acetylide ion attacks the positively polarized (and therefore electrophilic) carbon atom of bromomethane and pushes out bromide ion with the electron pair from the former C—Br bond, yielding propyne as product. We call such a reaction an **alkylation**, since a new alkyl group has become attached to the starting alkyne.

The nucleophilic acetylide anion uses its electron lone pair to form a bond to the positively polarized, electrophilic carbon atom of bromomethane. As the new C—C bond begins to form, the C—Br bond begins to break in the transition state.

$$H{-}C{\equiv}C{:}^{-}\ Na^{+} \quad + \quad H_3C{-}Br$$

$$\downarrow$$

$$\left[ H{-}C{\equiv}\overset{\delta^-}{C}\cdots CH_3 \cdots \overset{\delta^-}{Br} \ + \ Na^+ \right]^{\ddagger}$$

Transition state

$$\downarrow$$

The new C—C bond is fully formed and the old C—Br bond is fully broken at the end of the reaction.

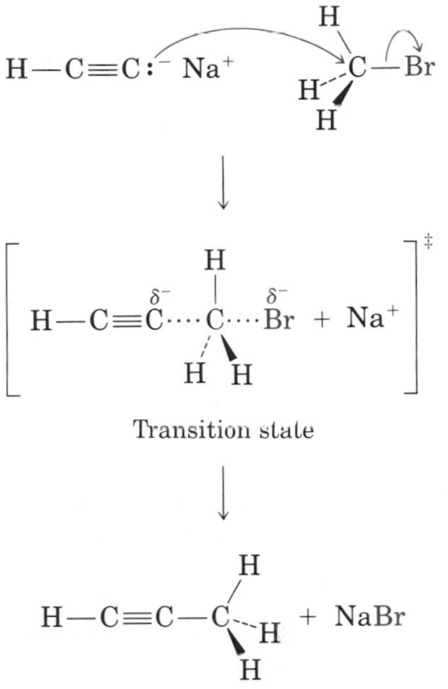

$$H{-}C{\equiv}C{-}CH_3 \ + \ NaBr$$

**Figure 8.7** A possible mechanism for the substitution reaction of acetylide anion with bromomethane

Alkyne alkylation is not limited to unsubstituted acetylide ion. *Any* terminal alkyne can be converted into its corresponding acetylide anion and alkylated by treatment with an alkyl halide to yield an internal alkyne product. For example, conversion of 1-hexyne into its anion, followed by reaction with 1-bromobutane, yields 5-decyne:

$$\underset{\text{1-Hexyne}}{CH_3CH_2CH_2CH_2C{\equiv}CH} \xrightarrow[\text{2. } CH_3CH_2CH_2CH_2Br,\ THF]{\text{1. } NaNH_2,\ NH_3} \underset{\text{5-Decyne (76\%)}}{CH_3CH_2CH_2CH_2C{\equiv}CCH_2CH_2CH_2CH_3}$$

Acetylide ion alkylation is limited to the use of primary alkyl bromides and iodides, $R{-}CH_2{-}X$, for reasons that will be discussed in more detail in Chapter 11. In addition to their reactivity as nucleophiles, acetylide ions are sufficiently strong bases to cause dehydrohalogenation instead of substitution when they react with secondary and tertiary alkyl halides. For example, reaction of bromocyclohexane with propyne anion leads to formation of the elimination product cyclohexene rather than the substitution product cyclohexylpropyne, as shown on page 244.

Bromocyclohexane (a secondary alkyl halide) + $CH_3C{\equiv}C:^-$ → Cyclohexene + $CH_3C{\equiv}CH$ + $:\ddot{Br}:^-$

(cyclohexylpropyne) Not formed

**PROBLEM**

**8.10** Show the terminal alkyne and alkyl halide starting materials from which the following products can be obtained. Where two routes look feasible, list both choices.

(a) $CH_3CH_2CH_2C{\equiv}CCH_3$

(b) $(CH_3)_2CHC{\equiv}CCH_2CH_3$

(c) cyclohexyl–$C{\equiv}CCH_3$

(d) 5-Methyl-2-hexyne

(e) 2,2-Dimethyl-3-hexyne

**PROBLEM**

**8.11** How would you prepare *cis*-2-butene, starting from 1-propyne, an alkyl halide, and any other reagents needed? (This problem can't be worked in a single step. You'll have to carry out more than one chemical reaction.)

## 8.9 Oxidative Cleavage of Alkynes

Alkynes, like alkenes, can be cleaved by reaction with powerful oxidizing agents such as potassium permanganate or ozone. A triple bond is generally less reactive than a double bond, however, and yields of cleavage products are sometimes low. The reaction is too complex mechanistically to discuss in detail, but the products obtained are carboxylic acids. If a terminal alkyne is oxidized, $CO_2$ is formed as one product.

An internal alkyne $\quad R{-}C{\equiv}C{-}R' \xrightarrow{KMnO_4 \text{ or } O_3} RCOOH + R'COOH$

A terminal alkyne $\quad R{-}C{\equiv}C{-}H \xrightarrow{KMnO_4 \text{ or } O_3} RCOOH + CO_2$

The major application of alkyne oxidation reactions is in structure determination. For example, when the substance called tariric acid was isolated from the Guatemalan plant *Picramnia tariri*, it was identified as an 18-carbon straight-chain acetylenic acid, but the position of the triple bond in the chain was unknown. Since oxidation of tariric acid with potassium permanganate gave a product identified as hexanedioic acid, the position of the triple bond could be defined.

$$CH_3(CH_2)_{10}C{\equiv}C(CH_2)_4COOH \xrightarrow[H_2O]{KMnO_4} CH_3(CH_2)_{10}COOH + HOOC(CH_2)_4COOH$$

6-Octadecynoic acid (Tariric acid) — Decanoic acid — 1,6-Hexanedioic acid (Adipic acid)

**PROBLEM**

**8.12** Propose structures for alkynes that give the following products on oxidative cleavage by $KMnO_4$:

(a) $C_6H_5COOH + CO_2$ (b) $2\ CH_3(CH_2)_7COOH + HO_2C(CH_2)_7COOH$

## 8.10 Preparation of Alkynes: Elimination Reactions of Dihalides

Perhaps the best method for the preparation of alkynes is the alkylation of acetylide anions. Remember that both terminal and internal alkynes can be prepared by suitable alkylation of primary alkyl halides:

$$H{-}C{\equiv}C{:}^-\ Na^+ + RCH_2Br \longrightarrow H{-}C{\equiv}C{-}CH_2R \quad \text{A terminal alkyne}$$

$$R{-}C{\equiv}C{:}^-\ Na^+ + R'CH_2Br \longrightarrow R{-}C{\equiv}C{-}CH_2R' \quad \text{An internal alkyne}$$

Alkynes can also be prepared by elimination of HX from alkyl halides in much the same manner as alkenes can (Section 7.12). Since an alkyne is doubly unsaturated, however, we have to eliminate *two* molecules of HX. Treatment of a 1,2-dihalide (a *vicinal* dihalide) with excess strong base such as KOH or $NaNH_2$ results in a twofold elimination of HX and formation of an alkyne. As with the elimination of HX to form an alkene, we'll defer a discussion of mechanism until Chapter 11.

The necessary vicinal dihalides are themselves readily available by addition of bromine or chlorine to alkenes. Thus, the overall sequence of halogenation–dehydrohalogenation provides an excellent method for going from an alkene to an alkyne. For example, diphenylethylene is converted into diphenylacetylene by reaction with bromine and subsequent base treatment:

$$C_6H_5CH{=}CHC_6H_5 \xrightarrow{Br_2,\ CCl_4} C_6H_5CHBrCHBrC_6H_5$$

1,2-Diphenylethylene (Stilbene) — 1,2-Dibromo-1,2-diphenylethane (a vicinal dibromide)

$$\xrightarrow[\text{Ethanol; } -2\ HBr]{2\ KOH} C_6H_5{-}C{\equiv}C{-}C_6H_5$$

Diphenylacetylene (85%)

Although different bases can be used for dehydrohalogenation, sodium amide ($NaNH_2$) is normally preferred since it usually gives higher yields. The twofold dehydrohalogenation takes place in discrete steps through a vinylic halide intermediate. This suggests that vinylic halides themselves should give alkynes when treated with strong base, which is indeed the case. For example,

$$\underset{\substack{\text{3-Chloro-2-buten-1-ol}\\\text{(a vinylic chloride)}}}{CH_3\underset{\substack{|\\Cl}}{C}{=}\overset{\substack{H\\|}}{C}CH_2OH} \xrightarrow[\text{2. } H_3O^+]{\text{1. 2 } NaNH_2} \underset{\text{2-Butyn-1-ol (85\%)}}{CH_3C{\equiv}CCH_2OH}$$

## 8.11 Organic Synthesis

The laboratory synthesis of an organic molecule from simple precursors might be carried out for many reasons. In the pharmaceutical industry, new organic molecules are designed and synthesized in the hope that some might be useful new drugs. In the chemical industry, synthesis is often undertaken to devise more economical routes to known compounds. In academic laboratories, the synthesis of highly complex molecules is sometimes done purely for the intellectual challenge involved in mastering so difficult a subject; the successful synthesis route is a highly creative work that is sometimes described by such subjective terms as *elegant* or *beautiful*.

In this book, too, we will often devise syntheses of molecules from simpler precursors. The purpose, however, is purely pedagogical. The ability to plan workable synthetic sequences demands a thorough knowledge of a wide variety of organic reactions. Furthermore, it requires more than a theoretical knowledge of reactions; it also requires a practical grasp for the proper fitting together of steps in a sequence such that each reaction does only what is desired. *Working synthesis problems is an excellent way to learn organic chemistry.*

Some of the syntheses we plan may appear trivial. Here's an example:

**PRACTICE PROBLEM**

Prepare octane from 1-pentyne.

*Solution* First alkylate the acetylide anion of 1-pentyne with 1-bromopropane, and then reduce the product using catalytic hydrogenation:

$$\underset{\text{1-Pentyne}}{CH_3CH_2CH_2C{\equiv}CH} \xrightarrow[\text{2. } BrCH_2CH_2CH_3\text{, THF}]{\text{1. } NaNH_2\text{, } NH_3} \underset{\text{4-Octyne}}{CH_3CH_2CH_2C{\equiv}CCH_2CH_2CH_3}$$

$$\downarrow H_2/Pd \text{ in ethanol}$$

$$\underset{\text{Octane}}{CH_3CH_2CH_2\overset{\substack{H\\|}}{\underset{\substack{|\\H}}{C}}{-}\overset{\substack{H\\|}}{\underset{\substack{|\\H}}{C}}CH_2CH_2CH_3}$$

Although the synthesis route just presented should work perfectly well, it has little practical value since a chemist can simply *buy* octane from several dozen chemical supply houses. The value of working the problem is that it makes us approach a chemical problem in a logical way, draw on our knowledge of chemical reactions, and organize that knowledge into a workable plan—it helps us *learn* organic chemistry.

There's no secret to planning organic syntheses. All it takes is a knowledge of the different reactions and lots of practice. But here's a hint: *Work backward.* Look at the final product and ask, "What was the *immediate* precursor of that product?" For example, if the end product is an alkyl halide, the immediate precursor might be an alkene (via HX addition). Having found an immediate precursor, proceed backward again, one step at a time, until a suitable starting material is found.

Let's work some examples of increasing complexity.

**PRACTICE PROBLEM**

Starting from 1-pentyne and any alkyl halide needed, synthesize *cis*-2-hexene. More than one step is required.

$$\underset{\text{1-Pentyne}}{CH_3CH_2CH_2C{\equiv}CH} + RX \overset{?}{\Rightarrow} \underset{\textit{cis}\text{-2-Hexene}}{(CH_3CH_2CH_2)HC{=}CH(CH_3)}$$

*Solution* First ask the question "What is an immediate precursor of a cis-disubstituted alkene?" We know that alkenes can be prepared from alkynes by partial reduction. The proper choice of experimental conditions will allow us to prepare either a trans-disubstituted alkene (using lithium in liquid ammonia) or a cis-disubstituted alkene (using catalytic hydrogenation over the Lindlar catalyst). Thus, reduction of 2-hexyne by catalytic hydrogenation using the Lindlar catalyst should yield *cis*-2-hexene:

$$\underset{\text{2-Hexyne}}{CH_3CH_2CH_2C{\equiv}CCH_3} \xrightarrow[\text{Lindlar catalyst}]{H_2} \underset{\textit{cis}\text{-2-Hexene}}{(CH_3CH_2CH_2)HC{=}CH(CH_3)}$$

Next ask "What is an immediate precursor of 2-hexyne?" We've seen that internal alkynes can be prepared by alkylation of terminal alkyne anions (acetylides). In the present instance, we are told to start with 1-pentyne. Thus, alkylation of the anion of 1-pentyne with iodomethane should yield 2-hexyne:

$$\underset{\text{1-Pentyne}}{CH_3CH_2CH_2C{\equiv}C{-}H} + NaNH_2 \xrightarrow{\text{In } NH_3} CH_3CH_2CH_2C{\equiv}C{:}^-\,Na^+$$

$$CH_3CH_2CH_2C{\equiv}C{:}^-\,Na^+ + CH_3I \xrightarrow{\text{In THF}} \underset{\text{2-Hexyne}}{CH_3CH_2CH_2C{\equiv}C{-}CH_3}$$

In three steps we have synthesized *cis*-2-hexene from the given starting materials:

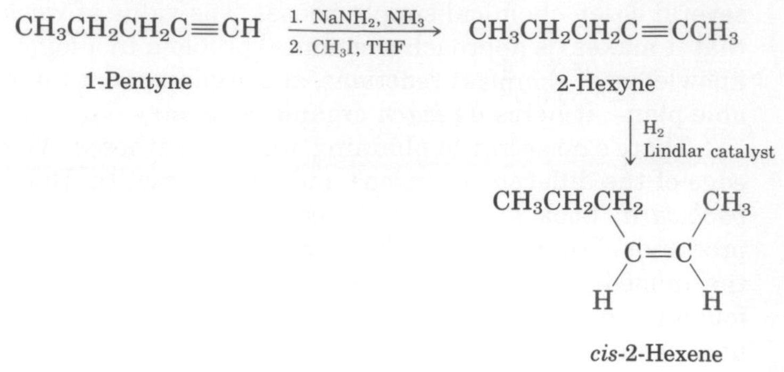

$$CH_3CH_2CH_2C{\equiv}CH \xrightarrow[\text{2. } CH_3I,\text{ THF}]{\text{1. } NaNH_2,\ NH_3} CH_3CH_2CH_2C{\equiv}CCH_3$$

1-Pentyne 2-Hexyne

$$\downarrow H_2,\ \text{Lindlar catalyst}$$

$$\begin{array}{ccc} CH_3CH_2CH_2 & & CH_3 \\ & C{=}C & \\ H & & H \end{array}$$

*cis*-2-Hexene

**PRACTICE PROBLEM**

Starting from acetylene and any alkyl halide needed, synthesize 2-bromopentane. More than one step is required.

$$HC{\equiv}CH + RX \overset{?}{\Rightarrow} CH_3CH_2CH_2CHBrCH_3$$

2-Bromopentane

*Solution* First ask "What is an immediate precursor of an alkyl halide?" Perhaps an alkene:

$$\begin{array}{c} CH_3CH_2CH_2CH{=}CH_2 \\ \text{or} \\ CH_3CH_2CH{=}CHCH_3 \end{array} \xrightarrow[\text{Ether}]{HBr} CH_3CH_2CH_2\overset{Br}{\overset{|}{C}}HCH_3$$

Of the two alkene possibilities, addition of HBr to 1-pentene looks like a better choice than addition to 2-pentene, since the latter would give a mixture of isomers.

Next ask "What is an immediate precursor of an alkene?" Perhaps an alkyne, which could be partially reduced:

$$CH_3CH_2CH_2C{\equiv}CH \xrightarrow[\text{Lindlar catalyst}]{H_2} CH_3CH_2CH_2CH{=}CH_2$$

Next ask "What is an immediate precursor of a terminal alkyne?" Perhaps sodium acetylide and an alkyl halide:

$$Na^+ \ {:}\bar{C}{\equiv}CH + BrCH_2CH_2CH_3 \longrightarrow CH_3CH_2CH_2C{\equiv}CH$$

In four steps we have synthesized the desired material from acetylene and 1-bromopropane:

$$\underset{\text{Acetylene}}{HC{\equiv}CH} \xrightarrow[\text{2. } CH_3CH_2CH_2Br,\ \text{THF}]{\text{1. } NaNH_2,\ NH_3} \underset{\text{1-Pentyne}}{CH_3CH_2CH_2C{\equiv}CH} \xrightarrow[\text{Lindlar catalyst}]{H_2} \underset{\text{1-Pentene}}{CH_3CH_2CH_2CH{=}CH_2}$$

$$\xrightarrow{\text{HBr, ether}} \underset{\text{2-Bromopentane}}{CH_3CH_2CH_2CH(Br)CH_3}$$

PRACTICE PROBLEM

Synthesize 1-hexanol from acetylene and any alkyl halide needed.

$$H{-}C{\equiv}C{-}H + RX \overset{?}{\Rightarrow} \underset{\text{1-Hexanol}}{HOCH_2CH_2CH_2CH_2CH_2CH_3}$$

*Solution* "What is an immediate precursor of a primary alcohol?" Perhaps an alkene, which could be hydrated by reaction with borane followed by oxidation with $H_2O_2$:

$$H_2C{=}CHCH_2CH_2CH_2CH_3 \xrightarrow[\text{2. } H_2O_2,\ HO^-]{\text{1. } BH_3} \underset{OH}{CH_2}{-}\underset{H}{CH}CH_2CH_2CH_2CH_3$$

"What is an immediate precursor of a terminal alkene?" Perhaps a terminal alkyne, which could be reduced:

$$H{-}C{\equiv}C{-}CH_2CH_2CH_2CH_3 \xrightarrow[\text{Lindlar catalyst}]{H_2} H_2C{=}CHCH_2CH_2CH_2CH_3$$

"What is an immediate precursor of 1-hexyne? Perhaps acetylene and 1-bromobutane:

$$H{-}C{\equiv}C{-}H \xrightarrow{NaNH_2} H{-}C{\equiv}C{:}^-\ Na^+ \xrightarrow{BrCH_2CH_2CH_2CH_3} H{-}C{\equiv}C{-}CH_2CH_2CH_2CH_3$$

We have completed the synthesis in three steps, by working backward.

PROBLEM

**8.13** Beginning with 4-octyne as your only source of carbon, and using any inorganic reagents necessary, how would you synthesize the following compounds?
(a) Butanoic acid (b) *cis*-4-Octene (c) 4-Bromooctane
(d) 4-Octanol (4-hydroxyoctane) (e) 4,5-Dichlorooctane

PROBLEM

**8.14** Beginning with acetylene and any alkyl halides needed, how would you synthesize the following compounds?
(a) Decane (b) 2,2-Dimethylhexane (c) Hexanal
(d) 2-Heptanone

## 8.12 Summary and Key Words

**Alkynes** are hydrocarbons that contain one or more carbon–carbon triple bonds. Alkyne carbon atoms are *sp* hybridized, and the triple bond is formed by one *sp–sp* sigma bond and two *p–p* pi bonds.

The chemistry of alkynes is dominated by **electrophilic addition reactions**, similar to those of alkenes. For example, alkynes react with HBr and HCl to yield **vinylic** halides, and with $Br_2$ and $Cl_2$ to yield 1,2-dihalides (**vicinal** dihalides). Alkynes can be hydrated (addition of $H_2O$) by either of two procedures. Reaction of an alkyne with aqueous sulfuric acid in the presence of mercuric ion catalyst leads to an intermediate **enol** that immediately isomerizes (**tautomerizes**) to yield a ketone. Since the addition reaction occurs with Markovnikov regiochemistry, a methyl ketone is produced from a terminal alkyne. Alternatively, addition of water can be effected by hydroboration of an alkyne, followed by oxidation with basic hydrogen peroxide. Disiamylborane, a sterically hindered dialkylborane, is often used in this reaction, allowing aldehydes to be prepared in good yield from terminal alkynes.

Alkynes can also be reduced to yield alkenes and alkanes. Complete reduction of the triple bond over a normal palladium hydrogenation catalyst yields an alkane, whereas **partial reduction** by catalytic hydrogenation over a **Lindlar catalyst** yields a cis alkene. Reduction of the alkyne with lithium in ammonia yields a trans alkene.

One of the most striking differences in chemical reactivity between alkynes and alkenes is due to their different acidities. Terminal alkynes contain an acidic hydrogen that can be removed by a strong base to yield an **acetylide anion**. Acetylide anions act as nucleophiles and can displace halide ion from primary alkyl halides (**alkylation** reaction). Acetylide anions are more stable than either alkyl anions or vinylic anions because their negative charge is in a hybrid orbital with much *s* character, allowing the charge to be closer to the nucleus.

There are relatively few general methods of alkyne synthesis. The two best are the alkylation of acetylide anions and the twofold elimination of HX from vicinal dihalides.

## 8.13 Summary of Reactions

1. Reactions of alkynes
   a. Addition of HX, where X = Br or Cl (Section 8.3)

$$R{-}C{\equiv}CH \xrightarrow[\text{Ether}]{HX} R{-}\overset{\overset{\large X}{|}}{C}{=}CH_2 \xrightarrow[\text{Ether}]{HX} R{-}\underset{\underset{\large X}{|}}{\overset{\overset{\large X}{|}}{C}}{-}CH_3$$

b. Addition of $X_2$, where X = Br or Cl (Section 8.3)

$$\text{R—C}\equiv\text{C—R}' \xrightarrow[\text{CCl}_4]{X_2} \underset{\text{R}\quad\quad\text{X}}{\overset{\text{X}\quad\quad\text{R}'}{\text{C=C}}} \xrightarrow[\text{CCl}_4]{X_2} \text{R—CX}_2\text{—CX}_2\text{—R}'$$

c. Mercuric sulfate-catalyzed hydration (Section 8.4)

$$\text{R—C}\equiv\text{CH} \xrightarrow[\text{HgSO}_4]{\text{H}_2\text{SO}_4,\ \text{H}_2\text{O}} \left[\text{R—}\overset{\text{OH}}{\overset{|}{\text{C}}}\text{=CH}_2\right] \longrightarrow \text{R—}\overset{\text{O}}{\overset{\|}{\text{C}}}\text{—CH}_3$$

A methyl ketone

d. Hydroboration–oxidation (Section 8.5)

$$\text{R—C}\equiv\text{C—H} \xrightarrow[\text{2. H}_2\text{O}_2,\ \text{NaOH}]{\text{1. Disiamylborane}} \text{R—CH}_2\text{—}\overset{\text{O}}{\overset{\|}{\text{C}}}\text{—H}$$

e. Reduction (Section 8.6)

(1) Catalytic hydrogenation

$$\text{R—C}\equiv\text{C—R} \xrightarrow{\text{H}_2,\ \text{Pd/C}} \text{R—CH}_2\text{CH}_2\text{R}$$

$$\text{R—C}\equiv\text{C—R} \xrightarrow[\text{Lindlar catalyst}]{\text{H}_2} \underset{\text{R}\quad\quad\text{R}}{\overset{\text{H}\quad\quad\text{H}}{\text{C=C}}}$$

A cis alkene

(2) Lithium/ammonia

$$\text{R—C}\equiv\text{C—R} \xrightarrow{\text{Li, NH}_3} \underset{\text{R}\quad\quad\text{H}}{\overset{\text{H}\quad\quad\text{R}}{\text{C=C}}}$$

A trans alkene

f. Acidity: Conversion into acetylide anions (Section 8.7)

$$\text{R—C}\equiv\text{C—H} \xrightarrow[\text{NH}_3]{\text{NaNH}_2} \text{R—C}\equiv\text{C:}^-\ \text{Na}^+ + \text{NH}_3$$

g. Acetylide ion alkylation (Section 8.8)

$$\text{H—C}\equiv\text{C:}^- + \text{RCH}_2\text{Br} \xrightarrow{\text{THF}} \text{R—CH}_2\text{—C}\equiv\text{CH} + :\ddot{\text{Br}}:^-$$

$$\text{R—C}\equiv\text{C:}^- + \text{R}'\text{—CH}_2\text{Br} \xrightarrow{\text{THF}} \text{R—C}\equiv\text{C—CH}_2\text{R}' + :\ddot{\text{Br}}:^-$$

h. Oxidative cleavage (Section 8.9)

$$R{-}C{\equiv}C{-}R' \xrightarrow[H_3O^+]{KMnO_4} RCOOH + R'COOH$$

2. Preparation of alkynes

a. Acetylide ion alkylation (Section 8.8)

$$R{-}C{\equiv}C{:}^- + R'CH_2Br \xrightarrow{THF} R{-}C{\equiv}C{-}CH_2R' + {:}\ddot{B}r{:}^-$$

b. Dehydrohalogenation of vicinal dihalides (Section 8.10)

$$R{-}CHBrCHBr{-}R' \xrightarrow[\text{or } NaNH_2,\ NH_3]{\text{2 KOH, Ethanol}} R{-}C{\equiv}C{-}R' + H_2O + KBr$$

$$R{-}C(Br){=}CHR' \xrightarrow[\text{or } NaNH_2,\ NH_3]{\text{KOH, Ethanol}} R{-}C{\equiv}C{-}R' + H_2O + KBr$$

## ADDITIONAL PROBLEMS

**8.15** Provide proper IUPAC names for these compounds:

(a) $CH_3CH_2C{\equiv}CC(CH_3)_3$ (b) $CH_3C{\equiv}CCH_2C{\equiv}CCH_2CH_3$

(c) $CH_3CH{=}C(CH_3)C{\equiv}CCH(CH_3)_2$ (d) $HC{\equiv}CC(CH_3)_2CH_2C{\equiv}CH$

(e) $H_2C{=}CHCH{=}CHC{\equiv}CH$

(f) $CH_3CH_2CH(CH_2CH_3)C{\equiv}CCH(CH_2CH_3)CH(CH_3)_2$

**8.16** Draw structures corresponding to these names:

(a) 3,3-Dimethyl-4-octyne (b) 3-Ethyl-5-methyl-1,6,8-decatriyne

(c) 2,2,5,5-Tetramethyl-3-hexyne (d) 3,4-Dimethylcyclodecyne

(e) 3,5-Heptadien-1-yne (f) 3-Chloro-4,4-dimethyl-1-nonen-6-yne

(g) 3-*sec*-Butyl-1-heptyne (h) 5-*tert*-Butyl-2-methyl-3-octyne

**8.17** The following names are incorrect. Draw the structures and give the correct names.

(a) 1-Ethyl-5,5-dimethyl-1-hexyne (b) 2,5,5-Trimethyl-6-heptyne

(c) 3-Methylhept-5-en-1-yne (d) 2-Isopropyl-5-methyl-7-octyne

(e) 3-Hexen-5-yne (f) 5-Ethynyl-1-methylcyclohexane

**8.18** These two hydrocarbons have been isolated from various plants in the sunflower family. Name them according to IUPAC rules.

(a) $CH_3CH{=}CHC{\equiv}CC{\equiv}CCH{=}CHCH{=}CHCH{=}CH_2$ (all trans)

(b) $CH_3C{\equiv}CC{\equiv}CC{\equiv}CC{\equiv}CC{\equiv}CCH{=}CH_2$

**8.19** Predict the products of these reactions:

$$C_6H_5{-}CH{=}CH{-}C{\equiv}C{-}H \xrightarrow{H_2,\ Pd/C} A$$

$$C_6H_5{-}CH{=}CH{-}C{\equiv}C{-}H \xrightarrow{H_2/\text{Lindlar}} B$$

**8.20** A hydrocarbon of unknown structure has the formula $C_8H_{10}$. On catalytic hydrogenation over the Lindlar catalyst, 1 equiv of $H_2$ is absorbed. On hydrogenation over a palladium catalyst, however, 3 equiv of $H_2$ are absorbed.
(a) How many rings/double bonds/triple bonds are present in the unknown?
(b) How many triple bonds are present?
(c) How many double bonds are present?
(d) How many rings are present?
Explain your answers and draw a possible structure that fits the data.

**8.21** Predict the products from reaction of 1-hexyne with these reagents:
(a) 1 equiv HBr
(b) 1 equiv $Cl_2$
(c) $H_2$, Lindlar catalyst
(d) $NaNH_2$ in $NH_3$, then $CH_3Br$
(e) $H_2O$, $H_2SO_4$, $HgSO_4$
(f) 2 equiv HCl

**8.22** Predict the products from reaction of 5-decyne with these reagents:
(a) $H_2$, Lindlar catalyst
(b) Li in $NH_3$
(c) 1 equiv $Br_2$
(d) $BH_3$ in THF, then $H_2O_2$
(e) $H_2O$, $H_2SO_4$, $HgSO_4$
(f) Excess $H_2$, Pd/C catalyst

**8.23** Predict the products from reaction of 2-hexyne with these reagents:
(a) 2 equiv $Br_2$
(b) 1 equiv HBr
(c) Excess HBr
(d) Li in $NH_3$
(e) $H_2O$, $H_2SO_4$, $HgSO_4$

**8.24** Acetonitrile, $CH_3CN$, contains a carbon–nitrogen triple bond. Sketch the orbitals involved in the bonding in acetonitrile and indicate the hybridization of each atom.

**8.25** How would you carry out these reactions?

(a) $CH_3CH_2C{\equiv}CH \xrightarrow{?} CH_3CH_2\overset{O}{\overset{\|}{C}}CH_3$

(b) $CH_3CH_2C{\equiv}CH \xrightarrow{?} CH_3CH_2CH_2CHO$

(c) $C_6H_5C{\equiv}CH \xrightarrow{?} C_6H_5C{\equiv}C{-}CH_3$

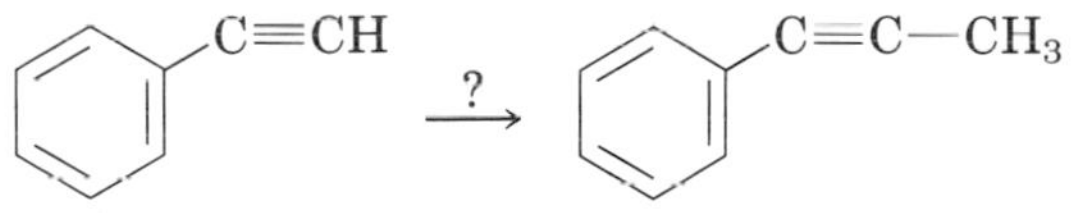

(d) $C_6H_5C{\equiv}CCH_3 \xrightarrow{?}$ (E)-$C_6H_5CH{=}CHCH_3$ [product drawn with H, H, $CH_3$ on the double bond]

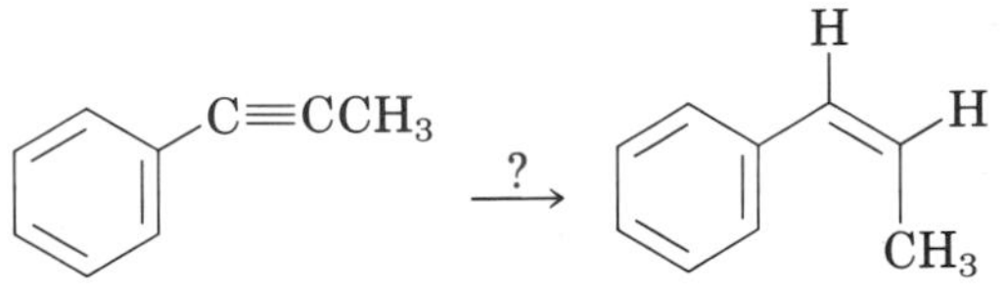

(e) $CH_3CH_2C{\equiv}CH \xrightarrow{?} CH_3CH_2COOH$

(f) $CH_3CH_2CH_2CH_2CH{=}CH_2 \xrightarrow{\text{(2 steps)}} CH_3CH_2CH_2CH_2C{\equiv}CH$

**8.26** Occasionally, chemists need to invert the stereochemistry of an alkene. That is, one might want to convert a cis alkene to a trans alkene, or vice versa. Although there is no one-step method for doing this alkene inversion, the transformation can be carried out by *combining* some of the reactions learned earlier in the proper sequence. How would you carry out these reactions?

(a) *trans*-5-Decene $\xrightarrow{?}$ *cis*-5-Decene

(b) *cis*-5-Decene $\xrightarrow{?}$ *trans*-5-Decene

**8.27** Propose structures for hydrocarbons that give the following products on oxidative cleavage by $KMnO_4$ or $O_3$.

(a) $CO_2 + CH_3(CH_2)_5COOH$

(b) $CH_3COOH$ + $C_6H_5COOH$ (benzoic acid)

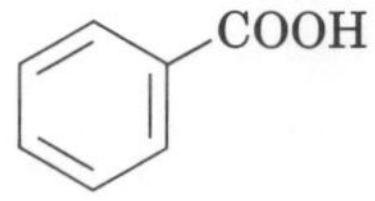

(c) $HOOC(CH_2)_8COOH$

(d) $CH_3CHO + CH_3COCH_2CH_2COOH + CO_2$

(e) $OHCCH_2CH_2CH_2CH_2COCOOH + CO_2$

**8.28** Each of the following syntheses requires more than one step. How would you carry them out?

(a) $CH_3CH_2CH_2C{\equiv}CH \xrightarrow{?} CH_3CH_2CH_2CHO$

(b) $(CH_3)_2CHCH_2C{\equiv}CH \xrightarrow{?}$ (E)-$(CH_3)_2CHCH_2CH{=}CHCH_2CH_3$ [H and $(CH_3)_2CHCH_2$ on one carbon; $CH_2CH_3$ and H on the other, trans]

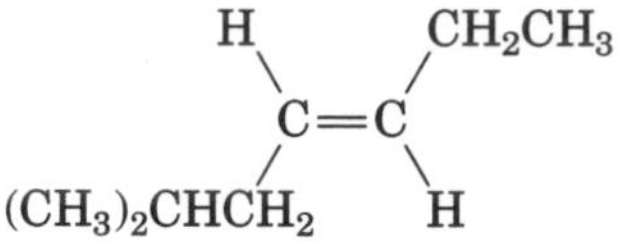

**8.29** How would you carry out the following transformation? More than one step is required.

$CH_3CH_2CH_2CH_2C{\equiv}CH \xrightarrow{?}$ cyclopropane with a $CH_2$ ring carbon, one ring carbon bearing H and $CH_3CH_2CH_2CH_2$, the other bearing H and $CH_3$ (cis)

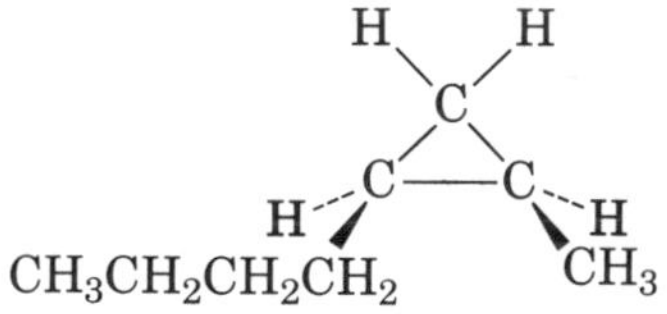

**8.30** How would you carry out the following conversion? More than one step is needed.

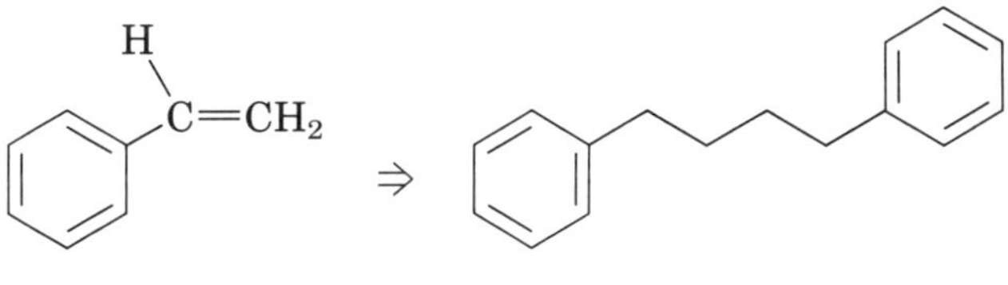

$C_6H_5CH{=}CH_2 \Rightarrow C_6H_5CH_2CH_2CH_2CH_2C_6H_5$

**8.31** Predict the products of these reactions:

(a) $CH_3(CH_2)_4C{\equiv}CH \xrightarrow[\text{2. } H_2O_2,\ ^-OH]{\text{1. } H{-}B[CH(CH_3)CH(CH_3)_2]_2,\ THF}$

(b) 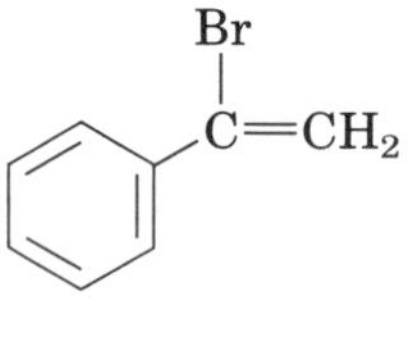 $C_6H_5CBr{=}CH_2 \xrightarrow[\text{2. } H_3O^+]{\text{1. } NaNH_2,\ NH_3}$

(c) 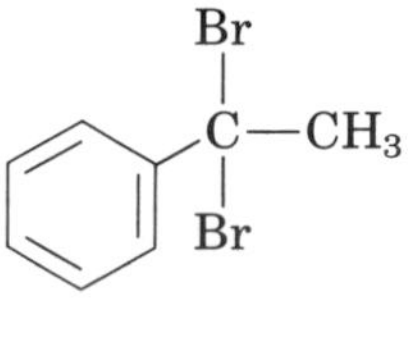 $C_6H_5CBr_2{-}CH_3 \xrightarrow[\text{2. } H_3O^+]{\text{1. } NaNH_2,\ NH_3}$

**8.32** Using 1-butyne as the only source of carbon, along with any inorganic reagents you need, synthesize the following compounds. More than one step may be needed.
(a) 1,1,2,2-Tetrachlorobutane (b) Octane (c) Butanal

**8.33** Using acetylene and any alkyl halides that have four or fewer carbons as starting materials, how would you synthesize the following compounds? More than one step may be required.

(a) $CH_3CH_2CH_2C{\equiv}CH$

(b) $CH_3CH_2C{\equiv}CCH_2CH_3$

(c) $(CH_3)_2CHCH_2CH{=}CH_2$

(d) $CH_3CH_2CH_2\overset{\overset{O}{\|}}{C}CH_2CH_2CH_2CH_3$

(e) $CH_3CH_2CH_2CH_2CH_2CHO$

**8.34** How would you carry out these reactions to introduce deuterium into organic molecules?

(a) $CH_3CH_2C{\equiv}CCH_2CH_3 \longrightarrow$ (D)(C₂H₅)C=C(D)(C₂H₅), both D on the same side (cis)

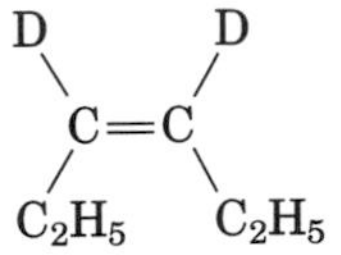

(b) $CH_3CH_2C{\equiv}CCH_2CH_3 \longrightarrow$ (D)(C₂H₅)C=C(C₂H₅)(D), D on opposite sides (trans)

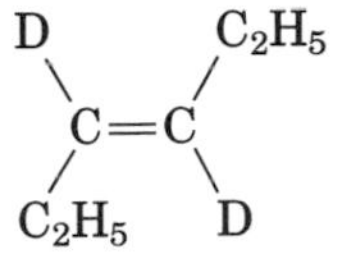

(c) $CH_3CH_2CH_2C{\equiv}CH \longrightarrow CH_3CH_2CH_2C{\equiv}CD$

(d) 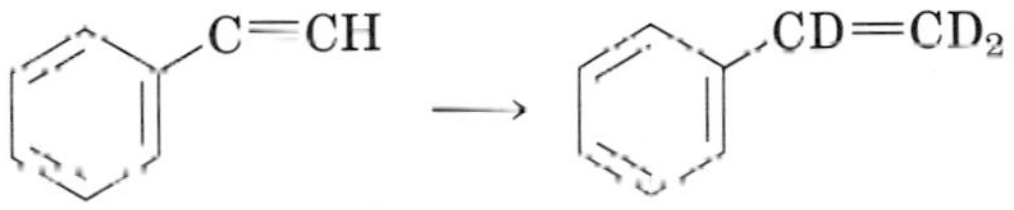 $C_6H_5C{\equiv}CH \longrightarrow C_6H_5CD{=}CD_2$

**8.35** A cumulene is a compound with three adjacent double bonds. Draw an orbital picture of a cumulene. What kind of hybridization do the two central carbon atoms have? What is the geometric relationship of the substituents on one end to the substituents on the other end? What kind of isomerism is possible? Make a model to help you see the answer.

$$R_2C{=}C{=}C{=}CR_2$$

A cumulene

**8.36** Although it is geometrically impossible for a triple bond to exist in a small ring, cycloalkynes with large rings are quite stable. How would you prepare cyclodecyne starting from acetylene and any alkyl halide needed?

**8.37** The sex attractant given off by the common housefly is a simple alkene named *muscalure*. Propose a synthesis of muscalure starting from acetylene and any alkyl halides. What is the IUPAC name for muscalure?

$$cis\text{-}CH_3(CH_2)_7CH{=}CH(CH_2)_{12}CH_3$$

Muscalure

**8.38** Compound A, with formula $C_9H_{12}$, absorbed 3 equiv of hydrogen on catalytic reduction over a palladium catalyst to give compound B, with formula $C_9H_{18}$. On ozonolysis, compound A gave, among other things, a ketone that was identified as cyclohexanone. On treatment with $NaNH_2$ in $NH_3$, followed by addition of iodomethane, compound A gave a new hydrocarbon, C, with formula $C_{10}H_{14}$. What are the structures of A, B, and C?

**8.39** Hydrocarbon A has the formula $C_{12}H_8$. It absorbs 8 equiv of hydrogen on catalytic reduction over a palladium catalyst. On ozonolysis, only two products are formed—oxalic acid, HOOC—COOH, and succinic acid, $HOOCCH_2CH_2COOH$. Formulate these reactions and propose a structure for A.

**8.40** Organometallic reagents such as sodium acetylide undergo an addition reaction with ketones, giving alcohols:

$$R-\overset{\overset{\displaystyle O}{\|}}{C}-R' \xrightarrow[\text{2. } H_3O^+]{\text{1. } Na^+ \; ^-{:}C\equiv CH} R-\underset{\underset{\displaystyle C\equiv CH}{|}}{\overset{\overset{\displaystyle OH}{|}}{C}}-R'$$

How might you use this reaction to prepare 2-methyl-1,3-butadiene, the starting material used in the manufacture of synthetic rubber?

**8.41** Erythrogenic acid, $C_{18}H_{26}O_2$, is an interesting acetylenic fatty acid that turns a vivid red on exposure to light. On catalytic hydrogenation over a palladium catalyst, 5 equiv of hydrogen are absorbed, and stearic acid, $CH_3(CH_2)_{16}COOH$, is produced. Ozonolysis of erythrogenic acid gives four products, which can be identified as follows: formaldehyde, $CH_2O$; oxalic acid, HOOC—COOH; azelaic acid, $HOOC(CH_2)_7COOH$; and the aldehyde acid $OHC(CH_2)_4COOH$. With this information, draw two possible structures for erythrogenic acid. Suggest a way to tell them apart by carrying out some simple reactions.

CHAPTER 9

# Stereochemistry

Are you right-handed or left-handed? Though most of us don't often think about it, handedness plays a surprisingly large role in our daily activities. Many musical instruments, such as oboes and clarinets, have a handedness to them; the last available softball glove always fits the wrong hand; left-handed people write in a "funny" way. The fundamental reason for these difficulties, of course, is that our hands aren't identical—they're **mirror images**. When you hold a *right* hand up to a mirror, the image you see looks like a *left* hand. Try it.

Handedness also plays a large role in organic chemistry as a direct consequence of the tetrahedral stereochemistry of $sp^3$-hybridized carbon. Let's see how handedness in organic molecules arises.

## 9.1 Optical Activity

The study of stereochemistry has its origins in the work of the French scientist Jean Baptiste Biot[1] in the early nineteenth century. Biot, a physicist, was investigating the nature of **plane-polarized light**. Now, a beam of ordinary light consists of electromagnetic waves that oscillate at right angles to the direction of light travel. Since ordinary light is unpolarized, this oscillation takes place in an infinite number of planes. When a beam of ordinary light is passed through a device called a **polarizer**, however, only the light waves oscillating in a *single* plane pass through. Light waves in all other planes are blocked out. The light that passes through the polarizer has its electromagnetic waves vibrating in a well-defined plane, hence

[1]Jean Baptiste Biot (1774–1862); b. Paris, physicist, Collège de France.

the name *plane-polarized light.* The polarization process is represented in Figure 9.1.

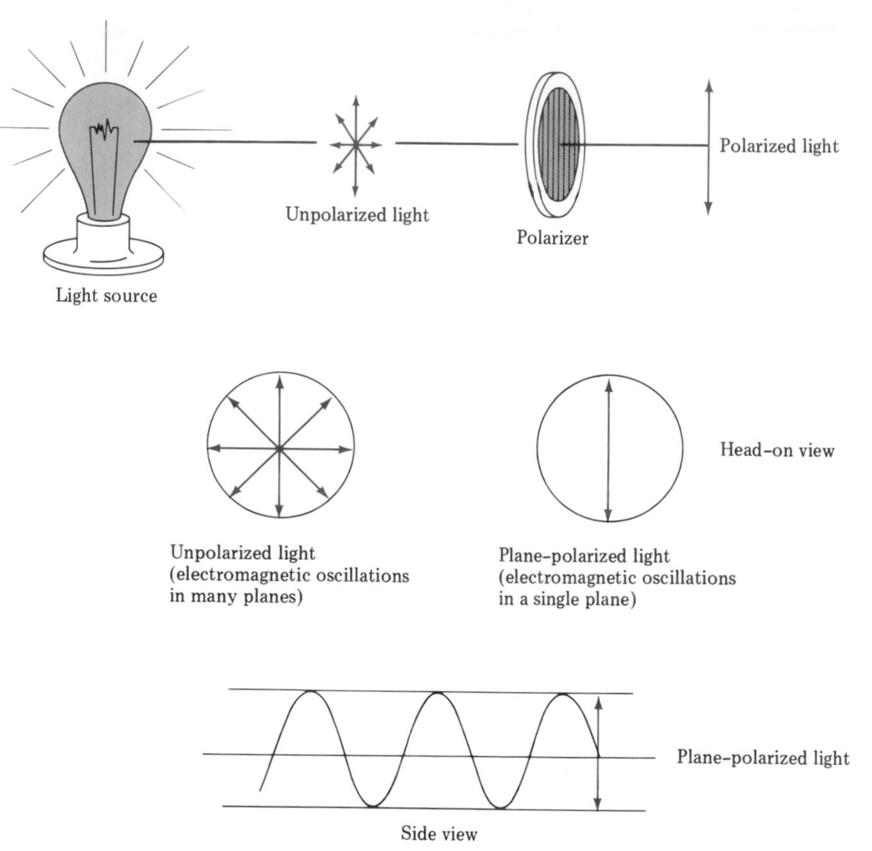


**Figure 9.1** Plane-polarized light

Biot made the remarkable observation that, when a beam of plane-polarized light is passed through solutions of certain organic molecules, such as sugar or camphor, the plane of polarization is *rotated.* Not all organic molecules exhibit this property, but those that do rotate plane-polarized light are said to be **optically active**.

The amount of rotation can be measured with an instrument known as a **polarimeter**, represented schematically in Figure 9.2. A solution of optically active organic molecules is first placed in a sample tube; plane-polarized light is passed through the tube; and rotation of the plane occurs. The light then goes through a second polarizer known as the **analyzer**. By rotating the analyzer until the light passes through it, we can find the new plane of polarization and can tell to what extent rotation has occurred. The amount of rotation is denoted $\alpha$ (Greek alpha) and is expressed in degrees.

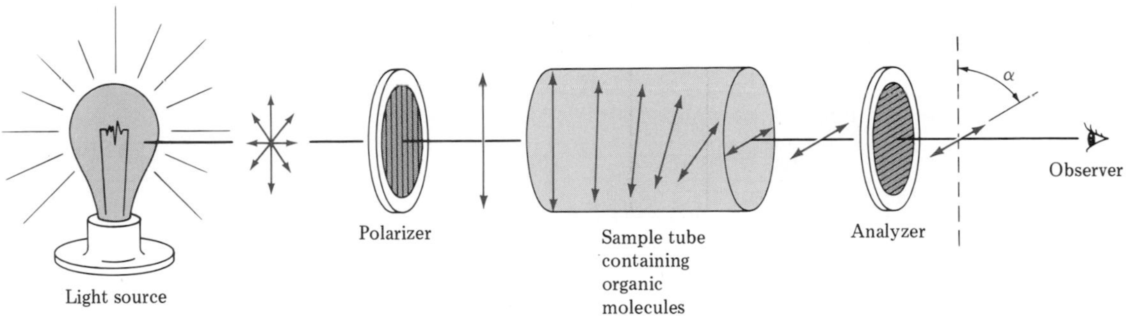


**Figure 9.2** Schematic representation of a polarimeter

In addition to determining the extent of rotation, we can also find out its direction. Some optically active molecules rotate polarized light to the left (counterclockwise) and are said to be **levorotatory**, whereas others rotate polarized light to the right (clockwise) and are said to be **dextrorotatory**. By convention, rotation to the left is given a minus sign (−), and rotation to the right is given a plus sign (+). For example, (−)-morphine is levorotatory, and (+)-sucrose is dextrorotatory.

## 9.2 Specific Rotation

Since the rotation of plane-polarized light is an intrinsic property of optically active organic molecules, it follows that the amount of rotation depends on the number of molecules that the light beam encounters. Thus, the exact amount of rotation observed is dependent both on sample concentration and on sample path length. If we double the concentration of sample in a tube, the observed rotation doubles. Similarly, if we keep the sample concentration constant but double the length of the sample tube, the observed rotation is doubled. It also turns out that the amount of rotation is dependent on the wavelength of the light used.

To express the data in a meaningful way so that comparisons can be made, we have to choose standard conditions. By convention, the **specific rotation, $[\alpha]_D$**, of a compound is defined as the observed rotation when light of 589 nanometers (1 nm = $10^{-9}$ m) wavelength is used with a sample path length $l$ of 1 decimeter (1 dm = 10 cm) and a sample concentration $C$ of 1 g/mL. (Light of 589 nm wavelength, the so-called sodium D line, is the yellow light emitted from common sodium lamps.)

$$[\alpha]_D = \frac{\text{Observed rotation, } \alpha}{\text{Path length, } l \text{ (dm)} \times \text{Concentration of sample, } C \text{ (g/mL)}}$$

$$= \frac{\alpha}{l \times C}$$

When optical rotation data are expressed in this standard way, the specific rotation, $[\alpha]_D$, is a physical constant that is characteristic of each optically active compound. Some examples are listed in Table 9.1.

**Table 9.1** Specific rotations of some organic molecules

| *Compound* | $[\alpha]_D$ *(degrees)* | *Compound* | $[\alpha]_D$ *(degrees)* |
|---|---|---|---|
| Camphor | +44.26 | Penicillin V | +223 |
| Morphine | −132 | Monosodium glutamate | +25.5 |
| Sucrose | +66.47 | Benzene | 0 |
| Cholesterol | −31.5 | Acetic acid | 0 |

PROBLEM

**9.1** Suppose you have a sucrose solution that appears to rotate plane-polarized light 90° to the right (dextrorotatory). How do you know for sure that the solution isn't rotating the plane of polarization to the *left* by 270° (levorotatory)? After all, the analyzer would be in exactly the same position in either case. [*Hint:* What effect would diluting the sample concentration have in each case?]

PROBLEM

**9.2** A 1.5 g sample of coniine, the toxic extract of poison hemlock, was dissolved in 10 mL ethanol and placed in a sample cell with a 5.0 cm path length. The observed rotation at the sodium D line was +1.2°. Calculate the specific rotation, $[\alpha]_D$, for coniine.

## 9.3 Pasteur's Discovery of Enantiomers

After Biot's discovery of optical activity in 1815, little was done until Louis Pasteur[2] entered the picture in 1849. Pasteur, who received his formal training in chemistry, had become interested in the subject of crystallography. He began work on crystalline salts of tartaric acid derived from wine and was repeating some measurements published a few years earlier when he made a surprising observation. When he recrystallized a concentrated solution of sodium ammonium tartrate below 28°C, two distinct kinds of crystals precipitated. Furthermore, the two kinds of crystals were nonsuperimposable *mirror images* of each other. That is, the crystals were not symmetrical, but were related to each other in exactly the same way that a right hand is related to a left hand.

Working carefully with a pair of tweezers, Pasteur was able to separate the crystals into two piles, one of "right-handed" crystals and one of "left-handed" crystals, like those shown in Figure 9.3. Although a solution of the original salt (a 50:50 mixture of right and left) was optically inactive, *solutions of the crystals in each of the individual piles were optically active*, and their specific rotations were equal in amount but opposite in sign.

[2]Louis Pasteur (1822–1895); b. Dôle, Jura, France; studied at Arbois, Besançon; professor, Dijon, Strasbourg (1849–1854), Lille (1854–1857), École Normale Supérieure (1857–1863).

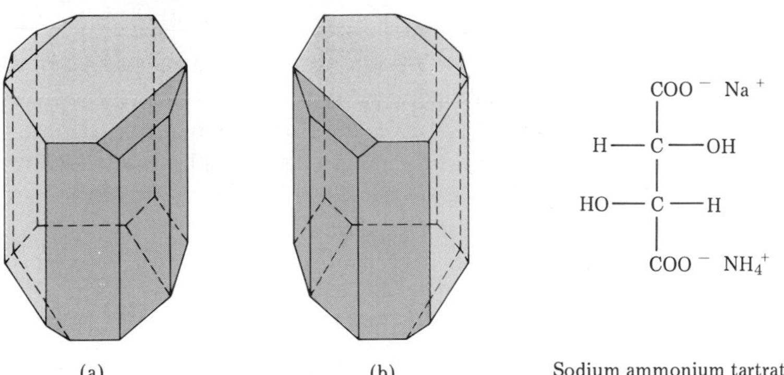


**Figure 9.3** Crystals of sodium ammonium tartrate: One of the crystals is "right-handed" and one is "left-handed." The drawings are taken from Pasteur's original sketches.

Pasteur was far ahead of his time. Although the structural theory of Kekulé had not yet been proposed, in explaining his results Pasteur spoke of the *molecules themselves*, saying: "It cannot be a subject of doubt that [in the *dextro* tartaric acid] there exists an asymmetric arrangement having a nonsuperimposable image. It is no less certain that the atoms of the *levo* acid possess precisely the inverse asymmetric arrangement." Pasteur's vision was extraordinary, for it was not until 25 years later that the theories of van't Hoff and Le Bel confirmed his ideas regarding the asymmetric carbon atom.

Today, we would describe Pasteur's work by saying that he had discovered the phenomenon of **optical isomerism**, or **enantiomerism**. **Enantiomers** (pronounced e-**nan**-tee-o-mers; from the Greek *enantio*, "opposite") are molecules that are mirror images of each other. The two "right-handed" and "left-handed" tartaric acid salts that Pasteur separated are identical in all respects except for their interaction with plane-polarized light. They have the same melting point, the same boiling point, the same solubilities, and the same spectroscopic properties. They are, however, related to each other as a right hand is to a left hand. Let's look further into the phenomenon of enantiomerism to see how it relates to the tetrahedral geometry of carbon.

## 9.4 Enantiomers and the Tetrahedral Carbon

By 1874 a sufficient number of pieces were available to complete the puzzle of stereochemistry, but they had not yet been assembled into a coherent picture. Let's see what facts were known:

1. Kekulé's structural theory indicated that carbon was always tetravalent.
2. Only *one* isomer of the general formula $CH_3X$ was known to be possible.

3. Only *one* isomer of the general formula $CH_2XY$ was known to be possible.
4. *Two* isomers of the general formula CHXYZ were known to be possible. Pasteur's (+)- and (−)-tartaric acids were the first examples, but by 1874 twelve other pairs of enantiomers had been found, including (+)-lactic acid from muscle tissue and (−)-lactic acid from sour milk.

Molecules of general formula CHXYZ:

$$Na^{+}\ {}^{-}OOCCH(OH)—\underset{\displaystyle OH}{\overset{\displaystyle H}{C}}—COO^{-}\ {}^{+}NH_4 \qquad H_3C—\underset{\displaystyle OH}{\overset{\displaystyle H}{C}}—COOH \qquad \left( X—\underset{\displaystyle Y}{\overset{\displaystyle H}{C}}—Z \right)$$

Sodium ammonium tartrate — Lactic acid

Starting with these four facts, Jacobus van't Hoff (Section 1.4) reasoned in the following way:

1. *The fact that there is only one isomer for formula $CH_3X$ indicates that all four valences on carbon are identical.* There is only one known $CH_3—Cl$, one known $CH_3—OH$, one known $CH_3—COOH$, and so on. If we imagine these molecules to be derived from methane by the replacement of one hydrogen atom with an X group, it doesn't matter which of the four methane hydrogens we replace. Replacement of any one of the four leads to the same $CH_3—X$ product, because all four hydrogens of methane are equivalent.

   What does the equivalence of methane hydrogens imply about the geometry of organic compounds? Van't Hoff reasoned that there were only three geometries in which all methane hydrogens could be equivalent—planar geometry, pyramidal geometry, and tetrahedral geometry. (You can convince yourself of the correctness of van't Hoff's reasoning by building molecular models; Figure 9.4 will help.)

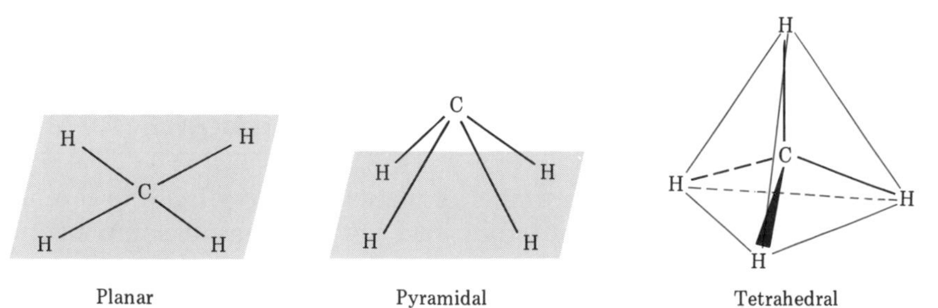


**Figure 9.4** Three possible geometries of methane in which all four hydrogens are equivalent

2. *The fact that there is only one known isomer of formula $CH_2XY$ indicates that the planar and pyramidal geometries are not correct.* Methane must therefore be tetrahedral. There is only one known $CH_3—CH_2—COOH$, one known $CH_3—CH_2—Br$, one known $CH_3CH_2—CH_2—OH$, and so on. If we imagine deriving these molecules by replacement of two hydrogens from planar methane or pyramidal methane, two isomers of each would be possible, as Figure 9.5 shows.

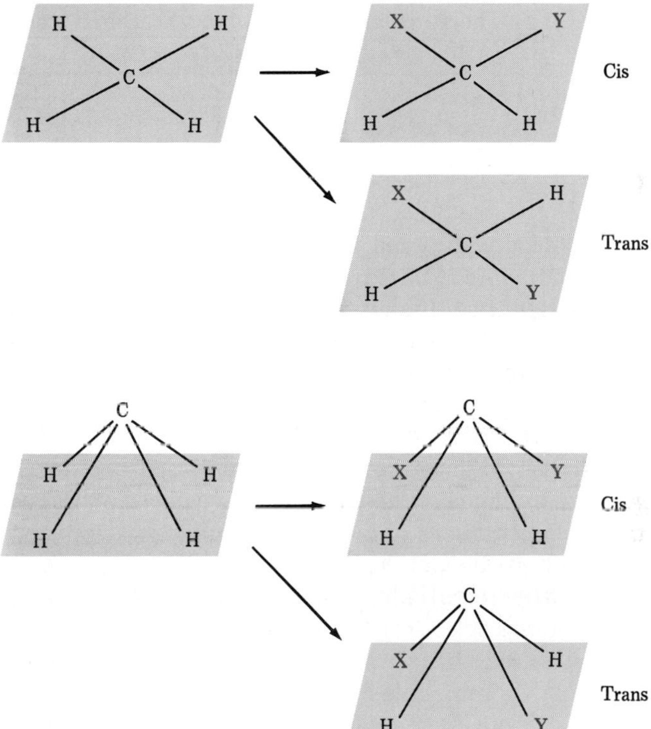


**Figure 9.5** Hypothetical cis–trans isomers of $CH_2XY$ that would result if methane were planar or pyramidal

Only tetrahedral methane allows for just one $CH_2XY$ isomer. We therefore conclude with van't Hoff that carbon has tetrahedral geometry. Again, you can convince yourself that this reasoning is sound by studying molecular models (Figure 9.6, page 264).

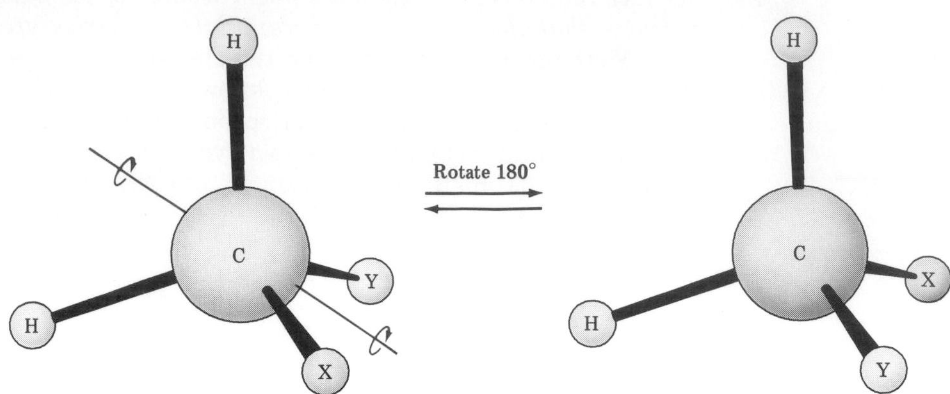


**Figure 9.6** The tetrahedral geometry of carbon: Only one molecule of general formula $CH_2XY$ is possible for tetrahedral carbon.

The logic that led van't Hoff to postulate the tetrahedral carbon is a remarkable piece of reasoning, but is still unsatisfying because its premises rest on *negative* evidence—the *lack* of a certain observation. Just because no one has yet observed two isomers with formula $CH_2XY$ doesn't mean that at some future time two isomers with that formula won't be found. Some form of *positive* evidence for the tetrahedral geometry of carbon—the positive observation of a predicted result—would be much more convincing. Fortunately, Pasteur's discovery of optical isomerism provides exactly this positive evidence. *The tetrahedral geometry of carbon predicts the existence of two enantiomers of formula CHXYZ.*

Look at the $CH_3X$, $CH_2XY$, and CHXYZ molecules shown in Figure 9.7 to see why tetrahedral geometry predicts the existence of mirror-image enantiomers for CHXYZ molecules. On the left of Figure 9.7 are three molecules; on the right are their images reflected in a mirror. The $CH_3X$ and $CH_2XY$ molecules are identical with their mirror images. If we were to make molecular models of each molecule and of its mirror image, we would find that we could superimpose one on top of the other.

The CHXYZ molecule, by contrast, is *not* identical with its mirror image. Try as we might, we can't superimpose a model of the molecule on top of a model of its mirror image for the same reason that we can't superimpose a left hand on a right hand. We might get *two* of the substituents superimposed, X and Y for example, but H and Z would be reversed. If the H and Z substituents were superimposed, X and Y would be reversed. Tetrahedral geometry predicts that a CHXYZ molecule can exist as a pair of enantiomers. Whenever a tetrahedral carbon is bonded to any four different substituents (one need not be H), optical activity can result.

To take an example of a molecule that is not identical with its mirror image, lactic acid, $CH_3CH(OH)COOH$, exists as a pair of enantiomers

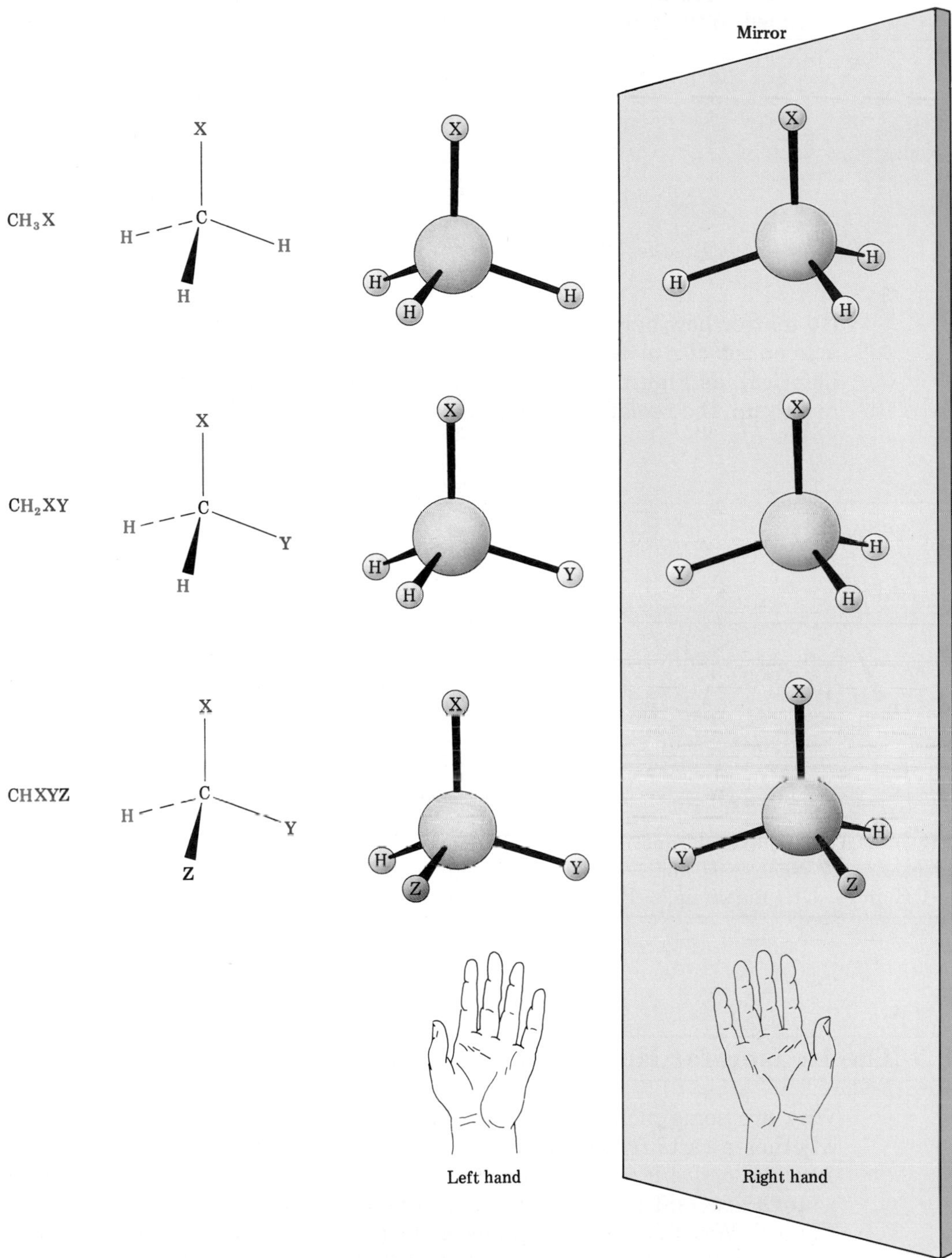


**Figure 9.7** Tetrahedral carbon atoms and their mirror images: Molecules of the type $CH_3X$ and $CH_2XY$ are identical to their mirror images, but a molecule of the type CHXYZ is not. A CHXYZ molecule is related to its mirror image in the same way that a right hand is related to a left hand.

because there are four different groups (—H, —OH, —$CH_3$, —COOH) attached to the central carbon atom:

Mirror

(+)-Lactic acid
$[\alpha]_D = +3.82°$

(−)-Lactic acid
$[\alpha]_D = -3.82°$

No matter how hard we try, we can't superimpose a molecule of (+)-lactic acid on top of a molecule of (−)-lactic acid; the two molecules simply aren't identical, as Figure 9.8 shows. If any two groups, say —H and —COOH, match up, the remaining two groups do not match.

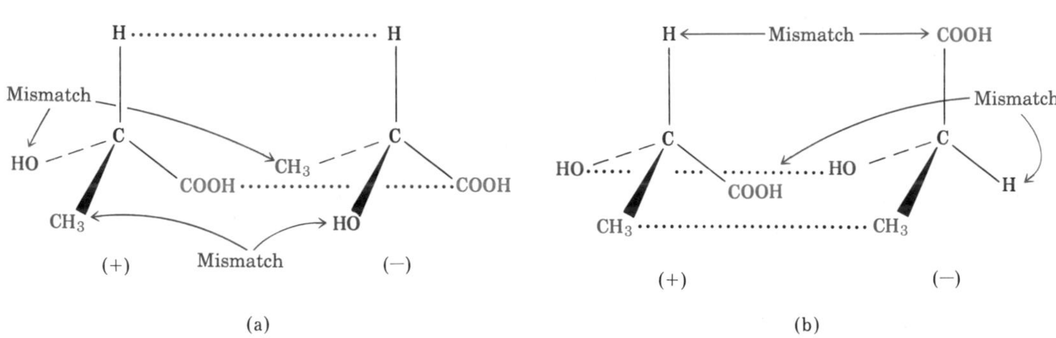


**Figure 9.8** Attempts at superimposing (+)-lactic acid and (−)-lactic acid to see if they are identical: (a) When —H and —COOH match up, —OH and —$CH_3$ don't match. (b) When —OH and —$CH_3$ match up, —H and —COOH don't match.

## 9.5 The Reason for Handedness in Molecules: Chirality

Why are some molecules handed but others aren't? How can we predict whether a certain compound is or is not optically active? *A compound is optically active if it is not superimposable on its mirror image.* Such compounds are said to be **chiral** (pronounced **ky**-ral, from the Greek *cheir*, "hand"). We can't take a chiral molecule and its mirror image (enantiomer) and place them on top of each other so that all atoms coincide.

*A compound cannot be chiral if it contains a plane of symmetry.* A **plane of symmetry** is an imaginary plane that cuts through an object (or molecule) in such a way that one half of the object is an exact mirror image of the other half. For example, an object like a flask has a plane of symmetry; if a plane were to cut the flask in half, one half would be an exact mirror image of the other half. An object like a hand, however, has no plane of symmetry. One "half" of a hand is not a mirror image of the other half (Figure 9.9).

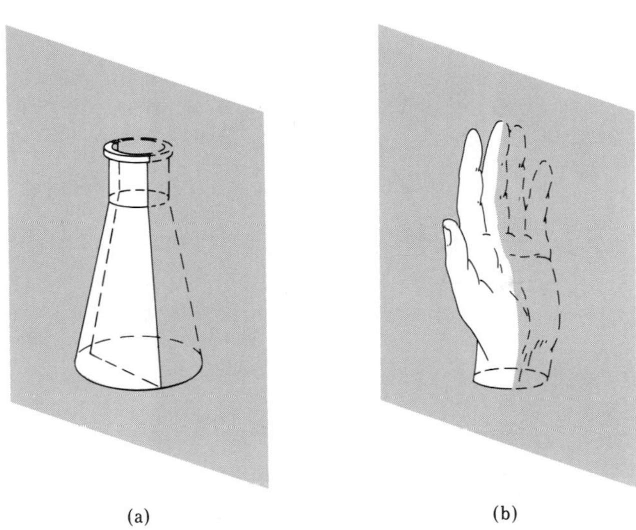


**Figure 9.9** The meaning of "symmetry plane": An object like a flask (a) has a symmetry plane cutting through it, making right and left halves mirror images. An object like a hand (b) has no symmetry plane; the right "half" of a hand is not a mirror image of the left "half."

Molecules that have planes of symmetry *must* be superimposable on their mirror images and hence must be nonchiral or **achiral** (pronounced a-**ky**-ral). Thus, hydroxyacetic acid contains a plane of symmetry and is achiral, whereas lactic acid (2-hydroxypropanoic acid) has no plane of symmetry and is chiral (Figure 9.10, page 268).

The most common (although not the only) cause of chirality in organic molecules is the presence of a carbon atom bonded to four different groups—for example, the central carbon atom in lactic acid. Such carbons are referred to as **chiral centers**. Detecting chiral centers in complex molecules is sometimes tricky because it's not always immediately apparent that four different groups are bonded to a given carbon. The differences don't necessarily appear right next to the chiral center. For example, 5-bromodecane is a chiral molecule because four different groups are bonded to C5, the chiral center (marked by an asterisk):

$$CH_3CH_2CH_2CH_2CH_2\overset{\displaystyle Br}{\underset{\displaystyle H}{\overset{|}{\underset{|}{C^*}}}}CH_2CH_2CH_2CH_3$$

5-Bromodecane (chiral)

*Substituents on carbon 5*

—H
—Br
—$CH_2CH_2CH_2CH_3$ (butyl)
—$CH_2CH_2CH_2CH_2CH_3$ (pentyl)

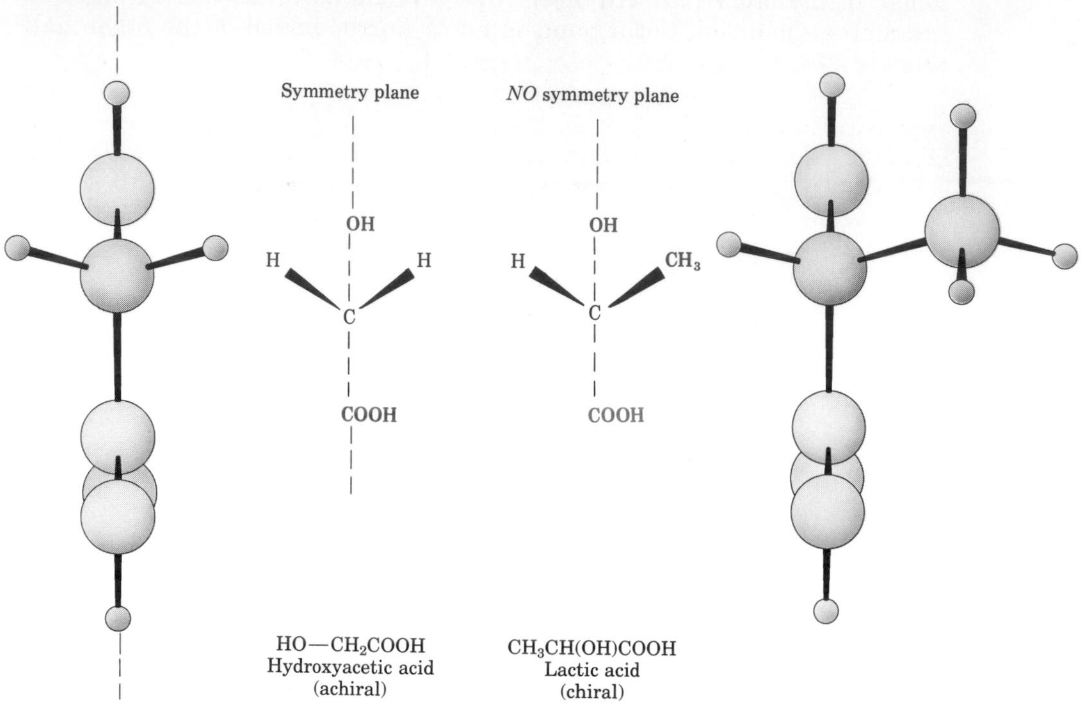


**Figure 9.10** The achiral hydroxyacetic acid molecule versus the chiral lactic acid molecule: Hydroxyacetic acid has a plane of symmetry that makes one side of the molecule a mirror image of the other side. Lactic acid, however, has no such symmetry plane.

A butyl substituent is *similar* to a pentyl substituent, but it isn't identical. The difference isn't apparent until four carbon atoms away from the chiral center, but there is still a difference.

As other examples, look at methylcyclohexane and 2-methylcyclohexanone. Are either of these molecules chiral?

H CH$_3$ \ / C1 ; 6 H$_2$C ; 2 CH$_2$ ; 5 H$_2$C ; 3 CH$_2$ ; 4 CH$_2$

O ‖ C1 ; 6 H$_2$C ; 2 *C(CH$_3$)(H) ; 5 H$_2$C ; 3 CH$_2$ ; 4 CH$_2$

Methylcyclohexane (achiral) 2-Methylcyclohexanone (chiral)

Methylcyclohexane is achiral because no carbon atom in the molecule is bonded to four different groups. We can immediately eliminate all —$CH_2$— carbons and the —$CH_3$ carbon from consideration, but what about C1 on the ring? The C1 carbon atom is bonded to a —$CH_3$ group, to an —H atom, and to C2 and C6 of the ring. Carbons 2 and 6 are equivalent, however, as are carbons 3 and 5. Thus, the C6—C5—C4 "substituent" is equivalent to the C2—C3—C4 "substituent," and methylcyclohexane is therefore achiral. (An alternative way of arriving at the same conclusion is to realize that methylcyclohexane has a symmetry plane passing through the methyl group and through carbons 1 and 4 of the ring. Make a molecular model to see this symmetry plane more clearly.)

The situation is different for 2-methylcyclohexanone. 2-Methylcyclohexanone has no symmetry plane and is chiral because C2 is bonded to four different groups: a —$CH_3$ group, an —H atom, a —COCH— ring bond (C1), and a —$CH_2CH_2$— ring bond (C3). Several more examples of chiral molecules appear in Figure 9.11. Check for yourself that the labeled centers are indeed chiral. [*Remember:* —$CH_2$—, —$CH_3$, and C=C centers *can't* be chiral, because they have at least two identical bonds.]

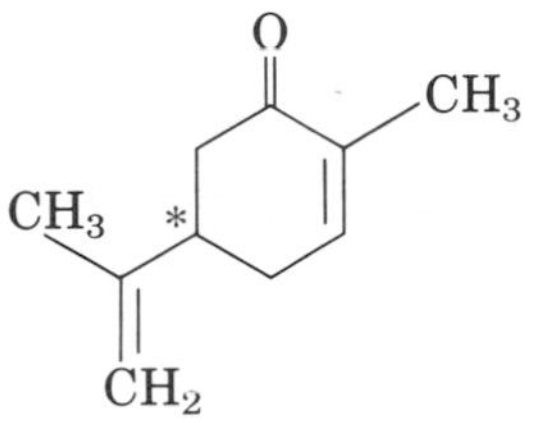

Carvone (spearmint oil)

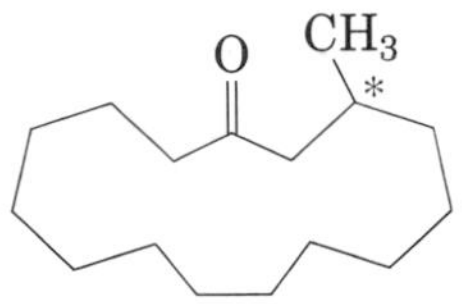

Muscone (musk oil)

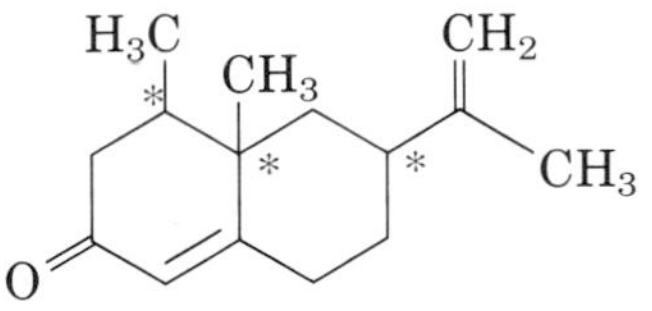

Nootkatone (grapefruit oil)

**Figure 9.11** Some chiral molecules. Note that nootkatone has three chiral centers.

PROBLEM

**9.3** Which of these objects are chiral (handed)?
(a) A screwdriver (b) A screw (c) A bean stalk
(d) A shoe (e) A hammer

PROBLEM

**9.4** Which of these compounds are chiral? Build molecular models if you need help seeing spatial relationships.

(a) Coniine (b) $HOCH_2CH(NH_2)COOH$ Serine (c) 1-Ethynylcyclohexanol

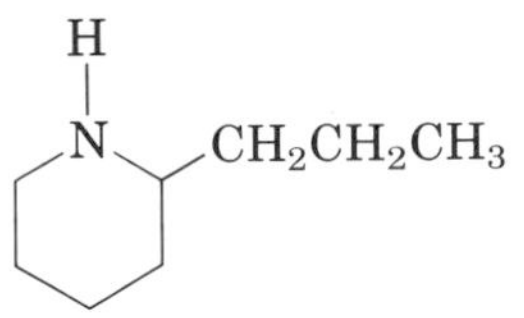

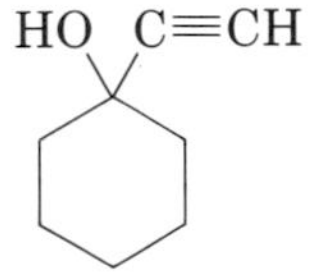

PROBLEM

**9.5** Place asterisks at all the chiral centers in these molecules:

(a) Menthol (b) Camphor (c) Dextromethorphan (a cough suppressant)

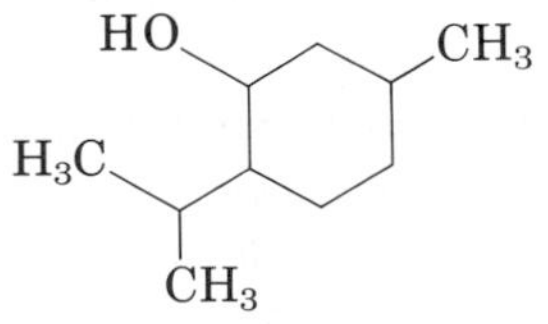

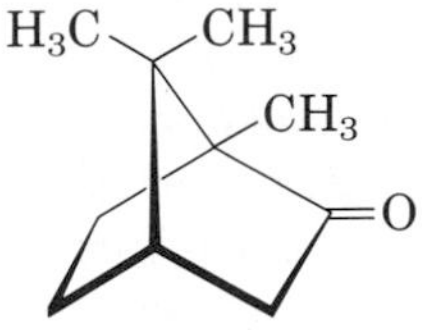

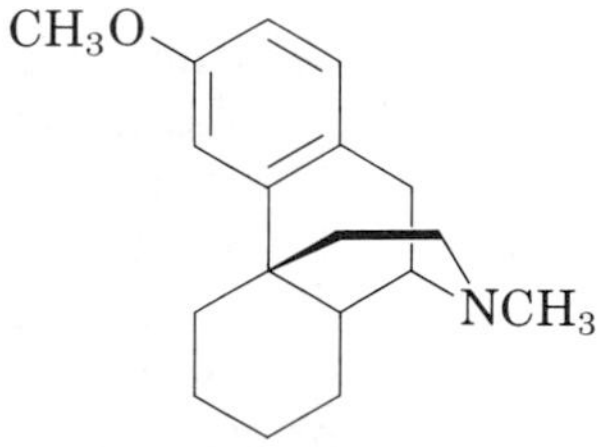

## 9.6 Sequence Rules for Specification of Configuration

Although drawings provide a pictorial representation of stereochemistry, they are difficult to translate into words. Thus, a verbal method for indicating the three-dimensional arrangement of atoms (the **configuration**) at a chiral center is also necessary. The method used employs the same sequence rules that we used in connection with the specification of alkene geometry ($Z$ versus $E$) in Section 6.6. Let's briefly review the Cahn–Ingold–Prelog sequence rules to see how they can be used for specifying the configuration of a chiral center. You should refer to Section 6.6 for an explanation of each rule.

1. Rank the atoms directly attached to the chiral center and assign priorities in order of decreasing atomic number. The group with highest atomic number is ranked first; the group with lowest atomic number is ranked fourth.

2. If a decision about priority can't be reached by applying rule 1, compare atomic numbers of the second atoms in each substituent, continuing on as necessary through the third or fourth atoms outward until the first point of difference.

3. Multiple-bonded atoms are considered as an equivalent number of singly bonded atoms. For example, a —CH=O substituent is equivalent to —CH—O—C.
   
   (with an O attached below the CH)

Following these sequence rules, we can assign priorities to the four substituent groups attached to a chiral carbon. To describe the stereochemical configuration around the chiral carbon, we mentally orient the molecule so that the group of lowest priority (fourth) is pointing directly back, away from us. We then look at the three remaining substituents, which now appear to radiate toward us like the spokes on a steering wheel (Figure 9.12). If a curved arrow drawn from the highest to second-highest to third-highest priority substituent ($1 \rightarrow 2 \rightarrow 3$) is *clockwise*, we say that the chiral center has the $R$ configuration (Latin *rectus*, "right"). If a curved arrow from the highest to second-highest to third-highest priority substituent ($1 \rightarrow 2 \rightarrow 3$) is *counterclockwise*, the chiral center has the $S$ configuration (Latin *sinister*, "left"). To remember these assignments, think of a car's steering wheel when making a right (clockwise) or left (counterclockwise) turn.

Look at (+)-lactic acid as an example of how configuration is assigned. The first step is to assign priorities to the four substituents. Sequence rule 1 says that —OH is first priority and —H is fourth priority, but it doesn't allow us to distinguish between $—CH_3$ and —COOH, since both groups have carbon as their first atom. Sequence rule 2 says that —COOH is higher priority than $—CH_3$, since oxygen outranks hydrogen (the second atom in each group).

OH
H—C—$CH_3$
COOH

(+)-Lactic acid

| *Priorities* | | |
|---|---|---|
| 4 | —H | (Low) |
| 3 | —$CH_3$ | |
| 2 | —C(=O)—OH | |
| 1 | —OH | (High) |

The next step is to orient the molecule so that the fourth-priority group (—H) is oriented toward the rear, away from the observer. Since a curved arrow from 1 (—OH) to 2 (—COOH) to 3 (—$CH_3$) is counterclockwise (left turn of the steering wheel), we assign the *S* configuration to (+)-lactic acid. Applying the same procedure to (−)-lactic acid should (and does) lead to the opposite assignment, as shown in Figure 9.12.

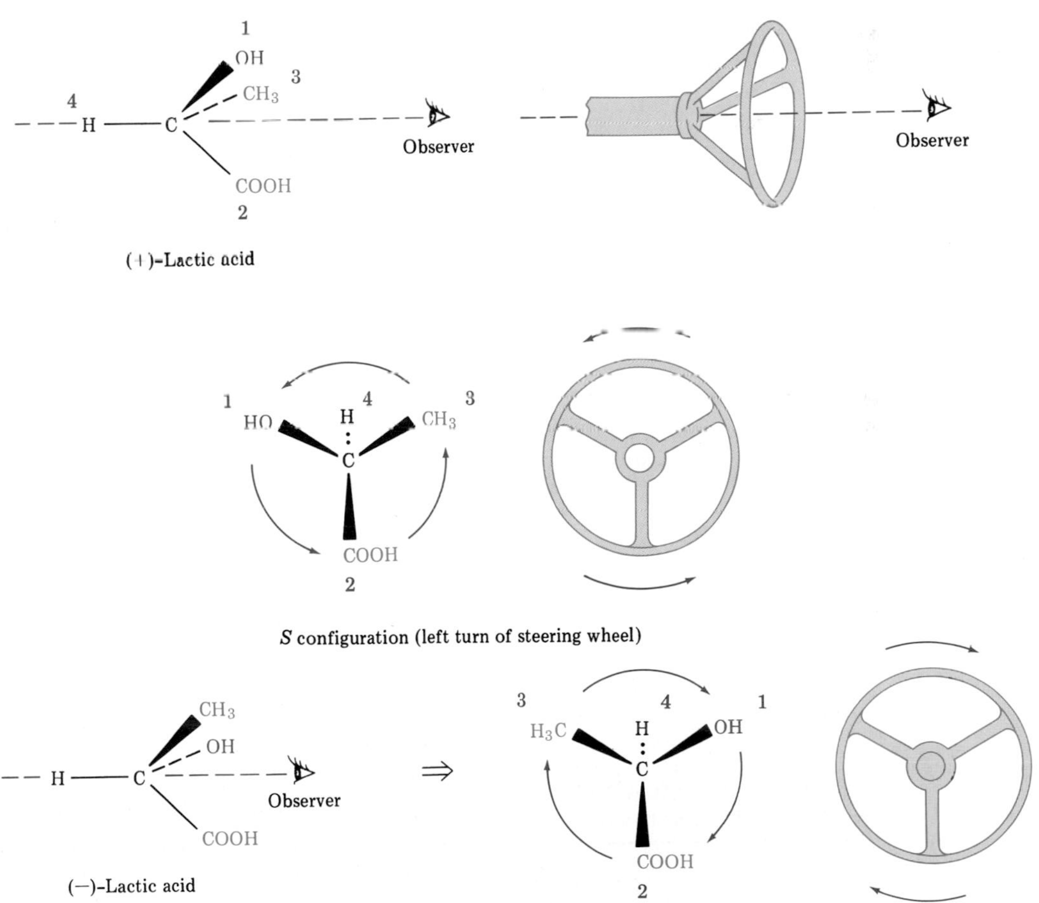


**Figure 9.12** Assignment of configuration to (*S*)-(+)-lactic acid and (*R*)-(−)-lactic acid

Further examples of how to assign configuration are provided by naturally occurring (+)-alanine and (−)-glyceraldehyde, which have the *S* configurations shown in Figure 9.13. Note that the sign of optical rotation, (+) or (−), is not related to the *R*,*S* designation. (*S*)-Alanine happens to be dextrorotatory (+), and (*S*)-glyceraldehyde happens to be levorotatory (−). There is no correlation between *R*,*S* configuration and direction or magnitude of optical rotation.

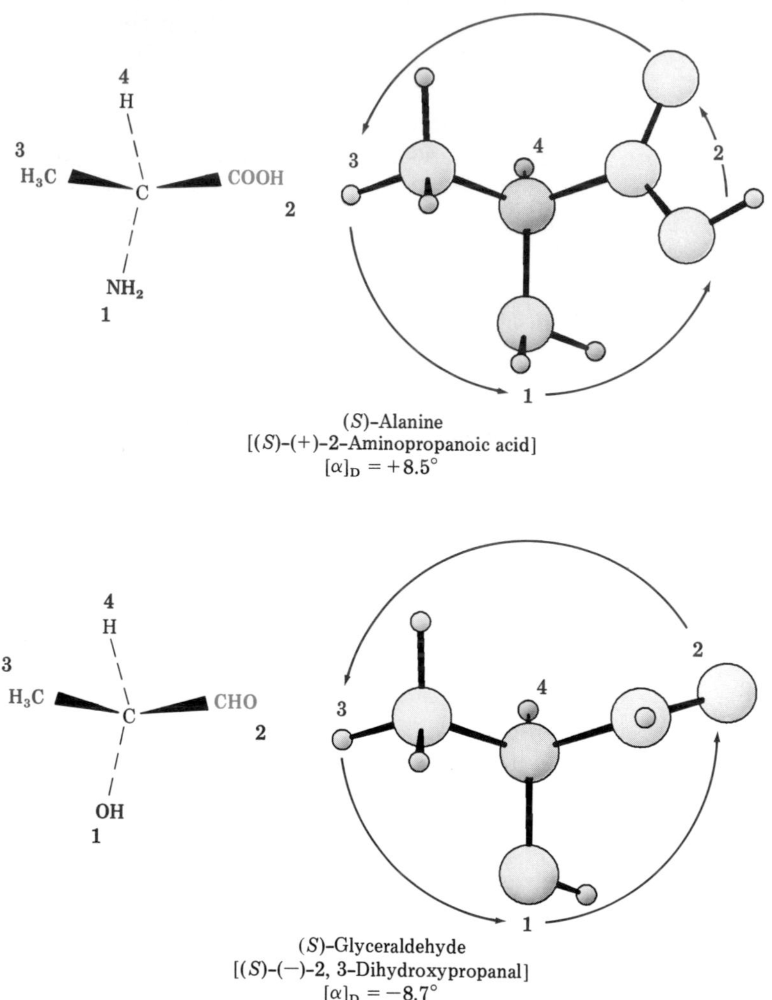


(*S*)-Alanine
[(*S*)-(+)-2-Aminopropanoic acid]
$[\alpha]_D = +8.5°$

(*S*)-Glyceraldehyde
[(*S*)-(−)-2, 3-Dihydroxypropanal]
$[\alpha]_D = -8.7°$

**Figure 9.13** Assignment of configuration to (+)-alanine and (−)-glyceraldehyde: Both happen to have the *S* configuration.

One further point bears mentioning—the matter of **absolute configuration**. How do we know that our assignments of *R*,*S* configuration are correct in an absolute, rather than a relative, sense? Since we can't see the molecules themselves, how do we know for certain that it is the dextrorotatory enantiomer of lactic acid that has the *R* configuration? This difficult question was not solved until 1951 when J. M. Bijvoet of the University of Utrecht reported an X-ray spectroscopic method for determining the absolute spatial arrangement of atoms in a molecule. Based on his results, we can say with certainty that our spatial representations of molecules are correct.

**PRACTICE PROBLEM**

Draw a tetrahedral representation of *R*-2-chlorobutane.

*Solution* The four substituents bound to the chiral carbon of *R*-2-chlorobutane can be assigned the following priorities: (1) —Cl, (2) —$CH_2CH_3$, (3) —$CH_3$, (4) —H. To draw a tetrahedral representation of the molecule, first orient the low-priority —H group toward the rear and imagine that the other three groups are coming out of the page toward you. Place the remaining three substituents in order such that the direction of travel from 1 → 2 → 3 is clockwise (right turn), and then tilt the molecule slightly to bring the rear hydrogen into view and end up with a standard tetrahedral representation:

Cl, H, $CH_2CH_3$, C, $CH_3$ = Cl, C, H, $H_3C$, $CH_2CH_3$

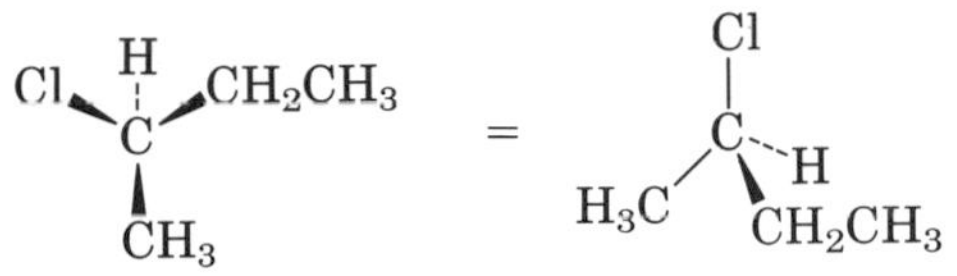

Using molecular models is a great help in working problems of this sort.

**PROBLEM**

**9.6** Assign priorities to these sets of substituents:
(a) —H, —Br, —$CH_2CH_3$, —$CH_2CH_2OH$
(b) —$CO_2H$, —$CO_2CH_3$, —$CH_2OH$, —OH
(c) —CN, —$CH_2NH_2$, —$CH_2NHCH_3$, —$NH_2$
(d) —Br, —$CH_2Br$, —Cl, —$CH_2Cl$

**PROBLEM**

**9.7** Assign *R*,*S* configurations to these molecules:

(a) 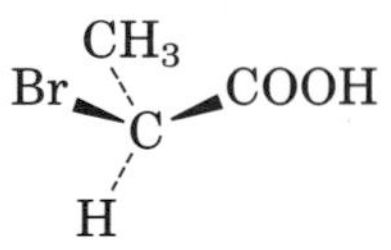 $CH_3$, Br, C, COOH, H (b) 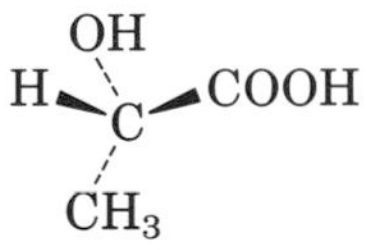 OH, H, C, COOH, $CH_3$ (c) $NH_2$, H, C, $CH_3$, CN

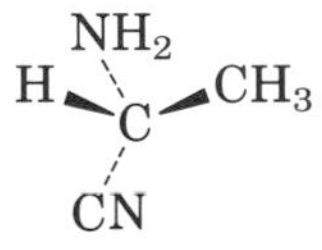

## 9.7 Diastereomers

Molecules such as lactic acid, alanine, and glyceraldehyde are relatively simple to deal with, since each has only one chiral center and can exist in only two enantiomeric configurations. The situation becomes more complex, however, for molecules that have more than one chiral center.

Look at the essential amino acid threonine (2-amino-3-hydroxybutanoic acid), for example. Since threonine has two chiral centers (C2 and C3), there are four possible stereoisomers, as shown in Figure 9.14. (Check for yourself that the *R*,*S* configurations are correct as shown.)

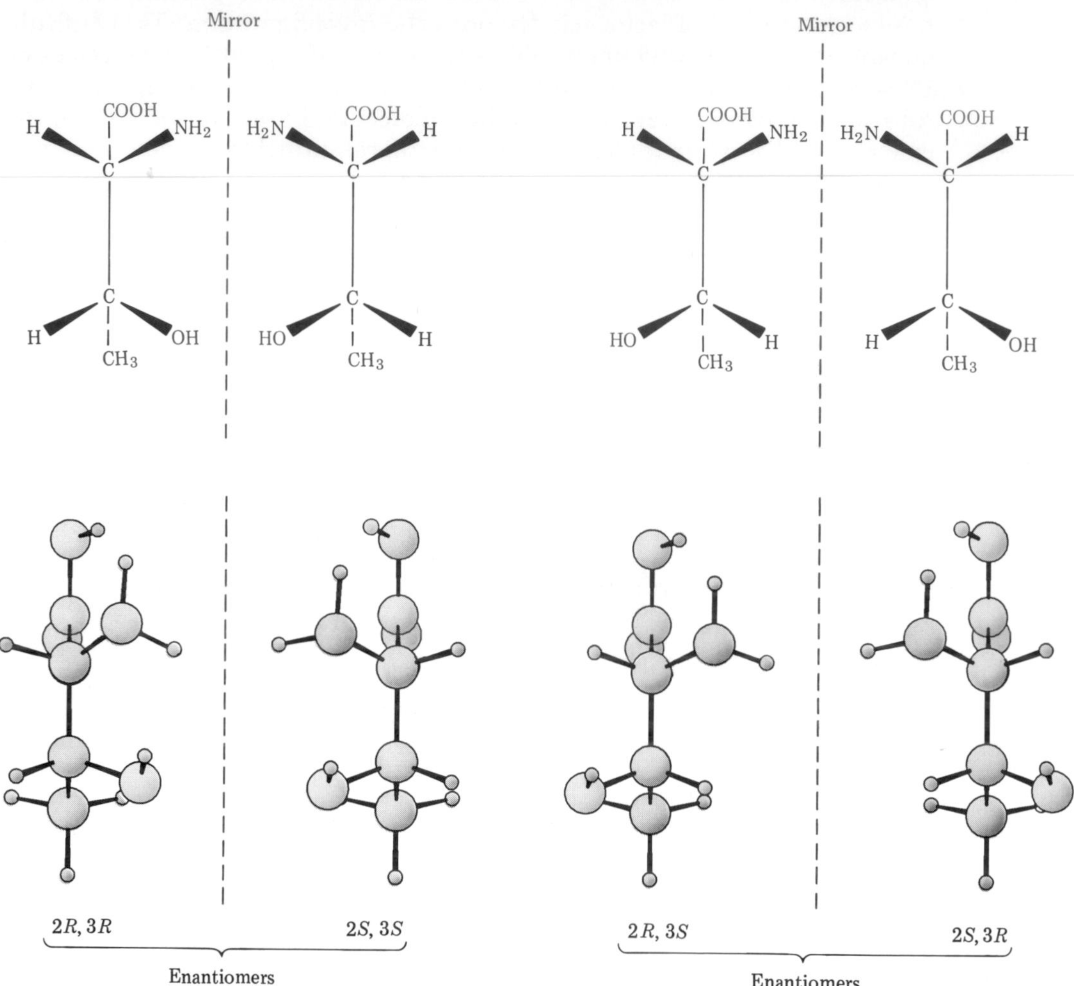


**Figure 9.14** The four stereoisomers of threonine (2-amino-3-hydroxybutanoic acid)

A careful look at the four threonine stereoisomers shows that they can be classified into two mirror-image pairs of enantiomers. The 2*R*,3*R* stereoisomer is the mirror image of 2*S*,3*S*, and the 2*R*,3*S* stereoisomer is the mirror image of 2*S*,3*R*. But what is the relationship between any two configurations that are not mirror images? What, for example, is the relationship between the 2*R*,3*R* compound and the 2*R*,3*S* compound? These two compounds are stereoisomers, yet they aren't superimposable and they aren't enantiomers. To describe such a relationship, we need a new term.

**Diastereomers** are stereoisomers that are not mirror images of each other. Since we used the right-hand/left-hand analogy to describe the relationship between two enantiomers, we might extend the analogy further by saying that diastereomers have a hand–foot relationship. Hands and feet look very *similar* but they aren't identical and they aren't mirror images.

The same is true of diastereomers; they're similar but not identical. Note carefully the difference between enantiomers and diastereomers: Enantiomers must have opposite (mirror-image) configurations at *all* chiral centers; diastereomers must have opposite configurations at *some* (one or more) chiral centers, but the same configurations at other chiral centers. A full description of the four threonine stereoisomers is given in Table 9.2.

**Table 9.2** Relationships between four stereoisomeric threonines

| *Stereoisomer* | *Enantiomeric with* | *Diastereomeric with* |
|---|---|---|
| 2*R*,3*R* | 2*S*,3*S* | 2*R*,3*S* and 2*S*,3*R* |
| 2*S*,3*S* | 2*R*,3*R* | 2*R*,3*S* and 2*S*,3*R* |
| 2*R*,3*S* | 2*S*,3*R* | 2*R*,3*R* and 2*S*,3*S* |
| 2*S*,3*R* | 2*R*,3*S* | 2*R*,3*R* and 2*S*,3*S* |

Of the four possible stereoisomers of threonine, only one, the 2*S*,3*R* isomer, $[\alpha]_D = -29.3°$, occurs naturally in plants and animals. Most biologically important molecules are chiral, and usually only a single stereoisomer is found in nature.

PROBLEM

**9.8** Assign *R*,*S* configurations to these molecules. Which are enantiomers and which are diastereomers?

(a) Br; H, $CH_3$; C; C; H, OH; $CH_3$

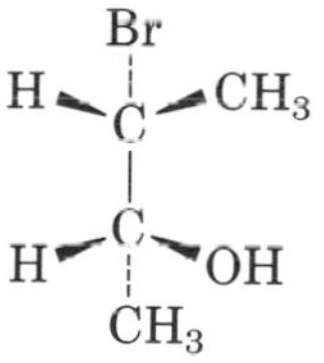

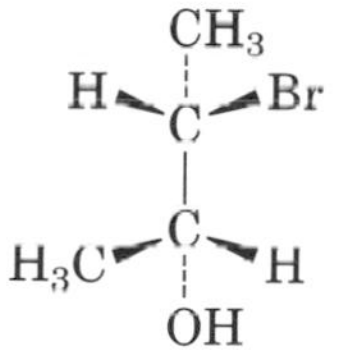

(b) $CH_3$; H, Br; C; C; $H_3C$, H; OH

(c) $CH_3$; Br, H; C; C; H, $CH_3$; OH

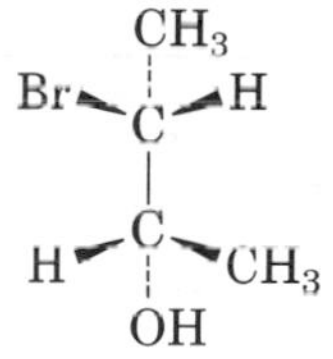

(d) H; Br, $CH_3$; C; C; $H_3C$, OH; H

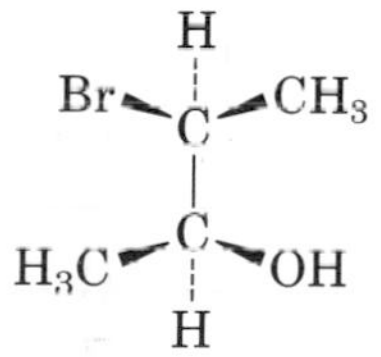

PROBLEM

**9.9** Chloramphenicol is a powerful antibiotic isolated in 1949 from the *Streptomyces venezuelae* bacterium. It is active against a broad spectrum of bacterial infections and is particularly valuable against typhoid fever. Assign *R*,*S* configurations to the chiral centers in chloramphenicol.

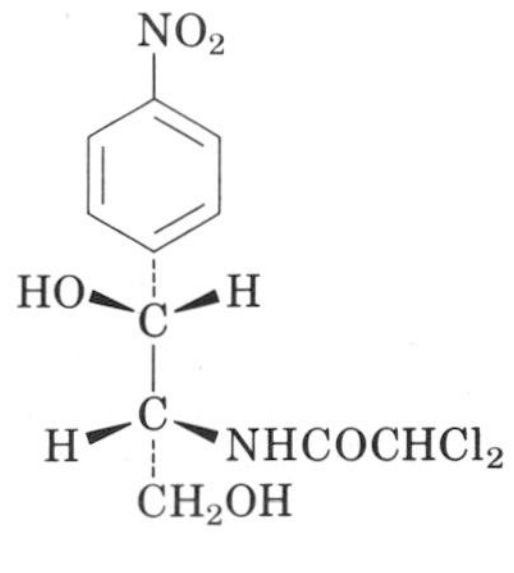

$NO_2$

HO, H; C; C; H, $NHCOCHCl_2$; $CH_2OH$

Chloramphenicol
$[\alpha]_D = +18.6°$

## 9.8 A Brief Review of Isomerism

We have seen several different kinds of isomers in the past few chapters, and it's probably a good idea at this point to see how they all relate to one another. As noted earlier, **isomers** are compounds that have the same chemical formula but have different structures. There are two fundamental types of isomerism, both of which we've now encountered: constitutional isomerism and stereoisomerism.

**Constitutional isomers** (Section 3.3) are compounds that have their atoms connected in a different order. Among the different kinds of constitutional isomers we've seen are skeletal, functional, and positional isomers.

Constitutional isomers (different connections between atoms):

**Skeletal isomers** (different carbon skeletons)

$CH_3-CH(CH_3)-CH_3$ and $CH_3-CH_2-CH_2-CH_3$

**Functional-group isomers** (different functional groups)

$CH_3-CH_2-OH$ (Ethyl alcohol) and $CH_3-O-CH_3$ (Dimethyl ether)

**Positional isomers** (different position of functional group)

$CH_3-CH(OH)-CH_3$ (2-Propanol) and $CH_3-CH_2-CH_2-OH$ (1-Propanol)

**Stereoisomers** (Section 3.10) are compounds that have their atoms connected in the same order but have a different geometry. Among the different kinds of stereoisomers we've seen are enantiomers, diastereomers, and cis–trans isomers (both in alkenes and in cycloalkanes). To be perfectly accurate, however, cis–trans isomers are really just a special category of diastereomers, since they meet the definition of being non-mirror-image stereoisomers.

Stereoisomers (same connections between atoms, but different geometry):

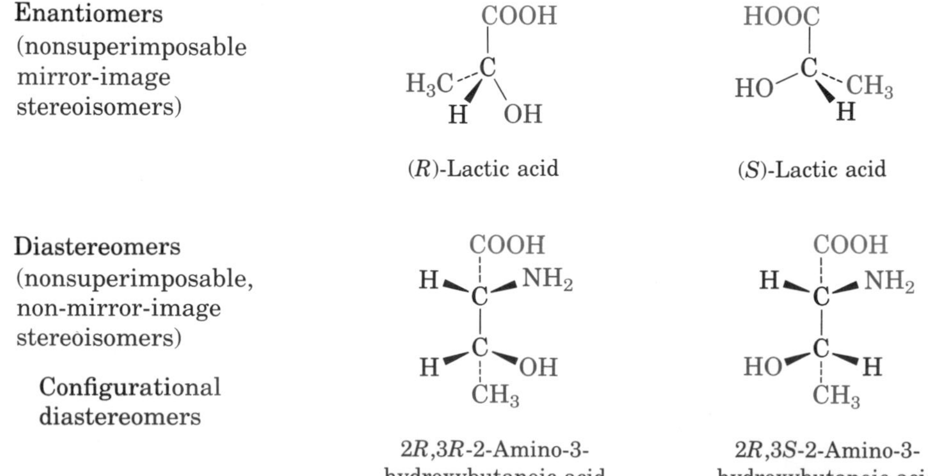

Cis–trans diastereomers (substituents on same side or opposite side of double bond or ring)

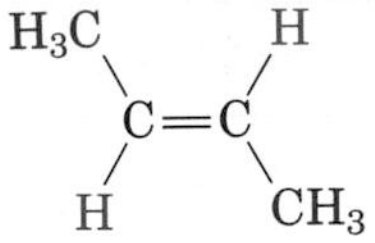

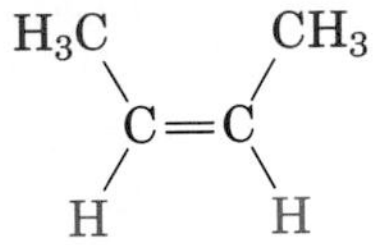

trans-2-Butene and cis-2-Butene

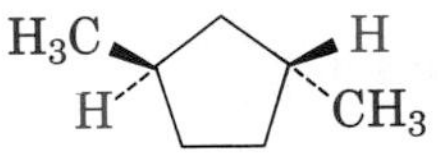

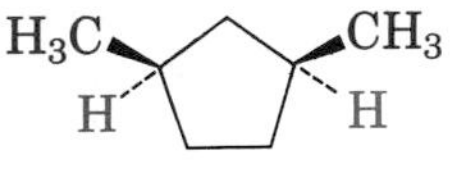

trans-1,3-Dimethylcyclopentane and cis-1,3-Dimethylcyclopentane

## 9.9 Meso Compounds

Let's look at one more example of a compound with two chiral centers—tartaric acid. We're already acquainted with tartaric acid for its role in Pasteur's discovery of optical activity, and we can now draw the four stereoisomers:

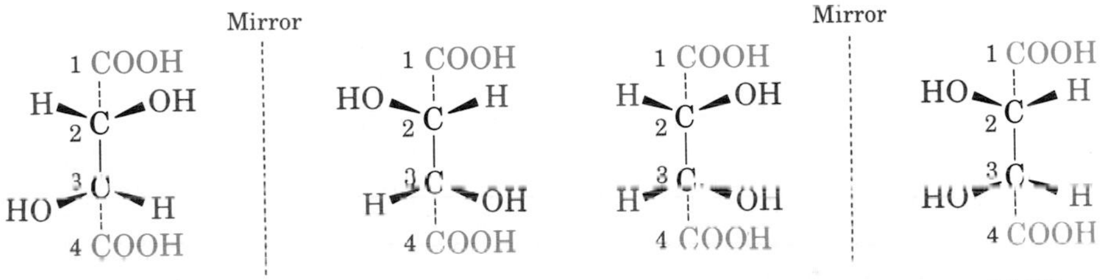

Mirror | Mirror

2*R*,3*R* | 2*S*,3*S* | 2*R*,3*S* | 2*S*,3*R*

The mirror-image 2*R*,3*R* and 2*S*,3*S* structures are not superimposable and are therefore a pair of enantiomers. A close look, however, reveals that the 2*R*,3*S* and 2*S*,3*R* structures are *identical*. We can see this readily by rotating one structure by 180°:

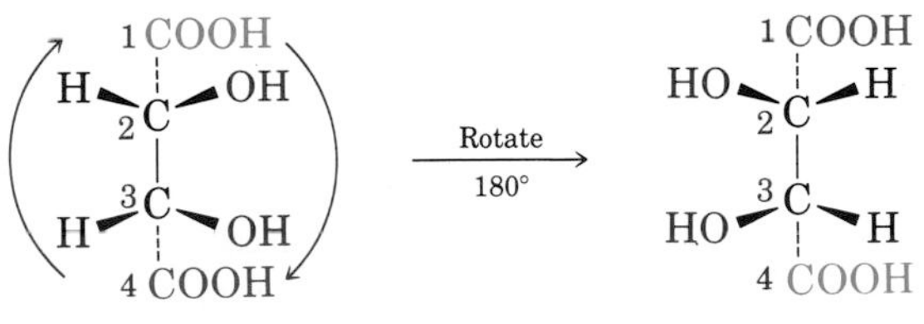

Rotate 180°

2*R*,3*S* | 2*S*,3*R*

Identical

The identity of the 2*R*,3*S* and 2*S*,3*R* structures results from the fact that the molecule has a plane of symmetry. The symmetry plane cuts through the C2—C3 bond, making one half of the molecule a mirror image of the other half (Figure 9.15, page 278).

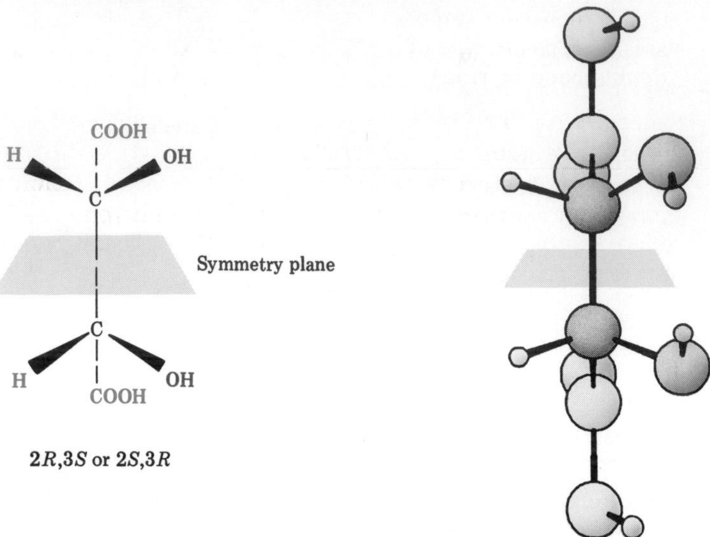


**Figure 9.15** A symmetry plane through the C2—C3 bond of *meso*-tartaric acid

Because of the plane of symmetry, the tartaric acid stereoisomer shown in Figure 9.15 must be achiral, despite the fact that it has two chiral centers. Compounds that are achiral by virtue of a symmetry plane, yet contain chiral centers, are called **meso compounds** (pronounced **me**-zo). Thus, tartaric acid exists in three stereoisomeric configurations: two enantiomers and one meso form.

PROBLEM

**9.10** Which, if any, of these structures represent meso compounds?

(a) H, OH, OH, H

(b) H, OH, OH, H

(c) $CH_3$, H

(d) H, Br, $CH_3$, C, C, $H_3C$, H, Br

PROBLEM

**9.11** Which of these substances have a meso form?
(a) 2,3-Dibromobutane (b) 2,3-Dibromopentane (c) 2,4-Dibromopentane

## 9.10 Molecules with More than Two Chiral Centers

We have seen how a single chiral center in a molecule gives rise to two stereoisomers (one pair of enantiomers), and how two chiral centers in a molecule give rise to a *maximum* of four stereoisomers (two pairs of enantiomers). In general, a molecule with $n$ chiral centers gives rise to a maximum of $2^n$ stereoisomers ($2^{n-1}$ pairs of enantiomers). For example, cholesterol contains eight chiral centers, making possible $2^8 = 256$ stereoisomers (128 enantiomeric pairs), although many would be too highly strained to exist. Only *one*, however, is produced in nature.

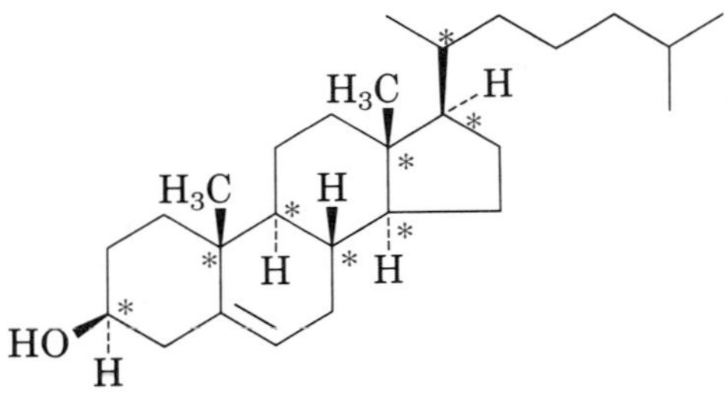

Cholesterol
(eight chiral centers)

PROBLEM

**9.12** How many chiral centers does morphine have? How many stereoisomers of morphine are possible in principle?

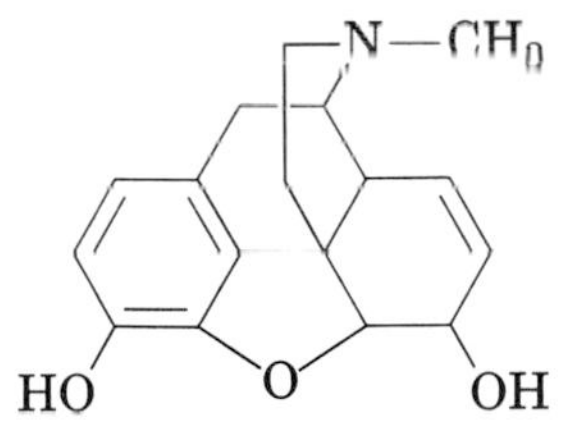

Morphine

## 9.11 Racemic Mixtures

To conclude our discussion of stereoisomerism, let's return to Pasteur's pioneering work. Pasteur took an optically inactive form of a tartaric acid salt and found that he could crystallize two optically active forms from it. These two optically active forms were the 2*R*,3*R* and 2*S*,3*S* configurations just discussed—but what was the optically inactive form he started with? It couldn't have been *meso*-tartaric acid, since *meso*-tartaric acid is a different chemical compound from the two chiral enantiomers, and can't interconvert with them without breaking and re-forming chemical bonds.

The answer is that Pasteur started with a 50:50 mixture of the two chiral enantiomers of tartaric acid. Such a mixture is called a **racemic** (pronounced ray-**see**-mic) **mixture**, or **racemate**, and is often denoted by the symbol (±). Racemic mixtures *must* show zero optical rotation because equal amounts of (+) and (−) forms are present, and the rotation from one enantiomer exactly cancels the rotation from the other enantiomer.

Through good fortune, Pasteur was able to **resolve** (±)-tartaric acid into its (−) and (+) enantiomers. Unfortunately, the method he used (fractional crystallization) does not work for most racemic mixtures, and other techniques are required. The method of resolution most often used is discussed in Section 25.5.

## 9.12 Physical Properties of Stereoisomers

We have seen how seemingly simple compounds such as tartaric acid can exist in several different stereoisomeric configurations. The question therefore arises whether the different stereoisomers of a compound have different physical properties. The answer is yes, they do.

Some physical properties of the three different stereoisomers of tartaric acid are listed in Table 9.3. As indicated, the (+)- and (−)-tartaric acids have identical melting points, solubilities, and densities, as they must since they are mirror images. They differ only in the sign of their rotation of plane-polarized light. The meso isomer, by contrast, is diastereomeric with the (+) and (−) forms. As such, it has no mirror-image relationship to (+)- and (−)-tartaric acids; it is a different compound altogether and has different physical properties.

The racemic mixture is different still. Though a mixture of enantiomers, racemates act as though they were pure compounds, different from either pure enantiomer. As the table shows, the physical properties of racemic tartaric acid differ from those of the two pure enantiomers and from those of the meso form.

**Table 9.3** Some properties of the stereoisomers of tartaric acid

| *Stereoisomer* | *Melting point (°C)* | $[\alpha]_D$ *(degrees)* | *Density* ($g/cm^3$) | *Solubility at 20°C* ($g/100\ mL\ H_2O$) |
|---|---|---|---|---|
| (+) | 168–170 | +12 | 1.7598 | 139.0 |
| (−) | 168–170 | −12 | 1.7598 | 139.0 |
| Meso | 146–148 | 0 | 1.6660 | 125.0 |
| (±) | 206 | 0 | 1.7880 | 20.6 |

## 9.13 Fischer Projections

When learning to visualize chiral molecules, it is best to begin by building molecular models. As more experience is gained, it becomes easier to draw pictures and work with mental images. To do this successfully, though, a standard method of representation is needed for depicting the three-dimensional arrangement of atoms on a flat page. In 1891, Emil Fischer suggested a convention based on the projection of a tetrahedral carbon atom onto a flat surface. These **Fischer projections** were soon adopted and are now a standard means of depicting stereochemistry at chiral centers.

A tetrahedral carbon atom is represented in a Fischer projection by two crossed lines. By convention, the horizontal lines represent bonds coming out of the page, and the vertical lines represent bonds going into the page. For example, (*R*)-lactic acid can be drawn as follows:

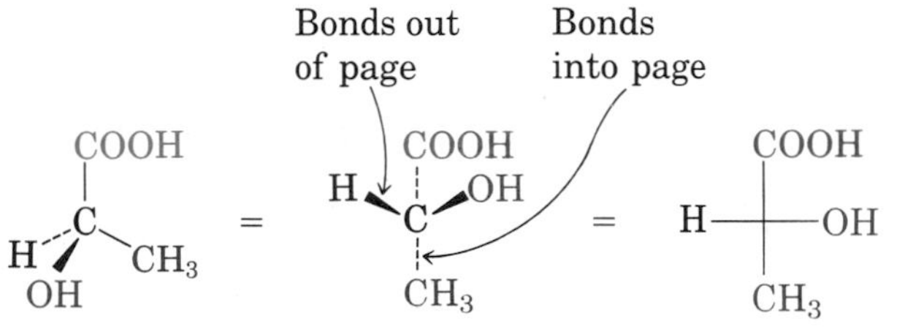

Bonds out of page

Bonds into page

COOH, H, OH, $CH_3$ = COOH, H, OH, $CH_3$ = H—┼—OH with COOH above and $CH_3$ below

(*R*)-Lactic acid

Fischer projection

Since a given molecule can be depicted in many different ways, it's often necessary to compare two different projections to see if they represent the same or different enantiomers. To test for identity, Fischer projections can be moved around on the paper, but care must be taken not to change the meaning of the projection inadvertently. Only two kinds of motions are allowed.

1. A Fischer projection can be rotated on the page by 180°, but *not by 90° or 270°*. The reason for this rule is simply that a 180° rotation obeys the Fischer convention by keeping the same two substituents on horizontal bonds always coming out of the plane and the same two substituents on vertical bonds always going into the plane. A 90° rotation, however, disobeys the Fischer convention. Substituents that were on horizontal bonds in one form become vertical after a 90° rotation, thus inverting their stereochemistry and changing the meaning of the projection. For example,

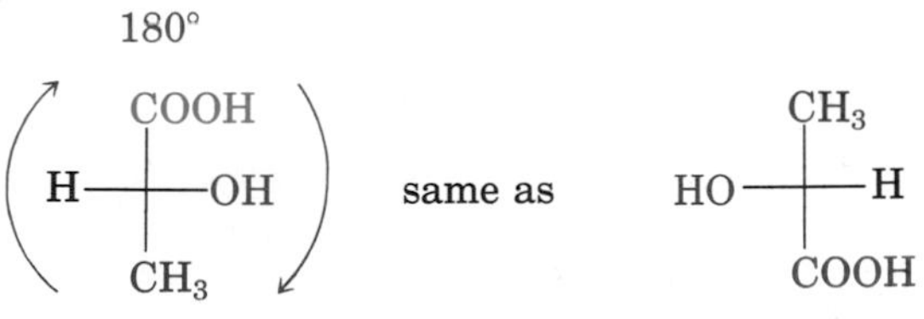

180°

H—┼—OH (COOH above, $CH_3$ below) same as HO—┼—H ($CH_3$ above, COOH below)

—COOH and —$CH_3$ go into plane of paper in both projections; —H and —OH come out of plane of paper in both projections.

but:

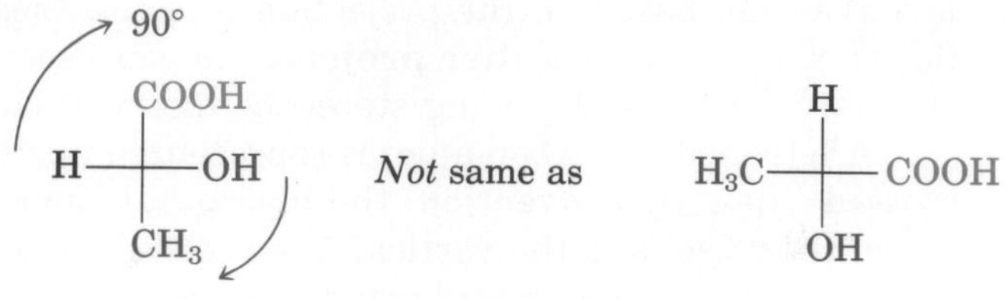

90°

COOH

H—+—OH  *Not* same as  $H_3C$—+—COOH

$CH_3$

H ... OH

—COOH and —$CH_3$ groups go into plane of paper in one projection but come out of plane of paper in other projection.

2. A Fischer projection can have one group held steady while the other three rotate in either a clockwise or counterclockwise direction. For example,

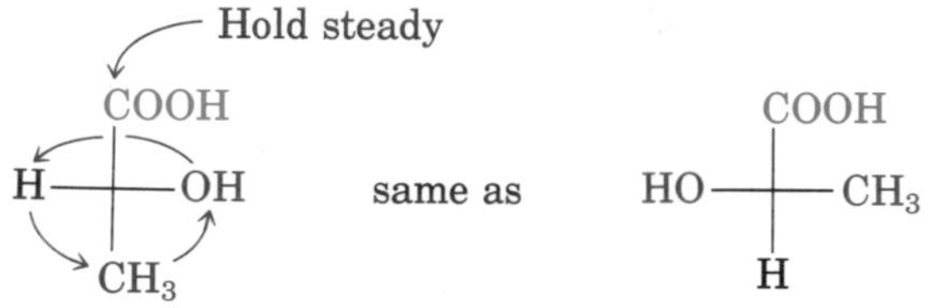

Hold steady

COOH

H—+—OH  same as  HO—+—$CH_3$

$CH_3$ ... H

These are the only kinds of motion allowed. Moving a Fischer projection in any other way inverts its meaning. Thus, if a Fischer projection of (*R*)-lactic acid is turned by 90°, a projection of (*S*)-lactic acid results (Figure 9.16).

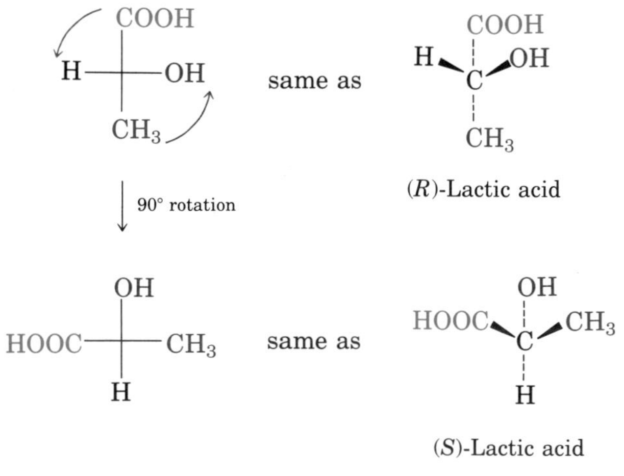

COOH

H—+—OH  same as  H, C, OH (COOH above, $CH_3$ below)

$CH_3$

(*R*)-Lactic acid

90° rotation

OH

HOOC—+—$CH_3$  same as  HOOC, C, $CH_3$ (OH above, H below)

H

(*S*)-Lactic acid

**Figure 9.16** Rotation of a Fischer projection by 90° inverts its meaning.

Knowing the two rules provides an easy way to superimpose molecules mentally. For example, three different Fischer projections of 2-butanol follow. Do they all represent the same enantiomer, or is one different?

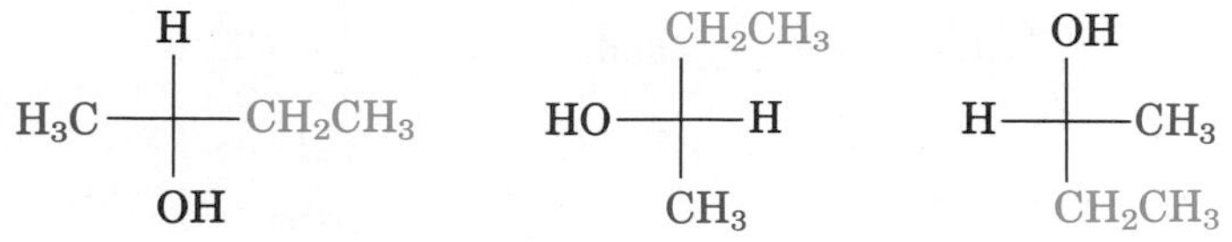

A: top H; left $H_3C$; right $CH_2CH_3$; bottom OH

B: top $CH_2CH_3$; left HO; right H; bottom $CH_3$

C: top OH; left H; right $CH_3$; bottom $CH_2CH_3$

The simplest way to determine if two Fischer projections represent the same enantiomer is to carry out allowed motions until *two* groups are superimposed. If the other two groups are also superimposed, then the Fischer projections are the same; if the other two groups are not superimposed, the Fischer projections are different.

Let's keep projection A unchanged and move B so that the $—CH_3$ and $—H$ substituents match up with those in A:

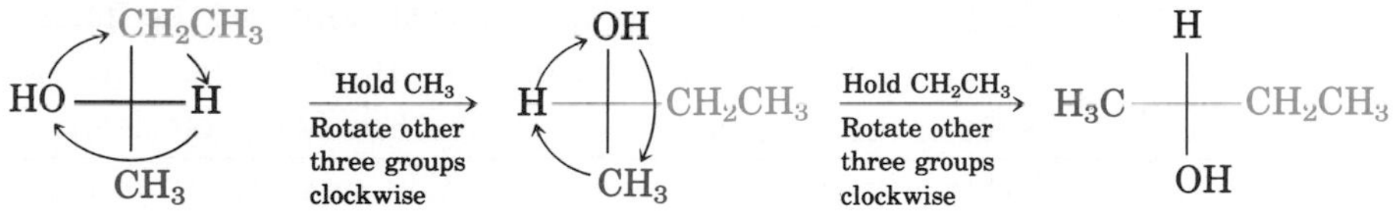

B (top $CH_2CH_3$; left HO; right H; bottom $CH_3$) → Hold $CH_3$, Rotate other three groups clockwise → (top OH; left H; right $CH_2CH_3$; bottom $CH_3$) → Hold $CH_2CH_3$, Rotate other three groups clockwise → A (top H; left $H_3C$; right $CH_2CH_3$; bottom OH)

By performing two allowed movements on projection B, we find that it is *identical* to projection A. Now let's do the same thing to projection C:

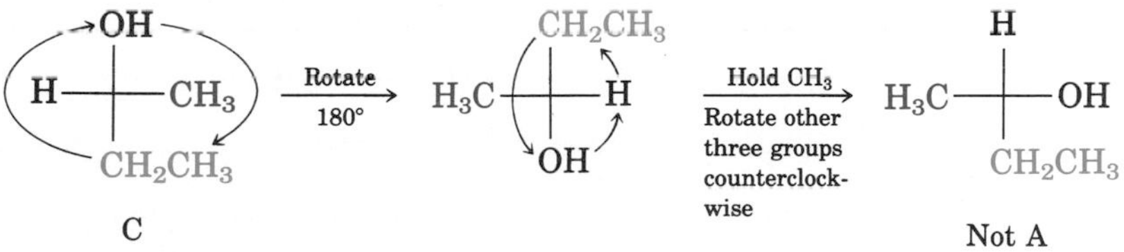

C (top OH; left H; right $CH_3$; bottom $CH_2CH_3$) → Rotate 180° → (top $CH_2CH_3$; left $H_3C$; right H; bottom OH) → Hold $CH_3$, Rotate other three groups counterclockwise → Not A (top H; left $H_3C$; right OH; bottom $CH_2CH_3$)

By performing two allowed movements on projection C, we can match up the $—H$ and $—CH_3$ substituents with those in A. After doing so, however, the $—OH$ and $—CH_2CH_3$ substituents *don't* match up. Thus, projection C must be enantiomeric with A and B.

**PROBLEM**

**9.13** Which of these Fischer projections of lactic acid represent the same enantiomer?

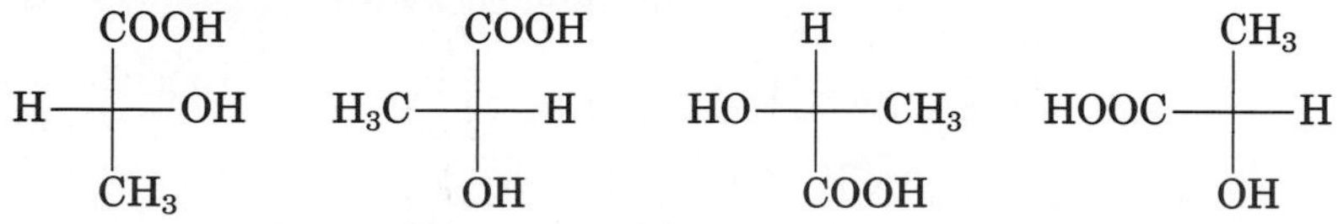

A: top COOH; left H; right OH; bottom $CH_3$

B: top COOH; left $H_3C$; right H; bottom OH

C: top H; left HO; right $CH_3$; bottom COOH

D: top $CH_3$; left HOOC; right H; bottom OH

PROBLEM

**9.14** Are these pairs of Fischer projections the same, or are they enantiomers?

(a)

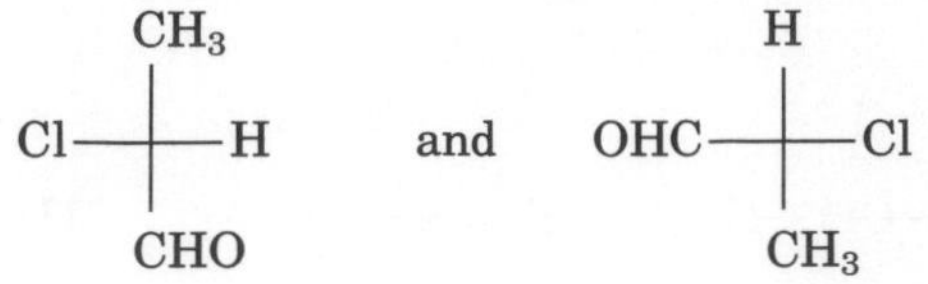

$CH_3$
Cl—+—H
CHO

and

H
OHC—+—Cl
$CH_3$

(b)

$CH_2OH$
H—+—OH
CHO

and

CHO
HO—+—$CH_2OH$
H

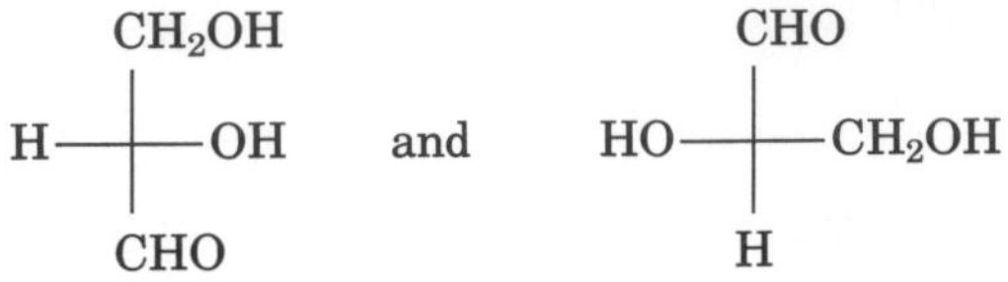

## 9.14 Assigning *R,S* Configurations to Fischer Projections

*R,S* stereochemical designations (Section 9.6) can be assigned to Fischer projections by following three steps:

1. Assign priorities to the four substituents in the usual way.
2. Perform one of the two allowed motions to place the group of lowest (fourth) priority at the top of the Fischer projection.
3. Determine the direction of rotation in going from priority 1 to 2 to 3, and assign *R* or *S* configuration as in the following practice problem.

PRACTICE PROBLEM

Assign *R* or *S* configuration to this Fischer projection of alanine, an amino acid:

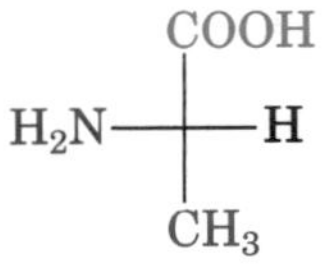

COOH
$H_2N$—+—H
$CH_3$

Alanine

*Solution* First, assign priorities to the four substituents on the chiral carbon. According to the sequence rules, the priorities are (1) $NH_2$, (2) —COOH, (3) —$CH_3$, and (4) —H. Next, perform one of the allowed motions on the Fischer projection to bring the group of lowest priority (—H) to the top. In the present instance, we might want to hold the —$CH_3$ group steady while rotating the other three groups counterclockwise:

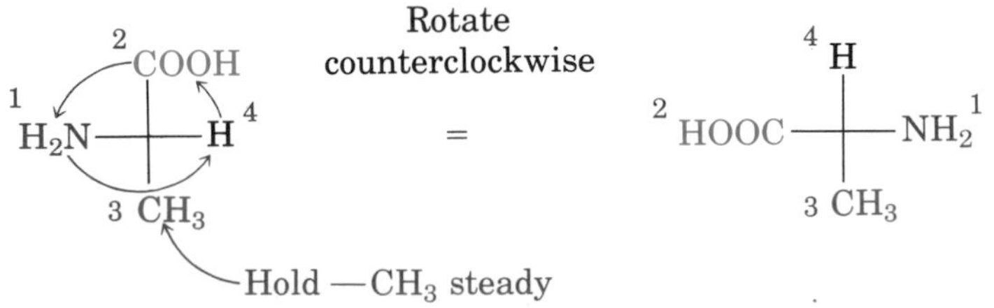

2 COOH
1 $H_2N$—+—H 4
3 $CH_3$

Hold —$CH_3$ steady

Rotate counterclockwise
=

4 H
2 HOOC—+—$NH_2$ 1
3 $CH_3$

Going now from first to second to third highest priority requires a counterclockwise (left-hand) turn, corresponding to *S* stereochemistry.

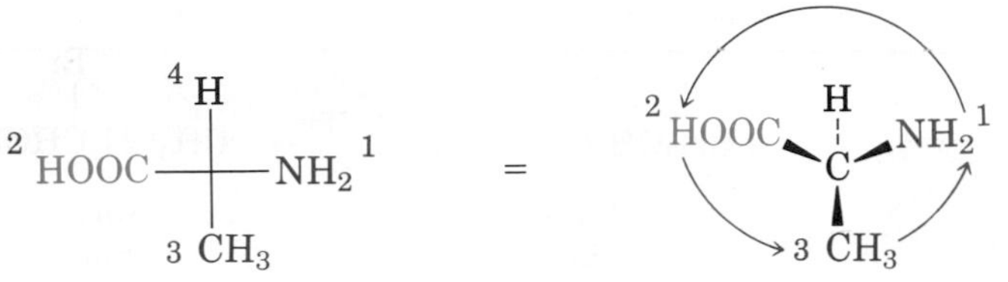

$$\begin{array}{c} {}^{4}\text{H} \\ {}^{2}\text{HOOC}-\!\!\!+\!\!\!-\text{NH}_2{}^{1} \\ {}^{3}\ \text{CH}_3 \end{array} \quad = \quad \text{(wedge/dash: } {}^{2}\text{HOOC},\ \text{H},\ \text{NH}_2{}^{1},\ {}^{3}\ \text{CH}_3\text{)}$$

*S* stereochemistry

Fischer projections can be used to specify more than one chiral center in a molecule simply by "stacking" the chiral centers one on top of the other. For example, threose, a simple four-carbon sugar, has the 2*S*,3*R* configuration:

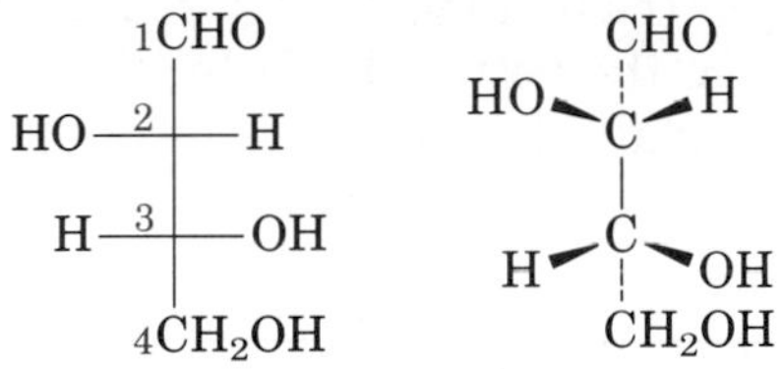

$$\begin{array}{c} {}_{1}\text{CHO} \\ \text{HO}\overset{2}{-\!\!+\!\!-}\text{H} \\ \text{H}\overset{3}{-\!\!+\!\!-}\text{OH} \\ {}_{4}\text{CH}_2\text{OH} \end{array}$$

Threose [(2*S*,3*R*)-2,3,4-Trihydroxybutanal]

Molecular models are particularly helpful in visualizing these structures.

PROBLEM

**9.15** Assign either *R* or *S* configuration to the chiral centers in these molecules:

(a) 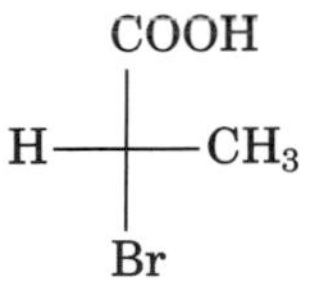 COOH at top, H left, $CH_3$ right, Br at bottom

(b) 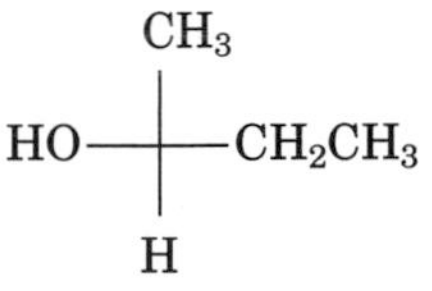 $CH_3$ at top, HO left, $CH_2CH_3$ right, H at bottom

(c) 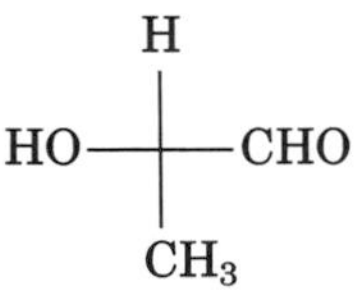 H at top, HO left, CHO right, $CH_3$ at bottom

## 9.15 Stereochemistry of Reactions: Addition of HBr to Alkenes

Many organic reactions, including some that we have studied, yield products with chiral centers. For example, HBr adds to 1-butene to yield 2-bromobutane, a chiral molecule. What predictions can we make about the stereochemistry of this chiral product? If a single enantiomer is formed, is it *R* or *S*? If a mixture of enantiomers is formed, how much of each is present?

In fact, the 2-bromobutane produced is a racemic mixture of *R* and *S* enantiomers.

$$\underset{\substack{\text{1-Butene} \\ \text{(achiral)}}}{CH_3CH_2CH{=}CH_2} \xrightarrow[\text{Ether}]{\text{HBr}} \underset{\substack{(\pm)\text{-2-Bromobutane} \\ \text{(chiral)}}}{CH_3CH_2\overset{\overset{\displaystyle Br}{|}}{\underset{*}{C}}HCH_3}$$

To understand *why* a racemic product results from reaction of HBr with 1-butene, think about what happens during the reaction. 1-Butene is first protonated by acid to yield an intermediate secondary (2°) carbocation. Since this ion is $sp^2$ hybridized, it has a plane of symmetry and is achiral. As a result, it can be attacked by bromide ion (also achiral) equally well from either the top or the bottom. Attack from the top leads to (*S*)-2-bromobutane, and attack from the bottom leads to (*R*)-2-bromobutane. Since both modes of attack are equally likely, a racemic mixture results (Figure 9.17).

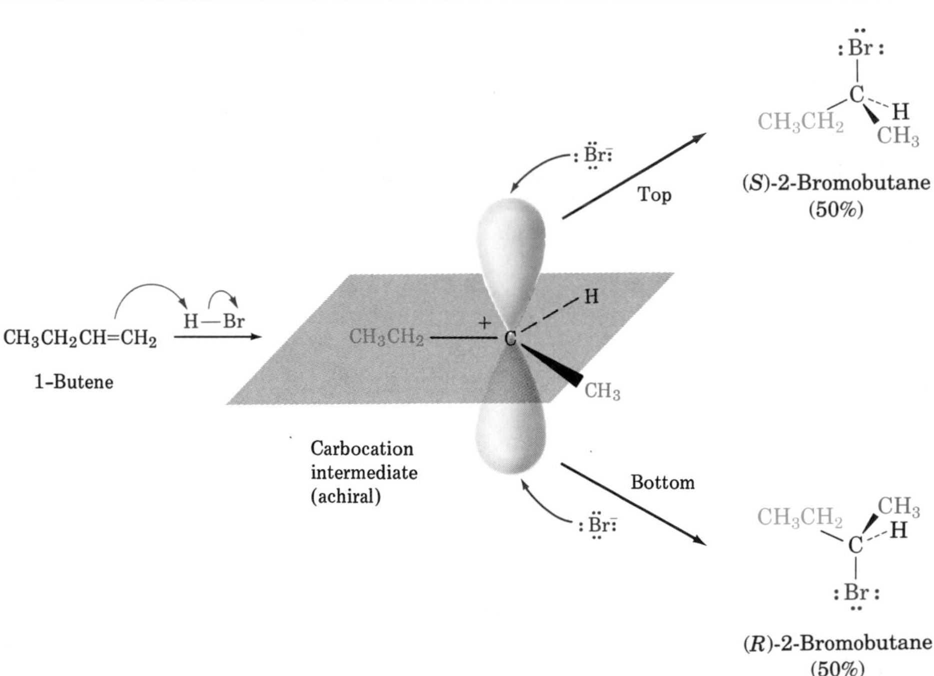


**Figure 9.17** Stereochemistry of the addition of HBr to 1-butene: The intermediate achiral carbocation is attacked equally well from both top and bottom, leading to a racemic product mixture.

Another way to think about the reaction is in terms of transition states. If the intermediate carbocation is attacked from the top, *S* product is formed through transition state 1 (TS 1) in Figure 9.18. If the cation is attacked from the bottom, *R* product is formed through TS 2. *The two transition states are mirror images.* They therefore have identical activation energies and are equally likely to occur. Both transition states form at identical rates, leading to a 50:50 mixture of *R* and *S* products.

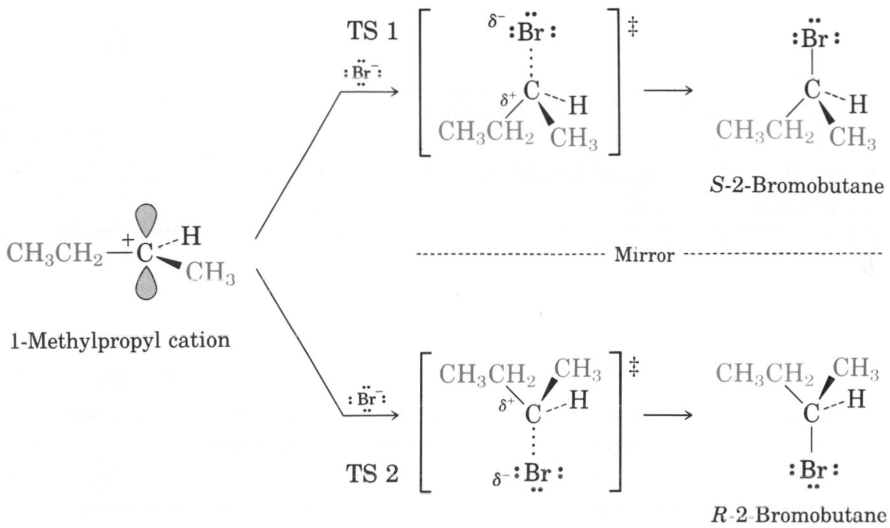

TS 1

1-Methylpropyl cation

S-2-Bromobutane

Mirror

TS 2

R-2-Bromobutane

**Figure 9.18** Attack of bromide ion on the 1-methylpropyl carbocation: Attack from the top leading to *S* product is the mirror image of attack from the bottom leading to *R* product. Since both are equally likely, racemic product is formed. The dotted C—Br bond in the transition state indicates partial bond formation.

## 9.16 Stereochemistry of Reactions: Addition of $Br_2$ to Alkenes

Addition of $Br_2$ to 2-butene leads to the formation of 2,3-dibromobutane and the generation of two chiral centers. What stereochemistry would we predict for such a reaction? If we begin with planar, achiral *cis*-2-butene, we would expect bromine to add to the double bond equally well from either the top or the bottom face to generate two intermediate bromonium ions. For the sake of simplicity, let's consider only the result of attack from the top face (Figure 9.19, page 288), keeping in mind that every structure we consider also has a mirror image.

The bromonium ion formed by top-face reaction of *cis*-2-butene can be attacked by bromide ion from either the right or the left side of the bottom face. Attack from the left (path a) leads to (2*S*,3*S*)-dibromobutane; attack from the right (path b) leads to (2*R*,3*R*)-dibromobutane. Since both modes

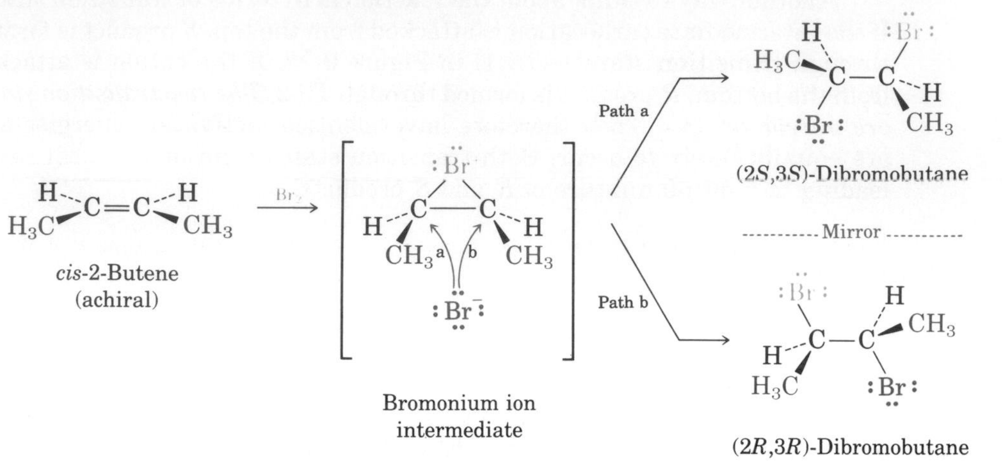

**Figure 9.19** Stereochemistry of the addition of $Br_2$ to *cis*-2-butene: A racemic mixture of 2*S*,3*S* and 2*R*,3*R* products is formed.

of attack on the symmetrical, achiral bromonium ion are equally likely, a 50:50 (racemic) mixture of the two enantiomeric products is formed. Thus, we obtain (±)-2,3-dibromobutane.

What about the addition of $Br_2$ to *trans*-2-butene? Is the same racemic product mixture formed? Perhaps surprisingly at first glance, the answer is no. *trans*-2-Butene reacts with bromine to form a bromonium ion, and, once again, we'll consider only top-face attack for simplicity (Figure 9.20). Attack of bromide ion on the bromonium ion intermediate takes place equally well from both right and left sides of the bottom face, leading to the formation of 2*R*,3*S* and 2*S*,3*R* products in equal amounts. A close look at the two products, however, shows that they are *identical*—both structures represent *meso*-2,3-dibromobutane.

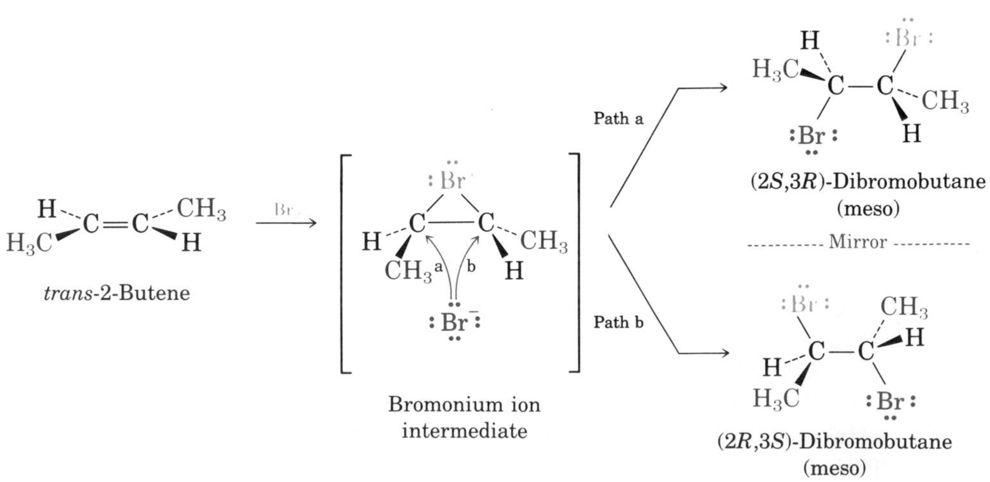

**Figure 9.20** Stereochemistry of the addition of $Br_2$ to *trans*-2-butene: A meso product is formed.

The key conclusion from all three of the addition reactions just discussed is that an optically inactive product has been formed. This is a general rule: *Any reaction between two achiral partners always leads to optically inactive products—either racemic or meso.* Put another way, optical activity can't come from out of nowhere; optically active products can't be produced from optically inactive starting materials.

PROBLEM

**9.16** What is the stereochemistry of the product that results from addition of $Br_2$ to cyclohexene? Is the product optically active? Explain.

PROBLEM

**9.17** Addition of $Br_2$ to an unsymmetrical alkene such as *cis*-2-hexene leads to racemic product, even though attack of bromide ion on the intermediate (unsymmetrical) bromonium ion is not equally likely at both ends. Make drawings of the intermediate and the products, and explain the observed stereochemical result.

PROBLEM

**9.18** Predict the stereochemical outcome of the reaction of $Br_2$ with *trans*-2-hexene. Show your reasoning.

## 9.17 Stereochemistry of Reactions: Addition of HBr to a Chiral Alkene

Both of the reactions considered in the two previous sections have involved additions to achiral alkene starting materials, and in both cases an optically inactive product was formed. What would happen, though, if we were to carry out a reaction on a single enantiomer of a *chiral* starting material? For example, what stereochemical result would be obtained from addition of HBr to a chiral alkene such as *R*-4-methyl-1-hexene? The product of this reaction, 2-bromo-4-methylhexane, has two chiral centers and might exist in any of four stereoisomeric configurations.

H $CH_3$ + HBr ⟶ $CH_3CH_2\overset{*}{C}H(CH_3)CH_2\overset{*}{C}H(Br)CH_3$

*R*-4-Methyl-1-hexene

2-Bromo-4-methylhexane

Let's think about the two chiral centers separately. What about the configuration at C4, the methyl-bearing carbon atom? Since C4 has the *R* configuration in the starting material, and since this chiral center is unaffected by the reaction, its configuration remains unchanged. Thus, the configuration of C4 in the product is also *R*.

What about the configuration at C2, the newly formed chiral center? As illustrated in Figure 9.21, the stereochemistry at C2 is established by bromide ion attack on a carbocation intermediate in the normal manner. *But this carbocation is not symmetrical; it is chiral because of the presence of the C4 center.* Since this intermediate carbocation has no plane of symmetry, it is not attacked equally well from top and bottom faces. One of the two faces is likely, for steric reasons, to be a bit more accessible than the other face, leading to a mixture of *R* and *S* product stereochemistries in some ratio other than 50:50. Thus, the two products, (2*R*,4*R*)-2-bromo-4-methylhexane and (2*S*,4*R*)-2-bromo-4-methylhexane, are formed in unequal amounts.

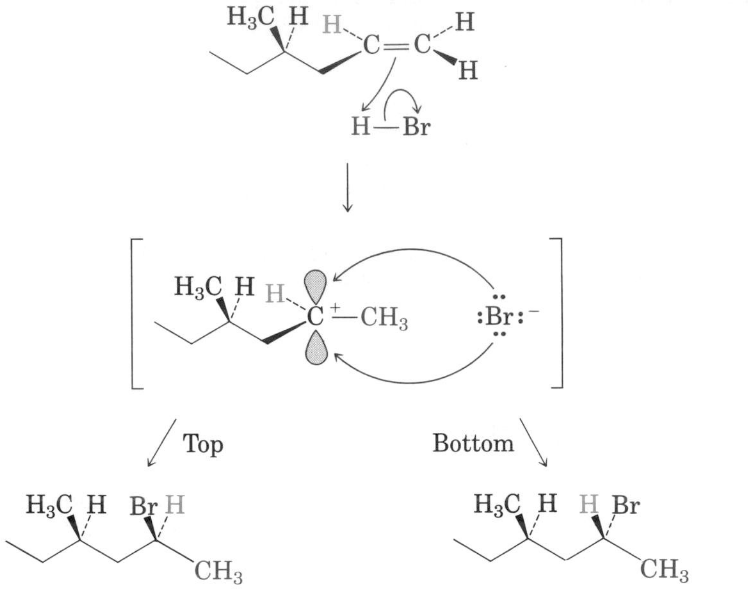

(2*S*,4*R*)-2-Bromo-4-methylhexane (2*R*,4*R*)-2-Bromo-4-methylhexane

**Figure 9.21** Stereochemistry of the addition of HBr to the chiral alkene *R*-4-methyl-1-hexene: A mixture of diastereomeric 2*R*,4*R* and 2*S*,4*R* products is formed in a non-50:50 ratio because attack on the unsymmetrical intermediate carbocation is not equally likely from both top and bottom.

The two products formed in the reaction of achiral HBr with chiral alkene are diastereomers, and both are optically active. This is a general rule: *Any reaction between an achiral reagent and a chiral reagent always leads to unequal amounts of diastereomeric products.* If the chiral starting material is optically active because only one enantiomer is used, then the products are also optically active.

PROBLEM

**9.19** What products would be formed and in what amounts from reaction of HBr with racemic (±)-4-methyl-1-hexene? Is the product mixture optically active?

PROBLEM

**9.20** What products would be formed and in what amounts from reaction of HBr with 4-methylcyclopentene?

## 9.18 Stereoisomerism and Chirality in Substituted Cyclohexanes

We saw in Section 4.13 that substituted cyclohexane rings adopt a chair-like geometry and that the conformation of a specific compound can be predicted by looking at steric interactions in the molecule. To complete a study of cyclohexane stereochemistry, we now need to examine the effect of conformation on stereoisomerism and chirality.

### 1,4-DISUBSTITUTED CYCLOHEXANES

1,4-Disubstituted cyclohexanes have no chiral centers by virtue of a symmetry plane passing through the substituents and through carbons 1 and 4 of the ring. Thus, only cis and trans stereoisomers (diastereomers) are possible. The symmetry plane is evident whether we use a flat view of the molecule or a three-dimensional view of the chair conformation (Figure 9.22).

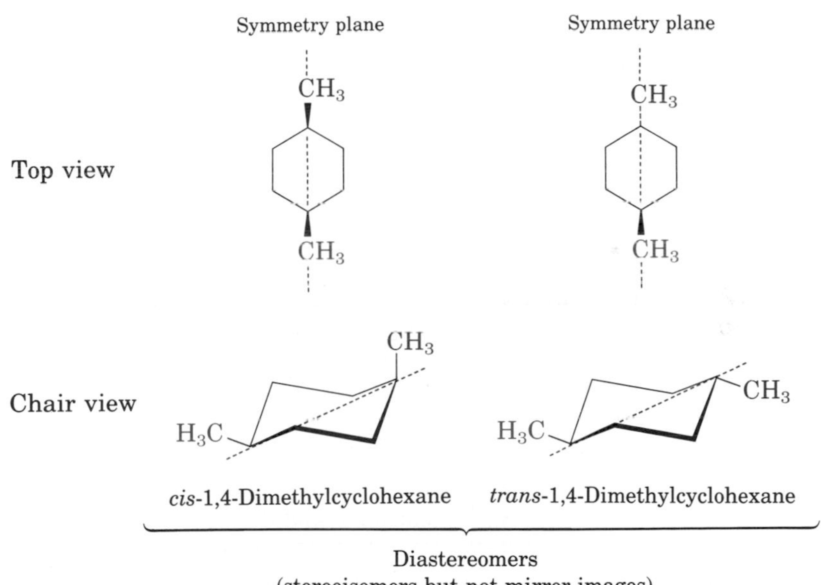


**Figure 9.22** The stereochemical relationships among 1,4-dimethylcyclohexanes: Both cis and trans isomers are achiral.

### 1,3-DISUBSTITUTED CYCLOHEXANES

1,3-Disubstituted cyclohexanes have two chiral centers, and a maximum of four stereoisomers is therefore possible. *cis*-1,3-Dimethylcyclohexane, however, has a symmetry plane and is thus a meso compound, whereas *trans*-1,3-dimethylcyclohexane has no symmetry plane and exists as a pair of enantiomers. Again, these symmetry properties are evident both from flat views and from conformational views (Figure 9.23).

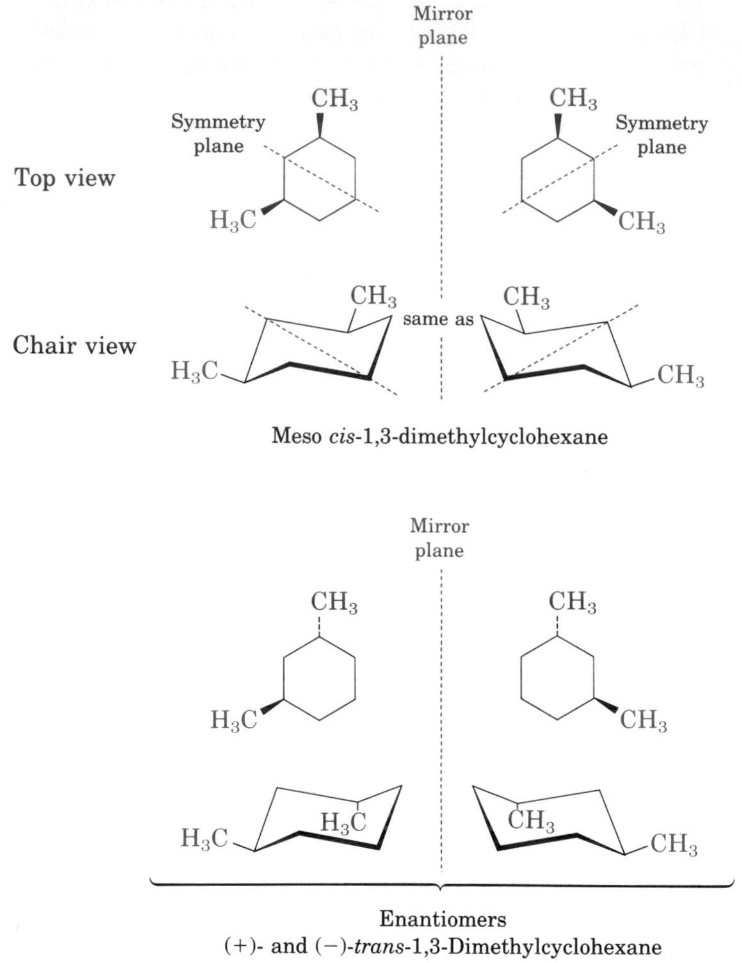

**Figure 9.23** The stereochemical relationships among 1,3-dimethylcyclohexanes: The cis isomer is a meso compound, and the trans isomer is a pair of enantiomers.

## 1,2-DISUBSTITUTED CYCLOHEXANES

1,2-Disubstituted cyclohexanes have two chiral centers, and four stereoisomers are again possible. The situation here is more complex, however. When we considered the stereoisomeric relationships in the 1,3- and 1,4-disubstituted cyclohexanes, we were able to look at flat structures to obtain the correct answers about stereoisomerism in these compounds. With 1,2-disubstituted cyclohexanes, however, we have to be more careful.

A top view of *trans*-1,2-dimethylcyclohexane shows that the molecule has no symmetry plane and must therefore exist as a pair of (+) and (−) enantiomers (Figure 9.24). A top view of *cis*-1,2-dimethylcyclohexane, however, shows an apparent symmetry plane and leads to the conclusion that the cis isomer is an optically inactive meso compound. If the cis isomer is viewed in chair conformation, though, the symmetry plane is no longer present because of the puckering of the ring, and we now predict that *cis*-1,2-dimethylcyclohexane exists as a pair of (+) and (−) enantiomers.

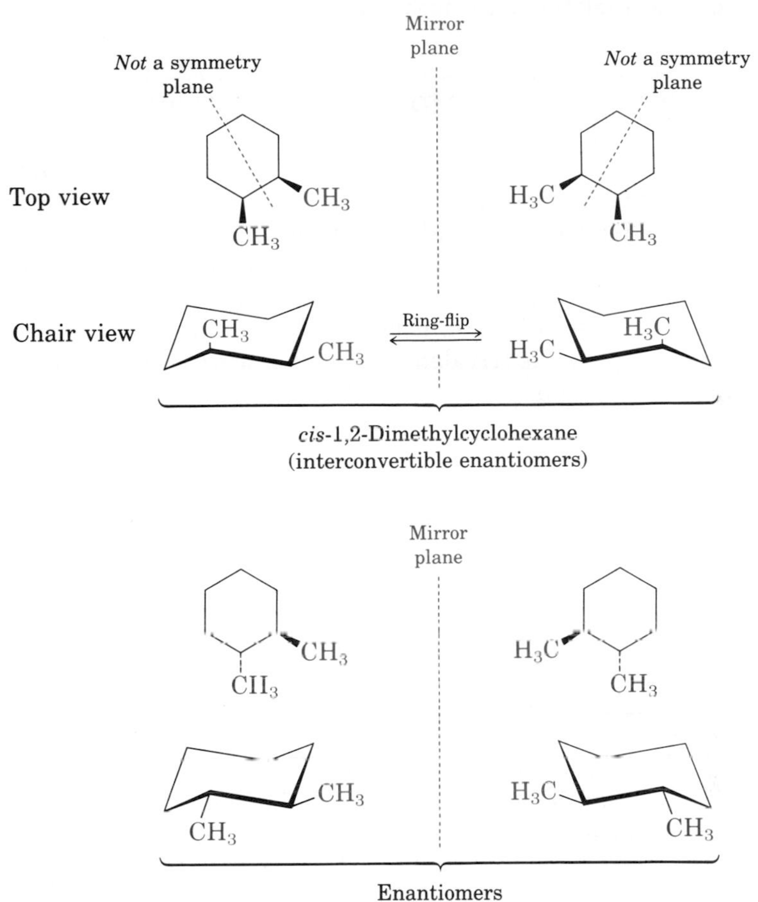


**Figure 9.24** Stereochemical relationships among the 1,2-dimethylcyclohexanes: The cis isomer exists as a pair of enantiomers that are interconvertible by a ring-flip. The trans isomer exists as a noninterconvertible pair of enantiomers.

Although the prediction that *cis*-1,2-dimethylcyclohexane exists as a pair of optically active enantiomers is true in principle, we observe no optical activity in practice because the two enantiomers can't be separated; they are interconverted by a ring-flip (Section 4.11). (This interconversion is much easier to see with molecular models.)

In general, it is possible to predict the presence or absence of optical activity in any substituted cycloalkane merely by looking at flat structures, without considering the exact three-dimensional chair conformations. The reasons for the lack of optical activity in a given case may be complex, however, as the situation present in *cis*-1,2-dimethylcyclohexane shows.

PROBLEM

**9.21** How many stereoisomers of 1-chloro-3,5-dimethylcyclohexane are there? Draw the most stable conformation.

PROBLEM

**9.22** How many 1,2-dimethylcyclopentane stereoisomers are there? What are the stereochemical relationships among them?

## 9.19 Chirality at Atoms Other than Carbon

Since the most common cause of chirality is the presence of four different substituents bonded to a tetrahedral atom, it follows that tetrahedral atoms other than carbon can also be chiral centers. Silicon, nitrogen, phosphorus, and sulfur are all commonly encountered in organic molecules, and all can be chiral centers under the proper circumstances. We know, for example, that trivalent nitrogen is tetrahedral and contains a lone pair of electrons (Section 1.12). Is trivalent nitrogen chiral? Does a compound such as ethylmethylamine exist as a pair of enantiomers?

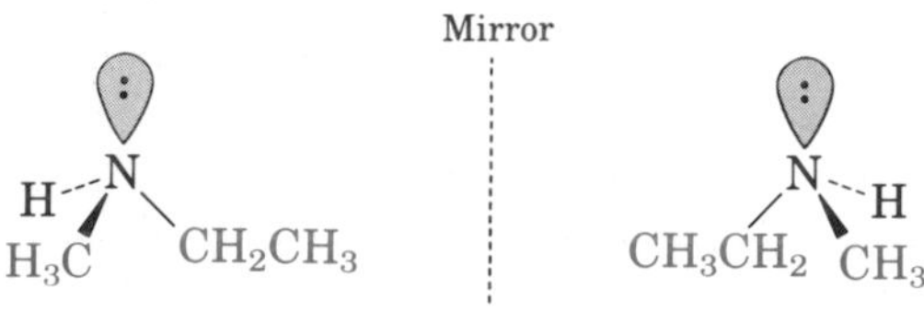


Ethylmethylamine

The answer is both yes and no. Yes in principle, but no in practice. Tetrahedral trivalent nitrogen compounds undergo a rapid "umbrella-like" inversion that interconverts enantiomers. We therefore can't separate or isolate enantiomers except in special cases, and their existence is of no great importance.

Mirror

H, $CH_3CH_2$, N, $H_3C$ ⇄ H, N, $CH_2CH_3$, $CH_3$

Rapid

## 9.20 Chirality in Nature

We saw in Section 9.12 that the different stereoisomeric forms of tartaric acid have different physical properties. It is usually the case that stereoisomers have different chemical and biological properties as well. One particularly dramatic example of how a simple change in chirality can affect the biological properties of a molecule is found in the amino acid, dopa. Dopa, more properly named 2-amino-3-(3,4-dihydroxyphenyl)propanoic acid, has a single chiral center and can thus exist in two stereoisomeric forms. Although the dextrorotatory enantiomer, D-dopa, has no physiological effect on humans, the levorotatory enantiomer, L-dopa, is widely used for its potent activity against Parkinson's disease, a chronic malady of the central nervous system.

HO, HO, COOH, H $NH_2$ — HOOC, $H_2N$ H, OH, OH

D-Dopa
(no biological effect)

L-Dopa
(anti-Parkinsonian agent)

Why do different stereoisomers have such dramatically different biological properties? In order to exert its biological action, a chiral molecule must fit into a chiral receptor at the target site, much as a hand fits into a glove. Just as a right hand can only fit into a right glove, so a particular stereoisomer can only fit into a receptor having the proper complementary shape. Any other stereoisomer will be a misfit like a right hand in a left glove. A schematic representation of the interaction between a chiral molecule and a chiral biological receptor is shown in Figure 9.25. One enantiomer fits the receptor perfectly, but the other doesn't.

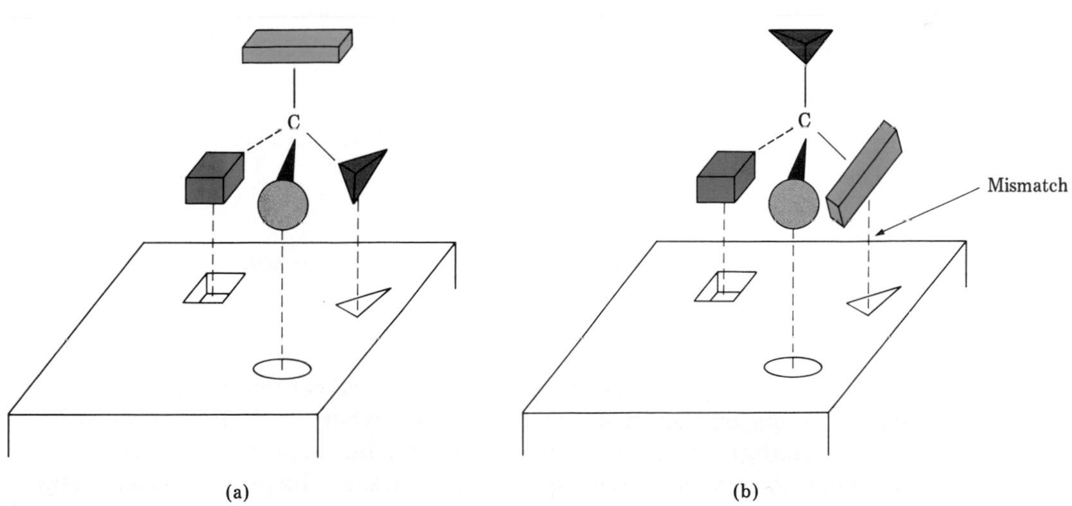


**Figure 9.25** (a) One enantiomer fits easily into a chiral receptor site to exert its biological effect, but (b) the other enantiomer can't fit into the same receptor and is thus without biological effect.

## 9.21 Summary and Key Words

When a beam of **plane-polarized light** is passed through a solution of certain organic molecules, the plane of polarization is rotated. Compounds that exhibit this behavior are called **optically active**. Optical activity is due to the asymmetric structure of the molecules themselves.

An object or molecule that is not superimposable on its mirror image is said to be **chiral**, meaning "handed." For example, a glove is chiral but a coffee cup is nonchiral, or **achiral**. A chiral molecule is one that does not contain a **plane of symmetry**—an imaginary plane that cuts through the molecule so that one half is a mirror image of the other half. The most common cause of chirality in organic molecules is the presence of a tetrahedral, $sp^3$-hybridized carbon atom bonded to four different groups. Compounds that contain such chiral carbon atoms exist as a pair of nonsuperimposable, mirror-image stereoisomers called **enantiomers**. Enantiomers are identical in all physical properties except for the direction in which they rotate plane-polarized light.

The stereochemical **configuration** of a chiral carbon atom can be depicted using **Fischer projections**, in which horizontal lines (bonds) are understood to come out of the plane of the paper and vertical bonds are understood to go back into the plane of the paper. The configuration can be specified as either *R* (*rectus*) or *S* (*sinister*) by using the Cahn–Ingold–Prelog sequence rules. This is done by first assigning priorities to the four substituents on the chiral carbon atom and then orienting the molecule so that the lowest-priority group points directly back away from the viewer. We then look at the remaining three substituents and let the eye travel from the group having the highest priority to second highest to third highest. If the direction of travel is clockwise, the configuration is labeled *R*; if the direction of travel is counterclockwise, the configuration is labeled *S*.

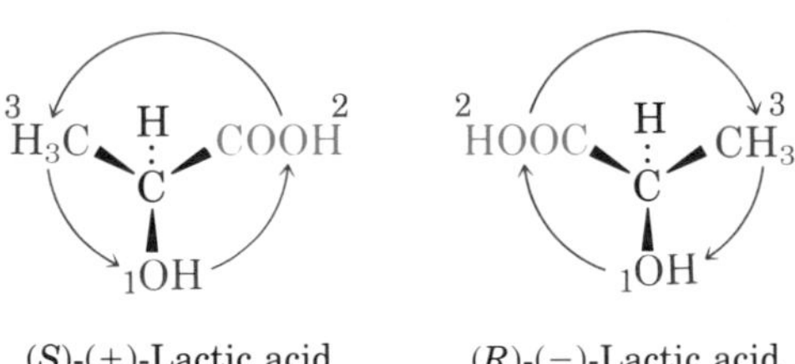


(*S*)-(+)-Lactic acid (*R*)-(−)-Lactic acid

Some molecules possess more than one chiral center. Enantiomers have opposite configuration at all chiral centers, whereas **diastereomers** have the same configuration in at least one center but opposite configurations at the others. A compound with $n$ chiral centers can have $2^n$ stereoisomers.

**Meso compounds** contain chiral centers, but are achiral overall because they contain a plane of symmetry. **Racemic mixtures**, or **racemates**, are 50:50 mixtures of (+) and (−) enantiomers. Racemic mixtures and individual diastereomers differ from each other in their physical properties such as solubility, melting point, and boiling point.

Most reactions give chiral products. If the starting materials are optically inactive, the products must also be optically inactive—either meso or racemic. If one or both of the starting materials is optically active, however, the product will also be optically active if the original chiral center remains.

## ADDITIONAL PROBLEMS

**9.23** Cholic acid, the major steroid found in bile, was observed to have a specific rotation of +2.22° when a 3.0 g sample was dissolved in 5.0 mL alcohol and the solution was placed in a sample tube with a 1.0 cm path length. Calculate $[\alpha]_D$ for cholic acid.

**9.24** Polarimeters for measuring optical rotation are quite sensitive and can measure rotations to 0.001°, an important fact when only small amounts of sample are available. Ecdysone, for example, is an insect hormone that controls molting in the silkworm moth. When 7.0 mg ecdysone was dissolved in 1.0 mL chloroform and the solution was placed in a 2.0 cm path-length cell, an observed rotation of +0.087° was found. Calculate $[\alpha]_D$ for ecdysone.

**9.25** What do these terms mean?
(a) Chirality (b) Chiral center (c) Optical activity
(d) Diastereomer (e) Enantiomer (f) Racemate

**9.26** Which of these compounds are chiral? Draw them and label the chiral centers.
(a) 2,4-Dimethylheptane (b) 3-Ethyl-5,5-dimethylheptane
(c) *cis*-1,4-Dichlorocyclohexane (d) 4,5-Dimethyl-2,6-octadiyne

**9.27** Draw chiral molecules that meet these descriptions:
(a) A chloroalkane, $C_5H_{11}Cl$ (b) An alcohol, $C_6H_{14}O$
(c) An alkene, $C_6H_{12}$ (d) An alkane, $C_8H_{18}$

**9.28** There are eight alcohols that have the formula $C_5H_{12}O$. Draw them. Which are chiral?

**9.29** Draw the nine chiral molecules that have the formula $C_6H_{13}Br$.

**9.30** Draw compounds that fit these descriptions:
(a) A chiral alcohol with four carbons
(b) A chiral carboxylic acid having the formula $C_5H_{10}O_2$
(c) A compound with two chiral centers
(d) A chiral aldehyde having the formula $C_3H_5BrO$

**9.31** Which of these objects are chiral?
(a) A basketball (b) A fork (c) A wine glass
(d) A golf club (e) A monkey wrench (f) A snowflake

**9.32** Penicillin V is an important broad-spectrum antibiotic that contains three chiral centers. Identify them with asterisks.

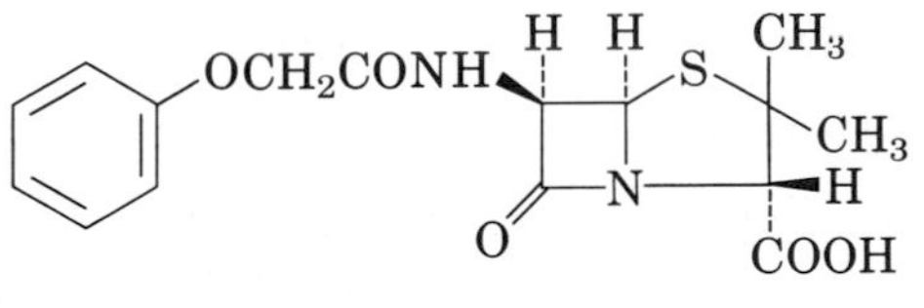

Penicillin V
(antibiotic)

**9.33** Draw examples of the following:
(a) A meso compound with the formula $C_8H_{18}$
(b) A meso compound with the formula $C_9H_{20}$
(c) A compound with two chiral centers, one *R* and the other *S*

**9.34** What is the relationship between the specific rotations of 2*R*,3*R*-dichloropentane and 2*S*,3*S*-dichloropentane? Between 2*R*,3*S*-dichloropentane and 2*R*,3*R*-dichloropentane?

**9.35** What is the stereochemical configuration of the enantiomer of 2*S*,4*R*-dibromooctane?

**9.36** What are the stereochemical configurations of the two diastereomers of 2*S*,4*R*-dibromooctane?

**9.37** Assign Cahn–Ingold–Prelog priorities to these sets of substituents:

(a) $-CH{=}CH_2$, $-CH(CH_3)_2$, $-C(CH_3)_3$, $-CH_2CH_3$

(b) $-C{\equiv}CH$, $-CH{=}CH_2$, $-C(CH_3)_3$, —(phenyl)

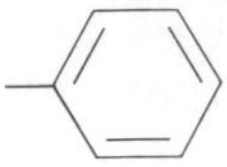

(c) $-CO_2CH_3$, $-COCH_3$, $-CH_2OCH_3$, $-CH_2CH_3$

(d) $-C{\equiv}N$, $-CH_2Br$, $-CH_2CH_2Br$, $-Br$

**9.38** Assign *R*,*S* configurations to the chiral centers in these molecules:

(a) 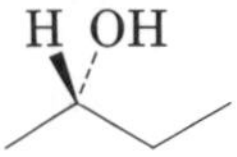 H OH

(b) 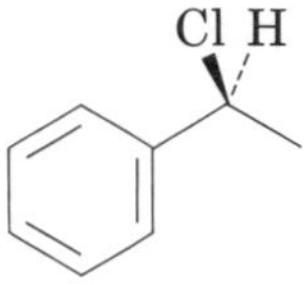 Cl H

(c) 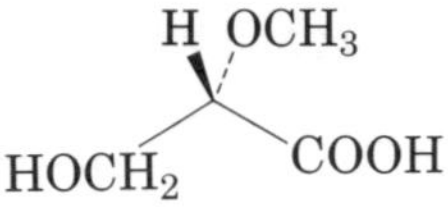 H $OCH_3$ $HOCH_2$ COOH

**9.39** Assign *R* or *S* configurations to each chiral center in these molecules:

(a) 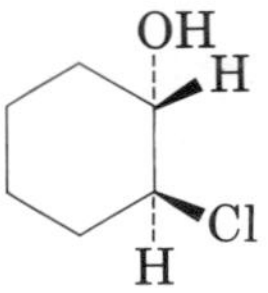 OH H Cl H

(b) 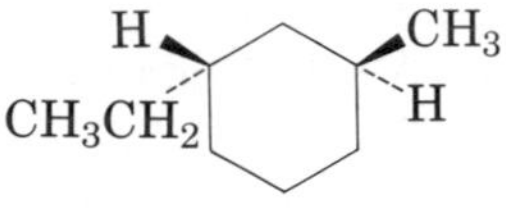 H $CH_3$ $CH_3CH_2$ H

**9.40** Draw tetrahedral representations of these molecules:
(a) (*S*)-2-Butanol, $CH_3CH_2CH(OH)CH_3$ (b) (*R*)-3-Chloro-1-pentene

**9.41** Draw tetrahedral representations of the two enantiomers of the amino acid cysteine, $HSCH_2CH(NH_2)COOH$, and identify each as *R* or *S*.

**9.42** Which of these pairs of Fischer projections represent the same enantiomer and which represent different enantiomers?

(a) 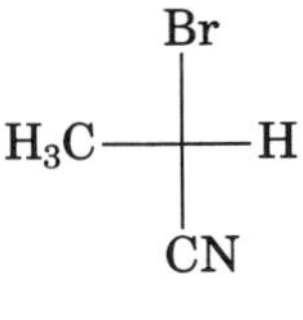 Br / $H_3C$—+—H / CN and CN / H—+—Br / $CH_3$

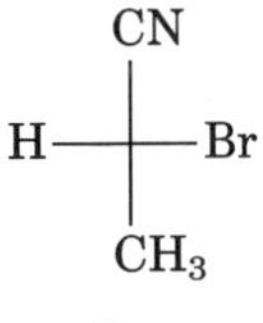

(b) 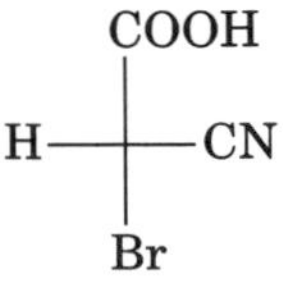 COOH / H—+—CN / Br and Br / H—+—COOH / CN

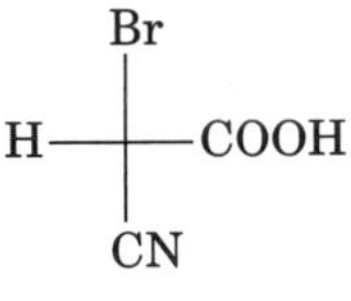

(c) 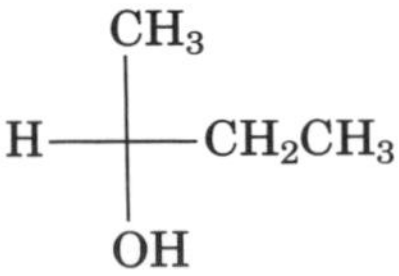 $CH_3$ / H—+—$CH_2CH_3$ / OH and OH / H—+—$CH_3$ / $CH_2CH_3$

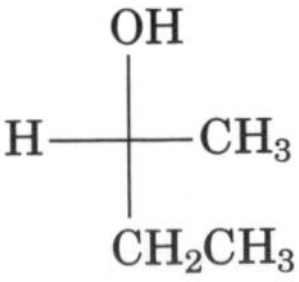

(d) Fischer projection: $CH_3$ (top), H (left), $NH_2$ (right), COOH (bottom) and Fischer projection: COOH (top), $H_3C$ (left), $NH_2$ (right), H (bottom)

**9.43** Assign *R,S* configurations to these Fischer projections.

(a) CN (top), H (left), Br (right), $CH_3$ (bottom)

(b) $CH{=}CH_2$ (top), H (left), $CH_2CH_3$ (right), $CO_2H$ (bottom)

(c) Br (top), H (left), phenyl (right), $CH_2CH_3$ (bottom)

**9.44** Assign *R* or *S* configurations to each chiral center in these molecules:

(a) H (top); $H_3C$—C—Br; Br—C—H; $CH_3$ (bottom)

(b) phenyl (top); $H_3C$—C—OH; $H_3C$—C—H; OH (bottom)

(c) $CO_2H$ (top); HO—C—H; H—C—OH; H—C—OH; $CH_2OH$ (bottom)

(d) $NH_2$ (top); H—C—$CO_2H$; H—C—OH; H—C—H; phenyl (bottom)

**9.45** Draw Fischer projections that fit these descriptions:
(a) The *S* enantiomer of 2-bromobutane
(b) The *R* enantiomer of alanine, $CH_3CH(NH_2)COOH$
(c) The *R* enantiomer of 2-hydroxypropanoic acid
(d) The *S* enantiomer of 3-methylhexane

**9.46** Assign *R* or *S* configurations to each chiral carbon atom in ascorbic acid (vitamin C).

O=C (ring to O); HO—C; ‖; HO—C; H—C (ring O); HO—C—H; $CH_2OH$

Ascorbic acid

9.47 Xylose is a common sugar found in many woods (maple, cherry). Because it is much less prone to cause tooth decay than sucrose, xylose has been used in candy and chewing gum. Assign *R*,*S* configurations to the chiral centers in xylose:

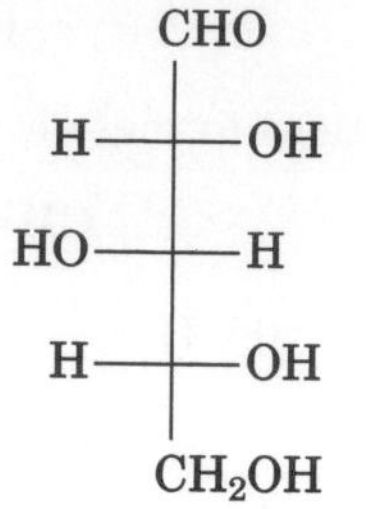

CHO
H—|—OH
HO—|—H
H—|—OH
$CH_2OH$

(+)-Xylose, $[\alpha]_D = +92°$

9.48 Hydroxylation of *cis*-2-butene with $OsO_4$ yields 2,3-butanediol. What stereochemistry do you expect for the product? (Review Section 7.7 if necessary.)

9.49 Hydroxylation of *trans*-2-butene with $OsO_4$ also yields 2,3-butanediol. What stereochemistry do you expect for the product?

9.50 Alkenes undergo reaction with peroxycarboxylic acids ($RCO_3H$) to give three-membered-ring cyclic ethers called *epoxides*. For example, 4-octene reacts with peroxyacids to yield 4,5-epoxyoctane:

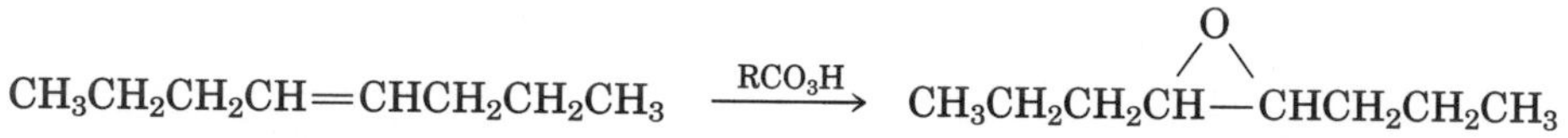

$$CH_3CH_2CH_2CH{=}CHCH_2CH_2CH_3 \xrightarrow{RCO_3H} CH_3CH_2CH_2CH\overset{O}{—}CHCH_2CH_2CH_3$$

4-Octene 4,5-Epoxyoctane

Assuming that this epoxidation reaction occurs with syn stereochemistry, draw the structure obtained from epoxidation of *cis*-4-octene. Is this epoxide chiral? How many chiral centers does it have? How would you describe the product stereochemically?

9.51 Answer Problem 9.50, assuming that the epoxidation reaction was carried out on *trans*-4-octene.

9.52 Write the products of the following reactions and indicate the stereochemistry obtained in each instance.

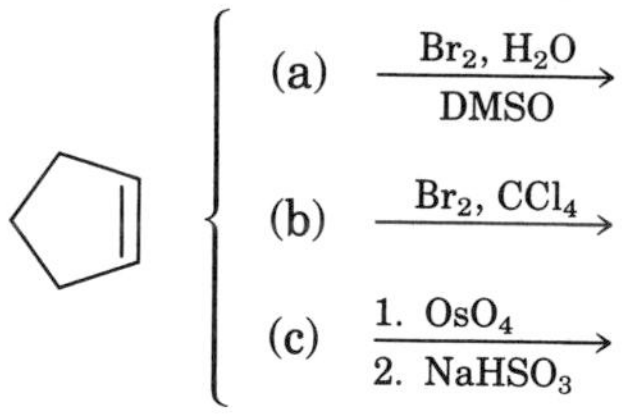

Cyclopentene:

(a) $\xrightarrow[DMSO]{Br_2,\ H_2O}$

(b) $\xrightarrow{Br_2,\ CCl_4}$

(c) $\xrightarrow[2.\ NaHSO_3]{1.\ OsO_4}$

9.53 Draw all possible stereoisomers of cyclobutane-1,2-dicarboxylic acid and indicate the interrelationships. Which, if any, are optically active? Do the same for cyclobutane-1,3-dicarboxylic acid.

9.54 Compound A, $C_7H_{12}$, was found to be optically active. On catalytic reduction over a palladium catalyst, 2 equiv of hydrogen were absorbed, yielding compound B, $C_7H_{16}$. On ozonolysis, two fragments were obtained. One fragment was identified as acetic acid. The other fragment, compound C, was an optically active carboxylic acid, $C_5H_{10}O_2$. Formulate the reactions, and draw structures for compounds A, B, and C.

**9.55** Compound A, $C_{11}H_{16}O$, was found to be an optically active alcohol. Despite its apparent unsaturation, no hydrogen was absorbed on catalytic reduction over a palladium catalyst. On treatment of compound A with dilute sulfuric acid, dehydration occurred, and an optically inactive alkene B, with formula $C_{11}H_{14}$, was produced as the major product. Alkene B, on ozonolysis, gave two products. One product was identified as propanal, $CH_3CH_2CHO$. Compound C, the other product, with formula $C_8H_8O$, was shown to be a ketone. How many multiple bonds and/or rings does compound A have? Formulate the reactions and identify compounds A, B, and C.

**9.56** Draw the structure of (*R*)-2-methylcyclohexanone.

**9.57** How many stereoisomers of 2,4-dibromo-3-chloropentane are there? Draw them and indicate which are optically active.

**9.58** The so-called tetrahedranes are an interesting class of compounds. The first member of this class was synthesized in 1979. Construct a model (carefully!) of tetrahedrane. Consider a substituted tetrahedrane with four different substituents. Is it chiral? Explain your answer.



A tetrahedrane

**9.59** *Allenes*, compounds with adjacent carbon–carbon double bonds, are well known. Many allenes are chiral, even though they don't contain chiral carbon atoms. Mycomycin, for example, a naturally occurring antibiotic isolated from the bacterium *Nocardia acidophilus*, is chiral and has $[\alpha]_D = -130°$. Why is mycomycin chiral? Making a molecular model should be helpful.

$$HC{\equiv}C{-}C{\equiv}C{-}CH{=}C{=}CH{-}CH{=}CH{-}CH{=}CH{-}CH_2COOH$$

Mycomycin
(an allene)

**9.60** Long before optically active allenes were known to exist (Problem 9.59), the resolution of 4-methylcyclohexylideneacetic acid into two enantiomers had been carried out. This was the first molecule to be resolved that was chiral yet did not contain a chiral center. Why is this molecule chiral? What geometric relation does this molecule have to allenes?



4-Methylcyclohexylideneacetic acid

**9.61** (*S*)-1-Chloro-2-methylbutane undergoes light-induced reaction with $Cl_2$ by a radical mechanism to yield a mixture of products. Among the products are 1,4-dichloro-2-methylbutane and 1,2-dichloro-2-methylbutane.

(a) Formulate the reaction, showing the correct stereochemistry of the starting material.

(b) One of the two products is optically active, but the other is optically inactive. Which is which?

(c) What can you conclude about the stereochemistry of radical chlorination reactions based on the results of this experiment?

**9.62** Grignard reagents, R—Mg—X, react with aldehydes to yield alcohols. For example, the reaction of methylmagnesium bromide with propanal yields 2-butanol:

$$\underset{\text{Propanal}}{CH_3CH_2—\overset{\overset{O}{\|}}{C}—H} \xrightarrow[2.\ H_3O^+]{1.\ CH_3MgBr} \underset{\text{2-Butanol}}{CH_3CH_2—\overset{\overset{OH}{|}}{\underset{\underset{H}{|}}{C}}—CH_3}$$

(a) Is the product chiral? Is it optically active?

(b) How many stereoisomers of butanol are formed, what are their stereochemical relationships, and what are their relative percentages?

**9.63** Imagine that another Grignard reaction similar to that in Problem 9.62 is carried out between methylmagnesium bromide and (*R*)-2-phenylpropanal to yield 3-phenyl-2-butanol:

$$\underset{(R)\text{-2-Phenylpropanal}}{C_6H_5—C(H)(CH_3)—\underset{\underset{H}{|}}{C}{=}O} \xrightarrow[2.\ H_3O^+]{1.\ CH_3MgBr} \underset{\text{3-Phenyl-2-butanol}}{CH_3—\underset{\underset{C_6H_5}{|}}{CH}—\overset{\overset{OH}{|}}{CH}—CH_3}$$

(a) Is the product chiral? Is it optically active?

(b) How many stereoisomers of 3-phenyl-2-butanol are formed, what are their stereochemical relationships, and what are their relative percentages?

CHAPTER 10

# Alkyl Halides

It would be difficult to study organic chemistry for long without becoming aware of halo-substituted alkanes. For example, alkyl halides such as tetrachloromethane, trichloromethane, and 1,1,1-trichloroethane are widely used as industrial solvents (although all are known to cause liver damage on chronic exposure); freons such as dichlorodifluoromethane (Freon 12) are used as refrigerants; and halogenated polymers such as polytetrafluoroethylene (Teflon) and poly-1,1-dichloroethylene (Saran) are used as plastics.

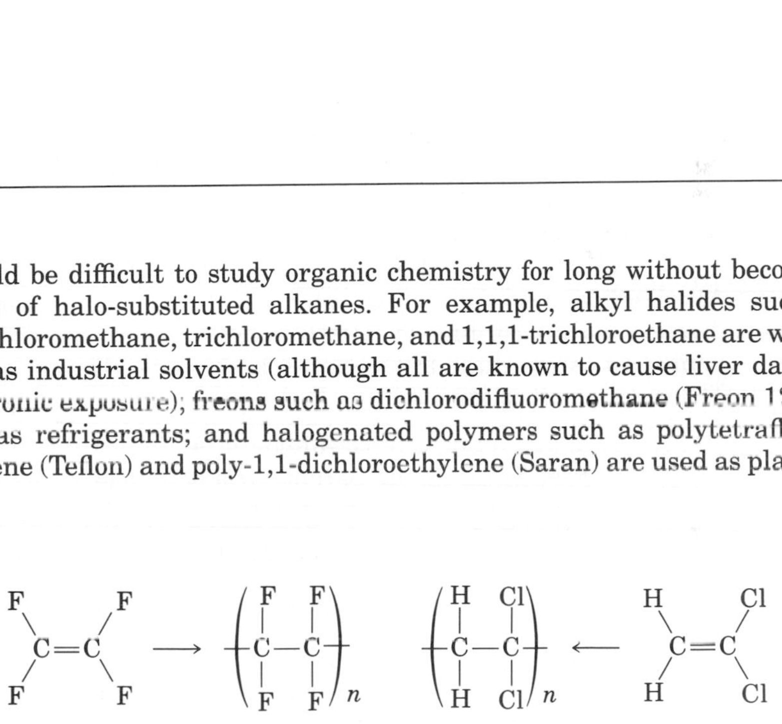

$$F_2C{=}CF_2 \longrightarrow \left(\begin{array}{c} F\;\;F \\ | \;\;\; | \\ -C-C- \\ | \;\;\; | \\ F\;\;F \end{array}\right)_n \qquad \left(\begin{array}{c} H\;\;Cl \\ | \;\;\; | \\ -C-C- \\ | \;\;\; | \\ H\;\;Cl \end{array}\right)_n \longleftarrow H_2C{=}CCl_2$$

Tetrafluoroethylene — Teflon — Saran — 1,1-Dichloroethylene

Alkyl halides also occur widely in nature, though mostly in marine rather than terrestrial organisms. The 1970s and 1980s have seen an explosive growth in the chemical investigation of marine organisms, and the structures of many naturally occurring halogenated molecules have been elucidated. Thus, simple halomethanes such as $CHCl_3$, $CCl_4$, $CBr_4$, $CH_3I$, and $CH_3Cl$ are constituents of the Hawaiian alga *Asparagopsis taxiformis*. In addition, many substances isolated from marine organisms exhibit interesting biological activity. For example, plocamene B, a trichlorocyclohexene derivative isolated from the red alga *Plocamium violaceum*, is similar in potency to DDT in showing insecticidal activity against mosquito larvae.

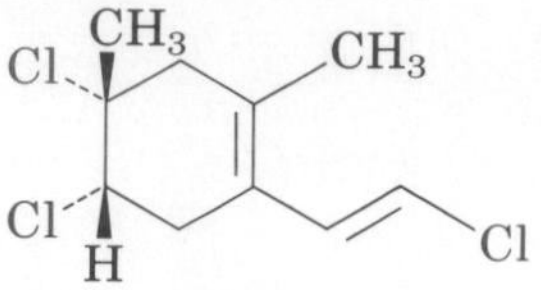

Plocamene B, a trichloride

Before discussing the chemistry of alkyl halides, it should be pointed out that we will be talking primarily about compounds having halogen atoms bonded to saturated, $sp^3$-hybridized carbon. Other classes of organohalides, such as aromatic (*aryl*) and alkenyl (*vinylic*) halides also exist, but much of their chemistry is different.

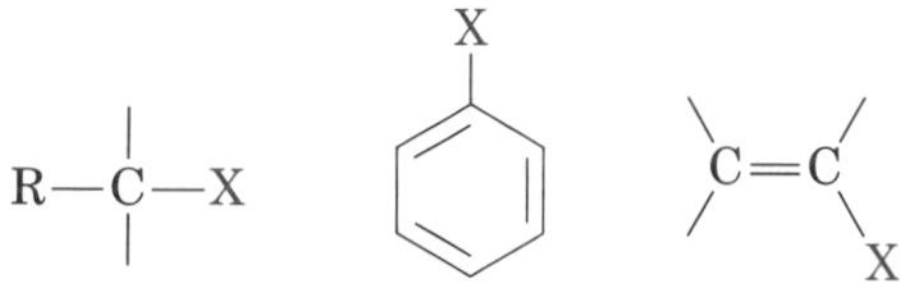

Alkyl halide Aryl halide Vinylic halide

## 10.1 Nomenclature of Alkyl Halides

Alkyl halides are named by an extension of the rules of alkane nomenclature (Section 3.4). According to IUPAC rules, halogens are considered substituents on a parent chain in the same sense that alkyl groups are substituents. Three rules suffice for naming alkyl halides:

1. Find and name the parent chain. As in naming alkanes, select the longest chain as the parent. (If a double or triple bond is present, however, the parent chain must contain it.)
2. Number the parent chain beginning at the end nearer the first substituent, regardless of whether it is alkyl or halo. For example,

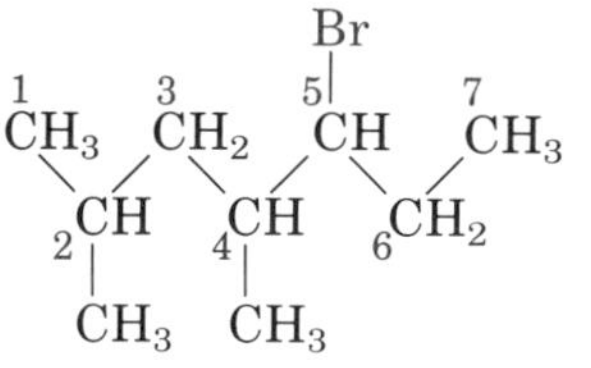

5-Bromo-2,4-dimethylheptane 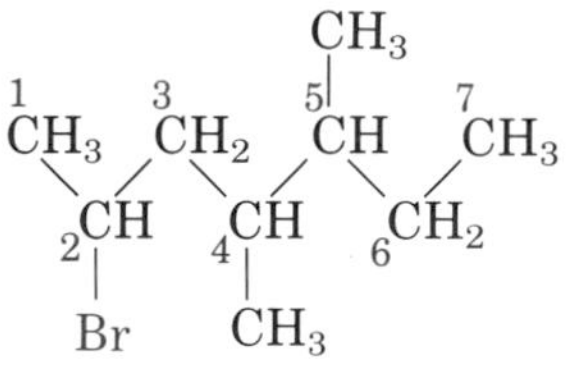 2-Bromo-4,5-dimethylheptane

a. If more than one of the same kind of halogen is present, number each and use one of the prefixes *di-*, *tri-*, *tetra-*, and so on. For example,

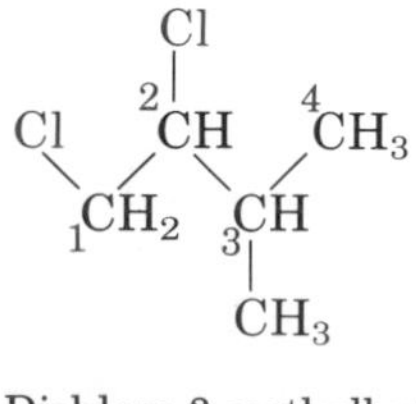

1,2-Dichloro-3-methylbutane

b. If different halogens are present, number each according to its position on the chain, but list all substituents in alphabetical order when writing the name. For example,

$$Br\overset{1}{C}H_2\overset{2}{C}H_2\overset{3}{C}H(Cl)\overset{4}{C}H(CH_3)\overset{5}{C}H_2\overset{6}{C}H_3$$

1-Bromo-3-chloro-4-methylhexane

3. If the parent chain can be properly numbered from either end by rule 2, begin at the end nearer the substituent (either alkyl or halo) that has alphabetical precedence. For example,

Br
$^{6}CH_3$ $^{4}CH_2$ $^{2}CH$
$_{5}CH$ $_{3}CH_2$ $_{1}CH_3$
$CH_3$

2-Bromo-5-methylhexane
(*not* 5-Bromo-2-methylhexane)

In addition to their systematic names, a small number of simple alkyl halides have alternative common names that are well entrenched in the chemical literature and in daily usage. Table 10.1 lists some of these common names, but they will not be used in this book.

**Table 10.1** Alternative names of some common alkyl halides

| *Structure* | *Systematic name* | *Common name* |
|---|---|---|
| $CH_3I$ | Iodomethane | Methyl iodide |
| $CH_2Cl_2$ | Dichloromethane | Methylene chloride |
| $CHCl_3$ | Trichloromethane | Chloroform |
| $CH_3CH_2Br$ | Bromoethane | Ethyl bromide |
| $CH_3CHClCH_3$ | 2-Chloropropane | Isopropyl chloride |
| $CH_3CH_2CHBrCH_3$ | 2-Bromobutane | *sec*-Butyl bromide |
| $(CH_3)_3CCl$ | 2-Chloro-2-methylpropane | *tert*-Butyl chloride |
| (cyclohexane ring)—Br | Bromocyclohexane | Cyclohexyl bromide |

PROBLEM

**10.1** Give the IUPAC names of these alkyl halides:

(a) $CH_3CH_2CH_2CH_2I$
(b) $(CH_3)_2CHCH_2CH_2Cl$
(c) $BrCH_2CH_2CH_2C(CH_3)_2CH_2Br$
(d) $(CH_3)_2CClCH_2CH_2Cl$
(e) $CH_3CHICH(CH_2CH_2Cl)CH_2CH_3$
(f) $CH_3CHBrCH_2CH_2CHClCH_3$

PROBLEM

**10.2** Draw structures corresponding to these IUPAC names:

(a) 2-Chloro-3,3-dimethylhexane
(b) 3,3-Dichloro-2-methylhexane
(c) 3-Bromo-3-ethylpentane
(d) 1,1-Dibromo-4-isopropylcyclohexane
(e) 4-*sec*-Butyl-2-chlorononane
(f) 1,1-Dibromo-4-*tert*-butylcyclohexane

## 10.2 Structure of Alkyl Halides

The carbon–halogen bond in an alkyl halide results from the overlap of a carbon $sp^3$ hybrid orbital with a halogen orbital. Thus, alkyl halide carbon atoms have an approximately tetrahedral geometry with H—C—X bond angles near 109°. Halogens increase in size going down the periodic table, an increase that is reflected in the bond lengths of the halomethane series (Table 10.2). Table 10.2 also indicates that C—X bond strengths decrease going down the periodic table.

**Table 10.2** Parameters for the C—X bond in halomethanes

| | | Bond strength | |
|---|---|---|---|
| *Halomethane* | *Bond length (Å)* | *(kcal/mol)* | *(kJ/mol)* |
| $H_3C—F$ | 1.39 | 109 | 456 |
| $H_3C—Cl$ | 1.78 | 84 | 351 |
| $H_3C—Br$ | 1.93 | 70 | 293 |
| $H_3C—I$ | 2.14 | 56 | 234 |

In an earlier discussion of bond polarity in functional groups (Section 5.5) we noted that halogens are electronegative with respect to carbon. The C—X bond is polar, with the carbon atom bearing a slight positive charge ($\delta^+$) and the halogen a slight negative charge ($\delta^-$):

$$\overset{\delta^+}{C}—\overset{\delta^-}{X}$$

where X = F, Cl, Br, or I (X is the standard abbreviation for a halogen).

We can get a rough idea of the amount of bond polarity by measuring the dipole moments of the halomethanes. As indicated in Table 10.3, the

**Table 10.3** Dipole moments of halomethanes

| *Halomethane* | *Dipole moment, μ (D)* |
|---|---|
| $CH_3F$ | 1.82 |
| $CH_3Cl$ | 1.94 |
| $CH_3Br$ | 1.79 |
| $CH_3I$ | 1.64 |

identity of the halogen has a rather small effect; all of the halomethanes have substantial dipole moments.

Since the carbon atom of alkyl halides is positively polarized, alkyl halides are good electrophiles (Section 5.5). We'll see in the next chapter that much of the chemistry alkyl halides undergo is dominated by their electrophilic behavior.

## 10.3 Preparation of Akyl Halides

We have already seen several methods of alkyl halide preparation. For example, we've seen that both HX and $X_2$ react with alkenes in electrophilic addition reactions to yield alkyl halides (Sections 6.9 and 7.1). HCl, HBr, and HI react with alkenes by a polar pathway to give the product of Markovnikov addition; HBr can also add by a radical pathway to give the non-Markovnikov product. Bromine and chlorine yield trans 1,2-dihalogenated addition products.

H CH$_3$

$X_2$ HBr radicals HX

H X X CH$_3$ H H Br CH$_3$ H X H CH$_3$

X = Cl or Br X = Cl, Br, or I

Another method of alkyl halide synthesis is the reaction of alkanes with chlorine by a radical chain-reaction pathway (Section 5.4). Although inert to most reagents, alkanes react readily with chlorine gas ($Cl_2$) in the presence of light to give chloroalkane substitution products. The reaction occurs by the radical mechanism shown in Figure 10.1 on page 308.

Recall from Section 5.4 that radical substitution reactions normally require three kinds of steps: **initiation**, **propagation**, and **termination**. Once an *initiation* step has started the process by producing radicals, the reaction continues in a self-sustaining cycle. The cycle requires two repeating *propagation* steps, in which a radical, the halogen, and the alkane yield alkyl halide product plus more radical to carry on the chain. The chain is occasionally *terminated* by the combination of two radicals.

Though interesting from a mechanistic point of view, alkane chlorination is a poor synthetic method for preparing different chloroalkanes. Let's see why this is so.

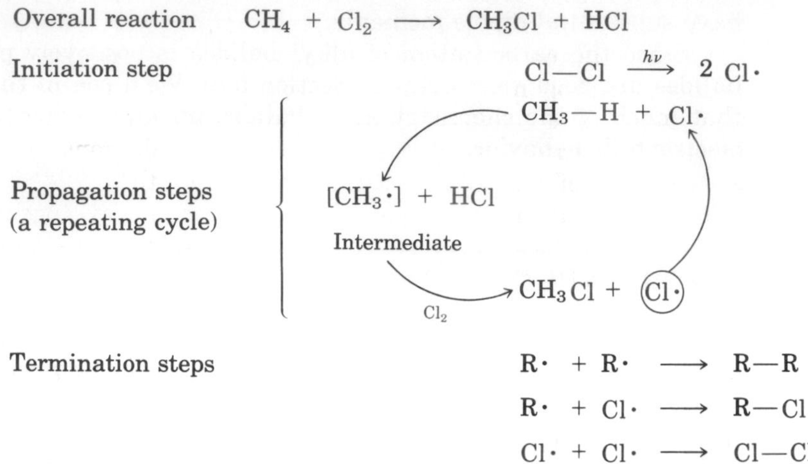


**Figure 10.1** Mechanism of the radical chlorination of methane (The symbol $h\nu$ shown in the initiation step is the standard way of indicating irradiation with light.)

## 10.4 Radical Chlorination of Alkanes

Alkane chlorination is a poor method of alkyl halide synthesis because mixtures of products invariably result. For example, chlorination of methane does not stop cleanly at the monochlorinated stage, but continues on to give a mixture of dichloro, trichloro, and even tetrachloro products:

$$CH_4 + n\,Cl_2 \longrightarrow CH_4 + CH_3Cl + CH_2Cl_2 + CHCl_3 + CCl_4 + n\,HCl$$

The situation is even worse for chlorination of alkanes that have more than one kind of hydrogen. For example, chlorination of butane gives two monochlorinated products in addition to dichlorobutane, trichlorobutane, and so on. Thirty percent of the monochloro product is 1-chlorobutane, and 70% is 2-chlorobutane:

$$\underset{\text{Butane}}{CH_3CH_2CH_2CH_3} + Cl_2 \xrightarrow{h\nu} \underbrace{\underset{\text{1-Chlorobutane}}{CH_3CH_2CH_2CH_2Cl} + \underset{\text{2-Chlorobutane}}{CH_3CH_2\overset{\displaystyle Cl}{\overset{|}{C}}HCH_3}}_{30:70} + \begin{matrix}\text{Dichloro-,}\\ \text{trichloro-,}\\ \text{tetrachloro-,}\\ \text{and so on}\end{matrix}$$

As another example, 2-methylpropane yields 2-chloro-2-methylpropane and 1-chloro-2-methylpropane in a ratio of 35:65, along with more highly chlorinated products:

$$\underset{\text{2-Methylpropane}}{CH_3-\overset{\displaystyle CH_3}{\overset{|}{\underset{\underset{\displaystyle CH_3}{|}}{C}}}-H} + Cl_2 \xrightarrow{h\nu} \underbrace{\underset{\text{2-Chloro-2-methylpropane}}{CH_3-\overset{\displaystyle CH_3}{\overset{|}{\underset{\underset{\displaystyle CH_3}{|}}{C}}}-Cl} + \underset{\text{1-Chloro-2-methylpropane}}{ClCH_2-\overset{\displaystyle CH_3}{\overset{|}{\underset{\underset{\displaystyle CH_3}{|}}{C}}}H}}_{35:65} + \begin{matrix}\text{Dichloro-,}\\ \text{trichloro-,}\\ \text{tetrachloro-,}\\ \text{and so on}\end{matrix}$$

From these and similar reactions, we can calculate a reactivity order toward chlorination for different types of hydrogen atoms in a molecule. Take the butane chlorination, for instance. Butane has six equivalent primary hydrogens ($—CH_3$) and four equivalent secondary hydrogens ($—CH_2—$). The fact that butane yields 30% of 1-chlorobutane product means that *each one* of the six primary hydrogens is responsible for 30%/6 = 5% of the product. Similarly, the fact that 70% of 2-chlorobutane is formed means that each one of the four secondary hydrogens is responsible for 70%/4 = 17.5% of the product. Thus, reaction of a secondary hydrogen happens 17.5/5 = 3.5 times as often as reaction of a primary hydrogen.

A similar calculation for the chlorination of 2-methylpropane indicates that each one of the nine primary hydrogens accounts for 65%/9 = 7.2% of the product, whereas the single tertiary hydrogen ($R_3CH$) accounts for 35% of the product. Thus, a tertiary hydrogen is evidently 35/7.2 = 5 times as reactive as a primary hydrogen.

| Relative reactivity toward chlorination | $R_3CH$ > $R_2CH_2$ > $RCH_3$ |
|---|---|
| | 5.0 > 3.5 > 1.0 |

What are the reasons for the observed reactivity order of alkane C—H bonds toward radical chlorination? A look at the bond dissociation energies given previously in Table 5.4 hints at the answer. As the data in Table 5.4 indicate, tertiary C—H bonds (91 kcal/mol; 380 kJ/mol) are weaker than secondary C—H bonds (95 kcal/mol; 397 kJ/mol), which are weaker in turn than primary C—H bonds (98 kcal/mol; 410 kJ/mol). Since less energy is needed to break a tertiary C—H bond than is needed to break a primary or secondary C—H bond, the resultant tertiary radical is more stable than a primary or secondary radical.

| Relative stability order of radicals | $R_3C\cdot$ > $R_2\dot{C}H$ > $R\dot{C}H_2$ |
|---|---|
| | 3° > 2° > 1° |

An explanation of the relationship between reactivity and bond strength in radical chlorination reactions is similar to that invoked earlier to explain why more stable carbocations form faster than less stable ones in alkene electrophilic addition reactions (Section 6.12). The reaction energy diagram for the formation of an alkyl radical during alkane chlorination looks like that in Figure 10.2. Although the hydrogen-abstraction step is slightly exothermic, there is nevertheless a certain amount of developing radical character in the transition state for this step. Thus, any factor (such as increased alkyl substitution) that stablizes a radical intermediate will also stabilize the transition state leading to that intermediate (lower its $\Delta G^{\ddagger}$). In other words, a more stable radical should form faster than a less stable one. Tertiary radicals are more stable, and tertiary hydrogen atoms are therefore more easily removed.

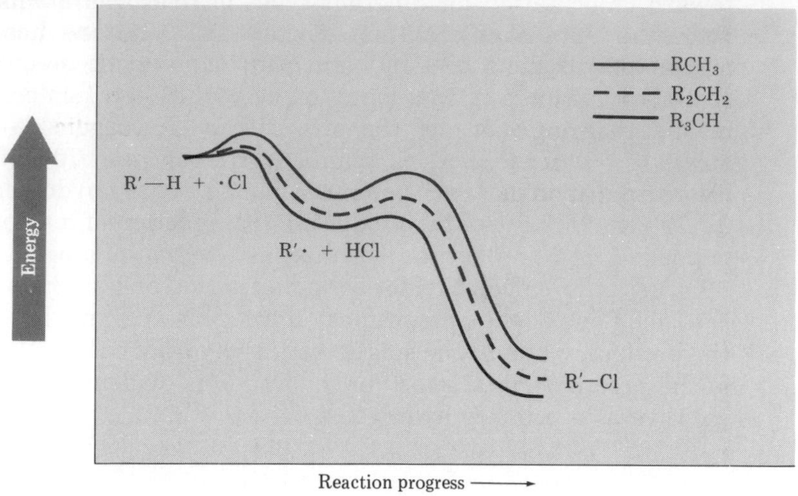


**Figure 10.2** Reaction energy diagram for alkane chlorination: The stability order of tertiary, secondary, and primary radicals is the same as their relative rate of formation.

PROBLEM

**10.3** Draw and name all monochloro products you would expect to obtain from radical chlorination of 2-methylpentane. Which, if any, are chiral?

PROBLEM

**10.4** Taking the known relative reactivities of 1°, 2°, and 3° hydrogen atoms into account, what product(s) would you expect to obtain from monochlorination of 2-methylbutane? What would be the approximate percentages of each product? (Don't forget to take into account the number of each type of hydrogen.)

## 10.5 Allylic Bromination of Alkenes

While repeating some earlier work of Wohl,[1] Karl Ziegler reported in 1942 that alkenes react with *N*-bromosuccinimide (NBS) in the presence of light to give products resulting from substitution of hydrogen by bromine at the position *next to* the double bond (the **allylic** position). Cyclohexene, for example, gives 3-bromocyclohexene in 85% yield:

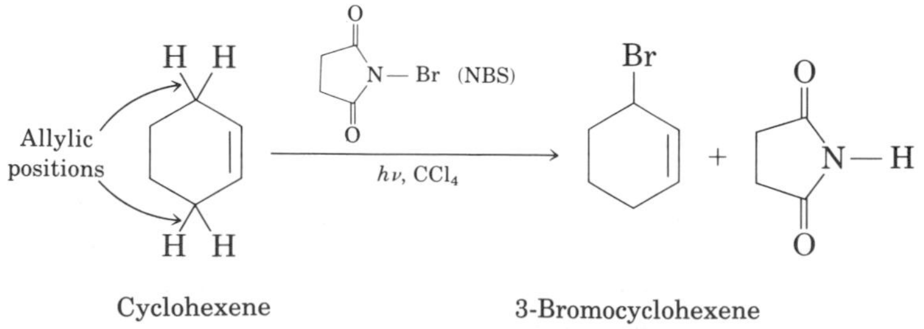


[1]Alfred Wohl (1863–1933); b. Graudentz; Ph.D., Berlin (Hofmann); professor, University of Danzig.

This allylic bromination with NBS looks very similar to the alkane chlorination reaction: In both cases, a C—H bond on a saturated carbon is broken and the hydrogen atom is replaced by halogen. Investigations have shown that allylic NBS brominations do in fact occur by a radical pathway. Although the exact mechanism of the reaction is complex, the crucial product-determining step involves abstraction of an allylic hydrogen atom and formation of the corresponding radical. Further reaction of this intermediate radical then yields the product.

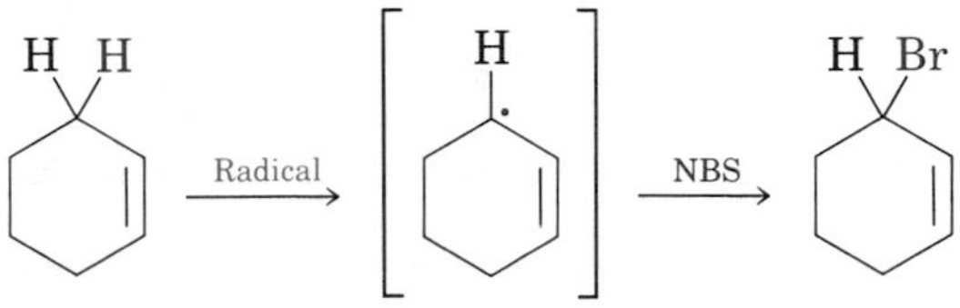

H H — Radical → [ H• ] — NBS → H Br

Intermediate allylic radical

Why does bromination occur exclusively at the position next to the double bond, rather than at one of the other positions? The answer, once again, is found by looking at bond dissociation energies to see the relative stabilities of various kinds of radicals.

There are three kinds of C—H bonds in cyclohexene, and Table 5.4 gives us an idea of their relative strengths. Although a typical secondary alkyl C—H bond has a strength of about 95 kcal/mol (397 kJ/mol), and a typical vinylic C—H bond has a strength of 108 kcal/mol (452 kJ/mol), an *allylic* C—H bond has a strength of only 87 kcal/mol (364 kJ/mol). An allylic radical is therefore more stable than a typical alkyl radical by about 7 kcal/mol (33 kJ/mol).

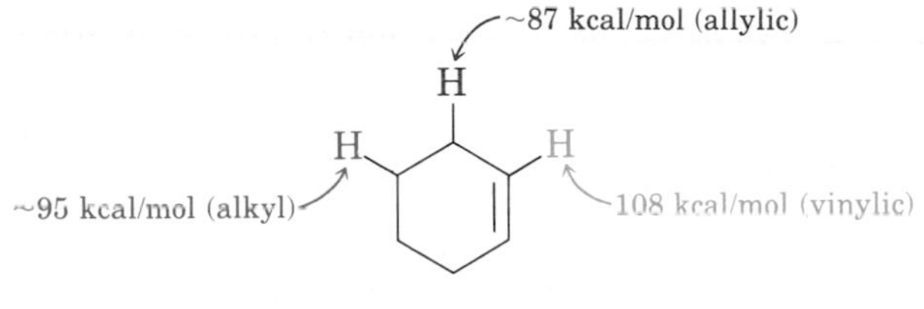

~87 kcal/mol (allylic)

~95 kcal/mol (alkyl)

108 kcal/mol (vinylic)

We can thus expand our stability ordering to include allylic and vinylic radicals:

Stability order of radicals $\quad$ Allylic > $R_3C\cdot$ > $R_2\dot{C}H$ > $R\dot{C}H_2$ > $\cdot CH_3$ > Vinylic

## 10.6 Stability of the Allyl Radical: Resonance

Why are allylic radicals so stable? To get an idea of the reason, look at the orbital picture of the allyl radical in Figure 10.3. The radical carbon atom next to the double bond can adopt $sp^2$ hybridization, placing the unpaired electron in a *p* orbital and giving a structure that is electronically symmetrical. The *p* orbital on the central carbon can overlap equally well with either of the two neighboring carbons.

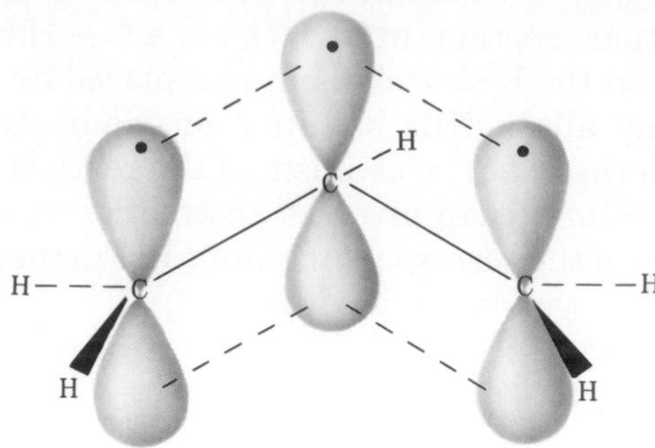


**Figure 10.3** Electronic structure of the allyl radical: The structure is electronically symmetrical.

Since the allyl radical is electronically symmetrical, there are two ways in which we can draw it: with the unpaired electron on the left and the double bond on the right, or with the unpaired electron on the right and the double bond on the left. *Neither structure is correct by itself; the true structure of the allyl radical is somewhere in between.*

$$H_2C{=}CH{-}\dot{C}H_2 \longleftrightarrow H_2\dot{C}{-}CH{=}CH_2$$

The two structures are called **resonance forms**, and their relationship is indicated by the double-headed arrow between them. The only difference between resonance forms is the placement of the bonding electrons. The atoms themselves don't move, but occupy exactly the same places in both resonance forms.

The best way to think about resonance is to realize that a species like the allyl radical is no different from any other organic substance. An allyl radical does *not* jump back and forth between two resonance forms, spending part of its time looking like one and the rest of its time looking like the other; rather, an allyl radical has a single unchanging structure that we call a **resonance hybrid**.

The real difficulty in visualizing the resonance concept is that we can't draw an accurate single picture of a resonance hybrid using familiar Kekulé structures, because the line-bond structures that serve so well to represent most organic molecules just don't work well for the allyl radical. The difficulty, however, lies with the *representation* of the structure rather than with the structure itself. We might try to represent the allyl radical by using a dotted line to indicate that the two C—C bonds are equivalent and that each is approximately $1\frac{1}{2}$ bonds, but such a drawing really doesn't help much and will be used infrequently in this book.

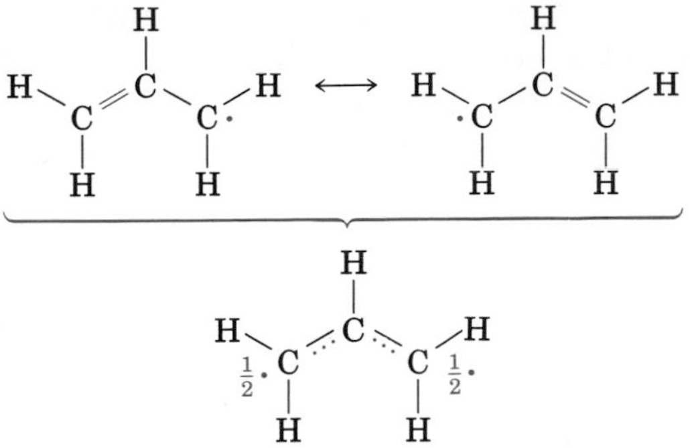

$$\mathrm{H_2C{=}CH{-}\dot{C}H_2 \longleftrightarrow H_2\dot{C}{-}CH{=}CH_2}$$

$$\mathrm{\tfrac{1}{2}\cdot CH_2 \cdots CH \cdots CH_2 \tfrac{1}{2}\cdot}$$

One of the most important postulates of resonance theory is that *the greater the number of possible resonance forms, the greater the stability of the compound.* For example, an allyl radical is a resonance hybrid of two Kekulé structures and is therefore more stable than a typical alkyl radical. In molecular orbital terms, the stability of the allyl radical is due to the fact that the unpaired electron can be *delocalized*, or spread out, over an extended pi orbital network rather than concentrated at only one site.

$$\mathrm{CH_2{=}CH{-}\dot{C}H_2 \;\updownarrow\; \dot{C}H_2{-}CH{=}CH_2}$$

Allyl radical (more stable)

$$\mathrm{CH_3CH_2\dot{C}H_2}$$

Propyl radical (less stable)

In addition to affecting stability, the delocalization of the unpaired electron in the allyl radical has chemical consequences. Since the unpaired electron is spread out over both ends of the pi orbital system, bromination of the allyl radical can occur *at either end.* Thus, allylic bromination of an unsymmetrical alkene often leads to a mixture of products. For example, bromination of 1-octene gives both 3-bromo-1-octene and 1-bromo-2-octene. Because the intermediate allylic radical in this case is unsymmetrical, reaction at the two ends is not equally likely and the two products are not formed in equal amounts.

$$\mathrm{CH_3(CH_2)_4CH_2CH{=}CH_2 \xrightarrow[CCl_4]{NBS} CH_3(CH_2)_4CH(Br)CH{=}CH_2 + CH_3(CH_2)_4CH{=}CHCH_2Br}$$

1-Octene

3-Bromo-1-octene (17%)

1-Bromo-2-octene (83%)
(53:47, trans:cis)

Since product mixtures are formed by bromination of unsymmetrical alkenes, allylic bromination is best carried out on symmetrical alkenes such as cyclohexene. The products of these reactions are particularly useful for conversion into dienes by dehydrohalogenation with base:

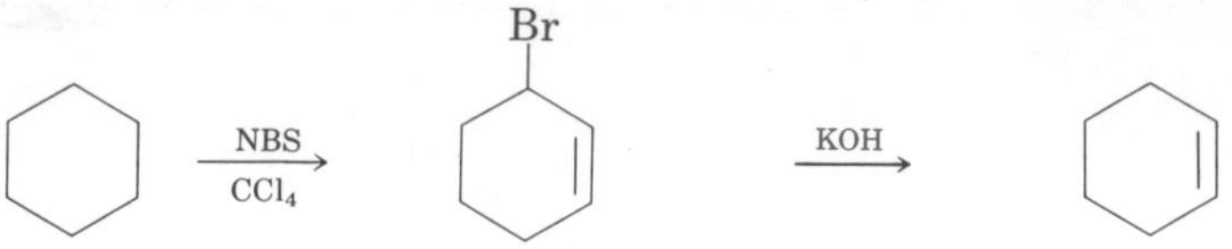

$$\text{Cyclohexene} \xrightarrow[CCl_4]{NBS} \text{3-Bromocyclohexene} \xrightarrow{KOH} \text{1,3-Cyclohexadiene}$$

PROBLEM

**10.5** The major product of the reaction of methylenecyclohexane and *N*-bromosuccinimide is 1-(bromomethyl)cyclohexene. How do you explain this?

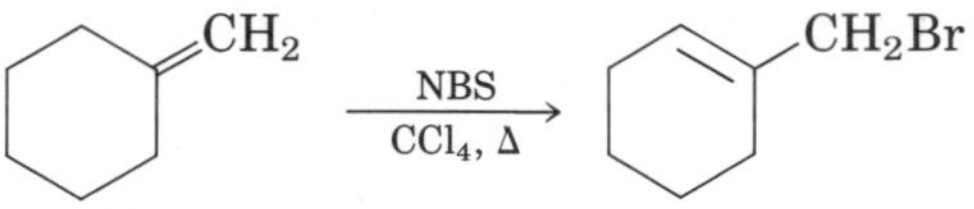

$$\text{Methylenecyclohexane } (=CH_2) \xrightarrow[CCl_4,\ \Delta]{NBS} \text{1-(bromomethyl)cyclohexene } (CH_2Br)$$

Major product

PROBLEM

**10.6** Alkylbenzenes such as toluene (methylbenzene) undergo reaction with NBS to give products in which bromine substitution has occurred at the position next to the aromatic ring. How can you account for this result? (Refer to Table 5.4 for a hint.)

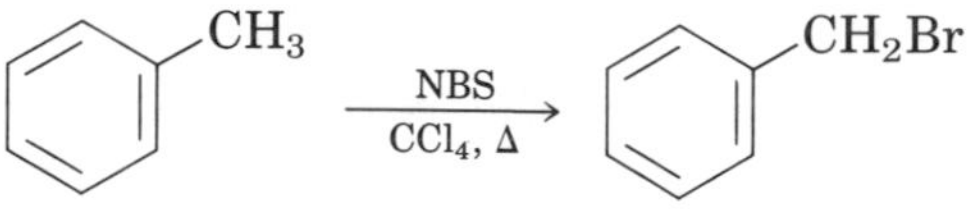

$$C_6H_5CH_3 \xrightarrow[CCl_4,\ \Delta]{NBS} C_6H_5CH_2Br$$

## 10.7 Drawing and Interpreting Resonance Forms

Resonance is an extremely useful concept in chemistry, and one that can be used to explain a variety of different phenomena. In inorganic chemistry, for example, the carbonate ion ($CO_3^{2-}$) is known to have identical bond lengths for its three C—O bonds. Although there is no *single* Lewis or line-bond structure that can account for this equality of C—O bonds, resonance theory accounts for it nicely. The carbonate ion is simply a resonance hybrid of three individual resonance forms. Each of the three oxygens shares the pi electrons and the negative charges equally:

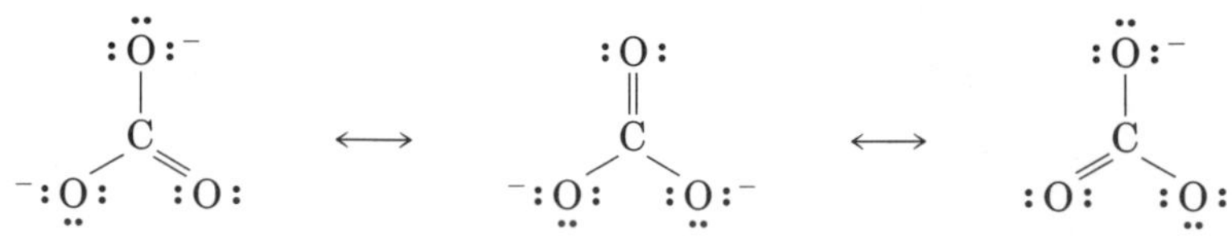

[Three resonance structures of the carbonate ion, each with one C=O double bond at a different oxygen and negative charges on the other two oxygens, connected by ⟷]

As an example from organic chemistry, we'll see in Chapter 15 that the six C—C bonds in aromatic compounds such as benzene are equivalent, because benzene is a resonance hybrid of two individual resonance forms. Though each individual form implies that benzene has alternating single and double bonds, neither form is correct by itself. The true benzene structure is a hybrid of the two individual forms:

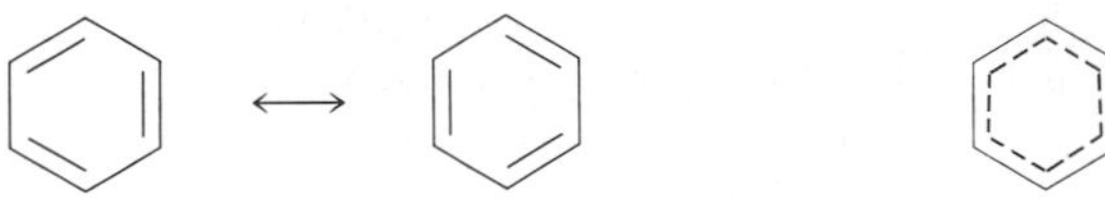

Two benzene resonance forms    Benzene resonance hybrid

When first dealing with resonance theory, it is often useful to have a set of guidelines that describe how to draw and interpret resonance forms. The following five rules should prove helpful.

*Rule 1* *Resonance forms are imaginary, not real.* The real structure is a composite or **resonance hybrid** of the different forms. Substances such as the allylic radical, the carbonate ion, and benzene are no different from any other substance; they have single, unchanging structures and do not switch back and forth between resonance forms. The only difference is in the way these molecules must be represented on paper. Although normal line-bond structures are adequate for representing *most* substances, they are inadequate for representing the bonding in some other substances.

*Rule 2* *Resonance forms differ only in the placement of pi electrons.* The positions of the atoms do not change from one resonance form to the next. In the carbonate ion, for example, the central carbon atom is $sp^2$ hybridized in all resonance forms, and the three oxygen atoms remain in exactly the same place in all resonance forms. Only the positions of the pi electrons differ from one form to another:

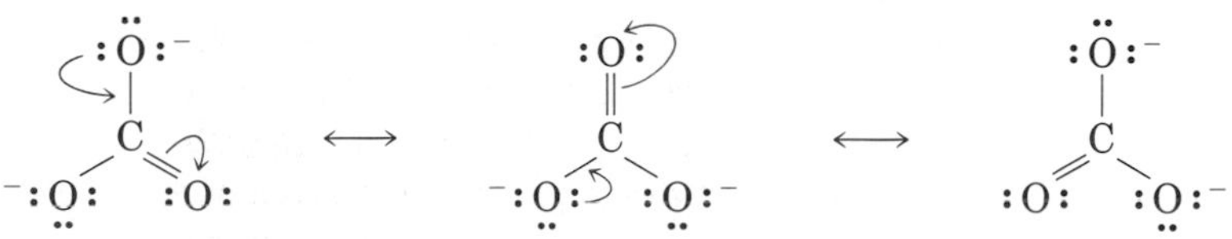

Similarly, in benzene the pi electrons in the double bonds move, but the carbon atoms remain in place:

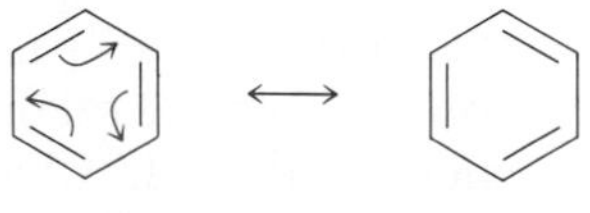

By contrast, two structures such as 1,3-cyclohexadiene and 1,4-cyclohexadiene are *not* resonance structures because they do not have their hydrogen atoms in the same position. Instead, the two dienes are simply constitutional isomers.

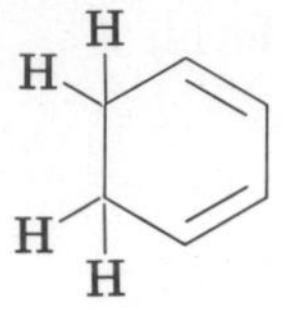

*Not* resonance forms

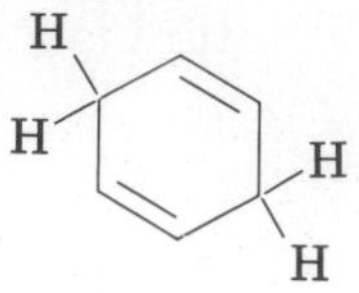

1,3-Cyclohexadiene

1,4-Cyclohexadiene

*Rule 3* *Different resonance forms of a substance do not have to be equivalent.* For example, bromination of 1-butene with NBS involves formation of an *unsymmetrical* allylic radical. One end of the delocalized pi electron system has a methyl substituent and the other end is unsubstituted. Even though the two resonance forms aren't equivalent, they both still contribute to the overall resonance hybrid.

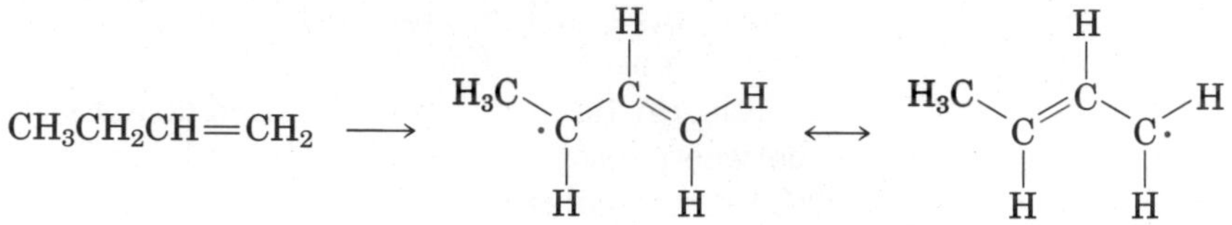

$CH_3CH_2CH{=}CH_2 \longrightarrow$

In general, when two resonance forms are nonequivalent, the actual structure of the resonance hybrid is closer to the more stable form than to the less stable form. Thus, we might expect the butenyl radical to look a bit more like a secondary radical than like a primary radical.

*Rule 4* *All resonance forms must obey normal rules of valency.* Resonance forms are like any other structure—the octet rule still holds. For example, one of the following structures for the carbonate ion is not a valid resonance form because the carbon atom has five bonds and ten electrons:

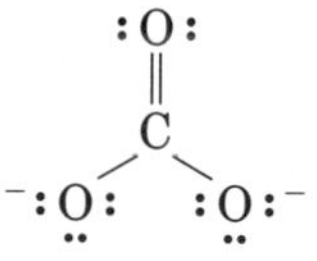

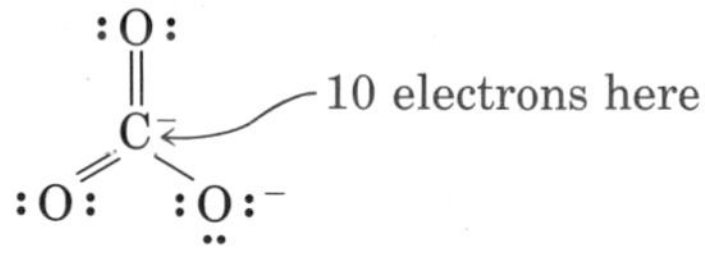

10 electrons here

Carbonate ion

*Not* a resonance form

*Rule 5* *The resonance hybrid is more stable than any single resonance form.* In other words, resonance leads to stability. The greater the number of resonance forms possible, the more stable the substance. We've already seen, for example, that an allylic radical is more stable than a normal radical. In a similar manner, we'll see in Chapter 15 that a benzene ring is more stable than a cyclic alkene.

PROBLEM

**10.7** Draw as many resonance structures as you can for these species:

(a) Nitromethane, $H_3C-\overset{+}{N}(=\ddot{O}:)-\ddot{O}:^-$

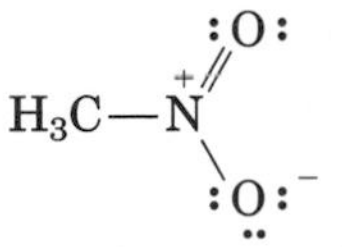

(b) Ozone, $:\ddot{O}=\overset{+}{\ddot{O}}-\ddot{O}:^-$

(c) Diazomethane, $H_2C=\overset{+}{N}=\overset{-}{\ddot{N}}:$

(d) $H_2C=CH-CH=CH-\dot{C}H_2$

PROBLEM

**10.8** Spectroscopic measurements indicate that the two oxygen atoms of sodium acetate are equivalent. Both C—O bonds have the same length (1.26 Å). Explain.

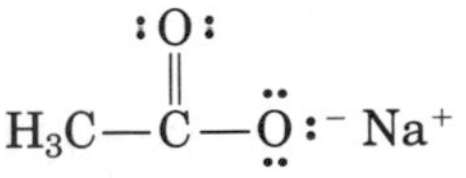

$$H_3C-C(=\ddot{O}:)-\ddot{O}:^-\ Na^+$$

Sodium acetate

PROBLEM

**10.9** In light of your answer to Problem 10.8, which would you expect to be more stable, acetate ion or methoxide ion, $H_3C-\ddot{O}:^-$? Which would you expect to be a stronger acid, $CH_3COOH$ or $CH_3OH$?

## 10.8 Preparing Alkyl Halides from Alcohols

The most valuable method for the preparation of alkyl halides is their synthesis from alcohols. A great many alcohols are commercially available, and we'll see later that a great many more can be obtained from carbonyl compounds. Because of the importance and generality of the reaction, a variety of different reagents have been used for transforming alcohols into alkyl halides.

The simplest (but also least generally useful) method for carrying out the conversion of an alcohol to an alkyl halide involves treating the alcohol with HCl, HBr, or HI:

$$R-OH + H-X \longrightarrow R-X + H_2O \qquad (X = Cl, Br, \text{ or } I)$$

For reasons to be discussed later (Section 11.14), the reaction works best when applied to tertiary alcohols, $R_3COH$. Primary and secondary alcohols also react, but at considerably slower rates and at considerably higher reaction temperatures. Although this is not a problem in simple cases, more complicated molecules are sometimes acid-sensitive and might be destroyed by the reaction conditions.

| Reactivity order of alcohols | $R_3COH$ | > | $R_2CHOH$ | > | $RCH_2OH$ | > | $CH_3OH$ |
|---|---|---|---|---|---|---|---|
| | 3° | > | 2° | > | 1° | > | Methanol |

The reaction of HX with a tertiary alcohol is so rapid that it's often carried out simply by bubbling the pure HX gas into a cold ether solution of the alcohol. Reaction is usually complete within a few minutes.

$$\text{1-Methylcyclohexanol} \xrightarrow[\text{Ether, 0°C}]{\text{HCl (gas)}} \text{1-Chloro-1-methylcyclohexane (90\%)} + H_2O$$

1-Methylcyclohexanol

1-Chloro-1-methylcyclohexane (90%)

Primary and secondary alcohols are best converted into alkyl halides by treatment with such reagents as thionyl chloride ($SOCl_2$) or phosphorus tribromide ($PBr_3$). These reactions, which normally take place readily under mild conditions, are less acidic and less likely to cause acid-catalyzed rearrangements than the HX method.

$$3\ CH_3CH_2CH(OH)CH_3 \xrightarrow[\text{Ether, 35°C}]{PBr_3} 3\ CH_3CH_2CHBrCH_3 + P(OH)_3$$

2-Butanol

2-Bromobutane (86%)

$$\text{Benzoin} \xrightarrow[\text{Pyridine}]{SOCl_2} \text{Desyl chloride} + SO_2 + HCl$$

Benzoin

Desyl chloride (86%)

As the above examples indicate, the yields of these $PBr_3$ and $SOCl_2$ reactions are generally high. Other functional groups such as ethers, carbonyls, and aromatic rings do not usually interfere. We'll look at the general mechanisms by which these substitution reactions take place in the next chapter.

**PROBLEM**

**10.10** How would you prepare these alkyl halides from the appropriate alcohols?

(a) 2-Chloro-2-methylpropane
(b) 2-Bromo-4-methylpentane
(c) $BrCH_2CH_2CH_2CH_2CH(CH_3)_2$
(d) $CH_3CH_2CH(CH_3)CH_2CCl(CH_3)_2$

## 10.9 Reactions of Alkyl Halides: Grignard Reagent Formation

Organic halides of widely varying structure—alkyl, aryl, and vinylic—react with magnesium metal in ether or tetrahydrofuran (THF) solvent to yield organomagnesium halides. These products, named **Grignard reagents** after their discoverer, Victor Grignard,[2] are examples of **organometallic compounds**, since they contain a carbon–metal bond.

$$R{-}X + Mg \xrightarrow[\text{or THF}]{\text{Ether}} R{-}Mg{-}X$$

where R = 1°, 2°, or 3° alkyl, aryl, or vinylic

X = Cl, Br, or I

For example,

Br (benzene ring) $\xrightarrow[\text{Ether}]{\text{Mg}}$ MgBr (benzene ring)

Phenylmagnesium bromide

$$(CH_3)_3CCl \xrightarrow[\text{Ether}]{\text{Mg}} (CH_3)_3C{-}MgCl$$

*t* Butylmagnesium chloride

All manner of organohalides form Grignard reagents. Steric hindrance in the halide doesn't appear to be a factor in the formation of Grignard reagents, since alkyl halides of all description, 1°, 2°, and 3°, react with ease. Aryl and vinylic halides also react, although it's best to use tetrahydrofuran as solvent for these cases. The halogen may be Cl, Br, or I, although chlorides are somewhat less reactive than bromides and iodides. Organofluorides rarely react with magnesium.

$$H_2C{=}C(CH_3)Br \xrightarrow[\text{THF}]{\text{Mg}} H_2C{=}C(CH_3)MgBr$$

2-Bromopropene — Isopropenylmagnesium bromide

As you might expect from the previous discussion of electronegativity and bond polarity (Section 5.5), the carbon–magnesium bond is highly polarized,

$$\overset{\delta^-}{C}{-}\overset{\delta^+}{MgBr}$$

making the carbon atom both nucleophilic and basic. In a formal sense, a Grignard reagent can be considered to be a carbon anion or **carbanion**—the magnesium salt of a hydrocarbon acid. It's more accurate, however, to view Grignard reagents as containing a highly polar covalent C—Mg bond, rather than an ionic bond between $C^-$ and $Mg^+$.

[2]François Auguste Victor Grignard (1871–1935); b. Cherbourg, France; professor, University of Nancy, Lyons; Nobel prize (1912).

Because of their nucleophilic/basic character, Grignard reagents react both with acids and with a wide variety of electrophiles. For example, Grignard reagents react with proton donors (Brønsted acids) such as $H_2O$, ROH, RCOOH, or $RNH_2$ to yield hydrocarbons. The overall sequence is a useful synthetic method for converting an alkyl halide into an alkane (R—X → R—H):

$$\underset{\text{Alkyl halide}}{\text{R—X}} \xrightarrow{\text{Mg}} \underset{\text{Grignard reagent}}{\text{R—Mg—X}} \xrightarrow{\text{H}_2\text{O}} \underset{\text{Alkane}}{\text{R—H}} + \text{HOMgX}$$

For example,

$$\underset{\text{1-Bromodecane}}{CH_3(CH_2)_8CH_2Br} \xrightarrow[\text{2. } H_2O]{\text{1. Mg}} \underset{\text{Decane (85\%)}}{CH_3(CH_2)_8CH_3}$$

PROBLEM

**10.11** How strong a base would you expect a Grignard reagent to be? Look at Tables 8.1 and 8.2 and then predict whether the following reactions are likely to occur.

(a) $CH_3MgBr + H—C{\equiv}C—H \longrightarrow CH_4 + H—C{\equiv}C—MgBr$

(b) $CH_3MgBr + NH_3 \longrightarrow CH_4 + H_2N—MgBr$

PROBLEM

**10.12** An important advantage of alkyl halide reduction via Grignard reagents is that the sequence can be used to introduce deuterium into a specific site in a molecule. How might you do this?

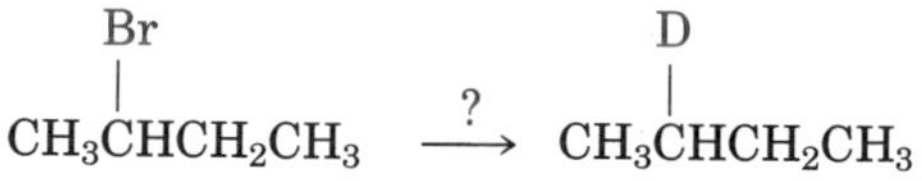

$$CH_3\overset{\overset{\text{Br}}{|}}{C}HCH_2CH_3 \xrightarrow{?} CH_3\overset{\overset{\text{D}}{|}}{C}HCH_2CH_3$$

PROBLEM

**10.13** Why do you suppose it is not possible to prepare a Grignard reagent from a bromo alcohol such as 4-bromo-1-pentanol?

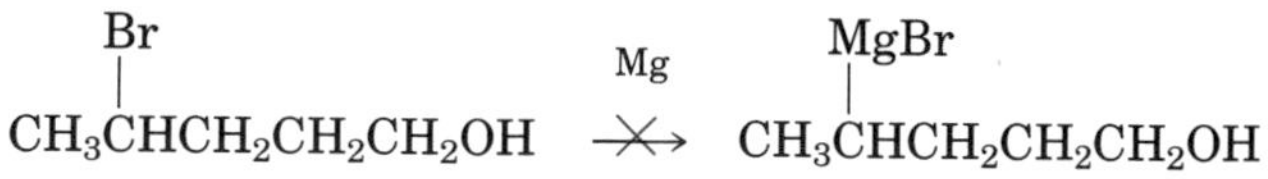

$$CH_3\overset{\overset{\text{Br}}{|}}{C}HCH_2CH_2CH_2OH \xrightarrow{\text{Mg}}\!\!\!\!\!\!\!\times\;\; CH_3\overset{\overset{\text{MgBr}}{|}}{C}HCH_2CH_2CH_2OH$$

Give another example of a molecule that is unlikely to form a Grignard reagent.

## 10.10 Organometallic Coupling Reactions

Several other organometallic reagents can be prepared in a manner similar to that of Grignard reagents. For example, **alkyllithium reagents** can be prepared by the reaction of an alkyl halide with lithium metal:

$$\underset{\text{1-Bromobutane}}{CH_3CH_2CH_2CH_2Br} \xrightarrow[\text{Pentane}]{\text{2 Li}} \underset{\text{Butyllithium}}{CH_3CH_2CH_2\overset{\delta^-}{C}H_2\overset{\delta^+}{Li}} + LiBr$$

Alkyllithiums are both nucleophiles and bases, and their chemistry is similar in many respects to that of the alkylmagnesium halides. One of the most valuable reactions of alkyllithiums is their use in preparing *lithium diorganocopper* reagents or **Gilman[3] reagents**.

$$\underset{\text{Bromomethane}}{CH_3Br} + 2\,Li \xrightarrow{\text{Pentane}} \underset{\text{Methyllithium}}{CH_3Li} + LiBr$$

$$\underset{\text{Methyllithium}}{2\,CH_3Li} + CuI \xrightarrow{\text{Ether}} \underset{\substack{\text{Lithium dimethylcopper}\\ \text{(a Gilman reagent)}}}{(CH_3)_2Cu^-\,Li^+} + LiI$$

Gilman reagents are easily prepared by reaction of an alkyllithium with cuprous iodide in ether solvent. Though rather unstable, they have the remarkable ability of undergoing organometallic *coupling* reactions with alkyl bromides and iodides (but not fluorides). One of the alkyl groups from the Gilman reagent replaces the halogen from alkyl halide, resulting in the formation of a hydrocarbon with a new carbon–carbon bond. Lithium dimethylcopper, for example, reacts with 1-iododecane to give undecane in 90% yield:

$$\underset{\text{Lithium dimethylcopper}}{(CH_3)_2CuLi} + \underset{\text{1-Iododecane}}{CH_3(CH_2)_8CH_2I} \xrightarrow[0^\circ C]{\text{Ether}} \underset{\text{Undecane (90\%)}}{CH_3(CH_2)_8CH_2CH_3} + LiI + CH_3Cu$$

This organometallic coupling reaction is highly versatile and is of great use in organic synthesis. As the following examples indicate, the coupling reaction can be carried out on aryl and vinylic halides as well as on alkyl halides.

$$\underset{\textit{trans}\text{-1-Iodo-1-nonene}}{(n\text{-}C_7H_{15})HC{=}CHI} + (n\text{-}C_4H_9)_2CuLi \longrightarrow \underset{\textit{trans}\text{-5-Tridecene (71\%)}}{(n\text{-}C_7H_{15})HC{=}CH(C_4H_9\text{-}n)} + n\text{-}C_4H_9Cu + LiI$$

$$\underset{\text{Iodobenzene}}{C_6H_5I} + (CH_3)_2CuLi \longrightarrow \underset{\text{Toluene (91\%)}}{C_6H_5CH_3} + CH_3Cu + LiI$$

[3]Henry Gilman (1893–1986); b. Boston; Ph.D. (1918) Harvard (Kohler); professor, Iowa State University (1923–1986).

An organocopper coupling reaction is even carried out on a commercial scale to synthesize *muscalure*, (9*Z*)-tricosene, the sex attractant secreted by the common housefly. Minute amounts of this insect hormone, or **pheromone**, greatly increase the lure of insecticide-treated fly bait and provide a species-specific means of insect control. Coupling of *cis*-1-bromo-9-octadecene with lithium dipentylcopper is used industrially to produce muscalure in 100 lb batches.

$$\underset{\textit{cis-1-Bromo-9-octadecene}}{(n\text{-}C_8H_{17})HC{=}CH(CH_2)_7CH_2Br} + \underset{\text{Lithium dipentylcopper}}{(n\text{-}C_5H_{11})_2CuLi} \longrightarrow \underset{\text{Muscalure [(9Z)-Tricosene] (99\%)}}{(n\text{-}C_8H_{17})HC{=}CH(C_{13}H_{27}\text{-}n)}$$

Although the details of the mechanism by which coupling occurs are not fully understood, radicals are probably involved. This coupling is *not* a typical nucleophilic substitution reaction of the sort considered in the next chapter.

PROBLEM

**10.14** How would you prepare these compounds using an organocopper coupling reaction at some point in the scheme? More than one step is required in each case.
(a) 3-Methylcyclohexene from cyclohexene (b) Octane from 1-bromobutane
(c) Decane from 1-pentene

## 10.11 Summary and Key Words

**Alkyl halides** are compounds containing halogen bonded to a saturated, $sp^3$-hybridized carbon atom. The C—X bond is polar, and alkyl halides can therefore behave as electrophiles.

Alkyl halides can be prepared by **radical chlorination** of alkanes, but this method is of little general value since mixtures of products always result. The reactivity order of alkanes toward chlorination is identical to the stability order of radicals: tertiary > secondary > primary. According to the Hammond postulate, the more stable radical intermediate is formed faster because the transition state leading to it is more stable.

Alkyl halides can also be prepared from alkenes. Alkenes add HX, and they react with NBS to give the product of **allylic bromination**. The NBS bromination of alkenes is a complex radical process that takes place through an intermediate allyl radical. Allyl radicals are stabilized by **resonance** and can be drawn in two different ways, neither of which is correct by itself. The true structure of the allyl radical is best described as a composite, or **resonance hybrid**, of the two individual resonance forms. The only difference between the two resonance structures is in the location of bonding electrons—the nuclei remain in the same places in both structures.

$$H_2C{=}CH{-}\dot{C}H_2 \longleftrightarrow H_2\dot{C}{-}CH{=}CH_2$$

Alcohols react with HX to form alkyl halides, but this works well only for tertiary alcohols, $R_3COH$. Primary and secondary alkyl halides are normally prepared from alcohols using either $SOCl_2$ or $PBr_3$. Alkyl halides react with magnesium in ether solution to form organomagnesium halides—**Grignard reagents**. Since Grignard reagents are both nucleophilic and basic, they react with Brønsted acids to form hydrocarbons. The overall result of Grignard formation and protonation is the conversion of an alkyl halide into an alkane (R—X → R—MgX → R—H).

Alkyl halides also react with lithium metal to form **organolithiums**. In the presence of CuI, these form diorganocoppers, or **Gilman reagents**. Gilman reagents react with alkyl halides by a radical process to yield coupled hydrocarbon products.

## 10.12 Summary of Reactions

1. Preparation of alkyl halides
   a. From alkenes by allylic bromination (Section 10.5)

$$>C{=}C{-}C(H)< \xrightarrow[CCl_4]{NBS} >C{=}C{-}C(Br)<$$

   b. From alkenes by addition of HBr and HCl (Sections 6.9 and 6.10)

$$>C{=}C< + HBr \longrightarrow -C(H)-C(Br)-$$

   c. From alcohols
      (1) Treatment with HX, where X = Cl, Br, or I (Section 10.8)

$$C{-}OH \xrightarrow[Ether]{HX} C{-}X$$

Reactivity order: 3° > 2° > 1°

      (2) Treatment of 1° and 2° alcohols with $SOCl_2$ in pyridine (Section 10.8)

$$H{-}C{-}OH \xrightarrow[Pyridine]{SOCl_2} H{-}C{-}Cl$$

      (3) Treatment of 1° and 2° alcohols with $PBr_3$ (Section 10.8)

$$H{-}C{-}OH \xrightarrow[Ether]{PBr_3} H{-}C{-}Br$$

2. Reactions of alkyl halides
   a. Grignard reagent formation (Section 10.9)

$$R—X \xrightarrow[\text{Ether}]{\text{Mg}} R—Mg—X$$

where X = Br, Cl, or I

R = 1°, 2°, 3° alkyl, aryl, or vinylic

   b. Diorganocopper (Gilman reagent) formation (Section 10.10)

$$R—X \xrightarrow[\text{Pentane}]{\text{2 Li}} R—Li + LiX$$

where *R* = 1°, 2°, 3° alkyl, aryl, or vinylic

$$2\,R—Li + CuI \xrightarrow{\text{In ether}} [R—Cu—R]^- \ Li^+$$

   c. Organometallic coupling (Section 10.10)

$$R_2CuLi + R'—X \xrightarrow{\text{In ether}} R—R' + RCu + LiX$$

   d. Conversion of alkyl halides to alkanes (Section 10.9)

$$R—X \xrightarrow[\text{Ether}]{\text{Mg}} R—Mg—X \xrightarrow{H_3O^+} R—H + HOMgX$$

## ADDITIONAL PROBLEMS

**10.15** Name these alkyl halides according to IUPAC rules:
(a) $(CH_3)_2CHCHBrCHBrCH_2CH(CH_3)_2$ (b) $CH_3CH{=}CHCH_2CHICH_3$
(c) $(CH_3)_2CBrCH_2CHClCH(CH_3)_2$ (d) $CH_3CH_2CH(CH_2Br)CH_2CH_2CH_3$
(e) $ClCH_2CH_2CH_2C{\equiv}CCH_2Br$

**10.16** Draw structures corresponding to these IUPAC names:
(a) 2,3-Dichloro-4-methylhexane (b) 4-Bromo-4-ethyl-2-methylhexane
(c) 3-Iodo-2,2,4,4-tetramethylpentane (d) *cis*-1-Bromo-2-ethylcyclopentane

**10.17** A chemist requires a large amount of 1-bromo-2-pentene as starting material for a synthesis. She finds a supply of 2-pentene in the stockroom, and decides to carry out an NBS allylic bromination reaction:

$$CH_3CH_2CH{=}CHCH_3 \xrightarrow[CCl_4]{\text{NBS}} CH_3CH_2CH{=}CHCH_2Br$$

What's wrong with this synthesis plan? What side products would form in addition to the desired product?

**10.18** What product(s) would you expect from the reaction of 1-methylcyclohexene with NBS? Would you use this reaction as part of a synthesis?

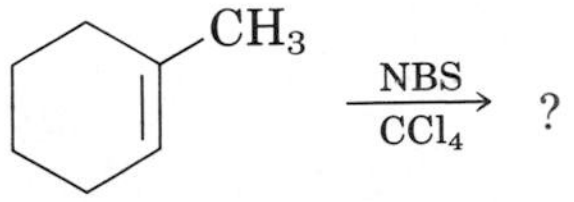

$$\text{1-methylcyclohexene} \xrightarrow[CCl_4]{NBS} ?$$

**10.19** How would you prepare the following compounds, starting with cyclopentene and any other reagents needed?

(a) Chlorocyclopentane
(b) Methylcyclopentane
(c) 3-Bromocyclopentene
(d) Cyclopentanol
(e) Cyclopentylcyclopentane
(f) 1,3-Cyclopentadiene

**10.20** Predict the product(s) of these reactions:

(a) 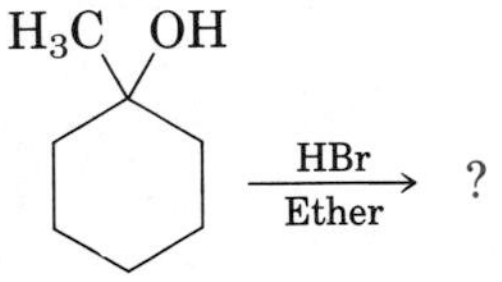 1-methylcyclohexanol ($H_3C$, OH) $\xrightarrow[Ether]{HBr}$ ?

(b) $CH_3CH_2CH_2CH_2OH \xrightarrow[Pyridine]{SOCl_2}$ ?

(c) 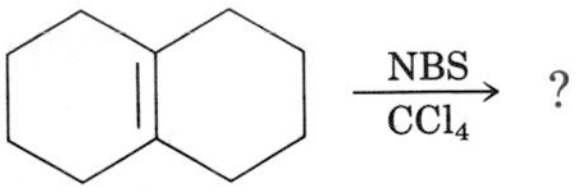 octahydronaphthalene $\xrightarrow[CCl_4]{NBS}$ ?

(d) 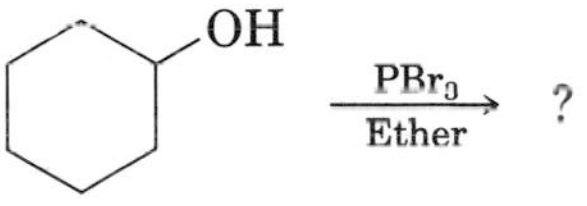 2-cyclohexenol (OH) $\xrightarrow[Ether]{PBr_3}$ ?

(e) $CH_3CH_2CHBrCH_3 \xrightarrow[Ether]{Mg} A \xrightarrow{H_2O} B$

(f) $CH_3CH_2CH_2CH_2Br \xrightarrow[Pentane]{Li} A \xrightarrow{CuI} B$

(g) $CH_3CH_2CH_2CH_2Br + (CH_3)_2CuLi \xrightarrow{Ether}$ ?

**10.21** Table 5.4 shows that the methyl C—H bond of toluene is 13 kcal/mol (54 kJ/mol) weaker than the C—H bond of ethane. In other words, the $C_6H_5CH_2\cdot$ (benzyl) radical is 13 kcal/mol more stable than the $\cdot CH_2CH_3$ radical. Draw as many resonance structures as you can for the benzyl radical.

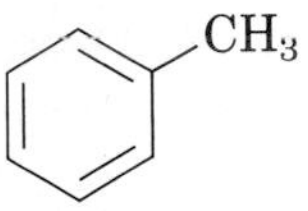

$CH_3$

Toluene

**10.22** What product(s) would you expect from the reaction of 1-phenyl-2-butene with NBS? Explain.

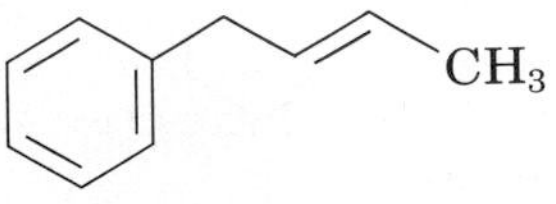

$CH_3$

**10.23** Which of the following pairs of structures represent resonance forms?

(a) 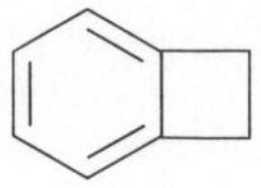 and 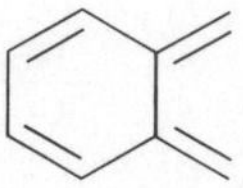 (b) 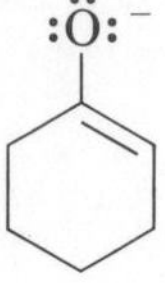 and

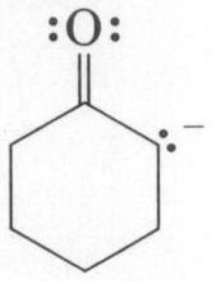

(c) 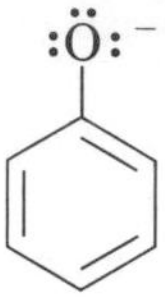 and 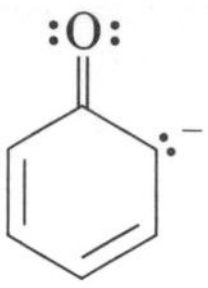 (d) 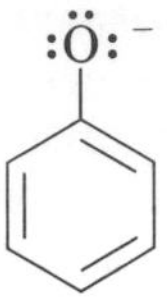 and

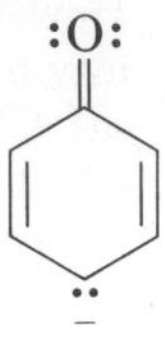

**10.24** Draw as many resonance structures as you can for these species:

(a) $H_3C{-}C({=}O){-}\ddot{C}H_2^-$ (b)

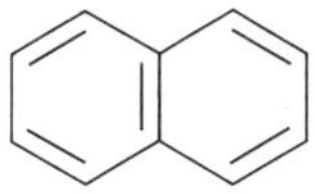

(c) $H_2\ddot{N}{-}C(NH_2){=}\overset{+}{N}H_2$ (d) $H_3C{-}\ddot{S}{-}\overset{+}{C}H_2$

(e) $H_2C{=}CH{-}CH_2^+$ (f) $H_2C{=}CH{-}CH{=}CH{-}\overset{+}{C}H{-}CH_3$

**10.25** Which of the following two resonance structures would you expect to contribute most to the resonance hybrid? Explain.

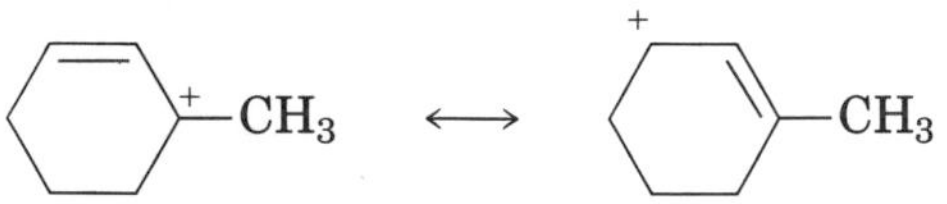

$\longleftrightarrow$

**10.26** How would you carry out these syntheses?
(a) Butylcyclohexane from cyclohexene
(b) Butylcyclohexane from cyclohexanol
(c) Butylcyclohexane from cyclohexane

**10.27** The syntheses shown here are unlikely to occur as written. What is wrong with each?

(a) $CH_3CH_2CH_2F \xrightarrow[2.\ H_3O^+]{1.\ Mg} CH_3CH_2CH_3$

(b) 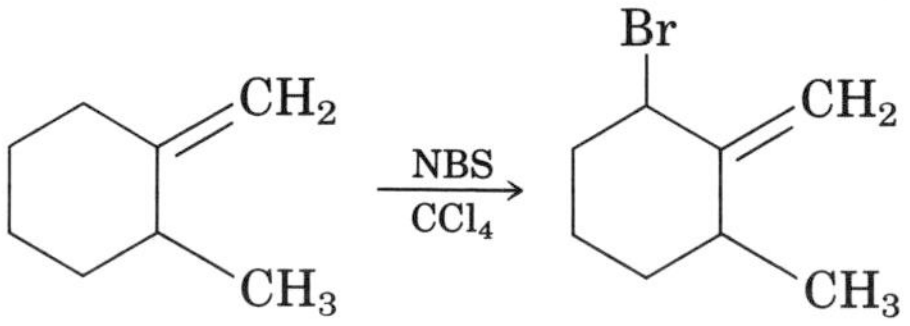 $\xrightarrow[CCl_4]{NBS}$

(c) 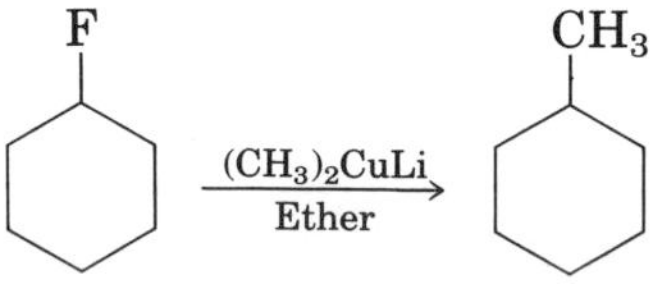 $\xrightarrow[Ether]{(CH_3)_2CuLi}$

**10.28** Which of the following pairs represent resonance structures?

(a) $CH_3C{\equiv}\overset{+}{N}{-}\ddot{\underset{..}{O}}{:}^-$ and $CH_3\overset{+}{C}{=}\ddot{N}{-}\ddot{\underset{..}{O}}{:}^-$

(b) $CH_3\overset{\overset{:O:}{\|}}{C}{-}\ddot{\underset{..}{O}}{:}^-$ and $\overset{-}{:}CH_2\overset{\overset{:O:}{\|}}{C}{-}\ddot{\underset{..}{O}}{-}H$

(c) $C_6H_5\overset{\overset{:O:}{\|}}{C}{-}\overset{+}{N}H_3$ and $C_6H_5\overset{\overset{+\ddot{O}-H}{\|}}{C}{-}\ddot{N}H_2$

(d) $CH_2{=}\overset{+}{N}(\ddot{\underset{..}{O}}{:}^-)_2$ and $\overset{-}{:}CH_2{-}\overset{+}{N}({=}\ddot{O}{:})(\ddot{\underset{..}{O}}{:}^-)$

CHAPTER 11

# Reactions of Alkyl Halides: Nucleophilic Substitutions and Eliminations

We saw in the last chapter that the carbon–halogen bond in alkyl halides is polar, and that the carbon atom is electron-poor. Thus, alkyl halides are electrophiles, and much of their chemistry involves polar reactions with electron-rich nucleophiles and bases.

$\overset{\delta^+}{C}$—$\overset{\delta^-}{X}$ Electrophilic carbon atom

Alkyl halides (R—X) do one of two things when they react with a nucleophile: Either they undergo *substitution* of the X group by the nucleophile, or they undergo *elimination* of H—X to yield an alkene.

Substitution: $Nu:^- + \text{—C—X} \longrightarrow \text{—C—Nu} + X:^-$

Elimination: $Nu:^- + \text{H—C—C—X} \longrightarrow \text{C=C} + \text{Nu—H} + X:^-$

Nucleophilic substitution and base-induced elimination are two of the most important reactions in organic chemistry. It's now time to take a close look at them to see how these reactions occur, what factors are involved, and how we can control the reactions for purposes of organic synthesis.

## 11.1 The Discovery of the Walden Inversion

In 1896, the German chemist Paul Walden[1] reported a remarkable discovery. He found that the pure enantiomeric (+)- and (−)-malic acids can be *interconverted* by a series of simple substitution reactions. When Walden treated (−)-malic acid with $PCl_5$, he isolated dextrorotatory (+)-chlorosuccinic acid. This, on treatment with wet silver oxide, gave (+)-malic acid. Similarly, reaction of (+)-malic acid with $PCl_5$ gave levorotatory (−)-chlorosuccinic acid, which was converted into (−)-malic acid when treated with wet silver oxide. The full cycle of reactions reported by Walden is shown in Figure 11.1.

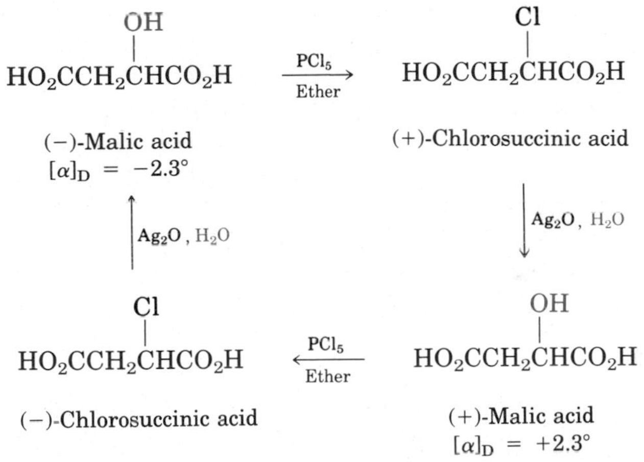

$HO_2CCH_2CH(OH)CO_2H$ (−)-Malic acid, $[\alpha]_D = -2.3°$ —($PCl_5$, Ether)→ $HO_2CCH_2CH(Cl)CO_2H$ (+)-Chlorosuccinic acid

(+)-Chlorosuccinic acid —($Ag_2O$, $H_2O$)→ $HO_2CCH_2CH(OH)CO_2H$ (+)-Malic acid, $[\alpha]_D = +2.3°$

(+)-Malic acid —($PCl_5$, Ether)→ $HO_2CCH_2CH(Cl)CO_2H$ (−)-Chlorosuccinic acid

(−)-Chlorosuccinic acid —($Ag_2O$, $H_2O$)→ (−)-Malic acid

**Figure 11.1** A Walden cycle, which interconverts (+)- and (−)-malic acids

At the time, the results were astonishing. The eminent chemist Emil Fischer called Walden's discovery "the most remarkable observation made in the field of optical activity since the fundamental observations of Pasteur." Since (−)-malic acid was being converted into (+)-malic acid, *some reactions in the cycle must have occurred with a change in configuration at the chiral center*. But which ones? (Recall that the direction of light rotation and the absolute configuration of a molecule aren't related. We can't tell by looking at the sign of rotation whether or not a change in configuration has occurred during a reaction.)

[1]Paul Walden (1863–1957); b. Latvia; Ph.D., Leipzig; student and professor, Riga Polytechnic, Russia (1882–1919); professor, University of Rostock, University of Tübingen, Germany.

Today, we would refer to the transformations taking place in Walden's cycle as **nucleophilic substitution reactions**, since each step involves the replacement (substitution) of one nucleophile (chloride ion, $:\ddot{\underset{..}{Cl}}:^-$, or hydroxide ion, $H\ddot{\underset{..}{O}}:^-$) by another.

## 11.2 Stereochemistry of Nucleophilic Substitution

Although Walden realized that changes in configuration must have taken place during his reaction cycle, he did not know at which steps the changes occurred. In the 1920s, however, Joseph Kenyon[2] and Henry Phillips began a series of investigations to elucidate the mechanism of nucleophilic substitution reactions and to find out how inversions of configuration occur. They recognized that the presence of the carboxylic acid group in Walden's work on malic acid may have led to complications, and they therefore carried out their own work on simpler cases. (In fact, the particular sequence of reactions studied by Walden *is* unusually complex for reasons we won't go into. The crucial point the sequence raises, however, is the idea that changes in three-dimensional configuration can evidently occur in organic reactions.)

Among the reactions studied by Kenyon and Phillips was one that interconverted the two enantiomers of 1-phenyl-2-propanol (Figure 11.2). Although this particular series of reactions involves nucleophilic substitution of an alkyl toluenesulfonate (a **tosylate**) rather than an alkyl halide, exactly the same type of reaction is involved as was studied by Walden. For all practical purposes, the *entire* tosylate group acts as if it were simply a halogen substituent:

$$\text{R—Y} + \text{Nu:}^- \longrightarrow \text{R—Nu} + \text{Y:}^-$$

where —Y = —Cl, —Br, —I, or —OTos $\left(\text{—O—}\overset{\displaystyle O}{\overset{\|}{\underset{\underset{\displaystyle O}{\|}}{S}}}\text{—}C_6H_4\text{—}CH_3\right)$ —OTos

Nu:⁻ = A nucleophile

In the three-step reaction sequence shown in Figure 11.2, (+)-1-phenyl-2-propanol is converted into its (−) enantiomer. Therefore, at least one of the three steps must involve an *inversion* (change) of configuration at the chiral center. The first step, formation of a toluenesulfonate, is known to occur by breaking the O—H bond of the alcohol rather than the C—O bond to the chiral carbon, and the configuration around carbon is therefore unchanged. Similarly, it can be shown (Section 21.7) that the third step, hydroxide ion cleavage of the acetate, also takes place without breaking the C—O bond at the chiral center; thus, inversion cannot occur in this step.

[2]Joseph Kenyon (1885–1961); b. Blackburn, England; D.Sc. (1914), London; British Dyestuffs Corp (1916–1920); Battersea Polytechnic, London (1920–1950).

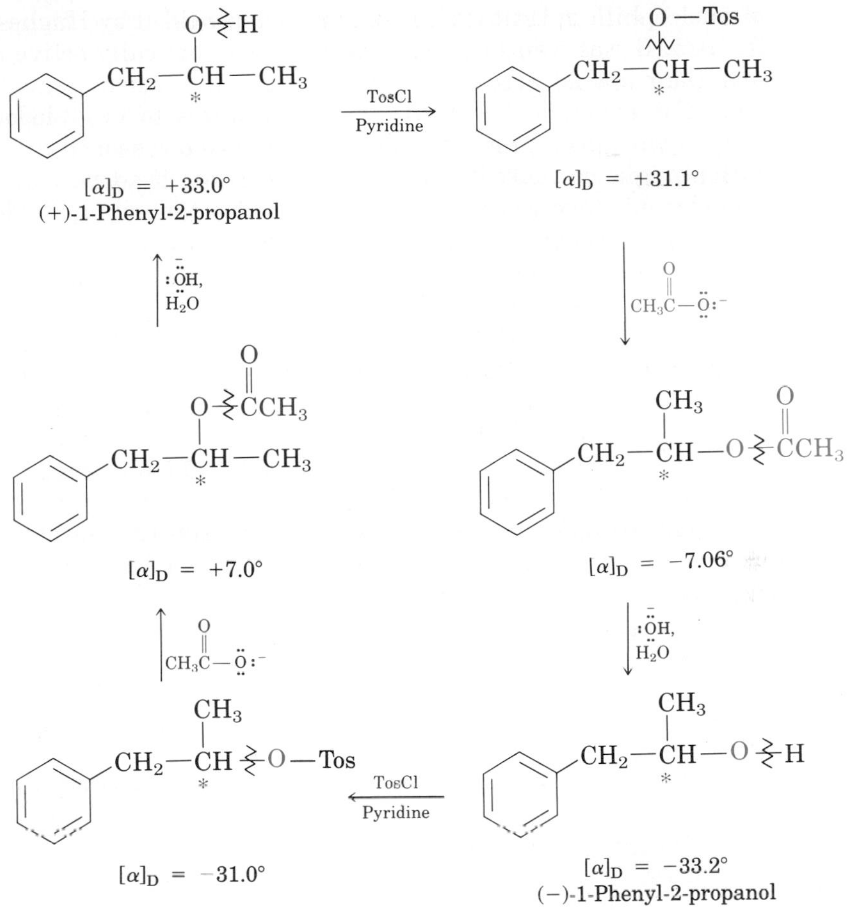

TosCl, Pyridine

$[\alpha]_D$ = +33.0°
(+)-1-Phenyl-2-propanol

$[\alpha]_D$ = +31.1°

$:\ddot{O}H^-$, $H_2O$

$CH_3C(=O)\ddot{O}:^-$

$[\alpha]_D$ = +7.0°

$[\alpha]_D$ = −7.06°

$CH_3C(=O)\ddot{O}:^-$

$:\ddot{O}H^-$, $H_2O$

TosCl, Pyridine

$[\alpha]_D$ = −31.0°

$[\alpha]_D$ = −33.2°
(−)-1-Phenyl-2-propanol

**Figure 11.2** A Walden cycle on 1-phenyl-2-propanol: Chiral centers are marked by asterisks, and the bonds broken in each reaction are indicated by wavy lines.

*The inversion of stereochemical configuration must therefore take place in the second step, the nucleophilic substitution of tosylate ion by acetate ion.*

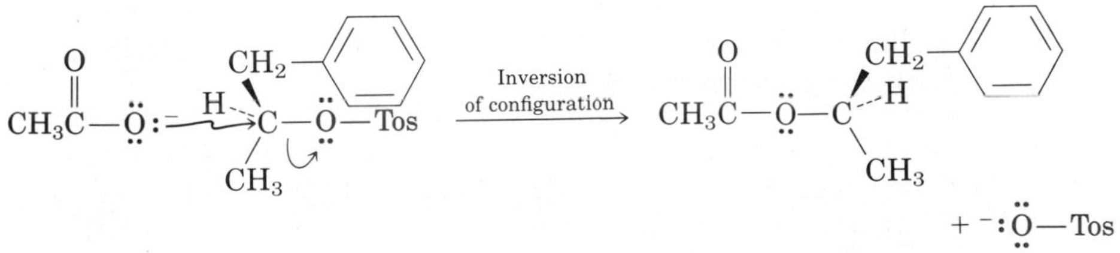

$CH_3C(=O)\ddot{O}:^-$ + H(CH₂Ph)(CH₃)C—Ö—Tos —Inversion of configuration→ $CH_3C(=O)$—Ö—C(CH₂Ph)(H)(CH₃) + $^-:\ddot{O}$—Tos

From this and nearly a dozen other series of reactions, Kenyon and Phillips concluded that the nucleophilic substitution reaction of primary and secondary alkyl halides and tosylates always proceeds with Walden inversion of configuration.

A second piece of stereochemical evidence concerning the mechanism of nucleophilic substitution reactions was provided by Hughes[3] and Ingold in 1935. It was well known at the time that optically active alkyl halides lose their optical activity and become racemic when treated with halide ions. For example, (−)-2-iodooctane racemizes to (±)-2-iodooctane when treated with lithium iodide. Hughes and Ingold reasoned that this racemization might be caused by a Walden inversion. If added iodide ion acts as a nucleophile and reacts with (*R*)-(−)-2-iodooctane via a Walden inversion, (*S*)-(+)-2-iodooctane will result.

$$^{-}:\ddot{I}: + (H)(CH_3)(n\text{-}C_6H_{13})C-\ddot{I}: \underset{\text{Walden inversion}}{\overset{\text{LiI}}{\rightleftharpoons}} :\ddot{I}-C(CH_3)(H)(n\text{-}C_6H_{13}) + :\ddot{I}:^{-}$$

(*R*)-(−)-2-Iodooctane (*S*)-(+)-2-Iodooctane

Hughes and Ingold reasoned that, if *every* substitution of iodide by iodide results in an inversion, then racemization will happen exactly twice as rapidly as iodide ion exchange. In other words, racemization will be complete when one-half of the starting (*R*)-2-iodooctane has been inverted by iodide exchange:

$$100\%\ (+)\ \text{molecules} \xrightarrow{50\%\ \text{inversion}} 50\%\ (+)\ \text{and}\ 50\%\ (-)\ \text{molecules}$$

Pure starting enantiomer Racemic mixture

This hypothesis was confirmed by an extremely clever experiment. The rate at which racemization occurs can be easily measured by using a polarimeter and determining how rapidly 2-iodooctane loses its optical activity. But how can the rate of iodide ion exchange be measured? Hughes and Ingold realized that, if they carried out the racemization reaction using *radioactive* iodide ion, $^{128}I^-$, and normal 2-iodooctane, $^{127}I$, they would be able to measure how rapidly the radioactive iodide became incorporated into the product and thus could tell how rapidly iodide exchange occurred.

The result of the experiment was exactly as predicted. When a large excess of radioactive $^{128}I^-$ was used, racemization was found to take place almost exactly twice as fast (1.93 ± 0.16) as iodide ion exchange. Thus, firm evidence was obtained to show that nucleophilic substitutions take place with *complete* inversion of configuration.

$$^{128}:\ddot{I}:^{-} + (H)(CH_3)(n\text{-}C_6H_{13})C-{}^{127}\ddot{I}: \longrightarrow {}^{128}:\ddot{I}-C(CH_3)(H)(n\text{-}C_6H_{13}) + {}^{127}:\ddot{I}:^{-}$$

[3]Edward David Hughes (1893–1970); b. Caernarvonshire, North Wales; Ph.D., Wales (Watson); D.Sc., London (Ingold); professor, University College, London (1930–1970).

## 11.3 Kinetics of Nucleophilic Substitution

We've referred several times up to this point about reactions being fast or slow. In fact, the speed at which a starting material reacts to give product is a quantity that can often be measured exactly. The determination of reaction rates, and the dependence of reaction rates on reagent concentrations, can be enormously useful in determining mechanisms. Let's see what can be learned about the nucleophilic substitution reaction by a study of reaction rates.

In all chemical reactions, there is a mathematical relationship between reaction rate and reagent concentrations. When we measure this relationship, we measure the **kinetics** of the reaction. For example, let's look at some factors that influence the rate of a simple nucleophilic substitution—the reaction of bromomethane with hydroxide ion to yield methanol:

$$H\ddot{\underset{..}{O}}{:}^- + CH_3{-}\ddot{\underset{..}{Br}}{:} \longrightarrow H\ddot{\underset{..}{O}}{-}CH_3 + {:}\ddot{\underset{..}{Br}}{:}^-$$

At a given concentration of reagents, the reaction occurs at a certain rate. If we double the concentration of hydroxide ion, the frequency of encounter between the reagents is also doubled, and we might therefore predict that the reaction rate will double. Similarly, if we double the concentration of bromomethane, we might expect that the reaction rate will again double. This behavior is exactly what is found. We call such a reaction, in which the rate is linearly dependent on the concentrations of two reagents, a **second-order reaction**. Mathematically, we can express this second-order dependence of the nucleophilic substitution reaction by setting up a **rate equation**:

$$\begin{aligned}\text{Reaction rate} &= \text{Rate of disappearance of starting material}\\ &= k \times [\text{RX}] \times [^-\text{OH}]\end{aligned}$$

$$\begin{aligned}\text{where } [\text{RX}] &= \text{CH}_3\text{Br concentration}\\ [^-\text{OH}] &= {}^-\text{OH concentration}\\ k &= \text{A constant value}\end{aligned}$$

This equation says that the reaction rate is the same as the rate of disappearance of starting material and is equal to a coefficient, $k$, times the alkyl halide concentration times the hydroxide ion concentration. The constant $k$ is called the **rate coefficient** for the reaction and is measured in units of liters per mole second (L/mol sec). The rate equation simply tells us that, as either [RX] or [$^-$OH] changes, the rate of the reaction changes accordingly. If the alkyl halide concentration is doubled, the reaction rate doubles; if the alkyl halide concentration is halved, the reaction rate is halved.

## 11.4 The $S_N2$ Reaction

At this point, two important pieces of information have been obtained about the nature of nucleophilic substitution reactions on primary and secondary alkyl halides and tosylates. What remains is to see how this information can be used to determine the mechanism of these reactions.

1. These reactions always occur with *complete* inversion of stereochemistry at the chiral carbon center.
2. These reactions show second-order kinetics and follow the rate law

$$\text{Rate} = k \times [\text{RX}] \times [\text{Nu}:^-]$$

How can we account for the observed reaction stereochemistry and second-order kinetics? A satisfactory explanation was first advanced in 1937 when Hughes and Ingold formulated the mechanism of what they called the **$S_N2$ reaction**—shorthand for substitution, nucleophilic, bimolecular. (*Bimolecular* means that two molecules—nucleophile and alkyl halide—are involved in the step whose kinetics are measured.)

The essential features of the $S_N2$ reaction mechanism formulated by Hughes and Ingold are that the reaction takes place in a single step without intermediates when the entering nucleophile attacks the substrate from a position 180° away from the leaving group. As the nucleophile comes in on one side and bonds to the chiral carbon, the leaving halide group departs from the other side and the stereochemical configuration of the molecule inverts (Figure 11.3).

The nucleophile $\text{Nu}:^-$ uses its lone-pair electrons to attack the alkyl halide 180° away from the halogen. This leads to a transition state with a partially formed C—Nu bond and a partially broken C—Y bond.

Nu:⁻ + C—Y

[Nu····C····Y] ‡ ($\delta^-$, $\delta^-$)

Transition state

The stereochemistry at carbon is inverted as the C—Nu bond forms fully and the halide departs with the electron pair from the original C—Y bond.

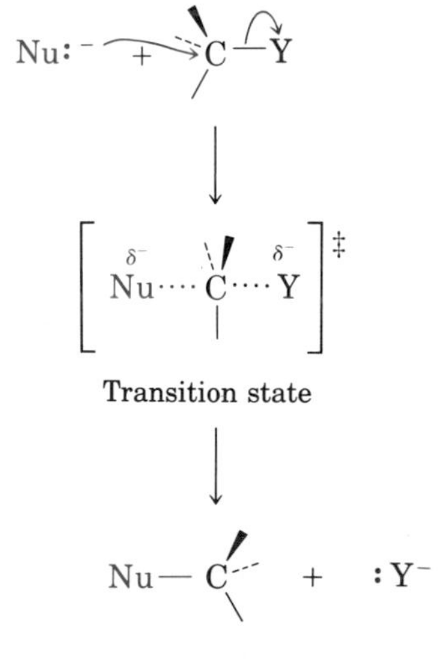

Nu—C + :Y⁻

**Figure 11.3** The mechanism of the $S_N2$ reaction: The reaction takes place in a single step when the incoming nucleophile approaches from a direction 180° away from the departing halide ion.

We can picture the reaction as occurring when an electron pair on the nucleophile, $Nu:^-$, forces out the leaving group, $Y:^-$, with its electron pair. This can occur through a transition state in which the new Nu—C bond is partially forming at the same time that the old C—Y bond is partially breaking, and in which the negative charge is shared by both the incoming nucleophile and the outgoing leaving group. The transition state for this inversion must have the remaining three bonds to carbon in a planar arrangement, as shown in Figure 11.4.

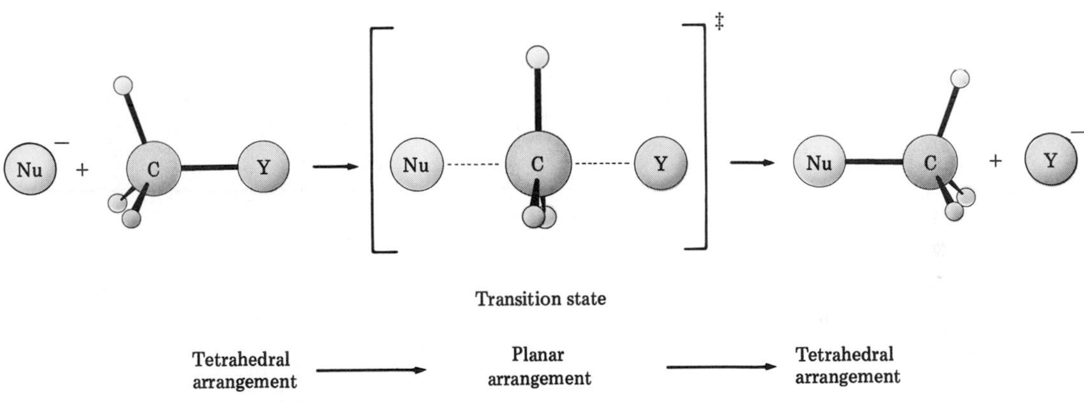


**Figure 11.4** The planar transition state of an $S_N2$ reaction

The mechanism proposed by Hughes and Ingold is consistent with experimental results, explaining both stereochemical and kinetic data. Thus, the requirement for back-side attack of the entering nucleophile from a direction 180° away from the departing Y group causes the stereochemistry of the substrate to invert, much like an umbrella turning inside out in the wind (Figure 11.5). The Hughes-Ingold mechanism also explains why second-order kinetics are found for $S_N2$ reactions. The reaction occurs in a single step that involves *both* alkyl halide and nucleophile. Two molecules are involved in the step whose rate is being measured.

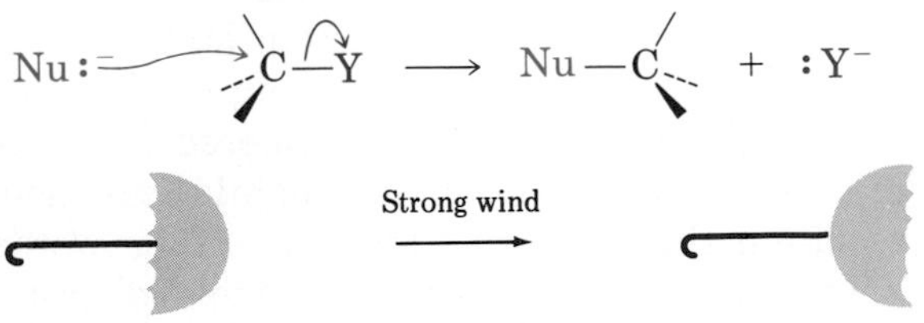

$Nu:^- \;\; C—Y \longrightarrow Nu—C \; + \; :Y^-$

Strong wind

**Figure 11.5** The inversion of a chiral center during an $S_N2$ reaction is similar to the inversion of an umbrella in a strong wind.

PROBLEM

11.1 What product would you expect to obtain from reaction of NaOH with (*R*)-2-bromobutane? Formulate the reaction showing the stereochemistry of both starting material and product.

PROBLEM

11.2 A further piece of evidence in support of requirement for back-side $S_N2$ displacement is the finding that the following alkyl bromide does not undergo a substitution reaction with hydroxide ion. Can you suggest a reason for the total lack of reactivity? Making a molecular model should be helpful.

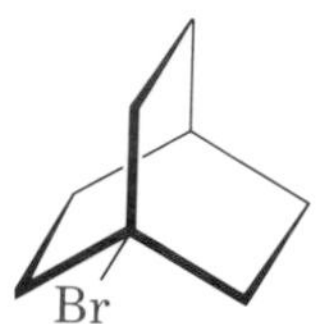


## 11.5 Characteristics of the $S_N2$ Reaction

We now have a good picture of how $S_N2$ reactions occur, but we also need to see how these substitutions can be used and what variables affect them. Some $S_N2$ reactions are fast and some are slow; some take place in high yield, and others in low yield. Understanding the different factors involved can be of tremendous value to chemists. Let's begin by reviewing what we've learned about reaction rates in general.

The rate of a chemical reaction is determined by $\Delta G^{\ddagger}$, the energy difference between reactant (ground state) and transition state. A change in reaction conditions can affect the magnitude of $\Delta G^{\ddagger}$ in two ways: either by changing the *reactant* energy level or by changing the *transition-state* energy level. If the reactant energy level is lowered, $\Delta G^{\ddagger}$ increases and reaction rate decreases; conversely, if the reactant energy level is raised, $\Delta G^{\ddagger}$ decreases and reaction rate increases, as can be seen in Figure 11.6(a).

Similarly, stabilization of the transition state lowers $\Delta G^{\ddagger}$ and raises reaction rate, whereas destabilization of the transition state raises $\Delta G^{\ddagger}$ and lowers reaction rate [Figure 11.6(b)]. We'll see examples of all these effects as we look at $S_N2$ reaction variables.

### THE SUBSTRATE: STERIC EFFECTS IN THE $S_N2$ REACTION

The first $S_N2$ reaction variable to look at is the steric bulk of the alkyl halide. We've said that the transition state for an $S_N2$ reaction involves partial bonding between the attacking nucleophile and the substrate. It therefore seems reasonable that a hindered, bulky substrate should prevent easy approach of an incoming nucleophile and should have a more difficult time reaching the transition state. In other words, sterically bulky substrates, in which the carbon atom is "shielded" from attack by the incoming nucleophile, react more slowly than less hindered substrates because the transition state for their reaction is sterically hindered and high in energy (Figure 11.7).

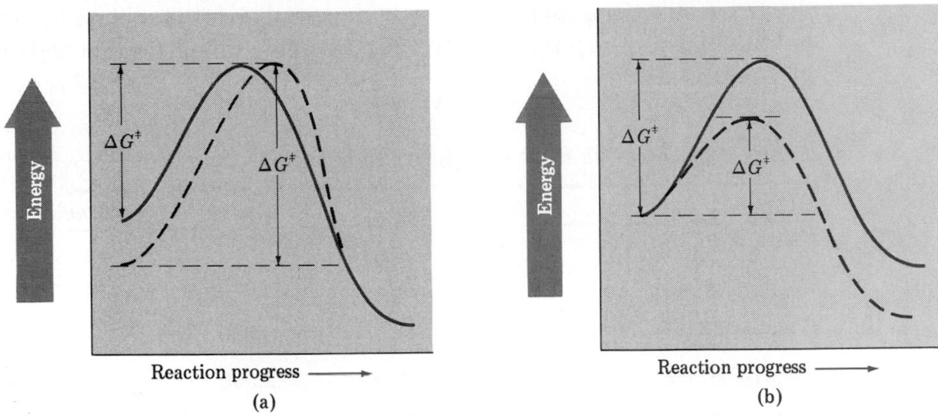


**Figure 11.6** The effect of changes in reactant and transition-state energy levels on reaction rate: (a) Higher reactant energy level corresponds to faster reaction (lower $\Delta G^{\ddagger}$). (b) Higher transition-state energy level corresponds to slower reaction (higher $\Delta G^{\ddagger}$).

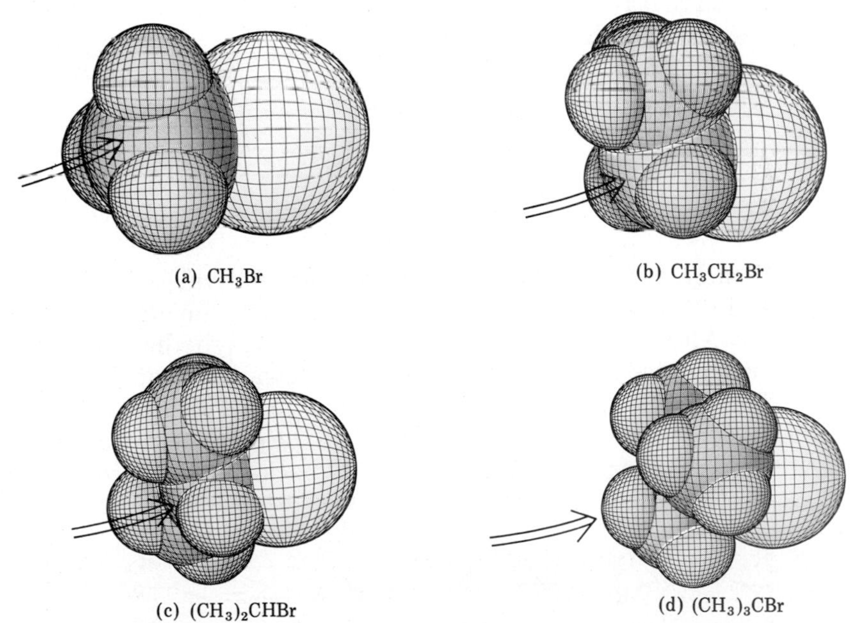


**Figure 11.7** Steric hindrance to the $S_N2$ reaction: As the computer-generated space-filling models indicate, the carbon atom in (a) bromomethane is readily accessible, resulting in a fast $S_N2$ reaction. The carbon atoms in (b) bromoethane (primary), (c) 2-bromopropane (secondary), and (d) 2-bromo-2-methylpropane (tertiary) are successively less accessible, resulting in successively slower $S_N2$ reactions.

As Figure 11.7 shows, the difficulty of nucleophilic attack increases as the three substituents bonded to the halo-substituted carbon atom increase in size. A list of relative reactivities for some different substrates is given in Table 11.1.

**Table 11.1** Relative rates of $S_N2$ reactions on alkyl halides

| *Alkyl halide* | *Type* | *Relative rate of reaction* |
|---|---|---|
| $CH_3—X$ | Methyl | 3,000,000 |
| $CH_3CH_2—X$ | 1° | 100,000 |
| $CH_3CH_2CH_2—X$ | 1° | 40,000 |
| $(CH_3)_2CH—X$ | 2° | 2,500 |
| $(CH_3)_3CCH_2—X$ | 1°, neopentyl | 1 |
| $(CH_3)_3C—X$ | 3° | ~0 |

Table 11.1 shows that methyl halides are by far the most reactive in $S_N2$ reactions, followed by primary alkyl halides such as ethyl and propyl. Alkyl branching next to the leaving group, as in isopropyl halides (2°), slows the reaction greatly, and further branching, as in *tert*-butyl halides (3°), effectively halts the reaction. Even branching that is one carbon removed from the leaving group, as in 2,2-dimethylpropyl (*neopentyl*) halides, greatly slows nucleophilic displacement.

From the information in Table 11.1, we can construct the following reactivity order for $S_N2$ reactions:

$$CH_3—X > RCH_2—X > R_2CH—X \gg R_3CCH_2—X > R_3C—X$$

$$\text{Methyl} > 1° > 2° \gg \text{Neopentyl} > 3°$$

Clearly, $S_N2$ reactions can only occur at relatively unhindered sites; methyl and unsubstituted primary substrates are by far the most reactive.

Although not shown in Table 11.1, vinylic halides ($R_2C{=}CRX$) and aryl halides are completely unreactive toward attempted $S_N2$ displacements. This lack of reactivity is probably due to steric factors, since the incoming nucleophile would have to approach in the plane of the carbon–carbon double bond in order to be able to carry out back-side displacements.

R C=C Cl R R Nu:⁻ ⟶✕ No reaction

Vinylic halide

:Nu⁻ Cl ⟶✕ No reaction

Aryl halide

Back-side approach of attacking nucleophiles is hindered, and $S_N2$ reactions cannot occur on these substrates.

## THE ATTACKING NUCLEOPHILE

The nature of the attacking nucleophile is a second variable that has a major effect on the $S_N2$ reaction. Any species, either neutral or negatively charged, can act as a nucleophile as long as it has an unshared pair of electrons (that is, is a Lewis base). If the nucleophile is negatively charged, the product is neutral, but if the nucleophile is neutral and an anion is displaced, the product is positively charged.

Negatively charged Nu:⁻ $\quad$ $Nu:^- + R-Y \longrightarrow R-Nu + Y:^-$

Neutral Nu: $\quad$ $Nu: + R-Y \longrightarrow R-Nu^+ + Y:^-$

Because of the wide scope of nucleophilic substitution reactions, a great many product types can be prepared from alkyl halides and tosylates. For example, Table 11.2 lists some common nucleophiles and the products of their reactions with bromomethane. Look carefully at the table to see the many kinds of products that can be made.

**Table 11.2** Some $S_N2$ reactions with bromomethane:

$Nu:^- + CH_3Br \longrightarrow Nu-CH_3 + :\ddot{Br}:^-$

| *Attacking nucleophile* | | *Product* | |
|---|---|---|---|
| *Formula* | *Name* | *Formula* | *Name* |
| $H:^-$ | Hydride | $CH_4$ | Methane |
| $CH_3\ddot{S}:^-$ | Methanethiolate | $CH_3S-CH_3$ | Dimethyl sulfide |
| $H\ddot{S}:^-$ | Hydrosulfide | $HS-CH_3$ | Methane thiol |
| $N\equiv C:^-$ | Cyanide | $N\equiv C-CH_3$ | Acetonitrile |
| $:\ddot{I}:^-$ | Iodide | $I-CH_3$ | Iodomethane |
| $H\ddot{O}:^-$ | Hydroxide | $HO-CH_3$ | Methanol |
| $CH_3\ddot{O}:^-$ | Methoxide | $CH_3O-CH_3$ | Dimethyl ether |
| $N=N=\ddot{N}:^-$ | Azide | $N_3-CH_3$ | Azidomethane |
| $:\ddot{Cl}:^-$ | Chloride | $Cl-CH_3$ | Chloromethane |
| $CH_3CO_2:^-$ | Acetate | $CH_3CO_2-CH_3$ | Methyl acetate |
| $H_3N:$ | Ammonia | $H_3\overset{+}{N}-CH_3\ Br^-$ | Methylammonium bromide |
| $(CH_3)_3N:$ | Trimethylamine | $(CH_3)_3\overset{+}{N}-CH_3\ Br^-$ | Tetramethylammonium bromide |

Although all the $S_N2$ reactions shown in Table 11.2 take place as indicated, some are much faster than others. What are the reasons for the reactivity differences? Why do some reagents appear to be much more "nucleophilic" than others?

The answers to these questions aren't straightforward and aren't yet fully understood. Part of the problem is that the very term *nucleophilic* is imprecise. Although most chemists use the term *nucleophilicity* to mean a measure of the affinity of a species for a carbon atom in the $S_N2$ reaction, the reactivity of a given nucleophile can change somewhat from one reaction to the next.

The exact nucleophilicity of a species in a given reaction depends on many factors, including the nature of the substrate, the identity of the solvent, and even the concentration of the reagents. In order to speak with any precision, we must study the relative reactivity of various nucleophiles on a *single* substrate in a *single* solvent system. Much work has been carried out on the $S_N2$ reactions of bromomethane in aqueous ethanol, and some quantitative results are listed in Table 11.3.

**Table 11.3** Relative nucleophilicities in $S_N2$ reactions on bromomethane:

$$Nu:^- + CH_3Br \longrightarrow Nu{-}CH_3 + :\ddot{Br}:^-$$

| *Nucleophile* | *Relative reactivity* | *Nucleophile* | *Relative reactivity* |
|---|---|---|---|
| $H\ddot{S}:^-$ | 125,000 | $C_6H_5\ddot{O}:^-$ (phenoxide) | 8,000 |
| $:CN^-$ | 125,000 | | |
| $:\ddot{I}:^-$ | 100,000 | $:\ddot{Cl}:^-$ | 1,000 |
| $CH_3CH_2\ddot{O}:^-$ | 25,000 | $(CH_3)_3N:$ | 700 |
| $H\ddot{O}:^-$ | 16,000 | $CH_3CO_2:^-$ | 500 |
| $:N_3^-$ | 10,000 | $H_2\ddot{O}:$ | 1 |

The data in Table 11.3 disclose a large range of reactivities. Although precise explanations for the observed nucleophilicities aren't known, there are some trends that can be detected in the data:

1. In comparing nucleophiles that have the same attacking atom, nucleophilicity roughly parallels basicity. Since "nucleophilicity" measures the affinity of a Lewis base for a carbon atom in the $S_N2$ reaction, and "basicity" measures the affinity of a base for a proton, it's easy to see why there might be a rough correlation between the two kinds of behavior. Table 11.4 provides a comparison between nucleophilicity and basicity for some oxygen nucleophiles. [Remember that basicity can be measured by obtaining the acidity ($pK_a$) of the conjugate acid. A strong base is derived from a weak acid (high $pK_a$), and a weak base is derived from a strong acid (low $pK_a$).]

**Table 11.4** Comparison of nucleophilicity and basicity for some oxygen nucleophiles; stronger bases are better nucleophiles

| *Nucleophile* | *Relative nucleophilicity toward bromomethane* | *Conjugate acid* | $pK_a$ |
|---|---|---|---|
| $CH_3CH_2\ddot{O}{:}^-$ | 25,000 | $CH_3CH_2OH$ | 16 |
| $H\ddot{O}{:}^-$ | 16,000 | $H_2O$ | 15.7 |
| $C_6H_5\ddot{O}{:}^-$ | 8,000 | $C_6H_5OH$ | 10 |
| $CH_3CO_2{:}^-$ | 500 | $CH_3CO_2H$ | 4.8 |
| $H_2\ddot{O}{:}$ | 1 | $H_3O^+$ | −1.7 |

Nucleophilicity

2. A second trend indicated in Table 11.3 is that nucleophilicity usually increases in going down a column of the periodic table. Thus $H\ddot{S}{:}^-$ is more nucleophilic than $H\ddot{O}{:}^-$, and the halide reactivity order is $:\ddot{I}:^- > :\ddot{Br}:^- > :\ddot{Cl}:^-$.

PROBLEM

**11.3** What products would you expect from reaction of 1-bromobutane with these reagents?
(a) NaI (b) KOH (c) H—C≡C—Na (d) $NH_3$

PROBLEM

**11.4** The tertiary amine base quinuclidine reacts with $CH_3I$ 50 times as fast as triethylamine. Can you suggest a reason for this difference?

$$R_3N{:} + CH_3I \longrightarrow R_3\overset{+}{N}—CH_3 \; :\ddot{I}:^-$$

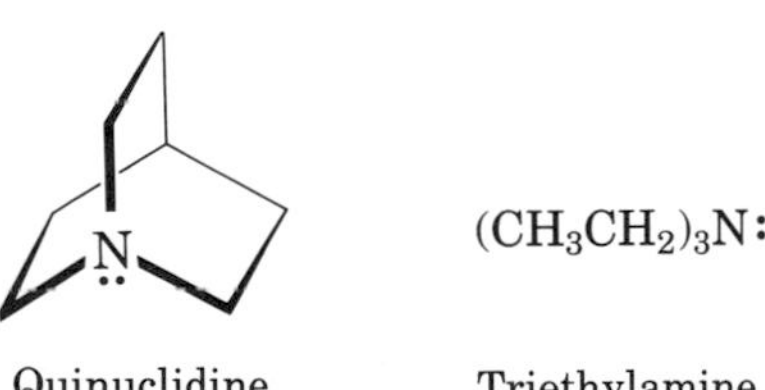


$(CH_3CH_2)_3N{:}$

Quinuclidine Triethylamine

PROBLEM

**11.5** Which reagent in each of the following pairs is more nucleophilic? Justify your choices.
(a) $(CH_3)_2\ddot{N}{:}^-$ and $(CH_3)_2\ddot{N}H$ (b) $(CH_3)_3B$ and $(CH_3)_3N{:}$
(c) $H_2\ddot{O}{:}$ and $H_2\ddot{S}{:}$

## THE LEAVING GROUP

A further variable that can strongly affect the $S_N2$ reaction is the nature of the species expelled by the attacking nucleophile—the **leaving group**.

Since the leaving group is expelled with a negative charge in most $S_N2$ reactions, we might expect the best leaving groups to be those that best stabilize the negative charge. Furthermore, since anion stability is related to basicity, we can also say that the best leaving groups should be the weakest bases.

The reason that stable anions (weak bases) make good leaving groups can be understood by looking at the transition state. In the transition state for an $S_N2$ reaction, the charge is distributed over both the attacking nucleophile and the leaving group. The greater the extent of charge stabilization by the leaving group, the more stable the transition state and the more rapid the reaction.

$$Nu:^- + C{-}Y \longrightarrow \left[ \overset{\delta^-}{Nu} \cdots C \cdots \overset{\delta^-}{Y} \right]^{\ddagger} \longrightarrow Nu{-}C + :Y^-$$

Transition state

Table 11.5 lists a variety of leaving groups in order of reactivity and shows their correlation with basicity. The weakest bases (anions derived from the strongest acids) are indeed the best leaving groups. The *p*-toluenesulfonate (tosylate) leaving group is the most easily displaced, although its basicity is out of line with others in the table for reasons that aren't well understood. Iodide and bromide ions are also excellent leaving groups, but chloride ion is much less effective.

**Table 11.5** Correlation of leaving-group ability with basicity; the anions of strong acids make good leaving groups in the $S_N2$ reaction

| *Leaving group* | $pK_a$ *of conjugate acid* | *Relative reactivity* |
|---|---|---|
| $CH_3{-}C_6H_4{-}SO_2{-}O:^-$ Tosylate | −6.5 | 60,000 |
| $:I:^-$ | −9.5 | 30,000 |
| $:Br:^-$ | −9 | 10,000 |
| $:Cl:^-$ | −7 | 200 |
| $:F:^-$ | 3.2 | 1 |
| $CH_3CO_2:^-$ | 4.8 | ~0 |
| $HO:^-$ | 15.7 | ~0 |
| $CH_3CH_2O:^-$ | 16 | ~0 |
| $H_2N:^-$ | 35 | ~0 |

Reactivity

It's just as important to know which are *poor* leaving groups as to know which are good, and Table 11.5 clearly indicates that *$F^-$, $RCOO^-$, $HO^-$, $RO^-$, and $H_2N^-$ are not displaced by nucleophiles.* In other words, alkyl fluorides, esters, alcohols, ethers, and amines do not undergo $S_N2$ reactions under normal circumstances.

**PROBLEM**

**11.6** Rank the following compounds in order of their expected reactivity toward $S_N2$ reaction:

$CH_3Br$, $CH_3OTos$, $CH_3COOCH_3$, $(CH_3)_3CCl$, $(CH_3)_2CHCl$

## THE SOLVENT

The rates of many $S_N2$ reactions are affected by the solvent used. **Protic solvents**—those that contain —OH groups—are generally the worst solvents for $S_N2$ reactions, whereas **polar aprotic solvents**—those that have strong dipoles but don't have —OH or —NH groups—are the best.

Protic solvents such as methanol and ethanol slow down $S_N2$ reactions by affecting the energy level of the nucleophilic *reactant* rather than the energy level of the transition state. Protic solvent molecules are able to form hydrogen bonds to negatively charged nucleophiles, orienting themselves into a "cage" around the nucleophile. This **solvation** strongly stabilizes the nucleophile, decreasing its reactivity toward electrophiles in the $S_N2$ reaction.

```
            OR
            |
            H
            :
RO—H····X:····H—OR
            :
            H
            |
            OR
```

A solvated anion
(reduced nucleophilicity due to enhanced ground-state stability)

In contrast to protic solvents, polar aprotic solvents favor $S_N2$ reactions. Particularly valuable are acetonitrile, $CH_3CN$; dimethylformamide, $(CH_3)_2NCHO$ (abbreviated DMF); dimethyl sulfoxide, $(CH_3)_2SO$ (abbreviated DMSO); and hexamethylphosphoramide, $[(CH_3)_2N]_3PO$ (abbreviated HMPA). These solvents are able to dissolve many salts because of their high polarity, and they tend to surround metal *cations* rather than nucleophilic anions. The unsolvated anions therefore have a far greater effective nucleophilicity in these solvents, and $S_N2$ reactions take place at correspondingly faster rates.

Rate increases of over a millionfold have been observed in going from methanol to hexamethylphosphoramide. Table 11.6 (page 344) gives the results of a study of the reaction of azide ion with 1-bromobutane in different solvents and shows that hexamethylphosphoramide is clearly preferred.

**Table 11.6** Relative rates for the $S_N2$ reaction of azide ion with 1-bromobutane:

$$CH_3CH_2CH_2CH_2Br + :N_3^- \longrightarrow CH_3CH_2CH_2CH_2N_3 + Br^-$$

| *Solvent* | *Relative rate* | *Solvent* | *Relative rate* |
|---|---|---|---|
| $CH_3OH$ | 1 | $(CH_3)_2NCHO$ (DMF) | 2,800 |
| $H_2O$ | 6.6 | $CH_3CN$ | 5,000 |
| $(CH_3)_2SO$ (DMSO) | 1,300 | $[(CH_3)_2N]_3PO$ (HMPA) | 200,000 |

Polar aprotic solvents increase the rate of $S_N2$ reactions by affecting the energy level of the nucleophilic reactant rather than the energy level of the transition state. They lower $\Delta G^\ddagger$ by *destabilizing* (raising) the ground-state energy level of the nucleophile (Figure 11.8).

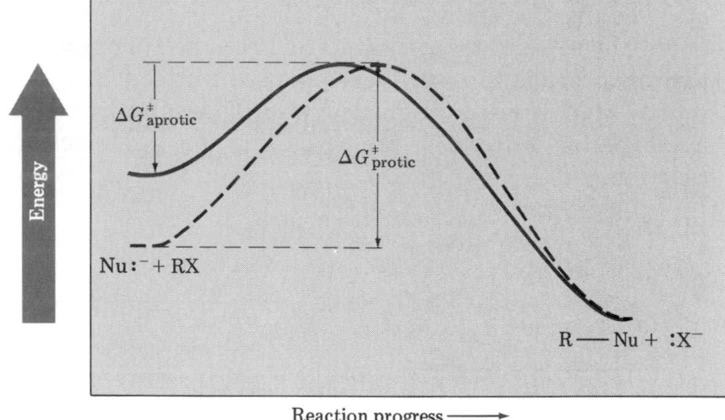


**Figure 11.8** The effect of solvent on $S_N2$ reactions: The nucleophile is stabilized by protic solvents but destabilized in polar aprotic solvents.

**PROBLEM**

**11.7** Normal organic solvents like benzene, ether, and chloroform are neither protic nor strongly polar. What effect would you expect these solvents to have on $S_N2$ reactions?

## 11.6 The $S_N1$ Reaction

We've seen that the $S_N2$ reaction occurs readily with primary alkyl halides and tosylates but that it is sensitive to steric factors. Secondary alkyl halides sometimes undergo $S_N2$ reactions, but tertiary halides are essentially inert to back-side attack by nucleophiles. For example, bromomethane reacts with acetate ion in 80% aqueous ethanol approximately 40 times faster than

2-bromopropane, and 2-bromo-2-methylpropane is unreactive toward $S_N2$ displacement by acetate ion.

Remarkably, however, a completely different picture emerges when different reaction conditions are used. When the alkyl bromides are heated with acetic acid, substitution of bromide ion by acetate occurs, and rate measurements show that 2-bromo-2-methylpropane reacts several thousand times *faster* than 2-bromopropane:

$$H_3C-\overset{\overset{O}{\|}}{C}-O-H + H_3C-\underset{\underset{CH_3}{|}}{\overset{\overset{CH_3}{|}}{C}}-Br \xrightarrow{\Delta} H_3C-\underset{\underset{CH_3}{|}}{\overset{\overset{CH_3}{|}}{C}}-O-\overset{\overset{O}{\|}}{C}-CH_3 + H-Br$$

The same trend is noted in many displacement reactions in which halides are heated with *neutral or nonbasic nucleophiles in protic solvents*: Tertiary alkyl halides react considerably faster than primary or secondary halides. For example, Table 11.7 gives the relative rates of reaction of some alkyl halides with water. 2-Bromo-2-methylpropane is more than *1 million times* as reactive as bromoethane.

**Table 11.7** Relative rates of reaction of some alkyl halides with water:

$$R-Br + H_2\ddot{O}: \longrightarrow R-\ddot{\underset{..}{O}}H + HBr$$

| *Alkyl halide* | *Type* | *Product* | *Relative rate of reaction* |
|---|---|---|---|
| $CH_3Br$ | Methyl | $CH_3OH$ | 1.0 |
| $CH_3CH_2Br$ | 1° | $CH_3CH_2OH$ | 1.0 |
| $(CH_3)_2CHBr$ | 2° | $(CH_3)_2CHOH$ | 12 |
| $(CH_3)_3CBr$ | 3° | $(CH_3)_3COH$ | 1,200,000 |

What are the reasons for this behavior? These reactions cannot be taking place by an $S_N2$ mechanism, and we must therefore conclude that *an alternative substitution mechanism* exists. This alternative mechanism is called the **$S_N1$ reaction** (for substitution, nucleophilic, unimolecular). Let's see what evidence is available concerning the $S_N1$ reaction.

## 11.7 Kinetics of the $S_N1$ Reaction

The reaction of acetic acid with 2-bromo-2-methylpropane seems analogous to the reaction of hydroxide ion with bromomethane, and we might therefore expect to observe second-order kinetics. In fact, we do not. We find instead that the reaction rate is dependent only on the alkyl halide concentration, and is independent of acetic acid concentration. In other words, the reaction is a **first-order process**. Only one molecule is involved in the step whose kinetics are measured. We can write the rate expression as follows:

$$\begin{aligned}\text{Reaction rate} &= \text{Rate of disappearance of alkyl halide}\\ &= k \times [RX]\end{aligned}$$

The rate of this $S_N1$ reaction is equal to a rate coefficient $k$ times the alkyl halide concentration. *Nucleophile concentration does not appear in the rate expression.* How can this result be explained? To answer this question, we must first learn more about kinetics measurements.

Many organic reactions are rather complex and occur in successive steps. One of these steps is usually slower than the others, and we call this the **rate-limiting step**. No reaction can proceed faster than its rate-limiting step, which acts as a kind of bottleneck. The overall reaction rate that we actually measure in a kinetics experiment is determined by the height of the highest *cumulative* energy barrier ($\Delta G^{\ddagger}$) between a low point and a subsequent high point in the energy diagram of the reaction. The hypothetical reaction energy diagrams in Figure 11.9 illustrate the idea of the rate-limiting step.

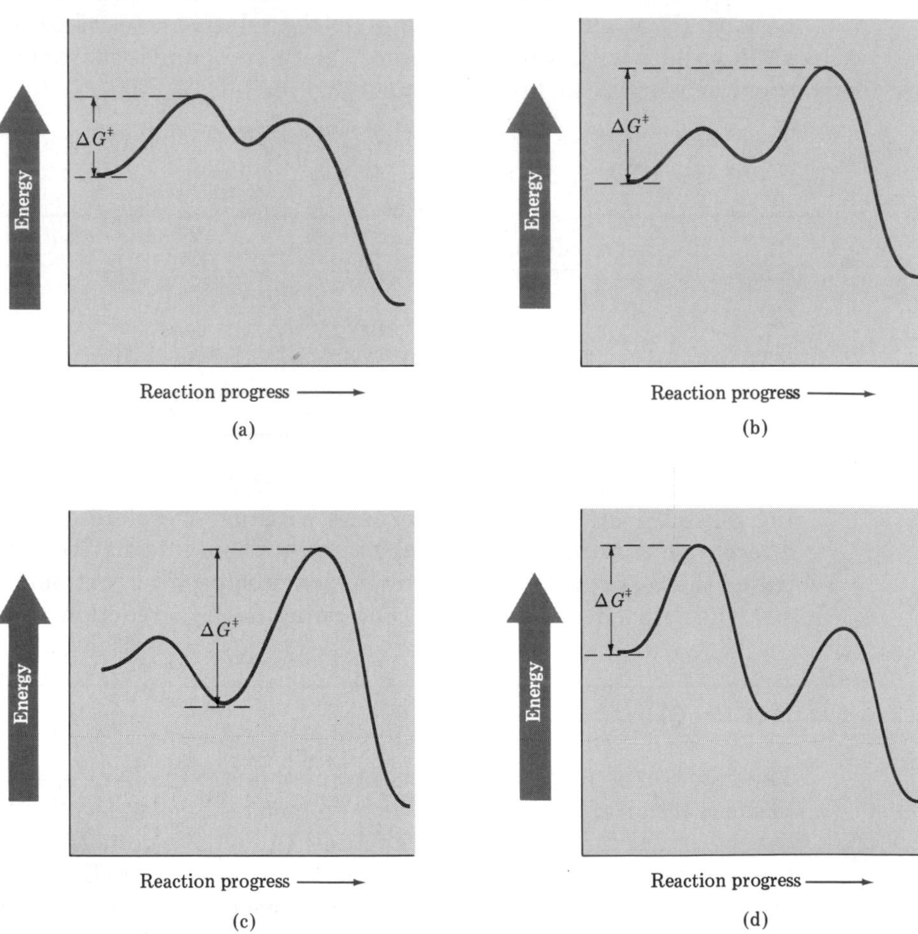


**Figure 11.9** Hypothetical reaction energy diagrams: The rate-limiting step in each is determined by the greatest cumulative difference in height between a low point and a subsequent high point.

In Figure 11.9(a), the rate-limiting step is simply the height of the barrier from the starting material to the first transition state, whereas in (b) the rate-limiting step represents the *cumulative* barrier from starting material to the highest-energy, *second* transition state. In (c) the step corresponding to the highest energy barrier is that from the stable intermediate to the second transition state, and in (d) the highest barrier is that from starting material to the first transition state. With this as background, let's look further into the kinetics of the $S_N1$ reaction.

The observation of first-order kinetics for the $S_N1$ reaction of 2-bromo-2-methylpropane with acetic acid indicates that the alkyl halide is involved in a unimolecular rate-limiting step. In other words, 2-bromo-2-methylpropane evidently undergoes some manner of spontaneous reaction *without assistance* from the nucleophile. The mechanism shown in Figure 11.10 can be postulated to account for the kinetic observations.

---

$(CH_3)_3C{-}\ddot{B}r{:}$

Spontaneous dissociation of the alkyl bromide occurs in a slow, rate-limiting step to generate a carbocation intermediate plus bromide ion.

Slow (rate-limiting) step

$[(CH_3)_3C^+] + {:}\ddot{B}r{:}^-$

Intermediate

The carbocation intermediate reacts with added nucleophile (acetate ion) in a fast step to yield neutral product.

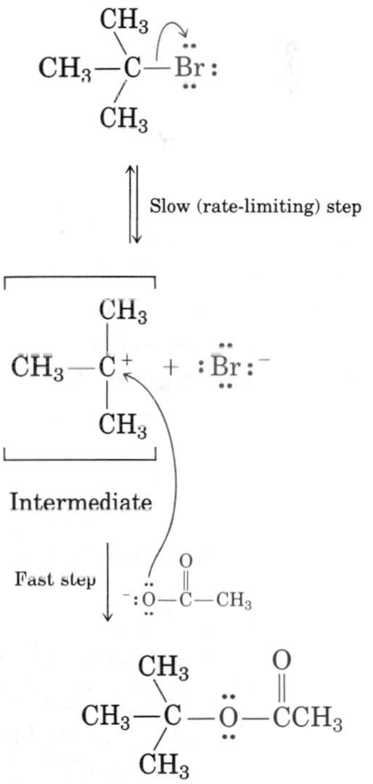

Fast step $\quad {}^-{:}\ddot{O}{-}C(=O){-}CH_3$

$(CH_3)_3C{-}\ddot{O}{-}C(=O)CH_3$

**Figure 11.10** The mechanism of the $S_N1$ reaction: Two steps are involved, with the first being rate-limiting.

---

If 2-bromo-2-methylpropane spontaneously dissociates to the *tert*-butyl carbocation plus bromide ion in a slow, rate-limiting step, and if the intermediate ion is immediately trapped by nucleophilic acetic acid in a fast step, then first-order kinetics will be obtained; *acetic acid plays no role in the step that is measured by kinetics*. The reaction energy diagram is shown in Figure 11.11 on page 348.

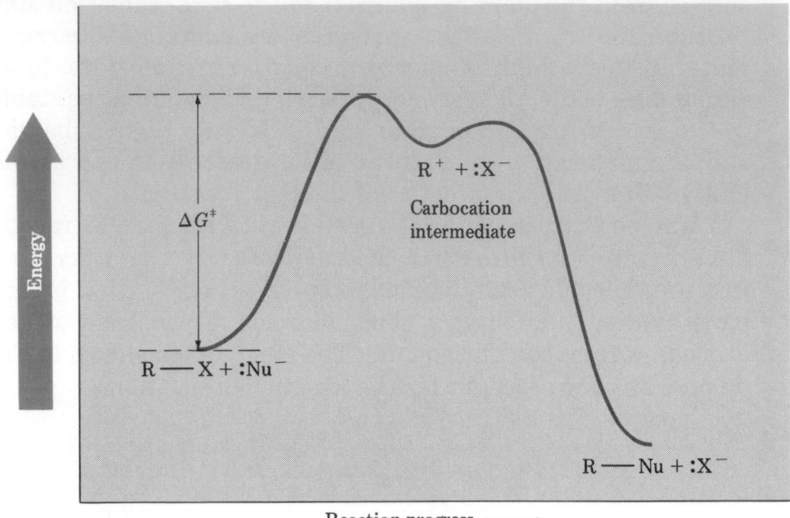


**Figure 11.11** A reaction energy diagram for an $S_N1$ reaction: The rate-limiting step is the spontaneous dissociation of the alkyl halide.

## 11.8 Stereochemistry of the $S_N1$ Reaction

If the postulate is correct that $S_N1$ reactions occur through carbocation intermediates, there should be clear stereochemical consequences. Carbocations are planar, $sp^2$-hybridized species and are therefore achiral. If we carry out an $S_N1$ reaction on a *chiral* starting material and go through an *achiral* intermediate, then the product must be optically inactive. In other words, the symmetrical intermediate carbocation can be attacked by a nucleophile equally well from either the right or the left side, leading to a 50:50 mixture of enantiomers—a racemic mixture (Figure 11.12).

The expectation that $S_N1$ displacements on chiral substrates should lead to racemic products has been amply borne out by experiment. Surprisingly, though, few $S_N1$ displacements occur with complete racemization. Most give a minor (0–20%) amount of inversion. For example, the reaction of optically active (*R*)-6-chloro-2,6-dimethyloctane with water leads to an alcohol product that is approximately 80% racemized and 20% inverted (80% *R*,*S* + 20% *S* is the same as 40% *R* + 60% *S*):

$$\underset{\text{(R)-6-Chloro-2,6-dimethyloctane}}{H_3C\text{-}C(C_2H_5)((CH_2)_3CH(CH_3)_2)\text{-}Cl} \xrightarrow[C_2H_5OH]{H_2\ddot{O}:} \underset{\substack{40\%\ R \\ \text{(retention)}}}{H_3C\text{-}C(C_2H_5)((CH_2)_3CH(CH_3)_2)\text{-}OH} + \underset{\substack{60\%\ S \\ \text{(inversion)}}}{HO\text{-}C(C_2H_5)(CH_3)((CH_2)_3CH(CH_3)_2)} + HCl$$

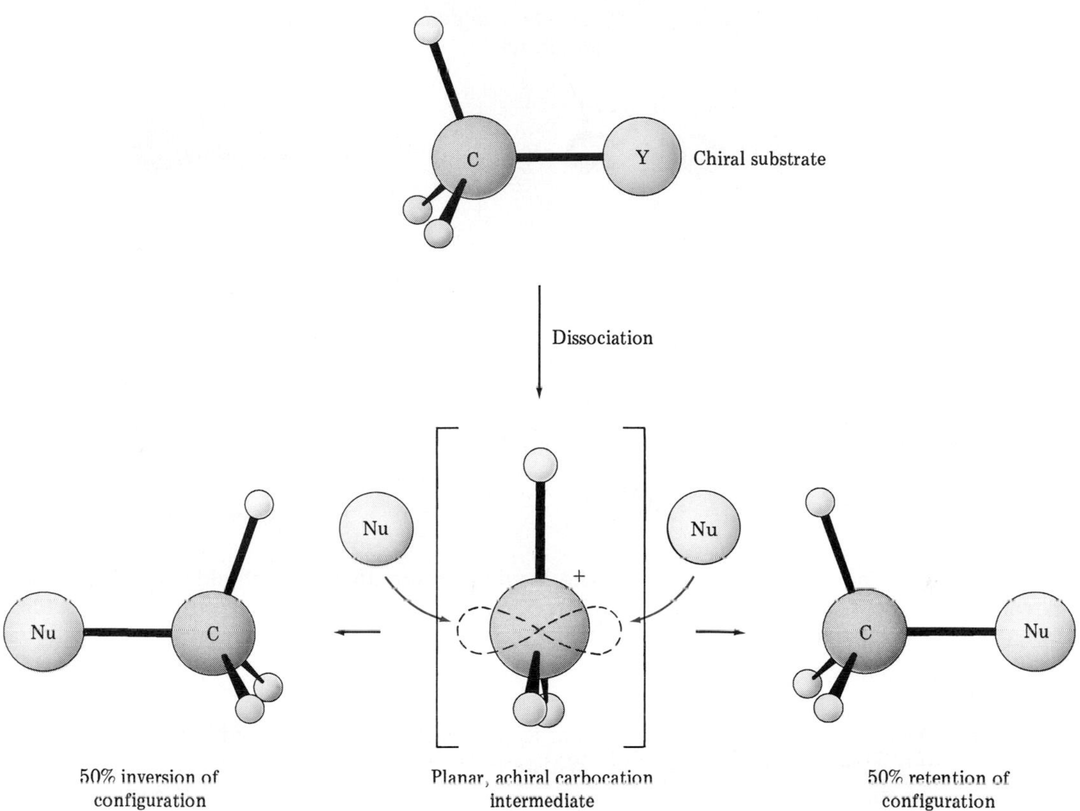


**Figure 11.12** Stereochemistry of the $S_N1$ reaction: An optically active starting material must give a racemic product.

The situation is complex, and the reasons for the lack of complete racemization in most $S_N1$ reactions aren't completely clear. An attractive suggestion, first proposed by Winstein,[4] is that **ion pairs** are involved. According to this suggestion, dissociation of the substrate occurs to give a complex in which the two ions are still loosely associated and in which the carbocation is effectively shielded from nucleophilic attack on one side by the departing ion. If a certain amount of substitution occurs before the two ions fully diffuse away from each other, then a net inversion of configuration will be observed (Figure 11.13, page 350).

**PROBLEM**

**11.8** What product(s) would you expect from reaction of (*S*)-3-chloro-3-methyloctane with acetic acid? Show the stereochemistry of both starting material and product.

[4]Saul Winstein (1912–1969); b. Montreal; Ph.D., California Institute of Technology (Lucas); professor, University of California, Los Angeles (1942–1969).

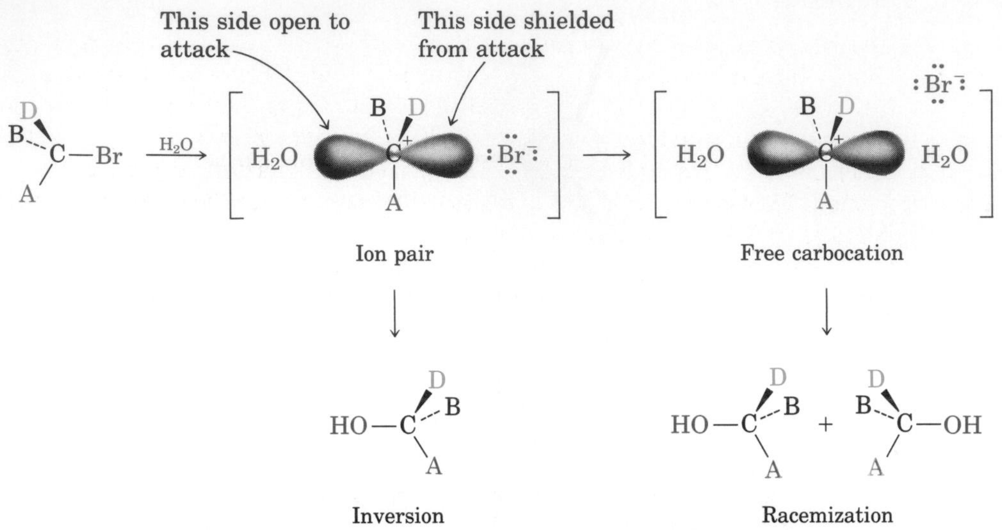


**Figure 11.13** The ion-pair hypothesis: The leaving group effectively shields one side of the developing carbocation intermediate from attack by added nucleophile.

PROBLEM

**11.9** Among the numerous examples of $S_N1$ reactions that occur with incomplete racemization is one reported by Winstein in 1952. The optically pure tosylate of 2,2-dimethyl-1-phenyl-1-propanol ($[\alpha]_D = -30.3°$) was heated in acetic acid to yield the corresponding acetate ($[\alpha]_D = +5.3°$). If complete inversion had occurred, the optically pure acetate would have had $[\alpha]_D = +53.6°$. What percentage racemization and what percentage inversion occurred in this reaction?

$$(CH_3)_3C{-}CH(OTos){-}C_6H_5 \xrightarrow[AcO^-]{HOAc} (CH_3)_3C{-}CH(OAc){-}C_6H_5 + HOTos$$

$[\alpha]_D = -30.3°$

Observed $[\alpha]_D = +5.3°$
(optically pure $[\alpha]_D = +53.6°$)

## 11.9 Characteristics of the $S_N1$ Reaction

Just as the $S_N2$ reaction is strongly influenced by such variables as solvent, leaving group, substrate structure, and nature of the attacking nucleophile, the $S_N1$ reaction is similarly influenced. Factors that lower $\Delta G^{\ddagger}$, either by stabilizing the transition state leading to carbocation formation or by raising the reactant energy level, favor faster $S_N1$ reactions. Conversely, factors that raise $\Delta G^{\ddagger}$, either by destabilizing the transition state leading to a carbocation or by lowering reactant energy level, slow down the $S_N1$ reaction.

## THE SUBSTRATE

According to the Hammond postulate, any factor that stabilizes a high-energy intermediate should also stabilize the transition state leading to that intermediate. Since the rate-limiting step in the $S_N1$ reaction is the spontaneous, unimolecular dissociation of the substrate, we would expect the reaction to be favored whenever stabilized carbocation intermediates are formed. This is exactly what is found—*the more stable the carbocation intermediate, the faster the $S_N1$ reaction.*

We've already seen (Section 6.11) that the stability order of alkyl carbocations is 3° > 2° > 1° > $—CH_3$. To this list we must now add the resonance-stabilized allyl and benzyl cations:

$$CH_2{=}CH{-}CH_2^+$$

Allyl carbocation

Benzyl carbocation ($C_6H_5CH_2^+$)

We saw in Sections 10.6 and 10.7 that allylic radicals have unusual stability because the unpaired electron can be delocalized over an extended pi orbital system. The same is true for allylic and benzylic carbocations. Delocalization of the positive charge over the extended pi orbital system of these carbocations results in unusual stability. As Figure 11.14 indicates,

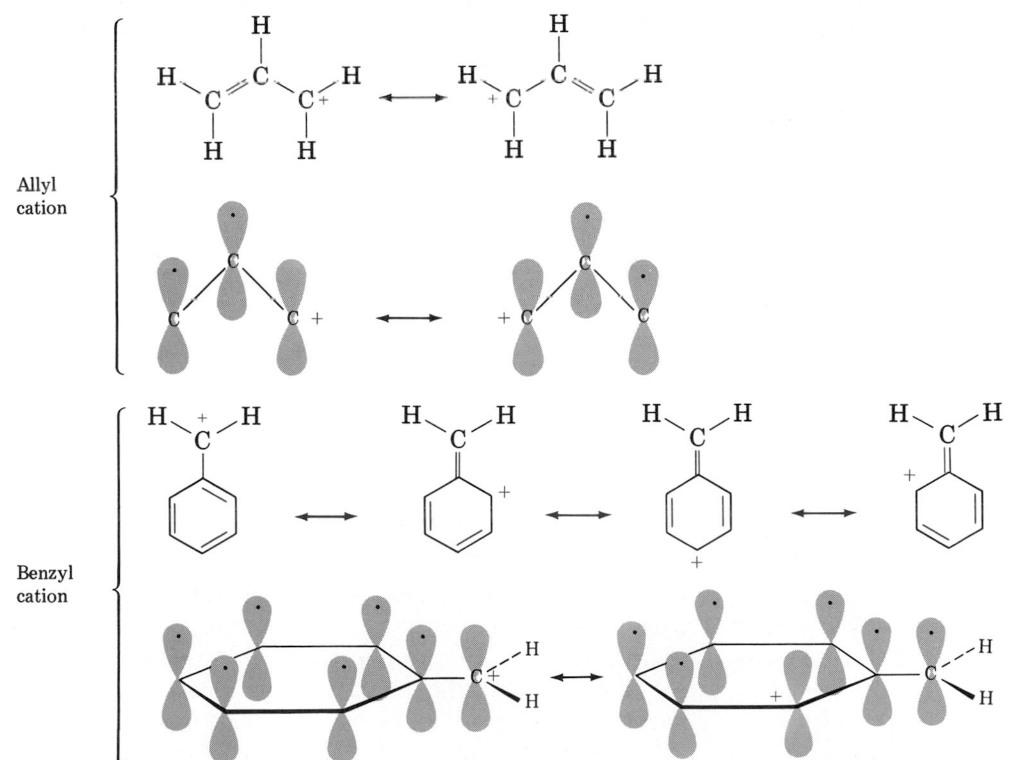


**Figure 11.14** Resonance forms of allylic and benzylic carbocations

an allylic cation has two equivalent resonance forms. In one form the double bond is on the "left," and in the other form the double bond is on the "right." A benzylic cation, however, has *four* resonance forms. These four forms are not equivalent, but all four make substantial contributions to the overall resonance hybrid.

As a result of the resonance stabilization in allylic and benzylic carbocations, carbocations have the following stability order. Note that a *primary* allylic or benzylic carbocation is approximately as stable as a *secondary* alkyl carbocation. (Similarly, a *secondary* allylic or benzylic carbocation is about as stable as a *tertiary* alkyl carbocation.)

| 3° | > | 2° | ≈ | Allyl | ≈ | Benzyl | > | 1° | > | Methyl |
|---|---|---|---|---|---|---|---|---|---|---|
| $R_3C^+$ | > | $R_2CH^+$ | ≈ | $H_2C{=}CH{-}CH_2^+$ | ≈ | $C_6H_5{-}CH_2^+$ | > | $RCH_2^+$ | > | $H_3C^+$ |

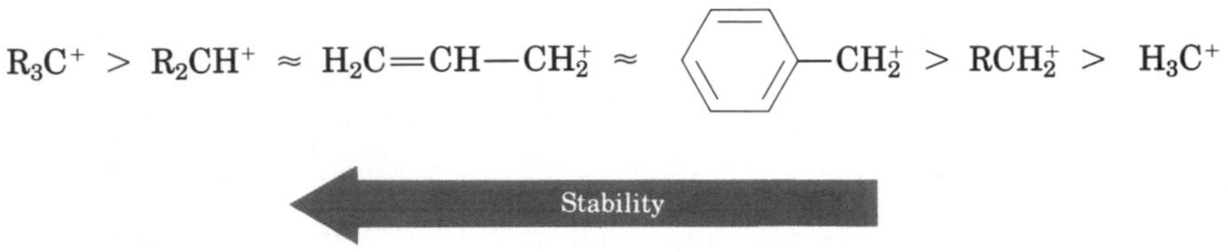

← Stability

The stability order of carbocations is exactly the order of $S_N1$ reactivity for alkyl halides and tosylates. Table 11.8, which shows the relative rates of reaction of some alkyl tosylates with ethanol, indicates a strong correlation between $S_N1$ rate and carbocation stabilities. Note particularly the extraordinary rate at which triphenylmethyl tosylate reacts.

**Table 11.8** Relative rates of reaction of some alkyl tosylates with ethanol at 25°C

| *Alkyl tosylate* | | *Product* | *Relative rate* |
|---|---|---|---|
| $CH_3CH_2OTos$ | ⟶ | $CH_3CH_2OCH_2CH_3$ | 1 |
| $(CH_3)_2CHOTos$ | ⟶ | $(CH_3)_2CHOCH_2CH_3$ | 3 |
| $H_2C{=}CHCH_2OTos$ | ⟶ | $H_2C{=}CHCH_2OCH_2CH_3$ | 35 |
| 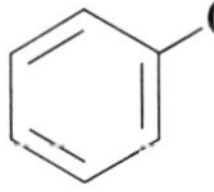 $C_6H_5CH_2OTos$ | ⟶ | 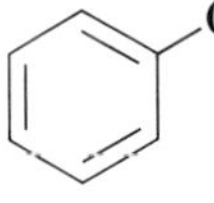 $C_6H_5CH_2OCH_2CH_3$ | 400 |
| $(C_6H_5)_2CHOTos$ | ⟶ | $(C_6H_5)_2CHOCH_2CH_3$ | $10^5$ |
| $(C_6H_5)_3COTos$ | ⟶ | $(C_6H_5)_3COCH_2CH_3$ | $10^{10}$ |

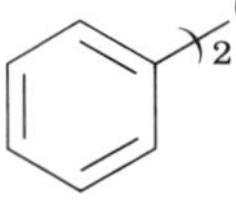

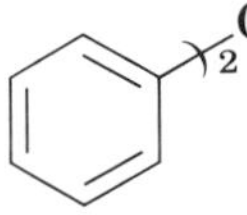

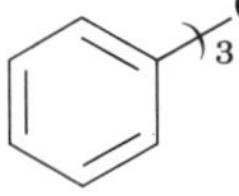

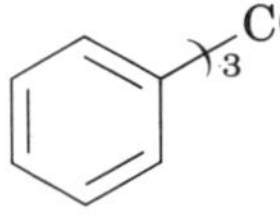

PROBLEM

**11.10** Rank the following alkyl halides in order of their expected $S_N1$ reactivity:

$CH_3CH_2Br$, $H_2C{=}CHCH(Br)CH_3$, $H_2C{=}CHBr$, $CH_3CH(Br)CH_3$

PROBLEM

**11.11** How can you account for the fact that 3-bromo-1-butene and 1-bromo-2-butene undergo $S_N1$ reaction at the same rate even though one is a secondary halide and the other is primary?

## THE LEAVING GROUP

During the discussion of $S_N2$ reactivity, we reasoned that the best leaving groups should be those that are most stable—that is, the conjugate bases of strong acids. An *identical* reactivity order is found for the $S_N1$ reaction, since the leaving group is intimately involved in the rate-limiting step. Thus, we find the $S_N1$ reactivity order to be

$$\text{Tosylate}^- > :\ddot{\underset{..}{I}}:^- > :\ddot{\underset{..}{Br}}:^- > :\ddot{\underset{..}{Cl}}:^- > H_2\ddot{O}:$$

Note that in the $S_N1$ reaction, which is often carried out under acidic conditions, neutral water can act as a leaving group. This is the case, for example, when an alkyl halide is prepared from a tertiary alcohol by reaction with HBr or HCl (Section 10.8). The alcohol is first protonated and then loses water to generate a carbocation. Reaction of the carbocation with halide ion then yields the alkyl halide (Figure 11.15).

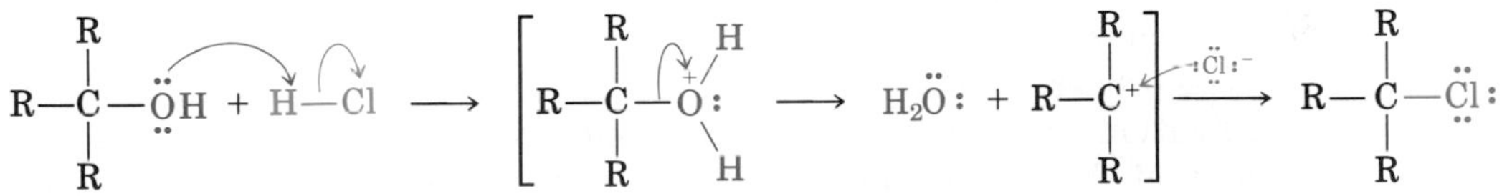

$$R_3C{-}\ddot{\underset{..}{O}}H + H{-}Cl \longrightarrow \left[R_3C{-}\overset{+}{\ddot{O}}H_2\right] \longrightarrow H_2\ddot{O}: + R_3C^+ \xrightarrow{:\ddot{\underset{..}{Cl}}:^-} R_3C{-}\ddot{\underset{..}{Cl}}:$$

**Figure 11.15** The $S_N1$ reaction of a tertiary alcohol with HCl to yield an alkyl halide: Neutral water is the leaving group.

Knowing that an $S_N1$ reaction is involved in the conversion of alcohols to alkyl halides makes it easier to understand why the reaction works well only for tertiary alcohols: Tertiary alcohols react fastest because they give the most stable carbocation intermediates.

## THE NUCLEOPHILE

The nature of the attacking nucleophile plays a major role in the $S_N2$ reaction. Should it play an equally major role in the $S_N1$ reaction? The answer is no. The $S_N1$ reaction, by its very nature, occurs through a rate-limiting step in which the added nucleophile plays no kinetic role. The nucleophile does not enter into the reaction until after rate-limiting dissociation has occurred. The reaction of 2-methyl-2-propanol with HX, for example, occurs at the same rate regardless of whether X is Cl, Br, or I:

$$\underset{\text{2-Methyl-2-propanol}}{(CH_3)_3COH} \xrightarrow[\text{Ethanol}]{HX} (CH_3)_3C{-}X + H_2O$$

PROBLEM

**11.12** How do you account for the fact that 1-chloro-1,2-diphenylethane reacts with the nucleophiles fluoride ion and triethylamine at exactly the same rate?

## THE SOLVENT

What about solvent? Does it have the same kind of effect in $S_N1$ reactions that it has in $S_N2$ reactions? The answer is both yes and no. Yes, solvents have a large effect on $S_N1$ reactions just as they do on $S_N2$ reactions; but no, the reasons for their effects are not the same. Solvent effects in the $S_N2$ reaction are due to stabilization or destabilization of the reactant nucleophile. Solvent effects in the $S_N1$ reaction, however, are due to stabilization of the transition state. Let's look again at the Hammond postulate to see this transition-state effect of solvent.

The Hammond postulate (Section 6.12) says that any factor stabilizing the intermediate carbocation should increase the rate of reaction. One factor that affects cation stability is structure, as previously discussed. Another factor is **solvation**—the interaction of the carbocation with solvent molecules. Solvation greatly lowers the energy of carbocation intermediates, thus lowering the transition-state energy for $S_N1$ reactions. Although the exact nature of carbocation stabilization by solvent is not easily defined, we might picture the solvent molecules orienting themselves around the cation in such a manner that the electron-rich ends of the solvent dipoles face the positive charge (Figure 11.16).

The properties of a solvent that contribute to its ability to stabilize ions by solvation aren't fully understood, but are undoubtedly related to the polarity of the solvent. Polar solvents such as water, methanol, dimethyl sulfoxide, and so forth are good at solvating ions, but most ether and hydrocarbon solvents are very poor at solvating ions.

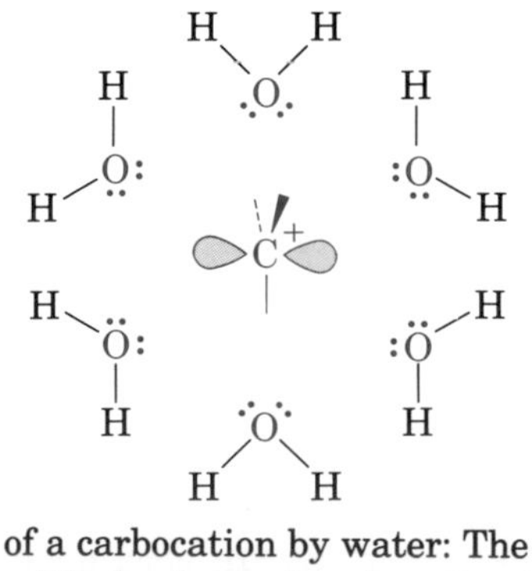

**Figure 11.16** Solvation of a carbocation by water: The electron-rich oxygen atoms of solvent surround the positively charged carbocation to stabilize it.

Solvent polarity is usually expressed in terms of **dielectric constants**, $\varepsilon$, which measure the ability of a solvent to act as an insulator of electric charges. In general, solvents of low dielectric constant such as hydrocarbons are nonpolar, whereas solvents of high dielectric constant such as water are polar. Table 11.9 lists the dielectric constants of some common solvents.

**Table 11.9** Dielectric constants of some common solvents

| *Name* | *Structure* | *Dielectric constant, ε* |
|---|---|---|
| **Aprotic (nonhydroxylic) solvents** | | |
| Hexane | $CH_3CH_2CH_2CH_2CH_2CH_3$ | 1.9 |
| Benzene | (benzene ring) | 2.3 |
| Diethyl ether | $CH_3CH_2—O—CH_2CH_3$ | 4.3 |
| Chloroform | $CHCl_3$ | 4.8 |
| Ethyl acetate | $CH_3C(=O)—O—CH_2CH_3$ | 6.0 |
| Acetone | $CH_3—C(=O)—CH_3$ | 20.7 |
| Hexamethylphosphoramide (HMPA) | $(CH_3)_2N—P(=O)(N(CH_3)_2)—N(CH_3)_2$ | 30 |
| Dimethylformamide (DMF) | $(CH_3)_2N—C(=O)—H$ | 38 |
| Dimethyl sulfoxide (DMSO) | $CH_3—S(=O)—CH_3$ | 48 |
| **Protic (hydroxylic) solvents** | | |
| Acetic acid | $CH_3C(=O)—OH$ | 6.2 |
| *tert*-Butyl alcohol | $(CH_3)_3COH$ | 10.9 |
| Ethanol | $CH_3CH_2OH$ | 24.3 |
| Methanol | $CH_3OH$ | 33.6 |
| Formic acid | $H—C(=O)—OH$ | 58 |
| Water | $H_2O$ | 80.4 |

All $S_N1$ reactions take place much more rapidly in highly polar solvents than in nonpolar solvents. Table 11.10 lists some relative rates measured for the reaction of 2-chloro-2-methylpropane with various solvents, and shows the magnitude of the rate differences due to solvent changes. A rate increase of 100,000 is observed on going from the polar solvent ethanol to the even more polar solvent water. The rate increases on going from hydrocarbon solvents to water are too large to measure accurately.

**Table 11.10** Relative rates for the reaction of 2-chloro-2-methylpropane with different solvents

| *Solvent* | *Relative rate* |
|---|---|
| Ethanol | 1 |
| Acetic acid | 2 |
| Aqueous ethanol (40%) | 100 |
| Aqueous ethanol (80%) | 14,000 |
| Water | $\sim 10^5$ |

The difference between effect of solvent on $S_N2$ and $S_N1$ reactions should be emphasized again. Although both reactions show large solvent effects, they do so for different reasons. $S_N2$ reactions (Section 11.5) are disfavored by protic solvents, since the *ground-state energy level* of the attacking nucleophile is lowered by solvation. $S_N1$ reactions, however, are favored by protic solvents, since the *transition-state energy level* leading to carbocation intermediate is lowered by solvation (Figure 11.17). (Compare Figure 11.17 to Figure 11.8, where the effect of solvent on the $S_N2$ reaction is illustrated.)

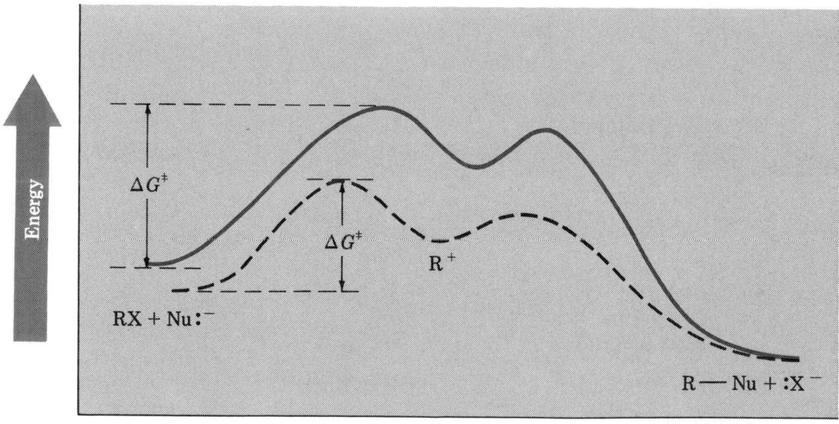


**Figure 11.17** The effect of solvent on $S_N1$ reactions: Both ground-state and transition-state energy levels are lowered by solvation in polar solvents, but the effect on transition state is far greater. (Solid curve, nonpolar solvent; dashed curve, polar solvent.)

PROBLEM

**11.13** As indicated in Problem 11.2, halides such as the one shown here are inert to $S_N2$ displacement. Perhaps more surprisingly, they are also unreactive to $S_N1$ substitution, even though they are tertiary. Considering the planar geometry of carbocations, can you suggest a reason for this low reactivity? (Molecular models may be helpful.)

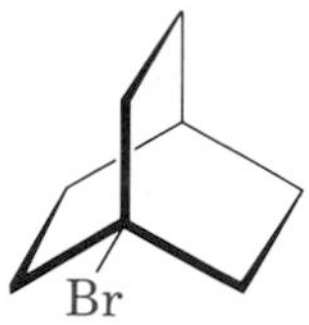

Br

## 11.10 Elimination Reactions of Alkyl Halides

We began this chapter by saying that two kinds of reactions are possible when a nucleophile/Lewis base attacks an alkyl halide. Often, the reagent will attack at carbon and substitute for the halide. Alternatively, though, attack at hydrogen can occur, resulting in elimination of HX to form an alkene.

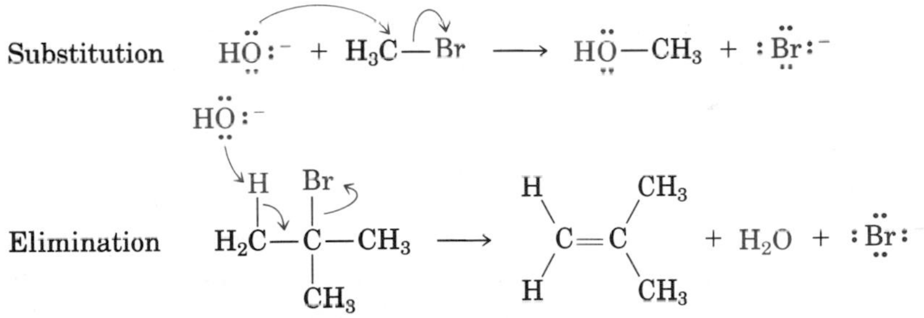

Substitution $$HO:^- + H_3C{-}Br \longrightarrow HO{-}CH_3 + :Br:^-$$

Elimination $$HO:^- + H_2C(H){-}C(Br)(CH_3){-}CH_3 \longrightarrow (H)_2C{=}C(CH_3)_2 + H_2O + :Br:^-$$

Elimination reactions are more complex than substitution reactions for several reasons. There is, for example, the problem of regiochemistry—what products result from dehydrohalogenation of unsymmetrical halides? In fact, elimination reactions almost always give *mixtures* of alkene products, and the best we can usually do is to predict which will be the major product.

According to a rule formulated by the Russian chemist Alexander Zaitsev,[5] base-induced elimination reactions generally give the more highly substituted alkene product—that is, the alkene with more alkyl substituents on the double-bond carbons. In the following two examples, **Zaitsev's rule** is clearly applicable. The more highly substituted alkene product predominates in both cases when sodium ethoxide in ethanol is used as the base.

[5]Alexander M. Zaitsev (1841–1910); b. Kasan, Russia (name also spelled *Saytzeff*, according to the German pronunciation).

$$\underset{\text{2-Bromobutane}}{CH_3CH_2CH(Br)CH_3} \xrightarrow[CH_3CH_2OH]{CH_3CH_2O^-Na^+} \underset{\substack{\text{2-Butene}\\(81\%)}}{CH_3CH{=}CHCH_3} + \underset{\substack{\text{1-Butene}\\(19\%)}}{CH_3CH_2CH{=}CH_2}$$

$$\underset{\text{2-Bromo-2-methylbutane}}{CH_3CH_2C(Br)(CH_3)CH_3} \xrightarrow[CH_3CH_2OH]{CH_3CH_2O^-Na^+} \underset{\substack{\text{2-Methyl-2-butene}\\(70\%)}}{CH_3CH{=}C(CH_3)_2} + \underset{\substack{\text{2-Methyl-1-butene}\\(30\%)}}{CH_3CH_2C(CH_3){=}CH_2}$$

The elimination of HX from alkyl halides is a general reaction that constitutes an excellent method for preparing alkenes. The subject is complex, however, and elimination reactions can take place through a variety of different mechanistic pathways. We'll consider two of them—the E1 and E2 reactions.

PROBLEM

11.14 What products would you expect from these elimination reactions? Which product will be major in each case?

(a) 1-Bromo-1-methylcyclohexane + $Na^+ \ {}^-OCH_2CH_3 \longrightarrow$ ?

(b) $CH_3CH_2C(Br)(CH_3)CH(CH_3)_2 + Na^+ \ {}^-OCH_2CH_3 \longrightarrow$ ?

## 11.11 The E2 Reaction

The **E2 reaction** (for elimination, bimolecular) is the most widely studied and commonly occurring pathway for elimination. It is closely analogous to the $S_N2$ reaction and may be formulated as shown in Figure 11.18.

The E2 reaction is a one-step process without intermediates. As the attacking base begins to abstract a proton from a carbon next to the leaving group, the C—H bond begins to break, a new carbon–carbon double bond begins to form, and the leaving group begins to depart, taking with it the electron pair from the C—X bond.

One of the most important pieces of evidence supporting this mechanism is the reaction kinetics. Since both base and alkyl halide enter into the rate-limiting step, E2 reactions show second-order kinetics. In other words, E2 reactions follow the rate law

$$\text{Rate} = k \times [\text{RX}] \times [\text{Base}]$$

A second and more compelling piece of evidence involves the stereochemistry of E2 eliminations. As indicated by a large amount of experimental data, E2 reactions are stereospecific. Elimination always occurs from a **periplanar** geometry, meaning that all four reacting atoms—the hydrogen, the two carbons, and the leaving group—lie in the same plane. There are two such geometries possible: *syn periplanar* geometry, in which the H and the X are on the *same* side of the molecule; and *anti periplanar* geometry,

Base (B:) attacks a neighboring C—H bond and begins to remove the H at the same time as the alkene double bond starts to form and the X group starts to leave.

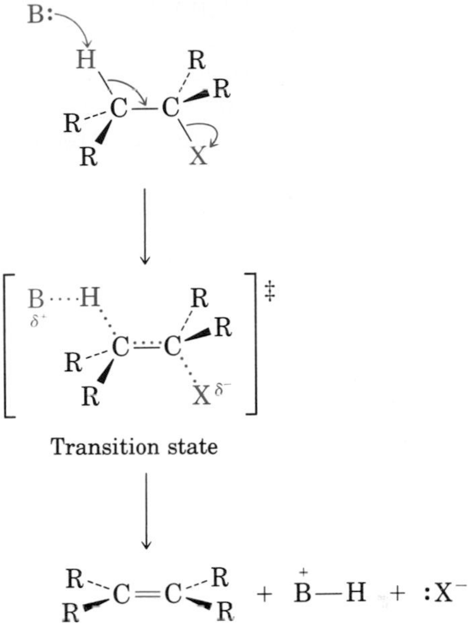

Transition state

Neutral alkene is produced when the C—H bond is fully broken and the X group has departed with the C—X bond electron pair.

**Figure 11.18** Mechanism of the E2 reaction of alkyl halides: The reaction takes place in a single step through a transition state in which the double bond begins to form at the same time the —H and —X groups are leaving.

in which the H and the X are on *opposite* sides of the molecule. Of the two choices, anti periplanar geometry is much preferred because it allows the two carbon centers to adopt a staggered relationship, whereas syn geometry requires the substituents on carbon to be eclipsed.

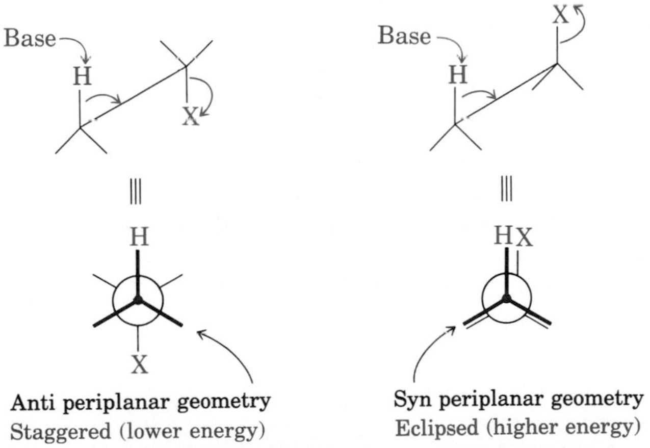

Anti periplanar geometry
Staggered (lower energy)

Syn periplanar geometry
Eclipsed (higher energy)

What's so special about periplanar geometry? A moment's reflection shows the answer. Since the original C—H and C—X $sp^3$ sigma orbitals in the starting material must overlap and become *p* pi orbitals in the alkene product, there must also be partial overlap in the transition state. This can occur only if all the orbitals are in the same plane to begin with—that is, if they're periplanar (Figure 11.19, page 360).

It might help to think of E2 elimination reactions with periplanar geometry as being similar to $S_N2$ reactions with 180° geometry. In an $S_N2$ reaction, an electron pair from the incoming nucleophile pushes out the leaving group on the opposite side of the molecule (back-side attack). In an E2 reaction, an electron pair from a neighboring C—H bond pushes out the leaving group on the opposite side of the molecule (anti periplanar).

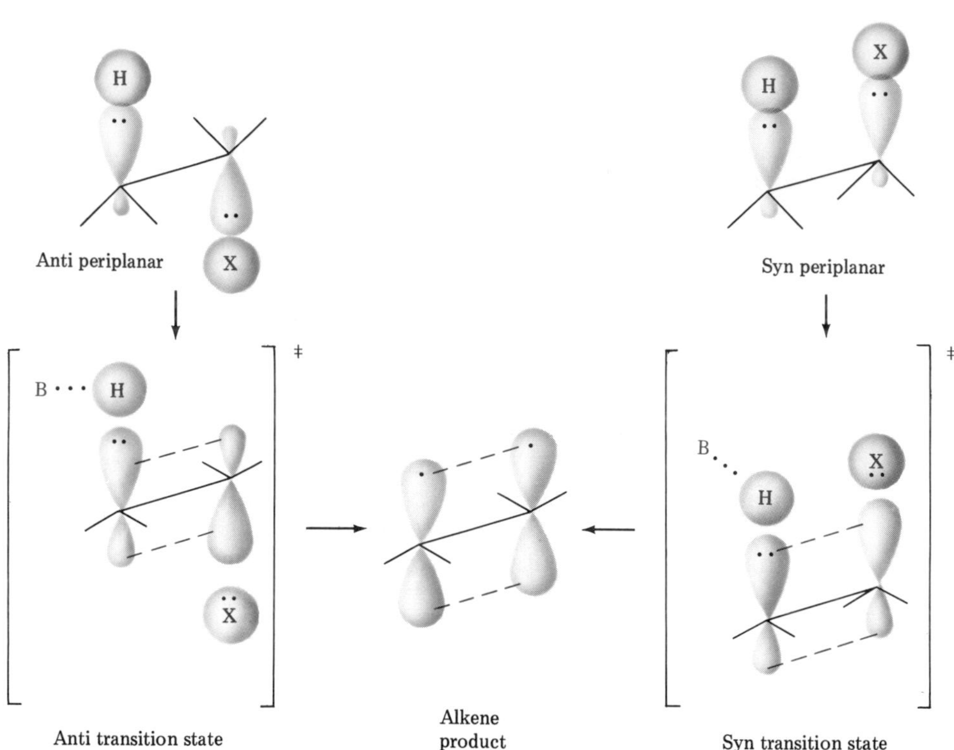


**Figure 11.19** The transition state for the E2 reaction of alkyl halides with base: Partial overlap of the developing *p* orbitals in the transition state requires periplanar geometry of the starting material.

Anti periplanar geometry for E2 eliminations has definite stereochemical consequences that provide strong evidence for the proposed mechanism. To take just one example, *meso*-1,2-dibromo-1,2-diphenylethane undergoes E2 elimination on treatment with base to give *only* the pure *E* alkene (phenyl groups cis). None of the isomeric *Z* alkene is formed because the transition state leading to *Z* alkene would have to have syn periplanar geometry.

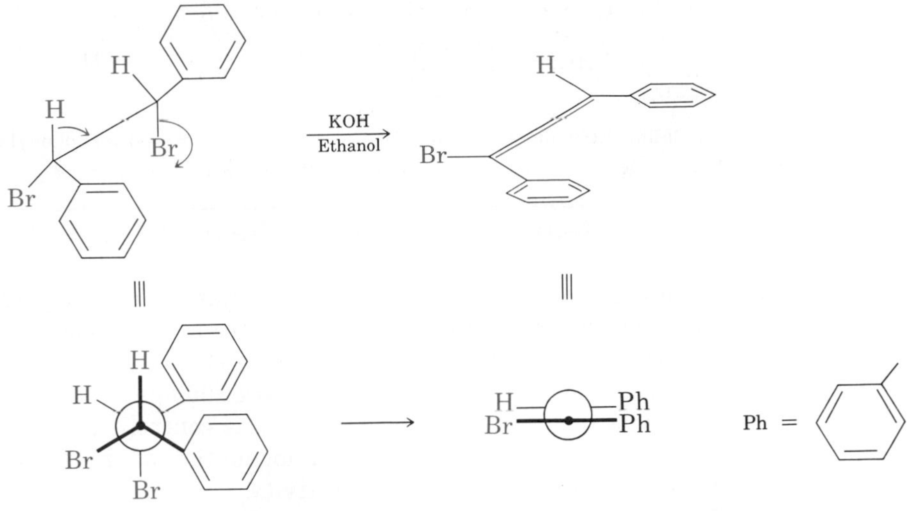

KOH / Ethanol

*meso*-1,2-Dibromo-1,2-diphenylethane (*E*)-1-Bromo-1,2-diphenylethylene

Ph =

**PRACTICE PROBLEM**

What stereochemistry would you expect for the alkene obtained by E2 elimination of (1*S*,2*S*)-1,2-dibromo-1,2-diphenylethane?

*Solution* First draw (1*S*,2*S*)-1,2-dibromo-1,2-diphenylethane so that you can see its stereochemistry and so that the —H and —Br groups to be eliminated are anti periplanar (molecular models are extremely helpful here). Keeping all substituents in approximately their same positions, eliminate HBr and see what alkene results. The product is (*Z*)-1-bromo-1,2-diphenylethylene.

(1*S*,2*S*)-1,2-Dibromo-1,2-diphenylethane (*Z*)-1-Bromo-1,2-diphenylethylene

**PROBLEM**

**11.15** What stereochemistry would you expect for the alkene obtained by E2 elimination of (1*R*,2*R*)-1,2-dibromo-1,2-diphenylethane? Draw a Newman projection of the reacting conformation.

**PROBLEM**

**11.16** 1-Chloro-1,2-diphenylethane can undergo E2 elimination to give either *cis*- or *trans*-1,2-diphenylethylene (stilbene). Draw Newman projections of the reactive conformations leading to both *cis*- and *trans*-1,2-diphenylethylene. Examine these two conformations and suggest a reason why the trans alkene is the major product, as shown on the next page.

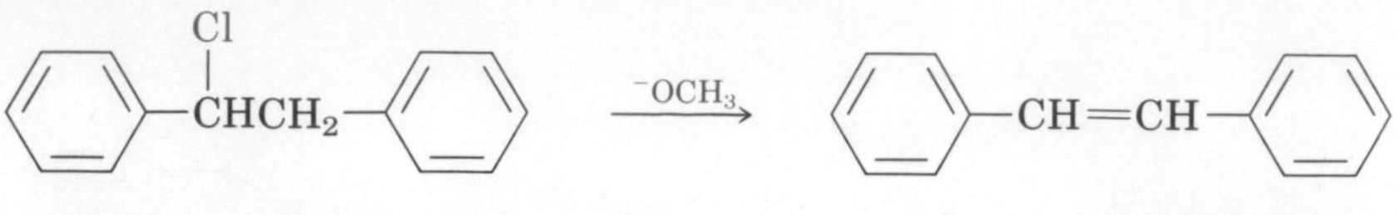

1-Chloro-1,2-diphenylethane $\xrightarrow{^{-}OCH_3}$ *trans*-1,2-Diphenylethylene

## 11.12 Elimination Reactions and Cyclohexane Conformation

Anti periplanar geometry for E2 reactions is particularly important in cyclohexane rings where rigid chair-like geometry forces a specific relationship between the substituents on neighboring carbon atoms (Section 4.10). As pointed out by Derek Barton[6] in a landmark 1950 paper, much of the chemical reactivity of substituted cyclohexanes is controlled by conformational effects. Let's look at the E2 dehydrohalogenation of chlorocyclohexanes to see the effect of conformation on reactivity.

The anti periplanar requirement for E2 reactions can be met in cyclohexanes only if the hydrogen and the halogen are trans diaxial (Figure 11.20). If either the halogen or hydrogen is equatorial, E2 elimination can't occur.

Axial chlorine: H and Cl are anti periplanar

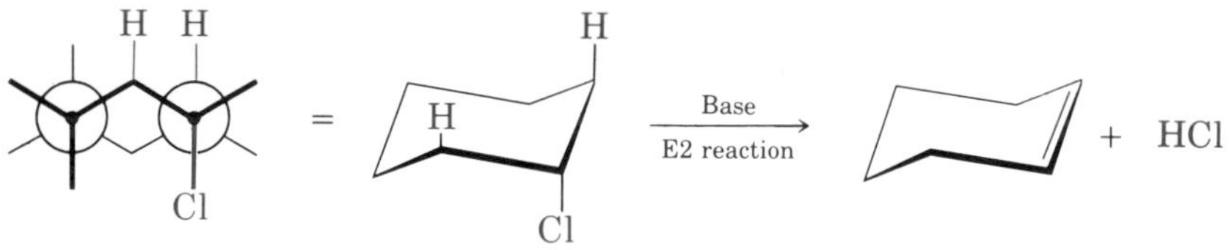

Base, E2 reaction → + HCl

Equatorial chlorine: H and Cl not anti periplanar

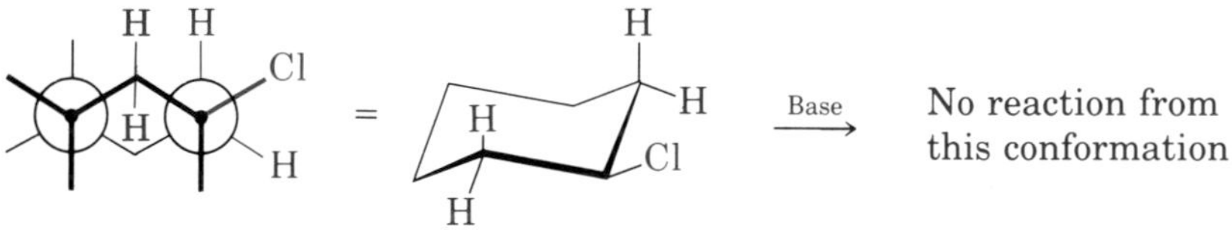

Base → No reaction from this conformation

**Figure 11.20** Geometric requirements for E2 reaction in cyclohexanes: The leaving group and the hydrogen must both be axial for anti periplanar elimination to occur.

The elimination of HCl from the menthyl chlorides (Figure 11.21) provides a good illustration of this trans-diaxial requirement. Neomenthyl chloride reacts with ethoxide ion 200 times as fast as menthyl chloride. Furthermore, neomenthyl chloride yields 3-menthene as the sole alkene product, whereas menthyl chloride yields 2-menthene.

[6]Derek H. R. Barton (1918– ); b. Gravesend, England; Ph.D. and D.Sc. London (Heilbron, E. R. H. Jones); professor, Birkbeck College, Harvard, Glasgow, Imperial College, London, Institut de Chimie des Substances Naturelles, Gif-sur-Yvette, France, Texas A and M; Nobel prize (1969).

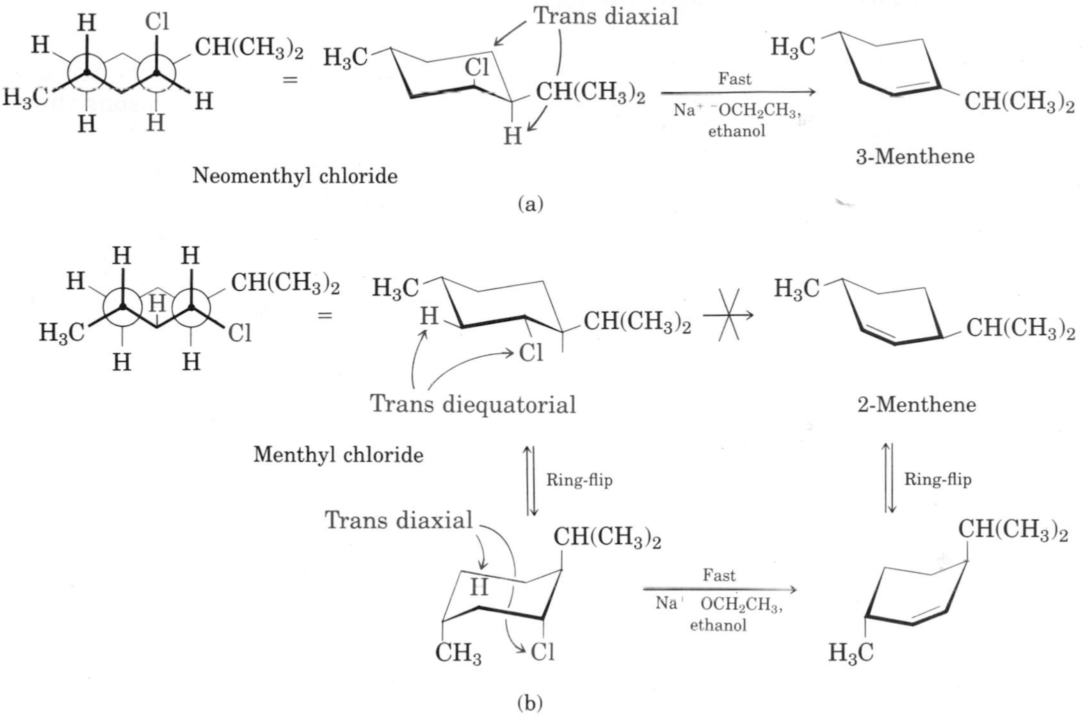

**Figure 11.21** Dehydrochlorination of menthyl and neomenthyl chlorides: Neomenthyl chloride can lose HCl from its more stable conformation, but menthyl chloride must first ring-flip before HCl loss can occur.

We can understand this difference in reactivity by looking at the most favorable chair conformations of the reactant molecules. Neomenthyl chloride should have the conformation shown in Figure 11.21(a), since the two large substituents, methyl and isopropyl, are equatorial. In this conformation, chlorine is rigidly held in an axial orientation—the perfect geometry for E2 elimination with loss of the hydrogen atom at C4 to yield the more substituted alkene product 3-menthene.

Menthyl chloride, by contrast, has the geometry shown in Figure 11.21(b), in which all three substituents are equatorial. Since the chlorine is equatorial, it can't undergo E2 elimination. In order to achieve the necessary geometry for elimination, menthyl chloride must first ring-flip to a highly energetic chair conformation, in which all three substituents are axial. Then E2 elimination occurs with loss of the only possible trans-diaxial hydrogen, leading to 2-menthene. The net effect of the simple change in chlorine stereochemistry is a 200-fold decrease in reaction rate and a complete change of product. The chemistry of the molecule is controlled by conformational effects.

PROBLEM

**11.17** Which isomer would you expect to react faster under E2 elimination conditions, *trans*-1-bromo-4-*tert*-butylcyclohexane or *cis*-1-bromo-4-*tert*-butylcyclohexane? Draw each molecule in its more stable chair conformation and explain your answer.

## 11.13 The Deuterium Isotope Effect

One final piece of evidence in support of the E2 mechanism is provided by a phenomenon known as the **deuterium isotope effect**. For reasons that we won't go into, a carbon–hydrogen bond is weaker by a small amount (about 1.2 kcal/mol) than a corresponding carbon–deuterium bond. Thus, a C—H bond is more easily broken than an equivalent C—D bond, and the rate of C—H bond cleavage is therefore faster. As an example of how this effect can be used to obtain mechanistic information, the base-induced elimination of HBr from 1-bromo-2-phenylethane proceeds 7.11 times as fast as the corresponding elimination of DBr from 1-bromo-2,2-dideuterio-2-phenylethane:

Faster reaction $C_6H_5\text{-}CH_2\text{-}CH_2Br \xrightarrow{\text{Base}} C_6H_5\text{-}CH{=}CH_2$

1-Bromo-2-phenylethane

Slower reaction $C_6H_5\text{-}CD_2\text{-}CH_2Br \xrightarrow{\text{Base}} C_6H_5\text{-}CD{=}CH_2$

1-Bromo-2,2-dideuterio-2-phenylethane

This result, which is fully consistent with our picture of the E2 reaction as a single-step process, tells us that the C—H (or C—D) bond is broken *in the rate-limiting step*. If it were otherwise, we couldn't measure a rate difference.

## 11.14 The E1 Reaction

Just as the $S_N2$ reaction has a close analog in the E2 reaction, the $S_N1$ reaction also has a close analog—the **E1 reaction** (for elimination, unimolecular). The E1 reaction can be formulated as shown in Figure 11.22 for 2-chloro-2-methylpropane.

E1 eliminations begin with the same spontaneous dissociation of a halide that we saw in the $S_N1$ reaction, but the dissociation is followed this time by loss of a proton from the intermediate carbocation. In fact, the E1 mechanism normally occurs in competition with $S_N1$ reaction when an alkyl halide is treated in a protic solvent with a nonbasic nucleophile. Thus, the best substrates are those that are also subject to $S_N1$ reaction, and mixtures

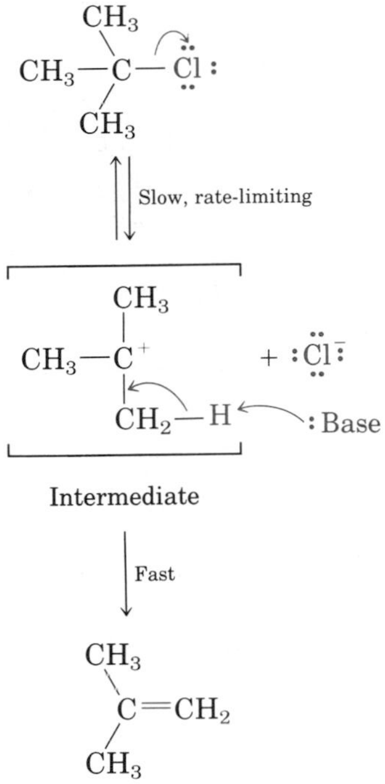

Spontaneous dissociation of the tertiary chloride yields an intermediate carbocation in a slow, rate-limiting step.

Abstraction of a neighboring $H^+$ by base in a fast step yields neutral alkene product. The C—H bond electron pair goes to form the alkene pi bond.

$$(CH_3)_3C{-}\ddot{Cl}: \underset{\text{Slow, rate-limiting}}{\rightleftharpoons} \left[ CH_3{-}\overset{CH_3}{\underset{CH_2-H}{C^+}} \right] + :\ddot{Cl}:^- \quad (\text{:Base})$$

Intermediate

$$\xrightarrow{\text{Fast}} (CH_3)_2C{=}CH_2$$

**Figure 11.22** Mechanism of the E1 reaction: Two steps are involved, the first of which is rate-limiting.

of products are almost always obtained. For example, when 2-chloro-2-methylpropane is warmed to 65°C in 80% aqueous ethanol, a mixture of 2-methyl-2-propanol ($S_N1$) and 2-methylpropene (E1) is obtained:

$$(CH_3)_3CCl \xrightarrow[65°C]{H_2O,\ CH_3CH_2OH} (CH_3)_3COH + (CH_3)_2C{=}CH_2 + HCl$$

2-Chloro-2-methylpropane — 2-Methyl-2-propanol (64%) — 2-Methylpropene (36%)

Much evidence has been obtained in support of the E1 mechanism. As expected, E1 reactions show first-order kinetics consistent with a spontaneous dissociation process:

$$\text{Rate} = k \times [\text{RX}]$$

A second piece of evidence involves the stereochemistry of elimination. Unlike the E2 reaction, where periplanar geometry is required, there is no geometric requirement on the E1 reaction; the intermediate carbocation can lose any available proton from a neighboring position. We might therefore expect to obtain the more stable (Zaitsev's rule) product from E1 reaction, which is just what we find. To return to a familiar example, menthyl chloride loses HCl under E1 conditions to give a mixture of alkenes in which the Zaitsev product, 3-menthene, predominates (Figure 11.23, page 366).

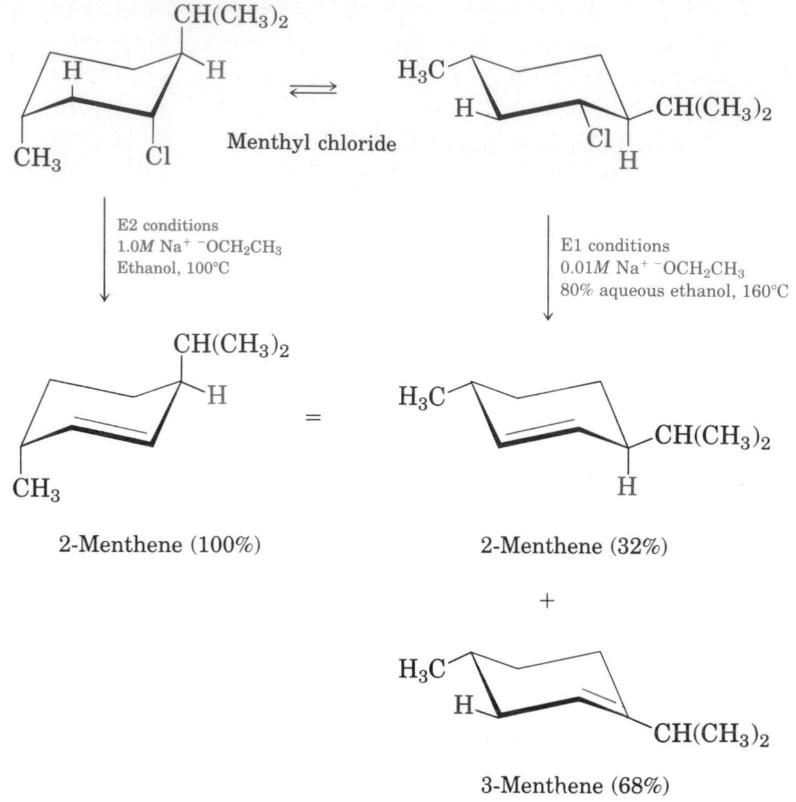

**Figure 11.23** Elimination reactions of menthyl chloride: E2 conditions (strong base) lead to 2-menthene, whereas E1 conditions (very dilute base) lead to a mixture of 2-menthene and 3-menthene.

One final piece of evidence is that *no* deuterium isotope effect is found for E1 reactions because rupture of the C—H (or C—D) bond occurs *after* the rate-limiting step, rather than during it. Thus, we can't measure a rate difference.

## 11.15 Summary of Reactivity: $S_N1$, $S_N2$, E1, E2

$S_N1$, $S_N2$, E1, E2—How can you keep it all straight? How can you predict what will happen in any given case? Will substitution or elimination occur? Will the reaction be bimolecular or unimolecular? There aren't any definitive answers to these questions, but it *is* possible to recognize some broad trends and make some generalizations about what to expect.

1. **Primary alkyl halides** react only by $S_N2$ or E2 mechanisms because they are relatively unhindered and their dissociation would give relatively unstable primary carbocations. If a good nucleophile

such as $R\ddot{S}:^-$, $:\ddot{I}:^-$, $^-:CN$, or $:\ddot{Br}:^-$ is used, nucleophilic substitution occurs to the virtual exclusion of elimination. Even strong bases such as hydroxide ion and ethoxide ion give almost entirely substitution product. Only when strong, bulky bases such as *tert*-butoxide are used do E2 eliminations with primary halides occur. In such cases, nucleophilic substitution is prevented by the steric bulk of the base, but elimination can still occur:

$$CH_3CH_2CH_2CH_2Br \xrightarrow[\text{Ethanol}]{CH_3CH_2\ddot{O}:^-} CH_3CH_2CH{=}CH_2 + CH_3CH_2CH_2CH_2OCH_2CH_3$$

1-Bromobutane → 1-Butene (10%) + Butyl ethyl ether (90%)

$$CH_3CH_2CH_2CH_2Br \xrightarrow{(CH_3)_3C\ddot{O}:^-} CH_3CH_2CH{=}CH_2 + CH_3CH_2CH_2CH_2OC(CH_3)_3$$

1-Bromobutane → 1-Butene (85%) + Butyl *tert*-butyl ether (15%)

2. **Secondary alkyl halides** can react by any of the four mechanisms, and we can often make one or the other pathway predominate by choosing appropriate experimental conditions. When a secondary halide is treated with a strong base such as ethoxide ion, hydroxide ion, or amide ion, E2 elimination normally occurs. But when the same secondary halide is treated in a polar aprotic solvent such as hexamethylphosphoramide with a good nucleophile or weak base, $S_N2$ substitution usually occurs. For example, 2-bromopropane undergoes different reactions when treated with ethoxide ion (strong base—E2) and with acetate ion (weak base—$S_N2$):

$$(CH_3)_2CHBr \xrightarrow[\text{(weak base)}]{CH_3CO_2:^-} (CH_3)_2CHOCOCH_3 + CH_3CH{=}CH_2$$

2-Bromopropane → Isopropyl acetate (100%) + Propene (0%)

$$(CH_3)_2CHBr \xrightarrow[\text{(strong base)}]{CH_3CH_2\ddot{O}:^-} (CH_3)_2CHOCH_2CH_3 + CH_3CH{=}CH_2$$

2-Bromopropane → Ethyl isopropyl ether (20%) + Propene (80%)

Secondary alkyl halides can also be made to undergo $S_N1$ and E1 reactions if weakly basic nucleophiles are used in protic solvents such as ethanol or acetic acid. Mixtures of products are usually obtained, though, and the reactions are of little use for synthesis.

3. **Tertiary halides** can react through three possible pathways—$S_N1$, E1, and E2—and one of the three can often be made to predominate by proper choice of reaction conditions. Steric hindrance at the tertiary carbon precludes $S_N2$ behavior. When a tertiary halide is treated with a strong base, E2 reaction predominates. For example, 2-bromo-2-methylpropane gives 97% elimination product when treated with ethoxide ion in ethanol. By contrast, reaction under $S_N1$ conditions (heating in pure ethanol) leads to a mixture of products in which substitution predominates:

$$(CH_3)_3CBr \xrightarrow[CH_3CH_2OH]{CH_3CH_2\ddot{O}:^-} (CH_3)_3COCH_2CH_3 + CH_3C(=CH_2)CH_3$$

2-Bromo-2-methylpropane → Ethyl *tert*-butyl ether (3%) + 2-Methylpropene (97%)

$$(CH_3)_3CBr \xrightarrow[\Delta]{CH_3CH_2OH} (CH_3)_3COCH_2CH_3 + CH_3C(=CH_2)CH_3$$

2-Bromo-2-methylpropane → Ethyl *tert*-butyl ether (80%) + 2-Methylpropene (20%)

Table 11.11 summarizes these generalizations.

**Table 11.11** Reactivity of alkyl halides toward substitution and elimination

| *Halide type* | *$S_N1$* | *$S_N2$* | *E1* | *E2* |
|---|---|---|---|---|
| Primary halide | Does not occur | Highly favored | Does not occur | Occurs when strong, hindered bases are used |
| Secondary halide | Can occur under solvolysis conditions in polar solvents | Favored by good nucleophiles in polar aprotic solvents | Can occur under solvolysis conditions in polar solvents | Favored when strong bases are used |
| Tertiary halide | Favored by nonbasic nucleophiles in polar solvents | Does not occur | Occurs under solvolysis conditions | Highly favored when bases are used |

**PROBLEM**

**11.18** Identify these reactions as to type:

(a) 1-Bromobutane + $NaN_3$ $\longrightarrow$ 1-Azidobutane

(b) Bromocyclohexane + NaOH $\longrightarrow$ Cyclohexene

(c) 2-Bromobutane + NaCN $\xrightarrow{\text{In HMPA}}$ 2-Methylbutanenitrile

## 11.16 Substitution Reactions in Synthesis

The reason we've discussed nucleophilic substitution reactions in such detail is because they are so important in organic chemistry. In fact, we've already seen a number of substitution reactions used in organic synthesis, although they weren't identified as such at the time.

One example is the alkylation of acetylide anions discussed in Section 8.8. We said that acetylide anions react well with primary alkyl bromides, iodides, and tosylates, to provide the internal alkyne product:

$$R{-}C{\equiv}C{:}^{-}\,Na^{+} + R'{-}CH_2{-}X \longrightarrow R{-}C{\equiv}C{-}CH_2{-}R' + NaX$$

where X = Br, I, or OTos.

Acetylide ion alkylation is, of course, an $S_N2$ reaction, and it is therefore understandable that only primary alkyl halides and tosylates react well. Since acetylide anion is a strong base as well as a good nucleophile, E2 elimination competes with $S_N2$ alkylation when a secondary or tertiary substrate is used. For example, reaction of sodio 1-hexyne with 2-bromopropane gives primarily the elimination product propene:

$$\underset{\text{Sodio 1-hexyne}}{CH_3(CH_2)_3C{\equiv}C{:}^{-}\,Na^{+}} + CH_3\overset{\displaystyle Br}{\overset{|}{C}}HCH_3 \longrightarrow \begin{cases} CH_3(CH_2)_3C{\equiv}C{-}CH(CH_3)_2 \\ \quad 7\%\ S_N2 \\ \quad + \\ CH_3(CH_2)_3C{\equiv}CH + CH_3CH{=}CH_2 \\ \quad 93\%\ E2 \end{cases}$$

Other substitution reactions we've seen include some of the various methods used for preparing alkyl halides from alcohols. We saw in Section 10.8, for example, that alkyl halides can be prepared by treating alcohols with HX—reactions now recognizable as nucleophilic substitutions of halide ions on the protonated alcohols. Tertiary alcohols react by an $S_N1$ pathway, whereas primary alcohols react by an $S_N2$ pathway (Figure 11.24).

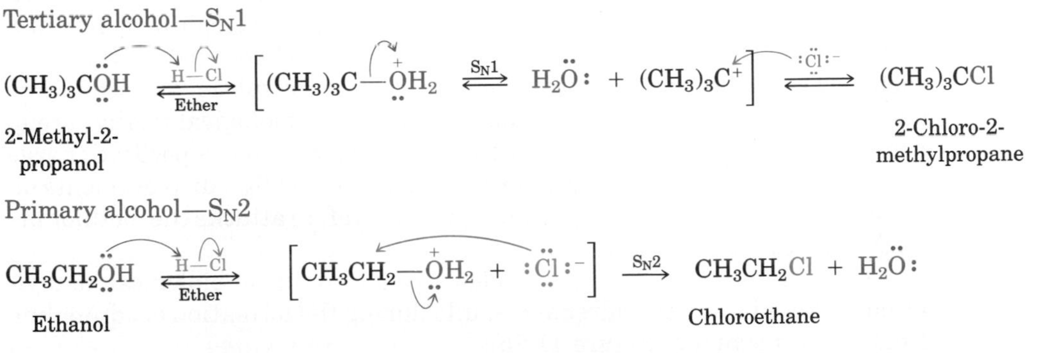

Tertiary alcohol—$S_N1$

$$\underset{\text{2-Methyl-2-propanol}}{(CH_3)_3C\ddot{O}H} \underset{\text{Ether}}{\overset{H-Cl}{\rightleftharpoons}} \left[(CH_3)_3C{-}\overset{+}{\ddot{O}}H_2 \overset{S_N1}{\rightleftharpoons} H_2\ddot{O}{:} + (CH_3)_3C^{+}\right] \overset{:\ddot{Cl}:^{-}}{\rightleftharpoons} \underset{\text{2-Chloro-2-methylpropane}}{(CH_3)_3CCl}$$

Primary alcohol—$S_N2$

$$\underset{\text{Ethanol}}{CH_3CH_2\ddot{O}H} \underset{\text{Ether}}{\overset{H-Cl}{\rightleftharpoons}} \left[CH_3CH_2{-}\overset{+}{\ddot{O}}H_2 + :\ddot{Cl}:^{-}\right] \overset{S_N2}{\longrightarrow} \underset{\text{Chloroethane}}{CH_3CH_2Cl} + H_2\ddot{O}{:}$$

**Figure 11.24** Mechanisms of reactions of HCl with tertiary and primary alcohols

Yet another substitution reaction we've seen is the conversion of primary and secondary alcohols into alkyl bromides by treatment with $PBr_3$

(Section 10.8). Although —OH is a poor leaving group and can't be displaced directly by nucleophiles, reaction with $PBr_3$ *activates* an alcohol toward displacement by transforming the hydroxyl into a better leaving group:

$$\underbrace{}_{\text{Poor leaving group}} \quad CH_3(CH_2)_4CH_2\text{—O—H} \xrightarrow[\text{Ether}]{PBr_3} \left[ CH_3(CH_2)_4CH_2\text{—}\overset{\overset{\text{H}}{|}}{\overset{+}{O}}\text{—}PBr_2 \right] \xrightarrow[S_N2]{:\ddot{Br}:^-} CH_3(CH_2)_4CH_2Br$$

Poor leaving group; Good leaving group

1-Hexanol → 1-Bromohexane

Alcohols react with $PBr_3$ to give dibromophosphites (R—O—$PBr_2$), which are highly reactive substrates in $S_N2$ reactions. Displacement by bromide ion then occurs rapidly on the primary carbon, and alkyl bromides are produced in good yield.

## 11.17 Substitution Reactions in Biological Systems

Many biological processes occur by reaction pathways analogous to those carried out in the laboratory. Thus, a number of reactions that occur in living organisms take place by nucleophilic substitution mechanisms.

### BIOLOGICAL METHYLATIONS

Perhaps the most common of all biological substitutions is the **methylation** reaction—the transfer of a methyl group from an electrophilic donor to a nucleophile. Although a laboratory chemist might choose iodomethane for such a reaction, living organisms operate in a more subtle way. The large and complex molecule *S*-adenosylmethionine is the biological methyl-group donor. Since the sulfur atom in *S*-adenosylmethionine has a positive charge (a *sulfonium* ion), it is an excellent leaving group for $S_N2$ displacements on the methyl carbon. A biological nucleophile therefore attacks the methionine methyl by an $S_N2$ reaction.

An example of the action of *S*-adenosylmethionine in biological methylations takes place in the adrenal medulla during the formation of adrenaline from norepinephrine (Figure 11.25).

After becoming used to dealing with simple halides such as bromomethane used for laboratory alkylations, it is something of a shock to encounter a molecule as complex as *S*-adenosylmethionine. From a chemical standpoint, however, $CH_3Br$ and *S*-adenosylmethionine do exactly the same thing—both transfer a methyl group by an $S_N2$ reaction. The same principles of reactivity apply to both.

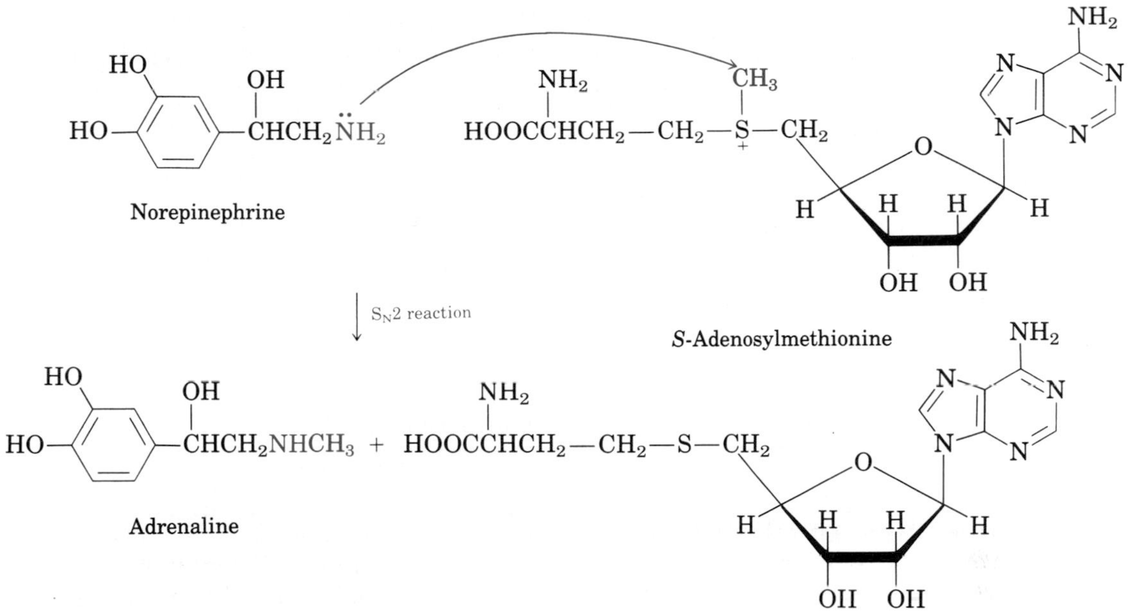

**Figure 11.25** The biological formation of adrenaline by reaction of norepinephrine with *S*-adenosylmethionine

## OTHER BIOLOGICAL ALKYLATIONS

Another example of a biological $S_N2$ reaction is involved in the response of organisms to certain toxic chemicals. Many reactive $S_N2$ substrates with deceptively simple structures are quite toxic to living organisms. For example, bromomethane is widely used as a fumigant to kill termites. The toxicity of these chemicals derives from their ability to transfer alkyl groups to nucleophilic amino groups ($—NH_2$) and mercapto groups ($—SH$) on enzymes. With enzymes modified by alkylation, normal biological chemistry is altered.

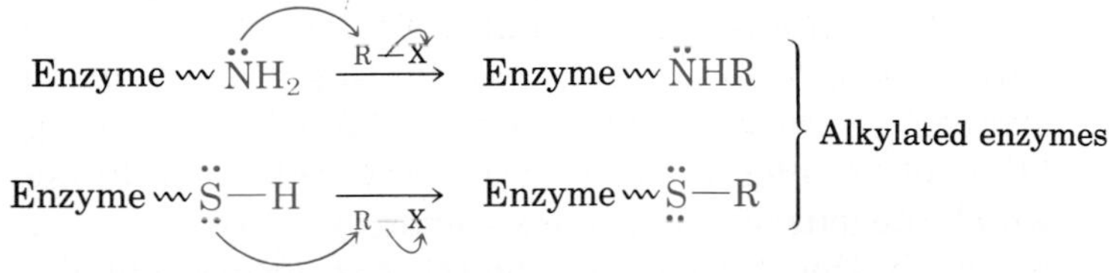

$$\text{Enzyme}\sim\ddot{N}H_2 \xrightarrow{R-X} \text{Enzyme}\sim\ddot{N}HR$$

$$\text{Enzyme}\sim\ddot{S}-H \xrightarrow[R-X]{} \text{Enzyme}\sim\ddot{S}-R$$

Alkylated enzymes

One of the best-known toxic alkylating agents is mustard gas, $ClCH_2CH_2SCH_2CH_2Cl$, which gained notoriety because of its use as a chemical-warfare agent during World War I. It has been estimated that some 400,000 casualties resulted from its use. Mustard gas, as a primary halide, is highly reactive toward $S_N2$ displacements by nucleophilic groups of proteins. It is thought to act through an intermediate sulfonium ion in much the same manner as *S*-adenosylmethionine (Figure 11.26, page 371).

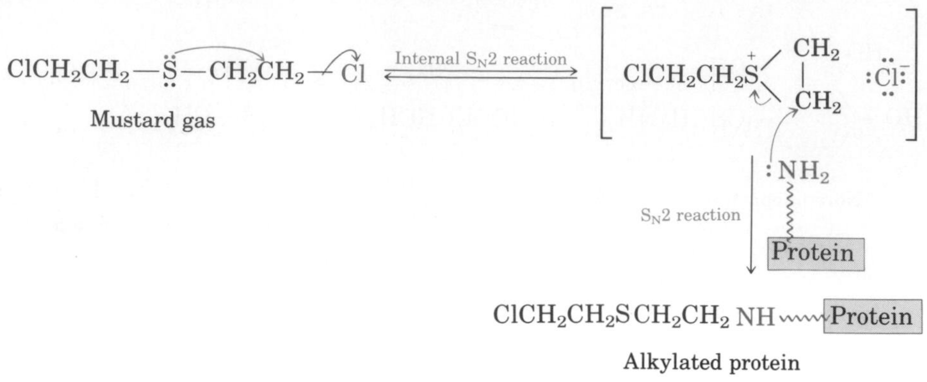

$ClCH_2CH_2—\ddot{S}—CH_2CH_2—Cl$ $\xrightarrow{\text{Internal } S_N2 \text{ reaction}}$ $\left[ ClCH_2CH_2\overset{+}{S}\langle CH_2 | CH_2 \quad :\ddot{Cl}:^- \right]$

Mustard gas

$:NH_2$ — Protein

$S_N2$ reaction

$ClCH_2CH_2SCH_2CH_2NH$—Protein

Alkylated protein

**Figure 11.26** The alkylation of a protein by mustard gas

## 11.18 Summary and Key Words

Reaction of an alkyl halide or tosylate with a nucleophile results either in **substitution** or in **elimination**. Both modes of reaction are extremely important in chemistry, and it's useful to be able to predict what will happen in specific cases.

Nucleophilic substitutions are of two types—**$S_N2$ reaction** and **$S_N1$ reaction**. In the $S_N2$ reaction, the entering nucleophile attacks the halide from a direction 180° away from the leaving group, resulting in an umbrella-like Walden **inversion of configuration** at the carbon atom. The reaction shows **second-order kinetics**, with rate $= k[RX][Nu]$, and is strongly inhibited by increasing steric bulk of the reagents. Thus, $S_N2$ reactions are favored only for primary and secondary substrates.

The $S_N1$ reaction occurs when the substrate spontaneously dissociates to a carbocation in a slow **rate-limiting step**, followed by a rapid attack of nucleophile. In consequence, $S_N1$ reactions show **first-order kinetics**, with rate $= k[RX]$, and take place with racemization of configuration at the carbon atom. They are most favored for tertiary substrates.

Eliminations of alkyl halides to yield alkenes also occur by two different mechanisms—**E2 reaction** and **E1 reaction**. In the E2 reaction, a base abstracts a proton at the same time the leaving group departs. The reaction takes place preferentially through an **anti periplanar** transition state in which the four reacting atoms—hydrogen, two carbons, and leaving group—are all in the same plane. The reaction shows second-order kinetics and a **deuterium isotope effect**, and occurs when a secondary or tertiary substrate is treated with a strong base. These elimination reactions usually give a mixture of alkene products in which the more highly substituted alkene predominates (**Zaitsev's rule**).

The E1 reaction takes place when the substrate spontaneously dissociates to yield a carbocation in the slow rate-limiting step before losing a neighboring proton. The reaction shows first-order kinetics and no deuterium isotope effect, and occurs when a tertiary substrate reacts in polar, nonbasic solution.

All four reactions—$S_N1$, $S_N2$, E1, and E2—are strongly influenced by many factors, summarized in Table 11.12.

**Table 11.12** Effects of reaction variables on substitution and elimination reactions

| *Reaction* | *Solvent* | *Nucleophile/base* | *Leaving group* | *Substrate structure* |
|---|---|---|---|---|
| $S_N1$ | Very strong effect; reaction favored by polar solvents | Weak effect; reaction favored by good nucleophile/ weak base | Strong effect; reaction favored by good leaving group | Strong effect; reaction favored by 3°, allylic, and benzylic substrates |
| $S_N2$ | Strong effect; reaction favored by polar aprotic solvents | Strong effect; reaction favored by good nucleophile/ weak base | Strong effect; reaction favored by good leaving group | Strong effect; reaction favored by methyl and 1° substrates |
| E1 | Very strong effect; reaction favored by polar solvents | Weak effect; reaction favored by weak base | Strong effect; reaction favored by good leaving group | Strong effect; reaction favored by 3°, allylic, and benzylic substrates |
| E2 | Strong effect; reaction favored by polar aprotic solvents | Strong effect; reaction favored by poor nucleophile/ strong base | Strong effect; reaction favored by good leaving group | Strong effect; reaction favored by 3° substrates |

## 11.19 Summary of Reactions

1. Nucleophilic substitutions
   a. $S_N1$ reaction; carbocation intermediate is involved (Sections 11.6–11.9)

$$R_3C{-}X \longrightarrow \left[R_3C^+\right] \xrightarrow{:Nu^-} R_3C{-}Nu + :X^-$$

Best for 3°, allylic, and benzylic halides and tosylates

b. $S_N2$ reaction; back-side attack of nucleophile occurs (Sections 11.4–11.5)

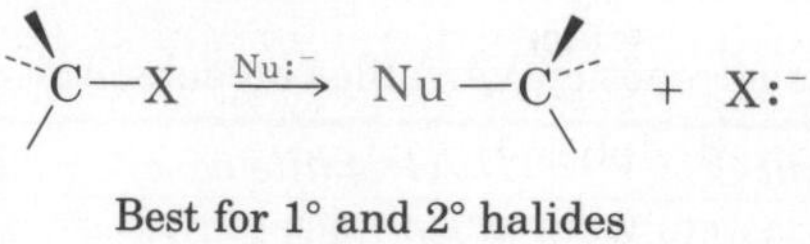

$$\text{C—X} \xrightarrow{\text{Nu:}^-} \text{Nu—C} + \text{X:}^-$$

Best for 1° and 2° halides

$Nu:^- = H:^-, ^-:CN, :\ddot{I}:^-, :\ddot{Br}:^-, :\ddot{Cl}:^-, H\ddot{O}:^-, ^-:\ddot{N}H_2, CH_3\ddot{O}:^-,$
$CH_3CO_2:^-, H\ddot{S}:^-, H_2\ddot{O}:, :NH_3$, and so on.

2. Eliminations

a. E1 reaction; more highly substituted alkene is formed (Section 11.14)

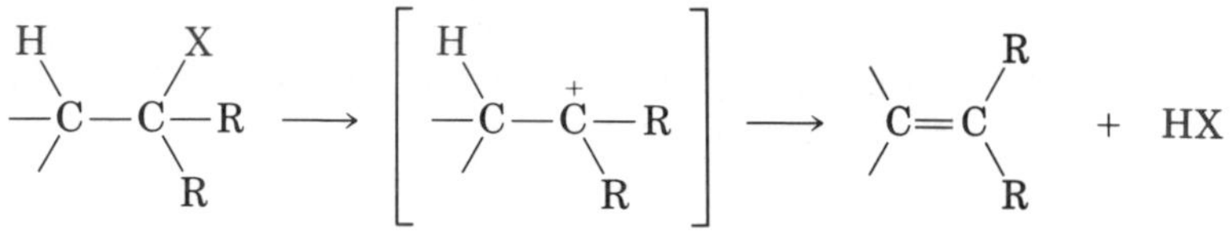

$$\text{H—C—C(X)(R)—R} \longrightarrow \left[\text{H—C—}\overset{+}{\text{C}}\text{(R)—R}\right] \longrightarrow \text{C=C(R)(R)} + \text{HX}$$

Best for 3° halides

b. E2; anti periplanar geometry is required (Section 11.11)

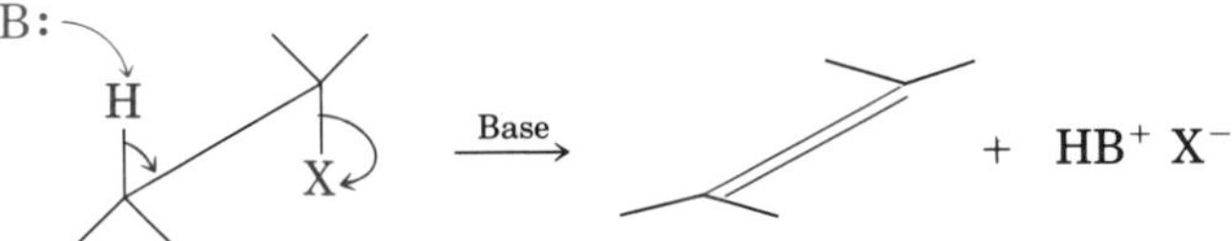


Best for 2° and 3° halides

## ADDITIONAL PROBLEMS

**11.19** Which reagent in each pair will react faster in an $S_N2$ reaction with hydroxide ion?
(a) $CH_3Br$ or $CH_3I$ (b) $CH_3CH_2I$ in ethanol or dimethyl sulfoxide
(c) $(CH_3)_3CCl$ or $CH_3Cl$ (d) $H_2C{=}CHBr$ or $H_2C{=}CHCH_2Br$

**11.20** How might you prepare each of the following molecules using a nucleophilic substitution reaction at some step?

(a) $CH_3C{\equiv}CCH(CH_3)_2$ (b) $CH_3CH_2CH_2CH_2CN$

(c) $H_3C{-}O{-}C(CH_3)_3$ (d) $CH_3CH_2CH_2NH_2$

(e) 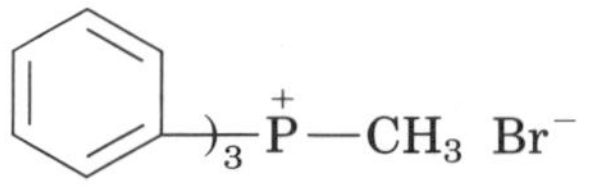 $(C_6H_5)_3\overset{+}{P}{-}CH_3\ Br^-$

A phosphonium salt

(f) 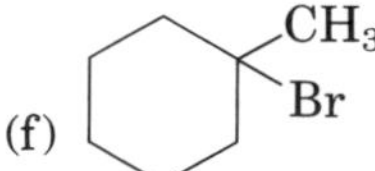

**11.21** Which reaction in each of these pairs would you expect to be faster?
(a) The $S_N2$ displacement by iodide ion on $CH_3Cl$ or on $CH_3OTos$
(b) The $S_N2$ displacement by acetate ion on bromoethane or on bromocyclohexane
(c) The $S_N2$ displacement on 2-bromopropane by ethoxide ion or by cyanide ion
(d) The $S_N2$ displacement by acetylide ion on bromomethane in benzene or in hexamethylphosphoramide

**11.22** What products would you expect from the reaction of 1-bromopropane with:

(a) $NaNH_2$ (b) $K^+ \ ^-O{-}C(CH_3)_3$ (c) NaI
(d) NaCN (e) $NaC{\equiv}CH$ (f) Mg, then $H_2O$

**11.23** Which reagent in each of the following pairs is more nucleophilic? Explain your answer.

(a) $^-{:}\ddot{N}H_2$ or $:NH_3$ (b) $H_2\ddot{O}{:}$ or $CH_3CO\ddot{O}{:}^-$
(c) $BF_3$ or $:\ddot{F}{:}^-$ (d) $(CH_3)_3P{:}$ or $(CH_3)_3N{:}$
(e) $:\ddot{I}{:}^-$ or $:\ddot{Cl}{:}^-$ (f) $^-{:}C{\equiv}N$ or $^-{:}\ddot{O}CH_3$

**11.24** Among the Walden cycles carried out by Phillips and Kenyon is the following series of reactions reported in 1923:

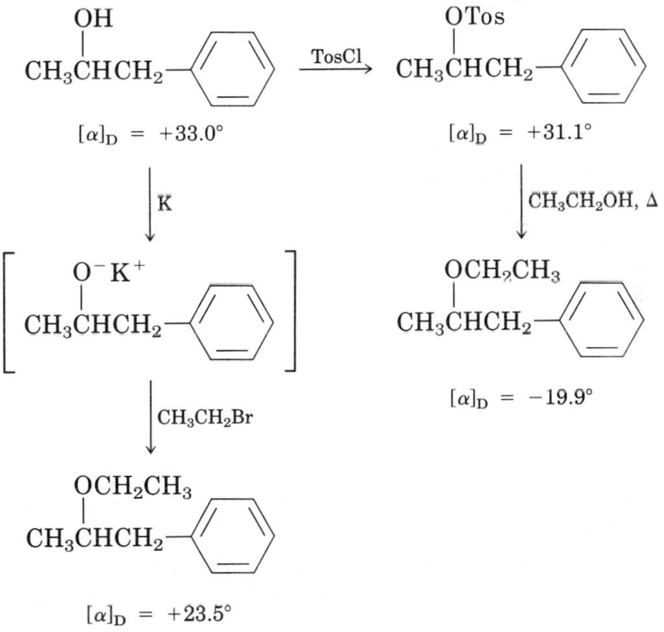

$CH_3CH(OH)CH_2C_6H_5$, $[\alpha]_D = +33.0°$ $\xrightarrow{TosCl}$ $CH_3CH(OTos)CH_2C_6H_5$, $[\alpha]_D = +31.1°$

$CH_3CH(OH)CH_2C_6H_5$ $\xrightarrow{K}$ $[CH_3CH(O^-K^+)CH_2C_6H_5]$ $\xrightarrow{CH_3CH_2Br}$ $CH_3CH(OCH_2CH_3)CH_2C_6H_5$, $[\alpha]_D = +23.5°$

$CH_3CH(OTos)CH_2C_6H_5$ $\xrightarrow{CH_3CH_2OH,\ \Delta}$ $CH_3CH(OCH_2CH_3)CH_2C_6H_5$, $[\alpha]_D = -19.9°$

Explain these results and indicate where Walden inversion is occurring.

**11.25** The synthetic sequences shown here are unlikely to occur as written. What is wrong with each?

(a) $CH_3CHBrCH_2CH_3 \xrightarrow[(CH_3)_3COH]{K^+ \ ^-OC(CH_3)_3} CH_3CH(O\text{-}t\text{-}Bu)CH_2CH_3$

(b) 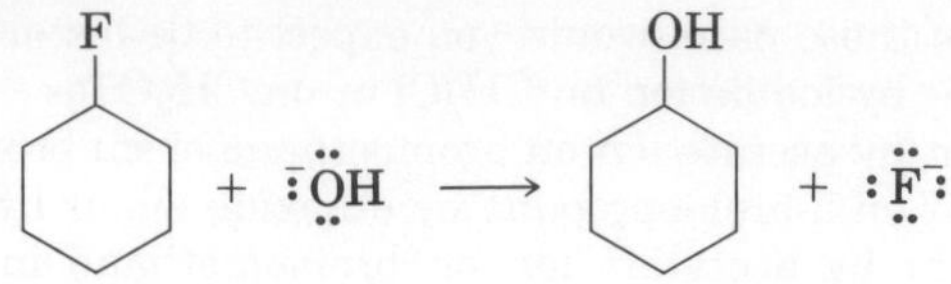 Fluorocyclohexane + $:\ddot{O}H^-$ ⟶ Cyclohexanol + $:\ddot{F}:^-$

(c) 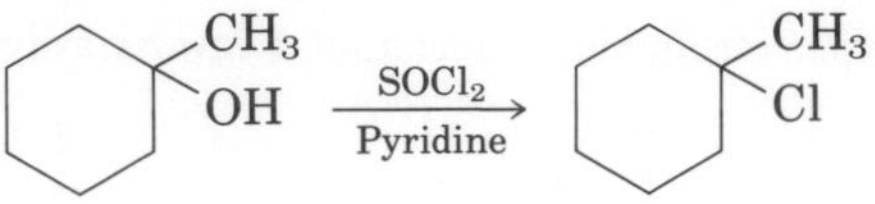 1-Methylcyclohexanol ($CH_3$, OH) $\xrightarrow[\text{Pyridine}]{SOCl_2}$ 1-Chloro-1-methylcyclohexane ($CH_3$, Cl)

**11.26** Order each set of compounds with respect to $S_N1$ reactivity:

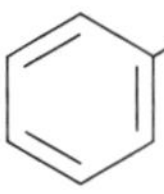

(a) $(CH_3)_3C—Cl$,  $C_6H_5C(CH_3)_2Cl$, $CH_3CH_2CH(NH_2)CH_3$

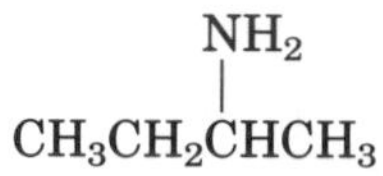

(b) $(CH_3)_3C—F$, $(CH_3)_3C—Br$, $(CH_3)_3C—OH$

(c)  $C_6H_5CH_2Br$, $C_6H_5CH(CH_3)Br$, $(C_6H_5)_3CBr$

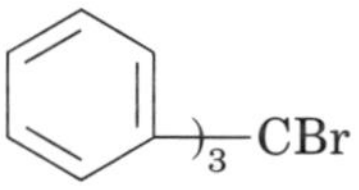

**11.27** Order each set of compounds with respect to $S_N2$ reactivity:
(a) $(CH_3)_3CCl$, $CH_3CH_2CH_2Cl$, $CH_3CH_2CHClCH_3$
(b) $(CH_3)_2CHCHBrCH_3$, $(CH_3)_2CHCH_2Br$, $(CH_3)_3CCH_2Br$
(c) $CH_3CH_2CH_2OCH_3$, $CH_3CH_2CH_2OTos$, $CH_3CH_2CH_2Br$

**11.28** Predict the product and give the stereochemistry resulting from reaction of the following nucleophiles with (*R*)-2-bromooctane:

(a) $^-:CN$ (b) $CH_3CO_2:^-$ (c) $CH_3\ddot{S}:^-$ (d) $:\ddot{Br}:^-$

**11.29** Describe in your own words the effects of the following variables on $S_N2$ and $S_N1$ reactions:
(a) Solvent
(b) Leaving group
(c) Attacking nucleophile
(d) Substrate structure

**11.30** Ethers can often be prepared by $S_N2$ reaction of alkoxide ions with alkyl halides. Suppose you wanted to prepare cyclohexyl methyl ether. Which of the two possible routes shown here would you choose? Explain.

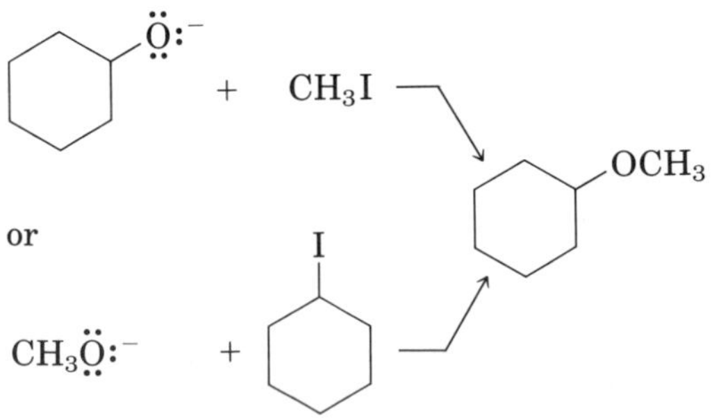

Cyclohexyl–$\ddot{O}:^-$ + $CH_3I$ ⟶ Cyclohexyl–$OCH_3$

or

$CH_3\ddot{O}:^-$ + Iodocyclohexane (I) ⟶ Cyclohexyl–$OCH_3$

**11.31** The $S_N2$ reaction can also occur *intramolecularly* (within the same molecule). What product would you expect from treatment of 4-bromo-1-butanol with base?

$$BrCH_2CH_2CH_2CH_2OH \xrightarrow{Na^+ \ ^-OCH_3} CH_3OH + [BrCH_2CH_2CH_2CH_2O^- \ Na^+] \longrightarrow ?$$

**11.32** In light of your answer to Problem 11.31, can you propose a synthesis of 1,4-dioxane starting only with a dihalide?

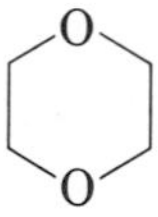

1,4-Dioxane

**11.33** Propose structures for compounds that fit these descriptions:
(a) An alkyl halide that gives a mixture of three alkenes on E2 reaction
(b) An alkyl halide that will not undergo nucleophilic substitution
(c) An alkyl halide that gives the non-Zaitsev product on E2 reaction
(d) An alcohol that reacts rapidly with HCl at 0°C

**11.34** Predict the major alkene product from each of these eliminations.

(a) 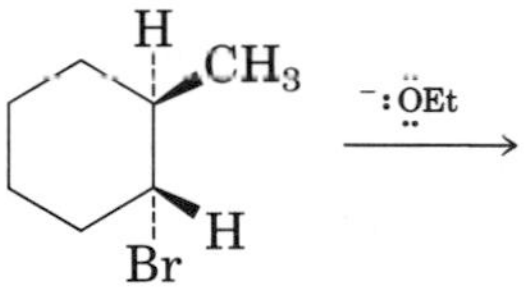 [cyclohexane with H, $CH_3$, Br, H substituents] $\xrightarrow{^-:\ddot{O}Et}$

(b) 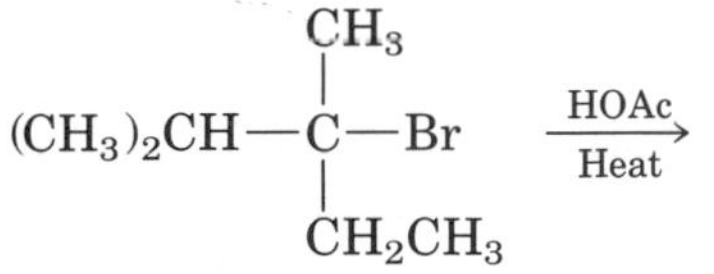 $(CH_3)_2CH—C(CH_3)(CH_2CH_3)—Br \xrightarrow[\text{Heat}]{\text{HOAc}}$

**11.35** The tosylate of (2*R*,3*S*)-3-phenyl-2-butanol undergoes E2 elimination on treatment with ethoxide ion to yield (*Z*)-2-phenyl-2-butene:

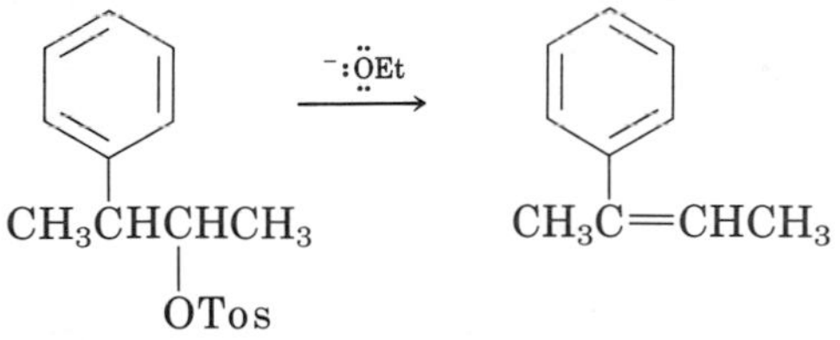

$$CH_3CH(C_6H_5)CH(OTos)CH_3 \xrightarrow{^-:\ddot{O}Et} CH_3C(C_6H_5){=}CHCH_3$$

Formulate the reaction, showing the proper stereochemistry. Explain the observed result by using Newman projections.

**11.36** In light of your answer to Problem 11.35, which alkene, *E* or *Z*, would you expect from the elimination reaction of the (2*R*,3*R*)-3-phenyl-2-butanol tosylate? Which alkene would result from E2 reaction on the (2*S*,3*R*) and (2*S*,3*S*) tosylates? Explain your answer.

**11.37** Alkynes can be made by dehydrohalogenation of vinylic halides in a reaction that is essentially an E2 process. In studying the stereochemistry of this elimination, it was found that (*Z*)-2-chloro-2-butene-1,4-dioic acid reacts 50 times as fast as the corresponding *E* isomer. What conclusion can you draw about the stereochemistry of eliminations in vinylic halides? How does this result compare with eliminations of alkyl halides?

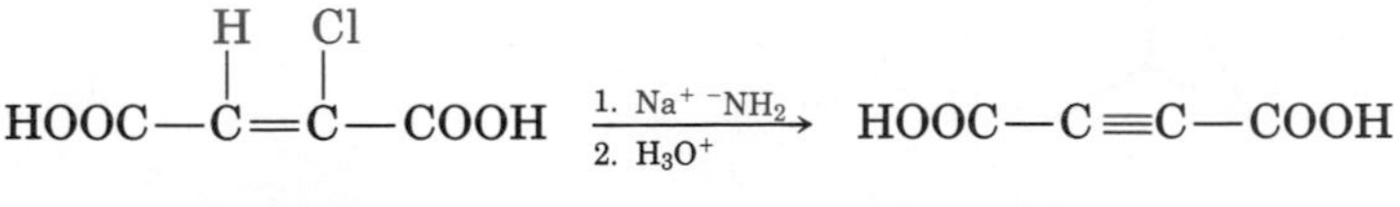

$$HOOC—C(H){=}C(Cl)—COOH \xrightarrow[2.\ H_3O^+]{1.\ Na^+ \ ^-NH_2} HOOC—C{\equiv}C—COOH$$

**11.38** Optically active 2-butanol slowly racemizes on standing in dilute sulfuric acid. Propose a mechanism to account for this observation.

$$CH_3CH_2CH(OH)CH_3$$

2-Butanol

**11.39** Suggest an explanation for the observation that reaction of HBr with (*R*)-3-methyl-3-hexanol leads to (±)-3-bromo-3-methylhexane.

$$CH_3CH_2CH_2C(OH)(CH_3)CH_2CH_3$$

3-Methyl-3-hexanol

**11.40** Explain the fact that treatment of 1-bromo-2-deuterio-2-phenylethane with strong base leads to a mixture of deuterated and nondeuterated phenylethylenes in which the deuterated product predominates by approximately 7:1

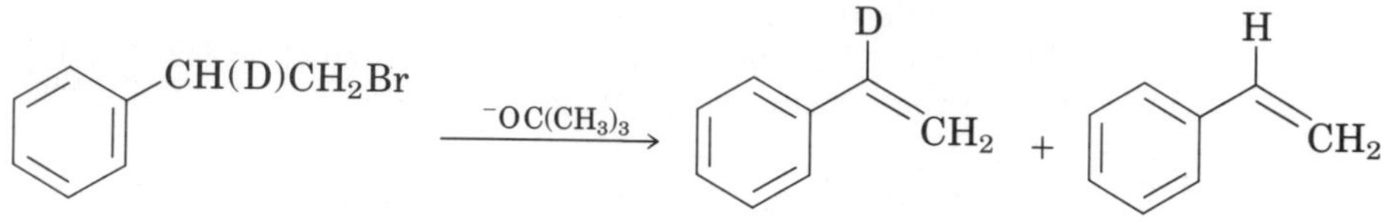

$C_6H_5CH(D)CH_2Br \xrightarrow{^-OC(CH_3)_3} C_6H_5C(D){=}CH_2 + C_6H_5C(H){=}CH_2$

7:1 ratio

**11.41** We've seen that allylic and benzylic halides are more reactive than saturated alkyl halides in the $S_N1$ reaction. It turns out that they're also more reactive in the $S_N2$ reaction. Look at the table of bond strengths (Table 5.4) and explain why allylic and benzylic halides are so readily displaced.

**11.42** Although anti geometry is preferred for E2 reactions, it isn't absolutely necessary. The deuterated bromo compound shown here reacts with strong base to yield an undeuterated alkene. Clearly, a syn elimination has occurred. Make a molecular model of the starting material to examine its geometry, and then explain the result.

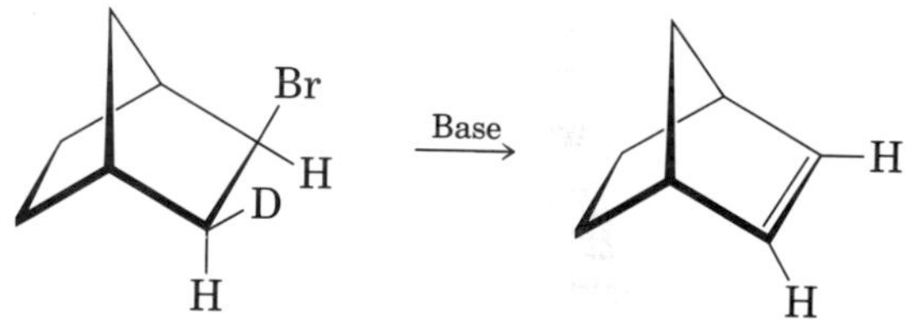

Br, H, D, H $\xrightarrow{\text{Base}}$ H, H

**11.43** In light of your answer to Problem 11.42, account for the observation that one of the following isomers undergoes E2 reaction approximately 100 times as fast as the other. Which isomer is the more reactive, and why?

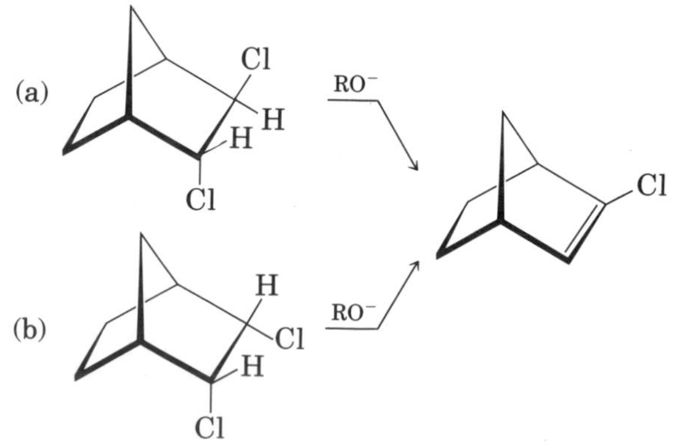

(a) Cl, H, H, Cl $\xrightarrow{RO^-}$ Cl

(b) H, Cl, H, Cl $\xrightarrow{RO^-}$

**11.44** There are eight possible diastereomers of 1,2,3,4,5,6-hexachlorocyclohexane. Draw them in their most stable chair conformations. One isomer loses HCl in an E2 reaction nearly 1000 times more slowly than the others. Which isomer reacts so slowly, and why?

**11.45** Consider the following methyl ester cleavage reaction:

$$\text{C}_6\text{H}_{11}\text{–C(=O)–O–CH}_3 \xrightarrow[\text{DMF}]{\text{LiI}} \text{C}_6\text{H}_{11}\text{–CO}_2^-\ \text{Li}^+ \quad + \quad \text{CH}_3\text{I}$$

The following evidence is available:

(a) The reaction occurs much faster in DMF than in ethanol.

(b) The corresponding ethyl ester cleaves approximately 10 times more slowly than the methyl ester.

Using this evidence, propose a mechanism for the reaction. What other kinds of experimental evidence could you gather to support your mechanistic hypothesis?

**11.46** Account for the fact that the rate of reaction of 1-chlorooctane with acetate ion to give octyl acetate is greatly accelerated by the presence of a small quantity of iodide ion.

**11.47** Compound X is optically inactive and has the formula $C_{16}H_{16}Br_2$. On treatment with strong base, X gives hydrocarbon Y, $C_{16}H_{14}$. Compound Y absorbs 2 equiv of hydrogen when reduced over a palladium catalyst, and reacts with ozone to give two fragments. One fragment, Z, is an aldehyde with formula $C_7H_6O$. The other fragment is glyoxal, CHOCHO. Formulate the reactions involved and suggest structures for X, Y, and Z. What is the stereochemistry of X?

**11.48** Propose a structure for an alkyl halide that gives (*E*)-3-methyl-2-phenyl-2-pentene and none of the Z isomer on E2 elimination. Make sure you indicate the stereochemistry.

CHAPTER 12

# Structure Determination: Mass Spectroscopy and Infrared Spectroscopy

Structure determination is central to organic chemistry. Every time a reaction is run, the products must be purified and identified. The desired products must be separated from solvents, from excess reagents, and from each other if more than one is formed, and their structures must be determined. Doing these things was a time-consuming process in the nineteenth and early twentieth centuries, but extraordinary advances have been made in the last few decades. Sophisticated (and usually expensive) instruments are now available that greatly simplify the problems of purification and structure determination. Use of these instruments doesn't guarantee good results—skill and patience are still required—but it does make things easier.

## 12.1 Purification of Organic Compounds

**Crystallization** is a simple yet effective method for purifying a solid product. The crude reaction product is dissolved in a minimum amount of a suitable hot solvent, and the solution is allowed to cool slowly. Pure crystals slowly form and precipitate, while impurities remain in solution. The crystalline product is then isolated by filtration.

**Distillation** is an equally simple and effective method for purifying a volatile liquid product. The crude liquid reaction product is heated to a boil, and the vapors are condensed into a receiver flask (Figure 12.1). Nonvolatile impurities remain in the sample flask.

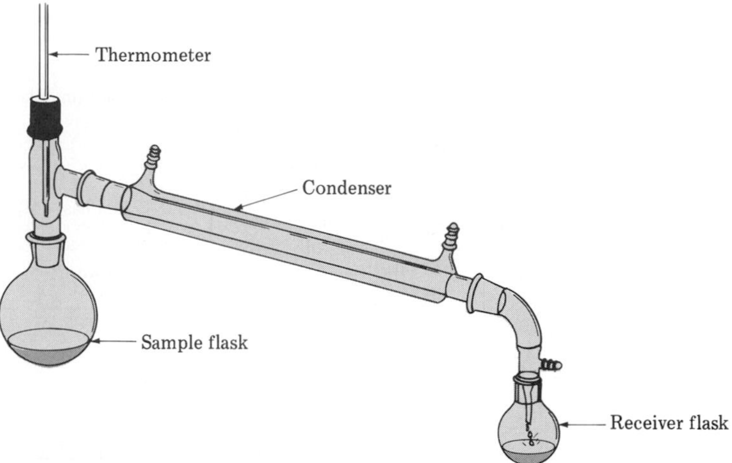


**Figure 12.1** A simple distillation apparatus: As the volatile sample is boiled, vapors rise, are condensed, and are collected in a receiver flask.

If the crude product is a mixture of two or more volatile compounds having different boiling points, **fractional distillation** can often effect a separation. The lower-boiling, more volatile component distills first, followed by the higher-boiling material. The theory of operation of distillation columns is complex, but as a practical matter good separation can usually be achieved in the laboratory if the components differ in boiling point by more than 10°C.

If neither crystallization nor distillation is effective, some form of **chromatography** is normally used to separate a mixture of organic compounds. The development of chromatography (literally, "color writing") as a separation technique dates back to the work of the Russian chemist Mikhail Tswett[1] in 1903. Tswett described the separation of the pigments in green leaves by dissolving the leaf extract and allowing the solution to run down through a vertical glass tube packed with chalk powder. Different pigments passed down the column at different rates, leaving a series of colored bands on the white chalk column. A simple chromatographic column is shown in Figure 12.2 on page 382.

The term *chromatography* is now used to refer to a variety of related separation techniques, all of which work on a common principle: The mixture to be separated is dissolved in a **mobile phase** and passed over an adsorbent **stationary phase**. Because different compounds adsorb to the stationary phase to differing extents, they migrate through the phase at different rates and are separated as they emerge from the end of the chromatography column. We will discuss briefly just three chromatographic techniques that are often used by organic chemists: liquid chromatography, high-performance liquid chromatography, and gas chromatography.

[1]Mikhail Semenovich Tswett (or Tsvett) (1872–1919); b. Asti, Italy; Russian subject; University of Warsaw (1901–1907); Institute of Technology, Warsaw (1908); in Moscow (1915); professor and director, Botanical Garden, Estonia (1917).

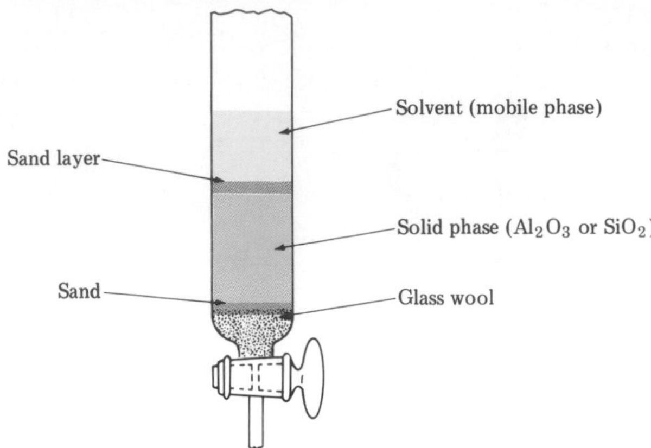


**Figure 12.2** A simple chromatography column: The glass tube is filled with adsorbent material, and a solution is allowed to drip through the column.

## 12.2 Liquid Chromatography

**Liquid chromatography** or **column chromatography** is one of the simplest and most often used chromatographic methods. As in Tswett's original experiments, a mixture of organic compounds is dissolved in a suitable solvent (the mobile phase) and is adsorbed onto a stationary phase such as alumina ($Al_2O_3$) or silica gel (hydrated $SiO_2$). More solvent is then passed down the column containing the stationary phase, and different compounds are removed (**eluted**) at different times.

The time at which a compound is eluted is strongly influenced by the polarity of the compound. As a general rule, molecules with polar functional groups (Section 5.5) are adsorbed more strongly and therefore migrate through the stationary phase more slowly than nonpolar molecules. For example, a mixture of an alcohol such as 1-heptanol and a related alkene such as 1-heptene can be easily separated by liquid chromatography. The relatively nonpolar alkene passes through the column much faster than the more polar alcohol.

**High-performance liquid chromatography** (**HPLC**) is a recent variant of the simple column technique. It has been found that the efficiency of column chromatography is vastly improved if the stationary phase is made up of very small, uniformly sized spherical particles. Small particle size ensures a large surface area for better adsorption, and a uniform spherical shape allows a tight, uniform packing. In practice, specially prepared silica microspheres of 10–25 micrometer ($\mu$m) size (1 $\mu$m = $10^{-6}$ m) are often used; 15 g of these microspheres have a surface area equivalent to the size of a football field.

High-pressure pumps are required to force solvent through a tightly packed HPLC column, and sophisticated detectors are required to monitor the appearance of material eluting from the column. These refinements are well worth it, however, for a good HPLC column can have up to several thousand times the separating power of a simple column. Figure 12.3 shows the results of HPLC analysis of a mixture of tetrachloromethane and five aromatic compounds, using silica microspheres as the stationary phase and hexane as the mobile phase. As each compound is eluted from the chromatography column, it passes through a detector, which registers its presence as a peak on a recorder chart.

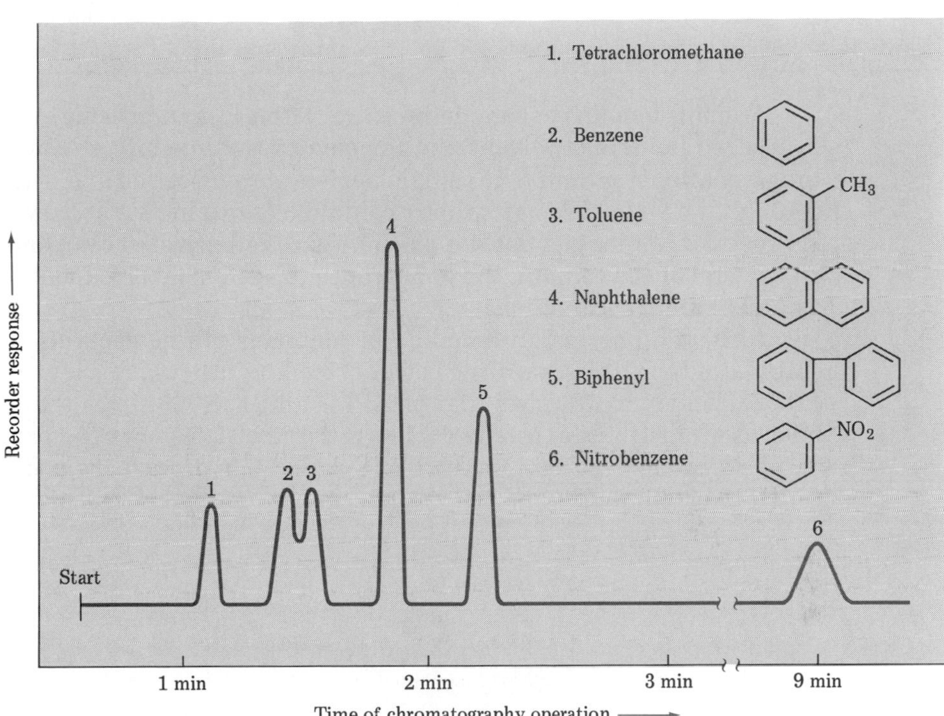


**Figure 12.3** The HPLC analysis of a mixture of aromatic compounds in tetrachloromethane: The sample was dissolved in hexane and forced through a 4 mm × 60 cm column under 2500 lb/in.$^2$ pressure.

## 12.3 Gas Chromatography

**Gas chromatography** differs from liquid chromatography in several respects. The most important difference is that a *gas* such as nitrogen or helium is used as the mobile phase, rather than a liquid solvent. Operationally, the technique employs an instrument known as a **gas chromatograph**, shown schematically in Figure 12.4 on page 384.

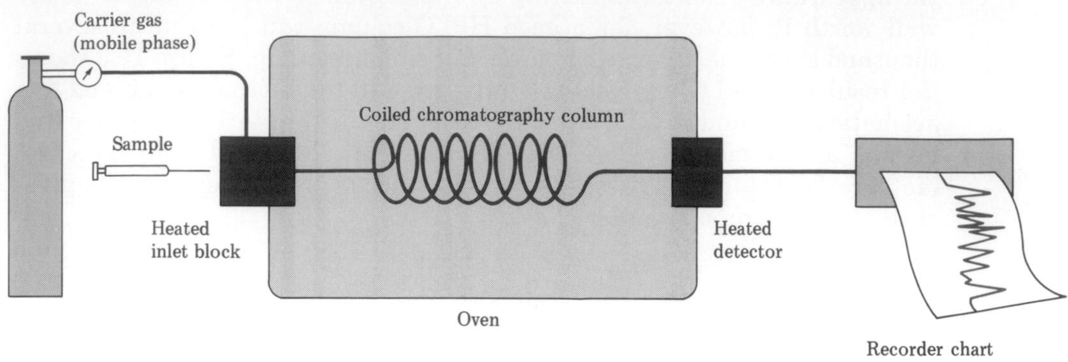


**Figure 12.4** Schematic representation of a gas chromatograph

A small amount of sample mixture (often less than 1 mg) is dissolved in a small volume of solvent and injected by syringe into a heated inlet of the gas chromatograph. The sample is instantly vaporized and is then swept through a heated chromatography column (containing stationary phase) by a stream of carrier gas (mobile phase). As pure separated components come off the end of the column, the appearance of each is detected and registered as a peak on a recorder chart.

Although only small amounts of material can be separated on a gas chromatograph, the separating power of modern instruments is phenomenal. Efficiencies up to 10 times that of HPLC and 100,000 times that of simple laboratory distillation can be achieved. Figure 12.5 shows the results of an analysis carried out on a mixture of $C_5$ and $C_6$ hydrocarbons, using a 10 m coiled-glass column.

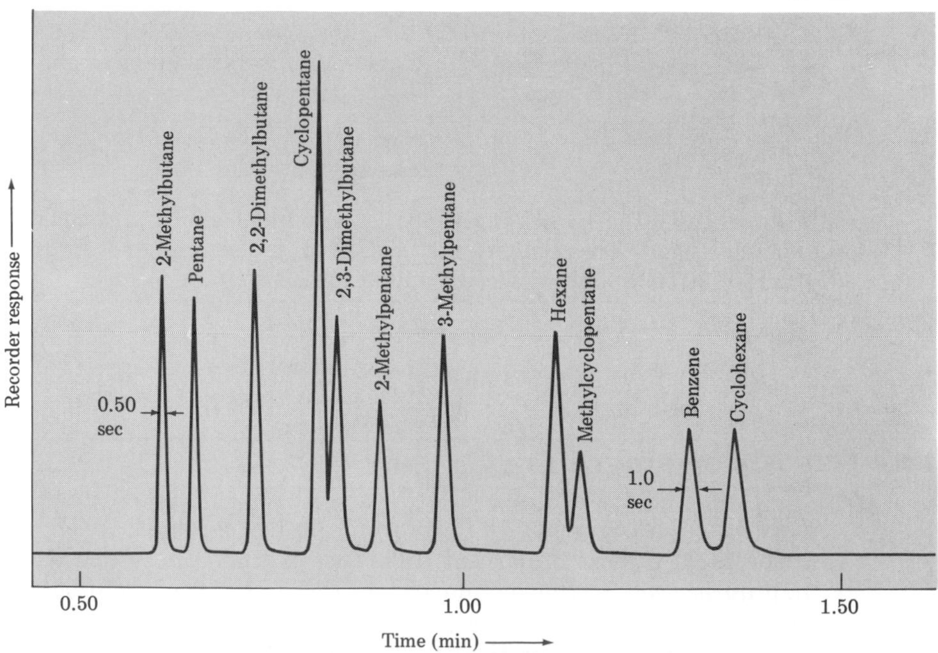


**Figure 12.5** Gas chromatographic analysis of a hydrocarbon mixture

## 12.4 Structure Determination: Spectroscopy

Once the products of a reaction have been separated and purified, the more difficult work begins. How do we identify what's been made? How do we know, for example, that the ionic addition of HBr to 1-butene gives 2-bromobutane and not 1-bromobutane?

$$CH_3CH_2CH{=}CH_2 \xrightarrow{HBr} CH_3CH_2\overset{\overset{\displaystyle Br}{|}}{C}HCH_3 \quad \text{or} \quad CH_3CH_2CH_2CH_2Br\,?$$

The answer is that we use several different kinds of **spectroscopy** to elucidate the structures of unknowns. In this and the next two chapters we'll look at four of the most useful spectroscopic techniques—mass spectroscopy (MS), infrared spectroscopy (IR), nuclear magnetic resonance spectroscopy (NMR), and ultraviolet spectroscopy (UV).

## 12.5 Mass Spectroscopy

At its simplest, **mass spectroscopy** (**MS**) is a technique that allows us to measure the mass (molecular weight) of a molecule. In addition, we can often gain valuable structural information about unknowns by measuring the masses of the fragments produced when high-energy molecules fly apart. There are several different kinds of **mass spectrometers** available, but one of the most common is the electron-impact, magnetic-sector instrument shown schematically in Figure 12.6.

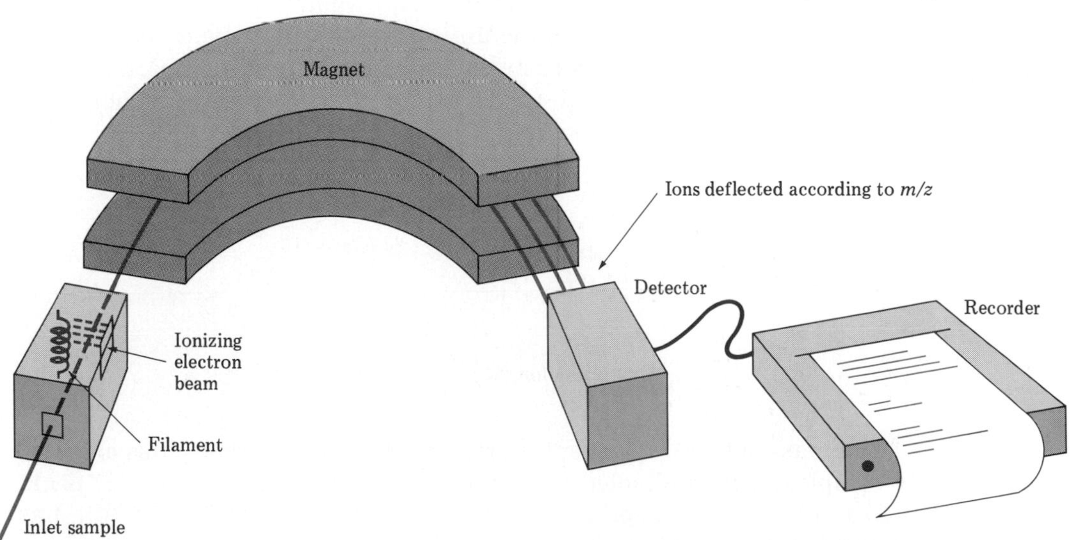


**Figure 12.6** Schematic representation of an electron-impact, magnetic-sector mass spectrometer

A small amount of sample is introduced into the mass spectrometer, where it is bombarded by a stream of high-energy electrons. The exact energy of the electron beam varies but is commonly around 70 electron volts (eV) or 1600 kcal/mol (6700 kJ/mol). When a high-energy electron strikes an organic molecule, it dislodges a valence electron from the molecule, producing a **cation radical** (*cation* because the molecule has lost a negatively charged electron; *radical* because the molecule now has an odd number of electrons).

$$\underset{\text{Organic molecule}}{RH} \xrightarrow{e^-} \underset{\text{Cation radical}}{RH^{+\cdot}} + e^-$$

Electron bombardment transfers such a large amount of energy to the sample molecules that the cation radicals *fragment* after ionization—they fly apart into numerous smaller pieces, some of which retain a positive charge, and some of which are neutral. The fragments then pass through a strong magnetic field, where they are deflected through a curved pipe according to their mass-to-charge ratio ($m/z$). Neutral fragments are not deflected by the magnetic field and are lost on the walls of the pipe, but positively charged fragments are sorted by the mass spectrometer onto a detector, which records them as peaks at the proper $m/z$ ratios. Since the number of charges, $z$, is usually 1, the peaks of ratio $m/z$ are simply $m$, the masses of the ions.

The **mass spectrum** of a compound is usually presented as a bar graph with unit masses ($m/z$ values) on the $x$-axis, and intensity (number of ions of a given $m/z$ striking the detector) on the $y$-axis. The highest peak, called the **base peak**, is arbitrarily assigned an intensity of 100%. Figure 12.7 shows the mass spectra of methane and propane.

The mass spectrum of methane is relatively simple, since few fragmentations are possible. As Figure 12.7(a) shows, the base peak has $m/z = 16$, which corresponds to the unfragmented methane cation radical, called the **parent peak**, or **molecular ion** (**$M^{+\cdot}$**). The mass spectrum also shows ions at $m/z = 15$ and 14, corresponding to cleavage of the molecular ion into $CH_3^+$ and $CH_2^{+\cdot}$ fragments.

$$CH_4 \xrightarrow{-e^-} \left[\begin{array}{c} H \\ H\!:\!\ddot{\underset{\cdot\cdot}{C}}\!\cdot\!H \\ H \end{array}\right]^{+\cdot} \longrightarrow [CH_3]^+ + H\cdot \quad (m/z = 15)$$

$$\longrightarrow [CH_2]^{+\cdot} + 2\,H\cdot \quad (m/z = 14)$$

$m/z = 16$
(molecular ion, $M^{+\cdot}$)

The mass spectral fragmentation patterns of larger molecules are usually complex, and the molecular ion is often *not* the highest (base) peak. For example, the mass spectrum of propane, shown in Figure 12.7(b), has a molecular ion at $m/z = 44$ that is only about 30% as high as the base peak at $m/z = 29$. In addition, many other fragment ions are observed.

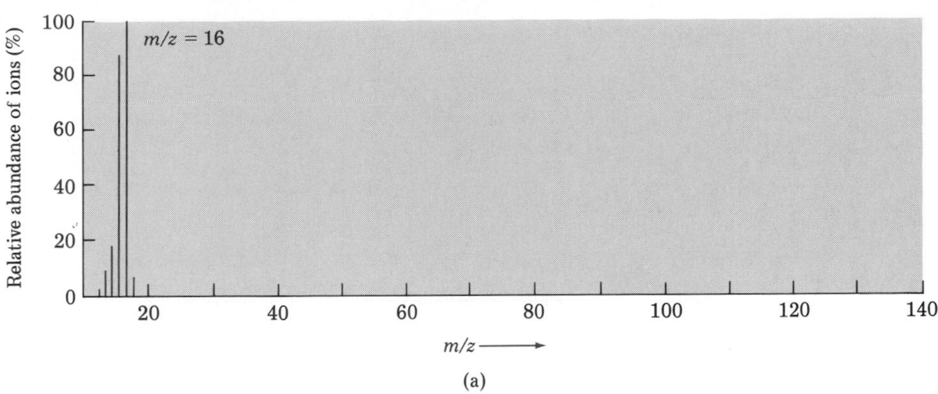


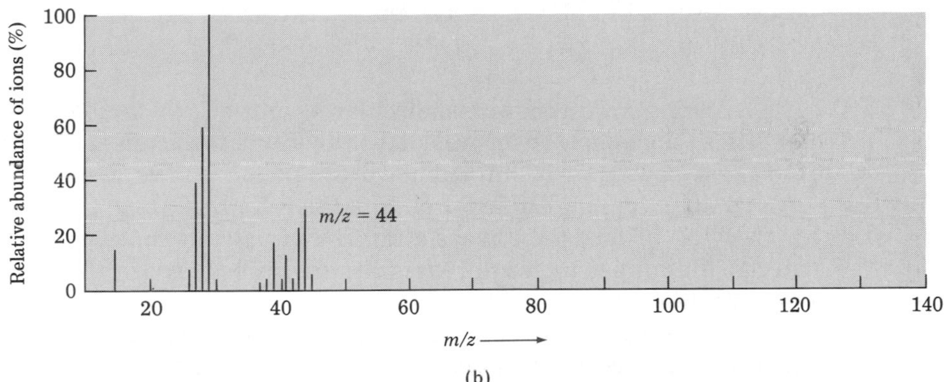


**Figure 12.7** Mass spectra of (a) methane ($CH_4$; mol wt = 16) and (b) propane ($C_3H_8$; mol wt = 44)

## 12.6 Interpreting Mass Spectra

What kinds of information can we get from studying the mass spectrum of a compound? Certainly the most obvious piece of information is the molecular weight (mol wt), which in itself can be invaluable. For example, if we were given three unlabeled bottles containing hexane (mol wt = 86), 1-hexene (mol wt = 84), and 1-hexyne (mol wt = 82), mass spectroscopy would readily distinguish between them.

Some instruments have such sophisticated detectors that they provide mass measurements accurate to 0.001 amu (atomic mass unit), allowing one to distinguish between two formulas with the same nominal mass. For example, although both $C_5H_{12}$ and $C_4H_8O$ have mol wt = 72, they differ slightly beyond the decimal point: $C_5H_{12}$ has an *exact mass* of 72.0939, whereas $C_4H_8O$ has an exact mass of 72.0575.

Unfortunately, not all compounds show a molecular ion in their mass spectrum. Although $M^{+\cdot}$ is usually easy to identify if it's abundant, some compounds such as 2,2-dimethylpropane fragment so readily that no molecular ion is observed (Figure 12.8, page 388).

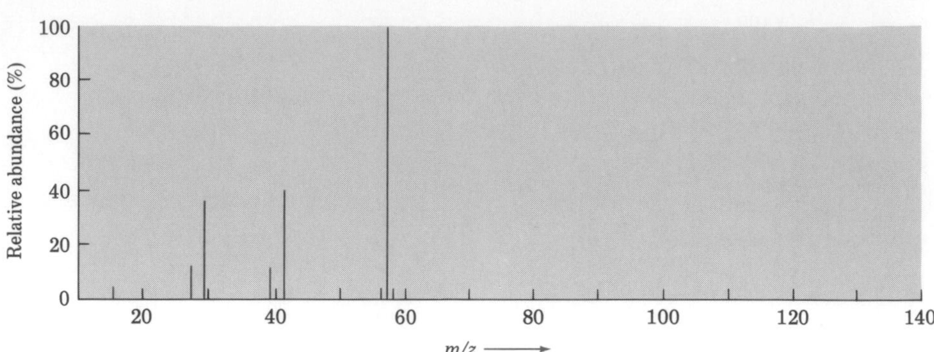


**Figure 12.8** Mass spectrum of 2,2-dimethylpropane ($C_5H_{12}$; mol wt = 72): No molecular ion is observed.

Knowing an unknown's molecular weight allows us to narrow the possibilities for molecular formula down to only a few choices. For example, if the mass spectrum of an unknown compound shows a molecular ion at $m/z$ = 110, the molecular formula is likely to be $C_8H_{14}$, $C_7H_{10}O$, $C_6H_6O_2$, or $C_6H_{10}N_2$. There are always a number of possible molecular formulas for all but the lowest molecular weights, and tables are available that list all possible choices.

A further point about mass spectroscopy can be seen by looking carefully at the mass spectra of methane and propane in Figure 12.7. Perhaps surprisingly, the molecular ions are *not* the highest mass peaks in the two spectra; there are small peaks in each spectrum at $(M + 1)^{+\cdot}$. These small peaks are due to the presence in the samples of small amounts of isotopically substituted molecules. Although $^{12}C$ is the most abundant carbon isotope, a small amount (1.11% natural abundance) of $^{13}C$ is also present. Thus, a certain percentage of the molecules analyzed in the mass spectrometer are likely to contain a $^{13}C$ isotope, giving rise to the observed $(M + 1)^{+\cdot}$ peak.

PROBLEM

**12.1** Write as many feasible molecular formulas as you can for compounds that have the following molecular ions in their mass spectra. (Assume that all of the compounds contain C and H, but that O may or may not be present.)
(a) $M^{+\cdot} = 86$ (b) $M^{+\cdot} = 128$ (c) $M^{+\cdot} = 156$

PROBLEM

**12.2** Nootkatone, one of the chemicals responsible for the odor and taste of grapefruit, shows a molecular ion at $m/z$ = 218 in the mass spectrum and is known to contain C, H, and O. Suggest several possible molecular formulas for nootkatone.

PROBLEM

**12.3** By knowing the natural abundances of minor isotopes, it's possible to calculate the relative heights of $M^{+\cdot}$ and $(M + 1)^{+\cdot}$ peaks. If $^{13}C$ has a natural abundance of 1.11%, what relative heights would you expect for $M^{+\cdot}$ and $(M + 1)^{+\cdot}$ peaks in the mass spectrum of benzene, $C_6H_6$?

PROBLEM

**12.4** The **nitrogen rule** of mass spectroscopy says that a compound containing an odd number of nitrogens has an odd-numbered molecular ion. Conversely, a compound containing an even number of nitrogens has an even-numbered $M^{+\cdot}$ peak. Why should this be so?

PROBLEM

**12.5** In light of the nitrogen rule mentioned in Problem 12.4, what is the molecular formula of pyridine, $M^{+\cdot} = 79$?

## 12.7 Interpreting Mass Spectral Fragmentation Patterns

Mass spectroscopy would be useful even if molecular weight and formula were the only information that could be obtained. In fact, though, we can get much more. For example, the mass spectrum of a compound serves as a kind of "molecular fingerprint." Each organic molecule fragments in a unique way depending on its structure, and the chance that two compounds will have identical mass spectra is small. Thus, it is sometimes possible to identify an unknown by computer-based matching of its mass spectrum to one of the more than 130,000 mass spectra recorded in a reference library established by the U.S. National Institutes of Health.

It is also possible to derive structural information about a sample by looking at its fragmentation pattern. Fragmentation occurs when the high-energy cation radical flies apart by spontaneous cleavage of a chemical bond. One of the two fragments retains the positive charge (that is, is a carbocation), whereas the other fragment is a neutral radical.

Not surprisingly, the positive charge usually remains with the fragment that is better able to stabilize it. In other words, the more stable carbocation is usually formed during fragmentation. For example, propane tends to fragment in such a way that the positive charge remains with the ethyl group rather than with the methyl because the ethyl carbocation is more stable than the methyl carbocation (Section 6.11). Propane therefore has a base peak at $m/z = 29$, and a barely detectable peak at $m/z = 15$ (Figure 12.7).

$$[CH_3CH_2CH_3]^{+\cdot} \longrightarrow CH_3CH_2^+ + \cdot CH_3$$

$M^{+\cdot} = 44$ $\qquad$ $m/z = 29$ $\qquad$ Neutral; not observed

Since mass spectral fragmentation patterns are usually complex, it is often difficult to assign definite structures to fragment ions. Most hydrocarbons fragment in many ways, as the mass spectrum of hexane, a typical alkane, shows (Figure 12.9, page 390).

The mass spectrum of hexane shows a moderately abundant molecular ion at $m/z = 86$, and fragment ions at $m/z = 71$, 57, 43, and 29. Since all of the carbon–carbon bonds of hexane are electronically similar, all of them break to a similar extent, giving rise to the observed ions. Figure 12.10 on page 390 shows how these fragments might arise.

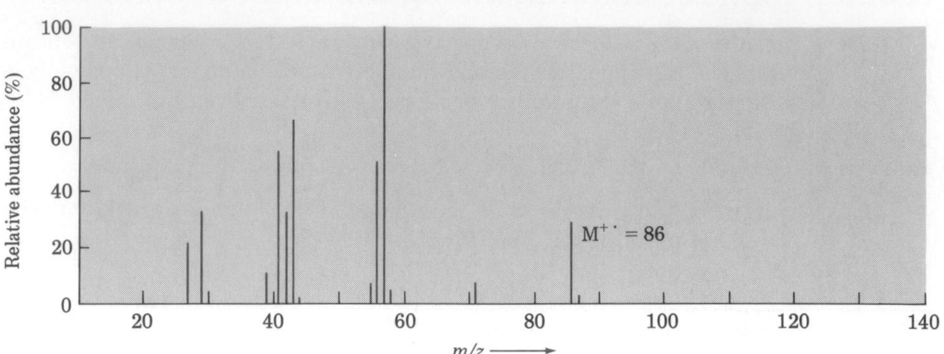


**Figure 12.9** Mass spectrum of hexane ($C_6H_6$; mol wt = 86): The base peak is at $m/z$ = 57.

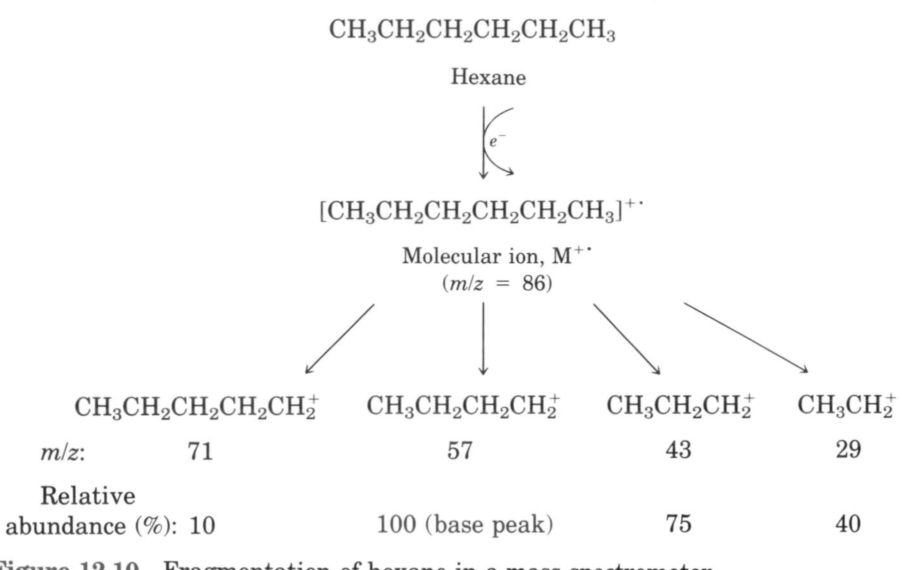


**Figure 12.10** Fragmentation of hexane in a mass spectrometer

The loss of a methyl radical from the hexane cation radical ($M^{+\cdot}$ = 86) gives rise to a fragment of mass 71; the loss of an ethyl radical accounts for a fragment of mass 57; the loss of a propyl radical accounts for a fragment of mass 43; and the loss of a butyl radical accounts for a fragment of mass 29. With skill and practice, chemists can learn to analyze the fragmentation patterns of unknown compounds and to work backward to a structure that is compatible with the available data.

Figure 12.11 shows how information from fragmentation patterns can be used. Assume that we have two unlabeled bottles, A and B. One contains methylcyclohexane, the other contains ethylcyclopentane, and we need to distinguish between them.

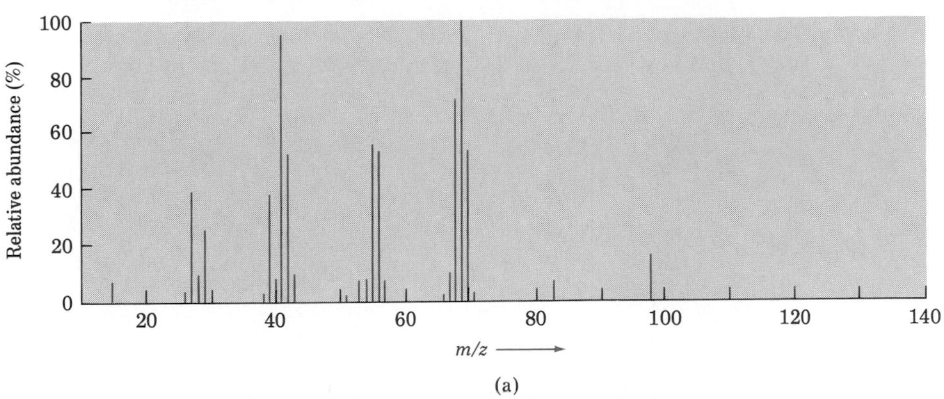


(a)

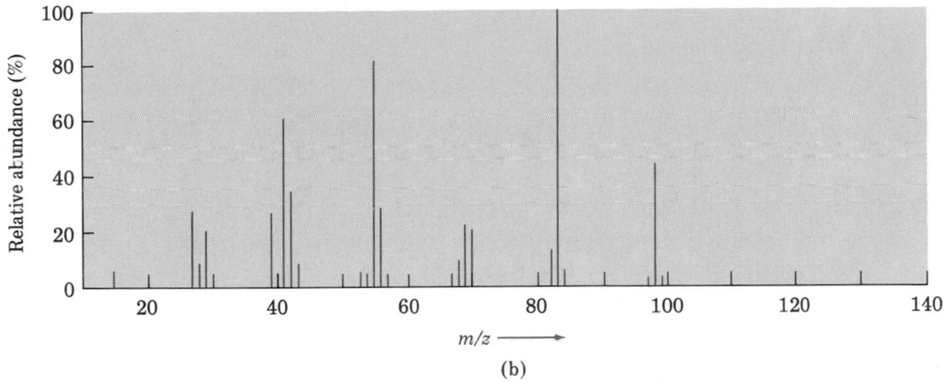


(b)

**Figure 12.11** Mass spectra of samples A and B

The mass spectra of both samples show molecular ions at $M^{+\cdot} = 98$, corresponding to $C_7H_{14}$. The two mass spectra differ considerably in their fragmentation patterns, however. Sample B has its base peak at $m/z = 83$, corresponding to the loss of a $\cdot CH_3$ group (15 mass units) from the molecular ion, but sample A has only a small peak at $m/z = 83$. Conversely, A has its base peak at $m/z = 69$, corresponding to the loss of a $\cdot CH_2CH_3$ group (29 mass units), but B has a rather small peak at $m/z = 69$. We can therefore be reasonably certain that B is methylcyclohexane and A is ethylcyclopentane.

This example is, of course, a simple one, but the principles used are broadly applicable for organic structure determination by mass spectroscopy. As we'll see in later chapters, specific functional-group families such as alcohols, ketones, aldehydes, and amines often show specific kinds of mass spectral fragmentations that can be interpreted to provide structural information.

PROBLEM

**12.6** The mass spectrum of 2,2-dimethylpropane (Figure 12.8) shows a base peak at $m/z = 57$. What molecular formula does this correspond to? Can you suggest a structure for the $m/z = 57$ fragment ion?

PROBLEM

**12.7** Two mass spectra are shown. One spectrum corresponds to 2-methyl-2-pentene; the other, to 2-hexene. Which do you think is which? Explain your choices.

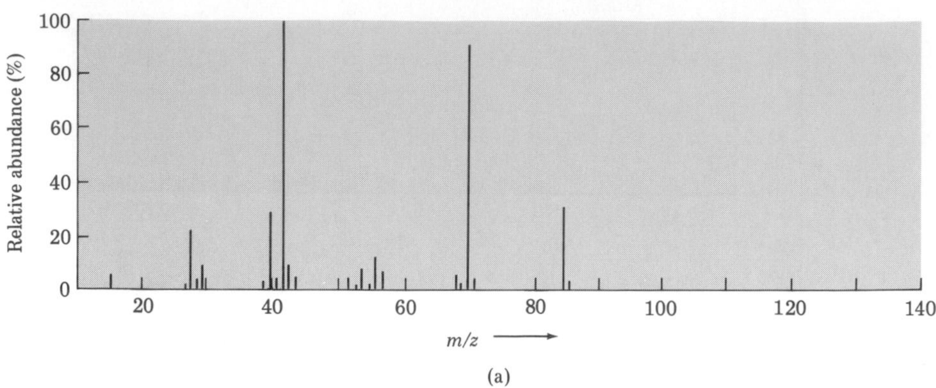


(a)

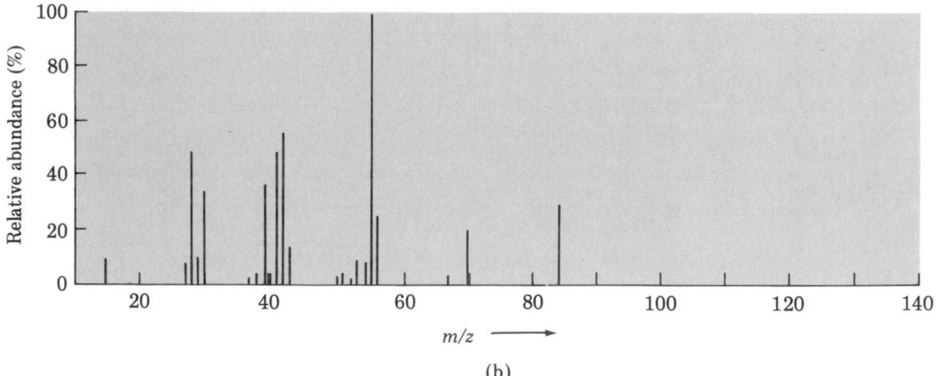


(b)

## 12.8 Spectroscopy and the Electromagnetic Spectrum

Infrared, ultraviolet, and nuclear magnetic resonance spectroscopies differ from mass spectroscopy in that they involve the interaction of molecules with electromagnetic energy rather than with a high-energy electron beam. Before beginning a study of these techniques, however, we need to look into the nature of radiant energy and the electromagnetic spectrum.

Visible light, X rays, microwaves, radio waves, and so forth are all different kinds of **electromagnetic radiation**. Collectively, they make up the **electromagnetic spectrum**, shown in Figure 12.12.

It's best to think of electromagnetic radiation as having dual behavior. In some respects it has the properties of a particle (called a **photon**), yet in other respects it behaves as a wave traveling at the speed of light. Electromagnetic waves can be described by their wavelength ($\lambda$) and frequency ($\nu$). **Wavelength** is simply the length of one complete wave cycle from trough to trough, and **frequency** is the number of wave cycles that travel past a fixed point in a certain unit of time (usually given in cycles per second, or **hertz, Hz**).

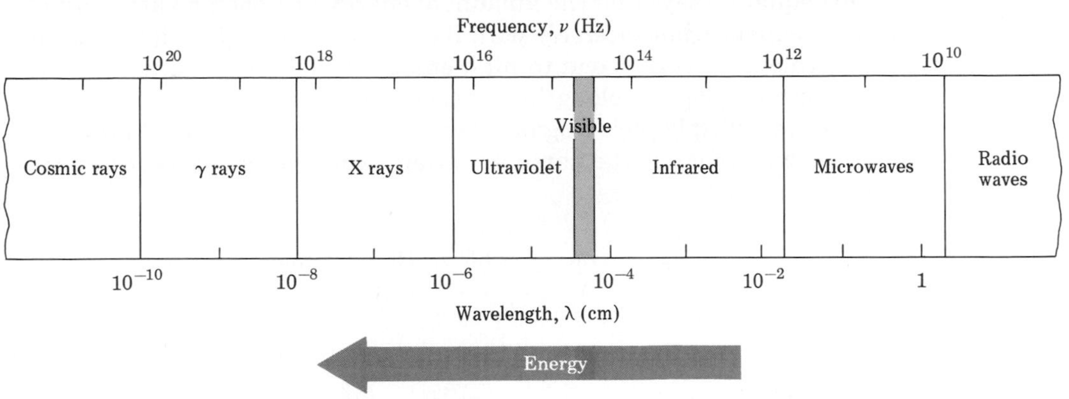


**Figure 12.12** The electromagnetic spectrum

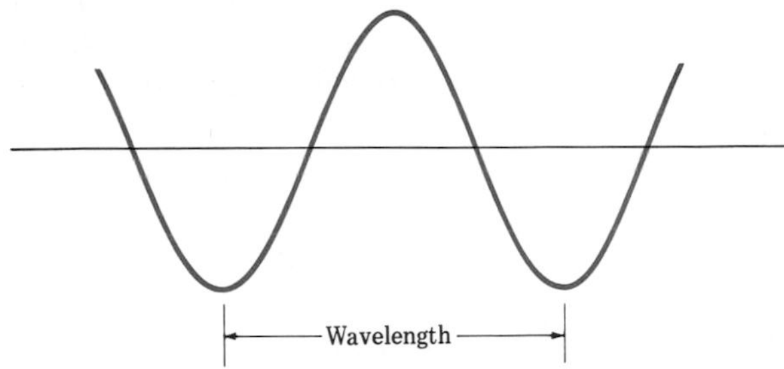


Wavelength and frequency are inversely related by the equation

$$\lambda = \frac{c}{\nu}$$

where $\lambda$ = Wavelength in centimeters

$c$ = Speed of light ($3 \times 10^{10}$ cm/sec)

$\nu$ = Frequency in hertz

As Figure 12.12 indicates, the electromagnetic spectrum is arbitrarily divided into various regions, with the familiar visible region accounting for only a small portion of the overall spectrum (from $3.8 \times 10^{-5}$ cm to $7.8 \times 10^{-5}$ cm in wavelength).

Electromagnetic energy is transmitted only in discrete energy packets, or **quanta**, where the amount of energy corresponding to 1 quantum of energy (or 1 photon) of a given frequency is expressed by the equation

$$\varepsilon = h\nu = \frac{hc}{\lambda}$$

where $\varepsilon$ = The energy of 1 photon (1 quantum)

$h$ = Planck's constant ($6.62 \times 10^{-34}$ J sec)

$\nu$ = Frequency in hertz

$\lambda$ = Wavelength in centimeters

$c$ = Speed of light ($3 \times 10^{10}$ cm/sec)

This equation says that the amount of energy in a photon varies *directly* with its frequency, but *inversely* with its wavelength; high frequencies and short wavelengths correspond to high-energy radiation (gamma rays); low frequencies and long wavelengths correspond to low-energy radiation (radio waves). If we multiply $\varepsilon$ by Avogadro's number, $N$, and convert to kilocalories per mole, we arrive at the same equation expressed in units familiar to organic chemists:

$$E = \frac{Nhc}{\lambda} = \frac{2.86 \times 10^{-3}\ \text{kcal/mol}}{\lambda\ (\text{cm})}$$

where $E$ represents the energy of Avogadro's number (a "mole") of photons of wavelength $\lambda$.

When a sample of an organic compound is exposed to electromagnetic radiation, energy of certain wavelengths is absorbed by the sample, but energy of other wavelengths passes through. Whether the light energy is absorbed or not absorbed depends both on the structure of the sample compound and on the wavelength of the radiation. If we irradiate the sample with energy of many different wavelengths and determine which wavelengths are absorbed and which pass through, we can determine the **absorption spectrum** of the compound.

Absorption spectra are usually displayed on graphs that plot wavelength versus amount of radiation transmitted. For example, the absorption spectrum of ethyl alcohol using infrared radiation is shown in Figure 12.13. The horizontal ($x$) axis records the wavelength, and the vertical ($y$) axis records the intensity of the various energy absorptions as they occur.

Since a molecule gains energy when it absorbs radiation, this energy gain must be distributed over the molecule in some way. For example, energy absorption might result in increased molecular motions, causing bonds to stretch, bend, or rotate. Alternatively, energy absorption might cause elec-

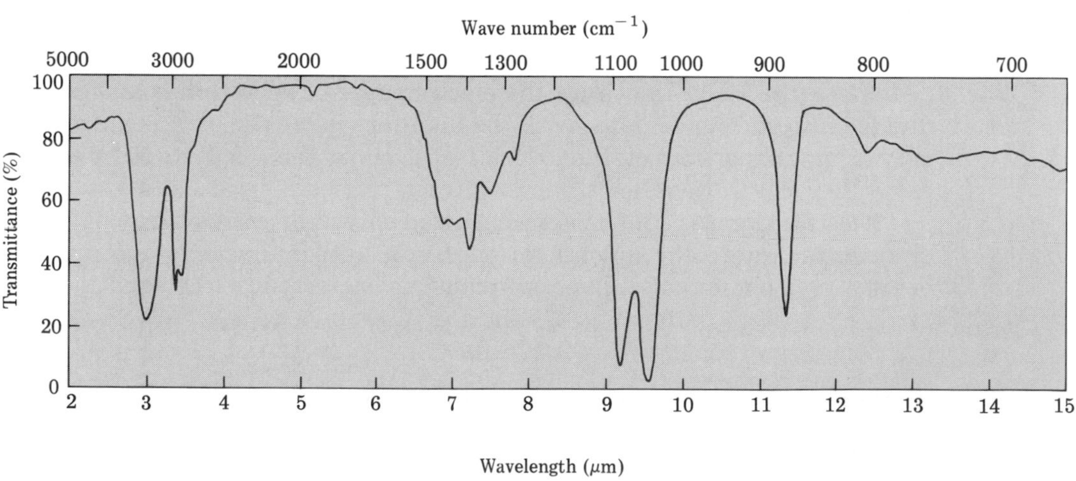


**Figure 12.13** The infrared spectrum of ethyl alcohol: A transmittance of 100% means that all the energy is passing through the sample, whereas a lower transmittance means that some energy is being absorbed. Thus, each downward spike corresponds to an energy absorption.

trons to be excited from a low-energy orbital to a higher one. Different radiation frequencies affect molecules in different ways, but each can provide structural information if we learn to interpret the results properly.

There are numerous kinds of spectroscopy, depending on which region of the electromagnetic spectrum is used. We'll look closely at just two, infrared spectroscopy and nuclear magnetic resonance spectroscopy, and have a brief introduction to a third, ultraviolet spectroscopy. Let's begin by seeing what happens when an organic sample absorbs infrared energy.

PROBLEM

**12.8** It is useful to develop an intuitive feeling for the amounts of energy that correspond to different parts of the electromagnetic spectrum. Using the relationships

$$E = \frac{2.86 \times 10^{-3}\ \text{kcal/mol}}{\lambda\ (\text{cm})} \quad \text{and} \quad \nu = \frac{c}{\lambda}$$

calculate the energies of the following:

(a) A gamma ray with $\lambda = 5 \times 10^{-9}$ cm (b) An X ray with $\lambda = 3 \times 10^{-7}$ cm
(c) Ultraviolet light with $\nu = 6 \times 10^{15}$ Hz
(d) Visible light with $\nu = 7 \times 10^{14}$ Hz
(e) Infrared radiation with $\lambda = 2 \times 10^{-3}$ cm
(f) Microwave radiation with $\nu = 10^{11}$ Hz
(g) Radio waves from the "Fat One," station KFAT in Gilroy, Ca., 102.5 MHz FM

## 12.9 Infrared Spectroscopy of Organic Molecules

The **infrared** (**IR**) region of the electromagnetic spectrum covers the range from just above the visible ($7.8 \times 10^{-5}$ cm) to approximately $10^{-2}$ cm, but only the midportion from $2.5 \times 10^{-3}$ cm to $2.5 \times 10^{-4}$ cm interests organic chemists (Figure 12.14). Specific wavelengths within the IR region are usually given in micrometers (1 $\mu$m = $10^{-4}$ cm), and frequencies are expressed in **wave numbers** ($\tilde{\nu}$) rather than in hertz. The wave number, expressed in units of $\text{cm}^{-1}$, is simply the reciprocal of the wavelength in centimeters:

$$\text{Wave number } (\tilde{\nu}) = \frac{1}{\lambda\ (\text{cm})}$$

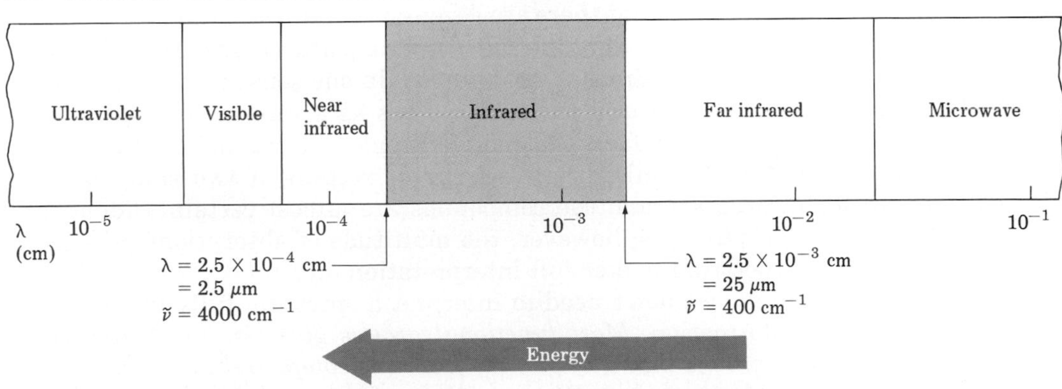


**Figure 12.14** The infrared region of the electromagnetic spectrum

Using the equation $E = (2.86 \times 10^{-3}$ kcal/mol$)/\lambda$, we can calculate that the energy levels of infrared radiation range from 1.13–11.3 kcal/mol (4.73–47.3 kJ/mol).

Why does an organic molecule absorb infrared radiation? It turns out that the amount of energy in infrared light corresponds exactly to the amount needed to increase certain molecular motions in organic compounds. Take bond lengths for example. Although we usually cite bond lengths as if they were fixed, the numbers we give are actually *averages*. In reality, bonds are constantly stretching and bending, lengthening and contracting. Thus, a typical C—H bond with an average bond length 1.10 Å is actually vibrating back and forth at a certain frequency, alternately stretching and compressing as if there were a spring connecting the two atoms. When the molecule is irradiated with electromagnetic radiation, *the vibrating bond will absorb energy from the light if the frequencies of the light and the vibration are the same.*

When a molecule absorbs infrared radiation, the molecular vibration with a frequency matching that of the light increases in intensity. In other words, the "spring" connecting the two atoms stretches and compresses a bit further. Since each light frequency absorbed by the molecule corresponds to a specific bond vibration, we can see what kinds of molecular vibrations a sample has by measuring its **infrared spectrum**. Then by working backward and interpreting the infrared spectrum, *we can find out what kinds of bonds (functional groups) are present in the molecule.*

PROBLEM

**12.9** There is some disagreement among chemists about the best way to refer to infrared data. Some people prefer to think in terms of micrometers, whereas others prefer wave numbers. To converse with both groups, it's useful to be able to interconvert the two systems of measurement rapidly. Do the following conversions:

(a) 3.1 $\mu$m to $cm^{-1}$
(b) 5.85 $\mu$m to $cm^{-1}$
(c) 2250 $cm^{-1}$ to $\mu$m
(d) 970 $cm^{-1}$ to $\mu$m

## 12.10 Interpreting Infrared Spectra

The full interpretation of an infrared spectrum is not easy. Most organic molecules are so large that there are dozens or hundreds of different possible bond stretching and bending motions, and an infrared spectrum therefore contains dozens or hundreds of absorptions. In one sense this complexity is valuable, since an infrared spectrum serves as a unique fingerprint of a specific compound. (In fact, the complex region of the infrared spectrum below 1500 $cm^{-1}$ is called the **fingerprint region**; if two samples have *identical* infrared spectra, the compounds are almost certainly identical.) For structural purposes, however, the multitude of absorptions present in an infrared spectrum makes full interpretation difficult.

Fortunately, we don't need to interpret a spectrum fully to get useful structural information. *Most functional groups give rise to characteristic infrared absorptions that change little from one compound to another.* Table 12.1 lists the characteristic IR absorptions of the most common functional groups.

**Table 12.1** Characteristic infrared absorptions of some functional groups

| *Functional group class* | *Band position* ($cm^{-1}$) | *Intensity of absorption* |
|---|---|---|
| Alkanes, alkyl groups | | |
| C—H | 2850–2960 | Medium to strong |
| Alkenes | | |
| =C—H | 3020–3100 | Medium |
| C=C | 1650–1670 | Medium |
| Alkynes | | |
| ≡C—H | 3300 | Strong |
| —C≡C— | 2100–2260 | Medium |
| Alkyl halides | | |
| C—Cl | 600–800 | Strong |
| C—Br | 500–600 | Strong |
| C—I | 500 | Strong |
| Alcohols | | |
| O—H | 3400–3640 | Strong, broad |
| C—O | 1050–1150 | Strong |
| Aromatics | | |
| C—H (aromatic ring) | 3030 | Medium |
| C=C (aromatic ring) | 1600, 1500 | Strong |
| Amines | | |
| N—H | 3310–3500 | Medium |
| C—N | 1030, 1230 | Medium |
| Carbonyl compounds[a] | | |
| C=O | 1670–1780 | Strong |
| Carboxylic acids | | |
| O—H | 2500–3100 | Strong, very broad |
| Nitriles | | |
| C≡N | 2210–2260 | Medium |
| Nitro compounds | | |
| $NO_2$ | 1540 | Strong |

[a]Acids, esters, aldehydes, and ketones.

For example, the C=O absorption of ketones is almost always in the range 1690–1750 $cm^{-1}$, the O—H absorption of alcohols is almost always in the range 3200–3600 $cm^{-1}$, and the C=C absorption of alkenes is almost always in the range 1640–1680 $cm^{-1}$. By learning to recognize where characteristic functional-group absorptions occur, we can gain valuable structural information from infrared spectra.

As an example of how infrared spectroscopy can be used, look at the IR spectra of hexane, 1-hexene, and 1-hexyne in Figure 12.15 (facing page). Although all three spectra contain many peaks, there are characteristic absorptions of the C=C and C≡C functional groups that allow the three compounds to be distinguished readily. Thus, 1-hexene shows a characteristic carbon–carbon double-bond peak at 1660 $cm^{-1}$ and a vinylic =C—H bond peak at 3100 $cm^{-1}$, whereas 1-hexyne exhibits a carbon–carbon triple-bond absorption at 2100 $cm^{-1}$ and a terminal alkyne ≡C—H bond absorption at 3300 $cm^{-1}$.

It helps in remembering the position of specific IR absorptions if we divide the infrared region from 4000 to 200 $cm^{-1}$ into four parts, as shown in Figure 12.16:

1. The region from 4000 to 2500 $cm^{-1}$ corresponds to absorptions caused by N—H, O—H, and C—H bond stretching motions. The N—H and O—H bonds absorb in the 3300–3600 $cm^{-1}$ range, whereas C—H bond stretching occurs near 3000 $cm^{-1}$.
2. The region from 2500 to 2000 $cm^{-1}$ is where triple-bond stretching occurs. Both nitriles (R—C≡N) and alkynes show peaks here.
3. The region from 2000 to 1500 $cm^{-1}$ is where double bonds of all kinds (C=O, C=N, and C=C) absorb. Carbonyl groups generally absorb in the range from 1670 to 1780 $cm^{-1}$, whereas alkene stretching normally occurs in a narrow range from 1640 to 1680 $cm^{-1}$.
4. The region below 1500 $cm^{-1}$ is the fingerprint portion of the IR range. A large number of absorptions due to a variety of C—C, C—O, C—N, and C—X single-bond vibrations occur here.

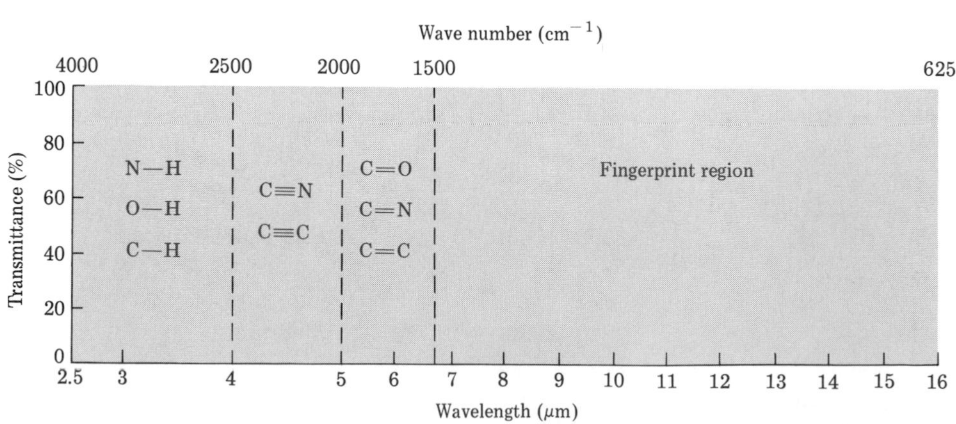


**Figure 12.16** Regions in the infrared spectrum

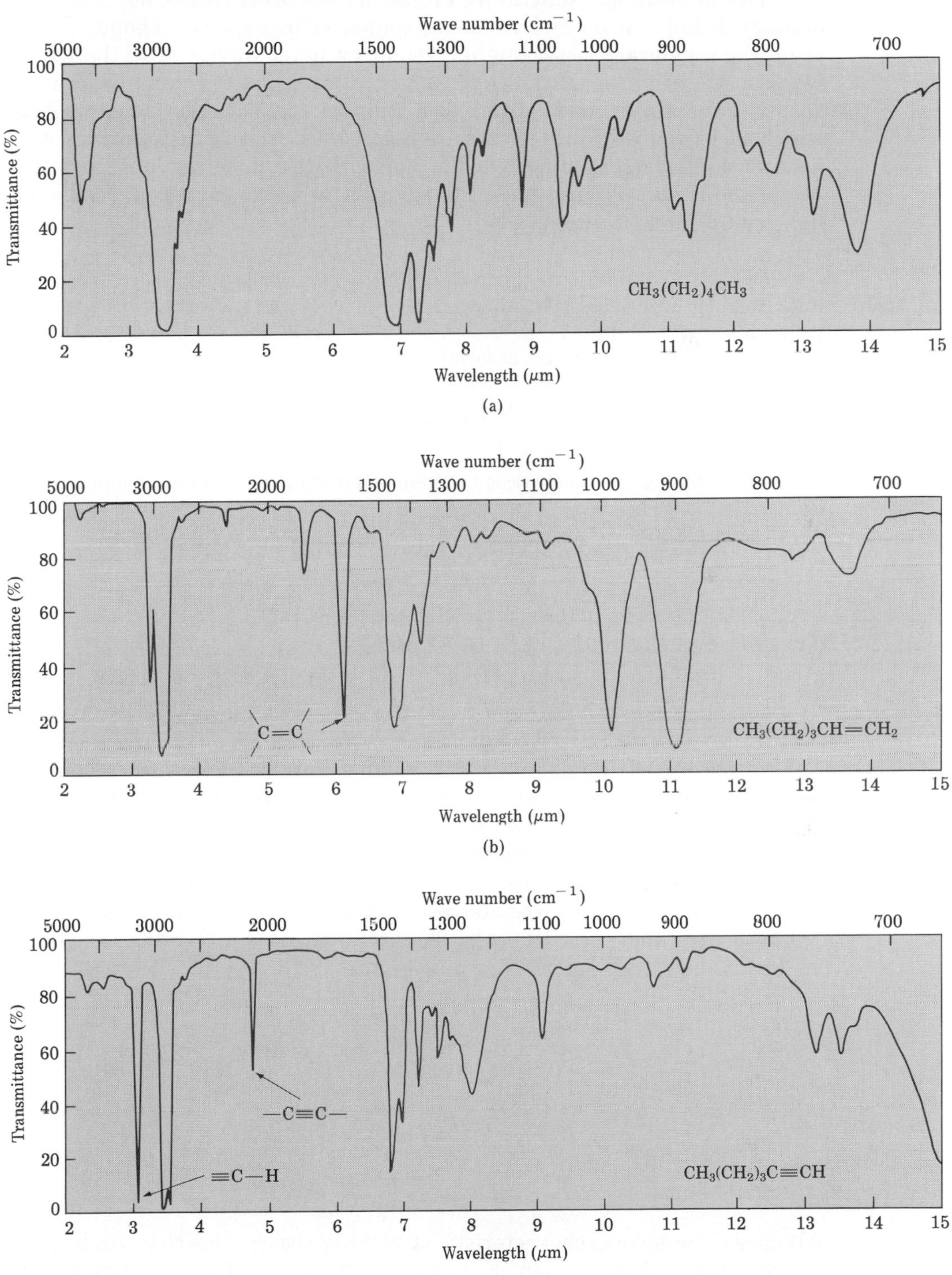


**Figure 12.15** Infrared spectra of (a) hexane, (b) 1-hexene, and (c) 1-hexyne: Spectra like these are easily obtained on small samples in a few minutes' time using commercially available instruments.

Why do different functional groups absorb where they do? The best analogy is that of two weights (atoms) connected by a spring (a bond). Short, strong bonds vibrate faster (higher frequency, lower wavelength) than long, weak bonds, just as a short, strong spring vibrates faster than a long, weak spring. Thus, triple bonds absorb at a higher frequency than double bonds, which in turn absorb higher than single bonds. In addition, springs connecting small weights vibrate faster than springs connecting large weights. Thus, C—H, O—H, and N—H bonds vibrate faster than bonds between heavier C, O, and N atoms.

**PROBLEM**

**12.10** Refer to Table 12.1, and make educated guesses as to what functional groups these molecules might contain:

(a) A compound with a strong absorption at 1710 $cm^{-1}$

(b) A compound with a strong absorption at 1540 $cm^{-1}$

(c) A compound with strong absorptions at 1720 $cm^{-1}$ and at 2500–3100 $cm^{-1}$

**PROBLEM**

**12.11** How might you use IR spectroscopy to help you distinguish between these pairs of isomers?

(a) $CH_3CH_2OH$ and $CH_3OCH_3$ (b) Cyclohexane and 1-hexene

(c) $CH_3CH_2COOH$ and $HOCH_2CH_2CHO$

## 12.11 Infrared Spectra of Hydrocarbons

### ALKANES

The infrared spectrum of an alkane is rather uninformative, since no functional groups are present and all absorptions are due to C—H and C—C bond stretching and bending. Alkane C—H bonds always show a strong absorption from 2850 to 2960 $cm^{-1}$, whereas saturated C—C bonds show a number of absorptions in the 800–1300 $cm^{-1}$ range. Since most organic compounds contain saturated alkane-like portions, most organic compounds have these characteristic IR peaks. These C—H and C—C peaks are clearly visible in the three spectra shown in Figure 12.15.

| | | |
|---|---|---|
| Alkanes | —C—H | 2850–2960 $cm^{-1}$ |
| | —C—C— | 800–1300 $cm^{-1}$ |

### ALKENES

Alkenes show several characteristic stretching absorptions that can be used for structural identification. For example, vinylic =C—H bonds absorb from 3020 to 3100 $cm^{-1}$, and alkene C=C bonds usually show an absorption near 1650 $cm^{-1}$, although in some cases this can be rather weak and difficult to see clearly. Both =C—H and C=C absorptions are diagnostic for alkenes, as the 1-hexene spectrum in Figure 12.15 shows.

Mono- and disubstituted alkenes have $=C-H$ bonds that give rise to specific out-of-plane bending absorptions in the 700–1000 $cm^{-1}$ range, thereby allowing the substitution pattern on a double bond to be determined. Monosubstituted alkenes such as 1-hexene show strong characteristic peaks at 910 and 990 $cm^{-1}$, and 2,2-disubstituted alkenes ($R_2C=CH_2$) have an intense band at 890 $cm^{-1}$.

| Alkenes | $=C-H$ | 3020–3100 $cm^{-1}$ |
|---|---|---|
| | $>C=C<$ | 1650–1670 $cm^{-1}$ |
| | $RCH=CH_2$ | 910 and 990 $cm^{-1}$ |
| | $R_2C=CH_2$ | 890 $cm^{-1}$ |

## ALKYNES

Alkynes exhibit a $C\equiv C$ stretching absorption at 2100–2260 $cm^{-1}$, a band that is much more intense for terminal alkynes than for internal alkynes. In fact, symmetrically substituted triple bonds like that in 3-hexyne show no absorption at all, for reasons we won't go into. Terminal alkynes such as 1-hexyne also have a characteristic $\equiv C-H$ stretch at 3300 $cm^{-1}$. This band is diagnostic for terminal alkynes, since it is fairly intense and quite sharp.

| Alkynes | $-C\equiv C-$ | 2100–2260 $cm^{-1}$ |
|---|---|---|
| | $\equiv C-H$ | 3300 $cm^{-1}$ |

One final point about infrared spectroscopy: It is also possible to derive much structural information from an infrared spectrum by noticing which characteristic absorptions are *not* present. For example, if the spectrum of an unknown does *not* contain absorptions at 3300 and 2150 $cm^{-1}$, it is not a terminal alkyne; if the spectrum has *no* absorption near 3400 $cm^{-1}$, the compound is not an alcohol, and so on.

**PROBLEM**

**12.12** Shown here is the infrared spectrum of ethynylcyclohexane. What absorption bands can you identify?

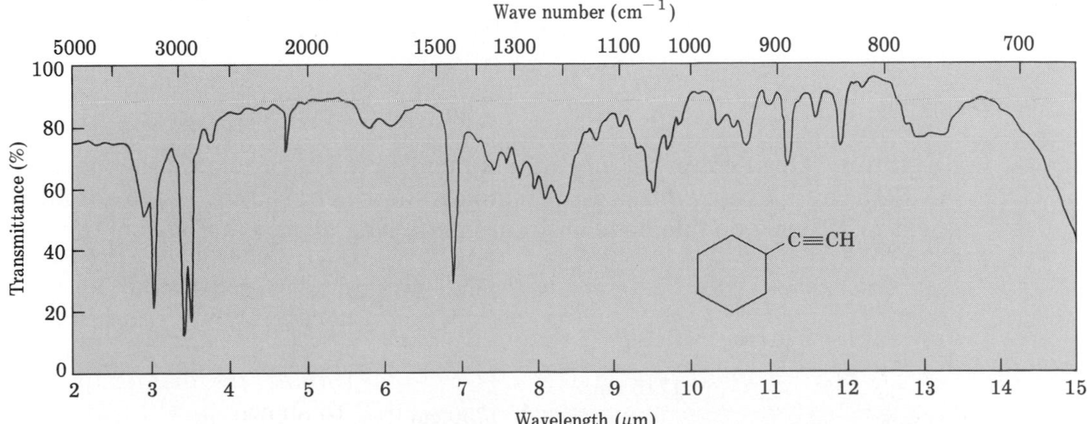

## 12.12 Infrared Spectra of Other Functional Groups

As each functional group is discussed in future chapters, the spectroscopic behavior of that group will be described. For the present, though, it's worthwhile to point out some distinguishing features of the more important functional groups.

### ALCOHOLS

The O—H functional group of alcohols is easy to spot in the IR. Alcohols have a characteristic band in the range 3300–3600 $cm^{-1}$ that is usually fairly broad and intense. If present, it's hard to miss this band or to confuse it with anything else.

Alcohols —O—H 3300–3600 $cm^{-1}$ (broad, intense)

### AMINES

The N—H functional group of amines is also easy to spot in the IR, with a characteristic absorption in the 3300–3500 $cm^{-1}$ range. Although alcohols also absorb in this range, an amine absorption is much sharper and less intense than a hydroxyl band.

Amines —N—H 3300–3500 $cm^{-1}$ (sharp, medium intensity)

### CARBONYL COMPOUNDS

Carbonyl functional groups are the easiest to identify of all IR absorptions, with a sharp, intense peak in the range 1670–1780 $cm^{-1}$. Most important, the *exact* position of absorption within the range can often be used to identify the exact kind of carbonyl functional group—aldehyde, ketone, ester, and so forth.

**Aldehydes** Saturated aldehydes absorb at 1730 $cm^{-1}$, whereas aldehydes next to either a double bond or an aromatic ring absorb at 1705 $cm^{-1}$.

Aldehydes R—C(=O)—H 1730 $cm^{-1}$; C₆H₅—C(=O)—H 1705 $cm^{-1}$

**Ketones** Open-chain ketones and six-membered-ring cyclic ketones absorb at 1715 $cm^{-1}$; five-membered-ring ketones absorb at 1750 $cm^{-1}$; and ketones next to either a double bond or an aromatic ring absorb at 1690 $cm^{-1}$.

Ketones R—C(=O)—R′ 1715 $cm^{-1}$; cyclopentanone 1750 $cm^{-1}$; C₆H₅—C(=O)—R 1690 $cm^{-1}$

**Esters** Saturated esters absorb at 1735 $cm^{-1}$, and esters next to either an aromatic ring or a double bond absorb at 1715 $cm^{-1}$.

Esters $R-\overset{\overset{O}{\|}}{C}-OR'$ 1735 $cm^{-1}$    $C_6H_5-\overset{\overset{O}{\|}}{C}-OR'$ 1715 $cm^{-1}$

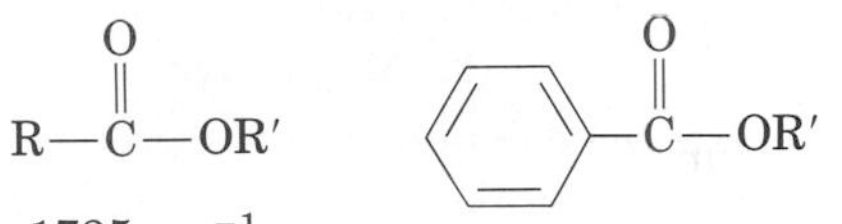

## 12.13 Summary and Key Words

Product purification and structure elucidation are two of the most difficult tasks facing laboratory chemists. If possible, purification is done by a direct method such as **crystallization** or **distillation**. Often, however, some form of **chromatography** must be used. The impure sample is applied to one end of a column containing an inert support material (the **stationary phase**), and the sample is carried along by a **mobile phase**, either gas or liquid. Separation of mixtures is effected because different compounds adsorb to the stationary phase in differing degrees and therefore pass along the column at different rates.

Once a reaction product has been purified, its structure must be determined. This is done using various spectroscopic methods such as mass spectroscopy and infrared spectroscopy. **Mass spectroscopy** gives information about the molecular weight and formula of unknown samples; **infrared spectroscopy** gives information about the functional groups present.

In mass spectroscopy, molecules are first ionized by collision with a high-energy electron beam. The ions then fragment into smaller pieces, which are magnetically sorted according to their mass-to-charge ratio (*m/z*). The ionized sample molecule is called the **molecular ion, $M^{+\cdot}$**, and measurement of its mass gives the molecular weight of the sample. Structural clues about unknown samples can be obtained by interpreting the **fragmentation pattern** of the molecular ion. Mass spectral fragmentations are usually complex, however, and interpretation is often difficult.

Infrared spectroscopy involves the interaction of a molecule with **electromagnetic radiation**. When an organic molecule is irradiated with infrared light, certain frequencies of light are absorbed by the molecule and others are not absorbed. The frequencies absorbed correspond to the amounts of energy needed to increase the amplitude of certain molecular vibrations such as bond stretchings and bendings. Each specific kind of functional group has a characteristic set of infrared absorptions. For example, the terminal alkyne $\equiv C-H$ bond absorbs infrared radiation of 3300 $cm^{-1}$ frequency, and the alkene $C=C$ bond absorbs in the range 1650–1670 $cm^{-1}$. By observing which frequencies of infrared radiation are absorbed by a molecule, and which are not, we can determine what functional groups a molecule contains.

## ADDITIONAL PROBLEMS

**12.13** Write as many molecular formulas as you can for hydrocarbons that show the following molecular ions in their mass spectra:
(a) $M^{+\cdot} = 64$ (b) $M^{+\cdot} = 186$ (c) $M^{+\cdot} = 158$ (d) $M^{+\cdot} = 220$

**12.14** In Section 6.2, we calculated the degree of unsaturation of molecules according to their molecular formulas. Write the molecular formulas of all hydrocarbons corresponding to the following molecular ions. How many degrees of unsaturation (double bonds and/or rings) are indicated by each formula?
(a) $M^{+\cdot} = 86$ (b) $M^{+\cdot} = 110$ (c) $M^{+\cdot} = 146$ (d) $M^{+\cdot} = 190$

**12.15** Draw the structure of a molecule that is consistent with the mass spectral data in each example:
(a) A hydrocarbon with $M^{+\cdot} = 132$ (b) A hydrocarbon with $M^{+\cdot} = 166$
(c) A hydrocarbon with $M^{+\cdot} = 84$

**12.16** Write as many molecular formulas as you can for compounds that show the following molecular ions in their mass spectra. Assume that the elements C, H, N, and O might be present.
(a) $M^{+\cdot} = 74$ (b) $M^{+\cdot} = 131$

**12.17** Camphor, a saturated monoketone from the Asian camphor tree is used, among other things, as a moth repellent and as a constituent of embalming fluid. If camphor has $M^{+\cdot} = 152$ in its mass spectrum, what is a reasonable molecular formula? How many rings does camphor have?

**12.18** Nicotine is a diamino compound that can be isolated from dried tobacco leaves. Nicotine has two rings and shows $M^{+\cdot} = 162$ in its mass spectrum. Propose a molecular formula for nicotine, and calculate the number of double bonds present. (There is no oxygen in nicotine.)

**12.19** Halogenated compounds are particularly easy to identify by their mass spectra because both chlorine and bromine occur naturally as mixtures of two abundant isotopes. Chlorine occurs as $^{35}Cl$ (75.5%) and as $^{37}Cl$ (24.5%), and bromine occurs as $^{79}Br$ (50.5%) and $^{81}Br$ (49.5%). At what masses do the molecular ion(s) occur for these formulas? What are the relative percentages of each molecular ion?
(a) Bromomethane, $CH_3Br$ (b) 1-Chlorohexane, $C_6H_{13}Cl$

**12.20** Molecular ions can be particularly complex for polyhalogenated compounds. Taking the natural abundance of Cl into account (Problem 12.19), calculate the masses of the molecular ions of the following formulas. What are the relative percentages of each ion?
(a) Chloroform, $CHCl_3$ (b) Freon 12, $CF_2Cl_2$ (Fluorine occurs only as $^{19}F$.)

**12.21** 2-Methylpentane ($C_6H_{14}$, mol wt 86) has the mass spectrum shown here. What peak represents $M^{+\cdot}$? Which is the base peak? Propose structures for fragment ions of $m/z$ = 71, 57, 43, and 29. Suggest a reason for the base peak's having the mass it does.

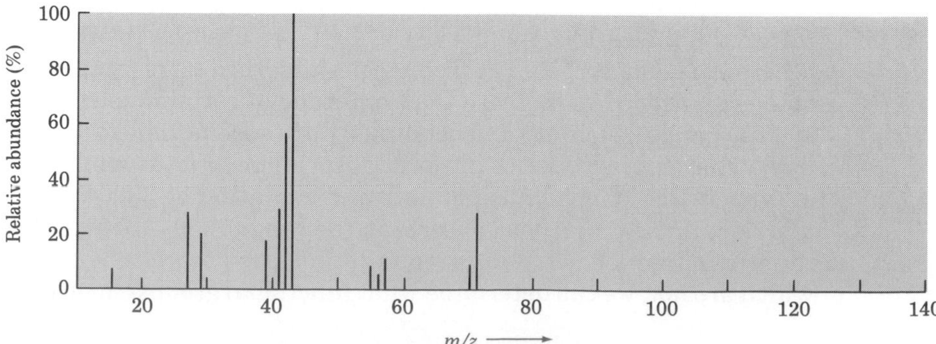

**12.22** The combined gas chromatograph/mass spectrometer is a sophisticated instrument that uses a mass spectrometer to detect compounds as they elute from a gas chromatograph. This technique allows one to inject a mixture of compounds onto a gas chromatography column and to determine automatically the mass spectrum of each compound present in the mixture. Assume that you are in the laboratory carrying out the catalytic hydrogenation of cyclohexene to cyclohexane. How could you use a gas chromatograph/mass spectrometer to determine when the reaction was complete?

**12.23** Convert the following infrared absorption values from micrometers to reciprocal centimeters:
(a) An alcohol, 2.98 $\mu m$ (b) An ester, 5.81 $\mu m$ (c) A nitrile, 4.93 $\mu m$

**12.24** Convert the following infrared absorption values from reciprocal centimeters to micrometers:
(a) A cyclopentanone, 1755 $cm^{-1}$ (b) An amine, 3250 $cm^{-1}$
(c) An aldehyde, 1725 $cm^{-1}$ (d) An acid chloride, 1780 $cm^{-1}$

**12.25** How might you use IR spectroscopy to distinguish between the three isomers 1-butyne, 1,3-butadiene, and 2-butyne?

**12.26** Would you expect two enantiomers such as (*R*)-2-bromobutane and (*S*)-2-bromobutane to have identical or different IR spectra? Explain.

**12.27** Would you expect two diastereomers such as *meso*-2,3-dibromobutane and (2*R*,3*R*)-dibromobutane to have identical or different IR spectra? Explain.

**12.28** Two infrared spectra are shown. One is the spectrum of cyclohexane, and the other is the spectrum of cyclohexene. Identify them and explain your answer.

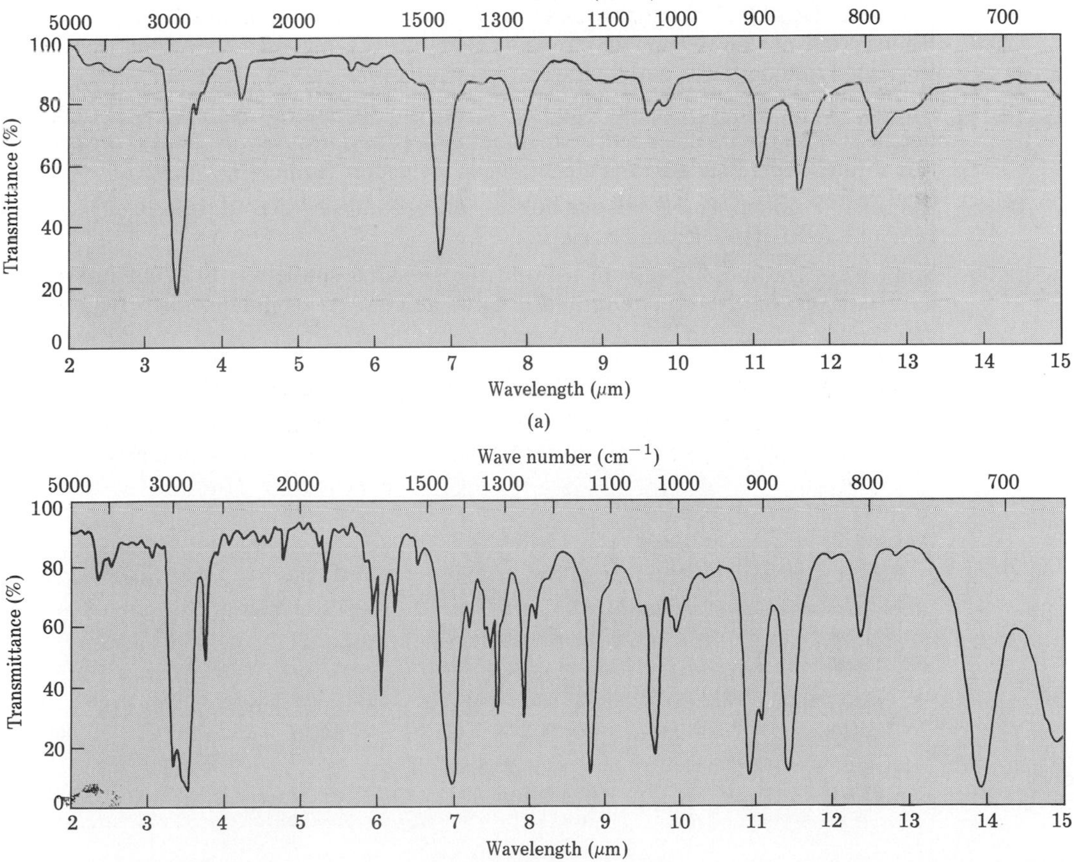

**12.29** How would you use infrared spectroscopy to distinguish between these pairs of constitutional isomers?

(a) $(CH_3)_3N$ and $CH_3CH_2NHCH_3$

(b) $CH_3CH_2\overset{O}{\overset{\|}{C}}CH_3$ and $CH_3CH{=}CHCH_2OH$

(c) $H_2C{=}CHOCH_3$ and $CH_3CH_2CHO$

**12.30** Assume you are carrying out the dehydration of 1-methylcyclohexanol to 1-methylcyclohexene. How could you use infrared spectroscopy to determine when the reaction was complete?

**12.31** Assume that you are carrying out the base-induced dehydrobromination of 3-bromo-3-methylpentane (Section 11.10). How could you use IR spectroscopy to tell which of two possible elimination products was formed?

**12.32** At what approximate positions might these compounds show IR absorptions?

(a) $CH_3CH_2\overset{O}{\overset{\|}{C}}CH_3$

(b) $(CH_3)_2CHCH_2C{\equiv}CH$

(c) $(CH_3)_2CHCH_2CH{=}CH_2$

(d) $CH_3CH_2CH_2COOCH_3$

(e) $C_6H_5{-}\overset{O}{\overset{\|}{C}}{-}CH_3$

**12.33** Which kind of C=O bond is stronger, that in an ester (1735 $cm^{-1}$) or that in a saturated ketone (1715 $cm^{-1}$)? Explain your answer.

**12.34** Carvone is an unsaturated ketone responsible for the odor of spearmint. If carvone has $M^{+\cdot} = 150$ in its mass spectrum, what molecular formulas are likely? If carvone has three double bonds and one ring, what molecular formula is correct?

**12.35** Carvone (Problem 12.34) has an intense infrared absorption at 1690 $cm^{-1}$. What kind of ketone does carvone contain?

**12.36** Shown are the mass spectrum (a) and the infrared spectrum (b) of an unknown hydrocarbon. Analyze the data and propose as many reasonable structures as you can.

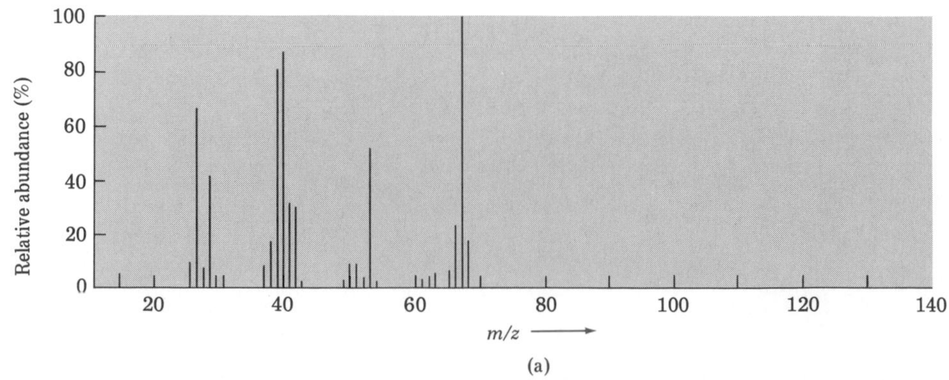


(a)

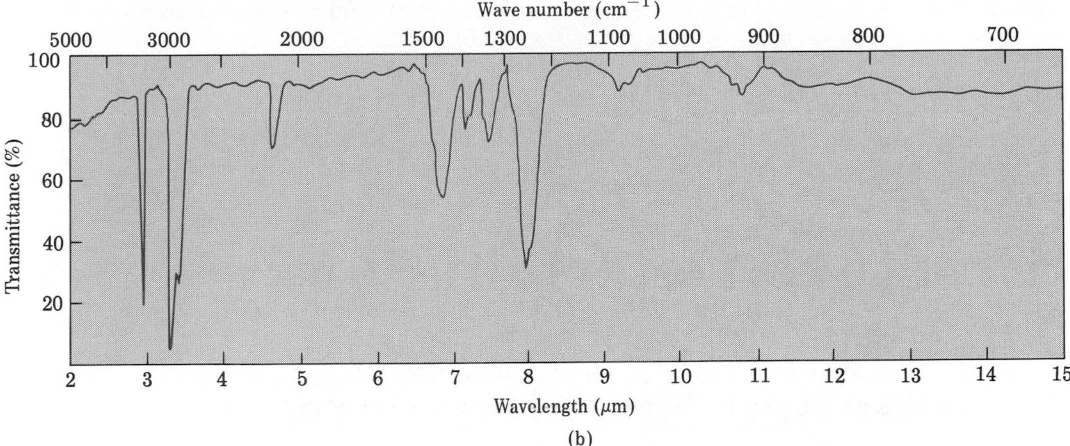


(b)

**12.37** Shown are the mass spectrum (a) and the infrared spectrum (b) of another unknown hydrocarbon. Analyze the data and propose as many reasonable structures as you can.

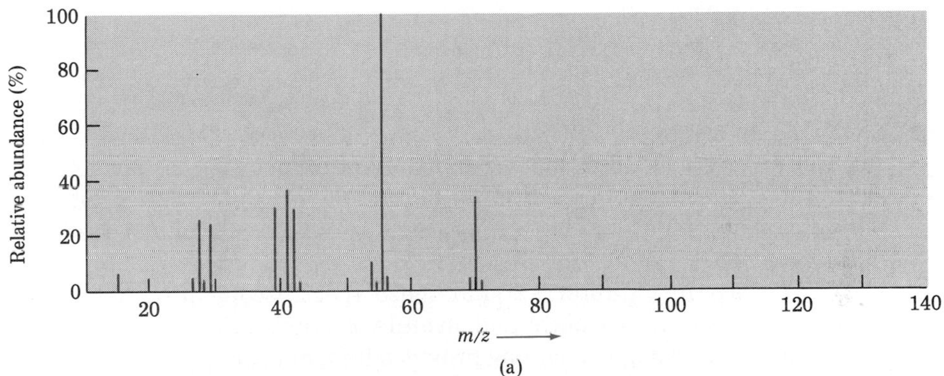


(a)

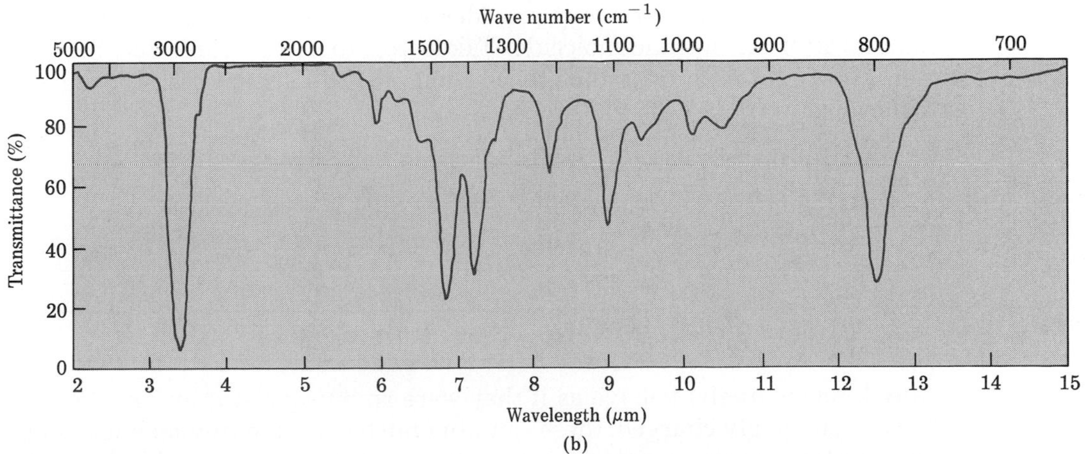


(b)

CHAPTER 13

# Structure Determination: Nuclear Magnetic Resonance Spectroscopy

Nuclear magnetic resonance (NMR) spectroscopy is perhaps the most valuable spectroscopic technique available to organic chemists. It is the method of structure determination to which chemists first turn for information.

We saw in Chapter 12 that mass spectroscopy provides information about the molecular weight and formula of a molecule of unknown structure, and that infrared spectroscopy provides information about the kinds of functional groups in the unknown. Nuclear magnetic resonance spectroscopy does not replace or duplicate either of these techniques; rather, it complements them. NMR spectroscopy provides a "map" of the carbon–hydrogen framework of an organic molecule. Taken together, the three techniques often allow us to obtain complete solutions for the structures of complex unknowns.

| | |
|---|---|
| Mass spectroscopy | Molecular size and formula |
| Infrared spectroscopy | Functional groups present |
| NMR spectroscopy | Map of carbon–hydrogen framework |

## 13.1 Nuclear Magnetic Resonance Spectroscopy

Many kinds of nuclei behave as if they were spinning about an axis. Since they are positively charged, these spinning nuclei act like tiny magnets and can therefore interact with an externally applied magnetic field (denoted $H_0$). Not all nuclei act this way but, fortunately for organic chemists, both the proton ($^1H$) and the $^{13}C$ nucleus do have spins. Let's see what the consequences of nuclear spin are, and how we can use the results.

In the absence of a strong external magnetic field, the nuclear spins of magnetic nuclei are oriented randomly. When these nuclei are placed between the poles of a strong magnet, however, they adopt specific orientations, much as a compass needle orients itself in the earth's magnetic field. A spinning $^1H$ or $^{13}C$ nucleus can orient so that its own tiny magnetic field is aligned either with (parallel to) or against (antiparallel to) the external field. These two possible orientations do not have the same energy and therefore are not present in equal amounts. The parallel orientation is slightly lower in energy, making this spin state slightly favored over the antiparallel orientation (Figure 13.1).

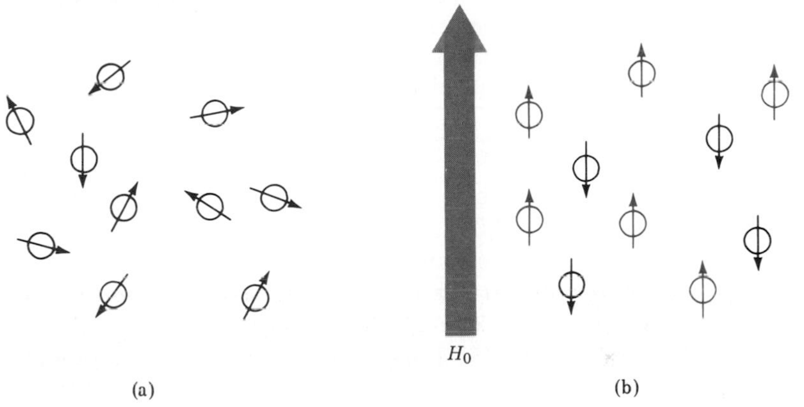


**Figure 13.1** Nuclear spins are oriented randomly in the absence of a strong external magnetic field (a), but have a specific orientation in the presence of an external field, $H_0$ (b). Note that some of the spins (red) are aligned parallel to the external field, whereas others are antiparallel. The parallel spin state is lower in energy.

If the oriented nuclei are now irradiated with electromagnetic radiation of the proper frequency, energy absorption occurs, and the lower energy state "spin-flips" to the higher energy state. When this spin-flip occurs, the nucleus is said to be in resonance with the applied radiation—hence the name, nuclear magnetic resonance.

The exact amount of radio-frequency (rf) energy necessary for resonance depends both on the strength of the external magnetic field and on the identity of the nucleus being irradiated. If a very strong magnetic field is applied, the energy difference between the two spin states is large, and higher-frequency (higher-energy) radiation is required for a spin-flip. If a weaker magnetic field is applied, less energy is required to effect the transition between nuclear spin states (Figure 13.2, page 410).

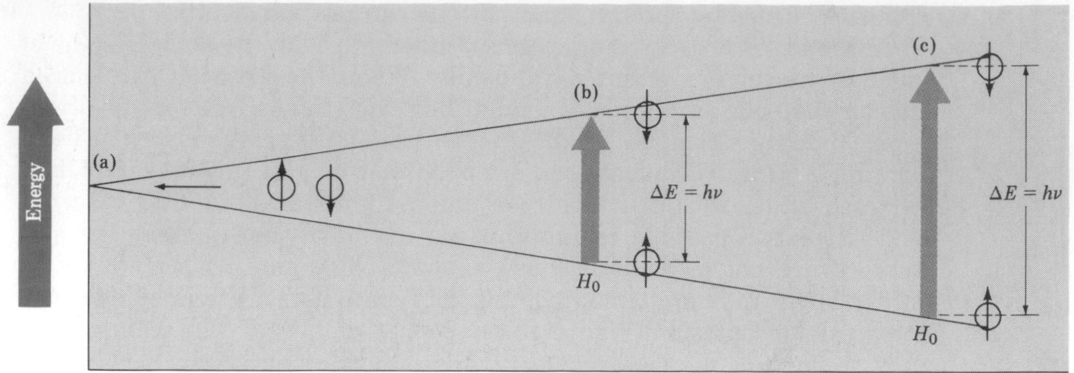


**Figure 13.2** The variation of energy difference between nuclear spin states as a function of the strength of the applied magnetic field: Absorption of rf energy of frequency $\nu$ converts a nucleus from a lower spin state to a higher spin state. (a) Spin states have equal energies in the absence of an applied magnetic field. (b) Spin states have unequal energies in the presence of a magnetic field. At $\nu = 60$ MHz, $\Delta E = 5.7 \times 10^{-6}$ kcal/mol. (c) The energy difference between spin states is greater at larger applied fields. At $\nu = 100$ MHz, $\Delta E = 9.5 \times 10^{-6}$ kcal/mol.

In practice, superconducting magnets that produce enormously powerful fields up to 140,000 gauss are sometimes used, but a field strength of 14,100 gauss is more common. At this magnetic field strength, rf energy in the 60 MHz range (1 MHz = 1 megahertz = 1 million cycles per second) is required to bring a $^{1}H$ nucleus into resonance, and rf energy of 15 MHz is required to bring a $^{13}C$ nucleus into resonance. These energy levels needed for NMR are much smaller than those required for infrared spectroscopy (1.1–11 kcal/mol). For example, 60 MHz rf energy corresponds to only $5.7 \times 10^{-6}$ kcal/mol ($2.4 \times 10^{-5}$ kJ/mol).

The $^{1}H$ and $^{13}C$ nuclei are not unique in their ability to exhibit the nuclear magnetic resonance phenomenon. All nuclei with odd-numbered masses, such as $^{1}H$, $^{13}C$, $^{19}F$, and $^{31}P$, show magnetic properties. Similarly, all nuclei with even-numbered masses but odd atomic numbers show magnetic properties ($^{2}H$ and $^{14}N$, for example). Nuclei having both even masses and even atomic numbers ($^{12}C$, $^{16}O$) do not give rise to magnetic phenomena (Table 13.1).

**Table 13.1** The NMR behavior of some common nuclei

| *Magnetic nuclei* | | *Nonmagnetic nuclei* | |
|---|---|---|---|
| $^{1}H$ | NMR observed | $^{12}C$ | No NMR observed |
| $^{13}C$ | | $^{16}O$ | |
| $^{2}H$ | | $^{32}S$ | |
| $^{14}N$ | | | |
| $^{19}F$ | | | |
| $^{31}P$ | | | |

PROBLEM ...................................................................................................

**13.1** The exact amount of energy required to spin-flip a magnetic nucleus depends not only on the strength of the external magnetic field but also on the intrinsic properties of the specific isotope. We saw earlier that, at a field strength of 14,100 gauss, rf energy of 60 MHz is required to bring a $^{1}H$ nucleus into resonance. At the same field strength, rf energy of 56 MHz will bring a $^{19}F$ nucleus into resonance. Use the equation given in Problem 12.8 to calculate the amount of energy required to spin-flip a $^{19}F$ nucleus. Is this amount greater or less than that required to spin-flip a $^{1}H$ nucleus?

PROBLEM ...................................................................................................

**13.2** Calculate the amount of energy required to spin-flip a proton in a spectrometer operating at 100 MHz. Does increasing the spectrometer frequency from 60 MHz to 100 MHz increase or decrease the amount of energy necessary for resonance?

## 13.2 The Nature of NMR Absorptions

From the description given thus far, you might expect all protons in a molecule to absorb rf energy at the same frequency and all $^{13}C$ nuclei to absorb at the same frequency. If this were true, we would observe only a single NMR absorption band in the $^{1}H$ or $^{13}C$ spectrum of an unknown, a situation that would be of little use for structure determination. In fact, the absorption frequency is not the same for all nuclei.

All nuclei in molecules are surrounded by electron clouds. When a uniform external magnetic field is applied to a sample molecule, the circulating electron clouds set up tiny local magnetic fields of their own. These local magnetic fields act in opposition to the applied field, so that the *effective* field actually felt by the nucleus is a bit smaller than the applied field.

$$H_{\text{effective}} = H_{\text{applied}} - H_{\text{local}}$$

In describing this effect, we say that the nuclei are **shielded** from the applied field by the circulating electron clouds. Since each kind of nucleus in a molecule is in a slightly different electronic environment, each nucleus is shielded to a slightly different extent. Thus, the effective magnetic field actually felt is not the same for each nucleus. If the NMR instrument is sensitive enough, the tiny differences in the effective magnetic fields experienced by different nuclei can be observed, and we can see different NMR signals for each nucleus.

*Each unique kind of proton and each unique kind of $^{13}C$ in a molecule give rise to a unique NMR signal.* Thus, the NMR spectrum of an organic compound provides us with a map of the carbon–hydrogen framework. With practice, we can learn how to read the map and thereby derive structural information about an unknown molecule.

Figure 13.3 shows both the $^{1}H$ and the $^{13}C$ NMR spectra of methyl acetate $CH_3COOCH_3$. The horizontal ($x$) axis shows the difference in effective field strength felt by the nuclei, and the vertical ($y$) axis indicates intensity of absorption of rf energy. Each peak in the NMR spectrum corresponds to a different kind of nucleus in a molecule. Note, however, that $^{1}H$ and $^{13}C$ spectra can't both be observed at the same time on the same

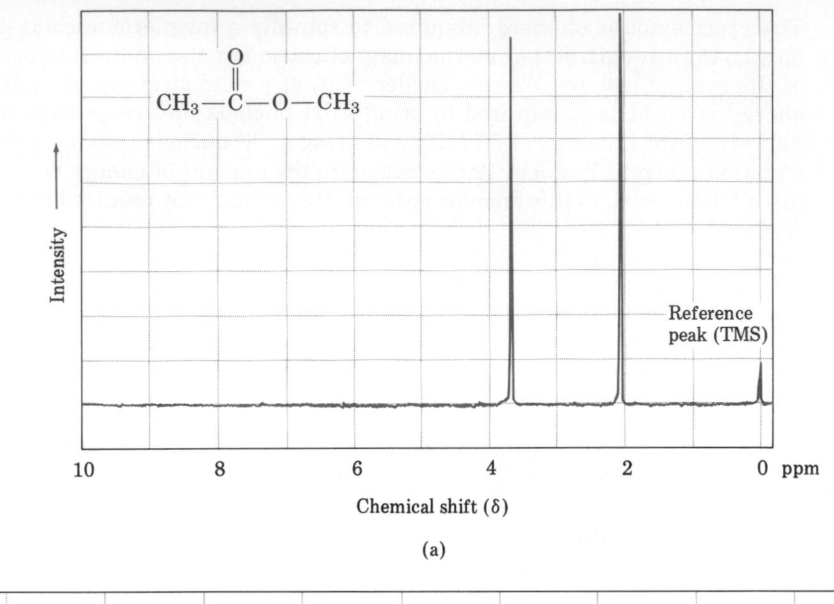


(a)

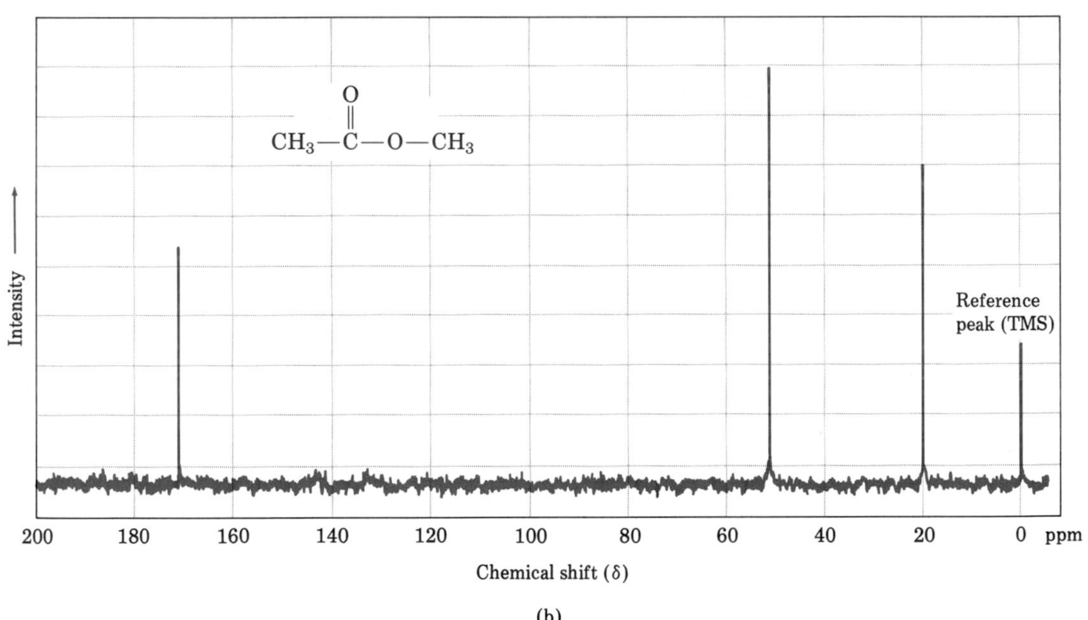


(b)

**Figure 13.3** (a) The $^1$H NMR spectrum and (b) the $^{13}$C NMR spectrum of methyl acetate, $CH_3COOCH_3$

spectrometer, since different amounts of energy are required to spin-flip the different kinds of nuclei. Each of the two kinds of spectra must be recorded separately.

The $^{13}$C spectrum of methyl acetate in Figure 13.3 shows three peaks, one for each of the three different carbon atoms present. The $^1$H NMR spectrum shows only *two* peaks, however, even though methyl acetate has six protons. One peak is due to the $CH_3CO$ protons, and the other to the $OCH_3$ protons. Since the three protons of each methyl group have the same chemical (and magnetic) environment, they are shielded to the same extent and show a single absorption. The two methyl groups themselves, however, are not equivalent; they therefore absorb at different positions.

The operation of an NMR spectrometer is illustrated schematically in Figure 13.4. An organic sample is dissolved in a suitable solvent and placed in a thin glass tube between the poles of a magnet. The strong magnetic field causes the $^{1}H$ and $^{13}C$ nuclei in the molecule to align in one of the two possible orientations, and the sample is then irradiated with rf energy. The exact amount of energy required depends both on the strength of the magnetic field and on the kind of nucleus we intend to observe. As noted before, the two most commonly observed nuclei, $^{1}H$ and $^{13}C$, absorb in quite different rf ranges, and we cannot observe both at the same time.

If the frequency of rf irradiation is held constant and the strength of the applied magnetic field is changed, each nucleus comes into resonance at a slightly different field strength. A sensitive detector monitors the absorption of rf energy, and the electronic signal is then amplified and displayed as a peak on a recorder chart.

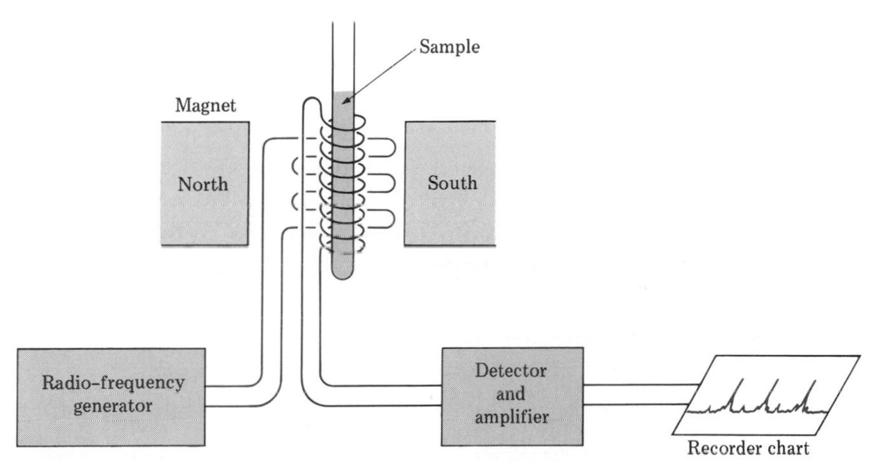


**Figure 13.4** Schematic operation of an NMR spectrometer

Nuclear magnetic resonance spectroscopy differs from infrared spectroscopy (Sections 12.8–12.12) in that the time scales of the two techniques are quite different. The absorption of infrared energy by a molecule giving rise to a change in vibrational states is an essentially instantaneous process (about $10^{-13}$ sec); the NMR process, however, requires far more time (about $10^{-3}$ sec).

The difference in time scales between IR and NMR spectroscopy can be compared to the difference between a camera operating at a very fast shutter speed and a camera operating at a very slow shutter speed. The fast camera (infrared) takes an instantaneous picture and "freezes" the action. If two rapidly interconverting species are present in a sample, infrared will record the spectrum of each. The slow camera (NMR), however, takes a blurred, "time-averaged" picture. If two species interconverting faster than $10^{3}$ times per second are present in a sample, NMR will record only a single, averaged spectrum, rather than separate spectra of the two discrete species.

Because of this "blurring" effect, NMR spectroscopy can sometimes be used to measure the rate of very fast processes such as conformational changes. The NMR spectrum of the molecule undergoing the fast process is recorded at varying temperatures until the point is found where blurring of peaks occurs. By then using the known frequency of radiation, it is possible to calculate the rate of the process under investigation.

PROBLEM

**13.3** How many signals would you expect each of the following molecules to have in its $^1H$ and $^{13}C$ spectra?

(a) $(CH_3)_2C{=}C(CH_3)_2$

(b) Cyclohexane

(c) $H_3COCH_3$

(d) $(CH_3)_3C\overset{O}{\overset{\|}{C}}OCH_3$

(e) $H_3C$–(benzene ring, para)–$CH_3$

(f) 1,1-Dimethylcyclopropane

## 13.3 Chemical Shifts

NMR spectra are displayed on charts that show the applied field strength increasing from left to right (Figure 13.5). Thus, the left part of the chart is the low-field (or **downfield**) side, and the right part is the high-field (or **upfield**) side. In order to define the position of absorptions, the NMR chart

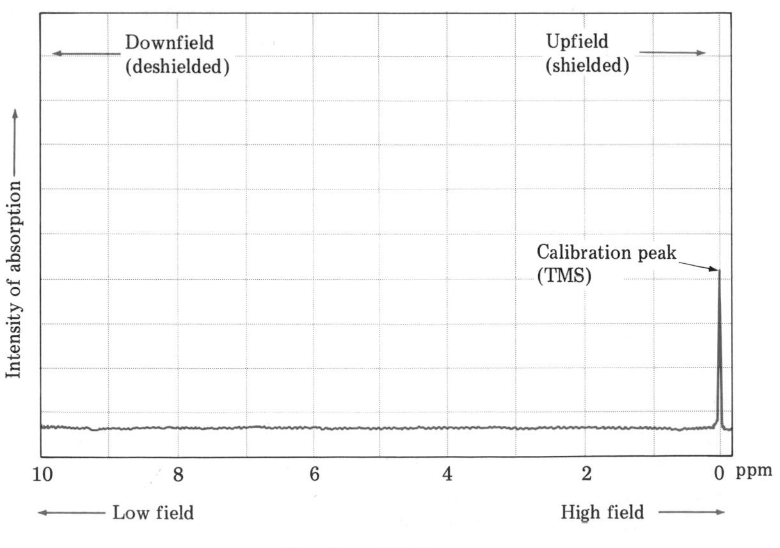


**Figure 13.5** The NMR chart

is calibrated, and a reference point is used. In practice, a small amount of tetramethylsilane [TMS, $(CH_3)_4Si$] is added to the sample so that a standard reference absorption line is produced when the spectrum is run. Tetramethylsilane is used as reference for both $^1H$ and $^{13}C$ measurements because it gives rise in both kinds of spectra to a single peak that occurs upfield (farther right on the chart) of all other absorptions normally found in organic compounds. The $^1H$ and $^{13}C$ spectra of methyl acetate in Figure 13.3 have the TMS reference peak indicated.

The exact place on the chart at which a nucleus absorbs is called its **chemical shift**. By convention, the chemical shift of TMS is arbitrarily set as the zero point, and all other absorptions normally occur downfield (to the left on the chart). For historical reasons, NMR charts are calibrated using an arbitrary scale called the **delta scale**. One delta unit ($\delta$) is equal to 1 part per million (ppm) of the spectrometer operating frequency. For example, if we were measuring the $^1H$ NMR spectrum of a sample using an instrument operating at 60 MHz, 1 $\delta$ (or 1 ppm of $60 \times 10^6$ Hz) would equal 60 Hz. Similarly, if we were measuring the spectrum using a 100 MHz instrument, then 1 $\delta$ = 100 Hz. The following equation can be used for any absorption:

$$\delta = \frac{\text{Observed chemical shift (number of hertz away from TMS)}}{\text{Spectrometer frequency in megahertz}}$$

Although this method of NMR calibration may seem needlessly complex, there is a good reason for it. As we saw earlier, the position of an NMR peak (chemical shift) depends on magnetic field strength, which in turn is proportional to spectrometer frequency. Since many different kinds of spectrometers, operating at many different magnetic field strengths and rf frequencies are available, chemical shifts given in hertz vary greatly from one instrument to another. By using a system of measurement in which NMR absorptions are expressed in *relative* terms (parts per million of spectrometer frequency) rather than in absolute terms (Hz), we can avoid much confusion. *The chemical shift of an NMR absorption given in ppm or $\delta$ units is constant, regardless of the operating frequency of the spectrometer.* A $^1H$ nucleus that absorbs at 2.0 $\delta$ on a 60 MHz instrument (2.0 ppm × 60 MHz = 120 Hz downfield from TMS) also absorbs at 2.0 $\delta$ on a 300 MHz instrument (2.0 ppm × 300 MHz = 600 Hz downfield from TMS).

The range in which most NMR absorptions occur is quite narrow. Almost all $^1H$ NMR absorptions occur 0–10 $\delta$ downfield from the proton absorption of TMS, and almost all $^{13}C$ absorptions occur 1–250 $\delta$ downfield from the carbon absorption of TMS. Thus, there is a considerable likelihood that accidental overlap of nonequivalent signals will occur. The advantage of using an instrument with high field strength (say, 300 MHz) rather than low field strength (60 MHz) is that different NMR absorptions are more widely separated at high field strength. The chances that two signals will accidentally overlap are also lessened, and interpretation of spectra becomes easier. For example, two signals that are only 6 Hz apart at 60 MHz (0.1 ppm) are 30 Hz apart at 300 MHz (still 0.1 ppm).

Let's now look more closely at the interpretation of NMR spectra to see how this tool can be used in organic structure determination. We'll begin by looking at $^{13}C$ NMR. For technical reasons, $^{13}C$ spectra are more difficult to obtain than $^1H$ spectra, but they are also much easier to interpret.

Although the complexities of spectrometer operation differ, the principles behind $^{13}C$ and $^{1}H$ NMR are the same. What we learn now about interpreting $^{13}C$ spectra will simplify the subsequent discussion of $^{1}H$ spectra.

PROBLEM

13.4 When the $^{1}H$ NMR spectrum of acetone, $CH_3COCH_3$, is recorded on an instrument operating at 60 MHz, a single sharp resonance line at 2.1 $\delta$ is observed.
(a) How many hertz downfield from TMS does the acetone resonance line correspond to?
(b) If the $^{1}H$ NMR spectrum of acetone were recorded at 100 MHz, what would be the position of the absorption in $\delta$ units?
(c) How many hertz downfield from TMS does this 100 MHz spectrum correspond to?

PROBLEM

13.5 The following $^{1}H$ NMR resonances were recorded on a spectrometer operating at 60 MHz. Convert each into $\delta$ units.
(a) $CHCl_3$, 436 Hz
(b) $CH_3Cl$, 183 Hz
(c) $CH_3OH$, 208 Hz
(d) $CH_2Cl_2$, 318 Hz

## 13.4 $^{13}C$ NMR

At first glance, it's surprising that carbon NMR is even possible. After all, $^{12}C$, the most abundant carbon isotope, has no nuclear spin. $^{13}C$ is the only naturally occurring carbon isotope with a spin, but its natural abundance is only about 1.1%. Thus, only about 1 out of every 100 carbons in organic molecules is observable by NMR. Although the low abundance of $^{13}C$ means that $^{13}C$ instrumentation must be far more sensitive (and expensive) than that required for $^{1}H$ NMR, these obstacles have been overcome through the use of improved electronics and computer techniques. Today $^{13}C$ NMR is a routine structural tool, and a $^{13}C$ NMR spectrum can often be obtained on 10 mg of sample in a few hours' time.

At its simplest, $^{13}C$ NMR allows us to count the number of carbons in a molecule of unknown structure. In addition, we can get information about the chemical (magnetic) environment of each carbon by observing its chemical shift.

Several different modes of operation are possible with $^{13}C$ NMR instruments. In the normal mode (called the **proton noise-decoupled mode**), a single sharp resonance line is observed for each unique (nonequivalent) kind of carbon atom present in a molecule. The spectrum of methyl acetate (Figure 13.3) illustrates this fact: There are three carbon atoms in methyl acetate and three peaks in its $^{13}C$ NMR spectrum.

Most $^{13}C$ resonances are between 0 and 220 ppm downfield from the TMS reference line, with the exact chemical shift of each $^{13}C$ resonance dependent on that carbon's environment within the molecule. Table 13.2 and Figure 13.6 show how environment and chemical shift may be correlated.

Table 13.2 Carbon-13 NMR chemical shift correlations

| Type of carbon | Chemical shift (δ) | Type of carbon | Chemical shift (δ) |
|---|---|---|---|
| C—I | 0–40 | =C | 100–150 |
| C—Br | 25–65 | C—O | 40–80 |
| C—Cl | 35–80 | C=O | 170–210 |
| —$CH_3$ | 8–30 | (aromatic ring carbon) | 110–160 |
| —$CH_2$— | 15–55 | C—N | 30–65 |
| >CH | 20–60 | | |
| ≡C | 65–85 | | |

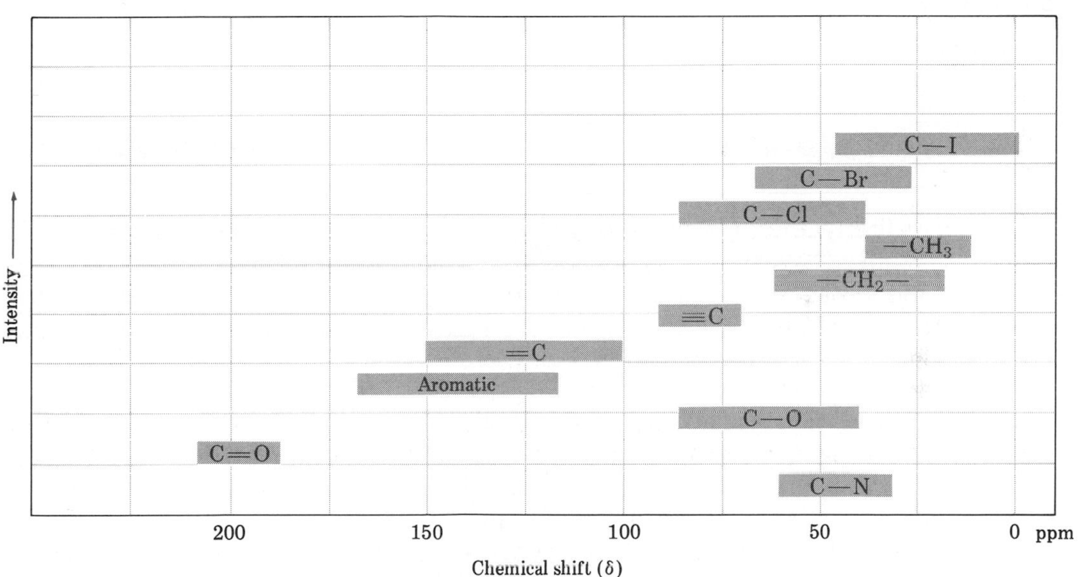


**Figure 13.6** Chemical shift correlations for $^{13}$C NMR

Although the factors involved in determining chemical shifts are complex, we can draw some generalizations from the information in Table 13.2 and Figure 13.6. One clear trend is that the chemical shift of a carbon is affected by the electronegativity of the atoms it is bonded to. Carbons bonded to electronegative atoms like oxygen, nitrogen, or halogen absorb downfield (to the left) of normal alkane carbons. This trend is easy to explain if you think for a moment about the causes of chemical shifts. Since electronegative atoms attract electrons, they pull electrons away from neighboring carbon atoms, causing the carbons to be deshielded and to come into resonance at a lower field than alkane carbons.

Another, less easily explained trend is that $sp^3$-hybridized carbons generally absorb in the 0–90 $\delta$ range and $sp^2$ carbons absorb in the 100–210 $\delta$ range. Carbonyl carbons (C=O) are particularly distinct in $^{13}$C NMR and are easily observed at the extreme low-field end of the spectrum in the range 170–210 $\delta$. Figure 13.7 shows the $^{13}$C NMR spectra of 2-butanone and *para*-bromoacetophenone and indicates the peak assignments. Note that the carbonyl carbons are at the left edge of the spectrum in each case.

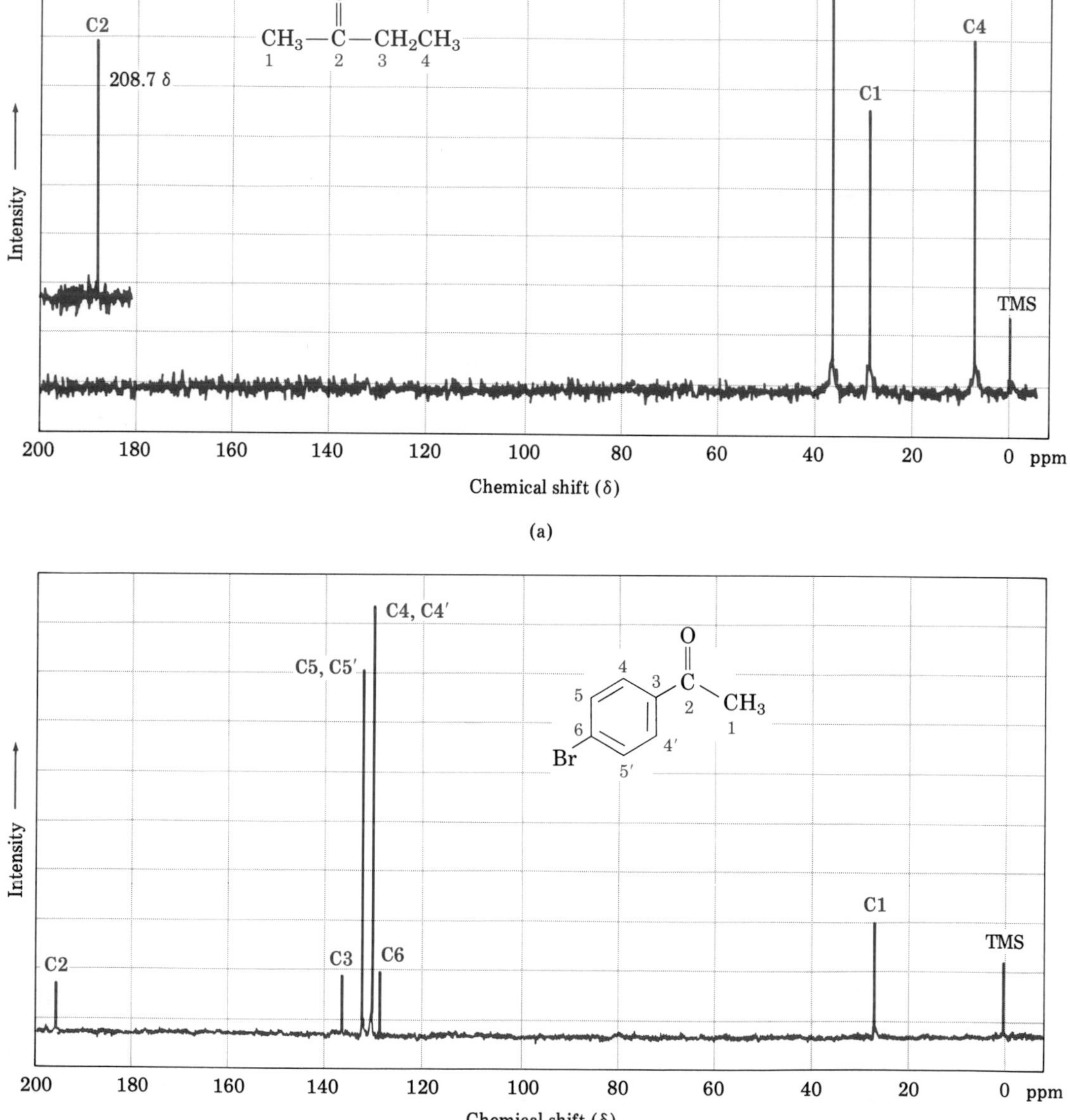


**Figure 13.7** Carbon-13 NMR spectra of (a) 2-butanone and (b) *para*-bromoacetophenone

The $^{13}C$ NMR spectrum of *para*-bromoacetophenone is interesting in several ways. Note particularly that only *six* carbon absorptions are observed even though the molecule contains eight carbons. This is because *para*-bromoacetophenone has a symmetry plane that makes ring carbons 4 and 4′, and carbons 5 and 5′, chemically and magnetically equivalent. They therefore have identical resonance frequencies.

Another interesting feature of both spectra is that the peaks aren't uniform in size. Some peaks appear larger than others even though all are one-carbon resonances (except for the two two-carbon peaks of *para*-bromoacetophenone). This difference in peak size is caused by complex factors that we won't go into, though it does have a relationship to the structure of the sample.

PROBLEM

**13.6** Assign the resonances in the following $^{13}C$ NMR spectrum of methyl propanoate, $CH_3CH_2CO_2CH_3$.

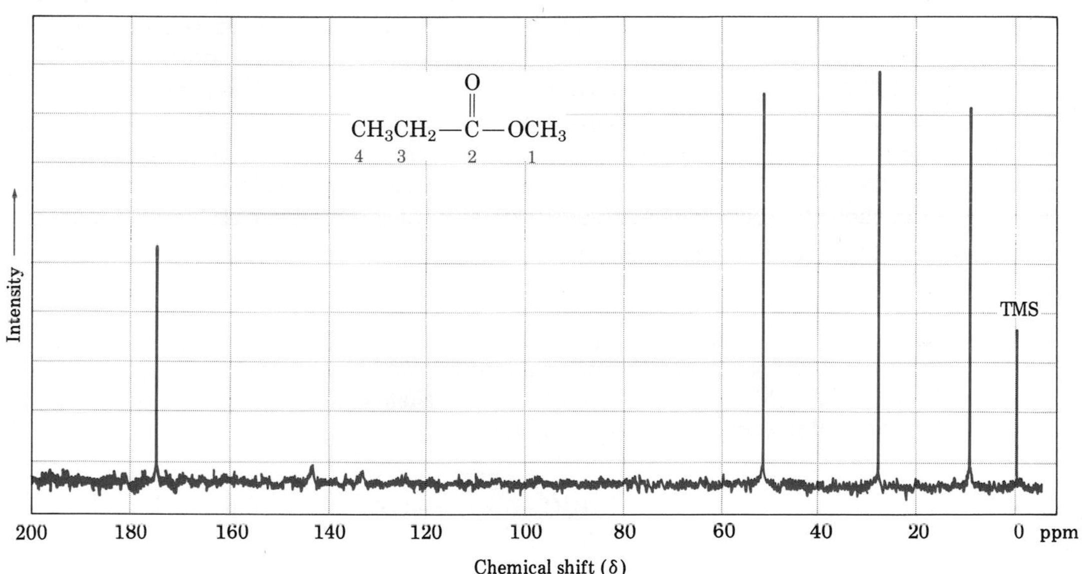


PROBLEM

**13.7** Predict the number of carbon resonance lines you would expect to observe in the $^{13}C$ NMR spectra of these compounds:

(a) Methylcyclopentane
(b) 1-Methylcyclohexene
(c) 1,2-Dimethylbenzene
(d) 2-Methyl-2-butene

PROBLEM

**13.8** Propose structures for compounds that fit these descriptions:

(a) A hydrocarbon with seven lines in its $^{13}C$ NMR spectrum
(b) A six-carbon compound with only five lines in its $^{13}C$ NMR spectrum
(c) A four-carbon compound with three lines in its $^{13}C$ NMR spectrum

## 13.5 Measurement of NMR Peak Areas: Integration

The relative sizes of the different peaks observed in a $^{13}C$ NMR spectrum depend to a considerable extent on the mode of spectrometer operation. When the spectrometer is operated in the normal mode, not all single-carbon peaks have the same area. When the NMR spectrometer is operated in an *integrating* mode (called the **gated-decoupled** mode), however, all single-carbon resonances have the same peak areas. By electronically measuring (integrating) the area under each peak, we can determine the relative number of carbon atoms each peak represents.

Integrated peak areas are presented on the chart in a "stair-step" fashion, with the height of each step proportional to the number of carbons causing that peak. To compare one peak size against another, you simply take out a ruler and measure the heights of the various steps. Figure 13.8 shows the integrated $^{13}C$ NMR spectra of 2-butanone and *para*-bromoacetophenone. Note that all four peaks in the 2-butanone spectrum have the same integrated peak area, while the *para*-bromoacetophenone spectrum shows four equal one-carbon resonance lines and two two-carbon resonance lines.

Integrated spectra contain more information than normal spectra, but this information comes at a price. The NMR spectrometer is two or three times less sensitive in the integrating mode than in the normal mode so that more sample and more time are required to obtain the spectrum. Since this price is not worth paying unless a particular ambiguity exists about the number of carbon atoms in a sample, integrated $^{13}C$ spectra are rarely obtained in practice. The concept of NMR spectra integration is an important one, though, and we'll soon see that peak integration is extremely useful in interpreting $^{1}H$ NMR spectra.

PROBLEM

**13.9** How many lines would you expect to see in the $^{13}C$ NMR spectrum of *para*-dimethylbenzene? What peak areas would you expect to find on integration of the spectrum? Refer to Table 13.2 for approximate values of chemical shifts and then sketch what the integrated spectrum might look like.

$H_3C$—⟨benzene ring⟩—$CH_3$

*para*-Dimethylbenzene

## 13.6 Spin–Spin Splitting of NMR Signals

For most purposes, the normal $^{13}C$ NMR spectrum is fully satisfactory. Obtaining the spectrum in this mode provides a carbon count of the sample molecule and gives information about the environment of each carbon. Occasionally, however, more detailed information is needed, and yet a third mode of spectrometer operation is employed. When the spectrometer is operated in the *spin-coupled* mode (called the **off-resonance mode**), single-carbon resonance lines can split into multiple lines. What is this signal splitting due to, and what use can we make of this phenomenon?

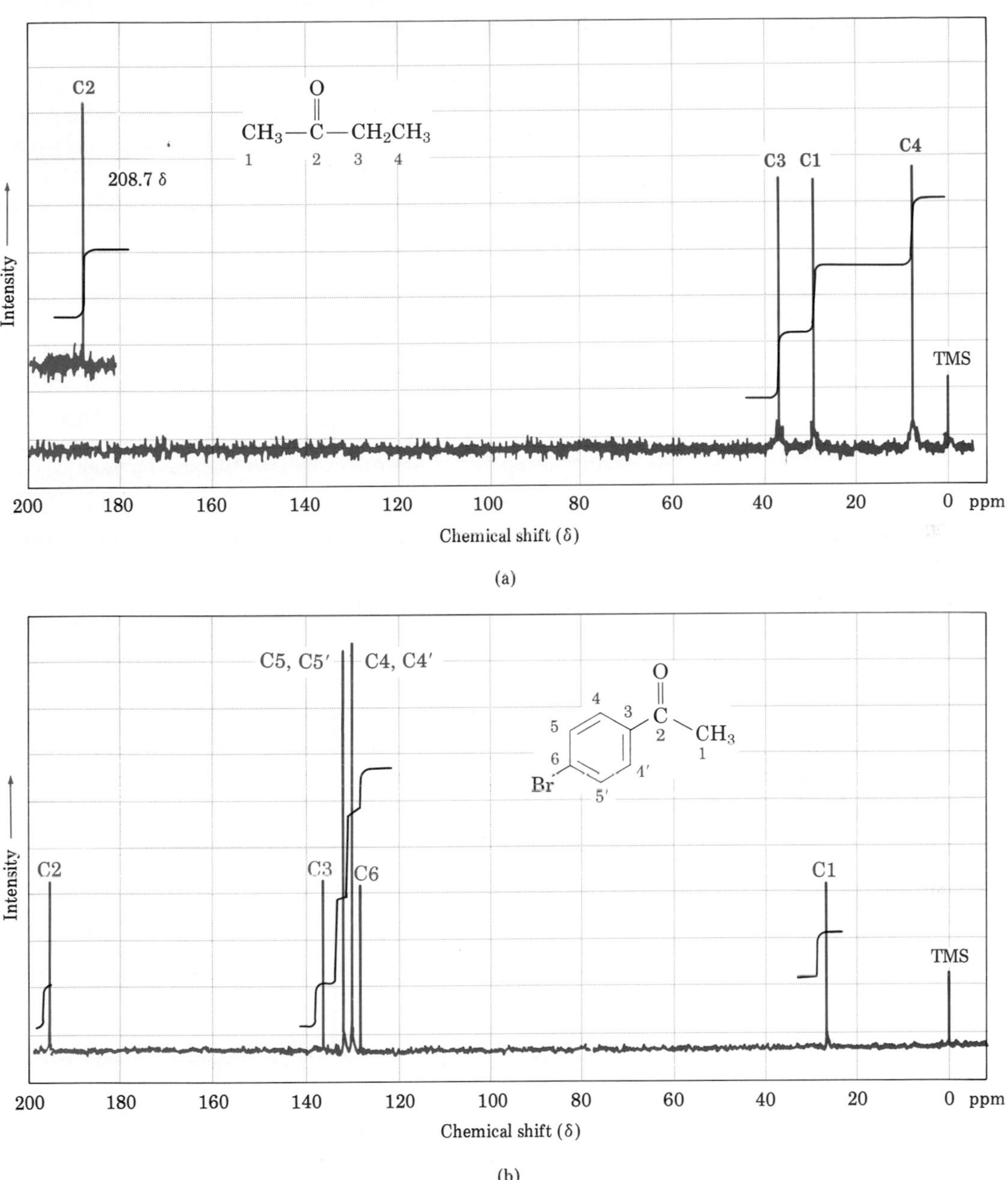


**Figure 13.8** Integrated $^{13}C$ NMR spectra of (a) 2-butanone and (b) *para*-bromoacetophenone

Look carefully at both the normal spectrum and the spin-coupled spectrum of dichloroacetic acid shown in Figure 13.9. When the two kinds of spectra are displayed on the same chart, it becomes clear that the carbonyl-carbon resonance at 170 δ remains a **singlet**, but the methine-carbon resonance ($R_3C—H$) at 64 δ splits into two peaks (a **doublet**). This phenomenon, known as **spin–spin splitting**, is due to the fact that the nuclear spin of one atom interacts, or **couples**, with the nuclear spin of a nearby atom.

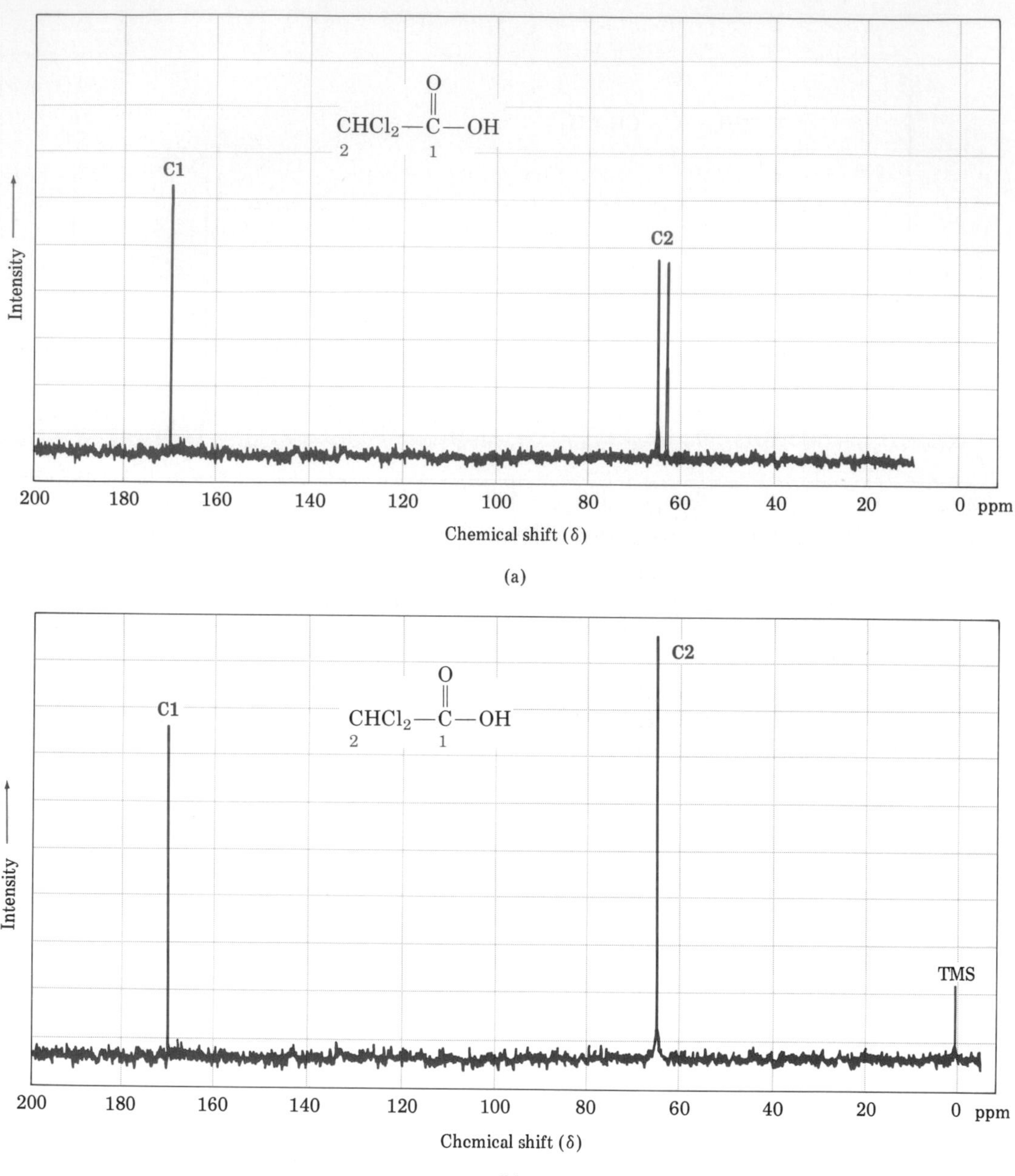


**Figure 13.9** (a) Spin-coupled spectrum and (b) normal spectrum of dichloroacetic acid, $CHCl_2COOH$

In other words, the tiny magnetic field of one nucleus affects the magnetic field felt by a neighboring nucleus.

How does spin–spin splitting arise? Let's review briefly. We've seen that, when a magnetic nucleus such as $^{13}C$ is placed in a strong magnetic field, the nucleus adopts one of two spin states. The $^{13}C$ spin lines up either with or against the applied magnetic field. In addition to orienting the nuclear spins, the applied magnetic field also causes electrons in the molecule to *shield* the nucleus by setting up tiny local magnetic fields that act in opposition to the applied field. Because of this shielding, the effective field actually felt by the nucleus is slightly less than the applied field. Differences in the extent of shielding at each nucleus account for the differences in chemical shifts between nuclei that we observe in $^{13}C$ NMR spectra.

In addition to being affected by electron shielding, the magnetic field felt by a nucleus is also affected by neighboring magnetic nuclei. For example, the C=O carbon of dichloroacetic acid is bonded only to nonmagnetic neighbor atoms, but the $—CHCl_2$ carbon is bonded to another magnetic nucleus—a proton, $^1H$. (Remember that the natural abundance of $^{13}C$ is only about 1.1%, and the likelihood of two $^{13}C$ atoms being adjacent in a molecule is therefore small. Thus, we don't observe $^{13}C–^{13}C$ magnetic interactions.)

When placed in a strong external magnetic field, the spin of the neighboring $^1H$ nucleus aligns either with or against the applied field. If the neighboring spin is aligned *with* the applied field, the total effective field at the neighboring carbon is slightly larger than it would otherwise be. Consequently, the applied field necessary to cause resonance is slightly reduced. Conversely, if the neighboring spin is aligned *against* the applied field, the effective field at the neighboring carbon is slightly smaller than it would otherwise be. Thus, the applied field needed to bring the carbon into resonance is slightly increased. Figure 13.10 shows schematically how spin–spin splitting arises.

The consequence of this coupling of adjacent nuclear spins is that the carbon comes into resonance at two slightly different values of the applied field. One resonance is a little above where it would be without coupling, and the other resonance is a little below where it would be without coupling. We therefore observe a doublet in the $^{13}C$ NMR spectrum of dichloroacetic acid.

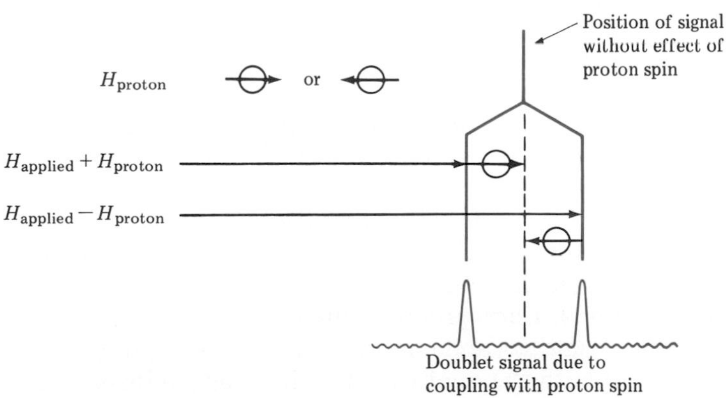


**Figure 13.10** Origin of spin–spin splitting: If the neighboring proton has its spin aligned *with* the applied field, then a lower field value is required for resonance. If the neighboring proton has its spin *against* the applied field, then a higher value is required. Resonance therefore occurs at two different values of the applied field, and a doublet absorption is observed.

It is important to realize that the signal from an *individual* carbon atom in an individual molecule is not a doublet; an individual carbon atom absorbs either at one position or the other depending on the orientation of its own adjacent hydrogen. We see a doublet in the NMR spectrum only because we are observing the summation of signals from a great many molecules.

If a carbon atom is bonded to more than one proton, more complex spin–spin splitting is observed. For example, Figure 13.11 shows both the normal spectrum and the spin-coupled spectrum of 2-butanone.

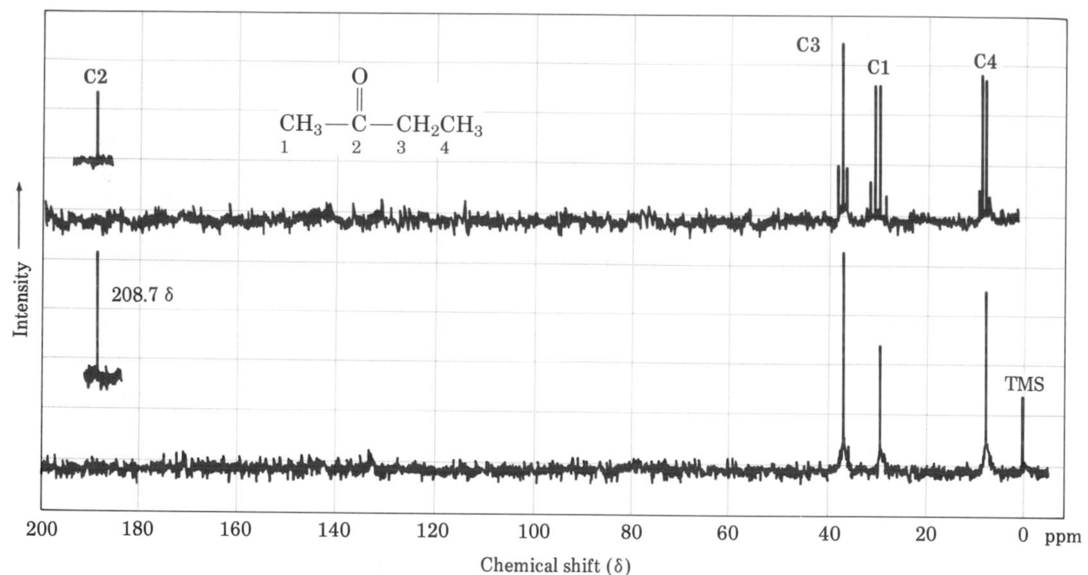


**Figure 13.11** Normal spectrum (lower scan) and spin-coupled spectrum (upper scan) of 2-butanone

The carbonyl-carbon (C2) resonance of 2-butanone remains a singlet in the spin-coupled spectrum at 208 δ because it is not bonded to any protons. The —$CH_2$— resonance at 36.8 δ splits into a **triplet**, however, and the two —$CH_3$ resonances at 6.6 δ and at 28.8 δ become **quartets**. According to the ***n* + 1 rule**, a carbon bonded to $n$ protons shows $n + 1$ peaks in its spin-coupled $^{13}C$ spectrum. How does this rule arise?

A methylene carbon (—$CH_2$—) is bonded to two other magnetic nuclei (protons), and the spins of these two protons can orient in three different combinations. Both proton spins can orient against the applied field; one can orient against and one with the applied field (two possibilities); or both can orient with the applied field. This leads to three peaks of relative intensity 1:2:1 in the spin-coupled spectrum of a —$CH_2$— carbon (Figure 13.12).

A similar analysis for —$CH_3$ carbons predicts that spin–spin splitting should lead to the observation of quartets of 1:3:3:1 intensity in the spin-coupled $^{13}C$ NMR spectrum. This is, in fact, observed (Figure 13.12). To repeat: A carbon bonded to $n$ protons gives a signal that is split into $n + 1$ peaks in the spin-coupled spectrum.

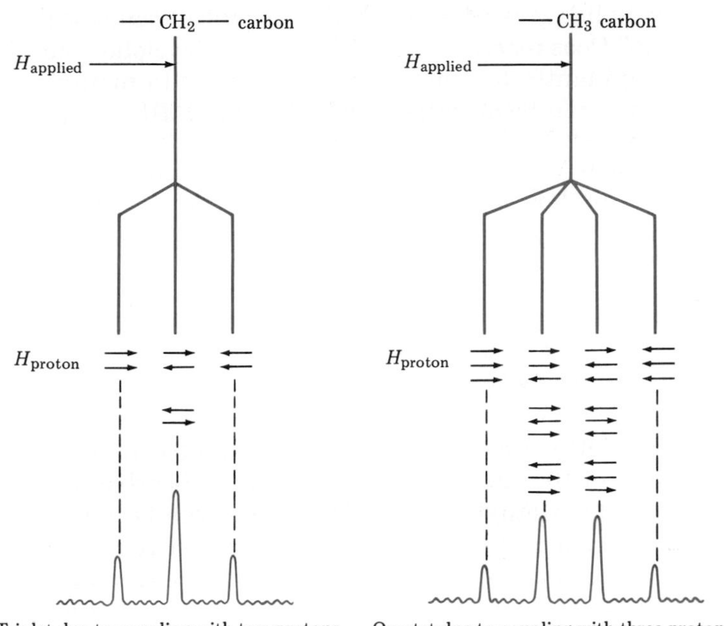


**Figure 13.12** Spin–spin splitting of a methylene carbon (—$CH_2$—) and a methyl carbon (—$CH_3$): The signal of a carbon bonded to $n$ hydrogens splits into $n + 1$ peaks (the $n + 1$ rule).

PROBLEM

**13.10** Sketch what the normal $^{13}$C NMR spectrum of propene might look like.
(a) How many carbon resonances are there?
(b) Into how many peaks does each carbon resonance split in the spin-coupled spectrum?

PROBLEM

**13.11** Sketch what the normal $^{13}$C NMR spectrum of cyclopentene might look like.
(a) How many carbon resonances are there?
(b) What further information would the integrated spectrum give you? Make a sketch of what the spin-coupled spectrum of cyclopentene might look like.

## 13.7 Summary of $^{13}$C NMR

The information derived from $^{13}$C NMR spectroscopy is extraordinarily useful for structure determination. Not only can we count the number of different kinds of carbon atoms in an unknown and deduce information about their chemical environments, we can also find how many protons are bonded to each carbon. This allows us to answer many structural questions that cannot be easily handled by infrared spectroscopy or mass spectroscopy.

Let's take an example. How might we prove that E2 elimination of an alkyl halide gives the more highly substituted alkene (Zaitsev's rule, Section 11.10)? Does reaction of 1-chloro-1-methylcyclohexane with strong base lead predominantly to 1-methylcyclohexene or to methylenecyclohexane? This question can be answered easily by $^{13}C$ NMR.

$H_3C$ Cl → (KOH, Ethanol) → $CH_3$ or $CH_2$ ?

1-Chloro-1-methylcyclohexane — 1-Methylcyclohexene — Methylenecyclohexane

1-Methylcyclohexene would be expected to show five $sp^3$-carbon resonances in the range 20–50 $\delta$ and two $sp^2$-carbon resonances in the range 100–150 $\delta$. Methylenecyclohexane, however, because of its symmetry, would be expected to have only three $sp^3$-carbon resonance peaks and two $sp^2$-carbon peaks. The spectrum of the actual reaction product shown in Figure 13.13 clearly identifies 1-methylcyclohexene as the substance formed in this E2 reaction.

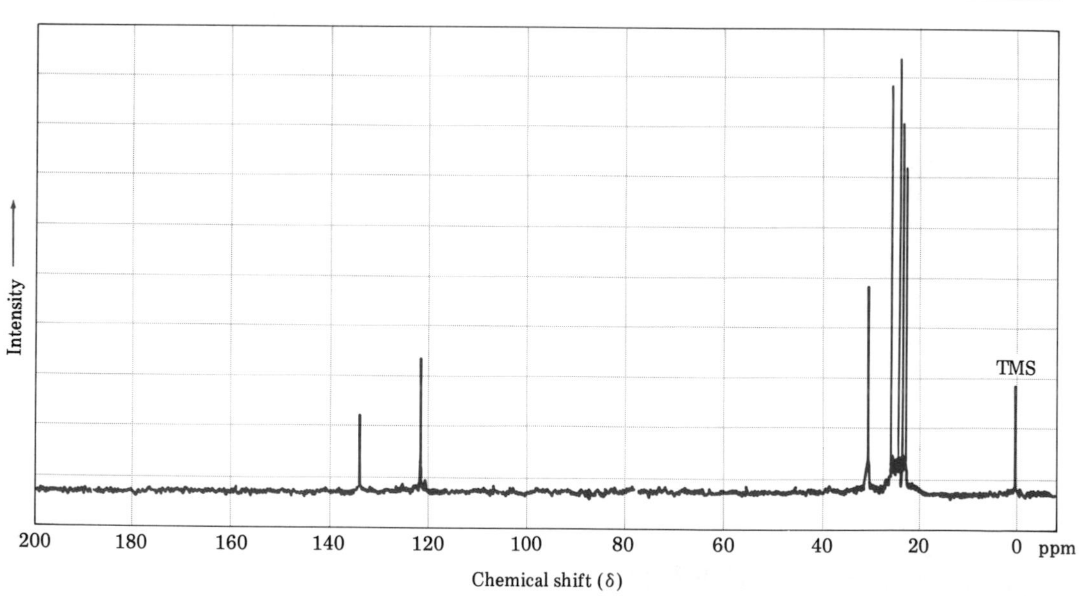


**Figure 13.13** Normal $^{13}C$ NMR spectrum of the elimination product from 1-chloro-1-methylcyclohexane

A summary of the operating modes of $^{13}C$ NMR spectrometers and of the kinds of information obtainable in each mode is given in Table 13.3.

Table 13.3 Summary of $^{13}$C NMR and the modes of spectrometer operation

| *Spectrometer operating mode* | *Structural information obtained* |
|---|---|
| Normal mode (proton noise-decoupled) | Gives one resonance line for each kind of nonequivalent carbon atom, allowing us to count the number of kinds of carbons and deduce their environments from chemical shifts. |
| Integrating mode (gated-decoupled) | All resonances are of equal peak area, allowing us to integrate the area under each and determine the relative number of carbons each peak represents. |
| Spin-coupled mode (off-resonance) | Spin–spin splitting causes carbon resonances to split into **multiplets**, allowing us to determine how many protons are bonded to each carbon (the $n + 1$ rule). |

PROBLEM

**13.12** We saw in Section 7.5 that addition of HBr to alkenes under radical conditions leads to the non-Markovnikov product—the bromine bonds to the less highly substituted carbon. How could you use $^{13}$C NMR to identify the product of radical addition of HBr to 2-methylpropene?

$$(CH_3)_2C{=}CH_2 \xrightarrow[\text{Radicals}]{\text{HBr}} ?$$

## 13.8 $^1$H NMR

We have seen four general features of NMR spectra in the previous pages:

1. *Number of NMR absorptions* Each nonequivalent $^{13}$C or $^1$H nucleus can give rise to a separate absorption peak.
2. *Chemical shifts* The exact position of an NMR absorption tells the chemical (magnetic) environment of a given nucleus.
3. *Integration of NMR absorptions* Electronic integration of the area under a peak tells how many nuclei cause that specific resonance peak.
4. *Spin–spin splitting* The splitting of resonance lines into multiplets by coupling of nuclear spins provides information about neighboring magnetic nuclei.

*All four of these features are just as applicable to $^1$H NMR spectra as they are to $^{13}$C NMR spectra.* In fact, they are even easier to observe for $^1$H spectra, since normal spectrometer operating conditions enable signal integration and spin–spin splitting. Proton NMR spectra are so easy to obtain that it usually takes only a few minutes' time and several milligrams of sample. Let's look at each of the four features in more detail.

## 13.9 Number of NMR Absorptions: Proton Equivalence

In $^{13}C$ NMR, each unique kind of carbon atom normally gives rise to a distinct peak in the spectrum. Exactly the same is true in $^{1}H$ NMR. Each unique kind of proton gives rise to an absorption peak, and we can use this information to determine how many different kinds of protons are present. For example, Figure 13.14 shows the $^{1}H$ NMR spectrum of methyl chloroacetate, $ClCH_2COOCH_3$. There are two kinds of protons present, Cl—$CH_2$— protons and —$CH_3$ protons; each gives rise to its own signal.

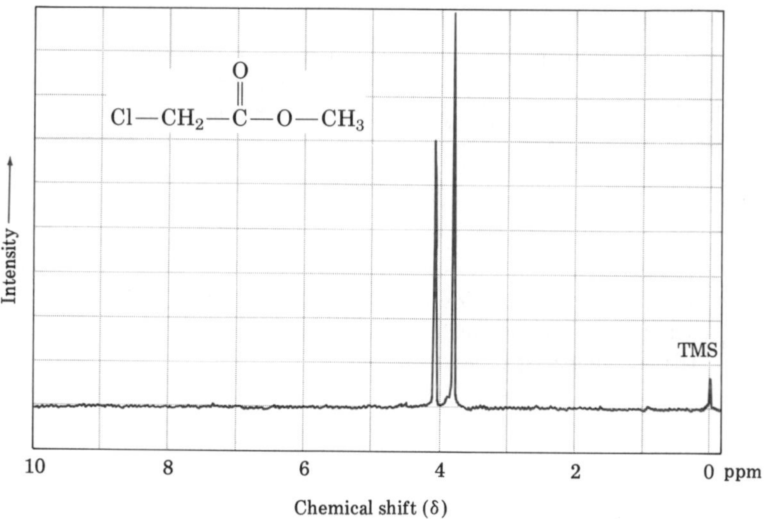


**Figure 13.14** The $^{1}H$ NMR spectrum of methyl chloroacetate, $ClCH_2COOCH_3$

Simple visual inspection of a structure is usually enough to decide how many kinds of nonequivalent protons are present. If doubt exists, however, the equivalence or nonequivalence of protons can be determined by asking whether or not we would get the same structure or different structures by mentally substituting an X group for one of the protons. If the protons are chemically equivalent, we would get the same product no matter which proton we substituted for. If they are not chemically equivalent, we would get different products on substitution.

For example, all 12 protons in 2,3-dimethyl-2-butene are equivalent. No matter which proton we mentally replace by an X group, we get the same structure. The 12 protons thus give rise to a single sharp $^{1}H$ NMR peak (Figure 13.15).

$(H_3C)_2C{=}C(CH_3)_2 \xrightarrow[\text{substitution}]{\text{Mental}} (H_3C)_2C{=}C(CH_3)(CH_2{-}X)$

All four methyl groups are equivalent.

Only one substitution product is possible.

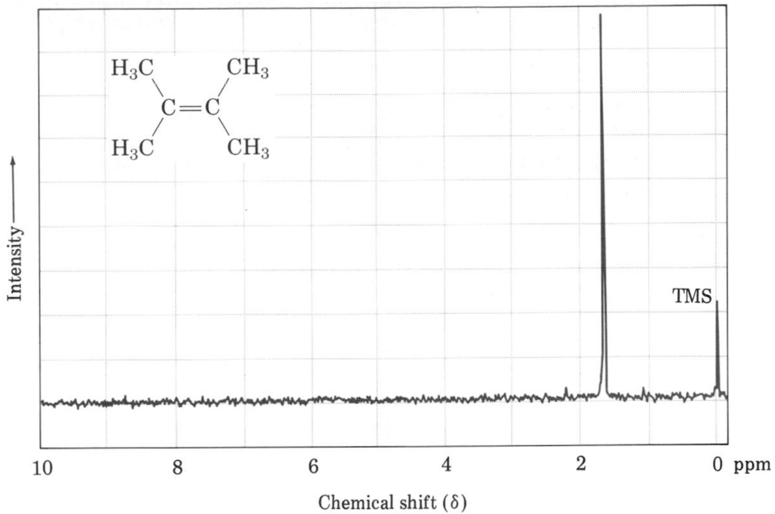


**Figure 13.15** The $^1H$ NMR spectrum of 2,3-dimethyl-2-butene

In contrast, the 10 protons of 2-methyl-2-butene are *not* all equivalent. There are three different kinds of methyl-group protons and one vinylic proton, leading to four different signals.

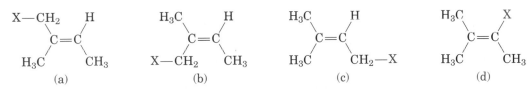

(a) $H_3C$ H (d)
C=C
(b) $H_3C$ $CH_3$ (c)

| (a) | (b) | (c) | (d) |
|---|---|---|---|
| $X{-}CH_2$, H / C=C / $H_3C$, $CH_3$ | $H_3C$, H / C=C / $X{-}CH_2$, $CH_3$ | $H_3C$, H / C=C / $H_3C$, $CH_2{-}X$ | $H_3C$, X / C=C / $H_3C$, $CH_3$ |

Four different substitution products

**PROBLEM**

**13.13** How many nonequivalent kinds of protons are present in these compounds?
(a) $CH_3CH_2Br$ (b) $CH_3OCH_2CH(CH_3)_2$ (c) $CH_3CH_2CH_2NO_2$
(d) Methylbenzene (e) 2-Methyl-1-butene (f) *cis*-3-Hexene

## 13.10 Chemical Shifts in $^1H$ NMR

We saw in $^{13}C$ NMR that differences in chemical shift are caused by small local magnetic fields due to electrons surrounding the various nuclei. Nuclei that are strongly shielded by electrons require a higher applied field to bring them into resonance (to the *right* on the NMR chart). Conversely, nuclei that are not strongly shielded absorb in the low-field region of the spectrum

Table 13.4 Correlation of $^1$H chemical shift with environment

| *Type of proton* | *Formula* | *Chemical shift (δ)* |
|---|---|---|
| Reference peak | $(CH_3)_4Si$ | 0 |
| Saturated primary | $—CH_3$ | 0.7–1.3 |
| Saturated secondary | $—CH_2—$ | 1.2–1.4 |
| Saturated tertiary | $\gt C—H$ (tertiary C–H) | 1.4–1.7 |
| Allylic primary | $\gt C{=}C—CH_3$ | 1.6–1.9 |
| Methyl ketones | $—C(=O)—CH_3$ | 2.1–2.4 |
| Aromatic methyl | $Ar—CH_3$ | 2.5–2.7 |
| Alkyl chloride | $Cl—C—H$ | 3.0–4.0 |
| Alkyl bromide | $Br—C—H$ | 2.5–4.0 |
| Alkyl iodide | $I—C—H$ | 2.0–4.0 |
| Alcohol, ether | $—O—C—H$ | 3.3–4.0 |
| Alkynyl | $—C{\equiv}C—H$ | 2.5–2.7 |
| Vinylic | $\gt C{=}C—H$ | 5.0–6.5 |
| Aromatic | $Ar—H$ | 6.5–8.0 |
| Aldehyde | $—C(=O)—H$ | 9.7–10.0 |
| Carboxylic acid | $—C(=O)—O—H$ | 11.0–12.0 |
| Alcohol | $—C—O—H$ | Extremely variable (2.5–5.0) |

(*left* on the NMR chart). Everything we have learned about $^{13}$C chemical shifts is also applicable to proton shifts. Proton chemical shifts tell a great deal about the chemical (magnetic) environments within a molecule.

Proton chemical shifts are expressed in $\delta$ units (recall 1 $\delta$ = 1 ppm of spectrometer frequency), just as for $^{13}$C shifts. In contrast to $^{13}$C shifts, however, proton chemical shifts fall within the rather narrow range of 0–10 $\delta$. The exact spot within this range is highly characteristic of environment, as shown in Table 13.4 and Figure 13.16.

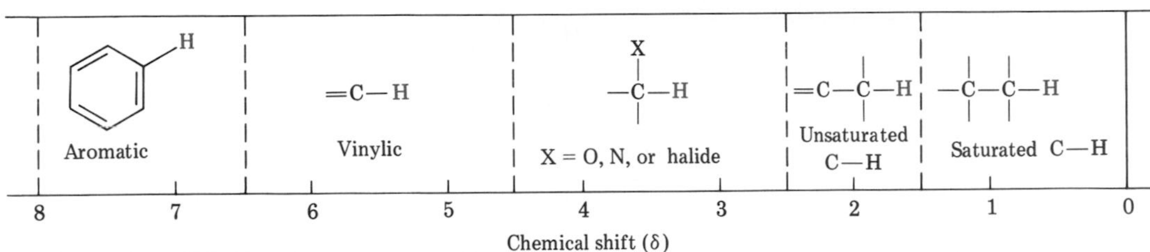


**Figure 13.16** Chemical shifts for different kinds of protons

The majority of $^1$H NMR absorptions occur from 0 to 8 $\delta$, a range that can be conveniently divided into five regions, as shown in Table 13.5. By memorizing the positions of these five regions, it's often possible to tell at a glance what general kinds of protons a molecule contains.

**Table 13.5** Regions of the $^1$H NMR spectrum

| *Region (δ)* | *Proton type* | *Comments* |
|---|---|---|
| 0–1.5 | —C—C—H | Protons on carbon next to saturated centers absorb in this region. Thus the alkane portions of most organic molecules show complex absorption here. |
| 1.5–2.5 | =C—C—H | Protons on carbon next to unsaturated centers (allylic, benzylic, next to carbonyl) show characteristic absorptions in this region, just downfield from other alkane resonance. |
| 2.5–4.5 | X—C—H | Protons on carbon next to electronegative atoms (halogen, O, N) are deshielded because of the electron-withdrawing ability of these atoms. Thus the protons absorb in this midfield region. |
| 4.5–6.5 | C=C—H | Protons on double-bond carbons (vinylic protons) are strongly deshielded by the neighboring pi bond and therefore absorb in this characteristic downfield region. |
| 6.5–8.0 | Aryl H (benzene ring—H) | Protons on aromatic rings (aryl protons) are strongly deshielded by the pi orbitals of the ring and absorb in this characteristic low-field range. |

**PROBLEM**

**13.14** Each of the following compounds exhibits a single $^1H$ NMR peak. Approximately where would you expect each compound to absorb?

(a) Cyclohexane (b) $CH_3COCH_3$ (c) Benzene

(d) Glyoxal, $H-\overset{O}{\overset{\|}{C}}-\overset{O}{\overset{\|}{C}}-H$ (e) $CH_2Cl_2$ (f) $(CH_3)_3N$

**PROBLEM**

**13.15** Identify the different kinds of protons in the following molecule, and tell where you would expect each to absorb.

H H H $CH_2CH_3$ C C $CH_3O$ H H H

## 13.11 Integration of NMR Absorptions: Proton Counting

Electronic integration of $^1H$ NMR peak areas tells us the relative numbers of protons responsible for those peaks. Peak integration was introduced during the discussion of $^{13}C$ NMR, but we remarked at that time that $^{13}C$ integration is not normally needed because each peak is usually due to a single carbon. By contrast, integration of $^1H$ NMR spectra is extremely useful since these spectra often show complicated patterns that are difficult to sort out.

An integrated $^1H$ spectrum is presented in a stair-step manner, with the height of each step proportional to the number of protons represented by that peak. For example, the integrated spectrum of methyl 2,2-dimethylpropanoate given in Figure 13.17 shows that the areas of the two absorptions

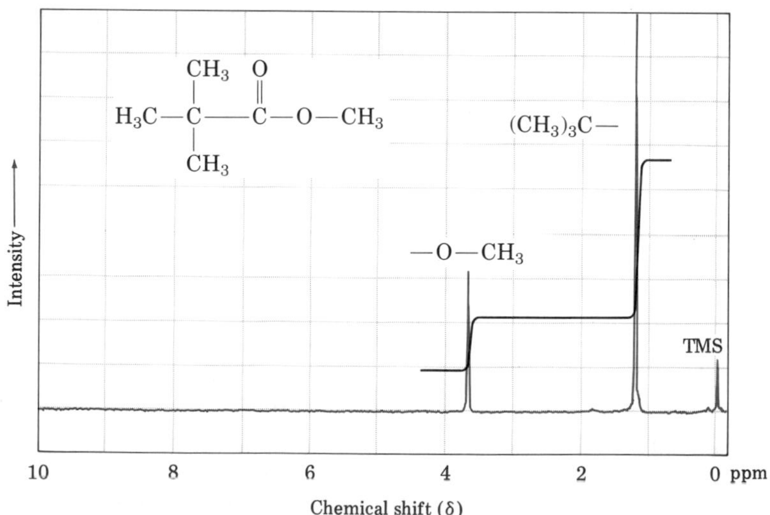


**Figure 13.17** The integrated $^1H$ NMR spectrum of methyl 2,2-dimethylpropanoate: The two peaks have a ratio of 3:9 or 1:3.

have a ratio of 1:3. This is just what we would expect, since the three —$OCH_3$ protons are equivalent and the nine $(CH_3)_3C$— protons are equivalent.

## 13.12 Spin–Spin Splitting in $^1H$ NMR Spectra

Spin–spin splitting of single absorption peaks into multiplets is due to the interaction or coupling of neighboring nuclear spins. For example, we've seen that a $^{13}C$ nucleus can couple with one or more nearby protons, leading to signal splitting when the $^{13}C$ off-resonance NMR spectrum is recorded. Spin–spin splitting is also observed in $^1H$ NMR.

Just as the spin of a $^{13}C$ nucleus can couple with the spins of neighboring protons, so the spin of one proton can couple with the spins of neighboring protons. (Because of the low natural abundance of $^{13}C$, the coupling of a proton's spin to a $^{13}C$ nucleus is of too low an intensity to be observed.) The resultant splitting patterns can be complex, but they can also provide much information. For example, Figure 13.18 shows the $^1H$ NMR spectrum of chloroethane. There are two distinct groups of peaks corresponding to the two different kinds of protons present, and we can account for the observed splitting pattern just as we accounted for splitting in $^{13}C$ spectra.

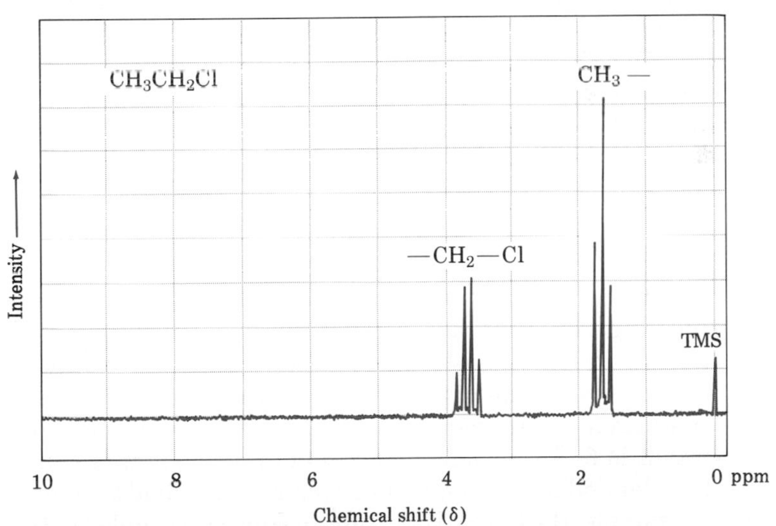


**Figure 13.18** The $^1H$ NMR spectrum of chloroethane, $CH_3CH_2Cl$

The two Cl—$CH_2$— protons at 3.5 $\delta$ are chemically equivalent and do not split each other's signals. Their signals *are*, however, split by the three protons on the neighboring carbon. Since each of the three neighboring protons can have its own spin aligned either with or against the applied field, there are four possible coupling combinations (remember the $n + 1$ rule).

Similarly, the $—CH_3$ protons of chloroethane are equivalent and do not split each other's signals. The methyl-proton signals *are* split by the two neighboring $Cl—CH_2—$ protons, however, and a triplet is therefore observed in the $^1H$ NMR spectrum (Figure 13.19).

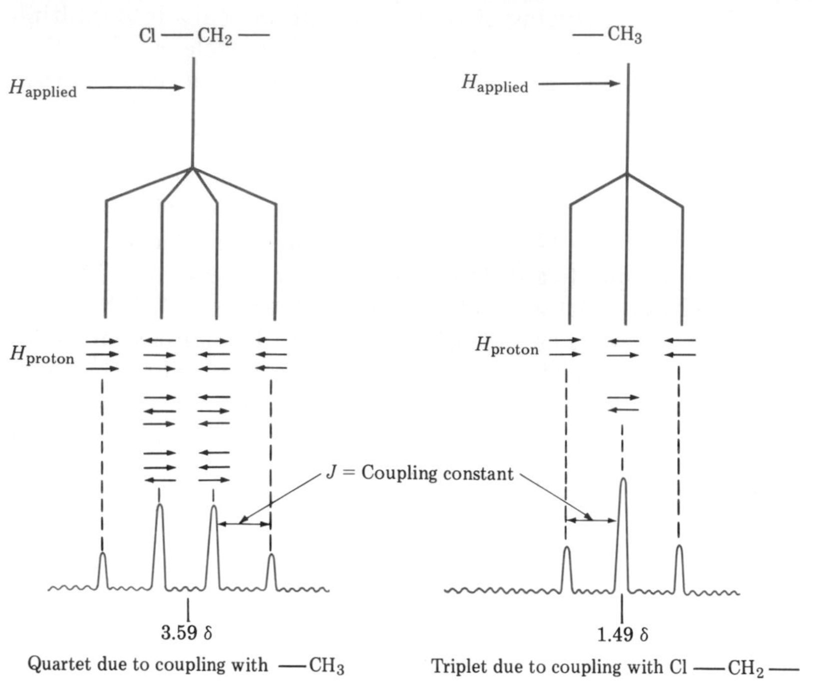


**Figure 13.19** Spin–spin splitting in chloroethane

The distance between individual peaks in the multiplets is called the **coupling constant**, denoted *J*. Coupling constants are measured in hertz and fall in the range 0–18 Hz. The exact value of the coupling constant between two groups of protons depends on several factors (such as geometric constraints on the molecule), but a typical value for an open-chain alkyl system is 6–8 Hz. Note that a coupling constant is shared by both groups of nuclei and is independent of spectrometer field strength. In chloroethane, for example, the $Cl—CH_2—$ proton spins are coupled with the $—CH_3$ proton spins and appear as a quartet with $J = 7$ Hz. The $—CH_3$ protons appear as a triplet with exactly the same coupling constant, $J = 7$ Hz.

Since coupling is a reciprocal interaction between the spins of two adjacent groups of protons, we can sometimes use this fact to tell which multiplets in a complex spectrum are related to each other. It often happens that an NMR spectrum contains many multiplets, and it is sometimes difficult to tell what is coupled with what. If two multiplets have exactly the same coupling constant, however, they are probably related, and the protons causing those multiplets are therefore adjacent in the molecule.

The chloroethane spectrum in Figure 13.19 illustrates three important rules about spin–spin splitting in $^1$H NMR:

1. Chemically equivalent protons do not exhibit spin–spin splitting. The equivalent protons may be on the same carbon or on different carbons, but their signals do not split.

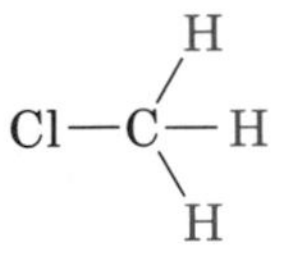

```
       H                 H     H
      /                   \   /
Cl—C—H              Cl—C—C—Cl
      \                   /   \
       H                 H     H
```

Three C—H protons are chemically equivalent; no splitting occurs.

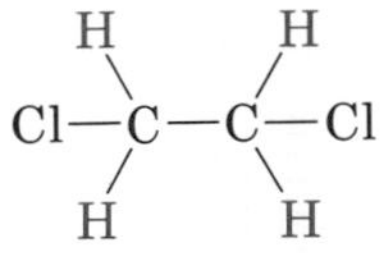

Four C—H protons are chemically equivalent; no splitting occurs.

2. A proton that has $n$ equivalent neighboring protons gives a signal that is split into a multiplet of $n + 1$ peaks with coupling constant $J$. Protons that are farther than two carbon atoms apart do not usually couple, although they sometimes show small coupling when they are separated by a pi bond.

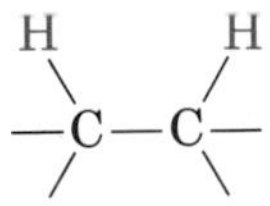

```
 H     H            H           H
  \   /              \    |    /
 —C—C—              —C—C—C—
  /   \              /    |    \
```

Splitting observed

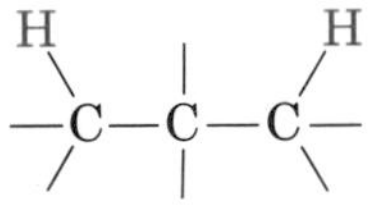

Splitting not usually observed

The most commonly observed coupling patterns are listed in Table 13.6, along with the relative intensities of the multiplet signals.

3. Two groups of protons coupled to each other must have the same coupling constant, $J$.

**Table 13.6** Some common spin multiplicities

| *Number of equivalent adjacent protons* | *Type of multiplet observed* | *Ratio of intensities* |
|---|---|---|
| 0 | Singlet | 1 |
| 1 | Doublet | 1:1 |
| 2 | Triplet | 1:2:1 |
| 3 | Quartet | 1:3:3:1 |
| 4 | Quintet | 1:4:6:4:1 |
| 5 | Sextet | 1:5:10:10:5:1 |
| 6 | Septet | 1:6:15:20:15:6:1 |

The spectra of 2-bromopropane and *para*-methoxypropiophenone in Figure 13.20 on page 436 further illustrate these three rules.

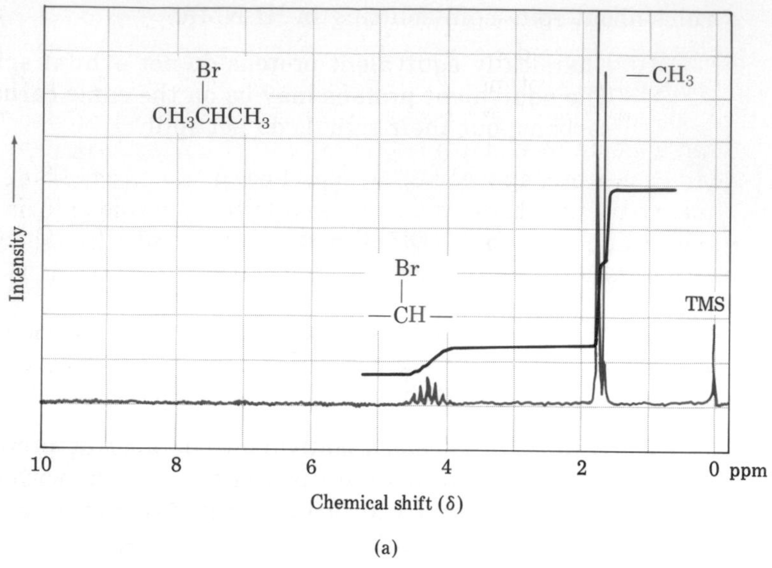

(a)

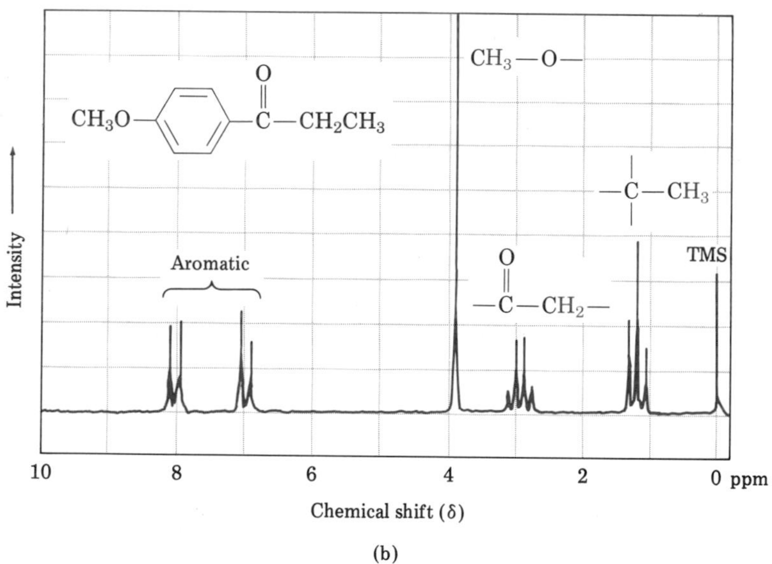

(b)

**Figure 13.20** The $^1$H NMR spectra of (a) 2-bromopropane and (b) *para*-methoxypropiophenone

The 2-bromopropane spectrum shows two groups of signals split into a doublet at 1.71 $\delta$ and a septet at 4.32 $\delta$. The downfield septet is due to splitting of the —CHBr— proton signal by six equivalent neighboring protons on the two methyl groups ($n = 6$ leads to $6 + 1 = 7$ peaks). The upfield doublet is due to signal splitting of the six equivalent methyl protons by the single —CHBr— proton ($n = 1$ leads to 2 peaks). Both multiplets have the same coupling constant, $J = 7$ Hz, and integration confirms the expected 6:1 ratio.

The *para*-methoxypropiophenone spectrum is more complex, but can nevertheless be interpreted in a straightforward way. The downfield absorptions at 6.98 and 8.0 $\delta$ are due to the four aromatic ring protons. There are two kinds of protons, each of which gives a signal that is split into a doublet by its neighbor. Thus we see two doublets. The $CH_3$—O— signal is unsplit and appears as a sharp singlet at 3.90 $\delta$. The —$CH_2$— protons next to the carbonyl group appear at 2.95 $\delta$ in the region expected for protons on carbon next to an unsaturated center, and their signal is split into a quartet due to coupling with the neighboring methyl group. The methyl-group protons appear as a triplet at 1.2 $\delta$ in the usual upfield region.

PROBLEM

**13.16** Predict the splitting patterns you would expect for each proton in these molecules:

(a) $CHBr_2CH_3$ (b) $CH_3OCH_2CH_2Br$

(c) $ClCH_2CH_2CH_2Cl$ (d) $CH_3CH_2O\overset{O}{\overset{\|}{C}}CH(CH_3)_2$

PROBLEM

**13.17** Draw structures for compounds that meet these descriptions:

(a) $C_2H_6O$; one singlet (b) $C_3H_7Cl$; one doublet and one septet

(c) $C_4H_8Cl_2O$; two triplets (d) $C_4H_8O_2$; one singlet, one triplet, and one quartet

PROBLEM

**13.18** The integrated $^1H$ NMR spectrum of a compound of formula $C_4H_{10}O$ is shown. Propose a structure consistent with the data.

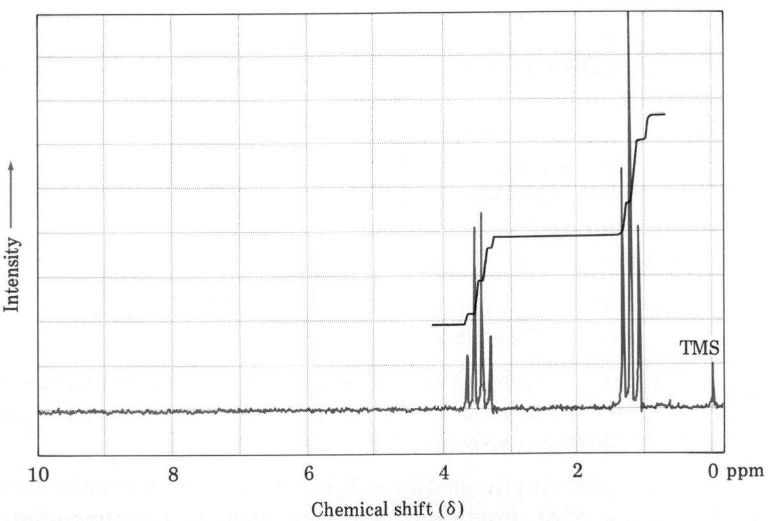


## 13.13 More Complex Spin–Spin Splitting Patterns

In all the NMR spectra seen so far, the chemical shifts of different protons have been quite distinct, and the spin–spin splitting patterns have been relatively simple. It often happens, however, that different kinds of protons have *overlapping* signals.

The spectrum of toluene (methylbenzene) in Figure 13.21, for example, shows that all five aromatic ring protons give a single overlapping absorption, even though they aren't all equivalent. This kind of accidental overlap of signals is something we must be aware of to avoid drawing false conclusions from spectra.

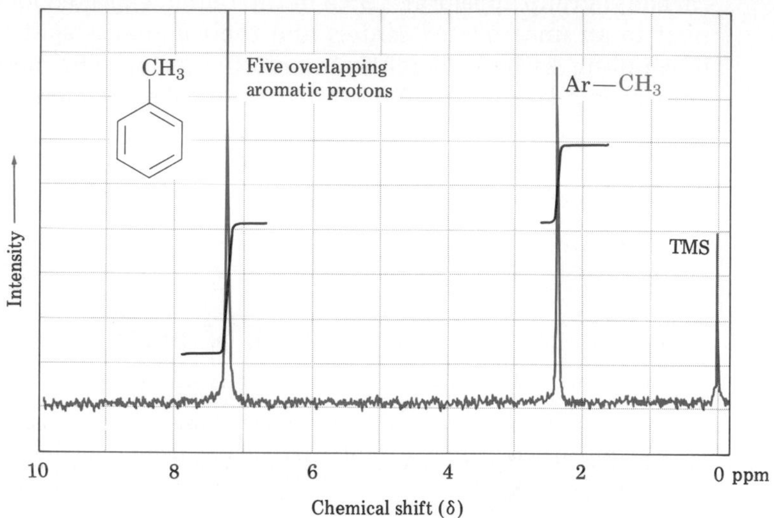


**Figure 13.21** The $^1$H NMR spectrum of toluene, showing the accidental overlap of the nonequivalent aromatic ring protons

Yet another complication can arise when a signal is split by two or more *nonequivalent* kinds of protons, as in the case of *trans*-cinnamaldehyde, isolated from oil of cinnamon (Figure 13.22). Although the $n + 1$ rule can be used to predict splitting due to *equivalent* protons, splittings due to nonequivalent protons are more complex.

The $^1$H NMR spectrum of *trans*-cinnamaldehyde is complex, but we can understand it if we isolate the different parts and look at each kind of proton individually:

1. The five aromatic proton signals overlap into a single broad resonance line at 7.45 $\delta$.
2. The aldehyde proton signal at C1 appears in the normal downfield position at $\delta = 9.67$ and is split into a doublet ($J = 7$ Hz) by the adjacent proton at C2.
3. The vinylic proton at C3 is next to the aromatic ring and is therefore shifted downfield from the normal vinylic region. This C3 proton signal appears at 7.42 $\delta$ and nearly overlaps the aromatic proton signals. Since it has one neighbor proton at C2, its signal is split into a doublet with $J = 15$ Hz.
4. The C2 vinylic proton signal appears at 6.66 $\delta$ and shows an interesting absorption pattern. It is coupled to the two nonequivalent protons at C1 and C3 with two different coupling constants: $J_{1-2} = 7$ Hz and $J_{2-3} = 15$ Hz.

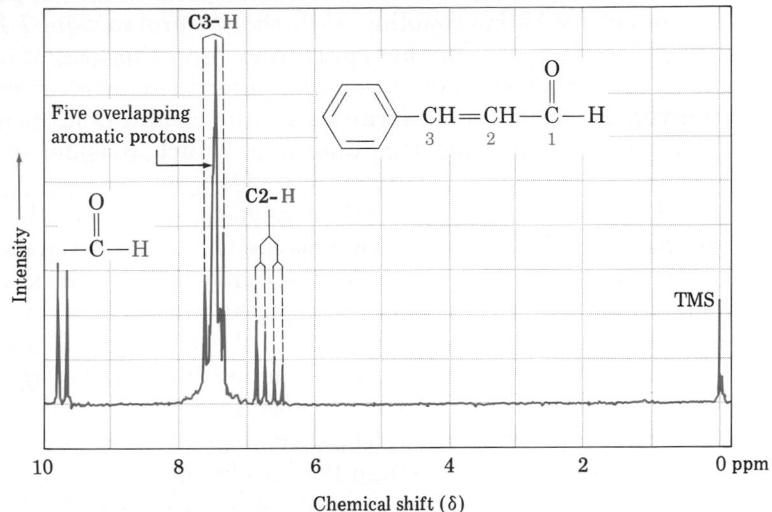


**Figure 13.22** The $^1$H NMR spectrum of *trans*-cinnamaldehyde

The best way to see the effect of multiple coupling is to draw a **tree diagram** like that shown in Figure 13.23. Tree diagrams show the individual effects of each coupling constant on the overall pattern.

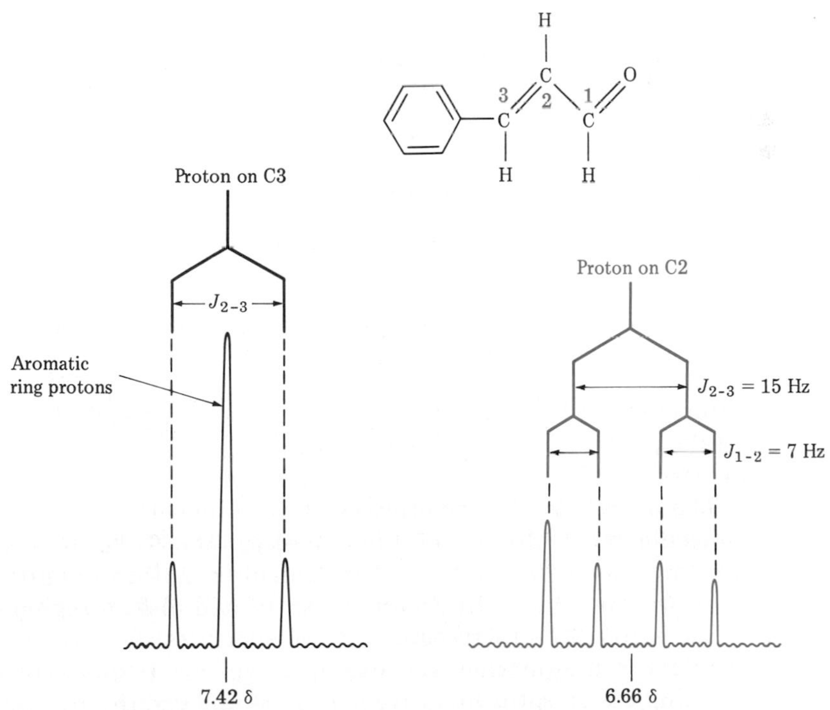


**Figure 13.23** A tree diagram for the C2 proton of *trans*-cinnamaldehyde

In *trans*-cinnamaldehyde, the signal due to the C2 proton is split into a doublet by 15 Hz coupling with the C3 proton. The 7 Hz coupling with the aldehyde proton further splits each leg of the doublet into new doublets. Thus we observe a four-line spectrum for the C2 proton of *trans*-cinnamaldehyde. Multiple coupling can look quite complex, but is usually amenable to simplification using tree diagrams, which consider each coupling separately.

One further point you might have noticed about the cinnamaldehyde spectrum is that the four peaks of the C2 proton resonance are not all the same size; the left-hand peaks are slightly larger. This size difference turns out to be due to the fact that the coupled nuclei have very similar chemical shifts. Whenever the chemical shifts of coupled nuclei are similar, there is a skewing of the multiplet, with the legs nearer the signal of the coupled partner becoming larger and the legs farther from the signal of the coupled partner becoming smaller. Thus, the legs of the C2 multiplet closer to the C3 absorption are larger than the legs farther away. This skewing effect on multiplets can sometimes be useful since it tells where to look in the spectrum to find the coupled partner: toward the direction of skewing.

PROBLEM

**13.19** 3-Bromo-1-phenyl-1-propene shows a complex NMR spectrum in which the vinylic proton at C2 is coupled with both the C1 vinylic proton ($J$ = 16 Hz) and the C3 methylene protons ($J$ = 8 Hz). Draw a tree diagram for the C2 proton signal and account for the fact that a five-line multiplet is observed.

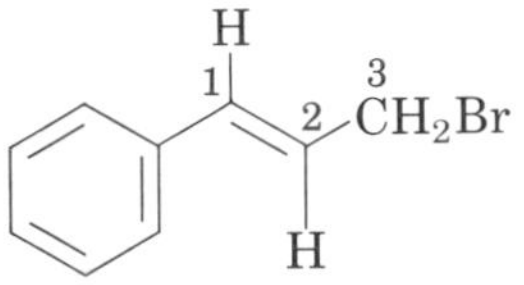

H 1 2 3 $CH_2Br$ H

3-Bromo-1-phenyl-1-propene

## 13.14 NMR Spectra of Larger Molecules

Digitoxigenin, the steroid responsible for the heart stimulant properties of digitalis preparations, has the formula $C_{22}H_{32}O_4$. Can a molecule this complex possibly give an interpretable NMR spectrum?

The answer is that we often can't *fully* interpret the spectrum of such a molecule, but NMR is useful nonetheless. The $^1H$ NMR spectrum of digitoxigenin taken on a 300 MHz instrument is shown in Figure 13.24. Even though we can't easily interpret the saturated alkane region between 1.5 and 2.0 $\delta$, we can still recognize many structural features that provide a great deal of information. For example, we can readily spot two methyl-group singlets on saturated carbon at 0.84 and 0.92 $\delta$. We can also see one vinylic proton singlet at 5.85 $\delta$, a pattern at 4.85 $\delta$ due to two protons on carbon next to oxygen, and a further broad absorption at 4.1 δ, due to one proton on carbon next to oxygen. Even in complex cases, NMR is an invaluable structural tool.

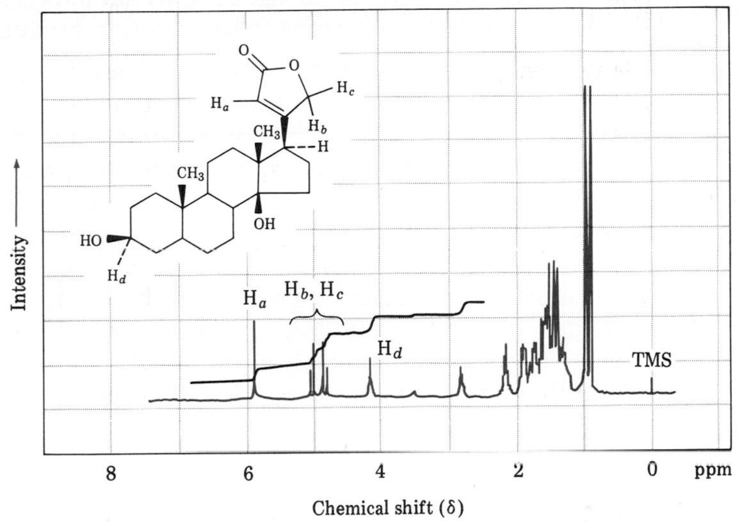


**Figure 13.24** The ¹H NMR spectrum of digitoxigenin taken at 300 MHz

## 13.15 Uses of ¹H NMR Spectra

NMR can be used to help identify the product of nearly every reaction run in the laboratory. For example, we said in Chapter 7 that addition of HCl to alkenes occurs with Markovnikov regiochemistry; that is, the more highly substituted alkyl chloride is formed. With the help of NMR, we can now prove this statement.

Does addition of HCl to 1-methylcyclohexene yield 1-chloro-1-methylcyclohexane or 1-chloro-2-methylcyclohexane?

$CH_3$ — 1-Methylcyclohexene $\xrightarrow[\text{Ether}]{\text{HCl}}$ $H_3C$ Cl — 1-Chloro-1-methylcyclohexane (Markovnikov) or $CH_3$ H H Cl ? — 1-Chloro-2-methylcyclohexane (non-Markovnikov)

The $^1H$ NMR spectrum of the reaction product is shown in Figure 13.25. The spectrum shows a large singlet absorption in the alkane methyl region at 1.5 $\delta$, indicating that the product has a methyl group bonded to a quaternary carbon, $R_3C—CH_3$, rather than to a tertiary carbon, $R_2CH—CH_3$. Furthermore, the spectrum shows *no* absorptions in the range 4–5 $\delta$, where we would expect the signal of an $R_2CHCl$ proton to occur. Thus, it is clear that 1-chloro-1-methylcyclohexane is the reaction product.

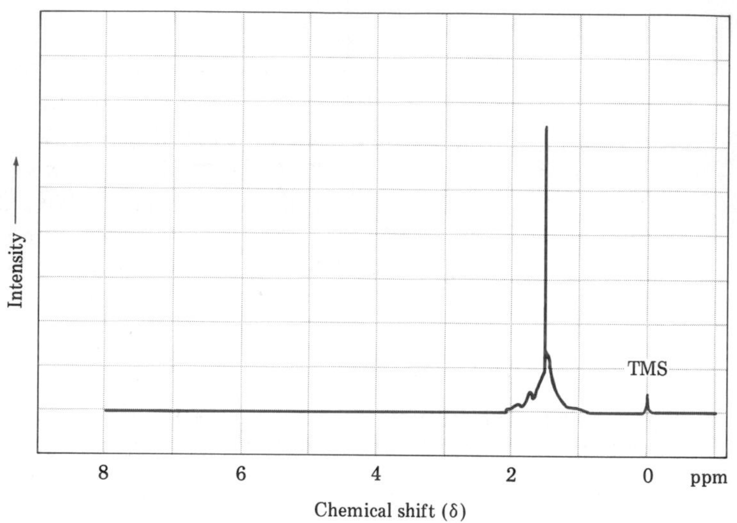


**Figure 13.25** The $^1H$ NMR spectrum of the reaction product from HCl and 1-methylcyclohexene

**PROBLEM**

**13.20** How could you use $^1H$ NMR to help you determine the regiochemistry of alkene hydroboration? Does hydroboration–oxidation of propene yield 1-propanol or 2-propanol?

$$CH_3CH{=}CH_2 \xrightarrow[2.\ H_2O_2,\ NaOH]{1.\ BH_3} CH_3CH_2CH_2OH \quad \text{or} \quad CH_3CH(OH)CH_3\ ?$$

## 13.16 Summary and Key Words

When $^1H$ and $^{13}C$ nuclei are placed in strong magnetic fields, their spins orient either with or against the applied field. On irradiation with radiofrequency (rf) waves, energy is absorbed and the nuclei "spin-flip" from the lower energy state to the higher energy state. This absorption of rf energy is detected, amplified, and displayed as a **nuclear magnetic resonance (NMR) spectrum**.

The NMR spectrum is obtained by irradiating a sample with a constant-frequency rf energy and slowly changing the value of the applied magnetic field. Different kinds of $^1H$ and $^{13}C$ nuclei come into resonance at slightly different applied fields, and we therefore see a different absorption line for each different kind of $^1H$ and $^{13}C$. The NMR chart is calibrated in **delta ($\delta$) units**, where 1 $\delta$ = 1 part per million (ppm) of spectrometer frequency. Tetramethylsilane (TMS) is used as a reference point to which other peaks are compared because it shows both $^1H$ and $^{13}C$ absorptions at unusually high values of the applied magnetic field. The TMS absorption occurs at the right hand (**upfield**) edge of the chart and is arbitrarily assigned a value of 0 $\delta$.

Both $^1H$ and $^{13}C$ NMR spectra display four general features:

1. *Number of resonance lines* Each different kind of $^1H$ or $^{13}C$ nucleus in a molecule can give rise to a different resonance line.
2. *Chemical shift* The exact position of each peak is its chemical shift. Chemical shifts are due to the effects of electrons setting up tiny local magnetic fields that **shield** a nearby nucleus from the applied field and therefore cause different nuclei to come to resonance at different places. By correlating chemical shifts with environment, we can learn about the chemical nature of each nucleus.
3. *Integration* The area under each NMR absorption peak can be electronically integrated to determine the relative number of nuclei ($^1H$ or $^{13}C$) responsible for each peak.
4. *Spin–spin splitting* Neighboring nuclear spins can **couple**, splitting NMR peaks into **multiplets**. The NMR signal of a $^{13}C$ nucleus bonded to $n$ protons splits into $n + 1$ peaks (the **$n$ + 1 rule**). Similarly, the NMR signal of a $^1H$ nucleus neighbored by $n$ equivalent adjacent protons splits into $n + 1$ peaks.

Most $^{13}C$ spectra are run in an operating mode (the **proton noise-decoupled mode**) that provides maximum sensitivity and gives a spectrum in which each nonequivalent carbon shows a single unsplit resonance line. Operating in the integrating (**gated-decoupled**) mode causes a loss of sensitivity, but provides a spectrum that can be electronically integrated to measure the number of carbon nuclei responsible for each peak. Operating in the spin-coupled (**off-resonance**) mode also causes a loss of sensitivity, but provides a spectrum in which spin–spin splitting is observed. Each carbon resonance is split into a multiplet depending on the number of protons to which it is bonded: Quaternary carbon resonances remain as singlets, tertiary carbons ($R_3CH$) appear as doublets, secondary carbons ($R_2CH_2$) appear as triplets, and primary carbons ($RCH_3$) appear as quartets.

Proton NMR spectra are even more useful than $^{13}C$ spectra. The sensitivity of $^1H$ instruments is high, and normal spectrometer operating conditions provide spectra that show spin–spin splitting and can be integrated. Proton resonances usually fall into the range 0–10 $\delta$ downfield from the TMS reference point.

Specific $^1H$ resonance peaks are often split into multiplets due to spin–spin splitting with the spins of protons on adjacent carbons. Equivalent protons do not split each other, but a proton with $n$ equivalent neighboring protons gives a signal that is split into $n + 1$ peaks with **coupling constant *J*.**

## ADDITIONAL PROBLEMS

**13.21** The following $^1H$ NMR absorptions were determined on a spectrometer operating at 60 MHz and are given in hertz downfield from the TMS standard. Convert the absorptions to $\delta$ units.
(a) 131 Hz (b) 287 Hz (c) 451 Hz (d) 543 Hz

**13.22** The following $^1H$ NMR absorptions given in $\delta$ units were obtained on a spectrometer operating at 80 MHz. Convert the chemical shifts from $\delta$ units into hertz downfield from TMS.
(a) 2.1 (b) 3.45 (c) 6.30 (d) 7.70

**13.23** When measured on a spectrometer operating at 60 MHz, chloroform ($CHCl_3$) shows a single sharp absorption at 7.3 $\delta$.
(a) How many parts per million downfield from TMS does chloroform absorb?
(b) How many hertz downfield from TMS would chloroform absorb if the measurement were carried out on a spectrometer operating at 360 MHz?
(c) What would be the position of the chloroform absorption in $\delta$ units when measured on a 360 MHz spectrometer?

**13.24** How many absorptions would you expect to observe in the normal $^{13}C$ NMR spectra of these compounds?
(a) 1,1-Dimethylcyclohexane (b) $CH_3CH_2OCH_3$
(c) *tert*-Butylcyclohexane (d) 3-Methyl-1-pentyne
(e) *cis*-1,2-Dimethylcyclohexane (f) 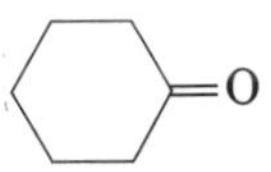 [cyclohexanone structure] =O

**13.25** Indicate the spin multiplicities you would expect to see for each carbon atom in the spin-coupled $^{13}C$ NMR spectra of the molecules shown in Problem 13.24.

**13.26** Why do you suppose accidental overlap of signals is much more common in $^1H$ NMR than in $^{13}C$ NMR?

**13.27** Tell what is meant by each of these terms:
(a) Chemical shift (b) Spin–spin splitting
(c) Applied magnetic field (d) Spectrometer operating frequency
(e) Coupling constant (f) Upfield/downfield

**13.28** How many types of nonequivalent protons are there in each of the following molecules?

(a) 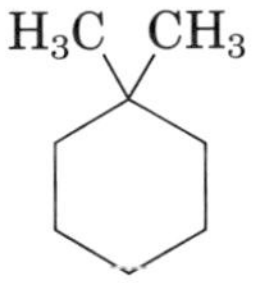 $H_3C$ $CH_3$ [1,1-dimethylcyclohexane structure]

(b) $CH_3CH_2CH_2OCH_3$

(c) 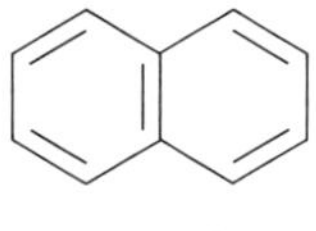 [naphthalene structure]
Naphthalene

(d) 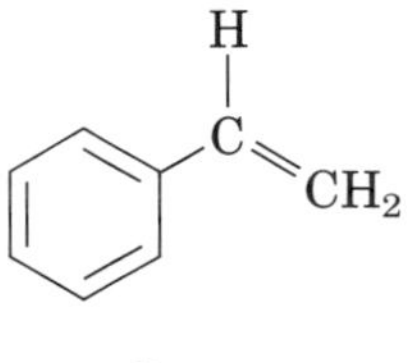 H, C, $CH_2$ [styrene structure]
Styrene

(e) 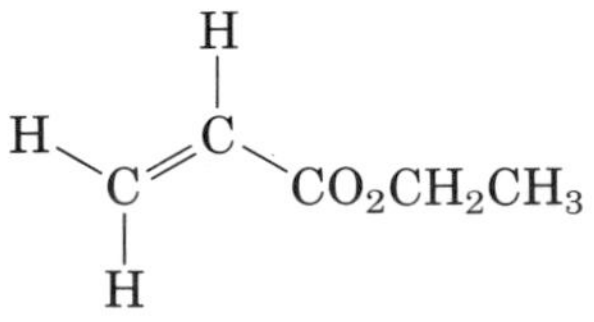 H, H, C, C, H, $CO_2CH_2CH_3$ [ethyl acrylate structure]
Ethyl acrylate

**13.29** The following compounds all show a single line in their $^1H$ NMR spectra. List them in expected order of increasing chemical shift.

$CH_4$; $CH_2Cl_2$; Cyclohexane; $CH_3COCH_3$; $H_2C{=}CH_2$; Benzene

**13.30** Predict the splitting pattern for each kind of hydrogen and each kind of carbon in these molecules:
(a) $(CH_3)_3CH$ (b) $CH_3CH_2COOCH_3$ (c) *trans*-2-Butene

**13.31** Predict the splitting pattern for each kind of hydrogen and each kind of carbon in isopropyl propanoate, $CH_3CH_2COOCH(CH_3)_2$.

**13.32** The acid-catalyzed dehydration of 1-methylcyclohexanol yields a mixture of two alkenes as product. After you separated them by chromatography, how would you use $^1H$ NMR to help you decide which was which?

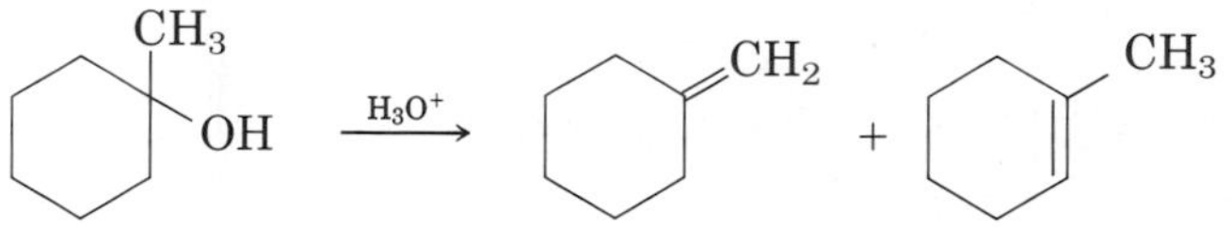

[1-methylcyclohexanol (ring with $CH_3$ and OH) $\xrightarrow{H_3O^+}$ methylenecyclohexane ($=CH_2$) + 1-methylcyclohexene ($CH_3$)]

**13.33** How would you use $^1H$ NMR to distinguish between these pairs of isomers?

(a) $CH_3CH{=}CHCH_2CH_3$ and $H_2C{-}CHCH_2CH_3$ (cyclopropane ring closed by a $CH_2$)

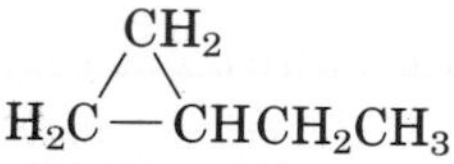

(b) $CH_3CH_2OCH_2CH_3$ and $CH_3OCH_2CH_2CH_3$

(c) $CH_3\overset{O}{\overset{\|}{C}}OCH_2CH_3$ and $CH_3CH_2\overset{O}{\overset{\|}{C}}OCH_3$

(d) $H_2C{=}C(CH_3)\overset{O}{\overset{\|}{C}}CH_3$ and $CH_3CH{=}CH\overset{O}{\overset{\|}{C}}CH_3$

**13.34** Propose structures for compounds with the following formulas that show only one peak in their $^1H$ NMR spectra.
(a) $C_5H_{12}$ (b) $C_5H_{10}$ (c) $C_4H_8O_2$

**13.35** Assume that you have a compound with formula $C_3H_6O$.
(a) How many double bonds and/or rings does your material contain?
(b) Propose as many structures as you can that fit the molecular formula.
(c) If your compound shows an infrared absorption peak at 1710 $cm^{-1}$, what inferences can you draw?
(d) If your compound shows a single $^1H$ NMR absorption peak at 2.1 $\delta$, what is its structure?

**13.36** How could you use $^1H$ and $^{13}C$ NMR to help you distinguish between the following isomeric compounds of formula $C_4H_8$?

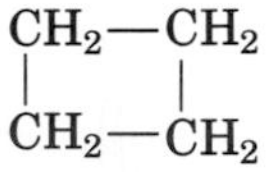

[cyclobutane: $CH_2{-}CH_2$ / $CH_2{-}CH_2$] $H_2C{=}CHCH_2CH_3$ $CH_3CH{=}CHCH_3$ $(CH_3)_2C{=}CH_2$

**13.37** How could you use $^1H$ and $^{13}C$ NMR to help you distinguish between these structures?

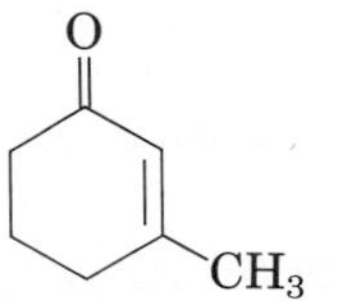

[structure: O, $CH_3$] 3-Methyl-2-cyclohexenone

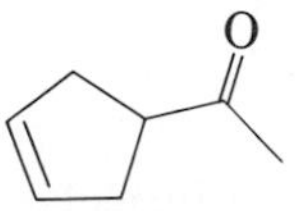

[structure: O] 4-Cyclopentenyl methyl ketone

**13.38** How could you use IR spectroscopy to help you distinguish between the two compounds shown in Problem 13.37?

**13.39** The compound whose $^1H$ NMR spectrum is shown here has the molecular formula $C_3H_6Br_2$. Propose a plausible structure.

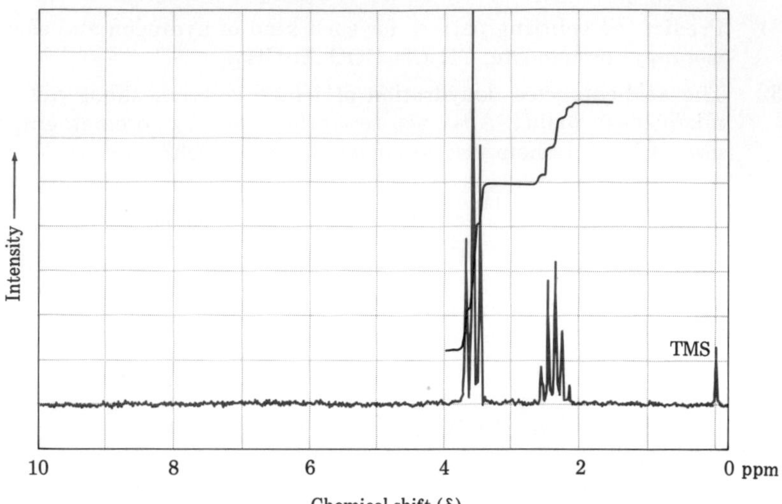


**13.40** Propose structures for compounds that fit the following $^1H$ NMR data:

(a) $C_5H_{10}O$
- 6 H doublet at 0.95 $\delta$, $J = 7$ Hz
- 3 H singlet at 2.10 $\delta$
- 1 H multiplet at 2.43 $\delta$

(b) $C_3H_5Br$
- 3 H singlet at 2.32 $\delta$
- 1 H broad singlet at 5.35 $\delta$
- 1 H broad singlet at 5.54 $\delta$

**13.41** The compound whose $^1H$ NMR spectrum is shown has the molecular formula $C_4H_7O_2Cl$ and shows an infrared absorption peak at 1740 $cm^{-1}$. Propose a plausible structure.

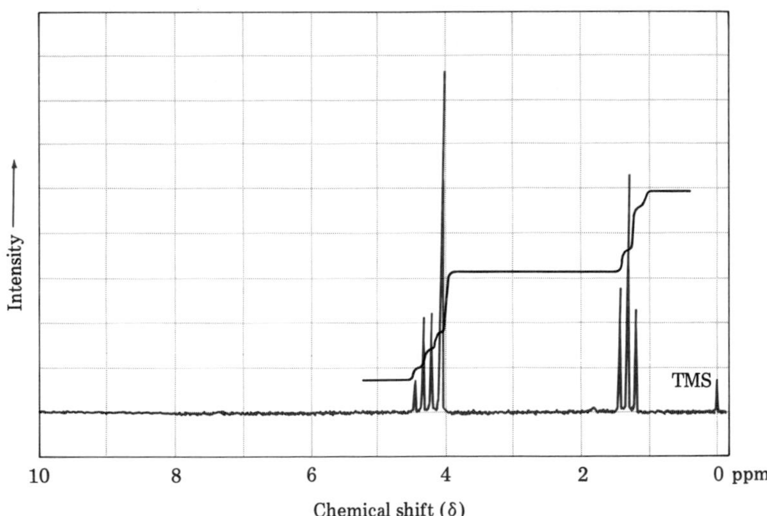


**13.42** Propose structures for compounds that fit the following $^1H$ NMR data:

(a) $C_4H_6Cl_2$
- 3 H singlet at 2.18 $\delta$
- 2 H doublet at 4.16 $\delta$, $J = 7$ Hz
- 1 H triplet at 5.71 $\delta$, $J = 7$ Hz

(b) $C_{10}H_{14}$
- 9 H singlet at 1.30 $\delta$
- 5 H singlet at 7.30 $\delta$

(c) $C_4H_7BrO$
3 H singlet at 2.11 $\delta$
2 H triplet at 3.52 $\delta$, $J$ = 6 Hz
2 H triplet at 4.40 $\delta$, $J$ = 6 Hz

(d) $C_9H_{11}Br$
2 H quintet at 2.15 $\delta$, $J$ = 7 Hz
2 H triplet at 2.75 $\delta$, $J$ = 7 Hz
2 H triplet at 3.38 $\delta$, $J$ = 7 Hz
5 H singlet at 7.22 $\delta$

**13.43** How might you use NMR (either $^1H$ or $^{13}C$) to differentiate between the following two isomeric structures?

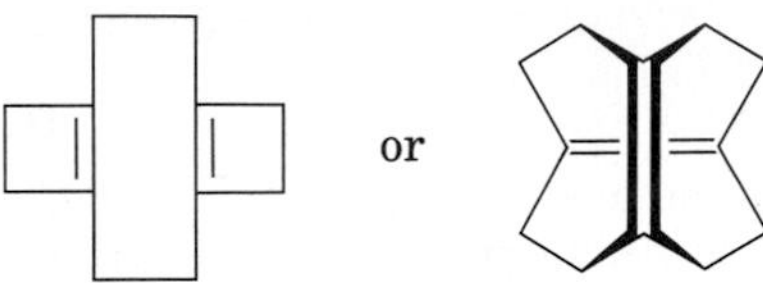


(You might want to build molecular models to help you examine the two structures more closely.)

**13.44** Propose plausible structures for the two compounds whose $^1H$ NMR spectra are shown.

(a) $C_4H_9Br$

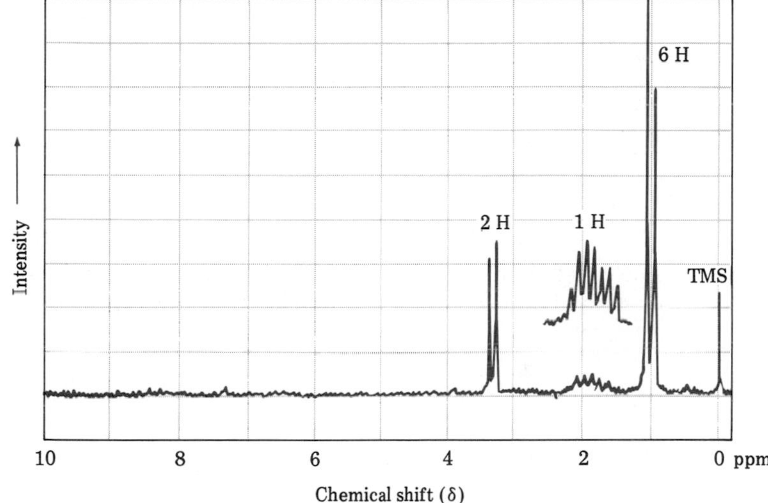


(b) $C_4H_8Cl_2$

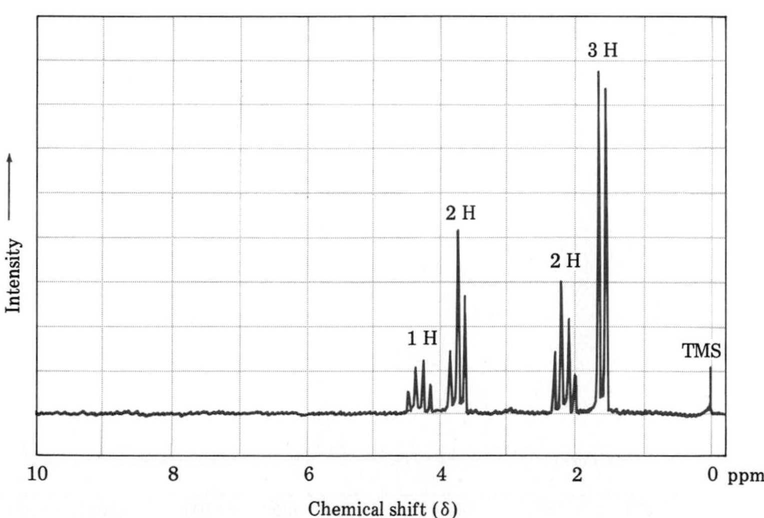

**13.45** We saw earlier that long-range coupling between protons more than two carbon atoms apart is sometimes observed when pi bonds intervene. One example of long-range coupling is found in 1-methoxy-1-buten-3-yne, whose $^1H$ NMR spectrum is shown.

$$
\begin{array}{c}
CH_3O \\
\diagdown \\
C{-}H_c \\
/\!/ \\
H_a{-}C{\equiv}C{-}C \\
\diagdown \\
H_b
\end{array}
$$

Not only does the acetylenic proton, $H_a$, couple with the vinylic proton $H_b$, it also couples with the vinylic proton $H_c$ (*four* carbon atoms away). The following coupling constants are observed:

$$J_{a-b} = 6\text{ Hz}; \qquad J_{a-c} = 2\text{ Hz}; \qquad J_{b-c} = 15\text{ Hz}$$

Construct tree diagrams that account for the observed splitting patterns of $H_a$, $H_b$, and $H_c$.

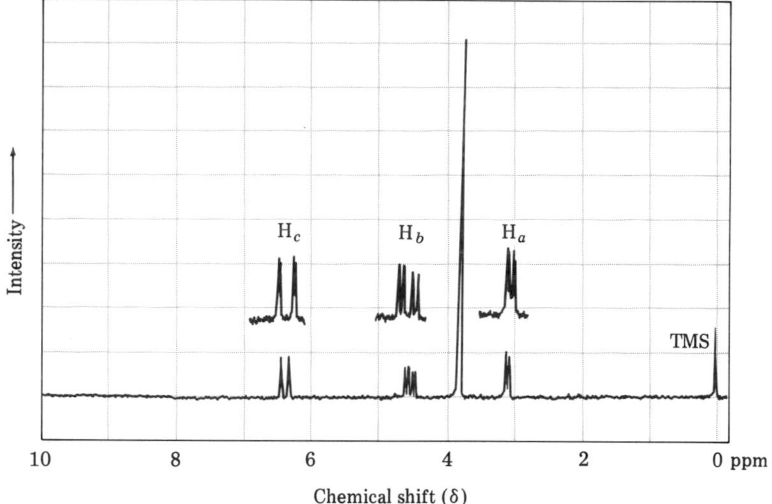


**13.46** Assign as many of the resonances as you can to specific carbon atoms in the $^{13}C$ NMR spectrum of ethyl benzoate shown here.

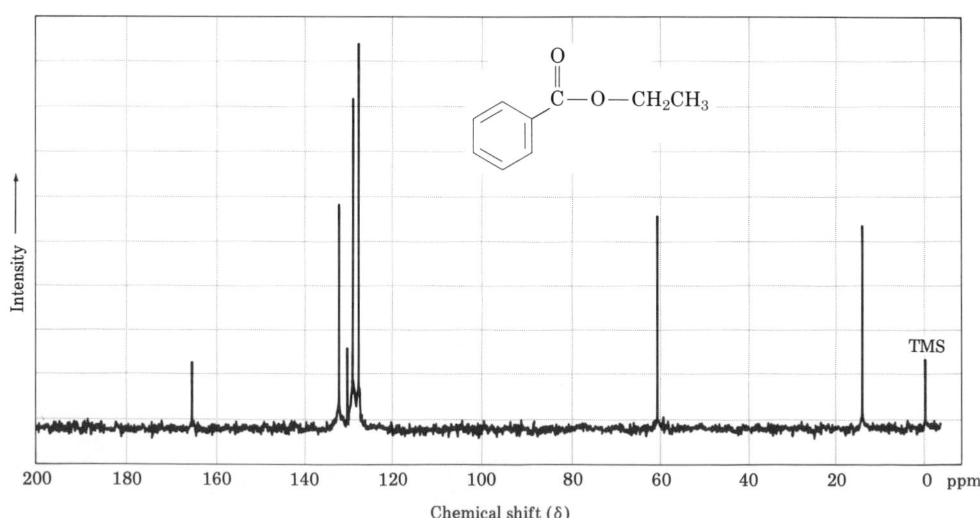

**13.47** The $^1H$ and $^{13}C$ NMR spectra of compound A, $C_8H_9Br$, are shown. Propose a possible structure for compound A, and assign peaks in the spectra to your structure.

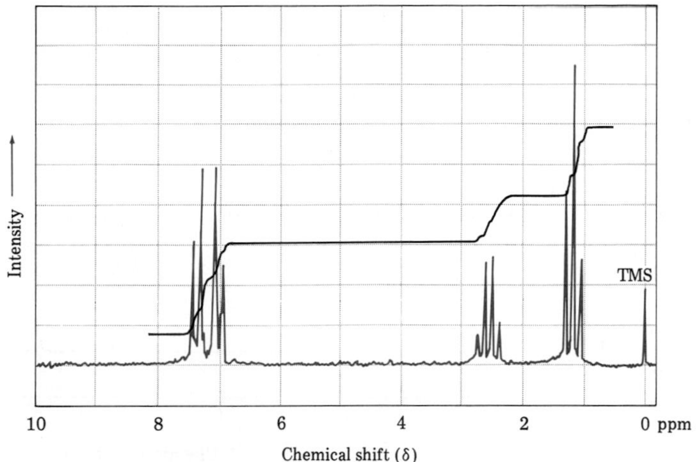


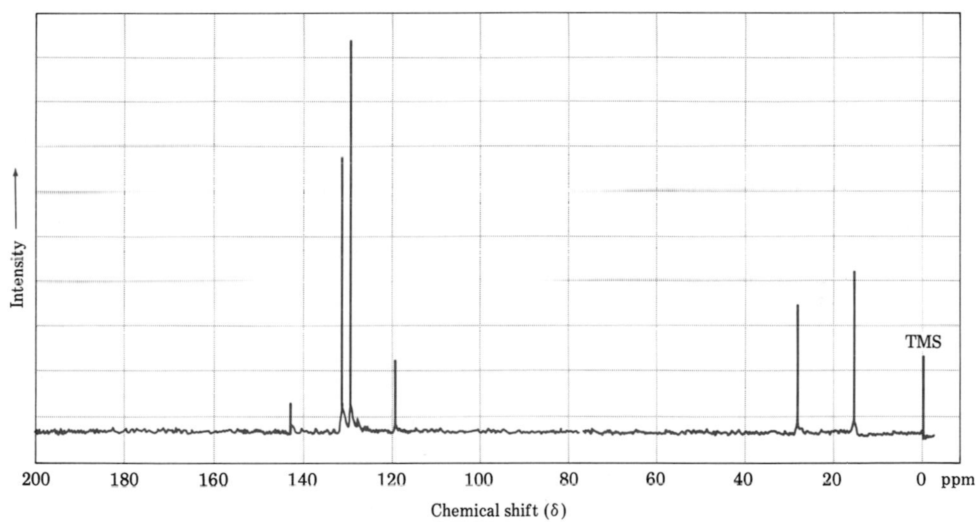

**13.48** Propose plausible structures for the three compounds whose $^{1}H$ NMR structures are shown.

(a) $C_4H_{10}O_2$

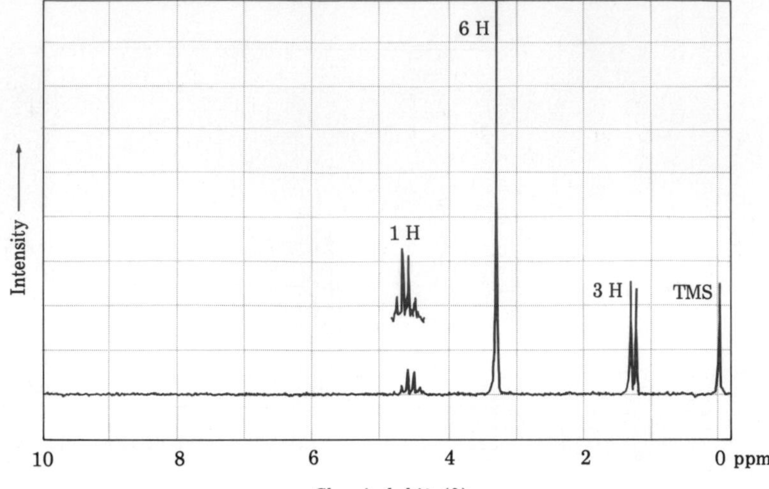


(b) $C_7H_7Br$

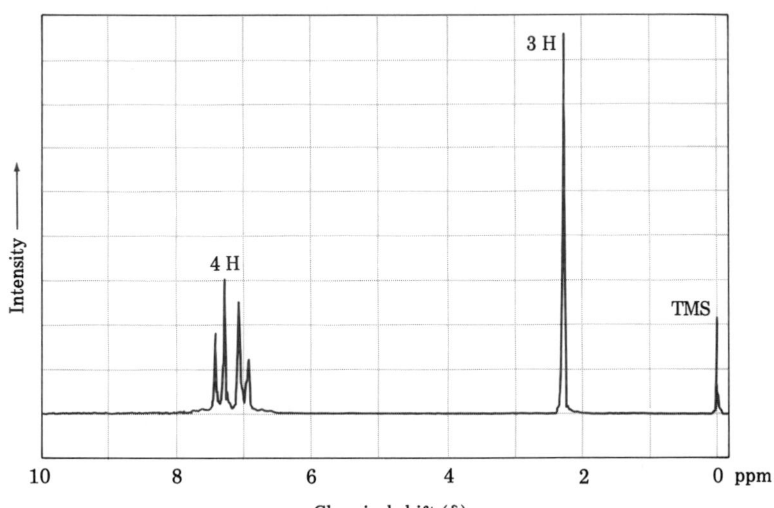


(c) $C_8H_9Br$

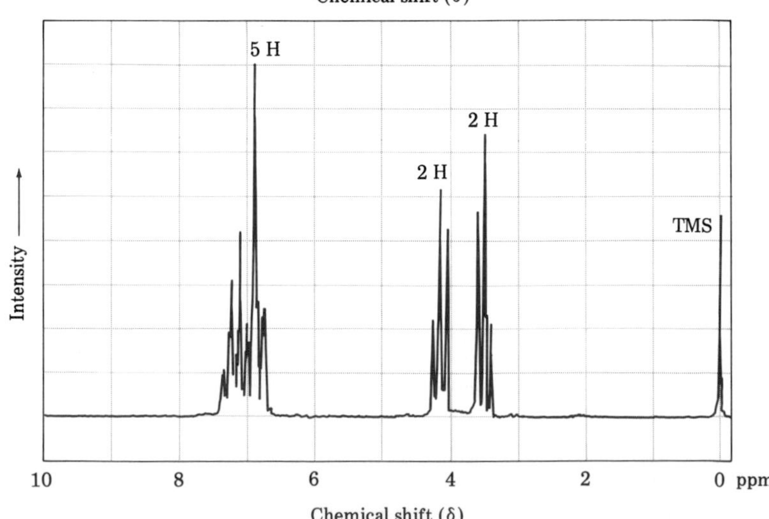

**13.49** The mass spectrum and $^{13}C$ NMR spectrum of a hydrocarbon are shown. Propose a suitable structure for this hydrocarbon and explain the spectral data.

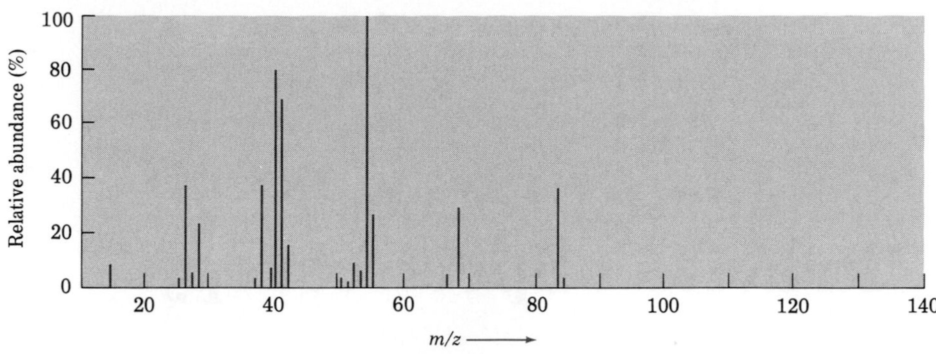


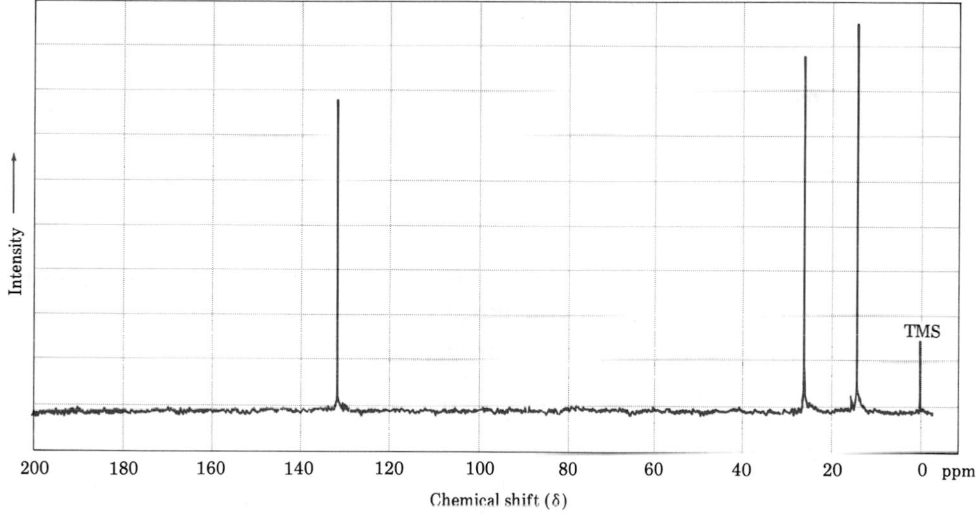

CHAPTER 14

# Conjugated Dienes and Ultraviolet Spectroscopy

Double bonds that alternate with single bonds are said to be **conjugated**. Thus, 1,3-butadiene is a **conjugated diene**, whereas 1,4-pentadiene is a nonconjugated diene with *isolated* double bonds.

$$H_2C{=}CH{-}CH{=}CH_2 \qquad H_2C{=}CH{-}CH_2{-}CH{=}CH_2$$

1,3-Butadiene (conjugated; alternating double and single bonds)

1,4-Pentadiene (nonconjugated; nonalternating double and single bonds)

There are other types of conjugated systems besides dienes, many of which play an important role in nature. For example, the pigments responsible for the brilliant reds and yellows of fruits and flowers are conjugated **polyenes** (*poly* = "many"); lycopene, the red pigment in tomatoes, is one such molecule. Conjugated **enones** (alk**ene** + ket**one**) are common structural features of important molecules such as progesterone, the "pregnancy hormone." Conjugated cyclic molecules such as benzene are a major field of study in themselves and will be considered in detail in the next chapter.

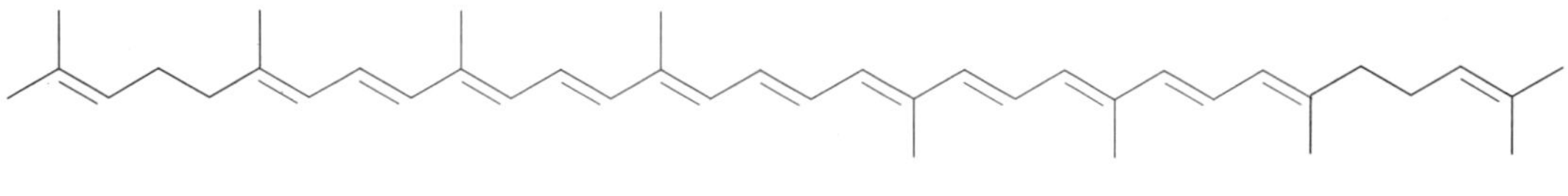

Lycopene

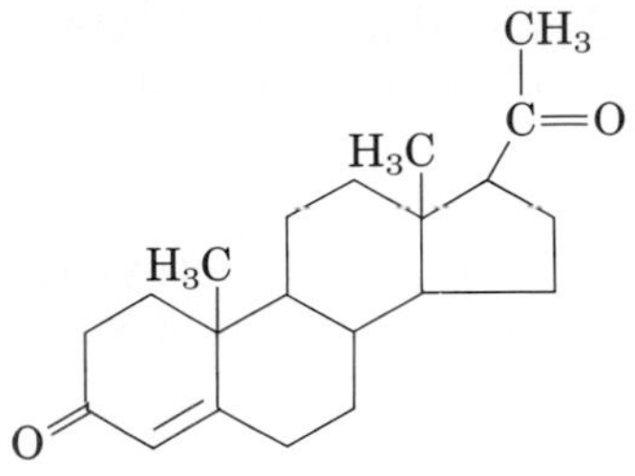

$CH_3$ C=O $H_3C$ $H_3C$ O

Progesterone

Benzene

PROBLEM

**14.1** Which of the following molecules contain conjugated systems? Circle the conjugated portion.

(a)

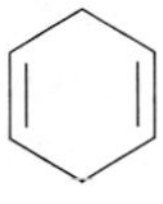

(b) 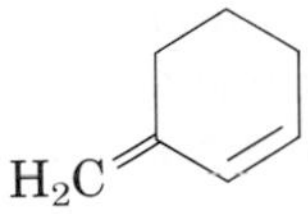 $H_2C$

(c) $H_2C{=}CH{-}C{\equiv}N$

(d) 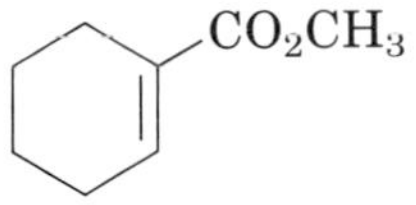 $CO_2CH_3$

(e) 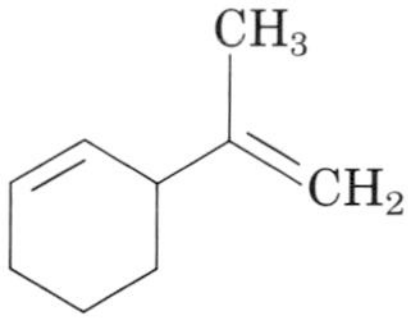 $CH_3$ $CH_2$

(f) 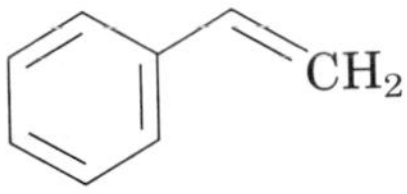 $CH_2$

## 14.1 Preparation of Conjugated Dienes

Conjugated dienes are generally prepared by the methods previously discussed for alkene synthesis. For example, the base-induced elimination of HX from an allylic halide produces a conjugated diene.

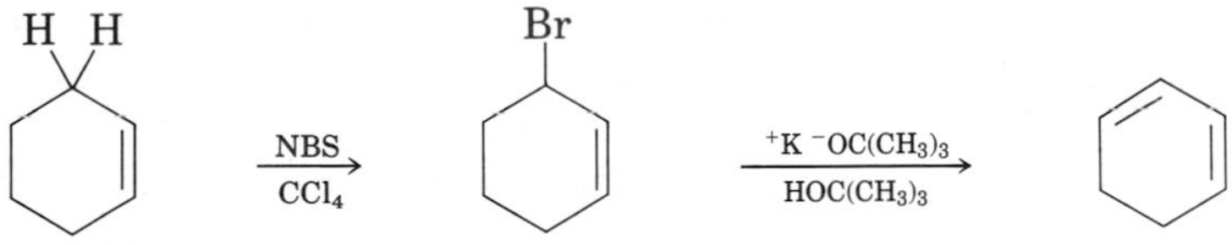

H H

$\xrightarrow[CCl_4]{NBS}$

Br

$\xrightarrow[HOC(CH_3)_3]{^+K\ ^-OC(CH_3)_3}$

Cyclohexene 3-Bromocyclohexene 1,3-Cyclohexadiene (76%)

1,3-Butadiene itself is prepared industrially on a vast scale for use in polymer synthesis. One industrial method involves thermal cracking of butane over a special chromium oxide–aluminum oxide catalyst, but this procedure is of no use in the laboratory.

$$CH_3CH_2CH_2CH_3 \xrightarrow[Catalyst]{600°C} H_2C{=}CHCH{=}CH_2 + 2\ H_2$$

Butane 1,3-Butadiene

Other simple conjugated dienes that have important uses in polymer synthesis include isoprene (2-methyl-1,3-butadiene) and chloroprene (2-chloro-1,3-butadiene). Isoprene has been prepared industrially by a number of methods, including the elimination of water (*dehydration*) from 2-methyl-3-buten-2-ol and the double dehydration of 3-methyl-1,3-butanediol over an alumina catalyst, but these methods are rarely used in the laboratory.

$$\underset{\text{3-Methyl-1,3-butanediol}}{\overset{\text{OH}\qquad\text{OH}}{CH_2CH_2C(CH_3)_2}} \xrightarrow[\Delta]{Al_2O_3} \underset{\substack{\text{Isoprene}\\ \text{(2-Methyl-1,3-butadiene)}}}{H_2C{=}CH{-}\overset{CH_3}{C}{=}CH_2} \xleftarrow[\Delta]{Al_2O_3} \underset{\text{2-Methyl-3-buten-2-ol}}{H_2C{=}CH\overset{OH}{C}(CH_3)_2}$$

## 14.2 Stability of Conjugated Dienes

Conjugated dienes are similar to isolated alkenes in much of their chemistry. There are, however, a few important differences, one of which is *stability*. Conjugated dienes are somewhat more stable than nonconjugated dienes.

Evidence for the extra stability of conjugated dienes comes from measurements of heats of hydrogenation (Table 14.1). We saw earlier in the

**Table 14.1** Heats of hydrogenation for some alkenes and dienes

| *Alkene* | | *Product* | $\Delta H^\circ_{hydrog}$ (*kcal/mol*) | $\Delta H^\circ_{hydrog}$ (*kJ/mol*) |
|---|---|---|---|---|
| $CH_3CH_2CH{=}CH_2$ 1-Butene | $\xrightarrow{H_2}$ | $CH_3CH_2CH_2CH_3$ | 30.3 | 127 |
| $CH_3CH_2\overset{CH_3}{C}{=}CH_2$ 2-Methyl-1-butene | $\xrightarrow{H_2}$ | $CH_3CH_2\overset{CH_3}{C}HCH_3$ | 26.9 | 113 |
| $H_2C{=}CHCH{=}CH_2$ 1,3-Butadiene | $\xrightarrow{H_2}$ | $CH_3CH_2CH{=}CH_2$ | 26.7 | 112 |
| $H_2C{=}CHCH{=}CH_2$ 1,3-Butadiene | $\xrightarrow{2\,H_2}$ | $CH_3CH_2CH_2CH_3$ | 57.1 | 239 |
| $H_2C{=}CH\overset{CH_3}{C}{=}CH_2$ 2-Methyl-1,3-butadiene | $\xrightarrow{2\,H_2}$ | $CH_3CH_2\overset{CH_3}{C}HCH_3$ | 53.4 | 223 |
| $H_2C{=}CHCH_2CH{=}CH_2$ 1,4-Pentadiene | $\xrightarrow{2\,H_2}$ | $CH_3CH_2CH_2CH_2CH_3$ | 60.8 | 254 |
| $H_2C{=}CHCH_2CH_2CH{=}CH_2$ 1,5-Hexadiene | $\xrightarrow{2\,H_2}$ | $CH_3CH_2CH_2CH_2CH_2CH_3$ | 60.5 | 253 |

discussion of alkene stabilities (Section 6.7) that alkenes of similar substitution pattern have remarkably similar $\Delta H^\circ_{\text{hydrog}}$ values. Monosubstituted alkenes such as 1-butene have values for $\Delta H^\circ_{\text{hydrog}}$ near 30 kcal/mol (125 kJ/mol), whereas disubstituted alkenes such as 2-methyl-1-butene show $\Delta H^\circ_{\text{hydrog}}$ values approximately 3 kcal/mol lower. We concluded from these data that highly substituted alkenes are more stable than less highly substituted ones. That is, substituted alkenes release less heat on hydrogenation because they contain less energy to start with. A similar conclusion can be drawn for conjugated dienes.

Table 14.1 shows that 1-butene, a monosubstituted alkene, has $\Delta H^\circ_{\text{hydrog}}$ = 30.3 kcal/mol (127 kJ/mol). We might therefore predict that a compound with two monosubstituted double bonds should have a $\Delta H^\circ_{\text{hydrog}}$ approximately twice this value, or 60.6 kcal/mol. This prediction is fully met by nonconjugated dienes such as 1,4-pentadiene ($\Delta H^\circ_{\text{hydrog}}$ = 60.8 kcal/mol) but is *not* met by the conjugated diene 1,3-butadiene ($\Delta H^\circ_{\text{hydrog}}$ = 57.1 kcal/mol). 1,3-Butadiene is approximately 3.7 kcal/mol (15 kJ/mol) more stable than predicted.

Confirmation of this unexpected stability comes from data on the partial hydrogenation of 1,3-butadiene. If 1,3-butadiene is partially hydrogenated to yield 1-butene, 26.7 kcal/mol (112 kJ/mol) energy is released. This is 3.6 kcal/mol less than we would expect for a normal isolated monosubstituted double bond. The same is true of other conjugated dienes.

| Compound | $\Delta H^\circ_{\text{hydrog}}$ *(kcal/mol)* | |
|---|---|---|
| $H_2C{=}CHCH_2CH{=}CH_2$ | 30.3 + 30.3 = 60.6 | Expected |
| 1,4-Pentadiene | 60.8 | Observed |
| | −0.2 | Difference |
| $H_2C{=}CHCH{=}CH_2$ | 30.3 + 30.3 = 60.6 | Expected |
| 1,3-Butadiene | 57.1 | Observed |
| | 3.5 | Difference |
| $H_2C{=}C(CH_3)CH{=}CH_2$ | 30.3 + 26.9 = 57.2 | Expected |
| 2-Methyl-1,3-butadiene | 53.4 | Observed |
| | 3.8 | Difference |

PROBLEM

**14.2** From the data in Table 14.1, calculate an expected heat of hydrogenation for allene, $H_2C{=}C{=}CH_2$. The measured value is 71.3 kcal/mol (298 kJ/mol). How stable is allene? Rank a conjugated diene, a nonconjugated diene, and an allene in order of stability.

## 14.3 Molecular Orbital Description of 1,3-Butadiene

Why do conjugated dienes have unexpected stability? Two different explanations have been advanced. One explanation says that 1,3-butadiene really does *not* have "unexpected" stability and that the problem is with our *expectations*, not with the diene. According to this view, our expectations are wrong because a nonconjugated diene such as 1,4-pentadiene is a poor choice for comparison with 1,3-butadiene.

In nonconjugated dienes, the carbon–carbon single bonds result from sigma overlap of an $sp^2$ orbital from one carbon with an $sp^3$ orbital from the neighboring carbon. In conjugated dienes, however, the carbon–carbon single bonds result from sigma overlap of $sp^2$ orbitals on both carbons. Since $sp^2$ orbitals have more *s* character than $sp^3$ orbitals, they form somewhat shorter, stronger bonds. Thus the "extra" stability of a conjugated diene has more to do with the hybridization of the orbitals forming the carbon–carbon single bond than with the double bonds themselves.

$H_2C{=}CH{-}CH_2{-}CH{=}CH_2$ $\qquad$ $H_2C{=}CH{-}CH{=}CH_2$

Bonds formed by overlap of $C_{sp^2}$ and $C_{sp^3}$ orbitals $\qquad$ Bond formed by overlap of $C_{sp^2}$ and $C_{sp^2}$ orbitals

The hybridization hypothesis makes a valid point, but another factor is also important: the interaction between pi orbitals of the conjugated diene system. To see how this interaction of pi orbitals arises, let's briefly review molecular orbital theory (Section 1.7). We've said that, when a covalent bond is formed by overlap of two atomic orbitals, the new orbitals that result are the property of the molecule, not of the individual atoms. The electrons in those orbitals are shared between atoms, rather than localized on one atom. For example, when two *p* atomic orbitals overlap to form a pi bond, the bonding electrons occupy a pi *molecular* orbital, rather than *p* atomic orbitals.

Since two *p* atomic orbitals are involved in forming a pi bond, two pi molecular orbitals are formed. One is lower in energy than the starting *p* orbitals and is therefore a **bonding molecular orbital**; the other is higher in energy and is an **antibonding orbital**. When we mentally construct the pi bond by assigning electrons to the orbitals, both electrons go into the low-energy bonding orbital, resulting in formation of a stable bond (Figure 14.1).

The pictorial representation of molecular orbital formation in Figure 14.1 shows how bonding arises. We saw earlier that *p* orbitals are dumbbell shaped, with the two lobes having different mathematical signs. When two *p* orbitals approach each other for overlap, they can orient in either of two ways. Overlap of lobes with identical signs is additive and corresponds to the low-energy bonding molecular orbital, whereas overlap of lobes with different signs is subtractive and corresponds to the high-energy antibonding orbital.

The change of sign between adjacent lobes in the antibonding orbital means that, if there were electrons in this orbital, they would not be shared between nuclei and there would be no bonding. Thus there is a region of **zero electron density** between the two nuclei, a region that we call a **node**. The bonding molecular orbital does not have a node between the nuclei. (There is also a node between the plus and minus lobes of each individual pi orbital, but only the nodes between nuclei affect bonding.)

Now let's bring two pi bonds together and allow four *p* atomic orbitals to interact, as occurs in a conjugated diene. In so doing, we generate a set of four molecular orbitals (Figure 14.2, page 458).

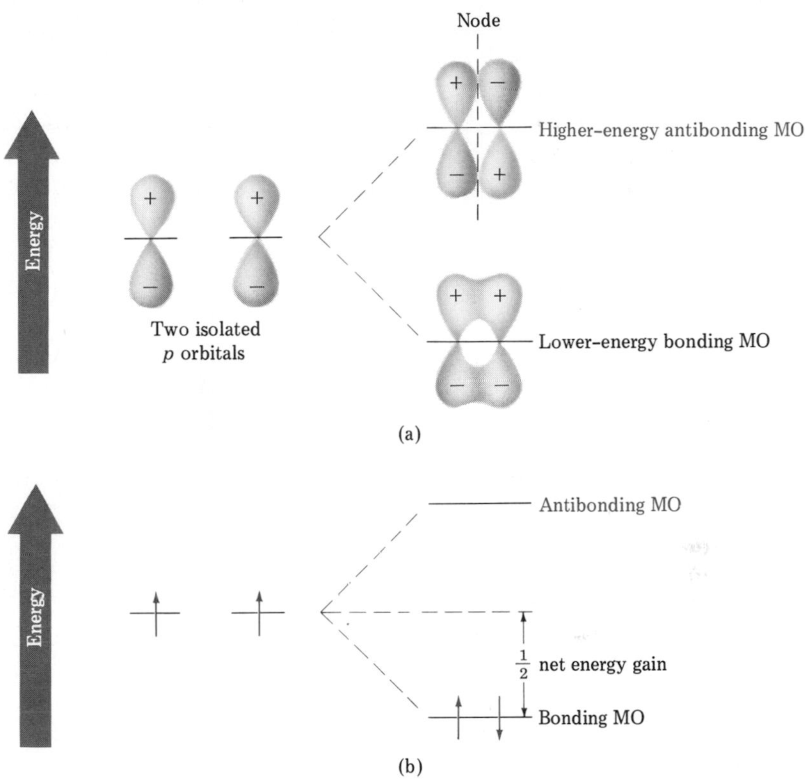


**Figure 14.1** (a) Two isolated $p$ orbitals combine to form two pi molecular orbitals. (b) When these orbitals are filled, both electrons occupy the low-energy bonding orbital, leading to formation of a stable bond.

The lowest-energy molecular orbital (denoted $\psi_1$, Greek psi) is a fully additive combination that has no nodes between the nuclei. It is therefore a bonding orbital and holds two electrons. The molecular orbital of next lowest energy, $\psi_2$, has one node between nuclei, is also a bonding orbital, and holds the remaining two electrons. Above $\psi_1$ and $\psi_2$ in energy are the two antibonding molecular orbitals $\psi_3^*$ and $\psi_4^*$. Of these, $\psi_3^*$ has two nodes and $\psi_4^*$, the highest-energy molecular orbital, has three nodes between nuclei. Note that the number of nodes increases as the energy level of the orbital increases.

Quantum mechanical calculations show that the sum of energy levels of the two bonding butadiene molecular orbitals is slightly lower than the sum of two isolated alkene molecular orbitals. In other words, placing the four electrons in the two bonding molecular orbitals results in a more stable arrangement than placing them in two isolated alkene orbitals (Figure 14.3). This "extra" stability is a consequence of a favorable bonding interaction across the C2—C3 bond of 1,3-butadiene.

In describing the 1,3-butadiene molecular orbitals, we say that the pi electrons are **delocalized** over the entire pi framework rather than localized between two specific nuclei. Electron delocalization always leads to lower-energy orbitals and greater stability.

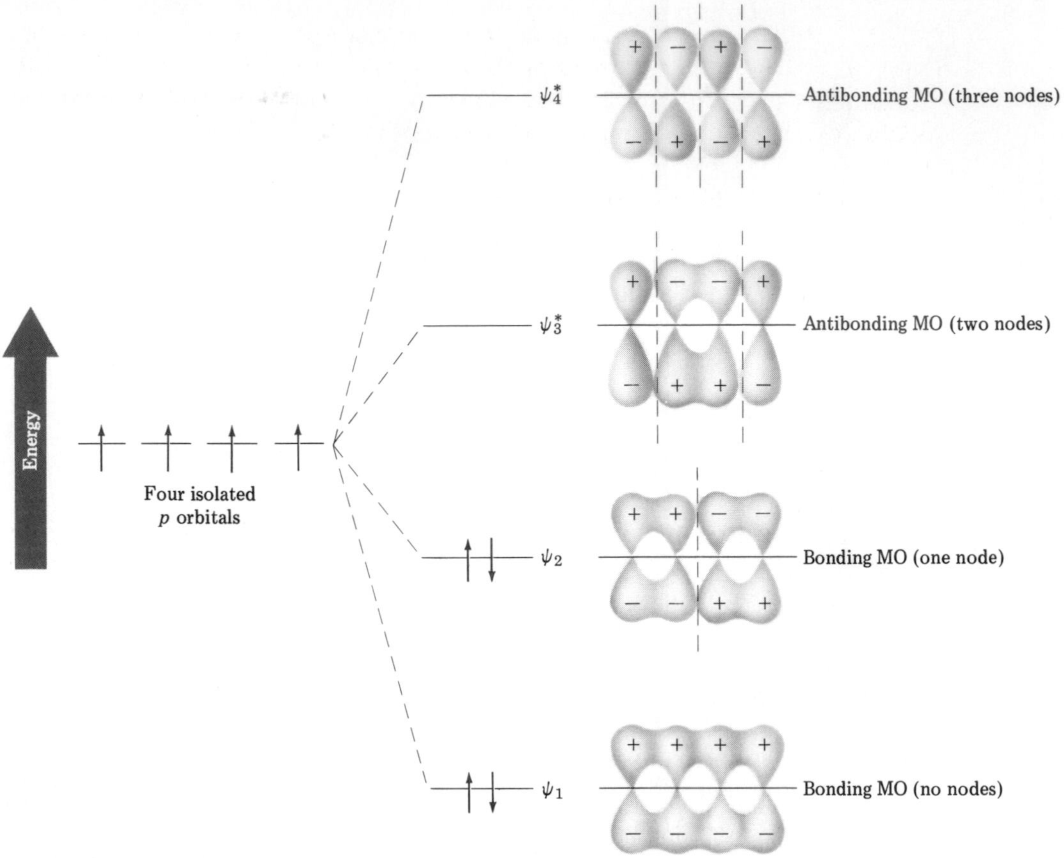

$\psi_4^*$ Antibonding MO (three nodes)

$\psi_3^*$ Antibonding MO (two nodes)

Energy

Four isolated *p* orbitals

$\psi_2$ Bonding MO (one node)

$\psi_1$ Bonding MO (no nodes)

**Figure 14.2** Pi molecular orbitals in 1,3-butadiene: The asterisk on $\psi_3^*$ and $\psi_4^*$ indicates antibonding orbitals.

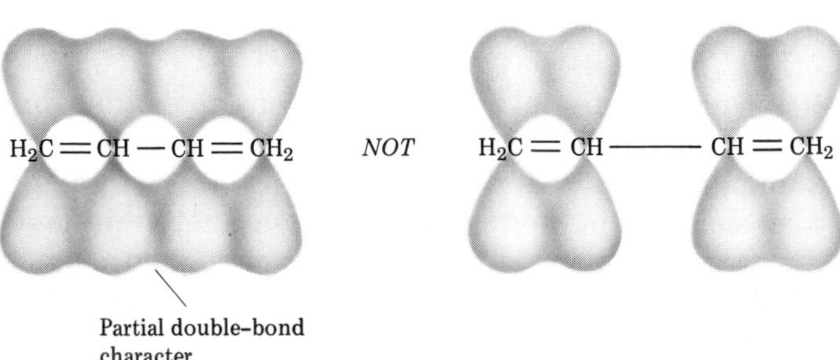


**Figure 14.3** An orbital view of 1,3-butadiene, showing the favorable pi bonding interaction between C2 and C3

## 14.4 Bond Lengths in 1,3-Butadiene

Further evidence for the special nature of conjugated dienes comes from bond length data. Measurements show that the C2—C3 single bond in 1,3-butadiene has a length of 1.48 Å and that the two equivalent carbon–carbon double bonds have lengths of 1.34 Å (Table 14.2).

**Table 14.2** Some carbon–carbon bond lengths

| *Bond* | *Bond length (Å)* | *Bond hybridization* |
|---|---|---|
| $CH_3—CH_3$ | 1.54 | $C_{sp^3}–C_{sp^3}$ |
| $H_2C{=}CH_2$ | 1.33 | $C_{sp^2}–C_{sp^2}$ |
| $H_2C{=}CH—CH{=}CH_2$ | 1.48 | $C_{sp^2}–C_{sp^2}$ |
| $H_2C{=}CHCH{=}CH_2$ | 1.34 | $C_{sp^2}–C_{sp^2}$ |

If we compare the length of the carbon–carbon single bond of 1,3-butadiene (1.48 Å) with that of ethane (1.54 Å), we find that the 1,3-butadiene single bond is shorter by 0.06 Å. Two different explanations have been advanced to account for this bond shortening. One explanation proposes that the shortening of the 1,3-butadiene single bond is due to the delocalization of pi electrons in the bonding molecular orbitals. According to this view, pi orbital overlap across the C2—C3 bond results in partial double-bond character (Figure 14.3) and consequent bond shortening to a value midway between a pure single bond (1.54 Å) and a pure double bond (1.33 Å). The partial double-bond character of the C2—C3 bond is sufficient to stabilize the molecule but not to prevent bond rotation from occurring.

Alternatively, it can be argued that the shortened 1,3-butadiene single bond is a natural consequence of the orbital hybridization involved. The C2—C3 bond results from sigma overlap of two carbon $sp^2$ orbitals, whereas a normal alkane bond results from overlap of two carbon $sp^3$ orbitals. Since overlap of $sp^2$ orbitals results in a single bond that has more *s* character than usual, the 1,3-butadiene single bond is a bit shorter and stronger than usual. Both explanations are probably valid, and both contribute to the bond shortening observed for 1,3-butadiene.

## 14.5 Electrophilic Additions to Conjugated Dienes: Allylic Carbocations

One of the most striking differences between the chemistry of conjugated dienes and isolated alkenes is in their electrophilic addition reactions.

As we've seen, the addition of electrophilic reagents to carbon–carbon double bonds is an important and general reaction (Section 6.8). Markovnikov regiochemistry is observed for these reactions because the more highly substituted (more stable) carbocation is involved as an intermediate. Thus, addition of HCl to 2-methylpropene yields 2-chloro-2-methylpropane rather than 1-chloro-2-methylpropane, and addition of 2 mol equiv of HCl to the nonconjugated diene 1,4-pentadiene yields 2,4-dichloropentane.

$$(CH_3)_2C{=}CH_2 \xrightarrow[\text{Ether}]{\text{HCl}} \left[(CH_3)_3C^+\right] \rightarrow (CH_3)_3CCl$$

2-Methylpropene → Tertiary carbocation intermediate → 2-Chloro-2-methylpropane

$(CH_3)_2CHCH_2Cl$ — 1-Chloro-2-methylpropane (*not formed*)

$$H_2C{=}CHCH_2CH{=}CH_2 \xrightarrow[\text{Ether}]{\text{2 HCl}} CH_3CH(Cl)CH_2CH(Cl)CH_3$$

1,4-Pentadiene (a nonconjugated diene) → 2,4-Dichloropentane

Conjugated dienes also undergo electrophilic addition reactions readily, but mixtures of products are invariably obtained. For example, addition of HBr to 1,3-butadiene yields a mixture of two products:

$$H_2C{=}CHCH{=}CH_2 \xrightarrow[0^\circ C]{\text{HBr}} H_2C{=}CHCH(Br)CH_2(H) + (H)CH_2CH{=}CHCH_2(Br)$$

1,3-Butadiene → 3-Bromo-1-butene (71%; 1,2 addition) + 1-Bromo-2-butene (29%; 1,4 addition)

3-Bromo-1-butene is the normal Markovnikov product (**1,2 addition**), but 1-bromo-2-butene appears unusual. The double bond in this product has moved to a position between carbons 2 and 3, while HBr has added to carbons 1 and 4 (**1,4 addition**).

Many other electrophiles besides HBr add to conjugated dienes, and mixtures of products are formed in all cases. For example, $Br_2$ adds to 1,3-butadiene to give a mixture of 1,4-dibromo-2-butene and 3,4-dibromo-1-butene.

$$H_2C{=}CH{-}CH{=}CH_2 \xrightarrow[25^\circ C]{Br_2} BrCH_2{-}CH{=}CH{-}CH_2Br \quad \text{1,4 addition}$$

1,3-Butadiene → 1,4-Dibromo-2-butene (45%)

+

$$BrCH_2{-}CH(Br){-}CH{=}CH_2 \quad \text{1,2 addition}$$

3,4-Dibromo-1-butene (55%)

How can we account for the formation of the 1,4-addition products? The answer is that **allylic carbocations** are involved as intermediates in the reactions. When the electron-rich pi bond of 1,3-butadiene is protonated, two carbocation intermediates are possible—a primary carbocation and a secondary allylic cation (recall that *allylic* means "next to a double bond"). Since an allylic cation is strongly stabilized (Section 11.9), it forms in preference to the less stable primary carbocation.

$$H_2C{=}CH{-}CH{=}CH_2 \xrightarrow{HBr} \left[ H_2C{=}CH{-}CH_2{-}\overset{+}{C}H_2 \right] Br^-$$

1,3-Butadiene → Primary carbocation (*not formed*)

$$H_2C{=}CH{-}CH{=}CH_2 \xrightarrow{HBr} \left[ H_2C{=}CH{-}\overset{+}{C}H{-}CH_3 \right] Br^-$$

Secondary, allylic carbocation

There are two ways to account for the stability of allylic carbocations. Resonance theory (Section 10.7) offers a pictorial representation of the situation through the use of different resonance forms. The more resonance forms that are possible, the more stable the compound is. Thus, an allylic cation has two resonance forms and is more stable than a nonallylic cation. Neither of the two resonance forms is correct by itself, of course; the true structure of the allylic cation is a combination, or resonance hybrid, of the two Kekulé structures. For example, the allylic carbocation produced by protonation of 1,3-butadiene is a resonance hybrid of the following two structures:

$$\overset{+}{C}H_2{-}CH{=}CH{-}CH_3 \longleftrightarrow CH_2{=}CH{-}\overset{+}{C}H{-}CH_3$$

Two resonance forms of an allylic carbocation

When the allylic cation reacts with bromide ion to complete the electrophilic addition reaction, attack can occur at either carbon 1 or carbon 3, since both share the positive charge. The result is a mixture of 1,2- and 1,4-addition products.

$$\overset{+}{C}H_2{-}CH{=}CHCH_3 \longleftrightarrow CH_2{=}CH{-}\overset{+}{C}HCH_3 \xrightarrow{:\ddot{Br}:^-} BrCH_2{-}CH{=}CHCH_3 + H_2C{=}CH{-}CH(Br)CH_3$$

1,4 addition (29%) 1,2 addition (71%)

An alternative way to account for the stability of allylic carbocations is to construct a molecular orbital description. When we allow three *p* orbitals to interact, three molecular orbitals are formed (Figure 14.4). One low-energy bonding orbital, one nonbonding orbital, and one high-energy antibonding orbital result. The two available pi electrons occupy the low-energy bonding orbital, indicating that a partial bond exists between carbons 2 and 3. Thus, the allylic cation is a conjugated system stabilized by electron delocalization in much the same way that 1,3-butadiene is stabilized by electron delocalization.

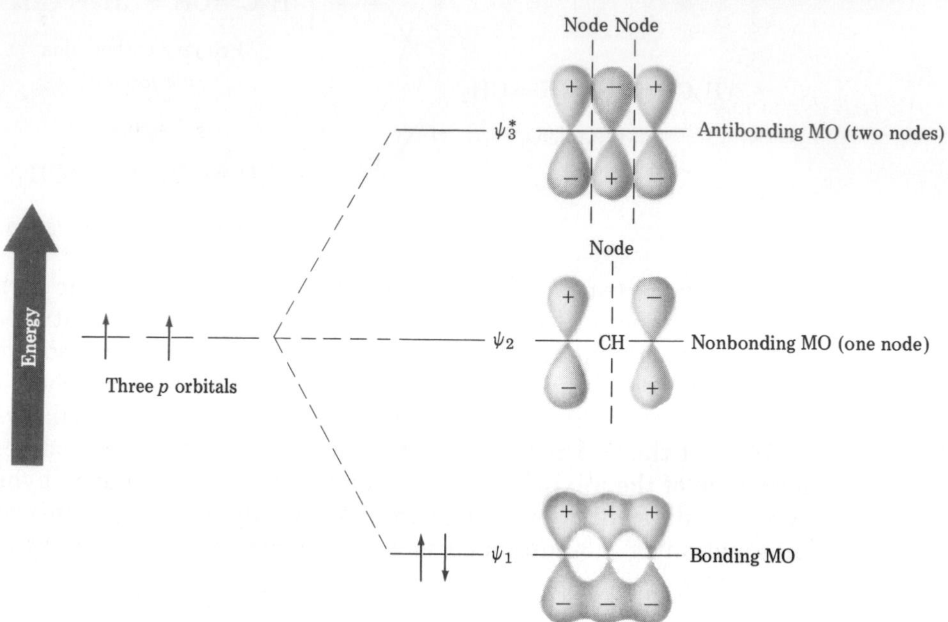


**Figure 14.4** Molecular orbitals of an allylic carbocation: The two available electrons occupy the low-energy bonding orbital, leading to a net stabilization.

It's important to realize that the two approaches just taken for describing allylic carbocations—the resonance approach and the molecular orbital approach—don't "compete" with each other. It's not that one approach is more "right" than the other; it's that both are really just alternative ways of saying the same thing: Both are pictorial ways of visualizing a phenomenon that is best handled mathematically. Sometimes one approach is more convenient, and sometimes the other is more convenient. We'll use both at various times.

PROBLEM

**14.3** Give the structures of the likely products from reaction of 1 equiv HCl with 1,3-pentadiene. Show both 1,2 and 1,4 adducts.

PROBLEM

**14.4** Examine the possible carbocation intermediates produced during addition of HCl to 1,3-pentadiene (Problem 14.3), and predict which of the 1,2 adducts predominates. Which 1,4 adduct would be expected to predominate?

PROBLEM

**14.5** Electrophilic addition of $Br_2$ to isoprene yields the following product mixture:

$$H_2C{=}C(CH_3)CH{=}CH_2 \xrightarrow{Br_2} \underset{(3\%)}{H_2C{=}C(CH_3)CHBrCH_2Br} + \underset{(21\%)}{BrCH_2CBr(CH_3)CH{=}CH_2} + \underset{(76\%)}{BrCH_2C(CH_3){=}CHCH_2Br}$$

Of the 1,2-addition products, explain why 3,4-dibromo-3-methyl-1-butene (21%) predominates over 3,4-dibromo-2-methyl-1-butene (3%).

PROBLEM

**14.6** The molecular orbital diagram for an allylic *radical*, $H_2C{=}CH{-}CH_2\cdot$ is similar to that for an allylic carbocation. Indicate which orbitals the three pi electrons occupy.

## 14.6 Kinetic versus Thermodynamic Control of Reactions

Addition of electrophiles to conjugated dienes at or below room temperature normally leads to a mixture of products in which the 1,2 adduct predominates over the 1,4 adduct. When the same reaction is carried out at higher temperatures, however, the product ratio often changes and the 1,4 adduct predominates. For example, addition of HBr to 1,3-butadiene at 0°C yields a 71:29 mixture of 1,2 and 1,4 adducts, but the same reaction carried out at 40°C yields a 15:85 mixture. Furthermore, when the product mixture formed at 0°C is heated to 40°C in the presence of more HBr, the ratio of adducts slowly changes from 71:29 to 15:85. How can we explain these observations?

$$H_2C{=}CHCH{=}CH_2 + HBr \xrightarrow[\text{40°C}]{\text{0°C}} \underset{\substack{\text{1,2 adduct} \\ (71\%)\ \text{at 0°C} \\ (15\%)\ \text{at 40°C}}}{H_2C{=}CHCHBrCH_3} + \underset{\substack{\text{1,4 adduct} \\ (29\%)\ \text{at 0°C} \\ (85\%)\ \text{at 40°C}}}{CH_3CH{=}CHCH_2Br}$$

To understand the reasons for the effect of reaction temperature on electrophilic addition reactions of conjugated dienes, we need to review what we've already learned about reactions and transition states. In principle, all reactions are reversible. In practice, however, it is sometimes difficult or impossible to reach equilibrium, and a nonequilibrium product distribution results. For example, imagine a reaction that can give either of two products depending on the reaction conditions used. Under mild, low-temperature reaction conditions, one product is formed, but under vigorous, high-temperature conditions, another product is formed:

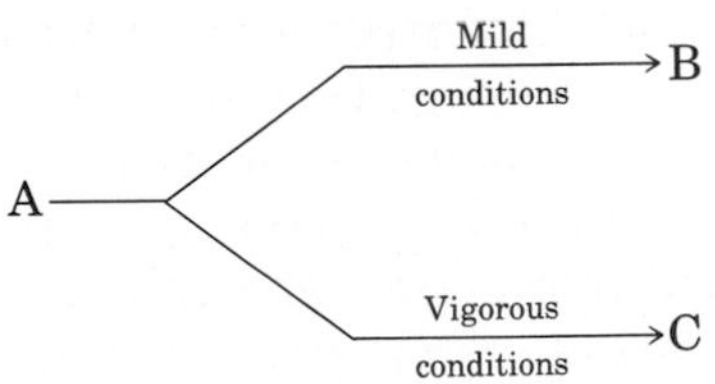


Let's assume that B forms faster than C (in other words, that $\Delta G^{\ddagger}_{B} < \Delta G^{\ddagger}_{C}$), but that C is a more stable product than B (in other words, that $\Delta G^{\circ}_{C} > \Delta G^{\circ}_{B}$). A reaction energy diagram for the two processes might look like that shown in Figure 14.5.

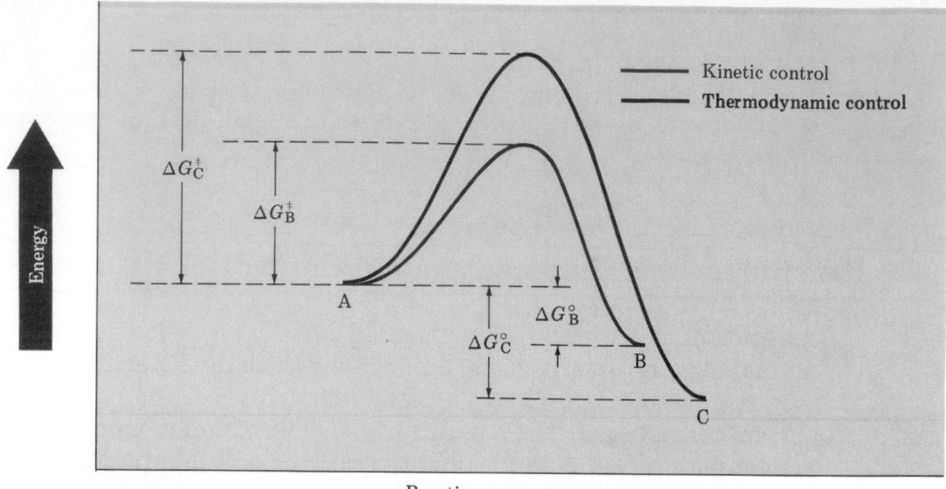


**Figure 14.5** A reaction energy diagram showing two competing reactions: The more stable product forms more slowly than the less stable product.

Let's first carry out the reaction under vigorous, high-temperature conditions such that both processes are reversible and an equilibrium is reached. Since C is more stable than B ($\Delta G^{\circ}_{C} > \Delta G^{\circ}_{B}$), C is the major product formed under the reaction conditions. It doesn't matter that C forms more slowly than B ($\Delta G^{\ddagger}_{B} < \Delta G^{\ddagger}_{C}$), because the reaction conditions have been chosen such that B and C are formed reversibly, and therefore interconvert rapidly. *All that matters in a reversible reaction is the thermodynamic stability of the products at equilibrium.* Such reactions are said to be under **thermodynamic control**, or equilibrium control.

$$B \rightleftharpoons A \rightleftharpoons C$$ Thermodynamic control (vigorous conditions; reversible)

Let's now carry out the same reaction under milder, low-temperature conditions such that both processes are *irreversible*, and an equilibrium is *not* reached. In other words, there is only enough energy present for the reactant A molecules to be able to climb the energy barrier separating A from B, but not enough energy for product B molecules to climb the hill in the reverse direction and go back to reactant A. Since B forms faster than C (has a lower $\Delta G^{\ddagger}$), B is the major product. It doesn't matter that C is more stable than B, because the reaction conditions have been chosen so that B and C are formed irreversibly and do not interconvert. *All that matters in an irreversible process is the reaction rate.* Such reactions are said to be under **kinetic control**.

$$B \longleftarrow A \longrightarrow C$$ Kinetic control (Mild conditions; irreversible)

We can now explain the effect of temperature on electrophilic addition reactions of conjugated dienes. Under mild, low-temperature conditions (0°C), HBr adds to 1,3-butadiene under kinetic control to give a 71:29 mixture of products with the 1,2 adduct predominating. Since these mild conditions don't allow the products to reach equilibrium, the product that forms faster predominates. Under more vigorous, high-temperature conditions (40°C), however, the reaction occurs reversibly under thermodynamic control to give a 15:85 mixture of products, with the more stable 1,4 adduct predominating. The higher temperature provides more energy for product molecules to climb the high energy barrier leading back to the allylic cation, and an equilibrium mixture of products therefore results. Figure 14.6 shows the situation on a reaction energy diagram.

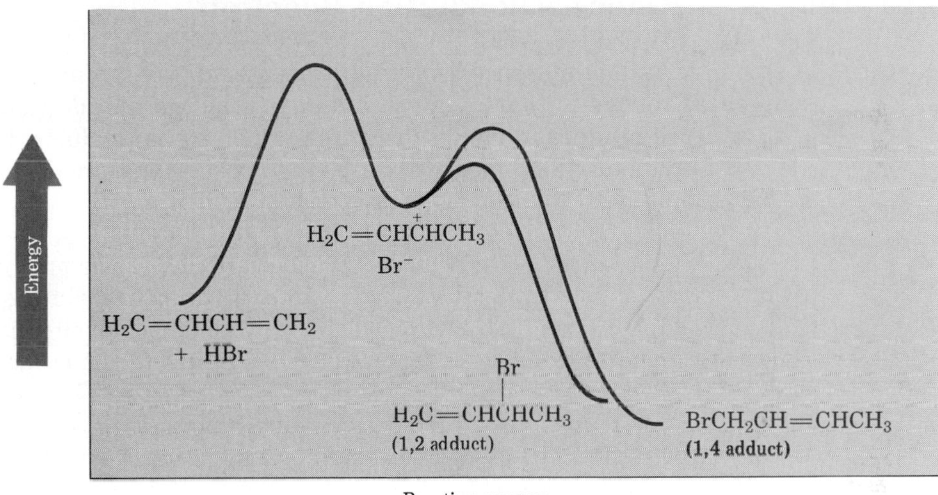


**Figure 14.6** Reaction energy diagram for the electrophilic addition of HBr to 1,3-butadiene: The 1,2 adduct is the kinetic product, and the 1,4 adduct is the thermodynamic product.

The electrophilic addition of HBr to 1,3-butadiene is an excellent example of how experimental conditions can determine the product of a reaction. The concept of thermodynamic control versus kinetic control is a valuable one that we can often use to advantage in the laboratory.

One more point: What about the Hammond postulate? We saw in Section 6.12 that more stable carbocations are formed *faster* than less stable ones because product stability is usually reflected in the transition states leading to the cations. Yet in the present case, the more stable product is formed *slower* than the less stable one.

The difference in the two cases is a consequence of the different natures of the product-determining steps. The formation of a carbocation by protonation of an alkene is an *endothermic* step; thus, the carbocation product is close in both energy and stability to the transition state. The formation of a 1,2 or 1,4 adduct by reaction of bromide ion with a high-energy carbocation is an *exothermic* step, however; thus, the alkyl halide product is far removed in both energy and stability from the transition state, and the Hammond postulate does not apply.

PROBLEM

**14.7** The 1,2 adduct and the 1,4 adduct of HBr with 1,3-butadiene are in equilibrium at 40°C. Propose a mechanism by which the interconversion of 3-bromo-1-butene and 1-bromo-2-butene takes place. [*Hint:* See Section 11.6.]

PROBLEM

**14.8** Why do you suppose 1,4 adducts of 1,3-butadiene are generally more stable than 1,2 adducts?

## 14.7 The Diels–Alder Cycloaddition Reaction

A second striking difference between the chemistries of conjugated and nonconjugated dienes is that conjugated dienes undergo an addition reaction with isolated alkenes to yield substituted cyclohexene products. For example, 1,3-butadiene and 3-buten-2-one give 3-cyclohexenyl methyl ketone in nearly 100% yield:

$$\text{H(CH}_2\text{=)C–C(=CH}_2\text{)H} + \text{H(COCH}_3\text{)C=CH}_2 \xrightarrow[\Delta]{\text{Benzene}} \text{3-cyclohexenyl–C(=O)CH}_3$$

1,3-Butadiene — 3-Buten-2-one — 3-Cyclohexenyl methyl ketone (100%)

This process, named the **Diels–Alder cycloaddition reaction** after its two discoverers, Otto Diels[1] and Kurt Alder,[2] is extremely useful in organic synthesis because it forms *two* carbon–carbon bonds in one step and is one of the few methods available for forming cyclic molecules. (As you might expect, a **cycloaddition reaction** is one in which two reactants *add* together to give a *cyclic* product.) The 1950 Nobel prize in chemistry was awarded to Diels and Alder in recognition of the importance of their discovery.

The mechanism of the Diels–Alder cycloaddition is quite different from all other reactions we have studied. It is neither a polar reaction nor a radical reaction; rather it is a pericyclic process. **Pericyclic reactions**, which we'll discuss in more detail in Chapter 30, are those processes that occur in a single step without intermediates and involve a cyclic redistribution of bonding electrons. The two reactants simply add together through a cyclic transition state in which both of the new carbon–carbon bonds form at the same time.

[1]Otto Diels (1876–1954); b. Hamburg; Ph.D. Berlin (E. Fischer); professor, University of Berlin (1906–1916), Kiel (1916–1948); Nobel prize (1950).

[2]Kurt Alder (1902–1958); b. Königshütte; Ph.D. Kiel (Diels); professor, University of Cologne (1940–1958); Nobel prize (1950).

We can picture a Diels–Alder addition as occurring by head-on (sigma) overlap of the two isolated-alkene *p* orbitals with the two *p* orbitals on carbons 1 and 4 of the diene (Figure 14.7). This is, of course, a *cyclic* orientation of the reactants.

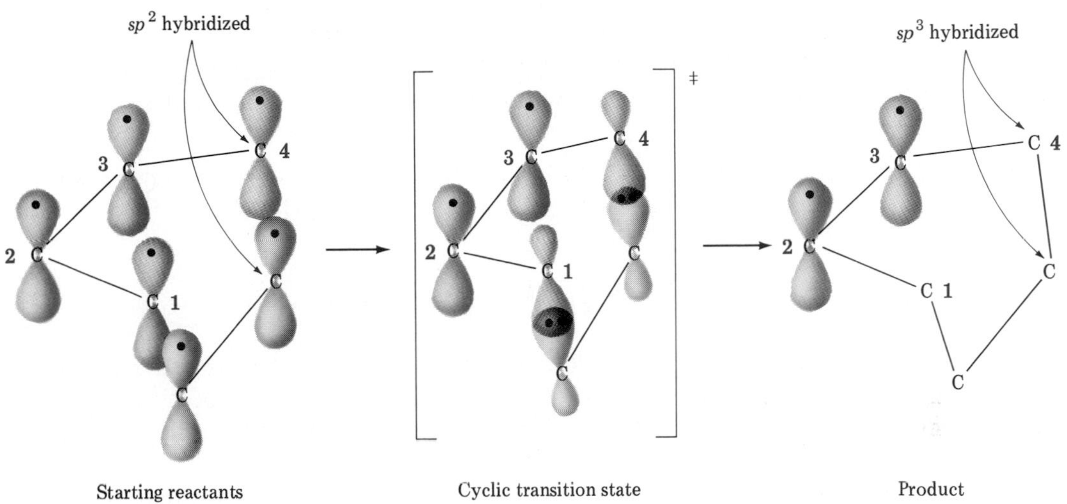


**Figure 14.7** Mechanism of the Diels–Alder cycloaddition reaction: The reaction occurs in a single step through a cyclic transition state in which the two new carbon–carbon bonds begin to form at the same time.

In the Diels–Alder transition state, the alkene carbons and carbons 1 and 4 of the diene rehybridize from $sp^2$ to $sp^3$ to form the two new single bonds. Carbons 2 and 3 of the diene remain $sp^2$ hybridized to form the new double bond in the cyclohexene product. We'll study this mechanism at greater length in Chapter 30 but will concentrate for the present on learning more about the chemistry of the Diels–Alder reaction.

## 14.8 Characteristics of the Diels–Alder Reaction

### THE DIENOPHILE

The Diels–Alder cycloaddition reaction takes place most rapidly and in highest yield if the alkene component, or **dienophile** ("diene lover"), is substituted by an electron-withdrawing group. Thus, ethylene itself is unreactive in the Diels–Alder reaction, but propenal, ethyl propenoate, maleic anhydride, benzoquinone, propenenitrile, and others are highly reactive (Figure 14.8). Note also that alkynes such as methyl propynoate can act as Diels–Alder dienophiles. In all of these cases, the dienophile double bond is next to the positively polarized carbon of a substituent that withdraws electrons.

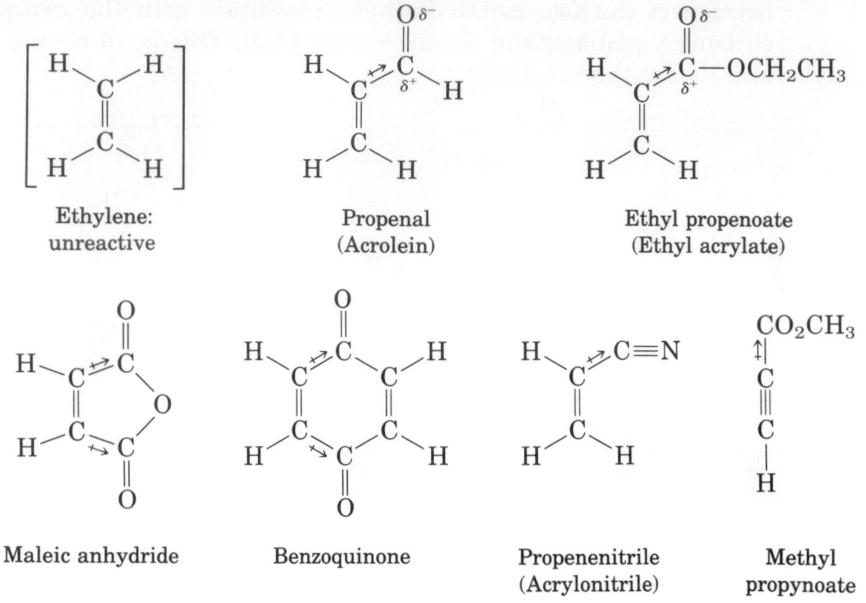

Ethylene: unreactive

Propenal (Acrolein)

Ethyl propenoate (Ethyl acrylate)

Maleic anhydride

Benzoquinone

Propenenitrile (Acrylonitrile)

Methyl propynoate

**Figure 14.8** Some Diels–Alder dienophiles: All contain electron-withdrawing substituents on the double bond.

One of the most important features of the Diels–Alder reaction is that it is *stereospecific*: The stereochemistry of the starting dienophile is maintained during the reaction. If we carry out the cycloaddition with a cis alkene such as methyl *cis*-2-butenoate, we produce only the cis-substituted cyclohexene. Conversely, Diels–Alder reaction with methyl *trans*-2-butenoate yields only the trans-substituted cyclohexene product.

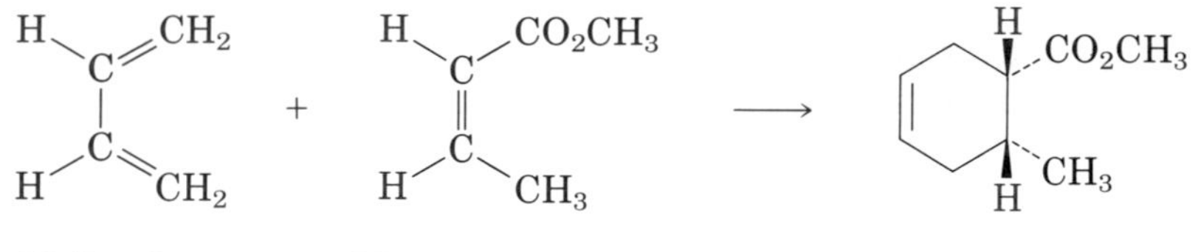

1,3-Butadiene + Methyl (*Z*)-2-butenoate ⟶ Cis product

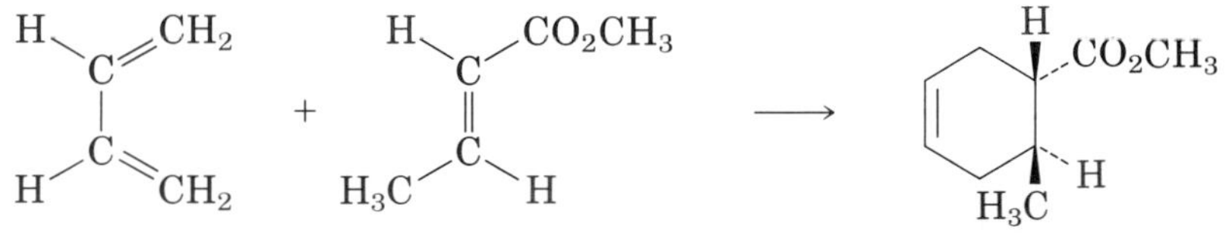

1,3-Butadiene + Methyl (*E*)-2-butenoate ⟶ Trans product

## THE DIENE

A diene must be able to adopt an **s-cis conformation** (a "cis-like" conformation about the *single* bond) in order to undergo the Diels–Alder reaction. Only in the *s*-cis conformation are carbons 1 and 4 of the diene close enough to react through a cyclic transition state to give a new ring. In the alternative *s*-trans conformation, the ends of the diene partner are too far apart to overlap the dienophile *p* orbitals successfully.

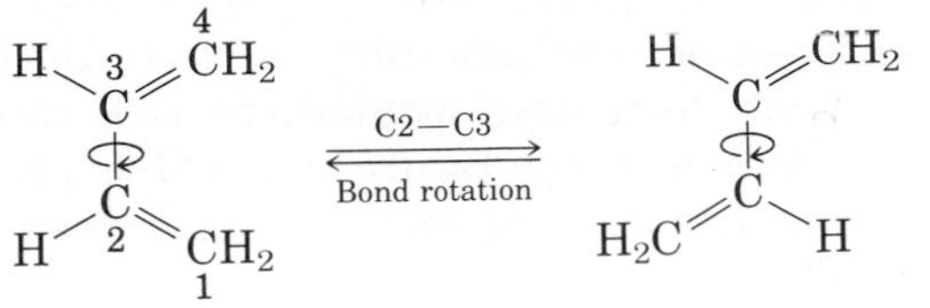

C2—C3 Bond rotation

*s*-Cis conformation

*s*-Trans conformation

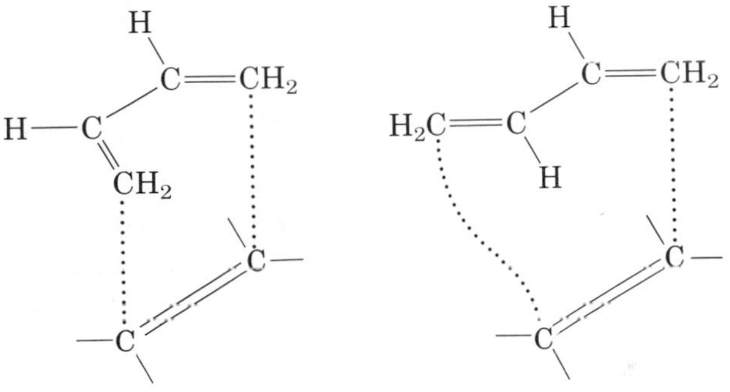

Successful reaction

No reaction (ends too far apart)

Examples of some dienes that cannot adopt *s*-cis conformation and therefore do not undergo Diels–Alder reaction are shown in Figure 14.9. In the case of the bicyclic (two-ring) diene, the double bonds are rigidly fixed in the *s*-trans arrangement by geometric constraints of the rings. In the case

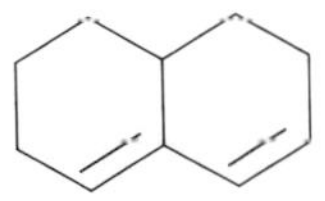

A bicyclic diene (rigid *s*-trans diene)

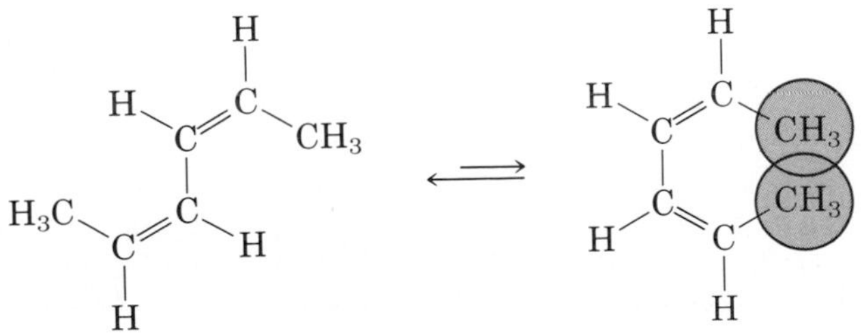

(2*Z*,4*Z*)-Hexadiene (*s*-trans, more stable)

Severe steric strain in *s*-cis form

**Figure 14.9** Two *s*-trans dienes that cannot undergo Diels–Alder reactions

of (2*Z*,4*Z*)-hexadiene, severe steric strain between the two methyl groups prevents the molecule from adopting *s*-cis geometry.

In contrast to the unreactive *s*-trans dienes, certain other dienes are rigidly fixed in the correct *s*-cis geometry and are therefore highly reactive in the Diels–Alder cycloaddition reaction. Such, for example, is the case with cyclopentadiene. Cyclopentadiene is so reactive, in fact, that it reacts with itself! At room temperature (about 25°C), cyclopentadiene **dimerizes**—one molecule acts as diene and another acts as dienophile.

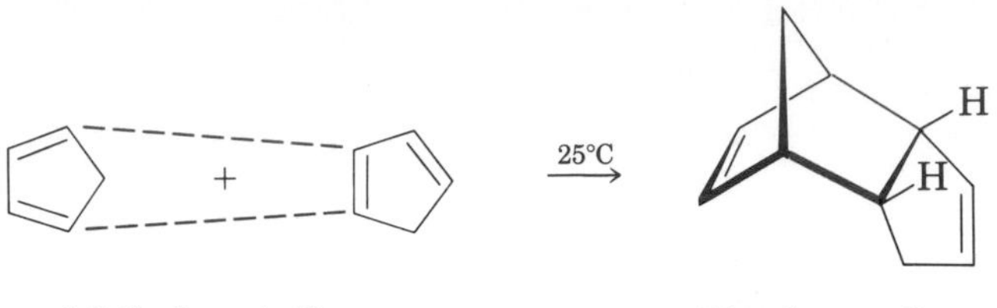

25°C

1,3-Cyclopentadiene (*s*-cis)

Bicyclopentadiene

This kind of Diels–Alder reaction is particularly important in the commercial production of a number of chlorinated insecticides such as chlordane. Thus, hexachlorocyclopentadiene is allowed to undergo Diels–Alder reaction with cyclopentadiene. Further addition of chlorine to the Diels–Alder product yields chlordane.

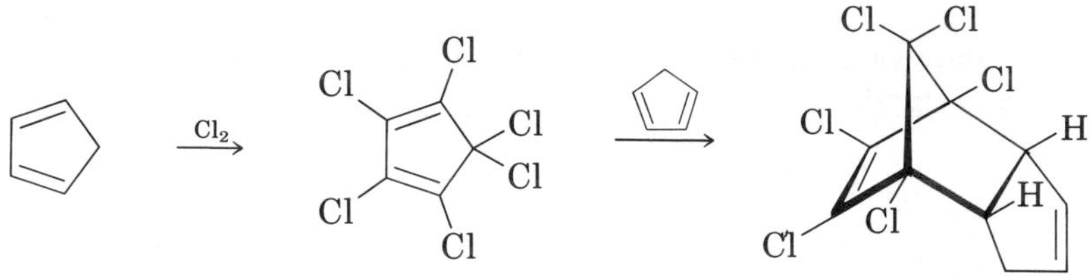

$Cl_2$

Cyclopentadiene

Hexachlorocyclopentadiene

Diels–Alder adduct

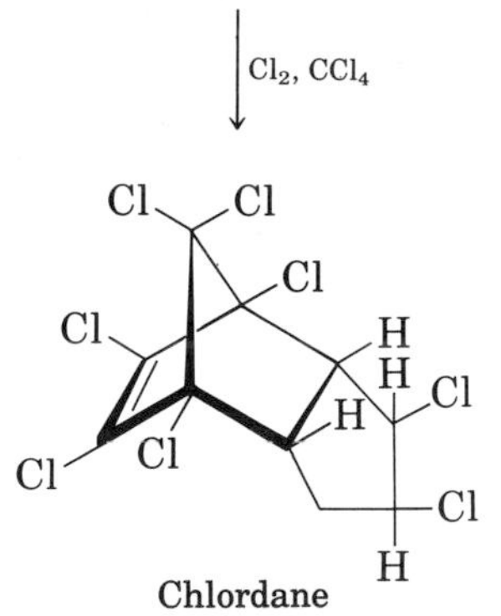

$Cl_2$, $CCl_4$

Chlordane (an insecticide)

**PROBLEM**

**14.9** Which of these alkenes would you expect to be good Diels–Alder dienophiles?

(a) $H_2C{=}CH\overset{O}{\overset{\|}{C}}Cl$ (b) $H_2C{=}CHCH_2CH_2\overset{O}{\overset{\|}{C}}OCH_3$ (c) 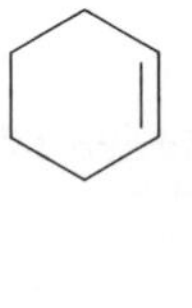 [cyclohexene]

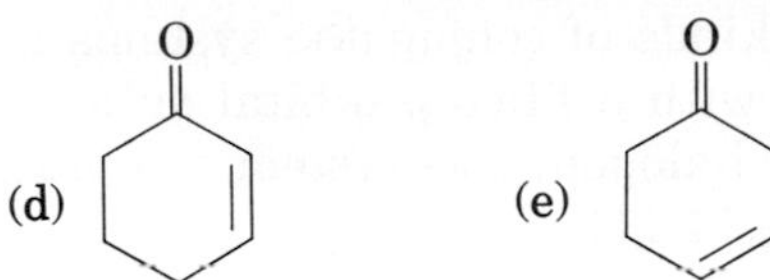


**PROBLEM**

**14.10** Which of the following dienes have an *s*-cis conformation and which an *s*-trans conformation? Of the *s*-trans dienes, which can readily rotate to *s*-cis?

(a) (b)

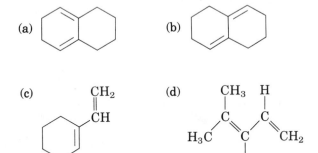

(c) $CH_2$ / CH (d) $CH_3$ H / C C / $H_3C$ C $CH_2$ / H

**PROBLEM**

**14.11** Although cyclopentadiene is highly reactive toward Diels–Alder cycloaddition reactions, 1,3-cyclohexadiene is less reactive, and 1,3-cycloheptadiene is nearly inert. Can you suggest a reason for this reactivity order? (Building molecular models should be helpful.)

## 14.9 Other Conjugated Systems

A conjugated system was defined earlier in this chapter as one that consists of alternating double and single bonds. After considering a molecular orbital description of 1,3-butadiene, however, we might now more accurately describe a conjugated system as one that consists of *an extended series of overlapping p orbitals* (Figure 14.10). Thus a 1,3-diene and an allylic cation are both examples of conjugated systems, but there are other kinds as well.

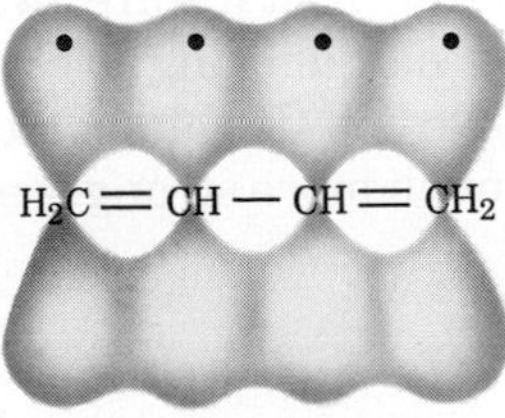


1,3-Butadiene
(four overlapping *p* orbitals)

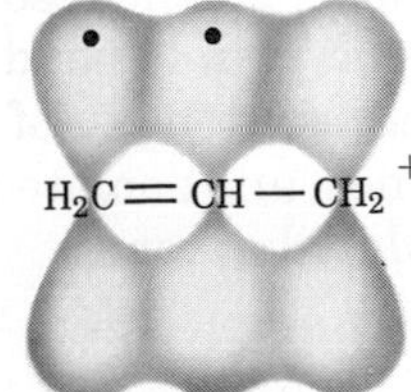


Allylic cation
(three overlapping *p* orbitals)

**Figure 14.10** Conjugated systems such as 1,3-butadiene and allylic cations contain an extended series of overlapping *p* orbitals.

Some of the most important kinds of conjugated systems result from overlap of double-bond *p* orbitals with a filled *p* orbital on a neighboring atom such as oxygen, nitrogen, or halogen. The system can either be negatively charged or neutral (Figure 14.11).

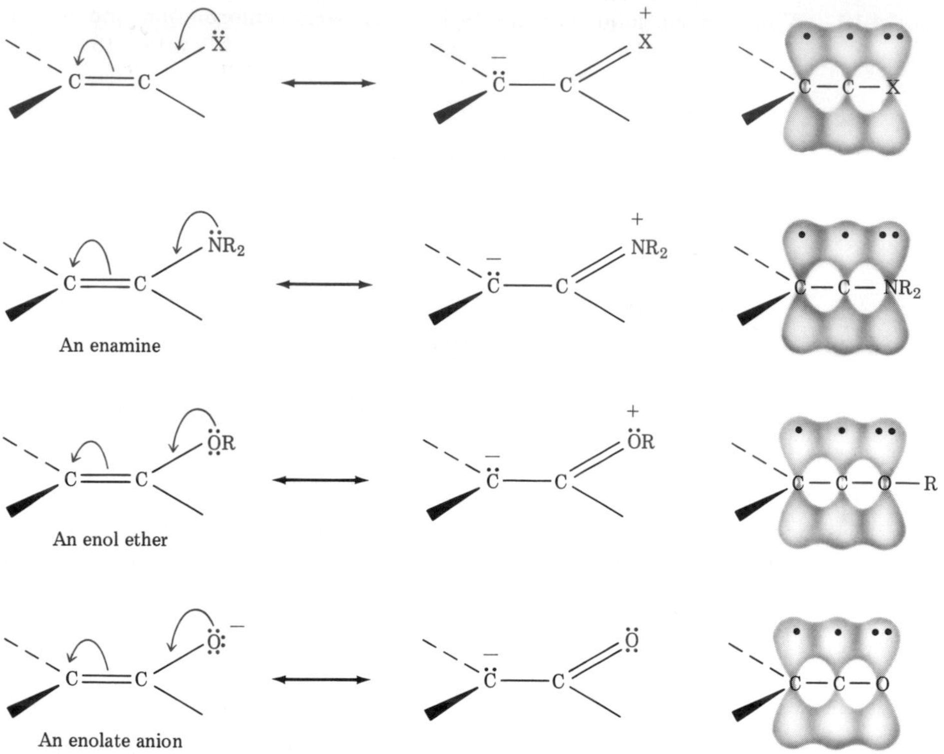

An enamine

An enol ether

An enolate anion

**Figure 14.11** Some conjugated systems resulting from overlap of double-bond *p* orbitals with filled *p* orbitals on neighboring atoms

These conjugated systems will be examined in more detail at a later point. For the present, it's sufficient to note that the electronic nature of the alkene pi bond is greatly affected by conjugation with a filled neighboring orbital. As the resonance forms shown in Figure 14.11 indicate, conjugation of the double bond with a filled neighboring orbital greatly increases the electron density of the carbon–carbon double bond. Thus enol ethers, enamines, and enolate ions are all much more strongly nucleophilic than normal isolated alkenes.

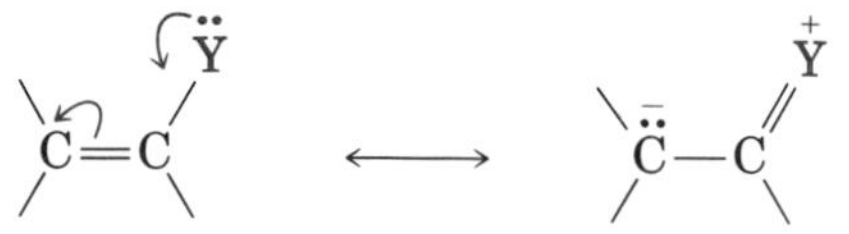

This resonance form puts extra electron density on *carbon,* making the carbon atom nucleophilic

## 14.10 Structure Determination of Conjugated Systems: Ultraviolet Spectroscopy

Infrared, nuclear magnetic resonance, and mass spectroscopy all help in the structure determination of conjugated systems. In addition to these three generally useful spectroscopic techniques, there's a fourth—**ultraviolet (UV) spectroscopy**—that is applicable solely to conjugated systems.

| | |
|---|---|
| 1. Infrared spectroscopy | Functional groups present |
| 2. Mass spectroscopy | Molecular size and formula |
| 3. Nuclear magnetic resonance spectroscopy | Carbon–hydrogen framework |
| 4. Ultraviolet spectroscopy | Nature of conjugated pi electron system |

Ultraviolet spectroscopy is less commonly used than the other three spectroscopic techniques because of the rather specialized information it gives. We'll therefore study it only briefly.

The ultraviolet region of the electromagnetic spectrum extends from the low wavelength end of the visible region ($4 \times 10^{-5}$ cm) down to $10^{-6}$ cm, but the narrow range from $2 \times 10^{-5}$ cm to $4 \times 10^{-5}$ cm is the portion of greatest interest to organic chemists. Absorptions in this region are usually measured in **nanometers**, nm (1 nm = $10^{-9}$ m = $10^{-7}$ cm). Thus, the ultraviolet range of interest is from 200 to 400 nm (Figure 14.12).

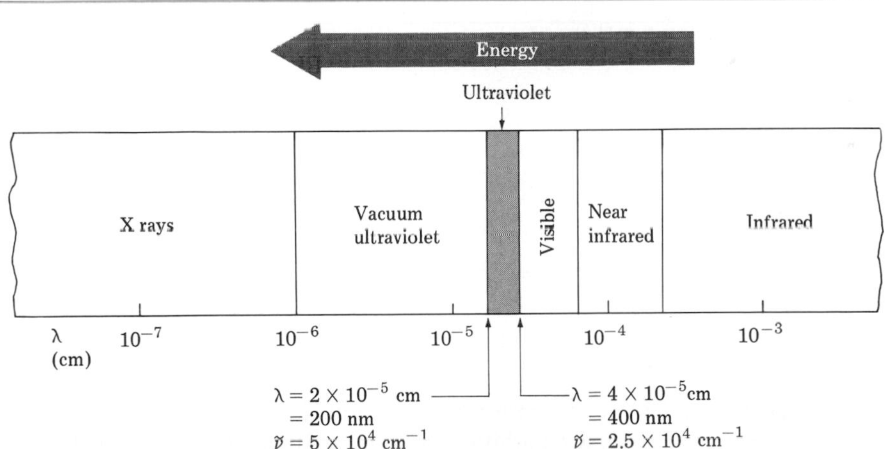


**Figure 14.12** The ultraviolet (UV) region of the electromagnetic spectrum

We saw in our discussion of infrared spectroscopy (Section 12.8) that, when an organic molecule is irradiated with electromagnetic waves, the radiation either is absorbed by the compound or passes through, depending on the exact energy of the waves. When infrared radiation is used, the energy absorbed corresponds to the amount necessary to increase molecular motions—bendings and stretchings—in functional groups. When ultraviolet radiation is used, the energy absorbed by a molecule corresponds to the amount necessary to excite electrons from one molecular orbital to another. Let's see what this means by looking first at 1,3-butadiene.

PROBLEM

**14.12** Calculate the energy range of electromagnetic radiation in the ultraviolet region of the spectrum from 200 to 400 nm wavelength. Recall

$$E = \frac{Nhc}{\lambda} = \frac{2.86 \times 10^{-3}\ \text{kcal/mol}}{\lambda\ \text{(cm)}}$$

PROBLEM

**14.13** How does the energy you calculated (Problem 14.12) for ultraviolet radiation compare with the values calculated previously for infrared spectroscopy and nuclear magnetic resonance spectroscopy?

## 14.11 Ultraviolet Spectrum of 1,3-Butadiene

1,3-Butadiene has four pi molecular orbitals. The two lower-energy bonding molecular orbitals are fully occupied in the ground state, and the two higher-energy antibonding molecular orbitals are unoccupied, as illustrated in Figure 14.13.

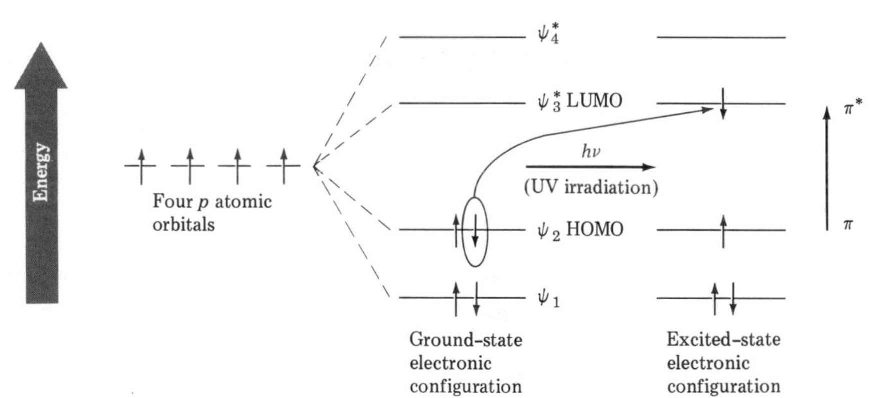


**Figure 14.13** Ultraviolet excitation of 1,3-butadiene results in the promotion of an electron from a bonding orbital ($\psi_2$) to an antibonding orbital ($\psi_3^*$).

On irradiation with ultraviolet light (denoted $h\nu$), 1,3-butadiene absorbs energy, and a pi electron is promoted from $\psi_2$, the highest occupied molecular orbital (**HOMO**), to $\psi_3^*$, the lowest unoccupied molecular orbital (**LUMO**). Since the electron is promoted from a bonding ($\pi$) molecular orbital to an antibonding ($\pi^*$) molecular orbital, we call this a $\pi \rightarrow \pi^*$ excitation (read as "pi to pi star"). The energy gap between the HOMO and the LUMO of

1,3-butadiene is such that ultraviolet light of 217 nm wavelength is required to accomplish the $\pi \rightarrow \pi^*$ electronic transition.

In practice, an ultraviolet spectrum is recorded by irradiating the sample with ultraviolet light of continuously changing wavelength. When the wavelength of light corresponds to the energy level required to excite an electron to a higher level, energy is absorbed. This absorption is detected and displayed on a chart that plots wavelength versus percent radiation absorbed (Figure 14.14). Note that UV spectra differ from IR spectra in the way they are recorded. IR spectra are usually displayed so that the "baseline" corresponding to zero absorption runs across the top of the chart and a valley indicates an absorption. UV spectra, however, are displayed with the baseline at the bottom of the chart so that a peak indicates an absorption.

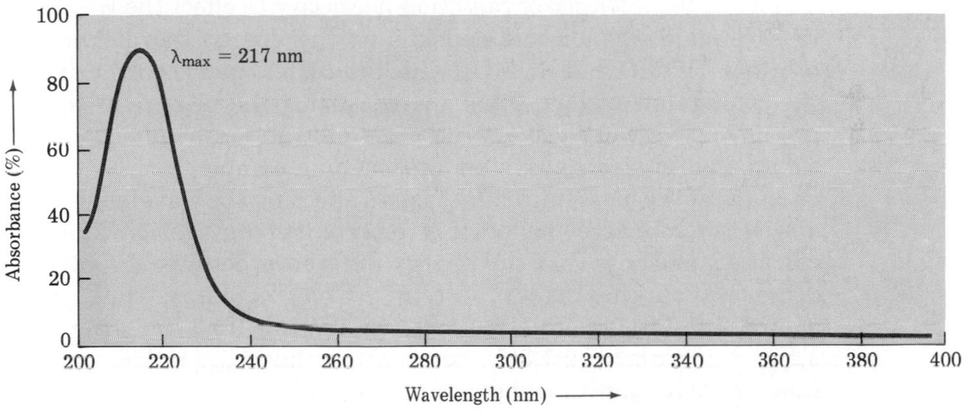


**Figure 14.14** The ultraviolet spectrum of 1,3-butadiene, $\lambda_{max} = 217$ nm

The exact amount of UV light absorbed is expressed as the sample's **molar absorptivity**, or **extinction coefficient**, $\varepsilon$, defined by the equation:

$$\text{Molar absorptivity } \varepsilon = \frac{A}{C \times l}$$

where $A$ = Percent absorbance

$C$ = Concentration in moles per liter

$l$ = Sample path length in centimeters

Molar absorptivity is a physical constant, characteristic of the particular molecule being observed, and thus characteristic of the particular pi electron system present in the sample. Typical values for conjugated dienes are in the range $\varepsilon = 10{,}000$–$25{,}000$.

Unlike infrared spectra and nuclear magnetic resonance spectra, which show many absorption lines for a given molecule, ultraviolet spectra are usually quite simple—often only a single peak is produced. The peak is usually broad, however, and we identify its position by noting the wavelength ($\lambda$) at the very top of the peak ($\lambda_{max}$, read as "lambda max").

PROBLEM

**14.14** Knowledge of molar absorptivities is particularly important in biochemistry where UV spectroscopy can provide an extremely sensitive method of analysis. For example, imagine that you wanted to determine the concentration of vitamin A in a sample. If pure vitamin A has $\lambda_{max} = 325$ ($\varepsilon = 50{,}100$), what is the vitamin A concentration in a sample whose absorbance at 325 nm is $A = 0.735$? Explain.

## 14.12 Interpreting Ultraviolet Spectra: The Effect of Conjugation

The exact wavelength of radiation necessary to effect the $\pi \rightarrow \pi^*$ transition in a conjugated molecule depends on the energy gap between molecular orbitals (HOMO and LUMO), which in turn depends on the exact nature of the conjugated system. Thus, by measuring the ultraviolet spectrum of an unknown, we can derive structural information about the nature of any conjugated pi electron system present in a sample.

One of the most important factors affecting the wavelength of ultraviolet absorption by a given molecule is the extent of conjugation. Molecular orbital calculations show that the energy difference between HOMO and LUMO decreases as the extent of conjugation increases; thus, 1,3-butadiene shows an absorption at $\lambda_{max} = 217$ nm, 1,3,5-hexatriene absorbs at $\lambda_{max} = 258$ nm, and 1,3,5,7-octatetraene has $\lambda_{max} = 290$ nm. [*Remember:* Longer wavelength means lower energy.]

Other kinds of conjugated systems besides dienes and polyenes also show ultraviolet absorptions. For example, conjugated enones and aromatic rings exhibit characteristic ultraviolet absorptions that aid in structure determination. More will be said about such compounds later when the functional groups are discussed in more detail, but the ultraviolet absorption maxima of some representative conjugated molecules are given in Table 14.3.

In addition to the $\pi \rightarrow \pi^*$ absorptions just discussed, other electronic transitions are observed in ultraviolet spectroscopy. Compounds with nonbonding electrons, such as the lone-pair electrons on oxygen, nitrogen, and halogen, also show weak ultraviolet absorption. In these cases, a nonbonding electron ($n$) is promoted to an antibonding orbital ($\pi^*$). In acetone, for example, a nonbonding lone-pair electron on oxygen is excited by ultraviolet irradiation into the carbonyl antibonding $\pi^*$ orbital. The resultant $n \rightarrow \pi^*$ transition shows an absorption peak at $\lambda_{max} = 272$ nm but is quite weak compared to the usual $\pi \rightarrow \pi^*$ absorption seen in conjugated systems.

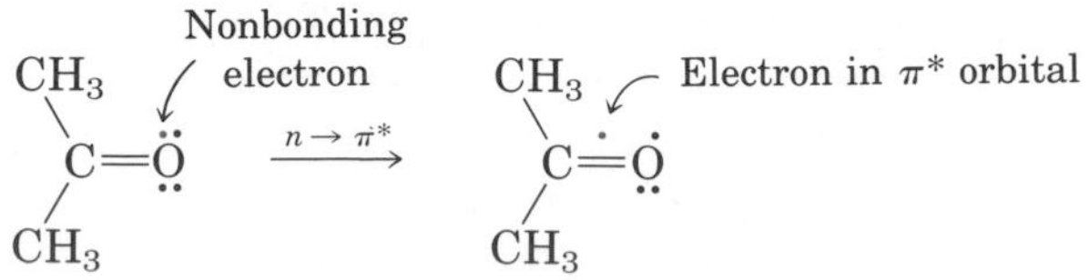

$$(CH_3)_2C{=}\ddot{\underset{\cdot\cdot}{O}} \xrightarrow{n \rightarrow \pi^*} (CH_3)_2C{=}\dot{\underset{\cdot\cdot}{O}}$$

Nonbonding electron

Electron in $\pi^*$ orbital

**Table 14.3** Ultraviolet absorption maxima of some conjugated molecules

| *Name* | *Structure* | $\lambda_{max}$ *(nm)* |
|---|---|---|
| Ethylene | $H_2C{=}CH_2$ | 171 |
| Cyclohexene | 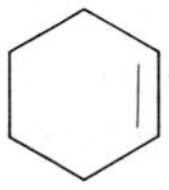 | 182 |
| 2-Methyl-1,3-butadiene | $H_2C{=}C(CH_3){-}CH{=}CH_2$ | 220 |
| 1,3-Cyclohexadiene |  | 256 |
| 1,3,5-Hexatriene | $H_2C{=}CH{-}CH{=}CH{-}CH{=}CH_2$ | 258 |
| 1,3,5,7-Octatetraene | $H_2C{=}CH{-}CH{=}CH{-}CH{=}CH{-}CH{=}CH_2$ | 290 |
| 2,4-Cholestadiene | 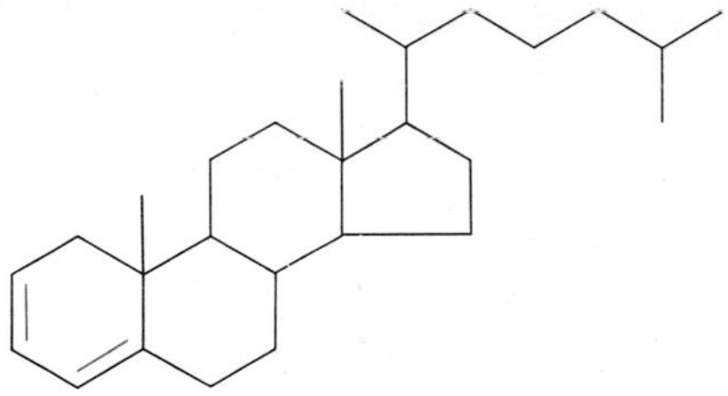 | 275 |
| 3-Buten-2-one | $H_2C{=}CH{-}C(CH_3){=}O$ | 219 |
| Benzene | 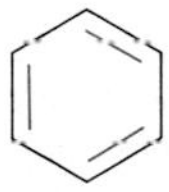 | 254 |
| Naphthalene | 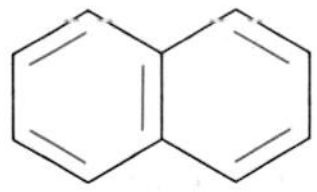 | 275 |

PROBLEM

**14.15** Which of the following compounds would you expect to show ultraviolet absorptions in the 200–400 nm range?

(a) 1,4-Cyclohexadiene (b) 1,3-Cyclohexadiene (c) $H_2C{=}CH{-}C{\equiv}N$

(d) 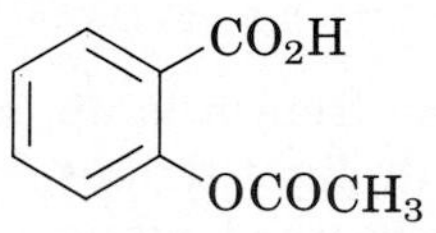 $CO_2H$, $OCOCH_3$

Aspirin

(e)

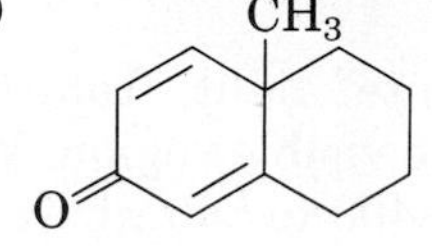


(f) 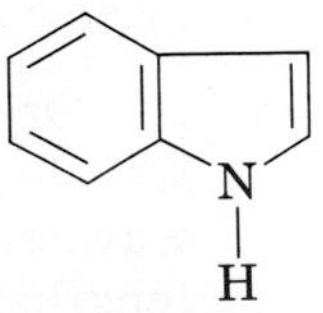 N H

Indole

## 14.13 Colored Organic Compounds and the Chemistry of Vision

Why are some organic compounds colored but others are not? Why is β-carotene (from carrots) orange but benzene is colorless? The answers have to do both with the structures of colored molecules and with the way we perceive light.

The visible region of the electromagnetic spectrum is adjacent to the ultraviolet region, extending from approximately 400 to 800 nm. Colored compounds such as β-carotene have such extended systems of conjugation that their "UV" absorptions actually extend out into the visible region. β-Carotene's absorption, for example, occurs at $\lambda_{max}$ = 455 nm (Figure 14.15).

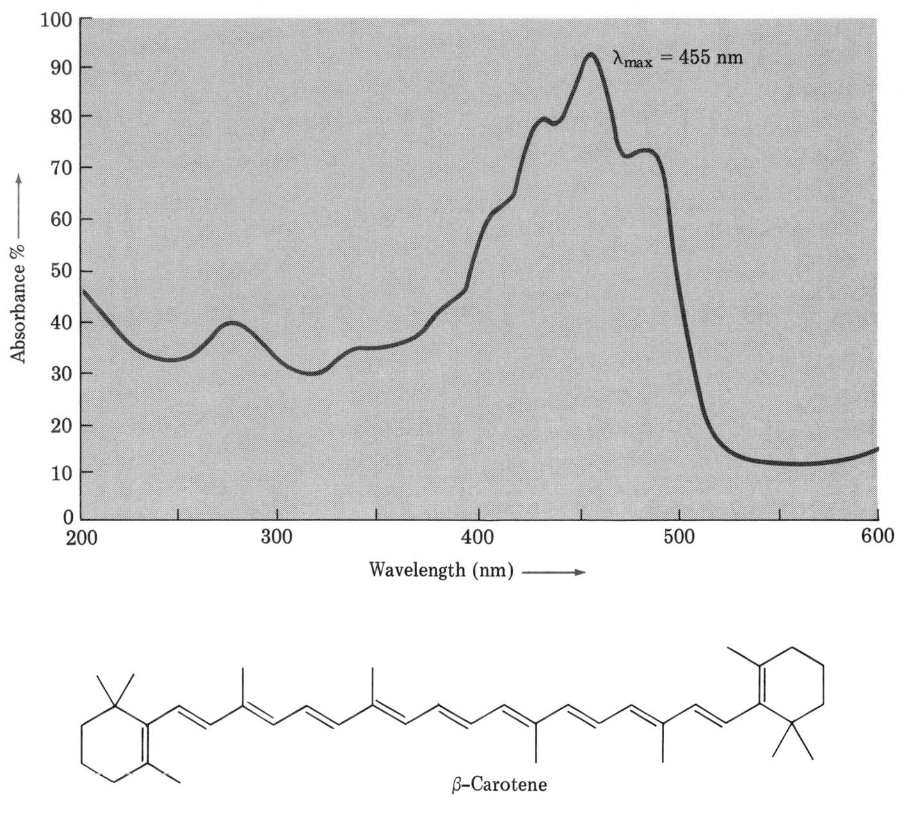


**Figure 14.15** Ultraviolet spectrum of β-carotene

Ordinary "white" light, from the sun or from a lamp, consists of all wavelengths in the visible region. When white light strikes β-carotene, the wavelengths from 400 to 500 nm (blue) are absorbed, while all other wavelengths reach our eyes. We therefore see the white light with the blue subtracted out, and we perceive a yellow-orange color for β-carotene. (The yellow-orange coloration accounts for the use of β-carotene as a food-coloring agent in margarine.)

What is true for β-carotene is true for all other colored organic compounds: All have an extended system of pi electron conjugation that gives rise to an absorption in the visible region of the electromagnetic spectrum.

## THE CHEMISTRY OF VISION

Conjugated molecules not only affect the color of a compound, they also make up the light-sensitive molecules on which the visual systems of all living things are based. 11-*cis*-Retinal, synthesized in the liver from dietary vitamin A, is the key substance.

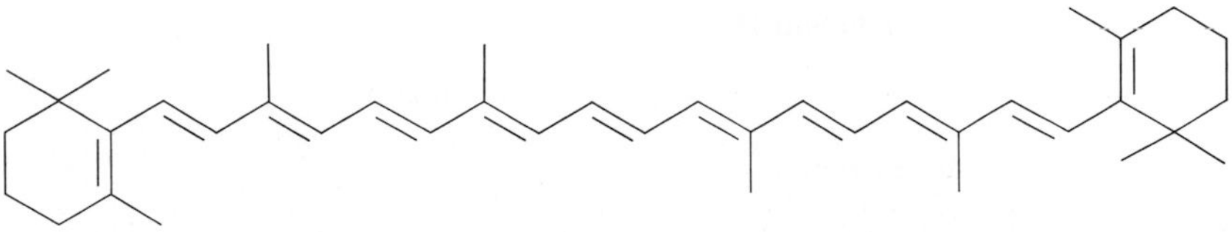

β-Carotene

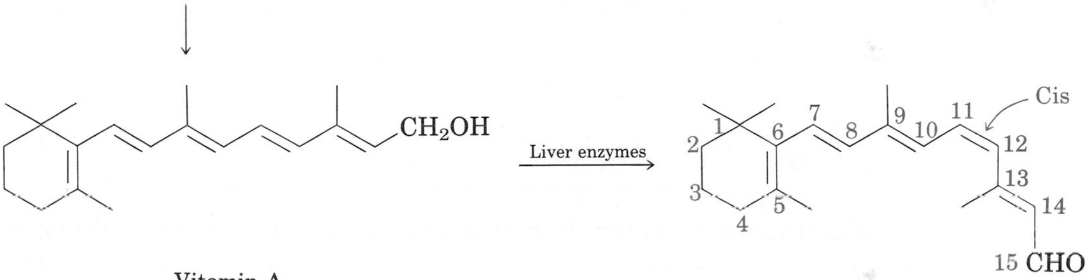

$CH_2OH$

Liver enzymes

1 2 3 4 5 6 7 8 9 10 11 12 13 14 15 CHO

Cis

Vitamin A

11-*cis*-Retinal

There are two types of light-sensitive receptor cells in the eye, *rod* cells and *cone* cells. Rod cells are primarily responsible for seeing in dim light, whereas cone cells are responsible for seeing in bright light and for the perception of bright colors. In the rod cells of the eye, 11-*cis*-retinal is converted into *rhodopsin*, a light-sensitive substance formed from the protein *opsin* and 11-*cis*-retinal. When light strikes the rod cell, isomerization of the C11—C12 double bond occurs and *trans*-rhodopsin, called metarhodopsin II, is produced. This cis–trans isomerization of rhodopsin is accompanied by a change in molecular geometry, which in turn causes a nerve impulse to be sent to the brain where it is perceived as vision.

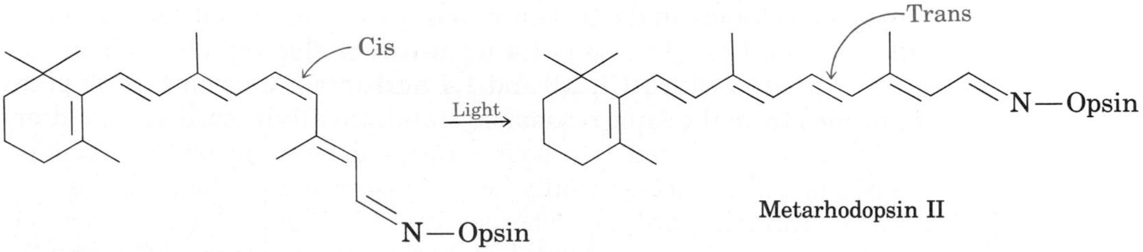

Cis

Light

Trans

N—Opsin

N—Opsin

Rhodopsin

Metarhodopsin II

Metarhodopsin II is then recycled back into rhodopsin by a multistep sequence involving cleavage into all-*trans*-retinal, conversion to vitamin A, cis–trans isomerization to 11-*cis*-vitamin A, and conversion back to 11-*cis*-retinal (Figure 14.16).

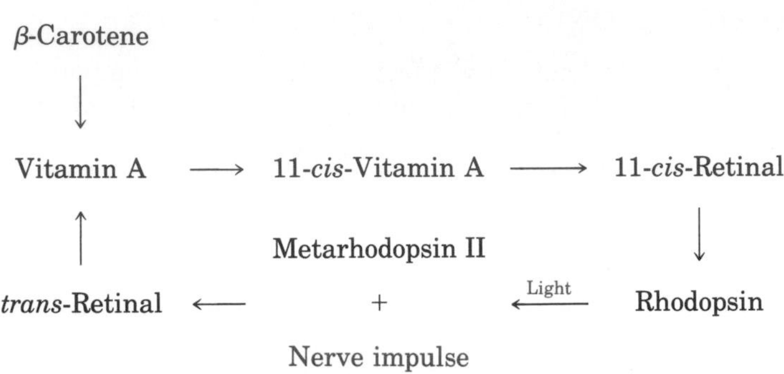


**Figure 14.16** The visual cycle: The actual series of events is more complicated than the diagram indicates, involving several intermediate steps in the light-induced conversion of rhodopsin into metarhodopsin II.

## 14.14 Summary and Key Words

A **conjugated** diene is one that contains alternating double and single bonds. Thus, 1,3-butadiene is conjugated, whereas 1,4-pentadiene is nonconjugated.

One important difference between conjugated and nonconjugated dienes is that conjugated dienes are somewhat more stable than we might expect. This unexpected stability can be explained by a molecular orbital description in which four *p* atomic orbitals overlap to form four **molecular orbitals**, denoted $\psi_1$, $\psi_2$, $\psi_3^*$, and $\psi_4^*$. Two of the molecular orbitals are **bonding** and two are **antibonding**. When four electrons are added, only the two bonding orbitals, $\psi_1$ and $\psi_2$, are filled. The two antibonding orbitals are unoccupied.

Calculations show that the sum of the energy levels of the two bonding 1,3-butadiene molecular orbitals is lower (that is, the system is more stable) than the sum of the energy levels of two isolated double bonds. The reason for this extra stability is a bonding (stabilizing) interaction between carbons 2 and 3, which introduces some partial double-bond character.

Conjugated dienes undergo two reactions not observed for nonconjugated dienes. The first of these is **1,4 addition** of electrophiles. When 1,3-butadiene is treated with HCl, **1,2** and **1,4 adducts** are formed. Both products are formed from the same resonance-stabilized allylic carbocation intermediate, and are produced in varying ratios depending on the reaction conditions. The 1,2 adduct is usually formed faster and predominates at low temperature (**kinetic control**), whereas the 1,4 adduct usually predominates when equilibrium is achieved at higher temperatures (**thermodynamic control**).

The second reaction unique to conjugated dienes is **Diels–Alder cycloaddition**. Conjugated dienes react with electron-poor alkenes (**dienophiles**) in a single step via a cyclic transition state to yield a cyclohexene product. This is an example of the general class of processes called **pericyclic reactions**, which have neither polar nor radical mechanisms. Diels–Alder reactions can occur only if the diene is able to adopt an ***s*-cis conformation**. For this reason, cyclic dienes such as cyclopentadiene are highly reactive.

**Ultraviolet (UV) spectroscopy** is a method of structure determination uniquely applicable to conjugated systems. When a conjugated molecule is irradiated with ultraviolet light, energy absorption occurs and a pi electron is promoted from the highest occupied molecular orbital (**HOMO**) to the lowest unoccupied molecular orbital (**LUMO**). For 1,3-butadiene, radiation of $\lambda_{max}$ = 217 nm is required. As a general rule, the greater the extent of conjugation, the less energy needed (longer wavelength radiation).

## ADDITIONAL PROBLEMS

**14.16** Provide IUPAC names for the following alkenes:

(a) $CH_3CH{=}C(CH_3)CH{=}CHCH_3$ (b) $H_2C{=}CHCH{=}CHCH{=}CHCH_3$

(c) $CH_3CH{=}C{=}CHCH{=}CHCH_3$ (d) $CH_3CH{=}C(CH_2CH_2CH_3)CH{=}CH_2$

**14.17** Circle any conjugated portions of these molecules:

(a) $CH_3CH{=}C{=}CHCH{=}CHCH_3$ (b) O, $OCH_3$

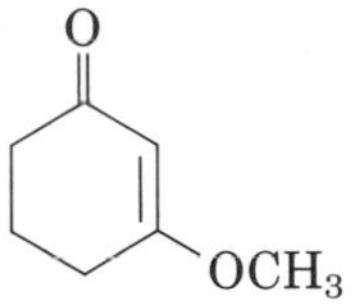

(c) $OCH_3$ (d) O

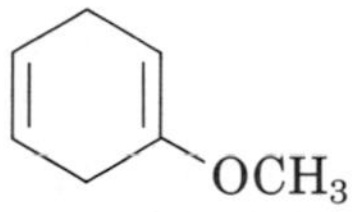

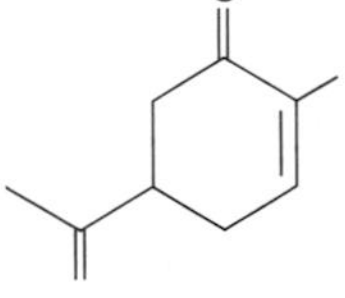

Carvone
(oil of spearmint)

(e) $CH_3O$ (f)

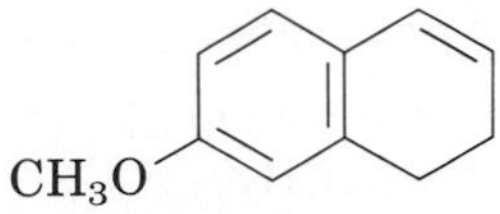

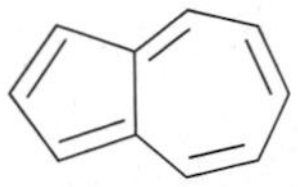

**14.18** What product(s) would you expect to obtain from reaction of 1,3-cyclohexadiene with each of the following?

(a) 1 mol $Br_2$ in $CCl_4$ (b) $O_3$ followed by Zn
(c) 1 mol HCl in ether (d) 1 mol DCl in ether
(e) 3-Buten-2-one ($H_2C{=}CHCOCH_3$) (f) Excess $OsO_4$, followed by $NaHSO_3$

**14.19** Draw and name the six possible diene isomers of formula $C_5H_8$. Which of the six are conjugated dienes?

**14.20** Treatment of 3,4-dibromohexane with strong base leads to loss of 2 equiv HBr and formation of a product with formula $C_6H_{10}$. Three possible products could be formed. Name each of the three and tell how you would use $^1H$ and $^{13}C$ NMR spectroscopy to help you identify the product. How would you use ultraviolet spectroscopy?

**14.21** Would you expect allene, $H_2C{=}C{=}CH_2$ to shown a UV absorption in the 200–400 nm range? Explain.

**14.22** Predict the products of these Diels–Alder reactions:

(a) 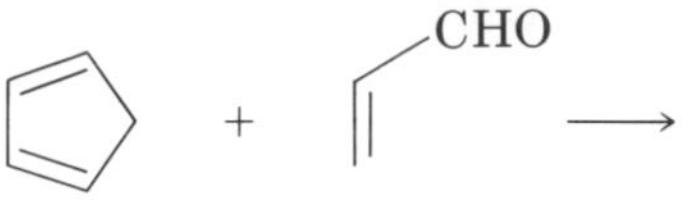 [cyclopentadiene] + $H_2C{=}CH{-}CHO$ ⟶ (b) 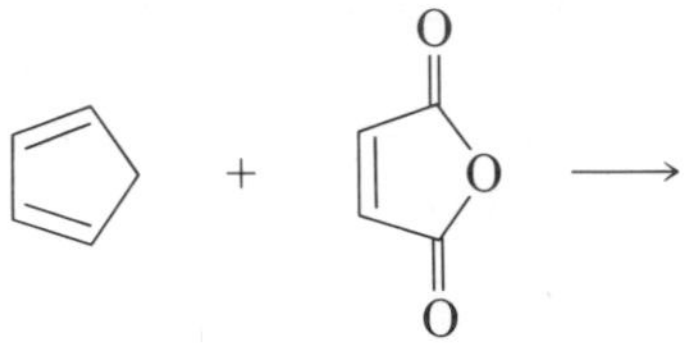 [cyclopentadiene] + [maleic anhydride] ⟶

(c) 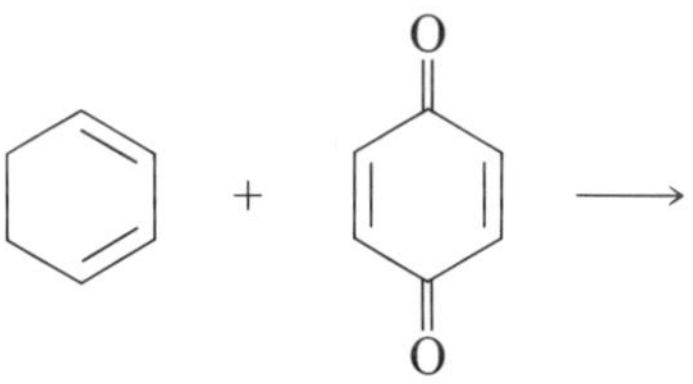 [1,3-cyclohexadiene] + [p-benzoquinone] ⟶ (d) 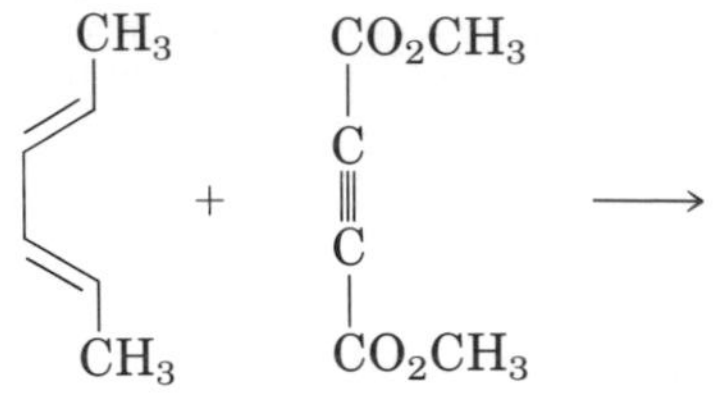 $CH_3CH{=}CH{-}CH{=}CHCH_3$ + $CH_3O_2C{-}C{\equiv}C{-}CO_2CH_3$ ⟶

**14.23** How do you account for the fact that *cis*-1,3-pentadiene is much less reactive than *trans*-1,3-pentadiene in the Diels–Alder reaction?

**14.24** Which of the following compounds would you expect to have $\pi \rightarrow \pi^*$ ultraviolet absorptions in the 200–400 nm range?

(a) 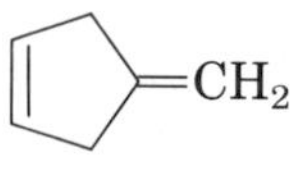 [cyclopentene ring]$=CH_2$ (b) 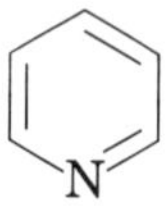 Pyridine (c) $(CH_3)_2C{=}C{=}O$ A ketene

**14.25** Would you expect a conjugated diyne such as 1,3-butadiyne to undergo Diels–Alder reaction with a dienophile? Explain.

**14.26** Propose a structure for a conjugated diene that gives the same product from both 1,2 and 1,4 addition of HBr.

**14.27** Draw the products resulting from addition of 1 mol HCl to 1-phenyl-1,3-butadiene. Which product or products would you expect to predominate, and why?

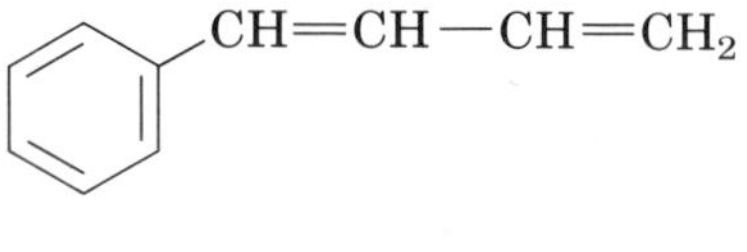

$C_6H_5{-}CH{=}CH{-}CH{=}CH_2$

1-Phenyl-1,3-butadiene

**14.28** Reaction of isoprene (2-methyl-1,3-butadiene) with ethyl propenoate gives a mixture of two Diels–Alder adducts. Show the structure of each and explain why a mixture is formed.

$$H_2C{=}C(CH_3){-}CH{=}CH_2 + H_2C{=}CHC(O)OCH_2CH_3 \longrightarrow ?$$

**14.29** Rank the following dienophiles in order of their expected reactivity in the Diels–Alder reaction. Explain.

$H_2C{=}CHCH_3$, $H_2C{=}CHCHO$, $(N{\equiv}C)_2C{=}C(C{\equiv}N)_2$, $(CH_3)_2C{=}C(CH_3)_2$

**14.30** How would you use Diels–Alder reactions to prepare these products? Show the starting dienes and dienophiles in each case.

(a)

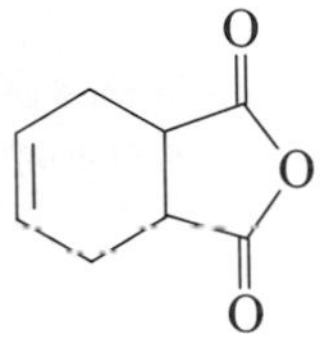

(b) H, CN

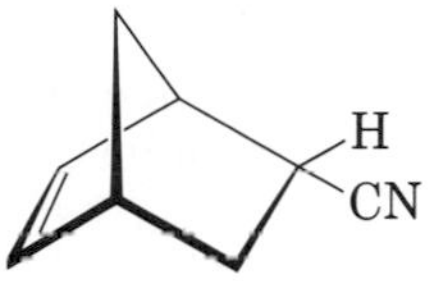

(c)

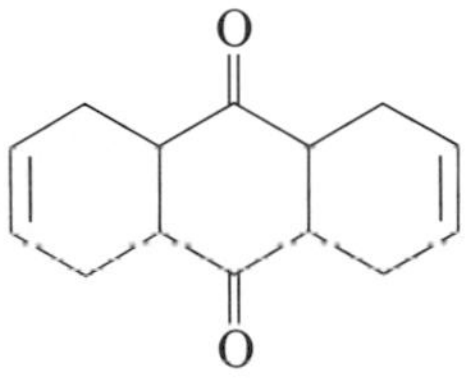

(d) $CO_2CH_3$

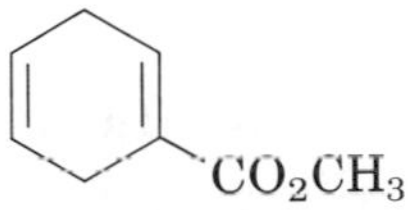

**14.31** We've seen that the Diels–Alder cycloaddition reaction is a pericyclic process that occurs in a concerted manner through a cyclic transition state. Depending on the exact energy levels of products and reactants, the Diels–Alder reaction can sometimes be made reversible. How can you account for the following reaction?

$$\xrightarrow{\Delta} \quad + \; H_2C{=}CH_2$$

**14.32** Propose a mechanism to explain the following reaction. [*Hint:* Consider the reversibility of the Diels–Alder reaction, Problem 14.31.]

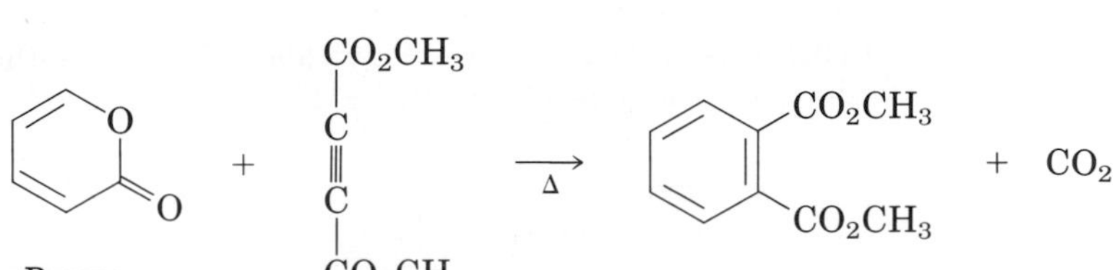

$$\alpha\text{-Pyrone} + CH_3O_2C{-}C{\equiv}C{-}CO_2CH_3 \xrightarrow{\Delta} C_6H_4(CO_2CH_3)_2 + CO_2$$

**14.33** The following ultraviolet absorption maxima have been measured:

| | $\lambda_{max}$ *(nm)* |
|---|---|
| 1,3-Butadiene | 217 |
| 2-Methyl-1,3-butadiene | 220 |
| 1,3-Pentadiene | 223 |
| 2,3-Dimethyl-1,3-butadiene | 226 |
| 2,4-Hexadiene | 227 |
| 2,4-Dimethyl-1,3-pentadiene | 232 |
| 2,5-Dimethyl-2,4-hexadiene | 240 |

What conclusion can you draw from these data concerning the effect of alkyl substitution on ultraviolet absorption maxima? Approximately what effect does each added alkyl group have?

**14.34** 1,3,5-Hexatriene has $\lambda_{max}$ = 258 nm. In light of your answer to Problem 14.33, approximately where would you expect 2,3-dimethyl-1,3,5-heptatriene to absorb? Explain.

**14.35** $\beta$-Ocimene is a pleasant-smelling hydrocarbon found in the leaves of certain herbs. It has the molecular formula $C_{10}H_{16}$ and exhibits an ultraviolet absorption maximum at 232 nm. On catalytic hydrogenation over palladium, 2,6-dimethyloctane is obtained. Ozonolysis of $\beta$-ocimene, followed by treatment with zinc and acetic acid, produces four fragments: acetone, formaldehyde, pyruvaldehyde, and malonaldehyde.

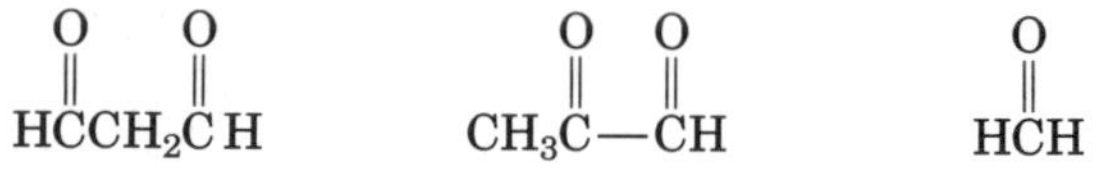

$$\overset{O}{\overset{\|}{H C}}CH_2\overset{O}{\overset{\|}{C}}H \qquad CH_3\overset{O}{\overset{\|}{C}}-\overset{O}{\overset{\|}{C}}H \qquad H\overset{O}{\overset{\|}{C}}H$$

Malonaldehyde Pyruvaldehyde Formaldehyde

(a) How many double bonds does $\beta$-ocimene have?
(b) Is $\beta$-ocimene conjugated or nonconjugated?
(c) Propose a structure for $\beta$-ocimene consistent with the observed data.
(d) Formulate all the reactions, showing starting material and products.

**14.36** Myrcene, $C_{10}H_{16}$, is found in oil of bay leaves and is isomeric with $\beta$-ocimene (Problem 14.35). It shows an ultraviolet absorption at 226 nm and can be catalytically hydrogenated to yield 2,6-dimethyloctane. On ozonolysis followed by zinc/acetic acid treatment, myrcene yields formaldehyde, acetone, and 2-oxopentanedial.

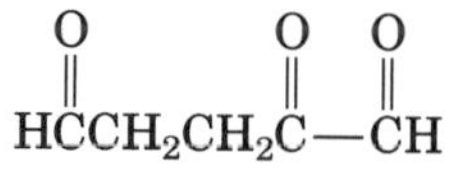

$$H\overset{O}{\overset{\|}{C}}CH_2CH_2\overset{O}{\overset{\|}{C}}-\overset{O}{\overset{\|}{C}}H$$

2-Oxopentanedial

Propose a structure for myrcene, and formulate all the reactions, showing starting material and products.

**14.37** Addition of HCl to 1-methoxycyclohexene yields 1-chloro-1-methoxycyclohexane as the sole product. Why is none of the other regioisomer formed?

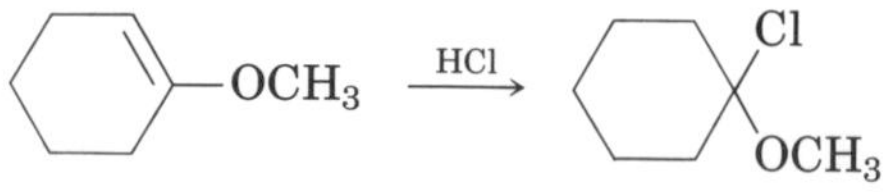

[1-methoxycyclohexene]$-OCH_3$ $\xrightarrow{HCl}$ [cyclohexane with Cl and $OCH_3$ on C1]

**14.38** Benzene has an ultraviolet absorption at $\lambda_{max} = 204$ nm and *para*-toluidine has $\lambda_{max} = 235$ nm. How do you account for this difference?

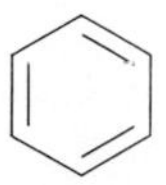

Benzene
($\lambda_{max} = 204$ nm)

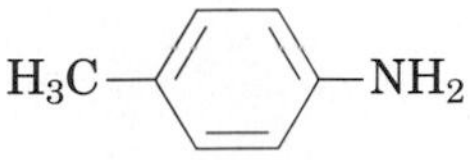

$H_3C$—⟨benzene ring⟩—$NH_2$

*para*-Toluidine
($\lambda_{max} = 235$ nm)

**14.39** When the ultraviolet spectrum of *para*-toluidine (Problem 14.38) is measured in the presence of a small amount of HCl, the $\lambda_{max}$ decreases to 207 nm, nearly the same value as for benzene. How do you account for the effect of acid?

**14.40** Phenol is a weak acid with p$K_a = 10.0$. In ethanol solution, phenol has an ultraviolet absorption at $\lambda_{max} = 210$ nm. When dilute NaOH is added to this solution, the absorption increases to $\lambda_{max} = 235$ nm. How do you account for this shift?

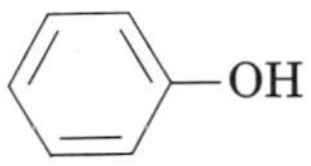

⟨benzene ring⟩—OH

Phenol

**14.41** Hydrocarbon A, $C_{10}H_{14}$, has an ultraviolet absorption at $\lambda_{max} = 236$ nm and gives hydrocarbon B, $C_{10}H_{18}$, on catalytic hydrogenation. Ozonolysis of A followed by zinc/acetic acid treatment yields the following diketo dialdehyde:

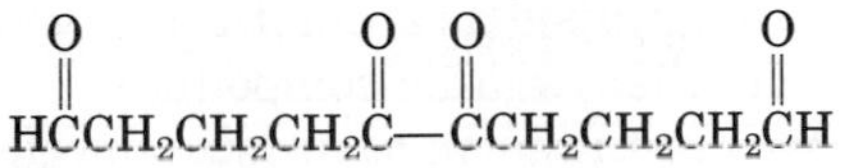

$$HC(=O)CH_2CH_2CH_2C(=O)—C(=O)CH_2CH_2CH_2CH(=O)$$

(a) Propose two possible structures for hydrocarbon A.
(b) Hydrocarbon A reacts with maleic anhydride to yield a Diels–Alder adduct. Which of your structures for A is correct?
(c) Formulate all reactions showing starting material and products.

**14.42** Adiponitrile, a starting material used in the manufacture of nylon, can be prepared in three steps from 1,3-butadiene. How would you carry out this synthesis?

$$H_2C{=}CH—CH{=}CH_2 \xrightarrow{\text{3 steps}} N{\equiv}CCH_2CH_2CH_2CH_2C{\equiv}N$$

Adiponitrile

CHAPTER 15

# Benzene and Aromaticity

In the early days of organic chemistry, the word *aromatic* was used to describe fragrant substances such as benzaldehyde (from cherries, peaches, and almonds) and toluene (from Tolu balsam). It was soon realized, however, that substances grouped as aromatic behaved in a chemically different manner from most other organic compounds.

Today, we use the term **aromatic** to refer to benzene and its structural relatives. We'll see in this and the next chapter that aromatic substances show chemical behavior quite different from that of the aliphatic substances we've studied to this point. Thus, chemists of the early nineteenth century were correct when they realized that a chemical difference exists between aromatic compounds and others, but the association of aromaticity with fragrance has long been lost.

Many compounds isolated from natural sources are aromatic in part. In addition to benzene, benzaldehyde, and toluene, complex compounds such as the female steroidal hormone, estrone, and the well-known analgesic, morphine, have aromatic rings. Most synthetic drugs used medicinally are also aromatic in part. The local anesthetic procaine and the tranquilizer diazepam (Valium) are two of many examples.

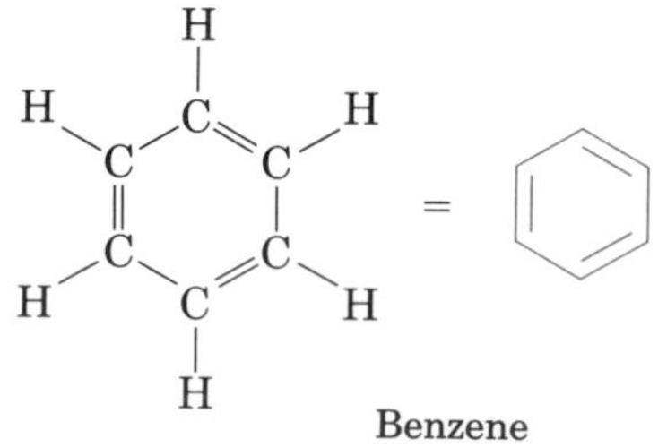

CHO

$CH_3$

Benzene

Benzaldehyde

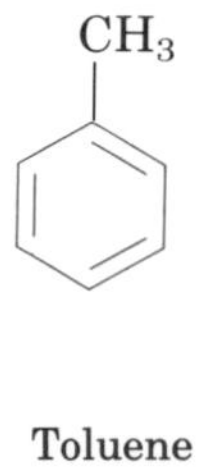

Toluene

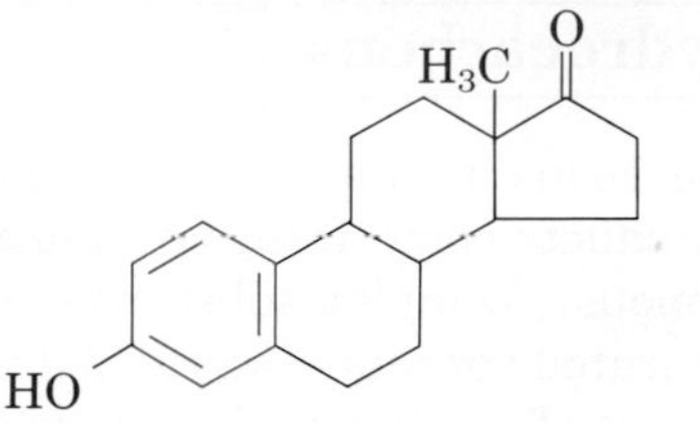

H$_3$C O HO

N—CH$_3$

HO HO

Estrone 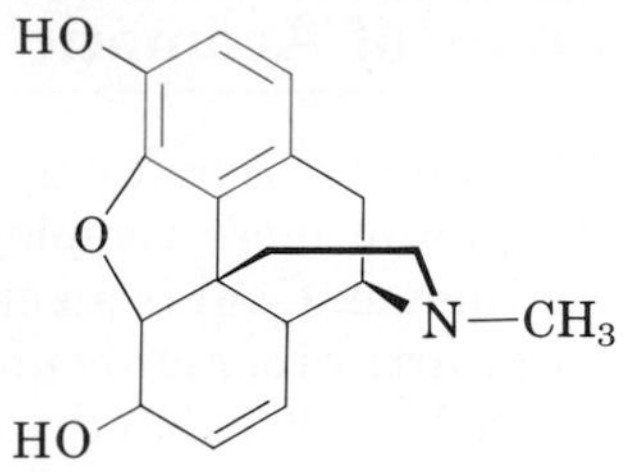 Morphine

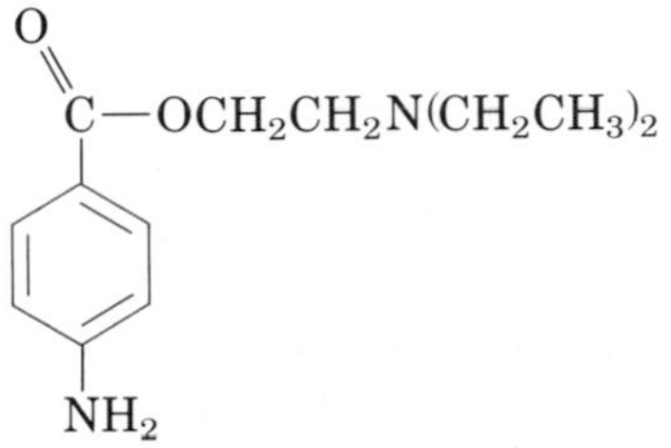

O

C—OCH$_2$CH$_2$N(CH$_2$CH$_3$)$_2$

CH$_3$

N O

Cl N

NH$_2$

Procaine
(a local anesthetic)

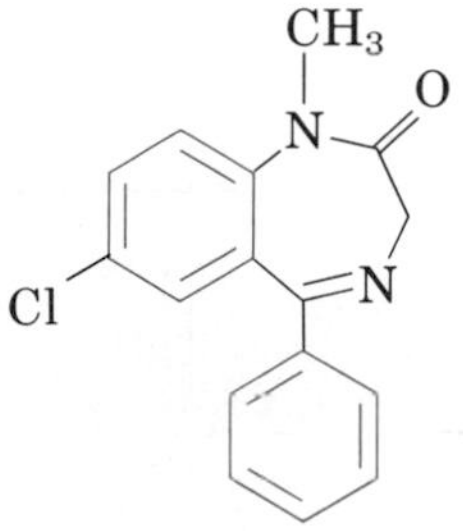

Diazepam
(Valium, a sedative hypnotic, often called a tranquilizer)

Benzene itself has been found to cause bone-marrow depression and consequent leukopenia (depressed white-blood-cell count) on prolonged exposure. Use of benzene as a laboratory solvent should therefore be avoided.

**PROBLEM**

**15.1** Circle the aromatic portions of these molecules:

(a) CH(OH)CH$_2$NHCH$_3$

OH

OH

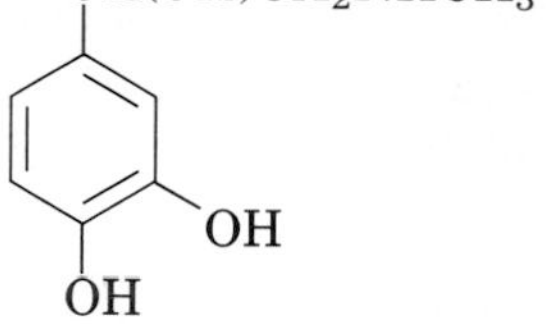

Adrenaline (Epinephrine)

(b) CH$_3$ CH$_3$ CH$_3$

H$_3$C O

(CH$_2$CH$_2$CH$_2$CH)$_3$CH$_3$

HO

CH$_3$

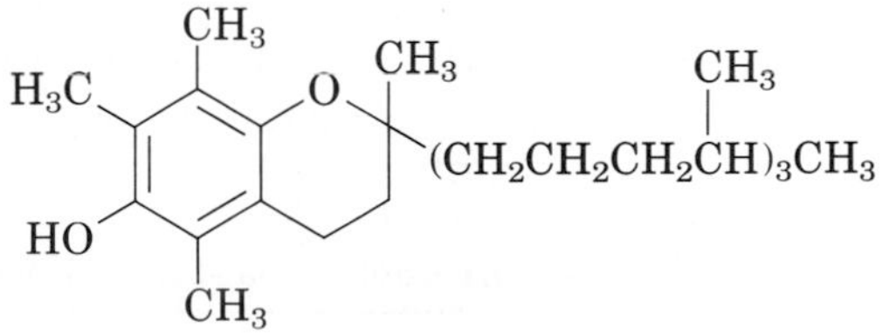

Vitamin E

(c) O

OCH$_2$C—NH S CH$_3$

CH$_3$

N

O CO$_2$H

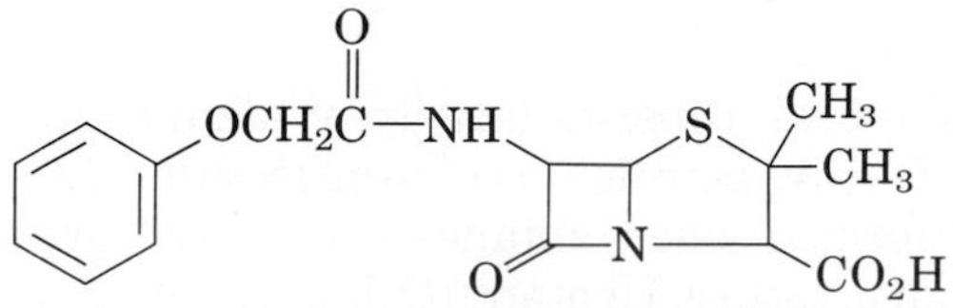

Penicillin V

## 15.1 Sources of Aromatic Hydrocarbons

The simple aromatic hydrocarbons used as starting materials for the preparation of more complex products come from two main sources, coal and petroleum. **Coal** is an enormously complex substance made up primarily of large arrays of highly unsaturated benzene-like rings linked together. When heated to 1000°C in the absence of air, thermal breakdown of coal molecules occurs and a mixture of volatile products, called **coal tar**, distills off. Further fractional distillation of coal tar yields benzene, toluene, xylene (dimethylbenzene), naphthalene, and a host of other aromatic compounds (Figure 15.1).

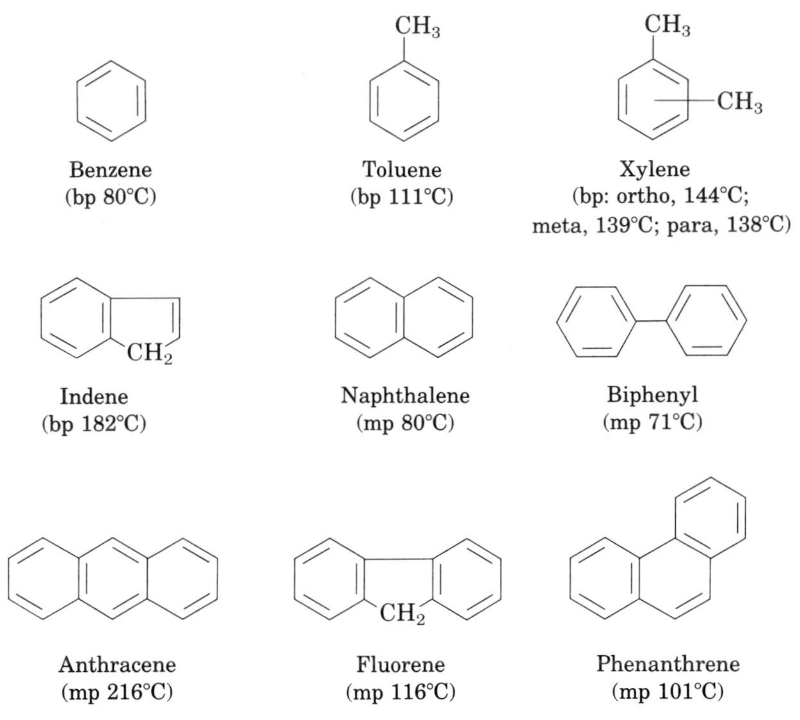


**Figure 15.1** Some coal-tar hydrocarbons

**Petroleum**, unlike coal, consists largely of alkanes and contains few aromatic compounds. During petroleum refining (Section 3.6), however, aromatic molecules are formed when alkanes are passed over a catalyst at about 500°C under high pressure. Heptane ($C_7H_{16}$), for example, is converted into toluene ($C_7H_8$) by dehydrogenation and cyclization.

## 15.2 Nomenclature of Aromatic Compounds

Aromatic substances, more than any other class of organic compounds, have acquired a large number of common, nonsystematic names. Although the use of such names is discouraged, IUPAC rules allow for some of the more widely used ones to be retained (Table 15.1). Thus, methylbenzene is known familiarly as *toluene*, hydroxybenzene as *phenol*, aminobenzene as *aniline*, and so on.

Table 15.1 Common names of some aromatic compounds

| *Formula* | *Name* | *Formula* | *Name* |
|---|---|---|---|
| $CH_3$ (on benzene ring) | Toluene (bp 110°C) | CHO (on benzene ring) | Benzaldehyde (bp 178°C) |
| OH (on benzene ring) | Phenol (mp 43°C) | COOH (on benzene ring) | Benzoic acid (mp 122°C) |
| $NH_2$ (on benzene ring) | Aniline (bp 184°C) | $CH_3$, $CH_3$ (on benzene ring, ortho) | *ortho*-Xylene (bp 144°C) |
| $H_3C$–CH–$CH_3$ (on benzene ring) | Cumene (bp 152°C) | $CH_3$, $CH_3$ (on benzene ring, meta) | *meta*-Xylene (bp 139°C) |
| CH=$CH_2$ (on benzene ring) | Styrene (bp 145°C) | $CH_3$, $CH_3$ (on benzene ring, para) | *para*-Xylene (bp 138°C) |

Monosubstituted benzene derivatives are systematically named in the same manner as other hydrocarbons, with *-benzene* used as the parent name. Thus, as shown at the top of the next page, $C_6H_5Br$ is bromobenzene, $C_6H_5NO_2$ is nitrobenzene, and $C_6H_5CH_2CH_2CH_3$ is propylbenzene.

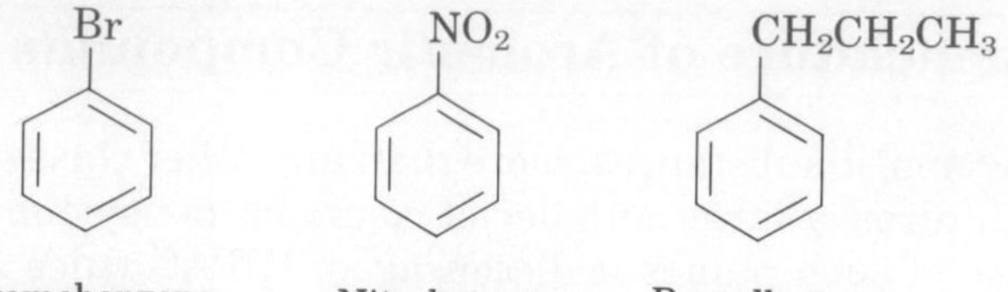

Br NO$_2$ $CH_2CH_2CH_3$

Bromobenzene Nitrobenzene Propylbenzene

Alkyl-substituted benzenes, sometimes referred to as **arenes**, are named in two different ways depending on the size of the alkyl group. If the alkyl substituent is small (six or fewer carbons), the arene is named as an alkyl-substituted benzene. If the alkyl substituent is larger than the ring (more than six carbons), the compound is named as a phenyl-substituted alkane. The name **phenyl**, pronounced **fen**-nil and often abbreviated as -Ph or $\phi$ (Greek phi), is used for the $-C_6H_5$ unit when the benzene ring is considered to be a substituent group. The word is derived from the Greek *pheno* ("I bear light"), commemorating the fact that benzene was discovered by Michael Faraday[1] in 1825 from the oily residue left by the illuminating gas used in London street lamps.

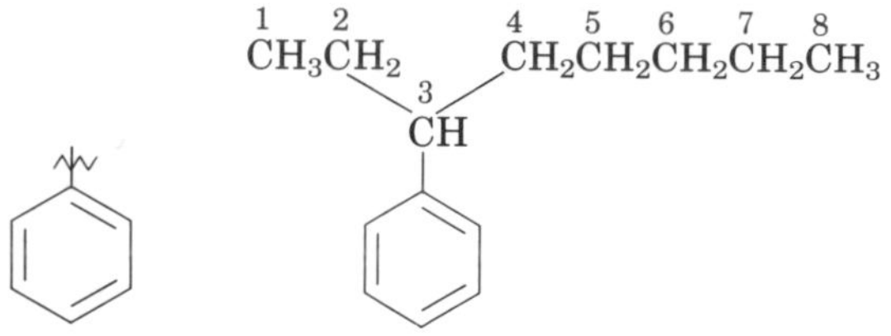

1 2 4 5 6 7 8
$CH_3CH_2$ $CH_2CH_2CH_2CH_2CH_3$
3
CH

A phenyl group 3-Phenyloctane

Disubstituted benzenes are named using one of the prefixes *ortho-*, *meta-*, or *para-*. An *ortho-* or *o*-disubstituted benzene has the two substituents in a 1,2 relationship on the ring; a *meta-* or *m*-disubstituted benzene has its two substituents in a 1,3 relationship; and a *para-* or *p*-disubstituted benzene has its substituents in a 1,4 relationship.

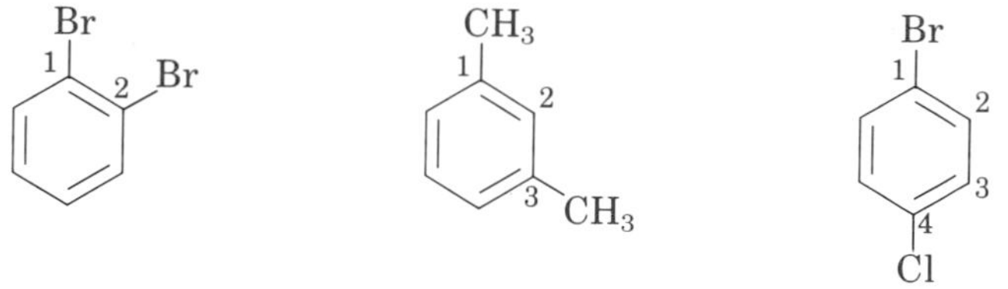

Br 1 2 Br $CH_3$ 1 2 3 $CH_3$ Br 1 2 3 4 Cl

*ortho*-Dibromobenzene, 1,2 disubstituted

*meta*-Dimethylbenzene (*m*-Xylene), 1,3 disubstituted

*para*-Bromochlorobenzene, 1,4 disubstituted

The ortho, meta, para system of nomenclature is also valuable when discussing reactions. For example, we might describe the reaction of bromine with toluene by saying "Reaction occurs in the para position"—in other words, at the position para to the methyl group already present on the ring.

[1]Michael Faraday (1791–1867); b. Newington Butts, Surrey, England; assistant to Sir Humphry Davy (1813); director, laboratory of the Royal Institution (1825); Fullerian Professor of Chemistry, Royal Institution (1833).

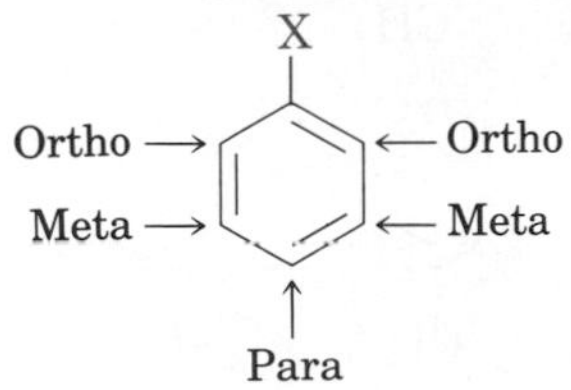

Ortho → ← Ortho
Meta → ← Meta
Para

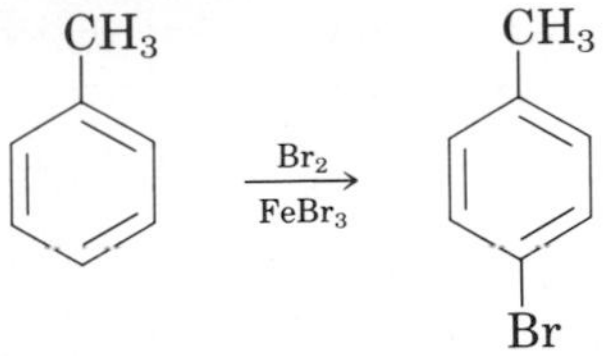

X

$CH_3$ $\xrightarrow[FeBr_3]{Br_2}$ $CH_3$, Br

Toluene p-Bromotoluene

Benzenes with more than two substituents must be named by numbering the position of each substituent on the ring. The numbering should be carried out in such a way that the lowest possible numbers are used. The substituents are listed alphabetically when writing the name.

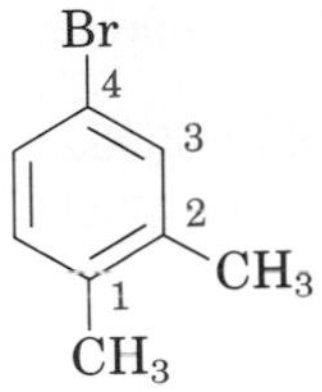

4-Bromo-1,2-dimethylbenzene 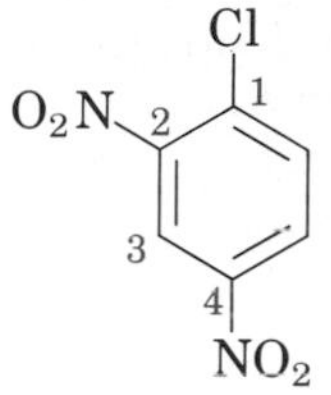 1-Chloro-2,4-dinitrobenzene 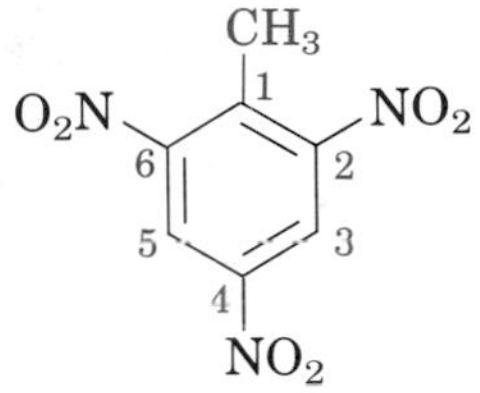 2,4,6-Trinitrotoluene (TNT)

Note in the third example shown that *-toluene* is used as the base name rather than *-benzene*. Any of the monosubstituted aromatic compounds shown in Table 15.1 can serve as a base name. In such cases the principal substituent (—$CH_3$ in toluene, for example) is assumed to be on C1. The following two examples further illustrate this rule (the base name is italicized in each case).

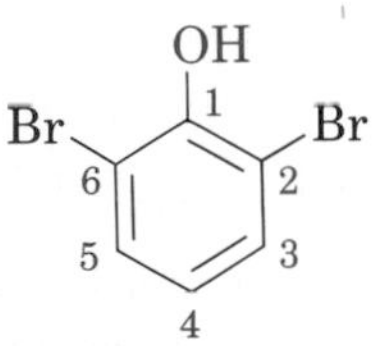

2,6-Dibromo*phenol* 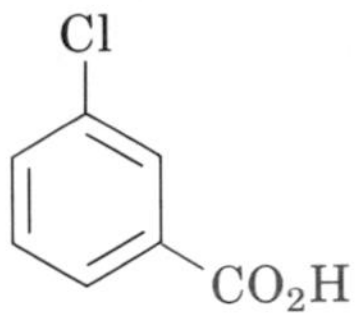 *m*-Chloro*benzoic acid*

PROBLEM

**15.2** Provide correct IUPAC names for these compounds:

(a) Cl, Br

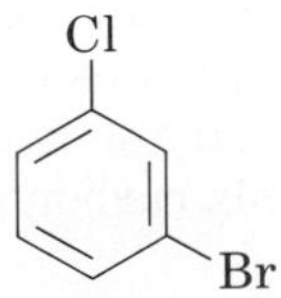

(b) $CH_2CH_2CH(CH_3)_2$

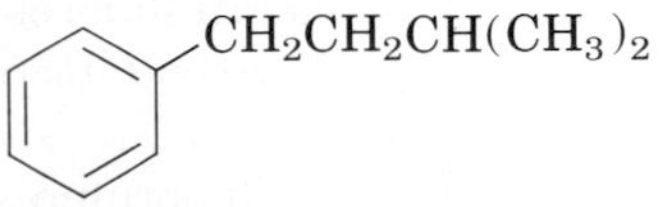

(c) 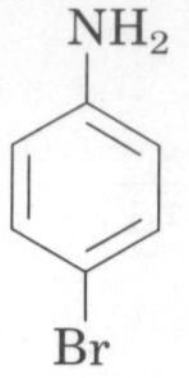 $NH_2$ … Br

(d) 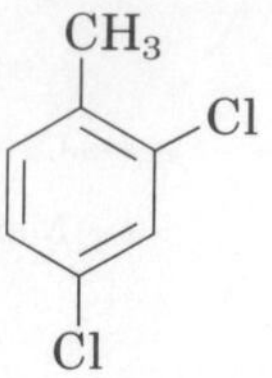 $CH_3$, Cl, Cl

(e) 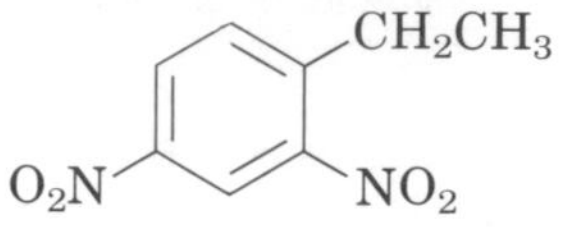 $CH_2CH_3$, $O_2N$, $NO_2$

(f) 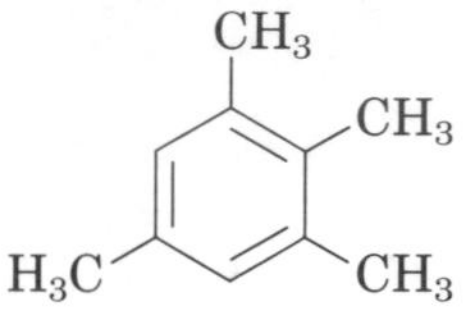 $CH_3$, $CH_3$, $H_3C$, $CH_3$

PROBLEM

**15.3** The following names are incorrect. Draw the structure represented by each, and provide correct IUPAC names.
(a) 2-Bromo-3-chlorobenzene
(b) 4,6-Dinitrotoluene
(c) 4-Bromo-1-methylbenzene
(d) 2-Chloro-*p*-xylene

## 15.3 Structure of Benzene: The Kekulé Proposal

By the mid-1800s, benzene was known to have the molecular formula $C_6H_6$, and its chemistry was being actively explored. The results, though, were puzzling. Although benzene is clearly "unsaturated"—a formula of $C_6H_6$ requires a combination of four multiple bonds/rings—it nevertheless fails to undergo reactions characteristic of alkenes. For example, it reacts with bromine in the presence of iron to give the *substitution* product $C_6H_5Br$, rather than the possible *addition* product $C_6H_6Br_2$. Furthermore, only one monobromo substitution product was known; no isomers of $C_6H_5Br$ had been prepared.

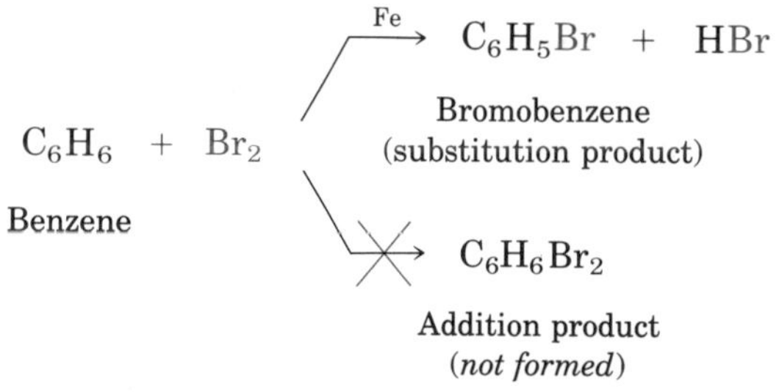

$$C_6H_6 + Br_2 \xrightarrow{Fe} C_6H_5Br + HBr$$

Benzene; Bromobenzene (substitution product)

$$C_6H_6 + Br_2 \not\rightarrow C_6H_6Br_2$$

Addition product (*not formed*)

On further reaction with bromine, disubstitution products are obtained, and three isomeric $C_6H_4Br_2$ compounds had been prepared. On the basis of these and similar results, Kekulé proposed in 1865 that benzene consists of a *ring* of carbon atoms and may be formulated as cyclohexatriene. Kekulé reasoned that this structure would readily account for the isolation of only a single monobromo substitution product, since all six carbon atoms and all six hydrogens in cyclohexatriene are equivalent.

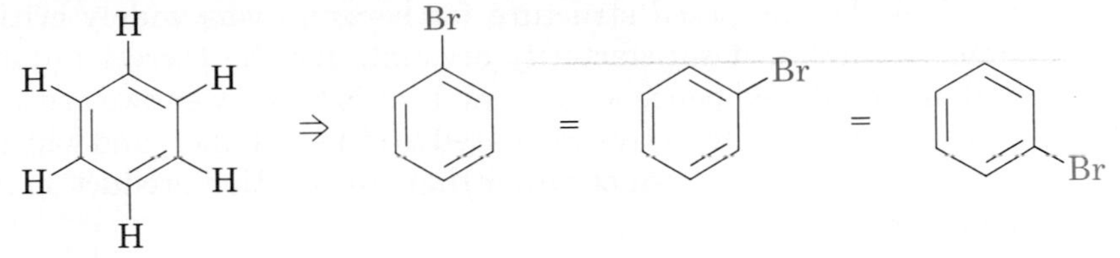

All six hydrogens of cyclohexatriene are equivalent

Only one possible monobromo product

The experimental observation that only three isomeric dibromo substitution products were known was more difficult to explain, since four structures can be written:

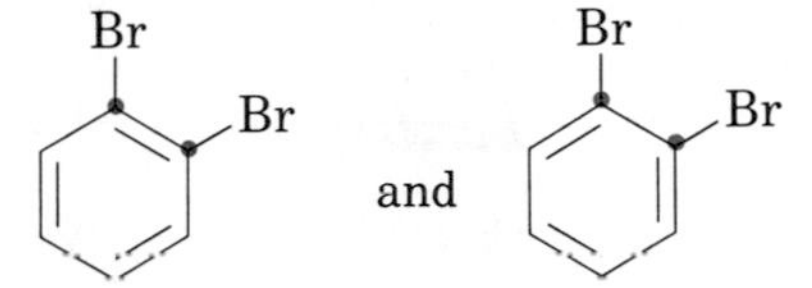

and

Two possible "1,2-dibromocyclohexatrienes"?

"1,3-Dibromocyclohexatriene" "1,4-Dibromocyclohexatriene"

Although there is only one possible 1,3 derivative and one possible 1,4 derivative, there appear to be two possible 1,2-dibromo substitution products, depending on the positions of the double bonds in the ring. Kekulé accounted for the formation of only three isomers by proposing that the double bonds in benzene rapidly "oscillate" between two positions. Thus, the two "1,2-dibromocyclohexatrienes" cannot be separated, according to Kekulé, because they interconvert too rapidly (Figure 15.2).

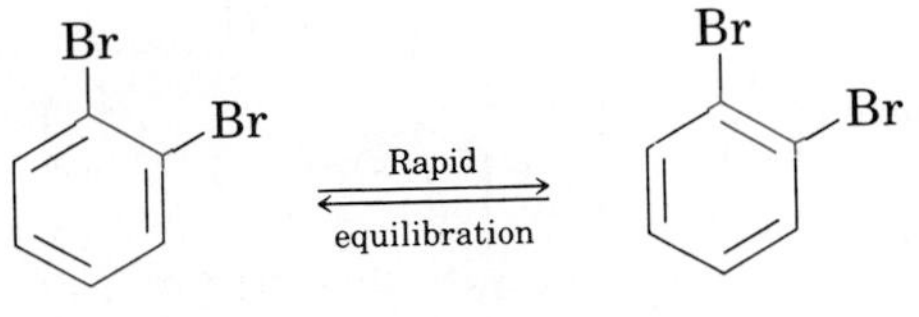

Rapid equilibration

**Figure 15.2** The Kekulé proposal: Benzene's double bonds rapidly oscillate back and forth between neighboring carbons. Distinct forms cannot be isolated.

Kekulé's proposed structure for benzene was widely criticized at the time. Although it satisfactorily accounts for the correct number of mono- and disubstituted benzene isomers, it fails to answer two critical questions: Why is benzene unreactive compared to other alkenes, and why does benzene give a substitution product rather than an addition product on reaction with bromine?

**PROBLEM**

**15.4** How many tribromo benzene derivatives are possible according to Kekulé's theory? Draw and name them.

**PROBLEM**

**15.5** The following structures with formula $C_6H_6$ were at one time suggested for benzene. If we assume that bromine can be substituted for hydrogen in these structures, how many monobromo derivatives are possible for each? How many dibromo derivatives?

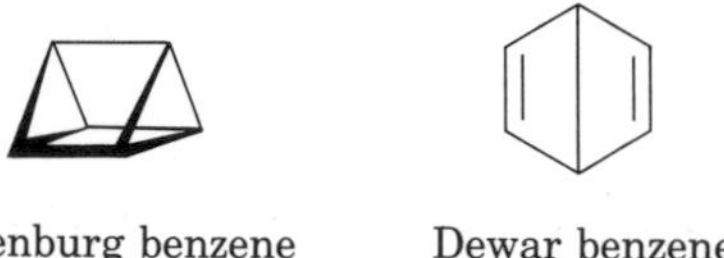


## 15.4 Stability of Benzene

The unusual stability of benzene was a great puzzle to early chemists. Although its formula, $C_6H_6$, indicates the presence of unsaturation, and although the Kekulé structure proposes three carbon–carbon double bonds, benzene shows none of the behavior characteristic of alkenes. For example, alkenes react readily with potassium permanganate to give cleavage products; they react rapidly with aqueous acid to give alcohols; and they react with gaseous HCl to give saturated alkyl chlorides. Benzene does none of these things (Figure 15.3). *Benzene does not undergo electrophilic addition reactions.*

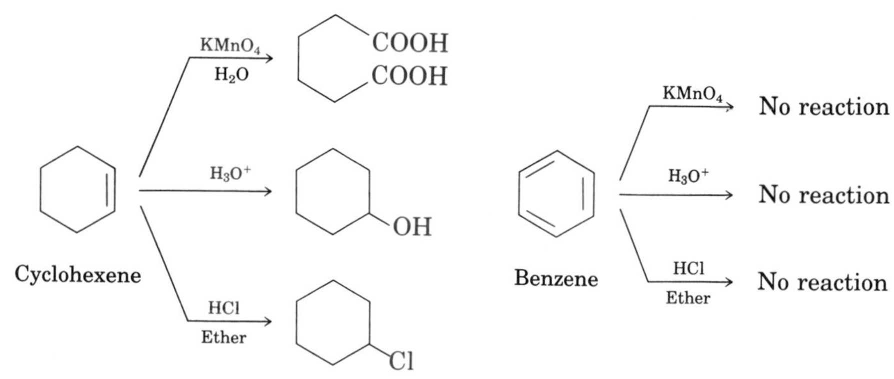


**Figure 15.3** A comparison of the reactivity of cyclohexene and benzene: Benzene does not undergo alkene electrophilic addition reactions.

We can get a quantitative idea of benzene's unusual stability by looking at data on heats of hydrogenation (Table 15.2). Cyclohexene, an isolated alkene, has $\Delta H^\circ_{\text{hydrog}} = 28.6$ kcal/mol (120 kJ/mol), and 1,3-cyclohexadiene, a conjugated diene, has $\Delta H^\circ_{\text{hydrog}} = 55.4$ kcal/mol (232 kJ/mol). As expected, the value for 1,3-cyclohexadiene is a bit less than twice the cyclohexene value since conjugated dienes are unusually stable (Section 14.2).

**Table 15.2** Heats of hydrogenation of cyclic alkenes

| | | $\Delta H^\circ_{\text{hydrog}}$ | |
|---|---|---|---|
| *Reactant* | *Product* | *(kcal/mol)* | *(kJ/mol)* |
| Cyclohexene | Cyclohexane | 28.6 | 120 |
| 1,3-Cyclohexadiene | Cyclohexane | 55.4 | 232 |
| Benzene | Cyclohexane | 49.8 | 208 |

Carrying the analogy one step further, we might predict that $\Delta H^\circ_{\text{hydrog}}$ for "cyclohexatriene" (benzene) should be about 86 kcal/mol, or three times the cyclohexene value. *The actual value is 49.8 kcal/mol (208 kJ/mol), some 36 kcal/mol less than expected.* Since 36 kcal/mol less heat than expected is released during hydrogenation of benzene, benzene must have 36 kcal/mol less energy than expected. In other words, benzene has 36 kcal/mol "extra" stability (Figure 15.4).

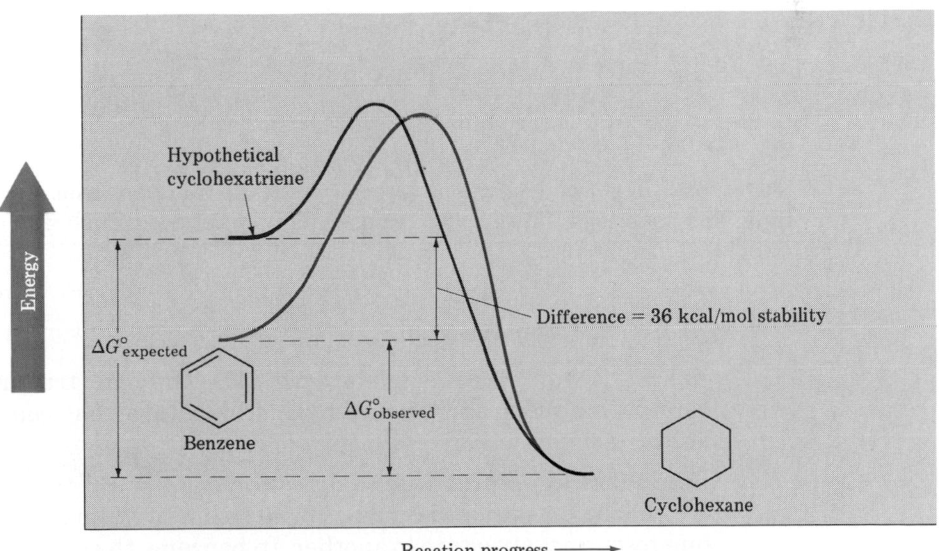


**Figure 15.4** Reaction energy diagram for the hydrogenation of benzene in comparison with a hypothetical cyclohexatriene

Further evidence for the unusual nature of benzene comes from spectroscopic studies showing that all carbon–carbon bonds in benzene have the same length, intermediate between normal single and double bonds. Although normal C—C single bonds are 1.54 Å long and normal C—C double bonds are 1.34 Å long, the carbon–carbon bonds in benzene are 1.39 Å long.

All C-C-C bond angles = 120°
All C—C bond lengths = 1.39 Å

Benzene

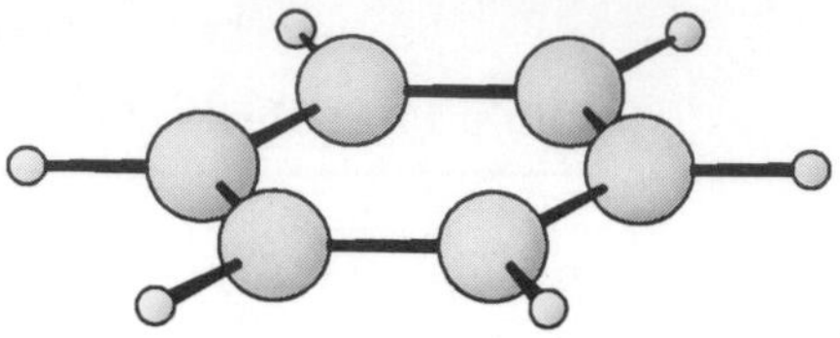

## 15.5 Representations of Benzene

How can we account for benzene's properties, and how can its structure best be represented? Resonance theory answers this question by saying that a *single* Kekulé structure is not satisfactory, and that benzene can best be described as a resonance hybrid of two equivalent Kekulé structures (Figure 15.5).

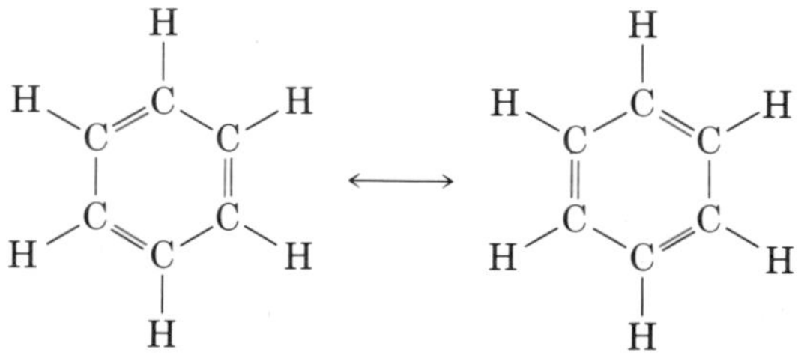

**Figure 15.5** The two equivalent resonance structures of benzene: Each carbon–carbon bond averages out to be 1.5 bonds—midway between single and double.

Recall from Section 10.7 that:

1. Resonance forms are imaginary, not real. The real structure of benzene is a single, unchanging hybrid structure that combines the characteristics of both resonance forms.
2. Resonance structures differ only in the positions of their electrons. Neither the position nor the hybridization of atoms changes from one resonance structure to another. In benzene, the six carbon nuclei form a regular hexagon while the pi electrons are shared equally between neighboring nuclei. Each carbon–carbon bond averages 1.5 electrons, and all bonds are equivalent.

3. Different resonance forms don't have to be equivalent. The more nearly equivalent the forms are, however, the more stable the molecule. Thus benzene, with two equivalent resonance forms is highly stable.
4. The more resonance structures there are, the more stable the molecule.

The true benzene structure can't be represented accurately by either single Kekulé structure and does not oscillate back and forth between the two. The true structure is somewhere in between the two extremes but is impossible to draw with our usual conventions. We might try to represent benzene as in Figure 15.6 by drawing it with either a full or dotted circle to indicate the equivalence of all carbon–carbon bonds, but these representations have to be used very carefully since they don't allow us to count the number of pi electrons in the ring. After all, how many electrons does a circle represent? In this book, benzene and other arenes will be represented by a single Kekulé structure. We will be able to keep count of all pi electrons this way, but we must still be aware of the limitations of the drawings.

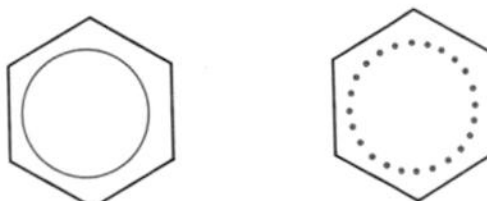

**Figure 15.6** Some alternative representations of benzene: These representations should be used with care because they don't specify the number of pi electrons in the ring.

There is a subtle yet important difference between Kekulé's representation of benzene and the resonance representation. Kekulé considered benzene as rapidly "oscillating" back and forth between two cyclohexatriene structures, whereas resonance theory considers benzene to be a single "resonance hybrid" structure:

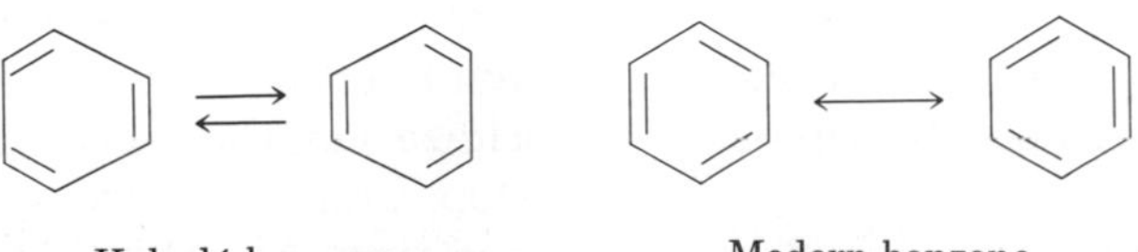


At any given instant, Kekulé's oscillating structures don't have all their carbon–carbon bonds the same length: Three bonds are short and three are long. This bond length difference implies that the carbon atoms must change position in oscillating from one structure to another, and thus the two structures are not the same as resonance forms.

The same reasoning that explains benzene's stability also explains why there is only one *ortho*-dibromobenzene, rather than two. The two possible

Kekulé structures of *o*-dibromobenzene are simply different resonance forms of a single compound whose true structure is intermediate between the two (Figure 15.7).

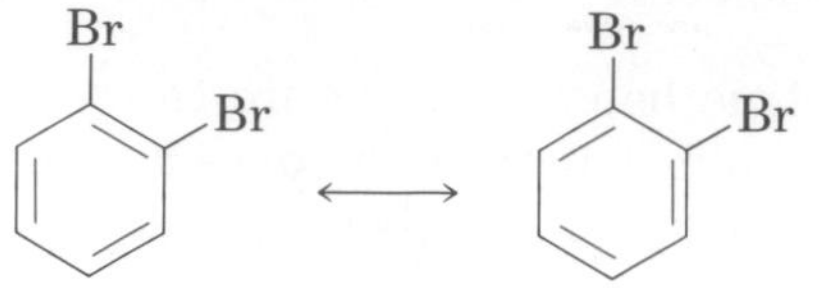

**Figure 15.7** Two equivalent resonance forms of *o*-dibromobenzene

**PROBLEM**

**15.6** In 1932, A. A. Levine and A. G. Cole studied the ozonolysis of *o*-xylene. They isolated three products: glyoxal (OHC—CHO), butane-2,3-dione ($CH_3COCOCH_3$), and pyruvaldehyde ($CH_3COCHO$):

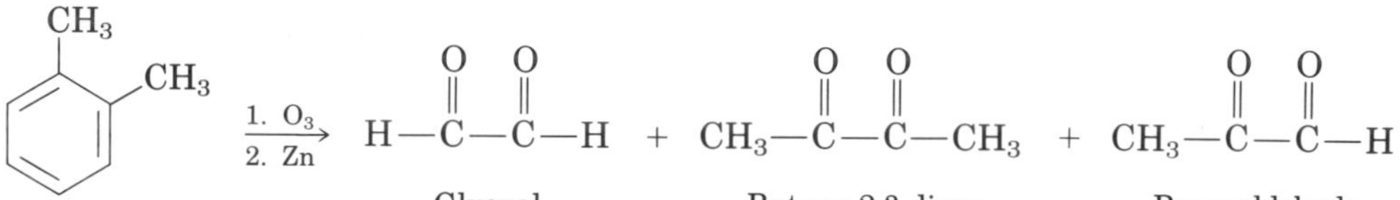

$$o\text{-xylene} \xrightarrow[2.\ Zn]{1.\ O_3} \underset{\text{Glyoxal}}{H{-}\overset{O}{\overset{\|}{C}}{-}\overset{O}{\overset{\|}{C}}{-}H} + \underset{\text{Butane-2,3-dione}}{CH_3{-}\overset{O}{\overset{\|}{C}}{-}\overset{O}{\overset{\|}{C}}{-}CH_3} + \underset{\text{Pyruvaldehyde}}{CH_3{-}\overset{O}{\overset{\|}{C}}{-}\overset{O}{\overset{\|}{C}}{-}H}$$

In what ratio would you expect these three products to be formed if *o*-xylene is a resonance hybrid of two Kekulé structures? The actual ratio found was 3 parts glyoxal, 2 parts pyruvaldehyde, and 1 part butane-2,3-dione. What conclusions can you draw about the structure of *o*-xylene?

## 15.6 Molecular Orbital Description of Benzene

Having just seen a resonance description of benzene, let's now see a molecular orbital description. MO theory describes benzene in a way that is, in many respects, superior to the simple resonance approach.

Benzene is a flat, hexagonal molecule with C-C-C bond angles of 120°. Since each carbon has $sp^2$ hybridization, the six carbon–carbon sigma bonds are formed by $C_{sp^2}$–$C_{sp^2}$ overlap, and there are six *p* orbitals perpendicular to the planar six-membered ring. An orbital picture looks like that in Figure 15.8.

Since all six *p* orbitals in benzene are equivalent, it is impossible to define three localized alkene pi bonds in which a given *p* orbital overlaps only one neighboring *p* orbital. Rather, each *p* orbital overlaps equally well with *both* neighboring *p* orbitals, leading to a picture of benzene in which the six pi electrons are completely delocalized around the ring. Benzene therefore has two doughnut-shaped clouds of electrons, one above and one below the ring.

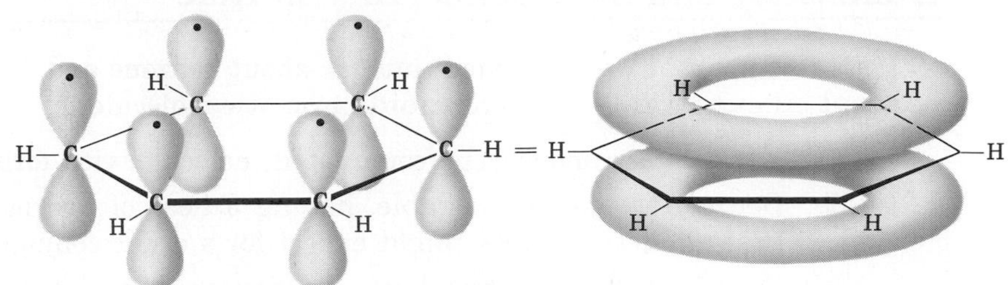


**Figure 15.8** An orbital picture of benzene

We can construct molecular orbitals for benzene just as we did for 1,3-butadiene in the preceding chapter. If we allow six *p* atomic orbitals to combine, six benzene molecular orbitals result, as shown in Figure 15.9. The three low-energy molecular orbitals, denoted $\psi_1$, $\psi_2$, and $\psi_3$, are lower in energy than the *p* atomic orbitals and are therefore bonding combinations. The three high-energy orbitals are antibonding. Note that two of the bonding orbitals, $\psi_2$ and $\psi_3$, have the same energy, as do the antibonding orbitals $\psi_4^*$ and $\psi_5^*$. Such orbitals are said to be **degenerate**. The six *p* electrons of benzene occupy the three bonding molecular orbitals and are therefore delocalized over the entire conjugated system, leading to the observed 36 kcal/mol stabilization energy of benzene.

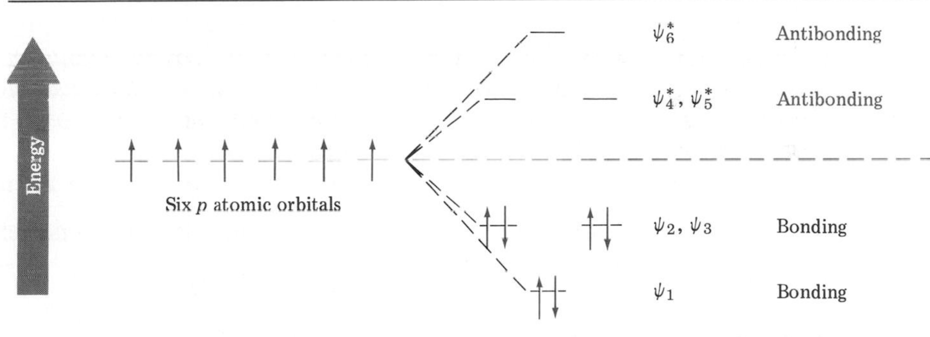


**Figure 15.9** The six molecular orbitals of benzene

**PROBLEM**

**15.7** Pyridine is a flat, symmetrical molecule with bond angles of 120°. It undergoes electrophilic substitution rather than addition and generally behaves as an aromatic substance. Draw an orbital picture of pyridine to explain its properties. Check your answer by looking ahead to Section 15.9.

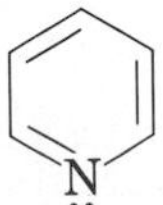


Pyridine

## 15.7 Aromaticity and the Hückel (4*n* + 2) Rule

Let's review what we've learned thus far about benzene and, by extension, about other benzene-like (**benzenoid**) aromatic molecules:

1. Benzene is a monocyclic conjugated molecule with formula $C_6H_6$.
2. Benzene is unusually stable, having a heat of hydrogenation 36 kcal/mol less than we might expect for a cyclic conjugated triene.
3. Benzene is a symmetrical, planar, hexagonal molecule. All C-C-C bond angles are 120°, and all C—C bonds are 1.39 Å in length.
4. Benzene undergoes substitution reactions that retain the cyclic conjugation, rather than electrophilic addition reactions that would destroy the conjugation.
5. Benzene can be described in terms of resonance theory as a hybrid whose structure is intermediate between two Kekulé structures:

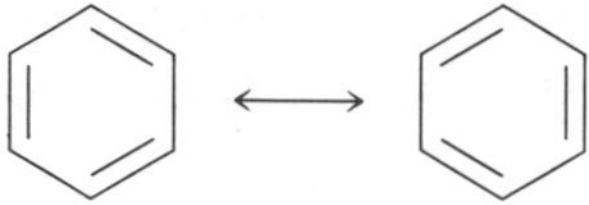

Taken together, these facts provide a description of benzene and of other aromatic molecules. But they aren't sufficient. Something else is needed to complete a description of aromaticity. According to theoretical calculations carried out by the German physicist Erich Hückel[2] in 1931, a molecule will be aromatic only if it has a planar, monocyclic system of conjugation with a *p* orbital on each atom, and *only if the p orbital system contains* $4n + 2$ *pi electrons*, where $n$ is an integer ($n = 0, 1, 2, 3, \ldots$). In other words, only molecules with 2, 6, 10, 14, 18, . . . pi electrons can be aromatic. Molecules with $4n$ pi electrons (4, 8, 12, 16, . . .) *cannot* be aromatic.

Let's look at some examples to see how the **Hückel (4*n* + 2) rule** works.

1. *Cyclobutadiene* has four pi electrons and is not aromatic:

Four pi electrons

Cyclobutadiene is a highly reactive molecule that shows none of the properties associated with aromaticity. A long history of attempts to synthesize the compound culminated in 1965 when cyclobutadiene was prepared, but not isolated, at low temperatures. Even at −78°C, however, cyclobutadiene dimerizes by Diels–Alder cycloaddition. One molecule behaves as a diene, whereas the other acts as dienophile:

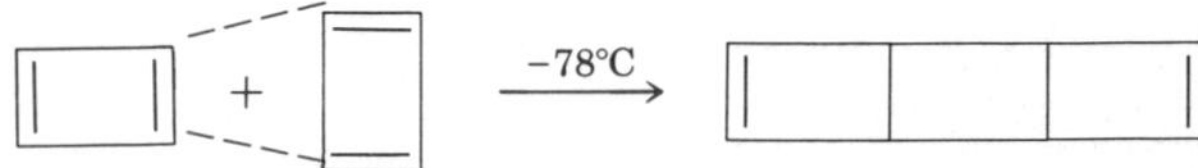


[2]Erich Hückel (1896–1980); b. Charlottenburg, Germany; Ph.D. Göttingen (Debye); professor at Stuttgart and Marburg.

2. *Benzene* has six pi electrons ($4n + 2$, where $n = 1$) and is, of course, a fully aromatic molecule:

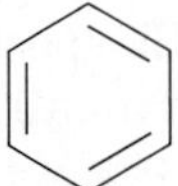

Six pi electrons

3. *Cyclooctatetraene* has eight pi electrons and is not aromatic:

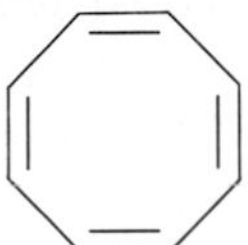

Eight pi electrons

In the early 1900s, chemists believed that the only requirement for aromaticity was the presence of a cyclic conjugated system. It was therefore expected that cyclooctatetraene, as a close analog of benzene, would also prove to be unusually stable. The facts proved otherwise. When cyclooctatetraene was first prepared in 1911 by the great German chemist Richard Willstätter,[3] it was found to resemble open-chain polyenes in its reactivity.

Cyclooctatetraene reacts readily with bromine, with potassium permanganate, and with hydrogen chloride, just as other alkenes do. We now know, in fact, that cyclooctatetraene is not even conjugated. It is tub-shaped, rather than planar, and has no cyclic conjugation because neighboring orbitals do not have the proper geometry for overlap (Figure 15.10). The pi

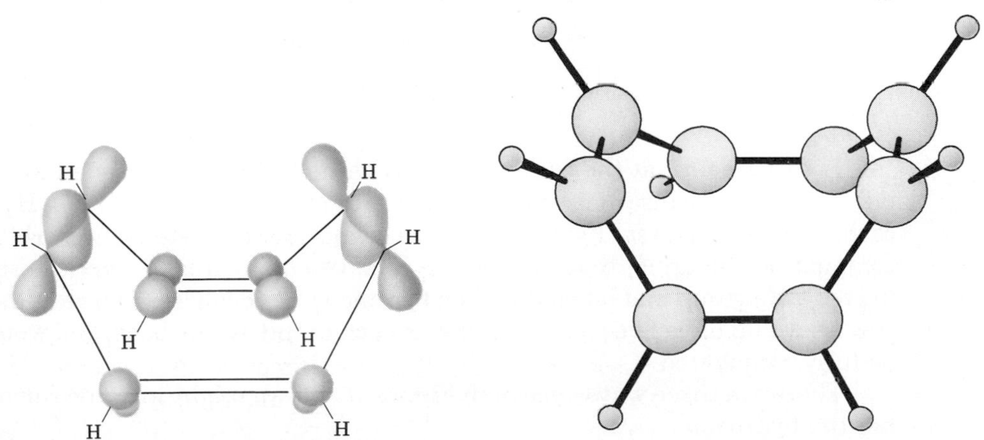


**Figure 15.10** An orbital view of cyclooctatetraene, a tub-shaped molecule that has no cyclic conjugation because its *p* orbitals are not aligned properly for overlap

[3]Richard Willstätter (1872–1942); b. Karlsruhe, Germany; Technische Hochschule, Munich (Einhorn) (1895); professor, Zurich, Dahlem, Munich; Nobel prize (1915).

electrons are localized in four distinct carbon–carbon double bonds rather than delocalized as in benzene. X-ray studies show that the carbon–carbon single bonds are 1.47 Å long, whereas the double bonds are 1.34 Å long. In addition, the $^1$H NMR spectrum shows a single sharp resonance line at 5.7 $\delta$, characteristic of an alkene rather than an aromatic molecule (Section 15.12).

PROBLEM

**15.8** In order to be aromatic, a molecule must be flat, so that $p$ orbital overlap can occur, and must have $(4n + 2)$ pi electrons. Cyclodecapentaene fulfills one of these criteria but not the other and is therefore nonaromatic. Explain. (Molecular models may be useful.)

Cyclodecapentaene (not aromatic)

## 15.8 Aromatic Ions

Look back at the Hückel definition of aromaticity in the previous section. In order to be aromatic, a molecule must have a cyclic system of conjugation, must have a $p$ orbital on each atom, and must have $(4n + 2)$ pi electrons. There is nothing in this definition that says the number of $p$ orbitals and the number of pi electrons has to be the same. In fact, they can be different. The $(4n + 2)$ rule is broadly applicable to many kinds of molecules, not just neutral hydrocarbons. For example, both the cyclopentadienyl anion and the cycloheptatrienyl cation are aromatic.

Cyclopentadienyl anion Cycloheptatrienyl cation
Six pi electrons, aromatic ions

Let's look first at the cyclopentadienyl anion. Cyclopentadiene itself is not aromatic because it is not a fully conjugated molecule; the $—CH_2—$ carbon in the ring is $sp^3$ hybridized, thus preventing complete cyclic $p$ orbital conjugation. Imagine, though, that we remove one hydrogen from the saturated $CH_2$ group and let that carbon become $sp^2$ hybridized. The resultant species would have five $p$ orbitals, one on each of the five carbons, and would be fully conjugated.

There are three ways, shown in Figure 15.11, we might imagine removing the hydrogen:

1. We could remove the hydrogen atom and *both* electrons from the C—H bond. Since the hydrogen removed must carry a negative charge, the cyclopentadienyl group that remains is positively charged.

2. We could remove the hydrogen and *one* electron from the C—H bond, leaving a cyclopentadienyl radical.
3. We could remove the hydrogen ion ($H^+$) with *no* electrons, leaving a cyclopentadienyl anion.

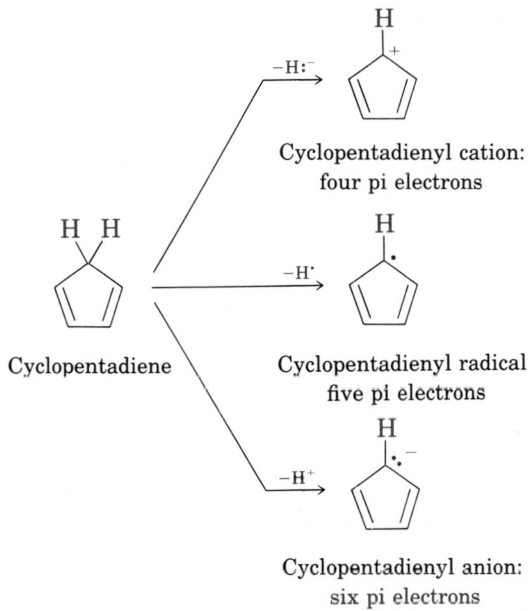


**Figure 15.11** Generating the cyclopentadienyl cation, radical, and anion by removing a hydrogen from cyclopentadiene

Resonance theory predicts that all three species should be highly stabilized, since it is possible to draw five equivalent resonance structures for each. Hückel theory, however, predicts that *only* the six-pi-electron anion should be aromatic. The four-pi-electron cyclopentadienyl carbocation and the five-pi-electron cyclopentadienyl radical are predicted not to be aromatic, even though they have cyclic conjugation.

In practice, both the cyclopentadienyl cation and the radical are highly reactive and difficult to prepare; neither species shows any sign of the unusual stability expected of an aromatic system. The six-pi-electron cyclopentadienyl anion, by contrast, is an easily prepared, remarkably stable **carbanion** (carbon anion). In fact, cyclopentadiene is one of the most acidic hydrocarbons known. Although most hydrocarbons have a $pK_a > 45$, cyclopentadiene has a $pK_a$ of 16, a value comparable to that of water! Cyclopentadiene is acidic because the anion formed by ionization is so stable. It doesn't matter that the cyclopentadienyl anion has only five *p* orbitals; all that matters is that there are six pi electrons, a Hückel number (Figure 15.12).

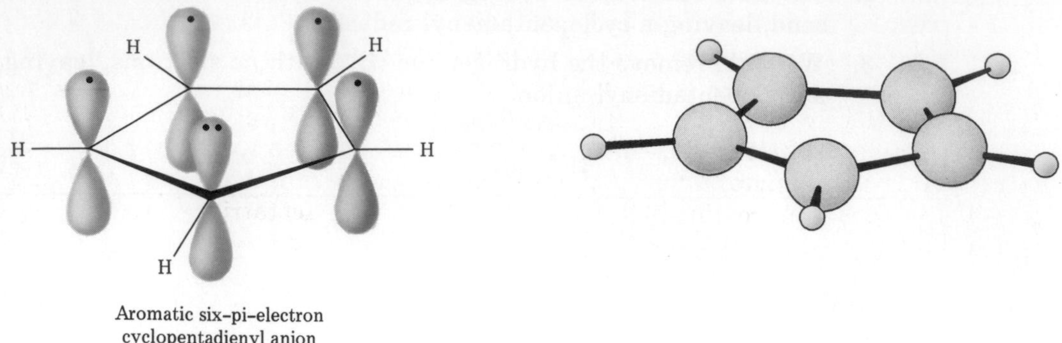


**Figure 15.12** An orbital view of the aromatic cyclopentadienyl anion, showing the cyclic conjugation and six pi electrons in five *p* orbitals

Similar arguments can be used to predict the stability of the cycloheptatrienyl cation, radical, and anion. Removal of a hydrogen from cycloheptatriene can generate either the six-pi-electron cation, the seven-pi-electron radical, or the eight-pi-electron anion (Figure 15.13). Resonance theory once again predicts a high level of stability for all three species, since seven equivalent resonance forms can be drawn for each. Hückel molecular orbital

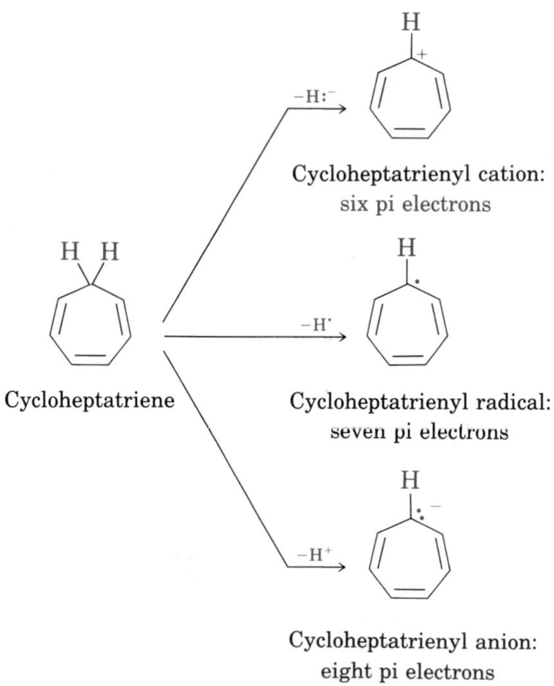


**Figure 15.13** Generation of the cycloheptatrienyl cation, radical, and anion: Only the six-pi-electron cation is aromatic.

theory, however, predicts that only the six-pi-electron cycloheptatrienyl cation should have aromatic stability.

Both the seven-pi-electron cycloheptatrienyl radical and the eight-pi-electron anion are highly reactive and difficult to prepare. The six-pi-electron cation, however, is extraordinarily stable. In fact, the cycloheptatrienyl cation was first prepared in 1891, although its structure was not recognized at the time. Later investigation showed that the salt-like material prepared by action of bromine on cycloheptatriene is cycloheptatrienylium bromide (Figure 15.14).

Cycloheptatriene $\xrightarrow{Br_2}$ HBr + $C_7H_7^+$ $Br^-$

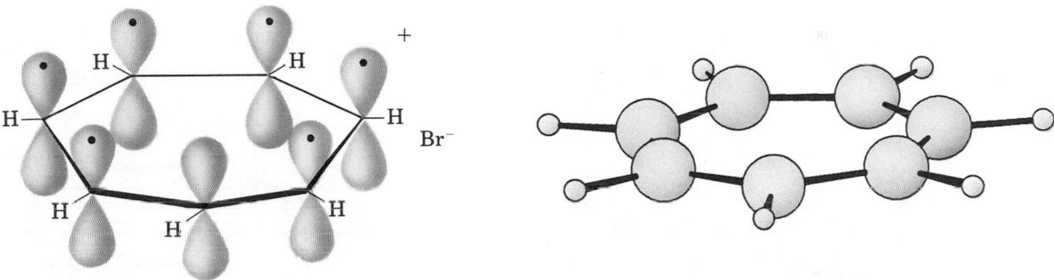


Cycloheptatrienylium bromide: six pi electrons

**Figure 15.14** Preparation of the aromatic cycloheptatrienyl anion by reaction of cycloheptatriene with bromine

**PROBLEM**

**15.9** Draw all possible resonance structures of the cyclopentadienyl anion. Are all carbon–carbon bonds equivalent? How many absorption lines would you expect to see in the $^1H$ NMR and $^{13}C$ NMR spectra of the anion?

**PROBLEM**

**15.10** Cyclooctatetraene readily accepts two electrons from potassium metal to form the cyclooctatetraene dianion, $C_8H_8^{2-}$. Why do you suppose this reaction occurs? What geometry would you expect for the cyclooctatetraene dianion?

$$C_8H_8 \xrightarrow{2\,K} 2\,K^+ \; [C_8H_8]^{2-}$$

## 15.9 Pyridine and Pyrrole: Two Aromatic Heterocycles

Look back once again at the definition of aromaticity: . . . a molecule that has a cyclic system of conjugation with a *p* orbital on each *atom* and that contains $(4n + 2)$ pi electrons. Nothing in this definition says that the atoms in the ring have to be *carbon*. In fact, **heterocyclic** compounds can also be aromatic. A **heterocycle** is a cyclic compound that has one or more atoms other than carbon in its ring. The heteroatom is often nitrogen or oxygen, but sulfur, phosphorus, and other elements are also found. Pyridine, for example, is a six-membered heterocycle with a nitrogen atom in its ring.

Pyridine is much like benzene in its pi electron structure. Each of the five $sp^2$-hybridized carbons has a *p* orbital perpendicular to the plane of the ring, and each contains one pi electron. The nitrogen atom also is $sp^2$ hybridized and has one electron in a *p* orbital, bringing the total to six pi electrons. The nitrogen lone-pair electrons are in an $sp^2$ orbital in the plane of the ring (perpendicular to the pi system) and are not involved with the aromatic pi system since they do not have the correct geometry for overlap (Figure 15.15).

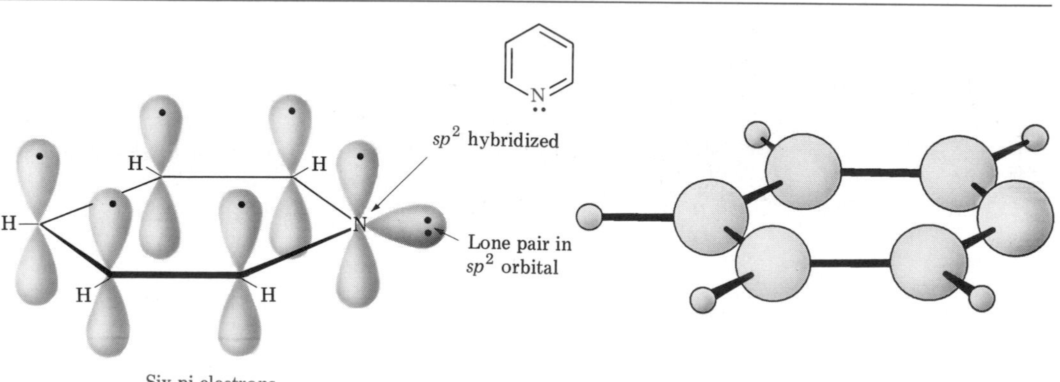


**Figure 15.15** Pyridine, an aromatic heterocycle, has a pi electron structure much like that of benzene.

The five-membered heterocycle, pyrrole (two *r*'s, one *l*), is also an aromatic six-pi-electron substance, with a pi electron system similar to that of the cyclopentadienyl anion. Each of the four $sp^2$-hybridized carbons has a *p* orbital perpendicular to the ring and each contributes one pi electron. The nitrogen atom is $sp^2$ hybridized, with its lone pair of electrons also occupying a *p* orbital. Thus, there is a total of six pi electrons, making pyrrole an aromatic molecule. An orbital picture of pyrrole is shown in Figure 15.16.

Note that the nitrogen atoms play different roles in pyridine and pyrrole even though both compounds are aromatic. The nitrogen atom in pyridine is part of a double bond and therefore contributes only *one* pi electron to the aromatic sextet, just as a carbon atom in benzene does. The nitrogen atom in pyrrole, however, is not part of a double bond. Like one of the carbons in the cyclopentadienyl anion, the pyrrole nitrogen atom contributes *two* pi electrons (the lone pair) to the aromatic sextet.

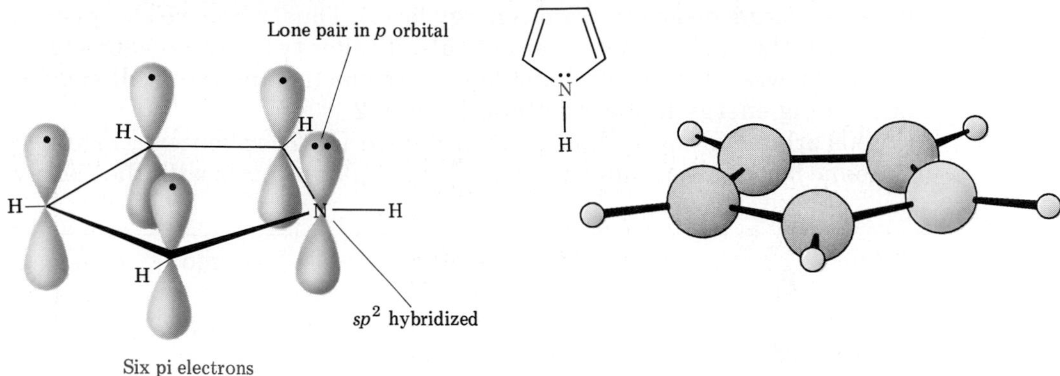


**Figure 15.16** Pyrrole, a five-membered aromatic heterocycle, has a pi electron structure much like that of the cyclopentadienyl anion.

Pyridine, pyrrole, and other aromatic heterocycles play an important role in many biochemical processes. Their chemistry will be discussed in more detail in Chapter 29.

PROBLEM

**15.11** The aromatic five-membered heterocycle imidazole is of great importance in a number of biological processes. One of its nitrogen atoms is "pyridine-like" in that it contributes one pi electron to the aromatic system, whereas the other nitrogen is "pyrrole-like" in that it contributes two pi electrons. Draw an orbital picture of imidazole to account for its aromaticity. How many pi electrons does each nitrogen contribute?

:N N̈—H

Imidazole

PROBLEM

**15.12** Assume that the oxygen atom in furan is $sp^2$ hybridized and draw an orbital picture to show how this heterocycle can be aromatic.

Ö

Furan

## 15.10 Why 4*n* + 2?

What's so special about (4*n* + 2) pi electrons? Why is it that having 2, 6, 10, 14, and so on, pi electrons leads to aromatic stability, but having other numbers of electrons does not? The answers to these questions have to do with the relative energy levels of the pi molecular orbitals.

When the energy levels of molecular orbitals for cyclic conjugated molecules are calculated, it turns out that there is always a *single* lowest-lying molecular orbital above which the molecular orbitals come in degenerate pairs (*degenerate* means the same energy level). Thus, when we assign electrons to fill the various molecular orbitals, it takes two electrons (one pair) to fill the lowest-lying orbital and four electrons (two pairs) to fill each of $n$ succeeding energy levels. The total is $4n + 2$.

This orbital filling is illustrated in Figure 15.17 for benzene, in which six atomic $p$ orbitals combine to give six molecular orbitals with the energy

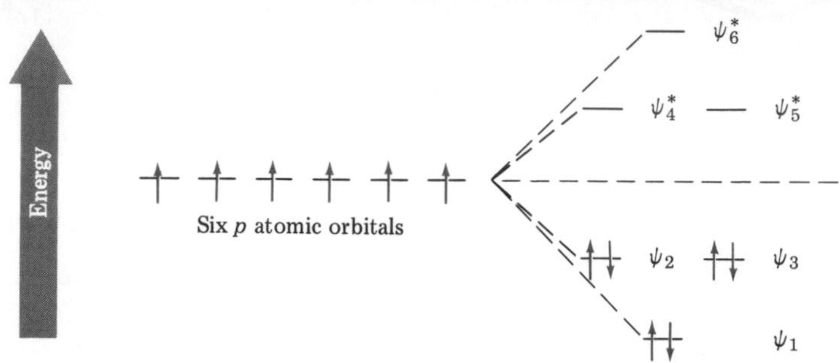


**Figure 15.17** Energy levels of benzene's six pi molecular orbitals

levels shown. Note that the lowest-energy molecular orbital, $\psi_1$, occurs singly and contains two electrons. The next two lowest-energy orbitals, $\psi_2$ and $\psi_3$, are degenerate, and it therefore takes four electrons to fill them. The result is a highly stabilized six-pi-electron aromatic molecule with filled bonding molecular orbitals.

A similar line of reasoning carried out for the cyclopentadienyl cation, radical, and anion leads to the conclusions illustrated in Figure 15.18. Once again, the five atomic $p$ orbitals combine to give five pi molecular orbitals in which there is a *single* lowest-energy orbital, and higher-energy degenerate pairs of orbitals. In the four-pi-electron cation, there are two electrons

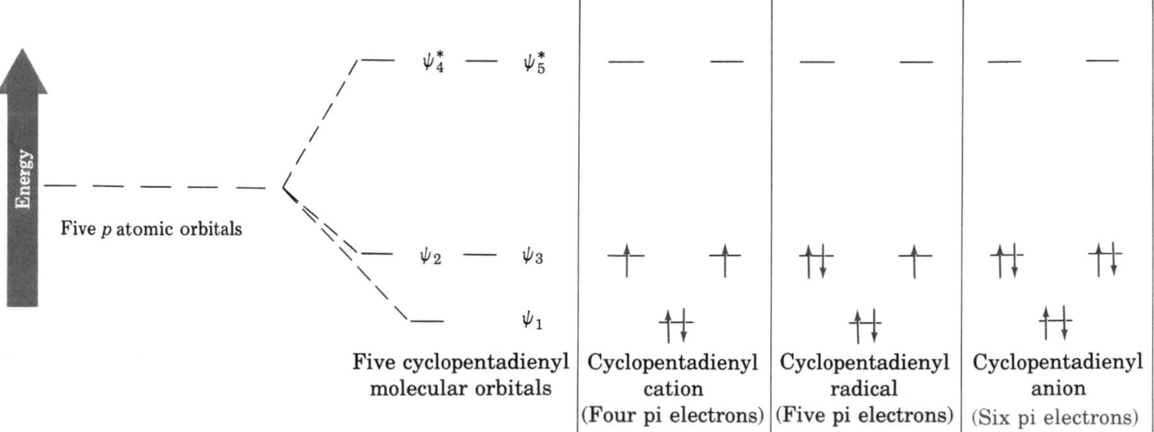


**Figure 15.18** Energy levels of the five cyclopentadienyl molecular orbitals: Only the six-pi-electron cyclopentadienyl anion has a filled-shell configuration leading to aromatic stability.

in $\psi_1$ but only one electron each in $\psi_2$ and $\psi_3$. Thus the cation has two orbitals that are only partially filled and is therefore not stabilized. In the five-pi-electron radical, $\psi_1$ and $\psi_2$ are filled, but $\psi_3$ is still only half-full. The radical is therefore not stabilized. Only in the six-pi-electron cyclopentadienyl anion are all of the bonding orbitals filled. Similar analyses can be carried out for all other aromatic species.

**PROBLEM**

**15.13** Show the relative positions of the seven pi molecular orbitals of the cycloheptatrienyl system. Indicate which of the seven orbitals are filled in the cation, radical, and anion, and account for the aromaticity of the cycloheptatrienyl cation.

## 15.11 Naphthalene: A Polycyclic Aromatic Compound

Although the Hückel rule is strictly applicable only to *monocyclic* aromatic compounds, the general concept of aromaticity can be extended beyond simple monocyclic compounds like benzene to include **polycyclic aromatic compounds**. Naphthalene, with two benzene-like rings fused together, is the simplest polycyclic aromatic molecule, but more complex substances such as anthracene, 1,2-benzpyrene, and coronene are known (Figure 15.19). 1,2-Benzpyrene is particularly interesting because it is one of the cancer-causing substances that has been isolated from tobacco smoke.

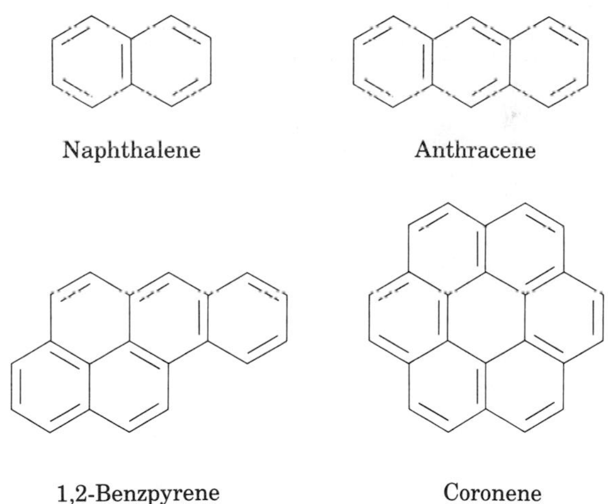


**Figure 15.19** Some polycyclic aromatic hydrocarbons

All polycyclic aromatic hydrocarbons can be represented by a number of different Kekulé structures. Naphthalene, for instance, has three resonance forms:

As was true for benzene with its two equivalent resonance forms, no single Kekulé structure is a true representation of naphthalene; the true structure of naphthalene lies somewhere in between the three resonance forms and is difficult to draw satisfactorily.

Naphthalene and other polycyclic hydrocarbons show many of the chemical properties we associate with aromaticity. Thus, measurements of heats of hydrogenation show an aromatic stabilization energy approximately 60 kcal/mol (250 kJ/mol) greater than might be expected if naphthalene had five isolated double bonds. Furthermore, naphthalene reacts slowly with electrophilic reagents such as bromine to give substitution products rather than double-bond addition products.

$$\text{Naphthalene} \xrightarrow[\Delta]{Br_2,\ Fe} \text{1-Bromonaphthalene (75\%)} + HBr$$

How can we explain the aromaticity of naphthalene? The orbital picture of naphthalene in Figure 15.20 shows a fully conjugated cyclic pi electron system, with *p* orbital overlap both around the 10-carbon periphery of the molecule and across the central bond. Since 10 pi electrons is a Hückel number, there is a high degree of pi electron delocalization and consequent aromaticity in naphthalene.

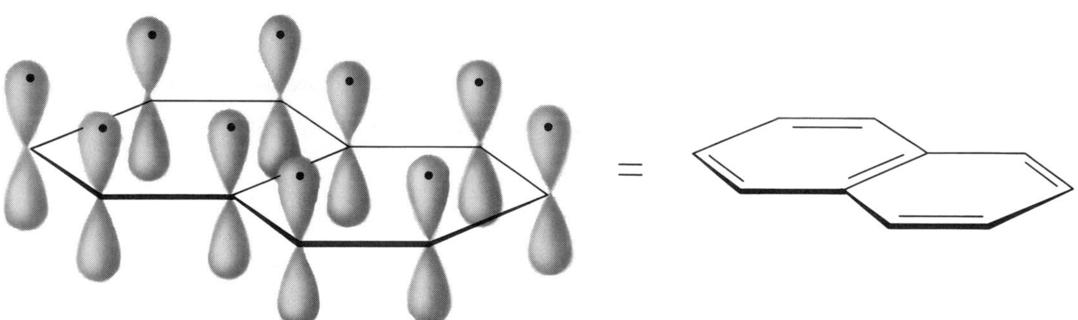

**Figure 15.20** An orbital picture of naphthalene, showing that the 10 pi electrons are fully delocalized throughout both rings

**PROBLEM**

**15.14** Look at the three resonance structures of naphthalene and then account for the fact that not all carbon–carbon bonds have the same length. The C1—C2 bond is 1.36 Å long, whereas the C2—C3 bond is 1.39 Å long.

**PROBLEM** ..............................................................................

**15.15** Naphthalene is sometimes represented with circles in each ring to represent aromaticity:

The difficulty with this representation is that it's not immediately apparent how many pi electrons are present. How many pi electrons are in each circle?

## 15.12 Spectroscopy of Aromatic Compounds

### INFRARED SPECTROSCOPY

Aromatic rings can be readily detected by infrared spectroscopy since they show a characteristic C—H stretching absorption at 3030 cm$^{-1}$ and a characteristic series of peaks in the 1450–1600 cm$^{-1}$ range. The aryl C—H band at 3030 cm $^{1}$ is generally of rather low intensity and occurs just to the left of a normal saturated C—H band. As many as four aryl absorptions are sometimes observed in the 1450–1600 cm$^{-1}$ region owing to complex molecular motions of the ring itself. Two bands, one at 1500 cm$^{-1}$ and one at 1600 cm$^{-1}$, are usually the most intense. In addition, aromatic compounds show strong absorptions in the 690–900 cm$^{-1}$ range due to C—H out-of-plane bending. The exact position of these absorptions is diagnostic of the substitution pattern of the aromatic ring:

| | | | |
|---|---|---|---|
| Monosubstituted: | 690–710 cm$^{-1}$<br>730–770 cm$^{-1}$ | *m*-Disubstituted: | 690–710 cm$^{-1}$<br>810–850 cm$^{-1}$ |
| *o*-Disubstituted: | 735–770 cm$^{-1}$ | *p*-Disubstituted: | 810–840 cm$^{-1}$ |

The infrared spectrum of toluene in Figure 15.21 shows these characteristic absorptions.

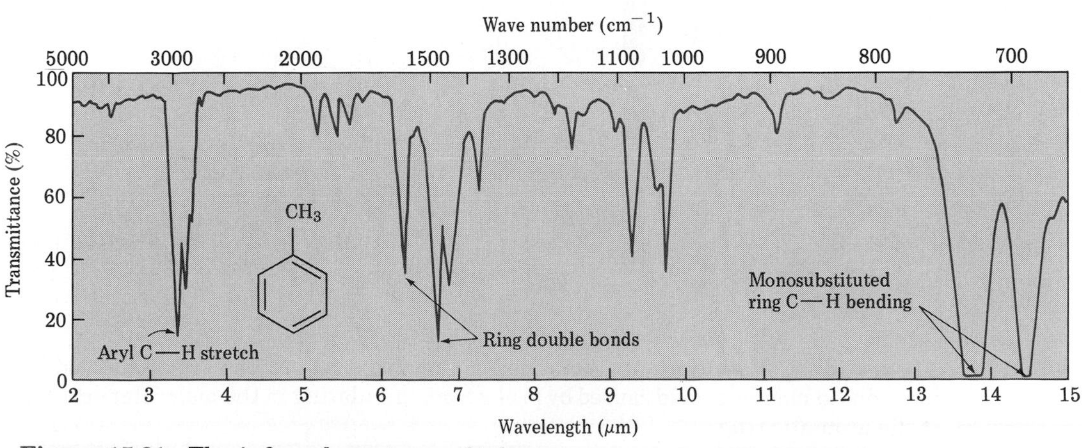


**Figure 15.21** The infrared spectrum of toluene

## ULTRAVIOLET SPECTROSCOPY

Aromatic rings are also detectable by ultraviolet spectroscopy since they contain a conjugated pi electron system. In general, aromatic compounds show a series of bands, with a fairly intense absorption near 205 nm and a less intense absorption in the 255–275 nm range. The presence of these bands in the ultraviolet spectrum of a molecule of unknown structure is a sure indication that an aromatic ring is present.

## NUCLEAR MAGNETIC RESONANCE SPECTROSCOPY

Hydrogens directly bonded to an aromatic ring are easily identifiable in the $^1H$ NMR spectrum. These hydrogens are strongly deshielded by the aromatic ring and absorb between 6.5 and 8.0 ppm downfield from the TMS standard. The spins of nonequivalent aryl protons on substituted rings often couple with each other, giving rise to spin–spin splitting patterns that can give information about the substitution pattern of the ring.

Much of the chemical shift difference between aryl protons (6.5–8.0 $\delta$) and vinylic protons (4.5–6.5 $\delta$) is due to a special property of aromatic rings called **ring current**. When an aromatic ring is oriented perpendicularly to a strong magnetic field, the pi electrons circulate around the ring in a direction such that they induce a tiny local magnetic field. This induced field opposes the applied field in the middle of the ring but *reinforces* the applied field outside the ring (Figure 15.22). Aryl protons are therefore deshielded; they experience an effective magnetic field greater than the applied field and thus come into resonance at a lower applied field.

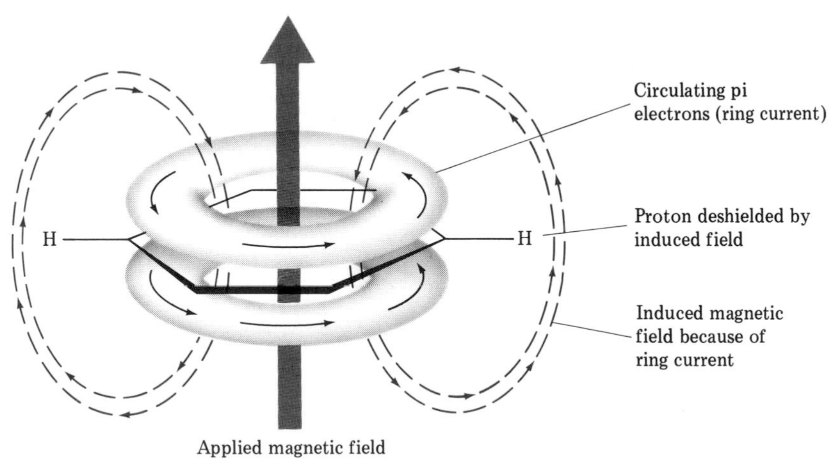


**Figure 15.22** The origin of aromatic ring current: Aryl protons are deshielded by the induced magnetic field caused by pi electrons circulating in the molecular orbitals of the aromatic ring.

Note that the existence of an aromatic ring current predicts different effects inside and outside the ring. If an aromatic ring were large enough to have both "inside" and "outside" protons, those protons on the outside should be deshielded and absorb at a field lower than normal, but those protons on the inside should be *shielded* and absorb at a field higher than normal. This prediction has been strikingly borne out by studies on [18]annulene, an 18-pi-electron cyclic conjugated polyene that contains a Hückel number of electrons ($4n + 2$, where $n = 4$) and is large enough to have both inside and outside protons:

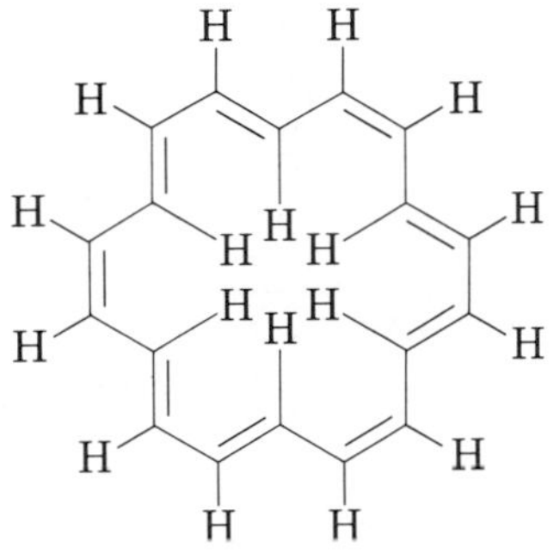

[18]Annulene

The $^1$H NMR spectrum of [18]annulene is just as predicted—the 6 inside protons are strongly shielded by the aromatic ring current and absorb at 3.0 ppm upfield from TMS, whereas the 12 outside protons are strongly deshielded and absorb in the typical aryl region at 9.3 ppm downfield from TMS.

The presence of a ring current is characteristic of all Hückel aromatic molecules and serves as an excellent test of aromaticity. For example, benzene, a 6-pi-electron aromatic molecule, absorbs at 7.37 δ, but cyclooctatetraene, an 8-pi-electron nonaromatic molecule, absorbs at 5.78 δ.

Protons on carbon next to aromatic rings also show distinctive absorptions in the NMR spectrum. Benzylic protons normally absorb downfield from other alkane protons in the region from 2.3 to 3.0 δ.

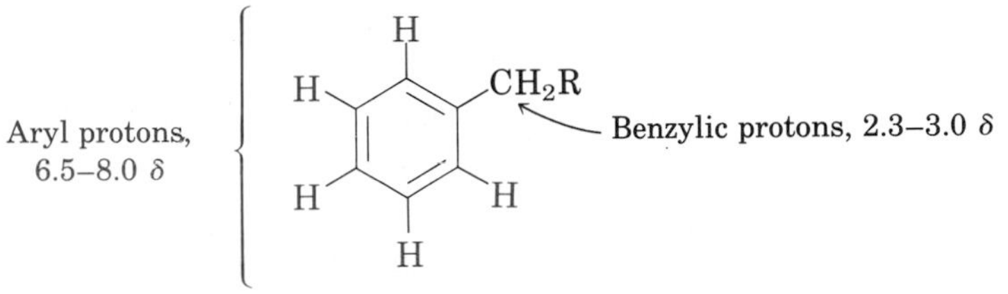

Aryl protons, 6.5–8.0 δ

$CH_2R$

Benzylic protons, 2.3–3.0 δ

The $^1$H NMR spectrum of *m*-bromotoluene shown in Figure 15.23 displays some of the features just discussed. The aryl protons absorb in a complex pattern from 6.9 to 7.6 δ, and the benzylic methyl protons absorb as a sharp singlet at 2.3 δ. Integration of the spectrum reveals the expected 4:3 ratio of peak areas.

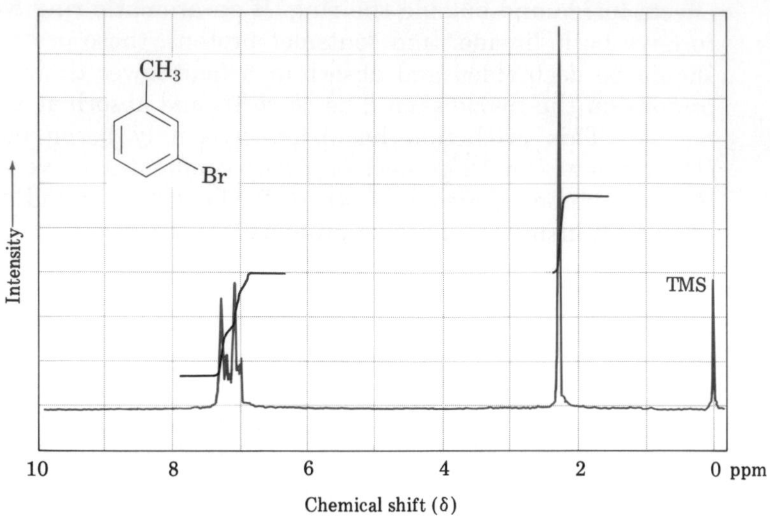


**Figure 15.23** The $^1H$ NMR spectrum of *m*-bromotoluene

Carbon atoms of an aromatic ring absorb in the range 110–160 $\delta$ in the $^{13}C$ NMR spectrum, as indicated by the examples in Figure 15.24. These resonances are easily distinguished from those of alkane carbons, but occur in the same range as alkene carbons. Thus, the presence of $^{13}C$ absorptions at 110–160 $\delta$ does not in itself establish the presence of an aromatic ring. Confirming evidence from infrared, ultraviolet, or $^1H$ NMR is needed.

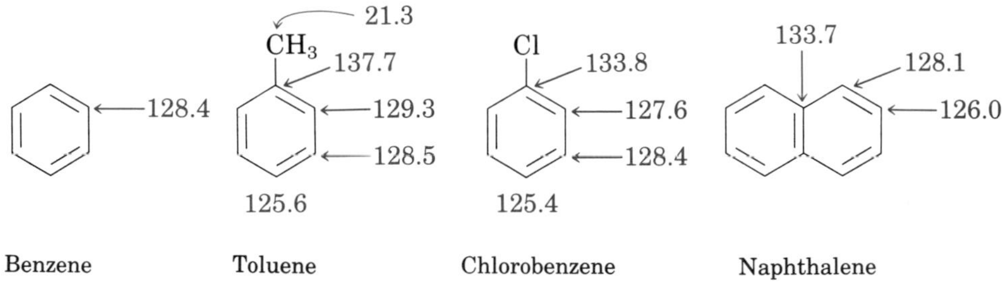

128.4
Benzene

21.3
$CH_3$ 137.7
129.3
128.5
125.6
Toluene

Cl 133.8
127.6
128.4
125.4
Chlorobenzene

133.7
128.1
126.0
Naphthalene

**Figure 15.24** Some $^{13}C$ NMR absorptions of aromatic compounds ($\delta$ units)

A summary of the kinds of information obtainable from different spectroscopic techniques is given in Table 15.3.

Table 15.3 Summary of spectroscopic information on aromatic compounds

| *Kind of spectroscopy* | *Absorption position* | *Interpretation* |
|---|---|---|
| Infrared ($cm^{-1}$) | 3030 | Aryl C—H stretch |
| | 1500 and 1600 | Two intense absorptions due to ring motions |
| | 690–900 | Intense C—H out-of-plane bending |
| Ultraviolet (nm) | 205 | Intense absorption |
| | 260 | Weak absorption |
| $^1$H NMR ($\delta$) | 2.3–3.0 | Benzylic protons |
| | 6.5–8.0 | Aryl protons |
| $^{13}$C NMR ($\delta$) | 110–160 | Aromatic ring carbons |

## 15.13 Summary and Key Words

The term **aromatic** is used for historical reasons to refer to the class of compounds related structurally to benzene. Many naturally occurring substances and most medicinally useful compounds contain aromatic (benzene-like) portions.

Aromatic compounds are systematically named according to IUPAC rules, but many common names are also used. Disubstituted benzenes are named as either **ortho** (1,2-disubstituted), **meta** (1,3-disubstituted), or **para** (1,4-disubstituted) derivatives. The —$C_6H_5$ unit itself is referred to as a **phenyl** group.

Benzene is described by resonance theory as a resonance hybrid of two equivalent Kekulé structures. Neither structure is correct by itself—the true structure of benzene is intermediate between the two:

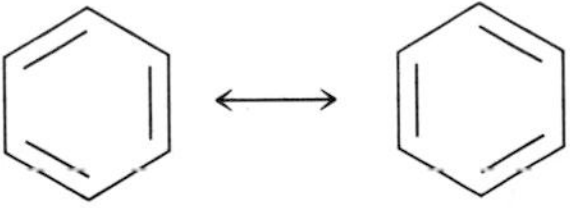

Benzene is described by molecular orbital theory as a cyclic conjugated molecule with six pi electrons. According to the **Hückel rule**, a cyclic conjugated molecule must have **($4n + 2$) pi electrons**, where $n = 0, 1, 2, 3$, and so on, in order to be aromatic.

Benzene has five key characteristics:

1. Benzene is a cyclic, planar, conjugated molecule.
2. Benzene is unusually stable; its heat of hydrogenation is approximately 36 kcal/mol less than we might expect for a cyclic triene.
3. Benzene reacts slowly with electrophiles to give substitution products in which cyclic conjugation is retained, rather than alkene addition products in which conjugation is destroyed.

4. Benzene is symmetrical. All carbon–carbon bonds are equivalent and have a length of 1.39 Å, a value that is intermediate between normal single- and double-bond lengths.
5. Benzene has a Hückel number of pi electrons, $(4n + 2)$, where $n = 1$. Thus, the pi electrons are delocalized over all six carbons, and there is a doughnut-shaped ring of electron density above and below the plane of the ring.

Other kinds of molecules besides benzenoid compounds can also be aromatic according to the Hückel $(4n + 2)$-pi-electron definition. For example, the cyclopentadienyl anion and the cycloheptatrienyl cation are both $(4n + 2)$-pi-electron aromatic ions. **Heterocyclic compounds**, which have atoms other than carbon in the ring, can also be aromatic. Pyridine, a six-membered nitrogen-containing heterocycle, resembles benzene electronically, whereas pyrrole, a five-membered heterocycle, resembles the cyclopentadienyl anion.

All of the spectroscopic techniques we have studied are applicable to the structure elucidation of aromatic compounds. Infrared, ultraviolet, and NMR spectroscopies all show characteristic aromatic absorption peaks.

## ADDITIONAL PROBLEMS

**15.16** Provide IUPAC names for these compounds:

(a) $CH_3$ $CH_3$ $CHCH_2CH_2CHCH_3$

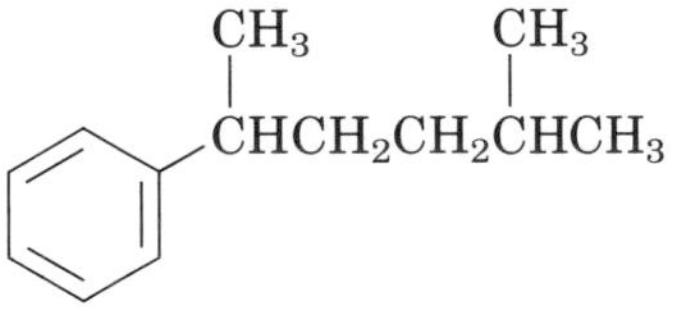

(b) $CO_2H$ Br

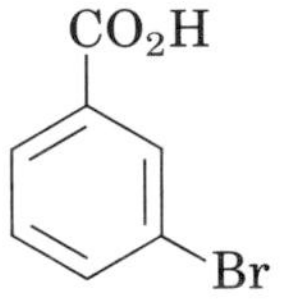

(c) Br $H_3C$ $CH_3$

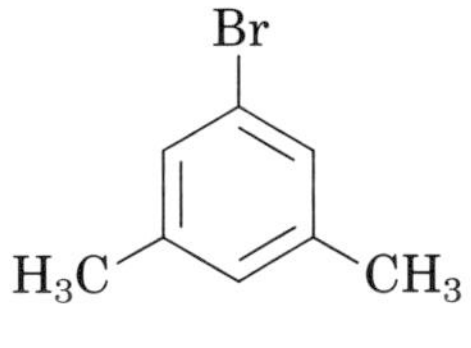

(d) Br $CH_2CH_2CH_3$

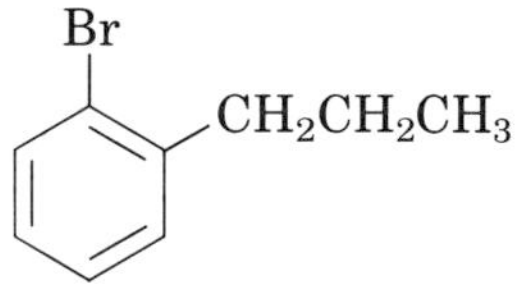

(e) F $NO_2$ $NO_2$

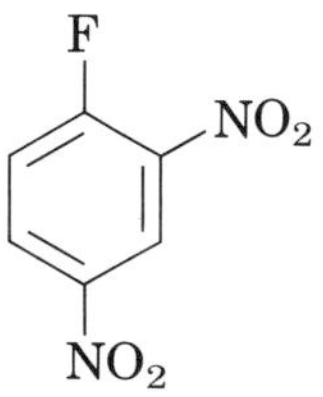

(f) $NH_2$ Cl

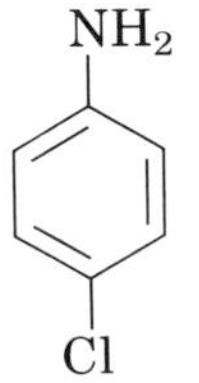

**15.17** Draw structures corresponding to these names:

(a) 3-Methyl-1,2-benzenediamine
(b) 1,3,5-Benzenetriol
(c) 3-Methyl-2-phenylhexane
(d) *o*-Aminobenzoic acid
(e) *m*-Bromophenol
(f) 2,4,6-Trinitrophenol (picric acid)
(g) *p*-Iodonitrobenzene

**15.18** Draw and name all possible isomeric:
(a) Dinitrobenzenes (b) Bromodimethylbenzenes (c) Trinitrophenols

**15.19** Draw and name all possible aromatic compounds with the formula $C_7H_7Cl$.

**15.20** Draw and name all possible aromatic compounds with the formula $C_8H_9Br$. (There are 14.)

**15.21** Propose structures for aromatic hydrocarbons that meet these descriptions:
(a) $C_9H_{12}$; gives only one $C_9H_{11}Br$ product on aromatic substitution by bromine
(b) $C_{10}H_{14}$; gives only one $C_{10}H_{13}Cl$ product on aromatic substitution by chlorine
(c) $C_8H_{10}$; gives three $C_8H_9Br$ products on aromatic substitution by bromine
(d) $C_{10}H_{14}$; gives two $C_{10}H_{13}Cl$ products on aromatic substitution by chlorine

**15.22** There are four resonance structures for anthracene, one of which is shown. Draw the other three.

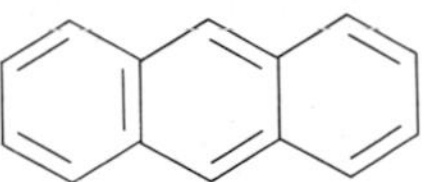

Anthracene

**15.23** There are five resonance structures of phenanthrene, one of which is shown. Draw the other four.

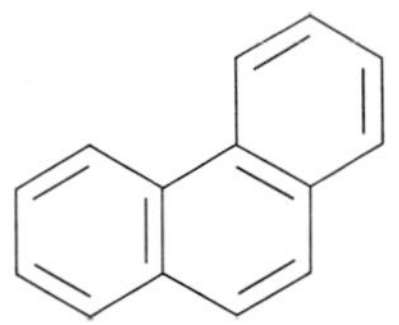

Phenanthrene

**15.24** Look at the five resonance structures for phenanthrene (Problem 15.23) and then predict which of its carbon–carbon bonds is shortest.

**15.25** Define these terms in your own words:
(a) Aromaticity (b) Conjugated
(c) Hückel ($4n + 2$) rule (d) Resonance hybrid

**15.26** Table 15.2 gives data on heats of hydrogenation for benzene, 1,3-cyclohexadiene, and cyclohexene. On the basis of these data, calculate the heats of hydrogenation for the partial hydrogenation shown here. Is the reaction exothermic or endothermic?

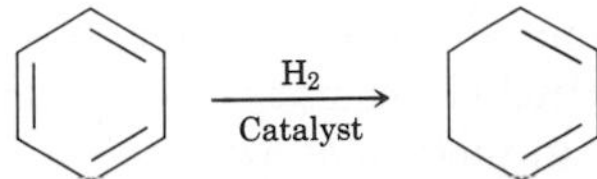


**15.27** 3-Chlorocyclopropene, on treatment with $AgBF_4$, gives a precipitate of AgCl and a stable solution of a species that shows only one absorption in the $^1H$ NMR spectrum at 11.04 δ. How do you explain this result? What is a likely structure for the product, and what is its relation to Hückel's rule?

3-Chlorocyclopropene

**15.28** Draw an energy diagram for the three molecular orbitals of the cyclopropenyl system (Problem 15.27). How are these three molecular orbitals occupied in the cyclopropenyl anion, cation, and radical? Which of these three is aromatic according to Hückel's rule?

**15.29** If we were to use the "circle" notation for aromaticity, we would draw the cyclopropenyl cation as shown here. How many pi electrons are represented by the circle in this instance?

**15.30** Cyclopropanone is a highly reactive molecule because of the large amount of angle strain it contains. Methylcyclopropenone, although more strained than cyclopropanone, is nevertheless quite stable and can even be distilled. How can you account for its stability? [*Hint:* Consider the polarity of the carbonyl group.]

O O

CH3

Cyclopropanone Methylcyclopropenone

**15.31** Cycloheptatrienone is a perfectly stable compound, but cyclopentadienone is so reactive that it can't be isolated. What do you suppose accounts for the stability difference between the two?

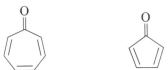


Cycloheptatrienone Cyclopentadienone

**15.32** Which would you expect to be most stable: cyclononatetraenyl radical, cation, or anion?

**15.33** Compound A, $C_8H_{10}$, yields three monobromo substitution products, $C_8H_9Br$, on reaction with bromine. Propose two possible structures. The $^1H$ NMR spectrum of compound A shows a complex four-proton multiplet at 7.0 $\delta$ and a six-proton singlet at 2.30 $\delta$. What is the correct structure of A?

**15.34** Azulene, $C_{10}H_8$, is a beautiful blue hydrocarbon that is isomeric with naphthalene. Unlike naphthalene, however, azulene has a large dipole moment ($\mu$ = 1.0 D).
(a) Is azulene a Hückel aromatic compound?
(b) Draw an orbital picture of azulene.
(c) How can you account for the observed dipole moment of azulene?

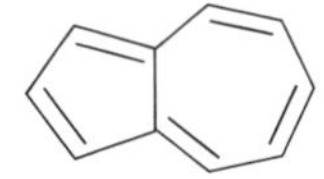

Azulene ($\mu$ = 1.0 D)

**15.35** What is the structure of a hydrocarbon that shows a molecular ion at $m/z$ = 120 in the mass spectrum and has the following $^1H$ NMR spectrum?

7.25 $\delta$, broad singlet, five protons

2.90 $\delta$, septet, $J = 7$ Hz, one proton

1.22 $\delta$, doublet, $J = 7$ Hz, six protons

**15.36** Propose structures for compounds that fit these descriptions:

(a) $C_{10}H_{14}$; $^1$H NMR: 7.18 $\delta$ (broad singlet, four protons), 2.70 $\delta$ (quartet, four protons, $J = 7$ Hz), 1.20 $\delta$ (triplet, six protons, $J = 7$ Hz); IR: 745 $cm^{-1}$.

(b) $C_{10}H_{14}$; $^1$H NMR: 7.0 $\delta$ (broad singlet, four protons), 2.85 $\delta$ (septet, one proton, $J = 8$ Hz), 2.28 $\delta$ (singlet, three protons), 1.20 $\delta$ (doublet, six protons, $J = 8$ Hz); IR: 825 $cm^{-1}$.

**15.37** Indole is an aromatic heterocycle that has a benzene ring fused to a pyrrole ring. Draw an orbital picture of indole.

(a) How many pi electrons are present?

(b) What is the electronic relationship of indole to naphthalene?

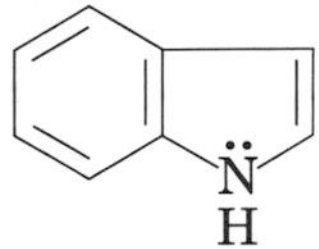


Indole

**15.38** On reaction with acid, 4-pyrone is protonated on the carbonyl-group oxygen to give a stable cationic product. Explain the stability of the protonated product.

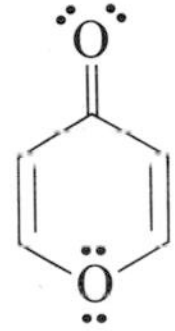


4-Pyrone

**15.39** Propose suitable structures for aromatic compounds that have the following $^1$H NMR spectra:

(a) $C_8H_9Br$;
IR: 820 $cm^{-1}$

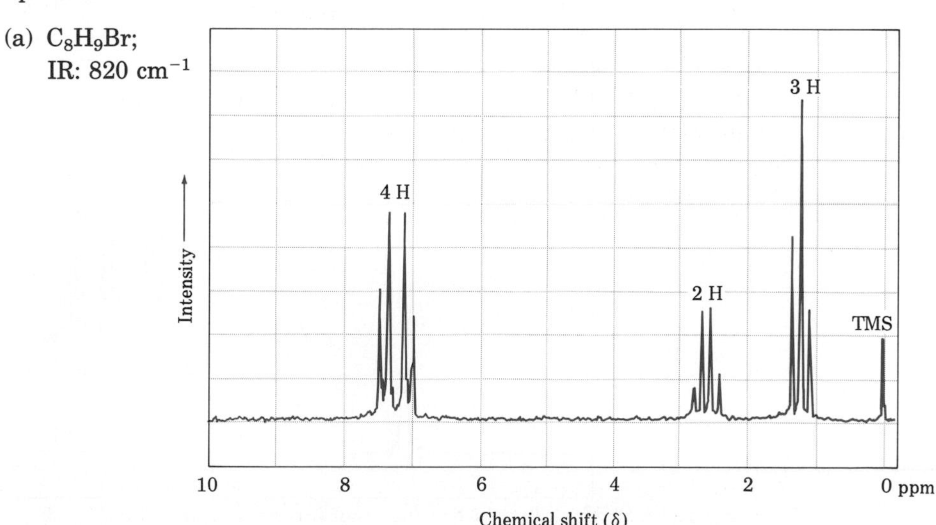

(b) $C_9H_{12}$;
IR: 750 $cm^{-1}$

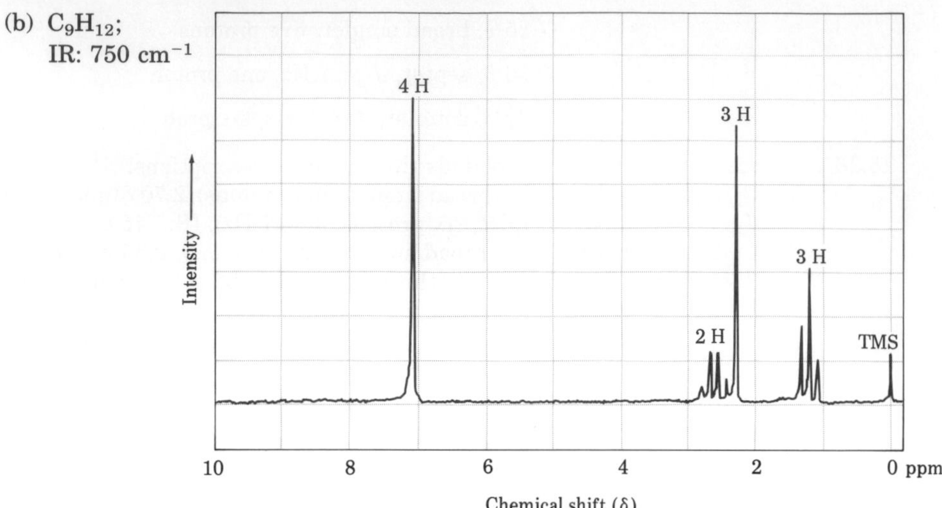


(c) $C_{11}H_{16}$;
IR: 820 $cm^{-1}$

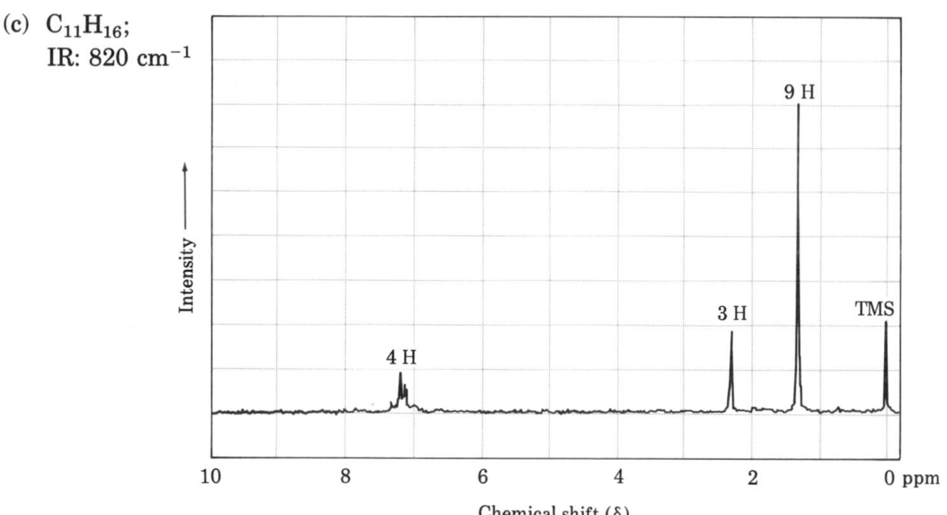


**15.40** Propose suitable structures for molecules that have the following $^1H$ NMR spectra:

(a) $C_9H_{11}Br$;
IR: 700,
760 $cm^{-1}$

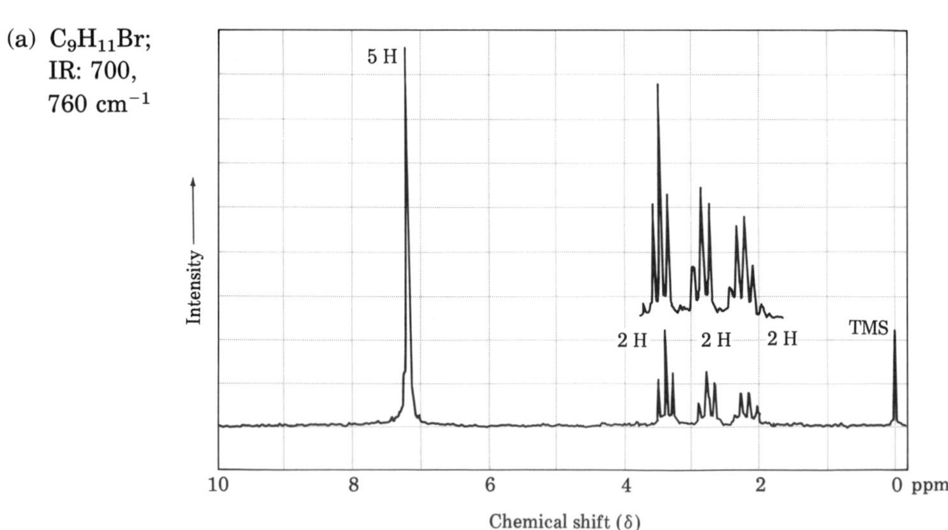

(b) $C_{14}H_{12}$;
IR: 700, 740, 890 $cm^{-1}$

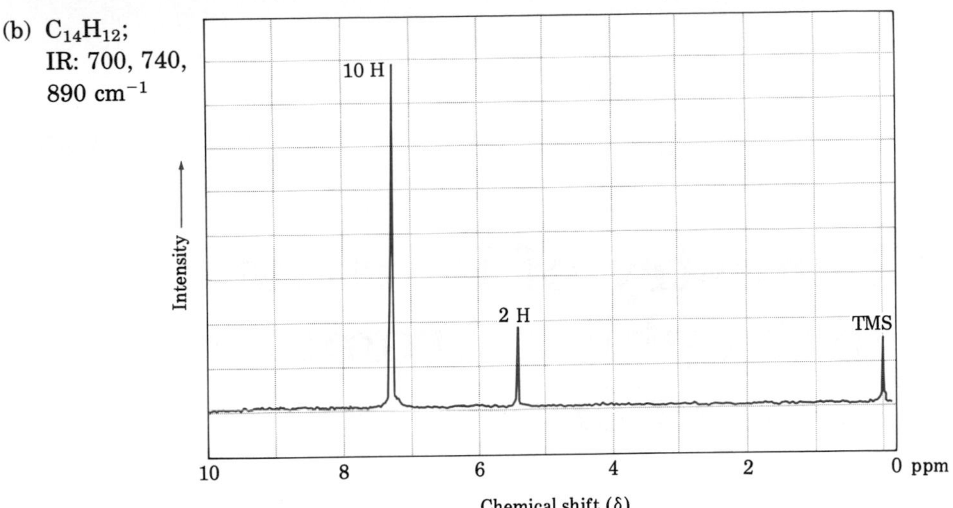


**15.41** Pentalene is a most elusive molecule that has never been isolated. The pentalene dianion, however, is well known and quite stable. Explain.

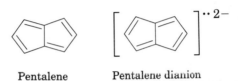

Pentalene    Pentalene dianion

**15.42** Purine is a heterocyclic aromatic compound that is a constituent of DNA and RNA. Why is purine considered to be aromatic? How many *p* electrons does each nitrogen donate to the aromatic pi system?

Purine

CHAPTER 16

# Chemistry of Benzene: Electrophilic Aromatic Substitution

The single most important reaction of aromatic compounds is **electrophilic substitution**. That is, an electrophile reacts with an aromatic ring and substitutes for one of the hydrogens (Figure 16.1).

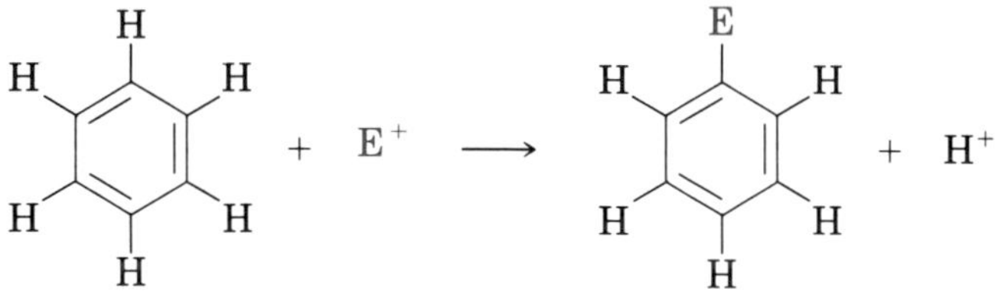

$$C_6H_6 + E^+ \longrightarrow C_6H_5E + H^+$$

**Figure 16.1** A generalized electrophilic aromatic substitution reaction: $E^+$ represents an electrophile.

Many different substituents can be introduced onto the aromatic ring using electrophilic substitution reactions. By choosing the proper conditions and reagents, we can **halogenate** (substitute a halogen: —F, —Cl, —Br, or —I), **nitrate** (substitute a nitro group: $—NO_2$), **sulfonate** (substitute a sulfonic acid group: $—SO_3H$), **alkylate** (substitute an alkyl group: —R), or **acylate** (substitute an acyl group: —COR) the aromatic ring. Starting from only a few simple materials, we can prepare many thousands of substituted aromatic compounds. Table 16.1 lists some of the possibilities.

Table 16.1 Some electrophilic aromatic substitution reactions

| *Name* | *Example* |
|---|---|
| Bromination | Ar—H + $Br_2$ $\xrightarrow{FeBr_3}$ Ar—Br (an aryl bromide) |
| Chlorination | Ar—H + $Cl_2$ $\xrightarrow{FeCl_3}$ Ar—Cl (an aryl chloride) |
| Nitration | Ar—H + $HNO_3$ $\xrightarrow{H_2SO_4}$ Ar—$NO_2$ (a nitro aromatic compound) |
| Sulfonation | Ar—H + $SO_3$ $\xrightarrow{H_2SO_4}$ Ar—$SO_3H$ (an aromatic sulfonic acid) |
| Friedel–Crafts alkylation | Ar—H + RCl $\xrightarrow{AlCl_3}$ Ar—R (an arene) |
| Friedel–Crafts acylation | Ar—H + $R\overset{O}{\overset{\|}{C}}Cl$ $\xrightarrow{AlCl_3}$ Ar—$\overset{O}{\overset{\|}{C}}R$ (an aryl ketone) |

All these reactions (and many more as well) take place by a similar mechanism. Let's begin a study of the process by examining one reaction in detail—the bromination of benzene.

## 16.1 Bromination of Aromatic Rings

Benzene, with six pi electrons in a cyclic conjugated system, is a site of electron density. Furthermore, the benzene pi electrons are sterically accessible to attacking reagents because of their location in circular clouds above and below the plane of the ring. Thus, benzene acts as an electron donor (a nucleophile) in most of its chemistry; most of the reactions of benzene take place with electron acceptors (electrophiles). For example, benzene reacts with bromine in the presence of $FeBr_3$ as catalyst to yield the substitution product bromobenzene.

Benzene + $Br_2$ $\xrightarrow{FeBr_3}$ Bromobenzene (80%) + HBr

Electrophilic substitution reactions are characteristic of all aromatic rings, not just of benzene and substituted benzenes. Indeed, the ability of a compound to undergo electrophilic substitution is an excellent test of aromaticity.

Before seeing how this electrophilic aromatic substitution reaction occurs, let's briefly recall what we learned in Chapter 6 about electrophilic *additions* to alkenes. When an electrophile such as HCl adds to an alkene, it approaches perpendicular to the plane of the double bond and forms a bond to one carbon, leaving a positive charge at the other carbon. This carbocation intermediate is then attacked by a nucleophile such as chloride ion to yield the addition product (Figure 16.2).

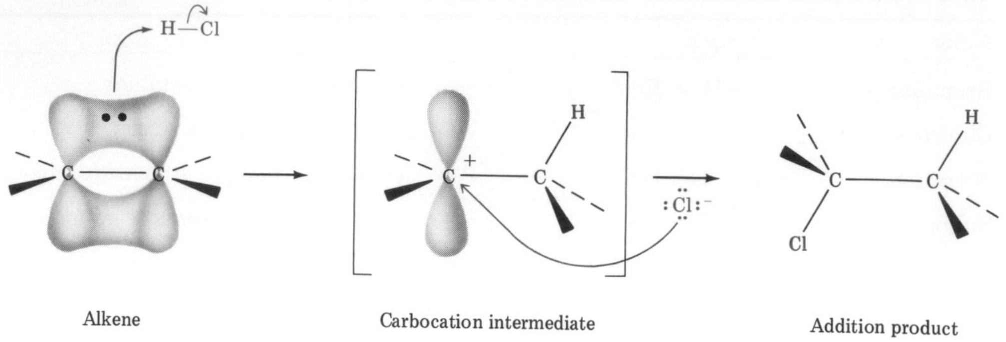

Alkene

Carbocation intermediate

Addition product

**Figure 16.2** The mechanism of alkene electrophilic addition reactions

An electrophilic aromatic substitution reaction begins in a similar way, but there are a number of differences. One difference is noticeable immediately: Aromatic rings are much less reactive than alkenes toward electrophiles. For example, bromine in $CCl_4$ solution reacts instantly with most alkenes but does not react with benzene. For bromination of benzene to take place, a catalyst such as $FeBr_3$ is needed. The catalyst exerts its effect by polarizing the $Br_2$ molecule, making it more electrophilic. The Lewis acid $FeBr_3$ complexes the bromine molecule, giving a polarized $FeBr_4^- \; Br^+$ species that reacts very much as if it were an electrophilic $Br^+$.

$$\overset{\delta^-}{Br}—\overset{\delta^+}{Br} \xrightarrow{FeBr_3} Br_3\overset{\delta^-}{Fe}\cdots\cdot Br \cdots\cdot \overset{\delta^+}{Br}$$

Bromine (a weak electrophile) — Polarized bromine (a strong electrophile)

The polarized bromine molecule is then attacked by the pi electron system of the nucleophilic benzene ring in a slow, rate-limiting step to yield a nonaromatic carbocation intermediate. This carbocation is doubly allylic (recall the allyl cation, Section 11.9) and can be written in three resonance forms:

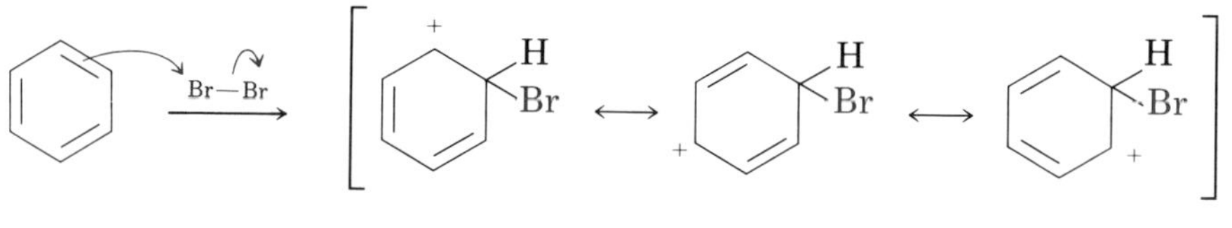

Although stable by comparison with most other carbocations, the intermediate in electrophilic aromatic substitution is nevertheless much less stable than the starting benzene ring itself with its 36 kcal/mol (152 kJ/mol) of aromatic stability. Thus, electrophilic attack on a benzene ring is highly endothermic, has a high activation energy ($\Delta G^{\ddagger}$), and is therefore a rather slow reaction. Figure 16.3 gives reaction energy diagrams comparing the reaction of an electrophile, $E^+$, with an alkene and with benzene.

The benzene reaction is slower (higher $\Delta G^{\ddagger}$) because the starting material is so much more stable.

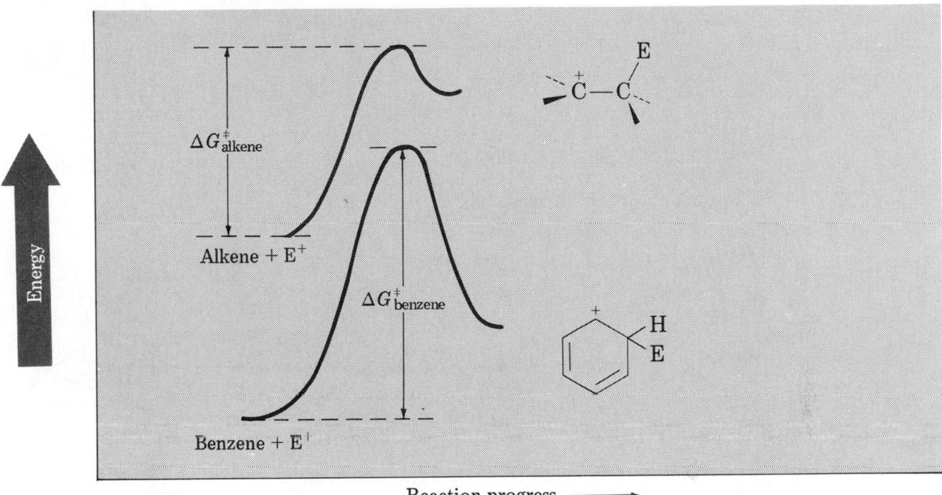


**Figure 16.3** A comparison of the reactions of an electrophile with an alkene and with benzene: $\Delta G^{\ddagger}_{\text{alkene}} \ll \Delta G^{\ddagger}_{\text{benzene}}$

A second major difference between alkene addition and aromatic substitution occurs after the electrophile has added to the benzene ring to give the carbocation intermediate. Although it would presumably be possible for a nucleophile such as bromide ion to attack the carbocation intermediate to yield the addition product dibromocyclohexadiene, this is not observed. Instead, the bromide ion (or some other base present in solution) abstracts a proton from the bromine-bearing carbon to yield the neutral, aromatic substitution product plus HBr. The net effect of reaction of $Br_2$ with benzene is the substitution of $H^+$ by $Br^+$. The overall mechanism is shown in Figure 16.4 (page 526).

Why does the reaction of an electrophile with benzene and other aromatic rings take a different course than its reaction with an alkene? The answer is quite simple: If *addition* occurred, the 36 kcal/mol stabilization energy of the aromatic ring would be lost and the overall reaction would be highly endothermic. By losing a proton, however, the carbocation intermediate reverts to an aromatic ring structure and regains aromatic stabilization. A reaction energy diagram for the overall process is shown in Figure 16.5 (page 527).

There are many other electrophilic aromatic substitutions besides bromination, and all are thought to occur by the same general mechanism. Let's see briefly what some of these other reactions are.

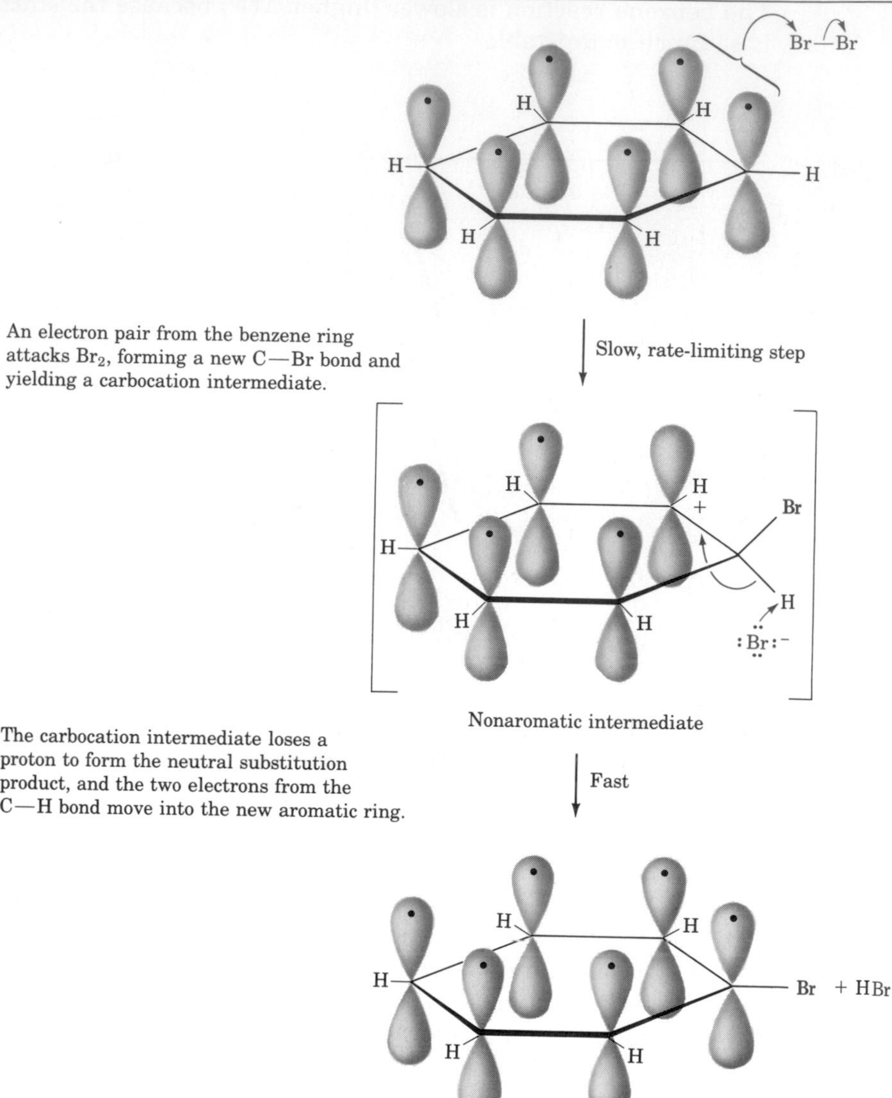
Br—Br
H
H
H
H
H
An electron pair from the benzene ring attacks $Br_2$, forming a new C—Br bond and yielding a carbocation intermediate.
Slow, rate-limiting step
H
H
+
Br
H
H
H
H
:Br:⁻
Nonaromatic intermediate
The carbocation intermediate loses a proton to form the neutral substitution product, and the two electrons from the C—H bond move into the new aromatic ring.
Fast
H
H
H
Br + HBr
H
H

**Figure 16.4** The mechanism of the electrophilic bromination of benzene

## 16.2 Other Aromatic Substitutions

### AROMATIC CHLORINATION AND IODINATION

Chlorine and iodine can be introduced into aromatic rings by electrophilic substitution reactions under the proper conditions, but fluorine is too reactive, and poor yields of monofluoroaromatic products are obtained. Chlorine reacts smoothly and gives excellent yields of chloro-substituted aromatic derivatives. As with bromination, a catalyst such as $FeCl_3$ is required to

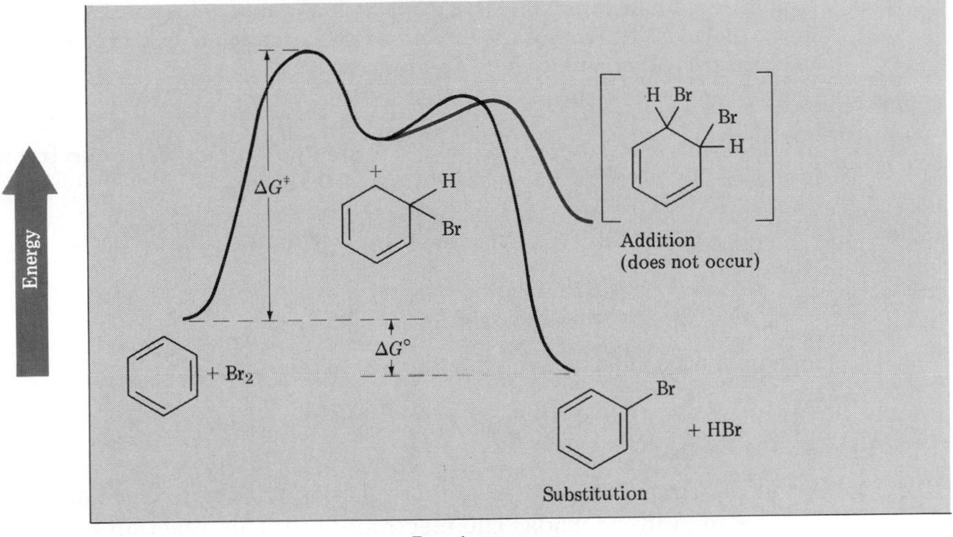


**Figure 16.5** A reaction energy diagram for the electrophilic bromination of benzene

increase the rate of reaction between chlorine and the aromatic ring. $FeCl_3$ acts by polarizing the chlorine molecule, making it more electrophilic.

$$FeCl_3 + Cl{-}Cl \longrightarrow \overset{\delta^-}{Cl_3Fe}\cdots Cl \cdots \overset{\delta^+}{Cl}$$

Benzene + Cl—Cl → [arenium ion intermediate, :Cl:⁻] → Chlorobenzene (86%) + HCl

Iodine by itself is unreactive toward aromatic rings, and a promoter is required to obtain a suitable reaction. The best promoters for aromatic iodination are oxidizing agents such as hydrogen peroxide, $H_2O_2$, or copper salts such as $CuCl_2$. These promoters are thought to work by oxidizing molecular iodine to a more powerful electrophilic species that reacts as if it were $I^+$. The aromatic ring then attacks $I^+$, yielding a normal substitution product.

$$I_2 + 2\,Cu^{2+} \longrightarrow 2\,\text{“}I^{+}\text{”} + 2\,Cu^{+}$$

Benzene + "I⁺" ($I_2 + CuCl_2$) → [arenium ion intermediate, :I:⁻] → Iodobenzene (65%) + HI

PROBLEM

**16.1** Aromatic iodination can be carried out with a number of reagents, including iodine monochloride, ICl. What is the direction of polarization of ICl? Propose a mechanism to account for the iodination of an aromatic ring.

PROBLEM

**16.2** Aryl fluorides can be prepared by a two-step process involving electrophilic **thallation** of the aromatic ring (substitution of thallium).

$$ArH + Tl(OCOCF_3)_3 \longrightarrow Ar{-}Tl(OCOCF_3)_2 \xrightarrow{KF,\ BF_3} ArF$$

Thallium tris(trifluoroacetate)

Propose a mechanism for the thallation of benzene.

## AROMATIC NITRATION

Aromatic rings can be nitrated by reaction with a mixture of concentrated nitric and sulfuric acids. The electrophile in this reaction is thought to be the **nitronium ion, $NO_2^+$**, which is generated from nitric acid by protonation and loss of water. The nitronium ion then reacts with benzene to yield a carbocation intermediate in much the same way that we discussed for $Br^+$. Loss of a proton from this intermediate gives the neutral substitution product nitrobenzene (Figure 16.6).

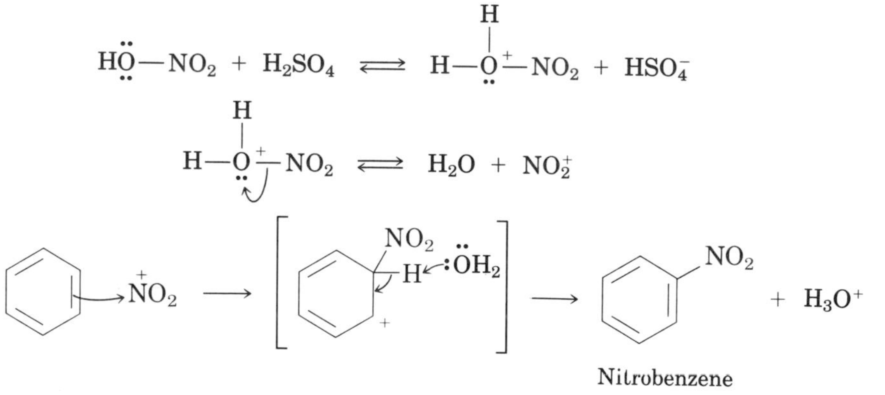

$$HO{-}NO_2 + H_2SO_4 \rightleftharpoons H{-}\overset{H}{\overset{|}{O^+}}{-}NO_2 + HSO_4^-$$

$$H{-}\overset{H}{\overset{|}{O^+}}{-}NO_2 \rightleftharpoons H_2O + NO_2^+$$



**Figure 16.6** Mechanism of the nitration of benzene

When, for some reason, nitration of an aromatic ring with $HNO_3–H_2SO_4$ is particularly sluggish, the reaction can also be carried out using a pure nitronium salt, $NO_2^+\ BF_4^-$. Nitronium tetrafluoroborate is a white crystalline material that will smoothly nitrate many aromatic compounds at room temperature or below.

Nitration of aromatic rings is a particularly important reaction because the product nitroarenes can be reduced by reagents such as iron or stannous chloride to yield aminoarenes (substituted anilines). We'll discuss this and other reactions of aromatic nitrogen compounds in Chapter 26.

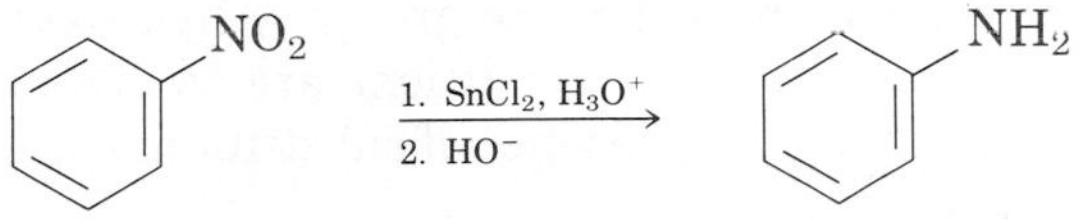

$NO_2$ — 1. $SnCl_2$, $H_3O^+$; 2. $HO^-$ → $NH_2$

Nitrobenzene Aniline (95%)

## AROMATIC SULFONATION

Aromatic rings can be sulfonated by reaction with sulfuric acid and sulfur trioxide, a mixture called fuming sulfuric acid ($H_2SO_4$ + $SO_3$):

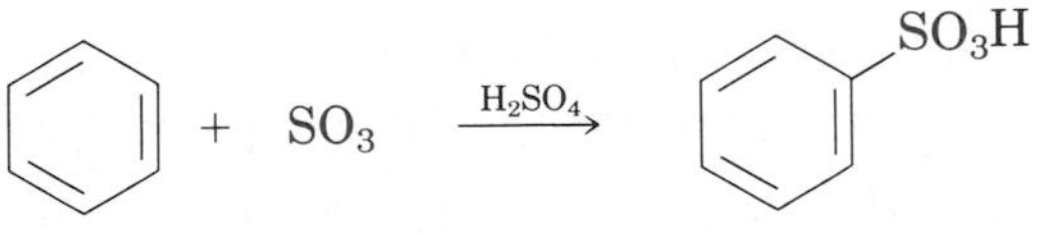

+ $SO_3$ $\xrightarrow{H_2SO_4}$ $SO_3H$

Benzene Benzenesulfonic acid (95%)

The reactive electrophile is either $HSO_3^+$ or neutral $SO_3$, depending on reaction conditions. Substitution occurs by the same two-step mechanism seen previously for bromination and nitration (Figure 16.7). Note, however, that the sulfonation reaction is reversible; it can occur either forward or backward, depending on the reaction conditions. Sulfonation is favored in strong acid, but desulfonation is favored in hot, dilute aqueous acid.

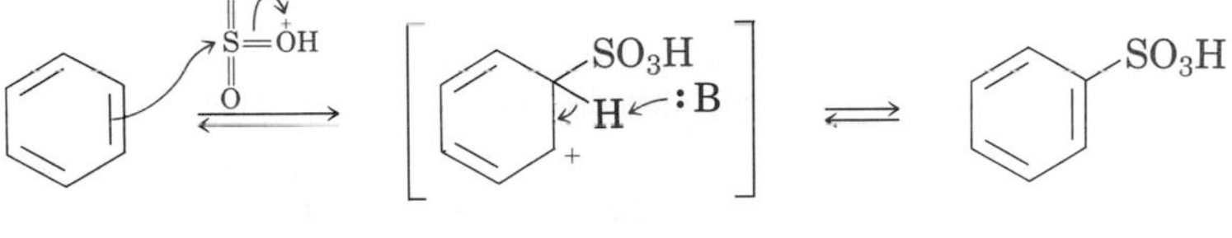

**Figure 16.7** Mechanism of the sulfonation of benzene

Aromatic sulfonic acids are valuable intermediates in the preparation of dyes and pharmaceuticals. For example, the sulfa drugs such as sulfanilamide were among the first useful antibiotics known. Although largely replaced today by more effective agents, sulfa drugs are still used in the treatment of meningitis. These drugs are prepared commercially by a process that involves aromatic sulfonation as the key step.

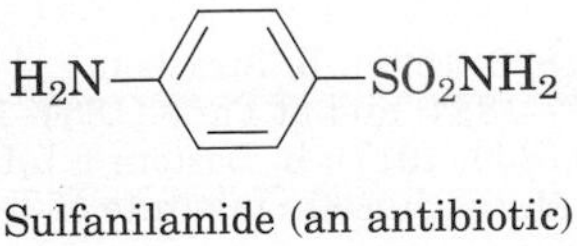

$H_2N$—⟨benzene⟩—$SO_2NH_2$

Sulfanilamide (an antibiotic)

Aromatic sulfonic acids are also valuable because of the further chemistry they undergo. Thus, **alkali fusion** of an arenesulfonic acid with NaOH at 300°C in the absence of solvent yields the corresponding phenol—a net replacement of the sulfonate group by hydroxyl. Yields in this process are generally good, but the conditions are so vigorous that the reaction is not compatible with the presence of substituents other than alkyl groups on the aromatic ring.

$H_3C-C_6H_4-SO_3H \xrightarrow[\text{2. } H_3O^+]{\text{1. NaOH, 300°C}} H_3C-C_6H_4-OH$

*p*-Toluenesulfonic acid — *p*-Cresol (72%) (a phenol)

PROBLEM

16.3 Show a detailed mechanism for the desulfonation reaction of benzenesulfonic acid to yield benzene. What is the electrophile in this reaction?

## 16.3 Alkylation of Aromatic Rings: The Friedel–Crafts Reaction

One of the most useful of all electrophilic aromatic substitution reactions is **alkylation**—the attachment of an alkyl group to the benzene ring. Charles Friedel[1] and James Crafts[2] reported in 1877 that benzene rings can be alkylated by reaction with an alkyl chloride in the presence of aluminum chloride as catalyst. For example, benzene reacts with 2-chloropropane and $AlCl_3$ to yield cumene (isopropylbenzene).

$C_6H_6 + (CH_3)_2CHCl \xrightarrow{AlCl_3} C_6H_5CH(CH_3)_2 + HCl$

Benzene — 2-Chloropropane — Cumene (85%)

The **Friedel–Crafts alkylation reaction** is an electrophilic aromatic substitution in which the aromatic ring attacks a *carbocation* electrophile. The carbocation is generated when $AlCl_3$ catalyst helps the alkyl halide to ionize, in much the same way that $FeCl_3$ catalyzes aromatic chlorinations by polarizing $Cl_2$ (Section 16.2). Loss of a proton then completes the reaction, as shown in Figure 16.8.

Though broadly useful for the synthesis of alkylbenzenes, Friedel–Crafts alkylations are nevertheless subject to certain limitations. One limitation is that only *alkyl* halides can be used. Alkyl fluorides, chlorides, bromides, and iodides all react well, but *aryl* halides and *vinylic* halides do not react. Aryl and vinylic carbocations are too unstable to form under Friedel–Crafts conditions.

[1]Charles Friedel (1832–1899); b. Strasbourg, France; studied at the Sorbonne; professor, École des Mines (1876–1884) and at Paris (1884–1899).

[2]James M. Crafts (1839–1917); b. Boston; L.L.D., Harvard (1898); professor, Cornell University (1868–1871); Massachusetts Institute of Technology (1871–1900).

$$(CH_3)_2CHCl + AlCl_3 \longrightarrow (CH_3)_2CH^+AlCl_4^-$$

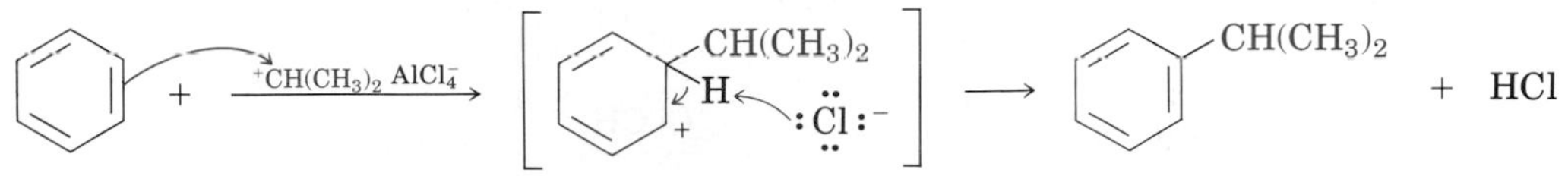

**Figure 16.8** Mechanism of the Friedel–Crafts alkylation reaction

An aryl halide    A vinylic halide
*Not reactive*

A second limitation is that Friedel–Crafts reactions do not succeed on aromatic rings that are substituted by strongly deactivating groups (Figure 16.9). We'll see in Section 16.5 that the presence of a substituent group on a ring can have a dramatic effect on that ring's subsequent reactivity toward further electrophilic substitution. Rings that contain any of the substituents listed in Figure 16.9 are simply not reactive enough to attack carbocations.

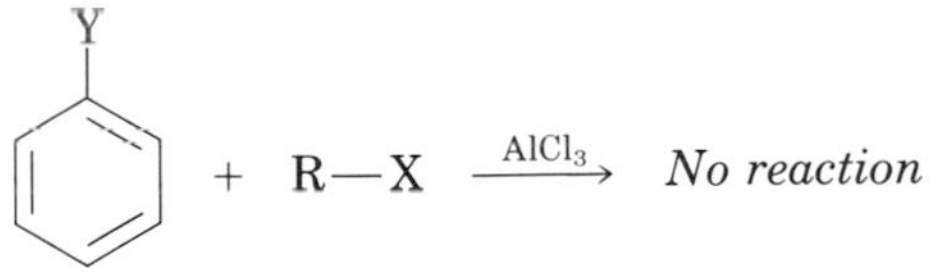

$$+ \text{R—X} \xrightarrow{AlCl_3} \textit{No reaction}$$

where $Y = -\overset{+}{N}R_3, -NO_2, -CN, -SO_3H, -CHO,$

$-COCH_3, -COOH, -COOCH_3$

$(-NH_2, -NHR, -NR_2)$

**Figure 16.9** Limitations on the aromatic substrate in Friedel–Crafts reactions

Yet a third fundamental limitation of the Friedel–Crafts alkylation is that it is often difficult to stop the reaction after a single substitution because the product is often more reactive than the starting material. Once the first group is on the ring, a second substitution reaction is facilitated for reasons we'll discuss in the next section. Thus, we often observe **polyalkylation**. For example, as shown at the top of the next page, reaction of benzene with 1 mol equiv of 2-chloro-2-methylpropane yields *p*-di-*tert*-butylbenzene as the major product, along with a small amount of *tert*-butylbenzene and unreacted starting material. High yields of monoalkylation product are obtained only when a large excess of benzene is used.

$C_6H_6$ + $(CH_3)_3CCl$ $\xrightarrow{AlCl_3}$ p-$C_6H_4[C(CH_3)_3]_2$ + $C_6H_5C(CH_3)_3$ + $C_6H_6$

Major product — Minor product

A final limitation to the Friedel–Crafts reaction is that skeletal *rearrangements* of the alkyl group sometimes occur during reaction, particularly when primary alkyl halides are used. The amount of rearrangement is variable and depends on catalyst, reaction temperature, and even reaction solvent. Thus, less rearrangement is usually found at lower reaction temperatures, but mixtures of products are often obtained. For example, treatment of benzene with 1-chlorobutane gives an approximately 2:1 ratio of rearranged (*sec*-butyl) to unrearranged (*n*-butyl) products when the reaction is carried out at 0°C using $AlCl_3$ as catalyst.

$C_6H_6$ + $CH_3CH_2CH_2CH_2Cl$ $\xrightarrow[0°C]{AlCl_3}$ $C_6H_5CH(CH_3)CH_2CH_3$ + $C_6H_5CH_2CH_2CH_2CH_3$

Benzene — 1-Chlorobutane — *sec*-Butylbenzene (~65%) — Butylbenzene (~35%)

These isomerizations take place by carbocation rearrangements of exactly the same sort we saw earlier during electrophilic additions to alkenes (Section 6.13). For example, the relatively unstable primary butyl carbocation produced by reaction of 1-chlorobutane with $AlCl_3$ rearranges to a more stable secondary carbocation by the shift of a hydrogen atom and its electron pair (a **hydride ion, H:⁻**) from C2 to C1:

$$CH_3CH_2CH(H)\overset{+}{C}H_2 \xrightarrow{\text{Hydride shift}} CH_3CH_2\overset{+}{C}HCH_2(H)$$

1° cation (less stable) — 2° cation (more stable)

Similarly, carbocation rearrangements can occur by *alkyl* shifts. For example, Friedel–Crafts alkylation of benzene with 1-chloro-2,2-dimethylpropane yields (1,1-dimethylpropyl)benzene as the sole product. The initially formed primary carbocation rearranges to a tertiary carbocation by shift of a methyl group and its electron pair from C2 to C1 (Figure 16.10).

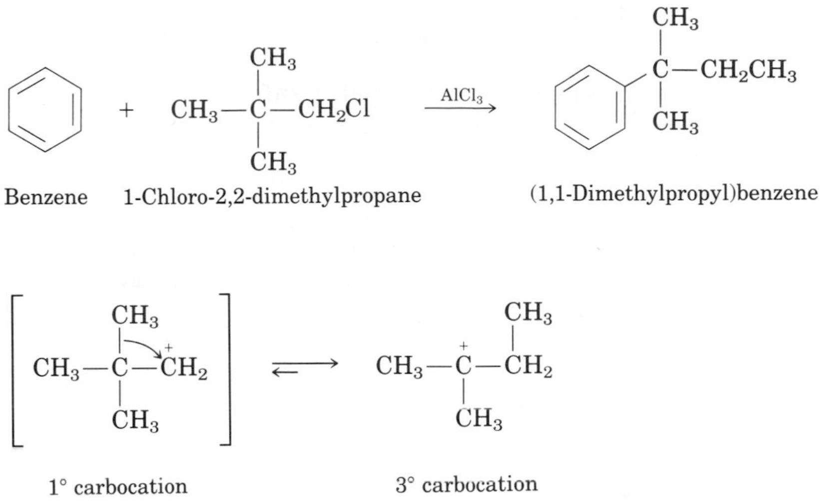

$$\text{Benzene} + CH_3-\underset{CH_3}{\overset{CH_3}{C}}-CH_2Cl \xrightarrow{AlCl_3} C_6H_5-\underset{CH_3}{\overset{CH_3}{C}}-CH_2CH_3$$

Benzene 1-Chloro-2,2-dimethylpropane (1,1-Dimethylpropyl)benzene

$$\left[ CH_3-\underset{CH_3}{\overset{CH_3}{C}}-\overset{+}{C}H_2 \right] \longrightarrow CH_3-\underset{CH_3}{\overset{+}{C}}-CH_2CH_3$$

1° carbocation 3° carbocation

**Figure 16.10** Rearrangement of a primary to a tertiary carbocation during Friedel–Crafts reaction of benzene with 1-chloro-2,2-dimethylpropane

PROBLEM

**16.4** What is the major monosubstitution product that you would expect to obtain from the Friedel–Crafts reaction of benzene and 1-chloro-2-methylpropane in the presence of $AlCl_3$?

PROBLEM

**16.5** The carbocation electrophile in a Friedel–Crafts reaction can be generated in ways other than by reaction of an alkyl chloride with $AlCl_3$. For example, reaction of benzene with 2-methylpropene in the presence of $H_3PO_4$ yields *tert*-butylbenzene. Formulate a mechanism for this reaction.

PROBLEM

**16.6** Which of the following alkyl halides would you expect to undergo Friedel–Crafts reaction *without* rearrangement? Explain.
(a) $CH_3CH_2Cl$ (b) $CH_3CH_2CH(Cl)CH_3$ (c) $CH_3CH_2CH_2Cl$
(d) $(CH_3)_3CCH_2Cl$ (e) Chlorocyclohexane

## 16.4 Acylation of Aromatic Rings

An **acyl** group, —COR (pronounced **ay**-sil), is introduced onto the ring when an aromatic compound is allowed to react with a carboxylic acid chloride, RCOCl, in the presence of $AlCl_3$. For example, reaction of benzene with acetyl chloride yields the ketone, acetophenone.

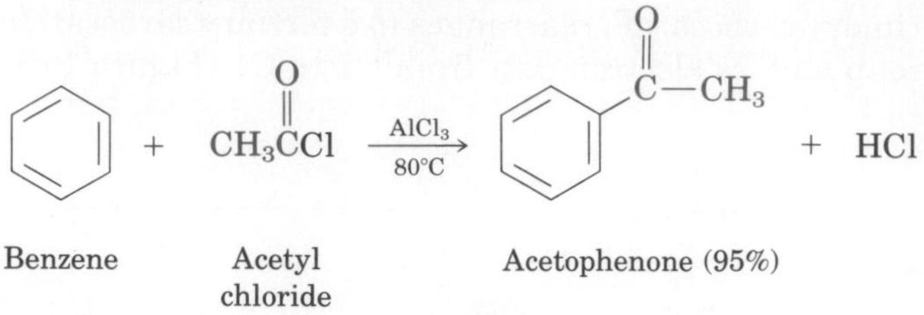

$$\text{Benzene} + CH_3CCl(=O) \xrightarrow[80^\circ C]{AlCl_3} \text{Acetophenone (95\%)} + HCl$$

Benzene | Acetyl chloride | Acetophenone (95%)

The mechanism of **Friedel–Crafts acylation** is similar to that of Friedel–Crafts alkylation. The reactive electrophile is a resonance-stabilized **acyl cation**, generated by reaction between the acyl chloride and $AlCl_3$ (Figure 16.11). An acyl cation is stabilized by overlap of the vacant orbital on carbon with a lone-pair orbital of the neighboring oxygen, as the resonance structures in Figure 16.11 indicate. Once formed, the acyl cation does not rearrange; rather, it is attacked by an aromatic ring to give unrearranged substitution product.

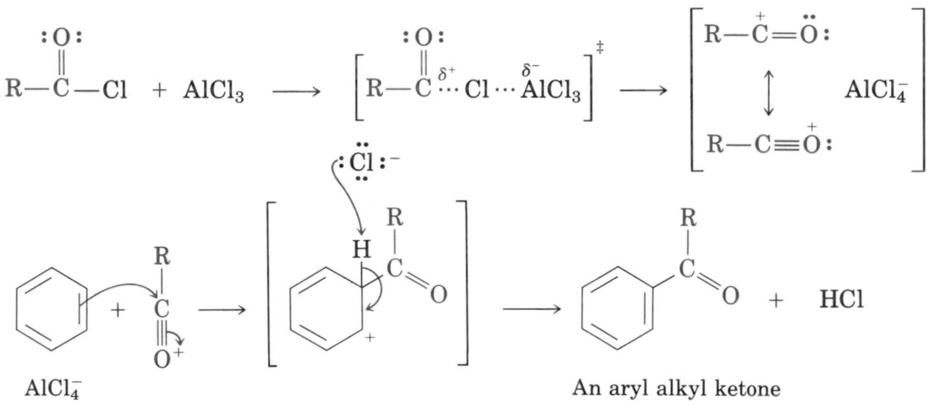

$AlCl_4^-$

An aryl alkyl ketone

**Figure 16.11** Mechanism of the Friedel–Crafts acylation reaction

Unlike the multiple substitutions that often occur in Friedel–Crafts alkylations, acylations never occur more than once on a ring, because the product acylbenzene is always less reactive than the nonacylated starting material. We'll account for these reactivity differences in the next section.

## 16.5 Reactivity of Aromatic Rings

Electrophilic substitution on benzene itself is straightforward. But what happens when we carry out a reaction on a ring that already has a substituent on it? Experiments show that substituents already present on the ring often have a profound effect on reactivity.

Substituents can be classified into two groups: those that *activate* the aromatic ring toward further electrophilic substitution, and those that *deactivate* it. Rings that contain an activating substituent are more reactive than benzene, whereas those with a deactivating substituent are less reactive than benzene. Figure 16.12 shows some groups in both categories.

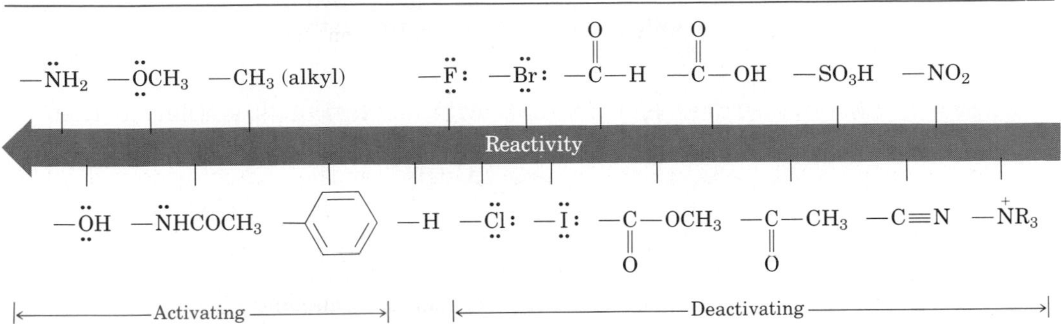


**Figure 16.12** Activating and deactivating substituents for electrophilic aromatic substitution

The common feature of all substituents within a category is that all activating groups are able to *donate* electrons to the ring, whereas all deactivating groups *withdraw* electrons from the ring. An aromatic ring with an electron-donating substituent is more electron-rich than benzene and therefore more reactive toward electrophiles; an aromatic ring with an electron-withdrawing substituent is less electron-rich than benzene and therefore less reactive toward electrophiles.

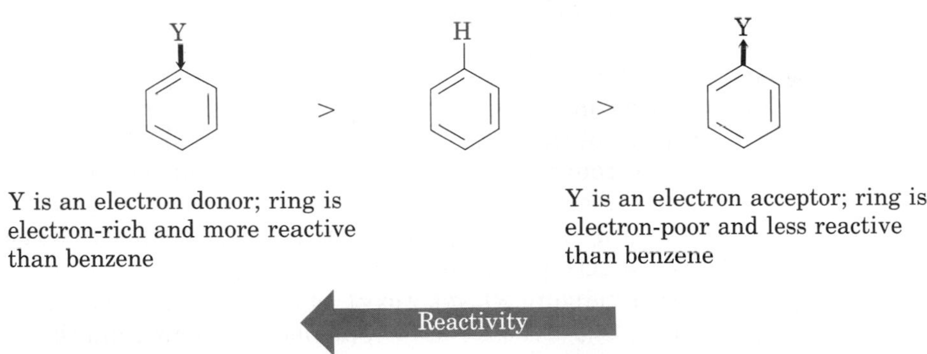


There are two ways by which a substituent group can donate or withdraw electrons from the aromatic ring—by inductive effects and by resonance effects. As we've seen previously (Sections 2.4 and 6.11), **inductive effects** are due to the intrinsic electronegativity of atoms and to bond polarity in functional groups. These effects operate by donating or withdrawing electrons through sigma bonds. For example, halogens, carbonyl groups, cyano groups, and nitro groups deactivate an aromatic ring by inductively withdrawing electrons from the ring through the sigma bond linking the substituent to the ring.

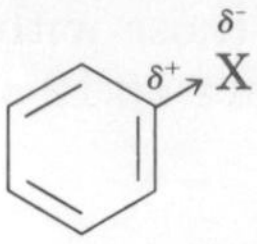

$\delta^-$ X, $\delta^+$

X = F, Cl, Br, or I; inductively electron-withdrawing because of electronegativity

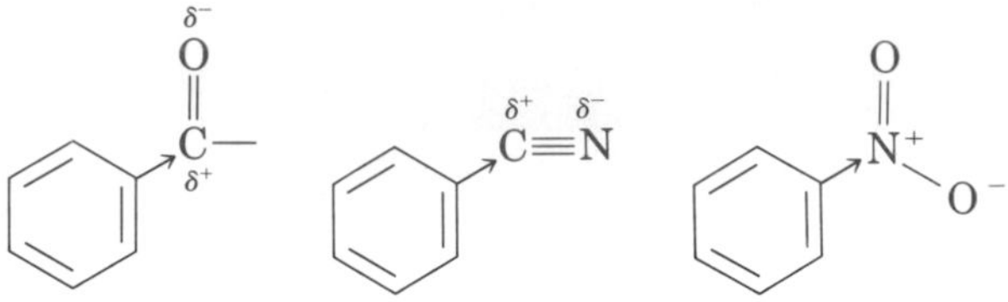

Carbonyl, cyano, nitro; inductively electron-withdrawing because of functional-group polarity

Alkyl groups are inductively electron-donating, and therefore activate the ring. The reasons for this are not fully understood but probably involve the same factors that cause alkyl substituents to stabilize alkenes (Section 6.7) and carbocations (Section 6.11).

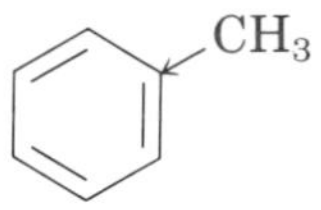

$CH_3$

Alkyl group; inductively electron-donating

**Resonance effects** operate by donating or withdrawing electrons through pi bonds when the substituent is connected to the aromatic ring by an atom that has a *p* orbital. For example, nitro, cyano, and carbonyl substituents are bonded to the aromatic ring through atoms that have *p* orbitals; the pi electrons of the aromatic ring can therefore be delocalized onto these substituents through *p* orbital overlap. As the resonance structures in Figure 16.13 indicate, electrons flow from the nitrobenzene and benzaldehyde rings onto the substituents, leaving a positive charge in the rings and deactivating them toward electrophilic attack.

Conversely, hydroxyl, methoxyl, and amino substituents *activate* the aromatic ring by resonance effects; pi electrons flow from the substituents to the ring as shown by the resonance structures in Figure 16.14. (Recall the discussion of conjugation in Section 14.9 where a similar effect was noted.)

It probably comes as a surprise to hear that hydroxyl, methoxyl, and amino groups activate an aromatic ring. After all, both oxygen and nitrogen are highly electronegative and would be expected to inductively *deactivate* a ring in the same way that halogens do. In fact, these groups *do* have an electron-withdrawing inductive effect. Nevertheless, the electron-*donating* resonance effect through pi orbitals is far stronger than the electron-withdrawing inductive effect through sigma orbitals, and these substituents therefore behave as ring activators.

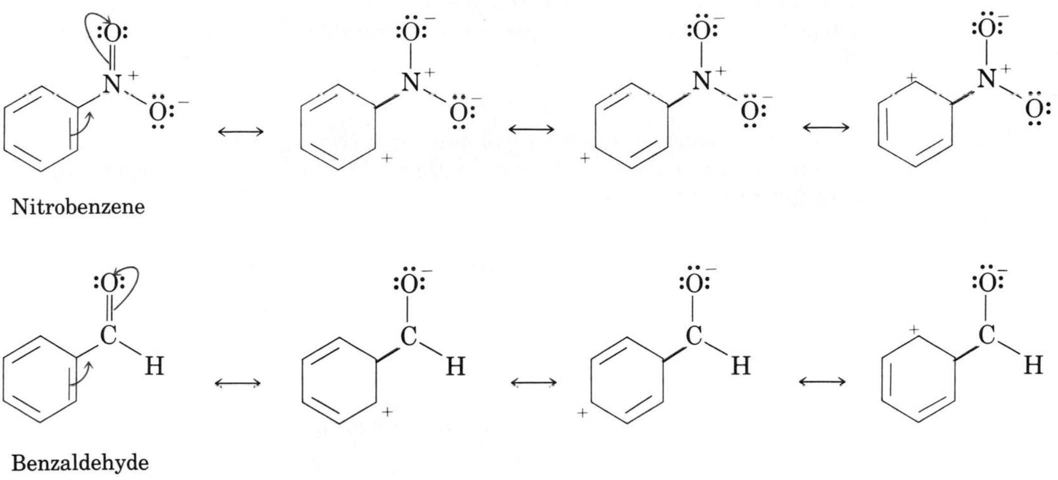

Nitrobenzene

Benzaldehyde

**Figure 16.13** Carbonyl, nitro, and similar substituents deactivate aromatic rings by resonance withdrawal of pi electrons. Note that the effect is felt most strongly at ortho and para positions.

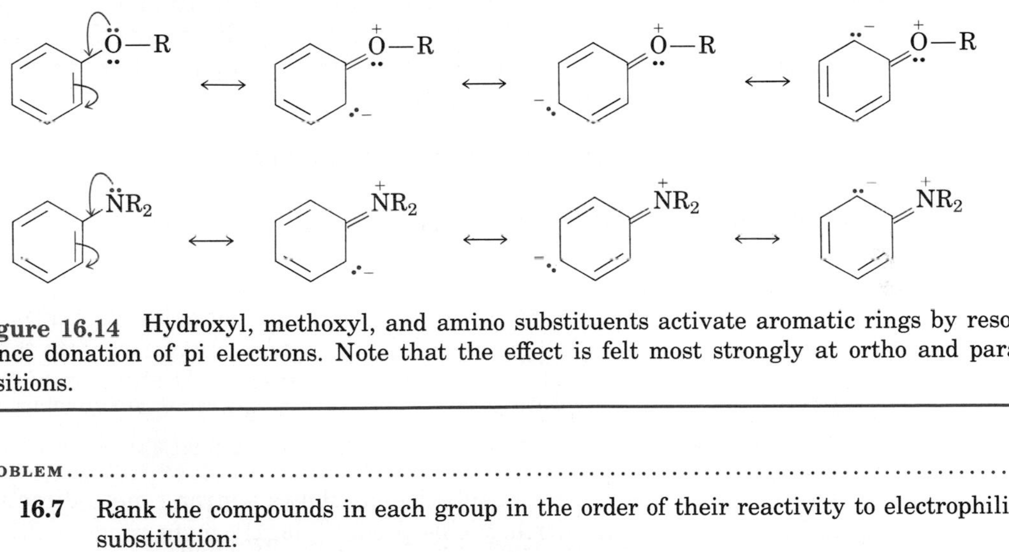

**Figure 16.14** Hydroxyl, methoxyl, and amino substituents activate aromatic rings by resonance donation of pi electrons. Note that the effect is felt most strongly at ortho and para positions.

PROBLEM

**16.7** Rank the compounds in each group in the order of their reactivity to electrophilic substitution:

(a) Nitrobenzene, phenol, toluene, benzene
(b) Phenol, benzene, chlorobenzene, benzoic acid
(c) Benzene, bromobenzene, benzaldehyde, aniline

PROBLEM

**16.8** Use Figure 16.12 to account for the fact that Friedel–Crafts *alkylations* often give polysubstitution but Friedel–Crafts *acylations* never give polysubstitution.

PROBLEM

**16.9** Write as many resonance forms as you can for the carbocation intermediates in these reactions. Be sure to consider the resonance effect of the hydroxyl substituent.

(a) Bromination of phenol at the para position

(b) Bromination of phenol at the meta position
(c) Bromination of phenol at the ortho position
At which position(s) does reaction appear most favorable? Which position(s) are least favored?

PROBLEM

**16.10** Write as many resonance forms as you can for nitration of the *N,N,N*-trimethylanilinium ion at ortho, meta, and para positions. At which position(s) does reaction look more favorable? Explain.

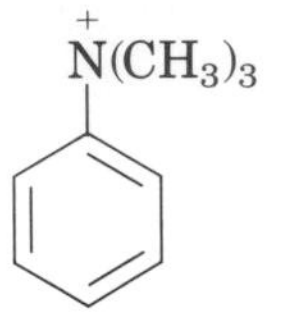

$\overset{+}{N}(CH_3)_3$

*N,N,N*-Trimethylanilinium ion

## 16.6 Orientation of Reactions on Substituted Aromatic Rings

In addition to affecting the reactivity of an aromatic ring, a substituent can also influence the position of further electrophilic substitution. For example, a methyl substituent shows a strong ortho- and para-directing effect. Nitration of toluene yields predominantly ortho (63%) and para (34%) products, along with only 3% of the meta isomer.

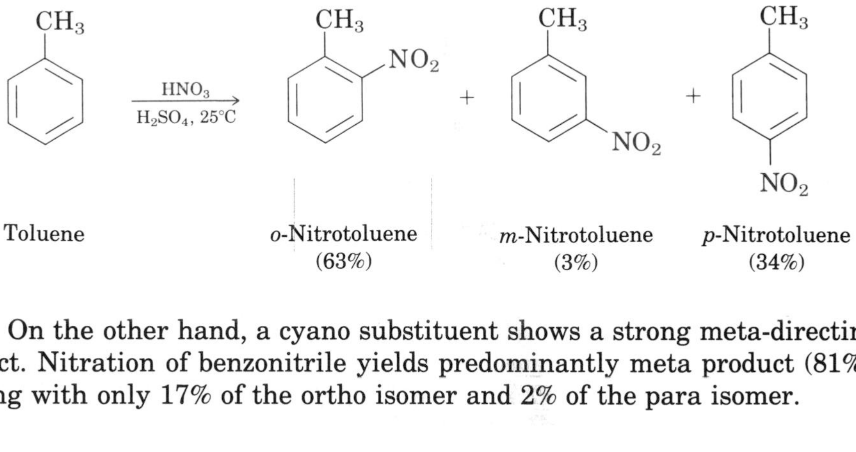

$CH_3$ → ($HNO_3$, $H_2SO_4$, 25°C) → $CH_3$, $NO_2$ + $CH_3$, $NO_2$ + $CH_3$, $NO_2$

Toluene — *o*-Nitrotoluene (63%) — *m*-Nitrotoluene (3%) — *p*-Nitrotoluene (34%)

On the other hand, a cyano substituent shows a strong meta-directing effect. Nitration of benzonitrile yields predominantly meta product (81%), along with only 17% of the ortho isomer and 2% of the para isomer.

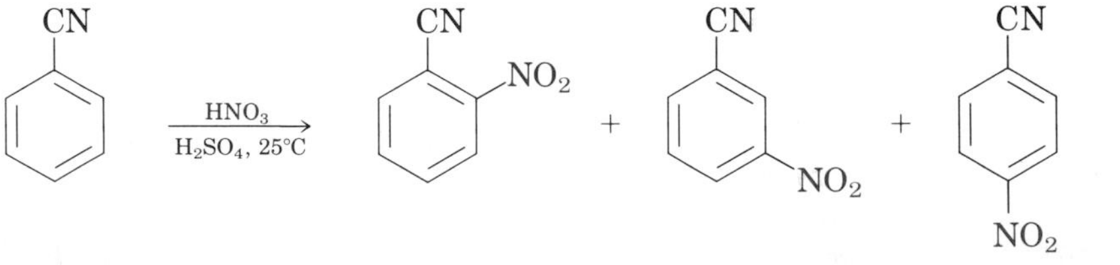

CN → ($HNO_3$, $H_2SO_4$, 25°C) → CN, $NO_2$ + CN, $NO_2$ + CN, $NO_2$

Benzonitrile — *o*-Nitrobenzonitrile (17%) — *m*-Nitrobenzonitrile (81%) — *p*-Nitrobenzonitrile (2%)

Substituents can be classified into three groups: ortho- and para-directing activators; ortho- and para-directing deactivators; and meta-directing deactivators. No meta-directing activators are known. Figure 16.15 shows some of the groups in each category, and Table 16.2 lists experimental results of the orientation of nitration in substituted benzenes. Compare Figure 16.15 with Figure 16.12 and notice how the groups fall into the three categories depending on their reactivities.

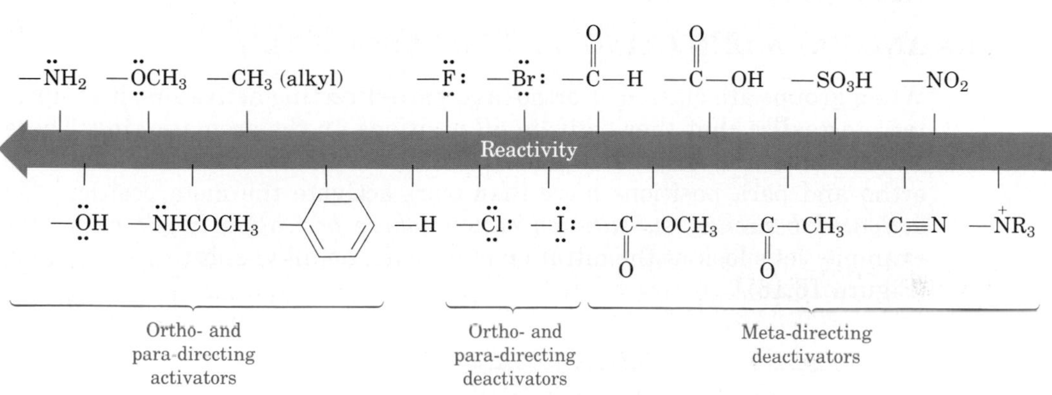


**Figure 16.15** Classification of directing effects for substituents

**Table 16.2** Orientation of nitration in substituted benzenes:

$$C_6H_5X \xrightarrow[H_2SO_4,\ 25^\circ C]{HNO_3} X\text{-}C_6H_4\text{-}NO_2$$

| X | Product (%) | | | X | Product (%) | | |
|---|---|---|---|---|---|---|---|
| | *Ortho* | *Meta* | *Para* | | *Ortho* | *Meta* | *Para* |
| **Meta-directing deactivators** | | | | **Ortho- and para-directing deactivators** | | | |
| $—\overset{+}{N}(CH_3)_3$ | 2 | 89 | 11 | —F | 13 | 1 | 86 |
| $—NO_2$ | 7 | 91 | 2 | —Cl | 35 | 1 | 64 |
| —COOH | 22 | 77 | 2 | —Br | 43 | 1 | 56 |
| —CN | 17 | 81 | 2 | —I | 45 | 1 | 54 |
| $—CO_2CH_2CH_3$ | 28 | 66 | 6 | **Ortho- and para-directing activators** | | | |
| $—COCH_3$ | 26 | 72 | 2 | $—CH_3$ | 63 | 3 | 34 |
| —CHO | 19 | 72 | 9 | $—\ddot{O}H$ | 50 | 0 | 50 |
| | | | | $—\ddot{N}HCOCH_3$ | 19 | 2 | 79 |

We can understand how substituents exert their directing influence by looking further at the consequences of resonance effects and inductive effects.

PROBLEM

16.11 Predict the major products of these reactions:
(a) Nitration of bromobenzene (b) Bromination of nitrobenzene
(c) Chlorination of phenol
(d) Nitration of bromobenzene, followed by $SnCl_2$ reduction
(e) Bromination of aniline

## ORTHO- AND PARA-DIRECTING ACTIVATORS: ALKYL

When groups are classed as ortho- and para-directing activators, it's important to realize that they activate *all* positions on the aromatic ring. These substituents are ortho- and para-directing only because they activate the ortho and para positions more than they activate the meta positions. *In activated benzenes, substitution occurs where activation is felt most.* For example, let's look at the nitration of toluene, an alkyl-substituted benzene (Figure 16.16).

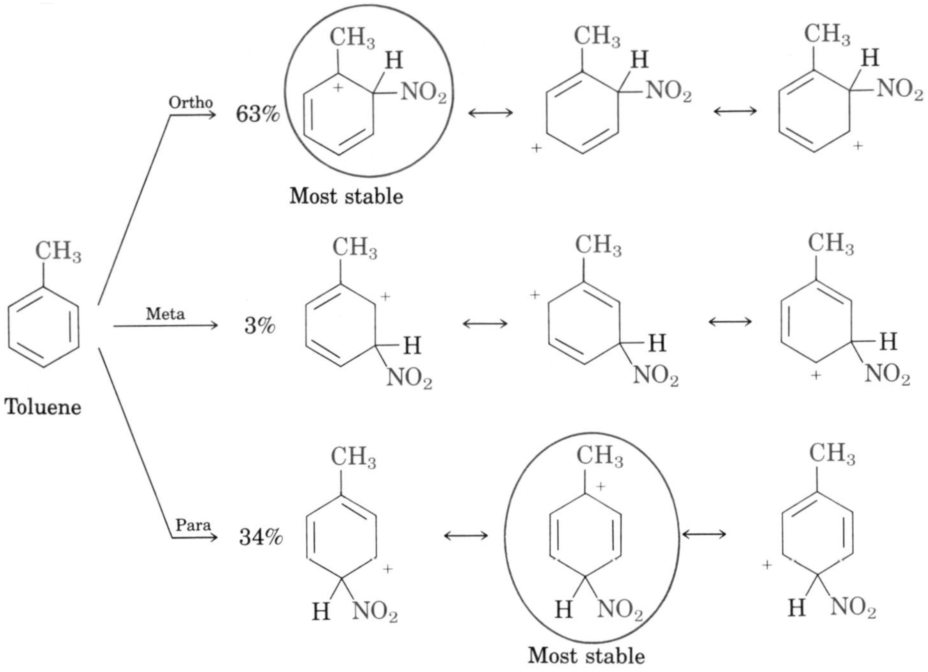

Toluene

Ortho 63%

Most stable

Meta 3%

Para 34%

Most stable

**Figure 16.16** Carbocation intermediates in the nitration of toluene: Ortho and para intermediates are more stable than meta intermediates.

Nitration of toluene at any one of the three positions leads to a resonance-stabilized carbocation intermediate, but the ortho and para intermediates are most stabilized. For both ortho and para (but not for meta) attack, a resonance form places the positive charge directly on the methyl-substituted carbon (a tertiary carbocation), where it can best be stabilized by the methyl inductive effect. The ortho and para intermediates are lower in energy than the meta intermediate and are therefore formed preferentially, as indicated by the reaction energy diagram in Figure 16.17.

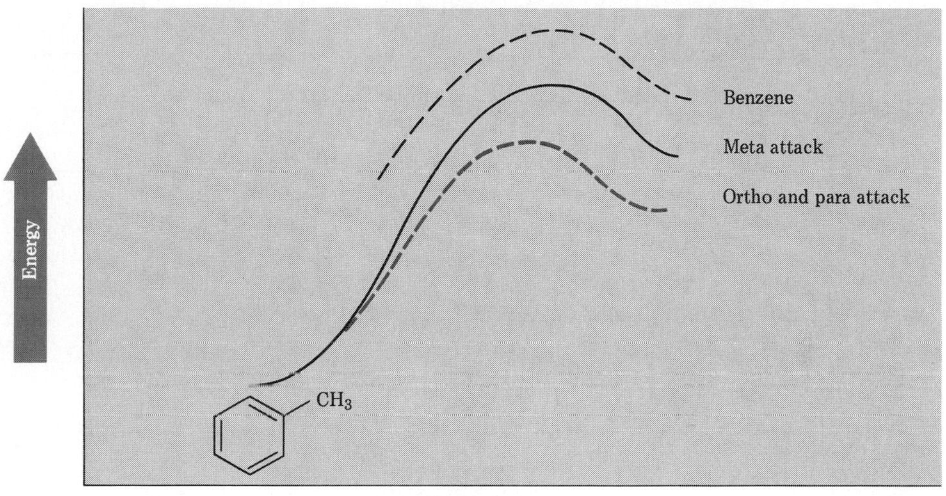


**Figure 16.17** Reaction energy diagram for the nitration of toluene: The carbocation intermediates formed by attack at ortho and para positions are more stable and are therefore formed more rapidly than the intermediate formed by attack at the meta position.

PROBLEM

**16.12** Which compound would you expect to be more reactive toward electrophilic substitution, toluene or (trifluoromethyl)benzene? Explain.

$CF_3$

(Trifluoromethyl)benzene

## ORTHO- AND PARA-DIRECTING ACTIVATORS: OH AND $NH_2$

Hydroxyl and amino groups (and their derivatives) are also ortho–para activators, but for a different reason than for alkyl groups. Hydroxyl and amino groups exert their activating influence through a strong resonance electron-donating effect, which is most pronounced at the ortho and para positions. The resonance forms of phenol shown previously in Figure 16.14 indicate how the oxygen lone-pair electrons can be shared only in the ortho and para positions, but not in the meta position.

When phenol is nitrated, the results shown in Figure 16.18 are obtained. Although three products are possible, only ortho and para attack is observed. All three of the possible carbocation intermediates are stabilized by resonance, but the intermediates from ortho and para attack are stabilized most. Only in ortho and para attack are there resonance forms in which the positive charge is stabilized by donation of an electron pair from oxygen. The product of meta attack has no such stabilization.

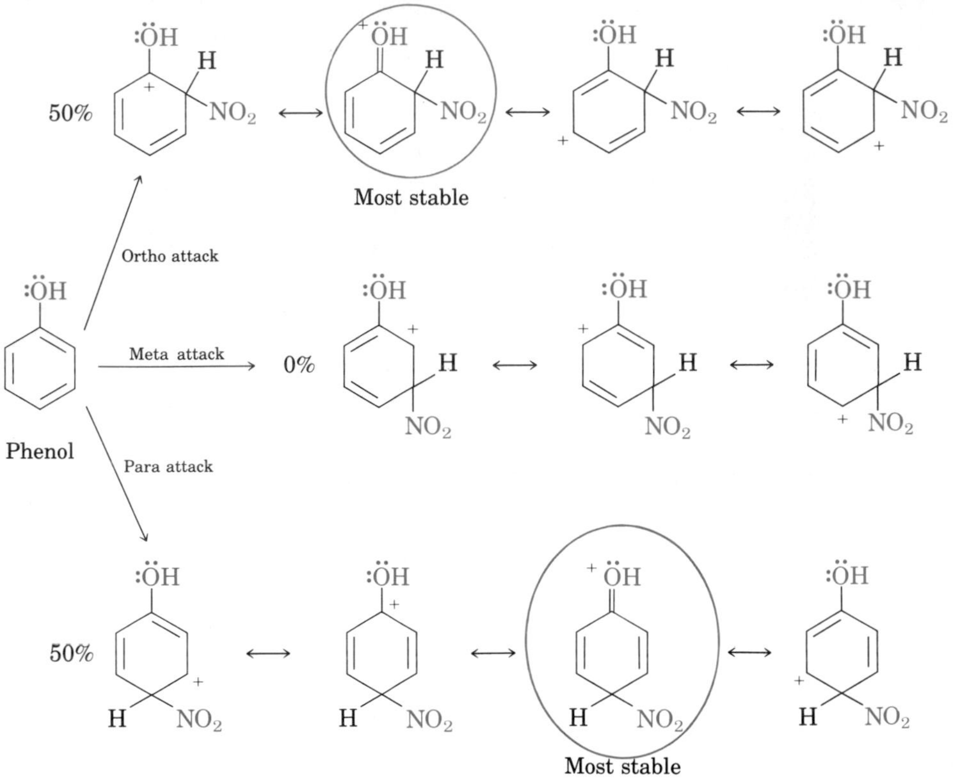

50% Ortho attack — Most stable

Phenol — Meta attack 0%

Para attack 50% — Most stable

**Figure 16.18** Carbocation intermediates in the nitration of phenol: The ortho and para intermediates are more stable than the meta intermediate because of resonance donation of the oxygen electrons.

PROBLEM

**16.13** Acetanilide is much less reactive toward electrophilic substitution than aniline. Explain.

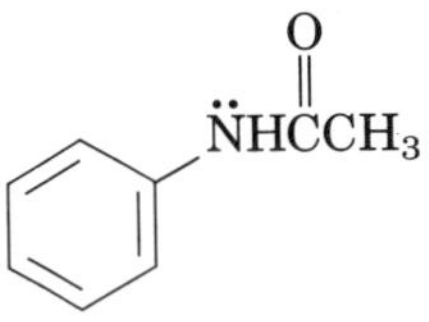

Acetanilide

PROBLEM

**16.14** What reactivity order toward electrophilic substitution would you expect for phenol, phenoxide ion, and phenyl acetate? Explain.

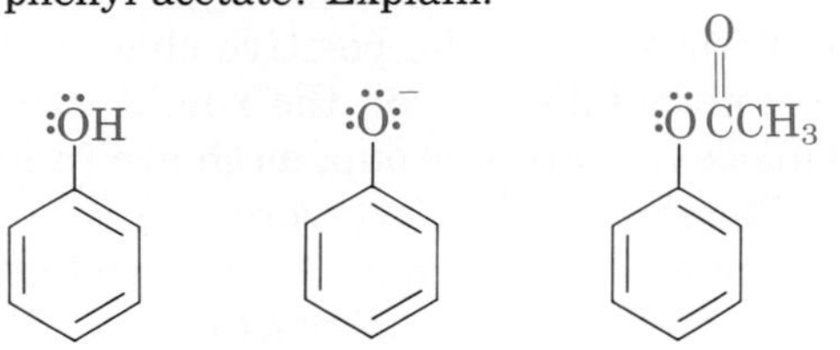

Phenol   Phenoxide ion   Phenyl acetate

## META-DIRECTING DEACTIVATORS

When groups such as carbonyl, nitro, and cyano are classified as deactivators, it's important to realize that these groups deactivate *all* positions on the ring. Meta directors exert their directing influence because they deactivate the meta position less than they deactivate the ortho and para positions. *In deactivated benzenes, substitution occurs at the least deactivated position.*

We can explain the influence of meta-directing deactivators by the same kinds of arguments just used for ortho- and para-directing activators. For example, let's look at the chlorination of benzaldehyde, shown in Figure 16.19. Of the three possible intermediates, the carbocations produced by reaction at ortho and para positions are least stable, and the cation produced by reaction at the meta position is most stable.

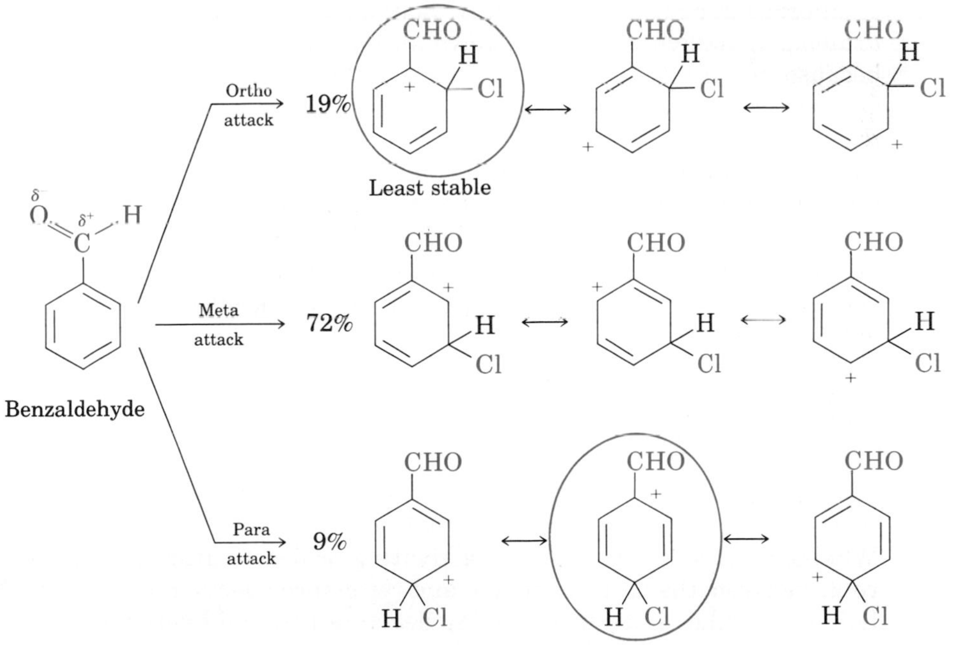

Ortho attack 19%   Least stable

Benzaldehyde   Meta attack 72%

Para attack 9%   Least stable

**Figure 16.19** Intermediates in the chlorination of benzaldehyde: The meta intermediate is more stable than the ortho or para intermediate.

Meta-directing deactivators such as carbonyl, nitro, and cyano exert their influence through a combination of inductive and resonance effects that reinforce each other. Inductively, both ortho and para intermediates are destabilized because the positive charge of the carbocationic intermediate can be placed directly on the ring carbon atom bearing the electron-withdrawing deactivating group, as shown in Figure 16.19. The meta intermediate is therefore the most favored.

Resonance electron withdrawal is also felt at the ortho and para positions, rather than at the meta position, as the dipolar resonance forms of benzaldehyde shown previously in Figure 16.13 indicate. Reaction with an electrophile therefore occurs at the meta position.

PROBLEM

**16.15** Draw resonance structures for the intermediates from attack of an electrophile at the ortho, meta, and para positions of nitrobenzene. Which intermediates are most favored?

## ORTHO- AND PARA-DIRECTING DEACTIVATORS: HALOGENS

Halogen substituents occupy a unique position since they are deactivating yet have an ortho- and para-directing effect. Why should this be?

As we saw in the previous section, halogen substituents are deactivating because of their strong electron-withdrawing inductive effect. Unlike other deactivating groups, however, halogens deactivate the ortho and para positions *less* than they deactivate the meta position. The reasons for this behavior can be seen by looking at the possible intermediates formed on nitration of chlorobenzene (Figure 16.20). Although halogens have an electron-*withdrawing* inductive effect, they have an electron-*donating* resonance effect because of *p* orbital overlap of lone-pair electrons on halogen with the pi orbitals of the aromatic ring. Thus, the halogen substituent can stabilize the positive charge of the carbocation intermediates from ortho and para attack in the same way that hydroxyl and amino substituents can. The meta intermediate, however, has no such stabilization.

In general, any substituent that has a lone pair of electrons on the atom bound to the aromatic ring has an electron-donating resonance effect and is thus an ortho–para director. Halogens, hydroxyl groups, and amino groups are therefore all ortho–para directors:

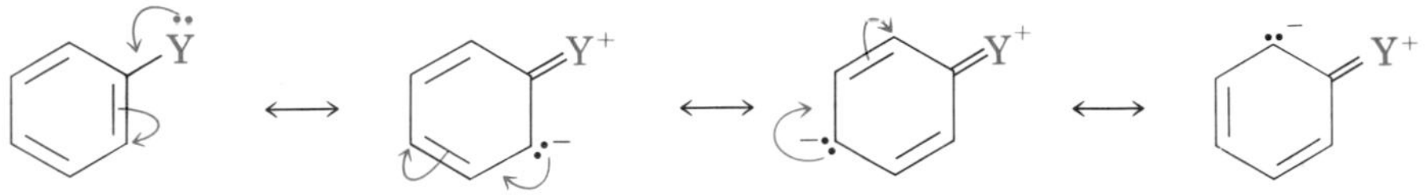

Y ⟷ $Y^+$ ⟷ $Y^+$ ⟷ $Y^+$

Whether a given substituent is activating or deactivating depends on the relative strengths of its inductive and resonance effects. For example, both halogens (ortho- and para-directing deactivators) and hydroxyl (ortho- and para-directing activators) show the same two effects—inductive electron withdrawal and resonance electron donation—but the relative strengths of the effects differ in each case. Halogens have a strong electron-withdrawing

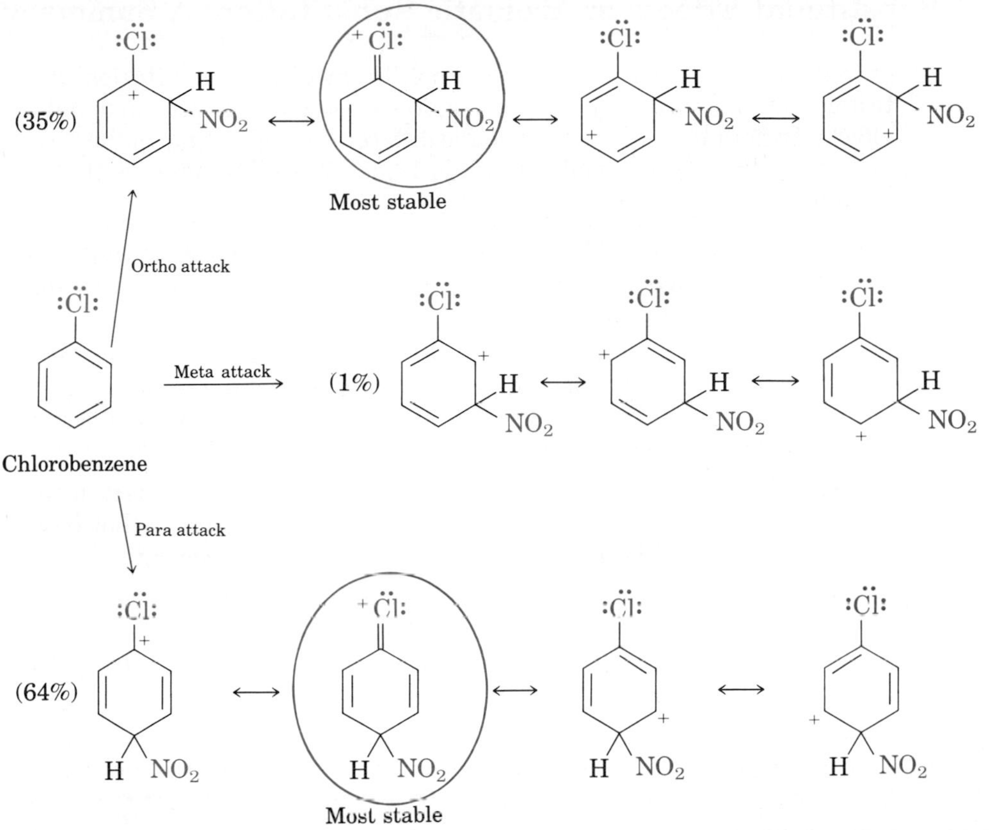

**Figure 16.20** Intermediates in the nitration of chlorobenzene: The ortho and para intermediates are more stable than the meta intermediate because of resonance electron donation of the halogen lone-pair electrons.

inductive effect but a weak electron-donating resonance effect and are thus deactivators. Hydroxyl and amino groups have weak electron-withdrawing inductive effects but have strong electron-donating resonance effects and are thus activators.

PROBLEM

**16.16** The nitroso group, —N=O, is one of the very few nonhalogens that is an ortho- and para-directing deactivating group. Draw resonance structures of intermediates in ortho and para electrophilic attack on nitrosobenzene, and explain why they are favored over the intermediate from meta attack.

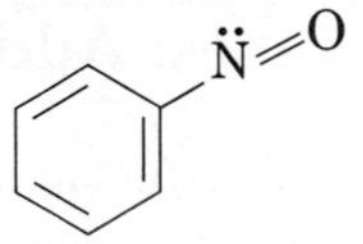

## 16.7 Substituent Effects in Aromatic Substitution: A Summary

Orientation and reactivity in electrophilic aromatic substitution are controlled by the interplay of two factors—resonance effects and inductive effects. Different substituents behave differently, depending on the direction and strength of the two effects (Table 16.3). We can summarize the results as follows:

1. *Alkyl groups* Moderate electron-donating inductive effect; no resonance effect. The net result is that alkyl groups are activating and ortho- and para-directing.
2. *Hydroxyl and amino groups (and derivatives)* Strong electron-donating resonance effect; moderate electron-withdrawing inductive effect. The net result is that these groups are activating and ortho- and para-directing.
3. *Halogens* Strong electron-withdrawing inductive effect; moderate electron-donating resonance effect. The net result is that halogens deactivate the ring but are ortho- and para-directing.
4. *Nitro, cyano, carbonyl, and similar groups* Strong electron-withdrawing resonance effect; strong electron-withdrawing inductive effect. The net result is that these groups are meta-directing and deactivating.

**Table 16.3** Substituent effects in electrophilic aromatic substitution

| *Substituent* | *Reactivity* | *Orientation* | *Inductive effect* | *Resonance effect* |
|---|---|---|---|---|
| $—CH_3$ | Activating | Ortho, para | Weak; electron-donating | None |
| $—\ddot{\underset{..}{O}}H$, $—\ddot{N}H_2$ | Activating | Ortho, para | Weak; electron-withdrawing | Strong; electron-donating |
| $—\ddot{\underset{..}{F}}:$, $—\ddot{\underset{..}{Cl}}:$, $—\ddot{\underset{..}{Br}}:$, $—\ddot{\underset{..}{I}}:$ | Deactivating | Ortho, para | Strong; electron-withdrawing | Weak; electron-donating |
| $—\overset{+}{N}(CH_3)_3$ | Deactivating | Meta | Strong; electron-withdrawing | None |
| $—NO_2$, $—CN$, $—CHO$, $—CO_2CH_3$, $—COCH_3$ | Deactivating | Meta | Strong; electron-withdrawing | Strong; electron-withdrawing |

## 16.8 Trisubstituted Benzenes: Additivity of Effects

Further electrophilic substitution of a disubstituted benzene is governed by the same resonance and inductive effects just discussed. The only difference is that now we must consider the *additive* effects of two different groups. In practice, this isn't as difficult as it sounds; three rules are usually sufficient to predict the results of a reaction:

1. If both groups direct substitution toward the same position, there is no problem. In *p*-nitrotoluene, for example, both the methyl and the nitro group direct further substitution to the same position (ortho to the methyl = meta to the nitro). A single product is thus formed during the reaction.

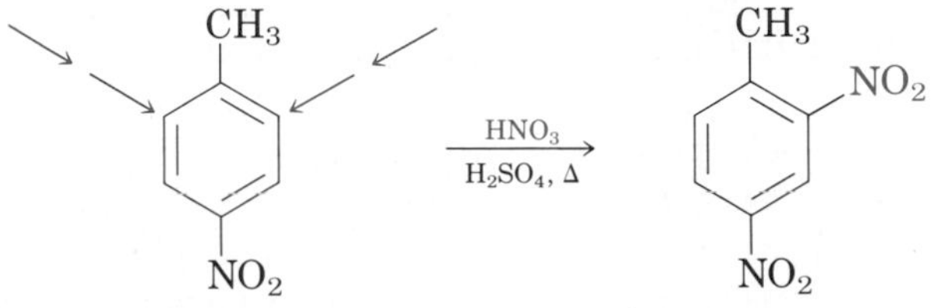

$CH_3$ $NO_2$ $\xrightarrow[H_2SO_4,\ \Delta]{HNO_3}$ $CH_3$ $NO_2$ $NO_2$

*p*-Nitrotoluene

2,4-Dinitrotoluene (sole product)

2. If the directing effects of the two groups *oppose* each other, the more powerful activating group has the dominant influence, but mixtures of products often result. For example, bromination of *p*-methylphenol yields largely 2-bromo-4-methylphenol, since hydroxyl is a more powerful activator than methyl.

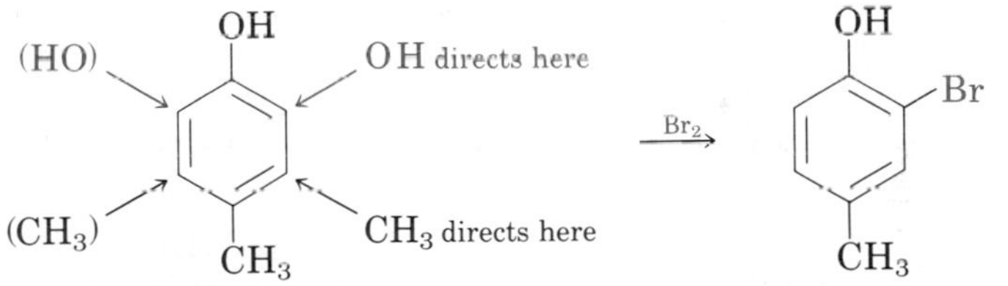

OH (HO) OH directs here ($CH_3$) $CH_3$ $CH_3$ directs here $\xrightarrow{Br_2}$ OH Br $CH_3$

*p*-Methylphenol (*p*-Cresol)

2-Bromo-4-methylphenol (major product)

3. Further substitution rarely occurs between the two groups in a meta-disubstituted compound because this site is too hindered for reaction to occur easily.

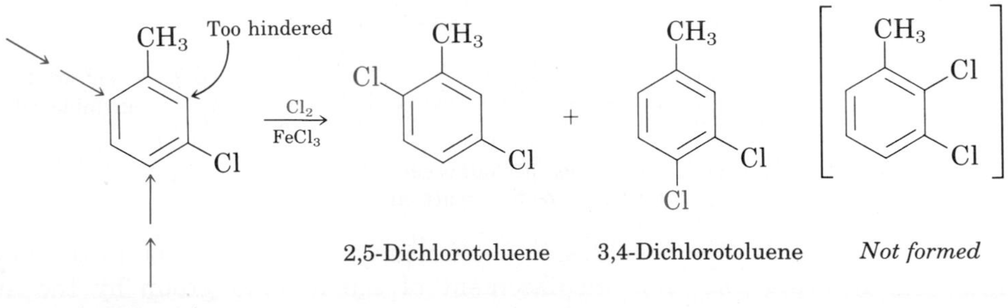

$CH_3$ Too hindered Cl $\xrightarrow[FeCl_3]{Cl_2}$ $CH_3$ Cl Cl + $CH_3$ Cl Cl [ $CH_3$ Cl Cl ]

*m*-Chlorotoluene

2,5-Dichlorotoluene

3,4-Dichlorotoluene

*Not formed*

PROBLEM

**16.17** Where would you expect electrophilic substitution to occur in these substances?

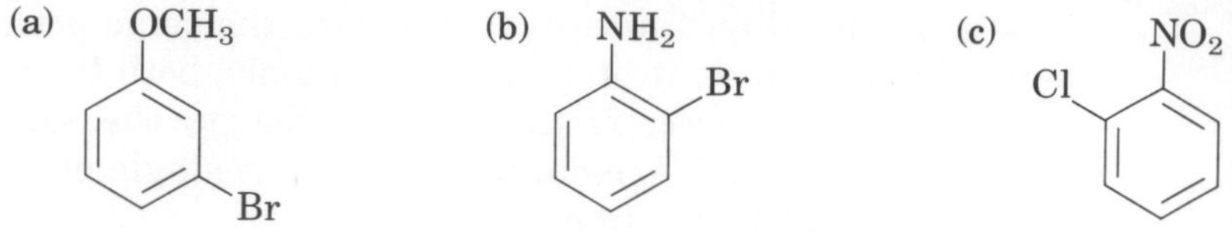

(a) $OCH_3$ / Br (b) $NH_2$ / Br (c) $NO_2$ / Cl

## 16.9 Nucleophilic Aromatic Substitution

The electrophilic substitutions just discussed are the most important and useful reactions of the aromatic ring. In certain cases, however, aromatic substitution can also occur by a nucleophilic mechanism. Nucleophilic substitutions aren't characteristic of *all* aromatic rings, though. Only aryl halides that have electron-withdrawing substituents undergo **nucleophilic aromatic substitution**. For example, 2,4,6-trinitrochlorobenzene reacts with aqueous sodium hydroxide at room temperature to give 2,4,6-trinitrophenol in 100% yield. The nucleophile, hydroxide ion, has substituted for chloride ion:

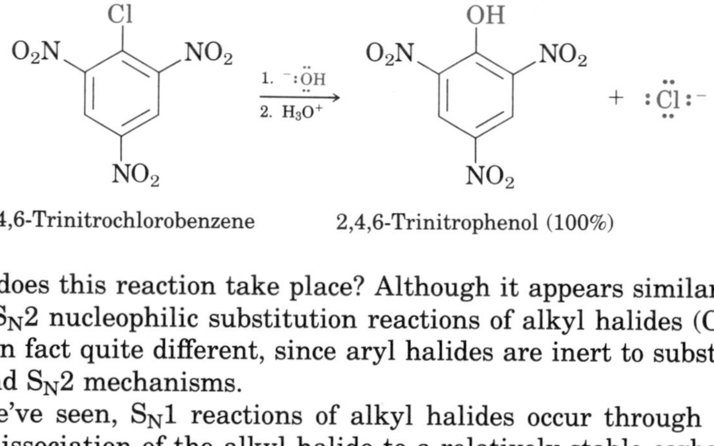

1. $^-$:ÖH 2. $H_3O^+$ → + :Cl:$^-$

2,4,6-Trinitrochlorobenzene 2,4,6-Trinitrophenol (100%)

How does this reaction take place? Although it appears similar to the $S_N1$ and $S_N2$ nucleophilic substitution reactions of alkyl halides (Chapter 11), it is in fact quite different, since aryl halides are inert to substitution by $S_N1$ and $S_N2$ mechanisms.

As we've seen, $S_N1$ reactions of alkyl halides occur through a rate-limiting dissociation of the alkyl halide to a relatively stable carbocation. Aryl halides, however, do not dissociate readily because aryl cations are relatively unstable in the same way that vinylic cations are unstable (Section 8.3). Thus aryl halides do not undergo $S_N1$ reactions.

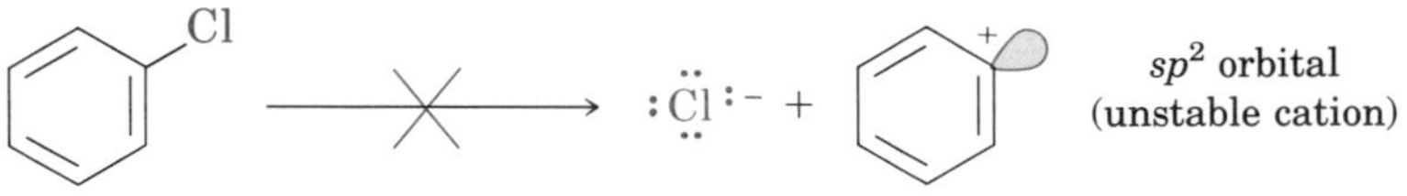

:Cl:$^-$ + $sp^2$ orbital (unstable cation)

Ionization does not occur; therefore, no $S_N1$ reaction

We've also seen that $S_N2$ reactions of alkyl halides occur through a rate-limiting back-side displacement of the leaving group by the attacking nucleophile. Aryl halides, however, are sterically shielded from back-side attack by the aromatic ring. In order for a nucleophile to attack an aryl halide, it would have to approach directly through the aromatic ring and invert the stereochemistry of the aromatic ring—a geometric impossibility. Nucleophilic aromatic substitutions must therefore occur by a different mechanism.

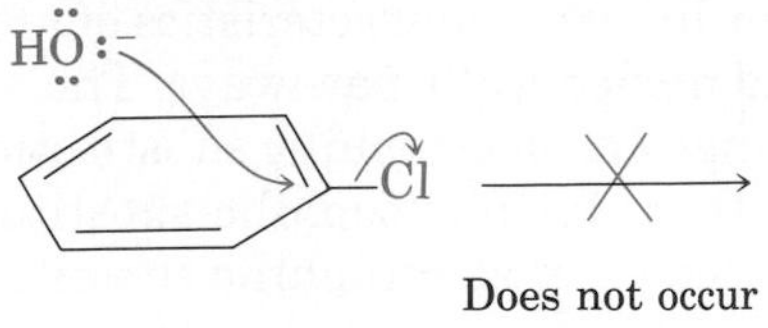

Does not occur

Nucleophilic aromatic substitutions proceed by the *addition–elimination* mechanism shown in Figure 16.21. The attacking nucleophile first *adds* to the electron-deficient aryl halide, forming a negatively charged intermediate (a **Meisenheimer[3] complex**), and halide ion is then *eliminated* in the second step.

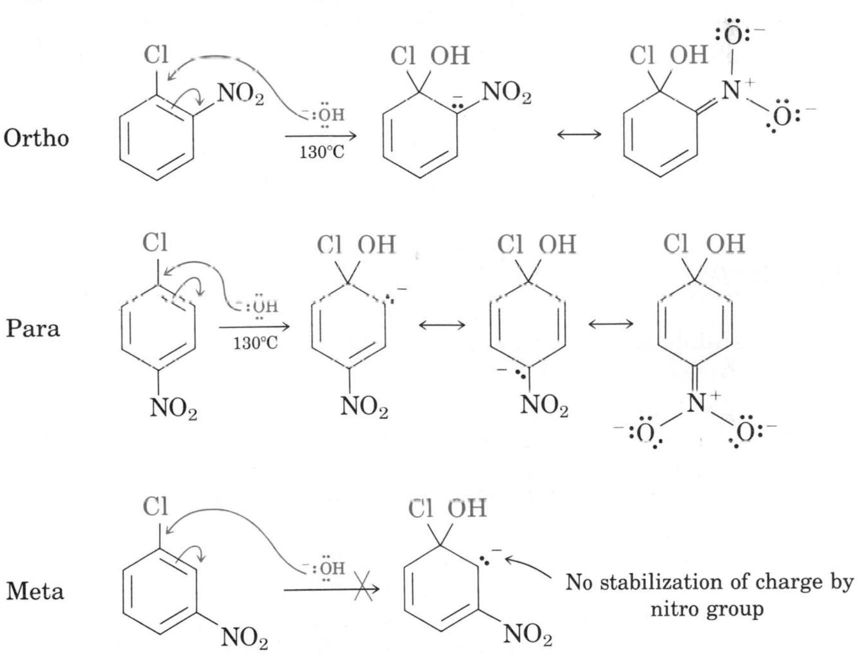

Ortho

Para

Meta

No stabilization of charge by nitro group

**Figure 16.21** Nucleophilic aromatic substitution on nitrochlorobenzenes: Only the ortho and para isomers undergo reaction.

Nucleophilic aromatic substitution occurs only if the halobenzene has electron-withdrawing substituents in the ortho and/or para positions; the more substituents there are, the faster the reaction goes. The reason for this requirement is that only ortho and para electron-withdrawing substituents can stabilize the anion intermediate through resonance. For example,

[3]Jacob Meisenheimer (1876–1934); b. Greisheim; Ph.D. Munich; professor, universities of Berlin, Greifswald, Tübingen.

*p*-chloronitrobenzene and *o*-chloronitrobenzene react with dilute hydroxide ion at 130°C to yield substitution products, but a meta substituent cannot offer resonance stabilization to the intermediate anion. *m*-Chloronitrobenzene is therefore inert to hydroxide ion (Figure 16.21).

Note the different characteristics of aromatic substitution through electrophilic and nucleophilic pathways. The electron-withdrawing groups that *deactivate* rings for electrophilic substitution (nitro, carbonyl, cyano, and so on) *activate* them for nucleophilic substitution. What's more, these groups are meta directors in electrophilic substitution, but ortho–para directors in nucleophilic substitution.

PROBLEM

**16.18** Propose a mechanism to account for the observation that 1-chloroanthraquinone reacts with methoxide ion to give the substitution product 1-methoxyanthraquinone.

1-Chloroanthraquinone $\xrightarrow{^-:\ddot{O}CH_3}$ 1-Methoxyanthraquinone + $:\ddot{Cl}:^-$

## 16.10 Benzyne

Halobenzenes without electron-withdrawing substituents are inert to nucleophiles under normal conditions. Under conditions of high temperature and pressure, however, even chlorobenzene can be forced to react. Scientists at the Dow Chemical Company announced in 1928 that phenol could be prepared on a large industrial scale by treatment of chlorobenzene with dilute aqueous sodium hydroxide at 340°C under 2500 psi (pounds per square inch) pressure.

Chlorobenzene $\xrightarrow[2.\ H_3O^+]{1.\ NaOH,\ H_2O,\ 340°C,\ 2500\ psi}$ Phenol + NaCl

This phenol synthesis is quite different from the other nucleophilic aromatic substitution reactions just studied. Experiments indicate that the reaction of chlorobenzene with hydroxide ion takes place by an **elimination–addition** mechanism. Strong base first causes the *elimination* of HX from halobenzene, yielding a highly reactive **benzyne** intermediate, and a nucleophile then *adds* to benzyne in a second step to yield the product. The two steps are the same as in other nucleophilic aromatic substitutions, but the order of steps is reversed (addition before elimination for the usual reaction versus elimination before addition for the benzyne reaction).

Chlorobenzene $\xrightarrow[\text{elimination}]{-HCl}$ [Benzyne] $\xrightarrow[\text{addition}]{H_2O}$ Phenol

Evidence in support of the benzyne mechanism has been obtained by studying the reaction between bromobenzene and the strong base potassium amide. When bromobenzene labeled with a radioactive $^{14}C$ carbon atom at the 1 position is used, the product has the label scrambled between positions 1 and 2. This result requires that the reaction proceed through a symmetrical intermediate in which positions 1 and 2 are equivalent—a requirement that only benzyne can meet:

Bromobenzene $\xrightarrow[NH_3\ (-HBr)]{:\ddot{N}H_2^-}$ Benzyne (symmetrical) $\xrightarrow{NH_3}$ 50% + 50% Aniline

Further evidence for a benzyne intermediate comes from trapping experiments. Although benzyne is far too reactive to be isolated as a pure compound, it can be intercepted as a Diels–Alder adduct if furan is added to the reaction. This is just the kind of behavior we would expect for so strained and reactive a species as benzyne.

Benzyne (a dienophile) + Furan (a diene) → Diels–Alder adduct

The electronic structure of benzyne, shown in Figure 16.22, can be compared to that of a highly distorted alkyne. Although a normal alkyne triple bond consists of a sigma bond formed by *sp*–*sp* overlap and two mutually perpendicular pi bonds formed by *p*–*p* overlap, the benzyne triple bond consists of a sigma bond formed by $sp^2$–$sp^2$ overlap, one pi bond formed by *p*–*p* overlap, and one pi bond formed by $sp^2$–$sp^2$ overlap. The latter pi bond is in the plane of the ring and is very weak because of poor orbital overlap.

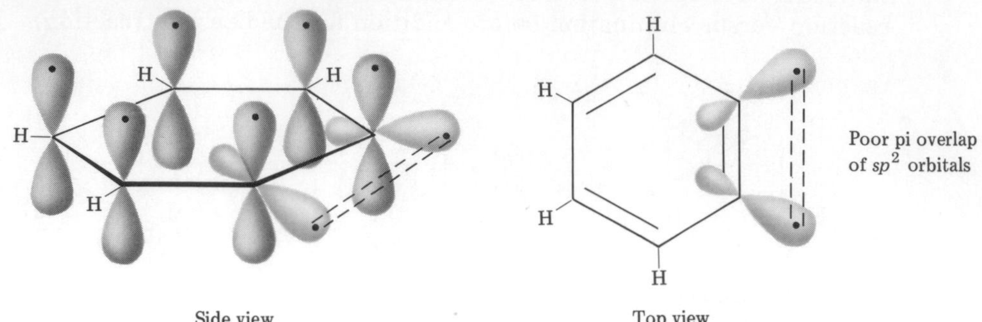


**Figure 16.22** An orbital picture of benzyne: The benzyne carbons are $sp^2$ hybridized, and the "third" bond results from weak overlap of two adjacent $sp^2$ orbitals.

PROBLEM

**16.19** Account for the fact that treatment of *p*-bromotoluene with NaOH at 300°C yields a mixture of two products, but treatment of *m*-bromotoluene with NaOH yields a mixture of three products.

## 16.11 Oxidation of Aromatic Compounds

### OXIDATION OF ALKYLBENZENE SIDE CHAINS

The benzene ring, despite its unsaturation, is normally inert to strong oxidizing agents such as potassium permanganate and sodium dichromate. (Recall that these reagents will cleave alkene carbon–carbon bonds; Section 7.8.) It turns out, however, that the presence of the aromatic ring has a dramatic effect on alkyl-group side chains. Alkyl side chains are readily attacked by oxidizing agents and are converted into carboxyl groups, —COOH. The net effect of side-chain oxidation is the conversion of an alkylbenzene into a benzoic acid, Ar—R → Ar—COOH. For example, both *p*-nitrotoluene and butylbenzene are oxidized by aqueous $KMnO_4$ in high yield to give the corresponding benzoic acids.

$CH_3$, $NO_2$ $\xrightarrow[H_2O,\ 95°C]{KMnO_4}$ $CO_2H$, $NO_2$

*p*-Nitrotoluene — *p*-Nitrobenzoic acid (88%)

$CH_2CH_2CH_2CH_3$ $\xrightarrow[H_2O]{KMnO_4}$ COOH

Butylbenzene — Benzoic acid (85%)

The exact mechanism of the reaction is not fully understood, but probably involves attack on side-chain C—H bonds at the position next to the aromatic ring to form intermediate benzylic radicals. *tert*-Butylbenzene has no benzylic hydrogens, however, and is therefore inert.

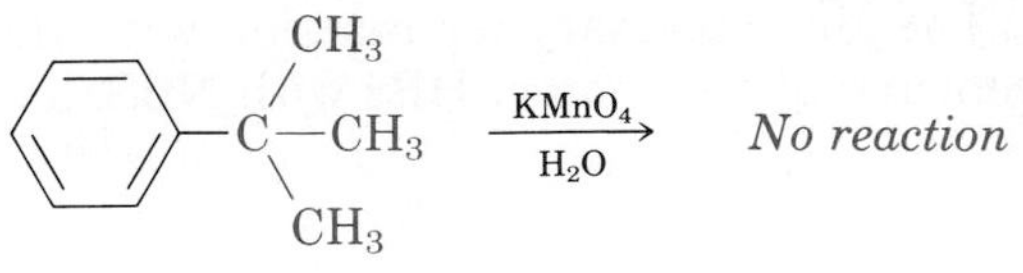

$$C_6H_5C(CH_3)_3 \xrightarrow[H_2O]{KMnO_4} \textit{No reaction}$$

*t*-Butylbenzene

A similar oxidation is employed industrially for the preparation of terephthalic acid, used in production of polyester fibers. Approximately 6 billion pounds per year of *p*-xylene are oxidized in this manner, using air as the oxidant and Co(III) salts as catalyst.

Industrial procedure

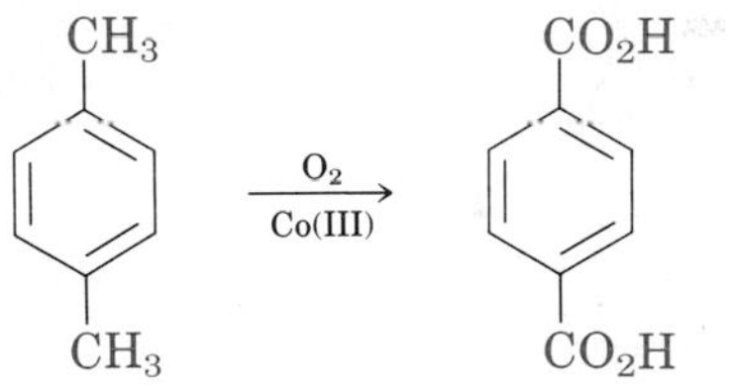

$$p\text{-}CH_3C_6H_4CH_3 \xrightarrow[Co(III)]{O_2} p\text{-}HO_2CC_6H_4CO_2H$$

*p*-Xylene — Terephthalic acid

PROBLEM

**16.20** What aromatic products would you expect to obtain from the $KMnO_4$ oxidation of these substances?

(a) Tetralin,

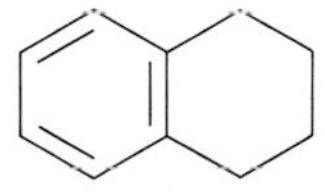

(b) *m*-Nitroisopropylbenzene

## BROMINATION OF ALKYLBENZENE SIDE CHAINS

Another kind of side-chain oxidation takes place when alkylbenzenes are treated with *N*-bromosuccinimide (NBS). *N*-Bromosuccinimide reacts with alkylbenzenes to brominate the benzylic position through a radical chain mechanism. For example, (3-bromopropyl)benzene gives (1,3-dibromopropyl)benzene in 99% yield on reaction with NBS in the presence of benzoyl peroxide, $(PhCO_2)_2$, as a radical initiator. Note that bromination occurs exclusively in the benzylic position and does not give a mixture of products.

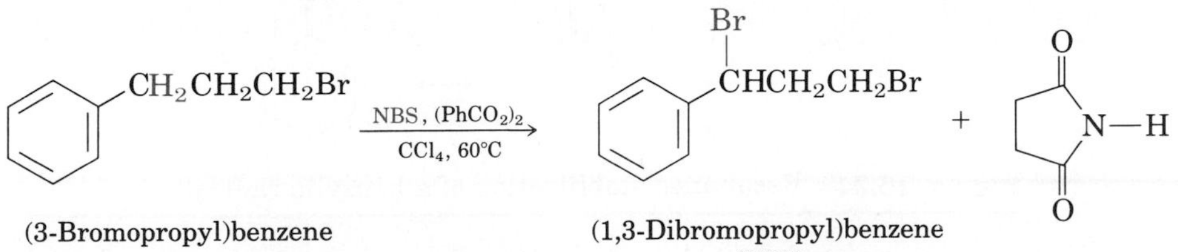

$$C_6H_5CH_2CH_2CH_2Br \xrightarrow[CCl_4,\ 60°C]{NBS,\ (PhCO_2)_2} C_6H_5CH(Br)CH_2CH_2Br + \text{succinimide (N—H)}$$

(3-Bromopropyl)benzene — (1,3-Dibromopropyl)benzene (99%)

The mechanism of benzylic bromination is similar to that seen previously for allylic bromination of alkenes (Section 10.5). Although the overall mechanism is somewhat complex, the critical step involves abstraction of a benzylic hydrogen atom of the alkylbenzene to generate an intermediate benzyl radical. The stabilized radical then reacts with $Br_2$ to yield product and a bromine radical, which cycles back into the reaction to carry on the chain. The $Br_2$ necessary for reaction with the benzyl radical is produced by a concurrent reaction of HBr with NBS, as shown in Figure 16.23.

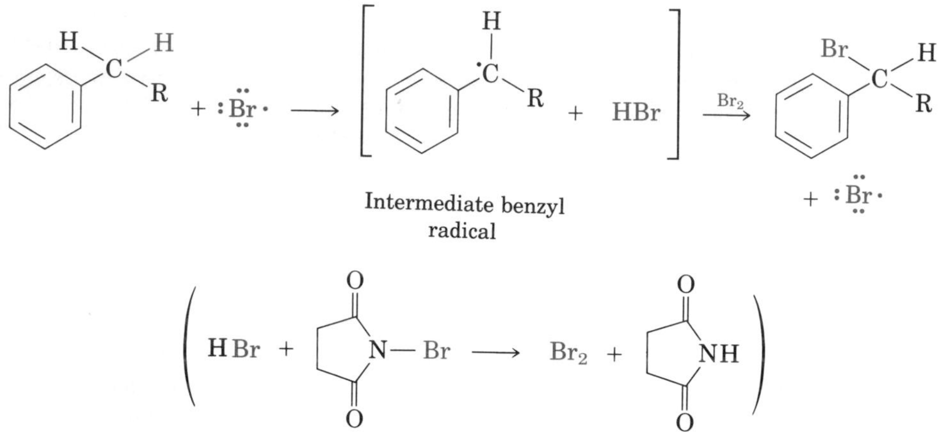

Intermediate benzyl radical

**Figure 16.23** Mechanism of benzylic bromination with *N*-bromosuccinimide

Reaction occurs exclusively at the benzylic position because the benzylic radical is highly stabilized by resonance. Figure 16.24 shows how this resonance stabilization arises, and Figure 16.25 shows an orbital view of the benzyl radical, indicating how the radical is stabilized by overlap of its *p* orbital with the ring pi electron system.

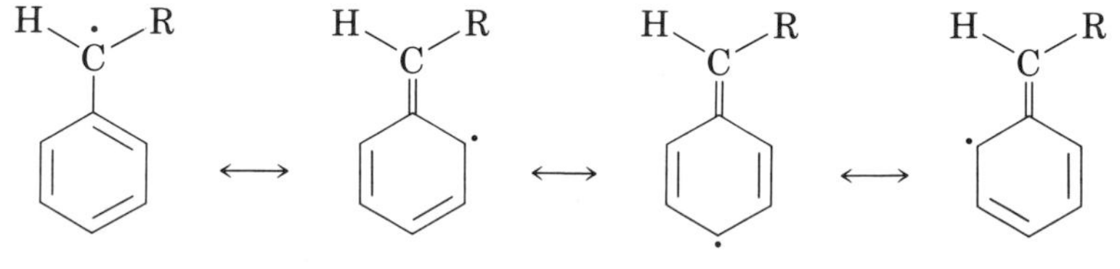

**Figure 16.24** Resonance stabilization of a benzylic radical

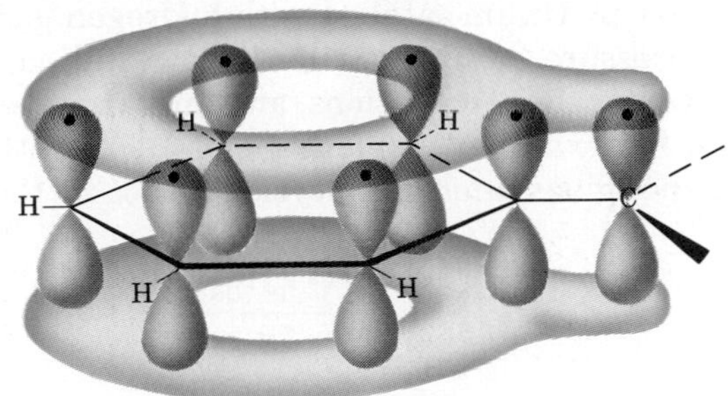


**Figure 16.25** An orbital picture of a benzylic radical, showing the overlap of a side-chain orbital with the aromatic ring orbitals

PROBLEM

**16.21** Styrene, the simplest alkenylbenzene, is prepared commercially for use in plastics manufacture by dehydrogenation of ethylbenzene over a special catalyst. How might you prepare styrene from benzene using reactions you have studied?

$C_6H_5$–CH=$CH_2$

Styrene

PROBLEM

**16.22** Refer to Table 5.4 for a quantitative idea of the stability of a benzyl radical. Approximately how much more stable (in kcal/mol) is the benzyl radical than a primary alkyl radical? How does a benzyl radical compare in stability to an allyl radical?

## 16.12 Reduction of Aromatic Compounds

### CATALYTIC HYDROGENATION OF AROMATIC RINGS

Just as aromatic rings are inert to oxidation under normal conditions, they are also inert to catalytic hydrogenation under conditions that reduce ordinary alkene double bonds. The usual platinum and palladium alkene hydrogenation catalysts don't affect aromatic rings under most conditions, and it is therefore possible to selectively reduce isolated double bonds in the presence of aromatic rings. For example, 4-phenyl-3-buten-2-one is selectively reduced to 4-phenyl-2-butanone when the reaction is carried out at room temperature and atmospheric pressure using a palladium catalyst. Neither the benzene ring nor the ketone carbonyl group is affected.

$$C_6H_5\text{–CH=CHCCH}_3\ (\text{C=O}) \xrightarrow[\text{Ethanol}]{H_2/Pd} C_6H_5\text{–CH}_2\text{–CH}_2\text{CCH}_3\ (\text{C=O})$$

4-Phenyl-3-buten-2-one → 4-Phenyl-2-butanone (100%)

In order to hydrogenate an aromatic ring, it is necessary either to use an ordinary platinum catalyst with hydrogen gas at several hundred atmospheres pressure or to use a more powerful catalyst such as rhodium on carbon. Under these conditions, aromatic rings are readily reduced to cyclohexanes. For example, *o*-xylene yields 1,2-dimethylcyclohexane, and 4-*tert*-butylphenol gives 4-*tert*-butylcyclohexanol in 100% yield.

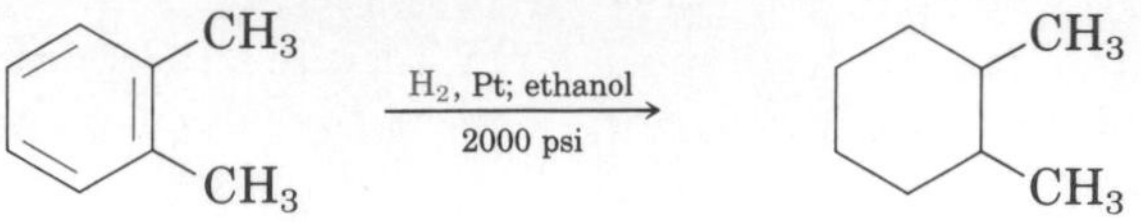

$CH_3$ $CH_3$ —$H_2$, Pt; ethanol / 2000 psi→ $CH_3$ $CH_3$

*o*-Xylene 1,2-Dimethylcyclohexane (100%)

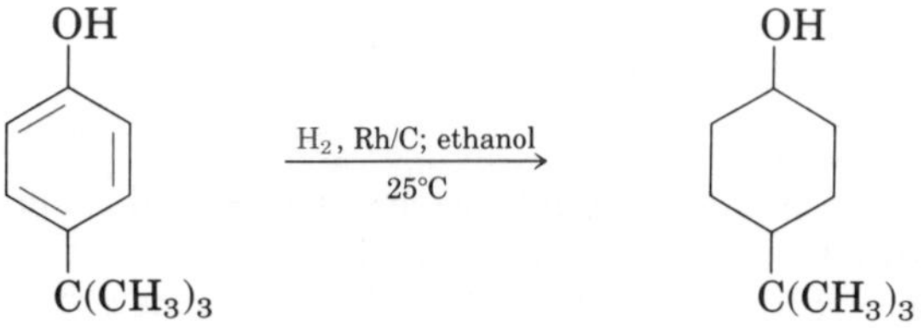

OH $C(CH_3)_3$ —$H_2$, Rh/C; ethanol / 25°C→ OH $C(CH_3)_3$

4-*tert*-Butylphenol 4-*tert*-Butylcyclohexanol (100%)

## REDUCTION OF ARYL ALKYL KETONES

Just as the presence of an aromatic ring activates a neighboring benzylic C—H position toward oxidation, it also activates a neighboring carbonyl group toward reduction. Thus, an aryl alkyl ketone prepared by Friedel–Crafts acylation of an aromatic ring can be converted into an alkylbenzene by catalytic hydrogenation over a palladium catalyst. For example, propiophenone is reduced to propylbenzene in 100% yield by catalytic hydrogenation. Since the net effect of Friedel–Crafts acylation followed by reduction is the preparation of a primary alkylbenzene, this two-step sequence of reactions allows us to circumvent the carbocation rearrangement problems associated with direct Friedel–Crafts alkylation using primary alkyl halides (Figure 16.26).

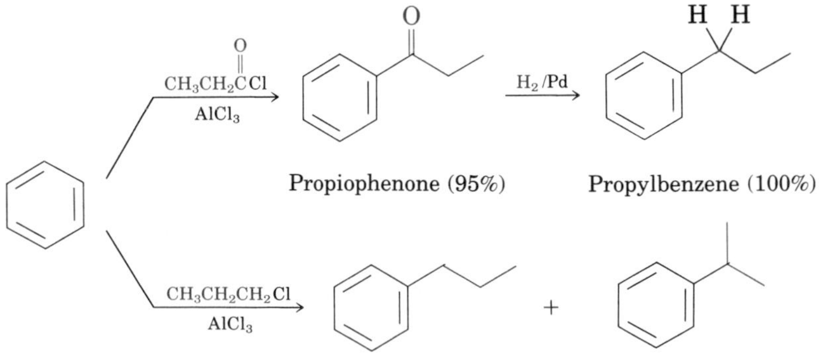

O

$CH_3CH_2\overset{\|}{C}Cl$ / $AlCl_3$ → O → $H_2$/Pd → H H

Propiophenone (95%) Propylbenzene (100%)

$CH_3CH_2CH_2Cl$ / $AlCl_3$ → +

Mixture of two products

**Figure 16.26** Use of the Friedel–Crafts acylation reaction to prepare a straight-chain alkylbenzene

Note that the conversion of a carbonyl group into a methylene ($C{=}O \rightarrow CH_2$) by catalytic hydrogenation is limited to *aryl* alkyl ketones. The presence of the neighboring aromatic ring is necessary to increase the reactivity of the carbonyl group toward hydrogenation; dialkyl ketones are not reduced under these conditions. It should also be pointed out that the catalytic reduction of aryl alkyl ketones is not compatible with the presence of a nitro substituent on the aromatic ring, since nitro groups are reduced to amino groups under the reaction conditions. We'll see a more general method for reducing ketone carbonyl groups to yield alkanes in Section 19.13.

$$\xrightarrow[\text{Pd/C; ethanol}]{H_2}$$

*m*-Nitroacetophenone — *m*-Ethylaniline

PROBLEM

**16.23** Show how you would prepare diphenylmethane, $(Ph)_2CH_2$, from benzene and an appropriate acid chloride.

## 16.13 Synthesis of Substituted Benzenes

One of the surest ways to acquire a command of organic chemistry is to work synthesis problems. The ability to plan a successful multistep synthesis of a complex molecule requires a working knowledge of the uses and limitations of many hundreds of organic reactions. Not only must you know *which* reactions to use, you must also know *when* to use them. The order in which reactions are carried out is often critical to the success of the overall scheme.

The ability to plan a sequence of reactions in the correct order is particularly valuable in the synthesis of substituted aromatic rings, where the introduction of one substituent is strongly affected by the directing effects of other substituents. Planning syntheses of substituted aromatic compounds is therefore an excellent way to gain facility with the many reactions learned in the past two chapters.

During our earlier discussion of the strategies that can be used in working synthesis problems (Section 8.11), we said that it's usually best to work problems backward. Look at the target molecule and ask the question "What is an immediate precursor of this compound?" Choose a likely answer and continue working backward, one step at a time, until you arrive at a simple starting material. Let's try some examples.

PRACTICE PROBLEM

Synthesize *p*-bromobenzoic acid from benzene.

*Solution* Ask yourself "What is an immediate precursor of *p*-bromobenzoic acid?"

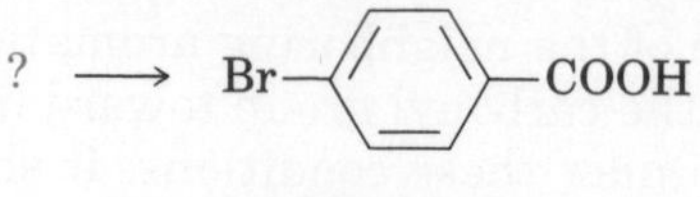

? ⟶ Br—C₆H₄—COOH

*p*-Bromobenzoic acid

There are only two substituents on the ring, a carboxyl group (—COOH), which is meta-directing, and a bromine, which is an ortho–para director. We can't brominate benzoic acid, because the wrong isomer (*m*-bromobenzoic acid) would be produced. We know, however, that oxidation of alkylbenzene side chains yields benzoic acids. Thus, an immediate precursor of our target molecule might be *p*-bromotoluene.

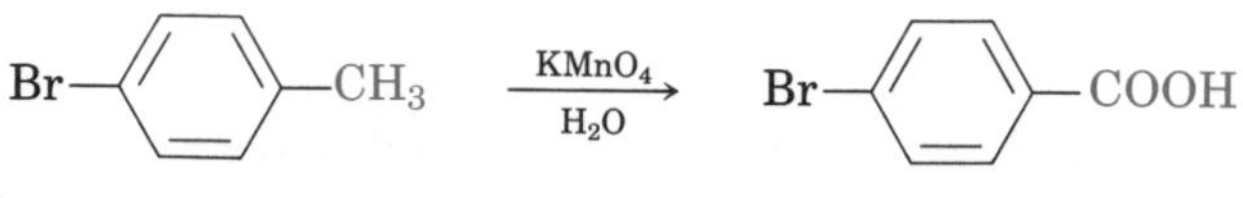

$$Br-C_6H_4-CH_3 \xrightarrow[H_2O]{KMnO_4} Br-C_6H_4-COOH$$

*p*-Bromotoluene *p*-Bromobenzoic acid

Next ask yourself "What is an immediate precursor of *p*-bromotoluene?" Perhaps toluene is an immediate precursor, since the methyl group directs bromination to the ortho and para positions, and we could then separate isomers. Alternatively, bromobenzene might be an immediate precursor, since we could carry out a Friedel–Crafts methylation and obtain para product. *Both answers are satisfactory*, although, in view of the difficulties often observed with polyalkylation in Friedel–Crafts reactions, bromination of toluene may well be the more efficient route.

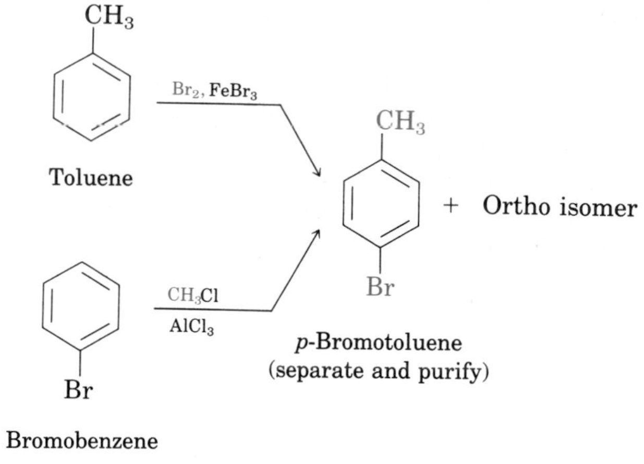

Toluene ($CH_3$) —$Br_2$, $FeBr_3$→ *p*-Bromotoluene ($CH_3$, Br) + Ortho isomer

Bromobenzene (Br) —$CH_3Cl$ / $AlCl_3$→ *p*-Bromotoluene

*p*-Bromotoluene
(separate and purify)

"What is an immediate precursor of toluene?" Benzene, which could be methylated in a Friedel–Crafts reaction.

$$\text{Benzene} \xrightarrow[AlCl_3]{CH_3Cl} \text{Toluene } (CH_3)$$

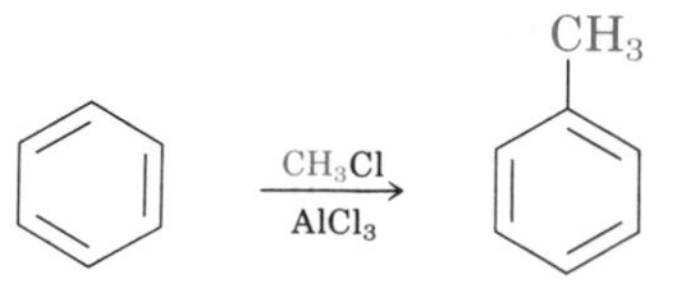

Benzene Toluene

Alternatively, "What is an immediate precursor of bromobenzene?" Benzene, which could be brominated.

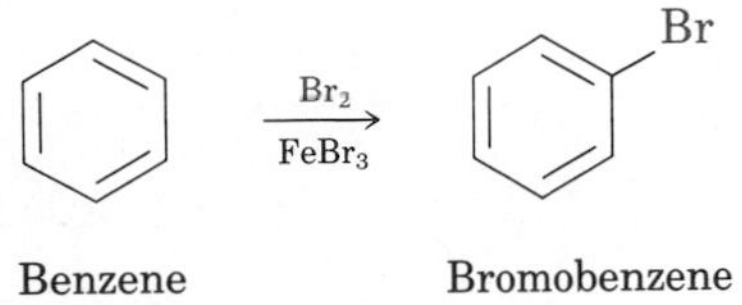

Br

$Br_2$ / $FeBr_3$

Benzene Bromobenzene

Our backward synthetic (**retrosynthetic**) analysis has provided two valid routes from benzene to *p*-bromobenzoic acid (Figure 16.27).

$Br_2$ / $FeBr_3$ ; $CH_3Cl$ / $AlCl_3$ ; Br ; $CH_3$ ; $CO_2H$ ; $KMnO_4$ ; Benzene ; $CH_3$ ; Br ; Br ; $CH_3Cl$ / $AlCl_3$ ; $Br_2$ / $FeBr_3$ ; *p*-Bromobenzoic acid

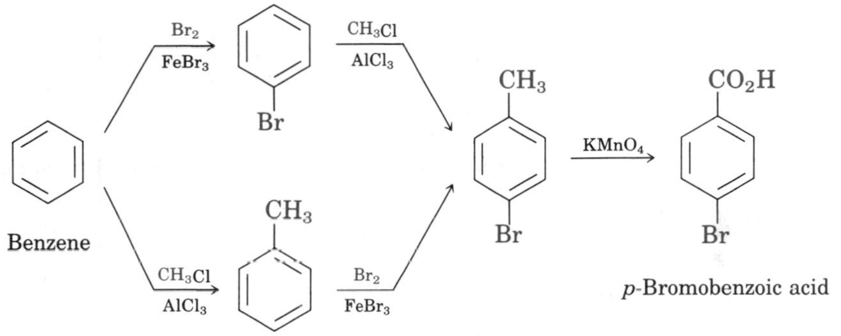

**Figure 16.27** Two routes for the synthesis of *p*-bromobenzoic acid from benzene

**PRACTICE PROBLEM**

Propose a synthesis of 4-chloro-1-nitro-2-propylbenzene from benzene.

*Solution* "What is an immediate precursor of the target?" Since the final step will involve introduction of one of three groups—chloro, nitro, or propyl—we have to consider three possibilities. Of the three, we know that chlorination of *o*-nitropropylbenzene can't be used because the reaction would occur at the wrong position. Similarly, a Friedel–Crafts reaction can't be used as the final step since these reactions don't work on nitro-substituted (deactivated) benzenes. Thus, the immediate precursor of our desired product is probably *m*-chloropropylbenzene, which can be nitrated. This nitration gives a mixture of product isomers, which must then be separated (Figure 16.28, page 560).

"What is an immediate precursor of *m*-chloropropylbenzene?" Since the two substituents have a meta relationship, the *first* substituent placed on the ring must be a meta director so that the second substitution will take place at the proper position. Furthermore, since primary alkyl groups such as propyl can't be introduced directly by Friedel–Crafts alkylation, the precursor of *m*-chloropropylbenzene is probably *m*-chloropropiophenone, which could undergo catalytic hydrogenation of the acyl group.

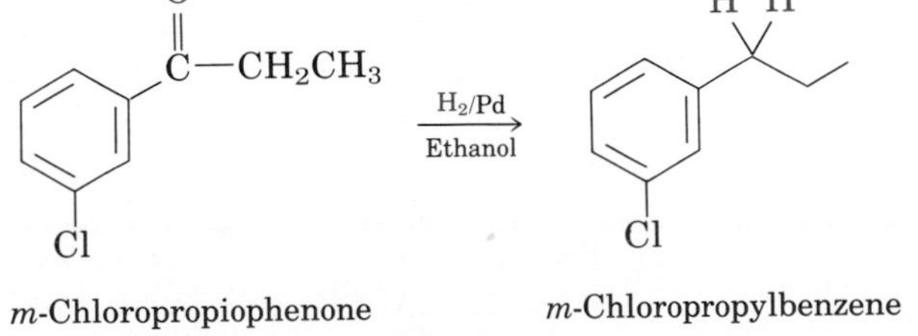

O ; C—$CH_2CH_3$ ; $H_2/Pd$ / Ethanol ; H H ; Cl ; Cl

*m*-Chloropropiophenone *m*-Chloropropylbenzene

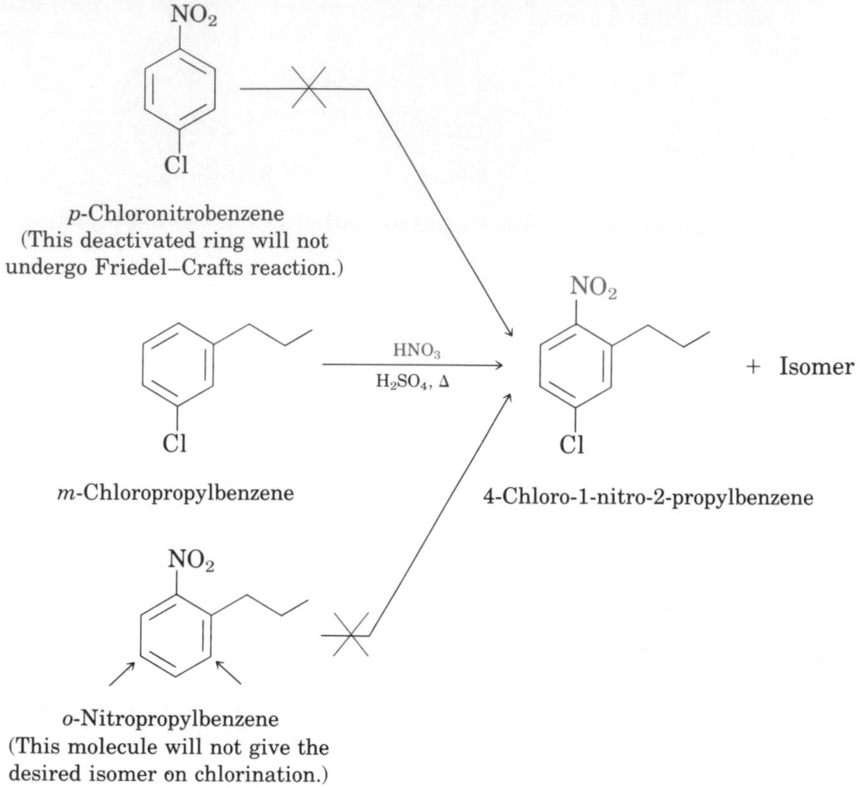

$NO_2$

Cl

*p*-Chloronitrobenzene
(This deactivated ring will not undergo Friedel–Crafts reaction.)

$HNO_3$

$H_2SO_4$, Δ

$NO_2$

+ Isomer

Cl

Cl

*m*-Chloropropylbenzene

4-Chloro-1-nitro-2-propylbenzene

$NO_2$

*o*-Nitropropylbenzene
(This molecule will not give the desired isomer on chlorination.)

**Figure 16.28** Possible routes for the synthesis of 4-chloro-1-nitro-2-propylbenzene

"What is an immediate precursor of *m*-chloropropiophenone?" Perhaps propiophenone, which could be chlorinated.

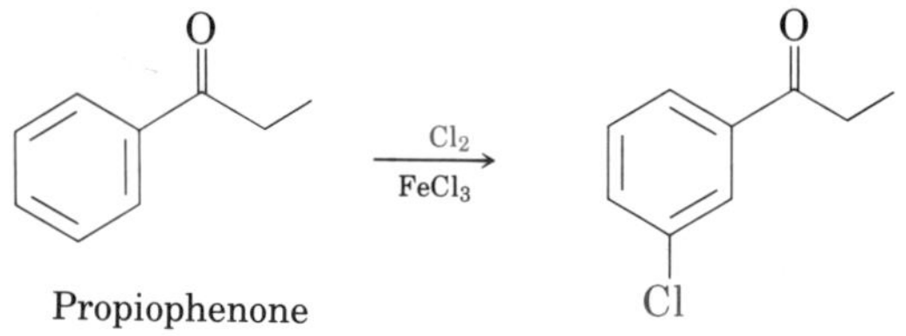

O

O

$Cl_2$

$FeCl_3$

Propiophenone

Cl

*m*-Chloropropiophenone

"What is an immediate precursor of propiophenone?" Benzene, which could undergo Friedel–Crafts acylation with propanoyl chloride and $AlCl_3$.

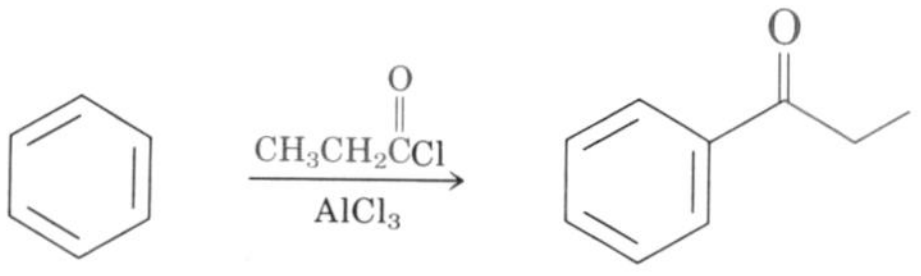

O

O

$CH_3CH_2\overset{\|}{C}Cl$

$AlCl_3$

Benzene

Propiophenone

Our final synthesis is a four-step route from benzene:

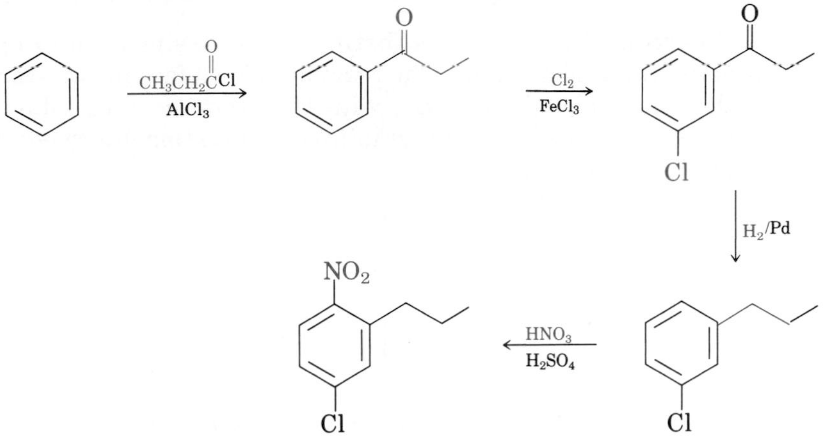

$CH_3CH_2\overset{O}{\overset{\|}{C}}Cl$, $AlCl_3$ → ; $Cl_2$, $FeCl_3$ → Cl ; $H_2/Pd$ ; $NO_2$ ... Cl ← $HNO_3$, $H_2SO_4$ ← Cl

..........

Planning organic syntheses has been compared to playing chess. There are no tricks; all that's required is a knowledge of the allowable moves (the organic reactions) and the discipline to work backward and to evaluate carefully the consequences of each move. Practicing is not always easy, but there is no surer way to learn organic chemistry.

PROBLEM ..........

**16.24** Propose syntheses of these substances from benzene:
(a) *m*-Chloronitrobenzene (b) *m*-Chloroethylbenzene
(c) *p*-Chloropropylbenzene

PROBLEM ..........

**16.25** In planning syntheses, it is as important to know what not to do as to know what to do. As written, the following reaction schemes have flaws that make their success unlikely. What is wrong with each one?

(a) CN → 1. $CH_3CH_2\overset{O}{\overset{\|}{C}}Cl$, $AlCl_3$; 2. $HNO_3$, $H_2SO_4$, $\Delta$ → CN, $NO_2$, $\underset{\underset{O}{\|}}{C}CH_2CH_3$

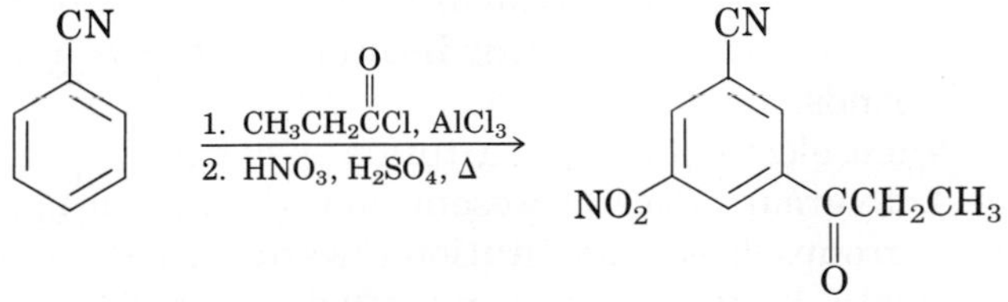

(b) Cl → 1. $CH_3CH_2CH_2Cl$, $AlCl_3$; 2. $Cl_2$, $FeCl_3$ → Cl, Cl, $CH_3CH_2CH_2$

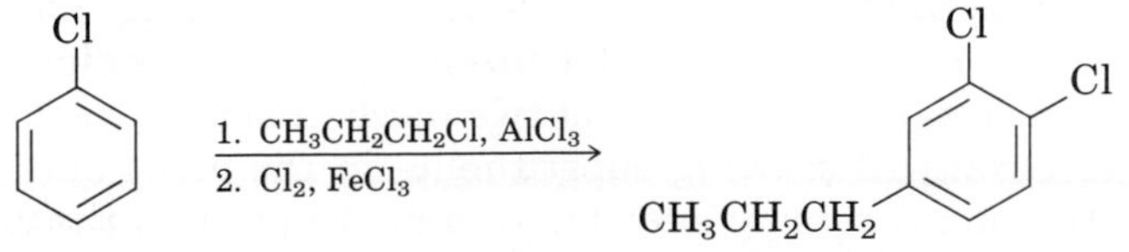

## 16.14 Summary and Key Words

**Electrophilic aromatic substitution** is the single most important reaction of aromatic compounds. The reaction takes place in two steps—initial reaction of an electrophile, $E^+$, with the aromatic ring, followed by loss of a proton from the resonance-stabilized carbocation intermediate to regenerate the aromatic ring:

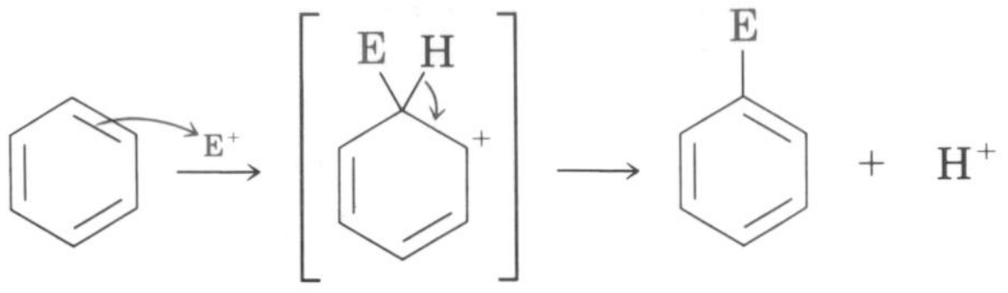

Many different substituents can be introduced onto the ring by this process. **Bromination, chlorination, iodination, nitration, sulfonation, alkylation,** and **acylation** can all be carried out with the proper choice of reagent.

**Friedel–Crafts alkylation** and **acylation**, which involve reaction of an aromatic with carbocation electrophiles, are particularly useful but are limited in several ways:

1. Only alkyl halides and acyl halides can be used. Vinylic and aryl halides do not react.
2. The aromatic ring must be at least as reactive as a halobenzene. Strongly deactivated rings do not react.
3. Polyalkylation often occurs in Friedel–Crafts alkylation, since the product alkylbenzene is more reactive than the starting material.
4. Carbocation rearrangements can occur during Friedel–Crafts alkylation, particularly when primary alkyl halides are used.

Substituents on the benzene ring affect both the reactivity of the ring toward further substitution, and the orientation of further substitution. Groups can be classified into three categories: **ortho- and para-directing activators, ortho- and para-directing deactivators,** and **meta-directing deactivators**.

Substituent effects are due to an interplay of resonance and inductive effects. **Resonance effects** are transmitted through *p* orbitals when the atom directly attached to the aromatic ring has a *p* orbital that can overlap the aromatic ring pi system; **inductive effects** are transmitted through sigma bonds.

When electrophilic substitution is carried out on a disubstituted benzene, both groups already present exert their orienting effects independently. If both groups direct substitution toward the same position, reaction occurs at that site. If the groups have conflicting directional effects, the more powerful activating substituent exerts a controlling influence.

In special cases, halobenzenes undergo **nucleophilic aromatic substitution** through either of two mechanisms. If the halobenzene has strong electron-withdrawing substituents in the ortho and/or para position, substitution occurs by addition of a nucleophile to the ring, followed by elimination of halide from the intermediate anion. If the halobenzene is not activated by electron-withdrawing substituents, nucleophilic substitution can occur by elimination of HX, followed by addition of a nucleophile to the intermediate **benzyne**.

The side chain of alkylbenzenes has unique reactivity because of the neighboring aromatic ring. Thus, the benzylic position can be brominated by reaction with *N*-bromosuccinimide, and the entire side chain can be degraded to a carboxylic acid by oxidation with aqueous potassium permanganate. Although aromatic rings are much less reactive than isolated alkene double bonds, they can nevertheless be reduced to cyclohexanes by hydrogenation over a platinum or rhodium catalyst. The neighboring aromatic ring also allows aryl alkyl ketones to be reduced to alkylbenzenes by hydrogenation over a platinum catalyst. The net effect of Friedel–Crafts acylation followed by catalytic hydrogenation is the synthesis of a straight-chain alkylbenzene.

## 16.15 Summary of Reactions

1. Electrophilic aromatic substitution

a. Bromination (Section 16.1)

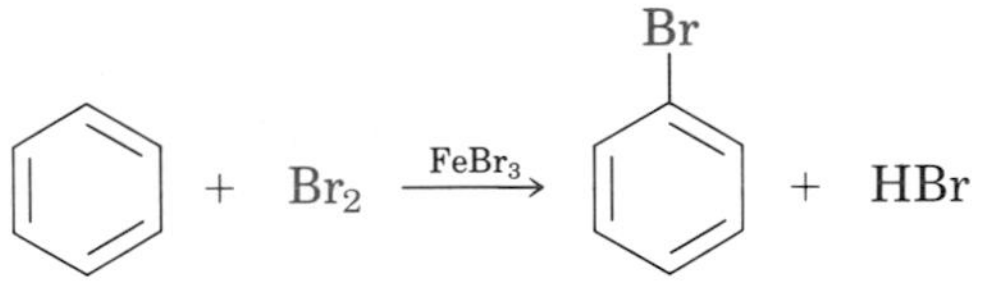

Br

$+ Br_2 \xrightarrow{FeBr_3}$ $+ HBr$

b. Chlorination (Section 16.2)

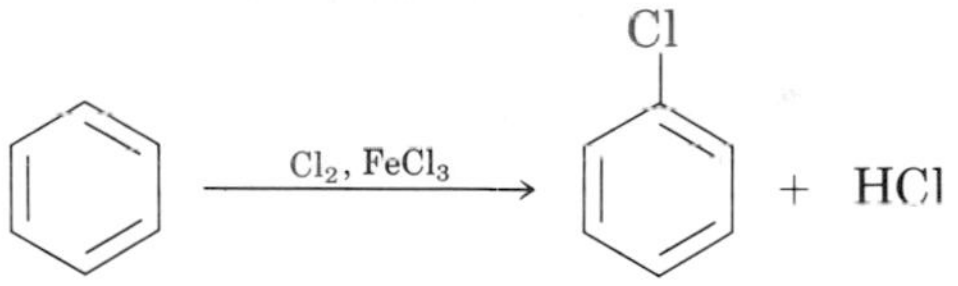

Cl

$\xrightarrow{Cl_2,\ FeCl_3}$ $+ HCl$

c. Iodination (Section 16.2)

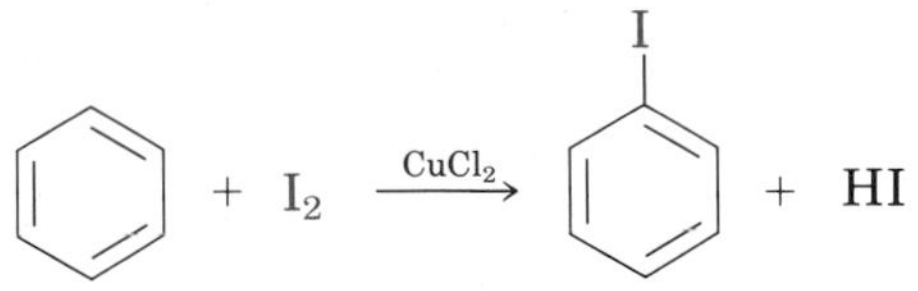

I

$+ I_2 \xrightarrow{CuCl_2}$ $+ HI$

d. Nitration (Section 16.2)

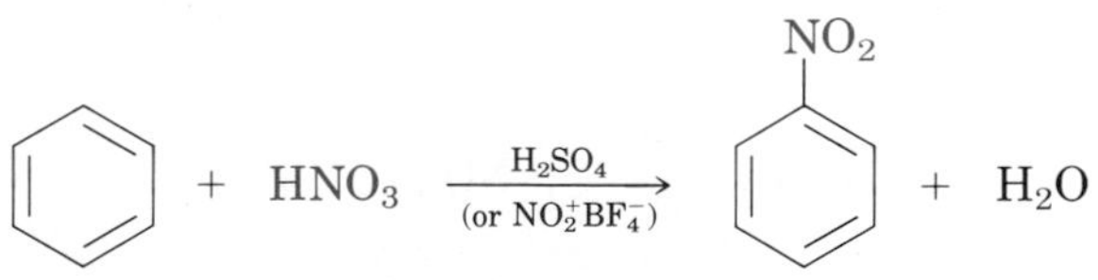

$NO_2$

$+ HNO_3 \xrightarrow[(or\ NO_2^+BF_4^-)]{H_2SO_4}$ $+ H_2O$

e. Sulfonation (Section 16.2)

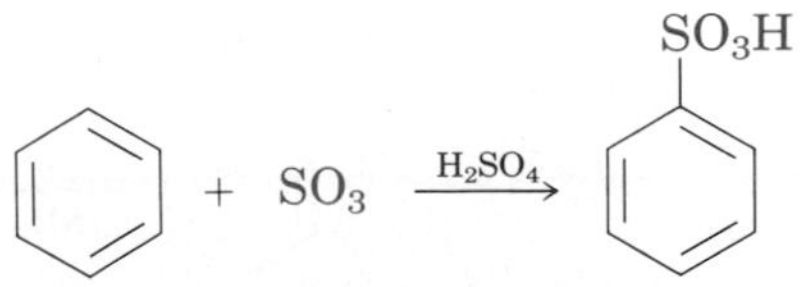

$SO_3H$

$+ SO_3 \xrightarrow{H_2SO_4}$

f. Friedel–Crafts alkylation (Section 16.3)

$$C_6H_6 + CH_3Cl \xrightarrow{AlCl_3} C_6H_5CH_3 + HCl$$

Aromatic ring: Must be at least as reactive as a halobenzene. Deactivated rings do not react.

Alkyl halide: Can be methyl, ethyl, 2°, or 3°; primary halides undergo carbocation rearrangement.

g. Friedel–Crafts acylation (Section 16.4)

$$C_6H_6 + CH_3\overset{O}{\overset{\|}{C}}Cl \xrightarrow{AlCl_3} C_6H_5\overset{O}{\overset{\|}{C}}CH_3 + HCl$$

2. Reduction of aromatic nitro groups (Section 16.2)

$$C_6H_5NO_2 \xrightarrow[2.\ HO^-]{1.\ SnCl_2,\ H_3O^+} C_6H_5NH_2$$

3. Alkali fusion of aromatic sulfonates (Section 16.2)

$$C_6H_5SO_3H \xrightarrow[2.\ H_3O^+]{1.\ NaOH} C_6H_5OH$$

4. Nucleophilic aromatic substitution

a. Via addition–elimination to activated aryl halides (Section 16.9)

$$\text{2,4,6-}(O_2N)_3C_6H_2Cl \xrightarrow[2.\ H_3O^+]{1.\ ^-{:}\ddot{O}H} \text{2,4,6-}(O_2N)_3C_6H_2OH + {:}\ddot{Cl}{:}^-$$

b. Via benzyne intermediate for unactivated aryl halides (Section 16.10)

$$C_6H_5Br \xrightarrow[NH_3]{^-{:}\ddot{N}H_2} C_6H_5NH_2 + {:}\ddot{Br}{:}^-$$

5. *N*-Bromosuccinimide bromination of alkylbenzenes (Section 16.11)

6. Oxidation of alkylbenzene side chain (Section 16.11)

Reaction occurs with 1° and 2°, but not 3°, alkyl side chains.

7. Catalytic hydrogenation of aromatic ring (Section 16.12)

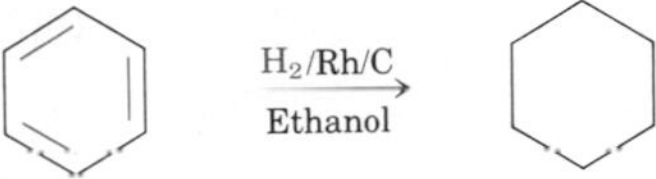


8. Reduction of aryl alkyl ketones (Section 16.12)

Reaction is specific for alkyl aryl ketones; dialkyl ketones are not affected.

## ADDITIONAL PROBLEMS

**16.26** Predict the major product(s) of mononitration of these substances. Which react faster, and which slower, than benzene?
(a) Bromobenzene (b) Benzonitrile (c) Benzoic acid
(d) Nitrobenzene (e) Benzenesulfonic acid (f) Methoxybenzene

**16.27** Rank the compounds in each group according to their reactivity toward electrophilic substitution:
(a) Chlorobenzene, *o*-dichlorobenzene, benzene
(b) *p*-Bromonitrobenzene, nitrobenzene, phenol
(c) Fluorobenzene, benzaldehyde, *o*-xylene
(d) Benzonitrile, *p*-methylbenzonitrile, *p*-methoxybenzonitrile

**16.28** Predict the major monoalkylation products you would expect to obtain from reaction of the following substances with chloromethane and $AlCl_3$:
(a) Bromobenzene (b) *m*-Bromophenol
(c) *p*-Chloroaniline (d) 2,4-Dichloronitrobenzene
(e) 2,4-Dichlorophenol (f) Benzoic acid
(g) *p*-Methylbenzenesulfonic acid (h) 2,5-Dibromotoluene

**16.29** Name and draw the major product(s) of electrophilic chlorination of these substances:
(a) *m*-Nitrophenol (b) *o*-Dimethylbenzene (c) *p*-Nitrobenzoic acid
(d) *p*-Bromobenzenesulfonic acid

**16.30** Predict the major product(s) you would obtain from sulfonation of these compounds:
(a) Fluorobenzene (b) *m*-Bromophenol (c) *m*-Dichlorobenzene
(d) 2,4-Dibromophenol

**16.31** Rank the following aromatic compounds in the expected order of their reactivity toward Friedel–Crafts alkylation. Which compounds are unreactive?
(a) Bromobenzene (b) Toluene (c) Phenol
(d) Aniline (e) Nitrobenzene (f) *p*-Bromotoluene

**16.32** Suggest a reason for the observation that bromination of biphenyl occurs at ortho and para positions rather than at meta. Use resonance structures of the intermediates to explain your answers.

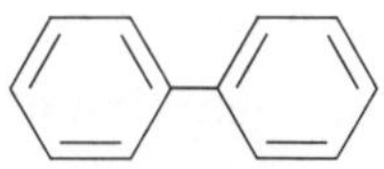

Biphenyl

**16.33** At what position, and on what ring, would you expect nitration of 4-bromobiphenyl to occur?

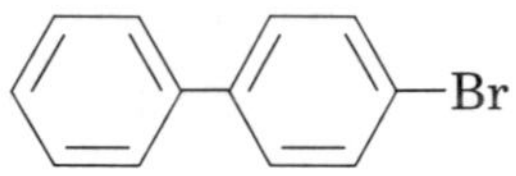

Br

4-Bromobiphenyl

**16.34** How do you explain the fact that electrophilic attack on 3-phenylpropanenitrile occurs at the ortho and para positions, whereas attack on 3-phenylpropenenitrile occurs at the meta position? Use resonance structures of the intermediates in your explanation.

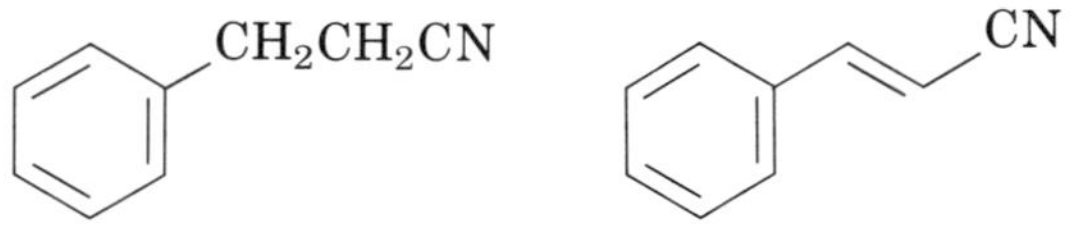

$CH_2CH_2CN$ CN

3-Phenylpropanenitrile 3-Phenylpropenenitrile

**16.35** Triphenylmethane can be prepared by reaction of benzene and chloroform in the presence of $AlCl_3$. Propose a mechanism for this reaction:

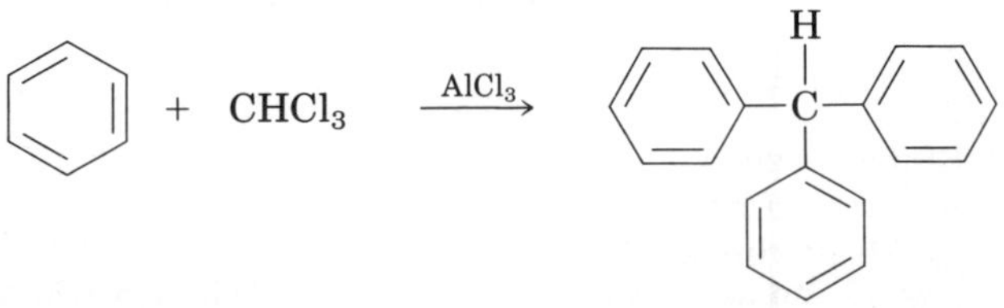

$$+ \ CHCl_3 \xrightarrow{AlCl_3}$$

H
C

**16.36** What product(s) would you expect to obtain from these reactions?

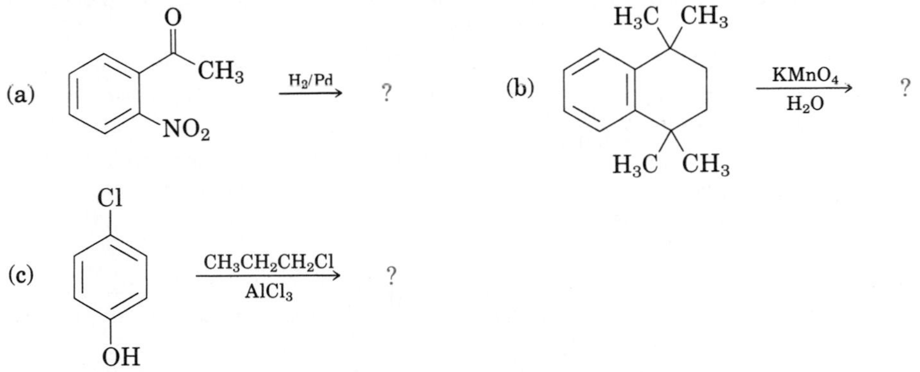

(a) O, $CH_3$, $NO_2$ $\xrightarrow{H_2/Pd}$ ?

(b) $H_3C$ $CH_3$, $H_3C$ $CH_3$ $\xrightarrow[H_2O]{KMnO_4}$ ?

(c) Cl, OH $\xrightarrow[AlCl_3]{CH_3CH_2CH_2Cl}$ ?

**16.37** How would you synthesize these substances, starting from benzene? Assume that ortho and para substitution products can be separated.
(a) *o*-Methylphenol (b) 2,4,6-Trinitrophenol
(c) 2,4,6-Trinitrobenzoic acid (d) *m*-Bromoaniline

**16.38** At what position, and on what ring, would you expect these substances to undergo electrophilic substitution?

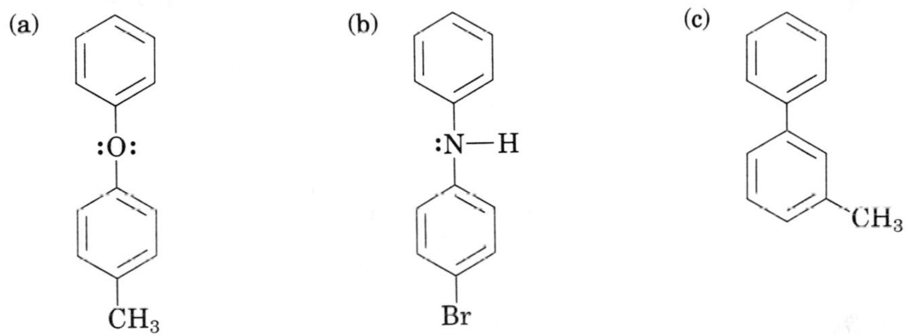


**16.39** At what position, and on what ring, would you expect bromination of benzanilide to occur? Explain your answer by drawing resonance structures of the intermediates.

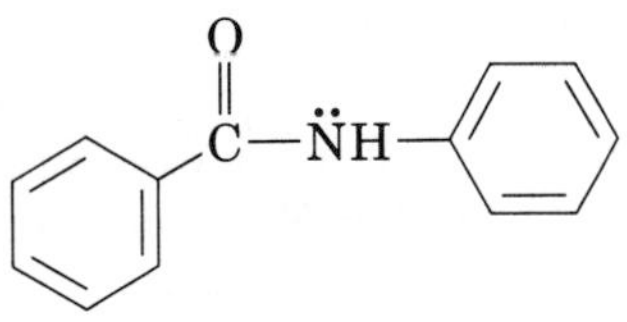

O, C—N̈H

Benzanilide

**16.40** Would you expect the Friedel–Crafts reaction of benzene with optically active 2-chlorobutane to yield optically active or racemic product? Explain your answer.

**16.41** Starting with benzene as your only source of aromatic compounds, how would you synthesize these substances? Assume that you can separate ortho and para isomers if necessary.
(a) *p*-Chlorophenol (b) *m*-Bromonitrobenzene
(c) *o*-Bromobenzenesulfonic acid (d) *m*-Chlorobenzenesulfonic acid

**16.42** Starting with either benzene or toluene, how would you synthesize these materials? Assume that ortho and para isomers can be separated.
(a) 2-Bromo-4-nitrotoluene (b) 1,3,5-Trinitrobenzene
(c) 2,4,6-Tribromoaniline (d) 2-Chloro-4-methylphenol

**16.43** As written, the following syntheses have certain flaws. What is wrong with each one?

(a) 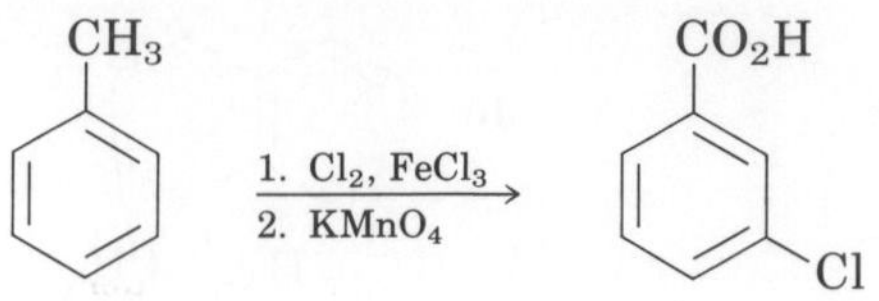 $CH_3$ (benzene ring) $\xrightarrow[\text{2. } KMnO_4]{\text{1. } Cl_2,\ FeCl_3}$ $CO_2H$, Cl (meta)

(b) 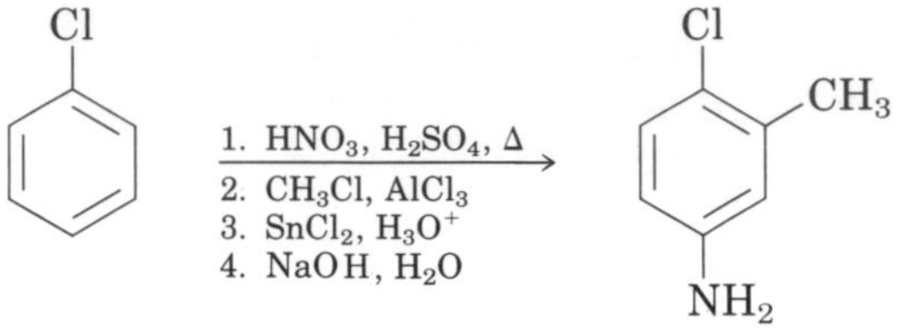 Cl (benzene ring) $\xrightarrow{\text{1. } HNO_3,\ H_2SO_4,\ \Delta}$ 2. $CH_3Cl$, $AlCl_3$ 3. $SnCl_2$, $H_3O^+$ 4. NaOH, $H_2O$ → Cl, $CH_3$, $NH_2$

(c) 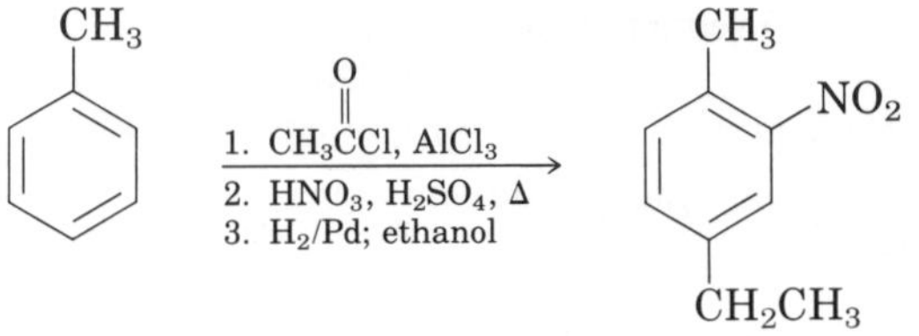 $CH_3$ (benzene ring) $\xrightarrow{\text{1. } CH_3CCl\ (C{=}O),\ AlCl_3}$ 2. $HNO_3$, $H_2SO_4$, Δ 3. $H_2$/Pd; ethanol → $CH_3$, $NO_2$, $CH_2CH_3$

**16.44** How would you synthesize these substances, starting from benzene?

(a) 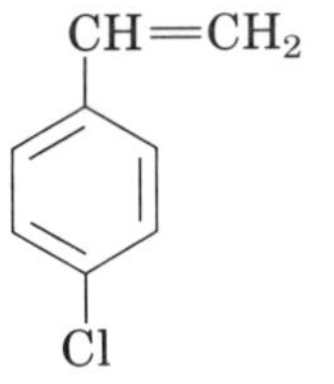 $CH{=}CH_2$ / Cl (para)

(b) 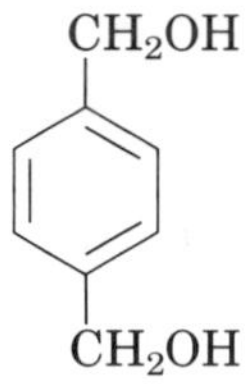 $CH_2OH$ / $CH_2OH$ (para)

(c) 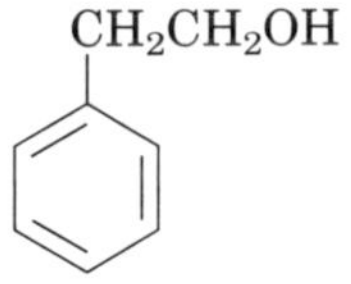 $CH_2CH_2OH$

**16.45** The compound MON-0585 is a nontoxic, biodegradable larvicide that is highly selective against mosquito larvae. How could you synthesize MON-0585 using only benzene as a source of the aromatic rings?

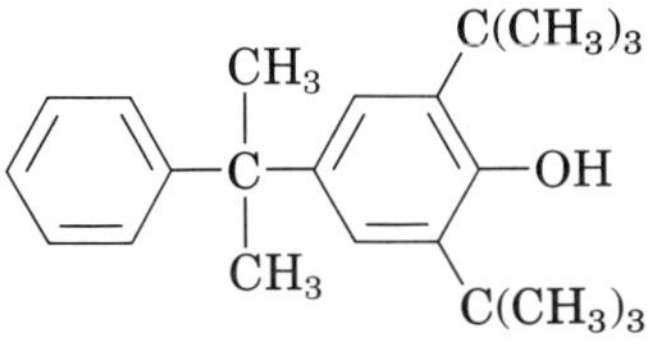

$CH_3$, C, $CH_3$, $C(CH_3)_3$, OH, $C(CH_3)_3$

MON-0585

**16.46** Hexachlorophene, a substance used in the manufacture of germicidal soaps, is prepared by reaction of 2,4,5-trichlorophenol with formaldehyde in the presence of concentrated sulfuric acid. Propose a mechanism to account for the reaction.

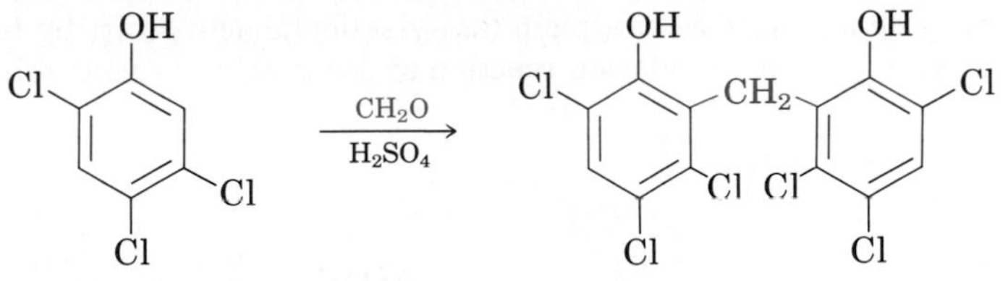

$$\text{2,4,5-Trichlorophenol} \xrightarrow[H_2SO_4]{CH_2O} \text{Hexachlorophene}$$

Hexachlorophene

**16.47** When heated, benzenediazonium carboxylate decomposes to yield $N_2$, $CO_2$, and a reactive organic substance that cannot be isolated. When benzenediazonium carboxylate is heated in the presence of furan, the following reaction is observed:

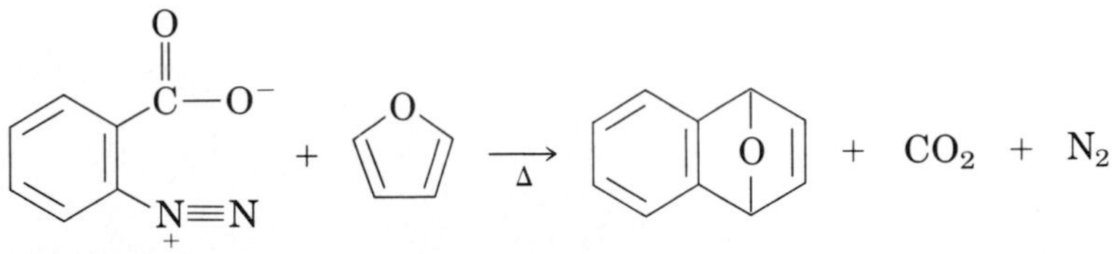

$$\text{Benzenediazonium carboxylate} + \text{Furan} \xrightarrow{\Delta} \text{Adduct} + CO_2 + N_2$$

What intermediate is involved in this reaction? Propose a mechanism for formation of this intermediate.

**16.48** Phenylboronic acid is nitrated to give 15% ortho substitution product and 85% meta. Account for the meta-directing effect of the —$B(OH)_2$ group.

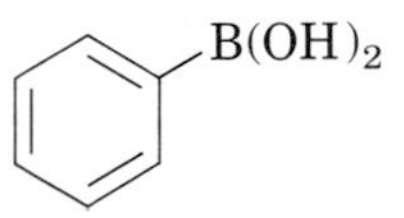

$B(OH)_2$

Phenylboronic acid

**16.49** Draw resonance structures of the intermediate carbocations in the bromination of naphthalene, and account for the fact that naphthalene undergoes electrophilic attack at C1 rather than C2.

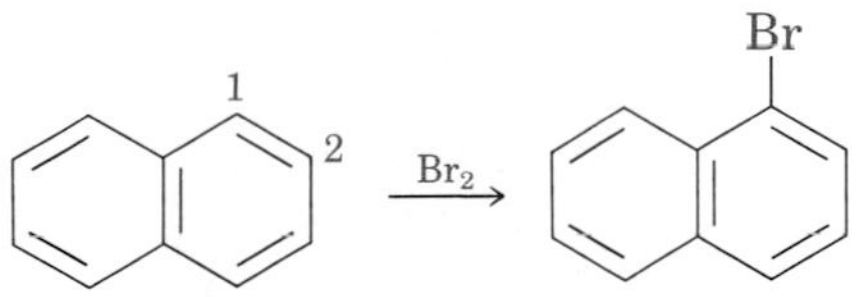

$$\text{Naphthalene} \xrightarrow{Br_2} \text{1-Bromonaphthalene}$$

**16.50** 4-Chloropyridine undergoes reaction with dimethylamine to yield 4-dimethylaminopyridine. Propose a mechanism to account for this result.

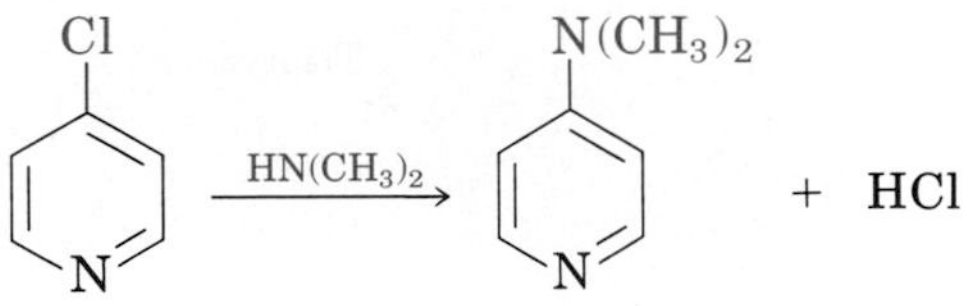

$$\text{4-Chloropyridine} \xrightarrow{HN(CH_3)_2} \text{4-Dimethylaminopyridine} + HCl$$

**16.51** How do you account for the fact that *p*-bromotoluene reacts with potassium amide to give a mixture of *m*- and *p*-methylaniline?

**16.52** Propose a synthesis of aspirin (acetylsalicylic acid) starting from benzene. You will need to use an acetylation reaction at some point in your scheme.

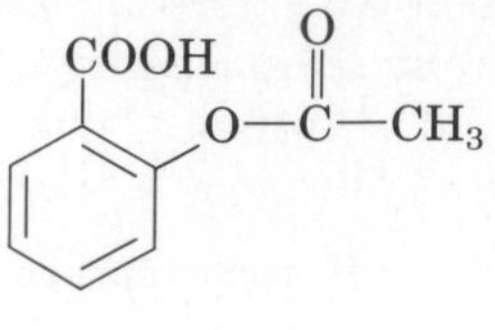

COOH, O—C(=O)—$CH_3$

Aspirin

$$\left[ ArOH \xrightarrow{CH_3C(=O)—O—C(=O)CH_3} ArOCOCH_3 \right]$$

An acetylation reaction

**16.53** Propose a mechanism to account for the following reaction of benzene with 2,2,5,5-tetramethyltetrahydrofuran.

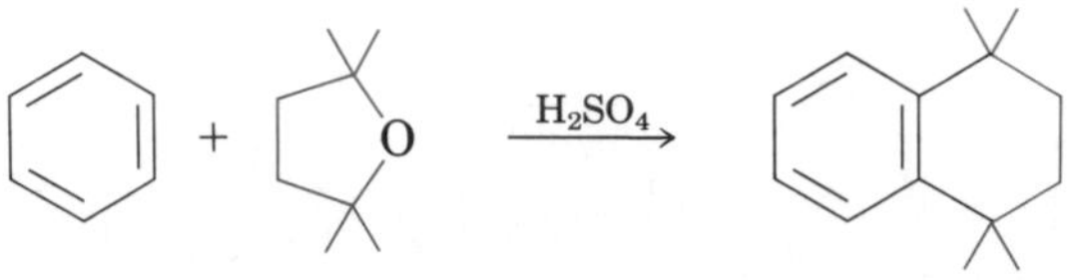

$\xrightarrow{H_2SO_4}$

**16.54** In the Gatterman–Koch reaction, a formyl group (—CHO) is introduced directly onto a benzene ring. For example, reaction of toluene with carbon monoxide and HCl in the presence of mixed $CuCl/AlCl_3$ gives *p*-methylbenzaldehyde in 55% yield. Propose a mechanism for this reaction.

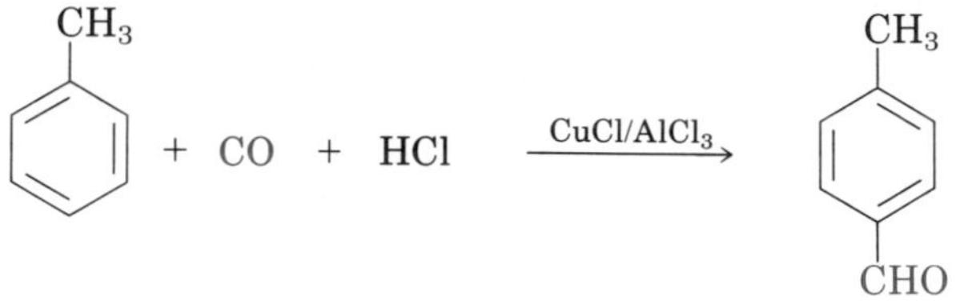

$CH_3$ + CO + HCl $\xrightarrow{CuCl/AlCl_3}$ $CH_3$, CHO

(55%)

**16.55** Triptycene is an unusual molecule that has been prepared by reaction of benzyne with anthracene. What kind of reaction is involved? Show the mechanism of the transformation.

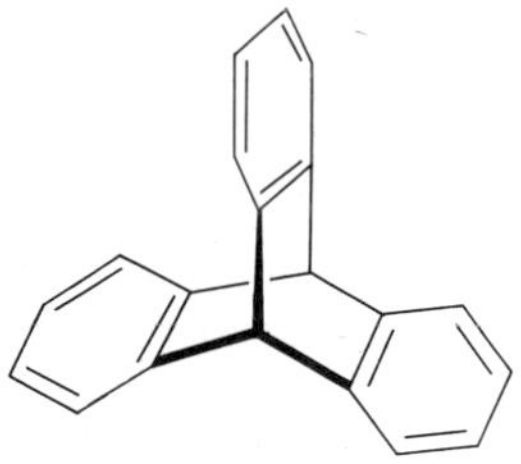

Triptycene

# Organic Reactions: A Brief Review

If you didn't believe it before you started, you believe it now—learning organic chemistry means memorizing a large number of reactions. The way to simplify the job, of course, is to organize the material. We said in Chapter 5 that organic reactions can be organized in two ways—by what kinds of reactions occur and by how they occur. Let's take a brief review of both organizational methods in light of what has been covered in the past several chapters.

## I. A Summary of the Kinds of Organic Reactions

There are four important kinds of reactions—additions, eliminations, substitutions, and rearrangements. We've now seen examples of all four, as summarized in Review Tables 1–4.

**Review Table 1** Some addition reactions

1. Additions to alkenes

   a. Electrophilic addition of HX (X = Cl, Br, I; Sections 6.9 and 6.10)

$$CH_3CH{=}CH_2 \xrightarrow[\text{Ether}]{\text{HCl}} CH_3\overset{\overset{\displaystyle Cl}{|}}{C}HCH_3$$

   b. Electrophilic addition of $X_2$ (X = Cl, Br; Section 7.1)

$$CH_3CH{=}CHCH_3 \xrightarrow[CCl_4]{Br_2} CH_3\overset{\overset{\displaystyle Br}{|}}{C}H{-}\overset{\overset{\displaystyle Br}{|}}{C}HCH_3$$

c. Electrophilic addition of HO—X (X = Cl, Br, I; Section 7.2)

$$CH_3CH{=}CH_2 \xrightarrow[NaOH,\ H_2O]{Br_2} CH_3CH(OH)CH_2Br$$

d. Electrophilic addition of water by hydroxymercuration (Section 7.3)

$$CH_3CH{=}CH_2 \xrightarrow[H_2O]{Hg(OAc)_2} CH_3CH(OH)CH_2HgOAc \xrightarrow{NaBH_4} CH_3CH(OH)CH_3$$

e. Addition of $BH_3$ (hydroboration; Section 7.4)

$$CH_3CH{=}CH_2 \xrightarrow{BH_3} CH_3CH(H)CH_2BH_2 \xrightarrow[NaOH]{H_2O_2} CH_3CH_2CH_2OH$$

f. Catalytic addition of $H_2$ (Section 7.6)

$$CH_3CH{=}CH_2 \xrightarrow[Pd/C\ catalyst]{H_2} CH_3CH(H){-}CH_2(H)$$

g. Hydroxylation with $OsO_4$ (Section 7.7)

$$CH_3CH{=}CH_2 \xrightarrow[2.\ NaHSO_3]{1.\ OsO_4} CH_3CH(OH)CH_2OH$$

h. Addition of carbenoids—cyclopropane formation (Section 7.10)

$$CH_3CH{=}CH_2 \xrightarrow{CH_2I_2,\ Zn(Cu)} CH_3\overset{CH_2}{CH{-}CH_2}\ \text{(cyclopropane)}$$

i. Cycloaddition to dienes—Diels–Alder reaction (Sections 14.7 and 14.8)

$$H_2C{=}CH{-}CH{=}CH_2 + H_2C{=}CHCOCH_3 \longrightarrow \text{4-acetylcyclohexene}$$

j. Radical addition of HBr (Section 7.5)

$$CH_3CH{=}CH_2 \xrightarrow[Radicals]{HBr} CH_3CH(H)CH_2Br$$

2. Additions to alkynes

a. Electrophilic addition of HX (X = Cl, Br, I; Section 8.3)

$$CH_3C{\equiv}CH \xrightarrow[Ether]{HBr} CH_3C(Br){=}CH_2$$

b. Electrophilic addition of $H_2O$ (Section 8.4)

$$CH_3C{\equiv}CH \xrightarrow[HgSO_4]{H_3O^+} CH_3\overset{OH}{\overset{|}{C}}{=}CH_2 \longrightarrow CH_3\overset{O}{\overset{\|}{C}}CH_3$$

c. Addition of $H_2$ (Section 8.6)

$$CH_3C{\equiv}CCH_3 \xrightarrow[\text{Lindlar catalyst}]{H_2} CH_3\overset{H}{\overset{|}{C}}{=}\overset{H}{\overset{|}{C}}CH_3$$

**Review Table 2** Some elimination reactions

1. Dehydrohalogenation of alkyl halides (Sections 11.10–11.12)

$$CH_3\overset{Br}{\overset{|}{C}}HCH_3 \xrightarrow{KOH} CH_3CH{=}CH_2 + KBr + H_2O$$

2. Dehydrohalogenation of vinylic halides (Section 8.10)

$$CH_3\overset{Br}{\overset{|}{C}}{=}CH_2 \xrightarrow{NaNH_2} CH_3C{\equiv}CH + NaBr + NH_3$$

3. Dehydrohalogenation of aryl halides—benzyne formation (Section 16.10)

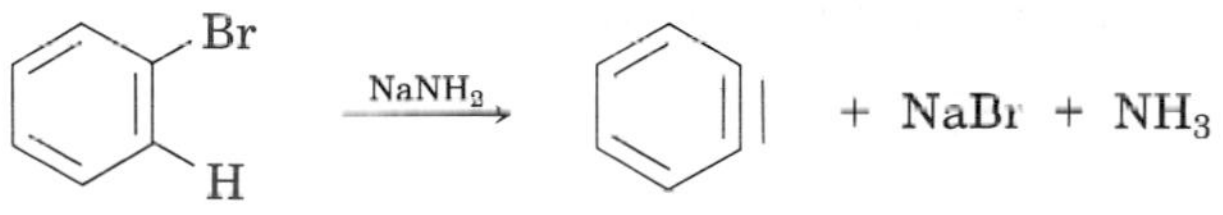

$$C_6H_4(Br)(H) \xrightarrow{NaNH_2} C_6H_4 \text{ (benzyne)} + NaBr + NH_3$$

4. Dehydration of alcohols (Section 7.12)

$$CH_3\overset{OH}{\overset{|}{C}}HCH_3 \xrightarrow{H_3O^+,\ \Delta} CH_3CH{=}CH_2 + H_2O$$

**Review Table 3** Some substitution reactions

1. $S_N2$ reactions of primary alkyl halides (Sections 11.2–11.5)

a. General reaction

$$CH_3CH_2CH_2X \xrightarrow{:Nu^-} CH_3CH_2CH_2Nu + :X^-$$

where X = Cl, Br, I, OTos

$:Nu^-$ = $CH_3O^-$, $HO^-$, $CH_3S^-$, $HS^-$, $CN^-$, $CH_3COO^-$, $NH_3$, $(CH_3)_3N$, etc.

b. Alkyne alkylation (Section 8.8)

$$CH_3C{\equiv}C{:}^-\ Na^+ \xrightarrow{CH_3I} CH_3C{\equiv}CCH_3 + Na^+\ I^-$$

2. $S_N1$ reactions of tertiary alkyl halides (Sections 11.6–11.9)
   a. General reaction

$$(CH_3)_3CX \xrightarrow{:NuH} (CH_3)_3CNu + HX$$

   b. Preparation of alkyl halides from alcohols (Section 10.8)

$$(CH_3)_3COH \xrightarrow{HBr} (CH_3)_3CBr + H_2O$$

3. Electrophilic aromatic substitution (Sections 16.1–16.4)
   a. Halogenation of aromatic compounds (Section 16.1)

$$C_6H_5{-}H \xrightarrow[FeBr_3]{Br_2} C_6H_5{-}Br + HBr$$

   b. Nitration of aromatic compounds (Section 16.2)

$$C_6H_5{-}H \xrightarrow[H_2SO_4]{HNO_3} C_6H_5{-}NO_2 + H_2O$$

   c. Sulfonation of aromatic compounds (Section 16.2)

$$C_6H_5{-}H \xrightarrow[H_2SO_4]{SO_3} C_6H_5{-}SO_3H$$

   d. Alkylation of aromatic rings (Section 16.3)

$$C_6H_5{-}H \xrightarrow[AlCl_3]{CH_3Cl} C_6H_5{-}CH_3 + HCl$$

   e. Acylation of aromatic rings (Section 16.4)

$$C_6H_5{-}H \xrightarrow[AlCl_3]{CH_3COCl} C_6H_5{-}COCH_3 + HCl$$

4. Nucleophilic aromatic substitution (Section 16.9)

$$O_2N{-}C_6H_4{-}Cl \xrightarrow[H_2O]{NaOH} O_2N{-}C_6H_4{-}OH + NaCl$$

5. Radical substitution reactions
   a. Chlorination of methane (Section 10.4)

$$CH_4 + Cl_2 \xrightarrow{Light} CH_3Cl + HCl$$

   b. NBS allylic bromination of alkenes (Section 10.5)

$$CH_3CH{=}CH_2 \xrightarrow[Radicals]{NBS} BrCH_2CH{=}CH_2$$

**Review Table 4** Some rearrangement reactions

**Carbocation rearrangements**

1. Rearrangement during electrophilic addition to alkenes (Section 6.13)

$$(CH_3)_2\overset{\displaystyle H}{\overset{|}{C}}CH{=}CH_2 + HCl \longrightarrow (CH_3)_2\overset{\displaystyle Cl}{\overset{|}{C}}CH_2CH_3$$

2. Rearrangement during Friedel–Crafts alkylation (Section 16.3)

$$C_6H_6 \xrightarrow[AlCl_3]{CH_3CH_2CH_2Cl} C_6H_5{-}CH(CH_3)_2$$

## II. A Summary of How Reactions Occur

The second method of organizing reactions is by how they occur—that is, by their *mechanisms*. We said in Chapter 5 that there are three fundamental reaction types—polar reactions, radical reactions, and pericyclic reactions. Having seen many different examples by now, let's see how this assertion stands up.

### A. POLAR REACTIONS

Polar reactions take place between electron-rich reagents (nucleophiles, or Lewis bases) and electron-poor reagents (electrophiles, or Lewis acids). These reactions are heterolytic processes and involve species with an even number of valence electrons. Bonds are made when a nucleophile donates an electron pair to an electrophile; bonds are broken when one product fragment leaves with an electron pair.

Heterogenic bond formation

$$\underset{\text{Electrophile}}{A^+} + \underset{\text{Nucleophile}}{:B^-} \longrightarrow A:B$$

Heterolytic bond cleavage

$$A:B \longrightarrow A^+ + :B^-$$

The polar reactions we've studied can be grouped into five general categories:

1. Electrophilic addition reactions
2. Elimination reactions
3. Nucleophilic alkyl substitution reactions
4. Electrophilic aromatic substitution reactions
5. Nucleophilic aromatic substitution reactions

**1. Electrophilic addition reactions** (Sections 6.9–6.11) Alkenes react with electrophiles such as HBr to yield saturated addition products. The reaction occurs in two steps: The electrophile first adds to the alkene double bond to yield a carbocation intermediate that reacts further to yield the addition product.

$$>C{=}C< \xrightarrow{HBr} \left[ -\overset{H}{\underset{|}{C}}-C^{+}< \;+\; :\ddot{\underset{..}{Br}}:^{-} \right] \longrightarrow -\overset{H}{\underset{|}{C}}-\overset{Br}{\underset{|}{C}}-$$

Alkene Carbocation Addition product

Many of the addition reactions listed in Review Table 1 take place by an electrophilic addition mechanism. The electrophile may be $H^+$ (HX addition; reactions 1a and 2a), $X^+$ (halogen addition and halohydrin formation; reactions 1b and 1c), or $Hg^{2+}$ (hydroxymercuration and alkyne hydration; reactions 1d and 2b), but the basic process is the same. The remaining addition reactions in Review Table 1 differ in that they occur without carbocation intermediates, but it's still convenient to group them together.

## 2. Elimination reactions

*a. E2 reaction* (Sections 11.10–11.12) Alkyl halides undergo elimination of HX to yield alkenes on treatment with base. When a strong base such as hydroxide ion ($HO^-$), alkoxide ion ($RO^-$), or amide ion ($NH_2^-$) is used, alkyl halides react by the E2 mechanism. E2 reactions occur in a single step involving removal by base of a neighboring hydrogen at the same time that the halide ion is leaving.

$$\underset{Br}{C}-\overset{H}{C} \xrightarrow{^{-}:\ddot{O}H\ Na^{+}} >C{=}C< \;+\; H_2O \;+\; NaBr$$

All the elimination reactions listed in Review Table 2 occur by the same E2 mechanism. Though they appear different, the elimination of an alkyl halide to yield an alkene (reaction 1), the elimination of a vinylic halide to yield an alkyne (reaction 2), and the elimination of an aryl halide to yield a benzyne (reaction 3) are all E2 reactions.

*b. E1 reaction* (Section 11.14) Tertiary alkyl halides undergo elimination by the E1 mechanism in competition with $S_N1$ substitution when a nonbasic nucleophile is used in a hydroxylic solvent. The reaction takes place in two steps: Spontaneous dissociation of the alkyl halide leads to a carbocation intermediate that then loses $H^+$.

$$H_3C-\overset{CH_3}{\underset{CH_3}{C}}-Br \longrightarrow \left[ H-\overset{H}{\underset{H}{C}}-C^{+}(CH_3)_2 \;+\; :\ddot{\underset{..}{Br}}:^{-} \right] \xrightarrow{H_2\ddot{O}:} \overset{H}{\underset{H}{>}}C{=}C\overset{CH_3}{\underset{CH_3}{<}} \;+\; HBr$$

Alkyl halide Carbocation Alkene product

### 3. Nucleophilic alkyl substitution reactions

*a.* $S_N2$ *reaction* (Sections 11.2–11.5) The nucleophilic alkyl substitution reaction is one of the most common reactions encountered in organic chemistry. As illustrated in Review Table 3 (reaction 1a), most primary halides and tosylates, and many secondary ones, undergo substitution reactions with a variety of different nucleophiles. The nucleophile might be hydroxide ion, alkoxide ion, ammonia, or many others. One particularly useful $S_N2$ reaction is the alkylation of a terminal alkyne anion (reaction 1b) to yield an internal alkyne product.

Mechanistically, $S_N2$ reactions take place in a single step involving attack of the incoming nucleophile from a direction 180° away from the leaving group. This results in an umbrella-like inversion of stereochemistry (Walden inversion).

$$\text{Br—CH}_3 \xrightarrow{:\ddot{O}H^-} \text{H}_3\text{C—}\ddot{\text{O}}\text{H} + :\ddot{\text{Br}}:^-$$

*b.* $S_N1$ *reaction* (Sections 11.6–11.9) Tertiary alkyl halides undergo nucleophilic substitution by the $S_N1$ mechanism, which occurs in two steps. Spontaneous dissociation of the alkyl halide into an anion and a carbocation intermediate takes place, followed by reaction of the carbocation with a nucleophile. The dissociation step is the slower of the two and is rate-limiting.

$$(CH_3)_3C\text{—Br} \longrightarrow \left[(CH_3)_3C^+ + :\ddot{Br}:^-\right] \xrightarrow{H_2\ddot{O}:} (CH_3)_3C\text{—}\ddot{O}H + HBr$$

Alkyl halide — Carbocation — Substitution product

Among the more useful $S_N1$ reactions is the conversion of a secondary or tertiary alcohol into an alkyl halide by reaction with HX (reaction 2b, Review Table 3).

**4. Electrophilic aromatic substitution reactions** (Sections 16.1–16.4) All of the electrophilic aromatic substitutions shown in Review Table 3 occur by a common two-step mechanism. The first step is similar to the first step in electrophilic addition to alkenes—an electron-poor reagent reacts with the electron-rich aromatic ring. The second step is identical to what happens during E1 elimination—a base present in solution attacks a hydrogen atom next to the positively charged carbon, and elimination of the proton occurs.

$$C_6H_6 \xrightarrow[FeBr_3]{Br\text{—}Br} \left[C_6H_6Br^+ \quad :\ddot{Br}:^-\right] \longrightarrow C_6H_5Br + HBr$$

Benzene — Carbocation — Bromobenzene

**5. Nucleophilic aromatic substitution reactions** (Section 16.9) Nucleophilic aromatic substitution (reaction 4, Review Table 3) is a polar reaction that must be placed in a unique category, since it is not related to the other general processes we have studied. In this reaction, a nucleophile attacks an electrophilic aromatic ring. The ring is made electrophilic, and hence reactive, only when substituted by strong electron-withdrawing groups such as nitro, cyano, and carbonyl. Although this electrophilic behavior of the aromatic ring is a reversal of its normal reactivity, the process still involves reaction between an electrophile and a nucleophile, as with all polar reactions.

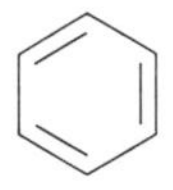

$NO_2$ Cl $O_2N$ $NO_2$

"Normal" aromatic ring; electron-rich and nucleophilic

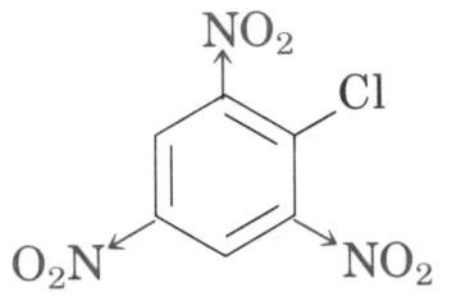

Electron-poor aromatic ring; electrophilic

## B. RADICAL REACTIONS

Radical reactions are homolytic processes that involve species with an odd number of electrons. Bonds are made when each reactant donates one electron, and bonds are broken when each product fragment leaves with one electron.

Homogenic bond formation $A\cdot + \cdot B \longrightarrow A:B$

Homolytic bond cleavage $A:B \longrightarrow A\cdot + \cdot B$

Since radical reactions are less common than polar reactions, we've only seen a few examples. Those we have studied can be classified as either radical addition reactions or radical substitution reactions. Radical additions such as the peroxide-catalyzed addition of HBr to alkenes (reaction 1j, Review Table 1) involve the multistep addition of a radical to an unsaturated substrate. As we'll see in Chapter 31, this kind of reaction is extremely important in the preparation of polymers like polystyrene and polypropylene.

Radical additions occur through three distinct kinds of steps, all of which involve odd-electron species: (1) initiation, (2) propagation, and (3) termination.

1. Initiation steps:
   a. $RO{-}OR \longrightarrow 2\,RO\cdot$
   b. $RO\cdot + H{-}Br \longrightarrow RO{-}H + \cdot Br$
2. Propagation steps:
   a. $>C{=}C< + \cdot Br \longrightarrow \cdot C{-}C(Br){-}$

b. $\cdot\text{C}{-}\text{C(Br)}{-} + \text{H}{-}\text{Br} \longrightarrow {-}\text{C(H)}{-}\text{C(Br)}{-} + \cdot\text{Br}$

3. Termination steps:

a. $\text{Br}\cdot + \cdot\text{Br} \longrightarrow \text{Br}{-}\text{Br}$

b. ${-}\text{C}\cdot + \cdot\text{Br} \longrightarrow {-}\text{C}{-}\text{Br}$

c. ${-}\text{C}\cdot + \cdot\text{C}{-} \longrightarrow {-}\text{C}{-}\text{C}{-}$

The reaction is initiated by thermal homolytic cleavage of a peroxide, which forms two radicals. These radicals abstract H· from HBr, yielding a Br· radical that then adds to the alkene, generating a new carbon radical and a carbon–bromine bond. Each reactant donates one electron to the new carbon–bromine bond. The reaction is completed by reaction of the carbon radical with HBr to yield neutral product and a bromine radical, which continues the chain. Note that the new carbon–hydrogen bond is also formed by donation of one electron from each reactant.

Radical substitution reactions, such as the light-induced chlorination of methane and the allylic bromination of alkenes with *N*-bromosuccinimide (reactions 5a and 5b; Review Table 3) are also common. The key feature of all these reactions is that one radical abstracts an atom from a neutral molecule, leaving a new radical.

## C. PERICYCLIC REACTIONS

Pericyclic reactions such as the Diels–Alder cycloaddition and the addition of a carbene to an alkene (reactions 1h and 1i, Review Table 1) involve neither radicals nor nucleophile–electrophile interactions. Rather, these processes take place in a single step by a reorganization of bonding electrons through a cyclic transition state. A fuller discussion of pericyclic reactions will be given in Chapter 30.

1,3-Butadiene + Methyl propenoate ($\text{H}_2\text{C}{=}\text{CH}{-}\text{CO}_2\text{CH}_3$) ⟶ [Cyclic transition state, $\text{CO}_2\text{CH}_3$]$^‡$ ⟶ cyclohexene-$\text{CO}_2\text{CH}_3$

Cyclic transition state

CHAPTER 17

# Alcohols and Thiols

**Alcohols** are compounds that have hydroxyl groups bonded to saturated, $sp^3$-hybridized carbon atoms. This definition purposely excludes phenols (hydroxyl groups bonded to an aromatic ring) and enols (hydroxyl groups bonded to a vinylic carbon), because the chemistry of the three types of compounds is quite different. Alcohols can be thought of as organic derivatives of water in which one of the water hydrogens is replaced by an organic group (H—O—H versus R—O—H).

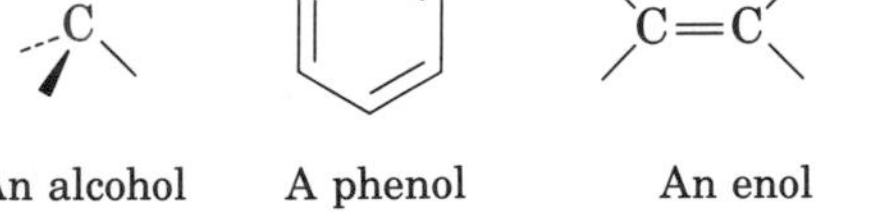

An alcohol A phenol An enol

Alcohols occur widely in nature and have a great many industrial and pharmaceutical applications. Ethanol, for example, is one of the simplest yet best known of all organic substances, finding use as a fuel additive, an industrial solvent, and a beverage; menthol, an alcohol isolated from peppermint oil, is widely used as a flavoring and perfumery agent; and cholesterol, a complicated-looking steroidal alcohol, has been implicated as a causative agent in heart disease.

$CH_3CH_2OH$

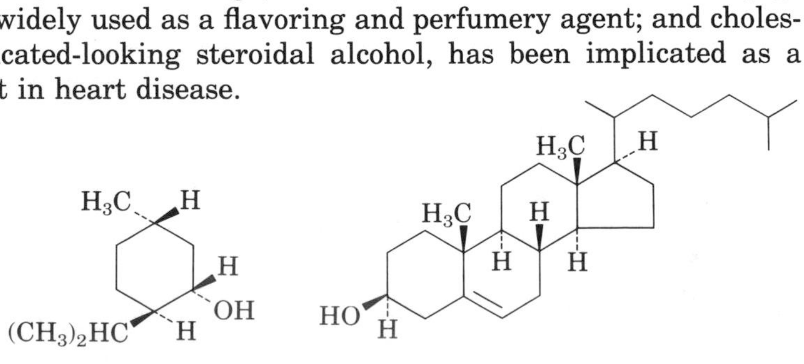

Ethanol Menthol Cholesterol (a steroid)

## 17.1 Nomenclature of Alcohols

Alcohols are classified as either primary (1°), secondary (2°), or tertiary (3°), depending on the number of carbon substituents bonded to the hydroxyl-bearing carbon.

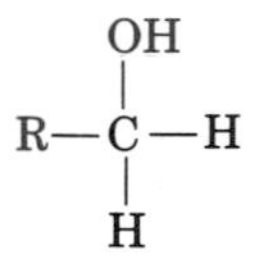

OH | R—C—H | H — A primary alcohol

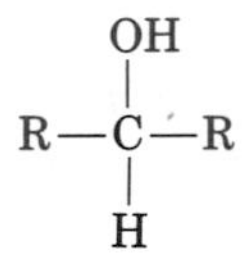

OH | R—C—R | H — A secondary alcohol

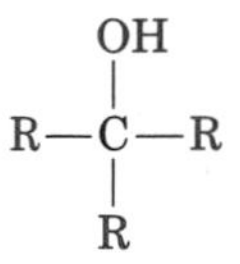

OH | R—C—R | R — A tertiary alcohol

Simple alcohols are named by the IUPAC system as derivatives of the parent alkane, using the suffix *-ol*:

1. Select the longest carbon chain *containing the hydroxyl group*, and derive the parent name by replacing the *-e* ending of the corresponding alkane with *-ol*.
2. Number the alkane chain beginning at the end nearer the hydroxyl group.
3. Number all substituents according to their position on the chain and write the name listing the substituents in alphabetical order.

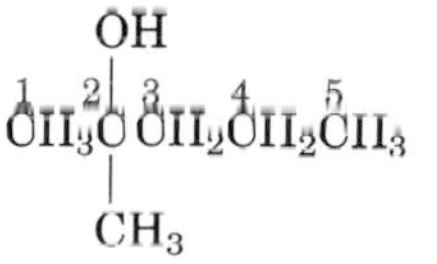

$\overset{1}{C}H_3\overset{2}{C}(OH)(CH_3)\overset{3}{C}H_2\overset{4}{C}H_2\overset{5}{C}H_3$

2-Methyl-2-pentanol

HO H (C-1), HO H (C-4), ring numbered 1–4

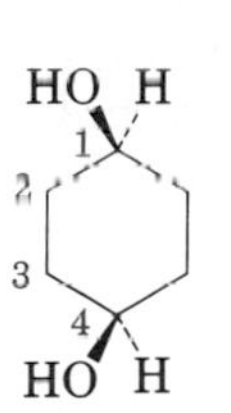

*cis*-1,4-Cyclohexanediol

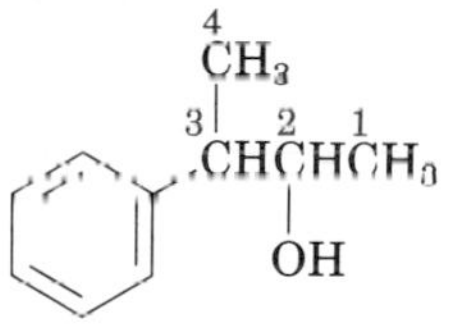

$C_6H_5\overset{3}{C}H(\overset{4}{C}H_3)\overset{2}{C}H(OH)\overset{1}{C}H_3$

3-Phenyl-2-butanol

Certain simple and widely occurring alcohols have common names that are accepted by IUPAC. For example,

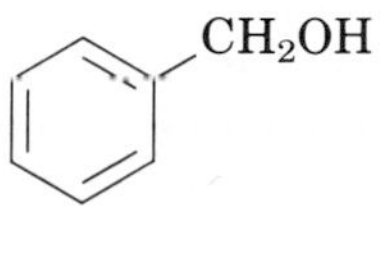

$C_6H_5CH_2OH$

Benzyl alcohol
(Phenylmethanol)

$H_2C{=}CHCH_2OH$

Allyl alcohol
(2-Propen-1-ol)

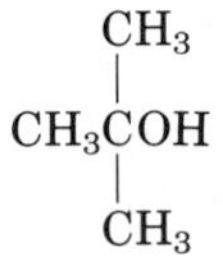

$CH_3C(CH_3)_2OH$

*tert*-Butyl alcohol
(2-Methyl-2-propanol)

$HOCH_2CH_2OH$

Ethylene glycol
(1,2-Ethanediol)

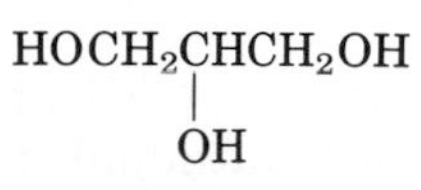

$HOCH_2CH(OH)CH_2OH$

Glycerol
(1,2,3-Propanetriol)

PROBLEM

**17.1** Provide IUPAC names for these compounds:

(a) $CH_3CH(OH)CH_2CH(OH)CH(CH_3)_2$

(b) 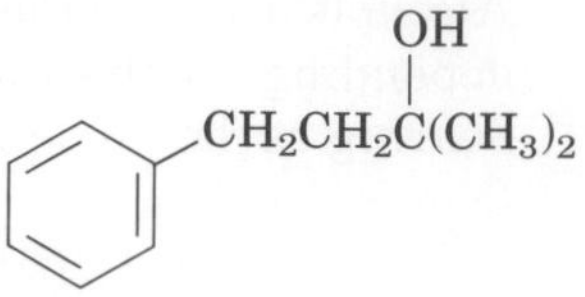 $C_6H_5CH_2CH_2C(OH)(CH_3)_2$

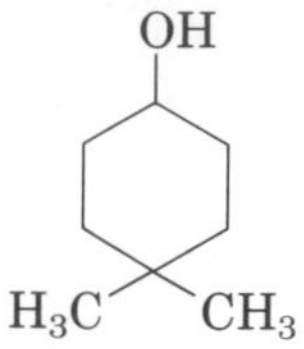

(c) OH, $H_3C$, $CH_3$

(d) 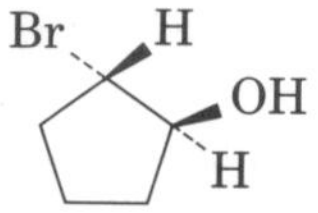 Br, H, OH, H

PROBLEM

**17.2** Draw structures corresponding to these IUPAC names:
(a) 2-Ethyl-2-buten-1-ol
(b) 3-Cyclohexen-1-ol
(c) *trans*-3-Chlorocycloheptanol
(d) 1,4-Pentanediol

## 17.2 Sources and Uses of Simple Alcohols

Methanol and ethanol are two of the most important of all industrial chemicals. Prior to the development of the modern chemical industry, methanol was prepared by heating wood in the absence of air and thus came to be called *wood alcohol*. Today, approximately 1.2 billion gallons of methanol are manufactured each year in the United States by catalytic reduction of carbon monoxide with hydrogen gas.

$$CO + 2\,H_2 \xrightarrow[\text{Zinc oxide/chromia}]{400°C} CH_3OH$$

Methanol is toxic to humans, causing blindness in low doses and death in larger amounts. Industrially, it is used both as a solvent and as a starting material for production of formaldehyde, $CH_2O$, and acetic acid, $CH_3COOH$.

Ethanol is one of the oldest known pure organic chemicals. Its production by fermentation of grains and sugars, and its subsequent purification by distillation, go back at least as far as the twelfth century AD. Fermentation is carried out by adding yeast to an aqueous sugar solution. Enzymes in the yeast break down carbohydrates into ethanol and $CO_2$.

$$\underset{\text{Glucose}}{C_6H_{12}O_6} \xrightarrow{\text{Yeast enzymes}} 2\,CH_3CH_2OH + 2\,CO_2$$

Only about 5% of the ethanol produced industrially comes from fermentation, although that figure may well change drastically in the next decade as demand for use in automobile fuel increases. Most ethanol is currently obtained by acid-catalyzed hydration of ethylene (Section 7.3).

Nearly 300 million gallons of ethanol a year are produced in the United States for use as a solvent or as a chemical intermediate in other industrial reactions.

$$H_2C{=}CH_2 + H_2O \xrightarrow{H_2SO_4} CH_3CH_2OH$$

## 17.3 Properties of Alcohols: Hydrogen Bonding

As mentioned earlier, alcohols can be thought of as organic derivatives of water in which one of the hydrogens has been replaced by an organic group. As such, alcohols have nearly the same geometry as water. The R—O—H bond angle has an approximately tetrahedral value (109° in methanol, for example), and the oxygen atom is $sp^3$ hybridized.

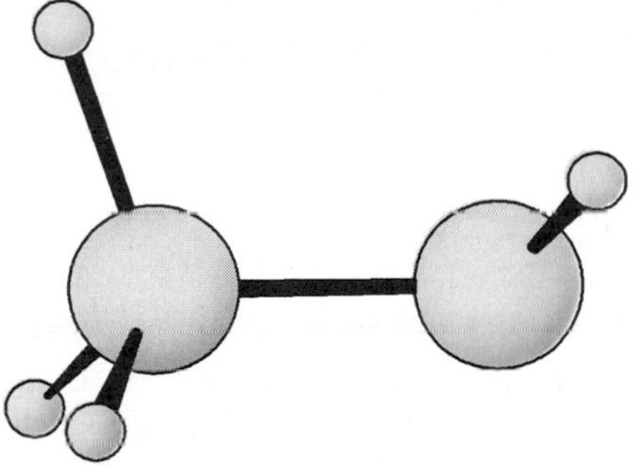

Methanol

Alcohols are quite different from the hydrocarbons and alkyl halides we've studied thus far. Not only is their chemistry much richer, their physical properties are also different. Table 17.1, which provides a comparison of the boiling points of some simple alcohols, alkanes, and chloroalkanes, shows that alcohols have much higher boiling points. For example, 1-propanol (mol wt = 60), butane (mol wt = 58), and chloroethane (mol wt = 65) are close in weight, yet 1-propanol boils at 97°C, compared to −0.5°C for the alkane and 12.5°C for the chloroalkane.

**Table 17.1** Boiling points of alkanes, chloroalkanes, and alcohols (°C)

| *Alkyl group,* R | *Alkane,* R—H | *Chloroalkane,* R—Cl | *Alcohol,* R—OH |
|---|---|---|---|
| $CH_3$— | −162 | −24 | 64.5 |
| $CH_3CH_2$— | −88.5 | 12.5 | 78.3 |
| $CH_3CH_2CH_2$— | −42 | 46.6 | 97 |
| $(CH_3)_2CH$— | −42 | 36.5 | 82.5 |
| $CH_3CH_2CH_2CH_2$— | −0.5 | 83.5 | 117 |
| $(CH_3)_3C$— | −12 | 51 | 83 |

The reason for their high boiling points is that alcohols, like water, are highly associated in solution because of the formation of **hydrogen bonds**. The positively polarized —O—H hydrogen atom from one molecule forms a weak hydrogen bond to the negatively polarized oxygen atom of another molecule (Figure 17.1). Although hydrogen bonds have a strength of only about 5 kcal/mol (20 kJ/mol), versus 103 kcal/mol (431 kJ/mol) for a typical O—H covalent bond, the presence of many hydrogen bonds means that extra energy must be added to break them during the boiling process.

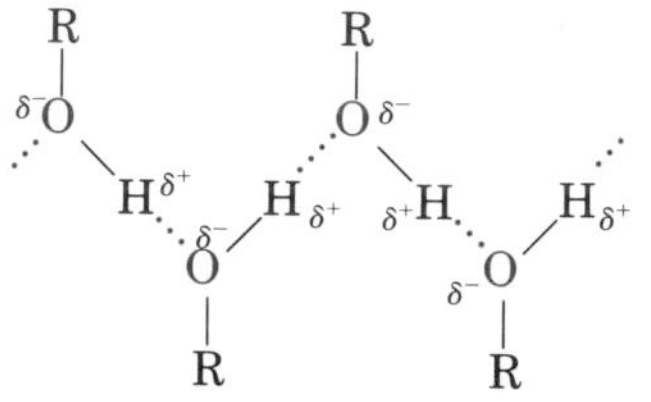

**Figure 17.1** Hydrogen bonding in alcohols

PROBLEM

**17.3** The following data for three isomeric four-carbon alcohols show that there is a decrease in boiling point with increasing substitution:

1-Butanol, bp 117.5°C

2-Butanol, bp 99.5°C

2-Methyl-2-propanol, bp 82.2°C

Propose an explanation to account for this trend.

## 17.4 Properties of Alcohols: Acidity

Like water, alcohols are weakly acidic. In dilute aqueous solution, alcohols dissociate by donating a proton to water.

$$R{-}\ddot{\underset{..}{O}}{-}H + H_2\ddot{O}: \rightleftharpoons R{-}\ddot{\underset{..}{O}}:^- + H_3O:^+$$

In our earlier discussion of acidity (Sections 2.6 and 2.7), we said that the strength of any acid HA in water can be defined by the expressions

$$K_a = \frac{[A^-][H_3O^+]}{[HA]} \qquad \text{and} \qquad pK_a = -\log K_a$$

where $K_a$ is the **acidity constant**. Compounds with a small $K_a$ (or high $pK_a$) are weakly acidic, whereas compounds with a larger $K_a$ (or smaller $pK_a$) are more strongly acidic.

The data presented in Table 17.2 show that alcohols are about as acidic as water. Structural effects, however, play a significant role in determining the exact acidity of a compound. For example, methanol and ethanol are similar to water in acidity, whereas *tert*-butyl alcohol is slightly less acidic. The effect of alkyl substitution on acidity is thought to be due to solvation. Water is able to surround the sterically accessible oxygen atom of unhindered alcohols and to stabilize the alkoxide by solvation, thus favoring its formation. Hindered alkoxides such as *tert*-butoxide, however, prevent solvation by their bulk and are therefore less stabilized.

**Table 17.2** Acidity constants of some alcohols

| *Alcohol* | $pK_a$ | |
|---|---|---|
| $(CH_3)_3COH$ | 18.00 | Weaker acid |
| $CH_3CH_2OH$ | 16.00 | ↓ |
| HOH (water)[a] | (15.74) | ↓ |
| $CH_3OH$ | 15.54 | ↓ |
| $CF_3CH_2OH$ | 12.43 | ↓ |
| $(CF_3)_3COH$ | 5.4 | ↓ |
| HCl (hydrochloric acid)[a] | (−7.00) | Stronger acid |

[a]Values for water and hydrochloric acid are shown for reference.

Inductive effects (Section 16.5) are also important in determining alcohol acidities. For example, electron-withdrawing halogen substituents stabilize an alkoxide anion by helping to spread out the charge over a large area, thus making the alcohol more acidic. This inductive effect can be observed by comparing the acidities of ethanol ($pK_a = 16$) and 2,2,2-trifluoroethanol ($pK_a = 12.43$), or of *tert*-butyl alcohol ($pK_a = 18$) and nonafluoro-2-methyl-2-propanol ($pK_a = 5.4$).

Electron-withdrawing groups stabilize alkoxide and lower $pK_a$

$(CF_3)_3C{-}O^-$ ($pK_a = 5.4$) versus $(CH_3)_3C{-}O^-$ ($pK_a = 18$)

Since alcohols are much weaker than carboxylic acids or mineral acids, they don't react with weak bases such as amines, bicarbonate ion, or metal hydroxides. Alcohols do, however, react with alkali metals and with strong bases like sodium amide ($NaNH_2$), sodium hydride (NaH), alkyllithium reagents (R—Li), and Grignard reagents (R—MgX). The metal salts of alcohols are themselves strong bases that are frequently used as reagents in organic chemistry.

$$CH_3OH + NaH \longrightarrow CH_3O^- Na^+ + H_2$$

Sodium methoxide

$$2\,(CH_3)_3COH + 2\,K \longrightarrow 2\,(CH_3)_3CO^- K^+ + H_2$$

Potassium *tert*-butoxide

$$C_6H_{11}\text{—OH} + CH_3MgBr \longrightarrow C_6H_{11}\text{—O}^- \,{}^+MgBr + CH_4$$

Bromomagnesium cyclohexoxide

PROBLEM

17.4 Which alcohol do you think would be more acidic, benzyl alcohol or *p*-nitrobenzyl alcohol? Explain.

## 17.5 Preparation of Alcohols

In many ways, alcohols occupy a central position in organic chemistry. They can be prepared from a variety of classes of functional groups (alkenes, alkyl halides, ketones, esters, and aldehydes, among others), and they can be transformed into a wide assortment of compound types (Figure 17.2).

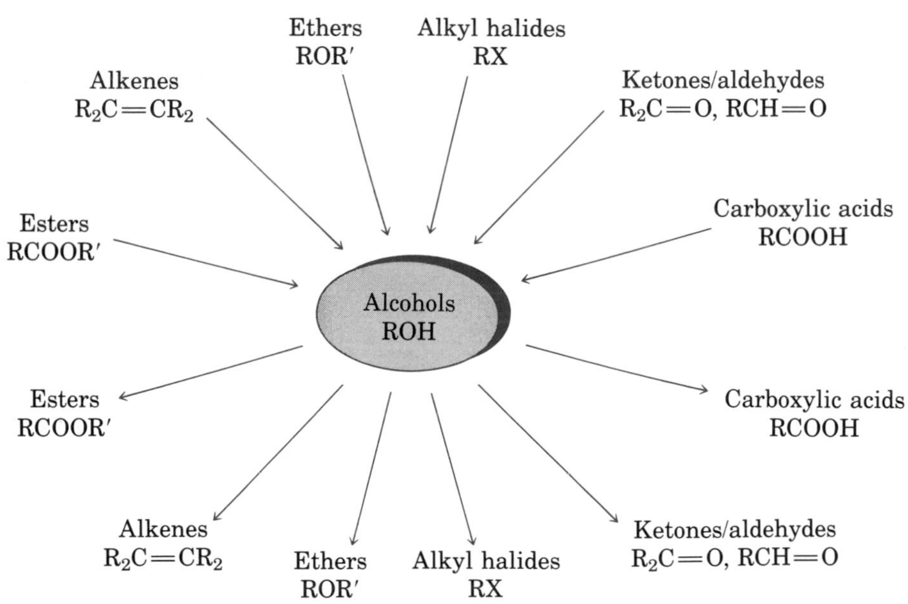


**Figure 17.2** The central position of alcohols in organic chemistry

Let's review briefly some of the methods of alcohol preparation we've already seen. Alcohols can be prepared by hydration of alkenes. Since the direct hydration of alkenes with aqueous acid is generally a poor reaction in the laboratory, two indirect methods are commonly used. A choice between the two is made depending on the product desired. Hydroboration–oxidation (Section 7.4) yields the product of syn non-Markovnikov hydration, whereas oxymercuration–reduction (Section 7.3) yields the product of Markovnikov hydration. Both reactions are mild and are generally applicable to most alkenes (Figure 17.3).

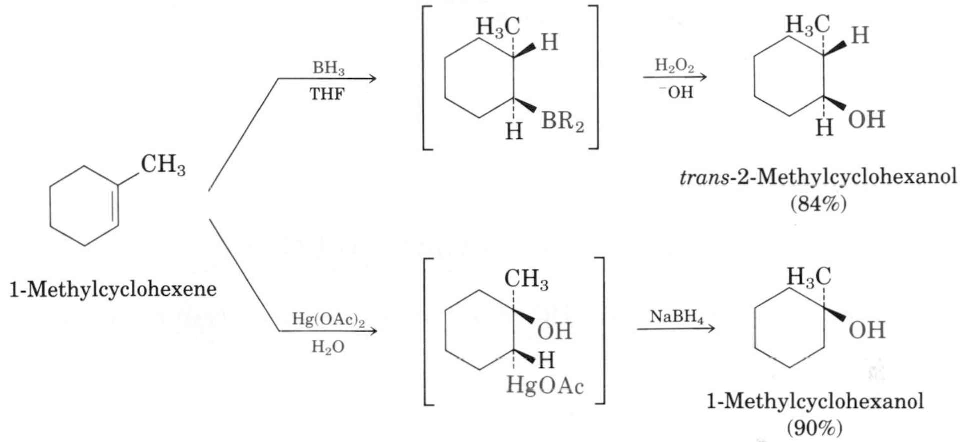

1-Methylcyclohexene

$BH_3$, THF → [intermediate with $H_3C$, H, H, $BR_2$] → $H_2O_2$, $^-OH$ → *trans*-2-Methylcyclohexanol (84%)

$Hg(OAc)_2$, $H_2O$ → [intermediate with $CH_3$, OH, H, HgOAc] → $NaBH_4$ → 1-Methylcyclohexanol (90%)

**Figure 17.3** Two complementary methods for the hydration of an alkene to yield an alcohol

1,2-Diols can be prepared by direct hydroxylation of an alkene with osmium tetraoxide followed by sodium bisulfite reduction (Section 7.7). The reaction takes place readily and is routinely used for the preparation of cis 1,2-diols (Figure 17.4). We'll see in the next chapter that 1,2-diols can also be prepared by acid-catalyzed hydrolysis of *epoxides*—compounds with a three-membered oxygen-containing ring. This method of opening epoxides is complementary to direct hydroxylation since it yields a trans 1,2-diol.

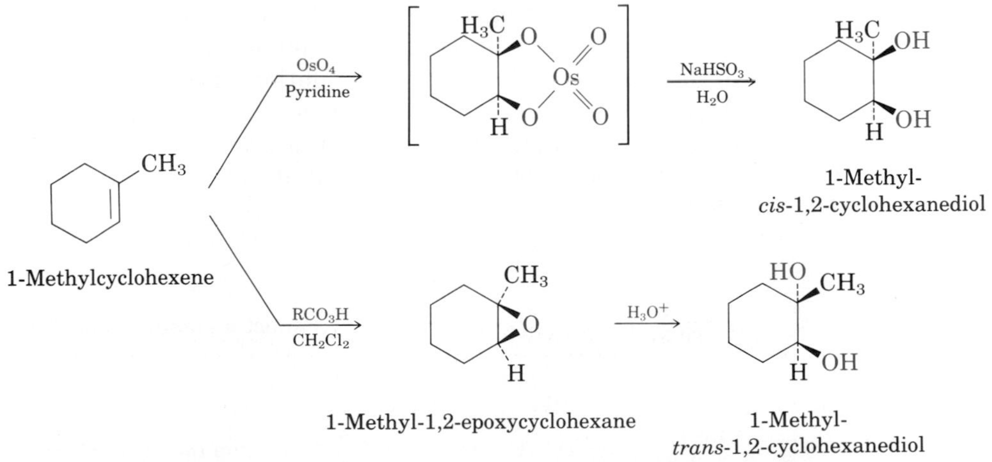

1-Methylcyclohexene

$OsO_4$, Pyridine → [osmate ester intermediate with $H_3C$, H, O, Os, O, O, O] → $NaHSO_3$, $H_2O$ → 1-Methyl-*cis*-1,2-cyclohexanediol

$RCO_3H$, $CH_2Cl_2$ → 1-Methyl-1,2-epoxycyclohexane → $H_3O^+$ → 1-Methyl-*trans*-1,2-cyclohexanediol

**Figure 17.4** Two complementary methods for the preparation of 1,2-diols

PROBLEM

**17.5** Predict the products of these reactions:

(a) 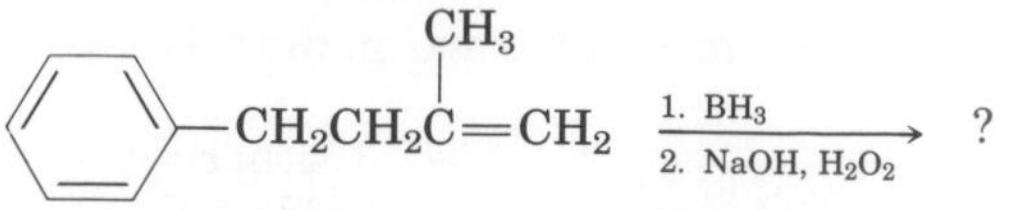 $C_6H_5-CH_2CH_2C(CH_3)=CH_2 \xrightarrow[\text{2. NaOH, } H_2O_2]{\text{1. } BH_3} ?$

(b) $CH_3CH_2CH=C(CH_3)_2 \xrightarrow[\text{2. } NaBH_4]{\text{1. Hg(OAc)}_2\text{, } H_2O} ?$

(c) Reaction of *cis*-5-decene with $OsO_4$, followed by $NaHSO_3$ reduction. Be sure to indicate the stereochemistry of the product.

## 17.6 Alcohols from Reduction of Carbonyl Groups

The most valuable method for preparing alcohols is by **reduction** of carbonyl compounds:

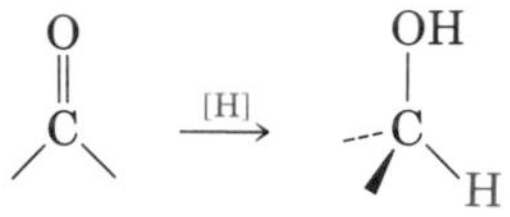

$$\text{C=O} \xrightarrow{[H]} \text{H-C-OH}$$

where [H] is a reducing agent.

In inorganic chemistry, reduction is defined as the gain of electrons by an atom, and oxidation is defined as the loss of electrons. In organic chemistry, however, it's often difficult to decide whether an atom gains or loses electrons during a reaction. Thus, the terms *oxidation* and *reduction* have less precise meanings. For our purposes, an **organic reduction** is a reaction that either *increases* the hydrogen content or *decreases* the oxygen, nitrogen, or halogen content of a molecule. Conversely, an **organic oxidation** is a reaction that either decreases the hydrogen content or increases the oxygen, nitrogen, or halogen content of a molecule.

For example, catalytic hydrogenation of an alkene is clearly a reduction, since two hydrogens are added to the starting material. Hydroxylation of an alkene with $OsO_4$, however, is clearly an oxidation since oxygen is added. Similarly, NBS allylic bromination of an alkene is an oxidation since a hydrogen is replaced by a halogen. Hydration of an alkene is neither an oxidation nor a reduction, since both hydrogen and oxygen are added to the alkene at the same time.

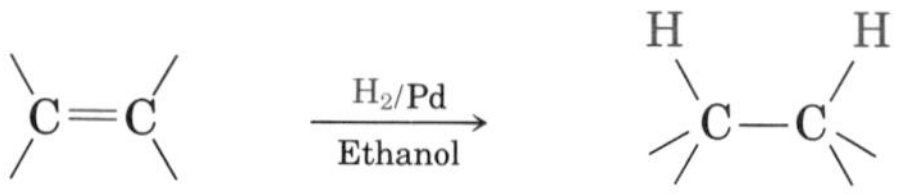

$$\text{C=C} \xrightarrow[\text{Ethanol}]{H_2/Pd} \text{H-C-C-H}$$ Reduction (addition of $H_2$)

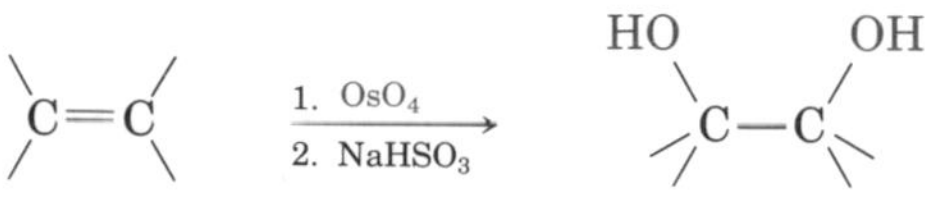

$$\text{C=C} \xrightarrow[\text{2. } NaHSO_3]{\text{1. } OsO_4} \text{HO-C-C-OH}$$ Oxidation (addition of O)

$$C_6H_5CH_3 \xrightarrow[CCl_4]{NBS} C_6H_5CH_2Br$$

Oxidation (removal of H and addition of Br)

$$\text{>C=C<} \xrightarrow{H_3O^+} \text{>C(H)—C(OH)<}$$

Neither oxidation nor reduction (addition of both H and —OH)

A list of classes of functional groups of increasing oxidation state is shown in Figure 17.5. Any reaction that converts a functional group from a lower state to a higher state is an oxidation; any reaction converting a functional group from a higher state to a lower state is a reduction; and any reaction that doesn't change the state is neither an oxidation nor a reduction.

| | | | | |
|---|---|---|---|---|
| | $H_2C{=}CH_2$ | $HC{\equiv}CH$ | | |
| $CH_3CH_3$ | $CH_3CH_2OH$ | $CH_3CH{=}O$ | $CH_3CO_2H$ | $CO_2$ |
| | $CH_3CH_2NH_2$ | $CH_3CII{=}NII$ | $CH_3C{\equiv}N$ | |
| | $CH_3CH_2Cl$ | $CH_3CHCl_2$ | $CH_3CCl_3$ | $CCl_4$ |

Low oxidation level → High oxidation level

**Figure 17.5** Oxidation states of some common functional groups

PROBLEM

**17.6** Rank the following series of compounds in order of increasing oxidation state:

(a) cyclohexane, benzene, cyclohexanone, 1-methoxycyclohexene ($OCH_3$)

(b) $CH_3CN$, $CH_3CH_2NH_2$, $NH_2CH_2CH_2NH_2$

PROBLEM

**17.7** Are these reactions oxidations, reductions, or neither?

(a) Bromocyclohexane + $NaNH_2$ ⟶ Cyclohexene

(b) Benzene + $Cl_2/FeCl_3$ ⟶ Chlorobenzene

(c) 1-Bromobutane + Mg, then $H_3O^+$ ⟶ Butane

(d) Benzene + $CH_3Cl/AlCl_3$ ⟶ Toluene

## REDUCTION OF ALDEHYDES AND KETONES

Aldehydes and ketones are easily reduced to yield alcohols. Aldehydes are converted into primary alcohols, and ketones are converted into secondary alcohols on reduction.

$$\mathrm{RC(=O)H} \xrightarrow{[H]} \mathrm{R{-}CH(OH){-}H}$$

An aldehyde — A primary alcohol

$$\mathrm{RC(=O)R'} \xrightarrow{[H]} \mathrm{R{-}C(OH)(R'){-}H}$$

A ketone — A secondary alcohol

Many reagents are available for reducing ketones and aldehydes to alcohols, but sodium borohydride, $NaBH_4$, is usually chosen because of its safety and ease of handling. Sodium borohydride is a white, crystalline solid that can be safely handled and weighed in the open atmosphere and that can be used either in water or in alcohol solution. High yields of alcohol products are usually obtained, as the following examples indicate.

$$\mathrm{CH_3CH_2CH_2C(=O){-}H} \xrightarrow[\text{2. } H_3O^+]{\text{1. NaBH}_4\text{, ethanol}} \mathrm{CH_3CH_2CH_2CH(OH){-}H}$$

Butanal — 1-Butanol (85%)

$m$-Hydroxybenzaldehyde $\xrightarrow[\text{2. } H_3O^+]{\text{1. NaBH}_4\text{, ethanol}}$ $m$-Hydroxybenzyl alcohol (93%)

Dicyclohexyl ketone $\xrightarrow[\text{2. } H_3O^+]{\text{1. NaBH}_4\text{, ethanol}}$ Dicyclohexylmethanol (88%)

Lithium aluminum hydride, $LiAlH_4$, a white powder soluble in ether and tetrahydrofuran, is another reducing agent that is sometimes used for reduction of ketones and aldehydes. Although more powerful and reactive than $NaBH_4$, $LiAlH_4$ is also far more dangerous and should be handled only by skilled persons. It reacts violently with water, decomposes explosively when heated above 120°C, and has even been known to explode when being ground with a mortar and pestle. Despite these drawbacks, $LiAlH_4$ is an extremely valuable reagent that is used daily in thousands of laboratories.

Lithium aluminum hydride is particularly useful for reducing $\alpha,\beta$-unsaturated ketones (ketones conjugated with carbon–carbon double bonds). Although $\alpha,\beta$-unsaturated ketones often undergo overreduction with $NaBH_4$ to give a mixture of both unsaturated alcohol and saturated alcohol, clean reduction to the allylic alcohol occurs with $LiAlH_4$. Thus, 2-cyclohexenone gives a 59:41 mixture of two products when $NaBH_4$ is used as a reducing agent, but gives largely one product when $LiAlH_4$ is used (Figure 17.6).

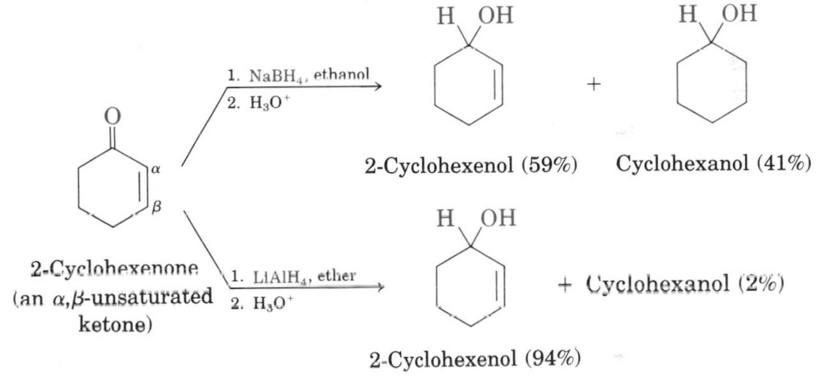

**Figure 17.6** The reduction of 2-cyclohexenone with $NaBH_4$ and with $LiAlH_4$

## REDUCTION OF ESTERS AND CARBOXYLIC ACIDS

Esters and carboxylic acids can be reduced to give primary alcohols:

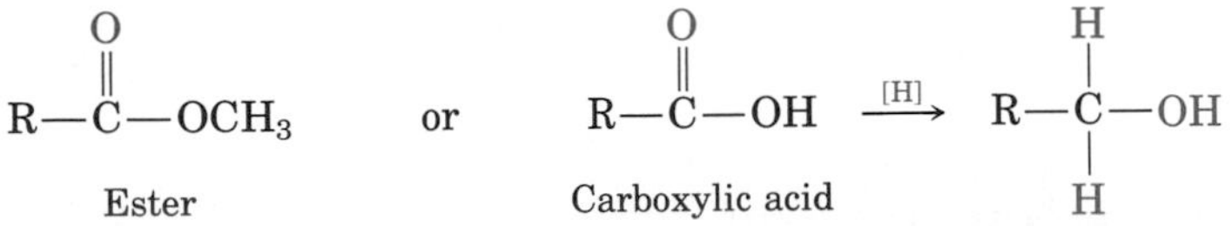

$$\underset{\text{Ester}}{R-\overset{O}{\overset{\|}{C}}-OCH_3} \quad \text{or} \quad \underset{\text{Carboxylic acid}}{R-\overset{O}{\overset{\|}{C}}-OH} \xrightarrow{[H]} R-\overset{H}{\overset{|}{\underset{H}{\underset{|}{C}}}}-OH$$

These reactions are more difficult than the corresponding reductions of aldehydes and ketones. For example, sodium borohydride only slowly reduces esters and does not reduce acids at all. Ester and carboxylic acid reductions are therefore usually carried out with lithium aluminum hydride. All carbonyl groups, including esters, acids, ketones, and aldehydes, are reduced by $LiAlH_4$ in high yield, as some of the following examples indicate. Note that *one* hydrogen atom is delivered to the carbonyl carbon atom during reductions of ketones and aldehydes, but that *two* hydrogens become bonded to the carbonyl carbon during ester and carboxylic acid reductions.

Ester reduction:

$$CH_3CH_2CH{=}CH\overset{O}{\overset{\|}{C}}OCH_2CH_3 \xrightarrow[\text{2. } H_3O^+]{\text{1. LiAlH}_4\text{, ether}} CH_3CH_2CH{=}CHCH_2OH$$

Ethyl 2-pentenoate — 2-Penten-1-ol (91%)

Ketone reduction:

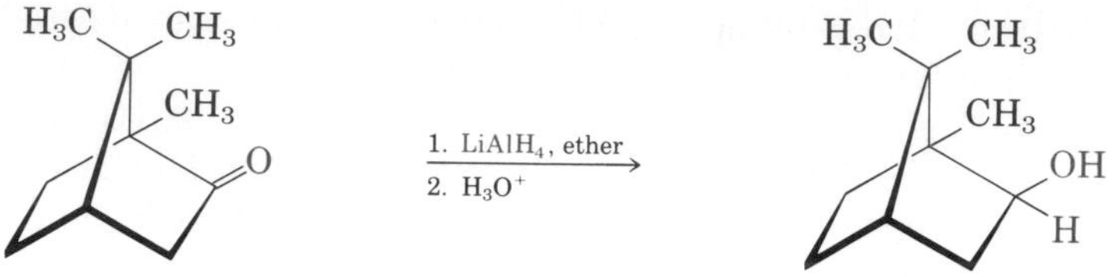

Camphor $\xrightarrow[\text{2. } H_3O^+]{\text{1. LiAlH}_4\text{, ether}}$ Borneol (95%)

Aldehyde reduction:

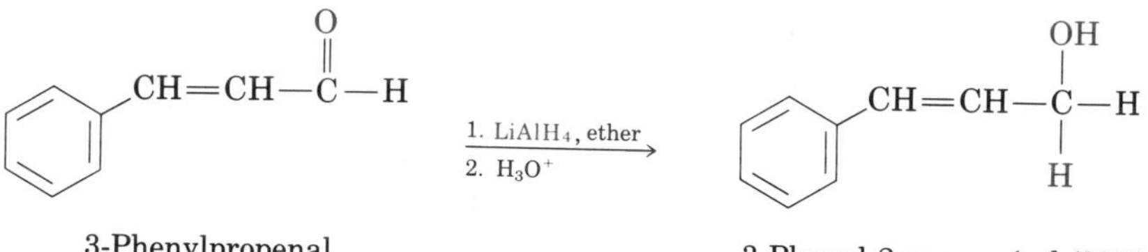

$$C_6H_5CH{=}CH{-}\overset{O}{\overset{\|}{C}}{-}H \xrightarrow[\text{2. } H_3O^+]{\text{1. LiAlH}_4\text{, ether}} C_6H_5CH{=}CH{-}CH(OH){-}H$$

3-Phenylpropenal — 3-Phenyl-2-propen-1-ol (90%)

Carboxylic acid reduction:

$$CH_3(CH_2)_7CH{=}CH(CH_2)_7CO_2H \xrightarrow[\text{2. } H_3O^+]{\text{1. LiAlH}_4\text{, THF}} CH_3(CH_2)_7CH{=}CH(CH_2)_7CH_2OH$$

Oleic acid — 9-Octadecen-1-ol (87%)

**PROBLEM**

**17.8** What reagent would you use to accomplish each of these reactions?

(a) $CH_3\overset{O}{\overset{\|}{C}}CH_2CH_2CO_2CH_3 \xrightarrow{?} CH_3\overset{OH}{\overset{|}{C}}HCH_2CH_2CO_2CH_3$

(b) $CH_3\overset{O}{\overset{\|}{C}}CH_2CH_2CO_2CH_3 \xrightarrow{?} CH_3\overset{OH}{\overset{|}{C}}HCH_2CH_2CH_2OH$

(c) Carvone (from spearmint oil) $\xrightarrow{?}$ corresponding alcohol (OH in place of C=O)

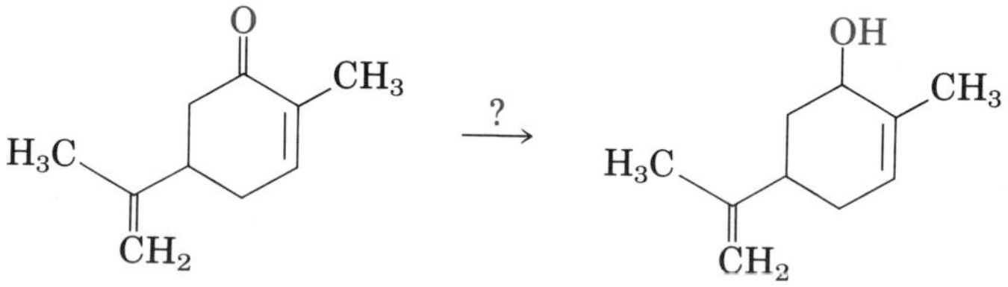

(d) 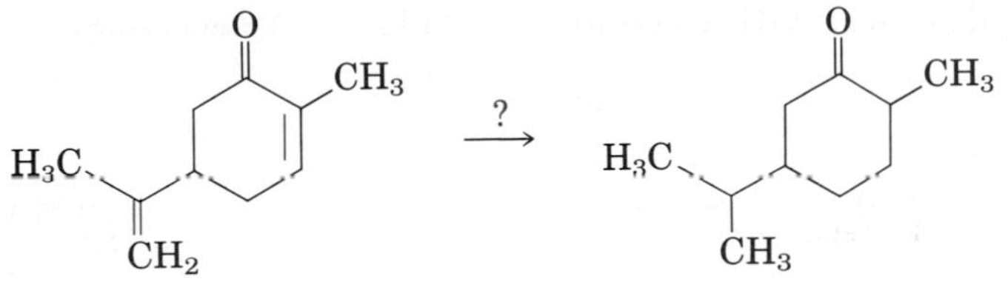 [structure: 2-methyl-5-(1-methylethenyl)cyclohex-2-enone] $\xrightarrow{?}$ [structure: 2-methyl-5-(1-methylethyl)cyclohexanone]

PROBLEM

**17.9** What carbonyl compounds give the following alcohols on reduction with $LiAlH_4$? Show all possibilities.

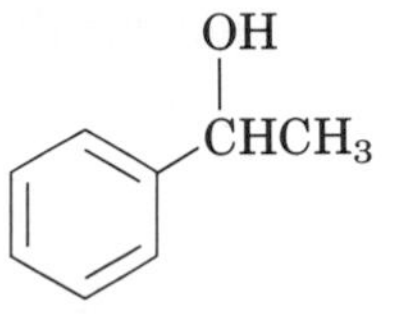

(a) 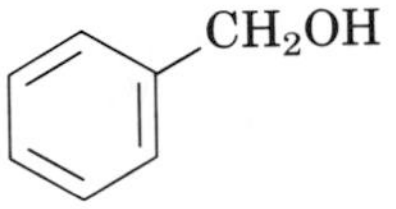 $C_6H_5CH_2OH$

(b) $C_6H_5CH(OH)CH_3$

(c) Cyclohexane bearing OH and H on the same carbon

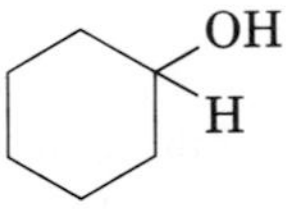

(d) $(CH_3)_2CHCH_2OH$

## 17.7 Alcohols from Addition of Grignard Reagents to Carbonyl Groups

Grignard reagents, RMgX, react with carbonyl compounds to yield alcohols in much the same manner that hydride reagents do. The result is a highly useful and general method of alcohol synthesis.

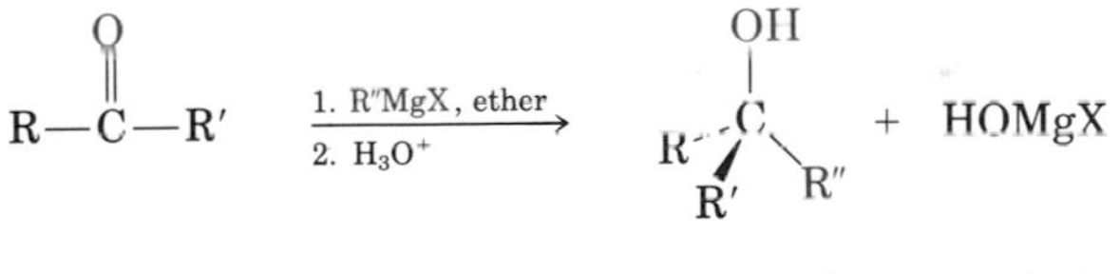

$$R{-}\overset{O}{\overset{\|}{C}}{-}R' \xrightarrow[2.\ H_3O^+]{1.\ R''MgX,\ ether} RR'R''C{-}OH + HOMgX$$

We saw in Section 10.9 that alkyl, aryl, and vinylic halides react with magnesium in ether or tetrahydrofuran solution to generate Grignard reagents, RMgX.

$$R{-}X + Mg \longrightarrow \overset{\delta^-}{R}{-}\overset{\delta^+}{MgX}$$

A Grignard reagent

where R = 1°, 2°, or 3° alkyl, aryl, or vinylic

X = Cl, Br, or I

A large number of alcohol products can be obtained from Grignard reactions, depending on the reagents used. For example, Grignard reagents react with formaldehyde, $CH_2{=}O$, to give primary alcohols:

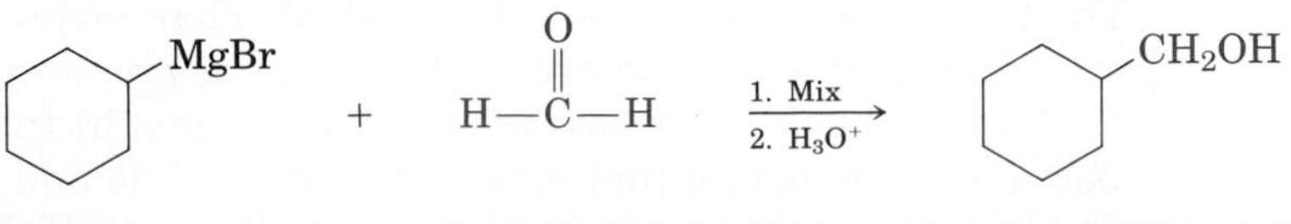

$$C_6H_{11}MgBr + H{-}\overset{O}{\overset{\|}{C}}{-}H \xrightarrow[2.\ H_3O^+]{1.\ Mix} C_6H_{11}CH_2OH$$

Cyclohexylmagnesium bromide + Formaldehyde → Cyclohexylmethanol (65%) (a 1° alcohol)

Aldehydes react with Grignard reagents to give secondary alcohols:

$$(CH_3)_2CHCH_2\overset{\overset{\displaystyle O}{\|}}{C}H \; + \; C_6H_5MgBr \xrightarrow[\text{then } H_3O^+]{\text{Ether}} (CH_3)_2CHCH_2\underset{\underset{\displaystyle OH}{|}}{C}H{-}C_6H_5$$

3-Methylbutanal — Phenylmagnesium bromide — 3-Methyl-1-phenyl-1-butanol (73%) (a 2° alcohol)

Ketones react similarly to yield tertiary alcohols:

$$\text{Cyclohexanone} \xrightarrow[\text{2. } H_3O^+]{\text{1. } CH_3CH_2{-}MgBr \text{ (ethylmagnesium bromide), ether}} \text{1-Ethylcyclohexanol (HO, } CH_2CH_3\text{)}$$

Cyclohexanone — 1-Ethylcyclohexanol (89%) (a 3° alcohol)

Esters react with Grignard reagents to yield tertiary alcohols in which two of the substituents bonded to the hydroxyl-bearing carbon have come from the Grignard reagent (just as $LiAlH_4$ reduction of esters adds *two* hydrogens). For example,

$$CH_3CH_2CH_2CH_2\overset{\overset{\displaystyle O}{\|}}{C}OCH_2CH_3 \xrightarrow[\text{2. } H_3O^+]{\text{1. } 2\,CH_3MgBr} CH_3CH_2CH_2CH_2\underset{\underset{\displaystyle CH_3}{|}}{\overset{\overset{\displaystyle OH}{|}}{C}}{-}CH_3 + CH_3CH_2OH$$

Ethyl pentanoate — 2-Methyl-2-hexanol (85%) (a 3° alcohol)

Carboxylic acids do not give addition products with Grignard reagents because the acidic carboxyl proton reacts with the Grignard reagent to produce a hydrocarbon and the magnesium salt of the acid. (We saw this reaction previously in Section 10.9 as a means of reducing alkyl halides to alkanes.)

$$R{-}Br + Mg \longrightarrow R{-}MgBr$$

$$R{-}MgBr + R'\overset{\overset{\displaystyle O}{\|}}{C}O{-}H \longrightarrow R{-}H + R'\overset{\overset{\displaystyle O}{\|}}{C}O{-}MgBr$$

Carboxylic acid — Hydrocarbon — Acid salt (unreactive)

The Grignard reaction, though useful, also has severe limitations. The major problem is that a Grignard reagent can't be prepared from an organohalide if there are other reactive functional groups in the same molecule. For example, a compound that is both an alkyl halide and a ketone will not form a Grignard reagent—instead, it reacts with itself. Similarly, a compound that is both an alkyl halide and a carboxylic acid, alcohol, or amine can't form a Grignard reagent because the acidic $RCO_2H$, ROH, or $RNH_2$ protons present in the same molecule simply react with the basic Grignard reagent as rapidly as it forms.

Generally speaking, Grignard reagents *cannot* be prepared from compounds with these functional groups (FG):

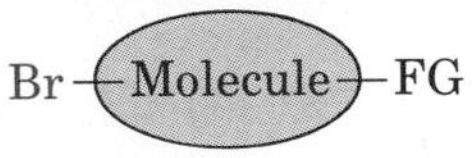

Br—(Molecule)—FG

where FG = —OH, —NH, —SH — The Grignard reagent is protonated by these groups.

FG = $—CO_2H$, $—NO_2$, —CHO, —COR, —CN, $—CONH_2$, or $—SO_2R$ — The Grignard reagent adds to these groups.

**PRACTICE PROBLEM**

How could you use the addition of a Grignard reagent to a ketone to synthesize 2-phenyl-2-propanol?

*Solution* First draw the structure of the product and identify the groups bonded to the alcohol carbon atom. In the present instance, there are two methyl groups ($—CH_3$) and one phenyl ($—C_6H_5$). One of the three will have come from a Grignard reagent, and the remaining two will have come from a ketone. Thus the possibilities are addition of methylmagnesium bromide to acetophenone and addition of phenylmagnesium bromide to acetone:

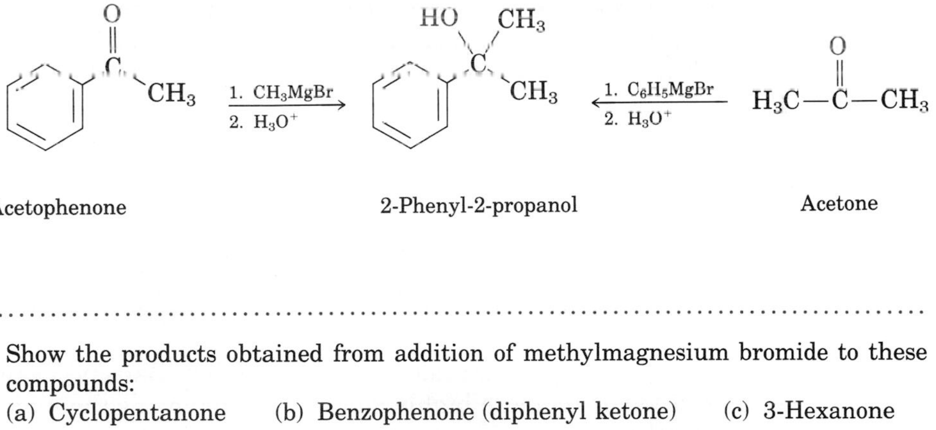

Acetophenone $\xrightarrow[2.\ H_3O^+]{1.\ CH_3MgBr}$ 2-Phenyl-2-propanol $\xleftarrow[2.\ H_3O^+]{1.\ C_6H_5MgBr}$ Acetone ($H_3C—C(=O)—CH_3$)

**PROBLEM**

**17.10** Show the products obtained from addition of methylmagnesium bromide to these compounds:
(a) Cyclopentanone (b) Benzophenone (diphenyl ketone) (c) 3-Hexanone

**PROBLEM**

**17.11** How could you use a Grignard addition reaction to prepare these alcohols?
(a) 2-Methyl-2-propanol
(b) 1-Methylcyclohexanol
(c) 3-Methyl-3-pentanol
(d) 2-Phenyl-2-butanol
(e) Benzyl alcohol

**PROBLEM**

**17.12** How can you explain the observation that treatment of 4-hydroxycyclohexanone with 1 equiv of methylmagnesium bromide yields none of the expected addition product, whereas treatment with an excess of Grignard reagent leads to a good yield of 1-methyl-1,4-cyclohexanediol?

## 17.8 Reactions of Alcohols

Reactions of alcohols can be divided into two groups—those that occur at the C—O bond and those that occur at the O—H bond:

$$-\overset{|}{\underset{|}{C}}-\ddot{O}-H \qquad \text{C—O reactions; O—H reactions}$$

Let's begin to look at reactions of both types by reviewing some of the alcohol reactions seen in previous chapters.

### DEHYDRATION OF ALCOHOLS TO YIELD ALKENES

One of the most important C—O bond reactions of alcohols is **dehydration** to give alkenes. The carbon–oxygen bond is broken, a neighboring C—H is broken, and an alkene pi bond is formed:

$$\text{H–C–C–OH} \longrightarrow \text{C=C} + H_2O \qquad \text{A dehydration reaction}$$

Because of the importance of the reaction, a number of alternative ways have been devised for carrying out dehydrations. One of the more common methods, which works particularly well for tertiary alcohols, is the acid-catalyzed method discussed earlier (Section 7.12). For example, when 1-methylcyclohexanol is warmed with aqueous sulfuric acid in a solvent such as tetrahydrofuran, loss of water occurs and 1-methylcyclohexene is formed:

$$\text{1-Methylcyclohexanol} \xrightarrow[50°C]{H_3O^+,\ THF} \text{1-Methylcyclohexene (91\%)}$$

Acid-catalyzed dehydrations normally follow Zaitsev's rule (Section 11.10) and yield the more highly substituted alkene as the major product. Thus, 2-methyl-2-butanol gives primarily 2-methyl-2-butene (trisubstituted double bond) rather than 2-methyl-1-butene (disubstituted double bond):

$$H_3C-C(CH_3)(OH)-CH_2CH_3 \xrightarrow[25°C]{H_3O^+,\ THF} (CH_3)_2C{=}CHCH_3 + CH_2{=}C(CH_3)-CH_2CH_3$$

2-Methyl-2-butanol → 2-Methyl-2-butene (trisubstituted), Major product + 2-Methyl-1-butene (disubstituted), Minor product

In normal laboratory practice, only tertiary alcohols are commonly dehydrated with acid. Secondary alcohols can be made to react, but the conditions are severe (75% $H_2SO_4$, 100°C) and sensitive molecules cannot survive.

$$\underset{\text{2-Butanol}}{CH_3CH_2\overset{\overset{\displaystyle OH}{|}}{C}HCH_3} \xrightarrow[100^\circ C]{75\%\ H_2SO_4} \underset{\text{2-Butene}}{CH_3CH{=}CHCH_3}$$

Primary alcohols are even less reactive than secondary ones, and very harsh conditions are necessary to cause dehydration (95% $H_2SO_4$, 150°C). Thus, the reactivity order for acid-catalyzed dehydrations is

$$R_3COH > R_2CHOH > RCH_2OH$$

The reasons for the observed reactivity order are best understood by looking at the mechanism of the reaction (Figure 17.7). As indicated, acid-catalyzed dehydrations are E1 reactions (Section 11.14), which occur by a

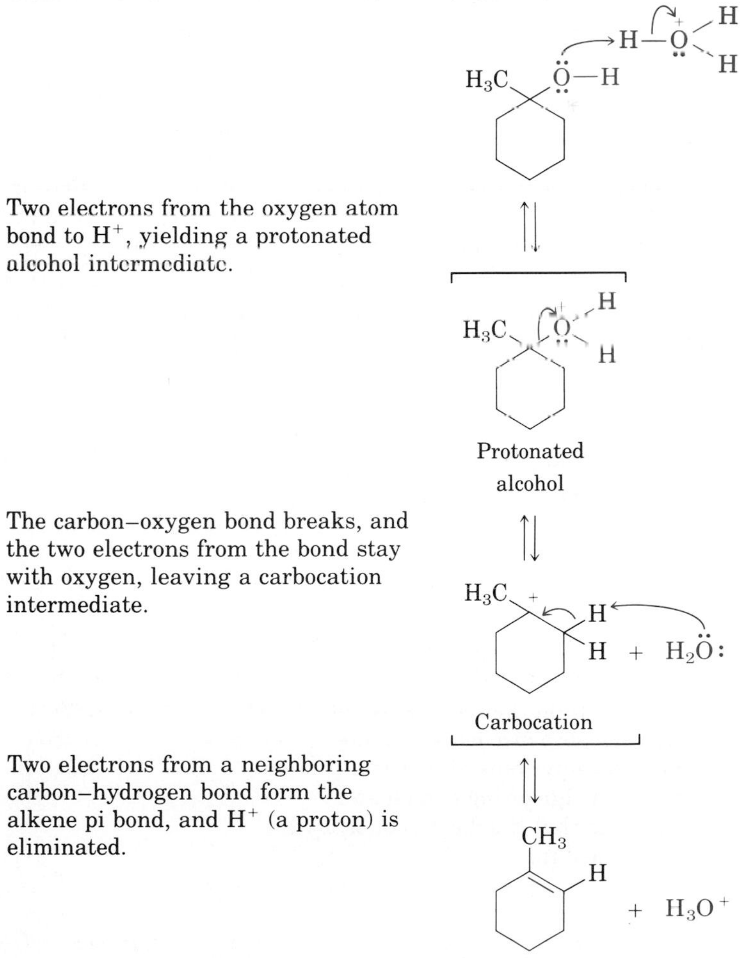

Two electrons from the oxygen atom bond to $H^+$, yielding a protonated alcohol intermediate.

Protonated alcohol

The carbon–oxygen bond breaks, and the two electrons from the bond stay with oxygen, leaving a carbocation intermediate.

$+ H_2\ddot{O}:$

Carbocation

Two electrons from a neighboring carbon–hydrogen bond form the alkene pi bond, and $H^+$ (a proton) is eliminated.

$+ H_3O^+$

**Figure 17.7** Mechanism of the acid-catalyzed dehydration of alcohols to yield alkenes

three-step mechanism involving protonation of the alcohol oxygen, loss of water to generate a carbocation intermediate, and final loss of a proton ($H^+$) from the neighboring carbon atom.

Once the acid-catalyzed dehydration is recognized to be an E1 reaction, the reason why tertiary alcohols react fastest becomes clear. Tertiary substrates *always* react fastest in E1 reactions because they lead to highly stabilized tertiary carbocation intermediates.

To circumvent the necessity for strong acids and allow the dehydration of secondary alcohols in a gentler way, other reagents have been developed that are effective under mild, basic conditions. One such reagent, phosphorus oxychloride ($POCl_3$), is often able to effect the dehydration of secondary and tertiary alcohols at 0°C in the basic amine solvent pyridine:

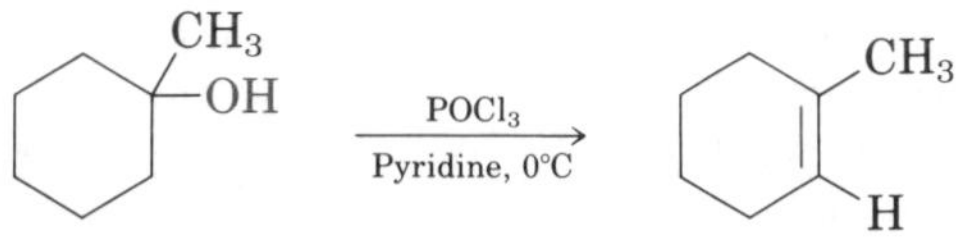

$CH_3$ OH —($POCl_3$ / Pyridine, 0°C)→ $CH_3$ H

1-Methylcyclohexanol → 1-Methylcyclohexene (96%)

Alcohol dehydrations carried out with $POCl_3$ take place by the mechanism shown in Figure 17.8. As indicated, the reaction is an E2 process. Since hydroxide ion is a poor leaving group (Section 11.5), direct E2 elimination of water from an alcohol does not occur. In the presence of $POCl_3$, however, the hydroxyl group is converted into a dichlorophosphate, which

The alcohol hydroxyl group reacts with $POCl_3$ to form a dichlorophosphate intermediate.

$\ddot{O}H$ + $O{=}PCl_3$

E2 elimination then occurs by the usual one-step mechanism as the amine base pyridine abstracts a proton from the neighboring carbon at the same time that the dichlorophosphate group is leaving.

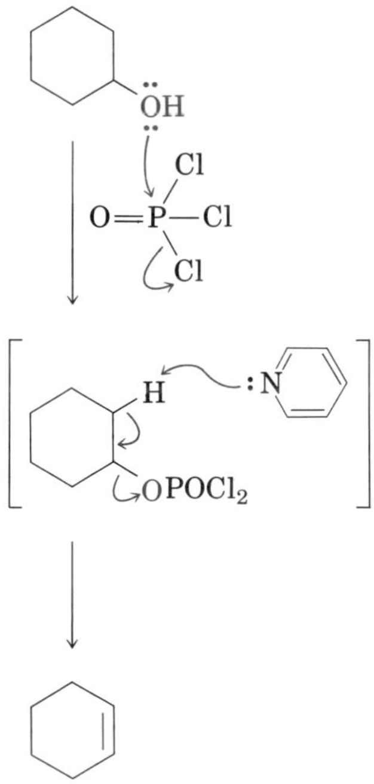

[H :N ; $OPOCl_2$]

**Figure 17.8** Mechanism of the dehydration of secondary and tertiary alcohols by reaction with $POCl_3$ in pyridine: The reaction is an E2 process.

is an excellent leaving group and is readily eliminated to yield an alkene. Pyridine, an organic amine, serves both as reaction solvent and as base to abstract a neighboring proton in the E2 elimination step.

PROBLEM

**17.13** What product(s) would you expect to obtain from dehydration of these alcohols with $POCl_3$ in pyridine? Indicate the major product in each case.

(a) $CH_3CH_2CH(OH)CH(CH_3)_2$ (OH on the third carbon)

(b) *trans*-2-Methylcyclohexanol

(c) *cis*-2-Methylcyclohexanol

PROBLEM

**17.14** Good evidence for the intermediacy of carbocations in the acid-catalyzed dehydration of alcohols comes from the observation that rearrangements sometimes occur. Propose a mechanism to account for the formation of 2,3-dimethyl-2-butene from 3,3-dimethyl-2-butanol. [*Hint:* See Section 6.13.]

$$(CH_3)_3CCH(OH)CH_3 \xrightarrow{H_2SO_4} (CH_3)_2C{=}C(CH_3)_2 + H_2O$$

## CONVERSION OF ALCOHOLS INTO ALKYL HALIDES

A second C—O bond reaction of alcohols is their conversion into alkyl halides (Section 10.8). Tertiary alcohols are readily converted into alkyl halides by treatment with either HCl or HBr at 0°C. Primary and secondary alcohols are much more resistant to acid, however, and are best converted into halides by treatment with either $SOCl_2$ or $PBr_3$.

As discussed in Section 11.16, the reaction of a tertiary alcohol with HX takes place by an $S_N1$ route. Acid protonates the hydroxyl oxygen atom, water is expelled to generate a carbocation, and the cation reacts with nucleophilic halide ion to give the alkyl halide product (Figure 17.9).

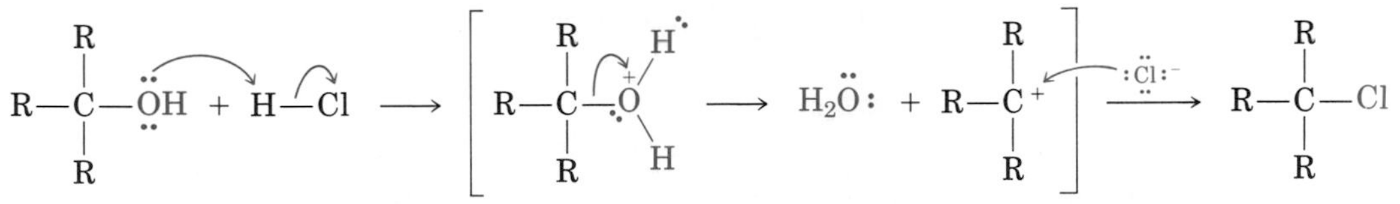

$$R_3C{-}\ddot{O}H + H{-}Cl \longrightarrow \left[R_3C{-}\overset{+}{O}H_2\right] \longrightarrow H_2\ddot{O}: + \left[R_3C^+\right] \xrightarrow{:\ddot{C}l:^-} R_3C{-}Cl$$

**Figure 17.9** The $S_N1$ reaction of a tertiary alcohol with HCl to yield an alkyl halide: Neutral water is the leaving group.

The reactions of primary and secondary alcohols with $SOCl_2$ and $PBr_3$ take place by $S_N2$ routes. Hydroxide ion itself is too poor a leaving group to be displaced by nucleophiles in $S_N2$ reactions, but reaction of an alcohol with $SOCl_2$ or $PBr_3$ converts the hydroxyl into a much better leaving group that is readily expelled by back-side nucleophilic attack (Figure 17.10).

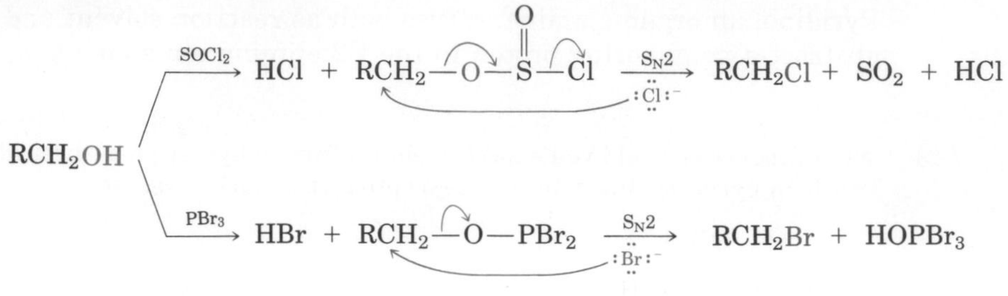

$$RCH_2OH \xrightarrow{SOCl_2} HCl + RCH_2{-}O{-}\overset{\overset{O}{\|}}{S}{-}Cl \xrightarrow[:\ddot{Cl}:^-]{S_N2} RCH_2Cl + SO_2 + HCl$$

$$RCH_2OH \xrightarrow{PBr_3} HBr + RCH_2{-}O{-}PBr_2 \xrightarrow[:\ddot{Br}:^-]{S_N2} RCH_2Br + HOPBr_3$$

**Figure 17.10** Conversion of a primary alcohol into alkyl halides by $S_N2$ reactions with $SOCl_2$ and $PBr_3$

## CONVERSION OF ALCOHOLS INTO TOSYLATES

We saw in Section 11.2 that alcohols react with *p*-toluenesulfonyl chloride (tosyl chloride, *p*-TosCl) in pyridine solution to yield alkyl tosylates, R—O—Tos. Only the O—H bond of the alcohol is broken in this reaction; the C—O bond remains intact, and no change of configuration occurs if the alcohol is chiral. The resultant alkyl tosylates behave much like alkyl halides in their chemistry, undergoing $S_N1$ and $S_N2$ substitution reactions with ease.

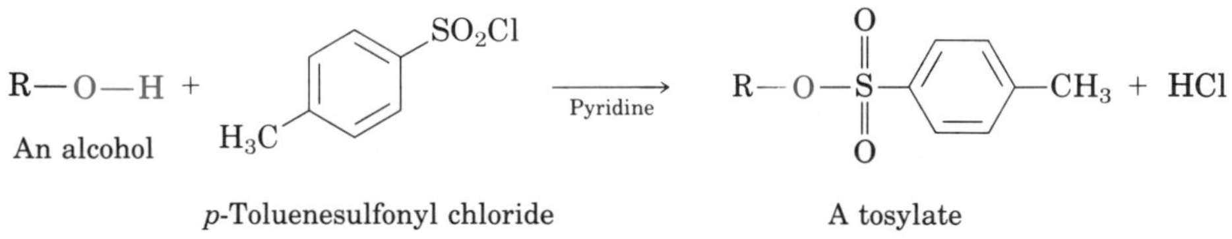

$$\underset{\text{An alcohol}}{R{-}O{-}H} + \underset{p\text{-Toluenesulfonyl chloride}}{H_3C{-}C_6H_4{-}SO_2Cl} \xrightarrow[\text{Pyridine}]{} \underset{\text{A tosylate}}{R{-}O{-}\overset{\overset{O}{\|}}{\underset{\underset{O}{\|}}{S}}{-}C_6H_4{-}CH_3} + HCl$$

One of the most important reasons for using tosylates instead of halides is stereochemical. In a sense, the two sequences of reactions

$$\text{Alcohol} \longrightarrow \text{Halide} \xrightarrow{S_N2\text{ reaction}} \text{Product}$$

and

$$\text{Alcohol} \longrightarrow \text{Tosylate} \xrightarrow{S_N2\text{ reaction}} \text{Product}$$

are stereochemically complementary. The $S_N2$ reaction via the halide proceeds with *two* Walden inversions—one to make the halide from the alcohol and one to substitute the halide—and yields a product with the same absolute stereochemistry as the starting material. The $S_N2$ reaction via the tosylate proceeds with only *one* Walden inversion and yields a product of opposite absolute stereochemistry from the starting material. Figure 17.11 gives a series of reactions on optically active 2-octanol that illustrates these stereochemical relationships.

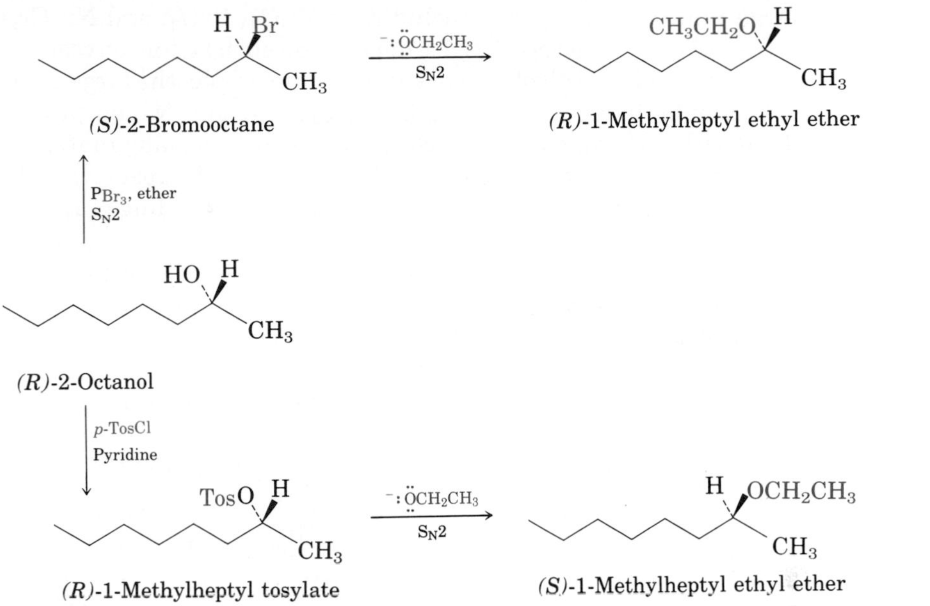

(S)-2-Bromooctane $\xrightarrow[S_N2]{^-:\ddot{O}CH_2CH_3}$ (R)-1-Methylheptyl ethyl ether

$PBr_3$, ether, $S_N2$ (from (R)-2-Octanol to (S)-2-Bromooctane)

(R)-2-Octanol

p-TosCl, Pyridine

(R)-1-Methylheptyl tosylate $\xrightarrow[S_N2]{^-:\ddot{O}CH_2CH_3}$ (S)-1-Methylheptyl ethyl ether

**Figure 17.11** Stereochemical consequences of some $S_N2$ reactions on derivatives of (R)-2-octanol

## 17.9 Oxidation of Alcohols

The most important reaction of alcohols is their oxidation to yield carbonyl compounds. Primary alcohols yield aldehydes or carboxylic acids; secondary alcohols yield ketones; and tertiary alcohols do not react with most oxidizing agents except under the most vigorous conditions.

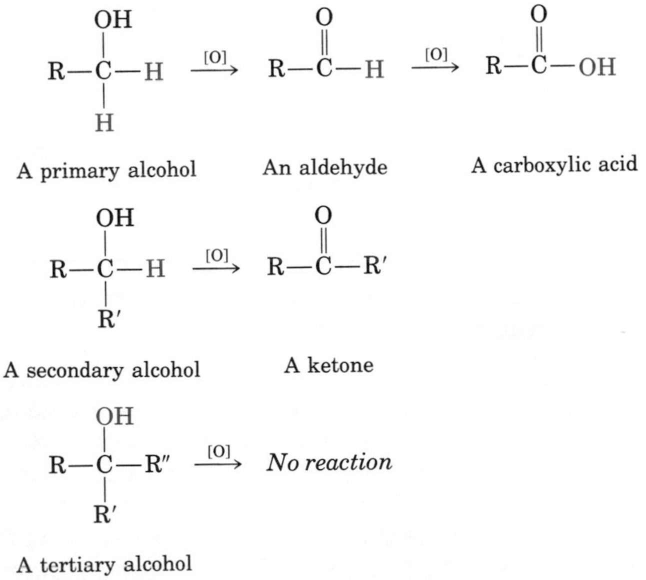

$$R\text{—}CH(OH)\text{—}H \xrightarrow{[O]} R\text{—}C(=O)\text{—}H \xrightarrow{[O]} R\text{—}C(=O)\text{—}OH$$

A primary alcohol — An aldehyde — A carboxylic acid

$$R\text{—}C(OH)(R')\text{—}H \xrightarrow{[O]} R\text{—}C(=O)\text{—}R'$$

A secondary alcohol — A ketone

$$R\text{—}C(OH)(R')\text{—}R'' \xrightarrow{[O]} \textit{No reaction}$$

A tertiary alcohol

where [O] = An oxidizing reagent

Oxidation of primary and secondary alcohols can be accomplished by a large number of reagents, including $KMnO_4$, $CrO_3$, and $Na_2Cr_2O_7$. Which reagent is used in a specific case depends on such factors as cost, convenience, reaction yield, and alcohol sensitivity. For example, the large-scale oxidation of a simple, inexpensive alcohol like cyclohexanol would probably best be done with a cheap oxidant such as potassium permanganate. On the other hand, the small-scale oxidation of a delicate and expensive polyfunctional alcohol would best be done with a mild and high-yielding reagent, regardless of cost.

Primary alcohols are oxidized either to aldehydes or to carboxylic acids, depending on the reagents chosen and on the conditions used. Probably the best method for preparing aldehydes from primary alcohols on a laboratory scale (as opposed to an industrial scale) is by use of pyridinium chlorochromate (PCC, $C_5H_6NCrO_3Cl$) in dichloromethane solvent.

$$CH_3(CH_2)_5CH_2OH \xrightarrow[CH_2Cl_2]{PCC} CH_3(CH_2)_5CHO$$

1-Heptanol → Heptanal (78%)

Citronellol (from rose oil) $\xrightarrow[CH_2Cl_2]{PCC}$ Citronellal (82%)

where PCC = $C_5H_5N^+{-}H\ CrO_3Cl^-$

Most other oxidizing agents, such as chromium trioxide ($CrO_3$) in aqueous sulfuric acid (**Jones' reagent**), oxidize primary alcohols to carboxylic acids. Aldehydes are involved as intermediates in the Jones oxidation but can't usually be isolated because they are further oxidized too readily.

$$CH_3(CH_2)_8CH_2OH \xrightarrow[\text{Acetone}]{\text{Jones' reagent } (CrO_3,\ H_2SO_4,\ H_2O)} CH_3(CH_2)_8CO_2H$$

1-Decanol → Decanoic acid (93%)

(1-Phenylcyclopentyl)methanol $\xrightarrow{\text{Jones' reagent}}$ 1-Phenylcyclopentanecarboxylic acid (85%)

Secondary alcohols are oxidized easily and in high yields to give ketones. For large-scale oxidations, an inexpensive reagent such as sodium dichromate in aqueous acetic acid is used.

4-*tert*-Butylcyclohexanol $\xrightarrow[H_2O,\ CH_3CO_2H,\ \Delta]{Na_2Cr_2O_7}$ 4-*tert*-Butylcyclohexanone (91%)

For more sensitive alcohols, pyridinium chlorochromate or Jones' reagent is often used, since these reactions are milder and occur at lower temperatures.

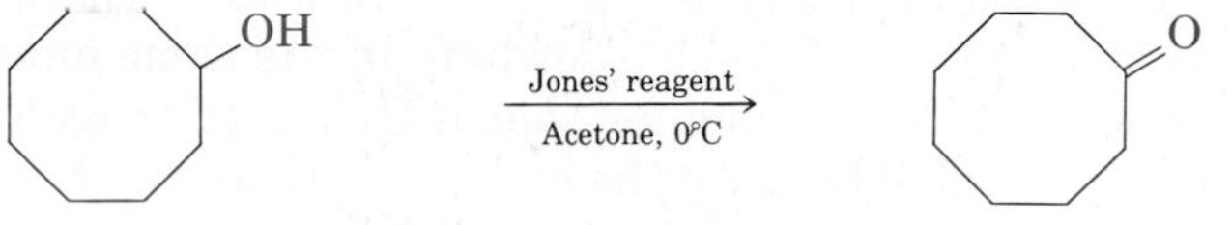

Jones' reagent
Acetone, 0°C

Cyclooctanol

Cyclooctanone (96%)

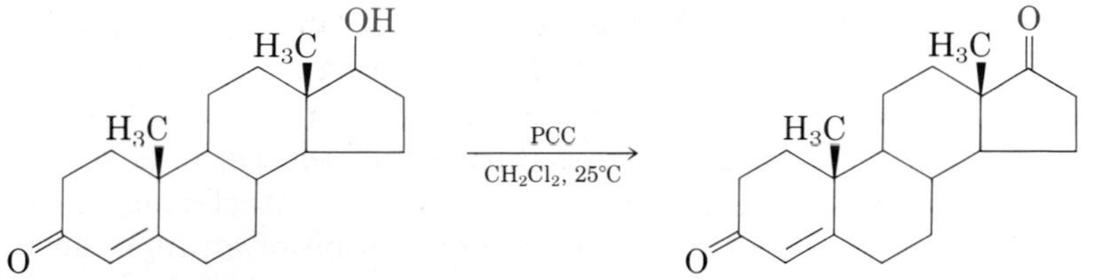

PCC
$CH_2Cl_2$, 25°C

Testosterone
(steroid; male sex hormone)

4-Androstene-3,17-dione (82%)

All these oxidations occur by a pathway closely related to the E2 reaction. The first step involves reaction between the alcohol and a chromium(VI) reagent to form an intermediate chromate. Bimolecular elimination then yields the carbonyl product.

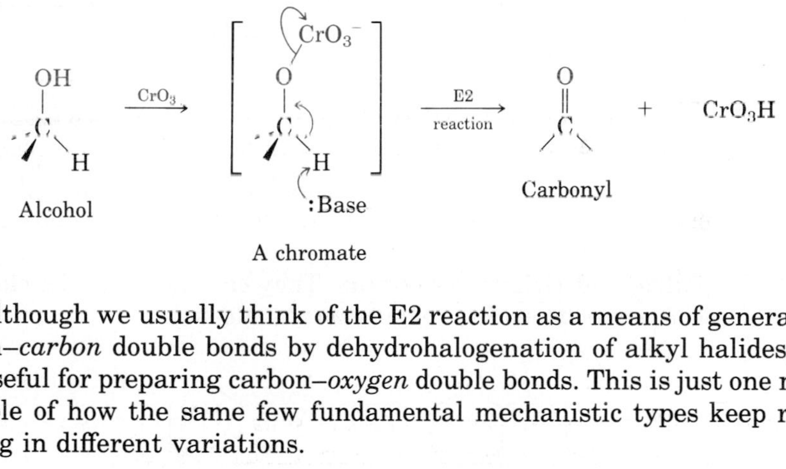

$CrO_3$

E2 reaction

+ $CrO_3H$

Alcohol

A chromate

:Base

Carbonyl

Although we usually think of the E2 reaction as a means of generating carbon–*carbon* double bonds by dehydrohalogenation of alkyl halides, it's also useful for preparing carbon–*oxygen* double bonds. This is just one more example of how the same few fundamental mechanistic types keep reappearing in different variations.

PROBLEM

**17.15** What alcohols would give these products on oxidation?

(a)

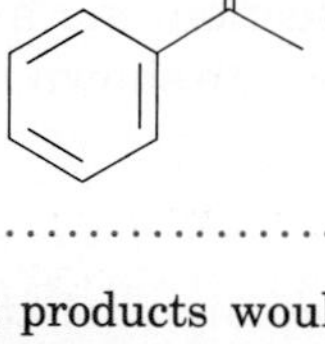

(b) $CH_3CH(CH_3)CHO$

(c)

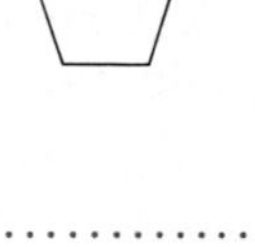

PROBLEM

**17.16** What products would you expect from oxidation of these compounds with Jones' reagent? With pyridinium chlorochromate?

(a) 1-Hexanol (b) 2-Hexanol (c) Hexanal

## 17.10 Protection of Alcohols

It often happens, particularly during the synthesis of complex molecules, that one functional group in a molecule interferes with an intended reaction on a second functional group elsewhere in the same molecule. For example, we saw earlier in this chapter that a Grignard reagent can't be prepared from a halo alcohol because the carbon–magnesium bond is not compatible with the presence of an acidic hydroxyl group in the same molecule.

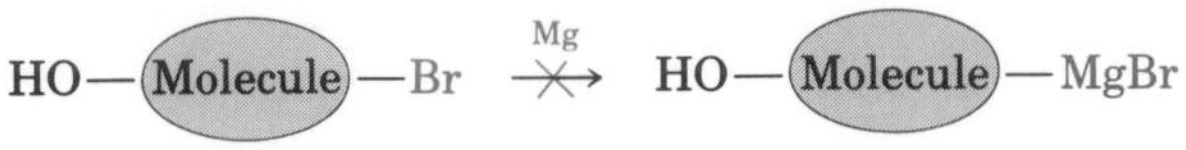

HO—(Molecule)—Br —Mg→ (crossed) HO—(Molecule)—MgBr

*Not formed*

When this kind of incompatibility arises, it is sometimes possible to circumvent the problem by *protecting* the interfering functional group. Protection involves three steps: (1) formation of an inert derivative, (2) carrying out the desired reaction, and (3) removal of the protecting group.

One common method of alcohol protection involves acid-catalyzed reaction with dihydropyran to yield a tetrahydropyranyl (THP) ether. The reaction is simply an acid-catalyzed electrophilic addition of the alcohol to the electron-rich double bond of dihydropyran:

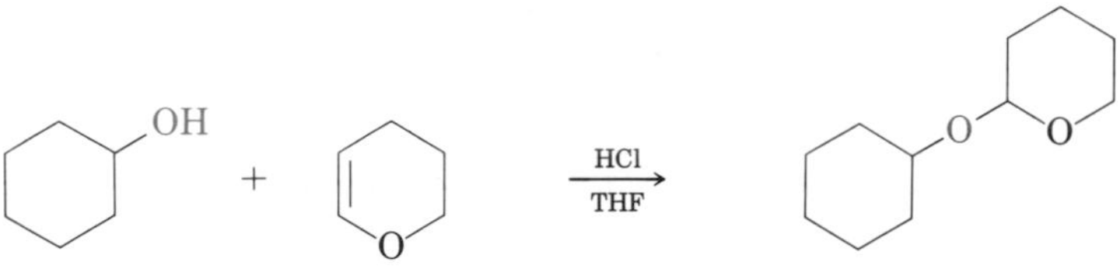

OH + O —HCl, THF→ O O

Cyclohexanol Dihydropyran Cyclohexyl 2-tetrahydropyranyl ether (95%)

Like most other ethers that we'll study in the next chapter, THP ethers are relatively unreactive. They have no acidic protons and are therefore protected against reaction with oxidizing agents, reducing agents, nucleophiles, and Grignard reagents. They can, however, be cleaved by reaction with aqueous acid to regenerate the alcohol.

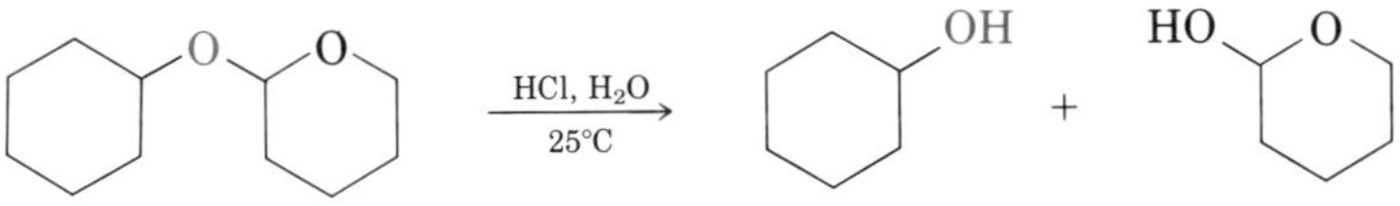

O O —HCl, $H_2O$, 25°C→ OH + HO O

Cyclohexyl THP ether Cyclohexanol Tetrahydropyran-2-ol

To complete the earlier example, it is possible to use a halo alcohol in a Grignard reaction by employing a three-step sequence: (1) protection, (2) Grignard formation and reaction, (3) deprotection. For example, we can add 3-bromo-1-propanol to acetaldehyde by the route shown in Figure 17.12.

PROBLEM

**17.17** Show the mechanism of THP ether formation by acid-catalyzed electrophilic addition of an alcohol to dihydropyran. How can you account for the regiochemistry of the reaction?

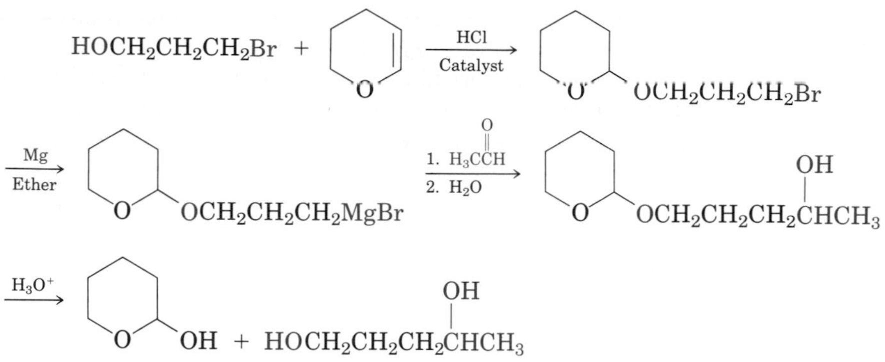

$HOCH_2CH_2CH_2Br$ + (dihydropyran) $\xrightarrow[\text{Catalyst}]{HCl}$ (THP)–$OCH_2CH_2CH_2Br$

$\xrightarrow[\text{Ether}]{Mg}$ (THP)–$OCH_2CH_2CH_2MgBr$ $\xrightarrow[2.\ H_2O]{1.\ H_3CCHO}$ (THP)–$OCH_2CH_2CH_2CH(OH)CH_3$

$\xrightarrow{H_3O^+}$ (THP)–$OH$ + $HOCH_2CH_2CH_2CH(OH)CH_3$

**Figure 17.12** Use of a THP-protected alcohol during a Grignard reaction

## 17.11 Spectroscopic Analysis of Alcohols

### INFRARED SPECTROSCOPY

Alcohols show a characteristic O—H stretching absorption at 3300–3600 $cm^{-1}$ in the infrared spectrum that simplifies their spectroscopic identification. The exact position of the absorption band depends on the extent of hydrogen bonding in the sample. Unassociated alcohols show a fairly sharp absorption near 3600 $cm^{-1}$, whereas hydrogen-bonded alcohols show a broader absorption in the 3300–3400 $cm^{-1}$ range. The hydrogen-bonded hydroxyl absorption is easily seen at 3350 $cm^{-1}$ in the spectrum of cyclohexanol (Figure 17.13). Alcohols also show a strong C—O stretching absorption near 1050 $cm^{-1}$.

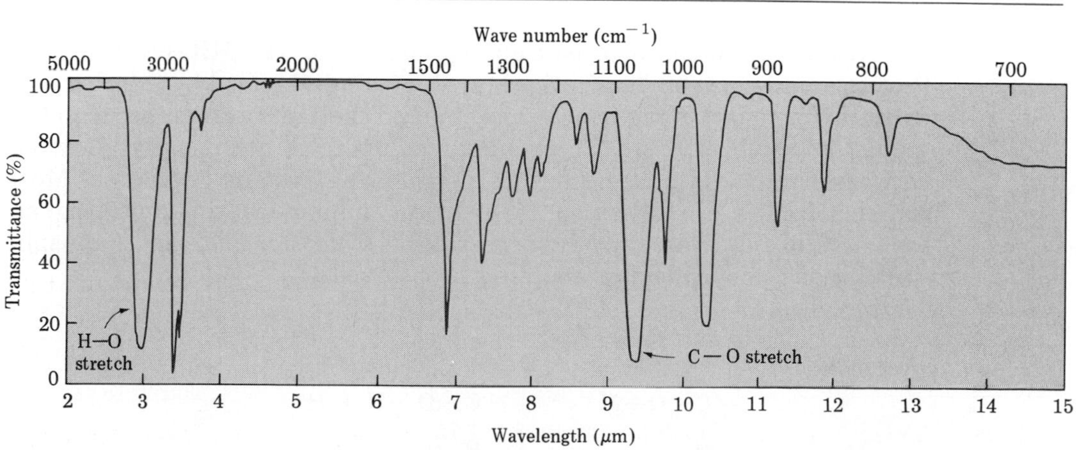


**Figure 17.13** Infrared spectrum of cyclohexanol

PROBLEM

**17.18** Let's assume that you needed to prepare 5-cholestene-3-one from cholesterol by Jones oxidation. How could you use infrared spectroscopy to tell if the reaction was successful? What differences would you look for in the infrared spectra of starting material and product?

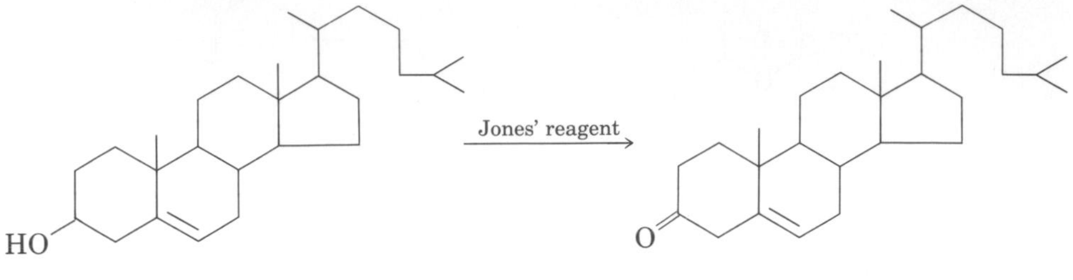

Jones' reagent

Cholesterol 5-Cholestene-3-one

## NUCLEAR MAGNETIC RESONANCE SPECTROSCOPY

Carbon atoms bearing hydroxyl substituents are somewhat deshielded and absorb at a lower field than do normal alkane carbon atoms in the $^{13}C$ NMR spectrum. Most alcohol carbon absorptions fall in the range 50–80 $\delta$, as the spectral data in Figure 17.14 illustrate for cyclohexanol.

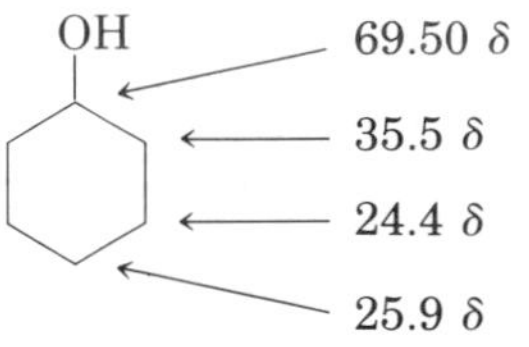

OH

69.50 $\delta$

35.5 $\delta$

24.4 $\delta$

25.9 $\delta$

**Figure 17.14** Carbon-13 NMR data for cyclohexanol

Alcohols show characteristic absorptions in the $^{1}H$ NMR spectrum also. Protons on the oxygen-bearing carbon atom are deshielded by the electron-withdrawing effect of the nearby oxygen, and their absorptions occur in the region from 3.5 to 4.5 $\delta$. Surprisingly, splitting is not usually observed between the O—H proton and the neighboring protons on carbon. Most samples contain small amounts of acidic impurities that catalyze an exchange of the hydroxyl proton on a time scale so rapid that spin–spin splitting is removed and no coupling is observed.

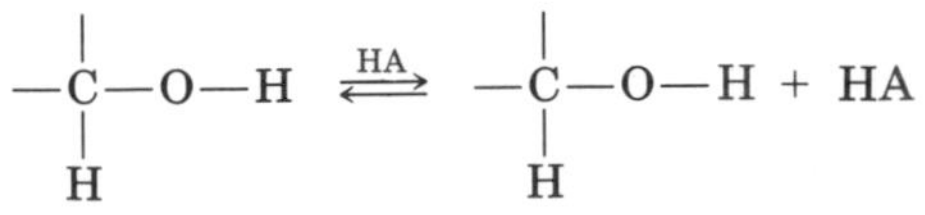

$$-\overset{|}{\underset{H}{\underset{|}{C}}}-O-H \overset{HA}{\rightleftharpoons} -\overset{|}{\underset{H}{\underset{|}{C}}}-O-H + HA$$

No NMR coupling observed

It is often possible to take advantage of this rapid exchange to identify the position of the O—H absorption. If a small amount of deuterated water, $D_2O$, is added to the NMR sample tube, the O—H proton is rapidly exchanged for deuterium, and the hydroxyl absorption disappears from the spectrum.

$$\text{—}\overset{\diagdown}{\underset{\diagup}{C}}\text{—O—H} \overset{D_2O}{\rightleftharpoons} \text{—}\overset{\diagdown}{\underset{\diagup}{C}}\text{—O—D} + \text{HDO}$$

Normal spin–spin splitting *is* observed between protons on the oxygen-bearing carbon and other neighbors. For example, the signal of the two $—CH_2O—$ protons in 1-propanol is split into a triplet by coupling with the neighboring $—CH_2—$ protons (Figure 17.15).

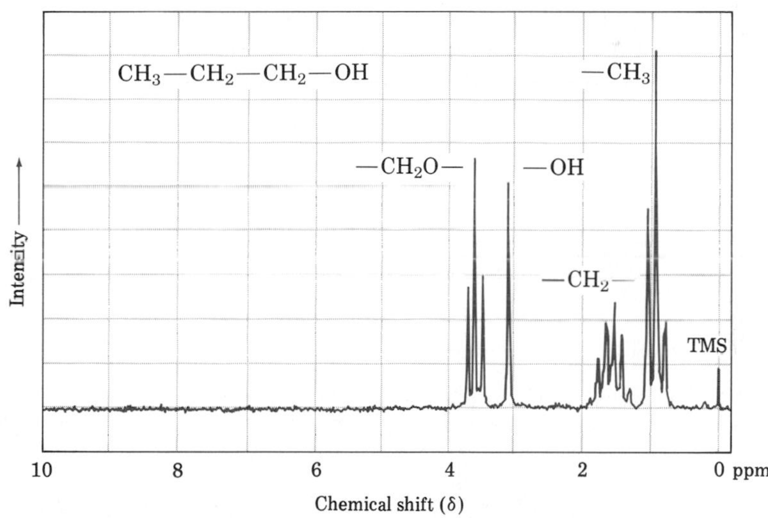


**Figure 17.15** Proton NMR spectrum of 1-propanol: The protons on the oxygen-bearing carbon are split into a triplet at 3.6 δ.

PROBLEM

**17.19** When the $^1H$ NMR spectra of alcohols are run in dimethyl sulfoxide (DMSO) solvent, exchange of the hydroxyl proton is slow, and spin–spin splitting is observed between the —O—H proton and C—H protons on the adjacent carbon. What spin multiplicities would you expect for the hydroxyl protons in these alcohols?

(a) 2-Methyl-2-propanol (b) Cyclohexanol (c) Ethanol
(d) 2-Propanol (e) Cholesterol (f) 1-Methylcyclohexanol

## MASS SPECTROSCOPY

Alcohols undergo fragmentation in the mass spectrometer by two characteristic pathways, **alpha cleavage** and **dehydration**. In the alpha-cleavage pathway, a carbon–carbon bond nearest the hydroxyl group is broken, yielding a neutral radical plus a charged oxygen-containing fragment:

$$\text{Alpha cleavage}\quad \left[\begin{array}{c} \text{OH} \\ | \\ \text{R—C} \nmid_{\alpha} \text{CH}_2\text{R} \\ | \\ \text{R} \end{array}\right]^{+\cdot} \longrightarrow \left[\begin{array}{c} \text{OH} \\ | \\ \text{R—C} \\ | \\ \text{R} \end{array}\right]^{+} + \cdot\text{CH}_2\text{R}$$

In the dehydration pathway, water is eliminated, yielding an alkene radical cation:

$$\text{Dehydration}\quad \left[\begin{array}{c} \text{H} \quad\quad \text{OH} \\ \diagdown \quad\quad \diagup \\ \text{—C—C—} \\ \diagup \quad\quad \diagdown \end{array}\right]^{+\cdot} \longrightarrow \text{H}_2\text{O} + \left[\begin{array}{c} \diagdown \quad \diagup \\ \text{C=C} \\ \diagup \quad \diagdown \end{array}\right]^{+\cdot}$$

Both of these characteristic alcohol-fragmentation modes are apparent in the mass spectrum of 1-butanol (Figure 17.16). The peak at $m/z = 56$ is due to loss of water from the molecular ion, whereas the peak at $m/z = 31$ is due to an alpha cleavage.

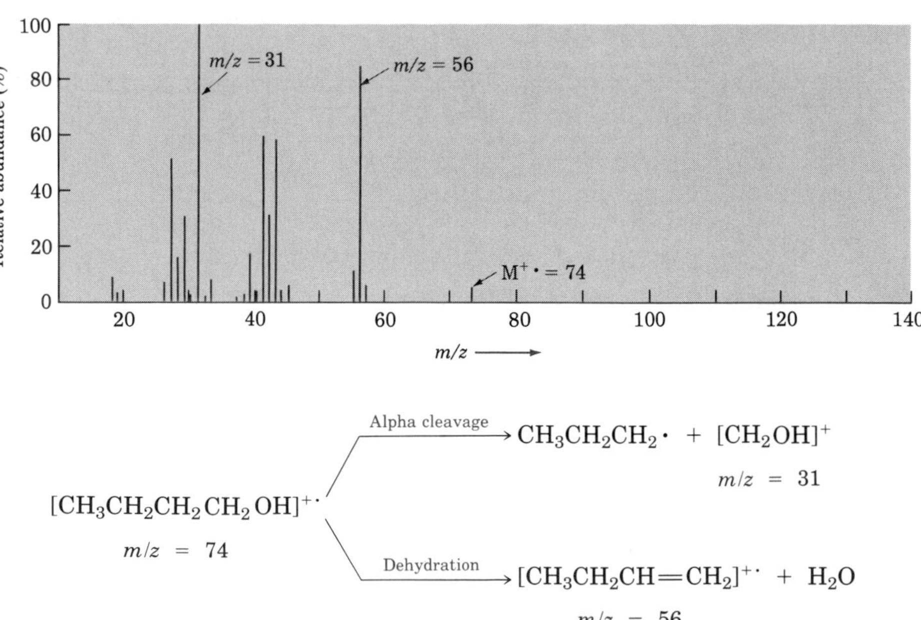


**Figure 17.16** Mass spectrum of 1-butanol: Dehydration gives a peak at $m/z = 56$; fragmentation by alpha cleavage gives a peak at $m/z = 31$.

## 17.12 Thiols

**Thiols, R—SH**, are sulfur analogs of alcohols. Thiols are named by the same system used for alcohols, with the suffix *-thiol* used in place of *-ol*. The —SH group itself is referred to as a **mercapto** group.

$CH_3CH_2SH$

Ethanethiol

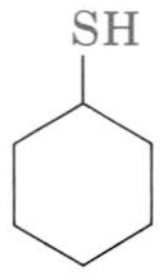

Cyclohexanethiol

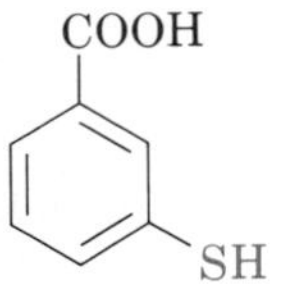

*m*-Mercaptobenzoic acid

The outstanding physical characteristic of thiols is their appalling odor. For example, skunk scent is caused primarily by the simple thiols, 3-methyl-1-butanethiol and 2-butene-1-thiol. Similarly, small amounts of low-molecular-weight thiols are added to natural gas to serve as an easily detectable warning in case of leaks.

Thiols are usually prepared from the corresponding alkyl halides by $S_N2$ displacement with a sulfur nucleophile such as hydrosulfide anion, $^-$:SH.

$$CH_3(CH_2)_6CH_2{-}Br + Na^+ \;^-{:}\ddot{S}H \longrightarrow CH_3(CH_2)_6CH_2SH + NaBr$$

1-Bromooctane — Sodium hydrosulfide — 1-Octanethiol

Yields are often poor in this reaction unless an excess of hydrosulfide anion is used, because the product thiol can undergo further reaction with alkyl halide, yielding a symmetrical sulfide, R—S—R, as a by-product. For this reason, thiourea, $(NH_2)_2C{=}S$, is often used as the nucleophile in the preparation of thiols from alkyl halides. The reaction occurs by displacement of the halide ion to yield an intermediate alkylisothiourea salt, followed by hydrolysis with aqueous base.

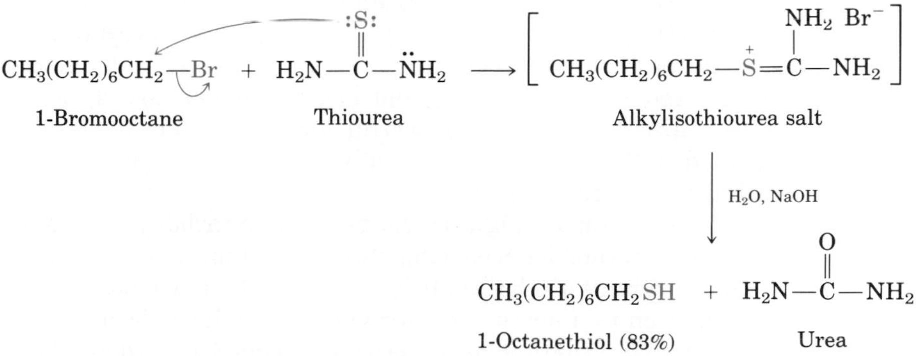

$$CH_3(CH_2)_6CH_2{-}Br + H_2N{-}C({=}S){-}NH_2 \longrightarrow \left[CH_3(CH_2)_6CH_2{-}\overset{+}{S}{=}C(NH_2){-}NH_2 \;\; Br^-\right]$$

1-Bromooctane — Thiourea — Alkylisothiourea salt

$$\xrightarrow{H_2O,\ NaOH} CH_3(CH_2)_6CH_2SH + H_2N{-}C({=}O){-}NH_2$$

1-Octanethiol (83%) — Urea

Thiols can be oxidized by mild reagents such as bromine or iodine to yield **disulfide** products. The reaction is easily reversed; disulfides can be reduced back to thiols by treatment with zinc and acid.

$$2\text{R—SH} \underset{\text{Zn, H}^+}{\overset{\text{Br}_2}{\rightleftharpoons}} \text{R—S—S—R} + 2\,\text{HBr}$$

A thiol — A disulfide

We'll see later that the thiol–disulfide interconversion is extremely important in biochemistry, where disulfide "bridges" form the cross-links between protein chains that help stabilize the three-dimensional conformations of proteins.

PROBLEM

**17.20** Name the following compounds by IUPAC rules:

(a) $CH_3CH_2CH(CH_3)SH$ (CH₃ on the CH carbon)

(b) $(CH_3)_3CCH_2CH(SH)CH_2CH(CH_3)_2$ (SH on the CH carbon)

(c) Cyclopentene ring bearing SH

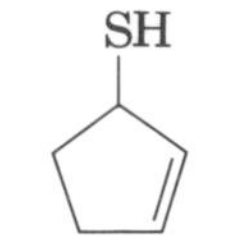

PROBLEM

**17.21** 2-Butene-1-thiol is one component of skunk spray. How would you synthesize this substance from methyl 2-butenoate? From 1,3-butadiene?

$$CH_3CH{=}CHCO_2CH_3 \overset{?}{\Longrightarrow} CH_3CH{=}CHCH_2SH$$

Methyl 2-butenoate — 2-Butene-1-thiol

## 17.13 Summary and Key Words

**Alcohols** are among the most versatile of all organic compounds. They occur widely in nature, are important industrially, and have an unusually rich chemistry. Alcohols can be prepared in a number of different ways, including osmium tetraoxide **hydroxylation** of alkenes, **hydroboration–oxidation** of alkenes, and **oxymercuration**–sodium borohydride reduction of alkenes.

The most important methods of alcohol synthesis involve carbonyl compounds. For example, aldehydes, ketones, esters, and carboxylic acids can be **reduced** by reaction with either sodium borohydride or lithium aluminum hydride. Aldehydes, esters, and carboxylic acids yield primary alcohols on reduction; ketones yield secondary alcohols. **Sodium borohydride** is safe to use and relatively unreactive. It reduces aldehydes and ketones rapidly, reduces esters slowly, and does not reduce carboxylic acids. **Lithium aluminum hydride** is a powerful and dangerous reagent that must be handled with great care. It rapidly reduces all types of carbonyl groups, including carboxylic acids.

The reaction of **Grignard reagents** with carbonyl compounds is another important method for preparing alcohols. Grignard addition to formaldehyde yields a primary alcohol, addition to an aldehyde yields a secondary alcohol, and addition to a ketone or ester yields a tertiary alcohol. Carboxylic acids do not give Grignard addition products. The Grignard synthesis of alcohols is limited by the fact that Grignard reagents cannot be prepared from alkyl halides that contain reactive functional groups in the same molecule. This problem can sometimes be avoided by **protecting** the interfering functional group. For example, alcohols are often protected by formation of **tetrahydropyranyl (THP) ethers**.

Alcohols undergo a great many different reactions. They can be **dehydrated** by treatment with $POCl_3$ and can be transformed into alkyl halides by treatment with $PBr_3$ or $SOCl_2$. Furthermore, alcohols are weakly acidic ($pK_a \approx 16–18$). They react with strong bases and with alkali metals to form **alkoxide anions**, which are much used in organic synthesis.

The most important reaction of alcohols is their **oxidation** to carbonyl compounds. Primary alcohols yield either aldehydes or carboxylic acids, depending on the conditions used; secondary alcohols yield ketones; and tertiary alcohols are not normally oxidized. Many oxidation methods are available, but only a few are commonly used. **Pyridinium chlorochromate (PCC)** in dichloromethane is often used for oxidizing primary alcohols to aldehydes and secondary alcohols to ketones. The conditions are mild (room temperature) and reaction is rapid (5–10 min). The **Jones reagent**, a solution of $CrO_3$ in aqueous sulfuric acid, is much used for oxidizing primary alcohols to carboxylic acids and secondary alcohols to ketones.

Alcohols show characteristic absorptions in both infrared and nuclear magnetic resonance spectra, making their spectroscopic analysis relatively straightforward.

**Thiols, RSH**, the sulfur analogs of alcohols, are usually prepared by $S_N2$ reaction of an alkyl halide with thiourea. Mild oxidation of a thiol yields a **disulfide, R—S—S—R**, and mild reduction of a disulfide gives back the thiol.

## 17.14 Summary of Reactions

1. Synthesis of alcohols
   a. Reduction of carbonyl compounds (Section 17.6)
      (1) Aldehydes

$$CH_3CH_2CH_2CHO \xrightarrow[\text{2. } H_3O^+]{\text{1. } NaBH_4\text{, ethanol (or } LiAlH_4\text{, ether)}} CH_3CH_2CH_2CH_2OH$$

85%

(2) Ketones

$$\text{Cyclohexyl methyl ketone} \xrightarrow[\text{2. } H_3O^+]{\text{1. } NaBH_4 \text{ or } LiAlH_4} \text{1-Cyclohexylethanol (H, OH)}$$

92%

(3) Esters

$$C_6H_5CO_2CH_3 \xrightarrow[\text{2. } H_3O^+]{\text{1. } LiAlH_4\text{, THF}} C_6H_5CH_2OH$$

89%

(4) Carboxylic acids

$$CH_3(CH_2)_7CH{=}CH(CH_2)_7CO_2H \xrightarrow[\text{2. } H_3O^+]{\text{1. LiAlH}_4\text{, THF}} CH_3(CH_2)_7CH{=}CH(CH_2)_7CH_2OH$$

87%

b. Grignard addition to carbonyl compounds (Section 17.7)

(1) Formaldehyde

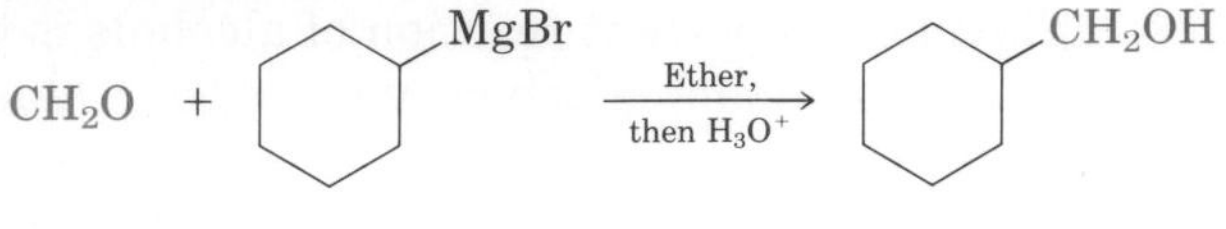

$$CH_2O + C_6H_{11}MgBr \xrightarrow[\text{then } H_3O^+]{\text{Ether,}} C_6H_{11}CH_2OH$$

65%

(2) Aldehydes

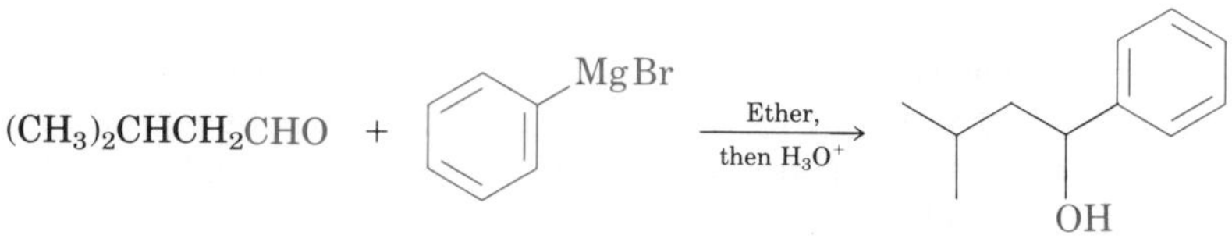

$$(CH_3)_2CHCH_2CHO + C_6H_5MgBr \xrightarrow[\text{then } H_3O^+]{\text{Ether,}} (CH_3)_2CHCH_2CH(OH)C_6H_5$$

73%

(3) Ketones

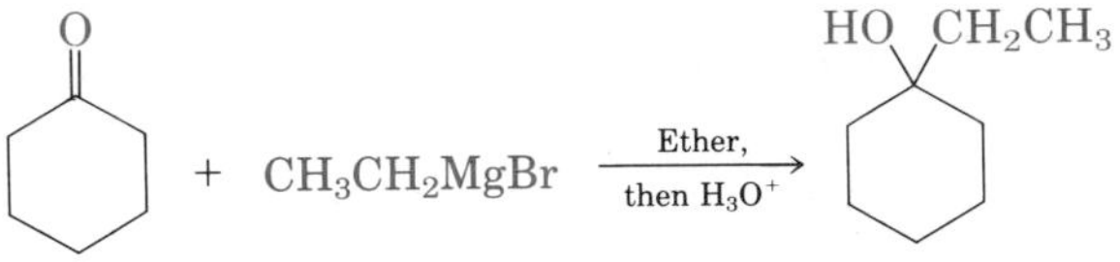

$$\text{Cyclohexanone} + CH_3CH_2MgBr \xrightarrow[\text{then } H_3O^+]{\text{Ether,}} \text{1-ethylcyclohexanol (HO, } CH_2CH_3\text{)}$$

89%

(4) Esters

$$CH_3CH_2CH_2CH_2CO_2CH_2CH_3 + CH_3MgBr \xrightarrow[\text{then } H_3O^+]{\text{Ether,}} CH_3CH_2CH_2CH_2C(OH)(CH_3)CH_3$$

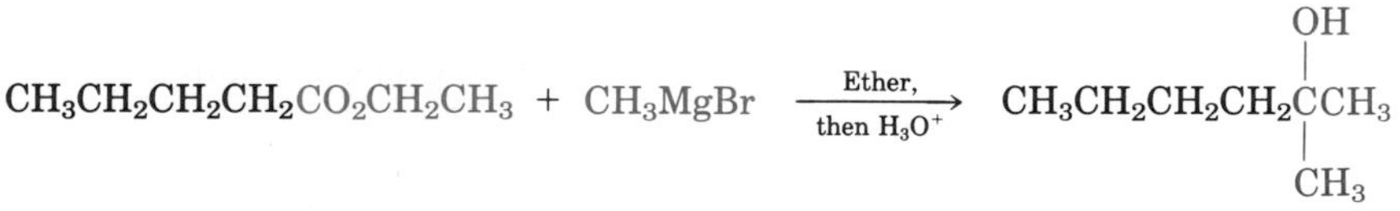

85%

2. Reactions of alcohols

a. Acidity (Section 17.4)

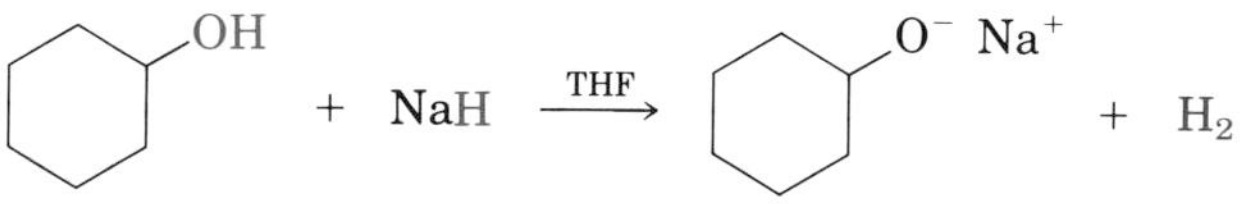

$$C_6H_{11}OH + NaH \xrightarrow{\text{THF}} C_6H_{11}O^- Na^+ + H_2$$

$$2\,(CH_3)_3COH + 2\,K \longrightarrow 2\,(CH_3)_3CO^-K^+ + H_2$$

b. Oxidation (Section 17.9)

(1) Primary alcohol

$$CH_3(CH_2)_5CH_2OH \xrightarrow[CH_2Cl_2]{\text{PCC,}} CH_3(CH_2)_5CHO$$

78%

$$CH_3(CH_2)_8CH_2OH \xrightarrow{\text{Jones' reagent}} CH_3(CH_2)_8CO_2H$$

93%

(2) Secondary alcohol

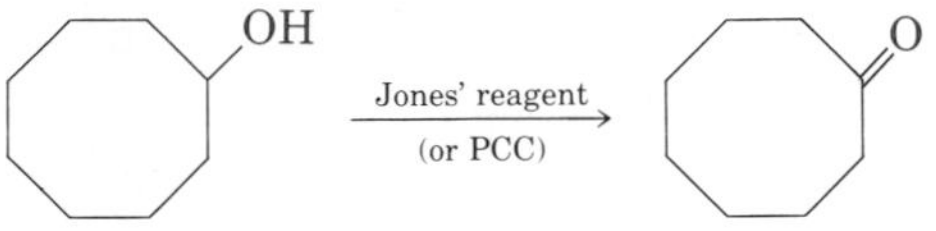

Cyclooctanol $\xrightarrow[\text{(or PCC)}]{\text{Jones' reagent}}$ cyclooctanone

96%

3. Synthesis of thiols (Section 17.12)

$$CH_3CH_2Br \xrightarrow[\text{2. }^-OH,\ H_2O]{\text{1. }(H_2N)_2C{=}S} CH_3CH_2SH$$

85%

4. Oxidation of thiols to disulfides (Section 17.12)

$$CH_3CH_2SH \xrightarrow{Br_2} CH_3CH_2S{-}SCH_2CH_3$$

## ADDITIONAL PROBLEMS

**17.22** Name the following compounds according to the IUPAC system:

(a) $HOCH_2CH_2CH(CH_3)CH_2OH$ (structure: $HOCH_2CH_2\overset{CH_3}{\overset{|}{C}}HCH_2OH$)

(b) $CH_3CH(OH)CHCH_2CH_3$ with $CH_2CH_2CH_3$ on the CH

(c) 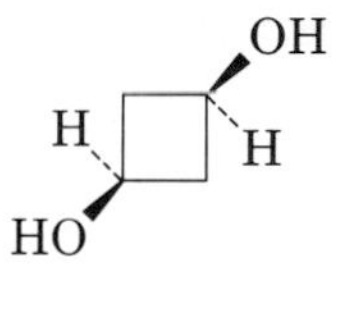 Cyclobutane bearing HO (wedge) and H at one carbon and OH (wedge) and H at the opposite carbon

(d) 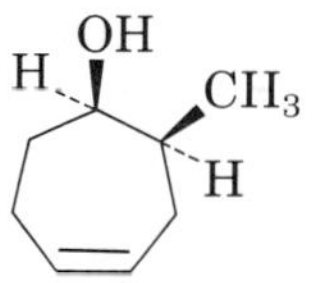 Cycloheptene ring bearing OH and H on one carbon and $CH_3$ (wedge) and H on the adjacent carbon

(e) 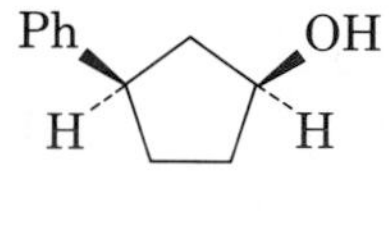 Cyclopentane bearing Ph and H on one carbon and OH and H on another carbon

(f) $(CH_3)_2CHC(SH)(CH_3)CH_2CH_2CH_3$

**17.23** Draw and name the eight isomeric alcohols with formula $C_5H_{12}O$.

**17.24** Which of the eight alcohols that you identified in Problem 17.23 react with Jones' reagent ($CrO_3$, $H_2O$, $H_2SO_4$)? Show the products you would expect from each reaction.

**17.25** How would you prepare the following compounds from 2-phenylethanol?

(a) Styrene
(b) Phenylacetaldehyde ($C_6H_5CH_2CHO$)
(c) Phenylacetic acid ($C_6H_5CH_2CO_2H$)
(d) Benzoic acid
(e) Ethylbenzene
(f) Benzaldehyde
(g) 1-Phenylethanol
(h) 2-Bromo-1-phenylethane

**17.26** How would you carry out these transformations?

(a) 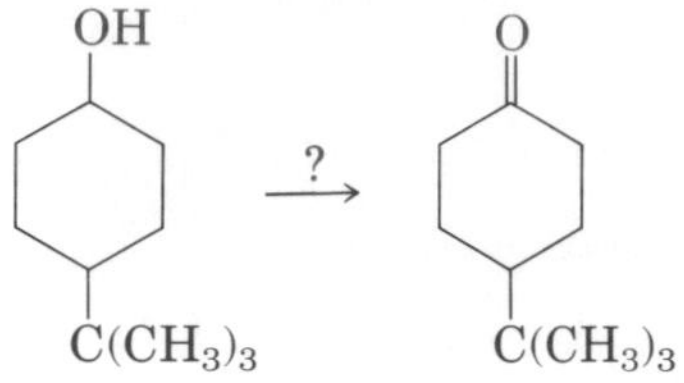 4-*tert*-butylcyclohexanol (OH; $C(CH_3)_3$) $\xrightarrow{?}$ 4-*tert*-butylcyclohexanone (O; $C(CH_3)_3$)

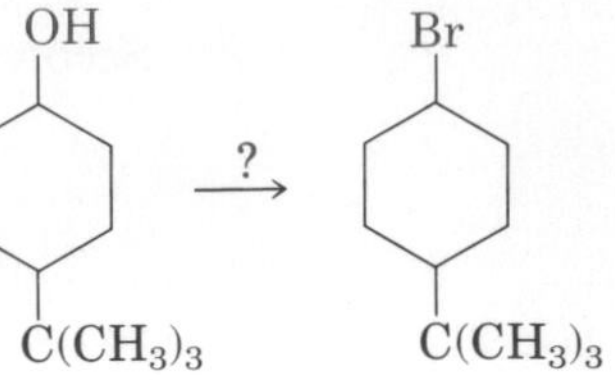

(b) 4-*tert*-butylcyclohexanol (OH; $C(CH_3)_3$) $\xrightarrow{?}$ 1-bromo-4-*tert*-butylcyclohexane (Br; $C(CH_3)_3$)

(c) 4-*tert*-butylcyclohexanol (OH; $C(CH_3)_3$) $\xrightarrow{?}$ 4-*tert*-butylcyclohexene ($C(CH_3)_3$)

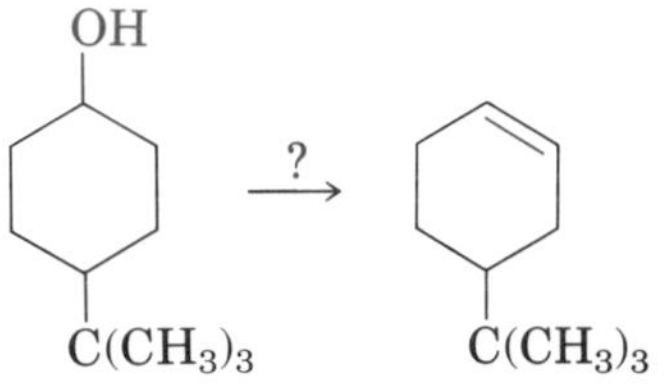

(d) $CH_3CH_2CH_2OH \xrightarrow{?} CH_3CH_2CHO$

(e) $CH_3CH_2CH_2OH \xrightarrow{?} CH_3CH_2CO_2H$

(f) $CH_3CH_2CH_2OH \xrightarrow{?} CH_3CH_2CH_2OTos$

(g) $CH_3CH_2CH_2OH \xrightarrow{?} CH_3CH_2CH_2Cl$

(h) $CH_3CH_2CH_2OH \xrightarrow{?} CH_3CH_2CH_2O^-\ Na^+$

**17.27** What Grignard reagent and what carbonyl compound might you start with to prepare these alcohols?

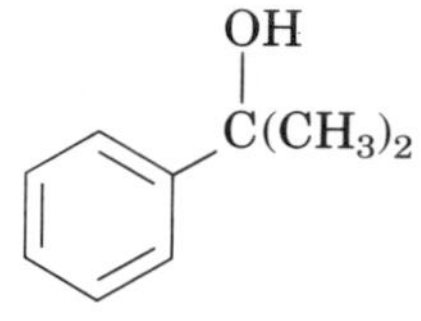

(a) 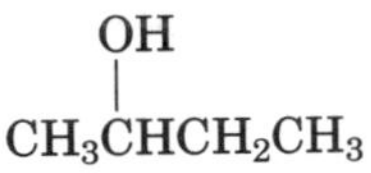 $CH_3CH(OH)CH_2CH_3$

(b) $C_6H_5C(OH)(CH_3)_2$ (benzene ring bearing $C(CH_3)_2$ with OH)

(c) $H_2C{=}C(CH_3){-}CH_2OH$

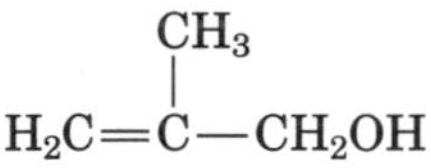

(d) $(Ph)_3COH$

(e)  $CH_3CH(OH)CH_2CH_2CH_2Br$

**17.28** Assume that you have been given a sample of (*S*)-2-octanol. How could you prepare (*R*)-2-chlorooctane? How could you prepare (*R*)-2-octanol?

**17.29** When 4-chloro-1-butanol is treated with a strong base such as sodium hydride, NaH, tetrahydrofuran is produced. Suggest a mechanism for this reaction.

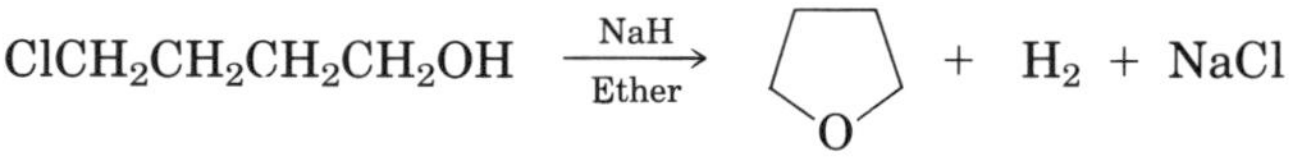

$$ClCH_2CH_2CH_2CH_2OH \xrightarrow[\text{Ether}]{\text{NaH}} \text{(tetrahydrofuran, ring O)} + H_2 + NaCl$$

**17.30** What carbonyl compounds would you reduce to prepare these alcohols? List all possibilities.

(a) 2,2-Dimethyl-1-hexanol

(b) 3,3-Dimethyl-2-butanol

(c) 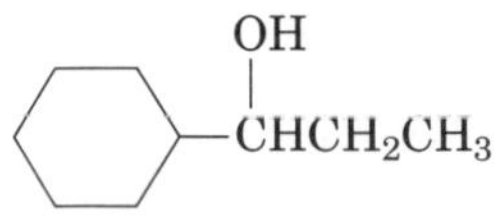 cyclohexyl$-CH(OH)CH_2CH_3$

(d) 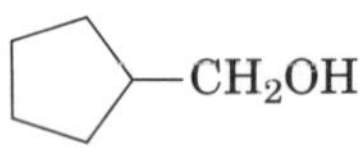 cyclopentyl$-CH_2OH$

**17.31** How would you carry out these transformations?

(a) 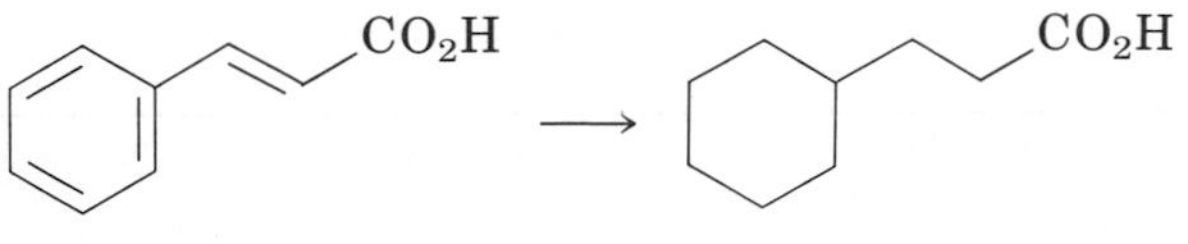 $CO_2H$ → $CO_2H$

(b) 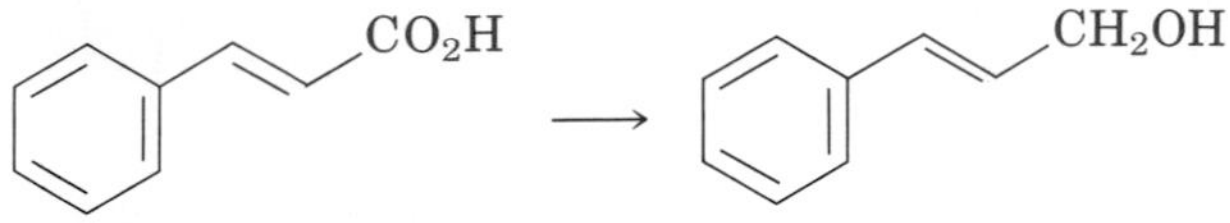 $CO_2H$ → $CH_2OH$

**17.32** What carbonyl compounds might you start with to prepare these compounds by Grignard reaction? List all possibilities.

(a) 2,2-Dimethyl-2-propanol (b) 1-Ethylcyclohexanol
(c) 3-Phenyl-3-pentanol (d) 2-Phenyl-2-pentanol

(e) 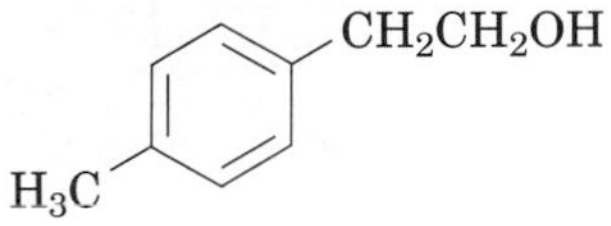 $H_3C$, $CH_2CH_2OH$ (f) 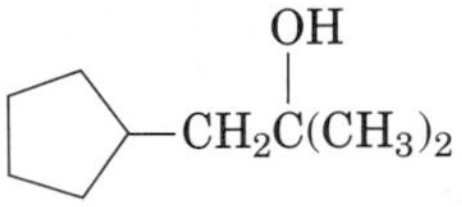 OH, $CH_2C(CH_3)_2$

**17.33** What products would you expect to obtain from reaction of 1-pentanol with these reagents?

(a) $PBr_3$ (b) $SOCl_2$ (c) $CrO_3$, $H_2O$, $H_2SO_4$ (d) PCC

**17.34** Acid-catalyzed dehydration of 2,2-dimethylcyclohexanol yields a mixture of 1,2-dimethylcyclohexene and isopropylidenecyclopentane. Propòse a mechanism to account for the formation of both products.

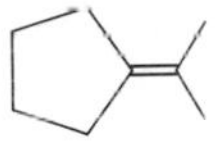

Isopropylidenecyclopentane

**17.35** How would you prepare these substances from cyclopentanol? More than one step may be required.

(a) Cyclopentanone (b) Cyclopentene
(c) 1-Methylcyclopentanol (d) *trans*-2-Methylcyclopentanol

**17.36** What products would you expect to obtain from reaction of 1-methylcyclohexanol with these reagents?

(a) HBr (b) NaH (c) $H_2SO_4$ (d) $Na_2Cr_2O_7$

**17.37** Testosterone is one of the most important male steroid hormones. When testosterone is dehydrated by treatment with acid, rearrangement occurs to yield the product indicated. Propose a mechanism to account for this reaction. [*Hint:* See Section 6.13.]

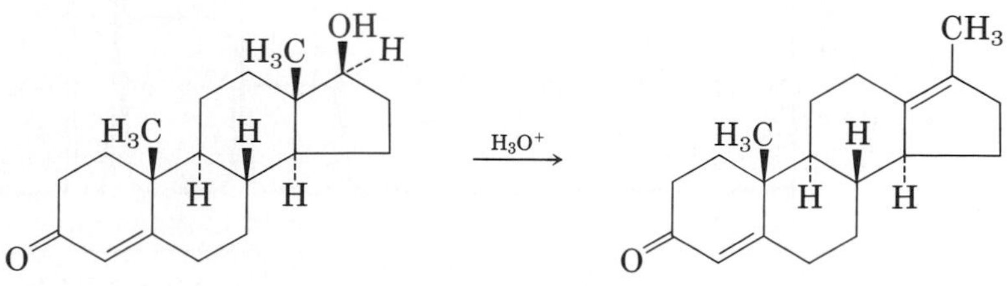


Testosterone

**17.38** Starting from testosterone (Problem 17.37), how would you prepare the following substances?

(a) (b)

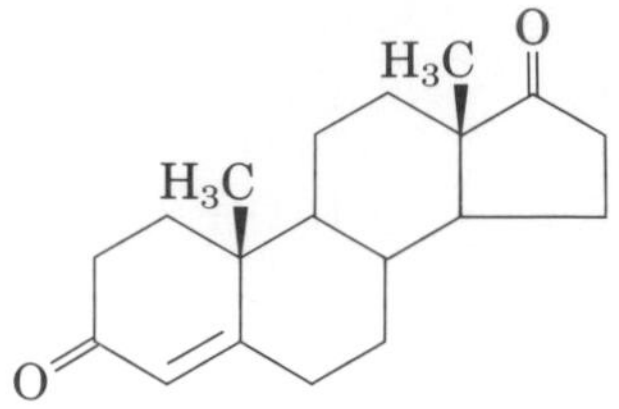

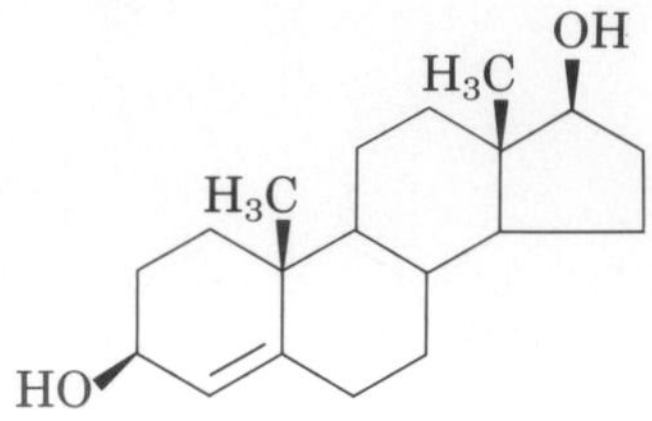

(c) 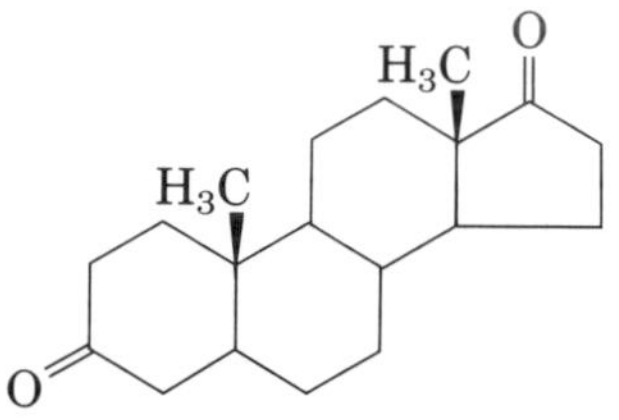 (d)

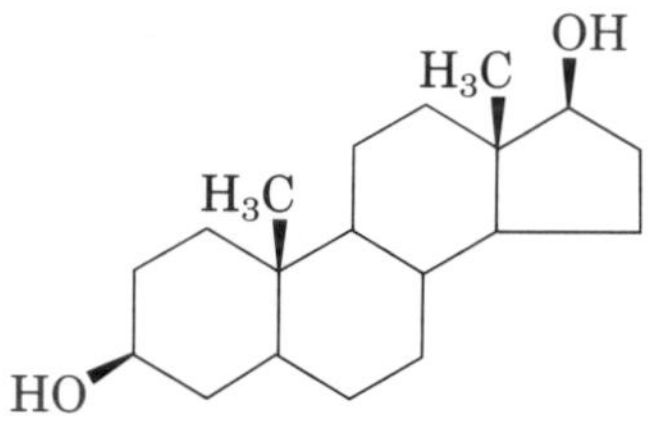

**17.39** Compound A, $C_{10}H_{18}O$, undergoes reaction with dilute $H_2SO_4$ at 25°C to yield a mixture of two alkenes, $C_{10}H_{16}$. The major alkene product, B, gives only cyclopentanone after ozone treatment followed by reduction with zinc in acetic acid. Formulate the reactions involved and identify A and B.

**17.40** Dehydration of *trans*-2-methylcyclopentanol with $POCl_3$ in pyridine yields predominantly 3-methylcyclopentene. What is the stereochemistry of this dehydration? Can you suggest a reason for formation of the observed product? (Make molecular models!)

**17.41** The $^1H$ NMR spectrum shown is that of 3-methyl-3-buten-1-ol. Assign all the observed resonance peaks to specific protons and account for the splitting patterns.

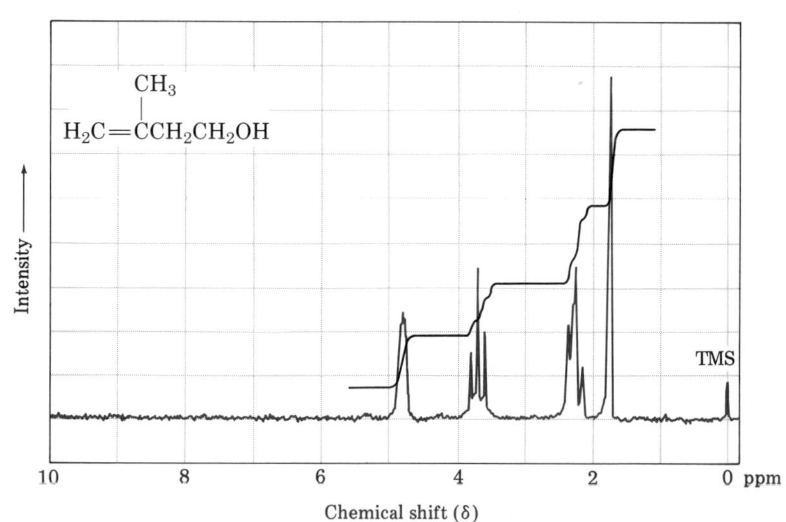

**17.42** Propose a structure consistent with the following spectral data for a compound of formula $C_8H_{18}O_2$:

Infrared: 3350 $cm^{-1}$

$^1$H NMR: 1.24 $\delta$ (12-proton singlet); 1.56 $\delta$ (4-proton singlet); 1.95 $\delta$ (2-proton singlet)

**17.43** The $^1$H NMR spectrum shown is of an alcohol of formula $C_8H_{10}O$. Propose a structure that is consistent with the observed spectrum.

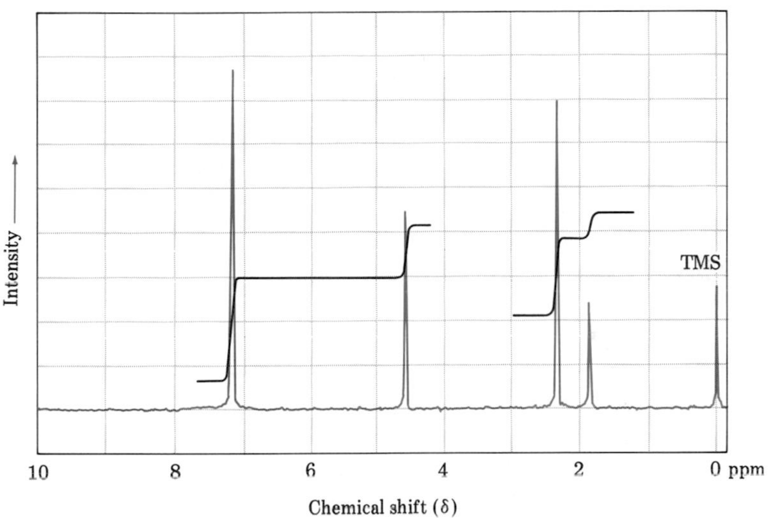


**17.44** 2,3-Dimethyl-2,3-butanediol has the trivial name *pinacol*. On heating with aqueous acid, pinacol rearranges to *pinacolone*, 3,3-dimethyl-2-butanone. Can you suggest a mechanism for this reaction? [*Hint:* See Section 6.13.]

$$\underset{\text{Pinacol}}{(CH_3)_2\overset{OH}{\overset{|}{C}}-\overset{OH}{\overset{|}{C}}(CH_3)_2} \xrightarrow{H_3O^+} \underset{\text{Pinacolone}}{CH_3\overset{O}{\overset{\|}{C}}C(CH_3)_3} + H_2O$$

**17.45** Compound A, $C_5H_{10}O$, is one of the basic building blocks of nature. All steroids and many other naturally occurring compounds are built up from compound A. Spectroscopic analysis of A yields the following information:

Infrared: 3400 $cm^{-1}$; 1640 $cm^{-1}$

$^1$H NMR: 1.63 $\delta$ (3-proton singlet); 1.70 $\delta$ (3-proton singlet); 3.83 $\delta$ (1-proton broad singlet); 4.15 $\delta$ (2-proton doublet, $J$ = 7 Hz); 5.70 $\delta$ (1-proton triplet, $J$ = 7 Hz)

(a) How many double bonds and/or rings does compound A have?
(b) From the infrared spectrum, what is the nature of the oxygen-containing functional group?
(c) What kinds of protons are responsible for the NMR absorptions listed?
(d) Propose a structure for compound A consistent with the $^1$H NMR data.

**17.46** Propose structures for alcohols that have the following $^1H$ NMR spectra:

(a) $C_3H_8O$

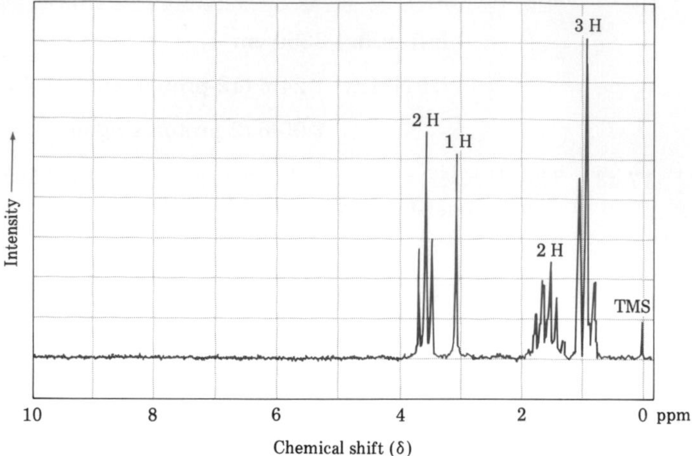


(b) $C_5H_{12}O$

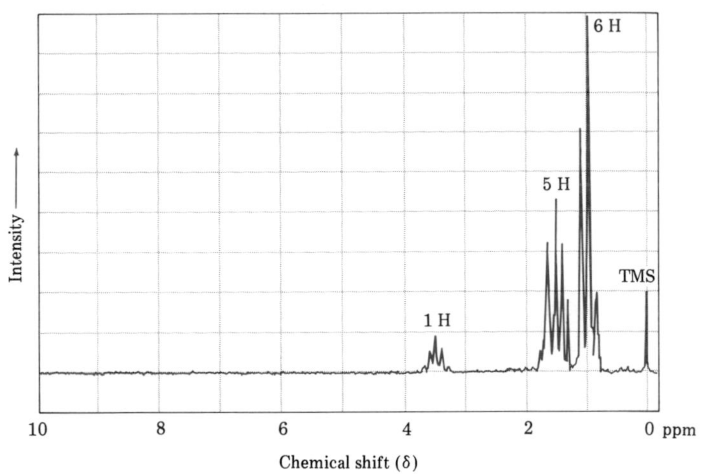


(c) $C_8H_{10}O$

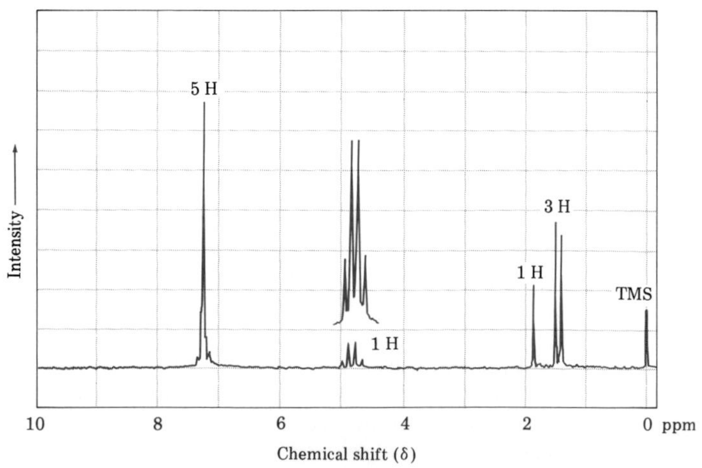

**17.47** Propose structures for alcohols that have the following $^1$H NMR spectra:

(a) $C_9H_{12}O$

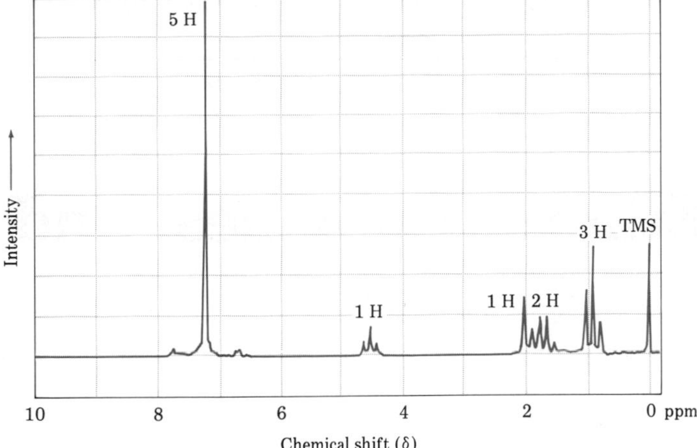


(b) $C_8H_{10}O_2$

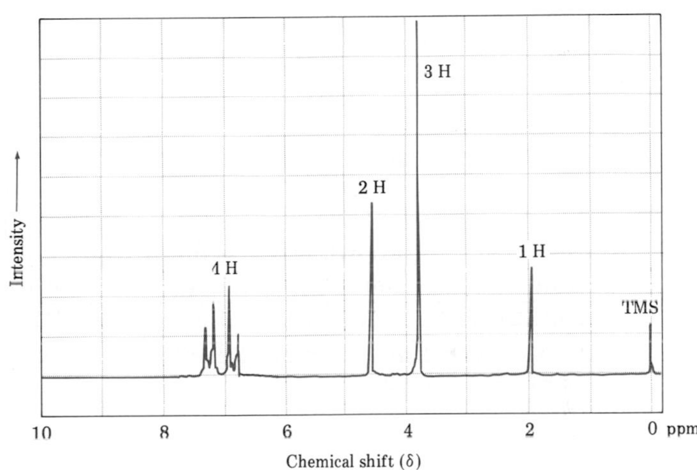


**17.48** As a general rule, axial alcohols oxidize somewhat faster than equatorial alcohols. Which would you expect to oxidize faster, *cis*-4-*tert*-butylcyclohexanol or *trans*-4-*tert*-butylcyclohexanol? Draw the more stable chair conformation of each molecule.

**17.49** Propose a synthesis of bicyclohexylidene, starting from cyclohexanone as the only source of carbon.

Bicyclohexylidene

**17.50** Since all hamsters look pretty much alike, attraction between sexes is controlled by chemical secretions. The sex attractant exuded by the female hamster has been shown to have the following spectral properties:

Mass spectrum: $M^{+\cdot} = 94$

Infrared: No characteristic absorptions above 1500 $cm^{-1}$

$^1$H NMR: 2.10 $\delta$ (singlet); no other absorptions

Propose a structure for the hamster sex attractant.

CHAPTER 18

# Ethers, Epoxides, and Sulfides

An ether is a substance that has two organic residues bonded to the same oxygen atom, R—O—R′. The organic residues may be alkyl, aryl, or vinylic, and the oxygen atom can be part of either an open chain or a ring. Perhaps the most well known ether is diethyl ether, a familiar substance that has been used medicinally as an anesthetic and is much used industrially as a solvent. Other useful ethers include anisole, a pleasant-smelling aromatic ether used in perfumery, and tetrahydrofuran (THF), a cyclic ether often used as a solvent.

$CH_3CH_2—\ddot{O}—CH_2CH_3$ Diethyl ether

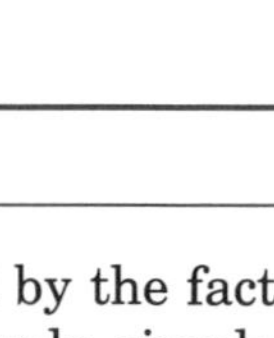

$:\ddot{O}—CH_3$ (on benzene ring) Anisole (Methyl phenyl ether)

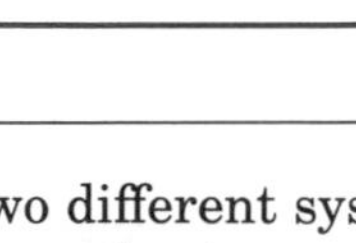

Tetrahydrofuran (a cyclic ether): ring of O, $H_2C$, $CH_2$, $H_2C—CH_2$

## 18.1 Nomenclature of Ethers

The naming of ethers is complicated by the fact that two different systems are allowed by IUPAC rules. Relatively simple ethers are best named by identifying the two organic residues and adding the word *ether*. For example,

| $CH_3OC(CH_3)_3$ | $CH_3CH_2OCH{=}CH_2$ | (phenyl–O–cyclopropyl) |
|---|---|---|
| *tert*-Butyl methyl ether | Ethyl vinyl ether | Cyclopropyl phenyl ether |

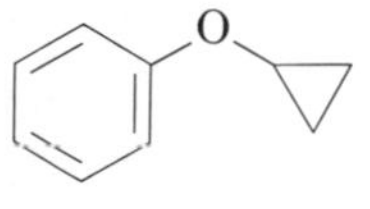

If more than one ether linkage is present in the molecule, or if other functional groups are present, the ether is named as an alkoxy-substituted parent compound. For example,

$CH_3O$—(benzene ring)—$OCH_3$

*p*-Dimethoxybenzene

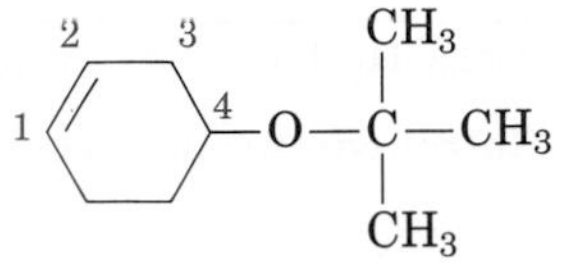

(cyclohexene ring, 1–4)—O—C($CH_3$)$_3$

4-*tert*-Butoxy-1-cyclohexene

**PROBLEM**

**18.1** Name these ethers according to IUPAC rules:

(a) $(CH_3)_2CHOCH(CH_3)_2$

(b) 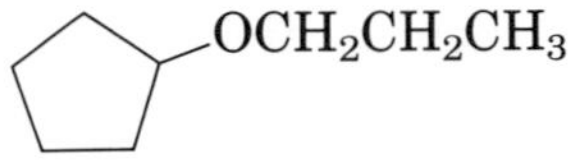 (cyclopentane)—$OCH_2CH_2CH_3$

(c) 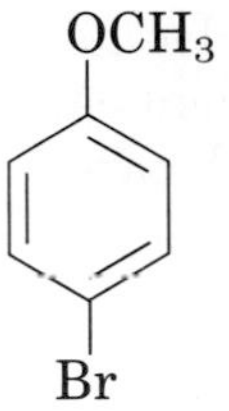 $OCH_3$ / (benzene ring) / Br

(d) 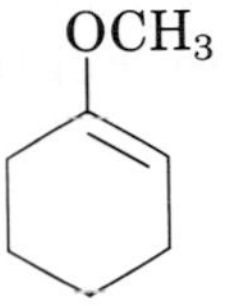 $OCH_3$ / (cyclohexene)

(e) $(CH_3)_2CHCH_2OCH_2CH_3$

(f) $H_2C{=}CHCH_2OCH{=}CH_2$

## 18.2 Structure and Properties of Ethers

Ethers can be thought of as organic derivatives of water in which the hydrogen atoms have been replaced by organic fragments, H—O—H versus R—O—R. As such, ethers have nearly the same geometry as water. The R—O—R bonds have an approximately tetrahedral bond angle (112° in dimethyl ether), and the oxygen atom is $sp^3$ hybridized.

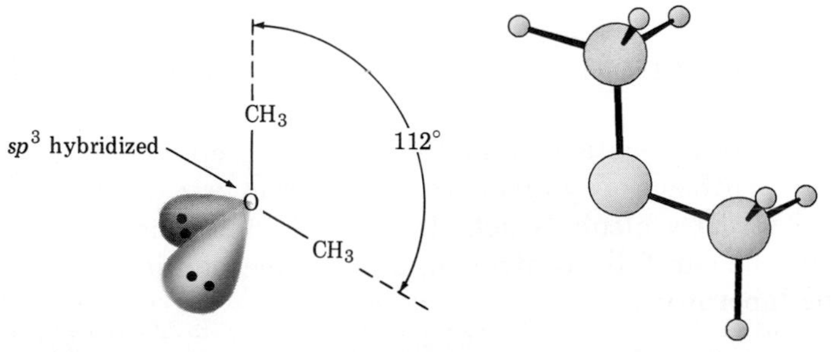

$sp^3$ hybridized

$CH_3$

112°

O

$CH_3$

The presence of the electronegative oxygen atom causes ethers to have a slight dipole moment, and the boiling points of ethers are therefore somewhat higher than the boiling points of comparable alkanes. Table 18.1 compares the boiling points of some common ethers with hydrocarbons of similar molecular weight.

**Table 18.1** Comparison of boiling points of ethers and hydrocarbons

| *Compounds* | *Boiling point (°C)* | |
|---|---|---|
| $CH_3OCH_3$ (versus $CH_3CH_2CH_3$) | −25 | (−45) |
| $CH_3CH_2OCH_2CH_3$ (versus $CH_3CH_2CH_2CH_2CH_3$) | 34.6 | (36) |
| Tetrahydrofuran (versus cyclopentane) | 65 | (49) |
| Anisole (versus ethylbenzene) | 158 | (136) |

Although ethers are unusually inert toward most reagents, certain ethers react slowly with air to give **peroxides**, compounds that contain oxygen–oxygen bonds. The peroxides from low-molecular-weight ethers such as diisopropyl ether and tetrahydrofuran are highly explosive and extremely dangerous, even in tiny amounts. This sensitivity to air, combined with their volatility and flammability, make ether solvents potentially dangerous unless handled by skilled persons. Ether solvents are very useful in the laboratory, but they must always be treated with care.

## 18.3 Industrial Preparation of Ethers

Diethyl ether and other simple symmetrical ethers are prepared industrially by the sulfuric acid-catalyzed dehydration of alcohols:

$$\underset{\text{Ethanol}}{2\ CH_3CH_2OH} \xrightarrow{H_2SO_4} \underset{\text{Diethyl ether}}{CH_3CH_2OCH_2CH_3} + H_2O$$

The reaction probably occurs by $S_N2$ displacement of water from a protonated ethanol molecule by the oxygen atom of a second ethanol:

$$CH_3CH_2\ddot{O}H + CH_2CH_3(\overset{+}{:}\ddot{O}H_2) \xrightarrow{S_N2\ \text{reaction}} CH_3CH_2\ddot{O}CH_2CH_3 + H_3O^+$$

This acid-catalyzed method of ether preparation is limited to the industrial synthesis of symmetrical ethers from primary alcohols, since secondary and tertiary alcohols dehydrate to yield alkenes. The reaction conditions must be carefully controlled, and the method is of little practical value in the laboratory.

PROBLEM ...........................................................................................

**18.2** Why do you suppose only symmetrical ethers can be prepared by the sulfuric acid-catalyzed dehydration procedure? What product(s) would you expect to get if ethanol and 1-propanol were allowed to react together? In what ratio would the products be formed if the two alcohols were of equal reactivity?

## 18.4 The Williamson Ether Synthesis

Metal alkoxides react with primary alkyl halides and tosylates by an $S_N2$ pathway to yield ethers, a process known as the **Williamson ether synthesis**. Discovered in 1850, the Williamson[1] synthesis is still the best method for the preparation of ethers, both symmetrical and unsymmetrical.

$$:\ddot{O}:^- K^+ \ + \ CH_3—I \xrightarrow[\text{solvent}]{\text{Ether}} \ :\ddot{O}CH_3 \ + \ KI$$

Potassium cyclopentoxide — Iodomethane — Cyclopentyl methyl ether (74%)

The alkoxide ions needed in the Williamson reaction are normally prepared by reaction of an alcohol with a strong base such as sodium hydride, NaH (Section 17.4). An acid–base reaction occurs between the alcohol and sodium hydride to generate the sodium salt of the alcohol:

$$R—O—H + NaH \longrightarrow R—O^- Na^+ + H_2$$

An important variation of the Williamson synthesis involves the use of silver oxide, $Ag_2O$, as base, rather than NaH. Under these conditions, the free alcohol reacts directly with alkyl halide, and there is no need to preform the metal alkoxide salt. For example, glucose reacts with iodomethane in the presence of $Ag_2O$ to generate a *pentaether* in 85% yield:

$$\alpha\text{-D-Glucose} \xrightarrow[Ag_2O]{CH_3I} \alpha\text{-D-Glucose pentamethyl ether} + AgI$$

α-D-Glucose (ring substituents: $CH_2OH$, HO, HO, HO, OH) — α-D-Glucose pentamethyl ether (85%) (ring substituents: $CH_2OCH_3$, $CH_3O$, $CH_3O$, $CH_3O$, $OCH_3$)

Mechanistically, the Williamson synthesis is simply an $S_N2$ displacement of halide ion by an alkoxide ion nucleophile. The Williamson synthesis is thus subject to all of the normal constraints on $S_N2$ reactions discussed previously (Section 11.5). Primary halides and tosylates work best, since competitive E2 elimination of HX can occur with more hindered substrates. For this reason, unsymmetrical ethers should be synthesized by reaction between the more hindered alkoxide partner and less hindered halide partner, rather than vice versa. For example, *tert*-butyl methyl ether is best prepared by reaction of *tert*-butoxide ion with iodomethane, rather than by reaction of methoxide ion with 2-chloro-2-methylpropane:

[1]Alexander W. Williamson (1824–1904); b. Wandsworth, England; Ph.D. (1846), Giessen; professor, University College, London (1849–1904).

$$\text{S}_\text{N}2\text{ reaction}\quad CH_3-\underset{CH_3}{\overset{CH_3}{C}}-\ddot{\underset{\cdot\cdot}{O}}:^- + CH_3-I \longrightarrow CH_3-\underset{CH_3}{\overset{CH_3}{C}}-\ddot{\underset{\cdot\cdot}{O}}-CH_3 + :\ddot{\underset{\cdot\cdot}{I}}:^-$$

*tert*-Butoxide ion — Iodomethane — *tert*-Butyl methyl ether

$$\text{E2 reaction}\quad CH_3\ddot{\underset{\cdot\cdot}{O}}:^- + \overset{H}{CH_2}-\underset{CH_3}{\overset{CH_3}{C}}-Cl \longrightarrow H_2C{=}C(CH_3)_2 + CH_3\ddot{\underset{\cdot\cdot}{O}}H + :\ddot{\underset{\cdot\cdot}{Cl}}:^-$$

Methoxide ion — 2-Chloro-2-methylpropane — 2-Methylpropene

**PROBLEM**

**18.3** Treatment of cyclohexanol with NaH gives an alkoxide ion that reacts with iodoethane to yield an ether. Write out the reaction showing all steps.

**PROBLEM**

**18.4** How would you prepare these compounds using a Williamson synthesis?
(a) Methyl propyl ether
(b) Anisole (methyl phenyl ether)
(c) Benzyl isopropyl ether
(d) Ethyl 2,2-dimethylpropyl ether

**PROBLEM**

**18.5** Rank the following halides in order of their expected reactivity in the Williamson synthesis:
(a) Bromoethane, 2-bromopropane, bromobenzene
(b) Chloroethane, bromoethane, 1-iodopropene

## 18.5 Alkoxymercuration–Demercuration of Alkenes

We saw in Section 7.3 that alkenes react with water in the presence of mercuric acetate to yield a **hydroxymercuration** product. Subsequent treatment of this addition product with sodium borohydride breaks the carbon–mercury bond and yields the alcohol. A similar **alkoxymercuration** reaction occurs when an alkene is treated with an *alcohol* in the presence of mercuric acetate. [Mercuric trifluoroacetate, $Hg(OOCCF_3)_2$, works even better.] Sodium borohydride-induced demercuration then yields an ether. As indicated by the following examples, the net result is Markovnikov addition of alcohol to the alkene:

$$C_6H_5CH{=}CH_2 \xrightarrow[CH_3OH]{Hg(O_2CCF_3)_2} \left[ C_6H_5\underset{}{CH}(OCH_3)CH_2HgO_2CCF_3 \right] \xrightarrow{NaBH_4} C_6H_5CH(OCH_3)CH_2H$$

Styrene — 1-Methoxy-1-phenylethane (97%)

Cyclohexene $\xrightarrow[\text{2. NaBH}_4]{\text{1. Hg(O}_2\text{CCF}_3)_2,\ \text{CH}_3\text{CH}_2\text{OH}}$ Cyclohexyl ethyl ether (100%) [cyclohexyl–$OCH_2CH_3$]

The mechanism of the alkoxymercuration–demercuration reaction is analogous to the mechanism described earlier (Section 7.3) for the hydroxymercuration procedure. The reaction is initiated by electrophilic addition of mercuric ion to the alkene, followed by reaction of the intermediate cation with alcohol. Displacement of mercury with sodium borohydride completes the process.

A wide variety of alcohols and alkenes can be used in the alkoxymercuration reaction. Primary, secondary, and even tertiary alcohols react smoothly, but ditertiary ethers cannot be prepared because of steric hindrance to reaction.

PROBLEM

**18.6** Show in detail the mechanism of the reaction between 1-methylcyclopentene, ethanol, and mercuric trifluoroacetate.

PROBLEM

**18.7** How would you prepare the following ethers? Use whichever method you think is most appropriate, the Williamson synthesis or the alkoxymercuration reaction.

(a) Butyl cyclohexyl ether
(b) Ethyl phenyl ether
(c) *tert*-Butyl *sec*-butyl ether
(d) Tetrahydrofuran

## 18.6 Reactions of Ethers: Acidic Cleavage

Ethers are unusually inert to most reagents used in organic chemistry, a property that accounts for their wide use as inert reaction solvents. Halogens, mild acids, bases, and nucleophiles have no effect on most ethers. In fact, ethers undergo only one reaction of general use—ethers are cleaved by strong acids.

The first example of acid-induced ether cleavage was observed in 1861 by Alexander Butleroff,[2] who found that 2-ethoxypropanoic acid reacts with aqueous HI at 100°C to yield iodoethane and lactic acid:

$$\underset{\text{2-Ethoxypropanoic acid}}{CH_3CH(OCH_2CH_3)CO_2H} + HI \xrightarrow[H_2O]{100°C} CH_3CH_2I + \underset{\text{Lactic acid}}{CH_3CH(OH)CO_2H}$$

Aqueous HI is still the preferred reagent for cleaving simple ethers, although HBr can also be used. HCl does not cleave ethers.

[2]Alexander M. Butleroff (1828–1886); b. Tschistopol, Russia; Ph.D. (1854), University of Moscow; professor, University of Kazan (1854–1867), University of St. Petersburg (1867–1880).

$$CH_3CH_2OCH(CH_3)_2 \xrightarrow[\text{Reflux}]{\text{HI, } H_2O} CH_3CH_2I + (CH_3)_2CHOH$$

Ethyl isopropyl ether — Iodoethane — Isopropyl alcohol

$$C_6H_5OCH_2CH_3 \xrightarrow[\text{Reflux}]{\text{HBr, } H_2O} C_6H_5OH + CH_3CH_2Br$$

Ethyl phenyl ether — Phenol — Bromoethane

$$C_6H_{11}OC(CH_3)_3 \xrightarrow[0°C]{\text{HBr, } H_2O} C_6H_{11}OH + (CH_3)_2C{=}CH_2$$

*tert*-Butyl cyclohexyl ether — Cyclohexanol — 2-Methylpropene

Acidic ether cleavages are typical nucleophilic substitution reactions that take place by either an $S_N1$ pathway or an $S_N2$ pathway, depending on the structure of the ether. Primary and secondary alkyl ethers react by an $S_N2$ pathway in which iodide or bromide ion attacks the protonated ether at the less highly substituted site. This usually results in a selective cleavage into a single alcohol and a single alkyl halide, rather than a mixture of products. For example, butyl isopropyl ether yields exclusively isopropyl alcohol and 1-iodobutane on cleavage by HI, since nucleophilic attack by iodide occurs at the less hindered primary site rather than at the more hindered secondary site. Similarly, anisole is cleaved to give phenol and iodomethane.

$$(CH_3)_2CH{-}\ddot{O}{-}CH_2CH_2CH_2CH_3 \xrightarrow[H_2O,\ 100°C]{\text{HI}} \left[(CH_3)_2CH{-}\overset{+}{\ddot{O}}(H){-}CH_2CH_2CH_2CH_3\right] \xrightarrow{:\ddot{I}:^-\ (S_N2\ \text{reaction})} (CH_3)_2CHOH + CH_3CH_2CH_2CH_2I$$

Butyl isopropyl ether — Isopropyl alcohol — 1-Iodobutane

$$C_6H_5OCH_3 \xrightarrow[H_2O,\ 100°C]{\text{HI}} C_6H_5OH + CH_3I$$

Anisole — Phenol — Iodomethane

Tertiary, benzylic, and allylic ethers tend to cleave by either an $S_N1$ or an E1 mechanism, since these substrates are capable of producing stabilized intermediate carbocations. These reactions are often fast and take place at moderate temperatures. *tert*-Butyl ethers, for example, can often be cleaved at room temperature or below. Trifluoroacetic acid, rather than HBr or HI, seems to be best for these reactions, as the following example shows:

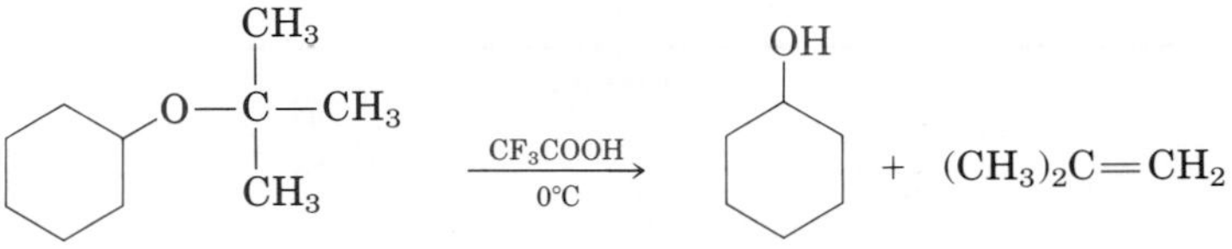

$$\text{tert-Butyl cyclohexyl ether} \xrightarrow[0^\circ C]{CF_3COOH} \text{Cyclohexanol (90\%)} + (CH_3)_2C{=}CH_2 \text{ (2-Methylpropene)}$$

tert-Butyl cyclohexyl ether — Cyclohexanol (90%) — 2-Methylpropene

**PROBLEM**

**18.8** Write the mechanism of the trifluoroacetic acid-catalyzed cleavage of a *tert*-butyl ether. Account for the fact that 2-methyl-2-propene is formed.

**PROBLEM**

**18.9** Suggest an explanation for the observation that HI and HBr are much more effective than HCl in cleaving ethers. [*Hint:* See Section 11.5.]

## 18.7 Cyclic Ethers: Epoxides

For the most part, cyclic ethers behave like acyclic ethers. The chemistry of the ether functional group is the same, whether it's in an open chain or in a ring. For example, common cyclic ethers such as tetrahydrofuran and dioxane are often used as solvents because of their inertness, yet they can be cleaved by strong acids.

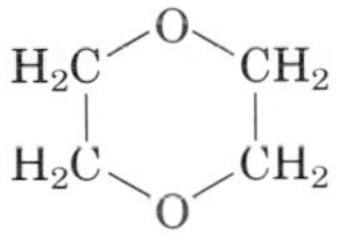

1,4-Dioxane 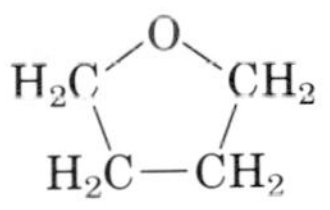 — Tetrahydrofuran

The one group of cyclic ethers that behaves differently is comprised of the three-membered, oxygen-containing rings called **epoxides**, or **oxiranes**. The strain of the three-membered ether ring makes epoxides highly reactive and confers unique chemical reactivity on them.

Ethylene oxide, the simplest epoxide, is an intermediate in the manufacture of both ethylene glycol (automobile antifreeze) and polyester polymers. More than 2.5 billion pounds of ethylene oxide are produced industrially each year by air oxidation of ethylene over a silver oxide catalyst at 300°C. This process is not of general utility for other epoxides, however, and is of little value in the laboratory.

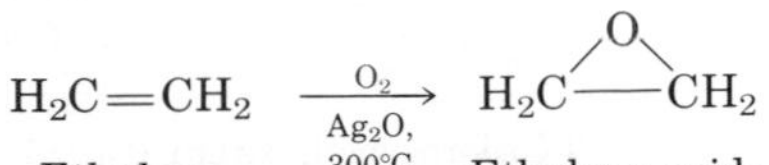

$$H_2C{=}CH_2 \xrightarrow[Ag_2O,\ 300^\circ C]{O_2} H_2C\overset{O}{\frown}CH_2$$

Ethylene — Ethylene oxide

In the laboratory, epoxides are normally prepared by treatment of an alkene with a **peroxyacid, $RCO_3H$**:

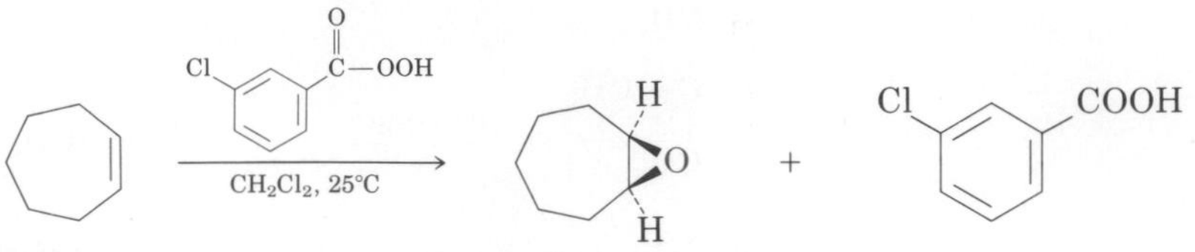

Cl C—OOH O H Cl COOH
$CH_2Cl_2$, 25°C O +
H

Cycloheptene 1,2-Epoxycycloheptane (78%) (Cycloheptene epoxide)

Many different peroxyacids can be used to accomplish epoxidation, but *m*-chloroperoxybenzoic acid is the preferred reagent on a laboratory scale. As opposed to most other peroxyacids, which are highly reactive and which readily decompose, *m*-chloroperoxybenzoic acid is a stable, crystalline, easily handled material.

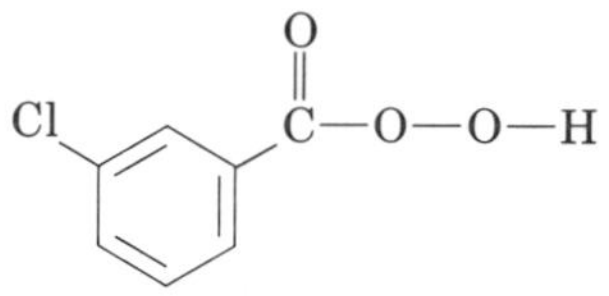

O
Cl C—O—O—H

*m*-Chloroperoxybenzoic acid

Peroxyacids are thought to transfer oxygen to an alkene through a complex one-step mechanism without intermediates. Thus, the epoxidation reaction is different from other alkene addition reactions we've studied. No carbocation intermediate is involved, and the details of the process aren't fully understood. There is good evidence to show, however, that the oxygen farthest from the carbonyl group is the one that is transferred.

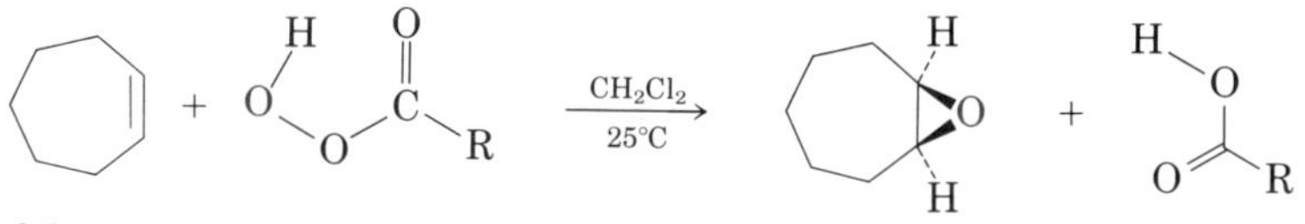

H O H H
+ O C $CH_2Cl_2$ O + O
O R 25°C H O R

Cycloheptene Peroxyacid 1,2-Epoxycycloheptane Acid

Another method for the synthesis of epoxides is through the use of halohydrins, prepared by electrophilic addition of HO—X to alkenes (Section 7.2). When halohydrins are treated with base, HX is eliminated, and an epoxide is produced.

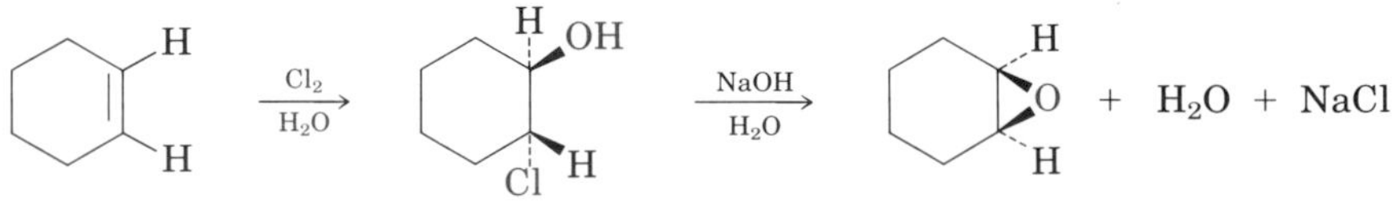

H H OH H
$Cl_2$ NaOH
$H_2O$ $H_2O$ O + $H_2O$ + NaCl
H Cl H H

Cyclohexene 2-Chlorocyclohexanol 1,2-Epoxycyclohexane (73%)

Note that the synthesis of epoxides by base treatment of halohydrins is actually an *intramolecular* Williamson ether synthesis. The nucleophilic oxygen atom and electrophilic carbon atom are in the same molecule, rather than in different molecules:

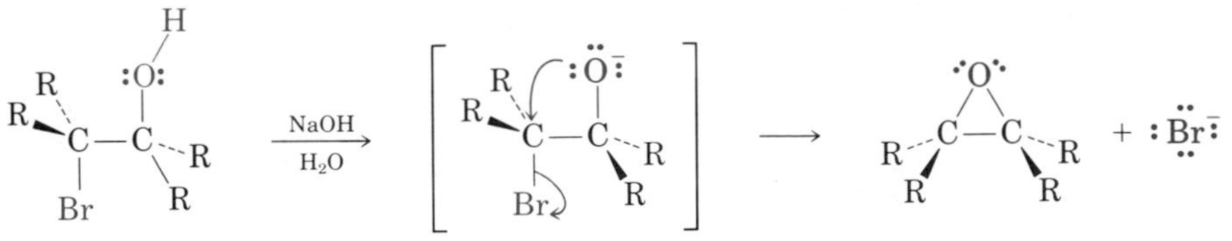

Bromohydrin — $\xrightarrow[H_2O]{NaOH}$ — Intramolecular substitution (within the same molecule) — $\longrightarrow$ Epoxide + $:\ddot{Br}:^-$

Recall the following:

$$CH_3\ddot{O}:^- + CH_3{-}Br \longrightarrow CH_3\ddot{O}CH_3 + :\ddot{Br}:^-$$

Intermolecular substitution
(between different molecules)

PROBLEM

**18.10** What product would you expect from reaction of *cis*-2-butene with *m*-chloroperoxybenzoic acid? Show the stereochemistry.

PROBLEM

**18.11** Reaction of *trans*-2-butene with *m*-chloroperoxybenzoic acid yields an epoxide different from that obtained by reaction of the cis isomer (Problem 18.10). Explain.

## 18.8 Ring-Opening Reactions of Epoxides

Epoxide rings can be cleaved by treatment with acid in much the same way as other ethers. The major difference is that epoxides react under much milder conditions because of ring strain. Dilute aqueous mineral acid at room temperature is sufficient to cause the hydrolysis of epoxides to 1,2-diols (also called **glycols**). Two million tons of ethylene glycol are produced each year by acid-catalyzed hydration of ethylene oxide. (Note that the name *ethylene glycol* is not a systematic one, since the *-ene* ending implies the presence of a double bond in the molecule. The name is used, however, because ethylene glycol is the glycol derived *from* ethylene by hydroxylation.)

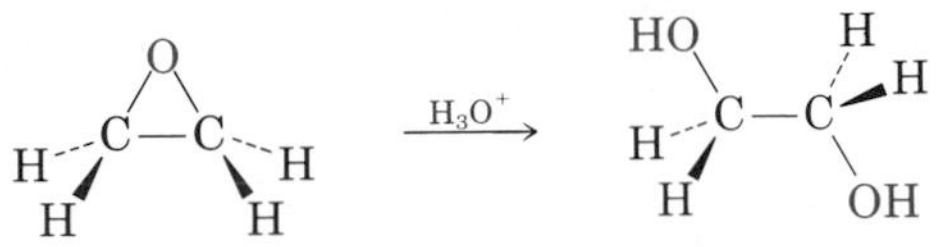

Ethylene oxide $\xrightarrow{H_3O^+}$ Ethylene glycol (1,2-Ethanediol)

Acid-induced epoxide ring opening takes place by $S_N2$ attack of a nucleophile on the protonated epoxide, in a manner analogous to the final step of alkene bromination, where a three-membered-ring bromonium ion is opened by nucleophilic attack (Section 7.1). When a cycloalkane epoxide is opened by aqueous acid, a trans 1,2-diol results (just as a trans 1,2-dibromide results from alkene bromination):

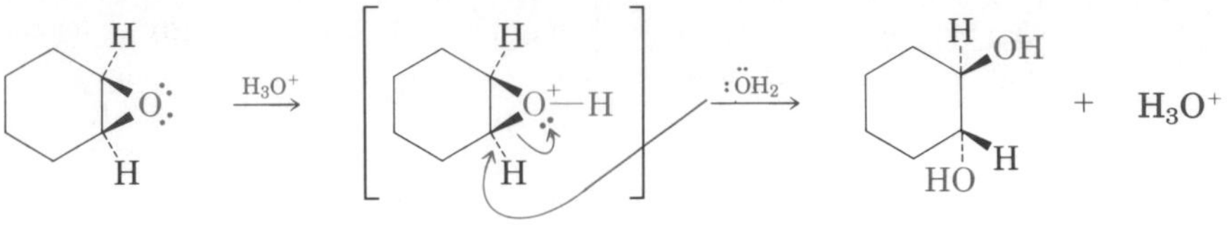

$H_3O^+$ ; $:\ddot{O}H_2$ ; $+ H_3O^+$

1,2-Epoxycyclohexane

*trans*-1,2-Cyclohexanediol (86%)

Recall the following reaction:

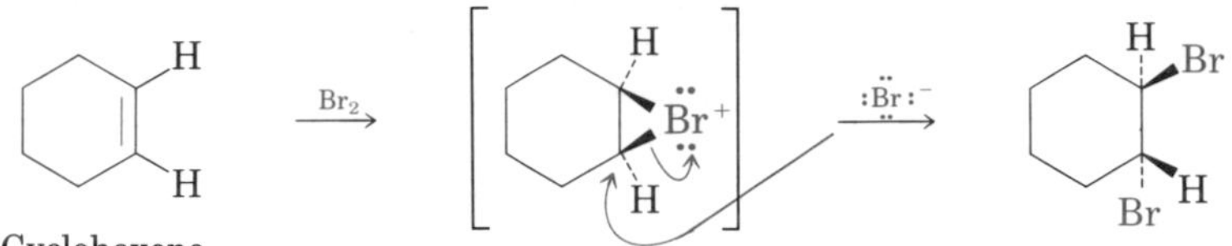

$Br_2$ ; $:\ddot{Br}:^-$

Cyclohexene

*trans*-1,2-Dibromocyclohexane

Protonated epoxides can also be opened by nucleophiles other than water. For example, if anhydrous HX is used, epoxides can be converted into trans halohydrins:

HX / Ether

A trans 2-halocyclohexanol

where X = F, Br, Cl, or I.

Epoxides, unlike other ethers, can also be cleaved by base. Although an ether oxygen is normally a very poor leaving group in an $S_N2$ reaction (Section 11.5), the reactivity of the three-membered ring is sufficient to cause epoxides to react with hydroxide ion at elevated temperatures.

$S_N2$ reaction

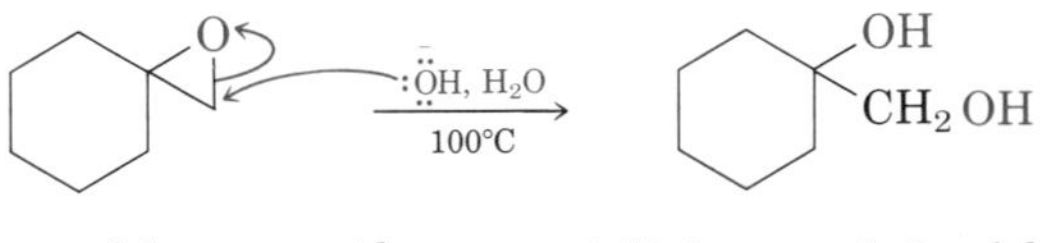

$:\ddot{O}H$, $H_2O$ ; 100°C

Methylenecyclohexane oxide

1-Hydroxymethylcyclohexanol (70%)

**PROBLEM**

**18.12** Show all the steps involved in the acidic hydrolysis of *cis*-5,6-epoxydecane. What is the stereochemistry of the product, assuming normal back-side $S_N2$ attack?

PROBLEM

**18.13** What is the stereochemistry of the product from acid-catalyzed hydrolysis of *trans*-5,6-epoxydecane? Is the product the same as or different from the one formed in Problem 18.12?

PROBLEM

**18.14** Acid-induced hydrolysis of a 1,2-epoxycyclohexane produces a trans diaxial 1,2-diol. What product would you expect to obtain from acidic hydrolysis of *cis*-3-*tert*-butyl-1,2-epoxycyclohexane? (Recall that the bulky *tert*-butyl group locks the cyclohexane ring into a specific conformation.)

## REGIOCHEMISTRY OF EPOXIDE RING OPENING

The direction in which unsymmetrical epoxide rings are opened depends on the conditions used. If a basic nucleophile is used in a typical $S_N2$ type of reaction, attack takes place at the less hindered epoxide carbon. For example, 1,2-epoxypropane is attacked by ethoxide ion exclusively at the less highly substituted, primary carbon to give 1-ethoxy-2-propanol.

$$CH_3-\overset{O}{CH-CH_2} \xrightarrow[CH_3CH_2OH]{^{-}:\ddot{O}CH_2CH_3} CH_3\overset{OH}{\overset{|}{C}}HCH_2OCH_2CH_3$$

No attack here (2°) — 1-Ethoxy-2-propanol (83%)

If acidic conditions are used, however, a different reaction course is followed and attack of the nucleophile occurs primarily at the *more* highly substituted carbon atom.

$$CH_3\overset{O}{CH-CH_2} \xrightarrow[Ether]{HCl} CH_3\underset{Cl}{\underset{|}{C}}H\overset{OH}{\overset{|}{C}}H_2 + CH_3\overset{OH}{\overset{|}{C}}HCH_2Cl$$

1,2-Epoxypropane — 2-Chloro-1-propanol (89%) — 1-Chloro-2-propanol (11%)

Acid-catalyzed epoxide opening is particularly interesting from a mechanistic viewpoint because it appears to be *midway* between a pure $S_N1$ reaction and a pure $S_N2$ reaction. Take the reaction of 1,2-epoxy-1-methylcyclohexane with HBr, for example. If this were a pure $S_N2$ reaction, bromide ion would attack the less highly substituted carbon atom, displacing the epoxide oxygen atom from the back side and yielding 2-bromo-1-methylcyclohexanol with the —Br and —OH groups trans to each other. On the other hand, if this were a pure $S_N1$ reaction, protonation of the epoxide oxygen atom followed by ring opening would yield a carbocation that could react with bromide ion from either side to give a mixture of two isomeric 2-bromo-2-methylcyclohexanols. One isomer would have the —Br and —OH groups cis; the other would have —Br and —OH trans (Figure 18.1).

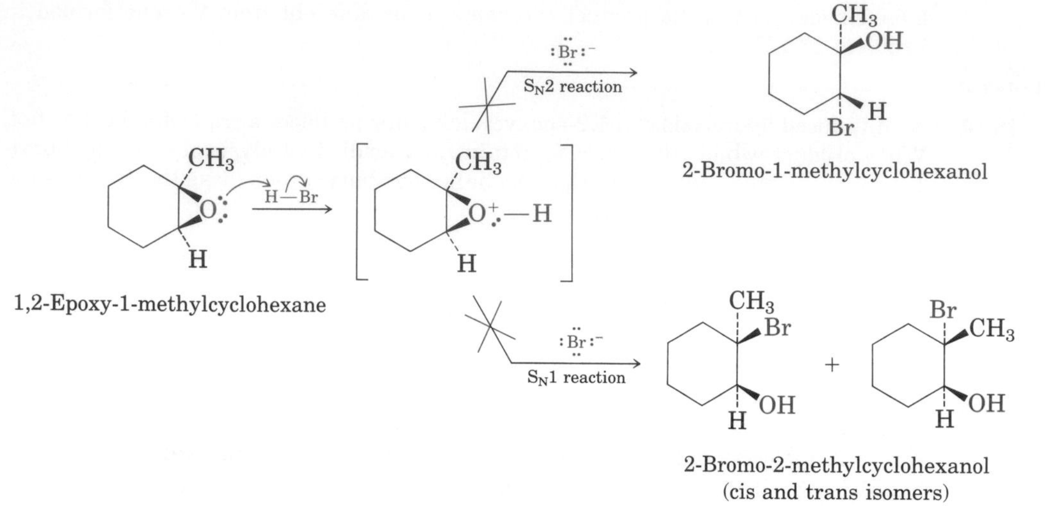

1,2-Epoxy-1-methylcyclohexane

$S_N2$ reaction

2-Bromo-1-methylcyclohexanol

$S_N1$ reaction

2-Bromo-2-methylcyclohexanol
(cis and trans isomers)

**Figure 18.1** Reaction products from hypothetical $S_N1$ and $S_N2$ ring openings of 1,2-epoxy-1-methylcyclohexane: In practice, neither reaction course is followed.

In fact, *neither* of the reaction courses shown in Figure 18.1 is followed. Instead, reaction of 1,2-epoxy-1-methylcyclohexane with HBr yields a single isomer of 2-bromo-2-methylcyclohexanol in which the bromo and hydroxyl groups are trans to each other. The observed regiochemistry can be accounted for by assuming that the reaction has characteristics of *both* $S_N1$ and $S_N2$ reactions. The fact that the single product formed has the entering bromine and the leaving oxygen on opposite sides of the ring is clearly an $S_N2$-like result. [*Remember:* $S_N2$ is the back-side displacement of the leaving group.] The fact that bromide ion attacks the tertiary side of the epoxide rather than the secondary side is clearly an $S_N1$-like result.

Both facts can be accommodated by postulating that the transition state for acid-induced epoxide ring opening has an $S_N2$-like geometry but a high degree of $S_N1$-like carbocationic character. Although the protonated epoxide is not a full carbocation, it is strongly polarized so that the positive charge is shared by the more highly substituted carbon atom. Thus, attack of a nucleophile occurs at the more highly substituted site (Figure 18.2).

PROBLEM

**18.15** Predict the major product of these reactions.

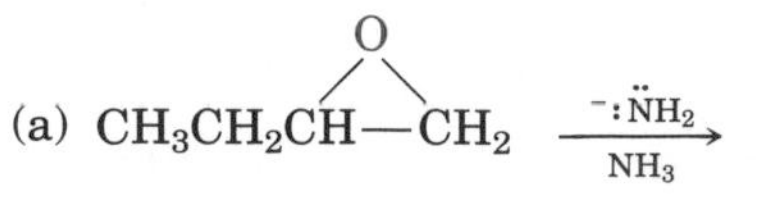

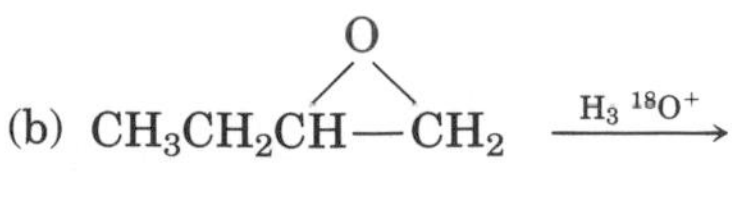

(a) $CH_3CH_2CH\overset{O}{—}CH_2 \xrightarrow[NH_3]{^-:\ddot{N}H_2}$ (b) $CH_3CH_2CH\overset{O}{—}CH_2 \xrightarrow{H_3\,^{18}O^+}$

1,2-Epoxy-1-methylcyclohexane

2° carbocation (*not formed*) → 2-Bromo-1-methylcyclohexanol

3° carbocation (more stable) → 2-Bromo-2-methylcyclohexanol

**Figure 18.2** Acid-induced ring opening of 1,2-epoxy-1-methylcyclohexane: There is a high degree of $S_N1$-like carbocationic character in the transition state, leading to exclusive formation of the isomer of 2-bromo-2-methylcyclohexanol that has —Br and —OH groups in the trans position.

## 18.9 Crown Ethers

**Crown ethers**, discovered in the early 1960s at the Du Pont Company, are a relatively recent addition to the ether family. Crown ethers are named according to the general format *x*-crown-*y*, where *x* is the total number of atoms in the ring and *y* is the number of oxygen atoms. Thus, 18-crown-6 ether is an 18-membered ring containing 6 ether oxygen atoms.

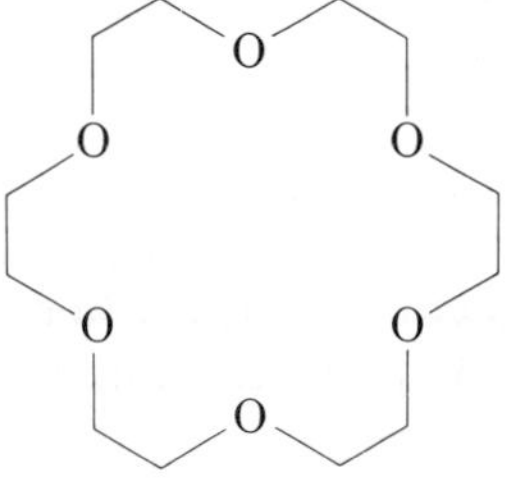


18-Crown-6 ether

The importance of crown ethers derives from their extraordinary ability to solvate metal cations by sequestering the metal in the center of the polyether cavity. Different crown ethers solvate different metal cations, depending on the match between ion size and cavity size. For example, 18-crown-6 is able to complex strongly with potassium ion.

Complexes between crown ethers and inorganic salts are soluble in nonpolar organic solvents, thus allowing many reactions to be carried out under aprotic conditions that would otherwise have to be carried out in aqueous solution. For example, the inorganic salt potassium permanganate actually dissolves in benzene in the presence of 18-crown-6. The resulting solution of "purple benzene" is a valuable reagent for oxidizing alkenes.

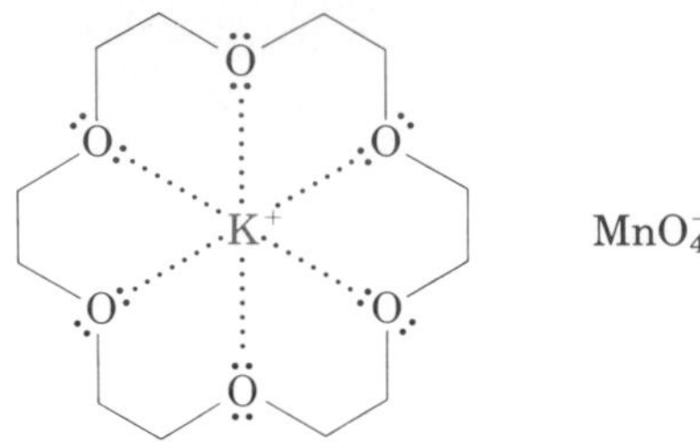


$KMnO_4$ solvated by 18-crown-6
(this solvate is soluble in benzene)

Many other inorganic salts, including KF, KCN, and $NaN_3$, can be made soluble in organic solvents by the use of crown ethers. The effect of using a crown ether to dissolve a salt in a hydrocarbon or ether solvent is similar to the effect of dissolving the salt in a polar aprotic solvent such as DMSO, DMF, or HMPA (Section 11.5). In both cases, the metal cation is strongly solvated, leaving the anion bare. Thus, the $S_N2$ reactivity of an anion is tremendously enhanced in the presence of a crown ether.

PROBLEM

**18.16** 15-Crown-5 and 12-crown-4 ethers complex $Na^+$ and $Li^+$, respectively. Make models of these crown ethers and compare the sizes of the cavities.

## 18.10 Spectroscopic Analysis of Ethers

### INFRARED SPECTROSCOPY

Ethers are difficult to distinguish by infrared spectroscopy. Although they show an absorption due to carbon–oxygen single-bond stretching in the range 1050–1150 $cm^{-1}$, many other kinds of absorptions occur in the same range. Figure 18.3 shows the infrared spectrum of diethyl ether and identifies the C—O stretch.

### NUCLEAR MAGNETIC RESONANCE SPECTROSCOPY

Protons on carbon next to the ether oxygen are shifted downfield from the normal alkane resonance and show $^1H$ NMR absorptions in the region 3.5–4.5 $\delta$. This downfield shift is clearly indicated in the spectrum of dipropyl ether shown in Figure 18.4.

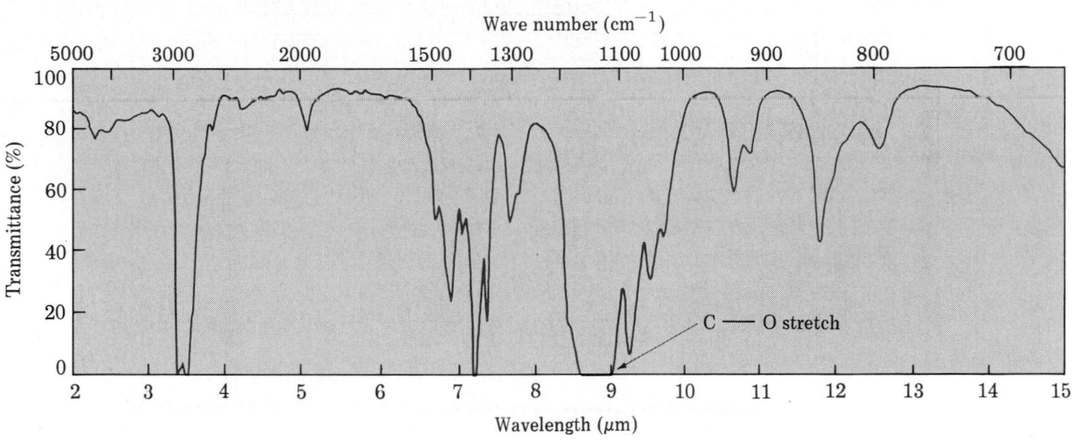


**Figure 18.3** The infrared spectrum of diethyl ether, $CH_3CH_2—O—CH_2CH_3$

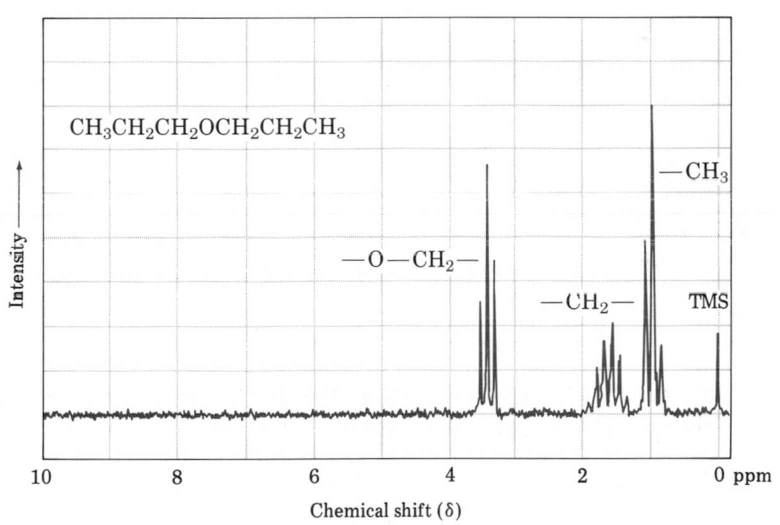


**Figure 18.4** The proton NMR spectrum of dipropyl ether: Protons on carbon next to oxygen are shifted downfield to 3.4 $\delta$.

Epoxides absorb at a slightly higher field than other ethers and show characteristic resonances at 2.5–3.5 $\delta$ in the $^1H$ NMR, as indicated for 1,2-epoxypropane in Figure 18.5, page 636.

Ether carbon atoms also exhibit a downfield shift in the $^{13}C$ NMR spectrum, where they usually absorb in the range 50–80 $\delta$. For example, the carbon atoms next to oxygen in methyl propyl ether absorb at 58.5 and 74.8 $\delta$ (Figure 18.6). Similarly, the methyl carbon in anisole absorbs at 54.8 $\delta$.

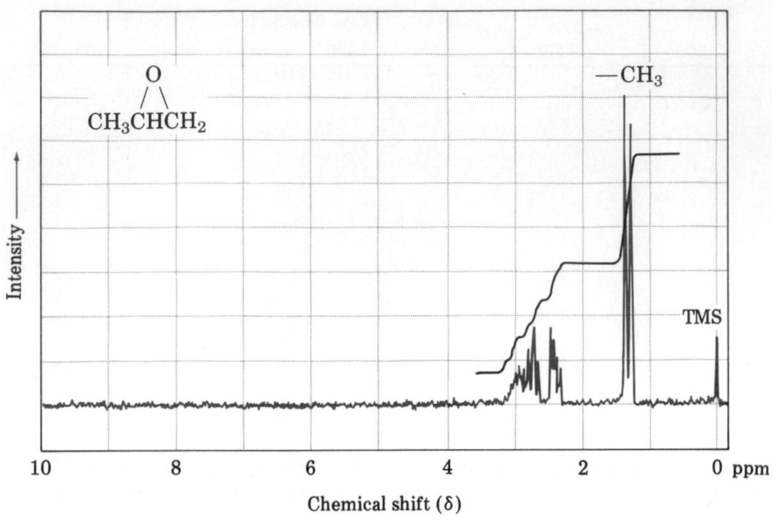


**Figure 18.5** The proton NMR spectrum of 1,2-epoxypropane

$CH_3—O—CH_2—CH_2—CH_3$: 58.5 ($CH_3$–O), 74.8 ($OCH_2$), 23.3 ($CH_2$), 10.7 ($CH_3$)

Anisole ($C_6H_5—O—CH_3$): 159.9 (C–O), 114.1, 129.5, 120.7, 54.8 ($OCH_3$)

**Figure 18.6** The $^{13}C$ NMR spectra of methyl propyl ether and anisole (δ units)

**PROBLEM**

**18.17** The $^1H$ NMR spectrum shown is that of a substance having the formula $C_4H_8O$. Propose a structure compatible with the observed spectrum.

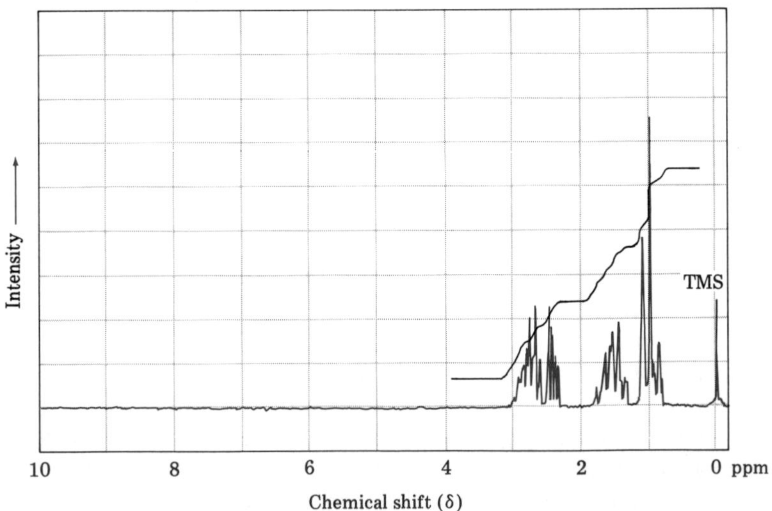

## 18.11 Sulfides

**Sulfides, R—S—R′**, are sulfur analogs of ethers. They are named by following the same rules used for ethers, with *sulfide* used in place of *ether* for simple compounds, and *alkylthio* used in place of *alkoxy* for more complex substances.

$CH_3—S—CH_3$

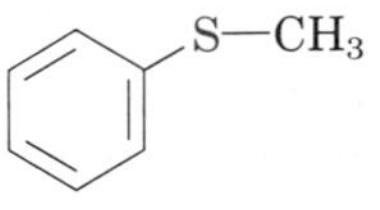

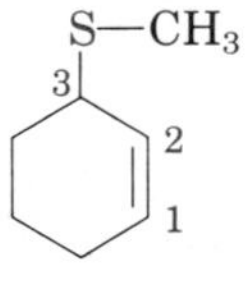

Dimethyl sulfide    Methyl phenyl sulfide    3-(Methylthio)cyclohexene

Sulfides are prepared by treatment of a primary or secondary alkyl halide with a **thiolate** anion, $R—\ddot{S}:^-$. The reaction occurs by an $S_N2$ mechanism, analogous to the Williamson synthesis of ethers (Section 18.4). Thiolate anions are among the best nucleophiles known (Section 11.5 and Table 11.3), and product yields are usually high in these substitution reactions.

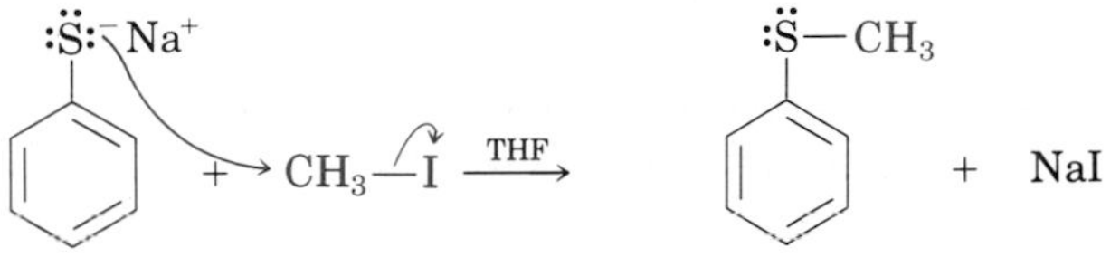

$$C_6H_5\ddot{S}:^-\ Na^+ + CH_3—I \xrightarrow{THF} C_6H_5\ddot{S}—CH_3 + NaI$$

Sodium benzenethiolate    Methyl phenyl sulfide (96%)

Since the valence electrons on sulfur are farther from the nucleus and are less tightly held than those on oxygen (3*p* electrons versus 2*p* electrons), there are some important differences between the chemistries of ethers and sulfides. For example, sulfur is more polarizable than oxygen, and sulfur compounds are thus more nucleophilic than their oxygen analogs. Unlike dialkyl ethers, dialkyl sulfides are good nucleophiles that react rapidly with primary alkyl halides by an $S_N2$ mechanism. The products of such reactions are **trialkylsulfonium salts, $R_3S^+$.**

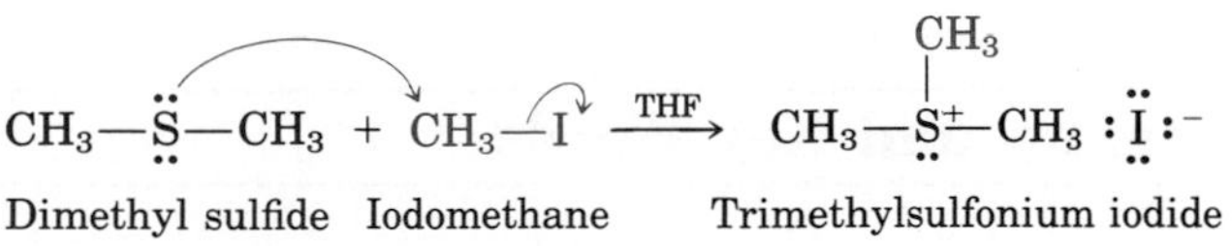

$$CH_3—\ddot{S}—CH_3 + CH_3—I \xrightarrow{THF} (CH_3)_2\overset{+}{S}—CH_3\ :\ddot{I}:^-$$

Dimethyl sulfide    Iodomethane    Trimethylsulfonium iodide

Once formed, trialkylsulfonium salts are themselves good alkylating agents, since a nucleophile can attack one of the groups bound to the positively charged sulfur, displacing a sulfide as leaving group. For example, nature makes extensive use of the trialkylsulfonium salt *S*-adenosylmethionine as a biological methylating agent (Section 11.17).

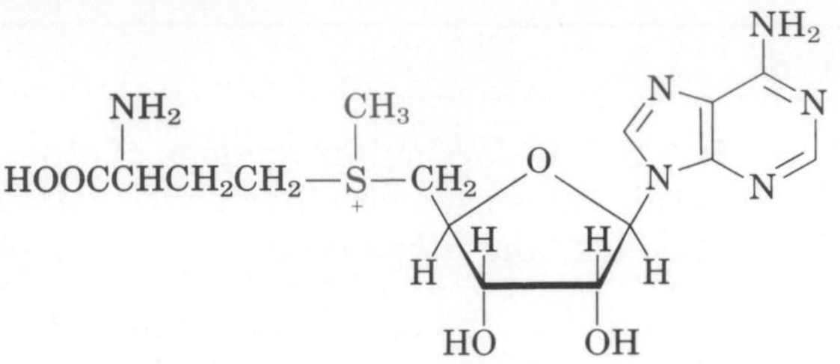

*S*-Adenosylmethionine (a sulfonium salt)

A second difference between sulfides and ethers is that sulfides are easily oxidized. Treatment of a sulfide with hydrogen peroxide, $H_2O_2$, at room temperature yields the corresponding **sulfoxide** (**$R_2SO$**), and further oxidation of the sulfoxide with a peroxyacid yields a **sulfone** (**$R_2SO_2$**).

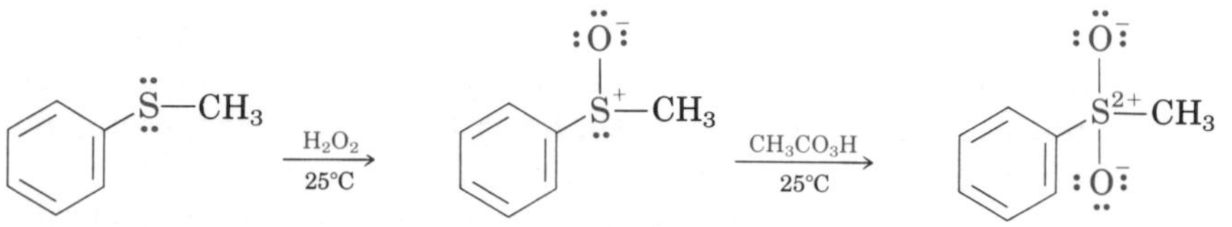

$$\xrightarrow[25°C]{H_2O_2} \qquad \xrightarrow[25°C]{CH_3CO_3H}$$

Methyl phenyl sulfide — Methyl phenyl sulfoxide — Methyl phenyl sulfone

Dimethyl sulfoxide (DMSO) is a particularly well known sulfoxide that finds wide use as a polar aprotic solvent. It must be handled with care, however, because it has a remarkable ability to penetrate the skin, carrying along whatever is dissolved in it.

PROBLEM

**18.18** Name the following compounds by IUPAC rules:

(a) $CH_3CH_2SCH_3$ (b) $(CH_3)_3CSCH_2CH_3$

(c) (d)

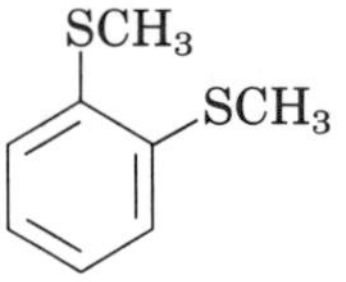

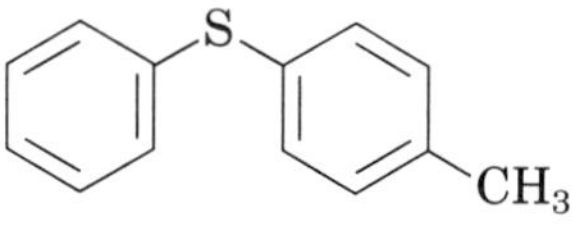

PROBLEM

**18.19** How can you account for the fact that dimethyl sulfoxide has a boiling point of 189°C and is miscible with water, whereas dimethyl sulfide has a boiling point of 37°C and is immiscible with water?

## 18.12 Summary and Key Words

**Ethers** are compounds that have two organic groups bonded to the same oxygen atom, R—O—R′. The organic groups may be alkyl, vinylic, or aryl, and the oxygen atom may be in a ring or in an open chain.

Ethers are normally prepared in the laboratory either by a Williamson synthesis or by an alkoxymercuration–demercuration sequence. The **Williamson ether synthesis** involves $S_N2$ attack of an alkoxide ion on a primary alkyl halide. The reaction is thus limited to the preparation of ethers in which one organic group is primary. The **alkoxymercuration–demercuration** sequence involves the formation of an intermediate alkoxy organomercurial compound, followed by sodium borohydride reduction of the carbon–mercury bond. The net result is Markovnikov addition of an alcohol to an alkene.

Ethers are inert to most common reagents but are attacked by strong acids to give cleavage products. Both HI and HBr are often used. The cleavage reaction takes place by an $S_N2$ mechanism if primary and secondary alkyl groups are bonded to the ether oxygen, but by an $S_N1$ mechanism if one of the alkyl groups bonded to oxygen is tertiary.

**Epoxides**, cyclic ethers with a three-membered oxygen-containing ring, differ from other ethers in their ease of cleavage. The high reactivity of the three-membered ether ring because of ring strain allows epoxide rings to be opened by nucleophilic attack of strong bases as well as by acids. Base-catalyzed epoxide ring opening occurs by $S_N2$ attack of a nucleophile at the less hindered epoxide carbon, whereas acid-induced epoxide ring opening occurs by $S_N1$-like attack at the more highly substituted epoxide carbon.

**Sulfides, R—S—R′**, are sulfur analogs of ethers. They are prepared by a Williamson type of $S_N2$ reaction between a thiolate anion and a primary or secondary alkyl halide. Sulfides are much more nucleophilic than ethers and can be oxidized to **sulfoxides ($R_2SO$)** and to **sulfones ($R_2SO_2$)**. Sulfides can also be alkylated by reaction with a primary alkyl halide to yield a **sulfonium salt, $R_3S^+$**.

Although ethers are not readily identifiable by infrared spectroscopy, they show characteristic downfield $^1H$ NMR absorptions that are easily detected. Carbon atoms bonded to oxygen show a similar downfield shift in the $^{13}C$ NMR.

## 18.13 Summary of Reactions

1. Preparation of ethers
   a. Williamson synthesis (Section 18.4)

$$CH_3CH_2\ddot{\underset{..}{O}}:^- + C_6H_5CH_2Cl \longrightarrow C_6H_5CH_2OCH_2CH_3 + :\ddot{\underset{..}{Cl}}:^-$$

   Alkyl halide should be primary.

   b. Alkoxymercuration–demercuration (Section 18.5)

$$C_6H_5CH{=}CH_2 + CH_3OH \xrightarrow[2.\ NaBH_4]{1.\ Hg(O_2CCF_3)_2} C_6H_5\underset{}{C}HCH_3(OCH_3)$$

   Markovnikov orientation is observed.

c. Epoxidation of alkenes with peroxyacids (Section 18.7)

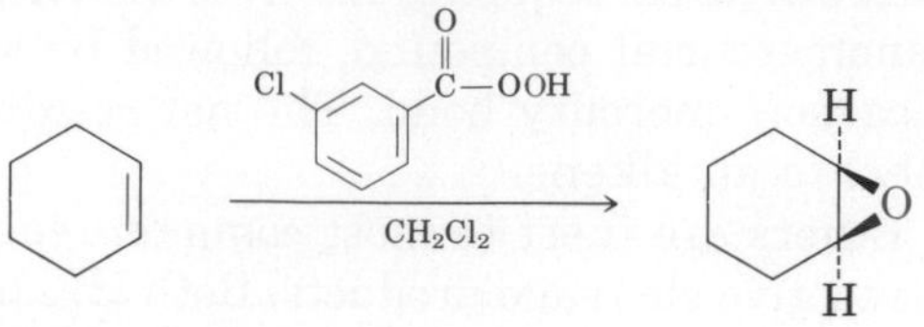

$$\text{cyclohexene} \xrightarrow[\text{CH}_2\text{Cl}_2]{m\text{-ClC}_6\text{H}_4\text{C(=O)}\text{—OOH}} \text{cyclohexene oxide (cis H, H)}$$

2. Reaction of ethers

a. Cleavage by HX, where HX = HBr or HI (Section 18.6)

$$CH_2CH_2OCH(CH_3)_2 \xrightarrow[H_2O]{HBr} CH_3CH_2Br + (CH_3)_2CHOH$$

b. Base-catalyzed epoxide ring opening (Section 18.8)

$$CH_3CH\overset{O}{—}CH_2 \xrightarrow[CH_3CH_2OH]{^-:\ddot{O}CH_2CH_3} CH_3\overset{OH}{\overset{|}{C}}HCH_2OCH_2CH_3$$

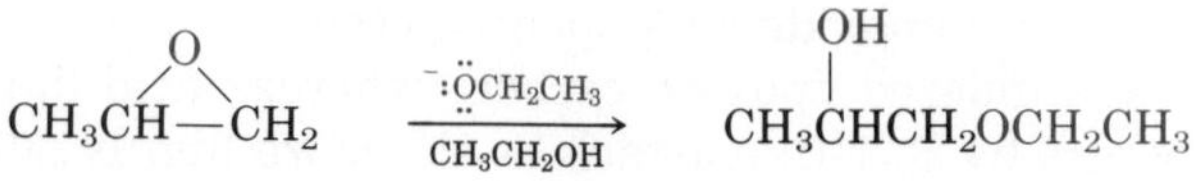

c. Acid-catalyzed hydrolysis of epoxides (Section 18.8)

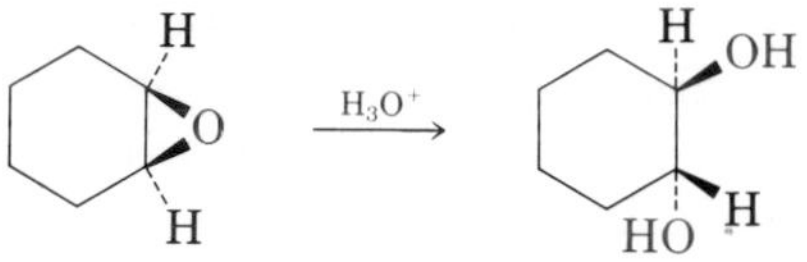

$$\text{cyclohexene oxide} \xrightarrow{H_3O^+} \text{trans-1,2-cyclohexanediol}$$

Trans 1,2-diols are produced.

d. Acid-induced epoxide ring opening (Section 18.8)

$$(CH_3)_2C\overset{O}{—}CH_2 \xrightarrow[H_2O]{HBr} (CH_3)_2\overset{Br}{\overset{|}{C}}CH_2OH$$

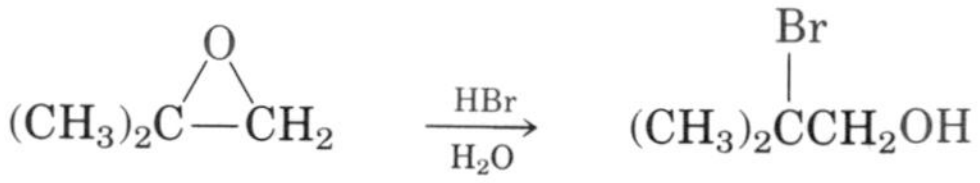

3. Preparation of sulfides (Section 18.11)

$$C_6H_5\ddot{\underset{\cdot\cdot}{S}}:^- + CH_3I \xrightarrow{THF} C_6H_5S—CH_3 + :\ddot{\underset{\cdot\cdot}{I}}:^-$$

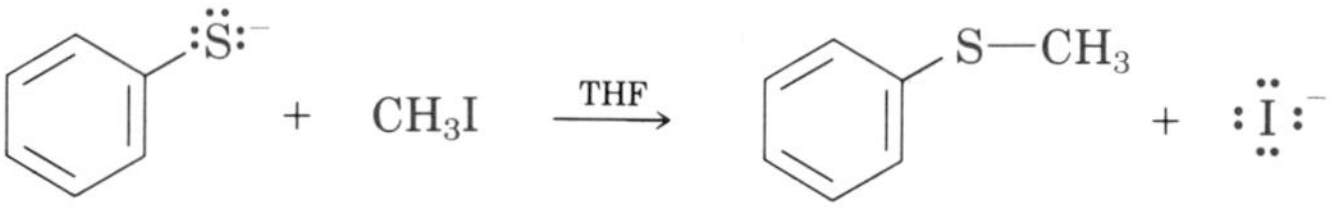

4. Oxidation of sulfides (Section 18.11)

a. Preparation of sulfoxides

$$CH_3—\ddot{\underset{\cdot\cdot}{S}}—CH_3 \xrightarrow{H_2O_2} CH_3—\overset{O^-}{\overset{|}{\underset{\cdot\cdot}{S^+}}}—CH_3$$

b. Preparation of sulfones

$$CH_3—\ddot{\underset{\cdot\cdot}{S}}—CH_3 \xrightarrow{RCO_3H} CH_3—\overset{O^-}{\overset{|}{\underset{\underset{O^-}{|}}{S^{2+}}}}—CH_3$$

## ADDITIONAL PROBLEMS

**18.20** Draw structures corresponding to these IUPAC names:
(a) Ethyl 1-ethylpropyl ether (b) Di(*p*-chlorophenyl) ether
(c) 3,4-Dimethoxybenzoic acid (d) Cyclopentyloxycyclohexane
(e) 4-Allyl-2-methoxyphenol (eugenol; from oil of cloves)

**18.21** Provide correct IUPAC names for these structures:

(a) 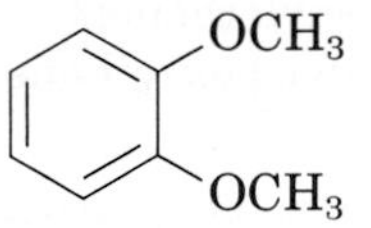 (b) 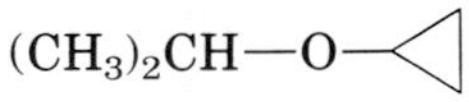 $(CH_3)_2CH{-}O{-}$cyclopropyl (c)

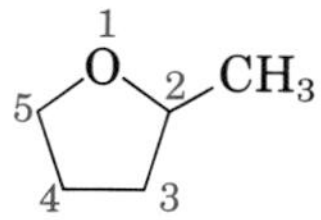

(d) 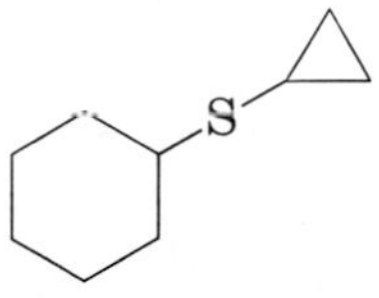 (e) $(CH_3)_2C(OCH_3)_2$ (f)

(g) $p\text{-}NO_2C_6H_4OCH_2CH_3$ (h)

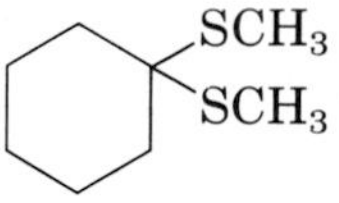

**18.22** When 2-methylpentane-2,5-diol is treated with sulfuric acid, dehydration occurs and 2,2-dimethyltetrahydrofuran is formed. Suggest a mechanism for this reaction. Which of the two oxygen atoms is most likely to be eliminated, and why?

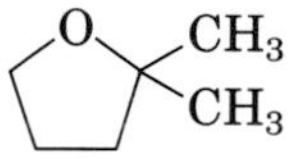

2,2-Dimethyltetrahydrofuran

**18.23** Predict the products of these ether cleavage reactions:

(a) 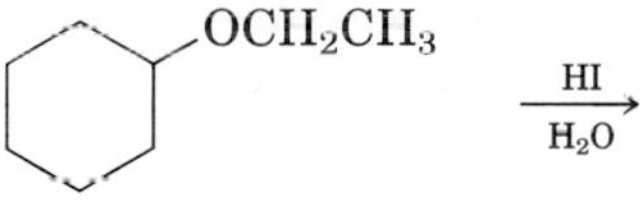 $\xrightarrow[H_2O]{HI}$ (b) 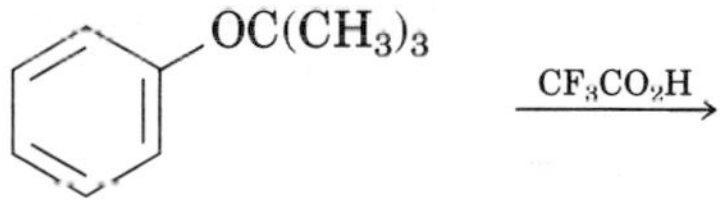 $\xrightarrow{CF_3CO_2H}$

(c) $H_2C{=}CHOCH_2CH_3 \xrightarrow[H_2O]{HI}$ (d) $(CH_3)_3CCH_2OCH_2CH_3 \xrightarrow[H_2O]{HI}$

**18.24** The **Zeisel method** is a classic analytical procedure for determining the number of methoxyl groups in a compound. A weighed amount of the compound is heated with concentrated HI, ether cleavage occurs, and the iodomethane formed is distilled off and passed into an alcohol solution of $AgNO_3$. The silver iodide that precipitates is then collected and weighed, and the percentage of methoxyl groups in the sample is thereby determined. For example, vanillin, the material responsible for the characteristic odor of vanilla, gives a positive Zeisel reaction; 1.06 g vanillin yields 1.60 g AgI. If vanillin has a molecular weight of 152, how many methoxyls does it contain?

**18.25** How would you prepare these ethers?

(a) 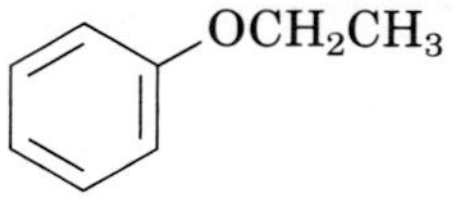 (b) 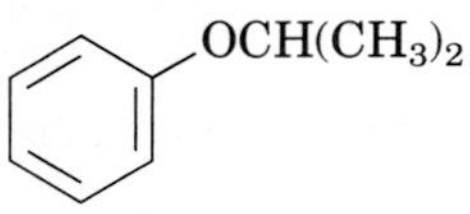 (c)

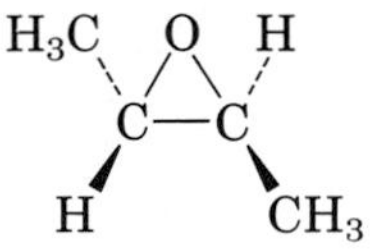

(d) 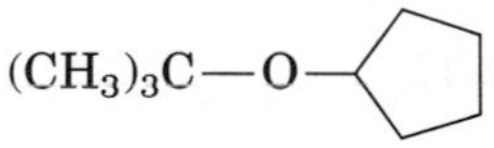 $(CH_3)_3C{-}O{-}$cyclopentyl (e) 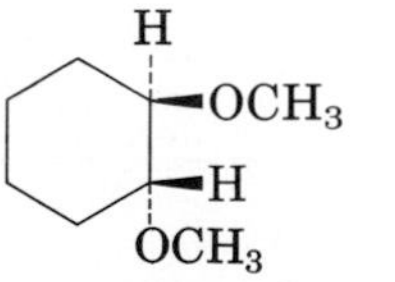 (f)

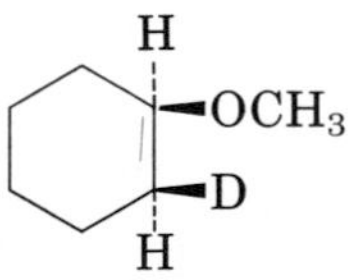

**18.26** **Meerwein's reagent**, triethyloxonium tetrafluoroborate, is a powerful ethylating agent that converts alcohols into ethyl ethers at neutral pH. Formulate the reaction of Meerwein's reagent with cyclohexanol and account for the fact that trialkyloxonium salts are much more reactive alkylating agents than alkyl iodides.

$$(CH_3CH_2)_3O^+ \ BF_4^-$$

Meerwein's reagent

**18.27** How would you prepare these compounds from 1-phenylethanol?
(a) Methyl 1-phenylethyl ether (b) Phenylepoxyethane
(c) *tert*-Butyl 1-phenylethyl ether

**18.28** How would you carry out these transformations? More than one step may be required.

(a) 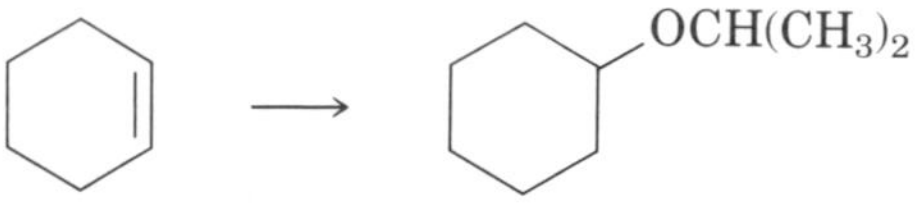 Cyclohexene ⟶ $OCH(CH_3)_2$ (cyclohexyl isopropyl ether)

(b) 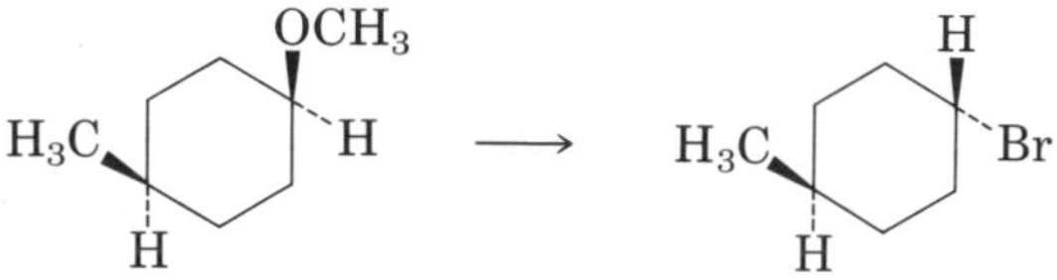 $H_3C$, H, $OCH_3$, H ⟶ $H_3C$, H, H, Br

(c) 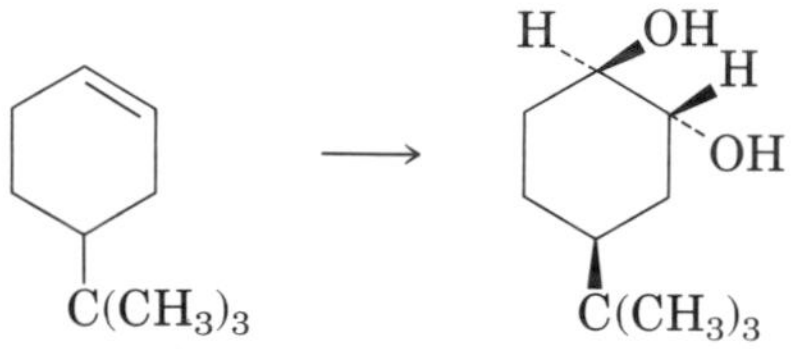 $C(CH_3)_3$ ⟶ H, OH, H, OH, $C(CH_3)_3$

(d) $CH_3CH_2CH_2CH_2C{\equiv}CH \longrightarrow CH_3CH_2CH_2CH_2CH_2CH_2OCH_3$

(e) $CH_3CH_2CH_2CH_2C{\equiv}CH \longrightarrow CH_3CH_2CH_2CH_2CH(OCH_3)CH_3$

**18.29** What product(s) would you expect from cleavage of tetrahydrofuran with HI?

**18.30** How could you prepare benzyl phenyl ether from benzene? More than one step is required.

**18.31** Methyl aryl ethers can be cleaved to iodomethane and a phenoxide anion by treatment with LiI in hot dimethylformamide solvent. Propose a mechanism for this reaction. What type of mechanism is involved?

**18.32** *tert*-Butyl ethers can be prepared by the reaction of an alcohol with 2-methylpropene in the presence of an acid catalyst. Propose a mechanism for this reaction.

**18.33** Safrole, isolated from oil of sassafras, is used as a perfumery agent. Propose a synthesis of safrole from catechol (1,2-benzenediol).

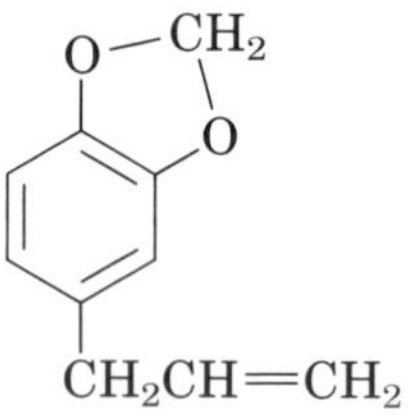

O—$CH_2$—O, $CH_2CH{=}CH_2$

Safrole

**18.34** Grignard reagents react with ethylene oxide to produce primary alcohols. Propose a mechanism for this reaction.

$$\text{Ethylene oxide } (CH_2{-}CH_2\text{ with O bridge}) \xrightarrow[\text{2. } H_3O^+]{\text{1. RMgX}} RCH_2CH_2OH$$

Ethylene oxide

**18.35** Grignard reagents also react with oxetane to produce primary alcohols, but the reaction is much slower than with ethylene oxide (Problem 18.34). Suggest a reason for the difference in reactivity between oxetane and ethylene oxide.

$$\text{Oxetane} \xrightarrow[\text{2. } H_3O^+]{\text{1. RMgX}} RCH_2CH_2CH_2OH$$

Oxetane

**18.36** Treatment of *trans*-2-chlorocyclohexanol with NaOH yields 1,2-epoxycyclohexane, but reaction of the cis isomer under the same conditions yields cyclohexanone. Propose mechanisms for both reactions and explain why the different results are obtained.

**18.37** Ethyl vinyl ether reacts with ethanol in the presence of an acid catalyst to yield 1,1-diethoxyethane, rather than 1,2-diethoxyethane. How can you account for the observed regioselectivity of addition?

**18.38** Anethole, $C_{10}H_{12}O$, a major constituent of the oil of anise, has the $^1H$ NMR spectrum shown here. On oxidation with hot sodium dichromate, anethole yields a compound identified as *p*-methoxybenzoic acid. What is the structure of anethole? Assign all peaks in the NMR spectrum and account for the observed splitting patterns.

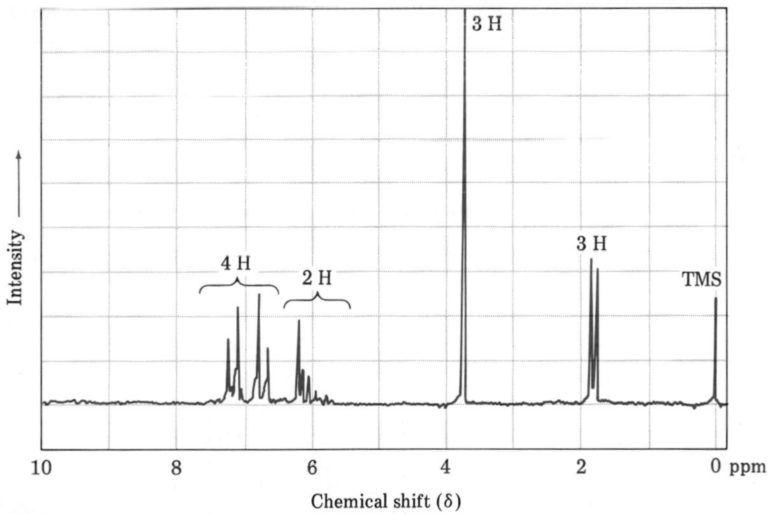


**18.39** How would you synthesize anethole (Problem 18.38) from benzene?

**18.40** The red fox (*Vulpes vulpes*) uses a chemical communication system based on scent marks in urine. Recent work has shown one component of fox urine to be a sulfide. Mass spectral analysis of the pure scent-mark component shows $M^{+\cdot} = 116$. Infrared spectroscopy shows an intense band at 890 $cm^{-1}$, and $^1H$ NMR spectroscopy reveals the following peaks: 3-proton singlet at 1.74 $\delta$; 3-proton singlet at 2.11 $\delta$; 2-proton triplet at 2.27 $\delta$, $J = 4.2$ Hz; 2-proton triplet at 2.57 $\delta$, $J = 4.2$ Hz; 2-proton broad peak at 4.73 $\delta$. Propose a structure consistent with these data. [*Note:* $(CH_3)_2S$ absorbs at 2.1 $\delta$ in the $^1H$ NMR.]

**18.41** How can you explain the fact that treatment of bornene with a peroxyacid yields a different epoxide from the one obtained by reaction of bornene with aqueous bromine, followed by base treatment?

$H_3C$ $CH_3$ … $CH_3$

Bornene $\xrightarrow{RCO_3H}$ Epoxide A

Bornene $\xrightarrow[\text{2. NaOH}]{\text{1. } Br_2/H_2O}$ Epoxide B

Bornene

Propose structures for epoxides A and B. [*Hint:* Build molecular models of bornene.]

**18.42** Disparlure, $C_{19}H_{38}O$, is a sex attractant released by the female gypsy moth, *Lymantria dispar*. The $^1H$ NMR spectrum of disparlure shows a large absorption in the alkane region, 1–2 $\delta$, and a triplet at 2.8 $\delta$. Treatment of disparlure, first with aqueous acid and then with $KMnO_4$, yields two carboxylic acids identified as undecanoic acid and 6-methylheptanoic acid. [$KMnO_4$ oxidatively cleaves 1,2-diols to yield carboxylic acids (Section 7.8).] Neglecting stereochemistry, propose a structure for disparlure consistent with these data. The actual compound is a chiral molecule with 7*R*,8*S* stereochemistry. Draw disparlure, showing the correct stereochemistry.

**18.43** How would you synthesize racemic disparlure (Problem 18.42) from compounds having 10 or fewer carbons?

**18.44** Treatment of 1,1-diphenyl-1,2-epoxyethane with aqueous acid yields diphenylacetaldehyde as the major product. Propose a mechanism to account for this reaction.

$$(Ph)_2C\overset{O}{—}CH_2 \xrightarrow{H_3O^+} (Ph)_2CHCHO$$

**18.45** Propose structures for ethers that have the following $^1H$ NMR spectra.

(a) $C_9H_{11}BrO$

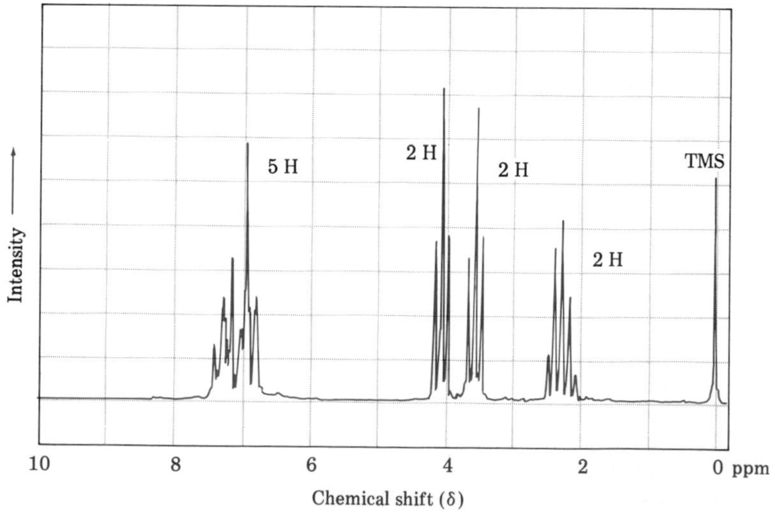

(b) $C_4H_{10}O_2$

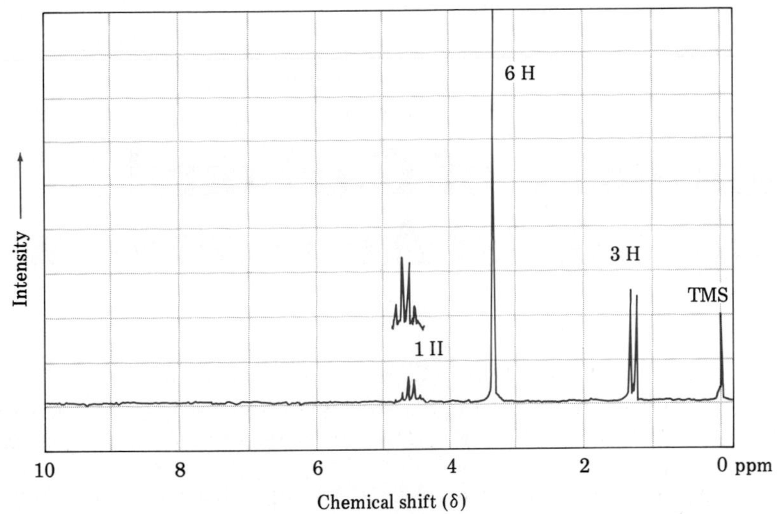


(c) $C_9H_{10}O$

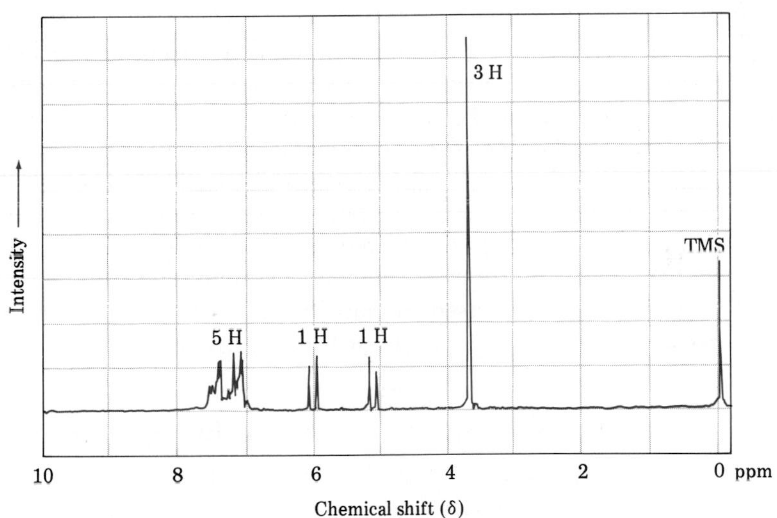

# Chemistry of Carbonyl Compounds: An Overview

In this and the next five chapters, we will discuss the most important functional group in organic chemistry—the **carbonyl group**, **C=O** (pronounced car-bo-**neel**).

Carbonyl compounds are everywhere in nature. The majority of biologically important molecules contain carbonyl groups, as do most pharmaceutical agents and many of the synthetic chemicals that touch our everyday lives. Acetic acid, the chief component of vinegar; acetaminophen, the active ingredient in many over-the-counter headache remedies; and Dacron, the polyester material used in clothing, all contain different kinds of carbonyl groups.

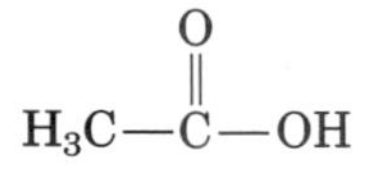

$$H_3C-\overset{\overset{\displaystyle O}{\|}}{C}-OH$$

Acetic acid
(a carboxylic acid)

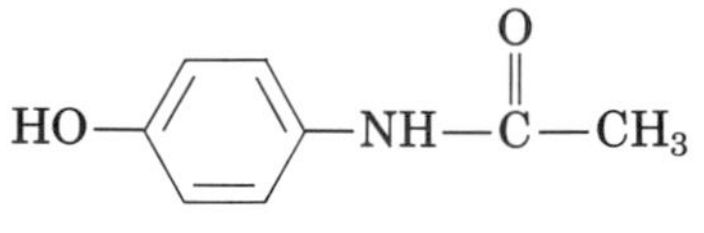

$$HO-C_6H_4-NH-\overset{\overset{\displaystyle O}{\|}}{C}-CH_3$$

Acetaminophen
(an amide)

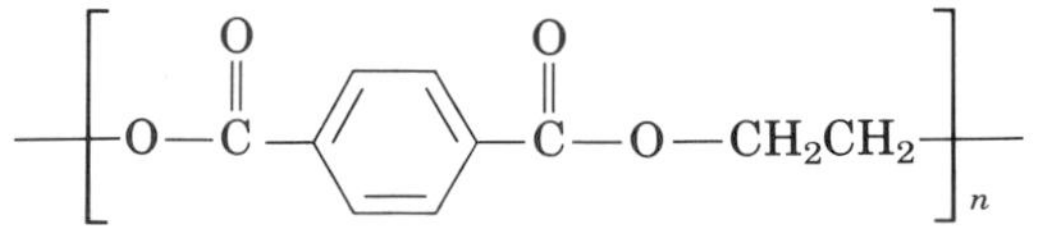

$$\left[-O-\overset{\overset{\displaystyle O}{\|}}{C}-C_6H_4-\overset{\overset{\displaystyle O}{\|}}{C}-O-CH_2CH_2-\right]_n$$

Dacron (a polyester)

## I. Kinds of Carbonyl Compounds

There are many different kinds of carbonyl compounds, depending on what groups are bonded to the C=O unit. The chemistry of carbonyl groups is quite similar, however, regardless of their exact structure.

Table 1 shows some of the many different kinds of carbonyl compounds. All contain an **acyl fragment**, $\mathbf{R{-}\overset{O}{\overset{\|}{C}}{-}}$, bonded to another residue. The R group of the acyl fragment may be alkyl, aryl, alkenyl, or alkynyl; the other residue to which the acyl fragment is bonded may be a carbon, hydrogen, oxygen, halogen, nitrogen, sulfur, or other substituent.

**Table 1** Some types of carbonyl compounds

| *Name* | *General formula* | *Name ending* |
|---|---|---|
| Aldehyde | $R{-}\overset{O}{\overset{\|}{C}}{-}H$ | *-al* |
| Ketone | $R{-}\overset{O}{\overset{\|}{C}}{-}R'$ | *-one* |
| Carboxylic acid | $R{-}\overset{O}{\overset{\|}{C}}{-}O{-}H$ | *-oic acid* |
| Acid chloride | $R{-}\overset{O}{\overset{\|}{C}}{-}Cl$ | *-yl* or *-oyl chloride* |
| Acid anhydride | $R{-}\overset{O}{\overset{\|}{C}}{-}O{-}\overset{O}{\overset{\|}{C}}{-}R'$ | *-oic anhydride* |
| Ester | $R{-}\overset{O}{\overset{\|}{C}}{-}O{-}R'$ | *-oate* |
| Lactone (cyclic ester) | $\diagdown\!\!\!\diagup C{-}\overset{O}{\overset{\|}{C}}{-}O$ (ring: C and O joined) | None |
| Amide | $R{-}\overset{O}{\overset{\|}{C}}{-}N\diagup\!\!\!\diagdown$ | *-amide* |
| Lactam (cyclic amide) | $\diagdown\!\!\!\diagup C{-}\overset{O}{\overset{\|}{C}}{-}N\diagup$ (ring: C and N joined) | None |

It turns out to be very useful to classify carbonyl compounds into two general categories, based on the kinds of chemistry they undergo:

Aldehydes ($RCH{=}O$)
Ketones ($R_2C{=}O$)

The acyl groups in these two families are bonded to substituents (—H and —R, respectively) that *can't stabilize a negative charge and therefore can't act as leaving groups*. Aldehydes and ketones behave similarly and undergo many of the same reactions.

| | |
|---|---|
| Carboxylic acids (RCOOH)<br>Esters (RCOOR′)<br>Acid chlorides (RCOCl)<br>Acid anhydrides (RCOOCOR′)<br>Amides ($RCONH_2$) | The acyl groups in carboxylic acids and their derivatives are bonded to substituents (oxygen, halogen, nitrogen) that *can stabilize a negative charge and can therefore serve as leaving groups in substitution reactions*. The chemistry of these compounds is therefore similar. |

## II. Nature of the Carbonyl Group

The carbon–oxygen double bond of carbonyl groups is similar in many respects to the carbon–carbon double bond of alkenes (Figure 1). The carbonyl carbon atom is $sp^2$ hybridized and forms three sigma bonds. The fourth valence electron remains in a carbon $p$ orbital and forms a pi bond to oxygen by overlap with an oxygen $p$ orbital. The oxygen atom also has two nonbonding pairs of electrons, which occupy its remaining two orbitals. (The oxygen atom is probably $sp^2$ hybridized, though there is some disagreement about this point.)

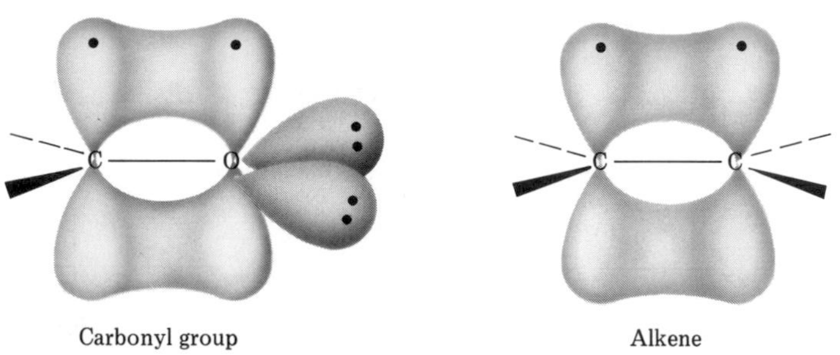


**Figure 1** Electronic structure of the carbonyl group

Like alkenes, carbonyl compounds are planar about the double bond and have bond angles of approximately 120°. Figure 2 shows the structure of acetaldehyde, and Table 2 indicates the experimentally determined bond lengths and angles. As you might expect, the carbon–oxygen double bond

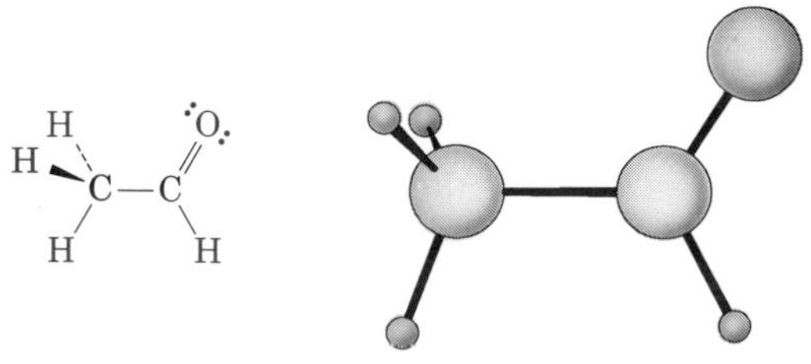

**Figure 2** Structure of acetaldehyde

**Table 2** Physical parameters of acetaldehyde

| *Bond angle (degrees)* | | *Bond length (Å)* | |
|---|---|---|---|
| H-C-C | 118 | C=O | 1.22 |
| C-C-O | 121 | C—C | 1.50 |
| H-C-O | 121 | OC—H | 1.09 |

is both shorter (1.22 Å versus 1.43 Å) and stronger [175 kcal/mol (732 kJ/mol) versus 92 kcal/mol (385 kJ/mol)] than a normal carbon–oxygen single bond.

Carbon–oxygen double bonds are polarized because of the high electronegativity of oxygen relative to carbon. Thus, all types of carbonyl compounds have substantial dipole moments, as shown in Table 3.

**Table 3** Dipole moments of some carbonyl compounds, $R_2C{=}O$

| *Carbonyl compound* | *Type of carbonyl group* | *Observed dipole moment (D)* |
|---|---|---|
| HCHO | Aldehyde | 2.33 |
| $CH_3CHO$ | Aldehyde | 2.72 |
| $(CH_3)_2CO$ | Ketone | 2.88 |
| $PhCOCH_3$ | Ketone | 3.02 |
| Cyclobutanone | Ketone | 2.99 |
| $CH_3COOH$ | Carboxylic acid | 1.74 |
| $CH_3COCl$ | Acid chloride | 2.72 |
| $CH_3CO_2CH_3$ | Ester | 1.72 |
| $CH_3CONH_2$ | Amide | 3.76 |
| $CH_3CON(CH_3)_2$ | Amide | 3.81 |

The most important consequence of carbonyl-group polarization is the chemical reactivity of the carbon–oxygen double bond. Since the carbonyl carbon carries a partial positive charge, it is an electrophilic site and is attacked by nucleophiles. Conversely, the carbonyl oxygen carries a partial negative charge and is a nucleophilic (basic) site. We'll see in the next five chapters that the majority of carbonyl-group reactions can be rationalized in terms of simple bond-polarization arguments.

Electrophilic carbon reacts with bases and nucleophiles → $\overset{\delta^+}{C}{=}\overset{\delta^-}{\ddot{O}}\!:$ ← Nucleophilic oxygen reacts with acids and electrophiles

## III. General Reactions of Carbonyl Compounds

Most reactions of carbonyl groups occur by one of four general reaction mechanisms—**nucleophilic addition, nucleophilic acyl substitution, alpha substitution**, and **carbonyl condensation**. These mechanisms have many variations, just as alkene electrophilic addition reactions and $S_N2$ reactions do, but the variations are much easier to learn when the fundamental features of the mechanisms are understood. Let's see what these four reaction mechanisms are and what kinds of chemistry carbonyl groups undergo.

### NUCLEOPHILIC ADDITION REACTIONS

The most common reaction of ketones and aldehydes is the **nucleophilic addition reaction**, which involves addition of a nucleophile to the electrophilic carbon of the carbonyl group. Since the nucleophile uses its electron pair to form a new bond to carbon, two electrons from the carbon–oxygen double bond must move onto the electronegative oxygen atom, where they can be stabilized as an alkoxide anion. The carbonyl carbon rehybridizes from $sp^2$ to $sp^3$ during the reaction, and the initial intermediate therefore has tetrahedral geometry.

:Nu⁻ (or :Nu—H)

Carbonyl compound ($sp^2$-hybridized carbon)

Tetrahedral intermediate ($sp^3$-hybridized carbon)

Once formed, and depending on the nature of the nucleophile, the tetrahedral intermediate can undergo either of the reactions shown in Figure 3. Often, the tetrahedral intermediate is simply protonated to form an alcohol, as happens during Grignard addition (Section 17.7). Alternatively, the tetrahedral intermediate can expel the oxygen completely to form a new double bond between the carbonyl-group carbon and the nucleophile. This second reaction is particularly common when amine nucleophiles are involved.

**Alcohol formation** The simplest reaction of a tetrahedral intermediate is protonation to yield an alcohol as final product. We have already seen two examples of this kind of process during reduction of ketones and aldehydes with hydride reagents such as $NaBH_4$ and $LiAlH_4$ (Section 17.6), and during Grignard reactions (Section 17.7). In the case of reduction, the nucleophile that adds to the carbonyl group is a **hydride ion, H:⁻**, whereas in the case of Grignard reaction, the nucleophile is a **carbanion, $R_3C$:⁻**.

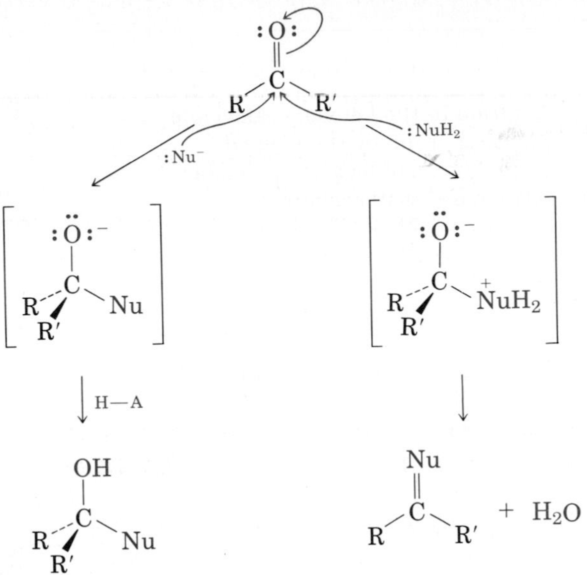

**Figure 3** The addition reaction of a nucleophile with a ketone or aldehyde

Hydride reduction

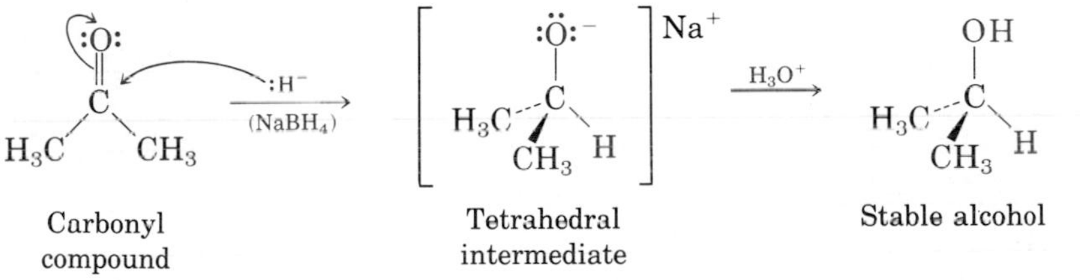

Carbonyl compound — Tetrahedral intermediate — Stable alcohol

Grignard reaction

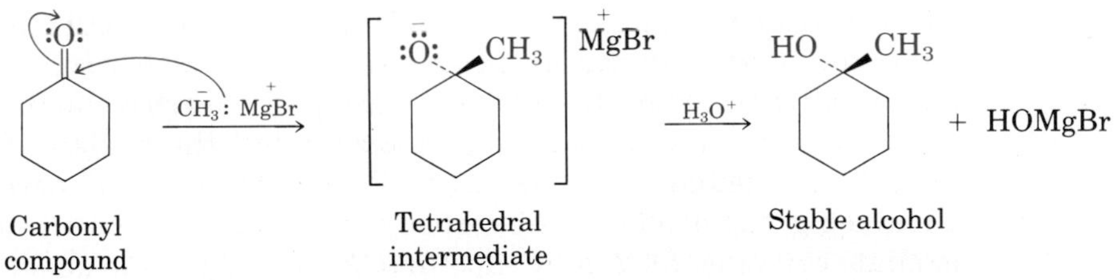

Carbonyl compound — Tetrahedral intermediate — Stable alcohol

**Formation of C=Nu** The second mode of nucleophilic addition, which often happens with amine nucleophiles, is that complete loss of oxygen can occur and a C=Nu double bond can form. For example, ketones and aldehydes react with primary amines, R—$NH_2$, to form **imines**, **$R_2C$=N—R′**. These reactions proceed through exactly the same kind of tetrahedral intermediate as that formed during hydride reduction and Grignard reaction, but the initially formed product is not isolated. Instead, it loses water to form an imine, as shown in Figure 4, page 652.

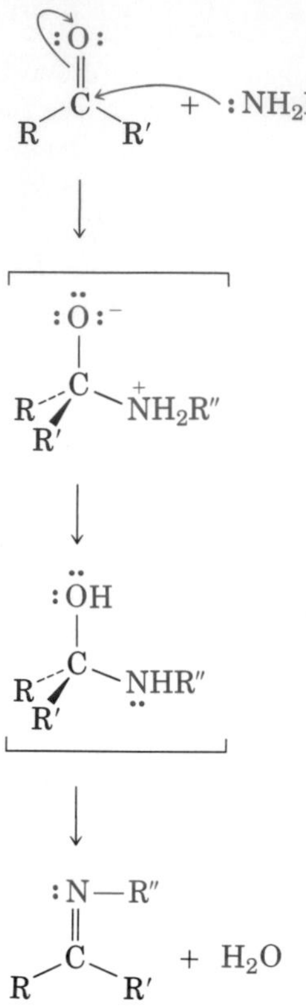

Addition to the ketone or aldehyde carbonyl group by the neutral amine nucleophile gives a dipolar tetrahedral intermediate.

Transfer of a proton from nitrogen to oxygen then yields a nonpolar amino alcohol intermediate.

Dehydration of the amino alcohol intermediate gives neutral imine plus water as final products.

**Figure 4** Mechanism of imine formation by reaction of an amine with a ketone or aldehyde

## NUCLEOPHILIC ACYL SUBSTITUTION REACTIONS

The second fundamental reaction of carbonyl compounds, **nucleophilic acyl substitution**, is related to the nucleophilic addition reaction just discussed, but occurs only with carboxylic acid derivatives rather than with ketones and aldehydes. When the carbonyl group of a carboxylic acid derivative reacts with a nucleophile, addition occurs, but the initially formed tetrahedral intermediate is not isolated. Since carboxylic acid derivatives have a leaving group bonded to the carbonyl-group carbon, the tetrahedral intermediate can react further by expelling the leaving group and forming a new carbonyl compound:

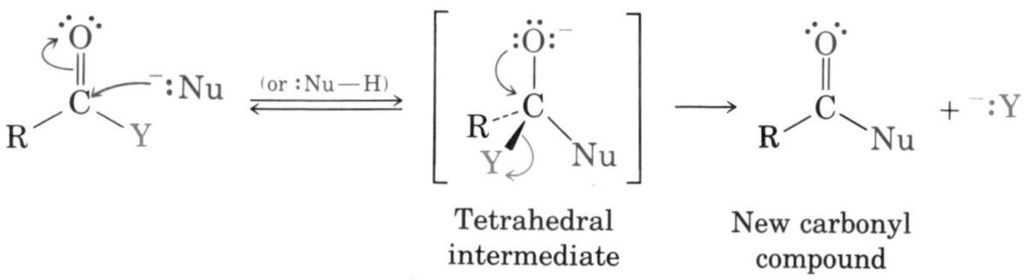

(or :Nu—H)

Tetrahedral intermediate

New carbonyl compound

where Y = —OR (ester), —Cl (acid chloride), $—NH_2$ (amide), or —OCOR′ (acid anhydride)

The net effect of nucleophilic acyl substitution is the replacement of the leaving group by the attacking nucleophile. For example, we'll see in Chapter 21 that acid chlorides are rapidly converted into esters by treatment with alkoxide ions. This reaction is simply a nucleophilic acyl substitution in which alkoxide replaces chloride (Figure 5).

Nucleophilic addition of alkoxide ion to an acid chloride yields a tetrahedral intermediate.

An electron pair from oxygen expels chloride ion and yields the substitution product, an ester.

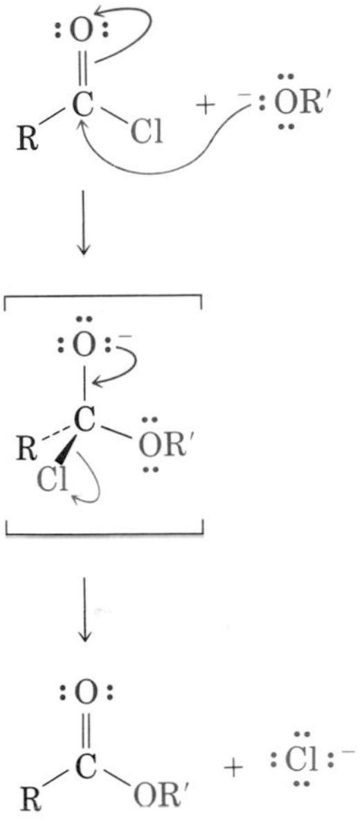

**Figure 5** Mechanism of the nucleophilic acyl substitution reaction of an acid chloride with an alkoxide ion to yield an ester

## ALPHA-SUBSTITUTION REACTIONS

The third major reaction of carbonyl compounds, **alpha substitution**, occurs at the position *next to* the carbonyl group, the alpha ($\alpha$) position. These processes, which involve substitution of an alpha hydrogen by some other group, take place by formation of intermediate **enols** or **enolate ions**:

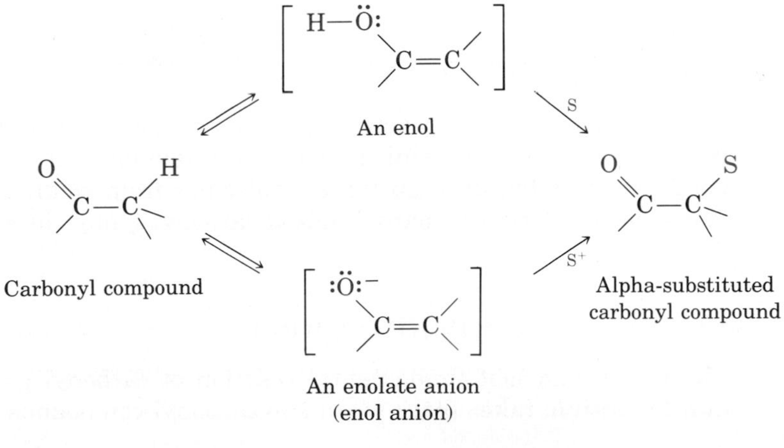

An enol

Carbonyl compound

Alpha-substituted carbonyl compound

An enolate anion (enol anion)

For reasons that we'll explore in Chapter 22, the presence of a carbonyl group renders the protons on the alpha carbon acidic. Carbonyl compounds therefore react with strong base to yield enolate ions.

Carbonyl compound ——:Base——→ [ An enolate ion ] + H—Base

What chemistry might we expect of enolate ions? Since they carry negative charges and have high electron density on the alpha carbon, enolate ions react as nucleophiles and take part in many of the reactions we have already studied. For example, enolates react with primary alkyl halides in the $S_N2$ reaction. The nucleophilic carbanion displaces halide ion and forms a new carbon–carbon bond:

Enolate + R—$CH_2$—X (Alkyl halide) ——$S_N2$ reaction——→ C—C—$CH_2$—R + :$X^-$

For example,

Cyclohexanone ⇌(Base) Cyclohexanone enolate ion ——$CH_3I$——→ 2-Methylcyclohexanone (65%) + :$I$:$^-$

The $S_N2$ alkylation reaction between an enolate ion and an alkyl halide is one of the most powerful methods available for making carbon–carbon bonds, thereby building up larger molecules from small precursors. We'll study the alkylation of many kinds of carbonyl groups in detail in Chapter 22.

## CARBONYL CONDENSATION REACTIONS

The fourth and last fundamental reaction of carbonyl groups, **carbonyl condensation**, takes place when two carbonyl compounds react with each other. For example, when acetaldehyde is treated with base, two molecules react with each other to yield the hydroxy aldehyde product known as *aldol* (*ald*ehyde + alcoh*ol*):

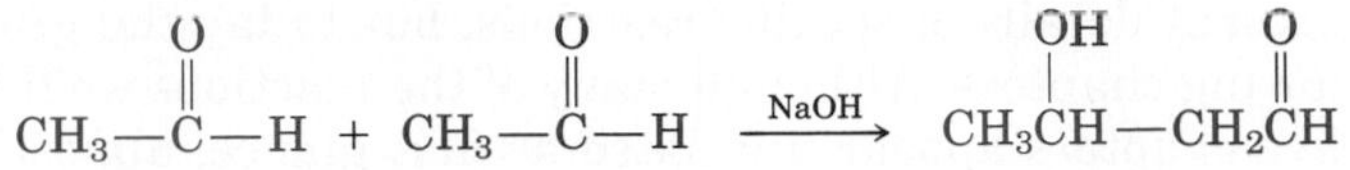

$$\underset{\text{Two acetaldehydes}}{CH_3{-}\overset{O}{\overset{\|}{C}}{-}H + CH_3{-}\overset{O}{\overset{\|}{C}}{-}H} \xrightarrow{NaOH} \underset{\text{Aldol}}{CH_3\overset{OH}{\overset{|}{C}H}{-}CH_2\overset{O}{\overset{\|}{C}}H}$$

Although the carbonyl condensation reaction appears different from the three general carbonyl-group processes already discussed, it's really quite similar. A carbonyl condensation reaction is simply a *combination* of a nucleophilic addition step and an alpha-substitution step. The initially formed enolate ion of acetaldehyde acts as a nucleophile and adds to the carbonyl group of another acetaldehyde molecule. Reaction occurs by the pathway shown in Figure 6.

---

$$\overset{H}{CH_2}{-}\overset{:O:}{\overset{\|}{C}}{-}H \quad {}^{-}:\ddot{O}H/H_2O$$

Hydroxide ion abstracts an acidic alpha proton from one molecule of acetaldehyde, yielding an enolate ion.

$$\rightleftharpoons$$

$$H_2C{=}\overset{:\ddot{O}:^-}{\overset{|}{C}}{-}H + H_2O$$

$$CH_3{-}\overset{:O:}{\overset{\|}{C}}{-}H$$

The enolate ion adds as a nucleophile to the carbonyl group of a second molecule of acetaldehyde, producing a tetrahedral intermediate.

$$\rightleftharpoons$$

$$CH_3{-}\underset{H}{\underset{|}{\overset{:\ddot{O}:^-}{\overset{|}{C}}}}{-}CH_2{-}\overset{:O:}{\overset{\|}{C}}{-}H$$

The intermediate is protonated by water solvent to yield the neutral aldol product and regenerate hydroxide ion.

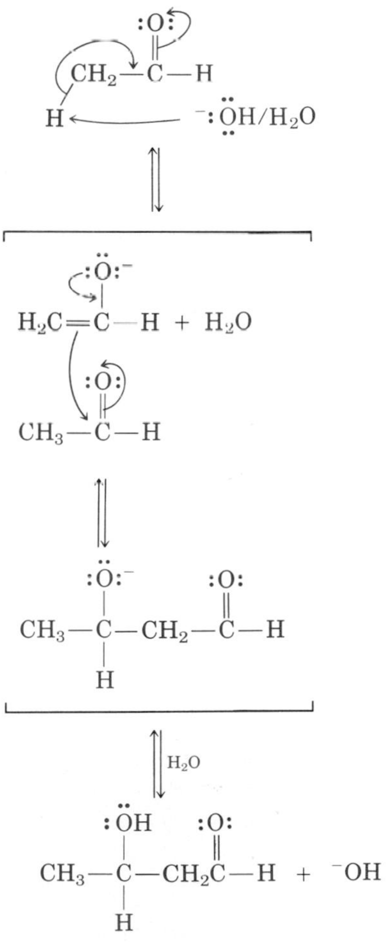

$$\xrightleftharpoons{H_2O}$$

$$CH_3{-}\underset{H}{\underset{|}{\overset{:\ddot{O}H}{\overset{|}{C}}}}{-}CH_2\overset{:O:}{\overset{\|}{C}}{-}H + {}^{-}OH$$

**Figure 6** Mechanism of a carbonyl condensation reaction between two molecules of acetaldehyde

## IV. Summary

The purpose of this short overview of carbonyl-group reactions is not to present details of specific reactions, but to lay the groundwork for the following chapters. Although many of the reactions we'll be seeing in the next five chapters appear unrelated at first glance, almost every reaction we'll encounter is either a **nucleophilic addition**, a **nucleophilic acyl substitution**, an **alpha substitution**, or a **carbonyl condensation**. Knowing where we'll be heading should help you to keep matters straight on this most important of all functional groups, the carbonyl group.

CHAPTER 19

# Aldehydes and Ketones: Nucleophilic Addition Reactions

Aldehydes and ketones are among the most important of all compounds, both in nature and in the chemical industry. In nature, many substances required by living systems are aldehydes or ketones. In the chemical industry, simple aldehydes and ketones are synthesized in great quantity for use both as solvents and as starting materials for a host of other products. For example, approximately 6.5 billion pounds per year of formaldehyde, $H_2C{=}O$, are manufactured for use in building insulation materials and in the adhesive resins that bind particle board and plywood. Concern over the possible toxicity of such materials may sharply curtail their uses, however. Acetone, $(CH_3)_2C{=}O$, is also widely used in industry, with some 2 billion pounds per year prepared for use as a solvent.

Formaldehyde is synthesized industrially by catalytic dehydrogenation (oxidation) of methanol, and one method of acetone preparation involves dehydrogenation of 2-propanol. These methods are of little use in the laboratory, however, because of the high temperatures required.

$$\underset{\text{Methanol}}{H_3C{-}OH} \xrightarrow[\Delta]{\text{Catalyst}} \underset{\text{Formaldehyde}}{H_2C{=}O} + H_2$$

$$\underset{\text{2-Propanol}}{CH_3CH(OH)CH_3} \xrightarrow[380°C]{\text{ZnO}} \underset{\text{Acetone}}{CH_3C(=O)CH_3} + H_2$$

## 19.1 Properties of Aldehydes and Ketones

As we saw in the previous overview, the carbonyl functional group is planar, and the carbon–oxygen double bond is polar.

R, $\delta^+$, $\delta^-$, C=O, R′, 120°

R = Alkyl, aryl, or alkenyl
R′ = Alkyl (ketone), or H (aldehyde)

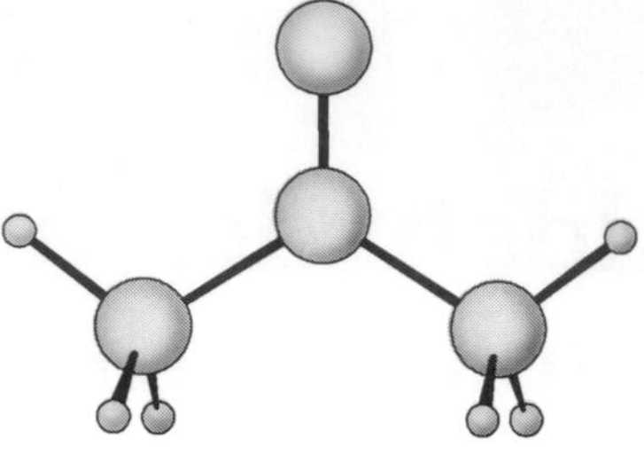

Acetone

One consequence of carbonyl bond polarity is that aldehydes and ketones are weakly associated and therefore have higher boiling points than alkanes of similar molecular weight. Since they cannot form hydrogen bonds, however, the boiling points of ketones and aldehydes are lower than those of the corresponding alcohols. Formaldehyde, the simplest aldehyde, is a gas at room temperature, but all other simple aldehydes and ketones are liquid (Table 19.1).

**Table 19.1** Physical properties of simple aldehydes and ketones

| *Compound name* | *Structure* | *Boiling point (°C)* | *Melting point (°C)* |
|---|---|---|---|
| Formaldehyde | HCHO | −21 | −92 |
| Acetaldehyde | $CH_3CHO$ | 21 | −121 |
| Propanal | $CH_3CH_2CHO$ | 49 | −81 |
| Butanal | $CH_3(CH_2)_2CHO$ | 76 | −99 |
| Pentanal | $CH_3(CH_2)_3CHO$ | 103 | −92 |
| Benzaldehyde | $C_6H_5CHO$ | 178 | −26 |
| Acetone | $CH_3COCH_3$ | 56 | −95 |
| 2-Butanone | $CH_3CH_2COCH_3$ | 80 | −86 |
| 2-Pentanone | $CH_3CH_2CH_2COCH_3$ | 102 | −78 |
| 3-Pentanone | $CH_3CH_2COCH_2CH_3$ | 102 | −40 |
| Cyclohexanone | ⬡=O | 156 | −16 |

## 19.2 Nomenclature of Aldehydes and Ketones

### ALDEHYDES

Aldehydes are named by replacing the terminal *-e* of the corresponding alkane name with *-al*. The longest chain selected to be the base name must contain the —CHO group, and the —CHO carbon is always numbered as carbon 1. For example,

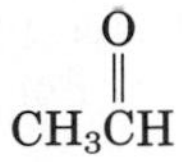

$CH_3CH{=}O$ — Ethanal (Acetaldehyde)

$CH_3CH_2CH{=}O$ — Propanal (Propionaldehyde)

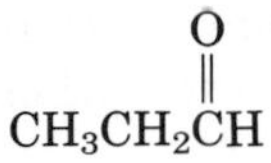

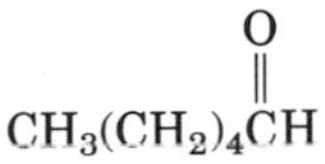

$CH_3(CH_2)_4CH{=}O$ — Hexanal

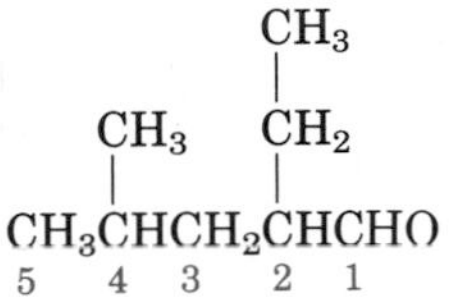

$CH_3CH(CH_3)CH_2CH(CH_2CH_3)CHO$ (carbons numbered 5 4 3 2 1) — 2-Ethyl-4-methylpentanal

Note that the longest chain in 2-ethyl-4-methylpentanal is a hexane, but this chain does not include the —CHO group and thus is not considered the parent.

For more complex aldehydes in which the —CHO group is attached to a ring, the suffix *-carbaldehyde* is used:

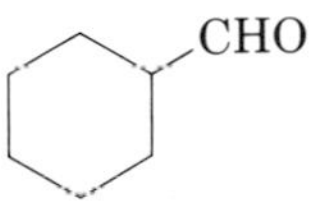

CHO — Cyclohexanecarbaldehyde

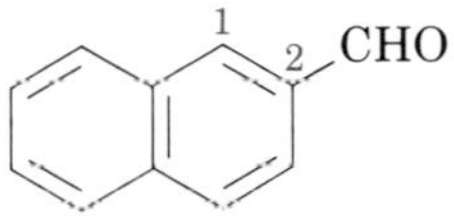

1, 2, CHO — 2-Naphthalenecarbaldehyde

Certain simple and well-known aldehydes have common names that are recognized by IUPAC. Some of the more important common names are given in Table 19.2.

**Table 19.2** Common names of some simple aldehydes

| *Formula* | *Common name* | *Systematic name* |
|---|---|---|
| HCHO | Formaldehyde | Methanal |
| $CH_3CHO$ | Acetaldehyde | Ethanal |
| $CH_3CH_2CHO$ | Propionaldehyde | Propanal |
| $CH_3CH_2CH_2CHO$ | Butyraldehyde | Butanal |
| $CH_3CH_2CH_2CH_2CHO$ | Valeraldehyde | Pentanal |
| $H_2C{=}CHCHO$ | Acrolein | 2-Propenal |
| 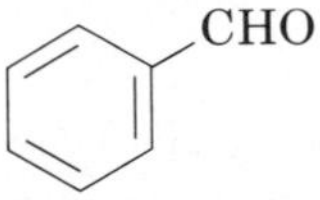 $C_6H_5$CHO (benzene ring–CHO) | Benzaldehyde | Benzenecarbaldehyde |

## KETONES

Ketones are named by replacing the terminal *-e* of the corresponding alkane name with *-one* (pronounced own). The chain selected for the base name is the longest one that contains the ketone group, and the numbering begins at the end nearer the carbonyl carbon. For example,

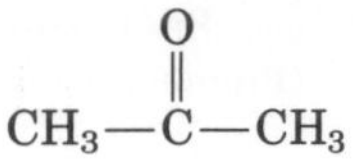

$$CH_3—\overset{O}{\overset{\|}{C}}—CH_3$$

Propanone
(Acetone)

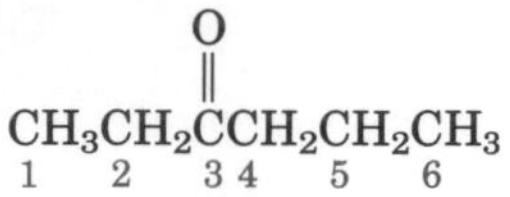

$$\underset{1}{CH_3}\underset{2}{CH_2}\underset{3}{\overset{O}{\overset{\|}{C}}}\underset{4}{CH_2}\underset{5}{CH_2}\underset{6}{CH_3}$$

3-Hexanone

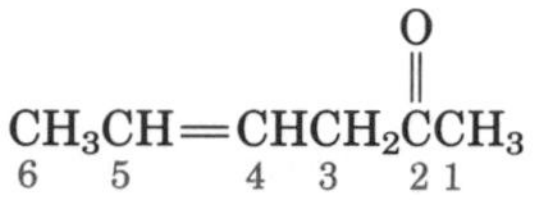

$$\underset{6}{CH_3}\underset{5}{CH}=\underset{4}{CH}\underset{3}{CH_2}\underset{2}{\overset{O}{\overset{\|}{C}}}\underset{1}{CH_3}$$

4-Hexen-2-one

$$\underset{6}{CH_3}\underset{5}{CH_2}\underset{4}{\overset{O}{\overset{\|}{C}}}\underset{3}{CH_2}\underset{2}{\overset{O}{\overset{\|}{C}}}\underset{1}{CH_3}$$

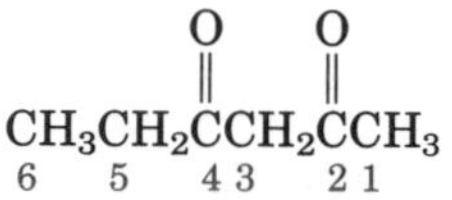

2,4-Hexanedione

Certain ketones are allowed by IUPAC to retain their common names, though these are few:

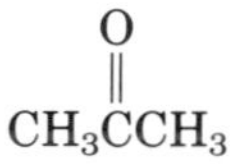

$$CH_3\overset{O}{\overset{\|}{C}}CH_3$$

Acetone

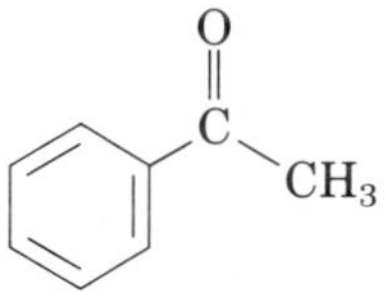

Acetophenone

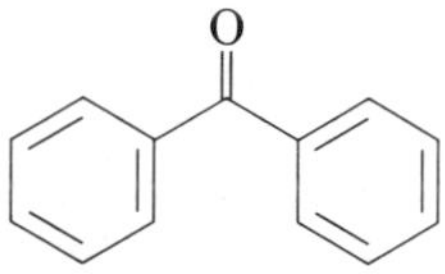

Benzophenone

When it becomes necessary to refer to the RCO— group as a substituent, the term **acyl** (pronounced **ay**-sil) is used. Similarly, —CHO is called a **formyl** group, and ArCO— is referred to as an **aroyl** group.

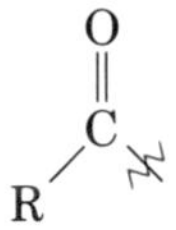

Acyl
(R = alkyl, alkenyl, or alkynyl)

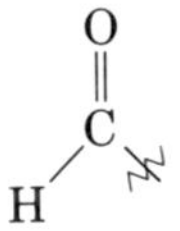

Formyl

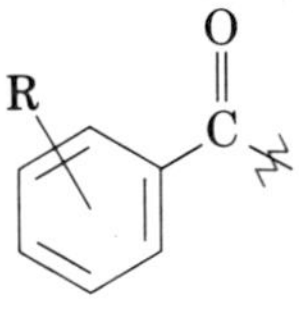

Aroyl

If other functional groups are present and the doubly bonded oxygen must be considered a substituent, the prefix *oxo-* is used. For example,

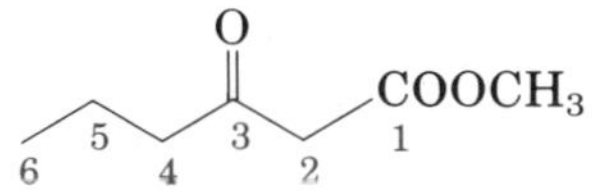

Methyl 3-oxohexanoate

PROBLEM

**19.1** Name these aldehydes and ketones according to IUPAC rules:

(a) $CH_3CH_2C(=O)CH(CH_3)_2$

(b) 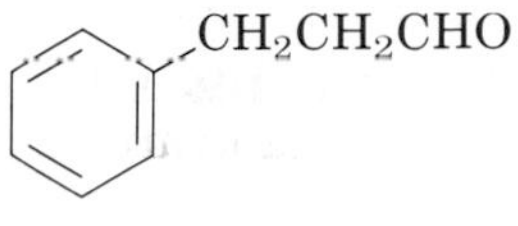 $C_6H_5CH_2CH_2CHO$

(c) $CH_3C(=O)CH_2CH_2CH_2C(=O)CH_2CH_3$

(d) 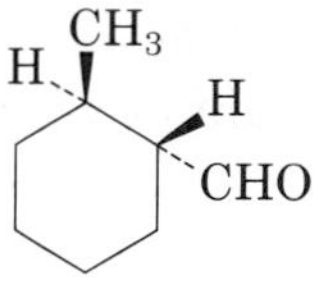 [cyclohexane ring with CH₃ and H on one carbon, H and CHO on the adjacent carbon]

(e) $OHCCH_2CH_2CH_2CHO$

(f) 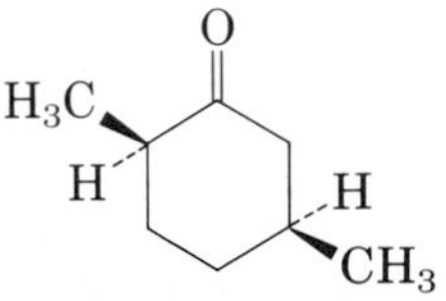 [cyclohexanone with $H_3C$/H on one carbon and H/$CH_3$ on another]

(g) $CH_3CH_2CH(CH_3)CH(CH_2CH_2CH_3)C(=O)CH_3$

(h) $CH_3CH{=}CHCH_2CH_2CHO$

PROBLEM

**19.2** Draw structures corresponding to these names:

(a) 3-Methylbutanal
(b) 4-Chloro-2-pentanone
(c) Phenylacetaldehyde
(d) *cis*-3-*tert*-Butylcyclohexanecarbaldehyde
(e) 3-Methyl-3-butenal
(f) 2-(1-Chloroethyl)-5-methylheptanal

## 19.3 Preparation of Aldehydes

We've already discussed two of the best methods of aldehyde synthesis—oxidation of primary alcohols and oxidative cleavage of alkenes—but let's review them briefly.

1. Primary alcohols can be oxidized to give aldehydes (Section 17.9). The reaction is often carried out using pyridinium chlorochromate (PCC) in dichloromethane solvent at room temperature, and high yields are usually obtained.

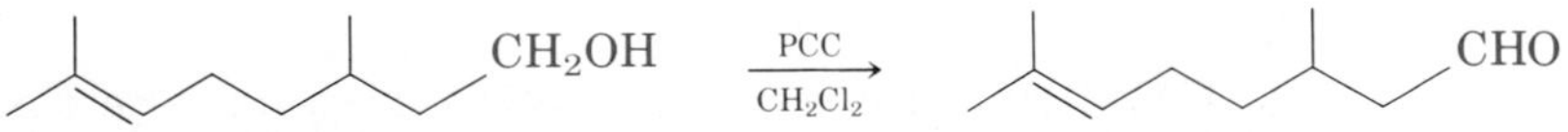

$$\text{Citronellol } (\ldots CH_2OH) \xrightarrow[CH_2Cl_2]{PCC} \text{Citronellal } (82\%) (\ldots CHO)$$

Citronellol — Citronellal (82%)

2. Alkenes that have at least one vinylic hydrogen undergo oxidative cleavage when treated with ozone to yield aldehydes (Section 7.8). If the ozonolysis reaction is carried out on cyclic alkenes, dicarbonyl compounds result.

$$\text{1-Methylcyclohexene} \xrightarrow[\text{2. Zn, } H_3O^+]{\text{1. } O_3} \text{6-Oxoheptanal (86\%)}$$

1-Methylcyclohexene 6-Oxoheptanal (86%)

Yet a third method of aldehyde synthesis is one that we will mention here just briefly and then return to for a more detailed explanation in Section 21.5. Certain carboxylic acid derivatives can be partially reduced to yield aldehydes:

$$R-\overset{\overset{\displaystyle O}{\|}}{C}-Y \xrightarrow{:H^-} R-\overset{\overset{\displaystyle O}{\|}}{C}-H + :Y^-$$

For example, acid chlorides can be hydrogenated over a catalyst of palladium on barium sulfate to yield aldehydes (the **Rosenmund reduction**). The reaction is particularly well suited for large-scale work because it is so simple and relatively inexpensive.

$$C_6H_{11}CH_2\overset{\overset{\displaystyle O}{\|}}{C}Cl \xrightarrow[\text{Ethyl acetate}]{H_2;\ Pd/BaSO_4} C_6H_{11}CH_2\overset{\overset{\displaystyle O}{\|}}{C}H$$

Cyclohexylacetyl chloride Cyclohexylacetaldehyde (70%)

Alternatively, the partial reduction of esters by diisobutylaluminum hydride (DIBAH) is an extremely important laboratory-scale method of aldehyde synthesis. The reaction is normally carried out at −78°C (dry-ice temperature) in toluene solution, and yields are often excellent.

$$CH_3(CH_2)_{10}\overset{\overset{\displaystyle O}{\|}}{C}OCH_3 \xrightarrow[\text{2. } H_3O^+]{\text{1. DIBAH, toluene, } -78^\circ C} CH_3(CH_2)_{10}\overset{\overset{\displaystyle O}{\|}}{C}H$$

Methyl dodecanoate Dodecanal (88%)

$$\text{where DIBAH} = (CH_3)_2CHCH_2-\overset{\overset{\displaystyle H}{|}}{Al}-CH_2CH(CH_3)_2$$

PROBLEM

**19.3** How would you prepare pentanal from these starting materials?
(a) 1-Pentanol (b) 1-Hexene (c) $CH_3CH_2CH_2CH_2COOCH_3$

## 19.4 Preparation of Ketones

For the most part, methods of ketone synthesis are analogous to those for aldehydes:

1. Secondary alcohols are oxidized by a variety of reagents to give ketones (Section 17.9). Jones' reagent ($CrO_3$ in aqueous sulfuric acid), pyridinium chlorochromate, and sodium dichromate in aqueous acetic acid are all effective, and the specific choice of reagent depends on factors such as reaction scale, cost, and acid or base sensitivity of the alcohol.

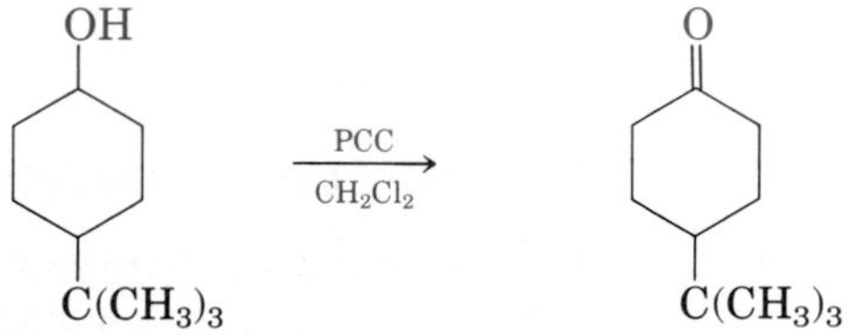

   OH ... $C(CH_3)_3$ $\xrightarrow[CH_2Cl_2]{PCC}$ O ... $C(CH_3)_3$

   4-*tert*-Butylcyclohexanol → 4-*tert*-Butylcyclohexanone (90%)

2. Ozonolysis of alkenes yields ketones if one of the unsaturated carbon atoms is disubstituted (Section 7.8):

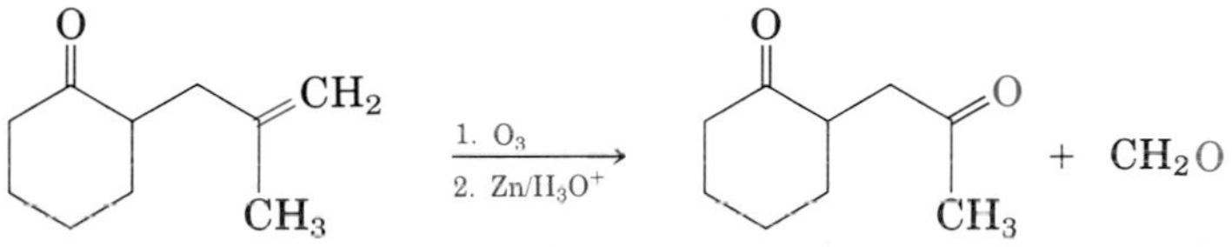

   O, $CH_2$, $CH_3$ $\xrightarrow[2.\ Zn/H_3O^+]{1.\ O_3}$ O, O, $CH_3$ + $CH_2O$

   70%

3. Aryl ketones are prepared by Friedel–Crafts acylation of an aromatic ring with an acid chloride in the presence of $AlCl_3$ catalyst (Section 16.4):

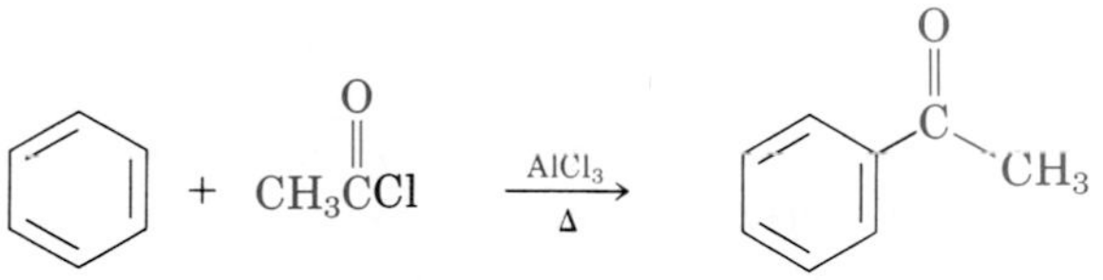

   Benzene + $CH_3\overset{O}{\overset{\|}{C}}Cl$ (Acetyl chloride) $\xrightarrow[\Delta]{AlCl_3}$ Acetophenone (95%)

4. Methyl ketones can be prepared by hydration of terminal alkynes catalyzed by mercuric ion (Section 8.4). Internal alkynes can also be hydrated, but a mixture of ketone products usually results.

   $$CH_3(CH_2)_3C{\equiv}CH \xrightarrow[Hg(OAc)_2]{H_3O^+} CH_3(CH_2)_3\overset{O}{\overset{\|}{C}}{-}CH_3$$

   1-Hexyne → 2-Hexanone (78%)

5. Ketones can be prepared from certain carboxylic acid derivatives, just as aldehydes can. We will discuss this subject in more detail in Section 21.5.

$$\mathrm{R{-}\overset{\overset{\large O}{\|}}{C}{-}Y} \xrightarrow{:R'^-} \mathrm{R{-}\overset{\overset{\large O}{\|}}{C}{-}R'} + :Y^-$$

Among the most useful syntheses of this type is the reaction between an acid chloride and a diorganocopper reagent. This process is reminiscent of the coupling reaction between alkyl halides and diorganocoppers that we saw earlier (Section 10.10):

$$\mathrm{CH_3(CH_2)_4\overset{\overset{\large O}{\|}}{C}Cl} + \mathrm{(CH_3)_2\overset{-}{Cu}Li^+} \longrightarrow \mathrm{CH_3(CH_2)_4\overset{\overset{\large O}{\|}}{C}CH_3}$$

Hexanoyl chloride — Dimethylcopper lithium — 2-Heptanone (81%)

PROBLEM

**19.4** How would you carry out these reactions? More than one step may be required.

(a) 3-Hexyne ⟶ 3-Hexanone

(b) Benzene ⟶ *m*-Bromoacetophenone

(c) Bromobenzene ⟶ Acetophenone

(d) 1-Methylcyclohexene ⟶ 2-Methylcyclohexanone

## 19.5 Oxidation of Aldehydes and Ketones

Aldehydes are readily oxidized to yield carboxylic acids, but ketones are unreactive toward oxidation except under the most vigorous conditions. This difference in behavior is a consequence of the structural difference between the two functional groups: Aldehydes have a —CHO proton that can be readily abstracted during oxidation, but ketones do not.

Many oxidizing agents such as hot nitric acid and potassium permanganate convert aldehydes into carboxylic acids, but the Jones reagent, $CrO_3$ in aqueous sulfuric acid, is a more common choice on a small laboratory scale (Section 17.9). Jones oxidations occur rapidly at room temperature and give good yields of product.

$$\mathrm{CH_3(CH_2)_4\overset{\overset{\large O}{\|}}{C}H} \xrightarrow[\text{Acetone, 0°C}]{\text{Jones' reagent}} \mathrm{CH_3(CH_2)_4\overset{\overset{\large O}{\|}}{C}OH}$$

Hexanal — Hexanoic acid (85%)

One drawback to the Jones oxidation is that the conditions under which it occurs are acidic, and sensitive molecules sometimes undergo acid-catalyzed decomposition. Where such decomposition is a possibility, aldehyde oxidations are often carried out using a dilute ammonia solution of silver oxide, $Ag_2O$ (**Tollens'[1] reagent**). Aldehydes are oxidized by the Tollens reagent in high yield without harming carbon–carbon double bonds or other functional groups in the molecule.

$CH_2$ ... CHO $\xrightarrow[NH_4OH,\ H_2O]{Ag_2O}$ $CH_2$ ... COOH + Ag

72%

A shiny mirror of metallic silver is deposited on the walls of the flask during a Tollens oxidation, forming the basis of a qualitative test for the presence of an aldehyde functional group in a molecule of unknown structure. A small sample of the unknown is dissolved in ethanol in a test tube, and a few drops of the Tollens reagent are added. If the test tube becomes silvery, the unknown is presumed to be an aldehyde.

Aldehyde oxidations are thought to occur through intermediate 1,1-diols or hydrates, which are formed by a reversible nucleophilic addition of water to the carbonyl group. Even though present in small amounts at equilibrium, the hydrate reacts like any normal primary or secondary alcohol and is oxidized to a carbonyl compound (Section 17.9).

$$R{-}\overset{O}{\overset{\|}{C}}{-}H \underset{}{\overset{H_2O}{\rightleftharpoons}} R{-}C(OH)(OH){-}H \xrightarrow[H_3O^+]{CrO_3} R{-}\overset{O}{\overset{\|}{C}}{-}OH$$

Aldehyde | Hydrate | Carboxylic acid

Ketones are inert to most common oxidizing agents but undergo a slow cleavage reaction when treated with hot alkaline $KMnO_4$. The carbon–carbon bond next to the carbonyl group is broken and carboxylic acid fragments are produced. The reaction is only useful for symmetrical ketones such as cyclohexanone, however, since product mixtures are formed from unsymmetrical ketones.

Cyclohexanone $\xrightarrow[\text{2. } H_3O^+]{\text{1. } KMnO_4,\ H_2O,\ OH^-,\ \Delta}$ Hexanedioic acid (79%)

[1]Bernhard Tollens (1841–1918); b. Hamburg, Germany; Ph.D. Göttingen; professor, University of Göttingen.

## 19.6 Nucleophilic Addition Reactions of Aldehydes and Ketones

The single most important reaction of ketones and aldehydes is the **nucleophilic addition reaction**. A nucleophile attacks the electrophilic carbon atom of the polar carbonyl group from a direction approximately perpendicular to the plane of the carbonyl $sp^2$ orbitals. Rehybridization of the carbonyl carbon from $sp^2$ to $sp^3$ then occurs, and a tetrahedral alkoxide ion intermediate is produced (Figure 19.1).

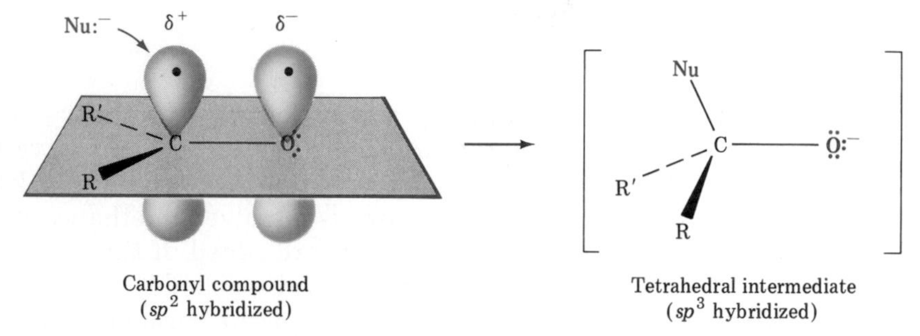


**Figure 19.1** A nucleophilic addition reaction

The attacking nucleophile can be either negatively charged ($Nu:^-$) or neutral (:Nu—H). If it is neutral, however, the nucleophile must be attached to a hydrogen atom that can subsequently be eliminated. For example:

Some negatively charged nucleophiles:
- $H\ddot{O}:^-$ (hydroxide ion)
- $H:^-$ (hydride ion)
- $R_3C:^-$ (a carbanion)
- $R\ddot{O}:^-$ (an alkoxide ion)
- $N{\equiv}C:^-$ (cyanide ion)

Some neutral nucleophiles:
- $H\ddot{O}H$ (water)
- $R\ddot{O}H$ (an alcohol)
- $H_3N:$ (ammonia)
- $R\ddot{N}H_2$ (an amine)

Nucleophilic addition to ketones and aldehydes usually has two variations, as shown in Figure 19.2: (1) The tetrahedral intermediate can be protonated to give an alcohol, or (2) the carbonyl oxygen atom can be expelled (as $HO^-$ or as $H_2O$) to give a new carbon–nucleophile double bond.

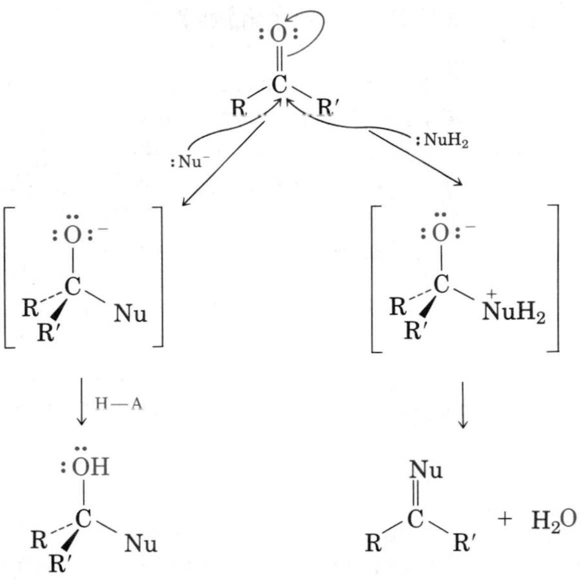

Figure 19.2 Two possible reaction pathways following addition of a nucleophile to a ketone or aldehyde

In the remainder of this chapter, we will examine some specific examples of nucleophilic addition reactions. In looking at the details of each reaction, we will be concerned with two key points—the *reversibility* of a given reaction, and the acid or base *catalysis* of that reaction. Some nucleophilic addition reactions take place without catalysis, but many others require acid or base to proceed.

PROBLEM

**19.5** What product would you obtain if cyanide ion ($^-$:C≡N) were added to acetone and the tetrahedral intermediate were protonated?

## 19.7 Relative Reactivity of Aldehydes and Ketones

Aldehydes are generally more reactive than ketones in nucleophilic addition reactions for both steric and electronic reasons. Sterically, the presence of two relatively large substituents in ketones versus only one large substituent in aldehydes means that attacking nucleophiles are able to approach aldehydes more readily. Thus, the transition state leading to the tetrahedral intermediate is less crowded and lower in energy for aldehydes than for ketones (Figure 19.3, page 668).

Electronically, aldehydes are more reactive than ketones because of the greater degree of polarity of aldehyde carbonyl groups. The easiest way to see this polarity difference is to recall the stability order of carbocations (Section 6.11). Primary carbocations are less stable than secondary ones because there is only one alkyl group inductively stabilizing the positive

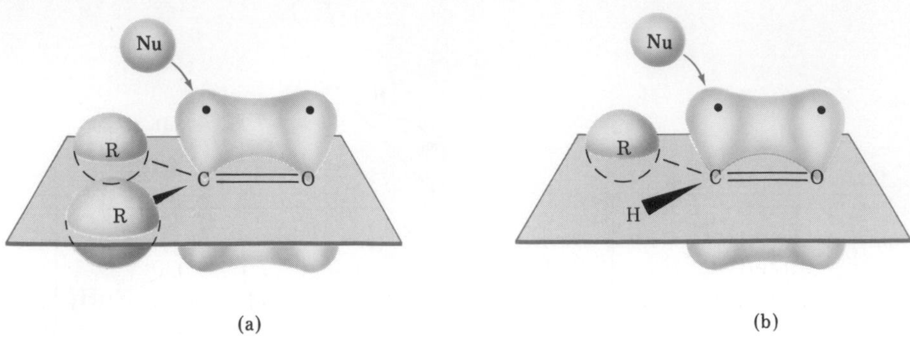


**Figure 19.3** Nucleophilic attack on a ketone (a) is sterically more hindered than attack on an aldehyde (b) because the ketone has two relatively large substituents attached to the carbonyl-group carbon.

charge, rather than two. For similar reasons, aldehydes are less stable (and therefore more reactive) than ketones because there is only one alkyl group inductively stabilizing the partial positive charge on the carbonyl carbon, rather than two.

| | |
|---|---|
| H–C(R)(H)$^{+}$ (R→C) | H–C(R)(R′)$^{+}$ (R→C←R′) |
| 1° carbocation (less stable, more reactive) | 2° carbocation (more stable, less reactive) |
| $O^{\delta-}$=$C^{\delta+}$(R)(H) (R→C) | $O^{\delta-}$=$C^{\delta+}$(R)(R′) (R→C←R′) |
| Aldehyde (less stabilization of $\delta^+$, more reactive) | Ketone (more stabilization of $\delta^+$, less reactive) |

PROBLEM

**19.6** Aromatic aldehydes such as benzaldehyde are less reactive in nucleophilic addition reactions than aliphatic aldehydes. How can you explain this observation?

PROBLEM

**19.7** Which would you expect to be more reactive toward nucleophilic additions, *p*-methoxybenzaldehyde or *p*-nitrobenzaldehyde? Explain.

## 19.8 Nucleophilic Addition of $H_2O$: Hydration

Aldehydes and ketones undergo reaction with water to yield 1,1-diols, or **geminal (gem) diols**. The hydration reaction is reversible; gem diols can eliminate water to regenerate ketones or aldehydes.

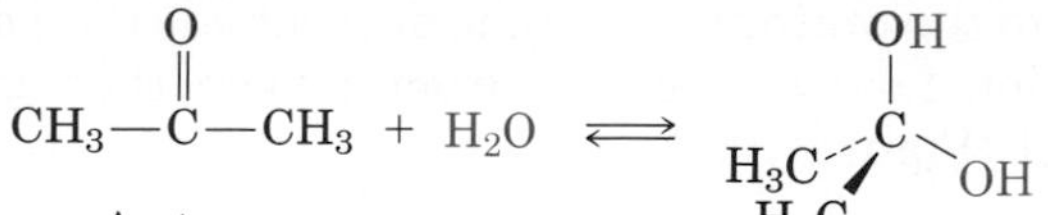

$$CH_3{-}\overset{\overset{\Large O}{\|}}{C}{-}CH_3 + H_2O \rightleftharpoons (H_3C)_2C(OH)_2$$

Acetone — Acetone hydrate (a gem diol)

The exact position of the equilibrium between gem diols and ketones/aldehydes depends on the structure of the carbonyl compound. Although the equilibrium strongly favors the carbonyl compound in most cases, the gem diol is favored for a few simple aldehydes. For example, an aqueous solution of acetone consists of about 0.1% gem diol and 99.9% ketone, whereas an aqueous solution of formaldehyde consists of 99.9% gem diol and 0.1% aldehyde.

The nucleophilic addition of water to ketones and aldehydes is rather slow in pure water, but is catalyzed by both acid and base. Although, like all catalysts, acid and base don't change the *position* of the equilibrium, they strongly affect the speed with which the hydration reaction occurs.

The base-catalyzed reaction takes place in several steps, as shown in Figure 19.4. The attacking nucleophile is the negatively charged hydroxide ion, rather than neutral water.

---

Hydroxide ion nucleophile adds to the ketone or aldehyde carbonyl group to yield an alkoxide ion intermediate.

The basic alkoxide ion intermediate abstracts a proton ($H^+$) from water to yield gem diol product and regenerate hydroxide ion catalyst.

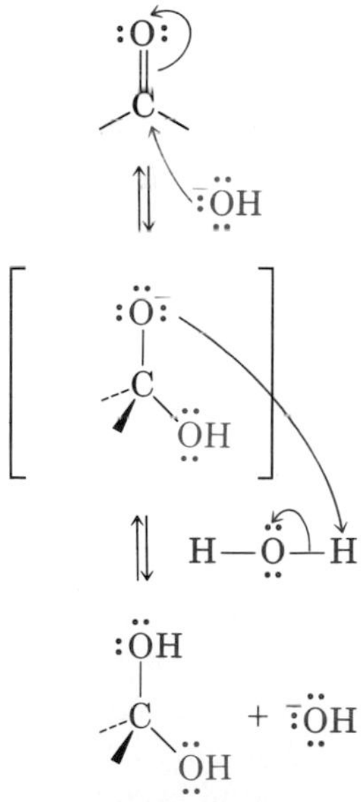

:O: (C=O) + :OH⁻ ⇌ [ :O:⁻ –C– OH ] ; + H—O—H ⇌ :OH –C– OH + ⁻:OH

**Figure 19.4** Mechanism of base-catalyzed hydration of a ketone or aldehyde: The purpose of the base is to serve as a far more reactive nucleophile than neutral water.

---

The acid-catalyzed hydration reaction also takes place in several steps. The acid catalyst first protonates the Lewis-basic oxygen atom of the carbonyl group, placing a positive charge on oxygen and thus making the carbonyl group far more electrophilic. Subsequent nucleophilic addition of water to the protonated ketone or aldehyde then occurs to yield a protonated gem diol. Loss of a proton from oxygen then gives the neutral gem diol product (Figure 19.5).

Acid catalyst protonates the basic carbonyl oxygen atom, making the ketone or aldehyde a much better acceptor of nucleophiles.

Nucleophilic addition of neutral water yields a protonated gem diol.

Loss of a proton regenerates the acid catalyst and gives neutral gem diol product.

$+ H_3O^+$

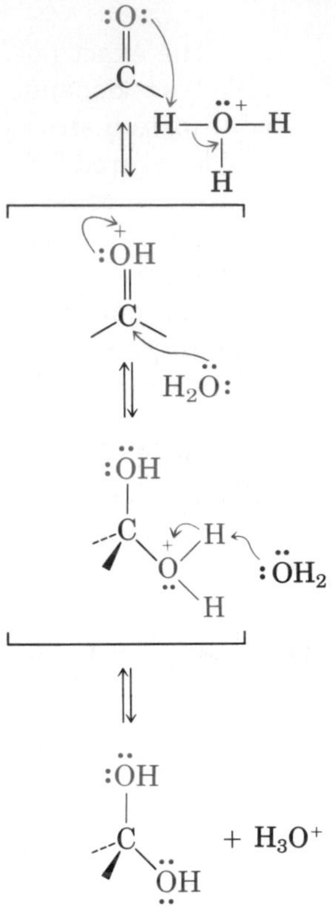

**Figure 19.5** Mechanism of acid-catalyzed hydration of a ketone or aldehyde: The purpose of the acid catalyst is to protonate the carbonyl starting material, thus making it more electrophilic and more reactive.

Note the important differences between the acid-catalyzed and base-catalyzed processes. The *base*-catalyzed reaction takes place rapidly because water is converted into hydroxide ion, a much better nucleophilic *donor*. The *acid*-catalyzed reaction takes place rapidly because the carbonyl compound is converted by protonation into a much better electrophilic *acceptor*.

The hydration reaction just described is typical of what happens when a ketone or aldehyde is treated with a nucleophile of the type H—Y, where the Y atom is electronegative (oxygen, halogen, or sulfur, for example). In such reactions, nucleophilic addition is reversible, with the equilibrium favoring the carbonyl starting material rather than the tetrahedral adduct. For example, treatment of ketones and aldehydes with reagents such as $H_2O$, HCl, HBr, or $H_2SO_4$ does not normally lead to addition products.

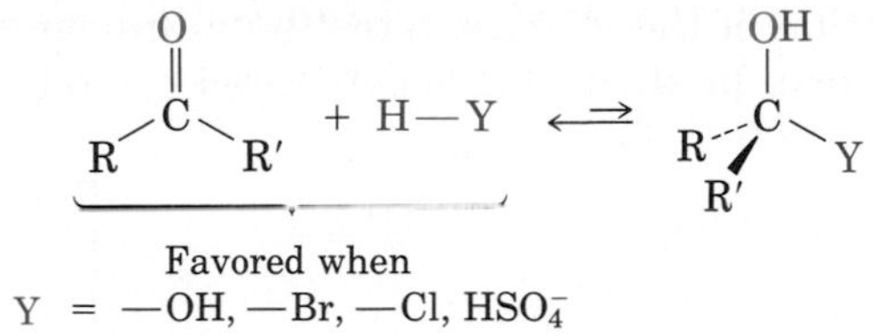

$$\mathrm{R{-}C({=}O){-}R'} + \mathrm{H{-}Y} \rightleftharpoons \mathrm{R{-}C(OH)(R'){-}Y}$$

Favored when
Y = —OH, —Br, —Cl, $HSO_4^-$

PROBLEM

**19.8** When dissolved in water, trichloroacetaldehyde (chloral, $CCl_3CHO$) exists primarily as the gem diol chloral hydrate, $CCl_3CH(OH)_2$ (better known by the non-IUPAC name "knock-out drops"). Show the structure of chloral hydrate.

PROBLEM

**19.9** The oxygen in water is primarily (99.8%) $^{16}O$, but water enriched with the heavy isotope $^{18}O$ is also available. When a ketone or aldehyde is dissolved in $^{18}O$-enriched water, the isotopic label becomes incorporated into the carbonyl group: $R_2C{=}O + H_2O^* \rightarrow R_2C{=}O^* + H_2O$ (where $O^* = {}^{18}O$). Explain.

PROBLEM

**19.10** How can you explain the observation that $S_N2$ reaction of (dibromomethyl)benzene, $C_6H_5CHBr_2$, with NaOH yields benzaldehyde rather than (dihydroxymethyl)benzene, $C_6H_5CH(OH)_2$?

## 19.9 Nucleophilic Addition of HCN: Cyanohydrins

Aldehydes and unhindered ketones react with HCN to yield **cyanohydrins, RCH(OH)C≡N**. For example, benzaldehyde gives the cyanohydrin mandelonitrile in 88% yield on treatment with HCN:

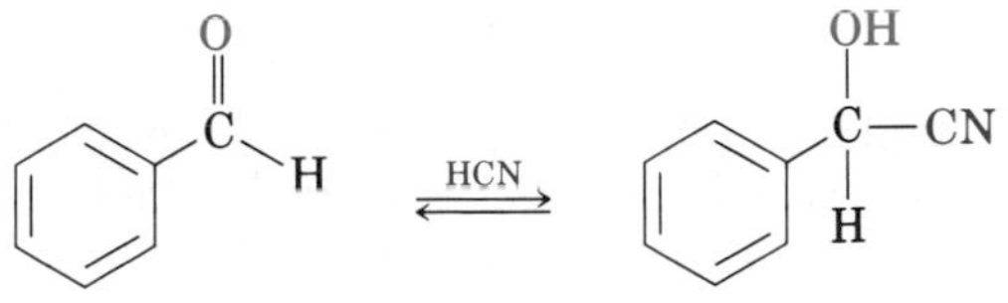

$$\mathrm{C_6H_5CHO} \xrightleftharpoons{\text{HCN}} \mathrm{C_6H_5CH(OH)CN}$$

Benzaldehyde

Mandelonitrile (88%)
(a cyanohydrin)

Detailed studies carried out by Arthur Lapworth[2] in the early 1900s showed that cyanohydrin formation is reversible and base catalyzed. Reaction occurs very slowly when pure HCN is used, but rapidly when a trace amount of base or cyanide ion is added. We can understand this result by recalling that HCN, a weak acid with $pK_a = 9.1$, is neither dissociated nor nucleophilic. Cyanide ion, however, is strongly nucleophilic, and addition to ketones and aldehydes occurs by a typical nucleophilic addition pathway.

[2]Arthur Lapworth (1872–1941); b. Galashiels, Scotland; D.Sc. City and Guilds Institute, London; professor, University of Manchester (1909–1941).

Protonation of the anionic tetrahedral intermediate yields the tetrahedral cyanohydrin product plus regenerated cyanide ion:

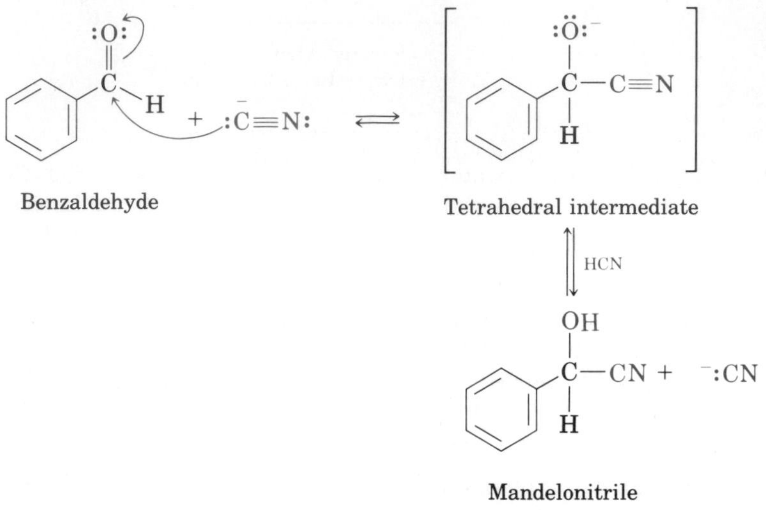

Benzaldehyde

Tetrahedral intermediate

HCN

Mandelonitrile

In order to avoid the dangers inherent in handling such a toxic gas as hydrogen cyanide, HCN is usually generated during the reaction by adding 1 equiv of mineral acid to a mixture of carbonyl compound and excess sodium cyanide.

Cyanohydrin formation is particularly interesting because it is one of the few examples of the addition of an acid to a carbonyl group. As noted earlier, acids such as HBr, HCl, $H_2SO_4$, and $CH_3COOH$ do not form carbonyl adducts because the equilibrium constant for reaction is unfavorable. With HCN, however, the equilibrium lies in favor of the adduct.

Cyanohydrin formation is useful because of the further chemistry that can be carried out. For example, nitriles (R—C≡N) can be reduced with $LiAlH_4$ to yield primary amines ($R—NH_2$) and can be hydrolyzed by aqueous acid to yield carboxylic acids. Thus, cyanohydrin formation provides a method for transforming a ketone or aldehyde into a different functional group while lengthening the carbon chain by one unit.

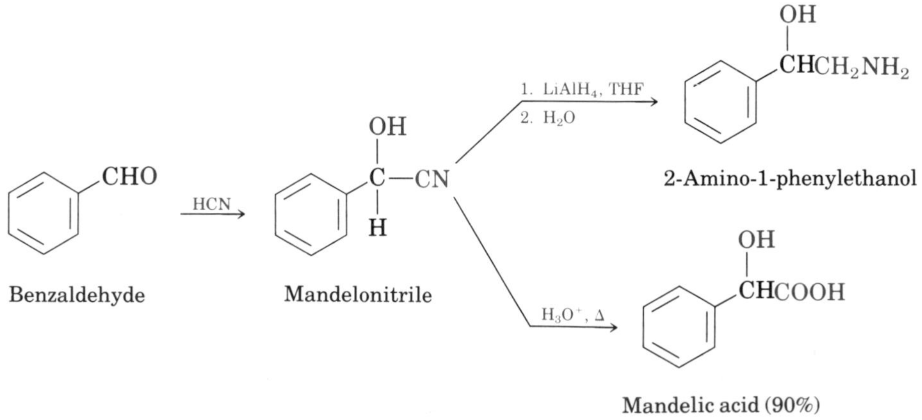

1. $LiAlH_4$, THF
2. $H_2O$

HCN

$H_3O^+$, Δ

Benzaldehyde

Mandelonitrile

2-Amino-1-phenylethanol

Mandelic acid (90%)

**PROBLEM**

**19.11** How can you account for the observation that cyclohexanone forms a cyanohydrin in good yield but that 2,2,6-trimethylcyclohexanone appears to be unreactive to HCN/KCN?

## 19.10 Nucleophilic Addition of Grignard Reagents: Alcohol Formation

Although we didn't point it out earlier, the synthesis of alcohols by reaction of Grignard reagents with ketones and aldehydes (Section 17.7) is simply a nucleophilic addition reaction. Unlike the nucleophilic addition of water, though, Grignard additions are generally irreversible.

Grignard reagents are nucleophiles because the carbon–magnesium bond is strongly polarized, with a high amount of electron density on carbon. Thus, Grignard reagents react as if they were carbanions, $R_3C:^-$. Acid–base complexation of magnesium ion with the carbonyl oxygen atom first serves to make the carbonyl group a better electrophile, and nucleophilic addition then produces a tetrahedrally hybridized magnesium alkoxide intermediate. Protonation of this intermediate by addition of aqueous acid yields the neutral alcohol (Figure 19.6).

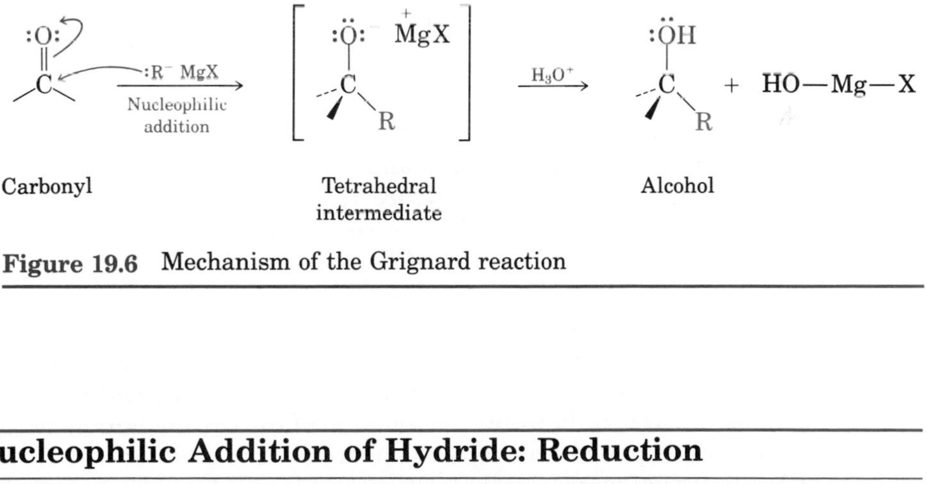

:O: C :R⁻ MgX → (Nucleophilic addition) [ :Ö:⁻ $\overset{+}{\text{MgX}}$ C R ] → ($H_3O^+$) :ÖH C R + HO—Mg—X

Carbonyl Tetrahedral intermediate Alcohol

**Figure 19.6** Mechanism of the Grignard reaction

## 19.11 Nucleophilic Addition of Hydride: Reduction

The reduction of ketones and aldehydes to yield alcohols (Section 17.6) is another nucleophilic addition reaction whose mechanism we didn't point out earlier. The exact details of carbonyl-group reduction by hydride agents such as $LiAlH_4$ and $NaBH_4$ are complex because there is no such species as a discrete hydride ion, $H:^-$. Nevertheless, the common reducing agents function as if they were "hydride ion equivalents," and the fundamental step in carbonyl-group reduction is a nucleophilic addition (Figure 19.7).

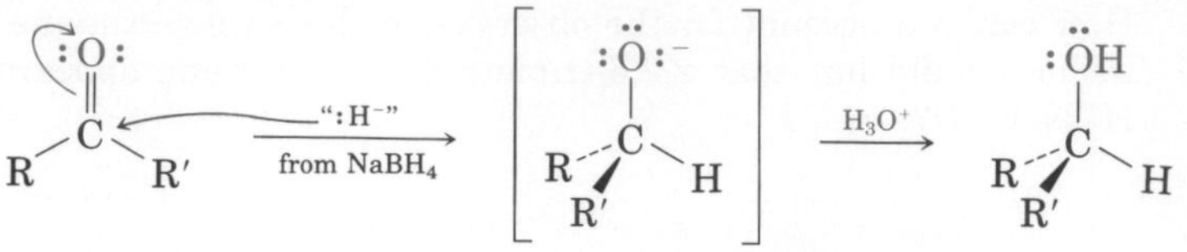

Figure 19.7 Mechanism of carbonyl-group reduction by nucleophilic addition of "hydride ion" from $NaBH_4$ or $LiAlH_4$

## 19.12 Nucleophilic Addition of Amines: Imine and Enamine Formation

Primary amines, $R—NH_2$, add to aldehydes and ketones to yield **imines**, $\mathbf{R_2C{=}NR}$. Secondary amines, $R_2NH$, add similarly to yield **enamines** (*-ene* + *amine*; unsaturated amine).

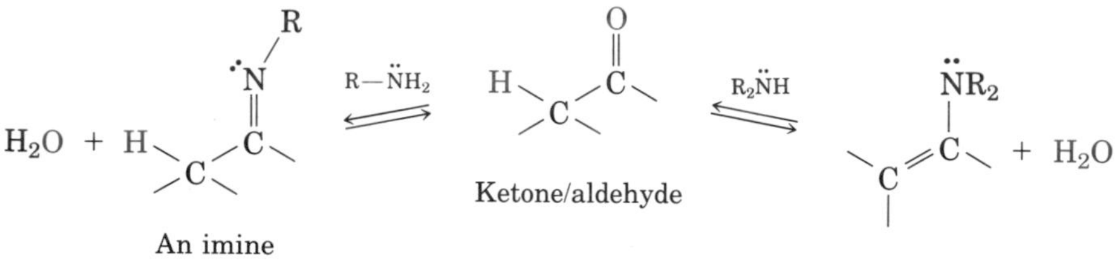

An imine

Ketone/aldehyde

An enamine

These two reactions appear to be different, since one leads to a carbon–nitrogen double-bond product and the other leads to a carbon–carbon double-bond product, but they are really quite similar. Both are typical examples of nucleophilic addition reactions in which the initially formed tetrahedral intermediate cannot be isolated. Instead, water is eliminated, and a new carbon-nucleophile double bond is formed.

Imines are formed by a reversible, acid-catalyzed process involving nucleophilic attack on the carbonyl group by the primary amine, followed by transfer of a proton from nitrogen to oxygen to yield a neutral amino alcohol (a **carbinolamine**). Protonation of the carbinolamine oxygen by the acid catalyst present converts the hydroxyl into a better leaving group, and E1-like loss of water produces an iminium ion. Loss of a proton then gives the final product and regenerates acid catalyst (Figure 19.8).

Imine formation is normally an acid-catalyzed process. Studies of this reaction have revealed a profile of pH versus reaction rate indicating that reaction is very slow at both high and low pH, but reaches a maximum rate at weakly acidic pH. For example, Figure 19.9 (page 676) shows the profile obtained for reaction between acetone and hydroxylamine, $H_2N—OH$, indicating that maximum reaction rate is obtained at pH 4.5.

Ketone/aldehyde

Nucleophilic attack on the ketone or aldehyde by the lone-pair electrons of an amine leads to a dipolar tetrahedral intermediate.

$:NH_2R$

A proton is then transferred from nitrogen to oxygen, yielding a neutral carbinolamine.

Proton transfer

Carbinolamine

Acid catalyst protonates the hydroxyl oxygen.

$H_3O^+$

The nitrogen lone-pair electrons expel water, giving an iminium ion.

$-H_2O$

Iminium ion

Loss of $H^+$ from nitrogen then gives the neutral imine product.

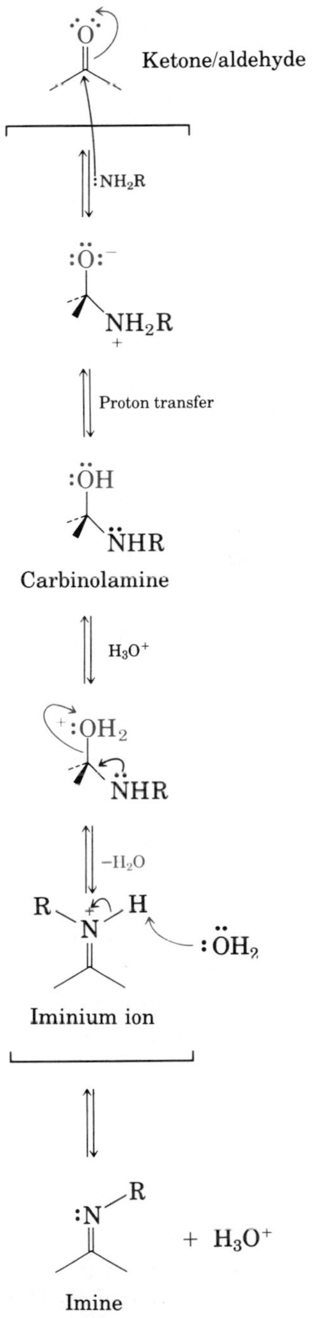

$+ H_3O^+$

Imine

**Figure 19.8** Mechanism of imine formation by reaction of a ketone or aldehyde with a primary amine

We can explain the observed maximum at pH 4.5 by looking at each individual step in the mechanism. As indicated in Figure 19.8, acid is

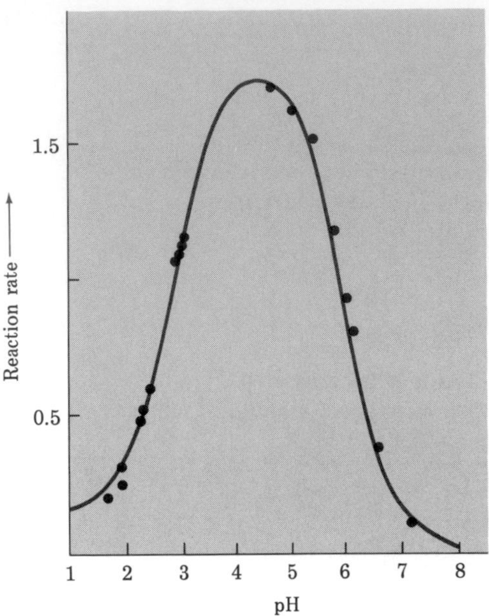


**Figure 19.9** Dependence on pH of the rate of reaction between acetone and hydroxylamine, $H_2NOH$:

$$CH_3{-}\overset{\displaystyle O}{\overset{\|}{C}}{-}CH_3 \xrightarrow{NH_2OH} CH_3{-}\overset{\displaystyle NOH}{\overset{\|}{C}}{-}CH_3 + H_2O$$

required to protonate the intermediate carbinolamine, thereby converting the hydroxyl into a better leaving group. Thus, reaction can't occur if there's not enough acid present (high pH). On the other hand, if too much acid is present (low pH), the attacking amine nucleophile is completely protonated and the initial nucleophilic addition step can't occur.

$$\underset{\text{Base}}{H_2\ddot{N}{-}OH} + \underset{\text{Acid}}{H{-}A} \longrightarrow \underset{\text{Nonnucleophilic}}{H_3\overset{+}{N}{-}OH} + A^-$$

Evidently, pH 4.5 represents a compromise between the need for *some* acid to catalyze the rate-limiting dehydration step, and the need for *not too much acid* to avoid complete protonation of the amine. Each individual nucleophilic addition reaction has its own specific requirements, and reaction conditions must often be controlled if maximum reaction rates are to be obtained.

Imine formation from such reagents as hydroxylamine ($NH_2OH$), semicarbazide ($NH_2NHCONH_2$), and 2,4-dinitrophenylhydrazine are particularly useful because the products of these reactions—**oximes**, **semicarbazones**, and **2,4-dinitrophenylhydrazones** (2,4-DNP's)—are usually crystalline and easy to handle. Such crystalline derivatives are sometimes prepared as a means of identifying liquid ketones or aldehydes.

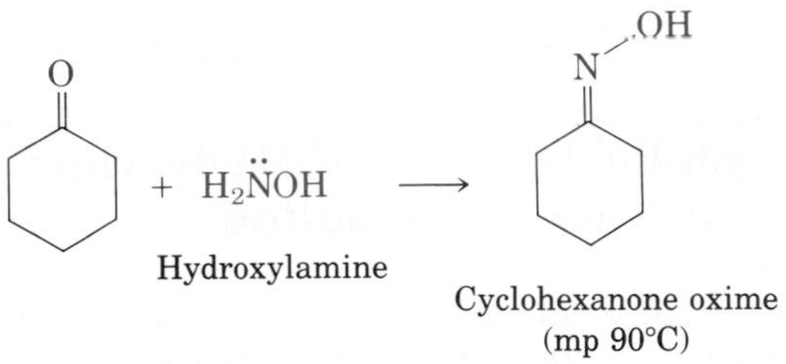

Cyclohexanone + $H_2\ddot{N}OH$ ⟶ Cyclohexanone oxime

Hydroxylamine

Cyclohexanone oxime (mp 90°C)

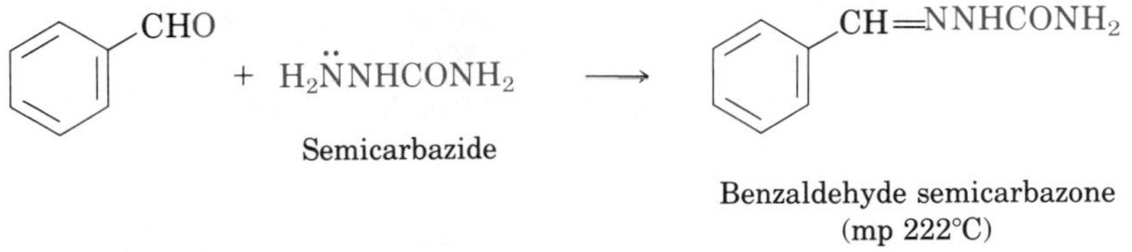

Benzaldehyde (CHO) + $H_2\ddot{N}NHCONH_2$ ⟶ $CH{=}NNHCONH_2$

Semicarbazide

Benzaldehyde semicarbazone (mp 222°C)

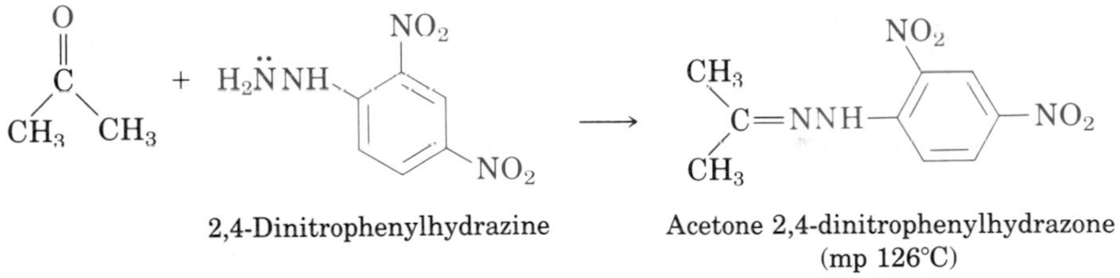

Acetone + $H_2\ddot{N}NH$–(2,4-dinitrophenyl) ⟶ $(CH_3)_2C{=}NNH$–(2,4-dinitrophenyl)

2,4-Dinitrophenylhydrazine

Acetone 2,4-dinitrophenylhydrazone (mp 126°C)

Enamines are formed when a ketone or aldehyde reacts with a secondary amine, $R_2NH$. The process is identical to imine formation up to the iminium ion stage, but at this point there is no proton on nitrogen that can be lost to yield neutral product. Instead, a proton is lost from the alpha-carbon atom, yielding an enamine (Figure 19.10).

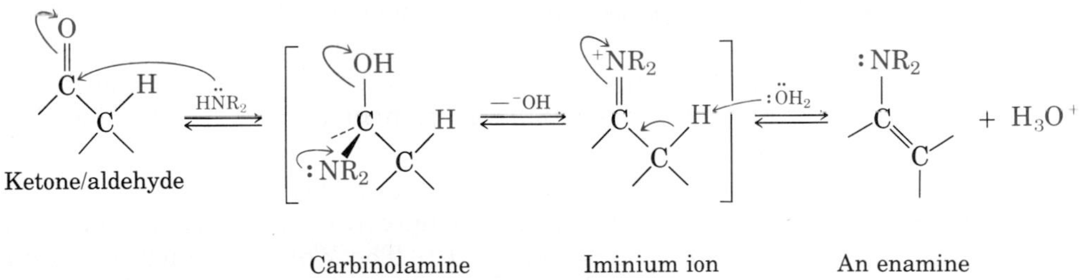

Ketone/aldehyde $\xrightarrow{H\ddot{N}R_2}$ [Carbinolamine $\xrightarrow{-\,^{-}OH}$ Iminium ion] $\xrightarrow{:\ddot{O}H_2}$ An enamine + $H_3O^+$

**Figure 19.10** Mechanism of enamine formation by reaction of a ketone or aldehyde with a secondary amine

PROBLEM

**19.12** Show the products you would obtain by reaction of cyclohexanone with ethylamine, $CH_3CH_2NH_2$, and with diethylamine, $(CH_3CH_2)_2NH$.

PROBLEM

19.13 Imine formation is reversible. Show all the steps involved in the hydrolysis of an imine to yield a ketone or aldehyde plus primary amine.

## 19.13 Nucleophilic Addition of Hydrazine: The Wolff–Kishner Reaction

An important variant of the imine formation just discussed involves the treatment of a ketone or aldehyde with hydrazine, $H_2NNH_2$, in the presence of KOH. This reaction, discovered independently in 1911 by Ludwig Wolff[3] in Germany and N. M. Kishner[4] in Russia, is an extremely valuable synthetic method for converting ketones or aldehydes into alkanes, $R_2C{=}O \rightarrow R_2CH_2$.

$H_2N{-}NH_2$; KOH, $H_2O$, $(HOCH_2CH_2)_2O$, 240°C

90%

Wolff–Kishner reaction

Cyclopropanecarbaldehyde → Methylcyclopropane (72%)

The **Wolff–Kishner reaction** is often carried out at 240°C in boiling diethylene glycol solvent, but a modification in which dimethyl sulfoxide is used as solvent allows the process to take place near room temperature.

The Wolff–Kishner reaction involves formation of a hydrazone intermediate, followed by base-catalyzed double-bond migration, loss of $N_2$ gas, and formation of alkane product (Figure 19.11). The double-bond migration takes place when base removes one of the weakly acidic N—H protons to generate a hydrazone anion. Since the hydrazone anion has an allylic resonance structure that places the double bond between nitrogens and the negative charge on carbon, protonation can occur on carbon to generate the double-bond migrated product. The final step—loss of nitrogen to give the alkane—is an unusual reaction, driven primarily by the enormous thermodynamic stability of the $N_2$ molecule.

Note that the Wolff–Kishner reduction accomplishes the same overall transformation that we saw earlier during catalytic hydrogenation of acyl benzenes to alkylbenzenes (Section 16.12). The Wolff–Kishner reduction is more general and more useful than catalytic hydrogenation, however, since it works well with both alkyl and aryl ketones.

In addition to the Wolff–Kishner reaction, there is a second process, called the **Clemmensen[5] reduction**, that also accomplishes the conversion

[3]Ludwig Wolff (1857–1919); b. Neustadt/Hardt; Ph.D. Strasbourg (Fittig); professor, University of Jena.

[4]N. M. Kishner (1867–1935); b. Moscow; Ph.D. Moscow (Markovnikov); professor, universities of Tomsk and Moscow.

[5]E. C. Clemmensen (1876–1941); b. Odense, Denmark; Ph.D. Copenhagen; Clemmensen Chemical Corp., Newark, NY.

Reaction of the ketone or aldehyde with hydrazine yields a hydrazone in the normal way.

Base then abstracts one of the weakly acidic protons from —$NH_2$, yielding a hydrazone anion. This anion has an "allylic" resonance form that places the negative charge on carbon and the double bond between nitrogens.

Protonation of the hydrazone anion takes place on carbon to yield a neutral intermediate.

Base-induced loss of nitrogen then gives a carbanion . . .

that is protonated to yield neutral alkane product.

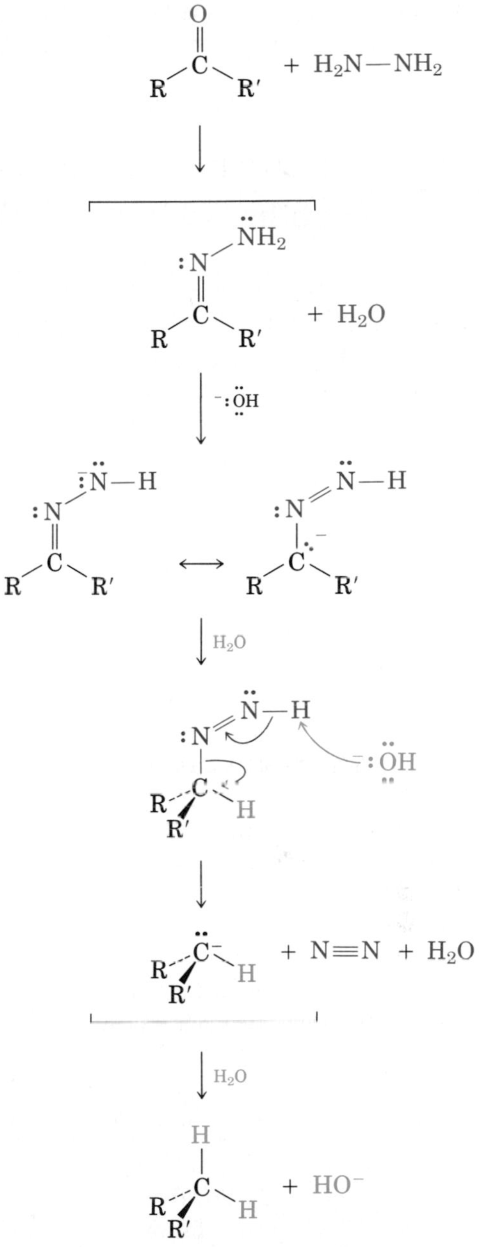

Figure 19.11 Mechanism of the Wolff–Kishner reduction of a ketone or aldehyde to yield an alkane

of a ketone or aldehyde to an alkane. The Clemmensen reduction, whose mechanism is complex and not fully understood, involves treatment of the carbonyl compound with amalgamated zinc, Zn(Hg), and concentrated aqueous HCl. It is used primarily when the ketone or aldehyde starting material is sensitive to the strongly basic conditions required by Wolff–Kishner reduction.

$$C_6H_5-\overset{O}{\overset{\|}{C}}-CH_2CH_3 \xrightarrow{Zn(Hg),\ HCl} C_6H_5-CH_2CH_2CH_3$$

Propiophenone — Propylbenzene (86%)

## 19.14 Nucleophilic Addition of Alcohols: Acetal Formation

Ketones and aldehydes react reversibly with alcohols in the presence of an acid catalyst to yield **acetals, $R_2C(OR')_2$**, also called *ketals* in older literature.

$$\overset{O}{\overset{\|}{C}} + 2\ R'OH \underset{catalyst}{\overset{Acid}{\rightleftharpoons}} C(OR')_2 + H_2O$$

Ketone/aldehyde — An acetal

Acetal formation is similar to the hydration reaction studied in Section 19.8. Like water, alcohols are relatively weak nucleophiles that add to ketones and aldehydes only slowly under neutral conditions. Under acidic conditions, however, the nucleophilic carbonyl oxygen is protonated, and the resultant protonated carbonyl compound is far more reactive than its neutral counterpart. Addition of alcohol therefore occurs rapidly.

$$\overset{\delta^-}{O}=\overset{\delta^+}{C} \underset{}{\overset{H-A}{\rightleftharpoons}} \overset{+}{O}(H)=C \longleftrightarrow O(H)-\overset{+}{C}$$

Neutral carbonyl group (moderately nucleophilic)

Protonated carbonyl group (strongly electrophilic and highly reactive toward nucleophiles)

The initial nucleophilic addition of alcohol to the carbonyl group yields a hydroxy ether called a **hemiacetal**. Hemiacetals are formed reversibly, with the equilibrium normally favoring the carbonyl compound. In the presence of acid, however, a further reaction can occur. Protonation of the hydroxyl group, followed by an E1-like loss of water, leads to a cation (an **oxonium ion, $R_3O^+$**) that then adds a second equivalent of alcohol to yield the acetal (Figure 19.12).

All the steps of acetal formation are reversible: The reaction can be driven either forward (from carbonyl compound to acetal) or backward (from acetal to carbonyl compound), depending on the conditions chosen. The forward reaction is accomplished by choosing conditions that remove water from the medium and thus drive the equilibrium to the right. In practice, this is often done by distilling off water as it forms. The reverse reaction is accomplished by treating the acetal with mineral acid and a large excess of water to drive the equilibrium to the left.

Protonation of the carbonyl oxygen strongly polarizes the carbonyl group and . . .

activates the carbonyl group for nucleophilic attack by oxygen lone-pair electrons from alcohol.

Loss of a proton yields a neutral hemiacetal tetrahedral intermediate.

Hemiacetal

Protonation of the hemiacetal hydroxyl converts it into a good leaving group.

Dehydration yields an intermediate oxonium ion.

Addition of a second equivalent of alcohol gives protonated acetal.

Loss of a proton yields neutral acetal product.

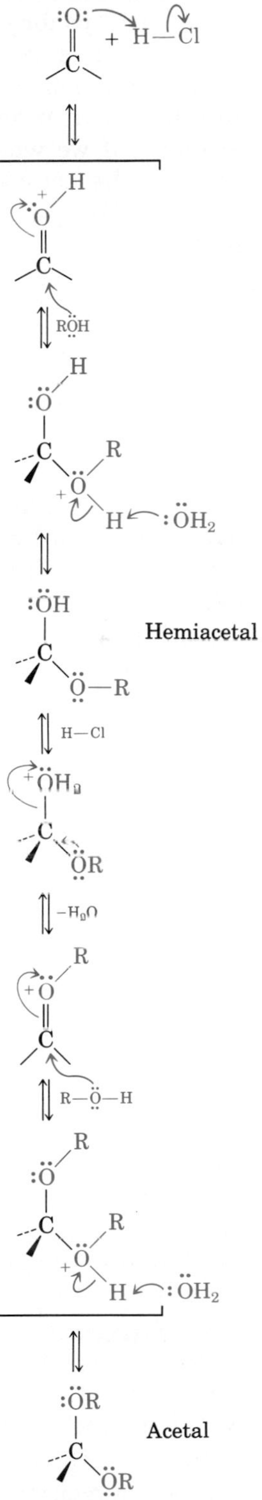

Acetal

**Figure 19.12** Mechanism of acid-catalyzed acetal formation by reaction of a ketone or aldehyde with an alcohol

Acetals are extremely useful compounds because they can serve as protecting groups for ketones and aldehydes in the same way that tetrahydropyranyl ethers serve as protecting groups for alcohols (Section 17.10). As we saw previously, it sometimes happens that one functional group may interfere with intended chemistry elsewhere in a complex molecule. For example, if we wanted to reduce only the ester group of ethyl 4-oxopentanoate, the ketone would interfere. Treatment of the starting keto ester with $LiAlH_4$ would reduce both the keto and the ester groups to give a diol product.

$$CH_3\overset{O}{\overset{\|}{C}}CH_2CH_2\overset{O}{\overset{\|}{C}}OCH_2CH_3 \xrightarrow{?} CH_3\overset{O}{\overset{\|}{C}}CH_2CH_2CH_2OH$$

Ethyl 4-oxopentanoate → 5-Hydroxy-2-pentanone

By protecting the keto group as an acetal, however, the problem can be circumvented. Like other ethers, acetals are inert to bases, hydride reducing agents, Grignard reagents, and catalytic reducing conditions, but they are acid-sensitive. Thus, we can accomplish the selective reduction of the ester group in ethyl 4-oxopentanoate by first converting the keto group to an acetal, then reducing the ester with $LiAlH_4$, and then removing the acetal by treatment with aqueous acid.

$$CH_3\overset{O}{\overset{\|}{C}}CH_2CH_2\overset{O}{\overset{\|}{C}}OCH_2CH_3 \xrightarrow[\text{done directly}]{\text{Cannot be}} CH_3\overset{O}{\overset{\|}{C}}CH_2CH_2CH_2OH$$

Ethyl 4-oxopentanoate

↓ Acid catalyst, $HOCH_2CH_2OH$ ; ↑ $H_3O^+$

$CH_2—CH_2$ (O, O ring) $CH_3CCH_2CH_2CO_2C_2H_5 \xrightarrow[\text{Ether}]{LiAlH_4}$ $CH_2—CH_2$ (O, O ring) $CH_3CCH_2CH_2CH_2OH$

In practice, it is convenient to use ethylene glycol as the alcohol and to form a cyclic acetal. The mechanism of cyclic acetal formation using 1 equiv of ethylene glycol is exactly the same as that using 2 equiv of methanol or other monoalcohol. The only difference is that both alcohol groups are now in the *same* molecule rather than in different molecules.

$$(CH_3)_3C\text{—C}_6H_{10}\text{=O} \xrightarrow[\text{Acid catalyst}]{HOCH_2CH_2OH} (CH_3)_3C\text{—C}_6H_{10}(OCH_2CH_2O) + H_2O$$

4-*tert*-Butylcyclohexanone → 4-*tert*-Butylcyclohexanone ethylene acetal (88%)

**PROBLEM**

**19.14** Show all the steps in the acid-catalyzed formation of a cyclic acetal from ethylene glycol and a ketone or aldehyde.

PROBLEM

**19.15** We saw in Section 19.7 that aldehydes are more reactive than ketones toward nucleophilic addition. How might you use this knowledge to carry out the following selective transformations? One of the two schemes requires a protection step.

(a) $CH_3COCH_2CH_2CH_2CHO \longrightarrow CH_3COCH_2CH_2CH_2CH_2OH$

(b) $CH_3COCH_2CH_2CH_2CHO \longrightarrow CH_3CH(OH)CH_2CH_2CH_2CHO$

## 19.15 Nucleophilic Addition of Thiols: Thioacetal Formation

Thiols, RSH, add to ketones and aldehydes by a reversible, acid-catalyzed pathway to yield **thioacetals, $R'_2C(SR)_2$**. As might be expected, the mechanism of thioacetal formation is identical in all respects to that of acetal formation (Figure 19.12) except that a thiol is used in place of an alcohol. Ethanedithiol is often chosen, and the resultant cyclic thioacetals form rapidly in high yield.

$H_3C$–(cyclohexane)=O $\xrightarrow[HCl]{HSCH_2CH_2SH}$ $H_3C$–(cyclohexane spiro-dithiolane, S, S) + $H_2O$

4-Methylcyclohexanone — A thioacetal (96%)

Thioacetals are useful because they undergo **desulfurization** when treated with a specially prepared nickel powder known as **Raney[6] nickel** (Raney Ni). Since desulfurization removes sulfur from a molecule, replacing it by hydrogen, thioacetal formation followed by Raney nickel desulfurization is an excellent method for reducing ketones or aldehydes to alkanes.

C=O $\longrightarrow$ C(S, S ring) $\xrightarrow[Ethanol]{Raney\ Ni}$ C(H)(H) + NiS

Aldehyde/ketone — Thioacetal — Alkane

For example,

$H_3C$–(cyclohexane spiro-dithiolane, S, S) $\xrightarrow{Raney\ Ni}$ $H_3C$–(cyclohexane, H, H) + NiS

Raney nickel desulfurization is a general method for reducing *any* R—S group to an R—H group. The mechanism of the process is not fully understood, but undoubtedly involves radical intermediates. The hydrogen atoms in the desulfurized products come from hydrogen gas, which is adsorbed onto the Raney nickel surface during preparation.

[6]Murray Raney (1885–1966); b. Tennessee; B.A. University of Kentucky; D.Sc. (Hon.) University of Kentucky; Gilman Paint and Varnish Company (1925–1950); Raney Catalyst Company (1950–1966).

Overall, the thioacetal desulfurization sequence is competitive with the Wolff–Kishner and Clemmensen reactions (Section 19.13) for accomplishing the reduction of a ketone or aldehyde to an alkane. Note, however, that each of the three methods has its own advantages. The Wolff–Kishner reaction occurs under basic conditions, the Clemmensen reaction occurs under acidic conditions, and the Raney nickel desulfurization reaction occurs under neutral conditions.

PROBLEM

**19.16** Show three different methods by which you might prepare cyclohexane from cyclohexanone.

PROBLEM

**19.17** How might you use Raney nickel desulfurization of a thioacetal to carry out the following reduction?

OHC–(cyclohexane)=O $\xrightarrow{?}$ $H_3C$–(cyclohexane)=O

## 19.16 Nucleophilic Addition of Phosphorus Ylides: The Wittig Reaction

Ketones and aldehydes are converted into alkenes by means of the **Wittig**[7] **reaction**. In this process, a phosphorus **ylide**, $R_2C{=}P(C_6H_5)_3$ (also called a **phosphorane**), adds to a ketone or aldehyde, yielding a dipolar intermediate called a **betaine**. (An ylide—pronounced **ill**-id—is a dipolar compound with adjacent plus and minus charges; a betaine—pronounced **bay**-ta-een—is a dipolar compound in which the charges are nonadjacent.)

The betaine intermediate in the Wittig reaction is not isolated; rather it decomposes at temperatures above 0°C to yield alkene and triphenylphosphine oxide. The net result is replacement of carbonyl oxygen by the organic fragment originally bonded to phosphorus (Figure 19.13).

The phosphorus ylides necessary for Wittig reaction are easily prepared by $S_N2$ reaction of primary and some secondary (but not tertiary) alkyl halides with triphenylphosphine, followed by treatment with base. Triorganophosphines are generally excellent nucleophiles in $S_N2$ reactions, and yields of crystalline tetraorganophosphonium salts are high. The proton on the carbon next to the positively charged phosphorus is weakly acidic and can be removed by a base such as sodium hydride or butyllithium (BuLi) to generate the neutral ylide. For example,

$$(C_6H_5)_3P\!: + CH_3{-}Br \xrightarrow{S_N2\text{ reaction}} (C_6H_5)_3\overset{+}{P}{-}CH_3\overset{-}{Br} \xrightarrow[\text{THF}]{\text{BuLi}} (C_6H_5)_3\overset{+}{P}{-}\overset{-}{C}H_2$$

Triphenylphosphine — Bromomethane — Methyltriphenylphosphonium bromide (99%) — Methylenetriphenylphosphorane

[7]Georg Wittig (1897– ); b. Berlin; Ph.D. Marburg (von Auwers); professor, universities of Freiburg, Tübingen, and Heidelberg; Nobel prize (1979).

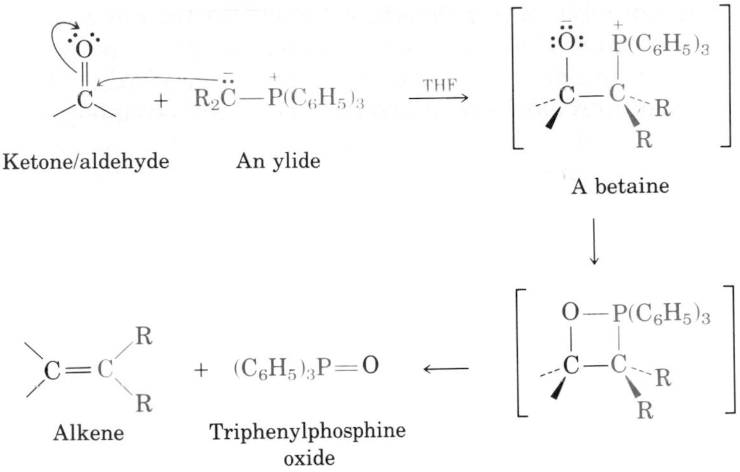

Ketone/aldehyde + An ylide $R_2\bar{C}—\overset{+}{P}(C_6H_5)_3$ $\xrightarrow{THF}$ A betaine

Alkene + $(C_6H_5)_3P{=}O$ Triphenylphosphine oxide

**Figure 19.13** The mechanism of the Wittig reaction between a ketone or aldehyde and a phosphorus ylide

The Wittig reaction is extremely useful, and a great many mono-, di-, and trisubstituted alkenes can be prepared from the appropriate combination of phosphorane and ketone or aldehyde. Tetrasubstituted alkenes cannot be prepared, however, presumably because of steric hindrance during the reaction.

The great value of the Wittig reaction is that pure alkenes of known structure are prepared—the alkene double bond is *always* exactly where the carbonyl group was in the precursor, and no product mixtures (other than $E,Z$ isomers) are formed. For example, addition of methylmagnesium bromide to cyclohexanone followed by dehydration with $POCl_3$ yields a mixture of two alkenes, 1-methylcyclohexene and methylenecyclohexane. Wittig reaction of cyclohexanone with methylenetriphenylphosphorane, however, yields only the single, pure alkene product methylenecyclohexane.

Cyclohexanone $\xrightarrow{1.\ CH_3MgBr \quad 2.\ POCl_3}$ 1-Methylcyclohexene + Methylenecyclohexane

Cyclohexanone $\xrightarrow{(C_6H_5)_3\overset{+}{P}—\bar{C}H_2,\ \text{THF solvent}}$ Methylenecyclohexane (84%) + $(C_6H_5)_3P{=}O$

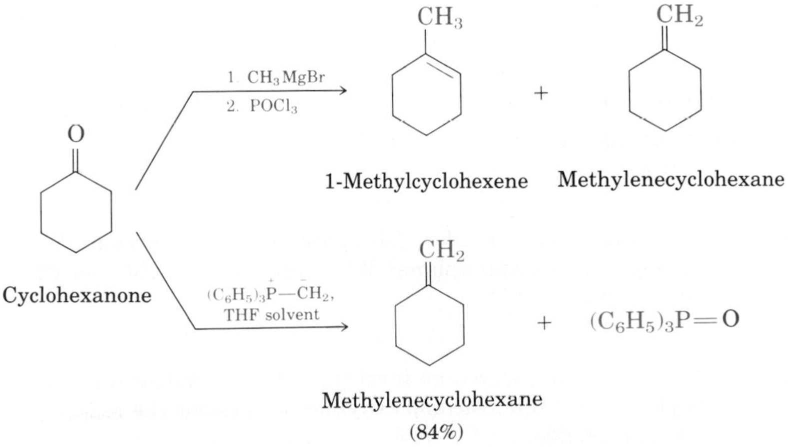

Wittig reactions are so important that they are even used commercially in a number of pharmaceutical applications. For example, the Swiss chemical firm of Hoffmann-LaRoche prepares β-carotene, a yellow food-coloring agent and dietary source of vitamin A, by Wittig reaction between retinal (vitamin A aldehyde) and retinylidenetriphenylphosphorane (Figure 19.14).

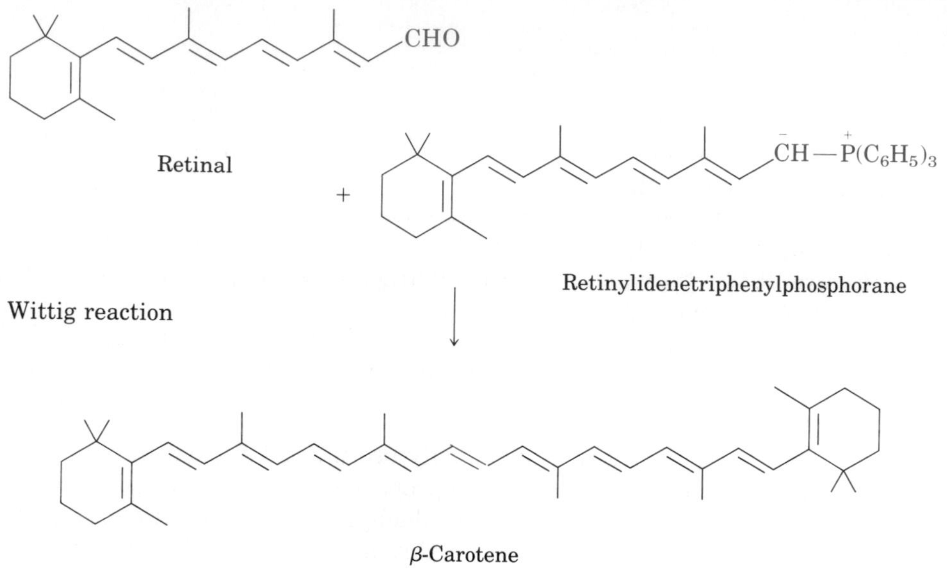

CHO

Retinal

+

$\overset{-}{C}H-\overset{+}{P}(C_6H_5)_3$

Retinylidenetriphenylphosphorane

Wittig reaction

β-Carotene

**Figure 19.14** Preparation of β-carotene using the Wittig reaction

PROBLEM

**19.18** What carbonyl compounds and what phosphorus ylides might you use to prepare these compounds?

(a) 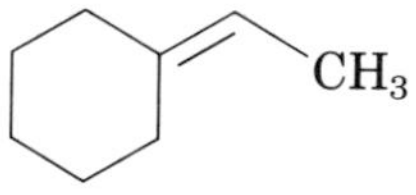 $CH_3$

(b) 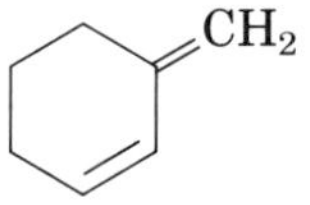 $CH_2$

(c) 2-Methyl-2-hexene

(d) $C_6H_5CH{=}CH(CH_3)_2$

(e) 1,2-Diphenylethylene

PROBLEM

**19.19** Why do you suppose tri*phenyl*phosphine is used to prepare Wittig reagents rather than, say, tri*methyl*phosphine? What problems might you run into if trimethylphosphine were used?

PROBLEM

**19.20** Another route to β-carotene involves a double Wittig reaction between 2 equiv of β-ionylideneacetaldehyde and a diylide. Formulate the reaction and show the structure of the diylide.

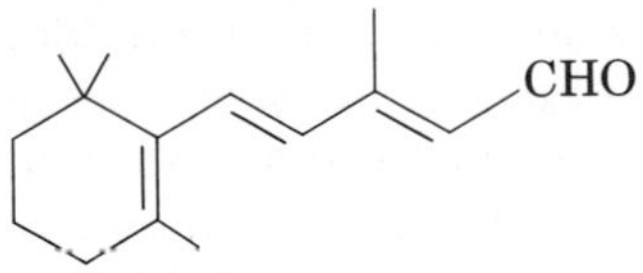

β-Ionylideneacetaldehyde

## 19.17 The Cannizzaro Reaction

We said in the overview of carbonyl chemistry that nucleophilic addition reactions are characteristic of ketones and aldehydes, but not of carboxylic acid derivatives. The reason for the difference is structural. As shown by the general reaction scheme in Figure 19.15, tetrahedral intermediates produced by addition of a nucleophile to a carboxylic acid derivative can eliminate a leaving group, leading to a net nucleophilic acyl substitution reaction. Tetrahedral intermediates produced by addition of a nucleophile to a ketone or aldehyde have only alkyl and hydrogen substituents, however, and thus cannot usually expel a leaving group. The **Cannizzaro[8] reaction**, discovered in 1853, is one exception to this rule.

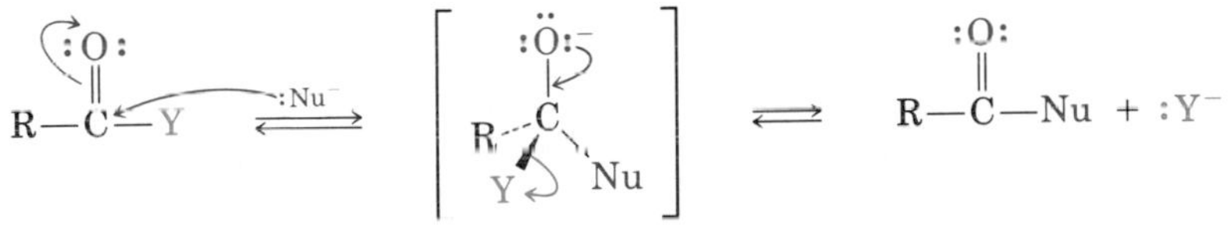

**Figure 19.15** A general reaction scheme for nucleophilic acyl substitution: Carboxylic acid derivatives have an electronegative group Y (—Br, —Cl, —OR′, —$NH_2$, and so forth) that can act as a leaving group. Ketones and aldehydes, however, have no such group.

When an aldehyde having no hydrogens on the carbon next to the —CHO group is heated with hydroxide ion, a disproportionation reaction occurs, yielding 1 equiv of carboxylic acid and 1 equiv of alcohol.

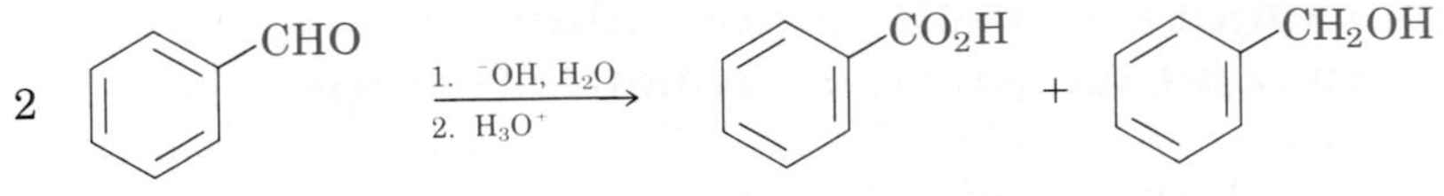

$$2\ C_6H_5CHO \xrightarrow[2.\ H_3O^+]{1.\ ^-OH,\ H_2O} C_6H_5CO_2H + C_6H_5CH_2OH$$

Benzaldehyde — Benzoic acid — Benzyl alcohol

[8]Stanislao Cannizzaro (1826–1910); b. Palermo, Italy; studied at Pisa (Piria); professor, universities of Genoa, Palermo, and Rome.

The Cannizzaro reaction takes place by nucleophilic addition of hydroxide ion to the aldehyde to give a tetrahedral intermediate, *which expels hydride ion as a leaving group*. A second equivalent of aldehyde then accepts the hydride ion in another nucleophilic addition step. The net result is that one molecule of aldehyde undergoes an acyl substitution of hydroxide for hydride and is thereby oxidized to an acid; a second molecule of aldehyde undergoes an addition of hydride and is thereby reduced to an alcohol.

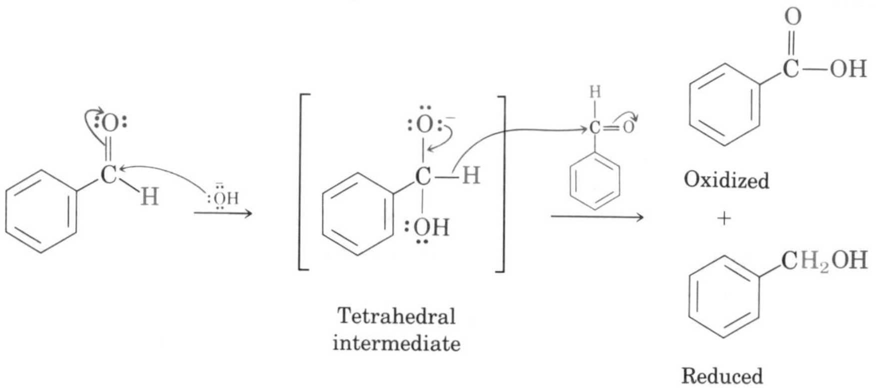

Tetrahedral intermediate

Oxidized

+

Reduced

The Cannizzaro reaction is primarily of mechanistic interest and has few practical applications. It is effectively limited to formaldehyde and substituted benzaldehydes, since aldehydes that have alpha protons undergo other processes involving enolization of their acidic alpha hydrogens.

PROBLEM

**19.21** When *o*-phthalaldehyde is treated with base, *o*-(hydroxymethyl)benzoic acid is formed. Propose a mechanism to account for this reaction.

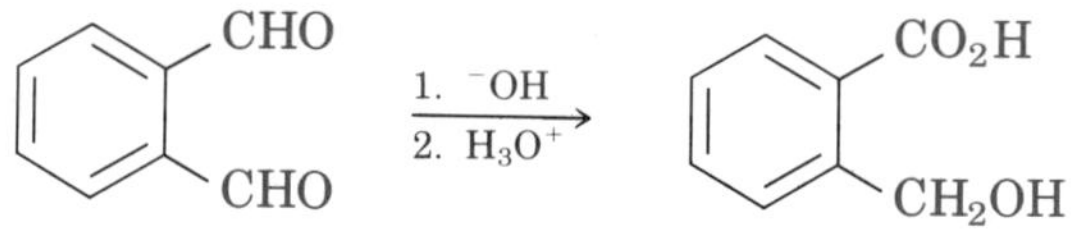

1. $^-OH$ 2. $H_3O^+$

*o*-Phthalaldehyde → *o*-(Hydroxymethyl)benzoic acid

## 19.18 Conjugate Nucleophilic Addition to α,β-Unsaturated Carbonyl Groups

Addition of nucleophiles to carbonyl groups is one of the most important reactions in organic chemistry. Closely related to the direct additions we've been discussing is the **conjugate addition** of nucleophiles to α,β-unsaturated ketones and aldehydes (Figure 19.16). (The two processes are often distinguished by calling them 1,2 addition and 1,4 addition, respectively.)

Direct addition

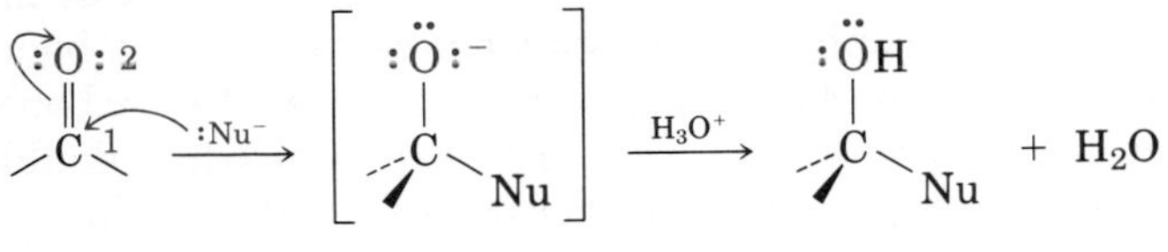

:O: 2 — C 1 — :Nu⁻ → [ :Ö:⁻ C Nu ] —$H_3O^+$→ :ÖH C Nu + $H_2O$

A direct addition product

Conjugate addition

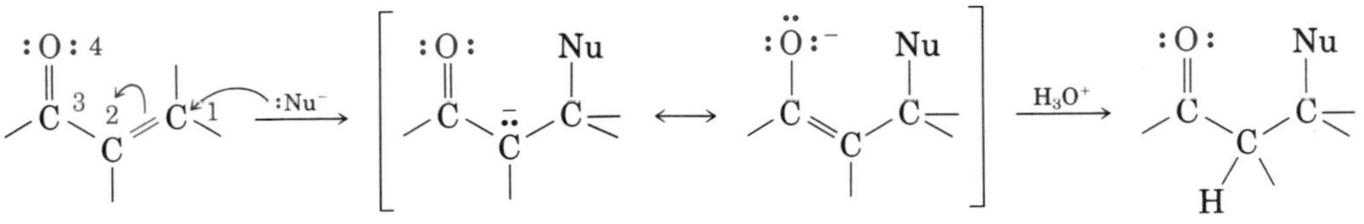

:O: 4 C 3 2 C 1 — :Nu⁻ → [ :O: C C̈ C Nu ⟷ :Ö:⁻ C C C Nu ] —$H_3O^+$→ :O: C C H C Nu

An α,β-unsaturated carbonyl group

An enolate ion

A conjugate addition product

**Figure 19.16** A comparison of direct (1,2) and conjugate (1,4) nucleophilic addition reactions

Conjugate addition of nucleophiles to α,β-unsaturated carbonyl groups is due to exactly the same electronic factors that are responsible for direct addition. As we've seen, a carbonyl group is polarized such that the carbonyl carbon is electropositive, and we can even draw a dipolar resonance structure to underscore the point:

Carbonyl group δ⁻ :O: C δ⁺ ⟷ :Ö:⁻ C +

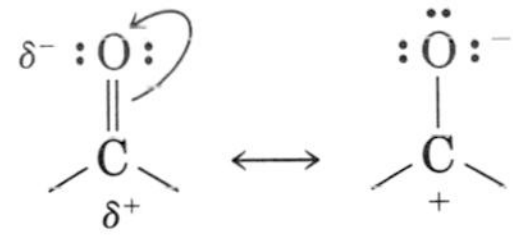

When we draw a similar resonance structure for an α,β-unsaturated carbonyl compound, however, the positive charge is part of an allylic cation and is shared by the beta carbon. In other words, the beta carbon of the enone is an electrophilic site that can react with nucleophiles:

α,β-Unsaturated carbonyl group δ⁻ :O: C δ⁺ α C β C ⟷ :Ö:⁻ C + C C ⟷ :Ö:⁻ C C C+

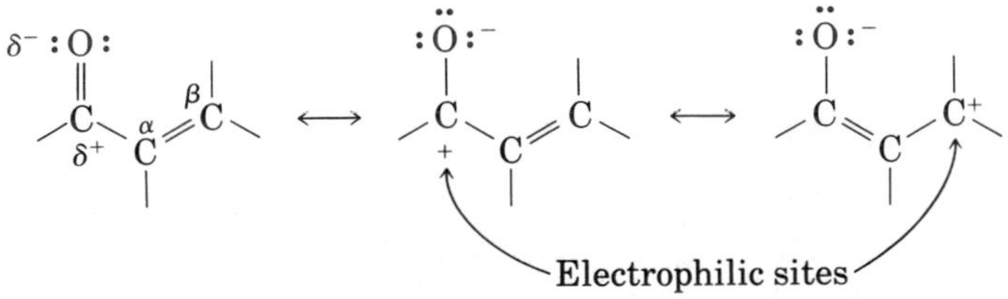

Electrophilic sites

Recall:

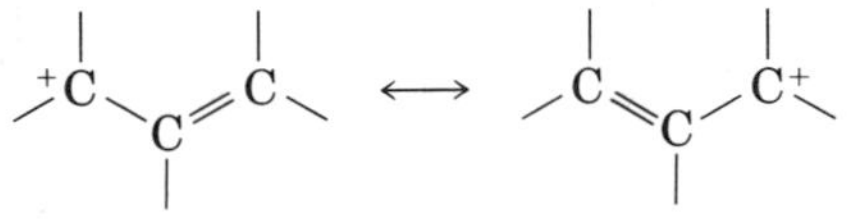

+C C C ⟷ C C C+

Allyl cation

Conjugate addition of a nucleophile to the beta carbon of an enone leads to an enolate ion intermediate, which is then protonated on the alpha carbon to give the saturated ketone product (Figure 19.16). The net effect of the reaction is addition of the nucleophile to the carbon–carbon double bond, with the carbonyl group itself unaffected. In fact, of course, the carbonyl group is crucial to the success of the reaction. The carbon–carbon double bond would not be polarized, and no reaction would occur, without the carbonyl group.

Activated double bond $\xrightarrow[\text{2. } H_3O^+]{\text{1. :Nu}^-}$ (Nu added to beta carbon, H to alpha carbon)

Unactivated double bond $\xrightarrow{:Nu^-}$ No reaction

## CONJUGATE ADDITION OF AMINES

Primary and secondary amines add to $\alpha,\beta$-unsaturated carbonyl compounds, yielding $\beta$-amino ketones and aldehydes. Reaction occurs rapidly under mild conditions, and yields are generally excellent. Note that, if only 1 equiv of amine is used, the conjugate addition product is obtained to the complete exclusion of the direct addition product.

$$CH_3\overset{O}{\overset{\|}{C}}CH{=}CH_2 + H\ddot{N}(CH_2CH_3)_2 \xrightarrow{\text{Ethanol}} CH_3\overset{O}{\overset{\|}{C}}CH_2CH_2N(CH_2CH_3)_2$$

3-Buten-2-one — Diethylamine — 4-*N*,*N*-Diethylamino-2-butanone (92%)

2-Cyclohexenone + $CH_3\ddot{N}H_2$ $\xrightarrow{\text{Ethanol}}$ 3-(*N*-Methylamino)cyclohexanone ($NHCH_3$)

2-Cyclohexenone — Methylamine — 3-(*N*-Methylamino)cyclohexanone

## CONJUGATE ADDITION OF HCN

The elements of HCN can be added to $\alpha,\beta$-unsaturated ketones and aldehydes, giving saturated keto nitriles:

Ketone/aldehyde $\xrightarrow{HCN}$ (CN added to beta carbon, H to alpha carbon)

Although this process can be carried out using sodium cyanide in aqueous alcohol, higher yields are obtained using a method introduced in 1966 by Wataru Nagata. The Nagata procedure for conjugate addition of HCN involves the use of diethylaluminum cyanide as the active reagent. Yields are generally excellent, as the following examples indicate. Note that only conjugate addition is observed; no cyanohydrin products of direct addition are formed.

$$CH_3\overset{O}{\overset{\|}{C}}CH{=}C(CH_3)_2 \xrightarrow[2.\ H_3O^+]{1.\ (C_2H_5)_2Al{-}CN\text{, toluene}} CH_3\overset{O}{\overset{\|}{C}}CH_2\overset{CN}{\overset{|}{C}}(CH_3)_2$$

4-Methyl-3-penten-2-one

2,2-Dimethyl-4-oxopentanenitrile (88%)

$$\xrightarrow[2.\ H_3O^+]{1.\ (C_2H_5)_2Al{-}CN}$$

89%

## CONJUGATE ADDITION OF ALKYL GROUPS: ORGANOCOPPER REACTIONS

Conjugate addition of an alkyl group to an $\alpha,\beta$-unsaturated ketone is one of the most important 1,4-addition reactions, just as direct addition of a Grignard reagent is one of the most important 1,2 additions.

$$\xrightarrow[2.\ H_3O^+]{1.\ \text{“}{:}R^-\text{”}}$$

$\alpha,\beta$-Unsaturated ketone/aldehyde

Conjugate addition of an alkyl group is carried out by treating the $\alpha,\beta$-unsaturated ketone with a lithium diorganocopper reagent (a **Gilman reagent**). As we saw in Section 10.10, a wide variety of diorganocopper reagents can be prepared by reaction between 1 equiv of cuprous iodide and 2 equiv of organolithium:

$$R{-}X \xrightarrow[\text{Pentane}]{2\ Li} R{-}Li + Li^+\ X^-$$

$$2\,R{-}Li \xrightarrow[\text{Ether}]{CuI} Li^+(R{-}\overset{-}{Cu}{-}R) + Li^+\ I^-$$

A lithium diorganocopper (Gilman reagent)

Primary, secondary, and even tertiary alkyl groups undergo the addition reaction, as do aryl and alkenyl groups. Alkynyl groups, however, react poorly in the conjugate addition process.

$$CH_3\overset{O}{\overset{\|}{C}}CH{=}CH_2 \xrightarrow[2.\ H_3O^+]{1.\ Li(CH_3)_2Cu,\ ether} CH_3\overset{O}{\overset{\|}{C}}CH_2CH_2CH_3$$

3-Buten-2-one → 2-Pentanone (97%)

2-Cyclohexenone $\xrightarrow[2.\ H_3O^+]{1.\ Li(H_2C{=}CH)_2Cu,\ ether}$ 3-Vinylcyclohexanone (65%)

2-Cyclohexenone $\xrightarrow[2.\ H_3O^+]{1.\ Li(C_6H_5)_2Cu,\ ether}$ 3-Phenylcyclohexanone (70%)

Diorganocopper reagents are unique in their ability to give conjugate addition products. Other organometallic reagents such as organomagnesiums (Grignard reagents) and organolithiums normally give direct carbonyl addition on reaction with $\alpha,\beta$-unsaturated ketones.

2-Cyclohexenone $\xrightarrow[2.\ H_3O^+]{1.\ CH_3MgBr,\ ether\ or\ CH_3Li}$ 1-Methyl-2-cyclohexen-1-ol (95%)

2-Cyclohexenone $\xrightarrow[2.\ H_3O^+]{1.\ Li(CH_3)_2Cu,\ ether}$ 3-Methylcyclohexanone (97%)

The mechanism of diorganocopper addition is not fully understood, but almost certainly involves radicals. The reaction is not a typical polar process like other nucleophilic additions.

PROBLEM

**19.22** Show how conjugate addition reactions of lithium diorganocopper reagents might be used to synthesize these compounds:

(a) 2-Heptanone (b) 3,3-Dimethylcyclohexanone

(c) 4-*tert*-Butyl-3-ethylcyclohexanone (d) [bicyclic ketone with $CH{=}CH_2$ at ring-fusion carbon]

## 19.19 Some Biological Nucleophilic Addition Reactions

Nature synthesizes the molecules of life using many of the same reactions that chemists use in the laboratory. This is particularly true of carbonyl-group reactions, where nucleophilic addition steps play an intimate role in the biosynthesis of many vital molecules.

For example, one of the pathways by which amino acids are made in the body involves nucleophilic addition of ammonia to $\alpha$-keto acids. To choose one specific case, alanine is synthesized from pyruvic acid and ammonia by bacterial enzymes from *Bacillus subtilis*:

$$\underset{\substack{\text{Pyruvic acid}\\(\text{an }\alpha\text{-keto acid})}}{CH_3\overset{O}{\overset{\|}{C}}COOH} + :NH_3 \xrightarrow{B.\ subtilis} \underset{\substack{\text{Alanine}\\(\text{an amino acid})}}{CH_3\overset{NH_2}{\overset{|}{C}}HCOOH}$$

The key step in this biological transformation is the nucleophilic addition of ammonia to the ketone carbonyl group of pyruvic acid. The tetrahedral intermediate loses water to yield an imine, which is further reduced (a second nucleophilic addition step) by an enzymatic reaction to yield alanine.

$$\underset{\text{Pyruvic acid}}{CH_3-\overset{O}{\overset{\|}{C}}-COOH} \xrightarrow{NH_3} \underset{\text{An imine}}{CH_3-\overset{NH}{\overset{\|}{C}}-COOH} \xrightarrow[\text{Reducing enzyme}]{[H]} \underset{\substack{\text{Alanine}\\(\text{an amino acid})}}{CH_3-\overset{NH_2}{\overset{|}{\underset{H}{\underset{|}{C}}}}-COOH}$$

Another nucleophilic addition reaction—this time in reverse—plays an interesting role in the chemical defense mechanisms used by certain plants and insects to protect themselves against predators. For example, when the millipede *Apheloria corrugata* is attacked by ants, it secretes mandelonitrile and an enzyme that catalyzes the decomposition of the cyanohydrin mandelonitrile into benzaldehyde and HCN, as shown on page 694. The millipede actually protects itself by discharging poisonous HCN at would-be attackers.

$$C_6H_5\text{–}CH(OH)\text{–}CN \xrightarrow[H_2O]{\text{Enzyme}} C_6H_5\text{–}CHO + HCN$$

Mandelonitrile
(from *Apheloria corrugata*)

## 19.20 Spectroscopic Analysis of Ketones and Aldehydes

### INFRARED SPECTROSCOPY

Ketones and aldehydes show a strong C=O bond absorption in the infrared region 1660–1770 $cm^{-1}$, as the spectra of benzaldehyde and cyclohexanone (Figures 19.17 and 19.18) demonstrate. In addition, aldehydes show two characteristic C—H absorptions in the range 2720–2820 $cm^{-1}$, due to stretching of the aldehyde —CO—H bond. The exact position of the C=O bond absorption varies slightly from compound to compound but is highly diagnostic of the exact nature of the carbonyl group. Table 19.3 shows the correlation between the infrared absorption maximum and carbonyl-group structure.

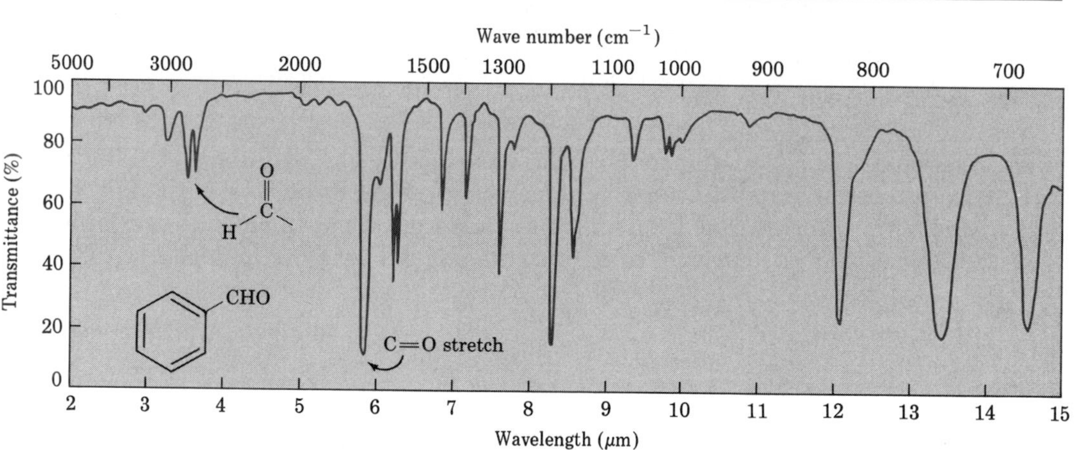


**Figure 19.17** Infrared spectrum of benzaldehyde

As the data in Table 19.3 indicate, saturated aldehydes usually show carbonyl absorptions near 1730 $cm^{-1}$ in the infrared spectrum, whereas conjugation of the aldehyde to an aromatic ring or a double bond lowers the absorption by 25 $cm^{-1}$ to near 1705 $cm^{-1}$. Saturated aliphatic ketones and cyclohexanones both absorb near 1715–1720 $cm^{-1}$, and conjugation with a double bond or aromatic ring again lowers the absorption by 30 $cm^{-1}$ to 1685–1690 $cm^{-1}$. Additional angle strain in the carbonyl group, caused by

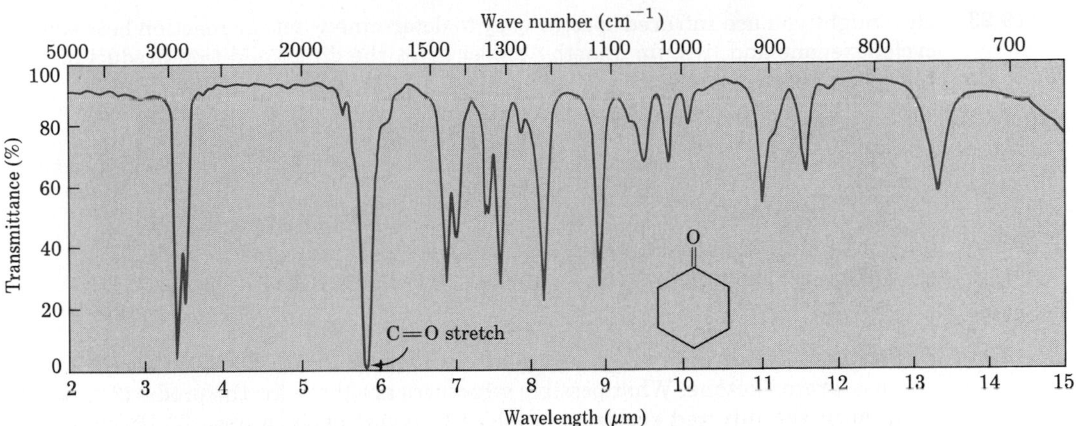


**Figure 19.18** Infrared spectrum of cyclohexanone

**Table 19.3** Infrared absorptions of some ketones and aldehydes

| *Carbonyl type* | *Example* | *Infrared absorption* ($cm^{-1}$) |
|---|---|---|
| Aliphatic aldehyde | Acetaldehyde | 1730 |
| Aromatic aldehyde | Benzaldehyde | 1705 |
| $\alpha,\beta$-Unsaturated aldehyde | $H_2C{=}CH{-}CHO$ | 1705 |
| Aliphatic ketone | Acetone | 1715 |
| Six-membered-ring ketone | Cyclohexanone | 1715 |
| Five-membered-ring ketone | Cyclopentanone | 1750 |
| Four-membered-ring ketone | Cyclobutanone | 1785 |
| Aromatic ketone | $C_6H_5C(=O)CH_3$ | 1690 |
| $\alpha,\beta$-Unsaturated ketone | $H_2C{=}CHC(=O)CH_3$ | 1685 |

reducing the ring size of cyclic ketones to four or five, results in a marked raising of the absorption position.

The values given in Table 19.3 are remarkably constant from one ketone to another or from one aldehyde to another. As a result, infrared spectroscopy is an extraordinarily powerful tool for diagnosing the nature and chemical environment of a carbonyl group in a molecule of unknown structure. An unknown that shows an infrared absorption at 1730 $cm^{-1}$ is almost certainly an aldehyde rather than a ketone; an unknown that shows an infrared absorption at 1750 $cm^{-1}$ is almost certainly a cyclopentanone, and so on.

PROBLEM

**19.23** How might you use infrared spectroscopy to determine whether reaction between 2-cyclohexenone and lithium dimethylcopper gives the direct addition product or the conjugate addition product?

PROBLEM

**19.24** Tell where you would expect each of these compounds to absorb in the infrared spectrum:

(a) 4-Penten-2-one (b) 3-Penten-2-one
(c) 2,2-Dimethylcyclopentanone (d) *m*-Chlorobenzaldehyde
(e) 3-Cyclohexenone (f) 2-Hexenal

PROBLEM

**19.25** Dehydration of 3-hydroxy-3-phenylcyclohexanone by treatment with acid leads to an unsaturated ketone. What possible structures are there for the product? At what position in the infrared spectrum would you expect each to absorb? If the actual product has an absorption at 1670 $cm^{-1}$, what is its structure?

## NUCLEAR MAGNETIC RESONANCE SPECTROSCOPY

Carbonyl-group carbon atoms show readily identifiable and highly characteristic $^{13}C$ NMR resonance peaks in the range 190–215 $\delta$. Since no other kinds of carbons absorb in this range, the presence of an NMR absorption near 200 $\delta$ is strong evidence for a carbonyl group. Isolated ketone or aldehyde carbons usually absorb in the region from 200 to 215 $\delta$, whereas $\alpha,\beta$-unsaturated carbonyl carbons absorb in the 190–200 $\delta$ region. Table 19.4 lists some specific examples.

**Table 19.4** Carbon-13 NMR absorptions of some ketones and aldehydes

| *Carbonyl compound* | *Carbon-13 NMR absorption of* $\rangle C{=}O$ ($\delta$) |
|---|---|
| Acetaldehyde | 201 |
| Benzaldehyde | 192 |
| 2-Butanone | 207 |
| Cyclohexanone | 211 |
| Acetophenone | 196 |

Proton NMR is also of considerable use for analysis of aldehydes, though less so for ketones. Aldehyde protons (R—CHO) absorb near 10 $\delta$ in the $^{1}H$ NMR spectrum and are highly distinctive, since no other kind of proton absorbs in this region. The aldehyde proton usually shows spin–spin coupling to neighbor protons, with coupling constant $J \approx 3$ Hz. Observation of the splitting pattern of the aldehyde proton enables us to tell the degree of substitution at the alpha position. Acctaldchyde, for example, shows a quartet at 9.8 $\delta$ for the aldehyde proton, indicating that there are three protons neighboring the —CHO group (Figure 19.19).

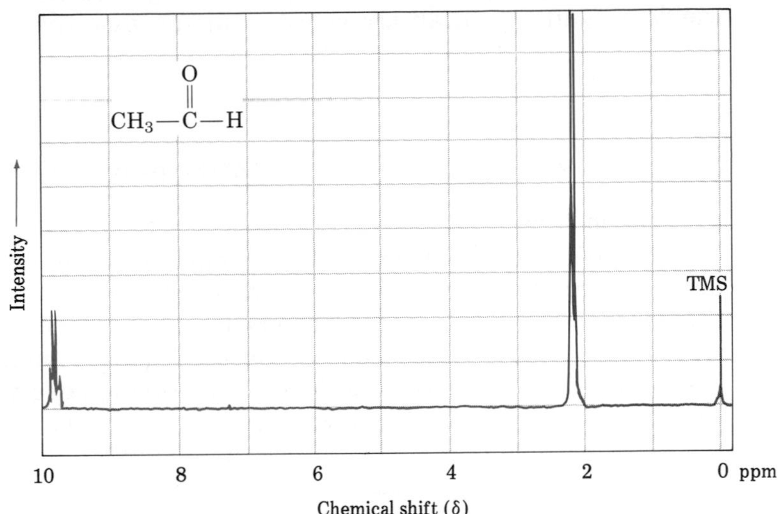


**Figure 19.19** Proton NMR spectrum of acetaldehyde

Protons on the carbon next to a carbonyl group are slightly deshielded and normally absorb near 2.0–2.3 $\delta$. (Note that the acetaldehyde methyl group in Figure 19.19 absorbs at 2.20 $\delta$). Methyl ketones are particularly distinctive, since they show a large sharp three-proton singlet near 2.1 $\delta$. Complex spin–spin splittings often obscure the absorption patterns of other ketones, however, and reduce the diagnostic usefulness of $^1$H NMR.

## MASS SPECTROSCOPY

Aliphatic ketones and aldehydes having hydrogens on their gamma-carbon atoms undergo a characteristic mass spectral cleavage called the **McLafferty[9] rearrangement**. In this rearrangement, a hydrogen atom is transferred from the gamma carbon to the carbonyl-group oxygen, the bond between the alpha and beta carbons is broken, and a neutral alkene fragment is produced. The charge remains with the oxygen-containing fragment.

$$\left[ R'{-}\overset{\gamma}{C}H(H){-}\overset{\beta}{C}H_2{-}\overset{\alpha}{C}H_2{-}C({=}O){-}R \right]^{+\cdot} \xrightarrow[\text{rearrangement}]{\text{McLafferty}} R'{-}\overset{\gamma}{C}H{=}\overset{\beta}{C}H_2 + \left[ \overset{\alpha}{C}H_2{=}C(OH){-}R \right]^{+\cdot}$$

In addition to fragmentation by the McLafferty rearrangement, ketones and aldehydes undergo cleavage of the bond between the alpha carbon and

[9]Fred Warren McLafferty (1923– ); b. Evanston, Ill.; Ph.D. (1950), Cornell University; Dow Chemical (1950–1964); professor, Purdue University (1964–1968), Cornell University (1968– ).

the carbonyl-group carbon (an alpha-cleavage reaction). Alpha cleavage yields a neutral radical; the charge remains with the oxygen-containing cation.

$$\left[RCH_2 \nmid \overset{\overset{\displaystyle O}{\|}}{C}-R'\right]^{+\cdot} \xrightarrow{\text{Alpha cleavage}} RCH_2\cdot + \left[\overset{\overset{\displaystyle O}{\|}}{C}-R\right]^{+}$$

Fragment ions resulting from both alpha cleavage and McLafferty rearrangement can be seen in the mass spectrum of 5-methyl-2-hexanone (Figure 19.20). Alpha cleavage occurs primarily at the more substituted side of the carbonyl group, leading to a $[CH_3CO]^+$ fragment with $m/z = 43$. McLafferty rearrangement and loss of 2-methylpropene yields a fragment with $m/z = 58$.

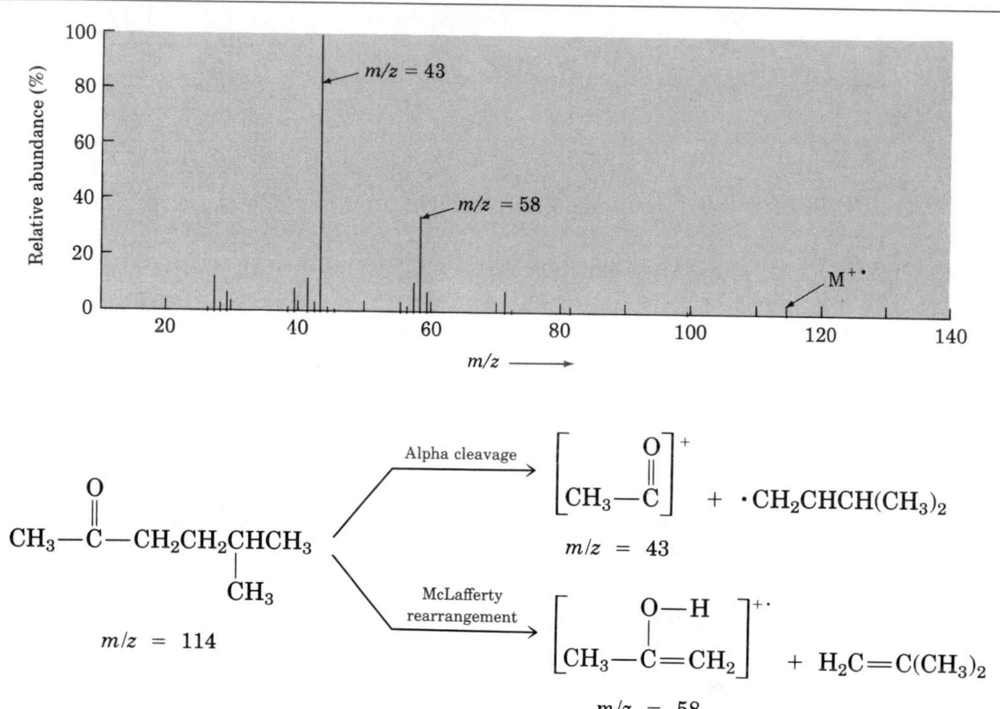


**Figure 19.20** Mass spectrum of 5-methyl-2-hexanone: The abundant peak at $m/z = 43$ is due to alpha cleavage at the more highly substituted side of the carbonyl group. The peak at $m/z = 58$ is due to McLafferty rearrangement. Note that the peak due to the molecular ion is very small.

PROBLEM

**19.26** How might you use mass spectroscopy to distinguish between these pairs of isomers?

(a) 3-Mcthyl-2-hexanone and 4-methyl-2-hexanone

(b) 3-Heptanone and 4-heptanone

(c) 2-Methylpentanal and 4-methylpentanal

## 19.21 Summary and Key Words

Aldehydes and ketones are among the most important of all compounds, both in biochemistry and in the chemical industry. Aldehydes are normally prepared in the laboratory by oxidative cleavage of alkenes, by oxidation of primary alcohols with pyridinium chlorochromate (PCC), or by partial reduction of acid chlorides or esters. Ketones are similarly prepared by oxidative cleavage of alkenes, by oxidation of secondary alcohols, or by addition of diorganocopper reagents to acid chlorides.

The **nucleophilic addition reaction** is the most important reaction of aldehydes and ketones. As shown in Figure 19.21, many different product types can be prepared by nucleophilic additions. The reactions are applicable to ketones and aldehydes, but aldehydes are generally more reactive for both steric and electronic reasons.

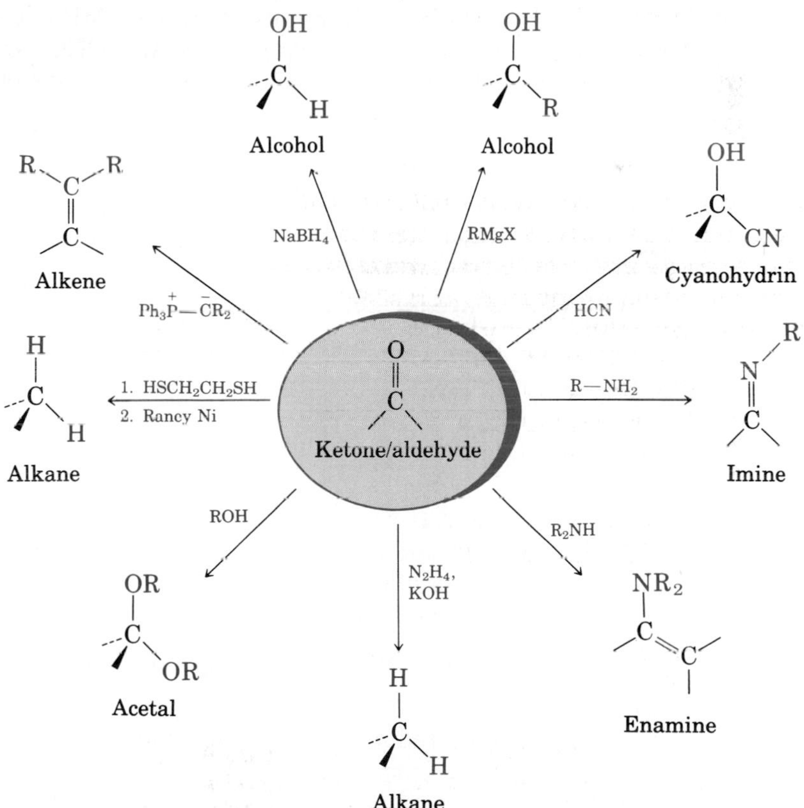


**Figure 19.21** Some nucleophilic addition reactions of ketones and aldehydes

Ketones and aldehydes are reduced by $NaBH_4$ or $LiAlH_4$ to yield secondary and primary alcohols, respectively. Addition of Grignard reagents to ketones and aldehydes also gives alcohols (tertiary and secondary, respectively), and addition of HCN yields cyanohydrins. Primary amines add to carbonyl compounds, yielding **imines**, and secondary amines yield **enamines**. Reaction of a ketone or aldehyde with hydrazine and base yields an alkane (the **Wolff–Kishner reaction**). Alcohols and thiols (RSH) add to carbonyl groups to yield **acetals** and **thioacetals**, respectively. Acetals are valuable as protecting groups, whereas thioacetals are useful because they can be desulfurized by Raney nickel treatment to produce alkanes. Phosphoranes add to ketones and aldehydes, giving alkenes (the **Wittig reaction**) in which the new C=C in the product is exactly where the C=O bond in the starting material was.

$\alpha,\beta$-Unsaturated ketones and aldehydes often react with nucleophiles to give the product of **conjugate addition**, or **1,4 addition**. Particularly important is the reaction with diorganocopper reagents, which results in the addition of alkyl, aryl, and alkenyl (but not alkynyl) groups.

Ketones and aldehydes can be analyzed by infrared and nuclear magnetic resonance spectroscopy. Carbonyl groups absorb in the range 1660–1770 $cm^{-1}$, with the exact infrared absorption position highly diagnostic of the precise kind of carbonyl group present in the molecule. Carbon-13 NMR spectroscopy is also useful for aldehydes and ketones since carbonyl carbons show resonances in the 190–215 $\delta$ range. Proton NMR is useful largely for analysis of aldehydes since aldehyde protons (R—CHO) absorb near 10 $\delta$. Ketones and aldehydes undergo two characteristic kinds of fragmentation in the mass spectrometer—**alpha cleavage** and **McLafferty rearrangement**.

## 19.22 Summary of Reactions

1. Preparation of aldehydes (Section 19.3)

   a. Oxidation of primary alcohols (Section 17.9)

$$R-\underset{H}{\overset{OH}{C}}-H \xrightarrow[CH_2Cl_2]{PCC} R-\overset{O}{\overset{\|}{C}}-H$$

   b. Ozonolysis of alkenes (Section 7.8)

$$RCH{=}CHR' \xrightarrow[2.\ Zn,\ CH_3COOH]{1.\ O_3} RCH{=}O + O{=}CHR'$$

   c. Rosenmund reduction of acid chloride (Section 19.3)

$$R-\overset{O}{\overset{\|}{C}}-Cl \xrightarrow[\text{Ethyl acetate}]{H_2,\ Pd/BaSO_4} R-\overset{O}{\overset{\|}{C}}-H$$

   d. Partial reduction of esters (Section 19.3)

$$R-\overset{O}{\overset{\|}{C}}-OR' \xrightarrow[2.\ H_3O^+]{1.\ \text{DIBAH, toluene}} R-\overset{O}{\overset{\|}{C}}-H$$

2. Preparation of ketones (Section 19.4)
   a. Oxidation of secondary alcohols (Section 17.9)

$$R{-}\overset{\displaystyle OH}{\underset{\displaystyle H}{\overset{|}{\underset{|}{C}}}}{-}R' \xrightarrow{\text{Cr(VI)}} R{-}\overset{\displaystyle O}{\overset{\|}{C}}{-}R'$$

   b. Ozonolysis of alkenes (Section 7.8)

$$R_2C{=}CR_2 \xrightarrow[\text{2. Zn, } CH_3COOH]{\text{1. } O_3} R_2C{=}O + O{=}CR_2$$

   c. Friedel–Crafts acylation (Section 16.4)

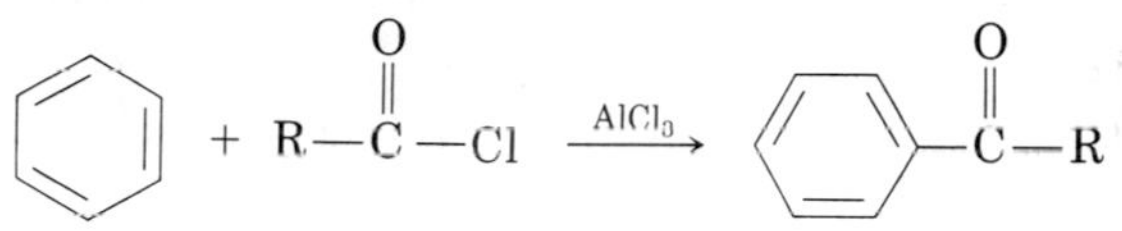

$$C_6H_6 + R{-}\overset{\displaystyle O}{\overset{\|}{C}}{-}Cl \xrightarrow{AlCl_3} C_6H_5{-}\overset{\displaystyle O}{\overset{\|}{C}}{-}R$$

   d. Alkyne hydration (Section 8.4)

$$R{-}C{\equiv}C{-}H \xrightarrow[H_2SO_4,\ H_2O]{Hg^{2+}} R{-}\overset{\displaystyle O}{\overset{\|}{C}}{-}CH_3$$

   e. Diorganocopper reaction with acid chlorides (Section 19.4)

$$R{-}\overset{\displaystyle O}{\overset{\|}{C}}{-}Cl + R'_2CuLi \xrightarrow{\text{Ether}} R{-}\overset{\displaystyle O}{\overset{\|}{C}}{-}R'$$

3. Reactions of aldehydes
   a. Oxidation (Section 19.5)

$$R{-}\overset{\displaystyle O}{\overset{\|}{C}}{-}H \xrightarrow[\text{or } Ag^+,\ NH_4OH]{\text{Jones' reagent}} R{-}\overset{\displaystyle O}{\overset{\|}{C}}{-}OH$$

   b. Cannizzaro reaction (Section 19.17)

$$2\ Ar{-}\overset{\displaystyle O}{\overset{\|}{C}}{-}H \xrightarrow[\text{2. } H_3O^+]{\text{1. } HO^-,\ H_2O} Ar{-}\overset{\displaystyle O}{\overset{\|}{C}}{-}OH + Ar{-}\overset{\displaystyle OH}{\underset{\displaystyle H}{\overset{|}{\underset{|}{C}}}}{-}H$$

4. Nucleophilic addition reactions of aldehydes and ketones

a. Addition of hydride: Reduction (Section 19.11)

$$R{-}\overset{\overset{\displaystyle O}{\|}}{C}{-}R' \xrightarrow[\text{2. } H_3O^+]{\text{1. NaBH}_4\text{, ethanol}} R{-}\underset{\underset{\displaystyle H}{|}}{\overset{\overset{\displaystyle OH}{|}}{C}}{-}R'$$

b. Addition of Grignard reagents (Section 19.10)

$$R{-}\overset{\overset{\displaystyle O}{\|}}{C}{-}R' \xrightarrow[\text{2. } H_3O^+]{\text{1. R''MgX, ether}} R{-}\underset{\underset{\displaystyle R''}{|}}{\overset{\overset{\displaystyle OH}{|}}{C}}{-}R'$$

c. Addition of HCN: Cyanohydrins (Section 19.9)

$$R{-}\overset{\overset{\displaystyle O}{\|}}{C}{-}R' \overset{\text{HCN}}{\rightleftharpoons} R{-}\underset{\underset{\displaystyle CN}{|}}{\overset{\overset{\displaystyle OH}{|}}{C}}{-}R'$$

d. Addition of primary amines: Imines (Section 19.12)

$$R{-}\overset{\overset{\displaystyle O}{\|}}{C}{-}R' \overset{\text{R''NH}_2}{\rightleftharpoons} R{-}\overset{\overset{\displaystyle N{-}R''}{\|}}{C}{-}R' + H_2O$$

For example:

Oximes, $R_2C{=}N{-}OH$

Semicarbazones, $R_2C{=}N{-}NHCONH_2$

2,4-Dinitrophenylhydrazones, $R_2C{=}N{-}NH{-}C_6H_4(NO_2)_2$

e. Addition of secondary amines: Enamines (Section 19.12)

$$R{-}\overset{\overset{\displaystyle O}{\|}}{C}{-}\underset{|}{\overset{|}{C}}{-}H \overset{\text{R}'_2\text{NH}}{\rightleftharpoons} (R'_2N)(R)C{=}C\langle + H_2O$$

f. Wolff–Kishner reaction (hydrazine addition) (Section 19.13)

$$R{-}\overset{\overset{\displaystyle O}{\|}}{C}{-}R' \xrightarrow[\text{KOH}]{H_2NNH_2} R{-}\underset{\underset{\displaystyle H}{|}}{\overset{\overset{\displaystyle H}{|}}{C}}{-}R' + N_2 + H_2O$$

g. Addition of alcohols: Acetals (Section 19.14)

$$\mathrm{R{-}\overset{\overset{\displaystyle O}{\|}}{C}{-}R' + 2\,R''OH \xrightarrow[\text{catalyst}]{\text{Acid}} R{-}\overset{\overset{\displaystyle OR''}{|}}{\underset{\underset{\displaystyle R'}{|}}{C}}{-}OR'' + H_2O}$$

h. Addition of thiols: Thioacetals (Section 19.15)

$$\mathrm{R{-}\overset{\overset{\displaystyle O}{\|}}{C}{-}R' + 2\,R''SH \xrightarrow[\text{catalyst}]{\text{Acid}} R{-}\overset{\overset{\displaystyle SR''}{|}}{\underset{\underset{\displaystyle R'}{|}}{C}}{-}SR'' + H_2O}$$

i. Desulfurization of thioacetals with Raney nickel (Section 19.15)

$$\mathrm{R{-}\overset{\overset{\displaystyle SR''}{|}}{\underset{\underset{\displaystyle R'}{|}}{C}}{-}SR'' \xrightarrow[\text{Ethanol}]{\text{Raney Ni}} R{-}\overset{\overset{\displaystyle H}{|}}{\underset{\underset{\displaystyle H}{|}}{C}}{-}R' + NiS}$$

j. Addition of phosphorus ylides: Wittig reaction (Section 19.16)

$$\mathrm{R{-}\overset{\overset{\displaystyle O}{\|}}{C}{-}R' + (C_6H_5)_3\overset{+}{P}{-}\overset{-}{C}HR'' \xrightarrow{\text{THF}} (R)(R')C{=}C(R'')(H) + (C_6H_5)_3P{-}O}$$

5. Conjugate additions to $\alpha,\beta$-unsaturated ketones and aldehydes (Section 19.18)

a. Addition of HCN

$$\mathrm{R{-}\overset{\overset{\displaystyle O}{\|}}{C}{-}\underset{|}{C}{=}C\langle \xrightarrow[\text{2. } H_3O^+]{\text{1. } (C_2H_5)_2AlCN} R{-}\overset{\overset{\displaystyle O}{\|}}{C}{-}\overset{\overset{\displaystyle H}{|}}{\underset{|}{C}}{-}\overset{|}{\underset{|}{C}}{-}CN}$$

b. Addition of amines

$$\mathrm{R{-}\overset{\overset{\displaystyle O}{\|}}{C}{-}\underset{|}{C}{=}C\langle \underset{}{\overset{R'NH_2}{\rightleftharpoons}} R{-}\overset{\overset{\displaystyle O}{\|}}{C}{-}\overset{\overset{\displaystyle H}{|}}{\underset{|}{C}}{-}\overset{|}{\underset{|}{C}}{-}NHR'}$$

c. Addition of alkyl groups: Diorganocopper reaction

$$\mathrm{R{-}\overset{\overset{\displaystyle O}{\|}}{C}{-}\underset{|}{C}{=}C\langle \xrightarrow[\text{2. } H_3O^+]{\text{1. } R'_2CuLi,\ ether} R{-}\overset{\overset{\displaystyle O}{\|}}{C}{-}\overset{\overset{\displaystyle H}{|}}{\underset{|}{C}}{-}\overset{|}{\underset{|}{C}}{-}R'}$$

## ADDITIONAL PROBLEMS

**19.27** Draw structures corresponding to these names:

(a) Bromoacetone
(b) 3,5-Dinitrobenzenecarbaldehyde
(c) 2-Methyl-3-heptanone
(d) 3,5-Dimethylcyclohexanone
(e) 2,2,4,4-Tetramethyl-3-pentanone
(f) 4-Methyl-3-penten-2-one
(g) Butanedial
(h) 3-Phenyl-2-propenal
(i) 6,6-Dimethyl-2,4-cyclohexadienone
(j) *p*-Nitroacetophenone
(k) (*S*)-2-Hydroxypropanal
(l) (2*S*,3*R*)-2,3,4-Trihydroxybutanal

**19.28** Draw and name the seven ketones and aldehydes having the formula $C_5H_{10}O$.

**19.29** Provide IUPAC names for these structures:

(a) O, $CH_3$

(b) CHO, H, C, OH, $CH_2OH$

(c) O

(d) $CH_3CH(CH_3)COCH_2CH_3$

(e) $CH_3CH(OH)CH_2CHO$

(f) CHO

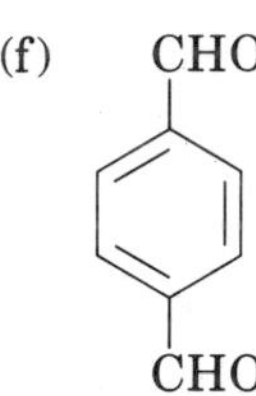

CHO

**19.30** Draw structures that fit these descriptions:

(a) An $\alpha,\beta$-unsaturated ketone, $C_6H_8O$
(b) An $\alpha$-diketone
(c) An aromatic ketone, $C_9H_{10}O$
(d) A diene aldehyde, $C_7H_8O$

**19.31** Predict the products of the reaction of phenylacetaldehyde with these reagents:

(a) $NaBH_4$, then $H_3O^+$
(b) Tollens' reagent
(c) Hydroxylamine, HCl
(d) Methylmagnesium bromide, then $H_3O^+$
(e) Methanol plus acid catalyst
(f) $H_2NNH_2$/KOH
(g) Methylenetriphenylphosphorane
(h) HCN, KCN

**19.32** Repeat Problem 19.31 for acetophenone.

**19.33** How would you prepare the following substances from 2-cyclohexenone? More than one step may be required.

(a)

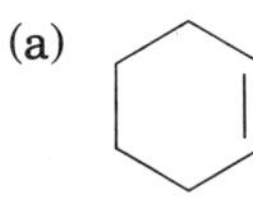

(b)

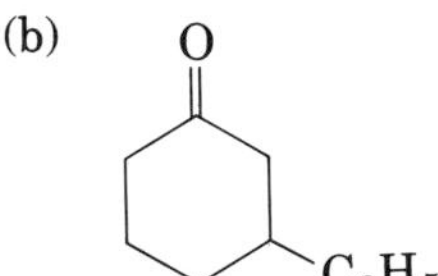


(c)

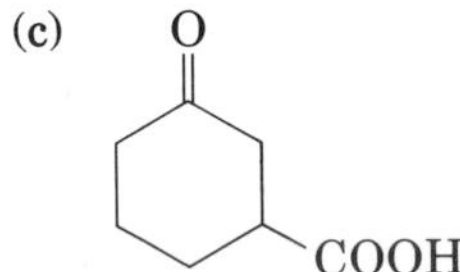


(d)

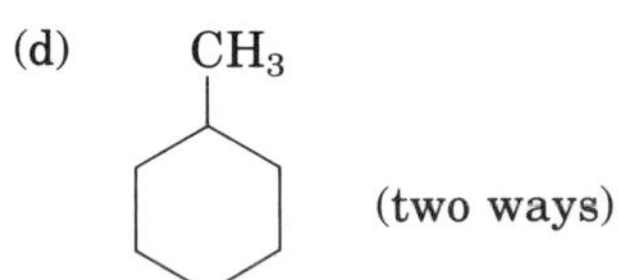

**19.34** How can you account for the fact that glucose reacts with the Tollens reagent to give a silver mirror, but glucose $\alpha$-methyl glycoside does not?

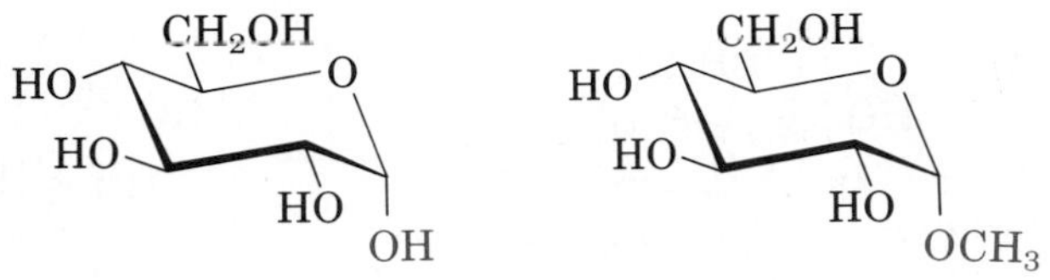

Glucose — Glucose $\alpha$-methyl glycoside

**19.35** Reaction of 2-butanone with $NaBH_4$ yields a chiral product. What stereochemistry does the product have? Is it optically active?

**19.36** Show how the Wittig reaction might be used to prepare these alkenes. Identify the alkyl halide and the carbonyl components that would be used.

(a) $C_6H_5CH{=}CH{-}CH{=}CHC_6H_5$

(b)

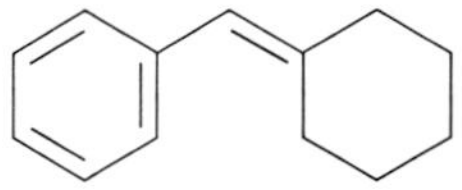

(c) 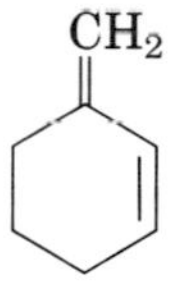 $CH_2$

(d) 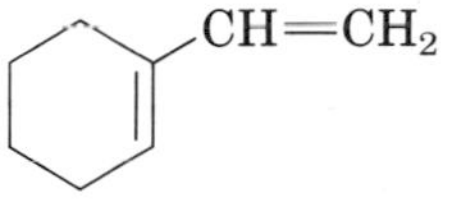 $CH{=}CH_2$

**19.37** How would you use a Grignard reaction on a ketone or aldehyde to synthesize these compounds?

(a) 2-Pentanol
(b) 1-Butanol
(c) 1-Phenylcyclohexanol
(d) Diphenylmethanol

**19.38** The Wittig reaction can be used to prepare aldehydes as well as alkenes. This is done by using (methoxymethylene)triphenylphosphorane as the Wittig reagent and hydrolyzing the product with acid. For example,

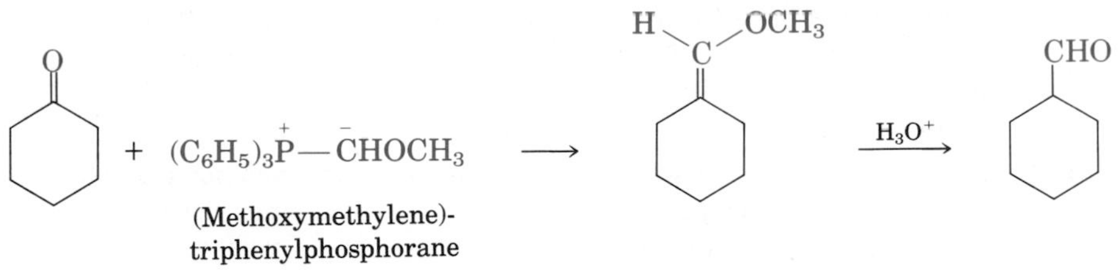

$$+ \ (C_6H_5)_3\overset{+}{P}{-}\overset{-}{C}HOCH_3 \longrightarrow \quad \xrightarrow{H_3O^+} \quad CHO$$

(Methoxymethylene)-triphenylphosphorane

(a) How would you prepare the required phosphorane?
(b) Propose a mechanism to account for the hydrolysis step.

**19.39** When 4-hydroxybutanal is treated with methanol in the presence of an acid catalyst, 2-methoxytetrahydrofuran is formed. Explain.

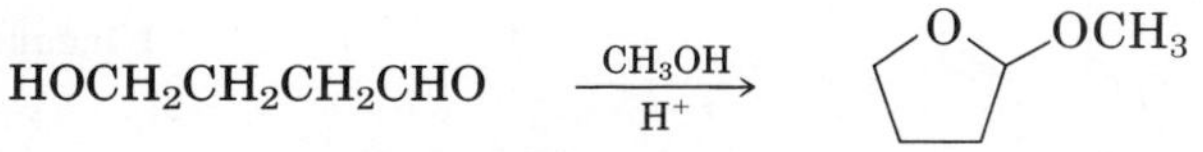

$$HOCH_2CH_2CH_2CHO \xrightarrow[H^+]{CH_3OH} \quad OCH_3$$

**19.40** When crystals of pure $\alpha$-glucose are dissolved in water, isomerization slowly occurs to produce $\beta$-glucose. Propose a mechanism to explain this isomerization.

$\alpha$-Glucose $\rightleftharpoons$ $\beta$-Glucose

**19.41** Give at least four methods for reducing a carbonyl group to a methylene group, $R_2C{=}O \rightarrow R_2CH_2$. What are the advantages and disadvantages of each?

**19.42** Carvone is the major constituent of spearmint oil. What products would you expect from reaction of carvone with the following reagents?

Carvone

(a) $(CH_3)_2Cu^+ Li^-$, then $H_3O^+$
(b) $LiAlH_4$, then $H_3O^+$
(c) $(C_2H_5)_2AlCN$, then $H_3O^+$
(d) $CH_3NH_2$
(e) $C_6H_5MgBr$, then $H_3O^+$
(f) $H_2$, Pd
(g) Jones' reagent
(h) $(C_6H_5)_3\overset{+}{P}{-}\overset{-}{C}HCH_3$
(i) $HSCH_2CH_2SH$, then Raney nickel
(j) $HOCH_2CH_2OH$, HCl

**19.43** Compound A (mol wt = 86) shows an infrared absorption at 1730 $cm^{-1}$ and a very simple $^1H$ NMR spectrum with peaks at 9.7 $\delta$ (1 H, singlet) and 1.2 $\delta$ (9 H, singlet). Propose a structure for compound A.

**19.44** Compound B is isomeric with compound A (Problem 19.43) and shows an infrared peak at 1720 $cm^{-1}$. The $^1H$ NMR spectrum of compound B has peaks at 2.4 $\delta$ (1 H, septet, $J$ = 7 Hz), 2.1 $\delta$ (3 H, singlet), and 1.2 $\delta$ (6 H, doublet, $J$ = 7 Hz). What is the structure of B?

**19.45** How would you synthesize these compounds from cyclohexanone?
(a) 1-Methylcyclohexene
(b) 2-Phenylcyclohexanone
(c) *cis*-1,2-Cyclohexanediol
(d) 1-Cyclohexylcyclohexanol

**19.46** At what position would you expect to observe infrared absorptions for the following molecules?

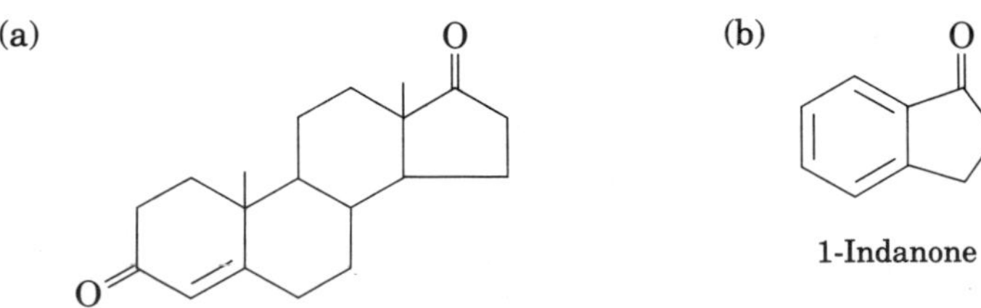

(a) 4-Androstene-3,17 dione

(b) 1-Indanone

(c) 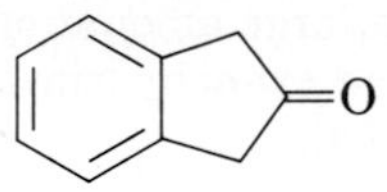 2-Indanone

(d) 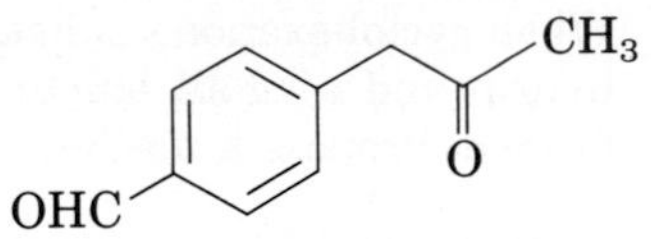 OHC—$C_6H_4$—$CH_2COCH_3$

**19.47** As written, each of the following reaction schemes contains one or more flaws. What is wrong in each case? How would you correct each scheme?

(a) 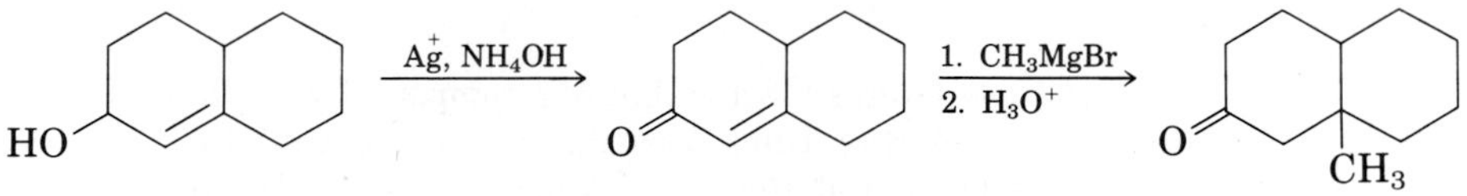 HO-substituted octahydronaphthalene $\xrightarrow{\text{Ag}^+,\ NH_4OH}$ enone $\xrightarrow{\text{1. } CH_3MgBr;\ \text{2. } H_3O^+}$ ketone with angular $CH_3$

(b) $C_6H_5CH{=}CHCH_2OH \xrightarrow{\text{Jones' reagent}} C_6H_5CH{=}CHCHO$

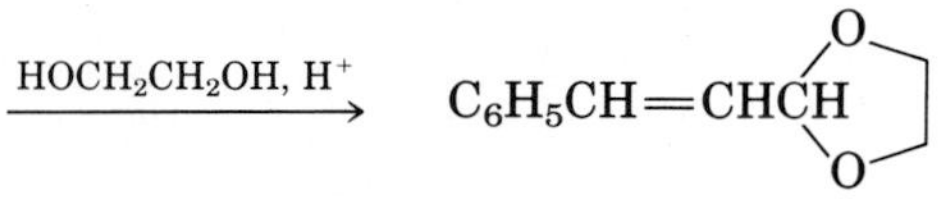

$\xrightarrow{HOCH_2CH_2OH,\ H^+} C_6H_5CH{=}CHCH$(cyclic acetal, $OCH_2CH_2O$)

(c) $CH_3COCH_3 \xrightarrow[KCN]{HCN} CH_3C(OH)(CN)CH_3 \xrightarrow{H_3O^+} CH_3C(OH)(CH_2NH_2)CH_3$

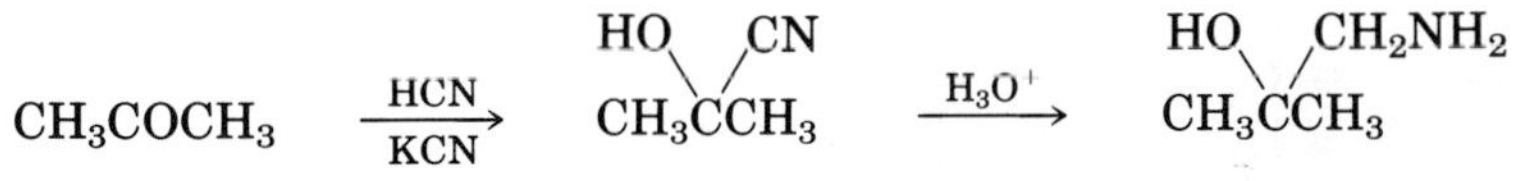

(d) 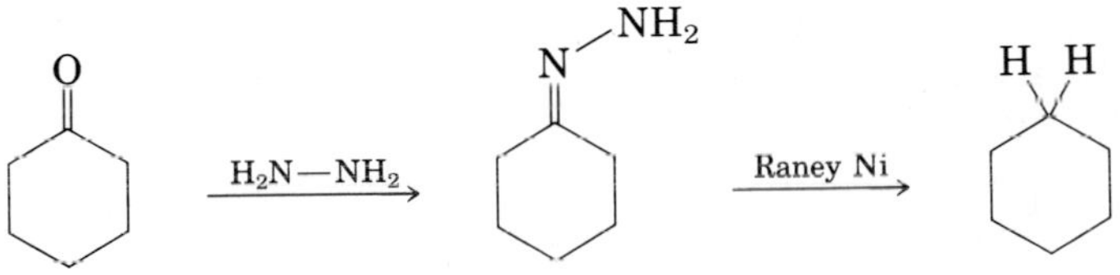 Cyclohexanone $\xrightarrow{H_2N{-}NH_2}$ cyclohexanone hydrazone ($N{-}NH_2$) $\xrightarrow{\text{Raney Ni}}$ cyclohexane (H H)

**19.48** 6-Methyl-5-hepten-2-one is a common constituent of many essential oils, particularly the lemongrass species. How could you synthesize this natural product from methyl 4-oxopentanoate?

$$CH_3\overset{O}{\overset{\|}{C}}CH_2CH_2COOCH_3$$

Methyl 4-oxopentanoate

**19.49** Ketones react with dimethylsulfonium methylide to yield epoxides. Suggest a mechanism for this reaction.

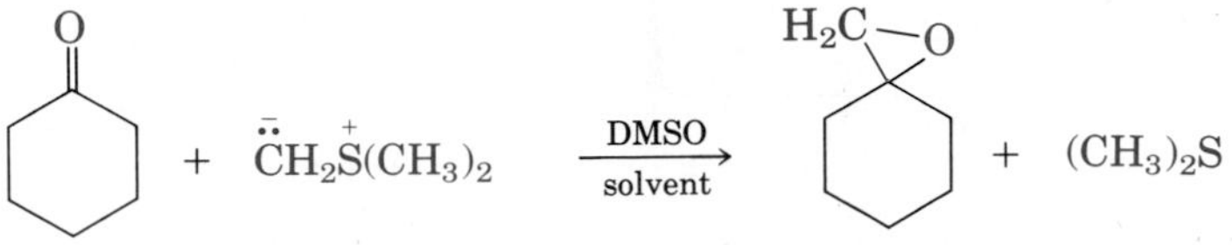

$$\text{Cyclohexanone} + \ddot{C}H_2\overset{+}{S}(CH_3)_2 \xrightarrow[\text{solvent}]{\text{DMSO}} \text{spiro epoxide } (H_2C{-}O) + (CH_3)_2S$$

Dimethylsulfonium methylide

**19.50** When cyclohexanone is heated in the presence of a large amount of acetone cyanohydrin and a small amount of base, cyclohexanone cyanohydrin and acetone are formed. Propose a mechanism for this transformation.

$$\text{(cyclohexanone)} + \underset{CH_3}{H_3C-\overset{OH}{\overset{|}{C}}-CN} \overset{^{-}OH}{\rightleftharpoons} \text{(cyclohexanone cyanohydrin: HO, CN)} + CH_3\overset{O}{\overset{\|}{C}}CH_3$$

**19.51** The NMR spectrum shown is that of a compound with formula $C_9H_{10}O$. How many double bonds and/or rings does this compound contain? If the unknown has an infrared absorption at 1690 cm$^{-1}$, what is a likely structure?

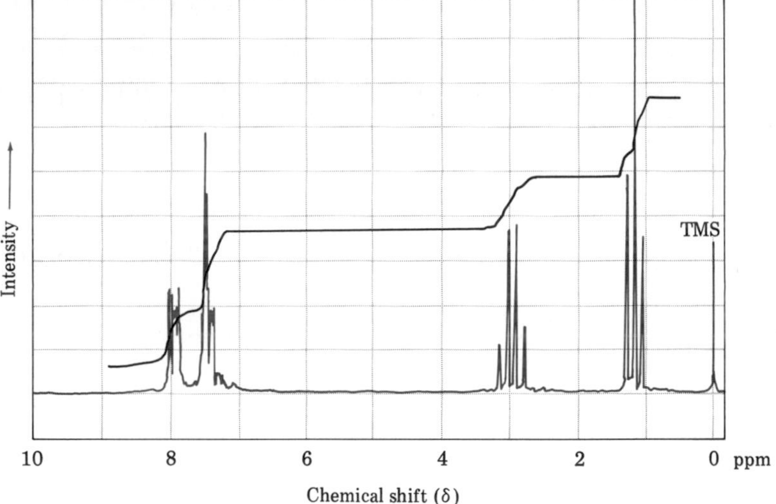


**19.52** The NMR spectrum shown is that of a compound isomeric with the one in Problem 19.51. This isomer has an infrared absorption at 1725 cm$^{-1}$. Propose a suitable structure.

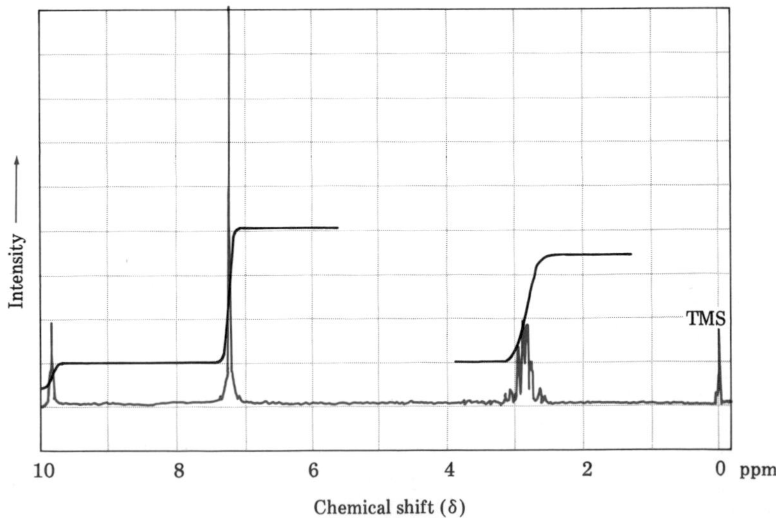

**19.53** Propose structures for molecules that meet the following descriptions, where s = singlet, d = doublet, t = triplet, and q = quartet.

(a) $C_6H_{12}O$;
IR: 1715 $cm^{-1}$;
proton-coupled $^{13}C$ NMR: 8.0 $\delta$ (q), 18.5 $\delta$ (q), 33.5 $\delta$ (t), 40.6 $\delta$ (d), 214.0 $\delta$ (s)

(b) $C_5H_{10}O$;
IR: 1725 $cm^{-1}$;
proton-coupled $^{13}C$ NMR: 22.6 $\delta$ (q), 23.6 $\delta$ (d), 52.8 $\delta$ (t), 202.4 $\delta$ (d)

(c) $C_6H_8O$;
IR: 1680 $cm^{-1}$;
proton-coupled $^{13}C$ NMR: 22.9 $\delta$ (t), 25.8 $\delta$ (t), 38.2 $\delta$ (t), 129.8 $\delta$ (d), 150.6 $\delta$ (d), 198.7 $\delta$ (s)

**19.54** Compound A, $C_8H_{10}O_2$, has an intense infrared absorption at 1750 $cm^{-1}$ and gives the $^{13}C$ NMR spectrum shown. Propose a suitable structure for compound A.

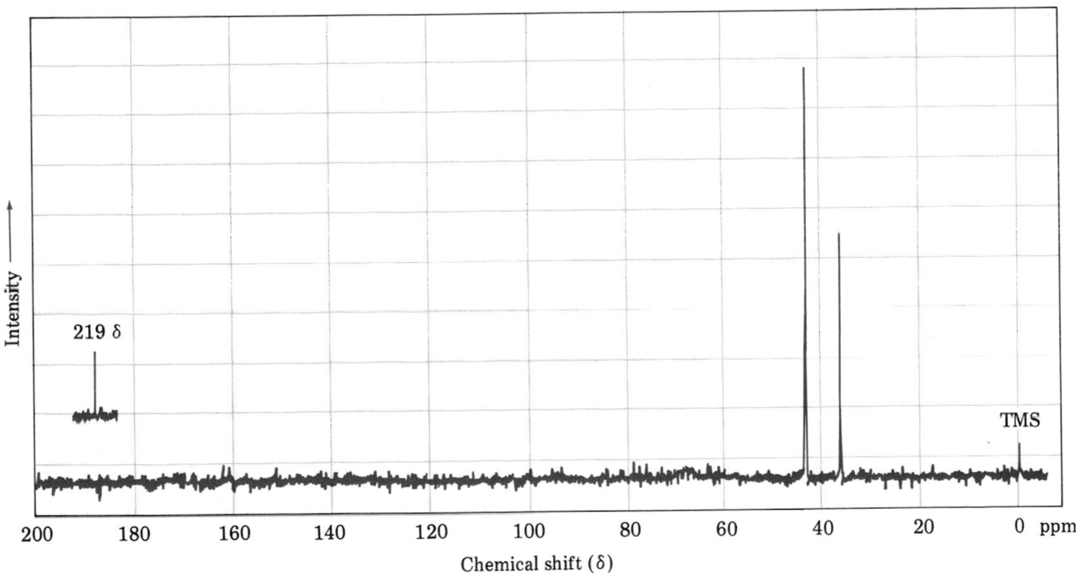


**19.55** The Meerwein–Ponndorf–Verley reaction involves reduction of a ketone by treatment with an excess of aluminum triisopropoxide. The mechanism of the process is closely related to the Cannizzaro reaction, in which a hydride ion serves as leaving group. Formulate a reasonable mechanism.

Cyclohexanone (O) $\xrightarrow[\text{2. } H_3O^+]{\text{1. } [(CH_3)_2CH-O]_3Al}$ Cyclohexanol (HO H) + $CH_3COCH_3$

**19.56** Propose structures for ketones or aldehydes that show the following $^1H$ NMR spectra:

(a) $C_5H_9ClO$;
IR: 1710 $cm^{-1}$

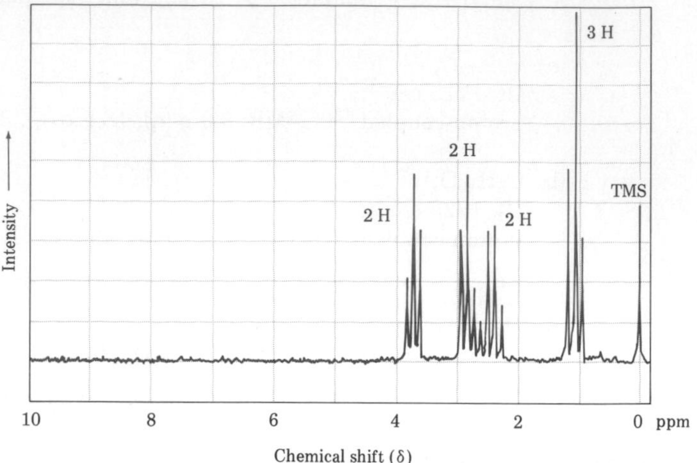


(b) $C_7H_{14}O$;
IR: 1710 $cm^{-1}$

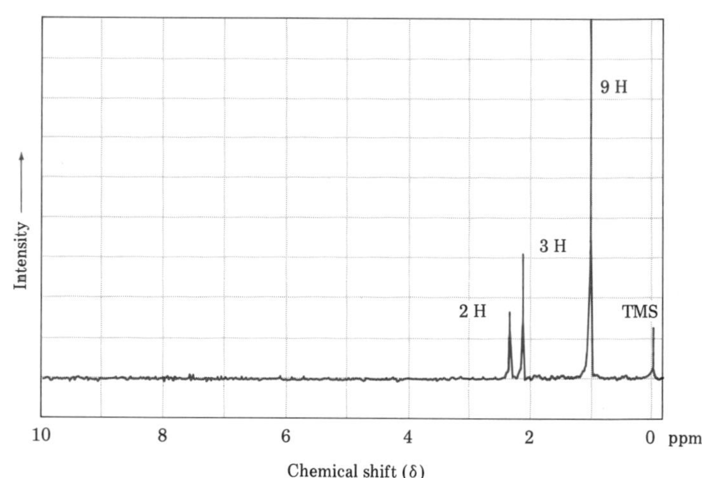


(c) $C_9H_{10}O_2$;
IR: 1695 $cm^{-1}$

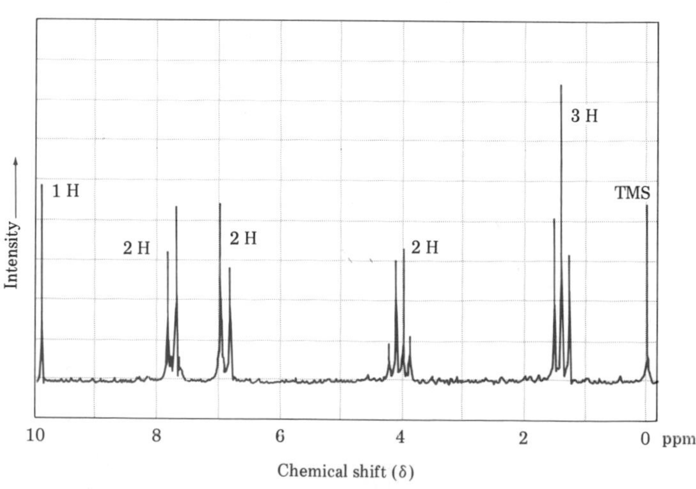

**19.57** Propose structures for ketones or aldehydes that have the following $^1H$ NMR spectra:

(a) $C_{10}H_{12}O$;
IR: 1710 $cm^{-1}$

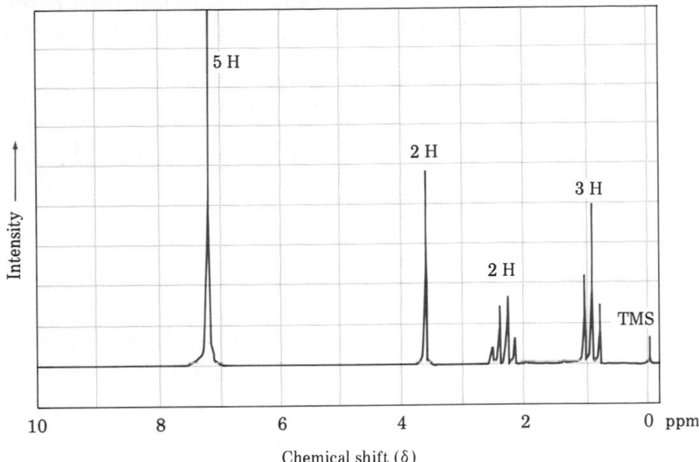


(b) $C_6H_{12}O_3$;
IR: 1715 $cm^{-1}$

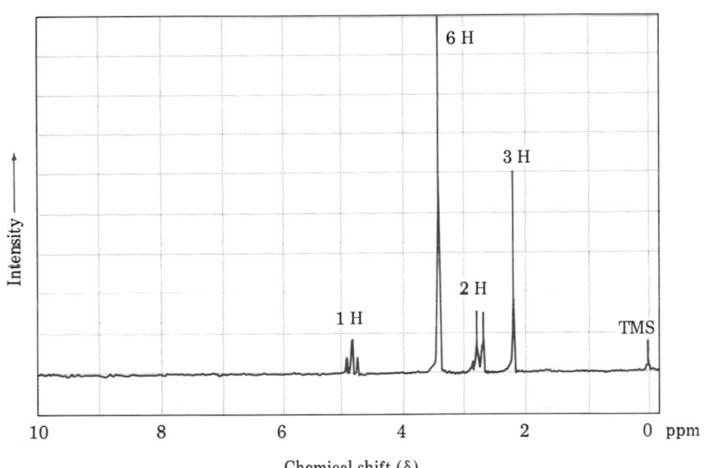


(c) $C_4H_6O$;
IR: 1690 $cm^{-1}$

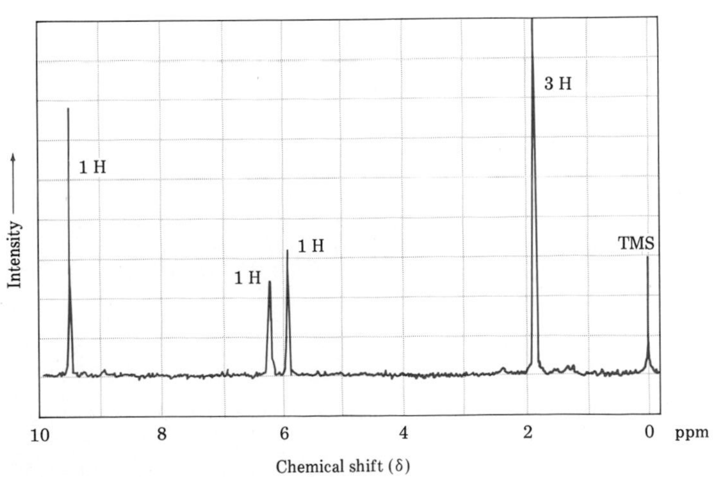

**19.58** Propose a mechanism to account for the formation of 3,5-dimethylpyrazole from hydrazine and 2,4-pentanedione. [*Hint:* Look carefully to see what has happened to each carbonyl carbon in going from starting material to product.]

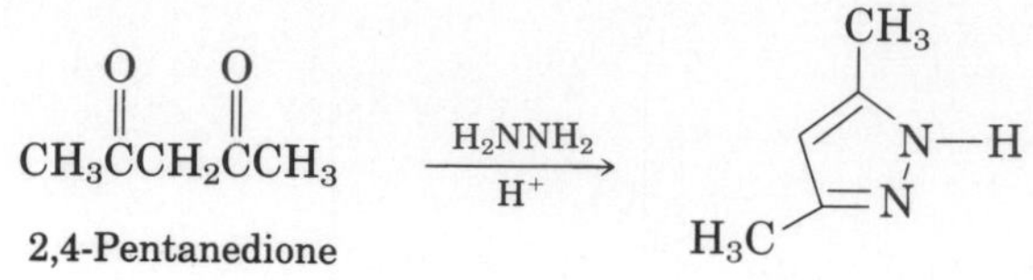

$$\underset{\text{2,4-Pentanedione}}{CH_3\overset{O}{\overset{\|}{C}}CH_2\overset{O}{\overset{\|}{C}}CH_3} \xrightarrow[H^+]{H_2NNH_2}$$

3,5-Dimethylpyrazole

**19.59** In light of your answer to Problem 19.58, propose a mechanism for the formation of 3,5-dimethylisoxazole from hydroxylamine and 2,4-pentanedione.

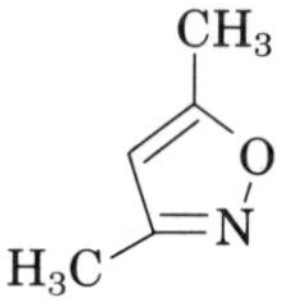

3,5-Dimethylisoxazole

CHAPTER 20

# Carboxylic Acids

Carboxylic acids occupy a central place among acyl derivatives. Not only are they important compounds themselves, they also serve as building blocks for preparing related derivatives such as esters and amides. Among important examples are cholic acid, a major component of human bile, and long-chain aliphatic acids such as oleic acid and linoleic acid, which are biological precursors of fats and other lipids.

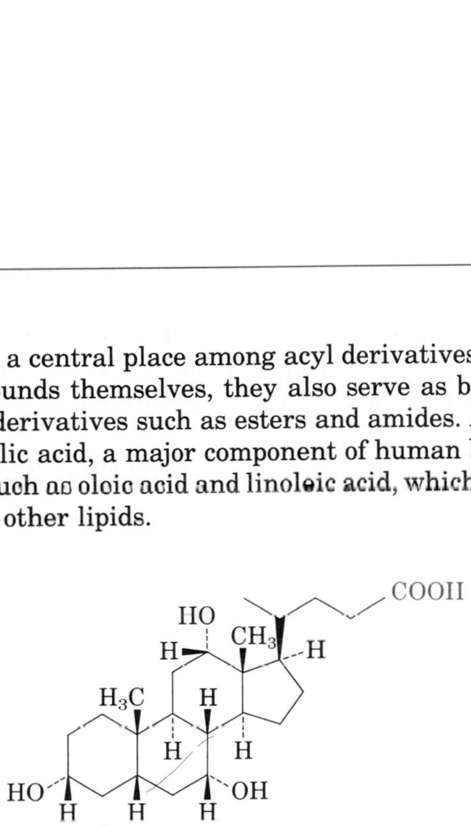

Cholic acid

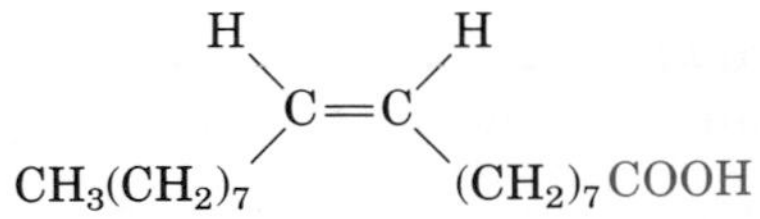

$CH_3(CH_2)_7CH=CH(CH_2)_7COOH$

Oleic acid

$CH_3(CH_2)_4CH=CHCH_2CH=CH(CH_2)_7COOH$

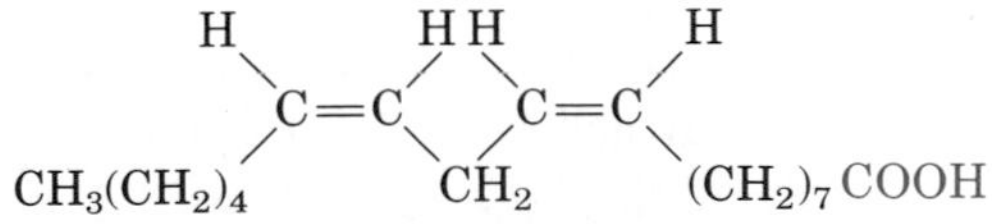

Linoleic acid

Many simple saturated carboxylic acids are also found in nature. For example, acetic acid, $CH_3COOH$, is the chief organic component of vinegar; butanoic acid, $CH_3CH_2CH_2COOH$, is responsible for the rancid odor of sour butter; and hexanoic acid (caproic acid), $CH_3(CH_2)_4COOH$, is partially responsible for the unmistakable aroma of goats and dirty gym socks (Latin *caper*, "goat").

Some 3.3 billion pounds per year of acetic acid are produced industrially for a variety of purposes, including use as a raw material for preparing the vinyl acetate polymer used in paints and adhesives. The industrial method of acetic acid synthesis involves a cobalt acetate-catalyzed air oxidation of acetaldehyde, but this method is not used in the laboratory.

$$CH_3CHO + O_2 \xrightarrow[80°C]{\text{Cobalt acetate}} CH_3COOH + H_2O$$

The Monsanto Company has developed an even more efficient synthesis involving a direct rhodium-catalyzed carbonylation of methanol:

$$CH_3OH + CO \xrightarrow{\text{Rh catalyst}} CH_3COOH$$

## 20.1 Nomenclature of Carboxylic Acids

IUPAC rules allow for two systems of nomenclature, depending on the complexity of the acid molecule. Carboxylic acids that are derived from open-chain alkanes are systematically named by replacing the terminal *-e* of the corresponding alkane name with *-oic acid*. The carboxyl carbon atom is always numbered C1 in this system.

$$\underset{6}{CH_3}\underset{5}{CH_2}\underset{4}{CH_2}\underset{3}{CH_2}\underset{2}{CH_2}\underset{1}{COOH}$$

Hexanoic acid
(Caproic acid)

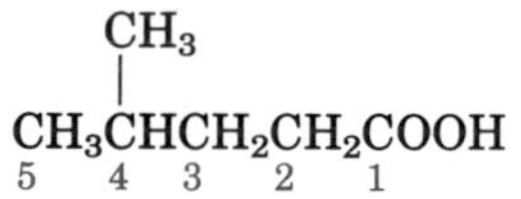

$$\underset{5}{CH_3}\underset{4}{\overset{\overset{CH_3}{|}}{C}H}\underset{3}{CH_2}\underset{2}{CH_2}\underset{1}{COOH}$$

4-Methylpentanoic acid

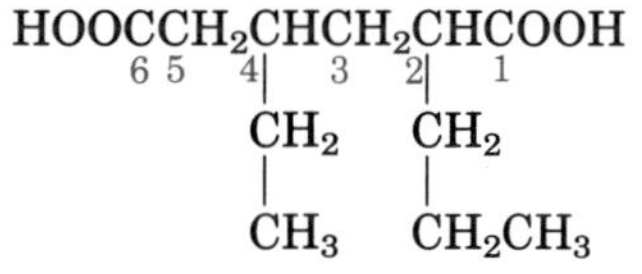

$$HOO\underset{6}{C}\underset{5}{CH_2}\underset{4}{\underset{\underset{\underset{CH_3}{|}}{CH_2}}{C}H}\underset{3}{CH_2}\underset{2}{\underset{\underset{\underset{CH_2CH_3}{|}}{CH_2}}{C}H}\underset{1}{COOH}$$

4-Ethyl-2-propylhexanedioic acid

Alternatively, compounds that have a —COOH group bonded to a ring are named using the suffix *-carboxylic acid*. The carboxylic acid carbon is attached to C1 and is not itself numbered in this system.

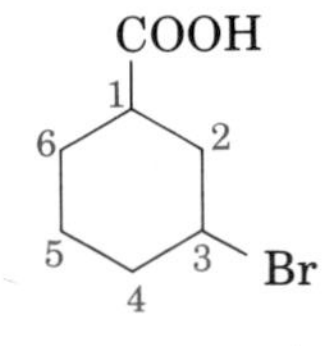

3-Bromocyclohexanecarboxylic acid

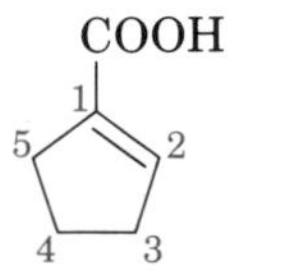

1-Cyclopentenecarboxylic acid

Many carboxylic acids were among the first organic compounds to be isolated and purified. Thus, for historical reasons, IUPAC rules make allowance for a large number of well-entrenched common names, some of which are given in Table 20.1. We will use systematic names in this book, with the exception of formic (methanoic) acid, HCOOH, and acetic (ethanoic) acid, whose names are so well known that it makes little sense to refer to them any other way. Also listed in Table 20.1 are the common names used for acyl groups derived from the parent acids.

**Table 20.1** Some common names of carboxylic acids and acyl groups

| *Carboxylic acid* | | *Acyl group* | |
|---|---|---|---|
| *Structure* | *Name* | *Name* | *Structure* |
| HCOOH | Formic | Formyl | HCO— |
| $CH_3COOH$ | Acetic | Acetyl | $CH_3CO$— |
| $CH_3CH_2COOH$ | Propionic | Propionyl | $CH_3CH_2CO$— |
| $CH_3CH_2CH_2COOH$ | Butyric | Butyryl | $CH_3(CH_2)_2CO$— |
| $CH_3CH_2CH_2CH_2COOH$ | Valeric | Valeryl | $CH_3(CH_2)_3CO$— |
| $(CH_3)_3CCOOH$ | Pivalic | Pivaloyl | $(CH_3)_3CCO$— |
| HOOCCOOH | Oxalic | Oxalyl | —OCCO— |
| $HOOCCH_2COOH$ | Malonic | Malonyl | —$OCCH_2CO$— |
| $HOOCCH_2CH_2COOH$ | Succinic | Succinyl | —$OC(CH_2)_2CO$— |
| $HOOCCH_2CH_2CH_2COOH$ | Glutaric | Glutaryl | —$OC(CH_2)_3CO$— |
| $HOOCCH_2CH_2CH_2CH_2COOH$ | Adipic | Adipoyl | —$OC(CH_2)_4CO$— |
| $H_2C{=}CHCOOH$ | Acrylic | Acryloyl | $H_2C{=}CHCO$— |
| $CH_3CH{=}CHCOOH$ | Crotonic | Crotonoyl | $CH_3CH{=}CHCO$— |
| $H_2C{=}C(CH_3)COOH$ | Methacrylic | Methacryloyl | $H_2C{=}C(CH_3)CO$— |
| $HC{\equiv}CCOOH$ | Propiolic | Propioloyl | $HC{\equiv}CCO$— |
| $HOOCCH{=}CHCOOH$ | *cis*-Maleic | Maleoyl | —$OCCH{=}CHCO$— |
| | *trans*-Fumaric | Fumaroyl | |
| 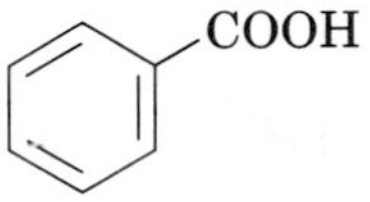 $C_6H_5$COOH (benzene ring with COOH) | Benzoic | Benzoyl | 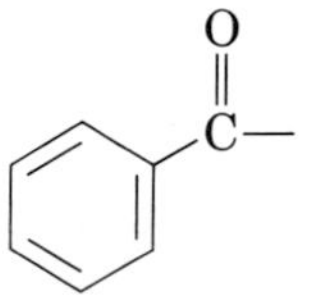 $C_6H_5$C(=O)— (benzene ring with C(=O)—) |
| 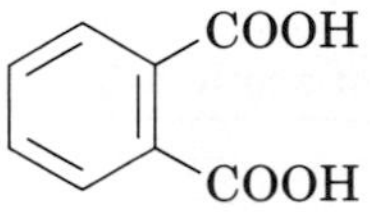 $C_6H_4$(COOH)$_2$ (benzene ring with two ortho COOH) | Phthalic | Phthaloyl | 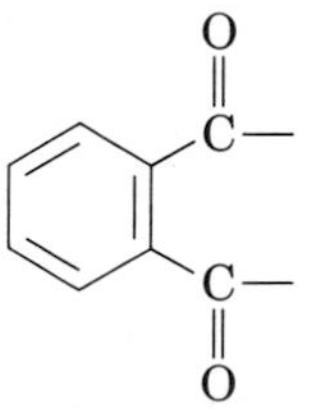 $C_6H_4$(C(=O)—)$_2$ (benzene ring with two ortho C(=O)—) |

PROBLEM

**20.1** Provide IUPAC names for these compounds:

(a) $(CH_3)_2CHCH_2COOH$

(b) $CH_3CHBrCH_2CH_2COOH$

(c) $CH_3CH{=}CHCH{=}CHCOOH$

(d) COOH
$CH_3CH_2\overset{|}{C}HCH_2CH_2CH_3$

(e) COOH, H, H, COOH

(f) $CH_3$, COOH

PROBLEM

**20.2** Draw structures corresponding to these IUPAC names:

(a) 2,3-Dimethylhexanoic acid

(b) 4-Methylpentanoic acid

(c) *trans*-1,2-Cyclobutanedicarboxylic acid

(d) *o*-Hydroxybenzoic acid

(e) (9*Z*,12*Z*)-9,12-Octadecadienoic acid

## 20.2 Structure and Physical Properties of Carboxylic Acids

Since carboxylic acid functional groups are structurally related to both ketones and alcohols, we might expect to see some familiar properties. Carboxylic acids are indeed similar to both ketones and alcohols in some ways, though there are also major differences. Like ketones, the carboxyl carbon is $sp^2$ hybridized. Carboxylic acid groups are therefore planar, with C-C-O and O-C-O bond angles of approximately 120°. Figure 20.1 shows the structure of acetic acid, and its physical parameters are given in Table 20.2.

**Table 20.2** Physical parameters of acetic acid

| *Bond angle (degrees)* | | *Bond length (Å)* | |
|---|---|---|---|
| C-C-O | 119 | C—C | 1.52 |
| C-C-OH | 119 | C=O | 1.25 |
| O-C-OH | 122 | C—OH | 1.31 |

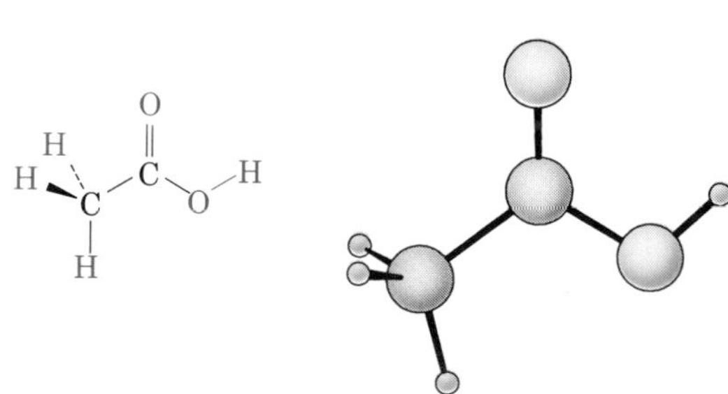


**Figure 20.1** Structure of acetic acid

Like alcohols, carboxylic acids are strongly associated because of hydrogen bonding between molecules. Studies have shown that most carboxylic acids exist as dimers held together by two hydrogen bonds:

```
       O····H—O
      //        \
R—C              C—R
      \         //
       O—H····O
```

A carboxylic acid dimer

This strong hydrogen bonding has a noticeable effect on boiling points, making carboxylic acids much higher boiling than the corresponding alcohols. Table 20.3 lists the observed properties of some common acids.

**Table 20.3** Physical constants of some carboxylic acids

| *Structure* | *Name* | *Melting point (°C)* | *Boiling point (°C)* |
|---|---|---|---|
| $HCOOH$ | Formic | 8.4 | 100.5 |
| $CH_3COOH$ | Acetic | 16.6 | 118 |
| $CH_3CH_2COOH$ | Propanoic | −22 | 141 |
| $CH_3CH_2CH_2COOH$ | Butanoic | −4.2 | 163 |
| $FCH_2COOH$ | Fluoroacetic | 35.2 | 165 |
| $BrCH_2COOH$ | Bromoacetic | 50 | 208 |
| $HOCH_2COOH$ | Glycolic | 80 | Decomposes |
| $H_2C{=}CHCOOH$ | Propenoic | 13 | 141 |
| $C_6H_5COOH$ | Benzoic | 122.4 | 249 |
| $HOOCCOOH$ | Oxalic | 189.5 | Decomposes |
| $HOOCCH_2COOH$ | Malonic | 135 | Decomposes |
| $HOOCCH_2CH_2COOH$ | Succinic | 188 | Decomposes |
| (*Z*)-$HOOCCH{=}CHCOOH$ | Maleic | 139 | Decomposes |

## 20.3 Dissociation of Carboxylic Acids

As their name implies, carboxylic acids are acidic. They therefore react with bases such as sodium hydroxide and sodium bicarbonate to give metal carboxylate salts. Although carboxylic acids with more than six carbons are only slightly soluble in water, alkali metal salts of carboxylic acids are generally quite water soluble because of their ionic nature. It's often possible to take advantage of this solubility to purify acids by extracting their salts into aqueous base, then reacidifying and extracting the pure acid back into an organic solvent.

$$R{-}\overset{\overset{\displaystyle O}{\|}}{C}{-}O{-}H + NaOH \xrightarrow{H_2O} R{-}\overset{\overset{\displaystyle O}{\|}}{C}{-}O^- \; Na^+ + H_2O$$

Like other Brønsted–Lowry acids (discussed in Section 2.6), carboxylic acids dissociate slightly in dilute aqueous solution to give $H_3O^+$ and carboxylate anion, $RCOO^-$.

$$RCOOH + H_2\ddot{O}: \rightleftharpoons RCOO^- + H_3O^+$$

As with all acids, we can define an acidity constant, $K_a$:

$$K_a = \frac{[RCOO^-][H_3O^+]}{[RCOOH]} \qquad \text{and} \qquad pK_a = -\log K_a$$

For most carboxylic acids, the acidity constant $K_a$ is on the order of $10^{-5}$. Acetic acid, for example, has $K_a = 1.8 \times 10^{-5}$, which corresponds to a $pK_a$ of 4.72. In practical terms, $K_a$ values near $10^{-5}$ mean that only about 1% of the molecules in a 0.1*M* solution are dissociated, as opposed to the 100% dissociation found with strong mineral acids such as HCl and $H_2SO_4$.

Although much weaker than mineral acids, carboxylic acids are nevertheless much stronger acids than alcohols. For example, the $K_a$ for ethanol is approximately $10^{-16}$, making ethanol a weaker acid than acetic acid by a factor of $10^{11}$.

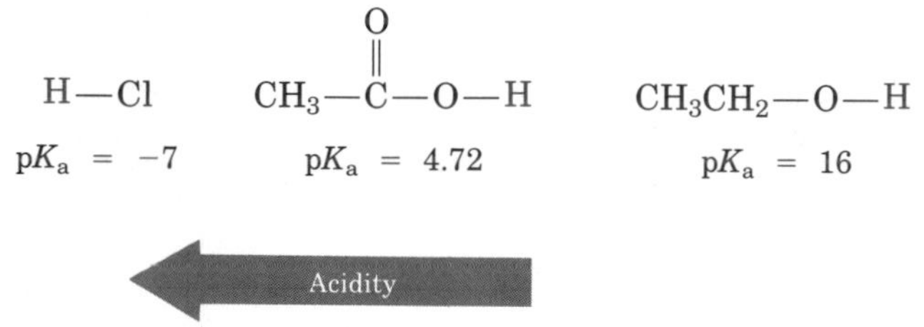

| H—Cl | $CH_3$—C(=O)—O—H | $CH_3CH_2$—O—H |
|---|---|---|
| $pK_a = -7$ | $pK_a = 4.72$ | $pK_a = 16$ |

← Acidity

Why do carboxylic acids dissociate to a greater extent than alcohols? The easiest way to answer this question is to look at the relative stability of carboxylate anions versus alkoxide anions. Alkoxides are oxygen anions in which the negative charge is localized on a single electronegative atom:

$$CH_3CH_2—\ddot{O}—H + H_2O \rightleftharpoons CH_3CH_2—\ddot{O}:^- + H_3O^+$$

Alcohol — Unstabilized alkoxide ion

Carboxylates are also oxygen anions, but the negative charge is delocalized over both oxygen atoms, resulting in stabilization of the ion. In resonance terms (Section 10.7), a carboxylate ion is a stabilized resonance hybrid of two equivalent Kekulé structures:

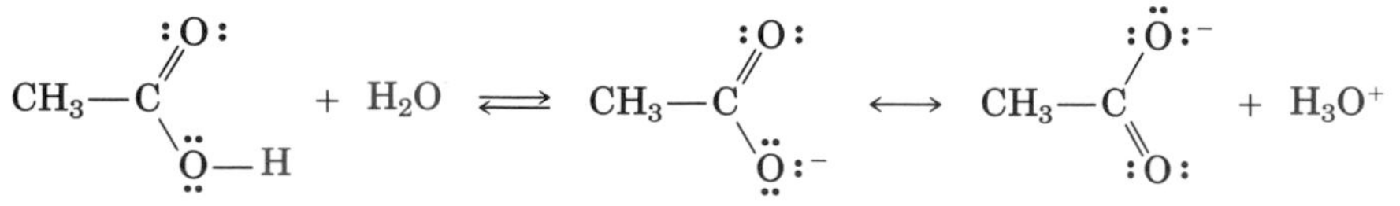

$$CH_3—C(=\ddot{O}:)—\ddot{O}—H + H_2O \rightleftharpoons CH_3—C(=\ddot{O}:)—\ddot{O}:^- \longleftrightarrow CH_3—C(—\ddot{O}:^-)=\ddot{O}: + H_3O^+$$

Carboxylic acid — Resonance-stabilized carboxylate ion (two equivalent resonance forms)

Since a carboxylate ion is more stable than an alkoxide ion, it is lower in energy and is present in greater amount at equilibrium. The situation is shown in Figure 20.2 using reaction energy diagrams. Dissociation of a carboxylic acid has a smaller $\Delta G°$ than dissociation of an alcohol, leading to a larger equilibrium constant, $K_a$.

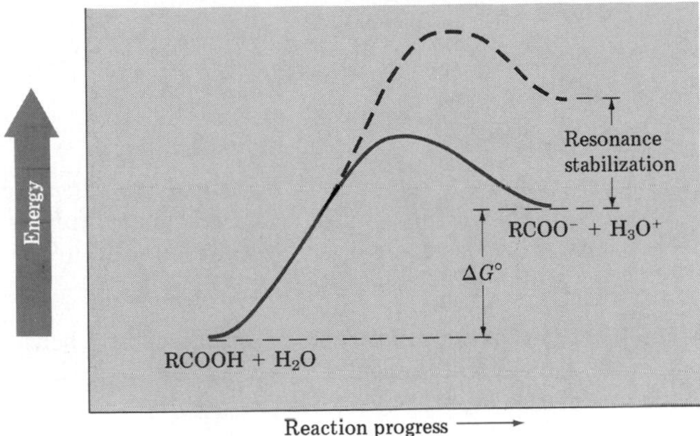


**Figure 20.2** A reaction energy diagram for the dissociation of an alcohol (dashed curve) and a carboxylic acid (solid curve): Resonance stabilization of the carboxylate anion lowers $\Delta G°$ for dissociation of the acid, leading to a larger $K_a$. (The starting energy levels of alcohol and acid are shown at the same point for ease of comparison.)

We can't really draw an accurate representation of the carboxylate resonance hybrid using Kekulé structures, but an orbital picture of acetate ion is helpful in making it clear that the carbon–oxygen bonds are equivalent and that each is intermediate between single and double bonds (Figure 20.3). The *p* orbital on the carboxylate carbon atom overlaps equally well with *p* orbitals from *both* oxygens, and the four *p* electrons are delocalized throughout the three-center pi electron system.

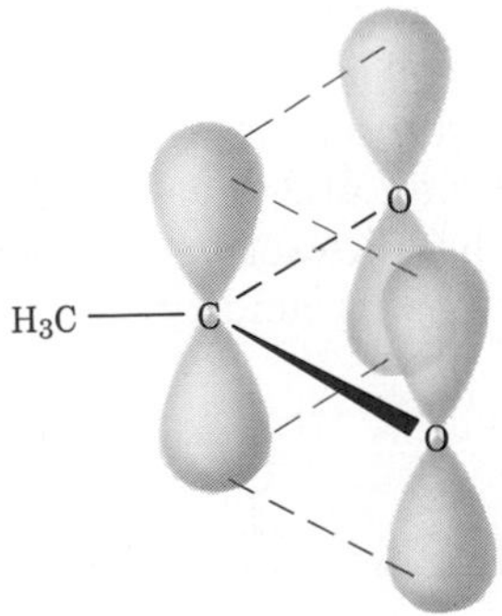


**Figure 20.3** An orbital picture of acetate ion, showing the equivalence of the two oxygen atoms

Physical evidence for the equivalence of the two carboxylate oxygens has been provided by X-ray studies on sodium formate. Both carbon–oxygen bonds are 1.27 Å in length, midway between the C═O double bond (1.20 Å) and C—O single bond (1.34 Å) of formic acid.

1.27Å
H—C(O)(O)$^{-}$ $Na^+$

Sodium formate

H—C(═O)—O—H 1.20Å 1.34Å

Formic acid

**PROBLEM**

**20.3** Use the expression $\Delta G° = -2.303\ RT \log K_a$ to calculate values of $\Delta G°$ for the dissociation of ethanol ($pK_a = 16.0$) and acetic acid ($pK_a = 4.72$) at 300 K (27°C). The gas constant $R$ has the value 1.98 cal/mol · K.

**PROBLEM**

**20.4** Assume that you had a mixture of naphthalene and benzoic acid that you wished to separate. How might you take advantage of the acidity of one component in the mixture to effect a separation?

**PROBLEM**

**20.5** The $K_a$ for dichloroacetic acid is $5.5 \times 10^{-2}$. Approximately what percentage of the acid is dissociated in a 0.1$M$ aqueous solution?

## 20.4 Substituent Effects on Acidity

The $pK_a$ values listed in Table 20.4 indicate a considerable difference in acidities for different carboxylic acids. For example, trifluoroacetic acid ($K_a = 0.59$) is more than 32,000 times as strong as acetic acid ($K_a = 1.8 \times 10^{-5}$). How can we account for such differences?

Since the dissociation of a carboxylic acid is an equilibrium reaction, any factor that stabilizes the carboxylate anion product relative to undissociated carboxylic acid should drive the equilibrium toward increased dissociation and result in increased acidity. Conversely, any factor that destabilizes carboxylate relative to undissociated acid should result in decreased acidity. For example, an electron-*withdrawing* group attached to the carboxyl should inductively withdraw electron density, thus stabilizing the carboxylate anion and increasing acidity. An electron-*donating* group, however, should have exactly the opposite effect by destabilizing the carboxylate and decreasing acidity.

Electron-withdrawing group stabilizes carboxylate and strengthens acid

Electron-donating group destabilizes carboxylate and weakens acid

**Table 20.4** Acidity of some carboxylic acids

| *Structure* | $K_a$ | p$K_a$ | |
|---|---|---|---|
| HCl (hydrochloric acid)[a] | ($10^7$) | (−7) | |
| $F_3CCOOH$ | 0.59 | 0.23 | |
| $Cl_3CCOOH$ | 0.23 | 0.64 | Strong acid |
| $Cl_2CHCOOH$ | $5.5 \times 10^{-2}$ | 1.26 | ↑ |
| $FCH_2COOH$ | $2.6 \times 10^{-3}$ | 2.59 | |
| $ClCH_2COOH$ | $1.4 \times 10^{-3}$ | 2.85 | |
| $BrCH_2COOH$ | $1.3 \times 10^{-3}$ | 2.89 | |
| $ICH_2COOH$ | $7.5 \times 10^{-4}$ | 3.12 | |
| HCOOH | $1.77 \times 10^{-4}$ | 3.75 | |
| $HOCH_2COOH$ | $1.5 \times 10^{-4}$ | 3.83 | |
| $ClCH_2CH_2COOH$ | $1.04 \times 10^{-4}$ | 3.98 | |
| $C_6H_5COOH$ | $6.46 \times 10^{-5}$ | 4.19 | |
| $H_2C{=}CHCOOH$ | $5.6 \times 10^{-5}$ | 4.25 | |
| $C_6H_5CH_2COOH$ | $4.9 \times 10^{-5}$ | 4.31 | Weak acid |
| $CH_3COOH$ | $1.8 \times 10^{-5}$ | 4.72 | |
| $CH_3CH_2COOH$ | $1.34 \times 10^{-5}$ | 4.87 | |
| $CH_3CH_2OH$ (ethanol)[a] | ($10^{-16}$) | (16) | |

[a]Values for hydrochloric acid and ethanol are shown for reference.

The p$K_a$ data of Table 20.4 show exactly the predicted effect. Highly electronegative substituents such as the halogens tend to make the carboxylate anion more stable by inductively withdrawing electrons. Thus, fluoroacetic, chloroacctic, bromoacctic, and iodoacetic acids are all stronger than acetic acid by factors of 50–150. Introduction of two electronegative substituents makes dichloroacetic acid some 3000-fold stronger than acetic acid, and introduction of three substituents makes trichloroacctic acid more than 12,000 times stronger (Figure 20.4).

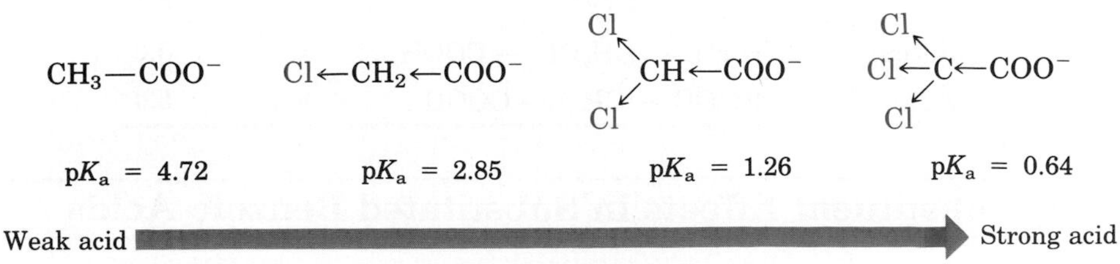

$CH_3{-}COO^-$    $Cl{\leftarrow}CH_2{\leftarrow}COO^-$    $Cl_2CH{\leftarrow}COO^-$    $Cl_3C{\leftarrow}COO^-$

p$K_a$ = 4.72    p$K_a$ = 2.85    p$K_a$ = 1.26    p$K_a$ = 0.64

Weak acid → Strong acid

**Figure 20.4** Relative strengths of chlorosubstituted acetic acids

Since inductive effects are strongly dependent on distance, the effect of halogen substitution decreases as the substituent is moved farther from the carboxyl. The chlorobutanoic acids show clearly what happens as the electronegative substituent is moved successively farther from the carbonyl group (Table 20.5). 2-Chlorobutanoic acid has a p$K_a$ of 2.86, the 3-substituted acid has a p$K_a$ of 4.05, and the 4-substituted acid, with a p$K_a$ of 4.52, has an acidity similar to that of butanoic acid itself.

**Table 20.5** Acidity of chlorosubstituted butanoic acids

| *Structure* | $K_a$ | p$K_a$ |
|---|---|---|
| $CH_3CH_2CH(Cl)COOH$ | $1.39 \times 10^{-3}$ | 2.86 |
| $CH_3CH(Cl)CH_2COOH$ | $8.9 \times 10^{-5}$ | 4.05 |
| $ClCH_2CH_2CH_2COOH$ | $3 \times 10^{-5}$ | 4.52 |
| $CH_3CH_2CH_2COOH$ | $1.5 \times 10^{-5}$ | 4.82 |

PROBLEM

**20.6** Rank the acids in the following groups in order of increasing acidity, without looking at a table of p$K_a$ values:
(a) $CH_3CH_2COOH$, $BrCH_2COOH$, $FCH_2COOH$
(b) Benzoic acid, *p*-nitrobenzoic acid, *p*-methoxybenzoic acid
(c) $CH_3CH_2OH$, $CH_3CH_2NH_2$, $CH_3CH_2COOH$

PROBLEM

**20.7** Dicarboxylic acids have two dissociation constants: one for the initial dissociation into a monoanion and one for the second dissociation into a dianion. For oxalic acid, HOOC—COOH, the first ionization constant has a p$K_1$ of 1.2, and the second ionization constant has a p$K_2$ of 4.2? Why is the second carboxyl group so much less acidic than the first?

PROBLEM

**20.8** Shown here are some p$K_a$ data for simple dibasic acids. How do you account for the fact that the difference between the first and second ionization constants decreases with increasing distance between the carboxyl groups?

| *Name* | *Structure* | p$K_1$ | p$K_2$ |
|---|---|---|---|
| Oxalic | HOOC—COOH | 1.2 | 4.2 |
| Succinic | HOOC—$CH_2CH_2$—COOH | 4.2 | 5.6 |
| Adipic | HOOC—$(CH_2)_4$—COOH | 4.4 | 5.4 |

## 20.5 Substituent Effects in Substituted Benzoic Acids

We saw during the discussion of electrophilic aromatic substitution (Section 16.5) that substituents on the aromatic ring play a large role in determining reactivity. Aromatic rings with electron-donating groups are activated toward further electrophilic substitution, and aromatic rings with electron-withdrawing groups are deactivated. Exactly the same effects are noticed on the acidity of substituted benzoic acids (Table 20.6).

**Table 20.6** Substituent effects on acidity of para-substituted benzoic acids

Y—$C_6H_4$—COOH

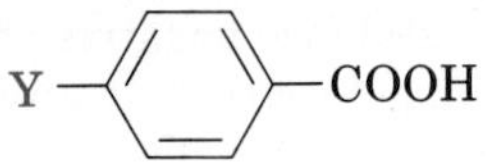

| | Y | $K_a$ | $pK_a$ | |
|---|---|---|---|---|
| Weak acid | —OH | $2.8 \times 10^{-5}$ | 4.55 | Activating groups |
| | —$OCH_3$ | $3.5 \times 10^{-5}$ | 4.46 | |
| | —$CH_3$ | $4.3 \times 10^{-5}$ | 4.34 | |
| ↓ | —H | $6.46 \times 10^{-5}$ | 4.19 | |
| | —Br | $1.1 \times 10^{-4}$ | 3.96 | Deactivating groups |
| | —Cl | $1.1 \times 10^{-4}$ | 3.96 | |
| Strong acid | —CHO | $1.8 \times 10^{-4}$ | 3.75 | |
| | —CN | $2.8 \times 10^{-4}$ | 3.55 | |
| | —$NO_2$ | $3.9 \times 10^{-4}$ | 3.41 | |

As Table 20.6 shows, electron-withdrawing (deactivating) groups increase acidity by stabilizing the carboxylate anion, and electron-donating (activating) groups decrease acidity by destabilizing the carboxylate anion. Thus, an activating group such as *p*-methoxy decreases the acidity of benzoic acid, but a deactivating group such as *p*-nitro increases the acidity.

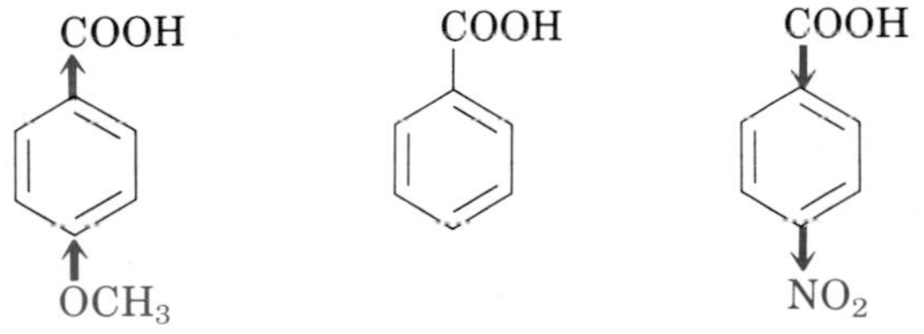

COOH / $OCH_3$ — *p*-Methoxybenzoic acid ($pK_a$ = 4.46)

COOH — Benzoic acid ($pK_a$ = 4.19)

COOH / $NO_2$ — *p*-Nitrobenzoic acid ($pK_a$ = 3.41)

Since it's much easier to measure the acidity of a substituted benzoic acid than it is to determine the relative electrophilic reactivity of a substituted benzene, the correlation between the two effects can be valuable in predicting reactivity. If we want to know the effect of a certain substituent on electrophilic reactivity, we can simply find the acidity of the corresponding benzoic acid.

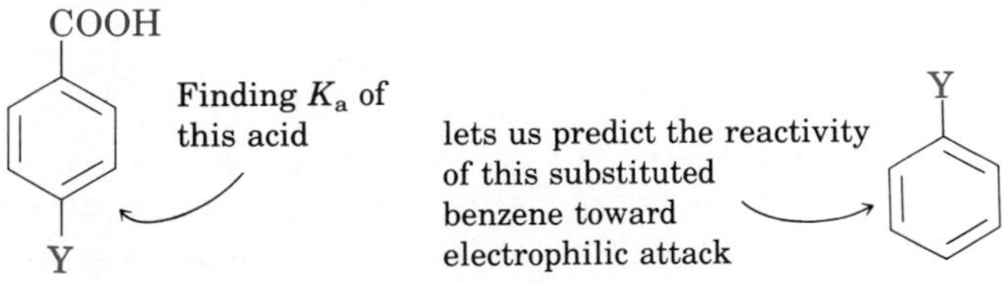

COOH / Y — Finding $K_a$ of this acid

lets us predict the reactivity of this substituted benzene toward electrophilic attack — Y

PRACTICE PROBLEM

The $pK_a$ of *p*-(trifluoromethyl)benzoic acid is 3.6. Would you expect the trifluoromethyl substituent to be an activating or deactivating group in the Friedel–Crafts reaction?

*Solution* A $pK_a$ of 3.6 means that *p*-(trifluoromethyl)benzoic acid is stronger than benzoic acid, whose $pK_a$ is 4.19. Thus, the trifluoromethyl substituent is favoring dissociation by helping to stabilize negative charge. Trifluoromethyl must therefore be an electron-withdrawing, deactivating group.

PROBLEM

**20.9** The $pK_a$ of *p*-cyclopropylbenzoic acid is 4.45. Is cyclopropylbenzene likely to be more reactive or less reactive than benzene toward electrophilic bromination? Explain.

PROBLEM

**20.10** Rank the compounds in the following groups in order of increasing acidity. Do not look at a table of $pK_a$ data to help with your answer.
(a) Benzoic acid, *p*-methylbenzoic acid, *p*-chlorobenzoic acid
(b) *p*-Nitrobenzoic acid, acetic acid, benzoic acid

## 20.6 Preparation of Carboxylic Acids

We've already seen most of the common methods for preparing carboxylic acids, but let's review them briefly:

1. Oxidation of substituted alkylbenzenes with potassium permanganate or sodium dichromate gives substituted benzoic acids (Section 16.11). Both primary and secondary alkyl groups can be oxidized in this manner, but tertiary groups are not affected.

$CH_3$, $NO_2$ $\xrightarrow[H_2O,\ 95°C]{KMnO_4}$ $COOH$, $NO_2$

*p*-Nitrotoluene — *p*-Nitrobenzoic acid (88%)

2. Oxidative cleavage of alkenes gives carboxylic acids if the alkene has at least one vinylic hydrogen (Section 7.8). The reaction can be carried out with sodium dichromate, potassium permanganate, or ozone.

$$CH_3(CH_2)_7CH{=}CH(CH_2)_7COOH \xrightarrow[H_2O,\ K_2CO_3]{KMnO_4} CH_3(CH_2)_7COOH + HOOC(CH_2)_7COOH$$

Oleic acid → Nonanoic acid + Nonanedioic acid

3. Oxidation of primary alcohols and aldehydes yields carboxylic acids (Sections 17.9 and 19.5). Primary alcohols are often oxidized with Jones' reagent ($CrO_3$, $H_2O$, $H_2SO_4$), and aldehydes are oxidized either with Jones' reagent or with basic silver oxide (Tollens' reagent). Both oxidations take place rapidly and in high yield.

$$\underset{\text{1-Decanol}}{CH_3(CH_2)_8CH_2OH} \xrightarrow[\text{reagent}]{\text{Jones'}} \underset{\text{Decanoic acid (93\%)}}{CH_3(CH_2)_8COOH}$$

$$\underset{\text{Hexanal}}{CH_3(CH_2)_4CHO} \xrightarrow[\text{reagent}]{\text{Tollens'}} \underset{\text{Hexanoic acid (85\%)}}{CH_3(CH_2)_4COOH}$$

## HYDROLYSIS OF NITRILES

Nitriles, R—C≡N, can be hydrolyzed by strong aqueous acid or base to yield carboxylic acids. Since nitriles themselves are most often prepared by $S_N2$ reaction between an alkyl halide and cyanide ion, the two-step sequence of cyanide displacement followed by nitrile hydrolysis is an excellent method for preparing carboxylic acids from alkyl halides.

$$RCH_2Br \xrightarrow[(S_N2)]{Na^+\ {}^-CN} RCH_2C{\equiv}N \xrightarrow{H_3O^+} RCH_2COOH + NH_3$$

The method works best with primary halides, since competitive E2 elimination reactions can occur when secondary and tertiary alkyl halides are used. Nevertheless, some unhindered secondary halides react well. A good example of the reaction occurs in the commercial synthesis of the antiarthritic drug fenoprofen. Note that this method yields a carboxylic acid product having one more carbon than the starting alkyl halide.

$$C_6H_5{-}O{-}C_6H_4{-}\underset{|\atop Br}{CHCH_3} \xrightarrow[\text{2. }{}^-OH/H_2O\text{ 3. }H_3O^+]{\text{1. NaCN}} C_6H_5{-}O{-}C_6H_4{-}\underset{|\atop CH_3}{CHCOOH}$$

Fenoprofen
(an antiarthritic agent)

## CARBOXYLATION OF GRIGNARD REAGENTS

Yet a further method for preparing carboxylic acids is the carboxylation of Grignard reagents. Both Grignard and organolithium reagents react with carbon dioxide to yield carboxylate salts, which can be protonated to give carboxylic acids. This **carboxylation** reaction is carried out either by pouring the Grignard reagent over dry ice (solid $CO_2$), or by bubbling a stream of dry $CO_2$ through the Grignard reagent solution. Grignard carboxylation generally gives good yields of acids from alkyl halides but is clearly limited in use to only those alkyl halides that can form Grignard reagents in the first place (Section 17.7).

1-Bromo-2,4,6-trimethylbenzene $\xrightarrow[\text{Ether}]{\text{Mg}}$ (2,4,6-trimethylphenyl)magnesium bromide $\xrightarrow[\text{2. } H_3O^+]{\text{1. } CO_2\text{, ether}}$ 2,4,6-Trimethylbenzoic acid (87%)

$$CH_3CH_2CH_2CH_2Cl \xrightarrow[\text{Ether}]{\text{Mg}} CH_3CH_2CH_2CH_2MgCl \xrightarrow[\text{2. } H_3O^+]{\text{1. } CO_2\text{, ether}} CH_3CH_2CH_2CH_2COOH$$

1-Chlorobutane → Pentanoic acid (73%)

The mechanism of Grignard carboxylation is similar to that of other Grignard reactions (Section 19.10). The organomagnesium halide adds to one of the C=O bonds of carbon dioxide in a typical nucleophilic addition reaction. Protonation of the carboxylate by addition of aqueous HCl then gives the free carboxylic acid product:

$$\overset{+}{X}Mg\ \overset{-}{R}{:} + O{=}C{=}O \longrightarrow R{-}C(=O)O^-\ \overset{+}{M}gX \xrightarrow{H_3O^+} RCOOH$$

Recall

$$\overset{+}{X}Mg\ \overset{-}{R}{:} + R_2C{=}O \longrightarrow R_2C(R){-}O^-\ \overset{+}{M}gX \xrightarrow{H_3O^+} R_2C(R){-}OH$$

**PROBLEM**

**20.11** We have seen two methods of converting an alkyl halide into a carboxylic acid having one more carbon atom: (1) substitution with cyanide ion followed by hydrolysis and (2) formation of a Grignard reagent followed by carboxylation. What are the strengths and weaknesses of the two methods? Under what circumstances might one method be better than the other?

**PROBLEM**

**20.12** In light of your answer to Problem 20.11, what methods would you use to prepare the following carboxylic acids from organohalides?

(a) Benzoic acid from bromobenzene
(b) $(CH_3)_3CCOOH$ from $(CH_3)_3CCl$
(c) $CH_3CH_2CH_2COOH$ from $CH_3CH_2CH_2Br$

## 20.7 Reactions of Carboxylic Acids

We commented earlier in this chapter that carboxylic acids are structurally similar to both alcohols and ketones. As you might expect, there are also chemical similarities. Like alcohols, carboxylic acids can be deprotonated to give anions, which are good nucleophiles in $S_N2$ reactions. Like ketones, carboxylic acids can undergo nucleophilic attack on their carbonyl group.

In addition, carboxylic acids undergo other reactions not characteristic of alcohols or ketones. Figure 20.5 shows some of the general types of reactions of carboxylic acids.

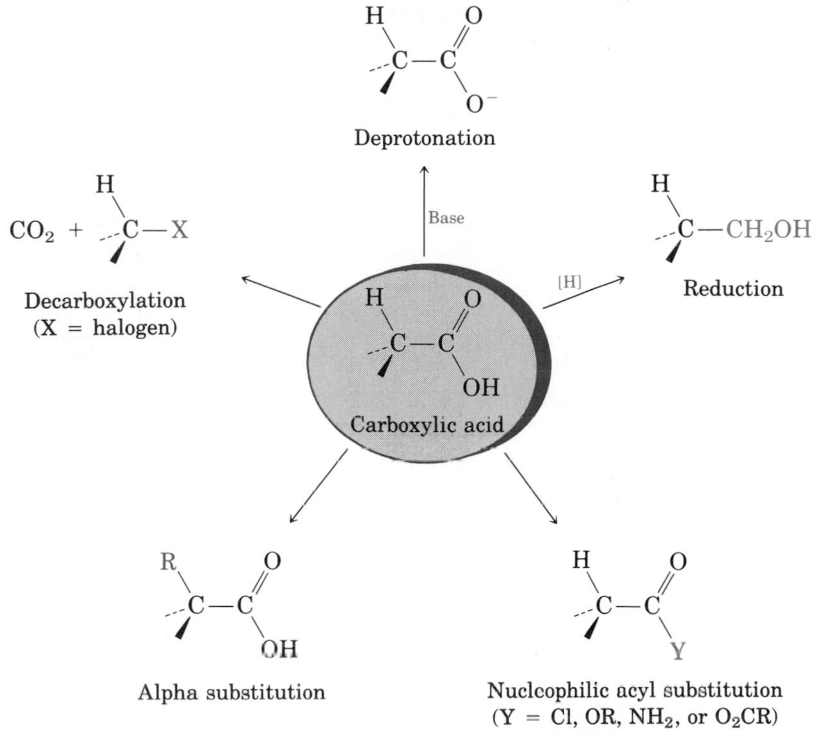

**Figure 20.5** Some general reactions of carboxylic acids

Reactions of carboxylic acids can be grouped into the five categories indicated in Figure 20.5. Of the five, we've already discussed the acidic behavior of carboxylic acids (Sections 20.3–20.5), and we'll discuss the decarboxylation reaction and reduction reaction later in this chapter. The remaining two general reactions of carboxylic acids—nucleophilic acyl substitution and alpha substitution—are two of the four fundamental carbonyl-group reaction mechanisms and will be discussed in detail in Chapters 21 and 22.

## 20.8 Reduction of Carboxylic Acids

Carboxylic acids are reduced by powerful hydride reagents like lithium aluminum hydride (but not $NaBH_4$) to yield primary alcohols (Section 17.6). The reaction is difficult, however, and often requires heating in tetrahydrofuran solvent to go to completion.

$$CH_3(CH_2)_7CH{=}CH(CH_2)_7COOH \xrightarrow[2.\ H_3O^+]{1.\ LiAlH_4,\ THF,\ \Delta} CH_3(CH_2)_7CH{=}CH(CH_2)_7CH_2OH$$

Oleic acid → *cis*-9-Octadecen-1-ol (87%)

Borane, $BH_3$, is also used for converting carboxylic acids into primary alcohols. Reaction of an acid with borane occurs rapidly at room temperature, and this procedure is often preferred to reduction with $LiAlH_4$ because of its relative ease and safety.

$$p\text{-}O_2N\text{-}C_6H_4\text{-}CH_2COOH \xrightarrow[\text{2. } H_3O^+]{\text{1. } BH_3,\ THF} p\text{-}O_2N\text{-}C_6H_4\text{-}CH_2CH_2OH$$

*p*-Nitrophenylacetic acid → 2-(*p*-Nitrophenyl)ethanol (94%)

## 20.9 Decarboxylation of Carboxylic Acids: The Hunsdiecker Reaction

Carboxylic acids lose carbon dioxide under certain conditions to give a product having one less carbon atom than the starting acid. The **Hunsdiecker**[1,2] **reaction**, which involves heating a heavy-metal salt of a carboxylic acid with bromine or iodine, is one example of such a decarboxylation process. Carbon dioxide is lost, and an alkyl halide having one less carbon atom than the starting acid is formed. The metal ion may be either silver, mercuric, or lead(IV)—all work equally well.

$$CH_3(CH_2)_{15}CH_2COOH \xrightarrow[CCl_4]{HgO,\ Br_2} CH_3(CH_2)_{15}CH_2Br + CO_2$$

Octadecanoic acid → 1-Bromoheptadecane (93%)

$$C_4H_7\text{-}COOH \xrightarrow[CCl_4]{Pb(IV),\ I_2} C_4H_7\text{-}I + CO_2$$

Cyclobutanecarboxylic acid → Iodocyclobutane (100%)

The Hunsdiecker reaction is thought to occur by the radical chain pathway shown in Figure 20.6. The initially formed acyl hypobromite undergoes homolytic cleavage of the weak O—Br bond to yield a carboxyl radical, which loses carbon dioxide to form an alkyl radical. The alkyl radical then propagates the chain by abstracting a bromine from acyl hypobromite.

[1]Heinz Hunsdiecker (1904– ); b. Cologne; Ph.D. Cologne (Wintgen); private laboratory, Cologne, Germany.

[2]Cläre Hunsdiecker (1903– ); b. Kiel; Ph.D. Cologne (Wintgen); private laboratory, Cologne, Germany.

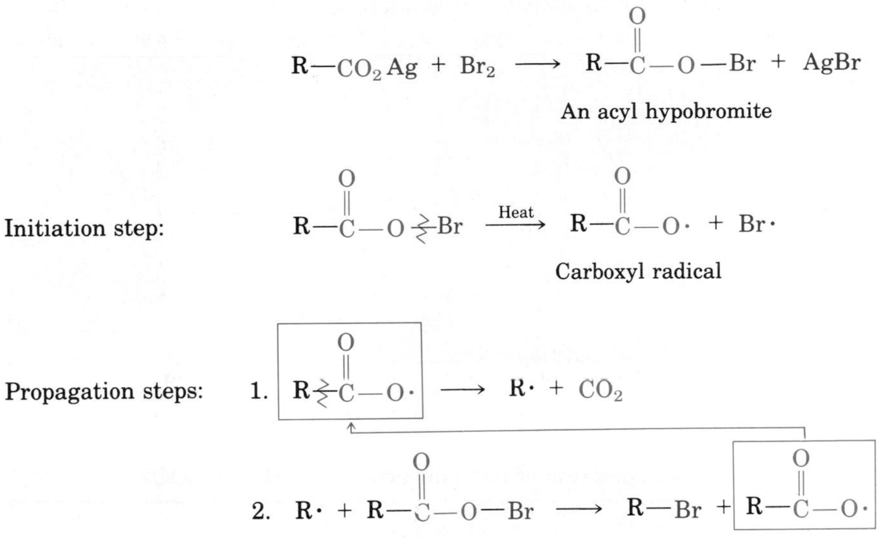

$$R{-}CO_2Ag + Br_2 \longrightarrow R{-}\overset{\displaystyle O}{\overset{\|}{C}}{-}O{-}Br + AgBr$$

An acyl hypobromite

Initiation step:

$$R{-}\overset{\displaystyle O}{\overset{\|}{C}}{-}O{\wr}Br \xrightarrow{\text{Heat}} R{-}\overset{\displaystyle O}{\overset{\|}{C}}{-}O\cdot + Br\cdot$$

Carboxyl radical

Propagation steps:

1. $R{\wr}\overset{\displaystyle O}{\overset{\|}{C}}{-}O\cdot \longrightarrow R\cdot + CO_2$

2. $R\cdot + R{-}\overset{\displaystyle O}{\overset{\|}{C}}{-}O{-}Br \longrightarrow R{-}Br + R{-}\overset{\displaystyle O}{\overset{\|}{C}}{-}O\cdot$

**Figure 20.6** The mechanism of the Hunsdiecker reaction

## 20.10 Spectroscopic Analysis of Carboxylic Acids

### INFRARED SPECTROSCOPY

Carboxylic acids show two highly characteristic absorptions in the infrared spectrum that make this functional group easily identifiable. The O—H bond of the carboxyl group gives rise to a very broad absorption over the range 2500–3300 $cm^{-1}$, and the C=O bond shows an absorption between 1710 $cm^{-1}$ and 1760 $cm^{-1}$. The exact position of carbonyl absorption depends both on the structure of the molecule and on whether the acid is free (monomeric) or associated (dimeric). Free carboxyl groups absorb at 1760 $cm^{-1}$, but the more commonly encountered associated carboxyl groups absorb at 1710 $cm^{-1}$.

Free carboxyl (uncommon), 1760 $cm^{-1}$

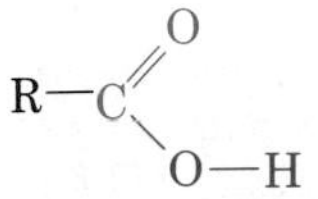

$$R{-}C{\scriptstyle\diagup\!\!\diagup}O,\ {\scriptstyle\diagdown}O{-}H$$

Associated carboxyl (usual case), 1710 $cm^{-1}$

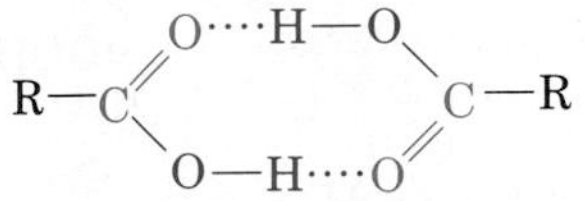

$$R{-}C(=O\cdots H{-}O)(O{-}H\cdots O=)C{-}R$$

The infrared spectrum of butanoic acid shown in Figure 20.7 has both the broad O—H absorption and the C=O absorption at 1710 $cm^{-1}$ (associated) identified.

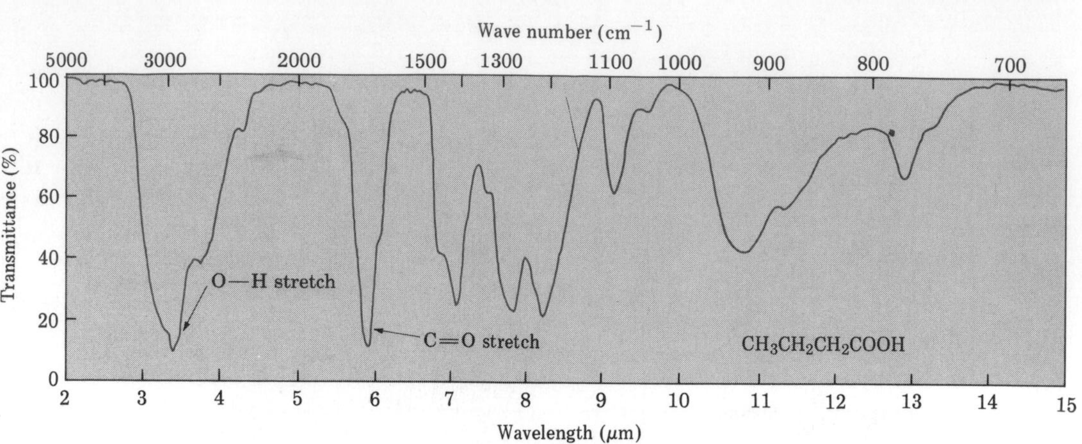


**Figure 20.7** Infrared spectrum of butanoic acid, $CH_3CH_2CH_2COOH$

## NUCLEAR MAGNETIC RESONANCE SPECTROSCOPY

Carboxylic acid groups can be detected by both $^1H$ and $^{13}C$ NMR spectroscopy. Carboxyl carbon atoms absorb in the range 165–185 $\delta$ in the $^{13}C$ NMR spectrum, with aromatic and $\alpha,\beta$-unsaturated acids near the upfield end of the range (~165 $\delta$) and saturated aliphatic acids near the downfield end (~185 $\delta$). The acidic —COOH proton normally absorbs near 12 $\delta$ in the $^1H$ NMR spectrum. Since the carboxylic acid proton has no neighbor protons, it is unsplit and occurs as a singlet. As with alcohols (Section 17.11), the —COOH proton can be replaced by deuterium upon addition of $D_2O$ to the sample tube, causing the —COOH absorption to disappear from the NMR spectrum.

Figure 20.8 indicates the positions of the $^{13}C$ NMR absorptions for several carboxylic acids, and Figure 20.9 shows the $^1H$ NMR spectrum of phenylacetic acid. Note that the carboxyl proton occurs at 10.6 $\delta$ and that all five aromatic ring protons show an accidental overlap as an apparent singlet at 7.3 $\delta$.

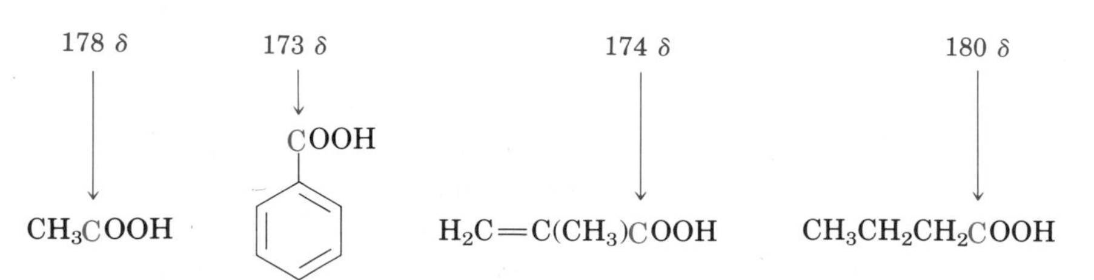


**Figure 20.8** Carbon-13 NMR absorptions of some carboxylic acids

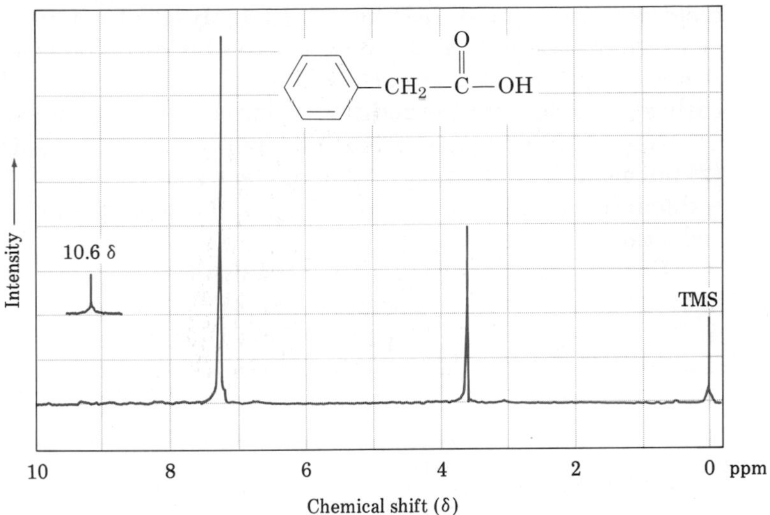


**Figure 20.9** Proton NMR spectrum of phenylacetic acid

## 20.11 Summary and Key Words

**Carboxylic acids** are among the most important building blocks for synthesizing other molecules, both in nature and in the chemical laboratory. They are named systematically by replacing the terminal *-e* of the corresponding alkane name with *-oic acid*. Like ketones and aldehydes, the carbonyl carbon atom is $sp^2$ hybridized; like alcohols, carboxylic acids are associated via hydrogen bonding and therefore have high boiling points.

The distinguishing characteristic of carboxylic acids is their acidity. Although weaker than mineral acids such as HCl, carboxylic acids nevertheless dissociate far more readily than alcohols. The reason for this difference lies in the stability of carboxylate ions: Carboxylate ions are stabilized by resonance between two equivalent forms:

$$\mathrm{R{-}C}\begin{matrix}\ddot{\text{O}}{:} \\ \| \\ \\ \ddot{\text{O}}{:}^- \end{matrix} \longleftrightarrow \mathrm{R{-}C}\begin{matrix}\ddot{\text{O}}{:}^- \\ \\ \| \\ \ddot{\text{O}}{:} \end{matrix}$$

Most alkanoic acids have p$K_a$ values near 5, but the exact acidity constant of a given acid is subject to considerable variation depending on structure. Carboxylic acids substituted by electron-withdrawing groups are more acidic (have a lower p$K_a$) because their carboxylate ions are stabilized. Carboxylic acids substituted by electron-donating groups are less acidic (have a higher p$K_a$) because their carboxylate ions are destabilized.

Methods of synthesis for carboxylic acids include: (1) **oxidation of alkylbenzenes** with sodium dichromate or potassium permanganate; (2)

**oxidative cleavage of alkenes**; (3) **oxidation of primary alcohols** or aldehydes with Jones' reagent; (4) **hydrolysis of nitriles**; and (5) reaction of Grignard reagents with $CO_2$ (**carboxylation**). The last two methods are particularly useful because they allow alkyl halides to be transformed into carboxylic acids with the addition of one carbon atom, R—Br → R—COOH.

General reactions of carboxylic acids include: (1) loss of the acidic proton; (2) nucleophilic acyl substitution at the carbonyl group; (3) loss of $CO_2$ via a **decarboxylation** reaction; (4) substitution on the alpha carbon; and (5) reduction.

Carboxylic acids are easily distinguished spectroscopically. They exhibit characteristic infrared absorptions at 2500–3300 $cm^{-1}$ (due to the O—H) and at 1710–1760 $cm^{-1}$ (due to the C=O). Acids also show $^{13}C$ NMR absorptions at 165–185 $\delta$ and $^{1}H$ NMR absorptions near 12 $\delta$.

## 20.12 Summary of Reactions

1. Preparation of carboxylic acids (Section 20.6)

   a. Oxidation of alkylbenzenes (Section 16.11)

$$C_6H_5-\underset{R}{\overset{H}{C}}-R \xrightarrow[H_2O,\ \Delta]{KMnO_4} C_6H_5-\overset{O}{\overset{\|}{C}}-OH$$

   b. Oxidative cleavage of alkenes (Section 7.8)

$$RCH{=}CHR' \xrightarrow{KMnO_4} R-\overset{O}{\overset{\|}{C}}-OH + R'-\overset{O}{\overset{\|}{C}}-OH$$

   c. Oxidation of primary alcohols (Section 17.9)

$$R-\underset{H}{\overset{OH}{C}}-H \xrightarrow[(CrO_3,\ H_3O^+)]{\text{Jones' reagent}} R-\overset{O}{\overset{\|}{C}}-OH$$

   d. Oxidation of aldehydes (Section 19.5)

$$R-\overset{O}{\overset{\|}{C}}-H \xrightarrow[\text{or } Ag^+,\ NH_4OH]{\text{Jones' reagent}} R-\overset{O}{\overset{\|}{C}}-OH$$

   e. Hydrolysis of nitriles (Section 20.6)

$$R-C{\equiv}N \xrightarrow[\text{or } NaOH,\ H_2O]{H_3O^+} R-\overset{O}{\overset{\|}{C}}-OH + NH_3$$

f. Carboxylation of Grignard reagents (Section 20.6)

$$\text{R—MgX} + \text{O=C=O} \xrightarrow[\text{2. } H_3O^+]{\text{1. Mix}} \text{R—}\overset{\overset{\displaystyle O}{\|}}{C}\text{—OH}$$

2. Reactions of carboxylic acids

a. Deprotonation (Section 20.3)

$$\text{R—}\overset{\overset{\displaystyle O}{\|}}{C}\text{—O—H} \xrightarrow{\text{Base}} \text{R—}\overset{\overset{\displaystyle O}{\|}}{C}\text{—}O^-$$

b. Reduction to primary alcohols (Section 20.8)

$$\text{R—}\overset{\overset{\displaystyle O}{\|}}{C}\text{—OH} \xrightarrow[\text{2. } H_3O^+]{\text{1. } LiAlH_4 \text{ or } BH_3} \text{R—}\underset{\underset{\displaystyle H}{|}}{\overset{\overset{\displaystyle OH}{|}}{C}}\text{—H}$$

c. Decarboxylation: Hunsdiecker reaction (Section 20.9)

$$\text{R—}\overset{\overset{\displaystyle O}{\|}}{C}\text{—OH} \xrightarrow[Br_2,\ CCl_4]{\text{HgO}} \text{R—Br} + CO_2$$

## ADDITIONAL PROBLEMS

**20.13** Provide IUPAC names for these compounds:

(a) $CH_3\underset{}{\overset{\overset{\displaystyle COOH}{|}}{C}}HCH_2CH_2\overset{\overset{\displaystyle COOH}{|}}{C}HCH_3$

(b) $(CH_3)_3CCOOH$

(c) $CH_3CH_2CH_2\underset{\underset{\displaystyle CH_2COOH}{|}}{\overset{\overset{\displaystyle CH_2CH_2CH_3}{|}}{C}}H$

(d) benzene ring bearing COOH and, para to it, $NO_2$

(e) cyclododecene ring with COOH on the double-bond carbon

(f) $BrCH_2CHBrCH_2CH_2COOH$

**20.14** Draw structures corresponding to these IUPAC names:

(a) *cis*-1,2-Cyclohexanedicarboxylic acid
(b) Heptanedioic acid
(c) 2-Hexen-4-ynoic acid
(d) 4-Ethyl-2-propyloctanoic acid
(e) 3-Chlorophthalic acid
(f) Triphenylacetic acid

**20.15** Acetic acid boils at 118°C, but its ethyl ester boils at 77°C. Why is the boiling point of the acid so much higher, even though it has the lower molecular weight?

**20.16** Draw and name the eight carboxylic acid isomers having the formula $C_6H_{12}O_2$.

**20.17** Order the compounds in each set with respect to increasing acidity:
(a) Acetic acid, oxalic acid, formic acid
(b) *p*-Bromobenzoic acid, *p*-nitrobenzoic acid, 2,4-dinitrobenzoic acid
(c) Phenylacetic acid, diphenylacetic acid, 3-phenylpropanoic acid
(d) Fluoroacetic acid, 3-fluoropropanoic acid, iodoacetic acid

**20.18** Arrange the compounds in each set in order of increasing basicity:
(a) Magnesium acetate, magnesium hydroxide, methylmagnesium bromide
(b) Sodium benzoate, sodium *p*-nitrobenzoate, sodium acetylide
(c) Lithium hydroxide, lithium ethoxide, lithium formate

**20.19** Account for the fact that phthalic acid (1,2-benzenedicarboxylic acid) has $pK_2 = 5.4$ but terephthalic acid (1,4-benzenedicarboxylic acid) has $pK_2 = 4.8$.

**20.20** How could you convert butanoic acid into the following compounds? Write out each step showing the reagents needed.
(a) 1-Butanol (b) 1-Bromobutane (c) Pentanoic acid
(d) 1-Butene (e) 1-Bromopropane (f) Octane

**20.21** How could you convert each of the following compounds into butanoic acid? Write out each step showing all reagents.
(a) 1-Butanol (b) 1-Bromobutane (c) 1-Butene
(d) 1-Bromopropane (e) 4-Octene (f) Pentanoic acid

**20.22** How would you prepare these compounds from benzene? More than one step is required in each case.
(a) *m*-Chlorobenzoic acid (b) *p*-Bromobenzoic acid
(c) Phenylacetic acid, $C_6H_5CH_2COOH$

**20.23** Calculate $pK_a$'s for these acids:
(a) Lactic acid, $K_a = 8.4 \times 10^{-4}$ (b) Acrylic acid, $K_a = 5.6 \times 10^{-6}$

**20.24** Calculate $K_a$'s for these acids:
(a) Citric acid, $pK_a = 3.14$ (b) Tartaric acid, $pK_a = 2.98$

**20.25** Predict the product of the reaction of *p*-methylbenzoic acid with each of the following reagents:
(a) $BH_3$, then $H_3O^+$ (b) N-Bromosuccinimide in $CCl_4$
(c) HgO, $I_2$ in $CCl_4$ (d) $CH_3MgBr$ in ether, then $H_3O^+$
(e) $KMnO_4$, $H_3O^+$ (f) $LiAlH_4$, then $H_3O^+$

**20.26** Using $^{13}CO_2$ as your only source of labeled carbon, how would you synthesize these compounds?
(a) $CH_3CH_2{}^{13}COOH$ (b) $CH_3{}^{13}CH_2COOH$

**20.27** Propose a structure for an organic compound, $C_6H_{12}O_2$, that dissolves in dilute sodium hydroxide and that shows the following $^1H$ NMR spectrum: 1.08 $\delta$ (9 H, singlet), 2.2 $\delta$ (2 H, singlet), and 11.2 $\delta$ (1 H, singlet).

**20.28** How would you carry out these transformations?

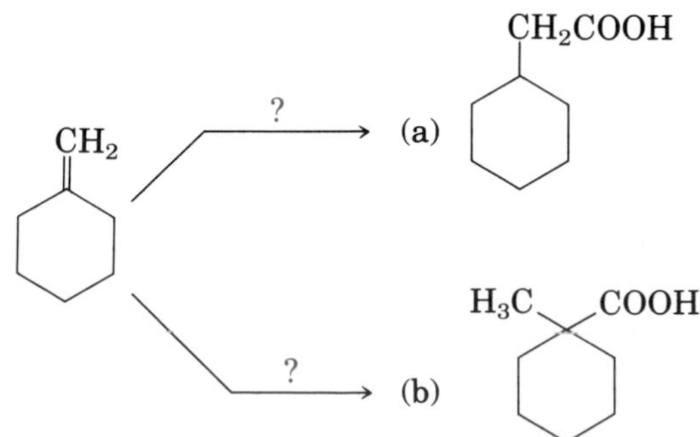

**20.29** What spectroscopic method might you use to distinguish among the following three isomeric acids? Tell exactly what characteristic features you would expect for each acid.

| $CH_3(CH_2)_3COOH$ | $(CH_3)_2CHCH_2COOH$ | $(CH_3)_3CCOOH$ |
|---|---|---|
| Pentanoic acid | 3-Methylbutanoic acid | 2,2-Dimethylpropanoic acid |

**20.30** Which method of acid synthesis—Grignard carboxylation or nitrile hydrolysis—would you use for each of the following reactions? Explain your choices.

(a) 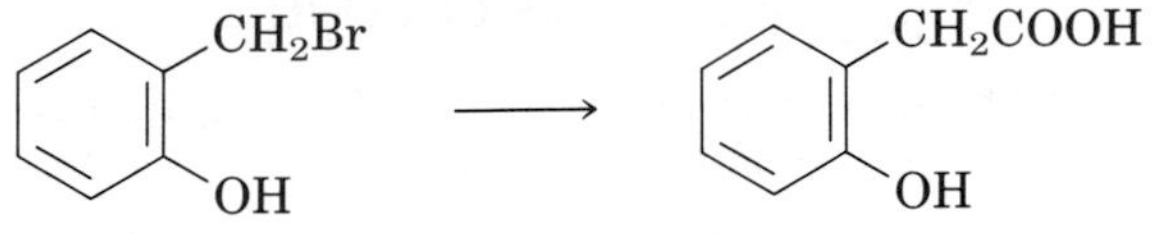 o-(Bromomethyl)phenol ($CH_2Br$, OH on benzene ring) $\longrightarrow$ o-hydroxyphenylacetic acid ($CH_2COOH$, OH on benzene ring)

(b) $CH_3CH_2CHBrCH_3 \longrightarrow CH_3CH_2CH(CH_3)COOH$

(c) $CH_3C(=O)CH_2CH_2CH_2I \longrightarrow CH_3C(=O)CH_2CH_2CH_2COOH$

(d) $HOCH_2CH_2CH_2Br \longrightarrow HOCH_2CH_2CH_2COOH$

**20.31** A chemist in need of 2,2-dimethylpentanoic acid decided to synthesize some by reaction of 2-chloro-2-methylpentane with NaCN followed by hydrolysis of the product. After carrying out the reaction sequence, however, none of the desired product could be found. What do you suppose went wrong?

**20.32** As written, the following synthetic schemes all have at least one flaw in them. What is wrong with each?

(a) 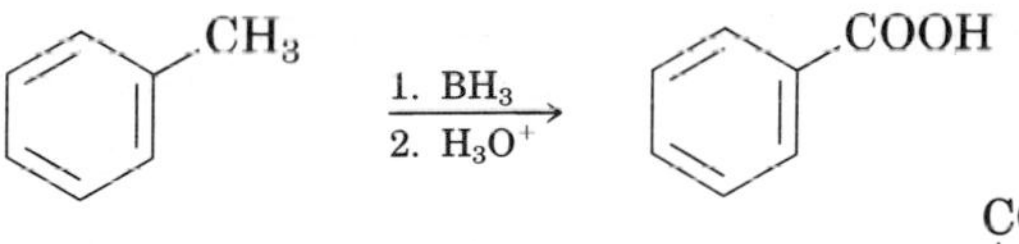 Toluene ($C_6H_5CH_3$) $\xrightarrow[2.\ H_3O^+]{1.\ BH_3}$ Benzoic acid ($C_6H_5COOH$)

(b) $CH_3CH_2CHBrCH_2CH_3 \xrightarrow[2.\ NaCN \quad 3.\ H_3O^+]{1.\ Mg} CH_3CH_2CH(COOH)CH_2CH_3$

(c) 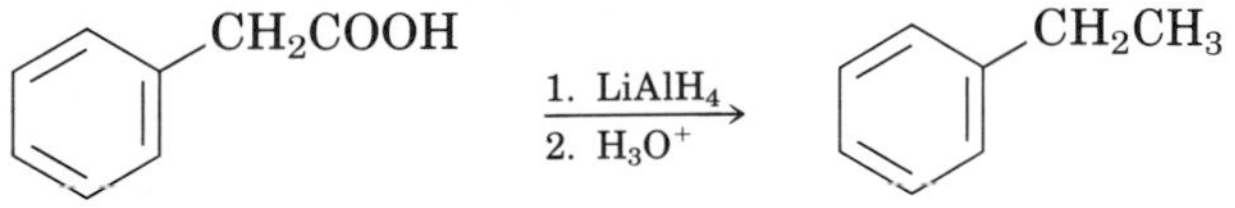 $C_6H_5CH_2COOH \xrightarrow[2.\ H_3O^+]{1.\ LiAlH_4} C_6H_5CH_2CH_3$

(d) $(CH_3)_2C(OH)CH_2CH_2Cl \xrightarrow[2.\ H_3O^+]{1.\ NaCN} (CH_3)_2C(OH)CH_2CH_2COOH$

**20.33** *p*-Aminobenzoic acid (PABA) is widely used as a sunscreen agent. Propose a synthesis of PABA starting from toluene.

**20.34** Lithocholic acid is a steroid found in human bile:

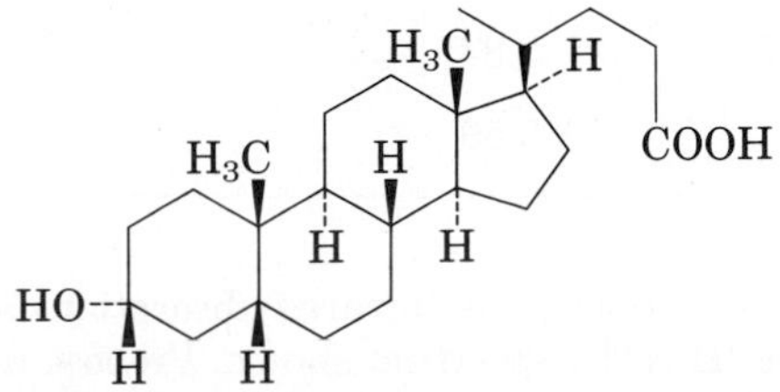

[Structure: steroid skeleton with HO, $H_3C$, H, and COOH labels]

Lithocholic acid

Predict the product of reaction of lithocholic acid with each of the following reagents. Don't worry about the size of the molecule; just concentrate on the functional groups.

(a) Jones' reagent
(b) Tollens' reagent
(c) $BH_3$, then $H_3O^+$
(d) Dihydropyran, acid catalyst
(e) $CH_3MgBr$, then $H_3O^+$
(f) $LiAlH_4$, then $H_3O^+$

**20.35** Propose a synthesis of the anti-inflammatory drug, Fenclorac from phenylcyclohexane.

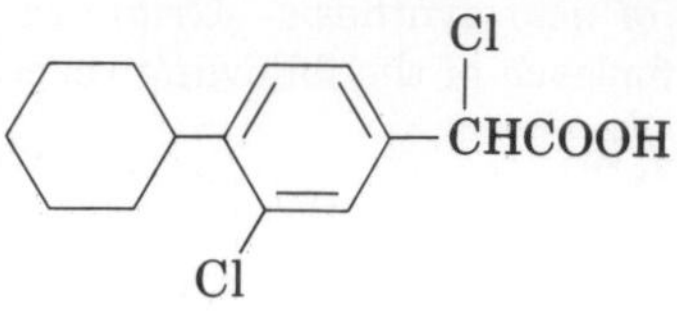

Fenclorac

**20.36** How would you use NMR (either $^{13}C$ or $^{1}H$) to distinguish between the following isomeric pairs?

(a) 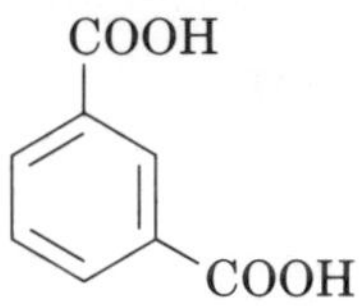 [m-benzenedicarboxylic acid] and [p-benzenedicarboxylic acid]

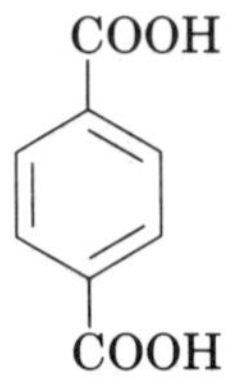

(b) $HOOCCH_2CH_2COOH$ and $CH_3CH(COOH)_2$
(c) $CH_3CH_2CH_2COOH$ and $HOCH_2CH_2CH_2CHO$

(d) $(CH_3)_2C{=}CHCH_2COOH$ and [cyclopentyl]—COOH

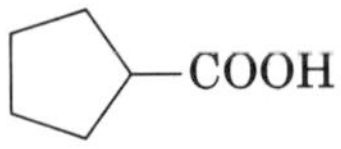

**20.37** The p$K_a$'s of five para-substituted benzoic acids (Y—$C_6H_4$—COOH) are given in the table. Rank the corresponding substituted benzenes (Y—$C_6H_5$) in order of their increasing reactivity toward electrophilic aromatic substitution. If benzoic acid has p$K_a$ = 4.20, which of the substituent groups are activators and which are deactivators?

| *Substituent* | p$K_a$ *of* 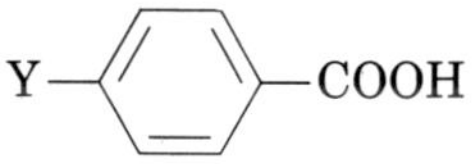 Y—$C_6H_4$—COOH |
|---|---|
| —$Si(CH_3)_3$ | 4.27 |
| —$CH{=}CHC{\equiv}N$ | 4.03 |
| —$HgCH_3$ | 4.10 |
| —$OSO_2CH_3$ | 3.84 |
| —$PCl_2$ | 3.59 |

**20.38** Compound A, $C_4H_8O_3$, has infrared absorptions at 1710 and 2500–3100 $cm^{-1}$, and exhibits the $^{1}H$ NMR spectrum shown. Propose a structure for A that is consistent with the data.

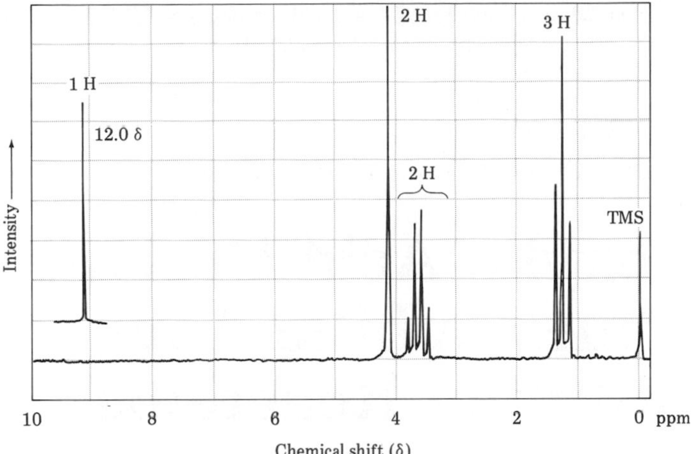


**20.39** The two $^1$H NMR spectra shown belong to crotonic acid (*trans*-$CH_3CH{=}CHCOOH$) and to methacrylic acid [$H_2C{=}C(CH_3)COOH$]. Which spectrum corresponds to which acid? Explain your answer.

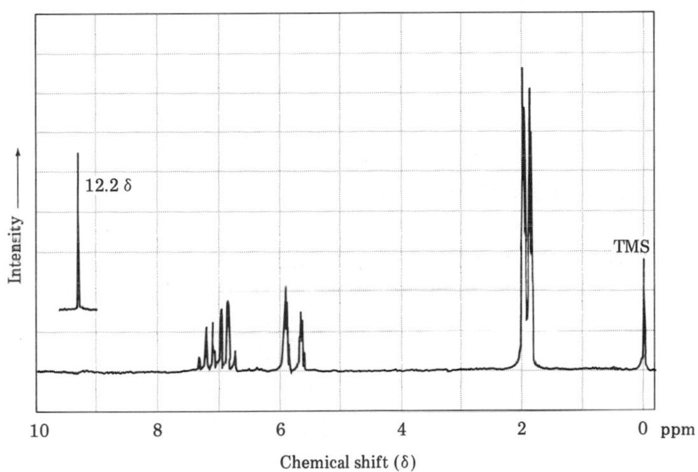


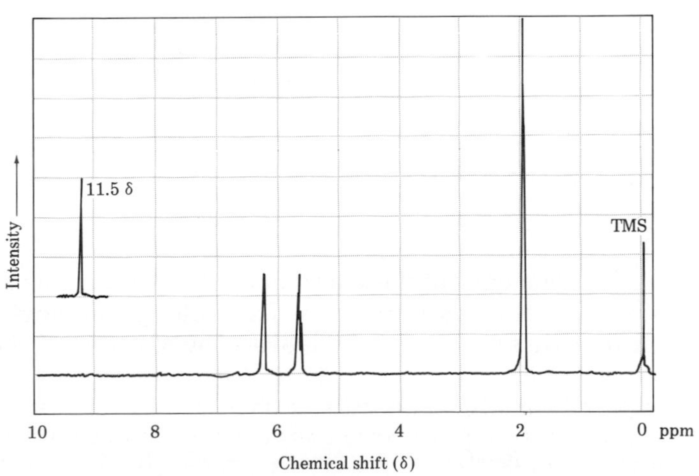

CHAPTER 21

# Carboxylic Acid Derivatives and Nucleophilic Acyl Substitution Reactions

Carboxylic acids are just one member of a class of **acyl derivatives, RCOY,** where the —Y substituent may be oxygen, halogen, nitrogen, or sulfur. Numerous acyl derivatives are possible, but we will be concerned only with four of the more common ones, in addition to carboxylic acids themselves: acid halides, acid anhydrides, esters, and amides. All these derivatives contain an acyl group, RCO—, bonded to an electronegative atom that can act as a leaving group in substitution reactions. Also in this chapter, we'll discuss **nitriles**, a class of compounds closely related to carboxylic acids.

$$\underset{\text{Carboxylic acid}}{R-\overset{O}{\overset{\|}{C}}-OH} \qquad \underset{\substack{\text{Acid halide}\\(X\,=\,F,\,Cl,\,Br,\,\text{or } I)}}{R-\overset{O}{\overset{\|}{C}}-X} \qquad \underset{\text{Acid anhydride}}{R-\overset{O}{\overset{\|}{C}}-O-\overset{O}{\overset{\|}{C}}-R'}$$

$$\underset{\text{Ester}}{R-\overset{O}{\overset{\|}{C}}-OR'} \qquad \underset{\text{Amide}}{R-\overset{O}{\overset{\|}{C}}-NH_2} \qquad \underset{\text{Nitrile}}{R-C\equiv N}$$

The chemistry of these acyl derivatives is similar and is dominated by a single general reaction type—the **nucleophilic acyl substitution reaction** that we saw briefly in the overview of carbonyl chemistry:

$$R-\overset{O}{\overset{\|}{C}}-Y + :Nu^- \longrightarrow R-\overset{O}{\overset{\|}{C}}-Nu + :Y^-$$

Let's first learn more about acyl derivatives and then explore the chemistry of acyl substitution reactions.

## 21.1 Nomenclature of Carboxylic Acid Derivatives

### ACID HALIDES: RCOX

Acid halides are named by identifying first the acyl group and then the halide. The acyl group name is derived from the carboxylic acid name by replacing the *-ic acid* ending with *-yl*, or the *-carboxylic acid* ending with *-carbonyl*. For example,

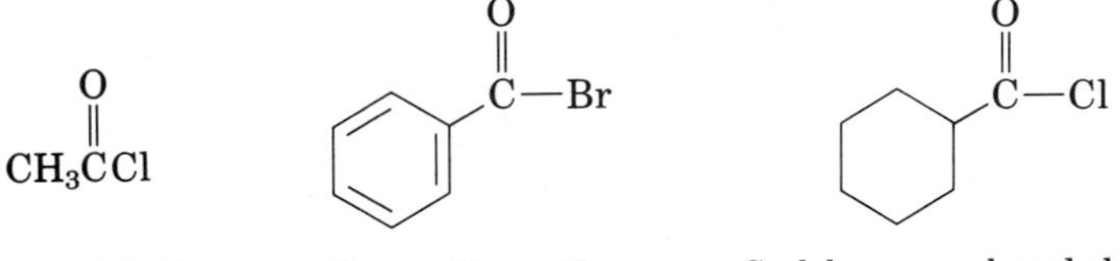


Acetyl chloride (from acetic acid) | Benzoyl bromide (from benzoic acid) | Cyclohexanecarbonyl chloride (from cyclohexanecarboxylic acid)

### ACID ANHYDRIDES: $RCO_2COR'$

Symmetrical anhydrides of straight-chain monocarboxylic acids, and cyclic anhydrides of dicarboxylic acids, are named by replacing the word *acid* with *anhydride*.

$CH_3\overset{O}{\overset{\|}{C}}-O-\overset{O}{\overset{\|}{C}}CH_3$

Acetic anhydride

$CH_3(CH_2)_5\overset{O}{\overset{\|}{C}}-O-\overset{O}{\overset{\|}{C}}(CH_2)_5CH_3$

Heptanoic anhydride

$H_3CO$ (3-methoxy-substituted benzene ring, positions 1–6, fused with C(=O)–O–C(=O) ring)

3-Methoxyphthalic anhydride

If the anhydride is derived from a substituted monocarboxylic acid, it is named by adding the prefix *bis-* (meaning two) to the acid name.

$ClCH_2\overset{O}{\overset{\|}{C}}-O-\overset{O}{\overset{\|}{C}}CH_2Cl$

Bis(chloroacetic) anhydride

### AMIDES: $RCONH_2$

Amides with an unsubstituted $-NH_2$ group are named by replacing the *-oic acid* or *-ic acid* ending with *-amide*, or by replacing the *-carboxylic acid* ending with *-carboxamide*.

$CH_3\overset{O}{\overset{\|}{C}}NH_2$

Acetamide (from acetic acid)

$CH_3(CH_2)_4\overset{O}{\overset{\|}{C}}NH_2$

Hexanamide (from hexanoic acid)

Cyclopentyl–$\overset{O}{\overset{\|}{C}}-NH_2$

Cyclopentanecarboxamide (from cyclopentanecarboxylic acid)

If the nitrogen atom is further substituted, the compound is named by first identifying the substituent groups and then citing the base name. The substituents are preceded by the letter *N* to identify them as being directly attached to nitrogen.

$CH_3CH_2C(=O)NHCH_3$

*N*-Methylpropanamide

$C_4H_7$–$C(=O)$–$N(CH_2CH_3)_2$

*N,N*-Diethylcyclobutanecarboxamide

## ESTERS: $RCO_2R'$

Systematic names for esters are derived by first citing the alkyl group attached to oxygen and then identifying the carboxylic acid. In so doing, the *-ic acid* ending is replaced by *-ate*.

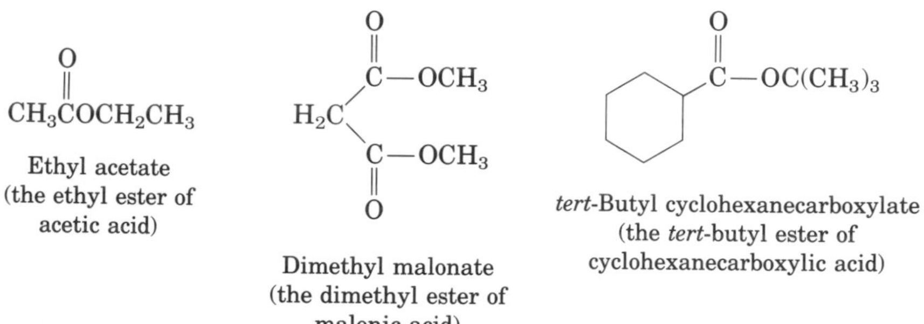


Ethyl acetate (the ethyl ester of acetic acid)

Dimethyl malonate (the dimethyl ester of malonic acid)

*tert*-Butyl cyclohexanecarboxylate (the *tert*-butyl ester of cyclohexanecarboxylic acid)

## NITRILES: $RC{\equiv}N$

Compounds containing the $-C{\equiv}N$ functional group are known as **nitriles**. Simple acyclic alkane nitriles are named by adding *-nitrile* as a suffix to the alkane name, with the nitrile carbon itself numbered as C1.

$$\overset{5}{CH_3}\overset{4}{CH}(CH_3)\overset{3}{CH_2}\overset{2}{CH_2}\overset{1}{C}N$$

4-Methylpentanenitrile

More complex nitriles are usually considered as derived from carboxylic acids, and are named either by replacing the *-ic acid* or *-oic acid* ending with *-onitrile*, or by replacing the *-carboxylic acid* ending with *-carbonitrile*. In this system, the nitrile carbon atom is attached to C1 but is not itself numbered.

$CH_3C{\equiv}N$

Acetonitrile (from acetic acid)

$C_6H_5$–$C{\equiv}N$

Benzonitrile (from benzoic acid)

1-CN, 2,2-($CH_3$)$_2$ cyclohexane (ring numbered 1–6)

2,2-Dimethylcyclohexanecarbonitrile (from 2,2-dimethylcyclohexanecarboxylic acid)

A summary of nomenclature rules for carboxylic acid derivatives is given in Table 21.1.

Table 21.1 Nomenclature of carboxylic acid derivatives

| *Functional group* | *Structure* | *Name ending* |
|---|---|---|
| Carboxylic acid | $R{-}C({=}O){-}OH$ | *-ic acid* (*-carboxylic acid)* |
| Acid halide | $R{-}C({=}O){-}X$ | *-yl halide* (*-carbonyl halide)* |
| Acid anhydride | $R{-}C({=}O){-}O{-}C({=}O){-}R$ | *anhydride* |
| Amide | $R{-}C({=}O){-}NH_2$ | *-amide* (*-carboxamide)* |
| Ester | $R{-}C({=}O){-}OR'$ | *-ate* (*-carboxylate)* |
| Nitrile | $R{-}C{\equiv}N$ | *-onitrile* (*-carbonitrile)* |

PROBLEM

**21.1** Provide IUPAC names for these structures:

(a) $(CH_3)_2CHCH_2CH_2COCl$

(b) cyclohexyl–$CH_2CONH_2$

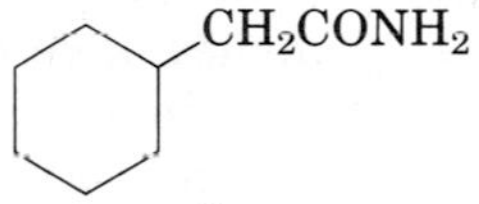

(c) $CH_3CH_2CH(CH_3)CN$

(d) $(C_6H_5C(=O))_2O$ (phenyl–$\overset{O}{\overset{\|}{C}}$$\!+_2O$)

(e) cyclopentyl–$CO_2CH(CH_3)_2$

(f) cyclopentyl–$O_2CCH(CH_3)_2$

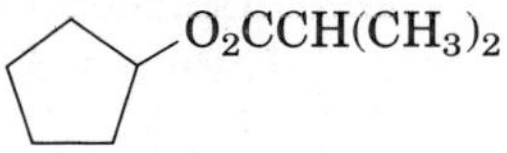

(g) $H_2C{=}CHCH_2CH_2CONH_2$

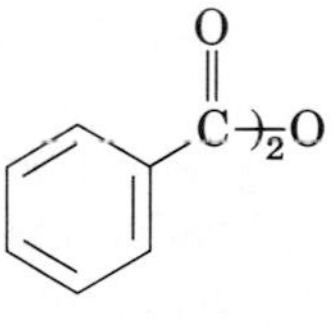

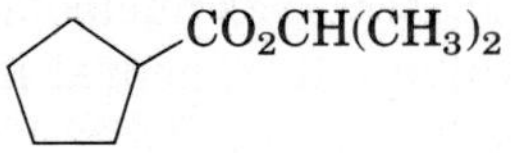

(h) $CH_3CH_2CH(CN)CH_2CH_3$

(i) $(CH_3)_2C{=}C(CH_3)COCl$

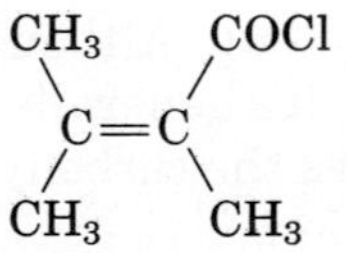

(j) $CF_3\overset{O}{\overset{\|}{C}}O\overset{O}{\overset{\|}{C}}CF_3$

PROBLEM

**21.2** The names shown for the following compounds are incorrect. Provide the correct names.

(a) $\overset{4}{C}H_3\overset{3}{C}H_2\overset{2}{C}H{=}\overset{1}{C}HCN$

1-Pentenenitrile

(b) $CH_3CH_2CH_2CONHCH_3$

Methylbutanamide

(c) $(CH_3)_2CHCH_2CH(CH_3)COCl$

2,4-Methylpentanoyl chloride

(d) Cyclohexane ring bearing $CH_3$ and $CO_2CH_3$ on the same carbon

Methyl-2-methylcyclohexane carboxylate

## 21.2 Nucleophilic Acyl Substitution Reactions

The addition of a nucleophile to the polar C=O bond is a general feature of carbonyl-group reactions and is the first step in two of the four major carbonyl-group reactions. When nucleophiles add to aldehydes and ketones, the initially formed tetrahedral intermediate can either be protonated to yield an alcohol, or it can eliminate the carbonyl oxygen, leading to a new C=Nu bond. When nucleophiles add to carboxylic acid derivatives, however, a different reaction course is followed. The initially formed tetrahedral intermediate expels one of the two substituents originally bonded to the carbonyl carbon, leading to a net nucleophilic acyl substitution (Figure 21.1).

What is the reason for the different behavior of ketones/aldehydes and carboxylic acid derivatives? The difference is simply a consequence of structure. Carboxylic acid derivatives have an acyl function bonded to a potential leaving group, Y, which can leave as a stabilized anion. As soon as the tetrahedral intermediate is formed, the negative charge on oxygen can readily expel this leaving group to generate a new carbonyl compound. Ketones and aldehydes have no such leaving group, however, and therefore do not undergo elimination.

As shown in Figure 21.2, the net effect of the two-step addition–elimination sequence is a substitution of the attacking nucleophile for the Y group originally bonded to the acyl carbon. Thus, the overall reaction is superficially similar to the kind of nucleophilic substitution that occurs during $S_N2$ reactions (Section 11.5). The *mechanisms* of the two reactions are completely different, however. $S_N2$ reactions occur in a single step by back-side displacement of the leaving group, whereas nucleophilic acyl substitutions take place in two steps and involve a tetrahedral intermediate.

## 21.3 Relative Reactivity of Carboxylic Acid Derivatives

Nucleophilic acyl substitution reactions take place in two steps—addition of the nucleophile and elimination of a leaving group. Although both steps can sometimes affect the overall rate of reaction, it's generally the first step that is rate-limiting. Thus, any factor that makes the carbonyl group more easily attacked by nucleophiles favors the reaction.

(a) (b)

**Figure 21.1** The reactions of carbonyl groups with nucleophiles: (a) Ketones and aldehydes undergo nucleophilic *addition* reactions to yield either alcohols or products with a C=Nu double bond, whereas (b) carboxylic acid derivatives undergo nucleophilic acyl *substitution* reactions.

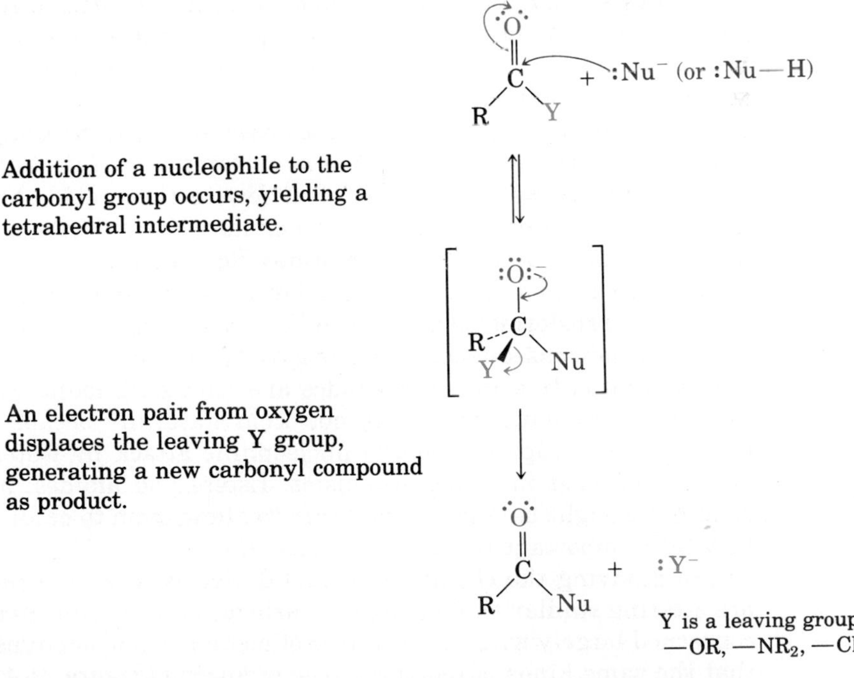

Addition of a nucleophile to the carbonyl group occurs, yielding a tetrahedral intermediate.
An electron pair from oxygen displaces the leaving Y group, generating a new carbonyl compound as product.
:Nu⁻ (or :Nu—H)
Y is a leaving group:
—OR, —NR₂, —Cl


**Figure 21.2** General mechanism of nucleophilic acyl substitution

Steric and electronic factors are both important in determining reactivity. Sterically, we find within a series of the same acid derivatives that unhindered, accessible carbonyl groups react with nucleophiles more readily than sterically hindered groups. For example, acetyl chloride, $CH_3COCl$, is much more reactive than 2,2-dimethylpropanoyl chloride $(CH_3)_3CCOCl$. Thus, we find a reactivity order:

$$\text{Less reactive} \quad R_3C\overset{O}{\overset{\|}{C}}— \;<\; R_2CH\overset{O}{\overset{\|}{C}}— \;<\; RCH_2\overset{O}{\overset{\|}{C}}— \;<\; CH_3\overset{O}{\overset{\|}{C}}— \quad \text{More reactive}$$

Electronically, we find that more strongly polarized acyl derivatives are attacked more readily than less polar derivatives. Thus, acid chlorides are the most reactive acyl derivatives because the electronegative chlorine atom polarizes the carbonyl group more strongly than alkoxy or amino groups. The observed reactivity order is

$$\text{Less reactive} \quad \underset{\text{Amide}}{R—\overset{O}{\overset{\|}{C}}—NH_2} \;<\; \underset{\text{Ester}}{R—\overset{O}{\overset{\|}{C}}—OR'} \;<\; \underset{\text{Anhydride}}{R—\overset{O}{\overset{\|}{C}}—O—\overset{O}{\overset{\|}{C}}—R'} \;<\; \underset{\text{Acid chloride}}{R—\overset{O}{\overset{\|}{C}}—Cl} \quad \text{More reactive}$$

The way that various substituents affect the polarization of a carbonyl group is similar to the way they affect the reactivity of an aromatic ring toward electrophilic substitution (Section 16.5). Thus, a chlorine substituent withdraws electrons from an aromatic ring in the same way that it withdraws electrons from a neighboring acyl group, whereas amino and methoxyl substituents donate electrons to aromatic rings in the same way that they donate electrons to acyl groups.

An important consequence of the observed reactivity differences is that *it is usually possible to transform a more reactive acid derivative into a less reactive one*. As we'll see in the next few sections, acid chlorides can be converted into anhydrides, esters, and amides, but amides can't be converted into esters, anhydrides, or acid chlorides. Remembering the reactivity order is therefore a way to keep track of a large number of reactions. Figure 21.3 shows the transformations that can be carried out.

A second consequence of the reactivity differences among acid derivatives is that only esters and amides are commonly found in nature. Acid halides and acid anhydrides do not occur naturally because they are too reactive. They rapidly undergo nucleophilic attack by water and are too reactive to exist in living organisms. Esters and amides, however, have exactly the right balance of reactivity to allow them to occur widely and to be vitally important in many life processes.

In studying the chemistry of acid derivatives, we will find that there are striking similarities among the various types of compounds. We will be concerned largely with the reactions of just a few nucleophiles, and will see that the same kinds of reactions keep occurring (Figure 21.4).

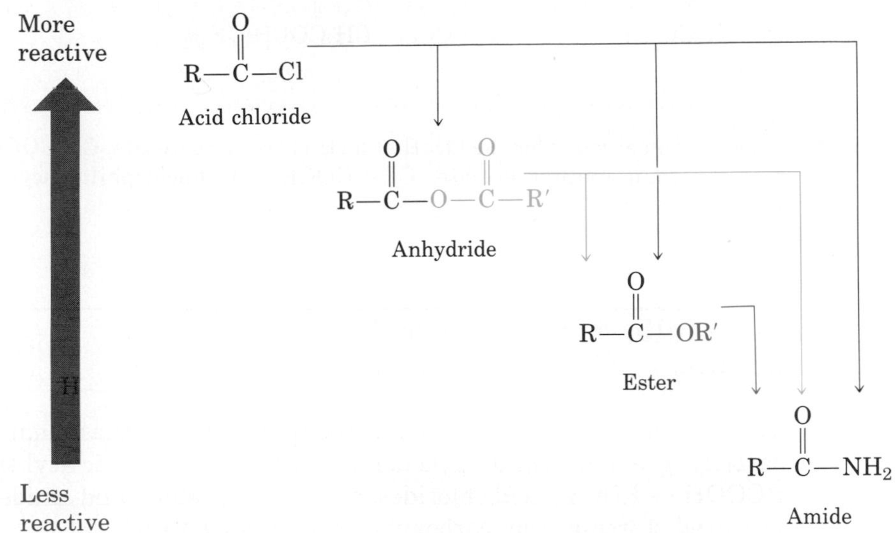


**Figure 21.3** Interconversions of carboxylic acid derivatives

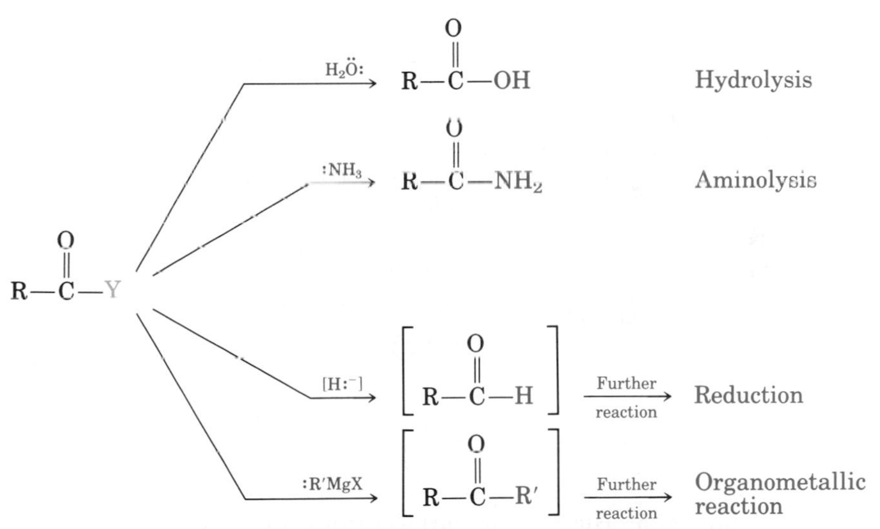


**Figure 21.4** General reactions of carboxylic acid derivatives

**PROBLEM**

**21.3** Rank the compounds in the following sets with regard to expected reactivity toward nucleophilic acyl substitution:

(a) $CH_3\overset{O}{\overset{\|}{C}}Cl$, $CH_3\overset{O}{\overset{\|}{C}}OCH_3$, $CH_3\overset{O}{\overset{\|}{C}}NH_2$

(b) $CH_3\overset{\overset{O}{\|}}{C}OCH_3$, $CH_3\overset{\overset{O}{\|}}{C}OCH_2CCl_3$, $CH_3\overset{\overset{O}{\|}}{C}OCH(CF_3)_2$

PROBLEM

21.4 How can you account for the fact that methyl trifluoroacetate, $CF_3COOCH_3$, is more reactive than methyl acetate, $CH_3COOCH_3$, in nucleophilic acyl substitution reactions?

## 21.4 Nucleophilic Acyl Substitution Reactions of Carboxylic Acids

The most important reactions of carboxylic acids are those that convert the carboxyl group into other acid derivatives by nucleophilic acyl substitution, RCOOH → RCOY. Acid chlorides, anhydrides, esters, and amides can all be prepared starting from carboxylic acids (Figure 21.5).

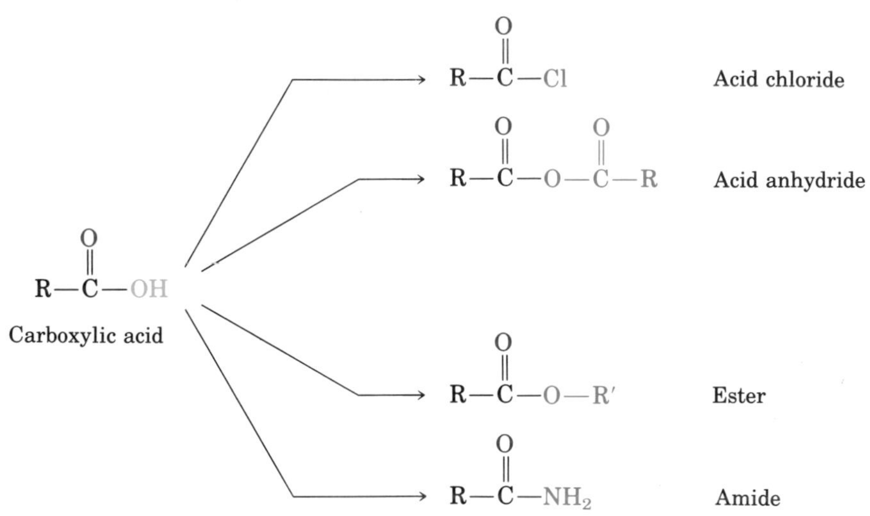


**Figure 21.5** Nucleophilic acyl substitution reactions of carboxylic acids

### CONVERSION INTO ACID CHLORIDES

Carboxylic acids are converted into carboxylic acid chlorides by treatment with thionyl chloride ($SOCl_2$), phosphorus trichloride ($PCl_3$), or oxalyl chloride (ClCOCOCl). Thionyl chloride is both cheap and convenient to use but is strongly acidic; only acid-stable molecules can survive the reaction conditions. Oxalyl chloride, on the other hand, is much more expensive but gives higher yields and reacts under milder conditions.

$$CH_3(CH_2)_7CH{=}CH(CH_2)_7\overset{O}{\overset{\|}{C}}OH \xrightarrow[\text{Benzene}]{ClCOCOCl} CH_3(CH_2)_7CH{=}CH(CH_2)_7\overset{O}{\overset{\|}{C}}Cl + HCl + CO + CO_2$$

Oleic acid → Oleyl chloride (97%)

$$\text{2,4,6-Trimethylbenzoic acid} \xrightarrow[CHCl_3]{SOCl_2} \text{2,4,6-Trimethylbenzoyl chloride (90\%)} + HCl + SO_2$$

These reactions occur by nucleophilic acyl substitution pathways in which the carboxylic acid is converted into a reactive derivative that is then attacked by a nucleophilic chloride ion. For example,

$$R{-}\overset{O}{\overset{\|}{C}}{-}OH + SOCl_2 \longrightarrow R{-}\overset{O}{\overset{\|}{C}}{-}O{-}\overset{O}{\overset{\|}{S}}{-}Cl + HCl$$

Carboxylic acid → A chlorosulfite

$$R{-}\overset{O}{\overset{\|}{C}}{-}O{-}\overset{O}{\overset{\|}{S}}{-}Cl \xrightarrow{:Cl:^-} \left[ R{-}\underset{Cl}{\overset{O^-}{C}}{-}O{-}\overset{O}{\overset{\|}{S}}{-}Cl \right] \longrightarrow R{-}\overset{O}{\overset{\|}{C}}{-}Cl + SO_2 + :Cl:^-$$

Chlorosulfite → Tetrahedral intermediate → Acid chloride

## CONVERSION INTO ACID ANHYDRIDES

Acid anhydrides are formally derived from two molecules of carboxylic acid by removing 1 equiv of water. Acyclic anhydrides are difficult to prepare directly from the corresponding acids, and only acetic anhydride is commercially available.

$$CH_3\overset{O}{\overset{\|}{C}}{-}O{-}\overset{O}{\overset{\|}{C}}CH_3$$

Acetic anhydride

Cyclic anhydrides of ring size five or six are readily obtained by high-temperature dehydration of the diacids.

$$\begin{array}{c} H_2C\text{—}COOH \\ | \\ H_2C\text{—}COOH \end{array} \xrightarrow{200^\circ C} \text{Succinic anhydride} + H_2O$$

Succinic acid Succinic anhydride

## CONVERSION INTO ESTERS

One of the most important reactions of carboxylic acids is their conversion into esters, RCOOH → RCOOR′. There are many excellent methods for accomplishing this transformation, one of which we've already studied—the $S_N2$ reaction between a carboxylate anion nucleophile and a primary alkyl halide (Section 11.5).

$$CH_3CH_2CH_2COO^-\ Na^+ + CH_3\text{—}I \xrightarrow[\text{reaction}]{S_N2} CH_3CH_2CH_2CO\text{—}CH_3 + NaI$$

Sodium butanoate Methyl butanoate, an ester (97%)

Alternatively, esters can be synthesized by a nucleophilic acyl substitution reaction between a carboxylic acid and an alcohol. Fischer[1] and Speier discovered in 1895 that esters result from simply heating a carboxylic acid in methanol or ethanol solution containing a small amount of mineral acid catalyst. Yields are good in the Fischer esterification reaction, but the need to use excess alcohol as solvent effectively limits the method to the synthesis of methyl, ethyl, and propyl esters.

$$C_6H_5CH(OH)COOH + CH_3CH_2OH \xrightarrow[\text{HCl}]{\text{Ethanol}} C_6H_5CH(OH)COOCH_2CH_3 + H_2O$$

Mandelic acid Ethyl mandelate (86%)

$$HOOC(CH_2)_4COOH + 2\ CH_3CH_2OH \xrightarrow[\text{HCl}]{\text{Ethanol}} CH_3CH_2OOC(CH_2)_4COOCH_2CH_3 + H_2O$$

Hexanedioic acid (Adipic acid) Diethyl hexanedioate (95%)

The **Fischer esterification reaction**, whose mechanism is shown in Figure 21.6, is a nucleophilic acyl substitution reaction carried out under acidic conditions. Although free carboxylic acids are not reactive enough to be attacked by most nucleophiles, they can be made much more reactive in the presence of a strong mineral acid such as HCl or $H_2SO_4$. The mineral acid acts by protonating the carbonyl-group oxygen atom, thereby giving

[1]Emil Fischer (1852–1919); b. Euskirchen, Germany; Ph.D. Strasbourg (Baeyer); professor, universities of Erlangen, Würzburg, and Berlin; Nobel prize (1902).

Protonation of the carbonyl oxygen activates the carbonyl group . . .

. . . toward nucleophilic addition by alcohol, yielding a tetrahedral intermediate.

Transfer of a proton from one oxygen to another yields a second tetrahedral intermediate and converts the hydroxyl into a good leaving group.

Loss of water yields a protonated ester.

Loss of a proton regenerates acid catalyst, and gives the free ester product.

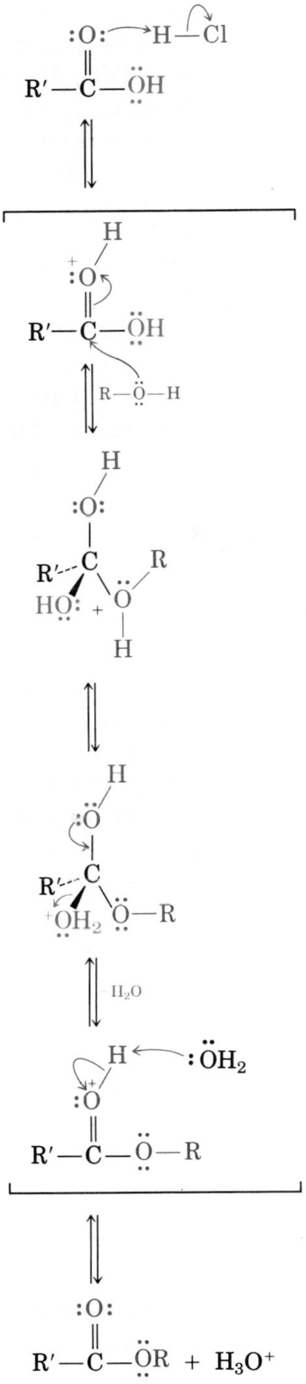

**Figure 21.6** Mechanism of Fischer esterification: The reaction is an acid-catalyzed nucleophilic acyl substitution process.

the carboxylic acid a positive charge and rendering it much more reactive toward nucleophilic attack by alcohol. Subsequent loss of water yields the ester product.

The net effect of Fischer esterification is substitution of an —OH group by —OR′. All steps are reversible, and the reaction can be driven in either direction by proper choice of reaction conditions. Ester formation is favored when a large excess of alcohol is used as solvent, but carboxylic acid formation is favored when a large excess of water is present.

One of the best pieces of evidence in support of the mechanism shown in Figure 21.6 comes from isotopic labeling experiments. When $^{18}O$-labeled methanol reacts with benzoic acid under Fischer esterification conditions, the methyl benzoate produced is found to be labeled with $^{18}O$, but the water produced is unlabeled. This experiment shows unequivocally that it is the CO—OH bond of the carboxylic acid that is cleaved, rather than the COO—H bond, and that it is the RO—H bond of the alcohol that is cleaved, rather than the R—OH bond.

$$C_6H_5\text{—}\overset{O}{\overset{\|}{C}}\text{—OH} + CH_3O\text{—H} \overset{HCl}{\rightleftharpoons} C_6H_5\text{—}\overset{O}{\overset{\|}{C}}\text{—OCH}_3 + HOH$$

A final method of ester synthesis is the reaction between a carboxylic acid and diazomethane, $CH_2N_2$, a reaction first described by von Pechmann[2] in 1894. The reaction takes place instantly at room temperature to give a high yield of the methyl ester. Though quite useful, this process does not involve a nucleophilic acyl substitution reaction, since it is the COO—H bond of the carboxylic acid that is broken.

$$C_6H_5\text{—COOH} + CH_2N_2 \xrightarrow{Ether} C_6H_5\text{—}\overset{O}{\overset{\|}{C}}\text{—O—CH}_3 + N_2$$

Benzoic acid — Methyl benzoate (100%)

The diazomethane method of ester synthesis is ideal, since it occurs under mild, neutral conditions and gives nitrogen gas as the only by-product. Unfortunately, diazomethane is both toxic and explosive, and should be handled only in small amounts by skilled persons.

PROBLEM

**21.5** Show how you would prepare these esters:
(a) Butyl acetate (b) Methyl butanoate

PROBLEM

**21.6** If 5-hydroxypentanoic acid is treated with acid catalyst, an intramolecular esterification reaction occurs. What is the structure of the product? (*Intramolecular* means within the same molecule.)

[2]Hans von Pechmann (1850–1902); b. Nuremberg; Ph.D. Greiswald; professor, universities of Munich and Tübingen.

PROBLEM

**21.7** The first step in the reaction of a carboxylic acid with diazomethane is a proton transfer to give the methyldiazonium cation ($CH_3N_2^+$) and carboxylate anion. What kind of mechanism accounts for the second step?

$$R{-}\overset{\overset{O}{\|}}{C}{-}O{-}H + :\overset{-}{C}H_2\overset{+}{N}{\equiv}N \longrightarrow R{-}\overset{\overset{O}{\|}}{C}{-}O^- + CH_3\overset{+}{N}{\equiv}N \xrightarrow{?} R{-}\overset{\overset{O}{\|}}{C}{-}O{-}CH_3 + N_2$$

## CONVERSION INTO AMIDES

Amides are carboxylic acid derivatives in which the acid hydroxyl group has been replaced by a nitrogen substituent, $-NH_2$, $-NHR$, or $-NR_2$. Amides are difficult to prepare directly from carboxylic acids, and high reaction temperatures are required. The reason for this lack of reactivity is that amines are bases (Section 25.4), which convert acidic carboxyl groups into their carboxylate anions. Since the carboxylate anion has a negative charge, it is no longer electrophilic and no longer likely to be attacked by nucleophiles.

$$R{-}\overset{\overset{O}{\|}}{C}{-}OH + :NH_3 \rightleftharpoons R{-}\overset{\overset{O}{\|}}{C}{-}O^- + NH_4^+$$

# 21.5 Chemistry of Acid Halides

## PREPARATION OF ACID HALIDES

Acid chlorides are prepared from carboxylic acids by reaction with thionyl chloride ($SOCl_2$), oxalyl chloride (ClCOCOCl), or phosphorus trichloride ($PCl_3$), as we saw in the previous section. Reaction of an acid with phosphorus tribromide ($PBr_3$) yields the acid bromide.

$$R{-}\overset{\overset{O}{\|}}{C}{-}OH \xrightarrow[\text{ClCOCOCl, or } PCl_3]{SOCl_2,} R{-}\overset{\overset{O}{\|}}{C}{-}Cl$$

$$R{-}\overset{\overset{O}{\|}}{C}{-}OH \xrightarrow[\text{Ether}]{PBr_3} R{-}\overset{\overset{O}{\|}}{C}{-}Br + HOPBr_2$$

## REACTIONS OF ACID HALIDES

Acid halides are among the most reactive of carboxylic acid derivatives and can therefore be converted into a variety of other compound types. For example, we have already seen the value of acid chlorides in preparing aryl alkyl ketones via the Friedel–Crafts reaction (Section 16.4).

$$Ar{-}H + R{-}\overset{\overset{O}{\|}}{C}{-}Cl \xrightarrow{AlCl_3} Ar{-}\overset{\overset{O}{\|}}{C}{-}R + HCl$$

Most acid halide reactions occur by nucleophilic acyl substitution mechanisms. As illustrated in Figure 21.7, the halogen can be replaced by —OH to yield an acid, by —OR to yield an ester, or by $—NH_2$ to yield an amide. In addition, the reduction of acid halides yields primary alcohols, and reaction with Grignard reagents yields tertiary alcohols. Although the reactions in Figure 21.7 are illustrated only for acid chlorides, they also take place with other acid halides.

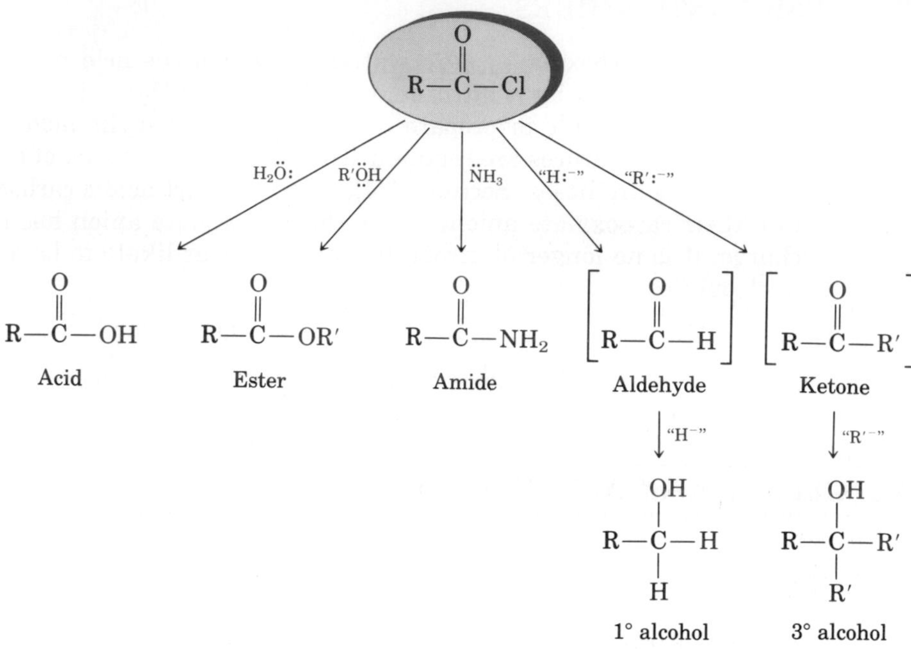


**Figure 21.7** Some nucleophilic acyl substitution reactions of acid chlorides

**Hydrolysis: Conversion of acid halides into acids** Acid chlorides react with water to yield carboxylic acids. This hydrolysis reaction is a typical nucleophilic acyl substitution process initiated by attack of water on the acid chloride carbonyl group. The initially formed tetrahedral intermediate undergoes elimination of chloride ion and loss of a proton to give the product carboxylic acid plus HCl.

$$\text{R—C(=O)—Cl} + :\ddot{O}H_2 \longrightarrow \left[\text{R—C(O}^-\text{)(Cl)—}\overset{+}{\ddot{O}}H_2 \longrightarrow \text{R—C(=O)—}\overset{+}{\ddot{O}}\text{—H (H)}\right] \xrightarrow{:\ddot{Cl}:^-} \text{R—C(=O)—OH} + \text{HCl}$$

An acid chloride

A carboxylic acid

Since HCl is generated during the hydrolysis, the reaction is often carried out in the presence of pyridine as a base to scavenge the HCl and prevent it from causing side reactions.

**Alcoholysis: Conversion of acid halides into esters** Acid chlorides react with alcohols to yield esters in a process analogous to their reaction with water to yield acids.

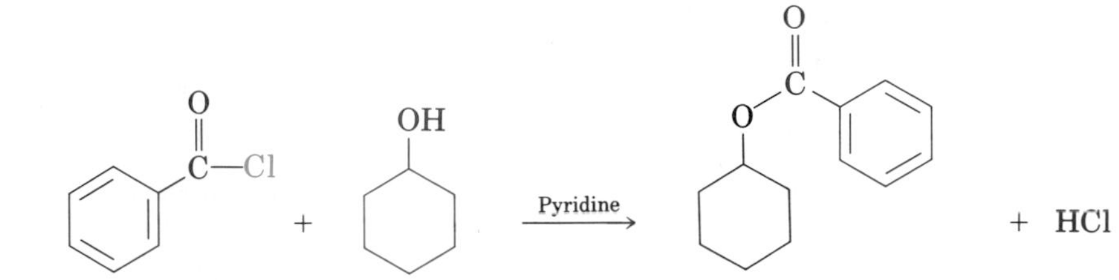

Benzoyl chloride + Cyclohexanol $\xrightarrow{\text{Pyridine}}$ Cyclohexyl benzoate (97%) + HCl

$$CH_3\overset{O}{\overset{\|}{C}}-Cl + CH_3CH_2CH_2CH_2OH \xrightarrow{\text{Pyridine}} CH_3\overset{O}{\overset{\|}{C}}\ OCH_2CH_2CH_2CH_3$$

Acetyl chloride 1-Butanol Butyl acetate (90%)

As with hydrolysis, alcoholysis reactions are usually carried out in the presence of pyridine to react with the HCl formed and prevent it from causing side reactions. If this were not done, the HCl might react with the alcohol to form an alkyl chloride, or it might add to a carbon–carbon double bond if one were present elsewhere in the molecule.

The esterification of alcohols with acid chlorides is strongly affected by steric hindrance. Bulky groups on either reaction partner slow down the rate of reaction considerably, resulting in a reactivity order among alcohols of primary > secondary > tertiary. As a result, it is often possible to esterify an unhindered alcohol selectively in the presence of a more hindered one. This can be important in complex synthesis where it is often necessary to distinguish chemically between similar functional groups. For example,

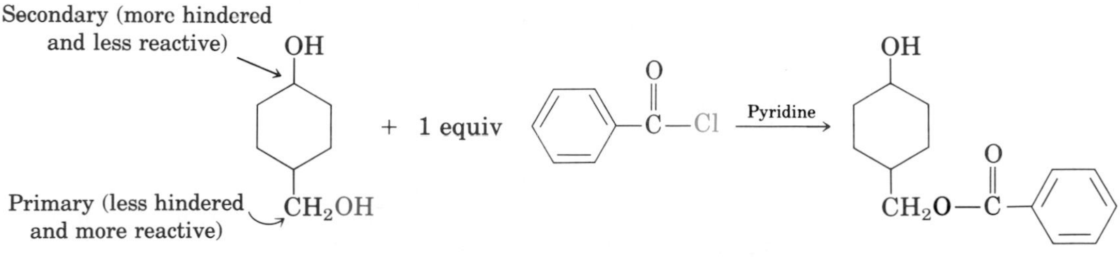

Secondary (more hindered and less reactive) OH

Primary (less hindered and more reactive) $CH_2OH$

+ 1 equiv $C_6H_5\overset{O}{\overset{\|}{C}}-Cl \xrightarrow{\text{Pyridine}}$ (OH; $CH_2O-\overset{O}{\overset{\|}{C}}-C_6H_5$) ~80%

**PROBLEM** ................................................................................

**21.8** How might you prepare these esters using a nucleophilic acyl substitution reaction of an acid chloride?

(a) $CH_3CH_2COOCH_3$ (b) $CH_3COOCH_2CH_3$ (c) Ethyl benzoate

PROBLEM

**21.9** Which method would you choose if you wanted to prepare cyclohexyl benzoate—Fischer esterification or reaction of an acid chloride with an alcohol? Explain.

**Aminolysis: Conversion of acid halides into amides** Acid chlorides react with ammonia and with amines to give amides. The reaction is rapid, and yields are usually excellent. Both mono- and disubstituted amines can be used, as well as ammonia.

$$(CH_3)_2CHC(=O){-}Cl + 2\,{:}NH_3 \xrightarrow{H_2O} (CH_3)_2CHC(=O){-}NH_2 + \overset{+}{N}H_4\overset{-}{Cl}$$

2-Methylpropanoyl chloride → 2-Methylpropanamide (83%)

$$C_6H_5C(=O){-}Cl + 2\,{:}NH(CH_3)_2 \longrightarrow C_6H_5C(=O){-}N(CH_3)_2 + \overset{+}{N}H_2(CH_3)_2\overset{-}{Cl}$$

Benzoyl chloride → *N,N*-Dimethylbenzamide (92%)

Since HCl is formed during the reaction, 2 equiv of the amine must be used; 1 equiv reacts with the acid chloride, and 1 equiv reacts with HCl to form an ammonium chloride salt. If, however, the amine component is valuable, amide synthesis is often carried out using 1 equiv of the desired amine plus 1 equiv of an inexpensive base such as NaOH.

Aminolysis reactions carried out with NaOH present are sometimes referred to as **Schotten–Baumann reactions** after their discoverers.[3] For example, the medically useful sedative trimetozine is prepared by reaction of 3,4,5-trimethoxybenzoyl chloride with the amine morpholine in the presence of 1 equiv NaOH.

$$(CH_3O)_3C_6H_2C(=O){-}Cl + H{-}\ddot{N}(CH_2CH_2)_2O \xrightarrow[H_2O]{NaOH} (CH_3O)_3C_6H_2C(=O){-}N(CH_2CH_2)_2O + NaCl$$

3,4,5-Trimethoxybenzoyl chloride + Morpholine → Trimetozine (an amide) + NaCl

PROBLEM

**21.10** Write the steps in the mechanism of the reaction between 3,4,5-trimethoxybenzoyl chloride and morpholine to form trimetozine.

PROBLEM

**21.11** Trisubstituted amines such as triethylamine can be used in place of NaOH to scavenge HCl during aminolysis reactions. Why doesn't triethylamine react with acid chlorides to yield amides?

[3]Carl Schotten (1853–1910); b. Marburg, Germany; Ph.D. Berlin (Hofmann); professor, University of Berlin.

PROBLEM

**21.12** How could you prepare these amides using an acid chloride and an amine or ammonia?
(a) $CH_3CH_2CONHCH_3$ (b) *N,N*-Diethylbenzamide (c) Propanamide

**Reduction: Conversion of acid chlorides into alcohols** Acid chlorides are reduced by lithium aluminum hydride to yield primary alcohols. The reaction is of little practical value, however, since the parent carboxylic acids are generally more readily available and are themselves reduced by $LiAlH_4$ to yield alcohols.

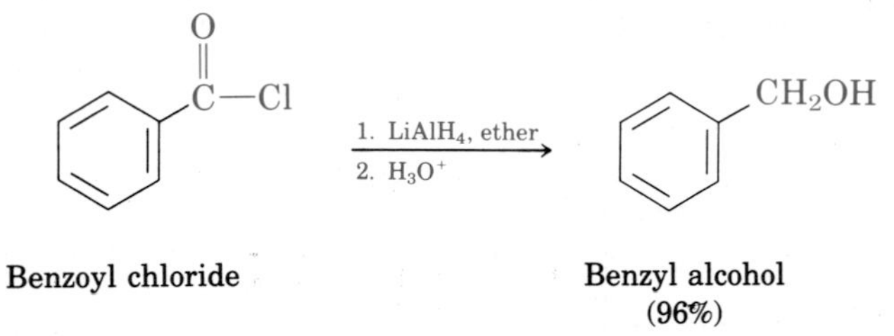

$$C_6H_5COCl \xrightarrow[2.\ H_3O^+]{1.\ LiAlH_4,\ ether} C_6H_5CH_2OH$$

Benzoyl chloride → Benzyl alcohol (96%)

Reduction occurs via a typical nucleophilic acyl substitution mechanism in which the reducing agent donates a hydride ion (H:$^-$) that attacks the carbonyl group, yielding a tetrahedral intermediate that expels chloride ion. The net effect is a substitution of —H for —Cl, yielding an aldehyde, which is then rapidly reduced by $LiAlH_4$ to yield the primary alcohol.

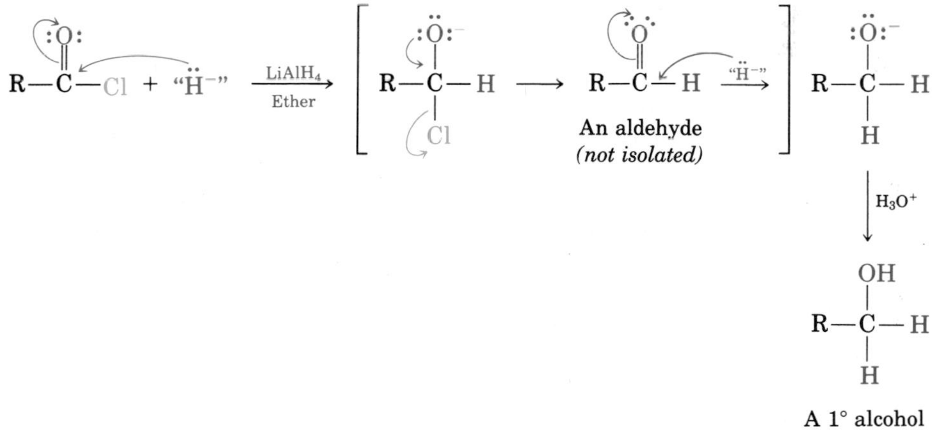

$$R{-}C({=}O){-}Cl + \text{"H}^-\text{"} \xrightarrow[Ether]{LiAlH_4} \left[ R{-}C(O^-)(H){-}Cl \longrightarrow R{-}C({=}O){-}H \right] \xrightarrow{\text{"H}^-\text{"}} R{-}CH_2{-}O^- \xrightarrow{H_3O^+} R{-}CH_2{-}OH$$

An aldehyde (*not isolated*)

A 1° alcohol

The aldehyde intermediate can be isolated if the less powerful hydride reducing agent, lithium tri-*tert*-butoxyaluminum hydride, is used. This reagent, which is obtained from reaction of $LiAlH_4$ with 3 equiv of *tert*-butyl alcohol, is particularly effective for carrying out the partial reduction of acid chlorides to aldehydes:

$$3\ (CH_3)_3COH + LiAlH_4 \longrightarrow Li^+\ {}^-AlH[OC(CH_3)_3]_3 + 3\ H_2$$

Lithium tri-*tert*-butoxyaluminum hydride

$$O_2N{-}C_6H_4{-}\overset{O}{\overset{\|}{C}}{-}Cl \xrightarrow[2.\ H_3O^+]{1.\ LiAlH[OC(CH_3)_3]_3,\ ether} O_2N{-}C_6H_4{-}\overset{O}{\overset{\|}{C}}{-}H$$

*p*-Nitrobenzoyl chloride

*p*-Nitrobenzaldehyde (81%)

Alternatively, the Rosenmund reduction of acid chlorides under catalytic hydrogenation conditions can be used, but this procedure is mechanistically quite different (Section 19.3).

$$C_6H_{11}{-}CH_2\overset{O}{\overset{\|}{C}}{-}Cl \xrightarrow[Ethyl\ acetate]{H_2,\ Pd/BaSO_4} C_6H_{11}{-}CH_2\overset{O}{\overset{\|}{C}}{-}H + HCl$$

Cyclohexylacetyl chloride

Cyclohexylacetaldehyde (70%)

**Reaction of acid chlorides with organometallic reagents** Grignard reagents react with acid chlorides to yield tertiary alcohols in which two of the substituents are identical:

$$RMgX + R'{-}\overset{O}{\overset{\|}{C}}{-}Cl \longrightarrow R'{-}\underset{R}{\underset{|}{\overset{OH}{\overset{|}{C}}}}{-}R + MgXCl$$

The mechanism for this Grignard reaction is similar to that for $LiAlH_4$ reduction. The first equivalent of Grignard reagent attacks the acid chloride. Loss of chloride ion from the tetrahedral intermediate then yields a ketone intermediate, which reacts with a second equivalent of the organometallic reagent to produce an alcohol:

$$C_6H_5{-}\overset{O}{\overset{\|}{C}}{-}Cl \xrightarrow[Ether]{CH_3MgBr} \left[C_6H_5{-}\overset{O}{\overset{\|}{C}}{-}CH_3\right] \xrightarrow[2.\ H_3O^+]{1.\ CH_3MgBr} C_6H_5{-}\underset{CH_3}{\underset{|}{\overset{OH}{\overset{|}{C}}}}{-}CH_3$$

Acid chloride

Ketone *(not isolated)*

3° alcohol

The ketone intermediate cannot usually be isolated in Grignard reactions, since addition of the second equivalent of organomagnesium reagent occurs so rapidly. Ketones can, however, be isolated from the reaction of acid chlorides with diorganocopper reagents (Section 19.4):

$$R-\overset{O}{\overset{\|}{C}}-Cl + R'_2Cu:^- Li^+ \longrightarrow R-\overset{O}{\overset{\|}{C}}-R'$$

Reactions between acid chlorides and a wide variety of lithium diorganocoppers, including dialkyl-, diaryl-, and dialkenylcoppers, are possible. Despite their apparent similarity to Grignard reactions, however, these diorganocopper reactions are almost certainly not typical nucleophilic acyl substitution processes. Rather, it is thought that diorganocopper reactions occur via a radical pathway. The reactions are generally carried out at −78°C in ether solution, and yields are often excellent. For example, manicone, a substance secreted by male ants to coordinate ant pairing and mating, has been synthesized by reaction of lithium diethylcopper with (*E*)-2,4-dimethyl-2-hexenoyl chloride:

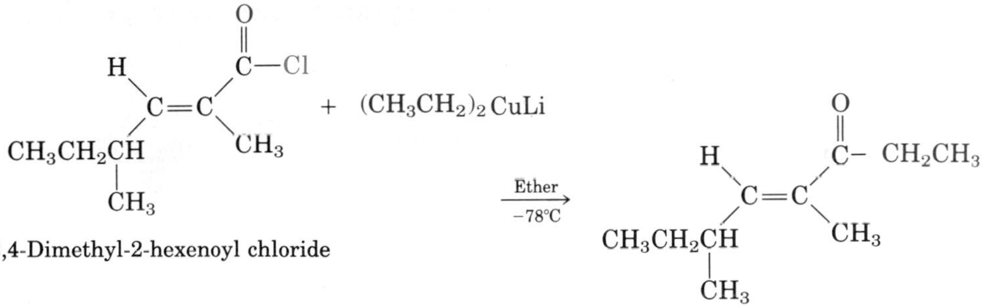

$$\underset{\text{2,4-Dimethyl-2-hexenoyl chloride}}{(H)(CH_3CH_2CH(CH_3))C=C(CH_3)(COCl)} + (CH_3CH_2)_2CuLi \xrightarrow[-78°C]{\text{Ether}} \underset{\text{Manicone (92\%)}}{(H)(CH_3CH_2CH(CH_3))C=C(CH_3)(COCH_2CH_3)}$$

Note that lithium diorganocoppers react only with acid chlorides. Acids, esters, anhydrides, and amides are inert to diorganocopper reagents.

PROBLEM

**21.13** Show how these ketones might be prepared by reaction of an acid chloride with a lithium diorganocopper reagent.

(a) 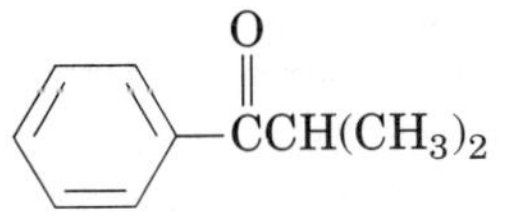 $C_6H_5-\overset{O}{\overset{\|}{C}}CH(CH_3)_2$ (b) $H_2C=CH\overset{O}{\overset{\|}{C}}CH_2CH_2CH_3$

PROBLEM

**21.14** It has been reported in the chemical literature that ketones can be prepared by reaction of acid chlorides with 1 equiv of Grignard reagent. It appears, however, that the Grignard reagent must be slowly added at low temperature to a solution of the acid chloride, rather than vice versa. Draw on your knowledge of the relative reactivity of ketones and acid chlorides to explain this observation.

## 21.6 Chemistry of Acid Anhydrides

### PREPARATION OF ACID ANHYDRIDES

The most general method of preparation of acid anhydrides is by nucleophilic acyl substitution reaction of an acid chloride with a carboxylate anion. Both symmetrical and unsymmetrical acid anhydrides can be prepared in this manner.

$$\underset{\text{Sodium formate}}{H{-}\overset{O}{\overset{\|}{C}}{-}O^-\ Na^+} + \underset{\text{Acetyl chloride}}{CH_3\overset{O}{\overset{\|}{C}}{-}Cl} \xrightarrow[25°C]{\text{Ether}} \underset{\substack{\text{Acetic formic anhydride}\\(64\%)}}{H\overset{O}{\overset{\|}{C}}O{-}\overset{O}{\overset{\|}{C}}CH_3} + NaCl$$

## REACTIONS OF ACID ANHYDRIDES

The chemistry of acid anhydrides is similar to that of acid chlorides. Although anhydrides react more slowly than acid chlorides, the kinds of reactions the two groups undergo are the same. Thus, acid anhydrides react with water to form acids, with alcohols to form esters, with amines to form amides, and with $LiAlH_4$ to form primary alcohols (Figure 21.8).

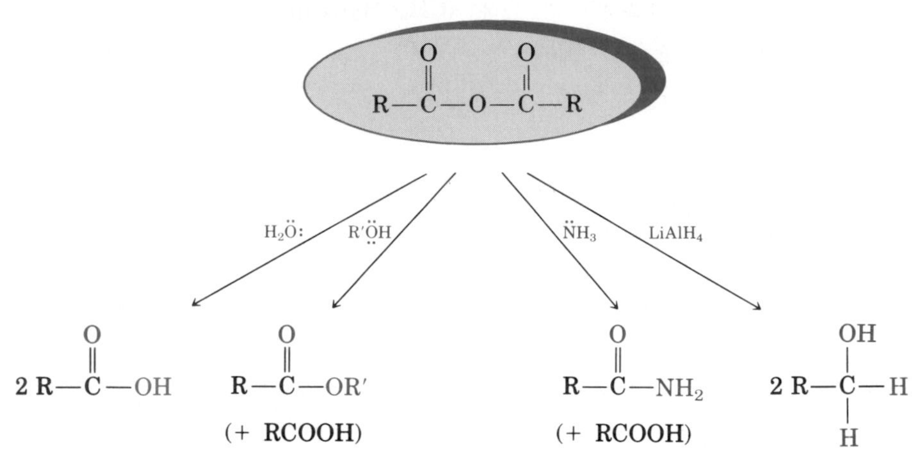


**Figure 21.8** Some reactions of acid anhydrides

Acetic anhydride is often used to prepare acetate esters of complex alcohols and to prepare substituted acetamides from amines. For example, phenacetin, a drug formerly used in headache remedies, can be prepared by reaction of *p*-ethoxyaniline with acetic anhydride. Aspirin (acetylsalicylic acid) is prepared similarly by the acetylation of *o*-hydroxybenzoic acid (salicylic acid) with acetic anhydride.

$$\underset{p\text{-Ethoxyaniline}}{:NH_2{-}C_6H_4{-}OCH_2CH_3} + \underset{\text{Acetic anhydride}}{CH_3\overset{O}{\overset{\|}{C}}O\overset{O}{\overset{\|}{C}}CH_3} \xrightarrow[H_2O]{NaOH} \underset{\text{Phenacetin (an amide)}}{H{-}N(\overset{O}{\overset{\|}{C}}CH_3){-}C_6H_4{-}OCH_2CH_3} + CH_3CO_2^-$$

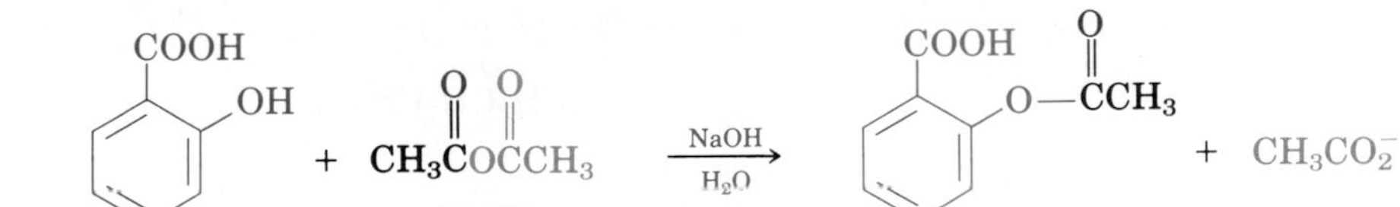

$$\text{Salicyclic acid (}o\text{-Hydroxybenzoic acid)} + CH_3\overset{O}{\overset{\|}{C}}O\overset{O}{\overset{\|}{C}}CH_3 \text{ (Acetic anhydride)} \xrightarrow[H_2O]{NaOH} \text{Aspirin (an ester)} + CH_3CO_2^-$$

Salicyclic acid (*o*-Hydroxybenzoic acid) — COOH, OH; Acetic anhydride; Aspirin (an ester) — COOH, $O-\overset{O}{\overset{\|}{C}}CH_3$

Notice in these two examples that only "half" of the anhydride molecule is used; the other half acts as the leaving group during the nucleophilic acyl substitution step and produces carboxylate anion as a by-product. Thus, anhydrides are inefficient to use, and acid chlorides are normally preferred for introducing acyl substituents other than acetyl groups.

PROBLEM

**21.15** What product would you expect to obtain from reaction of 1 equiv of methanol with a cyclic anhydride such as phthalic anhydride? What is the fate of the second "half" of the anhydride in such cases?

PROBLEM

**21.16** Write the steps involved in the mechanism of reaction between *p*-ethoxyaniline and acetic anhydride to prepare phenacetin.

PROBLEM

**21.17** Why is 1 equiv of a base such as sodium hydroxide required for the reaction between an amine and an anhydride to go to completion?

## 21.7 Chemistry of Esters

Esters are among the most important and most widespread of naturally occurring compounds. Many simple low-molecular-weight esters are pleasant-smelling liquids that are responsible for the fragrant odors of fruits and flowers. For example, methyl butanoate has been isolated from pineapple oil, and isopentyl acetate is a constituent of banana oil. The ester linkage is also present in animal fats and many biologically important molecules.

$$CH_3CH_2CH_2\overset{O}{\overset{\|}{C}}OCH_3$$

Methyl butanoate (from pineapples)

$$CH_3\overset{O}{\overset{\|}{C}}OCH_2CH_2CH(CH_3)_2$$

Isopentyl acetate (from bananas)

$$\begin{array}{l} CH_2OCOR \\ | \\ CHOCOR' \\ | \\ CH_2OCOR'' \end{array}$$

A fat (R, R′, R″ = $C_{12-18}$ chains)

The chemical industry uses esters for a variety of purposes. For example, ethyl acetate is a common solvent found in nail-polish remover, and dialkyl phthalates are used to keep plastics from turning brittle.

$$\text{C}_6\text{H}_4(\text{COOCH}_2\text{CH}_2\text{CH}_2\text{CH}_3)_2$$

Dibutyl phthalate (a plasticizer)

## PREPARATION OF ESTERS

Esters are usually prepared either from acids or from acid anhydrides by the methods already discussed. Thus, carboxylic acids are converted directly into esters either by $S_N2$ reaction of a carboxylate salt with a primary alkyl halide, by Fischer esterification of a carboxylic acid with a low-molecular-weight alcohol in the presence of a mineral acid catalyst, or by reaction of a carboxylic acid with diazomethane (Section 21.4). In addition, acid chlorides are converted into esters by treatment with an alcohol in the presence of base (Section 21.5).

$$R\text{—}\overset{O}{\overset{\|}{C}}\text{—}OH \xrightarrow[2.\ R'X]{1.\ \text{Salt formation}} R\text{—}\overset{O}{\overset{\|}{C}}\text{—}OR'$$

$$R\text{—}\overset{O}{\overset{\|}{C}}\text{—}OH \xrightarrow{R'OH,\ HCl} R\text{—}\overset{O}{\overset{\|}{C}}\text{—}OR'$$

$$R\text{—}\overset{O}{\overset{\|}{C}}\text{—}OH \xrightarrow{CH_2N_2,\ \text{ether}} R\text{—}\overset{O}{\overset{\|}{C}}\text{—}OCH_3$$

$$R\text{—}\overset{O}{\overset{\|}{C}}\text{—}Cl + R'OH \xrightarrow{\text{Pyridine}} R\text{—}\overset{O}{\overset{\|}{C}}\text{—}OR'$$

## REACTIONS OF ESTERS

Esters exhibit the same kinds of chemistry that we've seen for other acid derivatives, but they are less reactive toward nucleophiles than either acid chlorides or anhydrides. Figure 21.9 shows some general reactions of esters. All of these reactions are equally applicable to both acyclic and cyclic esters (**lactones**).

**Hydrolysis: Conversion of esters into carboxylic acids** Esters are hydrolyzed either by aqueous base or by aqueous acid to yield carboxylic acid plus alcohol:

$$R\text{—}\overset{O}{\overset{\|}{C}}\text{—}O\text{—}R' \xrightarrow[H_3O^+ \text{ or } ^-OH]{H_2O} R\text{—}\overset{O}{\overset{\|}{C}}\text{—}OH + R'OH$$

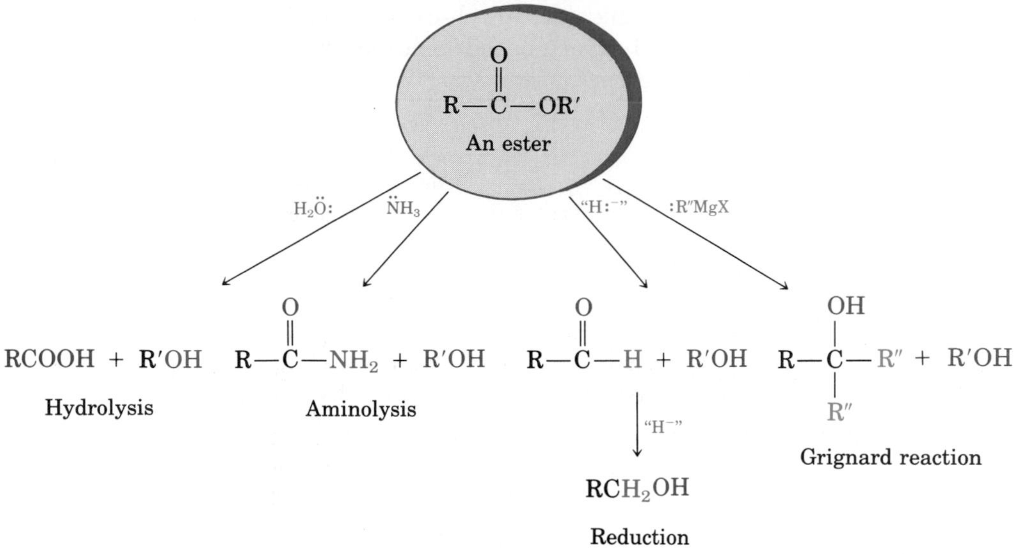

An ester: $R-\overset{O}{\overset{\|}{C}}-OR'$

$H_2\ddot{O}:$ → $RCOOH + R'OH$ — Hydrolysis

$\ddot{N}H_3$ → $R-\overset{O}{\overset{\|}{C}}-NH_2 + R'OH$ — Aminolysis

"H:⁻" → $R-\overset{O}{\overset{\|}{C}}-H + R'OH$ —("H⁻")→ $RCH_2OH$ — Reduction

:R″MgX → $R-\underset{R''}{\overset{OH}{C}}-R'' + R'OH$ — Grignard reaction

**Figure 21.9** Some reactions of esters

Ester hydrolysis in basic solution is called **saponification**, after the Latin *sapo*, "soap." As we'll see in Section 28.2, the boiling of wood ash extract with animal fat to make soap is indeed a saponification, since wood ash contains alkali and fats have ester linkages. Ester hydrolysis occurs through a typical nucleophilic acyl substitution pathway in which hydroxide ion nucleophile adds to the ester carbonyl group to give a tetrahedral intermediate. Loss of alkoxide ion then gives a carboxylic acid, which is deprotonated to give the carboxylate salt.

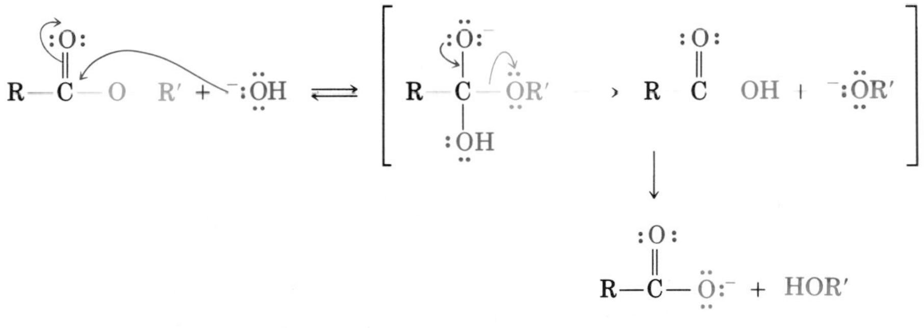

$$R-\overset{:O:}{\overset{\|}{C}}-O-R' + {}^{-}:\ddot{O}H \rightleftharpoons \left[ R-\underset{:\ddot{O}H}{\overset{:\ddot{O}:^{-}}{C}}-\ddot{O}R' \longrightarrow R-\overset{:O:}{\overset{\|}{C}}-OH + {}^{-}:\ddot{O}R' \right]$$

$$\downarrow$$

$$R-\overset{:O:}{\overset{\|}{C}}-\ddot{O}:^{-} + HOR'$$

One of the most elegant experiments in support of this mechanism involves isotope labeling. The Russian chemist D. N. Kursanov[4] showed that, when ethyl propanoate labeled with $^{18}O$ in the ether-type oxygen is hydrolyzed in aqueous sodium hydroxide, the $^{18}O$ label shows up exclusively in the ethanol product. None of the label remains with the propanoate salt,

[4]D. N. Kursanov (1899– ); graduate of Moscow University (1924); professor, Moscow Textile Institute (1930–1947); Institute of Organic Chemistry (1947–1953); Institute of Scientific Information, USSR Academy of Sciences (1953– ).

indicating that saponification occurs by cleavage of the acyl–oxygen bond (RCO—OR′) rather than the alkyl–oxygen bond (RCOO—R′). This result is just what we would expect, based on our knowledge of the nucleophilic acyl substitution mechanism.

$$CH_3CH_2\overset{O}{\overset{\|}{C}}-{}^{18}OCH_2CH_3 + {}^-:\ddot{O}H \xrightarrow{H_2O} CH_3CH_2\overset{O}{\overset{\|}{C}}-\ddot{O}:^- + H^{18}OCH_2CH_3$$

Acidic hydrolysis of esters can occur by more than one mechanism, depending on the structure of substrate. The usual pathway, however, is just the reverse of the Fischer esterification reaction (Section 21.4). The ester is first activated toward nucleophilic attack by protonation of the carboxyl oxygen atom. Nucleophilic attack by water, followed by transfer of a proton and elimination of alcohol, then yields the carboxylic acid (Figure 21.10).

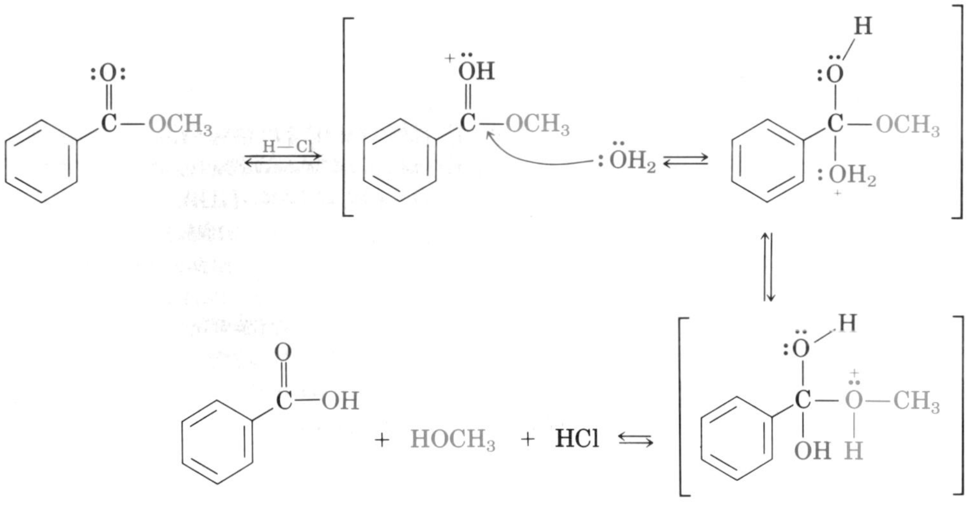

**Figure 21.10** Mechanism of acid-catalyzed ester hydrolysis: The forward reaction is a hydrolysis; the back-reaction is a Fischer esterification.

**PROBLEM**

**21.18** How would you synthesize the $^{18}O$-labeled ethyl propanoate used by Kursanov in his mechanistic studies? Assume that $^{18}O$-labeled acetic acid is your only source of isotopic oxygen.

**PROBLEM**

**21.19** Why do you suppose saponification of esters is irreversible? In other words, why doesn't treatment of a carboxylic acid with alkoxide ion lead to ester formation?

**Aminolysis: Conversion of esters into amides** Esters react with ammonia and amines via a typical nucleophilic acyl substitution pathway to yield amides. The reaction is not often used, however, since higher yields are normally obtained by aminolysis of acid chlorides (Section 21.5).

$$C_6H_5CO_2CH_3 \xrightarrow[\text{Ether}]{:NH_3} C_6H_5CONH_2 + CH_3OH$$

Methyl benzoate → Benzamide

**Reduction: Conversion of esters into alcohols** Esters are easily reduced by treatment with lithium aluminum hydride to yield primary alcohols (Section 17.6).

$$CH_3CH_2CH{=}CHCOCH_2CH_3 \xrightarrow[\text{2. } H_3O^+]{\text{1. } LiAlH_4\text{, ether}} CH_3CH_2CH{=}CHCH_2OH + CH_3CH_2OH$$

Ethyl 2-pentenoate → 2-Penten-1-ol (91%)

$$\text{(γ-valerolactone, } CH_3\text{)} \xrightarrow[\text{2. } H_3O^+]{\text{1. } LiAlH_4\text{, ether}} HOCH_2CH_2CH_2CH(OH)CH_3$$

1,4-Pentanediol (86%)

The mechanism of ester (and lactone) reductions is similar to that we saw earlier for acid chloride reduction. A hydride ion first adds to the carbonyl group, followed by elimination of alkoxide ion to yield an aldehyde intermediate. Further addition of hydride to the aldehyde gives the primary alcohol.

$$R{-}C({=}O){-}OR' + \text{“}:H^-\text{”} \xrightarrow{LiAlH_4} \left[ R{-}C(O^-)(H){-}OR' \longrightarrow R{-}C({=}O){-}H + R'O^- \right]$$

$$\xrightarrow{\text{“}:H^-\text{”}} \left[ R{-}C(O^-)(H){-}H \right] \xrightarrow{H_3O^+} R{-}C(OH)(H){-}H$$

The aldehyde intermediate can be isolated if DIBAH (diisobutylaluminum hydride) is used as the reducing agent instead of $LiAlH_4$. Great care must be taken—exactly 1 equiv of hydride reagent must be used, and the reaction must be carried out at $-78°C$. If these conditions are met, however, the DIBAH reduction of esters can be an excellent method of aldehyde synthesis.

$$CH_3(CH_2)_{10}\overset{O}{\overset{\|}{C}}OCH_2CH_3 \xrightarrow[2.\ H_3O^+]{1.\ \text{DIBAH in toluene}} CH_3(CH_2)_{10}\overset{O}{\overset{\|}{C}}H + CH_3CH_2OH$$

Ethyl dodecanoate — Dodecanal (88%)

where DIBAH = $[(CH_3)_2CHCH_2]_2AlH$

PROBLEM

**21.20** What product would you expect from the reaction of butyrolactone with DIBAH?

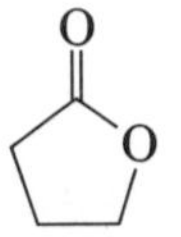

Butyrolactone

PROBLEM

**21.21** Show the products you would obtain by reduction of these esters with $LiAlH_4$.
(a) $CH_3CH_2CH_2CH(CH_3)COOCH_3$ (b) Phenyl benzoate

**Reaction of esters with Grignard reagents** Esters and lactones react with 2 equiv of Grignard reagent or organolithium reagent to yield tertiary alcohols (Section 17.7). The reactions occur readily and give excellent yields of products by the usual nucleophilic addition mechanism. For example,

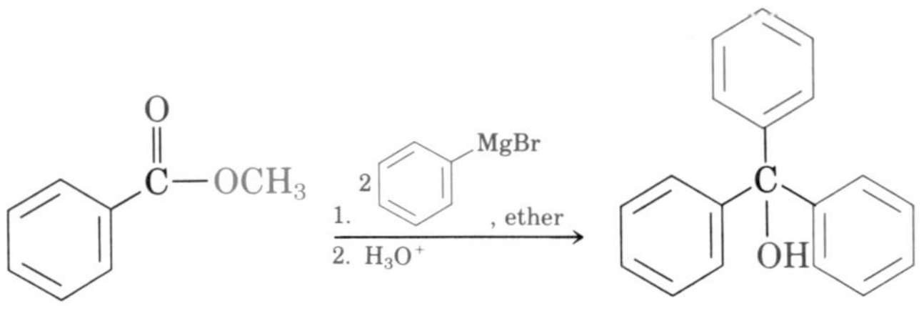

$$C_6H_5COOCH_3 \xrightarrow[2.\ H_3O^+]{1.\ 2\ C_6H_5MgBr,\ \text{ether}} (C_6H_5)_3COH$$

Methyl benzoate — Triphenylmethanol (96%)

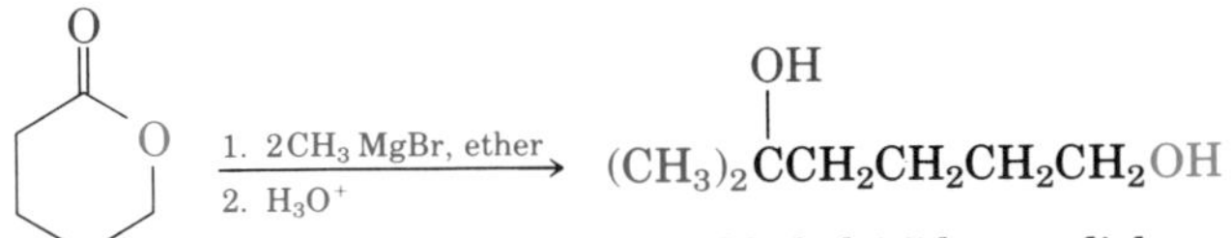

$$\text{Valerolactone} \xrightarrow[2.\ H_3O^+]{1.\ 2CH_3MgBr,\ \text{ether}} (CH_3)_2\overset{OH}{\overset{|}{C}}CH_2CH_2CH_2CH_2OH$$

Valerolactone — 5-Methyl-1,5-hexanediol

PROBLEM

**21.22** What ester and what Grignard reagent might you start with to prepare these alcohols?
(a) 2-Phenyl-2-propanol (b) 1,1-Diphenylethanol (c) 3-Ethyl-3-heptanol

## 21.8 Chemistry of Amides

### PREPARATION OF AMIDES

Amides are usually prepared by reaction of an acid chloride with an amine. Ammonia, monosubstituted amines, and disubstituted amines all undergo this reaction (Section 21.5).

$$R{-}\overset{O}{\overset{\|}{C}}{-}Cl \xrightarrow{:NH_3} R{-}\overset{O}{\overset{\|}{C}}{-}NH_2$$

$$R{-}\overset{O}{\overset{\|}{C}}{-}Cl \xrightarrow{R'\ddot{N}H_2} R{-}\overset{O}{\overset{\|}{C}}{-}NHR'$$

$$R{-}\overset{O}{\overset{\|}{C}}{-}Cl \xrightarrow{R'_2\ddot{N}H} R{-}\overset{O}{\overset{\|}{C}}{-}NR'_2$$

### REACTIONS OF AMIDES

Amides are much less reactive than acid chlorides, acid anhydrides, or esters. Thus, the amide linkage serves as the basic unit from which all proteins are made (Chapter 27).

$$\underset{\text{Amino acids}}{H_2N{-}\overset{R}{\overset{|}{C}H}{-}COOH} \Rightarrow \underset{\substack{\text{A protein} \\ \text{(a polyamide)}}}{\sim NH{-}\overset{R}{\overset{|}{C}H}{-}\underset{\underset{O}{\|}}{C}{-}NH{-}\overset{R'}{\overset{|}{C}H}{-}\underset{\underset{O}{\|}}{C}{-}NH{-}\overset{R''}{\overset{|}{C}H}{-}\underset{\underset{O}{\|}}{C}\sim}$$

Amides undergo hydrolysis to yield carboxylic acids plus amine upon heating in either aqueous acid or aqueous base. The conditions required for amide hydrolysis are more severe than those required for the hydrolysis of acid chlorides or esters, but the mechanisms are similar. The basic hydrolysis of amides yields an amine and a carboxylate ion as products, and occurs via nucleophilic addition of hydroxide to the amide carbonyl group, followed by elimination of amide ion ($^-{:}\ddot{N}H_2$). The acidic hydrolysis reaction occurs by nucleophilic addition of water to the protonated amide, followed by loss of ammonia.

Acidic hydrolysis

$$R{-}\overset{\ddot{O}:}{\overset{\|}{C}}{-}NH_2 \underset{}{\overset{HCl}{\rightleftharpoons}} \left[ R{-}\overset{^+\ddot{O}{-}H}{\overset{\|}{C}}{-}NH_2 \overset{:\ddot{O}H_2}{\rightleftharpoons} R{-}\overset{:\ddot{O}{-}H}{\overset{|}{\underset{\underset{^+OH_2}{|}}{C}}}{-}\ddot{N}H_2 \rightleftharpoons R{-}\overset{:\ddot{O}{-}H}{\overset{|}{\underset{\underset{:\ddot{O}H}{|}}{C}}}{-}\overset{+}{N}H_3 \right] \longrightarrow NH_3 + R{-}\overset{O}{\overset{\|}{C}}{-}OH$$

Basic hydrolysis

$$R{-}\overset{O}{\overset{\|}{C}}{-}NH_2 + {:}\ddot{O}H^- \rightleftharpoons \left[ R{-}\underset{:OH}{\overset{:O:^-}{C}}{-}\ddot{N}H_2 \longrightarrow R{-}\overset{O}{\overset{\|}{C}}{-}OH + {}^-{:}\ddot{N}H_2 \right] \longrightarrow R{-}\overset{O}{\overset{\|}{C}}{-}O^- + NH_3$$

Like other carboxylic acid derivatives, amides can be reduced by $LiAlH_4$. The product of this reduction, however, is an amine rather than an alcohol. The net effect of amide reduction reaction is to convert the amide carbonyl group into a methylene group ($C{=}O \rightarrow CH_2$). This kind of reaction is specific for amides and does not occur with other carboxylic acid derivatives.

$$CH_3(CH_2)_{10}\overset{O}{\overset{\|}{C}}NHCH_3 \xrightarrow[2.\ H_2O]{1.\ LiAlH_4,\ ether} CH_3(CH_2)_{10}\underset{H}{\overset{H}{C}}NHCH_3$$

*N*-Methyldodecanamide → Dodecylmethylamine (95%)

Amide reduction is thought to occur by initial nucleophilic addition of hydride ion to the amide carbonyl group, followed by expulsion of the oxygen atom as an aluminate anion to give an imine intermediate. The intermediate imine is then further reduced by $LiAlH_4$ to yield the amine.

$$R{-}\overset{O:}{\overset{\|}{C}}{-}NH_2 \xrightarrow[Ether]{:H^-\ LiAlH_3} \left[ R{-}\underset{H}{\overset{:\ddot{O}{-}\bar{A}lH_3}{C}}{-}\ddot{N}H_2 \longrightarrow R{-}\overset{^+NH_2}{\overset{\|}{C}}{-}H \rightleftharpoons R{-}\overset{\ddot{N}H}{\overset{\|}{C}}{-}H \right]$$

Iminium ion; Imine

$$\xrightarrow{:H^-} \left[ R{-}\underset{H}{\overset{^-{:}\ddot{N}H}{C}}{-}H \right] \xrightarrow{H_2O} R{-}\underset{H}{\overset{:NH_2}{C}}{-}H$$

Lithium aluminum hydride reduction is equally effective with both acyclic and cyclic amides (**lactams**). Lactam reductions provide cyclic amines in good yield, and constitute a valuable method of synthesis.

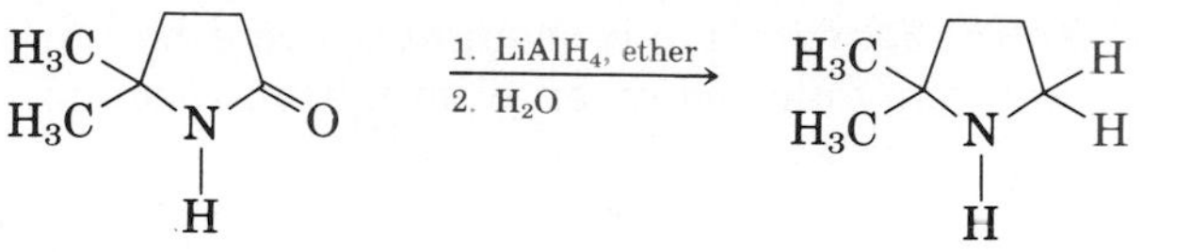

$$\text{A lactam (2,2-dimethyl-5-oxopyrrolidine: } H_3C, H_3C \text{ on C; N–H; C=O)} \xrightarrow[\text{2. } H_2O]{\text{1. } LiAlH_4\text{, ether}} \text{A cyclic amine (80\%)}$$

A lactam — A cyclic amine (80%)

PROBLEM

**21.23** How would you convert *N*-ethylbenzamide to these products?
(a) Benzoic acid (b) Benzyl alcohol (c) $C_6H_5CH_2NHCH_2CH_3$

PROBLEM

**21.24** How would you use the reaction between an amide and $LiAlH_4$ as the key step in going from bromocyclohexane to (dimethylaminomethyl)cyclohexane? Formulate all steps involved in the reaction sequence.

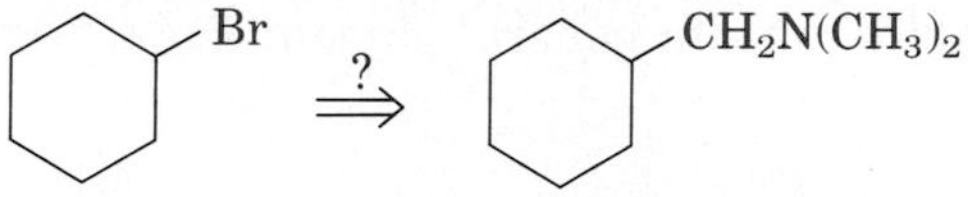

$$C_6H_{11}Br \overset{?}{\Longrightarrow} C_6H_{11}CH_2N(CH_3)_2$$

(Dimethylaminomethyl)cyclohexane

## 21.9 Chemistry of Nitriles

Nitriles are not related to carboxylic acids in the same sense that acyl derivatives are. Nevertheless, the chemistry of nitriles and carboxylic acids is so entwined that the two classes of compounds should be considered together.

### PREPARATION OF NITRILES

The simplest method of nitrile preparation is $S_N2$ reaction of cyanide ion with a primary alkyl halide, a reaction discussed in Section 20.6. This method is limited by the usual $S_N2$ steric constraints to the synthesis of $\alpha$-unsubstituted nitriles.

$$R\text{—}CH_2\text{—}Br + Na^+CN^- \xrightarrow[\text{reaction}]{S_N2} R\text{—}CH_2\text{—}CN + NaBr$$

Another excellent method for preparing nitriles is by dehydration of a primary amide. Thionyl chloride is often used to effect this reaction, although other dehydrating agents such as $P_2O_5$, $POCl_3$, and acetic anhydride can also be used.

$$CH_3CH_2CH_2CH_2\underset{\displaystyle CH_2CH_3}{\underset{|}{C}}H\overset{\displaystyle O}{\overset{\|}{C}}\text{—}NH_2 \xrightarrow[80^\circ C]{SOCl_2\text{, benzene}} CH_3CH_2CH_2CH_2\underset{\displaystyle CH_2CH_3}{\underset{|}{C}}HC{\equiv}N + SO_2 + 2\ HCl$$

2-Ethylhexanamide — 2-Ethylhexanenitrile (94%)

Amide dehydration is thought to occur by initial reaction on the amide oxygen atom, followed by an elimination reaction.

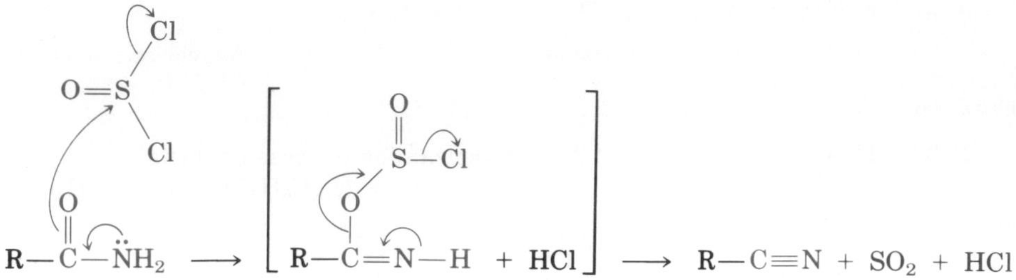

$$R{-}C(=O){-}\ddot{N}H_2 + O{=}SCl_2 \longrightarrow \left[ R{-}C(OS(=O)Cl){=}N{-}H + HCl \right] \longrightarrow R{-}C{\equiv}N + SO_2 + HCl$$

Although both methods of nitrile synthesis—$S_N2$ displacement by cyanide ion on an alkyl halide and amide dehydration—are useful, the synthesis from amides is more general since it is not limited by steric hindrance.

## REACTIONS OF NITRILES

The chemistry of nitriles is similar in many respects to the chemistry of carbonyl compounds. Like carbonyl groups, nitriles are strongly polarized making the carbon atom electrophilic. Thus, nitriles are attacked by nucleophiles to yield $sp^2$-hybridized intermediate imine anions in a reaction analogous to the formation of an $sp^3$-hybridized alkoxide ion intermediate by nucleophilic addition to a carbonyl group (Figure 21.11).

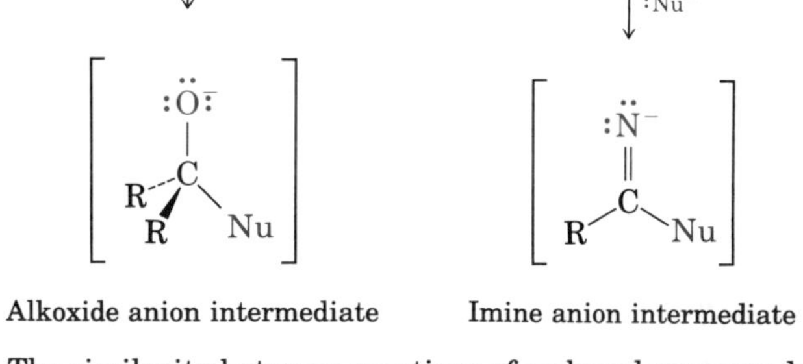

$$R_2C^{\delta+}{=}\ddot{O}^{\delta-} \text{ (Carbonyl compound)} \xrightarrow{:Nu^-} \left[ R_2C(Nu){-}\ddot{O}:^- \right] \text{ (Alkoxide anion intermediate)}$$

$$R{-}C^{\delta+}{\equiv}N^{\delta-} \text{ (Nitrile)} \xrightarrow{:Nu^-} \left[ R(Nu)C{=}\ddot{N}:^- \right] \text{ (Imine anion intermediate)}$$

**Figure 21.11** The similarity between reactions of carbonyl compounds and nitriles with nucleophiles

The two most important reactions of nitriles are hydrolysis and reduction. In addition, nitriles can be partially reduced and hydrolyzed to yield aldehydes, and can be treated with Grignard reagents to yield ketones (Figure 21.12).

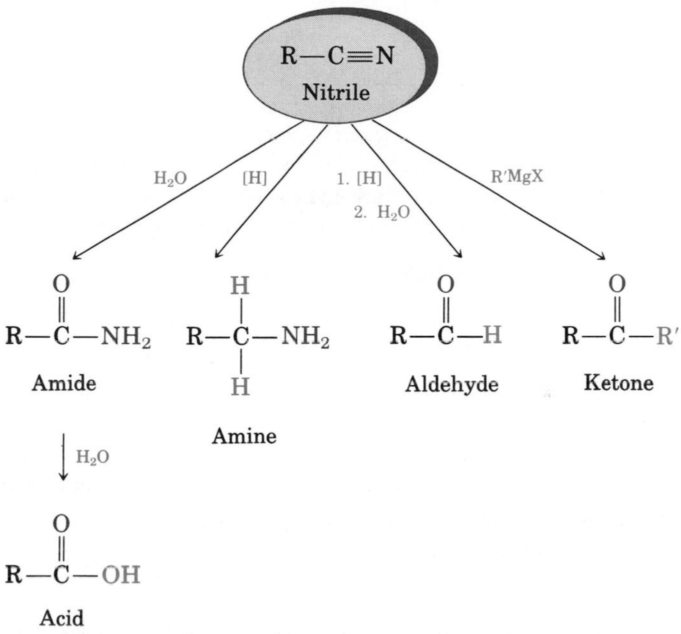


**Figure 21.12** Some reactions of nitriles

**Hydrolysis: Conversion of nitriles into acids** Nitriles are hydrolyzed in either acidic or basic aqueous solution to yield carboxylic acids and ammonia or an amine:

$$R{-}C{\equiv}N \xrightarrow{H_3O^+ \text{ or } ^-\!:OH} RCOOH + NH_3$$

The mechanism of the alkaline hydrolysis involves nucleophilic addition of hydroxide ion to the polar C≡N bond in a manner analogous to that of nucleophilic addition to a polar carbonyl C=O bond. The initial product is a hydroxy imine, which is rapidly converted to an amide in a step similar to the conversion of an enol to a ketone. Further hydrolysis of the amide then yields the carboxylic acid (Figure 21.13).

The conditions required for nitrile alkaline hydrolysis are severe (KOH, 200°C), and the amide intermediate can sometimes be isolated if milder conditions are used.

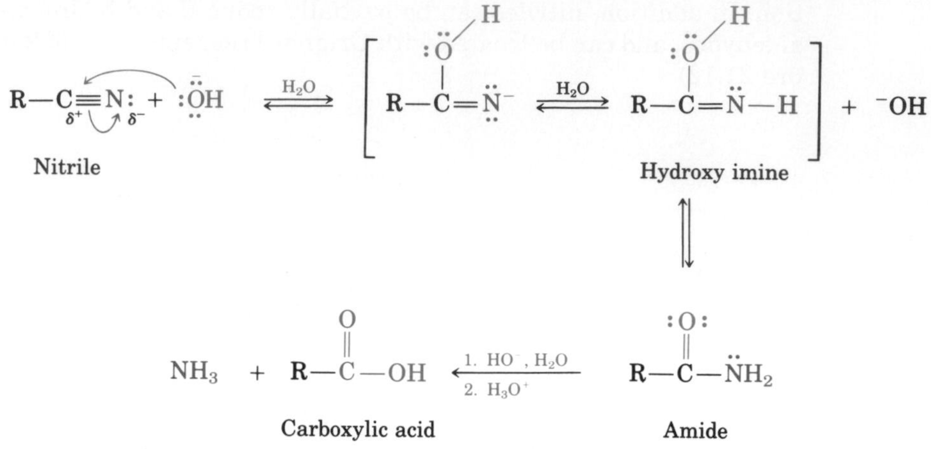

$$\mathrm{R{-}C{\equiv}N\!:} + \mathrm{:\ddot{O}H^-} \xrightleftharpoons{H_2O} \left[ \mathrm{R{-}C(OH){=}N^-} \xrightleftharpoons{H_2O} \mathrm{R{-}C(OH){=}N{-}H} \right] + {}^-\mathrm{OH}$$

Nitrile — Hydroxy imine

$$\updownarrow$$

$$\mathrm{NH_3} + \mathrm{R{-}C({=}O){-}OH} \xleftarrow[2.\ H_3O^+]{1.\ HO^-,\ H_2O} \mathrm{R{-}C({=}O){-}NH_2}$$

Carboxylic acid — Amide

Recall the following:

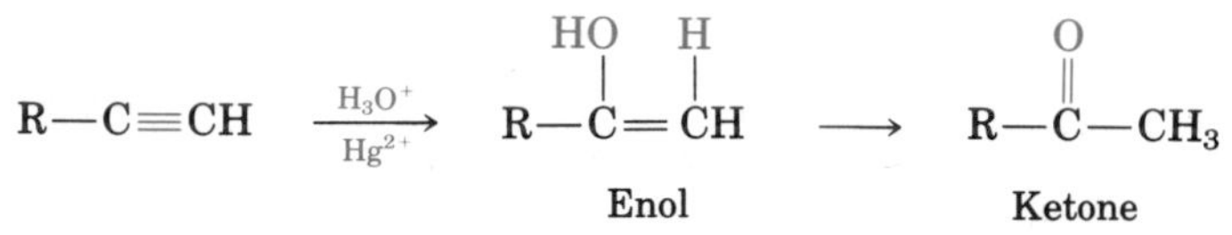

$$\mathrm{R{-}C{\equiv}CH} \xrightarrow[Hg^{2+}]{H_3O^+} \mathrm{R{-}C(OH){=}CH_2} \longrightarrow \mathrm{R{-}C({=}O){-}CH_3}$$

Enol — Ketone

**Figure 21.13** Mechanism of amide hydrolysis to yield a carboxylic acid

**PROBLEM**

**21.25** Acid-catalyzed nitrile hydrolysis occurs by initial protonation of the nitrile nitrogen atom, followed by nucleophilic addition of water. Show all the steps involved in the acidic hydrolysis of a nitrile to yield a carboxylic acid.

**Reduction: Conversion of nitriles into amines** Treatment of nitriles with lithium aluminum hydride gives primary amines in high yields. For example,

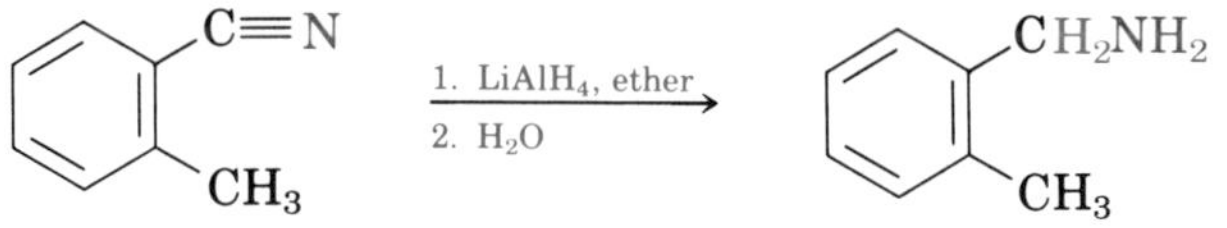

$$o\text{-}\mathrm{CH_3C_6H_4C{\equiv}N} \xrightarrow[2.\ H_2O]{1.\ LiAlH_4,\ ether} o\text{-}\mathrm{CH_3C_6H_4CH_2NH_2}$$

*o*-Methylbenzonitrile — *o*-Methylbenzylamine (88%)

The reaction occurs by nucleophilic addition of hydride ion to the polar C≡N bond, yielding an imine anion, which undergoes further addition of a second equivalent of hydride to give the final product. If, however, a less powerful reducing agent such as DIBAH is used, the second addition of hydride does not occur, and the imine intermediate can be hydrolyzed to yield an aldehyde:

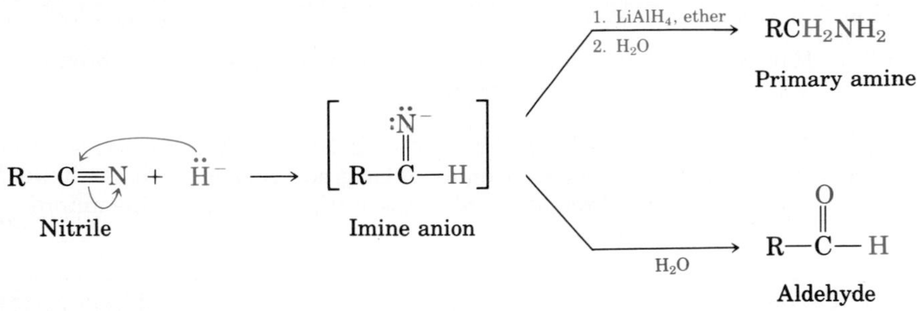

$$R{-}C{\equiv}N + \ddot{H}^- \longrightarrow \left[ R{-}\underset{}{C}(={\ddot{N}}^-){-}H \right] \begin{cases} \xrightarrow{\text{1. LiAlH}_4\text{, ether; 2. } H_2O} RCH_2NH_2 \text{ (Primary amine)} \\ \xrightarrow{H_2O} R{-}C(=O){-}H \text{ (Aldehyde)} \end{cases}$$

Nitrile — Imine anion

For example,

$$\text{2-(1-cyano-2-propyl)-1-isopropenylcyclohexane: } C_6H_{10}(CH(CH_3)CH_2C{\equiv}N)(C(CH_3){=}CH_2) \xrightarrow[\text{2. } H_2O]{\text{1. DIBAH, toluene, } -78^\circ C} C_6H_{10}(CH(CH_3)CH_2CHO)(C(CH_3){=}CH_2)$$

96%

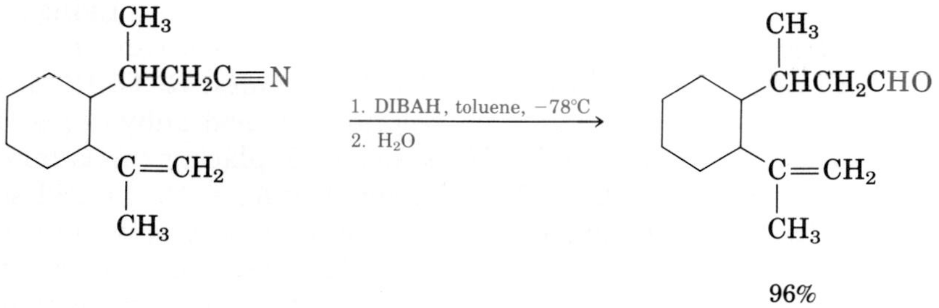

**Reaction of nitriles with organometallic reagents** Grignard reagents add to nitriles giving intermediate imine anions that can be hydrolyzed to yield ketones:

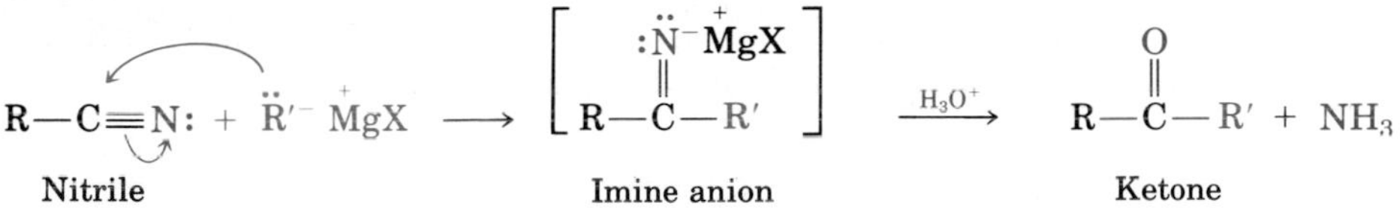

$$R{-}C{\equiv}N: + \ddot{R'}^- \overset{+}{Mg}X \longrightarrow \left[ R{-}C(={:}\ddot{N}^- \overset{+}{Mg}X){-}R' \right] \xrightarrow{H_3O^+} R{-}C(=O){-}R' + NH_3$$

Nitrile — Imine anion — Ketone

The reaction is similar to the DIBAH reduction of nitriles, except that the attacking nucleophile is a carbanion rather than a hydride ion. Yields are generally high. For example,

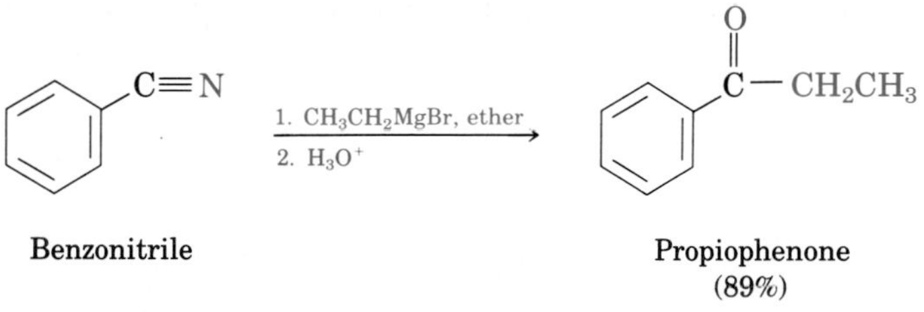

$$C_6H_5{-}C{\equiv}N \xrightarrow[\text{2. } H_3O^+]{\text{1. } CH_3CH_2MgBr\text{, ether}} C_6H_5{-}C(=O){-}CH_2CH_3$$

Benzonitrile — Propiophenone (89%)

PROBLEM

**21.26** Show how would you prepare these carbonyl compounds from an appropriate nitrile.

(a) $CH_3CH_2COCH_2CH_3$
(b) $CH_3CH_2COCH(CH_3)_2$
(c) $(CH_3)_2CHCHO$
(d) Acetophenone
(e) Dicyclohexyl ketone ($C_6H_{11}{-}C(=O){-}C_6H_{11}$)

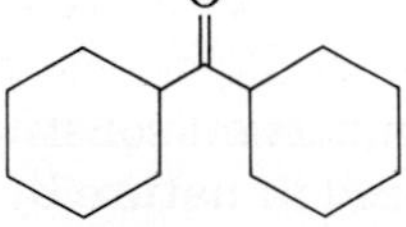

PROBLEM

**21.27** How would you prepare 1-phenyl-2-butanone from benzyl bromide, $C_6H_5CH_2Br$? More than one step is required.

PROBLEM

**21.28** Why do you suppose only 1 equiv of Grignard reagent adds to nitriles? Why doesn't a second Grignard reagent add to the initially formed imine anion?

## 21.10 Thiol Esters: Biological Carboxylic Acid Derivatives

Nucleophilic acyl substitution reactions take place in living organisms just as they take place in the chemical laboratory. The same principles apply in both cases. Nature, however, uses **thiol esters, RCOSR′**, as reactive acyl derivatives, rather than acid chlorides or acid anhydrides. The $pK_a$ of a typical alkane thiol (R—SH) is about 10, placing thiols midway between carboxylic acids ($pK_a \approx 5$) and alcohols ($pK_a \approx 16$) in acid strength. As a result, thiol esters are intermediate in reactivity between acid anhydrides and esters. They are not so reactive that they hydrolyze rapidly like anhydrides, yet they are more reactive than normal esters toward nucleophilic attack.

Acetyl coenzyme A (abbreviated acetyl CoA; Figure 21.14) is the most common thiol ester found in nature. Acetyl CoA is an enormously complex molecule by comparison with acetyl chloride or acetic anhydride, yet it serves exactly the same purpose. Nature uses acetyl CoA as a reactive acylating agent in nucleophilic acyl substitution reactions.

$CH_3—C(=O)—S—CH_2CH_2NHC(=O)CH_2CH_2NHC(=O)CH(OH)C(CH_3)_2CH_2OP(=O)(OH)—O—P(=O)(OH)—OCH_2$–(ribose, 3′-$OPO_3^-$, 2′-OH)–adenine ($NH_2$)

Thiol ester

$$CH_3—\overset{O}{\overset{\|}{C}}—\ddot{S}—CoA + :Nu^- \longrightarrow CH_3—\overset{O}{\overset{\|}{C}}—Nu + {}^-:\ddot{S}—CoA$$

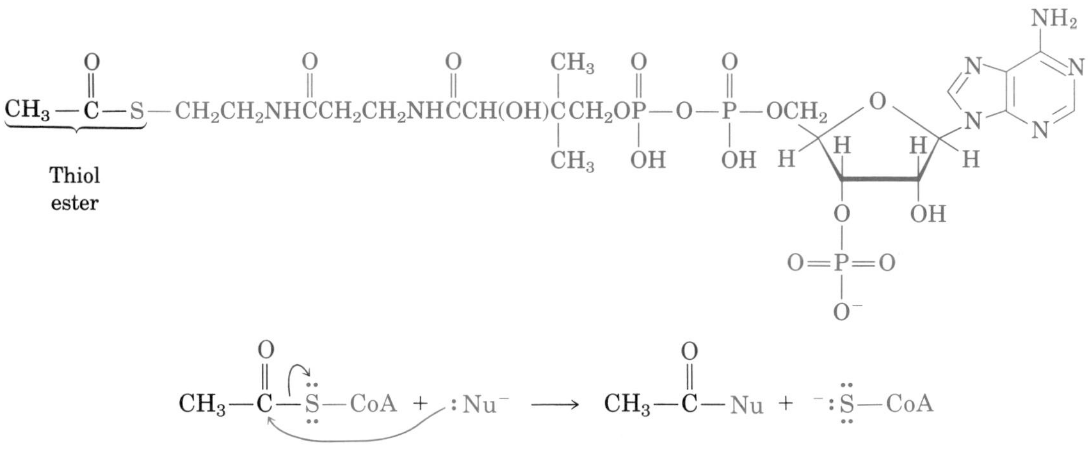

**Figure 21.14** The structure of acetyl coenzyme A, abbreviated acetyl CoA or $CH_3COSCoA$

For example, *N*-acetylglucosamine, an important constituent of surface membranes in mammalian cells, is synthesized in nature by an aminolysis reaction between glucosamine and acetyl CoA:

Glucosamine
(an amine)

+ $CH_3C(=O)-S-CoA$ ⟶ + H—S—CoA

*N*-Acetylglucosamine
(an amide)

## 21.11 Spectroscopy of Carboxylic Acid Derivatives

### INFRARED SPECTROSCOPY

All carbonyl-containing compounds have intense infrared absorptions in the range 1650–1850 $cm^{-1}$. As shown in Table 21.2, the exact position of the

Table 21.2 Infrared absorptions of some carbonyl compounds

| *Carbonyl type* | *Example* | *Infrared absorption ($cm^{-1}$)* |
|---|---|---|
| Aliphatic acid chloride | Acetyl chloride | 1810 |
| Aromatic acid chloride | Benzoyl chloride | 1770 |
| Aliphatic acid anhydride | Acetic anhydride | 1820, 1760 |
| Aliphatic ester | Ethyl acetate | 1735 |
| Aromatic ester | Ethyl benzoate | 1720 |
| Six-membered-ring lactone | | 1735 |
| Aliphatic amide | Acetamide | 1690 |
| Aromatic amide | Benzamide | 1675 |
| *N*-Substituted amide | *N*-Methylacetamide | 1680 |
| *N,N*-Disubstituted amide | *N,N*-Dimethylacetamide | 1650 |
| Aliphatic nitrile | Acetonitrile | 2250 |
| Aromatic nitrile | Benzonitrile | 2230 |
| Aliphatic aldehyde | Acetaldehyde | 1730 |
| Aliphatic ketone | Acetone | 1715 |
| Aliphatic carboxylic acid | Acetic acid | 1710 |

absorption provides information about the nature of the specific kind of carbonyl group. For comparison, the absorptions of ketones, aldehydes, and acids are included in the table, along with values for carboxylic acid derivatives.

As the data in the table indicate, acid chlorides are readily detected in the infrared by their characteristic carbonyl-group absorption near 1800 $cm^{-1}$. Acid anhydrides can be identified by the fact that they show two absorptions in the carbonyl region, one at 1820 $cm^{-1}$ and the second at 1760 $cm^{-1}$. Esters are detected by their absorption at 1735 $cm^{-1}$, a position somewhat higher than for either ketones or aldehydes. Amides, by contrast, absorb near the low end of the carbonyl region, with the degree of substitution on nitrogen affecting the exact position of the infrared band. Nitriles are easily recognized by the presence of an intense absorption near 2250 $cm^{-1}$. Since few other functional groups absorb in this region, infrared spectroscopy is highly diagnostic for nitriles.

PROBLEM

**21.29** What kinds of functional groups might compounds have if they exhibit the following infrared spectral properties?
(a) Absorption at 1735 $cm^{-1}$
(b) Absorption at 1810 $cm^{-1}$
(c) Absorptions at 2500–3300 $cm^{-1}$ and 1710 $cm^{-1}$
(d) Absorption at 2250 $cm^{-1}$
(e) Absorption at 1715 $cm^{-1}$

PROBLEM

**21.30** Propose structures for compounds having the following formulas and infrared absorptions:
(a) $C_3H_5N$, 2250 $cm^{-1}$
(b) $C_6H_{12}O_2$, 1735 $cm^{-1}$
(c) $C_4H_9NO$, 1650 $cm^{-1}$
(d) $C_4H_5ClO$, 1780 $cm^{-1}$

## NUCLEAR MAGNETIC RESONANCE SPECTROSCOPY

Although $^{13}C$ NMR is useful for determining the presence or absence of a carbonyl group in a molecule of unknown structure, precise information about the nature of the carbonyl group is difficult to obtain. Carbonyl carbon atoms show resonances in the range 160–210 $\delta$, as Table 21.3 shows.

**Table 21.3** Positions of $^{13}C$ NMR absorptions in some carbonyl compounds

| *Compound* | *Absorption* ($\delta$) | *Compound* | *Absorption* ($\delta$) |
|---|---|---|---|
| Acetic acid | 117.7 | Acetic anhydride | 166.9 |
| Ethyl acetate | 170.7 | Acetonitrile | 117.4 |
| Acetyl chloride | 170.3 | Acetone | 205.6 |
| Acetamide | 172.6 | Acetaldehyde | 201.0 |

Protons on the carbon next to a carbonyl group are slightly deshielded and absorb near 2 $\delta$ in the $^1H$ NMR spectrum. The exact nature of the carbonyl group cannot be distinguished by $^1H$ NMR, however, since all acyl derivatives absorb in the same range. Figure 21.15 shows the $^1H$ NMR spectrum of ethyl acetate.

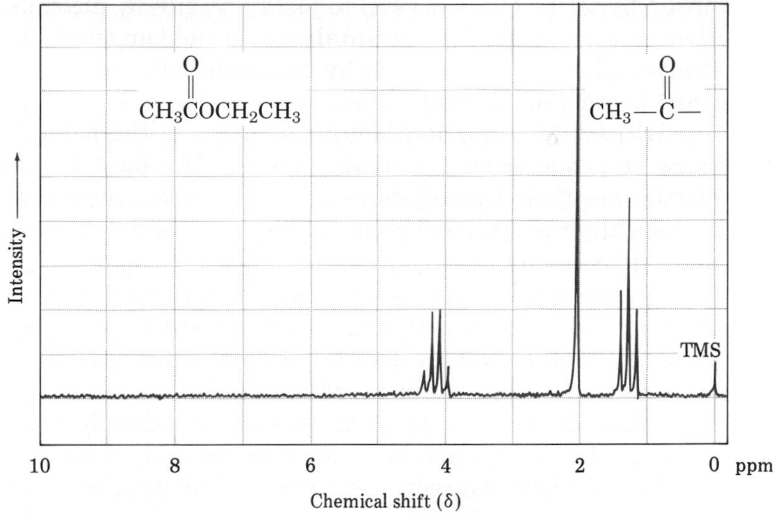


**Figure 21.15** Proton NMR spectrum of ethyl acetate

## 21.12 Summary and Key Words

Carboxylic acids can be transformed into a variety of acid derivatives in which the acid —OH group has been replaced by other substituents. **Acid chlorides, acid anhydrides, esters,** and **amides** are the most important derivatives.

The chemistry of the different acid derivatives is dominated by a single general reaction type—the **nucleophilic acyl substitution reaction.** Mechanistically, these substitutions take place by addition of a nucleophile to the polar carbonyl group of the acid derivative, followed by expulsion of a leaving group from the tetrahedral intermediate:

$$R{-}C({=}O){-}Y + :Nu^- \longrightarrow \left[ R{-}C(O^-)(Y){-}Nu \right] \longrightarrow R{-}C({=}O){-}Nu + :Y^-$$

Tetrahedral intermediate

where Y = Cl, Br, I (acid halide); OR (ester); OCOR (anhydride); or $NH_2$ (amide)

The reactivity of an acid derivative, RCOY, toward substitution depends both on the steric environment near the carbonyl group and on the electronic nature of the substituent, Y. Thus, we find a reactivity order:

Acid halide > Anhydride > Ester > Amide

The most important reactions of carboxylic acid derivatives are substitution by water (**hydrolysis**) to yield an acid, by alcohols (**alcoholysis**) to yield an ester, by amines (**aminolysis**) to yield an amide, by hydride (**reduction**) to yield an alcohol, and by organometallic reagents (**Grignard reaction**) to yield an alcohol.

Nitriles undergo nucleophilic addition to the polar C≡N bond in the same way that carbonyl compounds do. The most important reactions of nitriles are their hydrolysis to carboxylic acids, their reduction to primary amines, their partial reduction to aldehydes, and their reaction with organometallic reagents to yield ketones.

Nature employs nucleophilic acyl substitution reactions in the biosynthesis of many molecules, using **thiol esters** for the purpose. Acetyl coenzyme A (**acetyl CoA**) is a complex thiol ester that is employed in living systems to acetylate amines and alcohols.

Infrared spectroscopy is an extremely valuable tool for the structure analysis of acid derivatives. Acid chlorides, anhydrides, esters, amides, and nitriles all show characteristic infrared absorptions that can be used to identify these functional groups in unknowns. Carbon-13 NMR is useful for establishing the presence or absence of a carbonyl carbon atom, but does not usually allow exact identification of the different functional groups.

## 21.13 Summary of Reactions

1. Reactions of carboxylic acids

a. Conversion into acid chlorides (Section 21.4)

$$R-\overset{O}{\overset{\|}{C}}-OH \xrightarrow[CHCl_3]{SOCl_2} R-\overset{O}{\overset{\|}{C}}-Cl + SO_2 + HCl$$

The reaction also can be carried out using $PCl_3$ or $(COCl)_2$.

b. Conversion into cyclic acid anhydrides (Section 21.4).

$$(CH_2)_n\begin{matrix}\diagup COOH \\ \\ \diagdown COOH\end{matrix} \xrightarrow[200°C]{} (CH_2)_n\begin{matrix}\diagup C{=}O \\ \quad O \\ \diagdown C{=}O\end{matrix} + H_2O$$

where $n = 2$ or 3.

c. Conversion into esters (Section 21.4).

$$R-\overset{O}{\overset{\|}{C}}-\ddot{\underset{..}{O}}:^- + R'X \longrightarrow R-\overset{O}{\overset{\|}{C}}-O-R'$$

Via $S_N2$ reaction

$$R-\overset{O}{\overset{\|}{C}}-OH + R'OH \xrightarrow[HCl]{} R-\overset{O}{\overset{\|}{C}}-OR' + H_2O$$

Via Fischer esterification

$$R{-}\overset{\overset{\large O}{\|}}{C}{-}OH + CH_2N_2 \longrightarrow R{-}\overset{\overset{\large O}{\|}}{C}{-}OCH_3 + N_2$$

Via diazomethane

d. Conversion into amides (Section 21.4)

$$R{-}\overset{\overset{\large O}{\|}}{C}{-}OH + NH_3 \xrightarrow[200°C]{} R{-}\overset{\overset{\large O}{\|}}{C}{-}NH_2 + H_2O$$

2. Reactions of acid chlorides

a. Hydrolysis to yield acids (Section 21.5)

$$R{-}\overset{\overset{\large O}{\|}}{C}{-}Cl + H_2O \longrightarrow R{-}\overset{\overset{\large O}{\|}}{C}{-}OH + HCl$$

b. Alcoholysis to yield esters (Section 21.5)

$$R{-}\overset{\overset{\large O}{\|}}{C}{-}Cl + R'OH \xrightarrow{\text{Pyridine}} R{-}\overset{\overset{\large O}{\|}}{C}{-}OR' + HCl$$

c. Aminolysis to yield amides (Section 21.5)

$$R{-}\overset{\overset{\large O}{\|}}{C}{-}Cl + 2\,NH_3 \longrightarrow R{-}\overset{\overset{\large O}{\|}}{C}{-}NH_2 + NH_4Cl$$

d. Reduction to yield primary alcohols (Section 21.5)

$$R{-}\overset{\overset{\large O}{\|}}{C}{-}Cl \xrightarrow[\text{2. } H_3O^+]{\text{1. } LiAlH_4\text{, ether}} R{-}\underset{\underset{\large H}{|}}{\overset{\overset{\large OH}{|}}{C}}{-}H$$

e. Partial reduction to yield aldehydes (Section 21.5)

$$R{-}\overset{\overset{\large O}{\|}}{C}{-}Cl \xrightarrow[\text{2. } H_3O^+]{\text{1. } LiAlH(O\text{-}tert\text{-}Bu)_3\text{, ether}} R{-}\overset{\overset{\large O}{\|}}{C}{-}H$$

f. Rosenmund reduction to yield aldehydes (Section 21.5)

$$R{-}\overset{\overset{\large O}{\|}}{C}{-}Cl \xrightarrow[\text{Ethyl acetate}]{H_2\text{, Pd, }BaSO_4} R{-}\overset{\overset{\large O}{\|}}{C}{-}H$$

g. Grignard reaction to yield tertiary alcohols (Section 21.5)

$$R{-}\overset{\overset{\large O}{\|}}{C}{-}Cl \xrightarrow[\text{2. } H_3O^+]{\text{1. } 2\,R'MgX\text{, ether}} R{-}\underset{\underset{\large R'}{|}}{\overset{\overset{\large OH}{|}}{C}}{-}R'$$

h. Diorganocopper reaction to yield ketones (Section 21.5)

$$\mathrm{R{-}\overset{O}{\overset{\|}{C}}{-}Cl \xrightarrow[\text{Ether}]{R'_2CuLi} R{-}\overset{O}{\overset{\|}{C}}{-}R'}$$

3. Reactions of acid anhydrides

a. Hydrolysis to yield acids (Section 21.6)

$$\mathrm{R{-}\overset{O}{\overset{\|}{C}}{-}O{-}\overset{O}{\overset{\|}{C}}{-}R + H_2O \longrightarrow 2\,R{-}\overset{O}{\overset{\|}{C}}{-}OH}$$

b. Alcoholysis to yield esters (Section 21.6)

$$\mathrm{R{-}\overset{O}{\overset{\|}{C}}{-}O{-}\overset{O}{\overset{\|}{C}}{-}R + R'OH \xrightarrow{\text{Pyridine}} R{-}\overset{O}{\overset{\|}{C}}{-}OR' + RCOOH}$$

c. Aminolysis to yield amides (Section 21.6)

$$\mathrm{R{-}\overset{O}{\overset{\|}{C}}{-}O{-}\overset{O}{\overset{\|}{C}}{-}R + 2\,NH_3 \longrightarrow R{-}\overset{O}{\overset{\|}{C}}{-}NH_2 + RCO_2^{-}\,NH_4^{+}}$$

d. Reduction to yield primary alcohols (Section 21.6)

$$\mathrm{R{-}\overset{O}{\overset{\|}{C}}{-}O{-}\overset{O}{\overset{\|}{C}}{-}R \xrightarrow[\text{2. } H_3O^+]{\text{1. LiAlH}_4\text{, ether}} 2\,R{-}\underset{H}{\underset{|}{\overset{OH}{\overset{|}{C}}}}{-}H}$$

4. Reactions of amides and lactams

a. Hydrolysis to yield acids (Section 21.8)

$$\mathrm{R{-}\overset{O}{\overset{\|}{C}}{-}NH_2 \xrightarrow[\text{or } ^{-}OH]{H_3O^+} R{-}\overset{O}{\overset{\|}{C}}{-}OH + NH_3}$$

b. Reduction to yield amines (Section 21.8)

$$\mathrm{R{-}\overset{O}{\overset{\|}{C}}{-}NH_2 \xrightarrow[\text{2. } H_2O]{\text{1. LiAlH}_4\text{, ether}} R{-}\underset{H}{\underset{|}{\overset{H}{\overset{|}{C}}}}{-}NH_2}$$

c. Dehydration of primary amides to yield nitriles (Section 21.9)

$$\mathrm{R{-}\overset{O}{\overset{\|}{C}}{-}NH_2 \xrightarrow{SOCl_2} R{-}C{\equiv}N + SO_2 + HCl}$$

5. Reactions of esters and lactones
   a. Hydrolysis to yield acids (Section 21.7)

$$\mathrm{R{-}\overset{\overset{\displaystyle O}{\|}}{C}{-}OR' + H_2O \xrightarrow{H_3O^+ \text{or} \ ^-OH} R{-}\overset{\overset{\displaystyle O}{\|}}{C}{-}OH + HOR'}$$

   b. Aminolysis to yield amides (Section 21.7)

$$\mathrm{R{-}\overset{\overset{\displaystyle O}{\|}}{C}{-}OR' + NH_3 \longrightarrow R{-}\overset{\overset{\displaystyle O}{\|}}{C}{-}NH_2 + HOR'}$$

   c. Reduction to yield primary alcohols (Section 21.7)

$$\mathrm{R{-}\overset{\overset{\displaystyle O}{\|}}{C}{-}OR' \xrightarrow[\text{2. } H_3O^+]{\text{1. LiAlH}_4\text{, ether}} R{-}\underset{\underset{\displaystyle H}{|}}{\overset{\overset{\displaystyle OH}{|}}{C}}{-}H + R'OH}$$

   d. Partial reduction to yield aldehydes (Section 21.7)

$$\mathrm{R{-}\overset{\overset{\displaystyle O}{\|}}{C}{-}OR' \xrightarrow[\text{2. } H_3O^+]{\text{1. DIBAH, toluene}} R{-}\overset{\overset{\displaystyle O}{\|}}{C}{-}H + R'OH}$$

   e. Grignard reaction to yield tertiary alcohols (Section 21.7)

$$\mathrm{R{-}\overset{\overset{\displaystyle O}{\|}}{C}{-}OR' \xrightarrow[\text{2. } H_3O^+]{\text{1. 2 R''MgX, ether}} R{-}\underset{\underset{\displaystyle R''}{|}}{\overset{\overset{\displaystyle OH}{|}}{C}}{-}R'' + R'OH}$$

6. Reactions of nitriles
   a. Hydrolysis to yield carboxylic acids (Section 21.9)

$$\mathrm{R{-}C{\equiv}N + H_2O \xrightarrow{H_3O^+ \text{or } HO^-} R{-}\overset{\overset{\displaystyle O}{\|}}{C}{-}OH + NH_3}$$

   b. Partial hydrolysis to yield amides (Section 21.9)

$$\mathrm{R{-}C{\equiv}N + H_2O \xrightarrow{H_3O^+ \text{or} \ ^-OH} R{-}\overset{\overset{\displaystyle O}{\|}}{C}{-}NH_2}$$

   c. Reduction to yield primary amines (Section 21.9)

$$\mathrm{R{-}C{\equiv}N \xrightarrow[\text{2. } H_2O]{\text{1. LiAlH}_4\text{, ether}} R{-}\underset{\underset{\displaystyle H}{|}}{\overset{\overset{\displaystyle H}{|}}{C}}{-}NH_2}$$

d. Partial reduction to yield aldehydes (Section 21.9)

$$R{-}C{\equiv}N \xrightarrow[\text{2. } H_3O^+]{\text{1. DIBAH, toluene}} R{-}\overset{O}{\overset{\|}{C}}{-}H + NH_3$$

e. Reaction with Grignard reagents to yield ketones (Section 21.9)

$$R{-}C{\equiv}N \xrightarrow[\text{2. } H_3O^+]{\text{1. R'MgX, ether}} R{-}\overset{O}{\overset{\|}{C}}{-}R' + NH_3$$

## ADDITIONAL PROBLEMS

**21.31** Provide IUPAC names for these compounds:

(a) 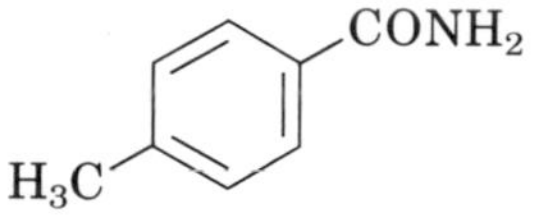 [4-methylbenzene ring with $CONH_2$; $H_3C$ substituent]

(b) $(CH_3CH_2)_2CHCH{=}CHCN$

(c) $CH_3O_2CCH_2CH_2CO_2CH_3$

(d) 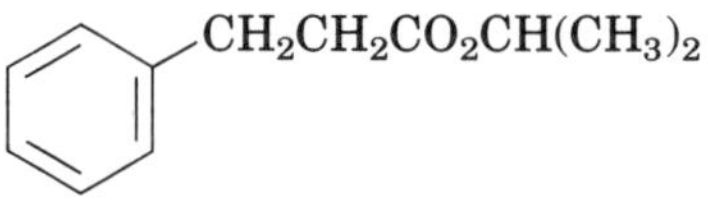 [benzene ring with $CH_2CH_2CO_2CH(CH_3)_2$]

(e) 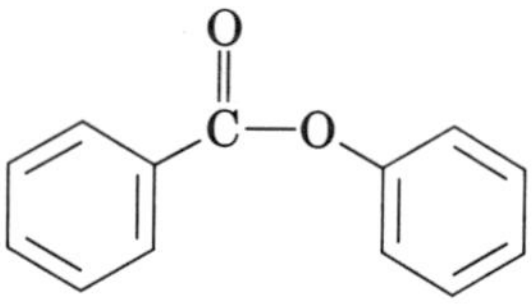 [benzene ring]$-\overset{O}{\overset{\|}{C}}-O-$[benzene ring]

(f) $CH_3CHBrCH_2CONHCH_3$

(g) 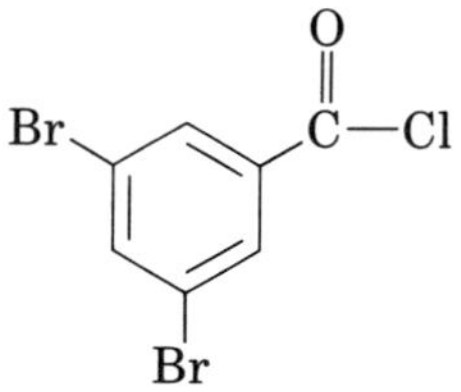 [benzene ring with Br, Br, and $\overset{O}{\overset{\|}{C}}-Cl$]

(h) 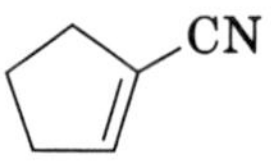 [cyclopentene ring with CN]

**21.32** Draw structures corresponding to these names:
(a) *p*-Bromophenylacetamide (b) *m*-Benzoylbenzonitrile
(c) 2,2-Dimethylhexanamide (d) Cyclohexyl cyclohexanecarboxylate
(e) 2-Cyclobutenecarbonitrile (f) 1,2-Pentanedicarbonyl dichloride

**21.33** Draw and name compounds meeting these descriptions:
(a) Three different acid chlorides having the formula $C_6H_9ClO$
(b) Three different amides having the formula $C_7H_{11}NO$
(c) Three different nitriles having the formula $C_5H_7N$

**21.34** The following reactivity order has been found for the saponification of alkyl acetates by aqueous hydroxide ion:

$$CH_3CO_2CH_3 > CH_3CO_2CH_2CH_3 > CH_3CO_2CH(CH_3)_2 > CH_3CO_2C(CH_3)_3$$

How can you explain this reactivity order?

**21.35** How can you explain the observation that attempted Fischer esterification of 2,4,6-trimethylbenzoic acid with methanol/HCl is unsuccessful? No ester is obtained and the acid is recovered unchanged. Can you suggest an alternative method of esterification that would be successful?

**21.36** How can you account for the fact that, when a carboxylic acid is dissolved in isotopically labeled water, the label rapidly becomes incorporated into *both* oxygen atoms of the carboxylic acid?

$$R-\overset{\overset{O}{\|}}{C}-OH + H_2{}^*O \rightleftharpoons R-\overset{\overset{^*O}{\|}}{C}-{}^*OH + H_2O$$

**21.37** Outline methods for the preparation of acetophenone (phenyl methyl ketone) from the following:
(a) Benzene (b) Bromobenzene (c) Methyl benzoate
(d) Benzonitrile (e) Styrene

**21.38** In the basic hydrolysis of para-substituted methyl benzoates, the following reactivity order has been found for Y: $NO_2 > Br > H > CH_3 > OCH_3$. How can you explain this reactivity order? Where would you expect Y = C≡N, Y = CHO, and Y = $NH_2$ to be in the reactivity list?

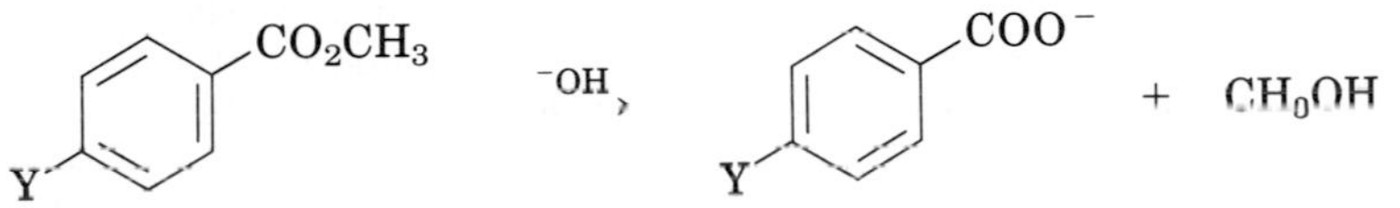

$$Y\text{-}C_6H_4\text{-}CO_2CH_3 \xrightarrow{^-OH} Y\text{-}C_6H_4\text{-}COO^- + CH_3OH$$

**21.39** How might you prepare these compounds from butanoic acid?
(a) 1-Butanol (b) Butanal (c) 1-Bromobutane
(d) Pentanenitrile (e) 1-Butene (f) *N*-Methylpentanamide
(g) 2-Hexanone (h) Butylbenzene

**21.40** What product would you expect to obtain from Grignard reaction of phenylmagnesium bromide with dimethyl carbonate, $CH_3OCOOCH_3$?

**21.41** When ethyl benzoate is heated in methanol containing a small amount of HCl, methyl benzoate is formed. Propose a mechanism for this reaction.

**21.42** In the iodoform reaction, a triiodomethyl ketone reacts with aqueous base to yield a carboxylate ion and iodoform (triiodomethane). Propose a mechanism for this reaction.

$$R-\overset{\overset{O}{\|}}{C}-CI_3 \xrightarrow[H_2O]{^-OH} R-COO^- + CHI_3$$

**21.43** Which of the two products shown would you expect from the reaction of methoxycarbonyl chloride with ammonia? Would you expect the nucleophile to replace the chlorine or the methoxy group? Explain your answer.

$$CH_3O-\overset{\overset{O}{\|}}{C}-Cl \xrightarrow{:NH_3} CH_3O-\overset{\overset{O}{\|}}{C}-NH_2 \quad \text{or} \quad H_2N-\overset{\overset{O}{\|}}{C}-Cl$$

**21.44** Which compound would you expect to have a more highly polarized carbonyl group, methoxycarbonyl chloride or *N*,*N*-dimethylaminocarbonyl chloride? Explain.

$$CH_3O{-}\overset{\overset{\large O}{\|}}{C}{-}Cl \qquad \text{or} \qquad (CH_3)_2N{-}\overset{\overset{\large O}{\|}}{C}{-}Cl$$

Methoxycarbonyl chloride — *N*,*N*-Dimethylaminocarbonyl chloride

**21.45** In light of your answer to Problem 21.44, which compound would you expect to react faster with nucleophiles—methoxycarbonyl chloride or *N*,*N*-dimethylaminocarbonyl chloride?

**21.46** *tert*-Butoxycarbonyl azide, an important reagent used in protein synthesis, is prepared by treating *tert*-butoxycarbonyl chloride with sodium azide. Propose a mechanism for this reaction.

$$(CH_3)_3COCOCl + NaN_3 \longrightarrow (CH_3)_3COCON_3 + NaCl$$

**21.47** Predict the product, if any, of reaction between propanoyl chloride and the following reagents:

(a) $(Ph)_2CuLi$ in ether (b) $LiAlH_4$, then $H_3O^+$ (c) $CH_3MgBr$, then $H_3O^+$
(d) $H_2$, $Pd/BaSO_4$ (e) $H_3O^+$ (f) Cyclohexanol
(g) Aniline (h) $CH_3COO^-\ {}^+Na$

**21.48** Answer Problem 21.47 for reaction of the listed reagents with methyl propanoate.

**21.49** Answer Problem 21.47 for reaction of the listed reagents with propanamide and with propanenitrile.

**21.50** A particularly mild method of esterification involves the use of trifluoroacetic anhydride. Treatment of a carboxylic acid with trifluoroacetic anhydride leads to a mixed anhydride that rapidly reacts with alcohol:

$$RCOOH + (CF_3CO)_2O \longrightarrow R{-}\overset{\overset{\large O}{\|}}{C}{-}O{-}\overset{\overset{\large O}{\|}}{C}{-}CF_3 \xrightarrow{R'OH} R{-}\overset{\overset{\large O}{\|}}{C}{-}OR' + CF_3COOH$$

(a) Propose a mechanism for formation of the mixed anhydride.
(b) Why is the mixed anhydride unusually reactive?
(c) Why does the mixed anhydride react specifically as indicated, rather than giving trifluoroacetate esters plus carboxylic acid?

**21.51** How would you accomplish the following transformations? More than one step may be required.

(a) $CH_3CH_2CH_2CH_2CN \longrightarrow CH_3CH_2CH_2CH_2CH_2NH_2$

(b) $CH_3CH_2CH_2CH_2CN \longrightarrow CH_3CH_2CH_2CH_2CH_2N(CH_3)_2$

(c) $CH_3CH_2CH_2CH_2CN \longrightarrow CH_3CH_2CH_2CH_2C(CH_3)_2OH$

(d) $CH_3CH_2CH_2CH_2CN \longrightarrow CH_3CH_2CH_2CH_2CH(OH)CH_3$

(e) $CH_3CH_2CH_2CH_2CN \longrightarrow CH_3CH_2CH_2CH_2CHO$

**21.52** List as many ways as you can think of for transforming cyclohexanol into cyclohexanecarbaldehyde (try to get at least four).

**21.53** Succinic anhydride yields succinimide when heated with ammonium chloride at 200°C. Propose a mechanism for this reaction. Why do you suppose such a high reaction temperature is required?

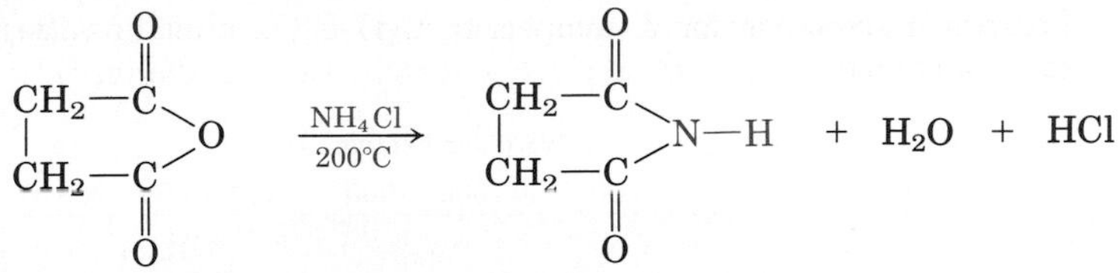

$$\begin{array}{c}CH_2{-}C(=O)\\ |\quad\quad\;\; \backslash O\\ CH_2{-}C(=O)\end{array} \xrightarrow[200^\circ C]{NH_4Cl} \begin{array}{c}CH_2{-}C(=O)\\ |\quad\quad\;\; \backslash N{-}H\\ CH_2{-}C(=O)\end{array} + H_2O + HCl$$

**21.54** Butacetin is an analgesic (pain-killing) agent that is synthesized commercially from *p*-fluoronitrobenzene. Propose a likely synthesis route.

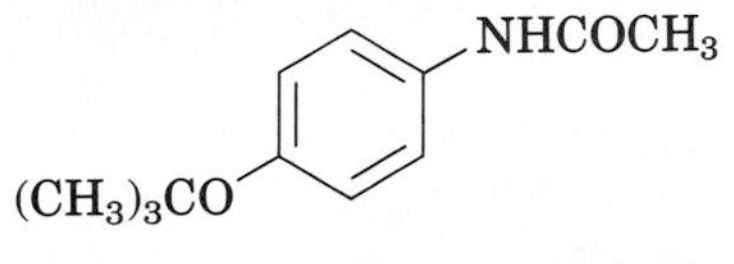

$NHCOCH_3$, $(CH_3)_3CO$

Butacetin

**21.55** Phenyl 4-aminosalicylate is a drug used in the treatment of tuberculosis. Propose a synthesis of this compound starting from 4-nitrosalicylic acid.

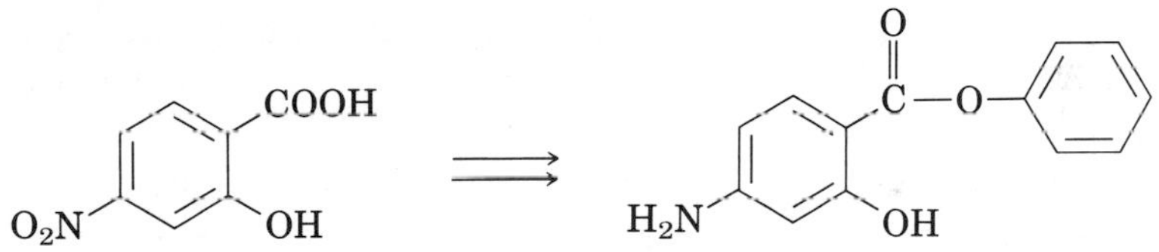

COOH, $O_2N$, OH $\Longrightarrow$ O, C—O—, $H_2N$, OH

4-Nitrosalicylic acid — Phenyl 4-aminosalicylate

**21.56** What spectroscopic technique would you use to distinguish between the following isomer pairs? Tell what differences you would expect to see.

(a) *N*-Methylpropanamide and *N*,*N*-dimethylacetamide
(b) 5-Hydroxypentanenitrile and cyclobutanecarboxamide
(c) 4-Chlorobutanoic acid and 3-methoxypropanoyl chloride
(d) Ethyl propanoate and propyl acetate

**21.57** *N*,*N*-Diethyl-*m*-toluamide (DEET) is the active ingredient in many insect-repellent preparations. How might you synthesize this substance from *m*-bromotoluene?

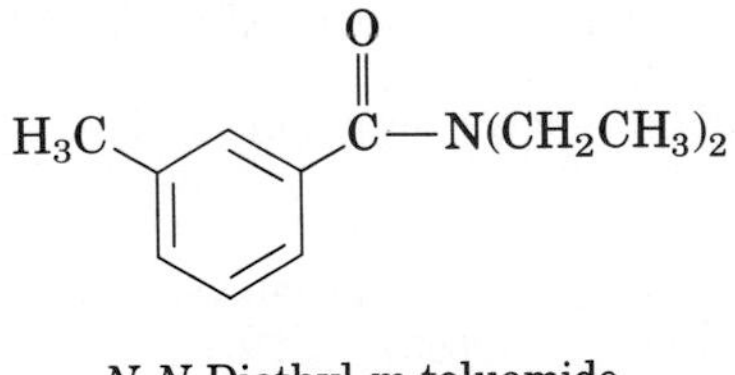

$H_3C$, O, $C{-}N(CH_2CH_3)_2$

*N*,*N*-Diethyl-*m*-toluamide

**21.58** Tranexamic acid, a drug useful against blood clotting, is prepared commercially from *p*-methylbenzonitrile. Formulate the steps likely to be used in the synthesis. (Don't worry about cis–trans isomers. Heating to 300°C interconverts the isomers.)

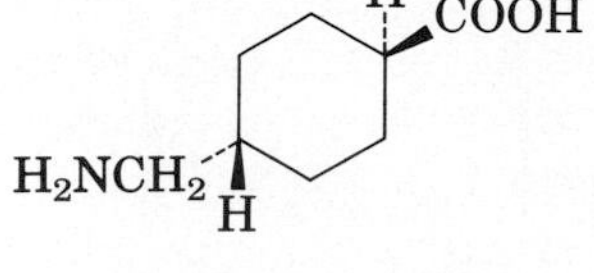

H, COOH, $H_2NCH_2$, H

Tranexamic acid

**21.59** Propose a structure for a compound, $C_4H_7ClO_2$, that has the infrared and NMR spectra shown.

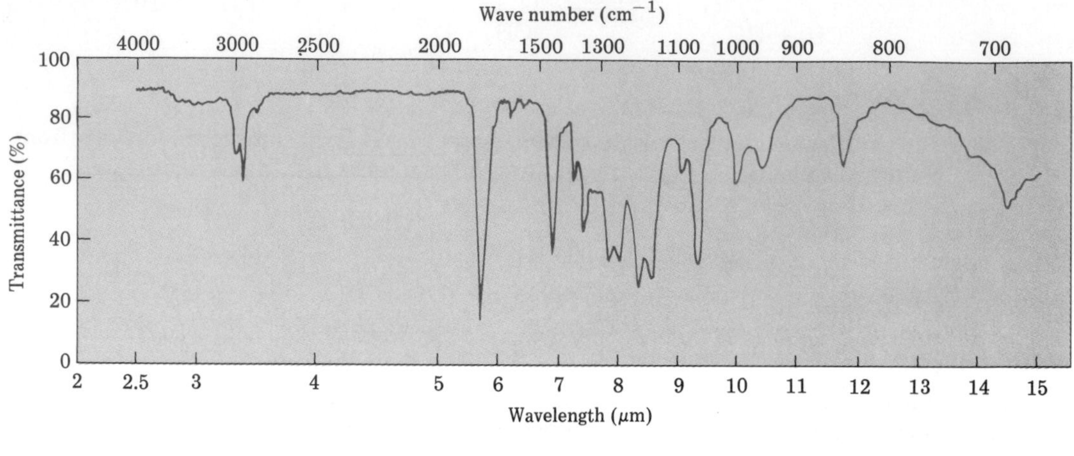


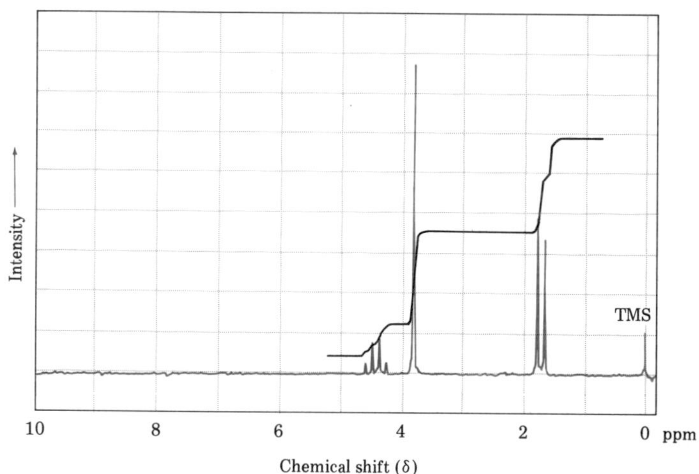


**21.60** Propose a structure for a compound, $C_4H_7N$, that has the infrared and NMR spectra shown.

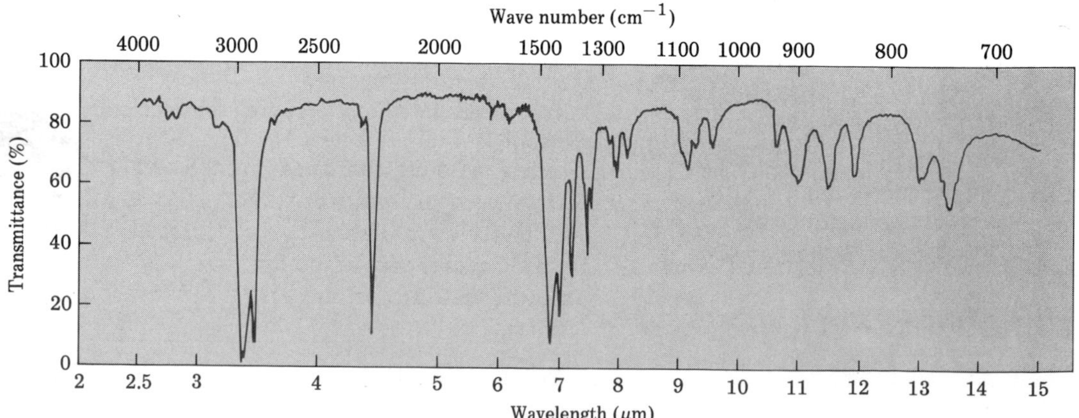

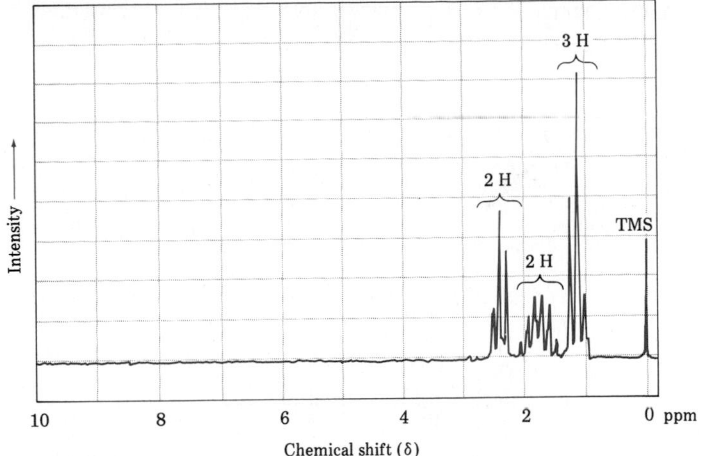


**21.61** The following $^1$H NMR spectrum is of ethyl propanoate, $CH_3CH_2COOCH_2CH_3$. Assign all peaks in the spectrum and account for the fact that the methyl groups appear as an evident *quartet* centered at 1.2 δ.

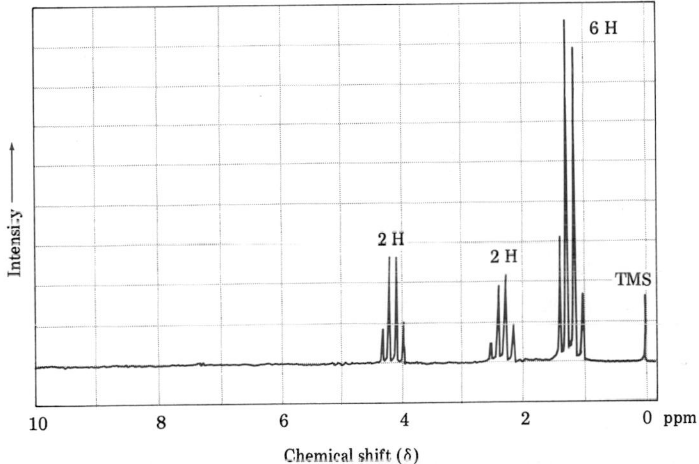


**21.62** Assign structures to compounds with the following $^1$H NMR spectra:

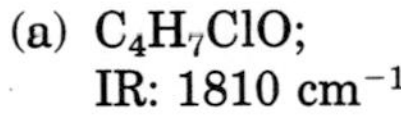

(a) $C_4H_7ClO$;
IR: 1810 $cm^{-1}$

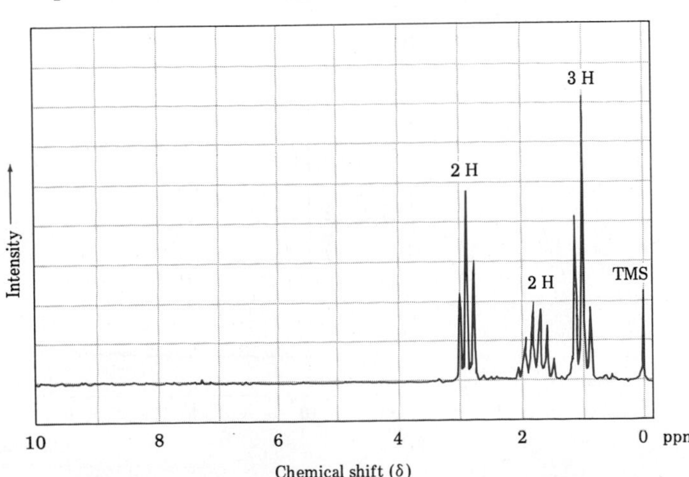

(b) $C_6H_9NO_2$;
IR: 2250, 1735 $cm^{-1}$

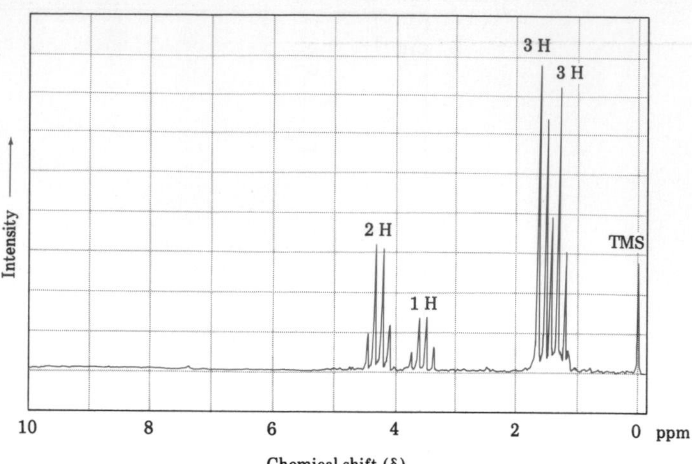


(c) $C_5H_{10}O_2$;
IR: 1735 $cm^{-1}$

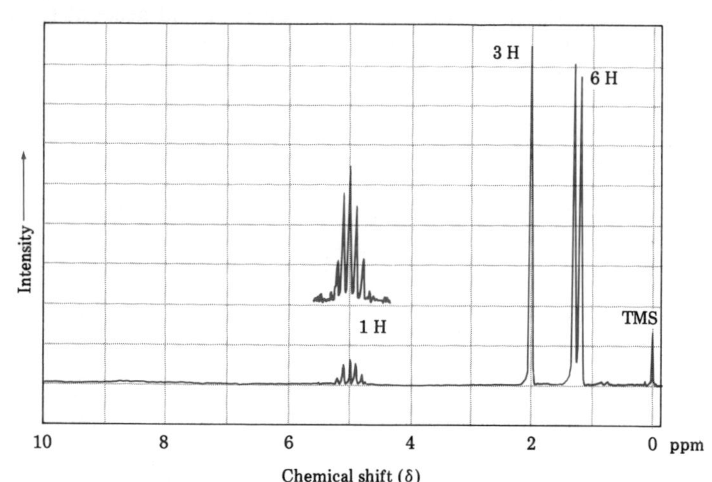


**21.63** Propose structures for compounds with the following $^1H$ NMR spectra:

(a) $C_5H_9ClO_2$;
IR: 1735 $cm^{-1}$

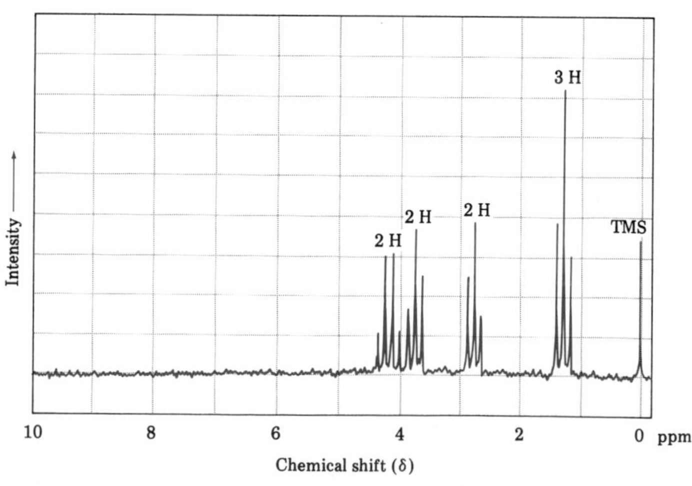

(b) $C_7H_{12}O_4$;
IR: 1735 $cm^{-1}$

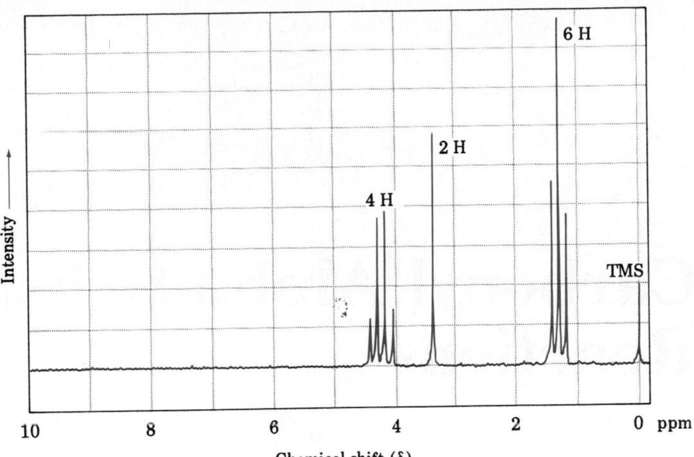


(c) $C_{11}H_{12}O_2$;
IR: 1710 $cm^{-1}$

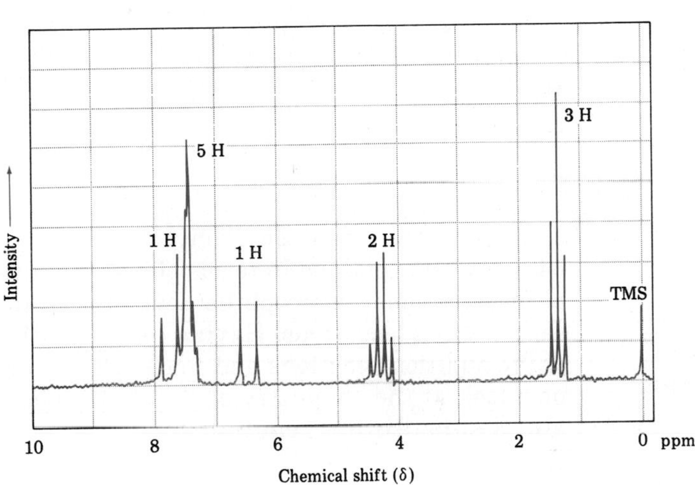

CHAPTER 22

# Carbonyl Alpha-Substitution Reactions

We said in the overview of carbonyl chemistry that most carbonyl-group chemistry can be explained in terms of just four fundamental reactions—nucleophilic additions, nucleophilic acyl substitutions, alpha substitutions, and carbonyl condensations. Having already looked at the chemistry of nucleophilic addition reactions and nucleophilic acyl substitution reactions, we'll now look at the chemistry of the third major carbonyl-group process—the **alpha-substitution reaction**.

R—C(=O)—Y + :Nu⁻ ⇌ [Tetrahedral intermediate: R, Y, Nu on C with :O:⁻]

→ R—C(=O)—Nu   Nucleophilic acyl substitution

→ R, Y, Nu on C with :OH   Nucleophilic addition

→ R—C(=Nu)—Y   Nucleophilic addition

Tetrahedral intermediate

H–Cα–C(=O) ⇌ [C=C–OH An enol or C=C–:O:⁻ An enolate anion] $\xrightarrow{E^+}$ E–C–C(=O)   Alpha substitution

$E^+$ = an electrophile

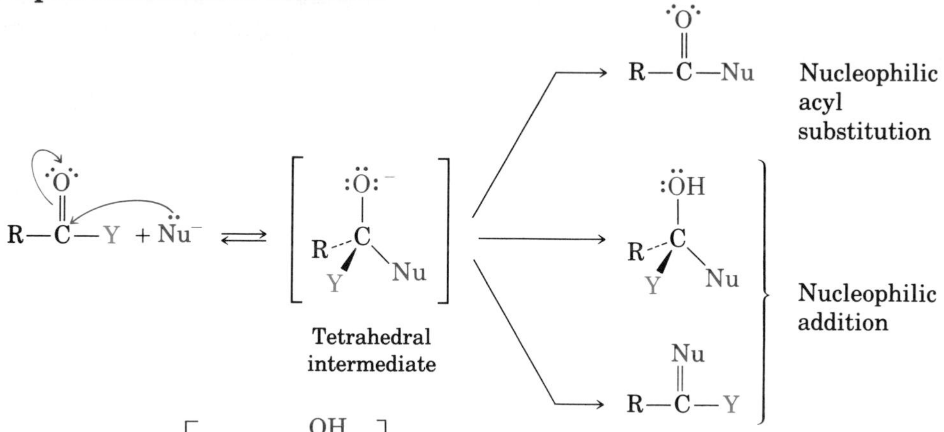

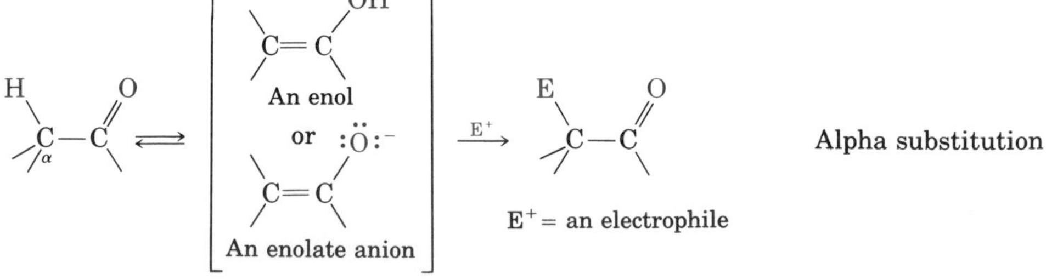

Alpha-substitution reactions occur at the position *next to* the carbonyl group—the **alpha (α) position**—and involve the substitution of an alpha hydrogen atom by some other group. They take place through the formation of either enol or enolate ion intermediates. Let's begin our study by learning more about these two species.

## 22.1 Keto–Enol Tautomerism

Carbonyl compounds that have hydrogen atoms on their alpha carbons are rapidly interconvertible with their corresponding **enols** (*ene* + *ol*, unsaturated alcohol). This rapid interconversion between two chemically distinct species is a special kind of isomerism known as **tautomerism** (taw-**tom**-er-ism; from the Greek *tauto*, "the same," and *meros*, "part"). Individual isomers are called **tautomers** (**taw**-toe-mers).

$$\text{H}-\underset{\alpha}{\text{C}}-\text{C}(=\text{O}) \underset{\text{equilibration}}{\overset{\text{Rapid}}{\rightleftharpoons}} \text{C}=\text{C}-\text{O}-\text{H}$$

Keto tautomer — Enol tautomer

Note that tautomerism requires the two different isomeric forms to be *rapidly* interconvertible. Thus, keto and enol carbonyl isomers are tautomers, but two isomeric alkenes such as 1-butene and 2-butene are not, since they do not rapidly interconvert under normal circumstances.

$$CH_3CH(H)-CH=CH_2 \underset{\text{happen}}{\overset{\text{Does not}}{\rightleftharpoons}} CH_3CH=CH-CH_2(H)$$

Most carbonyl compounds exist almost exclusively in the keto form at equilibrium, and it is difficult to isolate the pure enol form. For example, cyclohexanone contains only about 0.001% of its enol tautomer at room temperature, and acetone contains only about 0.0001% enol. The percentage of enol tautomer is even less for carboxylic acids and acyl derivatives such as esters and amides. Even though enols are difficult to isolate and are present only to a small extent at equilibrium, they are nevertheless extremely important and are involved in much of the chemistry of carbonyl compounds.

Cyclohexanone: cyclohexanone (99.999%) $\underset{}{\overset{H_3O^+ \text{ or } ^-OH}{\rightleftharpoons}}$ cyclohex-1-en-1-ol (0.001%)

Acetone: $H_3C-C(=O)-CH_3$ (99.9999%) $\overset{H_3O^+ \text{ or } ^-OH}{\rightleftharpoons}$ $H_3C-C(OH)=CH_2$ (0.0001%)

Keto–enol tautomerism of carbonyl compounds is catalyzed by both acids and bases. Acid catalysis involves protonation of the carbonyl oxygen atom (a Lewis base) to give an intermediate cation that can lose a proton from the alpha carbon to yield neutral enol (Figure 22.1). This proton loss from the cation intermediate is analogous to what occurs when a carbocation loses a proton to form an alkene in an E1 reaction (Section 11.14).

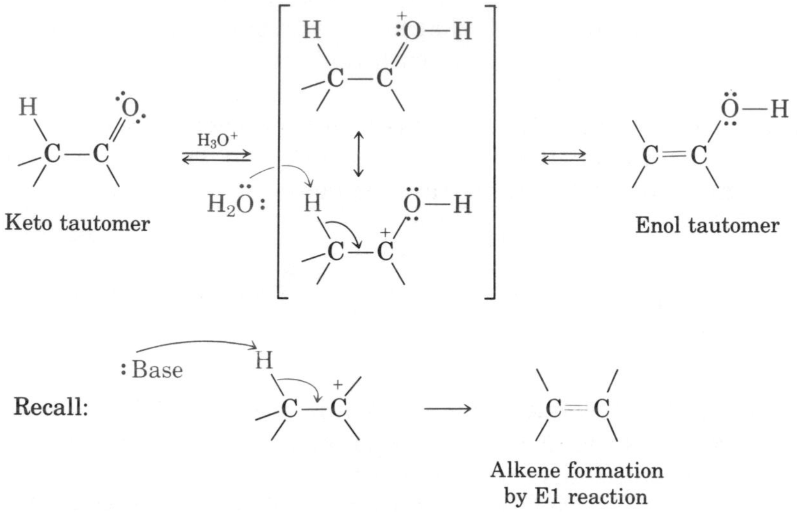

Keto tautomer ⇌ ($H_3O^+$) [intermediate cation; $H_2O$:] ⇌ Enol tautomer

Recall: :Base + H–C–C$^+$ ⟶ C=C

Alkene formation by E1 reaction

**Figure 22.1** Mechanism of acid-catalyzed enol formation

Base-catalyzed enol formation occurs by an acid–base reaction between catalyst and carbonyl compound. The carbonyl compound acts as a weak protic acid and donates one of its alpha hydrogens to the base. The resultant anion—an **enolate ion**—is then reprotonated to yield a neutral compound. Since the enolate ion is a resonance hybrid of two forms, it can be protonated either on carbon to regenerate the keto tautomer or on oxygen to give an enol tautomer (Figure 22.2).

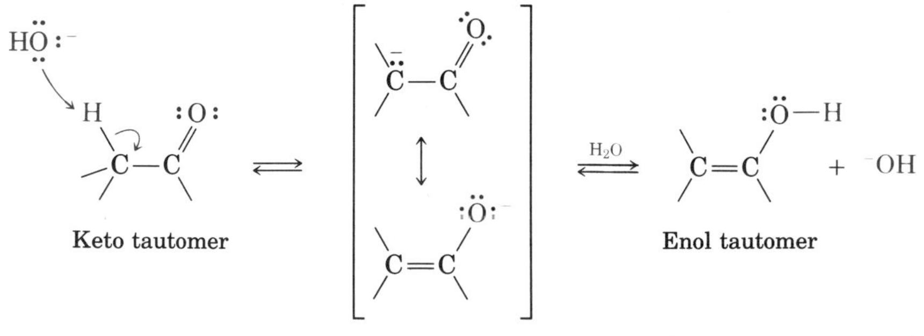

$HO^-$ + Keto tautomer ⇌ [enolate ion] ⇌ ($H_2O$) Enol tautomer + $^-OH$

**Figure 22.2** Mechanism of base-catalyzed enol formation: The intermediate enolate anion, a resonance hybrid of two forms, can be protonated either on carbon to regenerate starting ketone, or on oxygen to give an enol.

Note that only the protons on the *alpha* position of carbonyl compounds are acidic. Protons at beta, gamma, delta, and so on, are not acidic and cannot be removed by base. We'll account for this unique behavior of alpha protons shortly.

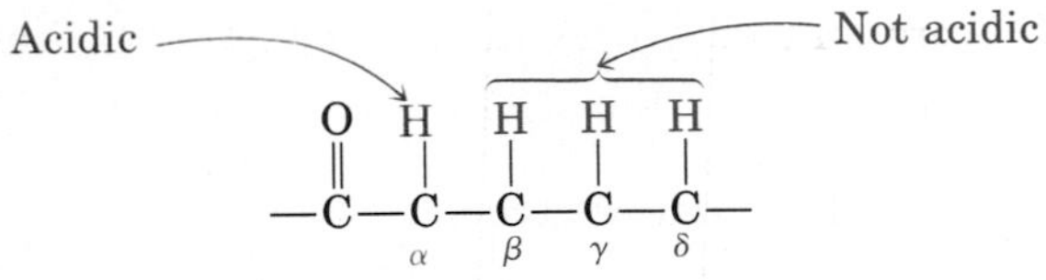

Acidic — Not acidic

$$-\overset{\overset{\displaystyle O}{\|}}{C}-\underset{\alpha}{\overset{\overset{\displaystyle H}{|}}{C}}-\underset{\beta}{\overset{\overset{\displaystyle H}{|}}{C}}-\underset{\gamma}{\overset{\overset{\displaystyle H}{|}}{C}}-\underset{\delta}{\overset{\overset{\displaystyle H}{|}}{C}}-$$

PROBLEM

**22.1** Draw structures for the enol tautomers of these compounds:
(a) Cyclopentanone (b) Acetyl chloride (c) Ethyl acetate
(d) Propanal (e) Acetic acid (f) Phenylacetone
(g) Acetophenone (methyl phenyl ketone)

PROBLEM

**22.2** How many acidic hydrogens does each of the molecules listed in Problem 22.1 have? Identify them.

PROBLEM

**22.3** Draw structures for the possible monoenol forms of 1,3-cyclohexanedione. How many enol forms are possible? Which would you expect to be most stable? Explain your answer.

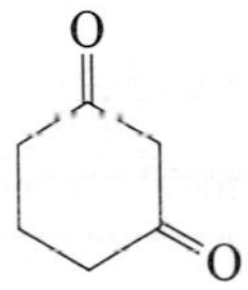

1,3-Cyclohexanedione

## 22.2 Reactivity of Enols: The Mechanism of Alpha-Substitution Reactions

What kind of chemistry should enols have? Since their double bonds are electron-rich, enols behave as nucleophiles and react with electrophiles in much the same way that alkenes do. Because of resonance electron donation of the lone-pair electrons on oxygen, however, enol double bonds are more reactive than alkenes (Section 14.9).

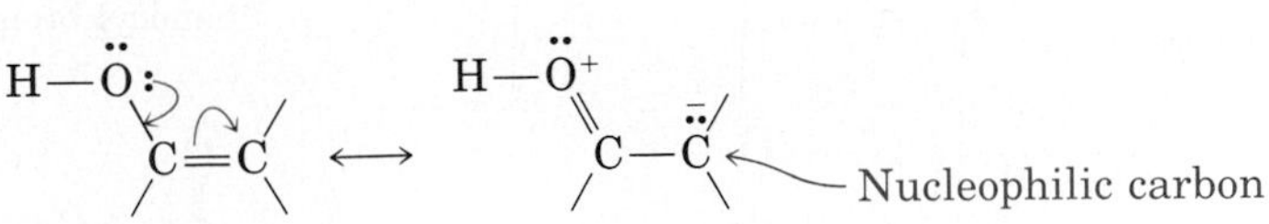

H—Ö: C=C ⟷ H—Ö⁺=C—C:⁻ ← Nucleophilic carbon

When an alkene reacts with an electrophile such as bromine, addition of $Br^+$ occurs to give an intermediate cation, which reacts with $Br:^-$. When an enol reacts with an electrophile, however, the intermediate cation loses the hydroxyl proton to regenerate a carbonyl compound. The net result of the reaction of an enol with an electrophile is alpha substitution. The general mechanism is shown in Figure 22.3.

Acid-catalyzed enolization occurs.

An electron pair from the enol attacks an electrophile, forming a new bond and leaving a positively charged intermediate that can be stabilized by two resonance forms.

Loss of a proton from oxygen yields the neutral alpha-substitution product, and the O—H bond electrons form a new C=O bond.

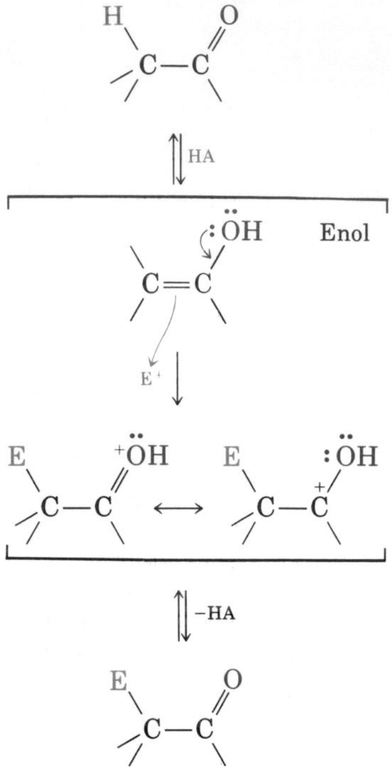

**Figure 22.3** General mechanism of a carbonyl alpha-substitution reaction

## 22.3 Alpha Halogenation of Ketones and Aldehydes

Ketones and aldehydes can be halogenated at their alpha positions by reaction with chlorine, bromine, or iodine in acidic solution. Bromine is most often used, and acetic acid is often employed as solvent.

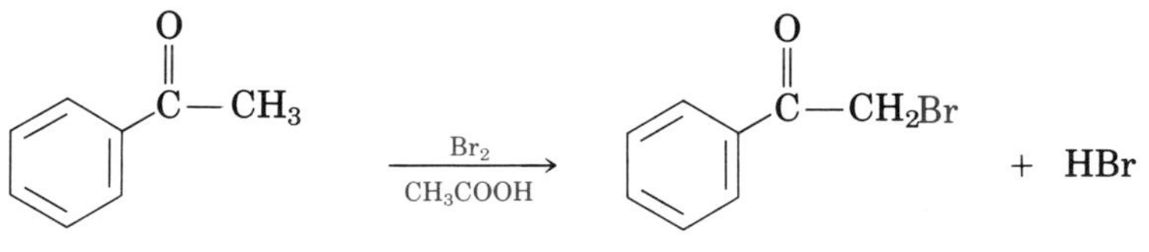

Acetophenone $\xrightarrow[CH_3COOH]{Br_2}$ Phenacyl bromide (72%) + HBr

Cyclohexanone $\xrightarrow[H_2O,\ HCl]{Cl_2}$ 2-Chlorocyclohexanone (66%) + HCl

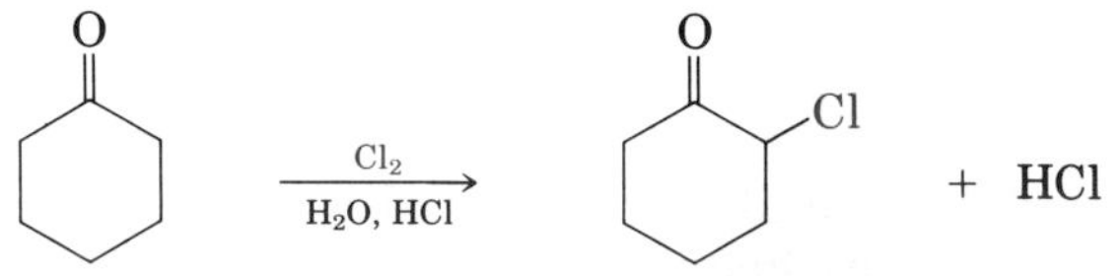

The alpha halogenation of ketones is a typical alpha-substitution reaction that proceeds by acid-catalyzed formation of enol intermediates, as shown in Figure 22.4.

Acid–base reaction between a catalyst and the carbonyl oxygen forms a protonated carbonyl compound.

This compound loses an acidic proton from the alpha carbon to yield an enol intermediate.

An electron pair from the enol attacks a positively polarized bromine atom, giving an intermediate cation that can be drawn as a resonance hybrid of two forms.

Loss of a proton then gives the alpha-halogenated product.

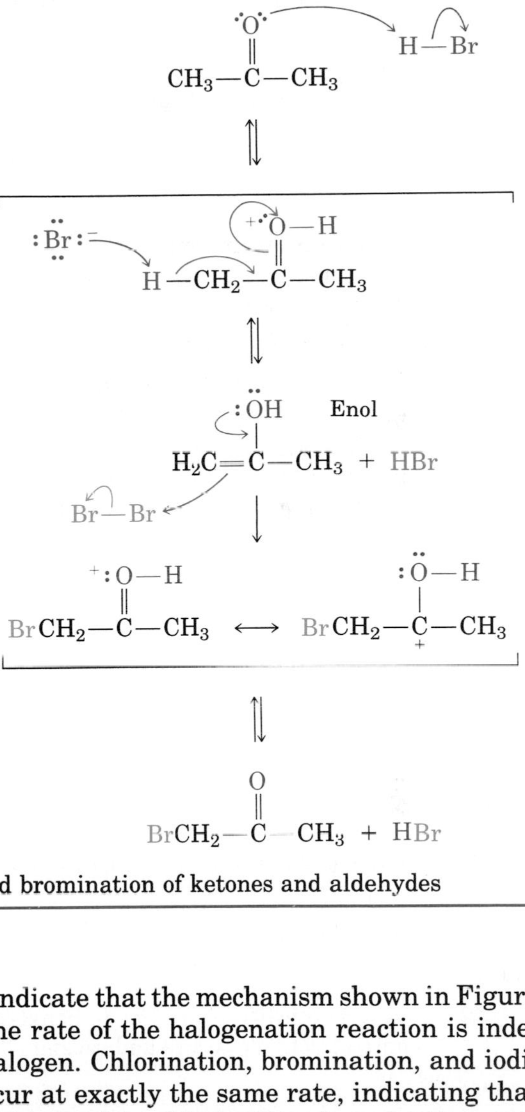

Figure 22.4 Mechanism of the acid-catalyzed bromination of ketones and aldehydes

There is much evidence to indicate that the mechanism shown in Figure 22.4 is correct. For example, the rate of the halogenation reaction is independent of the nature of the halogen. Chlorination, bromination, and iodination of a given ketone all occur at exactly the same rate, indicating that the same rate-limiting step is involved in chlorination, bromination, and iodination.

Additional evidence comes from measurements indicating that acid-catalyzed ketone halogenations exhibit second-order kinetics and follow the rate law:

$$\text{Reaction rate} = k[\text{Ketone}][\text{HA}]$$

This information tells us that ketone halogenations depend only on ketone and acid concentrations and are independent of halogen concentration. Thus, halogen is not involved in the rate-limiting step.

A final piece of evidence comes from deuteration experiments. If a ketone is treated with $D_3O^+$ instead of $H_3O^+$, the acidic alpha-hydrogen atoms are replaced by deuterium atoms. For a given ketone, the rate of deuteration is identical to the rate of halogenation, indicating that the same intermediate is involved in both deuteration and halogenation. The common intermediate that satisfies all the evidence just presented can only be an enol.

Ketone/aldehyde — $H_3O^+$ ($D_3O^+$), Slow, rate-limiting → Enol — $D_3O^+$, Fast / $X_2$, Fast

$X$ = Cl, Br, or I

$\alpha$-Bromo ketones are useful in organic synthesis because they can be dehydrobrominated by base treatment to yield $\alpha,\beta$-unsaturated ketones. For example, 2-bromo-2-methylcyclohexanone gives 2-methyl-2-cyclohexenone in 62% yield when refluxed in pyridine. The reaction, which takes place by an E2 elimination pathway (Section 11.11), is an excellent method of introducing carbon–carbon double bonds into molecules.

E2 reaction — 2-Bromo-2-methylcyclohexanone → (pyridine, Δ) 2-Methyl-2-cyclohexenone (62%) (an $\alpha,\beta$-unsaturated ketone) + HBr

**PROBLEM**

**22.4** Show in detail the mechanism of the deuteration of acetone on treatment with $D_3O^+$.

$$CH_3COCH_3 \xrightarrow{D_3O^+} CH_3COCH_2D$$

**PROBLEM**

**22.5** When optically active ($R$)-3-phenyl-2-butanone is exposed to aqueous acid, a loss of optical activity occurs and racemic 3-phenyl-2-butanone is produced. Explain.

**PROBLEM**

**22.6** In light of your answer to Problem 22.5, would you expect optically active ($R$)-3-methyl-3-phenyl-2-pentanone to be racemized by acid treatment? Explain.

## 22.4 Alpha Bromination of Carboxylic Acids: The Hell–Volhard–Zelinskii Reaction

Direct alpha bromination of carbonyl compounds by bromine in acetic acid is limited to ketones and aldehydes, since acids, esters, and amides do not enolize sufficiently for halogenation to take place. Carboxylic acids, however, can be α-brominated by a mixture of bromine and phosphorus tribromide—the **Hell–Volhard–Zelinskii (HVZ) reaction**:

$$\underset{\text{Heptanoic acid}}{CH_3CH_2CH_2CH_2CH_2\overset{\overset{\large H}{|}}{C}HCOOH} \xrightarrow[2.\ H_2O]{1.\ Br_2,\ PBr_3} \underset{\text{2-Bromoheptanoic acid (90\%)}}{CH_3CH_2CH_2CH_2CH_2\overset{\overset{\large Br}{|}}{C}HCOOH} + HBr$$

The first step in the Hell–Volhard–Zelinskii reaction takes place between $PBr_3$ and carboxylic acid to yield an intermediate acid bromide plus HBr (Section 21.5). The HBr produced in the first step next catalyzes enolization of the acid bromide, and the enol reacts rapidly with bromine. Hydrolysis of the α-bromo acid bromide by addition of water then gives the α-bromo carboxylic acid product.

$$R{-}\overset{\overset{\large H}{|}}{C}H{-}\overset{\overset{\large O}{\|}}{C}{-}OH \xrightarrow{PBr_3} \left[\underset{\text{Acid bromide}}{R{-}\overset{\overset{\large H}{|}}{C}H{-}\overset{\overset{\large O}{\|}}{C}{-}Br} \rightleftharpoons \underset{\text{Acid bromide enol}}{R{-}CH{=}\overset{\overset{\large OH}{|}}{C}{-}Br}\right]$$

$$\downarrow Br_2$$

$$R{-}\overset{\overset{\large Br}{|}}{C}H{-}\overset{\overset{\large O}{\|}}{C}{-}OH \xleftarrow{H_2O} \left[R{-}\overset{\overset{\large Br}{|}}{C}H{-}\overset{\overset{\large O}{\|}}{C}{-}Br + HBr\right]$$

The overall result of the Hell–Volhard–Zelinskii reaction is the transformation of an acid into an α-bromo acid. Note, however, that the key step involves bromination of an acid bromide enol in a manner analogous to that occurring during ketone bromination.

PROBLEM

**22.7** If an optically active carboxylic acid such as (*R*)-2-phenylpropanoic acid were brominated under Hell–Volhard–Zelinskii conditions, would you expect the product to be optically active or racemic? Explain.

PROBLEM

**22.8** If methanol rather than water is added at the end of the Hell–Volhard–Zelinskii reaction, an ester rather than an acid is produced. Propose a mechanism for this transformation.

$$RCH_2COOH \xrightarrow[2.\ CH_3OH]{1.\ Br_2,\ PBr_3} RCHBrCOOCH_3$$

## 22.5 Acidity of Alpha-Hydrogen Atoms: Enolate Ion Formation

During the discussion of base-catalyzed enol formation earlier in this chapter (Section 22.1), we said that carbonyl compounds act as weak protic acids. Strong bases can abstract acidic alpha protons from carbonyl compounds to yield enolate anions:

$$\text{:Base} \curvearrowright \text{H}-\overset{|}{\underset{|}{\text{C}}}-\text{C}(=\ddot{\text{O}}) \underset{}{\overset{\text{Base}}{\rightleftharpoons}} \overset{..}{\text{C}}^{-}-\text{C}(=\ddot{\text{O}}) \longleftrightarrow \text{C}=\text{C}-\ddot{\text{O}}:^{-} + \text{Base}-\text{H}^{+}$$

An enolate anion

Why are carbonyl compounds somewhat acidic? If we compare acetone, $pK_a \approx 20$, with ethane, $pK_a \approx 50$, we find that the presence of a neighboring carbonyl group increases the acidity of a ketone over an alkane by a factor of $10^{30}$.

$$CH_3-CH_2-H \quad \text{versus} \quad H_3C-\overset{O}{\overset{\|}{C}}-CH_2-H$$

Ethane, $pK_a \approx 50$ — Acetone, $pK_a \approx 20$

The reason for this increased acidity is best seen by viewing an orbital picture of the enolate ion (Figure 22.5). Proton abstraction from a carbonyl compound occurs when the alpha C—H sigma bond is oriented in the plane of the carbonyl-group *p* orbitals. The alpha-carbon atom of the enolate ion product is $sp^2$ hybridized and has a *p* orbital that overlaps the neighboring carbonyl-group *p* orbitals. Thus, the negative charge is shared by the electronegative oxygen atom, and the enolate ion is stabilized by resonance between two forms.

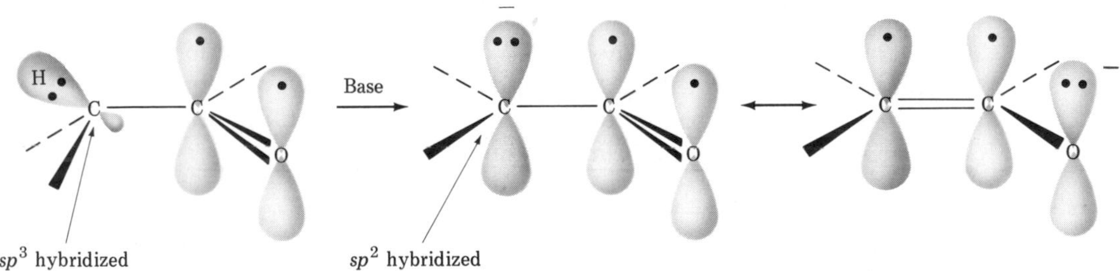


**Figure 22.5** Mechanism of enolate ion formation by abstraction of an alpha proton from a carbonyl compound

Carbonyl compounds are more acidic than alkanes for the same reason that carboxylic acids are more acidic than alcohols (Section 20.3). In both cases, the anions are stabilized by resonance. Enolate ions differ from carboxylate ions in that their two resonance forms are not equivalent—the form with the negative charge on oxygen is undoubtedly of lower energy than the form with the charge on carbon. Nevertheless, the principle behind resonance stabilization is the same in both cases.

$$CH_3-CH_2-H \quad \text{versus} \quad CH_3\overset{O}{\overset{\|}{C}}-CH_2-H \overset{H_2O}{\rightleftharpoons} CH_3\overset{:O:}{\overset{\|}{C}}-\overset{..}{C}H_2^{-} \longleftrightarrow CH_3\overset{:\ddot{O}:^{-}}{\overset{|}{C}}=CH_2 + H_3O^{+}$$

$pK_a \geq 50$ — $pK_a \approx 20$ — Nonequivalent resonance forms of ketone enolate ion

Recall the following:

$$CH_3-O-H \quad \text{versus} \quad CH_3\overset{\overset{\displaystyle O}{\|}}{C}-O-H \xrightleftharpoons{H_2O} CH_3\overset{\overset{\displaystyle :O:}{\|}}{C}-\ddot{\underset{\cdot\cdot}{O}}:^- \longleftrightarrow CH_3\overset{\overset{\displaystyle :\ddot{O}:^-}{|}}{C}=\ddot{\underset{\cdot\cdot}{O}} + H_3O^+$$

$pK_a \approx 16$ $\qquad$ $pK_a \approx 5$ $\qquad$ Equivalent resonance forms of carboxylate ion

Since alpha-hydrogen atoms of carbonyl compounds are only weakly acidic when compared with carboxylic acids, strong bases must be used to effect enolate ion formation. If an alkoxide ion such as sodium ethoxide is used as base, ionization takes place only to the extent of about 0.01%, since ethanol ($pK_a = 16$) is a stronger acid than acetone. If, however, a powerful base such as sodium hydride (NaH, the sodium "salt" of $H_2$), sodium amide ($NaNH_2$, the sodium salt of ammonia), or lithium diisopropylamide [$LiN(i\text{-}C_3H_7)_2$, the lithium salt of diisopropylamine] is used, a carbonyl compound can be completely converted into its enolate ion.

$$\text{Cyclohexanone} \xrightarrow[\text{THF}]{Li^+\ ^-N(i\text{-}C_3H_7)_2} \text{Cyclohexanone enolate } (:\ddot{O}:^-\ Li^+) + H-N(i\text{-}C_3H_7)_2$$

Cyclohexanone $\qquad$ Cyclohexanone enolate (100%)

Lithium diisopropylamide (LDA) is widely used as a base for preparing enolate ions from carbonyl compounds. It is easily prepared by reaction between butyllithium and diisopropylamine and has nearly ideal properties: It is an exceedingly powerful base since diisopropylamine has $pK_a \approx 40$; it is soluble in organic solvents such as THF; it is too bulky to add to carbonyl groups in nucleophilic addition reactions but is nevertheless extremely reactive, even at −78°C.

$$H-N[CH(CH_3)_2]_2 + n\text{-}C_4H_9Li \xrightarrow[\text{solvent}]{\text{THF}} Li^+\ :\bar{N}[CH(CH_3)_2]_2 + C_4H_{10}$$

Diisopropylamine $\qquad$ Lithium diisopropylamide (LDA)

All types of carbonyl compounds, including aldehydes, ketones, esters, acid chlorides, and amides have greatly enhanced alpha-hydrogen acidity compared to alkanes. Thus, all can be converted into enolate ions by reaction with LDA. Table 22.1 lists the approximate $pK_a$ values of different kinds of carbonyl compounds and shows how these values compare with other acidic substances we have seen.

**Table 22.1** Acidity constants for some organic compounds

| *Compound type* | *Compound* | $pK_a$ |
|---|---|---|
| Carboxylic acid | $CH_3COOH$ | 5 |
| 1,3-Diketone | $CH_2(COCH_3)_2$ | 9 |
| 1,3-Keto ester | $CH_3COCH_2CO_2C_2H_5$ | 11 |
| 1,3-Dinitrile | $CH_2(CN)_2$ | 11 |
| 1,3-Diester | $CH_2(CO_2C_2H_5)_2$ | 13 |
| Water | HOH | 16 |
| Primary alcohol | $CH_3CH_2OH$ | 16 |
| Acid chloride | $CH_3COCl$ | 16 |
| Aldehyde | $CH_3CHO$ | 17 |
| Ketone | $CH_3COCH_3$ | 20 |
| Ester | $CH_3CO_2C_2H_5$ | 25 |
| Nitrile | $CH_3CN$ | 25 |
| Dialkylamide | $CH_3CON(CH_3)_2$ | 30 |
| Ammonia | $NH_3$ | 35 |
| Dialkylamine | $HN(i\text{-}C_3H_7)_2$ | 40 |
| Alkyne | $HC{\equiv}CH$ | 25 |
| Alkene | $CH_2{=}CH_2$ | 49 |
| Alkane | $CH_3CH_3$ | 50 |

When a hydrogen atom is flanked by two carbonyl groups, the acidity is enhanced even more. Thus, as Table 22.1 indicates, compounds such as 1,3-diketones (known as ***β*-diketones**), 1,3-keto esters (***β*-keto esters**), and 1,3-diesters are much more acidic than water. The reason for the high acidity of these *β*-dicarbonyl compounds is that their enolate ions are highly stabilized by resonance stabilization of the charge by *both* neighboring carbonyl groups. For example, the enolate ion of 2,4-pentanedione has three resonance forms:

$$CH_3\overset{\overset{:O:}{\|}}{C}-\underset{\alpha}{\overset{\overset{H}{|}}{C}H}-\overset{\overset{:O:}{\|}}{C}CH_3 \xrightleftharpoons{\text{Base}} CH_3\overset{\overset{:O:}{\|}}{C}-\ddot{C}H^{-}-\overset{\overset{:O:}{\|}}{C}CH_3 \longleftrightarrow CH_3\overset{\overset{:\ddot{O}:^-}{|}}{C}=CH-\overset{\overset{:O:}{\|}}{C}CH_3 \updownarrow CH_3\overset{\overset{:O:}{\|}}{C}-CH=\overset{\overset{:\ddot{O}:^-}{|}}{C}CH_3$$

2,4-Pentanedione, a *β*-diketone ($pK_a$ = 9) ($\beta$ and $\alpha$ label the carbons)

Similar resonance forms can be drawn for other doubly stabilized enolate ions.

**PROBLEM**

**22.9** Identify the most acidic hydrogens in these molecules:
(a) $CH_3CH_2CHO$ (b) $(CH_3)_3CCOCH_3$ (c) $CH_3COOH$
(d) Benzamide (e) $CH_3CH_2CH_2CN$ (f) $CH_3CON(CH_3)_2$
(g) 1,3-Cyclohexanedione

**PROBLEM**

**22.10** When optically active (*R*)-2-methylcyclohexanone is treated with aqueous NaOH, racemic 2-methylcyclohexanone is produced. Propose a mechanism to account for this racemization.

**PROBLEM**

**22.11** Would you expect optically active (*S*)-3-methylcyclohexanone to be racemized on base treatment in the same way that 2-methylcyclohexanone is (Problem 22.10)? Explain.

## 22.6 Reactivity of Enolate Ions

Enolate ions are much more useful than enols for two reasons. The first is that pure enols cannot normally be isolated; they are usually generated only as fleeting intermediates in small concentration. By contrast, solutions of pure enolate ions are easily prepared from most carbonyl compounds by reaction with a strong base.

Second and more important is the fact that enolate ions are much more reactive than enols; they undergo many important reactions that enols don't. Whereas enols are neutral, enolate ions carry a full negative charge, making them much better nucleophiles. Thus, the alpha-carbon atom of an enolate ion is highly reactive toward electrophiles.

:ÖH
C=C

Enol: neutral, moderately reactive, very difficult to isolate

:Ö:⁻ ⟷ :O:
C=C ⟷ C—C

Enolate: negatively charged, very reactive, easily prepared

Since enolate ions are resonance hybrids of two nonequivalent resonance forms, they can be looked at either as vinylic alkoxides ($C{=}C{-}O^-$) or as $\alpha$-keto carbanions ($^-C{-}C{=}O$). Thus, enolate ions can react with electrophiles either on oxygen or on carbon. Reaction on oxygen would yield an enol derivative, whereas reaction on carbon would yield an alpha-substituted carbonyl compound (Figure 22.6). Both kinds of reactivity are known, but reaction on carbon is the more commonly observed pathway.

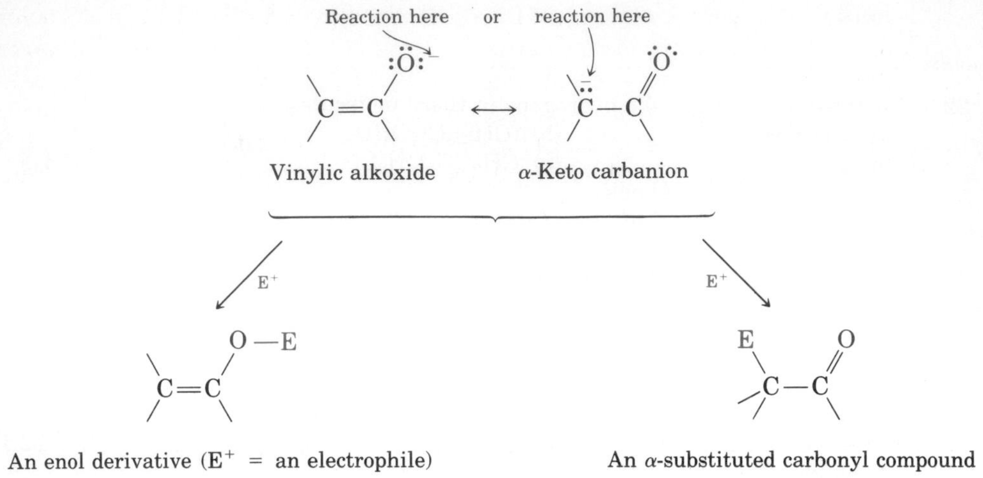

Reaction here or reaction here

Vinylic alkoxide ⟷ α-Keto carbanion

$E^+$ ↙ ↘ $E^+$

An enol derivative ($E^+$ = an electrophile) An α-substituted carbonyl compound

**Figure 22.6** Two modes of enolate ion reactivity: Reaction on carbon to yield an alpha-substituted carbonyl product is the more commonly followed path.

## 22.7 Halogenation of Enolate Ions: The Haloform Reaction

Halogenation of ketones is promoted by base as well as by acid. The base-promoted reaction occurs through enolate ion intermediates. Even relatively weak bases such as hydroxide ion are effective for halogenation, since it is not necessary to convert the ketone completely into its enolate ion. It is only necessary to generate a small amount of enolate at any one time because the reaction with halogen occurs as soon as the enolate ion is formed.

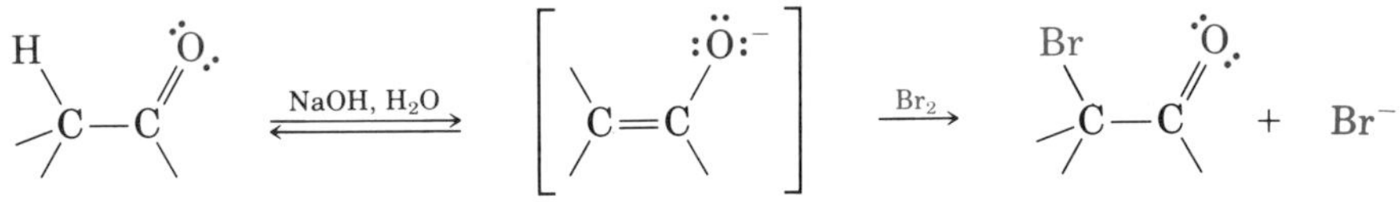

NaOH, $H_2O$ $Br_2$ + $Br^-$

Base-promoted halogenation of ketones is little used in practice because it is difficult to stop the reaction at the monosubstituted product. An alpha-halogenated ketone product is generally more acidic than the starting unsubstituted ketone because of the electron-withdrawing inductive effect of the halogen atom. Thus, the monohalo products are themselves rapidly turned into enolates and further halogenated.

If excess base and halogen are used, methyl ketones are trihalogenated and then cleaved by base in the **haloform reaction**:

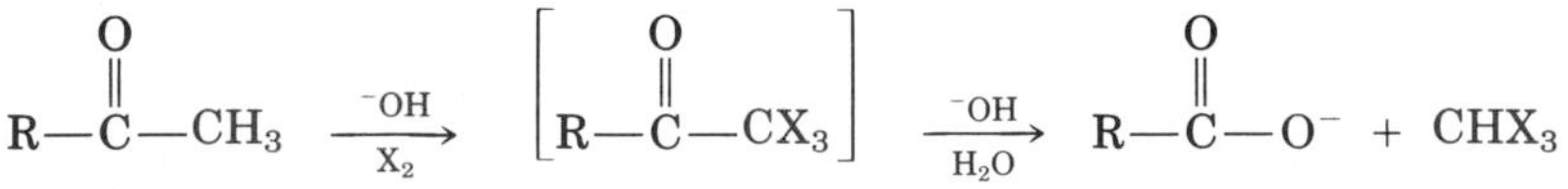

$$\mathrm{R{-}\overset{O}{\overset{\|}{C}}{-}CH_3} \xrightarrow[X_2]{^-OH} \left[\mathrm{R{-}\overset{O}{\overset{\|}{C}}{-}CX_3}\right] \xrightarrow[H_2O]{^-OH} \mathrm{R{-}\overset{O}{\overset{\|}{C}}{-}O^-} + \mathrm{CHX_3}$$

where X = Cl, Br, or I.

The haloform reaction, which converts a methyl ketone into a carboxylic acid plus a **haloform** (chloroform, $CHCl_3$; bromoform, $CHBr_3$; or iodoform, $CHI_3$), is the basis for a qualitative test for methyl ketones. A sample of unknown structure is dissolved in THF and placed in a test tube. Dilute solutions of aqueous sodium hydroxide and iodine are then added, and the test tube is observed. Formation of a yellow precipitate of iodoform, $CHI_3$, signals a positive test and indicates that the sample is a methyl ketone.

PROBLEM

**22.12** Base-promoted chlorination and bromination of a given ketone occur at the same rate. Explain.

PROBLEM

**22.13** Why do you suppose ketone halogenations in acidic media are referred to as being acid *catalyzed*, whereas halogenations in basic media are base *promoted*? Why is a full equivalent of base required for halogenation?

PROBLEM

**22.14** The second step of the haloform reaction is simply a nucleophilic acyl substitution—the replacement of $^{-}CX_3$ by $^{-}OH$. Show the mechanism of this transformation.

## 22.8 Selenenylation of Enolate Ions: Enone Synthesis

One of the most important recent advances in enolate ion chemistry occurred in the mid-1970s when it was found that carbonyl compounds can be **selenenylated**. That is, a *selenium* atom can be attached to the alpha position of a carbonyl compound. Selenenylation is accomplished by first allowing the carbonyl compound to react with a strong base such as LDA in THF solvent, followed by treatment of the enolate ion solution with 1 equiv of benzeneselenenyl bromide ($C_6H_5SeBr$). Immediate alpha-substitution reaction yields an $\alpha$-phenylseleno-substituted product.

The great value of the selenenylation reaction lies in the fact that the product can be easily converted into an $\alpha,\beta$-unsaturated carbonyl compound. On treatment with dilute hydrogen peroxide at room temperature, the selenium group is oxidized, elimination occurs, and an $\alpha,\beta$-unsaturated carbonyl compound is formed. The net result is introduction of a carbon–carbon double bond into the $\alpha,\beta$ position of the saturated carbonyl starting material. Yields are generally excellent, and the method is often superior to the alternative alpha-bromination–dehydrobromination route.

$$\text{H–C–C(H)–C(=O)} \xrightarrow[\text{2. } C_6H_5SeBr]{\text{1. LDA, THF}} \text{H–C–C(SeC}_6\text{H}_5\text{)–C(=O)} \xrightarrow{H_2O_2} \text{C=C–C(=O)} + C_6H_5SeOH$$

An $\alpha$-phenylseleno product

Mechanistically, the alpha-phenylselenenylation reaction is similar to halogenation except that the reactive enolate ion intermediate is prepared in a separate step prior to reaction, rather than during reaction as in halogenation. The overall process is shown in Figure 22.7.

The strong base LDA abstracts an acidic alpha-hydrogen atom from the carbonyl compound, yielding an enolate ion.

$Li^+ :\bar{N}:$ $i$-$C_3H_7$ $i$-$C_3H_7$

Enolate Li$^+$ + $HN(i\text{-}C_3H_7)_2$

The nucleophilic enolate ion then attacks electrophilic selenium and carries out a displacement of bromide, yielding the $\alpha$-phenylseleno product.

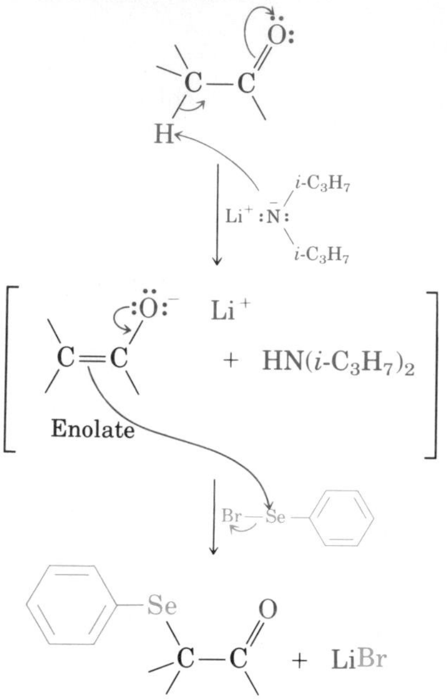

Br—Se

Se + LiBr

Figure 22.7 Mechanism of the reaction of an enolate ion with benzeneselenenyl bromide

Note that the key step in the alpha phenylselenenylation of carbonyl compounds is similar to an $S_N2$ reaction (Section 11.4). The alpha-carbon atom of the enolate ion is the nucleophile that displaces bromide by attack on selenium:

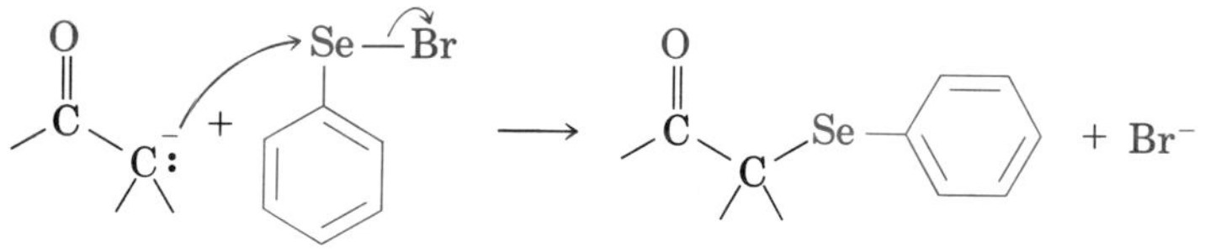

Se—Br + → Se + $Br^-$

Recall

$$Nu:^- + C—Br \xrightarrow[\text{reaction}]{S_N2} Nu—C + :\ddot{Br}:^-$$

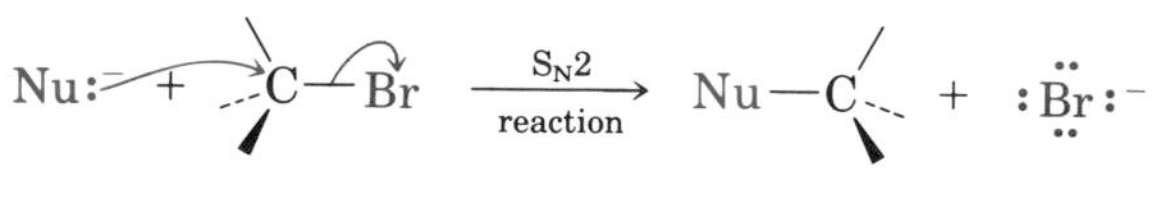

The elimination step involves oxidation of the phenylseleno intermediate to a phenyl selenoxide, which undergoes a spontaneous intramolecular elimination reaction. No added base is required (as it is in dehydrobromination), and the reaction occurs at room temperature under very mild conditions.

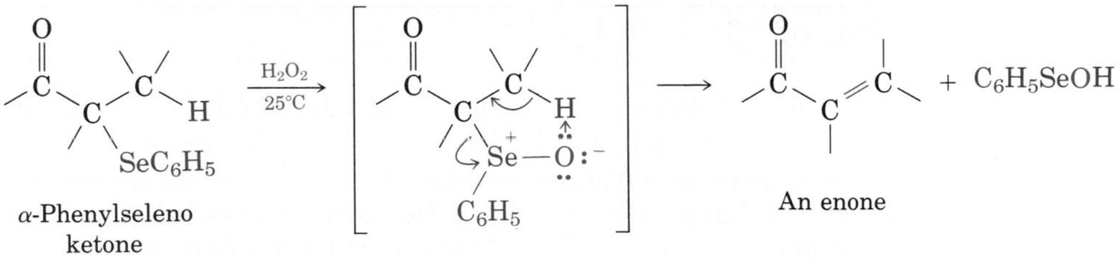

$$\alpha\text{-Phenylseleno ketone} \xrightarrow[25^\circ C]{H_2O_2} [\text{A selenoxide}] \longrightarrow \text{An enone} + C_6H_5SeOH$$

Ketones, esters, and nitriles all undergo the alpha-phenylselenenylation–elimination reaction in good yield. Aldehydes, however, often give product mixtures because their enolate ions are too reactive.

Ketone: Propiophenone ($C_6H_5COCH_2CH_3$) → 1. LDA in THF; 2. $C_6H_5SeBr$; 3. $H_2O_2$ → Phenyl vinyl ketone ($C_6H_5CO{-}CH{=}CH_2$) (80%)

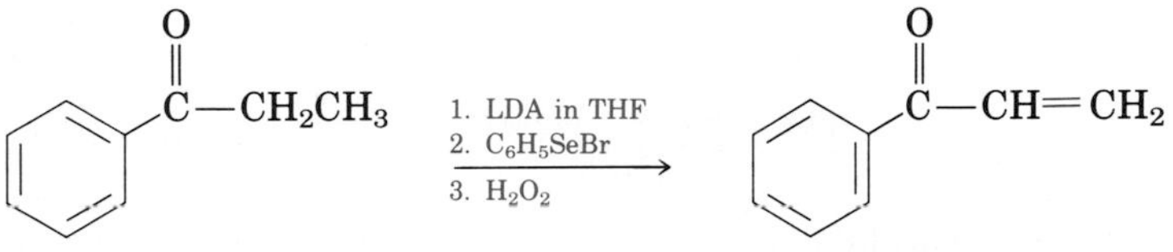

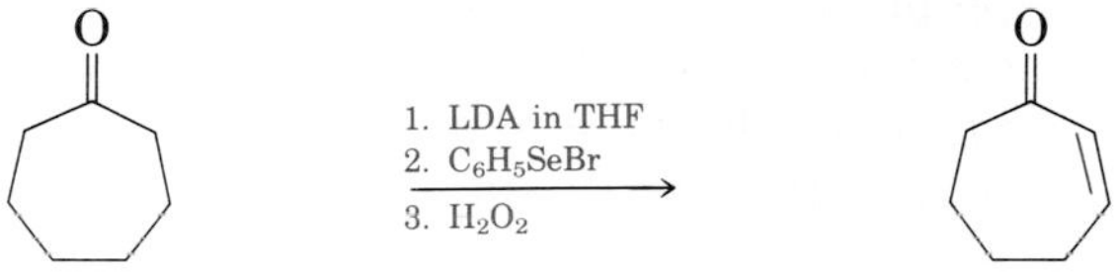

Ketone: Cycloheptanone → 1. LDA in THF; 2. $C_6H_5SeBr$; 3. $H_2O_2$ → 2-Cycloheptenone (55%)

Ester: Methyl cyclohexanecarboxylate ($C_6H_{11}{-}CO_2CH_3$) → 1. LDA in THF; 2. $C_6H_5SeBr$; 3. $H_2O_2$ → Methyl 1-cyclohexenecarboxylate (96%)

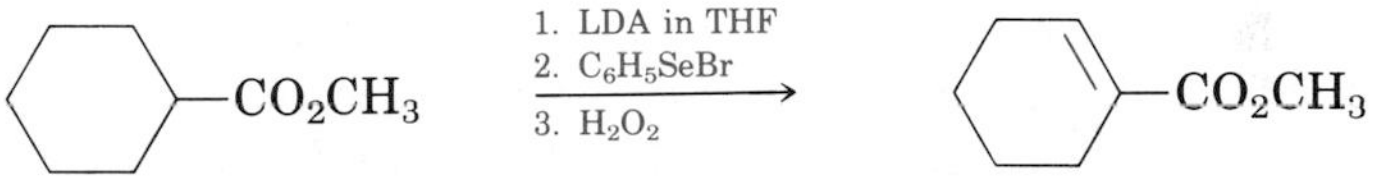

Nitrile: $CH_3(CH_2)_4CH_2CH_2CN$ (Octanenitrile) → 1. LDA in THF; 2. $C_6H_5SeBr$; 3. $H_2O_2$ → $CH_3(CH_2)_4CH{=}CHCN$ (2-Octenenitrile) (80%)

**PROBLEM**

**22.15** Show how you would prepare the following unsaturated molecules from saturated precursors.
(a) 1-Penten-3-one (b) 4-*tert*-Butyl-2-cyclohexenone
(c) 2-Methyl-2-pentenenitrile

**PROBLEM**

**22.16** Ketones react slowly with benzeneselenenyl chloride in the presence of HCl to yield $\alpha$-phenylseleno ketones. Propose a mechanism for this acid-catalyzed alpha-substitution reaction. To what other reaction is this process analogous?

## 22.9 Alkylation of Enolate Ions

One of the most important reactions of enolate ions is their **alkylation** by treatment with an alkyl halide. The alkylation reaction is extraordinarily useful for synthesis purposes because it allows the formation of a new carbon–carbon bond, thereby joining two small pieces into one larger molecule. Alkylation occurs when the nucleophilic enolate anion reacts with the electrophilic alkyl halide and displaces the leaving group by an $S_N2$ back-side attack.

Enolate ion + Alkyl halide $\xrightarrow[\text{reaction}]{S_N2}$ product + $^-$:Y

Alkylation reactions are subject to the same constraints that affect all $S_N2$ reactions (Section 11.4). Thus, the leaving group, —Y, in the alkylating agent can be chloride, bromide, iodide, or tosylate. The alkyl group, R—, must be primary or methyl and preferably should be allylic or benzylic. Secondary halides react poorly, and tertiary halides do not alkylate at all since competing E2 elimination of HX occurs instead. Vinylic and aryl halides are also unreactive, since back-side attack is sterically prevented.

R—Y

—Y: Tosylate > —I > —Br > —Cl

R—: Allylic ≈ Benzylic > $H_3C$— > $RCH_2$—

### THE MALONIC ESTER SYNTHESIS

The **malonic ester synthesis**, one of the oldest and best known carbonyl alkylation reactions, is an excellent method for preparing alpha-substituted acetic acids from alkyl halides:

$$\underset{\text{Alkyl halide}}{R—X} \xrightarrow[\text{ester synthesis}]{\text{Via malonic}} \underset{\alpha\text{-Substituted acetic acid}}{R—CH_2CO_2H}$$

Diethyl propanedioate, commonly called diethyl malonate or *malonic ester*, is more acidic than many other carbonyl compounds ($pK_a = 13$) because its alpha-hydrogen atoms are flanked by two carbonyl groups. Thus, malonic ester is easily converted into its enolate anion by reaction with sodium ethoxide in ethanol. The enolate ion, in turn, is a good nucleophile that reacts rapidly with alkyl halides, yielding alpha-substituted malonic esters.

$$\underset{\text{Malonic ester}}{H-\underset{H}{\overset{CO_2CH_2CH_3}{C}}-CO_2CH_2CH_3} \xrightarrow[CH_3CH_2OH]{Na^+\ ^-OCH_2CH_3} \underset{\text{Sodio malonic ester}}{Na^+\ ^-:\underset{H}{\overset{CO_2CH_2CH_3}{C}}-CO_2CH_2CH_3} \xrightarrow{RX} \underset{\text{An alkylated malonic ester}}{R-\underset{H}{\overset{CO_2CH_2CH_3}{C}}-CO_2CH_2CH_3} + NaX$$

The product of malonic ester alkylation has one acidic alpha-hydrogen atom left, and the alkylation process can therefore be repeated a second time to yield a dialkylated malonic ester:

$$R-\underset{H}{\overset{CO_2CH_2CH_3}{C}}-CO_2CH_2CH_3 \xrightarrow[2.\ R'X]{1.\ Na^+\ ^-OCH_2CH_3} \underset{\text{A dialkylated malonic ester}}{R-\underset{R'}{\overset{CO_2CH_2CH_3}{C}}-CO_2CH_2CH_3} + NaX$$

Once formed, the alkylated malonic esters can be hydrolyzed and decarboxylated when heated with aqueous acid (**decarboxylation** is the loss of carbon dioxide, $CO_2$). The product is a substituted monoacid.

$$R-\underset{R'}{\overset{CO_2CH_2CH_3}{C}}-CO_2CH_2CH_3 \xrightarrow[\Delta]{H_3O^+} R-\underset{H}{\overset{R'}{C}}-COOH + CO_2 + 2\ CH_3CH_2OH$$

Decarboxylation is not a general reaction of carboxylic acids, but is a unique feature of compounds like malonic acid that have a *second* carbonyl group two atoms away. That is, only $\beta$-keto acids and substituted malonic acids undergo loss of $CO_2$ on heating. The decarboxylation reaction occurs in two steps and involves initial acid-catalyzed hydrolysis of the diester to a diacid. The diacid then loses carbon dioxide by a cyclic mechanism.

$$\underset{\text{A malonic ester}}{R_2C(CO_2CH_2CH_3)_2} \xrightarrow[\Delta]{H_3O^+} \left[ \underset{\text{A diacid}}{R_2C(COOH)_2} \longrightarrow \underset{\text{An acid enol}}{R_2C{=}C(OH)_2} + CO_2 \right]$$

$$\downarrow$$

$$\underset{\text{A carboxylic acid}}{R_2CH\,COOH}$$

The malonic ester synthesis is an excellent method for converting alkyl halides into carboxylic acids while lengthening the carbon chain by two atoms. For example,

$$CH_3(CH_2)_2CH_2Br + Na^{+}\ {}^{-}\!:CH(CO_2C_2H_5)_2 \longrightarrow CH_3(CH_2)_3CH(CO_2C_2H_5)_2$$

1-Bromobutane — 84%

$$\xrightarrow{H_3O^+,\ \Delta} CH_3(CH_2)_4COOH$$

Hexanoic acid (75%)

$$CH_3CH_2CH_2CH_2CH(CO_2C_2H_5)_2 + CH_3I \xrightarrow[\text{Ethanol}]{Na^{+}\ {}^{-}OCH_2CH_3} CH_3CH_2CH_2CH_2C(CH_3)(CO_2C_2H_5)_2$$

$$\xrightarrow{H_3O^+,\ \Delta} CH_3CH_2CH_2CH_2CH(CH_3)COOH$$

2-Methylhexanoic acid (74%)

The malonic ester synthesis can also be used to prepare *cyclo*alkanecarboxylic acids by the proper choice of alkyl halide. For example, if 1,4-dibromobutane is treated with diethyl malonate in the presence of 2 equiv of sodium ethoxide base, the first alkylation occurs as expected, but the second alkylation step occurs *internally* to yield a five-membered-ring product. Hydrolysis and decarboxylation then lead to cyclopentanecarboxylic acid (Figure 22.8). Three-, four-, five-, and six-membered rings can be prepared in this manner, but yields decrease drastically for larger ring sizes.

**PRACTICE PROBLEM**

How would you prepare heptanoic acid via a malonic ester synthesis?

*Solution* The malonic ester synthesis converts an alkyl halide into a carboxylic acid having two more carbons in its chain. Thus, a seven-carbon acid chain must be derived from the five-carbon alkyl halide 1-bromopentane.

$$CH_3CH_2CH_2CH_2CH_2Br + CH_2(COOCH_2CH_3)_2 \xrightarrow[2.\ H_3O^+,\ \Delta]{1.\ Na^{+}\ {}^{-}OCH_2CH_3} CH_3CH_2CH_2CH_2CH_2CH_2COOH$$

**PROBLEM**

**22.17** What alkyl halide would you use to prepare these compounds by a malonic ester synthesis?
(a) Butanoic acid
(b) 5-Methylhexanoic acid

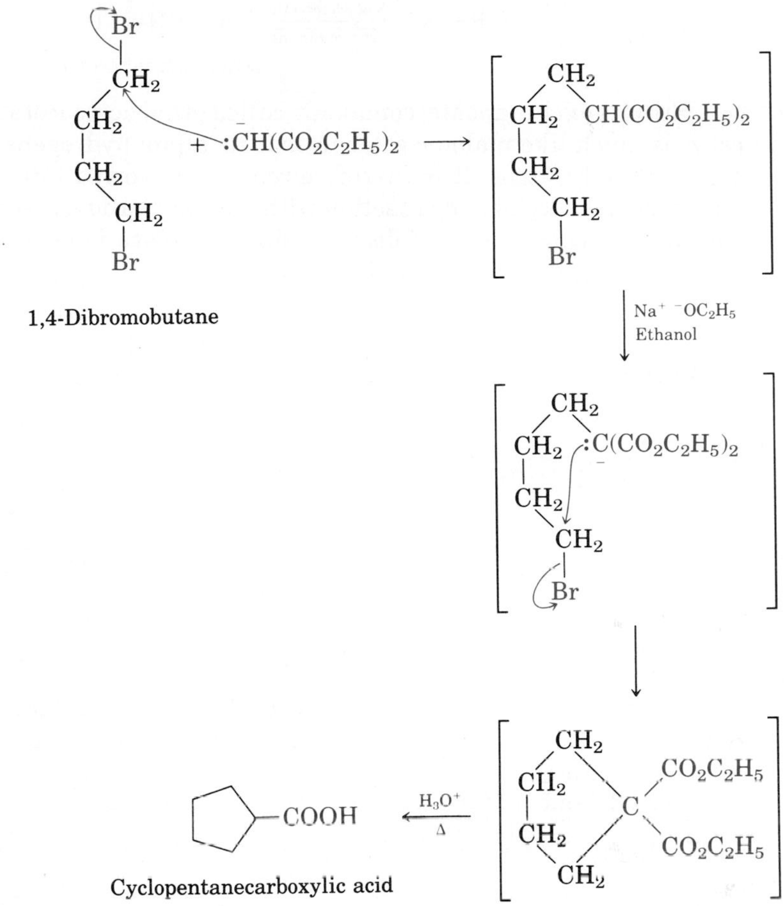

1,4-Dibromobutane + $:\overset{-}{C}H(CO_2C_2H_5)_2$ ⟶ $[BrCH_2CH_2CH_2CH_2CH(CO_2C_2H_5)_2]$

$Na^+ \ ^-OC_2H_5$, Ethanol ⟶ $[BrCH_2CH_2CH_2CH_2\overset{-}{C}:(CO_2C_2H_5)_2]$ ⟶ $[\text{cyclo-}(CH_2)_4C(CO_2C_2H_5)_2]$

$H_3O^+$, $\Delta$ ⟶ Cyclopentanecarboxylic acid (cyclopentane–COOH)

**Figure 22.8** Malonic ester synthesis of cyclopentanecarboxylic acid

PROBLEM

**22.18** How could you use a malonic ester synthesis to prepare these compounds? Show all steps.

(a) 3-Phenylpropanoic acid
(b) 2-Methylpentanoic acid
(c) 4-Methylpentanoic acid
(d) Ethyl cyclobutanecarboxylate

PROBLEM

**22.19** Monoalkylated and dialkylated acetic acids can be prepared by the malonic ester synthesis, but trialkylated acetic acids ($R_3CCOOH$) cannot be prepared. Explain.

## THE ACETOACETIC ESTER SYNTHESIS

The **acetoacetic ester synthesis** provides a method for preparing alpha-substituted acetone derivatives from alkyl halides in the same way that the malonic ester synthesis provides a method for preparing alpha-substituted acetic acids.

$$\text{R—X} \xrightarrow[\text{ester synthesis}]{\text{Via acetoacetic}} \text{R—CH}_2\overset{\overset{\text{O}}{\|}}{\text{C}}\text{CH}_3$$

α-Substituted acetone

Ethyl 3-oxobutanoate, commonly called ethyl acetoacetate or *acetoacetic ester*, is much like malonic ester in that its alpha hydrogens are flanked by two carbonyl groups. It is therefore readily converted into its enolate ion, which can be alkylated by reaction with an alkyl halide. A second alkylation can also be carried out, if desired, since acetoacetic ester has two acidic alpha protons that can be replaced.

$$\text{CH}_3\overset{\overset{\text{O}}{\|}}{\text{C}}\text{—}\underset{\text{H}}{\overset{\text{H}}{\text{C}}}\text{—}\overset{\overset{\text{O}}{\|}}{\text{C}}\text{OC}_2\text{H}_5 + \text{Na}^+ \ {}^-\text{OC}_2\text{H}_5 \xrightarrow{\text{Ethanol}} \left[\text{CH}_3\overset{\overset{\text{O}}{\|}}{\text{C}}\text{—}\underset{\text{H}}{\ddot{\text{C}}}\text{—}\overset{\overset{\text{O}}{\|}}{\text{C}}\text{OC}_2\text{H}_5\right]$$

Acetoacetic ester

$$\downarrow \text{R—X}$$

$$\text{CH}_3\text{—}\overset{\overset{\text{O}}{\|}}{\text{C}}\text{—}\underset{\text{R}'}{\overset{\text{R}}{\text{C}}}\text{—CO}_2\text{C}_2\text{H}_5 \xleftarrow[\text{2. R}'\text{X}]{\text{1. Na}^+ \ {}^-\text{OC}_2\text{H}_5} \text{CH}_3\overset{\overset{\text{O}}{\|}}{\text{C}}\text{—}\underset{\text{H}}{\overset{\text{R}}{\text{C}}}\text{—CO}_2\text{C}_2\text{H}_5$$

Upon heating with aqueous HCl, the alkylated acetoacetic ester is hydrolyzed and decarboxylated via a β-keto acid intermediate to yield an alpha-substituted acetone product. If a monoalkylated acetoacetic ester is hydrolyzed and decarboxylated, an alpha-monosubstituted acetone is formed; if a dialkylated acetoacetic ester is hydrolyzed and decarboxylated, an α,α-disubstituted acetone is formed.

$$\text{H(R)C(CO}_2\text{C}_2\text{H}_5\text{)(COCH}_3\text{)} \xrightarrow[\Delta]{\text{H}_3\text{O}^+} \left[\text{H(R)C(COOH)(C(=O)CH}_3\text{)} \longrightarrow \underset{\text{R}}{\overset{\text{H}}{}}\text{C=C}\underset{\text{CH}_3}{\overset{\text{O—H}}{}} + \text{CO}_2\right]$$

A monoalkylated acetoacetic ester

$$\downarrow$$

$$\text{RCH}_2\overset{\overset{\text{O}}{\|}}{\text{C}}\text{CH}_3$$

An α-monosubstituted acetone

$$\text{R—}\underset{\text{R}'}{\overset{\text{CO}_2\text{C}_2\text{H}_5}{\text{C}}}\text{—COCH}_3 \xrightarrow[\Delta]{\text{H}_3\text{O}^+} \text{RR}'\text{CH}\overset{\overset{\text{O}}{\|}}{\text{C}}\text{CH}_3 + \text{CO}_2 + \text{C}_2\text{H}_5\text{OH}$$

A dialkylated acetoacetic ester

An α,α-disubstituted acetone

For example,

$$\underset{\text{Acetoacetic ester}}{\overset{\displaystyle O \atop \|}{CH_2CCH_3} \atop |\ CO_2C_2H_5} + \underset{\text{1-Bromobutane}}{CH_3CH_2CH_2CH_2Br} \xrightarrow[\text{2. } H_3O^+,\ \Delta]{\text{1. } Na^+\ {}^-OC_2H_5} \underset{\text{2-Heptanone (65\%)}}{CH_3CH_2CH_2CH_2{-}CH_2\overset{O}{\overset{\|}{C}}CH_3}$$

The three-step sequence of (1) enolate ion formation, (2) alkylation, and (3) hydrolysis–decarboxylation is applicable to *all* $\beta$-keto esters with acidic alpha hydrogens, not just to acetoacetic ester itself. Thus, cyclic $\beta$-keto esters such as ethyl 2-oxocyclohexanecarboxylate can be alkylated and decarboxylated to give 2-substituted cyclohexanones in high yield. For example,

Ethyl 2-oxocyclohexanecarboxylate $\xrightarrow[C_2H_5OH]{Na^+\ {}^-OC_2H_5}$ [enolate, $COC_2H_5$] $\xrightarrow{C_6H_5CH_2Br}$ (2-oxo-1-$CH_2C_6H_5$-1-$CO_2C_2H_5$-cyclohexane) $\xrightarrow{H_3O^+,\ \Delta}$ 2-Benzylcyclohexanone (77%)

PRACTICE PROBLEM....................................................................

How would you prepare 2-pentanone via an acetoacetic ester synthesis?

*Solution* The acetoacetic ester synthesis yields a ketone product by adding three carbons to an alkyl halide:

$$\underbrace{CH_3\overset{O}{\overset{\|}{C}}CH_2}_{\text{These three carbons from acetoacetic ester}}{-}\underbrace{CH_2CH_3}_{\text{These carbons from alkyl halide}}$$

(This bond formed)

Thus, the acetoacetic ester synthesis of 2-pentanone would involve reaction of bromoethane:

$$CH_3\overset{O}{\overset{\|}{C}}CH_2COOCH_2CH_3 + CH_3CH_2Br \xrightarrow[\text{2. } H_3O^+]{\text{1. } Na^+\ {}^-OCH_2CH_3} CH_3\overset{O}{\overset{\|}{C}}CH_2CH_2CH_3$$

PROBLEM

**22.20** What alkyl halides would you use to prepare these ketones via an acetoacetic ester synthesis?

(a) 5-Methyl-2-hexanone (b) 5-Phenyl-2-pentanone

PROBLEM

**22.21** How would you prepare methyl cyclopentyl ketone, using an acetoacetic ester synthesis?

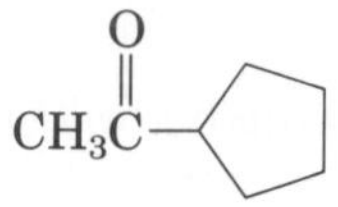

Methyl cyclopentyl ketone

PROBLEM

**22.22** Which of the following compounds cannot be prepared by an acetoacetic ester synthesis? Explain.

(a) 2-Butanone (b) Phenylacetone
(c) Acetophenone (d) 3,3-Dimethyl-2-butanone

## DIRECT ALKYLATION OF KETONES, ESTERS, AND NITRILES

The malonic ester synthesis and the acetoacetic ester synthesis are rather special, since both alkylation reactions take place at doubly carbonyl-activated centers. By contrast, it's also possible in certain cases to alkylate at the alpha position of *mono*ketones, *mono*esters, and nitriles.

The experimental conditions necessary to carry out the direct alkylation of carbonyl compounds are precise. Solvent and reaction temperature are both important, and the exact nature of the base used to generate the enolate ion is critical. The base must be sufficiently strong to convert a carbonyl compound ($pK_a \approx 20–25$) into its enolate anion quickly, yet it must also be sufficiently bulky that it will not add to the carbonyl group in a nucleophilic addition or substitution reaction. Research carried out in the 1970s showed that LDA in THF solvent is highly effective in promoting alkylation reactions of carbonyl compounds.

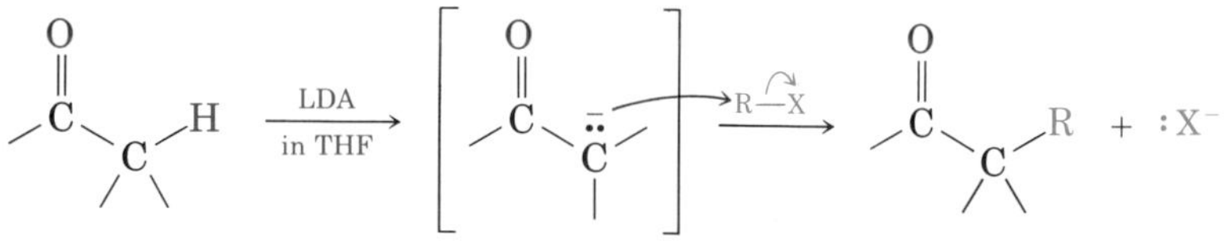

LDA in THF; R—X

Ketone/ester

α-Substituted ketone/ester

Ketones, esters, and nitriles can all be alkylated by using LDA or related dialkylamide bases in THF, but aldehydes rarely give high yields of pure products. (The problem is not that aldehydes aren't acidic enough, it's that aldehyde enolate ions undergo carbonyl condensation reactions instead of alkylation. We'll study this condensation reaction in the next chapter.) Some specific examples of alkylation reactions are shown.

Lactone

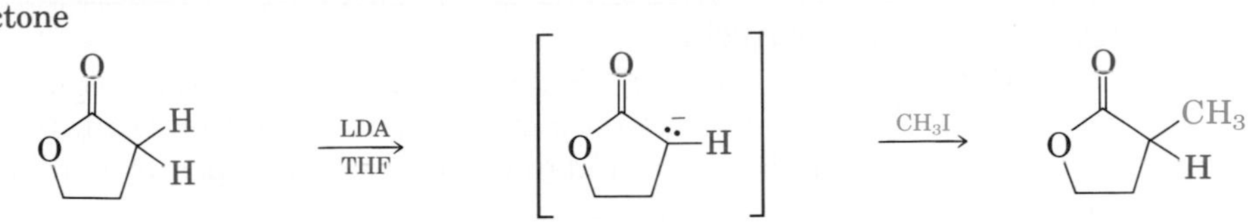

Butyrolactone $\xrightarrow[\text{THF}]{\text{LDA}}$ [enolate] $\xrightarrow{CH_3I}$ 2-Methylbutyrolactone (88%)

Ester

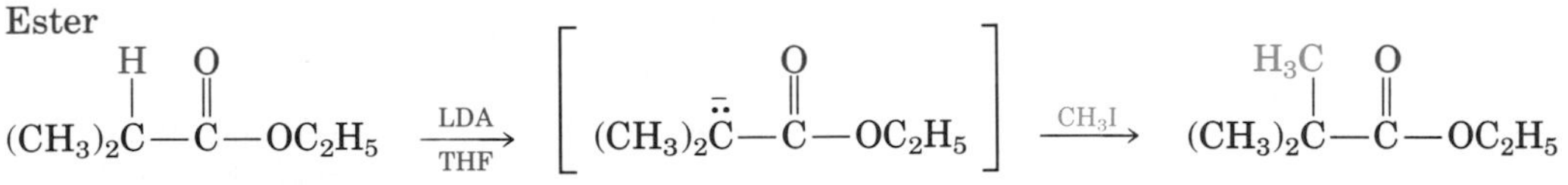

$(CH_3)_2CH-C(=O)-OC_2H_5$ $\xrightarrow[\text{THF}]{\text{LDA}}$ $[(CH_3)_2\ddot{C}^{-}-C(=O)-OC_2H_5]$ $\xrightarrow{CH_3I}$ $(CH_3)_2C(CH_3)-C(=O)-OC_2H_5$

Ethyl 2-methylpropanoate

Ethyl 2,2-dimethylpropanoate (87%)

Nitrile

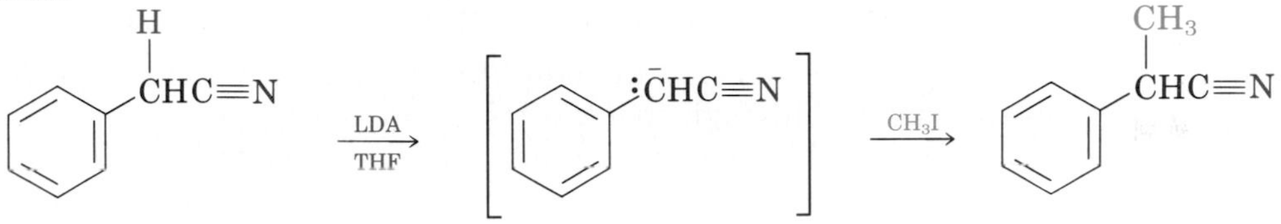

$C_6H_5CH_2C{\equiv}N$ $\xrightarrow[\text{THF}]{\text{LDA}}$ $[C_6H_5\ddot{C}^{-}HC{\equiv}N]$ $\xrightarrow{CH_3I}$ $C_6H_5CH(CH_3)C{\equiv}N$

Phenylacetonitrile

2-Phenylpropanenitrile (71%)

Ketone

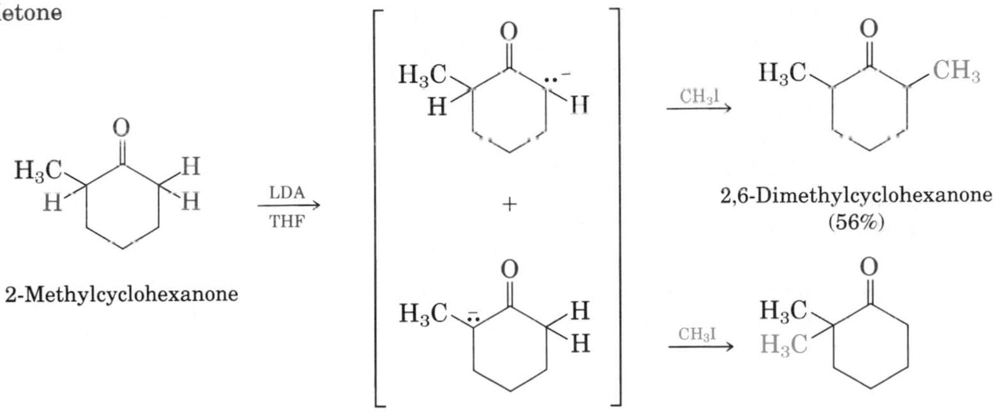

2-Methylcyclohexanone $\xrightarrow[\text{THF}]{\text{LDA}}$ [enolate at C-6 + enolate at C-2] $\xrightarrow{CH_3I}$ 2,6-Dimethylcyclohexanone (56%) + 2,2-Dimethylcyclohexanone (6%)

Note in the previous examples that alkylation of the unsymmetrically substituted ketone, 2-methylcyclohexanone, leads to a mixture of products, since both possible enolate ions are formed. In general, the major product in such cases is derived from alkylation at the less hindered, more accessible position. Thus, alkylation of 2-methylcyclohexanone occurs primarily at the secondary 6-position, rather than at the tertiary 2-position.

**PROBLEM**

**22.23** How might you prepare the following compounds, using an alkylation reaction as the key step?

(a) 3-Phenyl-2-butanone
(b) 2-Ethylpentanenitrile
(c) 2-Allylcyclohexanone
(d) 2,2,6,6-Tetramethylcyclohexanone

## 22.10 Summary and Key Words

The **alpha substitution** of carbonyl compounds via **enol** or **enolate ion** intermediates is one of the four fundamental reaction types in carbonyl-group chemistry.

Enol

Enolate

$\alpha$-Substituted product
($E^+$ = an electrophile)

All carbonyl compounds rapidly equilibrate with their enols, a process called **tautomerism**. Although enol tautomers are normally present to only a small extent at equilibrium and cannot usually be isolated in pure form, they nevertheless contain a highly nucleophilic double bond and react rapidly with a variety of electrophiles. For example, ketones and aldehydes are rapidly halogenated at the alpha position by reaction with chlorine, bromine, or iodine in acetic acid solution. Alpha bromination of carboxylic acids can be similarly accomplished by the **Hell–Volhard–Zelinskii reaction**, in which an acid is treated with $Br_2$ and $PBr_3$. The alpha-halogenated products can then undergo base-induced E2 elimination to yield $\alpha,\beta$-unsaturated carbonyl products.

Alpha-hydrogen atoms of carbonyl compounds are acidic and can be abstracted by strong bases to yield enolate ions. Ketones, aldehydes, esters, amides, and nitriles can all be deprotonated in this manner if a sufficiently powerful base such as **lithium diisopropylamide (LDA)** is used. Enolate ions are highly reactive as nucleophiles because of their negative charge. For example, enolates react with benzeneselenenyl bromide to yield $\alpha$-phenylselenenylated products. These, in turn, yield $\alpha,\beta$-unsaturated carbonyl compounds when treated with $H_2O_2$.

The most important reaction of enolates is their $S_N2$ **alkylation** by alkyl halides. The nucleophilic enolate ion attacks an alkyl halide from the back side and displaces the leaving halide group to yield an alpha-alkylated carbonyl product.

The **malonic ester synthesis**, which involves alkylation of diethyl malonate with an alkyl halide, provides a method for preparing monoalkylated or dialkylated acetic acids. Similarly, the **acetoacetic ester synthesis** provides a method for preparing monoalkylated or dialkylated acetone derivatives. Most important of all is that many carbonyl compounds, including ketones, esters, and nitriles, can be directly alkylated. These carbonyl alkylation reactions constitute what is perhaps the most important method in organic chemistry for synthesizing complex molecules.

## 22.11 Summary of Reactions

1. Ketone/aldehyde halogenation, where X = Cl, Br, or I (Section 22.3)

$$R-\overset{O}{\overset{\|}{C}}-\overset{H}{\overset{|}{\underset{|}{C}}}- \xrightarrow[CH_3COOH]{X_2} R-\overset{O}{\overset{\|}{C}}-\overset{X}{\overset{|}{\underset{|}{C}}}- + HX$$

2. Hell–Volhard–Zelinskii bromination of acids (Section 22.4)

$$-\overset{H}{\overset{|}{\underset{|}{C}}}-COOH \xrightarrow[2.\ H_2O]{1.\ Br_2\ +\ PBr_3} -\overset{Br}{\overset{|}{\underset{|}{C}}}-COOH$$

3. Dehydrobromination of $\alpha$-bromo ketones (Section 22.3)

$$R-\overset{O}{\overset{\|}{C}}-\overset{Br}{\overset{|}{\underset{|}{C}}}-\overset{H}{\overset{|}{\underset{|}{C}}}- \xrightarrow[\Delta]{Pyridine} R-\overset{O}{\overset{\|}{C}}-\underset{|}{C}=C\langle$$

4. Haloform reaction, where X = Cl, Br, or I (Section 22.7)

$$R-\overset{O}{\overset{\|}{C}}-CH_3 \xrightarrow[^-OH]{X_2} R-\overset{O}{\overset{\|}{C}}-O^- + HCX_3$$

5. Phenylselenenylation–elimination (Section 22.8)

$$R-\overset{O}{\overset{\|}{C}}-\overset{H}{\overset{|}{\underset{|}{C}}}-\overset{H}{\overset{|}{\underset{|}{C}}}- \xrightarrow[2.\ C_6H_5SeBr]{1.\ LDA\ in\ THF} R-\overset{O}{\overset{\|}{C}}-\overset{SeC_6H_5}{\overset{|}{\underset{|}{C}}}-\overset{H}{C}\langle \xrightarrow{H_2O_2} R-\overset{O}{\overset{\|}{C}}-\underset{|}{C}=C\langle$$

6. Alkylation of enolate ions (Section 22.9)
   a. Malonic ester synthesis

$$CH_2(CO_2C_2H_5)_2 + RX \xrightarrow[Ethanol]{NaOC_2H_5} R-CH(CO_2C_2H_5)_2 \xrightarrow{H_3O^+} R-CH_2COOH$$

$$RCH(CO_2C_2H_5)_2 + R'X \xrightarrow[Ethanol]{NaOC_2H_5} R'-\overset{R}{\overset{|}{C}}(CO_2C_2H_5)_2 \xrightarrow{H_3O^+} R'-\overset{R}{\overset{|}{C}}HCOOH$$

b. Acetoacetic ester synthesis

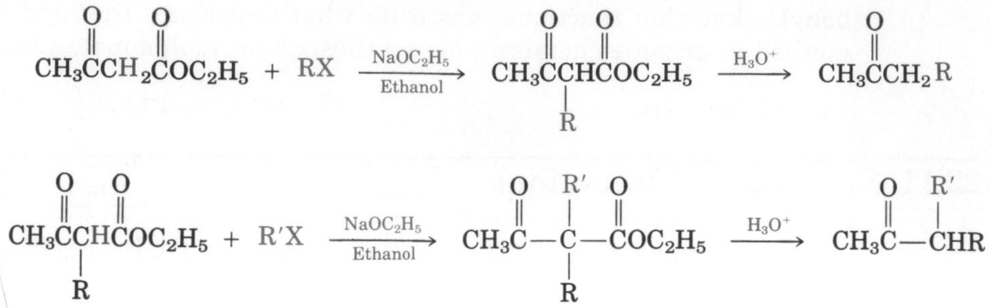

$$CH_3\overset{O}{\overset{\|}{C}}CH_2\overset{O}{\overset{\|}{C}}OC_2H_5 + RX \xrightarrow[\text{Ethanol}]{NaOC_2H_5} CH_3\overset{O}{\overset{\|}{C}}\underset{R}{\underset{|}{C}}H\overset{O}{\overset{\|}{C}}OC_2H_5 \xrightarrow{H_3O^+} CH_3\overset{O}{\overset{\|}{C}}CH_2R$$

$$CH_3\overset{O}{\overset{\|}{C}}\underset{R}{\underset{|}{C}}H\overset{O}{\overset{\|}{C}}OC_2H_5 + R'X \xrightarrow[\text{Ethanol}]{NaOC_2H_5} CH_3\overset{O}{\overset{\|}{C}}-\overset{R'}{\overset{|}{\underset{R}{\underset{|}{C}}}}-\overset{O}{\overset{\|}{C}}OC_2H_5 \xrightarrow{H_3O^+} CH_3\overset{O}{\overset{\|}{C}}-\overset{R'}{\overset{|}{C}}HR$$

c. Alkylation of ketones

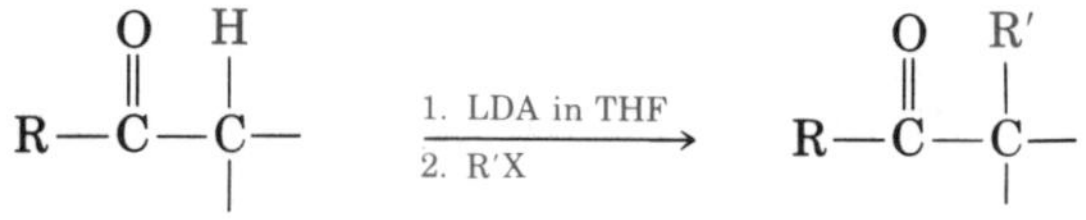

$$R-\overset{O}{\overset{\|}{C}}-\overset{H}{\overset{|}{\underset{|}{C}}}- \xrightarrow[\text{2. R'X}]{\text{1. LDA in THF}} R-\overset{O}{\overset{\|}{C}}-\overset{R'}{\overset{|}{\underset{|}{C}}}-$$

d. Alkylation of esters

$$-\overset{H}{\overset{|}{\underset{|}{C}}}-\overset{O}{\overset{\|}{C}}OC_2H_5 \xrightarrow[\text{2. RX}]{\text{1. LDA in THF}} R-\overset{|}{\underset{|}{C}}-\overset{O}{\overset{\|}{C}}-OC_2H_5$$

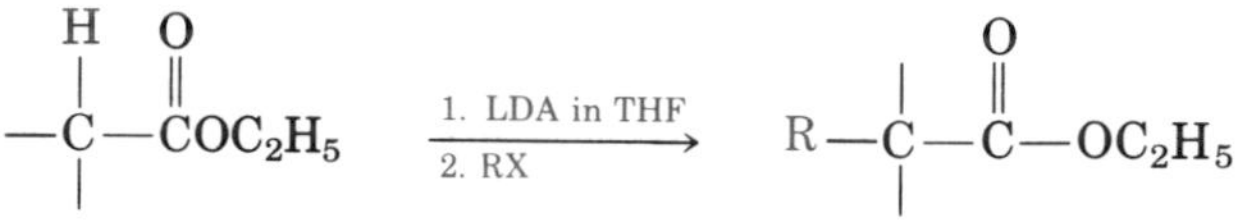

e. Alkylation of nitriles

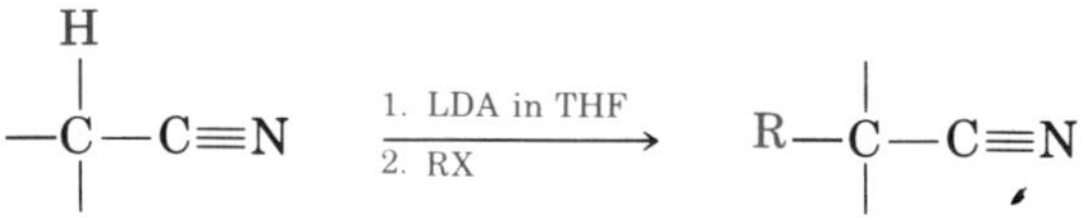

$$-\overset{H}{\overset{|}{\underset{|}{C}}}-C\equiv N \xrightarrow[\text{2. RX}]{\text{1. LDA in THF}} R-\overset{|}{\underset{|}{C}}-C\equiv N$$

## ADDITIONAL PROBLEMS

**22.24** Acetone is enolized only to the extent of about 0.0001% at equilibrium, whereas 2,4-pentanedione is 76% enolized. Explain.

**22.25** Write resonance structures for these anions:

(a) 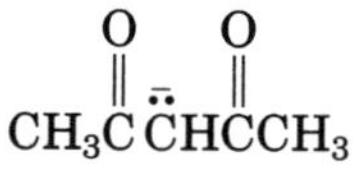 $CH_3\overset{O}{\overset{\|}{C}}\ddot{C}^{-}H\overset{O}{\overset{\|}{C}}CH_3$

(b) $:\bar{C}H_2C\equiv N$

(c) 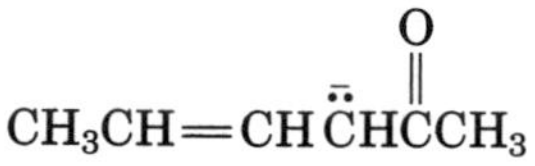 $CH_3CH=CH\ddot{C}^{-}H\overset{O}{\overset{\|}{C}}CH_3$

(d) $N\equiv C\ddot{C}^{-}HCO_2C_2H_5$

(e) 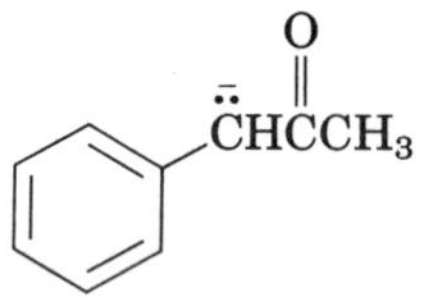 $C_6H_5\ddot{C}^{-}H\overset{O}{\overset{\|}{C}}CH_3$ (benzene ring attached to the carbanion carbon)

**22.26** Indicate all the acidic hydrogen atoms in the following structures:

(a) $HOCH_2CH_2\overset{\displaystyle O}{\overset{\|}{C}}CH_3$ (b) $HOCH_2CH_2\overset{\displaystyle O}{\overset{\|}{C}}C(CH_3)_3$

(c) 1,3-Cyclopentanedione (d) $CH_3CH{=}CHCHO$

**22.27** One way to determine the number of acidic hydrogens in a molecule is to treat the compound with NaOD in $D_2O$, isolate the product, and determine its molecular weight by mass spectroscopy. For example, if cyclohexanone is treated with NaOD in $D_2O$, the product has a molecular weight of 102. Explain how this method works.

**22.28** 2-Methylcycloheptanone and 3-methylcycloheptanone are nearly indistinguishable by spectroscopic techniques. How could you use the method outlined in Problem 22.27 to differentiate them?

**22.29** Rank the following compounds in order of increasing acidity:

(a) $CH_3CH_2COOH$ (b) $CH_3CH_2OH$ (c) $(CH_3CH_2)_2NH$

(d) $CH_3COCH_3$ (e) $CH_3\overset{\displaystyle O}{\overset{\|}{C}}CH_2\overset{\displaystyle O}{\overset{\|}{C}}CH_3$ (f) $CCl_3COOH$

**22.30** All attempts to isolate primary and secondary nitroso compounds result only in the formation of oximes. Tertiary nitroso compounds, however, are quite stable. Explain.

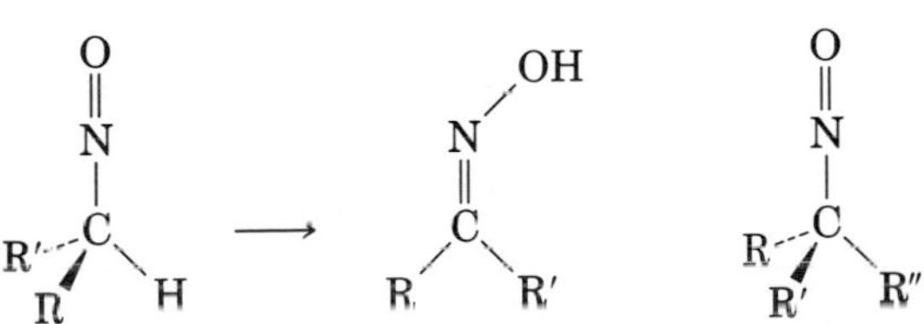


1° or 2° nitroso compound (unstable) Oxime 3° nitroso compound (stable)

**22.31** Which of these compounds can be prepared by a malonic ester synthesis? Show the alkyl halide you would use in each case.
(a) Ethyl pentanoate (b) Ethyl 3-methylbutanoate
(c) Ethyl 2-methylbutanoate (d) Ethyl 2,2-dimethylpropanoate

**22.32** How would you prepare these ketones using an acetoacetic ester synthesis?

(a) $(CH_3CH_2)_2CHCOCH_3$ (b) $CH_3CH_2CH_2\underset{\displaystyle CH_3}{\underset{|}{C}}HCOCH_3$

**22.33** How would you prepare these compounds using either an acetoacetic ester synthesis or a malonic ester synthesis?

(a) $(CH_3)_2C(CO_2C_2H_5)_2$ (b) 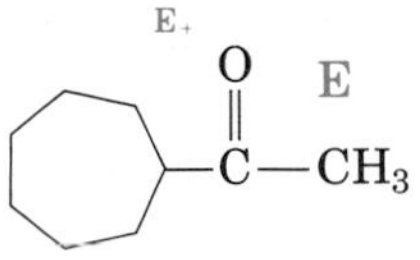 cycloheptyl$-\overset{\displaystyle O}{\overset{\|}{C}}-CH_3$

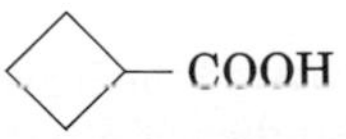

(c) cyclobutyl$-$COOH (d) $H_2C{=}CHCH_2CH_2COCH_3$

22.34 Predict the product(s) of these reactions:

(a) 1,1-cyclohexanedicarboxylic acid ($CO_2H$, $CO_2H$) $\xrightarrow{\Delta}$

(b) $(CH_3)_2CHCO_2C_2H_5 \xrightarrow[\text{2. } C_6H_5SeBr]{\text{1. LDA, THF}} A \xrightarrow{H_2O_2} B$

(c) 1,3-cyclopentanedione $\xrightarrow[\text{2. } CH_3I]{\text{1. } NaOC_2H_5}$

(d) $CH_3CH_2CH_2COOH \xrightarrow{Br_2,\ PBr_3} A \xrightarrow{H_2O} B$

(e) $C_6H_5$–C(=O)–$CH_3$ $\xrightarrow[I_2]{^-OH,\ H_2O}$

22.35 Nonconjugated $\beta,\gamma$-unsaturated ketones such as 3-cyclohexenone are in an acid-catalyzed equilibrium with their conjugated $\alpha,\beta$-unsaturated isomers. Propose a mechanism for the acid-catalyzed isomerization.

3-cyclohexenone ($\alpha$, $\beta$, $\gamma$) $\underset{}{\overset{H_3O^+}{\rightleftarrows}}$ 2-cyclohexenone ($\alpha$, $\beta$, $\gamma$)

22.36 The $\beta,\gamma$-to-$\alpha,\beta$ interconversion of unsaturated ketones (Problem 22.35) is also catalyzed by base. Explain.

22.37 One interesting consequence of the base-catalyzed $\beta,\gamma$-to-$\alpha,\beta$ isomerization of unsaturated ketones (Problem 22.35) is that 2-substituted 2-cyclopentenones can be interconverted with 5-substituted 2-cyclopentenones. Propose a mechanism to account for this isomerization.

2-methyl-2-cyclopentenone ($CH_3$) $\overset{^-OH}{\rightleftarrows}$ 5-methyl-2-cyclopentenone ($CH_3$)

22.38 Although 2-substituted 2-cyclopentenones are in a base-catalyzed equilibrium with their 5-substituted 2-cyclopentenone isomers (Problem 22.37), the analogous isomerization is not observed for 2-substituted 2-cyclohexenones. Explain.

2-methyl-2-cyclohexenone ($CH_3$) $\overset{^-OH}{\not\rightleftarrows}$ 6-methyl-2-cyclohexenone ($CH_3$)

22.39 At least as far back as the sixteenth century, the Incas chewed the leaves of the coca bush, *Erythroxylon coca*, to combat fatigue. Chemical studies of *Erythroxylon coca* by Friedrich Wöhler in 1862 resulted in the discovery of *cocaine* as the active component. It was soon found that basic hydrolysis of cocaine led to methanol, benzoic acid, and another compound called *ecgonine*. Chromium trioxide oxidation of ecgonine led to a keto acid that readily lost $CO_2$ on heating, giving tropinone.

$$\underset{\text{Cocaine}}{C_{17}H_{21}NO_4} \xrightarrow[H_2O]{^-OH} CH_3OH + C_6H_5COOH + \underset{\text{Ecgonine}}{C_9H_{15}NO_3}$$

$$\text{Ecgonine} \xrightarrow{CrO_3} \underset{\text{Keto acid}}{C_9H_{13}NO_3} \xrightarrow[\Delta]{} CO_2 + \text{(tropinone, } H_3C\text{–N bridged bicyclic ketone, =O)}$$

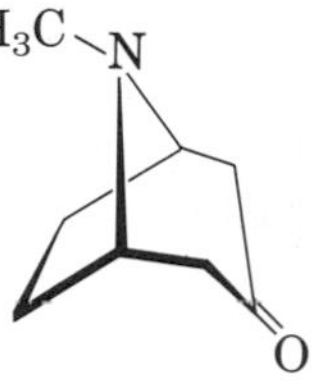

(a) What is a likely structure for the keto acid?
(b) What is a likely structure for ecgonine?
(c) What is a likely structure for cocaine?
(d) Formulate the reactions involved.

**22.40** Which of these substances would give a positive haloform reaction?

(a) $CH_3COCH_3$ (b) Acetophenone (c) $CH_3CH_2CHO$
(d) $CH_3COOH$ (e) $CH_3C{\equiv}N$

**22.41** Show how you might convert geraniol into either ethyl geranylacetate or geranylacetone.

$$\xrightarrow{?} \underset{\text{Ethyl geranylacetate}}{(CH_3)_2C{=}CHCH_2CH_2C(CH_3){=}CHCH_2CH_2CO_2C_2H_5}$$

$$\underset{\text{Geraniol}}{(CH_3)_2C{=}CHCH_2CH_2C(CH_3){=}CHCH_2OH}$$

$$\xrightarrow{?} \underset{\text{Geranylacetone}}{(CH_3)_2C{=}CHCH_2CH_2C(CH_3){=}CHCH_2CH_2COCH_3}$$

**22.42** How would you synthesize the following compounds from cyclohexanone? More than one step may be required.

(a) 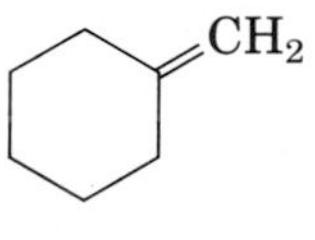 cyclohexane with $=CH_2$
(b) 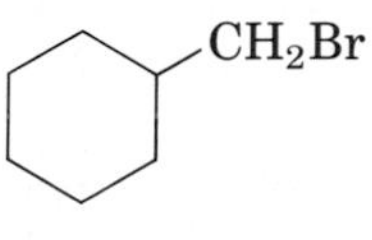 cyclohexane with $CH_2Br$
(c) 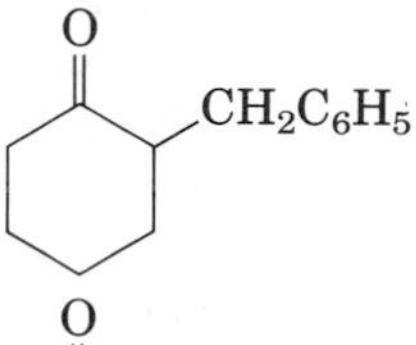 cyclohexanone with $CH_2C_6H_5$ (O)
(d) 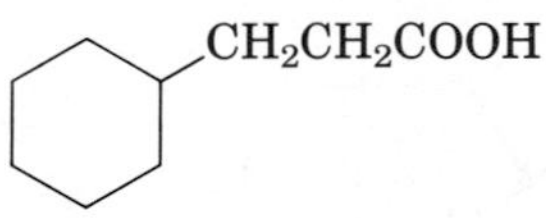 cyclohexane with $CH_2CH_2COOH$
(e) 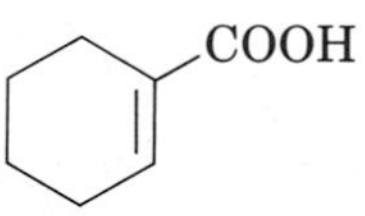 cyclohexene with $COOH$
(f) 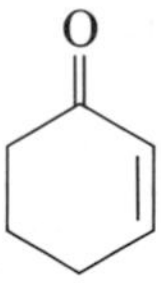 cyclohexenone (O)
(g) 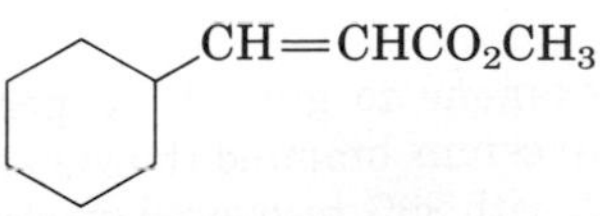 cyclohexane with $CH{=}CHCO_2CH_3$
(h) 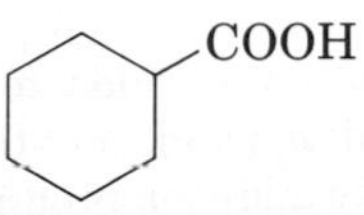 cyclohexane with $COOH$

**22.43** How can you account for the fact that *cis*- and *trans*-4-*tert*-butyl-2-methyl-cyclohexanone are interconverted by base treatment?

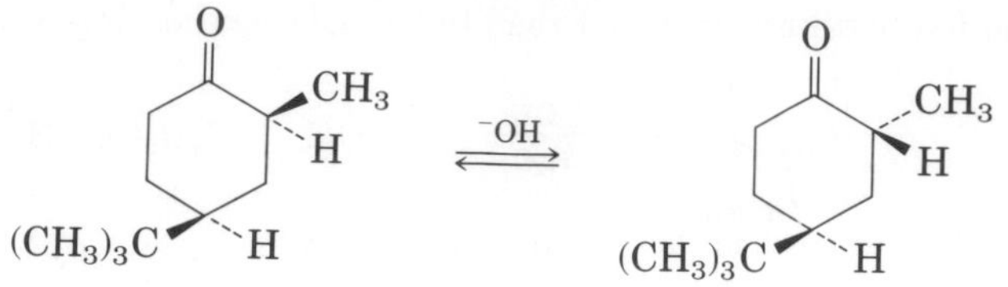

$(CH_3)_3C$ ... $CH_3$, H ... H $\xrightleftharpoons{^{-}OH}$ $(CH_3)_3C$ ... $CH_3$, H ... H

Which of the two isomers do you think is more stable, and why? Use molecular models to help formulate your answer.

**22.44** The following synthetic routes are incorrect as drawn. What is wrong with each?

(a) $CH_3CH_2CH_2CH_2CO_2CH_3 \xrightarrow[2.\ \Delta,\ pyridine]{1.\ Br_2,\ CH_3COOH} CH_3CH_2CH{=}CHCO_2CH_3$

(b) $CH_3CH(CO_2CH_2CH_3)_2 \xrightarrow[2.\ C_6H_5Br]{1.\ ^{-}OC_2H_5/C_2H_5OH} C_6H_5C(CH_3)(CO_2C_2H_5)_2$

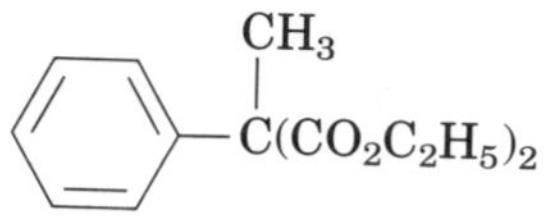

(c) $CH_3COCH_2CO_2C_2H_5 \xrightarrow[2.\ H_2C{=}CHCH_2Br;\ 3.\ H_3O^+]{1.\ ^{-}OC_2H_5/C_2H_5OH} H_2C{=}CHCH_2CH_2COOH$

(d) $CH_3CH_2CH_2CN \xrightarrow[2.\ C_6H_5SeBr;\ 3.\ \Delta,\ pyridine]{1.\ ^{-}OC_2H_5/C_2H_5OH} CH_3CH{=}CHCN$

(e) cyclohexanone $\xrightarrow[2.\ \Delta,\ pyridine]{1.\ Br_2,\ NaOH}$ 2-cyclohexenone

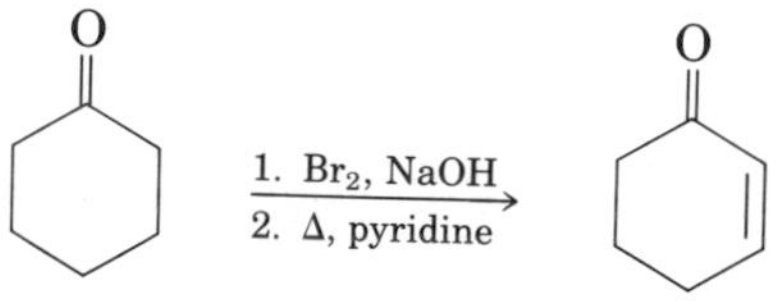

(f) $(CH_3)_2CHCOCH_3 \xrightarrow[2.\ C_6H_5CH_2Br]{1.\ LDA,\ THF} (CH_3)_2C(CH_2C_6H_5)COCH_3$

**22.45** Unlike most $\beta$-diketones such as 2,4-pentanedione, the $\beta$-diketone shown has no detectable enol content and is about as acidic as acetone. Explain this behavior. Molecular models should prove helpful.

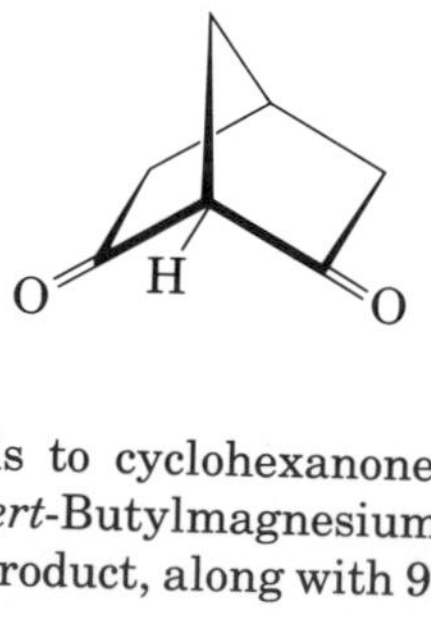

O, H, O

**22.46** Methylmagnesium bromide adds to cyclohexanone to give the expected tertiary alcohol product in high yield. *tert*-Butylmagnesium bromide, however, gives only about a 1% yield of the addition product, along with 99% recovered starting material.

Furthermore, if $D_3O^+$ is added to the reaction mixture after a suitable period, one deuterium atom is incorporated into the recovered cyclohexanone. Explain.

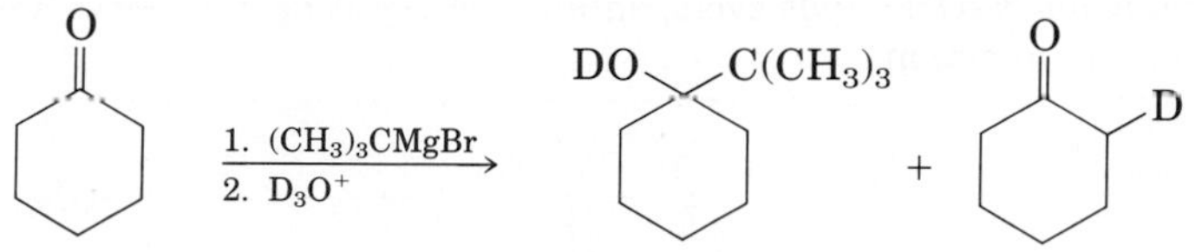

1. $(CH_3)_3CMgBr$
2. $D_3O^+$

DO C$(CH_3)_3$ + D

**22.47** The final step in an attempted synthesis of laurene, a hydrocarbon isolated from the alga *Laurencia glandulifera*, involved the Wittig reaction shown. The product obtained, however, was not laurene, but an isomer. Propose a mechanism to account for these unexpected results.

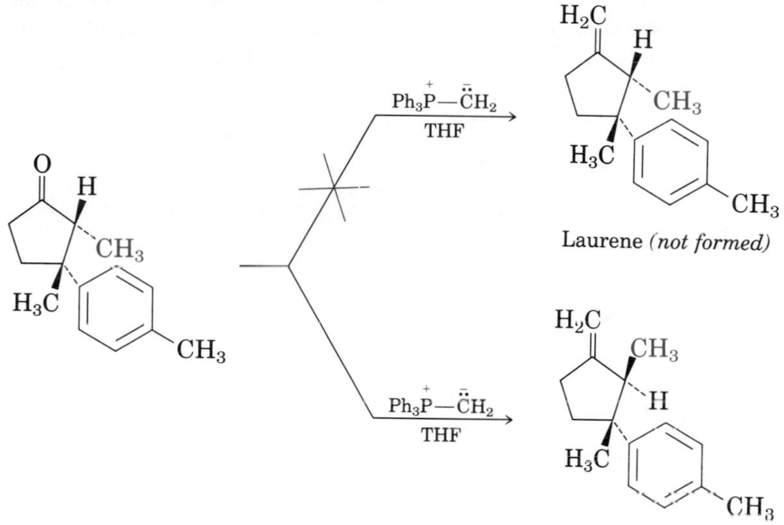

$Ph_3\overset{+}{P}—\overset{-}{\ddot{C}}H_2$ THF

Laurene *(not formed)*

**22.48** The key step in a reported laboratory synthesis of sativene, a hydrocarbon isolated from the mold *Helminthosporium sativum*, is shown. What kind of reaction is occurring? How would you complete the synthesis?

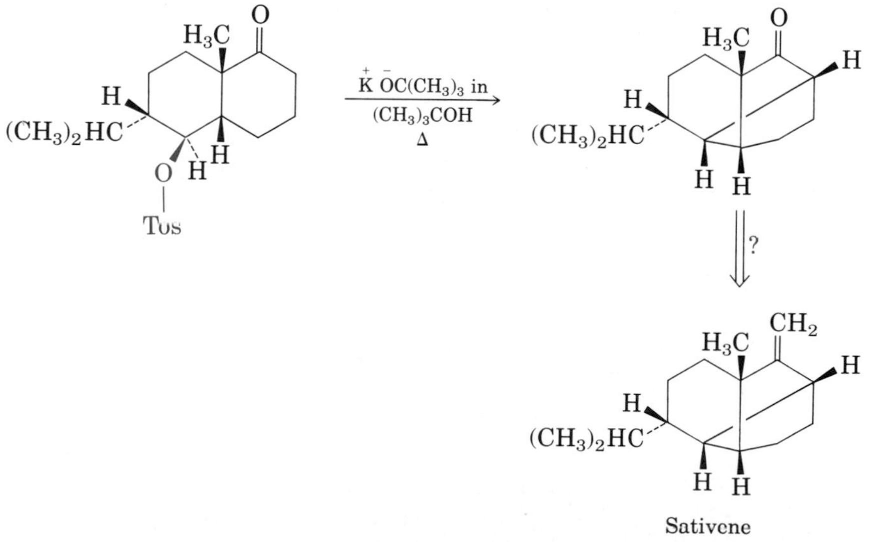

$\overset{+}{K}\ \overset{-}{O}C(CH_3)_3$ in $(CH_3)_3COH$ Δ

?

Sativene

**22.49** The Favorskii reaction involves treatment of an $\alpha$-bromo ketone with base to yield a ring-contracted product. For example, reaction of 2-bromocyclohexanone with aqueous NaOH yields cyclopentanecarboxylic acid. Propose a mechanism to account for this reaction.

$$\text{2-bromocyclohexanone} \xrightarrow[\text{2. } H_3O^+]{\text{1. NaOH}} \text{cyclopentane-COOH}$$

**22.50** Treatment of a cyclic ketone with diazomethane is a method for accomplishing a *ring-expansion reaction.* For example, treatment of cyclohexanone with diazomethane yields cycloheptanone. Propose a mechanism to account for this reaction.

$$\text{cyclohexanone} \xrightarrow[\text{Ether}]{:\overset{-}{C}H_2-\overset{+}{N}\equiv N} \text{cycloheptanone} + N_2$$

CHAPTER 23

# Carbonyl Condensation Reactions

We've seen three general kinds of carbonyl-group reactions in the past four chapters and have studied two general kinds of behavior. In nucleophilic addition reactions and nucleophilic acyl substitution reactions, the carbonyl group behaves as an electrophile by accepting electrons from an attacking nucleophile. In alpha substitution reactions, however, the carbonyl compound behaves as a nucleophile when it is converted into its enolate ion or enol tautomer. The **carbonyl condensation reactions** that we'll study in the present chapter involve *both* types of reactivity.

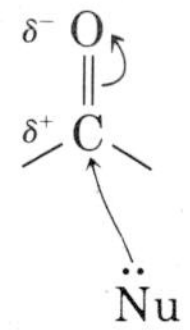

Electrophilic carbonyl is attacked by nucleophiles ($:Nu^-$)

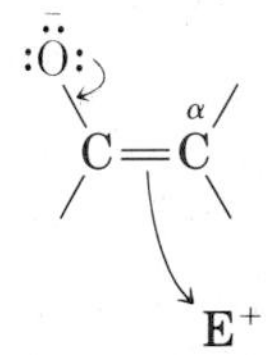

Nucleophilic alpha position of enolate attacks electrophiles ($E^+$)

## 23.1 General Mechanism of Carbonyl Condensation Reactions

Carbonyl condensation reactions take place between two carbonyl components and involve a *combination* of nucleophilic addition and alpha-substitution steps. One component acts as a nucleophilic electron donor in undergoing a nucleophilic addition, while the other component acts as an electrophilic electron acceptor in undergoing an alpha substitution. The general mechanism of a carbonyl condensation reaction is shown in Figure 23.1.

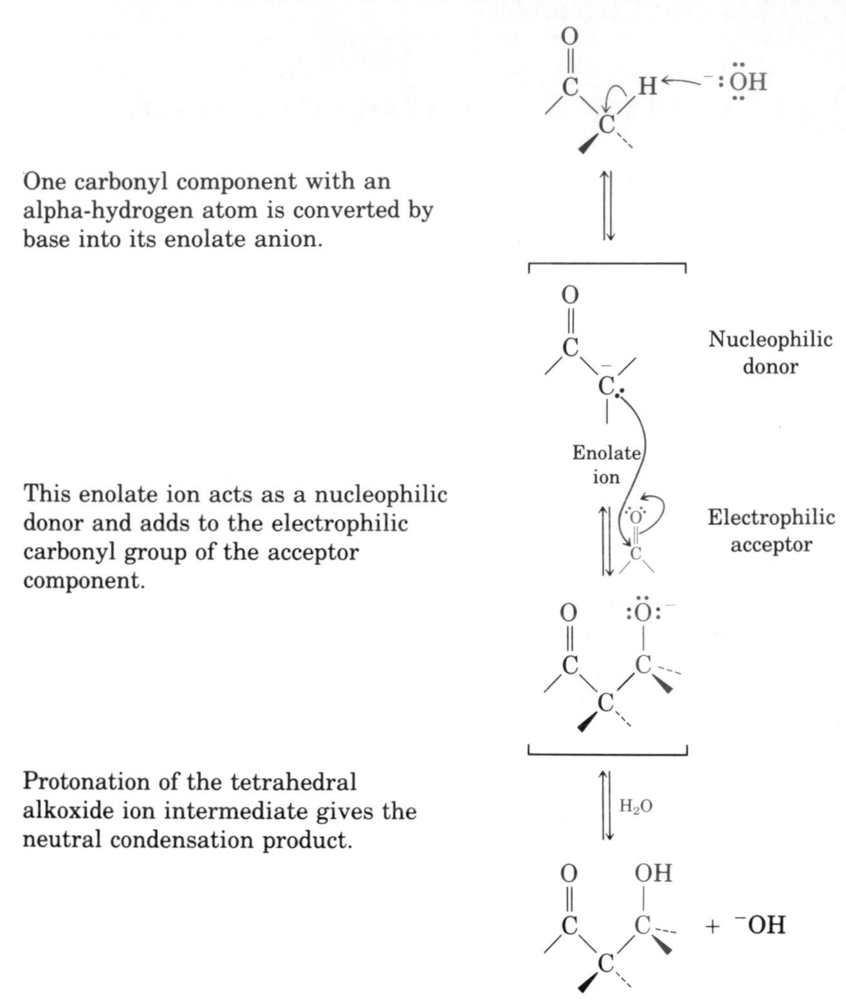

One carbonyl component with an alpha-hydrogen atom is converted by base into its enolate anion.
Nucleophilic donor
Enolate ion
This enolate ion acts as a nucleophilic donor and adds to the electrophilic carbonyl group of the acceptor component.
Electrophilic acceptor
Protonation of the tetrahedral alkoxide ion intermediate gives the neutral condensation product.
H2O
+ ⁻OH


**Figure 23.1** The general mechanism of a carbonyl condensation reaction: One component (the donor) acts as a nucleophile while the other component (the acceptor) acts as an electrophile.

Viewed from the side of the donor component, a carbonyl condensation reaction is simply an alpha-substitution process. Viewed from the side of the acceptor component, a carbonyl condensation reaction is a nucleophilic addition process. However one chooses to view the reaction, carbonyl condensations are among the most useful reactions in organic chemistry.

All manner of carbonyl compounds, including aldehydes, ketones, esters, amides, acid anhydrides, thiol esters, and nitriles, enter into condensation reactions. Nature uses carbonyl condensation reactions as key steps in the biosynthesis of many naturally occurring compounds, and chemists use the same reactions in the laboratory.

## 23.2 Condensations of Aldehydes and Ketones: The Aldol Reaction

When acetaldehyde is treated in a hydroxylic solvent with a basic catalyst such as sodium ethoxide or sodium hydroxide, a rapid and reversible condensation reaction occurs. The product is the $\beta$-hydroxy aldehyde known commonly as **aldol** (*alde*hyde + alco*hol*).

$$2\ CH_3CHO \xrightleftharpoons[\text{Ethanol}]{Na^+\,^-OC_2H_5} CH_3\overset{\overset{\displaystyle OH}{|\beta}}{\underset{\underset{\displaystyle H}{|}}{C}}\overset{\alpha}{-}CH_2CHO$$

Acetaldehyde

Aldol (a $\beta$-hydroxy aldehyde)

Known as the **aldol reaction**, base-catalyzed dimerization is a general reaction for all ketones and aldehydes having alpha-hydrogen atoms. If the ketone or aldehyde does not have an alpha-hydrogen atom, however, aldol condensation can't occur. As the following examples indicate, the aldol equilibrium generally favors condensation product in the case of monosubstituted acetaldehydes ($RCH_2CHO$), but favors starting material for disubstituted acetaldehydes ($R_2CHCHO$) and for most ketones. Steric factors are probably responsible for these trends, since increased substitution near the reaction site greatly increases steric congestion in the aldol product.

2 Cyclohexanone $\xrightleftharpoons{\text{NaOH, ethanol}}$ 2-(1-hydroxycyclohexyl)cyclohexanone (OH) 22%

Cyclohexanone

22%

$$2\ CH_3\overset{\overset{\displaystyle O}{\|}}{C}CH_3 \xrightleftharpoons{\text{NaOH}} (CH_3)_2\overset{\overset{\displaystyle OH}{|}}{C}-CH_2\overset{\overset{\displaystyle O}{\|}}{C}CH_3$$

Acetone

$\approx 5\%$

$$2\ C_6H_5CH_2CHO \xrightleftharpoons{\text{NaOH, ethanol}} C_6H_5CH_2\overset{\overset{\displaystyle OH}{|}}{C}H-\underset{\underset{\displaystyle C_6H_5}{|}}{C}HCHO$$

Phenylacetaldehyde

90%

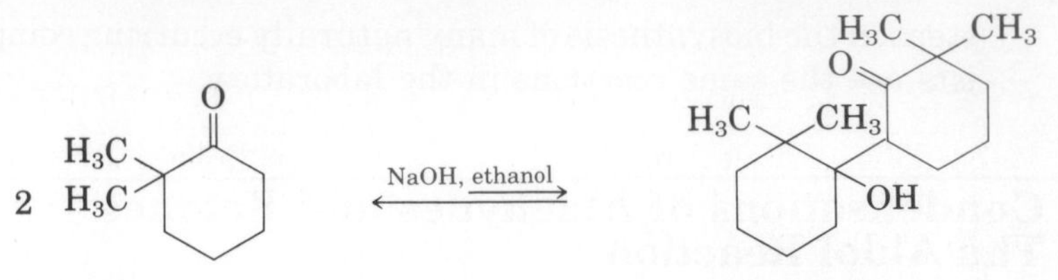

2 2,2-Dimethylcyclohexanone $\xrightleftharpoons{\text{NaOH, ethanol}}$ Low yield

$$2\ \mathrm{H{-}C(CH_3)_2{-}CHO} \xrightleftharpoons{\text{NaOH, ethanol}} \mathrm{(CH_3)_2CHCH(OH){-}C(CH_3)_2{-}CHO}$$

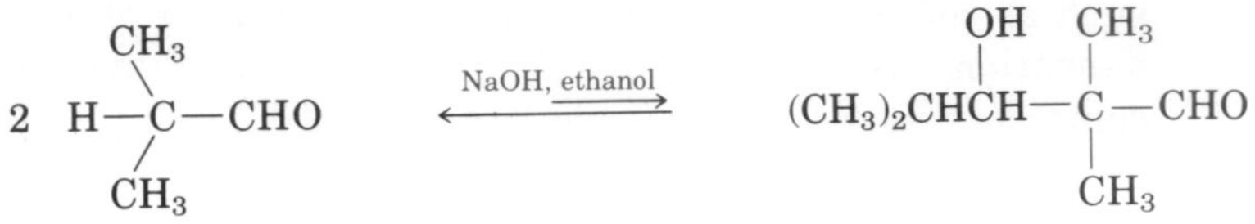

2-Methylpropanal — Low yield

Aldol reactions are typical carbonyl condensations. They occur by nucleophilic addition of the enolate ion of the donor molecule to the carbonyl group of the acceptor molecule, yielding a tetrahedral intermediate that is protonated to give the alcohol as final product. The reverse process occurs in exactly the opposite manner: Base abstracts the hydroxyl proton to yield an alkoxide, which fragments to give one molecule of enolate ion and one molecule of neutral carbonyl compound (Figure 23.2).

Base removes an acidic alpha hydrogen from one aldehyde molecule, yielding a resonance-stabilized enolate ion.

$$\mathrm{H_3C{-}CHO} + {}^-\!:\!\mathrm{OCH_3} \longrightarrow {}^-\!:\!\mathrm{CH_2{-}C(=O){-}H} + \mathrm{CH_3OH}$$

The enolate ion next attacks a second aldehyde molecule in a nucleophilic addition reaction to give a tetrahedral alkoxide ion intermediate.

$$\mathrm{H_3C{-}C(=O){-}H} \longrightarrow \mathrm{H_3C{-}CH(O^-){-}CH_2CHO}$$

Protonation of the alkoxide intermediate yields neutral aldol product and regenerates the base catalyst.

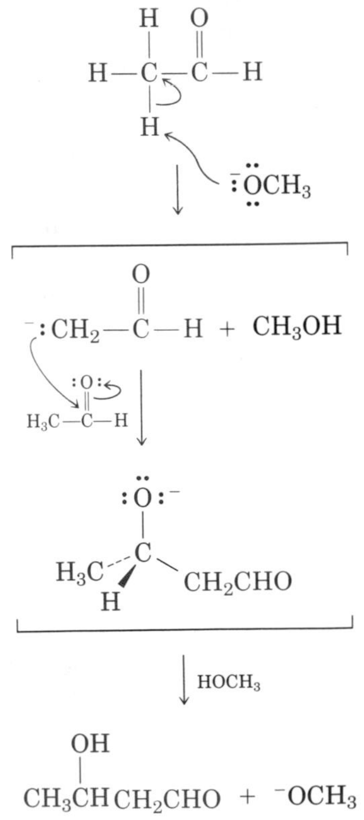

$$\xrightarrow{\mathrm{HOCH_3}} \mathrm{CH_3CH(OH)CH_2CHO} + {}^-\mathrm{OCH_3}$$

**Figure 23.2** Mechanism of the aldol reaction

**PRACTICE PROBLEM**

What is the structure of the aldol product derived from propanal?

*Solution* An aldol reaction combines two molecules of starting material, forming a bond between the alpha carbon of one partner and the carbonyl carbon of the second partner:

$$CH_3CH_2{-}\overset{\overset{\displaystyle O}{\|}}{C}{-}H + \underset{\underset{\displaystyle CH_3}{|}}{CH_2}{-}\overset{\overset{\displaystyle O}{\|}}{C}{-}H \xrightarrow{NaOH} CH_3CH_2{-}\underset{\underset{\displaystyle H}{|}}{\overset{\overset{\displaystyle OH}{|}}{C}}{-}\underset{\underset{\displaystyle CH_3}{|}}{CH}{-}\overset{\overset{\displaystyle O}{\|}}{C}{-}H$$

Bond formed here

**PROBLEM**

**23.1** Predict the product of aldol reaction of these compounds:
(a) Butanal (b) 2-Butanone (c) Cyclopentanone

**PROBLEM**

**23.2** The aldol reaction is catalyzed by acid as well as by base. What is the reactive nucleophilic species in the acid-catalyzed aldol reaction? Propose a possible mechanism for this acidic reaction.

**PROBLEM**

**23.3** Show by means of curved arrows how the base-catalyzed reverse aldol reaction of 4-methyl-4-hydroxy-2-pentanone to yield two molecules of acetone takes place.

## 23.3 Carbonyl Condensation Reactions versus Alpha-Substitution Reactions

Two of the four general carbonyl-group reactions—carbonyl condensation and alpha substitution—take place under basic conditions and involve the formation of enolate ion intermediates. Since reaction conditions for both processes are similar, how can we be sure which of the two possible reaction courses will be followed in a given case? When we generate an enolate ion with the intention of carrying out an alpha alkylation, how can we be sure that a carbonyl condensation reaction doesn't occur instead?

Although there are no simple answers to these questions, the experimental conditions usually have much to do with the result. Alpha-substitution reactions require a full equivalent of base, and are normally carried out in such a way that the carbonyl compound is rapidly and completely converted into its enolate ion at as low a temperature as possible. An electrophile is then added rapidly to ensure that the reactive enolate ion is quenched quickly.

In a malonic ester synthesis, for example, we might use 1 equiv of sodium ethoxide in ethanol solution at room temperature. Instant and complete generation of the malonic ester enolate ion would happen, and no unreacted starting material would be left so that no condensation reaction could occur. We would then immediately add an alkyl halide to complete the alkylation reaction.

$$CH_2(CO_2CH_2CH_3)_2 \xrightarrow[\text{Ethanol, 25°C (very fast reaction)}]{\text{1 equiv } NaOCH_2CH_3} [Na^+ \; ^-\!:CH(CO_2CH_2CH_3)_2]$$

100%

$\downarrow$ Add R—X

$$R—CH(CO_2CH_2CH_3)_2$$

Similarly, if we wanted to carry out the direct alkylation of a ketone such as 2-methylcyclohexanone, we might use 1 equiv of the strong base LDA in THF solution at very low temperature. Again, complete generation of the ketone enolate ion would occur rapidly, and no unreacted starting material would remain. Rapid addition of an alkyl halide would then let the desired alkylation (alpha-substitution) reaction occur.

H$_3$C, O — 1 equiv LDA / THF, −78°C → H$_3$C, O$^-$ Li$^+$ — R—X → H$_3$C, O, R

On the other hand, we might want to carry out a carbonyl condensation reaction. This, too, we could accomplish by selecting the proper reaction conditions. Since we need to generate only a small amount of the enolate ion in the presence of unreacted carbonyl compound, the aldol reaction requires only a *catalytic* amount of base, rather than a full equivalent. Once a condensation has occurred, the basic catalyst is regenerated.

To carry out an aldol reaction on propanal, for example, we might dissolve propanal in methanol, add 0.05 equiv of sodium methoxide, and then warm the mixture. A high yield of aldol product would result.

$$CH_3CH_2CHO \underset{CH_3OH,\ \Delta}{\overset{5\%\ NaOCH_3}{\rightleftharpoons}} [CH_3\ddot{C}HCHO]$$

Present in tiny amount

$\updownarrow$ $CH_3CH_2CHO$

$$^-OCH_3 + CH_3CH_2CH(OH)CH(CH_3)CHO \overset{HOCH_3}{\rightleftharpoons} [CH_3CH_2CH(O^-)CH(CH_3)CHO]$$

Regenerated catalyst

## 23.4 Dehydration of Aldol Products: Synthesis of Enones

The $\beta$-hydroxy ketones and $\beta$-hydroxy aldehydes formed in aldol reactions can be readily dehydrated to yield conjugated enones. In fact, it is this loss of water that gives the aldol *condensation* its name, since the enone product is more "condensed" than the hydroxy aldehyde/ketone.

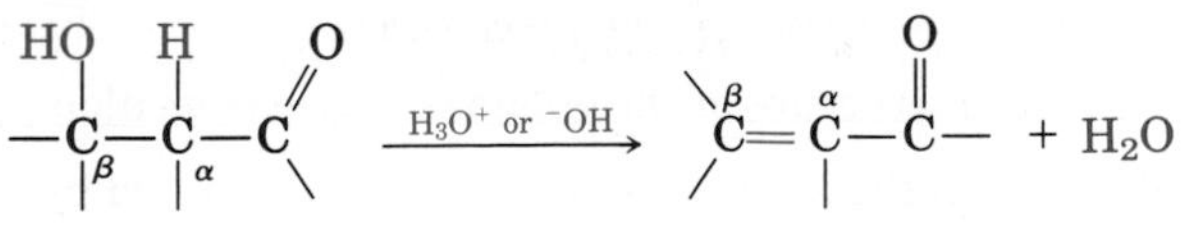

A β-hydroxy aldehyde/ketone $\xrightarrow{H_3O^+ \text{ or } ^-OH}$ A conjugated enone $+ H_2O$

Most alcohols are resistant to dehydration by dilute acid or base, and powerful reagents like $POCl_3$ must therefore be used (Section 17.8). Hydroxyl groups beta to a carbonyl group are special, however, because of the nearby carbonyl group. Under *basic* conditions, an acidic alpha hydrogen is abstracted, yielding an enolate ion from which the hydroxide ion leaving group is expelled. Under *acidic* conditions, an enol is formed, the hydroxyl group is protonated, and water is expelled.

Basic elimination $\xrightarrow{^-OH}$ Enolate ion $\rightleftharpoons$ $+ \ ^-OH$

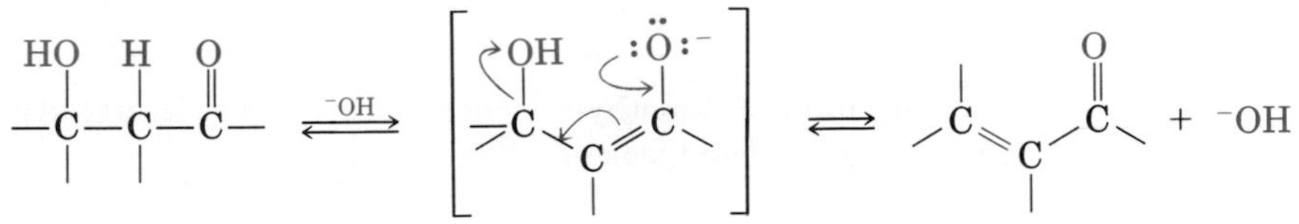

Acidic elimination $\xrightarrow{H_3O^+}$ Enol $\rightleftharpoons$ $+ H_3O^+$

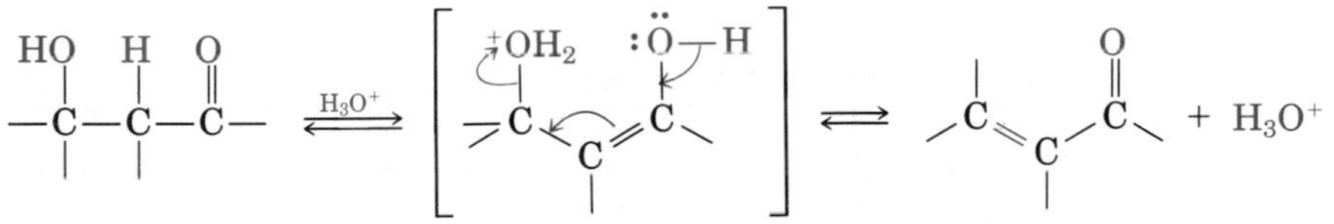

The conditions required to effect aldol dehydration reactions are often only a bit more vigorous (slightly higher temperature, for example) than the conditions required for the aldol dimerization itself. As a result, conjugated enones are often obtained directly from aldol reactions, and the intermediate β-hydroxy carbonyl compounds are not isolated.

Conjugated enones are more stable than nonconjugated enones for the same reasons that conjugated dienes are more stable than nonconjugated dienes (Section 14.2). Interaction between the pi electrons of the carbon–carbon double bond and the pi electrons of the carbonyl group leads to a molecular orbital description of conjugated enones that shows a partial delocalization of the pi electrons over all four atomic centers.

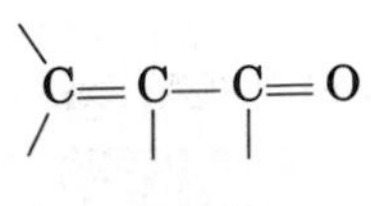

C=C—C=O
Conjugated enone (more stable)

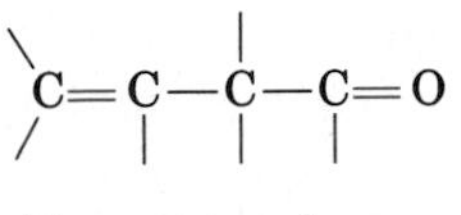

C=C—C—C=O
Nonconjugated enone (less stable)

Recall the following:

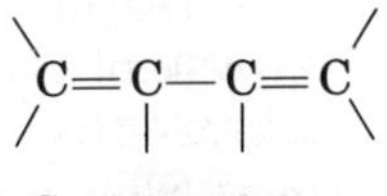

C=C—C=C
Conjugated diene (more stable)

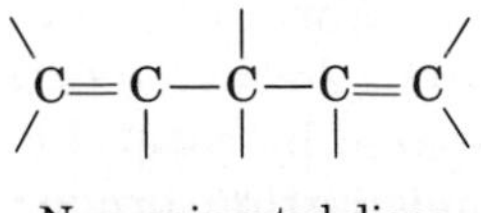

C=C—C—C=C
Nonconjugated diene (less stable)

**PRACTICE PROBLEM**

What is the structure of the enone obtained from aldol condensation of acetaldehyde?

*Solution* In the aldol reaction, $H_2O$ is eliminated by removing two hydrogens from the acidic alpha position of one partner and the oxygen from the second partner:

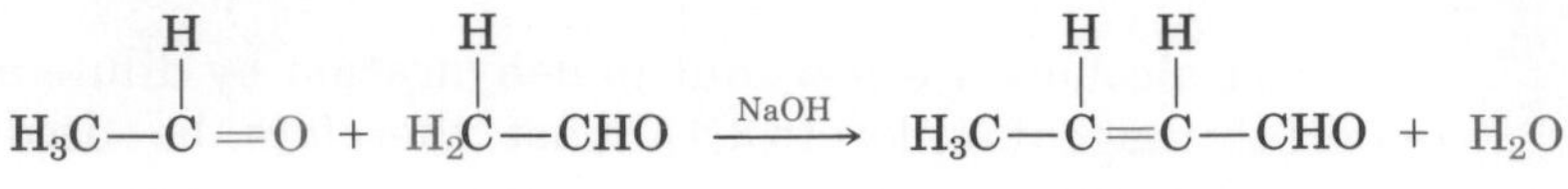

$$H_3C-\overset{\displaystyle H}{\overset{|}{C}}=O + H_2\overset{\displaystyle H}{\overset{|}{C}}-CHO \xrightarrow{NaOH} H_3C-\overset{\displaystyle H}{\overset{|}{C}}=\overset{\displaystyle H}{\overset{|}{C}}-CHO + H_2O$$

2-Butenal

**PROBLEM**

**23.4** What enone products would you expect from aldol condensation of the following compounds?
(a) Cyclopentanone (b) Acetophenone (c) 3-Methylbutanal

**PROBLEM**

**23.5** Aldol condensation of 3-methylcyclohexanone leads to a mixture of two products, not counting double-bond isomers. Draw them.

## 23.5 Recognizing Aldol Products

The aldol condensation reaction yields either $\beta$-hydroxy ketones/aldehydes or $\alpha,\beta$-unsaturated ketones/aldehydes, depending on the specific case and on the reaction conditions. By learning how to think *backward*, it's possible to predict when the aldol reaction might be useful in synthesis. Any time the target molecule contains either a $\beta$-hydroxy ketone/aldehyde or a conjugated enone functional group, it might come from an aldol reaction:

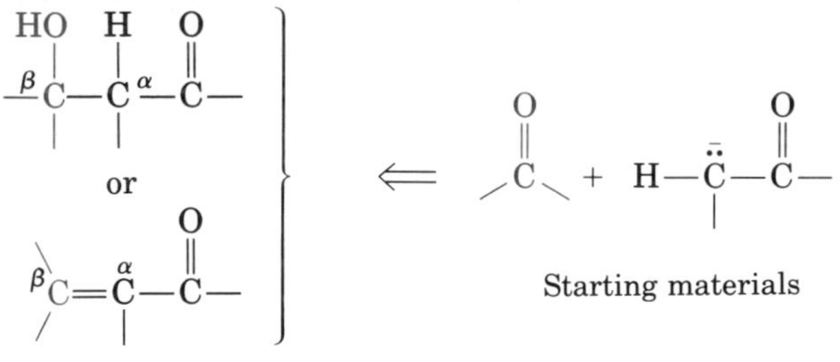

$$\left.\begin{matrix} -\overset{\displaystyle HO}{\overset{|}{\underset{|}{C}}}{}^{\beta}-\overset{\displaystyle H}{\overset{|}{\underset{|}{C}}}{}^{\alpha}-\overset{\displaystyle O}{\overset{\|}{C}}- \\ \text{or} \\ {}^{\beta}C=\overset{\alpha}{\underset{|}{C}}-\overset{\displaystyle O}{\overset{\|}{C}}- \end{matrix}\right\} \Longleftarrow \overset{\displaystyle O}{\overset{\|}{C}} + H-\overset{\cdot\cdot}{\underset{|}{C}}-\overset{\displaystyle O}{\overset{\|}{C}}-$$

Products | Starting materials

We can extend this kind of reasoning even further by considering that subsequent transformations might be carried out on the aldol products. For example, a saturated ketone might be prepared by catalytic hydrogenation of an enone product. A good example can be found in the industrial preparation of 2-ethyl-1-hexanol, an alcohol used in the synthesis of plasticizers.

Although 2-ethyl-1-hexanol bears little resemblance to an aldol product at first glance, it is in fact prepared commercially from butanal by an aldol reaction. Working backward, we can reason that 2-ethyl-1-hexanol might come from 2-ethylhexanal by a reduction. 2-Ethylhexanal, in turn, might be prepared by catalytic reduction of 2-ethyl-2-hexenal, which is the aldol self-condensation product of butanal. The reactions that follow show the sequence in reverse order.

$$\underset{\text{Target: 2-ethyl-1-hexanol}}{CH_3CH_2CH_2CH_2\underset{\displaystyle CH_2CH_3}{\underset{|}{C}}HCH_2OH} \xleftarrow[\text{(industrially, by } H_2/Pt)]{[H]} \underset{\text{Starting material: 2-ethylhexanal}}{CH_3CH_2CH_2CH_2\underset{\displaystyle CH_2CH_3}{\underset{|}{C}}HCHO}$$

$$\underset{\text{2-Ethylhexanal}}{CH_3CH_2CH_2CH_2\underset{\displaystyle CH_2CH_3}{\underset{|}{C}}HCHO} \xleftarrow{H_2/Pt} \underset{\text{2-Ethyl-2-hexenal}}{CH_3CH_2CH_2CH{=}\underset{\displaystyle CH_2CH_3}{\underset{|}{C}}CHO}$$

$$\underset{\text{2-Ethyl-2-hexenal}}{CH_3CH_2CH_2CH{=}\underset{\displaystyle CH_2CH_3}{\underset{|}{C}}CHO} \xleftarrow{KOH} \underset{\text{Butanal}}{2\ CH_3CH_2CH_2CHO}$$

**PROBLEM**

**23.6** Which of the following compounds are aldol self-condensation products? What is the ketone or aldehyde precursor of each?
(a) 2,2,3-Trimethyl-3-hydroxybutanal (b) 2-Methyl-2-hydroxypentanal
(c) 5-Ethyl-4-methyl-4-hepten-3-one

**PROBLEM**

**23.7** 1-Butanol is prepared commercially by a route that begins with the aldol reaction of acetaldehyde. Show the steps that are likely to be involved.

## 23.6 Mixed Aldol Reactions

Until now, we've considered only symmetrical aldol reactions, in which the two carbonyl components have been identical. What would happen, though, if we attempted to carry out a *mixed* aldol reaction between two different carbonyl partners?

In general, a mixed aldol reaction between two similar ketone or aldehyde components leads to a mixture of four possible products. For example, base treatment of a mixture of acetaldehyde and propanal gives a complex product mixture containing two "symmetrical" aldol products and two "mixed" aldol products. Clearly, such a reaction is of no practical value in the laboratory.

$$\underset{\text{Acetaldehyde}}{CH_3CHO} + \underset{\text{Propanal}}{CH_3CH_2CHO} \xrightarrow{\text{Base}} CH_3\overset{\displaystyle OH}{\overset{|}{C}}HCH_2CHO + CH_3\overset{\displaystyle OH}{\overset{|}{C}}H\underset{\displaystyle CH_3}{\underset{|}{C}}HCHO + CH_3CH_2\overset{\displaystyle OH}{\overset{|}{C}}H\underset{\displaystyle CH_3}{\underset{|}{C}}HCHO + CH_3CH_2\overset{\displaystyle OH}{\overset{|}{C}}HCH_2CHO$$

Symmetrical products: $CH_3CHOHCH_2CHO$ (upper left) and $CH_3CH_2CHOHCH(CH_3)CHO$ (lower left)

Mixed products: $CH_3CHOHCH(CH_3)CHO$ (upper right) and $CH_3CH_2CHOHCH_2CHO$ (lower right)

On the other hand, mixed aldol reactions *can* lead cleanly to a single product, if one of two conditions is met:

1. If one of the carbonyl components contains no alpha hydrogens (and thus cannot form an enolate ion to become a donor), but does contain a reactive carbonyl group that is a good acceptor of nucleophiles, then a mixed aldol reaction is likely to be successful. This is the case, for example, when benzaldehyde or formaldehyde is used as one of the carbonyl components:

$$\text{2-Methylcyclohexanone (donor)} + \text{Benzaldehyde (acceptor)} \xrightarrow[\text{Ethanol}]{Na^+ \ {}^-OC_2H_5} \text{2-benzylidene-6-methylcyclohexanone (78\%)} + H_2O$$

2-Methylcyclohexanone (donor) Benzaldehyde (acceptor) 78%

Benzaldehyde (or formaldehyde) can't form an enolate ion to condense with itself or with another partner, yet its carbonyl group is unhindered and reactive. Thus, in the presence of a ketone such as 2-methylcyclohexanone, the ketone enolate adds preferentially to benzaldehyde, giving the mixed aldol product.

2. If one of the carbonyl components is unusually acidic and easily transformed into its enolate ion, then a mixed aldol reaction is likely to be successful. This is the case, for example, when ethyl acetoacetate is used as one of the carbonyl components:

$$\text{Cyclohexanone} + CH_3\overset{O}{\overset{\|}{C}}CH_2\overset{O}{\overset{\|}{C}}OC_2H_5 \xrightarrow[\text{Ethanol}]{Na^+ \ {}^-OC_2H_5} (CH_3\overset{O}{\overset{\|}{C}})(CO_2C_2H_5)C{=}C_6H_{10} + H_2O$$

Cyclohexanone (acceptor) Ethyl acetoacetate (donor) 80%

Ethyl acetoacetate is completely converted into its enolate ion in preference to enolate ion formation from other carbonyl partners such as cyclohexanone. Aldol condensation therefore occurs preferentially to give the mixed product.

The situation can be summarized by saying that a mixed aldol reaction between two different carbonyl partners leads to a mixture of products unless one of the partners is either an unusually good nucleophilic donor (such as ethyl acetoacetate) or else has no alpha protons and is a good electrophilic acceptor (such as benzaldehyde).

PROBLEM

**23.8** Which of the following compounds can probably be prepared by a mixed aldol reaction?

(a) $C_6H_5CH{=}CHCCH_3$ (with C=O)

(b) $C_6H_5C(CH_3){=}CHCCH_3$ (with C=O)

(c) O, $CHCH_2CH_3$

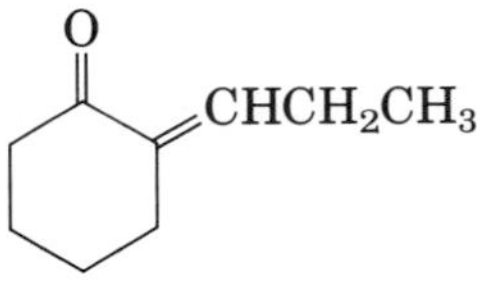

(d) 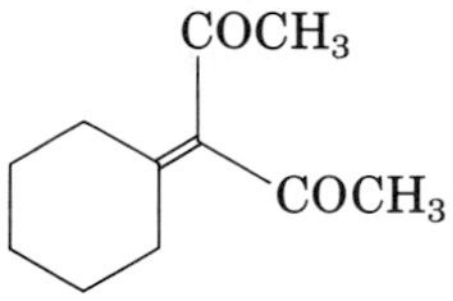 $COCH_3$, $COCH_3$

## 23.7 Intramolecular Aldol Reactions

The aldol reactions we've seen up to this point have been *inter*molecular. That is, they have taken place between two different molecules. When certain *di*carbonyl compounds are treated with base, however, *intra*molecular aldol reactions can occur, leading to the formation of cyclic products. For example, base treatment of a 1,4-diketone like 2,5-hexanedione yields a cyclopentenone product, and similar base treatment of a 1,5-diketone like 2,6-heptanedione yields a cyclohexenone product.

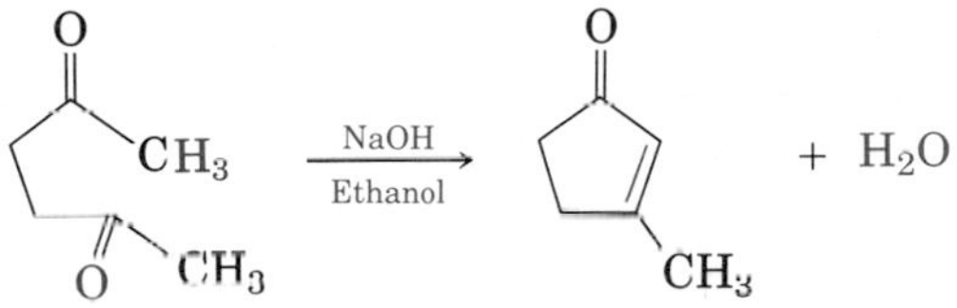

2,5-Hexanedione (a 1,4-diketone) $\xrightarrow[\text{Ethanol}]{\text{NaOH}}$ 3-Methyl-2-cyclopentenone + $H_2O$

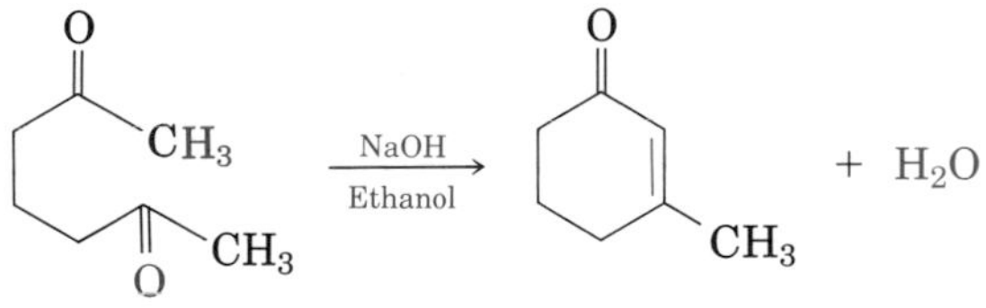

2,6-Heptanedione (a 1,5-diketone) $\xrightarrow[\text{Ethanol}]{\text{NaOH}}$ 3-Methyl-2-cyclohexenone + $H_2O$

The mechanism of these intramolecular aldol reactions is similar to the mechanism of the corresponding intermolecular reactions. The only difference is that both the nucleophilic carbonyl anion donor and the electrophilic carbonyl acceptor are in the same molecule.

In principle, intramolecular aldol reactions can lead to a mixture of products, depending on which enolate ion is formed. For example, 2,5-hexanedione might yield either the five-membered-ring product 3-methyl-2-cyclopentenone or the three-membered-ring product (2-methylcyclopropenyl)ethanone (Figure 23.3). In practicc, though, only the cyclopentenone is formed.

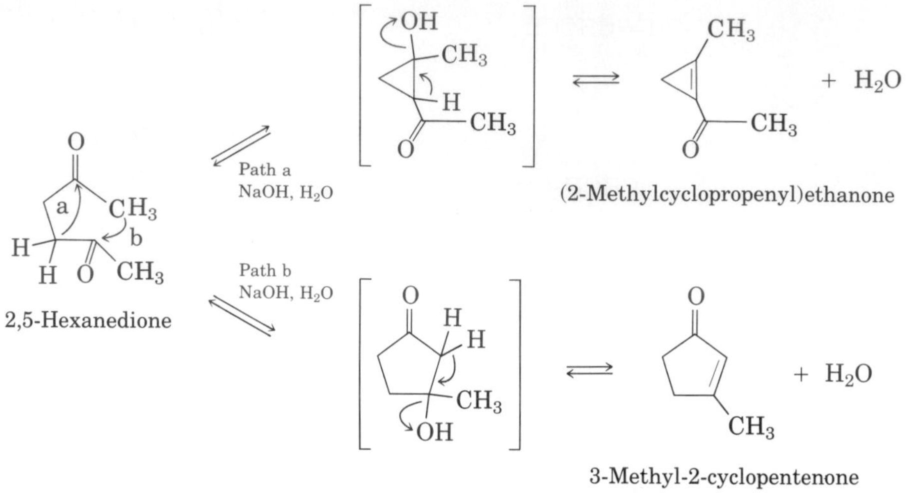

**Figure 23.3** Intramolecular aldol reaction of 2,5-hexanedione to yield 3-methyl-2-cyclopentenone

The reason for the observed product selectivity in intramolecular aldol reactions is that all steps in the mechanism are reversible and an equilibrium is reached. Thus, the relatively strain-free cyclopentenone product is considerably more stable than the highly strained cyclopropene alternative. For similar reasons, intramolecular aldol reactions of 1,5-diketones lead only to cyclohexenone products rather than to cyclobutenes.

PROBLEM

**23.9** Why do you suppose 1,3-diketones do not undergo internal aldol condensation to yield cyclobutenones?

PROBLEM

**23.10** What product would you expect to obtain from base treatment of 1,6-cyclodecanedione?

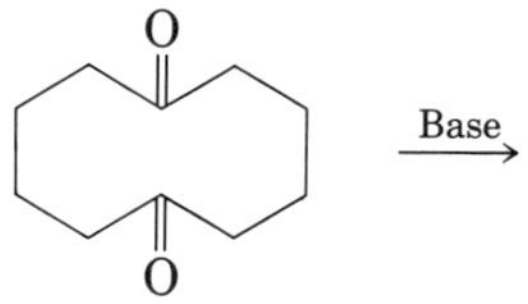

## 23.8 Reactions Similar to the Aldol Condensation

The aldol reaction is usually defined as the condensation of aldehydes and ketones. In a more general sense, however, there are many possible similar reactions in which a compound with acidic alpha hydrogens condenses with a second carbonyl component. For example, diethyl malonate can condense with aldehydes and unhindered ketones to yield $\alpha,\beta$-unsaturated diesters—the **Knoevenagel[1] reaction**. Like other malonates, these diesters can then be decarboxylated (Section 22.9).

$$C_6H_5CHO + CH_2(CO_2C_2H_5)_2 \xrightarrow[\text{Ethanol}]{Na^+\ ^-OCH_2CH_3} C_6H_5CH{=}C(CO_2C_2H_5)_2 \ (91\%)$$

Benzaldehyde — Diethyl malonate

$$\xrightarrow{H_3O^+} C_6H_5CH{=}CHCOOH + CO_2 + 2\ C_2H_5OH$$

Cinnamic acid

Many other possibilities exist, including condensations with nitriles, anhydrides, and nitro compounds as the acidic donor components. The many possibilities lead to an extremely rich and varied chemistry, but all are simply variations of the generalized aldol mechanism in which a nucleophilic carbon donor adds to an electrophilic carbonyl acceptor.

PROBLEM

**23.11** Show the mechanism of the Knoevenagel reaction of diethyl malonate and benzaldehyde.

PROBLEM

**23.12** In the **Perkin reaction**, an anhydride condenses with an aromatic aldehyde to yield a cinnamic acid. The reaction takes place by a mixed carbonyl condensation of the anhydride with the aldehyde to yield an $\alpha,\beta$-unsaturated intermediate that undergoes hydrolysis to yield the cinnamic acid. What is the structure of the unsaturated intermediate?

$$C_6H_5CHO + CH_3\overset{O}{\overset{\|}{C}}O\overset{O}{\overset{\|}{C}}CH_3 \xrightarrow[\text{2. } H_2O]{\text{1. } CH_3COONa,\ \Delta} C_6H_5CH{=}CHCOOH$$

Benzaldehyde — Acetic anhydride — Cinnamic acid (64%)

## 23.9 The Claisen Condensation Reaction

Esters, like aldehydes and ketones, are weakly acidic. When an ester having an alpha hydrogen is treated with 1 equiv of a base such as sodium ethoxide, a reversible condensation reaction occurs to yield a $\beta$-keto ester product. For example, ethyl acetate yields ethyl acetoacetate on treatment with base.

[1]Emil Knoevenagel (1865–1921); b. Linden/Hannover, Germany; Ph.D. Göttingen; professor, University of Heidelberg.

This reaction between two ester components is known as the **Claisen[2] condensation reaction**.

$$2\ CH_3\overset{O}{\overset{\|}{C}}OCH_2CH_3 \xrightarrow[2.\ H_3O^+]{1.\ Na^+\ ^-OCH_2CH_3,\ ethanol} CH_3\overset{O}{\overset{\|}{\underset{\beta}{C}}}-\underset{\alpha}{CH_2}\overset{O}{\overset{\|}{C}}OCH_2CH_3 + CH_3CH_2OH$$

Ethyl acetate

Ethyl acetoacetate, a $\beta$-keto ester (75%)

The mechanism of the Claisen reaction is similar to that of the aldol reaction. As shown in Figure 23.4, the Claisen condensation involves the nucleophilic addition of an ester enolate ion donor to the carbonyl group of a second ester molecule. Viewed from the side of the donor component, the reaction is simply an alpha substitution; viewed from the side of the acceptor component, the reaction is a nucleophilic acyl substitution.

The only difference between an aldol condensation and a Claisen condensation involves the fate of the initially formed tetrahedral intermediate. The tetrahedral intermediate in the aldol reaction is protonated to give an alcohol product—exactly the behavior previously seen for ketones and aldehydes (Section 19.6). The tetrahedral intermediate in the Claisen reaction expels an alkoxide leaving group to yield the acyl substitution product—exactly the behavior previously seen for esters (Section 21.7).

If the starting ester has more than one acidic alpha hydrogen, the product $\beta$-keto ester has a highly acidic, doubly activated hydrogen atom that can be abstracted by base. This deprotonation of the product requires that a full equivalent of base, rather than a catalytic amount, be used in the reaction. Furthermore, the deprotonation serves to drive the Claisen equilibrium completely to the product side so that high yields are often obtained.

**PRACTICE PROBLEM**

What product would you obtain from Claisen condensation of methyl propanoate?

*Solution* The Claisen condensation of an ester results in loss of one molecule of alcohol and formation of a product in which an acyl group of one reactant bonds to the alpha carbon of the second reactant:

$$CH_3CH_2\overset{O}{\overset{\|}{C}}-OCH_3 + H-\underset{CH_3}{\underset{|}{C}H}\overset{O}{\overset{\|}{C}}OCH_3 \xrightarrow[2.\ H_3O^+]{1.\ NaOCH_3} CH_3CH_2\overset{O}{\overset{\|}{C}}-\underset{CH_3}{\underset{|}{C}H}\overset{O}{\overset{\|}{C}}OCH_3 + CH_3OH$$

2 Methylpropanoate

Methyl 1-methyl-2-oxopentanoate

[2]Ludwig Claisen (1851–1930); b. Cologne; Ph.D. Bonn (Kekulé); professor, University of Bonn, Owens College (Manchester), universities of Munich, Aachen, Kiel, and Berlin; Godesberg (private laboratory).

Ethoxide base abstracts an acidic alpha hydrogen atom from an ester molecule, yielding an ester enolate ion.

$$CH_3\overset{O}{\overset{\|}{C}}OC_2H_5 \xrightleftharpoons{^-OC_2H_5} :\overset{-}{C}H_2\overset{O}{\overset{\|}{C}}OC_2H_5 + C_2H_5OH$$

Nucleophilic donor

This ion does a nucleophilic addition to a second ester molecule, giving a tetrahedral intermediate.

$$CH_3\overset{O}{\overset{\|}{C}}OC_2H_5$$

Electrophilic acceptor

$$\rightleftharpoons CH_3\overset{:\ddot{O}:^-}{\overset{|}{C}}(OC_2H_5)—CH_2\overset{O}{\overset{\|}{C}}OC_2H_5$$

The tetrahedral intermediate is not stable. It expels ethoxide ion to yield the new carbonyl compound, ethyl acetoacetate.

$$\rightleftharpoons CH_3\overset{O}{\overset{\|}{C}}CH_2\overset{O}{\overset{\|}{C}}OC_2H_5 + C_2H_5\ddot{O}:^-$$

But ethoxide ion is a base. It therefore converts the $\beta$-keto ester product into its enolate, thus shifting the equilibrium and driving the reaction to completion.

$$\rightleftharpoons CH_3\overset{O}{\overset{\|}{C}}\ddot{C}H\overset{O}{\overset{\|}{C}}OC_2H_5 + C_2H_5OH$$

Protonation by addition of acid yields the final product.

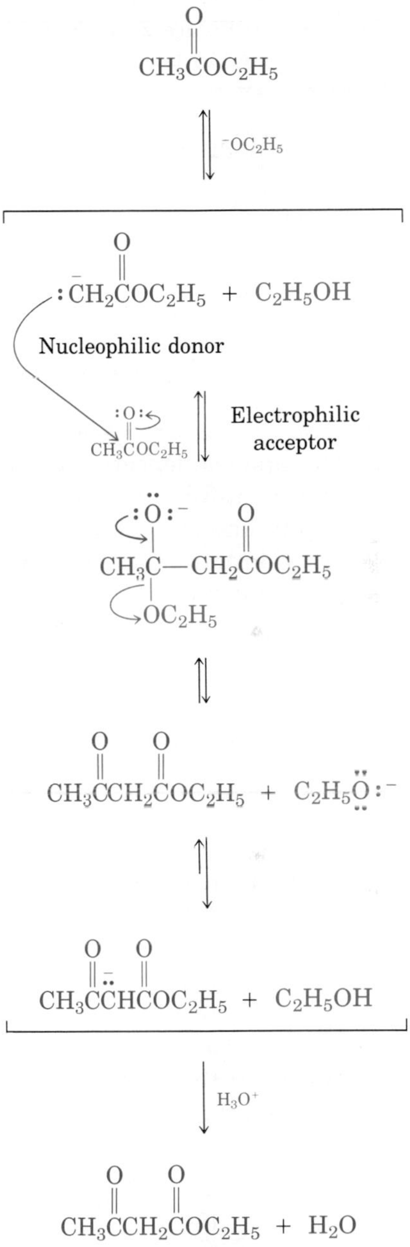

$$\xrightarrow{H_3O^+} CH_3\overset{O}{\overset{\|}{C}}CH_2\overset{O}{\overset{\|}{C}}OC_2H_5 + H_2O$$

**Figure 23.4** Mechanism of the Claisen condensation reaction

**PROBLEM**

**23.13** Show the products you would expect to obtain by Claisen condensation of these esters:

(a) $(CH_3)_2CHCH_2COOCH_3$ (b) Methyl phenylacetate

(c) Methyl cyclohexylacetate

PROBLEM

**23.14** As shown in Figure 23.4, the Claisen reaction is reversible. That is, a $\beta$-keto ester can be cleaved by base into two fragments. Show in detail the mechanism by which this cleavage occurs.

$$C_6H_5COCH_2CO_2C_2H_5 \xrightarrow[\text{Ethanol}]{\text{1 equiv NaOH}} C_6H_5COO^- + CH_3CO_2C_2H_5$$

## 23.10 Mixed Claisen Condensations

The mixed Claisen condensation of two different esters is similar to the mixed aldol condensation (Section 23.6). Mixed Claisen reactions are generally successful only when one of the two ester components has no alpha hydrogens and thus cannot form an enolate ion. For example, ethyl benzoate and ethyl formate cannot form enolate ions and thus cannot serve as donors. They can, however, act as the electrophilic acceptor components in reactions with other ester anions to give good yields of mixed $\beta$-keto ester products.

$$C_6H_5COOC_2H_5 + CH_3COOC_2H_5 \xrightarrow[\text{2. } H_3O^+]{\text{1. NaH/THF}} C_6H_5COCH_2COOC_2H_5 + C_2H_5OH$$

Ethyl benzoate (acceptor) + Ethyl acetate (donor) → Ethyl benzoylacetate

Mixed Claisen-like reactions can also be carried out between esters and ketones. The result is an excellent synthesis of $\beta$-diketones. The reaction works best when the ester component has no alpha hydrogens and thus cannot act as the nucleophilic donor. For example, ethyl formate gives particularly high yields in mixed Claisen condensations with ketones.

$$\text{2,2-Dimethylcyclohexanone} + HCOOC_2H_5 \xrightarrow[\text{2. } H_3O^+]{\text{1. } Na^+\ {}^-OC_2H_5\text{, ethanol}} \text{A } \beta\text{-keto aldehyde}$$

2,2-Dimethylcyclohexanone (donor) + Ethyl formate (acceptor) → A $\beta$-keto aldehyde (91%)

PROBLEM

**23.15** Would you expect diethyl oxalate, $C_2H_5O_2CCO_2C_2H_5$, to give good yields in mixed Claisen reactions? Explain your answer. What product would you expect to obtain from mixed Claisen reaction of ethyl acetate with diethyl oxalate?

PROBLEM

**23.16** What product would you expect from a mixed Claisen-like reaction of 2,2-dimethylcyclohexanone with diethyl oxalate (Problem 23.15)?

## 23.11 Intramolecular Claisen Condensations: The Dieckmann Cyclization

Intramolecular Claisen condensations can be carried out with diesters, just as intramolecular aldol condensations can be carried out with diketones (Section 23.7). Called the **Dieckmann[3] cyclization**, the reaction works best on 1,6-diesters and 1,7-diesters. Five-membered-ring cyclic $\beta$-keto esters result from Dieckmann cyclization of 1,6-diesters, and six-membered-ring $\beta$-keto esters result from cyclization of 1,7-diesters.

Dimethyl hexanedioate (a 1,6-diester) $\xrightarrow[\text{2. } H_3O^+]{\text{1. } Na^+ \ ^-OCH_3,\ CH_3OH}$ Methyl 2-oxocyclopentanecarboxylate (82%) + $CH_3OH$

Dimethyl heptanedioate (a 1,7-diester) $\xrightarrow[\text{2. } H_3O^+]{\text{1. } Na^+ \ ^-OCH_3,\ CH_3OH}$ Methyl 2-oxocyclohexanecarboxylate + $CH_3OH$

The mechanism of the Dieckmann cyclization, shown in Figure 23.5 on page 838, is analogous to the Claisen reaction. One of the two ester groups is converted into an enolate ion, which then carries out a nucleophilic attack on the second ester group at the other end of the molecule. A cyclic $\beta$-keto ester product results.

The products of the Dieckmann cyclization are cyclic $\beta$-keto esters that can be further alkylated and decarboxylated by a series of reactions analogous to the acetoacetic ester synthesis (Section 22.9). For example, alkylation and subsequent decarboxylation of methyl 2-oxocyclohexanecarboxylate yields a 2-alkylcyclohexanone, as shown at the top of page 839. The overall sequence of Dieckmann cyclization, $\beta$-keto ester alkylation, and decarboxylation is an excellent method for preparing 2-substituted cyclohexanones and cyclopentanones.

[3]Walter Dieckmann (1869–1925); b. Hamburg, Germany; Ph.D. Munich (Bamberger); professor, University of Munich.

Base abstracts an acidic alpha proton from the carbon atom next to one of the ester groups, yielding an enolate ion.

Intramolecular nucleophilic addition of the ester enolate ion to the carbonyl group of the second ester group at the other end of the chain then gives a cyclic tetrahedral intermediate.

Loss of alkoxide ion from the tetrahedral intermediate forms a cyclic $\beta$-keto ester.

Deprotonation of the acidic $\beta$-keto ester gives an enolate ion . . .

. . . which is protonated by addition of aqueous acid at the end of the reaction to generate the neutral $\beta$-keto ester product.

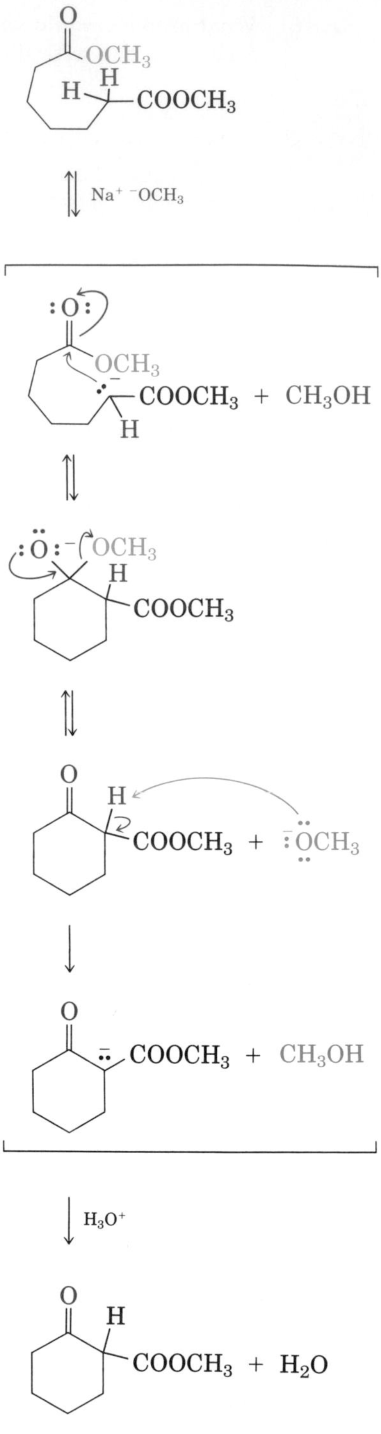

O
OCH3
H
H
COOCH3
Na+ −OCH3
:O:
OCH3
COOCH3 + CH3OH
H
:O:−
OCH3
H
COOCH3
O
H
COOCH3 + :OCH3
O
COOCH3 + CH3OH
H3O+
O
H
COOCH3 + H2O


**Figure 23.5** Mechanism of the Dieckmann cyclization of a diester to yield a cyclic $\beta$-keto ester product

$$\text{Methyl 2-oxocyclohexanecarboxylate} \xrightarrow[\text{2. } H_2C{=}CHCH_2Br]{\text{1. } Na^+\ {}^-OCH_3,\ CH_3OH} \text{2-}(CO_2CH_3)\text{-2-}(CH_2{-}CH{=}CH_2)\text{cyclohexanone} \xrightarrow[\Delta]{H_3O^+} CO_2 + CH_3OH + \text{2-Allylcyclohexanone}$$

Methyl 2-oxocyclohexanecarboxylate

2-Allylcyclohexanone
(83%)

PROBLEM

**23.17** What product would you expect on treatment of diethyl 4-methylheptanedioate with sodium ethoxide followed by acidification?

PROBLEM

**23.18** How can you account for the fact that Dieckmann cyclization of diethyl 3-methylheptanedioate gives a mixture of two $\beta$-keto ester products? What are their structures, and why is a mixture formed?

## 23.12 The Michael Reaction

We saw earlier (Section 19.18) that nucleophiles such as amines and cyanide ion can react with $\alpha,\beta$-unsaturated ketones and aldehydes to give the conjugate addition product, rather than the direct addition product:

$$H{-}Nu{:} + {>}C{=}C{-}\overset{O}{\overset{\|}{C}}{-} \longrightarrow Nu{-}\overset{|}{\underset{|}{C}}{-}\overset{H}{\underset{|}{C}}{-}\overset{O}{\overset{\|}{C}}{-}$$

Conjugate addition product

Exactly the same kind of conjugate addition can occur when a nucleophilic enolate ion reacts with an $\alpha,\beta$-unsaturated carbonyl compound—a process known as the **Michael[4] reaction.**

The best Michael reactions take place when particularly stable enolate ions such as those derived from $\beta$-keto esters or $\beta$-diesters (malonic esters) add to unhindered $\alpha,\beta$-unsaturated ketones. For example, ethyl acetoacetate

[4]Arthur Michael (1853–1942); b. Buffalo, New York; studied Heidelberg, Berlin, École de Médecine, Paris; professor, Tufts University (1882–1889 and 1894–1907), Harvard University (1912–1936).

reacts with 3-buten-2-one in 94% yield in the presence of sodium ethoxide catalyst to yield the conjugate addition product:

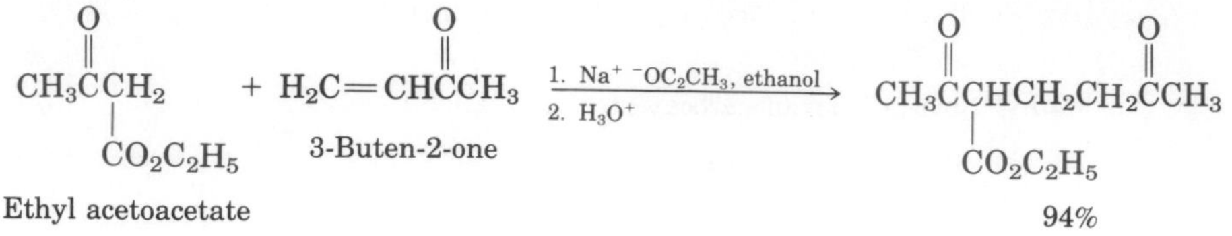

$$\underset{\text{Ethyl acetoacetate}}{\mathrm{CH_3\overset{O}{\overset{\|}{C}}\underset{\substack{|\\ CO_2C_2H_5}}{C}H_2}} + \underset{\text{3-Buten-2-one}}{\mathrm{H_2C{=}CH\overset{O}{\overset{\|}{C}}CH_3}} \xrightarrow[\text{2. } H_3O^+]{\text{1. } Na^+\ {}^-OC_2CH_3,\ \text{ethanol}} \underset{94\%}{\mathrm{CH_3\overset{O}{\overset{\|}{C}}\underset{\substack{|\\ CO_2C_2H_5}}{C}HCH_2CH_2\overset{O}{\overset{\|}{C}}CH_3}}$$

Michael reactions take place by addition of a nucleophilic enolate ion donor to the beta carbon of an $\alpha,\beta$-unsaturated carbonyl acceptor, as indicated in Figure 23.6.

---

The base catalyst removes an acidic alpha proton from the starting $\beta$-keto ester to generate a stabilized enolate nucleophile.

$$\mathrm{CH_3\overset{O}{\overset{\|}{C}}CH_2\overset{O}{\overset{\|}{C}}OC_2H_5}$$

$$\Big\updownarrow\ {}^-OC_2H_5,\ \text{ethanol}$$

The nucleophile takes part in a conjugate (Michael) addition to the $\alpha,\beta$-unsaturated ketone electrophile. The product anion is a ketone enolate.

$$\mathrm{CH_3\overset{O}{\overset{\|}{C}}\ddot{\bar{C}}H\overset{O}{\overset{\|}{C}}OC_2H_5}$$

$$\Big\downarrow\quad \mathrm{\overset{\delta^+}{H_2C}{=}\overset{\delta^-}{C}H{-}\underset{\delta^+}{\overset{O^{\delta^-}}{\overset{\|}{C}}}{-}CH_3}$$

$$\mathrm{CH_3\overset{O}{\overset{\|}{C}}\underset{\substack{|\\ CH_2\ddot{\bar{C}}H\overset{O}{\overset{\|}{C}}CH_3}}{C}H\overset{O}{\overset{\|}{C}}OC_2H_5}$$

The ketone enolate abstracts an available proton, either from solvent or from the starting $\beta$-keto ester, to yield the final product.

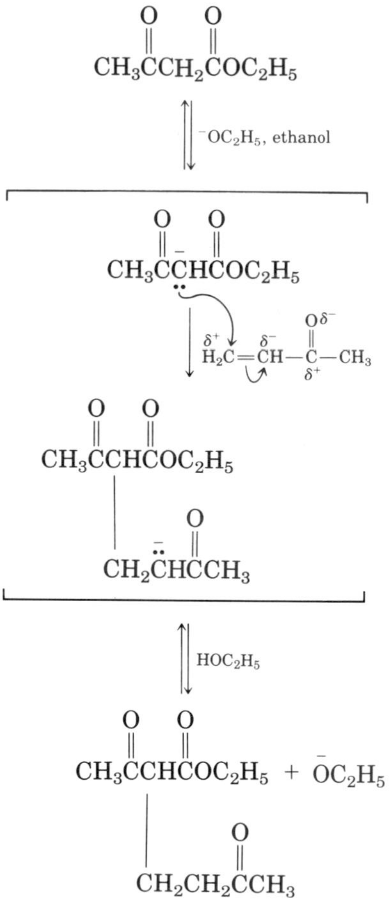

$$\Big\updownarrow\ HOC_2H_5$$

$$\mathrm{CH_3\overset{O}{\overset{\|}{C}}\underset{\substack{|\\ CH_2CH_2\overset{O}{\overset{\|}{C}}CH_3}}{C}H\overset{O}{\overset{\|}{C}}OC_2H_5} + {}^-OC_2H_5$$

**Figure 23.6** Mechanism of the Michael reaction between a $\beta$-keto ester and an $\alpha,\beta$-unsaturated ketone

---

The Michael reaction is general for a wide variety of $\alpha,\beta$-unsaturated carbonyl compounds, not just conjugated enones. Conjugated aldehydes, esters, nitriles, amides, and nitro compounds can all act as the electrophilic acceptor component in the Michael reaction (Table 23.1). Similarly, a variety of different donors can be used, including β-diketones, β-keto esters, malonic esters, β-keto nitriles, and nitro compounds.

**Table 23.1** Some Michael acceptors and Michael donors

| *Michael acceptors* | | *Michael donors* | |
|---|---|---|---|
| $H_2C{=}CHCHO$ | Propenal | $RCOCH_2COR'$ | β-Diketone |
| $H_2C{=}CHCO_2CH_3$ | Methyl propenoate | $RCOCH_2CO_2CH_3$ | β-Keto ester |
| $H_2C{=}CHC{\equiv}N$ | Propenenitrile | $CH_3O_2CCH_2CO_2CH_3$ | Malonic ester |
| $H_2C{=}CHCOCH_3$ | 3-Buten-2-one | $RCOCH_2C{\equiv}N$ | β-Keto nitrile |
| $H_2C{=}CHNO_2$ | Nitroethylene | $RCH_2NO_2$ | Nitro compound |
| $H_2C{=}CHCONH_2$ | Propenamide | | |

PROBLEM

**23.19** What products would you obtain from base-induced Michael reaction of 2,4-pentanedione with these α,β-unsaturated acceptors?
(a) 2-Cyclohexenone (b) Propenenitrile (c) Methyl 2-butenoate

PROBLEM

**23.20** What products would you obtain from base-induced Michael reaction of 3-buten-2-one with these nucleophilic donors?
(a) Diethyl malonate (b) [cyclopentanone ring with $COOCH_3$ substituent and $=O$] (c) Nitromethane

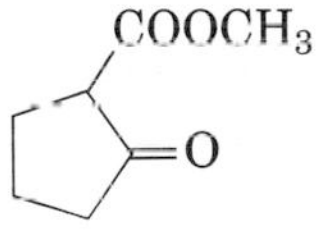

PROBLEM

**23.21** How might the following compounds be prepared using Michael reactions? Show the nucleophilic donor and the electrophilic acceptor in each case.

(a) $CH_3C(=O)CH(CO_2CH_3)CH_2CH_2COC_6H_5$

(b) $CH_3C(=O)CH_2CH_2CH_2C(=O)CH_3$

(c) $(CH_3O_2C)_2CHCH_2CH_2CN$

(d) $CH_3CH(NO_2)CH_2CH_2CO_2CH_3$

(e) $(CH_3O_2C)_2CHCH_2CH_2NO_2$

## 23.13 The Stork Enamine Reaction

Other kinds of carbon nucleophiles besides enolate ions add to α,β-unsaturated acceptors in the Michael reaction, greatly extending the usefulness and versatility of the process. Among the most important such nucleophiles are *enamines*. Recall from Section 19.12 that enamines are readily prepared by reaction between a ketone and a secondary amine:

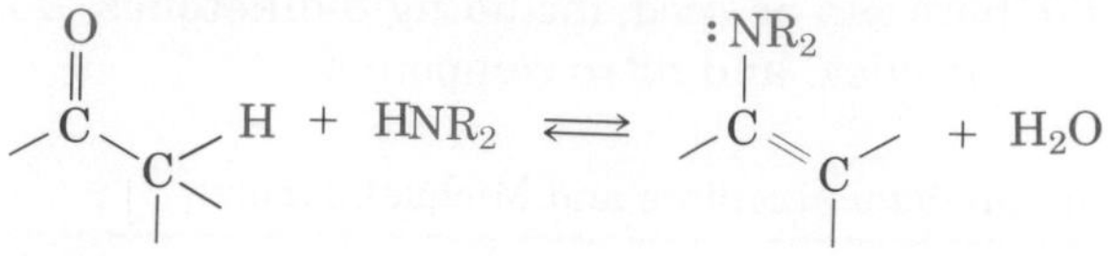

For example,

Cyclohexanone Pyrrolidine 1-Pyrrolidinocyclohexene (87%) (an enamine)

As the following resonance structures indicate, enamines are electronically similar to enolate ions. Overlap of the nitrogen lone-pair orbital with the double-bond $p$ orbitals leads to an increase in electron density on the alpha-carbon atom, making it strongly nucleophilic.

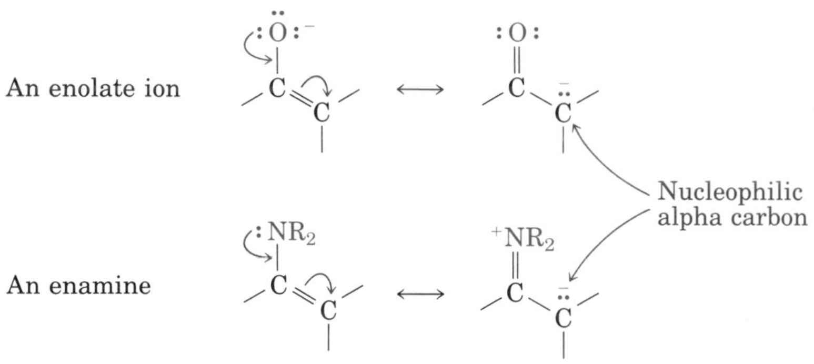

An enolate ion

An enamine

Nucleophilic alpha carbon

Enamines behave much the same as enolate ions in many respects and enter into many of the same kinds of reactions that enolate ions do. In the **Stork[5] enamine reaction**, for example, an enamine acts as a nucleophile in adding to an $\alpha,\beta$-unsaturated carbonyl acceptor in a Michael-type process. The initial product is then hydrolyzed by aqueous acid to yield a 1,5-dicarbonyl compound. The overall Stork enamine reaction is a three-step sequence:

1. Enamine formation from a ketone
2. Michael-type addition to an enone
3. Enamine hydrolysis back to a ketone

The net effect of the Stork enamine sequence is to carry out a Michael addition of a ketone to an $\alpha,\beta$-unsaturated acceptor. For example, cyclohexanone reacts with the cyclic amine pyrrolidine to yield an enamine; further reaction with an enone such as 3-buten-2-one yields a Michael-type adduct, and aqueous hydrolysis completes the sequence to provide a 1,5-diketone product, as shown in Figure 23.7.

[5]Gilbert Stork (1921– ); b. Brussels, Belgium; Ph.D. Wisconsin (McElvain); professor, Harvard University, Columbia University (1953– ).

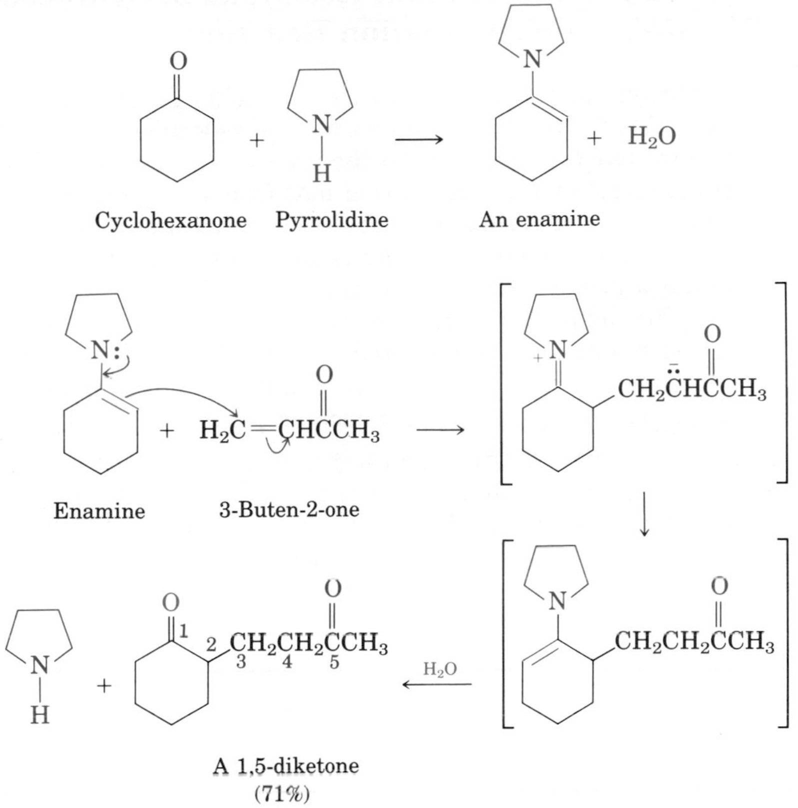

Cyclohexanone + Pyrrolidine ⟶ An enamine + $H_2O$

Enamine + 3-Buten-2-one ($H_2C{=}CHCCH_3$) ⟶ [$CH_2\ddot{C}HCCH_3$] ⟶ [$CH_2CH_2CCH_3$] $\xleftarrow{H_2O}$ Pyrrolidine + A 1,5-diketone (71%)

**Figure 23.7** A Stork enamine reaction between cyclohexanone and 3-buten-2-one: Cyclohexanone is first converted into an enamine; the enamine then adds to the $\alpha,\beta$-unsaturated ketone in a Michael reaction; and the initial product is hydrolyzed to yield a 1,5-diketone.

PROBLEM

**23.22** Draw the structures of the enamines you would obtain from reaction of pyrrolidine with these ketones:
(a) Cyclopentanone (b) 2,2-Dimethylcyclohexanone

PROBLEM

**23.23** What products would result from reaction of each enamine you prepared in Problem 23.22 with these $\alpha,\beta$-unsaturated acceptors? (Assume that the initial product is hydrolyzed.)
(a) Ethyl propenoate (b) Propenal (acrolein)

PROBLEM

**23.24** Show how you might use an enamine reaction to prepare these compounds:

(a) O, $CH_2CH_2CN$ (b) O, $CH_2CH_2CO_2CH_3$

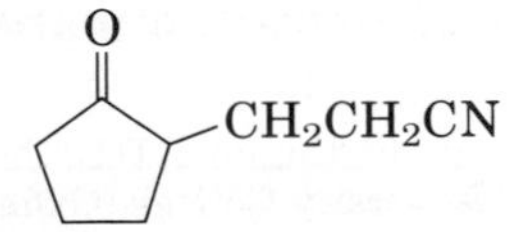

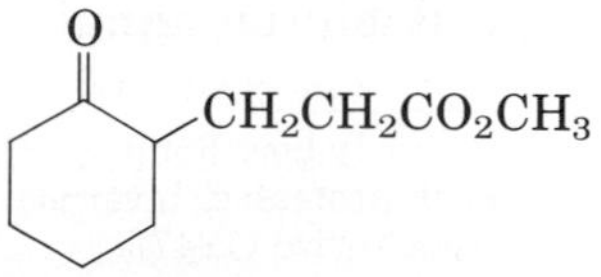

## 23.14 Carbonyl Condensation Reactions in Synthesis: The Robinson Annulation Reaction

Carbonyl condensation reactions are among the most valuable methods available for the synthesis of complex molecules. By putting a few fundamental reactions together in the proper sequence, some remarkably useful transformations can be carried out. One such example is the **Robinson[6] annulation reaction**, used for the synthesis of polycyclic molecules. [An *annulation* reaction (from the Latin *annulus*, meaning "ring") is one that builds a new ring onto a molecule.]

The Robinson annulation is a two-step process that combines a Michael reaction with an internal aldol reaction. It takes place between a Michael-type nucleophilic donor such as a $\beta$-keto ester or $\beta$-diketone and an $\alpha,\beta$-unsaturated ketone acceptor such as 3-buten-2-one. The product is a substituted 2-cyclohexenone:

$CO_2C_2H_5$

3-Buten-2-one + Ethyl acetoacetate

$\xrightarrow[\text{Michael reaction}]{Na^+\ ^-OC_2H_5,\ \text{ethanol}}$

[ Michael product ] ($CO_2C_2H_5$)

Aldol reaction ↓ $Na^+\ ^-OC_2H_5$, ethanol

Annulation product ($CO_2C_2H_5$) + $H_2O$

The first step of the Robinson annulation is simply a Michael reaction—a stabilized $\beta$-diketone or $\beta$-keto ester enolate ion effects a conjugate addition to an $\alpha,\beta$-unsaturated ketone, yielding a 1,5-diketone. But as we saw in Section 23.7, 1,5-diketones undergo intramolecular aldol condensation to yield cyclohexenones when treated with base. Thus, the final product contains a six-membered ring, and an annulation has been accomplished.

An excellent example of the practical importance of the Robinson annulation reaction occurs as a key step during the commercial synthesis of the female steroid hormone estrone (Figure 23.8).

In this example, 2-methyl-1,3-cyclopentanedione (a $\beta$-diketone) is used to generate the stabilized enolate ion required for Michael reaction, and an aryl-substituted $\alpha,\beta$-unsaturated ketone is used as the acceptor. Base-catalyzed Michael reaction between the two yields an intermediate triketone, which immediately cyclizes in an intramolecular aldol condensation to give a Robinson annulation product. Several further transformations are then

[6]Sir Robert Robinson (1886–1975); b. Rufford/Chesterfield, England; D.Sc. Manchester (Perkin); professor, Liverpool, Manchester (1922–1928), University College, Oxford (1930–1955); Nobel prize (1947).

required to complete the synthesis of estrone, but the Robinson annulation serves as the key step for assembling much of the molecule.

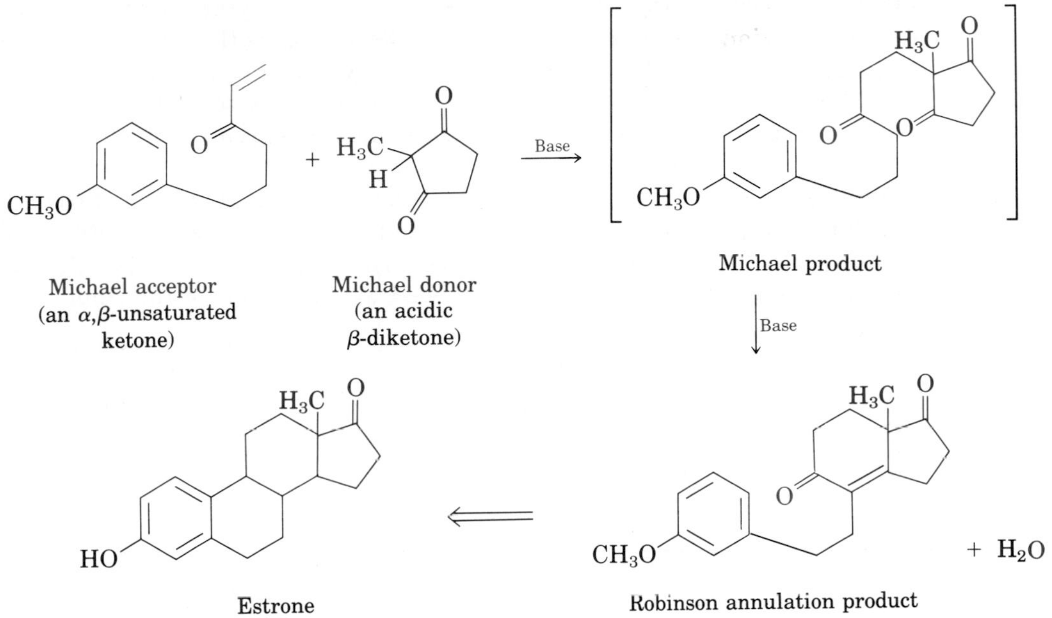

Michael acceptor (an $\alpha,\beta$-unsaturated ketone) + Michael donor (an acidic $\beta$-diketone) $\xrightarrow{\text{Base}}$ [Michael product] $\xrightarrow{\text{Base}}$ Robinson annulation product + $H_2O$ ⟹ Estrone

**Figure 23.8** A Robinson annulation reaction used in the commercial synthesis of the steroid hormone estrone

**PROBLEM**

**23.25** What product would you expect from a Robinson annulation reaction of 2-methyl-1,3-cyclopentanedione and 3-buten-2-one?

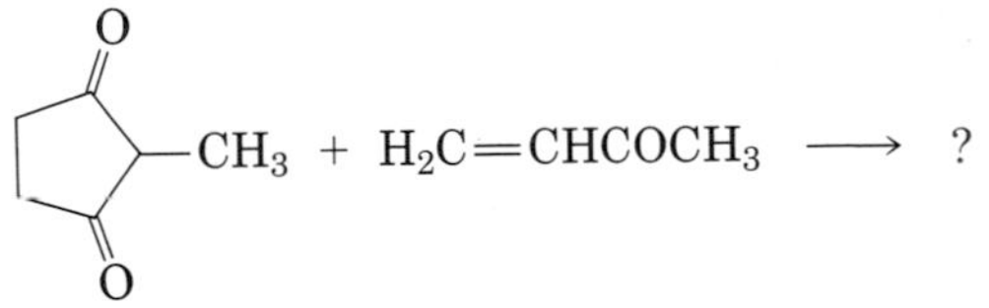

$$\text{(2-methyl-1,3-cyclopentanedione)} + H_2C{=}CHCOCH_3 \longrightarrow ?$$

**PROBLEM**

**23.26** How would you prepare the following compound using a Robinson annulation reaction between a $\beta$-diketone and an $\alpha,\beta$-unsaturated ketone acceptor? Draw the structures of both reactants you would use and the structure of the intermediate Michael addition product.

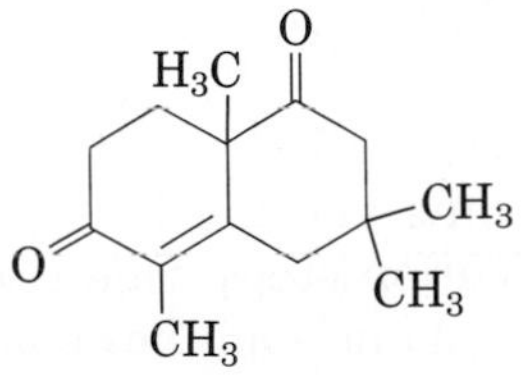

## 23.15 Biological Carbonyl Condensation Reactions

Carbonyl condensation reactions are used in nature for the biological synthesis of a great many different molecules. Fats, amino acids, steroid hormones, and many other kinds of compounds are all synthesized by plants and animals using carbonyl condensation reactions as the key step.

Nature uses the two-carbon acetate fragment of acetyl CoA as the major building block for synthesis. Not only can acetyl CoA serve as an electrophilic acceptor for attack of nucleophiles at the acyl carbon, it can also serve as a nucleophilic donor by loss of its acidic alpha proton to generate an enolate ion. The enolate ion of acetyl CoA can then add to another carbonyl group in a condensation reaction. For example, citric acid is biosynthesized by addition of acetyl CoA to the ketone carbonyl group of oxaloacetic acid in a kind of mixed aldol reaction, followed by hydrolysis of the thiol ester group.

$$CH_3\overset{O}{\overset{\|}{C}}SCoA \longrightarrow \text{“}:\overset{-}{C}H_2\overset{O}{\overset{\|}{C}}SCoA\text{”}$$

Acetyl CoA, a thiol ester

$$O{=}C(COOH)(CH_2COOH) + :\overset{-}{C}H_2\overset{O}{\overset{\|}{C}}SCoA \Longrightarrow HO{-}C(COOH)(CH_2COOH){-}CH_2COOH$$

Oxaloacetic acid → Citric acid

Acetyl CoA is also involved as a primary building block in the biosynthesis of steroids, fats, and other lipids, where the key step is a Claisen-like condensation reaction. We'll go into more of the details of this process in Section 28.4.

$$CH_3\overset{O}{\overset{\|}{C}}{-}SCoA + :\overset{-}{C}H_2\overset{O}{\overset{\|}{C}}SCoA \Longrightarrow CH_3\overset{O}{\overset{\|}{C}}{-}CH_2\overset{O}{\overset{\|}{C}}SCoA$$

Acetyl CoA → Acetoacetyl CoA

Acetoacetyl CoA ⇒ Fats, steroids, prostaglandins

## 23.16 Summary and Key Words

A **carbonyl condensation reaction** is one that takes place between two carbonyl components and involves a combination of nucleophilic addition and alpha-substitution steps. One carbonyl component (the donor) is converted by base into a nucleophilic enolate ion, which adds to the electrophilic carbonyl group of the second component (the acceptor). The donor molecule undergoes an alpha substitution while the acceptor molecule undergoes a nucleophilic addition.

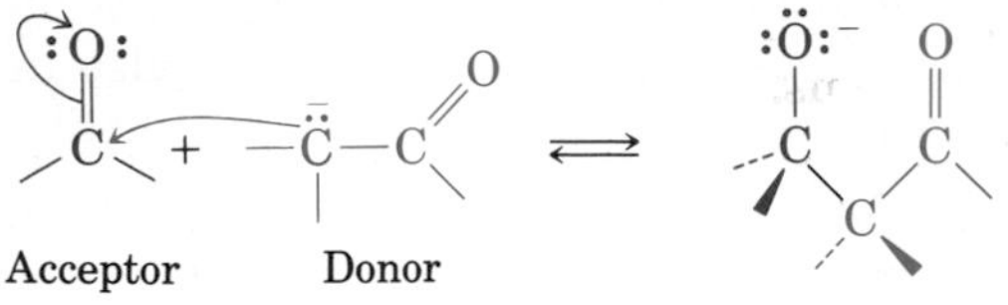

Acceptor Donor

The **aldol reaction** is a carbonyl condensation that occurs between two ketone or aldehyde components. Aldol reactions are reversible, leading first to $\beta$-hydroxy ketones (or aldehydes) and then to $\alpha,\beta$-unsaturated ketones.

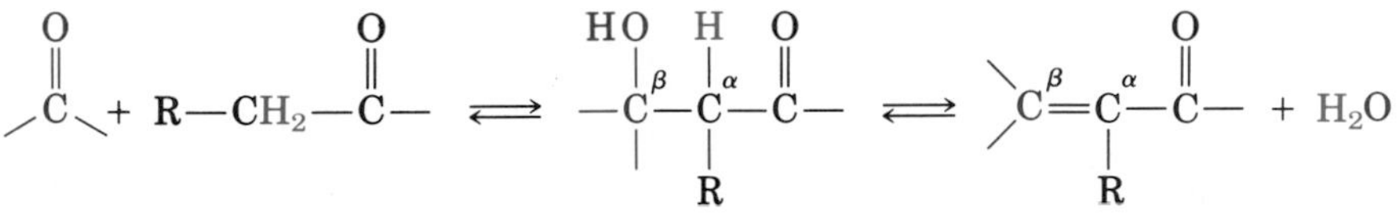

A $\beta$-hydroxy ketone An $\alpha,\beta$-unsaturated ketone

Mixed aldol condensations between two different ketones generally give a mixture of all four possible products. A mixed reaction can be successful, however, if one of the two components is an unusually good donor (as with ethyl acetoacetate), or if it can act only as an acceptor (as with formaldehyde and benzaldehyde). Intramolecular aldol condensations of 1,4- and 1,5-diketones are also successful and provide an excellent method for preparing five and six-membered enone rings.

The **Claisen reaction** is a carbonyl condensation that occurs between two ester components and leads to $\beta$-keto ester products:

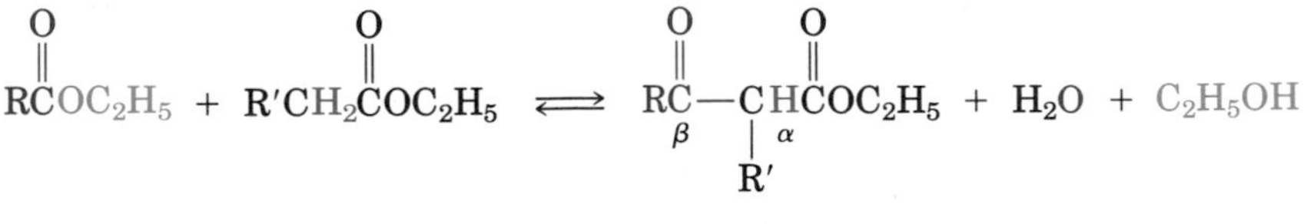

$$RCOC_2H_5 + R'CH_2COC_2H_5 \rightleftharpoons RC-CHCOC_2H_5 + H_2O + C_2H_5OH$$

A $\beta$-keto ester

Mixed Claisen condensations between two different esters are successful only when one of the two components has no acidic alpha hydrogens (as with ethyl benzoate and ethyl formate) and thus can function only as the acceptor component. Intramolecular Claisen condensations (**Dieckmann cyclization reactions**) provide excellent syntheses of five- and six-membered cyclic $\beta$-keto esters, starting from 1,6- and 1,7-diesters.

The conjugate addition of carbon nucleophiles to an $\alpha,\beta$-unsaturated acceptor is known as the **Michael reaction**. The best Michael reactions take place between unusually acidic donors ($\beta$-keto esters or $\beta$-diketones) and unhindered $\alpha,\beta$-unsaturated acceptors. Enamines, prepared by reaction of a ketone with a disubstituted amine, are also excellent Michael donors.

Carbonyl condensation reactions enjoy widespread use in synthesis. One example of their versatility is the **Robinson annulation reaction**, which leads to the formation of substituted cyclohexenone products. Treatment of a $\beta$-diketone or $\beta$-keto ester with an $\alpha,\beta$-unsaturated ketone leads first to a Michael addition, which is followed by intramolecular aldol cyclization of the product. Condensation reactions are also used widely in nature for the biosynthesis of such molecules as fats and steroids.

## 23.17 Summary of Reactions

1. Aldol reaction—a condensation between two ketones, two aldehydes, or one ketone and one aldehyde

   a. Ketones (Section 23.2)

$$2\ RCH_2\overset{O}{\overset{\|}{C}}R' \xrightleftharpoons{\text{NaOH, ethanol}} RCH_2\underset{R'}{\underset{|}{\overset{OH}{\overset{|}{C}}}}-\underset{R}{\underset{|}{CH}}\overset{O}{\overset{\|}{C}}R'$$

   b. Aldehydes (Section 23.2)

$$2\ RCH_2\overset{O}{\overset{\|}{C}}H \xrightleftharpoons{\text{NaOH, ethanol}} RCH_2\overset{OH}{\overset{|}{C}H}\underset{R}{\underset{|}{C}H}\overset{O}{\overset{\|}{C}}H$$

   c. Mixed aldol reaction (Section 23.6)

$$RCH_2\overset{O}{\overset{\|}{C}}R' + ArCHO \xrightleftharpoons{\text{NaOH, ethanol}} Ar\overset{OH}{\overset{|}{C}H}\underset{R}{\underset{|}{C}H}\overset{O}{\overset{\|}{C}}R'$$

$$RCH_2\overset{O}{\overset{\|}{C}}R' + CH_2O \xrightleftharpoons{\text{NaOH, ethanol}} HOCH_2\underset{R}{\underset{|}{C}H}\overset{O}{\overset{\|}{C}}R'$$

   d. Intramolecular aldol reaction (Section 23.7)

Diketone ($O$, $O$, $CH_2$, $R$, $R'$) $\xrightleftharpoons{\text{NaOH, ethanol}}$ substituted cyclohexenone ($O$, $R$, $R'$) $+ H_2O$

2. Dehydration of aldol products (Section 23.4)

$$-\overset{HO}{\overset{|}{\underset{|}{C}}}-\overset{H}{\overset{|}{\underset{|}{C}}}-\overset{O}{\overset{\|}{C}}- \xrightarrow[\text{or } H_3O^+]{\text{Base}} \rangle C=\underset{|}{C}-\overset{O}{\overset{\|}{C}}- + H_2O$$

3. Claisen reaction—the condensation between two esters, or one ester and one ketone (Section 23.9)

a.
$$2\ RCH_2CO_2R' \xrightleftharpoons[\text{ethanol}]{NaOCH_2CH_3} RCH_2\overset{\overset{O}{\|}}{C}-\underset{\underset{R}{|}}{C}HCO_2R' + HOR'$$

b. Mixed Claisen reaction (Section 23.10)

$$RCH_2CO_2R' + HCO_2R' \xrightleftharpoons{NaOCH_2CH_3,\ \text{ethanol}} H\overset{\overset{O}{\|}}{C}-\underset{\underset{R}{|}}{C}HCO_2R' + HOR'$$

4. Dieckmann cyclization; intramolecular Claisen condensation (Section 23.11)

$$RO_2C(CH_2)_4CO_2R \xrightleftharpoons{NaOCH_2CH_3,\ \text{ethanol}} \text{2-oxocyclopentane-}CO_2R + HOR$$

$$RO_2C(CH_2)_5CO_2R \xrightleftharpoons{NaOCH_2CH_3,\ \text{ethanol}} \text{2-oxocyclohexane-}CO_2R + HOR$$

5. Michael reaction (Section 23.12)

$$-\overset{\overset{O}{\|}}{C}CH_2\overset{\overset{O}{\|}}{C}- + >C=\underset{|}{C}-\overset{\overset{O}{\|}}{C}- \xrightarrow{NaOCH_2CH_3,\ \text{ethanol}} -\overset{\overset{O}{\|}}{C}\underset{\underset{\underset{|}{C=O}}{|}}{C}H-\overset{|}{\underset{|}{C}}-\overset{|}{C}H\overset{\overset{O}{\|}}{C}-$$

6. Enamine reaction (Section 23.13)

$$:NR_2\text{-}C=C< + >C=\underset{|}{C}-\overset{\overset{O}{\|}}{C}< \xrightarrow{H_2O} >\overset{\overset{O}{\|}}{C}-\overset{}{C}<-\overset{}{C}<-\overset{}{C}<(H)-\overset{\overset{O}{\|}}{C}<$$

7. Robinson annulation reaction (Section 23.14)

$$\text{enone} + \text{ketone} \xrightarrow{NaOCH_2CH_3,\ \text{ethanol}} \text{cyclohexenone}$$

## ADDITIONAL PROBLEMS

**23.27** Which of the following compounds would be expected to undergo aldol self-condensation?

(a) Trimethylacetaldehyde
(b) Cyclobutanone
(c) Benzophenone (diphenyl ketone)
(d) 3-Pentanone
(e) Decanal
(f) 3-Phenyl-2-propenal

**23.28** Show the product from each compound listed in Problem 23.27 that is capable of undergoing the aldol reaction.

**23.29** What product would you expect to obtain from aldol cyclization of hexanedial ($OHCCH_2CH_2CH_2CH_2CHO$)?

**23.30** How might you synthesize the following compounds using aldol reactions? In each case, show the structure of the starting ketone(s) or aldehyde(s) you would use.

(a) $C_6H_5CH{=}CHCOC_6H_5$

(b) 2-Cyclohexenone

(c)

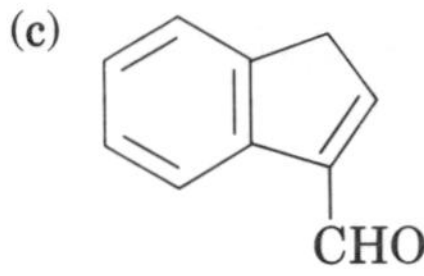


(d)

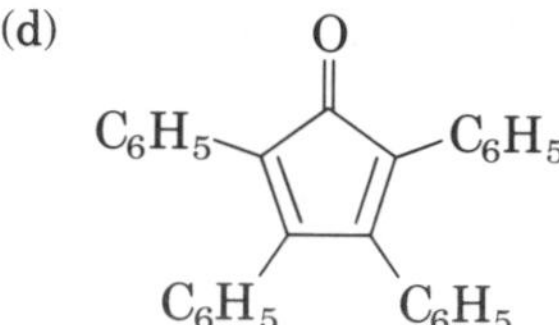


**23.31** How can you account for the fact that 2,2,6-trimethylcyclohexanone yields no detectable aldol product even though it has an acidic alpha hydrogen?

**23.32** Cinnamaldehyde, the aromatic constituent of cinnamon oil, can be synthesized by a mixed aldol condensation. Show the starting materials you would use and formulate the reaction.

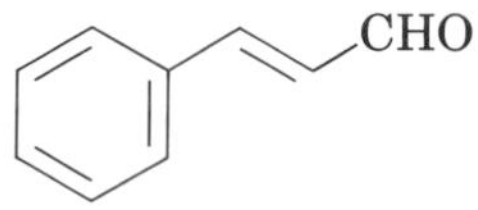

CHO

Cinnamaldehyde

**23.33** The so-called Wieland–Miescher ketone is a valuable starting material used in the laboratory synthesis of steroid hormones. How might you prepare it from 1,3-cyclohexanedione?

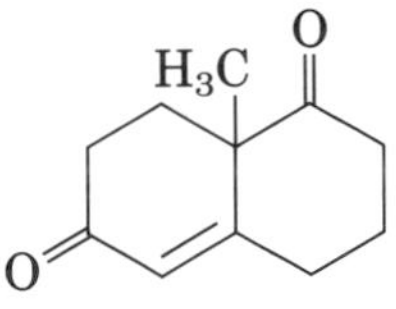

$H_3C$ O O

Wieland-Miescher ketone

**23.34** The bicyclic ketone shown does not undergo aldol self-condensation even though it has two alpha-hydrogen atoms. Explain. [*Hint:* Try to build a molecular model of the enolate ion intermediate.]

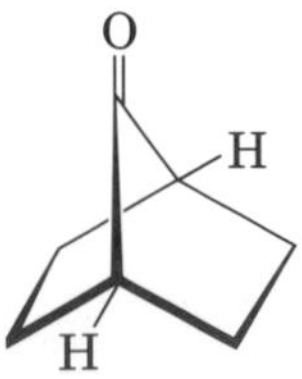

O H H

**23.35** What condensation products would you expect to obtain by treatment of these substances with sodium ethoxide in ethanol?

(a) Ethyl butanoate (b) Cycloheptanone
(c) 3,7-Nonanedione (d) 3-Phenylpropanal

**23.36** Give the structures of all the possible Claisen condensation products from these reactions. Tell which, if any, you would expect to predominate in each case.

(a) $CH_3CO_2CH_3 + CH_3CH_2CO_2CH_3$ (b) $C_6H_5CO_2CH_3 + C_6H_5CH_2CO_2CH_3$
(c) $CH_3OCO_2CH_3$ + Cyclohexanone (d) $C_6H_5CHO + CH_3CO_2CH_3$

**23.37** As written, the following reactions are unlikely to provide the desired product in high yield. What is wrong with each?

(a) $CH_3CHO + CH_3COCH_3 \xrightarrow[\text{Ethanol}]{Na^+ \ ^-OC_2H_5} CH_3CH(OH)CH_2COCH_3$

(b) $CH_2(CO_2C_2H_5)_2 + H_2C{=}CHCOCH_3 \xrightarrow{\text{Base}} H_2C{=}CHC(OH)(CH_3)-CH(CO_2C_2H_5)_2$

(c) 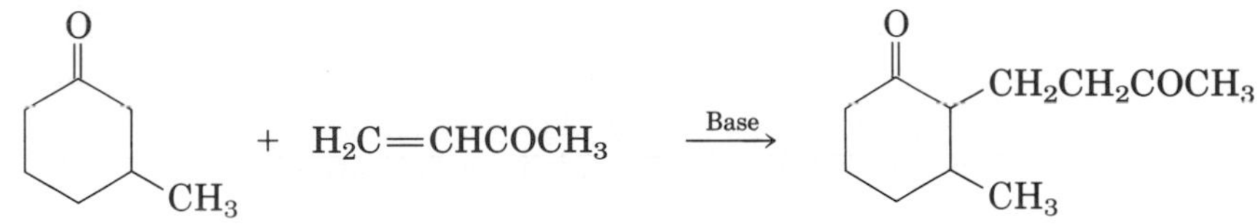 3-Methylcyclohexanone + $H_2C{=}CHCOCH_3 \xrightarrow{\text{Base}}$ 2-($CH_2CH_2COCH_3$)-3-($CH_3$)cyclohexanone

(d) 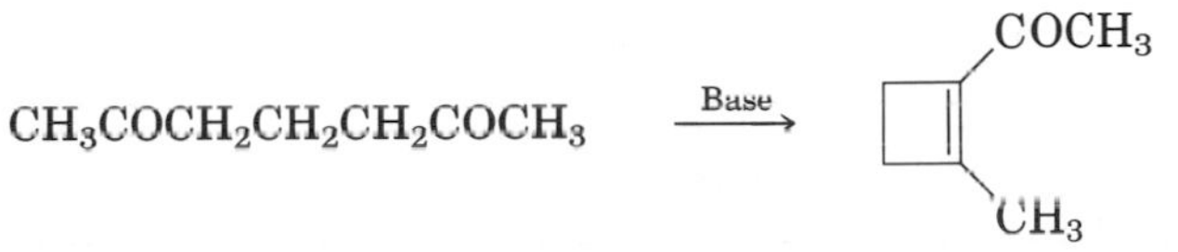 $CH_3COCH_2CH_2CH_2COCH_3 \xrightarrow{\text{Base}}$ 1-acetyl-2-methylcyclobutene ($COCH_3$, $CH_3$)

(e) 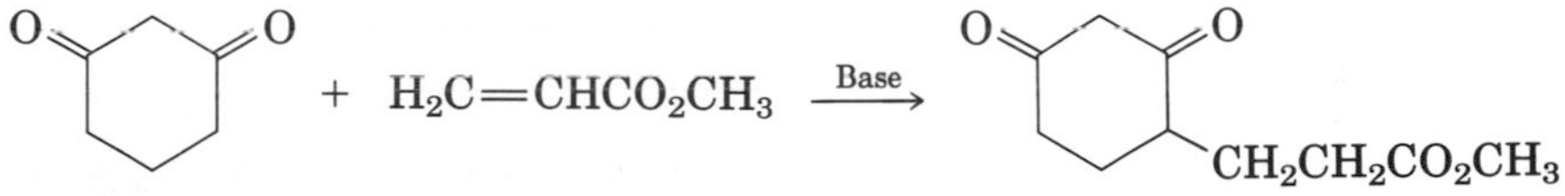 1,3-Cyclohexanedione + $H_2C{=}CHCO_2CH_3 \xrightarrow{\text{Base}}$ 4-($CH_2CH_2CO_2CH_3$)-1,3-cyclohexanedione

**23.38** In the mixed Claisen reaction of cyclopentanone with ethyl formate, a much higher yield of the desired product is obtained by first mixing the two carbonyl components and then adding base, rather than by first mixing base with cyclopentanone and then adding ethyl formate. Explain.

**23.39** Ethyl dimethylacetoacetate reacts instantly at room temperature when treated with ethoxide ion to yield two products, ethyl acetate and ethyl 2-methylpropanoate. Propose a mechanism for this cleavage reaction.

$$CH_3\overset{O}{\overset{\|}{C}}C(CH_3)_2COOCH_2CH_3 \xrightarrow[\text{Ethanol, 25°C}]{Na^+ \ ^-OC_2H_5} CH_3COOCH_2CH_3 + (CH_3)_2CHCOOCH_2CH_3$$

**23.40** In contrast to the rapid reaction shown in Problem 23.39, ethyl acetoacetate itself requires temperatures of over 150°C to undergo the same kind of cleavage reaction. How can you explain the difference in reactivity?

$$CH_3\overset{O}{\overset{\|}{C}}CH_2COOCH_2CH_3 \xrightarrow[\text{Ethanol, 150°C}]{Na^+ \ ^-OC_2H_5} 2\ CH_3COOCH_2CH_3$$

**23.41** The Darzens reaction involves a two-step base-catalyzed condensation of ethyl chloroacetate with a ketone to yield an epoxy ester. The first step is a carbonyl condensation reaction, and the second step is an $S_N2$ reaction. Formulate the complete mechanism by which this reaction occurs.

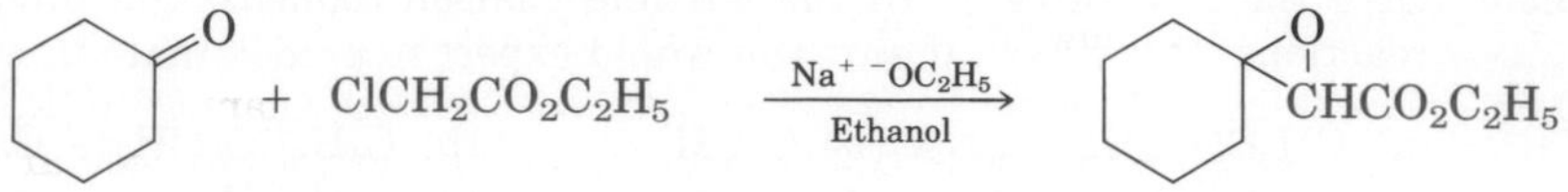

$$\text{Cyclohexanone} + ClCH_2CO_2C_2H_5 \xrightarrow[\text{Ethanol}]{Na^+\ ^-OC_2H_5} \text{(spiro epoxide)}\ CHCO_2C_2H_5$$

**23.42** How would you prepare these compounds from cyclohexanone?

(a) $C_6H_5CH$=(2,6-bis(benzylidene)cyclohexanone)=$CHC_6H_5$

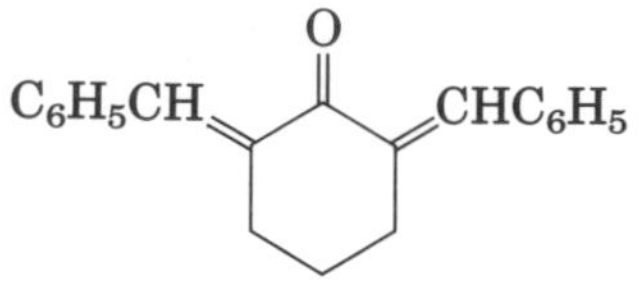

(b) 2-$CO_2C_2H_5$-cyclohexanone

(c) 2-$CH_2CH_2CN$-cyclohexanone

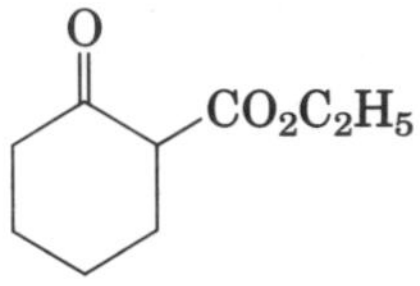

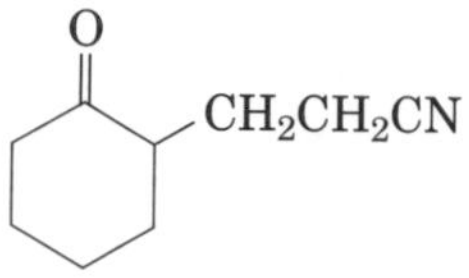

(d) 2-$CH_2CH{=}CH_2$-cyclohexanone

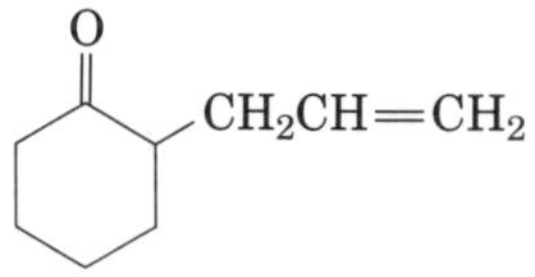

**23.43** Griseofulvin, an antibiotic produced by the mold *Penicillium griseofulvum* (Dierckx), has been synthesized by a route that employs a twofold Michael reaction as the key step. Propose a mechanism for this transformation.

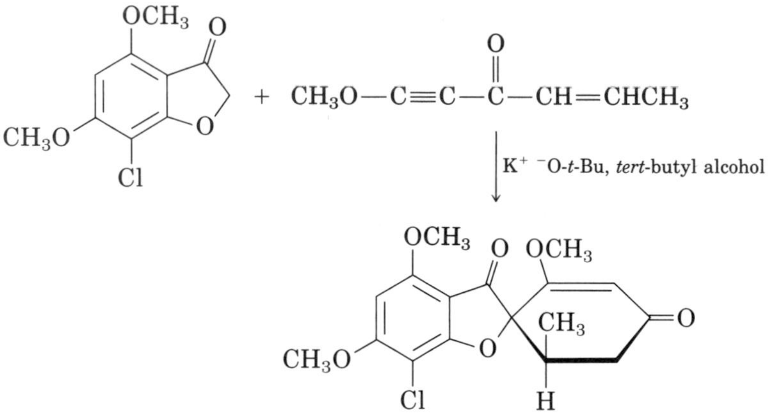

$$\text{(benzofuranone with } OCH_3, CH_3O, Cl) + CH_3O{-}C{\equiv}C{-}\overset{O}{\overset{\|}{C}}{-}CH{=}CHCH_3 \xrightarrow{K^+\ ^-O\text{-}t\text{-Bu},\ tert\text{-butyl alcohol}}$$

Griseofulvin (10%)

**23.44** The useful compound known as Hagemann's ester is prepared by treatment of a mixture of formaldehyde and ethyl acetoacetate with base, followed by acid-catalyzed decarboxylation:

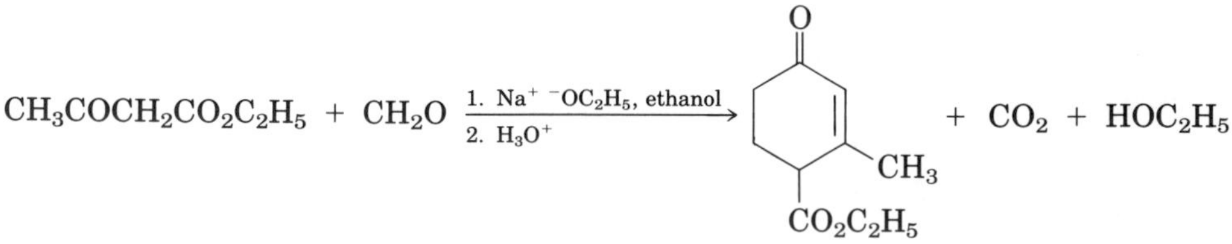

$$CH_3COCH_2CO_2C_2H_5 + CH_2O \xrightarrow[2.\ H_3O^+]{1.\ Na^+\ ^-OC_2H_5,\ \text{ethanol}} \text{(cyclohexenone with } CH_3,\ CO_2C_2H_5) + CO_2 + HOC_2H_5$$

Hagemann's ester

1. The first step in the reaction is an aldol-like condensation between ethyl acetoacetate and formaldehyde to yield an $\alpha,\beta$-unsaturated product. Write the reaction and show the structure of the product.
2. The second step in the reaction is a Michael reaction between a second equivalent of ethyl acetoacetate and the unsaturated product of step 1. Formulate the reaction and show the structure of the product.

**23.45** The third and fourth steps in the synthesis of Hagemann's ester from ethyl acetoacetate and formaldehyde (Problem 23.44) are an intramolecular aldol cyclization to yield a substituted cyclohexenone ring, and a decarboxylation reaction. Formulate both reactions and write the products of each step.

**23.46** When 2-methylcyclohexanone is converted into an enamine, only one product is formed, despite the fact that the starting ketone is unsymmetrical. Build molecular models of the two possible products and explain the fact that the sole product is the one with the double bond away from the methyl-substituted carbon.

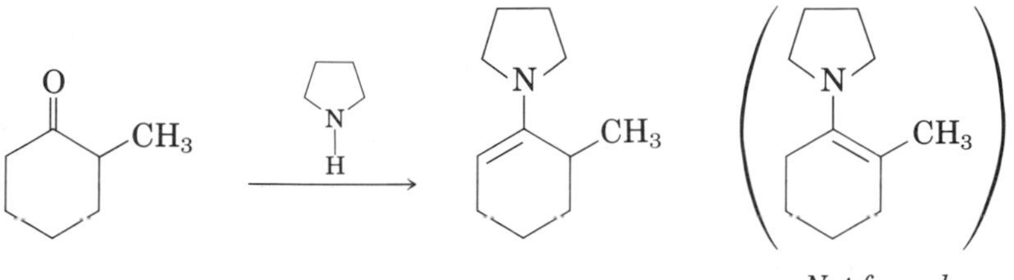

*Not formed*

**23.47** Intramolecular aldol cyclization of 2,5-heptanedione with aqueous NaOH yields a mixture of two products in the approximate ratio 9:1. Write their structures and show how each is formed.

**23.48** The major product formed by intramolecular aldol cyclization of 2,5-heptanedione (Problem 23.47) has two singlet absorptions in the $^1H$ NMR at 1.65 $\delta$ and 1.90 $\delta$, and has no absorptions in the range 3–10 $\delta$. What is the structure of this major product?

**23.49** Treatment of the minor product formed in the intramolecular aldol cyclization of 2,5-heptanedione (Problems 23.47 and 23.48) with aqueous NaOH converts it into the major product. Propose a mechanism to account for this base-catalyzed isomerization. [*Hint:* Remember that *all* steps in the aldol reaction are reversible.]

**23.50** The Stork enamine reaction and the intramolecular aldol reaction can be carried out in sequence to allow the synthesis of cyclohexenone rings. For example, reaction of the pyrrolidine enamine of cyclohexanone with 3-buten-2-one, followed by enamine hydrolysis and base treatment yields the product indicated. Show the mechanisms of the different steps.

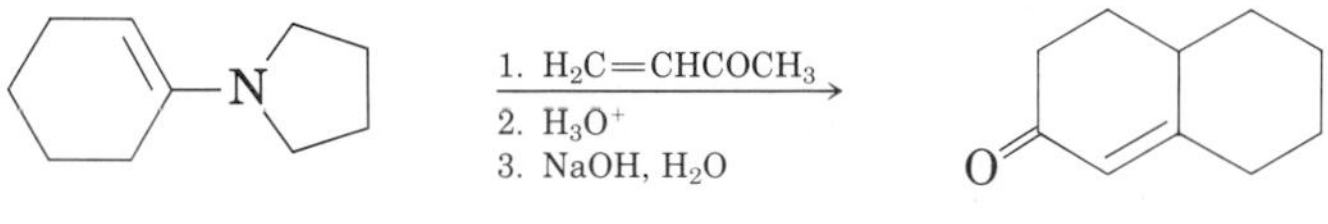

1. $H_2C{=}CHCOCH_3$
2. $H_3O^+$
3. NaOH, $H_2O$

**23.51** How could you prepare these cyclohexenones by combining a Stork enamine reaction with an intramolecular aldol condensation? [*Hint:* See Problem 23.46.]

(a)

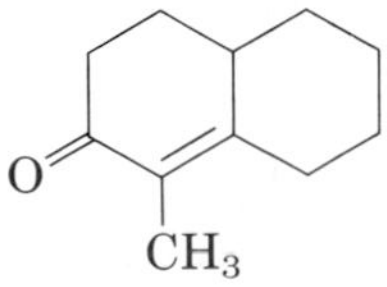

(b)

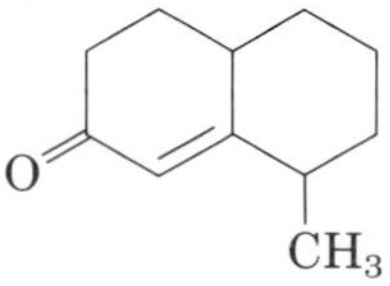

**23.52** Propose a mechanism to account for the following reaction:

$\xrightarrow[\text{Ethanol}]{\text{NaOH}}$ HO

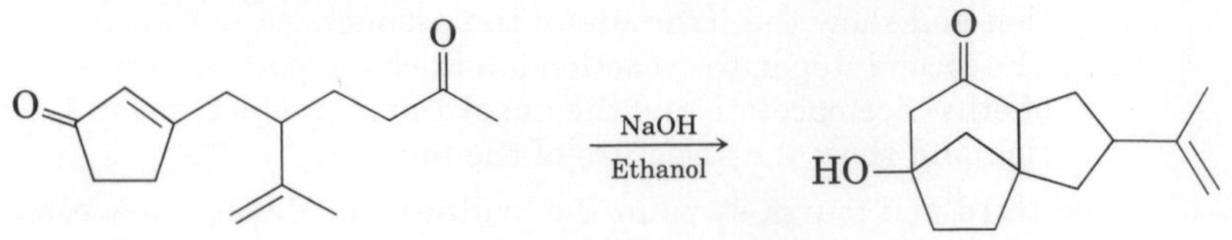

**23.53** Propose a mechanism to account for the following reaction:

$COOCH_3$

$-C{\equiv}C-COOCH_3$ $\xrightarrow{(CH_3)_2CuLi}$ $COOCH_3$, $CH_3$

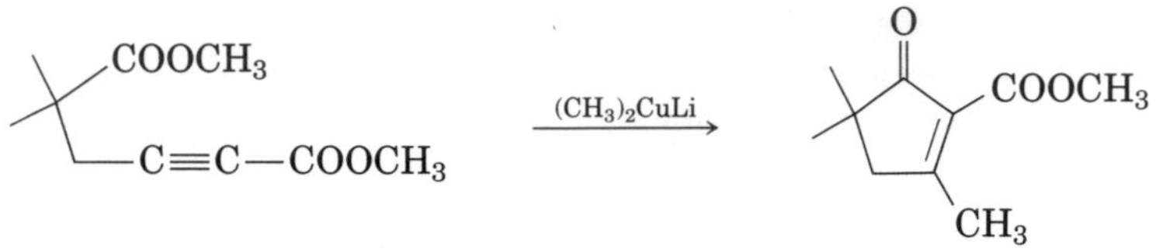

CHAPTER 24

# Carbohydrates

Carbohydrates are everywhere in nature. They occur in every living organism and are essential to life. The sugar and starch in food, and the cellulose in wood, paper, and cotton, are nearly pure carbohydrate. Some modified carbohydrates form part of the coating around living cells; others are found in the DNA that carries genetic information; and still others, such as gentamicin, are invaluable as medicines.

The word **carbohydrate** derives historically from the fact that glucose, the first simple carbohydrate to be purified, has the molecular formula $C_6H_{12}O_6$ and was originally thought to be a "hydrate of carbon," $C_6(H_2O)_6$. This view was soon abandoned, but the name persisted. Today, the term *carbohydrate* is used to refer loosely to the broad class of polyhydroxylated aldehydes and ketones commonly called *sugars*.

CHO
|
HCOH
|
HOCH
|
HCOH
|
HCOH
|
$CH_2OH$

Glucose (also called dextrose),
a pentahydroxyhexanal

Carbohydrates synthesized by green plants during photosynthesis are the chemical intermediaries by which solar energy is stored and used to support life. When broken down by metabolic processes in the cell, they provide the major source of energy required by living organisms.

Photosynthesis is a complex process in which carbon dioxide is converted into glucose. Many molecules of glucose are then chemically linked together for storage by the plant in the form of either cellulose or starch. It has been estimated that more than 50% of the dry weight of the earth's biomass—all plants and animals—consists of glucose polymers.

$$6\ CO_2 + 6\ H_2O \xrightarrow{\text{Sunlight}} 6\ O_2 + \underset{\text{Glucose}}{C_6H_{12}O_6} \longrightarrow \text{Cellulose, Starch}$$

When eaten, glucose can be either metabolized in the body to provide energy or stored by the body in the form of glycogen for use at a later time. Since humans and most other mammals lack the enzymes needed for digestion of cellulose, they require starch as their dietary source of carbohydrates. Grazing animals such as cows, however, contain in their rumen microorganisms that are able to digest cellulose. The energy stored in cellulose is thus moved up the biological food chain when these animals are used for food.

## 24.1 Classification of Carbohydrates

Carbohydrates are generally classed into two groups, simple and complex. **Simple sugars**, or **monosaccharides**, are carbohydrates like glucose and fructose that can't be hydrolyzed into smaller molecules. **Complex carbohydrates** are made of two or more simple sugars linked together. For example, sucrose (table sugar) is a **disaccharide** (two sugars) made up of one glucose molecule linked to one fructose molecule; cellulose is a **polysaccharide** (many sugars) made up of several thousand glucose molecules linked together. Hydrolysis of these polysaccharides breaks them down into their constituent monosaccharide units.

$$1\ \text{Sucrose} \xrightarrow{H_3O^+} 1\ \text{Glucose} + 1\ \text{Fructose}$$

$$\text{Cellulose} \xrightarrow{H_3O^+} {\sim}3000\ \text{Glucose}$$

Monosaccharides can be further classified as either **aldoses** or **ketoses**. The *-ose* suffix is used to designate a carbohydrate, and the *aldo-* and *keto-* prefixes designate the nature of the carbonyl group (aldehyde or ketone). The number of carbon atoms in the monosaccharide is given by using *tri-*, *tetr-*, *pent-*, *hex-*, and so forth as the base name. When prefix, base, and suffix are combined, a monosaccharide is fully classified. For example, glucose is an aldohexose (a six-carbon aldehydo sugar); fructose is a ketohexose (a six-carbon keto sugar); and ribose is an aldopentose (a five-carbon aldehydo sugar). Most of the commonly occurring sugars are either aldopentoses or aldohexoses.

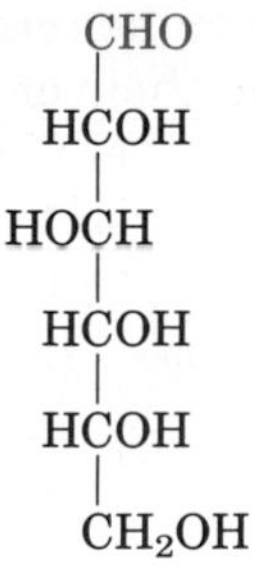

CHO
HCOH
HOCH
HCOH
HCOH
$CH_2OH$

Glucose
(an aldohexose)

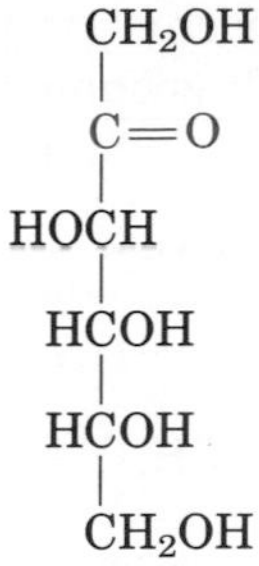

$CH_2OH$
C=O
HOCH
HCOH
HCOH
$CH_2OH$

Fructose
(a ketohexose)

CHO
HCOH
HCOH
HCOH
$CH_2OH$

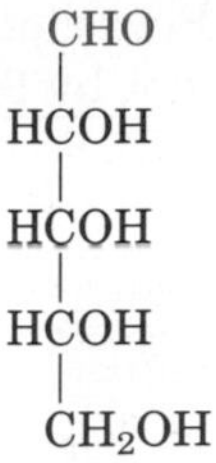

Ribose
(an aldopentose)

PROBLEM

**24.1** Classify each of the following monosaccharides:

(a) 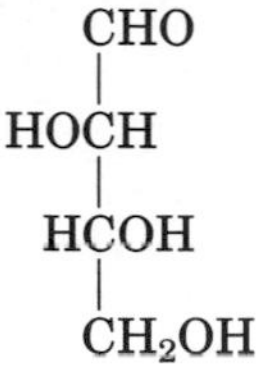 CHO
HOCH
HCOH
$CH_2OH$

Threose

(b) 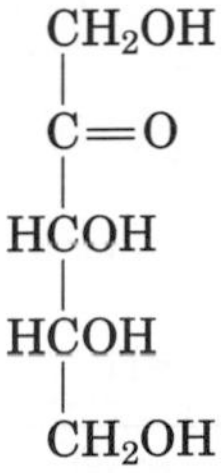 $CH_2OH$
C=O
HCOH
HCOH
$CH_2OH$

Ribulose

(c) 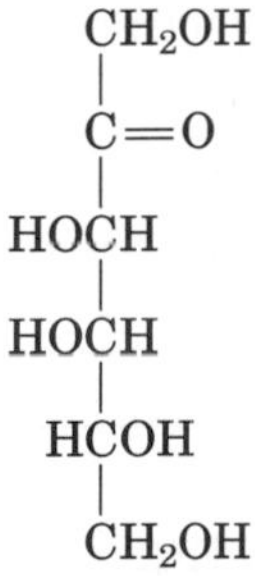 $CH_2OH$
C=O
HOCH
HOCH
HCOH
$CH_2OH$

Tagatose

(d) 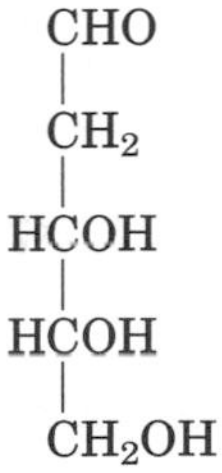 CHO
$CH_2$
HCOH
HCOH
$CH_2OH$

2-Deoxyribose

## 24.2 Fischer Projections for Depicting Carbohydrates

Since all carbohydrates have chiral carbon atoms, it was recognized long ago that a standard method of representation was needed to designate carbohydrate stereochemistry. The method most commonly used employs **Fischer projections** (Section 9.13) for depicting a chiral center on a flat page.

Recall that a tetrahedral carbon atom is represented in a Fischer projection by two crossed lines. By convention, the horizontal lines represent bonds coming out of the page, and the vertical lines represent bonds going into the page. For example, (*R*)-glyceraldehyde, the simplest monosaccharide, can be drawn as follows:

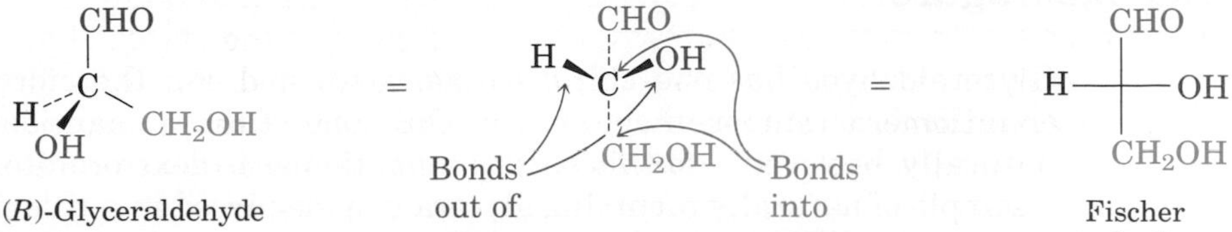

CHO, H, OH, C, $CH_2OH$

(*R*)-Glyceraldehyde

=

CHO, H, OH, C, $CH_2OH$

Bonds out of page

Bonds into page

=

CHO
H—+—OH
$CH_2OH$

Fischer projection

Recall also that Fischer projections can be rotated on the page by 180° (but not by 90° or 270°) without changing their meaning:

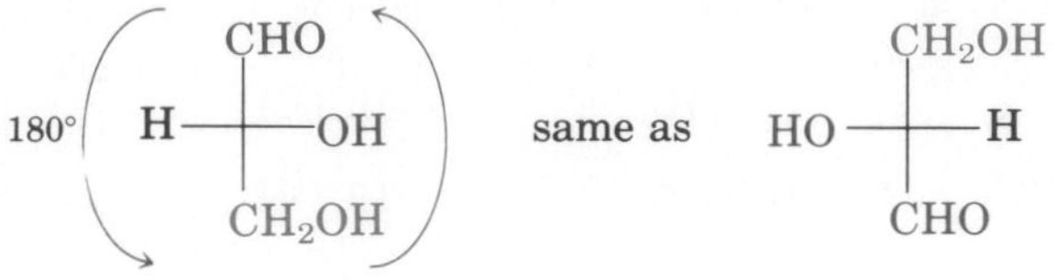

180° ( CHO / H—+—OH / $CH_2OH$ ) same as $CH_2OH$ / HO—+—H / CHO

*(R)*-Glyceraldehyde

Fischer projections can be used to depict more than one chiral center in a molecule simply by stacking the chiral centers one on top of the other. By convention, the carbonyl carbon is always placed either at the top or near the top when showing the Fischer projection of a carbohydrate. For example, glucose has four chiral centers stacked on top of each other in Fischer projection:

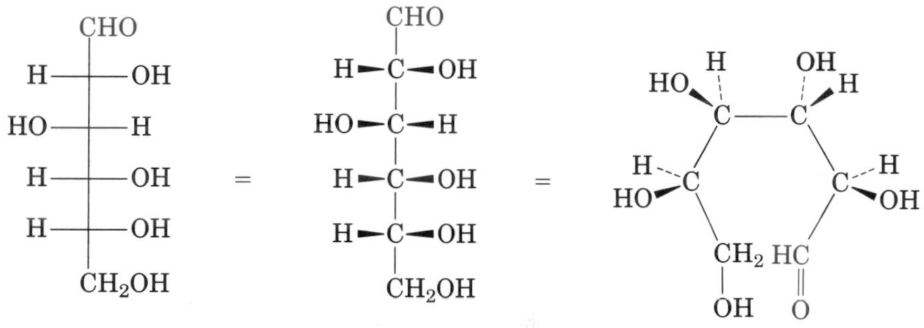

Glucose
(carbonyl group at top)

**PROBLEM**

**24.2** Which of these Fischer projections of glyceraldehyde represent the same enantiomer?

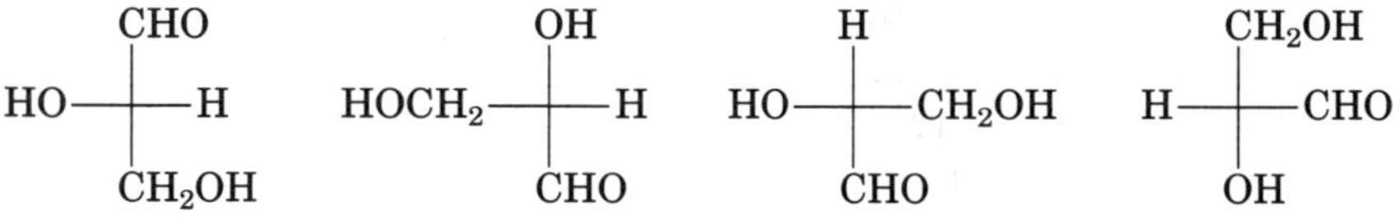

| A | B | C | D |
|---|---|---|---|
| CHO / HO—+—H / $CH_2OH$ | OH / $HOCH_2$—+—H / CHO | H / HO—+—$CH_2OH$ / CHO | $CH_2OH$ / H—+—CHO / OH |

## 24.3 D,L Sugars

Glyceraldehyde has one chiral carbon atom and can therefore have two enantiomeric (mirror-image) forms. Only one of these enantiomers occurs naturally, however, and this natural enantiomer is dextrorotatory. That is, a sample of naturally occurring glyceraldehyde placed in a polarimeter will rotate plane-polarized light in a clockwise direction, which we denote (+).

Since (+)-glyceraldehyde is known to have the *R* configuration at C2, we can represent it in Fischer projection as shown in Figure 24.1. For historical reasons dating back long before the adoption of the *R*,*S* system, (*R*)-(+)-glyceraldehyde is also referred to as D-*glyceraldehyde* (D from dextrorotatory). The nonnatural enantiomer, (*S*)-(−)-glyceraldehyde, is similarly known as L-*glyceraldehyde* (L from levorotatory).

For reasons having to do with the way monosaccharides are biosynthesized in nature, it turns out that glucose, fructose, and almost all other naturally occurring monosaccharides have the same stereochemical configuration as D-glyceraldehyde at the chiral carbon atom farthest from the carbonyl group. In Fischer projections, therefore, most naturally occurring sugars have the hydroxyl group at the lowest chiral carbon atom on the right (Figure 24.1). Such compounds are referred to as **D sugars**.

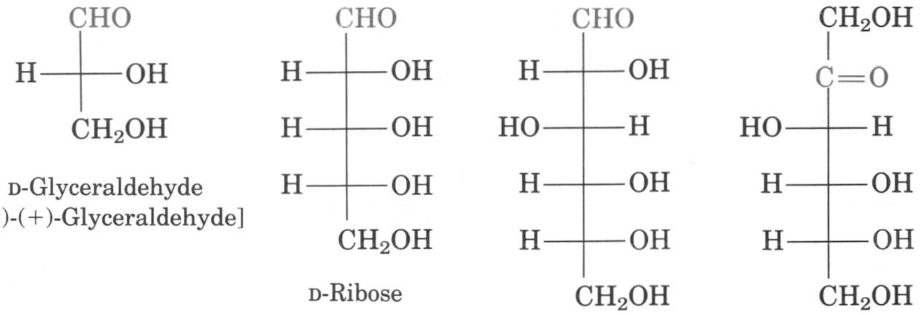

CHO
H—|—OH
$CH_2OH$

D-Glyceraldehyde
[(*R*)-(+)-Glyceraldehyde]

CHO
H—|—OH
H—|—OH
H—|—OH
$CH_2OH$

D-Ribose

CHO
H—|—OH
HO—|—H
H—|—OH
H—|—OH
$CH_2OH$

D-Glucose

$CH_2OH$
C=O
HO—|—H
H—|—OH
H—|—OH
$CH_2OH$

D-Fructose

**Figure 24.1** Some naturally occurring D sugars: The hydroxyl group at the chiral center farthest from the carbonyl group is on the right when the molecule is drawn in Fischer projection.

In contrast to D sugars, all **L sugars** have the hydroxyl group at the lowest chiral carbon atom on the *left* in Fischer projection. Thus, L sugars are mirror images (enantiomers) of D sugars.

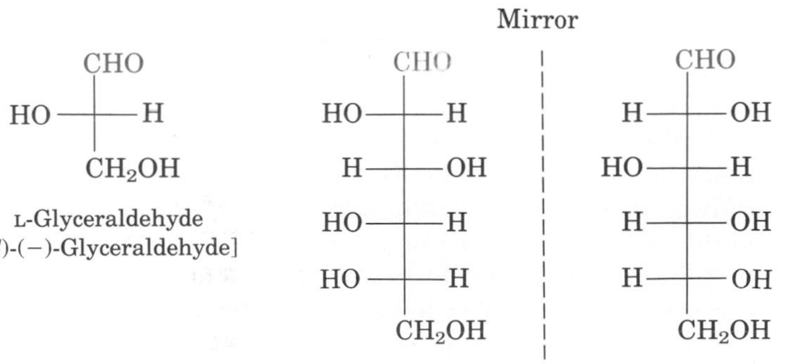

CHO
HO—|—H
$CH_2OH$

L-Glyceraldehyde
[(*S*)-(−)-Glyceraldehyde]

Mirror

CHO
HO—|—H
H—|—OH
HO—|—H
HO—|—H
$CH_2OH$

L-Glucose
(not naturally occurring)

CHO
H—|—OH
HO—|—H
H—|—OH
H—|—OH
$CH_2OH$

D-Glucose

Although widely used by carbohydrate chemists, the D and L notations have no relation to the direction in which a given sugar rotates plane-polarized light. A D sugar may be either dextrorotatory or levorotatory. The prefix D indicates only that the stereochemistry of the lowest chiral carbon atom is to the right in Fischer projection when the molecule is drawn in the standard way with the carbonyl group at or near the top.

The D,L system of carbohydrate nomenclature is of limited use, since it describes the configuration at only one chiral center and says nothing about other chiral centers that may be present. The advantage of the system, though, is that it allows a person to relate one sugar to another rapidly and visually.

PROBLEM

**24.3** Assign $R$ or $S$ configuration to each chiral carbon atom in the following sugars, and tell if each is a D sugar or an L sugar.

(a)
```
      CHO
HO ——+—— H
HO ——+—— H
      CH2OH
```

(b)
```
      CHO
 H ——+—— OH
HO ——+—— H
 H ——+—— OH
      CH2OH
```

(c)
```
      CH2OH
      C=O
HO ——+—— H
 H ——+—— OH
      CH2OH
```

PROBLEM

**24.4** (+)-Arabinose, an aldopentose that is widely distributed in plants, can be named systematically as (2$R$,3$S$,4$S$)-5-tetrahydroxypentanal. Draw a Fischer projection of (+)-arabinose and identify it as a D or L sugar.

## 24.4 Configurations of the Aldoses

Aldotetroses are four-carbon sugars that have two chiral centers. There are $2^2 = 4$ possible stereoisomers, or two D,L pairs of enantiomers. These enantiomeric pairs are called *erythrose* and *threose*.

Aldopentoses have three chiral centers, leading to a total of $2^3 = 8$ stereoisomers, or four D,L pairs of enantiomers. These four pairs are called *ribose, arabinose, xylose,* and *lyxose*. All except lyxose occur widely in nature. D-Ribose is an important constituent of RNA (ribonucleic acid); L-arabinose is found in many plants; and D-xylose is found in wood.

Aldohexoses have four chiral centers, for a total of $2^4 = 16$ stereoisomers, or eight D,L pairs of enantiomers. The names of the eight pairs are *allose, altrose, glucose, mannose, gulose, idose, galactose,* and *talose*. Of the eight, only D-glucose (from starch and cellulose) and D-galactose (from gums and fruit pectins) are found widely in nature. D-Mannose and D-talose also occur naturally, but in lesser abundance.

Fischer projections of the four-, five-, and six-carbon aldoses can be constructed as shown in Figure 24.2 for the D series. Starting from D-glyceraldehyde, we can construct the two D-aldotetroses by inserting a new chiral

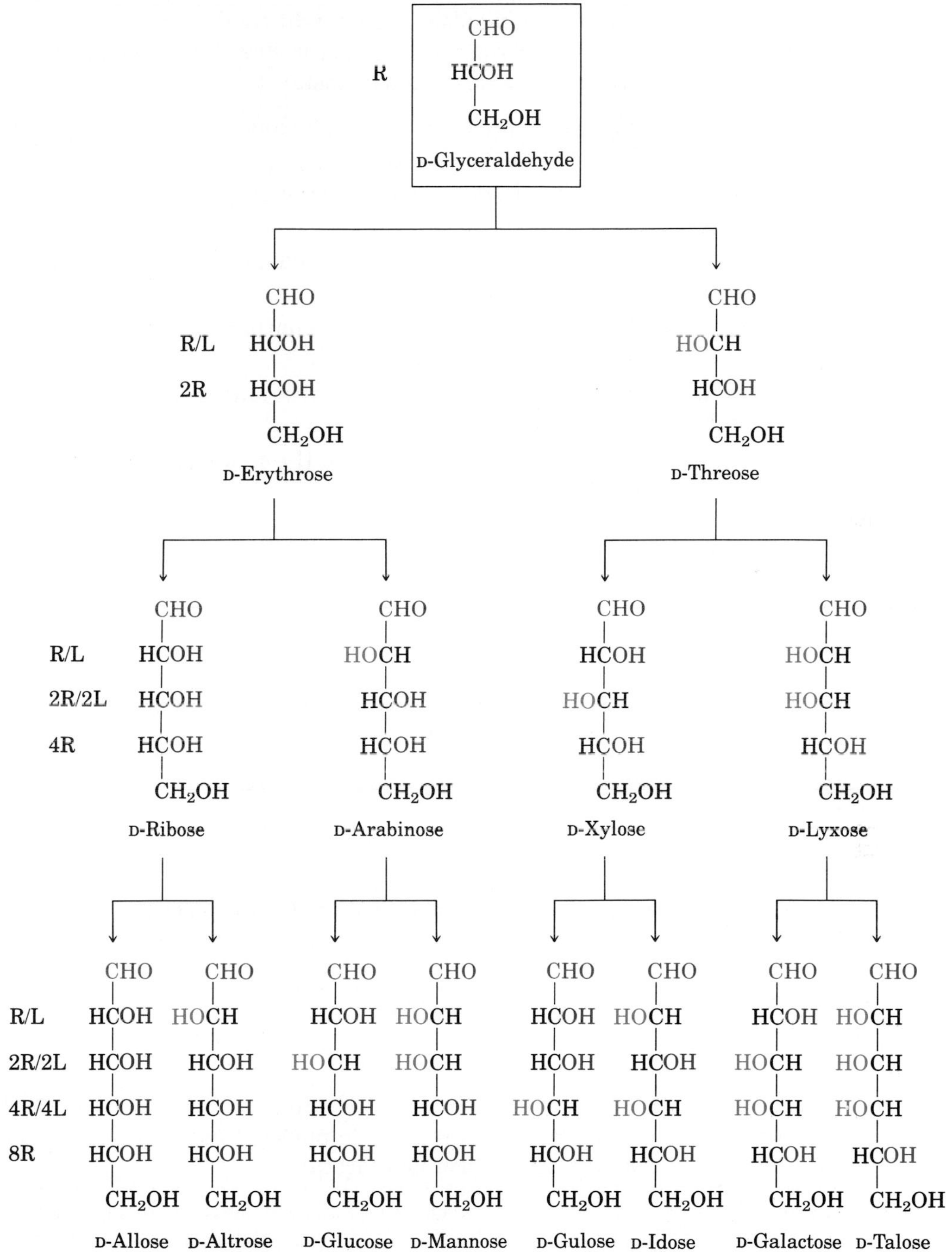


**Figure 24.2** Configurations of D aldoses: The structures are arranged in order from left to right so that the hydroxyl groups on C2 alternate right/left (R/L) in going across a series. Similarly, the hydroxyl groups at C3 alternate two right/two left (2R/2L); the hydroxyl groups at C4 alternate 4R/4L; and the hydroxyl groups at C5 are to the right in all eight (8R).

carbon atom just below the aldehyde carbon. Each of the two D-aldotetroses can then lead to two D-aldopentoses (four total), and each of the four D-aldopentoses can lead to two D-aldohexoses (eight total).

Louis Fieser[1] of Harvard suggested this procedure for remembering the names and structures of the eight D-aldohexoses:

1. Set up eight Fischer projections with the aldehyde group on top and the $CH_2OH$ group at the bottom.
2. Indicate stereochemistry at C5 by placing all eight hydroxyl groups to the right (D series).
3. Indicate stereochemistry at C4 by alternating four hydroxyl groups to the right and four to the left.
4. Indicate stereochemistry at C3 by alternating two hydroxyl groups to the right, two to the left, and so on.
5. Indicate stereochemistry at C2 by alternating hydroxyl groups right, left, right, left, and so on.
6. Name the eight isomers according to the mnemonic "All altruists gladly make gum in gallon tanks."

The four D-aldopentose structures can be generated in a similar way and can be named by the mnemonic "Ribs are extra lean."

**PROBLEM**

**24.5** Only the D sugars are shown in Figure 24.2. Draw Fischer projections for these L sugars:
(a) L-Xylose (b) L-Galactose (c) L-Glucose

**PROBLEM**

**24.6** How many aldoheptoses are there? How many are D sugars and how many are L sugars?

**PROBLEM**

**24.7** Draw Fischer projections for the two D-aldoheptoses whose stereochemistry at C3, C4, C5, and C6 corresponds to that of glucose at C2, C3, C4, and C5.

## 24.5 Cyclic Structures of Monosaccharides: Hemiacetal Formation

We said during our discussion of carbonyl-group chemistry (Section 19.14) that alcohols undergo a rapid and reversible nucleophilic addition reaction with ketones and aldehydes to form hemiacetals:

$$\mathrm{R{-}O{-}H} + \mathrm{R'{-}\overset{\overset{\displaystyle O}{\|}}{C}{-}H} \rightleftharpoons \mathrm{RO{-}\overset{\overset{\displaystyle OH}{|}}{C}(R')(H)}$$

A hemiacetal

[1]Louis F. Fieser (1899–1977); b. Columbus, Ohio; Ph.D. Harvard (Conant); professor, Bryn Mawr College, Harvard University.

If both the hydroxyl and the carbonyl group are in the same molecule, an *intramolecular* nucleophilic addition can take place, leading to the formation of a *cyclic* hemiacetal. Five- and six-membered cyclic hemiacetals are particularly stable, and many carbohydrates therefore exist in an equilibrium between open-chain and cyclic forms. For example, glucose exists in aqueous solution primarily as the six-membered **pyranose** ring formed by intramolecular nucleophilic addition of the hydroxyl group at C5 to the C1 aldehyde group. Fructose, on the other hand, exists to the extent of about 20% as the five-membered **furanose** ring formed by addition of the hydroxyl group at C5 to the C2 ketone. The words *pyranose* (six-membered ring) and *furanose* (five-membered ring) are derived from the names of the simple oxygen-containing cyclic compounds pyran and furan. The cyclic forms of glucose and fructose are shown in Figure 24.3.

Carbohydrate chemists often represent pyranose and furanose forms by using **Haworth[2] projections**, as shown in Figure 24.3, rather than Fischer

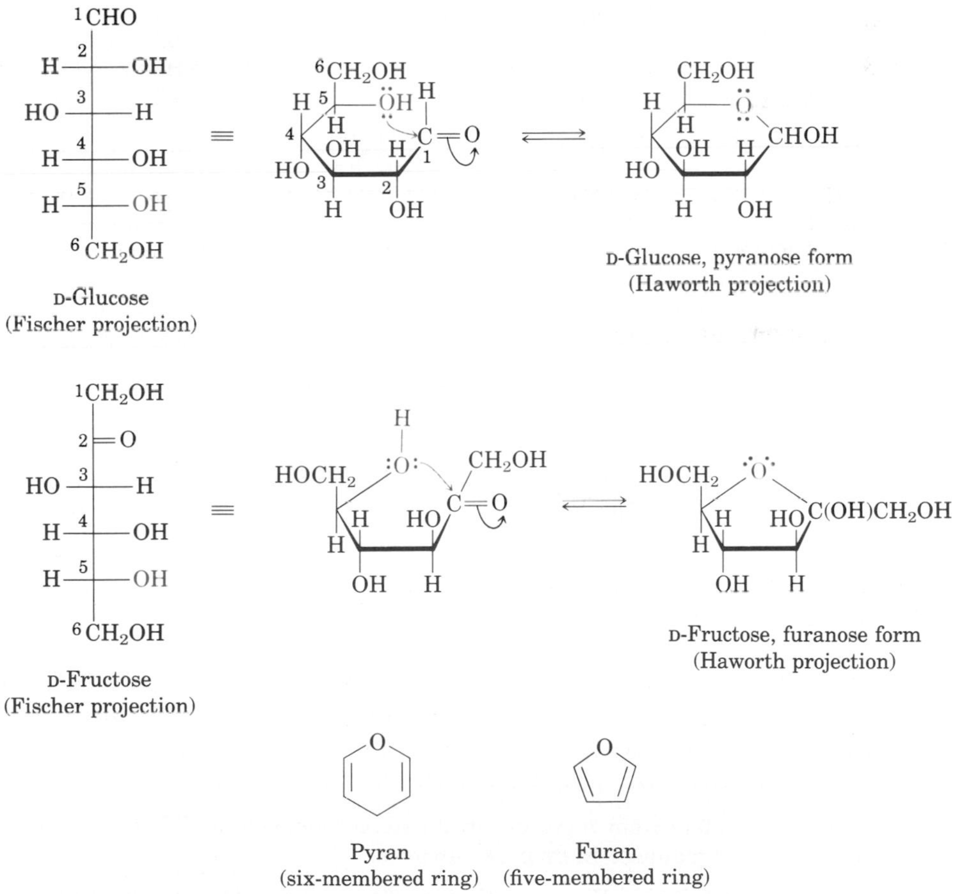

D-Glucose (Fischer projection)

D-Glucose, pyranose form (Haworth projection)

D-Fructose (Fischer projection)

D-Fructose, furanose form (Haworth projection)

Pyran (six-membered ring)

Furan (five-membered ring)

**Figure 24.3** Glucose and fructose in their cyclic furanose and pyranose forms

[2]Sir Walter Norman Haworth (1883–1950); b. Chorley, Lancashire; Ph.D. Göttingen; D.Sc. Manchester; professor, University of Birmingham; Nobel prize (1937).

projections. In a Haworth projection, the hemiacetal ring is drawn as if it were flat and is viewed edge-on with the oxygen atom at the upper right. Though convenient, this view is not really accurate since pyranose rings are actually chair shaped like cyclohexane (Section 4.9), rather than flat. Nevertheless, Haworth projections are widely used because they allow us to see at a glance the cis–trans relationships among hydroxyl groups on the ring.

When converting from one kind of projection to the other, remember that a hydroxyl on the *right* in a Fischer projection is *down* in a Haworth projection. Conversely, a hydroxyl on the *left* in a Fischer projection is *up* in a Haworth projection. For D sugars, the terminal —$CH_2OH$ group is always up in Haworth projections, whereas for L sugars the —$CH_2OH$ group is down. Figure 24.4 illustrates the conversion for D-glucose.

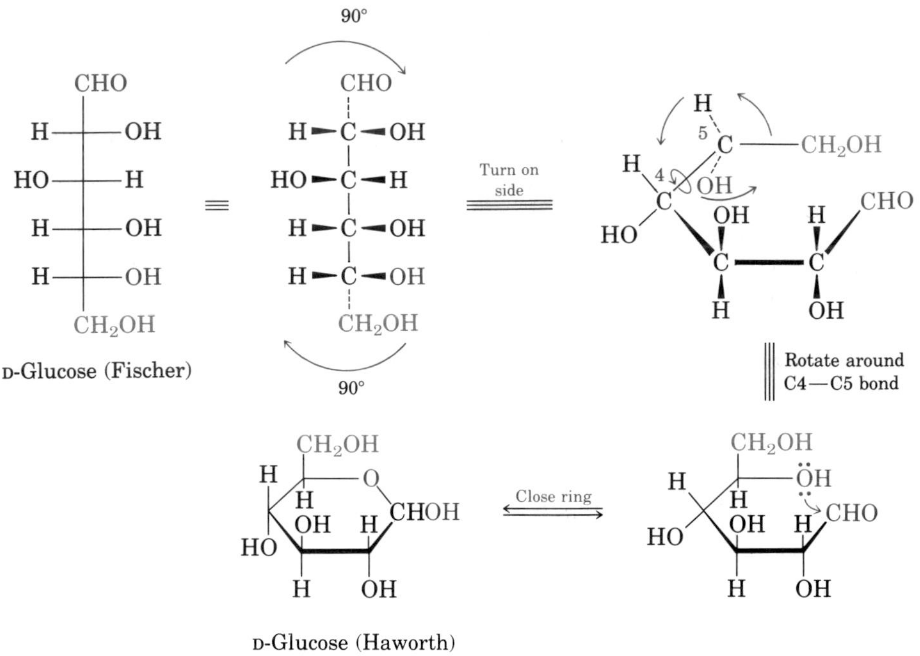

**Figure 24.4** Interconversion of Fischer and Haworth projections of D-glucose

**PRACTICE PROBLEM**

D-Mannose differs from D-glucose in its stereochemistry at C2. Draw a Haworth projection of D-mannose in its pyranose form.

*Solution* First draw a Fischer projection of D-mannose. Then lay it on its side, and curl it around so that the aldehyde group (C1) is toward the front and the $CH_2OH$ group (C6) is toward the rear. Now connect the hydroxyl at C5 to the C1 carbonyl group to form a pyranose ring.

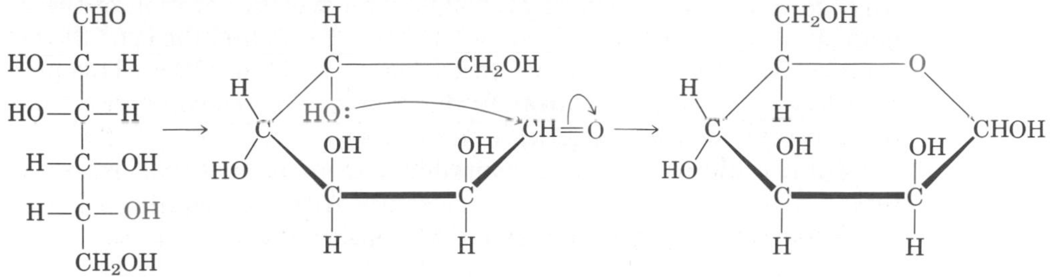

**PROBLEM**

**24.8** D-Galactose differs from D-glucose in its stereochemistry at C4. Draw a Haworth projection of D-galactose in its pyranose form.

**PROBLEM**

**24.9** Draw Haworth projections of L-glucose in its pyranose form and D-ribose in its furanose form.

## 24.6 Monosaccharide Anomers: Mutarotation

When an open-chain monosaccharide cyclizes to a furanose or pyranose form, a new chiral center is formed at what used to be the carbonyl carbon. The two diastereomers produced are called **anomers**, and the hemiacetal carbon atom is referred to as the **anomeric carbon**. For example, glucose cyclizes reversibly in aqueous solution to a 36:64 mixture of two anomers. The minor anomer with the C1—OH group trans to the $—CH_2OH$ substituent at C5 (down in a Haworth projection) is called the **alpha anomer**; its complete name is α-D-glucopyranose. The major anomer with the C1—OH group cis to the $—CH_2OH$ substituent at C5 (up in a Haworth projection) is called the **beta anomer**; its complete name is β-D-glucopyranose.

D-Glucose

Trans · Cis · Anomeric carbon

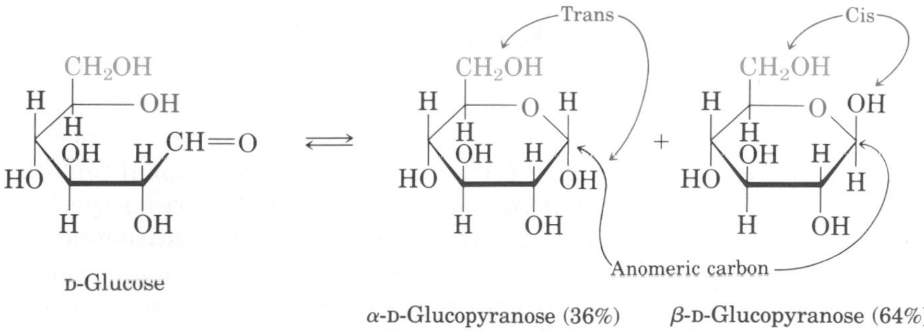

α-D-Glucopyranose (36%)
Alpha anomer: OH and $CH_2OH$ are trans

β-D-Glucopyranose (64%)
Beta anomer: OH and $CH_2OH$ are cis

Both anomers of D-glucopyranose can be crystallized and purified. Pure α-D-glucopyranose has a melting point of 146°C and a specific rotation, $[\alpha]_D$, of +112.2°; pure β-D-glucopyranose has a melting point of 148–155°C and a specific rotation of +18.7°. When a sample of either pure anomer is dissolved in water, however, the optical rotations slowly change and ultimately converge to a constant value of +52.6°. The specific rotation of the alpha-anomer solution decreases from +112.2° to +52.6°, and the specific rotation of the beta-anomer solution increases from +18.7° to +52.6°. This phenomenon, known as **mutarotation**, is due to the slow conversion of the *pure* anomers into the 36:64 equilibrium *mixture*.

Mutarotation occurs by a reversible ring opening of each anomer to the open-chain aldehyde, followed by reclosure. Although equilibration is slow at neutral pH, it is catalyzed by either acid or base.

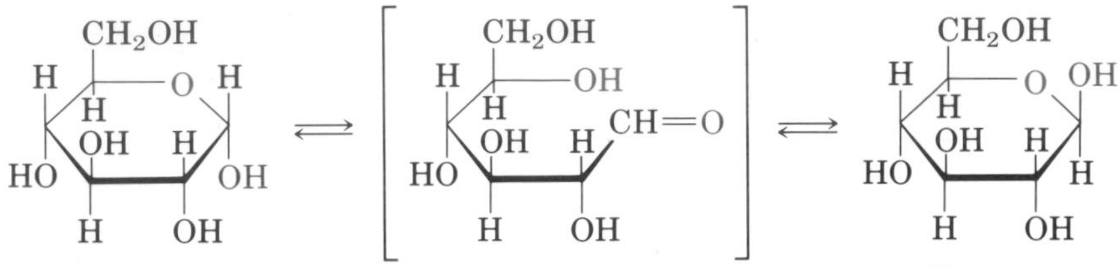

α-D-Glucose (36%)
$[\alpha]_D = +112.2°$

β-D-Glucose (64%)
$[\alpha]_D = +18.7°$

**PROBLEM**

**24.10** Knowing that the specific rotation of pure α-D-glucopyranose is +112.2° and that the specific rotation of pure β-D-glucopyranose is +18.7°, show how the equilibrium percentages of alpha and beta anomers can be calculated from the equilibrium specific rotation of +52.6°.

**PROBLEM**

**24.11** Many other sugars besides glucose exhibit mutarotation. For example, α-D-galactopyranose has $[\alpha]_D = +150.7°$, and β-D-galactopyranose has $[\alpha]_D = +52.8°$. If either anomer is dissolved in water and allowed to reach equilibrium, the specific rotation of the solution is +80.2°. What are the percentages of each anomer at equilibrium? Draw the pyranose forms of both anomers using Haworth projections.

## 24.7 Conformations of Monosaccharides

Although Haworth projections are relatively easy to draw and readily show cis–trans relationships between substituents on furanose and pyranose rings, they don't give an accurate three-dimensional picture of molecular conformation. Pyranose rings, like cyclohexane rings (Section 4.9), have a chair-like geometry with axial and equatorial substituents. Any substituent that's up in a Haworth projection is also up in a chair conformational formula, and any substituent that's down in a Haworth projection is down in the chair conformation. Haworth projections can be converted into chair representations by following three steps:

1. Draw the Haworth projection with the ring oxygen atom at the upper right.
2. Raise the leftmost carbon atom (C4) *above* the ring plane.
3. Lower the anomeric carbon atom (C1) *below* the ring plane. Figure 24.5 shows how this is done for α-D-glucopyranose and β-D-glucopyranose. Make molecular models to see the process more clearly.

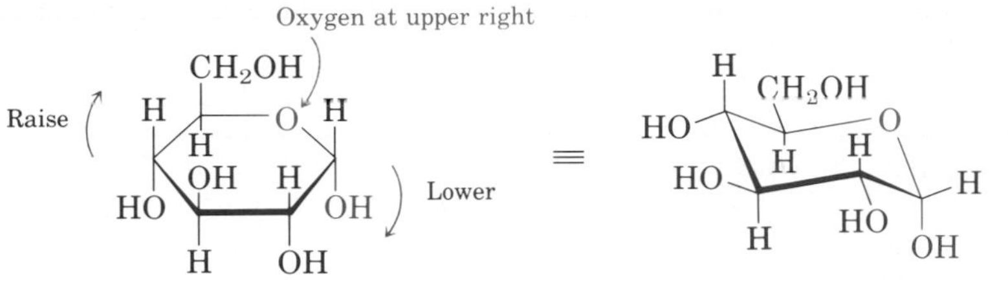

Oxygen at upper right

Raise Lower

α-D-Glucopyranose

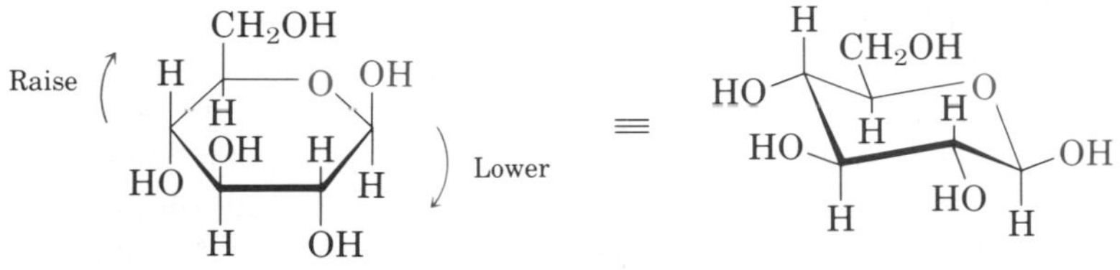

Raise Lower

β-D-Glucopyranose

Recall:

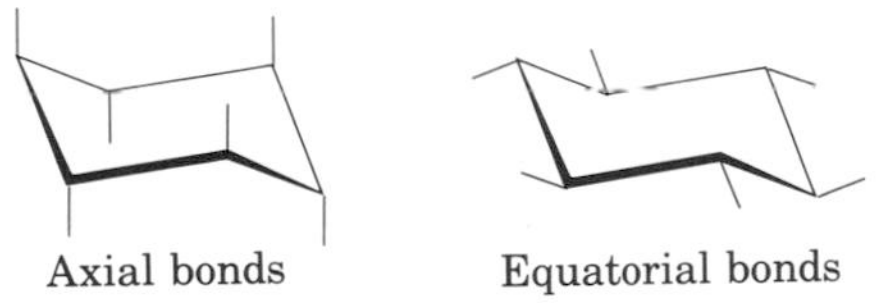

Axial bonds Equatorial bonds

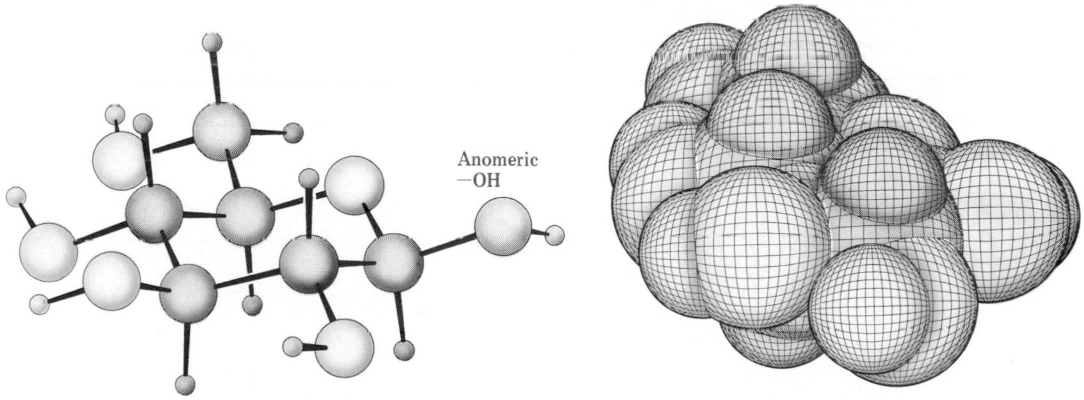

Anomeric —OH

**Figure 24.5** Chair representations of α-D-glucopyranose and β-D-glucopyranose: Computer-generated representations of β-D-glucopyranose are shown at the bottom of the figure.

Note that in β-D-glucopyranose, all the substituents on the ring are equatorial. Thus, β-D-glucopyranose is the least sterically crowded and most stable of the eight D-aldohexoses.

PROBLEM

**24.12** Draw chair conformations of β-D-galactopyranose and β-D-mannopyranose. Label the ring substituents as either axial or equatorial. Which would you expect to be more stable—galactose or mannose?

PROBLEM

**24.13** Draw a chair conformation of β-L-glucopyranose and label the substituents as either axial or equatorial.

## 24.8 Reactions of Monosaccharides

### ESTER AND ETHER FORMATION

Monosaccharides behave as simple alcohols in much of their chemistry. For example, carbohydrate hydroxyl groups can be converted into esters and ethers.

Esterification is normally carried out by treating the carbohydrate with an acid chloride or acid anhydride in the presence of a base. *All* the hydroxyl groups react, including the anomeric one. For example, β-D-glucopyranose is converted into its pentaacetate by treatment with acetic anhydride in pyridine solution.

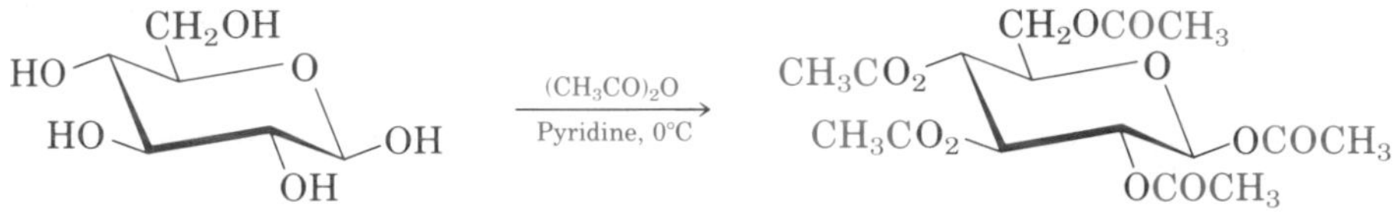


β-D-Glucopyranose — Penta-*O*-acetyl-β-D-glucopyranose (91%)

Carbohydrates can be converted into ethers by treatment with an alkyl halide in the presence of base (the Williamson ether synthesis, Section 18.4). Normal Williamson conditions using a strong base tend to degrade the sensitive sugar molecules, but Purdie[3] showed in 1903 that silver oxide works particularly well and that high yields of ethers are obtained. For example, α-D-glucopyranose is converted into its pentamethyl ether in 85% yield on reaction with iodomethane and silver oxide.

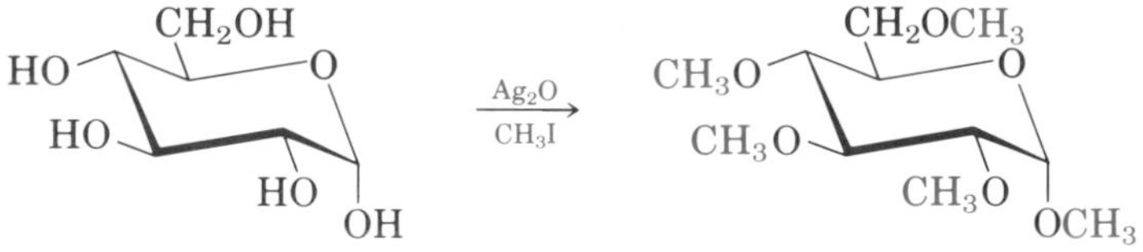


α-D-Glucopyranose — α-D-Glucopyranose pentamethyl ether (85%)

Ester and ether derivatives of carbohydrates are often prepared because they are easier to work with than the free sugars. Because of their many hydroxyl groups, monosaccharides are usually soluble in water but insoluble in organic solvents such as ether. They are also difficult to purify and have a tendency to form syrups rather than crystals when water is removed. Ester and ether derivatives, however, behave like most other organic compounds in that they tend to be soluble in organic solvents and to be readily purified and crystallized.

---

[3]Thomas Purdie (1843–1916); b. Biggar, Scotland; Ph.D. Würzburg; professor, St. Andrews University.

## GLYCOSIDE FORMATION

We saw in Section 19.14 that treatment of a hemiacetal with an alcohol and an acid catalyst yields an acetal:

$$\text{C(OH)(OR)} + \text{ROH} \xrightleftharpoons{\text{HCl}} \text{C(OR)(OR)} + H_2O$$

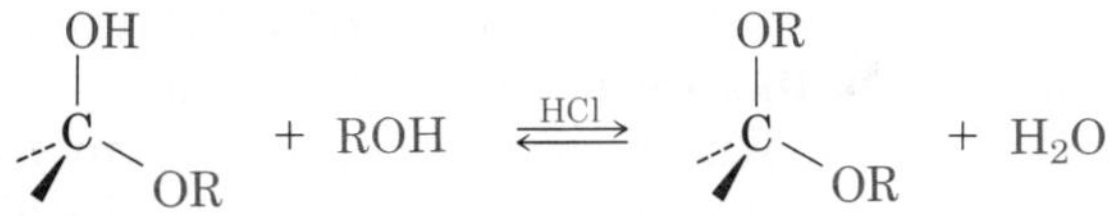

In the same way, treatment of a monosaccharide hemiacetal with an alcohol and an acid catalyst yields an acetal in which the anomeric hydroxyl has been replaced by an alkoxy group. For example, reaction of glucose with methanol gives methyl β-D-glucopyranoside:

$$\beta\text{-D-Glucopyranose} \xrightleftharpoons[H_3O^+]{CH_3OH,\ HCl} \text{Methyl } \beta\text{-D-Glucopyranoside} + H_2O$$

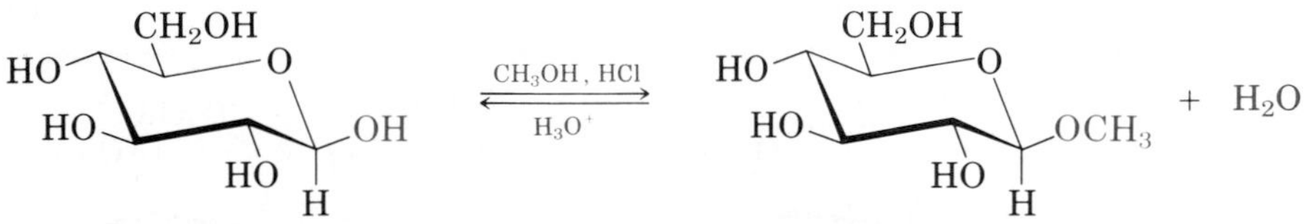

β-D-Glucopyranose (a hemiacetal)

Methyl β-D-Glucopyranoside (an acetal)

Carbohydrate acetals are called **glycosides**. They are named by citing the alkyl group and adding the *-oside* suffix to the name of the specific sugar. Note that glycosides, like all acetals, are stable to water. They are not in equilibrium with an open-chain form and they do not show mutarotation. They can, however, be converted back to the original monosaccharide by hydrolysis with aqueous acid.

Glycosides are widespread in nature, and a great many biologically important molecules contain glycosidic linkages. For example, digitoxin, the active component of the digitalis preparations used for treatment of heart disease, is a glycoside consisting of a complex steroid alcohol linked to a trisaccharide (Figure 24.6). Note also that the three sugars are linked to each other by glycosidic bonds.

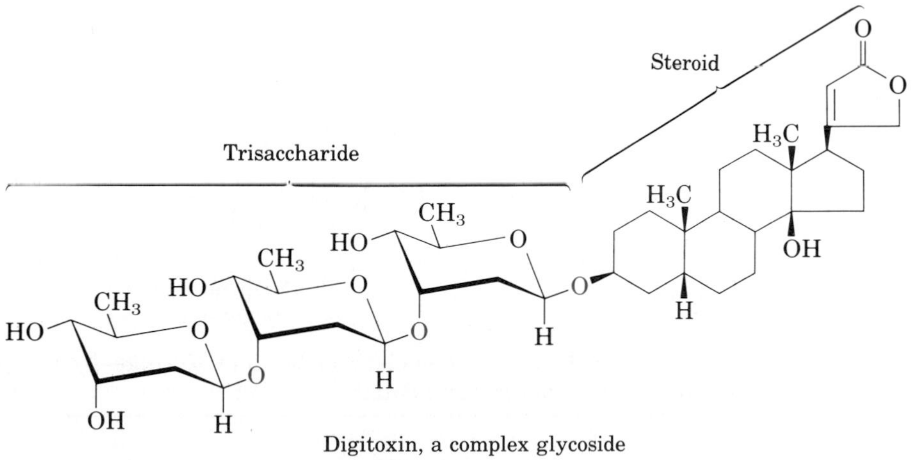

Trisaccharide

Steroid

Digitoxin, a complex glycoside

**Figure 24.6** The structure of digitoxin, a complex glycoside

The laboratory synthesis of glycosides is often difficult, and a successful reaction scheme is strongly dependent on the structures of both alcohol and saccharide components. One method (the **Koenigs–Knorr[4] reaction**) that is particularly suitable for preparation of glucose $\beta$-glycosides involves treatment of glucose pentaacetate with HBr, followed by addition of the appropriate alcohol in the presence of silver oxide.

The Koenigs–Knorr reaction sequence involves formation of a pyranosyl bromide, followed by nucleophilic substitution, yielding a $\beta$-glycoside. For example, methylarbutin, a glycoside found in pear leaves, has been prepared by reaction between *p*-methoxyphenol (hydroquinone monomethyl ether) and tetraacetyl-$\alpha$-D-glucopyranosyl bromide (Figure 24.7).

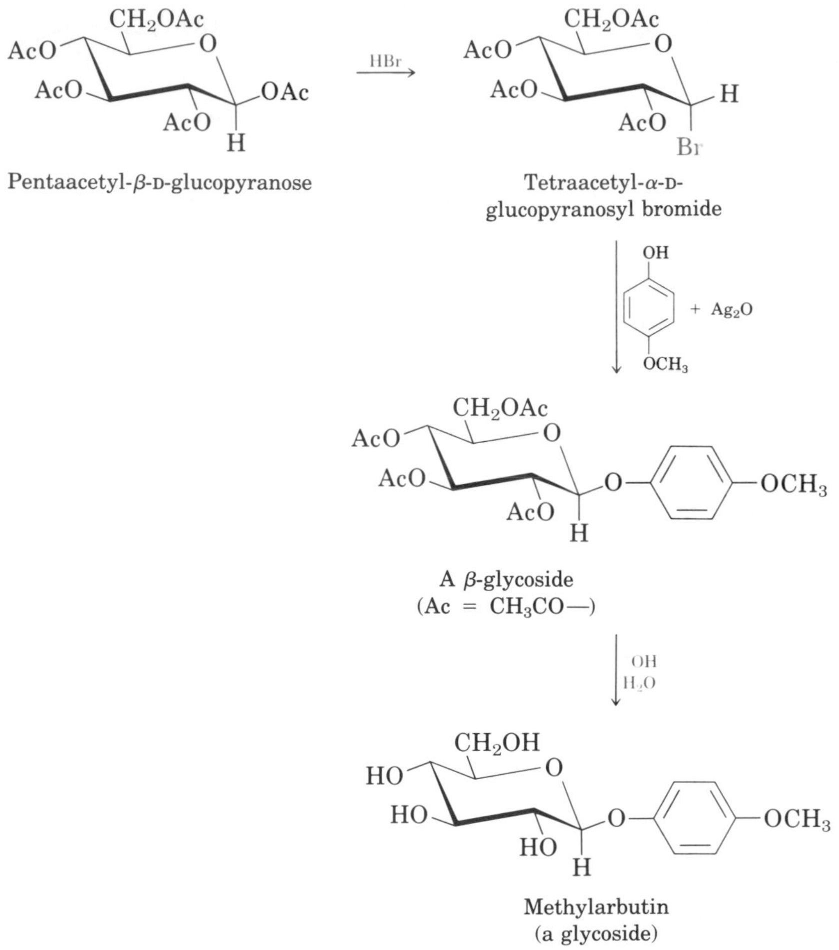

Pentaacetyl-$\beta$-D-glucopyranose $\xrightarrow{HBr}$ Tetraacetyl-$\alpha$-D-glucopyranosyl bromide

$\downarrow$ *p*-methoxyphenol (OH, $OCH_3$) + $Ag_2O$

A $\beta$-glycoside
($Ac = CH_3CO—$)

$\downarrow$ $^-OH$, $H_2O$

Methylarbutin
(a glycoside)

**Figure 24.7** Synthesis of the glycoside methylarbutin by Koenigs–Knorr reaction of pentaacetyl-$\beta$-D-glucopyranose with *p*-methoxyphenol

[4]Ludwig Knorr (1859–1921); Ph.D. Erlangen, 1882; professor, University of Jena.

Although the Koenigs–Knorr reaction appears to involve a simple back-side $S_N2$ displacement of bromide ion by alkoxide ion in a Williamson ether synthesis, the actual situation is more complex. Thus, both alpha and beta anomers of tetraacetyl-D-glucopyranosyl bromide give the *same* β-glycoside product, suggesting that both anomers react by a common pathway.

This observation is best explained by assuming that tetraacetyl-D-glucopyranosyl bromide (either alpha or beta anomer) undergoes a spontaneous $S_N1$ loss of bromide ion, followed by internal reaction of the cationic center at C1 with an oxygen atom of the ester group at C2 to form a stabilized oxonium ion intermediate. Since the ester group at C2 is on the bottom of the glucose ring, the new carbon–oxygen bond also forms from the bottom. An $S_N2$ displacement of the oxonium ion by back-side attack at C1 then occurs with the usual inversion of configuration, yielding a β-glycoside and regenerating the acetate ester group at C2 (Figure 24.8).

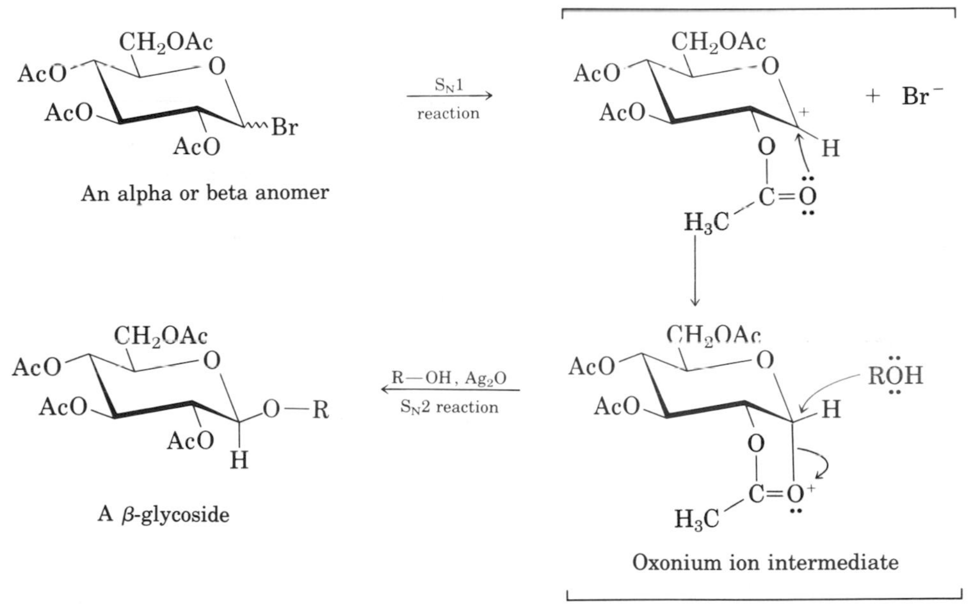


**Figure 24.8** Mechanism of the Koenigs–Knorr reaction, showing the neighboring-group effect of a nearby acetoxyl

This kind of participation by a nearby group, referred to as a **neighboring-group effect**, is a common occurrence in organic chemistry. Neighboring-group effects are usually noticeable only because they affect the rate or stereochemistry of a reaction; the nearby group itself does not undergo any evident change during the reaction.

## REDUCTION OF MONOSACCHARIDES

The carbonyl groups of monosaccharides undergo many reactions characteristic of simple ketones and aldehydes. For example, treatment of an aldose or ketose with $NaBH_4$ reduces it to a polyalcohol called an *alditol.* The reduction occurs by interception of the open-chain monosaccharide present in the aldehyde/ketone ⇄ hemiacetal equilibrium. Although only a small amount of open-chain form is present at any one time, that small amount is reduced, more is produced by opening of the pyranose form, that additional amount is reduced, and so on until the entire sample has undergone reaction.

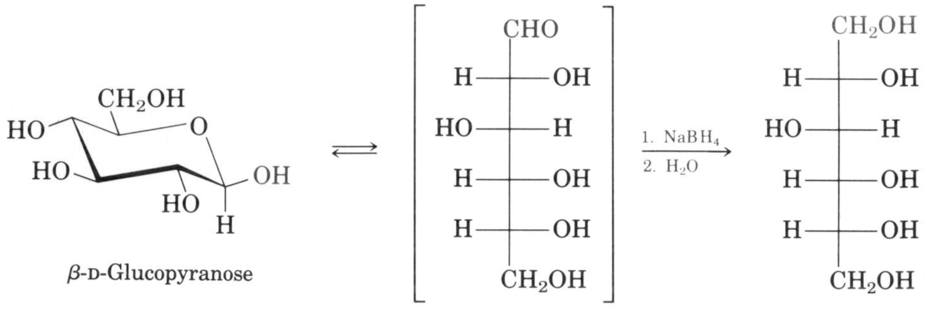

β-D-Glucopyranose ⇄ [D-Glucose] $\xrightarrow[\text{2. } H_2O]{\text{1. } NaBH_4}$ D-Glucitol (D-Sorbitol), an alditol

D-Glucitol, the alditol produced on reduction of D-glucose, is itself a natural product that has been isolated from many fruits and berries. It is used in many foods, under the name D-sorbitol, as an artificial sweetener and sugar substitute.

PROBLEM

**24.14** How can you account for the fact that reduction of D-galactose (Figure 24.2) with $NaBH_4$ leads to an alditol that is optically inactive?

PROBLEM

**24.15** Reduction of L-gulose with $NaBH_4$ leads to the same alditol (D-glucitol) as reduction of D-glucose. Explain.

## OXIDATION OF MONOSACCHARIDES

Like other aldehydes, aldoses are easily oxidized to yield carboxylic acids. Aldoses react with Tollens' reagent ($Ag^+$ in aqueous ammonia), Fehling's reagent ($Cu^{2+}$ in aqueous sodium tartrate), and Benedict's reagent ($Cu^{2+}$ in aqueous sodium citrate) to yield the oxidized sugar and a reduced metallic species. All three reactions serve as simple chemical tests for what are called **reducing sugars** (*reducing* because the sugar reduces the oxidizing agent).

If Tollens' reagent is used, metallic silver is produced as a shiny mirror on the walls of the reaction flask or test tube. If Fehling's or Benedict's reagent is used, a reddish precipitate of cuprous oxide signals a positive result. The diabetes self-test kits sold in drugstores for home use employ the Benedict test. As little as 0.1% glucose in urine gives a positive test.

All aldoses are reducing sugars since they contain an aldehyde carbonyl group, but some ketoses are reducing sugars as well. For example, fructose reduces Tollens' reagent even though it contains no aldehyde group. This occurs because fructose is readily isomerized to an aldose in basic solution by a series of keto ⇄ enol tautomeric shifts (Figure 24.9). Once formed, the aldose is oxidized normally. Glycosides, however, are nonreducing; they do not react with Tollens' reagent because the acetal group cannot open to an aldehyde under basic conditions.

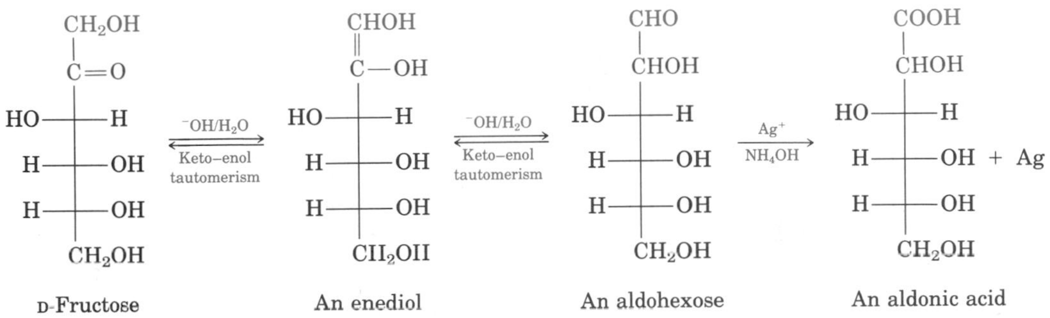

D-Fructose $\underset{\text{Keto–enol tautomerism}}{\overset{^{-}OH/H_2O}{\rightleftharpoons}}$ An enediol $\underset{\text{Keto–enol tautomerism}}{\overset{^{-}OH/H_2O}{\rightleftharpoons}}$ An aldohexose $\underset{NH_4OH}{\overset{Ag^+}{\longrightarrow}}$ An aldonic acid + Ag

**Figure 24.9** Fructose gives a positive Tollens test as a reducing sugar because it undergoes base-catalyzed keto–enol tautomerism that results in its conversion to an aldohexose.

Although the Tollens and Fehling reactions serve as useful tests for reducing sugars, they do not give good yields of carboxylic acid products because the alkaline conditions used cause decomposition of the carbohydrate skeleton. It has been found, however, that a buffered solution of aqueous bromine oxidizes aldoses to monocarboxylic acids called **aldonic acids**. The reaction is specific for aldoses; ketoses are not oxidized by bromine water.

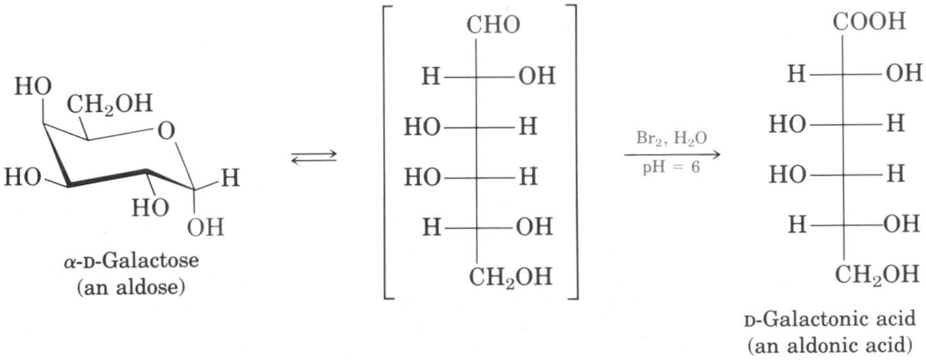

α-D-Galactose (an aldose) ⇌ [open-chain form] $\underset{pH = 6}{\overset{Br_2,\ H_2O}{\longrightarrow}}$ D-Galactonic acid (an aldonic acid)

If a more powerful oxidizing agent such as warm dilute nitric acid is used, aldoses are oxidized to dicarboxylic acids called **aldaric acids**. Both the —CHO group at C1 and the terminal —$CH_2OH$ group are oxidized in this reaction.

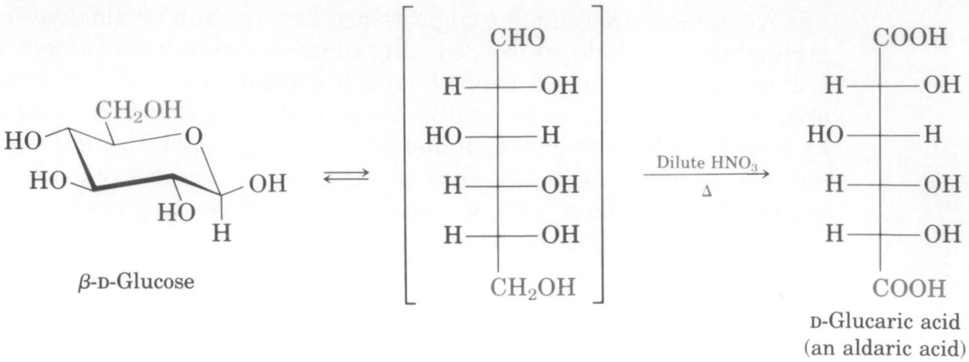

β-D-Glucore ⇌ Dilute $HNO_3$, Δ →

D-Glucaric acid
(an aldaric acid)

A summary of the various kinds of carbohydrate derivatives is shown in Figure 24.10.

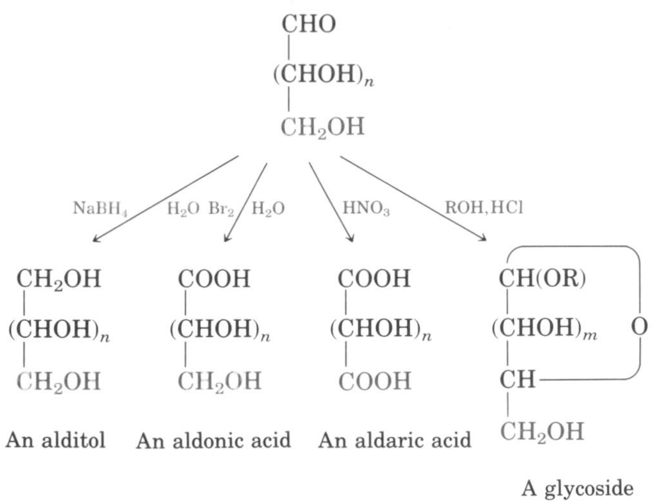

An alditol An aldonic acid An aldaric acid A glycoside

**Figure 24.10** Summary of carbohydrate derivatives

PROBLEM

**24.16** D-Glucose yields an optically active aldaric acid on treatment with nitric acid, but D-allose yields an optically inactive aldaric acid. Explain.

PROBLEM

**24.17** Which of the other six D-aldohexoses yield optically active aldaric acids on oxidation, and which yield meso aldaric acids? (See Problem 24.16.)

## CHAIN LENGTHENING: THE KILIANI–FISCHER SYNTHESIS

Much early activity in carbohydrate chemistry was devoted to unraveling the various stereochemical relationships among monosaccharides. One of the most important methods used was the **Kiliani–Fischer synthesis**, which results in the lengthening of an aldose chain by one carbon atom. For example, an aldo*pent*ose is converted by the Kiliani–Fischer synthesis into an aldo*hex*ose, as shown in Figure 24.11.

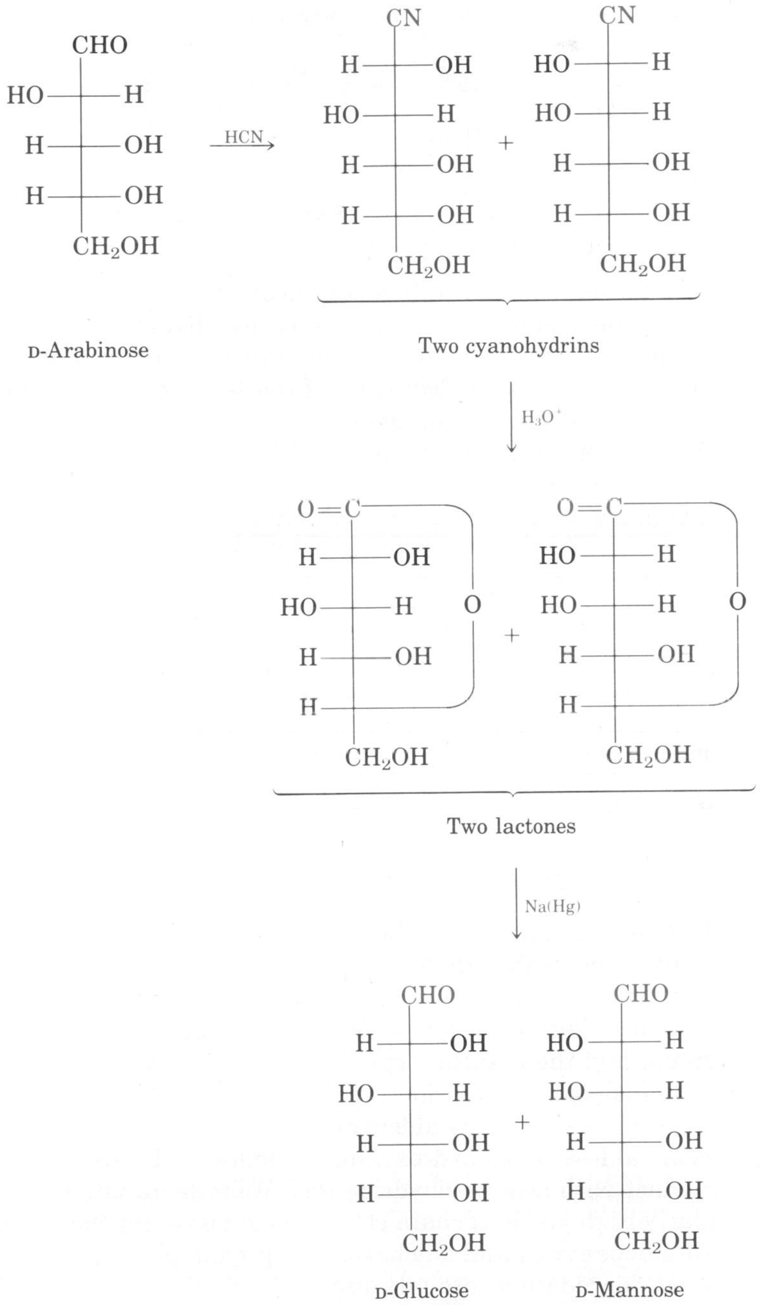

**Figure 24.11** Kiliani–Fischer chain lengthening of D-arabinose leads to a mixture of D-glucose and D-mannose.

Discovery of the chain-lengthening sequence was initiated by the observation of Heinrich Kiliani[5] in 1886 that aldoses react with HCN to form cyanohydrins (Section 19.9). Emil Fischer immediately realized the importance of Kiliani's discovery and in 1890 published a method for converting the cyanohydrin nitrile group into an aldehyde group.

$$\begin{array}{c} \mathrm{CHO} \\ | \\ \mathrm{CHOH} \\ \wr \end{array} \xrightarrow{\mathrm{HCN}} \begin{array}{c} \mathrm{CN} \\ | \\ \mathrm{CHOH} \\ | \\ \mathrm{CHOH} \\ \wr \end{array} \Longrightarrow \begin{array}{c} \mathrm{CHO} \\ | \\ \mathrm{CHOH} \\ | \\ \mathrm{CHOH} \\ \wr \end{array}$$

An aldose — A cyanohydrin — A chain-lengthened aldose

Conversion of the nitrile into an aldehyde is accomplished by first hydrolyzing the cyanohydrin intermediate to a hydroxy carboxylic acid and then forming a lactone ring by internal esterification with a hydroxyl group four carbon atoms away. Reduction of the lactone carbonyl group with sodium amalgam (an alloy of sodium and mercury) then yields the chain-lengthened aldose. Note that the initial cyanohydrin is a mixture of stereoisomers at the new chiral center. Thus, *two* new aldoses, differing only in their stereochemistry at C2, result from Kiliani–Fischer synthesis. For example, chain extension of D-arabinose yields a mixture of D-glucose and D-mannose (Figure 24.11).

PROBLEM

**24.18** What product(s) would you expect from Kiliani–Fischer reaction of D-ribose?

PROBLEM

**24.19** What aldohexose would give a mixture of L-gulose and L-idose on Kiliani–Fischer chain extension?

## CHAIN SHORTENING: THE WOHL DEGRADATION

Just as the Kiliani–Fischer synthesis lengthens an aldose chain by one carbon, the **Wohl[6] degradation** shortens an aldose chain by one carbon. The Wohl degradation is almost exactly the opposite of the Kiliani–Fischer sequence: The aldose aldehyde carbonyl group is first converted into a nitrile group, and the resulting cyanohydrin loses HCN under basic conditions (a retronucleophilic addition reaction).

Conversion of the aldehyde into a nitrile is accomplished by treatment of an aldose with hydroxylamine, followed by dehydration of the oxime product with acetic anhydride. The Wohl degradation does not give particularly high yields of chain-shortened aldoses, but the reaction is general for all aldopentoses and aldohexoses. For example, D-galactose is converted by Wohl degradation into D-lyxose:

[5]Heinrich Kiliani (1855–1945); b. Würzburg, Germany; Ph.D. Munich (Erlenmeyer); professor, University of Freiburg.

[6]Alfred Wohl (1863–1933); b. Graudenz, Poland; Ph.D. Berlin (Hofmann); professor, University of Danzig.

CHO
H—OH
HO—H
HO—H
H—OH
$CH_2OH$
D-Galactose

$\xrightarrow{H_2NOH}$

CH=NOH
H—OH
HO—H
HO—H
H—OH
$CH_2OH$
D-Galactose oxime

$\xrightarrow[NaO_2CCH_3]{(CH_3CO)_2O}$

C≡N
H—OH
HO H
HO—H
H—OH
$CH_2OH$
A cyanohydrin

$\xrightarrow{NaOCH_3}$

CHO
HO—H
HO—H + HCN
H—OH
$CH_2OH$
D-Lyxose (37%)

PROBLEM

**24.20** What two D-aldopentoses yield D-threose on Wohl degradation?

## 24.9 Stereochemistry of Glucose: The Fischer Proof

In the late 1800s, the stereochemical theories of van't Hoff and Le Bel on the tetrahedral geometry of carbon were barely a decade old. Modern chromatographic methods of product purification were unknown, and modern spectroscopic techniques of structure determination were undreamed of. Despite these obstacles, Emil Fischer published in 1891 what stands today as one of the finest examples of chemical logic ever recorded—a structure proof of the stereochemistry of glucose. Let's follow Fischer's logic and see how he arrived at his conclusions.

**FACT 1: (+)-Glucose is an aldohexose.** Glucose has four chiral centers and can therefore be any one of $2^4 = 16$ possible stereoisomers. These 16 possible stereoisomers consist of eight pairs of enantiomers. Since no method was available at the time for determining the *absolute* three-dimensional stereochemistry of a molecule, Fischer realized that the best he could do would be to limit his choices for the structure of glucose to a pair of enantiomers. He decided to simplify matters by considering only the eight enantiomers having the C5 hydroxyl group on the right in Fischer projections (D series). Fischer was well aware that this arbitrary choice of D-series stereochemistry had only a 50:50 chance of being right, but it was finally shown some 60 years later by the use of sophisticated X-ray techniques that the choice was indeed correct.

The four possible D-aldopentoses, and the eight possible D-aldohexoses derived from them by Kiliani–Fischer synthesis, are shown in Figure 24.12. One of the eight aldohexoses is glucose, but which one?

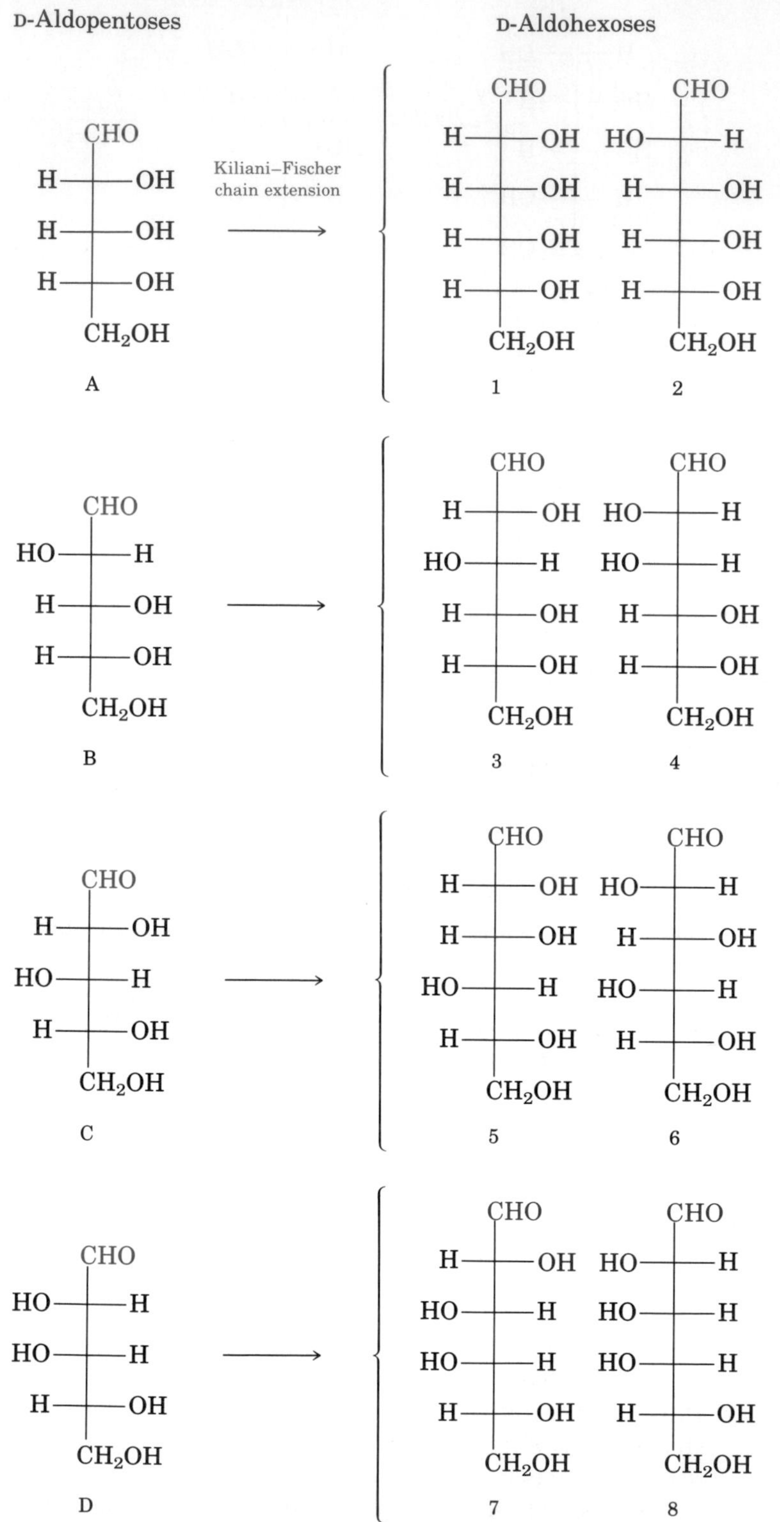


**Figure 24.12** The four D-aldopentoses and the eight D-aldohexoses

**FACT 2: Arabinose, an aldopentose, is converted by Kiliani–Fischer chain extension into a mixture of glucose and mannose.** This means that glucose and mannose have the same stereochemistry at C3, C4, and C5, and differ only at C2. Glucose and mannose are therefore represented by one of the pairs of structures 1 and 2, 3 and 4, 5 and 6, or 7 and 8 in Figure 24.12.

**FACT 3: Arabinose is converted by treatment with warm nitric acid into an optically active aldaric acid.** Of the four possible aldopentoses (A, B, C, and D in Figure 24.12), A and C give optically inactive meso aldaric acids, whereas B and D give optically active products. Thus, arabinose must be either B or D, and mannose and glucose must therefore be either 3 and 4 or 7 and 8 (Figure 24.13).

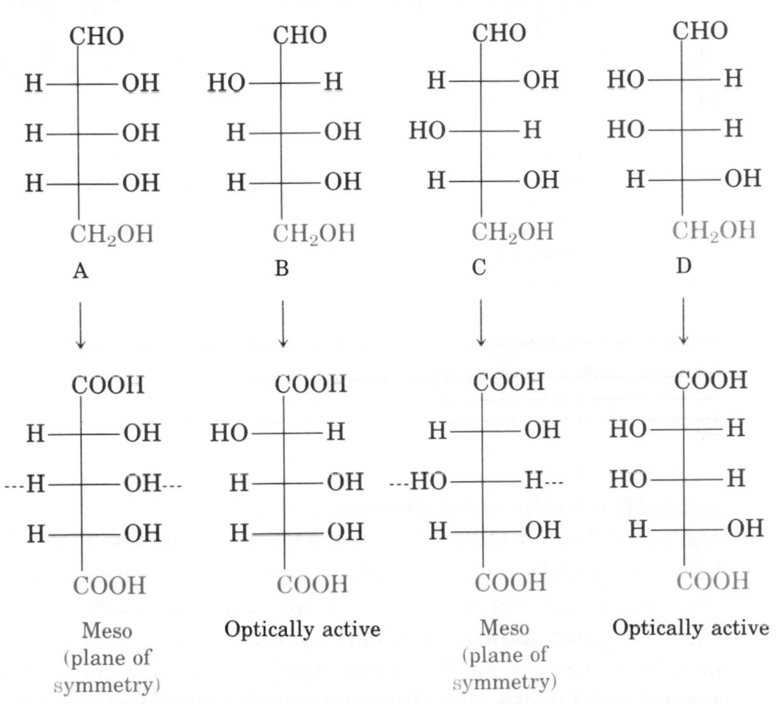


**Figure 24.13** Oxidation of aldopentoses to aldaric acids: Only structures B and D lead to optically active products.

**FACT 4: Both glucose and mannose are oxidized by warm nitric acid to optically active aldaric acids.** Of the possibilities left at this point, the pair represented by structures 3 and 4 would *both* be oxidized to optically active aldaric acids, but the pair represented by 7 and 8 would not *both* give optically active products. Compound 7 would give an optically inactive meso aldaric acid (Figure 24.14). Thus, glucose and mannose must be 3 and 4, though we can't yet tell which is which.

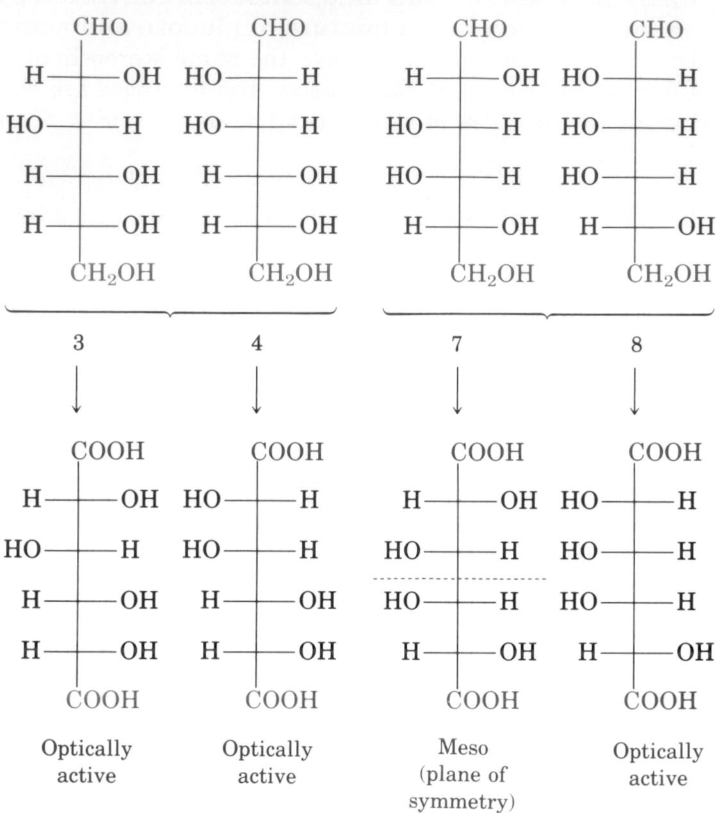


**Figure 24.14** Oxidation of aldohexoses to aldaric acids: Only the pair of structures 3 and 4 both give optically active products.

**FACT 5: One of the other 15 aldohexoses is converted by nitric acid oxidation to the same aldaric acid as that derived from glucose.** How can two different aldohexoses give the same aldaric acid? Since aldaric acids have —COOH groups at both ends of the carbon chain, there is no way to tell which was originally the —CHO end and which was the —$CH_2OH$ end. Thus, a given aldaric acid can have two different precursors. The aldaric acid from compound 3, for example, might also come from oxidation of a second aldohexose, and the aldaric acid from compound 4 might come from oxidation of a second aldohexose (Figure 24.15).

If we look carefully at the aldaric acids derived from compounds 3 and 4, we find that the aldaric acid derived from compound 3 could also come from oxidation of another aldohexose (L-gulose), but that the aldaric acid derived from compound 4 could not. The "other" aldohexose that could produce the same aldaric acid as the one from compound 4 is in fact identical to 4 by a simple 180° rotation. Thus, glucose must have structure 3 and mannose must have structure 4 (Figure 24.15).

Reasoning similar to that shown for glucose allowed Fischer to determine the stereochemistry of 12 of the 16 aldohexoses. For this remarkable achievement, he was awarded the 1902 Nobel prize in chemistry.

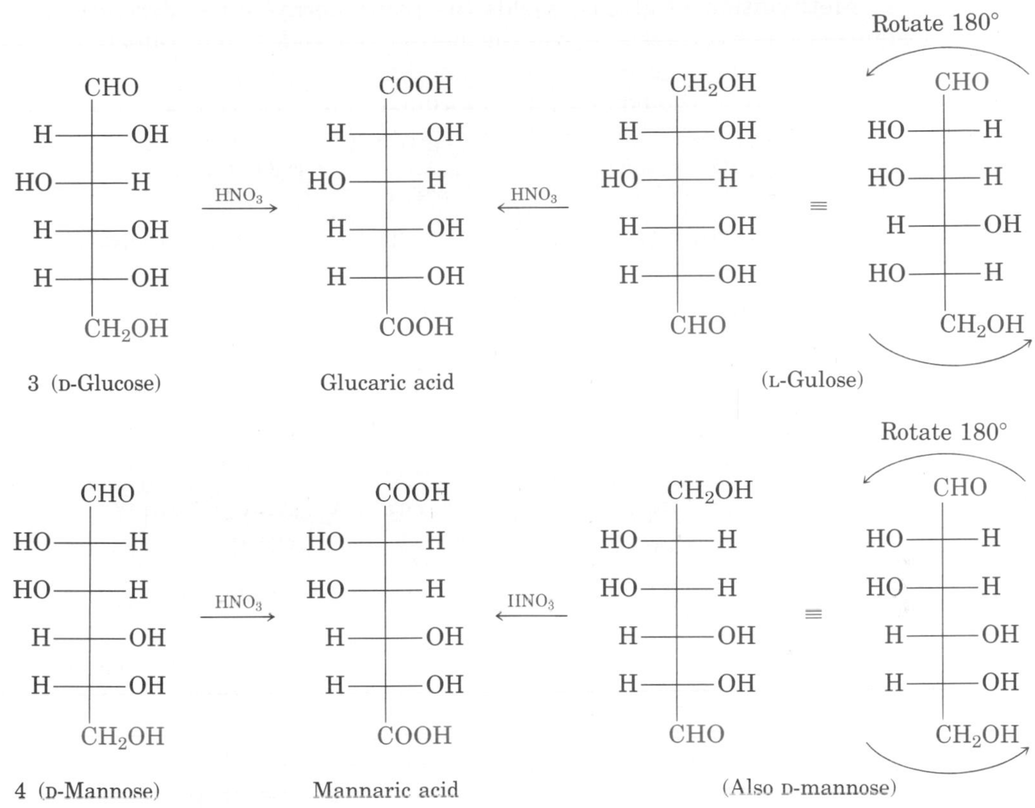


**Figure 24.15** There is another aldohexose (L-gulose) that can produce the same aldaric acid as compound 3, but there is no other aldohexose that can produce the same aldaric acid as compound 4. Thus, glucose has structure 3.

**PROBLEM**

**24.21** The structures of the four aldopentoses, A, B, C, and D, are shown in Figure 24.12. In light of fact 2 presented by Fischer, what is the structure of arabinose? In light of fact 3, what is the structure of lyxose, another aldopentose that yields an optically active aldaric acid?

**PROBLEM**

**24.22** The aldotetrose D-erythrose yields a mixture of D-ribose and D-arabinose on Kiliani–Fischer chain extension.

(a) What is the structure of D-ribose?
(b) What is the structure of D-xylose, the fourth possible aldopentose?
(c) What is the structure of D-erythrose?
(d) What is the structure of D-threose, the other possible aldotetrose?

## 24.10 Determination of Monosaccharide Ring Size

With the stereochemistry of glucose known, the only problem remaining is to determine the size of the cyclic hemiacetal ring. Is glucose a furanose or a pyranose? This problem was solved by Haworth and Hirst in 1926 by recourse to simple yet effective chemistry.

Methylation of glucose yields the pentamethyl ether derivative, and aqueous acid hydrolysis cleaves the methyl glycoside to a tetramethyl ether. In the ring-opened form of this tetramethyl ether, only the hydroxyl group that was part of the hemiacetal ring remains unmethylated.

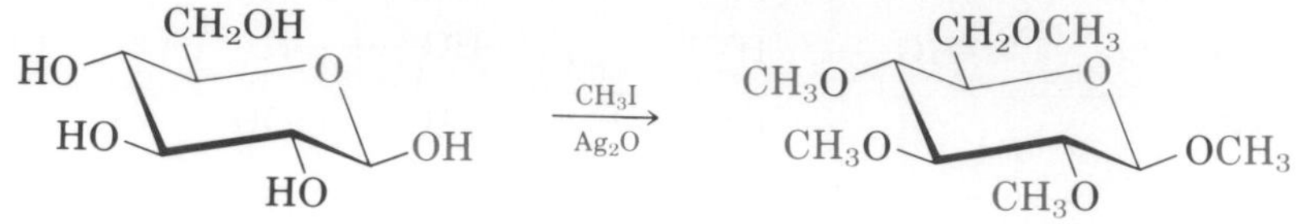

D-Glucose $\xrightarrow[Ag_2O]{CH_3I}$ D-Glucose pentamethyl ether $\xrightarrow{H_3O^+}$ D-Glucose tetramethyl ether

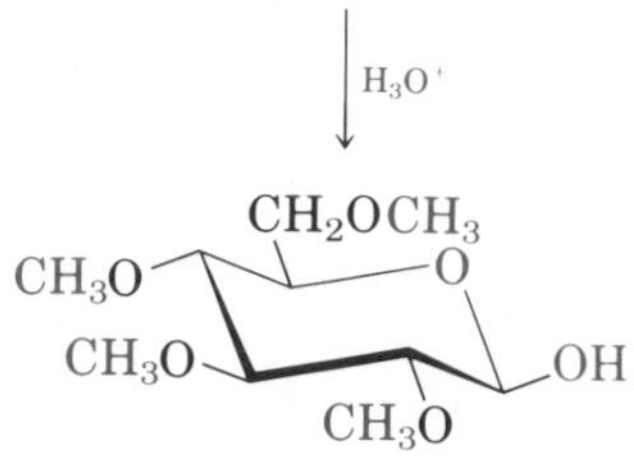

D-Glucose tetramethyl ether (pyranose) ⇌ open-chain form: CHO; H—OCH₃; CH₃O—H; H—OCH₃; H—OH ← Unmethylated; CH₂OCH₃

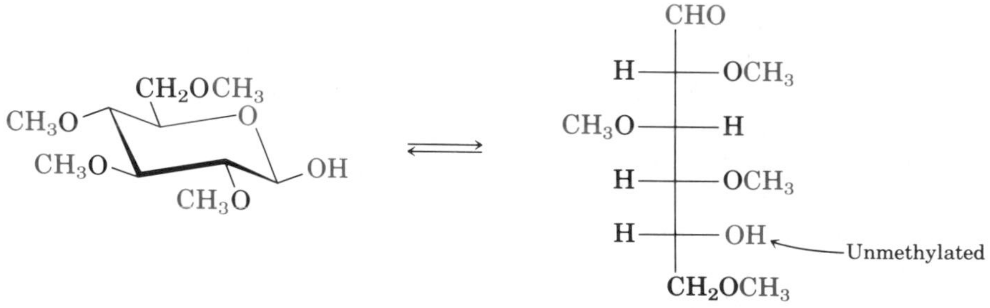

Haworth and Hirst determined the position of the free hydroxyl group by oxidation of the tetramethyl ether. Under the conditions used (hot nitric acid), the aldehyde group at C1 and the free hydroxyl were both oxidized, and cleavage occurred next to the ketone carbonyl group to yield a compound identified as dimethoxytartaric acid. This could happen only if the free hydroxyl group were at C5 and if glucose were therefore a pyranose. The structure of glucose was complete!

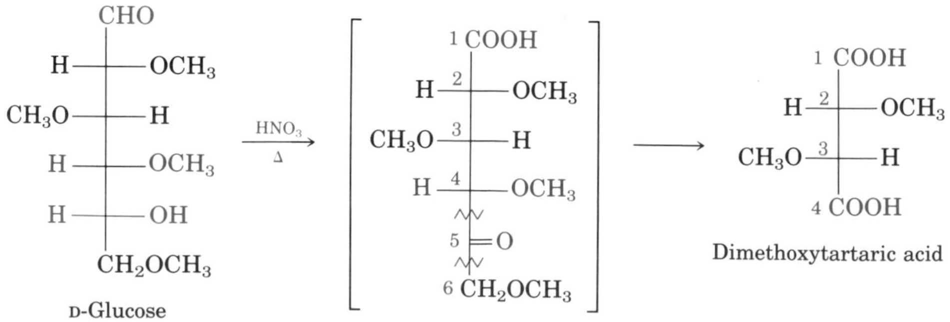

D-Glucose tetramethyl ether $\xrightarrow[\Delta]{HNO_3}$ [1 COOH; 2 H—OCH₃; 3 CH₃O—H; 4 H—OCH₃; 5 C=O; 6 CH₂OCH₃] ⟶ Dimethoxytartaric acid (1 COOH; 2 H—OCH₃; 3 CH₃O—H; 4 COOH) + $CO_2$ + Other products

PROBLEM

**24.23** What product would you expect to obtain from oxidation of glucose tetramethyl ether if glucose were a furanose?

PROBLEM

**24.24** Is the dimethoxytartaric acid obtained by degradation of glucose optically active or meso? Explain.

## 24.11 Disaccharides

We saw earlier that reaction of a monosaccharide hemiacetal with an alcohol yields a glycoside in which the anomeric hydroxyl group is replaced by an alkoxy substituent. If the alcohol is itself a sugar, however, the glycosidic product is a **disaccharide**.

### CELLOBIOSE AND MALTOSE

Disaccharides are compounds that contain a glycosidic acetal bond between C1 of one sugar and a hydroxyl group at *any* position on the other sugar. A glycosidic bond between C1 of the first sugar and C4 of the second sugar is particularly common but is by no means required. Such a bond is called a **1,4′ link** (read as "one, four-prime"). The prime superscript indicates that the 4′ position is on a different sugar than the nonprime 1 position.

A glycosidic bond to the anomeric carbon can be either alpha or beta. For example, cellobiose, the disaccharide obtained by partial hydrolysis of cellulose, consists of two D-glucopyranoses joined by a 1,4′-β-glycoside bond. Maltose, the disaccharide obtained by enzyme-catalyzed hydrolysis of starch, consists of two D-glucopyranoses joined by a 1,4′-α-glycoside bond.

Cellulose $\xrightarrow{H_3O^+}$

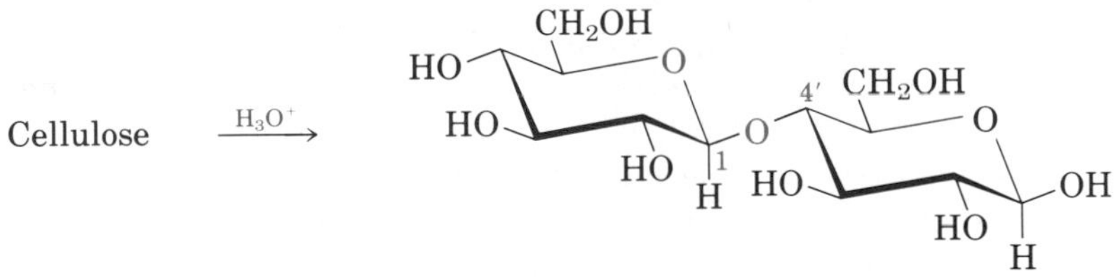

Cellobiose, a 1,4′-β-glycoside
[4-*O*-(β-D-Glucopyranosyl)-β-D-glucopyranose]

Starch $\xrightarrow{\text{Enzyme}}$

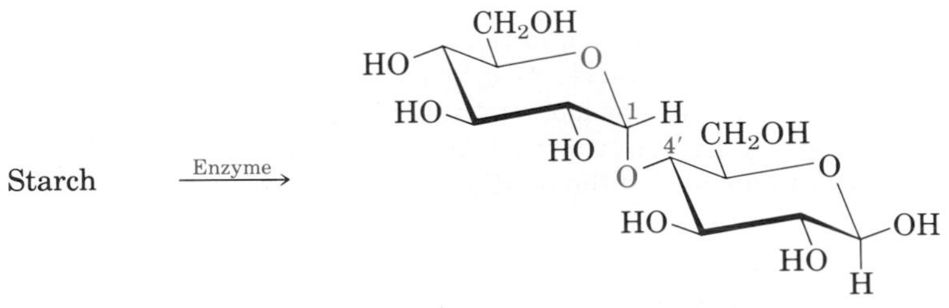

Maltose, a 1,4′-α-glycoside
[4-*O*-(α-D-Glucopyranosyl)-β-D-glucopyranose]

Both maltose and cellobiose are reducing sugars because the anomeric carbons on the right-hand sugar are part of a hemiacetal. Both are therefore in equilibrium with aldehyde forms, which can reduce Tollens' or Fehling's reagent. For a similar reason, both maltose and cellobiose exhibit mutarotation of alpha and beta anomers of the glucopyranose unit on the right (Figure 24.16).

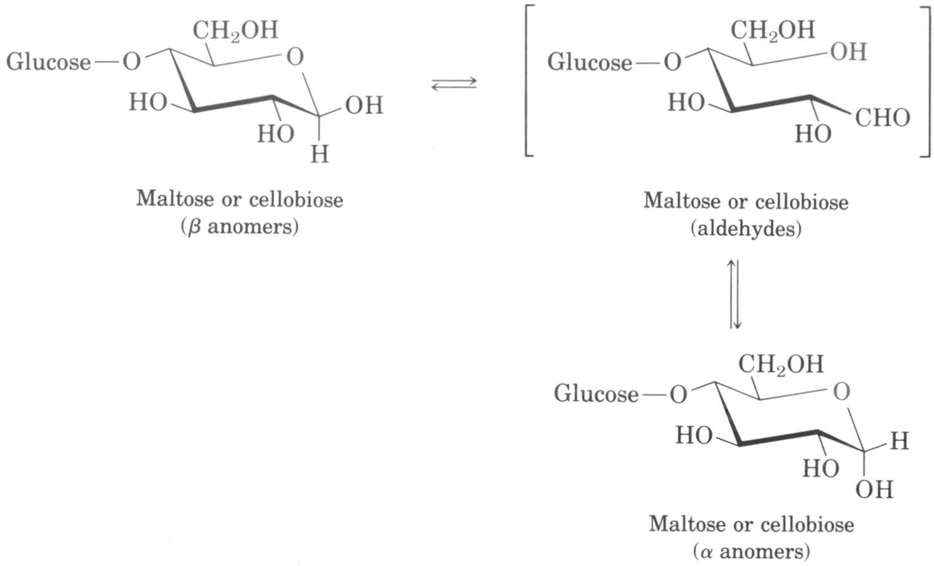

Maltose or cellobiose
($\beta$ anomers)

Maltose or cellobiose
(aldehydes)

Maltose or cellobiose
($\alpha$ anomers)

**Figure 24.16** Mutarotation of maltose and cellobiose

Despite the similarities of their structures, cellobiose and maltose are dramatically different biologically. Cellobiose cannot be digested by humans and cannot be fermented by yeast. Maltose, however, is digested without difficulty and is fermented readily.

PROBLEM

**24.25** Show the product you would obtain from the reaction of cellobiose with these reagents:
(a) $NaBH_4$ (b) $Br_2$, $H_2O$ (c) $CH_3COCl$, pyridine

PROBLEM

**24.26** The position of the glycosidic link in cellobiose can be determined by a modification of the method used by Haworth and Hirst to determine the ring size of glucose (Section 24.10). Reaction of cellobiose with iodomethane and $Ag_2O$ yields an octamethyl ether derivative. Acid hydrolysis of this octamethyl ether yields a tri-*O*-methyl-glucopyranose and a tetra-*O*-methylglucopyranose. What are the structures of these octamethyl, trimethyl, and tetramethyl ethers? How can you use this information to determine the position of the glycosidic link in cellobiose?

## LACTOSE

Lactose is a disaccharide that occurs naturally in both human and cow's milk. It is widely used in baking and in commercial infant-milk formulas.

Like cellobiose and maltose, lactose is a reducing sugar. It exhibits mutarotation and is a 1,4′-β-linked glycoside. Unlike cellobiose and maltose, however, lactose contains two *different* monosaccharide units. Acidic hydrolysis of lactose yields 1 equiv of D-glucose and 1 equiv of D-galactose; the two are joined by a β-glycoside bond between C1 of galactose and C4 of glucose.

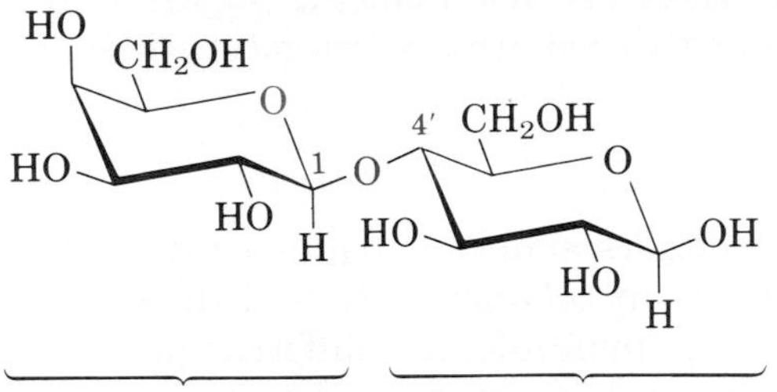

β-Galactopyranoside β-Glucopyranose

Lactose, a 1,4′-β-glycoside
[4-*O*-(β-D-Galactopyranosyl)-β-D-glucopyranose]

## SUCROSE

Sucrose—ordinary table sugar—is probably the single most abundant pure organic chemical in the world and the one most widely known to nonchemists. Whether from sugar cane (20% by weight) or sugar beets (15% by weight), and whether raw or refined, common sugar is still sucrose.

Sucrose is a disaccharide that yields 1 equiv of glucose and 1 equiv of fructose on acidic hydrolysis. This 1:1 mixture of glucose and fructose is often referred to as **invert sugar**, since the sign of optical rotation changes (inverts) during the hydrolysis from sucrose ($[\alpha]_D = +66.5°$) to a glucose–fructose mixture ($[\alpha]_D \approx -22.0°$). Certain insects, particularly honeybees, have enzymes called invertases that catalyze the hydrolysis of sucrose to a glucose–fructose mixture. Honey, in fact, is primarily a mixture of these three sugars.

Unlike most other disaccharides, sucrose is not a reducing sugar and does not exhibit mutarotation. These facts imply that sucrose has no hemiacetal linkages and that glucose and fructose must both be glycosides. This can happen only if the two sugars are joined by a glycoside link between C1 of glucose and C2 of fructose.

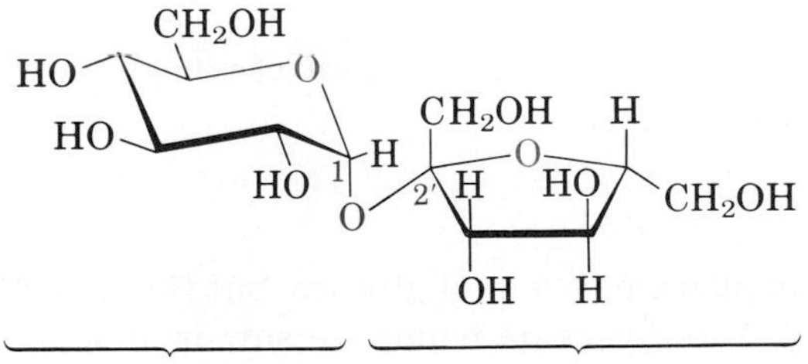

α-D-Glucopyranoside β-D-Fructofuranoside

Sucrose, a 1,2′-glycoside
[2-*O*-(α-D-Glucopyranosyl)-β-D-fructofuranoside]

## 24.12 Polysaccharides

**Polysaccharides** are carbohydrates in which tens, hundreds, or even thousands of simple sugars are linked together through glycoside bonds. Since they have no free anomeric hydroxyls (except for one at the end of the chain), polysaccharides are not reducing sugars and do not show mutarotation. Cellulose and starch are the two most widely occurring polysaccharides.

### CELLULOSE

Cellulose consists simply of D-glucose units linked by the 1,4′-β-glycoside bonds we saw in cellobiose. Several thousand glucose units are linked to form one large molecule, and different molecules can then interact to form a large aggregate structure held together by hydrogen bonds.

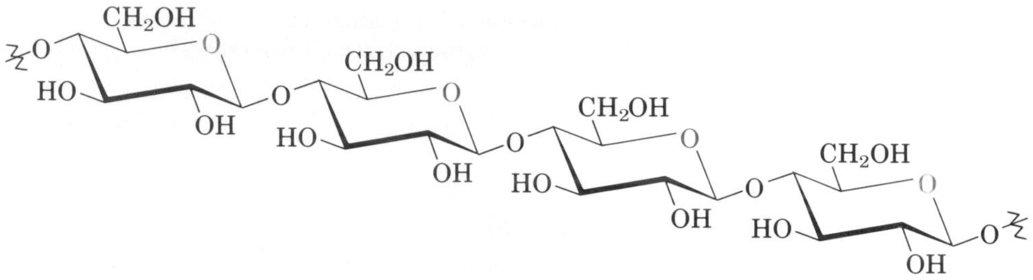

Cellulose, a 1,4′-*O*-(β-D-glucopyranoside) polymer

Nature uses cellulose primarily as a structural material to impart strength and rigidity to plants. Wood, leaves, grasses, and cotton are primarily cellulose. Cellulose also serves as raw material for the manufacture of cellulose acetate, which is widely used in clothing fiber.

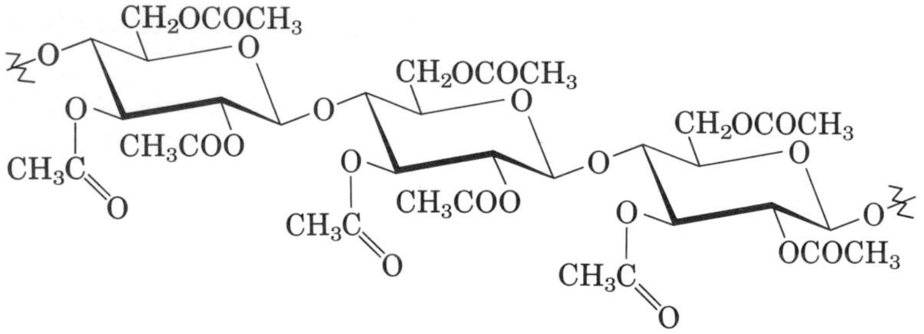

A segment of cellulose acetate

### STARCH

Starch is also a polymer of glucose, but the monosaccharide units are linked by the 1,4′-α-glycoside bonds we saw in maltose. Starch can be separated into two fractions—a fraction soluble in cold water, called *amylopectin*, and a fraction insoluble in cold water, called *amylose*. Amylose, which accounts for about 20% by weight of starch, consists of several hundred glucose molecules linked together by 1,4′-α-glycoside bonds.

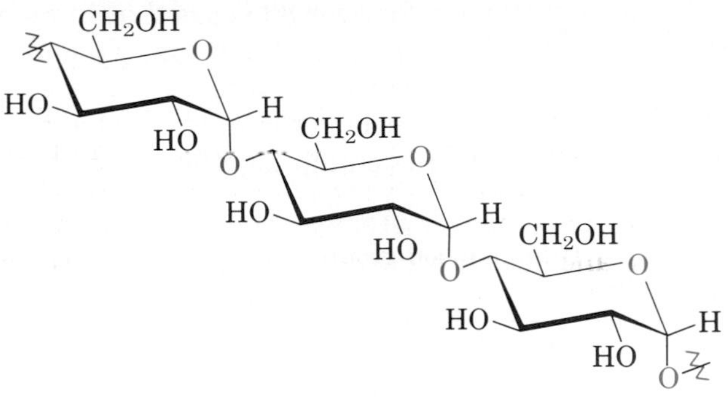

Amylose, a 1,4′-*O*-(α-D-glucopyranoside) polymer

Amylopectin, which accounts for the remaining 80% of starch, is more complex in structure than amylose. Unlike cellulose and amylose, which are linear or straight-chain polymers, amylopectin contains 1,6′-α-glycoside branches approximately every 25 glucose units. As a result, amylopectin has an exceedingly complex three-dimensional structure (Figure 24.17). Nature uses starch as the means by which plants store energy for later use. Potatoes, corn, and cereal grains contain large amounts of starch.

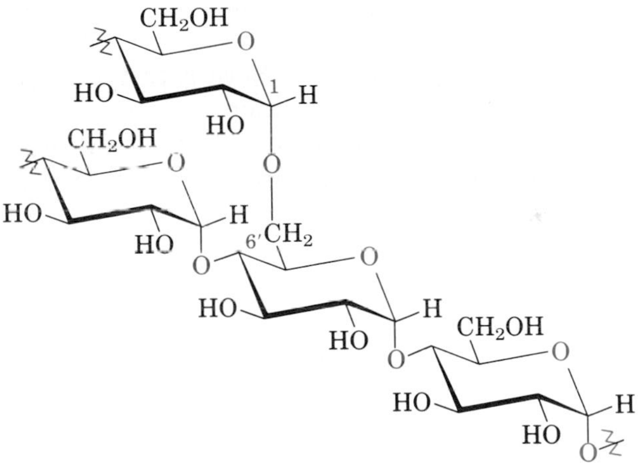

**Figure 24.17** A 1,6′-α-glycoside branch in amylopectin

## 24.13 Carbohydrates on Cell Surfaces

For many years, carbohydrates were thought to be rather dull compounds whose only biological purposes were to serve as structural materials and as energy sources. Although carbohydrates do indeed fill these two roles, recent research has shown that they perform many other important biochemical functions as well. For example, polysaccharides are known to be centrally involved in the critical process by which one cell type recognizes another. Small polysaccharide chains, covalently bound by glycosidic links to hydroxyl groups on proteins (**glycoproteins**), act as biochemical labels on cell surfaces, as exemplified by the human blood-group antigens.

It has been known for over 80 years that human blood can be classified into four blood-group types, A, B, AB, and O, and that blood from a donor of one type cannot be transfused into a recipient with another type unless the two types are compatible (Table 24.1). Should an incompatible mix be made, the red blood cells clump together, or *agglutinate*.

**Table 24.1** Human blood-group compatibilities

| *Donor blood type* | *Acceptor blood type* | | | |
|---|---|---|---|---|
| | *A* | *B* | *AB* | *O* |
| *A* | ○ | × | ○ | × |
| *B* | × | ○ | ○ | × |
| *AB* | × | × | ○ | × |
| *O* | ○ | ○ | ○ | ○ |

○ = compatible; × = incompatible

This agglutination of incompatible types of red blood cells, which indicates that the body's immune system has recognized the presence of foreign cells in the body and has formed antibodies against them, results from the presence of polysaccharide markers on the surface of the cells. Types A, B, and O red blood cells each have characteristic markers (**antigenic determinants**), and type AB cells have both type A and type B markers. The structures of all three blood-group determinants are shown in Figure 24.18.

Note that some rather unusual carbohydrates are involved. Thus, all three contain *N*-acetylamino sugars as well as the unusual monosaccharide L-fucose.

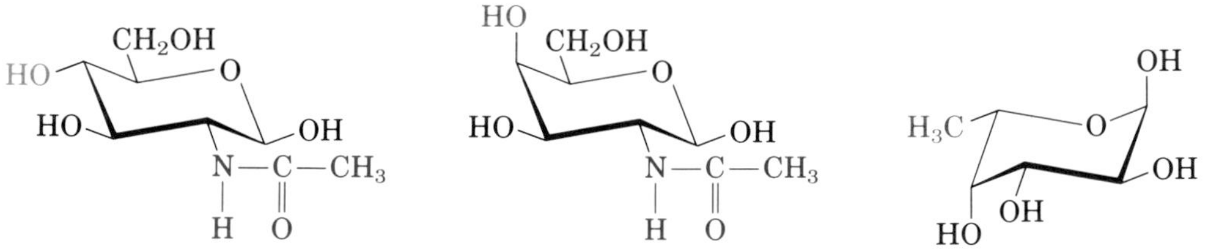

β-D-*N*-Acetylglucosamine (D-2-Acetamino-2-deoxyglucose)

β-D-*N*-Acetylgalactosamine (D-2-Acetamino-2-deoxygalactose)

α-L-Fucose (L-6-Deoxygalactose)

The antigenic determinant of blood group O is a trisaccharide, whereas the determinants of blood groups A and B have an additional saccharide attached at C3 of the galactose unit. The type A and B determinants differ only in the substitution of an acetylamino group (—$NHCOCH_3$) for a hydroxyl in the terminal galactose residue.

Elucidation of the role of carbohydrates in cell recognition is an exciting area of current research that offers hope of breakthroughs in the understanding of a wide range of diseases from bacterial infections to cancer. All stages of this work—isolation, purification, structure determination, and chemical synthesis of the carbohydrate cell markers—are extremely difficult, but the ultimate rewards and benefits may be enormous.

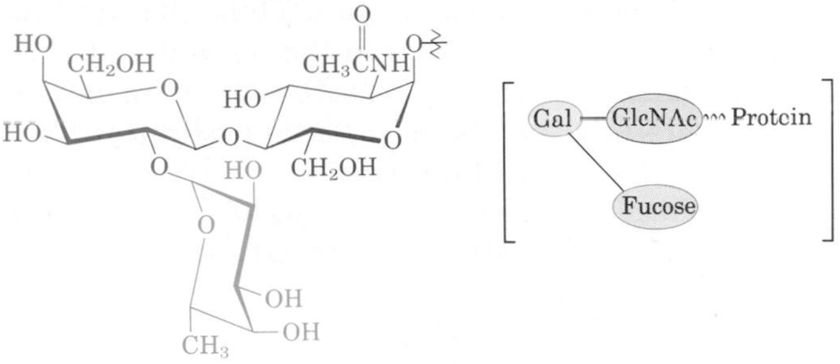

Blood group O

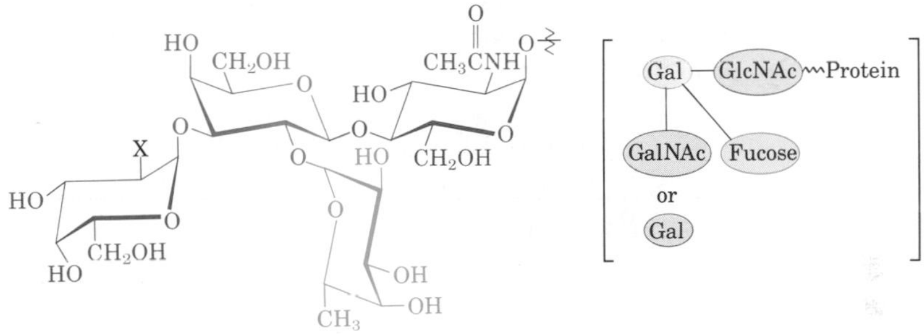

Blood group A, X = $NHCOCH_3$
Blood group B, X = OH

**Figure 24.18** Structures of the A, B, and O blood group antigenic determinants (Gal = D-galactose; GlcNAc = *N*-acetylglucosamine; GalNAc = *N*-acetylgalactosamine)

## 24.14 Summary and Key Words

**Carbohydrates** are polyhydroxy aldehydes and ketones. They can be classified according to the number of carbon atoms and the kind of carbonyl group they contain; thus, glucose is an **aldohexose**, a six-carbon aldehydo sugar. Monosaccharides are further classified as either **D** or **L sugars**, depending on the stereochemistry of the chiral carbon atom farthest from the carbonyl group.

Monosaccharides normally exist as **cyclic hemiacetals** rather than as open-chain aldehydes or ketones. The hemiacetal linkage results from reaction of the carbonyl group with a hydroxyl group three or four carbon atoms away. A five-membered-ring hemiacetal is called a **furanose**, and a six-membered-ring hemiacetal is called a **pyranose**. Cyclization leads to the formation of a new chiral center and production of two diastereomeric hemiacetals called **alpha** and **beta anomers**.

Stereochemical relationships among monosaccharides are portrayed in several ways. **Fischer projections** display chiral carbon atoms as a pair of crossed lines. These projections are useful in allowing us quickly to relate one sugar to another, but cyclic **Haworth projections** provide a more accurate view. Any group to the right in a Fischer projection is down in a Haworth projection.

Much of the chemistry of monosaccharides is the now-familiar chemistry of alcohols and aldehydes/ketones. Thus, the hydroxyl groups of carbohydrates form esters and ethers in the normal way. The carbonyl group of a monosaccharide can be reduced with sodium borohydride to form an **alditol**, oxidized with bromine water to form an **aldonic acid**, oxidized with warm nitric acid to form an **aldaric acid**, or treated with an alcohol in the presence of acid to form a **glycoside**. Monosaccharides can also be chain-lengthened by the multistep **Kiliani–Fischer synthesis**, and can be chain-shortened by the **Wohl degradation**.

**Disaccharides** are complex carbohydrates in which two simple sugars are linked by a glycoside bond between the anomeric carbon of one unit and a hydroxyl of the second unit. The two sugars can be the same, as in maltose and cellobiose, or different, as in lactose and sucrose. The glycosidic bond can be either $\alpha$ (maltose) or $\beta$ (cellobiose, lactose), and can involve any hydroxyl of the second sugar. [A 1,4′ link is most common (cellobiose, maltose), but others such as 1,2′ (sucrose) are also known.]

## 24.15 Summary of Reactions

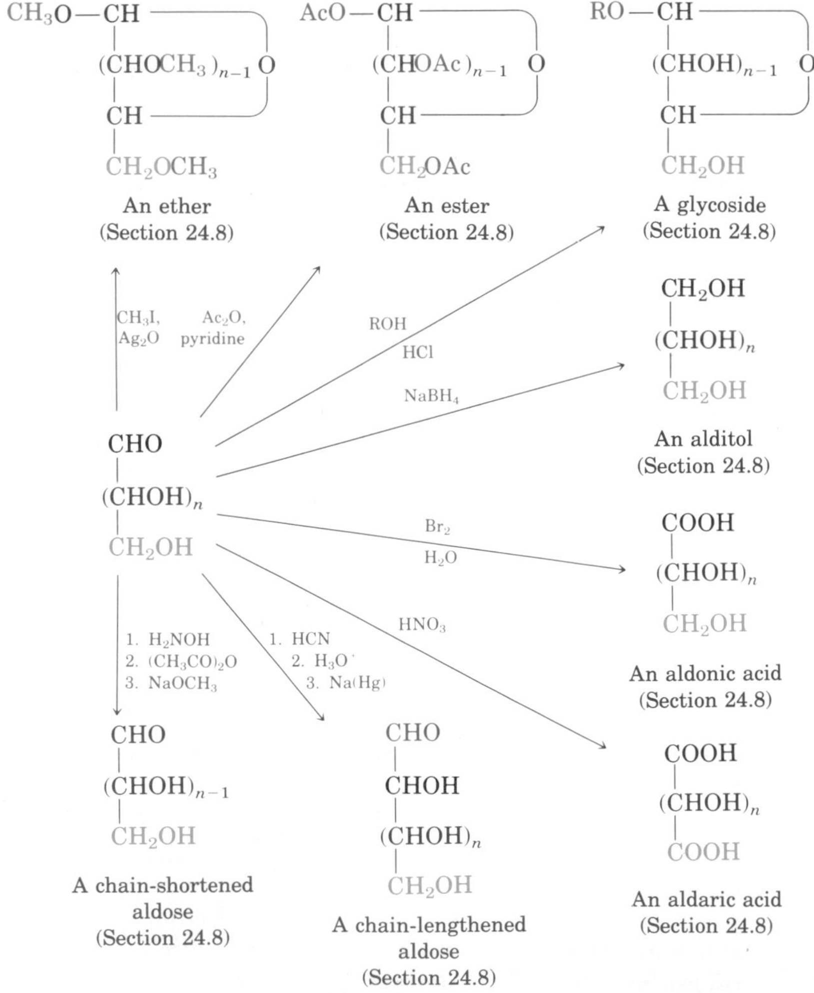

## ADDITIONAL PROBLEMS

**24.27** Classify the following sugars by type; for example, glucose is an aldohexose.

(a) $CH_2OH$–C=O–$CH_2OH$

(b) $CH_2OH$ / H–C–OH / C=O / H–C–OH / $CH_2OH$

(c) CHO / H–C–OH / HO–C–H / H–C–OH / HO–C–H / H–C–OH / $CH_2OH$

**24.28** Draw a Haworth projection of ascorbic acid (vitamin C).

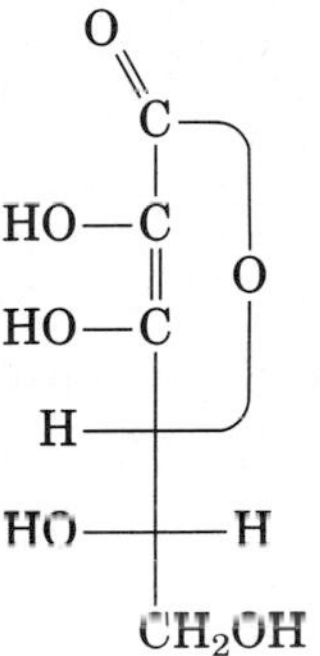

Ascorbic acid

**24.29** Draw structures for the products you would expect to obtain from reaction of $\beta$-D-talopyranose with each of the following reagents:
(a) $NaBH_4$ in $H_2O$ (b) Warm dilute $HNO_3$ (c) $Br_2$, $H_2O$
(d) $CH_3CH_2OH$, HCl (e) $CH_3I$, $Ag_2O$ (f) $(CH_3CO)_2O$, pyridine

**24.30** How many D-2-ketohexoses are possible? Draw them.

**24.31** One of the D-ketohexoses (Problem 24.30) is called *sorbose*. On treatment with $NaBH_4$, sorbose yields a mixture of gulitol and iditol. What is the structure of sorbose?

**24.32** Another D-2-ketohexose, *psicose*, yields a mixture of allitol and altritol when reduced with $NaBH_4$. What is the structure of psicose?

**24.33** Fischer prepared the L-gulose needed for his structure proof of glucose in the following way. D-Glucose was oxidized to D-glucaric acid, which can form two six-membered-ring lactones. These were separated and reduced with sodium amalgam to give D-glucose and L-gulose. What are the structures of the two lactones, and which one is reduced to L-gulose?

**24.34** What other D-aldohexose gives the same alditol as D-talose?

**24.35** Which of the eight D-aldohexoses give the same aldaric acids as their L enantiomers?

**24.36** Which of the other three D-aldopentoses gives the same aldaric acid as D-lyxose?

**24.37** Gentiobiose, a rare disaccharide found in saffron and gentian, is a reducing sugar and forms only glucose on hydrolysis with aqueous acid. Reaction of gentiobiose with iodomethane and silver iodide yields an octamethyl derivative, which can be hydrolyzed with aqueous acid to give 1 equiv of 2,3,4,6-tetra-*O*-methyl-D-glucopyranose and 1 equiv of 2,3,4-tri-*O*-methyl-D-glucopyranose. If gentiobiose contains a $\beta$-glycoside link, what is its structure?

**24.38** Amygdalin, or Laetrile, is a glycoside isolated in 1830 from almond and apricot seeds. It is known as a cyanogenic glycoside since acidic hydrolysis liberates HCN, along with benzaldehyde and 2 equiv of D-glucose. Structural studies have shown amygdalin to be a $\beta$-glycoside of benzaldehyde cyanohydrin with gentiobiose (Problem 24.37). Draw the structure of amygdalin.

**24.39** Trehalose is a nonreducing disaccharide that is hydrolyzed by aqueous acid to yield 2 equiv of D-glucose. Methylation followed by acidic hydrolysis yields 2 equiv of 2,3,4,6-tetra-*O*-methylglucose. How many possible structures are there for trehalose?

**24.40** Trehalose (Problem 24.39) is cleaved by enzymes that hydrolyze $\alpha$-glycosides but not by enzymes that hydrolyze $\beta$-glycosides. What is the structure and systematic name of trehalose?

**24.41** Isotrehalose and neotrehalose are chemically similar to trehalose (Problem 24.39) except for the fact that neotrehalose is hydrolyzed only by $\beta$-glycosidase enzymes, whereas isotrehalose is hydrolyzed by both $\alpha$- and $\beta$-glycosidase enzymes. What are the structures of isotrehalose and neotrehalose?

**24.42** Propose a scheme for the synthesis of gentiobiose methyl glycoside (Problem 24.37), starting from $\beta$-D-glucose and methyl 2,3,4-tri-*O*-acetyl-$\beta$-D-glucopyranoside.

**24.43** D-Glucose reacts with acetone in the presence of acid to yield the nonreducing 1,2:5,6-diisopropylidene-D-glucofuranose. Propose a mechanism for this reaction.

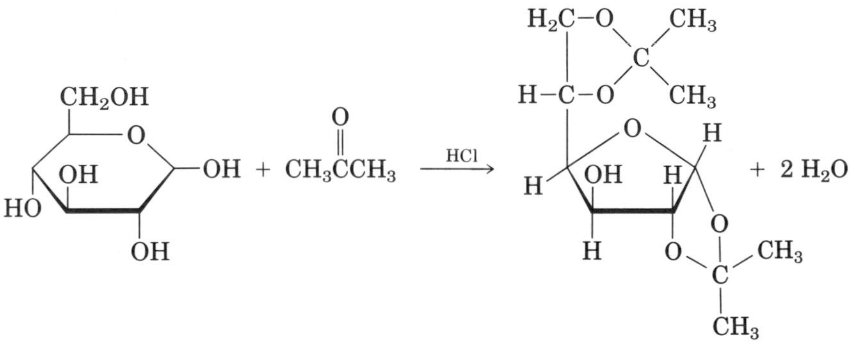

$$\text{D-glucose} + CH_3COCH_3 \xrightarrow{HCl} \text{1,2:5,6-diisopropylidene-D-glucofuranose} + 2\,H_2O$$

1,2:5,6-Diisopropylidene-D-glucofuranose

**24.44** D-Mannose reacts with acetone to give a diisopropylidene derivative that is still reducing toward Tollens' reagent. Propose a likely structure for this derivative.

**24.45** Propose a mechanism to account for the fact that D-gluconic acid and D-mannonic acid are interconverted when either is heated in pyridine solvent.

**24.46** The cyclitols are a group of carbocyclic sugar derivatives having the general formulation 1,2,3,4,5,6-cyclohexanehexol. How many stereoisomeric cyclitols are possible? Draw them in Haworth projection.

**24.47** Compound A is a D-aldopentose that can be oxidized to an optically inactive aldaric acid, compound B. On Kiliani–Fischer chain extension, compound A is converted into compounds C and D. Compound C can be oxidized to an optically active aldaric acid, E, but compound D is oxidized to an optically inactive aldaric acid, F. What are the structures of compounds A–F?

CHAPTER 25

# Aliphatic Amines

**Amines** are organic derivatives of ammonia in the same way that alcohols and ethers are organic derivatives of water. Amines are classified as either **primary ($RNH_2$), secondary ($R_2NH$),** or **tertiary ($R_3N$),** depending on the number of organic substituents attached to nitrogen. For example, methylamine, $CH_3NH_2$, is a primary amine; dimethylamine, $(CH_3)_2NH$, is a secondary amine; and trimethylamine, $(CH_3)_3N$, is a tertiary amine. Note that this usage of the terms *primary, secondary,* and *tertiary* is different from our previous usage. When we speak of a tertiary alcohol or alkyl halide, we refer to the degree of substitution at the alkyl carbon atom; when we speak of a tertiary amine, however, we refer to the degree of substitution at the nitrogen atom.

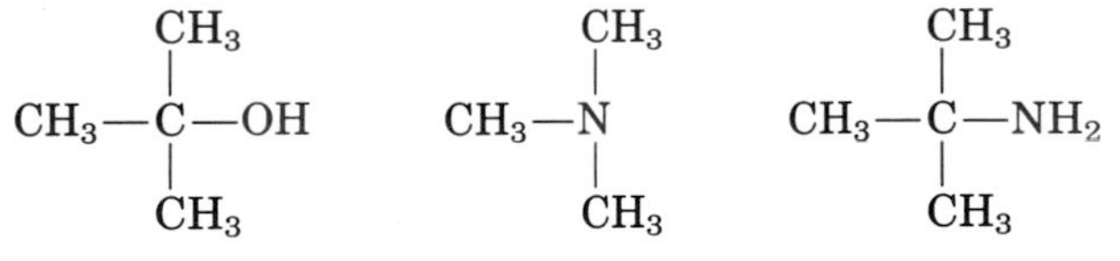

$$CH_3-\overset{\displaystyle CH_3}{\underset{\displaystyle CH_3}{\overset{|}{\underset{|}{C}}}}-OH \qquad CH_3-\overset{\displaystyle CH_3}{\underset{\displaystyle CH_3}{\overset{|}{\underset{|}{N}}}} \qquad CH_3-\overset{\displaystyle CH_3}{\underset{\displaystyle CH_3}{\overset{|}{\underset{|}{C}}}}-NH_2$$

*tert*-Butyl alcohol (a tertiary alcohol) | Trimethylamine (a tertiary amine) | *tert*-Butylamine (a primary amine)

Compounds with four groups attached to nitrogen are also possible, but the nitrogen atom must carry a positive charge. Such compounds are called **quaternary ammonium salts.**

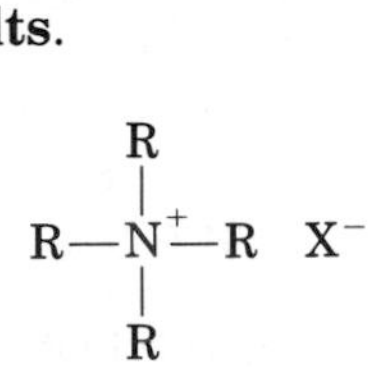

$$R-\overset{\displaystyle R}{\underset{\displaystyle R}{\overset{|}{\underset{|}{N^+}}}}-R \quad X^-$$

A quaternary ammonium salt

Amines can be either alkyl substituted or aryl substituted. Much of the chemistry of the two classes is similar, but there are sufficient differences that we will consider the classes separately. Arylamines will be discussed in Chapter 26.

$CH_3CH_2\ddot{N}H_2$ — Ethylamine (an aliphatic amine)

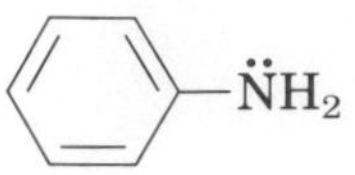

$C_6H_5-\ddot{N}H_2$ — Aniline (an arylamine)

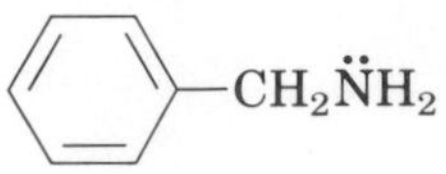

$C_6H_5-CH_2\ddot{N}H_2$ — Benzylamine (an aliphatic amine)

PROBLEM

**25.1** Classify these compounds as either primary, secondary, or tertiary amines, or as quaternary ammonium salts.

(a) $C_6H_{11}-NH_2$ (cyclohexyl)

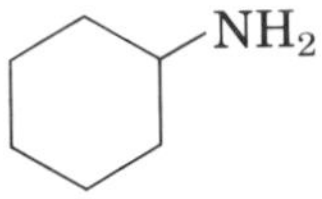

(b) $C_6H_{11}-N(CH_3)-CH_3$ (cyclohexyl)

(c) $C_6H_5-CH_2\overset{+}{N}(CH_3)_3\ ^{-}I$

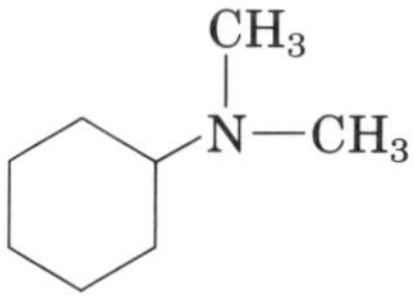

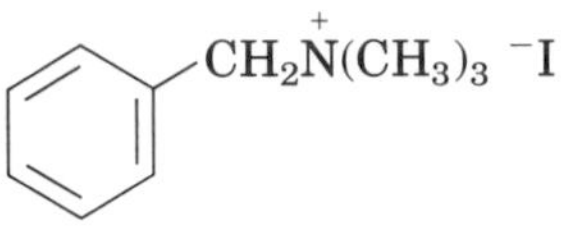

(d) $[(CH_3)_2CH]_2NH$

(e) pyrrolidine ring, N—H

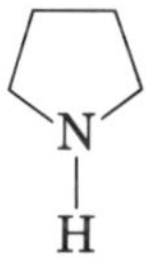

## 25.1 Nomenclature of Amines

Primary amines, $R-NH_2$, are named in the IUPAC system in several ways, depending on their structure. For relatively simple amines, the suffix *-amine* is added to the name of the alkyl substituent.

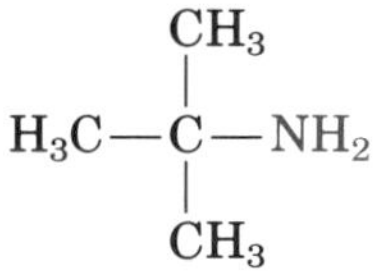

$H_3C-C(CH_3)_2-NH_2$ — *tert*-Butylamine

$C_6H_{11}-NH_2$ — Cyclohexylamine

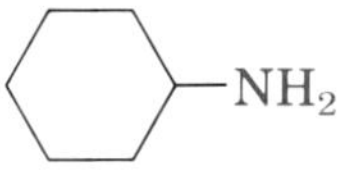

$H_2NCH_2CH_2CH_2CH_2NH_2$ — 1,4-Butanediamine

For more complex amines, it is also correct to add the suffix *-amine* in place of the final *-e* in the name of the parent compound.

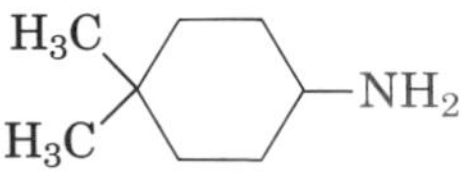

$(H_3C)_2C_6H_9-NH_2$

4,4-Dimethylcyclohexanamine

Amines having more than one functional group are named by considering the $-NH_2$ as an *amino* substituent on the parent molecule.

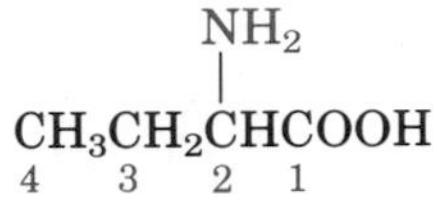
NH2
CH3CH2CHCOOH
4 3 2 1

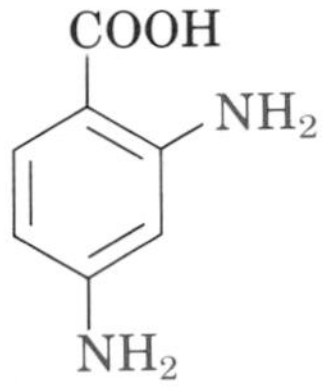
COOH
NH2
NH2

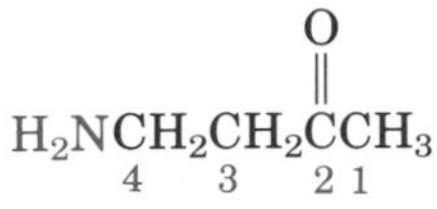
O
H2NCH2CH2CCH3
4 3 2 1

2-Aminobutanoic acid 2,4-Diaminobenzoic acid 4-Amino-2-butanone

Symmetrical secondary and tertiary amines are named by adding the prefix *di-* or *tri-* to the alkyl group.

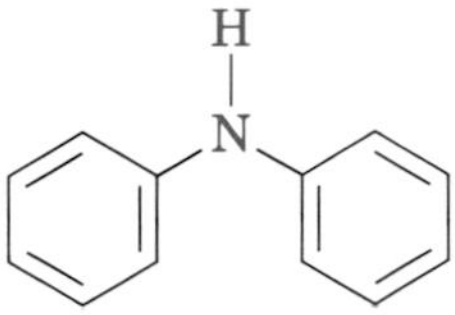
H
N

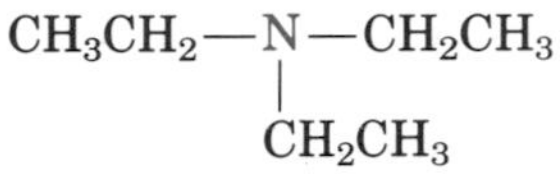
CH3CH2—N—CH2CH3
CH2CH3

Diphenylamine Triethylamine

Unsymmetrically substituted secondary and tertiary amines are named as *N*-substituted primary amines. The largest alkyl group is chosen as the parent name, and the other alkyl groups are considered *N*-substituents on the parent (*N* since they are attached to nitrogen).

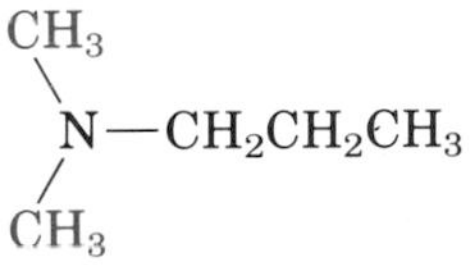
CH3
N—CH2CH2CH3
CH3

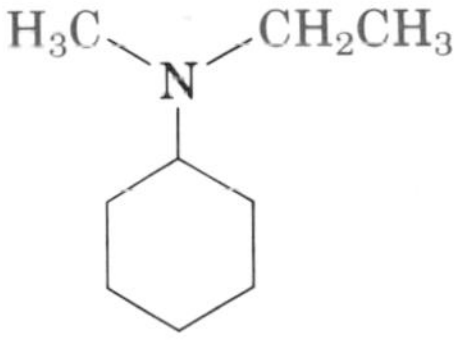
H3C
CH2CH3
N

*N,N*-Dimethylpropylamine (propylamine is the parent name; the two methyl groups are substituents on nitrogen)

*N*-Ethyl-*N*-methylcyclohexylamine (cyclohexylamine is the parent name; methyl and ethyl are *N*-substituents)

There are relatively few common names for simple amines, but IUPAC rules do recognize the names *aniline* and *toluidine* for aminobenzene and aminotoluene, respectively.

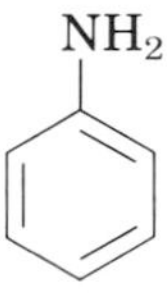
NH2

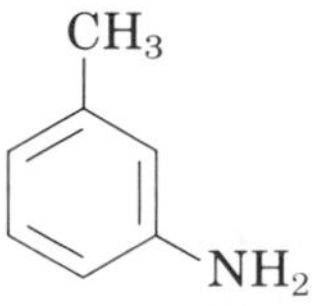
CH3
NH2

Aniline *m*-Toluidine

**Heterocyclic amines**—compounds in which the nitrogen atom occurs as part of a ring—are common, and each different heterocyclic ring system is given its own parent name. In all cases, the nitrogen atom is numbered as position 1.

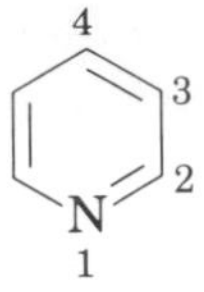

Pyridine 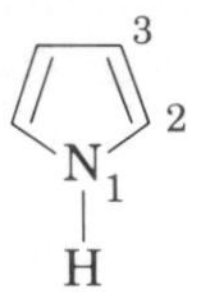    Pyrrole 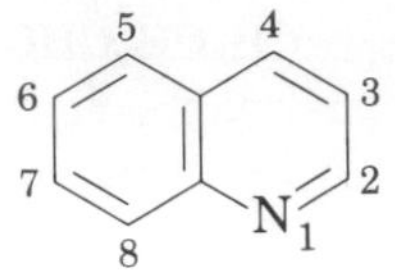    Quinoline

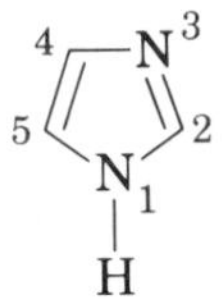

Imidazole 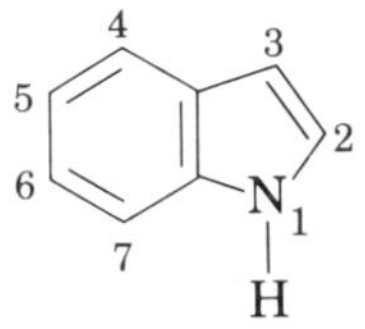    Indole 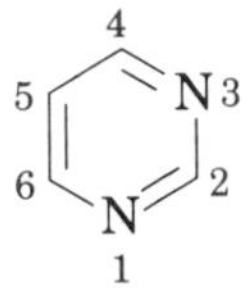    Pyrimidine

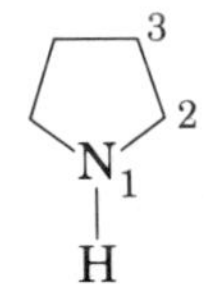

Pyrrolidine 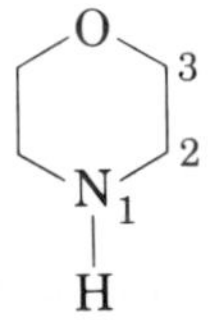    Morpholine 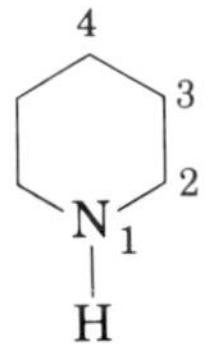    Piperidine

PROBLEM

**25.2** Name these compounds by IUPAC rules:

(a) $CH_3NHCH_2CH_3$

(b) 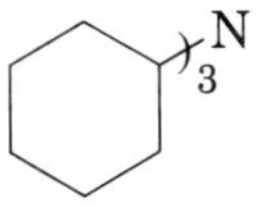 $(C_6H_{11})_3N$

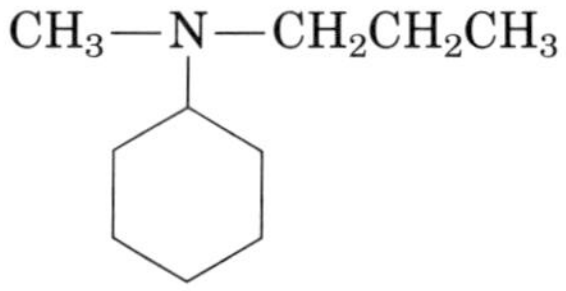

(c) $CH_3—N(C_6H_{11})—CH_2CH_2CH_3$

(d) 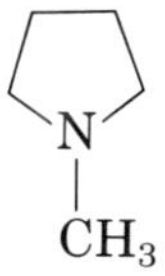 N-$CH_3$ pyrrolidine

(e) $[(CH_3)_2CH]_2NH$

(f) 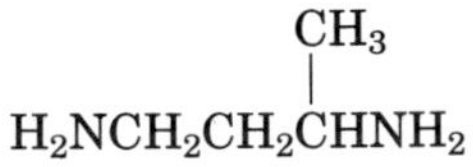 $H_2NCH_2CH_2CH(CH_3)NH_2$

PROBLEM

**25.3** Draw structures corresponding to these IUPAC names:

(a) Triethylamine
(b) Triallylamine
(c) *N*-Methylaniline
(d) *N*-Ethyl-*N*-methylcyclopentylamine
(e) *N*-Isopropylcyclohexylamine
(f) *N*-Ethylpyrrole

## 25.2 Structure and Bonding in Amines

Bonding in amines is similar to bonding in ammonia. The nitrogen atom is $sp^3$ hybridized, with the three substituents occupying three corners of a tetrahedron. The nitrogen's nonbonding lone pair of electrons occupies the fourth corner. As expected, the C-N-C bond angles are very close to the 109° tetrahedral value. For trimethylamine, the C-N-C bond angle is 108°, and the C—N bond length is 1.47 Å.

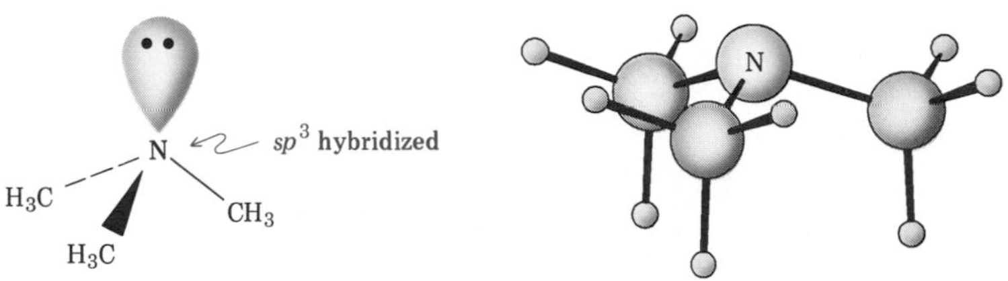


Trimethylamine

One consequence of tetrahedral geometry is that amines with three different substituents on nitrogen are chiral. Such an amine has no plane of symmetry and therefore is not superimposable on its mirror image. If we consider the lone pair of electrons to be the fourth substituent on nitrogen, these chiral amines are analogous to chiral alkanes with four different substituents attached to carbon:

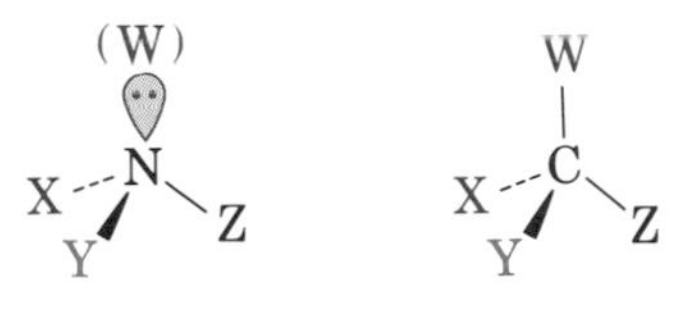


A chiral amine A chiral alkane

Unlike chiral alkanes, however, most chiral amines can't be resolved into their two enantiomers, because the two enantiomeric forms rapidly interconvert by a **pyramidal inversion**, much as an alkyl halide inverts in an $S_N2$ reaction. Pyramidal inversion occurs by a momentary rehybridization of the nitrogen atom to planar, $sp^2$ geometry, followed by rehybridization of the planar intermediate to tetrahedral, $sp^3$ geometry (Figure 25.1, page 898).

Spectroscopic studies have shown that the barrier to nitrogen inversion is about 6 kcal/mol (25 kJ/mol), a figure only twice as large as the barrier to rotation about a carbon–carbon single bond. Pyramidal inversion is therefore rapid at room temperature, and the two optically active forms cannot normally be isolated.

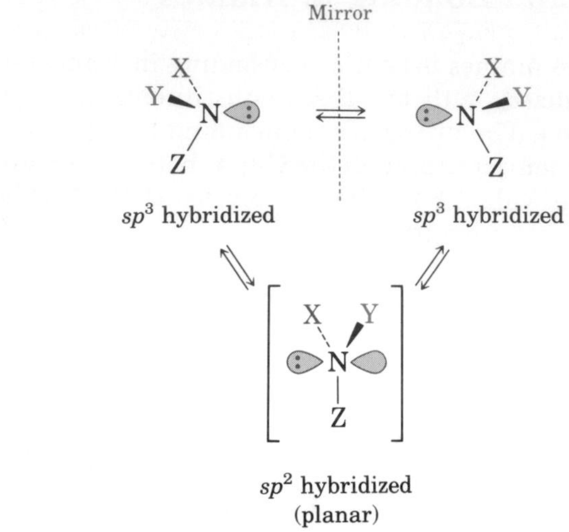

Mirror

X, Y, N, Z — $sp^3$ hybridized ⇌ X, Y, N, Z — $sp^3$ hybridized

[X, Y, N, Z] — $sp^2$ hybridized (planar)

**Figure 25.1** Pyramidal inversion of amines interconverts the two mirror-image forms.

PROBLEM

**25.4** Although rapid pyramidal inversion prevents chiral trialkylamines from being resolved, chiral tetraalkylammonium salts such as *N*-ethyl-*N*-methyl-*N*-propylbenzylammonium chloride are configurationally stable. Draw both the *R* and *S* enantiomers of this chiral ammonium salt and explain why they do not interconvert by pyramidal inversion.

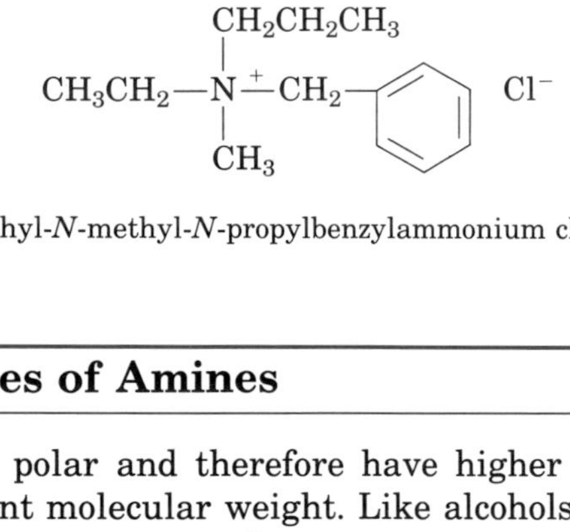

$$CH_3CH_2-\overset{CH_2CH_2CH_3}{\underset{CH_3}{N^+}}-CH_2-C_6H_5 \quad Cl^-$$

*N*-Ethyl-*N*-methyl-*N*-propylbenzylammonium chloride

## 25.3 Physical Properties of Amines

Amines are highly polar and therefore have higher boiling points than alkanes of equivalent molecular weight. Like alcohols, amines with fewer than five carbon atoms are generally water-soluble. Also like alcohols, primary and secondary amines form strong hydrogen bonds and are highly associated in the liquid state. The physical properties of some simple amines are given in Table 25.1.

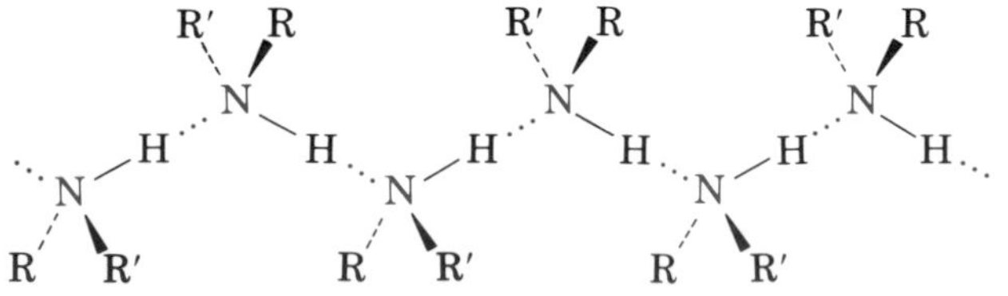

One characteristic of amines that does not show up in Table 25.1 is *odor*. All low-molecular-weight amines have a characteristic and distinctive fish-like aroma. Diamines such as putrescine (1,4-butanediamine) have names that are self-explanatory.

**Table 25.1** Physical properties of some simple amines

| *Name* | *Structure* | *Melting point (°C)* | *Boiling point (°C)* |
|---|---|---|---|
| Ammonia | $NH_3$ | −77.7 | −33.3 |
| **Primary amines** | | | |
| Methylamine | $CH_3NH_2$ | −94 | −6.3 |
| Ethylamine | $CH_3CH_2NH_2$ | −81 | 16.6 |
| *tert*-Butylamine | $(CH_3)_3CNH_2$ | −67.5 | 44.4 |
| Aniline (an arylamine) | $C_6H_5NH_2$ | −6.3 | 184.1 |
| **Secondary amines** | | | |
| Dimethylamine | $(CH_3)_2NH$ | −93 | 7.4 |
| Diethylamine | $(CH_3CH_2)_2NH$ | −48 | 56.3 |
| Diisopropylamine | $[(CH_3)_2CH]_2NH$ | −61 | 84 |
| Pyrrolidine | (ring) NH | 2 | 89 |
| **Tertiary amines** | | | |
| Trimethylamine | $(CH_3)_3N$ | −117 | 3 |
| Triethylamine | $(CH_3CH_2)_3N$ | −114 | 89.3 |
| *N*-Methylpyrrolidine | (ring) N—$CH_3$ | −21 | 81 |

## 25.4 Amine Basicity

The chemistry of amines is dominated by a single feature of their structure—the nitrogen lone pair of electrons. Because of the nitrogen lone pair, amines are both basic and nucleophilic. Amines react with Lewis acids to form acid/base salts, and they react with electrophiles in many of the polar reactions we've seen in past chapters.

$$\ddot{N}: \; + \; H{-}A \;\rightleftarrows\; \overset{+}{N}{-}H \; + \; :\bar{A}$$

An amine (a Lewis base) — An acid — A salt

Amines are much more basic than alcohols, ethers, or water, but not all amines are equal in base strength. Some amines are stronger and some are weaker. Just as we were able to measure the acid strength of a carboxylic acid (Section 20.3), we can also measure the base strength of an amine.

The most convenient way to measure the *basicity* of an amine is to look at the *acidity* ($pK_a$) of the corresponding ammonium salt. Since a strongly basic amine holds a proton tightly, its corresponding ammonium ion is relatively nonacidic (high $pK_a$). A weakly basic amine, however, holds a proton less tightly, and its corresponding ammonium ion is thus relatively acidic (low $pK_a$). This correlation between acidity and basicity is, of course, just another example of the general relationship we saw in Section 2.6: The conjugate base ($RNH_2$) that results from deprotonation of a weak acid ($RNH_3^+$) is a strong base, and the conjugate base of a strong acid is a weak base.

If this ammonium salt has a lower $pK_a$ (stronger acid), then this amine is a weaker base.

$$R{-}NH_3^+ + H_2O \rightleftharpoons R{-}\ddot{N}H_2 + H_3O^+$$

If this ammonium salt has a higher $pK_a$ (weaker acid), then this amine is a stronger base.

Table 25.2, which lists the measured $pK_a$'s of some common ammonium salts, indicates that there is relatively little effect of substitution on alkylamine basicity. The salts of most simple alkylamines have $pK_a$'s in the narrow range 10–11, regardless of their substitution pattern.

**Table 25.2** Basicity of some common alkylamines

| *Name* | *Structure* | $pK_a$ *of ammonium ion* |
|---|---|---|
| Ammonia | $:NH_3$ | 9.26 |
| **Primary alkylamine** | | |
| Methylamine | $CH_3\ddot{N}H_2$ | 10.64 |
| Ethylamine | $CH_3CH_2\ddot{N}H_2$ | 10.75 |
| **Secondary alkylamine** | | |
| Dimethylamine | $(CH_3)_2\ddot{N}H$ | 10.73 |
| Diethylamine | $(CH_3CH_2)_2\ddot{N}H$ | 10.94 |
| Pyrrolidine | (five-membered ring) :NH | 11.27 |
| **Tertiary alkylamine** | | |
| Trimethylamine | $(CH_3)_3N:$ | 9.79 |
| Triethylamine | $(CH_3CH_2)_3N:$ | 10.75 |

In contrast to amines, **amides** (**$RCONH_2$**) are completely nonbasic. Amides do not form salts when treated with aqueous acids; their aqueous solutions are neutral; and they are very poor nucleophiles. There are two main reasons for the difference in basicity between amines and amides. First, the ground state of an amide is stabilized by delocalization of the nitrogen lone-pair electrons through orbital overlap with the carbonyl group. In resonance terms, we can draw two contributing forms:

$$R{-}C({=}\ddot{O}{:})\text{–}\ddot{N}H_2 \longleftrightarrow R{-}C(\text{–}{:}\ddot{O}{:}^-){=}\overset{+}{N}H_2$$

Since this amide resonance stabilization is lost in the protonated product, protonation is disfavored (high $\Delta G°$). Second, a protonated amide is higher in energy than a protonated amine because the electron-withdrawing carbonyl group inductively destabilizes the neighboring positive charge.

Protonated amide (no resonance stabilization; inductive destabilization of positive charge)

$$R{-}\overset{\delta^+}{C}({=}\overset{\delta^-}{O}){-}\overset{+}{N}H_3$$

Both factors—increased stability of an amide versus an amine, and decreased stability of a protonated amide versus a protonated amine—lead to a large difference in $\Delta G°$ and a resultant large difference in basicity for amines and amides. Figure 25.2 shows these relationships.

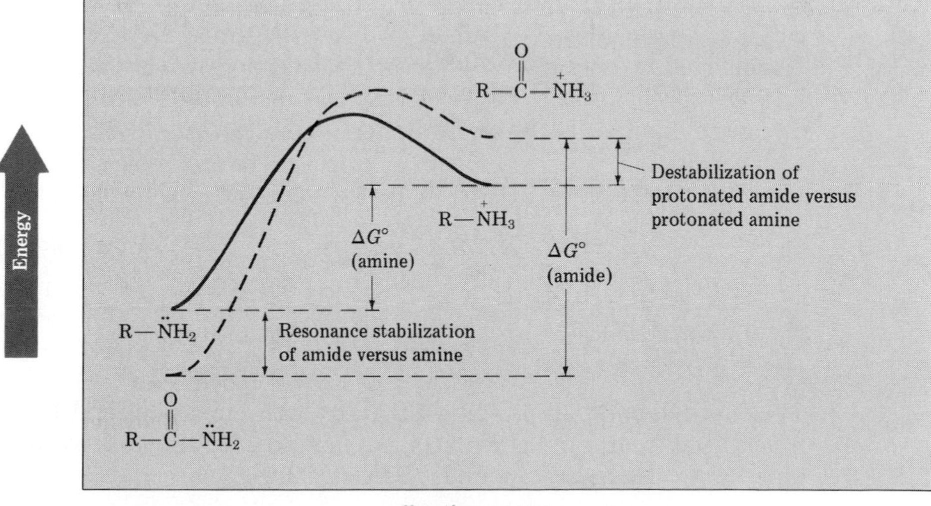


**Figure 25.2** Reaction energy diagram comparing the protonation of amines and amides: The $\Delta G°$ for amide protonation is larger because of increased reactant stability and decreased product stability.

We can often take advantage of the basicity of amines to purify them. For example, if we have a mixture of a basic amine and a neutral compound like a ketone, alcohol, or ether, we can simply dissolve the mixture in an organic solvent and extract with aqueous acid. The basic amine dissolves in the water layer as its protonated salt while the neutral compound remains in the organic solvent layer. Separation, basification, and extraction of the aqueous layer with organic solvent then allow us to recover pure amine (Figure 25.3).

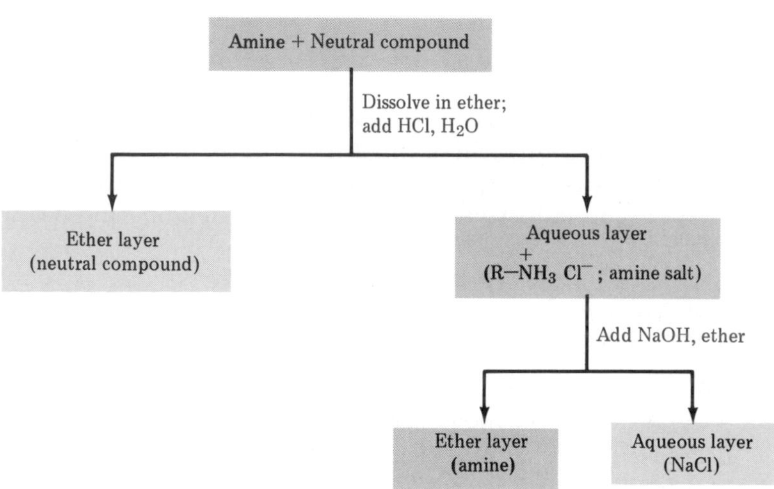


**Figure 25.3** Purification of an amine component from a mixture

In addition to their behavior as bases, primary and secondary amines can also be considered extremely weak *acids*, since their N—H protons can be removed by a sufficiently strong base. We've already seen, for example, how diisopropylamine ($pK_a \approx 40$) reacts with butyllithium to yield lithium diisopropylamide (LDA) and butane (Section 22.5):

$$[(CH_3)_2CH]_2\ddot{N}{-}H + n\text{-}C_4H_9Li \longrightarrow [(CH_3)_2CH]_2\ddot{N}:^-\ Li^+ + n\text{-}C_4H_{10}$$

Diisopropylamine ($pK_a \approx 40$) → Lithium diisopropylamide (LDA) + Butane ($pK_a \approx 50$)

Dialkylamide anions like LDA are extremely powerful bases that are much used in organic chemistry, particularly for the generation of enolate ions from carbonyl compounds (Section 22.9).

PROBLEM

**25.5** Which compound in each of the following pairs is more basic?
(a) $CH_3CH_2NH_2$ or $CH_3CH_2CONH_2$ (b) NaOH or $CH_3NH_2$
(c) $CH_3NHCH_3$ or $CH_3OCH_3$

PROBLEM

**25.6** Protonation of an amide actually occurs on oxygen rather than on nitrogen. Suggest a reason for this behavior.

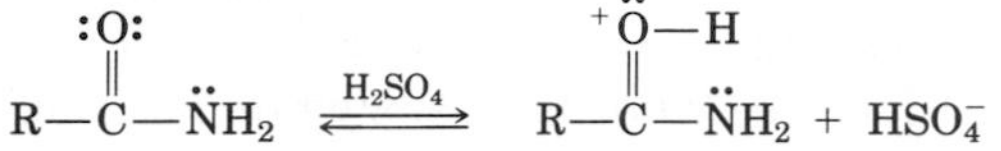

$$R{-}\overset{\overset{\displaystyle :O:}{\|}}{C}{-}\ddot{N}H_2 \xrightleftharpoons{H_2SO_4} R{-}\overset{\overset{\displaystyle {}^{+}\ddot{O}{-}H}{\|}}{C}{-}\ddot{N}H_2 + HSO_4^-$$

## 25.5 Resolution of Enantiomers via Amine Salts

We saw in the previous section that it's possible to take advantage of the basic properties of an amine to carry out its purification. We can also take advantage of amine basicity in another important way—to carry out the **resolution** of a racemic carboxylic acid into its two pure enantiomers. [Recall from Section 9.11 that a racemic mixture is a 50:50 mixture of (+) and (−) enantiomers.]

Historically, Louis Pasteur was the first person to resolve a racemic mixture when he was able to crystallize a salt of (±)-tartaric acid and to separate two different kinds of crystals by hand (Section 9.3). Pasteur's method is not generally applicable, however, since few racemic compounds crystallize into separate mirror-image forms. The most commonly used method of resolution makes use of an acid–base reaction between a racemic mixture of carboxylic acids and a chiral amine.

To understand how this method works, let's see what happens when a racemic mixture of chiral (+)- and (−)-lactic acids reacts with an achiral amine base like methylamine (Figure 25.4). Stereochemically, the situation is analogous to what happens when left and right hands (chiral) pick up a ball (achiral). Both left and right hands pick up the ball equally well, and the results—ball in right hand and ball in left hand—are mirror images.

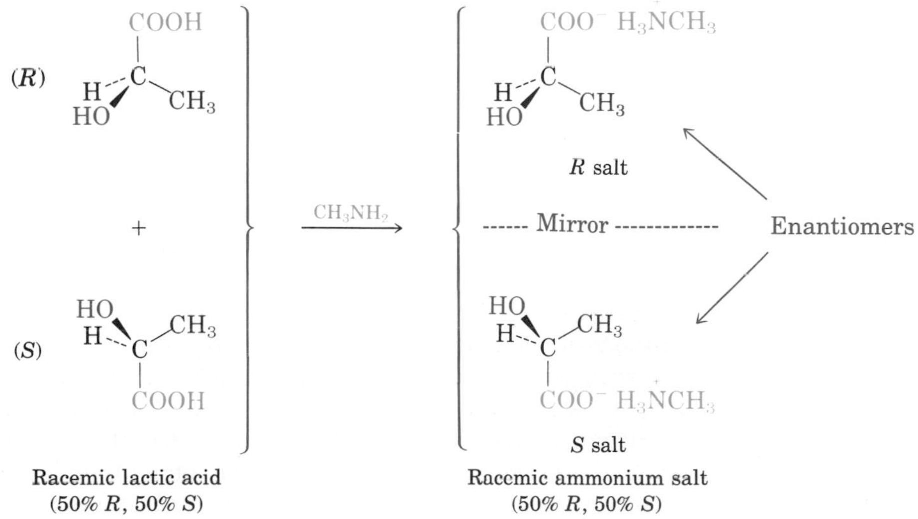

(R) COOH, H, HO, C, $CH_3$ + (S) HO, H, C, $CH_3$, COOH $\xrightarrow{CH_3NH_2}$ $COO^-\ H_3NCH_3$, H, HO, C, $CH_3$ — R salt; ------ Mirror ------ Enantiomers; HO, H, C, $CH_3$, $COO^-\ H_3NCH_3$ — S salt

Racemic lactic acid (50% R, 50% S) — Racemic ammonium salt (50% R, 50% S)

**Figure 25.4** Reaction of racemic lactic acid with achiral methylamine leads to a racemic mixture of ammonium salts.

In the same way, both (+)- and (−)-lactic acid react with methylamine equally well, and the product is a mixture of two salts: methylammonium (+)-lactate and methylammonium (−)-lactate. Just as with the chiral hands and the achiral ball, the two salts are mirror images and we still have a racemic mixture.

Now let's see what happens when the racemic mixture of (+)- and (−)-lactic acid reacts with a single enantiomer of a chiral amine base such as (*R*)-1-phenylethylamine (Figure 25.5). Stereochemically, the situation is analogous to what happens when a hand (chiral reagent) puts on a right-handed glove (also a chiral reagent). Left and right hands do not put on the *same* glove in the same way. The results—right hand in right glove and left hand in right glove—are not mirror images. They are altogether different.

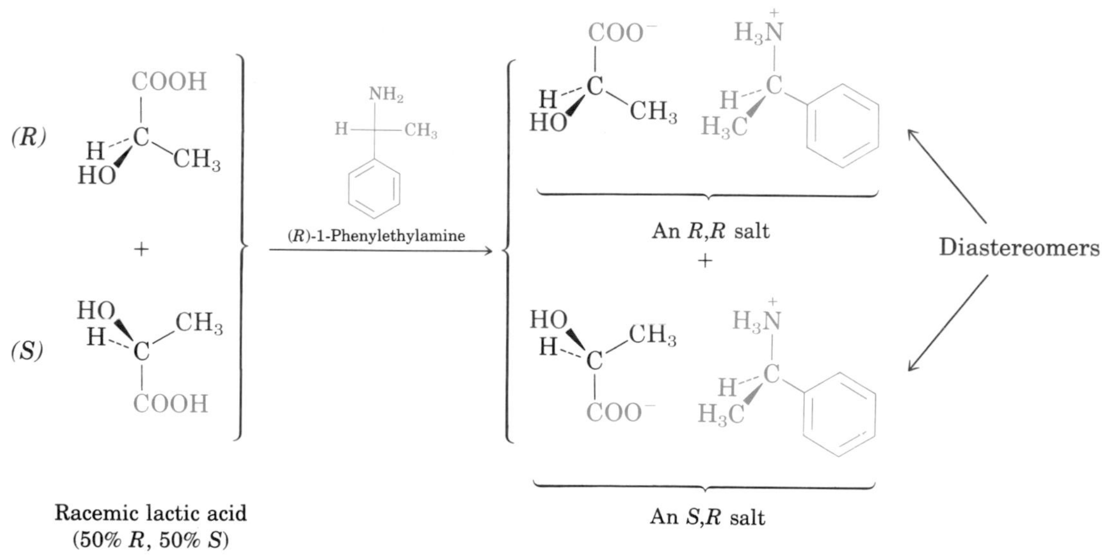

**Figure 25.5** Reaction of racemic lactic acid with pure (*R*)-1-phenylethylamine yields a mixture of diastereomeric ammonium salts.

In the same way, (+)- and (−)-lactic acid react with (*R*)-1-phenylethylamine to give two different products. (*R*)-Lactic acid reacts with (*R*)-1-phenylethylamine to give the *R*,*R* ammonium carboxylate salt, whereas (*S*)-lactic acid reacts with the same *R* amine to give the *S*,*R* salt. *These two salts are diastereomers*. They are different compounds and have different chemical and physical properties. It may therefore prove possible to separate them physically by fractional crystallization or by some other laboratory technique. Once separated, acidification of the two diastereomeric salts with mineral acid then allows us to isolate the two pure enantiomers of lactic acid and to recover the pure chiral amine for further use. A flow diagram for the overall process is shown in Figure 25.6.

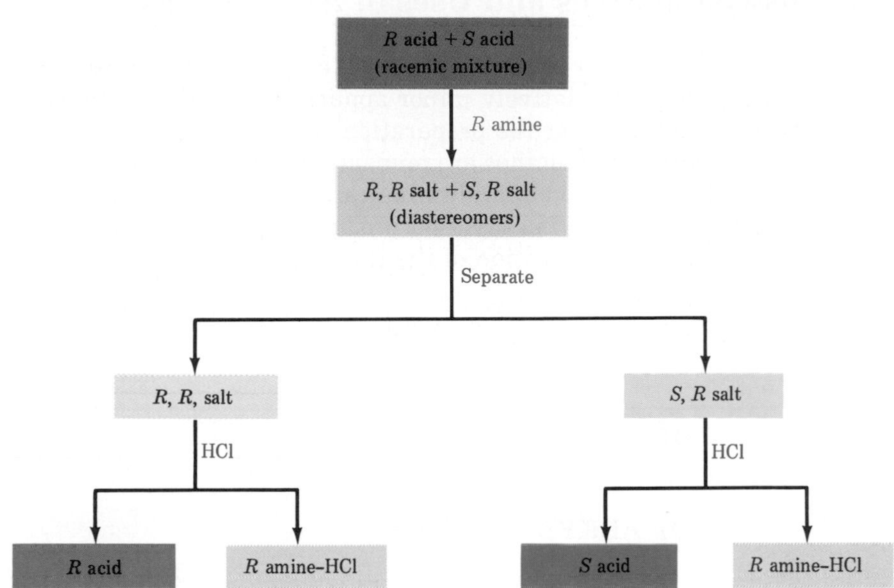


**Figure 25.6** Flow diagram for the resolution of a racemic carboxylic acid into its two pure enantiomers

The reaction between two enantiomers of a chiral acid and one enantiomer of a chiral base to yield diastereomeric salts is an example of a general rule: *Reaction between a racemic mixture and another chiral partner always yields a mixture of diastereomeric products.*

PROBLEM

**25.7** Suppose that racemic lactic acid reacts with methanol to yield the ester, methyl lactate. What stereochemistry would you expect the product(s) to have? What is the relationship of one product to another?

PROBLEM

**25.8** Suppose that a chiral acid such as (*S*)-lactic acid reacts with a chiral alcohol such as (*R*)-2-butanol:

$$(S)\ CH_3CH(OH)CO_2H \ + \ (R)\ CH_3CH_2CH(OH)CH_3$$

$$\xrightarrow[\text{catalyst}]{\text{Acid}} CH_3CH(OH)\overset{\overset{\displaystyle O}{\|}}{C}O\underset{\underset{\displaystyle CH_3}{|}}{C}HCH_2CH_3 \ + \ H_2O$$

What is the stereochemistry at the two chiral centers in the product? Draw the starting materials and product.

PROBLEM

**25.9** Suppose that racemic lactic acid reacts with (*S*)-2-butanol to form an ester. What stereochemistry does the product(s) have? What is the relationship of one product to another? How might you use this reaction to resolve (±)-lactic acid?

## 25.6 Industrial Sources and Uses of Alkylamines

Small alkylamines like methylamine, ethylamine, and dimethylamine all find a variety of relatively minor applications in the chemical industry as starting materials for the preparation of insecticides and pharmaceuticals. Simple methylated amines are prepared by reaction of ammonia with methanol in the presence of an alumina catalyst. The reaction yields a mixture of mono-, di-, and trimethylated products but is nonetheless useful industrially since the separation of the three products by distillation is easy.

$$NH_3 + CH_3OH \xrightarrow[450°C]{Al_2O_3} CH_3NH_2 + (CH_3)_2NH + (CH_3)_3N + H_2O$$

## 25.7 Synthesis of Amines

### $S_N2$ REACTIONS OF ALKYL HALIDES

Ammonia and other alkylamines are excellent nucleophiles in $S_N2$ reactions. As a result, the simplest method of amine synthesis is by $S_N2$ alkylation of ammonia or an alkylamine with an alkyl halide. If ammonia is used, a primary amine results; if a primary amine is used, a secondary amine results; and so on. Even tertiary amines react rapidly with alkyl halides to yield quaternary ammonium salts, $R_4N^+ X^-$.

| | | | | | |
|---|---|---|---|---|---|
| Ammonia | $\ddot{N}H_3 + R{-}X$ | $\longrightarrow$ | $RNH_3^+ \bar{X}$ | $\xrightarrow{NaOH} RNH_2$ | Primary |
| Primary | $R\ddot{N}H_2 + R{-}X$ | $\longrightarrow$ | $R_2NH_2^+ X^-$ | $\xrightarrow{NaOH} R_2NH$ | Secondary |
| Secondary | $R_2\ddot{N}H + R{-}X$ | $\longrightarrow$ | $R_3NH^+ X^-$ | $\xrightarrow{NaOH} R_3N$ | Tertiary |
| Tertiary | $R_3\ddot{N} + R{-}X$ | $\longrightarrow$ | $R_4N^+ X^-$ | Quaternary ammonium salt | |

$S_N2$ reaction

Unfortunately, these reactions do not stop cleanly after a single alkylation has occurred. Since primary, secondary, and even tertiary amines are all of similar reactivity, the initially formed monoalkylated product can undergo further reaction to yield a mixture of products. For example, treatment of 1-bromooctane with a twofold excess of ammonia leads to a mixture containing only 45% of the desired octylamine. A nearly equal amount of dioctylamine is produced by overalkylation, along with smaller amounts of trioctylamine and tetraoctylammonium bromide.

$$\underset{\text{1-Bromooctane}}{CH_3(CH_2)_6CH_2Br} + :NH_3 \longrightarrow \underset{\text{Octylamine (45\%)}}{CH_3(CH_2)_6CH_2\ddot{N}H_2} + \underset{\text{Dioctylamine (43\%)}}{[CH_3(CH_2)_6CH_2]_2\ddot{N}H}$$

$$+ \underset{\text{Trace}}{[CH_3(CH_2)_6CH_2]_3N:} + \underset{\text{Trace}}{[CH_3(CH_2)_6CH_2]_4\overset{+}{N}\overset{-}{Br}}$$

Higher yields of monoalkylated product can sometimes be obtained by using a large excess of the starting amine, but even so the reaction is a poor one, and better methods of synthesis are needed. Two such methods are often used—the **azide synthesis** and the **Gabriel synthesis**. Both methods are excellent ways to prepare primary amines from alkyl halides:

$$RCH_2Br \xrightarrow[\text{or Gabriel synthesis}]{\text{Via azide synthesis}} RCH_2NH_2$$

An alkyl halide — A primary amine

Azide ion, $N_3^-$, is a nonbasic, highly reactive nucleophile in $S_N2$ reactions. It displaces halide ion from primary and even secondary alkyl halides to give alkyl azides, $R{-}N_3$, in high yield. Since alkyl azides are not themselves nucleophilic, overalkylation cannot occur. Reduction of alkyl azides, either by catalytic hydrogenation over palladium or by reaction with $LiAlH_4$, leads to the desired primary amine. Yields are usually excellent, but the value of the process is tempered by the fact that low-molecular-weight alkyl azides are explosive and must be handled with great care.

$$C_6H_5CH_2CH_2Br \xrightarrow[S_N2\text{ reaction}]{NaN_3\text{, ethanol}} C_6H_5CH_2CH_2N{=}\overset{+}{N}{=}\overset{-}{N} + NaBr$$

1-Bromo-2-phenylethane — 2-Phenylethyl azide

$$\xrightarrow[2.\ H_2O]{1.\ LiAlH_4\text{, ether}} C_6H_5CH_2CH_2NH_2$$

2-Phenylethylamine (89%)

$$C_6H_{11}Br \xrightarrow{NaN_3} C_6H_{11}N{=}\overset{+}{N}{=}\overset{-}{N} + NaBr \xrightarrow[2.\ H_2O]{1.\ LiAlH_4\text{, ether}} C_6H_{11}NH_2$$

Bromocyclohexane — Cyclohexyl azide — Cyclohexylamine (54%)

The Gabriel[1] amine synthesis, introduced in 1887, makes use of an imide alkylation to provide an alternative means of preparing primary amines from alkyl halides. **Imides** (**—CONHCO—**) are structurally similar to ethyl acetoacetate in that the proton on nitrogen is flanked by two acidifying carbonyl groups. Thus, imides are readily deprotonated by bases

[1]Sigmund Gabriel (1851–1924); b. Berlin; Ph.D. University of Berlin (1874); assistant to A. W. von Hofmann; professor, University of Berlin.

such as KOH, and their resultant anions are readily alkylated in a reaction similar to the acetoacetic ester synthesis (Section 22.9). Basic hydrolysis of the *N*-alkyl phthalimide then yields a primary amine product. The reaction is best carried out in the highly polar solvent dimethylformamide, DMF [$HCON(CH_3)_2$]. Note that the imide hydrolysis step to yield amine plus phthalate ion is closely analogous to the hydrolysis of an amide (Section 21.8).

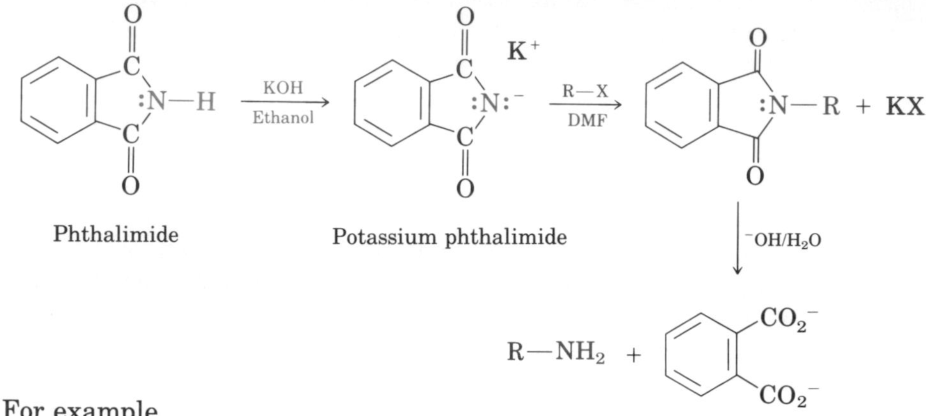

Phthalimide $\xrightarrow[\text{Ethanol}]{\text{KOH}}$ Potassium phthalimide ($K^+$) $\xrightarrow[\text{DMF}]{\text{R—X}}$ N-alkyl phthalimide + KX $\xrightarrow{^-OH/H_2O}$ $R—NH_2$ + phthalate ($CO_2^-$, $CO_2^-$)

For example,

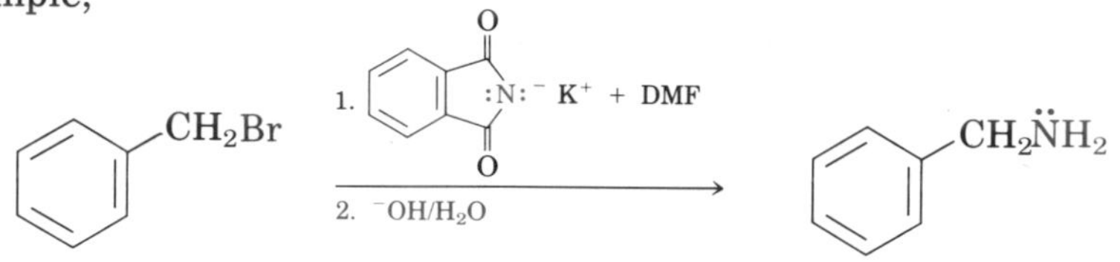

Benzyl bromide ($C_6H_5CH_2Br$) $\xrightarrow[\text{2. } ^-OH/H_2O]{\text{1. potassium phthalimide } (:N:^- K^+) \text{ + DMF}}$ Benzylamine ($C_6H_5CH_2NH_2$) (81%)

Recall the acetoacetic ester synthesis:

$$C_2H_5O—C(=O)—CH_2—C(=O)—CH_3 \xrightarrow[HOC_2H_5]{NaOC_2H_5} C_2H_5O—C(=O)—CH:^-—C(=O)—CH_3 \xrightarrow{RX} C_2H_5O—C(=O)—CH(R)—C(=O)—CH_3$$

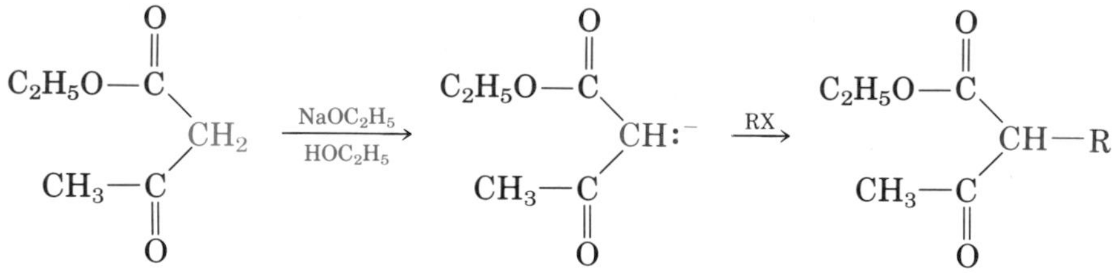

**PROBLEM**

**25.10** Propose a mechanism to account for the last step in the Gabriel amine synthesis, the base-promoted hydrolysis of an imide to yield an amine plus phthalate ion.

**PROBLEM**

**25.11** Show two methods for the synthesis of dopamine, a neurotransmitter involved in regulation of the central nervous system. Use any alkyl halide needed.

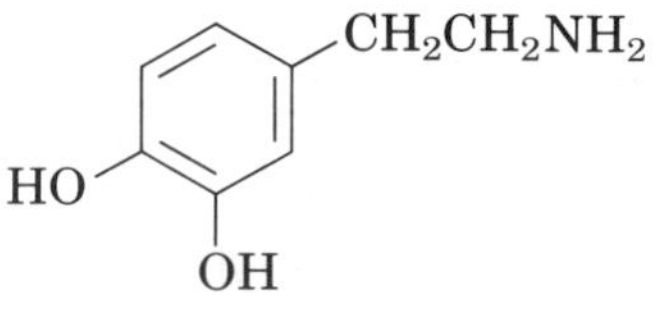

3,4-(HO)$_2C_6H_3CH_2CH_2NH_2$

Dopamine

## REDUCTION OF NITRILES AND AMIDES

We have already seen how amines can be prepared by reduction of nitriles (Section 21.9) and amides (Section 21.8) with $LiAlH_4$. Both reactions usually take place in high yield.

The two-step sequence of $S_N2$ displacement with cyanide ion followed by reduction provides an excellent method for converting alkyl halides into primary amines having one more carbon atom. Amide reduction provides an excellent method for converting carboxylic acids and their derivatives into amines:

$$\underset{\text{Alkyl halide}}{R{-}X} \xrightarrow{\text{NaCN}} R{-}CN \xrightarrow[\text{2. } H_2O]{\text{1. } LiAlH_4\text{, ether}} \underset{1^\circ\text{ amine}}{R{-}CH_2NH_2}$$

$$\underset{\text{Carboxylic acid}}{R{-}\overset{\overset{\displaystyle O}{\|}}{C}{-}OH} \xrightarrow[\text{2. } NH_3]{\text{1. } SOCl_2} R{-}\overset{\overset{\displaystyle O}{\|}}{C}{-}NH_2 \xrightarrow[\text{2. } H_2O]{\text{1. } LiAlH_4\text{, ether}} \underset{1^\circ\text{ amine}}{R{-}CH_2NH_2}$$

PROBLEM

**25.12** Propose structures of either a nitrile or an amide that might be a precursor of each of these amines.

(a) Propylamine
(b) Dipropylamine
(c) Benzylamine, $C_6H_5CH_2NH_2$
(d) *N*-Ethylaniline

## REDUCTIVE AMINATION OF KETONES AND ALDEHYDES

Amines can be synthesized in a single step by treatment of a ketone or aldehyde with ammonia or an amine in the presence of a reducing agent, a reaction called **reductive amination**. For example, amphetamine, a central nervous system stimulant, is prepared commercially by reductive amination of phenyl-2-propanone with ammonia, using hydrogen gas over a Raney nickel catalyst as reducing agent.

$$\underset{\text{Phenyl-2-propanone}}{C_6H_5CH_2\overset{\overset{\displaystyle O}{\|}}{C}CH_3} \xrightarrow[H_2/\text{Raney Ni}]{:NH_3} \underset{\text{Amphetamine}}{C_6H_5CH_2\overset{\overset{\displaystyle :NH_2}{|}}{C}HCH_3} + H_2O$$

Reductive amination takes place by the pathway shown in Figure 25.7 (page 910). As indicated, an imine intermediate is first formed by a nucleophilic addition reaction (Section 19.12), and the imine is then reduced.

Ammonia, primary amines, and secondary amines can all be used in the reductive amination reaction to yield primary, secondary, and tertiary amines, respectively.

Ammonia attacks the carbonyl group in a nucleophilic addition reaction to yield an intermediate carbinolamine.

$:NH_3$

The intermediate loses water to give an imine.

$-H_2O$

The imine is reduced catalytically over Raney nickel to yield the amine product.

$H_2$/Raney Ni

**Figure 25.7** Mechanism of reductive amination of a ketone to yield an amine

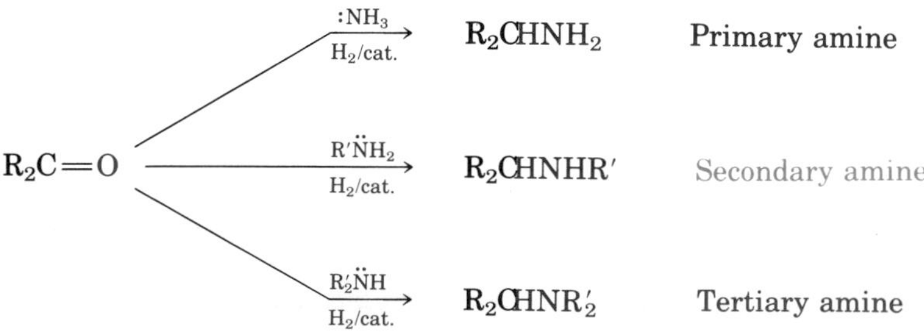


Many different reducing agents are effective, but the most common choice on a laboratory scale (as opposed to an industrial scale) is sodium cyanoborohydride, $NaBH_3CN$.

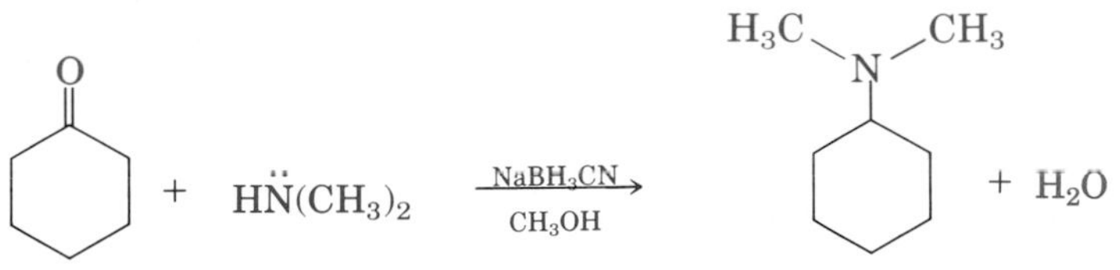

$$\text{Cyclohexanone} + H\ddot{N}(CH_3)_2 \xrightarrow[CH_3OH]{NaBH_3CN} N,N\text{-Dimethylcyclohexylamine (85\%)} + H_2O$$

Cyclohexanone

*N,N*-Dimethylcyclohexylamine (85%)

PROBLEM

**25.13** Show how these amines might be prepared by reductive amination of a ketone or aldehyde. Show all precursors if more than one is possible.
(a) $CH_3CH_2NHCH(CH_3)_2$ (b) *N*-Ethylaniline
(c) *N*-Methylcyclopentylamine

PROBLEM

**25.14** Show the mechanism of reductive amination of cyclohexanone and dimethylamine with $NaBH_3CN$. What intermediates are involved?

PROBLEM

**25.15** Ephedrine is an amino alcohol that is widely used for the treatment of bronchial asthma. Show how a reductive amination step might be used to synthesize ephedrine.

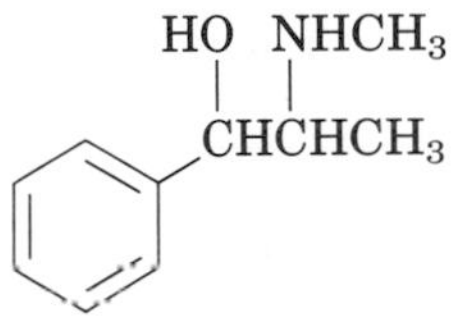

$C_6H_5CH(OH)CH(NHCH_3)CH_3$

Ephedrine

## HOFMANN AND CURTIUS REARRANGEMENTS OF AMIDES

Carboxylic acid derivatives can be converted into primary amines with loss of one carbon atom ($RCOY \rightarrow RNH_2$) by both the **Hofmann[2] rearrangement** and the **Curtius[3] rearrangement**. Although the Hofmann rearrangement involves a primary amide and the Curtius rearrangement involves an acyl azide, both proceed through similar mechanisms:

Hofmann rearrangement $$RCONH_2 \xrightarrow[H_2O]{^-OH,\ Br_2} RNH_2 + CO_2$$

Curtius rearrangement $$RCON_3 \xrightarrow[\Delta]{H_2O} RNH_2 + CO_2 + N_2$$

Hofmann rearrangement occurs when a primary amide, $RCONH_2$, is treated with halogen and base (Figure 25.8). Although the overall mechanism is lengthy, most of the individual steps have been encountered before. Thus, the bromination of an amide in steps 1 and 2 is analogous to the base-promoted bromination of a ketone enolate ion (Section 22.7), and the rearrangement of the bromoamide anion in step 4 is analogous to a carbocation

[2]August Wilhelm von Hofmann (1818–1892); b. Giessen, Germany; professor, Bonn, the Royal College of Chemistry, London (1845–1864), Berlin (1865–1892).
[3]Theodor Curtius (1857–1928); b. Duisberg; Ph.D. Leipzig; professor, universities of Kiel, Bonn, and Heidelberg (1898–1926).

Base abstracts an acidic amide proton, yielding an anion.

$$R{-}C(=O){-}NH_2 \xrightarrow[\text{①}]{:\ddot{O}H^-} R{-}C(=O){-}\ddot{N}H^- + H_2O$$

The anion reacts with bromine in an alpha-substitution reaction to give an *N*-bromoamide.

$$\xrightarrow[\text{②}]{Br{-}Br} R{-}C(=O){-}NHBr + :\ddot{B}r:^-$$

A bromoamide

Base abstraction of the remaining amide proton gives a bromoamide anion.

$$\xrightarrow[\text{③}]{:\ddot{O}H^-} R{-}C(=O){-}\ddot{N}^-{-}Br + H_2O$$

The bromoamide anion spontaneously rearranges as the R— group attached to the carbonyl carbon migrates to nitrogen at the same time the bromide ion leaves, yielding an isocyanate.

$$\xrightarrow{\text{④}} O{=}C{=}\ddot{N}{-}R + :\ddot{B}r:^-$$

An isocyanate

The isocyanate adds water in a nucleophilic addition step to yield a carbamic acid.

$$\xrightarrow[\text{⑤}]{:\ddot{O}H_2} H{-}O{-}C(=O){-}\ddot{N}H{-}R$$

A carbamic acid

The carbamic acid spontaneously loses $CO_2$, yielding the final product.

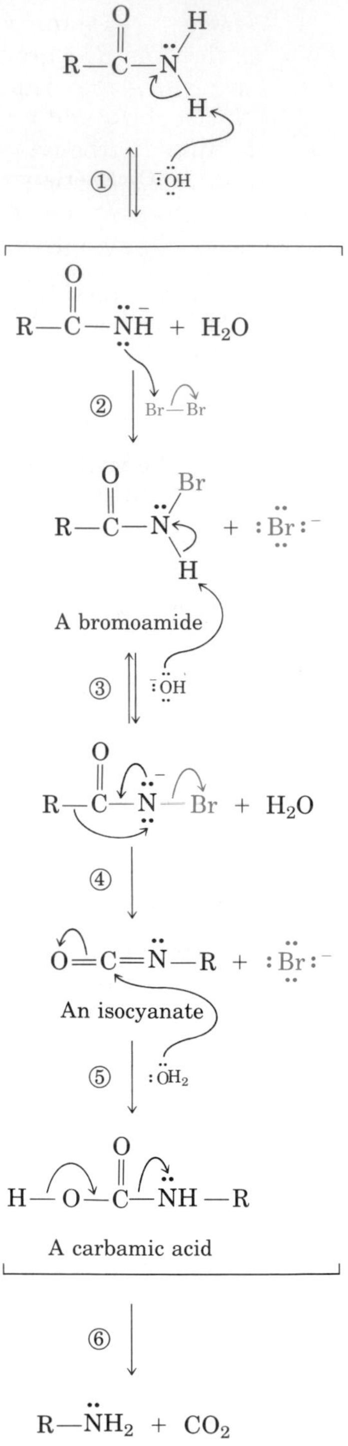

$$\xrightarrow{\text{⑥}} R{-}\ddot{N}H_2 + CO_2$$

**Figure 25.8** Mechanism of the Hofmann rearrangement of an amide to an amine: Each step is analogous to reactions studied previously.

rearrangement (Section 6.13). The primary difference between the migration step in a Hofmann rearrangement and that in a carbocation rearrangement is that the R— group begins its migration to the neighboring atom *at the same time* that the bromide ion is leaving, rather than after it has left. Nucleophilic addition of water to the isocyanate carbonyl group in step 5 is a typical carbonyl-group process (Section 19.8), as is the final decarboxylation step (Section 22.9).

Despite its mechanistic complexity, the Hofmann rearrangement often gives high yields of both aryl- and alkylamines. For example, the appetite-suppressing drug phentermine is prepared commercially by Hofmann rearrangement of a primary amide.

$$C_6H_5\text{—}CH_2\text{—}C(CH_3)_2\text{—}CONH_2 \xrightarrow[H_2O]{^-OH,\ Cl_2} C_6H_5\text{—}CH_2\text{—}C(CH_3)_2\text{—}NH_2$$

2,2-Dimethyl-3-phenylpropanamide — Phentermine

The Curtius rearrangement, like the Hofmann rearrangement, involves migration of an R— group from the carbonyl carbon atom to the neighboring nitrogen with concomitant loss of a leaving group. The reaction takes place on heating an acyl azide that is itself prepared by nucleophilic acyl substitution of an acid chloride:

$$R\text{—}\overset{O}{\overset{\|}{C}}\text{—}Cl \xrightarrow{NaN_3} R\text{—}\overset{O}{\overset{\|}{C}}\text{—}\ddot{N}^-\text{—}\overset{+}{N}{\equiv}N \xrightarrow{\Delta} N_2 + O{=}C{=}\ddot{N}\text{—}R \xrightarrow{H_2O} R\text{—}\ddot{N}H_2 + CO_2$$

Acid chloride — Acyl azide — Isocyanate — Amine

For example, the antidepressant drug tranylcypromine is prepared commercially by Curtius rearrangement of 2-phenylcyclopropanecarbonyl chloride.

$$\text{trans-}C_6H_5\text{-cyclopropane-COCl} \xrightarrow{1.\ NaN_3;\ 2.\ \Delta;\ 3.\ H_2O} \text{trans-}C_6H_5\text{-cyclopropane-}NH_2$$

*trans*-2-Phenylcyclopropanecarbonyl chloride — Tranylcypromine

**PROBLEM**

**25.16** Formulate the mechanism of the Curtius rearrangement of an acyl azide to an isocyanate, showing the origin and fate of all bonding electrons. Formulate also the mechanism of the addition of water to an isocyanate to yield a carbamic acid.

**PROBLEM**

**25.17** What starting materials would you use to prepare these amines by Hofmann and Curtius rearrangements?

(a) $(CH_3)_3CCH_2CH_2NH_2$

(b) $H_3C\text{—}C_6H_4\text{—}NH_2$ (para)

## 25.8 Reactions of Amines

We have already studied the two most important reactions of alkylamines—alkylation and acylation. As we saw earlier in this chapter, primary, secondary, and tertiary amines can all be alkylated by reaction with primary alkyl halides. Alkylations of primary and secondary amines are rather difficult to control and often give mixtures of products, but tertiary amines are cleanly alkylated to give quaternary (tetrasubstituted) ammonium salts.

Primary and secondary (but not tertiary) amines also can be acylated by reaction with acid chlorides or acid anhydrides (Sections 21.5 and 21.6) to yield amides.

$$NH_3 \xrightarrow[\text{Pyridine}]{\text{RCOCl}} R{-}\overset{O}{\overset{\|}{C}}{-}NH_2 + HCl$$

$$R'NH_2 \xrightarrow[\text{Pyridine}]{\text{RCOCl}} R{-}\overset{O}{\overset{\|}{C}}{-}NHR' + HCl$$

$$R'_2NH \xrightarrow[\text{Pyridine}]{\text{RCOCl}} R{-}\overset{O}{\overset{\|}{C}}{-}NR'_2 + HCl$$

If a **sulfonyl chloride ($RSO_2Cl$)** is used as the acylating agent, a sulfonamide ($R_2N{-}SO_2R'$) is produced. This reaction forms the basis of a classic laboratory test for distinguishing among primary, secondary, and tertiary amines (the **Hinsberg test**). A primary amine yields a sulfonamide product that has one remaining acidic N—H proton and can therefore be identified by its solubility in aqueous NaOH. A secondary amine yields a sulfonamide that has no acidic protons and can be identified by its lack of base solubility. A tertiary amine yields a product that reacts instantly with water to give back amine starting material. Thus no apparent reaction occurs with a tertiary amine.

Primary amine: $R\ddot{N}H_2 + C_6H_5{-}SO_2{-}Cl \xrightarrow{\text{Pyridine}} C_6H_5{-}SO_2{-}NHR$ — Base-soluble product

Benzenesulfonyl chloride — A sulfonamide

Secondary amine: $R_2\ddot{N}H + C_6H_5{-}SO_2{-}Cl \xrightarrow{\text{Pyridine}} C_6H_5{-}SO_2{-}NR_2$ — Base-insoluble product

$$\text{Tertiary amine} \quad R_3N: + C_6H_5SO_2Cl \xrightarrow{\text{Pyridine}} \left[ C_6H_5SO_2\overset{+}{N}R_3Cl^- \right] \xrightarrow{H_2O} R_3N: + C_6H_5SO_3H$$

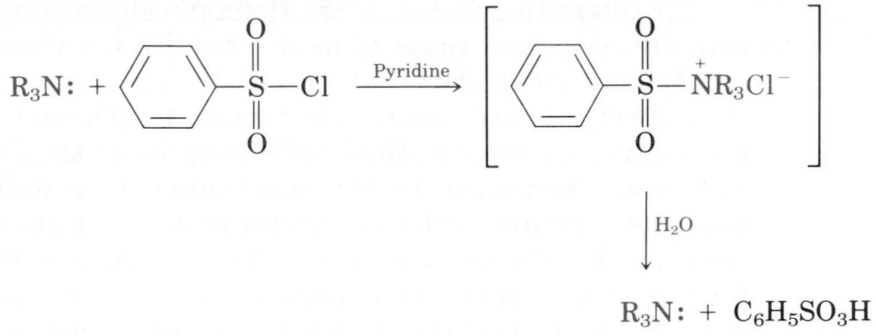

## HOFMANN ELIMINATION

Just as alcohols can be dehydrated to yield alkenes, amines can also be converted into alkenes under suitable conditions. In the **Hofmann elimination reaction**, an amine is first methylated with excess iodomethane to produce a quaternary ammonium iodide, which then undergoes an elimination reaction to give an alkene on heating with silver oxide. For example, 1-hexene is formed from hexylamine in 60% yield:

$$\underset{\text{Hexylamine}}{CH_3CH_2CH_2CH_2CH_2CH_2\ddot{N}H_2} \xrightarrow[\text{(excess)}]{CH_3I} \underset{\text{Hexyltrimethylammonium iodide}}{CH_3(CH_2)_3CH_2CH_2\overset{+}{N}(CH_3)_3\ I^-}$$

$$\xrightarrow{Ag_2O,\ H_2O} CH_3(CH_2)_3CH_2CH_2\overset{+}{N}(CH_3)_3\ \overset{-}{O}H + \mathbf{AgI}$$

$$\xrightarrow{\Delta} \underset{\text{1-Hexene (60\%)}}{CH_3CH_2CH_2CH_2CH{=}CH_2} + N(CH_3)_3 + H_2O$$

Silver oxide functions by exchanging hydroxide ion for iodide ion in the quaternary salt, thus providing the base necessary to effect elimination. The actual elimination step is an E2 reaction (Section 11.11) in which hydroxide ion removes a proton and the positively charged nitrogen atom acts as the leaving group (Figure 25.9).

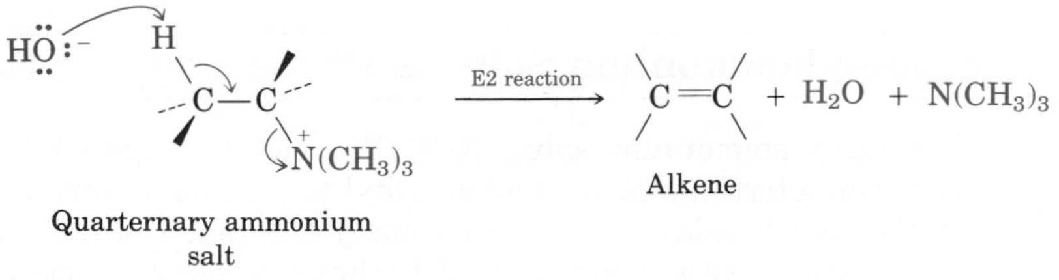

$$\underset{\text{Quarternary ammonium salt}}{H\ddot{O}:^- \quad H{-}C{-}C{-}\overset{+}{N}(CH_3)_3} \xrightarrow{\text{E2 reaction}} \underset{\text{Alkene}}{C{=}C} + H_2O + N(CH_3)_3$$

**Figure 25.9** Mechanism of the Hofmann elimination of an amine to yield an alkene

An interesting feature of the Hofmann elimination is that it gives products different from those of most other E2 reactions. Whereas the *more* highly substituted alkene product generally predominates in the E2 reaction of an alkyl halide (Zaitsev's rule; Section 11.11), the *less* highly substituted alkene predominates in the Hofmann elimination of a quaternary ammonium salt. The reasons for this selectivity are not well understood, but are probably steric in origin. Owing to the large size of the trialkylamine leaving group, hydroxide ion must abstract a hydrogen from the least hindered, most sterically accessible position. For example, (1-methylbutyl)trimethylammonium hydroxide yields 1-pentene and 2-pentene in a 94:6 ratio.

$$\underset{\text{(1-Methylbutyl)trimethylammonium hydroxide}}{CH_3CH_2CH_2\overset{\overset{+}{N}(CH_3)_3\ {}^{-}OH}{\overset{|}{C}}HCH_3} \xrightarrow[\Delta]{} \underbrace{\underset{\text{1-Pentene}}{CH_3CH_2CH_2CH{=}CH_2} + \underset{\text{2-Pentene}}{CH_3CH_2CH{=}CHCH_3}}_{\text{94:6 ratio}}$$

The Hofmann elimination reaction is important primarily because of its historical use as a degradative tool in the structure determination of many complex naturally occurring amines. The reaction is not often used today, since the product alkenes can usually be made more easily in other ways.

PROBLEM

**25.18** What products would you expect to obtain from Hofmann elimination of these amines. If more than one product is formed, indicate which is major.
(a) $CH_3CH_2CH_2CH(NH_2)CH_2CH_2CH_2CH_3$
(b) Cyclohexylamine
(c) $CH_3CH_2CH_2CH(NH_2)CH_2CH_2CH_3$
(d) *N*-Ethylcyclohexylamine

PROBLEM

**25.19** What product would you expect from Hofmann elimination of a cyclic amine such as piperidine? Formulate all the steps involved.

$$\text{Piperidine (NH)} \xrightarrow[\text{3. } \Delta]{\text{1. } CH_3I \text{ (excess)} \quad \text{2. } Ag_2O,\ H_2O} \ ?$$

## 25.9 Tetraalkylammonium Salts as Phase-Transfer Agents

Tetraalkylammonium salts, $R_4N^+\ X^-$, easily prepared by $S_N2$ reaction between a tertiary amine and an alkyl halide, have come to be widely used in the past decade as catalysts for many different kinds of organic reactions. For example, we saw in Section 7.10 that chloroform reacts with strong base to generate dichlorocarbene, which can then add to a carbon–carbon double bond to yield a dichlorocyclopropane.

Imagine an experiment where we dissolve cyclohexene in chloroform and stir the organic solution with 50% aqueous sodium hydroxide. Since the organic layer and the water layer are immiscible, the strong base in the aqueous phase is unable to come into contact with chloroform in the organic phase, and there is no reaction. If, however, we add a small amount of a tetraalkylammonium salt such as benzyltriethylammonium chloride to the two-phase mixture, an immediate reaction occurs, leading to formation of the dichlorocyclopropane product in 77% yield:

$$\text{cyclohexene} \xrightarrow[C_6H_5CH_2\overset{+}{N}(CH_2CH_3)_3\ Cl^-\ \text{(trace)}]{CHCl_3,\ 50\%\ NaOH/H_2O} \text{7,7-dichlorobicyclo[4.1.0]heptane (Cl, Cl)}$$

77%

No reaction occurs without $C_6H_5CH_2\overset{+}{N}(CH_2CH_3)_3Cl^-$

How does the tetraalkylammonium salt exert its catalytic effect? Benzyltriethylammonium ion, even though positively charged, is nevertheless soluble in organic solvents because of the four hydrocarbon substituents on nitrogen. When the positively charged tetraalkylammonium ion goes into the organic layer, a negatively charged counter-ion must also move into the organic layer to preserve charge neutrality. Hydroxide ion, present in far greater amount than chloride ion, is thus transferred into the organic phase where reaction with chloroform immediately occurs (Figure 25.10).

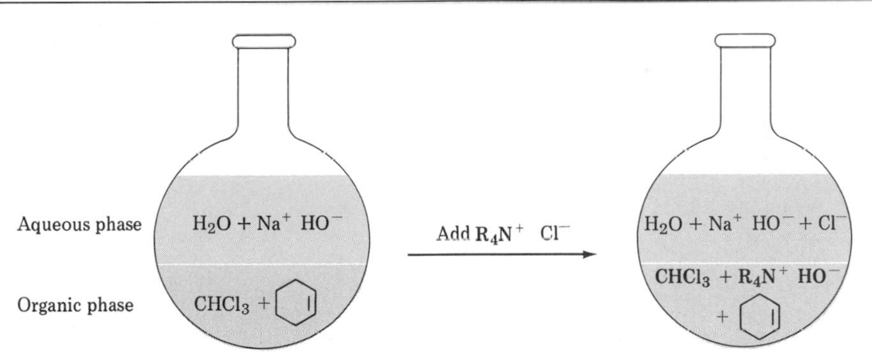


**Figure 25.10** Phase-transfer catalysis: Addition of a small amount of a tetraalkylammonium salt to a two-phase mixture allows an inorganic anion to be transferred from the aqueous phase into the organic phase, where a reaction can occur.

Transfer of an inorganic ion from one phase to another is called **phase transfer**, and the tetraalkylammonium salt is referred to as a **phase-transfer catalyst**. Many different kinds of organic reactions, including oxidations, reductions, carbonyl-group alkylations, and $S_N2$ reactions, are subject to phase-transfer catalysis, often with considerable improvements in yield. $S_N2$ reactions are particularly good candidates for phase-transfer catalysis, since inorganic nucleophiles can be transferred from an aqueous (protic) phase to an organic (aprotic) phase, where they are far more reactive. For example,

$$\underset{\text{1-Bromooctane}}{CH_3(CH_2)_6CH_2Br} + NaCN \xrightarrow[C_6H_5CH_2\overset{+}{N}(CH_2CH_3)_3\ Cl^-]{H_2O,\ \text{benzene}} \underset{\text{Nonanenitrile (92\%)}}{CH_3(CH_2)_6CH_2CN} + NaBr$$

## 25.10 Naturally Occurring Amines: Alkaloids

A great variety of amines are widely distributed among plants and animals. For example, trimethylamine occurs in animal tissues and is partially responsible for the distinctive odor of many fish; morphine is a powerful analgesic (painkiller) isolated from the opium poppy; quinine is an important antimalarial drug isolated from the bark of the South American *Cinchona* tree; and reserpine is a useful antihypertensive (blood-pressure-lowering) agent isolated from the Indian shrub *Rauwolfia serpentina*.

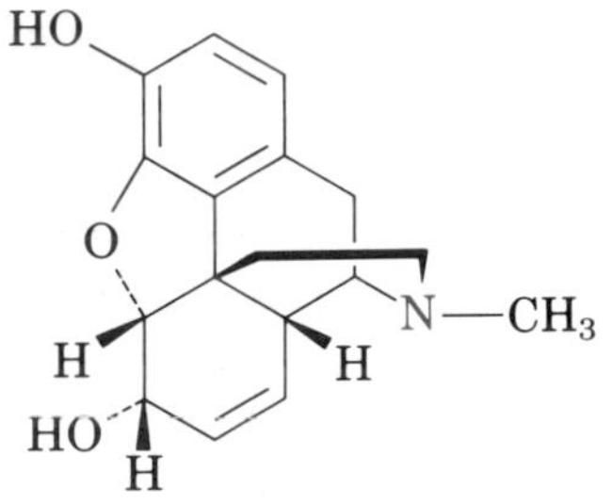

Morphine (analgesic)

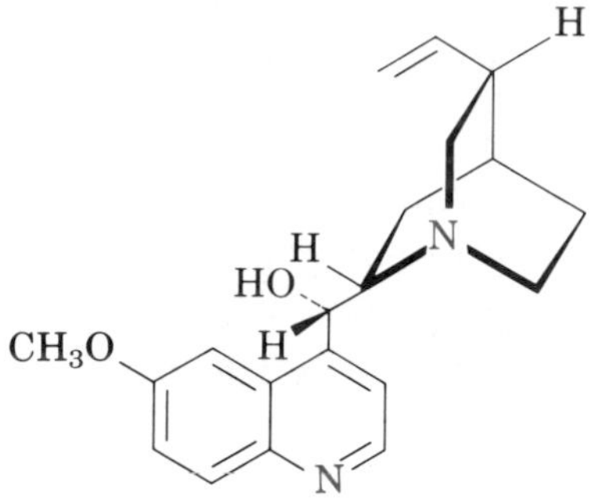

Quinine (antimalarial)

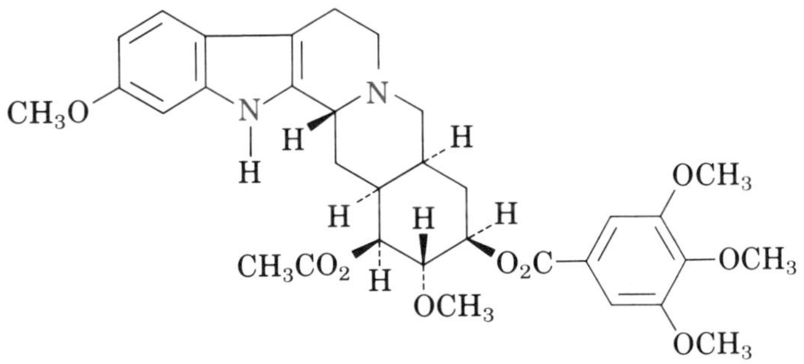

Reserpine (antihypertensive)

Naturally occurring amines derived from plant sources were once known as "vegetable alkali," since their aqueous solutions are basic, but they are now referred to as **alkaloids**. The study of alkaloids provided much of the impetus for the growth of organic chemistry in the nineteenth century and remains a fascinating area of research. Rather than attempt to classify the many kinds of alkaloids, let's look briefly at one particular group, the morphine alkaloids.

The medical uses of morphine alkaloids have been known at least since the seventeenth century, when crude extracts of the opium poppy, *Papaver somniferum*, were used for the relief of pain. Morphine was the first pure alkaloid to be isolated from the poppy, but its close relative, codeine, also occurs naturally. Codeine, which is simply the methyl ether of morphine, is used in prescription cough medicines. Heroin, another close relative of morphine, does not occur naturally but is synthesized by diacetylation of morphine.

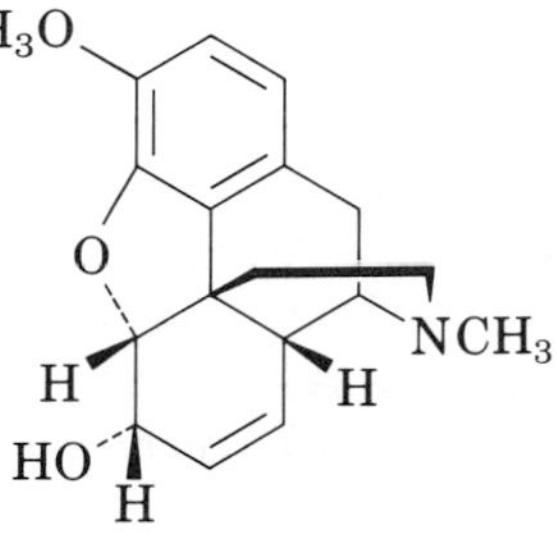

$CH_3O$ $NCH_3$ H H HO H O

Codeine

$CH_3\overset{O}{\overset{\|}{C}}O$ $NCH_3$ O H H $CH_3\underset{\underset{O}{\|}}{C}O$ H

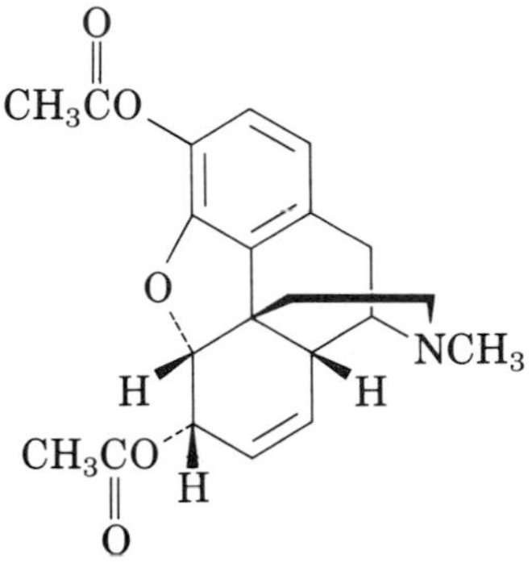

Heroin

Chemical investigations into the structure of morphine occupied some of the finest chemical minds of the nineteenth and early twentieth centuries, until the puzzle was finally solved by Robert Robinson in 1925. The key reaction used to establish structure was the Hofmann elimination.

Morphine and its relatives constitute a class of exceedingly useful pharmaceutical agents, yet they also pose a social problem of great proportion because of their addictive properties. Much effort has therefore gone into a search to understand the mode of action of morphine and to develop modified morphine analogs that retain the desired analgesic activity but do not cause physical dependence.

Our present understanding is that morphine appears to bind to opiate receptor sites in the brain. It does not interfere with or lessen the transmission of a pain signal to the brain, but rather changes the brain's reception of the signal. Much progress has been made in the search for modified morphine-like agents. For example, replacement of the morphine *N*-methyl group by an *N*-allyl substituent yields *nalorphine*, an analgesic agent that acts as a narcotic *antagonist* to reverse many of the undesirable side effects of morphine.

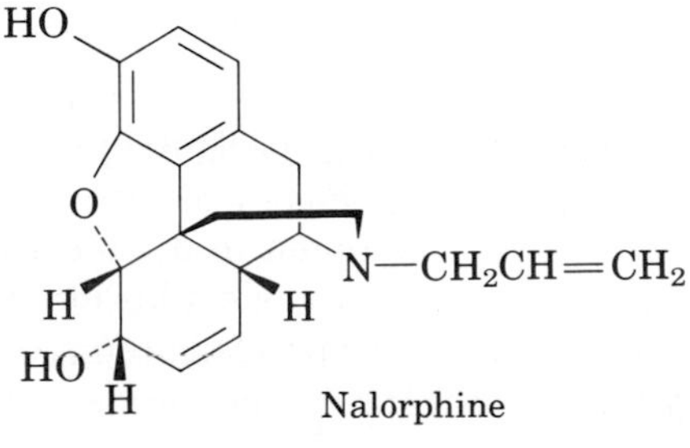

HO O $N{-}CH_2CH{=}CH_2$ H H HO H

Nalorphine

Studies have shown that not all of the complex tetracyclic framework of morphine is necessary for biological activity. According to the "morphine rule," biological activity requires: (1) an aromatic ring attached to (2) a quaternary carbon atom, and (3) a tertiary amine situated (4) two carbon atoms farther away. Thus, the bicyclic amine meperidine (Demerol) is widely used as a painkiller, and methadone has been used in the treatment of heroin addiction.

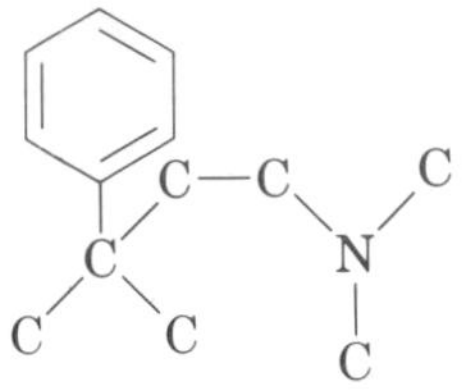

The morphine rule: an aromatic ring, a quaternary carbon, two carbons, a tertiary amine

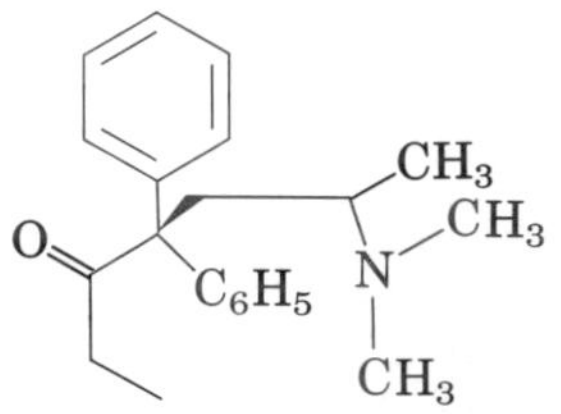

Methadone

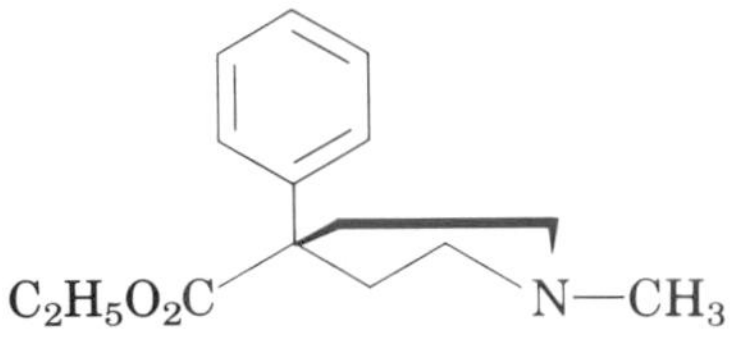

Meperidine

PROBLEM

**25.20** Show how the morphine rule fits the structure of dextromethorphan, a common constituent of cough remedies.

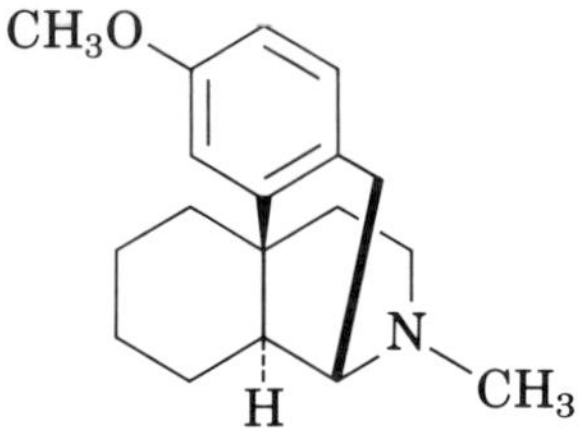

Dextromethorphan

## 25.11 Spectroscopy of Amines

### MASS SPECTROSCOPY

The "nitrogen rule" of mass spectroscopy says that compounds with an odd number of nitrogen atoms have odd-numbered molecular weights. Thus, we detect the presence of nitrogen in a molecule simply by observing its mass spectrum. An odd-numbered molecular ion usually means that the unknown compound has one or three nitrogen atoms, and an even-numbered molecular ion usually means that the compound has either zero or two nitrogen atoms.

The logic behind this rule derives from the fact that nitrogen is trivalent, thus requiring an odd number of hydrogen atoms in the molecule. For example, methylamine has the formula $CH_5N$ and a molecular weight of 31; morphine has the formula $C_{17}H_{19}NO_3$ and a molecular weight of 285.

Aliphatic amines undergo a characteristic alpha cleavage in the mass spectrometer, similar to the cleavage observed for aliphatic alcohols (Section 17.11). A carbon–carbon bond nearest the nitrogen atom is broken, yielding an alkyl radical and a nitrogen-containing cation:

$$\left[ RCH_2 \overset{}{\underset{\alpha}{\wr}} CH_2 - N \begin{matrix} R' \\ R' \end{matrix} \right]^{+\cdot} \xrightarrow{\text{Alpha cleavage}} RCH_2\cdot + \left[ CH_2 - N \begin{matrix} R' \\ R' \end{matrix} \right]^{+}$$

For example, the mass spectrum of *N*-ethylpropylamine shown in Figure 25.11 exhibits peaks at $m/z$ = 58 and $m/z$ = 72, corresponding to the two possible modes of alpha cleavage.

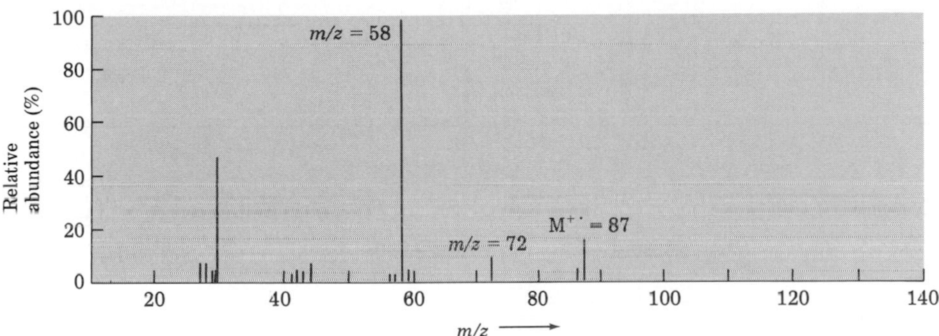


$$\left[ CH_3 \underset{\alpha}{\wr} CH_2 - \overset{H}{\overset{|}{N}} - CH_2 \underset{\alpha}{\wr} CH_2CH_3 \right]^{+\cdot} \quad m/z = 87$$

Alpha cleavage

$$\longrightarrow \cdot CH_3 + \left[ CH_2 = \overset{H}{\overset{|}{N}} - CH_2CH_2CH_3 \right]^{+} \quad m/z = 72$$

$$\longrightarrow \left[ CH_3CH_2 - \overset{H}{\overset{|}{N}} = CH_2 \right]^{+} + \cdot CH_2CH_3 \quad m/z = 58$$

**Figure 25.11** Mass spectrum of *N*-ethylpropylamine: The two possible modes of alpha cleavage occur, leading to the observed fragment ions at $m/z$ = 58 and $m/z$ = 72.

## INFRARED SPECTROSCOPY

Primary and secondary amines can be identified by characteristic N—H bond stretching absorptions in the 3300–3500 $cm^{-1}$ range of the infrared spectrum. Alcohols also absorb in this range (Section 17.11), but amine absorption bands are generally both sharper and less intense than hydroxyl bands. Primary amines show a pair of bands at about 3400 and 3500 $cm^{-1}$, and secondary amines show a single band at 3350 $cm^{-1}$. Tertiary amines show no absorption in this region, since they have no N—H protons. Representative infrared spectra of both primary and secondary amines are shown in Figure 25.12.

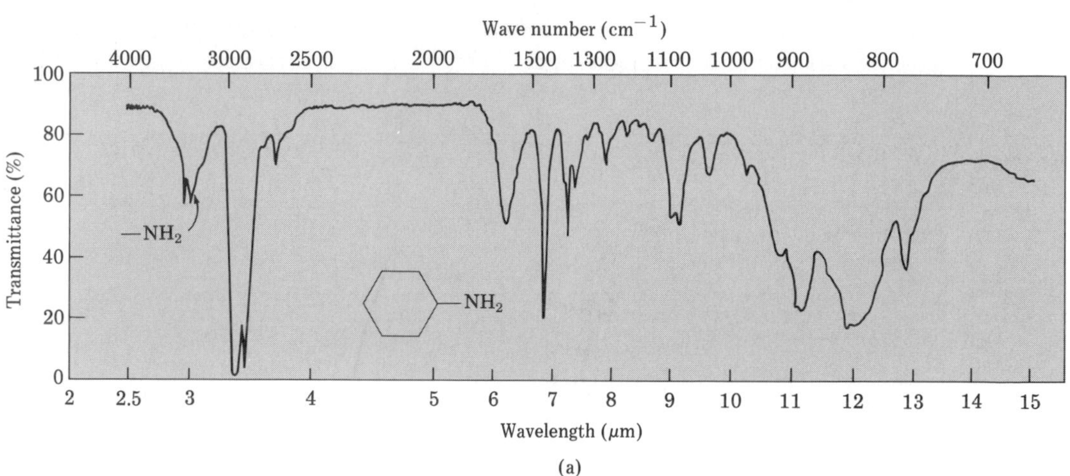


(a)

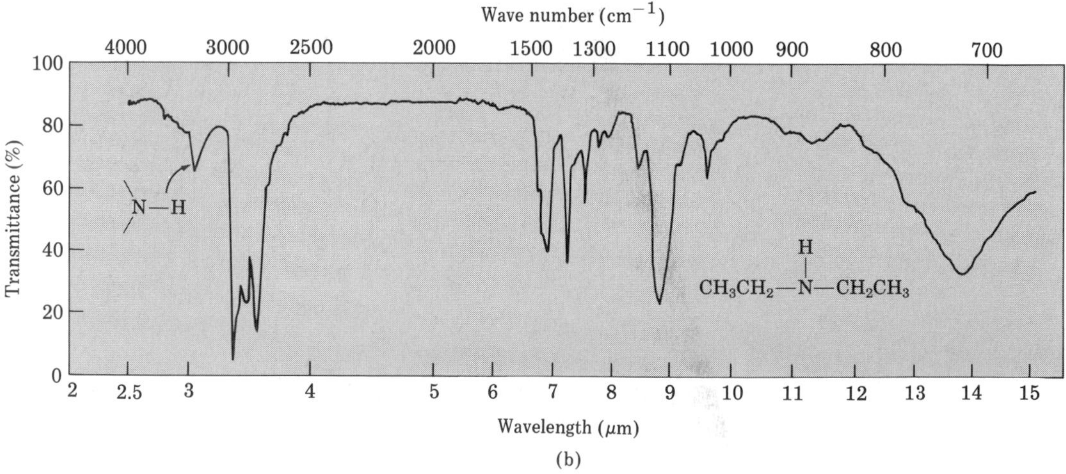


(b)

**Figure 25.12** Infrared spectra of (a) cyclohexylamine and (b) diethylamine

In addition to looking for characteristic N—H bands, there's also a simple trick that can be used to tell whether or not a given unknown is an amine. Addition of a small amount of mineral acid produces a broad and strong ammonium band in the 2200–3000 cm$^{-1}$ range if the sample contains an amino group. All protonated amines give rise to this readily observable absorption caused by the ammonium $R_3N—H^+$ bond. Figure 25.13 gives an example.

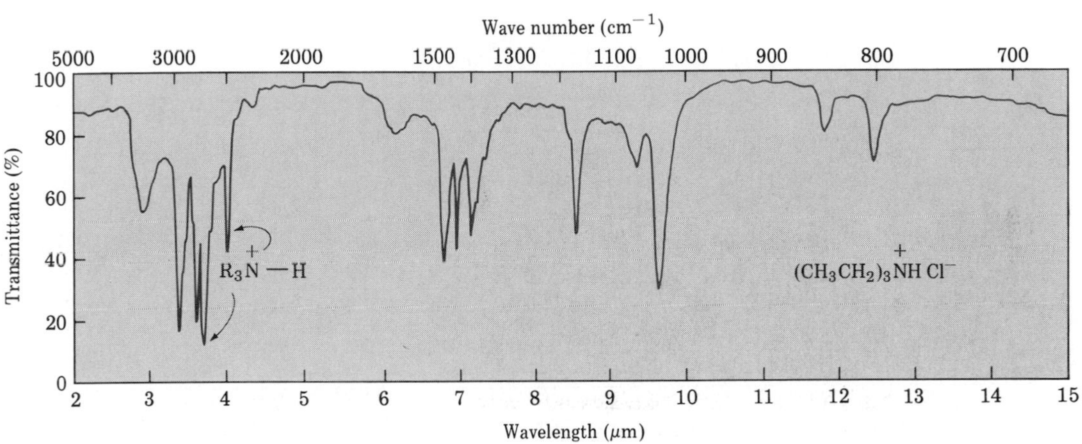


**Figure 25.13** Infrared spectrum of triethylammonium chloride

## NUCLEAR MAGNETIC RESONANCE SPECTROSCOPY

Amines are often difficult to identify solely by $^1$H NMR spectroscopy because N—H protons tend to appear as very broad resonances without clear-cut coupling to neighboring C—H protons. The situation is similar to that for hydroxyl protons (Section 17.11). As with hydroxyl O—H protons, amine N—H proton absorptions can appear over a wide range and are best identified by adding a small amount of $D_2O$ to the sample tube. Exchange of N—D for N—H occurs, and the N—H signal disappears from the NMR spectrum.

$$\rangle N—H \overset{D_2O}{\rightleftharpoons} \rangle N—D + HDO$$

Protons on the carbon next to nitrogen are somewhat deshielded because of the electron-withdrawing effect of the nitrogen, and they therefore absorb at lower field than alkane protons. *N*-Methyl groups are particularly distinctive, since they absorb as a sharp three-proton singlet at 2.2–2.6 $\delta$. The $N$-methyl resonance at 2.42 $\delta$ is easily seen in the $^1$H NMR spectrum of $N$-methylcyclohexylamine (Figure 25.14).

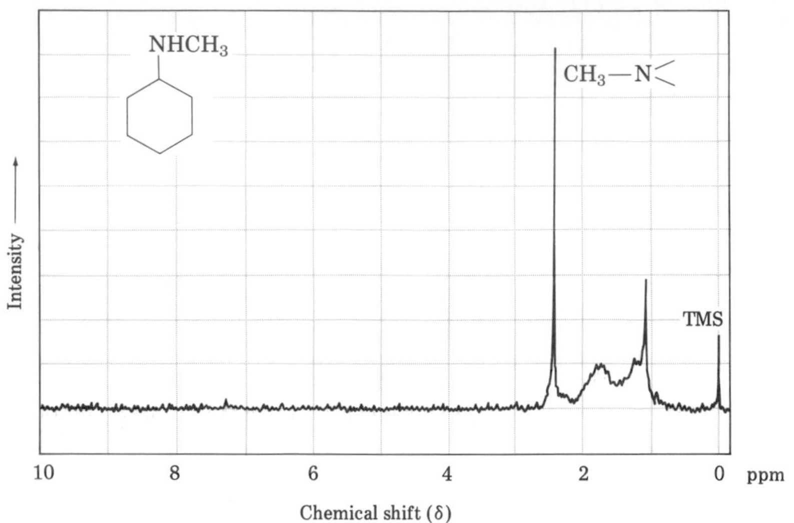


**Figure 25.14** Proton NMR spectrum of $N$-methylcyclohexylamine

Carbons next to amine nitrogens are slightly deshielded in the $^{13}$C NMR and absorb about 20 ppm downfield from where they would absorb in an alkane of similar structure. For example, in $N$-methylcyclohexylamine, the ring carbon to which nitrogen is attached absorbs at a position 24 ppm lower than that of any other ring carbon (Figure 25.15).

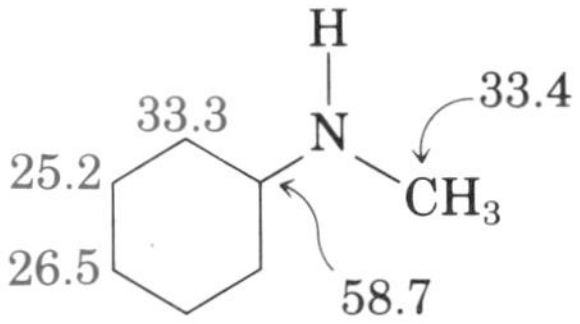

**Figure 25.15** Carbon-13 NMR absorptions (ppm) for $N$-methylcyclohexylamine

PROBLEM

**25.21** The infrared and $^1$H NMR spectra shown at the top of the next page are those of methamphetamine, which has a molecular ion in the mass spectrum at $m/z$ = 149. Propose a structure for methamphetamine and justify your answer. [*Note*: The signal at 1.2 $\delta$ disappears when $D_2O$ is added to the sample.]

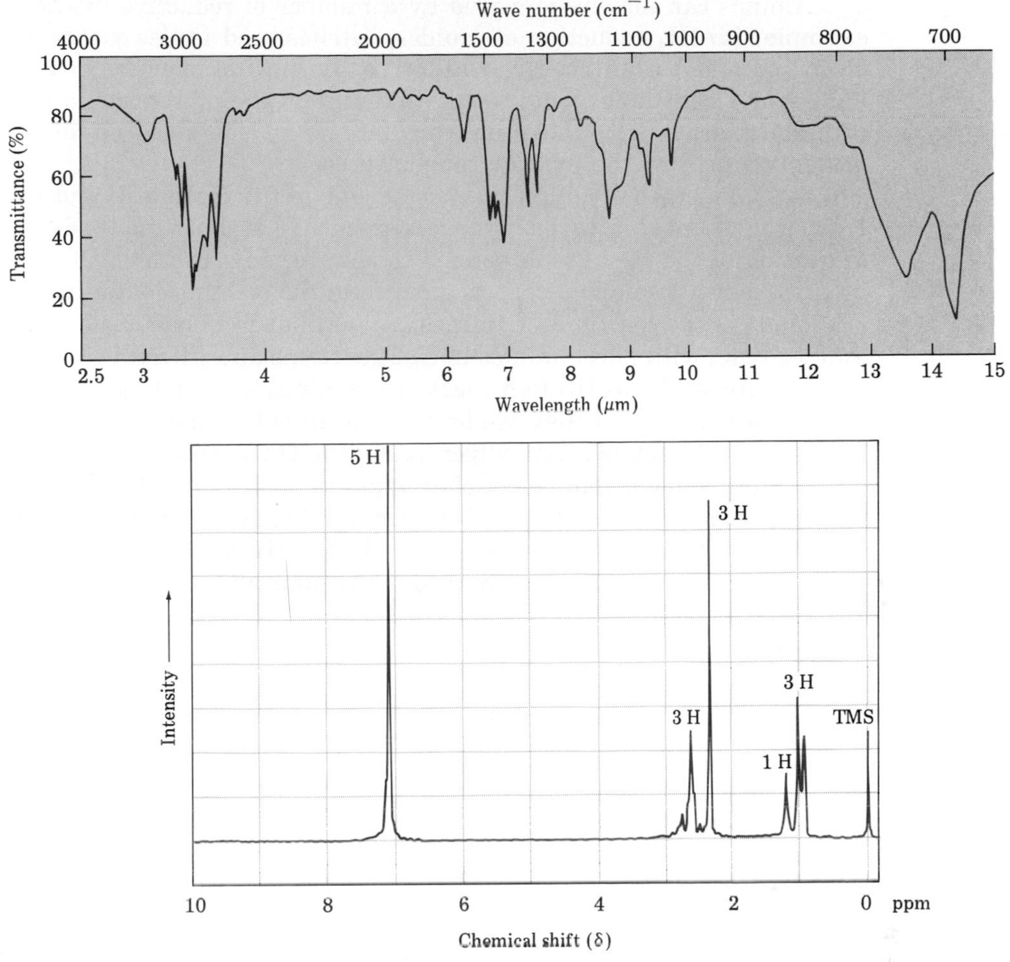


## 25.12 Summary and Key Words

**Amines** are organic derivatives of ammonia. They are named in the IUPAC system either by adding the suffix *-amine* to the names of the alkyl substituents or by considering the amino group as a substituent on a more complex parent molecule.

Bonding in amines is similar to that in ammonia. The nitrogen atom is $sp^3$ hybridized; the three substituents are directed to three corners of a tetrahedron; and the lone pair of nonbonding electrons occupies the fourth corner of the tetrahedron. An interesting feature of this tetrahedral structure is that amines undergo a rapid, umbrella-like pyramidal inversion, which interconverts mirror-image structures.

The chemistry of amines is dominated by the presence of the lone-pair electrons on nitrogen. Thus, amines are both basic and nucleophilic. The simplest method of amine synthesis involves $S_N2$ reaction of ammonia or an amine with an alkyl halide. Alkylation of ammonia yields a primary amine; alkylation of a primary amine yields a secondary amine; and so on. This method often gives poor yields, however, and an alternative such as the **Gabriel amine synthesis** is often preferred.

Amines can also be prepared by a number of reductive methods. For example, $LiAlH_4$ reduction of amides, nitriles, and azides yields amines. Even more important is the **reductive amination** reaction, whereby a ketone or aldehyde is treated with an amine in the presence of a reducing agent such as $NaBH_3CN$. An intermediate imine is formed and then immediately reduced by the hydride reagent present.

A final method of amine synthesis involves the **Hofmann** and **Curtius rearrangements** of carboxylic acid derivatives. Both methods involve migration of the R— group bonded to the carbonyl carbon and provide a product that has one less carbon atom than the starting material.

Many of the reactions of amines are familiar from past chapters. Thus, amines react with alkyl halides in $S_N2$ reactions and with acid chlorides in nucleophilic acyl substitution reactions. Amines also undergo E2 elimination to yield alkenes if they are first quaternized by treatment with methyl iodide and then heated with silver oxide (the **Hofmann elimination**).

Amines show a number of characteristic features that aid their spectroscopic identification. In the infrared spectrum, N—H absorptions are readily detectable at 3200–3500 $cm^{-1}$. In the $^1H$ NMR spectrum, protons on carbon next to nitrogen are deshielded and absorb in the range 2.2–2.6 $\delta$. In the mass spectrum, amines undergo a characteristic alpha cleavage.

## 25.13 Summary of Reactions

1. Preparation of amines (Section 25.7)

   a. The $S_N2$ alkylation of alkyl halides

| | | | | |
|---|---|---|---|---|
| Ammonia | $:NH_3 + RX \longrightarrow$ | $RNH_3^+ X^-$ | $\xrightarrow{NaOH} RNH_2$ | Primary |
| Primary | $:NH_2R + RX \longrightarrow$ | $R_2NH_2^+ X^-$ | $\xrightarrow{NaOH} R_2NH$ | Secondary |
| Secondary | $:NHR_2 + RX \longrightarrow$ | $R_3NH^+ X^-$ | $\xrightarrow{NaOH} R_3N$ | Tertiary |
| Tertiary | $:NR_3 + RX \longrightarrow$ | $R_4N^+ X^-$ | Quaternary ammonium salt | |

   b. Gabriel amine synthesis

$$\text{Phthalimide anion } (\ddot{N}:^-) + RX \xrightarrow[\text{solvent}]{\text{DMF}} \text{N-alkylphthalimide } (:N{-}R) \xrightarrow[H_2O]{^-OH} RNH_2$$

   c. Reduction of azides

$$RX + NaN_3 \xrightarrow[\text{solvent}]{\text{Ethanol}} RN_3 \xrightarrow[\text{2. } H_2O]{\text{1. } LiAlH_4\text{, ether}} RNH_2$$

   d. Reduction of nitriles

$$RX + NaCN \xrightarrow{\text{DMF}} RCN \xrightarrow[\text{2. } H_2O]{\text{1. } LiAlH_4\text{, ether}} RCH_2NH_2$$

e. Reduction of amides

$$RCONH_2 \xrightarrow[2.\ H_2O]{1.\ LiAlH_4,\ ether} RCH_2NH_2 \quad \text{Primary}$$

$$RCONHR \xrightarrow[2.\ H_2O]{1.\ LiAlH_4,\ ether} RCH_2NHR \quad \text{Secondary}$$

$$RCONR_2 \xrightarrow[2.\ H_2O]{1.\ LiAlH_4,\ ether} RCH_2NR_2 \quad \text{Tertiary}$$

f. Reductive amination of ketones/aldehydes

$$NH_3 + R_2C{=}O \xrightarrow[\text{Ethanol}]{NaBH_3CN} R_2CHNH_2 \quad \text{Primary}$$

$$R'NH_2 + R_2C{=}O \xrightarrow[\text{Ethanol}]{NaBH_3CN} R_2CHNHR' \quad \text{Secondary}$$

$$R_2'NH + R_2C{=}O \xrightarrow[\text{Ethanol}]{NaBH_3CN} R_2CHNR_2' \quad \text{Tertiary}$$

g. Hofmann rearrangement of amides

$$RCONH_2 \xrightarrow[H_2O,\ \Delta]{^-OH,\ Br_2} RNH_2 + CO_2$$

h. Curtius rearrangement of acyl azides

$$RCOCl + NaN_3 \xrightarrow{\text{Ethanol}} RCON_3 \xrightarrow[\Delta]{H_2O} RNH_2 + CO_2$$

2. Reactions of amines (Section 25.8)

a. Alkylation of alkyl halides [see reaction 1(a)]

b. Nucleophilic acyl substitution (see also Section 21.5)

$$\text{Ammonia} \quad NH_3 + RCOCl \xrightarrow{\text{Ether}} RCONH_2 + HCl$$

$$\text{Primary} \quad R'NH_2 + RCOCl \xrightarrow{\text{Ether}} RCONHR' + HCl$$

$$\text{Secondary} \quad R_2'NH + RCOCl \xrightarrow{\text{Ether}} RCONR_2' + HCl$$

c. Sulfonamides (Hinsberg test)

$$\text{Primary} \quad R'NH_2 + RSO_2Cl \longrightarrow RSO_2NHR' + HCl$$

$$\text{Secondary} \quad R_2'NH + RSO_2Cl \longrightarrow RSO_2NR_2' + HCl$$

$$\text{Tertiary} \quad R_3'N + RSO_2Cl \longrightarrow \text{Product hydrolyzes}$$

d. Hofmann elimination

$$HO^- \; H{-}\overset{|}{\underset{|}{C}}{-}\overset{|}{\underset{^+NR_3}{C}}{-} \xrightarrow[\Delta]{Ag_2O} \underset{\text{Alkene}}{>C{=}C<} + \underset{3^\circ\ \text{amine}}{R_3N} + H_2O$$

## ADDITIONAL PROBLEMS

**25.22** Classify each of the amine nitrogen atoms in these substances as either primary, secondary, or tertiary:

(a) Lysergic acid diethylamide

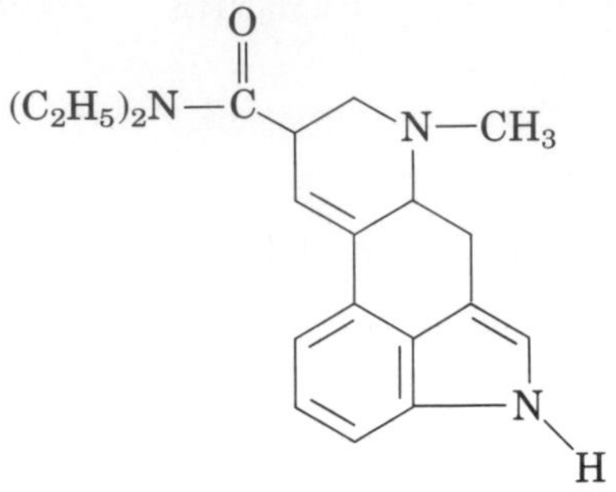

(b) Caffeine

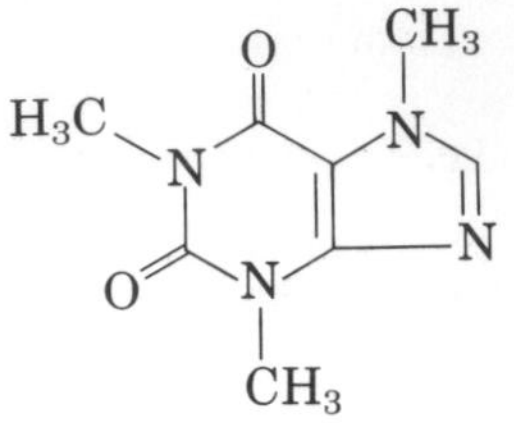

**25.23** Draw structures corresponding to these IUPAC names:

(a) *N*,*N*-Dimethylaniline
(b) (Cyclohexylmethyl)amine
(c) *N*-Methylcyclohexylamine
(d) (2-Methylcyclohexyl)amine
(e) 3-(*N*,*N*-Dimethylamino)propanoic acid
(f) *N*-Isopropyl-*N*-methylcyclohexylamine

**25.24** Name these compounds by IUPAC rules:

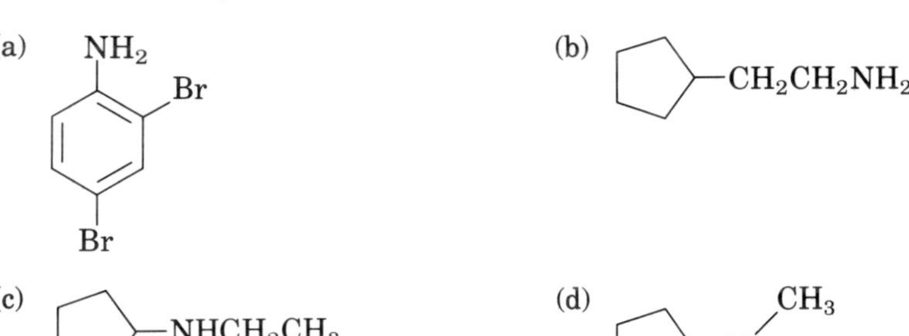

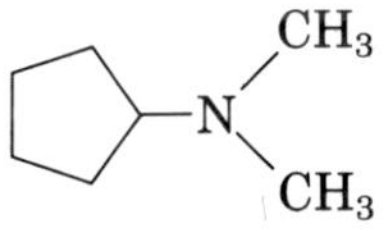

(e) $N—CH_2CH_2CH_3$

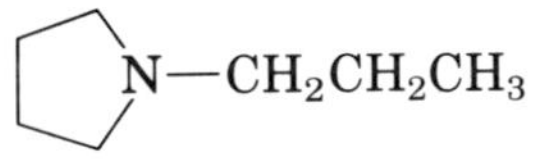

(f) $H_2NCH_2CH_2CH_2CN$

**25.25** How can you explain the fact that trimethylamine (bp 3°C) boils at a lower temperature than dimethylamine (bp 7°C) even though it has a higher molecular weight?

**25.26** How would you prepare these substances from 1-butanol?

(a) Butylamine
(b) Dibutylamine
(c) Propylamine
(d) Pentylamine
(e) *N*,*N*-Dimethylbutylamine
(f) Propene

**25.27** How would you prepare these substances from pentanoic acid?

(a) Pentanamide
(b) Butylamine
(c) Pentylamine
(d) 2-Bromopentanoic acid
(e) Hexanenitrile
(f) Hexylamine

**25.28** Treatment of bromoacetone with ammonia yields a compound having the formula $C_6H_{10}N_2$, rather than the expected 1-amino-2-propanone. What is a likely structure for the product?

**25.29** Propose structures for substances that fit these descriptions:
(a) A chiral quaternary ammonium salt
(b) A five-membered heterocyclic amine
(c) A secondary amine, $C_6H_{11}N$

**25.30** How might you prepare pentylamine from these starting materials?
(a) Pentanamide (b) Pentanenitrile (c) 1-Butene
(d) Hexanamide (e) 1-Butanol (f) 5-Decene
(g) Pentanoic acid

**25.31** Although most chiral trisubstituted amines cannot be resolved into enantiomers because nitrogen pyramidal inversion occurs too rapidly, one exception to this generalization is the substance known as Tröger's base. Make molecular models of Tröger's base and then explain why it is resolvable into enantiomers.

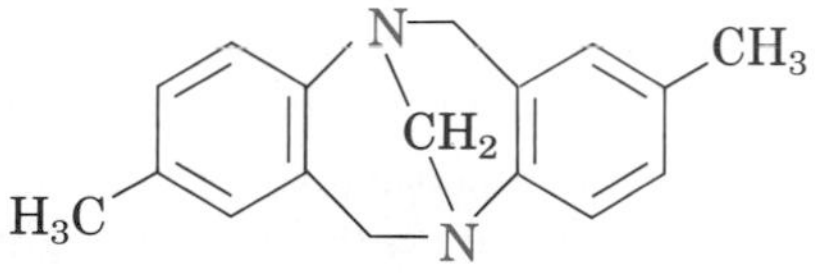

$CH_3$ $CH_2$ $H_3C$ N N

Tröger's base

**25.32** Predict the product(s) of these reactions. If more than one product is formed, tell which is major.

(a) 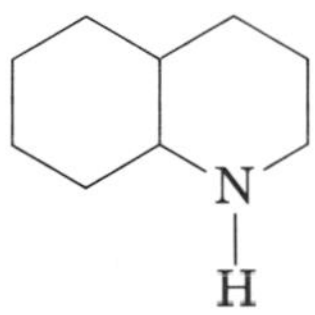 [decahydroquinoline, N–H] $\xrightarrow{CH_3I \text{ (excess)}}$ A $\xrightarrow{Ag_2O,\ H_2O}$ B $\xrightarrow[\Delta]{}$ C

(b) 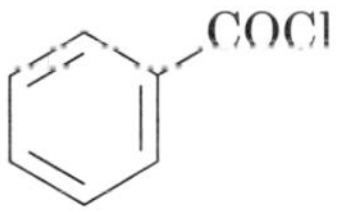 [benzene–COCl] $\xrightarrow{NaN_3}$ A $\xrightarrow[\Delta]{}$ B $\xrightarrow{H_2O}$ C

(c) 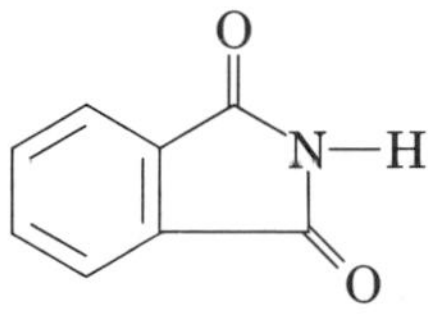 [phthalimide, N–H, O, O] $\xrightarrow{KOH}$ A $\xrightarrow{C_6H_5CH_2Br}$ B $\xrightarrow[H_2O]{KOH}$ C

(d) $BrCH_2CH_2CH_2CH_2Br$ + 1 equiv $CH_3NH_2$ $\xrightarrow[H_2O]{NaOH}$ ?

**25.33** Phthalimide, used in the Gabriel synthesis, is prepared by reaction of ammonia with phthalic anhydride (1,2-benzenedicarboxylic anhydride). Propose a mechanism for the reaction.

**25.34** The following syntheses are incorrect as written. What is wrong with each?

(a) $CH_3CH_2CONH_2$ $\xrightarrow{Br_2,\ NaOH,\ H_2O}$ $CH_3CH_2CH_2NH_2$

(b) 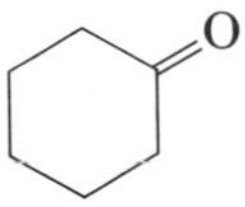 [cyclohexanone, O] + $(CH_3)_3N$ $\xrightarrow{NaBH_3CN}$ [cyclohexyl–N($CH_3$)($CH_3$)]

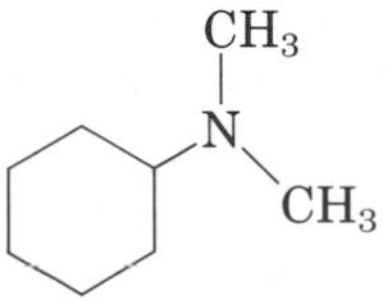

(c) $(CH_3)_3C{-}Br + NH_3 \longrightarrow (CH_3)_3C{-}NH_2$

(d) 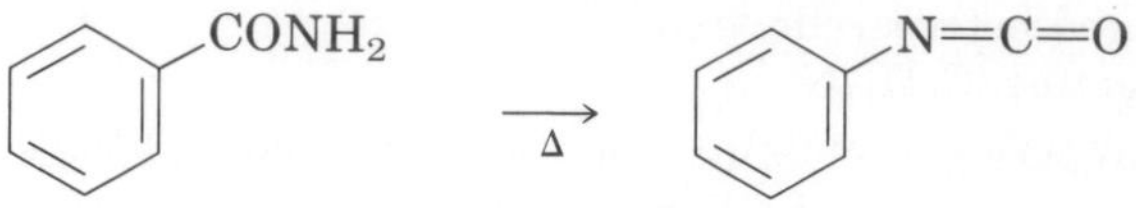 $C_6H_5CONH_2 \xrightarrow{\Delta} C_6H_5N{=}C{=}O$

(e) $CH_3CH_2CH_2CH(NH_2)CH_3 \xrightarrow[\text{2. } Ag_2O \quad \text{3. } \Delta]{\text{1. } CH_3I \text{ (excess)}} CH_3CH_2CH{=}CHCH_3$

**25.35** Coniine, $C_8H_{17}N$, is the toxic principle of poison hemlock drunk by Socrates. When subjected to Hofmann elimination, coniine yields 5-(*N*,*N*-dimethylamino)-1-octene. When subjected to the Hinsberg test, coniine yields a benzenesulfonamide derivative that is insoluble in base. What is the structure of coniine?

**25.36** Atropine, $C_{17}H_{23}NO_3$, is a poisonous alkaloid isolated from the leaves and roots of *Atropa belladonna*, commonly called deadly nightshade. In low doses, atropine acts as a muscle relaxant; 0.5 ng (nanogram, $10^{-9}$ g) is sufficient to cause pupil dilation. On basic hydrolysis, atropine yields tropic acid, $C_6H_5CH(CH_2OH)COOH$, and tropine, $C_8H_{15}NO$. Tropine is an optically inactive alcohol that yields tropidene on dehydration with $H_2SO_4$. Propose a structure for atropine.

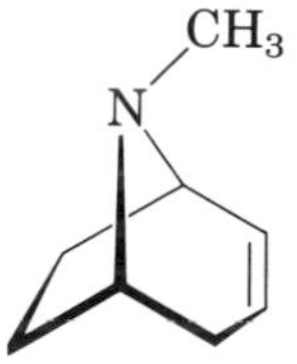

$CH_3$
N

Tropidene

**25.37** Tropidene (Problem 25.36) can be converted by a series of steps into tropilidene (1,3,5-cycloheptatriene). How would you accomplish this conversion?

**25.38** One problem with reductive amination as a method of amine synthesis is the fact that by-products are sometimes obtained. For example, reductive amination of benzaldehyde with methylamine leads to a mixture of methylbenzylamine and methyldibenzylamine. How do you suppose the tertiary amine by-product is formed? Propose a mechanism.

**25.39** What are the major products you would expect from Hofmann elimination of these amines?

(a) *N*-Methylcyclopentylamine
(b) $(CH_3)_2CHCH(NH_2)CH_2CH_2CH_3$
(c) *N*-Phenyl-*N*-(2-hexyl)amine

**25.40** Cyclopentamine is an amphetamine-like central nervous system stimulant. Propose a synthesis of cyclopentamine from materials of five carbons or less.

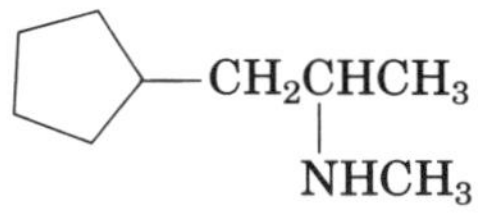

$C_5H_9{-}CH_2CH(NHCH_3)CH_3$

Cyclopentamine

**25.41** Prolitane is an antidepressant drug that is prepared commercially by a route that involves a reductive amination. What amine and what carbonyl precursors are used?

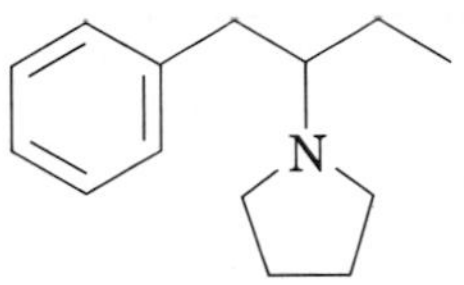

Prolitane

**25.42** Tetracaine is a substance used medicinally as a spinal anesthetic during lumbar punctures (spinal taps).

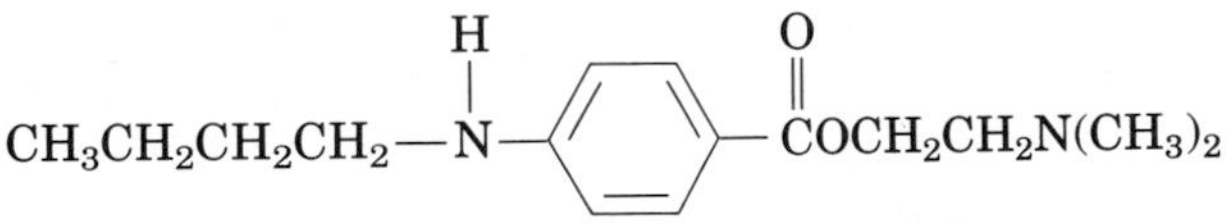

$CH_3CH_2CH_2CH_2{-}NH{-}C_6H_4{-}COCH_2CH_2N(CH_3)_2$ (C=O)

Tetracaine

(a) How would you prepare tetracaine from the corresponding aniline derivative, $ArNH_2$?
(b) How would you prepare tetracaine from *p*-nitrobenzoic acid?
(c) How would you prepare tetracaine from benzene?

**25.43** Propose a structure for the product of formula $C_9H_{17}N$ that results when 2-(2-cyanoethyl)cyclohexanone is reduced catalytically.

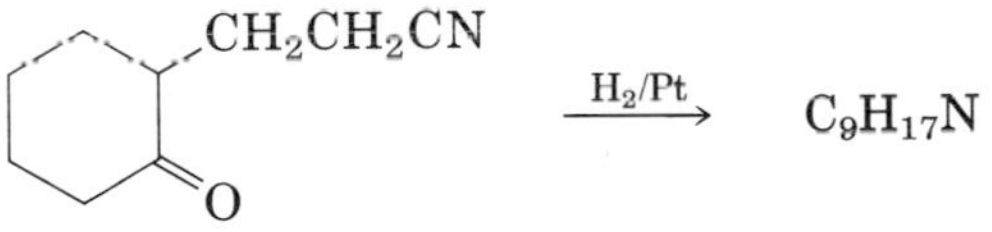

$$\text{2-}(CH_2CH_2CN)\text{cyclohexanone} \xrightarrow{H_2/Pt} C_9H_{17}N$$

**25.44** How would you synthesize coniine (Problem 25.35) from acrylonitrile ($H_2C{=}CHCN$) and ethyl 3-oxohexanoate ($CH_3CH_2CH_2COCH_2CO_2C_2H_5$)? [*Hint*: See Problem 25.43.]

**25.45** Although the barrier to nitrogen inversion is normally too low to allow isolation of one enantiomer of a chiral amine, (+)-1-chloro-2,2-diphenylaziridine has been prepared optically pure and has been shown to be stable for several hours at 0°C. Propose an explanation of this unusual stability.

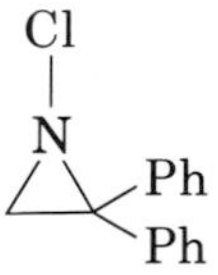

1-Chloro-2,2-diphenylaziridine

**25.46** The hydrolysis of phthalimides is often slow, and an alternative method is sometimes needed to liberate the primary amine in the last stage of a Gabriel synthesis. In the

Ing–Manske modification, reaction of an *N*-alkylphthalimide with hydrazine is employed. Propose a mechanism for this *hydrazinolysis*.

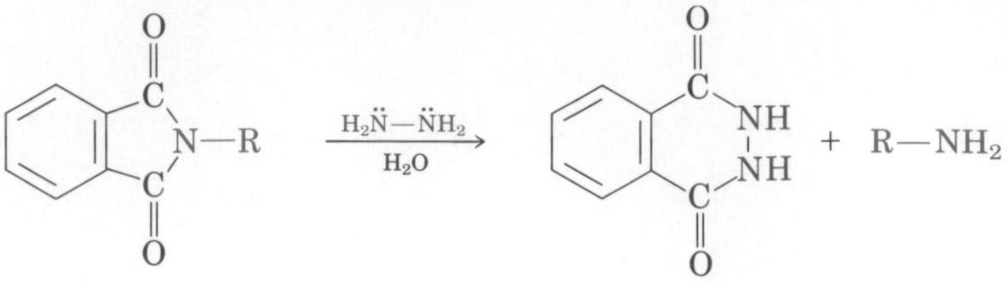

Phthalhydrazide

**25.47** Cyclooctatetraene was first synthesized in 1911 by a route that involved the following transformation:

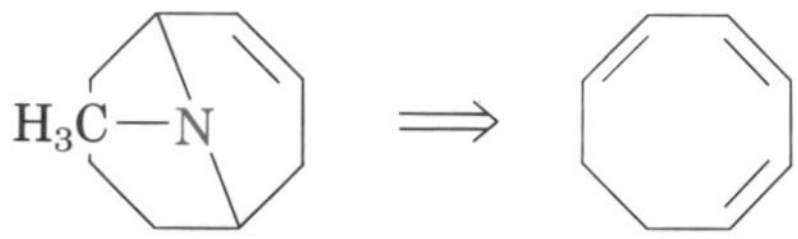

How might you use the Hofmann elimination to accomplish this reaction? How would you finish the synthesis by converting cyclooctatriene into cyclooctatetraene?

**25.48** Propose a mechanism for the following reaction:

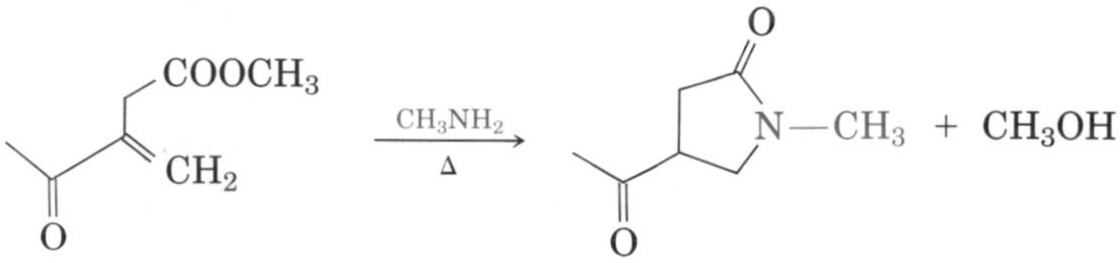

**25.49** Propose a mechanism for the following reaction:

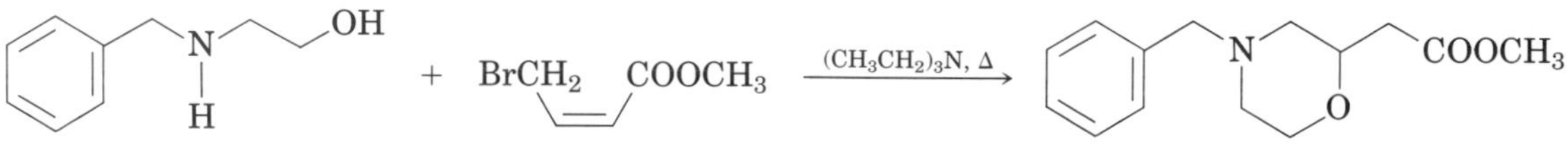

**25.50** Propose structures for amines with the following $^1$H NMR spectra. The peaks marked by an asterisk disappear when $D_2O$ is added to the sample.

(a) $C_4H_{11}N$

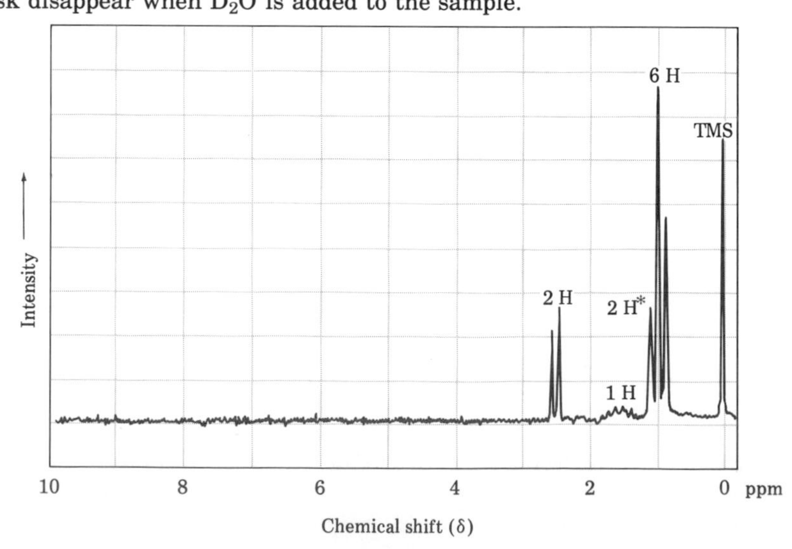

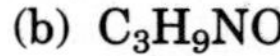
(b) $C_3H_9NO$

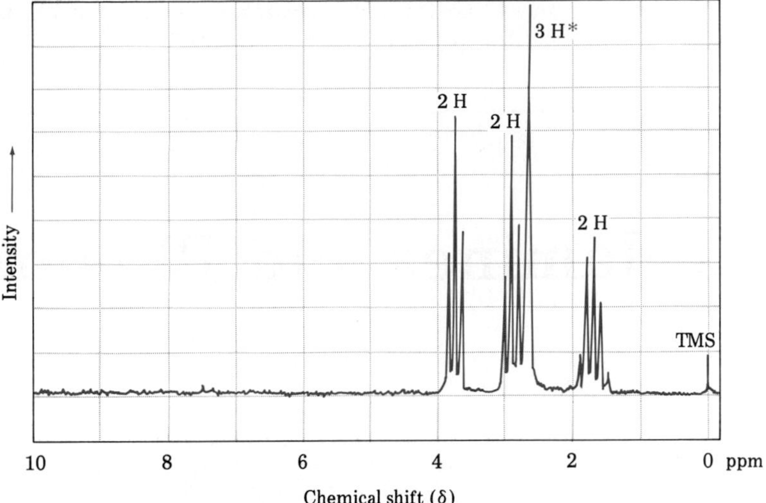

(c) $C_4H_{11}NO_2$

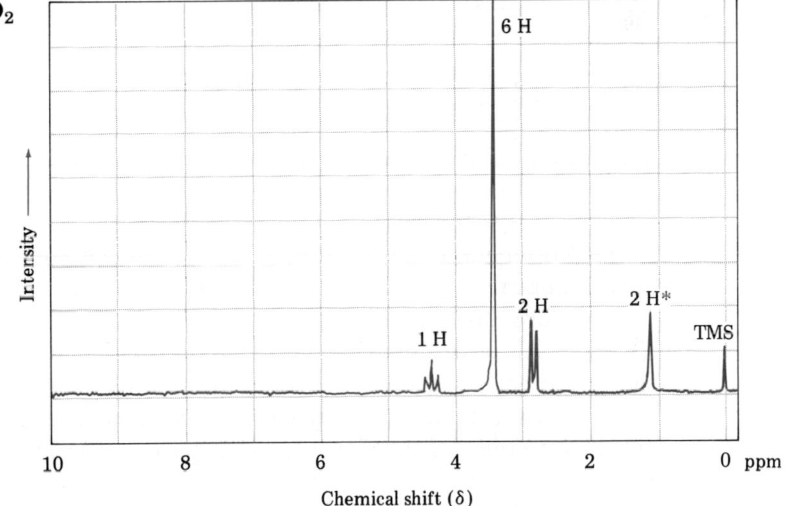

CHAPTER 26

# Arylamines and Phenols

## 26.1 Aniline and the Discovery of Synthetic Dyes

The founding of the modern organic chemical industry can be traced to the need for a single organic compound—aniline—and to the activities of one person—Sir William Henry Perkin.[1] Perkin, a student of Hofmann's at the Royal College of Chemistry in London, worked during the day on problems assigned him by Hofmann but spent his free time working on his own ideas in an improvised home laboratory. One day during Easter vacation in 1856, he decided to examine the oxidation of aniline with potassium dichromate. Although the reaction appeared unpromising at first, yielding a tarry black product, Perkin was able by careful extraction with methanol to isolate a small amount of a beautiful purple pigment that had the properties of a dye.

Since the only dyes known at the time were the naturally occurring vegetable dyes such as indigo, Perkin's synthetic purple dye, which he named *mauve*, created a sensation. Realizing the possibilities, Perkin resigned his post with Hofmann and, at the age of 18, formed a company to exploit his remarkable discovery.

No chemical industry existed at the time, since there had never before been a need for synthetic chemicals. Large-scale chemical manufacture was unknown, and Perkin's first task was to devise a procedure for preparing the needed quantities of aniline. He therefore worked out the techniques of manufacture and soon learned to prepare aniline on a large scale by nitration of benzene, followed by reduction of nitrobenzene with iron and hydrochloric acid. A similar procedure is used today to prepare some 350,000 tons of aniline annually in the United States, although the reduction step is now carried out by catalytic hydrogenation.

[1]Sir William Henry Perkin (1838–1907); b. London; studied at Royal College of Chemistry, London; industrial consultant, London.

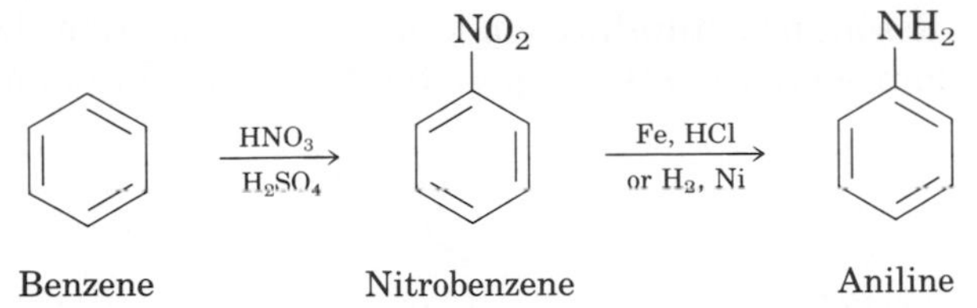

$$\text{Benzene} \xrightarrow[H_2SO_4]{HNO_3} \text{Nitrobenzene} \xrightarrow[\text{or } H_2,\ Ni]{Fe,\ HCl} \text{Aniline}$$

Benzene Nitrobenzene Aniline

Subsequent work showed that Perkin's original mauve was in fact not derived from aniline but from a toluidine (methylaniline) impurity in his starting material. Pure aniline yields a similar dye, however, which came to be marketed under the name *pseudomauveine.*

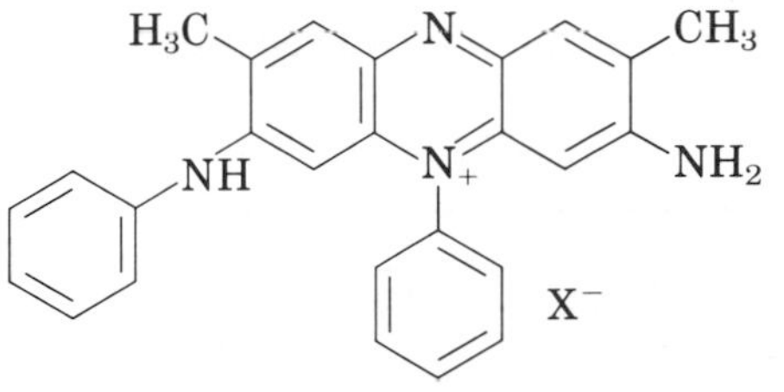

$H_3C$ N $CH_3$ NH $N^+$ $NH_2$ $X^-$

Perkin's mauve
(pseudomauveine has no methyl groups)

Today, dyestuff manufacture is a thriving and important part of the chemical industry, and many commonly used pigments are derived from aniline. Although aniline itself and several substituted anilines are available naturally from coal tar, synthesis from benzene is the major source.

## 26.2 Basicity of Arylamines

**Arylamines**, like their aliphatic counterparts, are basic; the lone pair of nonbonding electrons on nitrogen can bond to Lewis acids, yielding an arylammonium salt. The base strength of arylamines is generally lower than that of aliphatic amines, however. Thus, methylammonium ion has $pK_a$ = 10.64, whereas anilinium ion has $pK_a = 4.63$. [*Remember*: The base strength of an amine is inversely related to the acid strength of its corresponding ammonium ion (Section 25.4). A stronger base like methylamine corresponds to a less acidic ammonium ion (higher $pK_a$) whereas a weaker base like aniline corresponds to a more acidic ammonium ion (lower $pK_a$).]

Arylamines are less basic than alkylamines because the nitrogen lone-pair electrons are delocalized by orbital overlap with the aromatic ring pi electron system and are less available for bonding. In resonance terms, arylamines are stabilized relative to alkylamines because of the five contributing resonance structures that can be drawn:

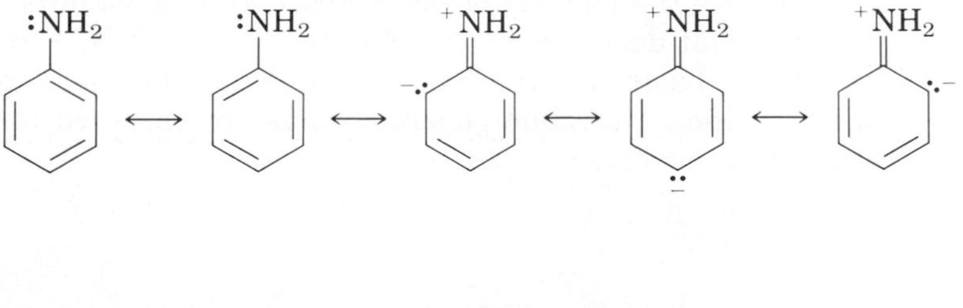

$:NH_2 \longleftrightarrow :NH_2 \longleftrightarrow {}^+NH_2 \longleftrightarrow {}^+NH_2 \longleftrightarrow {}^+NH_2$

Resonance stabilization is lost on protonation, however, since only two resonance structures are possible for the arylammonium ion:

$^+NH_3$ (benzene ring) $\longleftrightarrow$ $^+NH_3$ (benzene ring)

As a result, the energy difference ($\Delta G°$) between protonated and nonprotonated forms is higher for arylamines than it is for alkylamines, and arylamines are therefore less basic. Figure 26.1, which compares reaction energy diagrams for protonation of alkylamines and arylamines, illustrates the difference in $\Delta G°$ for the two reactions.

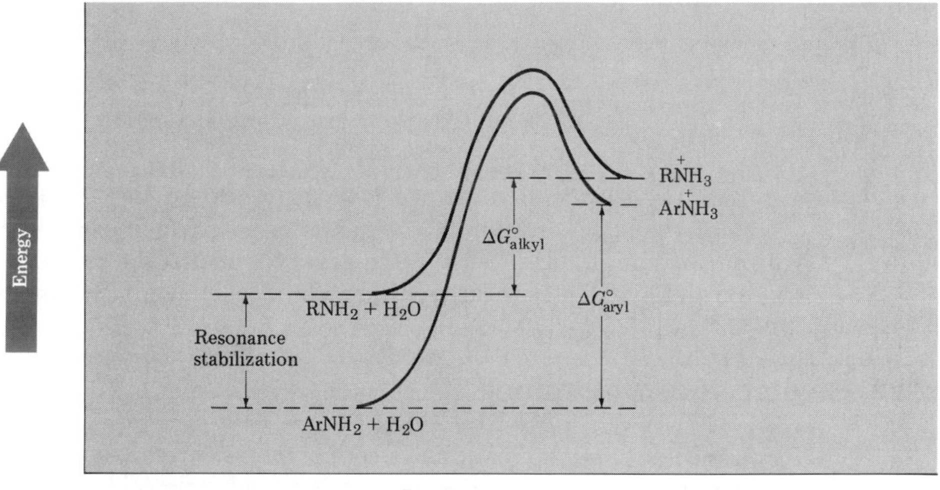


**Figure 26.1** Reaction energy diagrams for the protonation of alkylamines (black curve) and arylamines (red curve): Arylamines have a larger $\Delta G°$ and are therefore less basic than alkylamines, primarily because of resonance stabilization of their ground state.

Substituted arylamines can be either more basic or less basic than aniline, depending on the nature of the substituent. Table 26.1 presents data for a variety of para-substituted anilines.

As a general rule, substituents that increase the reactivity of an aromatic ring toward electrophilic substitution ($—CH_3$, $—NH_2$, $—OCH_3$) also increase the basicity of the corresponding arylamine. Conversely, substituents that decrease ring reactivity ($—Cl$, $—NO_2$, $—CN$) also decrease arylamine basicity. Although Table 26.1 considers only para-substituted anilines, the same general trends are observed for ortho and meta derivatives.

**Table 26.1** Base strength of para-substituted anilines

$$Y\text{–}C_6H_4\text{–}\ddot{N}H_2 + H_2O \rightleftharpoons Y\text{–}C_6H_4\text{–}\overset{+}{N}H_3 + {}^{-}OH$$

| | *Substituent*, Y | $pK_a$ | |
|---|---|---|---|
| Stronger base | $-NH_2$ | 6.15 | Activating groups |
| | $-OCH_3$ | 5.34 | |
| | $-CH_3$ | 5.08 | |
| | $-H$ | 4.63 | |
| | $-Cl$ | 3.98 | Deactivating groups |
| | $-Br$ | 3.86 | |
| Weaker base | $-C{\equiv}N$ | 1.74 | |
| | $-NO_2$ | 1.00 | |

The best way to understand the effect of substituent groups on arylamine basicity is to look at reaction energy diagrams for the amine protonation step (Figure 26.2). Activating substituents make the aromatic ring electron-rich, thereby increasing the stability of the positively charged nitrogen. Deactivating substituents, however, make the aromatic ring electron-poor, thereby decreasing the stability of the positively charged nitrogen. We therefore find a lower $\Delta G°$ for protonation of an activated arylamine than for protonation of a deactivated arylamine.

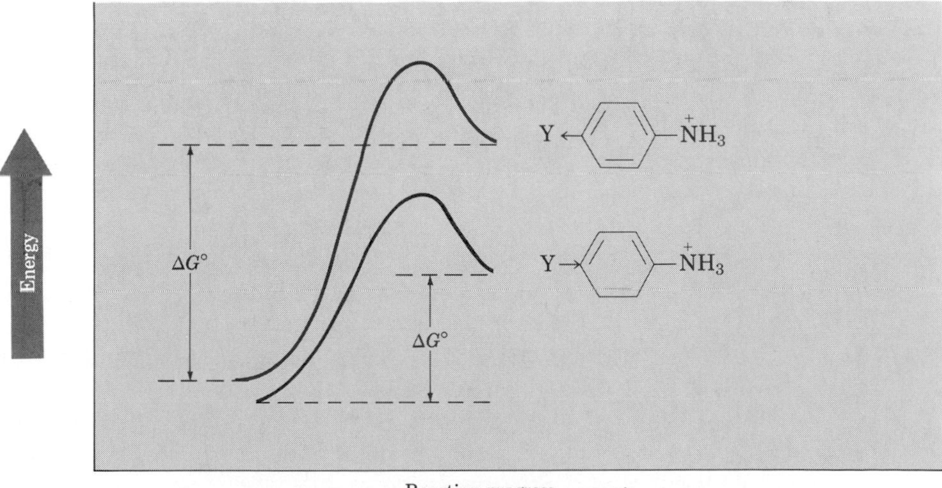


**Figure 26.2** Reaction energy diagrams for the protonation of substituted arylamines: An electron-donating substituent (red curve) stabilizes the ammonium salt much more than an electron-withdrawing substituent (black curve).

PROBLEM

**26.1** How do you account for the fact that *p*-nitroaniline (p$K_a$ = 1.0) is less basic than *m*-nitroaniline (p$K_a$ = 2.5) by a factor of 30? Draw resonance structures to support your answer. (The p$K_a$ values refer to the corresponding ammonium ions.)

PROBLEM

**26.2** Rank the following compounds in order of ascending basicity. Don't look at Table 26.1 to determine your answers.
(a) *p*-Nitroaniline, *p*-aminobenzaldehyde, *p*-bromoaniline
(b) *p*-Chloroaniline, *p*-aminoacetophenone, *p*-methylaniline
(c) *p*-(Trifluoromethyl)aniline, *p*-methylaniline, *p*-(fluoromethyl)aniline

## 26.3 Preparation of Arylamines

Arylamines are almost always prepared by nitration of an aromatic starting material, followed by reduction of the nitro group. No other method of synthesis approaches this nitration–reduction route with respect to versatility and generality.

The reduction step can be carried out in many different ways, depending on the circumstances. Catalytic hydrogenation over platinum is clean and gives high yields, but is often not compatible with the presence of other reducible groups (such as carbon–carbon double bonds or carbonyl groups) elsewhere in the molecule. Iron, zinc, tin metal, and stannous chloride ($SnCl_2$) are also effective when used in aqueous solution. Stannous chloride is particularly mild and is often used when other reducible functional groups are present.

$NO_2$ / $C(CH_3)_3$ —($H_2$/Pt, Ethanol)→ $NH_2$ / $C(CH_3)_3$

*p-tert*-Butylnitrobenzene → *p-tert*-Butylaniline (100%)

$NO_2$ / $NO_2$ / $CH_3$ —(1. Fe, HCl; 2. NaOH, $H_2O$)→ $NH_2$ / $NH_2$ / $CH_3$

2,4-Dinitrotoluene → Toluene-2,4-diamine (74%)

$NO_2$ / CHO —(1. $SnCl_2$, $H_3O^+$; 2. NaOH, $H_2O$)→ $NH_2$ / CHO

*m*-Nitrobenzaldehyde → *m*-Aminobenzaldehyde (90%)

## 26.4 Reactions of Arylamines

### ELECTROPHILIC AROMATIC SUBSTITUTION

Amino substituents are strongly activating, ortho- and para-directing groups in electrophilic aromatic substitution reactions (Section 16.5). The high reactivity of amino-substituted benzenes can be a drawback at times, since it is sometimes difficult to prevent polysubstitution. For example, reaction of aniline with bromine takes place rapidly to yield the 2,4,6-tribrominated product. The amino group is so strongly activating that it is not possible to stop cleanly at the monobromo stage.

$NH_2$ $\xrightarrow[H_2O]{Br_2}$ $NH_2$, Br, Br, Br

Aniline

2,4,6-Tribromoaniline
(100%)

Another drawback to the use of amino-substituted benzenes in electrophilic aromatic substitution reactions is that Friedel–Crafts reactions are not successful (Section 16.3). The amino group forms an acid–base complex with the aluminum chloride Friedel–Crafts catalyst, which prevents further reaction from occurring. Both of these drawbacks—high reactivity and amine basicity—can be overcome by carrying out electrophilic aromatic substitution reactions on the corresponding amide, rather than on the free amine.

As we saw in Section 21.6, treatment of an arylamine with acetic anhydride yields an *N*-acetylated product. Though still ortho- and para-directing and activating, amido substituents (—NHCOR) are much less strongly activating and much less basic than amino groups because their nitrogen lone-pair electrons are delocalized by overlap with the neighboring carbonyl group (Section 25.4). As a result, bromination of an *N*-arylamide occurs cleanly to give a monobromo product, which can be hydrolyzed by aqueous base to give the free bromoamine. For example, *p*-toluidine (4-methylaniline) can be acetylated, brominated, and hydrolyzed to yield 2-bromo-4-methylaniline in 79% yield. None of the 2,6-dibrominated product is obtained.

$:NH_2$, $CH_3$ $\xrightarrow[\text{Pyridine}]{(CH_3CO)_2O}$ $:NHCOCH_3$, $CH_3$ $\xrightarrow{Br_2}$ $:NHCOCH_3$, Br, $CH_3$ $\xrightarrow[H_2O]{^-OH}$ $:NH_2$, Br, $CH_3$ + $CH_3CO_2^-$

*p*-Toluidine

2-Bromo-4-methylaniline
(79%)

Friedel–Crafts alkylations and acylations of *N*-arylamides also proceed normally. For example, benzoylation of acetanilide (*N*-acetylaniline) under Friedel–Crafts conditions gives *p*-aminobenzophenone in 80% yield after hydrolysis:

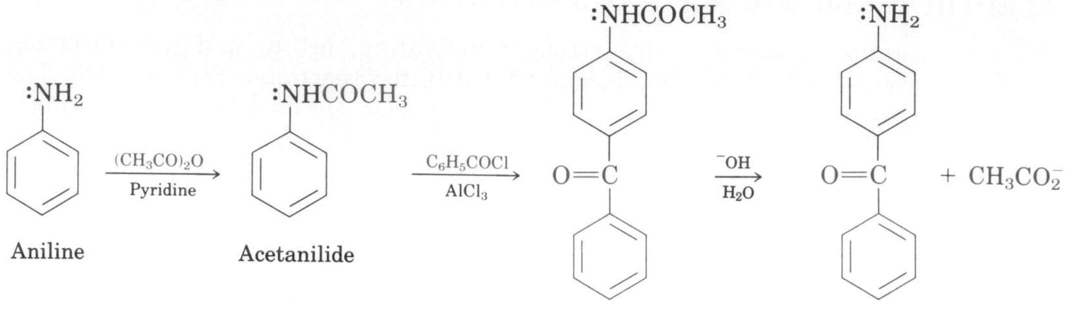

:$NH_2$ —[$(CH_3CO)_2O$, Pyridine]→ :$NHCOCH_3$ —[$C_6H_5COCl$, $AlCl_3$]→ :$NHCOCH_3$ (O=C) —[$^-OH$, $H_2O$]→ :$NH_2$ (O=C) + $CH_3CO_2^-$

Aniline

Acetanilide

*p*-Aminobenzophenone
(80%)

The simple chemical trick of modulating the reactivity of amino-substituted benzenes by forming an amide is an extremely useful one that allows many kinds of electrophilic aromatic substitutions to be carried out that would otherwise be impossible. A good example is the preparation of **sulfa drugs**.

Sulfa drugs such as sulfanilamide were among the first antibiotics to be used clinically against infection. Although they have largely been replaced today by safer and more powerful antibiotics, sulfa drugs were widely used in the 1940s and were credited with saving the lives of thousands of wounded during World War II. These compounds are prepared by chlorosulfonation of acetanilide, followed by reaction of *p*-(*N*-acetylamino)-benzenesulfonyl chloride with ammonia or some other amine to give a sulfonamide. Hydrolysis of the amide then yields the sulfa drug. (Note that this hydrolysis can be carried out in the presence of the sulfonamide group, because sulfonamides hydrolyze very slowly.)

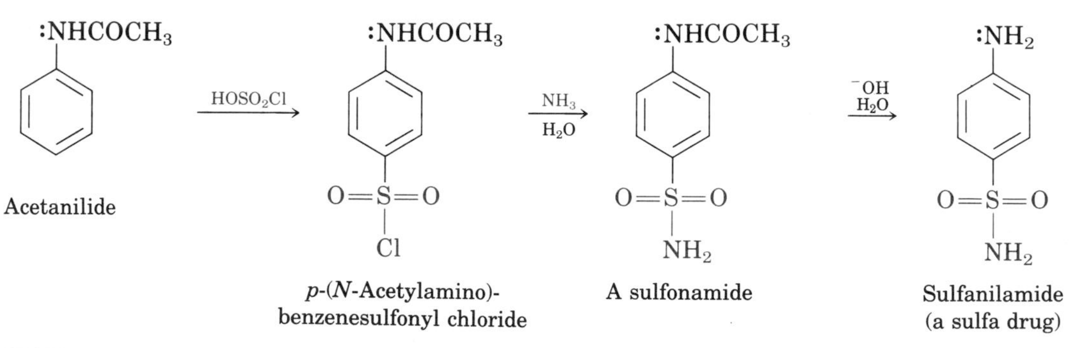


**PROBLEM**

**26.3** Propose a synthesis of sulfathiazole from benzene and any necessary amine.

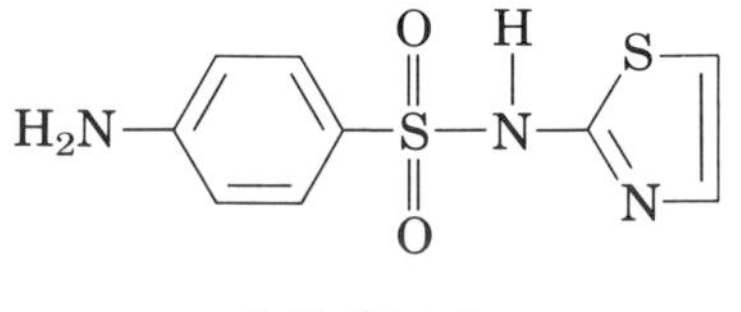

$H_2N$–$C_6H_4$–$SO_2$–NH–(thiazol-2-yl)

Sulfathiazole

PROBLEM

**26.4** Account for the fact that an amido substituent ($—\ddot{N}HCOR$) is ortho- and para-directing, by drawing resonance structures that share the nitrogen lone-pair electrons with the aromatic ring.

PROBLEM

**26.5** Propose syntheses of these compounds from benzene:
(a) *N,N*-Dimethylaniline (b) *p*-Chloroaniline
(c) *m*-Chloroaniline (d) 2,4-Dimethylaniline

## DIAZONIUM SALTS: THE SANDMEYER REACTION

Primary aromatic amines react with nitrous acid, $HNO_2$, to yield stable **arenediazonium salts, $Ar—\overset{+}{N}{\equiv}N\ X^-$**. This **diazotization** reaction is compatible with the presence of a wide variety of substituents on the aromatic ring.

$$Ar—NH_2 + HNO_2 + H_2SO_4 \longrightarrow Ar—\overset{+}{N}{\equiv}N\ HSO_4^- + 2\ H_2O$$

Alkylamines also react with nitrous acid, but the products of these reactions, alkanediazonium salts, are too reactive to isolate since they lose nitrogen instantly.

Arenediazonium salts are extremely useful in synthesis, because the diazonio group ($N_2^+$) can be replaced by nucleophiles:

$$Ar—\overset{+}{N}{\equiv}N + :Nu^- \longrightarrow Ar—Nu + N{\equiv}N$$

Many different nucleophiles react with arenediazonium salts, and many different substituted benzenes can be prepared with this reaction. The overall sequence of (1) nitration, (2) reduction, (3) diazotization, and (4) nucleophile replacement is probably the single most versatile method of aromatic substitution (Figure 26.3). Although the details are not fully understood, the mechanism by which these substitutions occur is probably a radical, rather than a polar, one.

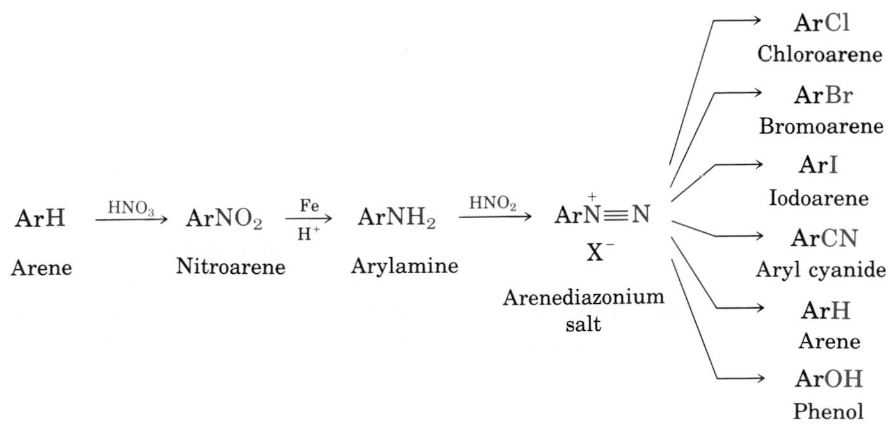


**Figure 26.3** Preparation of substituted aromatic compounds by diazonio replacement reactions

Aryl chlorides and bromides are prepared by reaction of an arenediazonium salt with the corresponding cuprous halide, CuX, a process called the **Sandmeyer[2] reaction**. Aryl iodides can be prepared by direct reaction with sodium iodide without using a cuprous salt. Yields generally fall in the 60–80% range.

$NH_2$ / $CH_3$ — $\xrightarrow[H_2SO_4]{HNO_2}$ — $N_2^+\ HSO_4^-$ / $CH_3$ — $\xrightarrow{CuBr}$ — Br / $CH_3$

*p*-Toluidine — *p*-Bromotoluene (73%)

$NH_2$ — $\xrightarrow[H_2SO_4]{HNO_2}$ — $N_2^+\ HSO_4^-$ — $\xrightarrow{NaI}$ — I

Aniline — Iodobenzene (67%)

Similar treatment of an arenediazonium salt with cuprous cyanide yields the arenenitrile ArCN. This reaction is particularly useful, since it allows the replacement of a nitrogen substituent by a carbon substituent. The nitrile can then be further converted into other functional groups such as carboxyl, —COOH. For example, hydrolysis of *o*-methylbenzonitrile, produced by Sandmeyer reaction of *o*-methylbenzenediazonium bisulfate with cuprous cyanide, yields *o*-methylbenzoic acid. This product could not be prepared from *o*-xylene by the usual side-chain oxidation route, since both methyl groups would be oxidized.

$NH_2$, $CH_3$ — $\xrightarrow[H_2SO_4]{HNO_2}$ — $N_2^+\ HSO_4^-$, $CH_3$ — $\xrightarrow{CuCN}$ — CN, $CH_3$ — $\xrightarrow{H_3O^+}$ — COOH, $CH_3$

*o*-Toluidine — *o*-Methylbenzenediazonium bisulfate — *o*-Methylbenzonitrile (70%) — *o*-Methylbenzoic acid

The diazonio group can also be replaced by —OH to yield phenols and by —H to yield arenes. Phenols are usually prepared by addition of the arenediazonium salt to hot aqueous acid. This reaction is especially important, since few other general methods exist for introducing an —OH group onto an aromatic ring.

[2]Traugott Sandmeyer (1854–1922); b. Wettingen, Switzerland; Ph.D. Heidelberg (Gattermann); Geigy Company, Basel, Switzerland.

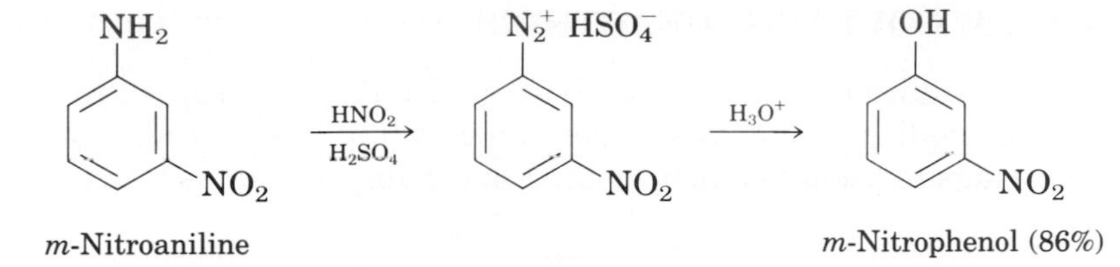

$NH_2$ / $NO_2$ —($HNO_2$, $H_2SO_4$)→ $N_2^+\ HSO_4^-$ / $NO_2$ —($H_3O^+$)→ OH / $NO_2$

*m*-Nitroaniline *m*-Nitrophenol (86%)

Arenes are produced by reduction of the diazonium salt with hypophosphorous acid, $H_3PO_2$. This reaction is not of great use, however, unless there is a particular need for temporarily introducing an amino substituent onto a ring to take advantage of its activating effect. The preparation of 3,5-dibromotoluene from *p*-toluidine illustrates how this can be done.

Dibromination of *p*-toluidine occurs ortho to the strongly directing amino substituent, and diazotization followed by treatment with hypophosphorous acid yields 3,5-dibromotoluene. This product cannot be prepared by direct bromination of toluene, however, since reaction would occur at positions 2 and 4.

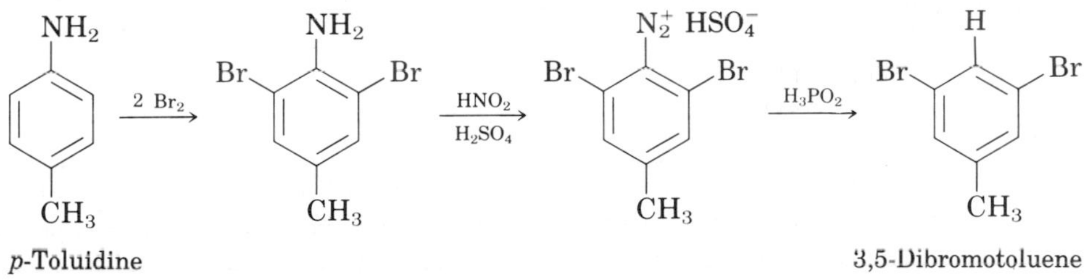

$NH_2$ / $CH_3$ —(2 $Br_2$)→ $NH_2$, Br, Br / $CH_3$ —($HNO_2$, $H_2SO_4$)→ $N_2^+\ HSO_4^-$, Br, Br / $CH_3$ —($H_3PO_2$)→ H, Br, Br / $CH_3$

*p*-Toluidine 3,5-Dibromotoluene

However,

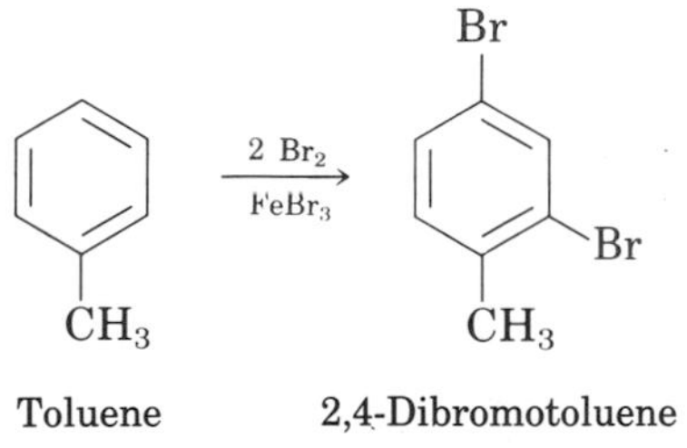

$CH_3$ —(2 $Br_2$, $FeBr_3$)→ Br, Br / $CH_3$

Toluene 2,4-Dibromotoluene

PROBLEM

**26.6** Alkanediazonium salts cannot be isolated because they spontaneously lose $N_2$ to yield a carbocation. Why do you suppose arenediazonium salts are more stable than alkanediazonium salts?

PROBLEM

**26.7** How would you prepare these compounds from benzene, using a diazonium replacement reaction at some point?

(a) *p*-Bromobenzoic acid
(b) *m*-Bromobenzoic acid
(c) *m*-Bromochlorobenzene
(d) *p*-Methylbenzoic acid
(e) 1,2,4-Tribromobenzene

## DIAZONIUM COUPLING REACTIONS

In addition to their reactivity in Sandmeyer-type substitution reactions, arenediazonium salts undergo a coupling reaction with activated aromatic rings to yield brightly colored **azo compounds, Ar—N=N—Ar′**:

$$C_6H_5-\overset{+}{N}\equiv N\ Cl^- + C_6H_5-\ddot{Y} \longrightarrow C_6H_5-N=N-C_6H_4-\ddot{Y}$$

An azo compound

where Y = —OH or —$NR_2$.

Diazonium coupling reactions are typical electrophilic aromatic substitution processes (Section 16.2) in which the positively charged diazonium ion is the electrophile that reacts with the electron-rich ring of a phenol or arylamine. Reaction almost always occurs at the para position, although ortho attack can take place if the para position is blocked.

$$C_6H_5-\overset{+}{N}\equiv\ddot{N}\ Cl^- \longleftrightarrow C_6H_5-\ddot{N}=\overset{+}{N}\text{:}\ Cl^- + C_6H_5-\ddot{O}H$$

Benzenediazonium chloride

Phenol

$$\downarrow\ :\ddot{O}H_2$$

$$\left[C_6H_5-\ddot{N}=\ddot{N}-C_6H_4(H)=\overset{+}{O}-H\right]$$

$$\downarrow$$

$$C_6H_5-\ddot{N}=\ddot{N}-C_6H_4-\ddot{O}H$$

*p*-Hydroxyazobenzene
(orange crystals, mp 152°C)

$$C_6H_5-\overset{+}{N}\equiv\ddot{N}\ Cl^- + C_6H_5-\ddot{N}(CH_3)_2 \longrightarrow C_6H_5-\ddot{N}=\ddot{N}-C_6H_4-\ddot{N}(CH_3)_2$$

Benzenediazonium chloride | *N,N*-Dimethylaniline | *p*-(Dimethylamino)azobenzene (yellow crystals, mp 127°C)

Azo-coupled products are widely used as dyes because their extended conjugated pi electron system causes them to absorb in the visible region of the electromagnetic spectrum. For example, *p*-(dimethylamino)azobenzene is bright yellow and was at one time used as a coloring agent in margarine. Another azo compound, Alizarin Yellow R, is used for dyeing wool.

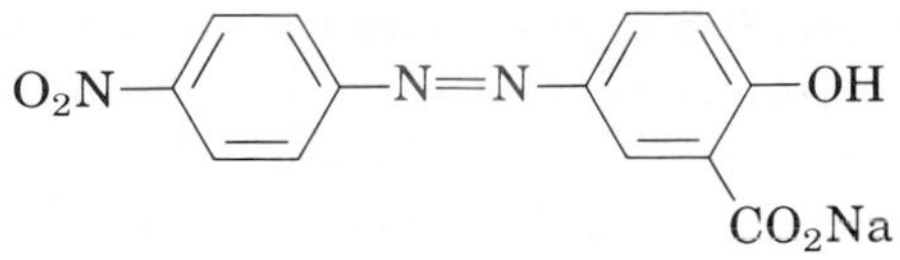

$O_2N-C_6H_4-N{=}N-C_6H_3(OH)(CO_2Na)$

Alizarin Yellow R

PROBLEM

**26.8** Propose a synthesis of *p*-(dimethylamino)azobenzene from benzene.

PROBLEM

**26.9** Methyl Orange is an azo dye that is widely used as a pH indicator. How would you synthesize Methyl Orange from benzene?

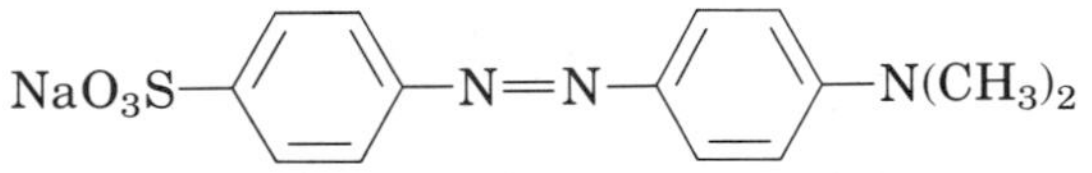

$NaO_3S-C_6H_4-N{=}N-C_6H_4-N(CH_3)_2$

Methyl Orange

## 26.5 Phenols

**Phenols** are compounds with a hydroxy group bonded directly to an aromatic ring, ArOH. They occur widely throughout nature, and they serve as intermediates in the industrial synthesis of products as diverse as adhesives and antiseptics. For example, phenol itself is a general disinfectant found in coal tar; methyl salicylate is a flavoring agent and liniment found in oil of wintergreen; and the urushiols are the main allergenic constituents of poison oak and poison ivy. (Note that the word *phenol* is the name both of a specific compound and of a class of compounds.)

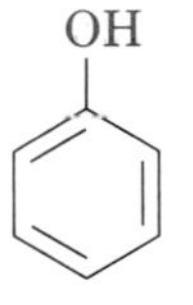

Phenol (also known as carbolic acid)

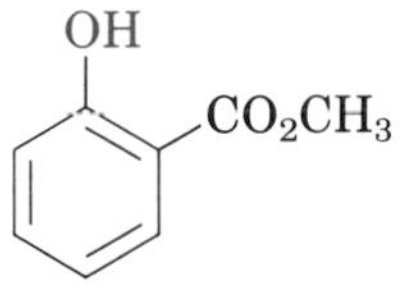

Methyl salicylate ($CO_2CH_3$)

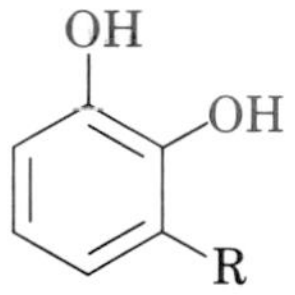

Urushiols ($R$ = different $C_{15}$ alkyl and alkenyl chains)

## 26.6 Industrial Uses of Phenols

The outbreak of World War I provided the stimulus for industrial preparation of large amounts of synthetic phenol, needed as raw material for the manufacture of the explosive picric acid (2,4,6-trinitrophenol). Today, approximately 1.5 million tons of phenol per year are manufactured in the United States for use in a variety of products, including Bakelite resin and adhesives for binding plywood.

For many years, phenol was manufactured by a process (Section 16.10) in which chlorobenzene reacts with sodium hydroxide at high temperature and pressure. Now, however, an alternative synthesis from isopropylbenzene (cumene) is in use. Cumene reacts with air at high temperature by a radical mechanism to form cumene hydroperoxide; acid treatment then gives phenol and acetone. This is a particularly efficient process, since two valuable chemicals are prepared at the same time.

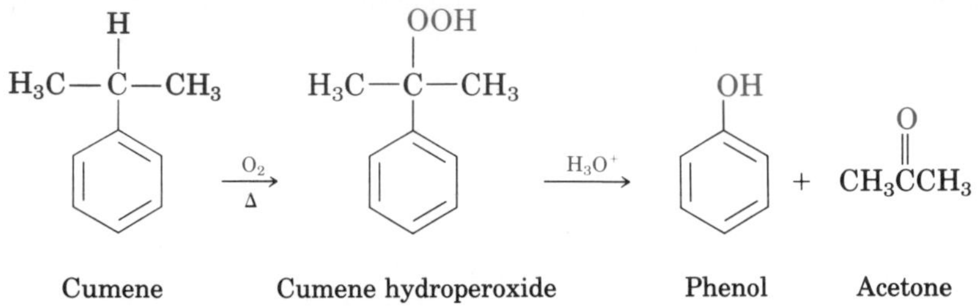

$$H_3C-\underset{C_6H_5}{\overset{H}{C}}-CH_3 \xrightarrow[\Delta]{O_2} H_3C-\underset{C_6H_5}{\overset{OOH}{C}}-CH_3 \xrightarrow{H_3O^+} C_6H_5OH + CH_3\overset{O}{\overset{\|}{C}}CH_3$$

Cumene Cumene hydroperoxide Phenol Acetone

The reaction occurs by protonation of oxygen, followed by rearrangement of the phenyl group from carbon to oxygen, and concurrent loss of water. Readdition of water then yields a hemiacetal, which breaks down to phenol and acetone (Figure 26.4).

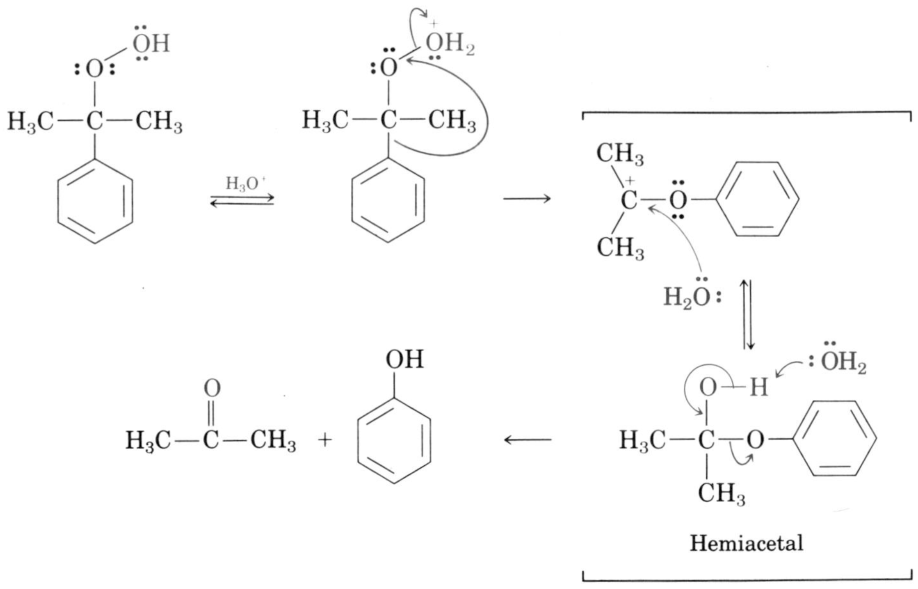

$$H_3C-\underset{C_6H_5}{\overset{:\ddot{O}-\ddot{O}H}{C}}-CH_3 \overset{H_3O^+}{\rightleftharpoons} H_3C-\underset{C_6H_5}{\overset{:\ddot{O}-\overset{+}{O}H_2}{C}}-CH_3 \longrightarrow (CH_3)_2\overset{+}{C}-\ddot{O}-C_6H_5 \xrightleftharpoons{H_2\ddot{O}:} H_3C-\underset{CH_3}{\overset{O-H}{C}}-O-C_6H_5 \;(:\ddot{O}H_2) \longrightarrow H_3C-\overset{O}{\overset{\|}{C}}-CH_3 + C_6H_5OH$$

Hemiacetal

**Figure 26.4** Mechanism of the formation of phenol by acid-catalyzed reaction of cumene hydroperoxide

In addition to its use in resins and adhesives, phenol also serves as starting material for the synthesis of chlorinated phenols and of the food preservatives BHT (butylated hydroxytoluene) and BHA (butylated hydroxyanisole). Thus, pentachlorophenol, a widely used wood preservative, is prepared by reaction of phenol with excess chlorine. The herbicide 2,4-D (2,4-dichlorophenoxyacetic acid) is prepared from 2,4-dichlorophenol, and the hospital antiseptic agent hexachlorophene is prepared from 2,4,5-trichlorophenol.

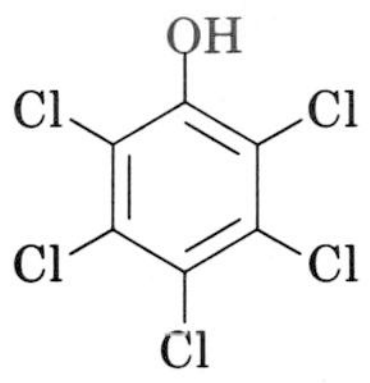

Pentachlorophenol (wood preservative)

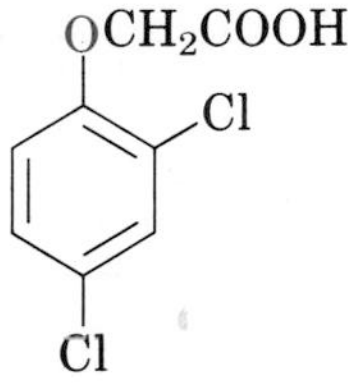

2,4-Dichlorophenoxyacetic acid, 2,4-D (herbicide)

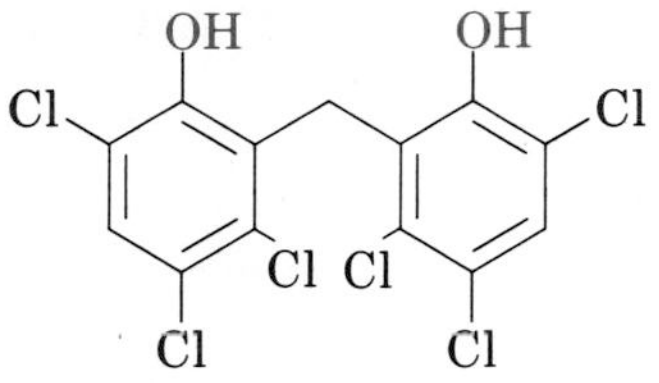

Hexachlorophene (antiseptic)

The food preservative BHT is prepared by Friedel–Crafts alkylation of *p*-methylphenol (*p*-cresol) with 2-methylpropene in the presence of acid; BHA is prepared similarly by alkylation of *p*-methoxyphenol.

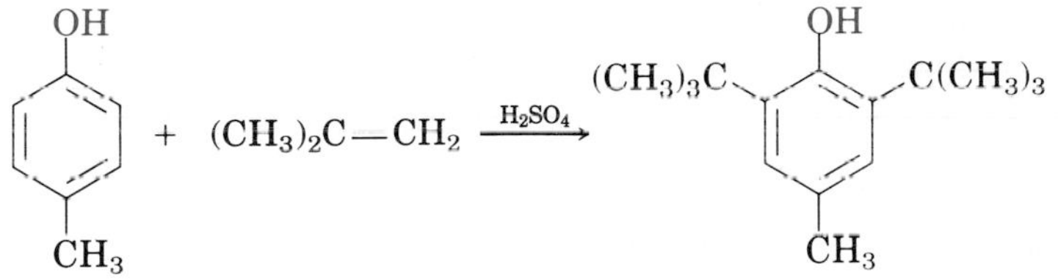

$$p\text{-Methylphenol} + (CH_3)_2C{=}CH_2 \xrightarrow{H_2SO_4} \text{BHT}$$

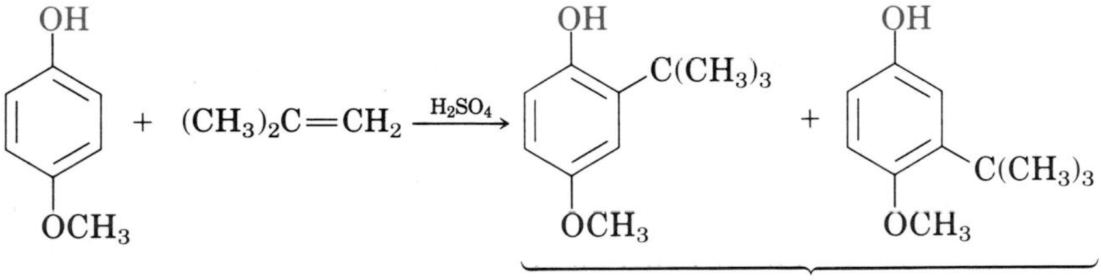

$$p\text{-Methoxyphenol} + (CH_3)_2C{=}CH_2 \xrightarrow{H_2SO_4} \text{BHA}$$

**PROBLEM**

**26.10** 2,4-Dichlorophenoxyacetic acid is prepared from phenol by a two-step sequence involving an electrophilic aromatic substitution followed by a Williamson ether synthesis. Formulate the reactions.

## 26.7 Properties of Phenols; Acidity

The properties of phenols are similar in many respects to those of alcohols. Thus, low-molecular-weight phenols are generally somewhat water soluble and are high boiling because of intermolecular hydrogen bonding. The most important property of phenols, however, is their acidity. Phenols are weak acids that can dissociate in aqueous solution to give $H_3O^+$ plus a phenoxide anion, $ArO^-$. Acidity values for some common phenols are given in Table 26.2.

**Table 26.2** Physical properties of some phenols

| *Phenol* | *Melting point (°C)* | *Boiling point (°C)* | p$K_a$ |
|---|---|---|---|
| Acetic acid [a] | | | 4.75 |
| 2,4,6-Trinitrophenol | 122 | — | 0.60 |
| *p*-Nitrophenol | 115 | — | 7.16 |
| *o*-Nitrophenol | 97 | — | 7.21 |
| *m*-Nitrophenol | 45 | 216 | 8.36 |
| *p*-Iodophenol | 94 | — | 9.30 |
| *p*-Bromophenol | 66 | 238 | 9.35 |
| *p*-Chlorophenol | 43 | 220 | 9.38 |
| Phenol | 43 | 182 | 10.00 |
| *p*-Methoxyphenol | 57 | 243 | 10.21 |
| *p*-Methylphenol (*p*-Cresol) | 35 | 202 | 10.26 |
| *p*-Aminophenol | 186 | — | 10.46 |
| Ethanol[a] | | | 16.00 |

[a]Values for acetic acid and ethanol are given for reference.

The data in Table 26.2 show that phenols are much more acidic than alcohols. Indeed, some phenols, such as the nitro-substituted ones, even approach or surpass the acidity of carboxylic acids. One practical consequence of this acidity is that phenols are soluble in dilute aqueous sodium hydroxide. Thus, a phenolic component can often be separated from a mixture of compounds simply by basic extraction into aqueous solution, followed by reacidification.

$$C_6H_5\text{—O—H} + NaOH \longrightarrow C_6H_5\text{—}O^-\ Na^+ + H_2O$$

Phenol Sodium phenoxide

Phenols are more acidic than alcohols because the phenoxide anion is resonance-stabilized by the aromatic ring. Sharing of the negative charge on oxygen with the ortho and para positions of the aromatic ring results in increased stability of the phenoxide anion relative to undissociated phenol, and in low $\Delta G°$ for the dissociation reaction.

Recall the following:

$$\mathrm{R{-}\overset{\overset{\displaystyle O}{\|}}{C}{-}O^-} \longleftrightarrow \mathrm{R{-}\overset{\overset{\displaystyle O^-}{|}}{C}{=}O}$$

Figure 26.5 compares the acidity of phenols and alcohols through reaction energy diagrams.

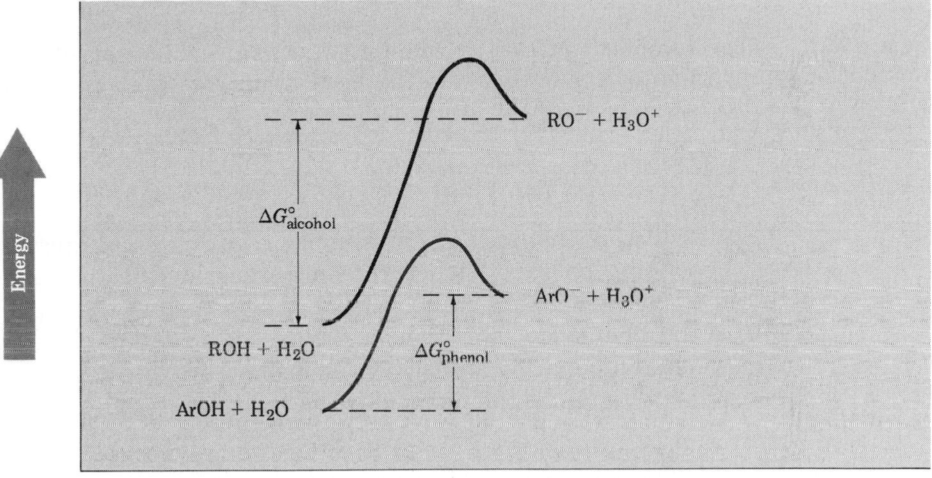


**Figure 26.5** A comparison of phenol and alcohol acidities: Phenols (red curve) are more acidic than alcohols (black curve) because a phenoxide ion is stabilized relative to free phenol more than an alkoxide ion is stabilized relative to free alcohol.

The arguments just used to account for phenol acidity are similar to those used in Section 26.2 to account for arylamine basicity, but the two situations are opposite. Arylamines are less basic than alkylamines because resonance stabilization of the free arylamine is greater than the stabilization of the arylammonium ion. Phenols, however, are more acidic than alcohols because resonance stabilization of the phenoxide ion is greater than that of the free phenol.

In general, any effect that stabilizes the starting material more than it stabilizes the product will raise $\Delta G°$ and disfavor the reaction. Conversely, any effect that stabilizes the product more than it stabilizes the starting material will lower $\Delta G°$ and favor the reaction. These effects are shown in Figure 26.6 (page 950).

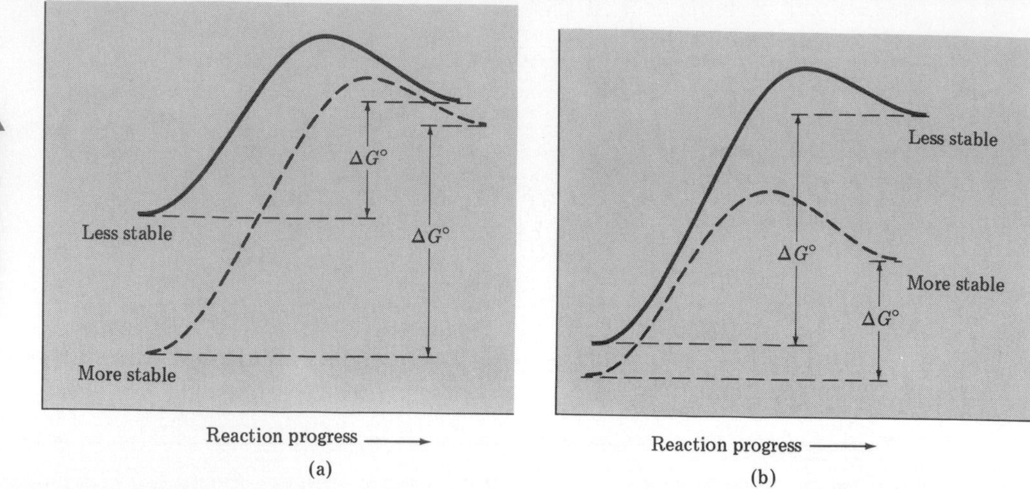


**Figure 26.6** Effect of stabilizing factors on equilibria ($\Delta G°$): (a) Stabilization of starting material relative to product raises $\Delta G°$ and disfavors the reaction, whereas (b) stabilization of product relative to starting material lowers $\Delta G°$ and favors the reaction.

Substituted phenols can be either more acidic or less acidic than phenol itself, depending on their structure. As a general rule, phenols with an electron-withdrawing substituent are more acidic, since these substituents *stabilize* the phenoxide ion by delocalizing the negative charge. Phenols with an electron-donating substituent, however, are less acidic, since these substituents *destabilize* the phenoxide ion by concentrating the charge.

EWG

Electron-withdrawing groups (EWG) stabilize phenoxide anion, resulting in increased phenol acidity

EDG

Electron-donating groups (EDG) destabilize phenoxide anion, resulting in decreased phenol acidity

The acidifying effect of an electron-withdrawing substituent is particularly noticeable for phenols having a nitro group at the ortho or para position because the phenoxide anion is strongly stabilized in these cases.

$NO_2$ ⟷ $NO_2$ ⟷ $NO_2$ ⟷ $^+N$ ⟷ $^+N$ ⟷ $NO_2$

Note that the effect of substituents on phenol acidity is the same as their effect on benzoic acid acidity (Section 20.5), but *opposite* to their effect on arylamine basicity (Table 26.3). The same resonance factors operate in all three cases.

**Table 26.3** Substituent effect on acidity of benzoic acids and phenols, and on basicity of arylamines

| *Type of compound* | *Effect of group* Y |
|---|---|
| 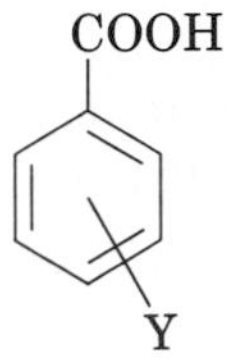 COOH (benzoic acid with substituent Y) | Electron-withdrawing; increased acidity<br>Electron-donating; decreased acidity |
| 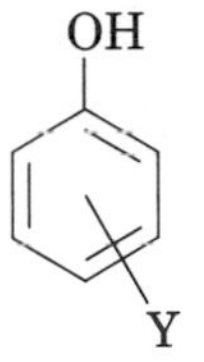 OH (phenol with substituent Y) | Electron-withdrawing; increased acidity<br>Electron-donating; decreased acidity |
| 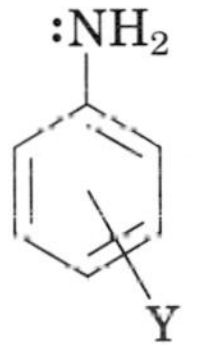 $:NH_2$ (arylamine with substituent Y) | Electron-withdrawing; decreased basicity<br>Electron-donating; increased basicity |

PROBLEM

**26.11** Rank the compounds in each group in order of increasing acidity:
(a) Phenol, *p*-methylphenol, *p*-(trifluoromethyl)phenol
(b) Benzyl alcohol, phenol, *p*-hydroxybenzoic acid
(c) *p*-Bromophenol, 2,4-dibromophenol, 2,4,6-tribromophenol

## 26.8 Preparation of Phenols

We've already seen the two best methods of phenol synthesis. To review briefly:

1. Alkali fusion of aromatic sulfonates takes place when an arenesulfonic acid is melted with sodium hydroxide at high temperature (Section 16.2). Few functional groups can survive such harsh conditions, however, and the reaction is therefore limited to the preparation of alkyl-substituted phenols.

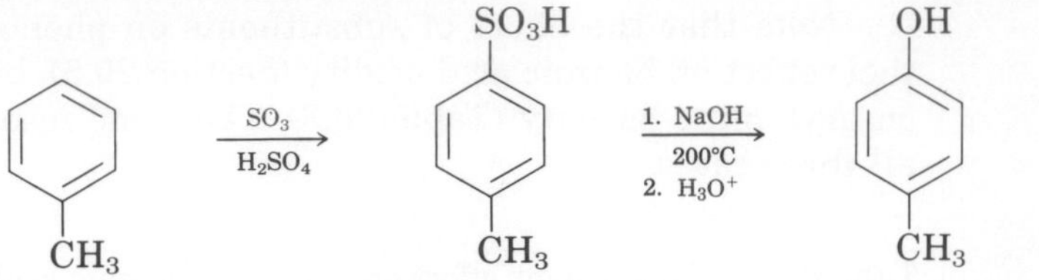

Toluene $\xrightarrow[H_2SO_4]{SO_3}$ p-Toluenesulfonic acid $\xrightarrow[\text{2. } H_3O^+]{\text{1. NaOH, 200°C}}$ p-Methylphenol (72%)

2. Hydrolysis of arenediazonium salts in a Sandmeyer-type replacement reaction is the most versatile and widely used laboratory method of phenol synthesis (Section 26.4). Most functional groups are compatible with the reaction conditions needed, and yields are generally good.

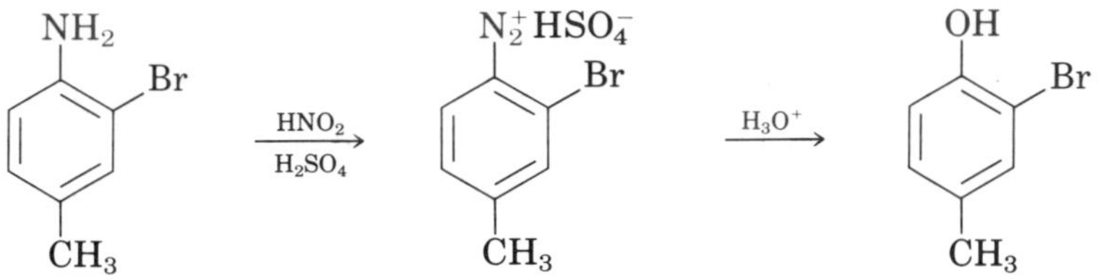

2-Bromo-4-methylaniline $\xrightarrow[H_2SO_4]{HNO_2}$ ($N_2^+ HSO_4^-$) $\xrightarrow{H_3O^+}$ 2-Bromo-4-methylphenol (92%)

PROBLEM

**26.12** *p*-Cresol (*p*-methylphenol) is used industrially as both an antiseptic and as a starting material to prepare the food additive BHT. How would you prepare *p*-cresol from benzene?

PROBLEM

**26.13** Carvacrol (5-isopropyl-2-methylphenol) is a natural product isolated from oregano, thyme, and marjoram. Propose two different syntheses of carvacrol from benzene.

## 26.9 Reactions of Phenols

### ALCOHOL-LIKE REACTIONS

The chemistry of phenols is similar in part to that of alcohols. Thus, phenols can be converted into esters by reaction with acid chlorides or acid anhydrides, and into ethers by reaction with alkyl halides in the presence of base (Williamson synthesis). Both reactions occur under relatively mild conditions since phenols are so much more acidic than alcohols, and since the reactive phenoxide ion intermediates are formed much more readily than alkoxide ions. Direct esterification by acid-catalyzed reaction between a phenol and a carboxylic acid, however, is not usually successful.

Ester formation

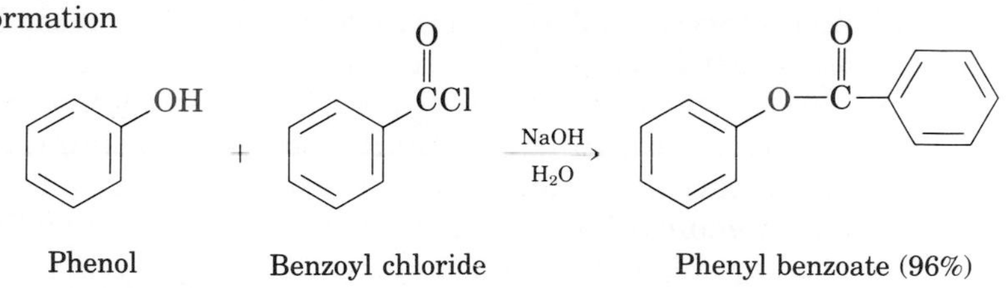

Phenol + Benzoyl chloride $\xrightarrow[H_2O]{NaOH}$ Phenyl benzoate (96%)

Ether formation

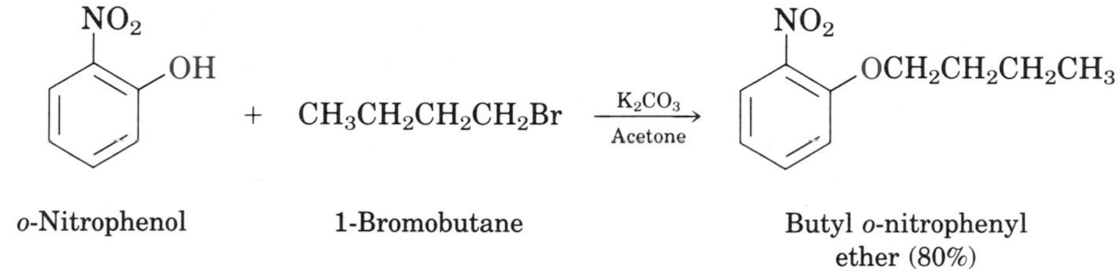

*o*-Nitrophenol + $CH_3CH_2CH_2CH_2Br$ (1-Bromobutane) $\xrightarrow[Acetone]{K_2CO_3}$ Butyl *o*-nitrophenyl ether (80%) ($OCH_2CH_2CH_2CH_3$, $NO_2$)

## ELECTROPHILIC AROMATIC SUBSTITUTION REACTIONS

The hydroxyl group is a strongly activating, ortho- and para-directing substituent in electrophilic aromatic substitution reactions (Section 16.5). As a result, phenols are highly reactive substrates for electrophilic halogenation, nitration, and sulfonation, as well as for coupling with diazonium salts to produce azo dyes.

Not surprisingly, phenoxide *ions* are even more reactive toward electrophilic aromatic substitution than neutral phenols because a full negative charge is present:

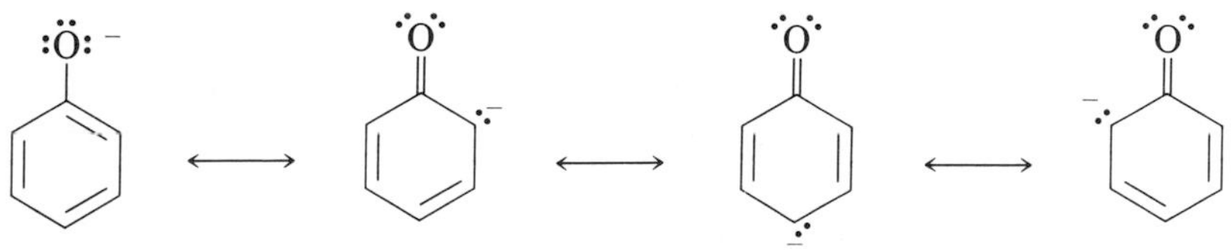

Phenoxide ion

Recall the following:

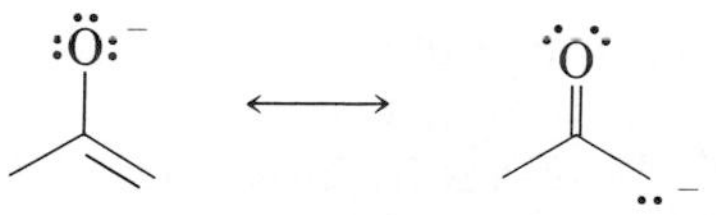

Ketone enolate ion

Resonance structures of phenoxide anion show a similarity to the resonance structures of a ketone enolate anion, which suggests the possibility that phenoxides might undergo alpha-substitution reactions similar to those of ketones (Section 22.6). In practice, phenoxide ions are less reactive than enolate ions because of the great stability of the benzene ring. Nevertheless, there are a number of examples of enolate-like reactivity of phenoxide ions. For example, in the **Kolbe–Schmitt**[3,4] **carboxylation reaction**, phenoxide ion adds to carbon dioxide under pressure to yield an intermediate keto acid anion that enolizes to give *o*-hydroxybenzoic acid (salicylic acid). This reaction is a key step in the industrial synthesis of aspirin (acetylsalicylic acid).

OH
NaOH
:Ö:⁻ Na⁺
O=C=O
O
H
COO⁻
Phenol
$H_3O^+$
$OCOCH_3$
COOH
$(CH_3CO)_2O$
OH
COOH
Acetylsalicylic acid (aspirin)
Salicylic acid

PROBLEM

**26.14** When sodium phenoxide is treated with allyl bromide in a Williamson ether synthesis, a mixture of phenyl allyl ether and *o*-allylphenol is formed. How can you account for the formation of *o*-allylphenol?

## OXIDATION OF PHENOLS: QUINONES

The susceptibility of phenols to electrophilic aromatic substitution is one consequence of the electron-rich nature of the phenol ring. Another consequence is the susceptibility of phenols to oxidation. Treatment of a phenol with any of a number of strong oxidizing agents yields a 2,5-cyclohexadiene 1,4-dione, or **quinone**. Older procedures employed sodium dichromate as oxidant, but Fremy's salt, potassium nitrosodisulfonate [$(KSO_3)_2NO$], is now preferred. The reaction takes place under mild conditions through a radical mechanism, and good yields are normally obtained. Arylamines are similarly oxidized to quinones.

[3]Herman Kolbe (1818–1884); b. Germany; Ph.D. Göttingen; professor, universities of Marburg and Leipzig.

[4]Rudolf Schmitt (1830–1898); b. Wippershain, Germany; Ph.D. Marburg; professor, University of Dresden.

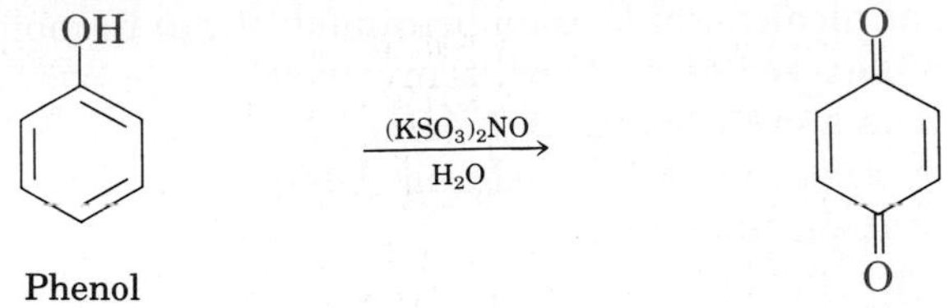

OH

$\xrightarrow[H_2O]{(KSO_3)_2NO}$

O

O

Phenol

Benzoquinone (79%)

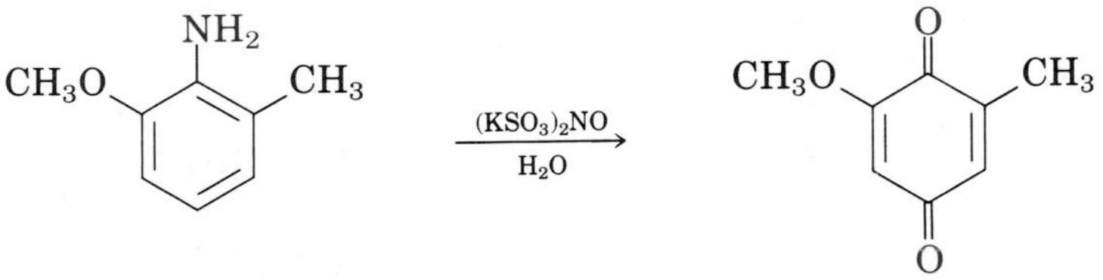

$NH_2$

$CH_3O$ $CH_3$

$\xrightarrow[H_2O]{(KSO_3)_2NO}$

O

$CH_3O$ $CH_3$

O

2-Methyl-6-methoxyaniline

2-Methyl-6-methoxybenzoquinone (96%)

Quinones are an interesting and valuable class of compounds because of their oxidation–reduction properties. They can be easily reduced to **hydroquinones** (*p*-dihydroxybenzenes) by reagents like $NaBH_4$ and $SnCl_2$, and hydroquinones can be easily reoxidized back to quinones by Fremy's salt.

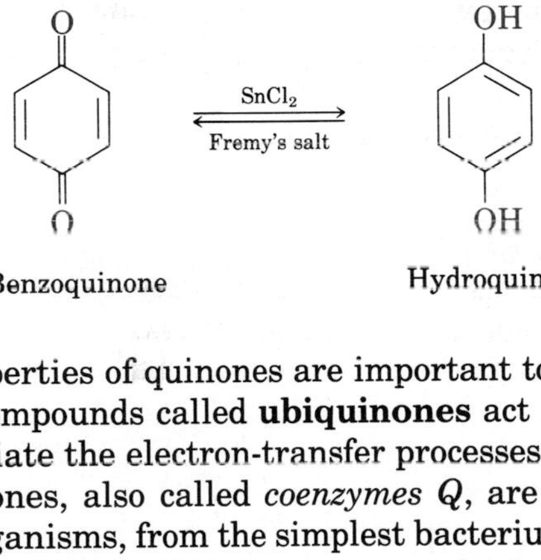

O

OH

$\underset{\text{Fremy's salt}}{\overset{SnCl_2}{\rightleftharpoons}}$

O

OH

Benzoquinone

Hydroquinone

These redox properties of quinones are important to the functioning of living cells, where compounds called **ubiquinones** act as biochemical oxidizing agents to mediate the electron-transfer processes involved in energy production. Ubiquinones, also called *coenzymes Q*, are components of the cells of all aerobic organisms, from the simplest bacterium to humans. They are so named because of their ubiquitous occurrence in nature.

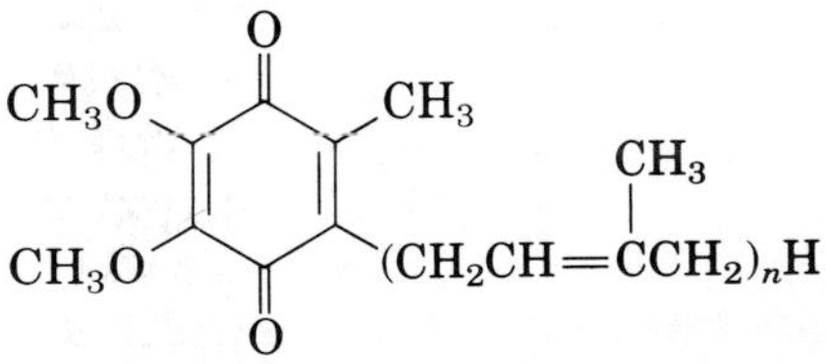

O

$CH_3O$ $CH_3$

$CH_3$

$CH_3O$ $(CH_2CH{=}CCH_2)_nH$

O

Ubiquinones ($n$ = 1–10)

Ubiquinones function within the mitochondria of cells as mobile electron carriers (oxidizing agents) to mediate the respiration process whereby electrons are transported from the biological reducing agent NADH (reduced form of nicotinamide adenine dinucleotide) to molecular oxygen. Although a complex series of steps is involved in the overall process, the ultimate result is a cycle whereby NADH is oxidized to NAD, oxygen is reduced to water, and energy is produced. Ubiquinone acts only as an intermediary and is itself unchanged.

*Step 1:*

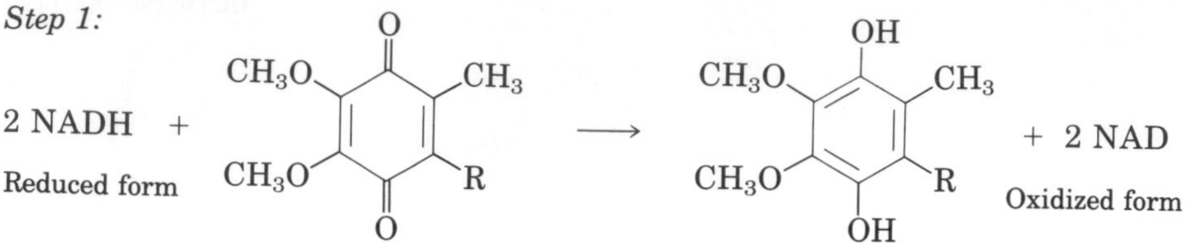

2 NADH + (ubiquinone: O, $CH_3O$, $CH_3$, $CH_3O$, R, O) ⟶ (hydroquinone: OH, $CH_3O$, $CH_3$, $CH_3O$, R, OH) + 2 NAD

Reduced form — Oxidized form

*Step 2:*

(hydroquinone: OH, $CH_3O$, $CH_3$, $CH_3O$, R, OH) + $\frac{1}{2}O_2$ ⟶ (ubiquinone: O, $CH_3O$, $CH_3$, $CH_3O$, R, O) + $H_2O$

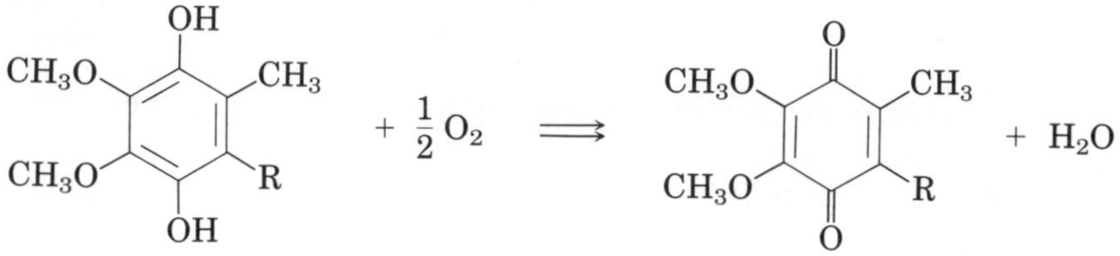

*Net change:*

$$2\ \text{NADH} + \frac{1}{2}O_2 \longrightarrow 2\ \text{NAD} + H_2O$$

PROBLEM

**26.15** Early work on the structural elucidation of ubiquinones was complicated by the fact that extraction of the compounds from cells was carried out using basic ethanol solution. Under these conditions, the ubiquinone methoxyl groups became exchanged for ethoxyls. Propose a mechanism to account for this exchange. (See Section 19.18.)

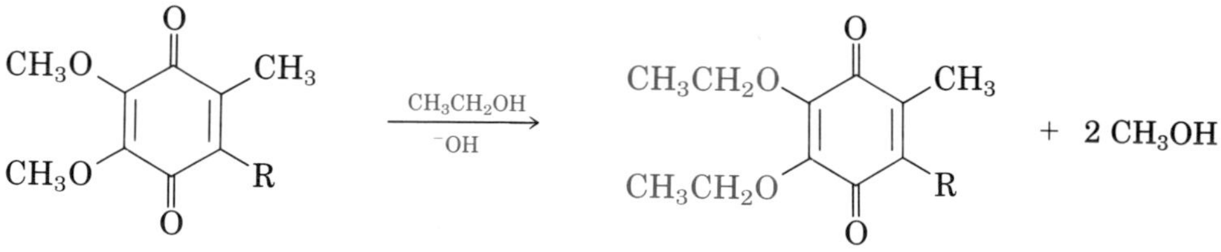

(ubiquinone: O, $CH_3O$, $CH_3$, $CH_3O$, R, O) $\xrightarrow[^{-}OH]{CH_3CH_2OH}$ (O, $CH_3CH_2O$, $CH_3$, $CH_3CH_2O$, R, O) + 2 $CH_3OH$

## CLAISEN REARRANGEMENT

Treatment of a phenoxide ion with 3-bromopropene results in a Williamson ether synthesis and production of an allyl phenyl ether. Heating the allyl phenyl ether to 200–250°C then effects rearrangement leading to an *o*-allylphenol (the **Claisen rearrangement**). The net effect is alkylation of the ortho position of the phenol.

Phenol + NaH $\xrightarrow[\text{solution}]{\text{THF}}$ Sodium phenoxide ($O^- Na^+$) $\xrightarrow{BrCH_2CH{=}CH_2}$ Allyl phenyl ether ($OCH_2CH{=}CH_2$)

Allyl phenyl ether $\xrightarrow[250°C]{\text{Claisen rearrangement}}$ *o*-Allylphenol ($CH_2CH{=}CH_2$)

Claisen rearrangement of allyl phenyl ethers to yield *o*-allylphenols is a general reaction that is compatible with the presence of many other substituents on the benzene ring. The reaction proceeds through a pericyclic mechanism in which a concerted reorganization of bonding electrons occurs via a cyclic six-membered-ring transition state. The 2-allylcyclohexadienone intermediate then tautomerizes to *o*-allylphenol.

Good evidence for this mechanism comes from the observation that the rearrangement takes place with an *inversion* of the allyl unit. For example, phenyl allyl ether containing a $^{14}C$ label on the allyl ether carbon atom yields *o*-allylphenol in which the label is on the terminal carbon. It would be very difficult to explain this result by any mechanism other than a pericyclic one, which we'll look at in more detail in Section 30.10.

Allyl phenyl ether → [Transition state]$^‡$ → [Intermediate (6-Allyl-2,4-cyclohexadienone)] → *o*-Allylphenol

PROBLEM

**26.16** What product would you expect to obtain from Claisen rearrangement of 2-butenyl phenyl ether?

$$\text{C}_6\text{H}_5\text{O—CH}_2\text{CH=CHCH}_3 \xrightarrow{250°\text{C}}$$

2-Butenyl phenyl ether

## 26.10 Spectroscopy of Arylamines and Phenols

### INFRARED SPECTROSCOPY

The infrared spectra of arylamines and phenols are little different from those of aliphatic amines and alcohols. Thus, aniline shows the usual N—H infrared absorptions at 3400 and 3500 $cm^{-1}$ characteristic of a primary amine, as well as a pair of bands at 1500 and 1600 $cm^{-1}$ characteristic of aromatic rings (Figure 26.7). Note that the infrared spectrum of aniline also shows the typical monosubstituted aromatic ring peaks at 690 and 760 $cm^{-1}$.

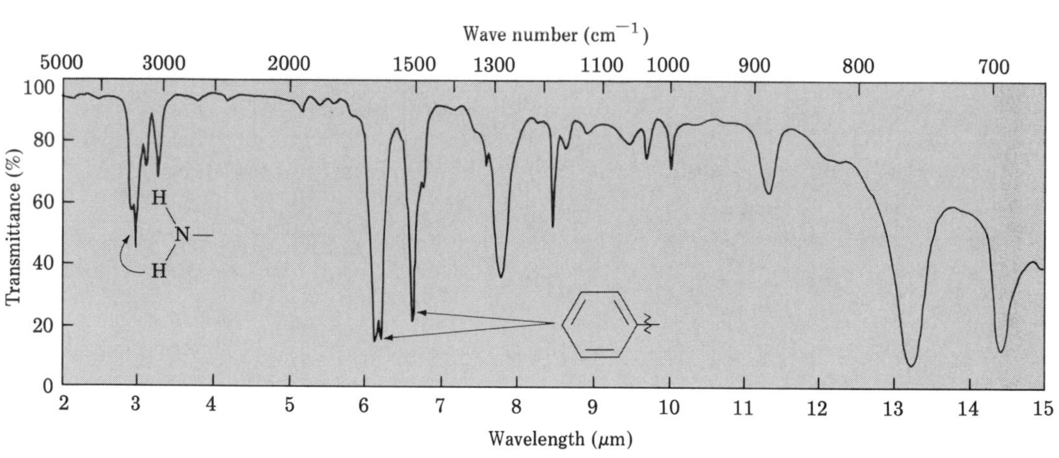


**Figure 26.7** Infrared spectrum of aniline

Phenol shows a characteristic broad absorption at 3500 $cm^{-1}$ due to the hydroxyl group, and the usual 1500 and 1600 $cm^{-1}$ aromatic bands (Figure 26.8). Here, too, the monosubstituted aromatic ring peaks at 690 and 760 $cm^{-1}$ are visible.

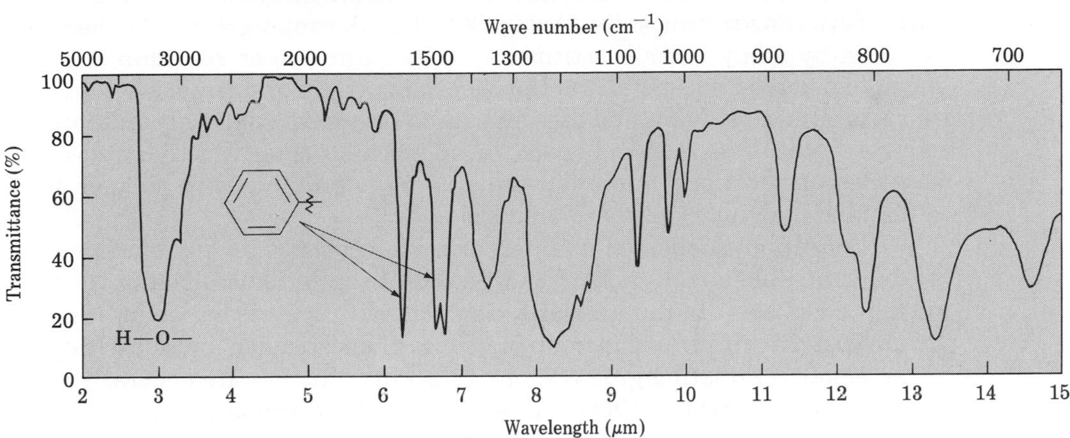


**Figure 26.8** Infrared spectrum of phenol

## NUCLEAR MAGNETIC RESONANCE SPECTROSCOPY

Arylamines and phenols, like all aromatic compounds, show $^1H$ NMR absorptions near 7–8 $\delta$, the expected position for aromatic ring protons. In addition, amine N—H protons usually absorb in the 2–3 $\delta$ range, and phenol O—H protons absorb at 2.5–6 $\delta$. In neither case are these absorptions uniquely diagnostic for arylamines or phenols, since other kinds of protons absorb in the same range. As a result, a combination of both NMR and infrared evidence is usually needed to assign structure.

As was true for alcohols (Section 17.11), the identity of the NMR peak due to N—H and O—H protons is easily determined by adding a small amount of $D_2O$ to the sample tube. Since the O—H and N—H protons rapidly exchange with added $D_2O$, their peaks disappear from the spectrum.

# 26.11 Summary and Key Words

**Arylamines**, like their aliphatic counterparts, are basic. The base strength of arylamines is generally lower than that of aliphatic amines, however, because the nitrogen lone-pair electrons are delocalized into the aromatic ring by orbital overlap with the aromatic pi system. As a general rule, electron-withdrawing substituents on the ring further weaken the basicity of a substituted aniline, whereas electron-donating substituents increase basicity.

Substituted anilines are almost always prepared by nitration of the appropriate aromatic ring, followed by reduction. An amine group is a strongly activating, ortho- and para-directing substituent, and electrophilic aromatic substitution is an important reaction of arylamines. If the amine group makes the ring too reactive, however, its reactivity can be modulated by converting it into a nonbasic amide.

The most important reaction of arylamines is conversion by nitrous acid into **arenediazonium salts, $ArN_2^+ X^-$**. The diazonio group can then be replaced by many other substituents in the **Sandmeyer reaction** to give a wide variety of substituted aromatic compounds. For example, aryl chlorides, bromides, iodides, and nitriles can be prepared from arenediazonium salts, as can arenes and phenols. In addition to their reactivity toward substitution reactions, diazonium salts undergo coupling with phenols and arylamines to give brightly colored azo dyes.

**Phenols** are aromatic counterparts of alcohols but are much more acidic, since phenoxide anions can be stabilized by delocalization of the negative charge into the aromatic ring. Substitution of the aromatic ring by an electron-withdrawing group increases phenol acidity, and substitution by an electron-donating group decreases acidity. Phenols are generally prepared by one of two methods: (1) alkali fusion of aromatic sulfonates or (2) hydrolysis of an arenediazonium salt.

Reactions of phenols can occur either at the hydroxyl group or on the aromatic ring. For example, a phenol hydroxyl can be converted into an ester or an ether group. Phenyl allyl ethers are particularly interesting since they undergo **Claisen rearrangement** to give *o*-allylphenols when heated to 250°C. The hydroxyl group strongly activates the aromatic ring toward electrophilic substitution reactions. In addition, phenols can be oxidized to **quinones** by reaction with Fremy's salt, potassium nitrosodisulfonate.

## 26.12 Summary of Reactions

1. Preparation of arylamines
   a. Reduction of nitrobenzenes (Section 26.3)

$$ArNO_2 + H_2 \xrightarrow[\text{Ethanol}]{\text{Pt}} ArNH_2$$

$$ArNO_2 + Fe \xrightarrow[\text{2. HO}^-]{\text{1. H}_3\text{O}^+} ArNH_2$$

$$ArNO_2 + SnCl_2 \xrightarrow[\text{2. HO}^-]{\text{1. H}_3\text{O}^+} ArNH_2$$

2. Reactions of arylamines
   a. Electrophilic aromatic substitution (Sections 16.2 and 26.4)

$$C_6H_5\ddot{N}H_2 + E^+ \longrightarrow p\text{-}E\text{-}C_6H_4\ddot{N}H_2$$

Ortho and para directing

b. Formation of arenediazonium salts (Section 26.4).

$$ArNH_2 + HNO_2 \xrightarrow{HX} ArN_2^+ X^-$$

c. Reaction of arenediazonium salts (Section 26.4)

(1) Aryl chlorides

$$ArN_2^+ X^- + CuCl \longrightarrow ArCl + CuX + N_2$$

(2) Aryl bromides

$$ArN_2^+ X^- + CuBr \longrightarrow ArBr + CuX + N_2$$

(3) Aryl iodides

$$ArN_2^+ X^- + NaI \longrightarrow ArI + NaX + N_2$$

(4) Arenenitriles

$$ArN_2^+ X^- + CuCN \longrightarrow ArCN + CuX + N_2$$

(5) Phenols

$$ArN_2^+ X^- + H_3O^+ \longrightarrow ArOH + HX + N_2$$

(6) Arenes

$$ArN_2^+ X^- + H_3PO_2 \longrightarrow ArH + HX + N_2$$

(7) Diazonium coupling

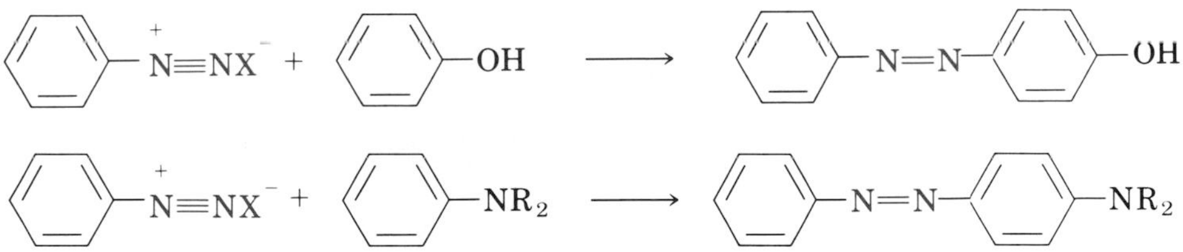

$$C_6H_5\text{-}\overset{+}{N}{\equiv}N\,X^- + C_6H_5\text{-}OH \longrightarrow C_6H_5\text{-}N{=}N\text{-}C_6H_4\text{-}OH$$

$$C_6H_5\text{-}\overset{+}{N}{\equiv}N\,X^- + C_6H_5\text{-}NR_2 \longrightarrow C_6H_5\text{-}N{=}N\text{-}C_6H_4\text{-}NR_2$$

d. Oxidation to quinones (Section 26.9)

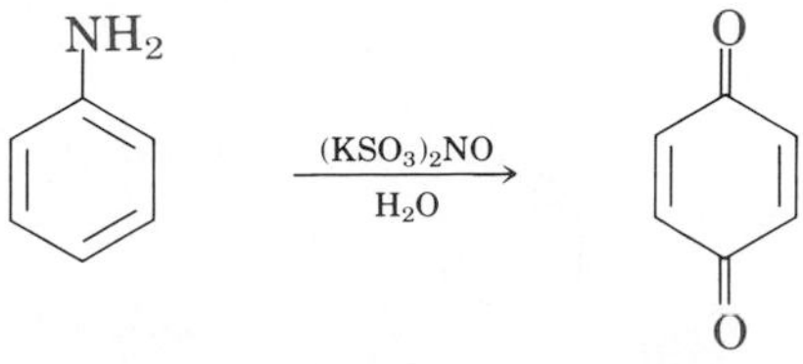

$$C_6H_5NH_2 \xrightarrow[H_2O]{(KSO_3)_2NO} \text{1,4-benzoquinone } (O{=}C_6H_4{=}O)$$

3. Preparation of phenols
   a. Alkali fusion of aryl sulfonates (Sections 16.2 and 26.8)

$$ArSO_3H + NaOH \xrightarrow{\Delta} ArOH + NaHSO_3$$

   b. Hydrolysis of arenediazonium salts (Section 26.4)

$$ArN_2^+ X^- + H_3O^+ \longrightarrow ArOH + N_2 + HX$$

4. Reactions of phenols
   a. Ester formation (Section 26.9)

$$ArOH + RCOCl \xrightarrow[H_2O]{NaOH} ArOCOR$$

   b. Williamson ether synthesis (Sections 18.4 and 26.9)

$$ArOH \xrightarrow[Ethanol]{NaOH} Ar\ddot{\underset{..}{O}}:^- \xrightarrow{RX} ArOR$$

   c. Kolbe–Schmitt carboxylation (Section 26.9)

$$C_6H_5OH \xrightarrow[\text{2. } CO_2,\ \Delta \quad \text{3. } H^+]{\text{1. NaOH}} o\text{-}HOC_6H_4COOH$$

   d. Oxidation to quinones (Section 26.9)

$$C_6H_5OH \xrightarrow[H_2O]{(KSO_3)_2NO} p\text{-benzoquinone } (O{=}C_6H_4{=}O)$$

   e. Claisen rearrangement (Sections 26.9 and 30.12)

$$C_6H_5OCH_2CH{=}CH_2 \xrightarrow{250°C} o\text{-}HOC_6H_4CH_2CH{=}CH_2$$

## ADDITIONAL PROBLEMS

**26.17** Provide IUPAC names for these compounds:

(a) [structure: benzene ring with OH and Br para]

(b) [structure: benzene ring with $NHCH_3$, Cl, Cl]

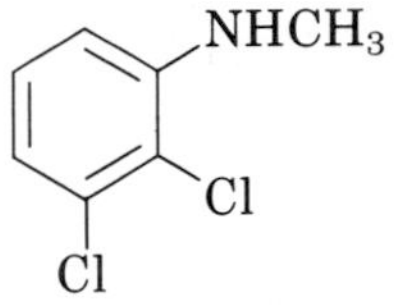

(c) [structure: benzene ring with $NH_2$, $CH_3$, $H_2N$]

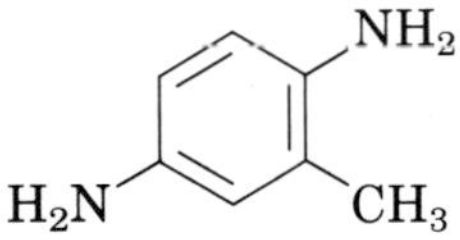

(d) [structure: benzene ring with OH, $H_3C$, $OCH_3$]

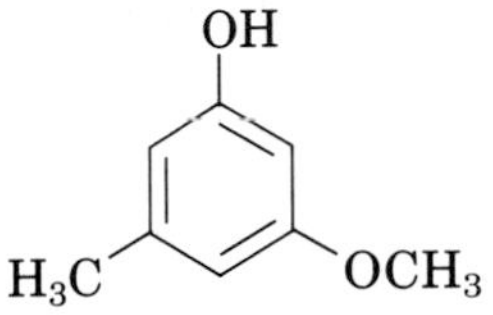

(e) [structure: benzene ring with OH, HO, OH]

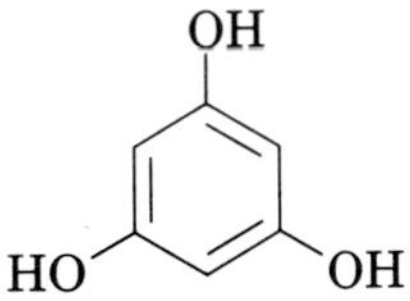

(f) [structure: benzene ring with $H_3C$, $N(CH_3)CH_2CH_3$, $CH_2CH_3$]

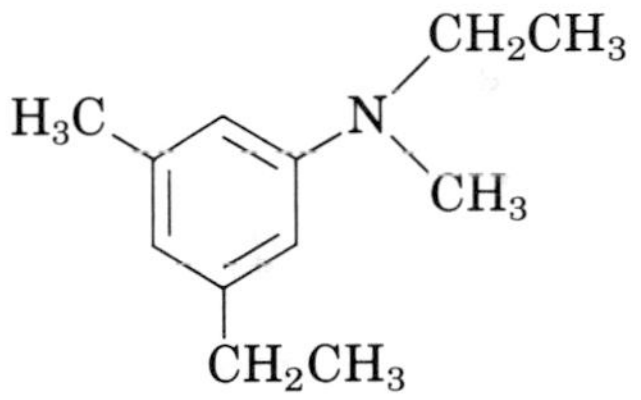

**26.18** How would you prepare aniline from these starting materials?
(a) Benzene (b) Benzamide (c) Toluene

**26.19** How would you convert aniline into each of the products listed in Problem 26.18?

**26.20** Suppose that you were given a mixture of toluene, aniline, and phenol and were asked to separate the mixture into its three pure components. Describe in detail how you would do this.

**26.21** Give the structures of the major organic products you would expect to obtain from reaction of *m*-toluidine (*m*-methylaniline) with these reagents:
(a) $Br_2$ (1 equiv)
(b) $(KSO_3)_2NO$
(c) $CH_3I$ (excess)
(d) $CH_3Cl + AlCl_3$
(e) $CH_3COCl$ in pyridine
(f) The product of part (e), then $HSO_3Cl$

**26.22** Benzoquinone is an excellent dienophile in the Diels–Alder reaction. What product would you expect to obtain from reaction of benzoquinone with 1 equiv of butadiene? From reaction with 2 equiv of butadiene?

**26.23** When the product, A, of the Diels–Alder reaction of benzoquinone and 1 equiv of butadiene (Problem 26.22) is treated with dilute acid or base, an isomerization occurs, and a new product, B, is formed. This new product shows a two-proton singlet in the $^1H$ NMR spectrum at 6.7 $\delta$ and an infrared absorption at 3500 $cm^{-1}$. What is the structure of the isomer B?

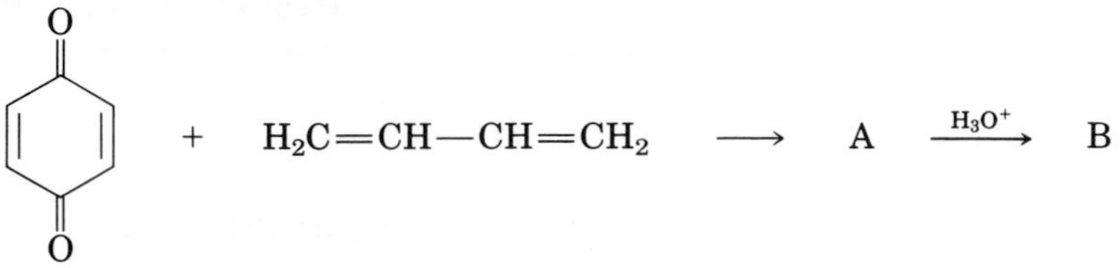

[benzoquinone] $+ \; H_2C{=}CH{-}CH{=}CH_2 \longrightarrow A \xrightarrow{H_3O^+} B$

**26.24** Tyramine is an alkaloid found, among other places, in mistletoe and ripe cheese. How would you synthesize tyramine from benzene? How would you synthesize it from toluene?

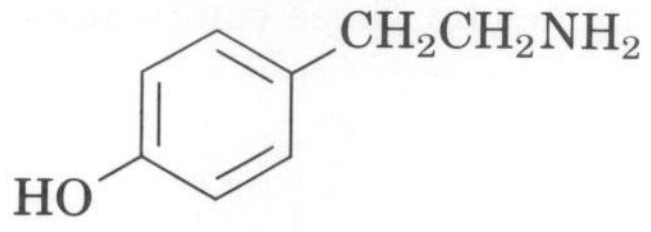

$CH_2CH_2NH_2$

HO

Tyramine

**26.25** How can you account for the fact that diphenylamine does not dissolve in dilute aqueous HCl and appears to be nonbasic?

**26.26** How would you prepare these compounds from toluene? A diazonio replacement reaction is needed in some instances.

(a) *p*-Methylaniline (b) *p*-Methylbenzylamine (c) 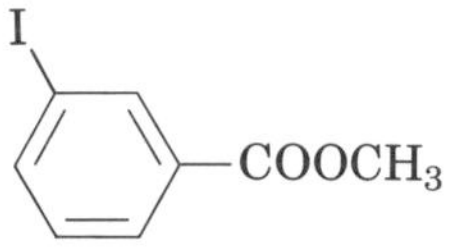 I, $COOCH_3$

**26.27** Show the products from reaction of *p*-bromoaniline with these reagents:

(a) Excess $CH_3I$
(b) HCl
(c) $NaNO_2$, $H_2SO_4$
(d) $CH_3COCl$
(e) $CH_3MgBr$
(f) $CH_3CH_2Cl$, $AlCl_3$
(g) Product of part (c) with CuCl
(h) Product of part (d) with $CH_3CH_2Cl$, $AlCl_3$

**26.28** Show the products from reaction of *o*-chlorophenol with these reagents:

(a) NaOH, then $CH_3I$
(b) $CH_3COCl$, pyridine
(c) Fremy's salt
(d) $CH_3CH_2CH_2Cl$, $AlCl_3$

**26.29** Reaction of anthranilic acid (*o*-aminobenzoic acid) with $NaNO_2$ and $H_2SO_4$ yields a diazonium salt that can be treated with base to yield a neutral diazonium carboxylate.

COOH

$NH_2$

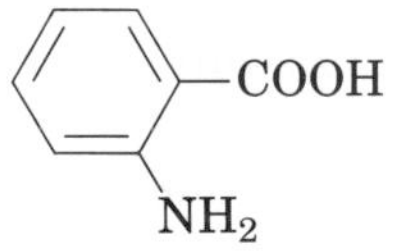

Anthranilic acid

(a) What is the structure of the neutral diazonium carboxylate?

(b) Heating the diazonium carboxylate results in the formation of $CO_2$, $N_2$, and a high-energy intermediate that reacts with 1,3-cyclopentadiene to yield the following organic product:

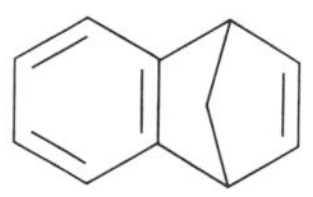

What is the structure of the reactive intermediate, and what kind of reaction does it undergo with cyclopentadiene?

**26.30** Mephenesin is a drug used as a muscle relaxant and sedative. Propose a synthesis of mephenesin from benzene and any other reagents needed.

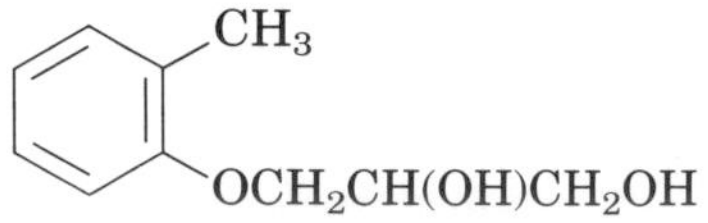

$CH_3$

$OCH_2CH(OH)CH_2OH$

Mephenesin

**26.31** How would you prepare these substances?
(a) *p*-Iodobenzoic acid from aniline (b) *o*-Iodobromobenzene from benzene

**26.32** Gentisic acid is a naturally occurring hydroquinone found in gentian. Its sodium salt is used medicinally as an antirheumatic agent. How would you prepare gentisic acid from benzene?

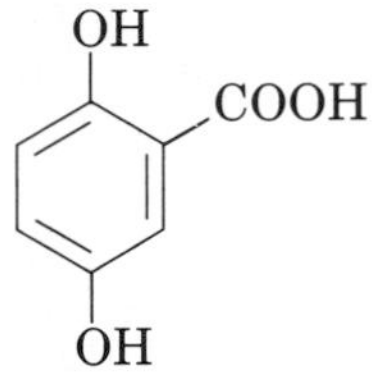

OH COOH OH

Gentisic acid

**26.33** Prontosil is an antibacterial azo dye that was once used for urinary tract infections. How would you prepare prontosil from benzene?

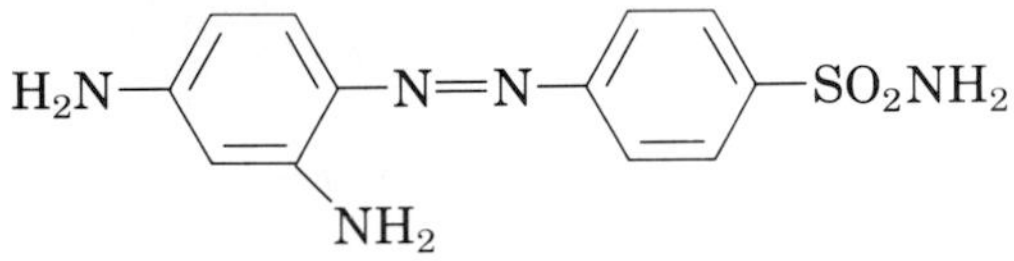

$H_2N$—N=N—$SO_2NH_2$ $NH_2$

Prontosil

**26.34** How would you synthesize the dye Orange II from benzene and $\beta$-naphthol?

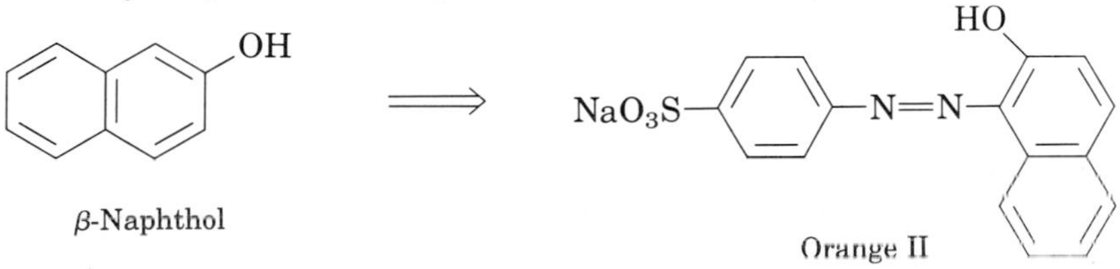

OH ⟹ $NaO_3S$—N=N— HO

$\beta$-Naphthol Orange II

**26.35** 2-Nitro-3,4,6-trichlorophenol is used as a lampricide—a compound toxic to lampreys—to combat the intrusion of sea lampreys into the Great Lakes. How would you synthesize this material from benzene?

**26.36** The germicidal agent hexachlorophene is prepared by condensation of two molecules of 2,4,5-trichlorophenol with one molecule of formaldehyde in the presence of sulfuric acid. Propose a mechanism to account for this reaction.

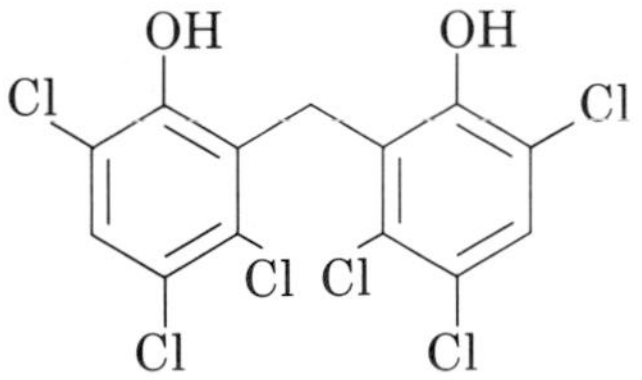

OH OH Cl Cl Cl Cl Cl Cl

Hexachlorophene

**26.37** Propose a route from benzene for the synthesis of the antiseptic agent trichlorosalicylanilide.

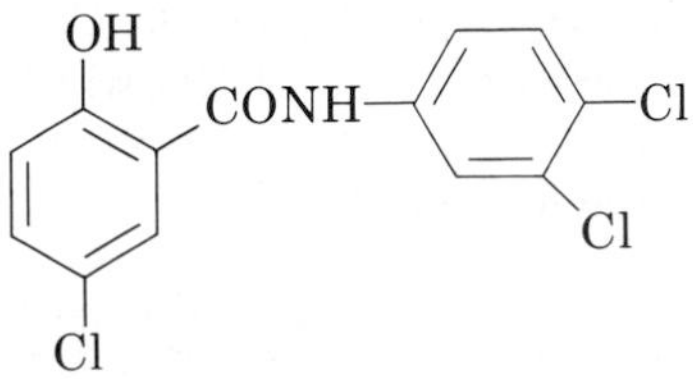

OH CONH— Cl Cl Cl

Trichlorosalicylanilide

**26.38** Compound A, $C_8H_{10}O$, has the infrared and $^1H$ NMR spectra shown. Propose a structure consistent with the observed spectral properties, and assign each peak in the NMR spectrum. Note that the absorption at 5.5 $\delta$ disappears when $D_2O$ is added.

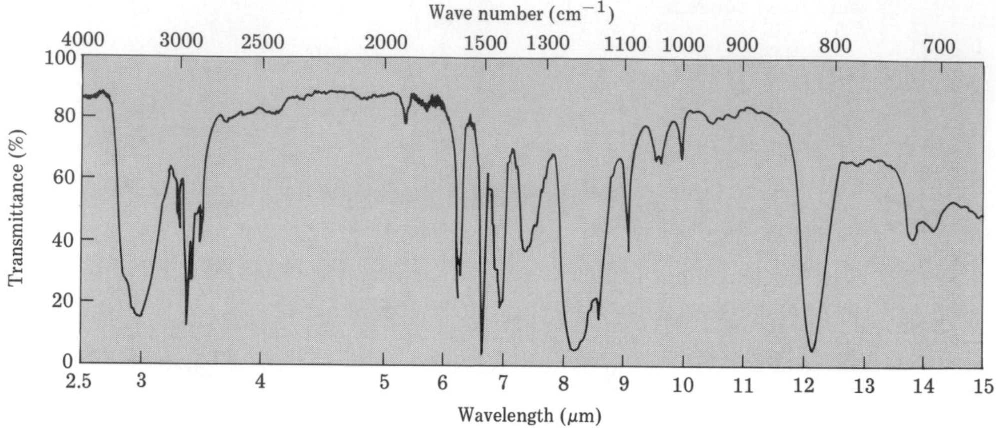


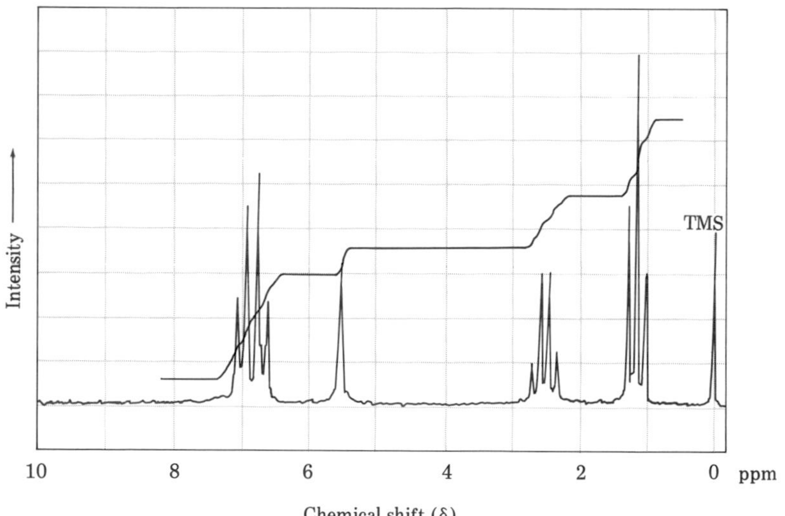


**26.39** Phenacetin, a substance formerly used in over-the-counter headache remedies, has the formula $C_{10}H_{13}NO_2$ and the infrared and NMR spectra shown. Phenacetin itself is neutral and does not dissolve in either acid or base. When warmed with aqueous hydroxide, phenacetin yields an amine, $C_8H_{11}NO$. When heated with HI, the amine is cleaved to an aminophenol $C_6H_7NO$, which, on treatment with Fremy's salt, yields benzoquinone. What is the structure of phenacetin, and what are the structures of the amine and the aminophenol?

$$\text{Phenacetin} \xrightarrow[\Delta]{\text{HO}^-} \text{Amine} \xrightarrow[\Delta]{\text{HI}} \text{Aminophenol} \xrightarrow[\text{salt}]{\text{Fremy's}} \text{Benzoquinone}$$

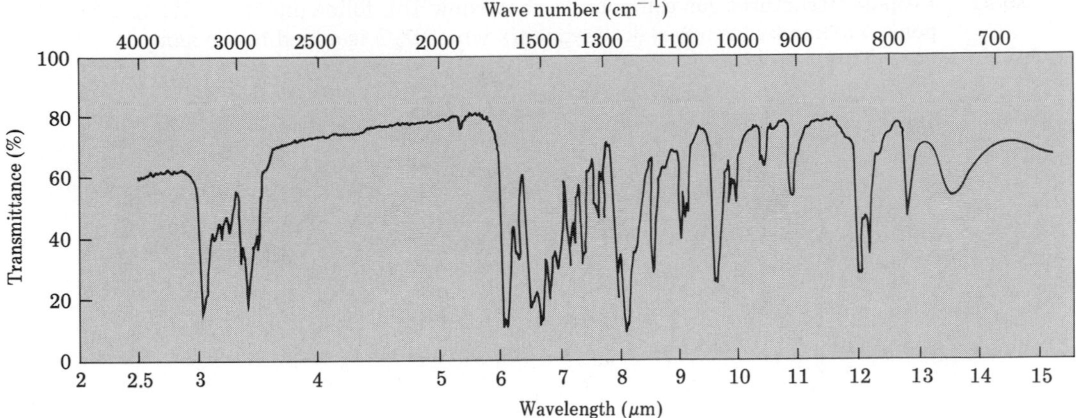


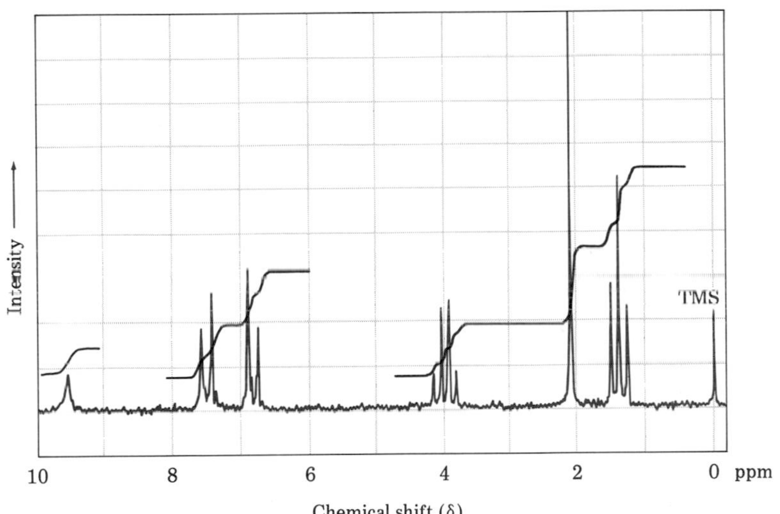


**26.40** In the Hoesch reaction, resorcinol (*m*-dihydroxybenzene) is treated with a nitrile in the presence of a Lewis acid catalyst. After hydrolysis, an acyl resorcinol is isolated. Propose a mechanism for the Hoesch reaction. To what other well-known reaction is this similar?

HO–$C_6H_4$–OH (resorcinol) + $CH_3C{\equiv}N$ $\xrightarrow[\text{2. } H_3O^+]{\text{1. } ZnCl_2\text{/ether}}$ HO, OH, $\underset{\displaystyle O}{\overset{\|}{C}}CH_3$ (2,4-dihydroxyacetophenone)

**26.41** Propose structures for compounds that show the following $^1H$ NMR spectra. The peak marked by an asterisk disappears when $D_2O$ is added to the sample.

(a) $C_{15}H_{24}O$

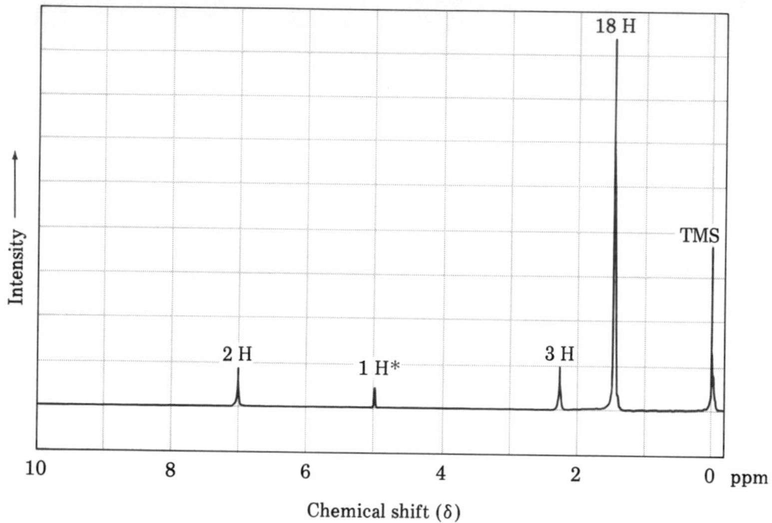


(b) $C_9H_{13}N$

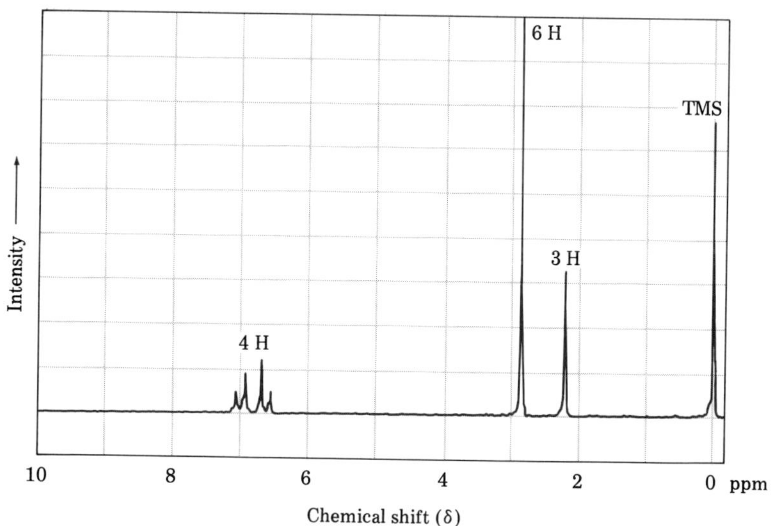

**26.42** Propose structures for compounds that show the following $^1H$ NMR spectra. The peak marked by an asterisk disappears when $D_2O$ is added to the sample.

(a) $C_{15}H_{17}N$

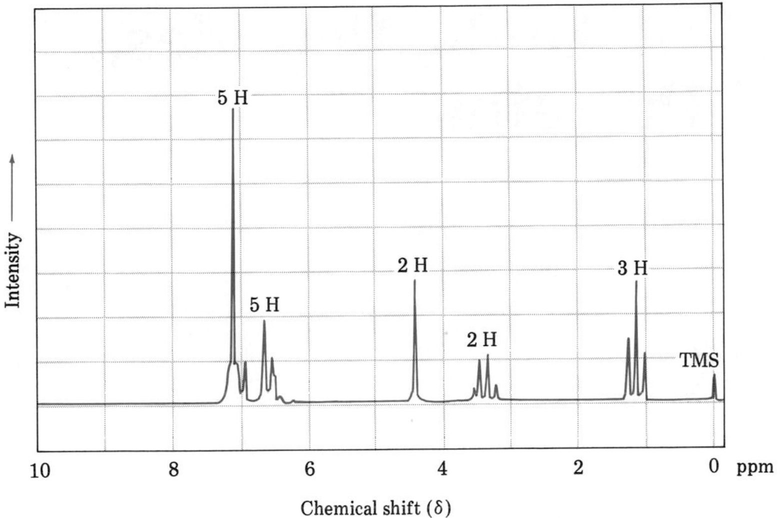


(b) $C_{10}H_{11}ClO_2$

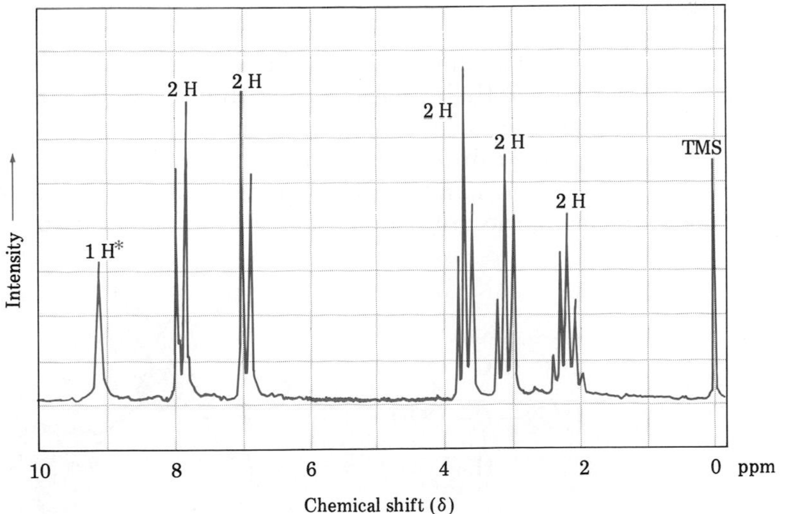

CHAPTER 27

# Amino Acids, Peptides, and Proteins

**Proteins** are large biomolecules that occur in every living organism. They are of many different types, and they serve many different biological roles. The keratin of skin and fingernails, the fibroin of silk and spider webs, and the collagen of tendons and cartilage are all **structural proteins**; the insulin that regulates glucose metabolism in the body is a **hormonal protein**; and the DNA polymerase and reverse transcriptase that serve as biological catalysts to carry out chemical reactions in the cell are proteins called **enzymes**.

Regardless of their appearance or their function, all proteins are chemically similar since all are made up of many amino acid units linked together in a long chain. **Amino acids** are the building blocks from which all proteins are made. As their name implies, amino acids are difunctional; they contain both a basic amino group and an acidic carboxyl group.

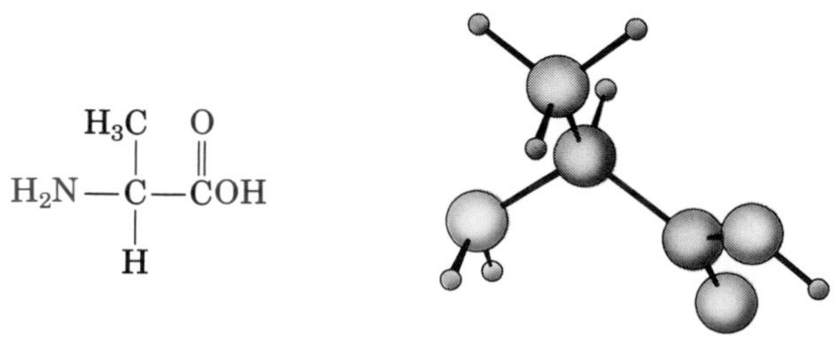

$$H_2N-\overset{\overset{H_3C}{|}}{\underset{\underset{H}{|}}{C}}-\overset{\overset{O}{\|}}{C}OH$$

Alanine, an amino acid

The great value of amino acids as biological building blocks stems from the fact that they can link together by forming amide, or **peptide**, bonds.

A **dipeptide** results when an amide bond is formed between the —$NH_2$ of one amino acid and the —COOH of a second amino acid; a **tripeptide** results from linkage of three amino acids via two amide bonds, and so on. Any number of amino acids can link together to form large chains. For classification purposes, chains with fewer than 50 amino acids are called **polypeptides**, whereas the term *protein* is reserved for larger chains.

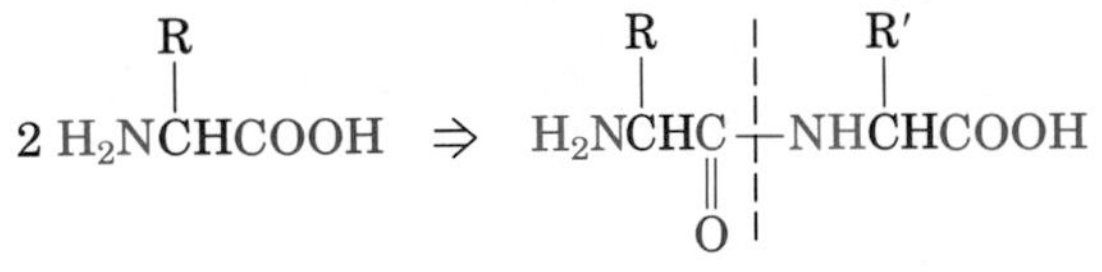

$$2\ H_2N\overset{R}{\overset{|}{C}}HCOOH \Rightarrow H_2N\overset{R}{\overset{|}{C}}H\underset{\underset{O}{\|}}{C}\text{—}NH\overset{R'}{\overset{|}{C}}HCOOH$$

A dipeptide (one amide bond)

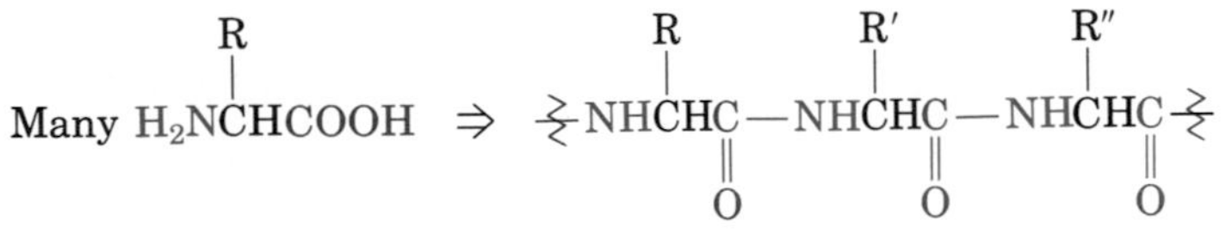

$$\text{Many } H_2N\overset{R}{\overset{|}{C}}HCOOH \Rightarrow \text{—}NH\overset{R}{\overset{|}{C}}H\underset{\underset{O}{\|}}{C}\text{—}NH\overset{R'}{\overset{|}{C}}H\underset{\underset{O}{\|}}{C}\text{—}NH\overset{R''}{\overset{|}{C}}H\underset{\underset{O}{\|}}{C}\text{—}$$

A polypeptide (many amide bonds)

## 27.1 Structures of Amino Acids

The structures of the 20 amino acids commonly found in proteins are shown in Table 27.1 (pages 972–973). All 20 are ***α*-amino acids**; that is, the amino group in each is a substituent on the carbon atom alpha ($\alpha$) to, or next to, the carbonyl group. The amino acid structures differ only in the nature of the side chains. Note that 19 of the 20 are primary amines, R—$NH_2$, but that proline is a secondary amine whose nitrogen and alpha-carbon atoms are part of a pyrrolidine ring. Proline can still form amide bonds like the other 19 $\alpha$-amino acids, however.

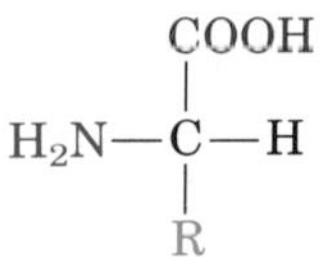

$$H_2N\text{—}\overset{COOH}{\underset{R}{C}}\text{—}H$$

Primary $\alpha$-amino acids (R = a side chain)

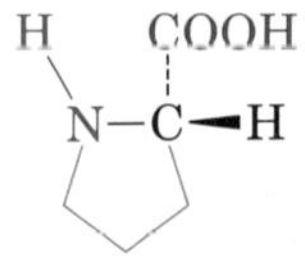

H COOH N—C—H (pyrrolidine ring)

Proline, a secondary $\alpha$-amino acid

Note also that each of the amino acids in Table 27.1 can be referred to by a mnemonic three-letter shorthand code: Ala for alanine, Gly for glycine, and so on. In addition, a new one-letter code is gaining popularity. This new code is shown in parentheses in the table.

With the exception of glycine, $H_2NCH_2COOH$, the alpha carbons of amino acids are chiral. Two different enantiomeric forms of each amino acid are therefore possible, but nature uses only a single enantiomer to construct proteins. In Fischer projections, naturally occurring amino acids are represented by placing the carboxyl group at the top as if drawing a carbohydrate (Section 24.2) and then placing the amino group on the left. Because

**Table 27.1** The twenty common amino acids found in proteins; essential amino acids are shown in blue

| *Name* | *Abbreviations* | | *Molecular weight* | *Structure* | *Isoelectric point* |
|---|---|---|---|---|---|
| **Neutral amino acids** | | | | | |
| Alanine | Ala | (A) | 89 | $CH_3CH(NH_2)COOH$ | 6.0 |
| Asparagine | Asn | (N) | 132 | $H_2NC(=O)CH_2CH(NH_2)COOH$ | 5.4 |
| Cysteine | Cys | (C) | 121 | $HSCH_2CH(NH_2)COOH$ | 5.0 |
| Glutamine | Gln | (Q) | 146 | $H_2NC(=O)CH_2CH_2CH(NH_2)COOH$ | 5.7 |
| Glycine | Gly | (G) | 75 | $CH_2(NH_2)COOH$ | 6.0 |
| Isoleucine | Ile | (I) | 131 | $CH_3CH_2CH(CH_3)CH(NH_2)COOH$ | 6.0 |
| Leucine | Leu | (L) | 131 | $(CH_3)_2CHCH_2CH(NH_2)COOH$ | 6.0 |
| Methionine | Met | (M) | 149 | $CH_3SCH_2CH_2CH(NH_2)COOH$ | 5.7 |
| Phenylalanine | Phe | (F) | 165 | $C_6H_5{-}CH_2CH(NH_2)COOH$ | 5.5 |
| Proline | Pro | (P) | 115 | $H_2C(CH_2)(CH_2)$ ring: $CH_2{-}CH{-}COOH$, $CH_2{-}N{-}H$ (pyrrolidine ring, $CH{-}COOH$ bonded to $N{-}H$) | 6.3 |
| Serine | Ser | (S) | 105 | $HOCH_2CH(NH_2)COOH$ | 5.7 |
| Threonine | Thr | (T) | 119 | $CH_3CH(OH)CH(NH_2)COOH$ | 5.6 |

| Name | Abbreviations | | Molecular weight | Structure | Isoelectric point |
|---|---|---|---|---|---|
| Tryptophan | Trp | (W) | 204 | 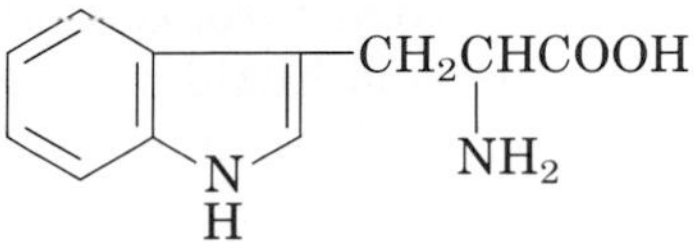 (indol-3-yl)$CH_2CH(NH_2)COOH$ | 5.9 |
| Tyrosine | Tyr | (Y) | 181 | 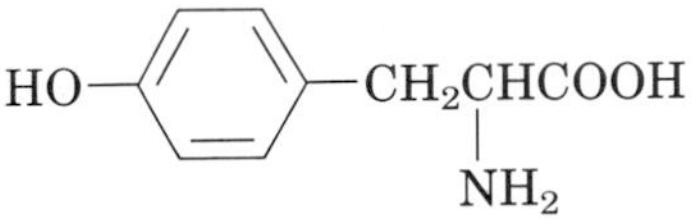 $HO-C_6H_4-CH_2CH(NH_2)COOH$ | 5.7 |
| Valine | Val | (V) | 117 | $(CH_3)_2CHCH(NH_2)COOH$ | 6.0 |
| **Acidic amino acids** | | | | | |
| Aspartic acid | Asp | (D) | 133 | $HOOCCH_2CH(NH_2)COOH$ | 3.0 |
| Glutamic acid | Glu | (E) | 147 | $HOOCCH_2CH_2CH(NH_2)COOH$ | 3.2 |
| **Basic amino acids** | | | | | |
| Arginine | Arg | (R) | 174 | $H_2NC(=NH)NHCH_2CH_2CH_2CH(NH_2)COOH$ | 10.8 |
| Histidine | His | (H) | 155 | 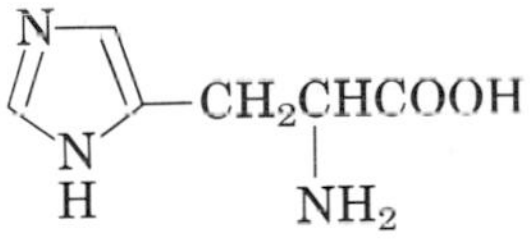 (imidazol-4-yl)$CH_2CH(NH_2)COOH$ | 7.6 |
| Lysine | Lys | (K) | 146 | $H_2NCH_2CH_2CH_2CH_2CH(NH_2)COOH$ | 9.7 |

of their stereochemical similarity to L sugars (Section 24.3), the naturally occurring α-amino acids are often referred to as L-amino acids.

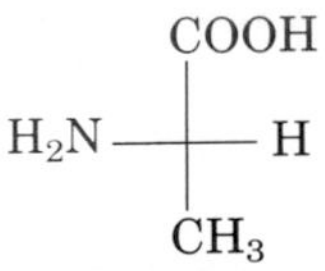

COOH; $H_2N$—C—H; $CH_3$
(*S*)-Alanine
(L-Alanine)

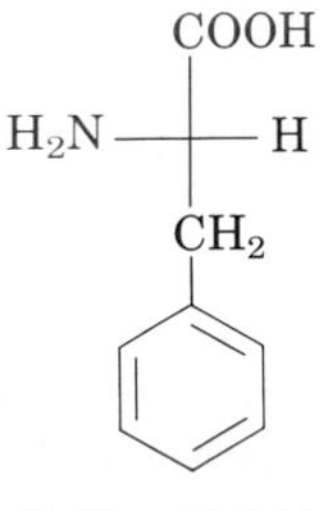

COOH; $H_2N$—C—H; $CH_2C_6H_5$
(*S*)-Phenylalanine
(L-Phenylalanine)

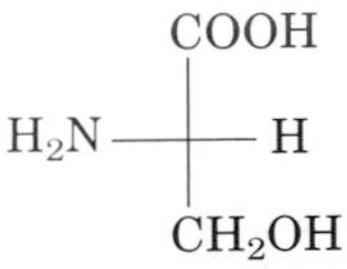

COOH; $H_2N$—C—H; $CH_2OH$
(*S*)-Serine
(L-Serine)

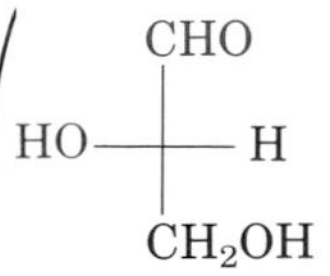

CHO; HO—C—H; $CH_2OH$
(Stereochemically similar to L-glyceraldehyde)

The 20 common amino acids can be further categorized as either neutral, acidic, or basic, depending on the nature of their specific side chains. Fifteen of the 20 have neutral side chains, but 2 (aspartic acid and glutamic acid) have an extra carboxylic acid group in their side chains, and three (lysine, arginine, and histidine) have basic amino groups in their side chains.

All 20 of the amino acids are required for protein synthesis, but humans are thought to be able to synthesize only 10 of the 20. The remaining 10 are called **essential amino acids** since they must be obtained from dietary sources. Failure to include an adequate dietary supply of these essential amino acids can lead to severe deficiency diseases.

PROBLEM

**27.1** Look carefully at the α-amino acids shown in Table 27.1. How many contain aromatic rings? How many contain sulfur? How many contain alcohols? How many contain hydrocarbon side chains?

PROBLEM

**27.2** Eighteen of the 19 L-amino acids have the *S* configuration at the alpha carbon. Cysteine is the only L-amino acid that has an *R* configuration. Explain.

PROBLEM

**27.3** The amino acid threonine, (2*S*,3*R*)-2-amino-3-hydroxybutanoic acid, has two chiral centers. Draw a Fischer projection of threonine.

PROBLEM

**27.4** Draw the Fischer projection of a threonine diastereomer, and label the chiral centers as *R* or *S* (see Problem 27.3).

## 27.2 Dipolar Structure of Amino Acids

Amino acids contain both acidic and basic groups in the same molecule. Thus, they undergo an intramolecular acid–base reaction and exist primarily in the form of a dipolar ion or **zwitterion** (German *zwitter*, "hybrid"):

$$H_2\ddot{N}-CH(R)-C(=O)-O-H \rightleftharpoons H_3\overset{+}{N}-CH(R)-C(=O)-O^-$$

A zwitterion

Amino acid zwitterions are a kind of internal salt and therefore have many of the physical properties we associate with salts. Thus, amino acids have large dipole moments; they are soluble in water but insoluble in hydrocarbons; and they are crystalline substances with high melting points. Furthermore, amino acids are **amphoteric**. They can react either as acids or as bases, depending on the circumstances. In aqueous acid solution, an amino acid zwitterion can *accept* a proton to yield a cation; in aqueous basic solution, the zwitterion can *lose* a proton to form an anion.

In acid solution
$$H_3\overset{+}{N}-\underset{}{\overset{R}{\overset{|}{C}}}H-CO_2^- + H_3O^+ \rightleftharpoons H_3\overset{+}{N}-\overset{R}{\overset{|}{C}}H-COOH + H_2O$$

In base solution
$$H-\underset{H}{\underset{|}{\overset{H}{\overset{|}{N^+}}}}-\overset{R}{\overset{|}{C}}H-CO_2^- + {}^-OH \rightleftharpoons H_2N-\overset{R}{\overset{|}{C}}H-CO_2^- + H-O-H$$

Note that it is the carboxylate anion, —COO$^-$, rather than the amino group that acts as the basic site and accepts the proton in acid solution. Similarly, it is the ammonium cation rather than the carboxyl group that acts as the acidic site and donates a proton in base solution. This behavior is simply another consequence of the zwitterionic structure of amino acids.

PROBLEM

27.5 Draw these amino acids in their zwitterionic forms:
(a) Phenylalanine (b) Serine (c) Proline

## 27.3 Isoelectric Point

In acid solution (low pH), an amino acid is protonated and exists primarily as a cation; in basic solution (high pH), an amino acid is deprotonated and exists primarily as an anion. Thus, at some intermediate point, the amino acid must be exactly balanced between anionic and cationic forms and exist primarily as the neutral, dipolar zwitterion. This pH is called the **isoelectric point**.

$$H_3\overset{+}{N}\overset{R}{\overset{|}{C}}HCOOH \underset{}{\overset{H_3O^+}{\rightleftharpoons}} H_3\overset{+}{N}\overset{R}{\overset{|}{C}}HCOO^- \overset{^-OH}{\rightleftharpoons} H_2N\overset{R}{\overset{|}{C}}HCOO^-$$

Low pH (protonated) — Isoelectric point (neutral zwitterion) — High pH (deprotonated)

The isoelectric point of a given amino acid depends on its structure, with values for the 20 most common amino acids given in Table 27.1. As indicated, the 15 amino acids with neutral side chains have isoelectric points near neutrality, in the pH range 5.0–6.5. (These values are not exactly at neutral pH = 7, because carboxyl groups are stronger acids in aqueous solution than amino groups are bases.) The two amino acids with acidic side chains have isoelectric points at lower (more acidic) pH to suppress dissociation of the extra —COOH in the side chain, and the three amino acids with basic side chains have isoelectric points at higher (more basic) pH to suppress protonation of the extra amino group. For example, aspartic acid has its isoelectric point at pH = 3.0, and lysine has its isoelectric point at pH = 9.7.

We can take advantage of the differences in isoelectric points to separate a mixture of amino acids (or a mixture of proteins) into its pure constituents.

In the technique known as **electrophoresis**, a solution of different amino acids is placed near the center of a strip of paper or gel. The paper or gel is moistened with an aqueous buffer of a given pH, and electrodes are connected to the ends of the strip. When an electric field is applied, those amino acids with negative charges (those that are deprotonated because their isoelectric points are below the pH of the buffer) migrate slowly toward the positive electrode. Similarly, those amino acids with positive charges (those that are protonated because their isoelectric points are above the pH of the buffer) migrate toward the negative electrode.

Different amino acids migrate at different rates, depending both on their isoelectric point and on the pH of the aqueous buffer. Thus, the different amino acids can be separated. Figure 27.1 illustrates this separation for a mixture of lysine (basic), glycine (neutral), and aspartic acid (acidic).

Paper strip

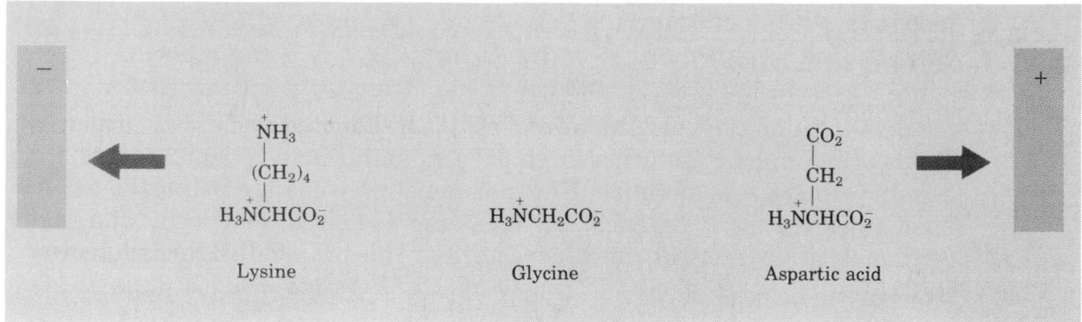


**Figure 27.1** Separation of an amino acid mixture by electrophoresis: At pH = 6.0, glycine molecules are mostly neutral and do not migrate; lysine molecules are largely protonated and migrate toward the negative electrode; and aspartic acid molecules are largely deprotonated and migrate toward the positive electrode. (Lysine has its isoelectric point at 9.7, glycine at 6.0, and aspartic acid at 3.0.)

PROBLEM

**27.6** Draw structures of the predominant forms of glycine at pH = 2.0, 6.0, and 10.0.

PROBLEM

**27.7** For the mixtures of amino acids indicated, predict the direction of migration of each component (toward anode or cathode) and relative rate of migration.
(a) Valine, glutamic acid, and histidine at pH = 7.6
(b) Glycine, phenylalanine, and serine at pH = 5.7
(c) Glycine, phenylalanine, and serine at pH = 5.5
(d) Glycine, phenylalanine, and serine at pH = 6.0

PROBLEM

**27.8** How can you account for the fact that tryptophan has a lower isoelectric point than histidine, even though both have five-membered-ring nitrogen atoms? Which nitrogen in the five-membered ring of histidine is more basic?

## 27.4 Synthesis of α-Amino Acids

α-Amino acids can be synthesized using some of the standard chemical methods already discussed. For example, one of the oldest methods of α-amino-acid synthesis begins with alpha bromination of a carboxylic acid by treatment with bromine and phosphorus (the Hell–Volhard–Zelinskii reaction, Section 22.4). Nucleophilic substitution ($S_N2$ reaction) of the α-bromo acid with ammonia (Section 25.7) then yields an α-amino acid.

$$(CH_3)_2CHCH_2CH_2COOH \xrightarrow[2.\ H_2O]{1.\ Br_2,\ P} (CH_3)_2CHCH_2\underset{\displaystyle Br}{\underset{|}{C}}HCOOH \xrightarrow{NH_3\,(excess)} (CH_3)_2CHCH_2\underset{\displaystyle NH_2}{\underset{|}{C}}HCOOH$$

4-Methylpentanoic acid — 2-Bromo-4-methylpentanoic acid — (*R,S*)-Leucine (45%)

Alternatively, higher product yields are obtained when the bromide displacement reaction is carried out by the Gabriel phthalimide method (Section 25.7), rather than by the ammonia method.

**PROBLEM**

**27.9** Show how you could prepare these α-amino acids starting from the appropriate carboxylic acids:
(a) Phenylalanine (b) Valine

**PROBLEM**

**27.10** Show how a Gabriel amine synthesis (Section 25.7) might be used to prepare isoleucine.

### THE STRECKER SYNTHESIS

Another method for preparing racemic α-amino acids is the **Strecker[1] synthesis**. Developed in 1850, this versatile two-step process involves treatment of an aldehyde with KCN and aqueous ammonia to yield an intermediate α-amino nitrile. Hydrolysis of the nitrile then gives an α-amino acid:

$$C_6H_5CH_2CHO \xrightarrow[H_2O]{NH_4Cl/KCN} C_6H_5CH_2\underset{\displaystyle NH_2}{\underset{|}{C}}HCN \xrightarrow{H_3O^+} C_6H_5CH_2\underset{\displaystyle NH_2}{\underset{|}{C}}HCOOH$$

Phenylacetaldehyde — An α-amino nitrile — (*R,S*)-Phenylalanine (53%)

The first part of the Strecker synthesis, formation of an α-amino nitrile by reaction of an aldehyde with ammonia and KCN, is simply a combination of two carbonyl-group reactions seen earlier in Chapter 19. The first step is

[1]Adolf Friedrich Ludwig Strecker (1822–1871); Ph.D. Giessen (1842); assistant to Liebig at Tübingen.

reaction of the aldehyde with ammonia to yield an imine (Section 19.12), which then adds HCN in a nucleophilic addition step similar to that involved in cyanohydrin formation (Section 19.9). The mechanism is shown in Figure 27.2.

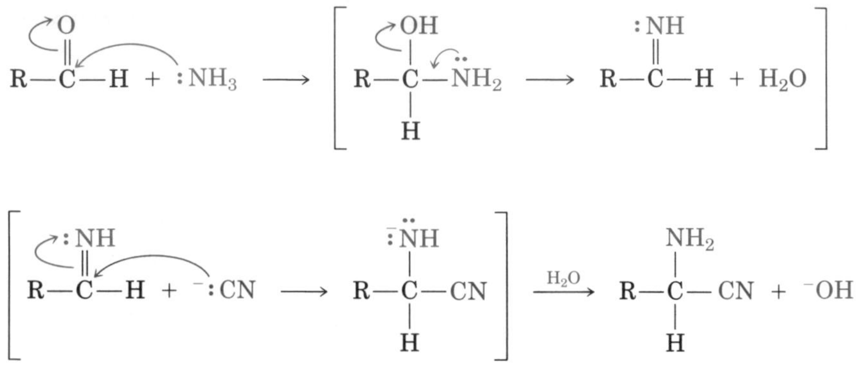

$$\mathrm{R{-}\overset{\overset{O}{\|}}{C}{-}H} + :NH_3 \longrightarrow \left[\mathrm{R{-}\underset{H}{\overset{OH}{C}}{-}\ddot{N}H_2} \longrightarrow \mathrm{R{-}\overset{\overset{:NH}{\|}}{C}{-}H} + H_2O\right]$$

$$\left[\mathrm{R{-}\overset{\overset{:NH}{\|}}{C}{-}H} + {}^{-}:CN \longrightarrow \mathrm{R{-}\underset{H}{\overset{:\ddot{N}H^{-}}{C}}{-}CN}\right] \xrightarrow{H_2O} \mathrm{R{-}\underset{H}{\overset{NH_2}{C}}{-}CN} + {}^{-}OH$$

**Figure 27.2** Mechanism of $\alpha$-amino nitrile formation in the Strecker amino acid synthesis

PROBLEM

**27.11** The rare amino acid L-dopa (3,4-dihydroxyphenylalanine) is useful as a drug against Parkinson's disease. Show how ($\pm$)-dopa might be synthesized from 3,4-dihydroxyphenylacetaldehyde.

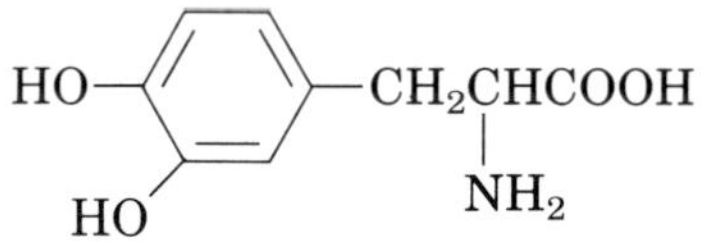

$$\text{(HO)}_2C_6H_3{-}CH_2\underset{NH_2}{CH}COOH$$

Dopa

## REDUCTIVE AMINATION OF $\alpha$-KETO ACIDS: BIOSYNTHESIS

Yet a third method for the synthesis of $\alpha$-amino acids is reductive amination of $\alpha$-keto acids (Section 25.7):

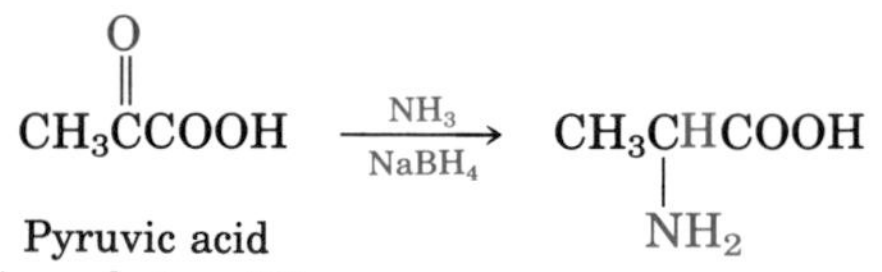

$$CH_3\overset{\overset{O}{\|}}{C}COOH \xrightarrow[NaBH_4]{NH_3} CH_3\underset{NH_2}{CH}COOH$$

Pyruvic acid (an $\alpha$-keto acid) → $(R,S)$-Alanine

Although not widely used, this reductive amination method is interesting because it is a close laboratory analogy of a pathway by which some amino acids are biosynthesized in nature. For example, the major route for glutamic acid synthesis in most organisms is reductive amination of α-ketoglutaric acid. The biological reducing agent is the rather complex molecule nicotinamide adenine dinucleotide (NADH), and the reductive amination step is catalyzed by an enzyme, L-glutamate dehydrogenase. Nevertheless, the fundamental chemical principles of this biosynthetic reaction are identical to those of the laboratory reaction.

$$\underset{\alpha\text{-Ketoglutaric acid}}{HOOCCH_2CH_2\overset{\overset{O}{\|}}{C}COOH} + NH_3 \xrightarrow[\text{L-Glutamate dehydrogenase}]{NADH} \underset{(S)\text{-Glutamic acid}}{HOOCCH_2CH_2\underset{NH_2}{\underset{|}{C}}HCOOH}$$

## THE AMIDOMALONATE SYNTHESIS

The most general method of preparation for α-amino acids is the **amidomalonate synthesis**. This route, a straightforward extension of the malonic ester synthesis (Section 22.9), involves initial conversion of diethyl acetamidomalonate into its enolate anion by treatment with base, followed by $S_N2$ reaction with a primary alkyl halide. Hydrolysis and decarboxylation occur when the alkylated product is warmed with aqueous acid, and a racemic α-amino acid results. For example, aspartic acid is prepared in good yield when diethyl acetamidomalonate is alkylated with ethyl bromoacetate, followed by hydrolysis and decarboxylation:

$$\underset{\text{Diethyl acetamidomalonate}}{CH_3\overset{\overset{O}{\|}}{C}NH-\overset{\overset{CO_2C_2H_5}{|}}{\underset{\underset{CO_2C_2H_5}{|}}{C}}-H} \xrightarrow[\text{Ethanol}]{Na^+\ ^-OC_2H_5} \left[CH_3\overset{\overset{O}{\|}}{C}NH-\overset{\overset{CO_2C_2H_5}{|}}{\underset{\underset{CO_2C_2H_5}{|}}{C}}:^-\right]$$

$$\xrightarrow{Br-CH_2CO_2C_2H_5} CH_3\overset{\overset{O}{\|}}{C}NH-\overset{\overset{CO_2C_2H_5}{|}}{\underset{\underset{CO_2C_2H_5}{|}}{C}}-CH_2CO_2C_2H_5 \xrightarrow{H_3O^+}$$

$$C_2H_5OH + CO_2 + 2\ C_2H_5OH + CH_3COOH + \underset{(R,S)\text{-Aspartic acid (55\%)}}{H_2N\overset{\overset{COOH}{|}}{C}HCH_2COOH}$$

Recall the malonic ester synthesis:

$$CH_2(CO_2C_2H_5)_2 \xrightarrow[\text{2. RX} \quad \text{3. } H_3O^+]{\text{1. } Na^+\ ^-OC_2H_5} R-CH_2COOH$$

PROBLEM

**27.12** Show the alkyl halides you would use to prepare these α-amino acids by the amidomalonate method.
(a) Leucine (b) Histidine (c) Tryptophan (d) Methionine

PROBLEM

**27.13** Serine can be synthesized by a simple variation of the amidomalonate method. Can you suggest how this might be done? [*Hint:* See Section 23.6.]

## 27.5 Resolution of *R,S* Amino Acids

The synthesis of chiral amino acids from achiral precursors by any one of the methods just described yields a racemic mixture—an equal mixture of *S* and *R* products. In order to use these synthetic amino acids for the laboratory synthesis of naturally occurring peptides, the racemic mixture must first be resolved into pure enantiomers.

Often this resolution can be done by the general method discussed earlier (Section 25.5), whereby the racemic mixture is converted into a mixture of diastereomeric salts by reaction with a chiral acid or base, and the different diastereomers are then separated by fractional crystallization.

Alternatively, biological methods of resolution can be used. **Enzymes** are chiral biological catalysts that often show an astounding selectivity toward one enantiomer of an *R,S* mixture. For example, the enzyme carboxypeptidase selectively catalyzes the hydrolysis of *S* amido acids but not *R* amido acids. We can therefore resolve an *R,S* mixture of amino acids by first allowing the mixture to react with acetic anhydride to form the *N*-acetyl derivatives. Selective hydrolysis of the *R,S* amido acid mixture with carboxypeptidase then yields a mixture of the desired *S* amino acid and the unchanged *N*-acetyl *R* amido acid, which can be separated by usual chemical techniques.

$$H_2N\overset{R}{\overset{|}{C}}HCOOH \xrightarrow{(CH_3CO)_2O} CH_3CONH\overset{R}{\overset{|}{C}}HCOOH$$

An *R,S* mixture of amino acids → An *R,S* mixture of amido acids

$\downarrow$ $H_2O$, Carboxypeptidase

$H_2N$—C(COOH)(R)—H + H—C(COOH)(R)—$NHCOCH_3$

An *S* enantiomer An *R* enantiomer

## 27.6 Peptides

**Peptides** are amino acid polymers in which the individual amino acid units, called **residues**, are linked together by amide, or peptide, bonds. An amino group from one residue forms an amide bond with the carboxyl of a second residue; the amino group of the second forms an amide bond with the carboxyl of a third, and so on. For example, alanylserine is the dipeptide formed when an amide bond is made between the alanine carboxyl and the serine amino group:

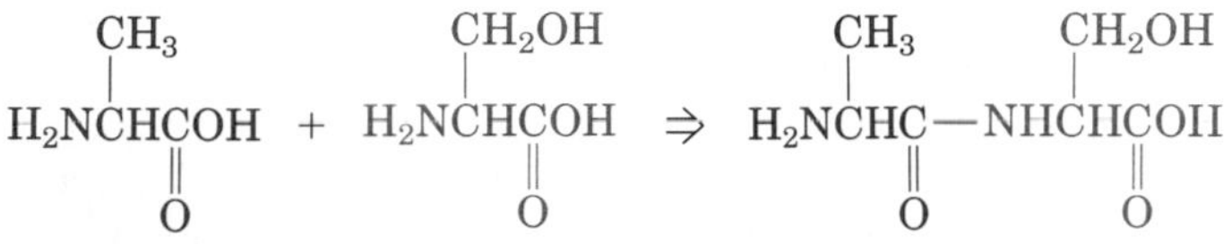

$$\underset{\substack{\text{Alanine}\\ \text{(Ala)}}}{H_2NCH(CH_3)C(=O)OH} + \underset{\substack{\text{Serine}\\ \text{(Ser)}}}{H_2NCH(CH_2OH)C(=O)OH} \Rightarrow \underset{\substack{\text{Alanylserine}\\ \text{(H-Ala-Ser-OH)}}}{H_2NCH(CH_3)C(=O){-}NHCH(CH_2OH)C(=O)OH}$$

Note that two dipeptides can result from reaction between alanine and serine, depending on which carboxyl group reacts with which amino group. If the alanine amino group reacts with the serine carboxyl, serylalanine results:

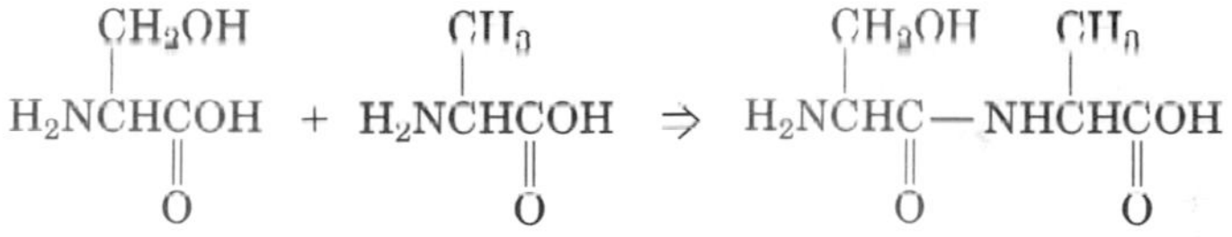

$$\underset{\substack{\text{Serine}\\ \text{(Ser)}}}{H_2NCH(CH_2OH)C(=O)OH} + \underset{\substack{\text{Alanine}\\ \text{(Ala)}}}{H_2NCH(CH_3)C(=O)OH} \Rightarrow \underset{\substack{\text{Serylalanine}\\ \text{(H-Ser-Ala-OH)}}}{H_2NCH(CH_2OH)C(=O){-}NHCH(CH_3)C(=O)OH}$$

By convention, peptides are always written with the **N-terminal amino acid** (the one with the free $-NH_2$ group) on the left, and the **C-terminal amino acid** (the one with the free $-COOH$ group) on the right. The name of the peptide is usually indicated by using the three-letter abbreviations listed in Table 27.1 for each amino acid. An H— is often appended to the abbreviation of the leftmost amino acid to underscore its position as the N-terminal residue, and an -OH is often appended to the abbreviation of the rightmost amino acid (C-terminal residue). For example, H-Gly-Val-Tyr-OH is the tripeptide glycylvalyltyrosine, whose structure is shown in Figure 27.3 (page 982).

The number of possible isomeric peptides increases rapidly as the number of amino acid units increases. There are six ways in which three amino acids can be joined, and more than 40,000 ways in which to join the eight amino acids present in the hormone angiotensin II, which regulates blood pressure (Figure 27.4).

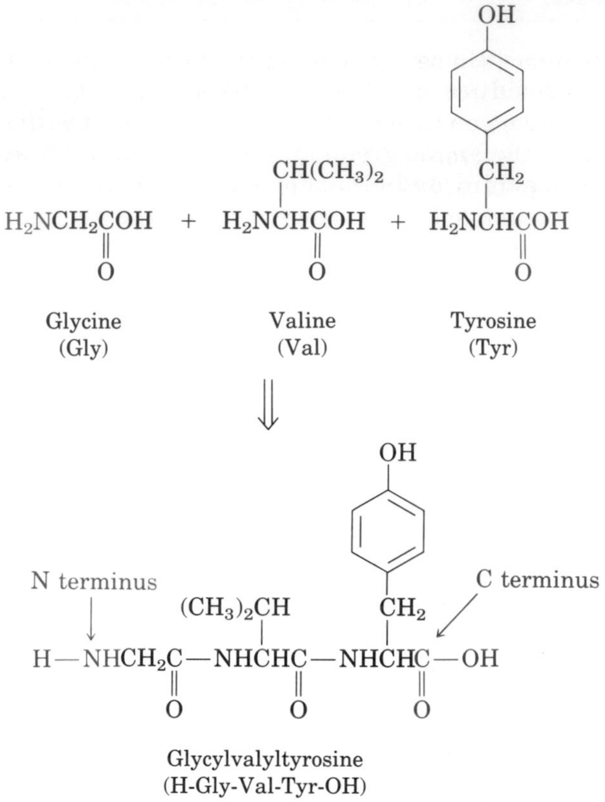

$$H_2NCH_2COOH + H_2NCH(CH(CH_3)_2)COOH + H_2NCH(CH_2C_6H_4OH)COOH$$

Glycine (Gly) Valine (Val) Tyrosine (Tyr)

$$\Downarrow$$

N terminus C terminus

$$H{-}NHCH_2C(=O){-}NHCH(CH(CH_3)_2)C(=O){-}NHCH(CH_2C_6H_4OH)C(=O){-}OH$$

Glycylvalyltyrosine
(H-Gly-Val-Tyr-OH)

**Figure 27.3** Structure of glycylvalyltyrosine, H-Gly-Val-Tyr-OH

H —Asp —Arg —Val —Tyr —Ile —His —Pro —Phe—OH

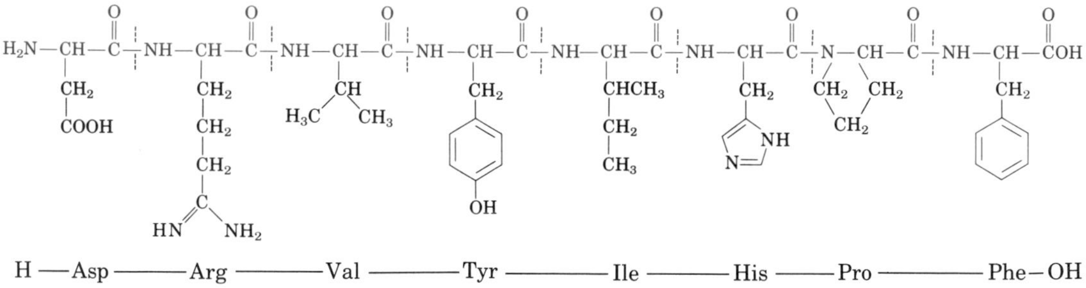

**Figure 27.4** The structure of angiotensin II, a hormone present in blood plasma that regulates blood pressure

PROBLEM

**27.14** Name the six possible isomeric tripeptides that contain valine, tyrosine, and glycine. Use the three-letter shorthand notation for each amino acid.

PROBLEM

**27.15** Draw the full structure of H-Met-Pro-Val-Gly-OH and indicate where the amide bonds are.

## 27.7 Covalent Bonding in Peptides

Amide bonds in peptides are similar to the simple amide bonds we've already discussed (Section 25.4). Amide nitrogens are nonbasic because their unshared electron pair is delocalized by orbital overlap with the carbonyl group. This overlap imparts a certain amount of double-bond character to the amide C—N bond and restricts its rotation (Figure 27.5).

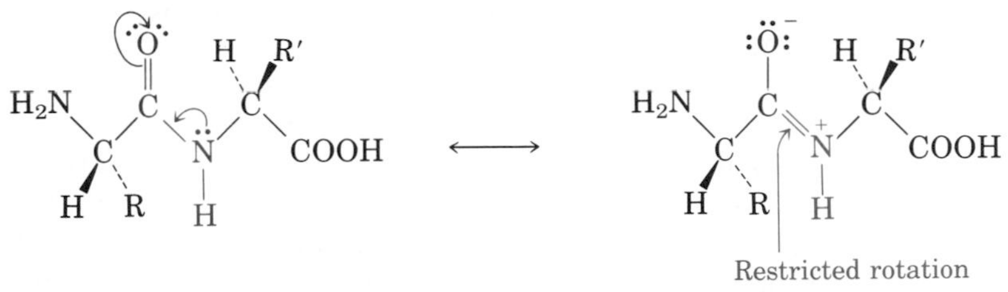

**Figure 27.5** Amide resonance causes restricted rotation around the C—N bond, giving it a certain amount of double-bond character.

A second kind of covalent bonding in peptides occurs when a disulfide linkage, R—S—S—R, is formed between two cysteine residues. As we've seen, disulfide bonds are easily formed by mild oxidation of thiols, R—SH, and are easily cleaved back to thiols by mild reduction (Section 17.12).

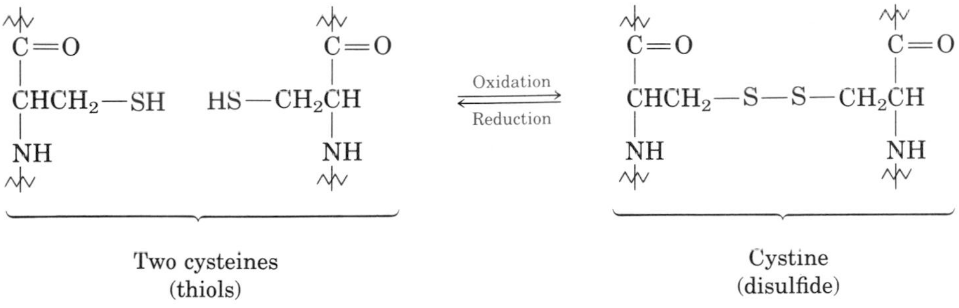

Disulfide bonds between cysteine residues in two different peptide chains can link the otherwise separate chains together. Alternatively, a disulfide bond between two cysteine residues within the same chain can cause a loop in the chain. Such is the case with the nonapeptide vasopressin, an antidiuretic hormone involved in controlling water balance in the body. Note also that the C-terminal end of vasopressin occurs as the primary amide, $—CONH_2$, rather than as the free acid.

Disulfide bridge

H-CyS-Tyr-Phe-Glu-Asn-CyS-Pro-Arg-Gly-$NH_2$

Vasopressin

## 27.8 Peptide Structure Determination: Amino Acid Analysis

Determining the structure of a peptide is a challenging task that requires finding the answers to three questions: What amino acids are present? How much of each is present? Where does each occur in the peptide chain? The answers to the first two questions are provided by a remarkable device, the **amino acid analyzer**.

The amino acid analyzer is an automated instrument based on analytical techniques worked out in the 1950s at the Rockefeller Institute by William Stein[2] and Stanford Moore.[3] The first step is to break the peptide down into its constituent amino acids by reducing all disulfide bonds and hydrolyzing all amide bonds with 6*M* HCl. Chromatography (Section 12.2) of the resultant amino acid mixture using a series of aqueous buffers as the mobile phase then effects a separation into component amino acids.

As each different amino acid elutes from the end of the chromatography column, it is allowed to mix with ninhydrin, a reagent that reacts with $\alpha$-amino acids to form an intense purple color. The purple color is detected by a spectrometer, and a plot of elution time versus spectrometer absorbance is obtained.

2 Ninhydrin + $H_2N\overset{R}{\overset{|}{C}}HCOOH$ (An $\alpha$-amino acid) ⟶ Purple color + RCHO + $CO_2$

Since the amount of time required for a given amino acid to elute from the chromatography column is reproducible from sample to sample, the identity of all amino acids in a peptide of unknown composition can be determined simply by noting the various elution times. The amount of each amino acid in a sample can be determined by measuring the intensity of the purple color resulting from its reaction with ninhydrin. Figure 27.6 shows the results of amino acid analysis of a standard equimolar mixture of 17 $\alpha$-amino acids and compares them to results obtained from analysis of methionine enkephalin, a pentapeptide with morphine-like analgesic activity.

**PROBLEM**

**27.16** Show the structures of the products obtained on reaction of valine with ninhydrin.

---

[2]William H. Stein (1911–1980); b. New York; Ph.D. Columbia; professor, Rockefeller Institute; Nobel prize (1972).

[3]Stanford Moore (1913–1982); b. Chicago; Ph.D. Wisconsin; professor, Rockefeller Institute; Nobel prize (1972).

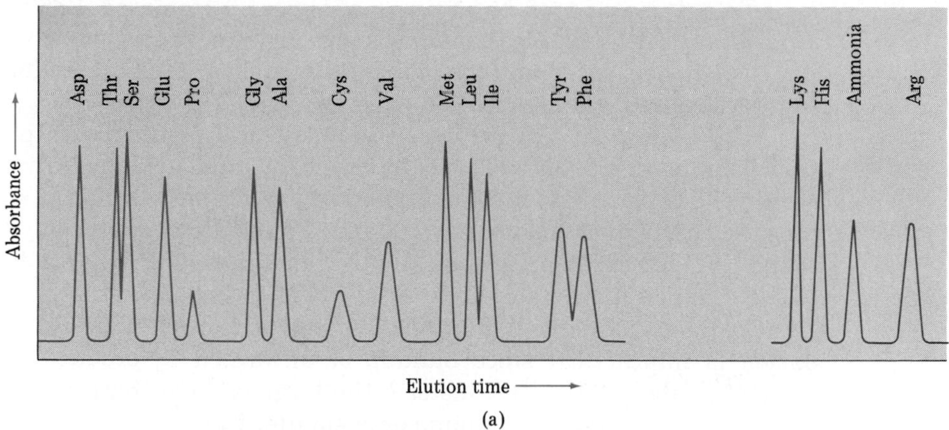


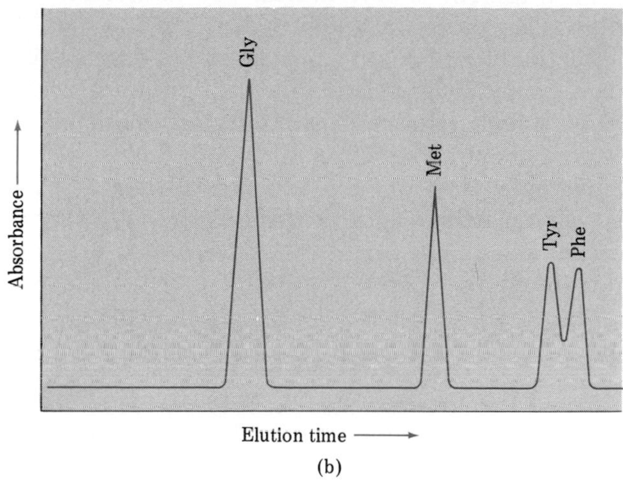


**Figure 27.6** Amino acid analysis of (a) an equimolar amino acid mixture and (b) methionine enkephalin (H-Tyr-Gly-Gly-Phe-Met-OH)

PROBLEM

**27.17** The data for amino acid analysis in Figure 27.6(a) indicate that proline is not easily detected by reaction with ninhydrin; only a very small peak is seen on the chromatogram. Explain.

## 27.9 Peptide Sequencing: The Edman Degradation

With the identity and amount of each amino acid known, the final task of structure determination is to **sequence** the peptide; that is, to find out in what order the amino acids are linked together. The general idea of peptide sequencing is to cleave one amino acid residue at a time from the end of the peptide chain (either N terminus or C terminus). That terminal amino acid is then separated and identified, and the cleavage reactions are repeated on the chain-shortened peptide until the entire peptide sequence is determined.

Almost all peptide sequencing is now done by **Edman[4] degradation**, an efficient method of N-terminal analysis. Automated Edman *protein sequenators* are available that allow a series of 20 or more repetitive sequencing steps to be carried out before a buildup of unwanted by-products begins to interfere with the results.

Edman degradation involves treatment of a peptide with phenyl isothiocyanate, $C_6H_5—N{=}C{=}S$, followed by mild acid hydrolysis. These steps yield a phenylthiohydantoin derivative of the N-terminal amino acid plus the chain-shortened peptide (Figure 27.7). The phenylthiohydantoin is then identified chromatographically by comparison with known derivatives of the common amino acids.

Complete sequencing of large peptides and proteins by Edman degradation is impractical since buildup of unwanted by-products limits the method to about 25 cycles. Instead, the large peptide chain is first cleaved by partial hydrolysis into a number of smaller fragments, and the sequence of each fragment is determined. The individual fragments are then fitted together like pieces in a jigsaw puzzle.

Partial hydrolysis of a peptide can be carried out either chemically with aqueous acid, or enzymatically with enzymes such as trypsin and chymotrypsin. Acidic hydrolysis is unselective and leads to a more or less random mixture of small fragments. Enzymic hydrolysis, however, is quite specific. For example, trypsin catalyzes hydrolysis only at the carboxyl side of the basic amino acids arginine and lysine; chymotrypsin cleaves only at the carboxyl side of the aryl-substituted amino acids phenylalanine, tyrosine, and tryptophan.

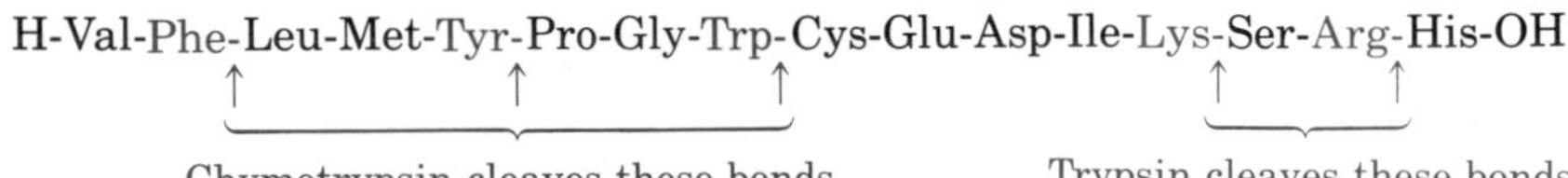
H-Val-Phe-Leu-Met-Tyr-Pro-Gly-Trp-Cys-Glu-Asp-Ile-Lys-Ser-Arg-His-OH
Chymotrypsin cleaves these bonds.
Trypsin cleaves these bonds.

To take an example of peptide sequencing, let's look at a hypothetical structure determination of angiotensin II, a hormonal octapeptide involved in controlling hypertension by regulating the sodium–potassium salt balance in the body.

1. Amino acid analysis of angiotensin II would show the presence of eight different amino acids: Arg, Asp, His, Ile, Phe, Pro, Tyr, and Val in equimolar amounts.
2. An N-terminal analysis by the Edman method would show that angiotensin II has an aspartic acid residue at the N terminus.
3. Partial hydrolysis of angiotensin II with dilute hydrochloric acid might yield the following fragments, whose sequences could be determined by Edman degradation:
   a. H-Asp-Arg-Val-OH
   b. H-Ile-His-Pro-OH
   c. H-Arg-Val-Tyr-OH
   d. H-Pro-Phe-OH
   e. H-Val-Tyr-Ile-OH

[4]Pehr Edman (1916– ); b. Stockholm; M.D. Karolinska Institute (E. Jorpes); professor, University of Lund.

Nucleophilic addition of the peptide terminal amino group to phenylisothiocyanate yields an *N*-phenylthiourea derivative.

Acid-catalyzed cyclization then yields a tetrahedral intermediate . . .

HCl

. . . which expels the chain-shortened peptide and forms a thiazolinone.

+ $H_2N$—Peptide

A thiazolinone

The thiazolinone rearranges in the presence of aqueous acid to yield the final *N*-phenylthiohydantoin derivative.

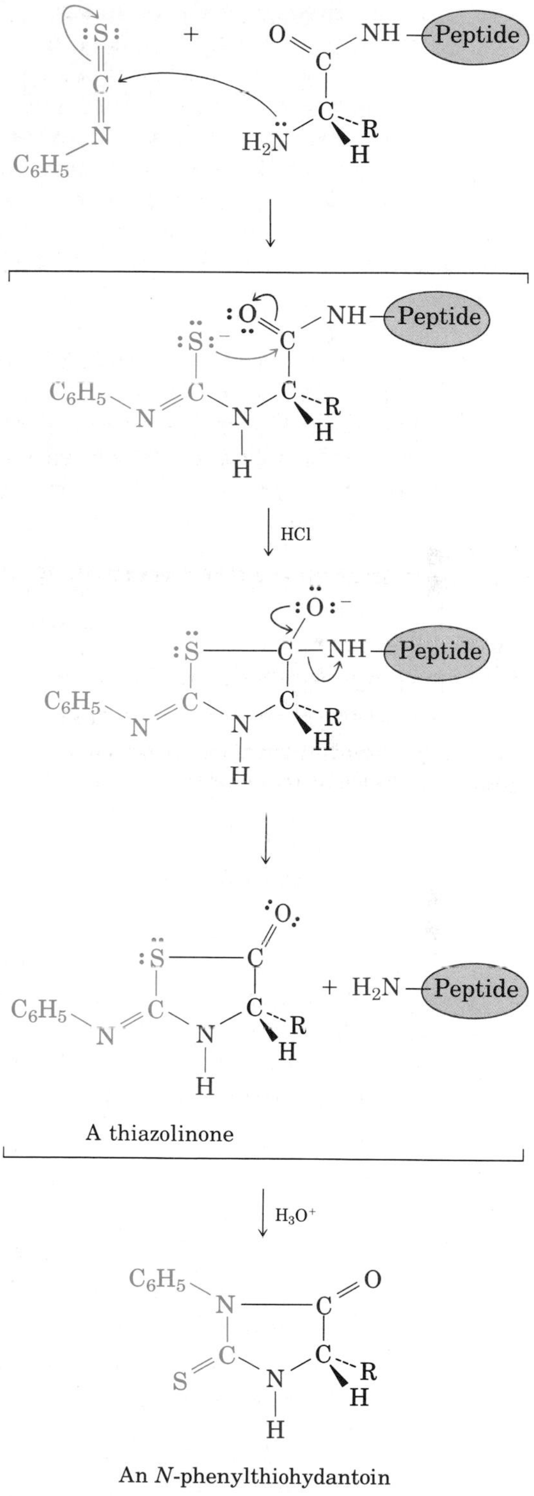

$H_3O^+$

An *N*-phenylthiohydantoin

**Figure 27.7** Mechanism of the Edman degradation for N-terminal analysis of peptides

4. Matching of overlapping fragment regions provides the full sequence of angiotensin II:

```
a. H-Asp-Arg-Val-OH
c.       H-Arg-Val-Tyr-OH
e.           H-Val-Tyr-Ile-OH
b.                   H-Ile-His-Pro-OH
d.                           H-Pro-Phe-OH
   H-Asp-Arg-Val-Tyr-Ile-His-Pro-Phe-OH
```

Angiotensin II

The structure of angiotensin II is relatively simple (the entire sequence could easily be done by a protein sequenator instrument), but the methods and logic used to solve this simple structure are the same as those used to solve more complex structures. Indeed, single protein chains with more than 400 amino acids have been sequenced by these methods.

PROBLEM

**27.18** What fragments would result if angiotensin II were cleaved with trypsin? With chymotrypsin?

PROBLEM

**27.19** Give the amino acid sequence of hexapeptides that produce these fragments on partial acid hydrolysis:
(a) Arg, Gly, Ile, Leu, Pro, Val gives H-Pro-Leu-Gly-OH, H-Arg-Pro-OH, H-Gly-Ile-Val-OH
(b) Asp, Leu, Met, Trp, $Val_2$ gives H-Val-Leu-OH, H-Val-Met-Trp-OH, H-Trp-Asp-Val-OH

## 27.10 Peptide Sequencing: C-Terminal Residue Determination

The Edman degradation is an excellent method of analysis for the N-terminal residue, but a complementary method of analysis for the C-terminal residue is also valuable. The best method currently available makes use of the enzyme carboxypeptidase specifically to cleave the C-terminal amide bond in a peptide chain.

$$\text{Peptide}-\text{NH}\underset{\underset{\displaystyle R'}{|}}{\text{CH}}\overset{\text{O}}{\underset{\|}{\text{C}}}-\text{NH}\underset{\underset{\displaystyle R}{|}}{\text{CH}}\text{COOH} \xrightarrow[H_2O]{\text{Carboxypeptidase}} \text{Peptide}-\text{NH}\underset{\underset{\displaystyle R'}{|}}{\text{CH}}\text{COOH} + H_2N\underset{\underset{\displaystyle R}{|}}{\text{CH}}\text{COOH}$$

The analysis is carried out by incubating the polypeptide with carboxypeptidase and watching for the appearance of the first free amino acid that appears in solution. Of course, further degradation also occurs, since a new C-terminus is produced when the first amino acid residue is cleaved off, until ultimately the entire peptide is hydrolyzed.

PROBLEM

**27.20** A hexapeptide with the composition Arg, Gly, Leu, $Pro_3$ is found to have proline at both C-terminal and N-terminal positions. Partial hydrolysis gives the following fragments:

H-Gly-Pro-Arg-OH  H-Arg-Pro-OH  H-Pro-Leu-Gly-OH

What is the structure of the hexapeptide?

PROBLEM

**27.21** Propose two structures for a tripeptide that gives Leu, Ala, and Phe on hydrolysis but does not react with carboxypeptidase and does not react with phenyl isothiocyanate.

## 27.11 Peptide Synthesis

Once the structure of a peptide has been determined, synthesis is often the next goal. This might be done either as a final proof of structure or as a means of obtaining larger amounts of a valuable peptide for biological evaluation.

Ordinary amide bonds are usually formed by reaction between amines and acylating agents (Section 21.8):

$$R'-\overset{\overset{\displaystyle O}{\|}}{C}-X + H_2NR \longrightarrow R'-\overset{\overset{\displaystyle O}{\|}}{C}-NHR + HX$$

An amide

Peptide synthesis is much more complex than simple amide synthesis, however, because of the requirement for specificity. Many different amide links must be formed, and they must be formed in a specific order, rather than at random. We can't expect simply to place a mixture of amino acids in a flask and obtain a single polypeptide product.

The solution to the specificity problem is *protection*. We can force a reaction to take only the desired course by protecting all of the amine and acid functional groups except for those we want to have react. For example, if we want to couple alanine with leucine to synthesize H-Leu-Ala-OH, we can protect the amino group of leucine and the carboxyl group of alanine to render them unreactive, then form the desired amide bond, and then remove the protecting groups, as shown at the top of the next page.

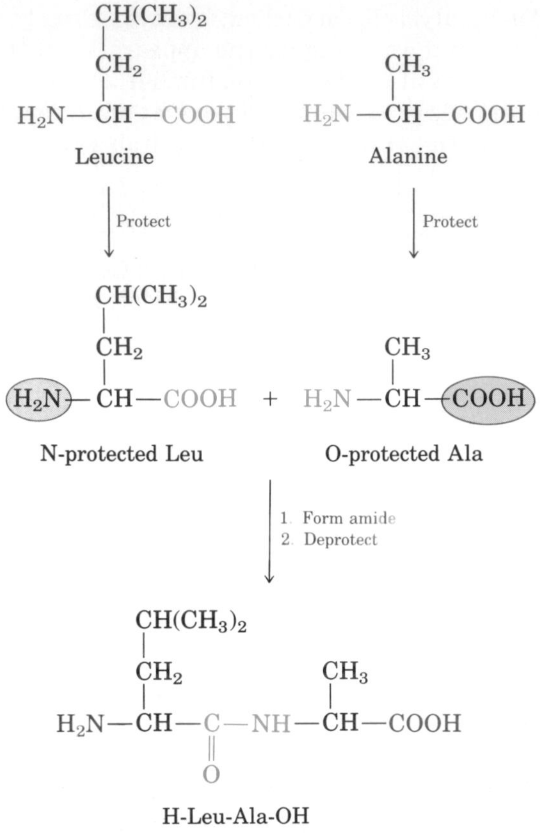

$$\underset{\text{Leucine}}{H_2N-\underset{\underset{CH(CH_3)_2}{|}}{\overset{CH_2}{}}CH-COOH} \qquad \underset{\text{Alanine}}{H_2N-\overset{CH_3}{\overset{|}{C}H}-COOH}$$

Leucine: $H_2N-CH(CH_2CH(CH_3)_2)-COOH$ — Protect → N-protected Leu

Alanine: $H_2N-CH(CH_3)-COOH$ — Protect → O-protected Ala

$$\text{(H}_2\text{N)}-CH(CH_2CH(CH_3)_2)-COOH \;+\; H_2N-CH(CH_3)-\text{(COOH)}$$

N-protected Leu    O-protected Ala

1. Form amide
2. Deprotect

$$H_2N-CH(CH_2CH(CH_3)_2)-\underset{\underset{O}{\|}}{C}-NH-CH(CH_3)-COOH$$

H-Leu-Ala-OH

Many different amino- and carboxyl-protecting groups have been devised but only a few are widely used. Carboxyl groups are often protected simply by converting them into methyl or benzyl esters. These groups are easily introduced by standard methods of ester formation and are easily removed by mild hydrolysis with aqueous sodium hydroxide. As Figure 27.8 shows, benzyl esters can also be cleaved by catalytic hydrogenolysis of the weak benzylic C—O bond ($ArCH_2—OCOR + H_2 \rightarrow ArCH_3 + RCOOH$).

Amino groups are often protected as their *tert*-butoxycarbonyl amide (BOC) derivatives. The BOC protecting group is easily introduced by reaction of the amino acid with di-*tert*-butyl dicarbonate (nucleophilic acyl substitution reaction; Section 21.6), and is removed by brief treatment with a strong acid such as trifluoroacetic acid, $CF_3COOH$.

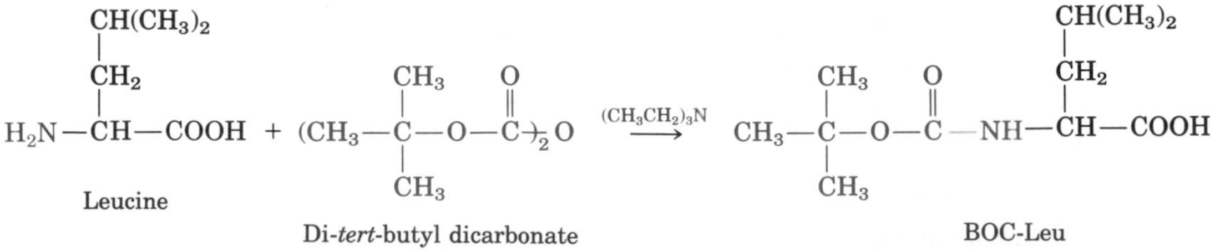

$$\underset{\text{Leucine}}{H_2N-CH(CH_2CH(CH_3)_2)-COOH} + \underset{\text{Di-}tert\text{-butyl dicarbonate}}{\left(CH_3-\underset{CH_3}{\overset{CH_3}{C}}-O-\overset{O}{\overset{\|}{C}}\right)_2O} \xrightarrow{(CH_3CH_2)_3N} \underset{\text{BOC-Leu}}{CH_3-\underset{CH_3}{\overset{CH_3}{C}}-O-\overset{O}{\overset{\|}{C}}-NH-CH(CH_2CH(CH_3)_2)-COOH}$$

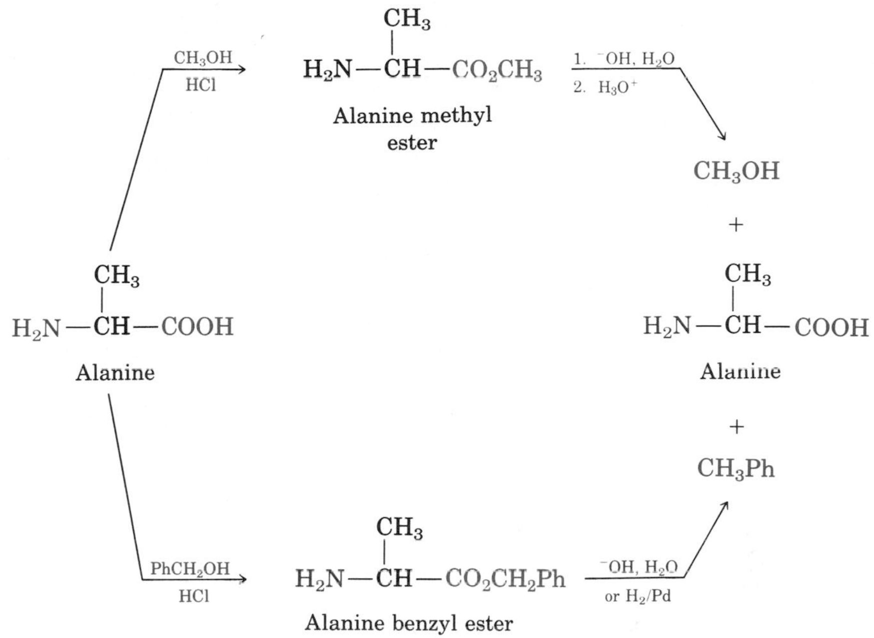


**Figure 27.8** Protection of an amino acid carboxyl group by ester formation

The step of peptide bond formation is usually accomplished by treating a mixture of protected acid and amine components with dicyclohexylcarbodiimide (DCC). As shown in Figure 27.9 on page 992, DCC functions by converting the carboxylic acid group into a reactive acylating agent that then undergoes a further nucleophilic acyl substitution with the amine.

We now have the knowledge needed to complete a synthesis of H-Leu-Ala-OH. Five separate steps are required:

1. Protect the amino group of leucine as the BOC derivative:

$$\text{H-Leu-OH} + (t\text{-BuO}\overset{\text{O}}{\overset{\|}{\text{C}}})_2\text{O} \longrightarrow \text{BOC-Leu-OH}$$

2. Protect the carboxyl group of alanine as the methyl ester:

$$\text{H-Ala-OH} + \text{CH}_3\text{OH} \xrightarrow[\text{catalyst}]{\text{Acid}} \text{H-Ala-OCH}_3 + \text{H}_2\text{O}$$

3. Couple the two protected amino acids using DCC:

$$\text{BOC-Leu-OH} + \text{H-Ala-OCH}_3 \xrightarrow{\text{DCC}} \text{BOC-Leu-Ala-OCH}_3$$

4. Remove the BOC protecting group by acid treatment:

$$\text{BOC-Leu-Ala-OCH}_3 \xrightarrow{\text{CF}_3\text{COOH}} \text{H-Leu-Ala-OCH}_3 + \text{CO}_2 + (\text{CH}_3)_2\text{C}{=}\text{CH}_2$$

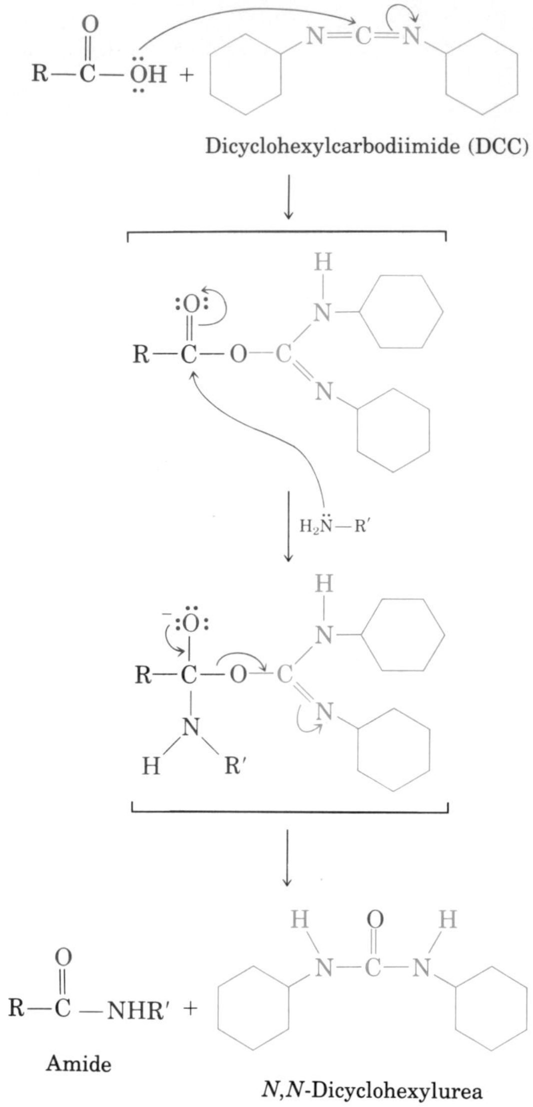

The carboxylic acid first adds to the carbodiimide reagent to yield a reactive acylating agent.

Dicyclohexylcarbodiimide (DCC)

Nucleophilic attack of the amine on the acylating agent gives a tetrahedral intermediate.

The intermediate loses dicyclohexylurea and yields the desired amide.

Amide

*N,N*-Dicyclohexylurea

**Figure 27.9** The mechanism of amide formation by reaction of a carboxylic acid and an amine with DCC (dicyclohexylcarbodiimide)

5. Remove the methyl ester by basic hydrolysis:

$$\text{H-Leu-Ala-OCH}_3 \xrightarrow[\text{H}_2\text{O}]{\text{NaOH}} \text{H-Leu-Ala-OH} + \text{HOCH}_3$$

These steps can be repeated to add one amino acid at a time to the growing chain or to link two peptide chains together. Many remarkable achievements in peptide synthesis have been reported, including a complete synthesis of human insulin. Insulin, the structure of which is shown in Figure 27.10, is composed of two chains totaling 51 amino acids linked by two disulfide bridges. Its structure was determined by Frederick Sanger,[5] who received the 1958 Nobel prize for his work.

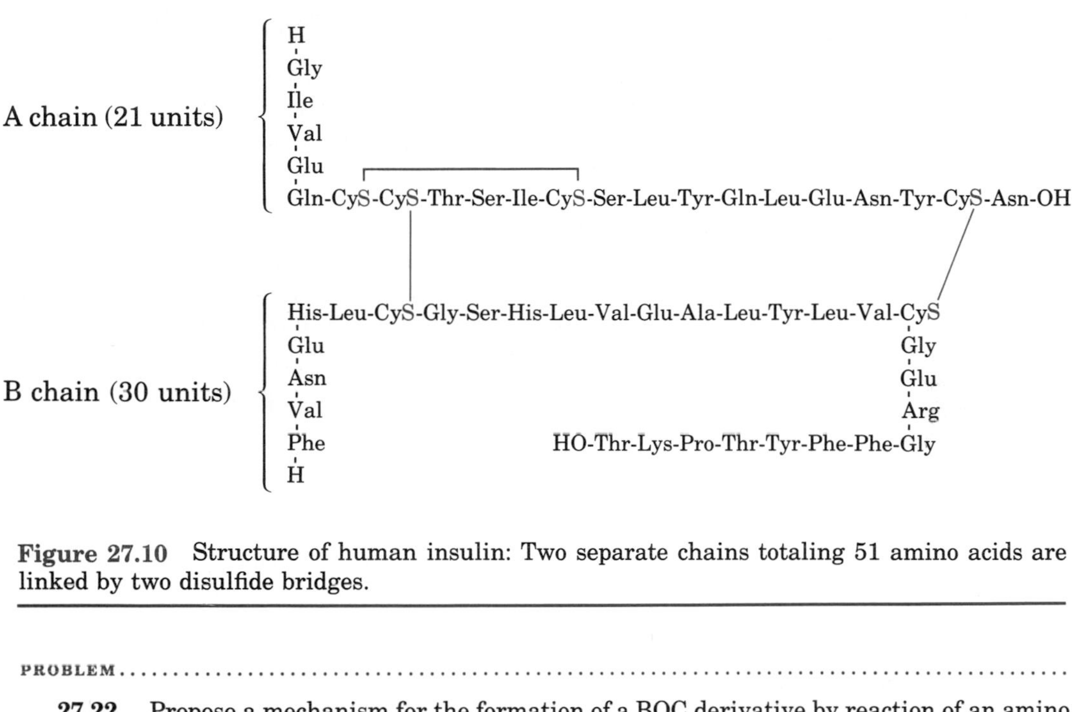


**Figure 27.10** Structure of human insulin: Two separate chains totaling 51 amino acids are linked by two disulfide bridges.

PROBLEM

**27.22** Propose a mechanism for the formation of a BOC derivative by reaction of an amino acid with di-*tert*-butyl dicarbonate.

PROBLEM

**27.23** Write all five steps required for the synthesis of H-Leu-Ala-OH from alanine and leucine.

PROBLEM

**27.24** How would you prepare these tripeptides?
(a) H-Leu-Ala-Gly-OH (b) H-Gly-Leu-Ala-OH

## 27.12 Automated Peptide Synthesis: The Merrifield Solid-Phase Technique

The synthesis of large peptide chains by sequential addition of one amino acid at a time is a long and arduous task. An immense simplification is possible, however, using the **solid-phase method** introduced by R. Bruce

[5]Frederick Sanger (1918– ); b. Gloucestershire, England; Ph.D. Cambridge; professor, Cambridge University; Nobel prize (1958, 1980).

Merrifield[6] at the Rockefeller University. In the Merrifield method, peptide synthesis is carried out on solid polymer beads of polystyrene, prepared so that one of every 100 or so benzene rings bears a chloromethyl ($-CH_2Cl$) group:

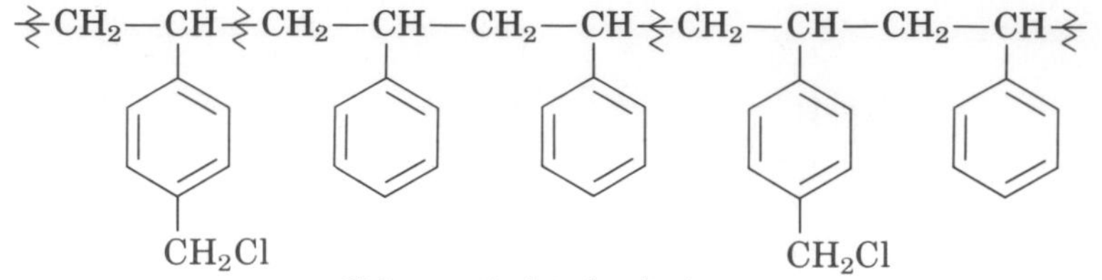

$CH_2-CH-CH_2-CH-CH_2-CH-CH_2-CH-CH_2-CH$ (each CH bearing a benzene ring; the first and fourth rings bear $CH_2Cl$)

Chloromethylated polystyrene

In the standard solution-phase method discussed in the previous section, a methyl ester was used to protect the carboxyl group during formation of the amide bond. In the solid-phase method, however, the solid polymer serves as the ester protecting group. Four steps are required in solid-phase peptide synthesis:

*Step 1* A covalent ester linkage is formed by reaction between a BOC-protected amino acid and the chloromethyl groups on the polystyrene polymer.

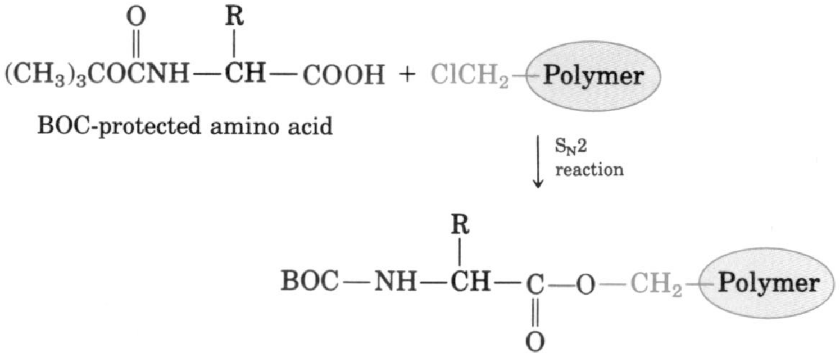

$$(CH_3)_3COC(=O)NH-CH(R)-COOH + ClCH_2-\text{Polymer} \xrightarrow{S_N2 \text{ reaction}} BOC-NH-CH(R)-C(=O)-O-CH_2-\text{Polymer}$$

BOC-protected amino acid

Ester link to polymer

*Step 2* After formation of the ester linkage is complete, the insoluble, polymer-bonded amino acid is washed free of excess reagents and treated with trifluoroacetic acid to remove the BOC group.

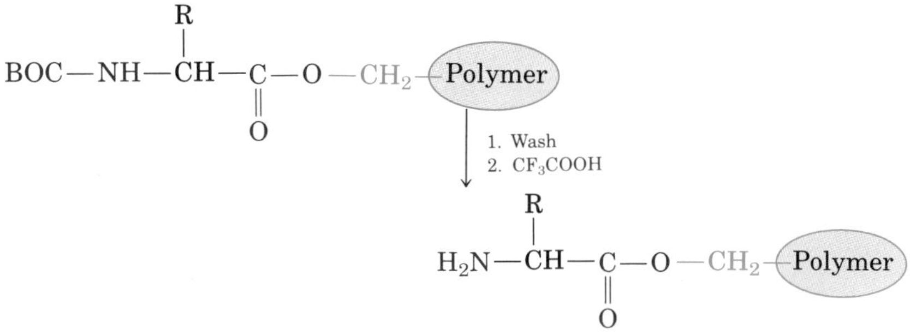

$$BOC-NH-CH(R)-C(=O)-O-CH_2-\text{Polymer} \xrightarrow{1.\ \text{Wash};\ 2.\ CF_3COOH} H_2N-CH(R)-C(=O)-O-CH_2-\text{Polymer}$$

Polymer-bonded amino acid

[6]Robert Bruce Merrifield (1921– ); Ph.D. University of California, Los Angeles (1949); professor, Rockefeller Institute; Nobel prize (1984).

*Step 3* A second BOC-protected amino acid is added along with the coupling reagent, DCC. A peptide bond forms, and excess reagents are then removed by washing the insoluble polymer:

$$\text{BOC—NH—}\underset{\displaystyle\text{R}'}{\text{CH}}\text{—}\underset{\displaystyle\text{O}}{\overset{\|}{\text{C}}}\text{—OH} + \text{H}_2\text{N—}\underset{\displaystyle\text{R}}{\text{CH}}\text{—}\underset{\displaystyle\text{O}}{\overset{\|}{\text{C}}}\text{—O—CH}_2\text{—Polymer}$$

$$\downarrow \text{1. DCC; 2. Wash}$$

$$\text{BOC—NH—}\underset{\displaystyle\text{R}'}{\text{CH}}\text{—}\underset{\displaystyle\text{O}}{\overset{\|}{\text{C}}}\text{—NH—}\underset{\displaystyle\text{R}}{\text{CH}}\text{—}\underset{\displaystyle\text{O}}{\overset{\|}{\text{C}}}\text{—O—CH}_2\text{—Polymer}$$

Polymer-bonded dipeptide

Step 2 is repeated to again remove a BOC group, and step 3 is repeated to add a third amino acid unit to the chain. In this way, dozens or even a hundred amino acid units can be efficiently and specifically linked to synthesize the desired polymer-bonded peptide.

*Step 4* After the proper number of coupling steps have been done and the desired peptide has been made, treatment with anhydrous hydrogen fluoride cleaves the ester bond to the polymer, yielding free peptide.

$$\text{BOC—NH—}\underset{\displaystyle\text{R}''}{\text{CH}}\text{—}\underset{\displaystyle\text{O}}{\overset{\|}{\text{C}}}\text{—(NH—}\underset{\displaystyle\text{R}'}{\text{CH}}\text{—}\underset{\displaystyle\text{O}}{\overset{\|}{\text{C}}}\text{)}_n\text{—NH—}\underset{\displaystyle\text{R}}{\text{CH}}\text{—}\underset{\displaystyle\text{O}}{\overset{\|}{\text{C}}}\text{—O—CH}_2\text{—Polymer}$$

$$\downarrow \text{HF}$$

$$\text{H}_2\text{N—}\underset{\displaystyle\text{R}''}{\text{CH}}\text{—}\underset{\displaystyle\text{O}}{\overset{\|}{\text{C}}}\text{—(NH—}\underset{\displaystyle\text{R}'}{\text{CH}}\text{—}\underset{\displaystyle\text{O}}{\overset{\|}{\text{C}}}\text{)}_n\text{—NH—}\underset{\displaystyle\text{R}}{\text{CH}}\text{—}\underset{\displaystyle\text{O}}{\overset{\|}{\text{C}}}\text{OH} + \text{HO—CH}_2\text{—Polymer}$$

Polypeptide

The solid-phase technique has now been automated. Peptide-growing machines are available for *automatically* repeating the coupling and deprotection steps with different amino acids as many times as desired. Each step occurs in extremely high yield, and mechanical losses are minimized since the peptide intermediates are never removed from the insoluble polymer until the final step. Among the many remarkable achievements recorded by Merrifield is the synthesis of bovine pancreatic ribonuclease, a protein containing 124 amino acid units. The entire synthesis required only 6 weeks and took place in 17% overall yield.

## 27.13 Classification of Proteins

Proteins can be classified into two major types according to their composition. **Simple proteins**, such as blood serum albumin, are those that yield only amino acids and no other organic compounds on hydrolysis. **Conjugated proteins**, such as are found in cell membranes, yield other compounds in addition to amino acids on hydrolysis.

Conjugated proteins are far more common than simple proteins and may be further classified according to the chemical nature of the non–amino acid portion. Thus, **glycoproteins** contain a carbohydrate part, **lipoproteins** contain a fatty part, **nucleoproteins** contain a nucleic acid part, and so on (Table 27.2). Glycoproteins are particularly widespread in nature and make up a large part of the membrane coating around living cells.

Table 27.2 Classification of some conjugated proteins

| *Class* | *Non–amino acid group* | *Weight of non–amino acid (%)* |
|---|---|---|
| **Glycoproteins** | | |
| $\gamma$-Globulin | Carbohydrate | 10 |
| Carboxypeptidase Y | Carbohydrate | 17 |
| Interferon | Carbohydrate | 20 |
| **Lipoproteins** | | |
| Plasma $\beta$ lipoprotein | Fats, cholesterol | 80 |
| **Nucleoproteins** | | |
| Ribosomal proteins | Ribonucleic acid | 60 |
| Tobacco mosaic virus | Ribonucleic acid | 5 |
| **Phosphoproteins** | | |
| Casein | Phosphate esters | 4 |
| **Metalloproteins** | | |
| Ferritin | Iron oxide | 23 |
| Hemoglobin | Iron | 0.3 |

Proteins can also be classified as either **fibrous** or **globular**, according to their three-dimensional shape. Fibrous proteins, such as collagen and $\alpha$-keratin, consist of polypeptide chains arranged side by side in long threads. Because these proteins are tough and insoluble in water, they are used in nature for structural materials like tendons, hooves, horns, and fingernails.

Globular proteins, by contrast, are usually coiled into compact, nearly spherical shapes. These proteins are generally soluble in water and are mobile within cells. Most of the 2000 or so known enzymes, as well as hormonal and transport proteins, are globular. Table 27.3 lists some common examples of both fibrous and globular proteins.

Table 27.3 Conformational classes of proteins

| Protein | Description |
|---|---|
| **Fibrous proteins (insoluble)** | |
| Collagen | Connective tissue, tendons |
| α-Keratin | Hair, horn, skin, nails |
| Elastin | Elastic connective tissue |
| **Globular proteins (soluble)** | |
| Insulin | Hormone controlling glucose metabolism |
| Lysozyme | Hydrolytic enzyme |
| Ribonuclease | Enzyme controlling RNA synthesis |
| Albumins | Proteins coagulated by heat |
| Immunoglobulins | Proteins involved in immune response |
| Myoglobin | Protein involved in oxygen transport |

## 27.14 Protein Structure

Proteins are so large in comparison to simple organic molecules that the word *structure* takes on a broader meaning when applied to these immense macromolecules. At its simplest, protein structure is the sequence in which amino acid residues are bound together. Called the **primary structure** of a protein, this is the most fundamental structural level.

There is, however, much more to protein structure than just amino acid sequence. The chemical properties of a protein are also dependent on higher levels of structure—on exactly how the peptide backbone is folded to give the molecule a specific three-dimensional shape. Thus, the term **secondary structure** refers to the way in which segments of the peptide backbone are oriented into a regular pattern; **tertiary structure** refers to the way in which the entire protein molecule is coiled into an overall three-dimensional shape; and **quaternary structure** refers to the way in which several protein molecules come together to yield large aggregate structures.

Let's look at three examples—α-keratin (fibrous), fibroin (fibrous), and myoglobin (globular)—to see how higher structure affects protein properties.

### α-KERATIN

α-Keratin is the fibrous structural protein found in wool, hair, nails, and feathers. Studies indicate that α-keratin is coiled into a right-handed secondary structure, as shown in Figure 27.11. This so-called **α-helix** is stabilized by hydrogen bonding between amide N—H groups and other amide carbonyl groups four residues away. Although the strength of a single hydrogen bond (about 5 kcal/mol) is only about 5% of the strength of a C—C or C—H covalent bond, the large number of hydrogen bonds made possible by helical winding imparts a great deal of stability to the α-helical structure. Each coil of the helix (the **repeat distance**) contains 3.6 amino acid residues; the distance between coils is 5.40 Å.

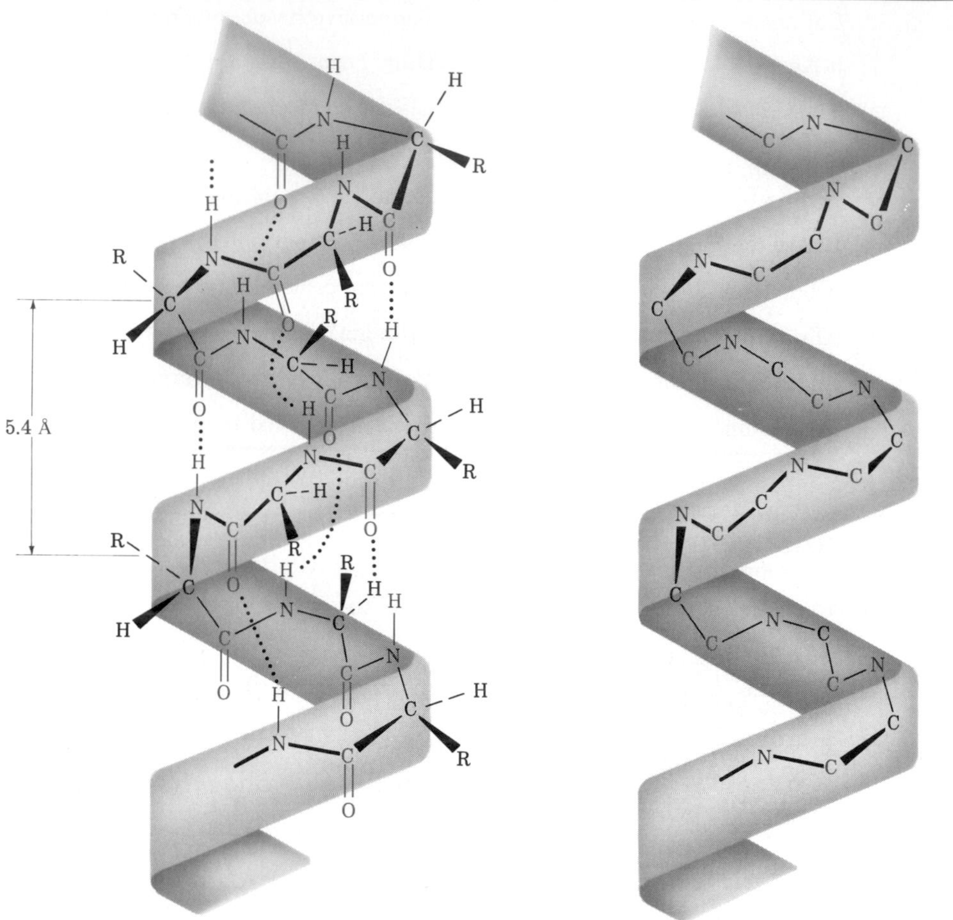


**Figure 27.11** The helical secondary structure of α-keratin

Further evidence suggests that the α-keratins of wool and hair also have a definite quaternary structure. The individual helices are themselves coiled about one another to form a *superhelix* that accounts for the threadlike properties and strength of these proteins.

## FIBROIN

Fibroin, the fibrous protein found in silk, has a secondary structure called a ***β*-pleated sheet**. In this pleated-sheet structure, polypeptide chains line up in a parallel arrangement held together by hydrogen bonds between chains (Figure 27.12). Although not as common as the $\alpha$-helix, small $\beta$-pleated-sheet regions are often found in proteins where sections of peptide chains double back on themselves.

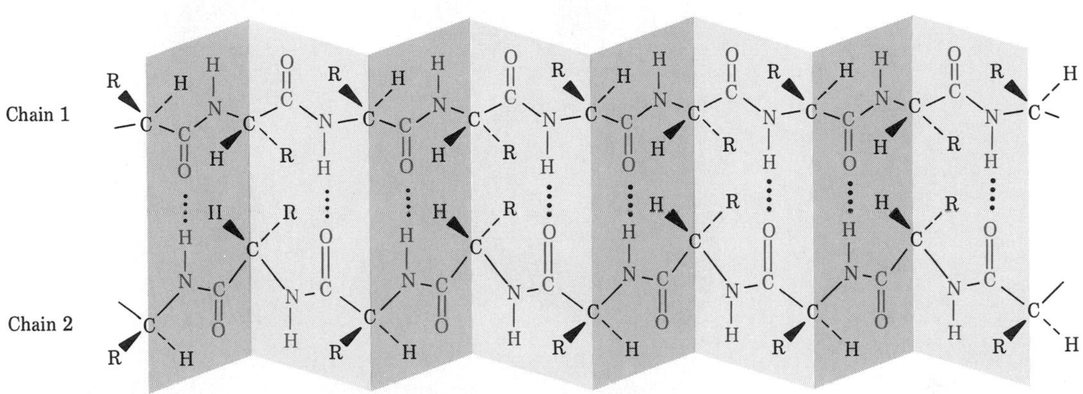


**Figure 27.12** The $\beta$-pleated-sheet structure in silk fibroin

## MYOGLOBIN

Myoglobin is a rather small globular protein containing 153 amino acid residues in a single chain. A relative of hemoglobin, myoglobin is found in the skeletal muscles of sea mammals, where it stores oxygen needed to sustain the animals during long dives. X-ray evidence obtained by Sir John Kendrew[7] and Max Perutz[8] has shown that myoglobin consists of eight straight segments, each of which adopts an $\alpha$-helical secondary structure. These helical sections are connected by bends to form a compact, nearly spherical, tertiary structure (Figure 27.13, page 1000). Although the bends appear to be irregular and the three-dimensional structure appears to be random, this is not the case. All myoglobin molecules adopt this same shape because it has a lower energy than any other possible shape.

Note that myoglobin is a conjugated protein that contains a covalently bound organic group (a **prosthetic group**) called *heme* (Figure 27.14). A great many proteins contain such prosthetic groups, which are crucial to their mechanism of action.

PROBLEM

**27.25** How can you account for the fact that proline is never present in a protein $\alpha$-helix? The $\alpha$-helical parts of myoglobin and other proteins stop whenever a proline residue is encountered in the chain.

---

[7]Sir John C. Kendrew (1917– ); b. Oxford, England; Ph.D. Cambridge; professor, Cambridge University; Nobel prize (1962).

[8]Max Ferdinand Perutz (1914– ); b. Vienna; universities of Vienna and Cambridge; professor, Cambridge University; Nobel prize (1962).

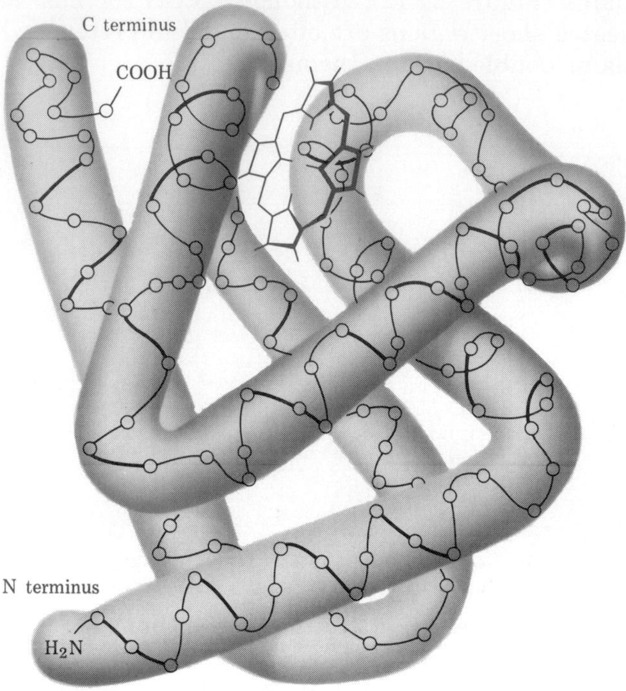


**Figure 27.13** Secondary and tertiary structure of myoglobin, a globular protein

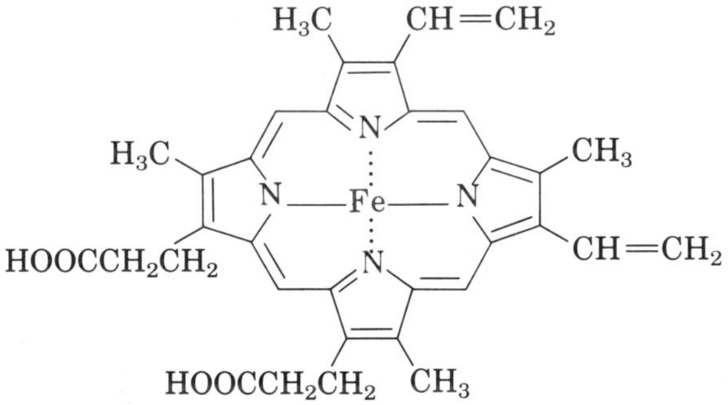

$H_3C$ $CH{=}CH_2$ $H_3C$ $CH_3$ $CH{=}CH_2$ $HOOCCH_2CH_2$ N Fe N N N $HOOCCH_2CH_2$ $CH_3$

**Figure 27.14** The structure of heme, a prosthetic group found in myoglobin

## 27.15 Protein Denaturation

The tertiary structure of globular proteins is delicately held together by weak intramolecular attractions. Often, a modest change in temperature or pH will disrupt the tertiary structure and cause the protein to become **denatured**. Denaturation occurs under such mild conditions that covalent bonds are not affected; the polypeptide primary structure remains intact, but the tertiary structure unfolds from a well-defined spherical shape to a randomly looped chain (Figure 27.15).

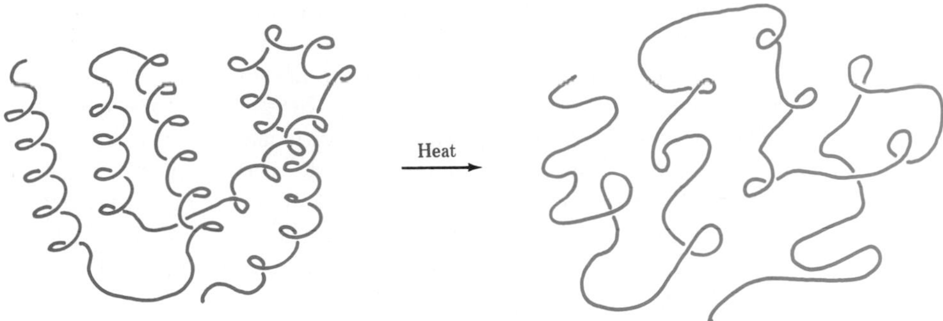


**Figure 27.15** Schematic representation of protein denaturation: A globular protein loses its specific three-dimensional shape and becomes randomly looped.

Denaturation is accompanied by changes in both physical and biological properties. Solubility is drastically decreased, as occurs when egg white is cooked and the albumins unfold and coagulate to an insoluble white mass. Most enzymes also lose all catalytic activity when denatured, since a precisely defined tertiary structure is required for their action.

Most, but not all, denaturation is irreversible. Eggs don't become uncooked when their temperature is lowered, and curdled milk doesn't become homogeneous. Many cases have now been found, however, where spontaneous **renaturation** of an unfolded protein occurs. Renaturation is accompanied by a full recovery of biological activity in the case of enzymes, indicating that the protein has completely returned to its stable tertiary structure.

## 27.16 Summary and Key Words

**Proteins** are large biomolecules made up of **α-amino acid residues** linked together by amide, or peptide, bonds. Twenty amino acids are commonly found in proteins; all are α-amino acids and all except glycine have stereochemistry similar to that of L sugars.

$$\mathrm{H_2N\overset{R}{\overset{|}{C}}H\underset{\underset{O}{\|}}{C}\!-\!\!\left(NH\overset{R'}{\overset{|}{C}}H\underset{\underset{O}{\|}}{C}\right)_{\!n}\!\!-NH\overset{R''}{\overset{|}{C}}H\underset{\underset{O}{\|}}{C}OH}$$

A polypeptide or protein

Amino acids can be synthesized by several methods, including ammonolysis of α-bromo acids, **reductive amination** of α-keto acids, **Strecker reaction** of aldehydes with $KCN/NH_4Cl$ followed by hydrolysis, and alkylation of diethyl **acetamidomalonate**. Resolution of the synthetic racemate then provides the optically active amino acid.

Determining the structure of a large polypeptide or protein is carried out in several steps. The identity and amount of each amino acid present in a peptide can be determined by **amino acid analysis**. The peptide is first hydrolyzed to its constituent $\alpha$-amino acids, which are then chromatographically separated and identified. Next, the peptide is **sequenced**. **Edman degradation** by treatment with phenyl isothiocyanate cleaves off one residue from the N terminus of the peptide and forms an easily identifiable derivative of the N-terminal amino acid. A series of sequential Edman degradations allows us to sequence peptide chains up to 25 residues in length.

Peptide synthesis is an equally challenging task, and one that must be solved by the use of selective **protecting groups**. An N-protected amino acid having a free carboxyl group is coupled with an O-protected amino acid having a free amino group in the presence of **dicyclohexylcarbodiimide (DCC)**. Amide formation occurs, the protecting groups are removed, and the sequence is repeated. Amines are usually protected as their *tert*-butoxy carbonyl (**BOC**) **derivatives**, and acids are protected as esters.

This synthetic sequence is often carried out by the **Merrifield solid-phase technique**, in which the peptide is esterified to an insoluble polymeric support.

Proteins are classified as either globular or fibrous, depending on their **secondary** and **tertiary structures**. **Fibrous proteins** such as $\alpha$-keratin are tough, rigid, and water insoluble and are used in nature for forming structures such as hair and nails. **Globular proteins** such as myoglobin are water soluble, roughly spherical in shape, and are mobile within cells. Most of the 2000 or so known enzymes are globular proteins.

## 27.17 Summary of Reactions

1. Amino acid synthesis
   a. From $\alpha$-bromo acids (Section 27.4)

$$\mathrm{RCH_2COOH} \xrightarrow[\mathrm{P}]{\mathrm{Br_2}} \mathrm{R\overset{\overset{\displaystyle Br}{|}}{C}HCOOH} \xrightarrow{\mathrm{NH_3}} \mathrm{R\overset{\overset{\displaystyle NH_2}{|}}{C}HCOOH}$$

   b. Strecker synthesis (Section 27.4)

$$\mathrm{RCHO} \xrightarrow[\mathrm{NH_4Cl}]{\mathrm{KCN}} \mathrm{R\overset{\overset{\displaystyle NH_2}{|}}{C}HCN} \xrightarrow{\mathrm{H_3O^+}} \mathrm{R\overset{\overset{\displaystyle NH_2}{|}}{C}HCOOH}$$

   c. Reductive amination (Sections 25.7 and 27.4)

$$\mathrm{R\overset{\overset{\displaystyle O}{\|}}{C}COOH} \xrightarrow[\mathrm{NaBH_4}]{\mathrm{NH_3}} \mathrm{R\overset{\overset{\displaystyle NH_2}{|}}{C}HCOOH}$$

d. Diethyl acetamidomalonate synthesis (Sections 22.9 and 27.4)

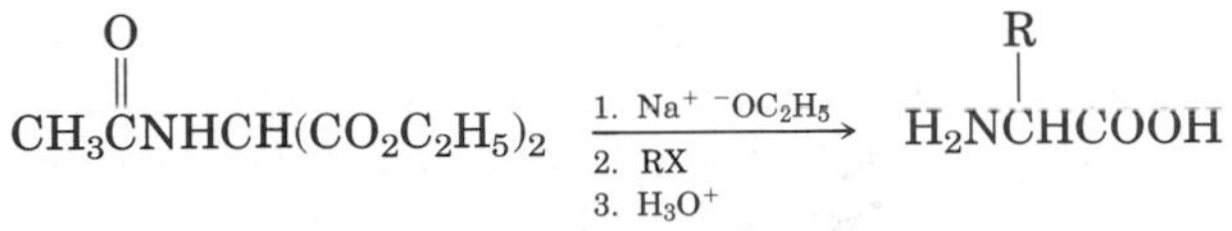

$$CH_3\overset{\overset{O}{\|}}{C}NHCH(CO_2C_2H_5)_2 \xrightarrow[\substack{2.\ RX \\ 3.\ H_3O^+}]{1.\ Na^+\ {}^-OC_2H_5} H_2N\overset{\overset{R}{|}}{C}HCOOH$$

2. Peptide synthesis

a. Nitrogen protection (Section 27.11)

$$H_2N\overset{\overset{R}{|}}{C}HCOOH + [(CH_3)_3CO\overset{\overset{O}{\|}}{C}]_2O \longrightarrow (CH_3)_3CO\overset{\overset{O}{\|}}{C}-NH\overset{\overset{R}{|}}{C}HCOOH$$

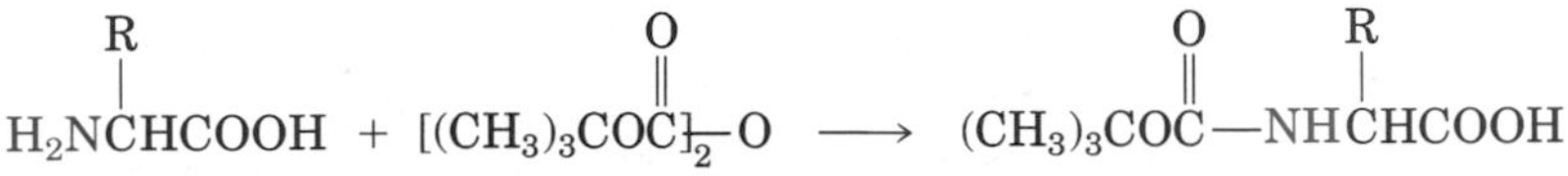

BOC-protected amino acid

The BOC protecting group can be removed by acid treatment:

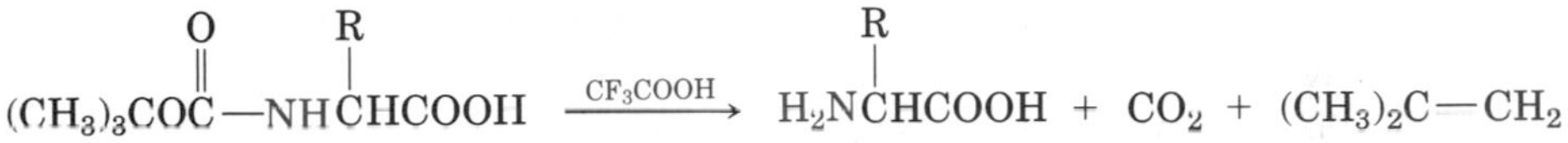

$$(CH_3)_3CO\overset{\overset{O}{\|}}{C}-NH\overset{\overset{R}{|}}{C}HCOOH \xrightarrow{CF_3COOH} H_2N\overset{\overset{R}{|}}{C}HCOOH + CO_2 + (CH_3)_2C{=}CH_2$$

b. Oxygen protection (Section 27.11)

$$H_2N\overset{\overset{R}{|}}{C}HCOOH + CH_3OH \xrightarrow{HCl} H_2N\overset{\overset{R}{|}}{C}HCOOCH_3$$

$$H_2N\overset{\overset{R}{|}}{C}HCOOH + C_6H_5CH_2OH \xrightarrow{HCl} H_2N\overset{\overset{R}{|}}{C}HCOOCH_2C_6H_5$$

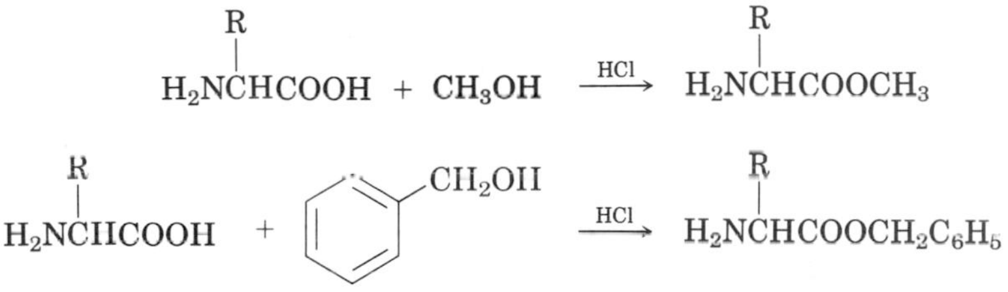

The ester protecting group can be removed by base hydrolysis:

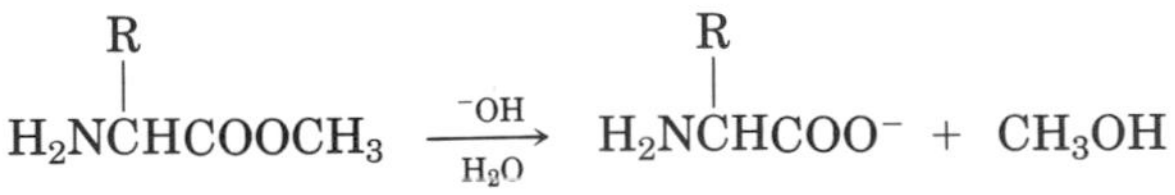

$$H_2N\overset{\overset{R}{|}}{C}HCOOCH_3 \xrightarrow[H_2O]{{}^-OH} H_2N\overset{\overset{R}{|}}{C}HCOO^- + CH_3OH$$

c. Amide bond formation (Section 27.11)

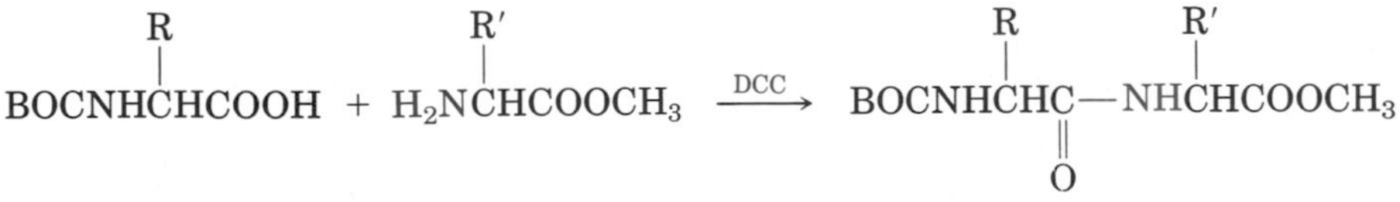

$$BOCNH\overset{\overset{R}{|}}{C}HCOOH + H_2N\overset{\overset{R'}{|}}{C}HCOOCH_3 \xrightarrow{DCC} BOCNH\overset{\overset{R}{|}}{C}H\underset{\underset{O}{\|}}{C}-NH\overset{\overset{R'}{|}}{C}HCOOCH_3$$

where DCC = $C_6H_{11}-N{=}C{=}N-C_6H_{11}$ (dicyclohexyl)

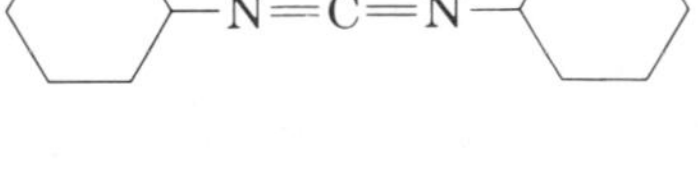

3. Peptide sequencing: Edman degradation (Section 27.9)

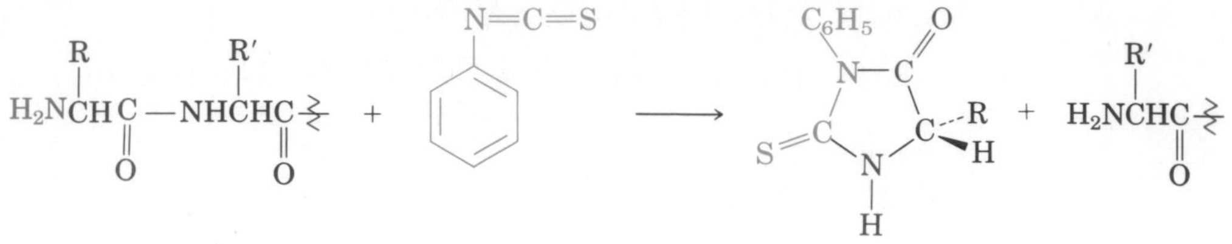

$$\underset{\substack{|\\R}}{H_2NCH}\overset{}{\underset{\|O}{C}}-NH\underset{R'}{CH}\underset{\|O}{C}\sim + C_6H_5N{=}C{=}S \longrightarrow \text{phenylthiohydantoin (R)} + H_2N\underset{R'}{CH}\underset{\|O}{C}\sim$$

## ADDITIONAL PROBLEMS

**27.26** Only *S* amino acids occur in proteins, but several *R* amino acids are also found in nature. Thus, (*R*)-serine is found in earthworms and (*R*)-alanine is found in insect larvae. Draw Fischer projections of (*R*)-serine and (*R*)-alanine.

**27.27** Draw a Fischer projection of (*S*)-proline.

**27.28** Show the structures of the amino acids corresponding to these three-letter codes:
(a) Trp (b) Ile (c) Cys (d) His

**27.29** Explain the observation that amino acids exist as dipolar zwitterions in aqueous solution, but exist largely as true amino carboxylic acids in chloroform solution.

$$\underset{\text{In } H_2O}{H_3\overset{+}{N}CH(R)CO_2^-} \rightleftharpoons \underset{\text{In } CHCl_3}{H_2NCH(R)COOH}$$

**27.30** At what pH would you carry out an electrophoresis experiment if you wanted to separate a mixture of histidine, serine, and glutamic acid? Explain.

**27.31** Define these terms:
(a) Amphoteric (b) Isoelectric point (c) Zwitterion

**27.32** Using the three-letter code names for amino acids, write the structures of all possible peptides containing these amino acids:
(a) Val, Ser, Leu (b) Ser, $Leu_2$, Pro

**27.33** Cytochrome *c* is an enzyme found in the cells of all aerobic organisms. Elemental analysis of cytochrome *c* shows that it contains 0.43% iron. What is the minimum molecular weight of this enzyme?

**27.34** Predict the product of the reaction of valine with these reagents:
(a) $CH_3CH_2OH$, acid (b) Di-*tert*-butyl dicarbonate
(c) KOH, $H_2O$ (d) $CH_3COCl$, pyridine; then $H_2O$

**27.35** Write out full structures for these peptides:
(a) H-Val-Phe-Cys-Ala-OH (b) H-Glu-Pro-Ile-Leu-OH

**27.36** Show the steps involved in a synthesis of H-Phe-Ala-Val-OH using the Merrifield procedure.

**27.37** Draw the structure of the phenylthiohydantoin product you would obtain by Edman degradation of these peptides:
(a) H-Ile-Leu-Pro-Phe-OH (b) H-Asp-Thr-Ser-Gly-Ala-OH

**27.38** The chloromethylated polystyrene resin used for Merrifield solid-phase peptide synthesis is prepared by treatment of polystyrene with chloromethyl methyl ether and a Lewis acid catalyst. Propose a mechanism for the reaction.

$$\text{-CH}_2\text{-CH(C}_6\text{H}_5\text{)-} \xrightarrow[\text{SnCl}_4]{\text{CH}_3\text{OCH}_2\text{Cl}} \text{-CH}_2\text{-CH(C}_6\text{H}_4\text{-}p\text{-CH}_2\text{Cl)-}$$

**27.39** Which amide bonds in the following polypeptide would be cleaved by trypsin? By chymotrypsin?

H-Phe-Leu-Met-Lys-Tyr-Asp-Gly-Gly-Arg-Val-Ile-Pro-Tyr-OH

**27.40** The synthesis of large peptides is much more efficient when done in a convergent manner. That is, higher overall yields of final products are obtained if several small chains are constructed and coupled, as opposed to constructing one long chain by stepwise addition of one amino acid residue at a time. For example, consider a synthesis of methionine enkephalin, H-Tyr-Gly-Gly-Phe-Met-OH, by two routes:

(a) Tyr + Gly $\longrightarrow$ Tyr-Gly $\xrightarrow{\text{Gly}}$ Tyr-Gly-Gly $\xrightarrow{\text{Phe}}$ Tyr-Gly-Gly-Phe $\xrightarrow{\text{Met}}$ Tyr-Gly-Gly-Phe-Met

(b) Tyr + Gly $\longrightarrow$ Tyr-Gly $\xrightarrow{\text{Gly}}$ Tyr-Gly-Gly

Phe + Met $\longrightarrow$ Phe-Met

Tyr-Gly-Gly + Phe-Met $\longrightarrow$ Tyr-Gly-Gly-Phe-Met

Assume a yield of 90% for each coupling step and calculate overall yields for the two routes.

**27.41** The Sanger end-group determination is sometimes used as an alternative to the Edman degradation. In the Sanger method, a peptide is allowed to react with 2,4-dinitrofluorobenzene, the peptide is hydrolyzed, and the N-terminal amino acid is identified by separation as its *N*-2,4-dinitrophenyl derivative:

$$\text{2,4-(NO}_2)_2\text{C}_6\text{H}_3\text{F} + \text{H}_2\text{NCH(R)C(=O)NH—Peptide} \longrightarrow \text{2,4-(O}_2\text{N)}_2\text{C}_6\text{H}_3\text{—NHCH(R)C(=O)NH—Peptide}$$

$$\xrightarrow{\text{H}_3\text{O}^+} \text{2,4-(O}_2\text{N)}_2\text{C}_6\text{H}_3\text{—NHCH(R)COOH} + \text{Amino acids}$$

Propose a mechanism to account for the initial reaction between peptide and dinitrofluorobenzene.

**27.42** Would you foresee any problems in using the Sanger end-group determination method (Problem 27.41) on a peptide such as H-Gly-Pro-Lys-Ile-OH? Explain.

**27.43** When $\alpha$-amino acids are treated with dicyclohexylcarbodiimide, DCC, 2,5-diketopiperazines result. Propose a mechanism for this reaction.

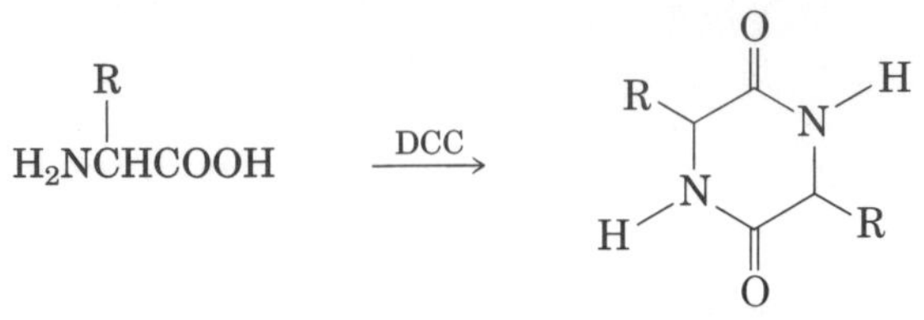

$$H_2N\underset{R}{C}HCOOH \xrightarrow{DCC}$$

A 2,5-diketopiperazine

**27.44** Arginine, which contains a guanidine functional group in its side chain, is by far the most basic of the 20 common amino acids. How can you account for this basicity? Use resonance structures to see how the protonated guanidino group is stabilized.

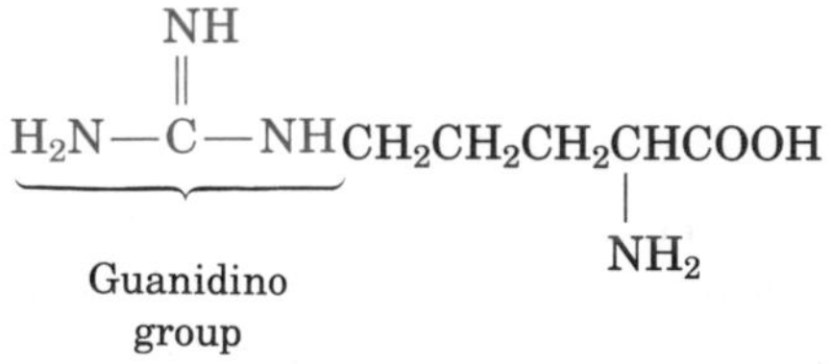

$$\underbrace{H_2N-\overset{NH}{\overset{\|}{C}}-NH}_{\text{Guanidino group}}CH_2CH_2CH_2\underset{NH_2}{\underset{|}{C}}HCOOH$$

Arginine

**27.45** Good evidence for restricted rotation around amide CO—N bonds comes from NMR studies. At room temperature, the $^1H$ NMR spectrum of *N,N*-dimethylformamide shows three peaks: 2.9 $\delta$ (singlet, 3 H), 3.0 $\delta$ (singlet, 3 H), 8.0 $\delta$ (singlet, 1 H). As the temperature is raised, however, the two singlets at 2.9 $\delta$ and 3.0 $\delta$ slowly merge. At 180°C, the $^1H$ NMR spectrum shows only two peaks: 2.95 $\delta$ (singlet, 6 H) and 8.0 $\delta$ (singlet, 1 H). Explain this temperature-dependent behavior.

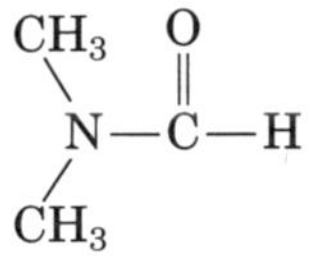

$$(CH_3)_2N-\overset{O}{\overset{\|}{C}}-H$$

*N,N*-Dimethylformamide

**27.46** An octapeptide shows the composition Asp, $Gly_2$, Leu, Phe, $Pro_2$, Val on amino acid analysis. Edman analysis shows a glycine N-terminal group, and carboxypeptidase cleavage produces leucine as the first amino acid to appear. Acidic hydrolysis gives the following fragments:

1. H-Val-Pro-Leu-OH
2. H-Gly-OH
3. H-Gly-Asp-Phe-Pro-OH
4. H-Phe-Pro-Val-OH

Propose a suitable structure for the starting octapeptide.

**27.47** Propose a mechanism to account for the reaction of ninhydrin with an $\alpha$-amino acid:

2 (ninhydrin) + $H_2N\overset{R}{C}HCOOH$ ⟶ (indanedione–N=indenone product, OH) + RCHO + $CO_2$

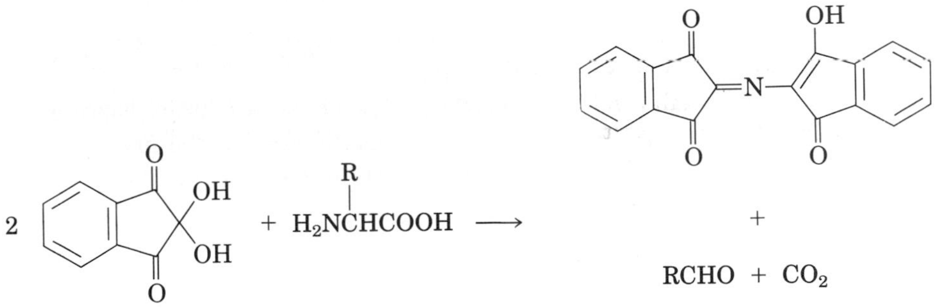

**27.48** Draw as many resonance forms as you can for the purple anion obtained by reaction of ninhydrin with an $\alpha$-amino acid (Problem 27.47).

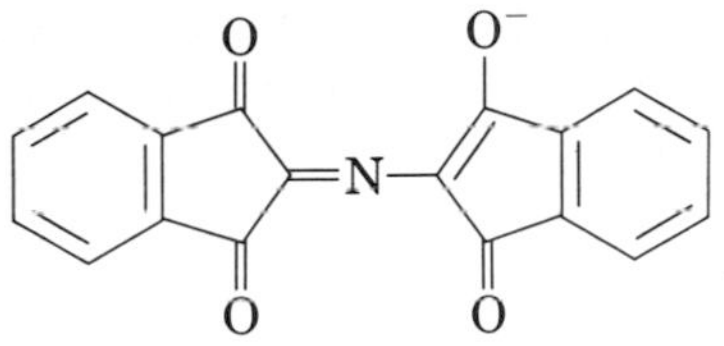

(purple anion structure: O, $O^-$, =N–, O, O)

**27.49** Look up the structure of human insulin (Figure 27.10) and indicate where in each chain the molecule would be cleaved by trypsin and chymotrypsin.

**27.50** What is the structure of a nonapeptide that gives the following fragments when cleaved?

Trypsin cleavage: H-Val-Val-Pro-Tyr-Leu-Arg-OH and H-Ser-Ile-Arg-OH

Chymotrypsin cleavage: H-Leu-Arg-OH and H-Ser-Ile-Arg-Val-Val-Pro-Tyr-OH

**27.51** Oxytocin, a nonapeptide hormone secreted by the pituitary gland, functions by stimulating uterine contraction and lactation during childbirth. Its sequence was determined from the following evidence:

1. Oxytocin is a cyclic compound containing a disulfide bridge between two cysteine residues.
2. When the disulfide bridge is reduced, oxytocin has the constitution Asn, $Cys_2$, Gln, Gly, Ile, Leu, Pro, Tyr.
3. Partial hydrolysis of reduced oxytocin yields seven fragments:

| | |
|---|---|
| H-Asp-Cys-OH | H-Ile-Glu-OH |
| H-Cys-Tyr-OH | H-Leu-Gly-OH |
| H-Tyr-Ile-Glu-OH | H-Glu-Asp-Cys-OH |
| H-Cys-Pro-Leu-OH | |

4. Gly can be shown to be the C-terminal group.
5. Both Glu and Asp are present as their side-chain amides (Gln and Asn) rather than as free side-chain acids.

On the basis of this evidence, what is the amino acid sequence of reduced oxytocin? What is the structure of oxytocin itself?

**27.52** *Aspartame*, a nonnutritive sweetener marketed under the trade name Nutra-Sweet, is the methyl ester of a simple dipeptide, H-Asp-Phe-$OCH_3$.
(a) Draw the full structure of aspartame.
(b) The isoelectric point of aspartame is 5.9. Draw the principal structure present in aqueous solution at this pH.
(c) Draw the principal form of aspartame present at physiological pH = 7.6.

**27.53** Refer to Figure 27.7 and propose a mechanism for the final step in the Edman degradation—the acid-catalyzed rearrangement of the thiazolinone to the *N*-phenylthiohydantoin.

CHAPTER 28

# Lipids

**Lipids** are naturally occurring organic molecules isolated from cells and tissues by extraction with nonpolar organic solvents. Since they usually have large hydrocarbon portions in their structures, lipids are insoluble in water but soluble in organic solvents. Note that this definition differs from the sort used for carbohydrates and proteins. Lipids are defined by *physical property* (solubility) rather than by structure.

Lipids can be further classified into two general types. **Complex lipids** such as fats and waxes contain ester linkages that can be hydrolyzed to yield smaller molecules. **Simple lipids** such as cholesterol and other steroids do not have ester linkages and cannot be hydrolyzed.

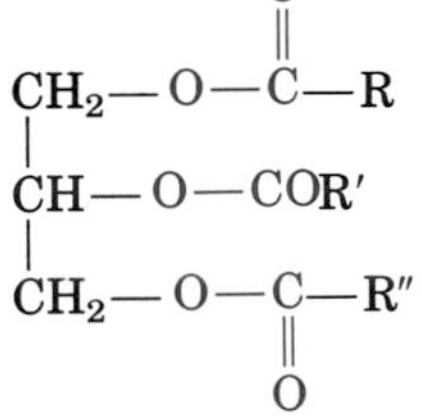

$$\begin{array}{l} CH_2-O-\overset{O}{\overset{\|}{C}}-R \\ | \\ CH-O-COR' \\ | \\ CH_2-O-\underset{O}{\underset{\|}{C}}-R'' \end{array}$$

Fat, a complex lipid
(R, R′, R″ = $C_{11}$–$C_{19}$ chains)

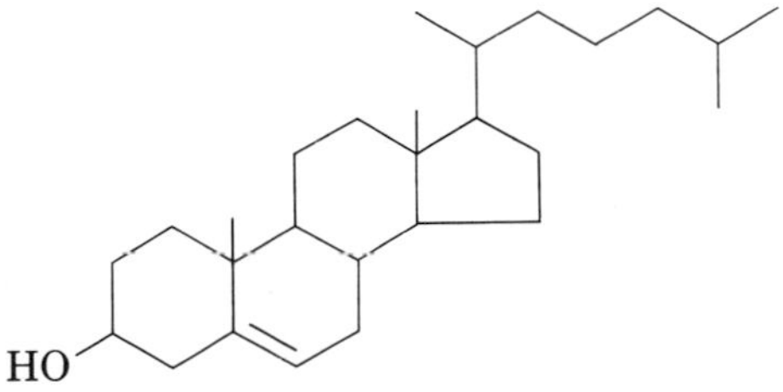

HO

Cholesterol, a simple lipid

$$CH_3(CH_2)_{20-24}-\overset{O}{\overset{\|}{C}}-O-(CH_2)_{27}CH_3$$

Beeswax, a complex lipid

## 28.1 Fats and Oils

Animal fats and vegetable oils are the most widely occurring lipids. Although they appear different—animal fats like butter and lard are solids, whereas vegetable oils like corn and peanut oil are liquid—their structures are closely related. Chemically, fats and oils are **triacylglycerols**; that is, triesters of glycerol with three long-chain carboxylic acids. Thus, hydrolysis of a fat with aqueous sodium hydroxide (**saponification**) yields glycerol and three **fatty acids**:

$$\begin{array}{l} CH_2O{-}\overset{\displaystyle O}{\overset{\|}{C}}{-}R \\ | \\ CHO{-}\overset{\displaystyle O}{\overset{\|}{C}}{-}R' \\ | \\ CH_2O{-}\overset{\displaystyle O}{\overset{\|}{C}}{-}R'' \\ \text{A fat} \end{array} \xrightarrow[2.\ H_3O^+]{1.\ ^-OH} \begin{array}{l} CH_2OH \\ | \\ CHOH \\ | \\ CH_2OH \\ \text{Glycerol} \end{array} + \begin{array}{l} RCOOH \\ \\ R'COOH \\ \\ R''COOH \\ \text{Fatty acids} \end{array}$$

The fatty acids obtained by hydrolysis of triacylglycerols are unbranched, contain an even number of carbon atoms between 12 and 20, and may be either saturated or unsaturated. If double bonds are present, they usually have *Z* (cis) geometry. The three fatty acids of a specific triacylglycerol molecule are not usually the same, and the fat or oil from a given source is likely to be a complex mixture of many different triacylglycerols. Table 28.1 lists some of the commonly occurring fatty acids, and Table 28.2 lists the approximate composition of some fats and oils from different sources.

**Table 28.1** Structures of some common fatty acids

| *Name* | *Carbons* | *Structure* | *Melting point (°C)* |
|---|---|---|---|
| **Saturated** | | | |
| Lauric | 12 | $CH_3(CH_2)_{10}COOH$ | 44 |
| Myristic | 14 | $CH_3(CH_2)_{12}COOH$ | 58 |
| Palmitic | 16 | $CH_3(CH_2)_{14}COOH$ | 63 |
| Stearic | 18 | $CH_3(CH_2)_{16}COOH$ | 70 |
| Arachidic | 20 | $CH_3(CH_2)_{18}COOH$ | 75 |
| **Unsaturated** | | | |
| Palmitoleic | 16 | $CH_3(CH_2)_5CH{=}CH(CH_2)_7COOH$ (cis) | 32 |
| Oleic | 18 | $CH_3(CH_2)_7CH{=}CH(CH_2)_7COOH$ (cis) | 4 |
| Ricinoleic | 18 | $CH_3(CH_2)_5CH(OH)CH_2CH{=}CH(CH_2)_7COOH$ (cis) | 5 |
| Linoleic | 18 | $CH_3(CH_2)_4CH{=}CHCH_2CH{=}CH(CH_2)_7COOH$ (cis,cis) | −5 |
| Arachidonic | 20 | $CH_3(CH_2)_4(CH{=}CHCH_2)_4CH_2CH_2COOH$ (all cis) | −50 |

**Table 28.2** Approximate fatty-acid composition of some fats and oils

| | *Saturated fatty acids (%)* | | | | *Unsaturated fatty acids (%)* | | |
|---|---|---|---|---|---|---|---|
| *Source* | $C_{12}$ *Lauric* | $C_{14}$ *Myristic* | $C_{16}$ *Palmitic* | $C_{18}$ *Stearic* | $C_{18}$ *Oleic* | $C_{18}$ *Ricinoleic* | $C_{18}$ *Linoleic* |
| **Animal fat** | | | | | | | |
| Lard | — | 1 | 25 | 15 | 50 | — | 6 |
| Butter | 2 | 10 | 25 | 10 | 25 | — | 5 |
| Human fat | 1 | 3 | 25 | 8 | 46 | — | 10 |
| Whale blubber | — | 8 | 12 | 3 | 35 | — | 10 |
| **Vegetable oil** | | | | | | | |
| Coconut | 50 | 18 | 8 | 2 | 6 | — | 1 |
| Corn | — | 1 | 10 | 4 | 35 | — | 45 |
| Olive | — | 1 | 5 | 5 | 80 | — | 7 |
| Peanut | — | — | 7 | 5 | 60 | — | 20 |
| Linseed | — | — | 5 | 3 | 20 | — | 20 |
| Castor bean | — | — | — | 1 | 8 | 85 | 4 |

The data listed in Table 28.1 show that unsaturated fatty acids generally have lower melting points than their saturated counterparts, a trend that also holds true for triacylglycerols. Since vegetable oils generally have a higher proportion of unsaturated to saturated fatty acids than animal fats (Table 28.2), they are lower melting.

This melting-point behavior is due to the fact that saturated fats have a uniform shape that allows them to pack together easily in a crystal lattice. Carbon–carbon double bonds in unsaturated vegetable oils, however, introduce bends and kinks into the hydrocarbon chains, making crystal formation difficult. The more double bonds there are, the harder it is for the molecules to crystallize, and the lower the melting point of the oil. Figure 28.1 (page 1012) illustrates this effect with space-filling molecular models.

The carbon–carbon double bonds present in vegetable oils can be reduced by catalytic hydrogenation (Section 7.6) to produce saturated solid or semisolid fats. Margarine and solid cooking fats such as Crisco are produced by hydrogenating soybean, peanut, or cottonseed oil until exactly the proper consistency is obtained.

PROBLEM

**28.1** Draw structures of these molecules. Which would you expect to have a higher melting point?
(a) Glyceryl tripalmitate (b) Glyceryl trioleate

PROBLEM

**28.2** Stearolic acid, $C_{18}H_{32}O_2$, yields stearic acid on catalytic hydrogenation and undergoes oxidative cleavage with ozone to yield nonanoic acid and nonanedioic acid. What is the structure of stearolic acid?

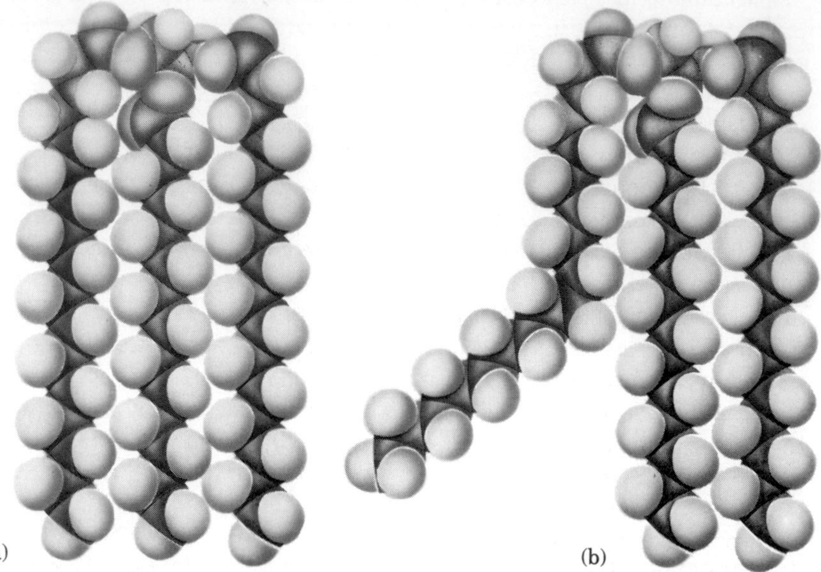


**Figure 28.1** Space-filling molecular models of (a) saturated and (b) unsaturated triacylglycerols: There is an unsaturated fatty acid in (b) that prevents the molecule from adopting a regular shape and crystallizing easily.

## 28.2 Soaps

Soap has been known since at least 600 BC, when the Phoenicians reportedly prepared a curdy material by boiling goat fat with extracts of wood ash. The cleansing properties of soap were not generally recognized, however, and the use of soap did not become widespread until the eighteenth century.

Chemically, soap is a mixture of the sodium or potassium salts of long-chain fatty acids produced by saponification of animal fat with alkali. Wood ash was used as a source of alkali until the mid-1800s, when the Leblanc process for producing $Na_2CO_3$ was invented, and NaOH thus became commercially available:

$$\begin{array}{l} CH_2{-}O{-}\overset{O}{\overset{\|}{C}}{-}R \\ | \\ CH{-}O{-}\overset{O}{\overset{\|}{C}}{-}R \\ | \\ CH_2{-}O{-}\overset{O}{\overset{\|}{C}}{-}R \\ \text{A fat} \end{array} \xrightarrow[H_2O]{NaOH} \underset{\text{Soap}}{3\,R\overset{O}{\overset{\|}{C}}O^-\ Na^+} + \begin{array}{l} CH_2OH \\ | \\ CHOH \\ | \\ CH_2OH \\ \text{Glycerol} \end{array}$$

where $R = C_{15}$–$C_{19}$ aliphatic chains.

Crude soap curds contain glycerol and excess alkali as well as soap, but purification can be effected by boiling with a large amount of water and adding NaCl to precipitate the pure sodium carboxylate salts. The smooth soap that precipitates is dried, perfumed, and pressed into bars for household use. Dyes are added if a colored soap is desired, antiseptics are added for medicated soaps, pumice is added for scouring soaps, and air is blown in for a soap that floats. Regardless of these extra treatments, and regardless of price, all soaps are basically the same.

Soaps exert their cleansing action because the two ends of a soap molecule are so different. The sodium salt end of the long-chain molecule is ionic; it is therefore **hydrophilic** (water-loving) and tries to dissolve in water. The long hydrocarbon chain portion of the molecule, however, is nonpolar; it is therefore **lipophilic** (fat-loving) and tries to dissolve in grease. The net effect of these two opposing tendencies is that soaps are attracted to both grease and water, and are therefore valuable as cleansers.

When soaps are dispersed in water, the long hydrocarbon tails cluster together in a lipophilic ball, while the ionic heads on the surface of the cluster stick out into the water layer. These spherical clusters, called **micelles**, are shown schematically in Figure 28.2. Grease and oil droplets are solubilized in water when they are coated by the nonpolar tails of soap molecules in the center of micelles. Once solubilized, the grease and dirt can be washed away.

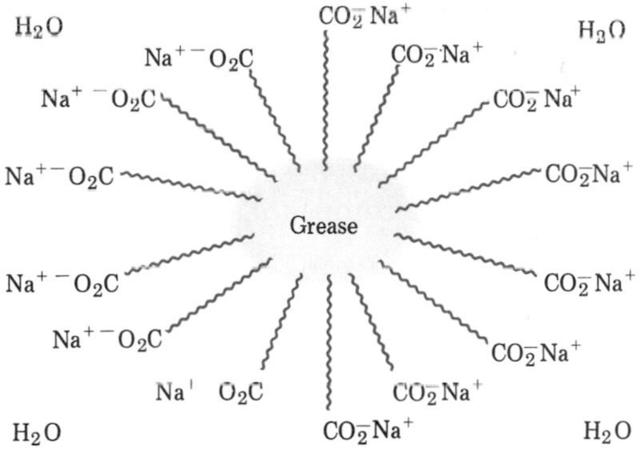


**Figure 28.2** A soap micelle solubilizing a grease particle in water

Soaps make life much more pleasant than it would otherwise be, but they have certain drawbacks. In hard water containing metal ions, soluble sodium carboxylates are converted into insoluble magnesium and calcium salts, leaving the familiar ring of scum around bathtubs and the "tattletale gray" on white clothes. Chemists have circumvented these problems by synthesizing a class of synthetic detergents based on salts of long-chain alkylbenzenesulfonic acids.

The principle by which synthetic detergents operate is identical to the principle of soaps—the alkylbenzene end of the molecule is lipophilic and attracts grease, but the sulfonate salt end is ionic and is attracted to water. Unlike soaps, however, sulfonate detergents do not form insoluble metal salts in hard water.

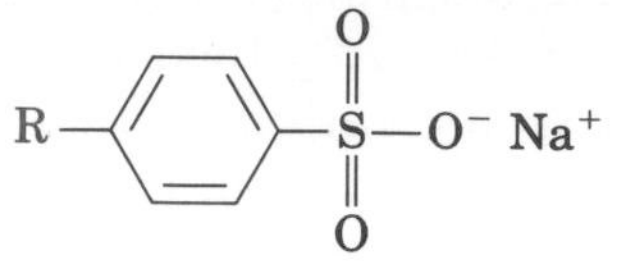

```
               O
               ‖
R—⟨benzene⟩—S—O⁻ Na⁺
               ‖
               O
```

A synthetic detergent

where R = a mixture of $C_{12}$ aliphatic chains.

**PROBLEM**

**28.3** Draw the structure of magnesium oleate, a component of bathtub scum.

## 28.3 Phospholipids

**Phospholipids** are esters of phosphoric acid, $H_3PO_4$. Most phospholipids are closely related to fats, since they contain a glycerol backbone linked by ester bonds to two fatty acids and one phosphoric acid. Although the fatty acid residues in these so-called **phosphoglycerides** may be any of the $C_{12}$–$C_{20}$ units normally present in fats, the acyl group at C1 is usually saturated, and that at C2 is usually unsaturated. The phosphate group at C3 is also bound by a separate ester link to an amino alcohol such as choline, $HOCH_2CH_2\overset{+}{N}(CH_3)_3$, or ethanolamine, $IIOCH_2CH_2NH_2$.

The most important phosphoglycerides are the *lecithins* and the *cephalins*. Note that these compounds are chiral and that they have the L or *R* configuration at C2.

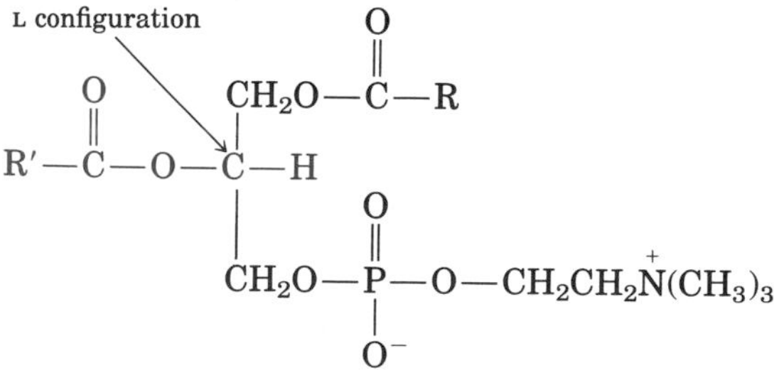

```
L configuration       O
                      ‖
   O        CH2O—C—R
   ‖         |
R′—C—O—C—H
             |      O
             |      ‖                 +
            CH2O—P—O—CH2CH2N(CH3)3
                    |
                    O⁻
```

Phosphatidylcholine, a lecithin

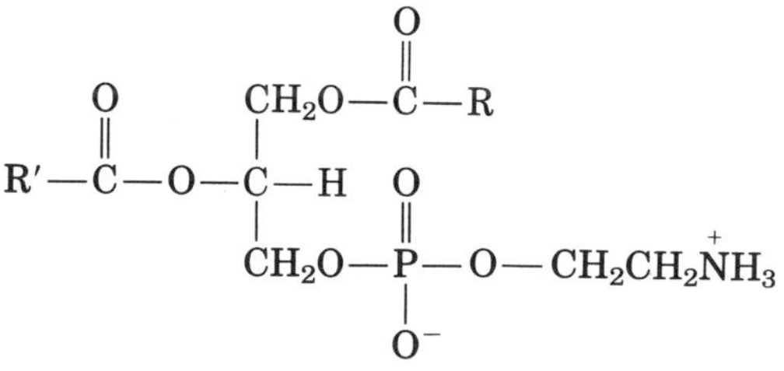

```
                      O
                      ‖
   O        CH2O—C—R
   ‖         |
R′—C—O—C—H   O
             |      ‖              +
            CH2O—P—O—CH2CH2NH3
                    |
                    O⁻
```

Phosphatidylethanolamine, a cephalin

where R is saturated and R′ is unsaturated.

Found widely in both plant and animal tissues, phosphoglycerides are the major lipid component of cell membranes (approximately 40%). Like soaps, phosphoglycerides have a long, nonpolar hydrocarbon tail bound to a polar ionic head (the phosphate group). Cell membranes are composed in large part of phosphoglycerides oriented into a **lipid bilayer** about 50 Å thick. As shown in Figure 28.3, the lipophilic tails aggregate in the center of the bilayer in much the same way that soap tails aggregate in the center of a micelle (Figure 28.2). This bilayer serves as an effective barrier to the passage of ions and other components into and out of the cell.

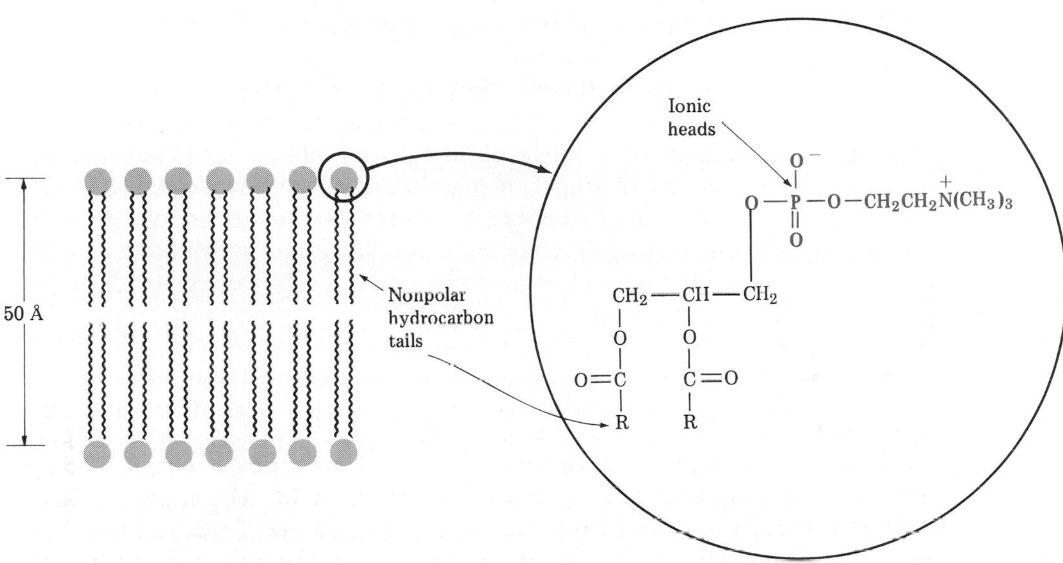


**Figure 28.3** Aggregation of phosphoglycerides into the lipid bilayer that composes cell membranes

The second major group of phospholipids are the **sphingolipids**. These complex lipids, which have *sphingosine* or a related dihydroxyamine as their backbones, are important constituents of plant and animal cell membranes. They are particularly abundant in brain and nerve tissue, where *sphingomyelins* are a major constituent of the coating around nerve fibers.

$$\begin{array}{l} CH_2OH \\ | \\ CHNH_2 \\ | \\ CHOH \\ | \\ CH{=}CH(CH_2)_{12}CH_3 \end{array}$$

Sphingosine

$$\begin{array}{l} CH_2O{-}\overset{\displaystyle O}{\overset{\|}{P}}{-}OCH_2CH_2\overset{+}{N}(CH_3)_3 \\ |\qquad\ \ \underset{\displaystyle O^-}{|} \\ CHNHCO(CH_2)_{16\text{–}24}CH_3 \\ | \\ CHOH \\ | \\ CH{=}CH(CH_2)_{12}CH_3 \end{array}$$

Sphingomyelin, a sphingolipid

## 28.4 Biosynthesis of Fatty Acids

One of the most striking features of the fatty acids shown in Table 28.1 is that all have an even number of carbon atoms. The reason for this characteristic is that all are derived biosynthetically from the simple two-carbon precursor, acetic acid. The pathway by which this is accomplished is shown in Figure 28.4.

The starting material for fatty acid synthesis is the thiol ester, acetyl CoA (Section 23.15), prepared in nature from acetic acid. The fatty acid synthetic pathway begins with several **priming reactions** that convert acetyl CoA into more reactive species.

The first two steps in the priming sequence convert acetyl CoA into *acetyl synthase*. This highly reactive acylating agent is capable of transferring the acetyl group to a nucleophile by nucleophilic acyl substitution reaction (Section 21.2). In step 1, a reaction catalyzed by the enzyme *acyl carrier protein (ACP) transferase* exchanges the thiol ester linkage of acetyl CoA for a different and somewhat more reactive thiol ester bond to ACP. Step 2 involves a further exchange of thiol ester linkages, resulting in the formation of acetyl synthase.

The next two steps in the priming sequence again start with acetyl CoA. In step 3, acetyl CoA is carboxylated by reaction with $CO_2$ and the enzyme acetyl CoA carboxylase to yield *malonyl CoA*. Step 4 is another thiol ester exchange reaction that converts malonyl CoA into the more reactive *malonyl ACP*.

The key carbon–carbon bond-forming reaction used to build the fatty acid chain occurs in step 5. This key step is simply a Claisen condensation (Section 23.9) between acetyl synthase as the electrophilic acceptor component and malonyl ACP as the nucleophilic donor component. An enolate ion derived from the doubly activated $—CH_2—$ group of malonyl ACP adds to the carbonyl group of acetyl synthase, yielding an intermediate $\beta$-keto acid, which loses carbon dioxide to give the four-carbon product, *acetoacetyl ACP*.

$$CH_3\overset{O}{\overset{\|}{C}}—S\ \text{synthase} \quad + \quad {}^{-}\!:\underset{COOH}{\underset{|}{C}H}\overset{O}{\overset{\|}{C}}SACP \longrightarrow \left[ CH_3\overset{O}{\overset{\|}{C}}—\underset{COOH}{\underset{|}{C}H}\overset{O}{\overset{\|}{C}}SACP \right]$$

$$\downarrow$$

$$CH_3\overset{O}{\overset{\|}{C}}—CH_2\overset{O}{\overset{\|}{C}}SACP + CO_2$$

Acetoacetyl ACP

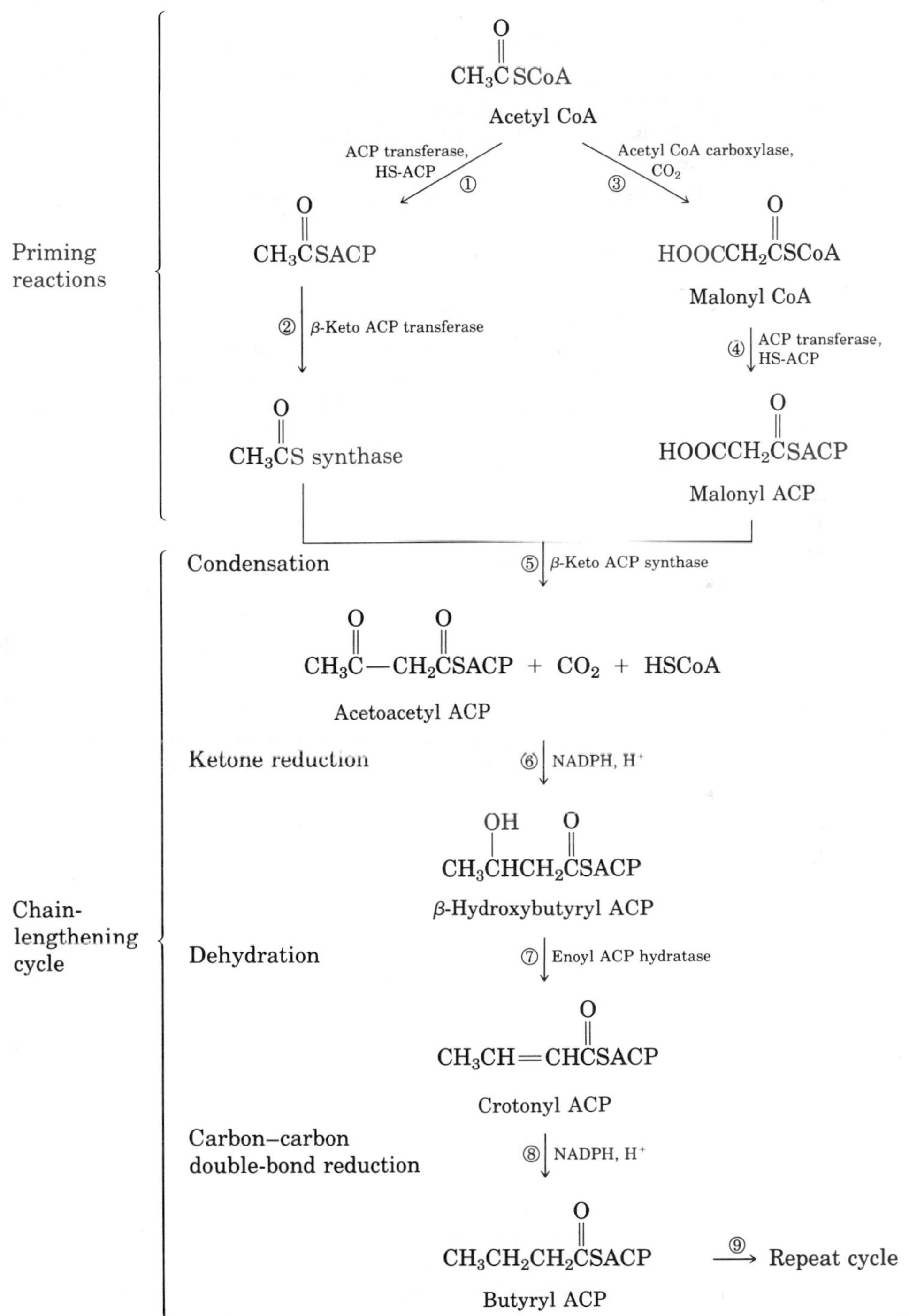


**Figure 28.4** Pathway for fatty acid biosynthesis

Step 6 involves the NADPH (nicotinamide adenine dinucleotide phosphate) enzyme-catalyzed reduction of the ketone carbonyl group in acetoacetyl ACP to an alcohol, *β-hydroxybutyryl ACP*, which is dehydrated in step 7 to yield *crotonyl ACP*. The carbon–carbon double bond of crotonyl ACP is further reduced by NADPH in step 8 to yield *butyryl ACP*. The net effect of these eight separate steps is to take two acetic acid units and combine them into a single four-carbon butanoic acid unit. Further repetition of the cycle by condensation of butyryl synthase with another malonyl ACP yields a six-carbon unit, and still further repetitions add two more carbon atoms to the chain each time.

Fatty acid biosynthesis is a lengthy process—some 64 separate steps are required for a cell to produce stearic acid—but all the individual steps are simple transformations of the same sort a chemist might do in the laboratory. We've seen them all before—nucleophilic acyl substitution, Claisen condensation, ketone reduction, dehydration, alkene hydrogenation. Clearly, biological chemistry and laboratory organic chemistry follow the same rules.

PROBLEM

**28.4** Show a likely reaction mechanism for the transformation of acetyl CoA into acetyl ACP.

$$CH_3\overset{\overset{O}{\|}}{C}{-}SCoA + HSACP \rightleftharpoons CH_3\overset{\overset{O}{\|}}{C}{-}SACP + HSCoA$$

PROBLEM

**28.5** Show a likely mechanism for the ready decarboxylation of the acylmalonyl ACP intermediate formed in step 5 of Figure 28.4.

$$\left[CH_3\overset{\overset{O}{\|}}{C}\underset{\underset{COOH}{|}}{CH}\overset{\overset{O}{\|}}{C}SACP\right] \longrightarrow CH_3\overset{\overset{O}{\|}}{C}CH_2\overset{\overset{O}{\|}}{C}SACP + CO_2$$

PROBLEM

**28.6** Evidence for the proposed role of acetate in fatty acid biosynthesis comes from isotope-labeling experiments. If acetate labeled with $^{14}C$ in the methyl group ($^{14}CH_3COOH$) were incorporated into fatty acids, at what positions in the fatty acid chain would you expect the $^{14}C$ label to appear?

## 28.5 Prostaglandins

Few compounds have caused as much excitement among medical researchers in the past decade as the **prostaglandins**. First isolated by Sune Bergstrom,[1] Bengt Samuelsson,[2] and their collaborators at the Karolinska Institute in Sweden, these simple lipids are synthesized in nature from the $C_{20}$ fatty acid, arachidonic acid. The name *prostaglandin* derives from the fact that these compounds were first thought to be produced by the prostate gland, but they have subsequently been shown to be present in small amounts in all body tissues and fluids.

[1]Sune Bergstrom (1916– ); M.D. Karolinska Institute; professor, Karolinska Institute; Nobel prize (1982).

[2]Bengt Samuelsson (1934– ); M.D. Lund; professor, Karolinska Institute; Nobel prize (1982).

The prostaglandins are simple in structure. All have a cyclopentane ring with two long side chains, though they differ in the number of oxygen atoms and the number of double bonds present. Prostaglandin $E_1$ ($PGE_1$) and prostaglandin $F_{2\alpha}$ ($PGF_{2\alpha}$) are representative structures.

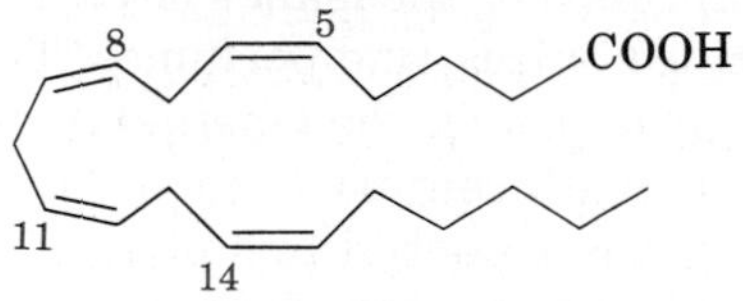

Arachidonic acid,
(5*Z*,8*Z*,11*Z*,14*Z*)-eicosatetraenoic acid

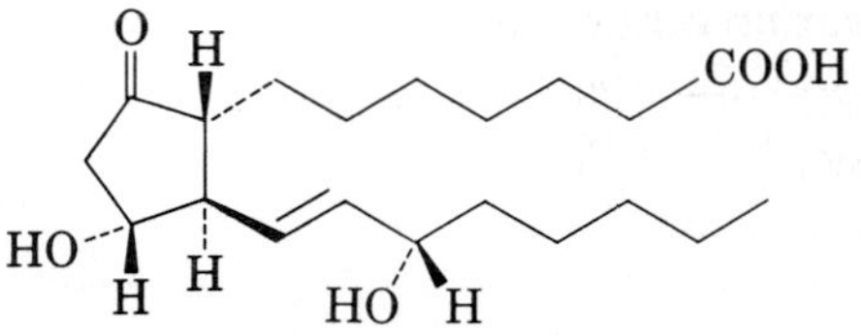

Prostaglandin $E_1$

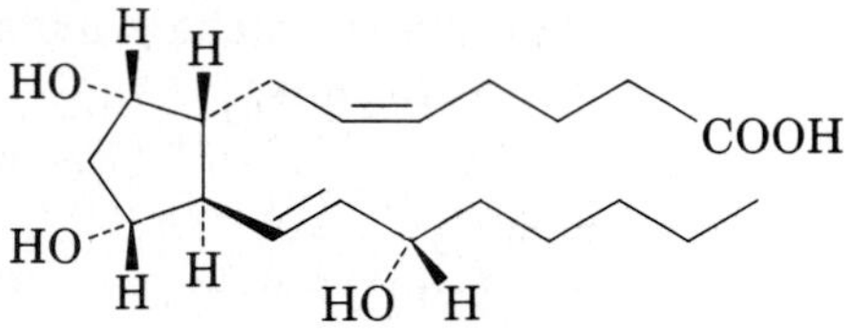

Prostaglandin $F_{2\alpha}$

The several dozen known prostaglandins have an extraordinarily wide range of biological activities. Among their known actions are their abilities to affect blood pressure, to affect blood-platelet aggregation during clotting, to affect gastric secretions, to control inflammation, to affect kidney function, to affect reproductive systems, and to stimulate uterine contractions during childbirth. In addition, compounds that are closely related to the prostaglandins have still other effects. Interest has centered particularly on the thromboxanes, on prostacyclin, and on the leukotrienes, whose release in the body appears to trigger the asthmatic response.

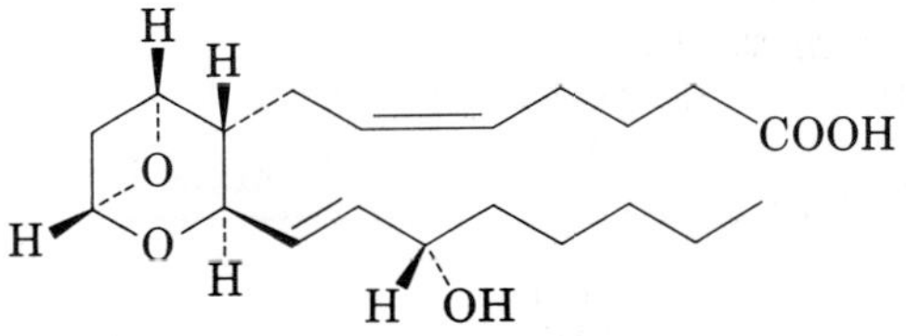

Thromboxane $A_2$

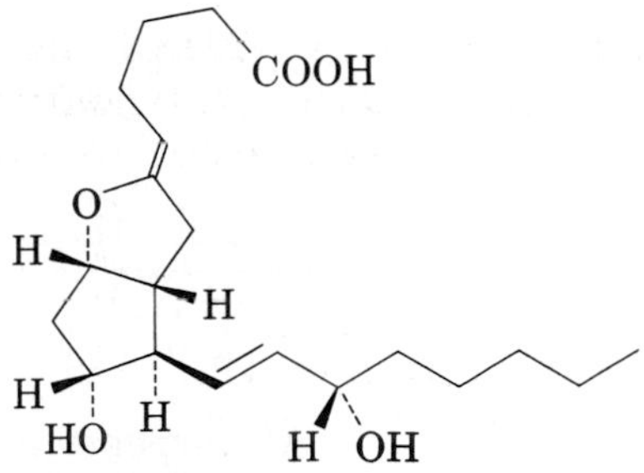

Prostacyclin

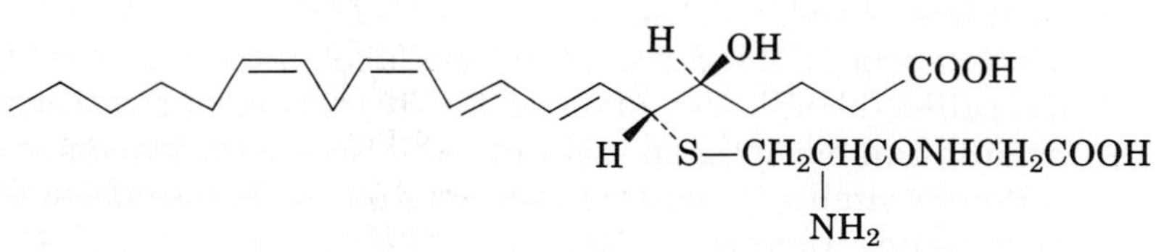

Leukotriene $D_4$

Prostaglandins are just beginning to be exploited in medicine, where their remarkable biological properties will surely lead to valuable new drugs.

## 28.6 Terpenes

It has long been known that codistillation of many plant materials with steam (**steam distillation**) produces a fragrant mixture of liquids called plant **essential oils**. For thousands of years, these plant extracts have been used as medicines, spices, and perfumes. The investigation of essential oils also played a major role in the emergence of organic chemistry as a science during the nineteenth century.

Chemically, the essential oils of plants consist largely of mixtures of simple lipids called **terpenes**. Terpenes are relatively small organic molecules that have an immense diversity of structure. Thousands of different terpenes are known; some are hydrocarbons, and others contain oxygen; some are open-chain molecules, and others contain rings. Figure 28.5 gives some examples.

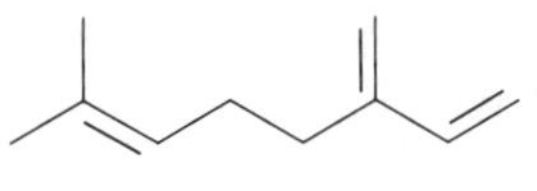

$CH_3$ = $H_3C$ $CH_3$

Myrcene (oil of bay) 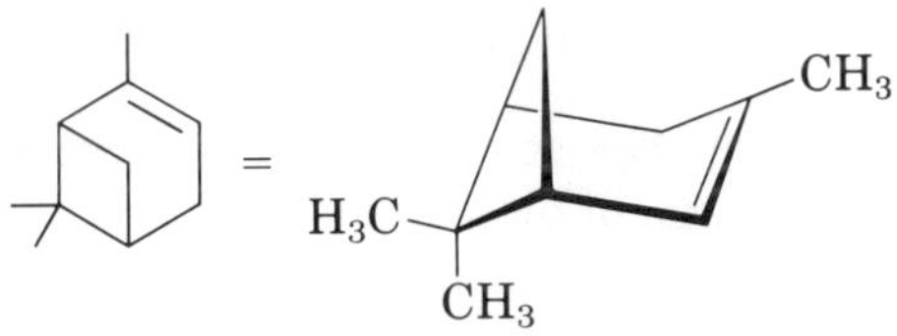 α-Pinene (oil of turpentine)

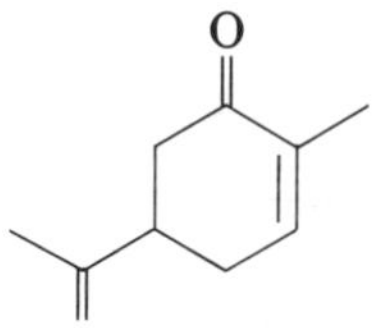

OH O $H_3C$ $H_3C$ $CH_3$ H $CH_3$

Carvone (oil of spearmint) 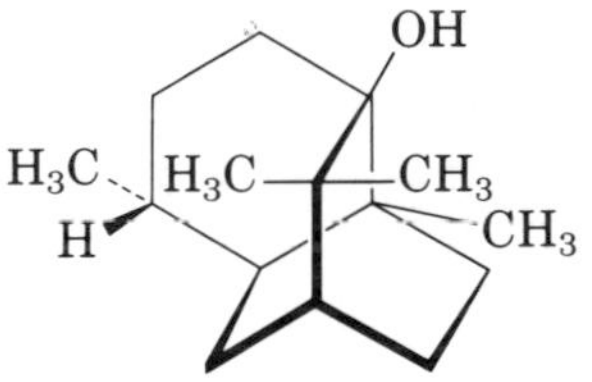 Patchouli alcohol (patchouli oil)

**Figure 28.5** The structures of some terpenes isolated from essential oils

All terpenes are related, regardless of their apparent structural differences. According to the **isoprene rule**, terpenes can be considered to arise from head-to-tail joining of simple five-carbon isoprene (2-methyl-1,3-butadiene) units. (Carbon 1 is called the head of the isoprene unit, and carbon 4 is called the tail.) For example, myrcene contains two isoprene units joined head to tail, forming an eight-carbon chain with two one-carbon branches. α-Pinene similarly contains two isoprene units assembled into a more complex cyclic structure.

Terpenes are classified into groups according to the number of 5-carbon isoprene units they contain. Thus, **monoterpenes** are 10-carbon substances biosynthesized from two isoprene units, **sesquiterpenes** are 15-carbon molecules from three isoprene units, and so on (Table 28.3).

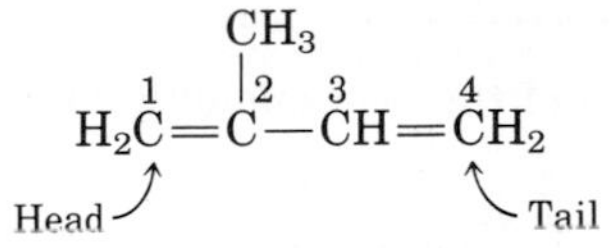

Isoprene (2-methyl-1,3-butadiene)

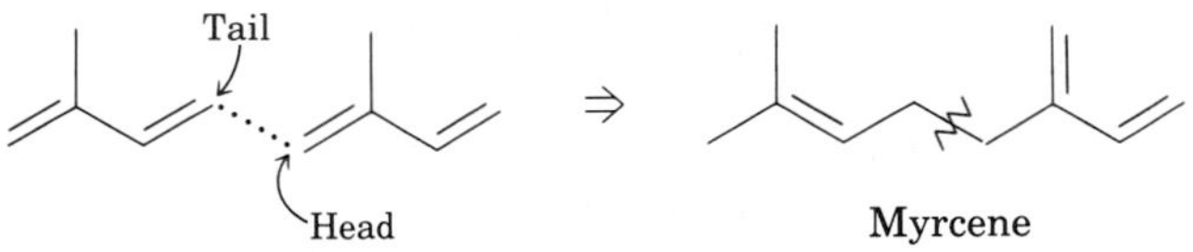

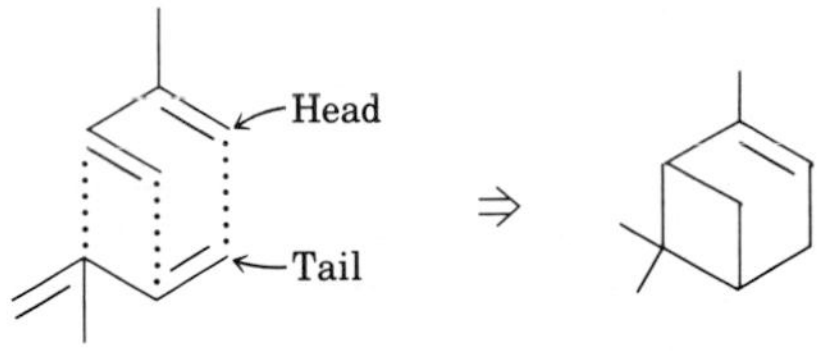

Two isoprenes

α-Pinene

**Table 28.3** Classification of terpenes

| *Carbon atoms* | *Isoprene units* | *Classification* |
|---|---|---|
| 10 | 2 | Monoterpenes |
| 15 | 3 | Sesquiterpenes |
| 20 | 4 | Diterpenes |
| 25 | 5 | Sesterterpenes |
| 30 | 6 | Triterpenes |
| 40 | 8 | Tetraterpenes |

Mono- and sesquiterpenes are found primarily in plants, but the higher terpenes occur in both plants and animals. Many of the higher terpenes have important biological roles. For example, the triterpene *lanosterol* is the precursor from which all steroid hormones are made in nature, and the tetraterpene *β-carotene* is a major dietary source of vitamin A.

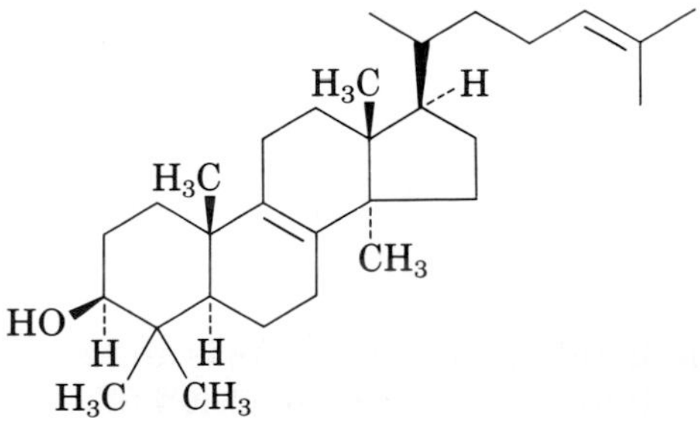

Lanosterol, a triterpene ($C_{30}$)

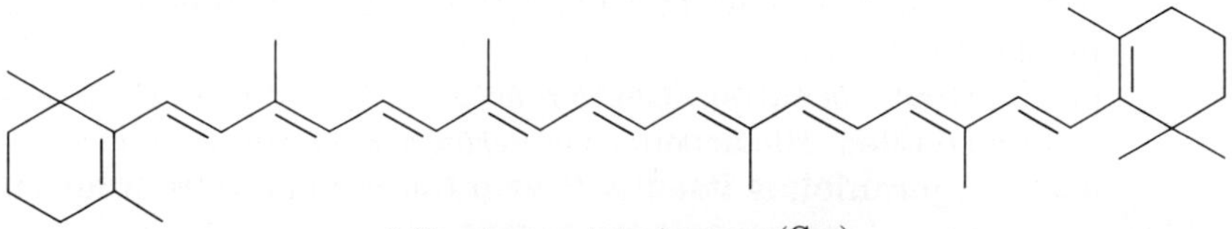

β-Carotene, a tetraterpene ($C_{40}$)

PROBLEM

**28.7** Show the positions of the isoprene units in these terpenes:

(a) Carvone (spearmint oil)

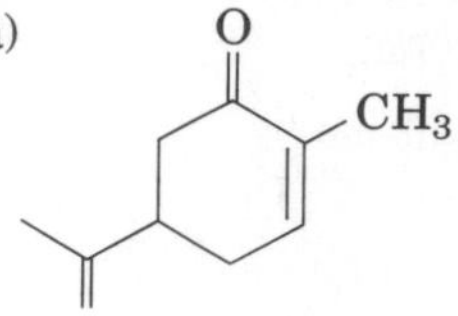

(b) Camphor

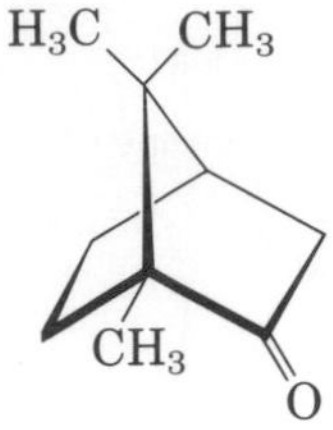

(c) Caryophyllene (cloves)

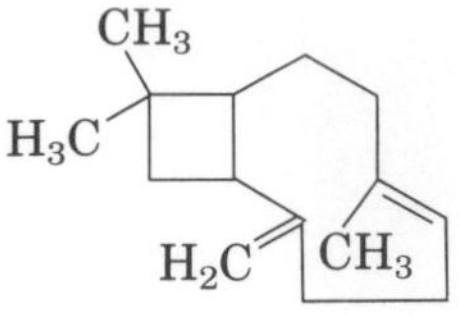

## 28.7 Biosynthesis of Terpenes

The isoprene rule is a convenient formalism for helping to determine new structures, but isoprene itself is not the biological precursor of terpenes. Nature instead uses two isoprene "equivalents," isopentenyl pyrophosphate and dimethylallyl pyrophosphate, for the biosynthesis of terpenes. These five-carbon molecules are themselves made from condensation of three acetyl CoA units.

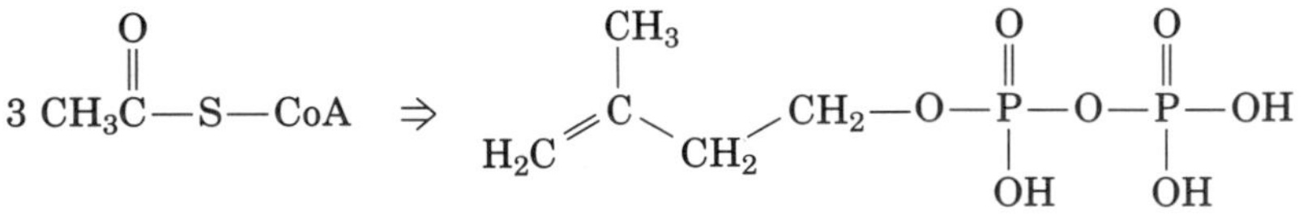

$$3\ CH_3\overset{O}{\overset{\|}{C}}—S—CoA \Rightarrow H_2C{=}C(CH_3)—CH_2—CH_2—O—\overset{O}{\overset{\|}{P}}(OH)—O—\overset{O}{\overset{\|}{P}}(OH)—OH$$

Isopentenyl pyrophosphate

and

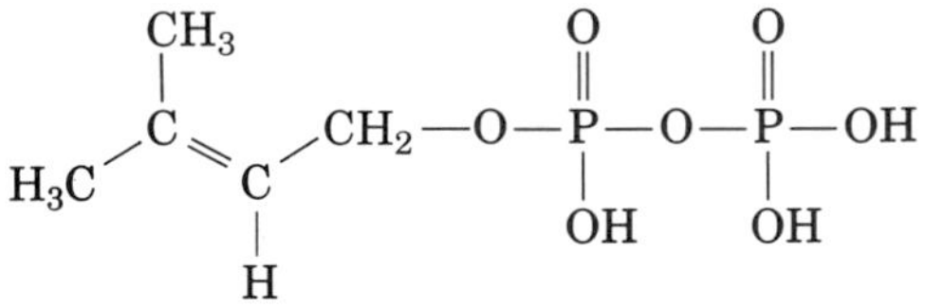

$$(CH_3)(H_3C)C{=}CH—CH_2—O—\overset{O}{\overset{\|}{P}}(OH)—O—\overset{O}{\overset{\|}{P}}(OH)—OH$$

Dimethylallyl pyrophosphate

Dimethylallyl pyrophosphate is an excellent alkylating agent since the primary, allylic pyrophosphate group (abbreviated OPP) is a good leaving group in nucleophilic substitution reactions (Section 11.5). We can therefore imagine a displacement (either $S_N1$ or $S_N2$ mechanism) of this leaving group by the nucleophilic double bond of isopentenyl pyrophosphate. Loss of a proton from the carbocationic reaction intermediate then leads to the head-to-tail coupled 10-carbon unit *geranyl pyrophosphate*. The corresponding alcohol, geraniol, is itself is a fragrant terpene that occurs in rose oil.

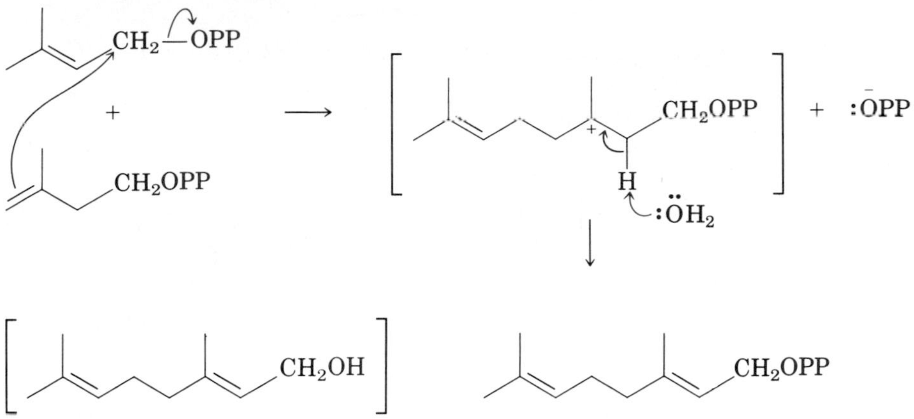

$CH_2$—OPP + $CH_2OPP$ ⟶ [ $CH_2OPP$ H :$\ddot{O}H_2$ ] + :$\bar{O}PP$

[ $CH_2OH$ ] Geraniol    $CH_2OPP$ Geranyl pyrophosphate

Geranyl pyrophosphate is the precursor of all other monoterpenes. The rough outlines of how the multitude of monoterpenes might arise can be rationalized by fundamental polar organic processes. For example, limonene, a monoterpene found in many citrus oils, can arise from geranyl pyrophosphate by a cis–trans double-bond isomerization followed by internal nucleophilic displacement of the pyrophosphate group and subsequent loss of a proton (Figure 28.6).

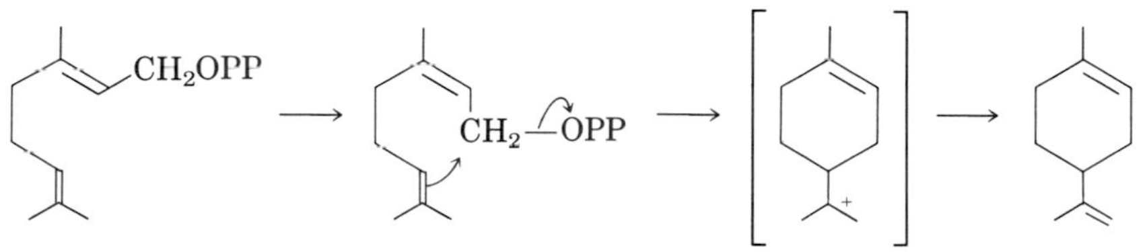

$CH_2OPP$ ⟶ $CH_2$—OPP ⟶ [ ] ⟶

Geranyl pyrophosphate    Neryl pyrophosphate    Limonene (from citrus oils)

**Figure 28.6** Biosynthesis of limonene from geranyl pyrophosphate

When geranyl pyrophosphate reacts with isopentenyl pyrophosphate, the 15-carbon *farnesyl pyrophosphate* results. Farnesyl pyrophosphate is the precursor of all sesquiterpenes. The corresponding alcohol farnesol is a terpene found in citronella oil and lemon oil.

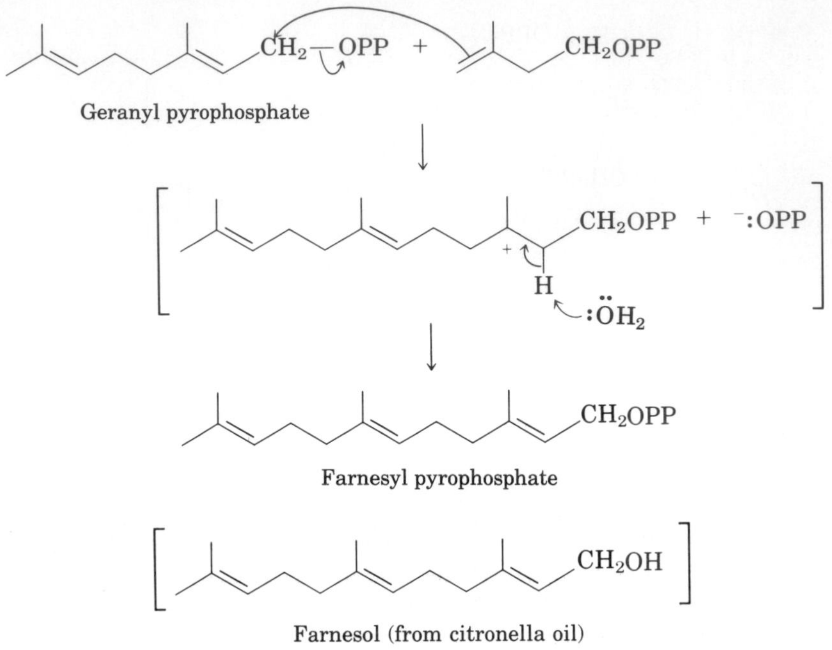

$CH_2$—OPP + $CH_2OPP$

Geranyl pyrophosphate

↓

[ $CH_2OPP$ + $^-$:OPP

H

:$\ddot{O}H_2$ ]

↓

$CH_2OPP$

Farnesyl pyrophosphate

[ $CH_2OH$ ]

Farnesol (from citronella oil)

Addition to farnesyl pyrophosphate of further isoprene units gives the 20-carbon (diterpene) and 25-carbon (sesterterpene) units. Triterpenes, however, arise biosynthetically by tail-to-tail coupling of two farnesyl pyrophosphates to give *squalene*, a 30-carbon hexaene (Figure 28.7). Squalene, found in high concentration in shark oil, is the precursor from which all other triterpenes and steroids arise.

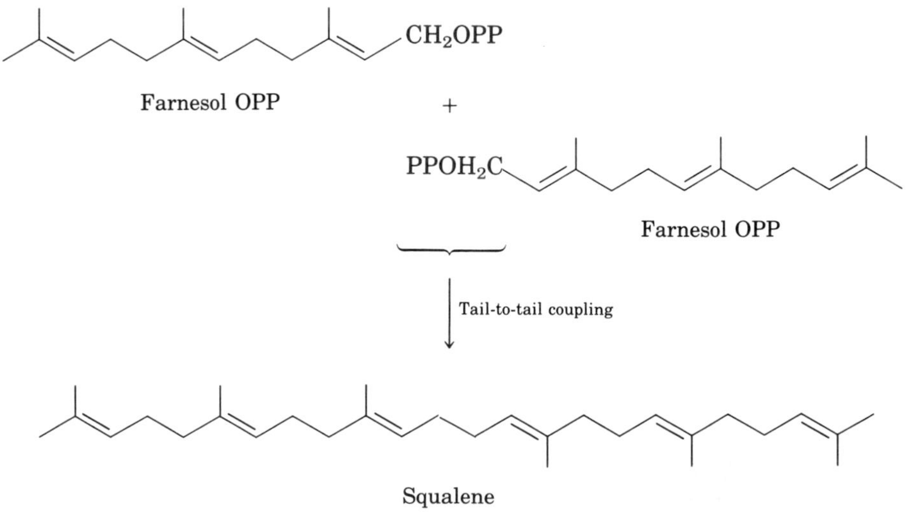

---

$CH_2OPP$

Farnesol OPP

+

$PPOH_2C$

Farnesol OPP

Tail-to-tail coupling

Squalene

**Figure 28.7** Biosynthesis of squalene, a $C_{30}$ precursor of triterpenes and steroids, by tail-to-tail coupling of two farnesyl pyrophosphate units

PROBLEM

**28.8** Propose a plausible pathway to account for the biosynthetic formation of γ-bisabolene from farnesyl pyrophosphate.

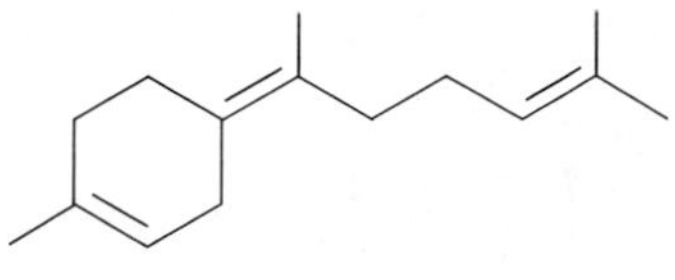

γ-Bisabolene

## 28.8 Steroids

In addition to fats, phospholipids, and terpenes, the lipid extracts of plants and animals also contain steroids. A **steroid** is an organic molecule whose structure is based on the tetracyclic ring system shown in Figure 28.8. The four rings are designated A, B, C, and D, beginning at the lower left, and the carbon atoms are numbered beginning in the A ring. Common examples are cholesterol, an animal steroid (and principal component of gallstones), and β-sitosterol, a ubiquitous plant steroid.

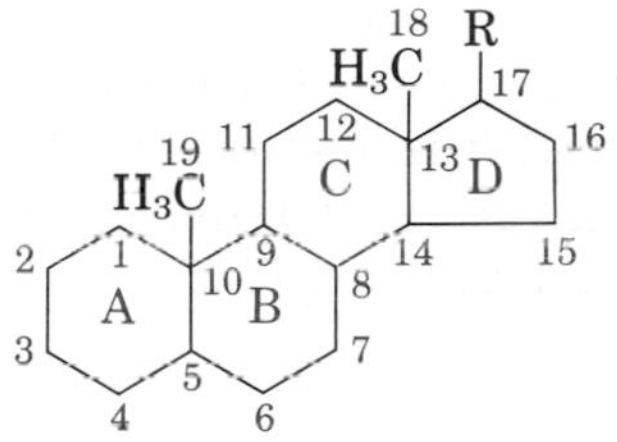

Steroid skeleton
(R = different side chains)

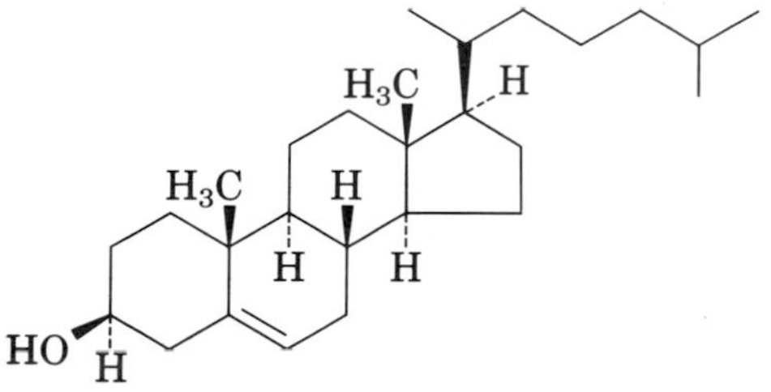

Cholesterol (animal sources)

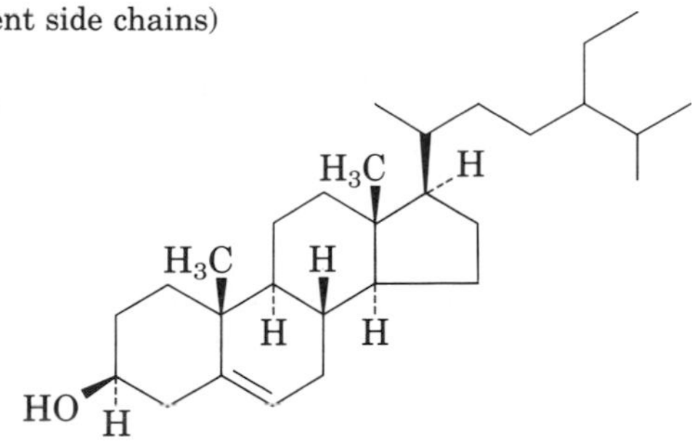

β-Sitosterol (plant sources)

**Figure 28.8** Some representative steroids

Steroids are widespread in both plant and animal kingdoms, and many have useful biological activity. For example, digitoxigenin, a plant steroid found in *Digitalis purpurea* (purple foxglove), is widely used medicinally as a heart stimulant; androsterone and estradiol are steroid sex hormones; and cortisone is a steroid hormone with anti-inflammatory properties.

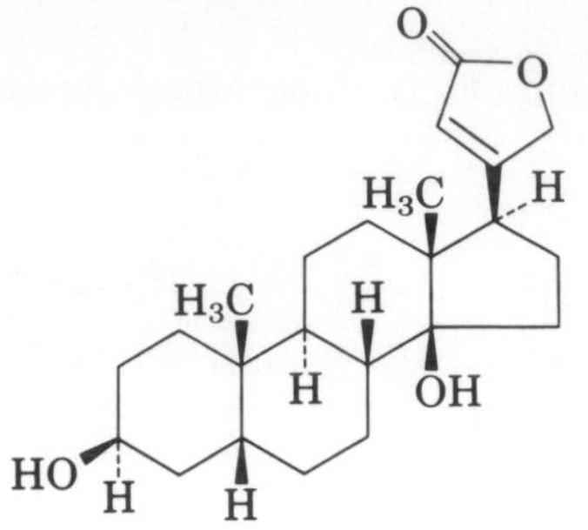

Digitoxigenin (heart stimulant)

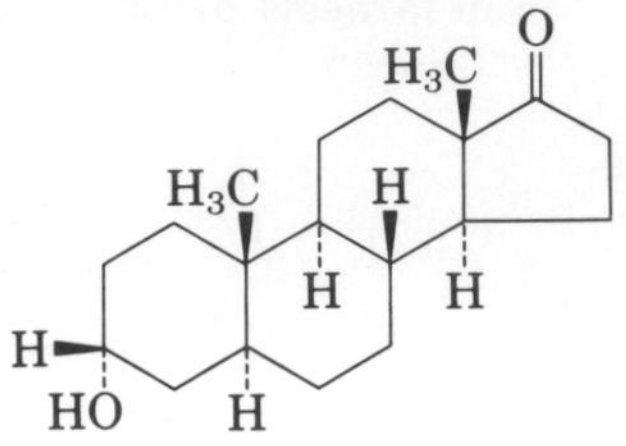

Androsterone (male sex hormone)

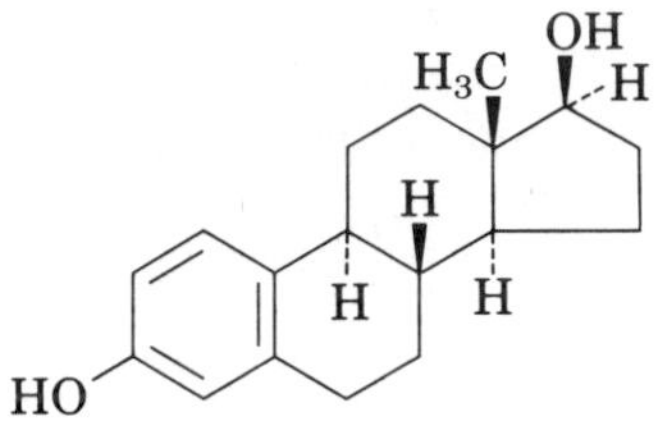

Estradiol (female sex hormone)

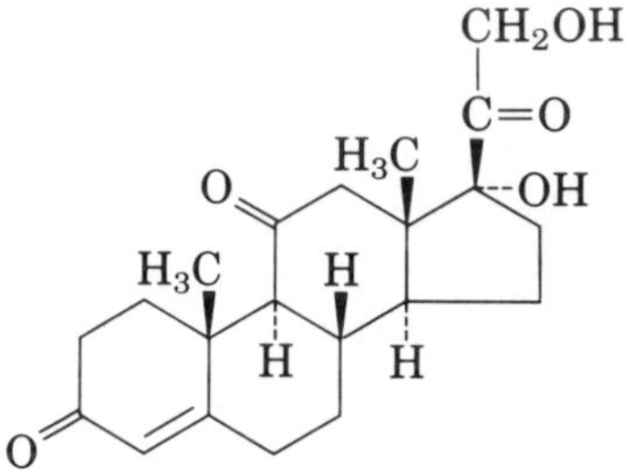

Cortisone (anti-inflammatory drug)

Many other steroids are produced synthetically by pharmaceutical companies. Even such nonnaturally occurring steroids as methandrostenolone (Dianabol, an anabolic or tissue-building steroid) and norethindrone (Norlutin, an oral contraceptive agent) have potent physiological effects.

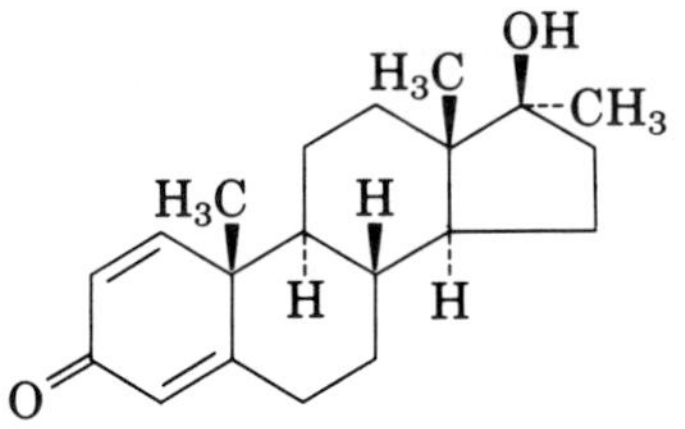

Methandrostenolone (anabolic)

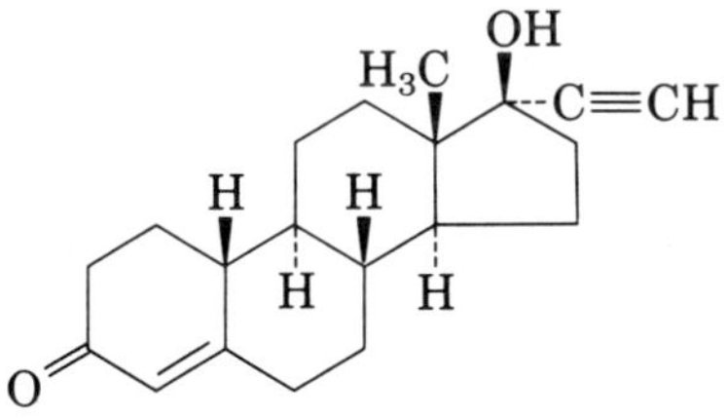

Norethindrone (oral contraceptive)

## 28.9 Stereochemistry of Steroids

The steroid skeleton is composed of four rings fused together with a specific stereochemistry. All three of the six-membered rings (rings A, B, and C) can adopt strain-free chair conformations, as indicated in Figure 28.9. Unlike simple cyclohexane rings, however, which can undergo chair–chair interconversions (Section 4.11), steroids are constrained to a rigid conformation and cannot undergo ring-flips.

An A,B trans steroid

An A,B cis steroid

**Figure 28.9** Steroid conformations: The three six-membered rings have chair conformations but are unable to undergo ring-flips.

Two cyclohexane rings can be joined in either a cis or a trans manner. In *cis*-decalin, both groups at the ring-junction positions (the *angular* groups) are on the same side of the two rings. In *trans*-decalin, however, the groups at the ring junctions are on opposite sides. These spatial relationships are best grasped by building molecular models of *cis*- and *trans*-decalin.

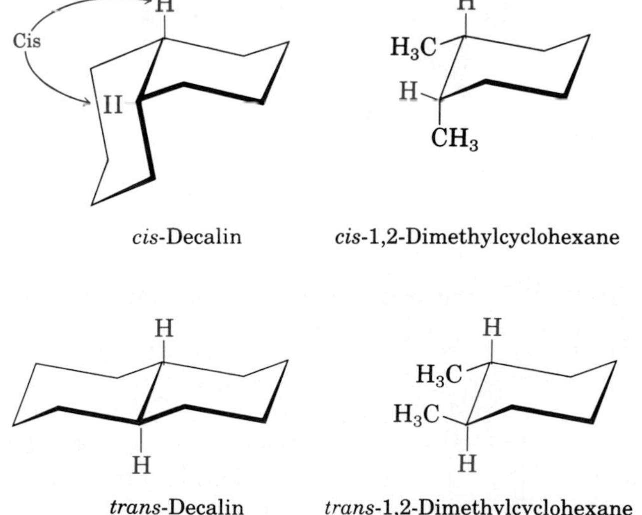

As indicated in Figure 28.9, steroids can have either a cis or a trans fusion of the A and B rings, but the other ring fusions (B–C and C–D) are usually trans. A,B trans steroids have the C19 angular methyl group "up" (denoted $\beta$) and the hydrogen atom at C5 "down" (denoted $\alpha$) on opposite sides of the molecule. By contrast, A,B cis steroids have both the C19 angular methyl group and the C5 hydrogen atom on the same side ($\beta$) of the molecule. Both kinds of steroids are relatively long, flat molecules that have their two methyl groups (C18 and C19) protruding axially above the ring system. The A,B trans steroids are by far the more common, though A,B cis steroids are found in liver bile.

Substituent groups on the steroid ring system may be either axial or equatorial. As was true for simple cyclohexanes (Section 4.12), equatorial substitution is generally more favorable than axial substitution for steric reasons. Thus, the hydroxyl group at C3 of cholesterol has the more stable equatorial orientation (Figure 28.10).

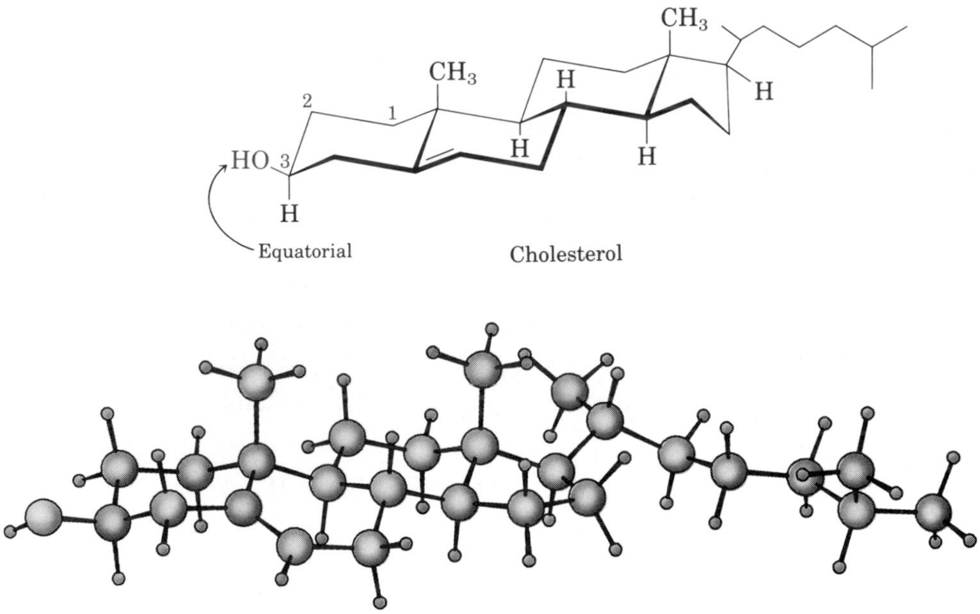

**Figure 28.10** The stereochemistry of cholesterol

**PROBLEM**

**28.9** Draw the following molecules in chair conformations, and indicate whether the ring substituents are axial or equatorial.

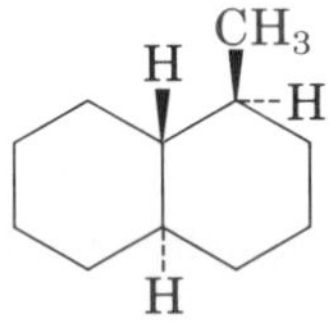

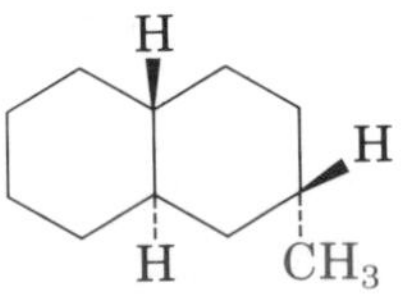

PROBLEM

**28.10** Lithocholic acid is an A,B cis steroid found in human bile. Draw lithocholic acid showing chair conformations as in Figure 28.9, and tell whether the hydroxyl group at C3 is axial or equatorial.

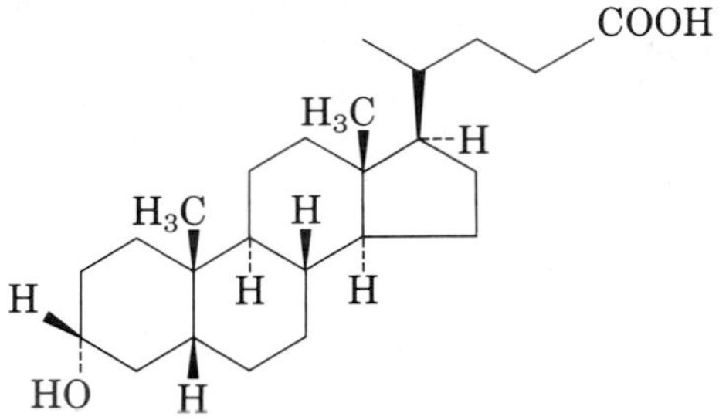

Lithocholic acid

## 28.10 Steroid Biosynthesis

Steroids, which can be thought of as heavily modified triterpenes, are biosynthesized in living organisms from squalene (Section 28.7). The exact pathway by which this remarkable transformation is accomplished is lengthy and complex, but most of the key steps have now been worked out, with notable contributions made by Konrad Bloch[3] and John Cornforth,[4] who received Nobel prizes for their accomplishments.

In essence, steroid biosynthesis occurs by enzyme-catalyzed epoxidation of squalene to yield squalene oxide, followed by acid-catalyzed cyclization and an extraordinary cascade of multiple carbocation rearrangements to yield lanosterol (Figure 28.11, page 1030). Lanosterol is then degraded by other enzymes to produce cholesterol, which is itself converted by other enzymes to produce a host of different steroids.

The exact series of carbocation rearrangements involved in the biosynthetic conversion of squalene to lanosterol involves the following steps:

*Step 1* The enzyme *squalene oxidase* selectively epoxidizes a terminal double bond of squalene to yield squalene oxide.

*Step 2* Squalene oxide is protonated, and the epoxide ring is opened by nucleophilic attack of the double bond six carbons away to yield a cyclic carbocation intermediate.

*Step 3* The tertiary carbocation intermediate produced in step 2 is attacked by another double bond six carbons away to yield a second carbocation intermediate.

*Step 4* A third cyclization occurs by attack of an appropriately positioned double bond on a carbocation.

*Step 5* A fourth and last cyclization takes place, this one giving a five-membered ring.

[3]Konrad E. Bloch (1912– ); b. Neisse, Germany; Ph.D. Columbia; professor, University of Chicago, Harvard University; Nobel prize in medicine (1964).

[4]John Warcup Cornforth (1917– ); b. Australia; Ph.D. Oxford (Robinson); National Institute of Medical Research (Great Britain); Nobel prize (1975).

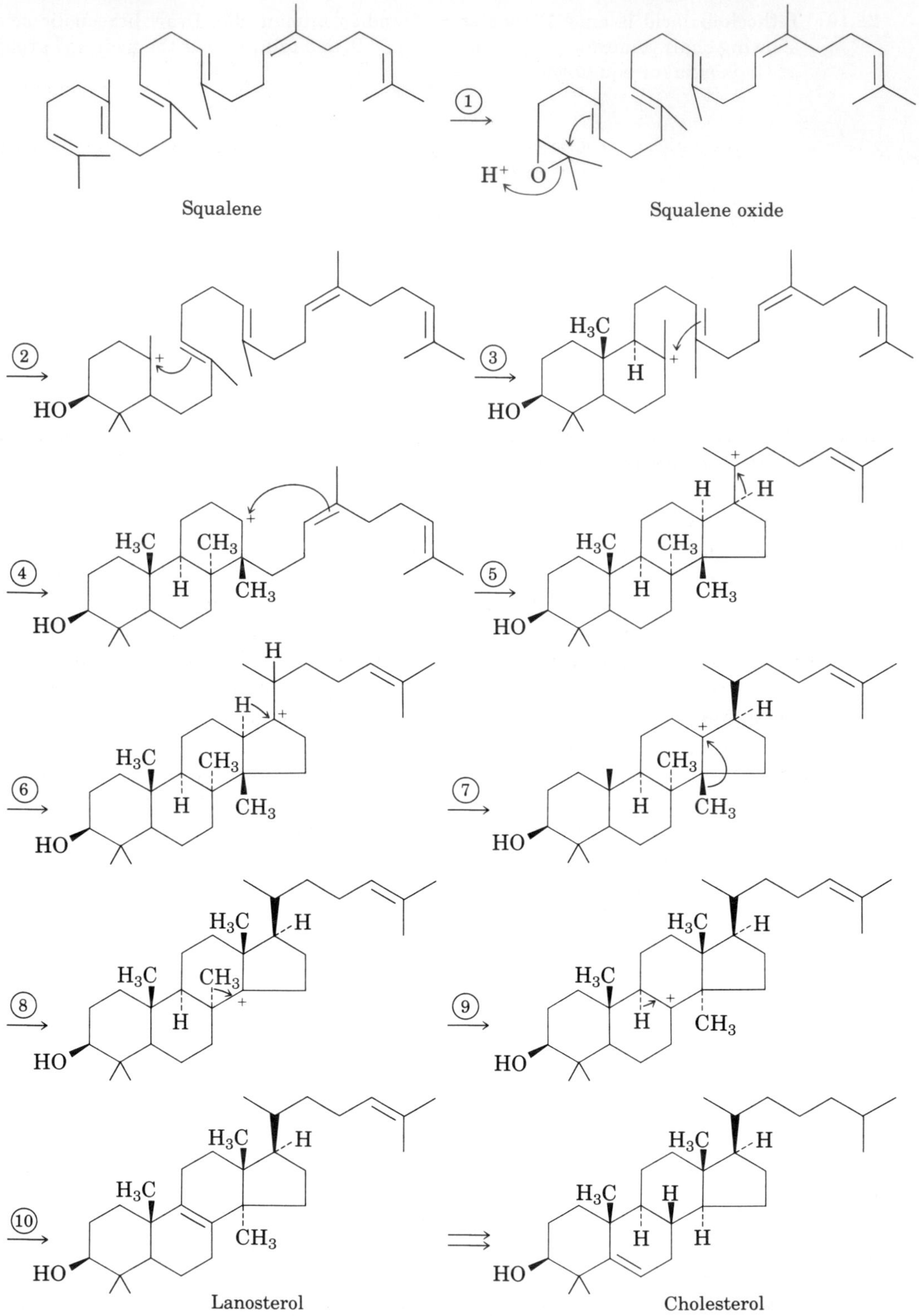

Squalene

Squalene oxide

Lanosterol

Cholesterol

**Figure 28.11** Biosynthesis of cholesterol from squalene

*Step 6* Carbocation rearrangement occurs by a hydride shift.

*Step 7* A second hydride shift gives still another carbocation.

*Step 8* Carbocation rearrangement occurs by shift of a methyl group.

*Step 9* A second methyl-group shift gives a final carbocation intermediate.

*Step 10* Loss of a proton (E1 reaction) from the carbon next to the cationic center gives lanosterol.

Although written in a stepwise way in Figure 28.11 for convenience, it is thought that the entire cyclization sequence (steps 2–5) takes place at one time without true intermediates being formed. Similarly, the carbocation rearrangements and proton loss (steps 6–10) take place at essentially the same time without intermediates.

PROBLEM

**28.11** Examine the structures of lanosterol and cholesterol and catalog the changes that have occurred in the transformation.

## 28.11 Summary and Key Words

**Lipids** are the naturally occurring materials isolated from plant and animal cells by extraction with organic solvents. They usually have large hydrocarbon portions in their structure. **Animal fats** and **vegetable oils** are the most widely occurring lipids. Both are triesters of glycerol with long-chain **fatty acids**, but animal fats are usually saturated, whereas vegetable oils usually have unsaturated fatty acid residues.

**Phosphoglycerides** such as **lecithin** and **cephalin** are closely related to fats. The glycerol backbone in these molecules is esterified to two fatty acids (one saturated and one unsaturated) and to one phosphate ester. **Sphingolipids**, another major class of phospholipids, have an amino alcohol such as sphingosine for their backbone. These compounds are important constituents of cell membranes.

Fatty acids are biosynthesized in nature by condensation of enzyme-bound two-carbon acetate units. The overall scheme is lengthy, but the individual steps are the fundamental organic reactions expected of the functional groups involved.

**Prostaglandins** and **terpenes** are still other classes of lipids. Prostaglandins, which are simple lipids found in all body tissues, have a wide range of physiological actions. Terpenes are often isolated from the essential oils of plants. They have an immense diversity of structure and are produced biosynthetically by head-to-tail coupling of the five-carbon "isoprene equivalents," isopentenyl pyrophosphate and dimethylallyl pyrophosphate.

**Steroids** are plant and animal lipids with a characteristic tetracyclic carbon skeleton. Like the prostaglandins, steroids occur widely in body tissue and have a large variety of physiological activities. Steroids are closely related to terpenes and arise biosynthetically from the triterpene precursor lanosterol. Lanosterol, in turn, arises from cyclization of the acyclic hydrocarbon squalene.

## ADDITIONAL PROBLEMS

**28.12** Fats can be either optically active or optically inactive, depending on their structure. Draw the structure of an optically active fat that yields 2 equiv of stearic acid and 1 equiv of oleic acid on hydrolysis. Draw the structure of an optically inactive fat that yields the same products.

**28.13** Show the products you would expect to obtain from reaction of glyceryl trioleate with the following:
(a) Excess $Br_2$ in $CCl_4$
(b) $H_2/Pd$
(c) $NaOH/H_2O$
(d) $O_3$, then $Zn/CH_3COOH$
(e) $LiAlH_4$, then $H_3O^+$
(f) $CH_3MgBr$, then $H_3O^+$
(g) $NaOH/H_2O$, then $CH_2N_2$ (diazomethane)

**28.14** Eleostearic acid, $C_{18}H_{30}O_2$, is a rare fatty acid found in the tung oil used for furniture finishing. On ozonolysis followed by treatment with zinc, eleostearic acid furnishes one part pentanal, two parts glyoxal (OHC—CHO), and one part 9-oxononanoic acid [$OHC(CH_2)_7COOH$]. What is the structure of eleostearic acid?

**28.15** Draw representative structures for:
(a) A fat
(b) A prostaglandin
(c) A steroid

**28.16** How would you convert oleic acid into these substances?
(a) Methyl oleate
(b) Methyl stearate
(c) Nonanal
(d) Nonanedioic acid
(e) 9-Octadecynoic acid (stearolic acid)
(f) 2-Bromostearic acid
(g) 18-Pentatriacontanone, $CH_3(CH_2)_{16}\overset{\overset{\displaystyle O}{\|}}{C}(CH_2)_{16}CH_3$

**28.17** How would you synthesize stearolic acid [Problem 28.16(e)] from 1-decyne and 1-chloro-7-iodoheptane?

**28.18** Vaccenic acid, $C_{18}H_{34}O_2$, is a rare fatty acid that gives heptanal and 11-oxoundecanoic acid [$OHC(CH_2)_9COOH$] on ozonolysis followed by zinc treatment. When allowed to react with $CH_2I_2/Zn(Cu)$, vaccenic acid is converted into lactobacillic acid. What are the structures of vaccenic and lactobacillic acids?

**28.19** Show the location of the isoprene units in these terpenes:

(a) 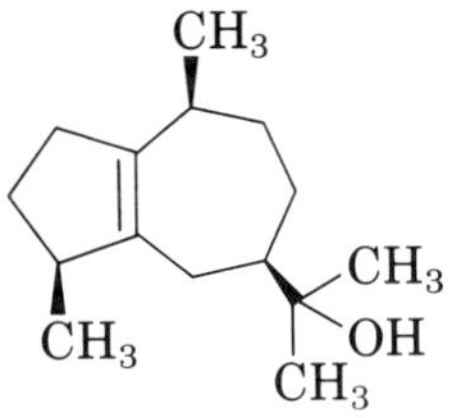


Guaiol

(b) 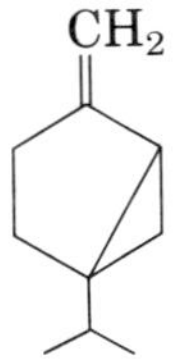


Sabinene

(c) 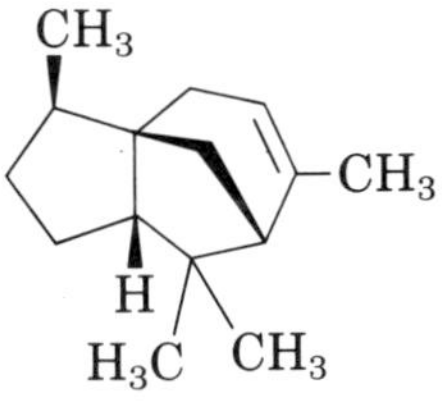


Cedrene

**28.20** Indicate by asterisks the chiral centers present in each of the three terpenes shown in Problem 28.19. How many stereoisomers of each are theoretically possible?

**28.21** Assume that the three terpenes in Problem 28.19 are derived biosynthetically from isopentenyl pyrophosphate and dimethylallyl pyrophosphate, each of which was isotopically labeled at the pyrophosphate-bearing carbon atom (C1). At what positions would the terpenes be isotopically labeled?

**28.22** Suggest a mechanistic pathway by which α-pinene might arise biosynthetically from geranyl pyrophosphate.

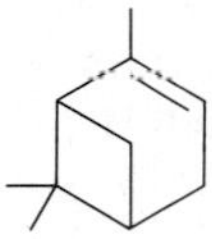

α-Pinene

**28.23** Suggest a mechanism by which ψ-ionone is transformed into β-ionone on treatment with acid.

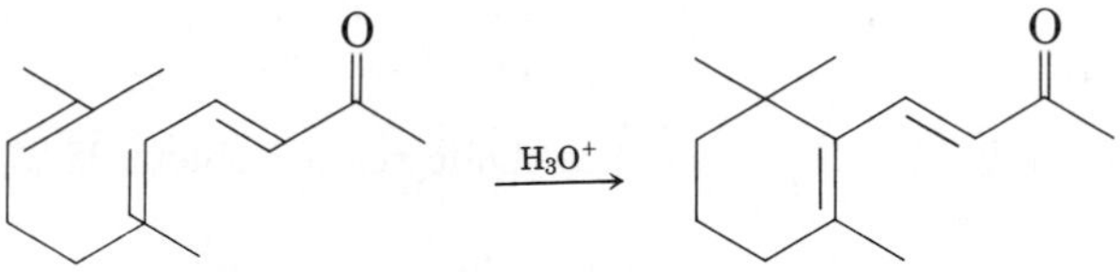

$H_3O^+$

ψ-Ionone β-Ionone

**28.24** Which isomer would you expect to be more stable—*cis*-decalin or *trans*-decalin? Explain. (You might want to review Section 4.13.)

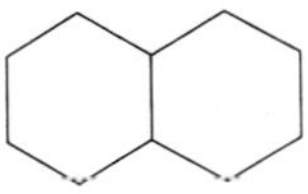

Decalin

**28.25** Draw the most stable chair conformation of dihydrocarvone.

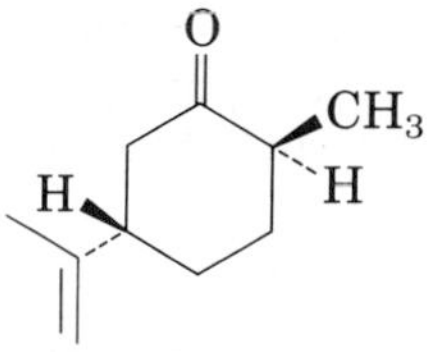

O $CH_3$ H H

Dihydrocarvone

**28.26** Draw the most stable chair conformation of menthol and label each substituent as axial or equatorial.

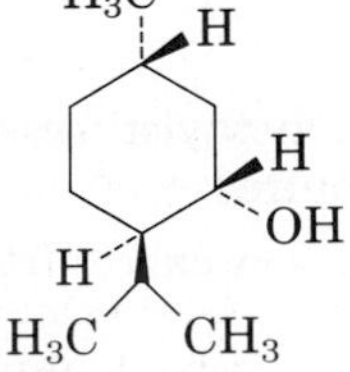

$H_3C$ H H OH H $H_3C$ $CH_3$

Menthol (from peppermint oil)

**28.27** Cholic acid, a major steroidal constituent of human bile, has the structure shown. Draw a conformational structure of cholic acid and label the three hydroxyl groups as axial or equatorial.

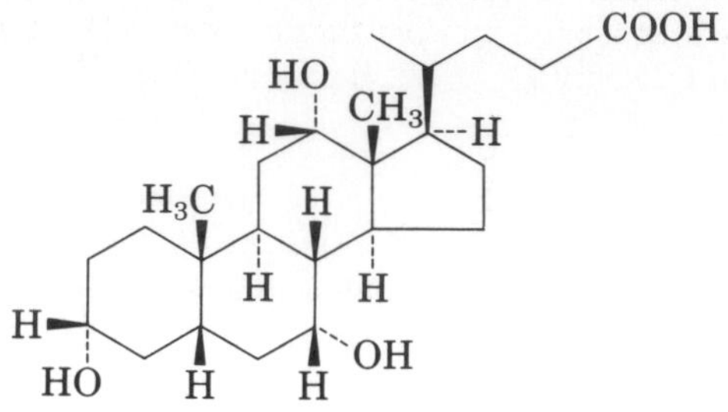

Cholic acid

**28.28** How many chiral centers does cholic acid (Problem 28.27) have? How many stereoisomers are possible?

**28.29** Show the products you would expect to obtain from reaction of cholic acid with these reagents:
(a) $C_2H_5OH$, HCl (b) Excess pyridinium chlorochromate in $CH_2Cl_2$
(c) $BH_3$ in THF, then $H_3O^+$

**28.30** As a general rule, equatorial alcohols are esterified more readily than axial alcohols. What product would you expect to obtain from reaction of these two compounds with 1 equiv of acetic anhydride?

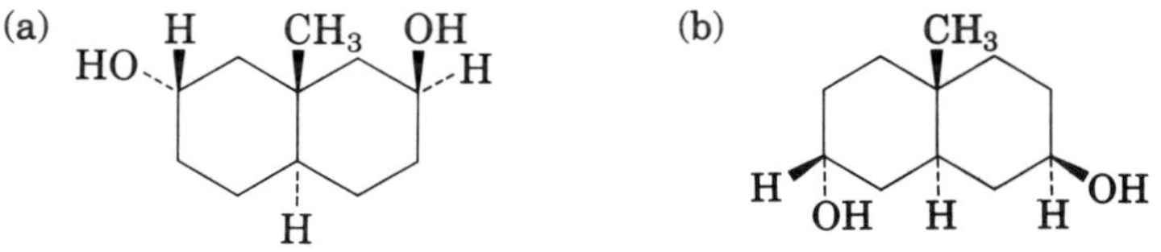

(a) (b)

**28.31** Diethylstilbestrol (DES) exhibits estradiol-like activity even though it is structurally unrelated to steroids. Once used widely as an additive in animal feed, DES has been implicated as a causative agent in several types of cancer. Look up the structure of estradiol (Section 28.8) and show how DES can be drawn so that it is sterically similar to estradiol.

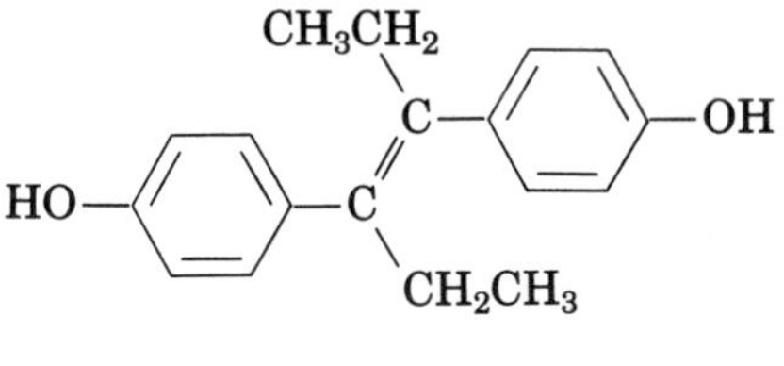

Diethylstilbestrol

**28.32** Propose a synthesis of diethylstilbestrol (Problem 28.31) from phenol and any other organic compound required.

**28.33** What products would you expect from reaction of estradiol (Problem 28.31) with these reagents?
(a) NaH, then $CH_3I$ (b) $CH_3COCl$, pyridine
(c) $Br_2$, $FeBr_3$ (d) Pyridinium chlorochromate in $CH_2Cl_2$

**28.34** Cembrene, $C_{20}H_{32}$, is a diterpene hydrocarbon isolated from pine resin. Cembrene has an ultraviolet absorption at 245 nm, but dihydrocembrene ($C_{20}H_{34}$), the product of hydrogenation with 1 equiv of hydrogen, has no ultraviolet absorption. On exhaustive hydrogenation, 4 equiv of hydrogen react, and octahydrocembrene, $C_{20}H_{40}$, is produced. On ozonolysis of cembrene, followed by treatment of the ozonide with zinc, four carbonyl-containing products are obtained:

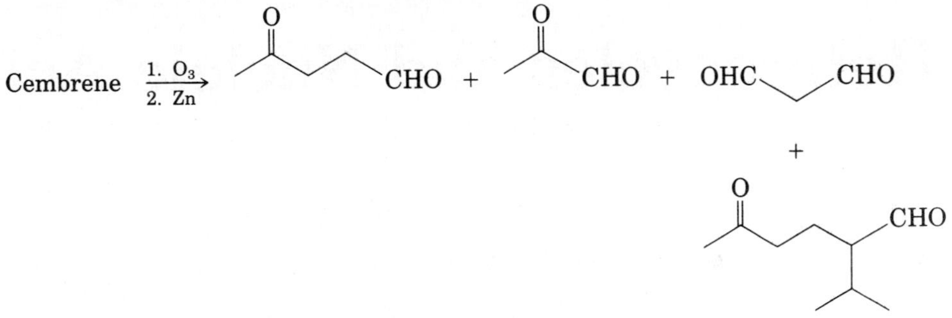

Cembrene $\xrightarrow[2.\ Zn]{1.\ O_3}$ O, CHO + O, CHO + OHC, CHO + O, CHO

Propose a suitable structure for cembrene that is consistent with the isoprene rule.

CHAPTER 29

# Heterocycles and Nucleic Acids

Cyclic organic compounds are classified either as **carbocycles** or as **heterocycles**. Carbocyclic rings contain only carbon atoms, but heterocyclic rings contain one or more different atoms in addition to carbon. Nitrogen, oxygen, and sulfur are the most common heteroatoms, but many others are also possible.

Heterocyclic compounds are common in organic chemistry, and many have important biological properties. For example, the antibiotic penicillin, the antiulcer agent cimetidine, the sedative phenobarbital, and the non-nutritive sweetener saccharin are all heterocycles.

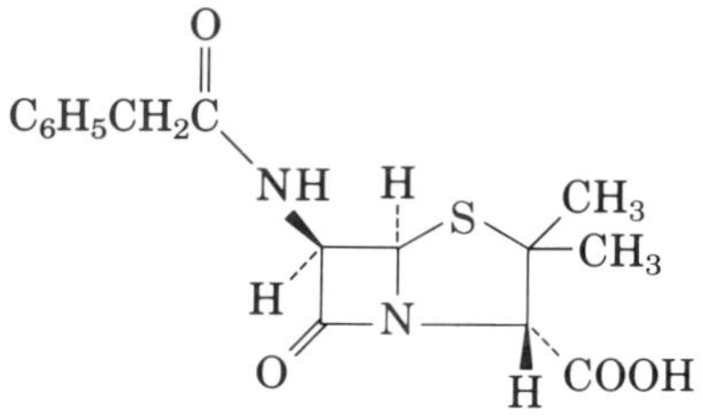

$C_6H_5CH_2C(=O)NH$ … $CH_3$, $CH_3$, S, N, $COOH$

Penicillin G
(an antibiotic)

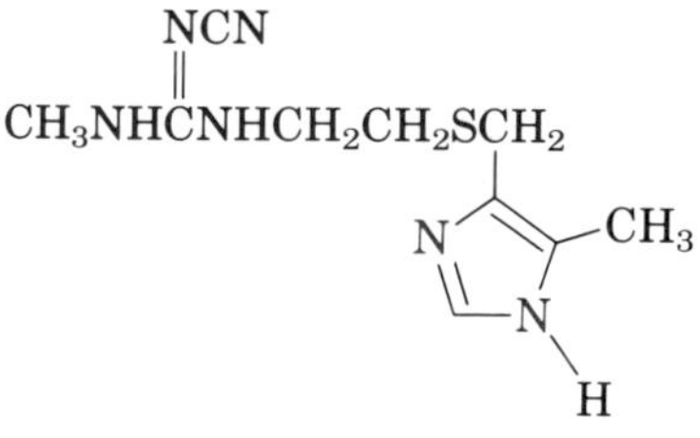

$CH_3NHC(=NCN)NHCH_2CH_2SCH_2$ … $CH_3$, N, N–H

Cimetidine
(an antiulcer agent)

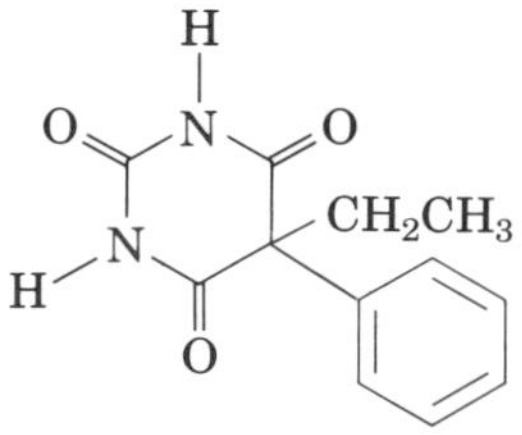

Phenobarbital
(a sedative)

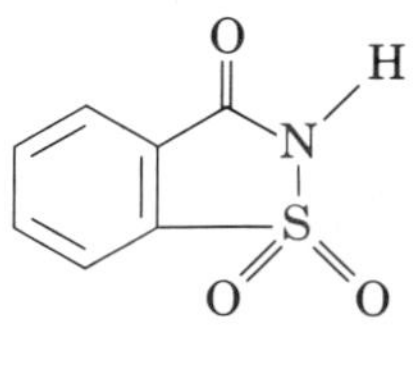

Saccharin
(an artificial sweetener)

Heterocycles aren't new to us; we've encountered them many times in previous chapters, usually without comment. Thus, epoxides (three-membered-ring ethers, Section 18.7), lactones (cyclic esters), and lactams (cyclic amides) are heterocycles, as are the solvents tetrahydrofuran (a cyclic ether) and pyridine (a cyclic amine). In addition, most carbohydrates exist as heterocyclic hemiacetals (Section 24.5).

Most heterocycles have the same chemistry as their open-chain counterparts: Lactones and acyclic esters behave similarly, lactams and acyclic amides behave similarly, and cyclic and acyclic ethers behave similarly. In certain cases, however, particularly when the ring is unsaturated, heterocycles have unique and interesting properties. Let's look first at the five-membered unsaturated heterocycles.

## 29.1 Five-Membered Unsaturated Heterocycles

Pyrrole, furan, and thiophene are the simplest five-membered unsaturated heterocycles. Each of the three has two double bonds and one heteroatom (N, O, or S).

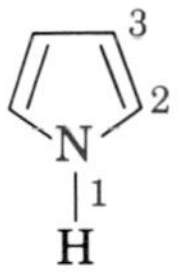

Pyrrole 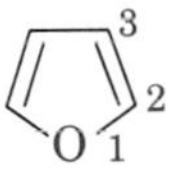 Furan 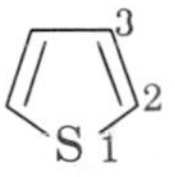 Thiophene

Pyrrole is obtained commercially either by distillation of coal tar or by treatment of furan with ammonia over an alumina catalyst at 400°C.

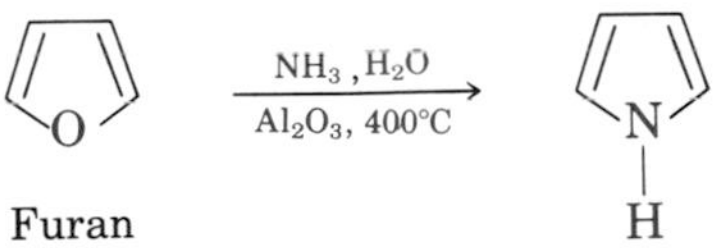

$$\text{Furan} \xrightarrow[Al_2O_3,\ 400°C]{NH_3,\ H_2O} \text{Pyrrole}$$

Furan is synthesized by catalytic loss of carbon monoxide (decarbonylation) from furfural, which is itself prepared by acidic dehydration of the pentoses found in oat hulls and corncobs.

$$\underset{\text{Pentose mixture}}{C_5H_{10}O_5} \xrightarrow{H_3O^+} \text{Furfural} \xrightarrow[280°C]{\text{Ni catalyst}} \text{Furan} + CO$$

Thiophene is found in small amounts in coal-tar distillates and is synthesized industrially by cyclization of butane or butadiene with sulfur at 600°C.

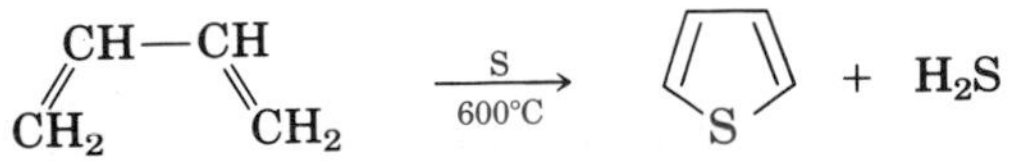

$$\underset{\text{1,3-Butadiene}}{CH_2{=}CH{-}CH{=}CH_2} \xrightarrow[600°C]{S} \text{Thiophene} + H_2S$$

All three of these heterocycles are liquid at room temperature, as indicated by the data on physical properties in Table 29.1.

**Table 29.1** Physical properties of some five-membered heterocycles

| *Name* | *Molecular weight* | *Melting point (°C)* | *Boiling point (°C)* |
|---|---|---|---|
| Furan | 68 | −85 | 31 |
| Pyrrole | 67 | −23 | 130 |
| Thiophene | 84 | −38 | |

The chemistry of these three heterocyclic ring systems contains some surprises. For example, pyrrole is both an amine and a conjugated diene, yet its chemical properties are not consistent with either of these structural features. Unlike most other amines, pyrrole is not basic; unlike most other conjugated dienes, pyrrole undergoes electrophilic substitution rather than addition reactions. The same is true of furan and thiophene: Both tend to react with electrophiles to give substitution products.

How can we explain these observations?

## 29.2 Structures of Pyrrole, Furan, and Thiophene

In fact, pyrrole, furan, and thiophene are *aromatic*. Each has six pi electrons ($4n + 2$, where $n = 1$) in a cyclic conjugated system. To choose pyrrole as an example, each of the four carbon atoms of pyrrole contributes one pi electron, and the $sp^2$-hybridized nitrogen atom contributes two (its lone pair). The six pi electrons occupy $p$ orbitals, with lobes above and below the plane of the ring, as shown in Figure 29.1. Overlap of the five $p$ orbitals forms aromatic molecular orbitals just as in benzene (Section 15.9).

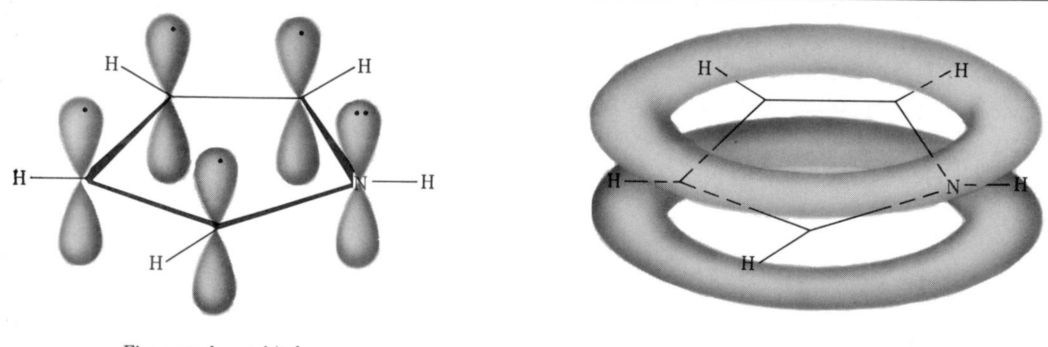


**Figure 29.1** Pi bonding in pyrrole, an aromatic heterocycle with six pi electrons

Note that the pyrrole nitrogen atom uses all five of its valence electrons in bonding. Three electrons are used in forming three sigma bonds (two to carbon and one to hydrogen), and the two lone-pair electrons are involved in aromatic pi bonding, as the following resonance structures indicate:

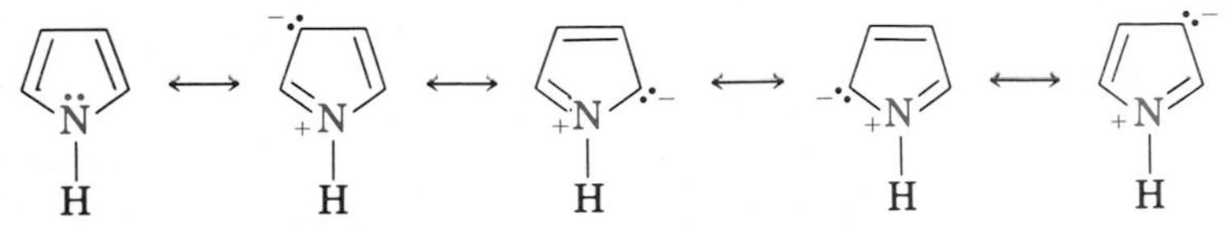

Because the nitrogen lone pair is a part of the aromatic sextet, it is less available for bonding to electrophiles, and pyrrole is therefore less basic ($pK_a$ of pyrrolinium ion $\approx$ 0) and less nucleophilic than aliphatic amines. By the same token, however, the carbon atoms of pyrrole are much *more* electron-rich and much more nucleophilic than typical double-bond carbon atoms. The pyrrole ring is therefore highly reactive toward electrophiles in the same way that activated benzene rings are reactive.

**PROBLEM**

**29.1** Draw an orbital picture of furan. Assume that the oxygen atom is $sp^2$ hybridized and show the orbitals that the two oxygen lone pairs are occupying.

**PROBLEM**

**29.2** Pyrrole has a dipole moment of 1.8 D. Look at the several possible resonance structures of pyrrole and predict which is the positive end and which is the negative end of the dipole.

## 29.3 Electrophilic Substitution Reactions of Pyrrole, Furan, and Thiophene

The chemistry of pyrrole, furan, and thiophene is similar to that of activated benzenoid aromatic rings. Like benzene, the five-membered aromatic heterocycles undergo electrophilic substitution rather than addition reactions. In general, these heterocycles are much more reactive toward electrophiles than benzene rings, and low-temperature reaction conditions are often necessary to prevent destruction of the starting material. Halogenation, nitration, sulfonation, and Friedel–Crafts acylation can all be accomplished if the proper reaction conditions are chosen. A reactivity order of furan > pyrrole > thiophene is normally found.

Bromination: Furan + $Br_2$ $\xrightarrow[0°C]{\text{Dioxane}}$ 2-Bromofuran (90%) + HBr

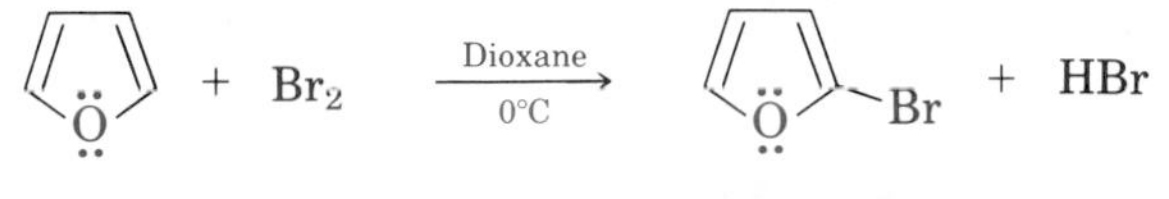

Nitration: Pyrrole + $HNO_3$ $\xrightarrow{\text{Acetic anhydride}}$ 2-Nitropyrrole (83%) + $H_2O$

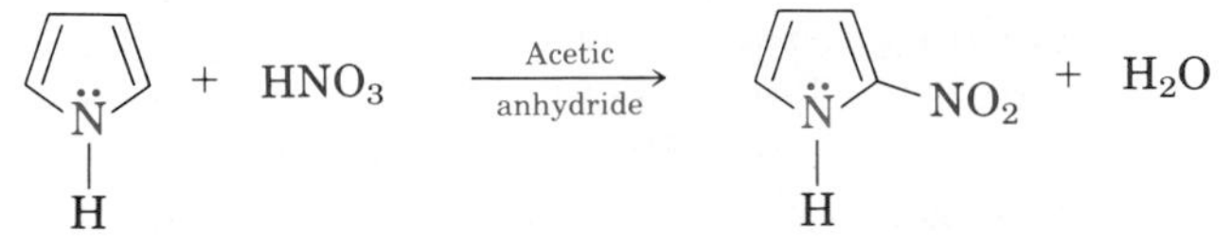

Friedel–Crafts acylation

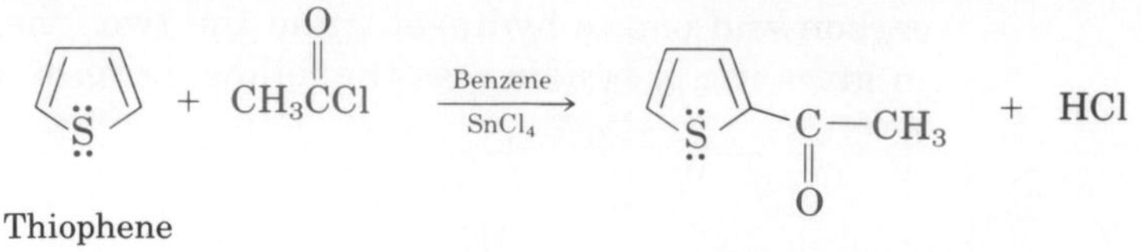

Thiophene + $CH_3CCl$ (O) $\xrightarrow[SnCl_4]{Benzene}$ 2-Acetylthiophene (83%) + HCl

Electrophilic substitution of these aromatic heterocycles normally occurs at C2, the position next to the heteroatom, because C2 is the most electron-rich (most nucleophilic) position on the ring. Another way of saying the same thing is to note that electrophilic attack at C2 leads to a more stable intermediate cation, with three resonance forms, than attack at C3, which results in a cation with only two resonance forms (Figure 29.2).

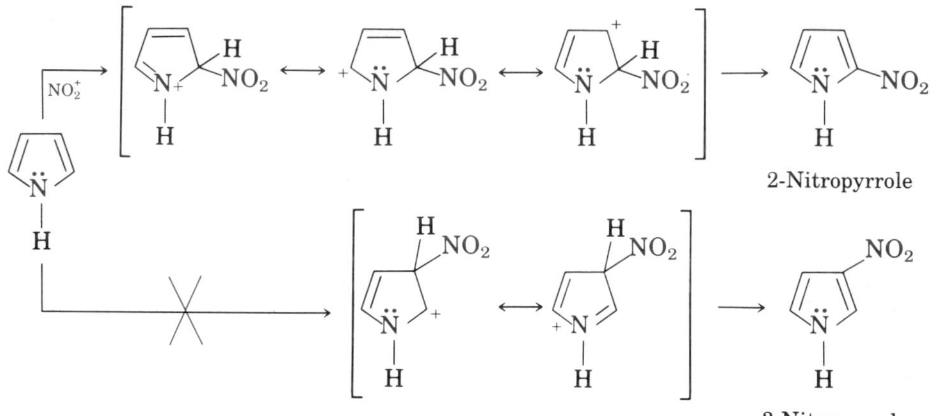

$NO_2^+$

2-Nitropyrrole

3-Nitropyrrole
*(not formed)*

**Figure 29.2** Electrophilic nitration of pyrrole: The intermediate produced by reaction at C2 is more stable than that produced by reaction at C3.

PROBLEM

**29.3** Propose a mechanism to account for the fact that treatment of pyrrole with deuteriosulfuric acid, $D_2SO_4$, leads to formation of 2-deuteriopyrrole.

## 29.4 Pyridine, a Six-Membered Heterocycle

Pyridine, obtained commercially by distillation of coal tar, is the nitrogen-containing heterocyclic analog of benzene. Like benzene, pyridine is aromatic. It is a flat molecule with bond angles of 120° and with carbon–carbon bond lengths of 1.39 Å, intermediate between normal single and double bonds. Each of the five carbon atoms contributes one pi electron, and the $sp^2$-hybridized nitrogen atom also contributes one pi electron to complete the aromatic sextet. Unlike the situation in pyrrole, however, the lone pair of electrons on the pyridine nitrogen atom is not involved in bonding but occupies an $sp^2$ orbital in the plane of the ring (Figure 29.3).

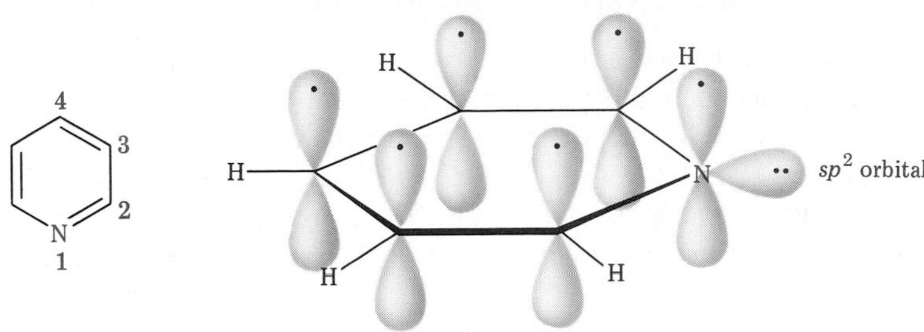


**Figure 29.3** Electronic structure of pyridine, a six-pi-electron nitrogen-containing analog of benzene

Pyridine is a stronger base than pyrrole ($pK_a = 5.25$ for pyridinium ion versus 0 for pyrrolinium ion) because of its electronic structure. The pyridine nitrogen's lone-pair electrons are not involved in aromatic pi bonding but are instead available for donation to a Lewis acid. Pyridine is a weaker base than aliphatic amines, however ($pK_a = 5.25$ versus 11), because of its hybridization. The lone-pair electrons in the $sp^3$-hybridized nitrogen orbital ($\frac{1}{4}$ $s$ character) of an aliphatic amine are held less closely to the positively charged nucleus than lone-pair electrons in an $sp^2$ hybrid orbital ($\frac{1}{3}$ $s$ character), with the result that $sp^3$-hybridized nitrogen is more basic.

Relative basicity: Pyrrolidine > Pyridine > Pyrrole

| Pyrrolidine | Pyridine | Pyrrole |
|---|---|---|
| ($pK_a = 11.27$) | ($pK_a = 5.25$) | ($pK_a = 0$) |

PROBLEM ..........

**29.4** Imidazolium ion has $pK_a = 6.95$. Draw an orbital structure of imidazole and indicate which nitrogen is more basic.

Imidazole

## 29.5 Electrophilic Substitution of Pyridine

The pyridine ring undergoes electrophilic aromatic substitution reactions only with great difficulty. Halogenation and sulfonation can be carried out under drastic conditions, but nitration occurs in very low yield, and Friedel–Crafts reactions are not successful. Reactions usually give the 3-substituted product.

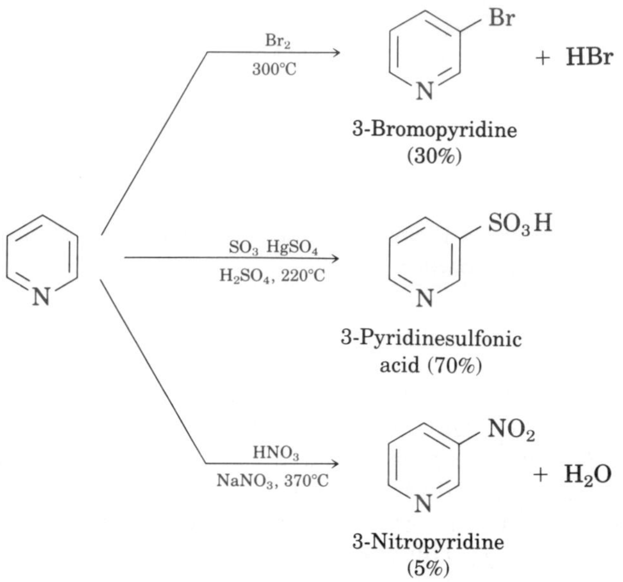

$Br_2$, 300°C → 3-Bromopyridine (30%) + HBr

$SO_3$ $HgSO_4$, $H_2SO_4$, 220°C → 3-Pyridinesulfonic acid (70%)

$HNO_3$, $NaNO_3$, 370°C → 3-Nitropyridine (5%) + $H_2O$

The low reactivity of pyridine toward electrophilic aromatic substitution is due to a combination of factors. Most important is that the electron density of the ring is decreased by the electron-withdrawing inductive effect of the electronegative nitrogen atom. Thus, pyridine has a dipole moment of 2.26 D, with the ring acting as the positive end of the dipole. Electrophilic attack on the positively polarized ring is therefore difficult.

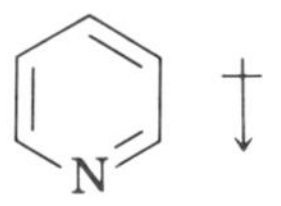

$\mu = 2.26$ D

A second factor decreasing the reactivity of the pyridine ring toward electrophilic attack is that acid–base complexation between the basic ring nitrogen atom and the attacking electrophile places a positive charge on the ring, further deactivating it.

**PROBLEM**

**29.5** Electrophilic aromatic substitution reactions of pyridine normally occur at C3. Draw the carbocation intermediates resulting from electrophilic attack at all possible positions, and explain the observed result.

## 29.6 Nucleophilic Substitution of Pyridine

In contrast to their lack of reactivity toward electrophilic substitution, certain substituted pyridine rings undergo *nucleophilic* aromatic substitution with relative ease. Both 2- and 4-halo-substituted (but not 3-substituted) pyridines react particularly well.

Cl → $\xrightarrow[C_2H_5OH]{Na^+ \ ^-:\ddot{O}C_2H_5}$ $OCH_2CH_3$ + $:\ddot{Cl}:^-$

4-Chloropyridine — 4-Ethoxypyridine (75%)

Br → $\xrightarrow[NH_3]{NaNH_2}$ $NH_2$ + $:\ddot{Br}:^-$

2-Bromopyridine — 2-Aminopyridine (67%)

These reactions are typical nucleophilic aromatic substitutions, analogous to those we saw earlier for halo-substituted benzenes (Section 16.9). Although a benzene ring needs to be further activated by the presence of electron-withdrawing substituents for nucleophilic substitution to occur, pyridines are already sufficiently activated. Reaction occurs by addition of the nucleophile to the C=N bond, followed by loss of halide ion from the anion intermediate.

N, Cl + $:\ddot{N}H_2^-$ ⟶ [ N, $NH_2$, Cl ] ⟶ N, $NH_2$ + $:\ddot{Cl}:^-$

This nucleophilic aromatic substitution is in some ways analogous to the nucleophilic acyl substitution of acid chlorides (Section 21.5). In both cases, the initial addition step is favored by the ability of the electronegative atom (nitrogen or oxygen) to stabilize the anion intermediate. The intermediate then expels chloride ion to yield the substitution product.

N, Cl $\xrightarrow{^-:\ddot{N}H_2}$ [ N, $NH_2$, Cl ] ⟶ N, $\ddot{N}H_2$

2-Chloropyridine — Stabilized anion — 2-Aminopyridine

O, Cl $\xrightarrow{^-:\ddot{N}H_2}$ [ :O, $\ddot{N}H_2$, Cl ] ⟶ O, $\ddot{N}H_2$

Acid chloride — Stabilized anion — Amide

PROBLEM

**29.6** Draw the anion intermediates expected from nucleophilic attack at C4 of a 4-halopyridine and at C3 of a 3-halopyridine. How can you account for the fact that substitution of the 4-halopyridine occurs readily, but the 3-halopyridine does not react?

PROBLEM

**29.7** If 3-bromopyridine is heated with $NaNH_2$ under forcing conditions, a mixture of 3- and 4-aminopyridine is obtained. Explain. [*Hint:* See Section 16.10.]

## 29.7 Fused-Ring Heterocycles

Quinoline, isoquinoline, and indole are **fused-ring heterocycles** that contain both a benzene ring and a heterocyclic aromatic ring. All three ring systems occur commonly in natural products, and many members of the class have pronounced biological activity. Thus, the quinoline alkaloid quinine is widely used as an antimalarial drug, and the indole alkaloid *N,N*-dimethyltryptamine is a powerful hallucinogen.

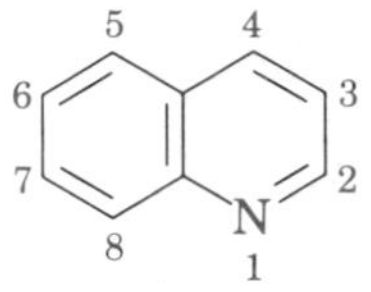

Quinoline

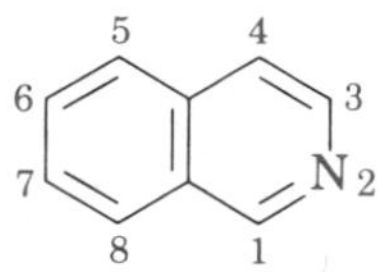

Isoquinoline

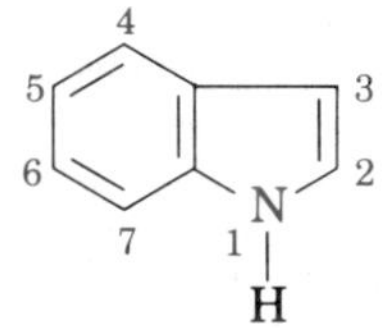

Indole

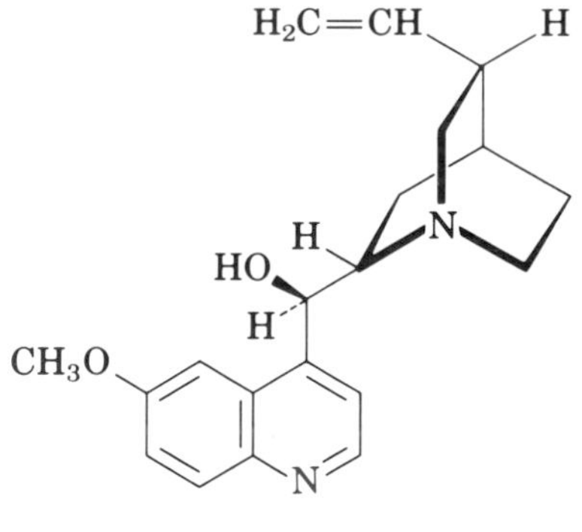

Quinine, an antimalarial drug (a quinoline alkaloid)

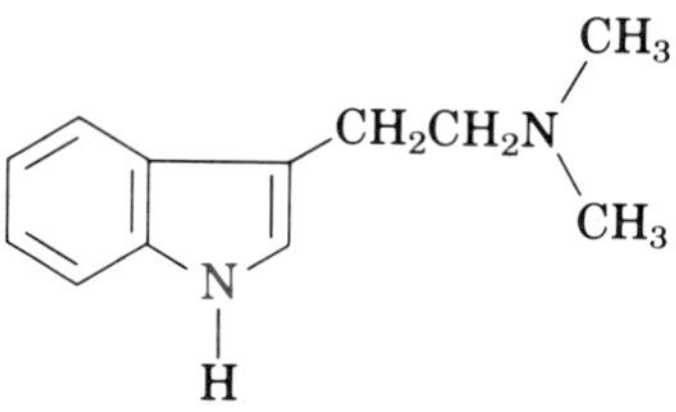

*N,N*-Dimethyltryptamine, a hallucinogen (an indole alkaloid)

The chemistry of these three classes of fused-ring heterocycles is just what we might expect from our knowledge of the simpler heterocycles pyridine and pyrrole. All three undergo electrophilic aromatic substitution reactions. Quinoline and isoquinoline both undergo electrophilic substitution more easily than pyridine but less easily than benzene, consistent with our previous observation that pyridine rings are deactivated compared to benzene. Note that reaction occurs on the benzene ring (not the pyridine ring) and that a mixture of C5 and C8 substitution products is obtained.

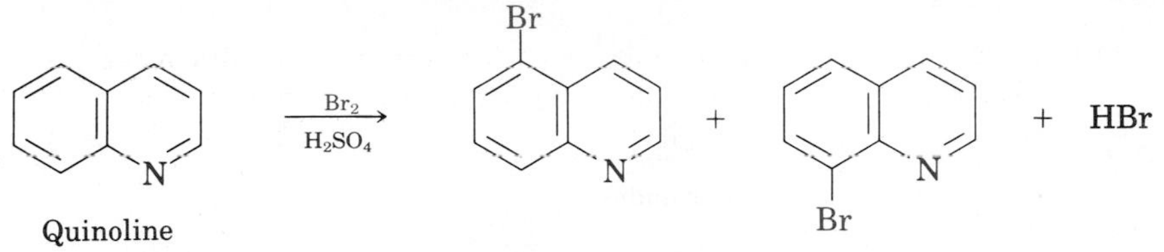

$$\text{Quinoline} \xrightarrow[H_2SO_4]{Br_2} \text{5-Bromoquinoline} + \text{8-Bromoquinoline} + HBr$$

A 51:49 ratio

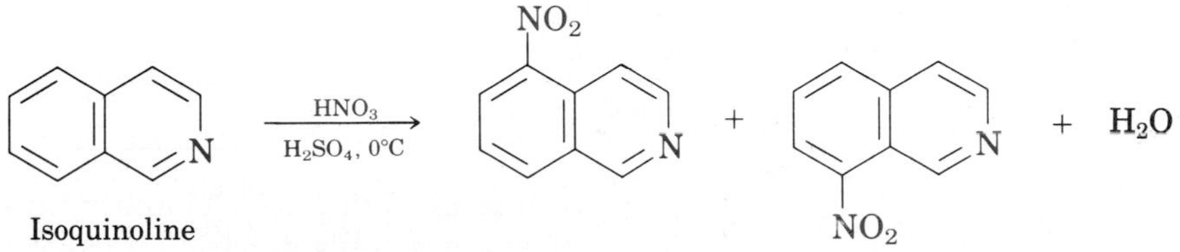

$$\text{Isoquinoline} \xrightarrow[H_2SO_4,\ 0°C]{HNO_3} \text{5-Nitroisoquinoline} + \text{8-Nitroisoquinoline} + H_2O$$

A 90:10 ratio

Indole undergoes electrophilic substitution more easily than benzene but less easily than pyrrole. Again, this is consistent with our previous observation that pyrrole rings are more strongly activated than benzene rings. Substitution occurs at C3 of the electron-rich pyrrole ring, rather than on the benzene ring, but reaction conditions must be chosen carefully to avoid destructive side reactions.

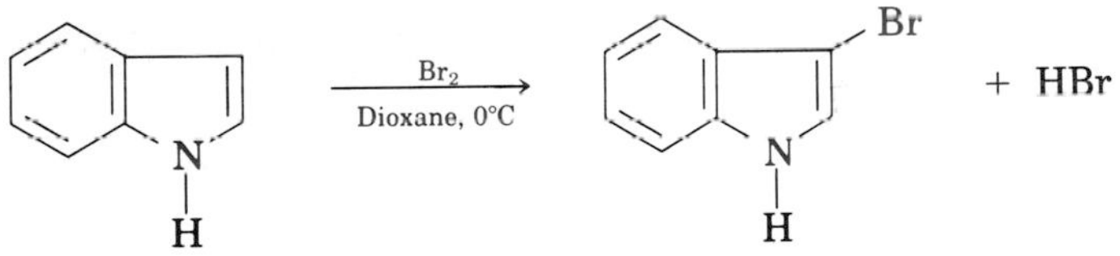

$$\text{Indole} \xrightarrow[\text{Dioxane, 0°C}]{Br_2} \text{3-Bromoindole} + HBr$$

PROBLEM

**29.8** Which nitrogen atom in *N,N*-dimethyltryptamine is more basic? Explain.

PROBLEM

**29.9** Indole reacts with electrophiles at C3 rather than at C2. Draw resonance forms of the intermediate cations resulting from attack at C2 and C3 and explain the observed results.

## 29.8 Pyrimidine and Purine

The most important heterocyclic ring systems from a biological viewpoint are **pyrimidine** and **purine**. Pyrimidine contains two nitrogens in a six-membered aromatic ring; purine has four nitrogens in a fused-ring structure.

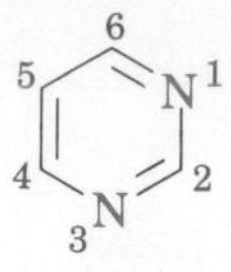

Pyrimidine

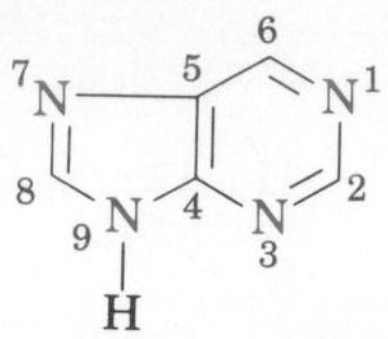

Purine

Both heterocycles are essential components of the last major class of biomolecules we will consider—the nucleic acids.

## 29.9 Nucleic Acids and Nucleotides

The nucleic acids—**deoxyribonucleic acid** (**DNA**) and **ribonucleic acid** (**RNA**)—are the chemical carriers of a cell's genetic information. Coded in a cell's DNA is all the information that determines the nature of the cell, controls cell growth and division, and directs biosynthesis of the enzymes and other proteins required for all cellular functions.

Like proteins, nucleic acids are polymers. Mild enzyme-catalyzed hydrolysis cleaves a nucleic acid into monomeric building blocks called **nucleotides**. Each nucleotide can be further cleaved by enzyme-catalyzed hydrolysis to give a **nucleoside** plus phosphoric acid, $H_3PO_4$, and each nucleoside can be hydrolyzed to yield a simple pentose sugar plus a heterocyclic purine or pyrimidine base.

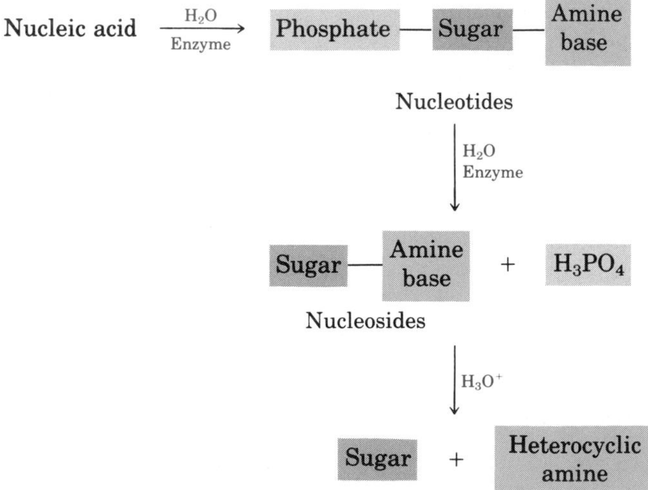


The sugar component in RNA is **ribose**, and the sugar in DNA is **2′-deoxyribose**. (The prefix *2′-deoxy* indicates that oxygen is missing from the 2′ position of ribose; numbers with a prime superscript refer to positions on the sugar component of a nucleotide, and numbers without a prime refer to positions on the heterocyclic amine base.)

Ribose 2′-Deoxyribose

Four different heterocyclic amine bases are found in deoxyribonucleotides. Two are substituted purines (**adenine** and **guanine**) and two are substituted pyrimidines (**cytosine** and **thymine**). Adenine, guanine, and cytosine also occur in RNA, but thymine is replaced in RNA by a different pyrimidine base called **uracil**.

Adenine Guanine Purines

Cytosine Uracil (RNA) Thymine (DNA) Pyrimidines

In both DNA and RNA, the heterocyclic amine base is bonded to C1′ of the sugar, whereas the phosphoric acid is bound by a phosphate ester linkage to the C5′ sugar position. Thus, nucleosides and nucleotides have the structure shown in Figure 29.4.

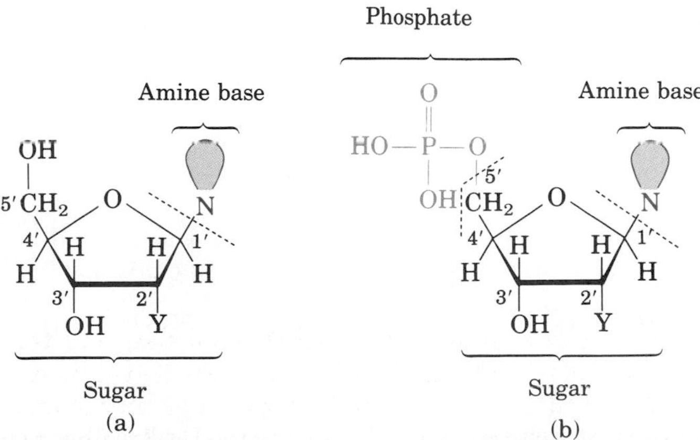


**Figure 29.4** (a) A nucleoside and (b) a nucleotide: When Y = H, the sugar is deoxyribose. When Y = OH, the sugar is ribose.

The complete names of the bases, nucleotides, and corresponding nucleosides are given in Table 29.2, and the complete structures of all four deoxyribonucleotides and all four ribonucleotides are shown in Figure 29.5.

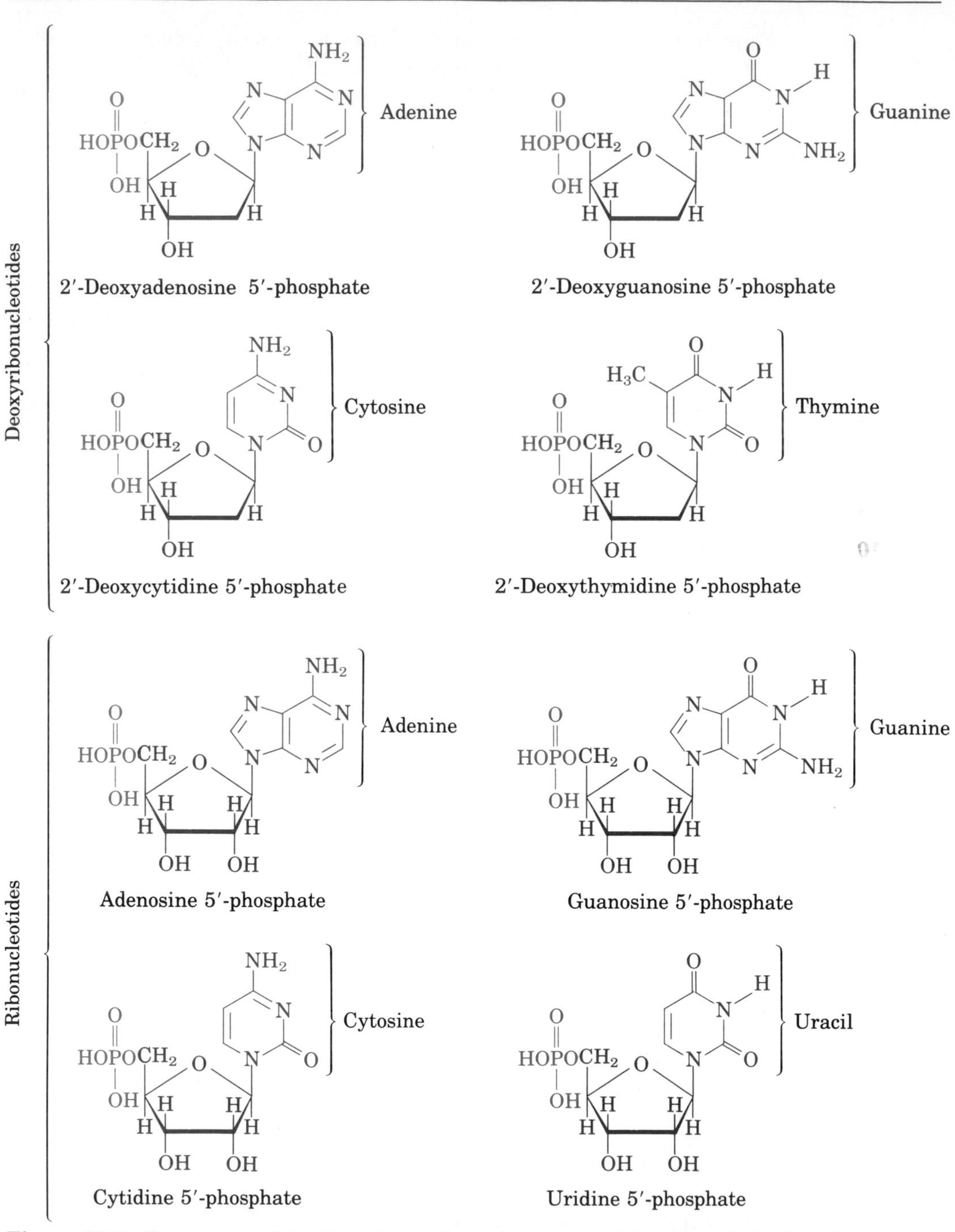


**Figure 29.5** Structures of the four deoxyribonucleotides and the four ribonucleotides

**Table 29.2** Names of bases, nucleosides, and nucleotides

| *Heterocyclic base* | *Source* | *Nucleoside* | *Nucleotide* |
|---|---|---|---|
| Adenine | RNA | Adenosine | Adenosine 5′-phosphate |
| | DNA | 2′-Deoxyadenosine | 2′-Deoxyadenosine 5′-phosphate |
| Guanine | RNA | Guanosine | Guanosine 5′-phosphate |
| | DNA | 2′-Deoxyguanosine | 2′-Deoxyguanosine 5′-phosphate |
| Cytosine | RNA | Cytidine | Cytidine 5′-phosphate |
| | DNA | 2′-Deoxycytidine | 2′-Deoxycytidine 5′-phosphate |
| Uracil | RNA | Uridine | Uridine 5′-phosphate |
| Thymine | DNA | 2′-Deoxythymidine | 2′-Deoxythymidine 5′-phosphate |

Though chemically similar, DNA and RNA are different in size and have different roles within the cell. Molecules of DNA are enormous. They have molecular weights of up to 50 *billion* and are found mostly in the nucleus of the cell. Molecules of RNA, by contrast, are much smaller (as low as 35,000 mol wt) and are found mostly outside the cell nucleus. We'll consider the two kinds of nucleic acids separately, beginning with DNA.

PROBLEM

**29.10** 2′-Deoxythymidine exists largely in the lactam form rather than in the tautomeric lactim form. Explain.

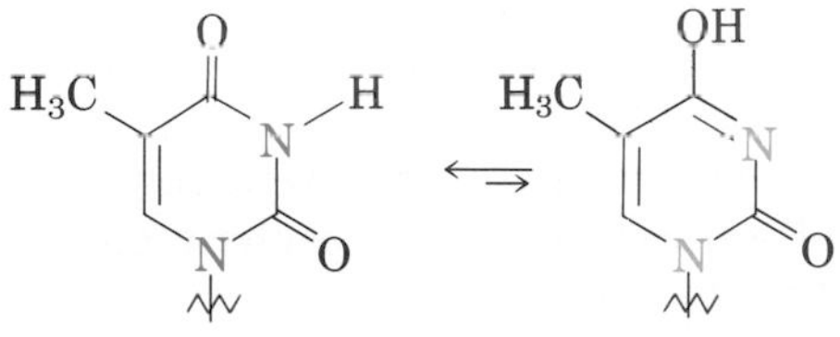

$H_3C$ O H N N O ⟷ $H_3C$ OH N N O

Lactam form    Lactim form

## 29.10 Structure of DNA

Nucleic acids consist of nucleotide units joined by a bond between the 5′-phosphate component of one nucleotide and the 3′-hydroxyl on the sugar component of another nucleotide (Figure 29.6, page 1050). One end of the nucleic acid polymer has a free hydroxyl at C3′ (called the **3′ end**) and the other end has a phosphoric acid residue at C5′ (the **5′ end**).

Just as the exact structure of a protein depends on the sequence in which individual amino acid residues are arranged, the exact structure of a nucleic acid depends on the sequence in which the individual nucleotides are arranged. To carry the analogy even further, just as a protein has a

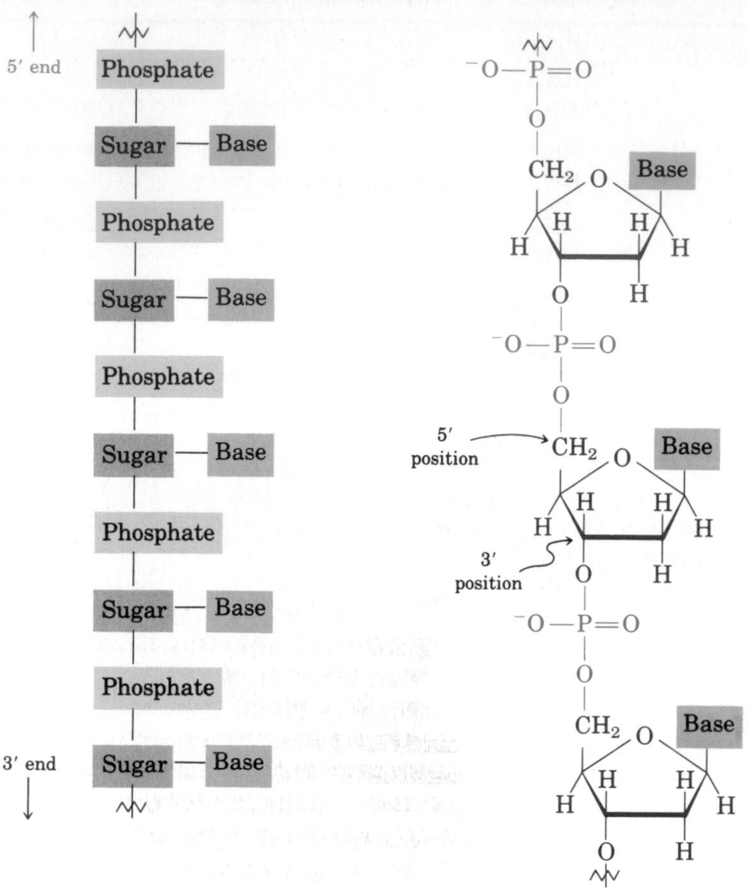


**Figure 29.6** Generalized structure of DNA

polyamide backbone with different side chains attached to it, a nucleic acid has an alternating sugar–phosphate backbone with different amine bases attached to it.

Different side chains

N terminus

C terminus

A protein

R1, R2, R3, R4, R5

—NH—CH(R1)—C(=O)—NH—CH(R2)—C(=O)—NH—CH(R3)—C(=O)—NH—CH(R4)—C(=O)—NH—CH(R5)—C(=O)—

Amide bonds

Different bases

5′ end

3′ end

A nucleic acid

Base1, Base2, Base3

—Phosphate—Sugar(Base1)—Phosphate—Sugar(Base2)—Phosphate—Sugar(Base3)—

Phosphate ester bonds

The sequence of nucleotides in a chain is described by starting at the 5′ end and identifying the bases in order of occurrence. Rather than write the full name of each nucleotide, however, it is more convenient to use simple abbreviations—A for adenosine, T for thymine, G for guanosine, and C for cytidine. Thus, a typical DNA sequence might be written as -T-A-G-G-C-T-.

Samples of DNA isolated from different tissues of the same species have the same proportions of heterocyclic bases, but samples from different species can have greatly differing proportions of bases. For example, human DNA contains about 30% each of adenine and thymine and about 20% each of guanine and cytosine. The bacterium *Clostridium perfringens*, however, contains about 37% each of adenine and thymine and only 13% each of guanine and cytosine. Note that in both of these examples the bases occur in pairs. Adenine and thymine are usually present in equal amounts, as are cytosine and guanine. The reason for this pairing of bases has much to do with the secondary structure of DNA.

In 1953, James Watson[1] and Francis Crick[2] made their now classic proposal for the secondary structure of DNA. According to the Watson–Crick model, DNA consists of two polynucleotide strands coiled around each other in a **double helix**. The two strands run in opposite directions and are held together by hydrogen bonds between specific pairs of bases. Adenine (A) and thymine (T) form strong hydrogen bonds to each other but not to other bases; guanine (G) and cytosine (C) form strong hydrogen bonds to each other but not to other bases.

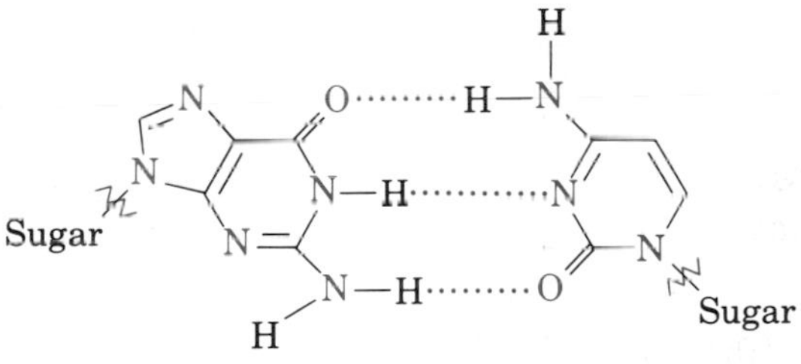

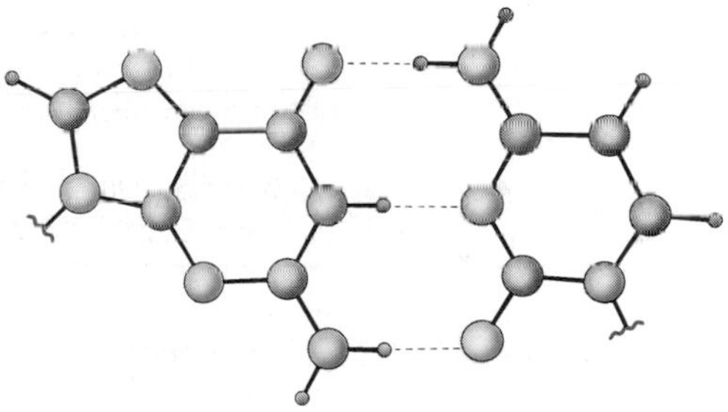

(Guanine) G : : : : : : C (Cytosine)

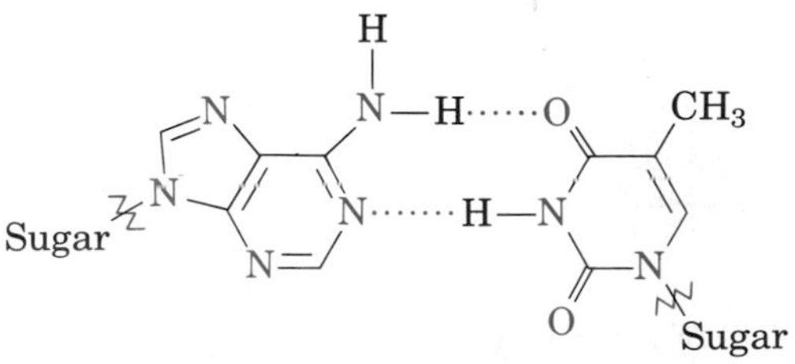

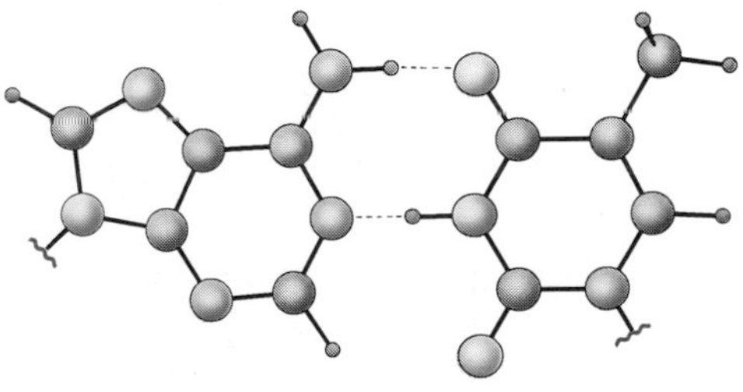

(Adenine) A : : : : : : T (Thymine)

[1]James Dewey Watson (1928– ); b. Chicago, Ill.; Ph.D. Indiana; professor, Harvard University; Nobel prize in medicine (1960).

[2]Francis H. C. Crick (1916– ); b. England; Ph.D. Cambridge; professor, Cambridge University; Nobel prize in medicine (1960).

The two strands of the DNA double helix are not identical; rather, they are complementary. Whenever a C base occurs in one strand, a G base occurs opposite it in the other strand. When an A base occurs in one strand, a T appears opposite it in the other strand. This complementary pairing of bases explains why A and T, and C and G, are always found in equal amounts. Figure 29.7 illustrates this base pairing, showing how the two complementary strands are coiled into the double helix. X-ray measurements show that the DNA double helix is 20 Å wide, that there are exactly 10 base pairs in each full turn, and that each turn is 34 Å in height.

A helpful mnemonic device to remember the nature of the hydrogen bonding between the four DNA bases is the simple phrase "Pure silver taxi":

| Pure | Silver | Taxi |
|---|---|---|
| Pur | Ag | TC |
| The purine bases, | A and G, | hydrogen bond to T and C. |

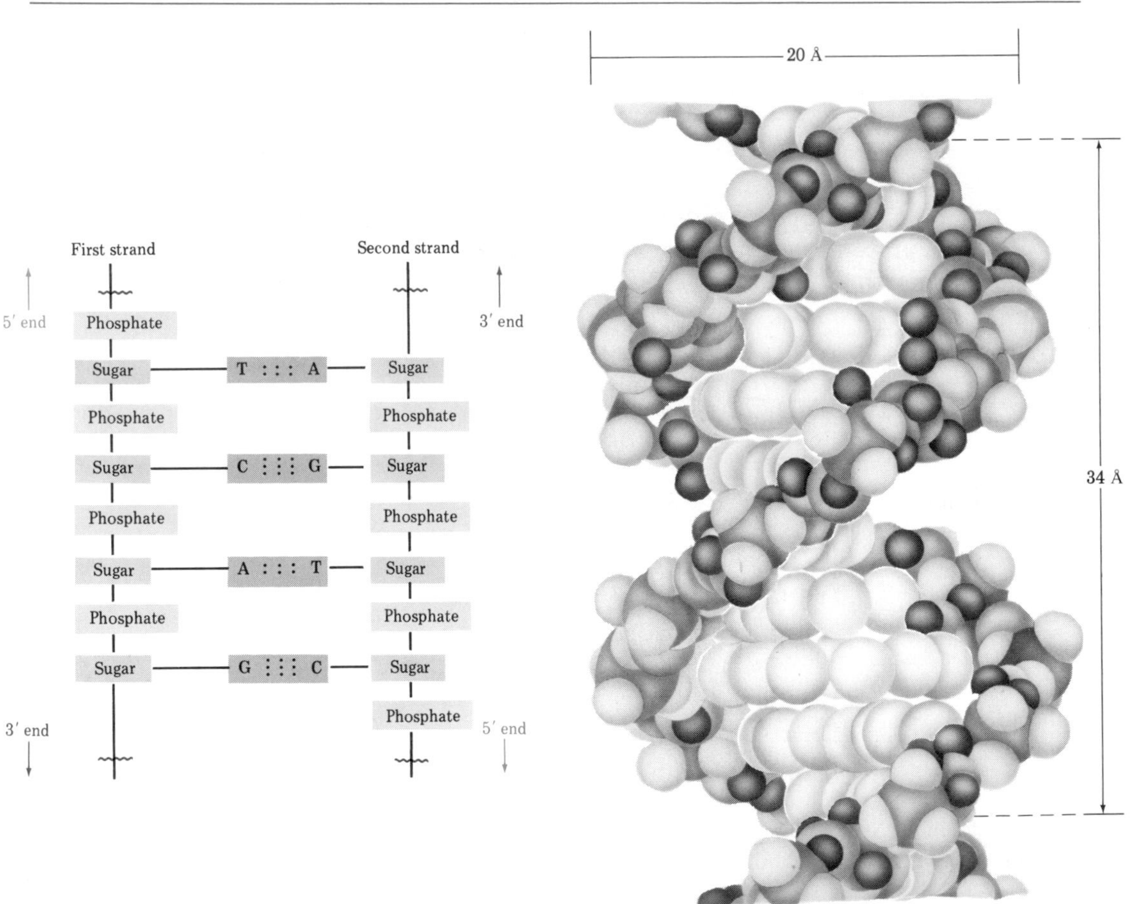


**Figure 29.7** Complementarity in base pairing in the DNA double helix: The sugar–phosphate backbone of DNA is shown in gray; the atoms of the amine bases are shown in color and lie inside the helix; the small black atoms are hydrogen.

PROBLEM

**29.11** What sequence of bases on one strand of DNA is complementary to the following sequence on another strand?

G-G-C-T-A-A-T-C-C-G-T

## 29.11 Nucleic Acids and Heredity

A DNA molecule is the chemical repository of an organism's genetic information, which is stored as a sequence of deoxyribonucleotides strung together in the DNA chain. For this information to be preserved, a mechanism must exist for the DNA molecule to be copied and passed on to succeeding generations. For this information to be used, a mechanism must exist for reading the DNA, for decoding the instructions contained therein, and for implementing those instructions to carry out the myriad biochemical processes necessary to sustain life.

What Crick has termed the *central dogma of molecular genetics* says that the function of DNA is to store information and pass it on to RNA at the proper time, whereas the function of RNA is to read, decode, and use the information received from DNA to make proteins. Each of the thousands of individual genes on each chromosome contains the instructions necessary to make a specific protein that is in turn needed for a specific biological purpose. By decoding the right genes at the right time in the right place, an organism can use genetic information to synthesize the many thousands of proteins necessary for carrying out the biochemical reactions required for smooth functioning.

$$\text{DNA} \longrightarrow \text{RNA} \longrightarrow \text{Proteins}$$

Three fundamental processes take place in the transfer of genetic information:

1. **Replication** is the process by which a replica, or identical copy, of DNA is made so that information can be preserved and handed down to offspring.
2. **Transcription** is the process by which the genetic messages contained in DNA are "read," or transcribed, and carried out of the nucleus to parts of the cell called ribosomes where protein synthesis occurs.
3. **Translation** is the process by which the genetic messages are decoded and used to build proteins.

## 29.12 Replication of DNA

The Watson–Crick model of DNA does more than just explain base pairing; it also provides a remarkably ingenious way for DNA molecules to reproduce exact copies of themselves. **Replication** of DNA is an enzyme-catalyzed process that begins by a partial unwinding of the double helix. As the strands separate and the bases are exposed, new nucleotides line up on each strand in an exactly complementary manner, A to T and C to G, and two new strands begin to grow. Each new strand is complementary to its old template strand, and two new identical DNA double helices are produced (Figure 29.8).

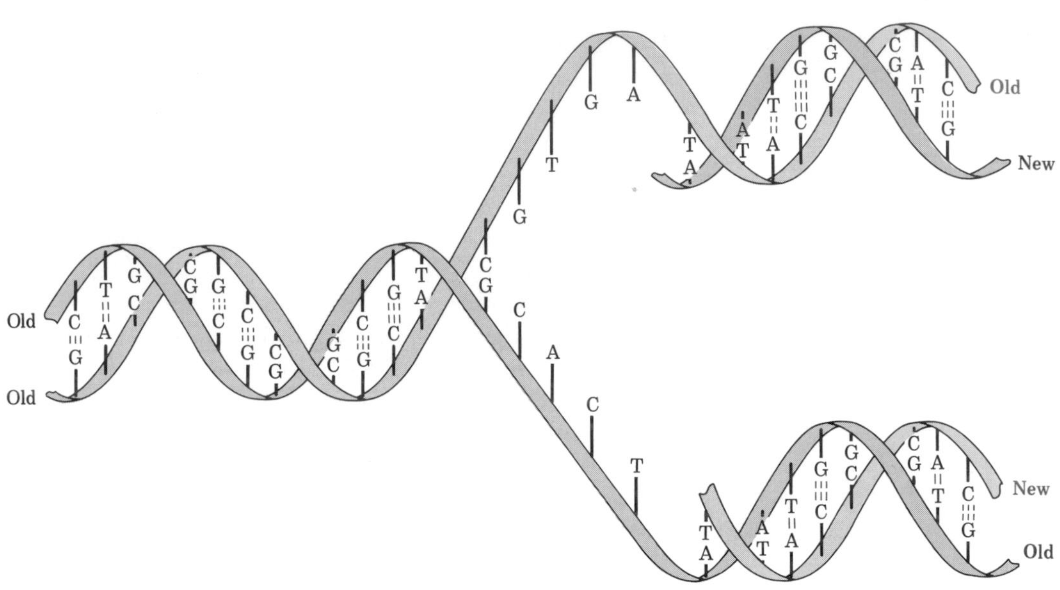


**Figure 29.8** Schematic representation of DNA replication

Crick probably described the process best when he used the analogy of the two DNA strands fitting together like a hand in a glove. The hand and glove separate; a new hand forms inside the glove; and a new glove forms around the hand. Two identical copies now exist where only one existed before.

The process by which the individual nucleotides are joined to create new DNA strands is complex, involving many steps and many different enzymes. Addition of new nucleotide units to the growing chain, which is catalyzed by the enzyme *DNA polymerase*, has been shown to occur by addition of a 5′-mononucleotide triphosphate to the free 3′-hydroxyl group of the growing chain, as indicated in Figure 29.9.

Both of the new DNA strands are synthesized in the same 5′-to-3′ direction, which implies that the two strands cannot be synthesized in exactly the same way. Since the two complementary DNA strands are lined up in opposite directions, one strand must have its 3′ end near the point of unraveling (the **replication fork**) whereas the other strand has its 5′ end near the replication fork. What evidently happens is that the complement of the original 3′ → 5′ strand is synthesized smoothly and in a single piece,

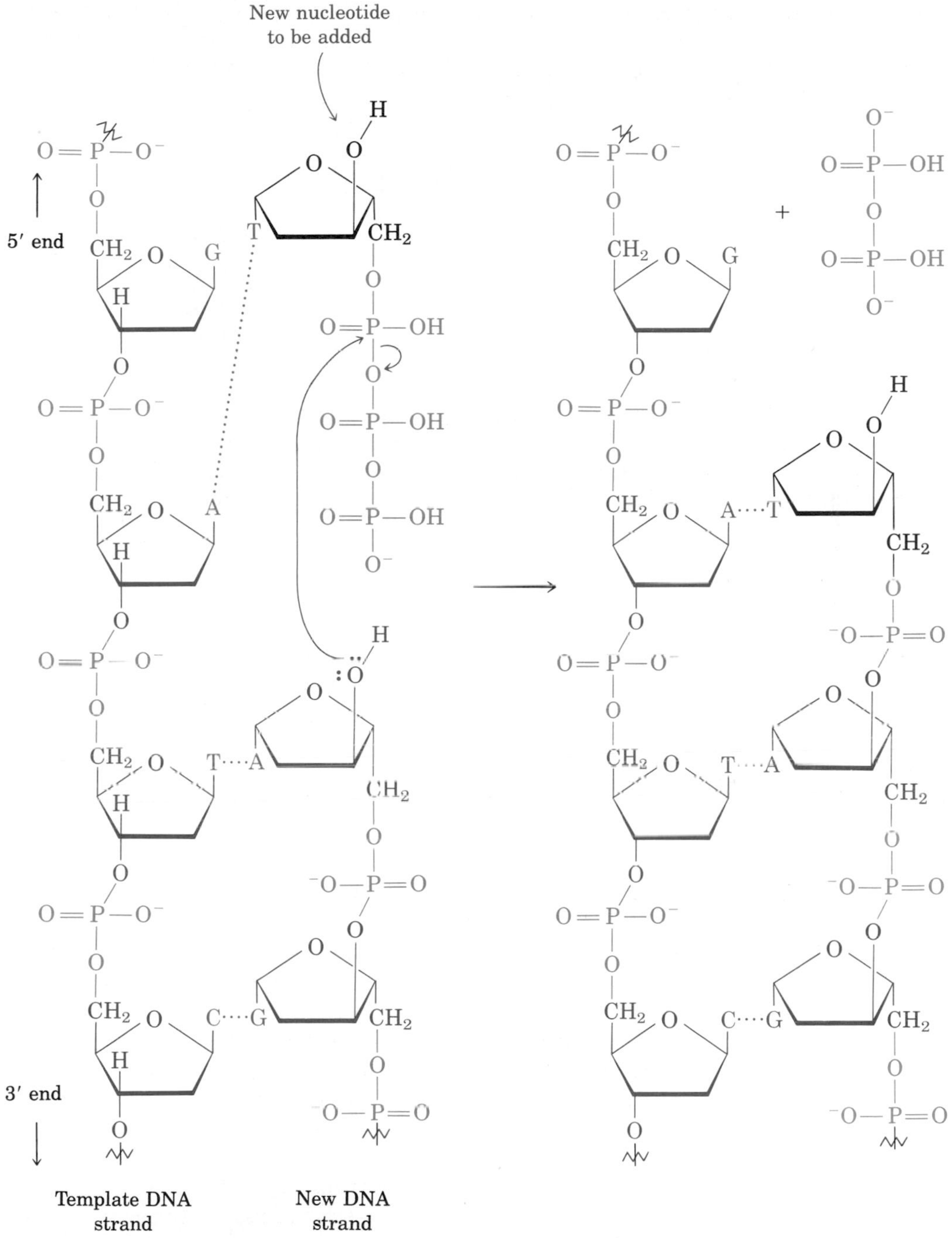

Figure 29.9 Addition of a new nucleotide to a growing DNA strand

but the complement of the original 5′ → 3′ strand is synthesized discontinuously in small pieces that are then linked at a later point by DNA ligase enzymes (Figure 29.10, page 1056).

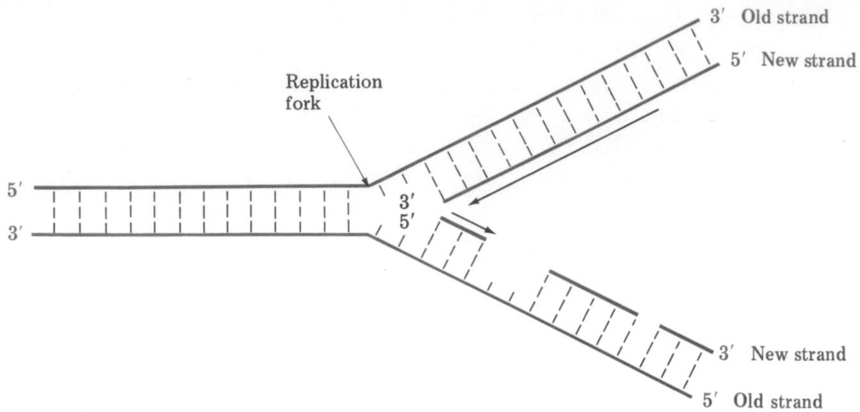


**Figure 29.10** Replication of DNA: Both new DNA strands are synthesized in a 5′-to-3′ direction. The original strand whose 3′ end is near the point of unraveling is replicated smoothly, but the original strand whose 5′ end is near the unraveling point is replicated in small pieces that are later joined. The arrowheads are at the growing ends of the chains.

It's difficult to conceive of the magnitude of the replication process. The total size of all genes in a human cell (the **genome**) is estimated to be approximately 3 billion base pairs. A single DNA chain might have a length of several centimeters and contain up to 75 million pairs of bases. Regardless of the size of these molecules, their base sequence is faithfully copied during replication. The copying process takes only minutes, and it's been estimated that an error occurs only about once each 10–100 billion bases.

## 29.13 Structure and Synthesis of RNA: Transcription

RNA is structurally similar to DNA. Both are sugar–phosphate polymers and both have heterocyclic bases attached. The only differences are that RNA contains ribose rather than deoxyribose, and uracil rather than thymine. Uracil in RNA forms strong hydrogen bonds to its complementary base, adenine, just as thymine does in DNA.

Uracil (in RNA) Thymine (in DNA)

There are three major kinds of ribonucleic acid, each of which serves a specific function: **messenger RNA (mRNA)** carries genetic messages from

DNA to *ribosomes*, small granular particles in the cytoplasm of a cell that can be thought of as "protein factories"; **ribosomal RNA (rRNA)** complexed with protein provides the physical makeup of the ribosomes; **transfer RNA (tRNA)** transports specific amino acids to the ribosomes where they are joined together to make proteins. All three kinds of RNA are structurally similar to DNA, but they are much smaller molecules than DNA and they remain single-stranded, rather than double-stranded like DNA.

Molecules of RNA are synthesized in the nucleus of the cell by **transcription** of DNA. A small portion of the DNA double helix unwinds, and the bases of the two strands are exposed. Ribonucleotides line up in the proper order by hydrogen bonding to their complementary bases on DNA, bond formation occurs in the 5′-to-3′ direction, and the completed RNA molecule then unwinds from DNA and migrates from the nucleus (Figure 29.11).

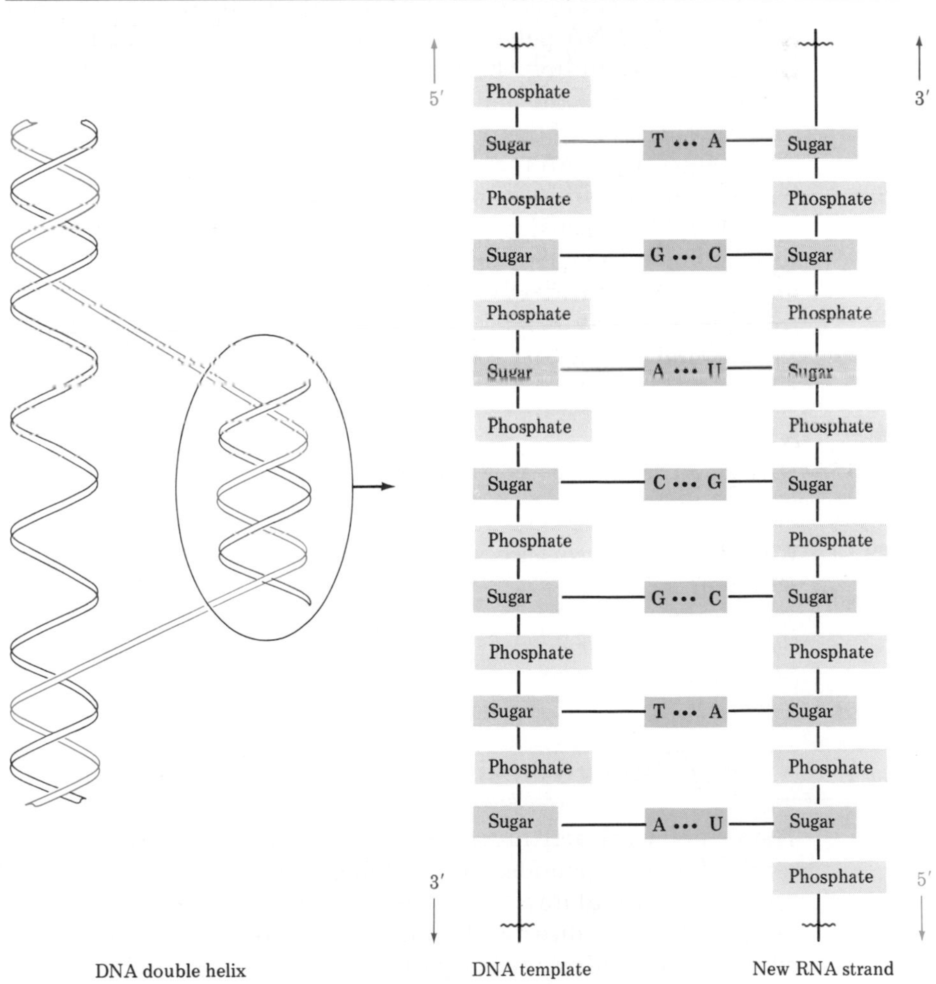


**Figure 29.11** Biosynthesis of RNA using a DNA segment as template

Unlike what happens in DNA replication, where both strands are copied, only one of the two DNA strands is transcribed into mRNA. The strand that contains the gene is called the **informational strand**, and its complement is called the **template strand**. Only the template strand is transcribed. Since the template strand and the informational strand are complementary, and since the template strand and the mRNA molecule are also complementary, it follows that *the messenger RNA molecule produced during transcription is a copy of the informational strand.* The only difference is that the mRNA molecule has a uracil everywhere the DNA informational strand has a thymine.

Transcription of DNA by the process just discussed raises as many questions as it answers. How does the DNA know where to unwind? Where along the chain does one gene stop and the next one start? How do the ribonucleotides know exactly the right place along the template strand to begin lining up and the right place to stop? Though these questions are extremely difficult to answer, the picture that has emerged in the last decade is that a DNA chain contains certain base sequences called **promoter sites** that bind the RNA polymerase enzyme that actually carries out RNA synthesis, thus signaling the beginning of a gene. Similarly there are other base sequences at the end of the gene that signal a stop.

Another part of the picture that has recently emerged is that genes are not necessarily continuous segments of the DNA chain. Often a gene will begin in one small section of DNA called an **exon**, then be interrupted by a seemingly nonsensical section called an **intron**, and then take up again further down the chain in another exon. The final mRNA molecule results only after the nonsense sections are cut out and the remaining pieces spliced together. Current evidence is that up to 90% of human DNA is made up of introns and only about 10% of DNA actually contains genetic instructions.

PROBLEM

**29.12** Show how uracil can form strong hydrogen bonds to adenine.

PROBLEM

**29.13** What RNA base sequence would be complementary to the following DNA base sequence?

G-A-T-T-A-C-C-G-T-A

## 29.14 RNA and Biosynthesis of Proteins: Translation

The primary cellular function of RNA is to direct biosynthesis of the thousands of diverse peptides and proteins required by an organism. These proteins in turn regulate all other biological processes. The mechanics of protein biosynthesis are directed by messenger RNA (mRNA) and take place on **ribosomes**, small granular particles in the cytoplasm of a cell that consist of about 60% ribosomal RNA and 40% protein. On the ribosome, mRNA serves as a template to pass on the genetic information it has transcribed from DNA.

The specific ribonucleotide sequence in mRNA forms a "code" that determines the order in which different amino acid residues are to be joined. Thus, each of the estimated 100,000 proteins in the human body is synthesized from a different mRNA that has been transcribed from a specific gene segment on DNA.

Each "word" or **codon** along the mRNA chain consists of a series of three ribonucleotides that is specific for a given amino acid. For example, the series cytosine–uracil–guanine (C-U-G) on mRNA is a codon directing incorporation of the amino acid leucine into the growing protein, and guanine–adenine–uracil (G-A-U) codes for aspartic acid. Of the $4^3 = 64$ possible triads of the four bases in RNA, 61 code for specific amino acids (certain amino acids are specified by more than one codon). In addition, 3 of the 64 codons specify chain termination. Table 29.3 shows the meaning of each codon.

**Table 29.3** Codon assignments of base triads

| *First base (5′ end)* | *Second base* | *Third base (3′ end)* | | | |
|---|---|---|---|---|---|
| | | *U* | *C* | *A* | *G* |
| U | U | Phe | Phe | Leu | Leu |
| | C | Ser | Ser | Ser | Ser |
| | A | Tyr | Tyr | Stop | Stop |
| | G | Cys | Cys | Stop | Trp |
| C | U | Leu | Leu | Leu | Leu |
| | C | Pro | Pro | Pro | Pro |
| | A | His | His | Gln | Gln |
| | G | Arg | Arg | Arg | Arg |
| A | U | Ile | Ile | Ile | Met |
| | C | Thr | Thr | Thr | Thr |
| | A | Asn | Asn | Lys | Lys |
| | G | Ser | Ser | Arg | Arg |
| G | U | Val | Val | Val | Val |
| | C | Ala | Ala | Ala | Ala |
| | A | Asp | Asp | Glu | Glu |
| | G | Gly | Gly | Gly | Gly |

The code expressed in mRNA is read by transfer RNA (tRNA) in a process called **translation**. There are at least 60 different transfer RNA's, one for each of the codons in Table 29.3. Each specific tRNA acts as a carrier to bring a specific amino acid into place so that it may be transferred to the growing protein chain.

A typical tRNA is roughly the shape of a cloverleaf, as shown in Figure 29.12 (page 1060). It consists of about 70–100 ribonucleotides and is bound to a specific amino acid by an ester linkage through the free 3′-hydroxyl on ribose at the 3′ end of the tRNA. Each tRNA also contains in its structure a segment called an **anticodon**, a sequence of three ribonucleotides complementary to the codon sequence. For example, the codon sequence C-U-G present on mRNA would be "read" by a leucine-bearing tRNA having the complementary anticodon base sequence G-A-C.

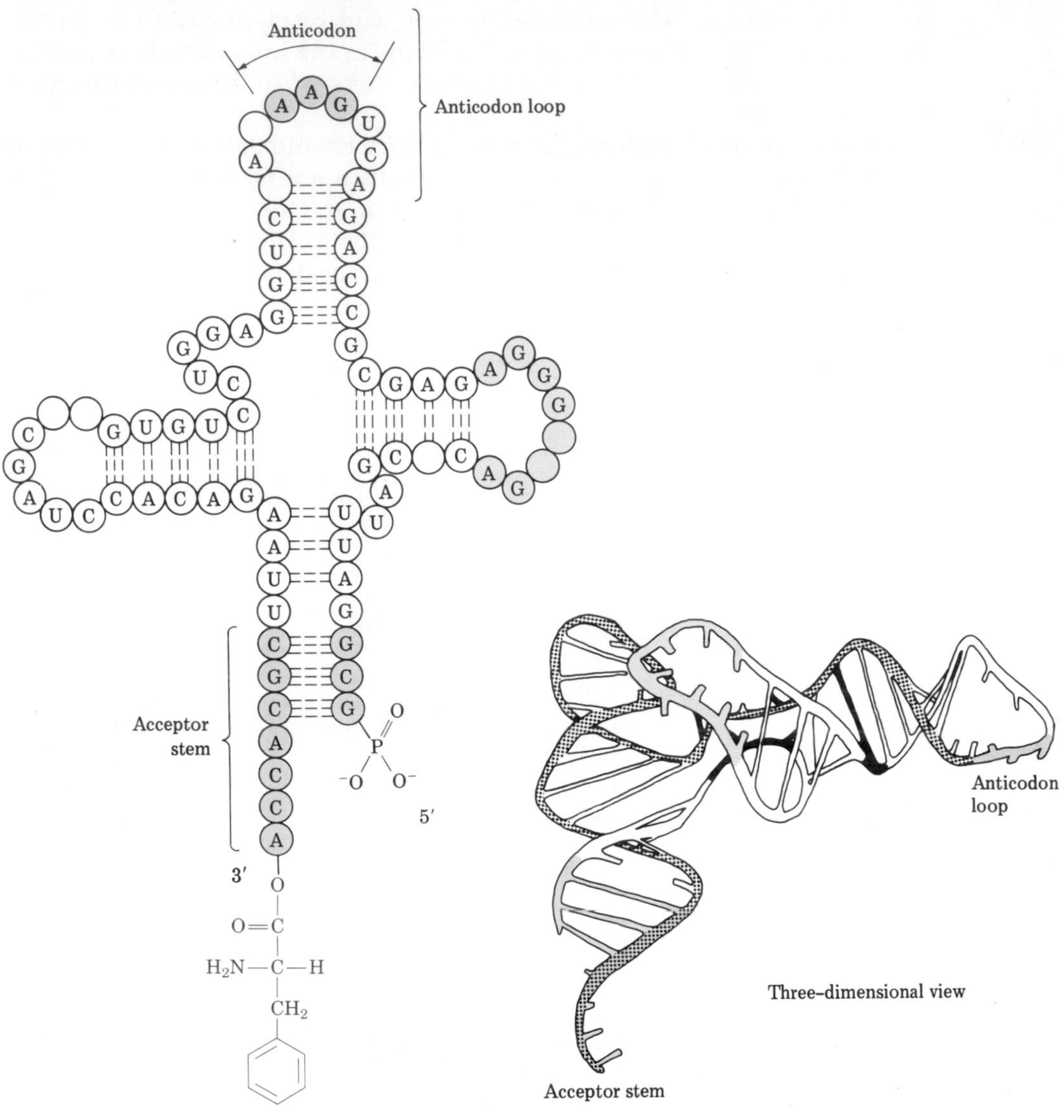


**Figure 29.12** Structure of a tRNA molecule: The tRNA is a roughly cloverleaf-shaped molecule containing an anticodon triplet on one "leaf" and a covalently attached amino acid unit at its 3′ end. The example shown is a yeast tRNA that codes for phenylalanine. The nucleotides not specifically identified are chemically modified analogs of the four normal nucleotides.

As each successive codon on mRNA is read, different tRNA's bring the correct amino acids into position for enzyme-mediated transfer to the growing peptide. When synthesis of the proper protein is completed, a "stop" codon signals the end, and the protein is released from the ribosome. The entire process of protein biosynthesis is illustrated schematically in Figure 29.13.

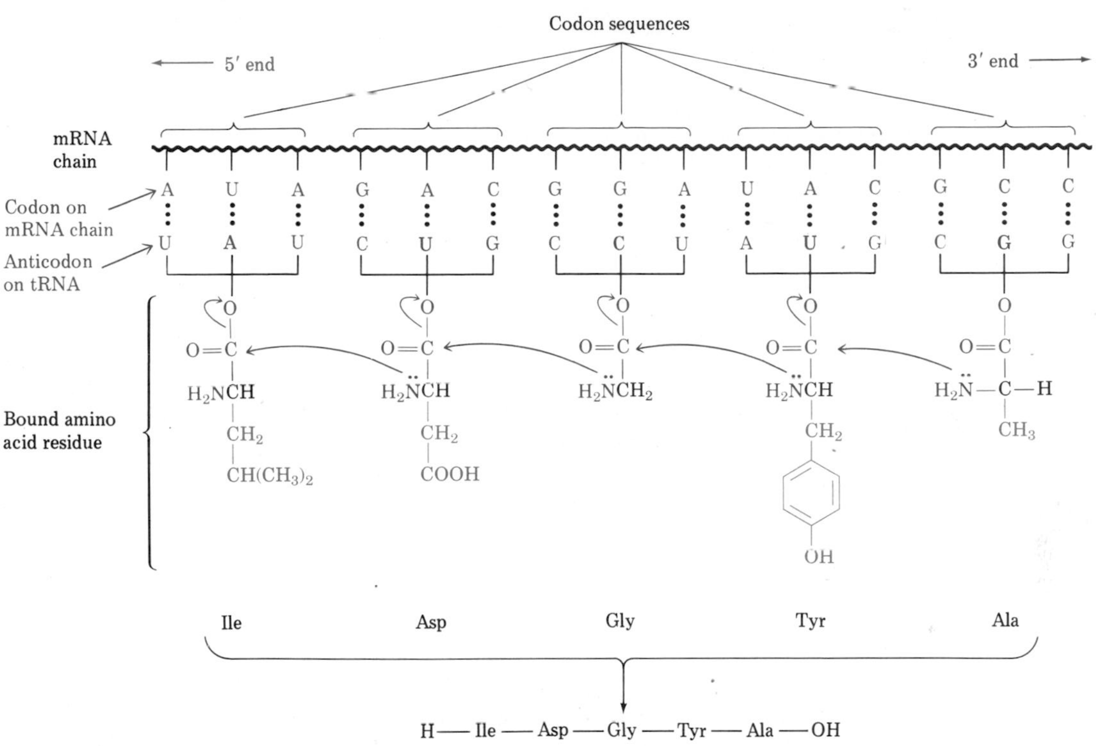


**Figure 29.13** Schematic representation of protein biosynthesis: The mRNA containing codon base sequences is read by tRNA containing complementary anticodon base sequences. Transfer RNA assembles the proper amino acids into position for incorporation into the growing peptide.

In summary, protein biosynthesis takes place in five discrete steps:

*Step 1* Messenger RNA, containing the genetic information transcribed from DNA, is synthesized in the nucleus and moves to the ribosomes.

*Step 2* Individual amino acids are activated by binding to specific tRNA's.

*Step 3* Transfer RNA's containing the correct anticodon sequences line up at the proper complementary codon sequences on one end of the mRNA chain and move the bound amino acid groups into position.

*Step 4* The polypeptide is produced as enzymes catalyze the addition of tRNA-bound amino acids to the growing peptide chain.

*Step 5* When the peptide is completed, a "stop" codon on the mRNA halts the biosynthesis, and the peptide is released from the ribosome.

PROBLEM

**29.14** List codon sequences for these amino acids:
(a) Ala (b) Phe (c) Leu (d) Tyr

PROBLEM

**29.15** List anticodon sequences on the tRNA's carrying the amino acids shown in Problem 29.14.

PROBLEM

**29.16** What amino acid sequence is coded for by the following mRNA base sequence?

CUU-AUG-GCU-UGG-CCC-UAA

PROBLEM

**29.17** What anticodon sequences of tRNA's are coded for by the mRNA in Problem 29.16?

PROBLEM

**29.18** What is the base sequence in the original DNA strand on which the mRNA sequence in Problem 29.16 was made?

## 29.15 Sequencing of DNA

> When we work out the structure of DNA molecules, we examine the fundamental level that underlies all processes in living cells. DNA is the information store that ultimately dictates the structure of every gene product, delineates every part of the organism. The order of the bases along DNA contains the complete set of instructions that make up the genetic inheritance. (Walter Gilbert, Nobel Prize Lecture, 1980)

DNA sequencing is now carried out by a remarkably efficient and powerful method developed in 1977 by Allan Maxam and Walter Gilbert.[3]

Since molecules of DNA are so enormous—some molecules of human DNA contain billions of base pairs—the first problem in DNA sequencing is to find a method for reproducibly and selectively cleaving the DNA chain at specific points to produce smaller, more manageable pieces. This problem has been solved by the use of enzymes called **restriction endonucleases**. Each different restriction enzyme, of which more than 200 are available, cleaves a DNA molecule between two nucleotides at well-defined points along the chain where specific base sequences occur. Since the required sequence is usually four or more nucleotides long, it is unlikely to occur very often in the overall DNA sequence.

By incubation of large DNA molecules with a given restriction enzyme, many different and well-defined segments of manageable length (100–200 nucleotides) are produced. For example, the restriction enzyme *Alu I* cleaves the linkage between G and C in the four-base sequence AG-CT; the enzyme *Hpa II* cleaves the C–C linkage in the four-base sequence C-CGG (Figure 29.14). If the original DNA molecule is incubated with another restriction enzyme having a different specificity for cleavage, still other segments are produced, whose sequences partially overlap those produced by the first

[3]Walter Gilbert (1932– ); b. Boston; Ph.D. Cambridge University (1957); professor, Harvard University (1958– ); Nobel prize (1980).

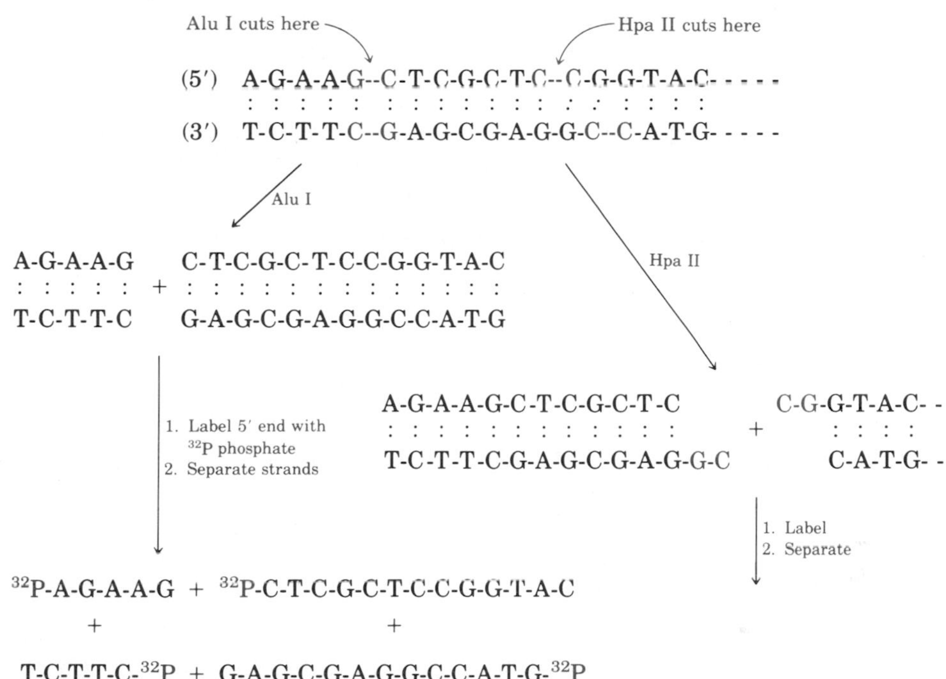


**Figure 29.14** Cleavage of a DNA molecule with restriction enzymes: The enzyme Alu I cleaves at the sequence AG-CT; the enzyme Hpa II cleaves at C-CGG. After cleavage of the double-stranded DNA, the fragments are isolated and each is labeled at its 5′ end by enzyme catalyzed formation of a radioactively labeled $^{32}P$-containing phosphate ester. The strands are then separated.

enzyme. Sequencing of all the segments, followed by identification of the overlapping sequences, then allows complete DNA sequencing in a manner similar to that used for protein sequencing (Section 27.9).

After restriction enzymes have cleaved DNA into smaller, more manageable pieces (restriction fragments) the various double-stranded fragments are isolated, and each is radioactively tagged by enzymatically incorporating a labeled $^{32}P$ phosphate group onto the 5′-hydroxyl group of the terminal nucleotide. The fragments are then separated into two strands by heating, and the strands are isolated.

The heart of the sequencing problem is finding reaction conditions for obtaining specific DNA chain breakage next to each of the four nucleotide bases so that the restriction fragments can be further degraded and ultimately sequenced. Maxam and Gilbert solved this problem using two reagents, dimethyl sulfate, $(CH_3O)_2SO_2$, and hydrazine, $H_2NNH_2$.

Treatment of a restriction fragment with dimethyl sulfate results in methylation ($S_N2$ reaction) of the purine bases A and G, but does not affect the pyrimidine bases C and T. Deoxyadenosine is methylated at $N_3$, and deoxyguanosine is methylated at $N_7$.

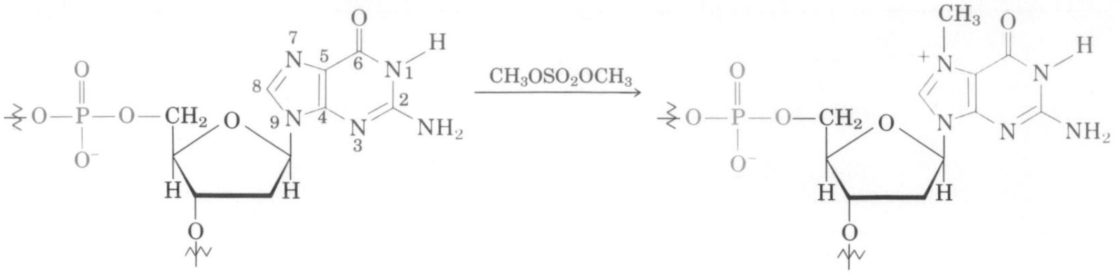

$\xrightarrow{CH_3OSO_2OCH_3}$

Deoxyguanosine

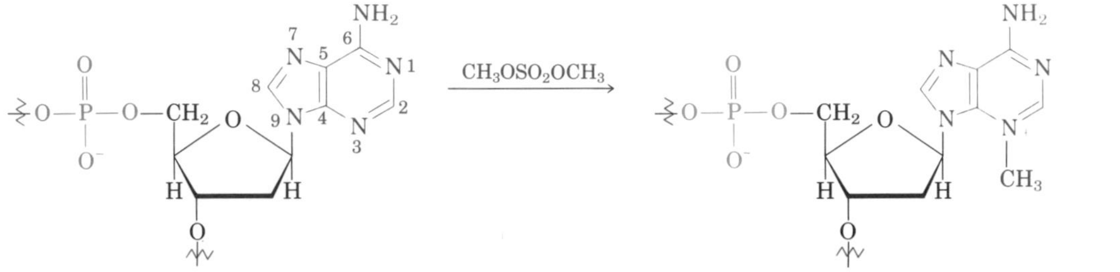

$\xrightarrow{CH_3OSO_2OCH_3}$

Deoxyadenosine

Treatment of methylated DNA with an aqueous solution of the secondary amine piperidine then brings about destruction of the methylated nucleotides and specific opening of the DNA chain at both the 3′ and the 5′ positions next to the methylated bases. The mechanism of the cleavage process is complex and involves a number of different steps, as shown in Figure 29.15 for deoxyguanosine breakage: (1) Hydrolysis occurs, opening the five-membered heterocycle; (2) hydrolysis of the aminoglycoside sugar linkage then yields an open-chain 2-deoxyribose; (3) formation of an enamine between piperidine and the 2-deoxyribose aldehyde group occurs; and (4) two eliminations of the 2-deoxyribose oxygen substituents at C3 and C5 take place. These two elimination reactions break open the DNA chain.

By working carefully, Maxam and Gilbert were able to find reaction conditions that are selective for cleavage either at A or at G. Thus, G methylates five times as rapidly as A but the hydrolytic breakdown of methylated A occurs more rapidly than that of methylated G if the product is first heated with dilute acid prior to base treatment.

Breaking the DNA chain next to both pyrimidine nucleotides C and T can be accomplished by treatment of DNA with hydrazine, followed by heating with aqueous piperidine. Although conditions that are selective for cleavage next to T have not been found, a selective cleavage next to C can be accomplished by carrying out the hydrazine reaction in 2*M* NaCl. The mechanism of deoxythymidine cleavage by hydrazine is shown in Figure 29.16 (page 1066).

Once again, the breakdown reaction involves numerous steps: (1) Hydrazine first undergoes a conjugate addition to thymine, followed by (2) an intramolecular nucleophilic acyl substitution reaction that opens the heterocyclic thymine ring; (3) breakage of the aminoglycoside linkage and (4) hydrolysis yields open-chain deoxyribose, which (5) forms an enamine by reaction with piperidine; (6) two elimination reactions then break open the DNA chain.

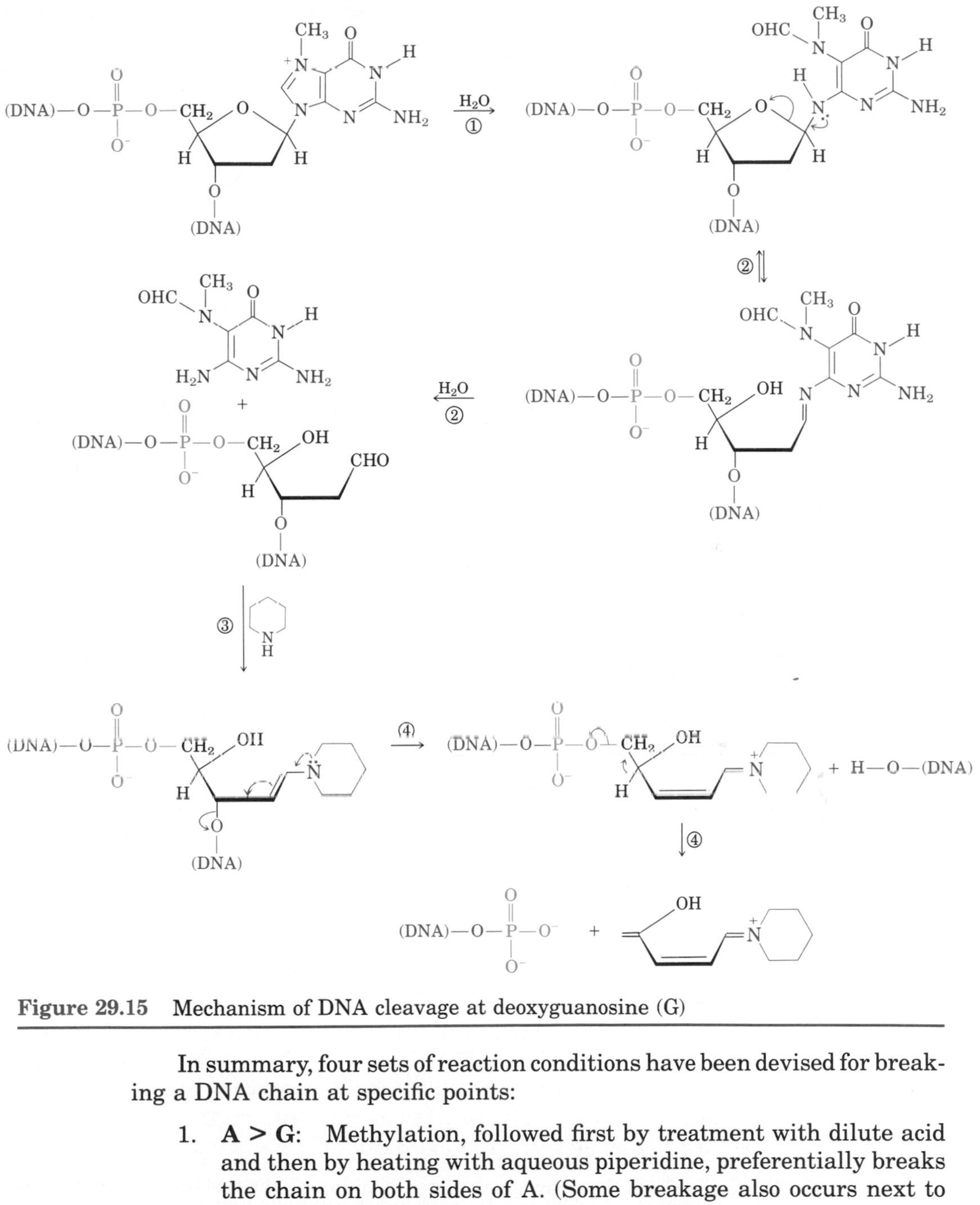

Figure 29.15 Mechanism of DNA cleavage at deoxyguanosine (G)

In summary, four sets of reaction conditions have been devised for breaking a DNA chain at specific points:

1. **A > G**: Methylation, followed first by treatment with dilute acid and then by heating with aqueous piperidine, preferentially breaks the chain on both sides of A. (Some breakage also occurs next to G.)
2. **G > A**: Methylation, followed by heating with aqueous piperidine, preferentially breaks the chain on both sides of G. (Some breakage also occurs next to A.)
3. **C**: Treatment with hydrazine in 2*M* NaCl, followed by heating with aqueous piperidine, breaks the chain on both sides of C.

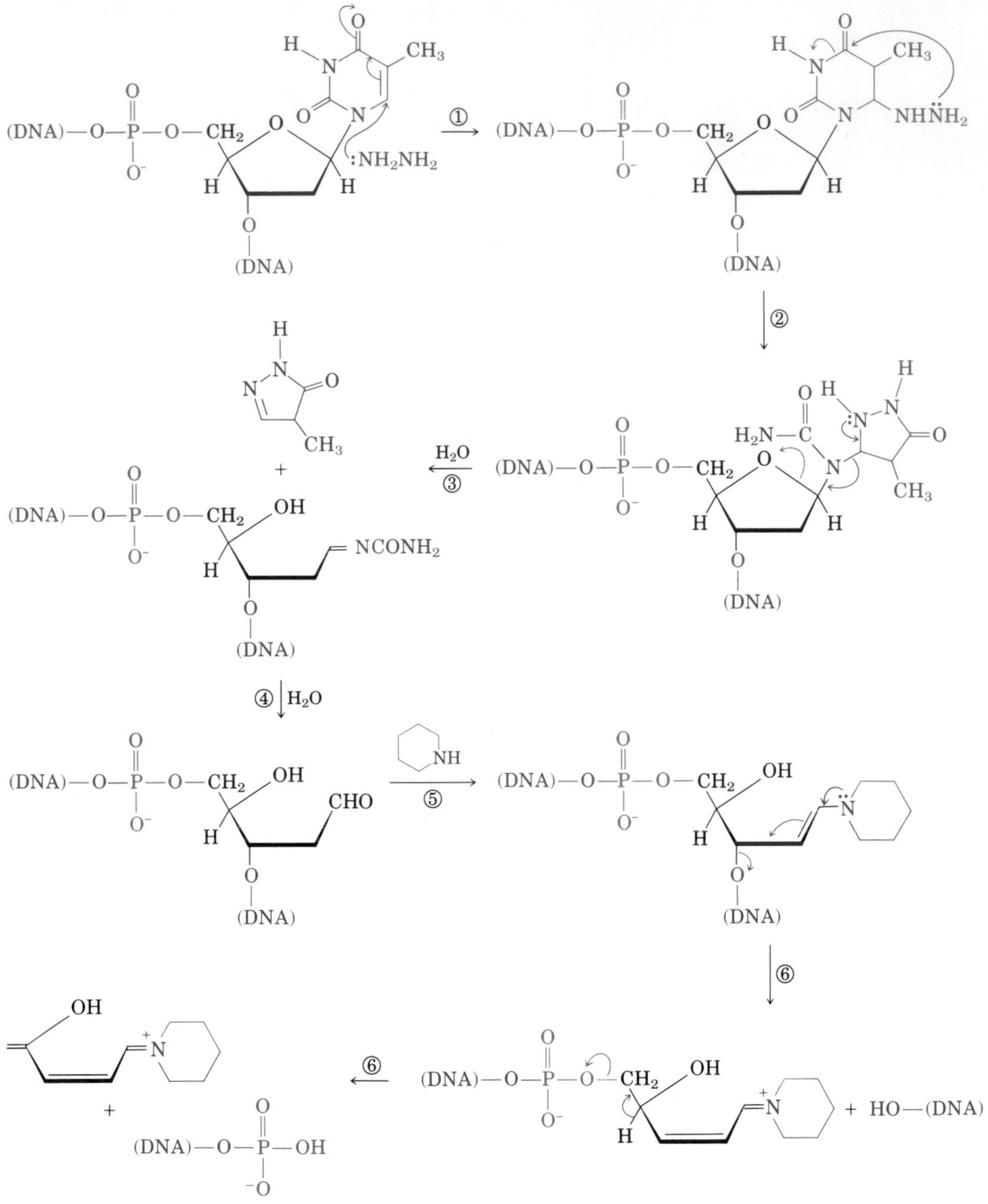

(DNA)—O—P—O—CH₂ … :NH₂NH₂ ① (DNA)—O—P—O—CH₂ … NHNH₂

② H₂N—C … CH₃

H₂O ③ NCONH₂ … CH₃

④ H₂O

CHO ⑤ NH

⑥ + HO—(DNA)

⑥ (DNA)—O—P—OH

**Figure 29.16** Mechanism of DNA cleavage at deoxythymidine (T) by reaction with hydrazine

4. **C + T**: Treatment with hydrazine in the absence of NaCl, followed by heating with aqueous piperidine, breaks the chain next to *both* C and T.

After the restriction fragment has been broken down by selective cleavage reactions into a mixture of smaller pieces, the mixture is separated by electrophoresis (Section 27.3). When the mixture of DNA pieces is placed at one end of a strip of buffered gelatinous polyacrylamide and a voltage difference is applied across the ends of the strip, electrically charged pieces move along the gel. Each piece moves at a rate that depends both on its size and on the number of negatively charged phosphate groups (the number of nucleotides) it contains; smaller pieces move rapidly, and larger pieces move more slowly. The technique is so sensitive that up to 250 DNA pieces, differing in size by only one nucleotide, can be separated.

Once separation of the pieces has been accomplished, the position on the gel of each radioactive $^{32}$P-containing piece is determined by exposing the gel to a photographic plate. Only the pieces containing the radioactively labeled 5′-end phosphate group are visualized; unlabeled pieces from the middle of the chain are present but do not appear on the photographic plate.

At this point, we've discussed methods for selectively degrading a DNA restriction fragment, for separating the degradation products, and for visualizing those products. How can these methods be used for sequencing? Let's follow a DNA fragment through the series of steps just discussed to see how they lead ultimately to a sequence for the DNA fragment.

*Step 1* A DNA molecule is incubated with a restriction enzyme, which cuts the chain at specific places and yields DNA fragments containing 100–200 nucleotide pairs.

*Step 2* The double-stranded DNA restriction fragments are radioactively labeled by incorporation of a $^{32}$P-containing phosphate group at the 5′-hydroxyl end of the terminal nucleotide.

*Step 3* The labeled restriction fragments are isolated, and each is separated into its two complementary strands. For example, imagine that we have now isolated a single-stranded DNA segment of approximately 100 nucleotides with the following partial structure:

(5′ end) $^{32}$P-C-T-C-A-G-T-A-C-C-G- - - - - - - - - (3′ end)

*Step 4* The labeled single-stranded DNA segment is subjected to four parallel sets of cleavage experiments under conditions that lead to (a) preferential splitting next to A, (b) preferential splitting next to G, (c) exclusive splitting next to C, and (d) splitting next to both T and C. *Mild reaction conditions are chosen so that only a few of the many possible splittings occur in each reaction.* In our example, the pieces shown in Table 29.4 (page 1068) would be produced.

*Step 5* Product mixtures from the four cleavage reactions are separated by gel electrophoresis, and the spots on the gel are visualized by exposing the gel to a photographic plate. The location

Table 29.4 Splitting a DNA restriction fragment under four sets of conditions[a]

| *Cleavage conditions* | *Pieces produced* | | |
|---|---|---|---|
| Original DNA segment | $^{32}$P-C-T-C-A-G-T-A-C-C-G- - - - | | |
| A > G | $^{32}$P-C-T-C | | |
| | $^{32}$P-C-T-C-A-G-T | + | Larger pieces |
| G > A | $^{32}$P-C-T-C-A | | |
| | $^{32}$P-C-T-C-A-G-T-A-C-C | + | Larger pieces |
| C | $^{32}$P-C-T | | |
| | $^{32}$P-C-T-C-A-G-T-A | | |
| | $^{32}$P-C-T-C-A-G-T-A-C | + | Larger pieces |
| C + T | $^{32}$P-C | | |
| | $^{32}$P-C-T | | |
| | $^{32}$P-C-T-C-A-G | | |
| | $^{32}$P-C-T-C-A-G-T-A | | |
| | $^{32}$P-C-T-C-A-G-T-A-C | + | Larger pieces |

[a]Only the pieces containing the radioactive end-label are considered. Other pieces are also produced but are not visualized.

of each radioactive piece appears as a dark band on the photographic plate; nonradioactive pieces are not visualized. The gel electrophoresis pattern shown in Figure 29.17 would be obtained in our hypothetical example.

*Step 6* The DNA sequence is then read directly from the gel. The band that appears farthest from the origin is the terminal mononucleotide (the smallest piece) and cannot be identified. Since the terminal mononucleotide appears only in the T + C column, however, it must have been produced by splitting *next to* a T or a C. Thus, the *second* nucleotide in the sequence is a T or a C. Since this smallest piece does not appear in the C column, however, the second nucleotide is not a C and must be a T.

The second farthest band from the origin is a dinucleotide that appears in both C and T + C columns. This dinucleotide is produced by splitting next to the third nucleotide, which must therefore be a C. The third farthest band appears mainly in the A > G column, which means that the fourth nucleotide is an A. (There is also a faint band in the G > A column, since the specificity of these splittings is not complete.)

Continuing in this manner, the entire sequence of the DNA can be read from the gel simply by noting in what columns the successively larger labeled polynucleotide pieces appear. Once read, the entire sequence can be checked by determining the sequence of the complementary strand. The identity of the 5′-

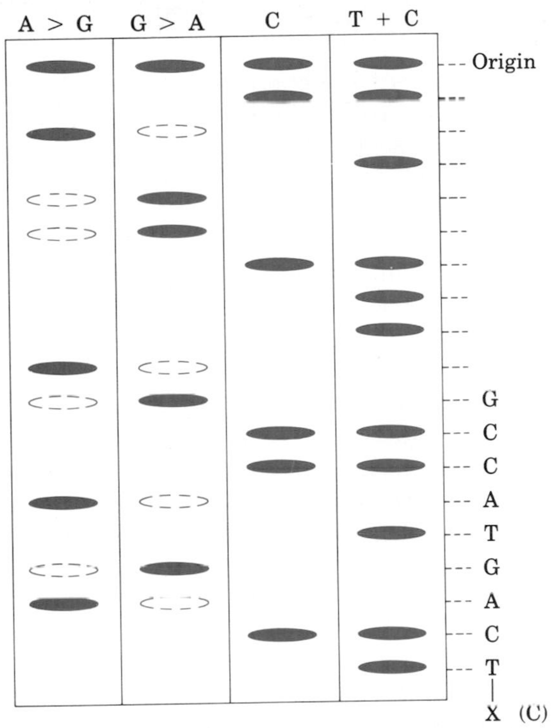


**Figure 29.17** Representation of a gel electrophoresis pattern: The products of the four parallel cleavage experiments are placed at the top of the gel and a voltage is applied between top and bottom. Smaller products migrate along the gel at a faster rate and thus appear at the bottom. The DNA sequence can be read from the positions of the radioactive spots.

terminal nucleotide can be determined by sequencing an overlapping segment produced by cleavage with another restriction enzyme.

The Maxam–Gilbert method of DNA sequencing is so efficient that sequencing rates of up to 200 nucleotides per day can be achieved. So powerful is the method, in fact, that some *protein* sequencing is now done by sequencing the DNA from which the protein's mRNA was transcribed. Impressive achievements are being recorded increasingly, including the sequencing of the Epstein–Barr virus having 170,000 base pairs.

PROBLEM

**29.19** Show the labeled cleavage products you would expect to obtain if the following DNA segment were subjected to each of the four cleavage reactions:

$^{32}$P-A-A-C-A-T-G-G-C-G-C-T-T-A-T-G-A-C-G-A

PROBLEM

**29.20** Sketch what you would expect the gel electrophoresis pattern to look like if the DNA segment in Problem 29.19 were sequenced.

PROBLEM

**29.21** Finish assigning the sequence to the pattern shown in Figure 29.17.

## 29.16 Laboratory Synthesis of DNA

The development of genetic engineering techniques in the 1970s brought with it an increased demand for efficient chemical methods for the synthesis of short DNA segments. Ideally, whole genes might be synthesized in the laboratory and inserted into the DNA of microorganisms, thereby directing the microorganisms to produce the specific protein coded for by that gene—perhaps insulin or some other valuable material.

The problems of DNA synthesis are similar to those of protein synthesis (Section 27.11) but are considerably more difficult because of the structural complexity of the deoxyribonucleotide monomers. Each nucleotide has several reactive sites that must be selectively protected and deprotected at the proper times, and coupling of the four nucleotides must be carried out in the proper sequence. Despite these difficulties, some extremely impressive achievements have been recorded, including the synthesis by Khorana[4] in 1979 of the tyrosine suppressor tRNA gene from the bacterium *Escherichia coli*. Some 207 base pairs were assembled in an effort that required 10 years of work.

More recently, automated "gene machines" have become available, which allow the fast and reliable synthesis of DNA sequences 10–20 nucleotides long. These DNA synthesizers, based on chemistry developed in the mid-1960s, operate on a principle similar to that of the Merrifield solid-phase peptide synthesizer (Section 27.12). In essence, a protected nucleotide is covalently bound to a solid support, and one nucleotide at a time is added to the chain by means of a coupling reagent. When the last nucleotide has been added, the protecting groups are removed, and the synthetic DNA is cleaved from the solid support.

The first step in DNA synthesis involves attachment of a protected nucleoside fragment to the polymer support by an ester linkage to the 3′ position of a deoxyribonucleoside. Both the 5′-hydroxyl and free amino groups on the heterocyclic bases must be protected to accomplish this attachment. Adenine and cytosine bases are protected by benzoyl groups, guanine is protected by an isobutyryl group, and thymine requires no protection. The deoxyribose 5′-hydroxyl group is protected as its DMT ether (*para*-dimethoxytrityl). Other protecting groups may also be used, but the ones mentioned here are common choices.

Step 2 involves deprotection of the 5′-deoxyribose hydroxyl by Lewis acid-catalyzed cleavage with $ZnBr_2$ to remove the DMT group.

Step 3 involves phosphate ester formation between the free 5′ position of the polymer-bound nucleoside and the 3′-phosphate group of a protected nucleotide. This ester-forming step involves use of a coupling agent such as 1-mesitylenesulfonyl-3-nitro-1,2,4-triazole (MSNT), a reagent that serves the same purpose in nucleotide synthesis that DCC serves in protein synthesis (Section 27.11). Note that one of the phosphate oxygens is protected as the *o*-chlorophenyl ester in this coupling step.

---

[4]Har Gobind Khorana (1922– ); b. Raipur, India; Ph.D. University of Liverpool; professor, Massachusetts Institute of Technology; Nobel prize in medicine (1968).

*Step 1.*

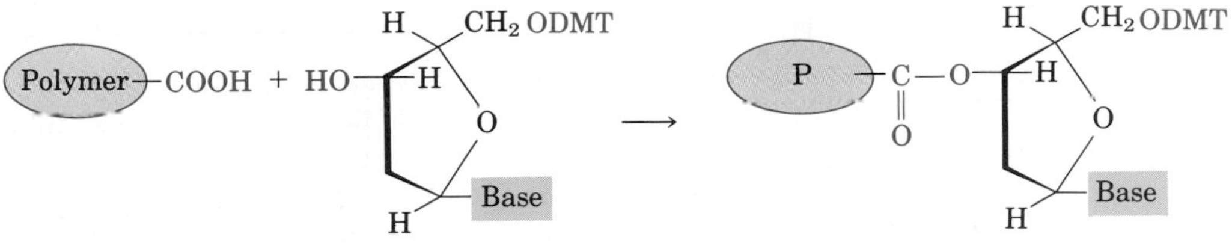

Polymer—COOH + HO—(deoxyribose, $CH_2ODMT$, Base) → P—C(=O)—O—(deoxyribose, $CH_2ODMT$, Base)

where DMT = $-C(C_6H_5)(C_6H_4OCH_3)_2$

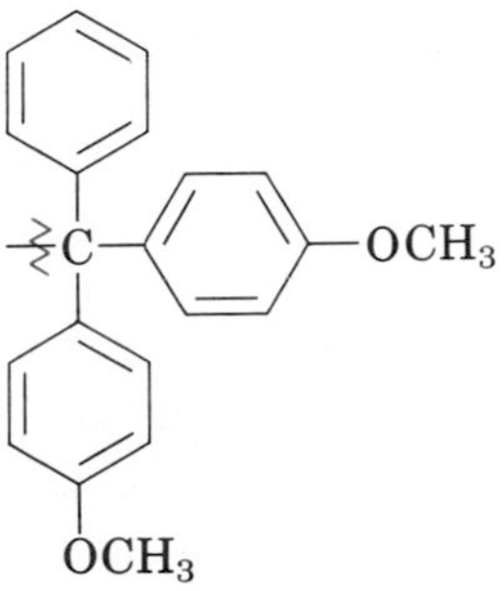

Base =

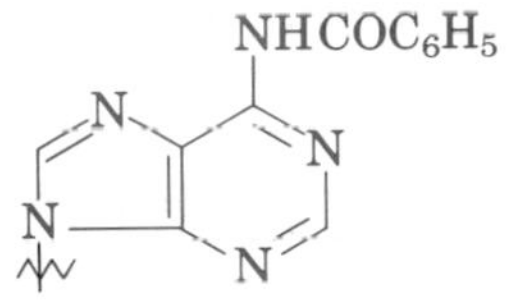

$NHCOC_6H_5$

N-protected adenine

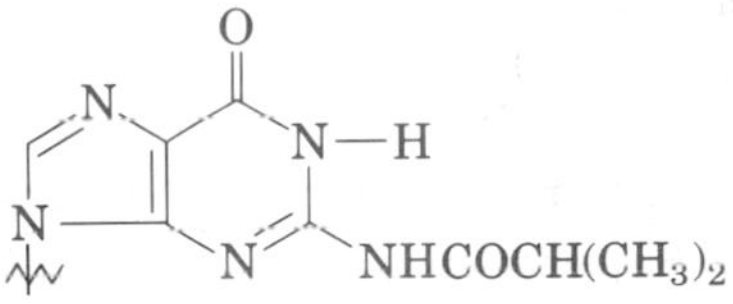

$NHCOCH(CH_3)_2$

N-protected guanine

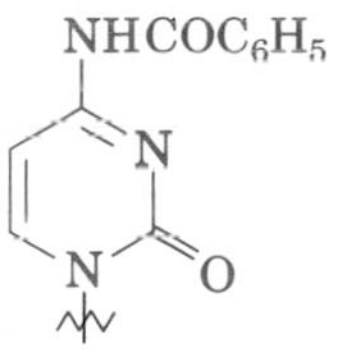

$NHCOC_6H_5$

N-protected cytosine

$H_3C$

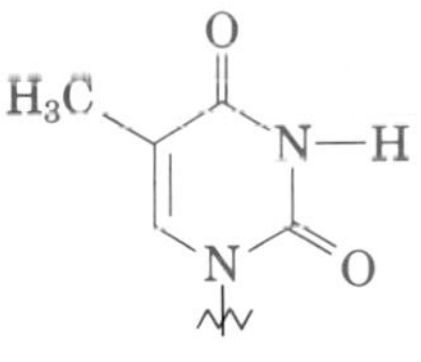

Thymine

*Step 2.*

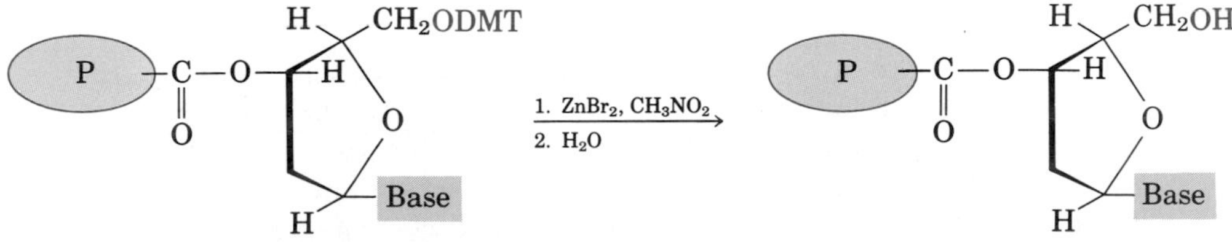


The cycle of steps 2 and 3 is then repeated until a polydeoxyribonucleotide chain of the desired length and sequence has been built. The final step is to cleave all protecting groups from the heterocyclic bases and from the phosphates and to cleave the ester bond holding the polynucleotide to the polymer. All these reactions can be effected by treatment with aqueous ammonia.

*Step 3.*

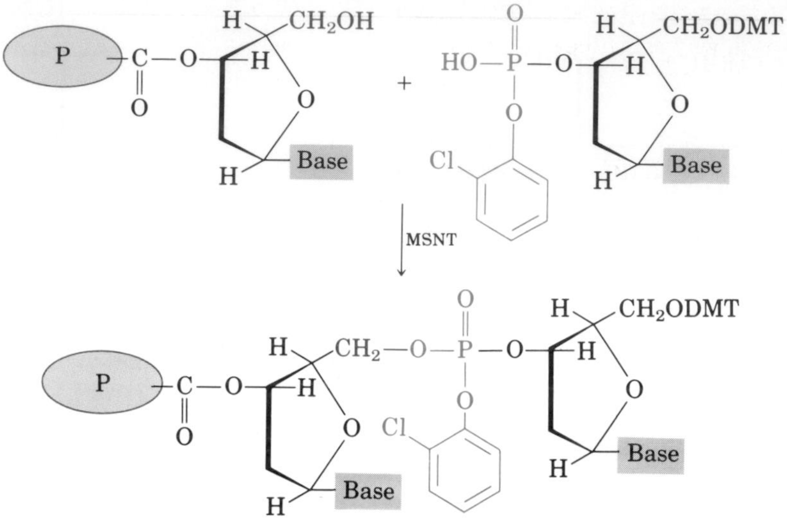

MSNT

where MSNT = $H_3C$—(2,6-dimethylphenyl)—$SO_2$—N(triazole)—$NO_2$

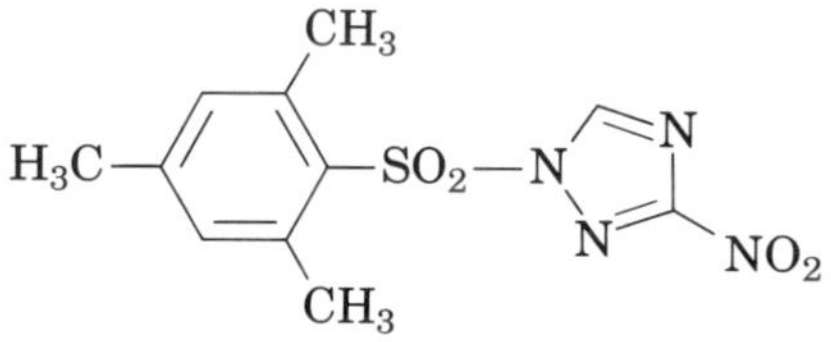

*Final step.*

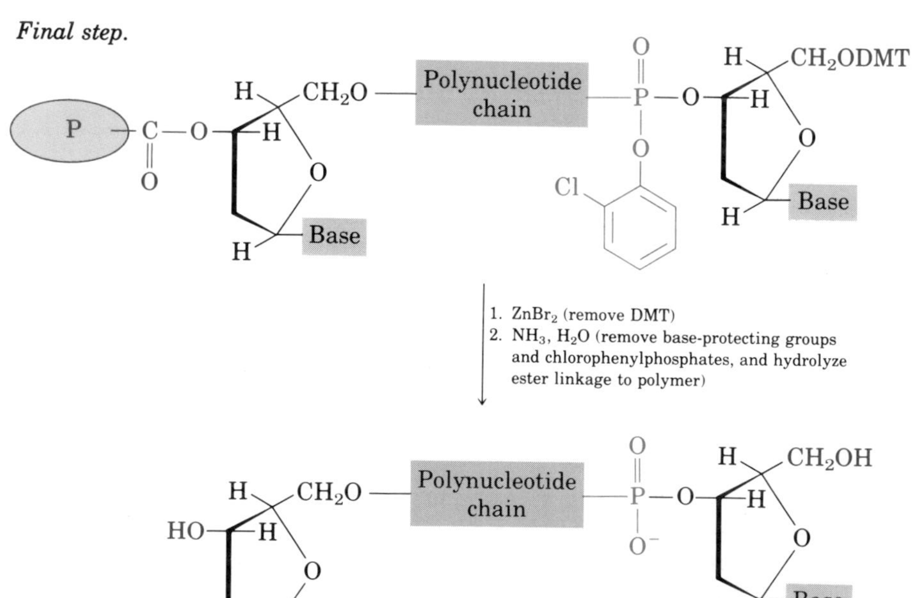


Synthetic DNA

PROBLEM

**29.22** *p*-Dimethoxytrityl (DMT) ethers are easily cleaved by mild Lewis acid. Suggest a reason why this ether cleavage is unusually easy.

## 29.17 Summary and Key Words

A **heterocycle** is a compound having a ring that contains more than one kind of atom. Nitrogen, oxygen, and sulfur are often found along with carbon in heterocyclic rings. Saturated heterocyclic amines, ethers, and sulfides usually display the same chemistry as their open-chain analogs, but unsaturated heterocycles often display aromaticity. Pyrrole, furan, and thiophene are the simplest five-membered aromatic heterocycles. All three are unusually stable, and all three undergo aromatic substitution when reacted with electrophiles. Reaction usually occurs at the highly activated position next to the heteroatom.

Pyridine is the six-membered-ring, nitrogen-containing heterocyclic analog of benzene. The pyridine ring is electron-poor and undergoes electrophilic aromatic substitution reactions with great difficulty. Nucleophilic aromatic substitutions of 2- or 4-halopyridines take place readily, however.

The nucleic acids—**DNA** (**deoxyribonucleic acid**) and **RNA** (**ribonucleic acid**)—are biological polymers that act as chemical carriers of an organism's genetic information. Enzyme-catalyzed hydrolysis of nucleic acids yields **nucleotides**, the monomer units from which RNA and DNA are constructed. Each nucleotide consists of a **purine** or **pyrimidine base** linked to C1′ of a simple pentose sugar (ribose in RNA and 2′-deoxyribose in DNA), with the sugar in turn linked through its C5′-hydroxyl to a phosphate group:

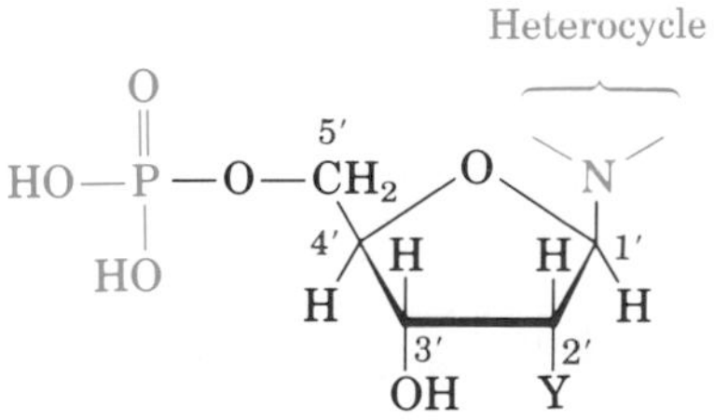

Heterocycle

O
||
HO—P—O—$CH_2$ (5′) — furanose ring O, N at 1′; H at 4′, 3′, 2′, 1′; 3′ OH; 2′ Y
|
HO

where Y = OH (a ribonucleotide)

= H (a deoxyribonucleotide)

The nucleotides are joined by phosphate links between the phosphate of one nucleotide and the 3′-hydroxyl group on the sugar of another nucleotide.

Molecules of DNA consist of two polynucleotide strands held together by hydrogen bonds between heterocyclic bases on the different strands and coiled into a **double-helix** conformation. **Adenine** and **thymine** form hydrogen bonds to each other, as do **cytosine** and **guanine**. The two strands of DNA are not identical but are complementary.

Three main processes take place in deciphering the genetic information of DNA:

1. **Replication** of DNA is the process by which identical DNA copies are made and genetic information is preserved. This occurs when the DNA double helix unwinds, complementary deoxyribonucleotides line up in order, and two new DNA molecules are produced.
2. **Transcription** is the process by which RNA is produced to carry genetic information from the nucleus to the ribosomes. This occurs when a segment of the DNA double helix unwinds, and complementary ribonucleotides line up to produce **messenger RNA** (**mRNA**).
3. **Translation** is the process by which mRNA directs protein synthesis. Each mRNA has segments called **codons** along its chain. These codons are ribonucleotide triads that are recognized by small amino acid carrying molecules of **transfer RNA** (**tRNA**), which then deliver the appropriate amino acids needed for protein synthesis.

Small DNA segments can be synthesized in the laboratory, and commercial "gene machines" are available for automating the work. Sequencing of DNA can be carried out rapidly and efficiently by the **Maxam–Gilbert method**.

## 29.18 Summary of Reactions

1. Electrophilic aromatic substitution in five-membered heterocycles (Section 29.3)

   a. Bromination

   Furan $\xrightarrow[\text{Dioxane}]{Br_2}$ 2-bromofuran + HBr

   b. Nitration

   Pyrrole $\xrightarrow[\text{Acetic anhydride}]{HNO_3}$ 2-nitropyrrole ($NO_2$) + $H_2O$

   c. Friedel–Crafts acylation

   Thiophene $\xrightarrow[SnCl_4]{CH_3COCl}$ 2-acetylthiophene ($\overset{O}{\overset{\|}{C}}CH_3$) + HCl

2. Nucleophilic aromatic substitution of halopyridines (Section 29.6)

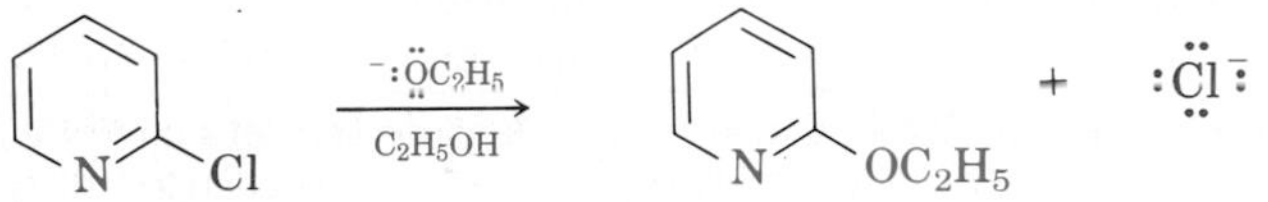

2-chloropyridine $\xrightarrow[C_2H_5OH]{^-:\ddot{O}C_2H_5}$ 2-ethoxypyridine ($OC_2H_5$) + $:\ddot{Cl}:^-$

## ADDITIONAL PROBLEMS

**29.23** Although pyrrole is a much weaker base than most other amines, it is a much stronger acid ($pK_a \approx 15$ for pyrrole versus 35 for diethylamine). The N—H proton is readily abstracted by base to yield the pyrrole anion, $C_4H_4N^-$. Explain.

**29.24** Oxazole is a five-membered aromatic heterocycle. Draw an orbital picture of oxazole, showing all *p* orbitals and all lone-pair orbitals. Would you expect oxazole to be more basic or less basic than pyrrole? Explain.

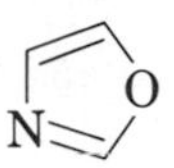

Oxazole

**29.25** Write the products of the reaction of furan with each of these reagents:
(a) $Br_2$, dioxane, 0°C (b) $HNO_3$, acetic anhydride (c) $CH_3COCl$, $SnCl_4$
(d) $H_2/Pd$ (e) $SO_3$, pyridine

**29.26** Nitrofuroxime is a pharmaceutical agent used in the treatment of urinary tract infections. Propose a synthesis of nitrofuroxime from furfural.

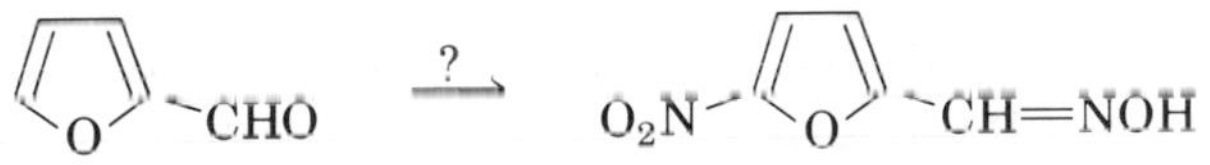

Furfural (furan–CHO) $\xrightarrow{?}$ Nitrofuroxime ($O_2N$–furan–CH=NOH)

**29.27** Substituted pyrroles are often prepared by treatment of the appropriate 1,4-diketone with ammonia. Suggest a mechanism by which this reaction occurs. (Review Section 19.12.)

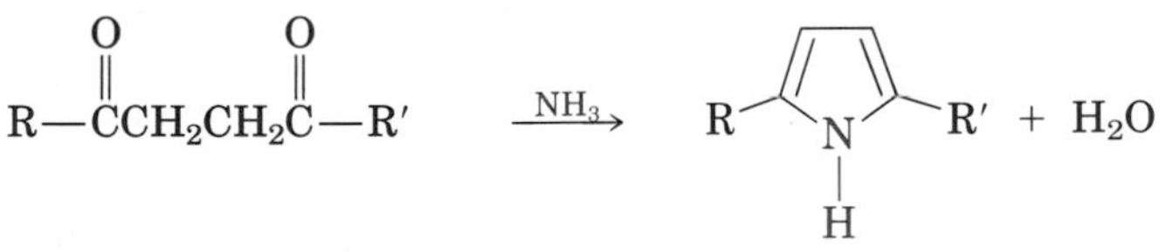

$$R{-}\overset{O}{\overset{\|}{C}}CH_2CH_2\overset{O}{\overset{\|}{C}}{-}R' \xrightarrow{NH_3} \text{2-R-5-R'-pyrrole (N–H)} + H_2O$$

**29.28** 3,5-Dimethylisoxazole is prepared by reaction of 2,4-pentanedione with hydroxylamine. Propose a mechanism for this reaction.

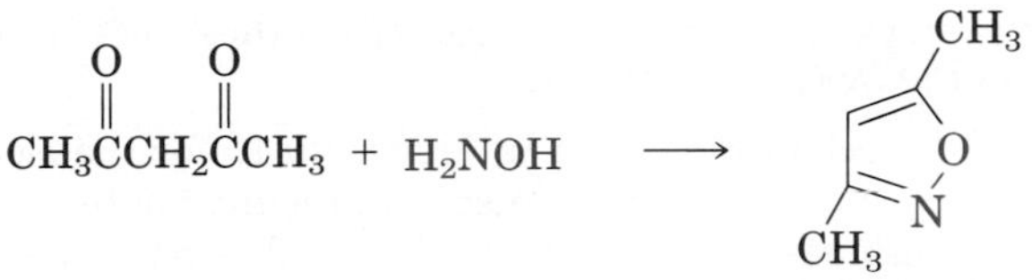

$$CH_3\overset{O}{\overset{\|}{C}}CH_2\overset{O}{\overset{\|}{C}}CH_3 + H_2NOH \longrightarrow \text{3,5-dimethylisoxazole } (CH_3, CH_3)$$

3,5-Dimethylisoxazole

**29.29** Define these terms:
(a) Heterocycle (b) DNA (c) Base pair
(d) Transcription (e) Translation (f) Replication (of DNA)
(g) Codon (h) Anticodon

**29.30** What amino acids do these ribonucleotide triplets code for?
(a) AAU (b) GAG (c) UCC (d) CAU

**29.31** From what DNA sequences were each of the mRNA codons in Problem 29.30 transcribed?

**29.32** What anticodon sequences of tRNA's are coded for by the codons in Problem 29.30?

**29.33** Draw the complete structure of the ribonucleotide codon U-A-C. For what amino acid does this sequence code?

**29.34** Draw the complete structure of the deoxyribonucleotide sequence from which the mRNA codon in Problem 29.33 was transcribed.

**29.35** Give an mRNA sequence that would code for synthesis of met-enkephalin:

H-Tyr-Gly-Gly-Phe-Met-OH

**29.36** Look up the structure of angiotensin II (Figure 27.4) and give an mRNA sequence that would code for its synthesis.

**29.37** What amino acid sequence is coded for by this mRNA base sequence?

CUA-GAC-CGU-UCC-AAG-UGA

**29.38** Isoquinolines are often synthesized by the Bischler—Napieralski cyclization of an *N*-acyl-2-phenylethyl amine with strong acid and $P_2O_5$, followed by oxidation of the initially formed dihydroisoquinoline. Suggest a mechanism by which this cyclization occurs.

NH, O=C, $CH_3$ $\xrightarrow[P_2O_5]{H_3PO_4}$ N, $CH_3$ $\xrightarrow{HNO_3}$ N, $CH_3$

A dihydroisoquinoline 1-Methylisoquinoline

**29.39** Quinolines are often prepared by the Skraup[5] synthesis, in which an aniline reacts with an $\alpha,\beta$-unsaturated aldehyde and the dihydroquinoline product is oxidized. Suggest a mechanistic pathway for the Skraup reaction.

$NH_2$ + $H_2C{=}CHCHO$ $\xrightarrow{HNO_3}$ N, H $\xrightarrow{HNO_3}$ N

1,2-Dihydroquinoline Quinoline

**29.40** Show the steps involved in a laboratory synthesis of the DNA fragment having the sequence C-T-A-G.

**29.41** Review the mechanism shown in Figure 29.15 for the cleavage of deoxyguanosine residues, and propose a mechanism to account for the similar cleavage of deoxyadenosine residues in a DNA chain. Recall that deoxyadenosine is first methylated at N3 prior to hydrolysis.

[5]Hans Zdenko Skraup (1850–1910); b. Austria; professor, Graz, Vienna, Austria.

CHAPTER 30

# Orbitals and Organic Chemistry: Pericyclic Reactions

Most organic reactions take place by polar mechanisms, in which an electron-rich nucleophile donates two electrons to form a bond to an electron-poor electrophile. Many other organic reactions take place by radical mechanisms, in which each of two reactants donates one electron to form a new bond. Although much remains to be learned about polar and radical reactions, the broad outlines of both classes have been studied for many years and seem relatively well understood.

By contrast, the fundamental principles of **pericyclic reactions**, the third major class of organic reaction mechanisms, have been understood only recently. Numerous individuals have made major contributions, but it was the work of Robert Woodward[1] and Roald Hoffmann[2] in the mid-1960s that made most chemists aware of the principles of pericyclic reactions.

## 30.1 Some Examples of Pericyclic Reactions

We previously defined a pericyclic reaction as one that occurs by a concerted process through a cyclic transition state. The word *concerted* means that all bonding changes occur *at the same time* and *in a single step*; no intermediates are involved. Rather than try to expand this definition now, let's look at some examples of the three major classes of pericyclic reactions.

The Diels–Alder reaction (Section 14.7) is an intermolecular pericyclic reaction between a diene and a dienophile to yield a cyclohexene. Three

[1]Robert Burns Woodward (1917–1979); b. Boston, Mass.; Ph.D. Massachusetts Institute of Technology (1937); professor, Harvard University (1941–1979); Nobel prize (1965).

[2]Roald Hoffmann (1937– ); b. Zloczow, Poland; Ph.D. Harvard University (1962); professor, Cornell University; Nobel prize (1981).

bonds are formed (two sigma bonds and one pi bond) and three bonds are broken (two diene pi bonds and one alkene pi bond) at the same time via a cyclic transition state. This is an example of a **cycloaddition reaction**, so named because the two reagents add together to yield a cyclic product. The product of a cycloaddition always has fewer pi bonds than the starting materials.

Diels-Alder reaction, a cycloaddition

A diene A dienophile Cyclic transition state A cyclohexene

The Claisen rearrangement (Section 26.9) is an intramolecular pericyclic reaction of an allyl aryl ether. Three new bonds are formed (one sigma bond and two pi bonds) and three bonds are broken (one sigma bond and two pi bonds) at the same time via a cyclic transition state. This is an example of a **sigmatropic rearrangement**, so named because the starting molecule rearranges by changing a sigma bond (Greek *tropos*; meaning "change" or "turn"). One sigma bond in the starting material breaks, and a new sigma bond forms in the product.

Claisen rearrangement, a sigmatropic rearrangement

An allyl aryl ether Cyclic transition state

Enolize

An *o*-allylphenol

A third kind of pericyclic process, called an **electrocyclic reaction**, occurs when polyenes are heated or irradiated with ultraviolet light. For example, heating (2*E*,4*Z*,6*E*)-octatriene leads exclusively to formation of *cis*-5,6-dimethyl-1,3-cyclohexadiene. Heating the (2*E*,4*Z*,6*Z*)-octatriene isomer, however, leads exclusively to formation of *trans*-5,6-dimethyl-1,3-cyclohexadiene. These electrocyclic reactions are so named because they involve a cyclic reorganization of electrons. The product of an electrocyclic reaction always has one more sigma bond than the starting material and one less pi bond.

Both of the above triene cyclizations, and many other pericyclic reactions as well, are **stereospecific**; that is, they yield only a *single* product stereoisomer rather than a mixture. Why should this be so?

Electrocyclic reactions

(2*E*,4*Z*,6*E*)-Octatriene → Cyclic transition state → *cis*-5,6-Dimethyl-1,3-cyclohexadiene

(2*E*,4*Z*,6*Z*)-Octatriene → Cyclic transition state → *trans*-5,6-Dimethyl-1,3-cyclohexadiene

The answer to this question is fundamental to the nature of pericyclic reactions and has to do with symmetry properties of the reactant and product orbitals. To understand how orbital symmetry affects reactivity, we need to look more deeply into the mathematical description of orbitals.

## 30.2 Atomic and Molecular Orbitals

Let's begin by briefly reviewing some of the concepts introduced in Chapter 1. We have defined an orbital as the region of space where a given electron is most likely to be found. In mathematical terms, orbitals are derived from quantum mechanical calculations involving electron wave functions. Although the exact nature of these calculations is not important for our purposes, the different lobes of an orbital turn out to have algebraic signs, + and −, when the wave functions are solved. For example, the two equivalent lobes of a *p* atomic orbital have plus and minus signs, and the two nonequivalent lobes of an $sp^3$ hybrid orbital also have plus and minus signs (Figure 30.1).

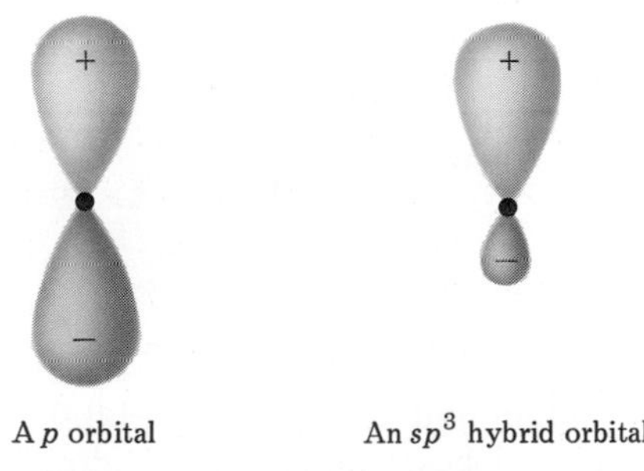


**Figure 30.1** Signs of lobes in *p* and $sp^3$ orbitals

When a bond is formed between two atoms by overlap of atomic orbitals, the two electrons in the bond no longer occupy atomic orbitals but instead occupy molecular orbitals (MO's). Overlap of two atomic orbitals leads to two molecular orbitals, one of which is lower in energy than either of the original two atomic orbitals, and one of which is higher in energy.

As shown in Figure 30.2, the lower-energy molecular orbital, denoted $\psi_1$ (Greek psi), is a **bonding molecular orbital** because it results from additive overlap of two lobes with the *same* algebraic sign. The higher-energy molecular orbital, denoted $\psi_2^*$, is an **antibonding molecular orbital** because it results from subtractive overlap of two lobes with the *opposite* algebraic sign. Note that the antibonding orbitals have nodes between nuclei—planes of zero electron density between lobes of opposite sign. Higher-energy orbitals always have more nodes between nuclei than lower-energy orbitals and thus have fewer favorable bonding interactions.

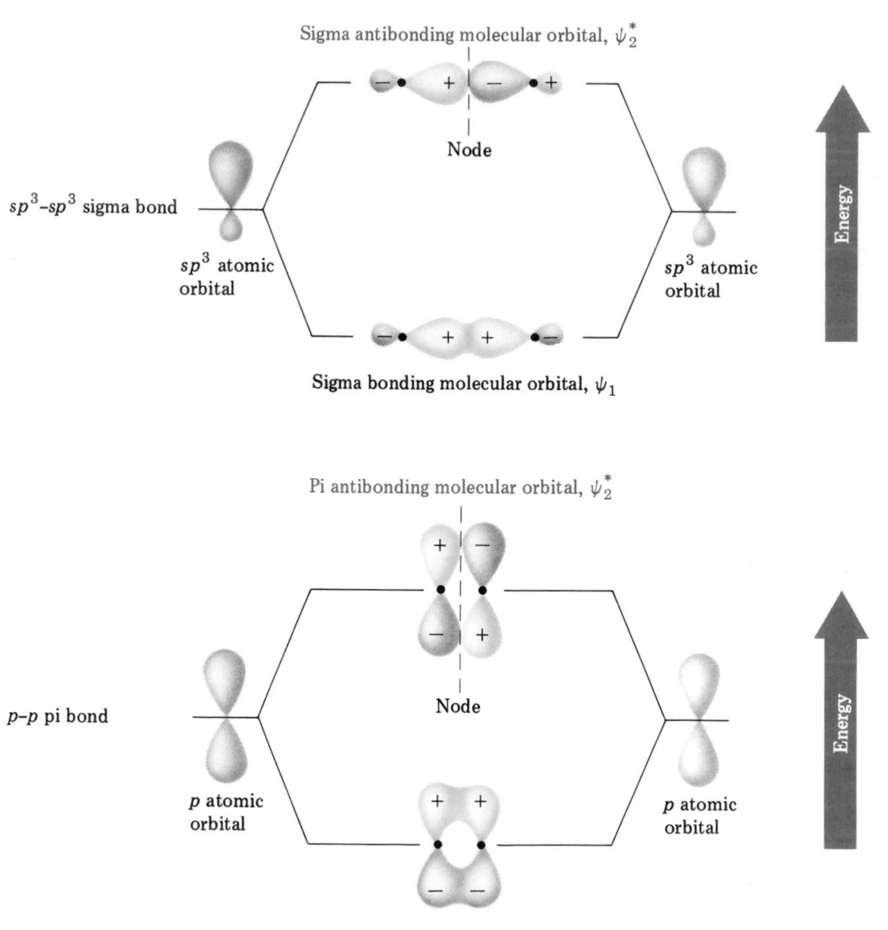


**Figure 30.2** Sigma and pi molecular orbitals: Bonding molecular orbitals result from overlap of like lobes, whereas antibonding molecular orbitals result from overlap of unlike lobes.

Having now constructed molecular orbitals, we can arrive at a complete description of sigma and pi bonds by assigning electrons to the orbitals. The method used is exactly the same as that used when we assigned electrons to atomic orbitals in describing the electronic configuration of atoms (the *aufbau* principle, Section 1.3). Each molecular orbital can hold two electrons of opposite spin.

The assignments are made by filling the lowest-energy molecular orbitals first and by filling each molecular orbital with two electrons before going on to the next higher orbital. For example, the ground-state electronic configuration of an alkene pi bond has $\psi_1$ filled and $\psi_2^*$ vacant. Irradiation with ultraviolet light, however, excites an electron from $\psi_1$ to $\psi_2^*$, leaving each orbital with one electron (Section 14.12). The electronic configurations of both ground and excited states of ethylene are shown in Figure 30.3.

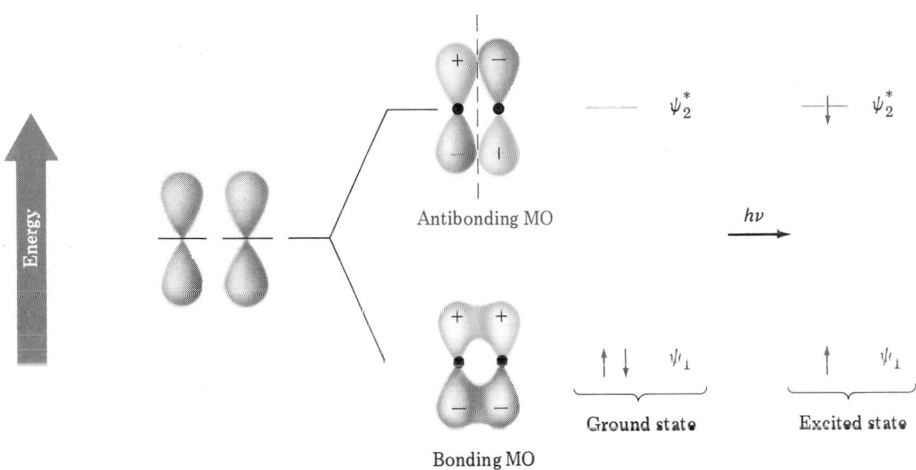


**Figure 30.3** Ground-state and excited-state electronic configurations of the ethylene pi bond

## 30.3 Molecular Orbitals of Conjugated Pi Systems

The molecular orbital description of a conjugated pi system is more complex than that of a simple alkene because the pi electrons are delocalized over more than two atoms. In 1,3-butadiene, for example, four $2p$ atomic orbitals combine into four diene molecular orbitals spanning the entire pi system. Two of these molecular orbitals are bonding (lower energy, $\psi_1$ and $\psi_2$), and two are antibonding (higher energy, $\psi_3^*$ and $\psi_4^*$), as shown in Figure 30.4 on page 1082. The two bonding orbitals in ground-state butadiene are occupied by four electrons, whereas the two antibonding orbitals are unoccupied. Note that, once again, the higher the energy of a molecular orbital, the more nodes it has between nuclei.

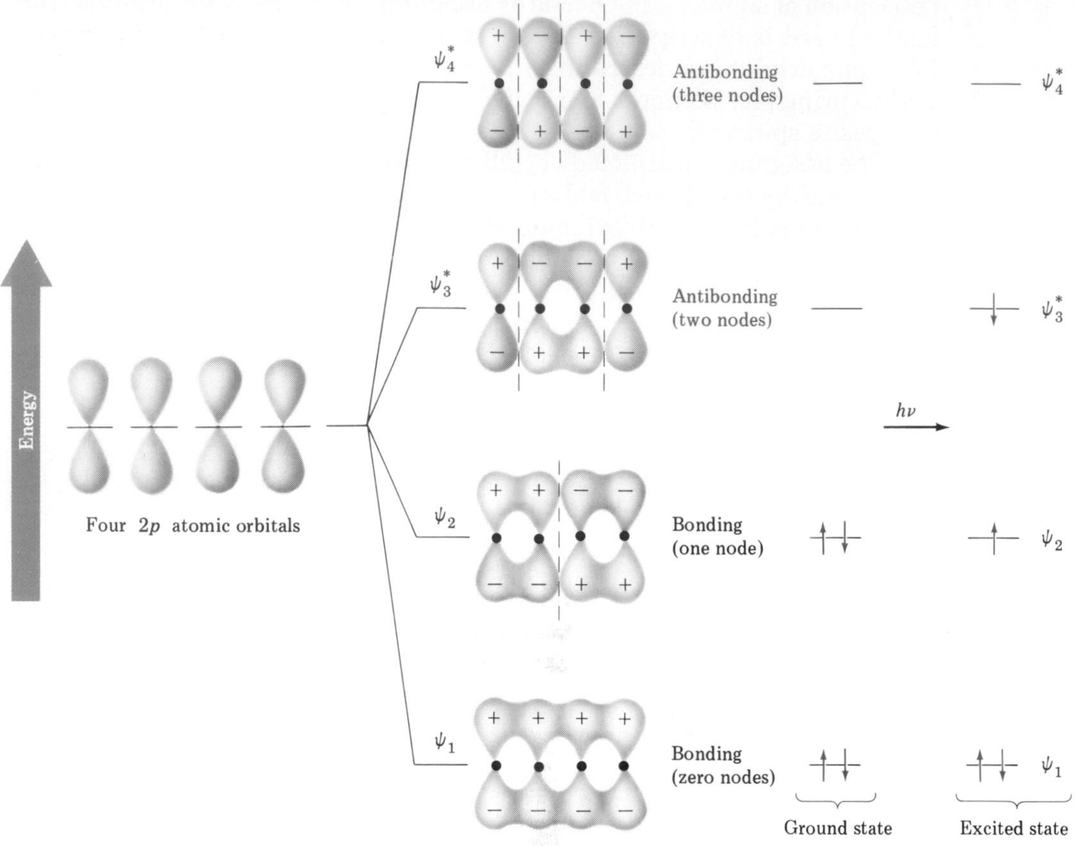


**Figure 30.4** Pi molecular orbitals of 1,3-butadiene: The two bonding orbitals are occupied, and the two antibonding orbitals are vacant in the ground state.

A similar sort of molecular orbital description can be derived for a conjugated triene, or for *any* conjugated pi electron system. The six molecular orbitals of 1,3,5-hexatriene are shown in Figure 30.5. Only the three bonding orbitals, $\psi_1$, $\psi_2$, and $\psi_3$, are occupied in the ground state, whereas $\psi_3$ and $\psi_4^*$ have one electron each in the excited state.

## 30.4 Molecular Orbitals and Pericyclic Reactions

What do molecular orbitals and the signs of their lobes have to do with pericyclic reactions? The answer is *everything*! According to a series of rules formulated by Woodward and Hoffmann, a pericyclic reaction can take place only if the symmetry of all reactant molecular orbitals is the same as the symmetry of the product molecular orbitals. In other words, the lobes of reactant molecular orbitals must be of the correct algebraic sign for bonding overlap to occur in the transition state leading to the product.

If the orbital symmetries of both reactant and product match up, or "correlate," the reaction is said to be **symmetry-allowed**. If the orbital

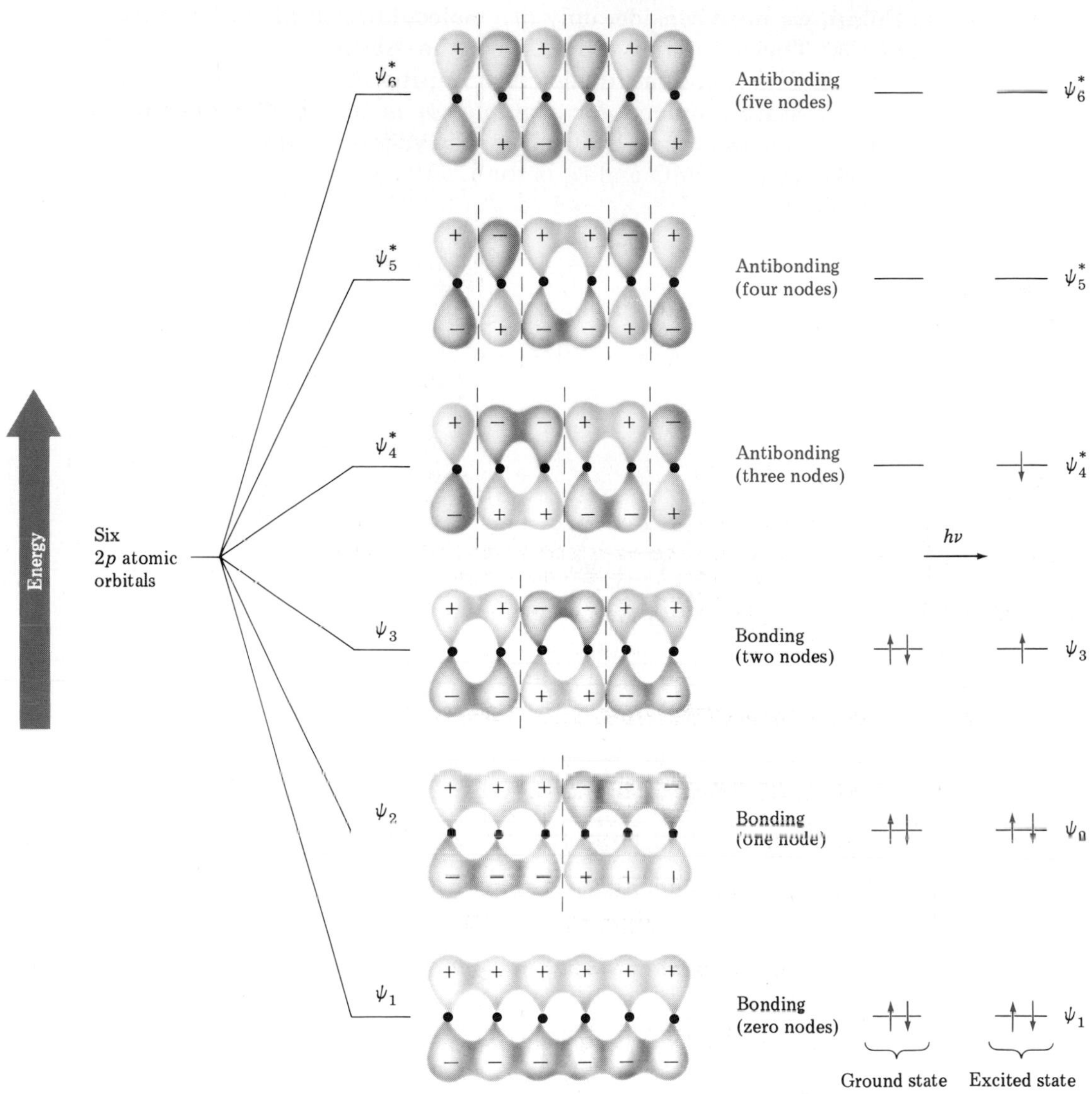


**Figure 30.5** Molecular orbitals of 1,3,5-hexatriene

symmetries of reactant and product do not correlate, however, the reaction is **symmetry-disallowed**. Symmetry-allowed reactions often occur under relatively mild conditions, but symmetry-disallowed reactions cannot occur by concerted paths. They take place either by nonconcerted pathways under forcing, high-energy conditions, or not at all.

The Woodward–Hoffmann rules for pericyclic reactions require an analysis of *all* reactant and product molecular orbitals, but Kenichi Fukui[3] at

[3]Kenichi Fukui (1918– ); b. Nara Prefecture, Japan; Ph.D. Kyoto University; professor, Kyoto University; Nobel prize (1981).

Kyoto University in Japan has introduced a simplified version. According to Fukui, we need consider only *two* molecular orbitals called the **frontier orbitals**. These frontier orbitals are the **highest occupied molecular orbital (HOMO)** and the **lowest unoccupied molecular orbital (LUMO)**. In ground-state ethylene, for example, $\psi_1$ is the HOMO since it has two electrons, and $\psi_2^*$ is the LUMO since it is vacant. In ground-state 1,3-butadiene, $\psi_2$ is the HOMO and $\psi_3^*$ is the LUMO (Figure 30.6).

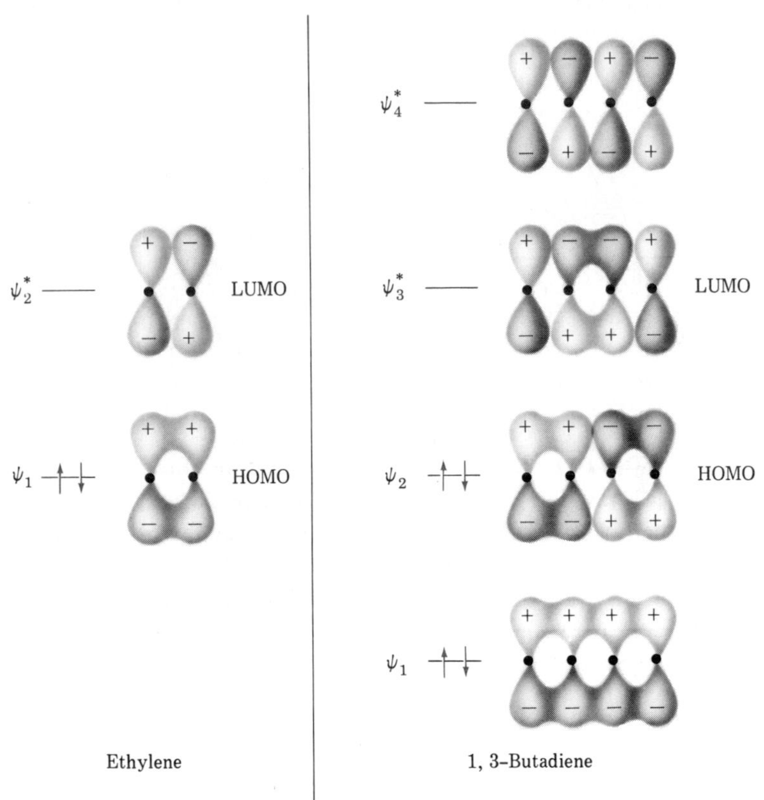


**Figure 30.6** The HOMO and LUMO orbitals of ground-state ethylene and 1,3-butadiene

The best way to understand how orbital symmetry affects pericyclic reactions is to look at some examples. Let's look first at some electrocyclic reactions of polyenes and then go on to cycloaddition reactions and sigmatropic rearrangements.

PROBLEM

**30.1** Refer to Figure 30.5 to find the molecular orbitals of a conjugated triene, and tell which molecular orbital is the HOMO and which is the LUMO for both ground and excited states.

## 30.5 Electrocyclic Reactions

**Electrocyclic reactions** are pericyclic processes that involve the cyclization of conjugated polyenes. One pi bond is broken, the other pi bonds change position, a new sigma bond is formed, and a cyclic compound results. For example, conjugated trienes can be converted into cyclohexadienes, and conjugated dienes can be converted into cyclobutenes.

Both of these reactions are reversible, and the position of the equilibrium depends on the specific case. In general, the triene ⇄ cyclohexadiene equilibrium favors the ring-closed product, whereas the diene ⇄ cyclobutene equilibrium favors the ring-opened product.

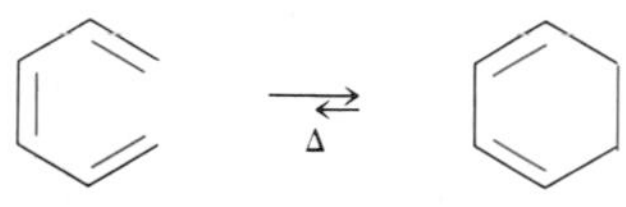

A conjugated triene ⇄ (Δ) A cyclohexadiene

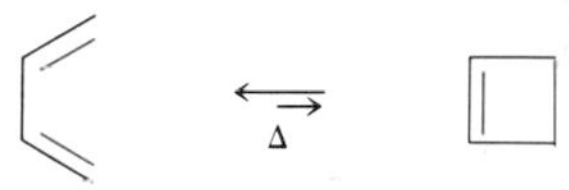

A conjugated diene ⇄ (Δ) A cyclobutene

The key feature of electrocyclic reactions is their stereochemistry. For example, (2*E*,4*Z*,6*E*)-octatriene yields only *cis*-5,6-dimethyl-1,3-cyclohexadiene when heated, and (2*E*,4*Z*,6*Z*)-octatriene yields only *trans*-5,6-dimethyl-1,3-cyclohexadiene. Remarkably, however, the stereochemical results change completely when the reactions are carried out under *photochemical*, rather than thermal, conditions. Thus, irradiation of (2*E*,4*Z*,6*E*)-octatriene with ultraviolet light yields *trans*-5,6-dimethyl-1,3-cyclohexadiene (Figure 30.7).

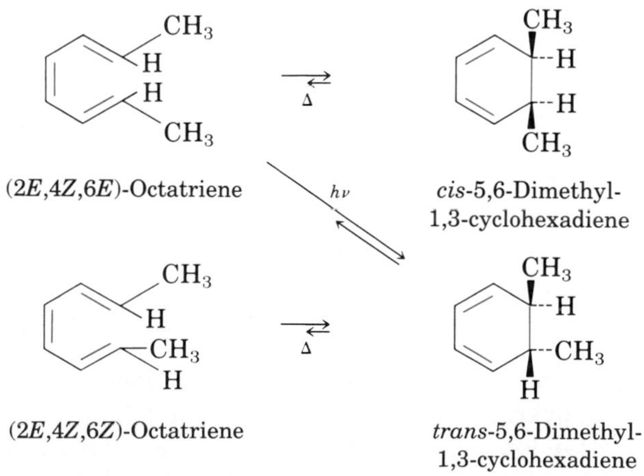

**Figure 30.7** Electrocyclic interconversions of 2,4,6-octatrienes and 5,6-dimethyl-1,3-cyclohexadienes

A similar result is obtained for the thermal electrocyclic ring opening of the 3,4-dimethylcyclobutenes. The trans isomer yields only (2*E*,4*E*)-hexadiene when heated, and the cis isomer yields only (2*E*,4*Z*)-hexadiene. On irradiation, however, the results are again different: Cyclization of the 2*E*,4*E* isomer under photochemical conditions yields cis product (Figure 30.8).

*cis*-3,4-Dimethylcyclobutene $\underset{175°C}{\rightleftarrows}$ (2*E*,4*Z*)-Hexadiene

*trans*-3,4-Dimethylcyclobutene $\underset{175°C}{\rightleftarrows}$ (2*E*,4*E*)-Hexadiene

(2*E*,4*E*)-Hexadiene $\xrightarrow{h\nu}$ *cis*-3,4-Dimethylcyclobutene

**Figure 30.8** Electrocyclic interconversions of 2,4-hexadienes and 3,4-dimethylcyclobutenes

All these stereospecific electrocyclic reactions can be accounted for by orbital-symmetry arguments. To do so, we need to look only at the symmetries of the two outermost lobes of the polyene. There are two possibilities: The lobes of like sign can be either on the same side or on opposite sides of the molecule.

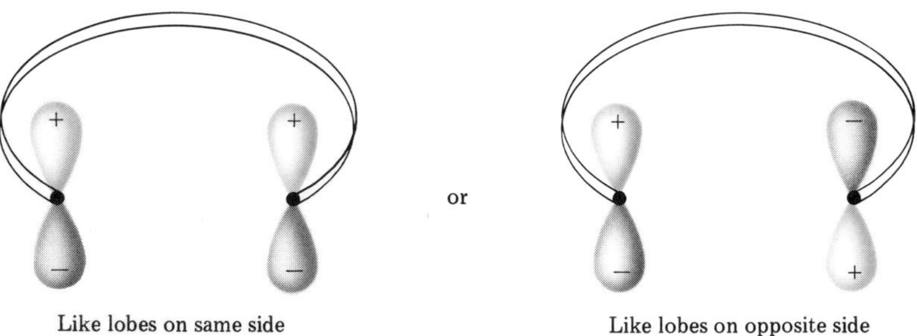


For a bond to form, the outermost pi lobes must rotate so that favorable bonding overlap is achieved—a positive lobe overlapping a positive lobe or a negative lobe overlapping a negative lobe. If two lobes of like sign are on the same side of the molecule, the two orbitals must rotate in *different* directions. One orbital must rotate clockwise and one must rotate counterclockwise. This kind of motion is referred to as **disrotatory**:

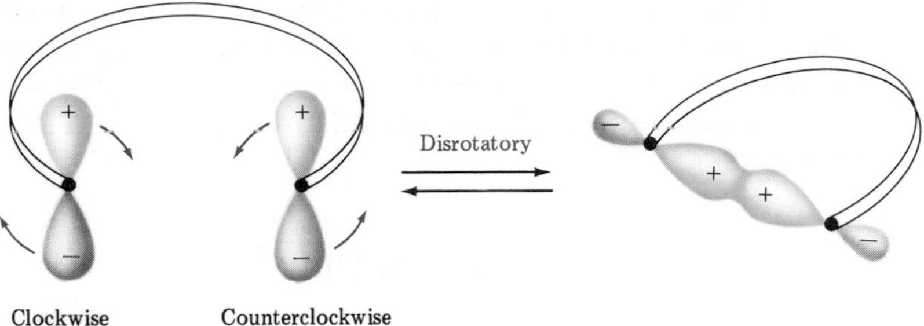


Conversely, if lobes of like sign are on opposite sides of the molecules, both orbitals must rotate in the *same* direction, either both clockwise or both counterclockwise. This kind of motion is called **conrotatory**:

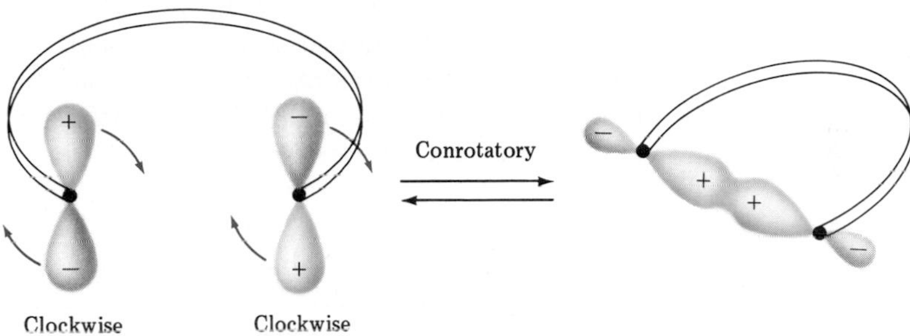


Now let's see what happens to substituents on the polyene carbon atoms when cyclization occurs. To choose a specific example, let's look at the thermal cyclization of (2*E*,4*Z*,6*E*)-octatriene. If a disrotatory cyclization were to occur, *cis*-5,6-dimethyl-1,3-cyclohexadiene would result. If a conrotatory cyclization were to occur, *trans*-5,6-dimethyl-1,3-cyclohexadiene would result (Figure 30.9).

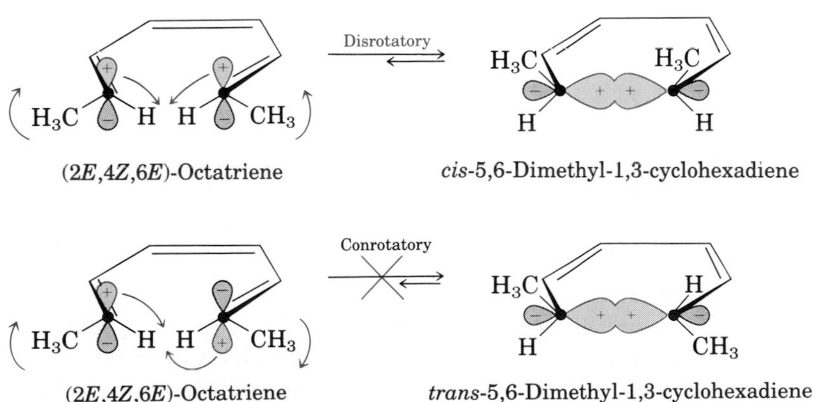


**Figure 30.9** Stereochemistry of disrotatory and conrotatory cyclization of (2*E*,4*Z*,6*E*)-octatriene

In fact, only the disrotatory cyclization of (2*E*,4*Z*,6*E*)-octatriene is observed to occur. We therefore conclude that the stereochemistry of an electrocyclic reaction is determined by the mode of ring closure. The mode of ring closure, in turn, is determined by the symmetry of reactant molecular orbitals.

## 30.6 Stereochemistry of Thermal Electrocyclic Reactions

How can we predict which mode of ring closure, conrotatory or disrotatory, will occur in a given case? How can we tell whether the terminal lobes of like sign will be on the same side or on opposite sides of the molecule?

According to frontier orbital theory, *the stereochemistry of an electrocyclic reaction is determined by the symmetry of the polyene's HOMO*. The electrons in the HOMO are the highest-energy, most loosely held electrons, and are therefore most easily moved during reaction. For thermal ring openings and closings, the ground-state electronic configuration is used to identify the HOMO; for photochemical ring openings and closings, the excited-state electronic configuration is used.

Let's look again at the thermal ring closure of conjugated trienes. According to Figure 30.5, the HOMO of a conjugated triene in its ground state has a symmetry that predicts a disrotatory ring closure:

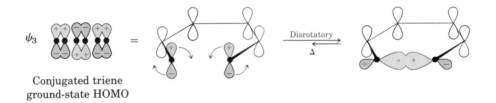

$\psi_3$ = Disrotatory, Δ

Conjugated triene ground-state HOMO

This disrotatory cyclization is exactly what is observed in the thermal cyclization of 2,4,6-octatriene. The 2*E*,4*Z*,6*E* isomer yields cis product; the 2*E*,4*Z*,6*Z* isomer yields trans product (Figure 30.10).

In a similar manner, the ground-state HOMO of a conjugated diene (Figure 30.4) has a symmetry that predicts conrotatory ring closure:

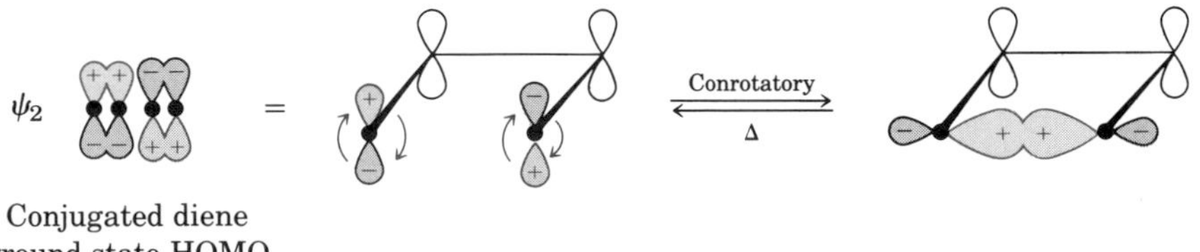


Conjugated diene ground-state HOMO

In practice, of course, the conjugated diene reaction can only be observed in the reverse direction (cyclobutene → butadiene) because of the position of the equilibrium. We therefore predict that the 3,4-dimethylcyclobutene ring will *open* in a conrotatory fashion, which is exactly what is observed.

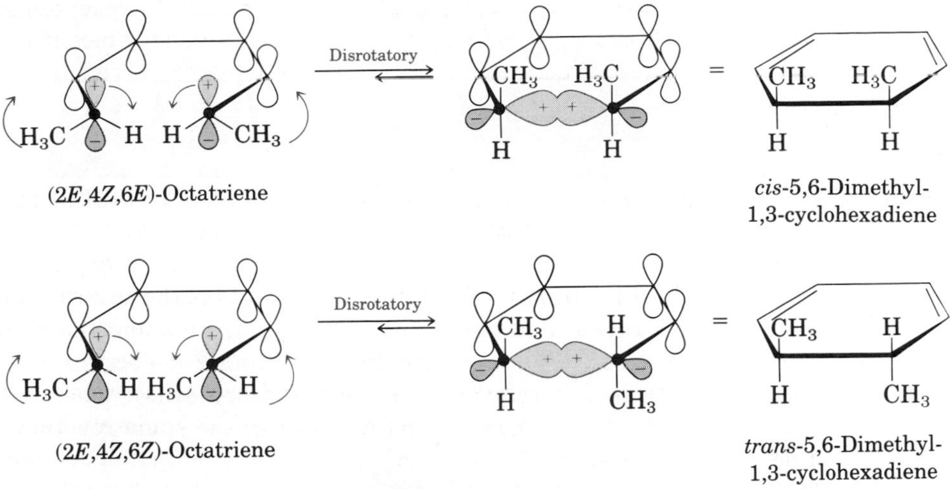

*cis*-5,6-Dimethyl-1,3-cyclohexadiene

*trans*-5,6-Dimethyl-1,3-cyclohexadiene

**Figure 30.10** Thermal disrotatory ring closure of 2,4,6-octatrienes

*cis*-3,4-Dimethylcyclobutene yields (2*E*,4*Z*)-hexadiene, and *trans*-3,4-dimethylcyclobutene yields (2*E*,4*E*)-hexadiene by conrotatory opening (Figure 30.11).

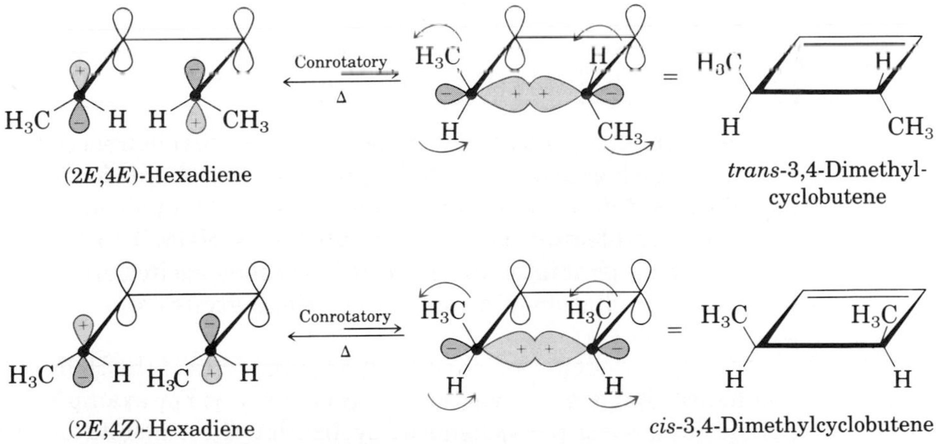

*trans*-3,4-Dimethylcyclobutene

*cis*-3,4-Dimethylcyclobutene

**Figure 30.11** Conrotatory ring opening of the *cis*- and *trans*-dimethylcyclobutenes

Note that a conjugated diene and a conjugated triene react in opposite stereochemical senses. The diene opens and closes by a conrotatory path, whereas the triene opens and closes by a disrotatory path. This difference is, of course, due to the different symmetries of the diene and triene HOMO's:

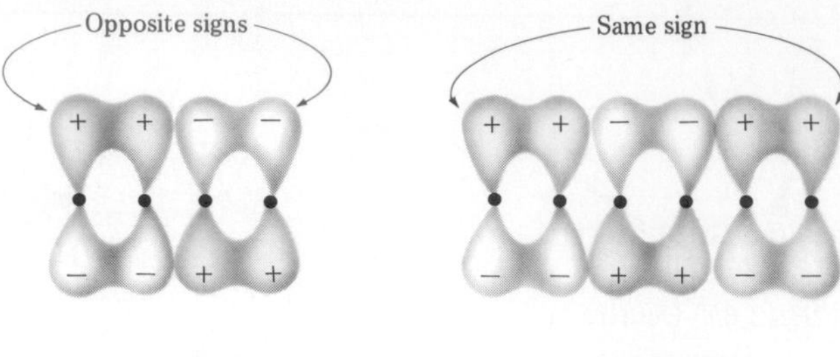


There is an alternating relationship between the number of electron pairs (double bonds) undergoing bond reorganization and the mode of ring closure (or opening). Polyenes with an even number of electron pairs undergo thermal electrocyclic reactions in a conrotatory sense, whereas polyenes with an odd number of electron pairs undergo the same reactions in a disrotatory sense.

**PROBLEM**

**30.2** Draw the products you would expect to obtain from conrotatory and disrotatory cyclizations of (2*Z*,4*Z*,6*Z*)-octatriene. Which of the two paths would you expect the thermal reaction to follow?

**PROBLEM**

**30.3** In theory, *trans*-3,4-dimethylcyclobutene can open by two conrotatory paths to give either (2*E*,4*E*)-hexadiene or (2*Z*,4*Z*)-hexadiene. Explain why both products are symmetry-allowed, and then account for the fact that only the 2*E*,4*E* isomer is obtained in practice.

## 30.7 Photochemical Electrocyclic Reactions

We noted previously that photochemical electrocyclic reactions take a different stereochemical course than their thermal counterparts. We can now explain this difference. Ultraviolet irradiation of a polyene causes an excitation of one electron from the ground-state HOMO to the ground-state LUMO. Thus, irradiation of a conjugated diene excites an electron from $\psi_2$ to $\psi_3^*$, and irradiation of a conjugated triene excites an electron from $\psi_3$ to $\psi_4^*$ (Figure 30.12).

Electronic excitation changes the symmetries of the HOMO and LUMO and hence changes the reaction stereochemistry. For example, (2*E*,4*E*)-hexadiene undergoes photochemical cyclization by a disrotatory path (recall that the thermal reaction is conrotatory):

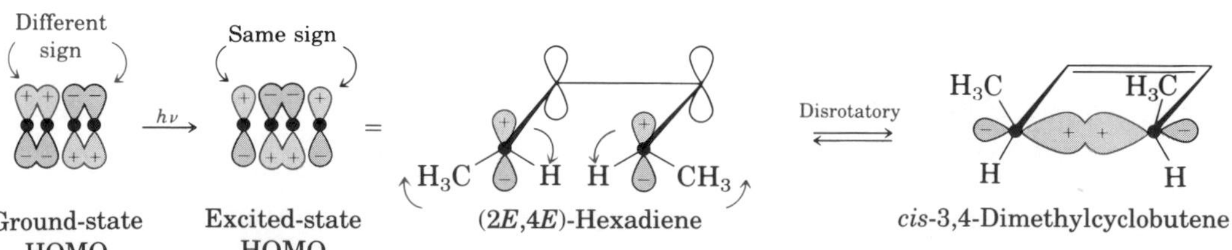

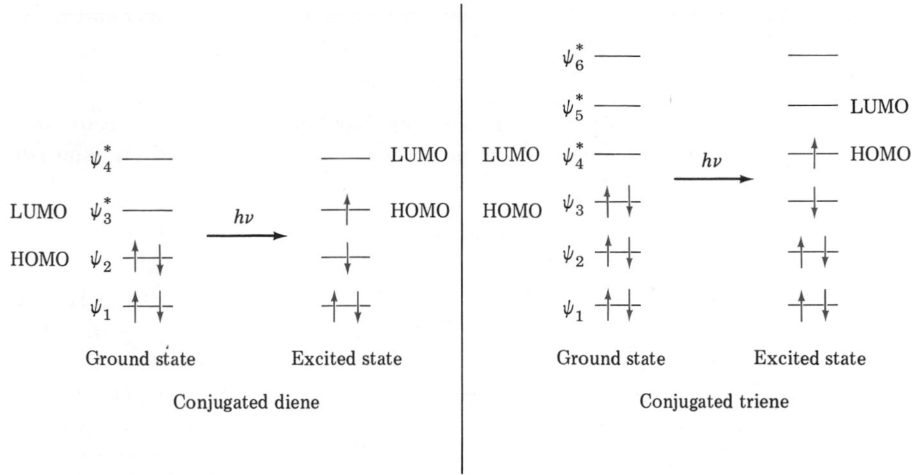


**Figure 30.12** Ground-state and excited-state electronic configurations of conjugated dienes and trienes

Similarly, (2*E*,4*Z*,6*E*)-octatriene undergoes photochemical cyclization by a conrotatory path (the thermal reaction is disrotatory):

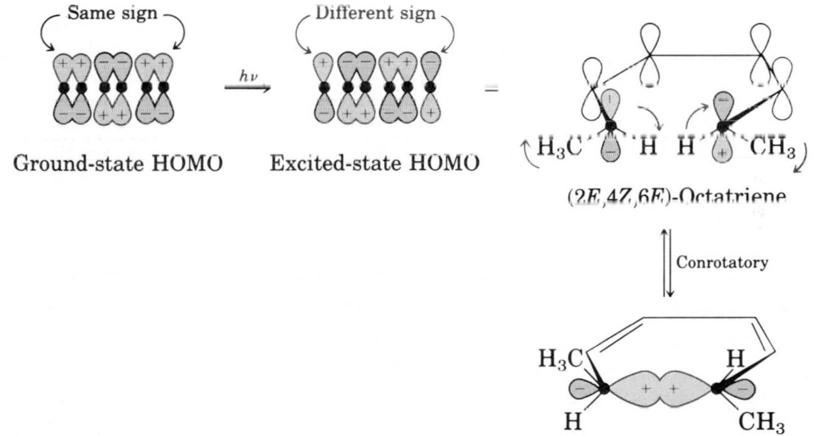


Thermal and photochemical electrocyclic reactions *always* take place with opposite stereochemistry, as the rules governing these processes indicate (Table 30.1). Learning these simple rules allows us to predict the stereochemistry of a large number of organic reactions.

**Table 30.1** Stereochemical rules for electrocyclic reactions

| *Electron pairs (double bonds)* | *Thermal reaction* | *Photochemical reaction* |
|---|---|---|
| Even number | Conrotatory | Disrotatory |
| Odd number | Disrotatory | Conrotatory |

PROBLEM

**30.4** What product would you expect to obtain from the photochemical cyclizations of (2*E*,4*Z*,6*E*)-octatriene? Of (2*E*,4*Z*,6*Z*)-octatriene?

PROBLEM

**30.5** The following thermal isomerization has been reported to occur under relatively mild conditions. Identify the pericyclic reactions involved and show how the rearrangement occurs.

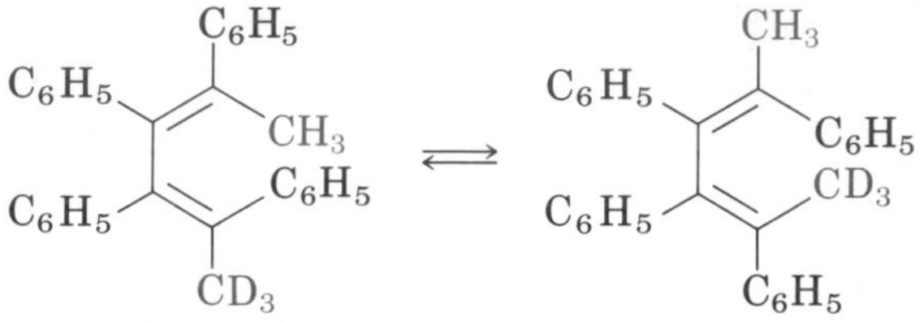

PROBLEM

**30.6** Would you expect the following reaction to proceed in a conrotatory or disrotatory manner? Show the stereochemistry of the cyclobutene product and explain your answer.

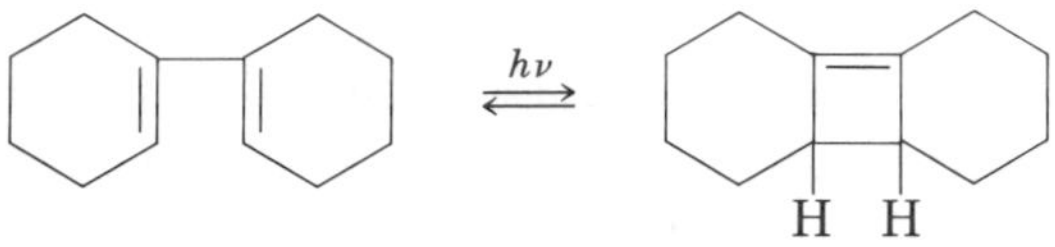

## 30.8 Cycloaddition Reactions

**Cycloaddition reactions** are intermolecular pericyclic processes in which two unsaturated molecules add to each other, yielding a cyclic product. As with electrocyclic reactions, cycloadditions are controlled by the orbital symmetry of the reactants. Symmetry-allowed processes often take place readily, but symmetry-disallowed processes take place with great difficulty, if at all, and then only by nonconcerted pathways. Let's look at two possible reactions to see how they differ.

The Diels–Alder cycloaddition reaction is a [4 + 2]-pi-electron process that takes place between a diene (4 pi electrons) and a dienophile (2 pi electrons) to yield a cyclohexene product. Thousands of examples of Diels–Alder reactions are known. They often take place under mild conditions (room temperature or slightly above), and they are stereospecific with respect to substituents. For example, room-temperature reaction between 1,3-butadiene and diethyl maleate (cis) yields exclusively cis-disubstituted cyclohexene product; reaction between 1,3-butadiene and diethyl fumarate (trans) yields exclusively trans-disubstituted product (Figure 30.13).

In contrast to the [4 + 2]-pi-electron Diels–Alder reaction, thermal cycloaddition between two alkenes (2 pi + 2 pi) does not occur. Photochemical [2 + 2]-pi-electron cycloadditions often take place readily, however, to yield cyclobutane products. How can we use orbital symmetry arguments to explain these results?

For a successful cycloaddition to take place, the terminal pi lobes of the two unsaturated reactants must have the correct symmetry for bonding

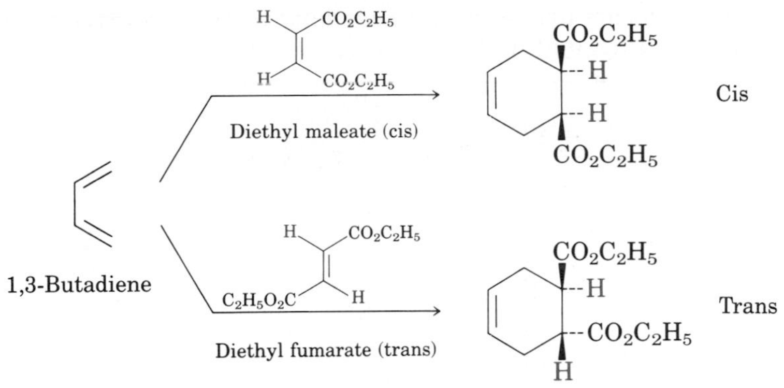

1,3-Butadiene

Diethyl maleate (cis) → Cis: $CO_2C_2H_5$, H, H, $CO_2C_2H_5$

Diethyl fumarate (trans) → Trans: $CO_2C_2H_5$, H, $CO_2C_2H_5$, H

**Figure 30.13** Diels–Alder cycloaddition reactions of diethyl maleate (cis) and diethyl fumarate (trans): The reactions are stereospecific.

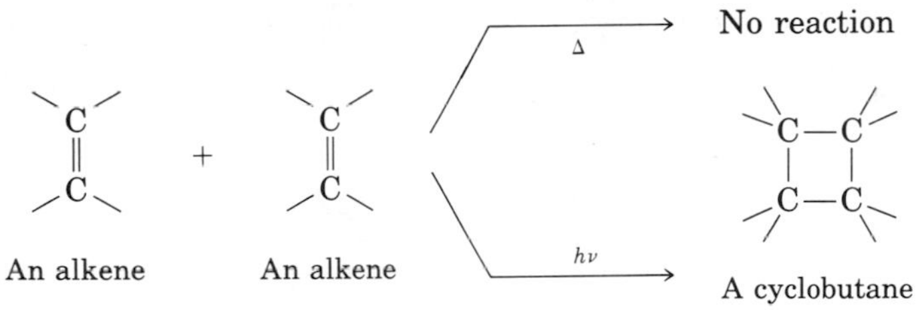

An alkene + An alkene

Δ → No reaction

$h\nu$ → A cyclobutane

overlap to occur. This can happen in either of two ways, designated suprafacial and antarafacial. **Suprafacial cycloadditions** occur when the orbital symmetries of the reactants are such that reactions occur between lobes on the *same* face of one component and lobes on the *same* face of the other component (Figure 30.14).

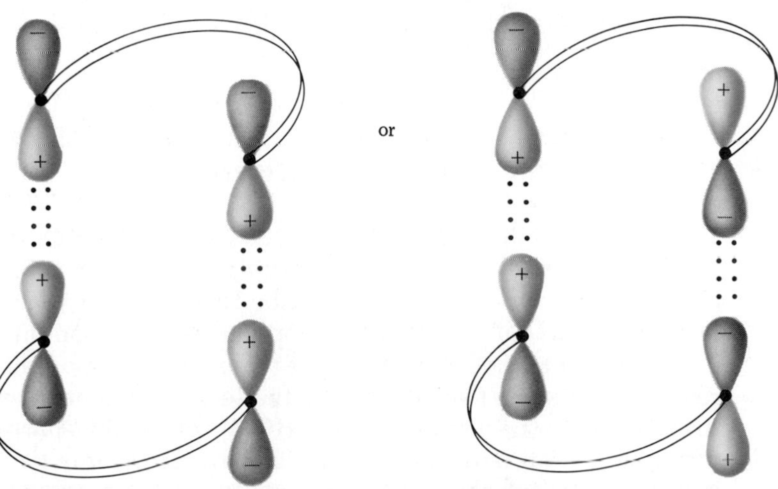


**Figure 30.14** Suprafacial cycloaddition occurs when there is bonding overlap between lobes on the same face of one reactant and lobes on the same face of a second reactant.

**Antarafacial cycloadditions** occur when the orbital symmetries of the reactants are such that reactions occur between lobes on the *same* face of one component and lobes on *opposite* faces of the other component (Figure 30.15).

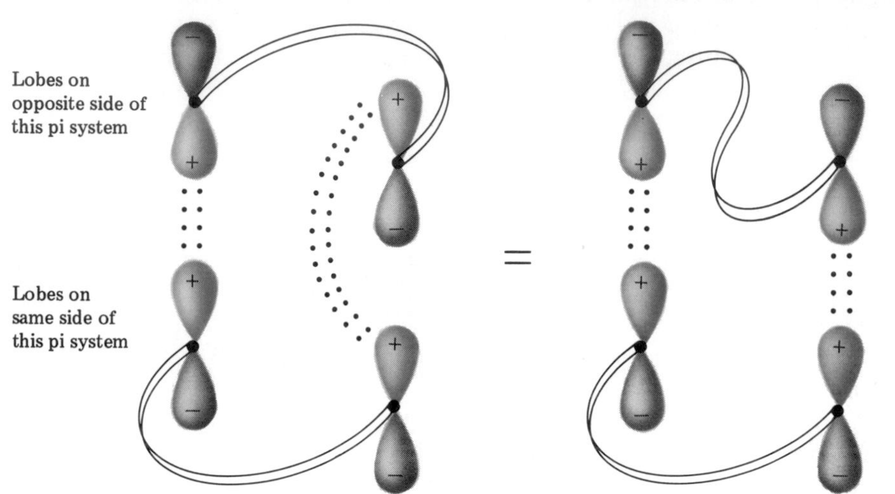


**Figure 30.15** Antarafacial cycloaddition occurs when there is bonding overlap between lobes on the same face of one reactant and lobes on opposite faces of the second reactant.

Note that both suprafacial and antarafacial cycloadditions are allowed on orbital-symmetry grounds. Geometric constraints often make antarafacial reactions difficult, however, since there must be twisting of the *p* orbital system. Thus, only suprafacial cycloadditions are possible for small pi systems.

## 30.9 Stereochemistry of Cycloadditions

How can we predict whether a given reaction will occur with suprafacial or with antarafacial geometry? According to frontier orbital theory, cycloaddition reactions take place when the HOMO of one reaction partner overlaps the LUMO of the other reaction partner in a bonding manner.

A good intuitive explanation of this rule is to imagine that one partner reacts by donating two electrons to the second partner. As with electrocyclic reactions, it is the electrons in the HOMO of the first partner that are least tightly held and most likely to be donated. Since only two electrons can be in any one orbital, these electrons must go into a vacant orbital of the second partner, and it is the LUMO that is lowest in energy and most likely to accept the electrons. Let's see how this rule applies to specific cases.

For the [4 + 2]-pi-electron cycloaddition (Diels–Alder reaction), let's arbitrarily select the diene LUMO and the alkene HOMO. (We could equally well use the diene HOMO and the alkene LUMO.) The symmetries of these two orbitals are such that bonding overlap of the terminal lobes can occur with suprafacial geometry (Figure 30.16). The Diels–Alder reaction therefore takes place readily under thermal conditions. Note that, as with electrocyclic reactions, we need be concerned only with the symmetries of the *terminal* lobes. For purposes of prediction, it doesn't matter whether or not the interior lobes have bonding or antibonding overlap in the product.

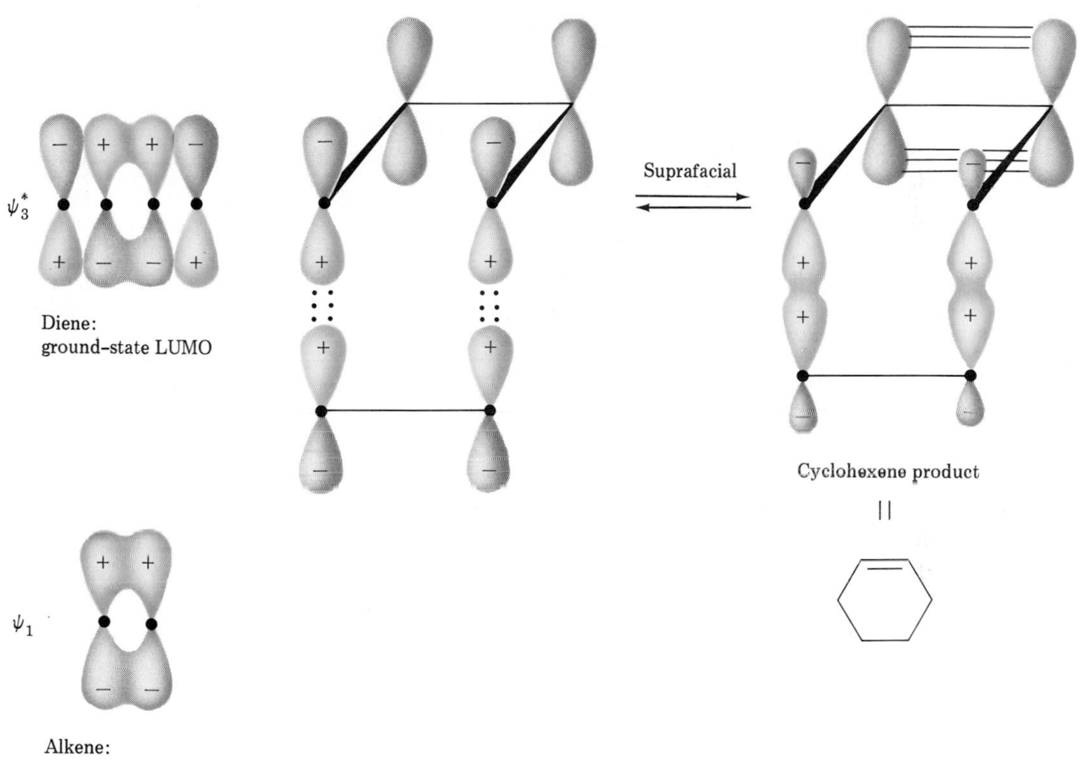


**Figure 30.16** Correlation of the diene LUMO and the alkene HOMO in a suprafacial [4 + 2] cycloaddition reaction (Diels–Alder reaction)

In contrast to the [4 + 2] Diels–Alder reaction, the [2 + 2] cycloaddition of two alkenes to yield a cyclobutane does not occur thermally but can only be observed photochemically. The explanation follows from orbital-symmetry arguments. Looking at the HOMO of one alkene and the LUMO of the second alkene, it is apparent that a thermal [2 + 2] cycloaddition must take place by an antarafacial pathway (Figure 30.17, page 1096). Geometric constraints make the antarafacial transition state impossible, however, and concerted thermal [2 + 2] cycloadditions are therefore not observed.

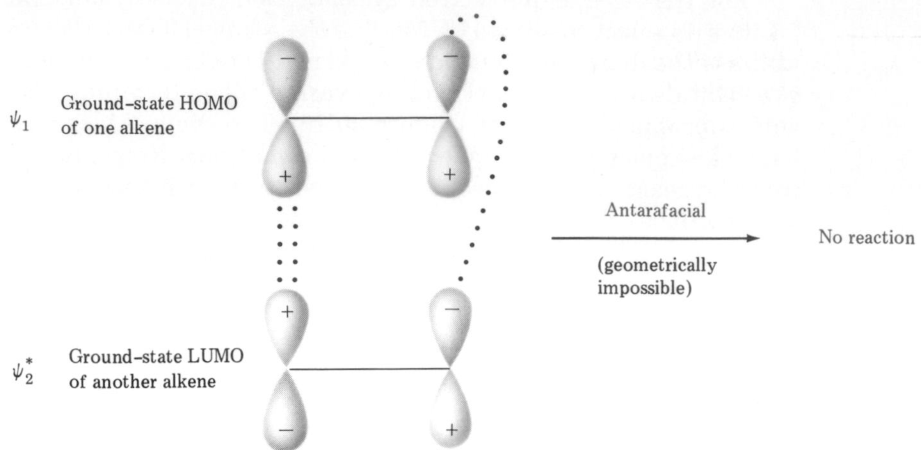


**Figure 30.17** Correlation of the HOMO and the LUMO in thermal [2 + 2] cycloaddition: The reaction does not occur because antarafacial geometry is too strained.

Photochemical [2 + 2] cycloadditions, however, *are* observed. Irradiation of an alkene with ultraviolet light excites an electron from $\psi_1$, the ground-state HOMO, to $\psi_2^*$, the excited-state HOMO. Correlation between the excited-state HOMO of one alkene and the LUMO of the second alkene indicates that a photochemical [2 + 2] cycloaddition reaction can occur by a suprafacial pathway (Figure 30.18).

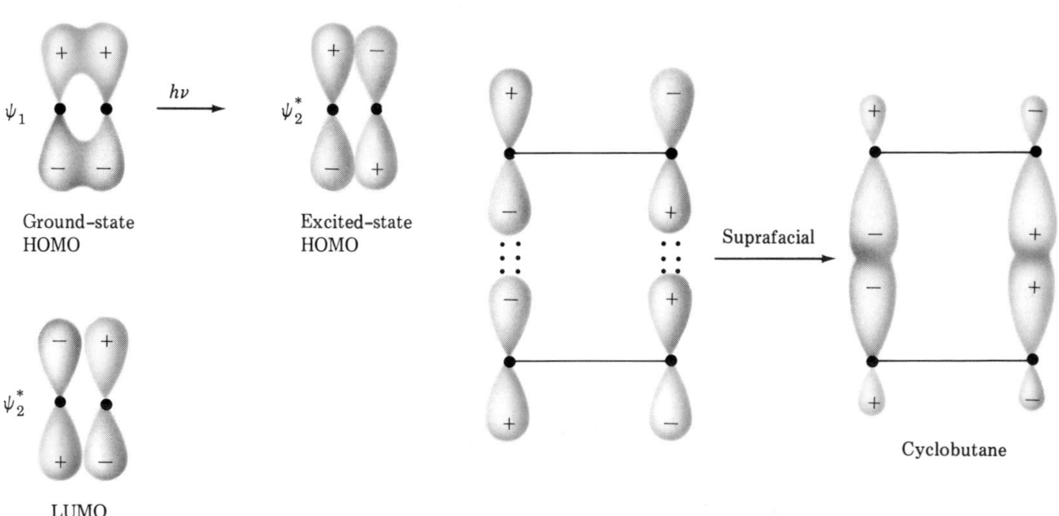


**Figure 30.18** Correlation of the HOMO and the LUMO in photochemical [2 + 2] cycloaddition reactions: The reaction occurs with suprafacial geometry.

The photochemical [2 + 2] cycloaddition reaction occurs smoothly and represents one of the best methods known for synthesizing cyclobutane rings. The reaction can take place either inter- or intramolecularly, as the following examples show:

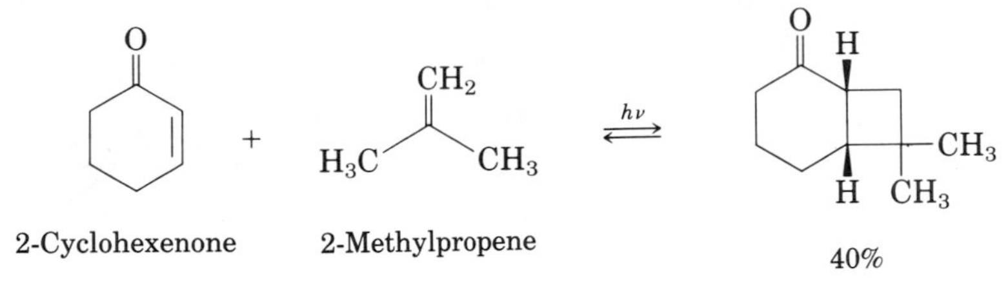

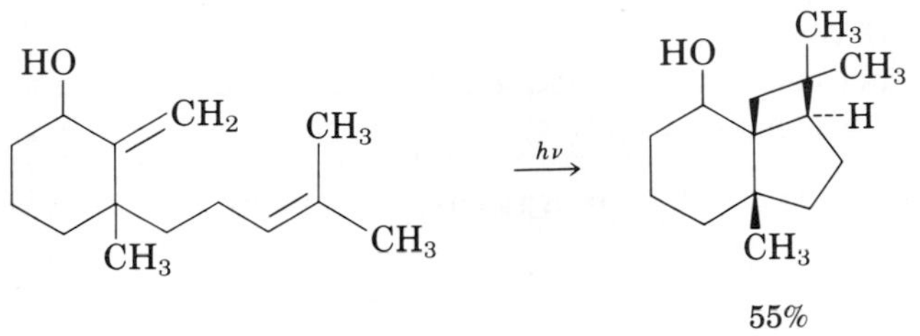

2-Cyclohexenone + 2-Methylpropene $\xrightleftharpoons{h\nu}$ 40%

$\xrightarrow{h\nu}$ 55%

Thermal and photochemical cycloaddition reactions always take place by opposite stereochemical pathways. As with electrocyclic reactions, we can categorize cycloadditions according to the total number of electron pairs (double bonds) involved in the rearrangement. Thus, a Diels–Alder [4 + 2] reaction between a diene and a dienophile involves an odd number (three) of electron pairs and takes place by a ground-state suprafacial pathway. A [2 + 2] reaction between two alkenes involves an even number (two) of electron pairs and must take place by a ground-state antarafacial pathway. The general rules governing such processes are given in Table 30.2.

**Table 30.2** Stereochemical rules for cycloaddition reactions

| *Electron pairs (double bonds)* | *Thermal reaction* | *Photochemical reaction* |
|---|---|---|
| Even number | Antarafacial | Suprafacial |
| Odd number | Suprafacial | Antarafacial |

It should be reiterated that both suprafacial and antarafacial cycloaddition pathways are symmetry-allowed processes. Only the geometric constraints inherent in twisting a conjugated pi electron system out of planarity make antarafacial reaction geometry difficult in many cases.

PROBLEM

**30.7** What stereochemistry would you expect for the product of the Diels–Alder reaction between (2*E*,4*E*)-hexadiene and ethylene? What stereochemistry would you expect if (2*E*,4*Z*)-hexadiene were used instead?

PROBLEM

**30.8** Cyclopentadiene reacts with cycloheptatrienone to give the product shown. Tell what kind of reaction is involved and explain the observed result. Is the reaction suprafacial or antarafacial?

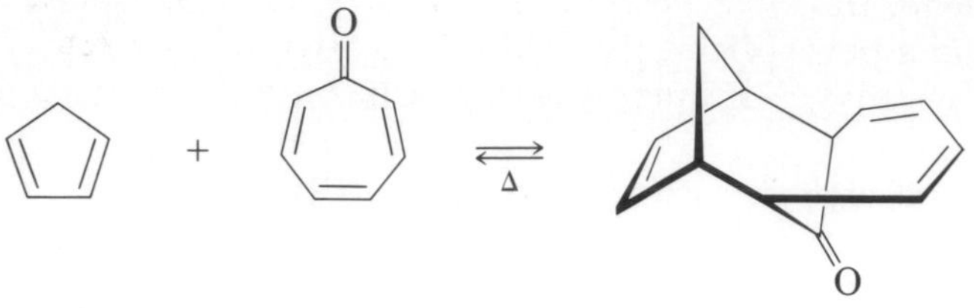

PROBLEM

**30.9** The following reaction takes place in two steps, one of which is a cycloaddition and the other of which is a *reverse* cycloaddition. Identify the two pericyclic reactions and show how they occur.

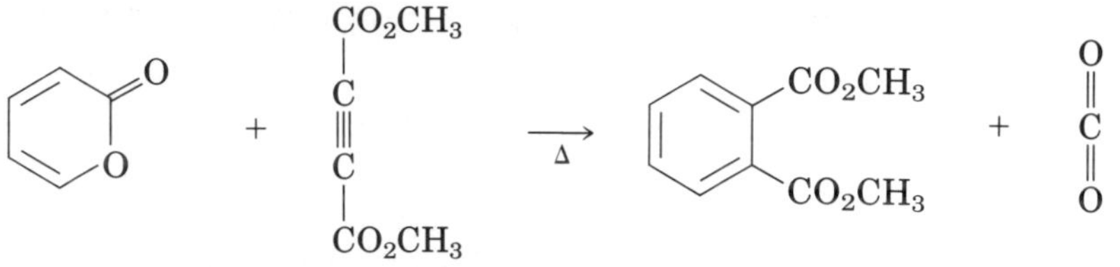

## 30.10 Sigmatropic Rearrangements

**Sigmatropic rearrangements** are pericyclic reactions in which a sigma-bonded substituent group (denoted here by a circled S) migrates across a pi electron system. One sigma bond is broken in the starting material, and a new sigma bond is formed in the product. The sigma-bonded group can be either at the end or in the middle of the pi system, as the following [1,3] and [3,3] rearrangements illustrate:

A [1,3] rearrangement

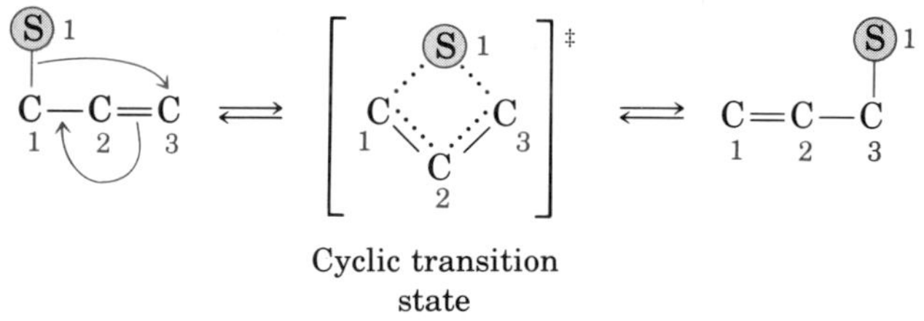

Cyclic transition state

A [3,3] rearrangement

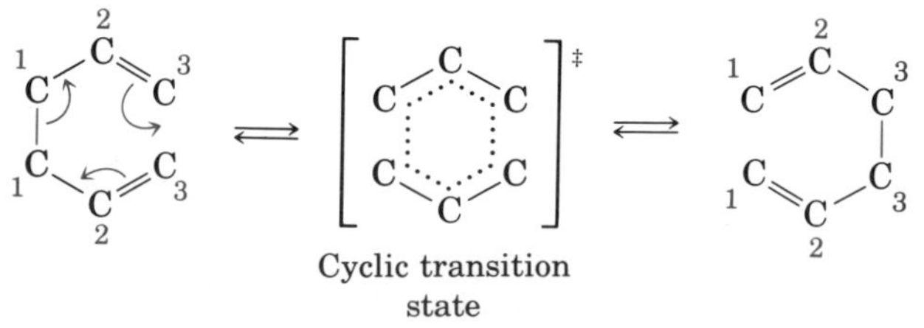

Cyclic transition state

The designations [1,3] and [3,3] describe the kind of rearrangement that has occurred. The two numbers in brackets refer to the two groups connected by the sigma bond and designate the positions in those groups *to which migration occurs*. For example, in the [1,5] sigmatropic rearrangement of a diene, the two groups connected by the sigma bond are a hydrogen atom and a pentadienyl fragment. Migration occurs to position 1 of the H group (the only possibility) and to position 5 of the pentadienyl group.

A [1,5] sigmatropic rearrangement

$$\overset{H_1}{\underset{1}{CH_2}}-\underset{2}{CH}=\underset{3}{CH}-\underset{4}{CH}=\underset{5}{CH_2} \rightleftharpoons \underset{1}{CH_2}=\underset{2}{CH}-\underset{3}{CH}=\underset{4}{CH}-\overset{H_1}{\underset{5}{CH_2}}$$

In the [3,3] Claisen rearrangement, the two groups connected by the sigma bond are an allyl group and a vinylic ether group. Migration occurs to position 3 of the allyl group and also to position 3 of the vinylic ether.

Claisen rearrangement, a [3,3] rearrangement

$$\text{O(1)}-\text{CH(2)}=\text{CH}_2\text{(3)};\ \text{O(1)}-\text{H}_2\text{C(1)}-\text{CH(2)}=\text{CH}_2\text{(3)} \longrightarrow \text{O(1)}=\text{CH(2)}-\text{CH}_2\text{(3)}-\text{CH}_2\text{(3)}-\text{CH(2)}=\text{H}_2\text{C(1)}$$

## 30.11 Stereochemistry of Sigmatropic Rearrangements

Sigmatropic rearrangements are more complex than either electrocyclic or cycloaddition reactions but are nonetheless controlled by orbital-symmetry considerations. There are two possible modes of reaction: Migration of a group across the same face of the pi system is called a suprafacial rearrangement, and migration of a group from one face of the pi system to the other face is an antarafacial rearrangement (Figure 30.19, page 1100).

The rules for sigmatropic rearrangements are identical to those for cycloaddition reactions, as summarized in Table 30.3. Both suprafacial and antarafacial sigmatropic rearrangements are symmetry-allowed processes, but suprafacial rearrangements are often easier for geometric reasons.

**Table 30.3** Stereochemical rules for sigmatropic rearrangements

| *Electron pairs (double bonds)* | *Thermal reaction* | *Photochemical reaction* |
|---|---|---|
| Even number | Antarafacial | Suprafacial |
| Odd number | Suprafacial | Antarafacial |

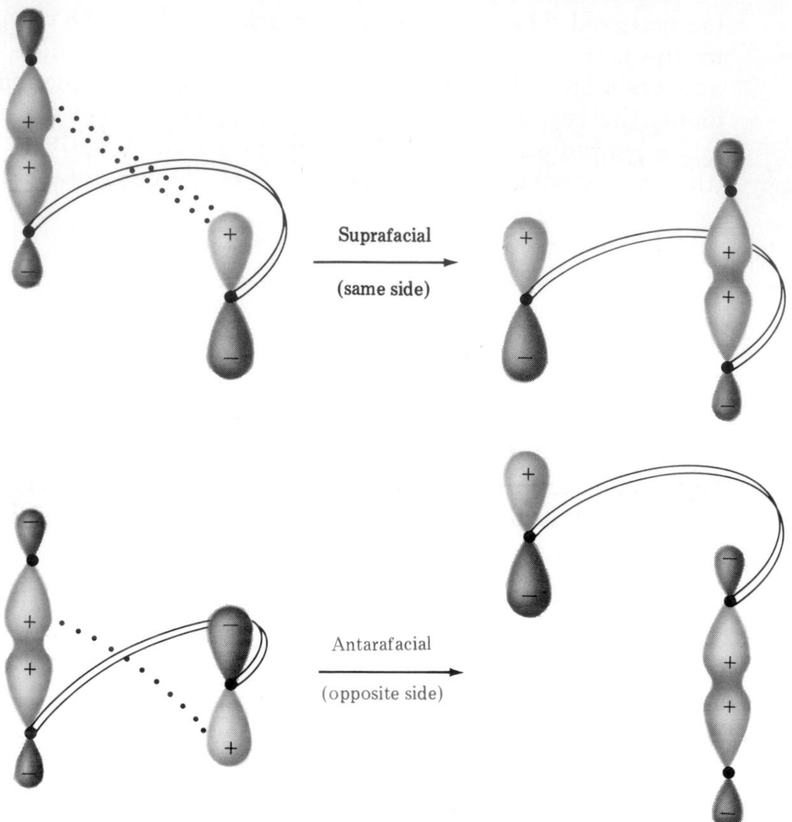


**Figure 30.19** Suprafacial and antarafacial sigmatropic rearrangements

## 30.12 Some Examples of Sigmatropic Rearrangements

Since a [1,5] sigmatropic rearrangement involves three electron pairs (two pi bonds and one sigma bond), the orbital-symmetry rules in Table 30.3 predict a suprafacial reaction. In fact, the [1,5] suprafacial shift of a hydrogen atom across two double bonds of a pi system is one of the most commonly observed of all sigmatropic rearrangements. For example, 5-methylcyclopentadiene rapidly rearranges at room temperature to yield a mixture of 1-methyl-, 2-methyl-, and 5-methyl-substituted products.

$H_3C$, H; 1, 2, 3, 4, 5 —[1,5] Shift, 25°C⇌— $CH_3$, H, H, H; 1, 2, 3, 4, 5 —[1,5] Shift, 25°C⇌— $CH_3$, H, H, H

5-Methylcyclopentadiene | 1-Methylcyclopentadiene | 2-Methylcyclopentadiene

As another example, heating 5,5,5-trideuterio-(1,3*Z*)-pentadiene causes scrambling of deuterium between positions 1 and 5.

$CD_3$ / $CH_2$ $\xrightleftharpoons[\Delta]{[1,5]\text{ D Shift}}$ $CD_2$ / $CH_2D$

Both of these [1,5] hydrogen shifts occur by a symmetry-allowed suprafacial rearrangement, as illustrated in Figure 30.20.

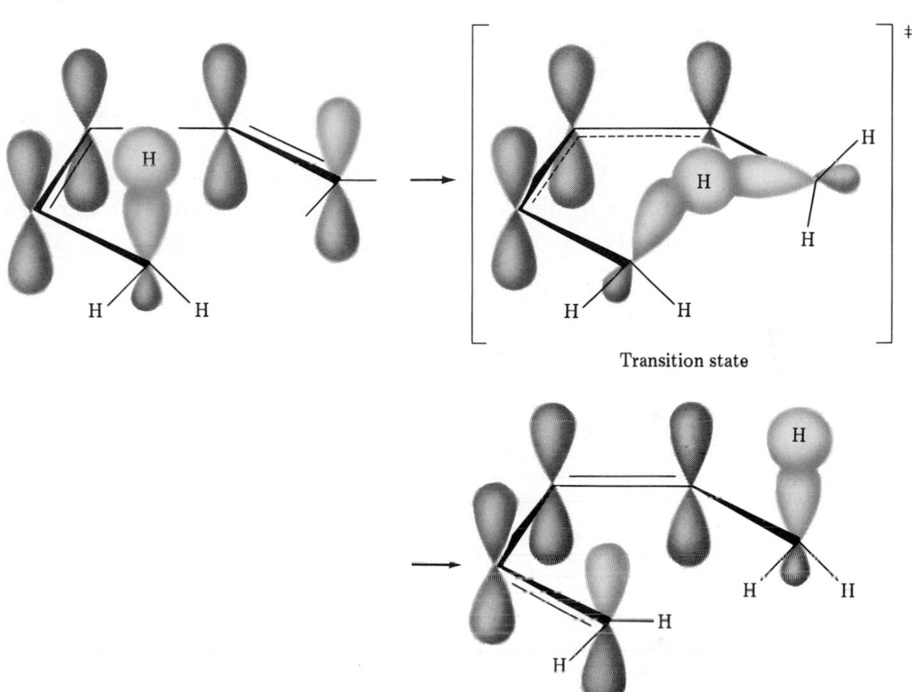


**Figure 30.20** An orbital view of a suprafacial [1,5] hydrogen shift

In contrast to the preceding two examples of [1,5] sigmatropic hydrogen shifts, thermal [1,3] hydrogen shifts are unknown. Were they to occur, they would have to proceed via an impossibly strained antarafacial reaction pathway.

Two other important sigmatropic reactions are the **Cope rearrangement** of a 1,5-hexadiene and the **Claisen rearrangement** of an allyl aryl ether (Section 26.9). These two, along with the Diels–Alder reaction, are the most useful pericyclic reactions for organic synthesis; many thousands of examples of all three are known. Note that the Claisen rearrangement works well with both allyl aryl ethers and with allyl vinylic ethers.

Claisen rearrangement

O $\xrightleftharpoons{[3,3]}$ O, H $\longrightarrow$ HO

An allyl aryl ether — An *o*-allylphenol

Claissen rearrangement

O [3,3] ⇄ O α β γ δ

An allyl vinylic ether — A γ,δ-unsaturated carbonyl compound

Cope rearrangement

$CH_3$ [3,3] ⇄ $CH_3$

A 1,5 diene — A new 1,5 diene

Since both Cope and Claisen rearrangements involve reorganization of an odd number of electron pairs (two pi bonds and one sigma bond), they are predicted to react by suprafacial pathways (Figure 30.21).

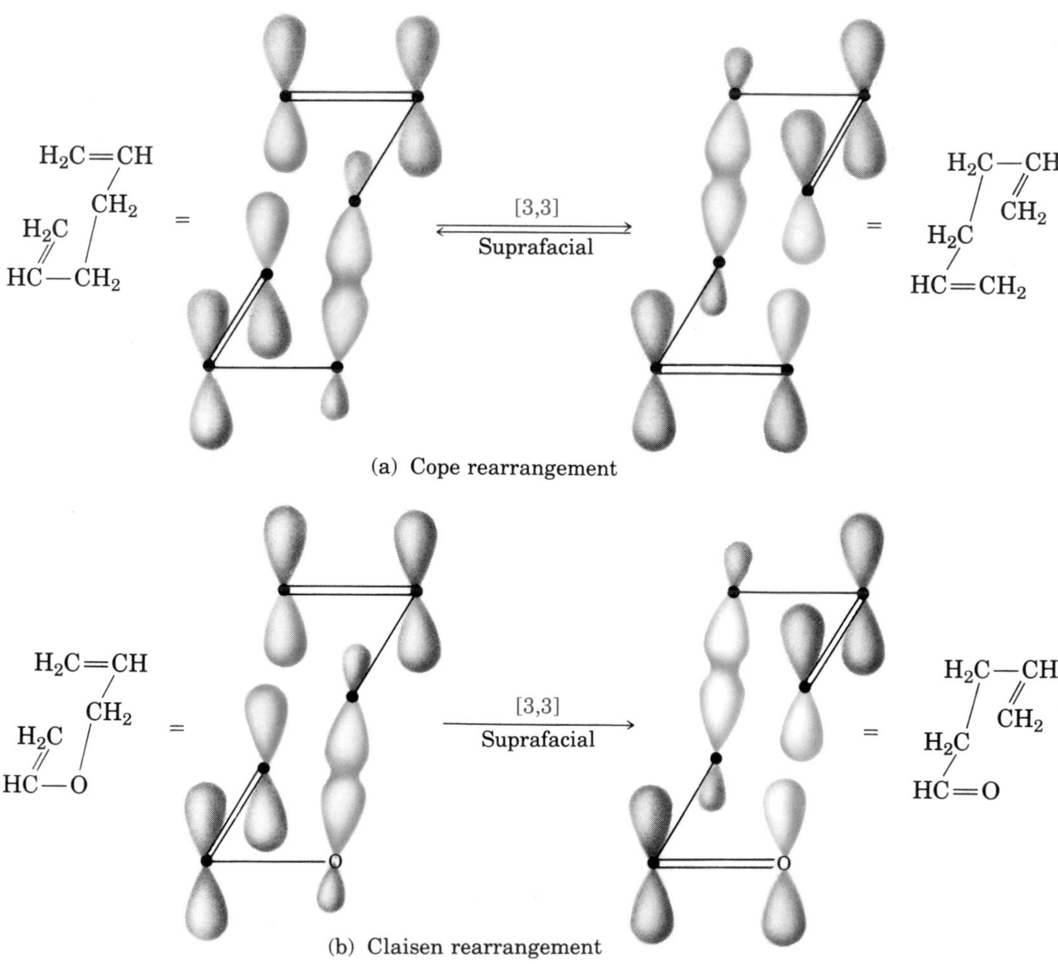


**Figure 30.21** Suprafacial [3,3] Cope and Claisen rearrangements

The ease with which these two rearrangements occur provides evidence for the correctness of orbital-symmetry predictions. For example, it has been estimated, based on spectroscopic measurements, that homotropilidene undergoes [3,3] Cope rearrangement *several hundred times each second* at room temperature.

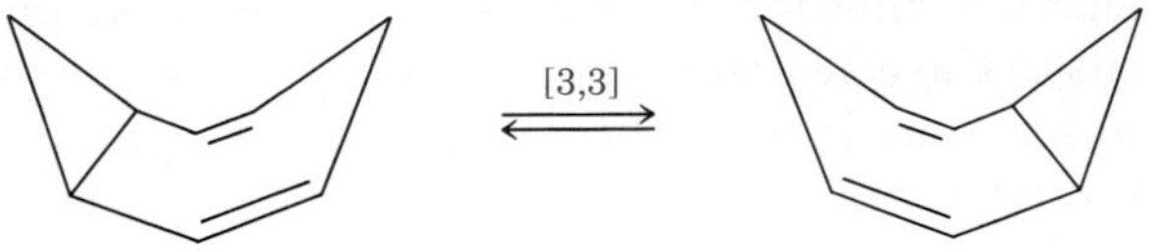

[3,3]

Homotropilidene

PROBLEM

**30.10** The $^{13}C$ NMR spectrum of homotropilidene taken at room temperature shows only three peaks. Explain.

PROBLEM

**30.11** How can you account for the fact that heating 1-deuterioindene scrambles the isotope label to all three positions on the five-membered ring?

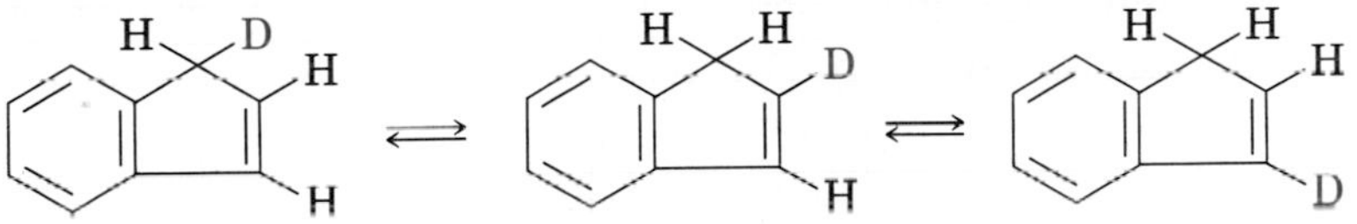

PROBLEM

**30.12** Classify the following sigmatropic reaction by order [*x*,*y*] and indicate whether you would expect it to proceed with suprafacial or antarafacial stereochemistry.

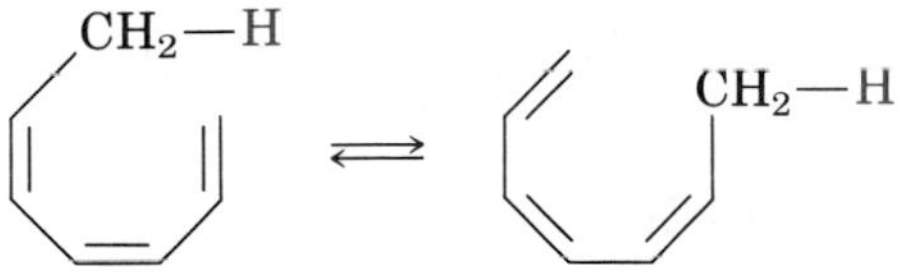

$CH_2$—H

$CH_2$—H

PROBLEM

**30.13** When a 2,6-disubstituted allyl phenyl ether is heated in an attempted Claisen rearrangement, migration occurs to give *p*-allyl product. Explain how this occurs.

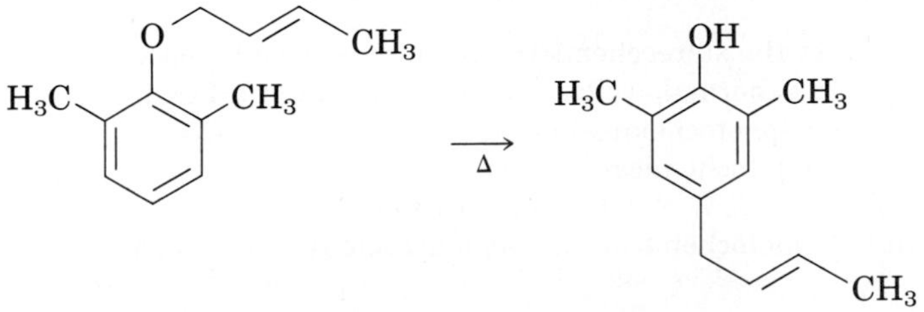

O $CH_3$ $H_3C$ $CH_3$ $\xrightarrow{\Delta}$ OH $H_3C$ $CH_3$ $CH_3$

## 30.13 A Summary of Rules for Pericyclic Reactions

Pericyclic, electrocyclic, cycloaddition, sigmatropic, conrotatory, disrotatory, suprafacial, antarafacial—how can we keep it all straight?

The information provided in Tables 30.1, 30.2, and 30.3 to summarize the selection rules for electrocyclic, cycloaddition, and sigmatropic reactions leads us to the conclusion that pericyclic processes can be grouped according to whether they involve the reorganization of an even or an odd number of electron pairs (bonds). All this information can be distilled into one simple phrase that, when memorized, provides an easy and accurate way to predict the stereochemical outcome of any pericyclic reaction:

> For a ground-state (thermal) pericyclic reaction, the groupings are *odd–supra–dis* and *even–antara–con.*

Cycloaddition and sigmatropic reactions involving an odd number of electron pairs (bonds) occur with suprafacial geometry; electrocyclic reactions involving an odd number of electron pairs occur with disrotatory stereochemistry. Conversely, pericyclic reactions involving an even number of electron pairs occur with either antarafacial geometry or conrotatory stereochemistry.

Once the selection rules for thermal reactions have been memorized, the rules for photochemical reactions are easily derived by simply remembering that they are the opposite of the thermal rules:

> For an excited-state (photochemical) pericyclic reaction, the groupings are *odd–antara–con* and *even–supra–dis.*

Both rules are summarized in Table 30.4. Memorizing this table will give you the ability to predict the stereochemistry of literally thousands of pericyclic reactions.

**Table 30.4** Generalized selection rules for pericyclic reactions

| *Electron state* | *Electron pairs* | *Stereochemistry* |
|---|---|---|
| Ground state (thermal) | Even number | Antara–con |
| | Odd number | Supra–dis |
| Excited state (photochemical) | Even number | Supra–dis |
| | Odd number | Antara–con |

**PROBLEM**

**30.14** Predict the stereochemistry of these pericyclic reactions:
(a) The thermal cyclization of a conjugated tetraene
(b) The photochemical cyclization of a conjugated tetraene
(c) A photochemical [4 + 4] cycloaddition
(d) A thermal [2 + 6] cycloaddition
(e) A photochemical [3,5] sigmatropic rearrangement

## 30.14 Summary and Key Words

A **pericyclic reaction** is one that takes place in a single step involving a cyclic transition state; no intermediates are involved. There are three major classes of pericyclic processes: **electrocyclic reactions**, **cycloaddition reactions**, and **sigmatropic rearrangements**. The stereochemistry of these reactions is controlled by the symmetry of the orbitals involved in bond reorganization.

Electrocyclic reactions involve the cyclization of conjugated polyenes. For example, 1,3,5-hexatriene cyclizes to 1,3-cyclohexadiene on heating. Electrocyclic reactions can occur by either **conrotatory** or **disrotatory** paths, depending on the symmetry of the terminal lobes of the pi system. Conrotatory cyclization requires that both lobes rotate in the same direction, whereas disrotatory cyclization requires that the lobes rotate in opposite directions (Figure 30.22). The reaction course for any specific case can be found by looking at the symmetry of the **highest occupied molecular orbital (HOMO)**.

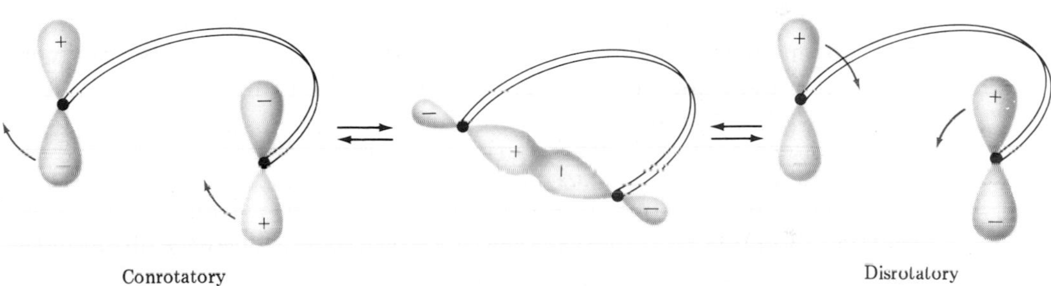


**Figure 30.22** Conrotatory and disrotatory motions during electrocyclic reactions

Cycloaddition reactions are those in which two unsaturated molecules add together to yield a cyclic product. For example, Diels–Alder reaction between a diene (four pi electrons) and a dienophile (two pi electrons) yields a cyclohexene. Cycloadditions can take place either by **suprafacial** or **antarafacial** pathways. Suprafacial cycloaddition involves reaction between lobes on the same face of one component and on the same face of the second component. Antarafacial cycloaddition involves reaction between lobes on the same face of one component and on opposite faces of the other component (Figure 30.23, page 1106). The reaction course in a specific case can be found by looking at the symmetry of the HOMO of one component and the **lowest unoccupied molecular orbital (LUMO)** of the other component.

Sigmatropic rearrangements involve the migration of a sigma-bonded group across a pi electron system. For example, **Claisen rearrangement** of an allyl vinylic ether yields an unsaturated carbonyl compound, and **Cope rearrangement** of a 1,5-hexadiene yields a new 1,5-hexadiene. Sigmatropic

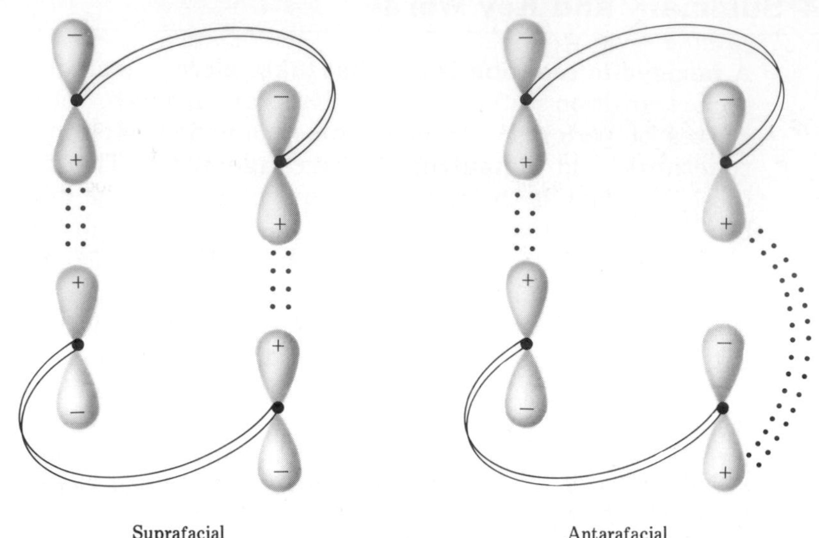


**Figure 30.23** Suprafacial and antarafacial cycloadditions

rearrangements can occur with either suprafacial or antarafacial stereochemistry; the selection rules for a given case are the same as those for cycloaddition reactions.

The stereochemistry of any pericyclic reaction can be predicted by counting the total number of electron pairs (bonds) involved in bond reorganization and then applying some simple rules. Thermal (ground-state) reactions involving an even number of electron pairs occur with either antarafacial or conrotatory stereochemistry (**even–antara–con**); thermal reactions involving an odd number of electron pairs occur with suprafacial or disrotatory stereochemistry (**odd–supra–dis**). Exactly the opposite rules apply to photochemical (excited-state) reactions.

## ADDITIONAL PROBLEMS

**30.15** Define these terms in your own words:

(a) Electrocyclic reaction
(b) Conrotatory motion
(c) Suprafacial
(d) Antarafacial
(e) Disrotatory motion
(f) Sigmatropic rearrangement

**30.16** Have the following reactions taken place in a conrotatory or in a disrotatory manner? Under what conditions, thermal or photochemical, would you carry out each reaction?

(a) 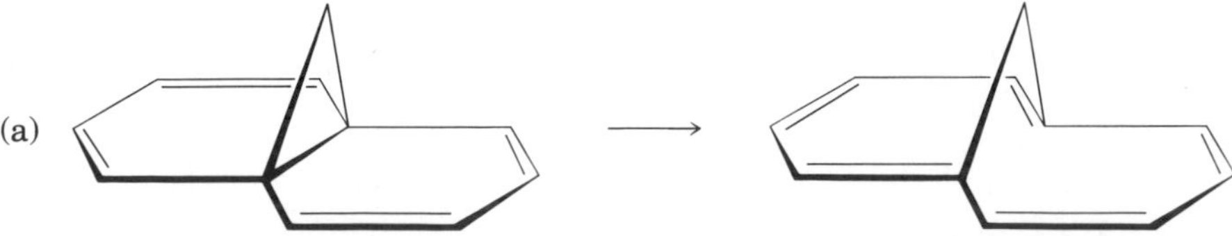

(b) 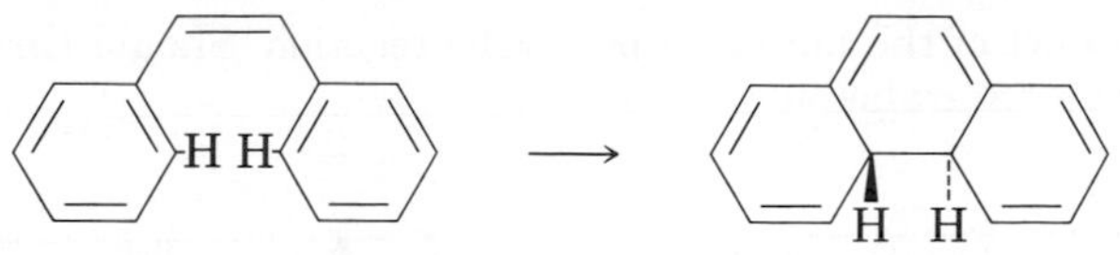 [structure] H H → [structure] H H

**30.17** (2*E*,4*Z*,6*Z*,8*E*)-Decatetraene has been cyclized to give 7,8-dimethyl-1,3,5-cyclooctatriene. Predict the manner of ring closure—conrotatory or disrotatory—for both thermal and photochemical reactions, and predict the stereochemistry of the product in each case.

**30.18** Answer Problem 30.17 for the thermal and photochemical cyclizations of (2*E*,4*Z*,6*Z*,8*Z*)-decatetraene.

**30.19** What stereochemistry would you expect to observe in these reactions?
(a) A photochemical [1,5] sigmatropic rearrangement
(b) A thermal [4 + 6] cycloaddition
(c) A thermal [1,7] sigmatropic rearrangement
(d) A photochemical [2 + 6] cycloaddition

**30.20** The cyclohexadecaoctaene shown isomerizes to two different isomers, depending on reaction conditions. Explain the observed results and indicate the conrotatory or disrotatory nature of each reaction.

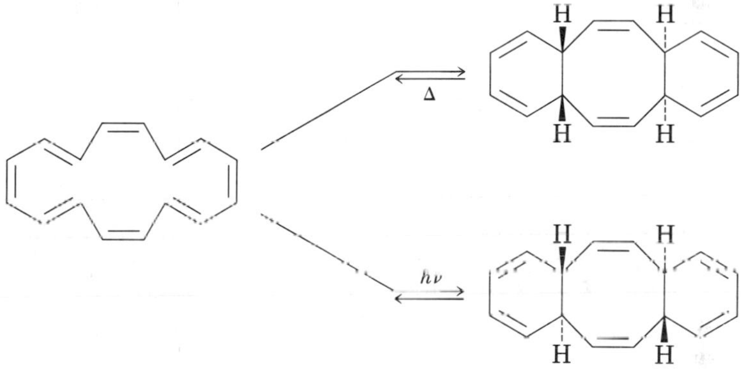

Δ; H H H H

hν; H H H H

**30.21** Which of the following two reactions is more likely to occur? Explain.

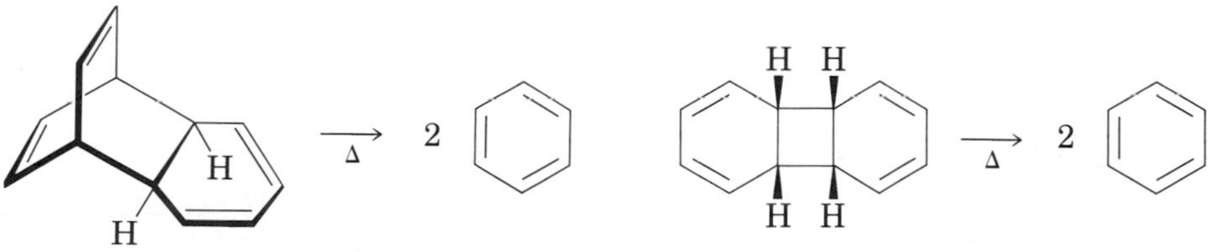

H H →(Δ) 2 [benzene]

H H H H →(Δ) 2 [benzene]

**30.22** The following thermal rearrangement involves two pericyclic reactions in sequence. Identify them and propose a mechanism to account for the observed result.

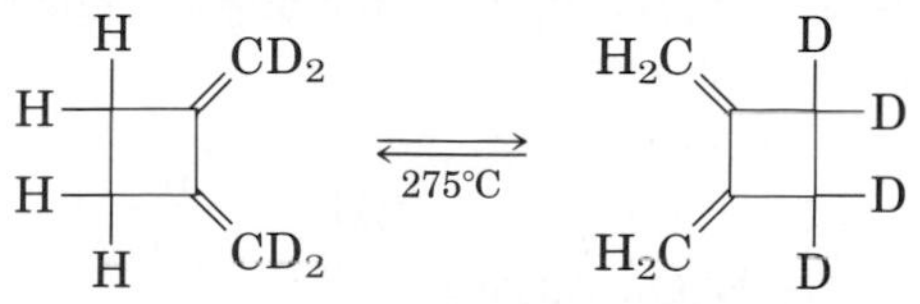

H, H, H, H, $CD_2$, $CD_2$ ⇌ (275°C) $H_2C$, $H_2C$, D, D, D, D

**30.23** Predict the product of the following pericyclic reaction. Should this [5,5] shift be a suprafacial or an antarafacial process?

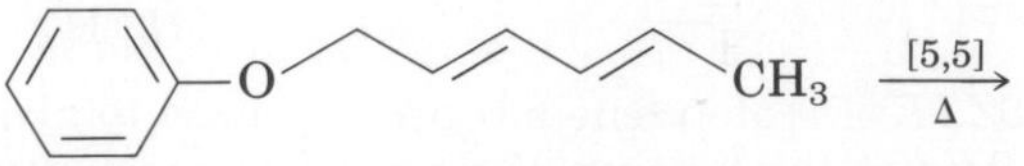

$CH_3$ $\xrightarrow[\Delta]{[5,5]}$

**30.24** How can you account for the fact that ring opening of the *trans*-cyclobutene isomer shown takes place at much lower temperatures than a similar ring opening of the *cis*-cyclobutene isomer? Identify the stereochemistry of each reaction as either conrotatory or disrotatory.

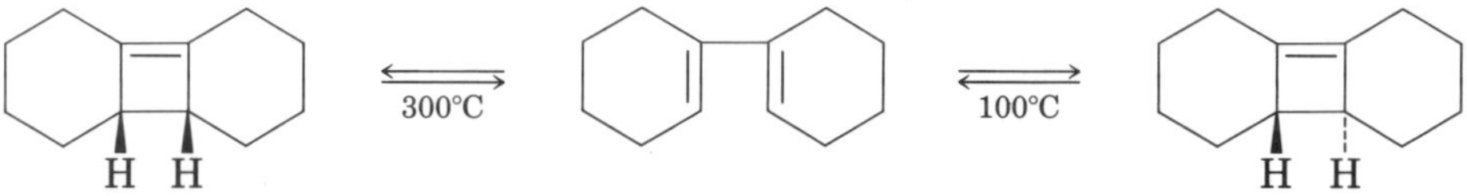

300°C

100°C

H H

H H

**30.25** Photolysis of the *cis*-cyclobutene isomer in Problem 30.24 yields *cis*-cyclododecaen-7-yne, but photolysis of the trans isomer yields *trans*-cyclododecaen-7-yne. Explain these results and identify the type and stereochemistry of the pericyclic reaction.

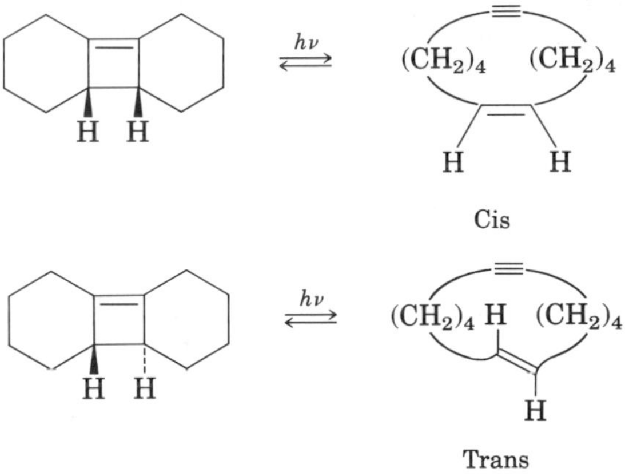

H H $\overset{h\nu}{\rightleftarrows}$ $(CH_2)_4$ $(CH_2)_4$ H H

Cis

H H $\overset{h\nu}{\rightleftarrows}$ $(CH_2)_4$ H $(CH_2)_4$ H

Trans

**30.26** Two pericyclic reactions are involved in the furan synthesis shown. Identify them and propose a mechanism for the transformation.

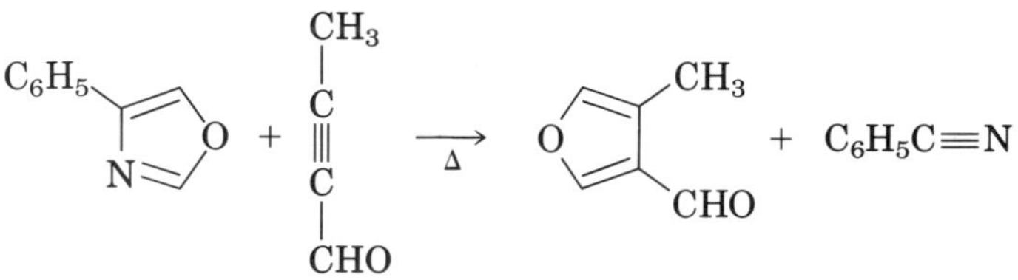

$C_6H_5$ O N + $CH_3-C\equiv C-CHO$ $\xrightarrow{\Delta}$ O $CH_3$ CHO + $C_6H_5C\equiv N$

**30.27** The following synthesis of dienones occurs readily. Propose a mechanism to account for the results, and identify the kind of pericyclic reaction involved.

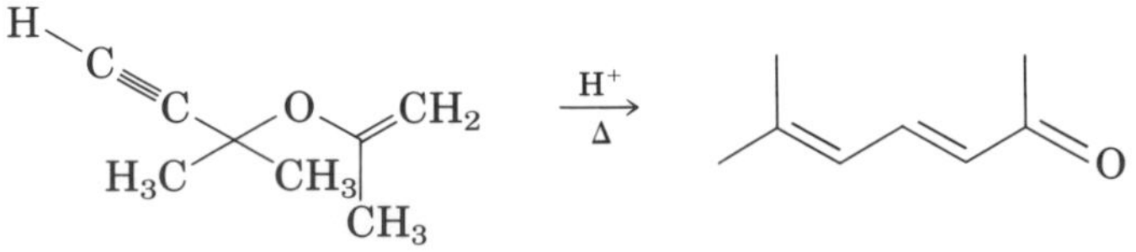

H C C $H_3C$ O $CH_3$ $CH_2$ $CH_3$ $\xrightarrow[\Delta]{H^+}$ O

**30.28** Karahanaenone, a compound isolated from oil of hops, has been synthesized by the thermal reaction shown. Identify the kind of pericyclic reaction and explain how karahanaenone is formed.

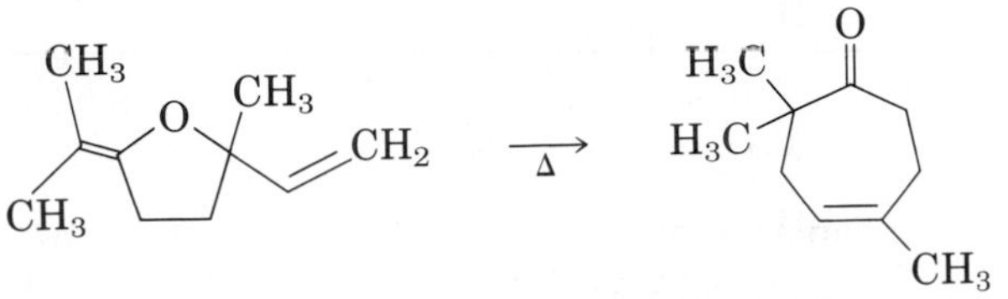

Karahanaenone

**30.29** The $^1$H NMR spectrum of bullvalene at 100°C consists only of a single peak at 4.22 δ. What conclusion can you draw from this? What explanation can you suggest to account for this result?

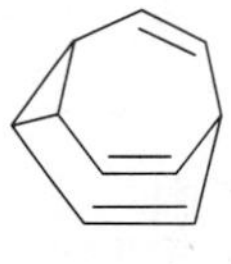

Bullvalene

**30.30** The rearrangement shown was devised and carried out to prove the stereochemistry of [1,5] sigmatropic hydrogen shifts. Explain how the observed result confirms the predictions of orbital symmetry.

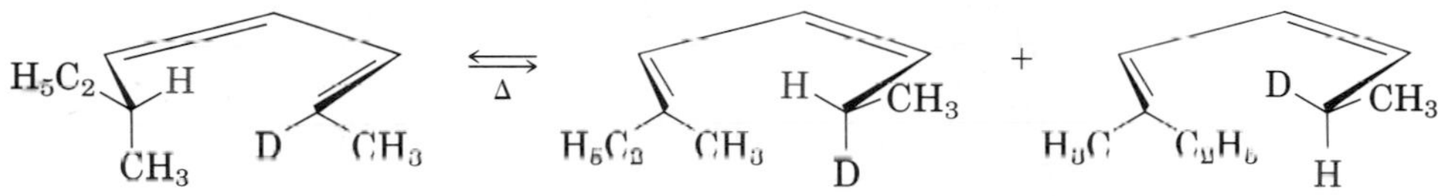

**30.31** The reaction shown is an example of a [2,3] sigmatropic rearrangement. Would you expect the reaction to be suprafacial or antarafacial? Explain.

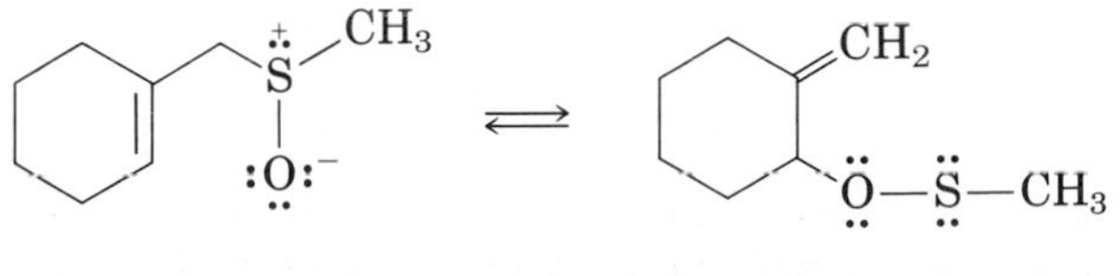

**30.32** When the compound having a cyclobutene fused to a five-membered ring is heated, (1*Z*,3*Z*)-cycloheptadiene is formed. When the related compound having a cyclobutene fused to an eight-membered ring is heated, however, (1*E*,3*Z*)-cyclodecadiene is formed. Explain these results and suggest a reason why the eight-membered ring opens at lower reaction temperatures.

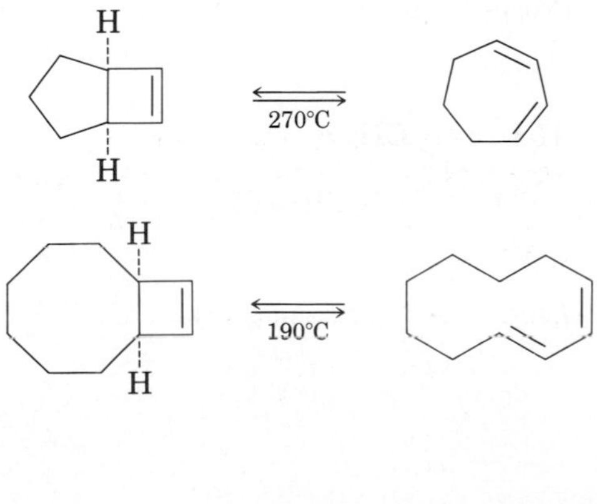

**30.33** In light of your answer to Problem 30.32, explain why a mixture of products occurs in the following reaction:

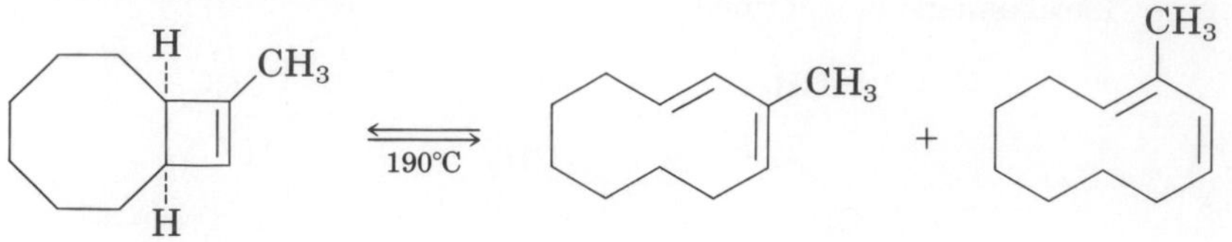

H, $CH_3$, H; 190°C; $CH_3$ + $CH_3$

**30.34** Estrone, a major female sex hormone, has been synthesized by a route that involves the following step. Identify the pericyclic reactions involved and propose a mechanism.

$H_3C$, O, $CH_3O$, H; Δ; $H_3C$, O, H, H, H, $CH_3O$

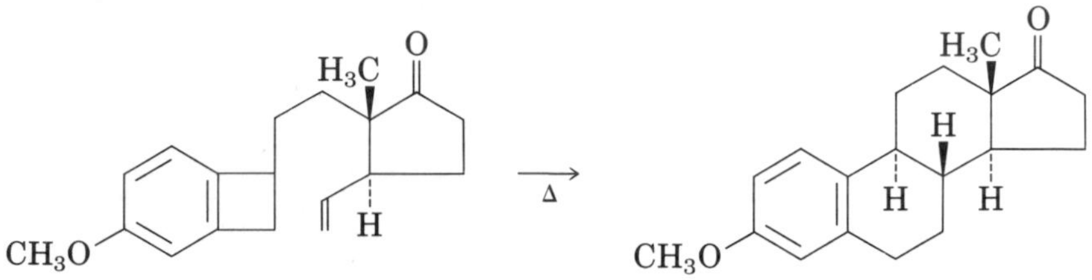

Estrone methyl ether

**30.35** Coronafacic acid, a bacterial toxin, was synthesized using a key step that involves three sequential pericyclic reactions. Identify them and propose a mechanism for the overall transformation. How would you complete the synthesis?

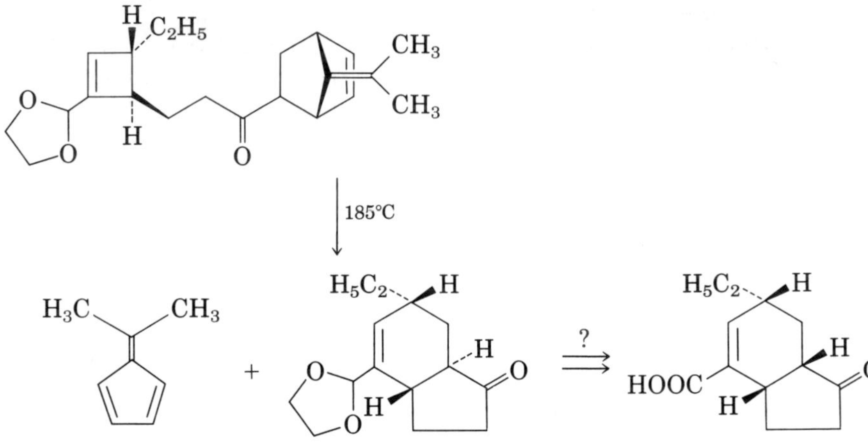

H, $C_2H_5$, O, O, H, O, $CH_3$, $CH_3$

185°C

$H_3C$, $CH_3$ + $H_5C_2$, H, O, O, H, H, O; ?; $H_5C_2$, H, HOOC, H, H, O

92%

Coronafacic acid

**30.36** The following rearrangement of *N*-allyl-*N*,*N*-dimethylanilinium ion has been observed. Propose a mechanism to account for the reaction.

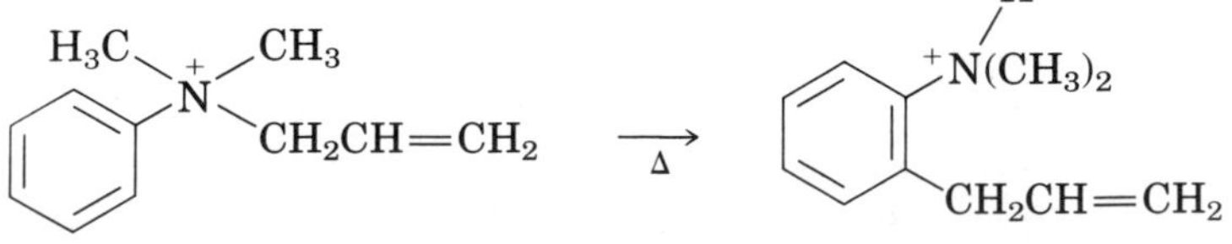

$H_3C$, $CH_3$, $\overset{+}{N}$, $CH_2CH{=}CH_2$; Δ; H, $^{+}N(CH_3)_2$, $CH_2CH{=}CH_2$

*N*-Allyl-*N*,*N*-dimethylanilinium ion

*o*-Allyl-*N*,*N*-dimethylanilinium ion

CHAPTER 31

# Synthetic Polymers

No other group of synthetic organic compounds has had as great an impact on our day-to-day living as the synthetic polymers. As plastics, adhesives, and paints, synthetic polymers have a multitude of uses, from the foam coffee cup to the life-saving artificial heart valve. Polymer synthesis is a major part of the chemical industry, and annual production figures for some of the more important polymers are shown in Table 31.1.

**Table 31.1** Production figures for some major polymers (1986)

| *Polymer* | *U.S. production (million lb/yr)* | *Polymer* | *U.S. production (million lb/yr)* |
|---|---|---|---|
| Polyethylene | | Acrylonitrile/butadiene/ styrene | 2,195 |
| Low-density | 8,903 | | |
| High-density | 7,182 | Acrylic fibers | 616 |
| Polypropylene | 5,812 | Epoxy adhesives | 398 |
| Polystyrene | 4,470 | Phenolic resins | 2,735 |
| Poly(vinyl chloride) | 7,256 | Urea–formaldehyde resins | 1,271 |
| Nylon | 2,514 | Styrene/butadiene rubber | 792 |
| Polyesters | 2,446 | Polybutadiene rubber | 325 |

Polymers aren't new to us. A **polymer** is simply a large molecule built up by repetitive bonding together of many smaller units called **monomers**. In fact, we've already studied the major classes of biopolymers: cellulose is a large carbohydrate polymer built of repeating glucose units; proteins are large polyamides built of repeating amino acid units; and nucleic acids are large molecules built of repeating nucleotide units.

Synthetic polymers are chemically much simpler than most biopolymers, since the repeating monomer units tend to be small, simple, and inexpensive molecules. There is, however, an immense diversity to the structure and properties of synthetic polymers, depending on the exact nature of the monomer and on the reaction conditions used for polymerization.

Polymers can be classified in many ways—by method of synthesis, by structure, by physical properties, or by end use, to name a few. Let's begin by looking at the organic chemistry of polymer synthesis and then see how polymer structure can be correlated with uses and physical properties.

## 31.1 General Classes of Polymers

Synthetic polymers can be classified by their method of synthesis as either chain-growth polymers or step-growth polymers. These categories are necessarily broad and imprecise but nevertheless provide a useful distinction. **Chain-growth polymers**, also called *addition polymers*, are produced by chain-reaction polymerization in which an initiator adds to a carbon–carbon double bond to yield a reactive intermediate. This intermediate reacts with a second molecule of monomer to yield a new intermediate, this new intermediate reacts with a third monomer unit, and so on. The polymer is built as more monomers add successively to the reactive end of the growing chain.

The initiator can be either an anion, a cation, or a radical, and the monomer unit can be any of a large number of substituted alkenes. Polyethylene, produced by radical-initiated polymerization of ethylene, is by far the most common example of a chain-growth polymer:

$$\underset{\text{A radical initiator}}{\text{In}\cdot} + CH_2{=}CH_2 \longrightarrow [\text{In}{-}CH_2CH_2\cdot] \xrightarrow{CH_2=CH_2} [\text{In}{-}CH_2CH_2CH_2CH_2\cdot]$$

Repeat many times

$$\text{In}{-}(CH_2{-}CH_2)_5{-}\cdots$$

A section of a polyethylene chain

**Step-growth polymers**, also called *condensation polymers*, are produced by processes in which the bond-forming step is one of the fundamental polar reactions we have studied. Reactions occur between two difunctional molecules, with each bond in the polymer being formed independently of the others. The polymer produced normally has the two monomers in an alternating order and usually has other atoms in addition to carbon in the main chain. Nylon, a polyamide formed by reaction of a diacid and a diamine, is the most common example of a step-growth polymer.

$$H_2N(CH_2)_nNH_2 + HOOC(CH_2)_mCOOH \xrightarrow{\Delta} \text{~}NH(CH_2)_nNH{-}\overset{O}{\overset{\|}{C}}(CH_2)_m\overset{O}{\overset{\|}{C}}\text{~} + H_2O$$

Nylon, a step-growth polymer
(a polyamide)

## 31.2 Radical Polymerization of Alkenes

Many low-molecular-weight alkenes undergo a rapid polymerization reaction when treated with small amounts of a radical initiator. For example, polyethylene, one of the first alkene chain-growth polymers to be manufactured commercially, has been produced since 1943. It has a current annual U.S. production volume of nearly 16 billion lb.

Polymerization of ethylene is usually carried out at high pressure (1000–3000 atm) and high temperature (100–250°C) with a radical catalyst such as benzoyl peroxide. The resultant polymer may have anywhere from a few hundred to a few thousand monomer units in the chain. As with all radical chain reactions, three kinds of steps are required: initiation steps, propagation steps, and termination steps.

*Step 1* Initiation occurs when trace amounts of radicals are generated by the catalyst:

$$C_6H_5{-}\overset{O}{\overset{\|}{C}}{-}O{-}O{-}\overset{O}{\overset{\|}{C}}{-}C_6H_5 \xrightarrow{\Delta} 2\ C_6H_5{-}\overset{O}{\overset{\|}{C}}{-}O\cdot = \text{“}In\cdot\text{”}$$

Benzoyl peroxide — Benzoyloxy radical

*Step 2* One of the benzoyloxy radicals produced in step 1 adds to ethylene to generate a new carbon radical, and the polymerization is off and running:

$$In\cdot + H_2C{=}CH_2 \longrightarrow In{-}CH_2{-}CH_2\cdot$$

*Step 3* Propagation occurs when the carbon radical adds to another ethylene molecule. Repetition of this step builds the polymer chain.

$$In{-}CH_2{-}CH_2\cdot + H_2C{=}CH_2 \longrightarrow In{-}CH_2CH_2CH_2CH_2\cdot$$

$$\xrightarrow[\text{many times}]{\text{Repeat}} In(CH_2CH_2)_nCH_2CH_2\cdot$$

*Step 4* Eventually, the polymer chain is terminated by reactions that consume the radical. Combination or disproportionation of two radicals are possible chain-terminating reactions:

$$2\ In(CH_2CH_2)_nCH_2CH_2\cdot \longrightarrow In(CH_2CH_2)_nCH_2CH_2{-}CH_2CH_2(CH_2CH_2)_nIn$$

$$2\ In(CH_2CH_2)_nCH_2CH_2\cdot \longrightarrow In(CH_2CH_2)_nCH{=}CH_2 + In(CH_2CH_2)_nCH_2CH_3$$

Many substituted ethylene monomers undergo radical reaction to yield polymers with substituent groups (denoted by a circled S) regularly spaced along the polymer backbone.

$$\underset{\text{Monomer}}{CH_2{=}\overset{\text{Ⓢ}}{\overset{|}{C}}H} \xrightarrow[\text{polymerization}]{\text{Radical}} \underset{\text{Polymer}}{\left( CH_2\overset{\text{Ⓢ}}{\overset{|}{C}}HCH_2\overset{\text{Ⓢ}}{\overset{|}{C}}HCH_2\overset{\text{Ⓢ}}{\overset{|}{C}}H \right)}$$

Table 31.2 shows some of the more important of these **vinyl monomers** and lists the industrial uses of the different polymers that result.

**Table 31.2** Some chain-growth polymers and their uses

| *Monomer name* | *Formula* | *Trade or common names of polymer* | *Uses* |
|---|---|---|---|
| Ethylene | $H_2C{=}CH_2$ | Polyethylene | Packaging, bottles, cable insulation, films and sheets |
| Propene (propylene) | $H_2C{=}CHCH_3$ | Polypropylene | Automotive moldings, rope, carpet fibers |
| Chloroethylene (vinyl chloride) | $H_2C{=}CHCl$ | Poly(vinyl chloride), Tedlar | Insulation, films, pipes |
| Styrene | $H_2C{=}CHC_6H_5$ | Polystyrene, Styron | Foam and molded articles |
| Tetrafluoroethylene | $F_2C{=}CF_2$ | Teflon | Valves and gaskets, coatings |
| Acrylonitrile | $H_2C{=}CHCN$ | Orlon, Acrilan | Fibers |
| Methyl methacrylate | $H_2C{=}\overset{CH_3}{\overset{|}{C}}CO_2CH_3$ | Plexiglas, Lucite | Molded articles, paints |
| Vinyl acetate | $H_2C{=}CHOCOCH_3$ | Poly(vinyl acetate) | Paints, adhesives |
| Vinyl alcohol | "$H_2C{=}CHOH$" | Poly(vinyl alcohol) | Fibers, adhesives |

Note that vinyl alcohol, the monomer corresponding to poly(vinyl alcohol), is an enol isomer that, if prepared, would tautomerize rapidly to acetaldehyde. Poly(vinyl alcohol) must therefore be made by polymerization of vinyl acetate ($H_2C{=}CHOCOCH_3$), followed by hydrolysis of poly(vinyl acetate):

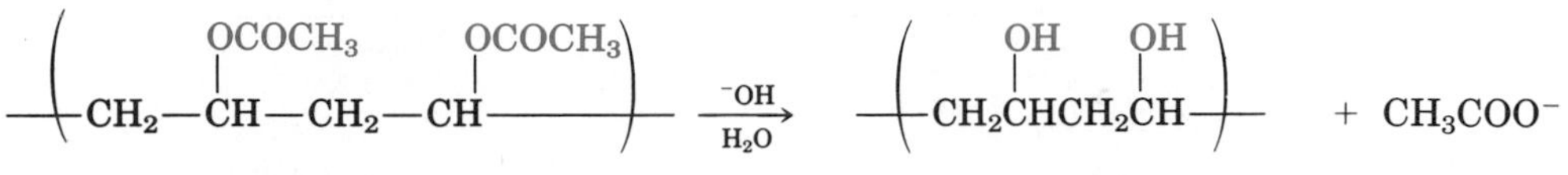

$$\left(\!\!-CH_2-\underset{\displaystyle OCOCH_3}{CH}-CH_2-\underset{\displaystyle OCOCH_3}{CH}-\!\!\right) \xrightarrow[H_2O]{^-OH} \left(\!\!-CH_2\underset{\displaystyle OH}{CH}CH_2\underset{\displaystyle OH}{CH}-\!\!\right) + CH_3COO^-$$

Poly(vinyl acetate) Poly(vinyl alcohol)

PROBLEM

**31.1** How might these polymers be prepared? Show the monomer units you would use.

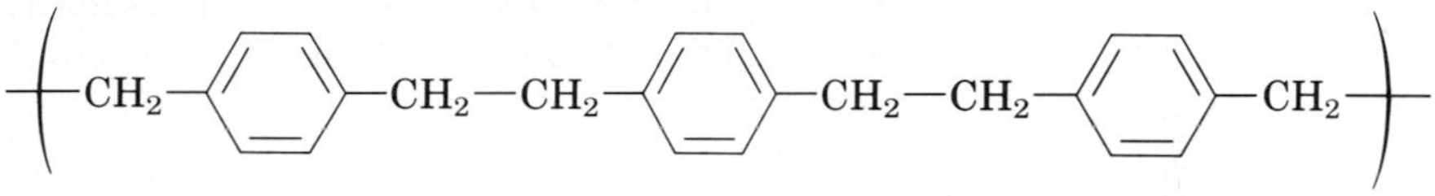

(a) $\left(\!\!-CH_2-\underset{\displaystyle OCH_3}{CH}-CH_2-\underset{\displaystyle OCH_3}{CH}-CH_2-\underset{\displaystyle OCH_3}{CH}-\!\!\right)$

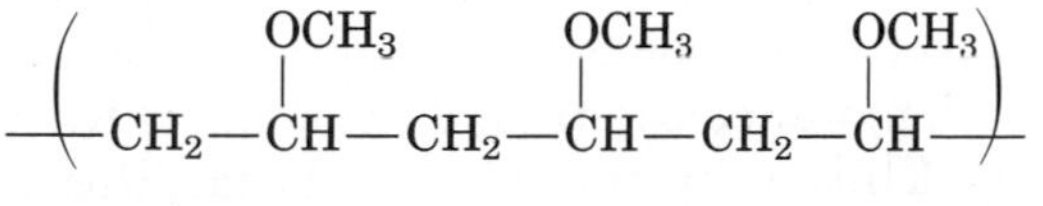

(b) $\left(\!\!-CH_2-C_6H_4-CH_2-CH_2-C_6H_4-CH_2-CH_2-C_6H_4-CH_2-\!\!\right)$

(c) $\left(\!\!-\underset{\displaystyle Cl}{CH}-\underset{\displaystyle Cl}{CH}-\underset{\displaystyle Cl}{CH}-\underset{\displaystyle Cl}{CH}-\underset{\displaystyle Cl}{CH}-\underset{\displaystyle Cl}{CH}-\!\!\right)$

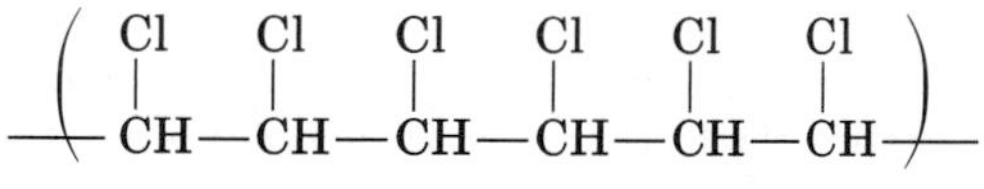

PROBLEM

**31.2** How can you account for the fact that radical polymerization of styrene yields a product in which the phenyl substituents are on alternate carbon atoms rather than on neighboring carbons?

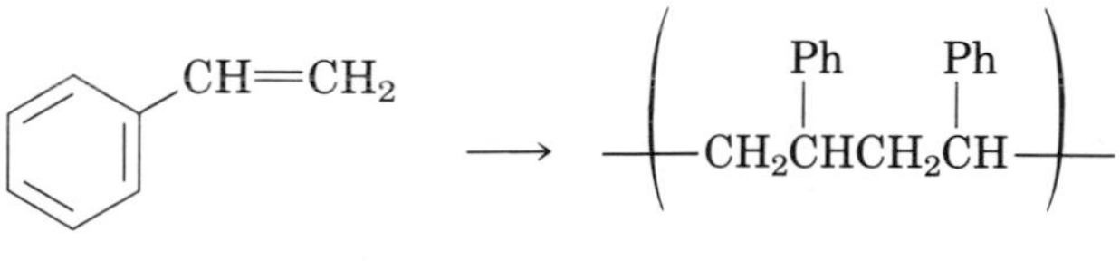

$$C_6H_5CH{=}CH_2 \longrightarrow \left(\!\!-CH_2\underset{\displaystyle Ph}{CH}CH_2\underset{\displaystyle Ph}{CH}-\!\!\right)$$

*not* $\left(\!\!-CH_2-\underset{\displaystyle Ph}{CH}-\underset{\displaystyle Ph}{CH}-CH_2-\!\!\right)$

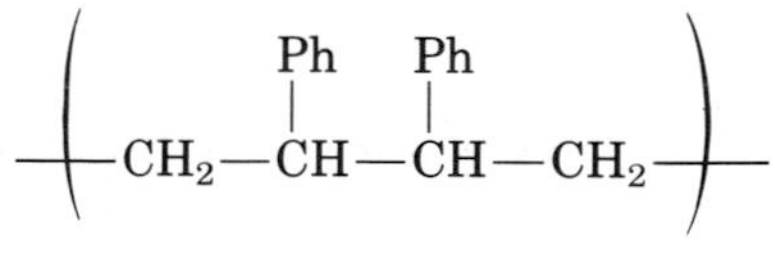

## 31.3 Cationic Polymerization

Certain alkene monomers can be polymerized by a cationic mechanism, as well as by a radical mechanism. Cationic polymerizations occur by a chain-reaction pathway and require the use of strong protic or Lewis acids as initiators. The key chain-carrying step is the electrophilic addition of a carbocation intermediate to the carbon–carbon double bond of another monomer unit.

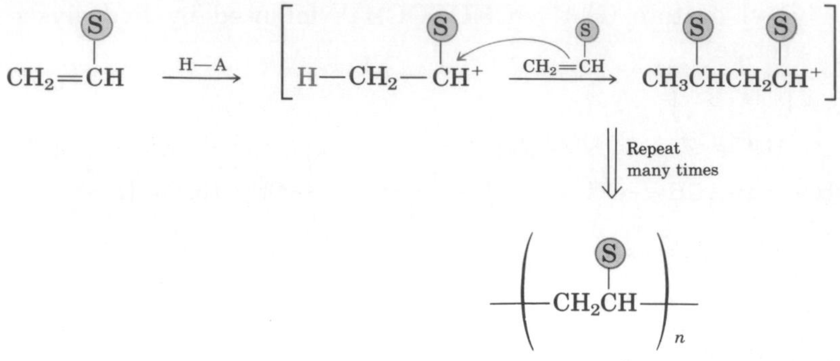

$$CH_2{=}CH(S) \xrightarrow{H-A} \left[ H-CH_2-CH^+(S) \xrightarrow{CH_2=CH(S)} CH_3CH(S)CH_2CH^+(S) \right]$$

Repeat many times

$$\left( CH_2CH(S) \right)_n$$

As we might expect, vinyl monomers with electron-donating substituents polymerize much more readily than do monomers with electron-withdrawing substituents. Thus, ethylene, vinyl chloride, and acrylonitrile do not polymerize easily under cationic conditions, but 2-methylpropene polymerizes rapidly. This difference in behavior simply reflects the difference in stability of the potential chain-carrying intermediate cations.

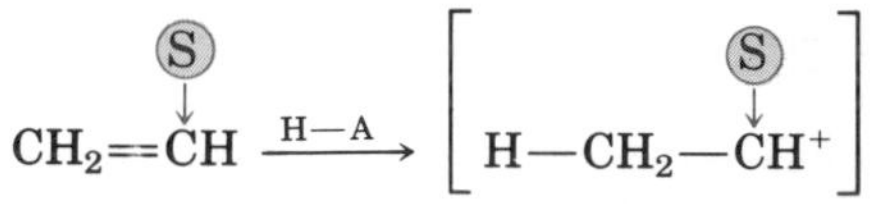

$$CH_2{=}CH(S) \xrightarrow{H-A} \left[ H-CH_2-CH^+(S) \right]$$

Electron-rich alkene; stabilized cation intermediate when substituent is electron-donating; good reaction

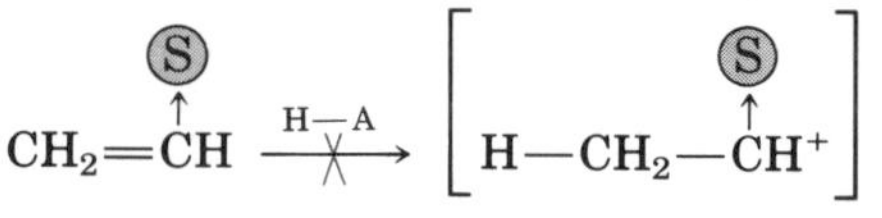

$$CH_2{=}CH(S) \not\xrightarrow{H-A} \left[ H-CH_2-CH^+(S) \right]$$

Electron-poor alkene; destabilized cation intermediate when substituent is electron-withdrawing; poor reaction

The most common commercial use of cationic polymerization occurs during the preparation of polyisobutylene by treatment of isobutylene (2-methylpropene) with $BF_3$ catalyst at $-80°C$. The product is used in the manufacture of inner tubes for truck and bicycle tires.

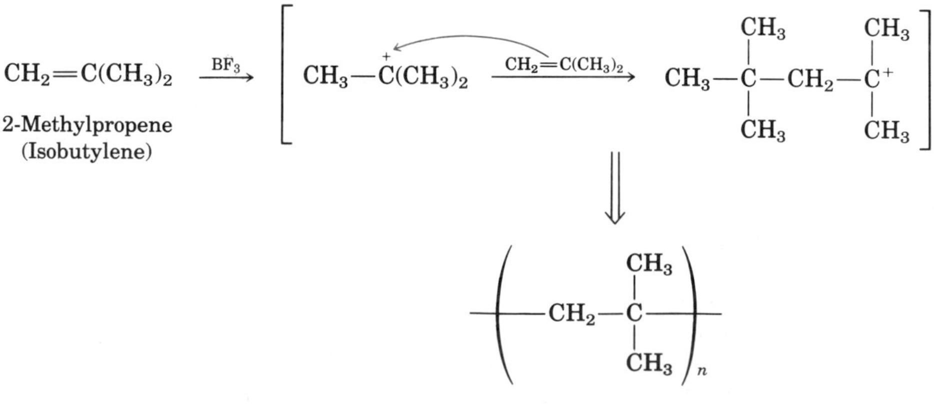

$$CH_2{=}C(CH_3)_2 \xrightarrow{BF_3} \left[ CH_3-\overset{+}{C}(CH_3)_2 \xrightarrow{CH_2=C(CH_3)_2} CH_3-C(CH_3)_2-CH_2-C^+(CH_3)_2 \right]$$

2-Methylpropene (Isobutylene)

$$\Downarrow$$

$$\left( CH_2-C(CH_3)_2 \right)_n$$

Polyisobutylene

PROBLEM

**31.3** List the expected reactivity order of the following monomers to cationic polymerization. Explain your ordering.

$H_2C{=}CHCH_3$, $H_2C{=}CHCl$, $H_2C{=}CH{-}C_6H_5$, $H_2C{=}CHCO_2CH_3$

## 31.4 Anionic Polymerization

Alkene monomers with electron-withdrawing (anion-stabilizing) substituents can be polymerized by anionic catalysts. A chain reaction occurs in which the key step is nucleophilic addition of an anion to the unsaturated monomer by a Michael-type reaction (Section 23.12). Acrylonitrile ($H_2C{=}CHCN$), methyl methacrylate [$H_2C{=}C(CH_3)COOCH_3$], and styrene ($H_2C{=}CHC_6H_5$) can all be polymerized anionically, although radical-initiated polymerization is preferred commercially.

$$CH_2{=}\overset{\text{S}}{\overset{|}{C}H} + Nu{:}^- \longrightarrow \left[ Nu{-}CH_2{-}\overset{\text{S}}{\overset{|}{C}H}{:}^- \longrightarrow Nu{-}CH_2\overset{\text{S}}{\overset{|}{C}H}CH_2\overset{\text{S}}{\overset{|}{C}H}{:}^- \right]$$

$$\Downarrow$$

$$\left( CH_2{-}\overset{\text{S}}{\overset{|}{C}H} \right)_n$$

where Nu = A nucleophilic initiator

S = An electron-withdrawing substituent

One particularly interesting example of anionic polymerization accounts for the remarkable properties of "super glue," one drop of which is claimed to support 2000 lb. Super glue is simply a solution of highly pure methyl $\alpha$-cyanoacrylate. Since the carbon–carbon double bond has two electron-withdrawing groups, anionic addition is particularly easy and particularly rapid. Trace amounts of water or bases on the surface of an object are sufficient to initiate polymerization of the cyanoacrylate and bind articles together. Skin is a good source of the necessary basic initiators, and many people have found their fingers stuck together after inadvertently touching super glue.

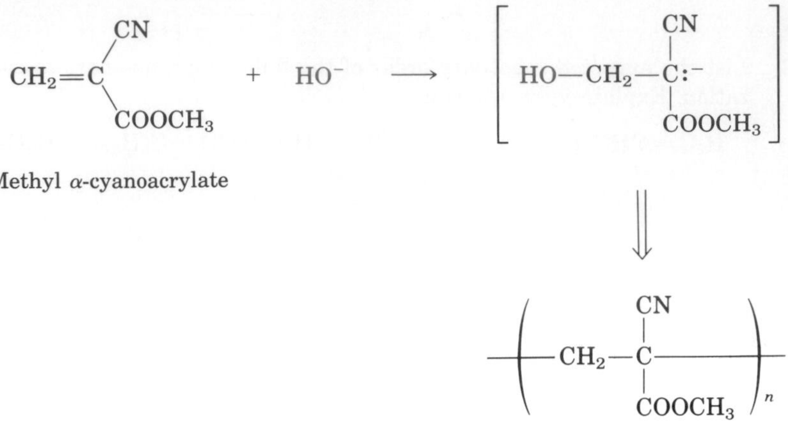

$$CH_2{=}C(CN)(COOCH_3) + HO^- \longrightarrow \left[ HO{-}CH_2{-}\overset{\displaystyle CN}{\underset{\displaystyle COOCH_3}{C}}{:}^- \right]$$

Methyl α-cyanoacrylate

$$\Downarrow$$

$$\left( CH_2{-}\overset{\displaystyle CN}{\underset{\displaystyle COOCH_3}{C}} \right)_n$$

Super glue

PROBLEM

**31.4** Order the following monomers with respect to their expected reactivity toward anionic polymerization. Explain your ordering.

$$H_2C{=}CHCH_3, \quad H_2C{=}CF_2, \quad H_2C{=}CHCN, \quad H_2C{=}CHC_6H_5$$

PROBLEM

**31.5** Poly(ethylene glycol), or Carbowax, can be made by base-induced polymerization of ethylene oxide. Propose a mechanism for this reaction.

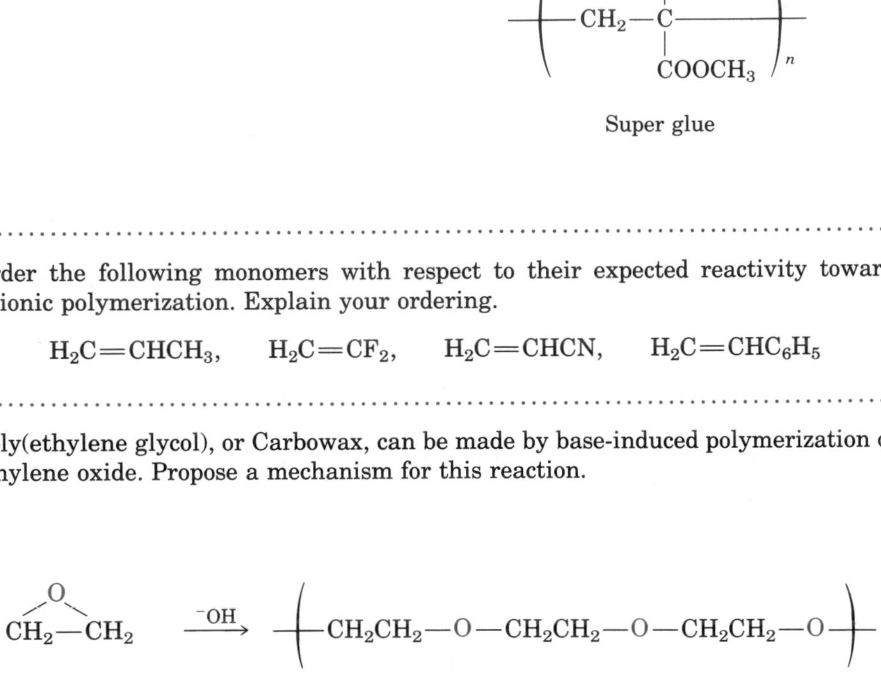

$$\underset{\text{Ethylene oxide}}{\overset{O}{CH_2{-}CH_2}} \xrightarrow{^-OH} \underset{\text{Carbowax}}{\left( CH_2CH_2{-}O{-}CH_2CH_2{-}O{-}CH_2CH_2{-}O \right)}$$

## 31.5 Chain Branching during Polymerization

The polymerization of unsaturated monomers is complicated in practice by several factors that greatly affect the properties of the product. One such problem is that radical polymerization yields a product that is not linear, but has numerous branches in it. Branches arise when the radical end of a growing chain abstracts a hydrogen atom from the middle of the chain to yield an internal radical site that continues the polymerization. The most common kind of branching, termed **short-chain branching**, arises from intramolecular hydrogen atom abstraction from a position four carbon atoms away from the chain end (Figure 31.1).

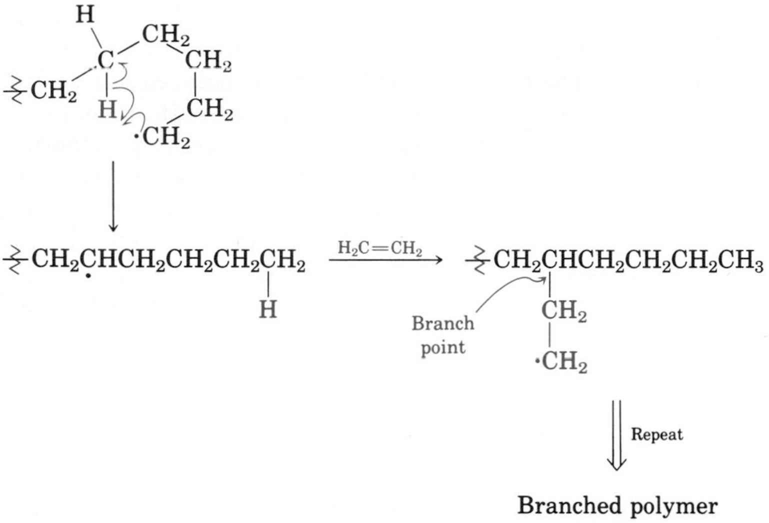

H
CH₂
C
CH₂
CH₂
H
CH₂
·CH₂

$$\text{-CH}_2\dot{\text{C}}\text{HCH}_2\text{CH}_2\text{CH}_2\text{CH}_2\text{-H} \xrightarrow{H_2C=CH_2} \text{-CH}_2\text{CHCH}_2\text{CH}_2\text{CH}_2\text{CH}_3 \text{ (with branch } \text{-CH}_2\text{-}\cdot\text{CH}_2\text{)}$$

Branch point

Repeat

Branched polymer

**Figure 31.1** Short-chain branching during polymerization of ethylene

Alternatively, intermolecular hydrogen atom abstraction can take place by reaction of the radical end of one chain with the middle of another chain. **Long-chain branching** results from this kind of reaction (Figure 31.2). Studies have shown that short-chain branching occurs about 50 times more often than long-chain branching.

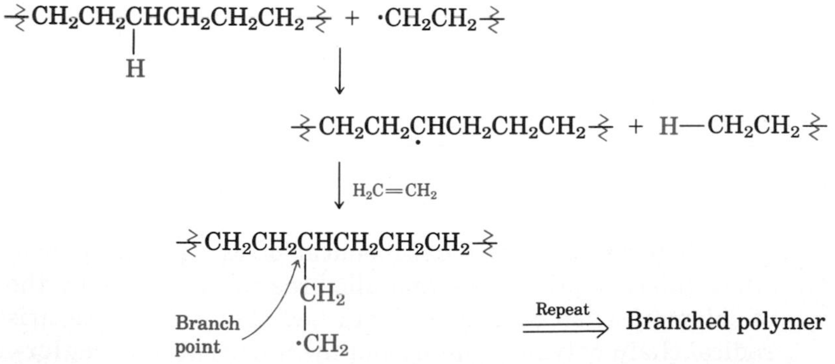

$$\text{-CH}_2\text{CH}_2\text{CH(H)CH}_2\text{CH}_2\text{CH}_2\text{-} + \cdot\text{CH}_2\text{CH}_2\text{-}$$

$$\downarrow$$

$$\text{-CH}_2\text{CH}_2\dot{\text{C}}\text{HCH}_2\text{CH}_2\text{CH}_2\text{-} + \text{H—CH}_2\text{CH}_2\text{-}$$

$$\downarrow H_2C=CH_2$$

$$\text{-CH}_2\text{CH}_2\text{CHCH}_2\text{CH}_2\text{CH}_2\text{-} \text{ (with branch } \text{-CH}_2\text{-}\cdot\text{CH}_2\text{)}$$

Branch point

Repeat ⟹ Branched polymer

**Figure 31.2** Long-chain branching during the polymerization of ethylene

Chain branching is a common occurrence during radical polymerizations and is not restricted to polyethylene. Polypropylene, polystyrene, and poly(methyl methacrylate) all contain branched chains.

## 31.6 Stereochemistry of Polymerization: Ziegler–Natta Catalysts

Yet another complication that arises during alkene polymerization has to do with stereochemistry. Although not pointed out earlier, polymerization of substituted alkenes leads to polymers with numerous chiral centers on the backbone. For example, propylene might polymerize with any of the three stereochemical outcomes shown in Figure 31.3. The conformation in which all methyl groups are on the same side of the zigzag backbone is called **isotactic**; that in which the methyl groups regularly alternate on opposite sides of the backbone is called **syndiotactic** (syn-**die**-oh-tac-tic); and the conformation in which the methyl groups are randomly oriented is called **atactic**.

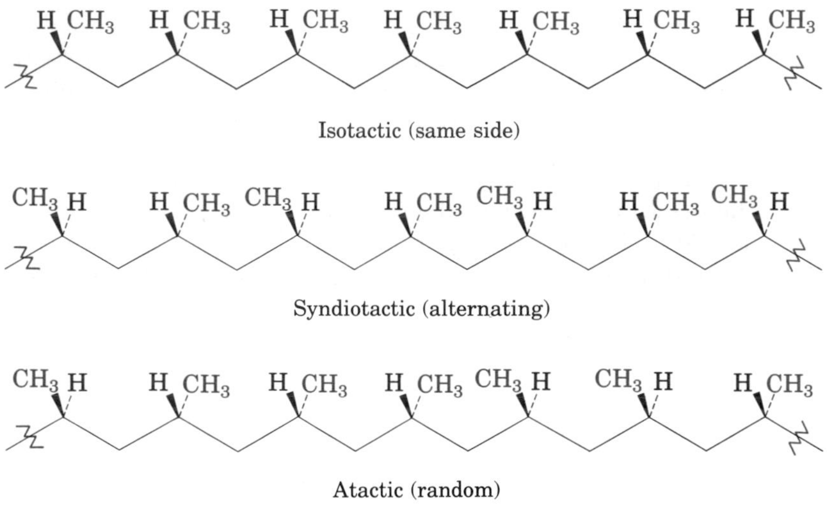

Isotactic (same side)

Syndiotactic (alternating)

Atactic (random)

**Figure 31.3** Isotactic, syndiotactic, and atactic forms of polypropylene

The three different stereochemical forms of polypropylene all have somewhat different properties, and all three can be made by the proper choice of polymerization conditions. Branched atactic polymers arise from normal radical chain polymerizations, but the use of special **Ziegler–Natta**[1,2] **catalysts**, allows preparation of isotactic and syndiotactic forms.

Ziegler–Natta catalysts are organometallic transition-metal complexes prepared by treatment of a trialkylaluminum with a titanium compound.

[1]Karl Ziegler (1889–1976); b. Helsa, near Kassel, Germany; Ph.D. Marburg University; director, Max Planck Institute for Coal Research, Mülheim-Ruhr, Germany; Nobel prize (1963).

[2]Giulio Natta (1903–1979); b. Imperia, Italy; D.C.E. Milan Polytechnic Institute; professor, Milan Polytechnic Institute; Nobel prize (1963).

Triethylaluminum and titanium trichloride form a typical preparation, although the precise structure of the active catalyst is still unknown:

$$(CH_3CH_2)_3Al + TiCl_3 \longrightarrow \text{Ziegler–Natta catalyst}$$

Introduced in 1953, Ziegler–Natta catalysts immediately revolutionized the field of polymer chemistry, largely because of two advantages:

1. Ziegler–Natta polymers are linear and have practically no chain branching.
2. Ziegler–Natta polymers are stereochemically regular. Either isotactic or syndiotactic forms can be produced, depending on the exact catalyst system used. All commercial polypropylene is now produced by the Ziegler–Natta process, since the product has greater strength, stiffness, and resistance to cracking than the branched atactic polypropylene prepared by radical polymerization.

Linear polyethylene produced by the Ziegler–Natta process (called *high-density polyethylene*) is a highly crystalline polymer with 500–1000 ethylene units per chain. High-density polyethylene has greater strength and heat resistance than the product of radical-induced polymerization (*low-density polyethylene*) and is used to produce plastic squeeze bottles and molded housewares.

The exact mechanism by which Ziegler–Natta catalysts operate is not clear, but the key chain-lengthening steps undoubtedly involve formation of alkyltitanium species, followed by coordination of alkene monomer to the titanium and insertion of coordinated alkene into the carbon–titanium bond.

$$\text{~}CH_2CH_2—Ti\langle \xrightarrow{CH_2=CH_2} \text{~}CH_2CH_2—Ti\langle(\cdots H_2C{=}CH_2) \longrightarrow \text{~}CH_2CH_2CH_2CH_2—Ti\langle$$

An alkyltitanium intermediate

Chain-extended alkyltitanium intermediate

⇓ Repeat

PROBLEM

**31.6** Account for the fact that vinylidene chloride, $H_2C{=}CCl_2$, does not polymerize in isotactic, syndiotactic, and atactic forms.

PROBLEM

**31.7** Polymers such as polypropylene contain a large number of chiral carbon atoms. Would you therefore expect samples of either isotactic, syndiotactic, or atactic polypropylene to rotate plane-polarized light? Explain.

## 31.7 Diene Polymers: Natural and Synthetic Rubbers

Although we've discussed only the polymerization of simple alkene monomers up to this point, the same principles apply to the polymerization of conjugated dienes. Diene polymers are structurally more complex, however, since double bonds remain every four carbon atoms along the chain. These double bonds may be either cis or trans, and the proper choice of Ziegler–Natta catalyst allows preparation of either geometry. Note that the polymerization reaction corresponds to 1,4 addition of the growing chain to each conjugated diene monomer (recall 1,4 ionic addition to dienes, Section 14.5).

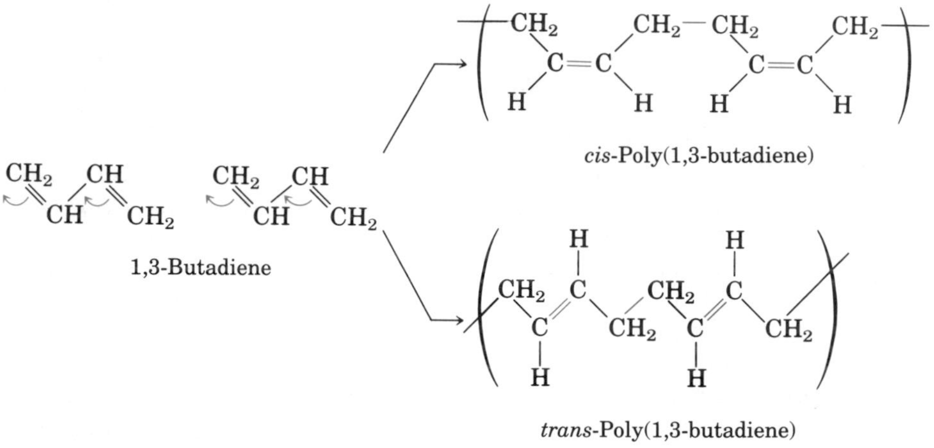

1,3-Butadiene

*cis*-Poly(1,3-butadiene)

*trans*-Poly(1,3-butadiene)

Natural rubber is a polymer of isoprene (2-methyl-1,3-butadiene) in which the double bonds have cis stereochemistry. Gutta-percha, the all-trans isomer of natural rubber, also occurs naturally as the exudate of certain trees but is less common. Gutta-percha is harder and more brittle than rubber but finds a variety of minor applications, including occasional use as the covering on golf balls.

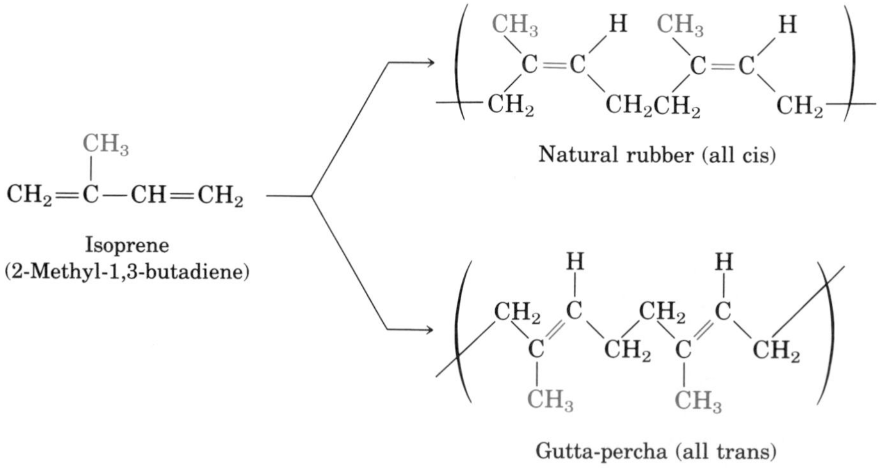

$$CH_2{=}C(CH_3){-}CH{=}CH_2$$

Isoprene
(2-Methyl-1,3-butadiene)

Natural rubber (all cis)

Gutta-percha (all trans)

A number of different synthetic rubbers are produced commercially by diene polymerization. Both *cis*- and *trans*-polyisoprene can be made under Ziegler–Natta conditions, and the synthetic rubber thus produced is quite similar to the natural material. Chloroprene (2-chloro-1,3-butadiene) is polymerized commercially to yield neoprene, an excellent, though expensive, synthetic rubber with good weather resistance. Neoprene is used in the production of industrial hoses and gloves, among other things.

$$CH_2{=}C(Cl){-}CH{=}CH_2 \xrightarrow[\text{catalyst}]{\text{Ziegler–Natta}} \left( CH_2{-}C(Cl){=}CH{-}CH_2{-}CH_2{-}C(Cl){=}CH{-}CH_2 \right)$$

Chloroprene
(2-Chloro-1,3-butadiene)

Neoprene (trans)

Both natural and synthetic rubbers are soft and tacky unless hardened by a process called **vulcanization**. Discovered in 1839 by Charles Goodyear (of subsequent tire fame), vulcanization involves heating the polymer with a few percent by weight of sulfur. The result is a much harder rubber with greatly improved resistance to wear and abrasion.

The chemistry of vulcanization is complex but involves formation of sulfur bridges or **cross-links** between polymer chains. Cross-linked polymers tend to be rigid, because the individual chains are locked together into immense single molecules that can no longer slip over one another (Figure 31.4). Note the similarity in structure between a vulcanized rubber and a peptide that has cysteine cross-links (Section 27.7).

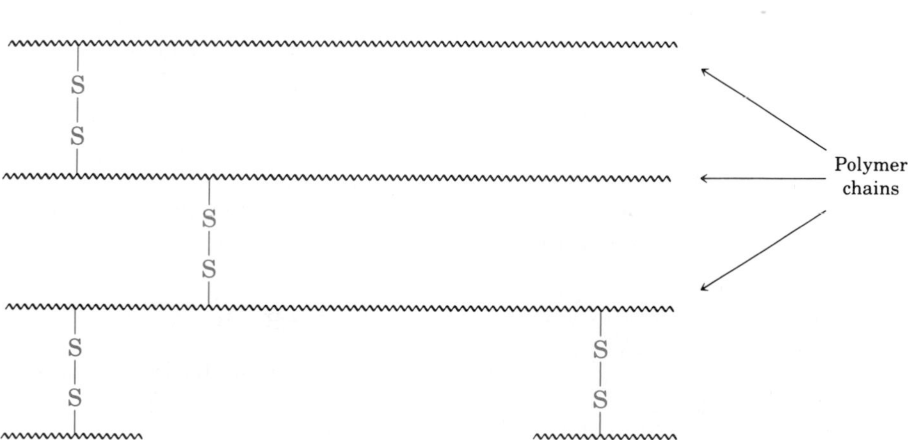


**Figure 31.4** Sulfur cross-linked chains resulting from vulcanization of rubber

PROBLEM

**31.8** Diene polymers contain occasional vinyl branches along the chain. How do you think these branches might arise?

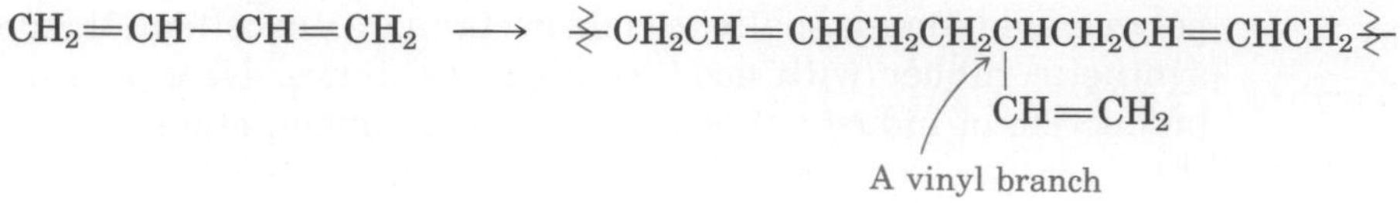

$$CH_2{=}CH{-}CH{=}CH_2 \longrightarrow \text{-}CH_2CH{=}CHCH_2CH_2\underset{\underset{CH{=}CH_2}{|}}{C}HCH_2CH{=}CHCH_2\text{-}$$

A vinyl branch

PROBLEM

**31.9** Radial tires, whose sidewalls are made of natural rubber, tend to crack and weather rapidly in areas around major cities where high levels of ozone and other industrial pollutants are found. Explain.

## 31.8 Copolymers

Up to this point we've discussed only **homopolymers**, polymers that are made up of identical repeating units. In practice, however, **copolymers** are more common and more important commercially. Copolymers are obtained when two or more different monomers are allowed to polymerize together. For example, copolymerization of vinyl chloride with vinylidene chloride (1,1-dichloroethylene) leads to the well-known polymer Saran:

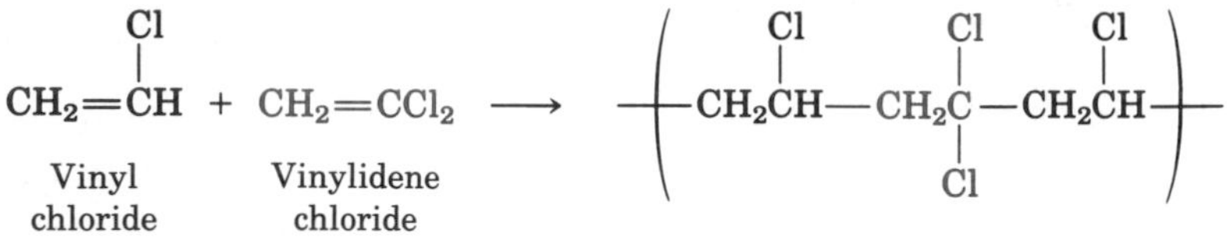

$$\underset{\text{Vinyl chloride}}{CH_2{=}\overset{\overset{Cl}{|}}{C}H} + \underset{\text{Vinylidene chloride}}{CH_2{=}CCl_2} \longrightarrow \left( CH_2\overset{\overset{Cl}{|}}{C}H{-}CH_2\overset{\overset{Cl}{|}}{\underset{\underset{Cl}{|}}{C}}{-}CH_2\overset{\overset{Cl}{|}}{C}H \right)$$

Saran

Copolymerization of monomer mixtures often leads to materials with properties quite different from those of either corresponding homopolymer, giving the polymer chemist a vast amount of flexibility for devising new materials. Table 31.3 lists some common copolymers and indicates their commercial applications.

Several different structural types of copolymers can be defined, depending on the distribution of monomer units in the chain. If we imagine, for example, that monomer A and monomer B are being copolymerized, the resultant product might have a random distribution of the two units throughout the chain, or it might have an alternating distribution:

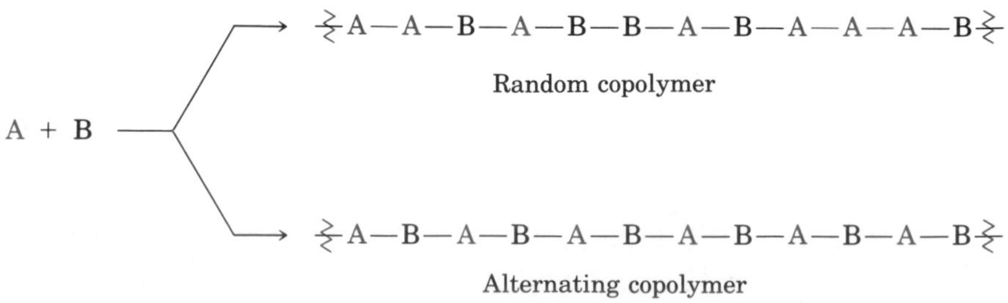

$$A + B \longrightarrow \text{-}A{-}A{-}B{-}A{-}B{-}B{-}A{-}B{-}A{-}A{-}A{-}B\text{-}$$

Random copolymer

$$A + B \longrightarrow \text{-}A{-}B{-}A{-}B{-}A{-}B{-}A{-}B{-}A{-}B{-}A{-}B\text{-}$$

Alternating copolymer

**Table 31.3** Some common copolymers and their uses

| *Monomer name* | *Formula* | *Trade or common name of polymer* | *Uses* |
|---|---|---|---|
| Vinyl chloride<br>Vinylidene chloride | $H_2C{=}CHCl$<br>$H_2C{=}CCl_2$ | Saran | Food wrapping, fibers |
| Styrene (25%)<br>Butadiene (75%) | $H_2C{=}CHC_6H_5$<br>$H_2C{=}CHCH{=}CH_2$ | SBR (styrene–butadiene rubber) | Tires |
| Hexafluoropropene<br>Vinylidene fluoride | $F_2C{=}CFCF_3$<br>$H_2C{=}CF_2$ | Viton | Gaskets, rubber articles |
| Acrylonitrile<br>Butadiene | $H_2C{=}CHCN$<br>$H_2C{=}CHCH{=}CH_2$ | Nitrile rubber | Latex, adhesives, gasoline hoses |
| Isobutylene<br>Isoprene | $H_2C{=}C(CH_3)_2$<br>$H_2C{=}C(CH_3)CH{=}CH_2$ | Butyl rubber | Inner tubes |
| Acrylonitrile<br>Butadiene<br>Styrene | $H_2C{=}CHCN$<br>$H_2C{=}CHCH{=}CH_2$<br>$H_2C{=}CHC_6H_5$ | ABS (initials of three monomers) | Pipes, high-impact applications |

The exact distribution depends on such factors as the proportion of the two reactant monomers used and their relative reactivities. In practice, neither perfectly random nor perfectly alternating copolymers are usually found. Most copolymers tend more toward the alternating form but have many random imperfections.

Two other special forms of copolymers that can be prepared under certain conditions are called **block copolymers** and **graft copolymers**. Block copolymers are those in which different blocks of identical monomer units alternate with each other; graft copolymers are those in which homopolymer branches of one monomer unit are "grafted" onto a homopolymer chain of another monomer unit.

$$\text{—A—A—A—A—A—A—A—A—B—B—B—B—B—B—B—B—}$$

Segment of a block copolymer

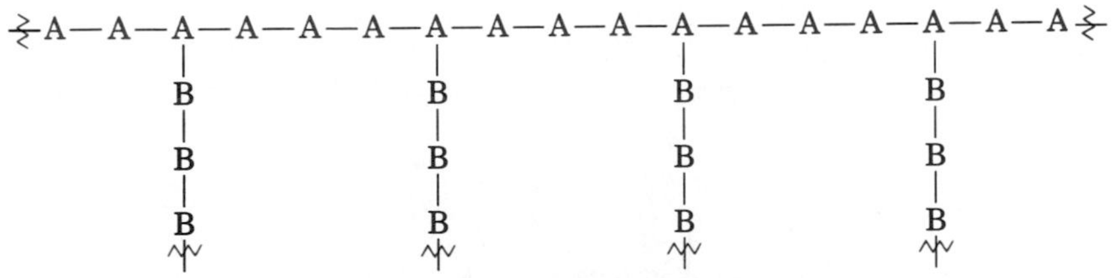

```
—A—A—A—A—A—A—A—A—A—A—A—A—A—A—A—A—A—
     |           |           |           |
     B           B           B           B
     |           |           |           |
     B           B           B           B
     |           |           |           |
     B           B           B           B
```

Segment of a graft copolymer

Block copolymers are prepared by initiating the radical polymerization of one monomer to grow homopolymer chains, followed by addition of an excess of the second monomer. Graft copolymers are made by gamma irradiation of a homopolymer chain in the presence of a second monomer. The high-energy irradiation knocks hydrogen atoms off the homopolymer chain at random points, thus generating radical sites that can initiate polymerization of the added monomer.

**PROBLEM**

**31.10** Draw the structure of an alternating segment of butyl rubber, a copolymer of 2-methyl-1,3-butadiene and 2-methylpropene prepared under cationic conditions.

**PROBLEM**

**31.11** One of the most important commercial applications of graft polymerization involves irradiation of polybutadiene, followed by addition of styrene. The product is used to make rubber soles for shoes. Draw the structure of a representative segment of this styrene–butadiene graft copolymer.

## 31.9 Step-Growth Polymers: Nylon

Step-growth polymers are produced by reactions between two difunctional molecules, as we saw earlier. Each bond is formed in a discrete step, independent of all other bonds in the polymer; chain reactions are not involved. The key bond-forming step is usually one of the fundamental polar reactions that we studied earlier, as opposed to a radical reaction:

$$\text{A}\sim\text{A} + \text{B}\sim\text{B} \longrightarrow \sim\text{A}\sim\text{A}{-}\text{B}\sim\text{B}{-}\text{A}\sim\text{A}{-}\text{B}\sim\text{B}\sim$$

where A and B are reactive functional groups.

A large number of different step-growth polymers have been made, with some of the more important ones shown in Table 31.4.

The best known step-growth polymers are the polyamides (**nylons**), first prepared by Wallace Carothers[3] at the Du Pont Company by heating diamines with diacids. For example, nylon 66 is prepared by reaction of the six-carbon adipic acid with the six-carbon hexamethylenediamine at 280°C:

$$\underset{\text{Adipic acid}}{HOOC(CH_2)_4COOH} + \underset{\text{Hexamethylenediamine}}{H_2N(CH_2)_6NH_2}$$

$$\downarrow$$

$$\underset{\text{Nylon 66}}{-\!\left(\overset{O}{\overset{\|}{C}}(CH_2)_4\overset{O}{\overset{\|}{C}}{-}NH(CH_2)_6NH\right)_{n}\!-} + n\, H_2O$$

[3]Wallace H. Carothers (1896–1937); b. Burlington, Iowa; Ph.D. Illinois (Adams); Du Pont Company.

**Table 31.4** Some common step-growth polymers and their uses

| *Monomer name* | *Formula* | *Trade or common name of polymer* | *Uses* |
|---|---|---|---|
| Adipic acid | $HOOC(CH_2)_4COOH$ | Nylon 66 | Fibers, clothing, tire cord, bearings |
| Hexamethylene diamine | $H_2N(CH_2)_6NH_2$ | | |
| Ethylene glycol | $HOCH_2CH_2OH$ | Dacron, Terylene, Mylar | Fibers, clothing, tire cord, film |
| Dimethyl terephthalate | 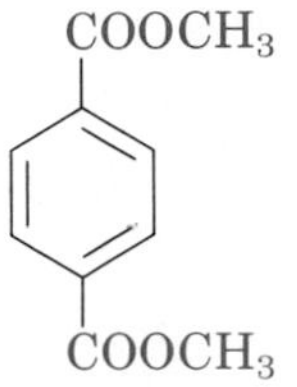 $COOCH_3$ (para-disubstituted benzene ring) $COOCH_3$ | | |
| Caprolactam | 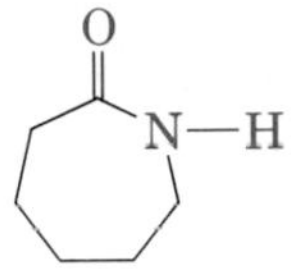 (seven-membered ring lactam: C=O, N—H) | Nylon 6, Perlon | Fibers, large cast articles |
| Diphenyl carbonate | $C_6H_5OCOOC_6H_5$ | Lexan, polycarbonate | Molded articles, machine housings |
| Bisphenol A | 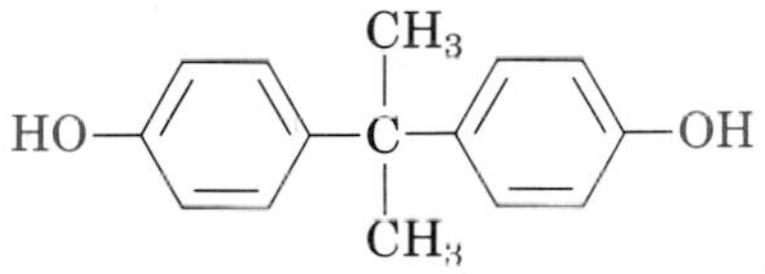 $HO-C_6H_4-C(CH_3)_2-C_6H_4-OH$ | | |
| Poly(2-butene-1,4-diol) | $HO\!+\!CH_2CH{=}CHCH_2\!+_{n}OH$ | Polyurethane, Spandex | Foams, fibers, coatings |
| Tolylene diisocyanate | 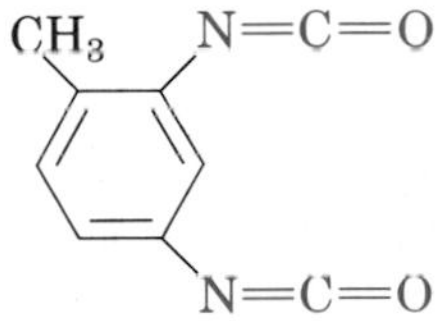 $CH_3$, $N{=}C{=}O$ (on benzene ring), $N{=}C{=}O$ | | |

Nylon 6, closely related in structure to nylon 66, is prepared by polymerization of caprolactam. Water is first added to hydrolyze caprolactam to 6-aminohexanoic acid, and strong heating then brings about dehydration and polymerization.

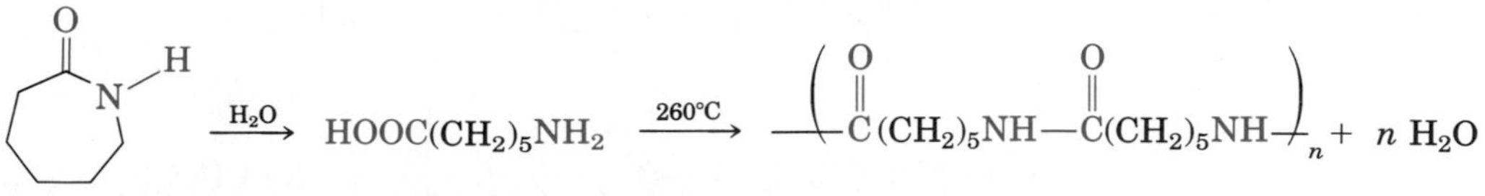

$$\text{Caprolactam} \xrightarrow{H_2O} \underset{\text{6-Aminohexanoic acid}}{HOOC(CH_2)_5NH_2} \xrightarrow{260^\circ C} \underset{\text{Nylon 6}}{\left( \overset{O}{\overset{\|}{C}}(CH_2)_5NH-\overset{O}{\overset{\|}{C}}(CH_2)_5NH \right)_n} + n\,H_2O$$

Caprolactam 6-Aminohexanoic acid Nylon 6

Nylons are used both in engineering applications and in making fibers. A combination of high impact strength and abrasion resistance makes nylon an excellent metal-substitute for bearings and gears. As fiber, nylon is used in a wide variety of applications, from clothing to tire cord to Perlon mountaineering ropes.

PROBLEM

**31.12** Nylon is far more easily damaged by accidental spillage of acid or base than are chain-growth polymers like polyethylene. In other words, wearing nylon stockings in a chemistry laboratory can be very expensive. Explain.

PROBLEM

**31.13** Draw structures of the step-growth polymers you would expect to obtain from these reactions:

(a) $BrCH_2CH_2CH_2Br + HOCH_2CH_2CH_2OH \xrightarrow{\text{Base}}$

(b) $HOCH_2CH_2OH + HOOCO(CH_2)_6COOH \xrightarrow{H_2SO_4 \text{ catalyst}}$

(c) $H_2N(CH_2)_6NH_2 + ClOC(CH_2)_4COCl \longrightarrow$

PROBLEM

**31.14** Kevlar, a nylon polymer prepared by reaction of 1,4-benzenedicarboxylic acid (terephthalic acid) with 1,4-diaminobenzene (*p*-phenylenediamine), is so strong that it is used to make bulletproof vests. Draw the structure of a segment of Kevlar.

## 31.10 Polyesters

Just as polyamides can be made by reaction between diacids and diamines, **polyesters** can be made by reaction between diacids and dialcohols. The most generally useful polyester is made by ester exchange reaction between dimethyl terephthalate and ethylene glycol. The product is widely used under the trade name Dacron to make clothing fiber and tire cord, and is used under the name Mylar to make plastic film. The tensile strength of poly(ethylene terephthalate) film is nearly equal to that of steel, and the film is unusually flex- and tear-resistant.

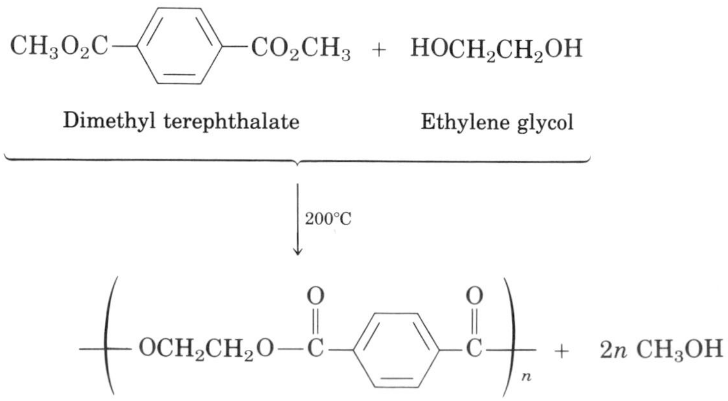

$$CH_3O_2C\text{—}C_6H_4\text{—}CO_2CH_3 + HOCH_2CH_2OH$$

Dimethyl terephthalate     Ethylene glycol

$$\xrightarrow{200°C}$$

$$\left[ OCH_2CH_2O\text{—}\overset{O}{\overset{\|}{C}}\text{—}C_6H_4\text{—}\overset{O}{\overset{\|}{C}} \right]_n + 2n\ CH_3OH$$

Polyester, Dacron, Mylar

Lexan, a polycarbonate prepared from diphenyl carbonate and bisphenol A, is another commercially valuable polyester. Lexan has an unusually high impact strength, making it valuable for use in machinery housings, telephones, and bicycle safety helmets.

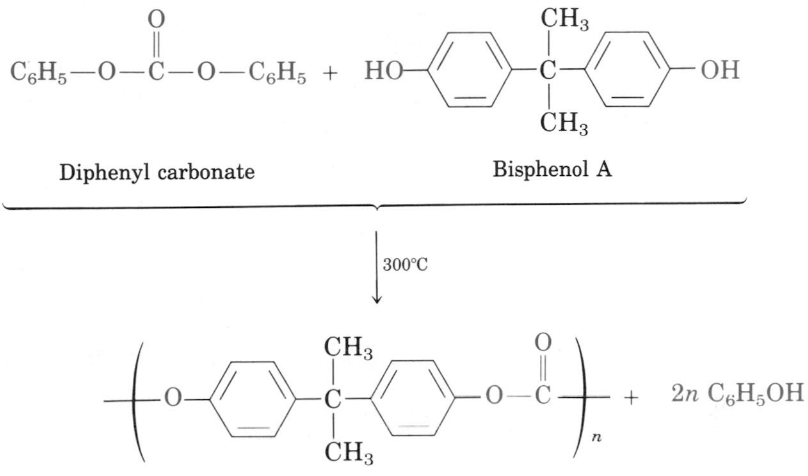

$$C_6H_5{-}O{-}\overset{\overset{\displaystyle O}{\|}}{C}{-}O{-}C_6H_5 \;+\; HO{-}C_6H_4{-}\underset{CH_3}{\overset{CH_3}{\overset{|}{\underset{|}{C}}}}{-}C_6H_4{-}OH$$

Diphenyl carbonate   Bisphenol A

$$\xrightarrow{300°C}$$

$$\left[{-}O{-}C_6H_4{-}\underset{CH_3}{\overset{CH_3}{\overset{|}{\underset{|}{C}}}}{-}C_6H_4{-}O{-}\overset{\overset{\displaystyle O}{\|}}{C}{-}\right]_n \;+\; 2n\ C_6H_5OH$$

Lexan

PROBLEM

**31.15** Draw the structure of the polymer you would expect to obtain from reaction of dimethyl terephthalate with a triol such as glycerol. What structural feature would this new polymer have that was not present in Dacron? How do you think this new feature would affect the properties of the polymer?

## 31.11 Polyurethanes

A **urethane** is a carbonyl-containing functional group in which the carbonyl carbon is bound both to an ether oxygen and to an amine nitrogen. As such, a urethane can be considered intermediate between a carbonate and a urea:

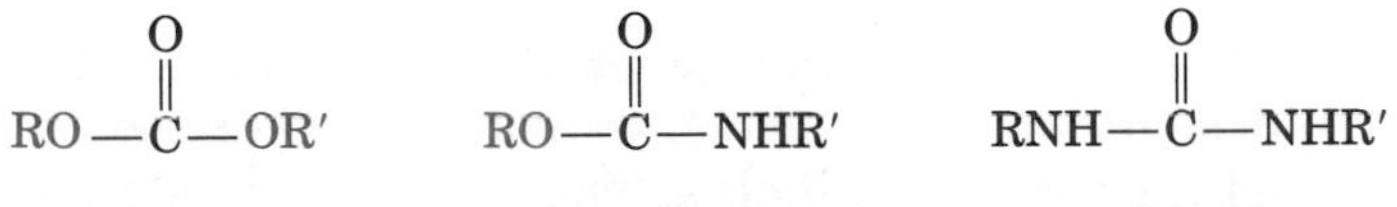

$$RO{-}\overset{\overset{\displaystyle O}{\|}}{C}{-}OR' \qquad RO{-}\overset{\overset{\displaystyle O}{\|}}{C}{-}NHR' \qquad RNH{-}\overset{\overset{\displaystyle O}{\|}}{C}{-}NHR'$$

A carbonate   A urethane   A urea

Urethanes are prepared by nucleophilic addition of alcohols to isocyanates:

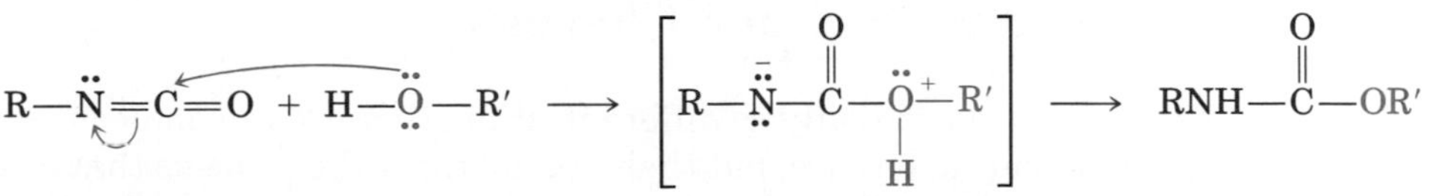

$$R{-}\ddot{N}{=}C{=}O \;+\; H{-}\ddot{O}{-}R' \longrightarrow \left[R{-}\overset{-}{\ddot{N}}{-}\overset{\overset{\displaystyle O}{\|}}{C}{-}\underset{H}{\underset{|}{\overset{+}{\ddot{O}}}}{-}R'\right] \longrightarrow RNH{-}\overset{\overset{\displaystyle O}{\|}}{C}{-}OR'$$

An isocyanate   A urethane

**Polyurethanes**, polymers containing urethane linkages, are prepared by reaction between a diol and a diisocyanate. The diol is itself a low-molecular-weight (mol wt ≈ 1000) polymer with hydroxyl end groups; the diisocyanate is often tolylene diisocyanate.

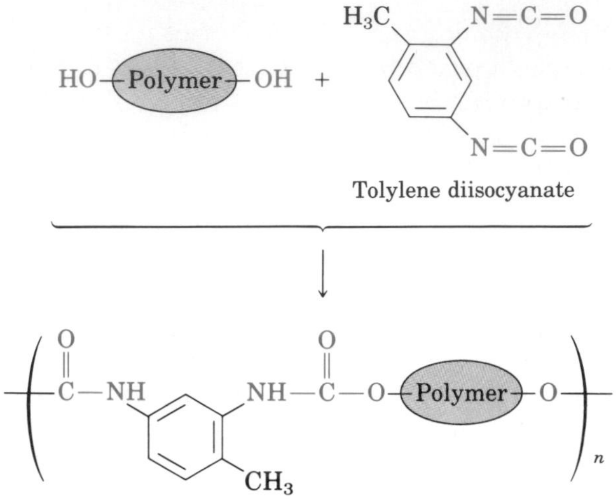

$$\text{HO-Polymer-OH} + \text{Tolylene diisocyanate } (H_3C\text{, } N{=}C{=}O\text{, } N{=}C{=}O \text{ on benzene ring})$$

Tolylene diisocyanate

$$\downarrow$$

$$\left( \overset{O}{\overset{\|}{C}}-NH-C_6H_3(CH_3)-NH-\overset{O}{\overset{\|}{C}}-O-\text{Polymer}-O \right)_n$$

A polyurethane

A number of different kinds of polyurethanes are produced, depending on the nature of the polymeric alcohol used and on the degree of cross-linking achieved. One major use of polyurethane is in the stretchable Spandex and Lycra fibers used for bathing suits and leotards. These polyurethanes have a rather low degree of cross-linking so that the resultant polymer is soft and elastic.

A second major use of polyurethanes is in foams. Foaming occurs when a small amount of water is added during polymerization. Water adds to isocyanate groups giving carbamic acids, which spontaneously lose $CO_2$, thus generating the foam bubbles.

$$R{-}N{=}C{=}O + H_2O \longrightarrow \left[ RNH-\overset{O}{\overset{\|}{C}}-OH \right] \longrightarrow RNH_2 + CO_2 \uparrow$$

A carbamic acid

Polyurethane foams generally have a higher amount of cross-linking than do polyurethane fibers, an amount that can be varied by using a polyalcohol (rather than a diol) as one of the reactive components. The result is a rigid but very light foam suitable for use as thermal insulation in building construction and in portable ice chests.

## 31.12 Polymer Structure and Chemistry

Polymers are not really so different from other organic molecules. They are much bigger, to be sure, but their chemistry is the same as that of analogous small molecules. The chemistry of polymers is the familiar chemistry of functional groups; molecular size plays little role. Thus, the ester linkages of a polyester such as Dacron are hydrolyzed by base; the aromatic rings of polystyrene undergo typical electrophilic aromatic substitution reactions; and the alkane chains of polyethylene undergo radical-initiated halogenation.

The major differences between small and very large organic molecules are in structure and in physical properties. Here too, though, the bulk structures and properties of polymers are the result of the same intermolecular forces that operate in small molecules.

The most important intermolecular forces between non-cross-linked polymer chains are **van der Waals forces**. These forces, which are due to weak attractive interactions between transient dipoles in nearby molecules (Section 3.7), are the same as those that act between small molecules in solution or in the solid state.

Since van der Waals forces operate only at close distances, they are strongest in those polymers like linear polyethylene in which chains can line up in a regular, close-packed way. Many polymers, in fact, have regions that are essentially crystalline. These regions, called **crystallites**, consist of highly ordered portions in which the zigzag polymer chains are bound together by van der Waals forces (Figure 31.5).

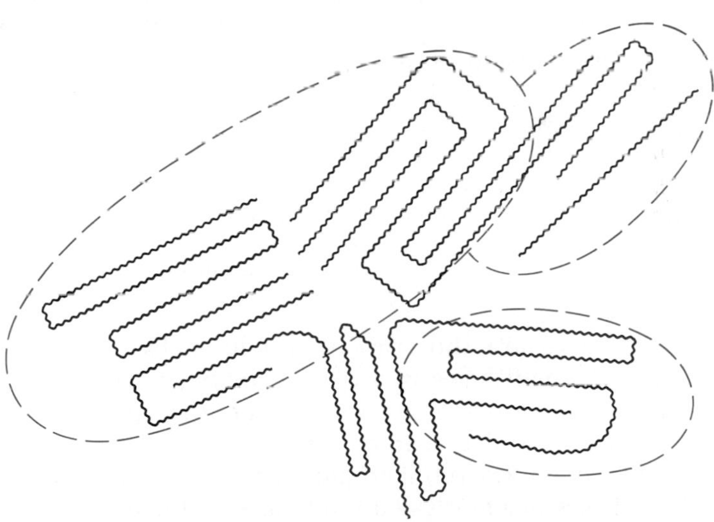

**Figure 31.5** Crystallites in linear polyethylene: The long polymer chains are arranged in parallel lines in the crystallite regions.

As we might expect, polymer crystallinity is strongly affected by the steric requirements of substituent groups on the chains. Thus, poly(methyl methacrylate) is noncrystalline because the chains cannot pack closely together in a regular way, but linear polyethylene is highly crystalline.

PROBLEM

31.16 What product would you expect to obtain from catalytic hydrogenation of natural rubber? How else might you obtain a similar polymer? Would the product be syndiotactic, atactic, or isotactic?

## 31.13 Polymer Structure and Physical Properties

Classification of synthetic polymers according to their physical properties is a useful exercise because it allows us to make a rough correlation between structure and property. In general, we can divide polymers into four major categories: thermoplastics, fibers, elastomers, and thermosetting resins.

**Thermoplastics** are the polymers most people think of when the word *plastic* is mentioned. These polymers are hard at room temperature but become soft and viscous when heated. As a result, they can be molded into toys, beads, telephone housings, or into any of a thousand other items. Because thermoplastics have little or no cross-linking, the individual chains can slip past one another on heating. Some thermoplastic polymers such as poly(methyl methacrylate) (Plexiglas) are amorphous and noncrystalline; others, such as polyethylene and nylon, are partially crystalline.

**Plasticizers**—small organic molecules that act as lubricants between chains—are usually added to plastics to keep them from becoming brittle at room temperature. Dialkyl phthalates are commonly used for this purpose and, in the past few decades, have become among the most widely dispersed of all environmental pollutants. Phthalate plasticizers have even been detected in the fat of antarctic penguins.

$COOCH_2CH_2CH_2CH_3$

$COOCH_2CH_2CH_2CH_3$

Dibutyl phthalate (a plasticizer)

**Fibers** are thin threads produced by extruding a molten polymer through small holes in a die, or *spinneret*. The fibers are then cooled and drawn out. Drawing has the effect of orienting the crystallite regions along the axis of the fiber, a process that adds considerable tensile strength (Figure 31.6). Nylon, Dacron, and polyethylene all have the semicrystalline structure necessary for drawing into oriented fibers.

**Elastomers** are amorphous polymers that have the ability to stretch out and spring back to their original shapes. These polymers must have a modest amount of cross-linking to prevent the chains from slipping over one another, and the chains must have an irregular shape to prevent crystallite formation. When stretched, the randomly coiled chains straighten out and orient along the direction of the pull. Van der Waals forces are too weak and too few to maintain this orientation, however, and the elastomer therefore reverts to its randomly coiled state when the stretching force is released (Figure 31.7).

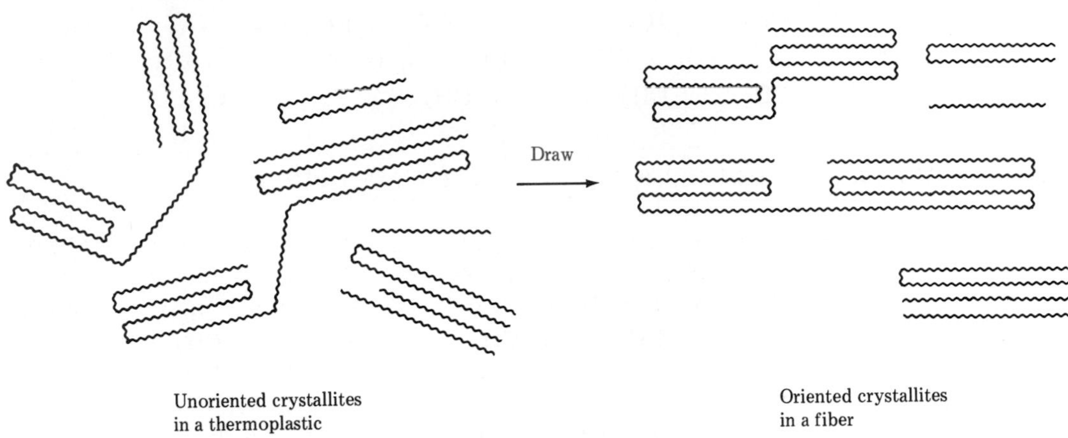


**Figure 31.6** Oriented crystallite regions in a polymer fiber

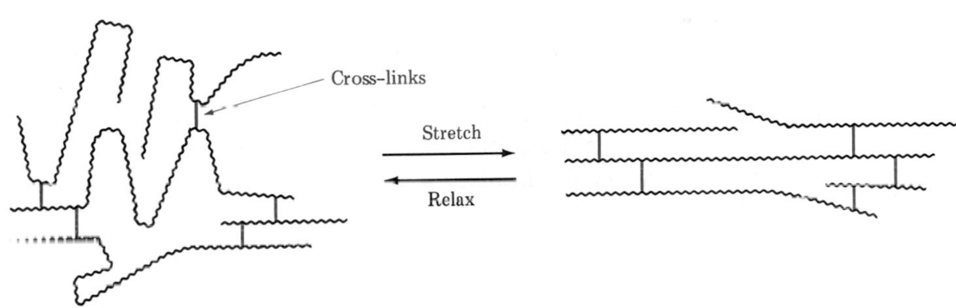


**Figure 31.7** Unstretched and stretched forms of an elastomer

Natural rubber is one example of an elastomer. Rubber has the long chains and occasional cross-links needed for elasticity, but its irregular geometry prevents close packing of the chains into crystallites. Gutta-percha, by contrast, is highly crystalline and is not an elastomer (Figure 31.8, page 1134).

**Thermosetting resins** are polymers that become highly cross-linked and solidify into a hard, insoluble mass when heated. We haven't paid particular attention to such polymers up to this point, but one example should suffice. *Bakelite*, a thermosetting resin first produced in 1907 by Leo Baekeland,[4] has been in commercial use longer than any other synthetic polymer. It is widely used for molded parts, for adhesives, for coatings, and even for high-temperature applications such as missile nose cones.

[4]Leo Hendrik Baekeland (1863–1944); b. Ghent, Belgium; founder and president, Bakelite Corp., United States (1910–1939).

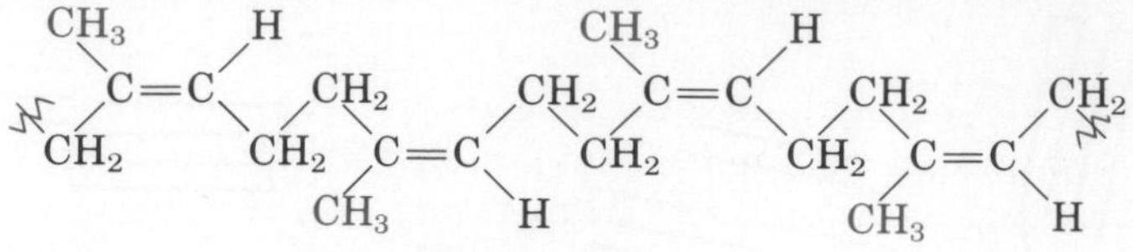

(a)

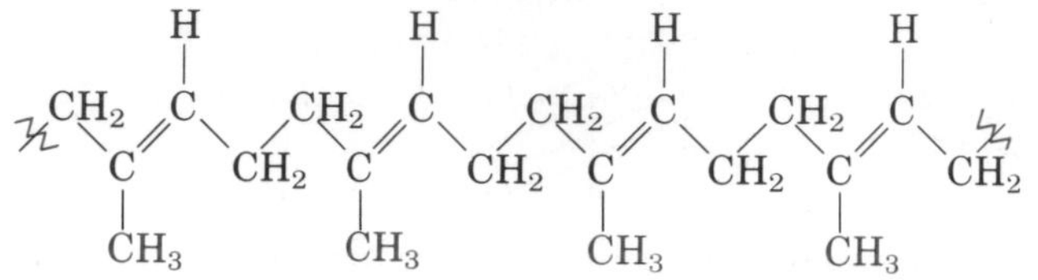

(b)

**Figure 31.8** (a) Natural rubber is elastic and noncrystalline because of its cis double-bond geometry, but (b) gutta-percha is nonelastic and crystalline because its geometry allows for better packing together of chains.

Chemically, Bakelite is a phenolic resin produced by reaction of phenol and formaldehyde. On heating, water is eliminated, many cross-links form, and the polymer sets into a rock-like mass. The cross-linking in Bakelite and other thermosetting resins is three-dimensional and is so extensive that we can't really speak of polymer "chains." A piece of Bakelite is essentially one large molecule.

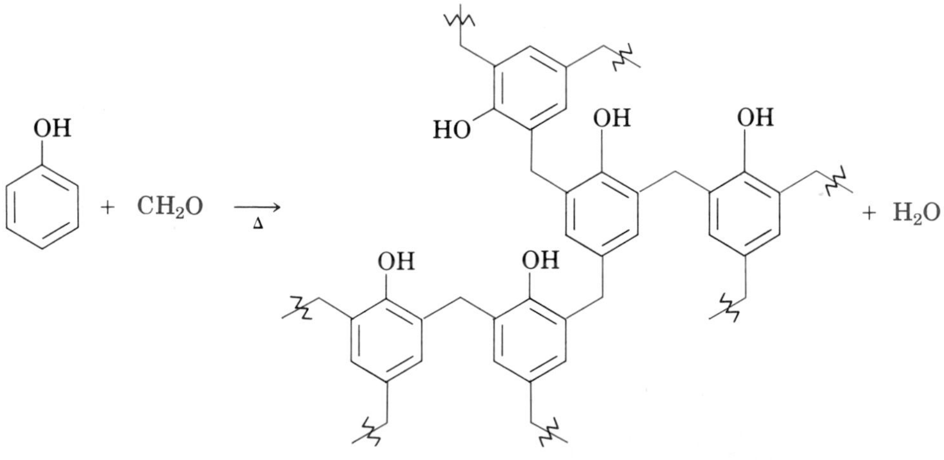

Bakelite

PROBLEM

**31.17** Propose a mechanism to account for the formation of Bakelite from acid-catalyzed polymerization of phenol and formaldehyde.

## 31.14 Summary and Key Words

Synthetic polymers can be classified as either chain-growth polymers or step-growth polymers. **Chain-growth polymers** are prepared by chain-reaction polymerization of unsaturated monomers in the presence of a radical, an anion, or a cation initiator. Radical polymerization is the most commonly used method, but alkenes such as 2-methylpropene that have electron-donating substituents on the double bond polymerize easily by a cationic route. Similarly, monomers such as methyl $\alpha$-cyanoacrylate that have electron-withdrawing substituents on the double bond polymerize by an anionic (Michael–type reaction) pathway.

Conjugated dienes such as 1,4-butadiene, isoprene, and chloroprene also undergo polymerization, giving products that have double bonds in their chains. Still another possibility is the copolymerization of two monomers to give a product that has properties different from those of either homopolymer. **Graft copolymers** and **block copolymers** are two special kinds of polymers whose physical properties can sometimes be controlled.

Alkene polymerization can be carried out in a much more controlled fashion if **Ziegler–Natta catalysts** are used. Ziegler–Natta polymerization minimizes the amount of chain branching in the polymer and leads to stereoregular chains—either **isotactic** (substituents on the same side of the chain) or **syndiotactic** (substituents on alternate sides of the chain), rather than **atactic** (substituents randomly disposed).

**Step-growth polymers**, the second major class of polymers, are prepared by reactions between two different difunctional molecules; the individual bonds in the polymer are formed independently of one another. **Polyamides** (nylons) are formed by step-growth polymerization between a diacid and a diamine; **polyesters** are formed from a diester and a diol; and **polyurethanes** are formed from a diisocyanate and a diol.

The chemistry of synthetic polymers is similar to the chemistry of small molecules with the same functional groups, but the physical properties of polymers are greatly affected by size. Polymers can be classified by physical property into four groups: **thermoplastics**, **fibers**, **elastomers**, and **thermosetting resins**. The properties of each group can be accounted for by the structure, the degree of crystallinity, the geometry, and the amount of cross-linking in the specific polymers.

## ADDITIONAL PROBLEMS

**31.18** Identify the monomer units in each of the following polymers.

(a) $\left(\!-CH_2\underset{}{\overset{NO_2}{\overset{|}{C}}}H-CH_2\overset{NO_2}{\overset{|}{C}}H-\right)$

(b) $\left(-CFCl-CF_2-CFCl-CF_2-\right)$, Kel-F

(c) $\left(-CH_2-O-CH_2-O-CH_2-O-\right)$, Delrin

(d) 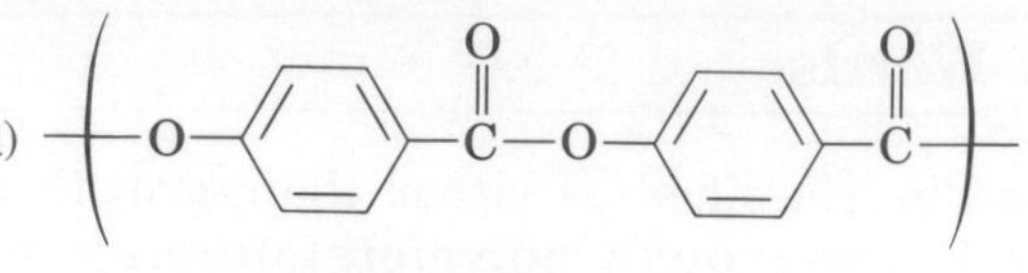 $\left( O - C_6H_4 - \overset{O}{\overset{\|}{C}} - O - C_6H_4 - \overset{O}{\overset{\|}{C}} \right)$

(e) $\left( CH(CH_2)_4CH{=}N(CH_2)_6N \right)$

**31.19** Categorize each polymer in Problem 31.18 as either a step-growth or a chain-growth polymer.

**31.20** Cyclopentadiene undergoes thermal polymerization to yield a polymer that has no double bonds in the chain. On strong heating, this polymer breaks down to regenerate cyclopentadiene. Propose a structure for the product. (See Section 14.7.)

**31.21** Draw a three-dimensional representation of segments of these polymers:
(a) Syndiotactic polyacrylonitrile (b) Atactic poly(methyl methacrylate)
(c) Isotactic poly(vinyl chloride)

**31.22** When styrene, $C_6H_5CH{=}CH_2$, is copolymerized in the presence of a few percent *p*-divinylbenzene, a hard, insoluble, cross-linked polymer is obtained. Show how this cross-linking of polystyrene chains occurs.

**31.23** One method for preparing the 1,6-hexanediamine needed in nylon production starts with 1,3-butadiene. How would you accomplish this synthesis?

$$H_2C{=}CH{-}CH{=}CH_2 \xrightarrow{?} H_2N(CH_2)_6NH_2$$

**31.24** Nitroethylene, $H_2C{=}CHNO_2$, is a sensitive compound that must be prepared with great care. Attempted purification of nitroethylene by distillation often results in low recovery of product and a white coating on the inner walls of the distillation apparatus. Explain.

**31.25** Poly(vinyl butyral) is used as the plastic laminate in the preparation of automobile windshield safety glass. How would you synthesize this polymer?

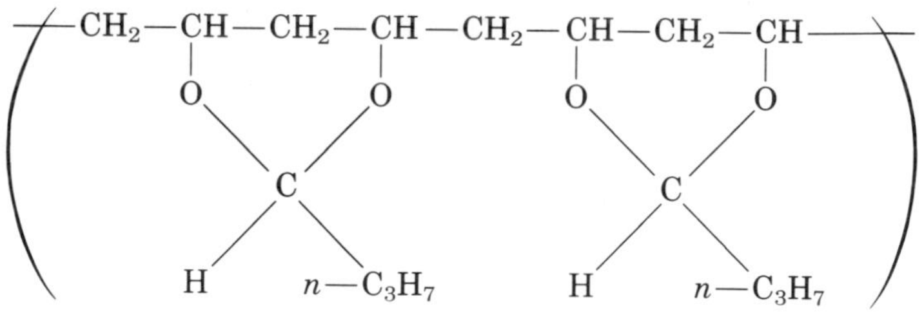

$$\left( CH_2{-}CH{-}CH_2{-}CH{-}CH_2{-}CH{-}CH_2{-}CH \right)$$
(each pair of adjacent CH groups bonded through O atoms to a C bearing H and $n{-}C_3H_7$)

Poly(vinyl butyral)

**31.26** Polyimides having the structure shown are used as coatings on glass and plastics to improve scratch resistance. How would you synthesize a polyimide?

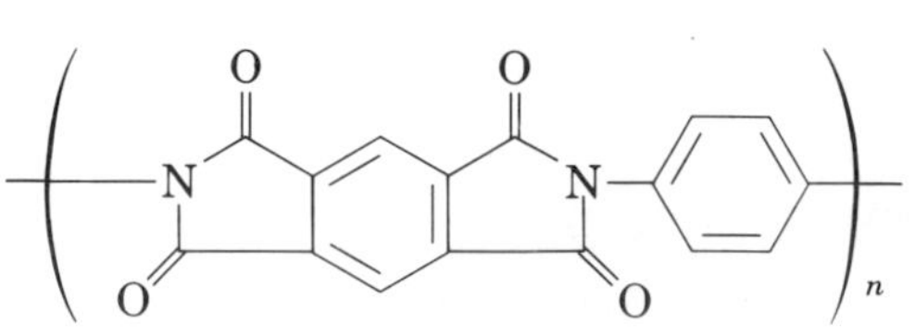

(pyromellitic diimide N—N linked to a *p*-phenylene unit, repeated $n$ times)

A polyimide

**31.27** *Qiana*, a polyamide fiber with a silk-like feel, has the structure indicated. What are the monomer units used in the synthesis of Qiana?

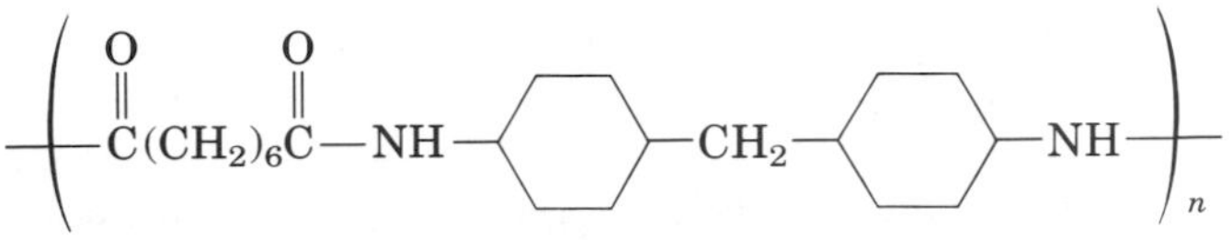

$$\left[ \overset{O}{\overset{\|}{C}}(CH_2)_6\overset{O}{\overset{\|}{C}}-NH-C_6H_{10}-CH_2-C_6H_{10}-NH \right]_n$$

Qiana

**31.28** Starting from aniline and formaldehyde, how would you synthesize the amine necessary for the preparation of Qiana (Problem 31.27)?

**31.29** What is the structure of the polymer produced by treatment of $\beta$-propiolactone with a small amount of hydroxide ion?

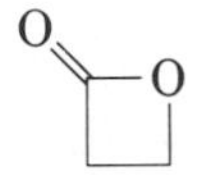

$\beta$-Propiolactone

**31.30** *Glyptal* is a highly cross-linked thermosetting resin produced by heating glycerol and phthalic anhydride (1,2-benzenedicarboxylic acid anhydride). Show the structure of a representative segment of glyptal.

**31.31** *Melmac*, a thermosetting resin often used to make plastic dishes, is prepared by heating melamine with formaldehyde. Propose a structure for Melmac. [*Hint:* See the structure of Bakelite, Section 31.13.]

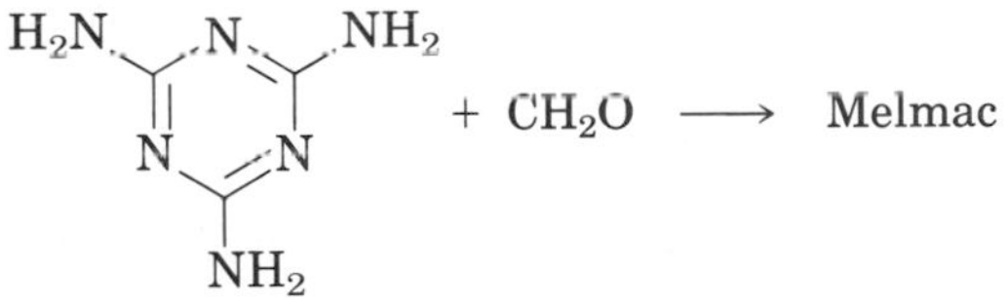

$$\text{Melamine} + CH_2O \longrightarrow \text{Melmac}$$

Melamine

**31.32** Epoxy adhesives are cross-linked resins prepared in two steps. The first step involves $S_N2$ reaction of the disodium salt of bisphenol A with epichlorohydrin to form a low-molecular-weight prepolymer. This prepolymer is then "cured" into a cross-linked resin by treatment with a triamine such as $H_2NCH_2CH_2NHCH_2CH_2NH_2$.

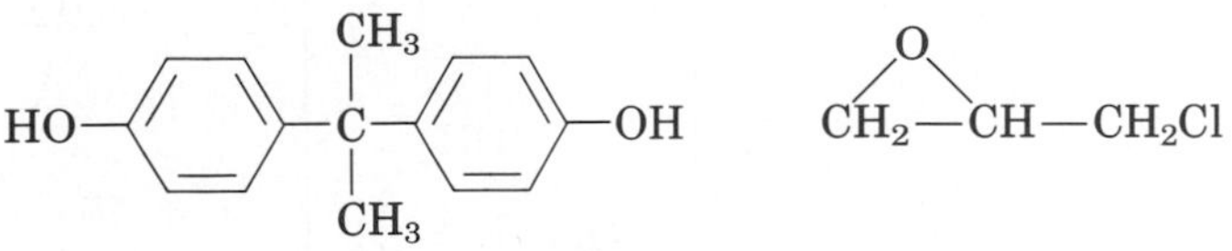

$$HO-C_6H_4-C(CH_3)_2-C_6H_4-OH \qquad \overset{O}{CH_2-CH}-CH_2Cl$$

Bisphenol A Epichlorohydrin

(a) What is the structure of the prepolymer?
(b) How does addition of the triamine result in cross-linking?

**31.33** The polyurethane foam used for home insulation uses methanediphenyldiisocyanate (MDI) as monomer. The MDI is prepared by reaction of aniline with formaldehyde, followed by treatment with phosgene, $COCl_2$. Propose mechanisms for the two steps.

$$2\ C_6H_5NH_2 \xrightarrow{CH_2O} H_2N{-}C_6H_4{-}CH_2{-}C_6H_4{-}NH_2 \xrightarrow{COCl_2} O{=}C{=}N{-}C_6H_4{-}CH_2{-}C_6H_4{-}N{=}C{=}O$$

MDI

**31.34** Write the structure of a representative segment of polyurethane prepared by reaction of ethylene glycol with MDI (Problem 31.33).

**31.35** Urea–formaldehyde resins have been prepared commercially for over 60 years for a variety of purposes, including use as adhesives and insulating foams. For example, the smoking salons of the great hydrogen-filled dirigibles of the 1930s were insulated with urea-formaldehyde foams, and urea–formaldehyde adhesives were used extensively in boats and aircraft during World War II. The structure of the urea–formaldehyde polymer is highly cross-linked, like that of Bakelite (Section 31.13). Propose a structure for this polymer.

$$\underset{\text{Urea}}{H_2N{-}\overset{O}{\overset{\|}{C}}{-}NH_2} + \underset{\text{Formaldehyde}}{CH_2O} \xrightarrow{\Delta}$$

**31.36** An improved polymeric resin used for Merrifield solid-phase peptide synthesis (Section 27.12) is prepared by treating polystyrene with *N*-(hydroxymethyl)phthalimide and trifluoromethanesulfonic acid, followed by reaction with hydrazine. Show how these steps occur.

$$\left(\!{-}CH_2{-}\underset{C_6H_5}{CH}{-}\!\right)_n \xrightarrow[CF_3SO_3H]{\text{phthalimide}{-}N{-}CH_2OH} \left[\left(\!{-}CH_2\underset{C_6H_4{-}CH_2{-}N(\text{phthalimide})}{CH}{-}\!\right)_n\right] \xrightarrow{H_2NNH_2} \left(\!{-}CH_2{-}\underset{C_6H_4{-}CH_2NH_2}{CH}{-}\!\right)_n$$

# Nomenclature of Polyfunctional Organic Compounds

Judging from the number of incorrect names that appear in the chemical literature, it's probably safe to say that relatively few practicing organic chemists are fully conversant with the rules of organic nomenclature. Simple hydrocarbons and monofunctional compounds present few difficulties since the basic rules for naming such compounds are logical and easy to understand. Problems, however, are often encountered with polyfunctional compounds. Whereas most chemists could correctly identify hydrocarbon **1** as 3-ethyl-2,5-dimethylheptane, rather few could correctly identify polyfunctional compound **2**. Should we consider **2** as an ether? As an ethyl ester? As a ketone? As an alkene? It is, of course, all four, but it has only one correct name: ethyl 3-(4-methoxy-2-oxo-3-cyclohexenyl)propanoate.

$$\begin{array}{c} CH_3CHCH_3 \\ | \\ CH_3CH_2CHCH_2CHCH_2CH_3 \\ | \\ CH_3 \end{array}$$

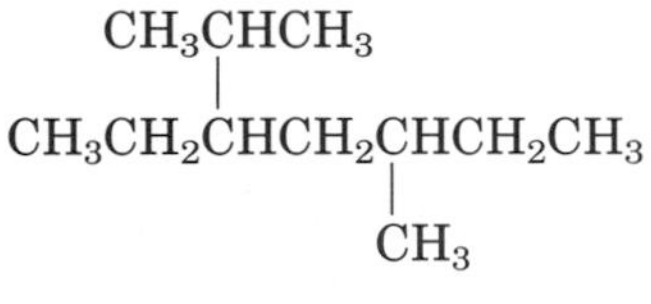

**1.** 3-Ethyl-2,5-dimethylheptane

O, $COOCH_2CH_3$, $CH_3O$

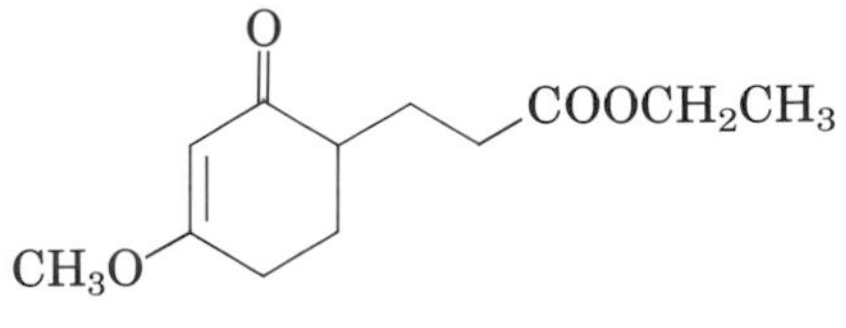

**2.** Ethyl 3-(4-methoxy-2-oxo-3-cyclohexenyl)propanoate

Naming polyfunctional organic compounds is really not much harder than naming monofunctional compounds. All that's required is a prior knowledge of monofunctional compound nomenclature and rigid application of a set of additional rules. In the following discussion, it is assumed that you have a good command of the rules of monofunctional compound nomenclature that were given throughout the text as each new functional group was introduced. A list of where these rules can be found is shown in Table A.1.

**Table A.1** Where to find nomenclature rules for simple functional groups

| *Functional group* | *Text section* | *Functional group* | *Text section* |
|---|---|---|---|
| Alkanes | 3.4 | Aldehydes | 19.2 |
| Cycloalkanes | 3.9 | Carboxylic acids | 20.1 |
| Alkenes | 6.3 | Acid halides | 21.1 |
| Alkynes | 8.2 | Acid anhydrides | 21.1 |
| Aromatic compounds | 15.2 | Amides | 21.1 |
| Alkyl halides | 10.1 | Esters | 21.1 |
| Ethers | 18.1 | Nitriles | 21.1 |
| Alcohols | 17.1 | Amines | 25.1 |
| Ketones | 19.2 | | |

The name of a polyfunctional organic molecule has four parts:

1. **Suffix**—the part that identifies the principal functional-group class to which the molecule belongs
2. **Parent**—the part that identifies the size of the main chain or ring
3. **Substituent prefixes**—parts that identify what substituents are located on the main chain or ring
4. **Locants**—numbers that tell where substituents are located on the main chain or ring

To arrive at the correct name for a complex molecule, the above four parts must be identified and then expressed in the proper order and format. Let's look at the four parts.

## THE SUFFIX—FUNCTIONAL-GROUP PRECEDENCE

A polyfunctional organic molecule may contain many different kinds of functional groups, but, for nomenclature purposes, we must choose just one suffix. It is not correct to use two suffixes. Thus, keto ester **3** shown below must be named either as a ketone with an *-one* suffix or as an ester with an *-oate* suffix, but can't be named as an *-onoate*. Similarly, amino alcohol **4** must be named either as an alcohol (*-ol*) or as an amine (*-amine*) but can't properly be named as an *-olamine*. The only exception to this rule is in naming compounds that have double or triple bonds. For example, $H_2C{=}CHCH_2COOH$ is 3-butenoic acid, and $HC{\equiv}CCH_2CH_2CH_2CH_2OH$ is 5-hexyn-1-ol.

$$\overset{\displaystyle O}{\overset{\|}{CH_3CCH_2CH_2COOCH_3}}$$

**3.** Named as an ester with a keto (oxo) substituent
Methyl 4-oxopentanoate

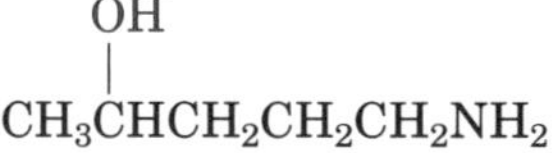


**4.** Named as an alcohol with an amino substituent
5-Amino-2-pentanol

How do we choose which suffix to use? Functional groups are divided into two classes, **principal groups** and **subordinate groups**, as shown in Table A.2. Principal groups are those that may be cited either as prefixes or as suffixes, whereas subordinate groups are those that may be cited only as prefixes. Within the principal groups, an order of precedence has been established. The proper suffix for a given compound is determined by identifying all of the functional groups present and then choosing the principal group of highest priority. For example, Table A.2 indicates

**Table A.2** Classification of functional groups for purposes of nomenclature[a]

| *Functional group class* | *Structure* | *Name when used as suffix* | *Name when used as prefix* |
|---|---|---|---|
| **Principal groups** | | | |
| Carboxylic acids | —COOH | -oic acid<br>-carboxylic acid | carboxy |
| Carboxylic anhydrides | —C(=O)—O—C(=O)— | -oic anhydride<br>-carboxylic anhydride | |
| Carboxylic esters | —COOR | -oate<br>-carboxylate | alkoxycarbonyl |
| Acyl halides | —COCl | -oyl halide<br>-carbonyl halide | halocarbonyl<br>(haloformyl) |
| Amides | $—CONH_2$ | -amide<br>-carboxamide | amido |
| Nitriles | —C≡N | -nitrile<br>-carbonitrile | cyano |
| Aldehydes | —CHO | -al<br>-carbaldehyde | formyl |
| | =O | | oxo (either aldehyde or ketone) |
| Ketones | =O | -one | oxo |
| Alcohols | —OH | -ol | hydroxy |
| Phenols | —OH | -ol | hydroxy |
| Thiols | —SH | -thiol | mercapto, sulfhydryl |
| Amines | $—NH_2$ | -amine | amino |
| Imines | =NH | -imine | imino |
| Alkenes | C=C | -ene | |
| Alkynes | C≡C | -yne | |
| Alkanes | C—C | -ane | |
| **Subordinate groups** | | | |
| Ethers | —OR | | alkoxy |
| Sulfides | —SR | | alkylthio |
| Halides | —F, —Cl, —Br, —I | | halo |
| Nitro | $—NO_2$ | | nitro |
| Azides | N=N=N | | azido |
| Diazo | =N=N | | diazo |

[a]The principal functional groups are listed in order of decreasing priority, but the subordinate functional groups have no established priority order. Principal functional groups may be cited either as prefixes or as suffixes; subordinate functional groups may be cited only as prefixes.

that keto ester **3** must be named as an ester rather than as a ketone, since an ester functional group is higher in priority than a ketone. Similarly, amino alcohol **4** must be named as an alcohol rather than as an amine. The correct name of **3** is methyl 4-oxopentanoate, and the correct name of **4** is 5-amino-2-pentanol. Further examples are shown below.

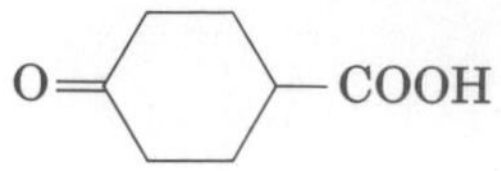

O═⬡—COOH

**5.** Named as a cyclohexanecarboxylic acid with an oxo substituent
4-Oxocyclohexanecarboxylic acid

$$HOOC-\overset{CH_3}{\underset{CH_3}{C}}-CH_2CH_2CH_2COCl$$

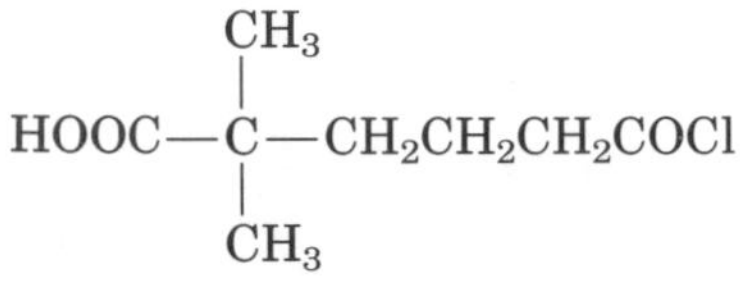

**6.** Named as a carboxylic acid with a chlorocarbonyl substituent
5-Chlorocarbonyl-2,2-dimethylpentanoic acid

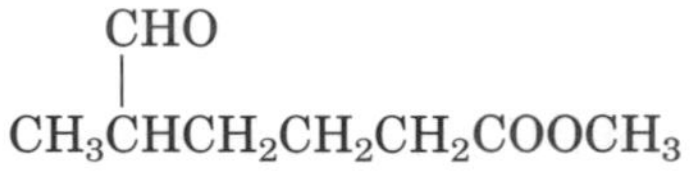

$$CH_3\overset{CHO}{CH}CH_2CH_2CH_2COOCH_3$$

**7.** Named as an ester with an oxo substituent
Methyl 5-methyl-6-oxohexanoate

## THE PARENT—SELECTING THE MAIN CHAIN OR RING

The parent or base name of a polyfunctional organic compound is usually quite easy to identify. If the group of highest priority is part of an open chain, we simply select the longest chain that contains the largest number of principal functional groups. If the highest-priority group is attached to a ring, we use the name of that ring system as the parent. For example, compounds **8** and **9** are isomeric aldehydo acids, and both must be named as acids rather than as aldehydes according to Table A.2. The longest chain in compound **8** has seven carbons, and the substance is therefore named 6-methyl-7-oxoheptanoic acid. Compound **9** also has a chain of seven carbons, but the longest chain that contains both of the principal functional groups has only three carbons. The correct name of this compound is 3-oxo-2-pentylpropanoic acid.

$$CH_3\overset{CHO}{CH}CH_2CH_2CH_2CH_2COOH$$

**8.** Named as a substituted heptanoic acid
6-Methyl-7-oxoheptanoic acid

$$CH_3CH_2CH_2CH_2CH_2\overset{CHO}{CH}COOH$$

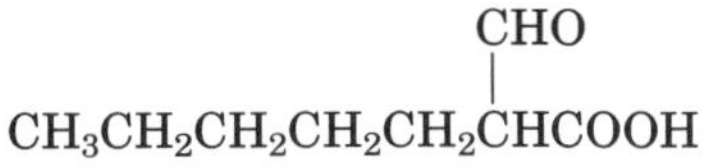

**9.** Named as a substituted propanoic acid
3-Oxo-2-pentylpropanoic acid

Similar rules apply for compounds **10–13**, which contain rings. Compounds **10** and **11** are keto nitriles, and both must be named as nitriles according to Table A.2. Substance **10** is named as a benzonitrile since the —CN functional group is a substituent on the aromatic ring, but substance **11** is named as an acetonitrile since the —CN functional group is on an open chain. The correct names are 2-acetyl-4-methylbenzonitrile (**10**) and (2-acetyl-4-methylphenyl)acetonitrile (**11**). Compounds **12** and **13** are both keto acids and must be named as acids. The correct names are 3-(2-oxocyclohexyl)propanoic acid (**12**) and 2-(3-oxopropyl)cyclohexanecarboxylic acid (**13**).

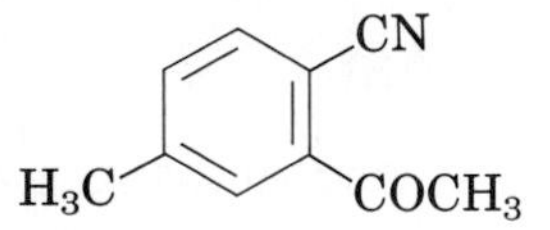

**10.** Named as a substituted benzonitrile
2-Acetyl-4-methylbenzonitrile

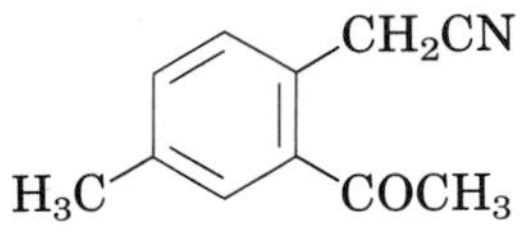

**11.** Named as a substituted acetonitrile
(2-Acetyl-4-methylphenyl)acetonitrile

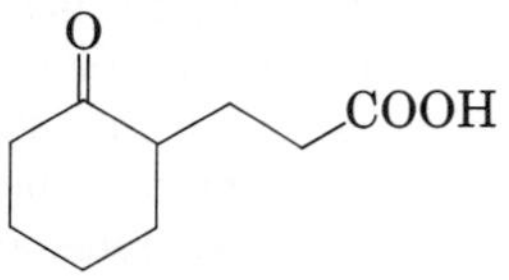

**12.** Named as a carboxylic acid
3-(2-Oxocyclohexyl)propanoic acid

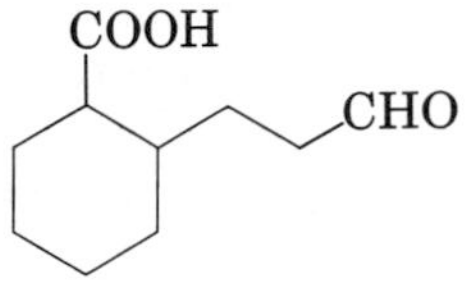

**13.** Named as a carboxylic acid
2-(3-Oxopropyl)cyclohexanecarboxylic acid

## THE PREFIXES AND LOCANTS

With the suffix and parent name established, the next step is to identify and number all substituents on the parent chain or ring. These substituents include all alkyl groups and all functional groups other than the one cited in the suffix. For example, compound **14** contains three different functional groups (carboxyl, keto, and double bond). Since the carboxyl group is highest in priority, and since the longest chain containing the functional groups is seven carbons long, **14** is a heptenoic acid. In addition, the main chain has an oxo (keto) substituent and three methyl groups. Numbering from the end nearer the highest-priority functional group, we find that **14** is 2,5,5-trimethyl-4-oxo-2-heptenoic acid. Note that the final *-e* of heptene is deleted in the word *heptenoic*. This deletion occurs only when the name would have two adjacent vowels (thus, *heptenoic* has the final *-e* deleted, but *heptenenitrile* retains the *-e*). Look back at some of the other compounds we have considered to see other examples of how prefixes and locants are assigned.

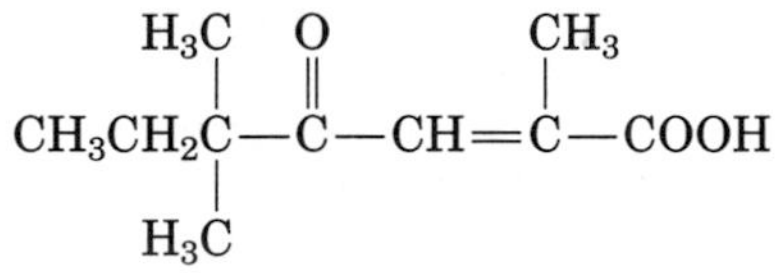

$$CH_3CH_2C(CH_3)_2—C(=O)—CH{=}C(CH_3)—COOH$$

**14.** Named as a heptenoic acid
2,5,5-Trimethyl-4-oxo-2-heptenoic acid

## WRITING THE NAME

Once the name parts have been established, the entire name is written out. Several additional rules apply:

1. *Order of prefixes* When the substituents have been identified, the main chain has been numbered, and the proper multipliers such as *di-* and *tri-* have been assigned, the name is written with the substituents listed in

alphabetical, rather than numerical, order. Multipliers such as *di-* and *tri-* are not used for alphabetization purposes, but the prefix *iso-* is used.

$$\begin{array}{c} \quad\quad\quad CH_3 \\ \quad\quad\quad | \\ H_2NCH_2CH_2CHCHCH_3 \\ \quad\quad\quad\quad\quad | \\ \quad\quad\quad\quad\quad OH \end{array}$$

**15.** 5-Amino-3-methyl-2-pentanol (*NOT* 3-Methyl-5-amino-2-pentanol)

2. *Single- and multiple-word names* The rule in such cases is to determine whether the principal functional group is itself an element or compound. If it is, then the name is written as a single word; if it is not, then the name is written as multiple words. For example, methylbenzene (one word) is correct because the parent, benzene, is itself a compound. Diethyl ether, however, is written as two words because the parent, ether, is a class name rather than a compound name. Some further examples are shown:

$H_3C—Mg—CH_3$

**16.** Dimethylmagnesium
(one word, since magnesium is an element)

$CH_3CHBrCOOH$

**17.** 2-Bromopropanoic acid
(two words, since "acid" is not a compound)

**18.** 4-(Dimethylamino)pyridine
(one word, since pyridine is a compound)

**19.** Methyl cyclopentanecarboxylate

3. *Parentheses* Parentheses are used to denote complex substituents when ambiguity would otherwise arise. For example, chloromethylbenzene has two substituents on a benzene ring, but (chloromethyl)benzene has only one complex substituent. Note that the expression in parentheses is not set off by hyphens from the rest of the name.

**20.** *p*-Chloromethylbenzene
(two substituents)

**21.** (Chloromethyl)benzene
(one complex substituent)

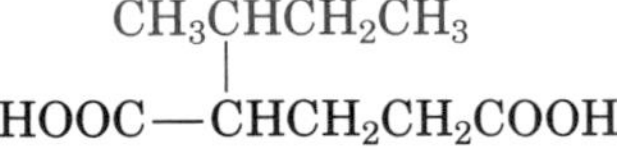

**22.** 2-(1-Methylpropyl)pentanedioic acid
(The 1-methylpropyl group is a complex substituent on C2 of the main chain.)

## ADDITIONAL READING

Further explanations of the rules of organic nomenclature can be found in the following references:

1. O.T. Benfey, "The Names and Structures of Organic Compounds," Wiley, New York, 1966.

2. J. H. Fletcher, O. C. Dermer, and R. B. Fox, "Nomenclature of Organic Compounds: Principles and Practice," Advances in Chemistry Series No. 126, American Chemical Society, Washington, DC, 1974.

3. "Nomenclature of Organic Chemistry, Sections A, B, C, D, E, F, and H," International Union of Pure and Applied Chemistry, Pergamon Press, Oxford, 1979.

4. J. G. Traynham, "Organic Nomenclature: A Programmed Introduction," Prentice-Hall, Englewood Cliffs, NJ, 1985.

APPENDIX B

# Summary of Functional Group Reactions

The following list summarizes the reactions of important functional groups. The functional groups are listed alphabetically, and references to the appropriate text sections are given.

### Acetals

1. Hydrolysis to yield a ketone or aldehyde plus alcohol (Section 19.14)

$$\mathrm{(RO)_2C\!<} \xrightarrow{H_3O^+} \mathrm{>C{=}O} + 2\,\mathrm{ROH}$$

### Acid anhydrides

1. Hydrolysis to yield a carboxylic acid (Section 21.6)

$$\mathrm{R{-}C(=O){-}O{-}C(=O){-}R} \xrightarrow[2.\ H_3O^+]{1.\ HO^-,\ H_2O} 2\,\mathrm{R{-}C(=O){-}OH}$$

2. Alcoholysis to yield an ester (Section 21.6)

$$\mathrm{R{-}C(=O){-}O{-}C(=O){-}R} \xrightarrow{R'OH,\ pyridine} \mathrm{R{-}C(=O){-}O{-}R'} + \mathrm{R{-}C(=O){-}OH}$$

3. Aminolysis to yield an amide (Section 21.6)

$$\mathrm{R{-}C(=O){-}O{-}C(=O){-}R} \xrightarrow{NH_3} \mathrm{R{-}C(=O){-}NH_2} + \mathrm{R{-}C(=O){-}OH}$$

4. Reduction to yield a primary alcohol (Section 21.6)

$$R{-}\overset{\displaystyle O}{\overset{\|}{C}}{-}O{-}\overset{\displaystyle O}{\overset{\|}{C}}{-}R \xrightarrow[\text{2. } H_2O]{\text{1. } LiAlH_4} 2\ R{-}CH_2OH$$

## Acid chlorides

1. Hydrolysis to yield a carboxylic acid (Section 21.5)

$$R{-}\overset{\displaystyle O}{\overset{\|}{C}}{-}Cl \xrightarrow[\text{2. } H_3O^+]{\text{1. } HO^-,\ H_2O} R{-}\overset{\displaystyle O}{\overset{\|}{C}}{-}OH + HCl$$

2. Alcoholysis to yield an ester (Section 21.5)

$$R{-}\overset{\displaystyle O}{\overset{\|}{C}}{-}Cl \xrightarrow{\text{R'OH, pyridine}} R{-}\overset{\displaystyle O}{\overset{\|}{C}}{-}OR' + HCl$$

3. Aminolysis to yield an amide (Section 21.5)

$$R{-}\overset{\displaystyle O}{\overset{\|}{C}}{-}Cl \xrightarrow{NH_3} R{-}\overset{\displaystyle O}{\overset{\|}{C}}{-}NH_2 + HCl$$

4. Reduction to yield a primary alcohol (Section 21.5)

$$R{-}\overset{\displaystyle O}{\overset{\|}{C}}{-}Cl \xrightarrow[\text{2. } H_2O]{\text{1. } LiAlH_4} R{-}CH_2OH + HCl$$

5. Partial reduction to yield an aldehyde (Section 21.5)

a. Reduction with $LiAlH(O\text{-}t\text{-}Bu)_3$

$$R{-}\overset{\displaystyle O}{\overset{\|}{C}}{-}Cl \xrightarrow{LiAlH(O\text{-}t\text{-}Bu)_3} R{-}\overset{\displaystyle O}{\overset{\|}{C}}{-}H$$

b. Rosenmund reduction

$$R{-}\overset{\displaystyle O}{\overset{\|}{C}}{-}Cl \xrightarrow{H_2/BaSO_4} R{-}\overset{\displaystyle O}{\overset{\|}{C}}{-}H$$

6. Grignard reaction to yield a tertiary alcohol (Section 21.5)

$$R{-}\overset{\displaystyle O}{\overset{\|}{C}}{-}Cl \xrightarrow[\text{2. } H_3O^+]{\text{1. } R'{-}MgX} R{-}\underset{\displaystyle R'}{\underset{|}{\overset{\displaystyle R'}{\overset{|}{C}}}}{-}OH + HCl$$

7. Reaction with lithium diorganocopper reagents to yield ketones (Section 21.5)

$$R{-}\overset{\displaystyle O}{\overset{\|}{C}}{-}Cl + R'_2CuLi \longrightarrow R{-}\overset{\displaystyle O}{\overset{\|}{C}}{-}R'$$

## Alcohols

1. Acidity (Section 17.4)

$$\mathrm{ROH} \xrightarrow{\mathrm{NaH}} \mathrm{RO^-\,Na^+} + \mathrm{H_2}$$

$$2\,\mathrm{ROH} \xrightarrow{\mathrm{K}} 2\,\mathrm{RO^-\,K^+} + \mathrm{H_2}$$

2. Oxidation (Section 17.9)

   a. Primary alcohol

$$\mathrm{RCH_2OH} \xrightarrow{\text{Pyridinium chlorochromate}} \mathrm{R{-}\overset{\overset{\large O}{\|}}{C}{-}H}$$

$$\mathrm{RCH_2OH} \xrightarrow[\text{Acetone}]{\mathrm{CrO_3,\ H_3O^+}} \mathrm{R{-}\overset{\overset{\large O}{\|}}{C}{-}OH}$$

   b. Secondary alcohol

$$\mathrm{R{-}\underset{\underset{\large H}{|}}{\overset{\overset{\large OH}{|}}{C}}{-}R'} \xrightarrow{\text{Pyridinium chlorochromate}} \mathrm{R{-}\overset{\overset{\large O}{\|}}{C}{-}R'}$$

3. Reaction with carboxylic acids to yield esters (Section 21.4)

$$\mathrm{ROH} \xrightarrow{\mathrm{R'COOH,\ H_2SO_4}} \mathrm{R'{-}\overset{\overset{\large O}{\|}}{C}{-}OR} + \mathrm{H_2O}$$

4. Reaction with acid chlorides to yield esters (Section 21.5)

$$\mathrm{ROH} \xrightarrow{\mathrm{R'COCl,\ pyridine}} \mathrm{R'{-}\overset{\overset{\large O}{\|}}{C}{-}OR} + \mathrm{HCl}$$

5. Dehydration to yield alkenes (Section 17.8)

$$\mathrm{R_2C(H){-}C(OH)R_2} \xrightarrow{\mathrm{H_2SO_4}} \mathrm{R_2C{=}CR_2} + \mathrm{H_2O}$$

6. Reaction with primary alkyl halides to yield ethers (Section 18.4)

$$\mathrm{ROH} \xrightarrow[\text{2. } \mathrm{R'CH_2X}]{\text{1. NaH}} \mathrm{ROCH_2R'} + \mathrm{NaX}$$

7. Conversion into alkyl halides (Section 17.8)

   a. Reaction of tertiary alcohols with HX

$$\mathrm{R{-}\underset{\underset{\large R}{|}}{\overset{\overset{\large R}{|}}{C}}{-}OH} \xrightarrow{\mathrm{HX}} \mathrm{R{-}\underset{\underset{\large R}{|}}{\overset{\overset{\large R}{|}}{C}}{-}X} + \mathrm{H_2O}$$

   where X = Cl, Br, I

b. Reaction of primary and secondary alcohols with $SOCl_2$

$$RCH_2OH \xrightarrow{SOCl_2} RCH_2Cl + HCl + SO_2$$

c. Reaction of primary and secondary alcohols with $PBr_3$

$$RCH_2OH \xrightarrow{PBr_3} RCH_2Br + P(OH)_3$$

## Aldehydes

1. Oxidation to yield carboxylic acids (Section 19.5)

$$R-\overset{\overset{\displaystyle O}{\|}}{C}-H \xrightarrow[\text{Acetone}]{CrO_3,\ H_3O^+} R-\overset{\overset{\displaystyle O}{\|}}{C}-OH$$

2. Nucleophilic addition reactions

a. Reduction to yield primary alcohols (Sections 17.6, 19.11)

$$R-\overset{\overset{\displaystyle O}{\|}}{C}-H \xrightarrow[2.\ H_2O]{1.\ NaBH_4} RCH_2OH$$

b. Reaction with Grignard reagents to yield secondary alcohols (Sections 17.7, 19.10)

$$R-\overset{\overset{\displaystyle O}{\|}}{C}-H \xrightarrow[2.\ H_2O]{1.\ R'-MgX} R-\underset{\underset{\displaystyle H}{|}}{\overset{\overset{\displaystyle R'}{|}}{C}}-OH$$

c. Grignard reaction of formaldehyde to yield primary alcohols (Section 17.7)

$$H-\overset{\overset{\displaystyle O}{\|}}{C}-H \xrightarrow[2.\ H_2O]{1.\ R'-MgX} R'CH_2OH$$

d. Reaction with HCN to yield a cyanohydrin (Section 19.9)

$$R-\overset{\overset{\displaystyle O}{\|}}{C}-H \xrightarrow{HCN} R-\underset{\underset{\displaystyle CN}{|}}{\overset{\overset{\displaystyle OH}{|}}{C}}-H$$

e. Wolff–Kishner reaction with hydrazine to yield alkanes (Section 19.13)

$$R-\overset{\overset{\displaystyle O}{\|}}{C}-H \xrightarrow{N_2H_4,\ KOH} RCH_3$$

f. Clemmensen reduction with zinc amalgam to yield alkanes (Section 19.13)

$$R-\overset{\overset{\displaystyle O}{\|}}{C}-H \xrightarrow{Zn(Hg),\ HCl} RCH_3$$

g. Reaction with alcohol to yield an acetal (Section 19.14)

$$R{-}\overset{O}{\overset{\|}{C}}{-}H \xrightarrow{HOCH_2CH_2OH,\ H^+} \text{(cyclic acetal: } R{-}C{-}H \text{ in a ring with O, O, } CH_2CH_2)$$

h. Reaction with a thiol to yield a thioacetal (Section 19.15)

$$R{-}\overset{O}{\overset{\|}{C}}{-}H \xrightarrow{HSCH_2CH_2SH,\ H^+} \text{(cyclic thioacetal: } R{-}C{-}H \text{ in a ring with S, S, } CH_2CH_2) \xrightarrow[\text{Ni}]{\text{Raney}} R{-}CH_3$$

i. Wittig reaction to yield alkenes (Section 19.16)

$$R{-}\overset{O}{\overset{\|}{C}}{-}H \xrightarrow{(Ph)_3\overset{+}{P}{-}\overset{-}{C}HR'} (R)(H)C{=}C(H)(R')$$

j. Reaction with an amine to yield an imine (Section 19.12)

$$R{-}\overset{O}{\overset{\|}{C}}{-}H \xrightarrow{R{-}NH_2} R{-}\overset{N-R}{\overset{\|}{C}}{-}H$$

3. Aldol reaction to yield a $\beta$-hydroxy aldehyde (Section 23.2)

$$RCH_2\overset{O}{\overset{\|}{C}}H \xrightarrow{NaOH} RCH_2\overset{OH}{\overset{|}{C}H}{-}\underset{R}{\underset{|}{C}}H\overset{O}{\overset{\|}{C}}H$$

4. Alpha bromination of aldehydes (Section 22.3)

$$RCH_2\overset{O}{\overset{\|}{C}}H \xrightarrow{Br_2,\ CH_2COOH} R\underset{Br}{\underset{|}{C}H}\overset{O}{\overset{\|}{C}}H + HBr$$

## Alkanes

1. Radical chlorination of methane (Section 10.4)

$$CH_4 + Cl_2 \xrightarrow{h\nu} CH_3Cl + HCl$$

## Alkenes

1. Electrophilic addition of HX to yield an alkyl halide (Sections 6.9–6.13)

$$\gt C{=}C\lt \xrightarrow{HX} \overset{H}{\overset{|}{C}}{-}\overset{X}{\overset{|}{C}}$$

where X = Cl, Br, I. Markovnikov regiochemistry is observed. H adds to the less highly substituted carbon, and X adds to the more highly substituted one.

2. Electrophilic addition of halogen to yield a 1,2-dihalide (Section 7.1)

$$\text{C}{=}\text{C} \xrightarrow{X_2} \text{XC}{-}\text{CX}$$

where X = Cl, Br. Anti stereochemistry is observed.

3. Oxymercuration to yield an alcohol (Section 7.3)

$$\text{C}{=}\text{C} \xrightarrow[\text{2. NaBH}_4]{\text{1. Hg(OAc)}_2\text{, H}_2\text{O}} \text{H}{-}\text{C}{-}\text{C}{-}\text{OH}$$

Markovnikov regiochemistry is observed, yielding the more highly substituted alcohol.

4. Hydroboration–oxidation to yield an alcohol (Section 7.4)

$$\text{C}{=}\text{C} \xrightarrow[\text{2. H}_2\text{O}_2\text{, NaOH}]{\text{1. BH}_3} \text{H}{-}\text{C}{-}\text{C}{-}\text{OH}$$

5. Hydrogenation of alkenes to yield alkanes (Section 7.6)

$$\text{C}{=}\text{C} \xrightarrow{\text{H}_2\text{, Pd}} \text{H}{-}\text{C}{-}\text{C}{-}\text{H}$$

6. Hydroxylation of alkenes to yield 1,2-diols (Section 7.7)

$$\text{C}{=}\text{C} \xrightarrow[\text{H}_2\text{O, NaOH}]{\text{KMnO}_4} \text{HO}{-}\text{C}{-}\text{C}{-}\text{OH}$$

7. Radical addition of HBr to yield an alkyl bromide (Section 7.5)

$$\text{C}{=}\text{C} \xrightarrow{\text{HBr radicals}} \text{H}{-}\text{C}{-}\text{C}{-}\text{Br}$$

8. Oxidative cleavage of alkenes to yield carbonyl compounds (Section 7.8)

$$\text{C}{=}\text{C} \xrightarrow[\text{H}_2\text{O}]{\text{KMnO}_4} \text{C}{=}\text{O} + \text{O}{=}\text{C}$$

9. Reaction with peroxyacids to yield epoxides (Section 18.7)

$$\text{C}{=}\text{C} \xrightarrow{\text{RCO}_3\text{H}} \text{C}{-}\text{C} \text{ (bridged by O)}$$

10. Simmons–Smith reaction with $CH_2I_2$ to yield a cyclopropane (Section 7.10)

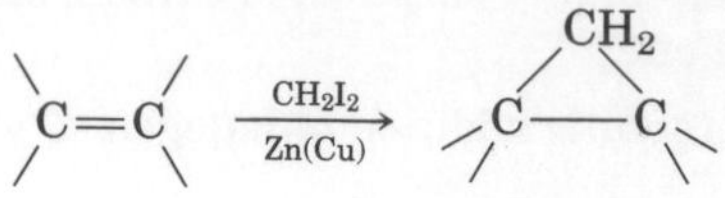

$$\text{>C=C<} \xrightarrow[\text{Zn(Cu)}]{CH_2I_2} \text{>C—C<} \text{ (bridged by } CH_2\text{, cyclopropane)}$$

## Alkynes

1. Electrophilic addition of HX to yield a vinylic halide (Section 8.3)

$$R—C{\equiv}C—R \xrightarrow{HX} RCH{=}CRX$$

where X = Cl, Br, I

2. Electrophilic addition of halogen to yield a dihalide (Section 8.3)

$$R—C{\equiv}C—R \xrightarrow{X_2} RCX{=}CRX$$

where X = Cl, Br, I

3. Mercuric sulfate-catalyzed hydration to yield a methyl ketone (Section 8.4)

$$R—C{\equiv}C—H \xrightarrow[HgSO_4]{H_3O^+} R—\overset{\overset{O}{\|}}{C}—CH_3$$

4. Hydroboration with disiamylborane to yield an aldehyde (Section 8.5)

$$R—C{\equiv}C—H \xrightarrow[\text{2. } H_2O_2\text{, NaOH}]{\text{1. } R_2B—H} R—CH_2—\overset{\overset{O}{\|}}{C}—H$$

5. Alkylation of alkyne anions (Section 8.8)

$$R—C{\equiv}C—H \xrightarrow[\text{2. } R'CH_2X]{\text{1. } NaNH_2} R—C{\equiv}C—CH_2R'$$

6. Reduction of alkynes (Section 8.6)

a. Hydrogenation to yield a cis alkene

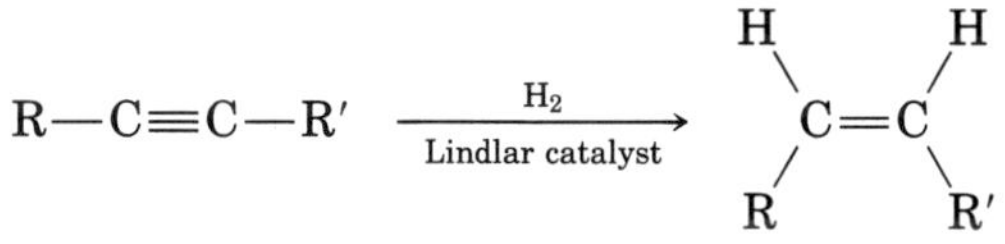

$$R—C{\equiv}C—R' \xrightarrow[\text{Lindlar catalyst}]{H_2} \text{(H)(R)C=C(H)(R') (cis: H, H on same side; R, R' on same side)}$$

b. Reduction to yield a trans alkene

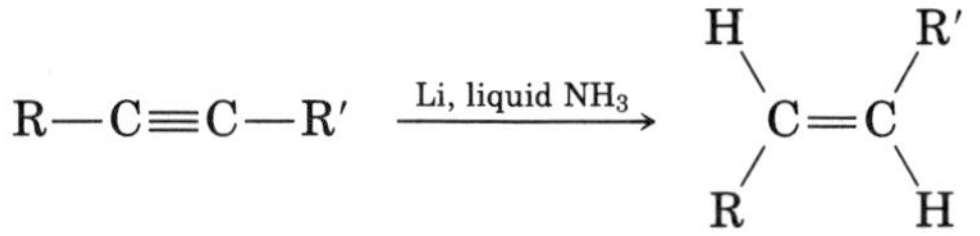

$$R—C{\equiv}C—R' \xrightarrow{\text{Li, liquid } NH_3} \text{(H)(R)C=C(R')(H) (trans: H opposite H; R opposite R')}$$

## Amides

1. Hydrolysis to yield carboxylic acids (Section 21.8)

$$R—\overset{\overset{O}{\|}}{C}—NH_2 \xrightarrow[\text{2. } H_3O^+]{\text{1. } HO^-\text{, } H_2O} R—\overset{\overset{O}{\|}}{C}—OH + NH_3$$

2. Reduction with $LiAlH_4$ to yield amines (Section 21.8)

$$R-\overset{\overset{O}{\|}}{C}-NH_2 \xrightarrow[2.\ H_2O]{1.\ LiAlH_4} RCH_2NH_2$$

3. Dehydration to yield nitriles (Section 21.9)

$$R-\overset{\overset{O}{\|}}{C}-NH_2 \xrightarrow{SOCl_2} RC\equiv N$$

## Amines

1. $S_N2$ alkylation of alkyl halides to yield amines (Section 25.7)

| | | | | |
|---|---|---|---|---|
| a. Ammonia | $NH_3 + RX$ | $\longrightarrow$ | $RNH_2$ | (Primary) |
| b. Primary | $RNH_2 + R'X$ | $\longrightarrow$ | $RNHR'$ | (Secondary) |
| c. Secondary | $R_2NH + R'X$ | $\longrightarrow$ | $R_2NR'$ | (Tertiary) |
| d. Tertiary | $R_3N + R'X$ | $\longrightarrow$ | $R_3\overset{+}{N}R'\ X^-$ | (Quaternary) |

2. Nucleophilic acyl substitution reactions
   a. Reaction with acid chlorides to yield amides (Section 21.5)

$$R-\overset{\overset{O}{\|}}{C}-Cl \xrightarrow{R'NH_2} R-\overset{\overset{O}{\|}}{C}-NHR' + HCl$$

   b. Reaction with acid anhydrides to yield amides (Section 21.6)

$$R-\overset{\overset{O}{\|}}{C}-O-\overset{\overset{O}{\|}}{C}-R \xrightarrow{R'NH_2} R-\overset{\overset{O}{\|}}{C}-NHR' + R-\overset{\overset{O}{\|}}{C}-OH$$

   c. Reaction with esters to yield amides (Section 21.7)

$$R-\overset{\overset{O}{\|}}{C}-OR'' \xrightarrow{R'NH_2} R-\overset{\overset{O}{\|}}{C}-NHR' + R''OH$$

3. Hofmann elimination to yield alkenes (Section 25.8)

$$\underset{/}{\overset{H}{\diagdown}}\!C-C\!\overset{NH_2}{\diagup}\!\diagdown \xrightarrow[2.\ \Delta,\ Ag_2O]{1.\ \text{Excess } CH_3I} \;\rangle C=C\langle\; + N(CH_3)_3 + H_2O$$

4. Formation of arenediazonium salts (Section 26.4)

$$C_6H_5-NH_2 \xrightarrow[H_2SO_4]{NaNO_2} C_6H_5-\overset{+}{N}\equiv N\ HSO_4^-$$

## Arenes

1. Oxidation of the alkylbenzene side chain (Section 16.11)

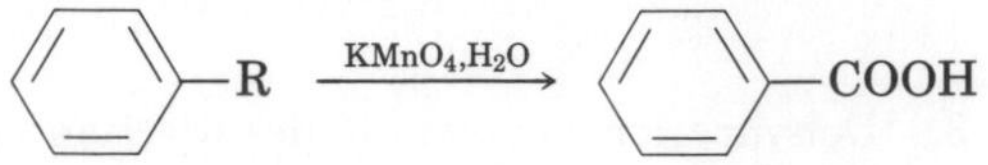

$$C_6H_5{-}R \xrightarrow{KMnO_4, H_2O} C_6H_5{-}COOH$$

2. Catalytic reduction to yield a cyclohexane (Section 16.12)

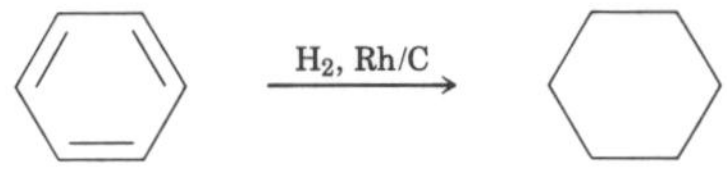

$$C_6H_6 \xrightarrow{H_2,\ Rh/C} C_6H_{12}$$

3. Reduction of aryl alkyl ketones (Section 16.12)

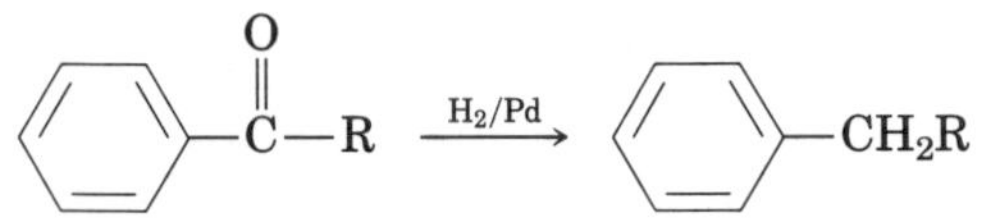

$$C_6H_5{-}\overset{O}{\overset{\|}{C}}{-}R \xrightarrow{H_2/Pd} C_6H_5{-}CH_2R$$

4. Electrophilic aromatic substitution

   a. Bromination (Section 16.1)

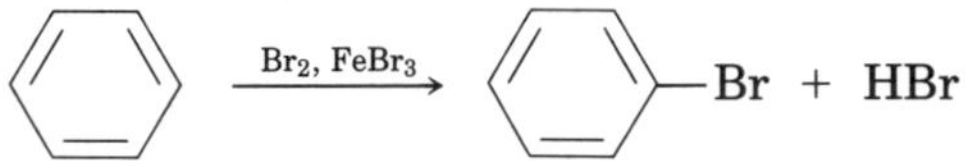

$$C_6H_6 \xrightarrow{Br_2,\ FeBr_3} C_6H_5{-}Br + HBr$$

   b. Chlorination (Section 16.2)

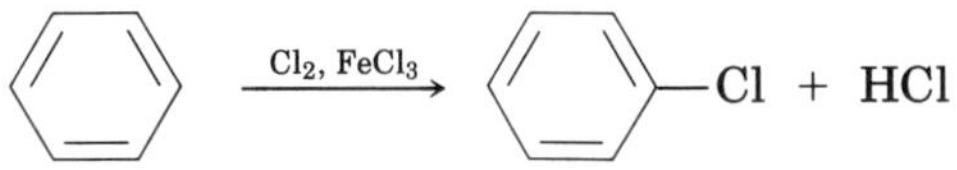

$$C_6H_6 \xrightarrow{Cl_2,\ FeCl_3} C_6H_5{-}Cl + HCl$$

   c. Iodination (Section 16.2)

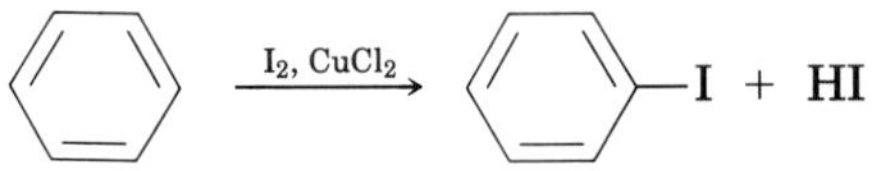

$$C_6H_6 \xrightarrow{I_2,\ CuCl_2} C_6H_5{-}I + HI$$

   d. Nitration (Section 16.2)

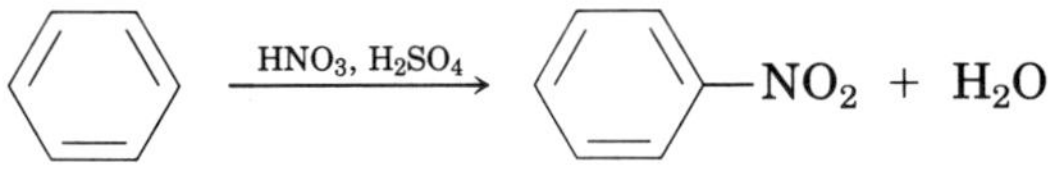

$$C_6H_6 \xrightarrow{HNO_3,\ H_2SO_4} C_6H_5{-}NO_2 + H_2O$$

   e. Sulfonation (Section 16.2)

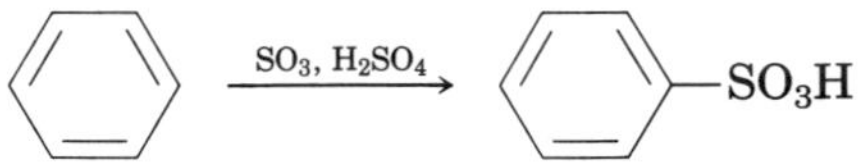

$$C_6H_6 \xrightarrow{SO_3,\ H_2SO_4} C_6H_5{-}SO_3H$$

   f. Friedel–Crafts alkylation (Section 16.3)

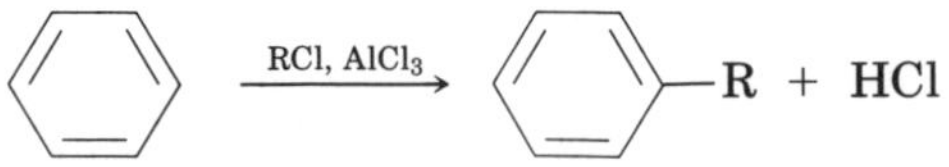

$$C_6H_6 \xrightarrow{RCl,\ AlCl_3} C_6H_5{-}R + HCl$$

Aromatic ring must be at least as reactive as a halobenzene; R must be methyl, ethyl, secondary, or tertiary.

g. Friedel–Crafts acylation (Section 16.4)

$$C_6H_6 \xrightarrow{RCOCl,\ AlCl_3} C_6H_5{-}\overset{O}{\overset{\|}{C}}{-}R + HCl$$

## Arenediazonium salts

1. Conversion into aryl chlorides (Section 26.4)

$$C_6H_5{-}\overset{+}{N}{\equiv}N\ HSO_4^- \xrightarrow{CuCl} C_6H_5{-}Cl + N_2$$

2. Conversion into aryl bromides (Section 26.4)

$$C_6H_5{-}\overset{+}{N}{\equiv}N\ HSO_4^- \xrightarrow{CuBr} C_6H_5{-}Br + N_2$$

3. Conversion into aryl iodides ((Section 26.4)

$$C_6H_5{-}\overset{+}{N}{\equiv}N\ HSO_4^- \xrightarrow{NaI} C_6H_5{-}I + N_2$$

4. Conversion into aryl cyanides (Section 26.4)

$$C_6H_5{-}\overset{+}{N}{\equiv}N\ HSO_4^- \xrightarrow{CuCN} C_6H_5{-}CN + N_2$$

5. Conversion into phenols (Section 26.4)

$$C_6H_5{-}\overset{+}{N}{\equiv}N\ HSO_4^- \xrightarrow{H_3O^+} C_6H_5{-}OH + N_2$$

6. Conversion into arenes (Section 26.4)

$$C_6H_5{-}\overset{+}{N}{\equiv}N\ HSO_4^- \xrightarrow{H_3PO_2} C_6H_5{-}H + N_2$$

## Arenesulfonic acids

1. Conversion into phenols (Section 16.2)

$$C_6H_5{-}SO_3H \xrightarrow{NaOH,\ \Delta} C_6H_5{-}OH$$

## Carboxylic acids

1. Acidity (Sections 20.3–20.5)

$$R{-}\overset{\overset{\displaystyle O}{\|}}{C}{-}OH \xrightarrow{NaOH} R{-}\overset{\overset{\displaystyle O}{\|}}{C}{-}O^-\ Na^+ + H_2O$$

2. Reduction to yield a primary alcohol (Sections 17.6, 20.8)

   a. Reduction with $LiAlH_4$

$$R{-}\overset{\overset{\displaystyle O}{\|}}{C}{-}OH \xrightarrow[2.\ H_3O^+]{1.\ LiAlH_4} RCH_2OH$$

   b. Reduction with $BH_3$

$$R{-}\overset{\overset{\displaystyle O}{\|}}{C}{-}OH \xrightarrow[2.\ H_3O^+]{1.\ BH_3} RCH_2OH$$

3. Nucleophilic acyl substitution reactions (Section 21.4)

   a. Conversion into an acid chloride

$$R{-}\overset{\overset{\displaystyle O}{\|}}{C}{-}OH \xrightarrow{SOCl_2} R{-}\overset{\overset{\displaystyle O}{\|}}{C}{-}Cl + HCl + SO_2$$

   b. Conversion into an acid anhydride

$$R{-}\overset{\overset{\displaystyle O}{\|}}{C}{-}OH \xrightarrow[2.\ R'COCl]{1.\ NaOH} R{-}\overset{\overset{\displaystyle O}{\|}}{C}{-}O{-}\overset{\overset{\displaystyle O}{\|}}{C}{-}R'$$

   c. Conversion into an ester

   (1) Fischer esterification

$$R{-}\overset{\overset{\displaystyle O}{\|}}{C}{-}OH \xrightarrow{R'OH,\ H_2SO_4} R{-}\overset{\overset{\displaystyle O}{\|}}{C}{-}OR' + H_2O$$

   (2) $S_N2$ reaction with an alkyl halide

$$R{-}\overset{\overset{\displaystyle O}{\|}}{C}{-}OH \xrightarrow[2.\ R'CH_2X]{1.\ NaOH} R{-}\overset{\overset{\displaystyle O}{\|}}{C}{-}OCH_2R'$$

   (3) Reaction with diazomethane

$$R{-}\overset{\overset{\displaystyle O}{\|}}{C}{-}OH \xrightarrow{CH_2N_2} R{-}\overset{\overset{\displaystyle O}{\|}}{C}{-}OCH_3$$

## Epoxides

1. Acid-catalyzed ring opening with HX to yield a halohydrin (Section 18.8)

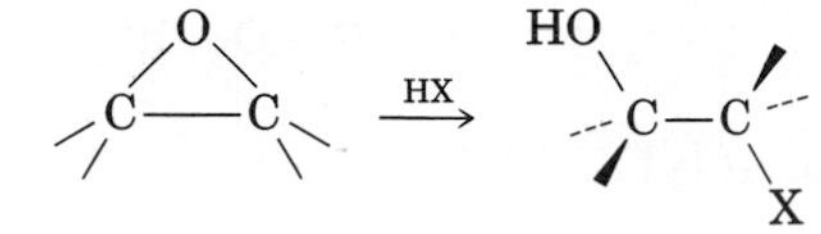

$$\text{epoxide (C—C with bridging O)} \xrightarrow{HX} \text{HO—C—C—X}$$

where X = Br, Cl, I

2. Ring opening with aqueous acid to yield a 1,2-diol (Section 18.8)

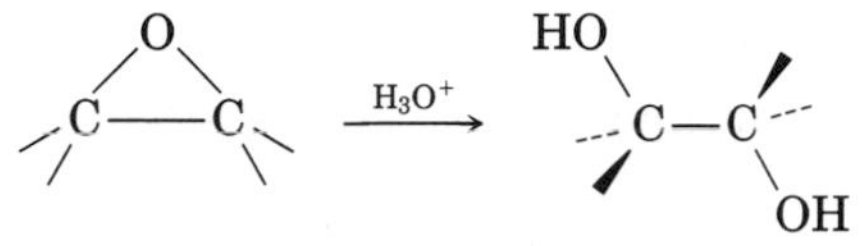

$$\text{epoxide (C—C with bridging O)} \xrightarrow{H_3O^+} \text{HO—C—C—OH}$$

## Esters

1. Hydrolysis to yield a carboxylic acid (Section 21.7)

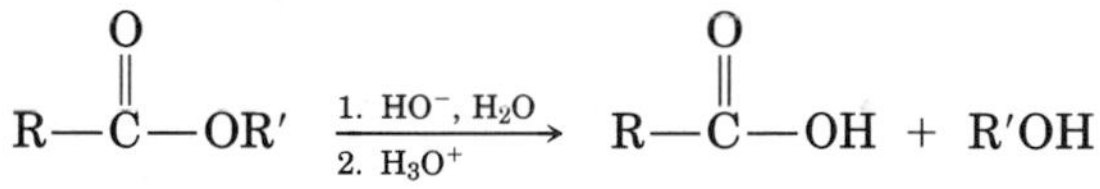

$$R—\overset{O}{\overset{\|}{C}}—OR' \xrightarrow[2.\ H_3O^+]{1.\ HO^-,\ H_2O} R—\overset{O}{\overset{\|}{C}}—OH + R'OH$$

2. Aminolysis to yield an amide (Section 21.7)

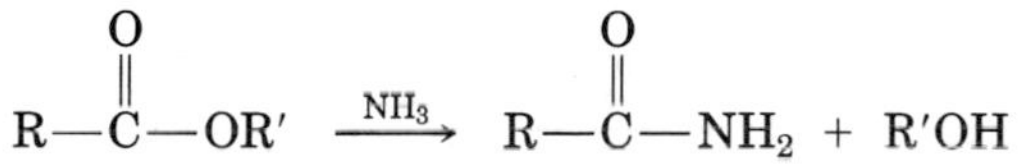

$$R—\overset{O}{\overset{\|}{C}}—OR' \xrightarrow{NH_3} R—\overset{O}{\overset{\|}{C}}—NH_2 + R'OH$$

3. Reduction to yield a primary alcohol (Section 17.6)

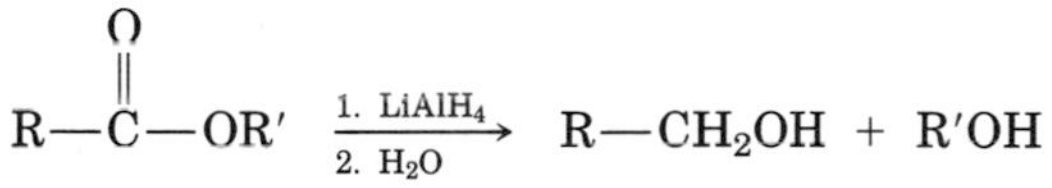

$$R—\overset{O}{\overset{\|}{C}}—OR' \xrightarrow[2.\ H_2O]{1.\ LiAlH_4} R—CH_2OH + R'OH$$

4. Partial reduction with DIBAH to yield an aldehyde (Section 21.7)

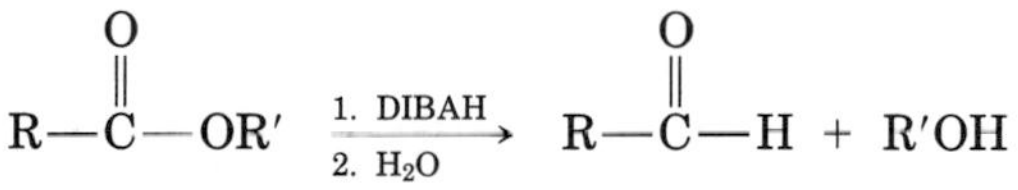

$$R—\overset{O}{\overset{\|}{C}}—OR' \xrightarrow[2.\ H_2O]{1.\ DIBAH} R—\overset{O}{\overset{\|}{C}}—H + R'OH$$

5. Grignard reaction to yield a tertiary alcohol (Section 17.7)

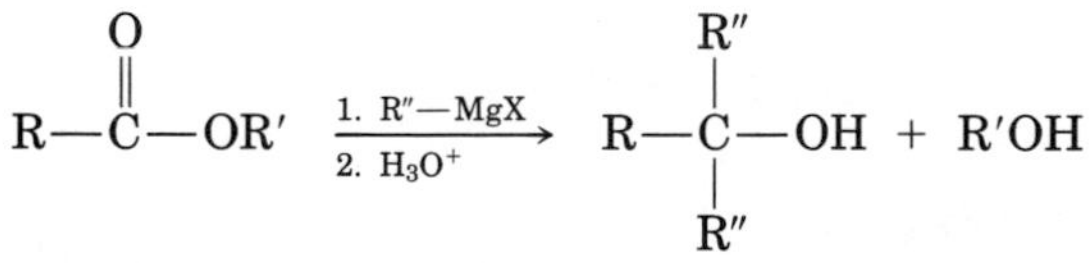

$$R—\overset{O}{\overset{\|}{C}}—OR' \xrightarrow[2.\ H_3O^+]{1.\ R''—MgX} R—\underset{R''}{\underset{|}{\overset{R''}{\overset{|}{C}}}}—OH + R'OH$$

6. Claisen condensation to yield a $\beta$-keto ester (Section 23.9)

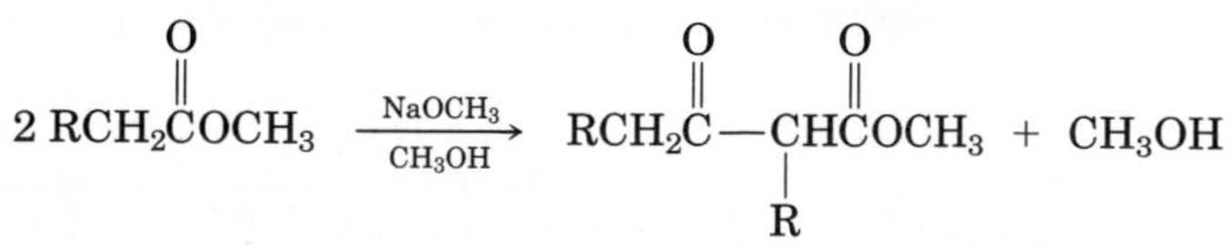

$$2\ RCH_2\overset{O}{\overset{\|}{C}}OCH_3 \xrightarrow[CH_3OH]{NaOCH_3} RCH_2\overset{O}{\overset{\|}{C}}—\underset{R}{\underset{|}{C}}H\overset{O}{\overset{\|}{C}}OCH_3 + CH_3OH$$

## Ethers

1. Acid-induced cleavage (Section 18.6)

$$R{-}O{-}R' \xrightarrow{HBr} ROH + BrR'$$

## Halides, alkyl

1. Reaction with magnesium to form Grignard reagents (Section 10.9)

$$RX + Mg \longrightarrow RMgX$$

where X = Br, Cl, I

2. Reduction to yield alkanes (Section 10.9)

$$RX \xrightarrow[2.\ H_3O^+]{1.\ Mg} RH$$

3. Nucleophilic substitution ($S_N1$ or $S_N2$) (Sections 11.1–11.9)

$$RX + :Nu^- \longrightarrow RNu + :X^-$$

where $Nu^-$ = $H^-$, $CN^-$, $I^-$, $Br^-$, $Cl^-$, $HO^-$, $NH_2^-$, $CH_3O^-$, $CH_3COO^-$, $HS^-$, $H_2O$, $NH_3$, and so on

X = Br, Cl, I, tosylate

4. Dehydrohalogenation (E1 or E2) (Sections 11.11, 11.14)

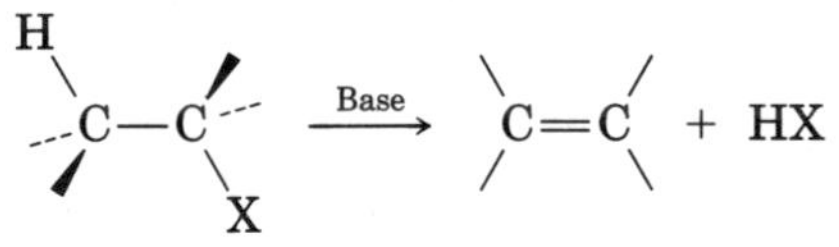

$$H{-}C{-}C{-}X \xrightarrow{Base} C{=}C + HX$$

where X = Br, Cl, I

## Halohydrins

1. Conversion into epoxides (Section 18.7)

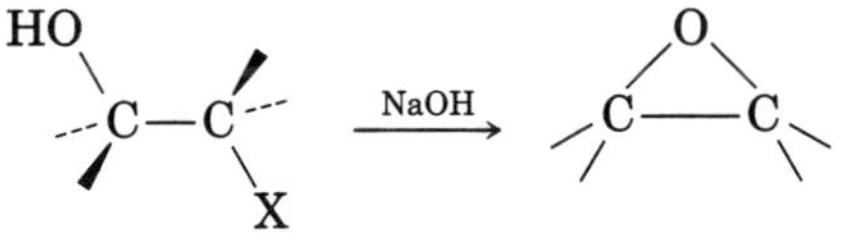

$$HO{-}C{-}C{-}X \xrightarrow{NaOH} C{-}C \text{ (epoxide, bridged by O)}$$

where X = Br, Cl, I

## Ketones

1. Nucleophilic addition reactions

a. Reduction to yield a secondary alcohol (Sections 17.6, 19.11)

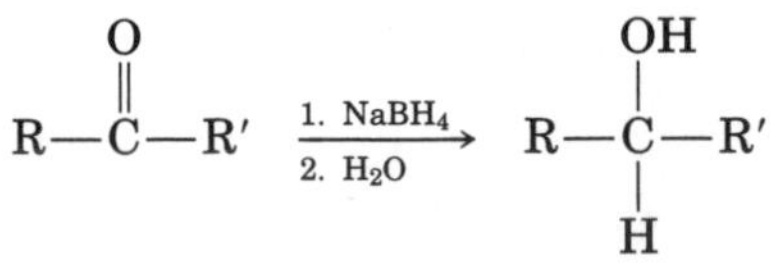

$$R{-}\overset{O}{\overset{\|}{C}}{-}R' \xrightarrow[2.\ H_2O]{1.\ NaBH_4} R{-}\overset{OH}{\overset{|}{\underset{H}{\underset{|}{C}}}}{-}R'$$

b. Reaction with Grignard reagents to yield a tertiary alcohol (Sections 17.7, 19.10)

$$\mathrm{R{-}\overset{O}{\overset{\|}{C}}{-}R''} \xrightarrow[\text{2. } H_2O]{\text{1. } R'{-}MgX} \mathrm{R{-}\underset{R''}{\underset{|}{\overset{R'}{\overset{|}{C}}}}{-}OH}$$

c. Wolff–Kishner reaction with hydrazine to yield an alkane (Section 19.13)

$$\mathrm{R{-}\overset{O}{\overset{\|}{C}}{-}R'} \xrightarrow{N_2H_4,\ KOH} \mathrm{R{-}CH_2{-}R'}$$

d. Reaction with HCN to yield a cyanohydrin (Section 19.9)

$$\mathrm{R{-}\overset{O}{\overset{\|}{C}}{-}R'} \xrightarrow{HCN} \mathrm{R{-}\underset{CN}{\underset{|}{\overset{OH}{\overset{|}{C}}}}{-}R'}$$

e. Reaction with alcohol to yield an acetal (Section 19.14)

$$\mathrm{R{-}\overset{O}{\overset{\|}{C}}{-}R'} \xrightarrow{HOCH_2CH_2OH,\ H^+} \text{(1,3-dioxolane: O, O ring)}\ \mathrm{R{-}C{-}R'}$$

f. Reaction with a thiol to yield a thioacetal (Section 19.15)

$$\mathrm{R{-}\overset{O}{\overset{\|}{C}}{-}R'} \xrightarrow{HSCH_2CH_2SH,\ H^+} \text{(1,3-dithiolane: S, S ring)}\ \mathrm{R{-}C{-}R'} \xrightarrow[\text{Ni}]{\text{Raney}} \mathrm{R{-}CH_2{-}R'}$$

g. Wittig reaction to yield an alkene (Section 19.16)

$$\mathrm{R{-}\overset{O}{\overset{\|}{C}}{-}R''} \xrightarrow{(Ph)_3\overset{+}{P}{-}\overset{-}{C}HR'} \mathrm{(R)(R'')C{=}C(H)(R')}$$

h. Reaction with an amine to yield an imine (Section 19.12)

$$\mathrm{R{-}\overset{O}{\overset{\|}{C}}{-}H} \xrightarrow{R{-}NH_2} \mathrm{R{-}\overset{N-R}{\overset{\|}{C}}{-}H}$$

2. Aldol reaction to yield a $\beta$-hydroxy ketone (Section 23.2)

$$\mathrm{RCH_2\overset{O}{\overset{\|}{C}}R'} \xrightarrow{NaOH} \mathrm{RCH_2\underset{R'}{\underset{|}{\overset{HO}{\overset{|}{C}}}}{-}\underset{R}{\underset{|}{C}}H\overset{O}{\overset{\|}{C}}R'}$$

3. Alpha bromination of ketones (Section 22.3)

$$\mathrm{RCH_2\overset{O}{\overset{\|}{C}}R'} \xrightarrow{\mathrm{Br_2,\ CH_3COOH}} \mathrm{R\underset{Br}{\underset{|}{C}}H\overset{O}{\overset{\|}{C}}R'} + \mathrm{HBr}$$

4. Clemmensen reduction with zinc amalgam to yield alkanes (Section 19.13)

$$\mathrm{R{-}\overset{O}{\overset{\|}{C}}{-}R'} \xrightarrow{\mathrm{Zn(Hg),\ HCl}} \mathrm{R{-}CH_2{-}R'}$$

## Nitriles

1. Hydrolysis to yield a carboxylic acid (Section 21.9)

$$\mathrm{RC{\equiv}N} \xrightarrow{\mathrm{H_3O^+}} \mathrm{R{-}\overset{O}{\overset{\|}{C}}{-}OH} + \mathrm{NH_3}$$

2. Reduction to yield a primary amine (Section 21.9)

$$\mathrm{RC{\equiv}N} \xrightarrow[\text{2. } \mathrm{H_2O}]{\text{1. } \mathrm{LiAlH_4}} \mathrm{R{-}CH_2NH_2}$$

3. Partial reduction with DIBAH to yield an aldehyde (Section 21.9)

$$\mathrm{RC{\equiv}N} \xrightarrow[\text{2. } \mathrm{H_2O}]{\text{1. DIBAH}} \mathrm{R{-}\overset{O}{\overset{\|}{C}}{-}H}$$

4. Reaction with Grignard reagents to yield ketones (Section 21.9)

$$\mathrm{RC{\equiv}N} \xrightarrow[\text{2. } \mathrm{H_2O}]{\text{1. RMgX}} \mathrm{R{-}\overset{O}{\overset{\|}{C}}{-}R'} + \mathrm{NH_3}$$

## Nitroarenes

1. Reduction by treatment with acid (Section 10.9)

$$\mathrm{C_6H_5{-}NO_2} \xrightarrow[\mathrm{H_3O^+}]{\text{Fe or Sn}} \mathrm{C_6H_5{-}NH_2}$$

## Organometallics

1. Reduction by treatment with acid (Section 10.9)

$$\mathrm{RMgX} \xrightarrow{\mathrm{H_3O^+}} \mathrm{RH}$$

2. Nucleophilic addition to carbonyl compounds (Section 17.7)

$$\mathrm{\overset{O}{\overset{\|}{C}}} \xrightarrow[\text{2. } \mathrm{H_3O^+}]{\text{1. RMgX}} \mathrm{\overset{:\ddot{O}:^-}{\overset{|}{C}}{-}R}$$

3. Conjugate addition of lithium diorganocopper reagents to $\alpha,\beta$-unsaturated ketones (Section 19.18)

$$\gt C{=}C(-)-\overset{O}{\overset{\|}{C}}-R \xrightarrow[2.\ H_3O^+]{1.\ R'_2CuLi} R'-\overset{|}{C}(-)-\underset{|}{CH}-\overset{O}{\overset{\|}{C}}-R$$

4. Coupling reaction of lithium diorganocopper reagents with alkyl halides (Section 10.10)

$$2\,RX + Li \longrightarrow RLi \xrightarrow{CuBr} R_2CuLi \xrightarrow{R'X} R{-}R'$$

5. Reaction with carbon dioxide to yield a carboxylic acid (Section 20.6)

$$R{-}MgX \xrightarrow[2.\ H_3O^+]{1.\ CO_2} R-\overset{O}{\overset{\|}{C}}-OH$$

## Phenols

1. Acidity (Section 26.7)

$$C_6H_5{-}OH \xrightarrow{NaOH} C_6H_5{-}O^-\ Na^+$$

2. Reaction with acid chlorides to yield esters (Section 26.9)

$$C_6H_5{-}OH \xrightarrow[2.\ RCOCl]{1.\ NaOH} C_6H_5{-}O-\overset{O}{\overset{\|}{C}}-R$$

3. Reaction with alkyl halides to yield ethers (Section 26.9)

$$C_6H_5{-}OH \xrightarrow[2.\ RX]{1.\ NaOH} C_6H_5{-}OR$$

4. Oxidation to yield quinones (Section 26.9)

$$C_6H_5{-}OH \xrightarrow{(KSO_3)_2NO} O{=}C_6H_4{=}O$$

## Quinones

1. Reduction to yield hydroquinones (Section 26.9)

$$O{=}C_6H_4{=}O \xrightarrow{NaBH_4} HO{-}C_6H_4{-}OH$$

## Sulfides

1. Reaction with alkyl halides to yield sulfonium salts (Section 18.11)

$$\text{R—S—R}' \xrightarrow{\text{R}''\text{X}} \text{R—}\overset{+}{\underset{\text{R}''}{\underset{|}{\text{S}}}}\text{—R}'\ \text{X}^-$$

2. Oxidation to yield sulfoxides (Section 18.11)

$$\text{R—S—R}' \xrightarrow{\text{H}_2\text{O}_2} \text{R—}\overset{\text{O}^-}{\overset{|}{\underset{+}{\text{S}}}}\text{—R}'$$

3. Oxidation to yield sulfones (Section 18.11)

$$\text{R—S—R}' \xrightarrow{\text{RCO}_3\text{H}} \text{R—}\overset{\text{O}}{\overset{\|}{\underset{\underset{\text{O}}{\|}}{\text{S}}}}\text{—R}'$$

## Thiols

1. Reaction with alkyl halides to yield sulfides (Section 17.12)

$$\text{R—S—H} \xrightarrow[\text{2. R}'\text{X}]{\text{1. NaH}} \text{R—S—R}'$$

2. Oxidation to yield disulfides (Section 17.12)

$$2\,\text{R—S—H} \xrightarrow{\text{Br}_2} \text{R—S—S—R} + 2\,\text{HBr}$$

APPENDIX C

# Summary of Functional Group Preparations

The following list summarizes the synthetic methods by which important functional groups can be prepared. The functional groups are listed alphabetically, followed by references to the appropriate text sections and a brief description of each synthetic method.

**Acetals, $R_2C(OR')_2$**

| | |
|---|---|
| Section 19.14 | From ketones and aldehydes by acid-catalyzed reaction with alcohols |

**Acid anhydrides, RCOOCOR′**

| | |
|---|---|
| Section 21.4 | From dicarboxylic acids by heating |
| Section 21.6 | From acid chlorides by reaction with carboxylate salts |

**Acid bromides, RCOBr**

| | |
|---|---|
| Section 21.5 | From carboxylic acids by reaction with $PBr_3$ |

**Acid chlorides, RCOCl**

| | |
|---|---|
| Section 21.4 | From carboxylic acids by reaction with either $SOCl_2$, $PCl_3$, or oxalyl chloride |

**Alcohols, ROH**

| | |
|---|---|
| Section 7.3 | From alkenes by oxymercuration–demercuration |
| Section 7.4 | From alkenes by hydroboration–oxidation |
| Section 7.7 | From alkenes by hydroxylation with either $OsO_4$ or $KMnO_4$ |
| Sections 11.4, 11.5 | From alkyl halides and tosylates by $S_N2$ reaction with hydroxide ion |
| Section 18.6 | From ethers by acid-induced cleavage |

| | |
|---|---|
| Section 18.8 | From epoxides by acid-catalyzed ring opening with either $H_2O$ or HX |
| Section 18.8 | From epoxides by base-induced ring opening |
| Sections 17.6, 19.11 | From ketones and aldehydes by reduction with $NaBH_4$ or $LiAlH_4$ |
| Sections 17.7, 19.10 | From ketones and aldehydes by addition of Grignard reagents |
| Section 20.8 | From carboxylic acids by reduction with either $LiAlH_4$ or $BH_3$ |
| Section 21.5 | From acid chlorides by reduction with $LiAlH_4$ |
| Section 21.5 | From acid chlorides by reaction with Grignard reagents |
| Section 21.6 | From acid anhydrides by reduction with $LiAlH_4$ |
| Sections 17.6, 21.7 | From esters by reduction with $LiAlH_4$ |
| Sections 17.7, 21.7 | From esters by reaction with Grignard reagents |

## Aldehydes, RCHO

| | |
|---|---|
| Section 7.8 | From disubstituted alkenes by ozonolysis |
| Section 7.9 | From 1,2-diols by cleavage with sodium periodate |
| Section 8.5 | From terminal alkynes by hydroboration with disiamylborane followed by oxidation |
| Sections 17.9, 19.3 | From primary alcohols by oxidation |
| Sections 19.3, 21.5 | From acid chlorides by catalytic hydrogenation (Rosenmund reduction) |
| Section 21.5 | From acid chlorides by partial reduction with $LiAl(O\text{-}t\text{-}Bu)_3H$ |
| Sections 19.3, 21.7 | From esters by reduction with DIBAH [$HAl(i\text{-}Bu)_2$] |
| Section 21.9 | From nitriles by partial reduction with DIBAH |

## Alkanes, RH

| | |
|---|---|
| Section 7.6 | From alkenes by catalytic hydrogenation |
| Section 10.9 | From alkyl halides by protonolysis of Grignard reagents |
| Section 10.10 | From alkyl halides by coupling with Gilman reagents |
| Section 19.13 | From ketones and aldehydes by Wolff–Kishner reaction |
| Section 19.13 | From ketones and aldehydes by Clemmensen reduction |
| Section 19.15 | From ketones and aldehydes by reduction of dithioacetals with Raney nickel |

## Alkenes, $R_2C{=}CR_2$

| | |
|---|---|
| Sections 7.12, 11.11 | From alkyl halides by treatment with strong base (E2 reaction) |
| Sections 7.12, 17.8 | From alcohols by dehydration |
| Section 8.5 | From alkynes by hydroboration followed by protonolysis |
| Section 8.6 | From alkynes by catalytic hydrogenation using the Lindlar catalyst |
| Section 8.6 | From alkynes by reduction with lithium in liquid ammonia |

| | |
|---|---|
| Section 19.16 | From ketones and aldehydes by treatment with alkylidenetriphenylphosphoranes (Wittig reaction) |
| Section 22.3 | From $\alpha$-bromo ketones by heating with pyridine |
| Section 22.8 | From saturated ketones by phenylselenenylation, followed by oxidative elimination |
| Section 25.8 | From amines by methylation and Hofmann elimination |

## Alkynes, RC≡CR

| | |
|---|---|
| Section 8.8 | From terminal alkynes by alkylation of acetylide anions |
| Section 8.10 | From dihalides by base-induced double dehydrohalogenation |

## Amides, $RCONH_2$

| | |
|---|---|
| Section 21.4 | From carboxylic acids by heating with ammonia |
| Sections 21.5, 25.8 | From acid chlorides by treatment with an amine or ammonia |
| Section 21.6 | From acid anhydrides by treatment with an amine or ammonia |
| Section 21.7 | From esters by treatment with an amine or ammonia |
| Section 21.9 | From nitriles by partial hydrolysis with either acid or base |
| Section 27.11 | From a carboxylic acid and an amine by treatment with dicyclohexylcarbodiimide (DCC) |

## Amines, $RNH_2$

| | |
|---|---|
| Section 25.7 | From primary alkyl halides by treatment with ammonia |
| Section 19.18 | From conjugated enones by addition of primary or secondary amines |
| Sections 21.8, 25.7 | From amides by reduction with $LiAlH_4$ |
| Sections 21.9, 25.7 | From nitriles by reduction with $LiAlH_4$ |
| Section 25.7 | From primary alkyl halides by Gabriel synthesis |
| Section 25.7 | From acid chlorides by Curtius rearrangement of acyl azides |
| Section 25.7 | From primary amides by Hofmann rearrangement |
| Section 25.7 | From primary alkyl azides by reduction with $LiAlH_4$ |
| Section 25.7 | From ketones and aldehydes by reductive amination with an amine and $NaBH_3CN$ |

## Amino Acids, $RCH(NH_2)COOH$

| | |
|---|---|
| Section 27.4 | From $\alpha$-bromo acids by $S_N2$ reaction with ammonia |
| Section 27.4 | From aldehydes by reaction with KCN and ammonia (Strecker synthesis) |
| Section 27.4 | From $\alpha$-keto acids by reductive amination |
| Section 27.4 | From primary alkyl halides by alkylation with diethyl acetamidomalonate |

## Arenes, Ar—R

| | |
|---|---|
| Section 16.3 | From arenes by Friedel–Crafts alkylation with a primary alkyl halide |

| | |
|---|---|
| Section 16.12 | From aryl alkyl ketones by catalytic reduction of the keto group |
| Section 26.4 | From arenediazonium salts by treatment with phosphorous acid |

### Arenediazonium salts, Ar—$N_2^+ X^-$

| | |
|---|---|
| Section 26.4 | From arylamines by reaction with nitrous acid |

### Arenesulfonic acids Ar—$SO_3H$

| | |
|---|---|
| Section 16.2 | From arenes by electrophilic aromatic substitution with $SO_3/H_2SO_4$ |

### Arylamines, Ar—$NH_2$

| | |
|---|---|
| Sections 16.2, 26.3 | From nitroarenes by reduction with either Fe, Sn, or $H_2$/Pd |

### Azides, R—$N_3$

| | |
|---|---|
| Sections 11.4, 25.7 | From primary alkyl halides by $S_N2$ reaction with azide ion |

### Carboxylic acids, RCOOH

| | |
|---|---|
| Section 7.8 | From mono- and 1,2-disubstituted alkenes by ozonolysis |
| Section 16.11 | From arenes by side-chain oxidation with $Na_2Cr_2O_7$ or $KMnO_4$ |
| Section 19.5 | From aldehydes by oxidation |
| Section 20.6 | From alkyl halides by conversion into Grignard reagents, followed by reaction with $CO_2$ |
| Sections 20.6, 21.9 | From nitriles by vigorous acid or base hydrolysis |
| Section 21.5 | From acid chlorides by reaction with aqueous base |
| Section 21.6 | From acid anhydrides by reaction with aqueous base |
| Section 21.7 | From esters by hydrolysis with aqueous base |
| Section 21.8 | From amides by hydrolysis with aqueous base |
| Section 22.7 | From methyl ketones by reaction with halogen and base (haloform reaction) |
| Section 26.9 | From phenols by treatment with $CO_2$ and base (Kolbe carboxylation) |

### Cyanohydrins, RCH(OH)CN

| | |
|---|---|
| Section 19.9 | From aldehydes and ketones by reaction with HCN |

### Cycloalkanes

| | |
|---|---|
| Section 16.12 | From arenes by rhodium-catalyzed hydrogenation |
| Section 7.10 | From alkenes by addition of dichlorocarbene |
| Section 7.10 | From alkenes by reaction with $CH_2I_2$ and Zn(Cu) (Simmons–Smith reaction) |

### Disulfides, RSSR′

| | |
|---|---|
| Section 17.12 | From thiols by oxidation with bromine |

**Enamines, $RCH{=}CRNR_2$**

| | |
|---|---|
| Section 19.12 | From ketones or aldehydes by reaction with secondary amines |

**Epoxides, $R_2C$—$CR_2$ (bridged by O)**

| | |
|---|---|
| Section 18.7 | From alkenes by treatment with a peroxyacid |
| Section 18.7 | From halohydrins by treatment with base |

**Esters, RCOOR′**

| | |
|---|---|
| Section 21.4 | From carboxylic acid salts by $S_N2$ reaction with primary alkyl halides |
| Section 21.4 | From carboxylic acids by acid-catalyzed reaction with an alcohol (Fischer esterification) |
| Section 21.4 | From carboxylic acids by reaction with diazomethane |
| Section 21.5 | From acid chlorides by base-induced reaction with an alcohol |
| Section 21.6 | From acid anhydrides by base-induced reaction with an alcohol |
| Section 22.9 | From alkyl halides by alkylation with diethyl malonate |
| Section 22.9 | From esters by treatment of their enolate ions with alkyl halides |
| Section 26.9 | From phenols by base-induced reaction with an acid chloride |

**Ethers, R—O—R′**

| | |
|---|---|
| Section 18.5 | From alkenes by alkoxymercuration–demercuration |
| Section 18.4 | From primary alkyl halides by $S_N2$ reaction with alkoxide ions (Williamson ether synthesis) |
| Section 16.9 | From activated haloarenes by reaction with alkoxide ions |
| Section 16.10 | From unactivated haloarenes by reaction with alkoxide ions via benzyne intermediates |
| Section 18.7 | From alkenes by epoxidation with peroxyacids |
| Section 26.9 | From phenols by reaction of phenoxide ions with primary alkyl halides |

**Halides, alkyl, $R_3C$—X**

| | |
|---|---|
| Section 6.9 | From alkenes by electrophilic addition of HX |
| Section 7.1 | From alkenes by addition of halogen |
| Section 7.2 | From alkenes by electrophilic addition of hypohalous acid (HOX) to yield halohydrins |
| Section 7.5 | From alkenes by radical-catalyzed addition of HBr |
| Section 8.3 | From alkynes by addition of halogen |
| Section 8.3 | From alkynes by addition of HX |
| Section 10.5 | From alkenes by allylic bromination with *N*-bromosuccinimide (NBS) |
| Section 10.8 | From alcohols by reaction with HX |
| Section 10.8 | From alcohols by reaction with $SOCl_2$ |
| Section 10.8 | From alcohols by reaction with $PBr_3$ |

| | |
|---|---|
| Sections 11.4, 11.5 | From alkyl tosylates by $S_N2$ reaction with halide ions |
| Section 16.11 | From arenes by benzylic bromination with *N*-bromosuccinimide (NBS) |
| Section 18.6 | From ethers by cleavage with HX |
| Section 20.9 | From carboxylic acids by treatment with $Ag_2O$ and bromine (Hunsdiecker reaction) |
| Section 22.3 | From ketones by alpha halogenation with bromine |
| Section 22.4 | From carboxylic acids by alpha halogenation with phosphorus and bromine (Hell–Volhard–Zelinskii reaction) |

**Halides, aryl, Ar—X**

| | |
|---|---|
| Sections 16.1, 16.2 | From arenes by electrophilic aromatic substitution with halogen |
| Section 26.4 | From arenediazonium salts by reaction with cuprous halides (Sandmeyer reaction) |
| Section 29.3 | From aromatic heterocycles by electrophilic aromatic substitution with halogen |

**Halohydrins, $R_2CXC(OH)R_2$**

| | |
|---|---|
| Section 7.2 | From alkenes by electrophilic addition of hypohalous acid (HOX) |
| Section 18.8 | From epoxides by acid-induced ring opening with HX |

**Imines, $R_2C{=}NR'$**

| | |
|---|---|
| Section 19.12 | From ketones or aldehydes by reaction with primary amines |

**Ketones, $R_2C{=}O$**

| | |
|---|---|
| Section 7.8 | From alkenes by ozonolysis |
| Section 7.9 | From 1,2-diols by cleavage reaction with sodium periodate |
| Section 8.4 | From alkynes by mercuric ion-catalyzed hydration |
| Section 8.5 | From alkynes by hydroboration–oxidation |
| Section 16.4 | From arenes by Lewis acid-catalyzed reaction with an acid chloride (Friedel–Crafts acylation) |
| Sections 17.9, 19.4 | From secondary alcohols by oxidation |
| Sections 19.4, 21.5 | From acid chlorides by reaction with lithium diorganocopper (Gilman) reagents |
| Section 19.18 | From conjugated enones by addition of lithium diorganocopper reagents |
| Section 21.9 | From nitriles by reaction with Grignard reagents |
| Section 22.9 | From primary alkyl halides by alkylation with ethyl acetoacetate |
| Section 22.9 | From ketones by alkylation of their enolate ions with primary alkyl halides |

**Nitriles, $R{-}C{\equiv}N$**

| | |
|---|---|
| Sections 11.5, 20.6 | From primary alkyl halides by $S_N2$ reaction with cyanide ion |
| Section 19.18 | From conjugated enones by addition of HCN |

| | |
|---|---|
| Section 21.9 | From primary amides by dehydration with $SOCl_2$ |
| Section 22.9 | From nitriles by alkylation of their alpha anions with primary alkyl halides |
| Section 26.4 | From arenediazonium ions by treatment with CuCN |

**Nitroarenes, Ar—$NO_2$**

| | |
|---|---|
| Section 16.2 | From arenes by electrophilic aromatic substitution with nitric/sulfuric acids |

**Organometallics, R—M**

| | |
|---|---|
| Section 10.9 | Formation of Grignard reagents from organohalides by treatment with magnesium |
| Section 10.10 | Formation of organolithium reagents from organohalides by treatment with lithium |
| Section 10.10 | Formation of lithium diorganocopper reagents (Gilman reagents) from organolithium reagents by treatment with cuprous halides |

**Phenols, Ar—OH**

| | |
|---|---|
| Sections 16.2, 26.8 | From arenesulfonic acids by fusion with KOH |
| Section 26.4 | From arenediazonium salts by reaction with aqueous acid |
| Section 16.9 | From aryl halides by nucleophilic aromatic substitution with hydroxide ion |

**Quinones,** O=C₆H₄=O

| | |
|---|---|
| Section 26.9 | From phenols by oxidation with Fremy's salt, $(KSO_3)_2NO$ |
| Section 26.9 | From arylamines by oxidation with Fremy's salt |

**Sulfides, R—S—R′**

| | |
|---|---|
| Section 18.11 | From thiols by $S_N2$ reaction of thiolate ions with primary alkyl halides |

**Sulfones, R—$SO_2$—R′**

| | |
|---|---|
| Section 18.11 | From sulfides or sulfoxides by oxidation with peroxyacids |

**Sulfoxides, R—SO—R′**

| | |
|---|---|
| Section 18.11 | From sulfides by oxidation with $H_2O_2$ |

**Thioacetals, $R_2C(SR')_2$**

| | |
|---|---|
| Section 19.15 | From ketones and aldehydes by acid-catalyzed reaction with thiols |

**Thiols, R—SH**

| | |
|---|---|
| Section 11.5 | From primary alkyl halides by $S_N2$ reaction with hydrosulfide anion |
| Section 17.12 | From primary alkyl halides by $S_N2$ reaction with thiourea followed by hydrolysis |

APPENDIX D

# Reagents in Organic Chemistry

The following list summarizes the uses of some important reagents in organic chemistry. The reagents are listed alphabetically, followed by brief descriptions of the uses of each and references to the appropriate text sections.

**Acetic acid, $CH_3COOH$**: Reacts with vinylic organoboranes to yield alkenes. The net effect of alkyne hydroboration followed by protonolysis with acetic acid is reduction of the alkyne to a cis alkene (Section 8.5).

**Acetic anhydride, $(CH_3CO)_2O$**: Reacts with alcohols to yield acetate esters (Sections 21.6, 24.8).

**Aluminum chloride, $AlCl_3$**: Acts as a Lewis acid catalyst in Friedel–Crafts alkylation and acylation reactions of aromatic ring compounds (Sections 16.3, 16.4).

**Ammonia, $NH_3$**: A solvent for the reduction of alkynes by lithium metal to yield trans alkenes (Section 8.6).

Reacts with acid chlorides to yield amides (Section 21.5).

**Borane, $BH_3$**: Adds to alkenes, giving alkylboranes that can be oxidized with alkaline hydrogen peroxide to yield alcohols (Section 7.4).

Adds to alkynes, giving vinylic organoboranes that either can be treated with acetic acid to yield cis alkenes (Section 8.5) or can be oxidized with alkaline hydrogen peroxide to yield aldehydes (Section 8.5).

Reduces carboxylic acids to yield primary alcohols (Section 20.8).

**Bromine, $Br_2$**: Adds to alkenes yielding 1,2-dibromides (Sections 7.1, 14.5).

Adds to alkynes yielding either 1,2-dibromoalkenes or 1,1,2,2-tetrabromoalkanes (Section 8.3).

Reacts with arenes in the presence of ferric bromide catalyst to yield bromoarenes (Section 16.1).

Reacts with ketones in acetic acid solvent to yield $\alpha$-bromo ketones (Section 22.3).

Reacts with carboxylic acids in the presence of phosphorus tribromide to yield $\alpha$-bromo carboxylic acids (Hell–Volhard–Zelinskii reaction; Section 22.4).

Reacts with methyl ketones in the presence of sodium hydroxide to yield carboxylic acids and bromoform (haloform reaction; Section 22.7).

Oxidizes aldoses to yield aldonic acids (Section 24.8).

***N*-Bromosuccinimide (NBS), $(CH_2CO)_2NBr$:** Reacts with alkenes in the presence of aqueous dimethylsulfoxide to yield bromohydrins (Section 7.2).

Reacts with alkenes in the presence of light to yield allylic bromides (Wohl–Ziegler reaction; Section 10.5).

Reacts with alkylbenzenes in the presence of light to yield benzylic bromides; (Section 16.11).

**di-*tert*-Butoxy dicarbonate, $(t\text{-BuOCO})_2O$:** Reacts with amino acids to give *t*-BOC-protected amino acids suitable for use in peptide synthesis (Section 27.11).

**Butyllithium, $CH_3CH_2CH_2CH_2Li$:** A strong base that reacts with alkynes to yield acetylide anions that can be alkylated (Section 8.7).

Reacts with dialkylamines to yield lithium dialkylamide bases such as LDA, lithium diisopropylamide (Section 22.5).

**Carbon dioxide, $CO_2$:** Reacts with Grignard reagents to yield carboxylic acids (Section 20.6).

Reacts with phenoxide anions to yield *o*-hydroxybenzoic acids (Kolbe–Schmitt carboxylation reaction; Section 26.9).

**Chlorine, $Cl_2$:** Adds to alkenes to yield 1,2-dichlorides (Sections 7.1, 14.5).

Reacts with alkanes in the presence of light to yield chloroalkanes by a radical chain reaction pathway (Section 10.4).

Reacts with arenes in the presence of ferric chloride catalyst to yield chloroarenes (Section 16.2).

***m*-Chloroperoxybenzoic acid, *m*-$ClC_6H_4CO_3H$:** Reacts with alkenes to yield epoxides (Section 18.7).

**Chromium trioxide, $CrO_3$:** Oxidizes alcohols in aqueous sulfuric acid (Jones reagent) to yield carbonyl-containing products. Primary alcohols yield carboxylic acids, and secondary alcohols yield ketones (Sections 17.9, 19.5, 20.6).

**Cuprous bromide, CuBr:** Reacts with arenediazonium salts to yield bromoarenes (Sandmeyer reaction; Section 26.4).

**Cuprous chloride, CuCl:** Reacts with arenediazonium salts to yield chloroarenes (Sandmeyer reaction; Section 26.4).

**Cuprous cyanide, CuCN:** Reacts with arenediazonium salts to yield substituted benzonitriles (Sandmeyer reaction; Section 26.4).

**Cuprous iodide, CuI:** Reacts with organolithiums to yield lithium diorganocopper reagents (Gilman reagents; Section 10.10).

**Diazomethane, $CH_2N_2$:** Reacts with carboxylic acids to yield methyl esters (Section 21.4).

**Dicyclohexylcarbodiimide (DCC), $C_6H_{11}{-}N{=}C{=}N{-}C_6H_{11}$:** Couples an amine with a carboxylic acid to yield an amide. DCC is often used in peptide synthesis (Section 27.11).

**Diethyl acetamidomalonate, $CH_3CONHCH(CO_2Et)_2$:** Reacts with alkyl halides in a common method of $\alpha$-amino acid synthesis (Section 27.4).

**Diethylaluminum cyanide, $(Et)_2AlCN$:** Reacts with $\alpha,\beta$-unsaturated ketones to yield $\beta$-keto nitriles (Section 19.18).

**Dihydropyran:** Reacts with alcohols in the presence of an acid catalyst to yield tetrahydropyranyl ethers that serve as useful hydroxyl-protecting groups (Section 17.10).

**Diiodomethane, $CH_2I_2$:** Reacts with alkenes in the presence of zinc–copper alloy to yield cyclopropanes (Simmons–Smith reaction; Section 7.10).

**Diisobutylaluminum hydride (DIBAH), $(i\text{-Bu})_2AlH$:** Reduces esters to yield aldehydes (Sections 19.3, 21.7).

Reduces nitriles to yield aldehydes (Section 21.9).

**2,4-Dinitrophenylhydrazine, 2,4-$(NO_2)_2C_6H_3NHNH_2$:** Reacts with ketones and aldehydes to yield 2,4-DNP's that serve as useful crystalline derivatives (Section 19.12).

**Disiamylborane, $[(CH_3)_2CHCH(CH_3)]_2BH$:** A hindered dialkylborane that adds to terminal alkynes, giving trialkylboranes that can be oxidized with alkaline hydrogen peroxide to yield aldehydes (Section 8.5).

**1,2-Ethanedithiol, $HSCH_2CH_2SH$:** Reacts with ketones or aldehydes in the presence of an acid catalyst to yield dithioacetals that can be reduced with Raney nickel to yield alkanes (Section 19.15).

**Ethylene glycol, $HOCH_2CH_2OH$:** Reacts with ketones or aldehydes in the presence of an acid catalyst to yield acetals that serve as useful carbonyl-protecting groups (Section 19.14).

**Ferric bromide, $FeBr_3$:** Acts as a catalyst for the reaction of arenes with bromine to yield bromoarenes (Section 16.1).

**Ferric chloride, $FeCl_3$:** Acts as a catalyst for the reaction of arenes with chlorine to yield chloroarenes (Section 16.2).

**Grignard reagent, RMgX:** Adds to carbonyl-containing compounds (ketones, aldehydes, esters) to yield alcohols (Section 19.10).

**Hydrazine, $H_2NNH_2$:** Reacts with ketones or aldehydes in the presence of potassium hydroxide to yield the corresponding alkanes (Wolff–Kishner reaction; Section 19.13).

**Hydrogen bromide, HBr:** Adds to alkenes to yield alkyl bromides. Markovnikov regiochemistry is observed (Sections 6.9, 14.5).

Adds to alkenes in the presence of a peroxide catalyst to yield alkyl bromides. Non-Markovnikov regiochemistry is observed (Section 7.5).

Adds to alkynes to yield either bromoalkenes or 1,1-dibromoalkanes (Section 8.3).

Reacts with alcohols to yield alkyl bromides (Sections 10.8, 17.8).

Cleaves ethers to yield alcohols and alkyl bromides (Section 18.6).

**Hydrogen chloride, HCl:** Adds to alkenes to yield alkyl chlorides. Markovnikov regiochemistry is observed (Sections 6.9, 14.5).

Adds to alkynes to yield either chloroalkenes or 1,1-dichloroalkanes (Section 8.3).

Reacts with alcohols to yield alkyl chlorides (Sections 10.8, 17.8).

**Hydrogen cyanide, HCN:** Adds to ketones and aldehydes to yield cyanohydrins (Section 19.9).

**Hydrogen iodide, HI:** Reacts with alcohols to yield alkyl iodides (Section 17.8).

Cleaves ethers to yield alcohols and alkyl iodides (Section 18.6).

**Hydrogen peroxide, $H_2O_2$:** Oxidizes organoboranes to yield alcohols. Used in conjunction with addition of borane to alkenes, the overall transformation effects syn Markovnikov addition of water to an alkene (Section 7.4).

Oxidizes vinylic boranes to yield aldehydes. Since the vinylic borane starting materials are prepared by hydroboration of a terminal alkyne with disiamyl borane, the overall transformation is the hydration of a terminal alkyne to yield an aldehyde (Section 8.5).

Oxidizes sulfides to yield sulfoxides (Section 18.11).

Reacts with $\alpha$-phenylselenenyl ketones to yield $\alpha,\beta$-unsaturated ketones (Section 22.8).

**Hydroxylamine, $NH_2OH$:** Reacts with ketones and aldehydes to yield oximes (Section 19.12).

Reacts with aldoses to yield oximes as the first step in the Wohl degradation of aldoses (Section 24.8).

**Hypophosphorous acid, $H_3PO_2$:** Reacts with arenediazonium salts to yield arenes (Section 26.4).

**Iodine, $I_2$:** Reacts with arenes in the presence of cupric chloride or hydrogen peroxide to yield iodoarenes (Section 16.2).

Reacts with carboxylic acids in the presence of lead tetraacetate to yield alkyl iodides (Hunsdiecker reaction; Section 20.9).

Reacts with methyl ketones in the presence of aqueous sodium hydroxide to yield carboxylic acids and iodoform (Section 22.7).

**Iodomethane, $CH_3I$:** Reacts with alkoxide anions to yield methyl ethers (Section 18.4).

Reacts with carboxylate anions to yield methyl esters (Section 21.7).

Reacts with enolate ions to yield $\alpha$-methylated carbonyl compounds (Section 22.9).

Reacts with amines to yield methylated amines (Section 25.7).

**Iron, Fe:** Reacts with nitroarenes in the presence of mineral acid to yield anilines (Section 26.3).

**Lead tetraacetate, $Pb(OCOCH_3)_4$:** Reacts with carboxylic acids in the presence of iodine to yield alkyl iodides and carbon dioxide (Hunsdiecker reaction; Section 20.9).

**Lindlar catalyst:** Acts as a catalyst for the hydrogenation of alkynes to yield cis alkenes (Section 8.6).

**Lithium, Li:** Reduces alkynes in liquid ammonia solvent to yield trans alkenes (Section 8.6).

Reacts with organohalides to yield organolithium compounds (Section 10.10).

**Lithium aluminum hydride, $LiAlH_4$:** Reduces ketones, aldehydes, esters, and carboxylic acids to yield alcohols (Section 17.6).

Reduces amides to yield amines (Section 21.8).

Reduces alkyl azides to yield amines (Section 25.7).

Reduces nitriles to yield amines (Sections 21.9, 25.7).

**Lithium diisopropylamide (LDA), $LiN(i\text{-}Pr)_2$:** Reacts with carbonyl compounds (aldehydes, ketones, esters) to yield enolate ions (Sections 22.5, 22.9).

**Lithium diorganocopper reagent (Gilman reagent), $LiR_2Cu$:** Couples with alkyl halides to yield alkanes (Section 10.10).

Adds to $\alpha,\beta$-unsaturated ketones to give 1,4-addition products (Section 19.18).

**Lithium tri-*tert*-butoxyaluminum hydride, $LiAl(O\text{-}t\text{-}Bu)_3H$:** Reduces acid chlorides to yield aldehydes (Section 21.5).

**Magnesium, Mg:** Reacts with organohalides to yield Grignard reagents (Section 10.9).

**Mercuric acetate, $Hg(OCOCH_3)_2$:** Adds to alkenes in the presence of water, giving $\alpha$-hydroxy organomercury compounds that can be reduced with sodium borohydride to yield alcohols. The overall reaction effects the Markovnikov hydration of an alkene (Section 7.3).

**Mercuric oxide, HgO:** Reacts with carboxylic acids in the presence of bromine to yield alkyl bromides and carbon dioxide (Hunsdiecker reaction; Section 20.9).

**Mercuric sulfate, $HgSO_4$:** Acts as a catalyst for the addition of water to alkynes in the presence of aqueous sulfuric acid, yielding ketones (Section 8.4).

**Mercuric trifluoroacetate, $Hg(OCOCF_3)_2$:** Adds to alkenes in the presence of alcohol, giving $\alpha$-alkoxy organomercury compounds that can be reduced with sodium borohydride to yield ethers. The overall reaction effects a net addition of an alcohol to an alkene (Section 18.5).

**1-Mesitylenesulfonyl-3-nitro-1,2,4-triazole (MSNT):** Acts as a coupling reagent for use in DNA synthesis (Section 29.16).

**Methyl sulfate, $(CH_3O)_2SO_2$:** A reagent used to methylate heterocyclic amine bases during Maxam–Gilbert DNA sequencing (Section 29.15).

**Nitric acid, $HNO_3$:** Reacts with arenes in the presence of sulfuric acid to yield nitroarenes (Section 16.2).

Oxidizes aldoses to yield aldaric acids (Section 24.8).

**Nitronium tetrafluoroborate, $NO_2BF_4$:** Reacts with arenes to yield nitroarenes (Section 16.2).

**Nitrous acid, $HNO_2$:** Reacts with amines to yield diazonium salts (Section 26.4).

**Osmium tetraoxide, $OsO_4$:** Adds to alkenes to yield 1,2-diols (Section 7.7).

Reacts with alkenes in the presence of periodic acid to cleave the carbon–carbon double bond, yielding ketone or aldehyde fragments (Section 7.9).

**Oxalyl chloride, ClCOCOCl:** Reacts with carboxylic acids, yielding acid chlorides (Section 21.4).

**Ozone, $O_3$:** Adds to alkenes to cleave the carbon–carbon double bond and give ozonides. The ozonides can then be reduced with zinc in acetic acid to yield carbonyl compounds (Section 7.8).

**Palladium on barium sulfate, $Pd/BaSO_4$:** Acts as a hydrogenation catalyst in the Rosenmund reduction of acid chlorides to yield aldehydes (Sections 19.3, 21.5).

**Palladium on carbon, Pd/C:** Acts as a hydrogenation catalyst in the reduction of carbon–carbon multiple bonds. Alkenes and alkynes are reduced to yield alkanes (Sections 7.6, 8.6).

Acts as a hydrogenation catalyst in the reduction of aryl ketones to yield alkylbenzenes (Section 16.12).

Acts as a hydrogenation catalyst in the reduction of nitroarenes to yield anilines (Section 26.3).

**Periodic acid, $HIO_4$:** Reacts with 1,2-diols to yield carbonyl-containing cleavage products (Section 7.9).

**Peroxyacetic acid, $CH_3CO_3H$:** Oxidizes sulfoxides to yield sulfones (Section 18.11)

**Phenyl isothiocyanate, $C_6H_5{-}N{=}C{=}S$:** A reagent used in the Edman degradation of peptides to identify *N*-terminal amino acids (Section 27.9).

**Phenylselenenyl bromide, $C_6H_5SeBr$:** Reacts with enolate ions to yield $\alpha$-phenylselenenyl ketones. On oxidation of the product with hydrogen peroxide, an $\alpha,\beta$-unsaturated ketone is produced (Section 22.8).

**Phosphorus oxychloride, $POCl_3$:** Reacts with secondary and tertiary alcohols to yield alkene dehydration products (Section 17.8).

**Phosphorus tribromide, $PBr_3$:** Reacts with alcohols to yield alkyl bromides (Section 10.8).

Reacts with carboxylic acids in the presence of bromine to yield $\alpha$-bromo carboxylic acids (Hell–Volhard–Zelinskii reaction; Section 22.4).

**Phosphorus trichloride, $PCl_3$:** Reacts with carboxylic acids to yield acid chlorides (Section 21.4).

**Platinum oxide (Adam's catalyst), $PtO_2$:** Acts as a hydrogenation catalyst in the reduction of alkenes and alkynes to yield alkanes (Section 7.6).

**Potassium *tert*-butoxide, KO-*t*-Bu**: Reacts with alkyl halides to yield alkenes (Sections 11.10-11.14).

Reacts with allylic halides to yield conjugated dienes in an elimination reaction (Section 14.1).

Reacts with chloroform in the presence of an alkene to yield a dichlorocyclopropane (Section 7.10).

**Potassium hydroxide, KOH**: Reacts with alkyl halides to yield alkenes by an elimination reaction (Sections 7.12, 11.10–11.14).

Reacts with 1,1- or 1,2-dihaloalkanes to yield alkynes by a twofold elimination reaction (Section 8.10).

**Potassium nitrosodisulfonate (Fremy's salt), $(KSO_3)_2NO$**: Oxidizes phenols and anilines to yield quinones (Section 26.9).

**Potassium permanganate, $KMnO_4$**: Oxidizes alkenes under alkaline conditions to yield 1,2-diols (Section 7.7).

Oxidizes alkenes under neutral or acidic conditions to give carboxylic acid double-bond cleavage products (Sections 7.8).

Oxidizes alkynes to give carboxylic acid triple-bond cleavage products (Section 8.9).

Oxidizes arenes to yield benzoic acids (Section 16.11).

**Potassium phthalimide, $C_6H_4(CO)_2NK$**: Reacts with alkyl halides to yield an *N*-alkylphthalimide that is hydrolyzed by aqueous sodium hydroxide to yield an amine (Gabriel amine synthesis; Section 25.7).

**Pyridine, $C_5H_5N$**: Reacts with $\alpha$-bromo ketones to yield $\alpha,\beta$-unsaturated ketones (Section 22.3).

Acts as a catalyst for the reaction of alcohols with acid chlorides to yield esters (Section 21.5).

Acts as a catalyst for the reaction of alcohols with acetic anhydride to yield acetate esters (Section 21.6).

**Pyridinium chlorochromate (PCC), $C_5H_6NCrO_3Cl$**: Oxidizes primary alcohols to yield aldehydes and secondary alcohols to yield ketones (Sections 17.9, 19.3, 19.4).

**Pyrrolidine, $C_4H_8N$**: Reacts with ketones to yield enamines for use in the Stork enamine reaction (Sections 19.12, 23.13).

**Raney nickel, Ni**: Reduces dithioacetals to yield alkanes by a desulfurization reaction (Section 19.15).

**Rhodium on carbon, Rh/C**: Acts as a hydrogenation catalyst in the reduction of benzene rings to yield cyclohexanes (Section 16.12).

**Silver oxide, $Ag_2O$**: Oxidizes primary alcohols in aqueous ammonia solution to yield aldehydes (Tollens oxidation; Sections 19.5, 20.6).

Acts as a catalyst for the reaction of alcohols with alkyl halides to yield ethers (Section 24.8).

**Sodium amide, $NaNH_2$**: Reacts with terminal alkynes to yield acetylide anions (Section 8.7).

Reacts with 1,1- or 1,2-dihalides to yield alkynes by a twofold elimination reaction (Section 8.10).

Reacts with aryl halides to yield anilines by a benzyne aromatic substitution mechanism (Section 16.10).

**Sodium azide, $NaN_3$**: Reacts with alkyl halides to yield alkyl azides (Section 25.7).

Reacts with acid chlorides to yield acyl azides. On heating in the presence of water, acyl azides yield amines and carbon dioxide (Section 25.7).

**Sodium bisulfite, $NaHSO_3$**: Reduces osmate esters, prepared by treatment of an alkene with osmium tetraoxide, to yield 1,2-diols (Section 7.7).

**Sodium borohydride, $NaBH_4$:** Reduces organomercury compounds, prepared by oxymercuration of alkenes, to convert the C—Hg bond to C—H (Section 7.3).
Reduces ketones and aldehydes to yield alcohols (Sections 17.6, 19.11).
Reduces quinones to yield hydroquinones (Section 26.9).

**Sodium cyanide, NaCN:** Reacts with alkyl halides to yield alkanenitriles (Sections 20.6, 21.9).

**Sodium cyanoborohydride, $NaBH_3CN$:** Reacts with ketones and aldehydes in the presence of ammonia to yield an amine by a reductive amination process (Section 25.7).

**Sodium dichromate, $Na_2Cr_2O_7$:** Oxidizes primary alcohols to yield carboxylic acids and secondary alcohols to yield ketones (Sections 17.9, 19.5).
Oxidizes alkylbenzenes to yield benzoic acids (Section 16.11).

**Sodium hydride, NaH:** Reacts with alcohols to yield alkoxide anions (Section 17.4).

**Sodium hydroxide, NaOH:** Reacts with arenesulfonic acids at high temperature to yield phenols (Section 16.2).
Reacts with aryl halides to yield phenols by a benzyne aromatic substitution mechanism (Section 16.10).
Catalyzes the reaction of acid chlorides with alcohols to yield esters (Schotten–Baumann reaction; Section 21.5).
Catalyzes the reaction of acid chlorides with amines to yields amides (Section 21.5).
Reacts with methyl ketones in the presence of iodine to yield carboxylic acids and iodoform (Section 22.7).

**Sodium iodide, NaI:** Reacts with arenediazonium salts to yield aryl iodides (Section 26.4).

**Stannous chloride, $SnCl_2$:** Reduces nitroarenes to yield anilines (Sections 16.2, 26.3).
Reduces quinones to yield hydroquinones (Section 26.9).

**Sulfur trioxide, $SO_3$:** Reacts with arenes in sulfuric acid solution to yield arenesulfonic acids (Section 16.2).

**Sulfuric acid, $H_2SO_4$:** Reacts with alcohols to yield alkenes (Sections 7.12, 17.8).
Reacts with alkynes in the presence of water and mercuric acetate to yield ketones (Section 8.4).
Catalyzes the reaction of nitric acid with aromatic rings to yield nitroarenes (Section 16.2).
Catalyzes the reaction of $SO_3$ with aromatic rings to yield arenesulfonic acids (Section 16.2).

**Thionyl chloride, $SOCl_2$:** Reacts with primary and secondary alcohols to yield alkyl chlorides (Section 10.8).
Reacts with carboxylic acids to yield acid chlorides (Section 21.4).

**Thiourea, $H_2NCSNH_2$:** Reacts with primary alkyl halides to yield thiols (Section 17.12).

***p*-Toluenesulfonyl chloride, *p*-$CH_3C_6H_4SO_2Cl$:** Reacts with alcohols to yield tosylates (Sections 11.2, 17.8).

**Trifluoroacetic acid, $CF_3COOH$:** Acts as a catalyst for cleaving *tert*-butyl ethers, yielding alcohols and 2-methylpropene (Section 18.6).
Acts as a catalyst for cleaving the *t*-BOC-protecting group from amino acids in peptide synthesis (Section 27.11).

**Triphenylphosphine, $(C_6H_5)_3P$:** Reacts with primary alkyl halides to yield the alkyltriphenylphosphonium salts used in Wittig reactions (Section 19.16).

**Zinc, Zn**: Reduces ozonides, produced by addition of ozone to alkenes, to yield ketones and aldehydes (Section 7.8).

Reduces disulfides to yield thiols (Section 17.12).

Reduces ketones and aldehydes in the presence of aqueous HCl to yield alkanes (Clemmensen reduction; Section 19.13).

**Zinc bromide, $ZnBr_2$**: Acts as a Lewis acid catalyst to cleave DMT ethers in DNA synthesis (Section 29.16).

**Zinc–copper alloy, Zn(Cu)**: Reacts with diiodomethane in the presence of alkenes to yield cyclopropanes (Simmons–Smith reaction; Section 7.10).

APPENDIX E

# Name Reactions in Organic Chemistry

The following list summarizes some reactions that are often referred to by name. The reactions are listed alphabetically, followed by references to the appropriate text sections.

**Acetoacetic ester synthesis** (Section 22.9): A multistep reaction sequence for converting a primary alkyl halide into a methyl ketone having three more carbon atoms in the chain:

$$RCH_2X + CH_3\overset{\overset{O}{\|}}{C}\overset{-}{C}H\overset{\overset{O}{\|}}{C}OEt \xrightarrow[2.\ H_3O^+,\ \Delta]{1.\ \Delta} RCH_2{-}CH_2\overset{\overset{O}{\|}}{C}CH_3 + CO_2 + EtOH$$

**Adams catalyst** (Section 7.6): $PtO_2$, a catalyst used for the hydrogenation of carbon–carbon double bonds.

**Aldol condensation reaction** (Section 23.2): The nucleophilic addition of an enol or enolate ion to a ketone or aldehyde, yielding a $\beta$-hydroxy ketone:

$$2\ R{-}\overset{\overset{O}{\|}}{C}{-}\underset{|}{\overset{|}{C}}{-}H \xrightarrow{\text{Base}} R{-}\overset{\overset{O}{\|}}{C}{-}\underset{|}{\overset{|}{C}}{-}\underset{\underset{R}{|}}{\overset{\overset{OH}{|}}{C}}{-}\underset{|}{\overset{|}{C}}{-}H$$

**Amidomalonate amino acid synthesis** (Section 27.4): A multistep reaction sequence, similar to the malonic ester synthesis, for converting a primary alkyl halide into an amino acid:

$$RCH_2X + {:}\overset{-}{C}H(NHAc)(COOEt)_2 \xrightarrow[2.\ H_3O^+,\ \Delta]{1.\ \Delta} RCH_2{-}\underset{\underset{NH_2}{|}}{C}H\overset{\overset{O}{\|}}{C}OH + CO_2 + 2\ EtOH$$

**Cannizzaro reaction** (Section 19.17): The disproportionation reaction that occurs when a nonenolizable aldehyde is treated with base:

$$2\ R_3C\overset{\overset{O}{\|}}{C}H \xrightarrow[2.\ H_3O^+]{1.\ HO^-} R_3C\overset{\overset{O}{\|}}{C}OH + R_3CCH_2OH$$

**Haloform reaction** (Section 22.7): The conversion of a methyl ketone to a carboxylic acid and haloform by treatment with halogen and base:

$$R-\overset{O}{\overset{\|}{C}}-CH_3 \xrightarrow[2.\ H_3O^+]{1.\ X_2,\ NaOH} R-\overset{O}{\overset{\|}{C}}-OH + CHX_3$$

**Hell–Volhard–Zelinskii reaction** (Section 22.4): The alpha bromination of carboxylic acids by treatment with bromine and phosphorus:

$$R-\underset{H}{\underset{|}{\overset{|}{C}}}-\overset{O}{\overset{\|}{C}}-OH \xrightarrow[2.\ H_2O]{1.\ Br_2,\ P} R-\underset{Br}{\underset{|}{\overset{|}{C}}}-\overset{O}{\overset{\|}{C}}-OH$$

**Hinsberg test** (Section 25.8): A chemical means of distinguishing among primary, secondary, and tertiary amines by observing their reactions with benzenesulfonyl chloride.

**Hofmann elimination** (Section 25.8): A method for effecting the elimination reaction of an amine to yield an alkene. The amine is first treated with excess iodomethane, and the resultant quaternary ammonium salt is heated with silver oxide:

$$\begin{matrix} R_2N & & H \\ & \diagdown\ \ \diagup & \\ -C & - & C- \\ \diagup & & \diagdown \end{matrix} \xrightarrow[2.\ Ag_2O,\ \Delta]{1.\ CH_3I} \begin{matrix} \diagdown & & \diagup \\ C & = & C \\ \diagup & & \diagdown \end{matrix}$$

**Hofmann rearrangement** (Section 25.7): The rearrangement of an *N*-bromoamide to a primary amine by treatment with aqueous base:

$$R-\overset{O}{\overset{\|}{C}}-NH_2 \xrightarrow[NaOH]{Br_2} R-\overset{O}{\overset{\|}{C}}-NHBr \longrightarrow RNH_2 + CO_2$$

**Hunsdiecker reaction** (Section 20.9): A reaction for converting a carboxylic acid into an alkyl halide by treatment with mercuric oxide and halogen:

$$R-\overset{O}{\overset{\|}{C}}-OH \xrightarrow[X_2]{HgO} RX + CO_2 + HX$$

**Jones' reagent** (Section 17.9): A solution of $CrO_3$ in acetone/aqueous sulfuric acid. This reagent oxidizes primary and secondary alcohols to carbonyl compounds under mild conditions.

**Kiliani–Fischer synthesis** (Section 24.8): A multistep sequence for chain-lengthening an aldose into the next higher homolog:

$$\underset{R}{\underset{|}{CHO}} \xrightarrow[3.\ Na-Hg]{1.\ HCN \quad 2.\ \Delta} \begin{matrix} CHO \\ | \\ CHOH \\ | \\ R \end{matrix}$$

**Koenigs–Knorr reaction** (Section 24.8): A method for synthesizing glycosides by reaction between an alcohol and a bromo-substituted carbohydrate:

[Structure: pyranose ring with $CH_2OAc$, OAc, AcO, AcO substituents and Br at anomeric carbon] $\xrightarrow{ROH,\ Ag_2O}$ [Structure: same ring with $CH_2OAc$, OAc, AcO, AcO substituents and OR at anomeric carbon]

**Kolbe–Schmitt carboxylation reaction** (Section 26.9): A method for introducing a carboxyl group in the ortho position of a phenol by treatment of the phenoxide anion with $CO_2$:

$$C_6H_5O^- \xrightarrow[2.\ H_3O^+]{1.\ CO_2} o\text{-}HOC_6H_4COOH$$

**Malonic ester synthesis** (Section 22.9): A multistep sequence for converting an alkyl halide into a carboxylic acid with the addition of two carbon atoms to the chain:

$$RCH_2X + :CH(COOEt)_2 \xrightarrow[2.\ H_3O^+,\ \Delta]{1.\ \Delta} RCH_2{-}CH_2\overset{O}{\overset{\|}{C}}OH + 2\ EtOH + 2\ CO_2$$

**Maxam–Gilbert DNA sequencing** (Section 29.15): A rapid and efficient method for sequencing long chains of DNA by employing selective cleavage reactions.

**McLafferty rearrangement** (Section 19.20): A general mass spectral fragmentation pathway for carbonyl compounds having a hydrogen three carbon atoms away from the carbonyl carbon:

$$\text{ketone with } \gamma\text{-H} \xrightarrow{e^-} \text{alkene} + H{-}O^{+\cdot}{=}C(CH_3)_2$$

**Meisenheimer complex** (Section 16.9): An intermediate formed in the nucleophilic aryl-substitution reaction of a base with a nitro-substituted aromatic ring:

$$2,6\text{-}(O_2N)_2C_6H_3Cl \xrightarrow{HO^-} [\text{Cl, OH adduct, } O_2N, NO_2, \text{ring anion } (-)]$$

**Merrifield solid-phase peptide synthesis** (Section 27.12): A rapid and efficient means of peptide synthesis in which the growing peptide chain is attached to an insoluble polymer support.

**Michael reaction** (Section 23.12): The 1,4-addition reaction of a stabilized enolate anion to an $\alpha,\beta$-unsaturated carbonyl compound:

$$R{-}\overset{O}{\overset{\|}{C}}{-}CH_2{-}\overset{O}{\overset{\|}{C}}{-}R' + H_2C{=}CH{-}\overset{O}{\overset{\|}{C}}{-}R'' \xrightarrow{\text{Base}} (R{-}\overset{O}{\overset{\|}{C}})(R'{-}\overset{O}{\overset{\|}{C}})CH{-}CH_2CH_2\overset{O}{\overset{\|}{C}}R''$$

**Nagata hydrocyanation reaction** (Section 19.18): A reaction for effecting the conjugate 1,4 addition of HCN to an $\alpha,\beta$-unsaturated ketone by treatment with diethylaluminum cyanide:

$$\rangle C{=}\overset{|}{C}{-}\overset{O}{\overset{\|}{C}}{-}R \xrightarrow[2.\ H_3O^+]{1.\ Et_2AlCN} N{\equiv}C{-}\overset{|}{\underset{|}{C}}{-}\overset{|}{\underset{|}{C}}{-}\overset{O}{\overset{\|}{C}}{-}R$$

**Raney nickel** (Section 19.15): A specially prepared form of nickel that is used for desulfurization of thioacetals:

$$\mathrm{(RS)(SR)C{<}} \xrightarrow{\text{Raney Ni}} \mathrm{(H)(H)C{<}} + 2\,\mathrm{RH}$$

**Robinson annulation reaction** (Section 23.14): A multistep sequence for building a new cyclohexenone ring onto a ketone. The sequence involves an initial Michael reaction of the ketone, followed by an internal aldol cyclization:

$$\text{methyl vinyl ketone} + \text{ketone} \xrightarrow{\text{Base}} \text{cyclohexenone}$$

**Rosenmund reduction** (Section 19.3): A method for conversion of an acid chloride into an aldehyde by catalytic hydrogenation:

$$\mathrm{R{-}\overset{O}{\overset{\|}{C}}{-}Cl} \xrightarrow{\mathrm{H_2/BaSO_4}} \mathrm{R{-}\overset{O}{\overset{\|}{C}}{-}H} + \mathrm{HCl}$$

**Sandmeyer reaction** (Section 26.4): A method for converting aryldiazonium salts into aryl halides by treatment with cuprous halide:

$$\mathrm{C_6H_5{-}\overset{+}{N}{\equiv}N\ HSO_4^-} + \mathrm{CuX} \longrightarrow \mathrm{C_6H_5{-}X} + \mathrm{N_2}$$

**Schotten–Baumann reaction** (Section 21.5): A method for preparing esters by treatment of an acid chloride with an alcohol in the presence of aqueous base:

$$\mathrm{R{-}\overset{O}{\overset{\|}{C}}{-}Cl} + \mathrm{R'OH} \xrightarrow{\mathrm{NaOH,\ H_2O}} \mathrm{R{-}\overset{O}{\overset{\|}{C}}{-}OR'}$$

**Simmons–Smith reaction** (Section 7.10): A method for preparing cyclopropanes by treating an alkene with $CH_2I_2$ and zinc–copper:

$$\mathrm{{>}C{=}C{<}} + \mathrm{CH_2I_2} \xrightarrow{\text{ZnCu}} \text{cyclopropane } (\mathrm{CH_2})$$

**Stork enamine reaction** (Section 23.13): A multistep sequence whereby ketones are converted into enamines by treatment with a secondary amine, and the enamines are then used in Michael reactions:

$$\mathrm{R{-}\overset{O}{\overset{\|}{C}}{-}\overset{|}{\underset{|}{C}}{-}H} \xrightarrow{\mathrm{HNR_2}} \mathrm{R{-}\overset{NR_2}{\overset{|}{C}}{=}C{<}} \xrightarrow[\text{2. } \mathrm{H_3O^+}]{\text{1. } \mathrm{H_2C{=}CHCOR'}} \mathrm{R{-}\overset{O}{\overset{\|}{C}}{-}\overset{|}{\underset{|}{C}}{-}CH_2CH_2\overset{O}{\overset{\|}{C}}{-}R'}$$

**Strecker amino acid synthesis** (Section 27.4): A multistep sequence for converting an aldehyde into an amino acid by initial treatment with ammonium cyanide, followed by hydrolysis:

$$\mathrm{R{-}\overset{O}{\overset{\|}{C}}{-}H} \xrightarrow{\mathrm{NH_3,\ KCN}} \mathrm{R{-}\overset{NH_2}{\overset{|}{C}H}{-}CN} \xrightarrow{\mathrm{H_3O^+}} \mathrm{R{-}\overset{NH_2}{\overset{|}{C}H}{-}COOH}$$

**Tollens test** (Section 19.5): A chemical test for detecting aldehydes by treatment with ammoniacal silver nitrate. A positive test is signaled by formation of a silver mirror on the walls of the reaction vessel.

**Walden inversion** (Section 11.1): The inversion of stereochemistry at a chiral center that occurs during $S_N2$ reactions:

$$Nu:^- + \text{C—X} \longrightarrow \text{Nu—C} + :X^-$$

**Williamson ether synthesis** (Section 18.4): A method for preparing ethers by treatment of a primary alkyl halide with an alkoxide ion:

$$RO:^- + R'CH_2X \longrightarrow R\text{—}O\text{—}CH_2R' + :X^-$$

**Wittig reaction** (Section 19.16): A general method of alkene synthesis by treatment of a ketone or aldehyde with an alkylidenetriphenylphosphorane:

$$R\text{—}\overset{O}{\overset{\|}{C}}\text{—}R' + \text{>C}{=}P(Ph)_3 \longrightarrow \overset{R}{\underset{R'}{>}}C{=}C< + (Ph)_3P{=}O$$

**Wohl degradation** (Section 24.8): A multistep reaction sequence for degrading an aldose into the next lower homolog:

$$\begin{matrix} CHO \\ | \\ CHOH \\ | \\ R \end{matrix} \xrightarrow[\text{3. NaOCH}_3]{\text{1. NH}_2\text{OH} \quad \text{2. Ac}_2\text{O}} \begin{matrix} CHO \\ | \\ R \end{matrix}$$

**Wohl–Ziegler reaction** (Section 10.5): A reaction for effecting allylic bromination by treatment of an alkene with *N*-bromosuccinimide:

$$>C{=}\overset{|}{C}\text{—}\overset{H}{\underset{|}{\overset{|}{C}}}\text{—} \xrightarrow{\text{NBS, CCl}_4} >C{=}\overset{|}{C}\text{—}\overset{Br}{\underset{|}{\overset{|}{C}}}\text{—}$$

**Wolff–Kishner reaction** (Section 19.13): A method for converting a ketone or aldehyde into the corresponding hydrocarbon by treatment with hydrazine and strong base:

$$\overset{O}{\overset{\|}{\diagup C \diagdown}} \xrightarrow{N_2H_4,\ KOH} \overset{H\ \ H}{\diagdown\diagup}\!\!\!\!\underset{\diagup\ \diagdown}{C}$$

**Woodward–Hoffmann orbital symmetry rules** (Section 30.13): A series of rules for predicting the stereochemistry of pericyclic reactions. Even-electron species react thermally through either antarafacial or conrotatory pathways, whereas odd-electron species react thermally through either suprafacial or disrotatory pathways (even–antara–con; odd–supra–dis).

**Ziegler-Natta polymerization** (Section 31.6): A method for carrying out the stereoregular polymerization of alkenes by using titanium–aluminum catalysts.

APPENDIX F

# Glossary

**Absolute configuration** (Section 9.6): The actual three-dimensional structure of a chiral molecule. Absolute configurations are specified verbally by the Cahn–Ingold–Prelog *R*,*S* convention and are represented on paper by Fischer projections.

**Absorption spectrum** (Section 12.8): A plot of wavelength of incident light versus amount of light absorbed. Organic molecules show absorption spectra in both the infrared and ultraviolet regions of the electromagnetic spectrum. By interpreting these spectra, useful structural information about the sample can be obtained. (*See* Infrared spectroscopy, Ultraviolet spectroscopy.)

**Acetal** (Section 19.14): A functional group consisting of two ether-type oxygen atoms bound to the same carbon, $R_2C(OR')_2$. Acetals are often used as protecting groups for ketones and aldehydes, since they are stable to basic and nucleophilic reagents but can be easily removed by acidic hydrolysis.

**Achiral** (Section 9.5): Having a lack of handedness. A molecule is achiral if it has a plane of symmetry and is thus superimposable on its mirror image. (*See* Chiral.)

**Activating group** (Section 16.5): An electron-donating group such as hydroxyl (—OH) or amino ($—NH_2$) that increases the reactivity of an aromatic ring toward electrophilic aromatic substitution. All activating groups are ortho- and para-directing.

**Activation energy, $\Delta E^{\ddagger}$** (Section 5.9): The difference in energy levels between ground state and transition state. The amount of activation energy required by a reaction determines the rate at which the reaction proceeds. The majority of organic reactions have activation energies of 10–25 kcal/mol.

**Acylation** (Section 16.4): The introduction of an acyl group, —COR, onto a molecule. For example, acylation of an alcohol yields an ester ($R'OH \rightarrow R'OCOR$), acylation of an amine yields an amide ($R'NH_2 \rightarrow R'NHCOR$), and acylation of an aromatic ring yields an alkyl aryl ketone ($ArH \rightarrow ArCOR$).

**Acylium ion** (Section 16.4): A resonance-stabilized carbocation in which the positive charge is located at a carbonyl-group carbon, $R—\overset{+}{C}{=}O \leftrightarrow R—C{\equiv}\overset{+}{O}$. Acylium ions are strongly electrophilic and are involved as intermediates in Friedel–Crafts acylation reactions.

**1,4 Addition** (Sections 14.5, 19.18): Addition of a reagent to the ends of a conjugated pi system. Conjugated dienes yield 1,4 adducts when treated with electrophiles such as HCl. Conjugated enones yield 1,4 adducts when treated with nucleophiles such as cyanide ion.

**Aldaric acid** (Section 24.8): The dicarboxylic acid resulting from oxidation of an aldose.

**Alditol** (Section 24.8): The polyalcohol resulting from reduction of the carbonyl group of a sugar.

**Aldonic acid** (Section 24.8): The monocarboxylic acid resulting from mild oxidation of an aldose.

**Alicyclic** (Section 3.8): Referring to an aliphatic cyclic hydrocarbon such as a cycloalkane or cycloalkene.

**Aliphatic** (Section 3.2): Referring to a nonaromatic hydrocarbon such as a simple alkane, alkene, or alkyne.

**Alkaloid** (Section 25.10): A naturally occurring compound that contains a basic amine functional group. Morphine is an example of an alkaloid.

**Alkylation** (Sections 8.8, 16.3, 18.4, 22.9): Introduction of an alkyl group onto a molecule. For example, certain aromatic rings can be alkylated to yield arenes (ArH → ArR), alkoxide anions can be alkylated to yield ethers ($R'O^- \rightarrow R'OR$), and enolate anions can be alkylated to yield alpha-substituted carbonyl compounds [$R'_2C{=}C(R')O^- \rightarrow RR'_2C{-}COR'$].

**Allylic** (Section 10.5): Used to refer to the position next to a double bond. For example, $CH_2{=}CHCH_2Br$ is an allylic bromide, and an allylic radical is a conjugated, resonance-stabilized species in which the unpaired electron is in a *p* orbital next to a double bond ($C{=}C{-}C\cdot \leftrightarrow \cdot C{-}C{=}C$).

**Angle strain** (Section 4.6): The strain introduced into a molecule when a bond angle is deformed from its ideal value. Angle strain is particularly important in small-ring cycloalkanes, where it results from compression of bond angles to less than their ideal tetrahedral values. For example, cyclopropane has approximately 22 kcal/mol angle strain owing to bond deformations from the 109° tetrahedral angle to 60°.

**Anomers** (Section 24.6): Cyclic stereoisomers of sugars that differ only in their configurations at the hemiacetal (anomeric) carbon.

**Antarafacial** (Section 30.8): A word used to describe the geometry of pericyclic reactions. An antarafacial reaction is one that takes place on opposite faces of the two ends of a pi electron system. (*See* Suprafacial.)

**Anti conformation** (Section 4.3): The geometric arrangement around a carbon–carbon single bond, in which the two largest substituents are 180° apart as viewed in a Newman projection.

**Anti stereochemistry** (Section 7.1): Referring to opposite sides of a double bond or molecule. An anti addition reaction is one in which the two ends of the double bond are attacked from different sides. For example, addition of $Br_2$ to cyclohexene yields *trans*-1,2-dibromocyclohexane, the product of anti addition. An anti elimination reaction is one in which the two groups leave from opposite sides of the molecule. (*See* Syn stereochemistry.)

**Antibonding orbital** (Section 1.7): A molecular orbital that is higher in energy than the atomic orbitals from which it is formed.

**Anticodon** (Section 29.14): A sequence of three bases on tRNA that reads the codons on mRNA and brings the correct amino acids into position for protein synthesis.

**Aromaticity** (Chapter 15): The special characteristics of cyclic conjugated pi electron systems that result from their electronic structures. These characteristics include unusual stability, the presence of a ring current in the $^1$H NMR spectrum, and a tendency to undergo substitution reactions rather than addition reactions on treatment with electrophiles. Aromatic molecules must be planar, cyclic, conjugated species that have $(4n + 2)$ pi electrons.

**Asymmetric center** (Section 9.5): *See* Chiral center.

**Atactic polymers** (Section 31.6): Chain-growth polymers that have a random stereochemical arrangement of substituents on the polymer backbone. These polymers result from high-temperature radical-initiated polymerization of alkene monomers.

**Aufbau principle** (Section 1.3): A guide for determining the ground-state electronic configuration of elements by filling the lowest-energy orbitals first.

**Axial bond** (Section 4.10): A bond to chair cyclohexane that lies along the ring axis perpendicular to the rough plane of the ring. (*See* Equatorial bond.)

Axial bonds

**Base peak** (Section 12.5): The most intense peak in a mass spectrum.

**Bent bonds** (Section 4.7): The bonds in small rings such as cyclopropane that bend away from the internuclear line and overlap at a slight angle, rather than head-on. Bent bonds are highly strained and highly reactive.

**Benzylic** (Sections 11.9, 16.11): Referring to the position next to an aromatic ring. For example, a benzylic cation is a resonance-stabilized, conjugated carbocation having its positive charge located on a carbon atom next to the benzene ring in a pi orbital that overlaps the aromatic pi system.

**Benzyne** (Section 16.10): An unstable intermediate having a triple bond in a benzene ring. Benzynes are implicated as intermediates in certain nucleophilic aromatic substitution reactions of aryl halides with strong bases.

**Betaine** (Section 19.16): A neutral dipolar molecule that has nonadjacent positive and negative charges. For example, the initial adducts of Wittig reagents with carbonyl compounds are betaines:

```
 O⁻          R
  \    |   +/
R—C—C—P—R
  /    |    \
 R           R
```

A Wittig betaine

**Bimolecular reaction** (Section 11.4): A reaction that occurs between two reagents.

**Block copolymer** (Section 31.8): A polymer consisting of alternating homopolymer blocks. Block copolymers are usually prepared by initiating chain-growth polymerization of one monomer, followed by addition of an excess of a second monomer.

**Boat cyclohexane** (Section 4.14): A three-dimensional conformation of cyclohexane that bears a slight resemblance to a boat. Boat cyclohexane has no angle strain, but has a large number of eclipsing interactions that make it less stable than chair cyclohexane:

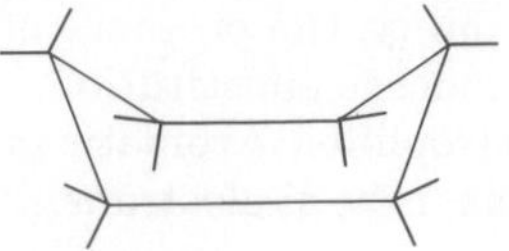

Boat cyclohexane

**Bond angle** (Section 1.8): The angle formed between two adjacent bonds.

**Bond-dissociation energy** (Section 5.8): The amount of energy needed to homolytically break a bond to produce two radical fragments.

**Bond length** (Section 1.7): The equilibrium distance between the nuclei of two atoms that are bonded to each other.

**Bond strength** (Section 1.7): *See* Bond-dissociation energy.

**Bonding orbital** (Section 1.7): A molecular orbital that is lower in energy than the atomic orbitals from which it is formed.

**Bromohydrin** (Section 7.2): The 1,2-disubstituted bromoalcohol that is obtained by addition of HOBr to an alkene.

**Brønsted acid** (Section 2.6): A substance that donates a hydrogen ion (proton) to a base.

**Carbanion** (Section 10.9): A carbon anion, or a substance that contains a trivalent, negatively charged carbon atom ($R_3C:^-$). Carbanions are $sp^3$ hybridized and have eight electrons in the outer shell of the negatively charged carbon.

**Carbene** (Section 7.10): A neutral substance that contains a divalent carbon atom having only six electrons in its outer shell ($R_2C:$).

**Carbinolamine** (Section 19.12): A molecule that contains the $R_2C(OH)NH_2$ functional group. Carbinolamines are produced as unstable intermediates during the nucleophilic addition of amines to carbonyl groups.

**Carbocation** (Section 6.9): A carbon cation, or a substance that contains a trivalent, positively charged carbon atom having six electrons in its outer shell ($R_3C^+$). Carbocations are planar and $sp^2$ hybridized.

**Carbocycle** (Section 15.9): A cyclic molecule that has only carbon atoms in the ring. (*See* Heterocycle.)

**Carbohydrate** (Chapter 24): A polyhydroxy aldehyde or polyhydroxy ketone. The name derives from the fact that glucose, the most abundant carbohydrate, has the formula $C_6H_{12}O_6$ and was originally thought to be a "hydrate of carbon." Carbohydrates can be either simple sugars such as glucose or complex sugars such as cellulose. Simple sugars are those that cannot be hydrolyzed to yield smaller molecules, whereas complex sugars are those that can be hydrolyzed to yield simpler sugars.

**Chain-growth polymer** (Section 31.1): A polymer produced by a chain-reaction procedure in which an initiator adds to a carbon–carbon double bond to yield a reactive intermediate. The chain is then built as more monomers add successively to the reactive end of the growing chain.

**Chain reaction** (Section 5.4): A reaction that, once initiated, sustains itself in an endlessly repeating cycle of propagation steps. The radical chlorination of alkanes is an example of a chain reaction that is initiated by irradiation with light and then continues in a series of propagation steps:

*Step 1* Initiation: $Cl_2 \longrightarrow 2\,Cl\cdot$

*Steps 2 and 3* Propagation: $Cl\cdot + CH_4 \longrightarrow HCl + \cdot CH_3$

$CH_3 + Cl_2 \longrightarrow CH_3Cl + Cl\cdot$

*Step 4* Termination: $R\cdot + R\cdot \longrightarrow R{-}R$

**Chair cyclohexane** (Section 4.9): A three-dimensional conformation of cyclohexane that resembles the rough shape of a chair. The chair form of cyclohexane, which has neither angle strain nor eclipsing strain, represents the lowest-energy conformation of the molecule:

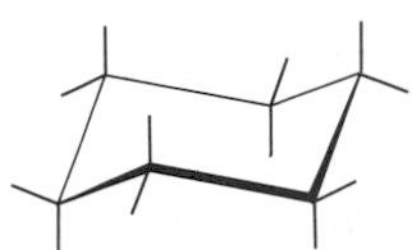

Chair cyclohexane

**Chemical shift** (Section 13.3): The position on the NMR chart where a nucleus absorbs. By convention, the chemical shift of tetramethylsilane (TMS) is arbitrarily set at zero, and all other absorptions usually occur downfield (to the left on the chart). Chemical shifts are expressed in delta units, $\delta$, where 1 $\delta$ equals 1 ppm of the spectrometer operating frequency. For example, 1 $\delta$ on a 60 MHz instrument equals 60 Hz. The chemical shift of a given nucleus is related to the chemical environment of that nucleus in the molecule, thus allowing one to obtain structural information by interpreting the NMR spectrum.

**Chiral** (Section 9.5): Having handedness. Chiral molecules are those that do not have a plane of symmetry and are therefore not superimposable on their mirror image. A chiral molecule thus exists in two forms, one right-handed and one left-handed. The most common (though not the only) cause of chirality in a molecule is the presence of a carbon atom that is bonded to four different substituents. (*See* Achiral.)

**Chiral center** (Section 9.5): An atom (usually carbon) that is bonded to four different groups and is therefore chiral. (*See* Chiral.)

**Chlorohydrin** (Section 7.2): The 1,2-disubstituted chloroalcohol that is obtained by addition of HOCl to an alkene.

**Chromatography** (Section 12.1): A technique for separating a mixture of compounds into pure components. Chromatography operates on a principle of differential adsorption whereby different compounds adsorb to a stationary support phase and are then carried along at different rates by a mobile phase.

**Cis–trans isomers** (Sections 3.10, 6.5): Special kinds of stereoisomers that differ in their stereochemistry about a double bond or on a ring. Cis–trans isomers are also called geometric isomers.

**Codon** (Section 29.14): A three-base sequence on the mRNA chain that encodes the genetic information necessary to cause specific amino acids to be incorporated into proteins. Codons on mRNA are read by complementary anticodons on tRNA.

**Concerted** (Section 30.1): Referring to a reaction that takes place in a single step without intermediates. For example, the Diels–Alder cycloaddition reaction is a concerted process.

**Configuration** (Section 9.6): The three-dimensional arrangement of atoms bonded to a chiral center relative to the stereochemistry of other chiral centers in the same molecule.

**Conformation** (Section 4.1): The exact three-dimensional shape of a molecule at any given instant, assuming that rotation around single bonds is frozen.

**Conformational analysis** (Section 4.13): A means of assessing the minimum-energy conformation of a substituted cycloalkane by totaling the steric interactions present in the molecule. Conformational analysis is particularly useful in assessing the relative stabilities of different conformations of substituted cyclohexane rings.

**Conjugate addition** (Section 19.18): Addition of a nucleophile to the $\beta$-carbon atom of an $\alpha,\beta$-unsaturated carbonyl compound. (*See* 1,4-Addition.)

**Conjugate base** (Section 2.6): The anion that results from dissociation of a Brønsted acid.

**Conjugation** (Section 14.1): A series of alternating single and multiple bonds with overlapping *p* orbitals. For example, 1,3-butadiene is a conjugated diene, 3-buten-2-one is a conjugated enone, and benzene is a cyclic conjugated triene.

**Conrotatory** (Section 30.5): A term used to indicate the fact that *p* orbitals must rotate in the same direction during electrocyclic ring opening or ring closure. (*See* Disrotatory.)

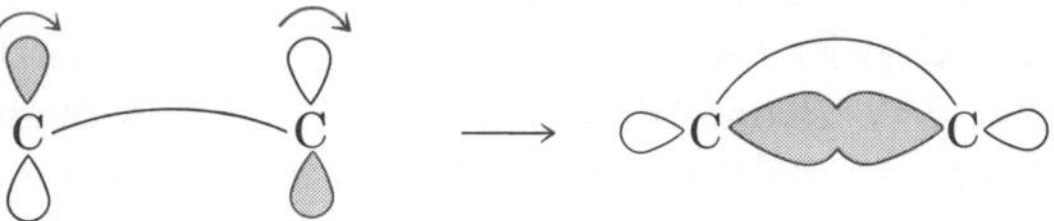

**Constitutional isomers** (Sections 3.3, 9.8): Isomers that have their atoms connected in a different order. For example, butane and 2-methylpropane are constitutional isomers.

**Copolymer** (Section 31.8): A polymer formed by chain-growth polymerization of a mixture of two or more different monomer units.

**Coupling constant** (Section 13.12): The magnitude (expressed in hertz) of the spin–spin splitting interaction between nuclei whose spins are coupled. Coupling constants are denoted *J*.

**Covalent bond** (Section 1.6): A bond formed by sharing electrons between two nuclei. (*See* Ionic bond.)

**Cracking** (Section 3.6): A process used in petroleum refining in which large alkanes are thermally cracked into smaller fragments.

**Cycloaddition** (Sections 14.7, 30.1): A pericyclic reaction in which two reactants add together in a single step to yield a cyclic product. The Diels–Alder reaction between a diene and a dienophile to give a cyclohexene is the best-known example of a cycloaddition.

**Deactivating group** (Section 16.5): An electron-withdrawing substituent that decreases the reactivity of an aromatic ring toward electrophilic aromatic substitution. Most deactivating groups, such as nitro, cyano, and carbonyl are meta-directors, but halogen substituents are ortho- and para-directors.

**Decarboxylation** (Sections 20.9, 22.9): A reaction that involves loss of carbon dioxide from the starting material. $\beta$-Keto acids decarboxylate particularly readily on heating.

**Degenerate orbitals** (Section 15.6): Two or more orbitals that have the same energy level.

**Dehydration** (Sections 7.12, 17.8): A reaction that involves loss of water from the starting material. Most alcohols can be dehydrated to yield alkenes, but aldol condensation products ($\beta$-hydroxy ketones) dehydrate particularly readily.

**Dehydrohalogenation** (Sections 7.12, 11.10): A reaction that involves loss of HX from the starting material. Alkyl halides undergo dehydrohalogenation to yield alkenes on treatment with strong base.

**Delocalization** (Section 10.6): A spreading out of electron density over a conjugated pi electron system. For example, allylic cations and allylic anions are delocalized because their charges are spread out by resonance stabilization over the entire pi electron system.

**Denaturation** (Section 27.15): The physical changes that occur in proteins when secondary and tertiary structures are disrupted. Denaturation is usually brought about by heat treatment or by a change in pH and is accompanied by a loss of biological activity.

**Deshielding** (Section 13.3): An effect observed in NMR that causes a nucleus to absorb downfield (to the left) of TMS standard. Deshielding is caused by a withdrawal of electron density from the nucleus and is responsible for the observed chemical shifts of vinylic and aromatic protons.

**Deuterium isotope effect** (Section 11.13): A tool for use in mechanistic investigations to establish whether or not a C—H bond is broken in the rate-limiting step of a reaction. Since carbon–deuterium bonds are stronger and less easily broken than carbon–protium bonds, one can measure the reaction rates of both protium- and deuterium-substituted substrates and see if they are the same. If they are different, a deuterium isotope effect is present, indicating that C—H bond breakage is rate-limiting.

**Dextrorotatory** (Section 9.1): A word used to describe an optically active substance that rotates the plane of polarization of plane-polarized light in a right-handed (clockwise) direction. The direction of rotation is not related to the absolute configuration of the molecule. (*See* Levorotatory.)

**Diastereomer** (Section 9.7): A term that indicates the relationship between non-mirror-image stereoisomers. Diastereomers are stereoisomers that have the same configuration at one or more chiral centers, but differ at other chiral centers.

**Diazotization** (Section 26.4): The conversion of a primary amine, $RNH_2$, into a diazonium salt, $RN_2^+$, by treatment with nitrous acid. Aryl diazonium salts are stable, but alkyl diazonium salts are extremely reactive and are rarely isolable.

**Dielectric constant** (Section 11.9): A measure of the ability of a solvent to act as an insulator of electric charge. Solvents that have high dielectric constants are highly polar and are particularly valuable in $S_N1$ reactions because of their ability to stabilize the developing positive charge of the intermediate carbocation.

**Dienophile** (Section 14.8): A compound containing a double bond that can take part in the Diels–Alder cycloaddition reaction. The most reactive dienophiles are those that have electron-withdrawing groups such as nitro, cyano, or carbonyl on the double bond.

**Dipolar aprotic solvent** (Section 11.5): Dipolar solvent that cannot function as a hydrogen ion donor. Dipolar aprotic solvents such as dimethyl sulfoxide (DMSO), hexamethylphosphoramide (HMPA), and dimethylformamide (DMF) are particularly useful in $S_N2$ reactions because of their ability to solvate cations.

**Dipole moment, $\mu$** (Section 2.5): A measure of the polarity of a molecule. A dipole moment arises when the centers of gravity of positive and negative charges within a molecule do not coincide.

**Disrotatory** (Section 30.5): A term used to indicate the fact that *p* orbitals rotate in opposite directions during electrocyclic ring opening or ring closing. (*See* Conrotatory.)

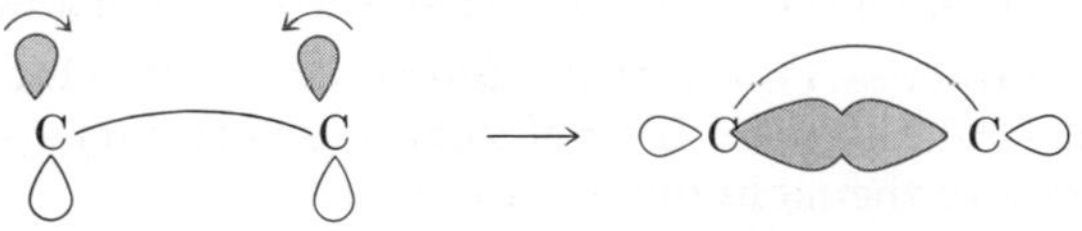

**d,l form** (Section 9.11): A shorthand way of indicating the racemic modification of a compound. (*See* Racemic mixture.)

**DNA** (Section 29.9): Deoxyribonucleic acid, the biopolymer consisting of deoxyribonucleotide units linked together through phosphate–sugar bonds. DNA, which is found in the nucleus of cells, contains an organism's genetic information.

**Doublet** (Section 13.6): A two-line NMR absorption caused by spin–spin splitting when the spin of the nucleus under observation couples with the spin of a neighboring magnetic nucleus.

**Downfield** (Section 13.3): Refers to the left-hand portion of the NMR chart. (*See* Deshielding.)

**Eclipsed conformation** (Section 4.1): The geometric arrangement around a carbon–carbon single bond in which the bonds to substituents on one carbon are parallel to the bonds to substituents on the neighboring carbon as viewed in a Newman projection. For example, the eclipsed conformation of ethane has the C—H bonds on one carbon lined up with the C—H bonds on the neighboring carbon:

Eclipsed conformation

**Eclipsing strain** (Section 4.1): The strain energy in a molecule caused by electron repulsions between eclipsed bonds. Eclipsing strain is also called torsional strain.

**Elastomers** (Section 31.13): Amorphous polymers that have the ability to stretch out and then return to their previous shape. These polymers have irregular shapes that prevent crystallite formation and have little cross-linking between chains.

**Electrocyclic reaction** (Section 30.1): A unimolecular pericyclic reaction in which a ring is formed or broken by a concerted reorganization of electrons through a cyclic transition state. For example, the cyclization of 1,3,5-hexatriene to yield 1,3-cyclohexadiene is an electrocyclic reaction.

**Electromagnetic spectrum** (Section 12.8): The range of electromagnetic energy, including infrared, ultraviolet, and visible radiation.

**Electron affinity** (Section 1.5): The measure of the tendency of an atom to gain an electron and form an anion. Elements on the right side of the periodic table, such as the halogens, have higher electron affinities than elements on the left side.

**Electronegativity** (Section 2.4): The ability of an atom to attract electrons and thereby polarize a bond. As a general rule, electronegativity increases in going across the periodic table from left to right and in going from bottom to top.

**Electrophile** (Section 5.5): An "electron-lover," or substance that accepts an electron pair from a nucleophile in a polar bond-forming reaction. (*See* Nucleophile.)

**Electrophoresis** (Section 27.3): A technique used for separating charged organic molecules, particularly proteins and amino acids. The mixture to be separated is placed on a buffered gel or paper, and an electric potential is applied across the ends of the apparatus. Negatively charged molecules migrate toward the positive electrode, and positively charged molecules migrate toward the negative electrode.

**Elution** (Section 12.2): The removal of a substance from a chromatography column.

**Empirical formula** (Section 2.9): A formula that gives the relative proportions of elements in a compound in smallest whole numbers.

**Enantiomers** (Section 9.3): Stereoisomers of a chiral substance that have a mirror-image relationship. Enantiomers must have opposite configurations at all chiral centers in the molecule.

**Endothermic** (Section 5.7): A term used to describe reactions that absorb energy and therefore have positive enthalpy changes. In reaction energy diagrams, the products of endothermic reactions have higher energy levels than the starting materials.

***Entgegen, E*** (Section 6.6): A term used to describe the stereochemistry of a carbon–carbon double bond. The two groups on each carbon are first assigned priorities according to the Cahn–Ingold–Prelog sequence rules, and the two carbons are then compared. If the high-priority groups on each carbon are on opposite sides of the double bond, the bond has $E$ geometry. (*See Zusammen.*)

**Enthalpy change, $\Delta H°$** (Section 5.7): The heat of reaction. The enthalpy change that occurs during a reaction is a measure of the difference in total bond energy between reactants and products.

**Entropy change, $\Delta S°$** (Section 5.7): The amount of disorder. The entropy change that occurs during a reaction is a measure of the difference in disorder between reactants and products.

**Enzyme** (Chapter 27): A biological catalyst. Enzymes are large proteins that catalyze specific biochemical reactions.

**Epoxide** (Section 18.7): A three-membered-ring ether functional group.

**Equatorial bond** (Section 4.10): A bond to cyclohexane that lies along the rough equator of the ring. (*See* Axial bond.)

Equatorial bonds

**Equilibrium constant** (Section 2.6): A measure of the equilibrium position for a reaction. The equilibrium constant, $K_{eq}$ for the reaction A + B → C + D is given by the expression

$$K_{eq} = \frac{[C][D]}{[A][B]}$$

where the letters in brackets refer to the molar concentrations of the reactants and products.

**Essential oil** (Section 28.6): The volatile oil that is obtained by steam distillation of a plant extract.

**Excited-state configuration** (Section 1.8): An electronic configuration having a higher energy level than the ground state. Excited states are normally obtained by excitation of an electron from a bonding orbital to an antibonding one, such as occurs during irradiation of a molecule with light of the proper frequency.

**Exothermic** (Section 5.7): A term used to describe reactions that release energy and therefore have negative enthalpy changes. On reaction energy diagrams, the products of exothermic reactions have energy levels lower than those of starting materials.

**Fat** (Section 28.1): A solid triacylglycerol derived from animal sources.

**Fibers** (Section 31.13): Thin threads produced by extruding a molten polymer through small holes in a die.

**Fibrous protein** (Section 27.13): Proteins that consist of polypeptide chains arranged side by side in long threads. These proteins are tough, insoluble in water, and occur in nature in structural materials such as hair, hooves, and fingernails.

**Fingerprint region** (Section 12.10): The complex region of the infrared spectrum from 1500 cm$^{-1}$ to 400 cm$^{-1}$. If two substances have identical absorption patterns in the fingerprint region of the IR, they are almost certainly identical.

**Fischer projection** (Section 9.13): A means of depicting the absolute configuration of chiral molecules on a flat page. A Fischer projection employs a cross to represent the chiral center; the horizontal arms of the cross represent bonds coming out of the plane of the page, whereas the vertical arms of the cross represent bonds going back into the plane of the page:

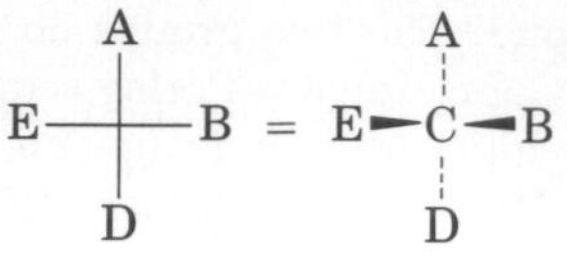

A, E, B, D = A, E, C, B, D

Fischer projection

**Formal charge** (Section 2.3): The difference in the number of electrons possessed by an atom in a molecule and by the same atom in its elemental state. The formal charge on an atom is given by the formula

$$\text{Formal charge} = \begin{pmatrix}\text{Number of outer-shell} \\ \text{electrons in a free atom}\end{pmatrix} - \begin{pmatrix}\text{Number of outer-shell} \\ \text{electrons in a bound atom}\end{pmatrix}$$

**Frequency** (Section 12.8): The number of electromagnetic wave cycles that travel past a fixed point in a given unit of time. Frequencies are usually expressed in units of cycles per second, or hertz.

**Functional group** (Section 3.1): An atom or group of atoms that is part of a larger molecule and has a characteristic chemical reactivity. Functional groups display the same chemistry in all molecules of which they are a part.

**Gated-decoupled mode** (Section 13.5): A mode of $^{13}C$ NMR spectrometer operation in which all one-carbon resonances are of equal intensity. Operating in this mode allows one to integrate the spectrum to find out how many of each kind of carbon atom is present.

**Gauche conformation** (Section 4.3): The conformation of butane in which the two methyl groups lie 60° apart as viewed in a Newman projection. This conformation has 0.9 kcal/mol steric strain:

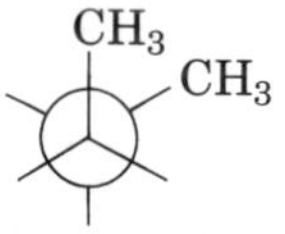

$CH_3$ $CH_3$

Gauche conformation

**Geometric isomers** (Sections 3.10, 6.5): *See* Cis–trans isomers.

**Gibbs free-energy change, $\Delta G°$** (Section 5.7): The total amount of free-energy change, both enthalpy and entropy, that occurs during a reaction. The standard Gibbs free-energy change for a reaction is given by the formula $\Delta G° = \Delta H° - T\Delta S°$.

**Globular protein** (Section 27.13): Proteins that are coiled into compact, nearly spherical shapes. These proteins, which are generally water-soluble and mobile within the cell, are the structural class to which enzymes belong.

**Glycol** (Section 7.7): A 1,2-diol such as ethylene glycol, $HOCH_2CH_2OH$.

**Glycoside** (Section 24.8): A cyclic acetal formed by reaction of a sugar with another alcohol.

**Graft copolymer** (Section 31.8): A copolymer that consists of homopolymer chains grafted onto a different homopolymer backbone. Graft copolymers are prepared by X-ray irradiation of a homopolymer to generate radical sites along the chain, followed by addition of a second monomer.

**Ground state** (Section 1.3): The most stable, lowest-energy electronic configuration of a molecule.

**Halohydrin** (Section 7.2): A 1,2-disubstituted haloalcohol such as is obtained on addition of HOBr to an alkene.

**Halonium ion** (Section 7.1): A species containing a positively charged, divalent halogen. Three-membered-ring bromonium ions are implicated as intermediates in the electrophilic addition of bromine to alkenes.

**Hammond postulate** (Section 6.12): A postulate stating that we can get a picture of what a given transition state looks like by looking at the structure of the nearest stable species. Exothermic reactions have transition states that resemble starting material, whereas endothermic reactions have transition states that resemble products.

**Haworth projection** (Section 24.5): A means of viewing stereochemistry in cyclic hemiacetal forms of sugars. Haworth projections are drawn so that the ring is flat and is viewed from an oblique angle with the hemiacetal oxygen at the upper right.

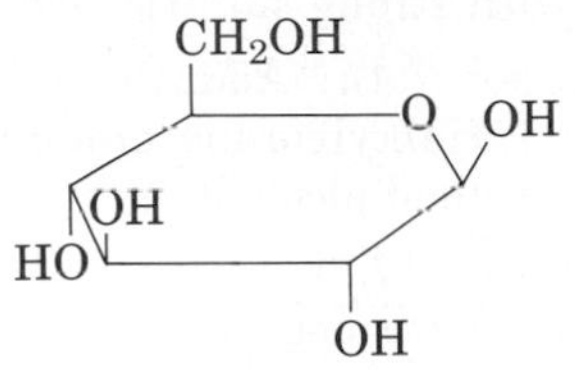

$CH_2OH$, O, OH, OH, HO, OH

Haworth projection of glucose

**Heat of combustion** (Section 4.5): The amount of heat released when a compound is burned in a calorimeter according to the equation

$$C_nH_m + O_2 \longrightarrow n\,CO_2 + \frac{m}{2}\,H_2O$$

**Heat of hydrogenation** (Section 6.7): The amount of heat released when a carbon–carbon double bond is hydrogenated. Comparison of heats of hydrogenation for different alkenes allows one to determine the stability of the different double bonds.

**Heterocycle** (Section 15.9, Chapter 29): A cyclic molecule whose ring contains more than one kind of atom. For example, pyridine is a heterocycle that contains five carbon atoms and one nitrogen atom in its ring.

**Heterogenic bond formation** (Section 5.2): What occurs when one partner donates both electrons in forming a new bond. Polar reactions always involve heterogenic bond formation:

$$A^+ + B{:}^- \longrightarrow A{:}B$$

**Heterolytic bond breakage** (Section 5.2): The kind of bond breaking that occurs in polar reactions when one fragment leaves with both of the bonding electrons, as in the equation

$$A{:}B \longrightarrow A^+ + B{:}^-$$

**HOMO** (Section 30.5): An acronym for highest occupied molecular orbital. The symmetries of the HOMO and LUMO are important in pericyclic reactions. (*See* LUMO.)

**Homogenic bond formation** (Section 5.2): What occurs in radical reactions when each partner donates one electron to the new bond:

$$A\cdot + B\cdot \longrightarrow A{:}B$$

**Homolytic bond breakage** (Section 5.2): The kind of bond breaking that occurs in radical reactions when each fragment leaves with one bonding electron according to the equation

$$A{:}B \longrightarrow A\cdot + B\cdot$$

**Homopolymer** (Section 31.8): A polymer made by chain-growth polymerization of a single monomer unit.

**Hückel's rule** (Section 15.7): A rule stating that monocyclic conjugated molecules having $(4n + 2)$ pi electrons ($n$ = an integer) show the unusual stability associated with aromaticity.

**Hybrid orbital** (Section 1.8): An orbital that is mathematically derived from a combination of ground-state ($s$, $p$, $d$) atomic orbitals. Hybrid orbitals, such as the $sp^3$, $sp^2$, and $sp$ hybrids of carbon, are strongly directed and form stronger bonds than ground-state atomic orbitals.

**Hydration** (Section 7.3): Addition of water to a molecule, such as occurs when alkenes are treated with strong sulfuric acid.

**Hydroboration** (Section 7.4): Addition of borane ($BH_3$) or an alkylborane to an alkene. The resultant trialkylborane products are useful synthetic intermediates that can be oxidized to yield alcohols.

**Hydrogen bond** (Section 17.3): A weak (5 kcal/mol) attraction between a hydrogen atom bonded to an electronegative element and an electron lone pair on another atom. Hydrogen bonding plays an important role in determining the secondary structure of proteins and in stabilizing the DNA double helix.

**Hydrogenation** (Section 7.6): Addition of hydrogen to a double or triple bond to yield the saturated product.

**Hyperconjugation** (Section 6.7): A weak stabilizing interaction that results from overlap of a $p$ orbital with a neighboring sigma bond. Hyperconjugation is important in stabilizing carbocations and in stabilizing substituted alkenes.

**Inductive effect** (Section 2.4): The electron-attracting or electron-withdrawing effect that is transmitted through sigma bonds as the result of a nearby dipole. Electronegative elements have an electron-withdrawing inductive effect, whereas electropositive elements have an electron-donating inductive effect.

**Infrared spectroscopy** (Section 12.8): A kind of optical spectroscopy that uses infrared energy. IR spectroscopy is particularly useful in organic chemistry for determining the kinds of functional groups present in molecules.

**Initiator** (Section 5.4): A substance with an easily broken bond that is used to initiate radical chain reactions. For example, radical chlorination of alkanes is initiated when light energy breaks the weak chlorine–chlorine bond to form chlorine radicals.

**Intermediate** (Section 5.10): A species that is formed during the course of a multistep reaction but is not the final product. Intermediates are more stable than transition states, but may or may not be stable enough to isolate.

**Intramolecular, intermolecular** (Section 23.7): Reactions that occur within the same molecule are intramolecular, whereas reactions that occur between two molecules are intermolecular.

**Ion pair** (Section 11.8): A loose complex between two ions in solution. Ion pairs are implicated as intermediates in $S_N1$ reactions in order to account for the partial retention of stereochemistry that is often observed.

**Ionic bond** (Section 1.5): A bond between two ions due to the electrical attraction of unlike charges. Ionic bonds are formed between strongly electronegative elements (such as the halogens) and strongly electropositive elements (such as the alkali metals). (*See* Covalent bond.)

**Ionization energy** (Section 1.5): The amount of energy required to remove an electron from an atom. Elements on the far right of the periodic table have high ionization energies, and elements on the far left of the periodic table have low ionization energies.

**Isoelectric point** (Section 27.3): The pH at which the number of positive charges and the number of negative charges on a protein or amino acid are exactly balanced.

**Isomers** (Section 3.3): Compounds that have the same molecular formula but different structures.

**Isoprene rule** (Section 28.7): An observation to the effect that terpenoids appear to be made up of isoprene (2-methyl-1,3-butadiene) units connected in a head-to-tail fashion. Monoterpenes have two isoprene units, sesquiterpenes have three isoprene units, diterpenes have four isoprene units, and so on.

**Isotactic polymer** (Section 31.6): A chain-growth polymer in which all substituents on the polymer backbone have the same three-dimensional orientation.

**Kekulé structure** (Sections 1.6, 2.1): A representation of molecules in which a line between atoms is used to represent a bond. (*See* Line-bond structure.)

**Kinetic control** (Section 14.6): Reactions that follow the lowest activation energy pathway are said to be kinetically controlled. The product formed in a kinetically controlled reaction is the one that is formed most rapidly, but is not necessarily the most stable. (*See* Thermodynamic control.)

**Kinetics** (Section 11.3): Referring to rates of reactions. Kinetics measurements can be extremely important in helping to determine reaction mechanisms.

**Leaving group** (Section 11.5): The group that is replaced in a substitution reaction. The best leaving groups in nucleophilic substitution reactions are those that form the most stable, least basic anions.

**Levorotatory** (Section 9.1): Used to describe an optically active substance that rotates the plane of polarization of plane-polarized light in a left-handed (counterclockwise) direction. (*See* Dextrorotatory.)

**Lewis acid** (Section 2.8): A substance having a vacant low-energy orbital that can accept an electron pair from a base. All electrophiles are Lewis acids, but transition metal salts such as $AlCl_3$ and $ZnCl_2$ are particularly good ones. (*See* Lewis base.)

**Lewis base** (Section 2.8): A substance that donates an electron lone pair to an acid. All nucleophiles are Lewis bases. (*See* Lewis acid.)

**Lewis structure** (Section 1.6): A representation of a molecule showing covalent bonds as a pair of electron dots between atoms.

**Line-bond structure** (Section 2.1): A representation of a molecule showing covalent bonds as lines between atoms. (*See* Kekulé structure.)

**Lipid** (Chapter 28): A naturally occurring substance isolated from cells and tissues by extraction with nonpolar solvents. Lipids belong to many different structural classes, including fats, terpenes, prostaglandins, and steroids.

**Lipophilic** (Section 28.2): Fat-loving. Long, nonpolar hydrocarbon chains tend to cluster together in polar solvents because of their lipophilic properties.

**Lone-pair electrons** (Section 1.12): Nonbonding electron pairs that occupy valence orbitals. It is the lone-pair electrons that are used by nucleophiles in their reactions with electrophiles.

**LUMO** (Section 30.6): An acronym for lowest unoccupied molecular orbital. The symmetries of the LUMO and HOMO are important in determining the stereochemistry of pericyclic reactions. (*See* HOMO.)

**Magnetic equivalence** (Section 13.9): Used to describe nuclei that have identical chemical and magnetic environments, and that therefore absorb at the same place in the NMR spectrum. For example, the six hydrogens in benzene are magnetically equivalent, as are the six carbons.

**Markovnikov's rule** (Section 6.10): A guide for determining the regiochemistry (orientation) of electrophilic addition reactions. In the addition of HX to an alkene, the hydrogen atom becomes bonded to the alkene carbon that has fewer alkyl substituents. A modern statement of this same rule is that electrophilic addition reactions proceed via the most stable carbocation intermediate.

**Mechanism** (Section 5.2): A complete description of how a reaction occurs. A mechanism must account for all starting materials and all products, and must describe the details of each individual step in the overall reaction process.

**Meso** (Section 9.9): A meso compound is one that contains chiral centers but is nevertheless achiral by virtue of a symmetry plane. For example, (2*R*,3*S*)-butanediol has two chiral carbon atoms, but is achiral because of a symmetry plane between carbons 2 and 3.

**Micelle** (Section 28.2): A spherical cluster of soap-like molecules that aggregate in aqueous solution. The ionic heads of the molecules lie on the outside where they are solvated by water, and the organic tails bunch together on the inside of the micelle.

**Mobile phase** (Section 12.1): The solvent (either gas or liquid) used in chromatography to move material along the solid adsorbent phase. (*See* Chromatography; Stationary phase.)

**Molecular formula** (Section 2.9): An expression of the total number of each kind of atom present in a molecule. The molecular formula must be a whole-number multiple of the empirical formula.

**Molecular ion** (Section 12.5): The cation produced in the mass spectrometer by loss of an electron from the parent molecule. The mass of the molecular ion corresponds to the molecular weight of the sample.

**Molecular orbital** (Section 1.7): An orbital that is the property of the entire molecule rather than of an individual atom. Molecular orbitals result from overlap of two or more atomic orbitals when bonds are formed and may be either bonding, nonbonding, or antibonding. Bonding molecular orbitals are lower in energy than the starting atomic orbitals, nonbonding MO's are equal in energy to the starting orbitals, and antibonding orbitals are higher in energy.

**Monomer** (Section 31.2): The simple starting units from which polymers are made.

**Multiplet** (Section 13.6): A symmetrical pattern of peaks in an NMR spectrum that arises by spin–spin splitting of a single absorption because of coupling between neighboring magnetic nuclei.

**Mutarotation** (Section 24.6): The spontaneous change in optical rotation observed when a pure anomer of a sugar is dissolved in water. Mutarotation is caused by the reversible opening and closing of the acetal linkage, which yields an equilibrium mixture of anomers.

***n* + 1 rule** (Sections 13.6, 13.12): A carbon bonded to $n$ hydrogens shows $n + 1$ peaks in its spin-coupled $^{13}C$ NMR spectrum, and a hydrogen with $n$ other hydrogens on neighboring carbons shows $n + 1$ peaks in its $^{1}H$ NMR spectrum.

**Neighboring-group effect** (Section 24.8): The effect on a reaction of a nearby functional group.

**Newman projection** (Section 4.1): A means of indicating stereochemical relationships between substituent groups on neighboring carbons. The carbon–carbon bond is viewed end-on, and the carbons are indicated by a circle. Bonds radiating from the center of the circle are attached to the front carbon, and bonds radiating from the edge of the circle are attached to the rear carbon:

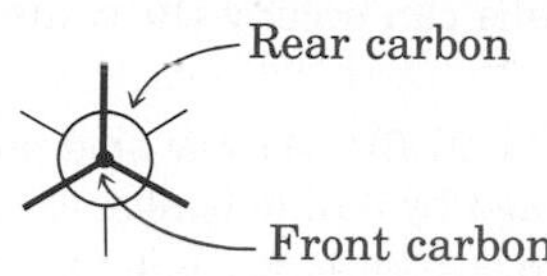


Newman projection

**Nitrogen rule** (Section 25.11): A rule stating that compounds having an odd number of nitrogens give rise to an odd-numbered molecular ion in the mass spectrum. Conversely, compounds with an even number of nitrogens give rise to even-numbered molecular ions.

**Node** (Sections 1.2, 14.3): The surface of zero electron density between lobes of orbitals. For example, a *p* orbital has a nodal plane passing through the center of the nucleus, perpendicular to the line of the orbital.

**Normal alkane** (Section 3.2): A straight-chain alkane, as opposed to a branched alkane. Normal alkanes are denoted by the suffix *n*, as in $n$-$C_4H_{10}$ ($n$-butane).

**Nuclear magnetic resonance, NMR** (Chapter 13): A spectroscopic technique that provides information about the carbon–hydrogen framework of a molecule. NMR works by detecting the energy absorption accompanying the transition between nuclear spin states that occurs when a molecule is placed in a strong magnetic field and irradiated with radio-frequency waves. Different nuclei within a molecule are in slightly different magnetic environments and therefore show absorptions at slightly different frequencies.

**Nucleophile** (Section 5.5): A "nucleus-lover," or species that donates an electron pair to an electrophile in a polar bond-forming reaction. Nucleophiles are also Lewis bases. (*See* Electrophile.)

**Nucleoside** (Section 29.9): A nucleic acid constituent, consisting of a sugar residue bonded to a heterocyclic purine or pyrimidine base.

**Nucleotide** (Section 29.9): A nucleic acid constituent, consisting of a sugar residue bonded both to a heterocyclic purine or pyrimidine base and to a phosphoric acid. Nucleotides are the monomer units from which DNA and RNA are constructed.

**Off-resonance mode** (Section 13.6): A mode of $^{13}C$ NMR spectrometer operation that allows for the observation of spin–spin splitting between carbons and their attached hydrogens. Carbons bonded to one hydrogen show a doublet; carbons attached to two hydrogens show a triplet; and carbons attached to three hydrogens show a quartet in the off-resonance NMR.

**Olefin**: An alternative name for an alkene.

**Optical isomers** (Section 9.3): *See* Enantiomers.

**Optically active** (Section 9.1): A substance that rotates the plane of polarization of plane-polarized light. Note that an optically active sample must contain chiral molecules, but all samples with chiral molecules are not optically active. Thus, a racemic sample is optically inactive even though the individual molecules are chiral. (*See* Chiral.)

**Orbital** (Section 1.2): The volume of space in which an electron is most likely to be found. Orbitals are described mathematically by wave functions, which delineate the behavior of electrons around nuclei.

**Ozonide** (Section 7.8): The product formed by addition of ozone to a carbon–carbon double bond. Ozonides are usually treated with a reducing agent such as zinc in acetic acid to produce carbonyl compounds.

**Paraffins** (Section 3.7): An alternative name for alkanes.

**Pauli exclusion principle** (Section 1.3): A statement of the fact that no more than two electrons can occupy the same orbital, and those two must have spins of opposite sign.

**Peptides** (Section 27.6): Amino acid polymers in which the individual amino acid residues are linked by amide bonds. (*See* Protein.)

**Pericyclic reaction** (Chapter 30): A reaction that occurs by a concerted reorganization of bonding electrons in a cyclic transition state.

**Periplanar** (Section 11.10): A conformation in which bonds to neighboring atoms have a parallel arrangement. In an eclipsed conformation, the neighboring bonds are syn periplanar; in a staggered conformation, the bonds are anti periplanar:

Anti periplanar    Syn periplanar

**Peroxide** (Section 18.2): A molecule containing an oxygen–oxygen bond functional group, R—O—O—R′ or R—O—O—H. The "peroxides" present as explosive impurities in ether solvents are usually of the latter type. Since the oxygen–oxygen bond is weak and easily broken, peroxides are often used to initiate radical chain reactions.

**Phase-transfer catalysts** (Section 25.9): Agents that cause the transfer of ionic reagents between phases, thus catalyzing reactions. Tetraalkylammonium salts, $R_4N^+\ X^-$, are often used to transport inorganic anions from the aqueous phase to the organic phase where the desired reaction then occurs. For example, permanganate ion is solubilized in benzene in the presence of tetraalkylammonium ions.

**Phospholipid** (Section 28.3): Lipids that contain a phosphate residue. For example, phosphoglycerides contain a glycerol backbone linked to two fatty acids and a phosphoric acid.

**Pi bond** (Section 1.7): The covalent bond formed by sideways overlap of atomic orbitals. For example, carbon–carbon double bonds contain a pi bond formed by sideways overlap of two *p* orbitals.

**Plane of symmetry** (Section 9.5): An imaginary plane that bisects a molecule such that one half of the molecule is the mirror image of the other half. Molecules containing a plane of symmetry are achiral.

**Plane-polarized light** (Section 9.1): Ordinary light that has its electric vectors in a single plane rather than in random planes. The plane of polarization is rotated when the light is passed through a solution of a chiral substance.

**Polar reaction** (Section 5.2): A reaction in which bonds are made when a nucleophile donates two electrons to an electrophile and bonds are broken when one fragment leaves with both electrons from the bond. Polar reactions are the most common class of reactions. (*See* Heterogenic bond formation; Heterolytic bond breakage.)

**Polarity** (Sections 2.4, 5.5): The unsymmetrical distribution of electrons in molecules that results when one atom attracts electrons more strongly than another.

**Polarizability** (Section 5.5): The measure of the change in the electron distribution in a molecule in response to changing electric interactions with solvents or ionic reagents.

**Polymer** (Chapter 31): A large molecule made up of repeating smaller units. For example, polyethylene is a synthetic polymer made from repeating ethylene units, and DNA is a biopolymer made of repeating deoxyribonucleotide units.

**Primary, secondary, tertiary, quaternary** (Section 3.5): Terms used to describe the substitution pattern at a specific site. A primary site has one organic substituent attached to it, a secondary site has two organic substituents, a tertiary site has three, and a quaternary site has four:

| | *Carbon* | *Hydrogen* | *Alcohol* | *Amine* |
|---|---|---|---|---|
| Primary | $RCH_3$ | $RCH_3$ | $RCH_2OH$ | $RNH_2$ |
| Secondary | $R_2CH_2$ | $R_2CH_2$ | $R_2CHOH$ | $R_2NH$ |
| Tertiary | $R_3CH$ | $R_3CH$ | $R_3COH$ | $R_3N$ |
| Quaternary | $R_4C$ | | | |

where R = Any organic substituent.

**Primary structure** (Section 27.14): The amino acid sequence in a protein. (*See* Secondary structure; Tertiary structure.)

**Principle of maximum overlap** (Section 1.7): The strongest bonds are formed when overlap between orbitals is greatest.

**Propagation step** (Section 5.4): The step or series of steps in a radical chain reaction that carry on the chain. The propagation steps must yield both product and a reactive intermediate to carry on the chain.

**Prostaglandin** (Section 28.5): A member of the class of lipids with the general carbon skeleton

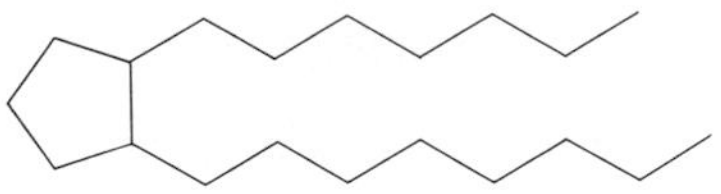

Prostaglandins are present in nearly all body tissues and fluids, where they serve a large number of important hormonal functions.

**Protecting group** (Section 17.10): A group that is introduced to protect a sensitive functional group from reaction elsewhere in the molecule. After serving its protective function, the group is then removed. For example, ketones and aldehydes are often protected as acetals by reaction with ethylene glycol, and alcohols are often protected as tetrahydropyranyl ethers.

**Protein** (Section 27.13): A large peptide, containing fifty or more amino acid residues. Proteins serve both as structural materials (hair, horns, fingernails) and as enzymes that control an organism's chemistry. (*See* Peptides.)

**Protic solvent** (Section 11.9): A solvent such as water or alcohol that can serve as a proton donor. Protic solvents are particularly good at stabilizing anions by hydrogen bonding, thereby lowering their reactivity. (*See* Dipolar aprotic solvent.)

**Proton noise-decoupled mode** (Section 13.4): The most common manner of $^{13}C$ NMR spectrometer operation, in which all nonequivalent carbon atoms in the sample show a single, unsplit resonance. Operating the spectrometer in this mode allows one to count the number of chemically different carbon atoms present in the sample molecule.

**Quartet** (Section 13.6): A set of four peaks in the NMR, caused by spin–spin splitting of a signal by three adjacent nuclear spins.

**Quaternary** (Section 3.5): *See* Primary.

**Quaternary structure** (Section 27.14): The highest level of protein structure, involving a specific aggregation of individual proteins into a larger cluster.

***R,S* convention** (Section 9.6): A method for defining the absolute configuration around chiral centers. The Cahn–Ingold–Prelog sequence rules are used to assign relative priorities to the four substituents on the chiral center, and the center is oriented such that the group of lowest (fourth) priority faces directly away from the viewer. If the three remaining substituents have a right-handed or clockwise relationship in going from first to second to third priority, then the chiral center is denoted *R* (*rectus*, right). If the three remaining substituents have a left-handed or counterclockwise relationship, the chiral center is denoted *S* (*sinister*, left). (*See* Sequence rules.)

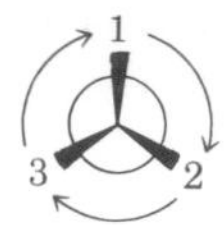


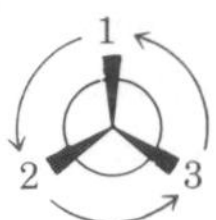


*R* configuration *S* configuration

**Racemic mixture** (Section 9.11): A mixture consisting of equal parts (+) and (−) enantiomers of a chiral substance. Even though the individual molecules are chiral, racemic mixtures are optically inactive.

**Racemization** (Section 9.11): The process whereby one enantiomer of a chiral molecule becomes converted into a 50:50 mixture of enantiomers, thus losing its optical activity. For example, this might happen during an $S_N1$ reaction of a chiral alkyl halide.

**Radical** (Section 5.2): When used in organic nomenclature, the word radical refers to a part of a molecule that appears in its name—for example, the "phenyl" in phenyl acetate. Chemically, however, a radical is a species that has an odd number of electrons, such as the chlorine radical, Cl·.

**Radical reaction** (Section 5.2): A reaction in which bonds are made by donation of one electron from each of two reagents and bonds are broken when each fragment leaves with one electron. (*See* Homogenic bond formation; Homolytic bond breakage.)

**Rate-limiting step** (Section 11.7): The slowest step in a multistep reaction sequence. The rate-limiting step acts as a kind of bottleneck in multistep reactions and is observed by kinetics measurements.

**Reaction energy diagram** (Section 5.9): A pictorial representation of the course of a reaction, in which potential energy is plotted as a function of reaction progress. Starting materials, transition states, intermediates, and final products are all represented, and their appropriate energy levels are indicated.

**Reducing sugar** (Section 24.8): Any sugar that reduces silver ion in the Tollens test or cupric ion in the Fehling or Benedict tests. All sugars that are aldehydes or can be readily converted into aldehydes are reducing. Glycosides, however, are not reducing sugars.

**Refining** (Section 3.6): The process by which petroleum is converted into gasoline and other useful products.

**Regiochemistry** (Section 6.10): A term describing the orientation of a reaction that occurs on an unsymmetrical substrate. Markovnikov's rule, for example, predicts the regiochemistry of electrophilic addition reactions.

**Regiospecific** (Section 6.10): A term describing a reaction that occurs with a specific regiochemistry to give a single product, rather than a mixture of products.

**Replication** (Section 29.12): The process by which double-stranded DNA uncoils and is replicated to produce two new copies.

**Resolution** (Section 25.5): The process by which a racemic mixture is separated into its two pure enantiomers. For example, a racemic carboxylic acid might be converted by reaction with a chiral amine base into a diastereomeric mixture of salts, which could be separated by fractional crystallization. Regeneration of the free acids would then yield the two pure enantiomeric acids.

**Resonance effect** (Section 16.5): The effect by which substituents donate or withdraw electron density through orbital overlap with neighboring pi bonds. For example, an oxygen or nitrogen substituent donates electron density to an aromatic ring by overlap of the O or N orbital with the aromatic ring *p* orbitals. A carbonyl substituent, however, withdraws electron density from an aromatic ring by *p* orbital overlap. These effects are particularly important in determining whether a given group is meta-directing or ortho- and para-directing in electrophilic aromatic-substitution reactions.

**Resonance hybrid** (Section 10.7): A molecule, such as benzene, that cannot be represented adequately by a single Kekulé structure but must instead be considered as an average of two or more resonance structures. The resonance structures themselves differ only in the positions of their electrons, not of their nuclei.

**Ring current** (Section 15.12): The circulation of pi electrons induced in aromatic rings by an external magnetic field. This effect accounts for the pronounced downfield shift of aromatic ring protons in the $^1H$ NMR.

**Ring-flip** (Section 4.11): The molecular motion that converts one chair conformation of cyclohexane into another chair conformation. The effect of a ring-flip is to convert an axial substituent into an equatorial substituent.

**RNA** (Section 29.9): Ribonucleic acid, the biopolymer found in cells that serves to transcribe the genetic information found in DNA and uses that information to direct the synthesis of proteins.

**Saccharide** (Section 24.1): A sugar.

**Saponification** (Section 21.7): An old term for the base-induced hydrolysis of an ester to yield a carboxylic acid salt.

**Saturated** (Section 3.2): A saturated molecule is one that has only single bonds and thus cannot undergo addition reactions. Alkanes, for example, are saturated, but alkenes are unsaturated.

**Sawhorse structure** (Section 4.1): A stereochemical manner of representation that portrays a molecule using a stick drawing and gives a perspective view of the conformation around single bonds.

Sawhorse structure

**Second-order reaction** (Section 11.3): A reaction whose rate-limiting step is bimolecular and whose kinetics are therefore dependent on the concentration of two reagents.

**Secondary** (Section 3.5): *See* Primary.

**Secondary structure** (Section 27.14): The level of protein substructure that involves organization of chain sections into ordered arrangements such as $\beta$-pleated sheets or $\alpha$-helices.

**Sequence rules** (Sections 6.6, 9.6): A series of rules devised by Cahn, Ingold, and Prelog for assigning relative priorities to substituent groups on a double-bond carbon atom or on a chiral center. Once priorities have been established, *E,Z* double-bond geometry and *R,S* configurational assignments can be made. (*See Entgegen*; *R,S* convention; *Zusammen*.)

**Shielding** (Section 13.2): An effect observed in NMR that causes a nucleus to absorb toward the right (upfield) side of the chart. Shielding is caused by donation of electron density to the nucleus. (*See* Deshielding.)

**Sigma bond** (Section 1.7): A covalent bond formed by head-on overlap of atomic orbitals.

**Sigmatropic reaction** (Section 30.1): A pericyclic reaction that involves the migration of a group from one end of a pi electron system to the other. For example, the [1,5] sigmatropic rearrangement of a hydrogen atom in cyclopentadiene is such a reaction.

**Skew conformation** (Section 4.3): Any conformation about a single bond that is intermediate between staggered and eclipsed. (*See* Staggered conformation; Eclipsed conformation.)

**Soap** (Section 28.2): The mixture of long-chain fatty acid salts obtained on base hydrolysis of animal fat.

**Solid-phase synthesis** (Section 27.12): A technique of synthesis whereby the starting material is covalently bound to a solid polymer bead and reactions are carried out on the bound substrate. After the desired transformations have been effected, the product is cleaved from the polymer and is isolated. This technique is particularly useful in peptide synthesis (Merrifield method).

***sp* orbital** (Section 1.11): A hybrid orbital mathematically derived from the combination of an *s* and a *p* atomic orbital. The two *sp* orbitals that result from hybridization are oriented at an angle of 180° to each other.

**$sp^2$ orbital** (Section 1.10): A hybrid orbital mathematically derived by combination of an *s* atomic orbital with two *p* atomic orbitals. The three $sp^2$ hybrid orbitals that result lie in a plane at angles of 120° to each other.

**$sp^3$ orbital** (Section 1.9): A hybrid orbital mathematically derived by combination of an *s* atomic orbital with three *p* atomic orbitals. The four $sp^3$ hybrid orbitals that result are directed toward the corners of a tetrahedron at angles of 109° to each other.

**Specific rotation, $[\alpha]_D$** (Section 9.2): The specific rotation of a chiral compound is a physical constant that is defined by the equation

$$[\alpha]_D = \frac{\text{Observed rotation}}{\text{Path length} \times \text{Concentration}} = \frac{\alpha}{l \times C}$$

where the path length of the sample solution is expressed in decimeters, and the concentration of the sample solution is expressed in grams per milliliter.

**Spin–spin splitting** (Section 13.6): The splitting of an NMR signal into a multiplet caused by an interaction between nearby magnetic nuclei whose spins are coupled. The magnitude of spin–spin splitting is given by the coupling constant, *J*.

**Staggered conformation** (Section 4.1): The three-dimensional arrangement of atoms around a carbon–carbon single bond in which the bonds on one carbon exactly bisect the bond angles on the second carbon as viewed end-on. (*See* Eclipsed conformation.)

Staggered conformation

**Stationary phase** (Section 12.1): The solid support used in chromatography. The molecules to be chromatographically separated adsorb to the stationary phase and are moved along by the mobile phase. Silica gel (hydrated $SiO_2$) and alumina ($Al_2O_3$) are often used as stationary phases in column chromatography of organic mixtures. (*See* Mobile phase.)

**Step-growth polymer** (Section 31.1): A polymer produced by a series of polar reactions between two difunctional monomers. The polymer normally has the two monomer units in alternating order and usually has other atoms in addition to carbon in the polymer backbone. Nylon, a polyamide produced by reaction between a diacid and a diamine, is an example of such a polymer.

**Stereochemistry** (Chapters 4, 9): The branch of chemistry concerned with the three-dimensional arrangement of atoms in molecules.

**Stereoisomers** (Section 9.3): Isomers that have their atoms connected in the same order but in different three-dimensional arrangements. The term stereoisomer includes both enantiomers and diastereomers, but does not include constitutional isomers.

**Stereospecific** (Section 7.10): A term indicating that only a single stereoisomer is produced in a given reaction, rather than a mixture.

**Steric strain** (Sections 4.3, 4.12): The strain imposed on a molecule when two groups are too close together and try to occupy the same space. Steric strain is responsible both for the greater stability of trans versus cis alkenes, and for the greater stability of equatorially substituted versus axially substituted cyclohexanes.

**Steroid** (Section 28.8): A lipid whose structure is based on the tetracyclic carbon skeleton:

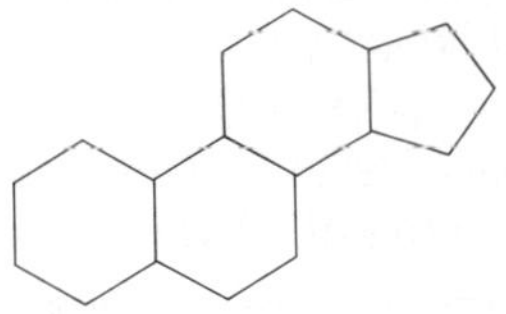

Steroids occur in both plants and animals and have a variety of important hormonal functions.

**Suprafacial** (Section 30.8): A word used to describe the geometry of pericyclic reactions. Suprafacial reactions take place on the same side of the two ends of a pi electron system. (*See* Antarafacial.)

**Symmetry-allowed, symmetry-disallowed** (Section 30.4): A symmetry-allowed reaction is a pericyclic process that has a favorable orbital symmetry for reaction through a concerted pathway. A symmetry-disallowed reaction is one that does not have favorable orbital symmetry for reaction through a concerted pathway.

**Syn stereochemistry** (Section 7.1): A syn addition reaction is one in which the two ends of the double bond are attacked from the same side. For example, $OsO_4$ induced hydroxylation of cyclohexene yields *cis*-1,2-cyclohexanediol, the product of syn addition. A syn elimination is one in which the two groups leave from the same side of the molecule. (*See* Anti stereochemistry.)

**Syndiotactic polymer** (Section 31.6): A chain-growth polymer in which the substituents on the polymer backbone have a regular alternating stereochemistry.

**Tautomers** (Section 8.4): Isomers that are rapidly interconverted. For example, enols and ketones are tautomers, since they are rapidly interconverted on treatment with either acid or base catalysts.

**Terpenes** (Section 28.6): Lipids that are formally derived by head-to-tail polymerization of isoprene units. (*See* Isoprene rule.)

**Tertiary** (Section 3.5): *See* Primary.

**Tertiary structure** (Section 27.14): The level of protein structure that involves the manner in which the entire protein chain is folded into a specific three-dimensional arrangement.

**Thermodynamic control** (Section 14.6): Equilibrium reactions that yield the lowest-energy, most stable product are said to be thermodynamically controlled. Although most stable, the product of a thermodynamically controlled reaction is not necessarily formed fastest. (*See* Kinetic control.)

**Thermoplastic** (Section 31.13): A polymer that is hard at room temperature but becomes soft and pliable when heated. Thermoplastics are used for the manufacture of a variety of molded objects.

**Thermosetting resin** (Section 31.13): A polymer that is highly cross-linked and sets into a hard, insoluble mass when heated. Bakelite is the best-known example of such a polymer.

**Torsional strain** (Section 4.1): The strain in a molecule caused by electron repulsion between eclipsed bonds. Torsional strain plays a major role in destabilizing boat cyclohexane relative to chair cyclohexane. Torsional strain is also called eclipsing strain.

**Transcription** (Section 29.13): The process by which the genetic information encoded in DNA is read and used to synthesize RNA in the nucleus of the cell. A small portion of double-stranded DNA uncoils, and complementary ribonucleotides line up in the correct sequence for RNA synthesis.

**Transition state** (Section 5.9): An imaginary activated complex between reagents, representing the highest-energy point on a reaction curve. Transition states are unstable complexes that cannot be isolated.

**Translation** (Section 29.14): The process by which the genetic information transcribed from DNA onto mRNA is read by tRNA and used to direct protein synthesis.

**Tree diagram** (Section 13.13): A diagram used in NMR to help sort out the complicated splitting patterns that can arise from multiple couplings.

**Triacylglycerol** (Section 28.1): Lipids such as animal fat and vegetable oil consisting chemically of triesters of glycerol with long-chain fatty acids.

**Triplet** (Section 13.6): A symmetrical three-line splitting pattern observed in the $^1$H NMR when a proton has two equivalent neighbor protons or in the $^{13}$C NMR when a carbon is bonded to two hydrogens.

**Ultraviolet (UV) spectroscopy** (Section 14.10): An optical spectroscopy employing ultraviolet irradiation. UV spectroscopy provides structural information about the extent of pi electron conjugation in organic molecules.

**Unimolecular reaction** (Section 11.7): A reaction that occurs by spontaneous transformation of the starting material without the intervention of other reagents. For example, the dissociation of a tertiary alkyl halide in the $S_N1$ reaction is a unimolecular process.

**Unsaturated** (Section 6.2): An unsaturated molecule is one that has multiple bonds and can undergo addition reactions. Alkenes and alkynes, for example, are unsaturated. (*See* Saturated.)

**Upfield** (Section 13.3): Used to refer to the right-hand portion of the NMR chart. (*See* Shielding.)

**Van der Waals forces** (Section 3.7): The attractive forces between molecules that are caused by dipole–dipole interactions. Van der Waals forces are one of the primary forces responsible for holding molecules together in the liquid state.

**Vicinal** (Section 8.10): A term used to refer to a 1,2-disubstitution pattern. For example, 1,2-dibromoethane is a vicinal dibromide.

**Vinylic** (Section 8.3): A term that refers to a substituent at a double-bond carbon atom. For example, chloroethylene is a vinylic chloride, and enols are vinylic alcohols.

**Vulcanization** (Section 31.7): A process for hardening rubber by heating in the presence of elemental sulfur. The sulfur functions by forming cross-links between polymer chains.

**Wave function** (Section 1.1): The mathematical expression that defines the behavior of an electron. The square of the wave function is the probability function that defines the shapes of orbitals.

**Wave number** (Section 12.9): The wave number is the reciprocal of the wavelength in centimeters. Thus, wave numbers are expressed in $cm^{-1}$.

**Wavelength** (Section 12.8): The length of a wave from peak to peak. The wavelength of electromagnetic radiation is inversely proportional to frequency and inversely proportional to energy. (*See* Frequency.)

**Ylide** (Section 19.16): A neutral dipolar molecule in which the positive and negative charges are adjacent. For example, the phosphoranes used in Wittig reactions are ylides.

**Zaitsev's rule** (Section 11.10): A rule stating that E2 elimination reactions normally yield the more highly substituted alkene as major product.

***Zusammen, Z*** (Section 6.6): A term used to describe the stereochemistry of a carbon–carbon double bond. The two groups on each carbon are assigned priorities according to the Cahn–Ingold–Prelog sequence rules, and the two carbons are compared. If the high-priority groups on each carbon are on the same side of the double bond, the bond has $Z$ geometry. (*See Entgegen*; Sequence rules.)

**Zwitterion** (Section 27.2): A neutral dipolar molecule in which the positive and negative charges are not adjacent. For example, amino acids exist as zwitterions, $H_3\overset{+}{N}$—CHR—$COO^-$. (Zwitterions are also called betaines.)

# Index

*The page references given in color refer either to entries in the text where terms are defined or to mini-biographies.*

# Periodic Chart

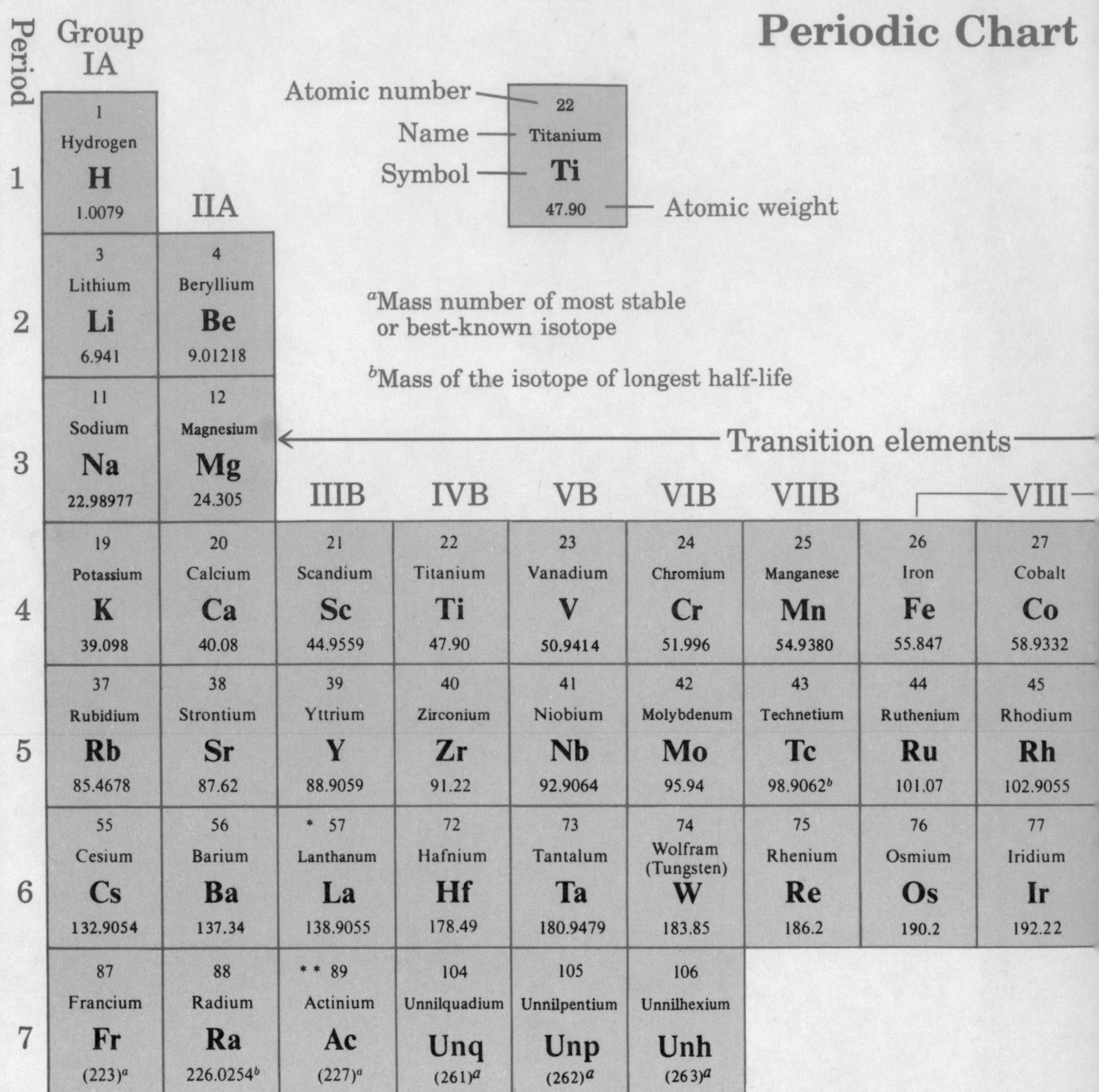

Atomic number — 22
Name — Titanium
Symbol — **Ti**
47.90 — Atomic weight

[a]Mass number of most stable or best-known isotope

[b]Mass of the isotope of longest half-life

← Transition elements →

| Period | Group IA | IIA | IIIB | IVB | VB | VIB | VIIB | VIII | |
|---|---|---|---|---|---|---|---|---|---|
| 1 | 1<br>Hydrogen<br>**H**<br>1.0079 | | | | | | | | |
| 2 | 3<br>Lithium<br>**Li**<br>6.941 | 4<br>Beryllium<br>**Be**<br>9.01218 | | | | | | | |
| 3 | 11<br>Sodium<br>**Na**<br>22.98977 | 12<br>Magnesium<br>**Mg**<br>24.305 | | | | | | | |
| 4 | 19<br>Potassium<br>**K**<br>39.098 | 20<br>Calcium<br>**Ca**<br>40.08 | 21<br>Scandium<br>**Sc**<br>44.9559 | 22<br>Titanium<br>**Ti**<br>47.90 | 23<br>Vanadium<br>**V**<br>50.9414 | 24<br>Chromium<br>**Cr**<br>51.996 | 25<br>Manganese<br>**Mn**<br>54.9380 | 26<br>Iron<br>**Fe**<br>55.847 | 27<br>Cobalt<br>**Co**<br>58.9332 |
| 5 | 37<br>Rubidium<br>**Rb**<br>85.4678 | 38<br>Strontium<br>**Sr**<br>87.62 | 39<br>Yttrium<br>**Y**<br>88.9059 | 40<br>Zirconium<br>**Zr**<br>91.22 | 41<br>Niobium<br>**Nb**<br>92.9064 | 42<br>Molybdenum<br>**Mo**<br>95.94 | 43<br>Technetium<br>**Tc**<br>98.9062[b] | 44<br>Ruthenium<br>**Ru**<br>101.07 | 45<br>Rhodium<br>**Rh**<br>102.9055 |
| 6 | 55<br>Cesium<br>**Cs**<br>132.9054 | 56<br>Barium<br>**Ba**<br>137.34 | * 57<br>Lanthanum<br>**La**<br>138.9055 | 72<br>Hafnium<br>**Hf**<br>178.49 | 73<br>Tantalum<br>**Ta**<br>180.9479 | 74<br>Wolfram (Tungsten)<br>**W**<br>183.85 | 75<br>Rhenium<br>**Re**<br>186.2 | 76<br>Osmium<br>**Os**<br>190.2 | 77<br>Iridium<br>**Ir**<br>192.22 |
| 7 | 87<br>Francium<br>**Fr**<br>(223)[a] | 88<br>Radium<br>**Ra**<br>226.0254[b] | ** 89<br>Actinium<br>**Ac**<br>(227)[a] | 104<br>Unnilquadium<br>**Unq**<br>(261)[a] | 105<br>Unnilpentium<br>**Unp**<br>(262)[a] | 106<br>Unnilhexium<br>**Unh**<br>(263)[a] | | | |

| | | | | | |
|---|---|---|---|---|---|
| * Lanthanide series 6 | 58<br>Cerium<br>**Ce**<br>140.12 | 59<br>Praseo-dymium<br>**Pr**<br>140.9077 | 60<br>Neodymium<br>**Nd**<br>144.24 | 61<br>Promethium<br>**Pm**<br>(145)[a] | 62<br>Samarium<br>**Sm**<br>150.4 |
| ** Actinide series 7 | 90<br>Thorium<br>**Th**<br>232.0381[b] | 91<br>Protactinium<br>**Pa**<br>231.0359[b] | 92<br>Uranium<br>**U**<br>238.029 | 93<br>Neptunium<br>**Np**<br>237.0482 | 94<br>Plutonium<br>**Pu**<br>(242)[a] |